RESEARCH AND MANAGEMENT TECHNIQUES FOR WILDLIFE AND HABITATS

Edited by

Theodore A. Bookhout

National Biological Service
Ohio Cooperative Fish and Wildlife Research Unit
The Ohio State University
Columbus, Ohio

The Wildlife Society
Bethesda, Maryland
1996

Research and Management Techniques for Wildlife and Habitats

This book is the fifth in a series on wildlife techniques published by The Wildlife Society

Editor, Henry S. Mosby
  Manual of Game Investigational Techniques
    (1) First Edition—May 1960
        Second Printing—February 1961
  Wildlife Investigational Techniques
    (2) Second Edition—May 1963
        Second Printing (Revised)—March 1965
        Third through Sixth Printing—March 1966 to September 1968

Editor, Robert H. Giles, Jr.
  Wildlife Management Techniques
    (3) Third Edition—June 1969
        Second Printing (Revised)—January 1971
        Third Printing—May 1972

Editor, Sanford D. Schemnitz
  Wildlife Management Techniques Manual
    (4) Fourth Edition—September 1980

Editor, Theodore A. Bookhout
  Research and Management Techniques for Wildlife and Habitats
    (5) Fifth Edition—January 1994
        Second Printing (Revised)—April 1996

————

Suggested citation formats:

Entire book

Bookhout, T. A., Editor. 1996. Research and management techniques for wildlife and habitats. Fifth ed., rev. The Wildlife Society, Bethesda, Md. 740pp.

Chapter of the book

Johnson, D. H. 1996. Population analysis. Pages 419–444 *in* T. A. Bookhout, ed. Research and management techniques for wildlife and habitats. Fifth ed., rev. The Wildlife Society, Bethesda, Md.

————

This book was produced on the Penta DeskTopPro/UX® and output to an AGFA SelectSet 7000 imagesetter. The text is Adobe Times Roman. The text paper is 50 pound Restorecote (50/10 recycled). The Roxite cloth cover was printed by Allen Press, Inc. and the case binding was done by Prizma Industries, Denver, Colorado. This book was printed on a Hantscho full-sized waterless web press by Allen Press, Inc.

ISBN 0-935868-81-X
Library of Congress Catalog Card Number: 93-61624

CONTENTS

DEDICATION

This book is dedicated to the memory of Dr. Willard D. Klimstra—
teacher, mentor, colleague, and friend

W. D. Klimstra
1919–1993

President, The Wildlife Society, 1973–74

Recipient, the Aldo Leopold Memorial Award, 1988

FOREWORD

There is little doubt that the combined impact of the first four editions of the *Wildlife Management Techniques Manual* has been greater than any other single publication in the field of wildlife management. Generations of wildlife professionals were exposed to the ''Techniques Manual'' in their college training, and they have used it in their daily work.

This fifth edition, *Research and Management Techniques for Wildlife and Habitats,* will do more than add to that proud history. This edition is more comprehensive than any of its predecessors, and its impact likely will exceed that of all the previous editions combined. The expanded title is reflective of the contents of this volume and the intention of the editor to put techniques into their proper context as means rather than ends.

The Wildlife Society has long sought to publish materials that are useful to practicing wildlife professionals. The Techniques Manual has proved itself over the years. By establishing standards and approaches to research and management, this work enables our profession to advance in our mission of protecting and managing the Earth's wild living resources. The fourth edition was translated into Spanish and it is likely that the fifth edition will be made available in other languages. These efforts strengthen The Wildlife Society and support professionals around the globe as they practice sound wildlife stewardship.

The success of the Techniques Manual is the result of the unselfish contribution of dozens of authors and reviewers who have volunteered their time and knowledge in completing this most important document. The members of the wildlife profession who will use this book owe those individuals a debt that they will repay by their future contributions of written works and an improved environment for all.

The Wildlife Society is extremely grateful to Editor Ted Bookhout, who has labored long and hard in working with authors and referees in the completion of their manuscripts and the final publication of the Manual. Ted has served The Wildlife Society in many capacities, including President, Editor of *The Journal of Wildlife Management,* and now Editor of this Manual. His continuing contributions set a standard that few of us will be able to match.

We already know that *Research and Management Techniques for Wildlife and Habitats* will be a best-seller. It will be on the shelves of most professional wildlife biologists within a short time, and few wildlife students will complete their college years without obtaining their own copy. The Wildlife Society is delighted to again sponsor this important contribution to the benefit of our World's wildlife resources and humankind.

W. Alan Wentz
President, 1992–93

PREFACE

When I accepted responsibility for editing this book, I believed strongly that readers would expect to find certain things when they took it off the shelf: how to conduct research on and how to manage wildlife and their habitats, and where to find more information on the subject. The questions quickly became, what research techniques, and what management methods? Previous editors faced the same dilemma: how to include the newer techniques and the newer literature, yet retain the older, relevant information without producing a volume that cannot be afforded nor lifted.

Soon after I became editor of the 5th edition, I sent a questionnaire to about 100 wildlife biologists—researchers, managers, and university faculty who teach a course in "wildlife techniques"—to solicit their views on what subjects they believed ought to be in the new manual. More than 70 individuals responded; although the range of ideas was wide, consensus about some subjects was obvious. The content of this 5th edition reflects some of that consensus, but the decision of which chapter titles to include in this book was mine.

In an early communication to lead authors I selected to be responsible for the chapters, I requested that they **not** write for comprehension by peers, but by college seniors. This was not to be construed as writing "down" to an intended audience; it simply meant that less familiarity with the subject matter was to be assumed. The goal was, and is, to produce a manual that will serve as a **useful** reference to wildlife biologists in field and laboratory in terms of "How do I do it?" and "Where can I find more information?" These chapters are overviews of the subjects, hence the presentations are not as exhaustive nor as detailed as some readers might like. But the literature for most subjects is extensive, and we provide a substantive number of citations, emphasizing the most recent ones.

I acknowledge the cooperation and assistance of several individuals who made my editing responsibility more bearable and who contributed directly to the final product, this volume. First, I am grateful to chapter authors for taking time from busy schedules to prepare the drafts, for responding positively to referee comments, and for producing a useful product for wildlife students, scientists, and managers. I encouraged lead authors to enlist co-authors to assist in preparing the chapters, leaving to them the selection of those individuals. The quality of the content of the chapters could not have been achieved without the careful scrutiny of nearly 90 referees, whose knowledge of the subjects resulted in improved text, tables, and figures and inclusion of many key references. These individuals are recognized elsewhere in this volume. Also assisting in the review process were R. D. Drobney, J. D. Erb, E. K. Fritzell, D. A. Haukos, J. D. Nichols, R. E. Reynolds, O. E. Rhodes, Jr., and G. W. Smith.

Nancy Pollack was copy editor for the 5th edition. She gave valuable input to decisions about editorial conventions, and she was an effective communicator with our printer, Allen Press; Nancy deserves much credit for the high quality of this book. I depended heavily on my secretary, Diane Rano, for assistance during the 7 years I worked on producing the 5th edition. She helped more times and in more ways than I can remember; I can only say, "Thanks, Diane." Harry Hodgdon, Executive Director of The Wildlife Society, was particularly helpful in communications with Allen Press, and he assisted in various other ways upon my request.

Graduate students in the Department of Zoology, The Ohio State University (OSU), critiqued several early draft chapters in a formal seminar in 1990, and their comments were relayed to authors as a reaction of students to chapter content. These students were: Brad Andres, Carol Bocetti, Joan Bradley, Nancy Buschhaus, Earl Campbell, David Cimprich, Jorge Coppen, Minna Hsu, Leslie Jackson, Joseph Robb, Thomas Kerr, and Reuven Yosef.

Verifying accuracy of literature references was enhanced by efforts of Melinda Ketring and Rene Auckerman, OSU Work Study students who spent many hours at computer terminals and at the bookshelves in the OSU libraries searching for publications and citations. Chapter authors ought to be as grateful as I for the many, many accurate citations that resulted from the work of these two undergraduate students.

Nancy Morin, Missouri Botanical Garden, and Richard Banks, U.S. Fish and Wildlife Service, were particularly helpful in providing advice about authorities for scientific names of plants and vertebrates, respectively. Dr. Banks and his colleagues at the U.S. National Museum kindly reviewed the scientific names of animals listed in the appendices. I am grateful for their assistance.

Brad Andres, graduate student, and David Stetson and Susan Earnst, OSU Department of Zoology, provided advice and assistance of many kinds that related to computer software and its use. I appreciate their patience toward an editor who needs more proficiency in using a PC.

The cover of this 5th edition was designed by David Dennis, Biological Illustrator, College of Biological Sciences, OSU. Dave also provided useful comments on several drafts I developed of the title page and first pages of chapters. I appreciate his critical eye.

Over the years that I worked at developing the 5th edition, many persons called or sent suggestions about improving the quality of subject material in this manual. I thank Dennis A. Demarchi, Robert A. Garrott, the late Gordon Gullion, Robert L. Hoover, and Monty Whiting for taking the time to be helpful. I acknowledge, too, the contributions of volunteer assistance by any other individuals whom I unintentionally have failed to mention.

Finally, I am grateful to the U.S. Fish and Wildlife Service for granting me the opportunity to accept and complete this responsibility. I particularly thank Rollin W. Sparrowe, John G. Rogers, Jr., Paul A. Vohs, W. Reid Goforth, M. Lynn Haines, and Edward T. LaRoe.

CHAPTER AUTHORS

Eric M. Anderson
College of Natural Resources
University of Wisconsin–Stevens Point
Stevens Point, WI 54481

Stanley H. Anderson
Wyoming Cooperative Fish and Wildlife Research Unit
University of Wyoming
Laramie, WY 82071

Morley W. Barrett
North American Waterfowl Association
Edmonton, Alberta T5M 3Z7

Jonathan R. Bart
Ohio Cooperative Fish and Wildlife Research Unit
Ohio State University
Columbus, OH 43210

Vernon C. Bleich[1]
Institute of Arctic Biology
University of Alaska Fairbanks
Fairbanks, AK 99775

Eric G. Bolen
Graduate School
University of North Carolina at Wilmington
Wilmington, NC 28403

Stephen J. Brady
U.S. Soil Conservation Service/Forest Service
Rocky Mountain Forest and Range Experiment Station
Ft. Collins, CO 80524

Robert L. Brownell, Jr.
U.S. Fish and Wildlife Service
National Ecology Research Center
Ft. Collins, CO 80525

Gary K. Clambey
Department of Botany/Biology
North Dakota State University
Fargo, ND 58105

William R. Clark
Department of Animal Ecology
Iowa State University
Ames, IA 50011

Richard N. Conner
U.S. Forest Service
Southern Forest Experiment Station
Nacogdoches, TX 75962

Lewis M. Cowardin
U.S. Fish and Wildlife Service
Northern Prairie Wildlife Research Center
Jamestown, ND 58401

Donald S. Davis
Department of Veterinary Pathology
Texas A&M University
College Station, TX 77843

Ralph W. Dimmick
Department of Forestry, Wildlife, and Fisheries
University of Tennessee
Knoxville, TN 37901

Richard A. Dolbeer
U.S. Department of Agriculture
Denver Wildlife Research Center
Sandusky, OH 44870

William J. Foreyt
Department of Veterinary Microbiology/Pathology
Washington State University
Pullman, WA 99164

Leigh H. Fredrickson
Gaylord Memorial Laboratory
University of Missouri
Puxico, MO 63960

Milton W. Friend
U.S. Fish and Wildlife Service
National Wildlife Health Research Center
Madison, WI 53711

Mark R. Fuller
U.S. Fish and Wildlife Service
Patuxent Wildlife Research Center
Laurel, MD 20708

Edward O. Garton
Department of Fish and Wildlife Resources
University of Idaho
Moscow, ID 83843

Kevin J. Gutzwiller
Department of Biology
Baylor University
Waco, TX 76798

John D. Harder
Department of Zoology
Ohio State University
Columbus, OH 43210

Harold J. Harju
Wyoming Game and Fish Department
Cheyenne, WY 82006

David L. Harlow
U.S. Fish and Wildlife Service
Nevada Ecological Services Field Office
Reno, NV 89502

---

[1]Present address: California Department of Fish and Game, Bishop, CA 93514.

Richard F. Harlow
Department of Forestry
Clemson University
Clemson, SC 29634

Jonathan B. Haufler
Department of Fisheries and Wildlife
Michigan State University
East Lansing, MI 48824

Donald W. Hawthorne
U.S. Department of Agriculture
Animal Damage Control, Western Region
Denver, CO 80225

Kenneth F. Higgins
South Dakota Cooperative Fish and Wildlife Research
Unit
South Dakota State University
Brookings, SD 57007

Nicholas R. Holler
Alabama Cooperative Fish and Wildlife Research Unit
Auburn University
Auburn, AL 36849

John Jacobson
Ducks Unlimited, Inc.
Memphis, TN 38120

Kurt J. Jenkins
Department of Wildlife and Fisheries Sciences
South Dakota State University
Brookings, SD 57007

Douglas H. Johnson
U.S. Fish and Wildlife Service
Northern Prairie Wildlife Research Center
Jamestown, ND 58401

John G. Kie
U.S. Forest Service
Pacific Southwest Forest & Range Experiment Station
Fresno, CA 93710

Roy L. Kirkpatrick
Department of Fisheries and Wildlife Sciences
Virginia Polytechnic Institute and State University
Blacksburg, VA 24061

Gregory T. Koeln
Earth Satellite Corporation
Rockville, MD 20852

Richard A. Lancia
Department of Forestry
North Carolina State University
Raleigh, NC 27695

Murray K. Laubhan
Gaylord Memorial Laboratory
University of Missouri
Puxico, MO 63960

John A. Litvaitis
Department of Natural Resources
University of New Hampshire
Durham, NH 03824

Louis N. Locke
U.S. Fish and Wildlife Service
National Wildlife Health Research Center
Madison, WI 53711

R. William Mannan
School of Renewable Natural Resources
University of Arizona
Tucson, AZ 85721

Bruce G. Marcot
U.S. Forest Service
Pacific Northwest Research Station
Portland, OR 97708

Keith R. McCaffery
Wisconsin Department of Natural Resources
Rhinelander, WI 54501

Alvin L. Medina
U.S. Forest Service
Rocky Mountain Forest & Range Experiment Station
Flagstaff, AZ 86001

Harvey W. Miller
Department of Range and Wildlife Management
Texas Tech University
Lubbock, TX 79409

Henry R. Murkin
Institute for Wetland and Waterfowl Research
% Ducks Unlimited Canada
Stonewall, Manitoba ROC 2ZO

Victor F. Nettles
Southeastern Cooperative Wildlife Disease Study
University of Georgia
Athens, GA 30602

James D. Nichols
U.S. Fish and Wildlife Service
Patuxent Wildlife Research Center
Laurel, MD 20708

Marie T. Nietfeld
Animal Sciences Division
Alberta Environmental Centre
Vegreville, Alberta T0B 4L0

William I. Notz
Department of Statistics
Ohio State University
Columbus, OH 43210

Bart W. O'Gara
Montana Cooperative Wildlife Research Unit
University of Montana
Missoula, MT 59812

John L. Oldemeyer
U.S. Fish and Wildlife Service
National Ecology Research Center
Fort Collins, CO 80525

James M. Peek
Department of Fish and Wildlife Resources
University of Idaho
Moscow, ID 83843

Michael R. Pelton
Department of Forestry, Wildlife and Fisheries
University of Tennessee
Knoxville, TN 37901

Kenneth H. Pollock
Department of Statistics
North Carolina State University
Raleigh, NC 27695

Daniel B. Pond
W. L. Gore & Associates
Flagstaff, AZ 86004

John T. Ratti
Department of Fish and Wildlife Resources
University of Idaho
Moscow, ID 83843

Frederic A. Reid
Ducks Unlimited, Inc.
Western Regional Office
Sacramento, CA 95827

Thomas J. Roffe
U.S. Fish and Wildlife Service
National Wildlife Health Research Center
Madison, WI 53711

Michael D. Samuel
U.S. Fish and Wildlife Service
National Wildlife Health Research Center
Madison, WI 53711

Gary J. San Julian
National Wildlife Federation
Washington, DC 20036

Sanford D. Schemnitz
Department of Fishery and Wildlife Sciences
New Mexico State University
Las Cruces, NM 88003

J. Michael Scott
Idaho Cooperative Fish and Wildlife Research Unit
University of Idaho
Moscow, ID 83844

Frederick A. Servello
Department of Wildlife
University of Maine
Orono, ME 04469

Mark L. Shaffer
The Wilderness Society
Washington, DC 20006

Henry L. Short
U.S. Fish and Wildlife Service
Office of Scientific Authority
Washington, DC 20240

Nova Silvy
Department of Wildlife and Fisheries Science
Texas A&M University
College Station, TX 77843

Loren M. Smith
Department of Range and Wildlife Management
Texas Tech University
Lubbock, TX 79409

Robert J. Stoll, Jr.
Ohio Division of Wildlife
Waterloo Wildlife Research Station
New Marshfield, OH 45766

M. Dale Strickland
Western EcoSystems Technology
Cheyenne, WY 82007

Laurence L. Strong
U.S. Fish and Wildlife Service
Northern Prairie Wildlife Research Center
Jamestown, ND 58401

Stanley A. Temple
Department of Wildlife Ecology
University of Wisconsin
Madison, WI 53706

Jack Ward Thomas
U.S. Forest Service
Pacific Northwest Research Station
LaGrande, OR 97850

Kimberly Titus
Alaska Department of Fish and Game
Division of Wildlife Conservation
Douglas, AK 99824

Dale E. Toweill
Wildlife Program Coordinator
Idaho Department of Fish and Game
Boise, ID 83707

Joe C. Truett
Truett Research
Glenwood, NM 88039

Larry W. VanDruff
College of Environmental Science and Forestry
State University of New York
Syracuse, NY 13210

Richard E. Warner
Center for Wildlife Ecology
Illinois Natural History Survey
Champaign, IL 61820

Gary C. White
Department of Fishery and Wildlife Biology
Colorado State University
Fort Collins, CO 80523

Samuel C. Williamson
U.S. Fish and Wildlife Service
National Ecology Research Center
Fort Collins, CO 80525

Dale A. Wrubleski
Institute for Wetland and Waterfowl Research
% Ducks Unlimited Canada
Stonewall, Manitoba ROC 2ZO

James D. Yoakum
U.S. Bureau of Land Management
Reno, NV 89520

CHAPTER REFEREES

Lowell W. Adams
National Institute for Urban Wildlife
Columbia, MD 21044

Paul R. Adamus
U.S. Environmental Protection Agency
EPA Research Laboratory
Corvallis, OR 97333

David R. Anderson
Colorado Cooperative Fish and Wildlife Research Unit
Colorado State University
Fort Collins, CO 80523

William H. Anderson
Center for Mapping
Ohio State University
Columbus, OH 43212

C. Davison Ankney
Department of Zoology
University of Western Ontario
London, Ontario N6A 5B7

I. Joseph Ball, Jr.
Montana Cooperative Wildlife Research Unit
University of Montana
Missoula, MT 59812

Gordon Beanlands
School for Resource and Environmental Studies
Dalhousie University
Halifax, Nova Scotia B3H 3E2

Louis B. Best
Department of Animal Ecology
Iowa State University
Ames, IA 50011

John A. Bissonette
Utah Cooperative Fish and Wildlife Research Unit
Utah State University
Logan, UT 84322

Kenneth P. Burnham
Colorado Cooperative Fish and Wildlife Research Unit
Colorado State University
Fort Collins, CO 80523

David E. Capen
School of Natural Resources
University of Vermont
Burlington, VT 05405

Len H. Carpenter
Colorado Division of Wildlife
Research Center
Fort Collins, CO 80523

Robert H. Chabreck
School of Forestry, Wildlife, and Fisheries
Louisiana State University
Baton Rouge, LA 70803

Richard N. Conner
U.S. Forest Service
Southern Forest Experiment Station
Nacogdoches, TX 75962

Michael J. Conroy
Georgia Cooperative Fish and Wildlife Research Unit
University of Georgia
Athens, GA 30602

Wayne L. Cornelius
Statistics Department
North Carolina State University
Raleigh, NC 27695

Scott Craven
Department of Wildlife Ecology
University of Wisconsin
Madison, WI 53706

Richard M. DeGraaf
U.S. Forest Service
University of Massachusetts
Amherst, MA 01003

Roger Edwards
Canadian Wildlife Service
Environment Canada
Edmonton, Alberta T6B 2X3

Craig Ely
U.S. Fish and Wildlife Service
Alaska Fish and Wildlife Research Center
Anchorage, AK 99503

Ned H. Euliss, Jr.
U.S. Fish and Wildlife Service
Northern Prairie Wildlife Research Center
Jamestown, ND 58401

Lester D. Flake
Department of Wildlife and Fisheries
South Dakota State University
Brookings, SD 57007

Robert A. Garrott
Department of Wildlife Ecology
University of Wisconsin
Madison, WI 53706

Thomas A. Gavin
Department of Natural Resources
Cornell University
Ithaca, NY 14853

James R. Gilbert
Department of Wildlife
University of Maine
Orono, ME 04469

Fred S. Guthery
Caeser Kleberg Institute
Texas A&I University
Kingsville, TX 78363

Jay B. Hestbeck
Massachusetts Cooperative Fish and Wildlife Research
  Unit
University of Massachusetts
Amherst, MA 01003

Kirk Horn
School of Forestry
University of Montana
Missoula, MT 59812

David A. Jessup
California Department of Fish and Game
Wildlife Investigations Laboratory
Rancho Cordova, CA 95670

Douglas H. Johnson
U.S. Fish and Wildlife Service
Northern Prairie Wildlife Research Center
Jamestown, ND 58401

Ron J. Johnson
Department of Forestry, Fisheries and Wildlife
University of Nebraska at Lincoln
Lincoln, NE 68583

John A. Kadlec
College of Natural Resources
Utah State University
Logan, UT 84322

Richard M. Kaminski
Department of Wildlife and Fisheries
Mississippi State University
Mississippi State, MS 39762

Cameron B. Kepler
U.S. Fish and Wildlife Service
University of Georgia
Athens, GA 30602

Paul R. Krausman
School of Renewable Natural Resources
University of Arizona
Tucson, AZ 85721

Charles J. Krebs
Department of Zoology
University of British Columbia
Vancouver, British Columbia V6T 1W5

William R. Lance
Wildlife Pharmaceuticals, Inc.
Fort Collins, CO 80524

L. Jack Lyon
U.S. Forest Service
Intermountain Research Station
Missoula, MT 59807

Richard J. Mackie
Department of Biology
Montana State University
Bozeman, MT 59717

R. Larry Marchinton
School of Forest Resources
University of Georgia
Athens, GA 30602

William C. McComb
Department of Forest Science
Oregon State University
Corvallis, OR 97331

Lyman McDonald
Western EcoSystems Technology
Littleton, CO 80122

Charles M. Nixon
Illinois Natural History Survey
Champaign, IL 61820

Tom Nudds
Department of Zoology
University of Guelph
Guelph, Ontario N1G 2W1

John L. Oldemeyer
U.S. Fish and Wildlife Service
National Ecology Research Center
Fort Collins, CO 80525

David Otis
South Carolina Cooperative Fish and Wildlife Research
  Unit
Clemson University
Clemson, SC 29634

Ray B. Owen, Jr.
Department of Wildlife
University of Maine
Orono, ME 04469

Edward D. Plotka
Marshfield Medical Research Foundation
Marshfield, WI 54449

Bryce Rickel
U.S. Forest Service
Southwestern Region
Albuquerque, NM 87102

Charles T. Robbins
Department of Natural Resources Science
Washington State University
Pullman, WA 99164

Robert E. Rolley[1]
Indiana Division of Fish and Wildlife
Bloomington, IN 47401

David M. Rosenberg
Department of Fisheries and Oceans
Freshwater Institute
Winnipeg, Manitoba R3T 2N6

---

[1]Present address: Bureau of Research, Wisconsin Department of Natural Resources, Monona, WI 53716.

Mark R. Ryan
School of Forestry, Fisheries, and Wildlife
University of Missouri
Columbia, MO 65221

Fred B. Samson
U.S. Forest Service
Alaska Region
Juneau, AK 99802

William M. Samuel
Department of Zoology
University of Alberta
Edmonton, Alberta T6G 2E9

Joseph M. Schaefer
Department of Wildlife and Range Science
University of Florida
Gainesville, FL 32611

Charles C. Schwartz
Moose Research Center
Alaska Department of Fish and Game
Soldotna, AK 99669

Ulysses S. Seal
Captive Breeding Specialists Group
Species Survival Commission, IUCN
Bloomington, MN 55420

Kieth Severson
U.S. Forest Service
Forestry Sciences Laboratory
Rapid City, SD 57701

Steven L. Sheriff
Missouri Department of Conservation
Fish and Wildlife Research Center
Columbia, MO 65201

Richard D. Slemons
Department of Veterinary Preventive Medicine
Ohio State University
Columbus, OH 43210

Norman S. Smith
Arizona Cooperative Fish and Wildlife Research Unit
University of Arizona
Tucson, AZ 85721

Dean F. Stauffer
Department of Fisheries and Wildlife Sciences
Virginia Polytechnic Institute and State University
Blacksburg, VA 24061

Gerald L. Storm
Pennsylvania Cooperative Fish and Wildlife Research
   Unit
Pennsylvania State University
University Park, PA 16802

Thomas C. Tacha
Caesar Kleberg Institute
Texas A&I University
Kingsville, TX 78363

Stanley A. Temple
Department of Wildlife Ecology
University of Wisconsin
Madison, WI 53706

Tom Thorne
Wyoming Game and Fish Department
Research Laboratory
Laramie, WY 82071

Nancy G. Tilghman
U.S. Forest Service
Ashville, NC 28802

Robert M. Timm
Hopland Field Station
University of California
Hopland, CA 95449

Robert E. Trost
U.S. Fish and Wildlife Service
Office of Migratory Bird Management
Washington, DC 20240

William J. Vander Zouwen
Wisconsin Department of Natural Resources
Madison, WI 53707

Michael R. Vaughan
Virginia Cooperative Fish and Wildlife Research Unit
Virginia Polytechnic Institute and State University
Blacksburg, VA 24061

Paul A. Vohs, Jr.
U.S. Fish and Wildlife Service
Office of Information Transfer
Fort Collins, CO 80525

James S. Wakeley
U.S. Army Corps of Engineers
Waterways Experiment Station Environmental
   Laboratory
Vicksburg, MS 39180

Robert J. Warren
School of Forest Resources
University of Georgia
Athens, GA 30602

Harmon P. Weeks, Jr.
Department of Forestry and Natural Resources
Purdue University
Lafayette, IN 47907

Ernie P. Wiggers
School of Natural Resources
University of Missouri
Columbia, MO 65211

Elizabeth S. Williams
Department of Veterinary Science
University of Wyoming
Laramie, WY 82070

John C. Wingfield
Department of Zoology
University of Washington
Seattle, WA 98195

Carle W. Wolfe, Jr.
Nebraska Game and Parks Commission
Lincoln, NE 68503

Vernon Wright
School of Forestry, Wildlife, and Fisheries
Louisiana State University
Baton Rouge, LA 70803

1

RESEARCH AND EXPERIMENTAL DESIGN

John T. Ratti and Edward O. Garton

INTRODUCTION

Wildlife biologists have a tremendous challenge and responsibility inherent in their profession. Wildlife habitat and populations are being lost at alarming rates, and the wildlife biologist will play an increasingly important role in the future of our wildlife resources. "To get some idea of the current rate of species extinction, consider that in one 3,000-year period of the Pleistocene during which great numbers of organisms perished, North America lost about 50 mammalian species and 40 birds—or about three species per hundred years. By way of contrast, since the arrival of the Puritans at Plymouth Rock in 1620, more than 500 species and subspecies of native animals and plants have become extinct. Another 170 U.S animals are today designated 'Endangered' by the U.S. Department of the Interior, and 1,867 plants have been proposed for endangered status. An additional 430 foreign animals receive protection from the Department of Interior's Endangered Species Program'' (Opler 1977:30). Western and Pearl (1989) estimated that 15–25% of *all* species may become extinct in response to human activity by the turn of the century. Most of the recent species loss is directly related to habitat loss or environmental degradation. Thus, our wildlife populations will be concentrated into fewer and smaller patches of habitat.

Many other problems will require skilled wildlife management. Nearly all populations subjected to hunting require research to assist proper management. Some popu-

lations are too abundant and require control, such as deer and elk in some parks. And, as world resources dwindle from greater use by higher human populations, utilization of some wildlife species for food and byproducts (e.g., hides) may become more important. To respond to these and other wildlife problems, management programs must be based on results of quality scientific investigations that produce objective, relevant information—and quality *science* is dependent upon carefully designed research.

Emergence of Rigor in Wildlife Science

Wildlife *science* is a term the wildlife profession has only recently nurtured. Our profession of wildlife conservation and management was built on natural history observations and conclusions from associations of wildlife population changes with environmental factors such as weather, habitat loss, or harvest. Thus, we have a long tradition of wildlife management based on "laws of association" rather than on experimental tests of specific hypotheses (Romesburg 1981).

Although Romesburg (1981) and others have been critical of wildlife research and resulting management, the wildlife biologist is confronted with a tremendous array of uncontrollable natural variation that might bias results and conclusions of an investigation. Scientists conducting experiments in the fields of physics and chemistry have the ability to *control* variables associated with an experiment, and they can repeat these experiments under the exact same conditions to confirm their results. They also have the ability to systematically alter the nature or form of specific variables to determine *cause and effect.*

The wildlife scientist, on the other hand, often conducts investigations in natural environments over large geographic areas. It is usually difficult to observe and accurately count or estimate density of the organism in question, and factors such as weather, habitat, predators, and competition change from one location to another, from one season to the next, and between years. Thus, rigorous scientific investigation in wildlife ecology is challenging and requires careful design. An early step that moved wildlife investigators from strictly descriptive natural-history studies (of the 1930s through 1950s) toward rigorous examination of processes was the application of statistics to wildlife data. In the past 30 years the profession has evolved from testing differences among sample means with simple *t*-tests to complex computer models and multivariate analyses. The application of statistical methods in wildlife research has made a rapid transition, and we now see the most sophisticated and modern statistical procedures used on data published in our journals.

However, the transition to thorough scientific inquiry has not been as rapid. As we will see in this chapter, scientific inquiry involves a systematic series of steps, and much of our wildlife research has taken only the initial steps in the process. Our scientific methods have been incomplete. Realization of this problem became more prominent after the publication of ''Wildlife science: gaining reliable knowledge,'' by Romesburg (1981). This publication will remain as a landmark in the transition from wildlife management with ''unreliable knowledge'' (Romesburg 1981:293) to management based on a series of carefully tested hypotheses and conclusions from sound scientific inquiry.

Experimental Versus Descriptive Research

To comprehend the process of wildlife research, we must understand the difference between and value of both descriptive and experimental investigations. Most wildlife research to date has been descriptive. Experimental research is the most powerful form of research, and it should be used much more in wildlife studies. However, in our quest for more critical science, we must not abandon descriptive natural-history studies. Descriptive research is usually an initial and essential phase of wildlife science, and it can produce answers to several important questions. Descriptive research often involves broad objectives rather than tests of specific hypotheses. For example, we might have a goal to describe and analyze gray partridge ecology. To do so, we might measure characteristics of nesting habitat, clutch size, hatching success, brood use of habitat, food habits, harvest rates, use of winter habitat, predation, and other mortality factors. From this information, we will learn a great deal about partridge biology that will be useful in our understanding and management of the species. However, we must accept these data for what they represent—descriptive observations with many limitations. If we observe that 90% of gray partridge nests are in habitat ''A,'' 10% in habitat ''B,'' and none in ''C'' and ''D,'' we are tempted to manage for habitat ''A'' to increase nesting density of partridge. However, many alternatives must be investigated. Possibly habitat ''A'' is the best *available* habitat, but partridge experience high nest mortality in this type. Maybe habitat ''X'' would be the best habitat for nesting, but it is simply not available on our study area. What habitat types are used by gray partridge in other regions? How do nest success and predation differ among regions and habitats? With answers to these questions we can begin to see that defining quality nesting habitat is complex. Nest success may be related not only to a specific habitat type, but to the spatial and proportional distribution of habitat types, the species of predators present, partridge density, and climatic conditions.

From these descriptive studies we might gain enough information to develop a general research hypothesis, or conceptual model, that attempts to explain the relationship between habitat and nesting success of partridge. Such descriptive models are very general, but they help us make specific predictions that can be tested to determine authenticity of the model. These predictions can be stated as *hypotheses.* We may define an experiment as a test made to examine the validity of a hypothesis. Although details of this process will be discussed, consider again the partridge study. Assume we have made a prediction about partridge nesting habitat and formulated the following hypothesis: gray partridge nesting density and nest success are higher in agricultural areas dominated (e.g., >75% of the available habitat) by pasture than in areas dominated by cultivated fields. To test this hypothesis we must establish a series of *control* study plots and *experimental* study plots. Our control plots will be randomly chosen from large blocks of cultivated land where agricultural practices have not changed in recent years and will not change throughout the duration of the study. On the experimental plots (established randomly within the same region as our control plots), the cultivated fields will be planted to pasture grass to test the validity of our hypothesis and predictions regarding the effect of habitat on partridge nesting. This process is difficult, because it requires large blocks of habitat, cooperation from landowners, several years to establish pasture grass on the experimental plots, and additional years of study to determine the response of birds to vegetative changes. At some point, the comparison between control plots and experimental plots will provide a basis to reject or fail to reject the hypothesis.

In summary, the wildlife profession has gained a tremendous body of information from 50 years of detailed descriptive studies of wildlife populations. Most of the management programs in place in North America are based on these data. However, advances in our ability to manage populations and to predict factors that cause population change have slowed tremendously in recent years. If our profession is to continue to advance with *management based on science,* we must use carefully designed experimental research to examine the many untested hypotheses that have evolved from descriptive studies. It is crucial that we know how to formulate testable questions and the methods or techniques that are available to examine these questions. We need to understand enough statistics to communicate with statisticians, and if a great loss of time and money is to be avoided, these communications must occur during the design phase prior to data collection.

SCIENTIFIC METHOD

One of the greatest contributions a biologist can make to the wildlife profession is advancement of *knowledge*. Kerlinger (1973) discussed four basic methods of knowing: (1) method of tenacity, (2) method of authority, (3) a priori method, and (4) method of science. We see the *method of tenacity* in practice when managers hold tightly to traditional beliefs even when there is no scientific basis to support these beliefs. The *method of authority* is practiced in a practical way when we contact an expert in a particular subject area for advice. The *a priori method* is used widely in the development of theory, especially quantitative theory, wherein a series of assumptions is made and logic, mathematics, or simulations are used to determine the consequences of these assumptions. The *method of science* is a circular process in which previous information is synthesized into a theory, the theory is stated explicitly in the form of hypotheses, predictions are deduced from these hypotheses, the predictions are tested through experimentation or observation, the theory is modified or expanded on the basis of the results of these tests, and the process starts again. Kerlinger (1973:6) noted that ''the scientific approach has one characteristic that no other method of attaining knowledge has: self-correction. There are built-in checks all along the way to scientific knowledge.''

One of the earliest papers published on the scientific method was by Chamberlin in *Science* in 1890 (reprinted in 1965). This method is commonly referred to as the hypothetico-deductive method and was formalized in classic contributions by Popper (1959, 1968). Platt (1964) reemphasized the importance of multiple hypotheses and proposed a systematic pattern of inquiry, referred to as strong inference, in which the investigator devises alternate hypotheses, develops an experimental design to reject as many hypotheses as possible, conducts the experiment to achieve clean results, and repeats the procedure on the remaining hypotheses. Other major works that provide highly detailed discussions of the scientific method include those of Dewey (1938), Bunge (1967), and Newton-Smith (1981).

The most widespread applications of the hypothetico-deductive method have come in the hard sciences (e.g., physics, chemistry, and physiology), in which experiments are relatively easy to conduct. The classic methods of natural history observation in wildlife biology and other natural sciences are expanding to include experimentation and hypothesis testing. James and McCulloch (1985:1) described this transition for avian biologists: ''traditional ornithologists accumulated facts but did not make generalizations or formulate causal hypotheses . . . modern ornithologists formulate hypotheses, make predictions, check the predictions with new data sets, perform experiments, and do statistical tests.'' This statement is equally applicable to wildlife research. In addition to James and McCulloch (1985), other excellent review papers on research design were written recently by Romesburg (1981), Quinn and Dunham (1983), Diamond (1986), Eberhardt and Thomas (1991), Murphy and Noon (1991), and Sinclair (1991). These papers should be read by any person engaged in professional wildlife research.

A systematic outline of the scientific method is pre-

Table 1. Systematic outline of sequential events in scientific research.

1. Identify the research problem.
2. Conduct a literature review of relevant topics.
3. Identify broad and basic research objectives.
4. Collect preliminary observations and data as necessary.
5. Conduct exploratory data analysis.
6. Formulate a research hypothesis (conceptual model).
7. Formulate predictions as testable hypotheses.
8. Design research and methodology for each hypothesis with assistance from a statistical consultant.
9. Prepare a written research proposal that reviews the problem, objectives, hypotheses, methodology, and procedures for data analysis.
10. Obtain peer review of the research proposal from experts on the research topic and revise if necessary.
11. Perform experiments or collect data.
12. Conduct data analysis.
13. Evaluate, interpret, and draw conclusions from the data.
14. Speculate on results and formulate new hypotheses.
15. Submit manuscript describing the research for peer-reviewed journal publication, agency publication, and/or presentation at scientific meetings.
16. Repeat the process with new hypotheses (starting at step 6 or 7).

sented in Table 1, and it provides an idealized but useful description of the process. The first step is a clear statement of the problem, which guides us in a careful review of literature on the topic and preliminary observations or data collection. The literature and preliminary data should be evaluated and synthesized by exploratory data analysis (Tukey 1977). These initial steps fall into the category of descriptive research. After careful consideration of this information, a conceptual model (or general research hypothesis) of the problem is developed. This conceptual model is essentially a broad theory that offers explanations and possible solutions to the problem and places the problem in a broader context. The next step is to develop predictions from the conceptual model, i.e., statements that would be true if the conceptual model were true. These predictions are then stated as testable hypotheses. The next step is to design research to test these hypotheses; ideally experimentation should be used whenever possible. The design should be reviewed by peers and by a statistician before data collection begins. Data analysis with appropriate statistical procedures leads to rejection of, or failure to reject, hypotheses. Evaluation and interpretation involve criticizing the actual conduct of the experiment or data-collection process and interpreting the meaning of the results. Final conclusions usually result in further speculation, modification of the original conceptual model and hypotheses, and formulation of new hypotheses. The publication process is the last, but essential, step, and peer-review comments should be considered carefully before research on new hypotheses is designed. These steps are discussed in detail later.

Problem Identification

The initial step in most wildlife research is problem identification. Most research falls into one of two categories, applied or basic. In wildlife science, applied research usually is related to a management problem, e.g., proper habitat management or protection for a certain species will require research on habitat requirements. Some biological problems that require research stem from political controversy or public demand. For example, we may study deer populations simply because the hunting public has demanded greater hunting success, but the biologist may not believe that a problem exists with the herd or that research is needed. However, research results should assist response to public concerns. Other applied studies may be politically supported due to projected loss of habitat by development or concerns over environmental problems such as contamination from agricultural chemicals. Unfortunately, few wildlife research projects are more basic in origin. We seldom have the luxury of studying wildlife populations to gain knowledge simply for the sake of knowledge and a more complete understanding of factors that affect behavior, reproduction, density, competition, mortality, habitat use, and population fluctuations. However, once there is political support for research on species of concern (e.g., declining raptor populations), we may enter a period when research funding is relatively abundant and allows for very basic studies to help solve a specific problem.

Most research should begin, after a thorough literature review, with a descriptive phase that has broad objectives. This phase is omitted only if much descriptive work already has been completed on the specific problem. Descriptive phases often include data collection on animal density (or relative density), habitat use, mortality factors, reproduction, behavior, population fluctuations, and environmental factors such as weather. In addition, political and social aspects of the problem should be investigated. An important aspect of this phase is exploratory data analysis (Tukey 1977, James and McCulloch 1985). During this process data are quantitatively analyzed in terms of means, medians, modes, standard deviations, and data distributions. Exploration of the data should be as complete and biologically meaningful as possible, which often includes comparison of data categories (e.g., mean values, proportions, ratios), multivariate analysis, correlation analysis, and regression. The "basic aim of exploratory data analysis is to look at patterns to see what the data indicate" (James and McCulloch 1985:21).

Hypothesis Formulation

Often data reveal patterns of association, or what Romesburg (1981) termed "laws of association" from the application of inductive reasoning (i.e., reasoning from the particular to the general). For example, if we studied pheasant brood production for 10 consecutive years and observed high brood production during dry years and few broods during years with above-average spring rain, we are tempted to make a direct association between brood survival and spring rainfall. Unfortunately, as noted by Romesburg (1981), most of the accepted principles of wildlife management are based on conclusions from this type of association. Such conclusions are termed "unreliable knowledge" (Romesburg 1981). They are unreliable because we have made premature judgments and have ignored many alternative explanations (i.e., correlation does not imply causality). For example, rainfall might

have had no effect on brood survival and the observed pattern of association was simply a coincidence. A second alternative explanation might be that spring rainfall altered the timing of hay mowing by farmers, which had a direct effect on nest destruction by farm machinery.

Exploratory data analysis and laws of association are most valuable when they lead to formulation of research hypotheses. Hypotheses are simply assertions subject to verification. Normally, our research hypothesis is what we initially consider to be the most likely explanation. However, we usually do not consider one hypothesis, but several alternate hypotheses that provide possible explanations or reasons for facts observed. For example, our primary research hypothesis might be:

H_1: Above-average spring rainfall reduces survival of ring-necked pheasant broods.

Alternative research hypotheses might be:

A_1: Above-average spring rainfall reduces egg viability of ring-necked pheasants.

A_2: Destruction of ring-necked pheasant nests by farm machinery is greatest during years of above-average spring rainfall.

A_3: The proportion of adult female ring-necked pheasants that nest is reduced by above-average spring rainfall.

A_4: Above-average spring rainfall results in above-average vegetative cover, which reduces the observability of ring-necked pheasant broods.

Note that hypothesis A_3 does not involve egg or brood mortality, but the number of birds that actually nest, which would affect the number of broods observed during our descriptive research phase. Hypothesis A_4 will explore the possibility that rainfall simply affects our ability to see birds during brood surveys. Other reasonable and testable hypotheses could be added to this list. Research designed to test our hypotheses will help us determine those factors related to brood production and production estimates and to explain *why* brood production varies.

From general research hypotheses (often called conceptual models), we usually formulate statistical hypotheses. The main difference is that research hypotheses represent theories, but statistical hypotheses represent predictions from theories (Dolby 1982, James and McCulloch 1985). An example of this difference was provided by James and McCulloch (1985:18):

"Suppose the model [research hypothesis] is that large harem size in red-winged blackbirds causes reproductive success to be higher. A prediction from this model is that the reproductive success of males with larger harems will be higher than the reproductive success of males with smaller harems. This statement can be translated into a statistical hypothesis if the hypothesis is stated as follows: The mean reproductive success (fledglings/season) of male red-winged blackbirds is a strictly increasing function of harem size."

If data were collected that supported our statistical hypothesis, we cannot conclude that our model is true, but only that it has not been rejected (James and McCulloch 1985). The central issue here is that we never *prove* a research hypothesis or theory to be correct, but the credibility of the hypothesis increases as more of its predictions are supported and alternate hypotheses are rejected.

We take an important step from descriptive natural history to science when we formulate hypotheses. Interpretation of exploratory data analysis and formulation of testable hypotheses are difficult aspects of science that require creativity, but they are essential to the future of wildlife science.

Hypothesis Testing and Data Analysis

We have combined our discussion of hypothesis testing with data analysis to emphasize that the research biologist should never proceed with data collection without careful consideration of how the data will be treated statistically. Sometimes only minor changes in methodology might mean the difference between obtaining a data set that provides relevant information and one that defies proper analysis and interpretation. Unfortunately, many hundreds of thousands of dollars have been wasted on poorly planned wildlife projects that resulted in relatively useless data sets.

Many different research options are available for the wildlife biologist to use to test statistical hypotheses (Fig. 1; also see Eberhardt and Thomas 1991). These research options differ dramatically in terms of two criteria: how certain are the conclusions reached and how widely applicable are the conclusions? No single option is perfect. The biologist must weigh the options carefully to find the best choice that fits within the constraints of time and resources.

Experiments consisting of manipulative trials are an underutilized approach in wildlife science. Laboratory experiments, in which most extraneous factors are controlled, provide the cleanest results with the most certainty, but the results generally have only narrow application to free-ranging wildlife populations (Fig. 1). By contrast, natural experiments (Diamond 1986), in which large-scale perturbations such as wildfires, disease outbreaks, and hurricanes manipulate populations naturally, provide only weak conclusions because of the lack of replication and our inability to control extrinsic factors through random assignment of treatments. Nevertheless, the results of natural experiments are applicable to a variety of populations. Field experiments, in which manipulative treatments are applied in the field, combine some of the advantages of laboratory and natural experiments (Fig. 1). Field experiments span a range (A in Fig. 1) from pseudoreplicated field experiments in which no true replication is possible and conclusions are not certain, to replicated field experiments in which conclusions are certain. Such field experiments provide conclusions that are broadly applicable to free-ranging wildlife populations.

Most wildlife studies fall somewhere along the continuum of field studies labeled B in Fig. 1. As we have seen before, case studies consisting of unreplicated natural-history descriptions are most useful at early stages in the development of the research process. At the other extreme of continuum B are replicated field studies wherein no manipulation or randomization of treatments occurs, but true replication is used and information is gathered to evaluate alternate hypotheses. Conclusions from such replicated field studies are broadly applicable, but the conclusions are always less certain than conclusions from rep-

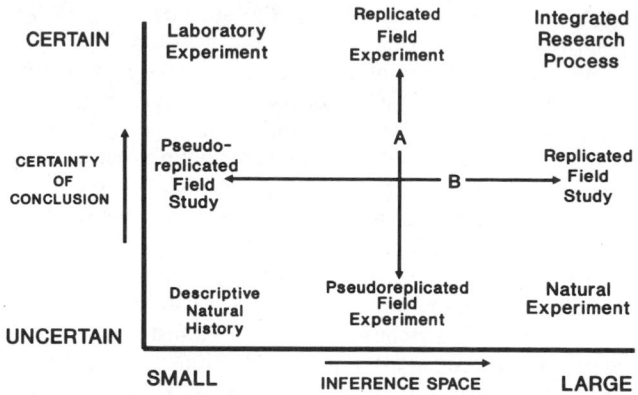

Fig. 1. The potential for wildlife study designs to produce conclusions with high certainty (few alternate hypotheses likely) and widespread applicability (a diversity of research populations where inferences apply).

licated field experiments. Some questions of importance in wildlife biology and management are not appropriate for experimentation. For example, we may be interested in the effects of weather on a particular animal population, but we cannot manipulate the weather. In addition, we may simply be interested in the *relative importance* of certain factors such as predation, habitat, and food limitations (Quinn and Dunham 1983).

Designing good field studies is more difficult than designing good experiments because of the potential for extraneous factors to invalidate the tests. One of the key steps for both experiments and field studies is designing a sampling procedure to draw observations (experimental units or sample units) from the research populations of interest. Only if this is done properly can the conclusions of the tests be applied to these research populations. The statistical specialty known as survey sampling (Cochran 1963) provides methods that are helpful in designing such sampling procedures. These methods are particularly important for field studies, but they are also useful in field experiments for drawing experimental units and subsamples (samples within one experimental unit). Survey sampling and experimentation are discussed in some detail later.

Once we have settled on a research option for each hypothesis, the actual testing process requires careful planning. For each hypothesis we must determine exactly what data will be collected, when, how, how much, and for how long. How will these data be treated statistically? Will the data meet the assumptions of the statistical test? Is the sample size adequate? Will the statistical hypothesis provide information directly related to the research hypothesis? Do biases exist in the data collection, research design, or data analysis that might lead to a spurious conclusion? These questions must be considered carefully for *each* hypothesis before fieldwork begins. Consulting a statistician is important, and the statistician should understand the basic biological problem, the overall objectives, and the research hypotheses. The biologist is cautioned not simply to ask, "How will I analyze these data?" This quick approach might lead to misunderstanding between the biologist and the statistician and, ultimately, a statistical test that does not address the specific research hypothesis.

We encourage one additional step at this point before data collection begins: obtain peer review of your proposed research from several people with expertise and experience with your research topic. Although many biologists are defensive of their work and find this critique process difficult, peer review will almost always improve a research design, and sometimes it will disclose serious problems that can be solved during the planning stage. This type of review process is highly constructive because it is a form of preventive medicine. Unfortunately, most peer review occurs after data collection when the final report or publication manuscript is written. At this point the research biologist has few, if any, corrective options. Hypothesis-testing options are discussed in more detail on p. 16.

Data Collection

After the researcher has identified specific objectives, statistical hypotheses, and methodology (i.e., research design, field procedures, and data analysis), data collection begins. Often the researcher becomes an organizational manager during this phase. Most research projects have limitations associated with budgets, the number of personnel, time required for travel and actual data collection, and lost field days caused by poor weather, illness, or accidents. These factors need to be monitored carefully to be sure that data collection proceeds on schedule and according to design. All data should be recorded on pre-printed data sheets. This approach ensures that each field person collects exactly the same data, and consistent organization of data simplifies analysis after data collection is complete. Data sheets should be duplicated after each field day (e.g., computer entry, photocopies, or transcribed) and stored in a separate location from the original data set. Any transcription of data (including computer data entry) must be followed by careful proofreading. All field personnel should receive careful instructions regarding data collection, and the principal researcher must check periodically to see that each person has similar skills and uses the same methods for observation, measurement, and recording (e.g., Kepler and Scott 1981). With regard to these factors, the principal researcher has responsibility for quality control. It has often been said that a chain is only as strong as its weakest link. In a similar fashion, the validity of research results is directly related to the quality of research design and data collection.

We offer one note of caution regarding data collection. Most novice research biologists are anxious to initiate data collection because of the attractiveness of working out-of-doors and the pleasure derived from observing wildlife-related phenomena. Avoid rushing the design phase simply to initiate fieldwork more quickly. Many successful research biologists spend about 40% of their time in design and planning phases, 20% in actual fieldwork, and 40% in data analysis and writing publications. Veteran research biologists can attest that data collection can be physically difficult, boring, and highly repetitive. Often the enjoyable and rewarding portion of research comes during the data-analysis phase when the biologist begins to see results from several years of planning and fieldwork.

Evaluation and Interpretation

As noted above, hypothesis formulation is a difficult and creative aspect of science compared to the mechanical and repetitive nature of data collection. In a similar fashion, evaluation and interpretation is also a creative phase, and the quality of conclusions drawn is dependent upon the biologist's past educational and professional experience as well as his or her willingness to consider standard as well as less-traditional interpretations. One great danger in wildlife science (and other fields) is that the researcher often has a conscious or unconscious expectation of results from the data. This bias might begin with development of the overall research objective and carry through to the interpretation phase. We must be careful not to bias field assistants with an expectation of results. This danger is so great that in some fields, such as medicine, experiments are performed with a double-blind approach in which neither the principal researcher nor assistants know the treatment and nontreatment groups. A good scientist must not design research or interpret data in a fashion that is more likely to support *preconceived* explanations of biological systems. This potential problem cannot be over-emphasized. Biologists who are consciously aware of their own biases and strive to keep an open mind to new ideas are most likely to make revolutionary discoveries.

The major objective of this phase is to organize clearly and concisely the results of the data collection, all of the exploratory data analysis, and results of specific statistical analyses. These results must be transformed from a collection of specific information into a synthesis explaining the biological system. Do specific statistical tests support one or more of the research hypotheses? Do the results provide a reasonable explanation of the biological system? Are there alternative explanations of the data and statistical tests? Are there specific problems with the data that should be identified, such as inadequate sample sizes or unusual variation in specific parameter measurements? What could have introduced bias into the estimates? Are additional data required?

During this phase, the biologist usually reaches some conclusions based on the data and results of statistical tests. The goal of science is knowledge, and in wildlife science we attempt to explain processes within biological systems and to predict how changes will affect specific wildlife populations. One relatively common problem is that conclusions often go beyond the data. Interpretation of research data must clearly separate conclusions based on data and *speculation*. For example, if we demonstrate that droppings from a forest grouse species are most abundant under lodgepole pine and Engelmann spruce, we can conclude that grouse use both tree species for some forms of behavior, but the type of behavior (e.g., roosting or feeding) is speculation without additional data (e.g., observations of feeding activity, and crop or fecal analyses).

Speculation and New Hypotheses

Rarely does a given research project provide the last word on any problem. More commonly, good research will generate more questions than it answers. Speculation, or inference based on inconclusive or incomplete evidence, is one of the most important aspects of science. As noted above, speculation must be identified as such and should not be confused with conclusions from data. But speculation is the fuel for future research. Many facts of nature have been discovered by accident—an unexpected result from some associated form of research. However, most research is directional, i.e., it attempts to support or falsify a theory reached by speculating from facts. Thus, the process of gaining knowledge is like construction of a building with bricks, and speculation is the first step to adding the next brick.

New hypotheses are basically a form of speculation, but they are verbalized in a more formal fashion and have a specific, testable format. For example, if we again consider our observations of forest grouse, we can formulate a basically untestable hypothesis that "Forest grouse have evolved a preference for use of lodgepole pine and Engelmann spruce trees." The statement is simply too vague and requires historical data that can never be collected. However, we can hypothesize that (1) forest grouse use lodgepole pine and Engelmann spruce trees for feeding, and (2) forest grouse use lodgepole pine and Engelmann spruce trees for roosting. From results designed to test these hypotheses we might learn that 80% of the forest grouse diet is lodgepole pine even though Engelmann spruce is more abundant. From these results we may hypothesize that "Needles from lodgepole pine have higher nutritional quality than needles from Engelmann spruce"—and the process goes on.

Publication

The final step of the scientific method is publication of the work. Unfortunately, many research dollars are wasted because the knowledge gained was not published, and the information is lost or hidden in file cabinets or boxes of data sheets.

The publication process is the most difficult phase for many biologists. Clear, concise scientific writing is tedious because most biologists have little formal training and inclination in that field. And peer review is often damaging to a person's ego, because we must subject our work to anonymous critiques that are used by editors to judge whether the manuscript is acceptable for publication.

In addition, agency administrators often do not encourage or reward employees for publishing their work, and in some instances it is discouraged. Some administrators are fairly shortsighted; they want an answer to an immediate problem and do not see the need for additional use of personnel time on the publication process. What these administrators fail to recognize is that peer review and publication will (1) correct errors and possibly lead to a better analysis, (2) help authors reach the most sound conclusions from their data, (3) help their personnel grow as scientists by responding to critical comments and careful consideration of past errors (that may have been overlooked without peer review), and (4) make a permanent contribution to wildlife management by placing results in a literature format that will be available to other agencies, researchers, and students.

The publication process is essential to science. Peer reviews almost always improve the quality of a manuscript, but some research is simply not suitable for publication.

On this latter point we emphasize the importance of the various topics discussed in this chapter. Rarely would any research effort that is properly planned, designed, and executed (including a well-written manuscript) be unpublishable. Simply stated, research is not complete and does not make a contribution to knowledge and the sound management of wildlife resources until the results are published in a fashion that effectively communicates to the scientific community and user groups, e.g., wildlife managers.

MAJOR COMPONENTS OF RESEARCH
Populations

Mayr (1970:424) defined populations as a group "of potentially interbreeding individuals at a given locality," and species as "a reproductively isolated aggregate of interbreeding populations." Thus, it is important to understand that a species is usually made up of many populations, and, conversely, a population is only one segment of a species. The relationships among species, subspecies, and populations are important to wildlife science, and the research biologist is encouraged to fully understand these concepts (for reviews see Mayr 1970, Selander 1971, Stebbins 1971, and Ratti 1980).

The wildlife profession usually deals with three types of populations: the biological population, the political population, and the research population. The biological population, as defined above, is an aggregation of individuals of the same species that occupy a specific locality, and often the boundaries can be described with accuracy. For example, the dusky Canada goose population breeds within a relatively small area on the Copper River Delta of Alaska and winters in the Willamette Valley near Corvallis, Oregon (Chapman et al. 1969). Between the breeding and wintering grounds of the dusky is the more restricted range of the relatively nonmigratory Vancouver Canada goose (Ratti and Timm 1979). Although these two populations are contiguous with no physical barriers between their boundaries, they remain reproductively isolated and independent. The Yellowstone National Park elk herds are additional examples of biological populations whose population boundaries have been described in detail (Houston 1982).

The political population has artificial constraints of political boundaries, often county, state, or international boundaries. For example, a white-tailed deer population within an intensively farmed agricultural region in the Midwest might be closely associated with a river drainage system due to the permanent riparian cover and food critical for winter survival when agricultural fields are void of vegetation. The biological population extends the entire length of the river drainage, and research and management should consider this population in its entirety. However, if the river flows through two states, the biological population is usually split into two political populations that are subjected to different management strategies and harvest regulations. Traditionally, this has been a common wildlife management problem. However, there has been more cooperation among adjacent governmental units with such situations in recent years. The research biologist should study such populations only if sound cooperation and compatible management programs exist between adjacent-administrative units. In addition, these situations should encourage cooperative studies, in which research personnel and funding resources can be pooled to benefit both agencies.

The most important population to our discussion here is the research population. We noted that biological populations are only a segment of a species; in a similar fashion, research populations are usually only a segment of a biological population. From this segment we take a sample. For this reason, the research population is commonly referred to as the *sample frame* (Scheaffer et al. 1986). In rare instances we might study an entire population that represents a species, e.g., endangered species with few individuals, such as whooping cranes. Or, our research population might represent an entire biological population, such as one of the Yellowstone elk herds. However, our research population usually is only a portion of the biological population. These restrictions are based on financial constraints, logistic problems, or our inability to obtain data from most individuals in the biological population. Thus, our sampling methodology is critical, for it provides the only link from our samples to the research population.

One important point is that conclusions from research are directly applicable only to the population from which the sample was drawn—the research population. However, research biologists usually have goals to obtain knowledge and solve problems regarding biological populations and species. The key questions that must be answered in this regard are: (1) Is the sample representative of the research population? (2) Is the research population representative of the biological population? and (3) If so, is the biological population representative of the species? These are difficult questions, and a positive answer should be reached with extreme caution (especially for #2 and #3). Variation of traits among segments of biological populations and among populations of a species are well documented. Thus, broad conclusions relative to a research hypothesis should be avoided until several projects from different populations and geographic locations provide similar results.

Experimental Versus Nonexperimental Hypotheses

We noted that descriptive natural-history studies are important for gathering baseline information and formulating research hypotheses. Statistical hypotheses test our predictions from research hypotheses, and we can examine the validity of these predictions in two fashions—through observations of natural events and through observations of the results of experimental manipulation, or *treatment*. To emphasize the difference, we will again use a hypothetical example.

Suppose we had a series of elk-feeding observations during summer that indicated elk consumed red-osier dogwood more than mountain-mahogany. Assume also that we had conducted analyses of feces that confirmed our field observations. We also measured food availability and observed that dogwood and mountain-mahogany were equally abundant in our study environment. From these observations we formulate a research hypothesis: "Elk select forage based on nutritional quality." From this hypothesis and our observations, we predict that "dogwood

has more digestible protein than mountain-mahogany," and we formulate a nonexperimental statistical null hypothesis: "Red-osier dogwood and mountain-mahogany have equal amounts of digestible protein." This is a nonexperimental hypothesis because we have provided no experimental manipulation or treatment. Note that the null hypothesis is simply a restatement of the statistical hypothesis such that there *is no difference.* If results of a laboratory analysis indicated greater protein levels in either species, we reject our statistical hypothesis.

Let us assume that dogwood had greater protein levels than mountain-mahogany. Then our research hypothesis is supported, but many alternative hypotheses have not been rejected. To build a convincing case for our research hypothesis, we must reject these alternate explanations (e.g., that elk select on the basis of color or some other factor unrelated to nutritional quality) and complete a causal link between nutritional quality and selection. For example, we must show (i.e., test the hypothesis) that "Elk are able to detect protein quality of forage." This could be the subject of experimentation, i.e., we can add experimental manipulation to our research design. Our experiment might involve feeding trials with 60 captive elk. First we will grind samples of dogwood and mountain-mahogany collected from our original study environment so that the texture and appearance of the forage are relatively similar. Each of our 60 captive elk will be subjected, simultaneously and with identical methodology, to feeding trials for 30 consecutive days. During these trials consumption of forage will be measured accurately. Our study will have three trials with 20 animals randomly assigned to each trial group. In trial I, we will provide each forage separately in feeding troughs with known quantities of dogwood and mountain-mahogany. This trial will serve as a *control* (see Controls [p. 11]). The two additional trials will be experimental and will be designed to test the predictive validity of our hypothesis. In trial II we again will provide both species of forage, but we will alter mountain-mahogany by increasing the protein level so that it is equal to the protein level of dogwood. In trial III, we again will provide both forage species, but we will increase the protein level of mountain-mahogany so that it is greater than that of dogwood. If our research hypothesis and resulting predictions are correct, we would expect the 20 animals in trial I to consume greater amounts of dogwood; we expect equal consumption of both forage species by the 20 experimental animals in trial II; and we expect greater consumption of mountain-mahogany by the 20 experimental animals in trial III. If results of all three trials are consistent with our predictions, we have fairly convincing support for our research hypothesis. However, we must be willing to critique our experimental design and consider alternate explanations. Might our protein additive have changed palatability or provided some additional nutrient to our treatment forage? Might animals have selected for these hidden changes rather than the changes in protein levels? If we answer yes to these questions, additional experimentation will be needed.

Pilot Study

A pilot study is simply a preliminary, short-term trial run through *all* phases of a research project. Pilot studies are an important but often neglected step in the research process. Information can be obtained that will help the researcher avoid potentially disastrous problems during or after the formal research phase. Benefits of a pilot study include:

1. *Cost Estimates.*—One necessity for good research planning is obtaining cost estimates, both direct-dollar costs for expenditures such as equipment and travel and indirect costs such as time commitments by personnel. Obtaining research grants and general funding for research is always a problem, and research biologists are constantly confronted with necessary compromises between what idealistically should be done and what is affordable under constraints of the research budget. Thus, accurate cost estimates are critical. Pilot studies often will disclose hidden costs or identify costs that were over- or underestimated.

2. *Methodology Problems.*—Research methodology will be specific to objectives or hypothesis tests. Often methods designed and planned at the desk simply do not work well when implemented. Unexpected discovery of poor methods during the formal research phase could cause complete loss of data collected for an entire sampling period. A pilot study also might reveal basic logistic problems, e.g., travel time among study plots might have been underestimated, or expectations for overall sample sizes might not be feasible without additional personnel and funding.

3. *Variance Estimates.*—Proper sample size is essential to statistically test hypotheses. Statistical procedures for estimating needed sample sizes require variance estimates of variables that will be measured, and these estimates often are available only from data gathered in a pilot study. These preliminary data might disclose that the variance of the population is large, and obtaining adequate sample sizes may be impossible. Although this might require complete reconsideration of the research design, it is far better to discover such problems before time, energy, personnel, and critical research dollars are committed to a formal research project.

If the research is part of an ongoing project, or if much research on the topic has been published, costs, methodology, and variance estimates may already be firmly established. However, going through the pilot-study steps on paper will still be beneficial because this forces one to evaluate these factors carefully.

Precision, Bias, and Accuracy

After a pilot study is completed (or estimates are obtained from other sources), how do we set the sample size for our formal study? We must decide how good our estimates need to be. But what does good mean? One measure of the quality of our estimates is their **precision.** Precision refers to the closeness to each other of repeated measurements of the same quantity (Cochran 1963, Zar 1984, Krebs 1989). Precision of an estimate is determined primarily by the variation in the population and the size of our sample. An indicator of the precision of an estimator is the confidence interval. Larger variation in the population leads to lower precision in an estimate, whereas a larger sample size produces higher precision in the estimator. Another measure of the quality of an estimator is termed **bias.** Bias describes how far the average value of the estimator is from the true population value. An

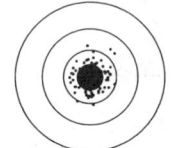 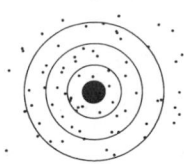

a. Unbiased and precise = accurate   b. Unbiased but not precise = not accurate

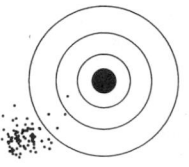 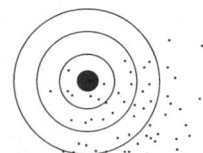

c. Biased but precise = not accurate   d. Biased and not precise = not accurate

Fig. 2. The concepts of bias, precision, and accuracy are illustrated with targets and a shot pattern (modified from Overton and Davis 1969 and White et al. 1982).

unbiased estimator centers about the true value for the population. If an estimate is both unbiased and precise, we say that it is **accurate** (defined here as an estimator with small mean-squared error, Cochran 1963). Accuracy refers to the size of deviations of the estimator from the true population value (Cochran 1963). Accuracy is the ultimate measure of the quality of an estimate. These concepts are illustrated graphically in Fig. 2 by analogy to sighting in a rifle. Let us look at their application to a typical population survey.

Suppose we were interested in estimating the density of elk on a large winter range. One approach we might take is to divide the area into a large number of count units of equal size and draw a sample of the units to survey from a helicopter. In this way we define our research population in terms of a geographic area rather than animals. The elements of our target population are count units, and we select a sample of these units using some objective sampling design. Using the helicopter we search each of the sampled units, attempting to count all of the elk present in each unit. We divide the number of elk counted in each unit by the size of that unit to obtain a density estimate for each unit. Typically we summarize the results of the sample in a frequency histogram as in Fig. 3A. The histogram suggests little variation in density on this winter range, for most of the units (80%) have densities between 1.5 and 2.3 elk/km^2. We need a single value that is representative of the entire winter range, and we choose the mean from our sample as the best estimate of the mean for the winter range. The variation from one unit to the next is small, so the mean from our sample is a fairly precise estimate. But suppose that instead of the results in Fig. 3A we had obtained the results in Fig. 3B. Now the variation from one unit to the next is great, and the sample mean is less precise and not as reliable as the previous estimate. Thus, for a given sample size, the former estimate is more precise because of less variation in the population.

Would our mean from the sample in Fig. 3A be an accurate estimate of the mean density of elk on this winter range? To answer this question, we must evaluate the bias in the estimate. If the winter range was partially forested

A

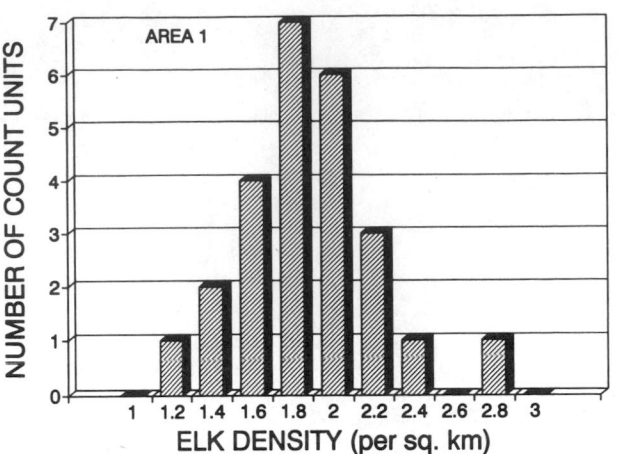

B

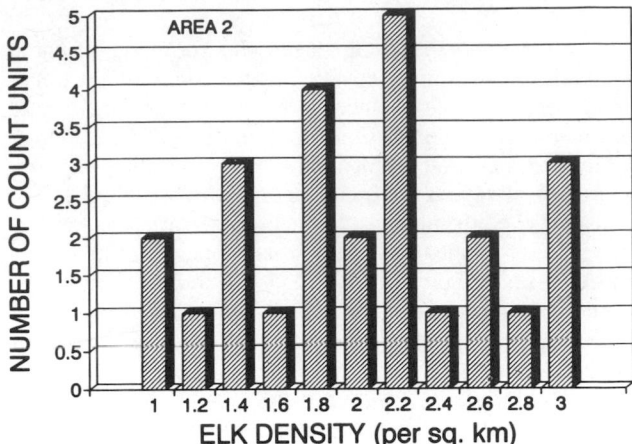

Fig. 3. Hypothetical example of elk counts and density estimates on Area 1 (A) and Area 2 (B).

or had tall brush capable of hiding elk from view, the aerial counts in each unit would be underestimates of the true number of elk present (Samuel et al. 1987). In this example the mean density from our sample would be a biased estimate of the density of elk on the winter range and therefore not highly accurate. On the other hand, if the winter range was a mixture of open brushfields and grasslands, where visibility bias would not occur, the mean density from the sample could be an accurate estimate of the density of elk on the entire winter range. We strive for accuracy in our estimates by selecting the approach with the least bias and most precision, applying a valid sampling or experimental design, and obtaining a large enough sample size to provide precise estimates.

Evaluating the bias in an estimate is difficult and usually has been done in the past on the basis of the researcher's biological knowledge and intuition. If the bias is constant, the estimate can be used to make relative comparisons and detect changes (Caughley 1977). Usually the bias is not constant, but the magnitude of the bias often can be measured so that a procedure to correct the estimates can be developed. For example, Samuel et al. (1987) measured the visibility bias in aerial surveys of elk from helicopters, and Steinhorst and Samuel (1989) developed a procedure to correct aerial surveys for this bias.

Replication

Sample size refers to the number of independent, random sample units drawn from the research population. In experiments, sample size is referred to as the number of replicates. Sample size must be distinguished from subsamples, which are the number of observations in a sampling unit. The precision of a statistic is measured by its standard error. Standard error is calculated from the variation in the original measurements and the sample size. These measurements must be true replicates, i.e., they must be independent random samples from the population. If they are not, the sample variance will underestimate the actual amount of variation in the population and the precision of our estimate will be overestimated. Let us look at an example to illustrate this point. Suppose we

wanted to evaluate the effect of prescribed fires on northern bobwhite habitat in a large valley (1,000 km^2 in size). We might conduct research on a habitat improvement project that involves burning 1 km^2 of grassland and brush (e.g., Wilson and Crawford 1979). We could place 10 permanent plots within the area to be burned and 10 more in an adjacent, unburned area. Measurements before and after the fire on the burned and unburned plots could be compared to measure the effect of the fire on bobwhite habitat. However, our 10 plots on the burned area are not really replicates but merely subsamples, or pseudoreplicates (Hurlbert 1984). In fact we have only one observation in this study because we have only one fire in one small area of our valley. We can see this by asking what would happen if we were to redesign the study to conduct 10 burns on 10 randomly chosen areas scattered throughout the valley. We would expect to see much more variation among these plots than among the 10 plots in a single burned area. The fallacy that the first design could produce is obvious. A statistical test would evaluate only whether the burned and unburned 1-km^2 areas differed and could lead to false conclusions about the effect of burning on bobwhite habitat in this valley. A more appropriate design would require randomly selecting 20 sites from throughout the valley and randomly assigning 10 of them to be burned (treatments) and 10 to be control sites. Each burned and control site would be sampled with five plots to measure bobwhite habitat before and after the treatment, and data would be analyzed by analysis of variance (ANOVA); the 20 sites are samples and the five plots per site are subsamples. In this way, the 10 sites of each type would be true replicates.

Sample Size and Power of a Test

In descriptive studies, sample size required to obtain an estimate of desired precision can be calculated after an estimate of the population variance is obtained from a pilot study or previous studies. Formulas for sample size are available for standard survey designs (e.g., Scheaffer et al. 1986).

In studies involving experiments or other types of comparisons, sample size is increased to improve the power of the test and to prevent our drawing erroneous conclusions. To help define power of a test, we consider the following example. Suppose we were using the fawn : doe ratio as an indicator of production for a mule deer herd (i.e., our research population). We want to know if the fawn : doe ratio had declined. Table 2 illustrates four possible outcomes from sampling the herd and testing for a decline in the fawn : doe ratio (the null hypothesis is that there is no change). We answer this question by comparing our test statistic to a tabled value for this statistic at our chosen level of significance (α). The level of significance represents the chance of concluding that the ratio changed when in fact it did not. An α of 0.05 means that we would make this error only five times in 100 tests. This is referred to as the Type I error. But we could make another error: we could conclude that the ratio had not changed when in fact it had declined. For the situation shown in Table 2 where we count 500 deer, we would fail to detect the decline in fawn : doe ratio 50% of the time. This second kind of error is referred to as the Type II error, and its likelihood is measured by β. When we perform a test, we typically set α low to minimize Type I errors. But Type II errors might be just as important (Alldredge and Ratti 1986), or even more important than Type I errors, as in Table 2. Obviously we want to detect a change when it occurs, and the probability of detecting a change is called the power of the test ($1-\beta$). The power of the test depends upon several factors—sample size, level of significance, variance in populations, the true change that occurred, and the efficiency of the particular test we are using. We cannot control the natural variation within the population or the actual change that occurred, but we have control of the other three factors. Parametric tests have the highest efficiency for normally distributed populations and for large samples. Nonparametric tests are superior when sample sizes are small (<30) and the populations are not normally distributed. The power of a test declines as the level of significance is made more stringent (decreasing α). In our example below (Table 2), this is a critical problem because the Type II error (failing to detect declining production) is the more serious error. It would be preferable to increase α so that the power of the test could be increased. In other instances the Type I error will be more serious and α must be kept low. Increasing sample size increases the power of the test. Calculating the sample size necessary for a desired level of power is

essential to designing a high-quality study (Toft and Shea 1983, Forbes 1990, Peterman 1990).

Controls

Observations on control sites are especially important in research design. In nonexperimental research, observations from randomly selected control sites can be compared with observations associated with a particular variable. For example, we may wish to know if habitat use by snowshoe hares is different than general habitat availability. To answer this question, we can make observations (e.g., measure vegetation) at habitat-use sites and compare those with observations from a series of random sites (controls) that we assume represent general habitat availability. If the use sites differ from the random control sites, we conclude habitat selection occurred (see Pietz and Tester 1983).

In experimental research, a control may be defined as parallel observations used to verify effects of experimental treatments. Control units are the same as experimental units except that they are not treated, and they are used to eliminate effects of confounding factors that could potentially influence conclusions or results. Creative use of controls would improve many wildlife studies. Experimental studies in wildlife that involve repeated measurements through time must always include controls because of the importance of weather and other factors that vary through time. Without adequate controls, distinguishing treatment effects from other sources of variation is often impossible. For example, in the bobwhite study described earlier, control sites are required to distinguish the effects of burning from the effects of rainfall and other weather characteristics that affect plant productivity. We might see an increase in grass production in the year following burning just because the rainfall was higher that year. Without control sites we cannot tell whether increased grass production resulted from increased rainfall, from the burning, or from a combination of both factors, and we cannot evaluate the relative importance of each factor.

SAMPLING

Most information gathered by wildlife biologists is used to meet descriptive rather than experimental objectives. Examples include estimates of population size, recruitment, herd composition, annual production of forage species, hunter harvest, and public attitudes. In such efforts we attempt to obtain estimates of characteristics that are

Table 2. Four possible outcomes of a statistical test for declining production in a deer herd. Counts of 500 antlerless deer (adult does and fawns) were obtained each year, and tests of the null hypothesis of no change by the fawn : doe ratio were performed at the 5% level of significance ($\alpha = 0.05$).

| | Fawns per 100 does | | | | | | | |
| | Actual herd values | | | Count values | | Conclusion | Result | Likelihood |
Case	1988	1989	Change	1988	1989	from test	of test	of this result
1	60	60	None	61	59	No change	No error	95% ($1 - \alpha$)
2	60	60	None	65	50	Declined	Type I error	5% (α)
3	65	50	Declined	65	50	Declined	No error	50% ($1 - \beta$)
4	65	50	Declined	62	57	No change	Type II error	50% (β)

Table 3. Survey design checklist.

Question	Example
1. What is the survey objective?	To estimate the percentage of successful hunters
2. What is the best technique or method?	A telephone survey of permit holders
3. To what population do we want to make inferences?	Everyone who has a permit for this hunting period
4. What will be the sample unit?	Individual permit holders
5. What is the size of the population to be sampled (N)?	$N = 350$ (for special permit hunt)
6. What sample design is best?	Simple random sample (Scheaffer et al. 1986)
7. How large should the sample be?[a]	$n = \dfrac{Np(1 - p)}{(N - 1)B^2/4 + p(1 - p)}$ where: N = population size (350) p = proportion of permit holders who hunted deer (from pilot survey = 0.24) B = bound on the estimate = 0.05 (we want an estimate with $P \pm 0.05$ confidence) Therefore $n = \dfrac{350(0.24)(1 - 0.24)}{(350 - 1)(0.05)^2/4 + 0.24(1 - 0.24)}$ $n = 159$, i.e., we should contact approximately 160 permit holders
8. Have you contacted a statistician to review design?	Yes!

[a]See Scheaffer et al. (1986:59).

important for management decisions. We want to obtain the best estimates possible within the constraints of our resources of time and money. A large body of statistical literature exists to help us do this. These types of studies are referred to as surveys, and the topic is known as survey sampling (Scheaffer et al. 1986). Sampling is also a critical part of experimental research and the test of formal statistical hypotheses. Choice of specific sampling methods is dependent on the objectives or hypotheses being addressed, the nature of the population being sampled, and many other factors such as species, weather conditions, topography, equipment, personnel, time constraints, and desired sample sizes. A variety of sampling designs is available for biologists to use in wildlife surveys and experimental research; some of the most valuable will be reviewed below.

Sampling Design

SIMPLE RANDOM

A simple random sample requires that every sample unit in the population have an equal chance of being drawn in the sample and that the procedure for selecting units be truly random. Strictly, this can be accomplished by assigning each member of the population a number and picking samples from a table of random numbers. For example, suppose we wanted to estimate the number of successful hunters in a special hunt for which a limited number of permits was issued. We might decide to contact a sample of permit buyers by telephone after the season to determine their hunting success. A survey design checklist (Table 3) helps us design such a survey properly. The population that we want to make statements about is all persons who obtained a permit. The list of the members of the population is usually called the sampling frame (Scheaffer et al. 1986). It is used to draw a random sample from the population. The sampling frame must be developed carefully or the resulting estimates may be biased. For example, if a portion of our permit buyers did not have telephones and we decided to drop them from the list, the results could be biased. To draw a random sample for our survey we could assign each person who purchased a permit a number and select the numbers to be contacted from a random numbers table or random number generator on the computer. In other types of surveys, obtaining a truly random sample of the population might be difficult. In such instances another method such as systematic sampling should be used (see below). A valid random sampling procedure must be independent of investigator decisions. For example, a widely used procedure to locate plots randomly in a study area consists of laying out an arbitrary baseline through the study area then picking random distances along the baseline and random distances off the baseline for actual plot location (Fig. 4A). This is a valid simple random sample because the randomization follows the arbitrary location of the baseline and removes its effects on plot location. Random-like methods referred to as haphazard or representative are sometimes used in place of truly random designs, but these should be avoided because they are subject to investigator bias. An example of these haphazard methods is the technique of facing in a random direction and throwing a pin over the shoulder to obtain the plot center for a vegetation plot. Although this sounds random, the odds of a field crew randomly facing away from a dense stand of thorny shrubs such as multiflora rose and throwing the pin into the middle of such a patch is practically zero. Seemingly minor losses of randomness can lead to substantial biases in the resulting estimates. Random sampling is not used widely because of the excessive time required to locate truly random samples. In addition, truly random samples occasionally produce poor estimates by chance due to poor spatial coverage of the area or population of interest.

SYSTEMATIC

A systematic sample is taken by selecting elements (sampling units) at regular intervals as they are encountered. This method is easier to perform and less subject to investigator errors than simple random sampling. For example, if we wanted to sample bird-watchers leaving a wildlife management area it would be difficult to draw a truly random sample. However, it would be fairly simple to draw a systematic sample of 10% of the population by sampling every tenth person leaving the area. Systematic

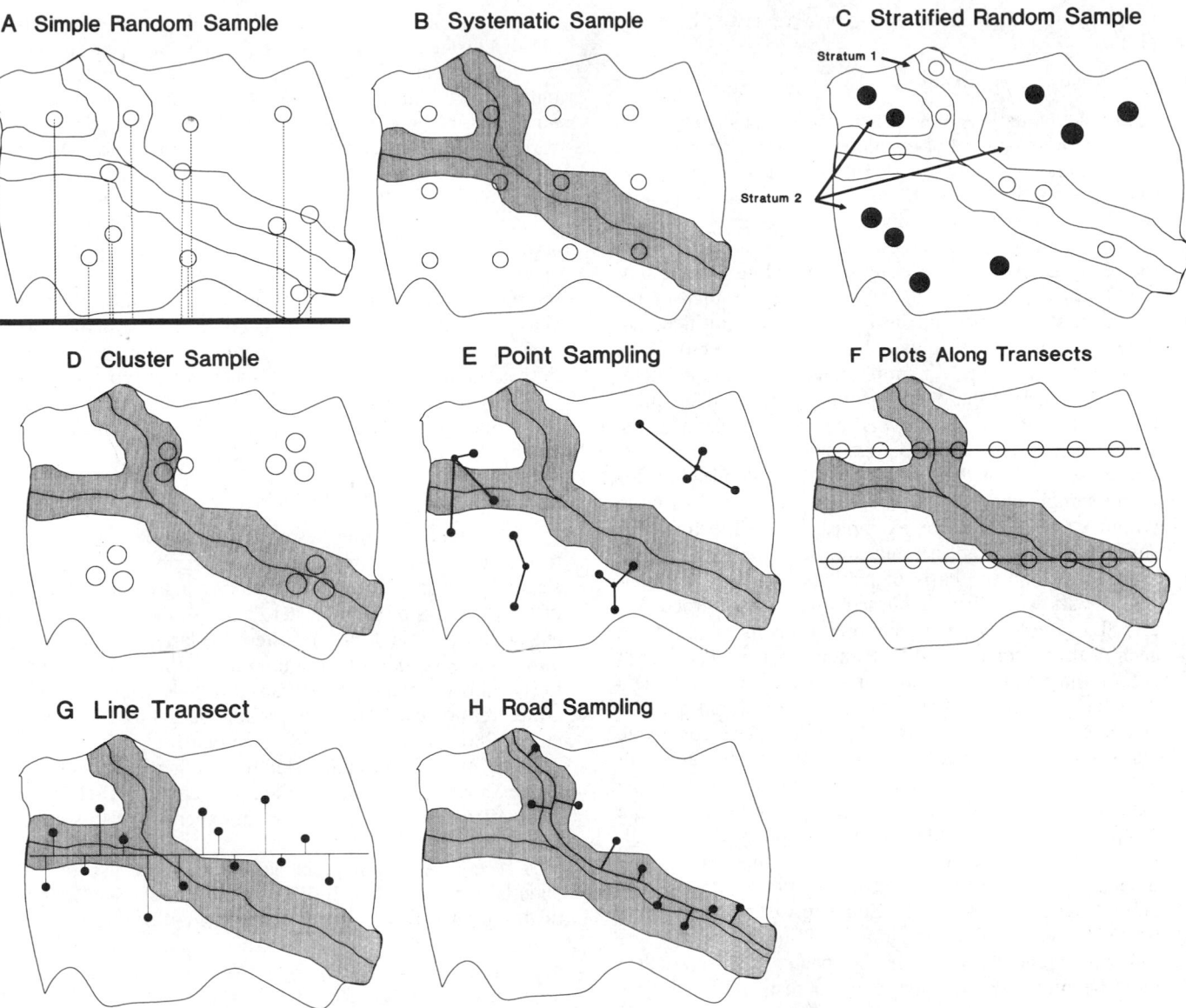

A Simple Random Sample

B Systematic Sample

C Stratified Random Sample

Stratum 1

Stratum 2

D Cluster Sample

E Point Sampling

F Plots Along Transects

G Line Transect

H Road Sampling

Fig. 4. Illustrative examples of sampling methods: A = simple random sample, B = systematic sample, C = stratified random sample, D = cluster sample, E = point sampling, F = plots along a transect, G = line transect, H = road sampling.

sampling is also used extensively in vegetation sampling because of its ease of use in the field. A valid application requires random placement of the first plot followed by systematic placement of the rest of the plots, usually along a transect or in a grid pattern (Fig. 4B). This approach often provides greater information per unit cost than simple random sampling because the sample is distributed uniformly over the entire population or study area. For random populations, systematic samples give estimates with the same variance as simple random samples.

The major danger with systematic samples is that they might give biased estimates with periodic populations (i.e., with nonuniform distributions). For example, if we were interested in estimating the number of people using a wildlife management area, we might set up a check station and take a systematic sample of days during the season. Such a procedure could yield extremely biased results if we chose to take a sample of one-seventh of the days. If our sample day fell during the week, we would get very different results than if it fell during the weekend. Additionally, our estimate of variance would be much too small, leading us to conclude that our estimate was much more precise than in reality. In this situation the population obviously is periodic, but in other situations the periodicity might be quite subtle. Thus, systematic sampling must be used with caution. The formal procedure is carried out by randomly selecting one of the first k elements to sample and every k^{th} element thereafter. For example, if we wanted to sample 10% of our population, k would equal 10. We would draw a random number between 1 and 10. Suppose we selected 3, then we would sample the third element and every tenth element thereafter (i.e., 13th, 23rd, 33rd, . . .). At a check station we might use this to sample 10% of the deer that came through the station. When laying out plots along a transect, we would randomly locate the starting point of the transect and then lay out plot centers at fixed intervals along the transect, such as every 100 m. Advantages and disadvantages of random and systematic sampling were reviewed by Krebs (1989).

STRATIFIED RANDOM

In many situations, obvious subpopulations exist within one total population. For example, tourists, bird-watchers, and hunters are readily divided into residents and nonresidents. A study area can be divided into habitats. A population of animals can be divided into age or sex groups. If the members of these subpopulations are similar in terms of the characteristics we are estimating and the subpopulations themselves differ one from another in the characteristic, a powerful design to use is stratified random sampling. The subpopulations are referred to as "strata," and we draw a simple random sample of members from each of these strata. Stratified random sampling is also indicated if we are particularly interested in the estimates for the subpopulations themselves. The strata are chosen so that they contain units of identifiably different sample characteristics, usually with lower variance within each stratum. For example, if the objective of a study of moose is to estimate moose density, we might define strata on the basis of habitats (e.g., bogs and riparian willow patches, unburned forest, and burned forest). To sample, we draw a simple random sample from each of these strata (Fig. 4C). If moose density is different among the strata, the variation in each stratum will be less than the overall variation. Thus, we will obtain a better estimate of moose density for the same or less cost. Note that if the strata are *not* different, stratified estimators will not be as precise as simple random estimators. In some instances, the cost of sampling is less for stratified random sampling than for simple random sampling. A final advantage of stratified random sampling is that separate estimates for each stratum, e.g., moose density in willows or in forests, are obtained at no extra cost. The formal procedure for stratified random sampling consists of the following three steps: (1) clearly specify the strata—they must be mutually exclusive and exhaustive, (2) classify all sampling units into their stratum, and (3) draw a simple random sample from each stratum. Formulas are available to determine the sample size and the optimal allocation of effort to the strata (Scheaffer et al. 1986).

CLUSTER SAMPLING

A cluster sample is a simple random sample in which each sample unit is a cluster or collection of observations (Fig. 4D). This approach has wide application in wildlife biology, because many birds and mammals occur in groups during all or part of the year. When we draw samples from such populations we draw clusters of observations, i.e., groups of animals. Likewise, many of the wildlife user groups (e.g., waterfowl hunters, tourists) occur in clusters (e.g., boats in wetlands, automobiles along highways). Cluster sampling is also useful where the cost of travel time from one sample unit to the next is a serious problem. This is commonly the situation in surveys of animals or habitat. The formal procedure for cluster sampling consists of the following three steps: (1) specify the appropriate clusters and make a list of all clusters, (2) draw a simple random sample of clusters, and (3) measure all elements within each cluster selected in the sample.

Making a formal list of clusters is rarely possible or essential. Instead, we emphasize obtaining a random sample of clusters. Cluster sampling of habitat is performed by picking a random sample of locations and then locating multiple plots in a cluster at each location. The optimal number of plots (cluster size) depends upon the pattern of variability in the habitat. If plots in a cluster tend to be similar (little variability within a cluster), cluster size should be small. If plots in a cluster tend to be heterogeneous (high variability within a cluster), cluster size should be large. For other types of cluster samples such as groups of animals (e.g., elk, deer, pheasants) or people in vehicles, the cluster size is not under control but is a characteristic of the population. For example, aerial surveys of elk and deer on winter ranges result in samples of these animals in clusters. Estimates of herd composition (e.g., fawn : doe or bull : cow ratios) are readily obtained from treating these data as cluster samples (Bowden et al. 1984).

OTHER SAMPLING METHODS

Many other sampling designs are available in addition to the four most common designs described here. Two-stage cluster sampling involves surveying only a portion of the members of each cluster drawn in the sample. This approach is efficient when clusters are large. Cluster sampling is one version of the more general method referred to as ratio estimation (Cochran 1963). Related methods are regression estimation and double sampling (Scheaffer et al. 1986), which have wide potential for application to wildlife research. The interested reader should consult one of the standard references on survey sampling (Scheaffer et al. 1986) and visit a statistician experienced in survey sampling.

A recent development in sampling is the use of sequential sampling, which differs from the classical statistical approach in that in sequential sampling the sample size is not fixed in advance. Instead, samples are drawn one at a time, and after each sample one decides whether a conclusion can be reached. Sampling is continued until either the null hypothesis is rejected or accepted with a specified level of certainty. This type of sampling is applicable to wildlife studies in which sampling is performed serially, i.e., the result of each sample is completed before the next sample is drawn (Krebs 1989). The major advantage of this approach is that it usually minimizes sample size and thereby saves time and money. For further details see Dixon and Massey (1983). Krebs (1989) provided an excellent overview of these methods with examples of their application to hypotheses concerning means and proportions.

Sampling Methodology

PLOTS

Plots are used widely to sample habitat characteristics, to count animal numbers, and to count animal sign. Used in this way, plots represent small geographic areas (circular, square, or rectangular) that are the elements of the geographically defined population. The population size is the number of these geographic areas (plots) that would cover the entire study area. Usually sufficient time, money, and personnel to study an entire area are not available, so a subset of plots is used with the assumption that it is representative of the area. Any of the survey designs can be applied (simple random, systematic, stratified random, and cluster), or more complicated designs such as two-stage designs may be applied (Cochran 1963). Selecting the best design requires insight into the characteristics and patterns of distribution of the species across the landscape. One of the advantages of this approach is that the size of the population is known and totals can be estimated (see Seber 1982). Selection of plot size and shape, also an important consideration, was reviewed in detail by Krebs (1989).

POINT SAMPLING

In point sampling a set of points is established throughout the population and measurements are taken from each sample point (Fig. 4E). A common measurement is distance from the point to a member of the population (e.g., plant or calling bird). Examples of this include the point quarter and nearest neighbor methods used widely to estimate density for trees and shrubs (Mueller-Dombois and Ellenberg 1974) and the variable circular plot method of estimating songbird density (Reynolds et al. 1980). Selection of sample points for plotless methods usually follows a systematic design, but other sample designs can be used as long as points are spaced widely enough that few members of the population are sampled more than once. Necessary sample size can be estimated from formulas if population size is assumed to be very large or unknown (see Zar 1984 for example).

TRANSECTS

A transect is a straight line or series of straight-line segments laid out in the area to be sampled. Transects are used for two general purposes: to organize or simplify establishment of a series of sample points or plots and as a sample unit themselves. Transects are used widely to obtain systematic samples of spatially distributed populations (e.g., plants). In these situations the plots placed along the transects are the actual sample units (Fig. 4F) and should be treated as described under systematic sampling. Plots can also be placed along transects at random intervals. When the transects themselves are used as sample units, they are commonly referred to as line transects (e.g., Burnham et al. 1980). Measurements of the perpendicular distance, or sighting distance and angle, to the sampled elements (e.g., flushing animals, groups of animals, carcasses, snags) are recorded as is the number of elements encountered along the line (Fig. 4G). Distances are used to estimate the effective width of the area sampled by the transect (Seber 1982). Each transect is treated as an independent observation, and transects should be laid out in a nonoverlapping manner according to one of the established sampling designs (i.e., simple random, systematic, stratified random). Transects are often easier to lay out in rough terrain than are plots, but they must be established carefully with compass or transit and measuring tape. Their use is becoming more widespread in aerial survey work because of the use of precise navigational systems such as LORAN-C (Patric et al. 1988). The critical assumptions for transect methods for sampling mobile objects such as animals (i.e., 100% detection directly on the line, no movement toward or away from the observer before detection) must be examined carefully before this sampling method is selected (Burnham et al. 1980). A strip transect is a similar method, but it is really a long, thin plot, because the method assumes all animals in the strip are counted (Krebs 1989).

ROAD SAMPLING

Sampling from roads is a widely used method for obtaining observations of species sparsely distributed over large areas, or for spreading observations of abundant species over a large geographic area. This sampling method is usually the basis for spotlight surveys for nocturnal species such as white-tailed deer (Boyd et al. 1986) and jack rabbits (Chapman and Willner 1986), for brood counts and call counts for upland game birds (Kozicky et al. 1952), for scent-station surveys (Nottingham et al. 1989), and for the U.S. Fish and Wildlife Service-sponsored continent-wide breeding-bird survey (Robbins et al. 1986). This approach involves drawing a sample from a population defined as that population occupying an area within a distance x of a road (Fig. 4H). The distance x is generally unknown and varies with any factor that would affect an animal's detectability, such as species conspicuousness, density and type of vegetation cover, or background noise for surveys based on aural cues. Roads rarely provide unbiased estimates for a region because they are generally placed along ridges or valleys and avoid steep or wet areas, and roads modify habitat for many species. Thus, sampling along roads rarely provides a representative sample of habitat. Although this bias is well known, unfortunately it is often ignored. As with all indices, every effort should be made to standardize counting conditions along fixed, permanently located routes (see Caughley 1977). The original sampling design for the location of these routes must follow one of the standard designs.

DEPENDENT (PAIRED) AND INDEPENDENT OBSERVATIONS

If we wish to make population comparisons, pairing observations is a powerful tool for sharpening our ability to detect differences. If there is a correlation between members of a pair, treating them as dependent or paired observations can improve the power of tests for differences. For example, to compare diets of adult female bighorn sheep and lambs, we might treat a ewe with a lamb as a pair and determine the diet of each animal by counting the number of bites of each plant they eat while foraging together. Treating these observations as pairs would sharpen the comparison between the age classes because it would compare animals that were foraging together and experiencing almost the same availability of plants. Pairing is a powerful technique in other contexts in which there is dependency between the observations that we want to remove. Pairing should be used only if an association really exists, otherwise the power of the comparison will be decreased.

Pairing also can be used to help answer a different question. For example, studies of habitat selection are often made by locating areas used by a species (i.e., nest sites or radio locations) and measuring the habitat characteristics at these use sites with sample plots. The available habitats are measured by selecting random sample plots from throughout the study area (Fig. 5A), and a

A

B

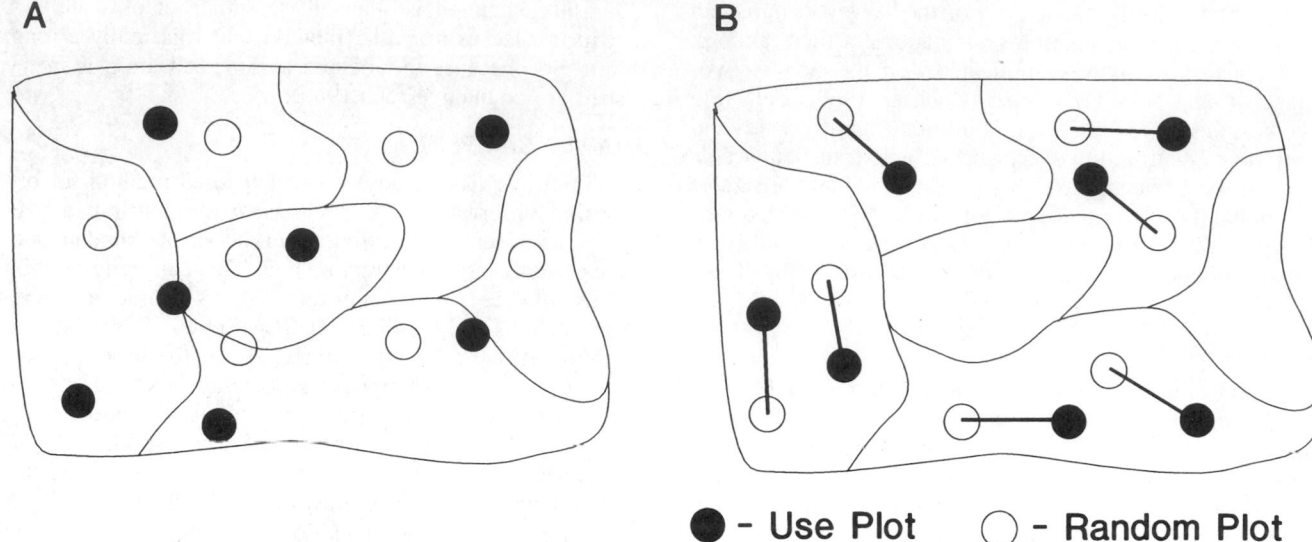

● - Use Plot   ○ - Random Plot

Fig. 5. Illustrative examples of use plots and random plots (A), and use plots paired with random plots (B).

comparison of these use and random plots will identify characteristics of areas selected by the species. An alternative approach involves pairing the use and random plots as follows: for each use plot select a random plot within a certain distance of the use plot (Fig. 5B). For analysis, the use and random plots are paired (i.e., random plot locations are dependent on the use sites). This comparison could produce quite different results from the unpaired comparison because it would be testing for habitat differences within areas used by the species (microhabitat selection), whereas the unpaired comparison (e.g., independent plots) would be testing for habitat differences within the general study area (macrohabitat selection).

HYPOTHESIS TESTING

Hypothesis testing incorporates one or more of four basic research options: field studies, natural experiments, field experiments, and laboratory experiments (Fig. 1). Wildlife field studies are common, but interpretation of the results has severe limitations. Experiments span a continuum from natural experiments over which we have no control to completely controlled laboratory experiments (Table 4).

Field Studies

Field studies are similar to experiments in that they are carried out to test hypotheses, but making inferences from field studies is difficult because we make *ex post facto* (Kerlinger 1973) or after-the-fact comparisons between groups, which already had the characteristic of interest. Drawing firm conclusions is difficult, because these groups also differ in many other aspects. For example, in a field study of dietary selection by Canada geese we might randomly select plots where flocks of geese have fed and those where they have not fed to determine if geese chose areas with vegetation that is more nutritious. If they did, a weak inference would be that geese are choosing nutritious food, but numerous alternative explanations remain untested (e.g., maybe geese preferred hilltop sites where visibility was good, and coincidentally these were also sites farmers fertilized most heavily to compensate for wind-soil erosion from previous years of tillage). The important aspect of a field study is that we have comparison groups (e.g., feeding plots versus nonfeeding plots), but we have no treatments. Well-designed field studies can make important contributions to wildlife science and management, but their limitations must not be overlooked.

Natural Experiments

Natural experiments are similar to field studies except that we study the effects of *uncontrolled treatments* such as wildfires, hurricanes, mass mortality from diseases, agricultural practices, and range expansions by animals or

Table 4. Strengths and weaknesses of different types of experiments (modified from Diamond 1986).

	Laboratory experiment	Field experiment	Natural experiment
Control of independent variables[a]	Highest	Medium	Low
Ease of inference	High	Medium	Low
Potential scale (time and space)	Lowest	Medium	Highest
Scope (range of manipulations)	Lowest	Medium	High
Realism	Low	High	Highest
Generality	Low	Medium	High

[a]Active regulation and/or site matching.

plants. A key problem in evaluating natural experiments is that we cannot assign treatments randomly. In natural experiments the treatment precedes the hypothesis and most comparisons must be made after the fact. In laboratory and field experiments the treatment follows the hypothesis. Many hypotheses of interest to wildlife biologists can be tested only with natural experiments, yet it is difficult to draw inferences from such experiments. The applied nature of wildlife management makes the realism and generality of natural experiments an important advantage. With our Canada goose example, a natural experiment might be to survey farmers in the region to locate pastures that have been fertilized and those that have not been fertilized in recent years. If our observations of feeding geese show more use of pastures that had been fertilized, we have more evidence indicating they are selecting forage that is more nutritious. However, many alternative explanations remain. For example, perhaps those pastures that were fertilized also were grazed later in the summer, and geese preferred fields with the shortest grass where ability to detect approaching predators was greatest.

Field Experiments

Field experiments offer advantages over natural experiments in terms of ease of inference and control but disadvantages of restricted scale and lower generality (Table 4). Compared to laboratory experiments, field experiments have greater scope and realism. The main advantage of field experiments is that we can randomly assign treatments. In field experiments, manipulations are carried out, but other factors are not subject to control (e.g., weather). In many situations in wildlife science, field experiments offer the best compromise between the limitations of laboratory experiments and natural experiments (Fig. 1; also see Wiens 1989). Let us continue with our example: a subsequent field experiment is to fertilize random plots in known foraging areas and determine if geese selected the fertilized plots more than nonfertilized control plots. If they did, a stronger inference about selection of nutritious foods could be made, because random assignment of plots to fertilization and control groups should have canceled the effects of extraneous confounding factors.

Laboratory Experiments

Drawing inferences from laboratory experiments is easy because of the high level of control, yet this advantage must be weighed against their disadvantages (Table 4) in terms of (1) scale (laboratory experiments are restricted to small spatial scales and short time periods), (2) scope (only a restricted set of potential manipulations is possible in the laboratory), (3) realism (the laboratory environment places many unnatural stresses and constraints on animals), and (4) generality (some laboratory results cannot be extrapolated to natural communities). A continuation of our example could be a logical laboratory experiment to determine if geese really can select the most nutritious forage when given several alternatives in a cafeteria-feeding trial (similar to the elk experiment described in MAJOR COMPONENTS OF RESEARCH [p. 7]).

Identifying one research option as best for all situations is not possible. All options should be considered as possibilities when a hypothesis test is designed. Sometimes the best evaluation of a hypothesis involves using a combination of field studies and various types of experiments. For example, Takekawa and Garton (1984) obtained field observations of birds feeding heavily on western spruce budworms during a budworm outbreak, which suggested that birds were a major source of budworm mortality. Field experiments were carried out to test this hypothesis by placing netting over trees to exclude birds. Survival of budworms on trees with netting was three to four times higher than on the control trees exposed to bird predation (Takekawa and Garton 1984). A second example is field observations by Ratti et al. (1984) that indicated spruce grouse fed exclusively on certain trees while ignoring numerous other similar trees of the same species. This led to a laboratory experiment with captive birds by Hohf et al. (1987) that tested the hypothesis that selected feeding trees had higher nutritional content than random trees, which they did. Diamond (1986) provided examples of the three types of experiments and excellent suggestions for improving each type. Other fine examples and discussion of experiments were provided by Cook and Campbell (1979), Milliken and Johnson (1984), Kamil (1988), and Hairston (1989).

Integrated Research Process

In wildlife science, proving research hypotheses with high certainty is unlikely because (1) free-ranging populations of animals are subjected to complex environmental variations that are often both uncontrolled and undetected, (2) wildlife are components of diverse and highly interactive communities, and (3) most changes observed result from multiple rather than single causes. Short-term studies cannot eliminate these problems, but longer-term studies that use an integrated research process (Fig. 1) can approach proof of a research hypothesis. The integrated research process builds on a solid base of natural-history observations. Field observations should lead to experiments, and the results of natural experiments should lead to field and laboratory experiments. The level of certainty increases as many predictions from the research hypothesis are supported and alternate hypotheses are rejected in successively more rigorous tests that use replicated research options. After such findings are repeated over broad geographic areas or throughout the range of the species, the research hypothesis may become a principle of wildlife science. The integrated research process should be the goal of wildlife science.

A Checklist for Experimental Design

The design of any experiment must be developed carefully or the conclusions reached will be subject to doubt. Four particularly critical elements in the design of a manipulative experiment are: specification of the research population, replication, proper use of controls, and random assignment of treatments to experimental units. The following experimental-design checklist provides a series of questions to assist in addressing these critical elements. Many of the questions also will be helpful with design of data-gathering efforts for studies involving nonexperimental hypothesis testing. Note that some experimental designs may address several hypotheses simultaneously; in other designs, each hypothesis may require independent experimental testing.

1. **What is the hypothesis to be tested?** The hypoth-

esis developed from the conceptual model must be stated clearly before any experiment can be designed. For example, an interesting hypothesis to test would be that nest predation on forest songbirds is higher at sharp, artificial edges, such as occur at typical forest clearcuts, than at feathered edges (partial timber removal), such as occur at the boundary of selectively logged areas (Ratti and Reese 1988).

2. **What is the dependent or response variable(s) and how should it be measured?** The dependent variable should be clear from the hypothesis (e.g., nest predation in the example above), but selecting the best technique to measure it might be more difficult. We must consider all possible methods and identify the one that will simultaneously maximize precision and minimize cost and bias. It is often helpful to contact others who have used the techniques, examine the assumptions of the techniques, and conduct a pilot study to test the potential techniques. In our example, we might choose to place artificial nests along forest edges and use a generalized Mayfield estimator (Heisey and Fuller 1985) of mortality rate. Choosing to use artificial nests would obviously be a compromise between the desirable goal of testing the hypothesis for naturally occurring nests of individual bird species on the one hand and the costs of finding a large number of such nests and all the confounding factors added by using natural nests on the other hand. Testing the hypothesis with artificial nests might be used as an early step in a major research program on the topic.

3. **What is the independent or treatment variable(s) and what levels of the variable(s) will we test?** The independent variable(s) should be clear from the hypothesis (sharp and feathered forest edges in our example), but selecting the levels to test will depend upon the population to which we want to make inferences. If we want to test the effects of our independent variable at any level, we must select the levels to test at random (random effects or Model II ANOVA, Zar 1984). If we are interested in only a few of the levels that our independent variable could take, we will use only those levels in our experiment and make inferences only to those levels tested (fixed effects or Model I ANOVA, Zar 1984). For example, if we wanted to evaluate the effects of forest edges of any type on predation rates, we would select our types of forest edges at random from all the types that occur. In this example we are interested only in the two types categorized as sharp and feathered. Additionally, our independent variable must be identified and classified clearly or measured precisely. Finally, how can we use controls to expand our understanding? In our example, comparing nest predation in undisturbed forest to predation at the two types of edges might be enlightening.

4. **To what population do we want to make inferences?** If the results of the experiment are to be applied to the real world, our experimental units must be drawn from some definable portion of that world, the research population. Likewise, the dependent and independent variables that we have chosen define the relationship(s) that we are examining and put constraints on the definition of this population. Finally, we must consider the impact of potential extraneous factors in selecting the population of interest. If the population is defined so broadly that many extraneous factors impact the results, the variation might be so large that we cannot test the hypothesis. Likewise, if the population is defined so narrowly that we have essentially a laboratory experiment, the application of the results might be severely limited (low generality). Reaching the proper balance between internal and external validity takes thought and insight. For example, we might want to compare nest-predation rates in sharp and feathered edges throughout the northern Rocky Mountains, but the logistics and cost of doing that would make the study impossible. Thus, we might restrict our population to one national forest in this region. Next we need to consider the types of forests. We might want to test the hypothesis for all the major forest types, but we know that the species of birds nesting in these forests and the predators on their nests differ from one forest type to another. We will need to restrict our population to one important type of forest such as Douglas-fir habitat types (Cooper et al. 1987) to remove the extraneous factors that could impact our results if we sampled a large variety of forest types. We need to ask what types of sharp and feathered edges occur and which ones we will sample. Sharp edges are commonly produced by clearcuts, powerline rights-of-way, and road rights-of-way. These three types differ dramatically in factors such as size, shape, human access, and disturbance after treatment. Additionally, our ability to design a true experiment involving random assignment of treatments is severely limited for all but the clearcuts. Therefore, we might restrict our populations to sharp edges created by clearcuts and feathered edges created by selective harvests.

5. **What will be our experimental unit?** What is the smallest unit that is independent of other units and to which we can randomly assign a treatment? This must be identified correctly or the resulting experiment might not have true replication, but instead have pseudoreplication (Hurlbert 1984). For example, we might erroneously decide that the experimental unit for our nest-predation study will be an individual nest. The resulting design might entail selecting three areas and randomly assigning one of them to be clearcut, one as a control, and the other to be selectively logged. Then 20 artificial nests could be placed along the edge of each area and monitored for predation. The resulting data would suggest 20 replicates of each treatment, but in fact only a single area was given each treatment. Only one area was randomly assigned each treatment, and the 20 nests were really subsamples. Such pseudoreplication restricts our potential inferences enormously. In effect we have sampled from populations consisting only of two logged areas and one unlogged area, and our inferences can be made only to those three areas, not to clearcuts, selective cuts, and undisturbed forests in general. In some situations, pseudoreplicated designs are unavoidable, but interpretation of their results is severely restricted because without replication the results could have been caused by confounding factors rather than the treatment. For example, in our nest predation experiment if one of the areas was within the home range of a raven and the other areas were not, this single confounding factor could determine the result regardless of which treatment fell in which area. A more reliable experiment would require that we identify several areas with potential to be logged, perhaps 15, that they be far enough apart that they are independent of each other, and that we

randomly assign five to be clearcut, five to be selectively harvested, and five to be controls. We would place several artificial nests in each area and they would be monitored. The nests in a single area would be correctly treated as subsamples and their overall success treated as the observation for that area. This approach attempts to remove the effect of confounding factors and thereby allow the development of a conclusion with general application to the populations sampled, i.e., edges created by clearcuts and selective cuts within this habitat type in this region.

6. **What experimental design is best?** A few of the most widely used designs are described below, but we advise consulting one of the excellent texts on experimental design and visiting a statistician before making the final selection. The choice depends primarily upon the type of independent and dependent variables (categorical or continuous), the number of levels of each, the ability to block experimental units together, and the type of relationship hypothesized (additive or with interactions). For our study of nest predation along two types of forest edges a simple, single-factor design would be appropriate.

7. **How large should the sample size be?** Estimating the sample size needed for proper analysis is essential. If the necessary sample size is too costly or impossible to obtain, it would be better to stop at this point and try to redesign the project or work on a different question that can be answered. The sample size depends upon the magnitude of the effect to be detected, variation in the populations, and the type of relationship that is hypothesized. Typically some preliminary data from a pilot test or from the literature are required to estimate variances. These estimates are used in the appropriate formulas available in statistical texts (e.g., Zar 1984).

8. **Have you contacted a statistician and received concurrence on your design?** Obtaining the input of a statistician before the data are gathered is essential. The statistician will not be able to help salvage an inadequate design afterwards, nor will he or she be very sympathetic at that point. Now is the time to get these comments!

Single-Factor Versus Multifactor Designs

Single-factor analyses are the simplest because they involve only comparisons between two or more levels of one factor. The appropriate statistical tests are described in Chapter 2. Evaluating the simultaneous effect of two or more independent variables (factors) at once requires the use of complicated statistical methods, which should be discussed with a statistician. Under many conditions we can test two factors at once without expending more effort than would be required to test either of the factors alone. A complicating issue is the potential for interaction between factors (for examples see Steel and Torrie 1980). An interaction occurs if the effects of one factor on the response variable are not the same at different levels of the second factor. For example, if we are interested in the effect of snow-melt date on nest success by arctic-nesting polymorphic snow geese, we might discover an interaction between color phase and the onset of spring snow melt, e.g., darker blue-phase birds are more cryptically colored during early snow-melt years and experience less nest predation, and during late snow-melt years white-phase birds are more cryptically colored and experience

less nest predation. In such situations many observations might be required to elucidate the relationships.

Dependent Experimental Units

Special forms of analysis have been developed for many types of dependency in experimental units. A common design involves pairing. In a paired design we match experimental units in pairs that are as similar as possible. The treatment is then applied to one member of each pair at random. If there is a confounding factor, which we succeed in matching in the pairs, this approach will lead to a much more powerful test than if pairing is not performed. For example, if we were studying the effects of spring burning on bobwhite habitat, we could lay out pairs of plots throughout our study area, being careful to place each pair in a homogeneous stand of vegetation. We would randomly assign one member of each pair to be burned in the spring. The analysis would then look at the differences between the members of a pair and test for a consistent improvement or decline in the burned portion of the pair. The pairing thus would remove the effects of vegetation difference from one part of the study area to another and result in a more sensitive experiment. If members of pairs are not more similar than members of the general population, the test will be less powerful because of the pairing. When more than two levels of a factor are compared, pairing is referred to as blocking. A block is a set of similar experimental units. Treatments are randomly assigned to units within each block, and the effectiveness of blocking can be tested during the analysis. For example, if we expanded our study of burning to include spring and autumn burning as treatments, a block design would be appropriate. Three adjacent plots would be laid out in homogeneous vegetation stands, and spring burning and autumn burning would be applied randomly to two of the three plots. The analysis would entail a randomized, block-design ANOVA. Another common form of dependency occurs when repeated measurements are taken on the same experimental unit through time. This is common in wildlife research wherein the effects of treatments may change over time and must be monitored over a series of years. For example, in our study of spring and autumn burning the effects may be different in the first, second, and third growing seasons after treatment. The plots should be monitored over several years to determine these effects. The measurements are repeated on the same plots, so they are not independent. This must be taken into account in the analysis by using repeated measures or multivariate analysis of variance (Johnson and Wichern 1988). Unless the biologist has extensive training in this topic, close cooperation with a consulting statistician is essential in designing and analyzing experiments involving such complicated designs.

Crossover Experiments

Crossover experiments provide a powerful tool to evaluate treatments that do not produce a long-lasting effect. A crossover experiment is performed by laying out pairs of experimental units and randomly assigning one member of each pair to be treated during the first treatment period. The second member of each pair serves as the control during this treatment period. In the second treatment period the control unit becomes the treatment and the former treatment becomes the control. In this way the effects of any underlying characteristics of experimental units are prevented from influencing the results. Obviously this technique is valid only if treatment effects do not persist into the second treatment period.

Consider the following example. Suppose we wanted to test the hypothesis that mowing hay before 4 July decreases pheasant nest success. We could test this by dividing our study area into five fairly homogeneous hay-field regions and then dividing each region into two portions. In one randomly selected portion of each region we could pay farmers not to mow their hay field until after 4 July (treatments). In the other portion of each region hay mowing will proceed as it does in most years, with the first cutting during mid-June, and these portions will serve as controls. To monitor nest success, we will locate nests by systematic field searches, being sure to search treatment and control areas with identical methodology, e.g., search intensity and seasonal timing. Nest success will be determined with standard techniques. After 1 year, we might measure significantly higher nesting success in the treatment portions, i.e., those areas with delayed hay mowing. However, the number of treatments is small, so we are not able to conclude with confidence if the higher success resulted from the treatment or from some undetected, inherent differences in the treated portions of each region, such as nest predators. At this point we implement the crossover experiment by switching portions in the second year so that the original control portions of the study regions now have mowing delayed until after 4 July (new treatments), and the original treatment portions revert to the standard practice of first cutting in mid-June (new controls). If the portions with late cutting treatments again have higher nest success, we have better evidence that delayed mowing is responsible for higher nest success than we had at the end of the first year (i.e., we have better evidence for a *cause-and-effect* relationship). If even stronger support for the hypothesis is desired, the crossover experiment might be repeated again in the same region and also in other farming regions with pheasant populations.

COMMON PROBLEMS

In this section we briefly review some previous topics and discuss some additional problems that are commonly associated with wildlife research.

Sample Size

The importance of sample size cannot be overemphasized. Often inadequate sample size is the result of (1) inadequate consideration of population variance, (2) inability to collect data (e.g., observe a rare species), or (3) insufficient funding, time, or personnel. Often a sample-size problem is overlooked initially because of failure to consider *sample-size reduction* throughout the study, i.e., we focus mostly on the initial sample size and not on the final sample size that represents the most important data for consideration of a hypothesis. We illustrate this problem by considering research designed to measure mallard brood movements from a nest site during the first 30 days of life after hatching. During the planning stage of this research we will likely give the most consideration to how many nests we can locate, and we assume this estimate is

somewhat representative of our final sample. Such assumptions can be misleading. Our example summarizes data from a 3-year study (J. J. Rotella and J. T. Ratti, unpubl. data) in which 258 mallard nests were located. To accomplish data collection on this question (brood movements after hatching), we used radiotelemetry techniques, because mallard broods are difficult to observe during the first 30 days. Thus, we located nests, fenced the nests with 5-cm-mesh poultry netting to protect them from ground predators such as striped skunks, trapped female mallards during incubation to attach radio transmitters, and recorded data on brood movements after hatching with standard radiotelemetry techniques. What one may not fully appreciate at the onset of this study is the magnitude of systematic sample-size reduction with each phase of the methodology, i.e., nests were lost at each stage to nest abandonment by laying hens, nest predation, our inability to trap some females, and nest abandonment caused by trapping efforts. Also, many broods experienced total mortality the first 2 weeks after hatching. Thus, after 3 years of intensive fieldwork, we located 258 mallard nests but obtained data on only 29 broods that lived 30 days. This represented an 89% sample-size reduction from nests located to actual brood data.

A comparable and probably more common problem is fairly large overall data sets that are not similar enough among years (or seasons) to combine, resulting in annual sample sizes that are too small for analysis. At the beginning of a research project we often set our desired sample size based on combining data collected over several continuous years. However, if the characteristic of interest is different among years of the study, combining the data would not be valid. For example, in a study of habitat selection by red foxes, habitat use might differ between a mild winter and a severe winter. In this example, combining the data would not be valid, yet the sample size in each year may be too small to detect selection (Alldredge and Ratti 1986).

Procedural Inconsistency

Procedural inconsistency is another common problem in research. Problems of this type occur from seemingly minor variations or alterations in methodology. For example, if a project is dependent upon field personnel to accurately identify songs of forest passerine birds, the data set might be biased by identification errors (e.g., Cyr 1981). In this situation, the magnitude of the bias will be directly related to the rate of errors by individuals, the difference in the rate of errors among individuals, and the relative proportion of data collected by each individual. Research methodology should be defined with great detail, and all individuals collecting data should have similar skills and knowledge of methods used (Kepler and Scott 1981). For another example, consider a research project in which you are estimating the density of a game-bird species using line-transect methods. After the first summer of data collection you receive a puppy as a gift. The following year you innocently take the dog along with you while walking transects, not realizing that the dog's presence changes flushing behavior and the average flushing distance of the game bird—factors that have an important effect on the mathematical model used to estimate density. One unfortunate aspect of biases of this type is that they are often overlooked (or ignored) as potential problems and are seldom reported in research publications. Thus, the data might be given unwarranted consideration for management and ecological interpretation.

Nonuniform Treatments

A third common bias is nonuniform treatments. This problem is illustrated by our considering two previous research examples. In the discussion of crossover experiments, we described a 2-year study in which mowing on the treatment areas was delayed until after 4 July. Assume that in the first year of this study, all treatment areas were cut between 4 and 7 July, as planned. But during year 2 of the study, a 3-day rainstorm began on 4 July, and the treatment areas were not cut until 9–12 July. Although this 5-day difference in mowing of the treatment areas may seem insignificant, the impact on the results and interpretation of our experiment is really unknown—and it may be serious. Thus, the second year of the experiment should be repeated.

For a second example, consider again the elk feeding trials described in MAJOR COMPONENTS OF RESEARCH (p. 7). In this experiment we were attempting to determine if elk could detect protein levels in forage and if they selected forage with higher protein levels. Uniform treatments in this research required that the protein additive placed in the forage was identical for all subsamples and all treatments. We noted that one alternative explanation for results of this experiment might be that the protein additive affected overall forage palatability. Although this possible bias must be addressed with additional research, nonuniform treatments would have hopelessly confused the results, e.g., if we altered forage with different protein additives so that we created different palatability that masked the effects of the intended treatment.

Pseudoreplication

Pseudoreplication occurs when sample or experimental units are not independent, i.e., they are really subsamples rather than replicates. This is a widespread problem in field ecology (Hurlbert 1984) that should be avoided wherever possible. In manipulative experiments, experimental units are independent only if we can randomly assign treatments to each unit. In field studies, a simple test for pseudoreplication is to ask if the values for two successive observations are more similar than the values for two observations drawn completely at random from the research population. If so, the successive observations are probably not true replicates and the research should be redesigned. There must be a direct tie between the sample or experimental unit and the research population. If the research population consists of one meadow in Yellowstone National Park, then two or more samples drawn from that meadow would be replicates. Note that in this example, our inferences or conclusions would apply only to that single meadow. If our research population consisted of all meadows in Yellowstone National Park, then two plots in the same meadow would not constitute true replicate samples. Also, repeated sampling of the same radio-collared animal often constitutes a form of pseudoreplication, e.g., if our research population consisted of moose in one ecoregion, repeated observations of habitat use by a single animal would not be true replicates. They

would have to be summarized into a single value such as the proportion of the observations in a certain habitat for statistical analysis. This would reduce our sample size to the number of radio-collared moose. Treating repeated observations as replicates is strictly justified only when the individual animal is the research population. In this situation, tests for serial correlation (Swihart and Slade 1985) should be conducted to assure that the observations are not repeated so frequently that they are still pseudoreplicates.

THE RESEARCH-MANAGEMENT CONNECTION

Three driving forces behind management programs of wildlife agencies are (1) public opinion, (2) politics, and (3) biology. We will restrict our comments here to the relationship between biological research and management, realizing that public-opinion and political influences on management programs are often substantial.

Wildlife management programs should be developed by the application of scientific knowledge, i.e., we should apply scientific facts and principles resulting from research on specific topics such as population ecology, habitat selection, or behavior. Initially, this is a sound practice for development of a new management program. In fact, the logic behind formulation of a management program is similar to formulation of a research hypothesis; both provide the opportunity for predictive statements. Our management prediction is that our plan of action will achieve a desired result. However, a major problem with nearly all wildlife management programs throughout the world is the lack of research on the effectiveness of programs (e.g., Macnab 1983, Gill 1985). Seldom is the question "Does our management lead to the desired result?" addressed in formal, well-designed, long-term research projects. For example, disparate sex ratios are common among North American mallard populations (i.e., more males than females, Bellrose et al. 1961), and our long-term management response to this fact (with monogamous species) is to set hunting regulations that direct more harvest pressure on the males. Initially this management plan seems appropriate, the assumption being that we shift harvest to the surplus segment of the population that adds little to overall recruitment. However, several important questions should be considered. Does reduction of excess males in the mallard population affect overall recruitment? For example, unpaired males often fertilize females attempting to renest. Is there an evolutionary adaptation to disparate sex ratios? With given levels of harvest, mallard population levels may not be maximized by disproportional harvest of the male segment. No research to date has adequately addressed these questions. If these basic biological questions cannot be answered, hunting regulations to increase male harvest may not be justified because of expensive public education and enforcement problems they create.

A second common example is prescribed burning as a management practice to increase deer and elk populations. The effectiveness of this management has not been addressed, and most follow-up evaluations have simply noted increases in browse forage species and changes in animal distributions. Increased population levels in response to prescribed burning have not been adequately documented or thoroughly studied (Peek 1989).

A third example is the use of population indices to monitor changes in population levels (e.g., ring-necked pheasant crowing counts). The primary assumption for use of a population index is that *the index is directly related to density.* Although nearly every wildlife management agency uses trend data from population indices for management decisions, only a few rare examples of index validation exist (e.g., Rotella and Ratti 1986, Crête and Messier 1987), and some studies have disclosed that index values are not related to density (e.g., see "PM calling data" of Rotella and Ratti 1986, Smith et al. 1984, Nottingham et al. 1989).

If wildlife agencies have the responsibility for management of wildlife populations, they also have the responsibility to conduct research on the effectiveness of management programs. Wildlife agency administrators should strive to develop long-term management-research programs as a basic component of annual agency operations.

SUMMARY

Carefully designed wildlife research will improve the reliability of knowledge that is the basis of wildlife management. Research biologists must rigorously apply the scientific method and make use of powerful techniques in survey sampling and experimental design. Much more effort must be dedicated to the design phase of research, including obtaining critiques from other biologists and statisticians, avoiding common problems such as insufficient sample sizes, procedural inconsistencies, nonuniform treatments, and pseudoreplication. Wherever possible, we must move from observational studies to experimental studies that provide a more reliable basis for interpretation and conclusions. Wildlife biologists have a tremendous responsibility associated with management of animal species experiencing increasing environmental-degradation problems, loss of habitat, and declining populations. We must face these problems armed with knowledge from quality scientific investigations.

Acknowledgments.—We thank J. R. Alldredge, J. H. Bassman, R. A. Black, W. R. Clark, F. W. Davis, R. A. Fischer, T. K. Fuller, G. D. Hayward, J. A. Kadlec, D. G. Miquelle, J. M. Peek, K. P. Reese, J. J. Rotella, J. M. Scott, R. K. Steinhorst, G. C. White, wildlife students at The Ohio State University and the University of Idaho, and three anonymous referees for valuable review comments. This is contribution 565, University of Idaho Forestry, Wildlife, and Range Experiment Station.

LITERATURE CITED

ALLDREDGE, J. R., AND J. T. RATTI. 1986. Comparison of some statistical techniques for analysis of resource selection. J. Wildl. Manage. 50:157–165.

BELLROSE, F. C., T. G. SCOTT, A. S. HAWKINS, AND J. B. LOW. 1961. Sex ratios and age ratios in North American ducks. Ill. Nat. Hist. Surv. Bull. 27:391–474.

BOWDEN, D. C., A. E. ANDERSON, AND D. E. MEDIN. 1984. Sampling plans for mule deer sex and age ratios. J. Wildl. Manage. 48:500–509.

BOYD, R. J., A. Y. COOPERRIDER, P. C. LENT, AND J. A. BAILEY. 1986. Ungulates. Pages 519–564 *in* A. Y. Cooperrider, R. J. Boyd, and H. R. Stuart, eds. Inventory and monitoring of wildlife habitat. U.S. Dep. Inter. Bur. Land Manage. Serv. Cent., Denver, Colo.

BUNGE, M. 1967. Scientific research. I: The search for system. Springer-Verlag, New York, N.Y. 536pp.

BURNHAM, K. P., D. R. ANDERSON, AND J. L. LAAKE. 1980. Estimation of density from line transect sampling of biological populations. Wildl. Monogr. 72. 202pp.

CAUGHLEY, G. 1977. Analysis of vertebrate populations. John Wiley & Sons, New York, N.Y. 234pp.

CHAMBERLIN, T. C. 1965. The method of multiple working hypotheses. Science 148:754–759.

CHAPMAN, J. A., C. J. HENNY, AND H. M. WIGHT. 1969. The status, population dynamics, and harvest of the dusky Canada goose. Wildl. Monogr. 18. 48pp.

———, AND G. R. WILLNER. 1986. Lagomorphs. Pages 453–473 *in* A. Y. Cooperrider, R. J. Boyd, and H. R. Stuart, eds. Inventory and monitoring of wildlife habitat. U.S. Dep. Inter. Bur. Land Manage. Serv. Cent., Denver, Colo.

COCHRAN, W. G. 1963. Sampling techniques. Second ed. John Wiley & Sons, New York, N.Y. 413pp.

COOK, T. D., AND D. T. CAMPBELL. 1979. Quasi experimentation: design and analysis issues for field studies. Houghton Mifflin, Boston, Mass. 405pp.

COOPER, S. V., K. E. NEIMAN, R. STEELE, AND D. W. ROBERTS. 1987. Forest habitat types of northern Idaho: a second approximation. U.S. For. Serv. Gen. Tech. Rep. INT-236. 135pp.

CRÊTE, M., AND F. MESSIER. 1987. Evaluation of indices of gray wolf, *Canis lupis,* density in hardwood-conifer forests of southwestern Quebec. Can. Field-Nat. 101:147–152.

CYR, A. 1981. Limitation and variability in hearing ability in censusing birds. Pages 327–333 *in* C. J. Ralph and J. M. Scott, eds. Estimating numbers of terrestrial birds. Stud. Avian Biol. 6.

DEWEY, J. 1938. Scientific method: induction and deduction. Pages 419–441 *in* J. Dewey, ed. Logic—the theory of inquiry. Holt and Co., New York, N.Y.

DIAMOND, J. R. 1986. Overview: laboratory experiments, field experiments and natural experiments. Pages 3–22 *in* J. R. Diamond and T. J. Case, eds. Community ecology. Harper & Row, New York, N.Y.

DIXON, W. J., AND F. J. MASSEY, JR. 1983. Introduction to statistical analysis. Fourth ed. McGraw-Hill, New York, N.Y. 678pp.

DOLBY, G. R. 1982. The role of statistics in the methodology of the life sciences. Biometrics 38:1069–1083.

EBERHARDT, L. L., AND J. M. THOMAS. 1991. Designing environmental field studies. Ecol. Monogr. 61:53–73.

FORBES, L. S. 1990. A note on statistical power. Auk 107:438–439.

GILL, R. B. 1985. Wildlife research—an endangered species. Wildl. Soc. Bull. 13:580–587.

HAIRSTON, N. G. 1989. Ecological experiments: purpose, design, and execution. Cambridge studies in ecology. Cambridge Univ. Press, New York, N.Y. 370pp.

HEISEY, D. M., AND T. K. FULLER. 1985. Evaluation of survival and cause-specific mortality rates using telemetry data. J. Wildl. Manage. 49:668–674.

HOHF, R. S., J. T. RATTI, AND R. CROTEAU. 1987. Experimental analysis of winter food selection by spruce grouse. J. Wildl. Manage. 51:159–167.

HOUSTON, D. B. 1982. The northern yellowstone elk—ecology and management. Macmillan Publ. Co., New York, N.Y. 474pp.

HURLBERT, S. H. 1984. Pseudoreplication and the design of ecological field experiments. Ecol. Monogr. 54:187–211.

JAMES, F. C., AND C. E. MCCULLOCH. 1985. Data analysis and the design of experiments in ornithology. Pages 1–63 *in* R. F. Johnston, ed. Current ornithology. Vol. 2. Plenum Press, New York, N.Y.

JOHNSON, R. A., AND D. W. WICHERN. 1988. Applied multivariate statistical analysis. Second ed. Prentice-Hall, Englewood Cliffs, N.J. 607pp.

KAMIL, A. C. 1988. Experimental design in ornithology. Pages 313–346 *in* R. F. Johnston, ed. Current ornithology. Vol. 5. Plenum Press, New York, N.Y.

KEPLER, C. B., AND J. M. SCOTT. 1981. Reducing bird count variability by training observers. Pages 366–371 *in* C. J. Ralph and J. M. Scott, eds. Estimating numbers of terrestrial birds. Stud. Avian Biol. 6.

KERLINGER, F. N. 1973. Foundations of behavioral research. Second ed. Holt, Rinehart and Winston, Inc., New York, N.Y. 741pp.

KOZICKY, E. L., G. O. HENDERSON, P. G. HOMEYER, AND E. B. SPEAKER. 1952. The adequacy of the fall roadside pheasant census in Iowa. Trans. North Am. Wildl. Nat. Resour. Conf. 17:293–305.

KREBS, C. J. 1989. Ecological methodology. Harper & Row, New York, N.Y. 654pp.

MACNAB, J. 1983. Wildlife management as scientific experimentation. Wildl. Soc. Bull. 11:397–401.

MAYR, E. 1970. Populations, species, and evolution. Belknap Press of Harvard Univ. Press, Cambridge, Mass. 453pp.

MILLIKEN, G. A., AND D. E. JOHNSON. 1984. Analysis of messy data: designed experiments. Vol. I. Van Nostrand Reinhold, New York, N.Y. 473pp.

MUELLER-DOMBOIS, D., AND H. ELLENBERG. 1974. Aims and methods of vegetation ecology. John Wiley & Sons, New York, N.Y. 547pp.

→MURPHY, D. D., AND B. D. NOON. 1991. Coping with uncertainty in wildlife biology. J. Wildl. Manage. 55:773–782.

NEWTON-SMITH, W. H. 1981. The rationality of science. Routledge and Kegan Paul, Boston, Mass. 294pp.

NOTTINGHAM, B. G., JR., K. G. JOHNSON, AND M. R. PELTON. 1989. Evaluation of scent-station surveys to monitor raccoon density. Wildl. Soc. Bull. 17:29–35.

OPLER, P. A. 1976. The parade of passing species: a survey of extinctions in the U.S. The Sci. Teacher 43:30–34.

OVERTON, W. S., AND D. E. DAVIS. 1969. Estimating the numbers of animals in wildlife populations. Pages 403–456 in R. H. Giles, ed. Wildlife management techniques. Third ed. The Wildl. Soc., Washington, D.C.

PATRIC, E. F., T. P. HUSBAND, C. G. McKIEL, AND W. M. SULLIVAN. 1988. Potential of LORAN-C for wildlife research along coastal landscapes. J. Wildl. Manage. 52:162–164.

PEEK, J. M. 1989. Another look at burning shrubs in northern Idaho. Pages 157–159 in D. M. Baumgartner, D. W. Breuer, and B. A. Zamora, eds. Proc. symp. prescribed fire in the intermountain region: forest site preparation and range improvement. Washington State Univ., Pullman.

PETERMAN, R. M. 1990. Statistical power analysis can improve fisheries research and management. Can. J. Fish. Aquatic Sci. 47:2–15.

PIETZ, P. J., AND J. R. TESTER. 1983. Habitat selection by snowshoe hares in north central Minnesota. J. Wildl. Manage. 47:686–696.

PLATT, J. R. 1964. Strong inference. Science 146:347–353.

POPPER, K. R. 1959. The logic of scientific discovery. Hutchinson and Co., London, U.K. 480pp.

———. 1968. Conjectures and refutations: the growth of scientific knowledge. Second ed. Harper & Row, New York, N.Y.

→QUINN, J. F., AND A. E. DUNHAM. 1983. On hypothesis testing in ecology and evolution. Am. Nat. 122:602–617.

RATTI, J. T. 1980. The classification of avian species and subspecies. Am. Birds 34:860–866.

———, D. L. MACKEY, AND J. R. ALLDREDGE. 1984. Analysis of spruce grouse habitat in north-central Washington. J. Wildl. Manage. 48:1188–1196.

———, AND K. P. REESE. 1988. Preliminary test of the ecological trap hypothesis. J. Wildl. Manage. 52:484–491.

———, AND D. E. TIMM. 1979. Migratory behavior of Vancouver Canada geese: recovery rate bias. Pages 208–212 in R. L. Jarvis and J. C. Bartonek, eds. Proc. Manage. Biol. Pacific Flyway geese. Northwest Sect., The Wildl. Soc., Portland, Oreg.

REYNOLDS, R. T., J. M. SCOTT, AND R. A. NUSSBAUM. 1980. A variable circular-plot method for estimating bird numbers. Condor 82: 309–313.

ROBBINS, C. S., D. BYSTRAK, AND P. H. GEISSLER. 1986. The breeding bird survey: its first 15 years, 1965–1979. U.S. Fish Wildl. Serv. Resour. Publ. 157. 154pp.

ROMESBURG, H. C. 1981. Wildlife science: gaining reliable knowledge. J. Wildl. Manage. 45:293–313.

ROTELLA, J. J., AND J. T. RATTI. 1986. Test of a critical density index assumption: a case study with gray partridge. J. Wildl. Manage. 50: 532–539.

SAMUEL, M. D., E. O. GARTON, M. W. SCHLEGEL, AND R. G. CARSON. 1987. Visibility bias during aerial surveys of elk in northcentral Idaho. J. Wildl. Manage. 51:622–630.

SCHEAFFER, R. L., W. MENDENHALL, AND L. OTT. 1986. Elementary survey sampling. Third ed. Duxbury Press, Boston, Mass. 324pp.

SEBER, G. A. F. 1982. The estimation of animal abundance and related parameters. Second ed. Charles Griffin, London, U.K. 600pp.

SELANDER, R. K. 1971. Systematics and speciation in birds. Pages 57–147 in D. S. Farner and J. R. King, eds. Avian biology. Vol. I. Academic Press, New York, N.Y.

SINCLAIR, A. R. E. 1991. Science and the practice of wildlife management. J. Wildl. Manage. 55:767–773.

SMITH, L. M., I. L. BRISBIN, JR., AND G. C. WHITE. 1984. An evaluation of total trapline captures as estimates of furbearer abundance. J. Wildl. Manage. 48:1452–1455.

STEBBINS, G. L. 1971. Processes of organic evolution. Second ed. Prentice-Hall, Englewood Cliffs, N.J. 193pp.

STEEL, R. G., AND J. H. TORRIE. 1980. Principles and procedures of statistics: a biometrical approach. Second ed. McGraw-Hill Book Co., New York, N.Y. 633pp.

STEINHORST, R. K., AND M. D. SAMUEL. 1989. Sightability adjustment methods for aerial surveys of wildlife populations. Biometrics 45: 415–425.

SWIHART, R. K., AND N. A. SLADE. 1985. Testing for independence of observations in animal movements. Ecology 66:1176–1184.

TAKEKAWA, J. Y., AND E. O. GARTON. 1984. How much is an evening grosbeak worth? J. For. 82:426–428.

TOFT, C. A., AND P. J. SHEA. 1983. Detecting community-wide patterns: estimating power strengthens statistical inference. Am. Nat. 122:618–625.

TUKEY, J. W. 1977. Exploratory data analysis. Addison-Wesley Publ. Co., Reading, Mass. 688pp.

WESTERN, D., AND M. C. PEARL. 1989. Conservation for the twenty-first century. Oxford Univ. Press, New York, N.Y. 365pp.

WHITE, G. C., D. R. ANDERSON, K. P. BURNHAM, AND D. L. OTIS. 1982. Capture-recapture and removal methods for sampling closed populations. Rep. LA-8787-NERP, UC-11, Los Alamos Natl. Lab., Los Alamos, N.M. 235pp.

WIENS, J. A. 1989. The ecology of bird communities: foundations and patterns. Vol. I. Cambridge Univ. Press, New York, N.Y. 539pp.

WILSON, M. M., AND J. A. CRAWFORD. 1979. Response of bobwhites to controlled burning in south Texas. Wildl. Soc. Bull. 7:53–56.

ZAR, J. H. 1984. Biostatistical analysis. Second ed. Prentice-Hall, Englewood Cliffs, N.J. 718pp.

2

ANALYSIS OF DATA

Jonathan Bart and William Notz

INTRODUCTION

This chapter describes general procedures for analyzing data collected during wildlife studies. Our goals are to provide some information about nearly any technique readers may need or want to use and to cover the most commonly used methods in greatest detail. In preparing the chapter we drew on our own background and a general literature review. We also scrutinized each article published during 1988 in *The Journal of Wildlife Management* and the *Wildlife Society Bulletin* to determine what types of analyses were conducted most frequently by wildlife biologists (Table 1). We relied heavily on this survey in deciding which techniques to discuss in greatest detail in this chapter.

We have tried to provide the level of quantitative detail about each topic that will be most useful to readers. For estimating single quantities or comparing two estimates, our experience is that biologists often carry out the calculations on a hand calculator or use simple programming commands on a computer (e.g., in a spreadsheet). For such users, it is essential that formulas be provided. Even if package programs are used, choosing among different

options is often necessary. Deciding whether to use a pooled or unpooled variance in carrying out a *t*-test is one example. Here, too, the reader is best served by having the detailed formulas needed for the calculations. In most multivariate analyses, however, the equations are too complex to be presented in detail here, and in any case they would be of little interest to many biologists; therefore we present only occasional equations in this section of the chapter and provide numerous references to more detailed accounts.

GRAPHICS
Introduction

One of the simplest yet most powerful ways of summarizing data is by means of charts, graphs, and tables. This is also one of the best techniques for misrepresenting data, either intentionally or unintentionally. Most articles in the wildlife literature contain at least one chart, graph, or table pertaining to the data collected or the subsequent statistical analyses. With the increasing popularity of desktop computers has come a proliferation of software, such as spreadsheets, graphics, and statistical software, for

Table 1. Number of articles reporting the use of different statistical methods in the 1988 volumes of *The Journal of Wildlife Management* (JWM) and The *Wildlife Society Bulletin* (WSB).

Subject	Method or data	JWM	WSB	Total
Sampling plans	Nonrandom	105	43	148
	Simple random	21	10	31
	Systematic	41	17	58
	Stratitifed	10	6	16
	Multiple-stage	58	15	73
Two-sample comparisons	Bivariate data	10	11	21
	Continuous data	61	25	86
Multiple comparisons	Bivariate data	13	11	24
	Continuous data	45	15	60
Regression correlation	Bivariate	33	12	45
	Multivariate	7	4	11
Other	Survival analysis	12	0	12
	Discriminant analysis	5	1	6
	Principal components	4	0	4
Number of articles		129	69	198

handling data. All such software has some capability for creating high-quality charts, graphs, and tables. Although certain types of charts, graphs, and tables are standard, there is room for creativity, and presenting complex data clearly is an art. Good graphics allow the reader to see the salient features of the data in a glance.

Before specific types of graphical displays are discussed, a few general principles are in order. First, always strive for clarity and honesty. This includes labelling displays clearly, avoiding distortion, and scaling properly. Second, always give the source of the data. Third, tables are generally preferable to graphics for displaying exact numerical values and for small data sets. For large data sets, where one wishes to display trends or overall relationships, graphics are preferable to tables. These principles and some educated common sense will carry one a long way. One of the best ways to decide how to make a chart, graph, or table of a set of data is to look for similar examples in the wildlife journals. Another excellent source of examples is the *Statistical Abstract of the United States*. This yearly compilation of data contains tables and graphs of every sort. Any library will have copies of it.

Charts

Among the most commonly used charts are pie charts, histograms, and bar charts. All are used to display characteristics of data that have been subdivided into various classes. The pie chart is simply a circle (pie) divided into slices and is used to display the relative sizes of the various classes. Each slice corresponds to a class, and the size of the slice represents the proportion of the data falling into that class. Alisauskas et al. (1988) used pic charts to display the composition of the winter diets (relative sizes of classes of plant structures comprising the diet) of lesser snow geese based on the type of region in which the geese fed (Fig. 1). It is immediately clear from the charts that one plant structure predominates in the diet of the geese, but this predominating structure is different in each type of region.

Histograms and bar charts convey information analo-

gous to that in a pie chart. A histogram is appropriate for numerical data and a bar chart for categorical (non-numerical) data, but in other respects these two types of graphs are nearly identical. A histogram is constructed by first dividing data into classes, each class consisting of a value or range of values of the data. Classes must be defined so that each data value belongs to one and only one class. A histogram is then constructed by drawing horizontal and vertical axes. The horizontal axis is marked off in the units of measurement of the data with a uniform scaling along the axis. The horizontal axis is then divided into segments corresponding to the range of values defining the different classes. The vertical axis represents frequencies (counts) or relative frequencies (proportions). Vertical bars are drawn above each class in such a way that the *area* of the bar is proportional to the frequency or relative frequency of observations in the class. If classes correspond to ranges of values all of equal length, one may draw the bars to have heights equal to frequencies or relative frequencies, and areas of bars will automati-

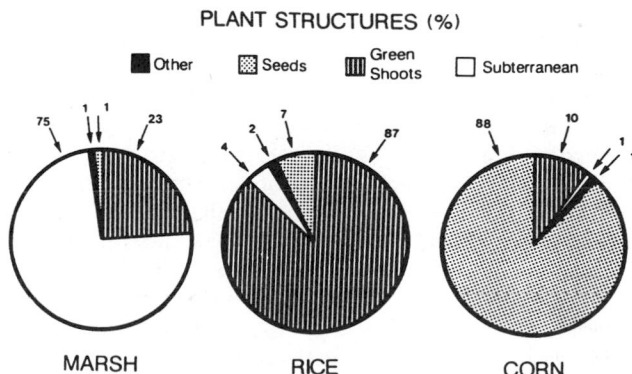

Fig. 1. Example of a pie-chart showing relative proportions (%) of a composition of winter diets of lesser snow geese on the basis of the plant structures consumed during 1983–84 in coastal marshes (La.)(MARSH), rice prairies (La. and Tex.)(RICE), and areas near the Missouri River Valley (La., Mo., and Kans.)(CORN) (from Alisauskas et al. 1988).

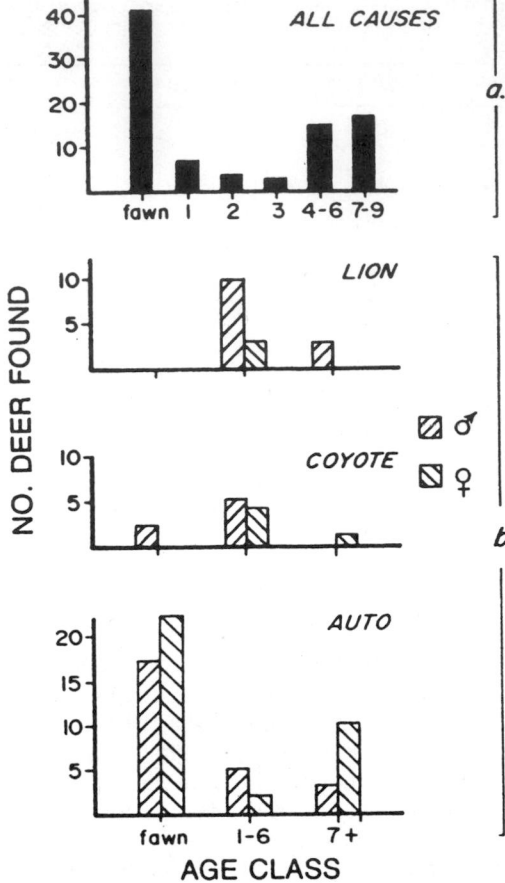

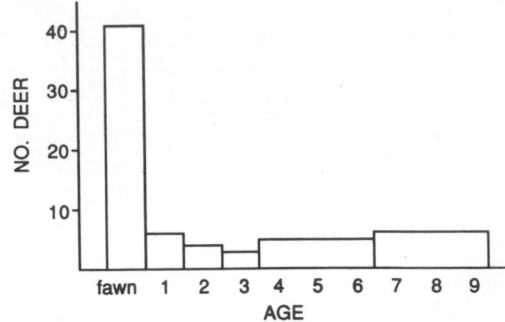

Fig. 3. A properly drawn histogram for the age structure of deer killed by mountain lions, coyotes, and automobiles combined. Notice the lack of vertical lines between the ages 4, 5, and 6. This suggests that these ages are not separated in the data. One might consider labelling the bar above 4, 5, and 6 as 4–6. Similar comments apply to the bar above 7, 8, and 9.

Fig. 2. Example of a bar chart showing the age structure of deer killed by mountain lions, coyotes, and automobiles combined (a) and separately (b) in western Montana during winters 1969–81 (from O'Gara and Harris 1988).

cally be in the correct proportions. If classes are of differing lengths, then, because areas of bars are equal to heights of bars times their widths, heights must be adjusted to offset discrepancies in the lengths and keep the overall area of the bars proportional to frequencies or relative frequencies. In this way, heights of bars represent frequencies or relative frequencies per unit length of the classes.

To illustrate how one might construct a histogram when classes are of differing lengths, we consider an example. O'Gara and Harris (1988) constructed charts of the number of deaths (frequencies) of deer in various age classes due to various causes (a separate graph for each cause) (Fig. 2). In these charts, the authors treated age as a categorical variable and so technically these charts are bar charts. Because age is actually a numerical variable, histograms might be more appropriate here. To see how to redraw these charts as histograms, consider the All Causes chart. The last two age categories are each three times as wide as the others. If we construct a proper histogram, the horizontal axis should be marked off in years and the bars for the last two categories should have heights that are one-third those displayed in the bar chart so that the areas of bars are proportional to frequencies (Fig. 3). The heights of the last two bars now represent the average number of deaths per year over the interval covered by

the class. Notice that by redrawing this bar chart we have altered the visual impression it makes. First, we see that the last two bars cover a wider age class than the other bars. Second, the original chart gives a false impression of a sudden large increase in the death rate when deer are >3 years of age. After adjustment in Fig. 3, the last two bars have heights comparable to the bars above the 1-, 2-, and 3-year classes. The increase in death rate does not appear as dramatic. For the remaining charts in Fig. 2, the appropriate adjustment for converting the Age Class charts to histograms would be to reduce the height of the bar above the 1–6 class by a factor of 6. In the 7+ age class, one would need to know the largest observed age to make the proper adjustment. The All Causes chart suggests this largest age is 9; if this is correct, bar height would be reduced by a factor of 3.

Bar charts are similar to histograms and are appropriate when data are categorical. The horizontal axis is now marked off with the category labels, and bars are drawn above each category. Additionally, the vertical axis can be used to display characteristics (such as means) of the classes instead of the frequencies or relative frequencies, if desired. Mason et al. (1989) used bar charts to represent the mean consumption of anthranilate and plain Purina Flight Bird Conditioner (PFBC) when starlings were given both as a choice. The classes correspond to different types of anthranilate used in the anthranilate PFBC. Heights of bars (Fig. 4) represent means rather than frequencies or relative frequencies. Notice that by displaying pairs of bars, the authors have essentially superimposed two separate bar charts. This allows the authors to display extra information in a single picture, and comparisons are readily made. The plain PFBC was clearly preferred to the anthranilate PFBC in all tests. Note one extra feature of this chart. The authors have also provided information about standard errors of the means by including capped vertical lines atop each bar. This feature is not unusual in bar charts representing means.

Although pie charts, histograms, and bar charts convey similar information, pie charts are probably best used, if at all, when classes are few and are merely categorical. Histograms and bar charts are always appropriate, especially when the number of classes is large. If the data are numerical and the classes represent ranges of values

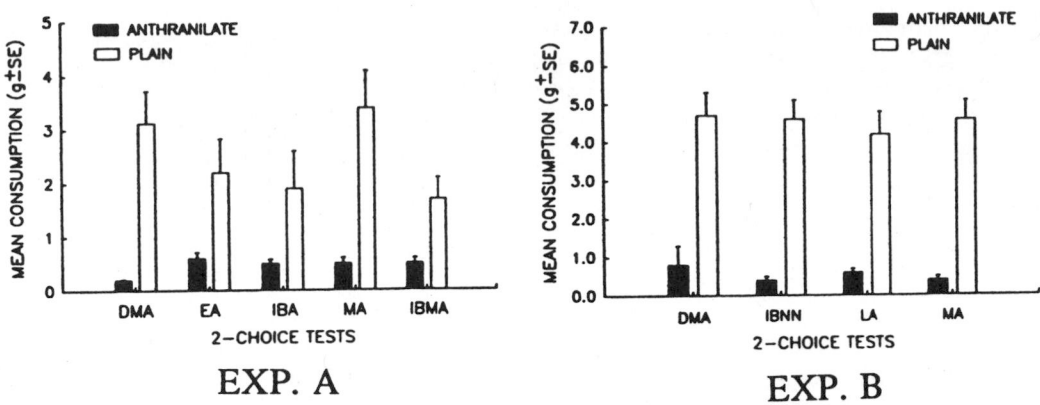

Fig. 4. Bar charts representing the consumption of anthranilate and plain Purina Flight Bird Conditioner (PFBC) in two experiments (A and B) when European starlings were given both as a choice. Dark bars represent consumption of anthranilate PFBC. Open bars represent consumption of plain PFBC. Capped vertical lines represent standard errors of the means. DMA = methyl-N-methyl anthranilate, EA = ethyl anthranilate, IBA = isobutyl anthranilate, MA = methyl anthranilate, IBMA = isobutyl methyl anthranilate, IBNN = isobutyl-N-N-dimethyl anthranilate, LA = linalyl anthranilate (from Mason et al. 1989).

of the variable, the use of a horizontal axis marked off in equal increments displays the ordering of the data. By displaying this ordering, the histogram conveys more information than the corresponding pie chart. Some authors (for example, Tufte 1983) advocate the use of tables instead of pie charts to display small data sets, in keeping with our third general principle. Nevertheless, pie charts remain a popular and familiar graphic.

Graphs

Perhaps the most common graphs one encounters are line graphs and scatterplots. In both types, the objective is to see how some measured response (usually plotted on the vertical axis of the graph) varies as a function of some independent variable (usually plotted on the horizontal axis). Scatterplots are described later (see Scatterplots and Correlation [p. 44].) Line graphs are generally used to display measurements that are taken in sequence, usually over a period of time. The vertical axis of the graph is marked off in the units of the measurements. The horizontal axis is marked off 1, 2, 3, 4, . . . if the measurements are merely ordered 1st, 2nd, 3rd, 4th, ... If the measurements are taken over time, the horizontal axis is scaled in units of time. Measurements are plotted as points on this graph, and consecutive measurements are connected with lines, hence the name line graph. Such a graph is particularly useful for seeing trends in measurements over time. Sæther and Gravem (1988) plotted temperature measurements (C) recorded over time (Nov–Apr) in Norway for the 1983/84 and 1984/85 winters (Fig. 5). Not surprisingly, temperatures dropped until January or February, after which they began to increase. Notice the 1984/85 winter was the colder of the two in the January–March period.

Two points are important to remember in the preparation of graphs. The first is to use a uniform scaling of the axes. Changing the scale along an axis will distort the appearance of the trend in the graph. The second involves location of zero points, particularly when the zero point represents the absence of some property. Generally one assumes that the intersection of the horizontal and vertical axes occurs at the zero point of both scales, and one may

unconsciously extrapolate trends in a graph to the zero point by extrapolating to the intersection of the axes. Thus one should consider making the intersection of the axes the zero point when feasible. Often, however, the range of measurements is far from zero and it becomes impossible to use a reasonable scaling of an axis while including the zero point and fitting the graph into a small space. In such a situation axes must be labelled clearly to alert the reader that the intersection of the axes is not the zero point. This is what Sæther and Gravem did in Fig. 5. Alternatively, one may still label the intersection of the axes as the zero point, but mark a break along the axis to inform the reader that a portion of the axis between the zero point and the beginning of the range of measurements has been omitted.

Tables

Tables are the most common means of displaying or summarizing information in the wildlife literature. No one style predominates, but all have elements in common. Typically rows (or groups of rows) correspond to classes, i.e., different categories or ranges of values of some variable. Columns give various summary information corre-

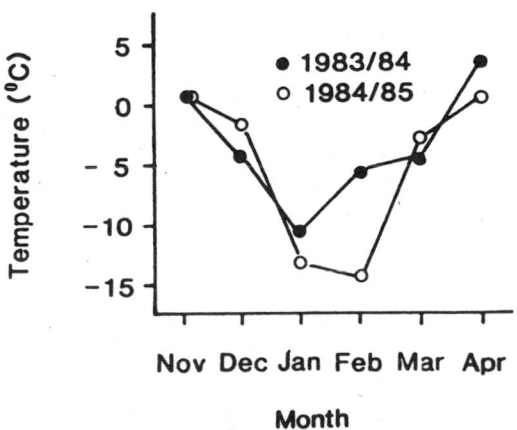

Fig. 5. Example of line graph showing temperatures in southeastern Norway during winters, 1984 and 1985 (Sæther and Gravem 1988).

Table 2. Number of observations, mean amount of kidney fat (g) of moose calves from southeastern Norway, SE of the mean, and range of the observations by sex for the winters 1984 and 1985 (from Sæther and Gravem 1988).

Winter	Sex	Kidney fat			
		n	$\bar{x}$	SE	Range
1984	M	9	2.9	0.34	1.1–4.4
	F	14	3.5	0.36	1.6–7.3
1985	M	7	3.2	0.31	2.5–4.5
	F	13	3.0	0.23	1.9–4.0
Both winters	M	16	3.0	0.23	1.1–4.5
	F	27	3.3	0.22	1.6–7.3

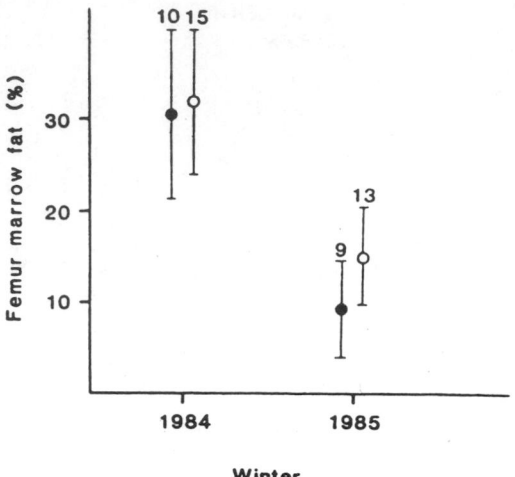

Fig. 6. Graphic representing the femur fat content (%) ($\bar{x} \pm$ SE) for male (black circles) and female (open circles) moose calves in southeastern Norway, 1984–85 (*n* indicated above SE) (from Sæther and Gravem 1988).

sponding to these classes. Sæther and Gravem (1988) provided summary statistics (number of animals and the mean, standard error, and range of kidney fat) for a sample of male and female moose for the winters 1984 and 1985 in southeastern Norway (Table 2).

The important principles to bear in mind when tables are constructed are, as for all graphics, to be clear (label clearly) and to provide the source of the data. In the example above, the authors make it clear how the data were collected in the text of the paper. The table itself provides sufficient information for the reader to understand what the numbers in the table refer to.

Perhaps the main drawback of tables is that they do not make as forceful an impression as other graphics. Consider the graphic in Fig. 6, also from Sæther and Gravem (1988), which gives information on fat content in moose femurs. This graphic presents the same kind of information as in Table 2. However, in this graphic one more readily sees patterns (such as the change from 1984 to 1985) than one would in a table.

Graphics Software

The above discussion is only a brief introduction to the use of graphics. More information along with numerous intriguing examples can be found in Tufte (1983). Additionally, many statistical software packages now have the capability for doing interactive graphics. Four that the authors are familiar with are the S language (for use on UNIX-based computers), SYSTAT™ and SYGRAPH™ (distributed by SYSTAT, Inc., and available both for PCs and Macintosh computers), Data Desk Professional (distributed by Odesta), and JMP (distributed by SAS™). The latter two are designed to run only on Macintosh computers and take advantage of the ease of use and multi-window capabilities of Macintosh computers. Several packages are available for PCs that allow interactive graphics. Besides enabling the user to "fiddle" interactively to produce the best possible graphic (for example, one could interactively construct histograms with greater or fewer bars, or scale axes differently to construct the histogram that most clearly displays important characteristics of the data), these programs allow the user to employ graphics as a means of "seeing" what the data are saying. This feature can be particularly useful for uncovering hidden patterns or trends in "messy" data sets. Because this "fiddling" is not truly formal analysis, any-

thing one uncovers should be taken as merely suggestive, rather than as strong evidence of characteristics or relationships. Follow-up experiments or studies can be designed to confirm any unusual features one finds through such interactive analyses. Another approach is to randomly divide one's data into two groups. Graphical displays can be used to investigate the data in one group, and the other group can be used to confirm any patterns found in the first group. This approach is sometimes called cross-validation. The interested reader can find more information about interactive graphical methods for data analysis in the manuals for SYGRAPH, Data Desk Professional, and JMP, as well as in Chambers et al. (1983).

QUANTITATIVE ANALYSIS

The remainder of this chapter is concerned with quantitative, largely statistical analysis (see Box 1). We have divided the material into three main sections: estimating a single quantity of interest, comparing two estimates, and investigating relationships between two or more quantities of interest. We begin by defining some terms that are used repeatedly in the rest of the chapter.

Population Units and Response Variables

The *population unit* is what we measure, or "the thing about which we record data." It is usually a plant or animal, a trap or recording device, or an area. The population unit often has a temporal dimension (e.g., a trap-night, an animal watched for 1 hour). In telemetry and behavior studies, instantaneous observations are often made on several animals. We will refer to this type of population unit as an "animal-time." The *sample units* are the population units selected for inclusion in the sample.

The *variables* are the measurements recorded on each sample unit. *Response variables*, also called *dependent variables*, are the attributes we are interested in estimating or studying. *Explanatory variables*, also known as *independent variables*, are used to obtain information about the response or dependent variables. The variables also may be classified according to how many different values they may take and by whether the values are simply labels or imply size or some other ordered relationship. *Continuous variables* take on an infinite number of values, such as all real numbers in an interval. *Dichotomous* or *binomial variables* can take only two values. Common examples in wildlife studies include male/female, young/adult, alive/dead, present/not present in a given habitat. Such data are often coded as 1s and 0s. The mean of the 1s and 0s equals the proportion of units that fall in one category. One reason for distinguishing between dichotomous and continuous variables is that different statistical methods are used for these two types of data. In practice, methods for continuous data are usually applied when the data take three or more different values. *Label* or *nominal* variables may take any number of values, but larger values do not imply larger size or any other ordering; they simply designate different conditions. Common examples include sex, which is also a dichotomous variable, study site, year, and age. Year and age might be used as either label or continuous variables according to whether trends through time or age are being estimated. The critical issue is whether the values of the variable are *ordered*.

Here are some examples illustrating these terms. Ely and Raveling (1989) collected geese wintering in the Sacramento Valley to estimate average body weight (among other objectives). The population unit in this study was one goose; the response variable was weight. The population consisted of all the geese in the study area; the sample was the geese they collected. Explanatory variables, used to predict or study weight, included such measures as sex, date, and location. Weight and perhaps date are continuous variables; sex is a dichotomous variable; location is a label variable. Santillo et al. (1989) studied the response of small mammals to herbicide treatment. For one of their objectives, they counted the number of hardwood stems on small, circular plots. The population unit, for this part of the study, was the area covered by one plot; the response variable was the number of stems, a continuous variable. The authors also measured small mammal abundance by setting snap traps and checking them once a day. The results were expressed as capture per trap-night. The population unit was one trap set for 24 hours (1 trap-night). The response variable was the number of captures, 0 or 1, a dichotomous variable.

Note that "population" in the small mammal example refers to the set of all possible trap-nights, not the set of all small mammals that might have been captured. Nearly any time animals are captured or surveyed, the statistical population is defined in terms of capture or survey effort (e.g., trap-nights, mist-net days, observer-hours, survey-miles). The animals captured or counted constitute the response variable. If this point seems confusing, it may be helpful to distinguish between the statistical population, defined as above, and the biological population, which is the group of animals being studied.

Types of Error

Error, in statistical terminology, refers to the difference between an estimate and the quantity being estimated, usually referred to as the *parameter*. Error may be divided into *sampling error*, caused by random selection of which items to include in the sample, and *bias*, a consistent tendency to over- or underestimate the parameter. More specifically, bias is any error that would have occurred even if we had selected a very large sample so that effects of random selection were negligible.

Measurement bias is the result of errors occurring as the data are recorded; *statistical bias* is caused by the way in which the data are analyzed. Measurement bias is possible anytime measurements are not recorded on some units selected for the sample or when measurements are recorded inaccurately. Statistical bias is usually zero or negligible when standard, widely recognized analytic methods are used. When newer methods, especially ones developed by the investigator, are used, careful study should be given to whether statistical bias exists. Some examples will make the three types of error clearer.

Leuschner et al. (1989) selected a random sample of hunters in the southeastern United States and asked them whether tax dollars should be spent on wildlife. The purpose was to estimate what proportion of all hunters in the study area would answer yes to this question. Sampling error was probably present in the study because, by chance, the proportion of respondents in the sample who believed that tax dollars should be spent on wildlife was probably not exactly equal to the proportion of all hunters who felt this way. Measurement bias could have been present, because 42% of the people selected for the sample were unreachable, gave unusable answers, or did not answer at all. These people might have felt differently, as a group, than those who did answer the question, and thus the proportion of yes answers in the responses obtained might have differed from the proportion that would have been obtained had everyone provided a usable answer. The authors used standard, widely accepted methods to analyze their results, so it is unlikely that any serious statistical bias was present in the estimates. Note that these three sources of error—sampling error, measurement bias, and statistical bias—are entirely separate from one another. Stating, as in the example above, that no statistical bias was present in the estimates does not reveal anything about the magnitude of sampling error or measurement bias.

Otis et al. (1978) developed methods for estimating population size when animals are captured, marked, and released, and then some of them are recaptured one or more times. The quantity of interest (the parameter) is the total number of animals in the population (assumed, in these particular models, to remain constant during the study). Sampling error would occur because the estimates depend on which animals are captured, and this in turn depends on numerous factors not under the biologists' control. Measurement bias would occur if animals lost their marks (this was assumed not to occur). The methods were relatively new, so the authors studied statistical bias with computer simulations. They reported little statistical bias under some conditions, but under others the estimates were consistently too high or too low. These errors occurred even if they assumed that no marks were lost and that the data were collected properly in all other ways (i.e., no measurement bias).

One reason for distinguishing among these types of error is that statistical analyses usually provide estimates of how serious sampling error is, but usually do not provide any information about the effects of bias. This point is often not stressed by introductory statistics texts because the authors assume that bias will be negligible. In many wildlife studies, however, this is not a reasonable assumption. For example, wildlife populations are often monitored with methods that do not detect all individuals in the sample plots. If the results are used to estimate density, the estimate is biased by the undercounting. Statistical analysis of the counts reveals the effects of sampling error but not the effects of the undercounting. Furthermore, if detection rates change, the counts may suggest a change in abundance when none has occurred or may miss a real change in abundance. Because some degree of bias is often unavoidable in wildlife studies, and because its effects are not estimated or accounted for in most statistical analyses, a great deal of effort in wildlife studies is directed towards placing upper bounds on the magnitude of bias.

SAMPLED POPULATION AND POPULATION OF INTEREST

Investigators usually wish to extrapolate their findings beyond the limits of the population that they studied. For example, most studies are confined to a relatively small study area and are carried out during a small number of years. Investigators hope, however, that their results are applicable to a much wider region and other years. Most questionnaire studies, for example, are made on smaller populations than the investigator is really interested in. Thus, in the study of attitudes toward spending tax money for wildlife, cited above, Leuschner et al. (1989) wished to determine the views of foresters as well as hunters. The sample was selected from the 1985 Virginia membership list of the Society of American Foresters, but the main population of interest included all foresters in the southeastern United States.

In examples such as these, distinguishing among the sampled population, the group from which samples were selected, and the larger population of interest is sometimes helpful. Two reasons exist for making this distinction. Conclusions about the sampled population have the force of statistical analysis behind them, whereas extrapolation to a larger population of interest must usually be justified primarily on nonstatistical grounds. In addition, the limits of the sampled population are usually well defined, whereas the limits of the population of interest usually are not. Thus, the term "forester" may be difficult to define rigorously, and the conclusions from a banding study probably apply best to birds that were in the area at the time the banding was done, less well to birds in the area in other years, and even less well to birds farther from the study area.

Point and Confidence Interval Estimates

By "point and confidence interval estimates" we mean, roughly, "What is our best guess about the parameter's value?" and "How accurate is our guess?" For example, Koehler and Hornocker (1989) studied habitat preferences of bobcats. They reported that their animals spent 10.3% of their time in alpine areas and that the 90% confidence interval (CI) for the estimate was 6.2–14.4%. To define the phrases best guess (or point estimate) and CI rigorously, we must imagine taking a large number of additional samples from the same population, using the same sampling plan (except that a different random sample would be drawn each time), and calculating the estimate and CI for each one. If we assume the estimates are unbiased, the average of all the point estimates would exactly equal the true value. This is the sense in which the method used to analyze the sample data provides our "best guess" about the parameter's value. Furthermore, if the assumptions of the analysis are met, 90% of the CIs obtained in a large series of samples would include the true value and 10% would not.

This interpretation of point and interval estimates relies heavily on the estimates being unbiased. If any bias exists, then, by definition, the average of all possible point estimates does not equal the parameter. The proportion of CIs including the parameter would probably be more or less than the nominal level, 90% in the bobcat example, and we would say that the interval estimate was also biased.

Construction of a CI begins with deciding how confident one wants to be that the parameter is included in the CI. The most common level of confidence is 95%, but any other level can be selected. The estimate can be nearly any quantity of interest. Typical examples include sample means, estimated population totals, survival rates, and estimates of relationship such as the slope from a regression analysis. The approach one then follows depends on several factors. For dichotomous data with small sample sizes, special tables or formulas may be needed; in most other situations a simple method is sufficient. The method requires the estimate, its standard error (SE), and the appropriate "t-value," a number that depends on the "degrees of freedom" (df) provided by the data. Formulas for the SE and df for the situations encountered most often in wildlife biology are given in the sections below and in the boxes.

In reporting estimates, one commonly provides the SE

rather than a CI. Readers then can construct their own confidence interval at whatever level of confidence they believe is appropriate. The SE also provides a more direct way to compare two point estimates than does a CI, as explained in later sections. The accuracy of an estimate can also be described by the quantity (SE/estimate). This quantity, called the coefficient of variation (CV), expresses the SE as a proportion of the estimate. Using the CV, readers can quickly determine how large the CI is compared to the estimate. For example, if the CV is 10%, and if we assume the *t*-value is 2 (in many studies its exact value is 1.96), we may infer that the 95% CI is the mean plus and minus 20% of the mean (because the CI = $\bar{y} \pm 2 \cdot SE = \bar{y}[1 \pm 2 \cdot SE/\bar{y}] = \bar{y}[1 \pm 2 \cdot CV]$). This gives us a very different picture of how accurate the estimate is than, say, a CV of 40%, which tells us that the 95% CI equals the mean plus and minus 80% of the mean. The CV also provides a useful way to describe several estimates at once. An investigator may report, for example, that all CVs were <15%.

Describing Variability in the Population

Notice that the CI, SE, and CV tell us how accurate the point estimate is, but they tell us little about how variable the measurements are. Sometimes we are interested in both the mean of the observations and in how variable they were. For example, Golightly and Hofstra (1989) studied the time required to immobilize elk with different chemicals. Investigators want this time to be as short as possible because animals may injure themselves or escape if the chemical requires a long time to take effect. In evaluating different chemicals, one must estimate the mean immobilization time but also must know how consistent the time is from animal to animal.

Two approaches are commonly used to analyze and describe how variable the measurements were in a random sample. The investigator may report the sample size and range, thereby informing readers of the extreme values. Golightly and Hofstra (1989) used this approach, reporting that the mean immobilization time for ketamine used on 23 elk was 8.7 minutes and the range was 5–14 minutes. The second approach is to calculate intervals that contained some proportion of the observations. Thus Golightly and Hofstra might have stated that two-thirds of the immobilization times were between 7.3 and 9.6 minutes or that 80% of them were between 5.8 and 12.7 minutes. A common variation of this approach is to determine whether the distribution of the measurements is about the same as a normal distribution. If it is, then the standard deviation (SD) may be used to calculate the width of any desired interval. For example, in a sample that has a normal distribution, 65% of the observations lie within 1 SD of the mean and 80% of them lie within 1.28 SD of the mean. This approach was used by Anthony and Isaacs (1989) in studying nest sites used by bald eagles. One of their objectives was to describe how much variation occurred in the heights of trees used as nest sites by the eagles. They measured 53 ponderosa pines that had been used by eagles as nest sites, determined that the distribution of the measurements was similar to a normal distribution, and reported that the mean height was 38.0 m and its SD was 5.3. Given this information, the reader can construct intervals that included (approximately) any de-

sired proportion of the measurements. For example, about 80% of the tree heights were probably between 31 and 45 m [38 ± (1.28)(5.3)]. Methods for determining whether observations follow a normal distribution and for calculating intervals estimated to contain any given proportion of the observations in the population are given in many statistics texts (e.g., Moore and McCabe 1988).

The CV, described above, can also be used to express variability in the observations. Its formula, for this purpose, is SD/mean rather than SE/mean. A report that the CV was 15% would tell us that approximately 65% of the measurements were within 15% of the mean and approximately 80% of them were within 19% (1.28 × 15%) of the mean.

Changing the Scale at Which Results are Reported

Investigators frequently record data at one scale of measurement and then need to report the results in a different scale. This is especially common when the population units are plots and the response value is a mean or count. For example, Wood (1988) studied the effects of prescribed fire on deer forage and nutrients. He measured the dry weight of vegetation in 1-m^2 plots before burning. Results were converted to weight per hectare for comparison with other variables.

Notice that conversion to the new scale, in this study, involved multiplying the original estimate by a constant. By the word constant, we mean a number that would have been the same regardless of what value had been obtained for the point estimate. When conversion is accomplished this way, the original SE is multiplied by the same constant to obtain the SE of the mean on the new scale. The constant in this example was 10,000 (the number of square meters in 1 ha), so the SE of the mean/ha would be 10,000 times the SE of the mean/m^2. CIs would be obtained in the usual way (mean/ha ± *t*-value × SE of the mean/ha).

This principle also provides a general way of obtaining estimated totals from estimated means. Suppose we have estimated that the mean number of animals per plot is 2.0, and the study area contains 100 plots. An obvious estimate of population size is then 100 × 2.0 = 200. The 100 in this example is a constant; its value would be the same regardless of what estimate we had obtained for the mean per plot. We can therefore apply the "change of scale" rule to obtain the SE, and thus CIs, for the estimated population total.

ESTIMATING ONE QUANTITY OF INTEREST

This section discusses methods for estimating means, proportions, and totals. Examples include average height, proportion of females that are lactating, and total number of animals present in a study area. Several variables can be measured on each population unit, but we assume they are analyzed one at a time. Thus we may estimate average weight, then average age, then proportion of individuals more than 2 years old, and so on. Methods for comparing estimates or studying the relationship between two or more variables are discussed in later sections.

Sometimes we record several quantities, but need only one from each population unit for the analysis. For example, condition indices may require several measurements from each animal. If we are estimating the average condition index, only one number per animal—its condition index—is needed for the estimate, so methods in this section may be used for the analysis.

The Finite Population Correction

The formulas for calculating SEs and CIs include a term called the finite population correction (fpc), which takes account of how much of the population is included in the sample. Formulas for the fpc vary with the sampling plan and are given in later sections. Including the fpc makes the SE smaller or, if only a tiny fraction of the population is included in the sample, has virtually no effect on the SE.

In most studies the sampled population is much larger than the sample, so the fpc can be ignored (and most of the formulas in standard statistics texts do not even include the fpc). Even if a substantial fraction of the sampled population is included in the sample, the fpc should still not be used if the *population of interest* is large. Thus, McAuley and Longcore (1988) estimated survival rates of young ring-necked ducks on three study areas in Maine. On each study area they found most of the broods, so their overall fpc was substantial, and, if employed in their statistical analysis, would have decreased their SEs by an appreciable amount. The purpose of their study, however, was to characterize survival rates of ring-necked ducks over a much larger area. The fpc was therefore not appropriate, and they did not include it.

Widely Used Sampling Methods

The formulas required for estimates and statistical tests vary according to the sampling plan used to collect the data. In most wildlife studies, the formulas developed for simple random sampling can be used.

SIMPLE RANDOM SAMPLING

Simple random sampling means that every unit in the population has the same probability of being chosen each time a new unit is selected for the sample. One method for selecting a simple random sample is to number all units in the population and then use a random numbers table to choose units for the sample. If units already in the sample cannot be chosen again, the plan is called simple random sampling without replacement. Wildlife studies nearly always use sampling without replacement, and throughout the chapter we assume this to be the case. True simple random sampling is not common in wildlife studies for two reasons. First, simple random sampling is difficult when the units are animals because the investigator is seldom able to identify all animals in the population and select them randomly. Second, when units are defined in space, time, or both, investigators usually prefer systematic sampling (see Chapter 1 and below). Only 31 of the 198 studies we reviewed employed simple random sampling, and in most studies the authors did so as part of a multiple-stage sampling design (see below) rather than as the sole method of sample selection.

Units also can be selected with unequal probabilities. For example, if our population unit was a landowner, we might select our sample by picking points on a map using simple random selection and then include in our sample the owners of the selected locations. With this plan, owners of larger farms would have more chance of being included in the sample than owners of smaller farms. Sampling with unequal probabilities is unusual in wildlife studies, so we do not discuss it in this chapter. If unequal probability sampling has been used, a statistician should be consulted before the analysis is undertaken, because some of the required analytic methods are complex (see Cochran 1977).

Formulas for calculating means, SEs, and CIs with simple random sampling are given in Boxes 2–4. The SE can be obtained easily on most hand calculators by entering the *n* sample values, asking for the SD, which may be identified as σ, and dividing this quantity by the square root of *n*. Note that SD in equation 2.3 has $(n - 1)$ in the denominator. Calculators generally offer the option of an SD calculated this way and another option also called SD that uses *n* in the denominator. Use the $n - 1$ version, and be sure to divide the result by the square root of *n* to obtain the SE. Students often confuse the issue of "*n* or $n - 1$ weighting" to obtain SDs with the requirement that the SD be divided by the square root of *n* to obtain the SE. The following hypothetical example may be used to check that the correct procedure is being used.

Ten plots, each comprising 1% of a study area, were surveyed for moose by helicopter. We assume that detection rates were 100% (and thus measurement bias is 0.0). The numbers recorded were 3, 3, 5, 7, 5, 4, 5, 6, 4, and 3. We wish to estimate the mean number of moose per plot and the total number of moose in the study area, and we want 95% CIs for each estimate. The mean is 4.5, the standard deviation is 1.354 so the SE is 0.428. With 10 plots, we have 9 degrees of freedom (see Box 2), so the *t*-value, for 95% confidence, is 2.262, and the 95% CI is $(4.5 - [2.262][0.428])$ to $(4.5 + [2.262][0.428])$ or 3.53 to 5.47 (see Box 3). Because each plot occupies 1% of the study area, there are 100 plots. The estimated total number of moose in the area is therefore 450, the SE for the estimated total is $100 \times 0.428 = 42.8$, and the 95% CI is therefore $(450 \pm [2.262][42.8])$, or 353 to 547. As noted above, one common mistake in calculating SEs is to obtain the SD, using "*n* weighting" and use it, instead of the correct SE, to calculate the CI. Had we done this, the improperly calculated SE would have been 1.28, and the CI for the total number of animals present would have been 161 to 740, a much wider interval than the correct one. Thus, incorrect estimation of SEs is likely to be a costly mistake.

NONRANDOM SAMPLING

As noted above, selecting units with a well-defined, random sampling plan is often not practical. For example, Hannon et al. (1988) studied nesting willow ptarmigan in northern Canada. Because of high travel costs, they delineated a relatively small study area and then attempted to find and study virtually every bird nesting in their area. Clearly there was no formal random selection of birds, nor could there have been. Brody and Pelton (1989) studied responses of black bears to roads in North Carolina. They captured 17 bears, attached radio transmitters, and estimated the home-range size and other quantities for each bear. As in Hannon's study, formal random sampling, in which every member of a well-defined population was given a number, and then a sample of these num-

---

Box 2. Estimating the population mean and the SE of the estimate with data from one-stage sampling or multiple-stage sampling with equal-size primary sampling units.

A. Definitions
1. For one-stage sampling, let

   n = number of units measured (sample size).

   y_i = measurement from (or value of) the ith unit.

2. For multiple-stage sampling with equal-sized primary units, let

   n = number of primary units.

   y_i = estimated mean for the ith primary unit.

B. The population mean, $\bar{y}$, the SE, SE($\bar{y}$), and degrees of freedom (df) are estimated as

$$\bar{y} = \Sigma\, y_i/n \qquad (2.1)$$

$$\mathrm{SE}(\bar{y}) = \mathrm{SD}(y_i)/\sqrt{n} \qquad (2.2)$$

$$\mathrm{SD}(y_i) = \sqrt{\Sigma\,(y_i - \bar{y})^2/(n-1)} \qquad (2.3)$$

$$= \sqrt{(\Sigma\, y_i^2 - n\bar{y}^2)/(n-1)} \qquad (2.4)$$

$$\mathrm{df} = n - 1,$$

where all sums are from $i = 1$ to n.

C. Notes
1. The formula for SE($\bar{y}$) assumes that population is "large." If this is not so, then multiply SE($\bar{y}$) (as calculated above) by the square root of $(1 - n/N)$, where N = number of units (primary units for multiple-stage sampling) in the population. This case seldom occurs in wildlife studies.
2. Some computer programs, especially spreadsheets, provide an easy means of calculating SDs, but use n rather than $n - 1$ in the denominator. To obtain the SE($\bar{y}$) from this SD, SD$_n$ say, divide SD$_n$ by the square root of $n - 1$.

---

bers was selected with a random numbers table, obviously was not feasible in this study. In our sample of wildlife studies, we found that nonrandom selection of population units—frequently animals—was used in about 75% of the studies (Table 1).

Although nonrandom sampling is often unavoidable, it can lead to problems in the analysis. With random selection we obtain three benefits: unbiased estimates of the parameter, unbiased estimates of the SE, and extrapolation of the results to a well-defined population. When random sampling is not possible, we may lose any or all of these benefits. Biased estimation of the parameter is especially likely when the population units are animals and are difficult to catch. The sampling method may tend to capture animals that are young or nonterritorial, for example, and these individuals may differ from older or territorial individuals with respect to the response variable. If all animals in one area are studied, they may be more like one another than a random sample from the region due to effects of habitat. In this situation, SEs, and thus CIs, will be too small. Instances in which the SE will be overestimated can also be imagined. Extrapolation of the results to a larger population of interest—the whole point of statistical analysis in these studies—is also often difficult if all the data come from a small area or time interval, or if capture methods are more effective for certain animals than for others. We may believe the analysis applies to *some* larger group, but deciding which areas, times, or animals the results characterize may be difficult, and different investigators may have different opinions.

Nonrandom samples are usually analyzed as though they are simple random samples. In some situations, however, one can make different assumptions about what stages of the sample selection were equivalent to random selection. These choices generally lead to different formulas for the SE, and the resulting estimates of precision may differ widely. A final problem with nonrandom sampling is that investigators may be tempted to explore different options and take the one producing the smallest SE. Nonrandom samples should thus be avoided whenever possible, and investigators who do use nonrandom sampling should be aware that rigorous statistical analysis of the resulting data often will be difficult.

SYSTEMATIC SAMPLING

Procedures for selecting a systematic sample were described in Chapter 1. This method was used in >25% of the studies we reviewed (Table 1). It is especially common for sampling in space or time because of its ease and efficiency.

Systematic samples are usually analyzed in exactly the same way that simple random samples are. This practice is so widespread, in fact, that investigators usually do not report that they used the formulas for simple random sampling. Systematic samples, however, are not equivalent to simple random samples, and one should be aware of a few facts about systematic sampling. First, the point estimates (of means, totals, or proportions) are unbiased if the first unit for the sample is selected randomly. Second, in many situations, systematic selection leads to estimates that are actually more precise (have smaller true SEs) than estimates obtained from simple random sampling, though occasionally the reverse is true. Third, a peculiar point, the *estimated* SEs (calculated with the simple random sampling formula) are often biased. In particular, if systematic selection gives a more precise estimate than simple random selection (because the systematic sample covers the population uniformly and does not coincide with any periodicity in the population), the simple random sampling formula overestimates the true SE. This leads to CIs that are larger than they should be. So if one has used systematic selection (and has followed the recommendations in Chapter 1), two options are available: use the simple random sampling formulas for analysis and accept

---

Box 3. CIs and tests for continuous data.

A. Notation

$\bar{y}$ = the estimate, typically a mean but may be any quantity (e.g., population total, slope from a regression analysis) as long as the normal approximation is appropriate.

$t_{\alpha,df}$ = value in Table A1, Appendix, with level of significance = α (usually 0.05) and degrees of freedom = df (see Boxes 2 or 5).

μ = the true value under the null hypothesis.

B. The $1 - \alpha$ CI for $\bar{y}$ is:

$$\bar{y} \pm t_{\alpha,df}[\mathrm{SE}(\bar{y})]. \qquad (3.1)$$

C. $\bar{y}$ is significantly different from μ if and only if

$$\frac{|\bar{y} - \mu|}{\mathrm{SE}(\bar{y})} \geq t_{\alpha,df}. \qquad (3.2)$$

---

the fact that the CIs will probably be too large, or consult a statistician for advice on whether some more complex method might be used to estimate the SE.

These three sample selection methods described above are conveniently referred to as one-stage sampling, the name referring to the fact that the population is not subdivided into groups prior to, or during, sampling. We now discuss two alternatives to one-stage sampling.

MULTIPLE-STAGE SAMPLING WITH EQUAL-SIZED PRIMARY SAMPLING UNITS

As noted in Chapter 1, multiple-stage sampling occurs when population units are first selected in groups, and then some or all of the units in each group are measured. We refer to the groups as primary sampling units, or just primary units. In habitat studies, a common design is to set out large plots and then sample on smaller subplots in each plot. The population unit in this example is a subplot, and the plots are the primary sampling units. In questionnaire studies, investigators often select a simple random sample of families and then determine the views of each member of the family. The population unit is a person, and the family is the primary sampling unit. In telemetry or behavior studies, investigators frequently mark several animals, and then record several observations on each animal. The population unit is thus an animal-time and the set of all possible measurements that might be recorded on one animal is the primary sampling unit. For simplicity, we often refer to each animal as a primary sampling unit.

In multiple-stage sampling the size of each primary unit is defined as the number of population units contained in each primary unit. In estimating means, proportions, or totals, one must know whether all of the primary units are

of equal size, and if they vary the size of each must be known. Frequently size is not well defined, but we know that each primary unit is the same size. For example, when animals are the primary units and the population unit is an instant in time, the "size" of the primary unit is not well defined, but as long as each animal is in the study area the same amount of time, the primary units are all of equal size. In habitat studies in which multiple-stage sampling is used, the primary units (plots) are usually the same size.

The fpc for multiple-stage sampling is defined as the proportion of primary units in the population of interest that are included in the sample. As with one-stage sampling, the fpc is nearly always negligible in wildlife studies.

Multiple-stage sampling is quite common in wildlife studies. It was used in more than one-third of the studies we reviewed. In virtually all of these studies, the primary sampling units were the same size, so we discuss this situation in greatest detail. We also assume that the fpc is negligible.

Analysis of data from a multiple-stage sample with equal-sized primary units is particularly easy because the simple random sampling formulas are used. With true simple random sampling, y_i is the observation on the ith unit in the sample, and n is the sample size. With multiple-stage sampling and equal-sized primary units, let y_i equal the estimate for the ith primary unit and n equal the number of primary units. Then use the formulas for simple random sampling in Box 2 to obtain means and SEs. No other special procedures are needed.

In most wildlife studies that use this method, selection of population units within primary sampling units is by one-stage sampling. Occasionally, however, a more complex sampling scheme may be used within primary units. In telemetry or behavioral studies, for example, observations may be selected in several stages (animal, season, day, hour, instant-in-time), and the stages may differ in length or size. Thus, the seasons may be pre-breeding, breeding, and post-breeding, and they may differ in length and in how much data were collected in each. In such situations, estimating the mean per animal may be complicated because the results from each stage of the sampling must be weighted properly. In making the calculations, keep in mind that the goal is to obtain unbiased estimates for each animal (or other primary unit). By unbiased, we mean that if a much larger sample had been taken, with the same sampling plan, the mean per animal would have equalled the true mean for that animal.

As explained in the boxes and sections below, general analytic methods can vary considerably according to whether the data are dichotomous or continuous. This distinction, however, is important only with one-stage sampling, because analysis of multiple-stage samples is invariably based on the estimates from each primary unit, and these estimates are continuous even if the original observations are dichotomous. For example, Bidwell et al. (1989) captured 41 turkeys and estimated the average proportion of time spent by each bird in each of several habitats. The observations were dichotomous (0 if the bird was not in the habitat, 1 if it was), but the sampling plan was clearly two-stage: selection of the birds, and then se-

Box 4. CIs and tests for proportions estimated with one-stage sampling.

A. Definitions

   p = the estimated proportion.

   $q = 1 - p$.

   n = the sample size.

   P = a hypothesized true value of p such as 0.5.

   $t_{\alpha,df}$ = the value in Table A1, Appendix, with level of significance α and degrees of freedom = df (see below).

B. Guidelines for when to use the normal approximation (modified from Cochran 1977:58).

p or $1 - p$, whichever is smaller	Minimum n for normal approximation to be used
0.5	30
0.4	50
0.3	80
0.2	200
0.1	600
0.05	1,400

C. Analysis with the normal approximation
   1. p is significantly different from P if and only if

   $$\frac{|p - P|}{\sqrt{PQ/n}} - \frac{c}{\sqrt{nPQ}} \geq t_{\alpha,\infty}, \quad (4.1)$$

   where c, the "correction for continuity" (Snedecor and Cochran 1980, Sec. 7.6), is defined as follows:
   a. For one-tailed tests, $c = 0.5$.
   b. For two-tailed tests, c depends on f = the fractional part of the quantity, $n|p - P|$. If $f > 0.5$, then $c = f$; otherwise, $c = f + 0.5$. Examples: If $n(p - P) = 7.9$ or -7.9, then $f = 0.9$ so $c = 0.9$. If $n(p - P) = 7.3$ or -7.3, then $f = 0.3$ so $c = 0.8$.
   2. The $1 - \alpha$ CI for p (Cochran 1977:57) is:

   $$p \pm t_{\alpha,n-1} \sqrt{\frac{pq}{n - 1} + \frac{0.5}{n}}. \quad (4.2)$$

   3. If the fpc is appropriate, then in eq. 4.1, replace "PQ/n" with "$(1 - n/N)PQ/n$" and in eq. 4.2 replace "pq" with "$(1 - n/N)pq$" where N = population size.
   4. Note that the significance test uses PQ/n for SE(p) because, under the null hypothesis, the variance is known. When it is unknown, as in constructing the CI, then $pq/(n - 1)$ is the appropriate formula (Cochran 1977:52, 57) though the incorrect expression, pq/n, is widely used.

D. Analysis when the normal approximation is not appropriate.
   1. Hypothesis tests are laborious to compute directly. An alternative (Steel and Torrie 1980:484) is to compute CIs (see below) and reject the null hypothesis if the CI does not include P.

   2. CIs
   a. Approximate limits may be determined with Fig. A1, Appendix, or by interpolation of values of binomial confidence limits (e.g., Steel and Torrie 1980:598).
   b. Exact limits may be calculated with a table of values from the F distribution (Table A2, Appendix). The f value depends on the level of significance and "numerator" and "denominator" degrees of freedom, symbolized below as ν_1, and ν_2, respectively.

   lower endpoint

   $$= [1 + f_{\alpha/2,\nu_1,\nu_2}(q + 1/n)/p]^{-1}, \quad (4.3)$$

   where $\nu_1 = 2(nq + 1)$

   $$\nu_2 = 2np$$

   upper endpoint

   $$= \left[1 + \frac{q}{(1/n + p)f_{\alpha/2,\nu_1,\nu_2}}\right]^{-1}, \quad (4.4)$$

   where $\nu_1 = 2(np + 1)$

   $$\nu_2 = 2nq.$$

lection of several times for each bird. The estimated means and SEs were therefore calculated from the means (of the 0s and 1s) for each bird, and these means are continuous. As noted above, continuous methods generally are used any time the random variables can take more than two values. Thus, even if only two observations were recorded for each bird, continuous methods would be used for the analysis because the estimate for each bird could be 0.0, 0.5, or 1.0.

STRATIFIED SAMPLING

In stratified sampling the population is divided into groups, just as in multiple-stage sampling, but we select a sample from each group, not from just some of them. In addition, the groups, which we call strata to distinguish them from primary sampling units, often differ in size, whereas this is unusual in multiple-stage sampling.

Stratified sampling often is used on wildlife surveys

that cover large areas. If the strata are delineated properly, stratified sampling often produces substantially smaller SEs than simple random sampling. Furthermore, use of stratification ensures that the sample is distributed throughout the area, permits heavier sampling in some areas, and provides separate estimates for portions of the study area (i.e., for individual strata). The Breeding Bird Survey, conducted by the U.S. Fish and Wildlife Service, is a typical example of stratified sampling. The survey area (much of North America) was first subdivided into 1°, latitude-longitude blocks. Survey routes were then drawn randomly from within these blocks. Waterfowl surveys to estimate abundance and production on the breeding grounds are designed the same way except the strata are larger, and survey lines are located systematically within each stratum. The distance between survey lines varies from stratum to stratum, thus the sample is not equivalent to a single, systematic sample covering the entire study area. National surveys of woodcock and mourning doves, and numerous state and provincial surveys, also use stratified sampling to select routes.

The analysis of data from a stratified sample varies according to the type of sampling plan used within each stratum. The procedure is: make a separate estimate of $\bar{y}_i$, the estimated mean per population unit in the ith stratum, and of $SE(\bar{y}_i)$, and then combine these to estimate the grand mean, $\bar{y}$, and $SE(\bar{y})$ using equations 5.1 and 5.2 (Box 5). In wildlife studies that employ stratified sampling, the sampling plan within strata is nearly always one-stage sampling or multiple-stage with equal-sized primary sampling units, so the $\bar{y}_i$ and $SE(\bar{y}_i)$ are easily obtained with the formulas for simple random sampling (Box 2).

One advantage of stratification is that sampling intensity can be varied among strata. Here is an example showing how valuable this technique can be. We once encountered a biologist studying habitat preferences of radio-collared gray foxes. He wanted to monitor their behavior throughout the 24-hour day. During the night he checked each animal once every 4 hours and felt comfortable with the resulting data. During the day, however, this sampling intensity seemed too labor-intensive because the foxes used the same resting areas each day and seldom moved during the day. The biologist was concerned, however, that he standardize procedures, so he continued to record locations every 4 hours during the day. If he had defined daytime and nighttime strata, he could have greatly reduced sampling intensity during the daytime period because there was virtually no variation in habitat use during the day. His $SE(\bar{y}_{day})$ would have been small despite the smaller sample size. Furthermore, if habitat use during the day and night differed, the SE of his overall estimate would probably have been even smaller than the estimate he actually obtained, because in stratified sampling the overall SE comes from the variance *within* strata.

OTHER DESIGNS

This category includes studies with large populations but unequal-sized primary units and studies in which the sample includes >10% of the primary units in the population. Neither category is common in wildlife work, but they do occur occasionally. For example, W. Butler (pers. commun.) and co-workers developed aerial surveys for nesting geese in western Alaska. The purpose was to estimate the number of geese per square kilometer. Their transects were each 1 km wide. They started at the coast and continued inland for variable distances, depending on how far inland the geese nested. The population unit was a square kilometer, the variable was number of geese, and each transect was a primary unit. Primary units were of unequal size because their length varied. The transects were 16 km apart and covered the entire study area, so the fpc was $\frac{1}{16}$. The authors calculated the mean number of geese/km^2 for each transect ($\bar{y}_i$) and then used equations 5.1 and 5.2 to calculate $\bar{y}$ and its SE. As explained in Box 5, the authors could have calculated the mean length of all possible transects in the study area (e.g., the average size of all the primary units in the population). This approach, however, probably would have been less precise.

EXAMPLES

Here are several examples of plans used in recently published wildlife studies showing how the data would be analyzed with the use of Boxes 2–5.

Serie and Sharp (1989) estimated the average body weight of adult female canvasbacks in Wisconsin during autumn migration. They collected 160 birds for the analysis. This is an example of nonrandom sampling. The population unit was a bird, and the variable was its weight. The equations in Box 2 were used to estimate the mean weight and its SE.

Lowney and Hill (1989) estimated the total number of wood duck cavities in a riparian study plot on the Noxubee National Wildlife Refuge by thoroughly searching randomly placed, long, narrow transects, each of which covered 0.5 ha. This is an example of simple random sampling. The population unit was a transect, and the variable was the number of wood duck cavities. The equations in Box 2 were used to estimate the mean number of nests per plot. The methods explained above for changing the scale at which results are expressed were used to estimate the total number of nests on the refuge and the SE of this estimate.

Reed et al. (1989) radio-collared 33 adult brant in four widely separated locations of northern North America and then searched for them during autumn migration at Izembek Lagoon in Alaska. The purpose was to determine whether a large fraction of the brant population passes through Izembek during autumn migration. This is an example of nonrandom sampling. The population unit was a radio-collared brant; the response variable was 1 if it was seen at Izembek and 0 otherwise. They recorded 79% of the 33 brant at Izembek, so their point estimate of the proportion was 0.79. Methods for constructing CIs for proportions are explained in Box 4. One of the choices is whether the normal approximation can be used. In this example the proportion and sample size are too small for the normal approximation to be appropriate, so graphs or small-sample procedures are used instead. The 95% CI is approximately 64–92%.

Arthur et al. (1989) collected 69 fisher scats to estimate the rate of occurrence in the scats of each of eight food types. This is an example of nonrandom sampling. Small mammals occurred in 22% of the scats. For a proportion of 0.22 and sample size of 69, the normal approximation

Box 5. Formulas for estimating the population mean and the SE of the estimate with data from a stratified sample or a multiple-stage sample with unequal-sized primary sampling units.

A. Notation

N = number of groups (strata or primary sampling units) in the population (sometimes indefinitely large).

n = number of groups sampled.

w_i = number of population units, or proportion of the population, in the ith group.

$\bar{w}$ = average of the w_i in the sample (except occasionally in C.2. below).

$\bar{y}$ = estimate of the population mean (i.e., the mean per population unit).

$\bar{y}_i$ = estimated mean per population unit in the ith group.

B. Estimate of the population mean, $\bar{y}$

$$\bar{y} = \frac{1}{n\bar{w}} \sum w_i \bar{y}_i. \qquad (5.1)$$

If w_i = proportion of the population in the ith group, then $n\bar{w} = 1$.

C. Formulas for SE($\bar{y}$) and its df

1. Stratified sampling, $n = N$ (modified from Cochran 1977:95).

$$\text{SE}(\bar{y}) = \sqrt{\frac{1}{(n\bar{w})^2} \sum w_i^2 [\text{SE}(\bar{y}_i)]^2} \quad (5.2)$$

If one-stage sampling is used within strata (the most common case), then eq. 2.2 (Box 2) can be used to obtain the SE($\bar{y}_i$) and

$$\text{df} = \frac{(\sum g_i)^2}{\sum (g_i/(n_i - 1))}, \qquad (5.3)$$

where $g_i = w_i^2 [\text{SE}(\bar{y}_i)]^2$ and n_i = sample size within the ith stratum.

If multiple-stage sampling is used within strata, then the SE($\bar{y}_i$) may be obtained with the formulas in Box 2 (equal-sized primary units and $n/N < 0.1$) or in C.2. below (all other cases). Consult a statistician for advice on calculating degrees of freedom.

2. Multiple-stage sampling, $n < N$ (modified from Cochran 1977, eq. 11.30). Use these formulas with unequal-sized primary units or $n/N \geq 0.1$. Otherwise, use Box 2.

a. Unequal-sized primary units and $n/N < 0.1$.

SE($\bar{y}$)

$$= \sqrt{\frac{1}{n\bar{w}^2} \sum w_i^2 (\bar{y}_i - \bar{y})^2/(n - 1)} \quad (5.4)$$

$$\text{df} = n - 1. \qquad (5.5)$$

Note: If the average size of the primary units in the entire population is known, then one may use it for $\bar{w}$ in calculating $\bar{y}$ (eq. 5.1). For SE($\bar{y}$), let $y_i = w_i \bar{y}_i/\bar{w}$ ($\bar{w}$ = population average) and calculate the SE as (Cochran 1977:303)

$$\text{SE}(\bar{y}) = \sqrt{\frac{\sum (y_i - \bar{y})^2}{n(n - 1)}} \qquad (5.6)$$

$$= \text{SD}(y_i)/\sqrt{n}. \qquad (5.7)$$

In most studies this SE($\bar{y}$) is larger than that in 3.4 above.

b. $n/N \geq 0.10$

SE($\bar{y}$)

$$= \sqrt{(1 - n/N)\text{SE}_1^2 + (n/N)\text{SE}_2^2}, \qquad (5.8)$$

where SE_1 is the SE($\bar{y}$) calculated with eq. 5.4 and SE_2 is the SE($\bar{y}$) calculated with eq. 5.2.

This formula may always be used for multiple-stage sampling, but with $n/N < 0.1$, the simpler versions (eqs. 2.2 or 3.4) yield nearly identical results. Consult a statistician to determine degrees of freedom.

is not appropriate (Box 4, B.). The 95% CI was approximately 15–34%. Notice that in analyzing these data, we consider one food type at a time and define the random variable y_i as 1 if the type occurs in the ith scat and 0 otherwise. We complete the analysis on that food type and then repeat the entire process. Studies of habitat preference and behavior are similar in that for each habitat or behavior we carry out a separate analysis.

Newton et al. (1989) studied the effects of herbicides on browse in clearcuts. They treated 18 1.0-ha plots with herbicides, and in each plot they measured several browse variables on each of 16 systematically placed, 0.004-ha circular subplots. This is an example of multiple-stage sampling with equal-sized primary units (1.0-ha plots). The small, circular subplots were the population units. The authors wanted to extrapolate their findings to a large area, so the fpc was not included. Because each plot was the same size, the authors calculated the means per plot and used these means as the y_i in Box 2 to obtain the grand mean and its SE. Conversion of the results from number of stems/0.004-ha plot to whatever scale they wished, for example stems/m^2 or stems/ha, was accomplished with the methods explained above for changing the scale. Separate analyses were carried out for each browse variable.

Rave and Baldassarre (1989) estimated the proportion of time that green-winged teals spent in each of several behaviors on their wintering grounds in Louisiana. During 1 month, they selected 161 teals and observed each one for 10 minutes, recording its behavior every 10 seconds. This is an example of multiple-stage sampling with equal-sized primary units. The population unit was a bird-time. Primary units were birds, or, more precisely, a primary unit was the set of all data that might have been recorded on one bird. Groups were the same size because each bird was assumed to be on the study area the same amount of time. The response variable, for a given behavior, was 1 if the bird was engaged in the behavior and 0 otherwise. The population size was the number of teals that used, or perhaps might have used, the study area and thus was treated as being indefinitely large. The authors calculated the proportion of time each bird spent in each behavior. For a given behavior, y_i was the proportion of time spent by the ith bird in the behavior. They used these y_i in Box 2 to obtain the grand mean and its SE.

Reed and Chagnon (1987) used stratified sampling to estimate the number of snow geese on an island in the Northwest Territories. They subdivided the island into strata on which they expected low, medium, and high densities of geese (based on maps of the island and knowledge of the species' habitat requirements). They selected plots in each stratum using simple random sampling and then surveyed each plot by helicopter. For the analysis, they used the equations in Box 2 to calculate $\bar{y}_i$ and its SE within each stratum and the equations in Box 5 to estimate the overall density and its SE. They did not report whether the fpc was used in calculating any of the SEs. They surveyed 52% of the plots in their high-density stratum, so the fpc might have been employed if their purpose was primarily to make inferences to the study area in the year of study. If they were primarily interested in extrapolation to a larger area or to the same area in other years, the fpc should not have been used.

Additional Points About Multiple-Stage and Stratified Sampling

MULTIPLE ANALYSES

When complex sampling designs have been used, the investigator commonly wants to obtain estimates for one or more ''populations'' (sets of population units). The definition of ''groups,'' and thus of the sampling plan, can change with each new question. For example, Hölzenbein and Schwede (1989) observed eight confined deer, recording several behaviors throughout the 24-hour day. They distinguished ''day'' and ''night'' periods and collected equal-sized samples in each period. The population unit was an animal-time, and the response variable was the behavior recorded, or more precisely 0 or 1 for a given behavior depending on whether the animal was engaged in the behavior. To estimate the mean for all deer, they regarded the eight animals as a sample from a large population. This is an example of multiple-stage sampling with equal-sized primary units. The authors calculated the mean of the 0s and 1s for each animal and used these

means as the y_i in Box 2 to obtain the grand mean and its SE.

Suppose they wanted to estimate the mean for a single animal. The animal—or all the observations that might have been collected on it—comprises the entire population, and the sampling plan involves dividing the population into two periods, day and night. Thus, for this question, we have stratified sampling with two strata. The means and SEs within the day, and within the night, periods would be calculated and used in Box 5 to obtain the grand mean (eq. 5.1) and its SE (eq. 5.2). Estimates of the SEs within periods (strata) would depend on the sampling plan used during each period.

PROBLEMS IN DEFINING PRIMARY SAMPLING UNITS

Sometimes deciding how primary sampling units should be defined is difficult. For example, suppose that we record territory size during the breeding season for 100 animals in each of 2 years. Should we view the 100 animals as comprising a one-stage sample from a large, hypothetical population, or should we consider the data set to be a two-stage sample in which years are the primary sampling units, and we have a sample of animals from each primary unit? The answer to this question is important because with the first plan our t-value (for a 5% level of significance) for tests or CIs will be 1.96 ($n = 200$), whereas if the data are viewed as a two-stage sample, we have only one df and the t-value will be 12.7 ($n = 2$). As a result the CI will probably be much wider and tests will have far less power in the second plan.

This is a difficult issue to resolve in specific instances and an even more difficult one to offer guidelines about. A few general comments can be made, however. If possible, situations with only a few primary units should be avoided. If this is not possible, consider carrying out a whole analysis in each primary unit and report the results for each primary unit separately. Some investigators test for a significant difference between primary unit means, or between the distribution of observations in primary units. If primary units are not significantly different, the data are treated as a simple random sample. If the units are significantly different, the multiple-stage sampling formulas are used for the analysis. This approach has the weakness that failing to find a difference does not mean that none exists. Furthermore, if one does exist, the investigator is more likely to detect it with a large sample than with a small sample. This leads to the paradoxical situation that one may be better off collecting only a small sample within primary units because then one is less likely to have to use the less powerful, multiple-stage sampling formulas. Despite these drawbacks, testing for significant differences does at least provide an objective decision-making process.

IMPORTANCE OF INDEPENDENT SAMPLING

The formulas for multiple-stage and stratified sampling are appropriate only if sampling is independent in different primary sampling units or strata. This means that a completely separate sampling effort is carried out in each group and that the selection of units in one group has no effect on which units are selected in any other group. Here is an example showing the importance of this assumption.

Andres (1989) surveyed shorebirds migrating through the delta of the Colville River in northern Alaska. He stratified the study area, using habitats to delineate strata borders. The population unit was a plot-time, and the response variable was number of shorebirds. Andres visited all of the habitats in $\frac{1}{5}$ of the study area on 1 day; the next day he visited all the habitats in the next $\frac{1}{5}$ of the study area, and so on. Sampling was therefore not independent (in time) in different strata (habitats); areas close together, but in different habitats, were counted on the same day far more frequently than would happen with independent selection. In some studies, this might not have mattered much, but shorebirds migrating through the area were strongly influenced by weather, so two plots in different strata that were sampled on the same day were likely to be much more similar to each other than the same two plots sampled on different days. It therefore became important to include the effects of sampling different strata at the same time. When mean density of birds was estimated and the lack of independence was ignored, the CV (mean/SE) was 0.06. When effects of the lack of independence were included, the CV increased to 0.11. Thus, the assumption of independence should be examined carefully, and when it is not satisfied a statistician should be consulted for assistance in developing the correct formula to estimate SEs. In addition, the assumption of independence should be met whenever possible, because the formula for the SE can be complex if the assumption is not met. Andres, for example, had to calculate more then 200 covariance terms in his analysis.

ESTIMATING QUANTITIES OTHER THAN MEANS, TOTALS, AND PROPORTIONS

The quantities of interest discussed thus far are all calculated from the means per sample or per group. Occasionally, more complex calculations are needed to obtain the point estimate—as in capture-recapture studies—and formulas for the SE may be quite complex. The SE of a diversity index provides another example of this problem. SEs can nearly always be calculated, though the advice of a statistician may be needed. If multiple-stage sampling with equal-sized primary units has been employed, however, and if the quantity of interest is simply the mean of the results for each primary unit, the calculations above apply for obtaining the SE of the final estimate. A few examples should make this point clear.

Thill and Martin (1989:541) calculated Kulcyznski's coefficient of similarity for the diets of "4 gentle cows and 3 tame deer." The authors calculated these coefficients during each of several periods. The coefficient is a complex quantity, and its SE could not be derived with methods in this section. The authors, however, were not primarily interested in the coefficient for any single period, rather they were interested in the mean of the coefficients from several periods. They therefore calculated the coefficient for each period and used these values as the y_i in Box 2 to calculate the average coefficient and its SE.

Guthery (1988) employed line-transect methods to estimate northern bobwhite density in each of eight areas (differing in size) in Texas. His purpose was primarily to evaluate the method, so he presented results for each area separately. To calculate the grand mean, assuming the eight areas were viewed as a simple random sample (of unequal-size primary units) from a larger region, he could have used the equations in Box 5 with the estimated densities from each area as the y_i and the sizes of the areas as the w_i.

COMPARING TWO ESTIMATES

Comparing two estimates is probably the most common statistical analysis carried out by wildlife biologists. More than one-half of the studies we reviewed contained at least one such comparison (Table 1). The general objective in such analyses is to make inferences about the parameters (i.e., the quantities of interest) in the two populations being studied. Two questions may be addressed in a two-sample comparison: do the data show that one of the parameters is larger than the other, and, if so, how much larger? The first question is answered by testing the hypothesis of no difference between the parameters; the second question is answered by constructing a CI for the estimated difference between the parameters. Procedures for carrying out the comparison are provided in Boxes 6–8. Several examples in which the hypothesis of no difference was tested in a two-sample comparison follow.

Quinn and Thompson (1987) studied lynx in Ontario. One of their objectives was to compare the physical condition of males and females of the same age. They collected carcasses from trappers and used renal fat as an index to condition. They considered the animals to be a random sample from a large population and tested the null hypothesis that in this population the renal fat indices of males and females were equal. They used a *t*-test to carry out the statistical analysis.

Cowan et al. (1987) studied the acceptance by European rabbits of baits containing Rhodamine B dye. The authors captured male and female rabbits during several periods after baits were placed in the field. Carcasses were examined for presence of the dye to determine what fraction of the rabbits had eaten the bait. One purpose was to determine whether males and females accepted the baits at different rates. They used ferrets to drive the rabbits into nets and considered the resulting catch of males and females to comprise random samples from large hypothetical populations of males and females. They tested the null hypothesis that the proportions of males and females consuming bait in these hypothetical populations were equal, using a chi-square analysis for the test.

Holl and Bleich (1987) compared the chemical content of soils from sites used as mineral licks by mountain sheep and from randomly selected sites to determine what chemicals might be important to the sheep. They collected soil from 12 mineral licks and 12 adjacent control sites, considered the pairs of samples to constitute a simple random sample from a large population, and tested the null hypothesis that the average concentrations of Na, K, Ca, Mg, and Cl were the same in the licks and control sites in the study area. They used a Wilcoxon signed-rank test for the analysis.

The main use of hypothesis testing in these and most wildlife studies is to help guard against unwarranted conclusions. One estimate may be substantially larger than the other, but statistical analysis may indicate that such a difference might readily occur even if the two populations were identical. In such situations, the data should not be

Box 6. Parametric methods (*t*-tests) for comparing two independent, continuous estimates.

A. Notation

$\bar{y}_1, \bar{y}_2$ = the estimates from populations 1 and 2.

$\bar{Y}_1, \bar{Y}_2$ = the true but unknown values (i.e., the parameters).

SE_1, SE_2 = the SEs of the estimates (obtained with the formulas in Boxes 2 or 5).

n_1, n_2 = sample sizes for the two estimates (see Boxes 2 and 5).

df_1, df_2 = the number of degrees of freedom for the estimates $\bar{y}_1$ and $\bar{y}_2$ (see Box 2, eq. 2.2 and 2.3).

$t_{\alpha,df}$ = the value in Table A2, Appendix, with level of significance α and degrees of freedom = df.

μ = value of $\bar{y}_1 - \bar{y}_2$ under the null hypothesis (usually 0.0).

B. General formulas for tests and CIs

1. $\bar{y}_1 - \bar{y}_2$ is significantly different from μ if and only if

$$\frac{|\bar{y}_1 - \bar{y}_2| - \mu}{SE(\bar{y}_1 - \bar{y}_2)} \geq t_{\alpha,df} \qquad (6.1)$$

$$df = df_1 + df_2, \qquad (6.2)$$

where df_1 and df_2 are the degrees of freedom for $\bar{y}_1$ and $\bar{y}_2$, respectively (see Boxes 2 and 5).

2. The $1 - \alpha$ CI on $\bar{y}_1 - \bar{y}_2$ is

$$\bar{y}_1 - \bar{y}_2 \pm t_{\alpha,df}[SE(\bar{y}_1 - \bar{y}_2)] \qquad (6.3)$$

$$df = \frac{(SE_1^2 + SE_2^2)^2}{SE_1^4/df_1 + SE_2^4/df_2}. \qquad (6.4)$$

C. Formulas for $SE(\bar{y}_1 - \bar{y}_2)$:

1. Hypothesis testing when the data are from one-stage samples or multiple-stage samples with equal-sized primary units and sample sizes, n_1 and n_2, are unequal.

$$\sqrt{\left(\frac{n_1 + n_2}{n_1 n_2}\right)\left(\frac{n_1(n_1 - 1)SE_1^2 + n_2(n_2 - 1)SE_2^2}{n_1 + n_2 - 2}\right)}. \qquad (6.5)$$

Note: Eq. 6.6 may be used for simplicity but usually yields a slightly larger value.

2. All other situations

$$\sqrt{SE_1^2 + SE_2^2}. \qquad (6.6)$$

taken as providing much evidence that the populations differ with respect to the quantity being estimated.

One- and Two-Tailed Tests

In most two-sample comparisons, the null hypothesis is that the two population means are equal, and the alternative hypothesis is that they are not equal. Under the alternative hypothesis, either population mean may be larger. Occasionally, however, one has more information about the difference between the parameters. For example, in comparing survival rates of animals with radio transmitters (test animals) and without radio transmitters (control animals), the investigator might assume that, at most, the survival rate of the test individuals might equal that of controls, but that test animals certainly would not survive at a higher rate than control animals. In such a situation, the observed difference (survival of test animals minus survival of control animals) should be close to zero or negative, but it should not be a large positive number. When positive (or negative) values of the true difference can be ruled out a priori, it is permissible to use a "one-tailed *t*-test" that recognizes the value of this information. Specifically, one adjusts the threshold value of the *t* statistic at which the null hypothesis will be rejected. This approach, however, is appropriate only if one is *certain* that the true difference cannot be positive (or negative), and for this reason one-tailed tests are rarely used in wildlife studies.

Cautions About the Two-Sample Comparison

Several points should be kept in mind when one makes comparisons such as the ones above. The hypothesis test is useful in determining which population has the larger mean (or other quantity of interest), but it does not provide any indication of how large the difference between populations is. Furthermore, it does not tell us whether the difference is large enough to be of biological importance.

Second, failure to detect a statistically significant difference does not mean that no difference exists. Population means are usually different; we may simply have too small a sample to detect the difference. In addition, failing to find a significant difference does not mean that the two populations are similar enough to each other that no difference of biological importance exists. It is entirely possible for two populations, for example males and females, to differ in biologically important ways, but for this difference to be undetectable from the data collected.

A final, frequently misunderstood point concerns deducing whether two estimates are significantly different by examination of plots that show the means (or other estimates) and error bars. The error bars usually represent 1 SE or the 95% CI, and biologists often assume that the estimates are not significantly different if the bars overlap and are different if the bars do not overlap. Neither of these claims is true. It is easy to produce data (or find published data) in which the means are significantly different even though the CIs overlap and to find data that are not significantly different even though the SE bars do not overlap. Only the *t*-test tells for certain (see Box 6). A rule of thumb that we have found occasionally useful, however, is that if the difference between estimates exceeds three times the length of the larger SE, the differ-

---

Box 7. Comparison of two proportions estimated with one-stage sampling (modified from Snedecor and Cochran 1980, Sec. 7.6).

A. Notation

   p_1, p_2 = the estimated proportions.
   n_1, n_2 = the sample sizes
       ($n_1 = n_2 = n$ for paired data).

B. Testing whether the proportions are significantly different.
   1. Paired data (estimates)
      p_1 and p_2 are significantly different if and only if

      $$\frac{|p_1 - p_2| - c}{SE(p_1 - p_2)} \geq t_{\alpha,\infty} \qquad (7.1)$$

      $c = 1/n$ if $n_1 = n_2$ or $0.5(1/n_1 + 1/n_2)$
          if $n_1 \neq n_2$

      $t_{\alpha,\infty}$ = value in Table A1 with level of significance α and degrees of freedom ∞ (infinity).

      Formulas for $SE(p_1 - p_2)$ are given in Part D below.

   2. Independent estimates
      a. Use Fisher's Exact Test whenever possible. Table A3, Appendix, explains the test and gives critical values for samples sizes of n_1, $n_2 \leq 15$. Many statistical packages provide the exact significance level for larger sample sizes.
      b. If sample sizes are too large to use Fisher's Test, use eq. 7.1.

C. CIs are difficult to calculate for small samples. The following large-sample approximation is recommended (Fleiss 1981) if the smallest of n_1p_1, n_1q_1, n_2p_2, and n_2, q_2 is < 5. It can be used for paired or independent estimates.
   The $1 - \alpha$ CI for $p_1 - p_2$ is

   $$p_1 - p_2 \pm t_{\alpha,\infty}[SE(p_1 - p_2)] + c. \qquad (7.2)$$

D. Formulas for $SE(p_1 - p_2)$

Data are	Analysis	$SE(p_1 - p_2)$
Paired[a]	Hypothesis test	$\sqrt{n_d/n}$
	CI	$\sqrt{(n_d - n_e^2)/n}$
Independent	Hypothesis test[b]	$\sqrt{pq/(1/n_1 + 1/n_2)}$
	CI	$\sqrt{\dfrac{p_1q_1}{n_1 - 1} + \dfrac{p_2q_2}{n_2 - 1}}$

---

[a] n_d = the number of pairs in which the observations were different, and n_e is defined as follows. Let pairs with different outcomes be coded y_{10} and y_{01} and let the numbers of y_{01} and y_{10} pairs be n_{10} and n_{01}. Then $n_e = n_{10} - n_{01}$.
[b] Let

$$p = \frac{n_1p_1 + n_2p_2}{n_1 + n_2} \qquad (7.3)$$

and $q = 1 - p$. If $n_1 = n_2$, then $p = (p_1 + p_2)/2$.

---

ence is significant at the 95% level. The rule holds only if the estimates are independent and the total sample size ($n_1 + n_2$) is at least 16.

Confidence Intervals

When a single comparison is of particular interest, calculating a CI on the difference is often useful. The meaning of the interval may be difficult to grasp at first, but it has exactly the same interpretation as the CI for a point estimate (see Point and Confidence Interval Estimates [p. 17]). Each of the two populations has a true but unknown mean (or other quantity of interest); we are estimating the difference between these means and may refer to it as the "true difference." A 95% CI on the differences tells us, with 95% probability, that the true difference lies within the computed interval. More precisely, if we were to repeat the sampling and calculations a large number of times, and if all assumptions are met, 95% of the computed CIs would include the true difference and 5% would not include it.

The CI has two practical uses. When the null hypothesis of no difference has been rejected, the CI tells us the largest and smallest value that is realistic for the true difference. Such a conclusion can be of great value. For example, suppose the mean weights of males and females

were 12 kg and 8 kg and the difference was statistically significant. Given only this information, a reader cannot determine whether the average weight of females is much smaller than the average for males or only slightly smaller. Now suppose that, in addition, we were told that the CI on the difference (4) was ± 3.5. This shows that the average weight of females may be only slightly less than the average for males (the difference may be as small as 0.5). In contrast, a CI of ± 0.7 would show that the average for females was probably (i.e., with 95% certainty) at least 3.3 kg smaller than the average for males. Thus, providing the CI helps the reader, and perhaps the investigator, evaluate the biological importance of the observed difference.

CIs can also be useful when the estimates are not significantly different. Suppose, for example, that the weights above were 12 kg for males and 11 kg for females and the result was not significant. Do these results show that males and females have about the same average weight? Again, the reader cannot answer this question without additional information. But if the CI on the difference was, say, ± 6 kg, the interval within which the true difference lies (with 95% certainty) would be −5–7 kg, and we might conclude that not a great deal had been learned about how similar the average weights of males and fe-

Box 8.   Nonparametric methods for comparing two estimates.

A. Paired data, Wilcoxon signed-rank test
  1. Rank the non-zero differences, ignoring sign, giving the smallest difference a rank of 1. If two or more of the differences are equal, assign to each of them the average of the ranks they would have received if they had differed slightly.
  2. Compute the sum of the ranks assigned to the positive differences and the sum of the ranks assigned to the negative differences.
  3. The test statistic, T, is the smaller sum and is used with the number of non-zero differences (n) in Table A4, Appendix, to determine whether the sample medians are significantly different.
  4. If $n > 20$, then the estimates are evaluated with a normal approximation. The medians are significantly different if and only if

$$\frac{[n(n + 1)/4] - T - 0.5}{n(n + 1)(2n + 1)/24} \geq t_{\alpha,\infty} \quad (8.1)$$

  where $t_{\alpha,\infty}$ is the value in Table A1, Appendix, with level of significance α and degrees of freedom ∞ (infinity).

B. Unpaired data, Mann-Whitney test
  1. Rank the observations, giving the smallest one rank 1. If there are ties, assign to each of them the average of the ranks they would have received if they had differed slightly.
  2. Calculate the test statistic, T.
    a. $n_1 = n_2$. Calculate the sum of the ranks received by the observations in each sample. $T =$ the smaller sum.
    b. $n_1 \neq n_2$. Calculate T_1, the sum of ranks for the sample with fewer observations, say n_1. Next calculate $T_2 = n_1(n_1 + n_2 + 1) - T_1$. $T =$ the smaller of T_1 and T_2.
  3. T is used with n_1 and n_2 in Table A5, Appendix, to determine whether the sample medians are significantly different.
  4. For values of n_1 and n_2 outside the limits of Table A5, the samples are evaluated with a normal approximation. The sample medians are significantly different if and only if

$$\frac{|n_1(n_1 + n_2 + 1)/2 - T| - 0.5}{n_1 n_2(n_1 + n_2 + 1)/12}$$
$$\geq t_{\alpha,\infty} \quad (8.2)$$

  where T is defined in 2.a. or 2.b. above and $t_{\alpha,\infty}$ is the value in Table A1, Appendix, with level of significance α and degrees of freedom ∞ (infinity).

males were. Conversely, if the CI was ± 0.7, the interval would be 0.3–1.7 kg and we might conclude that the average weights were fairly similar to each other. This analysis—construction of CIs when the null hypothesis is not rejected—is similar to carrying out a power calculation. It tells us how large a difference could have been detected with high probability given the data collected.

The CI and hypothesis test are similar in that if the CI on a difference does not cross 0.0, the significance test will always reject the null hypothesis. This rejection is reasonable, because the CI tells us the region within which it is plausible to assume the true difference falls. If this region does not include 0.0, one population parameter must be larger than the other. A test of significance often is not needed if a CI is to be constructed.

Care must be taken, however, in applying this principle of the equivalence of CIs and hypothesis tests. CIs and hypothesis tests derived under the same assumptions yield equivalent results. Discrepancies may occur, however, when the two are derived under different assumptions. For example, in testing for differences in populations by comparing means, one usually assumes that the populations have the same variance under the null hypothesis of no difference. If one rejects the hypothesis of no difference, one may also wish to drop the assumption that the populations have the same variance. In real data, when means differ, variances often differ, too. If one constructs CIs for differences in means and no longer assumes the populations have the same variance, the resulting CIs might include 0 and apparently contradict the results of the hy-

pothesis test! The problem is that we changed the assumptions underlying the procedures in "midstream." By estimating extra parameters for the CIs (two variances instead of a single common variance), we lose some power by reducing degrees of freedom.

In many studies several comparisons are made, and the main conclusions of the study hinge on the pattern of results, rather than on the difference observed in any single comparison. In such situations, investigators usually rely on hypothesis tests, rather than CIs. The general rationale is that if the differences are all, or nearly all, significant, one has good reason to conclude that if the study were repeated, the same pattern of results would be obtained. In contrast, if most of the differences are not significant, the data provide no grounds for such an expectation. A completely different pattern of results—perhaps leading to different biological conclusions—might be obtained if a new set of data were selected from the same population. When several comparisons are made, reporting which were significant is relatively simple, especially if they all were significant and if the direction of the differences were all consistent with a single biological explanation. In contrast, reporting and interpreting the sizes of the CIs for each comparison may be cumbersome.

Analyzing Paired Data With Parametric Tests

The phrase "paired data" means that two observations were collected on every population unit in the sample. The quantity of interest usually is the difference between the means of the two data sets. Typical examples in wildlife studies include recording the number of some event (such as activity level) before and after some treatment (such as presentation of food), and measuring the habitat at several sites of interest (such as dens) and at a randomly selected location near to each site of interest. In these examples, the data are clearly paired in the sense that we have only a single random sample from the population of interest, we collect two observations on each sampled item, and our goal is to estimate the difference between the means of the two populations.

Paired data fall naturally into two categories: dichotomous data from one-stage samples, and all other situations. Methods for the first category are explained in Box 7. For other types of samples, (1) create a new variable for each unit in the sample and define it as the difference between the observations for the unit, (2) calculate the mean and SE of this variable, and (3) use the guidelines in Boxes 2–5 to test the null hypothesis that the true mean is 0.0. If the null hypothesis is rejected, the two original samples are significantly different.

Occasionally, some of the observations are paired and others are not. For example, abundance may be estimated in 2 years with the same survey routes, but some routes might not be used each year, or data might be collected on animals in summer and winter to estimate effects of season, but some of the animals might die before winter. Numerous other examples could be cited in which the observations are largely, but not completely, paired. Several approaches are available for analyzing such data. We recommend consulting a statistician for advice.

t-Tests and Chi-Square Tests for Comparing Proportions

Proportions estimated with one-stage sampling with independent samples can be compared in two ways. One way is the *t*-test, described in Box 7. The other way is by means of a chi-square test. The two tests are algebraically identical (unless a different correction for continuity is employed). We prefer the *t*-test because it shows the relation of this particular two-sample comparison to other two-sample comparisons, and because it extends more easily than the chi-square approach to construction of CIs on the difference. One disadvantage of the *t*-test approach is that it does not extend easily to multiple comparisons, whereas the chi-square approach does (e.g., contingency tables). Most statistics texts (e.g., Snedecor and Cochran 1980) discuss the chi-square approach.

Multiple Comparisons

In most wildlife studies, investigators are interested in making several comparisons. Weight of males and females may be compared at each of several times, home-range size may be compared in several different cohorts (e.g., older vs. younger, paired vs. unpaired), density of a species may be compared before and after a treatment on each of several plots, and so on. When several comparisons are made, investigators must guard against reaching unwarranted conclusions simply because of the number of tests carried out. Thus, if no differences actually existed, we would nevertheless expect about one in 20 comparisons (made at the 5% level of significance) to be statistically significant. That is the meaning of the 5% significance level: even if populations are not different, there is a 5% probability of (incorrectly) rejecting the null hypothesis.

If all or nearly all of the differences of interest are statistically significant, no special precautions are needed because the probability of such an outcome if none of the populations really differ is extremely small (far less than 5%). When only a few of the outcomes are significantly different at the 5% level, additional measures are needed, and many different approaches have been suggested. We discuss two methods that are widely used in wildlife studies and recommend Snedecor and Cochran (1980, Sec. 2.12 and 2.13), Steel and Torrie (1980, Chapter 8), and Sokal and Rohlf (1981, Secs. 9.6 and 9.7) for additional information.

In many studies, the individual *t*-tests can be preceded by a comprehensive analysis to determine whether any of the individual comparisons is significant. For continuous data, the comprehensive test is usually an analysis of variance (ANOVA). For counts, or frequency data, a chi-square test is employed for the same purpose. The assumptions and procedures required for ANOVA or chi-square tests vary according to the particular data set being examined and will not be reviewed here. The general approach, however, is to compute the F or chi-square statistic and conclude that specific comparisons are statistically significant only if the result of the comprehensive test is also significant. When this approach is followed, the 5% (or other nominal) significance level is used in both the comprehensive and the individual pairwise tests.

The approach of employing a comprehensive test prior to the individual pairwise tests can also be used with nonparametric tests. The Kruskal-Wallis test is the most widely used nonparametric analogue to the ANOVA. Others are discussed in Hollander and Wolfe (1973) and Lehmann (1975).

The approach above is often not appropriate because the assumptions required for a comprehensive test are not met. One way to control the "multiple comparisons problem" in such situations is to adjust the level of significance required to reject the null hypothesis. If k comparisons are to be made, and the actual level of significance desired is α, then use t with significance level α/k. Thus, if five comparisons are being made, and a 5% level of significance is desired, use the column in the table of *t*-values labelled $\alpha = 0.01$ rather than the column labelled $\alpha = 0.05$. Interval estimates constructed with this larger *t*-value are often called Bonferoni CIs.

Choosing Between Parametric and Nonparametric Tests

In comparing two estimates based on continuous data, one has a choice of using *t*-tests or nonparametric tests. The *t*-test evaluates whether the population means are equal; the nonparametric equivalents test whether the population medians are equal. The nonparametric tests are generally regarded as requiring fewer assumptions (see below) and as having less power in some situations. The choice between the two approaches involves a number of issues, including personal preference. In many specific situations the outcome is nearly the same regardless of which test is used. In certain situations, however, one test may be much more appropriate than the other. The most commonly used nonparametric tests are the Wilcoxon signed-rank test for paired data and the Mann-Whitney rank sum test for independent estimates.

At least three issues warrant attention in choosing between parametric and nonparametric methods. First, and perhaps most important, t-tests lead naturally to estimates of the smallest likely difference between populations, either by construction of a CI or simply by inspection of the estimates and SEs. CI procedures are available for most nonparametric methods (Randles and Wolfe 1979), but they are complex and rarely used in wildlife studies. Thus, if one is interested in how large the difference between populations is—not just in showing which population is larger—then t-tests are probably a better choice than nonparametric tests. Many biologists are concerned that t-tests may be invalid because the observations are not normally distributed, but this problem is rarely serious enough to preclude the use of t-tests.

If the goal is restricted to determining whether the data demonstrate a difference between populations, and the sign of the difference, nonparametric tests may be quite useful. Keep in mind, however, that the nonparametric methods test whether the medians, not the means, are equal. This point is important because the mean from one population might be larger and the median from the other population might be larger. Thus, the conclusion about which population is "larger" could depend on whether a t-test or a nonparametric test was used.

One other point about nonparametric tests seems worth mentioning. With large samples, the distribution of the test statistic is approximately normal, so the tables used for parametric tests can be used to determine whether the result is significant. Biologists sometimes mistakenly believe that if these tables are used in the analysis, one must be assuming that the underlying population has a normal distribution, or, at least, it must be true that the nonparametric test is then equivalent to a t-test. Neither of these views is correct. No assumption about normality of the observations is required, and the nonparametric test is still testing the null hypothesis that medians—not means—are equal. Furthermore, the nonparametric test statistic, evaluated with the large sample normal approximation, is much less affected by extreme values than is the parametric test statistic.

STUDYING THE RELATIONSHIP BETWEEN TWO OR MORE VARIABLES

Introduction

When two or more variables are recorded for each sample unit, we are said to have *multivariate* data. One can, of course, calculate point estimates, construct CIs, or test hypotheses for each variable separately. However, one is often interested in exploring relationships between or among the variables measured in a multivariate data set. For example, Franzmann and Schwartz (1988) analyzed blood samples from 298 black bears, measuring 26 blood parameters in each of the summer, autumn, and winter seasons. Sex, age, and several "condition" variables also were recorded. The bears represented the sample units, and 75 measurements were recorded for each bear. The stated objective was to determine which blood parameters were useful in assessing the condition of the bears, and

this was to be accomplished by investigating the relation between the blood parameters and change in condition.

In studying relationships between or among variables, we must answer some preliminary questions. Are we interested in relationships between *categorical* variables (variables that are essentially labels indicating which of several classes, such as sex or race, a unit belongs to), between *quantitative* variables (variables that take numerical values and for which quantities such as means are meaningful), or between mixtures of categorical and quantitative variables? Different statistical techniques are appropriate for studying relationships between different types of variables. Another question we must ask is whether we simply wish to establish the existence of some relationship, often for purposes of prediction, or hope to show that changes in some of the variables, called *explanatory* or *independent* variables, explain or cause changes in other of the variables, called *response* or *dependent* variables. In the Franzmann and Schwartz study, interest seemed to be more in exploring the relation between blood parameters and change in condition (for purposes of prediction) rather than establishing a causal relation between blood parameters and changes in condition. Conover (1988) measured rye biomass in January and in April, and whether geese grazed on several plots (the sample units) of ground on which rye was being grown. Here the objective was to show that grazing by geese caused changes in rye biomass.

A final question that must be addressed concerns the purpose of the study. In some studies the purpose is to make inferences, i.e., test formally posed hypotheses about or estimate parameters that describe the underlying population from which the data were collected. In other studies the purpose is merely descriptive or exploratory. One is informally using the data to see if relationships appear to exist between variables, explore the possible form of such relationships, and gain some information about the population from which the data were drawn. Such informal study is useful for posing formal questions that are to be answered in a future, more carefully designed investigation.

Scatterplots and Correlation

Among the simplest and most commonly used methods for studying the relationship between two quantitative variables are scatterplots and correlations. A scatterplot is produced by plotting the pair of measurements on each sample unit as a point on a graph, the vertical axis representing one of the variables (the dependent variable if one of the variables is so designated) and the horizontal axis the other (the independent variable if one is so designated). The result is a plot that looks like a scattering of points. Gese et al. (1988) made the scatterplot in Fig. 7 of coyote home-range size (the dependent variable) versus percent available pinyon-juniper habitat (the independent variable) to investigate the relationship between these two variables. The scatterplot suggests that home-range size tends to decrease as percentage of available pinyon-juniper habitat increases.

When a categorical variable is used to explain changes in a quantitative variable, a scatterplot also can be constructed. The categories of the categorical variable are represented as equally spaced points on the horizontal axis,

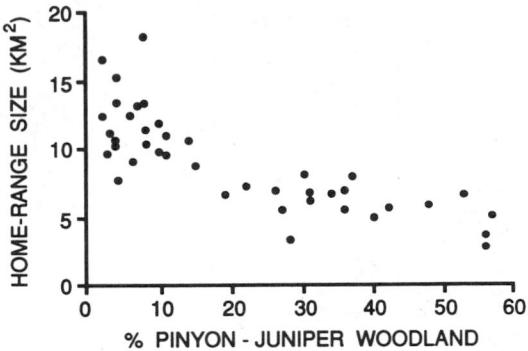

Fig. 7. Example of a scatterplot showing coyote home-range size versus percent available pinyon-juniper habitat (from Gese et al. 1988).

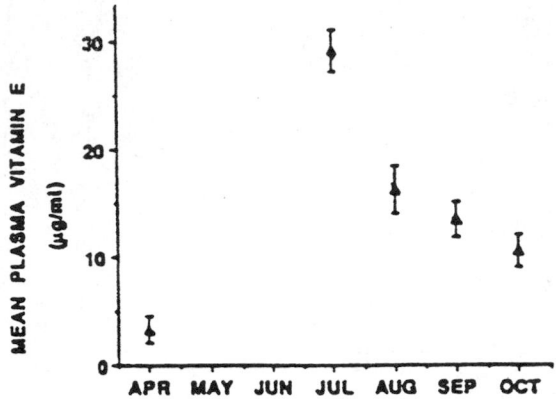

Fig. 8. Example of a scatterplot involving a categorical independent variable. The plot displays both the mean of plasma vitamin E per month and a bar representing the standard error of the means (from Dierenfeld et al. 1989).

and the resulting scatterplot looks like a series of vertical bars of points. One may wish to display the mean or median of the quantitative variable at each value of the qualitative variable on the plot. For example, Diernfeld et al. (1989) plotted the mean of plasma vitamin E (the dependent variable) of peregrine falcons versus the categorical independent variable month in the plot in Fig. 8. The individual data points should be suppressed, and only the mean and a bar representing the SE of the data points for a given month should be shown, to keep the plot from appearing too cluttered. Among other things, we see that in July mean plasma vitamin E is highest.

It is customary to attempt to quantify any relationships that appear in a scatterplot. The simplest possible relation is when the points in a scatterplot appear to be centered along a straight line. When above-average values of one variable tend to accompany above-average values of the other variable, we say the variables are *positively associated.* When above-average values of one variable tend to accompany below-average values of the other, and vice versa, the variables are said to be *negatively associated.* Scatterplots of variables that appear to be centered on a positively (negatively) sloping line are positively (negatively) associated. The more tightly the points appear to be clustered about a straight line, the more highly associated they are. A numerical measure of this degree of association is the (Pearson product moment) *correlation coefficient.* If we have a sample of n observations on two variables X and Y denoted

$$(X_1, Y_1), (X_2, Y_2), \ldots, (X_n, Y_n),$$

the correlation coefficient r is defined to be

$$r = \frac{1}{n-1} \sum_{i=1}^{n} \left(\frac{X_i - \bar{X}}{\mathrm{SD}(X)}\right)\left(\frac{Y_i - \bar{Y}}{\mathrm{SD}(Y)}\right),$$

where $\mathrm{SD}(X)$ and $\mathrm{SD}(Y)$ are the SDs of the X_i and the Y_i, respectively. If, whenever X takes on a value above its mean $\bar{X}$, the corresponding value of Y tends also to be above its mean $\bar{Y}$, the products of the terms in parentheses in the summation tend to be positive and r will be positive. Likewise if, whenever X takes on a value above its mean, the corresponding value of Y tends also to be below its mean, the products of the terms in parentheses in the summation tend to be negative and r will be negative. Positively (negatively) associated variables will therefore have a positive (negative) correlation. A mathematician

can show that the correlation coefficient must take on a value between -1 and $+1$, achieving the values ± 1 only if all the observations fall exactly on a straight line. Data that fall perfectly on a horizontal line are defined to have correlation 0 (Y has constant value independent of the value of X). The correlation coefficient is undefined for data that fall perfectly on a vertical line (X has constant value). In these latter examples, either X or Y remains constant and hence questions concerning how changes in one variable relate to changes in the other cannot be answered, because one of the variables does not change. Some sample scatterplots and the associated value of r are given in Fig. 9.

Two subtle points about the relation between scatterplots and the correlation coefficient are worth mentioning. First, the equation for the correlation coefficient actually implies that how tightly points are scattered about a straight line is determined by their *vertical* distance to the line rather than their perpendicular distance to the line. The discrepancy between these two distances can be large if the line is steep. Thus two scatterplots may appear to be equally tightly clustered (in the sense of perpendicular distance) about a line yet have quite different values of r because the lines have quite different slopes. Second, a

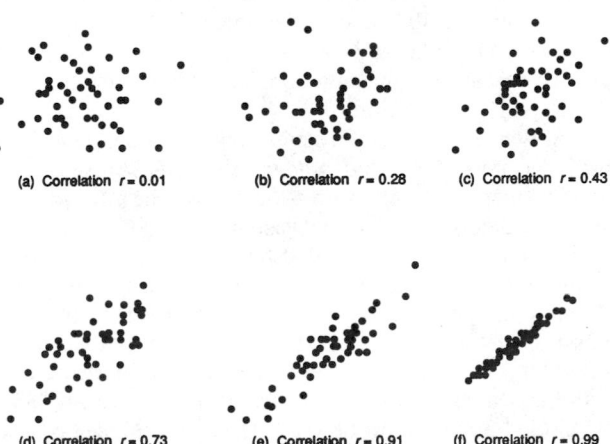

Fig. 9. Examples of scatterplots and their associated correlations (from Moore 1979).

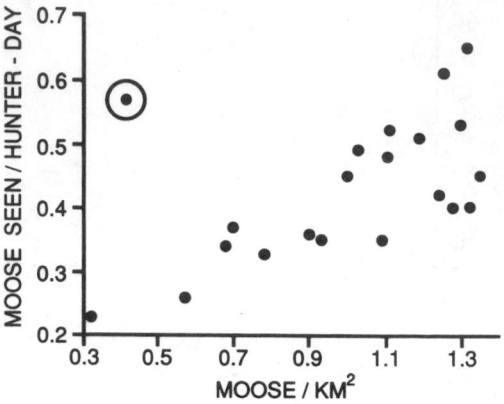

Fig. 10. Example of an outlier (circled) in a scatterplot. The plot displays moose density (per km²) versus moose seen per hunter-day (from Fryxell et al. 1988).

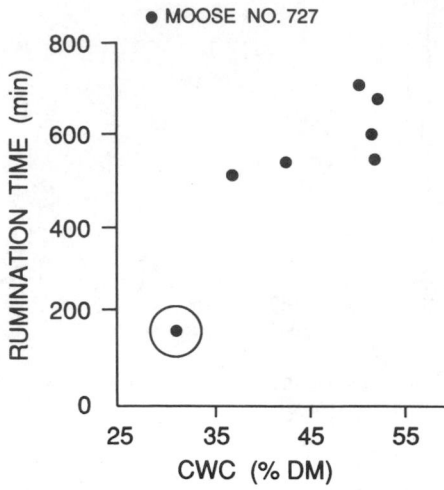

Fig. 11. Example of an influential observation (circled) in a scatterplot. The plot displays the time a female moose spent ruminating cell wall constituents (CWC) in the diet dry matter (DM) for selected times from December 1982 to January 1984 at the Ministik Wildlife Research Station, Alberta, Canada (from Renecker and Hudson 1989).

scatterplot can show a distinct trend or pattern and yet the correlation can be 0. This is because the correlation coefficient indicates only whether there is a *straight-line* relation between two variables. When a scatterplot displays a relationship other than a straight line, the ''strength'' of the relationship can be measured through other measures such as the coefficient of multiple determination in multiple regression (more on this later) or by investigating whether a straight-line relation exists between transformations (functions) of the two variables. An example of this latter method is given in the next section on simple linear regression.

A few words of caution concerning correlation need to be said at this point. **The presence of correlation between two variables, even a substantial correlation (near +1 or −1), does not imply that a cause-and-effect relationship exists between the two variables.** The correlation may be present because both variables are responding to changes in some third variable. For example, changes in the purchase of charcoal briquettes and the wearing of shorts may be positively correlated because both are responding to changes in outdoor temperature. Correlation may be present but impossible to interpret when a cause-and-effect relation exists between two variables, but this effect is ''mixed up'' or ''confounded'' with the fact that changes in several other factors are also causing changes in the two variables. Determining why an observed correlation is actually present can be difficult, and the researcher must be careful whenever attempting to interpret a correlation.

Additional features of scatterplots that one should be aware of are *outliers* and *influential* points. An outlier is a point that lies well above or below the ''band'' or ''cloud'' determined by the vast majority of the points. An influential point is one that has a strong effect on the impression that a trend is present in the data, i.e., removal of this point would have a significant effect on our impression of the trend present. Isolated points to the extreme left or right of a scatterplot are often influential.

In the scatterplot in Fig. 10, from Fryxell et al. (1988), the circled point would be considered an outlier, because it lies well above the ''band'' determined by the other points. In the scatterplot in Fig. 11, adapted from Renecker and Hudson (1989) (we have retained only the

points corresponding to moose number 727), the circled point is influential. The trend suggested by the plot with the point present is much steeper than when the point is removed.

When outliers are present, it is worth investigating whether the observation corresponding to the point is special in any way. If it is, analyzing it separately from the remaining data may be worthwhile. When a point is influential, the data should be analyzed twice, once with the point present and once with the point absent. Conclusions based on analysis with the point present must be regarded with caution if they differ from conclusions based on analysis with the point absent. Conclusions whose validity rests on a single observation cannot be made with confidence.

REGRESSION
Simple Linear Regression
FORMAL MODELS

If the scatterplot or correlation coefficient suggests the relation between *bivariate data* (i.e., data in which two variables are measured on each sample unit) is a straight-line trend (recall that the equation of a straight line is $Y = b_0 + b_1X$, where b_0 is the Y-intercept and b_1 is the slope), often one will want to explore this apparent relation further. The methodology for such exploration is called *simple linear regression*. The word ''regression'' was coined by the English scientist Francis Galton (1822–1911) and was based on his study of the relation between the heights of fathers and their sons. He observed that tall parents tended to produce offspring that were taller than average, but not as tall as their parents and called this phenomenon ''regression toward mediocrity.''

In the wildlife literature, the question most commonly of interest is whether the apparent straight-line trend indicates that a true straight-line trend exists in the population from which the data were drawn. To answer this question, one must think about the population from which the sample (X_1, Y_1), (X_2, Y_2), . . . , (X_n, Y_n) comes. If one thinks of the X_i as independent variables and the Y_i as dependent variables, a formal regression analysis assumes that for a given value of X, the distribution of the values of Y for all population units having the given value of X is normal with mean $\beta_0 + \beta_1 X$ and variance σ^2. Thus the population units are scattered around the line $Y = \beta_0 + \beta_1 X$, and σ^2 determines how tightly the points cluster around the line. In particular, the proportion of population units within a given band about the line $Y = \beta_0 + \beta_1 X$ is determined by the normal distribution. Notice that the variance σ^2 of the population of Y-values for a given X is independent of X. This is sometimes referred to as the assumption of *homogeneity of variance*. One makes no assumptions about the distribution of the values of X in the population. If neither the X_i nor Y_i is regarded as an independent variable, the population is regarded as being described by a *bivariate normal* distribution, wherein for a given value of X the distribution of the values of Y for all population units having the given value of X is normal with mean $\beta_0 + \beta_1 X$ and variance σ^2. In addition, the values of X are also assumed to be normally distributed in the population, and in the population of all units the correlation coefficient between the X and Y values is assumed to be ρ. Discussion of the bivariate normal distribution is beyond the scope of this chapter, and the interested reader is referred to a text on regression or multivariate analysis. See, for example, Neter et al. (1983) for additional discussion of regression in this context. Although this description of the population is rather complicated, the validity of any formal inference one makes depends on the extent to which this description holds. When this adequately describes the population, we see that the line $Y = \beta_0 + \beta_1 X$, the variance σ^2, and the correlation ρ (when the population is bivariate normal) describe the relation between the variables X and Y.

INFERENCE

In the wildlife literature, researchers usually wish to make inferences about the slope β_1, generally testing whether the slope is 0 (which is interpreted as equivalent to testing whether the relation between Y and X in the population is a straight line). One may also wish to make inferences about the intercept β_0, the variance σ^2, the correlation ρ (when the population is bivariate normal), and predictions of future values of Y for a given X based on the line $Y = \beta_0 + \beta_1 X$. The first step is to obtain estimates of the slope β_1 and intercept β_0. This is generally done by the *method of least squares*. This method seeks to find the equation of the straight line (called the least squares regression line) having the property that it, on the average, minimizes the square of the vertical distances of the individual data points from the line. If we have n observations on two variables X and Y, denoted (X_1, Y_1), (X_2,Y_2), $\ldots$, (X_n,Y_n), using calculus one can show that the least squares line has the equation $Y = b_0 + b_1 X$, where

$$ b_1 = \frac{\sum_{i=1}^{n} (X_i - \bar{X})(Y_i - \bar{Y})}{\sum_{i=1}^{n} (X_i - \bar{X})^2} $$

$$ b_0 = \bar{Y} - b_1 \bar{X} $$

where $\bar{X}_1$ and $\bar{Y}$ are the means of the X_i and Y_i, respectively. Notice in the denominator of b_1 that if all the X_i are the same, then all X_i will equal $\bar{X}$ and the denominator will be 0. Hence we must take observations at two or more different values of X_i (and hence at least two observations) to estimate the two quantities b_0 and b_1. It also turns out that at least three observations involving at least two different values of the X_i are necessary to estimate variances and do statistical inference. If the pairs (X_1,Y_1), (X_2,Y_2), $\ldots$, (X_n,Y_n) in the sample are independent (this will be true if sample units were selected from some population by simple random sampling), one can show that unbiased estimates of β_0 and β_1 are given by the least squares estimates b_0 and b_1.

An unbiased estimate of σ^2 is given by the mean square error (MSE)

$$ \frac{1}{n-2} \sum_{i=1}^{n} (Y_i - b_0 - b_1 X_i)^2 $$

and, if appropriate, ρ is estimated by the correlation coefficient r for the sample data. In addition, b_0 and b_1 have normal distributions with means β_0 and β_1 and SEs

$$ \sigma(b_0) = \sigma \sqrt{\frac{1}{n} + \frac{\bar{X}^2}{\sum_{i=1}^{n} (X - \bar{X})^2}} $$

$$ \sigma(b_1) = \frac{\sigma}{\sqrt{\sum_{i=1}^{n} (X_i - \bar{X})^2}}, $$

respectively. Estimates of these SEs, denoted $\text{SE}(b_0)$ and $\text{SE}(b_1)$, are obtained by replacing σ by its estimate $\sqrt{\text{MSE}}$. MSE, $\text{SE}(b_0)$, and $\text{SE}(b_1)$ all have two degrees of freedom associated with them. CIs and hypothesis tests for β_0 and β_1 based on normal theory are applicable, and the general procedures discussed previously can be used. For example, a $(1 - \alpha) \times 100\%$ CI for the true slope β_1 of the regression line is

$$ b_1 \pm t_{n-2,\frac{\alpha}{2}} \text{SE}(b_1) $$

and an α-level test of the hypotheses $H_0: \beta_1 = 0$ vs. $H_1: \beta_1 \neq 0$, i.e., a test of whether the slope differs from 0, is to reject H_0 if

$$ |b_1| > t_{n-2,\frac{\alpha}{2}} \text{SE}(b_1). $$

Rejection of H_0 implies that there is evidence of a straight-line relation between X and Y and hence that the correlation ρ is nonzero. In fact, one can show that the test of $H_0: \beta_1 = 0$ vs. $H_1: \beta_1 \neq 0$ is equivalent to testing $H_0: \rho = 0$ vs. $H_1: \rho \neq 0$. Let us return again to the study of coyote home range. Gese et al. (1988) reported the slope of the simple linear regression between coyote home-range size and the square root of percent available pinyon-juniper habitat to be -1.52 and that it was significant at $\alpha < 0.01$, i.e., the hypothesis $H_0: \beta_1 = 0$ would have been rejected at $\alpha = 0.01$.

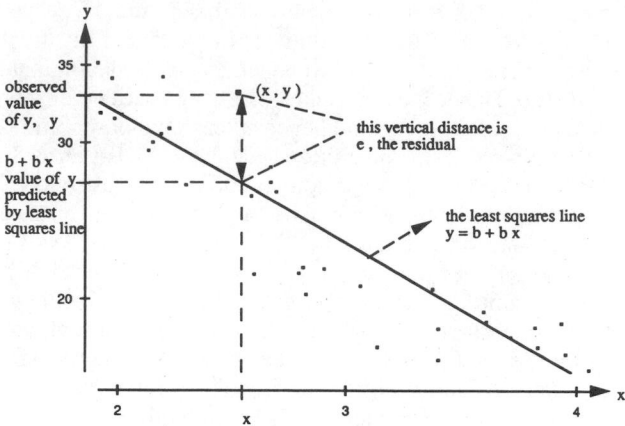

Fig. 12. The least squares line superimposed on a scatterplot. Also shown for a particular point are the observed value of y, the predicted value of y, and the residual as the vertical distance from a point to the regression line.

EXAMINING ASSUMPTIONS

The validity of any inference one makes depends on the extent to which our formal model describes the population from which the data were drawn. Thus a complete regression analysis should include examining the assumption that the data do indeed follow a normal distribution, have homogeneity of variance, do not follow a trend other than a straight line, and are independent. "Unusual" observations, such as outliers or influential points, are also identified and given careful scrutiny. Examination of assumptions is often carried out by examining the residuals $e_i = Y_i - b_0 - b_1 X_i$. Notice that the residual is simply the difference between the value of Y actually observed and the value Y would have if it fell exactly on the least squares line (Fig. 12).

Two ways in which the residuals are used to examine assumptions are the following. First, recall that the scatter of our population units (of which our data are a sample) about a line is determined by the normal distribution. In particular, the proportion within a given band about this line is determined by the normal distribution. One consequence of this is that the residuals, which measure how far a particular observation is from the least squares line, should behave approximately as though they have a normal distribution with mean 0. If one calculates all the residuals (many statistical software packages that do regression will calculate residuals), one can use statistical procedures to investigate whether the residuals, in fact, appear to follow a normal distribution. Second, the homogeneity of variance assumption implies that the population units should display the same magnitude of variability about a line. As a consequence, the magnitude of the residuals should not display any tendency to increase or decrease as the associated value of x increases or decreases. Such tendencies may indicate a violation of the homogeneity of variance assumption.

Corrective action may be necessary if the assumptions are not valid. For example, if the data do not appear normal or do not satisfy the homogeneity of variance assumption (violations of these two assumptions often occur together), one may try replacing (transforming) the values of $Y_1, \ldots, Y_n$ by some function of these values, i.e., by

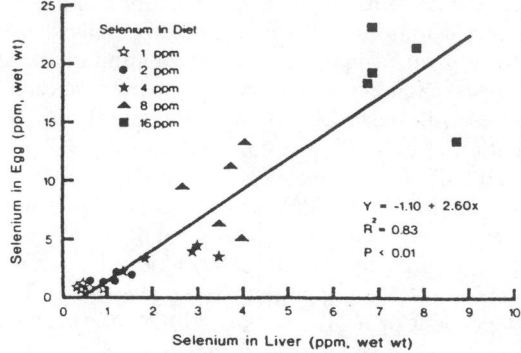

Fig. 13. Example of a violation of the homogeneity of variance assumption. The scatterplot displays the relationship between the concentration of selenium in the liver of female mallards and the concentration in the eighth eggs for females fed diets containing 1, 2, 4, 8, or 16 ppm selenium as selenomethionine (from Heinz et al. 1989).

$f(Y_1), \ldots, f(Y_n)$. Common functions of f are the logarithm, square root, reciprocal, or arcsin. Often, the transformed Y-values will satisfy the assumptions of normality or homogeneity of variance, and regression can proceed on the pairs $(X_1, f(Y_1)), \ldots, (X_n, f(Y_n))$. One must proceed with caution, however, because conclusions will now refer to the *transformed* data. For example, if a reciprocal transformation is used and one determines that the relation between $1/Y$ and X is $1/Y = -2X$, it is incorrect to conclude that increases in X produce decreases in Y. In fact, the relation between Y and X is $Y = -\frac{1}{2}X$ and Y increases (becomes less negative) as X increases.

Heinz et al. (1989), studying impaired reproduction of mallards fed an organic form of selenium, observed that data in their study that were measured as percentages were made to appear more normal by a transformation. Also, the scatterplot in Fig. 13 of the measured concentration of selenium in the eighth eggs of females fed diets of selenium versus the concentration in their liver suggests a violation of the homogeneity of variance assumption. Notice that variation about the regression line increases as selenium in the diet increases.

Although a formal discussion concerning the effects of violations of assumptions on the resulting inference is quite mathematical, a few comments can be made. If the expected value of Y at a given X is in fact $\beta_0 + \beta_1 X$ (i.e., on average the value of Y at a given X is $\beta_0 + \beta_1 X$), the least squares estimates of β_0 and β_1 are unbiased even if the assumptions of homogeneity of variance, normality, and independence do not hold. If the expected value of Y at a given X is in fact $\beta_0 + \beta_1 X$ and the assumptions of homogeneity of variance and independence hold, MSE is an unbiased estimate of the variance even if the assumption of normality does not hold. For testing hypotheses or constructing CIs all assumptions must hold, although these procedures are felt to be "robust" to departures from the assumption of normality, i.e., these inferences are still valid even if the assumption of normality is somewhat suspect.

The above discussion of simple linear regression is rather sketchy. The important thing to remember is that simple linear regression is a statistical tool for studying possible straight-line relationships between pairs of variables. A thorough discussion of simple linear regression, including checking of assumptions, can be found in any book on regression analysis (see, for example, Neter et al. 1983). Many introductory texts on statistics also contain discussion of simple linear regression (see, for example, Moore and McCabe 1988).

Multiple Regression

FORMAL MODELS

The techniques used in simple linear regression can be extended to provide methods for examining relationships other than straight-line relationships between sets of variables. The methodology for examining these relationships is called *multiple regression analysis* and applies to so-called *linear models*. Suppose a sample of n units is selected from some population, and for each of these units one records a dependent variable Y and p independent variables $X_1, \ldots, X_p$. One allows some of the X_i to be functions of the others; for example, one might allow X_2 to be the square of X_1. Let Y_i and $X_{1i}, \ldots, X_{pi}$ be the variables associated with unit i. One is said to have a linear model if the population from which the units are drawn and the method of selecting units are such that the relationship between Y and the Xs can be written as

$$Y_i = \beta_0 + \beta_1 X_{1i} + \beta_2 X_{2i} + \ldots + \beta_p X_{pi} + \epsilon_i.$$

The ϵ_i represent the cumulative effects of measurement error and independent variables not included in the model whose individual effects are considered to be small relative to those of the Xs. The Y_i are assumed to be independent (this is reasonable if units are selected by simple random sampling), and each Y_i is assumed to have mean $\beta_0 + \beta_1 X_{1i} + \beta_2 X_{2i} + \ldots + \beta_p X_{pi}$, have variance denoted by σ^2 (this is called the homogeneity of variance assumption because this variance is the same regardless of the values of the Xs), and be normally distributed. These last three assumptions concern the population from which the data are drawn. $\beta_0, \beta_1, \ldots, \beta_p$ are unknown constants (parameters) that one wishes to make inferences about. We may note that these assumptions imply that the ϵ_i are independent and that each ϵ_i has mean 0, has variance σ^2, and is normally distributed. Thus, a given value Y_i of the dependent variable will generally not equal $\beta_0 + \beta_1 X_{1i} + \beta_2 X_{2i} + \ldots + \beta_p X_{pi}$ exactly, but discrepancies will average to 0 in the long run, i.e., will yield 0 when averaged over all units in the population for which $X_1 = X_{i1}, \ldots, X_p = X_{pi}$. In mathematical language, one says the expected value of Y_i is $\beta_0 + \beta_1 X_{1i} + \beta_2 X_{2i} + \ldots + \beta_p X_{pi}$. $\beta_0 + \beta_1 X_{1i} + \beta_2 X_{2i} + \ldots + \beta_p X_{pi}$ is called the multiple regression function. As with simple linear regression, the validity of any inference one makes in multiple regression will depend on the extent to which the above assumptions hold, and an important component of any analysis involves checking that these assumptions are reasonable.

The β_i are often interpreted as the "effects" of the X_i in the sense that a unit change in X_k, holding the other X_i fixed, will produce a change of size β_k in Y, on average. This change may be directly caused by X_i or may occur because a change in X_i causes changes in other variables, which in turn cause a change in Y. Obviously if β_k is 0, changes in X_k, with the other X_i fixed, produce no change

in Y and so X_k has no "effect" on Y. Testing hypotheses as to whether β_k is 0 is considered one way to examine whether X_k has an "effect" on Y. In practice, however, one may encounter the following problem. In many experiments, the researcher has little or no control over the values of the X_i. This occurs in observational studies, for example, when units are selected from some population by simple random sampling and the X_i are characteristics of the unit (i.e., age, sex, weight). In such instances, the X_i are likely to be correlated and so a change in X_k is associated with changes in other of the X_i, which in turn affect Y through the other β_i. One never observes a set of units for which all the X_i, except, say, X_k, remain fixed and only X_k varies. Roughly, this means that we get no "direct" information about β_k, only "indirect" information that is subject to the additional variation in the other X_i. If the correlation among the X_i is large, this additional variation is large and one's uncertainty about β_k is increased. This state of affairs is called *multicollinearity* and can lead to rather uncertain inferences about β_k, i.e., estimates will have large SEs.

The term "linear" in linear model is borrowed from a branch of mathematics called linear algebra and refers to the fact that such an equation says that the expected value of Y is a so-called linear function of the β_i. This means the model is a sum of terms of the form

(parameter) $\times$ (some function of the independent variables).

It does not mean that Y is a straight-line function of the X_i. For example, if only a single independent variable X is measured, one can define $X_j = X^j$, so that the linear model is

$$Y_i = \beta_0 + \beta_1 X_i + \beta_2 X^2_i + \ldots + \beta_p X^p_i + e_i,$$

i.e., Y is a polynomial in X. Models that are not linear in the β_i, such as

$$Y_i = \beta_0 + \beta_1 X_i^{\beta_2}$$

are called *nonlinear* models.

INFERENCE AND INTERPRETATION

Analysis of a linear model proceeds in a manner analogous to that used for simple linear regression. The method of least squares can be used to obtain estimates of the β_i. A word of caution is needed here. Recall from the discussion of simple linear regression that one needs at least two different values of the independent variables to estimate the two parameters β_0 and β_1. In multiple regression a similar problem arises. One needs at least $p + 1$ different sets of values of the independent variables (and hence at least $p + 1$ observations) to be able to estimate the $p + 1$ parameters $\beta_0, \beta_1, \ldots, \beta_p$. If, in addition, estimates of variances are desired, at least $p + 2$ observations and at least $p + 1$ different sets of values of the independent variables are required. Furthermore, to evaluate the fit of the model, via a so-called lack-of-fit test (see Neter et al. 1983 for details), one needs at least one repeat observation at a fixed set of values of the independent variables. Thus to do inference in multiple regression, one needs a minimum of $p + 2$ observations at at least $p + 1$ different values of the independent variables with a repeated observation of at least one set. Of course, the more observations taken, the better, and, for purposes of inference, the wider the range of values of the independent variables at which observations are obtained (with several repeat observations at several sets of values of the independent variables), the better. Unfortunately, the cost or difficulty of obtaining observations may place severe limits on the number of observations one can obtain. Giving guidelines concerning how many observations to take is therefore difficult, and perhaps the best advice to give is to consult a statistician knowledgeable in design of experiments.

If the β_i can be estimated and if the errors e_i are assumed to be normal, formulas for the SEs of these estimates can be derived and used to construct CIs and test hypotheses with the normal methods discussed previously (see Box 4, the appropriate value for df here is $n - p - 1$). Unfortunately, these formulas are rather complex, and a knowledge of matrix algebra is necessary for their derivation (see, for example, Neter et al. 1983 for details). In practice these estimates and SEs are obtained with statistical software.

Interpretation of results can be complicated. The following example illustrates this. Bergerud and Ballard (1988) used multiple regression to study the effect of snow depth (mean depth in centimeters over an 8-month winter period), wolf numbers (in winter after birth), and total caribou numbers on an index of caribou recruitment in south-central Alaska. The index of caribou recruitment used was the percentage of 2.5-year-old caribou among all caribou ≥ 2.5 years old. Several multiple regression models were run. One result was

$$\text{recruitment} = 20.980 + 0.128 \text{ snow} - 0.064 \text{ wolves}.$$

For example, this model would predict that for a mean snow depth of 50 cm and 200 wolves in the winter after birth, recruitment would be $20.980 + 0.128 \times 50 - 0.064 \times 200 = 14.58\%$. Notice that this model has a positive coefficient for the snow-depth term, which would seem to suggest that increased snow depth (indicating a more severe winter) increases recruitment. This is counterintuitive and is undoubtedly due to multicollinearity, i.e., number of wolves and snow depth may be correlated and so the coefficients are somewhat difficult to interpret. The presence of multicollinearity is further indicated by the fact that a simple linear regression of recruitment on snow depth yielded the model

$$\text{recruitment} = 23.261 - 0.166 \text{ snow}.$$

In this model, the coefficient for snow is negative, i.e., less snow yields higher recruitment, which would seem more sensible. When the coefficient of a term in a multiple regression model changes sign or changes in size dramatically when other independent variables are added to the model, multicollinearity is often present, and interpretation of individual coefficients must be done with care, if at all.

In a multiple regression analysis one often reports more than simply least squares estimates and SEs of parameters. Several measures of how well the model fits the data and generalizations of the correlation coefficient discussed in simple linear regression are also reported. For the model

$$Y_i = \beta_0 + \beta_1 X_{1i} + \beta_2 X_{2i} + \ldots + \beta_p X_{pi} + e_i$$

these include the following:

1. The sum of squares total (SSTO). This measures the total variation in the dependent variable Y and is given by the formula

$$\text{SSTO} = \Sigma(Y_i - (\bar{Y}))^2,$$

where $(\bar{Y})$ is the mean of the Y_i.

2. The sum of squares for error (SSE). This measures how much the actual values of the dependent variable vary about the fitted multiple regression model $b_0 + b_1 X_1 + \ldots + b_p X_p$, where b_i is the least squares estimate of β_i. The formula for SSE is

$$\text{SSE} = \Sigma(Y_i - b_0 - b_1 X_{1i} - \ldots - b_p X_{pi})^2.$$

Additionally, one can define the mean sum of squares for error (MSE) to be $\text{SSE}/(n - p - 1)$, where n is the number of observations. MSE is an unbiased estimate of σ^2, the variance of the errors e_i, and provides a measure of how well the model fits the data; the smaller the MSE the better the fit. As pointed out above, the number of observations (n) must exceed $p + 1$ for MSE to be defined. If $n < p + 1$, one is dividing SSE by 0 or a negative number. Estimating variances by negative numbers makes no sense.

3. The sum of squares for regression (SSR). This measures how much of the variation in the dependent variable is accounted for by the multiple regression model and is given by the formula

$$\text{SSR} = \text{SSTO} - \text{SSE}.$$

4. The coefficient of multiple determination, denoted R^2. This measures the fraction of the variation in the dependent variable accounted for by the multiple regression model and is given by the formula

$$R^2 = \text{SSR}/\text{SSTO}.$$

R^2 is always between 0 and 1 and is similar to SSR in interpretation. One shortcoming of SSR as a measure of how well a multiple regression model fits a set of data is that whether SSR is sufficiently large depends on how large SSTO is, i.e., SSR must be interpreted relative to SSTO. R^2 accomplishes this automatically by taking the ratio of SSR and SSTO. As such it is a unitless quantity. In simple linear regression, one can compute R^2 and it turns out to equal the square of the usual correlation coefficient. For this reason R, the positive square root of R^2, is often perceived as the generalization of the correlation coefficient from simple linear regression to multiple regression and is called the *multiple correlation coefficient*.

These four measures are routinely reported by statistical software for regression and form the basis for comparing various multiple regression models. Such comparisons are formally conducted as follows. To determine whether the "full" regression model

$$Y_i = \beta_0 + \beta_1 X_{1i} + \beta_2 X_{2i} + \ldots + \beta_p X_{pi} + e_i$$

is necessary to explain the variation in the dependent variable Y, or if the "reduced" model

$$Y_i = \beta_0 + \beta_1 X_{1i} + \beta_2 X_{2i} + \ldots + \beta_p X_{qi} + e_i,$$

involving only the independent variables $X_1, X_2, \ldots, X_q$

($q < p$), which are a subset of $X_1, X_2, \ldots, X_p$, is adequate to explain the variation in the dependent variable, calculate SSR and SSE for the full model and for the reduced model. Let $\mathrm{SSR}(X_1, \ldots, X_p)$ and $\mathrm{SSE}(X_1, \ldots, X_p)$ denote SSR and SSE for the full model and $\mathrm{SSR}(X_1, \ldots, X_q)$ and $\mathrm{SSE}(X_1, \ldots, X_q)$ denote SSR and SSE for the reduced model. The quantity

$$\begin{aligned} \mathrm{SSR}(X_{q+1}, &\ldots, X_p \mid X_1, \ldots, X_q) \\ &= \mathrm{SSR}(X_1, \ldots, X_p) - \mathrm{SSR}(X_1, \ldots, X_q) \end{aligned}$$

is called the *extra sum of squares* and measures how much better the full model fits the dependent variable than does the reduced model. If this is sufficiently large—more precisely, if $(\mathrm{SSR}[X_{q+1}, \ldots, X_p \mid X_1, \ldots, X_q]/[p-q])/(\mathrm{SSE}[X_1, \ldots, X_p]/[n-p])$ exceeds the appropriate critical value of the F statistic with $p - q$ numerator and $n - p$ denominator df—one decides that the full model is necessary. Otherwise one decides that the reduced model is adequate. Formally this tests the hypothesis of whether the independent variables $X_{q+1}, \ldots, X_p$ have a significant effect on the dependent variable after accounting for the effects of the independent variables $X_1, \ldots, X_q$. This method is purely statistical and does not take into account the scientific "reasonableness" of the full or reduced model. The statistical decision must be modified by scientific considerations in any final decision concerning an appropriate model.

An informal test of the above hypothesis is often carried out by simply comparing the R^2 values for the full and reduced models and selecting the full model if the R^2 is appreciably higher, although how much higher is "appreciably higher" is rather subjective. The formal hypothesis test is probably the better way to make comparisons.

For the Bergerud and Ballard (1988) study, the two multiple regression models mentioned above,

$$\text{recruitment} = 23.261 - 0.166\ \text{snow}$$

and

$$\text{recruitment} = 20.980 + 0.128\ \text{snow} - 0.064\ \text{wolves},$$

have R^2 values of 0.10 and 0.79, respectively. These values would seem to suggest that snow depth is not a particularly significant predictor of recruitment, but wolf numbers, when added to a model containing snow depth, seem to be a significant predictor of recruitment. The authors also fit a model using only wolf numbers as an independent variable and obtained

$$\text{recruitment} = 24.379 - 0.057\ \text{wolves}$$

with $R^2 = 0.75$. This suggests that wolf numbers are a significant predictor of recruitment but that the addition of snow depth to a model containing wolf numbers is not particularly significant (R^2 increases only to 0.79). Unfortunately, no information about formal tests of hypotheses is mentioned in the paper, so these conclusions are somewhat subjective.

Several general observations can be made from this example. First, the value of R^2 increased in the above models when an additional independent variable was added. This always occurs in multiple regression, i.e., the addition of an independent variable will always cause R^2 to increase (or at worst stay the same). This is intuitively plausible, because the addition of independent variables provides extra information and cannot detract from our predictive ability. One could always ignore the extra information. Because R^2 can be inflated by adding independent variables, one must be careful to avoid adding extra independent variables simply to get a large R^2. A balance between reasonable R^2 and relatively few independent variables (simplicity of the model and hence ease in interpretation) is the goal. This balance is called *parsimony*.

Second, notice that interpretations were a bit awkward. For example, in comparing the model with snow depth and wolf numbers as independent variables to the model with only snow as an independent variable, we concluded that wolf numbers added predictive power to a model already containing snow depth as an independent variable. This "conditional" interpretation is a bit different than merely saying wolf numbers are a significant predictor of recruitment. This illustrates the kind of care that one must take in interpreting the results of a multiple regression analysis. Undoubtedly this sounds like nitpicking, but care is necessary if one is to interpret the results of statistical analyses properly.

Third, we mentioned that the model with only snow depth as an independent variable had an R^2 of 0.10, which did not seem to be particularly significant. Actually it is possible in multiple regression to have a very low R^2 (any value > 0, even 0.000001) and yet have statistical significance in a formal hypothesis test. Conversely, it is possible to have a large value of R^2 and not have statistical significance. For this reason it is good practice to conduct formal tests of hypotheses in addition to reporting R^2 values.

Fourth, again examining the model with only snow depth as an independent variable, we were tempted to conclude snow depth was not useful as a predictor of recruitment. Technically one can conclude only that a straight-line relationship between snow depth and recruitment does not exist. In theory, one might find that a multiple regression model like

$$\begin{aligned} \text{recruitment} = \text{constant} &+ b_1 \times \text{snow} + b_2 \times \text{snow}^2 \\ &+ b_3 \times \text{snow}^3 \end{aligned}$$

has a fairly high R^2 and is statistically significant, indicating snow depth is useful for predicting recruitment, but the prediction relation is more complicated (here a cubic polynomial) than a simple straight-line relation. Bergerud and Ballard (1988) reported that a three-way ANOVA was conducted and that snow depth as a main effect and in interactions was not significant. This kind of analysis does suggest that snow depth is not useful as a predictor of recruitment (although the authors do not make it clear exactly how the variable snow depth was categorized so as to make it a classification variable suitable for ANOVA). In general, multiple regression tends to provide information about the specific way in which an independent variable may be useful for predicting a dependent variable. ANOVA (or regression with indicator variables—see below) is more suitable for answering the question of whether an independent variable is useful in some way (no specific functional form specified) for prediction.

Fifth, notice that the dependent variable, being a percentage, is constrained to lie between 0% and 100%. For the model with snow depth as the only independent vari-

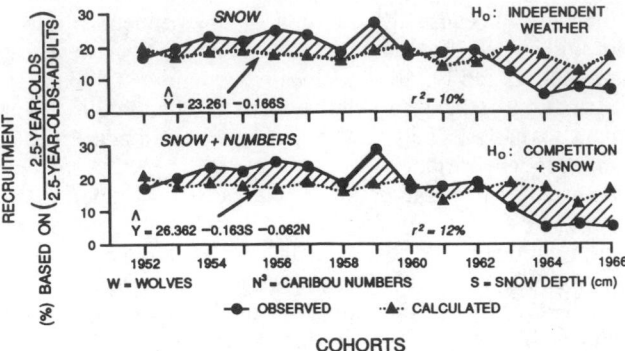

Fig. 14. Example of a possible violation of the assumptions of multiple regression. The plots display actual recruitment at 2.5 years of age in the Nelchina caribou herd, south-central Alaska, versus predicted recruitment using first snow depth and then both snow depth and caribou numbers as independent variables, for the years 1952–66 (from Bergerud and Ballard 1988).

able, a snow depth of 150 cm would predict recruitment at −1.639% which is, of course, nonsense. Examination of the authors' data shows that actual snow depth never exceeded 75 cm. Substituting a value of 150 cm, therefore, involves extrapolating to data outside the range used to estimate the multiple regression model. Such extrapolation should be avoided, and multiple regression models should be considered valid only for the range of data used to establish the model.

PARTIAL CORRELATION

The *coefficient of partial correlation* is often reported in multiple regression analyses. Consider once again the multiple regression model

$$Y_i = \beta_0 + \beta_1 X_{1i} + \beta_2 X_{2i} + \ldots + \beta_p X_{pi} + e_i.$$

The amount of additional variability explained by adding X_j to a model already containing the r variables $X_{k_1}, \ldots, X_{k_r}$ is called the *coefficient of partial determination* between Y and X_j given $X_{k_1}, \ldots, X_{k_r}$ and is defined to be

$$r^2_{j \cdot k_1 \ldots k_r}$$

$$= \mathrm{SSR}(X_j \mid X_{k_1}, \ldots, X_{k_r}) / \mathrm{SSE}(X_{k_1}, \ldots, X_{k_r}).$$

The corresponding coefficient of partial correlation is the square root of $r^2_{j \cdot k_1 \ldots k_r}$ having sign equal to that of b_j in the fitted model

$$Y = b_0 + b_j X_j + b_{k_1} X_{k_1} + \ldots + b_{k_r} X_{k_r}.$$

The relationship between the coefficient of partial determination and the coefficient of partial correlation is analogous to that between the coefficient of multiple determination (R^2) and the correlation coefficient (r) in regression. In particular, the coefficient of partial determination is easier to interpret than the coefficient of partial correlation. Compton et al. (1988) fitted a multiple regression model with number of deer (ND) observed at various locations along the lower Yellowstone River as the dependent variable; amount of riparian cover in hectares (RC) and amount of riparian cover with cattle in hectares (GR) were the independent variables. The fitted model was

$$ND = -3.69 + 0.92RC - 0.50GR$$

with an R^2 of 0.57. The coefficient of partial correlation of GR for a model already containing RC was −0.53. Notice the sign matches that of the coefficient of GR in the fitted model. The coefficient of partial determination is $(-0.53)^2 = 0.28$. We conclude that the addition of GR to a model already containing RC accounts for an additional 28% of the variance (SSE[RC]) still remaining.

EXAMINING ASSUMPTIONS

In any multiple regression one should thoroughly check whether the model assumptions seem reasonable, i.e., whether the errors are normally distributed with mean 0 and constant variance σ^2. For example, in Bergerud and Ballard (1988), the plot in Fig. 14 of the observed and calculated (from the fitted model) values of the dependent variable shows that the early data tended to have observed values above those predicted by the model, whereas in later years the observed values were below the predicted values. This suggests that the errors do not have mean 0, but a mean dependent on time. Time should probably be included in the model as an additional independent variable. This is good practice for any data collected over time and may require the use of time-series analysis for a thorough statistical investigation.

Because the dependent variable in the Bergerud and Ballard (1988) models is constrained to lie between 0% and 100%, it cannot technically be considered normally distributed. This problem may not be serious if the values of the dependent variable do not tend to cluster near the extremes of 0% or 100% (they do not seem to cluster near the extremes in the Bergerud and Ballard study) and appear approximately normal over the range of values observed. In such a situation, the multiple regression analysis is probably satisfactory.

A multiple regression analysis may be statistically valid in the sense that all assumptions seem reasonable and the calculations are done properly, but it may be criticized on other grounds. For example, Van Ballenberghe (1989) criticized the multiple regressions of Bergerud and Ballard on the ground that wolf numbers were obtained artificially and the apparent relation between recruitment and wolf numbers may be partly due to something in the artificial method of estimating wolf numbers rather than actual wolf numbers, which were not measured. This possibility deserves further consideration, and to address this issue it may be necessary to conduct a study in which actual wolf numbers are determined and compared to values based on this artificial method.

CATEGORICAL VARIABLES

Categorical variables can be incorporated into multiple regression models in a number of ways. As an illustration of this, suppose one records eye colors of human subjects as brown, blue, or other. Eye color is thus a categorical variable with three categories. One way to quantify this variable might be to denote it by the letter Z and write $Z = 1$ if eye color is brown, $Z = 2$ if eye color is blue, and $Z = 3$ if eye color is other. Suppose we now proceed to use multiple regression to determine the relation between eye color and blood pressure (Y). Treating eye color, Z, as the independent variable and blood pressure, Y, as the dependent variable, we would get a regression equation of the form

$$Y = b_0 + b_1 Z.$$

Unfortunately, this equation predicts that blood pressure for brown-eyed people is $b_0 + b_1$, that blood pressure for blue-eyed people is $b_0 + 2b_1$, and that blood pressure for other eye colors is $b_0 + 3b_1$. Regardless of the values of b_0 and b_1, our coding scheme used to define Z forces the predicted value of blood pressure Y, for blue-eyed individuals, as given by the regression equation, to take on a value between that for brown-eyed individuals and that for individuals with other eye colors, even if the data indicate otherwise. Furthermore, the difference in predicted blood pressure, based on the regression equation, between brown- and blue-eyed individuals is automatically the same as that between blue-eyed individuals and those with other eye colors. The way in which Z was defined automatically imposes these relations (possibly incorrect) between eye color and blood pressure as predicted by the regression equation. The above way of quantifying eye color leads to poor results in multiple regression.

A better way to quantify eye color in the example is to define two variables, Z_1 and Z_2, as follows. Let

$Z_1 = 1$ if the subject has brown eyes,
  $= 0$ if the subject does not have brown eyes,
and let

$Z_2 = 1$ if the subject has blue eyes,
  $= 0$ if the subject does not have blue eyes.

A variable such as Z_1 or Z_2 that takes on only the values 0 and 1, 1 if a certain characteristic is present and 0 if the characteristic is not present, is called an indicator variable. Notice that for a brown-eyed subject $Z_1 = 1$ and $Z_2 = 0$, for a blue-eyed subject $Z_1 = 0$ and $Z_2 = 1$, and for a subject with some other eye color $Z_1 = 0$ and $Z_2 = 0$. There is thus a unique pair of values for each eye color and hence no need to define a third variable Z_3. Fitting a multiple regression model as before yields an equation of the form

$$Y = b_0 + b_1 Z_1 + b_2 Z_2.$$

If a subject has brown eyes, the regression equation predicts a blood pressure of $b_0 + b_1$. If the subject has blue eyes, the regression equation predicts a blood pressure of $b_0 + b_2$. For subjects with other eye colors, the regression equation predicts a blood pressure of b_0. Notice b_1 and b_2, the coefficients of Z_1 and Z_2, respectively, represent the difference in the effects of brown and blue eyes, respectively, from the effect of other eye colors on blood pressure, and thus b_0, the effect of other eye colors, becomes a sort of reference value. Because b_0, b_1, and b_2 can take on any values, the equation has the flexibility to predict any blood pressures for the different eye colors.

The example above indicates that one must exercise care in quantifying the values of a categorical variable. The second method indicated is the better way to proceed. In general, if a categorical variable is an independent variable, it is quantified for use in multiple regression by means of indicator or 0-1 variables. If a categorical variable can take on c possible values, the c minus 1 indicator variables are defined

$Z_i = 1$ if the categorical variable has the ith possible value,

$= 0$ otherwise

for $i = 1, \ldots, c - 1$. If $Z_1, \ldots, Z_{c-1}$ are all 0, obviously the categorical variable has value c. There is no need to define Z_c because it is redundant. Notice the ith indicator variable "indicates" whether the categorical variable takes on the ith value. The $c - 1$ indicator variables are all added to the multiple regression equation to represent the (main) effects of the categorical variable. If the coefficient of any of these indicator variables in the fitted multiple regression model is found, in a hypothesis test, to be significantly different from 0, the effect of that value of the categorical variable differs significantly from that of the cth value. The cth value becomes the reference value. By clever use of indicator variables and their cross products, one can represent ANOVA models as multiple regression models and test all the standard hypotheses of ANOVA. Mixing quantitative independent variables with indicator variables allows one to represent analysis of covariance models as multiple regression models. Additional discussion of the regression approach to ANOVA and analysis of covariance can be found in Neter and Wasserman (1974). Use of indicator variables makes multiple regression models more general than might first appear and illustrates the fact that regression, ANOVA, and analysis of covariance have much in common. In fact, they are all special cases of general linear models for which an extensive theory exists.

If a categorical variable is the dependent variable, the assumption of normally distributed errors is clearly violated, and methods other than multiple regression are needed. Categorical dependent variables arise, for example, in problems of classification such as when one may wish to identify the sex of an animal based on a set of morphological measurements. Special sets of procedures exist for classification, including discriminant analysis, which will be discussed later, and logistic or loglinear models, which we now discuss briefly.

If the dependent variable takes on only two values, the natural tendency is to treat it as a binomial random variable. For such binomial dependent variables, logistic regression or loglinear models are typically used to analyze data. The basic idea is not to fit a regression model to the dependent variables directly, but instead to assume that a regression model exists that relates the independent variables to the probability p of observing a particular value of the dependent variable (or more precisely to some function of p such as $\log[p/(1 - p)]$, which is called the logit of p). Elementary discussion of these topics can be found in Neter et al. (1983). Cox and Snell (1989) provide more thorough treatment. Many computer packages have subroutines to analyze such data. If the dependent variable takes on more than two values, one again fits regression models to functions of the probability that the dependent variable will take on a particular value, using multinomial logit models. Computer packages that handle these models are somewhat less common, but a package called GLIM enables analysis of such models. In fact, GLIM analyzes so-called *generalized* linear models of which linear models, loglinear models, and multinomial logit models are special cases. For more information see McCullagh and Nelder (1989) or Aitkin et al. (1989).

Holm et al. (1988) studied the effectiveness of certain

chemicals in discouraging deer mice from feeding on corn. These repellents were applied to corn kernels that were then offered to deer mice for a number of days. Each kernel was categorized as having or not having sustained damage. This categorical variable was the dependent variable. The different chemicals were categorical independent variables. As part of the analysis, the authors also recorded when a kernel sustained damage, and a loglinear model was used to study how the probability of sustaining damage changed over time.

STEPWISE REGRESSION

Often in multiple regression many independent variables are measured. Some of these variables may be significantly correlated, and part of the goal of the analysis is to produce a model that makes scientific sense and fits the data well (has high R^2 or small value of MSE for example) while retaining only a relatively small number of independent variables. The best way to find such a model is simply to fit every possible regression model with some or all of the independent variables and to choose the one that strikes the desired balance among scientific sense, good fit, and small number of independent variables. Several rules of thumb are available for deciding what constitutes a desirable balance (see Neter et al. 1983 for a discussion of the balance between good fit and small number of independent variables), but ultimately the choice is somewhat subjective. For example, Nixon et al. (1988) wished to study the effect of 24 habitat variables (the independent variables) on the presence or absence of deer (the dependent variable). After examining all possible regression models on the basis of R^2, the authors decided a model involving only five of the independent variables was satisfactory. Notice here that because the dependent variable was categorical with two values, a logistic regression might have been more appropriate.

If one has p independent variables, there are $2^p - 1$ possible models involving at least one independent variable, so the number of models gets large very rapidly. For example, with $p = 24$, as in Nixon et al. (1988), one must examine $2^{24} - 1$, or 16,777,215 models. Even on modern computers, examining this many models is time consuming. For large p, therefore, algorithms have been developed that "cleverly" search for models with good fit while examining only a fraction of the possible models. These algorithms have been implemented on computer packages and are called *stepwise regressions*. The *forward stepwise regression* algorithm starts by trying all models with a single independent variable and selecting the one with highest R^2 or highest value of the F-statistic for testing whether the model fits. If this highest R^2 or value of F exceeds a prespecified cut-off, the algorithm accepts this model and continues. It now adds the independent variables not currently in the multiple regression equation to the one it has just accepted and finds the variable that increases R^2 the most or has highest value of F. If this exceeds the cut-off, the algorithm accepts this model and proceeds. The algorithm continues to add variables until there is inadequate improvement, at which point the computer stops and prints out the final model accepted as best. Changing the user-specified cut-off values can change the final model produced by the algorithm.

Backward stepwise regression works just the reverse of forward stepwise regression. It begins with all variables in the model and determines which one decreases R^2 or F the least. If this decrease does not exceed a user-specified cut-off, the variable is dropped from the model and the algorithm is repeated. This process continues until no more variables can be removed, at which point it ceases and the final model is printed. The model resulting from a backward stepwise regression may vary as one changes the cut-off values and it need not agree with the model produced by a forward stepwise regression.

The most popular stepwise procedure is the *full stepwise regression*, which alternates between a forward and backward stepwise approach. Variables added at a given stage may be removed at a later stage, and those removed at a given stage may be replaced later. The user must supply two cut-off values (one for the forward part and one for the backward part), and the choice will affect the final result. The result of a full stepwise regression need not agree with either a forward or backward stepwise regression. Johnson et al. (1989) used stepwise regression (presumably the full stepwise algorithm) to examine the effects of 15 land use variables on a variable measuring bird damage to grapefruits in southern Texas. The final model involved only three of the independent variables.

Although a stepwise regression will generally lead to a model with reasonably good fit, some words of caution are in order. These algorithms do not examine all possible models, so they may miss models with better fit and possibly fewer variables than those produced by the stepwise procedure. Models produced by these algorithms need not make scientific sense nor need they satisfy our regression assumptions. Any model produced by a stepwise procedure should therefore be investigated further. In addition to checking model assumptions, one may wish to add or delete variables to produce a model that achieves a better balance among scientific sense, good fit, and small number of independent variables. One may also wish to compare the model produced by a stepwise procedure to other models. Use of a stepwise procedure does not eliminate the need for additional investigation before one decides on a final regression model. Examination of all possible models is therefore recommended when feasible, i.e., when the number of independent variables is not too large. Stepwise procedures should be used only when this is not the case. For more information on stepwise procedures see Neter et al. (1983).

Additional topics in regression beyond the scope of the present discussion include the analysis of nonlinear models (Neter et al. 1983, Gallant 1987), inverse regression or calibration (Neter et al. 1983), and response surface methodology (Box and Draper 1987). All use some of the principles and techniques described above.

PATH ANALYSIS

When using regression, one can be tempted to (incorrectly) conclude that cause-and-effect relations exist between the independent and dependent variables simply because the resulting model was statistically significant and had a high value of R^2. Such a result could occur, however, because both the independent and dependent variables were caused by additional factors not observed by the investigator. **Good-fitting regression models need not imply cause and effect**. Many experimenters are nevertheless interested in determining if cause-and-effect relations exist, and a technique using regression, called *path analysis* (first introduced by Sewell Wright in the 1920s), can be exploited to help provide evidence of cause and effect and analyze cause-and-effect relations. In the following discussion we give a simplified overview of path analysis. Additional information can be found in Sokal and Rohlf (1981) and Namboodiri et al. (1975).

A path analysis begins with a specification of variables, say $X_1, \ldots, X_p$, whose interrelationships one wishes to study. For later ease, the X_i are usually assumed to be standardized by subtracting from each X_i its mean and dividing the result by its SD. One then writes down the cause-and-effect relationships that one believes exist between the X_i, perhaps using known sample correlations between the X_i to guide one in specifying these relationships. The cause-and-effect relationships are typically specified by means of a *path diagram* that displays the variables and joins with an arrow pairs thought to be related by cause and effect. If X_i is thought to *directly* cause X_j, then these two factors are connected with an arrow pointing from X_i to X_j. If X_i and X_j are directly related but it is not clear which causes which, X_i and X_j are connected with a two-headed arrow pointing at both.

In the Bergerud and Ballard (1988) study, suppose the interest had been in studying cause-and-effect relations among caribou recruitment, mean winter snow depth, and wolf numbers. Recall that, as pointed out by Van Ballenberghe (1989), the wolf numbers used by Bergerud and Ballard were not the true wolf numbers, but numbers based on artificial estimates. With this in mind, we might suggest the following possible description of the relationships among the standardized variables: greater snow depth directly causes reductions in both caribou recruitment and true wolf numbers; increased true wolf numbers directly cause decreases in caribou recruitment; and true wolf numbers directly cause the artificial wolf numbers recorded. Additionally, changes in caribou recruitment and true wolf numbers are directly caused by random factors (error). A path representing these relations might look like that given in Fig. 15.

After specifying cause-and-effect relationships, one next writes down a series of multiple regression equations, linear in the X_i (i.e., no $X_i^2, X_i^3, \ldots, X_i^n$ terms or cross-product terms) and with no intercept term, relating each X_j, in turn, to the other X_i thought to directly cause it as specified in the path diagram. This system of equations, plus any data one might have, is used to analyze the cause-and-effect relationships. This analysis might take several directions. One direction, common in the absence of data, is simply to analyze mathematically the system of equations to draw conclusions about:

(a) the logical consequences that can be deduced from the cause-and-effect relations that have been specified,
(b) the feasibility of estimating correlations or coefficients between the X_i that cannot be done directly (this is related to the question of whether the system of multiple regression equations resulting from the path diagram can be solved for certain unknowns), and
(c) whether the proposed cause-and-effect relations are reasonable, i.e., lead to reasonable conclusions.

Another direction one can pursue when data are present

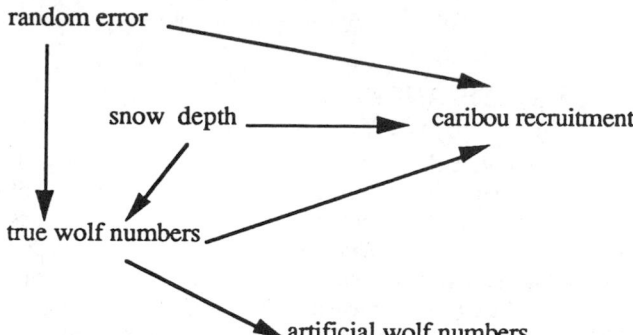

Fig. 15. Example of a path diagram of hypothetical relationships among snow depth, true wolf numbers, artificial wolf numbers, random error, and caribou recruitment. Arrows indicate directions of possible causal relations.

is to use the data (after they are standardized) to estimate the regression coefficients in each of the regression equations and then do (a), (b), or (c) above using the actual estimated equations. Some statistical inference on the estimates also can be done.

From the path diagram given in the previous example, we might write down the three regression equations (remember, these are in terms of the standardized variables, i.e., with the means subtracted and the result divided by the SD):

caribou recruitment
$$= b_1 \text{ snow depth} + b_2 \text{ true wolf numbers} + b_3 \text{ error,}$$

true wolf numbers
$$= c_1 \text{ snow depth} + c_2 \text{ error, and}$$

artificial wolf numbers
$$= d_1 \text{ true wolf numbers.}$$

Unfortunately, true wolf numbers were not measured and, of course, error is not directly observable and so cannot be measured. The data on caribou recruitment, snow depth, and artificial wolf numbers are not sufficient to fit any of these models by regression. Substituting the third equation into the first yields

caribou recruitment
$$= b_1 \text{ snow depth} + (b_2/d_1) \text{ artificial wolf numbers} + b_3 \text{ error,}$$

and multiple regression could be used to estimate b_1 and b_2/d_1 (but not b_3 since the actual errors are not observed) from the data collected. If b_2/d_1 was not significantly different from 0, this would suggest b_2 was not significantly different from 0 and hence there is not significant evidence that true wolf numbers cause changes in caribou recruitment in a linear manner. Note these conclusions depend on the validity of the path diagram.

The path diagram in a path analysis often involves variables that are not directly observable. Nevertheless, by borrowing techniques from factor analysis (to be discussed below), one can reach conclusions about the effects of these factors. The computer package LISREL VI (LISREL stands for linear statistical relationships) is available to carry out the necessary calculations. The procedures and results are rather complicated, and one should consult a statistician before proceeding with such an analysis.

FACTOR ANALYSIS AND PRINCIPAL COMPONENTS

Investigators collect multivariate data in essentially all real experiments, usually with a fairly large number of variables measured on each unit. Browsing through issues of *The Journal of Wildlife Management* or the *Wildlife Society Bulletin* will quickly convince one of this fact. For example, Cruz (1988) measured the frequency of use of 20 foraging categories for 11 species of birds. Thus on each species (basic experimental unit) there are 20 measured variables, resulting in 11 multivariate measurements.

When many variables are measured on each unit, often these variables are correlated with each other. In fact, "redundancies" may exist in the variables, i.e., the information in one variable is essentially contained in a subset of the other variables. In multiple regression terms, this one variable could be predicted fairly precisely with a multiple regression equation involving a subset of the other variables and hence eliminated from the data without much loss of information. Thus, for multivariate data, the following question may be worth investigating. Is it possible to replace the set of variables actually measured with a much smaller set of "artificial" variables without sacrificing much of the information contained in the original measurements? *Principal components analysis* provides a means of answering this question.

Suppose $X_1, \ldots, X_n$ are the n variables measured on each experimental unit. A *linear combination* or (linear) *score* of these variables is any function of the form $a_1 X_1 + a_2 X_2 + \ldots + a_n X_n$ where the a_i are known constants (some may be 0). Principal components analysis attempts to find the set of m linear combinations or scores of the X_i, say

$$a_{11}X_1 + \ldots + a_{1n}X_n$$

$$a_{21}X_1 + \ldots + a_{2n}X_n$$

$$\ldots$$

$$\ldots$$

$$\ldots$$

$$a_{m1}X_1 + \ldots + a_{mn}X_n$$

with m much smaller than n, which "best" replaces the original variables $X_1, \ldots, X_n$. Here "best" has the following meaning. The sum of the sample variances of these m linear combinations or scores is the largest possible for any set of m scores. The set with largest variance is said to account for as much of the variability in the original data as possible by use of only m scores, and hence "best" explains the variation in the original data. The goal of most studies is to understand why the data vary, so this set of m scores is the "best" way to replace the original variables by a smaller subset of m new variables for purposes of investigating the variation. These best m scores are called the *first m principal components*. The $a_{11}, a_{12}, \ldots, a_{mn}$ in the above linear combinations are called the *factor loadings*.

One reason for using principal components is to replace a large, unwieldy set of measurements by a much smaller set without sacrificing too much of the information in the original measurements. Smaller sets of data are generally easier to analyze and interpret. In Cruz (1988), it turned out to be important to reduce the number of variables. Recall in the discussion of multiple regression we mentioned that the method of least squares required at least $p + 1$ observations to estimate the $p + 1$ parameters and $p + 2$ observations to estimate variances in the multiple regression equation. The Cruz data consisted of 11 multivariate observations, each containing 20 variables. For reasons similar to those in multiple regression, one needs at least 21 (and preferably more) observations to jointly investigate correlations among the 20 variables without problems. With only 11 observations one is capable of examining correlations among only 10 variables. Thus in the Cruz data, the number of variables must be reduced before investigation of relationships among the variables can proceed. Principal components provides a method for doing this while retaining as much information as possible. In fact, Cruz reduced the number of variables from 20 to the best five scores, or first five principal components, using principal components analysis.

Determining m, i.e., the number of scores to reduce to, in a principal components analysis is somewhat subjective. The mathematics of a principal components analysis guarantees that if one chooses $m = n$, i.e., the number of scores equals the number of original variables (in this case one is said to have determined all n principal components), the resulting scores will not be the original variables but rather a set of uncorrelated scores that can be ordered so that the first one is the best single score (first principal component), the first two are the best pair of scores (first two principal components), the first three are the best set of three scores (first three principal components), and so on. The fact that these scores are uncorrelated suggests that they can be interpreted as being independent. Associated with each score is a number (called an eigenvalue) that is proportional to the additional variance explained by the score. The sum of all $m = n$ eigenvalues equals the total variance in the data, and so the eigenvalue of a particular score divided by this total indicates the fraction of the total variability explained by the score. The sum of the eigenvalues of the first j principal components divided by the total variability indicates the fraction of the total variability accounted for by these j scores. One wants to keep the number of scores that one reduces to fairly small while explaining a reasonable fraction of the total variability. Adding scores increases this fraction, so there is a tradeoff between these two objectives. In the Cruz paper, the first five principal components accounted for 0.75 of the total variability, and this was believed to be a reasonable compromise. Thompson and Capen (1988) measured 13 habitat variables on each of 24 species of birds. They used a principal components analysis to reduce the number of habitat variables and determined that the best three scores, or first three principal components, accounted for 0.886 of the variation. This was believed to be satisfactory. The ideal would be to have the first one or two scores account for 0.90 or greater of the variability, but this rarely happens in practice. Generally there is no point in adding additional principal components when the addition of the next one increases the fraction of the variability explained by only a small amount.

Having determined the first m principal components, one commonly attempts to attach subject matter interpretations to them. This is generally quite a subjective undertaking. For example, suppose one had 10 variables, say $X_1, \ldots, X_{10}$, and the first principal component was

$$0.87X_1 + 0.12X_2 + 0.09X_3 - 0.17X_4 + 0.75X_5$$
$$- 0.06X_6 + 0.84X_7 + 0.29X_8 + 0.18X_9$$
$$- 0.22X_{10}.$$

The coefficients or loadings of X_1, X_5, and X_7 are considerably larger in absolute value than the rest and are of the same order of magnitude. One might say, therefore, that the first principal component is essentially the mean or total of X_1, X_5, and X_7 because it is basically an equally weighted sum of these three. If X_1, X_5, and X_7 share some common characteristic and this characteristic is not shared by the other variables, this principal component might further be interpreted as measuring this characteristic. In the Cruz (1988) data, the first principal component had large, positive loadings (coefficients) for the variables "gleaning" and "probing," large negative loadings for several sally or hovering variables, and somewhat smaller loadings for all other variables. This component was therefore interpreted as a variable that distinguished between species that sally or hover (such as hummingbirds) and species that mainly glean or probe for food (such as woodpeckers).

If a small number of principal components can be found that explain a high fraction of the variability in the data, and if reasonable subject matter interpretations can be attached to these components, one is tempted to conclude that these principal components represent "unmeasurable" factors (i.e., quantities that cannot be directly measured) that are responsible for generating the values of the variables actually measured. In an education context, one might believe that grades in math courses and scores on math achievement tests are all basically driven by a single "unmeasurable" factor called innate math ability. Although the results of a principal components analysis may suggest that a large collection of variables can be "explained" by a relatively small number of "unmeasurable" factors, a formal attempt to answer such a question involves carrying out a *factor analysis*.

Discussion of factor analysis requires some advanced mathematics (a knowledge of linear algebra), and so it will not be presented in detail here. Johnson and Wichern (1988) contains detailed discussion of factor analysis as well as principal components analysis. A fairly readable, nontechnical discussion can also be found in Hair et al. (1987). In essence, a formal factor analysis involves writing a model, not unlike a regression model, that relates a large quantity of measured variables to a small quantity of (unmeasurable) factors. Some statistical method (maximum likelihood and least squares are two that are used) is used to fit the model (estimate parameters). The results are used to determine if the factor model adequately describes the data, if fewer factors are needed, or if additional factors are necessary. Formal statistical tests can be performed to answer these questions if the method of maximum likelihood is used and errors (discrepancies between the fitted model and the actual data) appear to be normally distributed. A typical factor analysis also involves attempts to find the model that relates the measured variables to the factors in the manner that makes interpretation of the factors most clear. This is done by means of so-called methods for *rotating* solutions. Fitting these models, rotating them, and carrying out statistical tests require using a statistical package such as SAS, SPSS, or LISREL. This, along with the difficulties of interpretation, suggests that one should not attempt to use factor analysis without first consulting a trained statistician. Rexstad et al. (1988) pointed out some of the pitfalls that can occur when these procedures are used without sufficient care.

In practice, many researchers use principal components analysis to identify a small number of "unmeasurable" factors that "explain" a large collection of measured variables. Although not a formal factor analysis, this use of principal components is a popular and acceptable substitute and is presented as factor analysis in some texts.

CLASSIFICATION AND DISCRIMINANT ANALYSIS

Occasionally, experiments are designed or data are collected for the purpose of determining how to classify an experimental unit into one of several groups on the basis of measurements on several independent variables. In the example of Nixon et al. (1988), discussed under stepwise regression, the authors wished to investigate how the 24 habitat variables could be used to classify sites into areas with deer during the winter versus areas with no deer during the winter. Crabtree et al. (1989) investigated the effectiveness of 15 habitat variables for classifying gadwall nests as likely or unlikely to be destroyed by mammalian predators.

The two basic objectives in a classification analysis are to determine which of several variables are useful for purposes of classification and to determine a classification rule based on these variables. Several methods are available for achieving these objectives, and all require that on each experimental unit one measures the independent variables of interest and knows how to classify the unit. Obviously not knowing to which group a unit belongs prevents determining the relation between the independent variables and groups for the unit, and thus the unit provides no useful information for the classification analysis.

The most widely used method for classification is *discriminant analysis*. For a discriminant analysis to be appropriate, all independent variables must be jointly normally distributed (more precisely, the independent variables must have a multivariate normal distribution). In addition, if the variances and correlations between the variables do not depend on the group from which the unit came (called homogeneity of covariance matrices or the assumption of equal covariance matrices), one can develop a classification rule as follows: (1) form a linear combination or score, as defined in principal components analysis, involving the independent variables; (2) for each unit, calculate this score from the observed data; and (3) run a one-way ANOVA using these scores, computing the F statistic for testing the hypothesis that the groups do not differ with respect to this score. As one changes the formula for the score, one produces different values of this F statistic. Using calculus one can determine the score that maximizes this F statistic and hence best displays differences in the groups. This score, called *Fisher's* linear discriminant function, best differentiates among the groups and can be used to do classification. If the F statistic associated with Fisher's linear discriminant function is sufficiently large (a formal hypothesis test can be constructed, see Johnson and Wichern 1988 for details), one concludes that the independent variables are useful for purposes of classification. This analysis is called a *discriminant analysis*. Hypothesis tests can also be constructed for testing whether a given variable significantly improves the discriminant function, and stepwise algorithms, not unlike those discussed in multiple regression, can be developed for determining a good subset of independent variables for purposes of classification. Such procedures have been implemented on many computer packages. Nixon et al. (1988) used a stepwise discriminant analysis and concluded that a linear discriminant function involving only 2 of the 24 habitat variables classified variables satisfactorily.

When the variances or correlations among variables depend on the groups, i.e., the assumption of homogeneity of covariance matrices fails, a variation on the above analysis, also called a discriminant analysis, can be carried out. Here techniques borrowed from advanced calculus and probability can be used to determine the rule for classifying units into groups that minimizes the number of incorrect classifications for the data collected (again see Johnson and Wichern 1988 for details). Because one knows to which groups the units in the sample belong, any candidate rule can be tried on the data, and its performance can be evaluated by comparing the results of the rule to the actual classifications. It turns out that this procedure actually yields Fisher's linear discriminant function when the covariance matrices are equal. Formal hypothesis tests concerning whether the classification rule is adequate are not available when the assumption of homogeneity of the covariance matrices fails. Adequacy is thus determined subjectively on the basis of the number of correct classifications when the rule is applied to the data. One drawback of this measure is that the rule was constructed to maximize the number of correct classifications in the data used to construct the rule and so, not surprisingly, it will appear to perform well on these data. For purposes of evaluating rules produced by this method, it is perhaps better to divide the data into two sets. One set is used to derive the discriminant function and the other set is used to evaluate it. This procedure will give a truer measure of the performance of the classification rule. The set used to construct the rule should generally be larger than that used to evaluate the rule, because one typically wants most of the information in the data to be used to construct the rule.

Many statistical software packages (for example, SAS, SPSS-X, and BMDP) do both of the above discriminant analyses, including testing for equality of covariance matrices. They also allow the user to test the resulting rule on a portion of the data set that has been set aside for that purpose.

When the independent variables are decidedly non-normal (this would occur if several of them were categorical), the above classification methods may not be appropriate (note, however, that if the assumption of homogeneity of covariance matrices holds, the analysis is thought to be somewhat insensitive to departures from normality). Other methods exist for doing classification analysis when discriminant analysis is inappropriate. One possibility is to use a categorical variable to represent the groups, treat this as the dependent variable, and do a logistic regression or fit a multinomial logit model to the data as mentioned in the discussion of multiple regression. Another possibility is to use Classification and Regression Trees (CART). This is a nonparametric method for determining which of a set of variables is useful for classification or prediction and for developing classification or prediction rules based on these variables. Because this method can be used for developing prediction equations, CART can be used in place of multiple regression when the assumptions of multiple regression are violated. For a description of CART see Breiman et al. (1984). Software is available for doing this analysis.

Discriminant analysis is by far the most common method of doing classification analysis, perhaps in part because other classification methods are not well known to nonstatisticians and software for doing these other analyses is not as widely available as that for discriminant analysis. Researchers should be aware, however, that discriminant analysis is not always appropriate for classification and that more appropriate methods do exist. The interested reader may wish to consult Hair et al. (1987) for a more detailed, readable description of discriminant analysis. Rexstad et al. (1988) discussed some of the pitfalls associated with discriminant analysis. Perhaps the best advice is for researchers to consult a trained statistician before doing classification analysis.

OTHER METHODS

Numerous additional statistical methods exist that have not been mentioned above. In the interest of space we mention some of these below with a brief, simplified description of the methods and references for additional information.

Canonical Correlation Analysis

The standard correlation coefficient allows one to measure correlations between pairs of variables. This is extended to R^2, the coefficient of multiple determination, in multiple regression and gives a measure of how a group of (independent) variables "correlates" with a single (dependent) variable. The next level of generalization would be to determine a measure of how one group of several variables correlates with another group of several variables. This is the goal of canonical correlation analysis.

Suppose one labels the variables in one set $X_1, \ldots, X_p$ and the variables in the other set $Y_1, \ldots, Y_q$. The *first canonical correlation* is defined to be the correlation between the pair of linear combinations or scores of the X variables and Y variables having the largest possible correlation coefficient. Because a linear combination of the X (Y) variables reduces the set of variables to a single variable, the two sets reduce to a pair of variables, and one can calculate the simple correlation coefficient between them. The scores producing the first canonical correlation are called the *first canonical variates*. The *second canonical correlation* is the largest correlation possible between a score in the X variables that is statistically independent of the first canonical variate in the Xs, and a score in the *Y* variables that is statistically independent of the first canonical variate in the *Y*s. The scores producing the second canonical correlation are called the *second canonical variates*. One proceeds to define additional canonical correlations and variates by determining the scores of the *X* and *Y* variables that are independent of all previous canonical variates and have maximum correlation. These are calculated with a computer, and most statistical software packages have routines for computing canonical correlations and variates.

The canonical correlations provide information concerning the strength of the relationship (''correlation'') between the set of *X* variables and the set of *Y* variables. One can actually test for statistical significance of the canonical correlations (i.e., do they differ from 0). The form of the scores making up the canonical variates may also provide subject-matter interpretations concerning why the sets are correlated. Interpretation of these scores is similar to the ''art'' of interpretation in a principal components analysis and is somewhat subjective.

Details of canonical correlation analysis were described in Johnson and Wichern (1988). A readable discussion of canonical correlation analysis was presented in Hair et al. (1987). Rexstad et al. (1988) discussed pitfalls associated with canonical correlation analysis.

Cluster Analysis

The basic objective of a cluster analysis is the following. Variables $X_1, \ldots, X_p$ are measured on each of *n* experimental units. One wishes to ''classify'' or ''cluster'' the *n* units into *k* groups. All units in a group are assumed to be ''similar'' or ''close'' to one another. Units in different groups are ''dissimilar'' or ''far apart.'' For example, suppose one measures two variables on each unit and makes a scatterplot of the data. If the points in the scatterplot separated into several ''clouds'' of points, as in Fig. 16 (we see four such clouds in Fig. 16), these ''clouds'' could be taken as the clusters. Unfortunately, this kind of graphical approach is possible only when one, two, or three variables are measured on each unit and the variables measured are quantitative. In general, *k* is usually not known in advance and is determined by the analysis. Also, there is no way of determining whether the groupings are ''correct.'' The groupings arrived at in a cluster analysis are decided on subjective grounds, and one is merely seeking to find evidence that some units are more alike than others.

To do a cluster analysis, one first needs a criterion for determining whether a pair of units is similar or close. Many measures of similarity or closeness are possible. Two common measures are the usual geometric or Euclidean distance and a ''statistical'' version of distance called the Mahalanobis distance. Whatever measure of distance is used, one must also decide how to use this quantity to measure the distance between two collections of observations. One possibility is to define this to be the distance between the means of the two collections. Many others are possible.

Having decided on a measure of closeness, one next needs to develop an algorithm for ''grouping'' the observations. If only one, two, or three variables are measured on each unit and the measure of distance is geometric distance, one can plot the measurements on each unit as

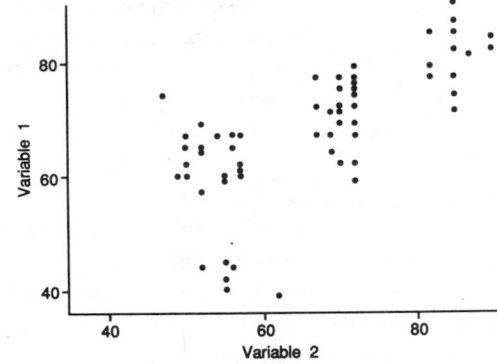

Fig. 16. Scatterplot displaying clustering. There appear to be four clusters in this plot.

points on a one-, two-, or three-dimensional graph, examine the graph, and decide on groupings from the way the plot looks. This fails when more than three measurements are taken on each unit. In this situation some algorithm that can be programmed on a computer is typically employed and the computer is allowed to determine groupings. Several algorithms are possible. One popular method is called hierarchical clustering. One begins by treating each unit (more precisely the set of measurements on the unit) as a separate cluster. A series of steps is now followed. In step 1, for every pair of existing clusters (initially there are *n* of these, one for each unit, and hence $n(n-1)/2$ pairs), the distance between each pair is calculated with the measure of distance decided upon. In step 2, the pair of clusters having the smallest distance (i.e., closest together) is merged to form a new cluster, thus reducing the number of existing clusters by one. This set of clusters is printed out and steps 1 and 2 are repeated. This continues until all units have been merged into a single cluster. The results at each stage are usually summarized in a graph called a *dendrogram*.

Cruz (1988) applied a hierarchical cluster analysis to see how 11 species might produce a cluster based on 20 foraging variables measured on each species. The analysis produced the dendrogram in Fig. 17. Notice that the first species joined into a single cluster were the striped-headed tanager and the Puerto Rican bullfinch. At the next step the Puerto Rican tanager was added to this group. Next, the red-legged thrush and pearly-eyed thrasher were joined into a single cluster. At the next-to-last stage the species were divided into two clusters. The emerald hummingbird and Puerto Rican tody (both sally-hoverers) formed one cluster and all other species formed the other.

Cluster analysis should be regarded as a descriptive method useful for identifying interesting features of the data. Because of the large (essentially infinite) number of possible distance measures and clustering algorithms, a set of data can be subjected to a large number of cluster analyses yielding different clusterings, none of which may be particularly dramatic or informative. One never knows if some untried measure might have yielded a clustering that would truly clarify the data or if the clustering obtained is biologically ''correct.'' Additional discussion of cluster analysis and related issues were given in Johnson and Wichern (1988), including the use of pictures to suggest clusters and a discussion of a technique called *multidi-*

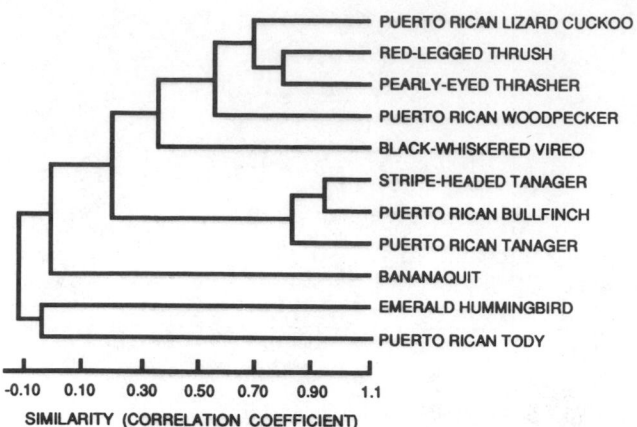

Fig. 17. Dendrogram indicating similarity in foraging relationships based on a correlation coefficient matrix of 11 species in the Cubuy Caribbean pine avifaunal assemblage, Luquilo Experimental Forest, Puerto Rico, March–August 1981–86 (from Cruz 1988).

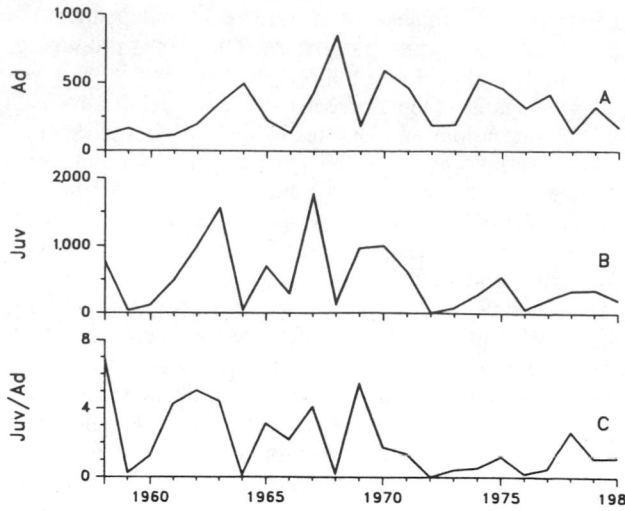

Fig. 18. Examples of time series, displaying numbers of California quail adults (A), juveniles (B), and computed juveniles/adults (C) from the annual counts in the Panoche Management Area, California, 1958–80 (from Botsford et al. 1988).

mensional scaling. Hair et al. (1987) provided a fairly readable introduction to cluster analysis and multidimensional scaling.

Related to cluster analysis is a large body of techniques termed spatial analysis. Rather than attempting to group units into similar clusters based on some measure of distance, the test determines if the units are uniformly separated or distributed in space with respect to the distance measure, or if they display some other interesting spatial distribution. In other instances, one simply estimates features of the spatial distribution. Buskirk et al. (1989) investigated the ecology of resting sites of martens in the central Rocky Mountains during 2 winters. Because these winter resting sites are often associated with coarse woody debris, the authors estimated spatial density of coarse woody debris on the forest floor.

In the interest of space, we shall not discuss spatial analysis further. The interested reader can consult Upton and Fingleton (1985) for additional discussion.

Time Series

Measurements taken over time often exhibit a periodic or "fluctuating" behavior. Everyday examples include meteorological and economic data. In such data, it may be of interest to study characteristics of the relationship between the measurements (dependent variable) and time (independent variable). Sometimes this can be accomplished with multiple regression. For purposes of studying periodicities (how often certain features repeat themselves) a more appropriate approach is to use time series analysis. Among other things, a time series analysis allows one to identify periodicities in data, determine what portion of the fluctuations in data is due to "noise" (random error) and what is due to more systematic sources, identify trends in data, and fit models to the data. The results can be used for forecasting (prediction) or simply for clarifying why the data behave as they do. To understand the methodology of time series, one must have some knowledge of mathematical analysis (including Fourier series) and stochastic processes. Many statistical software packages will do time series analyses. Two references

containing more information on time series are Box and Jenkins (1976) and Cryer (1986).

As a parting example, Botsford et al. (1988) plotted the numbers of adult, juvenile, and juvenile/adult California quail from 1958 to 1980. The resulting plots (Fig. 18) were time series. Notice the periodic or fluctuating behavior of the data. Peaks seem to repeat roughly every 4 or 5 years, and the height of the peaks appears to decline gradually over time. As part of the analysis, the authors tried to relate this behavior to precipitation (also time series data).

Directional Data

Occasionally experimenters are interested in recording a direction (angle from 0° to 359°) or day of year (1 to 365, leap years ignored) as part of their data. One oddity of such data is that the two most extreme values (0° and 359° or day 1 and day 365) are actually adjacent. In fact absolute differences in values do not necessarily indicate how far apart two measurements are. Diefenbach et al. (1988) recorded day of year of band recoveries for American black ducks and mallards. They were interested, among other things, in investigating how recoveries were distributed over the year.

Data such as described above have a cyclical structure and can be thought of as lying on a circle. Such data are called *directional data*, and several special procedures have been developed for analyzing such data in light of some of the peculiarities of these data. Further discussion is beyond the scope of this chapter, but the interested reader may consult Mardia (1972) for further information.

Miscellaneous Topics

Additional procedures one may encounter in the wildlife literature include MANOVA or multivariate analysis of variance (ANOVA methods applied to situations in which several dependent variables are measured on each unit, see Johnson and Wichern 1988 and Hair et al. 1987 for more information), survival analysis (methods for investigating the effects of factors on survival of experimental units, see Lee 1980), and Bayesian decision making and methodology (see Lindley 1971; Cohen 1988 has some discussion in a wildlife context). Bayesian methodology is perhaps being underutilized in experimental science in general. Of course, we have not mentioned numerous other statistical procedures. Perhaps the best advice we can give is to consult a trained statistician if designing or analyzing data from an experiment is contemplated.

LITERATURE CITED

AITKIN, M. A., D. ANDERSON, B. FRANCIS, AND J. HINDE. 1989. Statistical modelling in GLIM. Oxford Univ. Press, Oxford, U.K. 374pp.

ALISAUSKAS, R. T., C. D. ANKNEY, AND E. E. KLAAS. 1988. Winter diets and nutrition of midcontinental lesser snow geese. J. Wildl. Manage. 52:403–414.

ANDRES, B. 1989. Littoral zone use by post-breeding shorebirds on the Colville River delta, Alaska. M.S. Thesis, Ohio State Univ., Columbus. 116pp.

ANTHONY, R. G., AND F. B. ISAACS. 1989. Characteristics of bald eagle nest sites in Oregon. J. Wildl. Manage. 53:148–159.

ARTHUR, S. M., W. B. KROHN, AND J. R. GILBERT. 1989. Habitat use and diet of fishers. J. Wildl. Manage. 53:680–688.

BERGERUD, A. T., AND W. B. BALLARD. 1988. Wolf predation on caribou: the Nelchina herd case history, a different interpretation. J. Wildl. Manage. 52:344–357.

BIDWELL, T. G., S. D. SHALAWAY, O. E. MAUGHAN, AND L. G. TALENT. 1989. Habitat use by female eastern wild turkeys in southeastern Oklahoma. J. Wildl. Manage. 53:34–39.

BOTSFORD, L. W., T. C. WAINWRIGHT, J. T. SMITH, S. MASTRUP, AND D. F. LOTT. 1988. Population dynamics of California quail related to meteorological conditions. J. Wildl. Manage. 52:469–477.

BOX, G. E. P., AND N. R. DRAPER. 1987. Empirical model-building and response surfaces. John Wiley & Sons, New York, N.Y. 669pp.

———, AND G. M. JENKINS. 1976. Time series analysis. Holden-Day, San Francisco, Calif. 575pp.

BREIMAN, L. J., D. FRIEDMAN, R. OLSHEN, AND C. STONE. 1984. Classification and regression trees. Wadsworth Int. Group, Belmont, Calif. 358pp.

BRODY, A. J., AND M. R. PELTON. 1989. Effects of roads on black bear movements in western North Carolina. Wildl. Soc. Bull. 17:5–10.

BUSKIRK, S. W., S. C. FORREST, M. G. RAPHAEL, AND H. J. HARLOW. 1989. Winter resting site ecology of marten in the central Rocky Mountains. J. Wildl. Manage. 53:191–196.

CHAMBERS, J. M., W. S. CLEVELAND, B. KLEINER, AND P. A. TUKEY. 1983. Graphical methods for data analysis. Wadsworth Int. Group, Belmont, Calif. 395pp.

COCHRAN, W. G. 1977. Sampling techniques. John Wiley & Sons, New York, N.Y. 428pp.

COHEN, Y. 1988. Bayesian estimation of clutch size for scientific and management purposes. J. Wildl. Manage. 52:787–793.

COMPTON, B. B., R. J. MACKIE, AND G. L. DUSEK. 1988. Factors influencing distribution of white-tailed deer in riparian habitats. J. Wildl. Manage. 52:544–548.

CONOVER, M. R. 1988. Effect of grazing by Canada geese on the winter growth of rye. J. Wildl. Manage. 52:76–80.

COWAN, D. P., J. A. VAUGHAN, AND W. G. CHRISTER. 1987. Bait consumption by the European rabbit in southern England. J. Wildl. Manage. 51:386–392.

COX, D. R., AND E. J. SNELL. 1989. The analysis of binary data. Second ed. Chapman and Hall, London, U.K. 236pp.

CRABTREE, R. L., L. S. BROOME, AND M. L. WOLFE. 1989. Effects of habitat characteristics on gadwall nest predation and nest-site selection. J. Wildl. Manage. 53:129–137.

CRUZ, A. 1988. Avian resource use in a Caribbean pine plantation. J. Wildl. Manage. 52:274–279.

CRYER, J. D. 1986. Time series analysis. Duxbury, Boston, Mass. 286pp.

DIEFENBACH, D. R., J. D. NICHOLS, AND J. E. HINES. 1988. Distribution patterns of American black duck and mallard winter band recoveries. J. Wildl. Manage. 52:704–710.

DIERENFELD, E. S., C. E. SANDFORT, AND W. C. SATTERFIELD. 1989.

Influence of diet on plasma vitamin E in captive peregrine falcons. J. Wildl. Manage. 53:160–164.

ELY, C. R., AND D. G. RAVELING. 1989. Body composition and weight dynamics of wintering greater white-fronted geese. J. Wildl. Manage. 53:80–87.

FLEISS, J. L. 1981. Statistical methods for rates and proportions. Second ed. J. Wiley & Sons, New York, N.Y. 321pp.

FRANZMANN, A. W., AND C. C. SCHWARTZ. 1988. Evaluating condition of Alaskan black bears with blood profiles. J. Wildl. Manage. 52:63–70.

FRYXELL, J. M., W. E. MERCER, AND R. B. GELLATELY. 1988. Population dynamics of Newfoundland moose using cohort analysis. J. Wildl. Manage. 52:14–21.

GALLANT, A. R. 1987. Nonlinear statistical models. J. Wiley & Sons, New York, N.Y. 610pp.

GESE, E. M., O. J. RONGSTAD, AND W. R. MYTTON. 1988. Home range and habitat use of coyotes in southeastern Colorado. J. Wildl. Manage. 52:640–646.

GOLIGHTLY, R. T., JR., AND T. D. HOFSTRA. 1989. Immobilization of elk with a ketamine-xylazine mix and rapid reversal with yohimbine hydrochloride. Wildl. Soc. Bull. 17:53–58.

GUTHERY, F. S. 1988. Line transect sampling of bobwhite density on rangeland: evaluation and recommendations. Wildl. Soc. Bull. 16:193–203.

HAIR, J. F., JR., R. E. ANDERSON, AND R. L. TATHAM. 1987. Multivariate data analysis. Macmillan Publ. Co., New York, N.Y. 449pp.

HANNON, S. J., K. MARTIN, AND J. O. SCHIECK. 1988. Timing of reproduction in two populations of willow ptarmigan in northern Canada. Auk 105:330–338.

HEINZ, G. H., D. J. HOFFMAN, AND L. G. GOLD. 1989. Impaired reproduction of mallards fed an organic form of selenium. J. Wildl. Manage. 53:418–428.

HOLL, S. A., AND V. C. BLEICH. 1987. Mineral lick use by mountain sheep in the San Gabriel Mountains, California. J. Wildl. Manage. 51:383–385.

HOLLANDER, M., AND D. A. WOLFE. 1973. Nonparametric statistical methods. John Wiley & Sons, New York, N.Y. 503pp.

HOLM, B. A., R. J. JOHNSON, D. D. JENSEN, AND W. W. STROUP. 1988. Responses of deer mice to methiocarb and thiram seed treatments. J. Wildl. Manage. 52:497–502.

HÖLZENBEIN, S., AND G. SCHWEDE. 1989. Activity and movements of female white-tailed deer during the rut. J. Wildl. Manage. 53:219–223.

JOHNSON, D. B., F. S. GUTHERY, AND N. E. KOERTH. 1989. Grackle damage to grapefruit in the lower Rio Grande Valley. Wildl. Soc. Bull. 17:46–50.

JOHNSON, R. A., AND D. W. WICHERN. 1988. Applied multivariate statistical analysis. Second ed. Prentice-Hall, Inc., Englewood Cliffs, N.J. 594pp.

KOEHLER, G. M., AND M. G. HORNOCKER. 1989. Influences of seasons on bobcats in Idaho. J. Wildl. Manage. 53:197–202.

LEE, E. T. 1980. Statistical methods for survival data analysis. Lifetime Learning Publ., Belmont, Calif. 557pp.

LEHMANN, E. I. 1975. Nonparametrics: statistical methods based on ranks. McGraw-Hill, New York, N.Y. 457pp.

LEUSCHNER, W. A., V. P. RITCHIE, AND D. F. STAUFFER. 1989. Options on wildlife: responses of resource managers and wildlife users in the southeastern United States. Wildl. Soc. Bull. 17:24–29.

LINDLEY, D. V. 1971. Making decisions. Wiley-Interscience, New York, N.Y. 195pp.

LOWNEY, M. S., AND E. P. HILL. 1989. Wood duck nest sites in bottomland hardwood forests of Mississippi. J. Wildl. Manage. 53:378–382.

MARDIA, K. V. 1972. Statistics and directional data. Academic Press, London, U.K. 357pp.

MASON, J. R., M. A. ADAMS, AND L. CLARK. 1989. Anthranilate repellency to starlings: chemical correlates and sensory perception. J. Wildl. Manage. 53:55–64.

MCAULEY, D. G., AND J. R. LONGCORE. 1988. Survival of juvenile ring-necked ducks on wetlands of different pH. J. Wildl. Manage. 52:169–176.

MCCULLAGH, P., AND J. A. NELDER. 1989. Generalized linear models. Chapman and Hall, London, U.K. 511pp.

MOORE, D. S. 1979. Statistics: concepts and controversies. W.H. Freeman Co., San Francisco, Calif. 313pp.

———, AND G. P. MCCABE. 1988. Introduction to the practice of statistics. W.H. Freeman Co., San Francisco, Calif. 790pp.

NAMBOODIRI, N. K., L. F. CARTER, AND H. M. BLALOCK. 1975. Applied multivariate analysis and experimental designs. McGraw-Hill, New York, N.Y. 688pp.

NETER, J., AND W. WASSERMAN. 1974. Applied linear statistical models. Richard D. Irwin, Homewood, Ill. 842pp.

———, ———, AND M. H. KUTNER. 1983. Applied linear regression models. Richard D. Irwin, Homewood, Ill. 547pp.

NEWTON, M., E. C. COLE, R. A. LAUTENSCHLAGER, D. E. WHITE, AND M. L. McCORMACK, JR. 1989. Browse availability after conifer release in Maine's spruce-fir forests. J. Wildl. Manage. 53:643–649.

NIXON, C. M., L. P. HANSEN, AND P. A. BREWER. 1988. Characteristics of winter habitats used by deer in Illinois. J. Wildl. Manage. 52:552–555.

O'GARA, B. W., AND R. B. HARRIS. 1988. Age and condition of deer killed by predators and automobiles. J. Wildl. Manage. 52:316–320.

OTIS, D. L., K. P. BURNHAM, G. C. WHITE, AND D. R. ANDERSON. 1978. Statistical inference from capture data on closed animal populations. Wildl. Monogr. 62. 135pp.

QUINN, N. W. S., AND J. E. THOMPSON. 1987. Dynamics of an exploited Canada lynx population in Ontario. J. Wildl. Manage. 51:297–305.

RANDLES, R. H., AND D. A. WOLFE. 1979. Introduction to the theory of nonparametric statistics. John Wiley & Sons, New York, N.Y. 450pp.

RAVE, D. P., AND G. A. BALDASSARRE. 1989. Activity budget of green-winged teal wintering in coastal wetlands of Louisiana. J. Wildl. Manage. 53:753–759.

REED, A., AND P. CHAGNON. 1987. Greater snow geese on Bylot Island, Northwest Territories, 1983. J. Wildl. Manage. 51:128–131.

———, R. STEHN, AND D. WARD. 1989. Autumn use of Izembek Lagoon, Alaska, by brant from different breeding areas. J. Wildl. Manage. 53:720–725.

RENECKER, L. A., AND R. J. HUDSON. 1989. Seasonal activity budgets of moose in aspen-dominated boreal forests. J. Wildl. Manage. 53:296–302.

REXSTAD, E. A., D. D. MILLER, C. H. FLATHER, E. M. ANDERSON, J. W. HUPP, AND D. R. ANDERSON. 1988. Questionable multivariate statistical inference in wildlife habitat and community studies. J. Wildl. Manage. 52:794–798.

SAETHER, B.-E., AND A. J. GRAVEM. 1988. Annual variation in winter body condition of Norwegian moose calves. J. Wildl. Manage. 52:333–336.

SANTILLO, D. J., D. M. LESLIE, JR., AND P. W. BROWN. 1989. Responses of small mammals and habitat to glyphosate application on clearcuts. J. Wildl. Manage. 53:164–172.

SERIE, J. R., AND D. E. SHARP. 1989. Body weight and composition dynamics of fall migrating canvasbacks. J. Wildl. Manage. 53:431–441.

SNEDECOR, G. W., AND W. G. COCHRAN. 1980. Statistical methods. Seventh ed. Iowa State Univ. Press, Ames. 507pp.

SOKAL, R. R., AND F. J. ROHLF. 1981. Biometry. Second ed. W.H. Freeman, San Francisco, Calif. 859pp.

STEEL, R. G. D., AND J. H. TORRIE. 1980. Principles and procedures of statistics. Second ed. McGraw-Hill Book Co., New York, N.Y. 633pp.

THILL, R. E., AND A. MARTIN, JR. 1989. Deer and cattle diets on heavily grazed pine-bluestem range. J. Wildl. Manage. 53:540–548.

THOMPSON, F. R., III, AND D. E. CAPEN. 1988. Avian assemblages in seral stages of a Vermont forest. J. Wildl. Manage. 52:771–777.

TUFTE, E. R. 1983. The visual display of quantitative information. Graphics Press, Cheshire, Conn. 197pp.

UPTON, G. J. G., AND B. FINGLETON. 1985. Spatial data analysis by example. John Wiley & Sons, New York, N.Y. 394pp.

VAN BALLENBERGHE, V. 1989. Wolf predation on the Nelchina caribou herd: a comment. J. Wildl. Manage. 53:243–250.

WOOD, G. W. 1988. Effects of prescribed fire on deer forage and nutrients. Wildl. Soc. Bull. 16:180–186.

Appendix I.   Tables and figure referenced in the text.

A. Level of significance = 0.05.

B. Level of significance = 0.01.

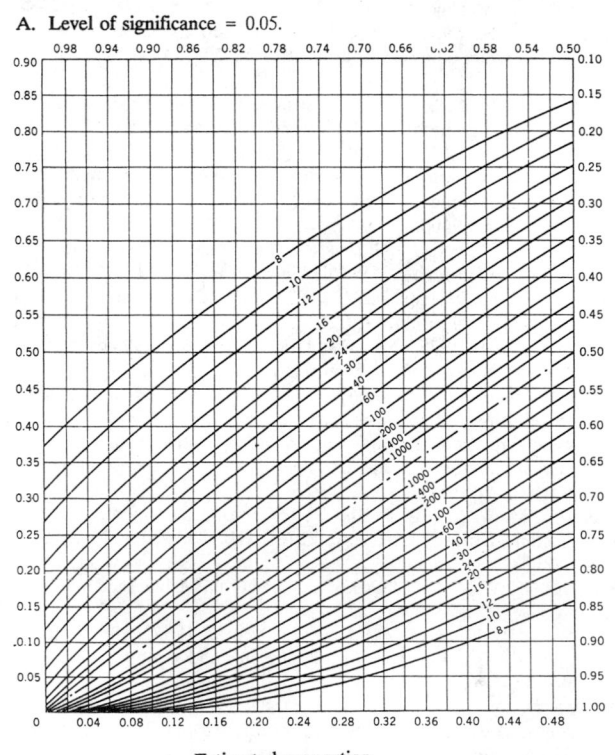

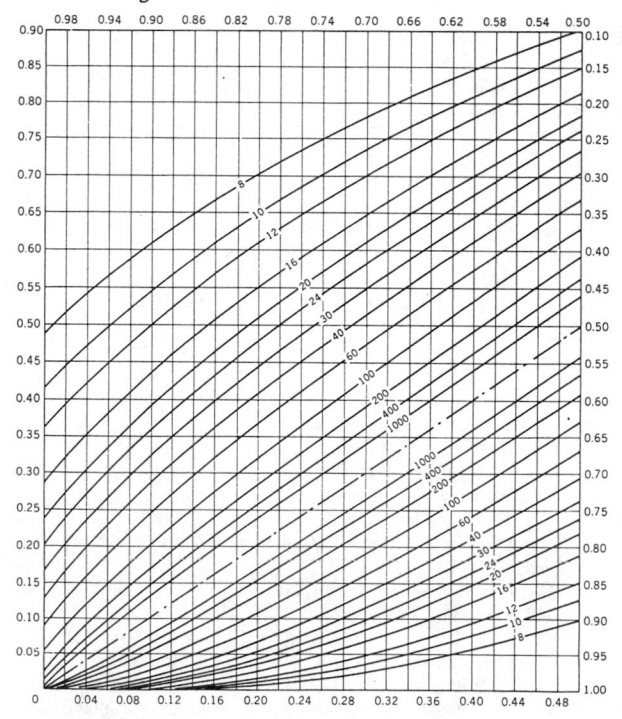

Estimated proportion

Estimated proportion

Fig. A1. Confidence limits for a proportion estimated using one-stage sampling (see Box 5). The numbers printed along the curves indicate the sample size. The upper and lower endpoints of the confidence interval are indicated along the *y* axis.

Table A1. Critical values for two-tailed *t*-tests.[a] Test statistics larger than the tabled values are statistically significant.

Degrees of freedom	Level of significance[a]								
	0.500	0.400	0.200	0.100	0.050	0.025	0.010	0.005	0.001
1	1.000	1.376	3.078	6.314	12.706	25.452	63.657		
2	0.816	1.061	1.886	2.920	4.303	6.205	9.925	14.089	31.598
3	0.765	0.978	1.638	2.353	3.182	4.176	5.841	7.453	12.941
4	0.741	0.941	1.533	2.132	2.776	3.495	4.604	5.598	8.610
5	0.727	0.920	1.476	2.015	2.571	3.163	4.032	4.773	6.859
6	0.718	0.906	1.440	1.943	2.447	2.969	3.707	4.317	5.959
7	0.711	0.896	1.415	1.895	2.365	2.841	3.499	4.029	5.405
8	0.706	0.889	1.397	1.860	2.306	2.752	3.355	3.832	5.041
9	0.703	0.883	1.383	1.833	2.262	2.685	3.250	3.690	4.781
10	0.700	0.879	1.372	1.812	2.228	2.634	3.169	3.581	4.587
11	0.697	0.876	1.363	1.796	2.201	2.593	3.106	3.497	4.437
12	0.695	0.873	1.356	1.782	2.179	2.560	3.055	3.428	4.318
13	0.694	0.870	1.350	1.771	2.160	2.533	3.012	3.372	4.221
14	0.692	0.868	1.345	1.761	2.145	2.510	2.977	3.326	4.140
15	0.691	0.866	1.341	1.753	2.131	2.490	2.947	3.286	4.073
16	0.690	0.865	1.337	1.746	2.120	2.473	2.921	3.252	4.015
17	0.689	0.863	1.333	1.740	2.110	2.458	2.898	3.222	3.965
18	0.688	0.862	1.330	1.734	2.101	2.445	2.878	3.197	3.922
19	0.688	0.861	1.328	1.729	2.093	2.433	2.861	3.174	3.883
20	0.687	0.860	1.325	1.725	2.086	2.423	2.845	3.153	3.850
21	0.686	0.859	1.323	1.721	2.080	2.414	2.831	3.135	3.819
22	0.686	0.858	1.321	1.717	2.074	2.406	2.819	3.119	3.792
23	0.685	0.858	1.319	1.714	2.069	2.398	2.807	3.104	3.767
24	0.685	0.857	1.318	1.711	2.064	2.391	2.797	3.090	3.745
25	0.684	0.856	1.316	1.708	2.060	2.385	2.787	3.078	3.725
26	0.684	0.856	1.315	1.706	2.056	2.379	2.779	3.067	3.707
27	0.684	0.855	1.314	1.703	2.052	2.373	2.771	3.056	3.690
28	0.683	0.855	1.313	1.701	2.048	2.368	2.763	3.047	3.674
29	0.683	0.854	1.311	1.699	2.045	2.364	2.756	3.038	3.659
30	0.683	0.854	1.310	1.697	2.042	2.360	2.750	3.030	3.646
35	0.682	0.852	1.306	1.690	2.030	2.342	2.724	2.996	3.591
40	0.681	0.851	1.303	1.684	2.021	2.329	2.704	2.971	3.551
45	0.680	0.850	1.301	1.680	2.014	2.319	2.690	2.952	3.520
50	0.680	0.849	1.299	1.676	2.008	2.310	2.678	2.937	3.496
55	0.679	0.849	1.297	1.673	2.004	2.304	2.669	2.925	3.476
60	0.679	0.848	1.296	1.671	2.000	2.299	2.660	2.915	3.460
70	0.678	0.847	1.294	1.667	1.994	2.290	2.648	2.899	3.435
80	0.678	0.847	1.293	1.665	1.989	2.284	2.638	2.887	3.416
90	0.678	0.846	1.291	1.662	1.986	2.279	2.631	2.878	3.402
100	0.677	0.846	1.290	1.661	1.982	2.276	2.625	2.871	3.390
120	0.677	0.845	1.289	1.658	1.980	2.270	2.617	2.860	3.373
∞	0.6745	0.8416	1.2816	1.6448	1.9600	2.2414	2.5758	2.8070	3.2905

[a]For one-tailed tests, double the level of significance, and then use the indicated column (i.e., for a one-tailed test at the 5% level, use the column labelled 0.10 to find the critical value).

Data Analysis

Table A2. Values from the F distribution. f_1 is the "numerator degrees of freedom" and f_2 is the "denominator degrees of freedom." For each combination of f_1 and f_2, the upper value in the table is for level of significance = 0.05. The lower value is for level of significance = 0.01.

							f_1					
f_2	1	2	3	4	5	6	7	8	9	10	11	12
1	161	200	216	225	230	234	237	239	241	242	243	244
	4,052	4,999	5,403	5,625	5,764	5,859	5,928	5,981	6,022	6,056	6,082	6,106
2	18.51	19.00	19.16	19.25	19.30	19.33	19.36	19.37	19.38	19.39	19.40	19.41
	98.49	99.00	99.17	99.25	99.30	99.33	99.36	99.37	99.39	99.40	99.41	99.42
3	10.13	9.55	9.28	9.12	9.01	8.94	8.88	8.84	8.81	8.78	8.76	8.74
	34.12	30.82	29.46	28.71	28.24	27.91	27.67	27.49	27.34	27.23	27.13	27.05
4	7.71	6.94	6.59	6.39	6.26	6.16	6.09	6.04	6.00	5.96	5.93	5.91
	21.20	18.00	16.69	15.98	15.52	15.21	14.98	14.80	14.66	14.54	14.45	14.37
5	6.61	5.79	5.41	5.19	5.05	4.95	4.88	4.82	4.78	4.74	4.70	4.68
	16.26	13.27	12.06	11.39	10.97	10.67	10.45	10.29	10.15	10.05	9.96	9.89
6	5.99	5.14	4.76	4.53	4.39	4.28	4.21	4.15	4.10	4.06	4.03	4.00
	13.74	10.92	9.78	9.15	8.75	8.47	8.26	8.10	7.98	7.87	7.79	7.72
7	5.59	4.74	4.35	4.12	3.97	3.87	3.79	3.73	3.68	3.63	3.60	3.57
	12.25	9.55	8.45	7.85	7.46	7.19	7.00	6.84	6.71	6.62	6.54	6.47
8	5.32	4.46	4.07	3.84	3.69	3.58	3.50	3.44	3.39	3.34	3.31	3.28
	11.26	8.65	7.59	7.01	6.63	6.37	6.19	6.03	5.91	5.82	5.74	5.67
9	5.12	4.26	3.86	3.63	3.48	3.37	3.29	3.23	3.18	3.13	3.10	3.07
	10.56	8.02	6.99	6.42	6.06	5.80	5.62	5.47	5.35	5.26	5.18	5.11
10	4.96	4.10	3.71	3.48	3.33	3.22	3.14	3.07	3.02	2.97	2.94	2.91
	10.04	7.56	6.55	5.99	5.64	5.39	5.21	5.06	4.95	4.85	4.78	4.71
11	4.84	3.98	3.59	3.36	3.20	3.09	3.01	2.95	2.90	2.86	2.82	2.79
	9.65	7.20	6.22	5.67	5.32	5.07	4.88	4.74	4.63	4.54	4.46	4.40
12	4.75	3.88	3.49	3.26	3.11	3.00	2.92	2.85	2.80	2.76	2.72	2.69
	9.33	6.93	5.95	5.41	5.06	4.82	4.65	4.50	4.39	4.30	4.22	4.16
13	4.67	3.80	3.41	3.18	3.02	2.92	2.84	2.77	2.72	2.67	2.63	2.60
	9.07	6.70	5.74	5.20	4.86	4.62	4.44	4.30	4.19	4.10	4.02	3.96
14	4.60	3.74	3.34	3.11	2.96	2.85	2.77	2.70	2.65	2.60	2.56	2.53
	8.86	6.51	5.56	5.03	4.69	4.46	4.28	4.14	4.03	3.94	3.86	3.80
15	4.54	3.68	3.29	3.06	2.90	2.79	2.70	2.64	2.59	2.55	2.51	2.48
	8.68	6.36	5.42	4.89	4.56	4.32	4.14	4.00	3.89	3.80	3.73	3.67
16	4.49	3.63	3.24	3.01	2.85	2.74	2.66	2.59	2.54	2.49	2.45	2.42
	8.53	6.23	5.29	4.77	4.44	4.20	4.03	3.89	3.78	3.69	3.61	3.55
17	4.45	3.59	3.20	2.96	2.81	2.70	2.62	2.55	2.50	2.45	2.41	2.38
	8.40	6.11	5.18	4.67	4.34	4.10	3.93	3.79	3.68	3.59	3.52	3.45
18	4.41	3.55	3.16	2.93	2.77	3.66	2.58	2.51	2.46	2.41	2.37	2.34
	8.28	6.01	5.09	4.58	4.25	4.01	3.85	3.71	3.60	3.51	3.44	3.37
19	4.38	3.52	3.13	2.90	2.74	2.63	2.55	2.48	2.43	2.38	2.34	2.31
	8.18	5.93	5.01	4.50	4.17	3.94	3.77	3.63	3.52	3.43	3.36	3.30
20	4.35	3.49	3.10	2.87	2.71	2.60	2.52	2.45	2.40	2.35	2.31	2.28
	8.10	5.85	4.94	4.43	4.10	3.87	3.71	3.56	3.45	3.37	3.30	3.23
21	4.32	3.47	3.07	2.84	2.68	2.57	2.49	2.42	2.37	2.32	2.28	2.25
	8.02	5.78	4.87	4.37	4.04	3.81	3.65	3.51	3.40	3.31	3.24	3.17
22	4.30	3.44	3.05	2.82	2.66	2.55	2.47	2.40	2.35	2.30	2.26	2.23
	7.94	5.72	4.82	4.31	3.99	3.76	3.59	3.45	3.35	3.26	3.18	3.12
23	4.28	3.42	3.03	2.80	2.64	2.53	2.45	2.38	2.32	2.28	2.24	2.20
	7.88	5.66	4.76	4.26	3.94	3.71	3.54	3.41	3.30	3.21	3.14	3.07
24	4.26	3.40	3.01	2.78	2.62	2.51	2.43	2.36	2.30	2.26	2.22	2.18
	7.82	5.61	4.72	4.22	3.90	3.67	3.50	3.36	3.25	3.17	3.09	3.03
25	4.24	3.38	2.99	2.76	2.60	2.49	2.41	2.34	2.28	2.24	2.20	2.16
	7.77	5.57	4.68	4.18	3.86	3.63	3.46	3.32	3.21	3.13	3.05	2.99
26	4.22	3.37	2.98	2.74	2.59	2.47	2.39	2.32	2.27	2.22	2.18	2.15
	7.72	5.53	4.64	4.14	3.82	3.59	3.42	3.29	3.17	3.09	3.02	2.96

Table A2. Extended.

f_1											
14	16	20	24	30	40	50	75	100	200	500	∞
245	246	248	249	250	251	252	253	253	254	254	254
6,142	6,169	6,208	6,234	6,261	6,286	6,302	6,323	6,334	6,352	6,361	6,366
19.42	19.43	19.44	19.45	19.46	19.47	19.47	19.48	19.49	19.49	19.50	19.50
99.43	99.44	99.45	99.46	99.47	99.48	99.48	99.49	99.49	99.49	99.50	99.50
8.71	8.69	8.66	8.64	8.62	8.60	8.58	8.57	8.56	8.54	8.54	8.53
26.92	26.83	26.69	26.60	26.50	26.41	26.35	26.27	26.23	26.18	26.14	26.12
5.87	5.84	5.80	5.77	5.74	5.71	5.70	5.68	5.66	5.65	5.64	5.63
14.24	14.15	14.02	13.93	13.83	13.74	13.69	13.61	13.57	13.52	13.48	13.46
4.64	4.60	4.56	4.53	4.50	4.46	4.44	4.42	4.40	4.38	4.37	4.36
9.77	9.68	9.55	9.47	9.38	9.29	9.24	9.17	9.13	9.07	9.04	9.02
3.96	3.92	3.87	3.84	3.81	3.77	3.75	3.72	3.71	3.69	3.68	3.67
7.60	7.52	7.39	7.31	7.23	7.14	7.09	7.02	6.99	6.94	6.90	6.88
3.52	3.49	3.44	3.41	3.38	3.34	3.32	3.29	3.28	3.25	3.24	3.23
6.35	6.27	6.15	6.07	5.98	5.90	5.85	5.78	5.75	5.70	5.67	5.65
3.23	3.20	3.15	3.12	3.08	3.05	3.03	3.00	2.98	2.96	2.94	2.93
5.56	5.48	5.36	5.28	5.20	5.11	5.06	5.00	4.96	4.91	4.88	4.86
3.02	2.98	2.93	2.90	2.86	2.82	2.80	2.77	2.76	2.73	2.72	2.71
5.00	4.92	4.80	4.73	4.64	4.56	4.51	4.45	4.41	4.36	4.33	4.31
2.86	2.82	2.77	2.74	2.70	2.67	2.64	2.61	2.59	2.56	2.55	2.54
4.60	4.52	4.41	4.33	4.25	4.17	4.12	4.05	4.01	3.96	3.93	3.91
2.74	2.70	2.65	2.61	2.57	2.53	2.50	2.47	2.45	2.42	2.41	2.40
4.29	4.21	4.10	4.02	3.94	3.86	3.80	3.74	3.70	3.66	3.62	3.60
2.64	2.60	2.54	2.50	2.46	2.42	2.40	2.36	2.35	2.32	2.31	2.30
4.05	3.98	3.86	3.78	3.70	3.61	3.56	3.49	3.46	3.41	3.38	3.36
2.55	2.51	2.46	2.42	2.38	2.34	2.32	2.28	2.26	2.24	2.22	2.21
3.85	3.78	3.67	3.59	3.51	3.42	3.37	3.30	3.27	3.21	3.18	3.16
2.48	2.44	2.39	2.35	2.31	2.27	2.24	2.21	2.19	2.16	2.14	2.13
3.70	3.62	3.51	3.43	3.34	3.26	3.21	3.14	3.11	3.06	3.02	3.00
2.43	2.39	2.33	2.29	2.25	2.21	2.18	2.15	2.12	2.10	2.08	2.07
3.56	3.48	3.36	3.29	3.20	3.12	3.07	3.00	2.97	2.92	2.89	2.87
2.37	2.33	2.28	2.24	2.20	2.16	2.13	2.09	2.07	2.04	2.02	2.01
3.45	3.37	3.25	3.18	3.10	3.01	2.96	2.98	2.86	2.80	2.77	2.75
2.33	2.29	2.23	2.19	2.15	2.11	2.08	2.04	2.02	1.99	1.97	1.96
3.35	3.27	3.16	3.08	3.00	2.92	2.86	2.79	2.76	2.70	2.67	2.65
2.29	2.25	2.19	2.15	2.11	2.07	2.04	2.00	1.98	1.95	1.93	1.92
3.27	3.19	3.07	3.00	2.91	2.83	2.78	2.71	2.68	2.62	2.59	2.57
2.26	2.21	2.15	2.11	2.07	2.02	2.00	1.96	1.94	1.91	1.90	1.88
3.19	3.12	3.00	2.92	2.84	2.76	2.70	2.63	2.60	2.54	2.51	2.49
2.23	2.18	2.12	2.08	2.04	1.99	1.96	1.92	1.90	1.87	1.85	1.84
3.13	3.05	2.94	2.86	2.77	2.69	2.63	2.56	2.53	2.47	2.44	2.42
2.20	2.15	2.09	2.05	2.00	1.96	1.93	1.89	1.87	1.84	1.82	1.81
3.07	2.99	2.88	2.80	2.72	2.63	2.58	2.51	2.47	2.42	2.38	2.36
2.18	2.13	2.07	2.03	1.98	1.93	1.91	1.87	1.84	1.81	1.80	1.78
3.02	2.94	2.83	2.75	2.67	2.58	2.53	2.46	2.42	2.37	2.33	2.31
2.14	2.10	2.04	2.00	1.96	1.91	1.88	1.84	1.82	1.79	1.77	1.76
2.97	2.89	2.78	2.70	2.62	2.53	2.48	2.41	2.37	2.32	2.28	2.26
2.13	2.09	2.02	1.98	1.94	1.89	1.86	1.82	1.80	1.76	1.74	1.73
2.93	2.85	2.74	2.66	2.58	2.49	2.44	2.36	2.33	2.27	2.23	2.21
2.11	2.06	2.00	1.96	1.92	1.87	1.84	1.80	1.77	1.74	1.72	1.71
2.89	2.81	2.70	2.62	2.54	2.45	2.40	2.32	2.29	2.23	2.19	2.17
2.10	2.05	1.99	1.95	1.90	1.85	1.82	1.78	1.76	1.72	1.70	1.69
2.86	2.77	2.66	2.58	2.50	2.41	2.36	2.28	2.25	2.19	2.15	2.13

Table A2. Continued.

f_2		1	2	3	4	5	6	7	8	9	10	11	12
	f_1												
27		4.21	3.35	2.96	2.73	2.57	2.46	2.37	2.30	2.25	2.20	2.16	2.13
		7.68	5.49	4.60	4.11	3.79	3.56	3.39	3.26	3.14	3.06	2.98	2.93
28		4.20	3.34	2.95	2.71	2.56	2.44	2.36	2.29	2.24	2.19	2.15	2.12
		7.64	5.45	4.57	4.07	3.76	3.53	3.36	3.23	3.11	3.03	2.95	2.90
29		4.18	3.33	2.93	2.70	2.54	2.43	2.35	2.28	2.22	2.18	2.14	2.10
		7.60	5.42	4.54	4.04	3.73	3.50	3.33	3.20	3.08	3.00	2.92	2.87
30		4.17	3.32	2.92	2.69	2.53	2.42	2.34	2.27	2.21	2.16	2.12	2.09
		7.56	5.39	4.51	4.02	3.70	3.47	3.30	3.17	3.06	2.98	2.90	2.84
32		4.15	3.30	2.90	2.67	2.51	2.40	2.32	2.25	2.19	2.14	2.10	2.07
		7.50	5.34	4.46	3.97	3.66	3.42	3.25	3.12	3.01	2.94	2.86	2.80
34		4.13	3.28	2.88	2.65	2.49	2.38	2.30	2.23	2.17	2.12	2.08	2.05
		7.44	5.29	4.42	3.93	3.61	3.38	3.21	3.08	2.97	2.89	2.82	2.76
36		4.11	3.26	2.86	2.63	2.48	2.36	2.28	2.21	2.15	2.10	2.06	2.03
		7.39	5.25	4.38	3.89	3.58	3.35	3.18	3.04	2.94	2.86	2.78	2.72
38		4.10	3.25	2.85	2.62	2.46	2.35	2.26	2.19	2.14	2.09	2.05	2.02
		7.35	5.21	4.34	3.86	3.54	3.32	3.15	3.02	2.91	2.82	2.75	2.69
40		4.08	3.23	2.84	2.61	2.45	2.34	2.25	2.18	2.12	2.07	2.04	2.00
		7.31	5.18	4.31	3.83	3.51	3.29	3.12	2.99	2.88	2.80	2.73	2.66
42		4.07	3.22	2.83	2.59	2.44	2.32	2.24	2.17	2.11	2.06	2.02	1.99
		7.27	5.15	4.29	3.80	3.49	3.26	3.10	2.96	2.86	2.77	2.70	2.64
44		4.06	3.21	2.82	2.58	2.43	2.31	2.23	2.16	2.10	2.05	2.01	1.98
		7.24	5.12	4.26	3.78	3.46	3.24	3.07	2.94	2.84	2.75	2.68	2.62
46		4.05	3.20	2.81	2.57	2.42	2.30	2.22	2.14	2.09	2.04	2.00	1.97
		7.21	5.10	4.24	3.76	3.44	3.22	3.05	2.92	2.82	2.73	2.66	2.60
48		4.04	3.19	2.80	2.56	2.41	2.30	2.21	2.14	2.08	2.03	1.99	1.96
		7.19	5.08	4.22	3.74	3.42	3.20	3.04	2.90	2.80	2.71	2.64	2.58
50		4.03	3.18	2.79	2.56	2.40	2.29	2.20	2.13	2.07	2.02	1.98	1.95
		7.17	5.06	4.20	3.72	3.41	3.18	3.02	2.88	2.78	2.70	2.62	2.56
55		4.02	3.17	2.78	2.54	2.38	2.27	2.18	2.11	2.05	2.00	1.97	1.93
		7.12	5.01	4.16	3.68	3.37	3.15	2.98	2.85	2.75	2.66	2.59	2.53
60		4.00	3.15	2.76	2.52	2.37	2.25	2.17	2.10	2.04	1.99	1.95	1.92
		7.08	4.98	4.13	3.65	3.34	3.12	2.95	2.82	2.72	2.63	2.56	2.50
65		3.99	3.14	2.75	2.51	2.36	2.24	2.15	2.08	2.02	1.98	1.94	1.90
		7.04	4.95	4.10	3.62	3.31	3.09	2.93	2.79	2.70	2.61	2.54	2.47
70		3.98	3.13	2.74	2.50	2.35	2.23	2.14	2.07	2.01	1.97	1.93	1.89
		7.01	4.92	4.08	3.60	3.29	3.07	2.91	2.77	2.67	2.59	2.51	2.45
80		3.96	3.11	2.72	2.48	2.33	2.21	2.12	2.05	1.99	1.95	1.91	1.88
		6.96	4.88	4.04	3.56	3.25	3.04	2.87	2.74	2.64	2.55	2.48	2.41
100		3.94	3.09	2.70	2.46	2.30	2.19	2.10	2.03	1.97	1.92	1.88	1.85
		6.90	4.82	3.98	3.51	3.20	2.99	2.82	2.69	2.59	2.51	2.43	2.36
125		3.92	3.07	2.68	2.44	2.29	2.17	2.08	2.01	1.95	1.90	1.86	1.83
		6.84	4.78	3.94	3.47	3.17	2.95	2.79	2.65	2.56	2.47	2.40	2.33
150		3.91	3.06	2.67	2.43	2.27	2.16	2.07	2.00	1.94	1.89	1.85	1.82
		6.81	4.75	3.91	3.44	3.14	2.92	2.76	2.62	2.53	2.44	2.37	2.30
200		3.89	3.04	2.65	2.41	2.26	2.14	2.05	1.98	1.92	1.87	1.83	1.80
		6.76	4.71	3.88	3.41	3.11	2.90	2.73	2.60	2.50	2.41	2.34	2.28
400		3.86	3.02	2.62	2.39	2.23	2.12	2.03	1.96	1.90	1.85	1.81	1.78
		6.70	4.66	3.83	3.36	3.06	2.85	2.69	2.55	2.46	2.37	2.29	2.23
1000		3.85	3.00	2.61	2.38	2.22	2.10	2.02	1.95	1.89	1.84	1.80	1.76
		6.66	4.62	3.80	3.34	3.04	2.82	2.66	2.53	2.43	2.34	2.26	2.20
∞		3.84	2.99	2.60	2.37	2.21	2.09	2.01	1.94	1.88	1.83	1.79	1.75
		6.64	4.60	3.78	3.32	3.02	2.80	2.64	2.51	2.41	2.32	2.24	2.18

Table A2. Extended.

					f_1						
14	16	20	24	30	40	50	75	100	200	500	∞
2.08	2.03	1.97	1.93	1.88	1.84	1.80	1.76	1.74	1.71	1.68	1.67
2.83	2.74	2.63	2.55	2.47	2.38	2.33	2.25	2.21	2.16	2.12	2.10
2.06	2.02	1.96	1.91	1.87	1.81	1.78	1.75	1.72	1.69	1.67	1.65
2.80	2.71	2.60	2.52	2.44	2.35	2.30	2.22	2.18	2.13	2.09	2.06
2.05	2.00	1.94	1.90	1.85	1.80	1.77	1.73	1.71	1.68	1.65	1.64
2.77	2.68	2.57	2.49	2.41	2.32	2.27	2.19	2.15	2.10	2.06	2.03
2.04	1.99	1.93	1.89	1.84	1.79	1.76	1.72	1.69	1.66	1.64	1.62
2.74	2.66	2.55	2.47	2.38	2.29	2.24	2.16	2.13	2.07	2.03	2.01
2.02	1.97	1.91	1.86	1.82	1.76	1.74	1.69	1.67	1.64	1.61	1.59
2.70	2.62	2.51	2.42	2.34	2.25	2.20	2.12	2.08	2.02	1.98	1.96
2.00	1.95	1.89	1.84	1.80	1.74	1.71	1.67	1.64	1.61	1.59	1.57
2.66	2.58	2.47	2.38	2.30	2.21	2.15	2.08	2.04	1.98	1.94	1.91
1.98	1.93	1.87	1.82	1.78	1.72	1.69	1.65	1.62	1.59	1.56	1.55
2.62	2.54	2.43	2.35	2.26	2.17	2.12	2.04	2.00	1.94	1.90	1.87
1.96	1.92	1.85	1.80	1.76	1.71	1.67	1.63	1.60	1.57	1.54	1.53
2.59	2.51	2.40	2.32	2.22	2.14	2.08	2.00	1.97	1.90	1.86	1.84
1.95	1.90	1.84	1.79	1.74	1.69	1.66	1.61	1.59	1.55	1.53	1.51
2.56	2.49	2.37	2.29	2.20	2.11	2.05	1.97	1.94	1.88	1.84	1.81
1.94	1.89	1.82	1.78	1.73	1.68	1.64	1.60	1.57	1.54	1.51	1.49
2.54	2.46	2.35	2.26	2.17	2.08	2.02	1.94	1.91	1.85	1.80	1.78
1.92	1.88	1.81	1.76	1.72	1.66	1.63	1.58	1.56	1.52	1.50	1.48
2.52	2.44	2.32	2.24	2.15	2.06	2.00	1.92	1.88	1.82	1.78	1.75
1.91	1.87	1.80	1.75	1.71	1.65	1.62	1.57	1.54	1.51	1.48	1.46
2.50	2.42	2.30	2.22	2.13	2.04	1.98	1.90	1.86	1.80	1.76	1.72
1.90	1.86	1.79	1.74	1.70	1.64	1.61	1.56	1.53	1.50	1.47	1.45
2.48	2.40	2.28	2.20	2.11	2.02	1.96	1.88	1.84	1.78	1.73	1.70
1.90	1.85	1.78	1.74	1.69	1.63	1.60	1.55	1.52	1.48	1.46	1.44
2.46	2.39	2.26	2.18	2.10	2.00	1.94	1.86	1.82	1.76	1.71	1.68
1.88	1.83	1.76	1.72	1.67	1.61	1.58	1.52	1.50	1.46	1.43	1.41
2.43	2.35	2.23	2.15	2.06	1.96	1.90	1.82	1.78	1.71	1.66	1.64
1.86	1.81	1.75	1.70	1.65	1.59	1.56	1.50	1.48	1.44	1.41	1.39
2.40	2.32	2.20	2.12	2.03	1.93	1.87	1.79	1.74	1.68	1.63	1.60
1.85	1.80	1.73	1.68	1.63	1.57	1.54	1.49	1.46	1.42	1.39	1.37
2.37	2.30	2.18	2.09	2.00	1.90	1.84	1.76	1.71	1.64	1.60	1.56
1.84	1.79	1.72	1.67	1.62	1.56	1.53	1.47	1.45	1.40	1.37	1.35
2.35	2.28	2.15	2.07	1.98	1.88	1.82	1.74	1.69	1.62	1.56	1.53
1.82	1.77	1.70	1.65	1.60	1.54	1.51	1.45	1.42	1.38	1.35	1.32
2.32	2.24	2.11	2.03	1.94	1.84	1.78	1.70	1.65	1.57	1.52	1.49
1.79	1.75	1.68	1.63	1.57	1.51	1.48	1.42	1.39	1.34	1.30	1.28
2.26	2.19	2.06	1.98	1.89	1.79	1.73	1.64	1.59	1.51	1.46	1.43
1.77	1.72	1.65	1.60	1.55	1.49	1.45	1.39	1.36	1.31	1.27	1.25
2.23	2.15	2.03	1.94	1.85	1.75	1.68	1.59	1.54	1.46	1.40	1.37
1.76	1.71	1.64	1.59	1.54	1.47	1.44	1.37	1.34	1.29	1.25	1.22
2.20	2.12	2.00	1.91	1.83	1.72	1.66	1.56	1.51	1.43	1.37	1.33
1.74	1.69	1.62	1.57	1.52	1.45	1.42	1.35	1.32	1.26	1.22	1.19
2.17	2.09	1.97	1.88	1.79	1.69	1.62	1.53	1.48	1.39	1.33	1.28
1.72	1.67	1.60	1.54	1.49	1.42	1.38	1.32	1.28	1.22	1.16	1.13
2.12	2.04	1.92	1.84	1.74	1.64	1.57	1.47	1.42	1.32	1.24	1.19
1.70	1.65	1.58	1.53	1.47	1.41	1.36	1.30	1.26	1.19	1.13	1.08
2.09	2.01	1.89	1.81	1.71	1.61	1.54	1.44	1.38	1.28	1.19	1.11
1.69	1.64	1.57	1.52	1.46	1.40	1.35	1.28	1.24	1.17	1.11	1.00
2.07	1.99	1.87	1.79	1.69	1.59	1.52	1.41	1.36	1.25	1.15	1.00

Table A3. Critical values for Fisher's Exact Test. Values of $n_2p_2 \geq$ the table entries indicate that the two estimated proportions are significantly different. n_1 and n_2 are the sample sizes; p_1 and p_2 are the estimated proportions.

n_1	n_2	n_1p_1	0.10	0.05	0.02	0.01
3	3	3	0	—	—	—
4	4	4	0	0	—	—
	3	4	0	—	—	—
5	5	5	1	1	0	0
		4	0	0	—	—
	4	5	1	0	0	—
		4	0	—	—	—
	3	5	0	0	—	—
	2	5	0	—	—	—
6	6	6	2	1	1	0
		5	1	0	0	—
		4	0	—	—	—
	5	6	1	0	0	0
		5	0	0	—	—
		4	0	—	—	—
	4	6	1	0	0	0
		5	0	0	—	—
	3	6	0	0	—	—
		5	0	—	—	—
	2	6	0	—	—	—
7	7	7	3	2	1	1
		6	1	1	0	0
		5	0	0	—	—
		4	0	—	—	—
	6	7	2	2	1	1
		6	1	0	0	0
		5	0	0	—	—
		4	0	—	—	—
	5	7	2	1	0	0
		6	1	0	0	—
		5	0	—	—	—
	4	7	1	1	0	0
		6	0	0	—	—
		5	0	—	—	—
	3	7	0	0	0	—
		6	0	—	—	—
	2	7	0	—	—	—
8	8	8	4	3	2	2
		7	2	2	1	0
		6	1	1	0	0
		5	0	0	—	—
		4	0	—	—	—
	7	8	3	2	2	1
		7	2	1	1	0
		6	1	0	0	—
		5	0	0	—	—
	6	8	2	2	1	1
		7	1	1	0	0
		6	0	0	0	—
		5	0	—	—	—
	5	8	2	1	1	0
		7	1	0	0	0
		6	0	0	—	—
		5	0	—	—	—

n_1	n_2	n_1p_1	0.10	0.05	0.02	0.01
4	8	8	1	1	0	0
		7	0	0	—	—
		6	0	—	—	—
	3	8	0	0	0	—
		7	0	0	—	—
	2	8	0	0	—	—
9	9	9	5	4	3	3
		8	3	3	2	1
		7	2	1	1	0
		6	1	1	0	0
		5	0	0	—	—
		4	0	—	—	—
	8	9	4	3	3	2
		8	3	2	1	1
		7	2	1	0	0
		6	1	0	0	—
		5	0	0	—	—
	7	9	3	3	2	2
		8	2	2	1	0
		7	1	1	0	0
		6	0	0	—	—
		5	0	—	—	—
	6	9	3	2	1	1
		8	2	1	0	0
		7	1	0	0	—
		6	0	0	—	—
		5	0	—	—	—
	5	9	2	1	1	1
		8	1	1	0	0
		7	0	0	—	—
		6	0	—	—	—
	4	9	1	1	0	0
		8	0	0	0	—
		7	0	0	—	—
		6	0	—	—	—
	3	9	1	0	0	0
		8	0	0	—	—
		7	0	—	—	—
	2	9	0	0	—	—
10	10	10	6	5	4	3
		9	4	3	3	2
		8	3	2	1	1
		7	2	1	1	0
		6	1	0	0	—
		5	0	0	—	—
		4	0	—	—	—
	9	10	5	4	3	3
		9	4	3	2	2
		8	2	2	1	1
		7	1	1	0	0
		6	1	0	0	—
		5	0	0	—	—
	8	10	4	4	3	2
		9	3	2	2	1
		8	2	1	1	0
		7	1	1	0	0
		6	0	0	—	—
		5	0	—	—	—

Table A3. Continued.

n_1	n_2	n_1p_1	0.10	0.05	0.02	0.01
	7	10	3	3	2	2
		9	2	2	1	1
		8	1	1	0	0
		7	1	0	0	—
		6	0	0	—	—
		5	0	—	—	—
	6	10	3	2	2	1
		9	2	1	1	0
		8	1	1	0	0
		7	0	0	—	—
		6	0	—	—	—
	5	10	2	2	1	1
		9	1	1	0	0
		8	1	0	0	—
		7	0	0	—	—
		6	0	—	—	—
	4	10	1	1	0	0
		9	1	0	0	0
		8	0	0	—	—
		7	0	—	—	—
	3	10	1	0	0	0
		9	0	0	—	—
		8	0	—	—	—
	2	10	0	0	—	—
		9	0	—	—	—
11	11	11	7	6	5	4
		10	5	4	3	3
		9	4	3	2	2
		8	3	2	1	1
		7	2	1	0	0
		6	1	0	0	—
		5	0	0	—	—
		4	0	—	—	—
	10	11	6	5	4	4
		10	4	4	3	2
		9	3	3	2	1
		8	2	2	1	0
		7	1	1	0	0
		6	1	0	0	—
		5	0	—	—	—
	9	11	5	4	4	3
		10	4	3	2	2
		9	3	2	1	1
		8	2	1	1	0
		7	1	1	0	0
		6	0	0	—	—
		5	0	—	—	—
	8	11	4	4	3	3
		10	3	3	2	1
		9	2	2	1	1
		8	1	1	0	0
		7	1	0	0	—
		6	0	0	—	—
		5	0	—	—	—
	7	11	4	3	2	2
		10	3	2	1	1
		9	2	1	1	0
		8	1	1	0	0
		7	0	0	—	—
		6	0	0	—	—
	6	11	3	2	2	1
		10	2	1	1	0
		9	1	1	0	0
		8	1	0	0	—
		7	0	0	—	—
		6	0	—	—	—
	5	11	2	2	1	1
		10	1	1	0	0
		9	1	0	0	0
		8	0	0	—	—
		7	0	—	—	—
	4	11	1	1	1	0
		10	1	0	0	0
		9	0	0	—	—
		8	0	—	—	—
	3	11	1	0	0	0
		10	0	0	—	—
		9	0	—	—	—
	2	11	0	0	—	—
		10	0	—	—	—
12	12	12	8	7	6	5
		11	6	5	4	4
		10	5	4	3	2
		9	4	3	2	1
		8	3	2	1	1
		7	2	1	0	0
		6	1	0	0	—
		5	0	0	—	—
		4	0	—	—	—
	11	12	7	6	5	5
		11	5	5	4	3
		10	4	3	2	2
		9	3	2	2	1
		8	2	1	1	0
		7	1	1	0	0
		6	1	0	0	—
		5	0	0	—	—
	10	12	6	5	5	4
		11	5	4	3	3
		10	4	3	2	2
		9	3	2	1	1
		8	2	1	0	0
		7	1	0	0	0
		6	0	0	—	—
		5	0	—	—	—
	9	12	5	5	4	3
		11	4	3	3	2
		10	3	2	2	1
		9	2	2	1	0
		8	1	1	0	0
		7	1	0	0	—
		6	0	0	—	—
		5	0	—	—	—
	8	12	5	4	3	3
		11	3	3	2	2
		10	2	2	1	1
		9	2	1	1	0
		8	1	1	0	0
		7	0	0	—	—
		6	0	0	—	—

Table A3. Continued.

n_1	n_2	$n_1 p_1$	0.10	0.05	0.02	0.01
	7	12	4	3	3	2
		11	3	2	2	1
		10	2	1	1	0
		9	1	1	0	0
		8	1	0	0	—
		7	0	0	—	—
		6	0	—	—	—
	6	12	3	3	2	2
		11	2	2	1	1
		10	1	1	0	0
		9	1	0	0	0
		8	0	0	—	—
		7	0	0	—	—
		6	0	—	—	—
	5	12	2	2	1	1
		11	1	1	1	0
		10	1	0	0	0
		9	0	0	0	—
		8	0	0	—	—
		7	0	—	—	—
	4	12	2	1	1	0
		11	1	0	0	0
		10	0	0	0	—
		9	0	0	—	—
		8	0	—	—	—
	3	12	1	0	0	0
		11	0	0	0	—
		10	0	0	—	—
		9	0	—	—	—
	2	12	0	0	—	—
		11	0	—	—	—
13	13	13	9	8	7	6
		12	7	6	5	4
		11	6	5	4	3
		10	4	4	3	2
		9	3	3	2	1
		8	2	2	1	0
		7	2	1	0	0
		6	1	0	0	—
		5	0	0	—	—
		4	0	—	—	—
	12	13	8	7	6	5
		12	6	5	5	4
		11	5	4	3	3
		10	4	3	2	2
		9	3	2	1	1
		8	2	1	1	0
		7	1	1	0	0
		6	1	0	0	—
		5	0	0	—	—
	11	13	7	6	5	5
		12	6	5	4	3
		11	4	4	3	2
		10	3	3	2	1
		9	3	2	1	1
		8	2	1	0	0
		7	1	0	0	0
		6	0	0	—	—
		5	0	—	—	—

n_1	n_2	$n_1 p_1$	0.10	0.05	0.02	0.01
10	13	13	6	6	5	4
		12	5	4	3	3
		11	4	3	2	2
		10	3	2	1	1
		9	2	1	1	0
		8	1	1	0	0
		7	1	0	0	—
		6	0	0	—	—
		5	0	—	—	—
9	13	13	5	5	4	4
		12	4	4	3	2
		11	3	3	2	1
		10	2	2	1	1
		9	2	1	0	0
		8	1	1	0	0
		7	0	0	—	—
		6	0	0	—	—
		5	0	—	—	—
8	13	13	5	4	3	3
		12	4	3	2	2
		11	3	2	1	1
		10	2	1	1	0
		9	1	1	0	0
		8	1	0	0	—
		7	0	0	—	—
		6	0	—	—	—
7	13	13	4	3	3	2
		12	3	2	2	1
		11	2	2	1	1
		10	1	1	0	0
		9	1	0	0	0
		8	0	0	—	—
		7	0	0	—	—
		6	0	—	—	—
6	13	13	3	3	2	2
		12	2	2	1	1
		11	2	1	1	0
		10	1	1	0	0
		9	1	0	0	—
		8	0	0	—	—
		7	0	—	—	—
5	13	13	2	2	1	1
		12	2	1	1	0
		11	1	1	0	0
		10	1	0	0	—
		9	0	0	—	—
		8	0	—	—	—
4	13	13	2	1	1	0
		12	1	1	0	0
		11	0	0	0	—
		10	0	0	—	—
		9	0	—	—	—
3	13	13	1	1	0	0
		12	0	0	0	—
		11	0	0	—	—
		10	0	—	—	—
2	13	13	0	0	0	—
		12	0	—	—	—

Table A3. Continued.

n_1	n_2	n_1p_1	Significance level 0.10	0.05	0.02	0.01
14	14	14	10	9	8	7
		13	8	7	6	5
		12	6	6	5	4
		11	5	4	3	3
		10	4	3	2	2
		9	3	2	2	1
		8	2	2	1	0
		7	1	1	0	0
		6	1	0	0	—
		5	0	0	—	—
		4	0	—	—	—
	13	14	9	8	7	6
		13	7	6	5	5
		12	6	5	4	3
		11	5	4	3	2
		10	4	3	2	2
		9	3	2	1	1
		8	2	1	1	0
		7	1	1	—	—
		6	1	0	—	—
		5	0	0	—	—
	12	14	8	7	6	6
		13	6	6	5	4
		12	5	4	4	3
		11	4	3	3	2
		10	3	3	2	1
		9	2	2	1	1
		8	2	1	0	0
		7	1	0	0	—
		6	0	0	—	—
		5	0	—	—	—
	11	14	7	6	6	5
		13	6	5	4	4
		12	5	4	3	3
		11	4	3	2	2
		10	3	2	1	1
		9	2	1	1	0
		8	1	1	0	0
		7	1	0	0	—
		6	0	0	—	—
		5	0	—	—	—
	10	14	6	6	5	4
		13	5	4	4	3
		12	4	3	3	2
		11	3	3	2	1
		10	2	2	1	1
		9	2	1	0	0
		8	1	1	0	0
		7	0	0	0	—
		6	0	0	—	—
		5	0	—	—	—
	9	14	6	5	4	4
		13	4	4	3	3
		12	3	3	2	2
		11	3	2	1	1
		10	2	1	1	0
		9	1	1	0	0
		8	1	0	0	—
		7	0	0	—	—
		6	0	—	—	—

n_1	n_2	n_1p_1	Significance level 0.10	0.05	0.02	0.01
8	14	14	5	4	4	3
		13	4	3	2	2
		12	3	2	2	1
		11	2	2	1	1
		10	2	1	0	0
		9	1	0	0	0
		8	0	0	0	—
		7	0	0	—	—
		6	0	—	—	—
7	14	14	4	3	3	2
		13	3	2	2	1
		12	2	2	1	1
		11	2	1	1	0
		10	1	1	0	0
		9	1	0	0	—
		8	0	0	—	—
		7	0	—	—	—
6	14	14	3	3	2	2
		13	2	2	1	1
		12	2	1	1	0
		11	1	1	0	0
		10	1	0	0	—
		9	0	0	—	—
		8	0	0	—	—
		7	0	—	—	—
5	14	14	2	2	1	1
		13	2	1	1	0
		12	1	1	0	0
		11	1	0	0	0
		10	0	0	—	—
		9	0	0	—	—
		8	0	—	—	—
4	14	14	2	1	1	1
		13	1	1	0	0
		12	1	0	0	0
		11	0	0	—	—
		10	0	0	—	—
		9	0	—	—	—
3	14	14	1	1	0	0
		13	0	0	0	—
		12	0	0	—	—
		11	0	—	—	—
2	14	14	0	0	0	—
		13	0	0	—	—
		12	0	—	—	—
15	15	15	11	10	9	8
		14	9	8	7	6
		13	7	6	5	5
		12	6	5	4	4
		11	5	4	3	3
		10	4	3	2	2
		9	3	2	1	1
		8	2	1	1	0
		7	1	1	0	0
		6	1	0	0	—
		5	0	0	—	—
		4	0	—	—	—

Table A3.   Continued.

n_1	n_2	n_1p_1	Significance level 0.10	0.05	0.02	0.01
	14	15	10	9	8	7
		14	8	7	6	6
		13	7	6	5	4
		12	6	5	4	3
		11	5	4	3	2
		10	4	3	2	1
		9	3	2	1	1
		8	2	1	1	0
		7	1	1	0	0
		6	1	0	—	—
		5	0	—	—	—
	13	15	9	8	7	7
		14	7	7	6	5
		13	6	5	4	4
		12	5	4	3	3
		11	4	3	2	2
		10	3	2	2	1
		9	2	2	1	0
		8	2	1	0	0
		7	1	0	0	—
		6	0	0	—	—
		5	0	—	—	—
	12	15	8	7	7	6
		14	7	6	5	4
		13	6	5	4	3
		12	5	4	3	2
		11	4	3	2	2
		10	3	2	1	1
		9	2	1	1	0
		8	1	1	0	0
		7	1	0	0	—
		6	0	0	—	—
		5	0	—	—	—
	11	15	7	7	6	5
		14	6	5	4	4
		13	5	4	3	3
		12	4	3	2	2
		11	3	2	2	1
		10	2	2	1	1
		9	2	1	0	0
		8	1	1	0	0
		7	1	0	0	—
		6	0	0	—	—
		5	0	—	—	—
	10	15	6	6	5	5
		14	5	5	4	3
		13	4	4	3	2
		12	3	3	2	2
		11	3	2	1	1
		10	2	1	1	0
		9	1	1	0	0
		8	1	0	0	—
		7	0	0	—	—
		6	0	—	—	—

n_1	n_2	n_1p_1	Significance level 0.10	0.05	0.02	0.01
9		15	6	5	4	4
		14	5	4	3	3
		13	4	3	2	2
		12	3	2	2	1
		11	2	2	1	1
		10	2	1	0	0
		9	1	1	0	0
		8	1	0	0	—
		7	0	0	—	—
		6	0	—	—	—
8		15	5	4	4	3
		14	4	3	3	2
		13	3	2	2	1
		12	2	2	1	1
		11	2	1	1	0
		10	1	1	0	0
		9	1	0	0	—
		8	0	0	—	—
		7	0	—	—	—
		6	0	—	—	—
7		15	4	4	3	3
		14	3	3	2	2
		13	2	2	1	1
		12	2	1	1	0
		11	1	1	0	0
		10	1	0	0	0
		9	0	0	—	—
		8	0	0	—	—
		7	0	—	—	—
6		15	3	3	2	2
		14	2	2	1	1
		13	2	1	1	0
		12	1	1	0	0
		11	1	0	0	0
		10	0	0	0	—
		9	0	0	—	—
		8	0	—	—	—
5		15	2	2	2	1
		14	2	1	1	1
		13	1	1	0	0
		12	1	0	0	0
		11	0	0	0	—
		10	0	0	—	—
		9	0	—	—	—
4		15	2	1	1	1
		14	1	1	0	0
		13	1	0	0	0
		12	0	0	0	—
		11	0	0	—	—
		10	0	—	—	—
3		15	1	1	0	0
		14	0	0	0	0
		13	0	0	—	—
		12	0	0	—	—
		11	0	—	—	—
2		15	0	0	0	—
		14	0	0	—	—
		13	0	—	—	—

Table A4. Critical values for the Wilcoxon signed rank test for paired data. Test statistics equal to or smaller than the tabled values are statistically significant.

Number of nonzero differences	Two-tailed tests: level of significance		One-tailed tests: level of significance	
	0.05	0.01	0.05	0.01
5	—	—	0	—
6	0	—	2	—
7	2	—	3	0
8	3	0	5	1
9	5	1	8	3
10	8	3	10	5
11	10	5	13	7
12	13	7	17	9
13	17	9	21	12
14	21	12	25	15
15	25	15	30	19
16	29	19	35	23
17	34	23	41	27
18	40	27	47	32
19	46	32	53	37
20	52	37	60	43
21	58	42	67	49
22	65	48	75	55
23	73	54	83	62
24	81	61	91	69
25	89	68	100	76

e A5. Critical values for the Mann-Whitney rank test for independent samples. Test statistics equal to or smaller than the tabled
ues are statistically significant. The table is for two-tailed tests. Define n_1 and n_2 so that $n_1 \leq n_2$.

Part One. Level of significance = 0.05.

n_2	2	3	4	5	6	7	8	9	10	11	12	13	14	15
4	ts[a]	ts	10											
5	ts	6	11	17										
6	ts	7	12	18	26									
7	ts	7	13	20	27	36								
8	3	8	14	21	29	38	49							
9	3	8	15	22	31	40	51	63						
10	3	9	15	23	32	42	53	65	78					
11	4	9	16	24	34	44	55	68	81	96				
12	4	10	17	26	35	46	58	71	85	99	115			
13	4	10	18	27	37	48	60	73	88	103	119	137		
14	4	11	19	28	38	50	63	76	91	106	123	141	160	
15	4	11	20	29	40	52	65	79	94	110	127	145	164	185
16	4	12	21	31	42	54	67	82	97	114	131	150	169	
17	5	12	21	32	43	56	70	84	100	117	135	154		
18	5	13	22	33	45	58	72	87	103	121	139			
19	5	13	23	34	46	60	74	90	107	124				
20	5	14	24	35	48	62	77	93	110					
21	6	14	25	37	50	64	79	95						
22	6	15	26	38	51	66	82							
23	6	15	27	39	53	68								
24	6	16	28	40	55									
25	6	16	28	42										
26	7	17	29											
27	7	17												
28	7													

(this area blank because n_1 must be $\leq n_2$)

(For combinations of n_1 and n_2 in this region, use the large sample approximation in Box 8.)

Part Two. Level of significance = 0.01.

n_2	2	3	4	5	6	7	8	9	10	11	12	13	14	15
5	ts	ts		15										
6			10	16	23									
7			10	17	24	32								
8			11	17	25	34	43							
9		6	11	18	26	35	45	56						
10		6	12	19	27	37	47	58	71					
11		6	12	20	28	38	49	61	74	87				
12		7	13	21	30	40	51	63	76	90	106			
13		7	14	22	31	41	53	65	79	93	109	125		
14		7	14	22	32	43	54	67	81	96	112	129	147	
15		8	15	23	33	44	56	70	84	99	115	133	151	171
16		8	15	24	34	46	58	72	86	102	119	137	155	
17		8	16	25	36	47	60	74	89	105	122	140		
18		8	16	26	37	49	62	76	92	108	125			
19	3	9	17	27	38	50	64	78	94	111				
20	3	9	18	28	39	52	66	81	97					
21	3	9	18	29	40	53	68	83						
22	3	10	19	29	42	55	70							
23	3	10	19	30	43	57								
24	3	10	20	31	44									
25	3	11	20	32										
26	3	11	21											
27	4	11												
28	4													

[a]Sample sizes are too small to detect a difference at the stated level of significance.

3

MICROCOMPUTER APPLICATIONS IN WILDLIFE MANAGEMENT AND RESEARCH

Gary C. White and William R. Clark

INTRODUCTION

Microcomputers in Wildlife Work

The use of computers in the management of wildlife populations and habitats has exploded in the last 2 decades, as the discipline has become more scientific and quantitative. Animals are counted, habitats are mapped and analyzed, and the data must be related to one another through statistical summaries and models. And all the information must be summarized for other professionals and transmitted to the general public. Computers have become an integral part of the wildlife profession, not only for doing the arithmetic of population management and graphically displaying the habitat changes, but for completing the daily responsibilities required of any professional.

Major advances in hardware (i.e., the computing machinery) and software (i.e., programs and computer codes) technologies that have occurred since the 1970s have had an impact on computer use by wildlife professionals. For example, 10 years ago applications of statistical analyses and population modeling generally were conducted on mainframe computers, i.e., centrally located machines in carefully regulated physical environments, by "computer experts." But development of microcomputers (e.g., desktop and home computers) in the early 1980s has influenced nearly all wildlife managers and researchers. Most offices now use a microcomputer word processor for routine correspondence, and mailing letters has a computerized counterpart as electronic mail. The graphics capabilities of microcomputers and workstations have stimulated habitat analyses through use of Geographic Information Systems. The portability and durability of microcomputers now make possible the recording of data electronically in the field.

Objectives of This Chapter

This chapter introduces a variety of present and future applications of computers in wildlife research and management. Our intention is to provide an understanding of how computers are used and how they have stimulated and interacted with the management and research challenges facing biologists. We have not discussed mainframe computers because most applications previously limited to mainframes are now available on microcomputers. In addition, the selection and operation of mainframes are usually out of the hands of the biologist. Microcomputers enable applications in situations where access to a mainframe is not available or is limited to dial-up modems over telephone lines. We emphasize systems compatible with the IBM PC and its disk operating system (DOS), because wildlife biologists have applied software developed for these machines, and these machines represent the largest user-base in the profession.

Specific objectives of this chapter are the following:

(1) Introduce the reader to current and potential applications of computers to problems encountered in wildlife management and research.
(2) Review specific software packages that will provide solutions to these problems.
(3) Provide the reader with advice on selection of hardware.

Overview of Chapter

We first present potential applications of software, because we believe these needs should direct acquisition of hardware. The first part of our discussion covers various commercial software applications: word processing, spreadsheets, graphics, database management, and statistical packages. Then we review software that solves problems specific to wildlife research and management. We address software relevant to field data collection, literature searches, and electronic mail. Finally, given these potential applications, we outline hardware available and discuss how well various configurations work with software packages. We suggest frequent reference to the glossary of terms because discussion of computer applications is full of jargon and acronyms. Some readers may first wish to look at the sections on Hardware Requirements (p. 87) and Deciding on System Requirements (p. 90) to gain an idea about the relationship between hardware and software.

Caution About Recommendations

It is not our intention to comparatively review software and hardware. Although we focus on specific software packages, readers should be aware that this software best meets our suggested criteria and is known to work well for specific applications based on the personal experience of the authors. Competition among software manufacturers assures that many other packages fulfill the specifications, and these might be preferred by others. Our preferences in no way imply endorsement by The Wildlife Society.

One recommendation is nearly inescapable, however: consider compatibility with other users and availability of help when selecting software. Using software compatible with the standard package used in your organization enhances productivity. For example, if 90% of the people in an office use WordPerfect™, your doing the same is advantageous. Files can be transferred among users, making it easy to develop documents with multiple authors. Local expertise helps a user learn applications of new software quickly and with less frustration.

A similar statement can be made about hardware. When selecting a system, consider compatibility with other hardware, especially printers and graphics devices.

An Aside on IBM and Macintosh Systems

We conspicuously have not considered systems such as the Apple Macintosh™ (Mac), which will disappoint some readers because they are powerful computing systems. Mac systems have gradually developed IBM compatibility, sometimes as a software package that runs like IBM hardware. At the same time, IBM PCs have developed more of the visual interface applauded by Macintosh users, i.e., software that uses a screen pointer to select options from a menu. This is particularly true with the release of operating systems like Microsoft Windows™ 3.0 and 3.1, which make the IBM systems very Mac-like. The distinctions that characterized the different hardware and software alternatives in the late 1980s have narrowed to the point that the perceived differences are more opinion than reality. In word processing, database management, and spreadsheet applications, the Mac and DOS platforms are essentially equal at present. In graphics the Mac still has an edge because so much time was required for IBM standard graphics to be improved. For the Mac there is more and better software—programs like Canvas™ and MacPaint™, which are bit-mapped graphics rather than object-oriented graphics. On the DOS side, programs like PCPaintbrush™ are equivalent. In applications software, particularly statistics and programming, DOS platforms still have the edge. DOS gives much more flexibility on level of operation, from 640Kb machines up to Windows applications with 8Mb of memory. We present more comparison of operating systems later in the chapter.

Because there is no competition on manufacture of Macintoshes, they will generally be $500–600 more expensive than DOS systems with similar configuration. Standard Macintosh systems have more memory and less hard-disk capacity than IBM-compatible systems. Also, fewer third-party companies write software for Macintosh than for DOS hardware.

In regard to networked systems, wherein network features rather than distributed CPU are important, DOS has the major advantage: DOS systems transmit information one order of magnitude faster than its counterparts for Apple communication. However, the DOS hardware and software are more expensive.

Another class of hardware that we mention only briefly is desktop mini-computers and networked workstations. Included in this class are the Unix-based systems that are commonly used for communications; GIS applications; and computing-intensive graphics, modeling, and statistical applications. This type of system is becoming more important as costs decline and common applications are converted to run on this architecture.

Finally, the most important reason that we did not emphasize Macintosh and workstation systems further is that most of the specialized applications discussed later will not run on these platforms. We will consider these alternatives further in the section on Deciding on System Requirements (p. 90).

SOFTWARE FOR INFORMATION PRESENTATION

Presentation of information is basic to everything wildlife professionals do, from disseminating new knowledge to communicating with the public. Documents must be polished and professional to get the message across most effectively. Microcomputers have revolutionized information transfer, and a variety of software is available for information processing. We first consider the preparation of documents and then preparation of visual aids and other graphical output.

Document Preparation

Word processing is the most common application of microcomputers in the wildlife profession; it includes everything from short memos to books. Short documents require only simple word processing capabilities, whereas a book-sized document requires more sophisticated software. We emphasize "every day" applications but also address special needs for lengthy scientific documents.

INCREASING PRODUCTIVITY

Most word processing involves simple documents such as memos, letters, and short reports. Most people are familiar with word processing programs that enable the entry, editing, formatting, and printing of documents. With a word processor, text is viewed on the screen and revised as desired before the document is printed on paper.

WYSIWYG describes some word processors: "What You See Is What You Get." This term applies to software packages whose document on the screen is the same as will be printed. Various degrees of this capability are available, depending on the resolution of the computer screen and the software package. The advantage of a WYSIWYG package is that editorial changes appear exactly as they will be printed. A disadvantage is that additional computer processing may be required, slowing entry and manipulation of text.

We recommend developing the skill of writing on the computer rather than on paper, rereading and revising as text is composed. An ability to type is helpful but not essential; software packages are available to assist learning at the keyboard. The software will center or emphasize text, change spacing, move blocks of text about, include parts of other documents, and use many other specialized features.

Good word processors have features that improve writing, including a spelling checker and thesaurus. With a single key stroke, an overworked word can be replaced with one that more precisely conveys your meaning. Spelling checkers not only find common errors but allow one to develop a library of technical terms or even scientific names.

Features such as mail merge and macro commands are useful for office work. Sending the same letter to a dozen people is easily done by automatically merging different addresses into the letter. Likewise it is useful to be able to "string together" commands to create a macro command that then can be used to perform a routine task; for example, a macro command that puts your address in the upper right corner of correspondence is desirable.

We have chosen WordPerfect as a package that meets our needs. A speller, a thesaurus, and a simple set of commands for text editing and manipulation make this package powerful, yet easy to learn and use. Graphics can be included in the text, and complex equations can be produced. This package is compatible with nearly all printers made, and a version is available to run in the Windows environment. Another powerful and nearly equally popular word processor is Microsoft Word™, especially the version Word for Windows. These two packages are also the most widely used software on the Macintosh platform.

Proofreading programs are available to check for grammatical errors such as disagreement of subject and verb, dangling participles, overuse of passive voice, jargon, and clichés. Most of these programs provide a measure of readability of the material, i.e., the grade level of the audience that would understand the material. Complexity of words and sentences contributes to the level of readability. Such programs have limitations, but using them can help break poor writing habits. Grammar and level of readability can be examined with programs such as RightWriter™ or Grammatik™.

SCIENTIFIC PAPERS AND TECHNICAL REPORTS

Scientific papers and reports require all basic word processing features, such as a spelling checker and a thesaurus, plus more advanced features. Incorporation of graphics into the document is particularly important, and most word processing packages allow graphics to be merged into documents. For example, a plot of data can be included in the body of the text, rather than as a drawing attached to the text. A WYSIWYG word processor intersperses the figure with the text. The increased quality demonstrates a desire to communicate in a professional manner.

An important feature of many word processors is the capability to automate outlining material before actual writing begins, and then fill in the outline to prepare the document. An outline is constructed and discussed with the other authors, then text is added. In addition, the capability to mark blocks of text that have been added to (redline) or deleted from (strikeout) the original draft, or to include comments to coauthors that will enhance discussion, is desirable.

A major limitation of some word processors is the amount of text that can be handled easily. Often, the length of a book-sized document requires specialized software to handle the material. Instead of reading the document sequentially, the material is "indexed" in a manner efficient for the computer to handle, enabling the writer to move readily within the document without having to wait for the computer to examine intervening material. WordPerfect handles a document sequentially, so that as the amount of text grows, the time required to locate the start of the last section in the document increases. Probably an upper limit of 200 pages for a single document is reasonable with WordPerfect, depending on the power of the computer. In contrast, some technical word processors are designed to handle documents of more than 700 pages without noticeable delays. Procedures to structure the document are used that allow moving directly to various levels of headings.

Mathematical notation and equations are often needed

for technical documents. The efficient formatting of equations requires some special capabilities usually not provided by microcomputer word processors. Some scientific word processors offer screen editing of equations and provide a WYSIWYG environment (or nearly so), but processing time may be slow enough that delays are noticeable when commands are given. Some software uses formatting codes for equations, so final equations are seen only when previewed on the screen or printed. The easiest software to use is that which allows specification of equations in simple, English-like phrases, but displays the result in a WYSIWYG environment.

Another special feature useful in technical writing is the capability of generating the literature cited section of a document from a few key words describing specific references. Word processing systems available now generally do not provide this capability, although stand-alone programs are available. However, most literature citation systems will not effectively search a database for citations based on a few key words taken from a document and generate the citation. Rather, the exact reference must be identified via author's names and publication date, or perhaps with a specific citation index key from the database. Several bibliographic retrieval systems are available that mesh, in varying degrees, with word processing packages. Papyrus™, Procite™, and Ref-11™ enable searches for a collection of citations and output selected items in a desired format (e.g., *The Journal of Wildlife Management* style), so they can be input directly to a document.

We do all of our scientific document preparation with the WordPerfect processor, which provides all the requirements discussed above except for a bibliographical retrieval capability. WordPerfect has the basic features of a good word processor, features a highly flexible equation system, incorporates graphics easily, and adequately handles larger documents. Although the desktop-publishing features are not as powerful as packages designed specifically for this purpose, the capabilities are more than adequate for production of technical documents with a professional appearance.

DESKTOP PUBLISHING

Desktop publishing, i.e., the production of documents that appear typeset, is often desirable. Producing publicity announcements or newsletters can be fairly simple, but producing long documents might require considerable time and experience. Quality of the final copy depends more on the printer than on the processing software. Laser printers now allow the production of documents that are nearly the quality of typeset material. Typesetters produce output at >1,000 dots per inch (dpi), whereas standard laser printers have a resolution of 300 dpi. Specialized capabilities available with software such as EROFF or TeX handle internal spacing of text, such as proportional spacing, kerning (different spacing between pairs of letters), rotated printing, and a variety of typefaces and fonts. All of these capabilities are needed to produce typeset documents with a professional appearance.

EROFF and TeX are designed to produce lengthy typeset documents. Figures and tables can be automatically floated to the top of the next page or to the bottom of the current page. In contrast, desktop publishing software such as Aldus Pagemaker™ or Ventura Publisher™ is useful for shorter documents such as a newsletter, because the user orchestrates the format of each page. However, for a lengthy book, the software should format the text and figures for each page automatically. All of these software packages require a significant investment in learning to use them effectively. Further, achieving a pleasing appearance with a typeset document requires artistic skill as well as awareness about combinations of typefaces and sizes that function well together.

Graphics

An advantage of microcomputers over mainframes is the easy access to sophisticated graphics they provide wildlife professionals. The old statement "a picture is worth a thousand words" has become even more true because of the simplicity of learning graphics on microcomputers and the savings of time and money to generate quality graphics. Easily used programs have standard formats for line graphs and bar and pie charts that can be modified for specific needs. Useful programs can take input data in forms compatible with other software. For example, a set of data graphed in a spreadsheet can be input to the graphics software, the text fonts changed, legends modified, and graphic details added to produce a high-quality visual aid. Graphics packages should be able to output figures in a format for inclusion directly into a word processor manuscript.

Output has been vastly improved by high-quality graphics printers. Not only can copies of the graph be printed on paper, the software can output for a film recorder to make a slide for an oral presentation or for a pen plotter to make overheads. Modifications can be made to improve the quality of the visual aid for each medium, so that a figure designed for a plot on paper can be redesigned for slides.

Most graphics packages can modify figures intended for printing to produce output appropriate for slides or transparencies. Users should consider the basics of presentation of material when converting among media to assure readability. Some criteria to think about are effective colors, text size, and boldness. Final production of transparencies and slides can be costly relative to viewing them on the screen, so your graphics package should allow previewing the image on the computer screen in the correct colors and relative size.

At least two packages meet the criteria described above. Although it was originally designed with business applications in mind, Harvard Graphics™ provides an easily learned system for presenting scientific data in charts and figures. The software easily makes title charts, histograms, data plots, multiple charts, and computerized slide shows. Lotus Freelance Plus™ also provides a menu system for preparing charts and figures, albeit not as sophisticated as Harvard Graphics. Although both provide free-hand drawing and manipulation capabilities, Freelance is somewhat more flexible than Harvard Graphics. We have used both packages to produce simple maps for presentation of spatial data, such as an animal's locations on a home-range map. Both programs can direct output to almost any desired medium, and versions of both programs are available to run in the Windows environment.

A hardware item worth mentioning at this point is the electronic scanner. Pictures or graphics can be optically scanned and converted to a format readable by graphics software. Then they can be modified, enhanced, or annotated with the software.

SOFTWARE FOR INFORMATION MANAGEMENT AND ANALYSIS
Database Management

Wildlife professionals often record data that can best be summarized and analyzed later with database management programs, most of which are interactive, thus allowing the user to enter data in a menu on the screen of the computer. Data are not limited to numeric type, but can be, e.g., a mailing list with information on licensed hunters, or literature citations. Good programs check items as they are entered, verifying that values are correct based on preset criteria. Data are then stored in a file, where they can be sorted according to contents of one or more items, or where portions of the data can be retrieved by defining selection criteria. Numeric data can be summarized with simple statistics such as means, standard deviations, and counts of frequencies. Some database managers have simple plotting and charting features as well.

NUMERIC AND ALPHANUMERIC DATA

The database language and file format that dominates the PC market today is dBase III™, and its successor, dBase IV™. Most other database packages are file compatible with dBase III, e.g., Paradox™. The dBase IV software provides a powerful programming language for applications such as data entry and summary. Therefore, dBase commonly is used to develop large-scale applications, such as state inventories of endangered species cross-referenced with habitat locations, but is more complex than necessary for routine summary of data. Various available software extends the capabilities of dBase, e.g., compilers speed the execution of the code and graphics programs are used for displaying bar and line charts. Most database managers do not include advanced statistical features, so the database file format should be read by a statistical package if more than routine statistical summaries are to be produced.

Borland's Paradox™ for Windows and Microsoft's Access™ are two database packages available for the Windows environment. Paradox for Windows provides power and flexibility for advanced programmers, whereas Access seems targeted at a broader range of less experienced database users. Both take advantage of the Windows interface and environment.

GEOGRAPHIC INFORMATION SYSTEMS

Geographic Information Systems (GIS) are a specialized form of databases whose application has become significant in wildlife management only since the middle 1980s. GIS packages provide manipulation and summary features for spatial data, that is, data collected in a format designed for maps (see Chapter 21). An example is a soils map for which soil type is known for all locations on a map, and for which specialized summaries such as the percentage of the map with a particular attribute are desired. Powerful analyses also can be performed by combining the data from two or more maps, i.e., all locations that have a soil type particular to wetlands, at least 1 ha

in size, in three specific townships. Analyses of changes in spatial characteristics over time also are possible.

Several public domain GIS packages have been developed through funding from the U.S. Government. Included in this category are MOSS (Map Overlay and Statistical System) and SAGIS (Systems Analysis Group Information System), originally developed by the U.S. Fish and Wildlife Service (and its contractors). Another is GRASS, developed by the U.S. Army Corps of Engineers. In contrast to the above GIS packages, commercially available packages such as ARC/INFO™ are often expensive (>$40,000). Some packages and versions of highly capable systems can be obtained for <$10,000.

The computer requirements of typical GIS applications preclude use on many personal computers because of the large files (hence large disks with at least 100Mb of storage needed) and the processing speed required to perform tasks in a reasonable amount of time. Most GIS software is run on high-resolution graphic workstations, which have a local CPU and hard disk but are also networked for sharing of software and files. Some software such as MOSS, pMAP (another public domain system), MIPS™, and PC ARC/INFO have been modified to operate on an IBM PC or compatible. Specialized hardware, principally fast processor speed and large rapid-access disk storage, will be needed when a microcomputer is purchased for GIS applications. It is beyond our goals to completely review GIS hardware and software, and readers are referred to Chapter 21.

Statistical Analysis

Statistical analysis was one of the earliest applications of computers in wildlife research and management. Statistical packages are still an important requirement for a wildlife microcomputer system.

REQUIREMENTS

We list a set of features that are appropriate for a statistical package for PCs. We were influenced by our use of several packages available for IBM PCs and by Dixon (1986), Geissler (1988), and Goldstein (1992).

Quality of Statistical Procedures

Before a statistical package is selected, we believe it is worth consulting a knowledgeable colleague, because many packages are limited in both the range and quality of statistical applications available. This may sound like an obvious point, but it is surprising how many of the packages first available for microcomputers lacked the necessary algorithms to operate on missing or extreme values, unbalanced designs, or large samples and render numerically correct results. Warnings about lack of numerical precision from rounding errors should be explicit, so that users do not misinterpret the output. The package should use the most current statistical procedure for analyses, providing the least biased estimators and most powerful tests. Error messages should make it clear to an inexperienced user if an analysis is incorrectly specified (e.g., zero degrees of freedom) or is not appropriate for the type of data supplied. Examples are calculating a chi-square statistic when expected frequencies are small or a pooled t-test when variances are unequal. Unfortunately, many unreliable and limited statistical packages are still marketed for microcomputers. The review by Dallal (1988) documents some problems encountered with some available packages.

Quantity of Statistical Procedures

A statistical package also should provide a variety of procedures meeting all the user's needs. Common difficulties are packages that include analysis of variance (ANOVA) for balanced designs but do not handle unbalanced designs, or provide univariate ANOVA but not multivariate ANOVA procedures. Another problem we often encounter is in the analysis of categorical data, e.g., frequency data such as the number of bucks and does counted during a deer sex-ratio survey. Because of rapid development in statistical theory in the last 20 years, even some of the major statistical packages lack the latest categorical data analysis procedures. Table 1 is a list of procedures that we believe should be available to analyze commonly encountered wildlife data sets.

Programming Language

A complete programming language for the transformation of data is an important feature of statistical packages. Manipulation of data before analysis is often crucial to statistical investigation. Therefore, a statistical package should have routine database management capabilities, such as sorting and subsetting. Other examples are the capability to sum variables over categories and then analyze the results, or to make simple transformations such as logarithms. An example of the increasing importance of programming in research is simulating data to test the robustness of a statistical procedure to violations of its assumptions.

User Interface

Useful statistical packages have a logical, easy-to-use interface and well-written manuals. Opinions differ regarding the type of interface that is easy to use, but many packages include on-line, menu-driven systems. As users learn the package or undertake larger, repetitive analyses, a menu system becomes tedious. Thus we suggest that a statistical package should include both interactive and noninteractive capabilities. At times there is a need to develop a series of commands and submit them for analysis, e.g., when the analyses will require more than a few minutes of time.

Another necessary component is on-line help. Simple questions about syntax of a command, or list of available options, should be available with a help request. Good programs also come with example data and analyses for a range of problems, which can be used to understand the full range of features of the system.

Graphics

Often we start statistical analyses with a graphical examination of the data, using simple *x-y* plots, histograms, and pie charts for frequencies and distributions of data and other kinds of plots to evaluate multivariate relationships. Graphics presentation within the statistical package can be simple, but output of graphics files should be possible in formats compatible with graphics packages, such as Lotus Freelance. With this type of interface, graphics

Table 1. Statistical procedures recommended in a statistical package to perform analyses commonly encountered with wildlife-related data.

Analysis	Application
Computation of simple summary statistics, such as mean, median, minimum, maximum, and standard deviation of a set of data	Summarize a set of numbers into a few summary statistics
Stem and leaf diagrams, box plots	Produce a graphical summary of a single set of data
Goodness-of-fit test for normality	Test a single set of data for goodness-of-fit to the normal distribution
Bivariate (*x* vs. *y*) plot	Produce a graphical summary of the relationship between two variables
Correlation	Measurement of degree of association between two variables
Linear regression, including step-wise procedures	Estimate and test linear relationships between two or more variables
Nonlinear regression	Estimate and test nonlinear relationships between two or more variables
t-tests	Test hypothesis of equality of two means when data are assumed normally distributed
Nonparametric equivalents to *t*-tests	Test hypothesis of equality of two sets of data when data are not assumed to be normally distributed
Unbalanced and multi-way ANOVA and MANOVA	Test hypotheses about means for complex, unbalanced designs, with multiple dependent variables
Chi-square tables	Compute chi-square tests of independence for two categorical variables
Log-linear analysis	Compute tests of independence for more than two categorical variables
Logistic regression	Compute tests of significance and estimate relationship between categorical and continuous variable
Survival estimation	Compute tests of significance and estimates for Kaplan-Meier survival procedures and Cox proportional hazard functions

can be manipulated to preferred form and used in documents and visual aids for presentations.

Compatibility With Other Systems

The statistical package must provide easy input of data, as text files and as files from database managers. The statistical package also must be able to export data to other packages, such as word processors, graphics packages, and spreadsheets (Hallahan 1992).

STATISTICAL PACKAGE RECOMMENDATION
Statistical Analysis System—SAS™

Our recommendation for a comprehensive statistical package is Statistical Analysis System, which meets all the above criteria. SAS (SAS Institute, Inc. 1985a) operates on a PC and has the largest array of statistical procedures available (SAS Institute, Inc. 1987). Reviews in the statistical and popular literature have shown that SAS procedures are numerically accurate and current (Wayner 1992).

Because SAS is so complete, it is somewhat intimidating to new users and requires a fully equipped PC to run. Depending on what modules are purchased, >130Mb of disk space may be required. However, the modular nature of SAS makes it easy to subset and configure for a particular user's needs. The DOS interface does not fully use the interactive capabilities of the PC environment but is still reasonably satisfactory. The latest release of SAS for a PC operates in the Windows environment, so the user interface is much improved. The manuals that accompany the software are reference manuals explaining how and when to use features, but not why the user would need the feature. An on-line tutorial comes with the software, and a videotape outlining use of the package and solving example problems is available from SAS.

Other Packages

Many other packages are available for the PC (Carpenter et al. 1984, Dixon 1986), including several adapted from mainframe environments. MINITAB/PC™ is useful as a teaching tool because it operates intuitively like a spreadsheet, but this sometimes makes it cumbersome for data manipulation and analysis. SPSS/PC+™ (Norušis 1986) and BMDP™ (Dixon 1983) are accurate and comprehensive systems; however, neither has the full data manipulation and programming capabilities of SAS that we feel are so important. SYSTAT™ includes a wide range of procedures, is fast and accurate (Carpenter et al. 1984), and is relatively easy to learn with a manual of examples. Geissler (1988) recommended SYSTAT over SAS for general use, although SYSTAT does not have all the options of SAS. SYSTAT is also available for Macintosh computers. Another package is STATGRAPHICS™, which lacks some complex statistical procedures but provides more extensive graphics capability. However, these graphics cannot be exported to other systems such as Lotus Freelance. Best and Morganstein (1991) reviewed six statistical packages for the Macintosh.

A difficulty in evaluating reviews of statistical packages is that reviewers place different emphasis on ease of use and graphics compared with the accuracy and currency of the statistical computations (Dallal 1988). Suggestions by Dixon (1986) concerning the selection of a statistical package are useful in making a decision. Furthermore, Dixon's (1986) warning that these packages change with time is also notable. Problems discussed in a review can be corrected in later versions of the software.

Spreadsheets
BUDGETS

Spreadsheet software can be thought of as a large accounting sheet that allows entries in rows and columns to be linked by formulas. Changing an entry in cell A1 automatically changes the total of the column, which might be in cell A20. Spreadsheets originally were designed for accounting applications, such as budgets, and economic models. These are common and important needs in the wildlife profession, and spreadsheets make the task of fiscal planning easier.

POPULATION MODELING

Beyond budgeting applications, spreadsheets have been used in wildlife for modeling populations. Rather than entries in a budget, visualize individuals in age classes in a population changing through time, based on survival and birth rates. Relationships among rates can be readily programmed into the formulas in a spreadsheet, and populations can be projected under different assumptions about the effect of management on the input rates.

GRAPHICS AND STATISTICAL COMPUTATIONS

Most spreadsheets can suffice as a simple statistical package if they include capabilities for computing means, variances, and linear regressions. They also may provide graphics, including x-y plots and bar graphs especially useful for quick examination of data. Graphics files can be generated for input to graphics packages that can be manipulated to produce the final figures, although the newer versions of most spreadsheet packages now allow extensive editing of graphs within the package (e.g., Quattro Pro™ for Windows).

DATABASE CAPABILITIES

We often use spreadsheets for routine data entry, similar to a database system, defining columns for each variable. This approach makes it easy to examine data or to look for outliers with plots. However, the amount of data that can be examined this way is limited, because all data normally must fit into memory. Another disadvantage is that data values are not checked by the program as they are entered, i.e., illogical values might be accidentally entered by the user. The spreadsheet must have the correct format for converting this information to a file compatible with the user's database program or statistical package. Usually, variable names must appear at the top of each column (within the first row), and the value in the second row of the spreadsheet defines the format for the variable, i.e., the entire column. This process often turns out to be intractable, so an alternative is to generate an ASCII file that can be read by a statistical package.

SOFTWARE RECOMMENDATIONS
Lotus 1-2-3™ or Compatible

Lotus 1-2-3 has dominated the spreadsheet market for the last few years, so most spreadsheets are Lotus-compatible. Therefore, we recommend that the spreadsheet chosen should be compatible with Lotus format. Lotus graph file format has become a standard, therefore many graphics packages can import files created in a spreadsheet stored in Lotus PIC format. The package we like is the Quattro Pro spreadsheet by Borland International. Quattro is Lotus-compatible, is available for the Windows environment, and can be purchased for a reasonable price by students.

Specialized Programs

Up to this point we have discussed general-purpose software packages that can be adapted to problems in the wildlife profession. However, much software has been designed specifically for analysis of problems in wildlife management and research. Our purpose here is to outline briefly the software appropriate and available for specific analyses. The reader should refer to the chapters by Lancia et al. (Chapter 9), Samuel and Fuller (Chapter 15), and Johnson (Chapter 16) for complete discussions of these methods.

SURVIVAL RATE ESTIMATION

Banding Data

Estimation of survival rates is an important step in managing any wildlife population. One approach to estimating survival rates is to band or tag animals and then recover information about the animal through return of the band, usually at the time of death. Primary programs for the estimation of survival from band returns are ESTIMATE, which estimates adult survival, and BROWNIE, which estimates adult and juvenile survival (Brownie et al. 1985). More sophisticated analyses are available with MULT (Conroy et al. 1989) or SURVIV (White 1983).

Capture-Recapture Data

Another approach to estimating survival from marked animals is to trap and release them alive or resight them. Methods developed by Jolly and Seber (Seber 1982) provide the statistical theory to estimate survival from this type of data. Program JOLLY (Pollock et al. 1990) provides the Jolly-Seber analysis for a single group of animals. POPAN-3 (Arnason and Baniuk 1978, Arnason and Schwarz 1987, Arnason and Miller 1988) also calculates Jolly-Seber estimates, allows extensive database manipulation by categories such as age and sex, and is now available for the PC (Arnason and Miller 1988). Program JOLLYAGE (Pollock et al. 1990) extends the Jolly-Seber analysis for two or more age classes. Program RELEASE (Burnham et al. 1987) enables comparison of survival rates for two or more groups of animals, e.g., a control group and a group treated with a chronic dose of a pesticide.

Telemetry Data

Biotelemetry now enables biologists to relocate individual animals on a regular basis, to determine survival status, and, sometimes, to identify cause of death. Many programs are available to help in analysis of these data. A recently refined approach to analyzing telemetry data is analogous to estimating survival rates in human populations, in which patients are contacted on a regular basis after receiving a treatment. Methods used to analyze human survival data, an application of statistical methods generally called failure-time analyses, might be applicable to wildlife studies if underlying assumptions can be met. Medical studies assume that survival rate is a function of the treatment and the age of individuals, but not of environmental conditions, which affect humans much less than wildlife populations. An important assumption of medical methods is the independence of censoring (loss of contact with the subject) and the cause of mortality, which might be violated in wildlife studies.

Various failure-time methods have been applied to ecological studies, especially the Kaplan-Meier estimator (Pollock et al. 1989a,b). This method is nonparametric, making no assumptions about constancy of the survival function. The Kaplan-Meier method is included in some statistical packages. SAS programs the procedure in PROC LIFETEST (SAS Institute, Inc. 1988), and BMDP includes the method in the P1L procedure (Dixon 1983). White and Garrott (1990) provided a SAS code that allows for staggered entry of subjects, and Pollock et al. (1989a) programmed it for Lotus 1-2-3.

MICROMORT.—Another procedure, presented by Heisey and Fuller (1985), assumes that daily survival is constant during intervals. Their approach is an extension of the familiar Trent and Rongstad (1974) method and the Mayfield method (Hensler and Nichols 1981). A stand-alone program, MICROMORT, carries out the calculations, including variance estimation and comparison of survival rates for pooling of time intervals. We emphasize that it is important to select intervals based on ecological data and to validate the assumption of constant survival rate.

SURVIV.—Survival also can be estimated from telemetry data by visualizing the data as a recovery matrix similar to a banding study. White (1983) wrote a general program, SURVIV, especially designed to test for differences in survival among groups of animals. The program requires that the user input the algebraic expressions for expected recoveries and be familiar with FORTRAN compilation. Although the program requires a knowledgeable user to analyze data, its flexibility makes it very powerful. This program can be used to estimate survival under band analysis, capture-recapture, and telemetry.

POPULATION ESTIMATION

Closed Populations—CAPTURE

Estimates of the number of animals in an area are one of the most fundamental requirements for managing wildlife populations. The most common method of estimating population size is to use capture-recapture experiments in a closed area and to analyze the data by assuming no births or deaths, immigration, or emigration. The simplest form of estimator is the familiar Lincoln-Petersen method, which assumes that each animal has the same capture probability on each occasion. Program CAPTURE implements extensions of this basic method (Otis et al. 1978, White et al. 1982). The methods in CAPTURE allow for differences between individuals (M_h), differences related to trap happiness or trap shyness (M_b), a combination of these two assumptions (M_{bh}), and an extension of the Lincoln-Petersen estimator to more than two occasions (M_t). Models are selected based on the variability of capture probabilities in the data by using the series of tests provided in CAPTURE, along with knowledge of the field methodology used to collect the data. A menu-driven system called 2CAPTURE has been developed to provide an easy method to generate input for CAPTURE (Rexstad and Burnham 1991).

Open Populations—Jolly-Seber Methods

Often the assumptions of complete demographic closure cannot be met because of births, deaths, or movements during the interval of interest. In this situation the Jolly-Seber model and its extensions (Seber 1982, Pollock et al. 1990) are appropriate. Open models lead to some confusion about the definition of the population that is being estimated (White et al. 1982), and we believe they might be more appropriate for estimating survival. JOLLY and JOLLYAGE (Pollock et al. 1990) provide programs that estimate population size from capture-recapture (or resighting) data for open populations. POPAN-3 (Arnason and Miller 1988) also can be used to calculate the Jolly-Seber estimates and extensions.

DENSITY ESTIMATION

Line Transects

In contrast to a population estimate, density is the number of animals per unit of area. The foremost approach to density estimation is the line-transect method, well summarized by Burnham et al. (1980). Program TRANSECT (Burnham et al. 1980) was developed to estimate density from grouped and ungrouped perpendicular distance data with five sighting functions. This approach is used most often with data collected on "flushing" transects, but it is applicable to a variety of problems, including estimating density of inanimate objects. In addition to perpendicular data, sighting distance and angle data can be analyzed with special models developed for these situations or by transforming the data to perpendicular distances. White et al. (1989) modified the original TRANSECT program to include an interactive interface. A successor to TRANSECT called DISTANCE is currently being developed by the Colorado Cooperative Fish and Wildlife Research Unit.

Other programs are available for line-transect analysis. Drummer (1987) wrote SIZETRAN, which extends the TRANSECT estimators to sighting data from groups of animals (Drummer and McDonald 1987), such as coveys of quail. Buckland (1985) developed codes for the application of the Hermite polynomial and hazard rate sighting functions for perpendicular distance data that are grouped into intervals. Gates (1979) provided estimators for a large collection of models. All of these codes operate on a PC.

A modification of the line-transect methods is to collect distances to the animals from a point. This technique, known as point transects (Burnham et al. 1980:195), can be applied with any of the line-transect programs described above. TRANSECT has been modified for point transect density estimates by squaring the observed distances and multiplying by π. However, the models specifically designed for point-transect data (Buckland 1985) appear better for this purpose.

Capture-Recapture

Density also can be estimated from capture-recapture data if locations of captures are recorded. Program CAPTURE enables estimation of density when a grid of traps is used (Otis et al. 1978, White et al. 1982). Another

method, using traps arranged in a "spider web" design, was developed by Anderson et al. (1983) and explored by Wilson and Anderson (1985a,b). Distances from the center of the web to each capture location are used as point-transect data and analyzed with Program TRANSECT. This approach requires large numbers of traps and captures but can produce precise estimates of density.

TRIANGULATION

A common problem in biotelemetry is estimating an animal's location by triangulation. Programs XYLOG and UTMTEL, developed by Dodge et al. (1986) and Dodge and Steiner (1986), are written in BASIC and run on laptop computers, such as the Radio Shack TRS 80 Model 100™. These programs have been extensively modified for DOS laptops like the light-weight TOSHIBA 2100, and later models (K. Kenow, U.S. Fish Wildl. Serv., LaCrosse, Wis., pers. commun.). White and Garrott (1984) provided a similar program (TRIANG) for the Model 100 and later translated the code to FORTRAN to operate on DOS laptop PCs. The PC version of TRIANG executes algorithms based on work by Lenth (1981) for maximum likelihood estimation of the triangulated location. This program has been modified for data collection from vehicles rather than fixed towers.

HOME-RANGE ESTIMATION

Often biologists wish to estimate home-range size from location data. The McPAAL program, developed by M. Stüwe and C. E. Blohowiak (Natl. Zool. Garden, Washington, D.C., pers. commun.), is the most comprehensive code now available for the PC. McPAAL includes the harmonic mean (Dixon and Chapman 1980), bivariate normal ellipse (Jennrich and Turner 1969), Fourier series (Anderson 1982), and minimum convex polygon (Mohr 1947) estimators. The interactive interface is easy to use for a small number of animals, but it is slow when a large number of animals must be processed. Other programs include DC80 (J. C. Carey, Univ. Wisconsin, Madison, pers. commun.), TELEM/PC (Koeln 1980), HOME RANGE (Samuel et al. 1985, Ackerman 1990), and HOMER (White and Garrott 1990). HOMER includes the multivariate Ornstein-Uhlenbeck estimator developed by Dunn and Gipson (1977) and programmed by Dunn (1978).

Hill and Fendley (1982) provided a set of routines in SAS to plot the minimum convex polygon with PROC MATRIX (SAS/IML, SAS Institute, Inc. 1985b). White and Garrott (1990) also provided SAS code, as part of the DATA step, which calculates the minimum convex polygon, Jennrich-Turner, and weighted ellipse (Samuel and Garton 1985) estimators.

MODELING APPLICATIONS

Dynamic systems models have been applied to a variety of problems in wildlife management (Gross et al. 1973, Molini et al. 1981, Starfield and Bleloch 1986). Most often models are written in higher programming languages (e.g., FORTRAN), although we note that population projection matrices can be easily programmed in Lotus 1-2-3. During the last 10 years researchers in the U.S. Fish and Wildlife Service have developed a stochastic model of mallard productivity (Johnson et al. 1987), for which a user interface has been written (T. Shaffer, U.S. Fish Wildl. Serv., Jamestown, N.D., pers. commun.). This model has been linked with GIS data and systems (in Chapter 21) and is becoming prominent in guiding management of prairie pothole habitat. It has also been modified for northern pintails by Carlson et al. (1993).

Applications of bioenergetics to management are also conspicuous in the literature. Hobbs (1989) linked energy balance to survival of mule deer and provided a readily accessible program for the microcomputer. REFMOD, a model of refuging waterfowl populations (Frederick et al. 1987), was used to assess consequences of alteration in agricultural practices and management of midcontinent refuges to snow geese. Recently, this model was extensively modified to simulate refuging Pacific white-fronted geese in the Klamath Basin of California and Oregon, including a menu-driven user interface, additional graphics display, and enhanced simulation of field feeding on various crops (Frederick et al. 1992). In both of these instances the main programs are written in FORTRAN for the microcomputer, and the user interface is written in BASIC. Clark et al. (REFMOD: user's guide for microcomputers, Iowa Coop. Fish. Wildl. Res. Unit, Ames, 1986) also prepared extensive documentation and a user's manual for REFMOD, an essential component enhancing utility and application of such complex software.

Habitat suitability indices (HSI models) have been developed by the U.S. Fish and Wildlife Service (1980*a,b*, 1981) for numerous species. These models now run on PCs using a menu interface that enables a biologist to select existing aquatic and terrestrial species models and to examine and modify functional relationships based on new data. Once a file is created with habitat conditions for an area of interest, overall habitat analyses and project comparisons can be completed.

OBTAINING WILDLIFE-SPECIFIC SOFTWARE

Many of the above specialized PC software packages can be obtained by downloading them from the TWS software exchange facility. A user must have a modem with the PC system. Documentation, the executable code, and some example input files are available at no cost other than phone charges. Details on use of the TWS bulletin board are provided in PUBLIC DATABASES (p. 37).

SOFTWARE FOR INFORMATION COLLECTION
Portable Data Collection

Field biologists should consider the use of portable computers for recording data in the field. As discussed by White and Garrott (1984) and Hensler et al. (1986), data checking can be finished on the spot, thus reducing errors. For example, when telemetry triangulation is done with three receiving stations, the three bearings should intersect. The program developed by White and Garrott (1984) plots bearings on the screen of the portable computer and thus verifies quality of the data. Simple programs can be written that prompt for data entry, ensure that all corresponding items are recorded, and check data to see that they are entered correctly, within specified bounds. If errors are detected by the program, the user can be prompted to make corrections immediately.

DEDICATED DEVICES

Foresters have used dedicated, hand-held field recorders for some time (Cooney 1985), but wildlife biologists have been slower to adopt these devices. With the advent of battery-operated portable computers, many new options exist. Unwin and Martin (1987) discussed recording animal behavior with a portable microcomputer. Hobbs (1988) discussed the decision of whether to purchase a portable PC as opposed to non-PC-compatible machines such as the Radio Shack TRS 80 Model 100. Advantages of the Model 100 are durability, low cost, long battery life (about 20 hr on 4 AA batteries), and cheap replacement of the power source. Portable PCs such as the TOSHIBA 2100H™, which have rechargeable NiCad batteries with operation time of 8 hours, require access to electric power to periodically recharge the system. We believe the availability of a fully compatible PC system, enabling the user to program more complex data-acquisition procedures, and to easily store and transfer large amounts of data, makes these systems preferable.

RADIO SHACK TRS 80

Several software packages for data collection have been adapted to the TRS 80 Model 100 and its replacement, the Model 200. We previously referred to the biotelemetry location systems of Dodge et al. (1986), Dodge and Steiner (1986), and White and Garrott (1984). C. Winchell (Pendleton Marine Base, Calif., pers. commun.) and S. Kovach (Western Div. Naval Facilities, Calif., pers. commun.) developed a general system for data entry called DDE (Direct Data Entry), and J. Ha (Colorado State Univ., pers. commun.) developed a system for recording animal behavior on a Model 100. P. F. Retief (Kruger Natl. Park, Republic of South Africa, pers. commun.) produced a similar system for recording animal feeding times and plant composition while the animal is observed.

PORTABLE PC CLONES

Portable or laptop computers are replacing the TRS 80 and non-PC compatibles as the former have become more reliable in the field. Kenow and Korschgen (U.S. Fish Wildl. Serv., La Crosse, Wis., pers. commun.) extensively modified XYLOG for use on laptop PCs, providing the full power of a PC in the field. The program includes an interface for input of UTM coordinates from a LORAN-C Unit, particularly useful in aerial telemetry. The current notebook computers are highly portable. Most come with at least 640Kb of RAM (expandable to 8Mb), a 3½" 1.4Mb disk drive, a full keyboard, both parallel (printer) and serial (communication) ports, and a liquid crystal display (LCD); they weigh about 1 kg. These machines operate on rechargeable NiCad batteries for up to 5 hours, can include an optional auto adaptor that enables them to run off a car battery, and have internal modems so data can be sent to other computers.

Analog to Digital Conversion

Computers also can be directly integrated with laboratory equipment to directly record data. For example, Takekawa (1987) integrated a microcomputer with an environmental chamber to automate physiological measurements of ducks. Measurements taken by the equipment were written directly to disk, so that errors that might have occurred during manual transcription were avoided. Laboratory equipment and the computer usually require special adaptation for this type of application, so it is advisable to consider options when equipment is purchased. An analog-to-digital board (to convert analog signals to a digital representation) or stand-alone unit is required for the computer, and accompanying software might dictate how frequently readings must be taken or how many variables can be recorded. Calibration must be checked initially, but savings in time and data recording errors often justify the time and expense.

Communicating With Other Computers

COMMERCIAL DATABASES

Online Reference Sources

Using the computer and modem to communicate with another computer is simply another form of data entry. Personal computers can be linked with a variety of commercial databases from which information such as literature citations or public-domain software can be obtained. Most university libraries have access to the DIALOG literature-searching facility, and the same databases can be searched with BRS After Dark™ on your PC. BRS is a commercial facility that charges users based on the number of references that are downloaded. This capability eliminates much of the need to manually search published lists of literature. Selected literature citations can be downloaded to one of the personal literature-filing systems we referred to above.

Other Information Sources

CompuServe is another example of a commercial database that provides information via a PC and modem, including airline schedules, research resources, and investment information. Forums on a variety of topics, including hardware and software, are available.

PUBLIC DATABASES

TWS Bulletin Board

Specialized software can be obtained with a PC through electronic bulletin boards. Many user groups have established bulletin boards to exchange the latest information and software. A bulletin board is supported by The Wildlife Society, making software available electronically to members of TWS. Members need only call the bulletin board to download software. The phone number is (301) 498-0402. Some software included on the bulletin board may have undergone a peer-review process through the Computer Software Exchange (CSE) facility (Samuel 1988), and published descriptions of software are available in the *Wildlife Society Bulletin.*

Most computer viruses and other destructive programs are obtained via unsecured bulletin boards. Computer viruses are programs that modify the executable file of other programs in such a way that the modified program appears to still function, but will perform a destructive action at some point in the future. A common act of destruction is to wipe out the user's hard disk, but less damaging activities include strange messages appearing on the screen and poor performance of your computer. Do not practice copying large amounts, because another user can upload a program without the program being checked by a bulletin board supervisor for a computer virus.

ELECTRONIC MAIL

Noncommercial Services

Many computer users are aware that they can exchange electronic mail with other users on a mainframe system, but are not familiar with electronic mail on a global basis. Among the noncommercial networks that operate worldwide are Internet (Krol 1992, LaQuey and Ryer 1993), a network operating on high-speed connections that developed from a U.S. military network, and BITNET, a network linking university computers. A PC user must use software that enables connection to and communication with a computer system on Internet or BITNET. BITNET and Internet can exchange mail via "gateways," which are machines that can communicate with either the Internet or BITNET networks. Internet provides gateways to both CompuServe and MCI mail, so mail can be exchanged among commercial patrons, governmental agencies, and most university mail systems. Software like NCSA Telnet (National Center for Super Computing Applications) enables the PC, operating with DOS, to communicate with Internet computers. Similar communications programs like Kermit or Procomm allow a PC with a modem to connect to another computer and exchange files. Electronic mail is an efficient way to exchange files as large as 300Kb. It is easiest to transfer ASCII files, but other formats are possible. If the recipient is monitoring the machine, mail can be delivered in 15 minutes. As an example, the authors, using Internet, exchanged the evolving WordPerfect file for this document between Colorado State and Iowa State many times during its development.

Commercial Services

Commercial networks providing international electronic mail service include MCI, CompuServe, and Western Union. Users enroll in an electronic mail service, selecting a particular service because one or more people with whom they want to communicate are also on this service. Fortunately, most commercial services now provide capabilities to transfer mail from their network to other networks, including BITNET and Internet.

Backing-Up Data

As software has become more complicated and the price of hard disks has declined, more and more users store large amounts of data and software on one or more hard disks. The need for maintaining a second copy of this information has led to several software packages that quickly copy the contents of a hard disk onto sequential floppy disks. Every user tends to ignore the task of backing up his/her hard disk until the information has been destroyed. Generally, we all like to think that we are protecting ourselves from a disk failure. In reality, the most likely way that the information will be lost is that we accidentally destroy the files ourselves!

We have used the Fastback™ and PCTools™ utilities to save the contents of our hard disks to floppies, but many competing packages are available. Price of the package, speed of copying information to floppies, and storage compression (efficiency of compressing the hard-disk information to fit on the minimum number of floppy disks) are criteria to consider when a product is purchased.

Specialized Utilities

As we said, the most common cause of lost files is the inevitable mistake of deleting files unintentionally. Utility software that supplements DOS (and was included with DOS 5.0) has therefore been developed that can "unerase" files and perform a variety of other tasks. For example, Norton Utilities™ includes programs that provide information about the organization of file directories and file attributes such as size and date of creation, and that enable the user to find, unerase, and protect files. The programs also can be used to reorganize the storage of files on the disk to speed access to the files. It is a good idea to routinely use such software to ensure optimum hard-disk performance and to update your own knowledge of the directories and files on the disk. Norton is designed to be customized for one's particular needs, whereas other utilities, such as PCTools, perform the basic tasks from a simple menu.

COMPUTER LANGUAGES
Languages Available

Many users were first exposed to computers by learning a high-level programming language such as FORTRAN. BASIC was the language first available on microcomputers. FORTRAN is still a common programming language in the wildlife science community, particularly for numerical processing applications. However, more applications are being written in PASCAL or C, newer, more structured computer languages. Most wildlife biologists will have no need to learn languages such as FORTRAN or PASCAL unless they desire to write customized applications packages for specialized problems.

When Are These Languages Needed?

Most biologists interested simply in processing their data can accomplish this task with the programming capabilities of a statistical package or a spreadsheet. Although computer languages such as FORTRAN were often emphasized in the past, modern software provides a powerful working environment that reduces the need to program an application in a language. Users can be efficient in programming problems in SAS or Lotus 1-2-3 because the software already includes specialized conventions for inputing, outputing, and manipulating data.

Nonetheless, biologists often develop or modify specialized applications by directly programming their computers. Developing skill with computer programming, whether it is in very high-level software like dBase or SAS, or only in BASIC, improves the ability to logically solve problems with the computer. For example, when SAS is used, some knowledge of programming is helpful to transform and organize data logically with FORTRAN-like commands. Sometimes writing a specialized program for a problem is more efficient, even though the task could be accomplished within a spreadsheet or statistical package. Now that laptop PCs are commonly used for field data collection, most biologists will have occasion to use BASIC at some time. Languages such as FORTRAN make more efficient use of machine time than do spreadsheets or SAS programs. Estimation programs such as CAPTURE (White et al. 1982) and models like REFMOD (Frederick et al. 1987) are written in FORTRAN. Although modern languages, such as C and PASCAL, are more efficient for a knowledgeable programmer, FORTRAN and BASIC are still most widely used in wildlife applications.

MICROCOMPUTER OPERATION
Operating Systems and Environments

The operating system of a computer is a combination of hardware and software that enables the user to execute commands. Generally the operating system can be viewed as the handyman who makes hardware work with software applications. Programs (like WordPerfect or a FORTRAN code) that will be executed generally are stored on the hard disk. To make it run, one must place a copy of the code in the computer's memory and then tell the processor to start executing the instructions that make up this code. Besides controlling this process and allocating memory, the operating system takes care of such things as requests from the program for access to data files that are stored on disks and sending information to the screen or printer. With most software the operating system is nearly transparent to the user.

For the IBM PCs and compatibles that we emphasize in this chapter, DOS is still the standard. DOS is a command-oriented operating system, which requires users to enter commands to examine or move files, for example. Because original versions of DOS included little help, users found it difficult to remember commands. Many users find other operating environments much more comfortable, most notably the graphical user interface (GUI) provided on Macintosh systems. Macintosh systems were designed for a GUI, but until about 1988 the computing power and graphics quality on IBM PCs were inadequate to operate well with a GUI. Now, operating systems such as Microsoft Windows enable much the same GUI environment as on Macintosh machines. There are tradeoffs, because a GUI requires much more of the hardware's processing power and time than a simple, command-oriented operating system. For example, running under DOS 5.0 the operating system will use about 90Kb of memory, whereas under Windows 3.0 the same machine will use about 1.1Mb, more than 10 times as much memory. Hardware with the additional memory, speed, and graphics quality for GUI will be more expensive.

One of the major advantages of the GUI for new users is the rapid initial learning. A part of the Mac's popularity can be explained by the fact that many users report they could operate the system effectively after a few hours of learning. For simple applications like word processing this is an advantage, but for more complex applications and programming, the rate of improvement becomes about the same for Mac and DOS systems. In fact, programming is much more difficult for a windows environment (whether on DOS or Mac systems) because of strict rules regarding memory space, management of files, and other dedicated features. Furthermore, many users find the menu-driven, mouse-executed operating system cumbersome as they become more proficient.

Because of the limitations of DOS and advances in

hardware technology, Microsoft Corporation and IBM jointly developed what they hoped would be DOS's successor, an operating system called OS/2™ (Operating System 2). OS/2 corrected many of the deficiencies of DOS and included a GUI. But OS/2 was slow in release and required more advanced hardware, and the PC-user community did not quickly adopt it. In 1990, Microsoft released Windows 3.0, which is software functioning much like an operating system for PC-compatible machines. It creates a user-friendly, GUI environment, manages memory, and enables multiple applications to be run simultaneously, yet retains DOS features. For example, using Windows one can write a letter with WordPerfect at the same time that SAS is executing a statistical analysis. It has become a replacement for the standard operating system, because the popularity and functionality of the GUI cause the IBM PC and Macintosh platforms to converge. Windows 3.1 was released in 1992.

Another operating system that can be run on a PC is Unix. Unix is an "old" system, dating back to the 1960s. Early versions of Unix were much like DOS (DOS is a descendent of Unix) and required much experience by the user. As with DOS, the quality of the Unix interface has advanced, and most 1990 Unix systems are typified by a GUI. Although not as well-known as DOS, Unix has a large repertoire of applications and a large following of users. Because it was specifically designed for multitasking, Unix is most commonly encountered on the desktop minicomputers, typified by Sun, Apollo, or DEC MicroVAX machines. Commonly, Unix systems now support a DOS window so that DOS software can be executed from within the Unix operating system.

A confusing issue for the naive user is that some software programs run only on certain operating systems, whereas other programs have been translated to run on two or more operating systems. WordPerfect is an example of a program that is distributed in versions that run on DOS, OS/2, Unix, and the Macintosh operating systems.

File Management

The operating system and applications software enable one to create and manipulate information by storing data in units consisting of related records called files. Whether you work with DOS or a GUI, you must have a means to keep track of the files you save or wish to retrieve. When you or software you are using issues any command that refers to a specific file, DOS searches the disk's directory to locate the file and, if necessary, copies the file's contents into a working area of memory. DOS expects file names to conform to certain conventions, usually a name and an extension such as FILEONE.TXT. DOS reserves certain common file extensions such as .BAT to refer to a batch file, .COM to refer to command files, .EXE to refer to executable files, and .DAT to refer to ASCII data files. Related files are usually stored in groups called directories (similar to folders on the Macintosh); for example, software like WordPerfect (i.e., all the component files) could be installed into a directory called WP51. DOS can be used to locate, move, copy, and delete files regardless of their contents or structure. However, most applications software uses additional conventions, including special characters and escape codes, when storing, re-

trieving, or manipulating data in files. For example, by default, Lotus stores files with an extension of .WK1, and only Lotus or compatible spreadsheet software can manipulate the data in such files. When additional codes are added, Lotus speeds access to and manipulation of the data but makes it more difficult to transfer data into other software. However, nearly all software can read and write ASCII files, data stored in standard code specifically designed for interchange.

HARDWARE REQUIREMENTS

With the explosion of personal computer use, organizations and individuals need information about how to select hardware. We have written this chapter from that perspective, suggesting that selection of a system first requires knowledge about the software. For example, SAS can be used most effectively with at least a 200Mb hard disk, and 350Mb would be even better. Because software developers enhance their products to take advantage of improved hardware, and further hardware improvements provide new possibilities, choosing hardware and software becomes an escalating spiral.

Keep several principles in mind when selecting a machine to meet personal needs:

(1) the latest versions of hardware generally have more capability than your knowledge and the software are able to access,
(2) software development lags behind hardware development by 1–3 years,
(3) a new generation of hardware has appeared as frequently as every 1–2 years, and
(4) skills and knowledge of potential applications will increase as software and hardware become more familiar.

These points lead to our recommendation of buying hardware that is more capable than presently used, because then software can be upgraded. To help make decisions about buying a microcomputer, we include the following sections on understanding the functions of the components and how these functions affect the speed and capability of the machine.

Processor

The "brain" of a PC is the microprocessor (the "microchip"). The first IBM PCs contained an Intel 8088 processor and operated at a speed of 4.7 megahertz (MHz). A revised version of this chip with the same capabilities was introduced later as the 8086 processor, although users did not notice any difference in the operation of the first two chips. The next generation of processor increased the speed by about two times (8 MHz) and was named the 80286 processor. This is the chip contained in IBM AT and compatible machines. In 1987, the first PCs appeared that contained the third-generation chip, the 80386, and these machines achieved another large increase in performance. Although the DOS operating system without Windows did not take full advantage of its features, machines with the 80386 have become the state-of-the-art for IBM PC-compatible machines. Intel Corporation now produces the 80486 chip, and PCs containing this chip appeared in 1989. The next generation chip has been designed, and PCs containing it will appear shortly.

Besides the fact that the design of processors has changed with time, i.e., 8086 through 80486, the speed of existing designs has been increased with improved manufacturing processes. Standard IBM AT chips started at 6MHz, i.e., 80286/6 Mhz, but manufacturers now provide 80286/12MHz, 80286/16MHz, and 80286/20MHz chips. The increase in speed is directly related to the ratio of the speeds (e.g., a 16-MHz chip is 1.33 times faster than a 12-Mhz chip). The fastest chips commonly available in 1993 are 80486/66, with 80486/100 chips expected in the near future.

MATH COPROCESSORS

The 8088, 8086, 80286, and 80386 chips use integer arithmetic, i.e., when 7 is divided by 2, the answer is 3, not 3.5. Floating-point arithmetic is done by software routines that compute the remainder and convert it to a decimal. Software is slower than microchips specifically designed to perform floating-point arithmetic. These chips are called math coprocessors and have been developed to assist the integer processors. The naming convention for PC math coprocessors simply renames the last digit, substituting a 7. For example, the 8088 and 8086 chips have a corresponding math coprocessor labeled 8087. Likewise, the 80287 and 80387 coprocessors work with the 80286 and 80386 chips, respectively. The 80486 processor contains its own math coprocessor internally, so there are no 80487 chips.

Math coprocessors enhanced the speed of 8088 and 8086 machines most because the use of software for floating-point arithmetic was slowest with these early versions. Numerical processing increased speed by about 10 times when an 8087 coprocessor was installed. Work that requires little floating-point arithmetic, such as word processing, is not affected by the installation of a math coprocessor. Although 80287 and 80387 coprocessors enhance speed, the increase is less because the primary chips are much faster. Note that some software packages with heavy numerical processing demands require math coprocessors to function. Graphics packages, in particular, require a math coprocessor for reasonable performance.

MEMORY REQUIREMENTS

Memory capacity is generally expressed in kilobytes (Kb) of Random Access Memory (RAM). Most machines are equipped with at least 640Kb (640,000 bytes, each byte approximately equivalent to one character) of usable RAM. Because 8088 and 8086 chips can access only 1 megabyte (i.e., 1Mb or 1,000,000 bytes) of memory, the DOS was designed for an upper limit for user's programs of 640Kb. The 80286 and 80386 chips can access 16Mb and 64Gb (gigabytes) of RAM, respectively, so they are considerably more powerful than what is supported directly by the DOS operating system.

The 80286, 80386, and 80486 processors can access memory above the 640Kb limit, known as extended memory. DOS software cannot use this extended memory unless the software has been designed specifically for these processors. Machines with extended memory often are used with virtual disk software that creates ''disk'' storage in memory, decreasing the time to access stored infor-

mation. This disk-caching software also takes advantage of extended memory, again speeding access to information contained on disks.

Modern software written for DOS uses large amounts of memory. For example, WordPerfect 5.1 uses 384Kb and SAS normally uses nearly all 640Kb of available memory. To access additional memory above the 640Kb limit, some DOS programs use protocols that ''trick'' the operating system and convert extended memory into what is referred to as expanded memory (EMS). Expanded memory is available to software that understands the Lotus-Intel-Microsoft (LIM) protocol. Examples of packages that can use expanded memory are Lotus 1-2-3, QUATTRO, and SAS. Microsoft Windows makes management of extended memory transparent for DOS applications.

Other operating systems such as OS/2 and Unix, which do not have the limit of 640Kb of DOS, can use the extended memory available with 80386 or 80486 processors. These operating systems generally require much more memory. For example, OS/2 requires a minimum of 2Mb to run, although 8Mb is more realistic.

Storage Devices

The distinction between memory, such as RAM, and disk storage is often confusing to new users. RAM is used by the processor (and coprocessor) to contain the operating system and program(s) being executed. When the task is finished, or the machine is turned off, the information in RAM is lost. In contrast, disks are used to store information for future use. Unlike memory, programs or data transferred to disk are not destroyed when the machine is turned off because a copy remains on the disk. Access to the information in RAM is several orders of magnitude faster than reading the same information from a disk.

HARD DISK

Data and software are stored on various kinds of media. Most PCs come with some type of disk drive with a removable disk, termed a floppy disk drive. In addition, a drive called a hard drive is included on all but the simplest computers, from which the magnetic media cannot be removed. Hard disk storage is generally at least 20Mb, more than an order of magnitude greater than the largest floppy disk, and can be >300Mb. More importantly, access time to hard disks is much faster than for floppy disks, because the media are fixed. As hard disks became common, software developers took advantage of the larger storage capacity and faster access time. Software executes much faster from a hard disk, and the user avoids the annoying requirement of removing and inserting floppy disks during execution. Most of the software discussed in this chapter is designed for hard disks, and most software systems are large enough that they no longer fit on a single floppy disk.

FLOPPY DISK

Floppy disks come in four standard formats. Originally, the most common format was a 5¼" disk that stores 360Kb of information, and these disks are still used. With the IBM AT machine, IBM introduced a 5¼" disk that stored 1.2Mb of information. A 1.2Mb floppy disk cannot be read in a 360Kb drive, but, fortunately, a 1.2Mb drive can read a 360Kb floppy disk and can even write files to the disk.

Two other floppy disk formats that have become the standard for PCs are the 3½" disks that can be formatted at either 720Kb or 1.44Mb. A major advantage of this type disk is that it is contained in a rigid plastic shell, with a metal window that opens, allowing access to the magnetic media inside. This extra protection makes the diskette easier to transport. The 3½" drives are standard in current lines of IBMs and clones, and most portable computers use the 3½" drives because of their smaller size.

REMOVABLE MASS STORAGE

Tape Drives

Because of the capacity of hard disks, copying information they contain to floppy disks can be a time-consuming process. A tape drive system can be installed to make backing up the hard disk easier. Tape drives can store 20Mb or more of information on a single tape, and about 5 minutes are required to copy that much information from the hard disk.

Because of the cost of tape drives (generally >$700), most users do not purchase such a sophisticated back-up system. If a user is judicious about copying files to floppy disks and keeps a master backup of the hard disk, restoration of files can be reasonable if a hard disk fails. Software designed to quickly back up the contents of a hard disk to a set of floppy disks was discussed earlier in this chapter.

Removable Hard Disks

As the processing power of personal computers has increased, the size of files has grown. Even floppy disks with 1.4Mb are no longer adequate for many PC files, so technology for large removable disks has developed. Now 20Mb, and even 150Mb, disks allow users to store large files and yet transport them like huge floppy disks.

Output Devices

Displaying words, quantities, or images on a computer screen or monitor has become synonymous with microcomputing. However, a computer monitor alone does not keep a permanent copy of information or transmit the information to someone else. Printers, plotters, and modems also are essential output devices for microcomputers.

MONITOR

Most computer screens are cathode ray tubes (CRT) that display 25 rows and 80 columns of text, in either color or monochrome. Monochrome monitors are generally less expensive, but color displays are more popular. Although monitors are generally used in "text mode," they also can display graphics. A monitor displays information by lighting a "pixel" (short for picture element). A text character is created when selected pixels in one of the 25 × 80 blocks are lit, representing the character. Text characters are seen immediately because they are built into the monitor hardware.

In "graphics mode," pixels are lit individually to display an image, without regard to the 25 × 80 text blocks. The typical resolution of a CRT might be 640 pixels across by 480 down. Time required to display a graphics image is considerably longer than for text because a list of which of the 640 × 480 pixels to light must be transmitted to the monitor. Monitors are driven by a graphics board, which holds specifically designed electronic components and is connected to the machine in a slot designed for such boards.

Color monitors and boards are available in a variety of resolution capabilities, characterized by both the pixel density and number of colors that can be displayed. The lowest resolution monitor is the Color Graphics Adaptor (CGA) monitor and board, with text resolution of 25 rows by 80 columns, and graphics resolution of 640 × 200 pixels with two colors or 320 × 200 pixels and four colors. Enhanced Graphics Adaptor (EGA) monitor systems can display at least 25 rows and 80 columns, sometimes up to 43 rows and 132 columns. Graphics resolution is 640 × 350 pixels with 16 colors. The current standard graphics system available for PCs is the Video Graphics Array (VGA) system originally developed by IBM for the System 2 line of computers. The graphics resolution of this system is 640 × 480 pixels with 16 colors, or 320 × 200 pixels with 256 colors. Advanced monitors for PCs available in 1993 are Super VGA monitors, capable of 1,024 × 768 pixel resolution.

PRINTERS

In addition to monitors, nearly all computer systems are equipped with a printer. Dot-matrix printers use a mechanical head that prints a series of dots and operate similarly to a monitor with a pixel on the paper represented by a dot of ink. Because dots can be printed anywhere on the page, these printers can reproduce graphics. Some printers are capable of high resolution, because they use many pixels to represent characters and graphics. Resolution of dot-matrix printers varies from 120 dots per inch (dpi) to >300 dpi. Early printers had low resolution with nine pins in the print head to form characters, whereas most dot matrix printers now use the higher resolution produced by 24 pins. Ink-jet printers use fine spray nozzles to disperse water-soluble ink onto the page to form the dots or pixels, instead of a mechanical system. These printers have similar capabilities as dot-matrix printers.

Laser printers produce the finest quality printed output and have quickly become the de facto standard for many applications. Laser technology is used to apply ink to the paper with an approach fundamentally similar to dot-matrix printers. However, laser printers usually print at least 300 dpi, so resolution is exceptionally good. Laser printing has made desktop publishing possible, because printers can now produce output nearly equivalent to that from a more expensive typesetter that operates at 1,000 dpi. Laser printers have large amounts of memory to store the complex instructions necessary to reproduce a wide array of type fonts, high-resolution graphics, and even color. They often have cartridges that can be inserted into the printer, in which microchips store the instructions permanently, so that printing complex images is faster. Because of its complex capabilities, software used to create laser images is more complicated than that used with other types of printers.

PEN PLOTTER AND FILM RECORDER

Graphics can be produced on other visual-aid media. Pen plotters driven by the microcomputer draw lines by moving a pen across paper or overhead transparencies. Images can be reproduced in color. Film recorders, driven by software such as Freelance or Harvard Graphics, record graphic images on film, either as 35-mm slides or as prints.

MODEM

Information is transmitted over a telephone line between two computers with modems. A modem on the sending machine converts the bits of a file into sounds that are transmitted across the line to the second computer. On the receiving end, another modem converts the sounds back into digital form for that computer. Modems can be used to send electronic mail to another user, or to download data from a portable computer in the field to another computer. More detail appears in the next section.

Input Devices

KEYBOARD, MOUSE, DIGITIZER, AND SCANNERS

The primary interface for input to a computer is still the keyboard, but other input devices are now used commonly. Most notably, mouse systems enable one to move the cursor around the screen and select items from a menu to execute various commands. As users demand more intuitive interaction with microcomputers and as hardware becomes more powerful, mouse systems are required. Manufacturers frequently develop systems and write software incorporating mouse devices and menu systems, e.g., Windows.

Digitizers are similar to mouse systems except that the digitizer provides absolute coordinates of the input, rather than relative coordinates. Most often they are used in graphics input, e.g., to input data from a map. Scanners are electronic devices similar in principle to photocopy machines, which scan a photograph and directly convert the image into digital form.

MODEM

Use of microcomputers does not remove the need to share information, and it presents particular problems for communication among users. If two machines are close together, they can be connected with a null modem cable. One machine reads the input while the other machine is transmitting across the cable. This is the common approach to downloading data from a field computer to a larger computer.

As we noted above, longer distances can be spanned with telephone communication lines and a modem. Modems can be external, connected to the serial (communications) port of the machine, or internal, wherein they use one of the internal slots on the system board (the primary circuit board that holds the processor and has slots for additional boards to be "plugged" into it). In either situation, a telephone line must be connected to the modem. Most modems allow the telephone line to be used for voice communication when the modem is not in use.

The most sophisticated connection of microcomputers is through the formation of a network. In a network, one machine acts as a server, and the contents of its hard disks are available to all machines connected on the network. Network software makes it appear as if the server hard disks are part of the user's machine. The foremost advantage of a network is that resources on the server are available to everyone, i.e., the same software is accessible to all users without putting copies on all machines. On most PC networks, machines hooked to the server cannot make use of the server's processor.

DECIDING ON SYSTEM REQUIREMENTS

This section is intended as a guide for both beginning users and professional wildlife biologists buying personal computer systems for themselves or for use within agencies and universities.

Document Processing

It is surprising how often people overlook consideration of how the keyboard feels. Check the resistance and sound of the keys and the general layout. Poor-quality keyboards lead to superfluous typing mistakes. Better-quality keyboards usually can be purchased for only a small increase in price. We think that a good-quality monitor is essential for word processing, and color has become the default choice. Colors help relieve eyestrain and can be varied by individual users. Color is used by most word processors (i.e., WordPerfect) to highlight types of text such as underlining, special characters, and size and font types. VGA quality text also reduces eyestrain during long periods of word processing.

For output, a good-quality dot-matrix or ink-jet printer is suitable for most letters and manuscripts, and is cost effective compared with a laser printer. However, for most groups, where high-quality output is needed, we recommend a laser printer. If desktop publishing or producing an assortment of graphics is the goal, a laser printer is essential.

Data Management

Here again consider the monitor, although color is not as much of a factor. However, most database managers use color to enhance the readability of data-entry screens. Consider how often the data file will be sorted and otherwise manipulated, and get a hard disk large enough for the database software and all the databases that will be stored (plus room for growth).

Data Analysis

Data analysis requires more computing power than word processing or database manipulation. Consider the size of files to be analyzed and the types of analyses. For example, multivariate analysis on large data sets will require more storage and speed than simple ANOVA. If numerically intensive jobs are foreseen, a 80486 processor is almost essential, with its math coprocessor part of the main processor.

Disk space also should be carefully considered with statistical analysis. For example, the full SAS/Windows system, including GRAPH, requires >130Mb. Storage of large data files or maintenance of additional software packages on the disk will easily fill a 200Mb hard disk.

Graphics

If only black-and-white graphics are needed, a monochrome monitor may suffice. However, because the trend is toward color for other applications, a high-resolution color monitor might be appropriate for graphics. Graphics applications generally require considerable processing power, so an 80486 processor is generally required.

Spreadsheets

For large spreadsheets, performance will be reduced without extended memory and a math coprocessor. Because most spreadsheet software uses color to denote differences in the types of cell entries (i.e., character versus numeric), a color monitor is useful. A graphics-quality monitor also should be a consideration.

Your Wish List

Here, we offer options among desirable items to purchase, depending on your budget. More expensive items should not be considered extravagant, but rather an investment toward the longer utility of the equipment. Buying hardware that provides capacity beyond current needs enables one to move ahead when the next revisions of the software are available.

PROCESSOR

We suggest that an 80386 processor is now a minimum requirement because more advanced operating systems will not run adequately with less capability. The extended memory capacity of the 80386 processor is needed with Windows. Software generally requires more power than older processors can supply.

For the best investment, consider the 80486 processor. The processing speed is improved enough over the 80386 processor that the useful life of the machine will be extended by 2–3 years over a 80386 machine. The difference in cost is generally <$200 and is recuperated over the life of the machine.

HARD DISK

Most computers are now equipped with at least a 200Mb hard disk, many with a 350Mb disk. For personal use we recommend a 300Mb disk and for an office we would recommend at least a 350Mb disk. Space for software can be reduced, but the current versions of the software we recommended require substantial disk space: WordPerfect, approximately 10Mb; Lotus 1-2-3, approximately 8Mb; Lotus Freelance, approximately 7Mb; dBase IV, approximately 6Mb; and SAS, approximately 130Mb. Reserve space for data files and documents, plus room for extras, such as Microsoft Windows (6Mb). Using SAS and other software will fill a 300Mb disk. Our experience is that empty disks always become full faster than expected, so plan accordingly.

MONITOR

All the software we discussed will run with any of the monitor systems we discussed. We recommend that a prospective buyer consider at least a VGA monitor and perhaps the SVGA system. Select the SVGA system if graphics expertise is important. We do not recommend a monochrome system because so many software packages use color capabilities.

PRINTER

A dot-matrix printer will allow printing both text and graphics nicely and an ink-jet printer will produce near-laser quality. Laser printers are now relatively inexpensive, especially the "personal" models (~$600). If the agency will do any desktop publishing or publication-quality graphics, purchase a laser printer.

ACCESSORIES

Two accessories will be desirable. A mouse has become the common input device for production of graphics and for menu-driven software, and is required for Windows. A modem is useful for communicating with bulletin boards and for electronic mail. Neither of these items was once a necessity, but we suggest that your needs will dictate these additions.

SOURCES OF HARDWARE AND SOFTWARE

Computer hardware and software can be purchased from a variety of sources, from a local computer store that sells major-brand hardware to mail-order catalogs. Equipment from computer stores will have excellent warranties, and the store will have technicians that can help set up and maintain your equipment. Catalog stores may sell nonbrand name equipment, with shorter warranty periods, and provide little to no help (perhaps phone-call support) in setting up your new system. However, the price difference between mail order and a computer store can be as much as 100%.

First check availability, service support, and prices provided by local sources with other knowledgeable users. Carefully consider the features of PC clones and their service support, because most are priced lower than well-known brands. Before following up on a good deal in a catalog, check with local people who have purchased equipment from the vendor or with computing bulletin boards for complaints about particular equipment or vendors.

PROJECTING THE FUTURE
Living With the Rapid Changes in Technology

Capabilities of personal computers and their application to wildlife research and management are expanding more rapidly than any other part of our profession. It is a challenge to envision ways in which the computer can make your work more productive and efficient and to stay current as new applications are developed. However, we can readily state that the full capability offered by microcomputers is not being used by most wildlife biologists.

How To Stay Current

We maintain contact with new ideas and industry trends by reading one of many computer magazines. We like *PC Magazine*, but many similar computer magazines are good sources for users interested in the applications we have described in this chapter. *PC Magazine* has columns on WordPerfect and Lotus 1-2-3 applications that appear in each issue. Competing software is compared, and prices and vendors are provided in these reviews. We also find the periodic comparisons and ratings of PC-compatible hardware based on performance and price useful when considering purchases.

An especially good way for new users to learn more about their hardware and software is to join a computer user's group or club. Most groups discuss problems and solutions with hardware, exchange software, and often obtain purchase discounts. If a group is not readily available, form a lunch group at work.

Obtaining information on applications specific to wildlife is more difficult. The TWS bulletin board provides one avenue to seeing what new software is available and to interacting with others using this software. The American Fisheries Society Computer Users Section (AFSCUS) publishes a newsletter suggesting tips on use of commercial software and lists software pertinent to the fisheries profession. Finally, conferences and workshops on computer applications in fisheries and wildlife are becoming common and provide the opportunity to interact with wildlife professionals who are using and developing software, to observe their applications, and to envision ways in which applications can be adapted to your work.

When we wrote the first draft of this chapter in 1989, IBM AT PCs (with 80286 chips) were commonly used and 80386 PCs were beginning to take over the market. In 1992, the 80486 with SVGA graphics became the standard. Exchanging information among software packages was difficult at best, and now we routinely copy graphics from Harvard Graphics into our WordPerfect documents. In fact, using the GUI of Windows, we can literally "drag" an image from one window and "drop" it into our manuscript. As the speed, capacity, graphics capability, and ease of use of computers have increased, wildlife researchers and managers are finding that they use the microcomputer to complete everything from the most routine tasks to simulating and displaying changing landscapes on their computer screens. Most of the tasks discussed in the other chapters of this manual are accomplished with microcomputers. In the future, when this edition of the manual faces revision, microcomputer systems will have advanced far beyond what we describe, and wildlife management and research will have developed ways to harness the technology in ways we are just beginning to imagine.

LITERATURE CITED

ACKERMAN, B., F. A. LEBAN, M. D. SAMUEL, AND E. O. GARTON. 1990. User's manual for program HOME RANGE. Second ed. Univ. Idaho For. Wildl. Range Exp. Stn. Tech. Rep. 15. 80pp.

ANDERSON, D. J. 1982. The home range: a new nonparametric estimation technique. Ecology 63:103–112.

ANDERSON, D. R., K. P. BURNHAM, G. C. WHITE, AND D. L. OTIS. 1983. Density estimation of small-mammal populations using a trapping web and distance sampling methods. Ecology 64:674–680.

ARNASON, A. N., AND L. BANIUK. 1978. POPAN-2: a data maintenance and analysis system for mark-recapture data. Charles Babbage Res. Cent., St. Norbert, Manit. 269pp.

——, AND D. W. MILLER. 1988. POPAN-PC: installation and user's manual for running POPAN-3 on IBM PC microcomputers. Charles Babbage Res. Cent., St. Norbert, Manit. 17pp.

——, AND C. J. SCHWARZ. 1987. POPAN-3: extended analysis and testing features for POPAN-2. Charles Babbage Res. Cent., St. Norbert, Manit. 83pp.

BEST, A. M., AND D. MORGANSTEIN. 1991. Statistics programs designed for the Macintosh: Data Desk, Exstatix, Fastat, JMP, StatView II, and SuperANOVA. Am. Stat. 45:318–338.

BROWNIE, C., D. R. ANDERSON, K. P. BURNHAM, AND D. S. ROBSON. 1985. Statistical inference from band recovery data—a handbook. Second ed. U.S. Fish Wildl. Serv. Resour. Publ. 156. 305pp.

BUCKLAND, S. T. 1985. Perpendicular distance models for line transect sampling. Biometrics 41:177–195.

BURNHAM, K. P., D. R. ANDERSON, AND J. L. LAAKE. 1980. Estimation of density from line transect sampling of biological populations. Wildl. Monogr. 72. 202pp.

——, ——, G. C. WHITE, C. BROWNIE, AND K. H. POLLOCK. 1987. Design and analysis methods for fish survival experiments based on release-recapture. Am. Fish. Soc. Monogr. 5. 437pp.

CARLSON, J. D., JR, W. R. CLARK, AND E. E. KLAAS. 1993. A model of the productivity of the northern pintail. U.S. Fish Wildl. Serv. Biol. Rep. 7. 20pp.

CARPENTER, J., D. DELORIA, AND D. MORGANSTEIN. 1984. Statistical software for microcomputers. Byte 1984(April):234–264.

CONROY, M. J., J. E. HINES, AND B. K. WILLIAMS. 1989. Procedures for the analysis of band-recovery data and user instructions for program MULT. U.S. Fish Wildl. Serv. Resour. Publ. 175. 61pp.

COONEY, T. M. 1985. Portable data collectors, and how they're becoming useful. J. For. 83:18–23.

DALLAL, G. E. 1988. Statistical microcomputing—like it is. Am. Stat. 42:212–216.

DIXON, K. R., AND J. A. CHAPMAN. 1980. Harmonic mean measure of animal activity areas. Ecology 61:1040–1044.

DIXON, P. 1986. Choosing a statistical package for a microcomputer. Bull. Ecol. Soc. 67:290–292.

DIXON, W. J., EDITOR. 1983. BMDP® statistical software. Univ. California Press, Berkeley. 734pp.

DODGE, W. E., AND A. J. STEINER. 1986. XYLOG: a computer program for field processing locations of radio-tagged wildlife. U.S. Fish Wildl. Serv. Tech. Rep. 4. 22pp.

——, D. S. WILKIE, AND A. J. STEINER. 1986. UTMTEL: a laptop computer program for location of telemetry "finds" using LORAN C. Massachusetts Coop. Wild. Res. Unit, Amherst. 21pp.

DRUMMER, T. 1987. Program documentation and user's guide for SIZE-TRAN. Math. Sci. Tech. Rep. MS-TR 87-1. Michigan Tech. Univ., Houghton. 26pp.

——, AND L. L. MCDONALD. 1987. Size bias in line transect sampling. Biometrics 43:13–21.

DUNN, J. E. 1978. Computer programs for the analysis of radio telemetry data in the study of home range. Stat. Lab. Tech. Rep. 7. Univ. Arkansas, Fayetteville. 73pp.

——, AND P. S. GIPSON. 1977. Analysis of radio telemetry data in studies of home range. Biometrics 33:85–101.

FREDERICK, R. B., W. R. CLARK, AND E. E. KLAAS. 1987. Behavior, energetics, and management of refuging waterfowl: a simulation model. Wildl. Monogr. 96. 35pp.

——, ——, AND J. Y. TAKEKAWA. 1992. Application of a computer simulation model to migrating white-fronted geese in the Klamath Basin. Pages 696–706 in D. R. McCullough and R. H. Barrett, eds. Wildlife 2001: populations. Elsevier Publ. Ltd., Essex, U.K.

GATES, C. E. 1979. LINETRAN user's guide. Inst. Statistics, Texas A&M Univ., College Station. 47pp.

GEISSLER, P. H. 1988. Criteria for evaluating microcomputer statistical packages. U.S. Fish Wildl. Serv. Res. Inf. Bull. 88-15. 2pp.

GOLDSTEIN, R. 1992. Editor's notes. Am. Stat. 46:48–49.

GROSS, J. E., J. E. ROELLE, AND G. L. WILLIAMS. 1973. Program ONE-POP and information processor: a systems modeling and communications project. Colorado Coop. Wildl. Res. Unit Program Rep., Ft. Collins. 327pp.

HALLAHAN, C. 1992. DBMS/COPY and DBMS/COPY Plus (version 2.0). Am. Stat. 46:49–52.

HEISEY, D. M., AND T. K. FULLER. 1985. Evaluation of survival and cause-specific mortality rates using telemetry data. J. Wildl. Manage. 49:668–674.

HENSLER, G. L., S. S. KLUGMAN, AND M. R. FULLER. 1986. Portable microcomputers for field collection of animal behavior data. Wildl. Soc. Bull. 14:189–192.

——, AND J. D. NICHOLS. 1981. The Mayfield method of estimating nesting success: a model, estimators and simulation results. Wilson Bull. 93:42–53.

HILL, H. S., AND T. T. FENDLEY. 1982. Animal movement analysis and home range determination package. Proc. Annu. Conf. Southeast. Assoc. Fish Wildl. Agencies 36:656–663.

HOBBS, N. T. 1988. Notebook computers in biological research: less technology, more productivity. Science Software 4:14–16.

———. 1989. Linking energy balance to survival of mule deer: development and test of a simulation model. Wildl. Monogr. 101. 39pp.

JENNRICH, R. I., AND F. B. TURNER. 1969. Measurement of non-circular home range. J. Theor. Biol. 22:227–237.

JOHNSON, D. H., D. W. SPARLING, AND L. M. COWARDIN. 1987. A model of the productivity of the mallard duck. Ecol. Model. 38: 257–275.

KOELN, G. T. 1980. A computer technique for analyzing radio-telemetry data. Pages 262–271 in J. M. Sweeney, ed. Proc. National Wild Turkey Symp. 4.

KROL, E. 1992. The whole Internet: user's guide and catalog. O'Reilly & Assoc., Sebastopol, Calif. 376pp.

LAQUEY, T., AND J. C. RYER. 1993. The Internet companion: a beginner's guide to global networking. Addison-Wesley, Reading, Mass. 196pp.

LENTH, R. V. 1981. On finding the source of a signal. Technometrics 23:149–154.

MOHR, C. O. 1947. Table of equivalent populations of North American small mammals. Am. Midl. Nat. 37:223–249.

MOLINI, J. J., R. A. LANCIA, J. BISHIR, AND H. E. HODGDON. 1981. A stochastic model of beaver population growth. Pages 1215–1245 in J. A. Chapman and D. Pursley, eds. Proc. Worldwide Furbearer Conf., Frostburg, Md.

NORUŠIS, M. J. 1986. SPSS/PC+™ for the IBM PC/XT/AT. SPSS Inc., Chicago, Ill. Var. pagin.

OTIS, D. L., K. P. BURNHAM, G. C. WHITE, AND D. R. ANDERSON. 1978. Statistical inference from capture data on closed animal populations. Wildl. Monogr. 62. 135pp.

POLLOCK, K. H., J. D. NICHOLS, C. BROWNIE, AND J. HINES. 1990. Statistical inference for capture-recapture experiments. Wildl. Monogr. 107. 97pp.

———, S. R. WINTERSTEIN, AND M. J. CONROY. 1989a. Estimation and analysis of survival distributions for radio-tagged animals. Biometrics 45:99–109.

———, ———, C. M. BUNCK, AND P. D. CURTIS. 1989b. Survival analysis in telemetry studies: the staggered entry design. J. Wildl. Manage. 53:7–15.

REXSTAD, E., AND K. BURNHAM. 1991. User's guide for interactive program CAPTURE. Colorado Coop. Fish Wildl. Res. Unit, Ft. Collins. 29pp.

SAMUEL, M. D. 1988. New feature for the *Wildlife Society Bulletin*. Wildl. Soc. Bull. 16:104.

———, AND E. O. GARTON. 1985. Home range: a weighted normal estimate and tests of underlying assumptions. J. Wildl. Manage. 49: 513–519.

———, D. J. PIERCE, E. O. GARTON, L. J. NELSON, AND K. R. DIXON. 1985. User's manual for program HOME RANGE. Second ed. For. Wildl. Range Exp. Stn. Tech. Rep. 15. Univ. Idaho, Moscow. 70pp.

SAS INSTITUTE INC. 1985a. SAS™ language guide for personal computers. Version 6 ed. SAS Inst., Inc., Cary, N.C. 429pp.

———. 1985b. SAS/IML™ user's guide for personal computers. Version 6 ed. SAS Inst., Inc., Cary, N.C. 243pp.

———. 1987. SAS/STAT™ guide for personal computers. Version 6 ed. SAS Inst., Inc., Cary, N.C. 1028pp.

———. 1988. SAS™ technical report P-179, additional SAS/STAT™ procedures. Release 6.03. SAS Inst., Inc., Cary, N.C. 255pp.

SEBER, G. A. F. 1982. The estimation of animal abundance and related parameters. Second ed. Griffin, London, U.K. 654pp.

SIPPL, C. J. 1985. MacMillan dictionary of microcomputing. MacMillan Press, London, U.K. 473pp.

STARFIELD, A. M., AND A. L. BLELOCH. 1986. Building models for conservation and wildlife management. Macmillan Publ. Co., New York, N.Y. 253pp.

TAKEKAWA, J. Y. 1987. Energetics of canvasbacks staging on an Upper Mississippi River pool during fall migration. Ph.D. Thesis, Iowa State Univ., Ames. 189pp.

TRENT, T. T., AND O. J. RONGSTAD. 1974. Home range and survival of cottontail rabbits in southwestern Wisconsin. J. Wildl. Manage. 38:459–472.

UNWIN, D. M., AND P. MARTIN. 1987. Recording behaviour using a portable microcomputer. Behaviour 101:87–100.

U.S. FISH AND WILDLIFE SERVICE. 1980a. Habitat as a basis for environmental assessment. Ecol. Serv. Man. 101. U.S. Fish Wildl. Serv., Washington, D.C. 28pp.

———. 1980b. Habitat evaluation procedures (HEP). Ecol. Serv. Man. 102. U.S. Fish Wildl. Serv., Washington, D.C. 84pp.

———. 1981. Standards for the development of suitability index models. Ecol. Serv. Man. 103. U.S. Fish Wildl. Serv., Washington, D.C. 68pp.

WAYNER, P. 1992. Ample waves of data: five tools to help you stay afloat. Byte 17:259–270.

WHITE, G. C. 1983. Numerical estimation of survival rates from band-recovery and biotelemetry data. J. Wildl. Manage. 47:716–728.

———, D. R. ANDERSON, K. P. BURNHAM, AND D. L. OTIS. 1982. Capture-recapture and removal methods for sampling closed populations. LA-8787-NERP, Los Alamos Natl. Lab., Los Alamos, N.M. 235pp.

———, R. M. BARTMANN, L. H. CARPENTER, AND R. A. GARROTT. 1989. Evaluation of aerial line transects for estimating mule deer densities. J. Wildl. Manage. 53:625–635.

———, AND R. A. GARROTT. 1984. Portable computer system for field processing biotelemetry triangulation data. Colo. Div. Wildl. Game Inf. Leafl. 110. 4pp.

———, AND ———. 1990. Analysis of wildlife radio-tracking data. Academic Press, New York, N.Y. 383pp.

WILSON, K. R., AND D. R. ANDERSON. 1985a. Evaluation of a density estimator based on a trapping web and distance sampling theory. Ecology 66:1185–1194.

———, AND ———. 1985b. Evaluation of two density estimators of small mammal population size. J. Mammal. 66:13–21.

GLOSSARY OF TERMS

Many definitions were adapted from Sippl (1985).

analog
: analog computers work on the physical quantity of voltage instead of numbers; analog data are continuously variable rather than discrete.

applications program
: any program that does actual problem solving.

ASCII
: American Standard Code for Information Interchange; a standard data-transmission code that was created to achieve compatibility among data devices.

BASIC
: an acronym for Beginners All-purpose Symbolic Instruction Code, the simple language most widely used on microcomputers.

batch file
: a batch or sequence of DOS commands placed in a file to be executed (e.g., AUTOEXEC.BAT).

bit
: binary digit, the basic unit of memory, which is either on or off.

board
: a modular computer component containing electronic circuits that is specifically designed to control a device such as a hard disk or monitor.

boot
: to pull the computer operating system up by its "bootstraps," from the ROM bootstrap loader.

byte
: a cell in memory that can store 8 bits of memory, approximately equivalent to one character.

CGA
: Color Graphics Adapter; a monitor and associated board that display colors.

chip
: a microchip or integrated circuit, a solid-state device containing many circuit elements formed on a single chip of semiconductor material.

clock rate
: the rate at which a word or parts of characters (bits) are transferred from one internal element to another, measured in MHz.

code
: a system of characters and rules for representing information; most often used in reference to a specific computer program.

command
: an instruction to DOS entered at the prompt.

command file
: file of binary code executable under DOS.

compatible
: capability of direct interconnection or translation; usually implies that requirements for code, or physical characteristics like speed, have been matched.

coprocessor
: a specialized microchip designed to do floating-point arithmetic.

CPU
: Central Processing Unit; conducts and controls the flow of information through the computer; it includes the microchips and associated memory.

database
: a collection of data stored in standard format that can be efficiently searched or sorted.

digital
: using numbers 0–9 to represent all the variables involved in a calculation.

digitizer
: a device designed to convert analog signals, such as a line traced on a pad, into digital data.

documentation
: orderly presentation and communication accompanying computer software.

DOS
: Disk Operating System; originally developed by IBM Corporation in cooperation with Microsoft Corporation.

dpi
: dots per inch.

EGA
: Enhanced Graphics Adapter; see CGA.

executable file
: a binary file of machine instructions that is loaded into memory to be executed.

extension
: the appended part of a file name (i.e., .BAT); it may be supplied by the user or by default by the software.

file
: a working compilation of data, such as a word processor document, that is stored and cataloged by name.

floating-point arithmetic
: arithmetic in which the location of the decimal point for each number is defined prior to the operation; using floating-point arithmetic makes programming complex arithmetic operations easier.

floppy disk
: a storage device made of a flexible material somewhat like magnetic tape.

FORTRAN
: acronym for FORmula TRANslator, a common scientific programming language.

Gb
: gigabyte; approximately 1 billion bytes, see Kb.

GIS
: Geographic Information System.

GUI
: Graphical User Interface, like Microsoft Windows.

hard disk
: a rigid storage medium, contained in a protected space in the computer, that can be rotated rapidly for fast access to data; sometimes called a Winchester disk.

hardware
: mechanical, magnetic, electronic, and electrical devices that form a microcomputer system.

input	an adjective referring to a device that transmits data to a computer or a verb referring to the process of transmission.
interactive input	input process in which the user is prompted for responses or selects choices from a menu.
Kb	kilobyte, approximately 1,000 bytes; actually 1,024 bytes, which is a round number in binary notation; 64Kb is 64 units of 1,024, or exactly 65,536.
LIM	Lotus-Intel-Microsoft protocol for converting extended memory into expanded, addressable memory.
Mb	megabyte, approximately 1 million bytes.
memory	the internal hardware in the computer that stores information for future use.
menu	a list, usually displayed graphically and often accessed with a mouse pointer, from which commands or functions may be selected.
MHz	a measure of frequency, 1 million cycles per second.
multitasking	often referring to operating systems designed to execute more than one task simultaneously.
network	a system of interconnected PCs or workstations.
online	equipment, devices, or software in direct interactive communication with the CPU.
output	an adjective referring to a device that displays data transmitted from a computer or a verb referring to the process of transmission from a computer.
parallel	processing or transmitting digits on a separate channel or line simultaneously.
pixel	picture element.
public domain	a term describing software available at no cost to the user because it was developed with funding from a government agency, or because the developer decided not to charge for the software. Another slang term for free software is freeware.
RAM	Random Access Memory, i.e., the memory available to the computer's CPU, whose contents can be read or written on regardless of memory location; DOS can access only 640Kb.
ROM	Read Only Memory; this is programmed during manufacture and cannot be changed.
serial	a method of data transfer between a computer and a peripheral device in which data are transmitted bit by bit over a single circuit.
shareware software	programs and computer code that work with the hardware of a computer.
slot	a plug in the circuit board of the computer into which various hardware boards can be connected.
Unix	an operating system, originally developed by AT&T Corporation, that is especially well-suited for multitasking and communications.
VGA	Video Graphics Array.
window	a portion of the video display that simultaneously shows a menu or operating software over other video output.
word processor	a software program designed to permit original entry, editing, and printing of text.
workstation	a specialized microcomputer or terminal connected within a network; GIS applications are most frequently conducted from graphics workstations.
WYSIWYG	"What you see is what you get"; term used to describe computer software that displays an image on the screen that will be identical (or nearly so) to what will be produced when the image is printed on paper.

4

GUIDELINES FOR PROPER CARE AND USE OF WILDLIFE IN FIELD RESEARCH

Milton Friend, Dale E. Toweill, Robert L. Brownell, Jr., Victor F. Nettles, Donald S. Davis, and William J. Foreyt

INTRODUCTION

Philosophy

Scientists do not operate in a vacuum, but rather in an arena with responsibilities to the organisms they study and to society. Professional scientists must consider the effects of their activities on the organisms under study, on the validity of study results, and on the use of these organisms by other segments of society. The Wildlife Society recognizes these relationships and supports the sound application of responsible methods for the conduct of animal research in all field and laboratory investigations. This position reflects our ethical and moral concerns regarding human interactions with each other and with other species, and recognizes the scientific benefits of investigations that are not compromised by the manner in which animals are handled or maintained. These concerns are the foundation for our philosophy that responsible methods of animal investigations must include all animal species. Wildlife professionals are urged to apply high standards of animal care and maintenance, and responsible methods of experimental procedures, in conducting each animal investigation.

Purpose

These guidelines are intended for field research involving wild animals. The variety of wild vertebrates investigated and of conditions encountered precludes provision of specific information applicable to each situation. Lists of useful references for those seeking more specific information are provided in the Appendices.

BACKGROUND

The Animal Welfare Act (7 U.S.C. 2131, et seq.) was enacted on 23 December 1985, with amendments including Parts 1, 2, and 3 (9CFR); *Fed. Register* 4(168) 36112–36163, effective 30 October 1989. The Act established definitions of terms (Part 1) used in the regulations (Part 2) and standards (Part 3) for the humane handling, care, treatment, and transportation of regulated animals used for research or exhibition purposes, sold as pets, or transported in commerce. Excluded from the provisions of the Act are cold-blooded vertebrates, birds, rats (*Rattus*) and mice (*Mus*) bred for use in research, horses and other farm animals used or intended for use as food and fiber, and livestock and poultry used or intended for use in improving animal nutrition, breeding, management, or production efficiency, or for improving the quality of food or fiber. Also excluded are field studies as defined by the Act, i.e., "any study conducted on free-living wild animals in their natural habitat, which does not involve an invasive procedure, and which does not harm or materially alter the behavior of the animals under study." Collection of blood samples, ear-notching, branding, and collection of routine weight and measurement data are examples of exempted activities.

Exclusion of animal species under the Act removes reporting requirements and reduces oversight by the U.S. Department of Agriculture, but does not negate coverage of these species under guidelines established by other agencies. Thus, fish, amphibians, reptiles, birds, and mammals are covered by the National Science Foundation (NSF) and the National Institutes of Health (NIH) guidelines. This coverage is extended to research grants funded by these agencies and to federal agencies, such as the U.S. Fish and Wildlife Service, that function under the guidelines of the Interagency Research Animal Care Committee.

ROLE OF INSTITUTIONAL ANIMAL CARE AND USE COMMITTEES

A major requirement of the Animal Welfare Act and NIH/NSF guidelines is establishment of institutional/facility Animal Care and Use Committees (ACUCs). The function of ACUCs is critical to the conduct of scientific investigations. Each ACUC must consist of at least three members, one of whom is the attending veterinarian of the research facility (or another veterinarian with delegated program responsibility) and one of whom is not affiliated in any way with the facility other than as a committee member. The purpose of the ACUC is to evaluate the care, treatment, housing, and use of animals and to certify compliance with the Act. This process involves evaluation of experimental protocols to ensure that animal pain and distress are minimized. ACUC oversight includes laboratory and field studies. Consensus recommendations on effective ACUCs for laboratory animals were provided by Orlans et al. (1987). Differences between laboratory and field studies (Orlans 1988) do not negate the need for application of responsible methods for care and use of animals during field research activities. ACUCs and field investigators must work together in reaching agreement on appropriate protocols and methods for specific circumstances of the field research to be undertaken. "Standards for humane treatment of wild vertebrates must continue to be constantly developed, applied, and re-examined. Practices that are acceptable today may well prove unacceptable to tomorrow's scientific community, and/or to society in general" (Canadian Council on Animal Care 1984:192). Wildlife professionals are strongly encouraged to serve on ACUCs and contribute their specific knowledge about the needs of free-living wildlife to help guide Committee actions involving protocol reviews for field investigations. Wildlife professionals also are encouraged to publish manuscripts that document the proper care and maintenance of free-living wildlife species during field investigations. Development of this information by knowledgeable field biologists provides specific species information for guiding ACUC decisions involving protocol reviews.

Irrespective of the species or circumstances involved, wildlife professionals should satisfy the following conditions for all field research studies. Written assurance that these conditions will be met is a prerequisite for project consideration and funding by many granting agencies. These conditions also are principal points for evaluation by the ACUC.

1. Procedures employed should avoid or minimize distress to animals consistent with sound research design.
2. Procedures that may cause more than momentary or slight distress to animals should be performed with appropriate sedation, analgesia, or anesthesia, except when justified for scientific reasons in writing by the investigator in advance.
3. Animals that otherwise would experience severe or chronic distress that cannot be relieved will be euthanized at the end of the procedure or, if appropriate, during the procedure.
4. Methods of euthanasia will be consistent with recommendations of the American Veterinary Medical Association (AVMA) Panel on Euthanasia (Andrews et al. 1993) unless deviation is justified for scientific reasons in writing by the investigator. However, species differences must be considered. As noted elsewhere, "The AVMA recommendations cannot be taken rigidly for ectotherms; the methods suggested for endotherms are often not applicable to ectotherms with significant anaerobic capacities" (American Society of Ichthyologists and Herpetologists [ASIH], the Herpetologists' League [HL], and the Society for the Study of Amphibians and Reptiles [SSAR] 1987:2).
5. Living conditions of animals held in captivity at field sites should be appropriate for that species and contribute to their health and well-being. Specific considerations include appropriate standards of hygiene, nutrition, group composition and numbers, provisions for refuge and seclusion, and protection from weather and other forms of environmental stress. The housing, feeding, and nonmedical care of these animals must be directed by a scientist trained and experienced in the proper care, handling, and use of the species being maintained or studied. Some experiments (e.g., competition studies) will require the housing of mixed species, possibly in the same enclosure. Mixed housing also is appropriate for holding or displaying certain species.

WILDLIFE OBSERVATIONS AND COLLECTIONS
General

Before initiating field research, investigators must be familiar with the target species and its response to disturbance, sensitivity to capture and restraint, and, if necessary, requirements for captive maintenance to the extent that these factors are known and applicable.

To the extent feasible, animals with dependent young should not be removed from the wild unless the young also are collected or removed alive and provided for in a manner that facilitates their survival beyond the period of dependency. Whenever possible, voucher specimens of animals, their tissues, and parasitic and microbial fauna collected during field investigations should be deposited in catalogued scientific collections available to others within the scientific community, to provide for maximum use of animals collected.

The number of animals required for investigations depends on questions being investigated, but provision of adequate sample size is essential to assure scientific validity of results and avoid unnecessary repetition of studies. Removal of animals from a population (either for translocation or by lethal means) should be restricted to the fewest animals necessary to achieve established goals, but should never jeopardize the population's well-being.

Investigator Disturbance and Impacts

Potential gains in knowledge from field investigations must be balanced against the potential adverse consequences associated with the conduct of the study (Animal Behavior Society/Animal Society for Animal Behavior 1986). A high level of sensitivity to the potential, indirect effects of investigator presence and study procedures must be maintained, and appropriate steps must be taken to minimize these effects. Examples of secondary impacts associated with field investigations may include nest desertion, abandonment of young, increased vulnerability to predation, traumatic injuries and mortality resulting from panic escape response, cessation of breeding activities, increased energy use by disrupted species, altered feeding behavior, habitat abandonment, long-term marring of fragile habitats, increased vulnerability to hunting, introduction of disease, and spread of disease. These effects may impact either research (target) or other (nontarget) species. Investigators should use available information on secondary impacts as a basis for taking appropriate precautions to minimize known potential impacts.

Such factors as frequency and timing of investigator presence can influence greatly research effects on target and nontarget species. When applicable, remote methods of data collection can be used to minimize disturbance. Also, habitat conservation should be practiced rigorously during all field investigations, and every reasonable effort should be made to leave the study area and access to it as undisturbed as possible.

Museum Collections and Other Killed Specimens

Collection of animals often is an essential component of field investigations. These collections may involve systematic zoology, comparative anatomy, disease assessments, food preference studies, environmental contaminant evaluations, and numerous other justifiable causes and scientific needs.

Assessment of the need should involve appropriate evaluations to determine that the proposed collections will provide scientific data that are not duplicative of information already available in the scientific literature (unless confirmation of these data is needed), or that are presently available in accessible scientific collections and repositories. These evaluations also should assess whether suitable information can be obtained from alternative methods that do not require taking live animals. Methods of collection must be responsible, minimize the potential for the taking of nontarget species, and not compromise the purpose of the study. In some instances it is possible and practical to capture animals and then apply approved euthanasia methods (see Andrews et al. 1993). However, for many field studies the only practical means of animal collection are those involving direct killing as the initial step in the collection process. Under these conditions, methods of vertebrate collection must be as species or age-class specific as possible. Methods must not be employed that compromise data evaluation. Appropriate provisions also must be made for proper collection and preservation of biological materials associated with the purpose of the study. Improperly collected or preserved specimens that fail as useful and valid sources of scientific information negate the purpose of collecting the animals.

When shooting is the collection method, the firearm and ammunition should be appropriate for the species and purpose of the study. The shooter should be sufficiently skilled to be able to kill the animal cleanly. If an animal is wounded, immediate attention must be given to appropriate follow-up actions to kill it quickly. Attention also must be given to the animal's location to assure it can be killed cleanly and that it will be readily accessible for retrieval and data collection.

Kill traps, with attendant baits and attractants, are acceptable and effective for animal collection when used in a manner that minimizes the potential for collecting nontarget species. All traps should be checked regularly, at least daily, to prevent specimen loss from scavengers and predators and should be rendered nonfunctional when not in use.

Live traps for nocturnal species should be set before dusk, checked as soon as possible after dawn, and closed during the day to prevent capture of nontarget species. Live traps for diurnal species should be shaded or positioned to avoid full exposure to the sun. Live traps for nonfossorial mammals should enclose a volume of space adequate for movement within the trap; for fossorial mammals, trap diameter should approximate that of the burrow. The live-trap mechanism should not cause serious injury to the animal, and trap doors should be effective in preventing the captive animal from becoming stuck or partially held in the door opening (Ad Hoc Committee on Acceptable Field Methods in Mammalogy 1987). Pitfalls used as live traps should contain adequate food to last until the next trap check and should be covered to keep out rain or punctured to permit drainage.

Blood and Tissue Collections

Only properly trained individuals proficient in the required techniques should attempt to take tissue samples from live animals. Collection of tissue samples requires proper animal restraint to avoid traumatic injuries to the animal and to the investigator taking the samples. Use of anesthetics is required when the sample procedure will cause more than slight or momentary pain. The institution/facility ACUC is the proper source for evaluating collection methods and use of anesthetics for noninvasive and invasive procedures for tissue collections from live animals.

Blood is the most common tissue sampled from live animals. A conservative rule of thumb is that the amount of blood drawn at one time from a healthy animal that is to be kept alive should be no more than 1% of its body weight. However, the amount of blood taken should be limited to actual needs, rather than the maximum amount that can be safely taken, to reduce stress from handling. Appropriate equipment (e.g., needle size) and sample site should be selected to provide the amount of blood needed for the species involved.

The three most common sites for bleeding birds are the right jugular vein of the neck, medial-metatarsal vein of the leg, and brachial vein of the wing. The jugular is preferred for bleeding most birds because of its accessibility and size and because large samples can be taken relatively easily. The medial-metatarsal vein is not recommended for use in raptors, nor is the brachial vein in large birds such as cranes. Feathers should not be plucked to locate these veins. Birds also can be bled from a variety of other sites including the heart and occipital venous sinus. However, there is seldom reason to assume the risk associated with these sites for nonlethal sampling, even though successful application of these techniques has been demonstrated.

Multiple sites also are available for drawing blood samples from mammals. Venipuncture of the cephalic, femoral, or jugular vein, the orbital sinus, or various venous plexuses are common procedures. In some instances cardiac bleeding also is acceptable. Need for anesthesia with any of these procedures depends upon methods of restraint, species being bled, physical condition of the animal, and volume of blood needed.

RESTRAINT AND HANDLING
General

Safety of both wild animals and scientists who are studying them should be the primary consideration when physical contact between them is judged to be necessary and unavoidable. Nondomesticated animals almost without exception will try to elude capture, handling, and restraint. The means by which a particular animal may try to prevent capture will vary with the species, sex, physiologic condition, and temperament of the individual. In attempts to elude capture, wild animals are capable of inflicting severe damage to themselves and their potential captors.

Behavioral characteristics of wild animals often may be used to assist the potential captor. For instance, animals in a small pen or cage often voluntarily will enter a smaller container to hide and evade capture. If that container provides adequate restraint, the potentially dangerous work of securing the animal can be accomplished more easily. Every effort involving contact between wild animals and humans should be carefully conceived and skillfully executed. Personnel involved must know the habits and behaviors of the animal to be handled; the plan must have suitable alternatives; and a genuine regard for the physical, physiological, and psychological welfare of the animal must be of deep concern to those actually handling the animals. If the planned and alternate procedures do not appear to be satisfactory, the responsible thing to do is cease immediately and return to the planning stage. Trying to enforce unworkable procedures in a particular situation is a virtual guarantee of injury to either the animals or the humans involved.

Physical Restraint

For many situations physical restraint is the most appropriate method of animal handling, because of risks from chemical immobilization to the animal and humans when potentially toxic drugs are used. When physical restraint is selected, an adequate number of sufficiently trained and equipped personnel must be available to complete the task safely. Location and type of capture, as well as procedures to be performed and time required to accomplish them, will influence the particular type of physical restraint. Gloves, catch poles, ropes, nets, body bags, holding boxes, corrals, squeeze chutes, or more sophisticated mechanical holding devices may be required for specific situations.

For some highly excitable or anatomically fragile species, prolonged physical restraint without some chemical tranquilization may result in self-inflicted trauma, physiological disturbances, or, occasionally, death. Investigators have an obligation to make every effort to avoid physical restraint procedures that result in cardiogenic shock, capture myopathy, and other stress-induced causes of mortality in their animal subjects. Stress-related damage may not be immediately apparent but may lead to debility or death after release.

Chemical Restraint

Use of chemicals or drugs to render a wild and potentially dangerous animal safe to handle has many applications in wildlife research and management (see Chapter 6). Use of anesthetics, analgesics, and sedatives is mandatory for the control of pain and distress before potentially painful procedures such as surgery are performed on animals. Use of drugs and "tranquilizer guns," however, is not the panacea to wild-animal restraint. Chemicals used for tranquilization and immobilization, if not correctly handled and delivered, may be dangerous to the target animals and humans. In addition, during the drug induction phase or during recovery, an unrestrained animal may be subject to increased potential for accidental injury or death including predation. While under the effects of the drug the animal may become hyper- or hypothermic, depending on chemicals used and ambient temperature, it may vomit and aspirate the vomitus, or pregnant females may abort. A darted animal may be able to elude its captors and hide before being completely anesthetized, a particularly acute hazard when chemicals are employed that require administration of an antidote. All of these circumstances and possibilities must be understood and evaluated by the researcher before a chemical is selected as the best method of restraint in a given instance.

If chemical restraint is selected, it is imperative for all members of the capture team to have a working knowledge of the chemical or drugs being used, even if they are to be handled and delivered by a veterinarian. It also is the responsibility of researchers to know the effects, side effects, advantages, and disadvantages of the drugs being used, and to have knowledge of such factors as the minimum and maximum induction times and potential for adverse drug reactions. This type of information is necessary to evaluate the danger to target animals, and to humans that might be exposed to the drugs. Researchers should be capable of monitoring the condition of anesthetized animals and be able to apply resuscitative routines in a life-threatening emergency. Specific recommendations for drug use and their dosage, drug delivery systems, and physical restraint techniques applicable to the specific species are available in the published literature (see also Chapter 6). Information on use of these methods exists in guidelines on acceptable field techniques by various professional societies (Appendix I).

ANIMAL MARKING

Developing means of reliably identifying individual animals to achieve field research objectives often is necessary. In addition to requiring individual identification, researchers may need information on nonconspicuous aspects of physiology or movements, or other aspects of animal ecology that can be determined directly or indirectly through specially designed markers. However, before initiating any marking procedure for wild animals, researchers must resolve the following questions to determine whether marking is required and appropriate for the particular situation.

(1) Do naturally occurring differences in the morphology of the animals under consideration provide sufficient identification to achieve research objectives?
(2) How many animals must be individually identifiable?
(3) If animals must be physically marked, can a sufficient number of animals be marked in the time available?
(4) Are the risks (to both the animal and researcher) associated with capture, handling, and marking, and subsequent well-being, minimal and acceptable in both responsible and scientific contexts?

If the marking process causes pain or distress, as defined by the Animal Welfare Act, appropriate analgesics or anesthetics should be used.

Criteria for Marking

When answers to the four initial questions lead to a decision to initiate an animal-marking program, researchers must search among a wide array of potential techniques with varying strengths and weaknesses to select the method(s) most suited to their particular project (see Chapter 7 for details of marking techniques). Technological and methodological constraints and available resources can vary widely from project to project and will require each researcher to examine each potential marking technique in terms of a standard set of criteria. Specific criteria relate to impacts of marking on the organism, validity of the study, and other constraints such as legal requirements. The following are essential criteria for evaluation:

(1) Marks should have minimal effect on the anatomy and physiology of the organism, i.e., no immediate or long-term physical hindrance.
(2) Marks should not influence the organism's behavior, i.e., they should not reduce an organism's ability to secure food or inhibit breeding activity (unless the marks are intended as a reproductive inhibitor).
(3) Marks that make an organism more conspicuous must be evaluated carefully to ensure that they neither cause others of the same species to react differently to it than to other conspecifics nor subject it to increased selection by potential predators (unless this is a purpose of the study).

(4) Marks should be retained for the minimal period required to achieve project goals.
(5) Unambiguous marks that are quick and easy to apply should be selected to avoid extensive handling or error potential.
(6) Marks must comply with federal, state, and other agency rules and regulations.

The first three of these criteria focus on the well-being of the organism being studied and the potential for marks to influence research results by affecting the fitness or behavior of the organisms. Criteria 4 and 5 may affect the validity of the research design, and criterion 6 reflects other constraints placed upon the researcher. Violation of any of the first five criteria may result in biased research results, so researchers should specifically address these criteria in any evaluation of research resulting from a sample of marked organisms.

Although marks that may be applied to organisms are commonly perceived as passive and visual, markers also exist that are active and visual (lights), that are auditory, that feature radiotelemetry, or that rely on chemical detection. A vast literature exists of techniques and potential concerns regarding the marking of organisms from insects to whales, and it has been summarized in detail elsewhere (see Appendix I, Day et al. 1980, Orlans 1988).

Other Professional and Ethical Considerations

Many organisms of interest to wildlife professionals are free-ranging and may be enjoyed by other segments of society in many ways, from observation or photography to harvest as meat or trophies. Professional ethics dictate that those other potential uses of organisms be considered and accommodated insofar as possible. Wild animals and birds are valued in part *because* they are wild, and the presence of human-caused marks may detract from that value. Accordingly, short-lived and inconspicuous marks should be selected whenever they can meet the objectives of proposed research. Scientists have an ethical responsibility to attempt to remove collars or other external markers at the conclusion of their research if possible and feasible. Furthermore, professional and ethical considerations dictate that permanent markers that injure or change the appearance of an animal (e.g., toe-clipping, branding, and tattooing) be employed only under the most humane conditions and when alternate methods are not available to achieve desired research objectives.

HOUSING AND MAINTENANCE OF FIELD SITES
General

Proper care and responsible treatment of incarcerated animals must depend on scientific and professional judgement, on concern for the animal, on knowledge of animal behavior and animal husbandry, and on familiarity with the species. Investigators working with species unfamiliar to them should obtain all pertinent information before confining those animals. It also may be necessary to test and compare several methods of housing to determine the most appropriate one for the well-being of the animal and the purpose of the study. Findings should be part of a permanent record system and animal logbook associated with the study and the maintenance facility.

Housing

Housing for wild vertebrates should approximate natural conditions as closely as possible. Housing should provide safety and comfort for the animal as well as meet the study objectives. Method of housing should provide for behavioral needs, safety, adequate exercise and rest, and conditions for the general well-being of the animal. Considerations depend on the animal involved and include isolation or refuge areas, natural materials, dust and water baths, natural foods, sunlight, and fresh air. Housing should incorporate as many aspects of natural living as possible, such as brushy areas for escape, resting cover, shade and protection from environmental elements, a natural stream traversing the pen, rocky areas for hoofed animals that need to wear down their hooves, and social groups of animals kept together. Housing of compatible species in a common pen also will provide for social interaction. Frequency of cleaning should be a compromise between level of cleanliness necessary to prevent disease and amount of stress imposed by cleaning.

In general, housing must be of adequate size to allow for the physical and behavioral needs of the animals, while allowing scientists to collect appropriate data. For many housing situations, the pen can be large and natural, with a smaller internal or attached catch pen to restrain animals for experimental techniques. Pen construction materials must provide for the safety of the animals, as well as prevent the animals from escaping. Materials should be of sufficient durability to last for the intended period of confinement. When long-term confinement (weeks or longer) is necessary, or pens are to be reused, materials with impervious surfaces should be used to facilitate sanitation and minimize the potential for survival of animal pathogens. All animals that are inherently dangerous, are environmentally injurious, or have a propensity for escape require special attention. Double walls or double enclosures, covered tops of enclosures, and construction with metal bars or chain link may be required, depending on the species. Mesh size and spacing between fencing materials must be small enough to prevent the head of an animal from extending through the fence. Smaller fencing mesh also is more visible to animals. Colored flagging material may be necessary for animals to visualize fencing until they become accustomed to it. Animals should be released into the housing in a calm and unstressed manner so that initial mortality and morbidity from fence encounters are minimal. A small dose of tranquilizer often will reduce the immediate flight response when an animal is released into the housing and may help prevent initial injuries. Once animals have investigated the limits of the housing, injury occurrence is minimized if investigators do not cause undo flight reactions.

Adequacy of housing often can be judged on normal behavior patterns, weight gains and growth, survival rates, reproductive success, and physical appearance of the animals involved in the research project. Established guidelines for housing laboratory and farm animals were provided by the Canadian Council on Animal Care (1980, 1984). Additional guidelines for housing requirements of fish, amphibians, reptiles, wild birds, and small mammals were reported by the appropriate professional societies and appear in the Animal Welfare Act (see also Appendix I).

Nutrition

Nutrition must meet the needs of the animal unless deviations are an approved purpose of the investigation. Researchers are responsible for determining the appropriate nutritional needs of study animals prior to placing them in confinement and for obtaining adequate food supplies to sustain the animals during the period of confinement. Feeding and watering should be under the direct supervision of an individual trained and experienced in animal care for the species being maintained. Animal care personnel must be familiar with the animals being studied so abnormalities in appearance and behavior that may be indicative of nutritional deficiencies can be recognized quickly.

TRANSPORTATION
General Considerations

A variety of vehicles such as conventional motor vehicles, all-terrain vehicles, snow machines, rotary and fixed-wing aircraft, and boats is used to transport wild animals. The species involved, method of transportation selected, and length of time an animal is to be transported are important factors regarding the type of care and conditions of containment required to maintain the animal in a state of well-being. To the extent possible, selection of transportation vehicles should take into account maintenance of the animal in a comfortable environment. Veterinary assistance may be required to prescribe and administer appropriate tranquilizers or other drugs when conditions of transportation are likely to result in a high level of stress to the animal due to its behavioral and physiological characteristics, restrictions of confinement, engine noise, and rigors of the trip. The transportation process should be as brief as possible. This can be expedited by proper and adequate planning to assure that transportation vehicles and housing units in appropriate numbers and size are available and ready for use as needed; that food, water, bedding, and other needs to provide for the animals also are available; that individuals involved in the transportation process are trained in the procedures to be used in containment and transportation of the animals; and that all permits, health certificates, and other paperwork have been completed to the extent possible.

When interstate movement of animals or shipment by commercial carriers is involved, scheduling of transportation segments to minimize the number of transfers and delays between transfers, having someone involved with the project meet the shipment at each transfer point, and, when appropriate, arranging for prompt clearance of animals by veterinary and customs inspectors can result in major reductions in transit time. The receiving party should be on-site when the animals reach their destination.

For some species, periodic rest periods are required to allow the animals to feed undisturbed. Other species are best transported when they are normally inactive and do not feed. Ventilation within the housing unit and transportation vehicle should provide for adequate air movement to keep animals comfortable and avoid buildup of exhaust gases. Subdued lighting and visual barriers between animals and humans and between animals and their transportation environment should be provided to help keep the animals calm. The U.S. Fish and Wildlife Service has published rules for the *Humane and Healthful Transport of Wild Animals and Birds to the United States* (see Fed. Regul. 50 CFR Part 14).

Confinement During Shipping

Animal containers should be inspected to assure they have no sharp edges, protrusions, or rough surfaces that could cause injury during transport. When appropriate, containers also should be padded to help prevent injury. The floor of shipping containers should allow reasonable footing to prevent falling due to a slippery surface. Also, containers should not have coatings or be constructed of materials that are toxic and could be consumed by the animal through licking or chewing during transportation. In general, housing units of porous materials, such as cardboard boxes, should not be reused; all other containers used to house animals should be suitably disinfected between uses. That portion of the transportation vehicle used to contain the housing units also should be disinfected.

Grouping or separation of animals being transported at the same time should take into consideration the species, age, and other appropriate factors. Direct contact generally should be maintained between females and their dependent young, particularly if abandonment may result (unless the young are to be maintained by some other means). Birds should be isolated in separate cells within the shipping container; if this cannot be done, each individual should have sufficient space to assume normal postures and engage in comfort and maintenance activities unimpeded by other birds (Ad Hoc Committee on the Use of Wild Birds in Research 1988).

Health Aspects

For short-term transportation (<30 min), basic considerations are to prevent pain, injury, and undue stress. Thermoregulation capabilities of the species must be considered when an animal is removed from its existing environment and placed in the transportation environment. Transported animals should be protected from exposure to inclement weather, harsh environmental conditions, and major temperature fluctuations and extremes.

Bedding, feed, and water should be provided, as appropriate, and the animals should be observed periodically to determine their state of well-being during transportation. On-site veterinary assistance may be warranted to monitor animals and to provide life-support assistance should a medical emergency occur during transportation or at the release or field study site. Selection of veterinary assistance should focus on the individual's knowledge and experience with the wildlife species involved. Any animals that die during transit should be removed as soon as practical from the sight and olfactory detection of other animals being transported. These carcasses should be retained for pathological examinations regarding cause of death. Similarly, animals that become severely injured or clinically ill should be removed and responsibly euthanized. Euthanasia should not take place in the presence of other live animals. Sick animals disposed of in this manner also should be retained for pathological assessments. Determinations of cause of death are needed to assess whether the remaining animals are at risk from pathogens associated with the dead animals.

SURGICAL AND MEDICAL PROCEDURES

Wildlife field research can involve surgical and medical procedures such as implanting radio transmitters and surgical sex determination in birds. Incorporation of such techniques into a research protocol should follow these guidelines:

1. Surgical and medical techniques used should be based on accepted protocols for the studied species *or* for the most closely related domesticated species. The Canadian Council on Animal Care's (1984) *Guide to the Care and Use of Experimental Animals,* Volume 2, is a good source of such information.
2. Protocols should be developed and, if possible, implemented in collaboration with a qualified veterinarian. Only properly trained personnel, conversant in all techniques necessary, should conduct the procedures.
3. Protocols must be reviewed carefully by the ACUC with special attention paid to limiting pain during the actual procedure and post-procedure period.
4. Adequate anesthesia and/or analgesia must be provided.

Minor Procedures

Minor medical procedures such as collection of blood, administration of drugs intravenously or intramuscularly, biopsies of superficial structures such as skin, and sutured attachment of radio transmitters usually can be performed safely and responsibly in the field without complicated equipment. However, it is the researcher's responsibility to choose the least invasive and least painful technique, minimize the duration of the procedure, use the most appropriate equipment and aseptic technique, and provide analgesia or sedation when indicated.

Major Procedures

As defined by the Animal Welfare Act, major operative procedures are (p. 36121) "any surgical intervention that penetrates and exposes a body cavity or any procedure which produces permanent impairment of physical or physiological functions." Major surgical procedures, when survival of the animal is intended, should be performed only under proper anesthesia and with sterile technique. Examples of major procedures used in wildlife research include laparotomy, surgical flight restraint, and sterilization. These procedures should be performed only in a clean space set aside for sterile surgery, with surgical instruments and drapes of the proper type, and with anesthesia protocols judged to be safe and responsible for the species involved. Necessary equipment and trained personnel to deal with surgery or anesthesia-related emergencies (i.e., severe blood loss, cessation of breathing or cardiac function, severe hypo- or hyperthermia, acid-base imbalances) should be available at all times. This will maximize the success and subsequent scientific return from those often costly procedures and, therefore, minimize the number of animals needed and amount of animal distress.

Medical Considerations

Wildlife field researchers should have access to veterinary consultation and take responsibility to prepare themselves to deal with any health problems that might arise in their study population. Sometimes intervention and control of a natural disease process may not be advisable and may interfere with the study's goals. However, if the health problem arises due to the researcher's work, or if it will interfere with the study, the researcher must be ready to respond. Preparations should include gaining familiarity with the common diseases and health problems of the species under study, establishing a contact with a veterinary consultant, and having appropriate treatment or control equipment and drugs on hand or easily accessible. The researcher also is responsible for evaluating the possible impact of disease in the study animals on the larger population or ecosystem as a whole, and for making the maintenance of their welfare a priority as decisions are made. This is especially true when release or translocation of animals is part of a study; disease must be considered in evaluating the advisability of the program.

Euthanasia

Euthanasia is defined under the Animal Welfare Act as (p. 36121) "the humane destruction of an animal accomplished by a method that produces rapid unconsciousness and subsequent death without evidence of pain or distress, or a method that utilizes anesthesia produced by an agent that causes painless loss of consciousness and subsequent death." Euthanasia may not be an approved component of a field study, but it may become a necessary health care option in a study involving capture, restraint, or surgical procedures. Therefore, all wildlife researchers involved in invasive studies must be familiar with the approved euthanasia methods for their study species (Andrews et al. 1993) and have the appropriate equipment/drugs on hand so euthanasia can be performed quickly.

DISEASE CONSIDERATIONS

Field investigators need to be fully aware of disease concepts so they may avoid introduction of new disease problems into animal populations or the spread of disease to other populations and locations as a result of their studies. Disease introductions and spread occur as a result of animals brought to the field research site to serve as biological sentinels, as decoys to lure and capture other animals, for species introductions or releases to supplement existing populations, for behavioral studies, for assistance in tracking or retrieving animals, and for other purposes. All of these uses of animals involve acceptable methods for scientific research and wildlife management. However, under no circumstances should the well-being of free-ranging wildlife populations be unduly jeopardized by disease risks associated with animal use in field research. Field investigators have ethical and professional obligations to take appropriate actions for minimizing the introduction of the following: (a) new disease agents, (b) vectors (e.g., ticks and internal parasites) capable of efficiently transmitting indigenous, dormant diseases or those not currently being effectively transmitted, and (c)

species that can serve as amplification hosts for transmitting indigenous diseases to other species.

In addition, animals that are highly susceptible to diseases indigenous to the study location should not be released into the wild without using applicable prophylactic measures, unless these animals are to serve as biological sentinels for disease investigations. Biological sentinels should be monitored closely and euthanized by approved, responsible methods as soon as is practical after study objectives have been met.

Disease introduction and spread can result from mechanical means such as contaminated personnel, supplies, and equipment in addition to the biological processes identified above. Steps taken to address disease prevention are far more cost effective than disease-control activities initiated after a problem has developed. Protection of free-ranging wildlife from disease is aided by the following actions:

(1) Appropriate health certification should be required for all animals being brought to the site of field investigations. State veterinary officials should be contacted to determine what specific testing must be done when animals are moved into their jurisdiction.
(2) Appropriate disinfection procedures should be used for investigators and their equipment when disease risks are present.
(3) Prior knowledge of disease activity at the study site should be obtained to guide actions involving the research study.
(4) Source for any animals being brought to a field investigation site (captive-reared and relocated wild stock) should be evaluated for inherent disease problems, and appropriate steps should be taken to avoid disease introductions.
(5) To the extent possible, animals should be held under surveillance for 15–30 days prior to their release into the wild, and only healthy animals should be released. These animals should not be mixed with other species during transportation and should be isolated from other animals during the surveillance period.
(6) Any animals that die should be examined by a disease diagnostic laboratory having competency for determining cause of death in the species involved; these findings should be used to guide appropriate actions.
(7) Animals that become clinically ill should be examined by disease specialists, and their counsel should be used to protect the well-being of other animals within the study area.

ANIMAL DISPOSITION AT COMPLETION OF STUDY

When live animals are in the possession of investigators or under their control at the time of study completion, an evaluation must be made as to whether these animals can be released to a free-ranging existence, should be maintained under controlled conditions, or should be euthanized. As a general rule, field-captured animals should be released only:

(a) At the site of the original capture, unless conservation efforts or safety considerations dictate otherwise. Prior approval for releases at noncapture sites should be obtained from appropriate state/federal agencies. Relocation release sites should be within the native range of the species, or established range for introduced species, and be in habitat suitable for species survival;

(b) When the released animal can be reasonably expected to function normally within the population;

(c) When local and seasonal conditions are conducive to survival;

(d) When the ability to survive in nature has not been irreversibly impaired; and

(e) When release is not likely to spread pathogens or contribute to disease processes in other ways.

The decision of whether to release captive-reared animals into the wild after completion of a field research project demands more rigorous evaluation than for field-captured animals. In addition to evaluating the future well-being of the animal being released, impacts on other animals of the same species and competition and risks for other species sharing that environment also must be considered. Rarely, if ever, will releases of captive-reared animals at the completion of research studies be justified on the basis of animal welfare considerations.

When animals are to be released, efforts should be made to enhance their chances of survival. Animals should be in good physical condition and released when weather conditions are favorable, at a time of day when they are able to locate food and cover that meet survival needs.

Animals that cannot be released should be considered for distribution to other scientists for further study. However, if the animal was subject to a major invasive procedure, it may not be appropriate for additional experimentation. Animals not suitable for research may be suitable display animals that can be donated to a zoo or other type of educational institution.

When animals must be euthanized, responsible methods appropriate for the species and circumstances must be used. Care must be taken to assure that the animal is dead before disposal of the carcass. Also, disposal procedures must prevent carcasses containing toxic substances or drugs from the research investigations or euthanasia procedures to enter the food web of other animals. To the extent feasible, euthanized animals should be properly preserved and used as voucher specimens or for teaching purposes.

SAFETY CONSIDERATIONS

Researchers working with free-ranging wildlife are subject to enhanced levels of exposure to wildlife diseases transmissible to humans. Disease transmission may in-volve direct contact with infected animals such as those with rabies, contact with disease vectors such as ticks transmitting Lyme disease, or contact with contaminated environments such as bird roosts and histoplasmosis. Field investigators should become familiar with the common diseases of wildlife species they are working with and the relative prevalence of those diseases in the populations they are studying. Consultation with a physician regarding immunization or other preventative treatment is advised when serious diseases for humans commonly occur in the populations being studied. Investigators who become ill should seek medical assistance and advise their physicians of their exposure to potentially hazardous animals, diseases, and environmental conditions.

Acknowledgments.—These guidelines were prepared by a committee of The Wildlife Society appointed by J. G. Teer during his Presidency. The committee acknowledges the contributions of F. J. Dein for his review of these guidelines and valuable input provided in enhancing the final content.

LITERATURE CITED

AD HOC COMMITTEE ON ACCEPTABLE FIELD METHODS IN MAMMALOGY. 1987. Acceptable field methods in mammalogy: preliminary guidelines approved by the American Society of Mammalogists. J. Mammal. 68(4, Suppl.). 18pp.

AD HOC COMMITTEE ON THE USE OF WILD BIRDS IN RESEARCH. 1988. Guidelines for use of wild birds in research. Auk 105(1, Suppl.). 41pp.

AMERICAN SOCIETY OF ICHTHYOLOGISTS AND HERPETOLOGISTS (ASIH), THE HERPETOLOGISTS' LEAGUE (HL), AND THE SOCIETY FOR THE STUDY OF AMPHIBIANS AND REPTILES (SSAR). 1987. Guidelines for the use of live amphibians and reptiles in field research. J. Herpetol. 4(Suppl.):1–14.

ANDREWS, E. J., ET AL. 1993. Report of the AVMA panel on euthanasia. J. Am. Vet. Med. Assoc. 202:229–249.

ANIMAL BEHAVIOR SOCIETY/ANIMAL SOCIETY FOR ANIMAL BEHAVIOR. 1986. ABS/ASAB guidelines for the use of animals in research. Anim. Behav. Soc. Newsletter 31:7–8.

CANADIAN COUNCIL ON ANIMAL CARE. 1980. Guide to the care and use of experimental animals. Vol. 1. Can. Counc. Anim. Care, Ottawa, Ont. 120pp.

———. 1984. Guide to the care and use of experimental animals. Vol. 2. Can. Counc. Anim. Care, Ottawa, Ont. 208pp.

DAY, G. I., S. D. SCHEMNITZ, AND R. D. TABER. 1980. Capturing and marking wild animals. Pages 61–88 *in* S. D. Schemnitz, ed. Wildlife management techniques manual. Fourth ed., rev. The Wildl. Soc., Washington, D.C.

ORLANS, F. B., EDITOR. 1988. Field research guidelines: impact on animal care and use committees. Sci. Cent. Anim. Welfare, Bethesda, Md. 23pp.

———, R. C. SIMMONDS, AND W. J. DODDS, EDITORS. 1987. Effective animal care and use committees. Sci. Cent. Anim. Welfare, Bethesda, Md. 178pp.

Appendix I. Professional society guidelines for use of live animals in field research.

Ad Hoc Committee on Acceptable Field Methods in Mammalogy. 1987. Acceptable field methods in mammalogy: preliminary guidelines approved by the American Society of Mammalogists. J. Mammal. 68(4, Suppl.). 18pp.

Ad Hoc Committee on the Use of Wild Birds in Research. 1988. Guidelines for use of wild birds in research. Auk 105(1, Suppl.). 41pp.

American Society of Ichthyologists and Herpetologists (ASIH), American Fisheries Society (AFS), and the American Institute of Fisheries Research Biologists (AIFRB). 1987. Guidelines for use of fishes in field research. Copeia 1987(Suppl.). 12pp.

American Society of Ichthyologists and Herpetologists (ASIH), the Herpetologists' League (HL), and the Society for the Study of Amphibians and Reptiles (SSAR). 1987. Guidelines for the use of live amphibians and reptiles in field research. J. Herpetol. 4(Suppl.):1–14.

Appendix II. Sources of assistance for technical information implementation and interpretation of the Animal Welfare Act.

Animal Welfare Information Center
National Agricultural Library
Beltsville, MD 20705
(301) 344-3212

National Library of Medicine
Bethesda, MD 20209
(301) 496-6097

Scientists Center for Animal Welfare
4805 St. Elmo Avenue
Bethesda, MD 20814
(301) 654-6390

Sector Supervisors
Regulatory Enforcement and Animal Care
Animal and Plant Health Inspection Service
USDA
Room 206
6505 Becrest Road
Hyattsville, MD 20782
(301) 436-6491

University of Illinois
Laboratory Animal Welfare Project
1301 W. Gregory Drive
Urbana, IL 61801
(217) 244-5802

5

CAPTURING AND HANDLING WILD ANIMALS

Sanford D. Schemnitz

INTRODUCTION

The art of capturing animals for food is as old as human existence on earth. However, few animals are captured solely for food today. Most animals are captured alive to implement various management and research studies. Successful capture programs result from the efforts of experienced wildlife biologists who have planned, studied, and tested various techniques before beginning any new program. Researchers must have valid trapping and banding permits before undertaking wildlife capture.

TRAP BAITS AND SCENTS

The success of most animal trapping operations depends on a suitable bait or scent to attract animals into traps. Numerous native foods, commercial foods, artificial lures, and prepared scents have been used as attractants. Unfortunately, no universal attractant exists that works successfully on all species. Consequently, wildlife biologists may have to test several baits or scents before they find ones that attract different species in their geographical areas.

Baits

Domestic livestock foods are probably the most common baits used in big-game trapping. Prebaiting with these foods is an important prerequisite to any trapping program. Howard and Engelking (1974) reported that apple and pear baits were best for trapping mule deer in New Mexico; alfalfa hay and cottonseed were good, and salt and corn were least successful. However, others have used salt and corn as bait to trap white-tailed deer (Hawkins et al. 1967, Ramsey 1968, Mattfeld et al. 1972).

Native browse plants are not used as frequently as in the past because of the time and effort required in gathering them. However, Mattfeld et al. (1972) trapped deer effectively with browse plants in the winter and with salt in the summer. In the arid climate of the Southwest, water has been an effective bait to trap desert bighorns (Papez and Tsukamoto 1970).

Various grains such as corn, milo, wheat, and oats are used extensively to attract game birds. Gullion (1961) observed that corn dyed orange, red, blue, or purple was more accepted by ruffed grouse than undyed, yellow shelled corn. He soaked the dyed corn in water, which caused it to swell and look like wild fruit.

A mixture of peanut butter and oatmeal has been used as a rodent bait for many years. Anderson and Ohmart (1977) recommended adding dimethylphthalate to bait to repel ants. Chabreck et al. (1986) tested Dursban, Diazinon, and RARK as ant repellents and observed that the number of small mammals captured did not differ between treated and untreated traps. Getz and Prather (1975) used a short-fiber cotton with peanut butter that was heated to about 65 C. This bait provided odor and taste of peanut butter, but the cotton made removal by insects difficult. Most rodent baits are applied to the trap by hand, but Johnson (1969) used a caulking gun to dispense bait more rapidly in live traps.

Scents

For many years fur trappers have used curiosity scents to attract furbearing mammals into traps. The major ingredients in scent mixtures are similar except for minor variations. Dobie (1949) reported the important items in

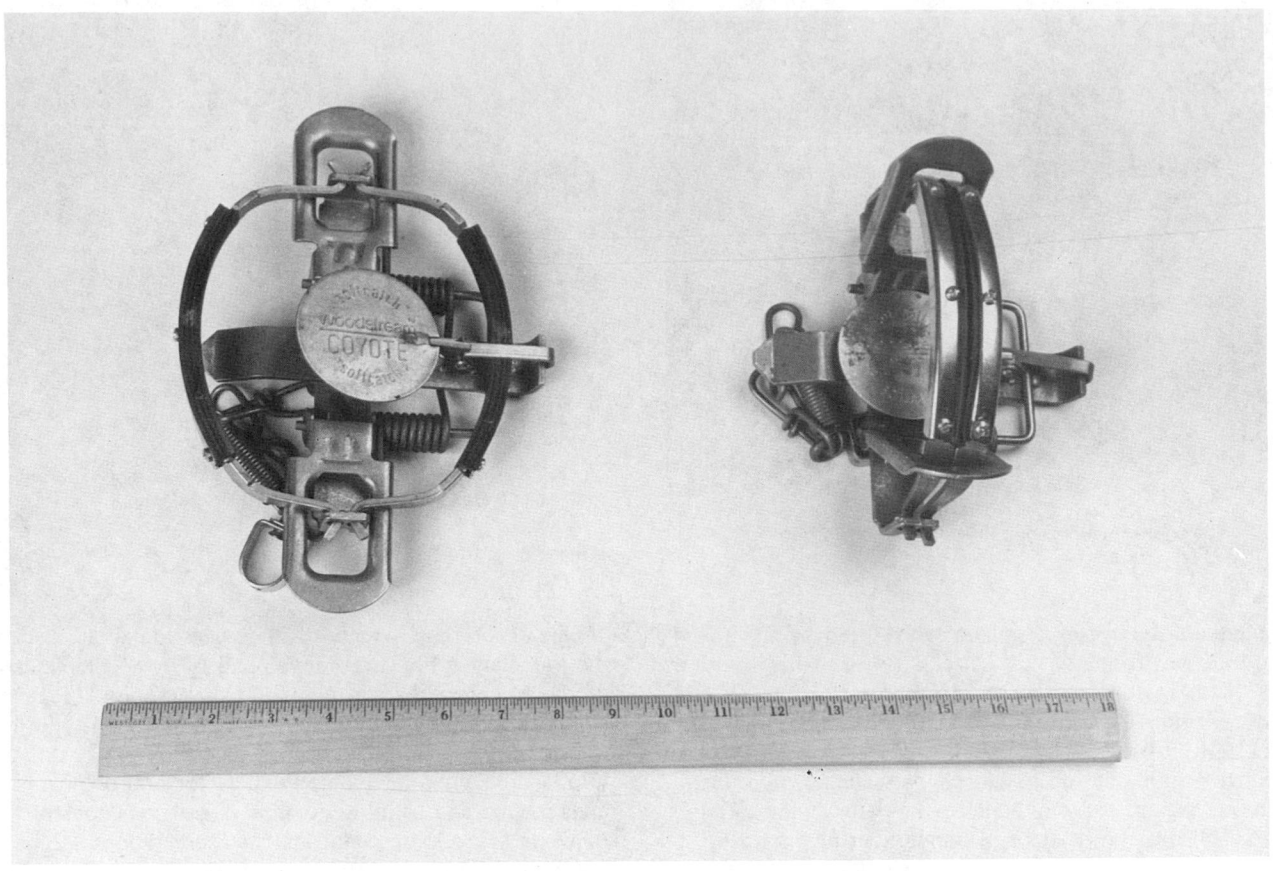

Fig. 1.   Victor softcatch leg-hold trap. Left, set; right, sprung.

coyote scent are coyote urine and anal glands, fish oil, and glycerine as a preservative. Carnivores are sometimes attracted to snares and traps by the scent of meat "stink" baits that are prepared from fish, poultry, or beef. For example, holes can be punched in cans of fish or meat to make a long-lasting, smelly bait. Scents comprising rotten eggs, decomposed meat, and fish oil also have been used for coyote trapping. Other items such as seal oil, Siberian musk oil, beaver castor, and skunk musk are widely used in scents. Fermented egg product and a liquid synthetic fermented egg product are effective canid attractants (Roughton 1982, Turkowski et al. 1983). Scrivner et al. (1987) and Graves and Boddicker (1988) reported tri-methylammonium decanoate (TMAD) to be an effective coyote attractant.

Often plant extractions are added to scents. The root of the Asiatic plant asafetida imparts a strong, persistent odor to scents. The oils from the herbs anise and valerian also have been added to scent mixtures. Some trappers like to add small amounts of cheap commercial perfumes to their scents.

Scents are used primarily to attract carnivores, but other mammals are attracted to them. Pedersen and Adams (1977) successfully trapped elk in the summer with anise oil and salt.

CAPTURING MAMMALS

Field biologists have several new techniques available to capture small rodents and large herbivores. Some techniques are either improved or modified versions of old capture methods. Animals are captured by hand, mechanical devices, remote injection of drugs, or drugs administered orally in baits. Major emphasis in this chapter will be on equipment other than drugs used for capture.

Steel and Snap Traps

For many years fur trappers and predator control agents have used various commercial steel traps to kill or capture animals. Palmisano and Dupuie (1975) compared leg-hold and conibear steel traps for taking animals in Louisiana. They concluded that long-spring leg-hold traps were more effective in taking nutria and raccoons, whereas conibears appeared superior to leg-holds for muskrats. Leg-hold traps with offset and padded jaws (Victor softcatch) (Fig. 1) also have been used to live-capture carnivores safely for marking and radio instrumentation. Erickson (1957), Black (1958), and Jonkel and Cowan (1971) used modified Newhouse #150 steel traps with lengthened trap chains to capture black bears for marking. Wolves were captured for study with #4 Newhouse steel traps (Van Ballenberghe et al. 1975). Storm et al. (1976) captured foxes at dens with #1 and #2 steel traps. However, Nellis (1968) concluded that steel traps were unacceptable for live coyote capture (and probably other canids) because of foot injuries. Berchielli and Tullar (1980) reported no significant difference in trap-related injuries between leg snare and standard leg-gripping traps, but the leg snare was significantly less effective than the leg-gripping trap in capturing animals.

The efficiency of padded-jaw leg-hold traps has been

Fig. 2. Bailey live trap set to capture beavers (photo from R. N. Conner, U.S. For. Serv.).

Fig. 3. Stephenson live trap with wooden drop doors used to capture white-tailed deer (photo by Ed Cleary, USDA, APHIS/ADC).

evaluated for coyotes (Linhart et al. 1986, Skinner and Todd 1990, Linhart and Dasch 1992), fisher (Arthur 1988), and other furbearers (Linscombe and Wright 1988). Tullar (1984), Olsen et al. (1986), and Onderka et al. (1990) observed that padded jaw traps caused less damage to the feet of foxes, coyotes, and raccoons and were no less effective in capture rate than Victor leg-hold traps. Balser (1965) attached tranquilizer tabs by wire to the trap jaw to reduce leg injury to wild carnivores. Diazepam is now a controlled substance and has been replaced for this use by propiopromazine hydrochloride. A suitable dose for coyotes is 600 mg (Linhart et al. 1981, Zemlicka and Bruce 1991). Various pan tension devices were tested by Turkowski et al. (1984). They reported that the curved leaf spring was most effective in avoiding the capture of nontarget animals.

Bailey (Fig. 2) and Hancock live traps or modified versions are used primarily to capture beavers. Buech (1983) modified the Bailey trap release and lengthened the lock mechanisms to curtail "misses" and increase trapping success. Northcott and Slade (1976) modified the Hancock trap to capture river otters along slides and pathways.

Bangs (1981) soldered a piece of #22 stainless-steel wire over the striker bar of Museum Special snap traps to increase capture rate of shrews. West (1985) compared old and new model Museum Special snap traps and observed that the old models had stronger springs than the new models, resulting in more *Tamias townsendi* and *Microtus oregoni* captures and more *Sorex trowbridgii* captures in the new trap model. The higher shrew capture rate was attributed to the smaller force required to release the trap treadle of the new model.

Leg Snares

Leg snares are useful in capturing large carnivores with a minimum of injury. Van Ballenberghe (1984) reported less injury to wolves captured with leg snares than with steel traps. Novak (1981) reported similar results with leg snares and coyotes, foxes, skunks, and raccoons. Aldrich leg snares have been widely used to capture bears. Johnson and Pelton (1980) added a heavy spring on the snare cable to act as a shock absorber to help minimize injury.

Box Traps

One of the most widely used devices to capture deer is the Stephenson box trap, developed by J. H. Stephenson of the Michigan Department of Conservation. McBeath (1941) first reported the use of the Stephenson-type live trap to capture white-tailed deer. The basic design of the trap has remained unchanged except for improvements in construction and tripping devices. The trap is constructed of wood (Fig. 3) or metal and measures about 1.2 m $\times$ 1.2 m $\times$ 3.7 m with drop gates at both ends. Trap gates can be tripped by a string or wire, but Webb (1943) used a steel trap as the release device. Williams and Pelton (1972) and Foreyt and Glazener (1979) modified the basic box trap design to increase portability for capturing deer and European and feral hogs. Runge (1972) described an improved tripping device made of monofilament fish line and steel rods. Masters (1978) described a simple trigger assembly that consisted of a gate hook, band of iron, and monofilament line.

Large box traps (Fig. 4) made of corrugated steel culverts have been used to capture bears (Erickson 1957, Black 1958, Troyer et al. 1962). Bear traps can be made of steel plates or 14-gauge culvert sections, 1.8–2.4 m long and 1.2 m wide, with a steel drop gate at one or both ends. The weight of these traps restricts their use to sites along roads. Most culvert traps are mounted on trailer frames to make them more portable. These traps are useful in capturing nuisance bears, which can be hauled within the trap to a new release site.

A basic, widely used portable net trap (Fig. 5) was designed by Clover (1954). Improvements were made by Sparrowe and Springer (1970), Roper et al. (1971), and McCullough (1975). Thompson et al. (1989) modified the Clover trap by making it collapsible to restrain and safely handle elk with a small field crew and without use of drugs or hobbles.

Apparatuses that resemble box traps have been constructed to restrain captive animals for blood collection and drugging. Sauer et al. (1969) used a plywood restraint chute with folding doors to control and handle mule deer. A similar device modified with movable padded side panels and a padded plunger successfully restrained white-tailed deer (Mautz et al. 1974). Masters (1978) used a

Fig. 4. Culvert trap used to capture bears (photo by Ariz. Game and Fish Dep.). ———

Fig. 5. Clover trap.

padded, plywood holding box to restrain captured deer for weighing and marking. Deer captured in box traps were forced into the smaller holding box where their necks were held between movable catch-bars on the front of the box.

Smaller mammals also can be captured by a variety of commercial "box-like" traps, such as Havahart, Longworth, National, Sherman, and Victor. However, many field-workers design and construct their own traps of wood, metal, wire, or plastic. A wooden box trap designed by Mosby (1955) was improved by Ludwig and Davis (1975) to increase catches and reduce escapes of woodchucks. They increased the trap length from 51 cm to 61 cm, added a transom latch to secure the door, lined the door with metal to discourage chewing, and used a wooden post instead of wire for a trigger. Cushwa and Burnham (1974) designed an inexpensive live trap for snowshoe hares. Brown et al. (1969) constructed a simple, small, lightweight, inexpensive live trap for small mammals from plastic (PVC) water pipe.

Keller et al. (1982) described a covered-lid, plastic-bucket shelter for trapping rodents in snow, sleet, and rain with live traps. Arboreal small mammals can be captured effectively with conventional live traps tied to a V-shaped frame attached to a limb or supported by a pulley system to sample at various heights in the tree canopy (Malcolm 1991).

Maly and Cranford (1985) reported that large and small Sherman live traps were equally effective in capturing smaller mammal species (mean weight 4–18 g), but the larger species (mean weight 40–47 g) were more readily caught in large traps.

Zoellick and Smith (1986) combined an enclosure and a box trap to capture kit foxes at dens. Foreyt and Rubenser (1980) used a similar trap for multiple captures of coyote pups at dens. Storm and Dauphin (1965), Storm et al. (1976), and Berchielli and Tullar (1981) used wire ferrets (Fig. 6) and net bags to capture red foxes at dens.

Layne (1987) designed a wire enclosure for protecting small-mammal traps from disturbance without modifying the probability of capture. Warner and Chesemore (1985)

held Sherman live traps in place with inexpensive, arch-shaped croquet wickets to avoid trap failure from high wind and livestock disturbance. Jackson and Hutchison (1985) painted live traps with camouflage paint to reduce human interference near urban areas. Barry et al. (1989) developed an inexpensive, accurate timer that can be attached to Sherman live traps to time the capture of small mammals.

Literature is contradictory on the effects of odors of previous occupants on the trap response of animals that subsequently may enter the trap. Stoddart (1982a,b) observed that traps that had previously been occupied deterred subsequent *Microtus agrestis* from entering. Montgomery (1979) reported that *Apodemus sylvaticus, A. flavicollis,* and *Clethrionomys glareolus* preferred traps that had previously caught conspecifics. Tew (1987) found no difference between the capture rate of "clean" and "dirty" traps.

Hayes (1982) attached a telemetry transmitter device to modified Clover traps to monitor capture activity in re-

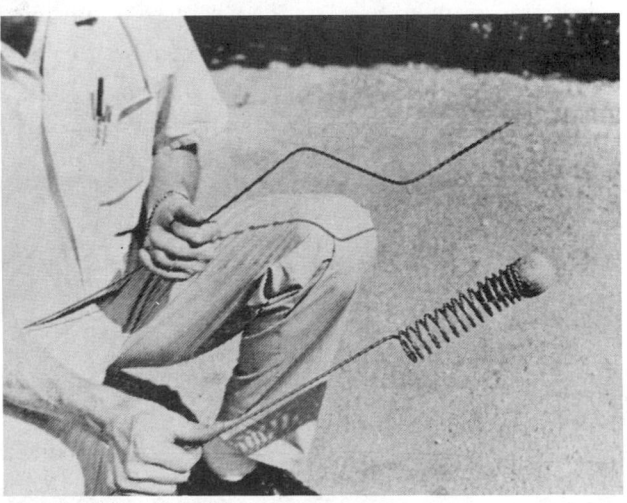

Fig. 6. The two ends of a wire ferret used to help capture red foxes. Above, the handle; below, the coiled spring (photo from G. L. Storm, U.S. Fish Wildl. Serv.).

Fig. 7. Above, bighorn sheep feeding on bait under a drop net. Note mule deer on far right. Below, net dropped with bighorn sheep entangled (photo by D. L. Reed, Colo. Div. Wildl.).

mote areas. Much time was saved, and unnecessary trips to check inactive traps were avoided.

Corral Traps

Corral traps have been used to capture several big-game species. Animals are lured into these traps by various baits, particularly native browse, alfalfa hay, apples, salt, and water. Generally, corral traps are permanent structures made of wire and lumber, although many are made from log poles found near the trap sites. Specifications for the corral trap at the Jackson Hole National Elk Refuge were described by Taber and Cowan (1969). R. Wilson (Wyo. Game and Fish Dep., pers. commun.) suggested the following modifications to the elk trap: (1) cover the top of each chute to prevent elk from rearing up, and (2) make one side of the chute completely solid. This seems to calm the elk and force them to the open side where tags and collars can be affixed.

A portable corral trap was used to capture Roosevelt elk (Mace 1971). This trap could be assembled by three persons in less than 1 day. Rempel and Bertram (1975) used a corral trap with seven panels (2.4 m × 2.4 m) to capture deer at a salt lick. The corral trap was modified to accommodate two Clover traps (Clover 1954, 1956). These Clover traps captured the deer as they attempted to escape from the corral, which greatly reduced the injuries associated with corral traps. Sugden (1956) designed a technique to close gates on big-game live traps by remote

control. Two blasting caps were detonated at 548 m to sever ropes holding trap gates. An unbaited drive corral with an exterior of chicken-wire fencing and an interior gill net were used effectively to capture jack rabbits (Henke and Demarais 1990).

A manual or automatic weighted plastic sheet enclosure has been widely used in southern Africa to capture groups of various ungulates. The unfolded plastic sheeting extends between metal poles and drops vertically when released. Bait or water is used to entice the animals to the trap site (M. D. Kock, Zimbabwe Dep. Natl. Parks Wildl. Manage., Harare, pers. commun.).

Net Traps

Cannon nets (Hawkins et al. 1968) and rocket nets have been used to capture deer. Drop nets (Fig. 7) are effective for capturing deer (Ramsey 1968, Conner et al. 1987) and bighorn sheep with a minimum of capture myopathy.

Drive nets and helicopters are a safe but efficient, labor-intensive method for capturing ungulates (Beasom et al. 1980). The nets are supported by paired wooden poles notched at the end and in an A-frame configuration at 3-m to 5-m intervals. Animals are moved into the net by a hovering helicopter and subdued by a ground crew. A low rate of mortality, 1.4%, was tallied in a sample of 430 deer captured in southern Texas by Sullivan et al. (1991). Kattel and Alldredge (1991) modified the drive-net capture technique by using people to direct musk deer in forested habitat into locally woven nets.

McCabe and Elison (1986) described success in capturing muskrats with nightlights and long-handled nets at a rate of 10–12 muskrats/hour.

Net Gun

The hand-held net gun (Fig. 8) fired from a helicopter has been widely used to capture a variety of wildlife including golden eagles (O'Gara and Getz 1986), waterfowl (Mechlin and Shaiffer 1980), caribou (Valkenburg et al. 1983), pronghorns (Firchow et al. 1986), desert bighorns (Krausman et al. 1985), white-tailed deer, Dall sheep (Barrett et al. 1982), and coyotes (Gese et al. 1987). Barrett et al. (1982) recommended the net gun as a highly portable, all-season tool for capture of specific animals but recommended practice to gain experience and discouraged use to capture species susceptible to capture myopathy such as pronghorns. Mortality from capture myopathy can be minimized by intraperitoneal injection of sodium bicarbonate.

Andryk et al. (1983) compared helicopter darting versus net gunning and concluded the net gun was more efficient. Kock et al. (1987) also reported the net gun to be superior to the drop net, drive net, and chemical immobilization for capturing bighorn sheep. Advantages included rapid and accurate deployment, resulting in short capture and processing time with less stress-induced mortality. De-Young (1988) concluded the net gun was more efficient (3.5 person hours per white-tail buck captured vs. 8.7 for the drive-net technique).

Oral Drugs

Oral drugs for animal capture and restraint should have the following qualities to be effective: (1) be readily taken in food or water; (2) have a wide safety margin because of the difficulty of controlling the amount ingested; (3) be fast-acting so drugged animals will not be able to move out of view; and (4) not be injurious to other animals consuming the bait. Unfortunately, few drugs on the market fit all of these requirements; consequently, reliable capture of animals with oral drugs remains largely in the future.

Austin and Peoples (1967) captured feral swine with alpha-chloralose, a sugar compound of chloral hydrate. They mixed 2 g of drug in 0.24 L of shelled corn. Stafford and Williams (1968) captured 9 of 17 black bears that ate bait containing alpha-chloralose. LeCount (1983) immobilized bears caught in culvert traps with a dosage of 1 g of alpha-chloralose per 15 kg of body weight mixed with 115 g of honey.

Miscellaneous Methods

Williams and Braun (1983) observed that pitfall traps caught greater numbers and diversity of small mammals than did conventional snap traps. Mengak and Guynn (1987) experienced similar capture rates with pitfall traps during winter and summer. Pitfall traps are more efficient than snap traps for capturing shrews. Boonstra and Rodd (1984) concluded that Longworth live traps were more efficient than pitfalls for capturing *Microtus pennsylvanicus*. The results were contradictory to results from live trap versus pitfall trap capture rates for *M. townsendii* (Boonstra and Krebs 1978). Pitfalls can be used as effective live traps, or as kill traps by adding water. Pitfall traps also allow the sampling of herptiles. Arrays of pitfall traps along 5-m-long drift fences were recommended by Bury and Corn (1977). Gibbons and Semlitsch (1981) recommended 50-cm-high aluminum flashing as a drift fence with 20-L buried plastic buckets as pitfall traps.

Successful capture of mice, rats, and squirrels with glue on metal or plastic sheets was reported by Srivastava and Srivastava (1985). Pagels and French (1987) suggested the examination of discarded bottles as a source of small-mammal distribution data to augment conventional collecting methods.

Eagle and Sargeant (1985) used live mink decoys placed in sprung live traps to attract mink to adjacent, set live traps at excavated den sets. They also used barrier tunnel sets. Barriers along marsh shorelines or in road culverts were made with rocks or boards, e.g., and double-door live traps formed tunnels through the barrier.

A transmitter was developed by Nolan et al. (1984) to monitor Aldrich leg snares set for grizzly bears. Hawley et al. (1985) perfected a clock for trap-monitoring transmitters to record the precise time of trap triggering.

Observation of distinctive behavior patterns and postures of doe deer facilitated the capture of cryptic fawns (Downing and McGinnes 1969, White et al. 1972, Huegel et al. 1985). Rubber gloves should be worn when fawns are handled so that subsequent predation is minimized.

Kunz and Kurta (1988) described and discussed bat capture methods in detail. They recommended headlights that can be adjusted for intensity. They mentioned hand capture; hand, mist, and canopy nets; and bucket, bag, funnel, and harp traps (Tuttle 1974) as useful methods and equipment for capturing bats.

Fig. 8. Four-barrel net gun loaded with 7-cm-diameter, large-mesh net (photo by D. H. Ellis, U.S. Fish Wildl. Serv.).

CAPTURING BIRDS

The *North American Bird Banding Manual* (U.S. Fish and Wildlife Service and Canadian Wildlife Service 1977), *Guide to Waterfowl Banding* (Addy 1956), and *Bird Trapping and Bird Banding* (Bub 1990) contain much more useful detail on trap design than can be presented in this chapter. Also, Wilbur (1967) and Reeves et al. (1968) are excellent sources on capture methods for upland game birds. Wilbur (1967) highlighted the importance of federal and state regulations that apply to trapping and marking game species. Special permits are required, and certain techniques may be prohibited. Anyone anticipating a trapping program should have full knowledge of the pertinent regulations.

Baited Traps

Many species of birds, especially the gregarious seed-eaters, can be captured in baited traps. Traps differ, aside from size and shape, mainly in the type of entrance. The simplest trap, often used for small birds, consists merely of a mesh box, supported on one end by a stick. When the birds feed under the box, the hidden operator pulls a string attached to the stick and the box falls, imprisoning the birds. This principle was used for capturing band-tailed pigeons by Wooten (1955), who used a 4.8-m-square wooden frame, tightly covered with 3.8-cm-mesh net, supported on one side by a 2.1-m pole. A pull-wire ran from the pole to a blind. Weiland (1964) described

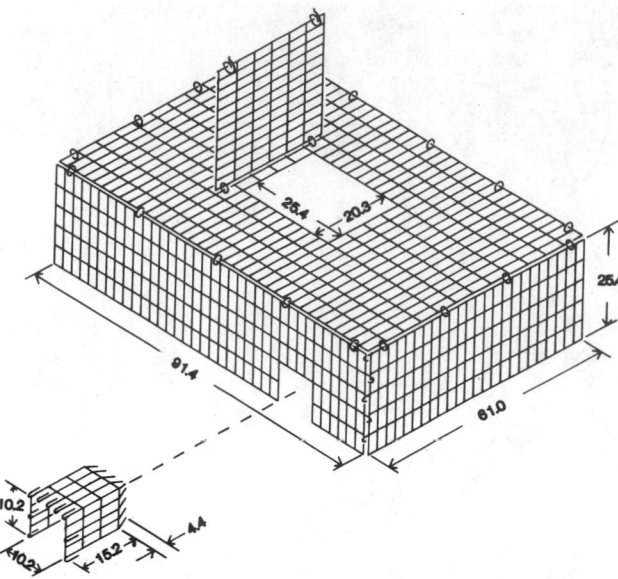

Fig. 9. Collapsible quail trap (measurements are in centimeters) (from Smith et al. 1981).

Fig. 10. Excavated tunnel trap beneath baited board. Tunnel opens into center of quail trap.

the use of a solenoid for instantaneous springing of such traps. Braun (1976) described a baited drop trap for band-tailed pigeons; when a cord was pulled, the trap fell off four wooden blocks onto the ground.

The funnel trap is constructed so that the wide portion of the funnel entrance is flush with the outer side of the trap and the narrow, inner opening projects well into the interior of the trap. Birds that enter seek a way out around the inner face of the box and usually overlook the funnel. Usually, horizontal wires are left to form a fringe around the inner entrance of the funnel to discourage use of the funnel as an exit. When traps are set on a fluctuating body of water, such as tidal marshes, the funnel is constructed tall enough to accommodate a bird at any level. The funnel entrance is most commonly used on waterfowl traps, being suitable for almost all species. If a trap is constructed of weak metal mesh, or fishnet, the funnel should be constructed of stout wire so that it will hold its shape. Smith et al. (1981) described an inexpensive, lightweight, and collapsible quail trap (Fig. 9).

The *swinging wire* or *bob* entrance is often used in traps designed for taking birds such as pheasants that are accustomed to walking through heavy vegetation.

Entrance to the *tip-top* is a door in the upper surface of the trap (which may be buried in the ground). The door is balanced with a light spring, so that the weight of the bird will cause it to open. When the bird drops off within the trap, the door springs shut. This trap is not widely used, but it is particularly successful for prairie grouse.

Tunnel (Fig. 10) and *ladder* traps (Fig. 11) were used successfully by Schemnitz (1961) to capture scaled quail.

The *swinging* or *sliding door* entrance is most often used in songbird traps. The door is supported by a device that is sprung when the weight of the bird depresses a bar or pan. This door is best used on a trap that is intended to catch only one bird at a time.

If the body of the trap is large, a smaller catch-box usually is added. The birds are driven into this box and

then can be removed through a door in the box. Trapped birds are vulnerable to crowding, exposure, and predation. The best protection against trap mortality is prompt processing of birds.

Traps set in water should be placed on a firm bottom, so that the bait will not sink in the mud and so that trapped birds will not churn the bottom into a soupy mud that will mat their feathers. A swift current of water through the trap will carry away bait, and a lack of current will permit water to freeze readily. Usually, vegetation should be removed from the vicinity of the entrance to waterfowl traps.

Szymczak and Corey (1976) modified materials used in the construction of the Salt Plains duck trap, effective for dabbling ducks and described by Addy (1956). A large waterfowl trap on a concrete foundation at a permanent site was designed by Arthur and Kennedy (1972). A "lily-lead" portable trap useful for diving ducks where water levels fluctuate daily was described by Hunt and Dahlka (1953). A portable, rectangular trap with metal bait-pans and an improved gathering box (McCall 1954) has been used for dabbling ducks. Floating raft traps readily capture waterfowl (Sugden and Poston 1970, Thornsberry and Cowardin 1971). Spitzkeit et al. (1987) described a portable and inexpensive swim-in trap for waterfowl. Haramis et

Fig. 11. Ladder trap for scaled quail.

al. (1987) devised a bait trap with a corral entrance for diving ducks that often catches 50–75 ducks within a few minutes.

Bacon (1987) described a hanging cylinder with multiple funnel entrances that used thistle as bait to capture American goldfinches and pine siskins.

Engle and Young (1989) evaluated several techniques for capturing the wary common raven in southwestern Idaho. They caught 23 of 24 ravens with padded leg-hold traps.

A variety of traps has been described for upland birds: by Aldous (1936) for white-necked ravens; by Schultz (1950) for northern bobwhites; by Chambers and English (1958), Edwards (1961), and Gullion (1961) for ruffed grouse; and by Wilbur (1967) for various upland game birds.

Trap size usually is a compromise between efficiency and portability. If the trapped birds are drawn from a population that is migrating through the area, so that new individuals are constantly being caught, the trap should be large and permanent. Often, however, a trap must be moved frequently to catch birds previously uncaught. The top of the trap should be made of some soft material, such as fishnet, so that birds jumping or flying upward will not be injured.

Net Traps

PROJECTILES AND DROP NETS

For birds that are too wary to enter an enclosure readily, such as the wild turkey, a baited net trap may be used. The cannon or rocket trap (Fig. 12) is used widely for wild turkeys and waterfowl. It consists of a large, light net that is carried over the baited birds by mortar projectiles or rockets. This is an especially good trap for geese, which are otherwise difficult to trap except during their flightless period (Dill and Thornsberry 1950, Salyer 1955 [U.S. Fish Wildl. Serv. Branch Wildl. Res., Wildl. Manage. Ser. 12]), and it also has been used successfully on sandhill cranes (Wheeler and Lewis 1972, Urbanek et al. 1991) and bald eagles (Grubb 1988). Each rocket is aimed about 20° from horizontal, and the corner rockets are aimed outward at about 45° from the edge of the net and set 5 m or 6 m in from the end of the net.

Firing is accomplished by use of regular electric cable and a blasting machine or radio (Grieb and Sheldon 1956) or by a 12-V battery. Two-way radio units may generate sufficient static electricity to cause accidental discharge of electric blasting caps. Blasting machines may fail to operate on cold days and should be warmed before use. The blind should be built first to accustom the birds to it; usually it is located 46–91 m from the net with a line of vision parallel to the folded net. The area under the net should be free of debris, and inconspicuous markers should be placed to show the location of the leading edge when the net is extended. Before the net is fired, its edges should be lapped under or staked. If the net is shot over water, net and birds should be pulled ashore before birds are removed. Up to 2 weeks may elapse before geese can again be lured into trapping position.

Giesen et al. (1982) described a vehicle-mounted rocket net to catch juvenile sage grouse. Sharp and Lokemoen (1980) developed a simple, remote-controlled radio-signal

Fig. 12. Rocket net for wild turkeys. Above, set; below, detonated (photo from D. Moreland, La. Dep. Wildl. Fish.).

device to trigger a rocket net. Garrott and Hayes (1984) devised a radio-controlled device for triggering drop nets at distances up to 1 km.

Nets and Lights at Night

Large ground-roosting birds, such as the ring-necked pheasant and sage grouse, can be taken at night on relatively flat terrain. An automobile, equipped with a strong searchlight (Fig. 13) and seats on the front for the netters,

Fig. 13. Night spotlighting rig for sage grouse (from J. Connelly, Id. Fish Game Dep.).

Fig. 14. Mist net with birds entangled (photo from T. A. Bookhout, U.S. Fish Wildl. Serv.).

frequently is used. When the birds are "fixed" in the light, the netters can pick them up with long-handled nets and swing them backward to the bed of the vehicle, where they are removed and placed in crates. Lights are more effective if a rather steady, loud vehicle and generator noise is maintained throughout the operation. This method can be used on waterfowl as well as ground-roosting birds. Labisky (1959, 1968) reported greatest success after birds had been roosting for 3–4 hours; he caught ring-necked pheasants (primarily), soras, Virginia rails, common barn-owls, screech owls, common coturnix, northern bobwhites, and greater prairie-chickens. Drewien et al. (1967) described a portable backpack unit for night-lighting upland game and waterfowl. This combination of light and a loud, steady noise has been applied successfully for capturing water birds by net from a boat at night (Cummings and Hewitt 1964). Brown (Ariz. Game and Fish Dep. P-R Proj. W-78-R-15, 1975) was aided by a pointing dog in the night capture of ground-roosting Mearns quail. Giesen et al. (1982) concluded that spotlights were the best method to capture adult sage grouse in spring and summer. Shuler et al. (1986) combined taped recordings of woodcock vocalizations and night-lights to capture woodcock with long-handled nets.

Mitchell (1963) described a successful floodlight trap for capturing large numbers of blackbirds and starlings at roosts. Kautz and Malecki (1990) caught pigeons by hand in barns at night in the dark. They used a headlight at intervals to help see where they were going while climbing a ladder.

Mist Nets

Mist netting (Fig. 14) long has been practiced in Asia and the Mediterranean area to catch birds for the market. McClure (1956) described the effective methods used in Japan. As described by Low (1957), the mist net is a fine, black, silk or nylon net, usually from 0.9 m to 2.1 m wide and 9.0 m to 11.6 m long. Mesh size determines which birds can be caught. A combination of 30-mm and 36-mm mesh sizes is best for capturing land birds weighing 5–100 g (Heimerdinger and Leberman 1966), but larger mesh is necessary to hold ducks, hawks, or ring-necked pheasants. A taut frame of stout twine crossed by hori-

zontal braces called "shelfstrings" is used in conjunction with the mist net. The net and shelfstrings are supported by poles at the ends; the shelfstrings are tight, but the net is loose. The excess netting is arranged in a loose bag or pocket 7.6 cm or 10.2 cm deep below each shelfstring except the topmost one. A bird striking the net from either side carries the net beyond the shelfstring and hangs in the pocket of the net. A net properly hung, with four shelves, is about 1.8 m high. When nets are used, a dark background is helpful. Wind interferes with netting. Capture rates generally peak at midmorning (0800–0930) and again in late afternoon (1600–1730) (Ralph 1976). Birds should not be left in the net more than 1 hour, and less if the net is in full sunlight. Nets should be closed immediately if rain begins. Jewell (1978) devised a simple, battery-operated microswitch device connected by wire to a bulb that lighted when a bird hit the net.

Phalaropes have been taken with a weighted mist net suspended horizontally over the water and dropped when they were underneath (Johns 1963). Dorio et al. (1978) captured incubating upland sandpipers and Wilson's phalaropes by lowering a mist net over the bird as it flushed. Hicklin et al. (1989) developed a simple, durable, and inexpensive "Fundy pull trap" (Fig. 15) to capture shorebirds with a minimum of handling time and mortality.

Frequently one or more persons can successfully drive birds, such as snipe, into a series of mist nets set in a group (Fogarty 1969). Otnes (1991) described a "swoop-netting" method of holding a mist net horizontally between two persons and then raising the net to a vertical position as shorebirds flew by. Blue grouse (Schladweiler and Mussehl 1969) and sage grouse (Browers and Connelly 1986) can be driven into a single net or two nets together.

Keyes and Grue (1982) reviewed in detail the capture of birds with mist nets with emphasis on general mist-netting procedures, including nets and poles, site selection

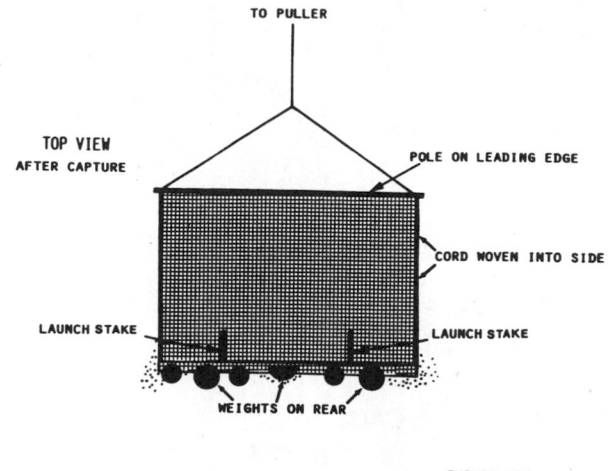

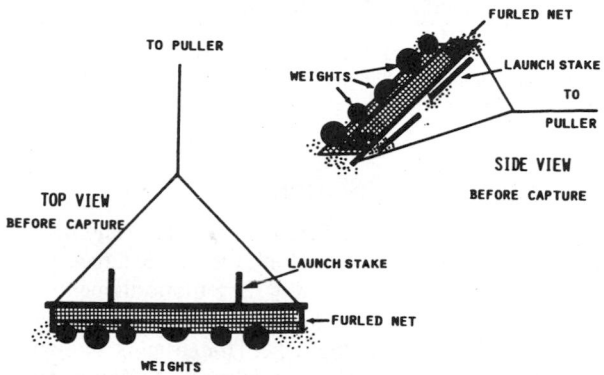

Fig. 15. Top and side view of Fundy pull trap before and after capture (from Hicklin et al. 1989).

and net placement, weather and time of day, and care of netted birds.

Sykes (1985) described a simple procedure for the construction of portable net poles and PVC pipe transport containers for the net poles. Nesbitt et al. (1982) used mist nets with taped calls of conspecifics near nest sites to capture red-cockaded woodpeckers. Barrentine (1984) developed a mist-netting technique for use with low bridges and deep water. DeJonghe and Cornuet (1983) described a system for capturing birds as high as 50 m above the ground by use of elevated mist nets without poles in mountainous areas. Karr (1979) developed a system to allow operation of nets up to 12 m above ground in forested situations. He used pulleys and guy lines to erect poles and nets. Munn (1991) devised a giant slingshot to shoot lines into tall, tropical trees to elevate and suspend mist nets (Fig. 16).

Mist netting is particularly useful for those species that will not come to bait. It is also useful as a sampling device, taking all species in proportion to their abundance. Special permission to use mist nets to capture migratory birds must be obtained from the U.S. Fish and Wildlife Service in the United States and from the Canadian Wildlife Service in Canada.

Drive and Drift Traps

Waterfowl molt all flight feathers at once during the summer, and while flightless they may be driven into traps. Currently, a fishnet is used in the pothole country for trapping flightless ducks and their half-grown young.

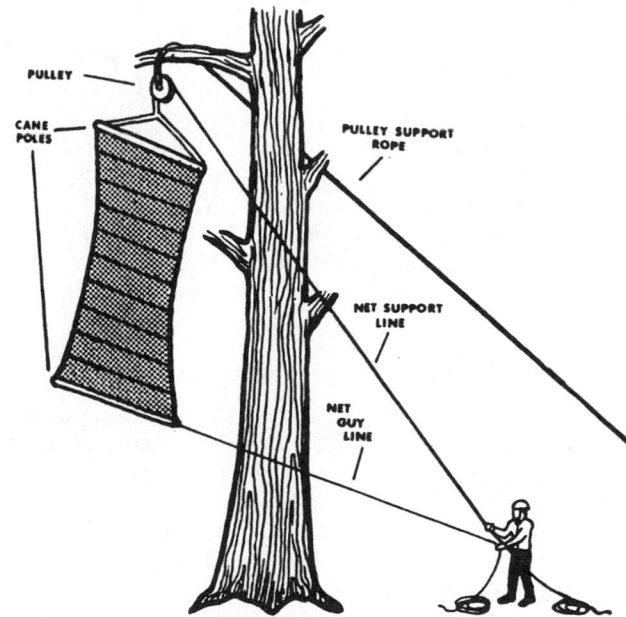

Fig. 16. Above, giant slingshot for shooting weighted lines over tall tree limbs to elevate mist net into canopy (from Munn 1991); below, vertical aerial mist-net rig.

Flightless geese may be taken by herding them into a corral trap (Cooch 1953). Heyland (1970) and Timm and Bromley (1976) described the advantages of the use of a helicopter to drive geese into a corral. Johnson (1972) modified a drive trap to capture flightless young goldeneyes by having part of the net submerged to entangle the ducklings as they dived to escape. The nets must be elevated rapidly to avoid drowning. Certain upland game birds, although capable of flight, generally run when herded. Large numbers of scaled quail have been taken by herding them into wings that lead to a tubular-enclosed cage (Schemnitz 1961). Tomlinson (1963) described catching blue grouse hens and broods by driving them into a portable trap with wings. Many upland game birds are reluctant to fly if frightened by hawks; imitation of a hawk

Fig. 17. Unbaited shorebird walk-in, drift trap with lead set for wood-cock. Netting, rather than chicken wire, is preferred for the top to min-imize injury to captured birds.

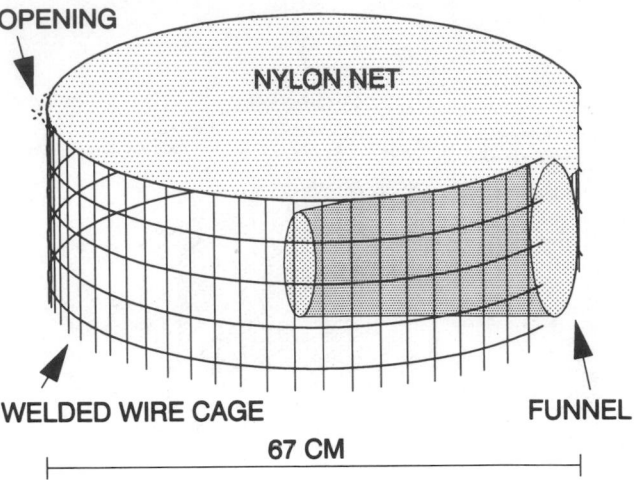

Fig. 18. Funnel design and configuration of the funnel of a welded wire trap with nylon net top for greater prairie-chickens (from Schroeder and Braun 1991).

call might be used in inhibiting their tendency to flush while being driven into traps.

Rails and shorebirds have been taken with drift traps that consist of long, chicken-wire leads directed into fun-nel-like openings (Fig. 17). The leads cross the feeding areas, so that walking birds encounter them and are guided into the funnels. These funnels face in both directions and lead into the box of the trap (Low 1935, Stewart 1951, Serventy et al. 1962). Toepfer et al. (1988) and Schroeder and Braun (1991) used a drift fence with several funnel entrances (Fig. 18) to capture prairie grouse hens on dis-play grounds.

Decoy and Enticement Lures

Various live animals and devices have been used suc-cessfully to lure animals for capture. Probably the most successful technique is the bal-chatri trap (Fig. 19) for capturing raptors (Berger and Mueller 1959). This trap is a chicken-wire cage that holds a bird or rodent as a lure. On the top of the trap are numerous monofilament nooses that can snag the talons of hawks that attack the caged animals. Berger and Hamerstrom (1962) used these traps to protect game-bird trapping stations from feeding hawks. Yosef and Lohrer (1992) combined a treadle and bal-chatri trap to capture loggerhead shrikes. Meng (1971) described the Swedish goshawk trap that is baited with live pigeons. Hamerstrom (1963) and Phillips (1978) de-scribed a vertical mesh-net trap, a Dho-Gaza, used to cap-ture hawks and baited with a live, tethered horned owl. Bryan (1988) perfected a radio-activated bow-net baited with a live house sparrow to capture American kestrels. Wegner (1981) modified the conventional live-baited noose trap by using carrion bait to capture American kes-trels. Clark (1981) described a modified and improved Dho-Gaza breakaway net to capture small raptors. Scharf (1985) used a noose system and a tethered magpie to trap territorial magpies. Owls are readily captured with live bait and mist nets (Bull 1987). Barn-owls have been cap-tured by several methods, including hoop and mist nets, trap doors, noose carpets, and hand capture (Colvin and Hegdal 1986). Anderson et al. (1980) described an im-proved spring-door decoy trap for diving ducks, and Sharp and Lokemoen (1987) devised a decoy trap for mallards.

Snares

A perch snare was perfected by Prevost and Baker (1984) for capturing ospreys. Dunk (1991) designed a se-lective pole trap for raptors that has monofilament line and a 1-m-long fishing rod tip. The monofilament noose closes around the bird's leg when the trap is manually triggered. A nylon noose inside a rubber tube can be used to capture woodpecker nestlings <10 days old (Jackson 1982). Schroeder (1986) described a modification for shortening the noosing pole to allow it to be carried in a backpack. Noosing poles have been used to capture com-mon nighthawks (McNicholl 1983), nestling bank swal-lows (Kramer 1988), and cormorants (Hogan 1985). Bar-rentine and Ewing (1988) built an inexpensive, compact, lightweight noose carpet of 1.3-cm-mesh hardware cloth with 20 monofilament nooses attached to capture burrow-ing owls. Winchell and Turman (1992) attached the noos-es to weighted dowel rods set at owl burrow entrances. The noose carpets were placed on a mound near the bur-row entrances. Bull (1987) described a noosing pole and a tethered deer mouse to capture spotted owls. A snare

Fig. 19. Cooper's hawk snared by noose on bal-chatri trap baited with a cowbird (photo by D. H. Ellis, U.S. Fish Wildl. Serv.).

mounted on a long, hand-held pole was useful in safely capturing spruce and blue grouse (Zwickel and Bendell 1967). Hoglund (1968) developed a foot snare set in gaps in vegetative fences to capture willow ptarmigan.

Nest Traps

Ground-nesting birds, such as most waterfowl, can be caught on the nest with a manually operated drop-net (Sowls 1955). A bow-net is a semicircular frame, hinged at the ends, that flips over the sitting bird when a cord is pulled quickly, carrying the net with it. Doty and Lee (1974) used a similar system to capture mallards incubating on nesting baskets. Coulter (1958) developed a trap consisting of a circular frame staked to the ground around the nest. A cylindrical net was fastened to the frame, and the open end of the net was laced with cord to form a purse-net closure when the cord was pulled. The cord led upward to a stake or limb and then to the blind. When the trap was open, the net was concealed around the frame; a pull of the cord raised and closed the net simultaneously. Guide rods prevented the hen from disarranging the net during nest-building and also ensured that the net would not close until it passed over the hen.

Miller (1962) used a nest trap with an automatic thermal-activating mechanism. An alarm clock also has been used; the turning key on the alarm springs a mousetrap that releases a rubber cord. The clock has the advantage of springing the trap at a set time, and the operator can promptly check to reduce possible harm to bird or nest. Shaiffer and Krapu (1978) further refined the nest-trap triggering device, using a telemetered remote-control system. An iron frame, covered with 2.5-cm-mesh netting, can be propped over the nest with a stick. Lead weights fastened to the front bar ensure a good fall when the trap is tripped with a long string.

A circular throw-net has been used successfully in capturing laying and incubating females. Also, a lightweight cotton net, 2.4 m × 2.4 m, stretched between two poles (3.6 m) has been thrown over a previously marked nest. This "clap-net" requires two persons for its operation. Leasure and Holt (1991) stretched a mist net attached to poles horizontally over the nests of short-eared owls, then flushed and captured incubating females. A blanket-net trap, measuring 3.6 m × 3.6 m with 2.5-cm mesh, can be suspended over the vegetation around a nest. The edges are brought down loosely and attached. After the set has been in place for one-half day, two persons approach from opposite sides and rush the nest. The hen usually flies straight up and hits the net (Addy 1956). Incubating diving ducks can be caught by setting a one-entrance funnel trap over the nest, the entrance pointing along the path the female uses in coming to the nest (Addy 1956). Kagarise (1978) readily captured nesting Wilson's phalaropes by placing a long-handled fisherman's landing net over the nest. The net was propped up 8 cm on vegetation to allow the bird to crawl under. This flush-net procedure also was used by Martin (1969) and Weaver and Kadlec (1970). Gartshore (1978) placed several fine, monofilament slip nooses attached to a wire ring over the nests of ground-nesting birds to catch incubating adults. Jewell and Bancroft (1991) used a welded wire nest trap for egrets and herons. They recommended trapping in the morning to minimize nest failure.

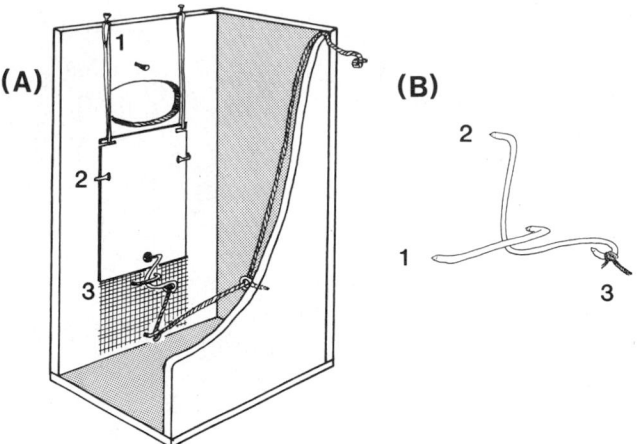

Fig. 20. Nest trap for capturing cavity-nesting ducks (photo from M. Zicus, Minnesota Dep. Nat. Resour.).

Mourning doves nesting in trees have been caught on the nest by manually operated traps (Swank 1952) and with automatic traps (Stewart 1954). Nolan (1961) used a small hoop net to catch birds at open nests in trees and shrubs.

Various methods for capturing birds nesting in tree cavities (Fischer 1944, Jackson 1977, Bull and Pedersen 1978) and nest boxes (DeHaven and Guarino 1969, Kibler 1969, Dhondt and Van Outryve 1971, Stewart 1971, Klimkiewicz and Jung 1977) involve boards or nets on poles and mousetrap and rat trap triggering devices. A simple trap for catching birds in nest boxes that requires <20 seconds to install was described by Stutchbury and Robertson (1986). Lombardo and Kemly (1983) described a complex radio-control method for trapping birds in nest boxes to facilitate the capture of a specific bird. Zicus (1989) installed an automatic, inconspicuous trap to capture cavity-nesting waterfowl in wooden nest boxes. The trap consisted of a sliding wooden door, spring-loaded with rubber bands, inside the nest box (Fig. 20). Jackson and Parris (1991) substituted a clear plastic bag for netting attached with duct tape to a fisherman's landing net. This device is held over the roost to capture the bird as it leaves at dawn.

Wilson and Wilson (1989) devised a remote-controlled sedative injector that they placed in the nest of breeding seabirds. They administered an intramuscular sedative by telemetry.

A tree-trunk trap, Grave's tree trap, is used to capture bark-foraging birds such as creepers, chickadees, nuthatches, titmice, and woodpeckers (Peters 1986).

Oral Drugs

Oral drugs are particularly useful in capturing large, wary flocking birds such as wild turkeys and sandhill cranes (Williams and Phillips 1973). Care must be exercised to prepare the proper dosage to avoid mortality. To prevent overdosing, Williams (1966) made a small incision in the crops of wild turkeys to wash out and remove excess drugs. Other workers used a turkey baster or large syringe to wash out and remove excess drug from the crop. Stouffer and Caccamise (1991) added 0.035 g of

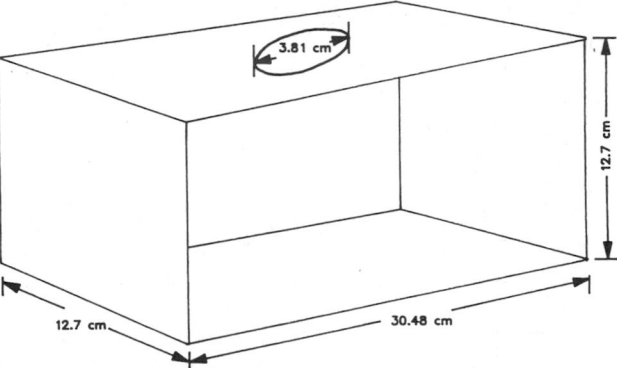

Fig. 22. Holding device for restraining birds. Above, northern bob-white in holder (note placement of clothespin for securing the bird); below, plans for a bobwhite holder (from S. DeMaso, Okla. Dep. Wildl. Conserv.).

Fig. 21. Pitfall trap with cover and drift fence. Above, placement of fence leading to trap; below, trap with weighted cover (photos by B. Tomberlin and T. Snell).

alpha-chloralose per fresh egg to capture wary and difficult-to-trap American crows.

Smith (1967) used tribromoethanol to capture seabirds. He put the drug in glycerine capsules that were inserted in fish and seal meat. Captured birds were forced to regurgitate drugged bait to reduce mortality losses. This was done by giving 1.5 L of warm water to each bird and then shaking it for 20–30 seconds while holding its legs and bill. Tribromoethanol has been tested on a variety of birds, but the major limitation is the inability to prevent partially narcotized birds from leaving the bait site (Williams and Phillips 1972, Evans et al. 1975, Krapu 1976).

Holbrook and Vaughan (1985) summarized past use of oral anaesthetics on wild turkeys. They used alpha-chloralose to capture 55 turkeys with a 5% adult mortality. If the bait site was near water, caution was necessary to avoid drowning. Nesbitt (1984) reported no serious detrimental effects of using alpha-chloralose to capture sandhill cranes.

Cline and Greenwood (1972) tested several anesthetic agents on captive mallard ducks and reported that alpha-chloralose was poisonous at high doses and had a long induction time. They concluded that tribromoethanol was the most satisfactory drug tested. Hofman and Weaver (1980) recommended a dosage of 40 mg of alpha-chloral-

ose per kilogram of body weight to capture northern pintails and mallards. Immobilization lasted for at least 5 hours.

CAPTURING REPTILES

Reptiles are captured by a variety of devices and methods, many of which are similar to those used for mammal capture. Balgooyen (1977) and Jones (1986) reviewed methods for collecting small reptiles. These methods include box traps, funnel traps with or without drift fences, snap traps, pitfall traps with drift fences (Fig. 21), hand snares, pole nooses, and rubber-band guns.

Recht (1981) blocked burrows of tortoises after their departure to promote capture. Knight (1986) devised a humane method for capturing snakes with glueboards. Captured snakes were released unharmed by pouring common cooking oil on them. The oil broke down the glue.

Jones (1965) described three methods for capturing alligators in Florida. He used a long-handled dip net to capture small individuals and a snare pole for those 1 m to 2 m in length. The most effective method was a harpoon made of a #8/0 fishhook embedded in a 10-cm length of tubing. About 0.6 cm of the hook shaft extended below the barb. A ring was fastened to the tube, which held a nylon cord 7.6 m long. A 4-L plastic bottle was attached to the cord as a float. The harpoon was delivered into the neck of the animals by a 3.7-m wooden pole.

A baited snare trap was used by Murphy and Fendley

Fig. 23. Handling a live red squirrel in a flexible, wire-rod cone. Above, squirrel taken from live trap with a canvas handling bag is transferred into the wire-rod handling cone; below, immobilized red squirrel is held by wire rods of the handling cone (from Halvorson 1972).

(1973) for capturing alligators. This trap consisted of two 1-m × 30-m plywood boards and anchor stakes. The boards were placed in a V-shape perpendicular to the shoreline. The boards guided the animals to baits, where they triggered a flexible pole snare. The snare was made of 0.6-m nylon rope attached to a tree on shore.

Webb and Messel (1977) reviewed several techniques used to capture crocodiles in Australia. They preferred tongs, harpoons, and rope traps for these animals. Mazzotti and Brandt (1988) described a simple wire-noose device to capture crocodiles.

HANDLING CAPTURED ANIMALS

Jessup et al. (1989) compiled a comprehensive and detailed handbook on wildlife restraint. Schmitt et al. (1983) decribed a safe squeeze chute to efficiently handle captive deer. McCullough et al. (1986) developed an inexpensive squeeze box device to restrain river otters. Layton and Cheal (1985) described an inexpensive restraint for small animals that was made of plexiglass with straps of Velcro. McCown et al. (1990) curtailed injuries to treed panthers by using a portable cushion and net during live capture.

Passmore (1979) restrained birds by wrapping them with a single strip of Velcro with a short overlap. Tweit (1982) used 1.85-L milk cartons to construct a simple holding box for birds. DeMaso and Peoples (1993) de-

vised a simple restraint for handling game birds (Fig. 22). Morrow et al. (1987) built a device of galvanized tin flashing attached to a 3.8-m-length pole to return nestling mourning doves, aged 9–12 days, to nests in trees after banding. Erickson (1981) described a safe transport case for incubated eggs.

A unique method of inducing sleep in birds was achieved by placing a small, flat stone weighing 7–20 g on the ear of birds (Tehsin 1988). The weight produced a hypnotic effect.

Ketamine hydrochloride, in combination with xylazine, is an effective tranquilizer for canids, mustelids, rodents, and opossums (see Chapter 6); it has a wide tolerance range and a relatively short recovery period of 40–120 minutes. It is recommended for restraint of small mammals in the field (Wright 1983). Methoxyflurane, an inhalant anaesthetic, was used with a restraint bag by King et al. (1990) to restrain cottontails and reduce nest abandonment of gray partridge (Smith et al. 1980) and mallards (Rotella and Ratti 1990). Halvorson (1972) developed a safe, simple, and effective handling technique and device for red squirrels (Fig. 23) featuring a weighing bag and wire-rod handling cone.

LITERATURE CITED

ADDY, C. E. 1956. Guide to waterfowl banding. U.S. Fish Wildl. Serv., Laurel, Md. 164pp.

ALDOUS, S. E. 1936. A cage trap useful in the control of white-necked ravens. U.S. Bur. Biol. Surv. Wildl. Res. Manage. Leafl. BS-27. 5pp.

ANDERSON, B. W., AND R. D. OHMART. 1977. Rodent bait additive which repels insects. J. Mammal. 58:242.

ANDERSON, M. G., R. D. SAYLER, AND A. D. AFTON. 1980. A decoy trap for diving ducks. J. Wildl. Manage. 44:217–219.

ANDRYK, T. A., L. R. IRBY, D. L. HOOK, J. J. McCARTHY, AND G. OLSON. 1983. Comparison of mountain sheep capture techniques: helicopter darting versus net-gunning. Wildl. Soc. Bull. 11:184–187.

ARTHUR, G. C., AND D. D. KENNEDY. 1972. A permanent site waterfowl trap. J. Wildl. Manage. 36:1257–1261.

ARTHUR, S. M. 1988. An evaluation of techniques for capturing and radiocollaring fishers. Wildl. Soc. Bull. 16:417–421.

AUSTIN, D. H., AND J. H. PEOPLES. 1967. Capturing hogs with alpha-chloralose. Proc. Annu. Conf. Southeast. Assoc. Game and Fish Comm. 21:202–205.

BACON, B. R. 1987. A hanging cylinder funnel trap. North Am. Bird Bander 12:46–47.

BALGOOYEN, T. G. 1977. Collecting methods for amphibians and reptiles. U.S. Bur. Land Manage. Tech. Note T/N 299. 12pp.

BALSER, D. S. 1965. Tranquilizer tabs for capturing wild carnivores. J. Wildl. Manage. 29:438–442.

BANGS, E. E. 1981. A modified museum special snap trap. J. Wildl. Manage. 45:1079.

BARRENTINE, C. D. 1984. A mist-netting technique for use with low bridges and deep water. North Am. Bird Bander 9:11–12.

———, AND K. D. EWING. 1988. A capture technique for burrowing owls. North Am. Bird Bander 13:107.

BARRETT, M. W., J. W. NOLAN, AND L. D. ROY. 1982. Evaluation of a hand-held net-gun to capture large mammals. Wildl. Soc. Bull. 10:108–114.

BARRY, R. E., JR., A. A. FRESSOLA, AND J. A. BRUSEO. 1989. Determining the time of capture for small mammals. J. Mammal. 70:660–662.

BEASOM, S. L., W. EVANS, AND L. TEMPLE. 1980. The drive net for capturing western big game. J. Wildl. Manage. 44:478–480.

BERCHIELLI, L. T., JR., AND B. F. TULLAR, JR. 1980. Comparison of a leg snare with a standard leg-gripping trap. N.Y. Fish Game J. 27:63–71.

———, AND ———. 1981. A technique for excavating red fox dens. N.Y. Fish Game J. 28:40–48.

BERGER, D. D., AND F. HAMERSTROM. 1962. Protecting a trapping station from raptor predation. J. Wildl. Manage. 26:203–206.

——, AND H. C. MUELLER. 1959. The bal-chatri: a trap for the birds of prey. Bird-Banding 30:18–26.

BLACK, H. C. 1958. Black bear research in New York. Trans. North Am. Wildl. Conf. 23:443–461.

BOONSTRA, R., AND C. Z. KREBS. 1978. Pitfall trapping of *Microtus townsendii*. J. Mammal. 59:136–148.

——, AND F. H. RODD. 1984. Efficiency of pitfalls versus live traps in enumeration of populations of *Microtus pennsylvanicus*. Can. J. Zool. 62:758–765.

BRAUN, C. E. 1976. Methods for locating, trapping and banding band-tailed pigeons in Colorado. Colo. Div. Wildl. Spec. Rep. 39. 20pp.

BROWERS, H. W., AND J. W. CONNELLY. 1986. Capturing sage grouse with mist nets. Prairie Nat. 18:185–188.

BROWN, E. B., II, W. R. SAATELA, AND W. D. SCHMID. 1969. A compact, lightweight live trap for small mammals. J. Mammal. 50:154–155.

BRYAN, J. R. 1988. Radio controlled bow-net for American kestrels. North Am. Bird Bander 13:30–31.

BUB, H. 1990. Bird trapping and bird banding. Cornell Univ. Press, Ithaca, N.Y. 448pp.

BUECH, R. R. 1983. Modification of the Bailey live trap for beaver. Wildl. Soc. Bull. 11:66–68.

BULL, E. L. 1987. Capture techniques for owls. Pages 291–293 *in* Biology and conservation of northern forest owls. U.S. For. Serv. Gen. Tech. Rep. RM-142.

——, AND R. J. PEDERSEN. 1978. Two methods of trapping adult pileated woodpeckers at their nest cavities. North Am. Bird Bander 3:95–99.

BURY, R. B., AND P. S. CORN. 1987. Evaluation of pitfall trapping in northwestern forests: trap arrays with drift fences. J. Wildl. Manage. 51:112–119.

CHABRECK, R. H., V. V. CONSTANTIN, AND R. B. HAMILTON. 1986. Use of chemical ant repellents during small mammal trapping. Southwest. Nat. 31:109–110.

CHAMBERS, R. E., AND P. F. ENGLISH. 1958. Modifications of ruffed grouse traps. J. Wildl. Manage. 22:200–202.

CLARK, W. S. 1981. A modified Dho-Gaza trap for use at a raptor banding station. J. Wildl. Manage. 45:1043–1044.

CLINE, D. R., AND R. J. GREENWOOD. 1972. Effect of certain anesthetic agents on mallard ducks. J. Am. Vet. Med. Assoc. 161:624–633.

CLOVER, M. R. 1954. A portable deer trap and catch-net. Calif. Fish Game 40:367–373.

——. 1956. Single-gate deer trap. Calif. Fish Game 42:199–201.

COLVIN, B. A., AND P. L. HEGDAL. 1986. Techniques for capturing common barn-owls. J. Field Ornithol. 57:200–207.

CONNER, M. C., E. C. SOUTIERE, AND R. A. LANCIA. 1987. Drop-netting deer: costs and incidence of capture myopathy. Wildl. Soc. Bull. 15:434–438.

COOCH, G. 1953. Techniques for mass capture of flightless blue and lesser snow geese. J. Wildl. Manage. 17:460–465.

COULTER, M. W. 1958. A new waterfowl nest trap. Bird-Banding 29:236–241.

CUMMINGS, G. E., AND O. H. HEWITT. 1964. Capturing waterfowl and marsh birds at night with light and sound. J. Wildl. Manage. 28:120–126.

CUSHWA, C. T., AND K. P. BURNHAM. 1974. An inexpensive live trap for snowshoe hares. J. Wildl. Manage. 38:939–941.

DEHAVEN, R. W., AND J. L. GUARINO. 1969. A nest-box trap for starlings. Bird-Banding 40:48–50.

DEJONGHE, J. F., AND J. F. CORNUET. 1983. A system of easily manipulated, elevated mist nets. J. Field Ornithol. 54:84–88.

DEMASO, S. J., AND A. D. PEOPLES. 1993. A restraining device for handling northern bobwhites. Wildl. Soc. Bull. 21:45–46.

DEYOUNG, C. A. 1988. Comparison of net-gun and drive-net capture for white-tailed deer. Wildl. Soc. Bull. 16:318–320.

DHONDT, A. A., AND E. J. VAN OUTRYVE. 1971. A simple method for trapping breeding adults in nesting boxes. Bird-Banding 42:119–121.

DILL, H. H., AND W. H. THORNSBERRY. 1950. A cannon-projected net trap for capturing waterfowl. J. Wildl. Manage. 14:132–137.

DOBIE, J. F. 1949. The voice of the coyote. Little, Brown and Co., Boston, Mass. 386pp.

DORIO, J. C., J. JOHNSON, AND A. H. GREWE. 1978. A simple technique for capturing upland sandpipers. Inl. Bird-Banding News 50:57–58.

DOTY, H. A., AND F. B. LEE. 1974. Homing to nest baskets by wild female mallards. J. Wildl. Manage. 38:714–719.

DOWNING, R. L., AND B. S. MCGINNES. 1969. Capturing and marking white-tailed deer fawns. J. Wildl. Manage. 33:711–714.

DREWIEN, R. C., H. M. REEVES, P. F. SPRINGER, AND T. L. KUCK. 1967. Back-pack unit for capturing waterfowl and upland game by night-lighting. J. Wildl. Manage. 31:778–783.

DUNK, J. E. 1991. A selective pole trap for raptors. Wildl. Soc. Bull. 19:208–210.

EAGLE, T. C., AND A. B. SARGEANT. 1985. Use of den excavations, decoys, and barrier tunnels to capture mink. J. Wildl. Manage. 49:40–42.

EDWARDS, M. G. 1961. New use of funnel trap for ruffed grouse broods. J. Wildl. Manage. 25:89.

ENGEL, K. A., AND L. S. YOUNG. 1989. Evaluation of techniques for capturing common ravens in southwestern Idaho. North Am. Bird Bander 14:5–8.

ERICKSON, A. W. 1957. Techniques for live-trapping and handling black bears. Trans. North Am. Wildl. Conf. 22:520–543.

ERICKSON, R. C. 1981. Transport case for incubated eggs. Wildl. Soc. Bull. 9:57–60.

EVANS, R. R., J. W. GOERTZ, AND C. T. WILLIAMS. 1975. Capturing wild turkeys with tribromoethanol. J. Wildl. Manage. 39:630–634.

FIRCHOW, K. M., M. R. VAUGHAN, AND W. R. MYTTON. 1986. Evaluation of the hand-held net gun for capturing pronghorns. J. Wildl. Manage. 50:320–322.

FISCHER, R. B. 1944. Suggestions for capturing hole-nesting birds. Bird-Banding 15:151–156.

FOGARTY, M. J. 1969. Capturing snipe with mist nets. Proc. Annu. Conf. Southeast. Assoc. Game and Fish Comm. 23:78–84.

FOREYT, W. J., AND W. C. GLAZENER. 1979. A modified box trap for capturing feral hogs and white-tailed deer. Southwest. Nat. 24:377–380.

——, AND A. RUBENSER. 1980. A live trap for multiple capture of coyote pups from dens. J. Wildl. Manage. 44:487–488.

GARROTT, R. A., AND R. W. HAYES. 1984. A radio-controlled device for triggering traps. Wildl. Soc. Bull. 12:320–324.

GARTSHORE, M. E. 1978. A noose trap for catching nesting birds. North Am. Bird Bander 3:1–2.

GESE, E. M., O. J. RONGSTAD, AND W. R. MYTTON. 1987. Manual and net-gun capture of coyotes from helicopters. Wildl. Soc. Bull. 15:444–445.

GETZ, L. L., AND M. L. PRATHER. 1975. A method to prevent removal of trap bait by insects. J. Mammal. 56:955.

GIBBONS, J. W., AND R. D. SEMLITSCH. 1981. Terrestrial drift fences with pitfall traps: an effective technique for quantitative sampling of animal populations. Brimleyana 7:1–16.

GIESEN, K. M., T. J. SCHOENBERG, AND C. E. BRAUN. 1982. Methods for trapping sage grouse in Colorado. Wildl. Soc. Bull. 10:224–231.

GRAVES, G. E., AND M. L. BODDICKER. 1988. Field evaluation of olfactory attractants and strategies used to capture depredating coyotes. Pages 195–204 *in* U.S. For. Serv. Gen. Tech. Rep. RM-154.

GRIEB, J. R., AND M. G. SHELDON. 1956. Radio-controlled firing device for the cannon-net trap. J. Wildl. Manage. 20:203–205.

GRUBB, T. G. 1988. A portable rocket-net system for capturing wildlife. U.S. For. Serv. Res. Note RM-484. 8pp.

GULLION, G. W. 1961. A technique for winter trapping of ruffed grouse. J. Wildl. Manage. 25:428–430.

HALVORSON, C. H. 1972. Device and technique for handling red squirrels. U.S. Fish Wildl. Serv. Spec. Sci. Rep. Wildl. 159. 10pp.

HAMERSTROM, F. 1963. The use of great horned owls in catching marsh hawks. Proc. Int. Ornithol. Congr. 13:866–869.

HARAMIS, G. M., E. L. DERLETH, AND D. G. MCAULEY. 1987. A quick-catch corral trap for wintering canvasbacks. J. Field Ornithol. 58:198–200.

HAWKINS, R. E., D. C. AUTRY, AND W. D. KLIMSTRA. 1967. Comparison of methods used to capture white-tailed deer. J. Wildl. Manage. 31:460–464.

——, L. D. MARTOGLIO, AND G. G. MONTGOMERY. 1968. Cannon-netting deer. J. Wildl. Manage. 32:191–195.

HAWLEY, A. W. L., M. W. BARRETT, AND C. D. MEWIS. 1985. Clocks for trap-monitoring transmitters. Wildl. Soc. Bull. 13:561–563.

HAYES, R. W. 1982. A telemetry device to monitor big game traps. J. Wildl. Manage. 46:551–553.

HEIMERDINGER, M. A., AND R. C. LEBERMAN. 1966. The comparative efficiency of 30 and 36 mm mesh in mist nets. Bird-Banding 37:280–285.

HENKE, S. E., AND S. DEMARAIS. 1990. Capturing jackrabbits by drive corral on grasslands in west Texas. Wildl. Soc. Bull. 18:31–33.

HEYLAND, J. D. 1970. Aircraft-supported Canada goose banding operations in arctic Quebec. Trans. Northeast Sect. Fish Wildl. Conf. 27:187–198.

HICKLIN, P. W., R. G. HOUNSELL, AND G. H. FINNEY. 1989. Fundy pull trap: a new method of capturing shorebirds. J. Field Ornithol. 60:94–101.

HOFMAN, D. E., AND H. WEAVER. 1980. Immobilization of captive mallards and pintails with alpha-chloralose. Wildl. Soc. Bull. 8:156–158.

HOGAN, G. G. 1985. Noosing adult cormorants for banding. North Am. Bird Bander 10:76–77.

HOGLUND, N. H. 1968. A method of trapping and marking willow grouse in winter. Viltrevy Swed. Wildl. 5:95–101.

HOLBROOK, H. T., AND M. R. VAUGHAN. 1985. Capturing adult and juvenile wild turkeys with adult dosages of alpha-chloralose. Wildl. Soc. Bull. 13:160–163.

HOWARD, V. W., JR., AND C. T. ENGELKING. 1974. Bait trials for trapping mule deer. J. Wildl. Manage. 38:946–947.

HUEGEL, C. N., R. B. DAHLGREN, AND H. L. GLADFELTER. 1985. Use of doe behavior to capture white-tailed deer fawns. Wildl. Soc. Bull. 13:287–289.

HUNT, G. S., AND K. J. DAHLKA. 1953. Live trapping of diving ducks. J. Wildl. Manage. 17:92–95.

JACKSON, J. A. 1977. A device for capturing tree cavity roosting birds. North Am. Bird Bander 2:14–15.

———. 1982. Capturing woodpecker nestlings with a noose—a technique and its limitations. North Am. Bird Bander 7:90–92.

———, AND S. D. PARRIS. 1991. A simple, effective net for capturing cavity roosting birds. North Am. Bird Bander 16:30–31.

JACKSON, M. H., AND W. M. HUTCHISON. 1985. The effect of camouflage on the vandalism and efficiency of Longworth small mammal traps. J. Zool. Ser. A (Lond.) 207:623–626.

JESSUP, D. A., W. E. CLARK, AND D. HUNTER. 1989. Wildlife restraint handbook. Calif. Dep. Fish Game, Sacramento. 151pp.

JEWELL, D. G. 1978. Building the better bird trap. North Am. Bird Bander 3:156.

JEWELL, S. D., AND J. T. BANCROFT. 1991. Effects of nest-trapping on nesting success of *Egretta* herons. J. Field Ornithol. 62:78–82.

JOHNS, J. E. 1963. A new method of capture utilizing the mist net. Bird-Banding 34:209–213.

JOHNSON, K. G., AND M. R. PELTON. 1980. Prebaiting and snaring techniques for black bears. Wildl. Soc. Bull. 8:46–54.

JOHNSON, L. L. 1972. An improved capture technique for flightless young goldeneyes. J. Wildl. Manage. 36:1277–1279.

JOHNSON, W. W. 1969. Dispensing bait with a caulking gun. J. Mammal. 50:149.

JONES, F. K. 1965. Techniques and methods used to capture and tag alligators in Florida. Proc. Annu. Conf. Southeast. Assoc. Game and Fish Comm. 19:98–101.

JONES, K. B. 1986. Amphibians and reptiles. Pages 267–290 in A. Y. Cooperrider, R. J. Boyd, and H. R. Stuart, eds. Inventory and monitoring of wildlife habitat. U.S. Bur. Land Manage. Serv. Cent., Denver, Colo.

JONKEL, C. J., AND I. McT. COWAN. 1971. The black bear in the spruce-fir forest. Wildl. Monogr. 27. 57pp.

KAGARISE, C. M. 1978. A simple trap for capturing nesting Wilson's phalaropes. Bird-Banding 49:281–282.

KARR, J. R. 1979. On the use of mist nets in the study of bird communities. Inl. Bird-Banding News 51:1–10.

KATTEL, B., AND A. W. ALLDREDGE. 1991. Capturing and handling of the Himalayan musk deer. Wildl. Soc. Bull. 19:397–399.

KAUTZ, J. E., AND R. A. MALECKI. 1990. Effects of harvest on feral rock dove survival, nest success and population size. U.S. Fish Wildl. Serv. Fish Wildl. Tech. Rep. 31. 22pp.

KELLER, B. L., C. R. GROVES, E. J. PITCHER, AND M. J. SMOLEN. 1982. A method to trap rodents in snow, sleet, or rain. Can. J. Zool. 60:1104–1106.

KEYES, B. E., AND C. E. GRUE. 1982. Capturing birds with mist nets: a review. North Am. Bird Bander 7:2–14.

KIBLER, L. F. 1969. The establishment and maintenance of a bluebird nest-box project. Bird-Banding 40:114–129.

KING, S. L., H. L. STRIBLING, AND D. W. SPEAKE. 1990. Use of methoxyflurane and a funnel bag as a cottontail rabbit restraint system. J. Wildl. Manage. 54:409–411.

KLIMKIEWICZ, M. K., AND P. D. JUNG. 1977. A new banding technique for nesting adult purple martins. North Am. Bird Bander 2:3–6.

KNIGHT, J. E. 1986. A humane method for removing snakes from dwellings. Wildl. Soc. Bull. 14:301–303.

KOCK, M. D., D. A. JESSUP, R. K. CLARK, C. E. FRANTI, AND R. A. WEAVER. 1987. Capture methods in five subspecies of free-ranging bighorn sheep—an evaluation of drop net, drive net, chemical immoblization and the net gun. J. Wildl. Dis. 23:634–640.

KRAMER, D. L. 1988. A noose apparatus and its usefulness in capturing nestling bank swallows. North Am. Bird Bander 13:66–67.

KRAPU, G. L. 1976. Experimental responses of mallards and Canada geese to tribromoethanol. J. Wildl. Manage. 40:180–183.

KRAUSMAN, P. R., J. J. HERVERT, AND L. L. ORDWAY. 1985. Capturing deer and mountain sheep with a net-gun. Wildl. Soc. Bull. 13:71–73.

KUNZ, T. H., AND A. KURTA. 1988. Capture methods and holding devices. Pages 1–30 in T. H. Kunz, ed. Ecological and behavioral methods for the study of bats. Smithsonian Inst. Press, Washington, D.C.

LABISKY, R. F. 1959. Night-lighting: a technique for capturing birds and mammals. Ill. Nat. Hist. Surv. Biol. Notes 40. 11pp.

———. 1968. Nightlighting: its use in capturing pheasants, prairie chickens, bobwhites, and cottontails. Ill. Nat. Hist. Surv. Biol. Notes 62. 12pp.

LAYNE, J. N. 1987. An enclosure for protecting small mammal traps from disturbance. J. Mammal. 68:666–668.

LAYTON, C. M., AND M. CHEAL. 1985. An inexpensive restraint for small animals. Physiol. Behav. 28:1115–1116.

LEASURE, S. M., AND D. W. HOLT. 1991. Technique for locating and capturing nesting short-eared owls (*Asio flammeus*). North Am. Bird Bander 16:32–33.

LeCOUNT, A. L. 1983. Immobilization of culvert-trapped black bears with alpha-chloralose. Ariz. Game and Fish Dep. Wildl. Digest Abstr. 14. 7pp.

LINHART, S. B., AND G. J. DASCH. 1992. Improved performance of padded jaw traps for capturing coyotes. Wildl. Soc. Bull. 20:63–66.

———, ———, C. B. MALE, AND R. M. ENGEMAN. 1986. Efficiency of unpadded and padded steel foothold traps for capturing coyotes. Wildl. Soc. Bull. 14:212–218.

———, ———, AND F. J. TURKOWSKI. 1981. The steel leg-hold trap: techniques for reducing foot injury and increasing selectivity. Pages 1560–1578 in J. A. Chapman and D. Pursley, eds. Proc. Worldwide Furbearer Conf., Frostburg, Md.

LINSCOMBE, R. G., AND V. L. WRIGHT. 1988. Efficiency of padded foothold traps for capturing terrestrial furbearers. Wildl. Soc. Bull. 16:307–309.

LOMBARDO, M. P., AND E. KEMLY. 1983. A radio-control method for trapping birds in nest boxes. J. Field Ornithol. 54:194–195.

LOW, S. H. 1935. Methods of trapping shore birds. Bird-Banding 6:16–22.

———. 1957. Banding with mist nets. Bird-Banding 28:115–128.

LUDWIG, J., AND D. E. DAVIS. 1975. An improved woodchuck trap. J. Wildl. Manage. 39:439–442.

MACE, R. U. 1971. Trapping and transplanting Roosevelt elk to control damage and establish new populations. Proc. Annu. Conf. West. Assoc. Game and Fish Comm. 51:464–470.

MALCOLM, J. R. 1991. Comparative abundances of neotropical small mammals by trap height. J. Mammal. 72:188–192.

MALY, M. S., AND J. A. CRAWFORD. 1985. Relative capture efficiency of large and small Sherman live traps. Acta Theriol. 30:165–167.

MARTIN, S. G. 1969. A technique for capturing nesting grassland birds with mist nets. Bird-Banding 40:233–237.

MASTERS, R. 1978. Deer trapping, marking and telemetry techniques. State Univ. New York Coll. Environ. Sci. For., Adirondack Ecol. Cent., Newcomb, N.Y. 72pp.

MATTFELD, G. F., J. E. WILEY, AND D. F. BEHREND. 1972. Salt versus browse—seasonal baits for deer trapping. J. Wildl. Manage. 36:996–998.

MAUTZ, W. W., R. P. DAVISON, C. E. BOARDMAN, AND H. SILVER. 1974. Restraining apparatus for obtaining blood samples from white-tailed deer. J. Wildl. Manage. 38:845–847.

MAZZOTTI, J. J., AND L. A. BRANDT. 1988. A method of live-trapping wary crocodiles. Herpetol. Rev. 19:40–41.

MCBEATH, D. Y. 1941. Whitetail traps and tags. Mich. Conserv. 10(11):6–7,11, 10(12):6–7.

MCCABE, T. R., AND G. ELISON. 1986. An efficient live-capture technique for muskrats. Wildl. Soc. Bull. 14:282–284.

MCCALL, J. D. 1954. Portable live trap for ducks, with improved gathering box. J. Wildl. Manage. 18:405–407.

MCCLURE, H. E. 1956. Methods of bird netting in Japan applicable to wildlife management problems. Bird-Banding 27:67–73.

MCCOWN, J. W., D. S. MAEHR, AND J. ROBOSKI. 1990. A portable cushion as a wildlife capture aid. Wildl. Soc. Bull. 18:34–36.

MCCULLOUGH, C. R., L. D. HEGGEMANN, AND C. H. CALDWELL. 1986. A device to restrain river otters. Wildl. Soc. Bull. 14:177–180.

MCCULLOUGH, D. R. 1975. Modification of the Clover deer trap. Calif. Fish Game 61:242–244.

MCNICHOLL, M. K. 1983. Use of a noosing pole to capture common nighthawks. North Am. Bird Bander 8:104–105.

MECHLIN, L. M., AND C. W. SHAIFFER. 1980. Net firing gun for capturing breeding waterfowl. J. Wildl. Manage. 44:895–896.

MENG, H. 1971. The Swedish goshawk trap. J. Wildl. Manage. 35:832–835.

MENGAK, M. T., AND D. C. GUYNN, JR. 1987. Pitfalls and snap traps for sampling small mammals and herpetofauna. Am. Midl. Nat. 118:284–288.

MILLER, W. R. 1962. Automatic activating mechanism for waterfowl nest trap. J. Wildl. Manage. 26:402–404.

MITCHELL, R. T. 1963. The floodlight trap: a device for capturing large numbers of blackbirds and starlings at roosts. U.S. Fish Wildl. Serv. Spec. Sci. Rep. Wildl. 77. 14pp.

MONTGOMERY, W. I. 1979. Trap-revealed home range in sympatric populations of *Apodemus sylvaticus* and *A. flavicollis*. J. Zool. (Lond.) 189:535–540.

MORROW, M. E., N. W. ATHERTON, AND N. J. SILVY. 1987. A device for returning nestling birds to their nests. J. Wildl. Manage. 51:202–204.

MOSBY, H. S. 1955. Live trapping objectionable animals. Virginia Polytechnic Inst. Agric. Ext. Serv. Circ. 667. 4pp.

MUNN, C. A. 1991. Tropical canopy netting and shooting lines over tall trees. J. Field Ornithol. 62:454–463.

MURPHY, T. M., JR., AND T. T. FENDLEY. 1973. A new technique for live-trapping of nuisance alligators. Proc. Annu. Conf. Southeast. Assoc. Game and Fish Comm. 27:308–311.

NELLIS, C. H. 1968. Some methods for capturing coyotes alive. J. Wildl. Manage. 32:402–405.

NESBITT, S. A., B. A. HARRIS, R. W. REPENNING, AND C. B. BROWNSMITH. 1982. Notes on red-cockaded woodpecker study techniques. Wildl. Soc. Bull. 10:160–163.

NESBITT, S. N. 1984. Effects of an oral tranquilizer on survival of sandhill cranes. Wildl. Soc. Bull. 12:387–388.

NOLAN, J. W., R. H. RUSSELL, AND F. ANDERKA. 1984. Transmitters for monitoring Aldrich snares set for grizzly bears. J. Wildl. Manage. 48:942–945.

NOLAN, V., JR. 1961. A method of netting birds at open nests in trees. Auk 78:643–645.

NORTHCOTT, T. H., AND D. SLADE. 1976. A livetrapping technique for river otters. J. Wildl. Manage. 40:163–164.

NOVAK, M. 1981. The foot-snare and the leg-hold traps: a comparison. Pages 1671–1685 in J. A. Chapman and D. Pursley, eds. Proc. Worldwide Furbearer Conf., Frostburg, Md.

O'GARA, B. W., AND D. C. GETZ. 1986. Capturing golden eagles using a helicopter and net gun. Wildl. Soc. Bull. 14:400–402.

OLSEN, G. H., S. B. LINHART, R. A. HOLMES, G. J. DASCH, AND C. B. MALE. 1986. Injuries to coyotes caught in padded and unpadded steel foothold traps. Wildl. Soc. Bull. 14:219–223.

ONDERKA, D. K., D. L. SKINNER, AND A. W. TODD. 1990. Injuries to coyotes and other species caused by four models of foot-holding devices. Wildl. Soc. Bull. 18:175–182.

OTNES, G. L. 1991. An alternate method of netting shorebirds in the Canadian subarctic. North Am. Bird Bander 15:139–140.

PAGELS, J. F., AND T. W. FRENCH. 1987. Discarded bottles as a source of small mammal distribution data. Am. Midl. Nat. 118:217–219.

PALMISANO, A. W., AND H. H. DUPUIE. 1975. An evaluation of steel traps for taking fur animals in coastal Louisiana. Proc. Annu. Conf. Southeast. Assoc. Game and Fish Comm. 29:342–347.

PAPEZ, N. J., AND G. K. TSUKAMOTO. 1970. The 1969 sheep trapping and transplant program in Nevada. Trans. Desert Bighorn Counc. 14:43–50.

PASSMORE, M. F. 1979. Use of Velcro for handling birds. Bird-Banding 50:369.

PEDERSEN, R. J., AND A. W. ADAMS. 1977. Summer elk trapping with salt. Wildl. Soc. Bull. 5:72–73.

PETERS, W. D. 1986. An improved Grave's tree trap. North Am. Bird Bander 11:10.

PHILLIPS, B. 1978. Hanging a Dho-Gaza. Inl. Bird-Banding News 50:211–217.

PREVOST, Y. A., AND J. M. BAKER. 1984. A perch snare for catching ospreys. J. Wildl. Manage. 48:991–993.

RALPH, C. J. 1976. Standardization of mist net captures for quantification of avian migration. Bird-Banding 47:44–47.

RAMSEY, C. W. 1968. A drop-net deer trap. J. Wildl. Manage. 32:187–190.

RECHT, M. A. 1981. A burrow-occluding trap for tortoises. J. Wildl. Manage. 45:557–559.

REEVES, H. M., A. D. GEIS, AND F. C. KNIFFIN. 1968. Mourning dove capture and banding. U.S. Fish. Wildl. Serv. Spec. Sci. Rep. Wildl. 117. 63pp.

REMPEL, R. D., AND R. C. BERTRAM. 1975. The Stewart modified corral trap. Calif. Fish Game 61:237–239.

ROPER, L. A., R. L. SCHMIDT, AND R. B. GILL. 1971. Techniques of trapping and handling mule deer in northern Colorado with notes on using automatic data processing for data analysis. Proc. Annu. Conf. West. Assoc. Game and Fish Comm. 51:471–477.

ROTELLA, J. J., AND J. T. RATTI. 1990. Use of methoxyflurane to reduce nest abandonment of mallards. J. Wildl. Manage. 54:627–628.

ROUGHTON, R. D. 1982. A synthetic alternative to fermented eggs as a canid attractant. J. Wildl. Manage. 46:230–234.

RUNGE, W. 1972. An efficient winter live-trapping technique for white-tailed deer. Saskatchewan Dep. Nat. Resour. Tech. Bull. 1. 16pp.

SAUER, B. W., H. A. GORMAN, AND R. J. BOYD. 1969. A new technique for restraining antlerless mule deer. J. Am. Vet. Med. Assoc. 155:1080–1084.

SCHARF, C. S. 1985. A technique for trapping territorial magpies. North Am. Bird Bander 10:34–36.

SCHEMNITZ, S. D. 1961. Ecology of the scaled quail in the Oklahoma panhandle. Wildl. Monogr. 8. 47pp.

SCHLADWEILER, P., AND T. W. MUSSEHL. 1969. Use of mist-nets for recapturing radio-equipped blue grouse. J. Wildl. Manage. 33:443–444.

SCHMITT, S. M., T. M. COOLEY, L. D. SCHRADER, AND M. A. BRADLEY. 1983. A squeeze chute to restrain captive deer. Wildl. Soc. Bull. 11:387–389.

SCHROEDER, M. A. 1986. A modified noosing pole for capturing grouse. North Am. Bird Bander 11:42.

———, AND C. E. BRAUN. 1991. Walk-in traps for capturing greater prairie chickens on leks. J. Field Ornithol. 62:378–385.

SCHULTZ, V. 1950. A modified Stoddard quail trap. J. Wildl. Manage. 14:243.

SCRIVNER, J. H., W. E. HOWARD, AND R. TERANISHI. 1987. Effectiveness of a lure called "coyote control." Wildl. Soc. Bull. 15:272–274.

SERVENTY, D. L., D. S. FARNER, C. A. NICHOLLAS, AND N.E. STEWART. 1962. Trapping and maintaining shore birds in captivity. Bird-Banding 33:123–130.

SHAIFFER, C. W., AND G. L. KRAPU. 1978. A remote controlled system for capturing nesting waterfowl. J. Wildl. Manage. 42:668–669.

SHARP, D. E., AND J. T. LOKEMOEN. 1980. A remote-controlled firing device for cannon net traps. J. Wildl. Manage. 44:896–898.

———, AND ———. 1987. A decoy trap for breeding-season mallards in North Dakota. J. Wildl. Manage. 51:711–715.

SHULER, J. F., D. E. SAMUEL, B. P. SHISSLER, AND M. R. ELLINGWOOD. 1986. A modified nightlighting technique for male American woodcock. J. Wildl. Manage. 50:384–387.

SKINNER, D. L., AND A. W. TODD. 1990. Evaluating efficiency of footholding devices for coyote capture. Wildl. Soc. Bull. 18:166–175.

SMITH, H. D., F. A. STORMER, AND R. D. GODFREY, JR. 1981. A collapsible quail trap. U.S. For. Serv. Res. Note RM-400. 3pp.

SMITH, L. M., J. W. HUPP, AND J. T. RATTI. 1980. Reducing abandonment of nest-trapped gray partridge with methoxyflurane. J. Wildl. Manage. 44:690–691.

SMITH, N. G. 1967. Capturing seabirds with avertin. J. Wildl. Manage. 31:479–483.

SOWLS, L. K. 1955. Prairie ducks: a study of their behavior, ecology and management. Wildl. Manage. Inst., Washington, D.C., and Stackpole Co., Harrisburg, Pa. 193pp.

SPARROWE, R. D., AND P. F. SPRINGER. 1970. Seasonal activity patterns of white-tailed deer in eastern South Dakota. J. Wildl. Manage. 34:420–431.

SPITZKEIT, J. W., J. R. NAWROT, AND W. B. KLIMSTRA. 1987. A portable swim-in trap for waterfowl. Wildl. Soc. Bull. 15:189–191.

SRIVASTAVA, V., AND R. C. SRIVASTAVA. 1985. Trapping rodents with glue. Indian J. Agric. Sci. 55:385–386.

STAFFORD, S. K., AND L. E. WILLIAMS, JR. 1968. Data on capturing black bears with alpha-chloralose. Proc. Annu. Conf. Southeast. Assoc. Game and Fish Comm. 22:161–165.

STEWART, P. A. 1954. Combination substratum and automatic trap for nesting mourning doves. Bird-Banding 25:6–8.

———. 1971. An automatic trap for use on bird nesting boxes. Bird-Banding 42:121–122.

STEWART, R. E. 1951. Clapper rail populations of the Middle Atlantic states. Trans. North Am. Wildl. Conf. 16:421–430.

STODDART, D. M. 1982a. Demonstration of olfactory discrimination by the short-tailed vole, *Microtus agrestis* L. Anim. Behav. 30: 293–301.

———. 1982b. Does trap odour influence estimation of population size of the short-tailed vole, *Microtus agrestis*? J. Anim. Ecol. 51:375–386.

STORM, G. L., R. D. ANDREWS, R. L. PHILLIPS, R. A. BISHOP, D. B. SINIFF, AND J. R. TESTER. 1976. Morphology, reproduction, dispersal, and mortality of midwestern red fox populations. Wildl. Monogr. 49. 82pp.

———, AND K. P. DAUPHIN. 1965. A wire ferret for use in studies of foxes and skunks. J. Wildl. Manage. 29:625–626.

STOUFFER, P. C., AND D. F. CACCAMISE. 1991. Capturing American crows using alpha-chloralose. J. Field Ornithol. 62:450–453.

STUTCHBURY, B. J., AND R. J. ROBERTSON. 1986. A simple trap for catching birds in nest boxes. J. Field Ornithol. 57:64–65.

SUGDEN, L. G. 1956. A technique for closing gates on big-game live traps by remote control. J. Wildl. Manage. 20:467.

———, AND H. J. POSTON. 1970. A raft trap for ducks. Bird-Banding 41:128–129.

SULLIVAN, J. B., C. A. DEYOUNG, S. L. BEASOM, J. R. JEFFELFINGER, S. P. COUGHLIN, AND M. W. HELLICKSON. 1991. Drive netting: incidence of mortality. Wildl. Soc. Bull. 19:393–396.

SWANK, W. G. 1952. Trapping and marking of adult nesting doves. J. Wildl. Manage. 16:87–90.

SYKES, P. W., JR. 1985. Construction of portable net poles and transport containers. North Am. Bird Bander 10:115–116.

SZYMCZAK, M. R., AND J. F. COREY. 1976. Construction and use of the Salt Plains duck trap in Colorado. Colo. Div. Wildl. Rep. 6. 13pp.

TABER, R. D., AND I. McT. COWAN. 1969. Capturing and marking wild animals. Pages 277–317 in R. H. Giles, ed. Wildlife management techniques. Third ed. The Wildl. Soc., Washington, D.C.

TEHSIN, R. H. 1988. Inducing sleep in birds. J. Bombay Nat. Hist. Soc. 85:435–436.

TEW, T. 1987. A comparison of small mammal responses to clean and dirty traps. J. Zool. (Lond.) 212:361–364.

THOMPSON, M. J., R. E. HENDERSON, T. O. LEMKE, AND B. A. STERLING. 1989. Evaluation of a collapsible Clover trap for elk. Wildl. Soc. Bull. 17:287–290.

THORNSBERRY, W. H., AND L. M. COWARDIN. 1971. A floating bail trap for capturing individual ducks in spring. J. Wildl. Manage. 35:837–839.

TIMM, D. E., AND R. G. BROMLEY. 1976. Driving Canada geese by helicopter. Wildl. Soc. Bull. 4:180–181.

TOEPFER, J. E., J. A. NEWELL, AND J. MONARCH. 1988. A method for trapping prairie grouse hens on display grounds. Pages 21–31 in Prairie chickens on the Sheyenne National Grasslands. U.S. For. Serv. Gen. Tech. Rep. RM-159.

TOMLINSON, R. E. 1963. A method for drive-trapping dusky grouse. J. Wildl. Manage. 27:563–566.

TROYER, W. A., R. J. HENSEL, AND K. E. DURLEY. 1962. Livetrapping and handling of brown bears. J. Wildl. Manage. 26: 330–331.

TULLAR, B. F., JR. 1984. Evaluation of a padded leg-hold trap for capturing foxes and raccoons. N.Y. Fish Game J. 31:97–103.

TURKOWSKI, F. J., A. R. ARMISTEAD, AND S. B. LINHART. 1984. Selectivity and effectiveness of pan tension devices for coyote foothold traps. J. Wildl. Manage. 48:700–708.

———, M. L. POPELKA, AND R. W. BULLARD. 1983. Efficacy of odor lures and baits for coyotes. Wildl. Soc. Bull. 11:136–145.

TUTTLE, M. D. 1974. An improved trap for bats. J. Mammal. 55:475–477.

TWEIT, R. C. 1982. A holding box for birds. North Am. Bird Bander 7:49.

URBANEK, R. P., J. L. McMILLEN, AND T. A. BOOKHOUT. 1991. Rocket-netting of greater sandhill cranes on their breeding grounds at Seney National Wildlife Refuge. Pages 241–245 in Proc. 1987 Int. Crane Workshop. Int. Crane Found., Baraboo, Wis.

U.S. FISH AND WILDLIFE SERVICE AND CANADIAN WILDLIFE SERVICE. 1977. North American bird banding manual. Vol. II. U.S. Dep. Inter., Washington, D.C. Var. pagin.

VALKENBURG, P., R. D. BOERTJE, AND J. L. DAVIS. 1983. Effects of darting and netting on caribou in Alaska. J. Wildl. Manage. 47: 1233–1237.

VAN BALLENBERGHE, V. 1984. Injuries to wolves sustained during live-capture. J. Wildl. Manage. 48:1425–1429.

———, A. W. ERICKSON, AND D. BYMAN. 1975. Ecology of the timber wolf in northeastern Minnesota. Wildl. Monogr. 43. 43pp.

WARNER, D. R., AND D. L. CHESEMORE. 1985. A technique to secure small mammal livetraps against disturbance. Calif. Fish Game 71: 184–185.

WEAVER, D. K., AND J. A. KADLEC. 1970. A method for trapping breeding adult gulls. Bird-Banding 41:28–31.

WEBB, G. J. W., AND H. MESSEL. 1977. Crocodile capture techniques. J. Wildl. Manage. 41:572–575.

WEBB, W. L. 1943. Trapping and marking white-tailed deer. J. Wildl. Manage. 7:346–348.

WEGNER, W. A. 1981. A carrion-baited noose trap for American kestrels. J. Wildl. Manage. 45:248–250.

WEILAND, E. C. 1964. Methods of tripping traps with a solenoid. Inl. Bird-Banding News 36:3–4,7,9.

WEST, S. D. 1985. Differential capture between old and new models of the Museum Special snap trap. J. Mammal. 66:798–800.

WHEELER, R. H., AND J. C. LEWIS. 1972. Trapping techniques for sandhill crane studies in the Platte River Valley. U.S. Fish Wildl. Serv. Resour. Publ. 107. 19pp.

WHITE, M., F. F. KNOWLTON, AND W. C. GLAZENER. 1972. Effects of dam-newborn fawn behavior on capture and mortality. J. Wildl. Manage. 36:897–906.

WILBUR, S. R. 1967. Live-trapping North American upland game birds. U.S. Fish Wildl. Serv. Spec. Sci. Rep. Wildl. 106. 37pp.

WILLIAMS, D. F., AND S. E. BRAUN. 1983. Comparison of pitfall and conventional traps for sampling small mammal populations. J. Wildl. Manage. 47:841–845.

WILLIAMS, L. E., JR. 1966. Capturing wild turkeys with alpha-chloralose. J. Wildl. Manage. 30:50–56.

———, AND R. W. PHILLIPS. 1972. Tests of oral anesthetics to capture mourning doves and bobwhites. J. Wildl. Manage. 36:968–971.

———, AND ———. 1973. Capturing sandhill cranes with alpha-chloralose. J. Wildl. Manage. 37:94–97.

WILLIAMSON, M. J., AND M. R. PELTON. 1972. New design for a large portable mammal trap. Proc. Annu. Conf. Southeast. Assoc. Game and Fish Comm. 25:315–322.

WILSON, R. P., AND M. T. J. WILSON. 1989. A minimal-stress bird-capture technique. J. Wildl. Manage. 53:77–80.

WINCHELL, C. S., AND J. W. TURMAN. 1992. A new trapping technique for burrowing owls: the noose rod. J. Field Ornithol. 63:66–70.

WOOTEN, W. A. 1955. A trapping technique for band-tailed pigeons. J. Wildl. Manage. 19:411–412.

WRIGHT, J. M. 1983. Ketamine hydrochloride as a chemical restraint for selected small mammals. Wildl. Soc. Bull. 11:76–79.

Yosef, R., and F. E. Lohrer. 1992. A composite treadle/bal-chatri trap for loggerhead shrikes. Wildl. Soc. Bull. 20:116–118.

Zemlicka, D. E., and K. J. Bruce. 1991. Comparison of handmade and molded rubber tranquilizer tabs for delivering tranquilizing materials to coyotes captured in steel leg-hold traps. Proc. Great Plains Wildl. Damage Workshop 10:52–56.

Zicus, M. C. 1989. Automatic trap for waterfowl using nest boxes. J. Field Ornithol. 60:109–111.

Zoellick, B. W., and N. S. Smith. 1986. Capturing desert kit foxes at dens with box traps. Wildl. Soc. Bull. 14:284–286.

Zwickel, F. C., and J. F. Bendell. 1967. A snare for capturing blue grouse. J. Wildl. Manage. 31:202.

6

CHEMICAL IMMOBILIZATION OF LARGE MAMMALS

Daniel B. Pond and Bart W. O'Gara

INTRODUCTION

The days when anyone can or should simply buy a capture gun, darts, and drugs and learn by trial and error how to immobilize and care for wild mammals is long past. Any person contemplating capture of a mammal by using an immobilizing drug should be guided by the following statement: "Use of 'Cap-Chur' guns or darts to shoot a sedative into the flank of a large mammal requires knowledge of proper dosage and adequate logistical support to track the mammal until the sedative takes effect. Unless the investigator has considerable experience in the use of this capture method, it is recommended that the advice of a wildlife veterinarian be obtained. Location of the mammal and time required for sedation should be considered to avoid injury or drowning of sedated mammals. Sedated mammals should be monitored closely and should not be released until they recover normal locomotor capabilities" (Ad Hoc Committee on Acceptable Field Methods in Mammalogy 1987:9).

Not only should advice be sought from a wildlife veterinarian, but a qualified veterinarian should be a member of the capture team. The Code of Federal Regulations 9 contains Uniform Rules of Practice for the Department of Agriculture, which enforces the Animal Welfare Act. The Code requires that research facilities, animal dealers, and animal exhibitors shall have an attending veterinarian to provide adequate guidance to persons involved in the care and use of animals regarding handling, immobilization, anesthesia, tranquilization, and euthanasia. The Code directs that handling of all animals shall be done as expe-

diently and carefully as possible in a manner that does not cause trauma, overheating, excessive cooling, behavioral stress, or unnecessary discomfort. The Code is revised at least once each calendar year, so anyone handling animals must be constantly aware of changes in regulations.

In the United States, the Food and Drug Administration (FDA) determines the species of mammals on which a drug may be used. Restrictions that users should memorize and comply with are printed on drug containers. The FDA also lists the level of controlled substances for which a registration certificate is needed for possession or administration of each drug. Registration certificates can be obtained from the U.S. Drug Enforcement Administration (DEA), P.O. Box 28083, Central Station, Washington, DC 20005. Anyone possessing or administering restricted drugs must have the proper registration certificate or work under the supervision of someone who has such a certificate. The holder of a registration certificate is responsible for record keeping, proper storage and use of drugs, safety of users, and welfare of animals. States also have veterinary practice codes that should be known and followed by anyone administering drugs to wildlife.

A person who immobilizes an animal is assuming a responsibility that should not be considered lightly. Each immobilization has some effect on the behavior, other activities, or life of an animal. From a humane and moral standpoint, minimum restraint should be used. Each time immobilizing an animal is proposed, two questions should be asked: What procedure will produce the least hazard? Who is most qualified to accomplish the task in the least amount of time and with the least stress to the animal?

Table 1. Recommended dosage ranges for chemical restraint agents. (All dosages are mg/kg unless otherwise noted; K = ketamine HCl, C = carfentanil citrate, X = xylazine HCl.)

Common name	Carfentanil citrate[a] and carfentanil citrate : xylazine HCl	Tiletamine HCl and zolazepam HCl[a]	Ketamine HCl[b]	Xylazine HCl[b]	Xylazine HCl : ketamine HCl
Wolf		4	5.6	5–10	2.2X:4.4K
Coyote		4	5.6	3	
Black bear	0.009–0.020	4	6–12	4–15	2X:4K
Grizzly bear	0.015–0.025	4	8–15		2X:4–5K
Badger			6–15		
Skunk		5–14	3–10		
Raccoon		0.8–25	3–10		
Mountain lion		2–4	5–11	3–10	0.5–2X:11K
Bobcat		1.3–3	5–11	4–15	
Lynx			5–11	4–15	
Deer	0.001–0.002	8–15		1–6	
Elk	0.006–0.014	9.2		1–6	
Moose	0.006–0.014	3–8			
Bighorn sheep	0.003–0.010	4.4–5.5			
Bison	0.004	3–4.5			0.5X:0.7K
Pronghorn	2C:33X[c]	13		45[c]	30X:20K[c]

[a]Available from distributors and Wildlife Laboratories, Inc., 1401 Duff Drive, Suite 600, Ft. Collins, CO 80524.
[b]Available from most veterinarians.
[c]Total dosage.

To be successful in immobilizing animals, one must understand their behavioral characteristics and have a working knowledge of the tools of restraint. Complete understanding is gained only through experience, preferably with the guidance of qualified individuals.

The general principles of chemical restraint will be discussed and tables will be presented to give current usage of the drugs we consider best to use for various species as well as drug reactions and complications. This chapter is intended to provide quick, general ''how to'' information; it is by no means complete because of the limited space available. Common trade names for drugs are presented in parentheses after generic names to aid persons looking up drugs they have heard referred to by trade names. For more complete information, consult the references or ask someone who has extensive experience. The tables can be copied and included as a reference outline in a drugging kit.

No single drug meets all requirements of safety and effectiveness for immobilizing mammals. However, certain drugs meet most of the qualifications for individual species. Dosages for all drugs discussed are listed in Table 1. The wide range of dosages reflects the fact that these drugs are used under varying conditions. The lower dose range generally is for animals adapted to confinement. Higher dosages usually are necessary for wild animals or animals that are highly excited.

An ideal drug has a high therapeutic index (lethal dose/effective dose), allowing for individual differences in physiology and error in estimating body weight. Drugs should be physically and chemically compatible with other useful drugs. Often, use of drugs in combination decreases the required dose for each and increases the effectiveness of immobilization.

Most drugs used in wildlife fieldwork are administered intramuscularly. An ideal drug causes a minimum of localized pain. It should be in high enough concentrations to allow the use of small volumes and thereby decrease the chance of muscle tearing and bruising as the liquid is injected under pressure.

As an example of minimizing drug volumes, ketamine hydrochloride (hydrochloride = HCl) and xylazine HCl are used commonly for immobilizing large mammals. The dosage rate, however, necessitates use of large volumes. Volumes can be reduced by freeze-drying the drugs to their crystalline state and remixing with sterile water at a higher concentration. If ketamine HCl and xylazine HCl are to be used together, a good practice is to freeze-dry the ketamine HCl and use xylazine HCl to remix the ketamine HCl at the required concentration and in the proper ratio. In this manner, these drugs are mixed routinely in concentrations of 200 mg ketamine HCl with 100 mg xylazine HCl per milliliter of solution. In the field, such mixtures should be carried in an inside pocket or other warm place until used. Solutions higher than 250 mg/ml of ketamine HCl must be kept heated for the drug to remain in solution.

A short induction time is desirable. No drug causes instant immobilization upon injection. Most drugs in routine use require 5–20 minutes after intramuscular injection before the animal is immobilized. This is a serious drawback for capturing free-ranging animals because these animals may run a great distance before being immobilized (if at all). As a result, one may not be able to locate the animal or it might run over a precipice or into water.

The ideal drug should have an antagonist (antidote). An antagonist allows for a short downtime (time the animal is immobilized) and allows reversing the drug's effect if problems arise.

Only drugs that are commonly used for wildlife work are included in this chapter, with a brief discussion of the action and effects of each drug. Each technical term is defined at its initial appearance.

NARCOTIC ANALGESICS

Analgesic drugs play an important role in wildlife immobilization. Within the last decade, analgesics such as etorphine HCl and carfentanil citrate have been introduced for use in large mammals. These agents are important in the alleviation of pain, and they are valuable in facilitating restraint and handling. Carfentanil citrate has largely replaced etorphine HCl for immobilization of wildlife because of the small volume of drug required; consequently, etorphine HCl is no longer produced in the United States. Thus, only carfentanil citrate will be discussed here; information on etorphine HCl may be obtained from Wallach et al. 1967, Hatch et al. 1976, Ballard et al. 1982, Franzman et al. 1984, Seal et al. 1985, Boever 1986, and Karesh et al. 1986.

Carfentanil Citrate (Carfentanil)

Carfentanil citrate is intended for use in wild, exotic, or free-ranging species; however, it must never be used unless an adequate supply of antagonist (naloxone HCl) is immediately available. When using carfentanil citrate, one must be prepared to employ procedures to maintain an animal under anesthesia. Users must have necessary equipment (especially to assist during respiratory depression), supplies, and experienced personnel to handle situations that may occur during or following immobilization to minimize possible injury to the animal or personnel. Be advised that accidental human exposure to carfentanil can cause death. Never work alone, and always wear eye protection and rubber gloves. All members of the drugging team should be trained in using cardiopulmonary resuscitation (CPR), in recognizing the effects of narcosis, and in administering the antidote.

Carfentanil citrate is supplied at a concentration of 3 mg/ml and has approximately 30,000 times the analgesic potency of morphine (analgesia = loss of pain without producing sleep). Hyperthermia is frequently observed, especially if the drug is used alone. Renarcotization (an animal is again affected by the drug after it has apparently recovered) may occur, especially in large mammals such as moose and polar bears.

ADMINISTRATION

Carfentanil citrate is administered intramuscularly (often in combination with xylazine HCl), and immobilization occurs 2–10 minutes following administration (Meuleman et al. 1984). The lower end of the dose range is suggested for animals that are quiet, confined, or in poor physical condition. The upper dose range is suggested for animals that are excited, after extensive pursuit, or when an extremely short induction time is required (Jessup et al. 1984).

SIDE EFFECTS

Side effects that are manifested after administration of carfentanil citrate are varied. Those reported include excitement during induction, tremors, excessive salivation, tachycardia (rapid heart beat), tachyapnea (rapid, shallow breathing), rumen regurgitation, lingual paralysis, and delayed renarcotization.

Fatal hyperthermia may occur. Extreme care must be used during times of high environmental temperatures and high humidity, after extended pursuit, or during any other activity that produces elevated body temperatures. Equipment to shade and cool the animal should be available when carfentanil citrate is used during hot weather.

Carfentanil citrate is dangerous to humans. It can produce narcosis through contact with mucous membranes or breaks in skin. Accidental human exposure may produce central nervous system (CNS) depression, resulting in respiratory depression or failure followed by coma or death. Depending on its route of administration, effects may be noted in 2–30 minutes. Treatment should start immediately by administering naloxone HCl intravenously or intramuscularly, maintaining an open airway, and being prepared to provide CPR. Naloxone HCl effects are transient, therefore additional doses may be necessary. Trained personnel and first-aid equipment should be available during immobilizing procedures with carfentanil citrate. Rubber gloves and glasses should be worn while darts and syringes are loaded (Parker and Haigh 1982).

ANTAGONIST

Reversal of carfentanil citrate narcosis is accomplished by administering naloxone HCl at 80–100 times the carfentanil dose (Parker and Haigh 1982, Jessup et al. 1984). Intravenous administration causes almost immediate results. Intramuscular injections require 2–10 minutes to effect reversal.

INDICATIONS

Carfentanil citrate has been used on a variety of wild and exotic species. It is the drug of choice for ungulates except Perissodactyla, especially when rapid induction times are desirable. Carfentanil citrate must not be used on animals intended for human consumption. Many veterinarians and research personnel interpret this to mean it must not be used on game species during or for 30 days before hunting seasons.

DISSOCIATIVE ANESTHETICS

Three drugs in this category are phencyclidine HCl, ketamine HCl, and tiletamine HCl with zolazepam HCl. The latter two are currently the most important. A cataleptic-like state referred to as "dissociative anesthesia" is typical when these drugs are used and is accompanied by marked analgesia in most species. The mechanism of action of these anesthetics is not fully understood. Generally, all exhibit the same pharmacological effects.

The conventional stages of anesthesia commonly described for anesthetics do not fit cataleptoid anesthetics. The term cataleptoid refers to waxy rigidity or skeletal muscle hypertonia (muscle contraction, stiffness). In discussions of cataleptoid anesthesia, the following categorization of phases may be used:

(1) Induction: time from intramuscular injection to loss of righting reflex (LRR).

(2) Cataleptoid anesthesia: from LRR until animal lifts head (HL).
(3) Surgical anesthesia: that portion of cataleptoid anesthesia during which the animal fails to show outward manifestation (limb withdrawal, shoulder twitching, blinking, or vocalization) in response to a strong stimulus (e.g., loud noise or pain).
(4) Emergence: from HL to standing.
(5) Recovery: time from standing until return to normal pretreatment state.

Phencyclidine Hydrochloride (Sernylan)

Phencyclidine HCl has numerous federal restrictions and is not readily available. For further information, review Seal and Erickson 1969, Hornocker and Wiles 1972, Bush et al. 1980, Ballard et al. 1982, and Kreeger 1987.

Ketamine Hydrochloride (Ketaset)

Ketamine HCl is easily obtained by any veterinarian and is inexpensive. It is available at concentrations of 20, 50, and 100 mg/ml.

Animals usually retain normal to somewhat diminished pharyngeal-laryngeal (swallowing and coughing) reflexes, minimizing the chances of inhaling food, vomitus, or saliva. Induction is characterized by uncoordinated movement, and the eyes become fixed and dilated. The animal lies down and displays a characteristic licking motion (serpentine tongue). In initial stages, animals become hypersensitive to sounds, but they rapidly become insensitive to external stimulation. Lateral nystagmus (rapid oscillation of the eyes) appears, then disappears with increased depth of anesthesia (Booth 1982).

Ketamine HCl is not a muscle relaxant. Catatonia (involuntary muscle contraction) is common. Marked analgesia is rapidly produced in the body except in the peritoneum (lining of the abdomen), but it may be less than necessary for abdominal surgery. Eyelids remain open with a ''fixed'' expression of the eyes. Palpebral reflexes (blinking reflex produced by touching the inside corner of the eyelid) usually remain. Pupils are dilated (Beck et al. 1971).

ADMINISTRATION

In wildlife work, intramuscular or subcutaneous injections are the usual routes of administration. Dose variations among species are tremendous (Bush et al. 1980).

First signs of induction appear 3–5 minutes after injection, and complete immobilization occurs in 5–10 minutes. Downtime varies with species and dosage: 15–30 minutes are usually available to complete procedures (Hash and Hornocker 1980). Most animals are fully ambulatory within 1 hour if no CNS depressant drugs are given in addition to ketamine HCl.

Recovery is usually smooth and uneventful. Long recovery periods of up to 5 hours are common. Domestic cats sometimes show slight depression for 24 hours after anesthesia.

SIDE EFFECTS

Convulsions occur in a small percentage of felids given ketamine HCl alone. The chances of convulsions can be reduced by administering diazepam (0.1–0.25 mg/kg) or acepromazine maleate (0.50 mg/kg) intramuscularly with the ketamine HCl. Once convulsions have begun, diazepam given intravenously may be required to control them. If seizures occur, body heat may be increased.

Prolonged apnea (slow, shallow breathing) occasionally occurs in large cats such as mountain lions, and assisting respiration may be necessary (Logan et al. 1986). Salivation that is excessive and not controlled by automatic swallowing can be readily controlled with atropine sulfate (Ramsden et al. 1976).

Hypothermia is not a common side effect of ketamine HCl. However, prolonged deep anesthesia during cold weather may result in hypothermia.

Accidental oral ingestion or injection can produce anesthesia in humans. Hallucinations may occur. Respiration should be monitored and assisted if necessary. Prompt hospitalization is required.

ANTAGONIST

Yohimbine (Hatch and Ruch 1974, Jessup et al. 1983) is the antagonist most commonly used.

INDICATIONS

Ketamine HCl is ideal for small carnivores (Wright 1983). In combination with xylazine HCl (discussed later), it is excellent for large carnivores, satisfactory for deer (Kreeger et al. 1986*a*), but not suitable for many ungulates. When ketamine HCl or a ketamine HCl/xylazine HCl combination is used on large animals, freeze-drying the drugs and remixing at higher concentrations will allow the use of smaller volumes. This is especially advantageous when darting equipment is used.

Tiletamine Hydrochloride and Zolazepam Hydrochloride (Telazol or CI-744)

Tiletamine HCl and zolazepam HCl combine to produce a non-narcotic, nonbarbiturate, injectable anesthetic, and a controlled substances registration certificate is required to obtain them. Tiletamine HCl and zolazepam HCl are supplied in sterile vials containing 500 mg of active drug (250 mg of each). Reconstitution with 5 ml of sterile water results in a 100-mg/ml solution. Concentrations up to 500 mg/ml may be obtained if less water is added, facilitating use of smaller projectile syringes.

Tiletamine HCl produces profound analgesia and cataleptoid anesthesia characterized by rapid induction and normal pharyngeal-laryngeal reflexes. Used alone, it does not provide adequate muscle relaxation for abdominal surgical procedures. When it is combined with zolazepam HCl, good muscle relaxation is attained. Low dosages produce a tranquil state, permitting or expediting procedures not requiring complete anesthesia. Higher dosages are required for excited animals (Kaufman and Hahnenberger 1975, Taylor et al. 1989).

At low dosage, the palpebral reflex remains. The eyes are not closed and immobilization cannot be monitored by this reflex. Application of a bland eye ointment and shade are necessary to prevent drying and sun damage. Pinnal (involuntary reflex of the ear when the ear canal is touched) and pedal (withdrawal of foot when pain is felt between toes) reflexes remain. The pinnal and pedal reflexes are markedly reduced or abolished at higher dosages. Pupils are slightly dilated. Nystagmus and serpentine tongue also occur.

ADMINISTRATION

Tiletamine HCl and zolazepem HCl may be given intravenously or intramuscularly. Most immobilizations will require intramuscular injections. Dosages are quite variable depending on species (Schobert 1987).

First signs of induction appear 2–6 minutes after injection, and immobilization is complete in 5–10 minutes. Downtime and recovery time are variable among species. Immobilization time is usually 30–60 minutes, but full recovery may require up to 5 hours.

During induction, motor paralysis proceeds from the rear limbs to the forelimbs (Clausen et al. 1984). Upon recovery, these effects are reversed.

SIDE EFFECTS

Excessive salivation is often present unless premedication with atropine sulfate has taken place. Salivation is usually not a problem because pharyngeal-laryngeal reflexes are maintained. Atropine sulfate at a rate of 0.04–0.1 mg/kg will control salivation. At higher dosages, chronic seizures may occur, especially involving facial and limb muscles. An irregular respiratory rate frequently tending toward an inspiratory breath-holding pattern may be seen. Irregular breathing may be stabilized by administering doxapram HCl (Hsu et al. 1985). Administration of doxapram HCl may significantly shorten the arousal time (recovery), but it should be used only when absolutely necessary because doxapram HCl is a powerful stimulant that increases oxygen requirements of tissues and can produce brain damage or death. Mild to severe ataxia may occur. Some ataxia may be observed after recovery. Hypothermia may occur if prolonged, deep anesthesia is maintained during cold weather.

ANTAGONIST

Doxapram HCl may shorten arousal time (Hatch et al. 1984, 1985; Hsu et al. 1985), but arousal may be brief.

INDICATIONS

The combination of tiletamine HCl and zolazepam HCl has a wide safety margin in a variety of species. This drug is especially suited for the larger carnivores (Haigh et al. 1985, Taylor et al. 1989, Kreeger et al. 1990).

NON-NARCOTIC SEDATIVES
Xylazine Hydrochloride (Rompun)

Xylazine HCl is easily obtained by veterinarians in concentrations of 20 mg/ml and 100 mg/ml. It is a non-narcotic sedative, analgesic, and muscle relaxant. Animals given this drug appear to be sleeping (Booth 1982). Stimulation during induction may prevent sedation. When approached too rapidly, a seemingly sedated animal may rouse explosively, which can be dangerous to the researcher as well as to the animal.

ADMINISTRATION

Xylazine HCl may be given intravenously or intramuscularly. Dosage varies widely among species and routes of administration. Immobilization occurs within 3–5 minutes of intravenous injection and 10–20 minutes or longer of intramuscular injection. Analgesia lasts 10–30 minutes,

but the sleep-like state may persist for hours or even days (Addison and Kolenosky 1979, Boever 1986).

SIDE EFFECTS

Muscle tremors, apnea, bradycardia (slowing of the heart rate), and respiratory depression occasionally occur at standard doses (Clark et al. 1982, Kolata and Rawlings 1982). Explosive response to stimuli can result in injury (Knight 1980). Waiting 5–10 minutes after an animal goes down is advisable before it is approached slowly and quietly. This will allow the animal to go into a deeper sleep. We have sedated free-ranging and captive deer, elk, and bison so they would sleep if left alone, but they could get up and walk with help. This is helpful to persons trying to move a large animal or load it into a trailer. These same animals, if startled during induction, may never go down, even with multiple doses.

Xylazine HCl produces additive effects when combined with tranquilizers and barbiturates (Cronin et al. 1983). If drugs are combined, dosages should be reduced and caution should be exercised (Karesh et al. 1986).

ANTAGONIST

Yohimbine HCl (Hatch et al. 1985, Jessup et al. 1985*b*) is the antagonist most commonly used.

INDICATIONS

Xylazine HCl can be used for mild sedation or complete immobilization (Bauditz 1972). It can be used singly or in combination with other drugs (carfentanil citrate, ketamine HCl) in a variety of species. It is not the drug of choice when animals are in a stressful situation (i.e., elk being darted from a helicopter), with herd animals such as pronghorns, or any time rapid immobilization is desired. Xylazine HCl in combination with ketamine HCl provides excellent immobilization of bears, wolves (Ballard et al. 1982, Kreeger et al. 1986*b*), and mountain lions (Logan et al. 1986). When a combination is used, dose and volume are reduced. Xylazine HCl balances other drugs, induction time is decreased, and recovery (though slowed) is not violent.

TRANQUILIZERS
Acepromazine Maleate (Acepromazine)

Acepromazine maleate is a phenothiazine tranquilizer available by prescription in tablets or injectable liquid. It depresses the CNS, leading to muscle relaxation and reduced physical activity.

ADMINISTRATION

Intramuscular injection is the most common method of administration; intravenous and subcutaneous routes are less common. Tablets may be given orally. When it is given intramuscularly, full effects are not seen for 15–30 minutes (Pusateri et al. 1982). Intravenous injections require 1–2 minutes for full effect, and oral doses require 30–60 minutes.

SIDE EFFECTS

Caution should be exercised when acepromazine maleate is used in conjunction with drugs that cause a reduction in blood pressure.

ANTAGONIST

There is no known antagonist.

INDICATIONS

Acepromazine maleate is rarely used alone as an immobilizing agent. Its muscle relaxing characteristics will complement the effects of phencyclidine HCl, ketamine HCl, tiletamine HCl and zolazepam HCl, and carfentanil citrate.

Diazepam (Valium)

Diazepam is a prescription tranquilizer that cannot be mixed with aqueous solutions. It is available in injection or capsule formulation (Pusateri et al. 1982).

ADMINISTRATION

Diazepam can be given intramuscularly, intravenously (Booth 1982), or orally (Montgomery and Hawkins 1967). Intramuscular injections require 10–35 minutes to take effect, depending on dosage. Effects may last up to 2 hours (Thomas et al. 1967) or longer with multiple doses.

SIDE EFFECTS

Diazepam is incompatible with most immobilizing agents and should not be mixed (Booth 1982). This chemical incompatibility precludes mixing diazepam with other drugs in darts or syringes, and it should not be administered intravenously soon after other intravenous injections have occurred.

ANTAGONIST

There is no known antidote.

INDICATIONS

Convulsive side effects of ketamine HCl and phencyclidine HCl can be controlled with diazepam. Intravenous injection can effectively control convulsions in progress.

ANTAGONISTS
Naloxone Hydrochloride (Naloxone or Narcan)

Naloxone HCl is available only to individuals having a controlled substances registration certificate.

ADMINISTRATION

Intravenous and intramuscular injections are most commonly used. Reversal effects are seen in 0.5–2 minutes after intravenous injections and 5–15 minutes after intramuscular injections (Jessup et al. 1985a). An intermediate reversal time will occur if the dosage is divided between intravenous and intramuscular injections. Dosage is 10–50 times the dosage of etorphine HCl and 80–100 times the dosage of carfentanil citrate (Fowler 1978).

SIDE EFFECTS

Renarcotization of the animal may occur because naloxone HCl is metabolized rapidly.

INDICATIONS

Naloxone HCl is used to reverse the effects of etorphine HCl and carfentanil citrate. Naloxone HCl must be on hand when the narcotic immobilizing agents are used. If human narcotization occurs with etorphine HCl or carfentanil citrate, naloxone HCl may be given to reverse those effects.

Yohimbine Hydrochloride (Yohimbine)

Yohimbine HCl is available in concentrations of 2 and 5 mg/ml and has been approved by the FDA for use in Cervidae. This drug has been used to antagonize the effects of xylazine HCl and ketamine HCl combinations (Jessup et al. 1983, Jacobson and Kollias 1984, Ramsay et al. 1985, Kreeger and Seal 1986).

ADMINISTRATION

Yohimbine HCl is given intravenously or intramuscularly (for a slower effect) at dosages of 0.1–0.35 mg/kg (Hatch et al. 1983, Hsu and Shulaw 1984). Full reversal effects are seen 2–5 minutes after intravenous injection.

SIDE EFFECTS

Animals injected with yohimbine HCl exhibit increased respiratory and heart rates (Hsu et al. 1985).

INDICATIONS

Yohimbine HCl is the antagonist of choice when xylazine HCl and xylazine HCl/ketamine HCl mixtures are used as immobilizing agents.

Doxapram Hydrochloride (Doxapram)

In the U.S., doxapram HCl is FDA approved for use on humans, dogs, cats, and horses, but not on animals intended for human consumption. It is available from veterinarians. It stimulates respiratory activity during the recovery stage of anesthesia. The principal effect on respiration is an increase in tidal volume. After administration the respiratory minute volume may increase more than 200% within 1 minute. However, this increase diminishes in 5–6 minutes when doxapram HCl is used alone. The overall improvement in respiration is characterized by changes in the acid-base status of the blood and in the oxygen tension of arterial blood (Hsu et al. 1985).

ADMINISTRATION

Doxapram HCl can be given intravenously at a rate of 0.4–5.0 mg/kg. The intravenous route will elicit the most rapid response. Injection into the base of the tongue will also produce rapid results.

SIDE EFFECTS

Convulsions may occur at high dosages (>50 times greater than therapeutic dose).

INDICATIONS

Doxapram HCl aids in reversing the effect of immobilization induced by ketamine HCl and ketamine HCl/xylazine HCl (MacKintosh and Van Reenen 1984, Hatch et al. 1985). It is of value primarily in treating respiratory depression from barbiturates and xylazine HCl.

MISCELLANEOUS
Atropine Sulfate

Atropine sulfate is obtainable from veterinarians. The drug is available at concentrations of 0.4–0.6 mg/ml. It decreases salivation, sweating, gut motility, bladder tone, and gastric and respiratory secretions and stabilizes heart rate (Klide et al. 1975).

ADMINISTRATION

Atropine sulfate can be given orally, intravenously, or intramuscularly at 0.04 mg/kg.

SIDE EFFECTS

Animals that dissipate excess heat and moisture by sweating may develop hyperthermia because of sweat inhibition (Hsu et al. 1985). Pupils are dilated and must be protected from sunlight. Because gastric motility is decreased, care should be taken to avoid bloat when atropine sulfate is administered to ruminants.

INDICATIONS

Atropine sulfate diminishes excess salivation caused by etorphine HCl, carfentanil citrate, ketamine HCl, phencyclidine HCl, and xylazine HCl.

Succinylcholine Chloride (Sucostrin or Anectine)

Succinylcholine chloride, a paralyzing agent with no anesthetic or analgesic effects, has been used for many years to immobilize a variety of species. Because it has no anesthetic or analgesic effect, we cannot recommend the use of succinylcholine. However, some researchers may want additional information, and we recommend reading Miller (1968), Allen (1970), Jacobson et al. (1976), and Amstrup and Segerstrom (1981).

CALCULATING DRUG DOSAGE

The responsibility for the correct use and storage of immobilization drugs rests entirely with the user and, in some states, the veterinarian who has delegated their use. Proper storage (out of sunlight and heat) and care (maintaining sterility) will ensure maximum shelf life. When removing drugs from vials, the user should refer to the manufacturer's instructions for storage and use sterile needles and syringes.

Table 1 lists suggested dosages for various species. Average adult dosages are given along with dosage by body weight (mg/kg). It is best to estimate each individual's body weight and calculate its dosage. This is usually more accurate and is safer for the animal than is using a standard dosage for all animals of the same sex, age group, and species.

Dosages should be adjusted according to the size, age, time of year, condition, and emotional status of the animal. We found that bighorn sheep rams in rut required two to three times the normal dosage of xylazine HCl needed during other times of the year. Because of excitement and stress, deer and elk tranquilized from a helicopter usually require a higher dose than do animals in traps.

Some drugs are available in different concentrations. When immobilizing large mammals by remote injection, use the highest concentration possible. A smaller volume of drug is required, and a smaller, more accurate dart can be used. Smaller volumes produce less injection trauma.

An example of dosage calculation is shown in Box 1. As one gains experience, estimating weights and correctly adjusting dosages becomes easier.

> **Box 1. Calculating the dosage of a drug to be used to immobilize a wild ungulate.**
>
> Estimated weight of the animal: 100 kg
> Dosage of drug: 2.6 mg/kg
> Concentration of drug (found on bottle): 100 mg/ml
> Dosage for animal (100 kg body weight $\times$ 2.6 mg/kg): 260 mg
> Volume of drug needed (260 mg dosage $\div$ 100 mg/ml concentration): 2.6 ml
> 1 ml = 1 cc

DRUG DELIVERY SYSTEMS

Many manufacturers produce high-quality injection equipment that employs various methods of drug delivery. The following is a general description of the basic methods used by manufacturers. Selection of appropriate equipment is often a matter of personal preference.

Hand-Held Syringes

The use of hand-held syringes in wildlife fieldwork is limited to animals that are physically restrained. We have used a pole noose to pin to the ground coyotes and badgers in leg-hold traps, making injecting drugs by hand easier. We have also injected deer, elk, mountain goats, and bighorn sheep that were tangled in nets. Metal or plastic syringes are best because they are not as likely as glass to break under pressure or if dropped. A large-gauge needle (16–18 gauge) should be used to deliver the drug quickly. The needle on the syringe should be tightened. Luer-Lok syringes are ideal.

Jab Stick, Pole Syringe

Various homemade and commercial jab sticks act as an extension of the hand for administering drugs to dangerous animals confined in small areas (i.e., bears in culvert traps). All work on the principle of injection immediately upon insertion of the needle. A quick jab is necessary to effect administration. Pressure against the animal must be maintained until all of the drug has been injected; if the animal jumps away, however, a second injection is necessary. Multiple injections with jab sticks are often difficult because the animal is "wise" to the jab stick after the first try. Carnivores often bite at the syringe, damaging it. More often than not, bears in culvert traps bite off the syringe on the first attempt. Hoofed animals are likely to kick, and the quick movement may bend or break the needle. Most commercially made jab sticks have devices that protect the needle and support the hub. Sharp, large-bore needles should be used for their added strength, easy insertion, and rapid injection of the drug. Needle length is dependent on the size of the animal. Needles must be long enough to penetrate muscles, but if they are too long they may bend, and the animal may jump away before injection.

Projected Darts or Syringes

Modern chemical restraint requires equipment capable of projecting a syringe some distance and discharging the

Table 2. Sources of supply of darting equipment.

BALLISTIVET, Inc.
4434 Centerville Road
White Bear Lake, MN 55127

CAP-CHUR Equipment
NASCO
901 Janesville Ave.
Fort Atkinson, WI 53538-0901

CAP-CHUR Equipment
NASCO WEST
1524 Princeton Ave.
Modesto, CA 95352-3837

IDEAL INSTRUMENTS
607 N. Western Avenue
Chicago, IL 60612

PAXARMS Equipment
Telonics
932 E. Impala Ave.
Mesa, AZ 85204-6699

PNEU-DART Inc.
P.O. Box 1415
Williamsport, PA 17703

TELINJECT Equipment
Telinject U.S.A., Inc.
16133 Ventura Blvd., Suite 635
Encino, CA 91436

WILDLIFE TECHNOLOGIES
3118 N. Park Drive
Flagstaff, AZ 86004

Wiley & Sons, Inc.
Exotic Game & Gun Ranch
Rt. 1, Box 303
Wills Point, TX 75169

contents upon impact. Several companies manufacture darts and dart guns (Table 2). Whatever make of equipment is used, the manufacturer's instructions must be thoroughly understood.

Most projection systems use modified firearms or pellet guns. Some short-range projectors are CO_2-powered pistols with a range of about 10 m or CO_2-powered rifles with a range of about 25 m. CO_2-powered projectors usually have the disadvantage of reduced power during cold weather. Carbon dioxide cartridges lose power before becoming depleted. One should be familiar with the projector and replace CO_2 cartridges before this happens.

Long-range projectors use .22-caliber powder charges. The maximum range is about 70 m, but accuracy is generally poor beyond 35 m. The powder charges are available in varying strengths for some projectors; others use one powder charge but have bleeder valves to reduce power for close shots. Darts also can be pushed down the barrels of some projectors to further reduce power.

When a powder-charged gun is used, the large, heavy darts travel at high velocity, resulting in a high-impact force. This impact force causes bruising of tissue and can break bones. Darts have been fired into and completely through the bodies of deer, sheep, elk, and polar bears. Care must be taken when small-bodied animals like coy-

otes or bobcats are darted, especially at close range. However, these are still the best systems available for capturing free-ranging mammals.

Darts used with projection systems usually inject drugs by a powder charge or compressed air. The powder charge is dependable and quick to load, but it has disadvantages. As the dart hits an animal, the drug is discharged in about 0.001 second no matter what its volume. This builds up tremendous pressure inside the injection site. As a result, the dart can "rocket" back out. Barbed needles are usually necessary to prevent the dart from coming out and the drug being lost. Barbed needles should be surgically removed by making a small incision in the skin. Barbs should be modified so that they are as small as possible.

The pressure buildup as the drug discharges also causes trauma to the muscles. Little imagination is required to visualize what happens to muscle tissue when 5–10 ml of a drug are injected in 0.001 second. Bruising and tearing of tissues are accompanied by considerable pain. Besides the risks of injuring or crippling an animal, a painful injection may cause the animal to flee, resulting in the animal's never being immobilized due to the increased excitement, or the animal's being lost.

After a hard impact, or if the wrong size of internal charge is used, the barrels of Cap-Chur darts may expand to the point that they will jam in the gun barrel. Because of this possibility and for greater accuracy, pushing the loaded dart through the gun barrel with a cleaning rod before loading and firing is advisable. Besides verifying that the dart will pass through the barrel, the push-through cleans the barrel the same amount for each shot. A clean barrel throws a dart higher than does a dirty barrel. Paper wadding from the powder charges will also reduce power after 5–10 shots. We advise cleaning the adapter after every fifth shot. When darting from a helicopter, one should carry extra adapters to save valuable time that would be used cleaning the adapter, or to replace one that is dropped and cannot be recovered readily.

Air-charged darts used in long-range projectors are similar in operation to blowgun darts (see description under blowguns). Such darts cause less tissue damage than powder-charged darts and require smaller or no barbs on the needles. Plastic darts may crack upon impact. If this happens, the power setting should be lowered.

Any dart may wobble in flight if its tailpiece is not symmetrical. Fabric tailpieces should be trimmed so they are symmetrical and should be clean. If a dart has greater capacity than the volume of drug to be used, the plunger should be pushed forward until the correct volume is achieved or the volume should be increased with saline solution (the drug should reach the threads in the nose of Cap-Chur darts).

To adjust sights on capture guns and to determine distances that are practical with different power loads, we suggest shooting water-filled darts of the size that will be used. Choose an appropriate target like a piece of carpeting over an open frame to mimic a mammal's skin. The needle should penetrate the target, but the nose of the dart should never penetrate. Insufficient impact force may fail to fire the injector charge. The force of impact needed can be determined while the sights are set.

Blowgun

The blowgun is popular because it is silent, causes little trauma on impact, is adaptable for use on small animals, is easily sighted, and has no mechanical parts to wear out (Wentges 1975). However, it can be unwieldy in confined spaces, and the range is short. Blowguns can be purchased commercially, but electrical conduit, copper, stainless steel, plastic, or other tubing of the correct diameter can be used. The inside of the tube must be smooth to minimize friction. A mouthpiece on the tube helps increase air pressure.

The length of blowguns varies from 1 m to 2 m and is determined by the distance to the target. A longer tube permits greater accuracy. Maximum range varies with the length of tubing and operator skill, but 10 m is average for a level shot. Firing into a tree is difficult.

Various commercial and homemade air- or butane-powered darts are used with blowguns and projectors. The drug is inserted into the front compartment and a plugged needle with a lateral hole drilled into the side is placed on the dart. A piece of sliding plastic tubing fits over the lateral hole. The rear compartment is filled with compressed air or liquid butane, putting the sliding plunger between the two compartments and the drug under pressure. As the needle penetrates the skin, the plastic tubing slides back, allowing the drug to be released. Complete injection requires 0.5–2 seconds, depending on the size of the dart. Blow-dart barrels are generally clear or translucent plastic, and one can see if the plunger has moved to the front of the dart and injected the drug.

Haigh and Hopf (1976) provided instructions for constructing economical butane-powered darts from plastic syringes. We have used such darts on most of the big game in Montana plus bison, coyotes, bobcats, and lynx with no difficulties. Mammals that were hit with these darts generally did not jump on impact and showed no excitement. Homemade blow-darts and their needles should be pressure-checked before use to be sure they do not leak. These darts sometimes break at the needle hub upon impact. Care should be taken to assure direct hits. If a dart strikes at an angle to the surface of the skin, breakage is more likely.

Because of the difficulty in shooting into trees with the blowgun, the senior author designed a simple barrel conversion for a pump-up pellet gun. With the conversion, the gun can shoot darts up to 10 cc in volume 15 m into a tree or 30 m on the level. The power can be varied because the gun is of the pump-up variety. CO_2 pistols are commercially available for blowdarts.

Precautions

Successful immobilization of a wild animal is an art. Many factors are involved. One must consider not only which equipment to use and the animal's condition, but personal ability as well. One not skilled in the use of a particular dart gun will have difficulty in hitting the target.

All drugging equipment should be in working order, cleaned, and lubricated. Needles should be inspected for plugs of skin, cleaned, sharpened, and sterilized. The most common operator fault is missing the target entirely or making an injection at an inappropriate site. Suitable injection sites are major muscle masses. The muscles of the upper hindquarter and the shoulder are preferred sites because they present the largest target areas. Proper site selection is dependent on the animal species and operator skill. The neck is an undesirable target site. The cervical vertebrae lie close to the surface in the midportion of the neck. The trachea, esophagus, and major blood vessels traverse the lower neck. If neck shots are necessary or desirable, use low-impact darts (blow-darts) and short needles. Injection at the wrong site can also result in the needle striking bone, causing a fracture or breaking the needle.

Restraint drugs are not absorbed equally well or at the same rate from all tissues. Injection into fat deposits, connective tissue, abdomen, or the skin may result in prolonged induction or failure of effect. Injections into the abdomen, chest, or a vein may accelerate induction and could lead to an overdose.

With sufficient impact, the whole syringe may be driven through an animal. This is likely to happen if too great a charge is used, if the dart is too heavy, or if distance is misjudged. Such trauma is not only aesthetically and morally undesirable, but it also results in decreased absorption of the drug. Delivery of a syringe by blowgun is much more gentle.

The operation of most discharge mechanisms in darts requires a certain degree of impact, necessitating that the needle enter the skin at a perpendicular angle. If the needle strikes at too acute an angle, the syringe may fail to discharge or glance off the animal.

Infection sometimes develops at the injection site. Cleaning the skin or applying antiseptic prior to injection is impossible (unless the animal is already physically restrained), and the needle passes through dirty hair and skin, carrying surface bacteria into the muscle. Syringes, needles, and all paraphernalia used to load syringes should be clean and sterile to minimize post-injection infections. Wound infections occur most frequently with extensive trauma at the impact site.

APPROACHING THE ANIMAL

For helicopter darting, a pilot with wildlife experience is very important. Such a pilot knows when to hold back and drift an animal toward suitable terrain and when to make a rapid approach for the shot. A time limit should be set (usually 5 min or less) for rapid pursuit. Usually, only the person who will administer the drug should stalk a free-ranging animal or one in a trap. All others should remain out of sight, sound, and smell of the animal. For large carnivores, having an armed backup person in addition to the person who administers the drugs may be wise. One should slowly approach the animal from downwind, staying out of sight as much as possible and remaining quiet. Movements should be kept to a minimum, and sudden movements should be avoided.

Once the drugs have been administered, one should withdraw to a vantage point to avoid detection, so the animal can be observed. The rule of thumb is to eliminate all stimuli that may frighten an animal before it is fully immobilized. Even after the animal is immobilized, keeping quiet is good practice. One should be quiet and unobtrusive when working with animals.

Once an animal is down, do not assume it is unconscious. Wait another few minutes before approaching. Fi-

Table 3. Chemical induction signs and effects of selected immobilizing drugs.

Drug	Induction signs	Side effects
Carfentanil citrate	Blank look Aimless walking, stumbling	Excitement Inhibits respiration Hyperthermia Tachycardia
Ketamine HCl	Fixed expression in the eyes	May cause tonic or clonic convulsions
Tiletamine HCl and zolazepam HCl	Pupils dilated, eyes open, nystagmus Swallowing Serpentine tongue Ataxia	Apnea Salivation Hyperthermia Hypothermia Catatonia
Xylazine HCl	Animals slow down, lower head and ears Lie down and appear sleeping	Muscle tremors Bradycardia Explosive response to stimuli

nal approach should be made from the rear in a slow, quiet manner. Knowing when it is safe to approach an immobilized animal requires skill. If one remembers the induction signs for the drugs being used (Table 3) and uses them as a "post-injection checklist" as the animal goes down, the ability to determine when to approach will develop.

POST-IMMOBILIZATION CARE

Once an animal has been immobilized, the first thing to do is be certain that the airways of the animal are open and clear. Foreign matter and vomitus should be removed from the mouth and nose. The tongue should be forward, the neck should be forward without stretching, and the head should be in a normal position. Nothing should be resting on the neck or chest that may restrict breathing. Researchers sometimes sit across the chest of fully immobilized elk, deer, and bears when they are putting on radio collars. This practice should be avoided unless the animal needs to be physically restrained. Normal respiration and heart rate should be verified. Proper circulation can be evaluated by examining the gums for pink coloration and rapid capillary refill. Capillary refill is determined by pressing on the gums with a finger so the gum turns white under the pressure. Upon release, the gum tissue should return to pink immediately. Poor capillary refill indicates shock.

Once a proper airway has been established, the animal

should be protected against injury, bad weather, and direct sunlight. Eyes often remain open and must be protected with a bland eye ointment to prevent drying. The eyes should be covered with a clean blindfold. During summer, the animal should remain in the shade. Regardless of the season, if the animal has undergone excessive physical exertion, it should be checked for elevated body temperature; dowsing the animal with water or snow from time to time to keep it cool may be advisable. In winter, insulation should be placed around and under the animal. These recommendations depend on the weather conditions, how long the animal will be down, and what drugs are used. Temperature must be monitored—not estimated.

As soon as possible, the animal should be placed in the proper position. All ruminants are placed in sternal recumbency (on their brisket) with their legs flexed naturally under them and the head and neck lifted. The mouth should be slightly lower than the neck to allow drainage. This position should be maintained, even if someone has to hold the animal up, to prevent inhalation of regurgitated rumen contents, to allow belching, to reduce the chances of bloat, and to maintain a proper airway.

Other species can be maintained in lateral recumbency (on their side) with their head and neck in a normal position relative to the body. If excessive salivation occurs, the animal should be positioned with its head downhill. To prevent fluid buildup in the lungs, one should turn the animal over every 20–30 minutes until recovery begins.

One word of caution should be given about turning animals over. Ruminants should be turned onto their stomachs, not onto their backs, and then over. Abdominal muscles may be sufficiently relaxed to allow the digestive tract to undergo torsion. The rumen is so heavy that, as the animal is rolled over onto its back and then to the other side, the rumen remains stationary, resulting in a twisting of the esophagus and intestines. If this happens, the animal will die within a few hours or days. No immediate outward appearances other than bloat will be evident if the animal has undergone torsion, and the problem cannot be corrected in the field.

Talking around an immobilized animal should be avoided. Physical handling should be kept to a minimum. Excessive stimuli may initiate defensive reflexes, increase

Table 4. Normal heart and respiration rates and body temperatures of large mammals.

	Heart rate (beats/min)	Respiration (per min)	Temperature (C)
Canids	70–120	10–30	38.6
Small felids	90–130	20–30	38.6
Mountain lions	55–65	18–22	38.6
Bears	55–90	20–30	37.8
Deer	70–80	16–20	38.3
Elk	60–70	8–12	38.3
Bighorn sheep	90–120	12–20	37.2

Table 5. Signs and possible medical problems encountered during restraint of large mammals.

Convulsions	Bloat (stomach or intestines)
Anoxia	Improper body position
Hyperthermia	Twisted intestine
Hypoxia, pneumonia, struggling, cardiac failure	Loose stool
Hypercalcemia	Fright
Catatonia (drug reaction)	Parasitism
Concussion	Response to drugs
Acidosis	Previous enteric disease
Fracture of cervical vertebrae	Regurgitation
Tetany (tonic muscle spasms)	Response to drugs (reverse peristalsis)
Hypocalcemia	Pressure on chest or abdomen
Hypothermia (shivering)	Improper body position
Elevated body temperature	Carrying a limb
Latent infection	Dislocation
Increased muscle activity	Bone fracture
Drug effects on CNS	Severe sprain
Restraint practices preventing heat dissipation	Unable to stand on legs
Catatonia	Nerve damage
Convulsions	Ruptured tendon or ligaments
Decreased temperature	Capture myopathy
Drug effects on CNS	Fracture, dislocation
Cold environment	Unable to use hindquarters
Prolonged anesthesia	Broken back, fractured pelvis
Pale mucous membranes	Capture myopathy
Shock	Delayed drug toxicity
Hemorrhage	Inability to use all four limbs
Dark mucous membranes	Broken neck
Normal pigmentation in that species	Concussion
Hypoxia (strangulation, pneumonia)	Urination and frequent defecation
	Fright
	Response to drugs
	Diuresis (xylazine HCl)

stress, or cause an early recovery. All unnecessary people should be kept away, and picture taking should be kept to a minimum.

A researcher's primary obligation is to ensure the physical well-being of the immobilized animal. One should concentrate on monitoring the vital life signs of an animal for the entire time that animal is immobilized or assign an assistant for this specific task. Changes in the vital signs will be the first indication that the animal is deteriorating and support treatment is necessary. Deteriorating conditions must be detected and reversed as early as possible for best results. Table 4 gives some normal heart and respiration rates and body temperatures. Some characteristics change or are not present with some drugs. The drug description or Table 3 describes these side effects.

If vital signs are normal, the animal should be given a physical exam to determine the presence of injuries, including immobilization-related trauma and signs of disease or illness. The animal should be examined in a systematic way from the head to the forelegs, body, hindlegs, and tail. Treatment of injuries like abscesses will depend on one's experience. If unsure of what to do next, leave the animal alone. Once the health of the animal is deter-

mined, whether to continue with the planned procedure can be decided. Sick animals may not be suitable for radio collaring, but one may want to use ear tags or color-marking or take samples to determine the nature of the illness. A record should be kept of each animal's condition and whether any abnormalities are immobilization-related. Such a record will help in continuing education in immobilization and problems that arise.

Before leaving an animal, one should retrieve the dart, treat the dart wound, and assure there is no excessive hemorrhaging. The wound should be left open and treated with a topical antibiotic. Giving a small injection of a long-acting antibiotic is also a good idea. This will help prevent stress-related invasions of bacteria like *Pasteurella* spp. Larger doses may be given to animals that have an active infection.

Numerous medical problems can develop during restraint procedures. Signs (symptoms) that may be observed and probable causes for those signs are listed in Table 5. Clinical conditions and treatments are described in Table 6. General dosage ranges for selected tranquilizers and antagonists are listed in Table 7.

An animal should never be left alone until it has re-

Table 6.   Medical problems encountered during restraint.

Condition	Clinical signs	Therapy

Respiratory distress	Rapid breathing
	Pale mucous membranes
	Slow capillary refill
	   Check by pressing on mucous
	   membranes and releasing. Color should return in 1 second.
	   Color: pink = normal; blue = oxygen deficient; white = shock/death
		Discontinue immobilization drugs and cool the animal.
		Open airways by clearing mouth and pulling tongue forward.
		Make sure the body is positioned correctly.
		Check that bloat or restraining measures aren't causing pressure.
		If breathing has stopped, tickle the back of the throat with a pen, stick, or finger. This may cause spontaneous inhalation.
		Administer antagonist for narcotics or xylazine HCl.
		Ventilate the animal:
		   Manual compression of the chest with the animal in lateral recumbency. Place the flat of your hand on the ribs just below the "armpit" and compress ribs 2.5–5.0 cm at eight compressions/minute. Not as efficient as mouth-to-nose resuscitation.
		   Mouth-to-nose resuscitation. Hold mouth shut by hand sealing the lips. Place mouth completely over nostrils, blowing steadily and deeply (15 breaths/minute).
		   Mechanical resuscitation with an AMBU bag and nose cone using same procedure as mouth to nose resuscitation.

Bloat	Distended abdomen Tympanitic sounds heard on percussion Marked dyspnea and cyanosis (lack of oxygen) and rapid pulse	Correctly position the animal on its brisket with head up. Reduce pressure by passing a stomach tube or inserting a large needle (12 gauge) or trocar and cannula through the left side into the rumen to release gas. If a needle or trocar is unavailable, a small knife may be inserted and the blade twisted sideways to allow gas passage. Administer antagonist and get animal on its feet if possible.
Circulatory failure	Absence of pulse and heart beat No respiration White mucous membranes Pupils may be dilated but remain active	Restore respiration and circulation immediately. Administer epinephrine and antagonist. Place the animal on its side, clear airways. Perform CPR: One person starts mouth–nose respiration at 15 breaths/minute. Another person starts chest compressions as previously described with the following modifications: place the hands, one on top of the other, in a position about ⅓ of the way above the brisket and just behind the shoulder. Compress 7 cm at 60 compressions/minute, allowing for 1 respiration every 4 compressions. One-person CPR: Place hands in the middle of the rib cage and compress at 60 compressions/minute. Air transfer is not as efficient as two-person CPR. After 5 minutes, check for pulse and pupil reactions. Continue CPR until pupils no longer react to light.
Acidosis	Excessive muscular activity Respiratory distress Listlessness, confusion Lapse into coma and/or convulsions Dehydration (noted from skin turgor) Rapid breathing	Prevent excessive muscular activity before and during restraint procedures. Open airways. Ventilate to help get rid of CO_2 in blood. Intravenous or subcutaneous administration of sodium bicarbonate (4–10 mEq/kg) in saline or dextrose solution. Intravenous administration is preferable.
Shock	Decreased blood pressure, pale mucous membranes, depression, cool skin, weakness, coma, rapid breathing, rapid–weak pulse, dilated pupils, decreased body temperature	Eliminate cause of shock; provide oxygen (ventilate); restore circulatory blood volume; open airways and ventilate. Give sodium bicarbonate IV at 0.03 mEq/ml in saline or dextrose at a rate of 1 L/hour. This will increase blood volume and prevent acidosis. If the animal recovers, give high doses of antibiotics.

Table 6. Continued.

Condition	Clinical signs	Therapy
Shock (continued)	Neurogenic shock caused by psychological stress—slow pulse rate, decreased blood pressure, skin warm and flushed	In the field, shock with complete circulatory failure will generally result in death. There are numerous therapies but these must be performed by a veterinarian in a hospital situation. NOTE: Shock results from severe physical or psychological insults and is the terminal result of traumatic or metabolic disorders associated with restraint. The following are important classifications of shock: Cardiac tamponade—heart failure due to pressure in the heart. Hemorrhage—whole blood loss. —plasma lost in contusions and burns. Dehydration—exercise, hyperthermia. Decreased blood volume—capillary dilation (pooling of blood). Neurogenic response—fear, anger, pain. Toxins—drugs. Best therapy—prevent shock, reduce *all* stress factors.
Capture myopathy	May appear up to 14 days after restraint. Predisposing factors: fear, anxiety, overexertion, repeated handling, failure to allow exhausted animals to recover before shipment, prolonged shipment, constant muscle tension. Cardiac and skeletal muscle necrosis. Painful, stiff movements of hind legs. Major muscle masses swollen, hard, and hot. Difficult or labored breathing and tachycardia in acute cases. Necropsy: light grayish streaks and hemorrhages in muscles	Prevent all predisposing factors. If severe stress does occur, treat as for acidosis. Keep animal well oxygenated. Once muscle necrosis has occurred, prognosis is poor.

Table 7. Dosage ranges for selected tranquilizers and antagonists for large mammals.

Name	Dosage and method
Tranquilizers	
Acepromazine maleate[a]	0.25–2.0 mg/kg IM
Diazepam[a]	0.5–3.5 mg/kg IM, IV
Antagonists	
Naloxone HCl[b]	10–50 × M-99 dose IM, IV 80–100 × Carfentanil dose IM, IV
Yohimbine HCl[b]	0.1–0.30 mg/kg IV
Doxapram HCl[b]	0.4–5.0 mg/kg IV

[a]Available from most veterinarians. Note: diazepam is chemically incompatible with most immobilizing agents. DO NOT MIX in syringes or give IV if other drugs were given IV.

[b]Available from distributors and Wildlife Laboratories, Inc., 1401 Duff Drive, Suite 600, Ft. Collins, CO 80524.

covered. Remaining in a safe position to observe the animal until it is safely on its way will ensure that the animal does not stumble and have an accident (e.g., fall off a cliff or fall asleep in a creek) or others of the same species do not harass it. As an example, we immobilized several elk to take blood samples. Two bulls were immobilized at the same time. We took the blood samples and went on to other animals, still keeping an eye on the animals that were down. One bull recovered before the other and when it became oriented, it promptly went to the other bull and mortally gored it. This was an unfortunate situation in which no one was close enough to the immobilized bull to scare off the other. It illustrates the need to stay and protect an immobilized animal until it has recovered and to use drugs that have appropriate antagonists. Certain species are especially difficult to keep from harming a fallen rival. More than one person is sometimes required to keep a pronghorn buck from goring an immobilized buck. Leaving an immobilized animal alone is also an invitation to predators and scavengers. Ravens are especially quick to locate a helpless animal and peck its eyes out.

LITERATURE CITED

ADDISON, E. M., AND G. M. KOLENOSKY. 1979. Use of ketamine hydrochloride and xylazine hydrochloride to immobilize black bears (*Ursus americanus*). J. Wildl. Dis. 15:253–258.

AD HOC COMMITTEE ON ACCEPTABLE FIELD METHODS IN MAMMALOGY. 1987. Acceptable field methods in mammalogy: preliminary guidelines approved by the American Society of Mammalogists. J. Mammal. 68(4, Suppl.). 18pp.

ALLEN, T. J. 1970. Immobilization of white-tailed deer with succinylcholine chloride and hyaluronidase. J. Wildl. Manage. 34:207–209.

AMSTRUP, S. C., AND T. B. SEGERSTROM. 1981. Immobilizing free-ranging pronghorns with powdered succinylcholine chloride. J. Wildl. Manage. 45:741–745.

BALLARD, W. B., A. W. FRANZMANN, AND C. L. GARDNER. 1982. Comparison and assessment of drugs used to immobilize Alaskan gray wolves (*Canis lupus*) and wolverines (*Gulo gulo*) from a helicopter. J. Wildl. Dis. 18:339–342.

BAUDITZ, R. 1972. Sedation, immobilization and anesthesia with Rompun in captive and free-living wild animals. Vet. Med. Rev. 9:204–226.

BECK, C. C., R. W. COPPOCK, AND B. S. OTT. 1971. Evaluation of Vetalar (ketamine HCl): a unique feline anesthetic. Vet. Med. Small Anim. Clin. 66:993–996.

BOEVER, W. J. 1986. Artiodactylids: restraint, handling, and anesthesia. Pages 940–952 *in* M. E. Fowler, ed. Zoo and wild animal medicine. W. B. Saunders Co., Philadelphia, Pa.

BOOTH, N. H. 1982. Nonnarcotic analgesics. Pages 297–320 *in* N. H. Booth and L. E. McDonald, eds. Veterinary pharmacology and therapeutics. Iowa State Univ. Press, Ames.

BUSH, M., R. S. CUSTER, AND E. E. SMITH. 1980. Use of dissociative anesthetics for the immobilization of captive bears: blood gas, hematology, and biochemistry values. J. Wildl. Dis. 16:481–489.

CLARK, D. M., R. A. MARTIN, AND C. A. SHORT. 1982. Cardiopulmonary responses to xylazine/ketamine anesthesia in the dog. J. Am. Anim. Hosp. Assoc. 18:815–821.

CLAUSEN, B., P. HJORT, H. STRANDGAARD, AND P. L. SOERENSEN. 1984. Immobilization and tagging of muskoxen (*Ovibos mosochatus*) in Jameson Lane, Northeastern Greenland. J. Wildl. Dis. 20:141–145.

CODE OF FEDERAL REGULATIONS 9. 1990. Parts 1 to 199, Revised as of January 1, 1990. Off. Fed. Register, Natl. Adm., U.S. Gov. Printing Off., Washington, D.C.

CRONIN, M. F., N. H. BOOTH, R. C. HATCH, AND J. BROWN. 1983. Acepromazine-xylazine combination in dogs: antagonism with 4-aminopyridine and yohimbine. Am. J. Vet. Res. 44:2037–2042.

FOWLER, M. E. 1978. Restraint and handling of wild and domestic animals. Iowa State Univ. Press, Ames. 332pp.

FRANZMANN, A. W., C. C. SCHWARTZ, D. C. JOHNSON, AND J. B. FARO. 1984. Immobilization of moose with carfentanil. Alces 20:259–282.

HAIGH, J. C., AND H. C. HOPF. 1976. The blowgun in veterinary practice: the uses and preparation. J. Am. Vet. Assoc. 169:881–883.

———, L. J. LEE, AND R. R. SCHWEINSBURG. 1985. Immobilization of polar bears with carfentanil. J. Wildl. Dis. 21:140–144.

HASH, H. S., AND M. G. HORNOCKER. 1980. Immobilizing wolverines with ketamine hydrochloride. J. Wildl. Manage. 44:713–715.

HATCH, R. C., ET AL. 1976. Immobilization of adult bull bison with etorphine. Proc. Ia. Acad. Sci. 83:67–70.

———, N. H. BOOTH, J. V. KITZMAN, B. M. WALLNER, AND J. D. CLARK. 1983. Antagonism of ketamine anesthesia in cats by 4-aminopyridine and yohimbine. Am. J. Vet. Res. 44:417–423.

———, J. D. CLARK, A. D. JENIGAN, AND C. H. TRACY. 1984. Searching for a safe, effective antagonist to Telazol overdose. Univ. Georgia Vet. Med. Exp. Pap. 2550. 6pp.

———, J. V. KITZMAN, AND M. ZAHNER. 1985. Antagonism of xylazine sedation with yohimbine, 4-aminopyridine, and doxapram in dogs. Am. J. Vet. Res. 46:371–375.

———, AND T. RUCH. 1974. Experiments on antagonism of ketamine anesthesia in cats given adrenergic, serotonergic, and cholinergic stimulants alone and in combination. Am. J. Vet. Res. 35:35–39.

HORNOCKER, M. G., AND W. V. WILES. 1972. Immobilizing pumas (*Felis concolor*) with phencyclidine hydrochloride. Int. Zoo Yearb. 12:220–222.

HSU, W. H., Z. X. LU, AND F. B. HEMBROUGH. 1985. Effect of xylazine on heart rate and arterial blood pressure in conscious dogs, as influenced by atropine, 4-aminopyridine, doxapram, and yohimbine. J. Am. Vet. Med. Assoc. 186:153–156.

———, AND W. P. SHULAW. 1984. Effect of yohimbine on xylazine-induced immobilization in white-tailed deer. J. Am. Vet. Med. Assoc. 185:1301–1303.

JACOBSEN, N. K., W. P. ARMSTRONG, AND A. N. MOEN. 1976. Seasonal variation in succinylcholine immobilization of captive white-tailed deer. J. Wildl. Manage. 40:447–453.

JACOBSON, E. R., AND G. V. KOLLIAS. 1984. Yohimbine antagonism of ketamine/xylazine tranquilization and immobilization in hoofstock. Proc. Am. Assoc. Zoo. Vet. 1984:57.

JESSUP, D. A., W. E. CLARK, P. A. GULLETT, AND K. R. JONES. 1983. Immobilization of mule deer with ketamine and xylazine, and reversal of immobilization with yohimbine. J. Am. Vet. Med. Assoc. 183:1339–1340.

———, AND K. R. JONES. 1984. Immobilization of captive mule deer with carfentanil. J. Zoo Anim. Med. 15:8–10.

———, ———, ———, R. CLARK, AND W. R. LANCE. 1985a. Immobilization of free-ranging desert bighorn sheep, tule elk and wild horses using carfentanil and xylazine: reversal with naloxone, diprenorphine and yohimbine. J. Am. Vet. Med. Assoc. 187:1253–1254.

———, K. JONES, R. MOHR, AND T. KUCERA. 1985b. Yohimbine antagonism to xylazine in free-ranging mule deer and desert bighorn sheep. J. Am. Vet. Med. Assoc. 187:1251–1253.

KARESH, W. B., D. L. JANSSEN, AND J. E. OOSTERHUIS. 1986. A comparison of carfentanil and etorphine/xylazine immobilization of axis deer. J. Zoo Anim. Med. 17:58–61.

KAUFMAN, P. L., AND R. HAHNENBERGER. 1975. CI-744 anesthesia for ophthalmological examination and surgery in monkeys. Invest. Ophthalmol. 14:788–791.

KLIDE, R. J., H. W. CALDERWOOD, AND L. R. SOMA. 1975. Cardiopulmonary effects of xylazine in dogs. Am. J. Vet. Res. 36:931–935.

KNIGHT, A. P. 1980. Xylazine. J. Am. Vet. Med. Assoc. 176:454–455.

KOLATA, R. J., AND C. A. RAWLINGS. 1982. Cardiopulmonary effects of intravenous xylazine, ketamine, and atropine in the dog. Am. J. Vet. Res. 43:2196–2198.

KREEGER, T. J., G. D. DELGUIDICE, U. S. SEAL, AND P. D. KARNS. 1986a. Immobilization of white-tailed deer with xylazine hydrochloride and ketamine hydrochloride and antagonism by tolazoline hydrochloride. J. Wildl. Dis. 22:407–412.

———, AND U. S. SEAL. 1986. Immobilization of coyotes with xylazine hydrochloride-ketamine hydrochloride and antagonism by yohimbine hydrochloride. J. Wildl. Dis. 22:604–606.

———, ———, M. CALLAHAN, AND M. BECKEL. 1990. Physiological and behavioral responses of grey wolves (*Canis lupus*) to immobilization with tiletamine and zolazepam. J. Wildl. Dis. 26:90–94.

———, ———, AND A. M. FAGGELLA. 1986b. Xylazine hydrochloride-ketamine hydrochloride immobilization of wolves and its antagonism by tolazoline hydrochloride. J. Wildl. Dis. 22:397–402.

LOGAN, K. A., E. T. THORNE, L. L. IRWIN, AND R. SKINNER. 1986. Immobilizing wild mountain lions (*Felis concolor*) with ketamine hydrochloride and xylazine hydrochloride. J. Wildl. Dis. 22:97–104.

MACKINTOSH, G. G., AND G. VAN REENEN. 1984. Comparison of yohimbine, 4-aminopyridine and doxapram antagonism of xylazine sedation in deer (*Cervus elaphus*). N.Z. Vet. J. 32:181–184.

MEULEMAN, T., J. D. PORT, T. H. STANLEY, K. F. WILLIARD, AND J. KIMBALL. 1984. Immobilization of elk and moose with carfentanil. J. Wildl. Manage. 48:258–262.

MILLER, F. L. 1968. Immobilization of free-ranging black-tailed deer with succinylcholine chloride. J. Wildl. Manage. 32:195–197.

MONTGOMERY, G. G., AND R. E. HAWKINS. 1967. Diazepam bait for capture of white-tailed deer. J. Wildl. Manage. 31:464–468.

PARKER, J. R. B., AND J. C. HAIGH. 1982. Human exposure to immobilizing agents. Pages 119–136 *in* L. Nielsen, J. C. Haigh, and M. E. Fowler, eds. Chemical immobilization of North American wildlife. Wis. Humane Soc., Milwaukee.

PUSATERI, F. M., C. P. HIBLER, AND T. M. POJAR. 1982. Oral administration of diazepam and promazine hydrochloride to immobilize pronghorn. J. Wildl. Dis. 18:9–16.

RAMSAY, M. A., I. STIRLING, L. O. KNUTSEN, AND E. BROUGHTON. 1985. Use of yohimbine hydrochloride to reverse immobilization of polar bears by ketamine hydrochloride and xylazine hydrochloride. J. Wildl. Dis. 21:396–400.

RAMSDEN, R. O., P. F. COPPIN, AND D. H. JOHNSON. 1976. Clinical observations on the use of ketamine hydrochloride in wild carnivores. J. Wildl. Dis. 12:221–224.

SCHOBERT, E. 1987. Telazol use in wild and exotic animals. Vet. Med. 24:1080–1084.

SEAL, U. S., AND A. W. ERICKSON. 1969. Immobilization of carnivora and other mammals with phencyclidine and promazine. Fed. Proc. 28:1410–1419.

———, AND T. J. KREEGER. 1987. Chemical immobilization of furbearers. Pages 191–215 *in* M. Novak, J. A. Baker, M. E. Obbard, and B. Malloch, eds. Wild furbearer management and conservation. Min. Nat. Resour., Toronto, Ont.

———, S. M. SCHMITT, AND R. O. PETERSON. 1985. Carfentanil and xylazine for immobilization of moose (*Alces alces*) on Isle Royale. J. Wildl. Dis. 21:48–51.

TAYLOR, W. P., JR., H. V. REYNOLDS, III, AND W. B. BALLARD. 1989. Immobilization of grizzly bears with tiletamine hydrochloride and zolazepam hydrochloride. J. Wildl. Manage. 53:979–981.

THOMAS, J. W., R. M. ROBINSON, AND R. G. MARBURGER. 1967. Use of diazepam in the capture and handling of cervids. J. Wildl. Manage. 31:686–692.

WALLACH, J. D., R. FRUEH, AND M. LENTZ. 1967. The use of M-99 as an immobilizing and analgesic agent in captive wild animals. J. Am. Vet. Med. Assoc. 151:870–876.

WENTGES, H. 1975. Medicine administered by blowpipe. Vet. Rec. 97:281.

WRIGHT, J. M. 1983. Ketamine hydrochloride as a chemical restraint for selected small mammals. Wildl. Soc. Bull. 11:76–79.

7

WILDLIFE MARKING TECHNIQUES

Marie T. Nietfeld, Morley W. Barrett, and Nova Silvy

INTRODUCTION

Studies of free-ranging animals often require that individuals be marked to obtain detailed information on population dynamics, movement patterns, and behavior. Techniques vary widely in relation to species studied and the nature and objective of the study. Marking techniques continue to evolve as new technologies and materials are developed, and as creative thinkers strive to meet the challenges of future studies. Despite this, a number of basic considerations, outlined in the first part of this chapter, remain fundamental to the selection and implementation of all marking techniques. Subsequently, techniques generally available for marking mammals, birds, and amphibians and reptiles are addressed and described. We have included information on retention time, visibility, and adverse effects of individual techniques where available. It is not within the scope of this chapter to document each marking technique in detail but rather to provide a summary of available techniques and an overview of factors to consider in the selection of an appropriate procedure.

Mark Selection

Numerous marking techniques are available for use. The mark selection process will depend on evaluating several interacting criteria to identify the most appropriate procedure. The marking technique selected must allow the study objectives to be accomplished while meeting acceptable professional and humane standards. Before a particular technique is selected, the following should be considered (Barclay and Bell 1988): the period of time the mark must persist, the distance at which marked animals may be identified, the need for individual identity (and, if so, the number of distinct marks required), how quickly the animals must be marked, time available for identifying marked individuals, and the effect of the mark on survival or behavior of the animal. A desirable marking technique would have the following criteria (Ferner 1979, Marion and Shamis 1977): (1) involve minimal pain or stress, (2) produce no adverse effects on survival and behavior, (3) display good retention and durability characteristics, (4) be easy to recognize, (5) be easy to apply, (6) be easy to obtain and assemble, and (7) be relatively inexpensive. A single marking technique will be unlikely to satisfy all these criteria, but all should be prioritized and considered during the mark selection process.

MARKER RETENTION

Normally, greater efficiency in accomplishing study objectives is derived from marks that persist over the duration of the study. Markers can be divided into three main types in relation to retention time: temporary, semipermanent, and permanent. Temporary marks generally are used when the duration of the study is short relative to the life-span of the animal, when more permanent marks would adversely affect the animal, or when no other marks are available. Streamers, adhesive tapes, nocturnal lights, and marks that are lost during subsequent molts, such as feather and fur-clipping, dyes, and paints, are included within this category. Semipermanent marks include a variety of tags and collars, some of which can persist for the life of the animal. The length of time these marks remain on the animal is influenced by the durability of the material from which they are constructed. Nonpermanent marks may be applied so as to fall off over time, or they may have to be removed by the investigator. Permanent marks include branding, tattoos, ear notching, toe clipping, and other mutilation techniques. Although these marks last a lifetime, scarring, tearing, or aging may reduce their effectiveness. Nonpermanent marks, which are generally more visible, can be used in addition to permanent marks to temporarily increase the amount of data that can be collected without sacrificing long-term recognition of an individual.

MARKER RECOGNITION

Most markers fall into two recognition categories: those that require the animal to be recaptured or killed for future identification, and those that allow the animal to be identified at a distance. Small, inconspicuous marks, such as tattoos or monel ear tags, are sufficient for the first category and provide information on population dynamics and estimates as well as general movement and distribution patterns. Larger, conspicuous marks are required for the second category. The size of the mark will be limited by the physical attributes of the animal, as well as the distance from which accurate observations must be made. Color and code marks frequently are employed to enhance recognition. When color is used for marking, the following should be chosen: bright colors that contrast the coloration of the animal, as few colors as possible, and only highly contrasting colors to avoid error in identification. Only colorfast materials should be used, because fading may result in indistinguishable colors and misidentification. Color markers may increase vulnerability to predation (Kessler 1964) and may affect behavioral interactions in some species (Burley et al. 1982). Information on a marker can be increased by the use of characters and symbols. These should contrast with the base color of the marker or animal and should be made with durable materials, paints, and inks. The use of a few distinct codes will enhance clarity and reduce confusion when read. Large size and adequate spacing of codes will extend the distance at which they can be identified effectively.

Marks can be used for individual or group recognition. Individual identity allows for detailed information to be obtained on such parameters as home-range size, social organization, and reproductive behavior. It is particularly useful when variation among individuals is important.

SPECIES-SPECIFIC ATTRIBUTES

Behavioral and anatomical features of a species are important to consider when a marking technique is chosen. Whether the animal is diurnal or nocturnal, is conspicuous or secretive, inhabits open country or densely vegetated areas, and is or is not easily approachable will influence which marker will be most appropriate. Using a visual marker on animals inhabiting dense forests and rarely seen would not be an efficient means of obtaining movement patterns. Furthermore, the anatomy of the animal may influence the choice of techniques and, for example, whether the mark could become obscured by fur or feathers during the functional period. Researchers must be certain that the mark will not significantly impact the social behavior of the species or its ability to function physically. Finally, the capability of the species to remove the mark,

or indirect mark loss due to species-specific behavior, such as digging or fighting, must be evaluated.

ADVERSE EFFECTS

Marks commonly affect the physiology or behavior of an animal, at least temporarily. History has shown that the magnitude of the effect can range from mild irritation resulting in increased grooming to infection, increased vulnerability, or other factors that eventually could lead to death. The adverse effects of marking may be immediately evident or may surface long after marking. Neck tags developed for marking game birds in the 1940s–50s (Taber 1949) were unsatisfactory due to potential injury from surgical pin and wire attachments. Severe icing on nasal tags and neck collars of ducks and geese may occur during inclement weather and affect physical condition and survival of birds (Zicus et al. 1983). Investigators should be aware of all effects of marking on particular animals, including those that restrict feeding or movement, disrupt breeding or social interactions, alter distribution or migration patterns, or cause direct physical damage. When selecting the appropriate technique, one should consider the following questions (Young and Kochert 1987): (1) Can study objectives still be met, or will the data be biased? (2) What are the limitations of the data caused by these effects, and how will these be handled during data analysis? (3) Does the information obtained outweigh the adverse effects on the marked population if study objectives are met? When a mark affects the particular parameter being studied, a more appropriate technique should be sought. Some mortality associated with the capture and marking of most species is inevitable, and researchers must assume responsibility to ensure it is minimal. When rare or endangered species are involved, this issue is particularly sensitive.

Proper design, fitting, and testing of markers on captive animals prior to a study can eliminate many potential problems. As well, advance training of field personnel will ensure that marks are applied in a proficient manner.

Marking Permits

Appropriate federal and state or provincial permits and authorizations must be obtained before most species of wildlife can be captured and marked. Permits for most resident populations are regulated by local state/provincial wildlife agencies, and permit requirements vary. Applications for the appropriate permits should be initiated long before field operations are scheduled.

Banding of migratory birds in North America is regulated by the federal governments, and prospective banders are advised to obtain the *North American Bird Banding Manual,* Volumes I and II (U.S. Fish and Wildlife Service and Canadian Wildlife Service 1986, 1991). These manuals contain extensive guidelines on the permits required and on the capture, handling, marking, band size, and record keeping for a wide range of migratory species. Extensive portions of these manuals have been revised or expanded since their publication, and requests for the manual and updates or questions related to the banding of migratory birds should be directed to the Canadian Bird Banding Office (Canadian Wildlife Service, Ottawa, Ontario, Canada K1A 0H3) or the Bird Banding Laboratory

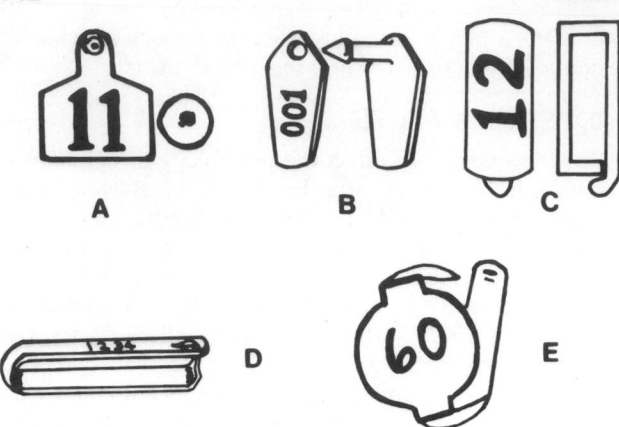

Fig. 1. Examples of ear/web/flipper tags used to mark mammals: (A) plastic All-Flex ear tags with button fastener, (B) plastic interlocking deer tags, (C) self-closing plastic Temple tags, (D) monel metal or steel self-piercing tags, and (E) aluminum interlocking button tags.

(Office of Migratory Bird Management, U.S. Fish and Wildlife Service, Laurel, MD 20708-9619).

MARKING TECHNIQUES FOR MAMMALS
Tags

EAR, WEB, AND FLIPPER TAGS

Tags made from metals and plastics in a variety of shapes, sizes, and colors and stamped with an identifying number are a common method for marking mammals. Tag-closing mechanisms can be interlocking, self-locking, or a rivet design that cannot be easily pried apart when the rivet is flattened (Fig. 1). Tags may be self-piercing or inserted through a punched hole or a knife slit. Ear tags usually are placed on the lower, inner region of the ear where there is heavier cartilage and the tag is more protected. This placement reduces tearing of the ear or the chance of tags being pulled out. Tags should be loose enough not to interfere with blood circulation, and puncture marks should be treated appropriately to prevent infection and ensure healing. Tags used to mark ears also have been used to mark foot webs and interdigital webbing of the hind and foreflippers. For seal species, the use of a knife slit rather than a round hole punch to attach the tags reduces tag loss. Tag losses increase with time since tagging (Hubert et al. 1976, Alt et al. 1985) and may result from infection, wear, grooming, or fighting. Thus the use of tags in both ears or webs, and in conjunction with a more permanent marking method (e.g., tattoo) where possible, will minimize the chance of losing identity of an animal over a long time period. The duration of the study and the required visibility of the tag are factors that will largely influence tag choice. Many tags require recapture of the animal for positive identification, and they may be missed in resightings due to their location and small size. For some species, such as seals, an ideal tag is not yet available because tag loss, codability, and color change remain problems.

Harper and Lightfoot (1966) and Downing and McGinnes (1969) reported good retention rates of aluminum ear tags on deer. These tags also exhibit good retention on the hind foot of sea otters but are bulky and difficult to read after the paint wears off (Miller 1979). Day (1973) considered Tamp-R-Pruf tags as the most satisfactory ear

tag available for medium to large animals. Rudge and Joblin (1976) reported steel Hasco cattle tags remained on feral goats for at least 5 years if the puncture marks healed.

Monel metal ear tags are suitable for long retention on black bears (Johnson and Pelton 1980, Alt et al. 1985) and may be better retained on red foxes than larger aluminum button tags (Hubert et al. 1976). Small monel tags placed on sea otter ears were retained for at least 2 years with no ill effects (Ames et al. 1983), and they have been successful markers on Steller's sea lions (Scheffer 1950). For marking the foreflippers of seals, monel metal tags are more durable than plastic tags, although they may be less visible on marked animals and still exhibit significant rates of loss (Hobbs and Russell 1979). Aluminum tags, which wear and corrode easily, are regarded as inferior to steel or monel metal tags for species inhabiting seawater.

Fingerling ear tags have been used to mark bats since the 1930s (Mohr 1934). These tags may not be suitable for large-eared bats or species that exhibit rapid ear movement synchronized with their echolocation emissions, or for medium- and large-sized bats due to poor retention (Stebbings 1978). Self-piercing fingerling and monel metal tags also have been used to mark the hind foot web of snowshoe hares (Keith et al. 1968) and nutria (Evans et al. 1971). Retention of tags on both feet of hares was 82% for 6 months, and the only losses on nutria over a 3-year period were due to improper crimping.

Plastic ear markers commonly used on ungulates include Roto tags and All-Flex tags. Beasom and Burd (1983) used the 1000 Series, Lonestar plastic livestock tags on both ears of deer as a discernible visual marker. Tag retention was 100% over 16 months, and 95% over 2 years. Round All-Flex tags and Roto tags also have been used with good results on a variety of seal species (Hobbs and Russell 1979), though Warneke (1979) reported that plastic tags became stained and fell off fur seals within 1–2 years. Miller (1979) used highly visible, colored, plastic Temple cattle ear tags placed systematically on each hind foot to individually mark sea otters, with no apparent damage to the animal. Ames et al. (1983) developed a technique of double anchoring Temple tags, resulting in a significant improvement in their retention. Delrin button tags are satisfactory for marking polar bears (Larsen 1971, Stirling 1979) and the hind foot of sea otters (Johnson 1979); although they have good retention, they are readable only in the hand.

NECK COLLARS AND RELATED BANDS

A variety of neck collars and bands has been designed for field identification of free-ranging animals (Fig. 2). Collars may be fixed in size or expansible to allow for growth. Properly fit collars should not restrict feeding, circulation, or breathing or cause entanglement. Collars are usually highly visible (Fig. 3), but their longevity depends on the material used, climate, and behavior and sex of the animal involved.

Beale and Smith (1973) and Keister et al. (1988) designed self-adjusting collars for young ungulates. Hawkins et al. (1967) constructed expansible and nonexpansible collars from vinyl plastic for male and female white-tailed deer, respectively. Plastic rope and polyethylene rope with numbered or coded color flaps and flags

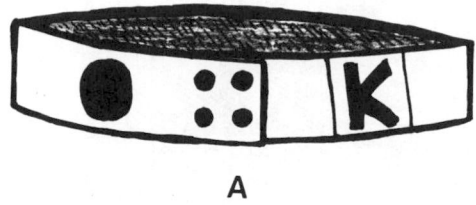

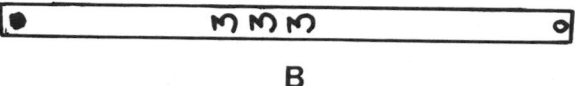

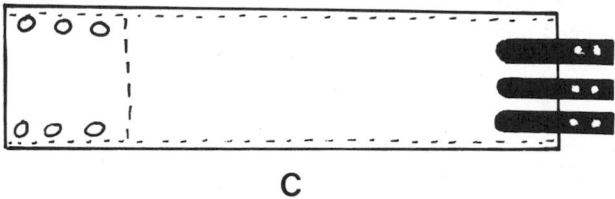

Fig. 2. (A) Vinyl collar with character codes secured by rivets, (B) plastic polyvinyl chloride band secured with bolts, and (C) nylon-coated fabric collar secured by either rivets or straps and buckles.

Fig. 3. Neck collar and ear streamer on white-tailed deer.

were used to mark deer and elk (Lightfoot and Maw 1963, Harper and Lightfoot 1966). Vinyl plastic and vinyl-coated nylon fabrics (Saflag, Armor-tite, Herculite, Sterkolite) were used with success, marking animals for up to 5 years (Knight 1966, Craighead et al. 1969, Phillips and Nicholls 1970, Smuts 1972). Hanks (1969) reported the use of plastic sewn onto canvas with a celluloid strip in between for stiffness, as well as a variety of PVC plastics and PVC-impregnated fabrics. Brooks (1981) designed a collar made from rubberized machine belting onto which stainless steel and colored Sterkolite symbols were attached. Over a 2-year period, 53–64% of such collars used on a variety of African ungulate species remained identifiable. Automatic collaring devices and snare collars have been designed for marking deer (Verme 1962, Siglin 1966, Taylor 1969) and pronghorns (Beale 1966).

Lentfer (1968) and Jones and Bush (1988) used nylon webbing collars on adult polar bears and arboreal monkeys, respectively. Rubber-impregnated machine belting was used for elephant collars (Hanks 1969), though tail collars of thin tin sheeting (Hanks 1969) and PVC plastic held in position by pop rivets (Viljoen 1986) may be more visible for this species. Rudge and Joblin (1976) used collars made of link galvanized steel chain, accompanied with number tag, address disc, and plasticized nylon flashers, to mark feral goats. The chain showed no wear after 2 years, although other parts were chewed. Bead-clasp "keychain" neck collars (Wilkinson 1985) and spiral bird rings mounted on the neck (Moran 1985) are successful alternatives to placing bands on forearms of some bats. Rubber neck bands were used to mark northern fur seals (Scheffer 1950), and peduncle straps made of machine belting in surgical tubing (Irvine and Scott 1984), ty rap and Velcro rubber peduncle belts, and Nasco multi-loc cattle leg belts (White et al. 1981) are potential markers for manatees and cetaceans.

ARMBANDS

The attachment of bands to the forearms has been the most widely used technique for marking bats (Stebbings 1978, Hooper 1983, Barclay and Bell 1988) (Fig. 4). Several different band types are available, including serially numbered metal bands, color-anodized aluminum bands, numbered and unnumbered colored plastic bands, and celluloid rings. Bird bands and bands designed specifically for bats have been used. Bonaccorso et al. (1976) recommended aluminum bands be used on bats <10 g, plastic bands on bats 10–50 g, and anodized aluminum and steel bands on bats >50 g. Plastic bands (Morrison 1978) and anodized aluminum bands (Davis 1963, Cockrum 1969) may lessen the problem of some bats, especially frugivores (Bonaccorso and Smythe 1972), chewing on aluminum bands. Bands attached to the hind leg or pollex of the bat are not effective markers due to band loss (Moran 1985). Reflective tape can be applied to metal or plastic bands to aid in nocturnal identification (Racey and Swift 1985, Bell et al. 1986). Injury from bands is a common problem, caused by the motion of the forearms during flight. Chafing, inflammation, and overgrowth of flesh at the banding site are frequent (Bradbury 1977, Hooper 1983, Phillips 1985). A small incision in the membranes of the forearm, allowing the band to slip through and completely encircle the forearm, may reduce injuries (Bateman and Vaughan 1974, Bonaccorso et al. 1976). LaVal et al. (1977) reported few band injuries from celluloid rings after 18 months. Perry and Beckett (1966) observed that banding neonatal bats resulted in damage to the developing bones of the forearm and manus, thus bands used should be large enough to allow for growth. Banding of temperate bats during hibernation, an energetically critical period, has been implicated as one of the major causes of population declines (Keen and Hitchcock 1980, Hooper 1983) and therefore should be avoided. Several countries have restricted or prohibited banding of bats in an effort to stop these apparent declines (Barclay and Bell 1988).

DORSAL FIN, BACK, AND RELATED TECHNIQUES

Discovery marks, small stainless steel projectiles with identifying information stamped on them, have been used to mark commercially valuable species of whales since

Fig. 4. Anodized aluminum colored band used to mark the forearm of bats.

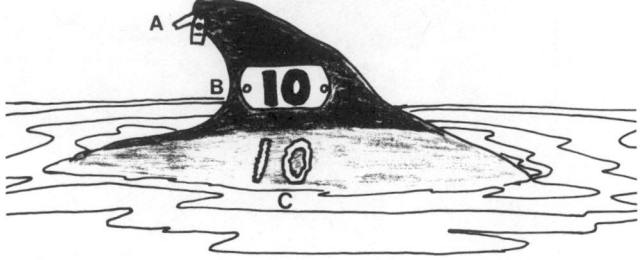

Fig. 5. Visual marking technique for small cetaceans include: (A) roto tags, (B) double bolt tags, and (C) freeze brand.

the 1920s (Brown 1978). A 23-cm-long projectile shot from a 12-gauge, shoulder-held shotgun has typically been used to mark large whales, and a modified, smaller mark propelled from a .410 shotgun has been used extensively to mark smaller whales (Clarke 1971). Marks are subsequently recovered in harvested whales and have provided information on movement and growth. Loss of discovery marks by ejection from the whale and incorrect tagging assessment are factors that could affect population estimates (de la Mare 1985). Discovery marks are not visible on living whales, so their use will decline with reduction in whaling activity (Leatherwood et al. 1976). Mitchell and Kozicki (1975) designed a prototype "spaghetti tag" visual marker for whales modified from the discovery tag. A strip of vinyl was attached to an anchor rivet behind the head of the mark by a length of braided Dacron, Teflon-coated line; the line and streamer are coiled in the tube and released when the mark is fired. Miyashita and Rowlett (1985) developed .410 streamer markers, and Kasamatsu et al. (1986) improved the design. Evans et al. (1972) concluded a similar spaghetti streamer tag was the best method for marking a large number of small cetaceans without the need to capture them, and Irvine and Scott (1984) used a spaghetti tag to mark manatees.

Other visual tags have been tested on cetaceans. Norris and Pryor (1970) attached plastic deer tags to the dorsal fin and reported them to be potentially durable markers. Bolt tags (rectangular fiberglass tags held in place on the dorsal fin by one or two Teflon bolts with stainless steel washers and cotter pins) provide easy identification from a distance and are a relatively durable tag (Irvine et al. 1982) (Fig. 5). Migration of the tags, injury to the dorsal fin, and covering of the tag with algae were problems associated with dorsal fin tags (Irvine et al. 1982, Tomilin et al. 1983).

Tomilin et al. (1983) designed a special clamp to hold a stencil on either side of the areas of the dorsal fin, caus-

ing the epithelium under the pressurized areas to be exfoliated and replaced by demelanized skin that remained distinct for at least 2 years. This procedure, however, required 4 days for the depigmented tissue to be produced, thus limiting its value as a field marker.

TAPES, STREAMERS, AND BELLS

Colored streamers made from plastic, nylon, and nylon-coated fabrics (Herculite, Saflag, and Armor-tite) have been used to visibly mark ungulate species by attachment to ears (Knowlton et al. 1964, Harper and Lightfoot 1966), horns (Jonkel et al. 1975, Reynolds and Garner 1983), achilles tendons (Queal and Hlavachick 1968), or to other marking devices (Downing and McGinnes 1969, Panagis and Stander 1989). Nylon-coated fabric streamers have been retained for several months to years (Harper and Lightfoot 1966, Downing and McGinnes 1969). Knowlton et al. (1964) observed some reluctance of does to accept ear-marked fawns, though their survival rates were not different from the estimated rates of unmarked fawns. Lentfer (1968) used colored flagging tape fastened with a metal band as an ear marker for polar bears. Different lengths and color codes provided a means of individual identification at a distance.

Many types of streamers and flags made from material such as fluorescent plastic, polypropylene, polyurethane, hypalon, orthoplast, nylon-coated vinyl, and vinyl tubing have been tested as markers for cetaceans and dolphins (Evans et al. 1972, Mitchell and Kozicki 1975, White et al. 1981). Steel barbs, nylon darts with adjoining flukes, umbrella anchors, and anchor rivets have been used to secure such markers to the animal. Duration of streamers and flags varies from a few days to several months. Difficulty in anchoring the marker into the tissue, water friction, behavior of marked animals, and tissue trauma are factors that contribute to the shedding of these markers.

Highly reflective plastic tape strips (Williams et al. 1966) and plastic-covered tape with coded numbers (Daan 1969) were glued to the head fur of bats as temporary individual markers. Colored, plastic, adhesive tape was used as a durable, visual marker on the horns of mountain sheep (Day 1973) and as a temporary marker on the quills of porcupines (Pigozzi 1988).

Bells have been used in conjunction with other individual marking methods (color-coded ear tags and collars) to facilitate locating and monitoring movements of deer (Schneegas and Franklin 1972) and collared peccaries (Ellisor and Harwell 1969). Ellisor and Harwell (1969) reported that periods of auditory observation of peccaries provided movement data comparable to those gained from telemetry. This technique also allowed for the activity and habitat use of the animal to be determined. The attraction of predators is a potential concern when this marker is used.

TRANSPONDERS

Passive integrated transponder tags have been developed as a permanent marker for animals and have been tested on sea otters (Thomas et al. 1987) and black-footed ferrets (Fagerstone and Johns 1987). The tags consist of an electromagnetic coil and a custom-designed microchip that emits an analog signal when excited by a scanning wand with electromagnetic energy. The transponder chip is uniquely programmed with an alpha or numeric code, and >34 billion combinations are available. It is implanted subcutaneously with a spring-loaded syringe. The chip is activated only when energized; therefore the life of the transponder should be indefinite, making reuse of retrieved tags possible. A reader presents an activated code on a liquid screen, and information may be stored in the reader until retrieved.

No adverse effects of transponders were observed in animals 4–6 months post-implant (Fagerstone and Johns 1987, Thomas et al. 1987). The major disadvantage of this system is that the reader must be close (≤7.5 cm) to the animal to record the code, and this may necessitate rehandling of the animal (Fagerstone and Johns 1987). Potential exists for the collection of remote readings; a reader tube can be inserted into burrows or nesting cavities, or along travel routes, reading the transponder number each time the marked animal passes it.

Mutilation

BRANDING

Branding provides an inexpensive, permanent, and visible means of marking animals. Historically, hot branding has been used to mark horns of mountain sheep (Aldous and Craighead 1958) and to brand a variety of African ungulates (Hanks 1969). Homestead et al. (1972) developed a variation of the conventional hot-branding technique and used an explosive branding device to permanently mark four species of seals. Conventional hot-iron branding has been a common method for permanently marking seals (Summers and Witthames 1978).

Hot branding has almost no role in modern wildlife management, and when used on soft tissue it is considered a form of barbarity by some persons (Ryder 1978). In addition to inducing extreme pain, hot branding often produces open wounds that lead to infection. This method now is rarely used for zoo animals except to mark horns of bovids (Ashton 1978).

Freeze branding, or cryo-branding, a technique originally developed for livestock by R. K. Farrell (Washington State Univ. Anim. Health Notes 6:4–5, 1966), is a useful, more humane marking system than hot branding for a selective range of wildlife (Figs. 5, 6). Branding irons are supercooled, most commonly in a mixture of dry ice and 95% methanol (−67 C to −77 C) or liquid nitrogen (−196 C), and placed on a shaved and washed area of the skin. The epidermis is temporarily frozen (about 20–30 sec), destroying the pigment-producing melano-

Fig. 6. Freeze branding mark on a Thomson's gazelle.

cytes in the hair follicles and causing regrowth of white instead of colored hair. Dry ice and liquid nitrogen are difficult to maintain and handle in remote field operations, but pressurized dichlorodifluoromethane (Lazarus and Rowe 1975, Miller et al. 1983) or canned freon (Russell 1981) has been used successfully to freeze brand a variety of rodents. Hadow (1972) summarized the technique as being safe and reliable and producing permanent marks that are identifiable from a distance. Freeze branding has particular value for permanently marking long-lived species (e.g., marine mammals), captive wildlife herds, or small animal colonies. Standardized, individual animal identification and classification systems have been proposed for freeze-branded domestic livestock (Farrell et al. 1969) and laboratory animals (Farrell and Johnston 1973), but no generalized system is in place for wildlife.

Freeze branding has been used successfully to mark a variety of wildlife including white-tailed deer (Newsom and Sullivan 1968), seals (Hobbs and Russell 1979), and beavers (Pfeifer et al. 1984). Irvine et al. (1982) evaluated several methods for marking bottle-nosed dolphins and reported that freeze branding was the most readable, long-lasting, and least harmful procedure.

TATTOOS

Tattoos provide a simple, efficient means of permanently marking a wide range of species. Best results are achieved by tattooing any lightly pigmented area that is clean and essentially hairless (Fig. 7). Standard or rotary pliers or electric tattooing pencils may be used, and strongly contrasting dye (e.g., green or black often recommended) must be applied liberally. Even mammals as small as rats and mice can be effectively ear tattooed with modern equipment (Honma et al. 1986). Tattoos add no weight to an animal and are inconspicuous to predators but have the major disadvantage of being unreadable without having the animal in hand (Brady and Pelton 1976). Tattoos often are used in conjunction with more conspicuous marking systems for field study of wildlife.

Tattooing the inside of the ear is a standard marking procedure for field studies of cottontail rabbits (Brady and Pelton 1976) and snowshoe hares (Keith et al. 1968). Although less commonly used in cervids, ear tattooing of white-tailed deer fawns (Downing and McGinnes 1969)

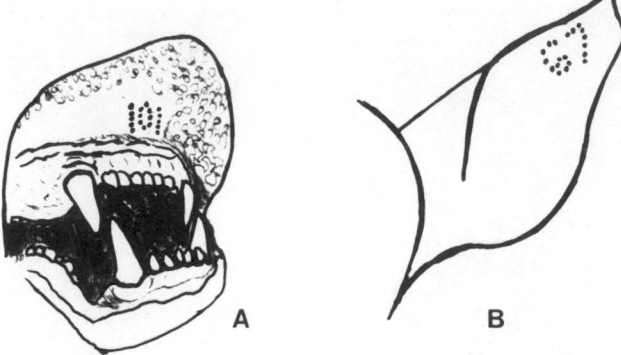

Fig. 7. Numeric characters tattooed on (A) the inside upper lip of a bear and (B) the inner ear of a deer.

and Père David's deer (Carnio and Killmar 1983) provided a satisfactory permanent marker. Bears are typically tattooed on the upper lip, axilla, or groin as part of the marking regime for these long-lived species (Lentfer 1968, Johnson and Pelton 1980).

Griffin (1934) successfully tattooed the wing membranes of bats but disliked the slowness of the procedure. Cheeseman and Harris (1982) used an electrically powered pen to tattoo badgers in the inguinal area of the abdomen. Soderquist and Dickman (1988) used a temporary tattooing method to mark marsupial pouch young. They tattooed small quantities of fluorescent pigments into the pinnae, the marks being highly visible under ultraviolet (UV) light for at least 6 months. Tattooing has been a proposed method of marking dolphinids (White et al. 1981), though an effective technique has not been developed. Geraci et al. (1986) reported tattooing to be an unsatisfactory method for marking beluga whales alone or in conjunction with freeze branding, because the tattooed tissue was sloughed off and replaced with new tissue within a few weeks.

TISSUE REMOVAL

Toe clipping is widely used to individually mark small mammals. The nail and first joint of the toe are removed with sterile dissecting scissors. The technique is inexpensive, rapid, and permanent, but sometimes clipped toes cannot be distinguished from toe loss in traps. Toe clipping has been most commonly used on small mammals, though a variety of species including fur seal pups (Gentry and Holt 1982) has been marked by this method. Standard methods were described for marking small mammals by toe clipping (Baumgartner 1940, Melchior and Iwen 1965). By clipping only two toes, one per foot, 98 animals may be marked, although an additional 106 may be marked if hyphenated numbers are also used. Removing no more than one toe per foot allows 899 animals to be marked. Kumar (1979) developed a toe-clipping code for identification of up to 9,999 animals, using no more than two digits clipped per foot. No direct adverse effects of toe clipping were reported for small mammals (Fullagar and Jewel 1965, Korn 1987), even when two toes were clipped per foot (Kumar 1979). However, toe clipping caused a temporary reduction in capture rates (Smal and Fairley 1982) and indirectly lowered the life-span of male meadow voles (Pavone and Boonstra 1985). Toe clipping

is not advised for bats because the toes are essential for roosting and grooming (Barclay and Bell 1988). In nursery roosts, however, Stebbings (1978) cut claws off juvenile bats that were too young to be banded; the marks lasted only a few weeks.

Toe clipping also has been used to mark hares (Dell 1957) and coyotes (Andelt and Gipson 1980) for the purpose of identifying tracks of marked individuals. A toe-end is disarticulated and surgically removed. Infection is a potential problem, and appropriate measures (antibiotics, convalescence time) should be taken to reduce the risk. Suitable conditions (e.g., snow) are required for track identification.

The ears of many small mammals can be marked by punching or clipping out small portions in a variety of coded systems, as outlined by Blair (1941) and Honma et al. (1986). Riley and Gwilliam (1981) developed an instrument that cuts a neat hole and completely removes the skin, reducing tearing of the skin and tissue regrowth into the hole. Large-eared ungulates, carnivores, and primates have been marked by cutting one or two notches at predetermined coded sites on the margin of the ear, allowing for a number of marking combinations. Ear notching larger species of mammals permits identification of marked animals at a distance. Notches usually last longer than tags, although they can be distorted by infection, growth, or damage (Ashton 1978). Ear notching is not advisable for mammals with highly specialized ears, such as bats, which use their ears for orientation and prey location (Barclay and Bell 1988), and seals, which have valve-like ears that function during deep sea dives (Scheffer 1950).

Punched holes or slits cut into foot webs have been used to mark beavers (Aldous 1940) and nutria (Davis 1963) and could be used for other web-footed species such as muskrats. The marks are permanent, but unclean cutting may produce a small scar rather than a hole. Holes punched in the web of the hind flipper of northern fur seals were distinct after 2 years, although the combination marks were difficult to observe on living, moving seals (Scheffer 1950). Identifying numbers punched through the wing membrane of bats with a tapered needle or a tattooing outfit produced a white scar in about 10 days (Bonaccorso and Smythe 1972) that persisted 1–5 months (Bonacoorso et al. 1976, Stebbings 1978). No injury to marked bats was reported, but animals must be rehandled to read the marks, as is generally the case with this technique.

The removal of some fur in a unique pattern is a nonpermanent, humane means of marking mammals. The marked animal is generally identifiable until the next molt, or, in fur seals, longer (Gentry 1979). Hair may be removed with mechanical clippers, a chemical substance, or heat, allowing for recognition at a distance or individual recognition. Clipping is especially useful on neonate seals that are too young to survive branding and tagging (Gentry 1979). A depilatory paste has been used to mark numbers on rats (Chitty and Shorten 1946), although depilatory agents may be extremely irritating to the skins of phocids (Gentry 1979). Hair burning, or "hair branding," produces a sharp, highly visible mark on fur seals, but a series of irons and fire is required for the marking (Gentry 1979). Unlike hot branding, this procedure does not burn skin tissue.

Dyes and Paints

Dyes and paints have been used as temporary external markers to identify mammals at a distance. They have been applied directly to immobilized or trapped animals or from a distance with paint-pellet pistols (Jonkel et al. 1975), modified Cap-Chur darts (Turner 1982), compressed spray tanks activated from blinds (Hansen 1964, Simmons and Phillips 1966), treadle-type spray devices (Clover 1954), and spraying devices used from aircraft (Simmons 1971). Paints and spray paints applied to the hides and pelage of terrestrial species, including bats (McCracken 1984) and elephants (Pienaar et al. 1966), and to the horns of antelopes and bovines (Hanks 1969, Clausen et al. 1984) persisted for a few weeks to several months. Marine and aquatic mammals have been marked less successfully with paints (Watkins and Schevill 1976, Gentry and Holt 1982). However, a promising aquatic marker is a fluorescent paste made from fluorescent pigment, vehicle binder, and solvent. It has visibly marked northern fur seals ≤2 years with no adverse behavioral effects or tissue abnormalities (Griben et al. 1984). Paint-stiks, oil-based, crayon-like markers, also may have potential as a temporary marker for manatees (Irvine and Scott 1984). Evans et al. (1971) reported codit white reflective liquid adequately marked nutria ≤30 days.

Nyanzol, Rhodamine B, and picric acid dissolved in ethyl or isopropyl alcohol (Hansen 1964, Brady and Pelton 1976) are dyes commonly used to mark terrestrial mammals with good results (Fig. 8). Color retention lasts 5–7 months or longer with no adverse effects. Red and orange aniline dyes (Day 1973) and clothing dyes (Simmons 1971) also have been used with success. Pinnipeds have been marked with nyanzol, but the subject must be dry and remain out of water at least 15 minutes for a clear mark to develop (Gentry 1979). Wool-lite, a pelage dye, was used to mark harbor seals a red color for ≤4 months (Pitcher 1979). Human hair dyes and lighteners, containing peroxide bleaches, were applied successfully to monk seals (Johnson et al. 1981) and northern fur seals (Gentry and Holt 1982), producing identifying marks that remained visible for ≤2 years. Bleaches are best applied to dry animals that remain dry for 30 minutes after application (Gentry 1979). Fur bleaching also has potential for short-term studies of bats (Bradbury 1977) and rodents (Hurst 1988).

Rhodamine B consumed orally acts as an internal marker, coloring gallbladder, gut, feces, urine, and oral and urogenital openings (Ellenton and Johnston 1975, Lindsey 1983). It has been used as a nonquantitative method for tracing bait consumption in opossums (Morgan 1981) and European rabbits (Cowan et al. 1984) and could be used in population estimation and determination of home range. Codit white reflective liquid was a satisfactory fecal tracer for nutria (Evans et al. 1971). Sudan black, orally consumed, stained fat deposits in rats for at least 112 days (Taylor and Quy 1973), but it was not effective as a bait consumption marker in European rabbits (Cowan et al. 1984).

Rhodamine B, in addition to its value as an external and internal marker, has been employed as a systemic marker, producing fluorescent banding of claws and hair of coyotes (Johns and Pan 1981) and mountain beavers

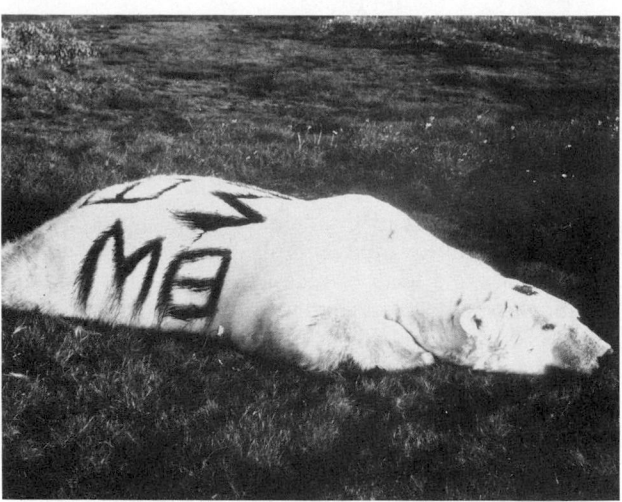

Fig. 8. Polar bear dye-marked with nyanzol for individual recognition from a distance (courtesy of Ian Stirling, Can. Wildl. Serv.).

(Lindsey 1983) that is visible under UV light. Pocket gophers also showed bands in whiskers and claws after oral dosing, although no marks were detected in body hairs (Lindsey 1983). Marks may be visible within 24 hours of dosing and may persist for several weeks. Scanning for Rhodamine B does not require necropsy, and the use of portable UV lamps allows entrapped animals to be examined and released immediately, reducing handling stress. The use of Rhodamine B as a systemic marker may be limited to certain periods of the year, because banding may occur only in actively growing tissue. No adverse physiological effects of Rhodamine B as a marker for wildlife have been reported.

Particle Markers

Lemen and Freeman (1985) and Mullican (1988) used fluorescent pigments for following movements of small mammals at night. Live-trapped animals were dusted with the pigment and released. The following night a fluorescent trail could be traced with ultraviolet lamps. The amount of vegetation cover, precipitation, and ambient light influenced trail detection (Mullican 1988). This technique enables detailed information on home range, movement patterns, and habitat use to be obtained within a few days and is easier to use than radiotelemetry. The pigment is detectable on the animal the second night, but the animal no longer leaves a trail. Boonstra and Craine (1986) located the nests of voles, and Dickman (1988) investigated social interactions of small mammals using similar techniques.

Microtaggants, small, plastic particles that are coded by means of colored layers, have been tested and proposed as a means of identifying acute toxicants in digestive tracts and baits without chemical analysis (Johns and Thompson 1979). Microtaggants do not cause bait aversion, remain intact, and, due to their fluorescent and magnetic properties, can be readily recovered from gut or fecal samples. Fluorescent acetate floss fibers also have been tested for measuring bait consumption (Cowan et al. 1984). As with microtaggants, floss fibers are a quantitative, nonpersistent marker. Floss fibers do not affect bait palatability and are more economic than microtaggants. Randolph (1973) put fluorescent acetate floss fibers in bait as a fecal marker for comparing individual home ranges of small mammals. Powdered aluminum pigment in baits was used as a fecal tracer in nutria (Evans et al. 1971).

Chemical Markers

Certain members of the tetracycline group of antibiotics, given orally or intravenously, combine with calcium in the bones and teeth of mammals to produce a characteristic yellow fluorescence under UV light. Tetracyclines are persistent, quantitative markers that can cross the placental barrier (Owen 1961). Marking captive coyotes with demethylchlorotetracycline (DMCT) (Linhart and Kennelly 1967) was most apparent on the mandible, and the greatest intensity and quantity of fluorescence were in younger animals. Marks were visible for at least 5 months after dosing. Crier (1970) used single oral doses of DMCT at 50 mg/kg of body weight to label the lower jaws of laboratory rats for up to 6 months, making DMCT a promising marker for field studies of rodents. Tetracycline was applied successfully as a time-specific marker to such species as northern fur seals (Yagi et al. 1963), dolphins (Best 1976), and beluga whales (Geraci et al. 1986), the marker appearing as fluorescent rings in the dentine of teeth. Taylor and Lee (1994) used tetracycline in a similar manner to obtain mark-recapture population estimates for polar bears. Nelson and Linder (1972) used chicken eggs containing DMCT to determine the percentage of raccoons and skunks that consumed baits, though DMCT may be ineffective for marking some species, and may affect bait acceptance.

Johns and Pan (1981) tested quinacrine dehydrochloride as a fluorescent chemical marker in rats and detected it in the blood with fluorometric and chromatographic analytical techniques. Larson et al. (1981) tested iophenoxic acid, an iodine-containing compound, and mirex, an organochlorine compound, as blood and tissue markers for bait-consuming animals. Baseline iodine was elevated for at least 7–8 weeks after oral dosing with 5 mg of iophenoxic acid, and mirex was detectable at ≥ 0.005 ppm at 8 weeks after dosing with 25–75 mg/kg of body weight. Baer et al. (1985) and Follmann et al. (1987) used iophenoxic acid as a serum marker for dogs or foxes fed rabies vaccine in the wild to determine how many had consumed baits. This marker could be used with other vaccines for the same purpose.

Radioactive Markers

In the selection of an appropriate radioactive marker, characteristics of the isotope to consider are its availability, type of radiation, energy levels emitted, physical and biological half-life, radiotoxicity, and metabolic characteristics (Pendleton 1956). The physical half-life of an isotope will influence radiation exposure to humans and animals and environmental contamination. Radioactivity causes tissue damage, so the smallest detectable dose should be used.

Radioactive tracers generally have been used to identify and acquire information on the behavior of specific small mammals. The three main methods of marking mammals with radioisotopes are inert implants, external attachments, and metabolizable radionucleotides (Table 1). Inert implants are suitable for monitoring specific movements such as nest visits of small mammals in the field with a manual or automatic detector (Bailey et al. 1973, Linn 1978). Radioactive wires, pins, and capsules containing isotopes have been inserted subcutaneously in small rodents and bats as inert implants. Radioactive material can be attached to external leg rings and forearm tags, or they can be made radioactive. Radioactive material also can be fed, injected, or implanted into the animal in a metabolizable form. Such materials may be incorporated into the tissues of the animal, passed on to offspring, or voided in feces and urine and can thus be used for many purposes other than tracking (Linn 1978). Dickman et al. (1983) injected ^{35}S into lactating small mammals. The radioisotope was passed to the young via the mother's milk. Different amounts of radioisotope resulted in different levels of radioactivity in the hair of juveniles, allowing filial relationships to be determined. Tamarin et al. (1983) injected pregnant females with one or more gamma-emitting radionuclides and accurately determined mother-offspring relatedness; Scott and Tan (1985) used a similar technique to establish mating success of males and estimate reproductive success in natural populations of marsupial mice.

The use of metabolizable radioisotope marking has been extended to larger mammals. Zinc^{65} (^{65}Zn) fed and injected into rabbits, opossums, foxes, and bobcats was detectable in feces from these animals for >1 year after injection (Nellis et al. 1967). Pelton and Marcum (1975) injected radioisotopes ^{65}Zn and ^{54}Mn into captured black bears, collected and analyzed scats for the isotope, and estimated population abundance using the ratio of tagged to untagged scats. This technique has since been used to estimate population abundance of a number of species including white-tailed deer (Kinningham et al. 1980), coyotes, and bobcats (Conner 1982).

Nocturnal Tracking Lights

Light sources attached to animals allow them to be visually tracked at night, providing information on movements and foraging behavior. Chemical, electrical, and radioactive types can be used alone or in conjunction with radiotelemetry. Evidence suggests that the use of optical light sources does not increase predation of marked individuals or adversely affect their behavior, although the potential for this does exist. Conversely, marked predators might have less success capturing prey. Barclay and Bell (1988) suggested that a constant light source may cause undue stress in bats.

Buchler (1976) used a chemical light source, Cyalume, to monitor the activity of bats. The light was obtained by mixing dibutyl phthalate and dimethyl phthalate liquids and sealing them in small glass spheres. The spheres were glued to the ventral fur of bats. The brightness and duration of light emission could be controlled by varying the proportions of the mixture. Light was emitted ≤ 3 hours and was visible from 225 m to 475 m with the aid of binoculars. LaVal et al. (1977) reported the potential for use of high-intensity Cyalume, which may be seen at distances of up to 1,500 m with binoculars but fades within 1 hour.

Barbour and Davis (1969) taped small "pinlights" with attached batteries to the fur of bats to follow their movement at night. Carpenter et al. (1977) put battery-operated

Table 1. Radioisotopes used for marking mammals.

Isotope	Half-life	Toxicity	Mammal(s)	Method of use	Reference
Cobalt-60	5.25 yr	Medium upper	Small mammals	Implanted or used in capsule on rings	Linn and Shillito 1960, Barbour 1963, Schnell 1968
Cadmium-115	43 days	—	Raccoon	Injected	Conner and Labisky 1985
Gold-198	2.7 days	Medium upper	Harvest mice	Implanted, biologically inert	Kaye 1960
Iodine-131	8.04 days	Medium upper	Mammals	Injected, used in capsules on rings, implanted, or injected in bait	Gifford and Griffin 1960, Johanningsmeier and Goodnight 1962
Magnesium-54	312 days	—	Black bear	Injected	Pelton and Marcum 1975
Phosphorus-32	14.3 days	Medium lower	Voles	Injected	Miller 1957
Sulphur-35	87.2 days	Medium lower	Small mammals	Injected	Dickman et al. 1983
Tantalum-182	115 days	Medium lower	Small mammals	Implanted, biologically inert	Graham and Ambrose 1967, Schnell 1968
Zinc-65	245 days	Medium lower	Small mammals, opossum, rabbit, fox, European badger, bobcat, black bear	Injected, fed	Nellis et al. 1967, Gentry et al. 1971, Pelton and Marcum 1975, Kruuk et al. 1980, Conner 1982

neon lights on neck collars for nocturnal observations of mule deer. Light intensity or blinking sequence was varied for individual identification. Wolcott (1977) designed and constructed a miniaturized light-emitting diode (LED) source and flasher. The device produced consistently timed flashes that could be used for individual identification. Batchelor and McMillan (1980) developed a similar system with individually programmable flashes, a light-sensitive flasher, and an optional attachment of radiotransmitter equipment to the same circuit. The size of the battery and the intensity of the light source influence life-span and visibility of the marker. The use of binoculars or night vision scopes greatly increases the distance at which these markers can be seen.

Betalights are a radioactive light source that consists of phosphor excited by tritium gas in glass capsules. The capsules can be produced in any shape and size, and different colors of Betalights are available. The useful range varies from about 50 m to 1 km depending on shape, size, and viewing method. The life-span of Betalights is about 15–20 years. Acceptable radiation levels should be determined when these light sources are used. Kruuk (1978) incorporated Betalights onto radio transmitters carried by badgers to assist in observation of located animals. Davey et al. (1980) glued Betalights to chicken wing-tabs with a cold-curing epoxy resin and attached them to the ears of rabbits. Some colors were used at different intensities to increase the number of individuals marked. Thompson (1982) glued Betalights to the top of the head of rodents after clipping fur from the area.

Natural Markings

Identification systems that depend on variations in natural markings, size, shape, or other features are useful in field studies when capture and marking of individuals may be impractical or undesirable. Photographs or drawings of individuals provide a means for later field identification.

For example, identification systems that use ear markings, shape of horn, wrinkle patterns, sex, and size have been applied to black rhinoceroses (Mukinya 1976), coat patterns to giraffes (Foster 1966), facial features in combination with other identifying features to primate groups (Kummer 1968), and fin notches, color patterns, scars, and callosities to dolphins and whales (Würsig and Würsig 1977, Irvine et al. 1982). Natural marking is best when modest numbers of individuals occupy a well-defined area and minimal immigration occurs (Pennycuick 1978).

When a natural marking system is evaluated, the benefits of not having to catch or mark individuals must be weighed against the costs of a reliable, artificial marking system. Natural marking systems are better for individuals with rare patterns, and the larger the population size, the smaller the proportion of individuals that can be reliably identified (Pennycuick 1978). The possibility of two or more identically marked individuals (Ashton 1978) and the amount of information required to accurately identify individuals within a population (Pennycuick 1978) must be realized. Some natural markings may be altered by further accidents or age, or observer changes may introduce errors into natural identification systems (Carnio and Killmar 1983). Natural identification, in many situations, is best employed in conjunction with or as a supplement to an artificial marking system.

MARKING TECHNIQUES FOR BIRDS
Leg Bands

Metal leg bands bearing an identification number and return address are the oldest and most common method of marking captured birds (Fig. 9). Although states and provinces are required to use their own bands for resident game birds, bands for migratory birds are issued by the U.S. Fish and Wildlife Service and the Canadian Wildlife Service, which require the bander to follow a common

　　　　　　　　　　　　　　　　Marking Wildlife

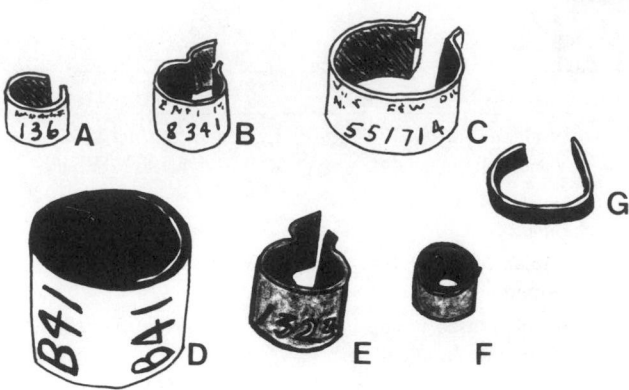

Fig. 9. Standard bird leg bands: (A) butt-end, (B) lock-on, (C) rivet, (D) laminated plastic wrap-around, (E) anodized aluminum colored, (F) soft, plastic, wrap-around, and (G) steel flipper band for marking penguins (not drawn to scale).

nomenclature (Appendix I). The butt-end or split-ring type of band is widely used for most species. Lock-on bands are of thinner metal than butt-end bands and are used on raptors and other birds capable of removing butt-end bands (Environment Canada 1984). Rivet bands are used for eagles, which are capable of removing butt-end (Berger and Mueller 1960, Robson 1986) and lock-on (Young and Kochert 1987) bands. A close-ring band is often used to mark birds raised in captivity (Spencer 1978).

Standard bands made of pure aluminum are sufficient for marking many species. Aluminum, however, is relatively weak and can be more easily damaged by abrasion and corrosion than heavier metals. As a result, other metals such as monel, incoloy, stainless steel, and titanium have been placed on long-lived birds, especially those inhabiting marine areas. These heavier metals are more difficult to apply and remove, and they are naturally darker than aluminum and are more inconspicuous on a bird. Colored anodized aluminum leg bands have been used with variable success to facilitate individual identification in many species (e.g., Cohen 1969, Godfrey 1975).

Bands should fit properly, allowing movement up and down and turning freely. The band should not be so loose that it can be pulled over the toes. Needle-nosed pliers or special banding pliers are useful for opening bands for placement on legs and for ensuring a tight closure of the bands. Bands flattened during securing should be removed and replaced, ensuring that the round shape of the band is retained. Care should be taken not to damage the bird's legs by bending, twisting, or pulling them during the banding process. Ducklings may be ringed with the aid of florist's wax or plasticine; as the tarsus grows, the plasticine or wax on the inner band yields (Spencer 1978). Banding the tibia rather than the tarsus can increase longevity of the ring and legibility of the code for seabirds (Perdeck and Wassenaar 1981, Zmud 1985), though this is an uncommon practice for most species. Turkey vultures, which excrete down their legs, should not be leg banded because excrement loading of the band can cause swelling, leading to loss of the leg or foot (Henckel 1976). Ice build-up on legs of banded passerines in cold climates also may cause impairment of leg movement or leg loss (Elmes 1955, Dunbar 1959, MacDonald 1961).

Birds can mutilate (Young 1941) and remove (Young

and Kochert 1987) bands, and loss of bands has occurred in nestlings (Kaczynski and Kiel 1963). The main causes of loss of leg bands, however, are abrasion and corrosion from saltwater and defecation. Rates of leg band wear and loss are related to the type of metal used, band size, location on the leg, characteristics of the species banded (e.g., habitat selection, foraging behavior), and even gender (Mills 1972). Rates of leg band loss are of concern to researchers and managers and can lead to inflated estimates of mortality and errors in population size, especially for long-lived species (Kadlec 1975, Nelson et al. 1980).

Colored bands made from plexiglass, vinylite, PVC, Darvic, Lynnply, Gravoply, and Saflag have been used alone or in conjunction with metal bands to mark individuals of a variety of species. Colored bands are primarily intended to permit rapid identification of individuals without recapturing and handling, thus being unsuitable for species whose behavioral or physical characteristics tend to obscure the bands (Spencer 1978, Forsman 1983). Color bands deteriorate quicker than metal bands and are best used in short-term studies (Seguin and Cooke 1983, Ottaway et al. 1984). Color bands, however, may increase the rate of recovery of dead birds by making them more conspicuous (Goss-Custard et al. 1982, Shedden et al. 1985). Durability and color retention are lowest in soft, plastic, wrap-around bands (Reese 1980, Anderson 1981), increasing in laminated, wrap-around bands (Lumsden et al. 1977, Anderson 1981) and plexiglass, butt-end bands, which require special equipment to fit in the field (Balham and Elder 1953). Seguin and Cooke (1983) reported that wide PVC bands had lower rates of band loss than narrow PVC bands. Strong et al. (1987) molded color bands to conform to the laterally compressed legs of loons.

Colored bands have caused severe leg abrasion (Reed 1953) and damage from constriction (Atherton et al. 1982), and may cause crippling in web-footed species as the result of band displacement (Colclough and Ross 1987). Color bands hindered social interactions of cranes (Wheeler and Lewis 1972), although survival did not appear to be affected by the banding (Hoffman 1985). Color bands also may influence mate selection in some species such as zebra finches (Burley et al. 1982).

Colored tags have been used in conjunction with bands to improve band retention and field recognition of birds. Scotch-brand, pressure-sensitive tapes (Carrick and Murray 1970) and labelled strips of Velcro (Willsteed and Fetterolf 1986) were used to mark nestlings too young for a leg band. Plasticized PVC tape, Saflag, and Herculite tags were attached to legs with a falconer's jessed knot, a slot and notch system, and pop-rivets (Downing and Marshall 1959, Arnold and Coon 1971, Platt 1980). Leg streamers and markers were attached to aluminum bands by pop-rivets (Frentress 1976), between the flanges of a rivet band (Cline and Clark 1981), and through slits in the marker (Thomas and Marburger 1964, Royall et al. 1974). Painted bands are of limited use because abrasion or paint removal by birds results in rapid marker loss (Childs 1952). Tags can cause entanglement (Royall et al. 1974), and if large enough they may hinder flight (Guarino 1968). Long markers on falcons may be mistaken for prey items by other falcons (Platt 1980).

Wing Markers

Wing markers are commonly used on birds. They are generally made from flexible PVC-coated nylon fabric, rigid PVC, and upholstery plastics, and they either wrap around the wing (Morgenweck and Marshall 1977, Kochert et al. 1983) or are pierced to the patagium by a stainless steel or nylon pin (Knowlton et al. 1964, Wallace et al. 1980), by a pop-rivet (Seel et al. 1982, Stiehl 1983), or by the marker itself (Baker 1983, Sweeney et al. 1985). Cummings (1987) attached wing tags to blackbirds with nylon fasteners and a fastener gun. Wing tags and streamers should be large enough for observational purposes but not so large as to hinder flight (Wallace et al. 1980) (Fig. 10). Durability and colorfastness are functions of material composition and manufacturing (Nesbitt 1979, Young and Kochert 1987), some materials lasting ≤10 years (Kochert et al. 1983). Wing markers with good visibility have been placed on a variety of bird species, allowing observers to collect more information on life histories of marked birds than could be gathered with leg bands. Tag loss is generally low the first year (Patterson 1978, Stiehl 1983), gradually increasing in subsequent years (Patterson 1978). Mudge and Ferns (1978), however, estimated that 25% of the wing tags attached to herring gulls were lost within 1 year. Double pinning the tags reduces marker loss (Hart and Hart 1987).

Wing markers often have no consistent effect on birds, although an initial adjustment period may range from a few days to 2 weeks (Bartelt and Rusch 1980, Sweeney et al. 1985). Light feather wear and patagium callousing have been commonly noted. Severe abrasion has been observed occasionally with some species (Harmata 1984) and consistently with falcons (Kochert et al. 1983). Abnormal replacement of feathers may occur (Howe 1980) and flight may be affected (Tacha 1979), although double pinning greatly reduces feather abrasion and callousing (Hart and Hart 1987). The reported effects of wing markers on reproductive and social behavior have been variable. Many species have shown no significant effect on fledging successs when at least one adult was marked (summarized in Young and Kochert 1987). However, brood size was slightly lower in marked ring-billed gulls compared to those unmarked (Southern and Southern 1983), and Jackson (1982) observed that the mean re-nesting interval for female red-winged blackbirds was lengthened in wing-marked birds versus leg-banded birds. Wing-marked cranes became outcasts or subordinates (Tacha 1979), although Wallace et al. (1980) and Sweeney et al. (1985) observed no differences in winners of aggressive encounters between marked and unmarked black and turkey vultures. Patagial marking can interfere with migration (Howe 1980, Southern and Southern 1985), alter habitat selection (Szymczak and Ringelman 1986), and contribute to mortality (Howe 1980, Southern and Southern 1985). Saunders (1988) contended that patagial tags should not be used on rare, vulnerable, or endangered species unless the reason for their use is compelling.

Flipper bands, made initially of aluminum and more recently from monel metal and stainless steel, have been used to mark the foreflipper of penguins. Cooper and Morant (1981) recommended the more durable stainless steel bands. A flipper band designed without safety fas-

Fig. 10. Wing markers on trumpeter swan cygnet (courtesy of Len Shandruk, Can. Wildl. Serv.).

teners (Sladen 1952) and which allows for flipper expansion during the molt is less likely to cause injury or death (Sallaberry and Valencia 1985).

Neckbands and Collars

Colored neckbands marked with letters and numbers have been used extensively to mark geese and swans (Fig. 11), sandhill cranes (Huey 1965), ring-necked pheasants, and scaled quail (Taber and Cowan 1963). However, they are not an appropriate marker for ducks due to low retention and high marker-related mortality (Idstrom and Lindmeier 1956). Collars have been constructed from flexible vinylite (Koerner et al. 1974), flexible plastic (Fjetland 1973), rigid plastic (acrylic resin) (Ballou and Martin 1964), and aluminum (MacInnes et al. 1969). Pirkola and Kalinainen (1984) used a double-layered, UV-protected plastic, a material that does not break down under natural circumstances and maintains its color and visibility codes even after 5 years of exposure. Maltby (1977) designed a durable, lightweight, aluminum collar that was riveted together. Craighead and Stockstad (1956) used plasticized polyvinyl chloride tape as a strap-type marker. Collars made from flexible materials are generally lighter and easier to apply than rigid collars. Collars should be loose enough not to hinder movement or foraging, though too small for the head to slip through.

Collars are usually highly visible and allow for more information to be gained than if only metal bands or colored leg bands are used. Collar retention rates of 76–100% the first year and 60–100% the second year were reported for geese (Ballou and Martin 1964, Sherwood 1966, Craven 1979), and some collars were retained for 11 years on Canada geese (Zicus and Pace 1986). A small loss of collars on geese results from entanglement in duck traps (Sherwood 1966), although the major cause is due to synthetic collars becoming old and brittle (Fjetland 1973). Collars should not be used on goslings <2 months of age because few are retained (Sherwood 1966).

Many studies report no or insignificant adverse effects of neck collars on breeding-related activities, social behavior, and physical damage beyond minor feather wear

Fig. 12. Nasal saddle on the bill of juvenile male blue-winged teal (courtesy of Ducks Unlimited).

Fig. 11. Trumpeter swan marked with coded neckband (courtesy of Len Shandruk, Can. Wildl. Serv.).

and irritation. However, neckbands disrupted pair bonds in black brants (Lensink 1968), lowered success in agonistic encounters in black brants (Abraham et al. 1983) and tundra swans (Hawkins and Simpson 1985), and contributed to starvation of female lesser snow geese (Ankney 1975). Zicus et al. (1983) observed severe icing on marked geese where strong winds, subzero temperatures, and snow or blowing snow were factors, resulting in an estimated 30–68% mortality. Icing is not a problem with aluminum neckbands, perhaps due to their conductive properties (MacInnes et al. 1969). MacInnes and Dunn (1988) indicated that collars may cause important non-hunting mortality in Canada geese or result in birds moving away from the breeding area.

Tags

NASAL TAGS

Nasal discs and saddles have been used extensively to mark waterfowl. Nasal tags are generally made from rigid or flexible PVC or nylon, marked with patterns or numbers, and attached by a short nylon or stainless steel pin through the nares (Fig. 12). Nasal discs were retained well on ducks for 1 year (Bartonek and Dane 1964), although tag loss was high for Canada geese (Sherwood 1966). Discs snagged on vegetation and tangled in nets during trapping operations and were thought to increase mortality of diving ducks (Erskine 1962). Proper design reduced such hazards (Sugden and Poston 1968), and nasal saddles have been used successfully on many diving and dabbling ducks. Davey and Fullagar (1985) constructed PVC saddles, held in place by a self-locking pin, to fit the size and shape of the bill of particular waterfowl species. They were also notched at the edge, producing a binary code by which individuals could be identified. Lokemoen and Sharp (1985) reported nylon nasal saddles to be nearly indestructible and color stable compared to lynply plastic and PVC markers, respectively.

Small ducks may have difficulty with the nasal saddle due to its large size and the shape of bill and nares (Joyner 1975, Koob 1981). Occasional entanglement in fences and traps has resulted in tag loss in mallards (Evrard 1986).

Greenwood and Blair (1974) reported icing on the nasal saddles of mallards, and Byers (1987) inferred that mortality of mallards from icing during unusually severe winter weather could reach 10%.

BACK TAGS

Tags designed to lie on the back have been used frequently to mark upland game birds and waterfowl, although Southern (1964) marked bald eagles and Frankel and Baskett (1963) marked mourning doves this way. Back tags are generally made from flexible PVC or PVC-coated nylon fabric and are attached by a leather or nylon cord harness whose straps pass around each wing base. Average retention time of leather-strapped back tags varied from 6.5 months for ruffed grouse (Gullion et al. 1962) to ≥12 months for coots (Anderson 1963), Hungarian partridges (Blank and Ash 1956), and ring-necked pheasants (Labisky and Mann 1962). Tags with nylon straps lasted ≥2 years. Back tag markers modified into ponchos were worn successfully by sharp-tailed grouse, sage grouse, and Hungarian partridges (Pyrah 1970). Although Hester (1963) reported back tagging too cumbersome for small birds, Furrer (1979) developed a tag that protruded from the bird's back, making it more visible, and used it to mark large thrushes, finches, and starling-sized birds. Cuthbert and Southern (1975) glued a circular numbered tag to the synsacrum region of recently hatched gull chicks, and Baltosser (1978) glued back tags to hummingbirds. Tags glued onto feathers are lost during the following molt.

WEB TAGS

Grice and Rogers (1965) developed a technique to mark wood ducks that were too young to take a leg band. They attached consecutively numbered fingerling fish tags to the center of one of the foot webs of newly hatched young, using a pair of light pliers to tighten the tag. Alliston (1975) applied this technique to marking ducklings in pipped eggs of ground-nesting species. Part of the shell and membrane of an egg were removed, a foot was extracted, tagged and replaced, and the hole was covered by masking tape. Web-tagging did not affect hatching success or survival after nest exodus. However, this technique should not be attempted before pipping is far enough advanced or bleeding may occur (Alliston 1975). Haramis and Nice (1980) modified a pair of needle-nosed pliers to accurately position and cleanly cut slots in the web, reducing the risk of the tag tearing out. These pliers can be used for either cavity-nesting ducks or ducklings in pipped eggs. Tag loss ranged from <1% to 3% in these studies. Web tags were also placed on gull chicks (Ryder and Ryder 1981).

Mutilation

FEATHER CLIPPING

Portions of vanes are clipped in different sizes and shapes from the shaft of several adjacent feathers, creating unique holes in the wings or tail that are used as identifying features. The clipping should be performed so as not to affect flight. This technique, most suitable for gliding species, is of limited value for sedentary species because the marks cannot be observed on perching birds. As well, the number of combinations of effective marks is limited. Snelling (1970) and Gargett (1973) used feather clipping to individually identify large African raptors, as did Enderson (1964) for prairie falcons and Garnett (1987) for captive frigate birds. Geis and Elbert (1956) clipped tail feathers on ring-necked pheasants, making a visible mark though reducing the breeding success of the birds.

TISSUE REMOVAL

Richdale (1951) used a leather punch to puncture holes in three precise places in each web on both feet of penguins, allowing for a large number of birds to be uniquely marked. Although some marks changed from injury or healing processes, the loss of identity of birds was negligible, and this method was more practical than using leg bands. Conversely, Reuther (1968) used web punching for zoo birds and reported that fighting and other causes often destroyed the identification marks. Regardless of lost marks, a major disadvantage to web punching is that the birds must be recaptured for the web holes to be read.

Burger et al. (1970) successfully marked nestling mallards by clipping off the alula, allowing subsequent identification in the hand. No difference in growth rate, behavior, or flight capability was noted between marked and unmarked birds.

Toenail clipping, similar to the technique of toe clipping in small mammals, was used to mark individual nestlings of eastern kingbirds and eastern phoebes (Murphy 1981), and tree swallows and house wrens (St. Louis et al. 1989). St. Louis et al. (1989) clipped toenails within 3 days of hatching, reporting no nestling mortality or toe loss, nor was ossification or growth of cartilaginous material within the nail impaired. Clipped toenails remained blunt enough at the tip to be distinguished throughout an 18-day nestling period when birds were too young to be banded, although the nails eventually grew back.

TATTOOING

Tattooing is a long-lasting method of identification which, to date, has been used most successfully as a method to mark captive birds, particularly birds of prey (Havelka 1983). A tattoo is usually put on the underside of the wing on the skin close to the body and is therefore visible only when the wings are opened.

Ricklefs (1973), however, used a syringe filled with black India ink to tattoo unique combinations of dots on the abdomens of nestling starlings that were too young to be marked satisfactorily by most tagging methods. No discomfort or infection was associated with tattooing, nor was development of the young affected. The tattoos became illegible after feathers grew in.

FREEZE BRANDING

Greenwood (1975) freeze branded mallard duckling feather tracts without success. Freeze branding of the premaxillae, however, was successful and was suggested as a potential temporary marker for ducklings.

Dyes, Paints, and Inks

Dyes, paints, and inks have been used to mark a variety of bird species for many investigative purposes. They are temporary markers and are generally lost during the following molt. Many methods have been designed to spray dye or ink on birds from a distance (Moffit 1942, Siegfried 1971, Moseley and Mueller 1975, Rodgers 1986), thus eliminating the need for capture.

Waterproof dyes should give an easily recognizable color, be nontoxic, resist fading, be harmless to the plumage, be capable of use with a wetting agent or solvent to ensure quick penetration and coverage, and be fast acting in a cool solution (Patterson 1978). Dyes are most effective when applied in a 33% alcohol/66% water solution (Wadkins 1948). Picric acid, Rhodamine B Extra, and Malachite Green are dyes that have strong color and exhibit good penetration and retention (Wadkins 1948, Handel and Gill 1983, Underhill and Hofmeyer 1987). Species with light plumage are more effectively marked with dyes (Paton and Pank 1986, Paullin and Kridler 1988) than species with dark plumage (Moffitt 1942). Dyes have been applied by dipping, brushing, and spraying. Caution should be used when dyes are applied in cool weather, because birds may undergo hypothermia (Kozlik et al. 1959). Birds captured for dye marking should be thoroughly dried before they are released.

Birds have also been marked with dyes in an indirect fashion. Evans (1951) injected food dyes into duck eggs just before hatching, and Rotterman and Monnett (1984) applied this technique to passerines. The newly hatched young are marked for a few days, permitting continuous identification of individual birds. Short-term survival and condition of the young do not appear to be affected, although care must be taken to avoid infection of the embryo. Mossman (1960) sprinkled thief detection powder on eggs and nests of glaucous-winged gulls, marking the adults as they incubated their eggs. Paton and Pank (1986) marked nesting cattle egrets by adding Rhodamine B powder to an oil-based silica gel and placing it on the upper surfaces of eggs within clutches. Incubating adults were marked for 2–6 months and could be observed from 200 m with spotting scopes. No clutches failed, although the effect of this dyeing technique on egg survival was not investigated.

Ellis and Ellis (1975) used human hair dyes to mark golden eagles, and White et al. (1980) and Malacarne and Griffa (1987) used a hair lightener powder mixed with hydrogen peroxide to bleach plumage on blackbirds and swifts, respectively. Visibility of marked birds ranged from 500 m to 2,000 m for golden eagles with lightened or darkened plumage, and ≤1,000 m for bleached blackbirds. Feather damage may occur if feathers are bleached at too high a temperature or for too long a period, thus the tips of primaries are generally not marked. White et al. (1980) also reported blackbirds were susceptible to hypo- and hyperthermia during the bleaching process.

Model airplane paint and spray paints also have been used to color mark birds. Marking is most effective on flight feathers, because subsequent preening may result in feather loss or mark deterioration (Swank 1952) and in matting of breast feathers (Dickson et al. 1982). Paints should be allowed to dry before a bird is released. Ruffed grouse (Bendell and Fowle 1950) and cattle egrets (Siegfried 1971) were marked with printer's ink, and marks lasted for at least 12 months on the latter species.

No adverse physiological effects have been reported for these markers when they are properly applied. Although they generally have no observable effects on behavior other than temporarily increased preening, certain markings may disrupt pair bonding (Goforth and Baskett 1965) or cause temporary shunning by mates (Butts 1930). Altering intraspecific recognition mechanisms in birds may severely alter social interactions (Rohwer 1977).

Feather Imping and Similar Techniques

The insertion of a colored feather into the clipped shaft of a bird's feather ("imping"), wherein a double-ended needle (Wright 1939) or cement (Hamerstrom 1942) secures the feather, has been used as a technique to mark birds. Rectrices are usually used, although remiges can be imped if the replacement feather closely matches the one cut off. Hester (1963) concluded that feather imping is unsatisfactory for passerines due to poor visibility of the marker, few color combinations, and the time required. Sowls (1950) noted that imping was a less effective marker for waterfowl than painting with airplane paint.

Dyed feathers also can be attached to natural feathers or wired to the rachis of natural feathers whose vanes have been clipped off (Edminster 1938), or glued to plumage in unnatural, conspicuous patterns (Neal 1964). Dickson et al. (1982) and Ritchison (1984) cut the barbs off a proportion of centrally located rectrices and applied colored tape to the rachis of the clipped area as a marker. These marking techniques are temporary, although nestling gulls have been marked by grafting the pollex to the skin of the head, resulting in alula feathers growing from the head region (Coppinger and Wentworth 1966).

Particle, Chemical, and Radioactive Markers

Bendell and Fowle (1950) placed aluminum and bronze dust in nesting places of ruffed grouse and found it later on shed feathers. Otis et al. (1986) used an aerial application of a liquid fluorescent pigmented material, visible under UV light, to mark large numbers of roosting blackbirds. The liquid dries, adhering to the feathers. Subsequent collection of marked birds provides data on dispersal and population dynamics. Microtaggants, small, color-coded plastic particles (Johns and Thompson 1979), reviewed in the mammalian section of this chapter, can be used to identify bait-consuming birds and have potential in raptor management.

Haramis et al. (1983) and Eadie et al. (1987) injected tetracycline as a marker to trace maternity in wood ducks and Barrow's goldeneyes. When injected intraperitoneally, tetracycline chelates with calcium ions in the forming eggshell and is detected by fluorescence under UV light. Tetracycline was detected in eggs up to 20 days after injection in some females. Haramis et al. (1983) did not detect any adverse side effects of tetracycline, but Eadie et al. (1987) reported egg-laying rate decreased and cautioned against its use in estimating the reproductive success of individual females.

Johns and Pan (1981) used quinacrine dehydrochloride as a fluorescent chemical marker in starlings. Larson et al. (1981) tested two chemicals (iophenoxic acid and mirex) for marking bait-consuming birds. Mirex was detectable in blood or tissue samples for 8 weeks following oral dosing at 75 mg/kg body weight. Iophenoxic acid, administered at 5 mg/kg body weight, was not an effective marker.

Lindsey (1983) observed that oral consumption of Rhodamine B produced fluorescent banding of feathers in domestic chickens and believed that it could be used as a marker for other bird species. Bands were most evident in the primary and secondary feathers. Bands persisted for 1–2 weeks in birds receiving 5 mg/kg body weight of Rhodamine B and 15–26 weeks at concentrations of 15 and 30 mg/kg body weight. No adverse physiological effects were noted. This type of marker should not affect the social structure of birds, because the bands are invisible in ambient light. However, its use may be limited to certain periods of the year due to growth characteristics of the feathers.

Radioisotopes have received little attention as a means of marking birds. Griffin (1952) placed radioactive leg bands on semipalmated plovers and recorded nest visits by marked birds with an automatic monitoring device.

Nocturnal Tracking Lights

Nocturnal tracking lights, similar to those described in the mammalian section, have also been used on birds. Radioactive Betalights were used with radiotelemetry to aid in nocturnal observations of foraging boreal owls (Hayward 1987). The most effective location of the Betalight was on the radio antenna, away from the body of the owl. Betalights did not increase mortality of radio-marked owls, although hunting success could potentially be affected. DeLong (1982) used a battery-powered, light-emitting diode (LED) to study the behavior of long-eared owls at their nest sites, and Clayton et al. (1978) marked black skimmers with this technique. They also used Cyalume to mark black skimmers. The mixture was injected into a plastic bulb, sealed with epoxy, and attached to the feathers of the spinal tract. The marker was visible at 600 m with the naked eye and 1.5 km with binoculars after 4 hours. After 12 hours, the marker was still visible at 80 m with the naked eye.

Natural Markings

Natural markings to identify individuals have been used less commonly for birds than for mammals. Particular plumage or bill patterns can be used as distinguishing features, although few individuals within a population may exhibit unique features and this may change with molt or age. Scott (1978) developed a system of identifying Bewick's swans by bill patterns and other body features, although bill features changed with age. The accuracy of these identifying features was questionable in relation to the size of the population studied (Pennycuik 1978). The potential for natural marking systems of such types is limited, but they have application on a short-term basis and in conjunction with other markers in some species.

Egg Markers

Hayward (1982) used 5 × 5-mm labels of colored plastic tape to mark eggs of ring-billed gulls. The tape label was firmly applied to the egg near the apex, and a different color or color combination was used for each egg laid within a clutch. No loss of this marker before hatching was recorded. Boss (1963) and Olsen et al. (1982) used marking pens to number eggs within clutches to determine the incubation times of American coots, Australian kestrels, brown falcons, brown goshawks, and welcome swallows. Olsen et al. (1982) observed no harmful effects from the marking, although they cautioned that marking pens should be used with discretion until possible embryotoxic effects are tested.

MARKING METHODS FOR AMPHIBIANS AND REPTILES

Marking methods for amphibians and reptiles were reviewed by Woodbury (1956), Thomas (1977), Swingland (1978), and Ferner (1979). Spellerberg and Prestt (1978) and Fitch (1987) reviewed the literature on methods for marking snakes. Because of the wide range in diversity in species of amphibians and reptiles, no specific approved methods of marking them for field research are practical or desirable (Society for the Study of Amphibians and Reptiles 1987; also see Chapter 4). Further, the ultimate responsibility for the ethical and scientific validity of methods used will rest with the investigator. Natural marks have the least adverse effect on animals and should be used whenever possible, whereas mutilation techniques have the potential for the greatest adverse effect on the individual animals marked. However, most techniques require capture, recapture, and handling of animals that may in themselves affect the animals' behavior.

Natural Markings

Perhaps the most ideal method of recognizing individuals is to use their own naturally occurring variation in color patterns. Hagstrom (1973) photographed belly patterns to recognize individuals of warty newts and smooth newts. Healy (1975) observed that variation in the dorsal spot pattern of eastern newts was useful in identifying individuals.

Stamps (1973) could recognize individuals of anoles by their distinctive patterns coupled with tail regenerations in various stages. Carlstrom and Edelstam (1946) successfully used photographs to record unique individual dorsal patterns in viviparous lizards and throat patterns in slow worm lizards.

Carlstrom and Edelstam (1946) photographed the ventral pattern of grass snakes and reported it was constant during the life history of each individual. Henley (1981) used a distinctive characteristic on the exuvia, removed that portion of the skin, and placed it on the specimen's data card for later identification of the snake. Shine et al.

(1988) recorded the number and position of divided versus entire subcaudal scales to separate individual snakes.

Tilley (1980) photographed dorsal color patterns of mountain dusky salamanders for a capture-recapture study. He detected recaptures by comparing the color photographs of specimens taken on different occasions (Tilley 1977). Tilley (1980) noted that the changes in the dorsal pattern proceeded slowly and that adults taken early in the study could be readily recognized as much as 7 years later.

External Marks

BRANDING

Branding sometimes causes infections and may possibly affect the animal's behavior or survival. Freeze branding, if properly done, rarely results in infection. However, freezing the skin for too long a time causes scab formation or tissue necrosis, resulting in the formation of new cells with intact melanocytes, which creates an indistinct mark. A major disadvantage of freeze branding is that the brand cannot be read until after the animal sheds. Tattoos usually cause the least problem; however, the major disadvantage with the technique is the animal must be recaptured for identification.

Hot Branding

Gulf Coast toads marked with heated branding irons could be identified for up to 21.5 months (Clark 1971); bullfrogs and northern leopard frogs also were marked successfully in this manner. Data suggested branding was similar to toe clipping in its effects on survival. Taber et al. (1975) branded hellbenders and noted most marks remained clear through a 2-year study. Woodbury and Hardy (1948) branded the carapace plates of desert tortoises. If scutes were burned too deeply, complete regeneration took place, and if they were burned too lightly the scar wore off in a few years. Woodbury (1956) added that using a 12- or 14-gauge brand yielded the best results. Clark (1971) branded the plastron of one common slider into the underlying bone and saw no evidence of regeneration or infection during the 36 days the animal was observed. Clark (1971) used hot branding successfully on green anoles and Texas horned lizards and preferred it to toe clipping. Weary (1969) hot branded redbelly and common garter snakes and reported no regeneration over a 2-year period for these snakes. Clark (1971) hot branded several species of snakes and noted the brand formed a scab within a few days and did not produce an open sore. He believed survivorship to be less influenced by branding than by scale clipping.

Chemical Branding

Anurans were branded with silver nitrate (Thomas 1975), and adult green treefrogs were branded with silver nitrate applicators (75% silver nitrate and 25% potassium nitrate). The silver nitrate caused a brown mark to form immediately, and within about 2 weeks the dark mark faded into a light mark. Thomas (1975) recommended the method for dark-colored amphibians; the marks persisted for >9 weeks.

Freeze Branding

Daugherty (1976) used irons cooled in chipped dry ice to freeze brand tailed frogs. Brands were placed on the ventral surfaces, could be read within 1 day, and were legible for >2 years in the field; however, the mark gradually lost pigment, so that after 1 year the viscera were often visible through the integument. Bull et al. (1983) evaluated freeze-branding application times for salamanders and concluded that a 0.75-second application produced the most readable marks. According to Lewke and Stroud (1974), freeze branding was used successfully on green turtles, but no details were given. They used a dry ice-alcohol coolant to mark two western rattlesnakes and seven pine snakes. Marks were observed for about 2 years with no adverse effects; however, all snakes molted within 3 weeks of marking. Ferner (1979) quoted R. K. Farrell (pers. commun.) as having had excellent results in freeze marking iguanas, the marks persisting through several skin sheddings.

Laser Marking

Ferner (1979) quoted R. K. Farrell (pers. commun.) as being unsuccessful in marking a turtle with a laser. Farrell also used a ruby laser to mark king snakes and rattlesnakes.

Tattoos

Kaplan (1958) described a marking technique for frogs that involved the incorporation of India ink into scarified skin of the venter. Numeralized grooves were etched into the skin with a hypodermic needle and then filled with ink. He also reported on an electric tattooing technique as an improvement upon the scarification method. He used a small electric tattoo marker and Higgins India ink mixed with a drop of glycerin to aid its spreading into the skin. Woodbury (1956) reported Kaplan's tattooing technique for snakes was "permanent" and conferred no adverse effects to the snakes after recovery from the tattooing operation. Chabreck (1965) tested tattooing as a possible marking method for American alligators. Numbers were written on the light-colored skin on the bottom of the tail. The numbers were clear when new but faded with age and were barely legible after several months.

TISSUE REMOVAL

The effect of most tissue-removal marking methods on survival and fitness is not adequately known and is a topic worth investigating (Society for the Study of Amphibians and Reptiles 1987). If behavior or survival of the animal is impaired, alternate marking methods should be used.

Toe/Tail Clipping

Clipping notches from the tail fin of amphibian tadpoles is a traditional method for marking (Turner 1960). However, Guttman and Creasey (1973) noted fin clipping caused a higher mortality than did staining techniques.

Although several toe-clipping methods are proposed in the literature, the method of Martof (1953) is most widely used. A modification of this technique by Carpenter (1954) has not been cited in recent literature. Martof (1953) cut off the toes of green frogs with scissors and observed no regeneration of the digits. Regeneration may be more common in primitive frogs, but Richards et al. (1975) also reported it in advanced forms such as the Kenyan reed frog; they observed that newly metamorphosed reed frogs completely regenerated amputated toes, including the structurally complex digital pad. They also noted that ecologists generally avoid toe clipping tree frogs because of their regenerative capabilities. However, Jameson (1957) noted only slight toe regeneration in Pacific chorus frogs after 1 year. Brown and Alcala (1970) also reported slow or nonexistent regeneration of toes in a red-eared frog during a 4-year field study. Briggs and Strom (1970) avoided clipping the thumbs of cascades frogs due to their importance in amplexus and usefulness in sexing individuals. Dole and Durant (1974) also did not clip thumbs and the smallest toes of harlequin frogs. The most serious criticism to toe clipping anurans was raised by Clarke (1972), who reported the probability of recapture of Fowler's toads decreased as the number of digits excised increased. However, his technique involved the removal of one or two toes per foot, complete removal on the front feet, but only to the webbing in back. Hero (1989) developed a simple code for toe clipping anurans that minimized the number of toes clipped. Daugherty (1976) reported a problem with weight loss in northern leopard frogs that were toe clipped.

At the time of Woodbury's (1956) review, toe clipping was the only successful marking technique reported for salamanders. Usually something similar (Twitty 1966) to Martof's (1953) system of marking anurans is employed. Regeneration needs to be considered, but whether it is a deterrent to using toe clipping for salamanders depends upon the species. In *Taricha* spp. the time interval for regeneration is several years (Twitty 1966); however, Efford and Mathias (1969) reported no regeneration in the roughskin newt if the animals were kept below or at 10 C. Heatwole (1961) observed that regeneration time for eastern red-backed salamanders averaged 7 months for animals kept in a laboratory. Wells and Wells (1976) noted a regeneration time of at least 2 years for slimy salamanders, and Hall and Stafford (1972) observed 50% of Wehrle's salamanders regenerated toes after 100 days; however, regenerated digits were identifiable by lack of pigmentation. Hillis and Bellis (1971) noted hellbenders averaged 1 year for regeneration, and Hendrickson (1954) observed that regenerated toes of slender salamanders were slow to regenerate but recognizable.

Healy (1974) marked post-larval metamorphs of the eastern newt by amputating one limb at the middle of the zeugopodium; however, few individuals were recaptured. Juvenile dusky salamanders have toes too small for clipping, but they were successfully marked by clipping small pieces of the tail (Orser and Shure 1972). This allowed juveniles to be distinguished for at least 1 month before regeneration became a problem with identification.

Cagle (1939) used a numbering system similar to that for lizards and clipped the first phalanx of toes on young turtles whose shells had not ossified. Toe clipping is the most popular technique for marking lizards (Ferner 1979). The most commonly cited method involves clipping no more than four toes, but no more than two per foot and never adjacent ones (Tinkle 1967). Medica et al. (1971) described a different numbering system similar to that used for salamanders and frogs. Woodbury (1956) suggested lettering the feet, numbering the toes, and giving the sex for an individual identification code. Minnich and Shoemaker (1970) toe clipped desert iguanas. None of the extensive studies of recent years documented any harmful effect on toe clipping lizards.

Chabreck (1965) observed toe clipping was successful and permanent but somewhat limited for individual recognition of American alligators. However, when combined with notching of tail scutes, it provided combinations for >3,000 distinct marks. Ferner (1979) quoted S.J. Gorzula (pers. commun.) as having successfully marked spectacled caiman by clipping triangular pieces from the ventral tail scutes. The marks were identifiable for 2 years, but individuals usually required remarking after 1 year because of regeneration of the scales. Although toe-clipping amphibians and reptiles confers disadvantages, it is still the most common marking technique used for anurans (Ferner 1979).

Skin Transplantation

Raginski (1977) removed pieces of skin from the orange venter and the bluish to brown-black dorsum from alpine newts. These grafts were exchanged and needed no adhesives. After transplantation, the newts were kept isolated out of water for 1 hour. He reported a 95% success rate and retention of grafts for ≥3 years.

Shell Notching

The most commonly used marking technique for turtles is to notch the shell. Cagle (1939) developed a coding system for notching that has had wide use. He cautioned that marks on young specimens may not be permanent. Ernst (1971) notched painted turtles and proposed a new coding system; he suggested that marginals at the bridge or junction of the plastron and carapace should not be notched so as not to weaken the shell.

Scale Clipping and Related Techniques

Scale clipping with scissors or clippers is the most commonly used method of marking snakes (Ferner 1979). Many investigators still follow the methods of Blanchard and Finster (1933). They cut pieces from the subcaudals, which leaves ''permanent'' scars, and numbered the subcaudals on each side, beginning at the proximal end of the tail. They reported no adverse effects on the snakes. Carlstrom and Edelstam (1946) criticized the technique of Blanchard and Finster (1933) because regeneration was a problem and marking hatchlings of grass snakes was difficult. Woodbury (1956) cited regeneration of clipped scales in 4–5 years as reported by Conant (1948) for rat snakes. Weary (1969) also noted three problems with the technique when redbelly and common garter snakes were marked: several minutes were required to remove a complete caudal scale, blood was frequently drawn, and clipping was difficult on young snakes <10 cm long.

Brown and Parker (1976) described a ventral scale clipping system to mark snakes. The ventrals are larger and easier to clip than subcaudals, and scars in that area cannot be lost by tail breakage. Their serial enumeration system is far less confusing than the code of Blanchard and Finster (1933). The authors did not observe direct adverse effects on snakes; however, scale clipping or handling caused increased over-winter mortality and weight loss in newly marked striped whipsnakes. They observed no effects on racers or pine snakes. Brown and Parker (1976) reported marks persisted 4 years, and 92% of the time shed skin from clipped racers could be precisely identified.

Tagging

JAW TAGGING

Raney (1940) developed a technique for jaw tagging toads. Stille (1950) reported loss of tags from jaw-tagged toads was significant, and Woodbury (1956) noted such tags were often lost and caused irritation to toads. Hirth (1966:8) tagged adult striped whipsnakes and racers with numbered monel tags that were ''clamped into the corner of the mouth.'' No details were given and the technique has not been widely used by others.

NECK COLLARS AND RELATED BANDS

Chabreck (1965) concluded that a collar made from vinyl plastic tape and placed around an American alligator's neck was satisfactory for short-term use. The collar provided a method for field identification and could be seen at a distance. However, because alligators grew at a rapid rate, material thin enough to break when the animal outgrew the collar was required.

WEB/FLIPPER MARKERS

Sea turtles are usually marked with numbered monel tags attached to the trailing edge of a foreflipper. These have been used on loggerhead (LeBuff and Beatty 1971), green (Pritchard 1976), leatherback (Bacon 1973), and olive ridley turtles (Pritchard 1976). Pritchard (1976) reported significant tag loss through metal corrosion and tissue necrosis, particularly with leatherback turtles. Frazer (1983) applied individually numbered plastic and monel stainless steel livestock tags to the trailing edge of the front flippers of loggerhead turtles. The incidence of tag loss from marine turtles has been high (Bjorndal 1980, Pritchard 1980). However, Balazs (1985) reported good retention of commercially available monel flipper tags on hatchling sea turtles. Eckert and Eckert (1989) observed greater retention of monel than of plastic tags attached to flippers of sea turtles.

LEG BANDS

Aluminum butt-end bird bands have been used on toes of frogs (Kaplan 1958). The bands were tightened so as to allow adequate circulation, but did pierce the webbing of the foot. Tags remained fixed indefinitely and caused no apparent problems. Rao and Rajabai (1972) tagged Sita's and bloodsucker lizards with different shaped colored aluminum rings. The rings were placed around the thigh and caused no apparent problems. Paulissen (1986) marked six-lined racerunners by gluing colored plastic birds bands to their tails. Average ''band life'' was 26.4 days (range 4–63 days).

BODY TAGS AND RELATED TECHNIQUES

Nace and Manders (1982) developed a technique for marking amphibians with colored beads. Fisher and

Muth (1989) modified the technique for marking lizards. Emlen (1968) used a nylon waistband in a behavioral study of male bullfrogs. Waistbands were 13 mm wide and painted with black numerals for individual identification. Waistbands were recognizable from maximum distances of 8–12 m with headlamp and binoculars, but numbers were visible only from 4 m to 6 m. Emlen reported no differences in behavior, mortality, or other problems; however, soiling and staining of the bands required seasonal replacement. Nickerson and Mays (1973) successfully marked hellbenders with Floy T-tags but gave no details on procedures.

Kaplan (1958) tagged turtles with numbered aluminum bands fastened through holes drilled in the carapace. Lonke and Obbard (1977) tagged snapping turtles with aluminum plates anchored in holes drilled through the marginal scutes just to the side of the tail. The tags could be observed at a distance and persisted for at least 3 years. Pough (1970) used a tool for fastening buttons to clothing to mark young turtles. Froese and Burghardt (1975) similarly marked snapping turtles and noted a problem when the plugs were too short to allow free movement in the holes. Layfield et al. (1988) placed wire rings through the posterior marginal scute of hatchling and adult snapping turtles. Aldabra tortoises were marked with numbered titanium disks fixed in depressions drilled into the keratin shield (Gaymer 1973). The tags were held in the depression by a metal-resin adhesive and had high retention. Graham (1986) warned against the use of Petersen disc tags in turtle studies; the tags entangled the animals in the mesh of traps used to recapture the animals and caused the turtles to drown. He also reported that turtles marked with Petersen tags became fouled in discarded monofilament fishing line. Ward et al. (1976) marked the carapace of spotted turtles with an "adhesive tag" that bore an identification number. Davis and Sartor (1975) tagged snapping turtles by putting a small wooden dowel through a hole drilled in the nuchal scute. To reduce the hinderance this caused turtles moving in thick vegetation, they connected two shorter dowels with a piece of rubber tubing.

Pough (1970) also used his button-fastening technique on snakes. Attachment was made by inserting a plug into the caudal musculature through the lateral region of a subcaudal scute. Hudnall (1982) threaded colored glass beads onto the tails of snakes using monofilament suture-line mounted on a surgical needle. Pendlebury (1972) tagged western rattlesnakes with a pair of colored vinyl discs fastened through the dorsal lobe of the second proximal segment of the rattle. The tag was observable ≤30 m with binoculars. The number of molts that had occurred since tagging could be determined by (N − 1), where N was the number of segments craniad to the tagged one. No adverse effects were reported for this technique. Stark (1984) used stainless steel wire to attach numbered colored sequins covered with a thin layer of epoxy between the proximal rattle segment and the body of rattlesnakes.

Chabreck (1965) attached a monel tag to a dorsal tail scute to mark American alligators. This tag was the self-piercing type (size 681, National Band and Tag Co., Newport, KY 41072) and was imprinted with a number and return address.

TAPES, STREAMERS, AND BELLS

Minnich and Shoemaker (1970) marked desert iguanas with colored Mystik cloth tape. This technique also was used to mark lizards (Minnich and Shoemaker 1970). Zwickel and Allison (1983) marked a small New Guinea skink with pressure sensitive rip-stop nylon tape. Robertson (1984) attached reflective tape to heads of bullfrogs with fast-setting cyanoacrylate tissue cement. The reflective squares remained attached for 16–41 days.

Chabreck (1965) tested streamers and internal anchor tags on alligators. Either a tag or a barb was placed beneath the skin, usually on the side of the tail. This was attached to a flexible chain or plastic strip that extended through the skin so as to be visible. However, openings through the skin were slow healing, making this method undesirable for long-term studies. Henderson (1974) tagged green iguanas by tying small bells around their necks with fishing line.

Trailing Devices

To study movements of northern leopard frogs, Dole (1965) glued a bobbin to an elastic band that was secured around the waist of the frog. A small stake was used to mark the point of capture and the trailing nylon sewing thread was tied to it. The 50 m of thread on the bobbin lasted from 1 hour to 7 days. The loaded trailing device weighed about 8.5 g. The device may have shortened the frog's jumping ability, and no individuals <60 mm long were trailed. Frogs had some difficulty swimming and entering crevices, and the waistband occasionally caused skin irritation. Grubb (1970) modified the technique by using 200 m of cotton thread.

Whitford and Massey (1970) tagged tiger salamanders by suturing a numbered plastic float through the tail with a monofilament line. The line was made long enough to allow the salamanders to move through the deepest portion of the lake.

Stickel (1950) used a trailing device on box turtles. She attached a wooden spool and thread with its housing to the carapace with waterproof adhesive tape. She found no evidence the device changed the turtle's behavior. LemKau (1970) attached a "thread trailer" radio transmitter packet to the carapace of a box turtle. A modification of Stickel's (1950) technique was developed by Reagan (1974) to avoid interference with mating. He used 35-mm film canisters to hold a wooden spool and thread. The unit was attached to the caudal end of the carapace with waterproof adhesive tape. This method was tangle-proof and weatherproof and weighed only 12 g as compared to 55 g for Stickel's (1950) device. Scott and Dobie (1980) developed a low-friction thread-release mechanism similar to those employed in spincast fishing reels for use on turtles. Carr et al. (1974) attached fiberglass-coated styrofoam floats to 24-m lines for observing movements of green turtles. A 3-V flashlight bulb attached to each float was powered by batteries imbedded in the styrofoam. A fiberglass mast also was attached and was topped by an orange pennant. No adverse effects were reported for this flotation device.

The fringe-toed lizard was tagged with a small piece of foil attached to 30 cm of light string placed around the lower abdomen (Deavers 1972). This tag allowed the measurements of burial depth of the lizards at night. Judd (1975) used a similar tag to locate buried keeled earless lizards for body temperature readings.

Dyes and Paints

Herreid and Kinney (1966) used neutral red to stain tadpoles of wood frogs. Guttman and Creasey (1973) similarly stained tadpoles of bullfrogs; these tadpoles had an immediate mortality of 8.7% (N = 567); survivors retained the stain for at least 10 days. All staining methods appear to be time limited (Ferner 1979). Travis (1981) concluded that staining with neutral red affected the growth of treefrogs. Treated tadpoles grew more slowly than control animals at low density, but grew no differently than controls at high density. He concluded that staining had a residual effect on the animal's growth well after the stain disappeared.

Woolley (1962) used a black felt ink pen to mark salamanders. Dilute acetic acid or ammonium hydroxide was used to remove slime on the salamander's tail before the ink was applied, and the marks were visible for at least 1 month. Burger and Montevecchi (1975) marked the plastrons of nesting diamondback terrapins with washable ink so they would not be recounted in a census while they were on land. In a nesting study, the eggs of turtles were marked with a permanent ink felt pen (Burger 1976). Ireland (1991) concluded that fluorescent pigments were useful for short-term marking of terrestrial salamanders.

Woodbury and Hardy (1948) concluded that colored paints were less permanent than branding for marking carapace scutes of desert tortoises. However, they found this to be less permanent than branding. Medica et al. (1975), studying the same species, painted the last vertebral scutes and then repainted them with a different color each year. Bennett et al. (1970) painted numbers on the carapace of mud turtles, chicken turtles, and common sliders. Bayless (1975) reported that numbers painted on the carapace of painted turtles had to be repainted each year because they were shed with the carapace scutes.

Tinkle (1967) painted adult side-blotched lizards with colored insignia to identify individuals without recapture. Shedding of skin made necessary the recapture and repainting of the lizards. Young were marked only with a small spot of paint on the back between the hind legs that did not allow individual recognition. Model paint was used to mark lizards (e.g., Tinkle 1973, Fox 1978), and Jenssen (1970) painted spots with quick drying paint on the backs of anoles. Vincgar (1975) marked the tails of fence lizards with model airplane paint. Stebbins and Cohen (1973) used a purple indelible pencil on the hind legs of western fence lizards to distinguish control and experimental individuals. Henderson (1974) placed numbers on the sides of green iguanas with a felt-tip pen and reported the marks persisted several weeks. How colored marks affected the behavior of individuals marked in these species, for which colors have seasonal social significance, is unknown. Jones and Ferguson (1980) studied the effects of paint marking on mortality of fence lizards. About one-half of the lizards captured were toe clipped only and the other half were also marked with a spot of model airplane paint at the base of the tail. No significant difference was noted in the proportions of the two groups recovered.

Pough (1966) marked three species of rattlesnakes by painting an identifying number on the basal rattle segment with quick-drying waterproof paint. Brown et al. (1984) used different colors of waterproof paint on the basal segments of the rattle and distal portion of the tail of several species of rattlesnakes. Striped whipsnakes were painted on the head and neck with a color code for individual recognition (Bennion and Parker 1976, Parker 1976).

Internal Markers

STAINS

Seale and Boraas (1974) developed a more permanent marking method for frog and salamander larvae. They used a 21:20 (by weight) ratio of mineral oil to petroleum jelly and organic biological stains ("Oil Red O" and "Oil Blue M") to dye the mineral oil. Variations in color and placement of marks were used to identify groups of individuals and groups. Marks were placed in the dorsal or ventral tail fin cavity with a hypodermic syringe and 22-gauge needle. No mortality, infection, impairment of mobility, or retardation of growth was reported. During metamorphosis, the organic dye-organic solvent mark was reabsorbed with the tail with no ill effects.

Woolley (1973) marked cave salamanders and long-tailed salamanders with a subcutaneous injection of two parts Liquitex Acrylic Polymer to one part distilled water. The mixture was injected into the lateral proximal caudal region, leaving a mark 7–10 mm in diameter. He observed no adverse effects and found only slight fading in a few individuals.

PARTICLE MARKERS

Ireland (1973) marked salamander larvae with fine grained fluorescent pigments. Four colors of fluorescent pigments in a melamine-sulfonamide-formaldehyde resin were mixed with acetone to make a paste. This paste was administered to the larvae on their mid-dorsal surface with a heated probe. The probe burned the outer epithelial layers, leaving a small scar (1 mm). The epithelium regenerated within 15 days and incorporated the pigments. In the laboratory, 50% of the many-ribbed salamanders and 47% of the ringed salamanders lost the fluorescent tag in 70 days. Loss of tags in the field was less than this, and no detrimental effects of the tag on the larvae were observed.

RADIOACTIVE MARKERS

The major disadvantage of using radioactive tags is the restriction imposed by state or federal regulations. These tags can cause damage to or death of the animals carrying them, can be lost, and can constitute a hazard to other animals or humans.

Karlstrom's (1957) study of the Yosemite toad was the first major discussion of the use of radioactive tagging (cobalt, Co) for amphibians. Breckenridge and Tester (1961) placed radioactive tags on Canadian toads. Salamander larvae were marked for short periods with radioactive sodium with no observed increase in mortality (Shoop 1971). Dusky salamanders were tagged with radioactive Co (Barbour et al. 1969*b*, Ashton 1975), and Appalachian salamanders were marked with radioactive tantalum (Ta) (Madison and Shoop 1970). In these and other studies of plethodontids, researchers observed local

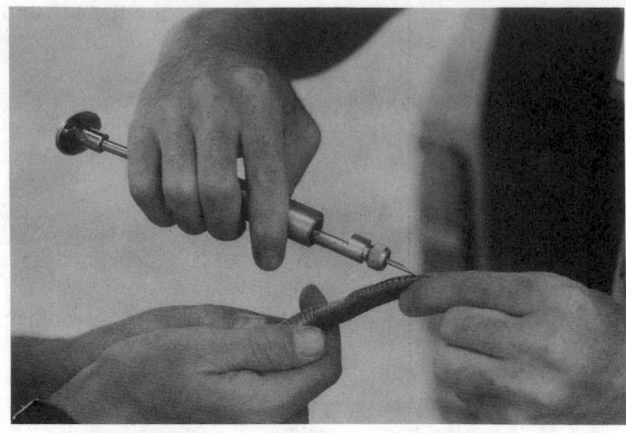

Fig. 13. Implanting a transponder into a snake with a metal syringe.

ulceration that eventually opened, exposing the tag. This was not a problem with chorus frogs, waterdogs, or reptiles. Bennett et al. (1970) marked mud turtles, chicken turtles, and common sliders with radioactive Ta, and Ward et al. (1976) used this isotope to mark spotted turtles. O'Brien et al. (1965) tagged one northern fence lizard with radioactive gold placed in polyethylene tubing and tied around the waist of the lizard. The tag was detectable from distances of 1.5–3.7 m for about 1 week, when recapture was necessary to attach a new tag. Ferner (1979) quoted H. Fitch (pers. commun.) as successfully using radioactive Ta wires on five-lined skinks, ground skinks, slender glass lizards, ring-necked snakes, and worm snakes. Barbour et al. (1969a) studied worm snakes with radioactive Co tags.

TRANSPONDERS

Camper and Dixon (1988) tested a passive, integrated transponder (PIT) for marking amphibians and reptiles. The PIT, encased in glass, encoded with a 10-space, alpha-numeric code and measuring 10 × 2.1 mm, was implanted with a modified metal syringe and a 12-gauge canula (needle) (Fig. 13). Twenty frogs and toads, one alligator, 20 snakes, 23 lizards, and 31 turtles (including eight sea turtles) were implanted with PITs. Only one PIT failed (cracked glass cover), and time trials of successful PIT readings on the first pass of the wand revealed a successful first-pass rate of 91.6%. Migration of the PIT from its implant site occurred in 60% of the amphibians and 36% of the reptiles. PIT movement affected the ''first-pass'' reading in only one lizard and one turtle. Despite some problems with the PIT tagging system, the authors believed it was superior to most systems for marking amphibians and reptiles.

LITERATURE CITED

ABRAHAM, K. F., C. D. ANKNEY, AND H. BOYD. 1983. Assortative mating by brant. Auk 100:201-203.

ALDOUS, M. C., AND F. C. CRAIGHEAD, JR. 1958. A marking technique for bighorn sheep. J. Wildl. Manage. 22:445–446.

ALDOUS, S. E. 1940. A method of marking beavers. J. Wildl. Manage. 4:145–148.

ALLISTON, W. G. 1975. Web-tagging ducklings in pipped eggs. J. Wildl. Manage. 39:625–628.

ALT, G. L., C. R. MCLAUGHLIN, AND K. H. POLLOCK. 1985. Ear tag loss by black bears in Pennsylvania. J. Wildl. Manage. 49:316–320.

AMES, J. A., R. A. HARDY, AND F. E. WENDELL. 1983. Tagging materials and methods for sea otters, *Enhydra lutris*. Calif. Fish Game 69:243–252.

ANDELT, W. F., AND P. S. GIPSON. 1980. Toe-clipping coyotes for individual identification. J. Wildl. Manage. 44:293–294.

ANDERSON, A. 1963. Patagial tags for waterfowl. J. Wildl. Manage. 27:284–288.

———. 1981. Making polyvinyl chloride (PVC) colored legbands. J. Wildl. Manage. 45:1067–1068.

ANKNEY, C. D. 1975. Neckbands contribute to starvation in female lesser snow geese. J. Wildl. Manage. 39:825–826.

ARNOLD, K. A., AND D. W. COON. 1971. A technique modification for color-marking birds. Bird-Banding 42:49–50.

ASHTON, D. G. 1978. Marking zoo animals for identification. Pages 24–34 *in* B. Stonehouse, ed. Animal marking: recognition marking of animals in research. The MacMillan Press Ltd., London, U.K.

ASHTON, R. E. 1975. A study of movement, home range, and winter behavior of *Desmognathus fuscus* (Rafinesque). J. Herpetol. 9:85–91.

ATHERTON, N. W., M. E. MORROW, A. E. BIVINGS, IV, AND N. J. SILVY. 1982. Shrinkage of spiral plastic leg bands with resulting leg damage to mourning doves. Proc. Annu. Conf. Southeast. Assoc. Game and Fish Agencies 36:666–670.

BACON, P. R. 1973. The orientation circle in the beach ascent crawl of the leatherback turtle, *Dermochelys coriacea*, in Trinidad. Herpetologica 29:343–348.

BAER, G. M., J. H. SHADDOCK, D. J. HAYES, AND P. SAVARIE. 1985. Iophenoxic acid as a serum marker in carnivores. J. Wildl. Manage. 49:49–51.

BAILEY, G. N. A, I. J. LINN, AND P. J. WALKER. 1973. Radioactive marking of small mammals. Mammal. Rev. 3:11–23.

BAKER, W. W. 1983. A non-clamp patagial tag for use on red-cockaded woodpeckers. Proc. Red-Cockaded Woodpecker Symp. 2:110–111.

BALAZS, G. H. 1985. Retention of flipper tags on hatchling sea turtles. Herpetol. Rev. 16:43–45.

BALHAM, R. W., AND W. H. ELDER. 1953. Colored leg bands for waterfowl. J. Wildl. Manage. 17:446–449.

BALLOU, R. M., AND F. W. MARTIN. 1964. Rigid plastic collars for marking geese. J. Wildl. Manage. 28:846–847.

BALTOSSER, W. H. 1978. New and modified methods for color-marking hummingbirds. Bird-Banding 49:47–49.

BARBOUR, R. W. 1963. *Microtus*: a simple method of recording time spent in the nest. Science 141:41.

———, AND W. H. DAVIS. 1969. Bats of America. Univ. Press of Kentucky, Lexington. 286pp.

———, M. J. HARVEY, AND J. W. HAUDIN. 1969a. Home ranges, movements and activity of the eastern worm snake, *Carphophis amoenus amoenus*. Ecology 50:470–476.

———, J. W. HAUDIN, J. P. SHAKER, AND M. J. HARVEY. 1969b. Home range, movements, and activity of the dusky salamander, *Desmognathus fuscus*. Copeia 1969:293–297.

BARCLAY, R. M. R., AND G. P. BELL. 1988. Marking and observational techniques. Pages 59–76 *in* T.H. Kunz, ed. Ecological and behavioral methods for the study of bats. Smithsonian Inst. Press, Washington, D.C.

BARTELT, G. A., AND D. H. RUSCH. 1980. Comparison of neck bands and patagial tags for marking American coots. J. Wildl. Manage. 44:236–241.

BARTONEK, J. C., AND C. W. DANE. 1964. Numbered nasal discs for waterfowl. J. Wildl. Manage. 28:688–692.

BATCHELOR, T. A., AND J. R. MCMILLAN. 1980. A visual marking system for nocturnal animals. J. Wildl. Manage. 44:497–499.

BATEMAN, G. C., AND T. A. VAUGHAN. 1974. Nightly activities of mormoopid bats. J. Mammal. 55:45–65.

BAUMGARTNER, L. L. 1940. Trapping, handling, and marking fox squirrels. J. Wildl. Manage. 4:444–450.

BAYLESS, L. E. 1975. Population parameters for *Chrysemys picta* in a New York pond. Am. Midl. Nat. 93:168–176.

BEALE, D. M. 1966. A self-collaring device for pronghorn antelope. J. Wildl. Manage. 30:209–211.

———, AND A. D. SMITH. 1973. Mortality of pronghorn antelope fawns in western Utah. J. Wildl. Manage. 37:343–352.

BEASOM, S. L., AND J. D. BURD. 1983. Retention and visibility of plastic ear tags on deer. J. Wildl. Manage. 47:1201–1203.

BELL, G. P., G. A. BARTHOLOMEW, AND K. A. NAGY. 1986. The roles of energetics, water economy, foraging behavior, and geothermal refugia in the distribution of the bat, *Macrotus californicus.* J. Comp. Physiol. B 156:441–450.

BENDELL, J. F. S., AND C. D. FOWLE. 1950. Some methods for trapping and marking ruffed grouse. J. Wildl. Manage. 14:480–482.

BENNETT, D. H., J. W. GIBBONS, AND J. C. FRANSON. 1970. Terrestrial activity in aquatic turtles. Ecology 51:738–840.

BENNION, R. S., AND W. S. PARKER. 1976. Field observations on courtship and aggressive behavior in desert striped whipsnakes, *Masticophis t. taeniatus.* Herpetologica 32:30–35.

BERGER, D. D., AND H. C. MUELLER. 1960. Band retention. Bird-Banding 31:90–91.

BEST, P. B. 1976. Tetracycline marking and the rate of growth layer formation in the teeth of a dolphin (*Lagenorhynchus obscurus*). South Afr. J. Sci. 72:216–218.

BJORNDAL, K. A. 1980. Demography of the breeding population of the green turtle, *Chelonia mydas,* at Tortuguero, Costa Rica. Copeia 1980:525–530.

BLAIR, W. F. 1941. Techniques for the study of mammal populations. J. Mammal. 22:148–157.

BLANCHARD, F. N., AND E. B. FINSTER. 1933. A method of marking living snakes for future recognition, with a discussion of some problems and results. Ecology 14:334–347.

BLANK, T. H., AND J. S. ASH. 1956. Marker for game birds. J. Wildl. Manage. 20:328–330.

BONACCORSO, F. J., AND N. SMYTHE. 1972. Punch-marking bats: an alternate to banding. J. Mammal. 53:389–390.

———, ———, AND S. R. HUMPHREY. 1976. Improved techniques for marking bats. J. Mammal. 57:181–182.

BOONSTRA, R., AND I. T. M. CRAINE. 1986. Natal nest location and small mammal tracking with a spool and line technique. Can. J. Zool. 64:1034–1036.

BOSS, A. S. 1963. Aging the nests and young of the American coot. M.S. Thesis, Univ. Minnesota, St. Paul. 62pp.

BRADY, J. R., AND M. R. PELTON. 1976. An evaluation of some cottontail rabbit marking techniques. J. Tenn. Acad. Sci. 51:89–90.

BRADBURY, J. W. 1977. Lek mating behavior in the hammer-headed bat. Z. Tierpsychol. 45:225–255.

BRECKENRIDGE, W. J., AND J. R. TESTER. 1961. Growth, local movements and hibernation of the Manitoba toad, *Bufo hemiophrys.* Ecology 42:637–646.

BRIGGS, J. L., AND R. M. STROM. 1970. Growth and population structure of the cascade frog, *Rana cascadae* Slater. Herpetologica 26:283–300.

BROOKS, P. M. 1981. Comparative longevity of a plastic and a new machine-belting collar on large African ungulates. South Afr. J. Wildl. Res. 11:143–145.

BROWN, S. G. 1978. Whale marking techniques. Pages 71–80 *in* B. Stonehouse, ed. Animal marking: recognition marking of animals in research. The MacMillan Press Ltd., London, U.K.

BROWN, W. C., AND A. C. ALCALA. 1970. Population ecology of the frog, *Rana erythraea,* in southern Negros, Philippines. Copeia 4:611–622.

BROWN, W. S., V. P. J. GANNON, AND D. M. SECOY. 1984. Paint-marking the rattle of rattlesnakes. Herpetol. Rev. 15:75–76.

———, AND W. S. PARKER. 1976. A ventral scale clipping system for permanently marking snakes (Reptila, Serpentes). J. Herpetol. 10:247–249.

BUCHLER, E. R. 1976. A chemiluminescent tag for tracking bats and other small nocturnal animals. J. Mammal. 57:173–176.

BULL, E. L., R. WALLACE, AND D. H. BENNETT. 1983. Freeze-branding: a long term marking technique on long-toed salamanders. Herpetol. Rev. 14:81–82.

BURGER, G. V., R. J. GREENWOOD, AND R. C. OLDENBURG. 1970. Alula removal technique for identifying wings of released waterfowl. J. Wildl. Manage. 34:137–146.

BURGER, J. 1976. Temperature relationships in nests of the northern diamondback terrapin, *Malaclemys terrapin terrapin.* Herpetologica 32:412–418.

———, AND W. A. MONTEVECCHI. 1975. Nests site selection in the terrapin, *Malaclemys terrapin.* Copeia 1975:113–119.

BURLEY, N., G. KRANTZBERG, AND P. RADMAN. 1982. Influence of colour-banding on the conspecific preferences of zebra finches. Anim. Behav. 30:444–455.

BUTTS, W. K. 1930. A study of the chickadee and white-breasted nuthatch by means of marked individuals. Part I: methods of marking birds. Bird-Banding 1:149–168.

BYERS, S. M. 1987. Extent and severity of nasal saddle icing on mallards. J. Field Ornithol. 58:499–504.

CAGLE, F. R. 1939. A system of marking turtles for future identification. Copeia 1939:170–173.

CAMPER, J. D., AND J. R. DIXON. 1988. Evaluation of a microchip marking system for amphibians and reptiles. Texas Parks Wildl. Dep. Res. Publ. 7100-159. 22pp.

CARLSTROM, D., AND C. EDELSTAM. 1946. Methods of marking reptiles for identification after recapture. Nature 158:748–749.

CARNIO, J., AND L. KILLMAR. 1983. Identification techniques. Pages 39–52 *in* B. B. Beck and C. Wemmer, eds. The biology and management of an extinct species—Père David's deer. Noyes Publ., Park Ridge, N.J.

CARPENTER, C. C. 1954. A study of amphibian movement in the Jackson Hole Wildlife Park. Copeia 3:197–200.

CARPENTER, L. H., D. W. REICHERT, AND F. WOLFE, JR. 1977. Lighted collars to aid night observations of mule deer. U.S. For. Ser. Res. Note RM-338. 4pp.

CARR, A., P. ROSS, AND S. CARR. 1974. Internesting behavior of the green turtle, *Chelonida mydas,* at a mid-ocean island breeding ground. Copeia 1974:703–706.

CARRICK, R., AND M. D. MURRAY. 1970. Readable band numbers and "Scotchlite" colour bands for silver gull. Aust. Bird Bander 8:51–56.

CHABRECK, R. H. 1965. Methods of capturing, marking and sexing alligators. Proc. Annu. Conf. Southeast. Assoc. Game and Fish Comm. 17:47–50.

CHEESEMAN, C. L., AND S. HARRIS. 1982. Methods of marking badgers (*Meles meles*). J. Zool. (Lond.) 197:289–292.

CHILDS, H. E., JR. 1952. Color bands. Western Bird Banding Assoc. News 27:4.

CHITTY, D., AND M. SHORTEN. 1946. Techniques for the study of the Norway rat *Rattus norvegicus.* J. Mammal. 27:63–78.

CLARK, D. R., JR. 1971. Branding as a marking technique for amphibians and reptiles. Copeia 1971:148–151.

CLARKE, R. 1971. The possibility of injuring small whales with the standard discovery whale mark. Int. Whaling Comm. Rep. Comm. 21:106–108.

CLARKE, R. D. 1972. The effect of toe clipping on survival in Fowler's toad (*Bufo woodhousei fowleri*). Copeia 1972:182–185.

CLAUSEN, B., P. HJORT, H. STRANDGAARD, AND P. L. SOERENSEN. 1984. Immobilization and tagging of muskoxen (*Ovibos moschatus*) in Jameson Land, northeastern Greenland. J. Wildl. Dis. 20:141–145.

CLAYTON, D. H., C. L. HARTLEY, AND M. GOCHFELD. 1978. Two optical tracking devices for nocturnal field studies of birds. Proc. Colonial Waterbird Group 1978:79–83.

CLINE, K. W., AND W. S. CLARK. 1981. Chesapeake Bay bald eagle banding project: 1981 report and five year summary. Natl. Wildl. Fed., Raptor Inf. Cent., Washington, D.C. 38pp.

CLOVER, M. R. 1954. A portable deer trap and catch-net. Calif. Fish Game 40:367–373.

COCKRUM, E. L. 1969. Migration of the guano bat, *Tadarida brasiliensis.* Univ. Kansas Mus. Nat. Hist. Misc. Publ. 51:303–336.

COHEN, R. 1969. Color-banded house finches. Eastern Bird Banding Assoc. News 32:81–82.

COLCLOUGH, J. H., AND G. J. B. ROSS. 1987. Colour band loss in cape gannets. Safring News 16:35–37.

CONANT, R. 1948. Regeneration of clipped subcaudal scales in a pilot black snake. Nat. Hist. Misc. 13:1–2.

CONNER, M. C. 1982. Determination of bobcat (*Lynx rufus*) and raccoon (*Procyon lotor*) population abundance by radioisotope tagging. M.S. Thesis, Univ. Florida, Gainesville. 55pp.

———, AND R. F. LABISKY. 1985. Evaluation of radioisotope tagging for estimating abundance of raccoon populations. J. Wildl. Manage. 32:698–711.

COOPER, J., AND P. D. MORANT. 1981. The design of stainless steel flipper bands for penguins. Ostrich 52:119–123.

COPPINGER, R. P., AND B. C. WENTWORTH. 1966. Identification of experimental birds with the aid of feather autografts. Bird-Banding 37:203–205.

COWAN, D. P., J. A. VAUGHAN, K. J. PROUT, AND W. G. CHRISTER. 1984. Markers for measuring bait consumption by the European wild rabbit. J. Wildl. Manage. 48:1403–1409.

CRAIGHEAD, J. J., M. G. HORNOCKER, M. W. SHOESMITH, AND R. I. ELLIS. 1969. A marking technique for elk. J. Wildl. Manage. 33:906–909.

————, AND D. S. STOCKSTAD. 1956. A colored neckband for marking birds. J. Wildl. Manage. 20:331–332.

CRAVEN, S. R. 1979. Some problems with Canada goose neckbands. Wildl. Soc. Bull. 7:268–273.

CRIER, J. K. 1970. Tetracyclines as a fluorescent marker in bones and teeth of rodents. J. Wildl. Manage. 34:829–834.

CUMMINGS, J. L. 1987. Nylon fasteners for attaching leg and wing tags to blackbirds. J. Field Ornithol. 58:265–269.

CUTHBERT, F. J., AND W. E. SOUTHERN. 1975. A method for marking young gulls for individual identification. Bird-Banding 46:252–253.

DAAN, S. 1969. Frequency of displacements as a measure of activity of hibernating bats. Lynx 10:13–18.

DAUGHERTY, C. H. 1976. Freeze-branding as a technique for marking anurans. Copeia 4:836–838.

DAVEY, C. C., AND P. J. FULLAGAR. 1985. Nasal saddles for Pacific black duck *Anas superciliosa* and austral teal. Corella 9:123–124.

————, ————, AND C. KOGON. 1980. Marking rabbits for individual identification and a use for Betalights. J. Wildl. Manage. 44:494–497.

DAVIS, R. A. 1963. Feral coypus in Britain. Proc. Assoc. Appl. Biol., Great Britain 51:345–348.

DAVIS, W., AND G. SARTOR. 1975. A method of observing movements of aquatic turtles. Herpetol. Rev. 6:13–14.

DAVIS, W. H. 1963. Anodizing bat bands. Bat Banding News 4:12–13.

DAY, G. I. 1973. Marking devices for big game animals. Ariz. Game and Fish Dep. Res. Abstr. 8:1–7.

DEAVERS, D. R. 1972. Water and electrolyte metabolism in the arenicolous lizard *Uma notata notata*. Copeia 1972:109–122.

DE LA MARE, W. K. 1985. Some evidence for mark shedding with discovery whale marks. Int. Whaling Comm. Rep. Comm. 35:477–486.

DELL, J. 1957. Toe clipping varying hares for track identification. N.Y. Fish Game J. 4:61–68.

DELONG, T. R. 1982. Effect of ambient conditions on nocturnal nest behavior in long-eared owls. M.S. Thesis, Brigham Young Univ., Provo, Ut. 24pp.

DICKMAN, C. R. 1988. Detection of physical contact interactions among free-living mammals. J. Mammal. 69:865–868.

————, D. H. KING, D. C. D. HAPPOLD, AND M. J. HOWELL. 1983. Identification of the filial relationships of free-living small mammals by ^{35}sulfur. Aust. J. Zool. 31:467–474.

DICKSON, J. G., R. N. CONNER, AND J. H. WILLIAMSON. 1982. An evaluation of techniques for marking cardinals. J. Field Ornithol. 53:420–421.

DOLE, J. W. 1965. Summer movements of adult leopard frogs, *Rana pipiens* Schreber, in northern Michigan. Ecology 46:236–255.

————, AND P. DURANT. 1974. Movements and seasonal activity of *Atelopus oxyrhynchus* (Anura: Atelopodidae) in a Venezuelan cloud forest. Copeia 1974:230–235.

DOWNING, R. L., AND C. M. MARSHALL. 1959. A new plastic tape marker for birds and mammals. J. Wildl. Manage. 23:223–224.

————, AND B. S. MCGINNES. 1969. Capturing and marking whitetailed deer fawns. J. Wildl. Manage. 33:711–714.

DUNBAR, I. K. 1959. Leg bands in cold climates. East. Bird Banding News 22:37.

EADIE, J. MCA., K. M. CHENG, AND C. R. NICHOLS. 1987. Limitations of tetracycline in tracing multiple maternity. Auk 104:330–333.

ECKERT, K. L., AND S. A. ECKERT. 1989. The application of plastic tags to leatherback sea turtles, *Deumochelys eumochelys corfacea*. Herpetol. Rev. 20:90–91.

EDMINSTER, F. C. 1938. The marking of ruffed grouse for field identification. J. Wildl. Manage. 2:55–57.

EFFORD, I. E., AND J. A. MATHIAS. 1969. A comparison of two salamander populations in Marion Lake, British Columbia. Copeia 1969:723–736.

ELLENTON, J. A., AND O. H. JOHNSTON. 1975. Oral biomarkers of calciferous tissues in carnivores. Pages 60–67 in R. E. Chambers, ed. Trans. eastern coyote workshop. Northeast Fish Wildl. Conf., New Haven, Conn.

ELLIS, D. H., AND C. H. ELLIS. 1975. Color marking golden eagles with human hair dyes. J. Wildl. Manage. 39:445–447.

ELLISOR, J. E., AND W. F. HARWELL. 1969. Mobility and home range of collared peccary in southern Texas. J. Wildl. Manage. 33:425–427.

ELMES, R. 1955. Loss of rings. Bird Study 2:153.

EMLEN, S. T. 1968. A technique for marking anuran amphibians for behavioral studies. Herpetologica 24:172–173.

ENDERSON, J. H. 1964. A study of the prairie falcon in the central Rocky Mountain region. Auk 81:332–352.

ENVIRONMENT CANADA. 1984. North American bird banding. Environ. Conserv. Serv. 1:1–3.

ERNST, C. H. 1971. Population dynamics and activity scales of *Chrysemys picta* in southeastern Pennsylvania. J. Herpetol. 5:151–160.

ERSKINE, A. J. 1962. Nasal disc method of color-marking waterfowl. In Abstr. Pap. 13th Int. Ornithol. Congr., Ithaca, N.Y. 84pp.

EVANS, C. D. 1951. A method of color marking young waterfowl. J. Wildl. Manage. 15:101–103.

EVANS, J., J. O. ELLIS, R. D. NASS, AND A. L. WARD. 1971. Techniques for capturing, handling, and marking nutria. Proc. Annu. Conf. Southeast. Assoc. Game and Fish Comm. 25:295–315.

EVANS, W. E., J. D. HALL, A. B. IRVINE, AND J. S. LEATHERWOOD. 1972. Methods for tagging small cetaceans. Fish. Bull. 70:61–65.

EVRARD, J. O. 1986. Loss of nasal saddle on mallard. J. Field Ornithol. 57:170–171.

FAGERSTONE, K. A., AND B. E. JOHNS. 1987. Transponders as permanent identification markers for domestic ferrets, black-footed ferrets, and other wildlife. J. Wildl. Manage. 51:294–297.

FARRELL, R. K., AND S. D. JOHNSTON. 1973. Identification of laboratory animals: freeze marking. Lab. Anim. Sci. 23:107–110.

————, G. A. LAISNER, AND T. S. RUSSELL. 1969. An international freeze-mark animal identification system. J. Am. Vet. Med. Assoc. 154:1561–1572.

FERNER, J. W. 1979. A review of marking techniques for amphibians and reptiles. Soc. Stud. Amphib. Reptiles Herpetol. Circ. 9. 42pp.

FISHER, M., AND A. MUTH. 1989. A technique for permanently marking lizards. Herpetol. Rev. 20:45–46.

FITCH, H. S. 1987. Collecting and life-history techniques. Pages 143–164 in R. A. Seigel, J. T. Collins, and S. S. Novak, eds. Snakes: ecology and evolutionary biology. Macmillan Publ. Co., New York, N.Y.

FJETLAND, C. A. 1973. Long-term retention of plastic collars on Canada geese. J. Wildl. Manage. 37:176–178.

FOLLMANN, E. H., P. J. SAVARIE, D. G. RITTER, AND G. M. BAER. 1987. Plasma marking of arctic foxes with iophenoxic acid. J. Wildl. Dis. 23:709–712.

FORSMAN, E. D. 1983. Methods and materials for locating and studying spotted owls. U.S. For. Serv. Gen. Tech. Rep. PNW-162. 8pp.

FOSTER, J. B. 1966. The giraffe of Nairobi National Park: home range, sex ratios, the herd, and food. E. Afr. Wildl. J. 4:139–148.

FOX, S. F. 1978. Natural selection on behavioral phenotypes of the lizard *Uta stansburiana*. Ecology 59:834–847.

FRANKEL, A. I., AND T. S. BASKETT. 1963. Color marking disrupts pair bonds of captive mourning doves. J. Wildl. Manage. 27:124–127.

FRAZER, N. B. 1983. Survivorship of adult female loggerhead sea turtles, *Caretta caretta,* nesting on Little Cumberland Island, Georgia, USA. Herpetologica 39:436–447.

FRENTRESS, C. 1976. "Pop" rivet fasteners for color markers. Inland Bird Banding Assoc. News 47:3–9.

FROESE, A. D., AND G. M. BURGHARDT. 1975. A dense natural population of the common snapping turtle (*Chelydra s. serpentina*). Herpetologica 31:204–208.

FULLAGAR, P. J., AND P. A. JEWELL. 1965. Marking small rodents and the difficulties of using leg rings. J. Zool. 147:224–228.

FURRER, R. K. 1979. Experiences with a new back-tag for open-nesting passerines. J. Wildl. Manage. 43:245–249.

GARGETT, V. 1973. Marking black eagles in the Matopos. Honeyguide 76:26–31.

GARNETT, S. 1987. Feather-clipping: a nauruan technique for short-term recognition of individual birds. Corella 11:30–31.

GAYMER, R. 1973. A marking technique for giant tortoises and field trials in Aldabra. J. Zool. (Lond.) 169:393–401.

GEIS, A. D., AND L. H. ELBERT. 1956. Relation of the tail length of cock ring-necked pheasants to harem size. Auk 73:289.

GENTRY, R. L. 1979. Adventitious and temporary marks in pinniped studies. Pages 39–43 in L. Hobbs and P. Russell, eds. Report on the pinniped tagging workshop, 18-19 January 1979, Seattle, Wash.

————, AND J. R. HOLT. 1982. Equipment and techniques for handling northern fur seals. NOAA Tech. Rep. NMFS Spec. Sci. Rep. Fish. 758. 18pp.

————, M. H. SMITH, AND R. J. BEYERS. 1971. Use of radioactively tagged bait to study movement patterns in small mammals. Ann. Zool. Fenn. 8:17–21.

GERACI, J. R., G. J. D. SMITH, AND T. G. FRIESEN. 1986. Assessment of marking techniques for beluga whale. Final Rep. to World Wildlife Fund Canada. Dep. Pathol., Univ. Guelph, Guelph, Ont. 94pp.

GIFFORD, C. E., AND D. R. GRIFFIN. 1960. Notes on homing and migratory behavior of bats. Ecology 41:378–381.

GODFREY, G. A. 1975. Home range characteristics of ruffed grouse broods in Minnesota. J. Wildl. Manage. 39:287–298.

GOFORTH, W. R., AND T. S. BASKETT. 1965. Effects of experimental color marking on pairing of captive mourning doves. J. Wildl. Manage. 29:543–553.

GOSS-CUSTARD, J. D., S. E. A. LE V. DIT DURELL, H. P. SITTERS, AND R. SWINFEN. 1982. Age-structure and survival of a wintering population of oystercatchers. Bird Study 29:83–98.

GRAHAM, T. W. 1986. A warning against the use of Petersen disc tags in turtle studies. Herpetol. Rev. 17:42–43.

GRAHAM, W. J., AND H. W. AMBROSE, III. 1967. A technique for continuously locating small mammals in field enclosures. J. Mammal. 48:639–642.

GREENWOOD, R. J. 1975. An attempt to freeze-brand mallard ducklings. Bird-Banding 46:204–206.

———, AND W. C. BAIR. 1974. Ice on waterfowl markers. Wildl. Soc. Bull. 2:130–134.

GRIBEN, M. R., H. R. JOHNSON, B. B. GALLUCCI, AND V. F. GALLUCCI. 1984. A new method to mark pinnipeds as applied to the northern fur seal. J. Wildl. Manage. 48:945–949.

GRICE, D., AND J. P. ROGERS. 1965. The wood duck in Massachusetts. Mass. Div. Fish. Game Final Rep., Fed. Aid Proj. W-19-R. 96pp.

GRIFFIN, D. R. 1934. Marking bats. J. Mammal. 15:202–207.

———. 1952. Radioactive tagging of animals under natural conditions. Ecology 33:329–335.

GRUBB, J. C. 1970. Orientation in post-reproductive Mexican toads, *Bufo valliceps*. Copeia 1970:674–680.

GUARINO, J. L. 1968. Evaluation of a colored leg tag for starlings and blackbirds. Bird-Banding 39:6–13.

GULLION, G. W., R. L. ENG, AND J. J. KUPA. 1962. Three methods for individually marking ruffed grouse. J. Wildl. Manage. 26:404–407.

GUTTMAN, S. I., AND W. CREASEY. 1973. Staining as a technique for marking tadpoles. J. Herpetol. 7:388–390.

HADOW, H. H. 1972. Freeze-branding: a permanent marking technique for pigmented mammals. J. Wildl. Manage. 36:645–649.

HAGSTROM, T. 1973. Identification of newt specimens (*Urodela, Trurus*) by recording the belly pattern and a description of photographic equipment for such registrations. Br. J. Herpetol. 7:321–326.

HALL, R. J., AND D. P. STAFFORD. 1972. Studies in the life history of Wehrles salamander, *Plethodon wehrlei*. Herpetologica 28:300–309.

HAMERSTROM, F. 1942. Dominance in winter flocks of chickadees. Wilson Bull. 54:32–42.

HANDEL, C. M., AND R. E. GILL, JR. 1983. Yellow birds stand out in a crowd. North Am. Bird Bander 8:6–9.

HANKS, J. 1969. Techniques for marking large African mammals. Puku 5:65–86.

HANSEN, C. G. 1964. A dye spraying device for marking desert bighorn sheep. J. Wildl. Manage. 28:584–587.

HARAMIS, G. M., W. G. ALLISTON, AND M. E. RICHMOND. 1983. Dump nesting in the wood duck traced by tetracycline. Auk 100:729–730.

———, AND A. D. NICE. 1980. An improved web-tagging technique for waterfowl. J. Wildl. Manage. 44:898–899.

HARMATA, A. R. 1984. Bald eagles of the San Luis Valley, Colorado: their winter ecology and spring migration. Ph.D. Thesis, Montana State Univ., Bozeman. 222pp.

HARPER, J. A., AND W. C. LIGHTFOOT. 1966. Tagging devices for Roosevelt elk and mule deer. J. Wildl. Manage. 30:461–466.

HART, A., AND A. D. M. HART. 1987. Patagial tags for herring gulls: improved durability. Ringing & Migr. 8:19–26.

HAVELKA, P. 1983. Registration and marking of captive birds of prey. Int. Zoo Yearb. 23:125–132.

HAWKINS, L. L., AND S. G. SIMPSON. 1985. Neckband a handicap in an aggressive encounter between tundra swans. J. Field Ornithol. 56:182–184.

HAWKINS, R. E., W. D. KLIMSTRA, G. FOOKS, AND J. DAVIS. 1967. Improved collar for white-tailed deer. J. Wildl. Manage. 31:356–359.

HAYWARD, G. D. 1987. Betalights: an aid in the nocturnal study of owl foraging habitat and behavior. J. Raptor Res. 21:98–102.

HAYWARD, J. L., JR. 1982. A simple egg-marking technique. J. Field Ornithol. 53:173.

HEALY, W. R. 1974. Population consequences of alternative life histories in *Notophthalmus v. viridescens*. Copeia 1974:221–229.

———. 1975. Terrestrial activity and home range in efts of *Notophthalmus viridescens*. Am. Midl. Nat. 93:131–138.

HEATWOLE, H. 1961. Inhibition of digital regeneration in salamanders and its use in marking individuals for field studies. Ecology 42:593–594.

HENCKEL, R. E. 1976. Turkey vulture banding problem. North Am. Bird Bander 1:126.

HENDERSON, R. W. 1974. Aspects of the ecology of the juvenile common iguana (*Iguana iguana*). Herpetologica 30:327–332.

HENDRICKSON, J. R. 1954. Ecology and systematics of salamanders of the genus *Batrochoseps*. Univ. Calif. Publ. Zool. 54:1–46.

HENLEY, G. B. 1981. A new technique for recognition of snakes. Herpetol. Rev. 12:56.

HERO, J. 1989. A simple code for toe clipping anurans. Herpetol. Rev. 20:66–67.

HERREID, C. F., AND S. KINNEY. 1966. Survival of Alaskan woodfrog (*Rana sylvatica*) larvae. Ecology 47:1039–1041.

HESTER, A. E. 1963. A plastic wing tag for individual identification of passerine birds. Bird-Banding 34:213–217.

HILLIS, R. E., AND E. D. BELLIS. 1971. Some aspects of the ecology of the hellbender, *Cryptobranchus alleganiensis alleganiens*. Herpetologica 5:121–126.

HIRTH, H. G. 1966. Weight changes and mortality of three species of snakes during hibernation. Herpetologica 22:8–12.

HOBBS, L., AND P. RUSSELL. 1979. Report on the pinniped tagging workshop. Pinniped Tagging Workshop, Seattle, Wash. 48pp.

HOFFMAN, R. H. 1985. An evaluation of banding sandhill cranes with colored leg bands. North Am. Bird Bander 10:46–49.

HOMESTEAD, R., B. BECK, AND D. E. SERGEANT. 1972. A portable, instantaneous branding device for permanent identification of wildlife. J. Wildl. Manage. 36:947–949.

HONMA, M., S. IWAKI, A. KAST, AND H. KREUZER. 1986. Experiences with the identification of small rodents. Exp. Anim. 35:347–352.

HOOPER, J. H. D. 1983. The study of horseshoe bats in Devon caves: a review of progress 1947–1982. Stud. Speleol. 4:59–70.

HOWE, M. A. 1980. Problems with wing tags: evidence of harm to willets. J. Field Ornithol. 51:72–73.

HUBERT G. F., JR., G. L. STORM, R. L. PHILLIPS, AND R. D. ANDREWS. 1976. Ear tag loss in red foxes. J. Wildl. Manage. 40:164–167.

HUDNALL, J. A. 1982. New methods for measuring and tagging snakes. Herpetol. Rev. 13:97–98.

HUEY, W. S. 1965. Sight records of color-marked sandhill cranes. Auk 83:640–643.

HURST, J. L. 1988. A system for the individual recognition of small rodents at a distance, used in free-living and enclosed populations of house mice. J. Zool. 215:363–367.

IDSTROM, J. M., AND J. P. LINDMEIER. 1956. Some tests of the rubber styrene neck bands for marking waterfowl. Minn. Dep. Conserv. Q. Prog. Rep. Wildl. Res. 16:134–137.

IRELAND, P. H. 1973. Marking larval salamanders with fluorescent pigments. Southwest. Nat. 18:252–253.

———. 1991. A simplified fluorescent marking technique for identification of terrestrial salamanders. Herpetol. Rev. 22:21–22.

IRVINE, A. B., AND M. D. SCOTT. 1984. Development and use of marking techniques to study manatees in Florida. Fla. Sci. 47:12–26.

———, R. S. WELLS, AND M. D. SCOTT. 1982. An evaluation of techniques for tagging small odontocete cetaceans. Natl. Oceanic and Atmos. Adm. Fish. Bull. 80:135–143.

JACKSON, J. J. 1982. Effect of wing tags on renesting interval in red-winged blackbirds. J. Wildl. Manage. 46:1077–1079.

JAMESON, D. L. 1957. Population structure and homing responses in the Pacific tree frog. Copeia 3:221–228.

JENSSEN, T. A. 1970. The ethoecology of *Anolis nebulosus* (Sauria, Iguanidae). J. Herpetol. 4:1–38.

JOHANNINGSMEIER, A. G., AND C. J. GOODNIGHT. 1962. Use of iodine-131 to measure movements of small animals. Science 138:147–148.

JOHNS, B. E., AND H. P. PAN. 1981. Analytical techniques for fluorescent chemicals used as systemic or external wildlife markers. Am. Soc. Testing Materials, Vertebr. Pest Control Manage. Materials 3:86–93.

———, AND R. D. THOMPSON. 1979. Acute toxicant identification in whole bodies and baits without chemical analysis. Pages 80–88 *in* E. E. Kenega, ed. Avian and mammalian wildlife toxicology. ASTM STP 693, Am. Soc. Testing Materials, Philadelphia, Pa.

JOHNSON, A. M. 1979. Factors contributing to difficulties in the analysis of mark-recapture data. Pages 27–29 *in* L. Hobbs and P. Russell, eds. Report on the pinniped tagging workshop, 18-19 January 1979, Seattle, Wash.

JOHNSON, K. G., AND M. R. PELTON. 1980. Marking techniques for black bears. Proc. Annu. Conf. Southeast. Assoc. Fish Wildl. Agencies 34:557–562.

JOHNSON, P. A., B. W. JOHNSON, AND L. T. TAYLOR. 1981. Interisland movement of a young Hawaiian monk seal between Laysan Island and Maro Reef. Elepaio 41:113–114.

JONES, S. M., AND G. W. FERGUSON. 1980. The effect of paint marking on mortality in a Texas population of *Sceloporus undulatus*. Copeia 1980:850–854.

JONES, W. T., AND B. B. BUSH. 1988. Darting and marking techniques for an arboreal forest monkey, (*Cerocophthecus ascanius*). Am. J. Primatol. 14:83–99.

JONKEL, C. J., D. R. GRAY, AND B. HUBERT. 1975. Immobilizing and marking wild muskoxen in Arctic Canada. J. Wildl. Manage. 39:112–117.

JOYNER, D. E. 1975. Nest parasitism and brood-related behavior of the ruddy duck (*Oxyura jamaicensis rubida*). Ph.D. Thesis, Univ. Nebraska, Lincoln. 152pp.

JUDD, F. W. 1975. Activity and thermal ecology of the keeled earless lizard, *Holbrookia propinqua*. Herpetologica 31:137–150.

KACZYNSKI, C. F., AND W. H. KIEL, JR. 1963. Band loss by nestling mourning doves. J. Wildl. Manage. 27:271–279.

KADLEC, J. A. 1975. Recovery rates and loss of aluminum, titanium, and incoloy bands on herring gulls. Bird-Banding 46:230–235.

KAPLAN, H. M. 1958. Marking and banding frogs and turtles. Herpetologica 14:131–132.

KARLSTROM, E. L. 1957. The use of Co(60) as a tag for recovering amphibians in the field. Ecology 38:187–195.

KASAMATSU, F., S. NISHIWAKI, AND M. SATO. 1986. Results of the test firing of improved .410 streamer marks, February 1985. Int. Whaling Commn. Rep. Comm. 36:201–204.

KAYE, S. V. 1960. Gold-198 wires used to study movements of small mammals. Science 131:824.

KEEN, R., AND H. B. HITCHCOCK. 1980. Survival and longevity of the little brown bat (*Myotis lucifugus*) in southeastern Ontario. J. Mammal. 61:1–7.

KEISTER, G. P., JR., C. E. TRAINER, AND M. J. WILLIS. 1988. A self-adjusting collar for young ungulates. Wildl. Soc. Bull. 16:321–323.

KEITH, L. B., E. C. MESLOW, AND O. J. RONGSTAD. 1968. Techniques for snowshoe hare population studies. J. Wildl. Manage. 32:801–812.

KESSLER, F. W. 1964. Avian predation on pheasants wearing differently colored plastic markers. Ohio J. Sci. 64:401–402.

KINNINGHAM, J. J., M. R. PELTON, AND D. C. FLYNN. 1980. Use of the pellet count technique for determining densities of deer in the southern Appalachians. Proc. Annu. Conf. Southeast. Assoc. Fish Wildl. Agencies 34:508–514.

KNIGHT, R. R. 1966. Effectiveness of neckbands for marking elk. J. Wildl. Manage. 30:845–846.

KNOWLTON, F. F., E. D. MICHAEL, AND W. C. GLAZENER. 1964. A marking technique for field recognition of individual turkeys and deer. J. Wildl. Manage. 28:167–170.

KOCHERT, M. N., K. STEENHOF, AND M. Q. MORITSCH. 1983. Evaluation of patagial markers for raptors and ravens. Wildl. Soc. Bull. 11:271–281.

KOERNER, J. W., T. A. BOOKHOUT, AND K. E. BEDNARIK. 1974. Movements of Canada geese color-marked near southwestern Lake Erie. J. Wildl. Manage. 38:275–289.

KOOB, M. D. 1981. Detrimental effects of nasal saddles on male ruddy ducks. J. Field Ornithol. 52:140–143.

KORN, H. 1987. Effects of live-trapping and toe-clipping on body weight of European and African rodent species. Oecologia (Berl.) 71:597–600.

KOZLIK, F. M., A. W. MILLER, AND W. C. RIENECKER. 1959. Color-marking white geese for determining migration routes. Calif. Fish Game 45:69–82.

KRUUK, H. 1978. Spatial organization and territorial behavior of the European badger *Meles meles*. J. Zool. (Lond.) 184:1–19.

———, M. GORMAN, AND T. PARRISH. 1980. The use of 65Zn for estimating populations of carnivores. Oikos 34:206–208.

KUMAR, R. K. 1979. Toe-clipping procedure for individual identification of rodents. Lab. Anim. Sci. 29:679–680.

KUMMER, H. 1968. Social organization of hamadryas baboons. Bibl. Primatol. 6, Karger, Basee.

LABISKY, R. F., AND S. H. MANN. 1962. Backtag markers for pheasants. J. Wildl. Manage. 26:393–399.

LARSEN, T. 1971. Capturing, handling, and marking polar bears in Svalbard. J. Wildl. Manage. 35:27–36.

LARSON, G. E., P. J. SAVARIE, AND I. OKUNO. 1981. Iophenoxic acid and mirex for marking wild, bait-consuming animals. J. Wildl. Manage. 45:1073–1077.

LAVAL, R. K., R. L. CLAWSON, M. L. LAVAL, AND W. CAIRE. 1977. Foraging behavior and nocturnal activity patterns of Missouri bats, with emphasis on the endangered species *Myotis grisescens* and *Myotis sodalis*. J. Mammal. 58:592–599.

LAYFIELD, J. A., D. A. GALBRAITH, AND R. J. BROOKS. 1988. A simple method to mark hatchling turtles. Herpetol. Rev. 19:78–79.

LAZARUS, A. B., AND F. P. ROWE. 1975. Freeze-marking rodents with a pressurized refrigerant. Mammal Rev. 5:31–34.

LEATHERWOOD, S., D. K. CALDWELL, AND H. E. WINN. 1976. Whales, dolphins, and porpoises of the western North Atlantic—a guide to their identification. NOAA Tech. Rep. NMFS Spec. Sci. Rep. Fish. 396. 176pp.

LEBUFF, C. R., AND R. W. BEATTY. 1971. Some aspects of nesting of the loggerhead turtle, *Caretta caretta caretta* (Linne) on the Gulf Coast of Florida. Herpetologica 27:153–156.

LEMEN, C. A., AND P. W. FREEMAN. 1985. Tracking mammals with fluorescent pigments: a new technique. J. Mammal. 66:134–136.

LEMKAU, P. J. 1970. Movements of the box turtle, *Terrapene c. carolina* (Linnaeus), in unfamiliar territory. Copeia 1970:781–783.

LENSINK, C. J. 1968. Neckbands as an inhibitor of reproduction in black brant. J. Wildl. Manage. 32:418–420.

LENTFER, J. W. 1968. A technique for immobilizing and marking polar bears. J. Wildl. Manage. 32:317–321.

LEWKE, R. R., AND R. K. STROUD. 1974. Freeze branding as a method of marking snakes. Copeia 1974:997–1000.

LIGHTFOOT, W. D., AND V. MAW. 1963. Trapping and marking mule deer. Proc. West. Assoc. State Game and Fish Comm. 43:138–142.

LINDSEY, G. D. 1983. Rhodamine B: a systemic fluorescent marker for studying mountain beavers (*Aplodontia rufa*) and other animals. Northwest Sci. 57:16–21.

LINHART, S. B., AND J. J. KENNELLY. 1967. Fluorescent bone labeling of coyotes with demethylchlortetracycline. J. Wildl. Manage. 31:317–321.

LINN, I. J. 1978. Radioactive techniques for small mammal marking. Pages 177–191 *in* B. Stonehouse, ed. Animal marking: recognition marking of animals in research. The MacMillan Press Ltd., London, U.K.

———, AND J. SHILLITO. 1960. Rings for marking very small mammals. Proc. Zool. Soc. Lond. 134:489–495.

LOKEMOEN, J. T., AND D. E. SHARP. 1985. Assessment of nasal marker materials and designs used on dabbling ducks. Wildl. Soc. Bull. 13:53–56.

LONKE, D. J., AND M. E. OBBARD. 1977. Tag success, dimensions, clutch size and nesting site fidelity for the snapping turtle, *Chelydra serpentina* (Reptilia, Testudines, Chelydridae) in Algonquin Park, Ontario, Canada. J. Herpetol. 11:243–244.

LUMSDEN, H. G., V. W. McMULLEN, AND C. L. HOPKINSON. 1977. An improvement in fabrication of large plastic leg bands. J. Wildl. Manage. 41:148–149.

MACDONALD, R. N. 1961. Injury to birds by ice-coated bands. Bird-Banding 72:59.

MACINNES, C. D., AND E. H. DUNN. 1988. Effects of neck bands on Canada geese nesting at the McConnell River. J. Field Ornithol. 59:239–246.

———, J. P. PREVETT, AND H. A. EDNEY. 1969. A versatile collar for individual identification of geese. J. Wildl. Manage. 33:330–335.

MADISON, D. M., AND C. R. SHOOP. 1970. Homing behavior, orientation, and home range of salamanders tagged with Tantalum-182. Science 168:1484–1487.

MALACARNE, G., AND M. GRIFFA. 1987. A refinement of Lack's method of swift studies. Sitta 1:175–177.

MALTBY, L. S. 1977. Techniques used for the capture, handling and marking of brant in the Canadian High Arctic. Can. Wildl. Serv. Prog. Notes 72. 6pp.

MARION, W. R., AND J. D. SHAMIS. 1977. An annotated bibliography of bird marking techniques. Bird-Banding 48:42–61.

MARTOF, B. S. 1953. Territoriality in the green frog, *Rana clamitans*. Ecology 34:165–174.

McCRACKEN, G. F. 1984. Communal nursing in Mexican free-tailed bat maternity colonies. Science 223:1090–1091.

MEDICA, P. A., R. B. BURY, AND F. B. TURNER. 1975. Growth of the desert tortoise (*Gopherus agassizi*) in Nevada. Copeia 1975:639–643.

——, G. A. HODDENBACH, AND J. R. LANNOM. 1971. Lizard sampling techniques. Rock Valley Misc. Publ. 1. 55pp.

MELCHIOR, H. R., AND F. A. IWEN. 1965. Trapping, restraining, and marking Arctic ground squirrels for behavioral observations. J. Wildl. Manage. 29:671–678.

MILLER, D. J. 1979. Sea otter capture and tagging in California. Pages 11–12 *in* L. Hobbs and P. Russell, eds. Report on the pinniped tagging workshop, 18-19 January 1979, Seattle, Wash.

MILLER, D. S., J. BERGLUND, AND M. JAY. 1983. Freeze-mark techniques applied to mammals at the Santa Barbara Zoo. Zoo Biol. 2:143–148.

MILLER, L. S. 1957. Tracing vole movements by radioactive excretory products. Ecology 38:132–136.

MILLS, J. A. 1972. A difference in band loss from male and female red-billed gulls, *Larus novaehollandiae scopulinus.* Ibis 114:252–255.

MINNICH, J. E., AND V. H. SHOEMAKER. 1970. Diet, behavior and water turnover in the desert iguana, *Depsosaurus dorsalis.* Am. Midl. Nat. 84:496–509.

MITCHELL, E., AND V. M. KOZICKI. 1975. Prototype visual mark for large whales modified from "Discovery" tag. Int. Whaling Comm. Rep. Comm. 25:236–239.

MIYASHITA, T., AND R. A. ROWLETT. 1985. Test-firing of .410 streamer marks. Int. Whaling Comm. Rep. Comm. 35:305–308.

MOFFITT, J. 1942. Apparatus for marking wild animals with colored dyes. J. Wildl. Manage. 6:312–318.

MOHR, C. E. 1934. Marking bats for later recognition. Proc. Pa. Acad. Sci. 8:26–30.

MORAN, S. 1985. Banding fruit bats. Israel J. Zool. 33:91–93.

MORGAN, D. R. 1981. Monitoring bait acceptance in brush-tailed possum populations: development of a tracer technique. N.Z. J. For. Sci. 11:271–277.

MORGENWECK, R. O., AND W. H. MARSHALL. 1977. Wing marker for American woodcock. Bird-Banding 48:224–227.

MORRISON, D. W. 1978. Foraging ecology and energetics of the frugivorous bat *Artibeus jamaicensis.* Ecology 59:716–723.

MOSELEY, L. J., AND H. C. MUELLER. 1975. A device for color-marking nesting birds. Bird-Banding 46:341–342.

MOSSMAN, A. S. 1960. A color marking technique. J. Wildl. Manage. 24:104.

MUDGE, G. P., AND P. N. FERNS. 1978. Durability of patagial tags on herring gulls. Ringing & Migr. 2:42–45.

MUKINYA, J. G. 1976. An identification method for black rhinoceros (*Diceros bicornis* Linn. 1758). E. Afr. Wildl. J. 14:335–338.

MULLICAN, T. R. 1988. Radio telemetry and fluorescent pigments: a comparison of techniques. J. Wildl. Manage. 52:627–631.

MURPHY, M. T. 1981. Growth and aging of nestling eastern kingbirds and eastern phoebes. J. Field Ornithol. 52:309–316.

NACE, G. W., AND E. K. MANDERS. 1982. Marking individual amphibians. J. Herpetol. 16:309–311.

NEAL, W. 1964. Extra white feather makes bird important. Inland Bird Banding Assoc. News 36:69–71.

NELLIS, D. W., J. H. JENKINS, AND A. D. MARSHALL. 1967. Radioactive zinc as a feces tag in rabbits, foxes, and bobcats. Proc. Annu. Conf. Southeast. Assoc. Game and Fish Comm. 21:205–207.

NELSON, L. J., D. R. ANDERSON, AND K. P. BURNHAM. 1980. The effect of band loss on estimates of annual survival. J. Field Ornithol. 51:30–38.

NELSON, R. L., AND R. L. LINDER. 1972. Percentage of raccoons and skunks reached by egg baits. J. Wildl. Manage. 36:1327–1329.

NESBITT, S. A. 1979. An evaluation of four wildlife marking materials. Bird-Banding 50:159.

NEWSOM, J. D., AND J. S. SULLIVAN, JR. 1968. Cryo-branding—a marking technique for white-tailed deer. Proc. Annu. Conf. Southeast. Assoc. Game and Fish Comm. 22:128–133.

NICKERSON, M. A., AND C. E. MAYS. 1973. A study of the Ozark hellbender, *Cryptobranchus alleganiensis bishopi.* Ecology 54:1154–1165.

NORRIS, K. S., AND K. W. PRYOR. 1970. A tagging method for small cetaceans. J. Mammal. 51:609–610.

O'BRIEN, G. P., H. K. SMITH, AND J. R. MEYER. 1965. An activity study of a radioisotope-tagged lizard, *Sceloporus undulata hyacinthimus* (Sauria, Iguanidae). Southwest Nat. 10:179–187.

OLSEN, J., T. BILLETT, AND P. OLSEN. 1982. A method for reducing illegal removal of eggs from raptor nests. Emu 82:225.

ORSER, P. N., AND D. J. SHURE. 1972. Effects of urbanization on the salamander *Desmognathus fuscus fuscus.* Ecology 53:1148–1154.

OTIS, D. L., C. E. KNITTLE, AND G. M. LINZ. 1986. A method for estimating turnover in spring blackbird roosts. J. Wildl. Manage. 50:567–571.

OTTAWAY, J. R., R. CARRICK, AND M. D. MURRAY. 1984. Evaluation of leg bands for visual identification of free-living silver gulls. J. Field Ornithol. 55:287–308.

OWEN, L. N. 1961. Fluorescence of tetracyclines in bone tumours, normal bone and teeth. Nature 190:500–502.

PANAGIS, K., AND P. E. STANDER. 1989. Marking and subsequent movement patterns of springbok lambs in the Etosha National Park, South West Africa/Namibia. Madoqua 16:71–73.

PARKER, W. S. 1976. Population estimates, age structure, and denning habits of whipsnakes, *Masticophis t. taeniatus,* in a northern Utah *Atriplex-Sarcobatus* community. Herpetologica 32:53–57.

PATON, P. W. C., AND L. PANK. 1986. A technique to mark incubating birds. J. Field Ornithol. 57:232–233.

PATTERSON, I. J. 1978. Tags and other distant-recognition markers for birds. Pages 54–62 *in* B. Stonehouse, ed. Animal marking: recognition marking of animals in research. The MacMillan Press Ltd., London, U.K.

PAULISSEN, M. A. 1986. A technique for marking teiid lizards in the field. Herpetol. Rev. 17:6,17.

PAULLIN, D. G., AND E. KRIDLER. 1988. Spring and fall migration of tundra swans dyed at Malheur National Wildlife Refuge, Oregon. Murrelet 69:1–9.

PAVONE, L. V., AND R. BOONSTRA. 1985. The effects of toe clipping on the survival of the meadow vole (*Microtus pennsylvanicus*). Can. J. Zool. 63:499–501.

PELTON, M. R., AND L. C. MARCUM. 1975. The potential use of radioisotopes for determining densities of black bears and other carnivores. Pages 221–236 *in* R. L. Phillips and C. Jonkel, eds. Proc. 1975 predator symposium. Mont. For. Conserv. Exp. Stn., Univ. Montana, Missoula.

PENDLEBURY, G. B. 1972. Tagging and remote identification of rattlesnakes. Herpetologica 28:349–350.

PENDLETON, R. C. 1956. Uses of marking animals in ecological studies: labelling animals with radioisotopes. Ecology 37:686–689.

PENNYCUICK, C. J. 1978. Identification using natural markings. Pages 147–159 *in* B. Stonehouse, ed. Animal marking: recognition marking of animals in research. The MacMillan Press Ltd., London, U.K.

PERDECK, A. C., AND R. D. WASSENAAR. 1981. Tarsus or tibia: where should a bird be ringed? Ringing & Migr. 3:149–157.

PERRY, A. E., AND G. BECKETT. 1966. Skeletal damage as a result of band injury in bats. J. Mammal. 47:131–132.

PFEIFER, S., F. H. WRIGHT, AND M. DONCARLOS. 1984. Freeze-branding beaver tails. Zoo Biol. 3:159–162.

PHILLIPS, R. S., AND T. H. NICHOLLS. 1970. A collar for marking big game animals. U.S. For. Serv. Res. Note NC-103. 4pp.

PHILLIPS, W. R. 1985. The use of bird bands for marking tree-dwelling bats a preliminary appraisal. Macroderma 1:17–21.

PIENAAR, U. DE V., J. W. VAN NIEKERK, E. YOUNG, P. VAN WYK, AND N. FAIRALL. 1966. The use of oripavine hydrochlorine (M.99) in the drug immobilisation and marking of the wild African elephant (*Loxodonta africana* Blumenbach) in the Kruger National Park. Koedoe 9:108–124.

PIGOZZI, G. 1988. Quill-marking: a method to identify crested porcupines individually. Acta Theriol. 33:138–142.

PIRKOLA, M. K., AND P. KALINAINEN. 1984. Use of neckbands in studying the movements and ecology of the bean goose *Anser fabalis.* Ann. Zool. Fenn. 21:259–263.

PITCHER, K. 1979. Pinniped tagging in Alaska. Pages 3–4 *in* L. Hobbs and P. Russell, eds. Report on the pinniped tagging workshop, 18–19 January 1979, Seattle, Wash.

PLATT, S. W. 1980. Longevity of herculite leg jess color markers on the prairie falcon (*Falco mexicanus*). J. Field Ornithol. 51:281–282.

POUGH, F. H. 1966. Ecological relationships of rattlesnakes in southeastern Arizona with notes on other species. Copeia 1966:676–683.

——. 1970. A quick method for permanently marking snakes and turtles. Herpetologica 26:428–430.

PRITCHARD, P. C. H. 1976. Post-nesting movements of marine turtles (*Cheloniidae* and *Dermochelyidae*) tagged in the Guianas. Copeia 1976:749–754.

————. 1980. The conservation of sea turtles: practices and problems. Am. Zool. 20:609–617.

PYRAH, D. 1970. Poncho markers for game birds. J. Wildl. Manage. 34:466–467.

QUEAL, L. M., AND B. D. HLAVACHICK. 1968. A modified marking technique for young ungulates. J. Wildl. Manage. 32:628–629.

RACEY, P. A., AND S. M. SWIFT. 1985. Feeding ecology of *Pipistrellus pipistrellus* (Chiroptera: Vespertilionidae) during pregnancy and lactation. I. Foraging behavior. J. Anim. Ecol. 54:205–215.

RAGINSKI, J. N. 1977. Autotransplantation as a method for permanent marking of urodele amphibians (Amphibia, Urodela). J. Herpetol. 11:241–242.

RANDOLPH, S. E. 1973. A tracking technique for comparing individual home ranges of small mammals. J. Zool. 170:509–520.

RANEY, E. C. 1940. Summer movements of a bullfrog, *Rana catesbeiana* Shaw, as determined by the jaw tag method. Am. Midl. Nat. 23:733–745.

RAO, M. V., AND B. S. RAJABAI. 1972. Ecological aspects of the agamid lizards *Sitana ponticeriana* and *Calotes nemoricola* in India. Herpetologica 28:285–289.

REAGAN, D. P. 1974. Habitat selection in the three-toed box turtle, *Terrapene carolina triunguis.* Copeia 1974:512–527.

REED, P. C. 1953. Danger of leg mutilation from the use of metal color bands. Bird-Banding 24:65–67.

REESE, K. P. 1980. The retention of colored plastic leg bands by black-billed magpies. North Am. Bird Bander 5:136–137.

REUTHER, R. T. 1968. Marking animals in zoos. Int. Zoo Yearb. 8:388–390.

REYNOLDS, P. E., AND G. W. GARNER. 1983. Immobilizing and marking muskoxen in the Arctic National Wildlife Refuge, Alaska. Proc. Alaska Sci. Conf. (Abstr.) 34:71.

RICHARDS, C. M., B. M. CARLSON, AND S. L. ROGERS. 1975. Regeneration of digits and forelimbs in the Kenyan reed frog, *Hyperolius viridiflavus ferniquei.* J. Morphol. 146:431–436.

RICHDALE, L. E. 1951. Banding and marking penguins. Bird-Banding 22:47–54.

RICKLEFS, R. E. 1973. Tattooing nestlings for individual recognition. Bird-Banding 44:63.

RILEY, J., AND R. GWILLIAM. 1981. A new ear-punch for small rodents. J. Inst. Anim. Technicians 32:53–55.

RITCHISON, G. 1984. A new marking technique for birds. North Am. Bird Bander 9:8.

ROBERTSON, J. G. 1984. A technique for individually marking frogs in behavioral studies. Herpetol. Rev. 15:56–57.

ROBSON, J. E. 1986. Ring "fit" on blackbreasted snake eagle. Safring News 15:56.

RODGERS, J. A., JR. 1986. A field technique for color-dyeing nestling wading birds without capture. Wildl. Soc. Bull. 14:399–400.

ROHWER, S. 1977. Status signaling in Harris sparrows: some experiments in deception. Behaviour 61:107–129.

ROTTERMAN, L. M., AND C. MONNETT. 1984. An embryo-dyeing technique for identification through hatching. Condor 86:79–80.

ROYALL, W. C., J. L. GUARINO, AND O. E. BRAY. 1974. Effects of color on retention of leg streamers by red-winged blackbirds. West. Bird Bander 49:64–65.

RUDGE, M. R., AND R. J. JOBLIN. 1976. Comparison of some methods of capturing and marking feral goats (*Capra hircus*). N.Z. J. Zool. 3:51–55.

RUSSELL, J. K. 1981. Patterned freeze-brands with canned freon. J. Wildl. Manage. 45:1078.

RYDER, P. L., AND J. P. RYDER. 1981. Reproductive performance of ring-billed gulls in relation to nest location. Condor 83:57–60.

RYDER, R. D. 1978. Postscript: towards humane methods of identification. Pages 229–234 *in* B. Stonehouse, ed. Animal marking: recognition marking of animals in research. The MacMillan Press Ltd., London, U.K.

SALLABERRY, A. M., AND D. J. VALENCIA. 1985. Wounds due to flipper bands on penguins. J. Field Ornithol. 56:275–277.

SAUNDERS, D. A. 1988. Patagial tags: do benefits outweigh risks to the animal? Aust. Wildl. Res. 15:565–569.

SCHEFFER, V. B. 1950. Experiments in the marking of seals and sea-lions. U.S. Fish Wildl. Serv. Spec. Sci. Rep. Wildl. 4. 33pp.

SCHNEEGAS, E. R., AND G. W. FRANKLIN. 1972. The Mineral King deer herd. Calif. Fish Game 58:133–140.

SCHNELL, J. H. 1968. The limiting effects of natural predation on experimental cotton rat populations. J. Wildl. Manage. 32:698–711.

SCOTT, A. F., AND J. L. DOBIE. 1980. An improved design for a thread trailing device used to study terrestrial movements of turtles. Herpetol. Rev. 11:106–107.

SCOTT, D. K. 1978. Identification of individual Bewick's swans by bill patterns. Pages 160–168 *in* B. Stonehouse, ed. Animal marking: recognition marking of animals in research. The MacMillan Press Ltd., London, U.K.

SCOTT, M. P., AND T. N. TAN. 1985. A radiotracer technique for the determination of male mating success in natural populations. Behav. Ecol. Sociobiol. 17:29–33.

SEALE, D., AND M. BORAAS. 1974. A permanent mark for amphibian larvae. Herpetologica 30:160–162.

SEEL, D. C., A. G. THOMPSON, AND G. H. OWEN. 1982. A wing-tagging system for marking larger passerine birds. Bangor Occas. Pap. 14: 1–6.

SEGUIN, R. J., AND F. COOKE. 1983. Band loss from lesser snow geese. J. Wildl. Manage. 47:1109–1114.

SHEDDEN, C. B., P. MONAGHAN, K. ENSOR, AND N. B. METCALFE. 1985. The influence of colour-rings on recovery rates of herring and lesser black-backed gulls. Ringing & Migr. 6:52–54.

SHERWOOD, G. A. 1966. Flexible plastic collars compared to nasal discs for marking geese. J. Wildl. Manage. 30:853–855.

SHINE, C., N. SHINE, R. SHINE, AND D. SLIP. 1988. Use of subcaudal scale anomalies as an aid in recognizing individual snakes. Herpetol. Rev. 19:79–80.

SHOOP, C. R. 1971. A method for short-term marking of amphibians with 24-sodium. Copeia 1973:264–272.

SIEGFRIED, W. R. 1971. Communal roosting of the cattle egret. Trans. R. Soc. South Africa 39:419–443.

SIGLIN, R. J. 1966. Marking mule deer with an automatic tagging device. J. Wildl. Manage. 30:631–633.

SIMMONS, N. M. 1971. An inexpensive method of marking large numbers of Dall sheep for movement studies. Trans. North Am. Wild Sheep Conf. 1:116–126.

————, AND J. L. PHILLIPS. 1966. Modifications of a dye-spraying device for marking desert bighorn sheep. J. Wildl. Manage. 30: 208–209.

SLADEN, W. J. L. 1952. Notes on methods of marking penguins. Ibis 94:541–543.

SMAL, C. M., AND J. S. FAIRLEY. 1982. The dynamics and regulation of small rodent populations in the woodland ecosystems of Killarney, Ireland. J. Zool. (Lond.) 196:1–30.

SMUTS, G. L. 1972. Seasonal movements, migration and age determination of Burchell's zebra (*Equus burchell: antiquorum,* H. Smith, 1841) in Kruger National Park. M.S. Thesis, Univ. Pretoria, Pretoria, South Africa.

SNELLING, J. C. 1970. Some information obtained from marking large raptors in the Kruger National Park, Republic of South Africa. Ostrich Suppl. 8:415–427.

SOCIETY FOR THE STUDY OF AMPHIBIANS AND REPTILES. 1987. Guidelines for use of live amphibians and reptiles in field research. J. Herpetol. Suppl. 4. 14pp.

SODERQUIST, T. R., AND C. R. DICKMAN. 1988. A technique for marking marsupial pouch young with fluorescent pigment tattoos. Aust. Wildl. Res. 15:561–563.

SOUTHERN, L. K., AND W. E. SOUTHERN. 1983. Responses of ring-billed gulls to cannon-netting and wing-tagging. J. Wildl. Manage. 47:234–237.

————, AND ————. 1985. Some effects of wing tags on breeding ring-billed gulls. Auk 102:38–42.

SOUTHERN, W. E. 1964. Additional observations on winter bald eagle populations: including remarks on biotelemetry techniques and immature plumages. Wilson Bull. 76:121–137.

SOWLS, L. K. 1950. Techniques for waterfowl-nesting studies. Trans. North Am. Wildl. Conf. 15:478–487.

SPELLERBERG, I. P., AND I PRESTT. 1978. Marking snakes. Pages 133–141 *in* R. Stonehouse, ed. Animal marking. Univ. Park Press, Baltimore, Md.

SPENCER, R. 1978. Ringing and related durable methods of marking birds. Pages 45–53 *in* B. Stonehouse, ed. Animal marking: recognition marking of animals in research. The MacMillan Press Ltd., London, U.K.

ST. LOUIS, V. L., J. C. BARALOW, AND J. R. A. SWEERTS. 1989. Toenail-clipping: a simple technique for marking individual nidicolous chicks. J. Field Ornithol. 60:211–215.

STAMPS, J. A. 1973. Displays and social organization in female *Anolis aeneus.* Herpetologica 30:160–162.

STARK, M. A. 1984. A quick, easy and permanent tagging technique for rattlesnakes. Herpetol. Rev. 15:110.

STEBBINGS, R. E. 1978. Marking bats. Pages 81–94 *in* B. Stonehouse, ed. Animal marking: recognition marking of animals in research. The MacMillan Press Ltd., London, U.K.

STEBBINS, R. C., AND N. W. COHEN. 1973. The effect of parietalectomy on the thyroid and gonads of free-living western fence lizards, *Sceloporus occidentalis*. Copeia 1973:663–668.

STICKEL, L. F. 1950. Populations and home range relationships of the box turtle, *Terrapene c. carolina* (Linnaeus). Ecol. Monogr. 20: 351–358.

STIEHL, R. B. 1983. A new attachment method for patagial tags. J. Field Ornithol. 54:326–328.

STILLE, W. T. 1950. The loss of jaw tags by toads. Chicago Nat. Hist. Mus., Nat. Hist. Misc. Publ. 74. 2pp.

STIRLING, I. 1979. Tagging Weddell and fur seals and some general comments on long-term marking studies. Pages 13–14 *in* L. Hobbs and P. Russell, eds. Report on the pinniped tagging workshop, 18–19 January 1979, Seattle, Wash.

STRONG, P. I. V., S. A. LaVALLEY, AND R. C. BURKE, II. 1987. A colored plastic leg band for common loons. J. Field Ornithol. 58: 218–221.

SUGDEN, L. G., AND H. L. POSTON. 1968. A nasal marker for ducks. J. Wildl. Manage. 32:984–986.

SUMMERS, C. F., AND S. R. WITTHAMES. 1978. The value of tagging as a marking technique for seals. Pages 63–70 *in* B. Stonehouse, ed. Animal marking: recognition marking of animals in research. The MacMillan Press Ltd., London, U.K.

SWANK, W. G. 1952. Trapping and marking of adult nesting doves. J. Wildl. Manage. 16:87–90.

SWEENEY, T. M., J. D. FRASER, AND J. S. COLEMAN. 1985. Further evaluation of marking methods for black and turkey vultures. J. Field Ornithol. 56:251–257.

SWINGLAND, I. R. 1978. Marking reptiles. Pages 119–132 *in* R. Stonehouse, ed. Animal marking. Univ. Park Press, Baltimore, Md.

SZYMCZAK, M. R., AND J. K. RINGELMAN. 1986. Differential habitat use of patagial-tagged female mallards. J. Field Ornithol. 57:230–232.

TABER, C. A., R. F. WILKINSON, AND M. S. TOPPING. 1975. Age and growth of hellbenders in the Niangua River, Missouri. Copeia 1975: 633–639.

TABER, R. D. 1949. A new marker for game birds. J. Wildl. Manage. 13:228–231.

———, AND I. McT. COWAN. 1963. Capturing and marking wild animals. Pages 250–283 *in* H. S. Mosby, ed. Wildlife investigational techniques. Second ed. Edwards Brothers, Inc., Ann Arbor, Mich.

TACHA, T. C. 1979. Effects of capture and color markers on behavior of sandhill cranes. Pages 177–179 *in* J. C. Lewis, ed. Proc. 1978 Crane Workshop. Colorado State Univ. Printing Serv., Ft. Collins.

TAMARIN, R. H., M. SHERIDAN, AND C. K. LEVY. 1983. Determining matrilineal kinship in natural populations of rodents using radionuclides. Can. J. Zool. 61:271–274.

TAYLOR, K. D., AND R. J. QUY. 1973. Marking systems for the study of rat movements. Mammal. Rev. 3:30–34.

TAYLOR, M., AND J. LEE. 1994. Tetracycline as a biomarker for polar bears. Wildl. Soc. Bull. (In press).

TAYLOR, R. H. 1969. Self-attaching collars for marking red deer in New Zealand. Deer 1:404–407.

THOMAS, A. E. 1975. Marking anurans with silver nitrate. Herpetol. Rev. 6:12.

THOMAS, J. A., L. H. CORNELL, B. E. JOSEPH, T. D. WILLIAMS, AND S. DREISCHMAN. 1987. An implanted transponder chip used as a tag for sea otters (*Enhydra lutris*). Mar. Mammal Sci. 3:271–274.

THOMAS, J. W., AND R. G. MARBURGER. 1964. Colored leg markers for wild turkeys. J. Wildl. Manage. 28:552–555.

THOMAS, R. A. 1977. Selected bibliography of certain vertebrate techniques. U.S. Bur. Land Manage. Tech. Note, Denver, Colo. 88pp.

THOMPSON, S.D. 1982. Microhabitat utilization and foraging behavior of bipedal and quadrupedal heteromyid rodents. Ecology 63:1303–1312.

TILLEY, S. G. 1977. Studies of life histories and reproduction in North American plethodontid salamanders. Pages 1–41 *in* D. H. Taylor and S. I. Guttman, eds. The reproductive biology of amphibians. Plenum Press, New York, N.Y.

———. 1980. Life histories and comparative demography of two salamander populations. Copeia 1980:806–821.

TINKLE, D. W. 1967. The life and demography of the side-blotched lizard, *Uta stansburiana*. Univ. Michigan Mus. Zool. Publ. 132. 182pp.

———. 1973. A population analysis of the sagebrush lizard, *Sceloporus graciosus* in southern Utah. Copeia 1973:284–296.

TOMILIN, A. G., Y. I. BLIZNYUK, AND A. V. ZANIN. 1983. A new method for marking small cetaceans. Int. Whaling Commn. Rep. Comm. 33:643–645.

TRAVIS, J. 1981. The effect of staining on the growth of *Hyla gratiosa* tadpoles. Copeia 1981:193–196.

TURNER, F. B. 1960. Population structure and dynamics of the western spotted frog, *Rana p. pretiosa* Baird and Girard, in Yellowstone Park, Wyoming. Ecol. Monogr. 30:251–278.

TURNER, J. C. 1982. A modified Cap-Chur dart and dye evaluation for marking desert sheep. J. Wildl. Manage. 46:553–557.

TWITTY, V. C. 1966. Of scientists and salamanders. W. H. Freeman Co., San Francisco, Calif. 178pp.

UNDERHILL, L., AND J. HOFMEYER. 1987. Experience with colour-dyed common terns. Safring News 16:29–30.

U.S. FISH AND WILDLIFE SERVICE AND CANADIAN WILDLIFE SERVICE. 1986. North American bird banding manual. Vol. II. U.S. Dep. Inter., Washington, D.C. Var. pagin.

———. 1991. North American bird banding manual. Vol. I. U.S. Dep. Inter., Washington, D.C. Var. pagin.

VERME, L. J. 1962. An automatic tagging device for deer. J. Wildl. Manage. 26:387–392.

VILJOEN, P. J. 1986. A plastic tail collar for marking wild elephants. South Afr. J. Wildl. Res. 16:158–159.

VINEGAR, M. B. 1975. Life history phenomena in two populations of the lizard *Sceloporous undulatus* in southwestern New Mexico. Am. Midl. Nat. 93:388–402.

WADKINS, L. A. 1948. Dyeing birds for identification. J. Wildl. Manage. 12:388–391.

WALLACE, M. P., P. G. PARKER, AND S. A. TEMPLE. 1980. An evaluation of patagial markers for cathartid vultures. J. Field Ornithol. 51:309–314.

WARD, F. P., C. J. HOHMANN, J. F. ULRICH, AND S. E. HILL. 1976. Seasonal microhabitat selections of spotted turtles (*Clemmys guttata*) in Maryland elucidated by radioisotope tracking. Herpetologica 32:60–64.

WARNEKE, B. M. 1979. Marking of Australian fur seals, 1966–1977. Pages 7–8 *in* L. Hobbs and P. Russell, eds. Report on the pinniped tagging workshop, 18–19 January 1979, Seattle, Wash.

WATKINS, W. A., AND W. A. SCHEVILL. 1976. Underwater paint marking of porpoises. Fish. Bull. 74:687–689.

WEARY, G. C. 1969. An improved method of marking snakes. Copeia 1969:854–855.

WELLS, K. D., AND R. A. WELLS. 1976. Patterns of movement in a population of the slimy salamander, *Plethodon glutinosus*, with observations on aggregations. Herpetologica 32:156–162.

WHEELER, R. H., AND J. C. LEWIS. 1972. Trapping techniques for sandhill crane studies in the Platte River Valley. U.S. Fish Wildl. Serv. Resour. Publ. 107. 19pp.

WHITE, M. J., JR., J. G. JENNINGS, W. F. GANDY, AND L. H. CORNELL. 1981. An evaluation of tagging, marking, and tattooing techniques for small dolphinids. NOAA Tech. Memo. NMFS 16:1–142.

WHITE, S. B., T. A. BOOKHOUT, AND E. K. BOLLINGER. 1980. Use of human hair bleach to mark blackbirds and starlings. J. Field Ornithol. 51:6–9.

WHITFORD, W. G., AND M. MASSEY. 1970. Responses of a population of *Ambystoma tigrinum* to thermal and oxygen gradients. Herpetologica 26:372–376.

WILKINSON, G. S. 1985. The social organization of the common vampire bat. I. Pattern and cause of association. Behav. Ecol. Sociobiol. 17:111–121.

WILLIAMS, T. C., J. M. WILLIAMS, AND D. R. GRIFFIN. 1966. The homing ability of the neotropical bat, *Phyllostomus hastatus*, with evidence for visual orientation. Anim. Behav. 14:468–473.

WILLSTEED, P. M., AND P. M. FETTEROLF. 1986. A new technique for individually marking gull chicks. J. Field Ornithol. 57:310–313.

WOLCOTT, T. G. 1977. Optical tracking and telemetry for nocturnal field studies. J. Wildl. Manage. 41:309–312.

WOODBURY, A. M. 1956. Uses of marking animals in ecological studies: marking amphibians and reptiles. Ecology 37:670–674.

———, AND R. HARDY. 1948. Studies of the desert tortoise, *Gopherus agassizii*. Ecol. Monogr. 18:145–200.

WOOLLEY, H. P. 1973. Subcutaneous acrylic polymer injections as a marking technique for amphibians. Copeia 1973:340–341.

WOOLLEY, P. 1962. A method of marking salamanders. Mo. Speleol. 4:69–70.

WRIGHT, E. G. 1939. Marking birds by imping feathers. J. Wildl. Manage. 3:238–239.

WÜRSIG, B., AND M. WÜRSIG. 1977. The photographic determination of group size, composition, and stability of coastal porpoises (*Tursiops truncatus*). Science 198:755–756.

YAGI, T., M. NISHIWAKI, AND M. NAKAJIMA. 1963. A preliminary study on the method of time marking with lead salt and tetracycline on the teeth of fur seals. Sci. Rep. Whales Res. Inst. 7:191–195.

YOUNG, J. B. 1941. Unusual behavior of a banded cardinal. Wilson Bull. 53:197–198.

YOUNG, L. S., AND M. N. KOCHERT. 1987. Marking techniques. Natl. Wildl. Fed. Sci. Tech. Ser. 10:125–156.

ZICUS, M. C., AND R. M. PACE, III. 1986. Neckband retention in Canada geese. Wildl. Soc. Bull. 14:388–391.

———, D. F. SCHULTZ, AND J. A. COOPER. 1983. Canada goose mortality from neckband icing. Wildl. Soc. Bull. 11:286–290.

ZMUD, M. E. 1985. Marking of the redshank *Tringa totanus* in the north-western Pricernomorije. The Ring 11(122-123):7–15.

ZWICKEL, F. C., AND A. ALLISON. 1983. A back marker for individual identification of small lizards. Herpetol. Rev. 14:82.

Appendix I. Standardized nomenclature and procedure for keeping records of all banded migratory birds.

New Birds—birds newly banded.

Returns—birds recaptured at the original banding station ≥90 days from the time of banding or last date of recapture.

Recoveries—birds captured by someone else at another location.

Repeats—birds recaptured at the same banding station where they were banded <90 days before; includes ''repeat'' birds originally banded at a different station.

Experimentals—birds held for more than a few hours after removal from the trap, released other than in the immediate vicinity of capture, marked in another fashion than legbands, or subject to anything more than being immediately released at the point of capture.

Hand-Reared—birds reared from eggs hatched in an incubator or under a setting hen, reared by domestic stock, reared at any stage in captivity, or reared by held, clipped, or pinioned parents.

Sick or Injured—birds not in good normal condition when obtained for banding; includes birds banded after recovery from sickness, such as botulism.

Wild Birds—healthy, normal birds raised in the wild.

L (local birds)—young birds, incapable of sustained flight.

HY (hatching year)—a bird capable of sustained flight and known to have hatched during the calendar year in which it was banded.

AHY (after hatching year)—a bird known to have hatched before the calendar year of banding with the year of hatch otherwise unknown.

SY (second year)—a bird known to have hatched in the calendar year preceding the year of banding and in its second calendar year of life.

ASY (after second year)—a bird known to have hatched in the calendar year preceding the year of banding and in its second calendar year of life.

TY (third year)—a bird known to have hatched in the calendar year preceding the year before the year of banding, and now in the third calendar year of life.

ATY (after third year)—a bird in at least its fourth calendar year of life.

8

CRITERIA OF SEX AND AGE

Ralph W. Dimmick and Michael R. Pelton

INTRODUCTION

Ascertaining the sex and age of an individual animal is the first step toward defining the sex ratio and age structure of a species' population. These characteristics may provide important insight into a population's recent history, current status, and likely immediate future trend. Being able to identify the sex and age of a bird or mammal is an interesting and desirable skill in its own right, but for a population ecologist, animal behaviorist, or wildlife manager, it often is essential.

The purposes of this chapter are to: (1) define the basic terminology associated with sexing and aging wildlife, (2) describe broad approaches to identifying sex and age classes of wildlife, and (3) delineate specific sex and age characteristics for species or groups of species of North American birds and mammals.

Many different techniques are used to distinguish sex and age among the various species of wildlife. Some of these are widely known and easily applied. Adult male deer, for example, typically display antlers; females do not. Breeding male birds typically are adorned with brightly colored plumage and exhibit sexual displays such as singing, dancing, strutting, or drumming. Their female counterparts are cryptically colored and generally more secretive. Beyond these obvious examples, however, external criteria of sex or age may be obscure or impossible to detect under most conditions in the field. This is especially true for sexually monomorphic species of birds, for immature individuals of most species of birds and mammals, and for mammals lacking obvious secondary sex characters such as antlers or horns. The presence of external genitalia permits easy sexing of most, but not all, mammals in hand. Birds lack external genitalia, but some, waterfowl for example, can be sexed by examining the cloaca.

Competent biologists can verify the sex of a bird or mammal by dissection and subsequent examination of the genitalia. However, determining the sex or age of living individuals in hand or in the field often requires knowledge specific to that species; in fact, assessing the age of animals in hand, live or dead, is frequently quite difficult even within broad categories. Furthermore, biologists often have available only portions of animals (e.g., wings or jawbones) rather than entire specimens.

In this chapter we provide information on sexing and aging birds and mammals that is not generally provided in field guides and manuals. Ideally, sex and age criteria will permit a biologist to:

1. Assign the individual animal to a sex or age group with a minimal amount of subjectivity.
2. Segregate sexes or age classes with little or no overlap.

Additionally, the most useful criteria are those that can be used alone, rather than in combination, and that can be identified on parts of animals easily stored and willingly contributed by hunters (e.g., wings, tails, jawbones, and premolars). A wealth of biological information can be obtained from materials contributed by hunters, but few are willing to part with edible portions or trophies.

SEX AND AGE CHARACTERISTICS OF BIRDS
Subcellular Techniques for Aging or Sexing

Genetic sexing of birds can be accomplished by chromosome analysis of cells from blood feathers. Van Turner and Valentine (1983) reported this technique is 100% reliable at any age, provided the sex chromosomes are successfully located. Specimens of white-naped cranes and whooping cranes were easily sexed from culture of single feathers. The process uses blood from rapidly dividing

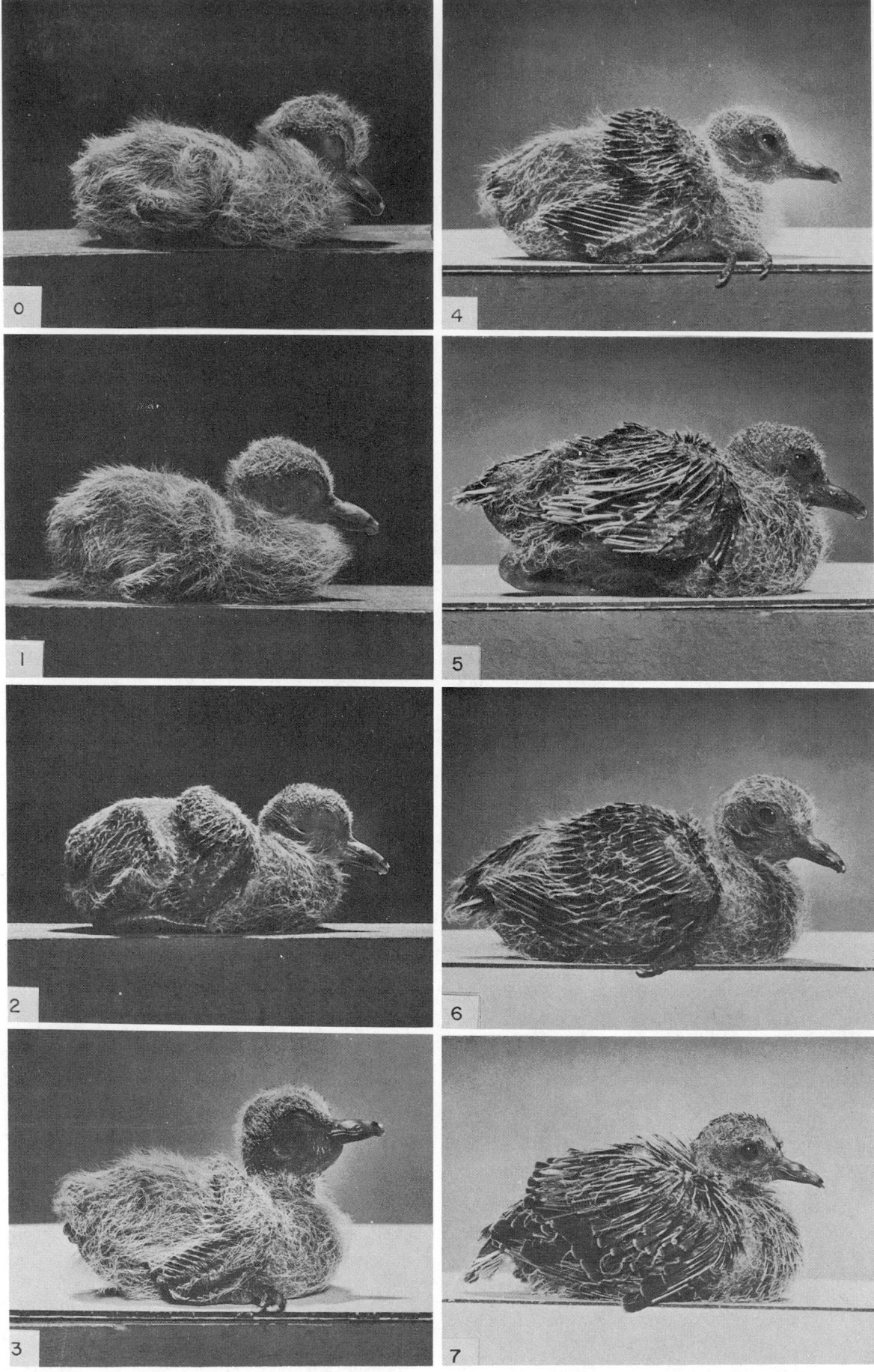

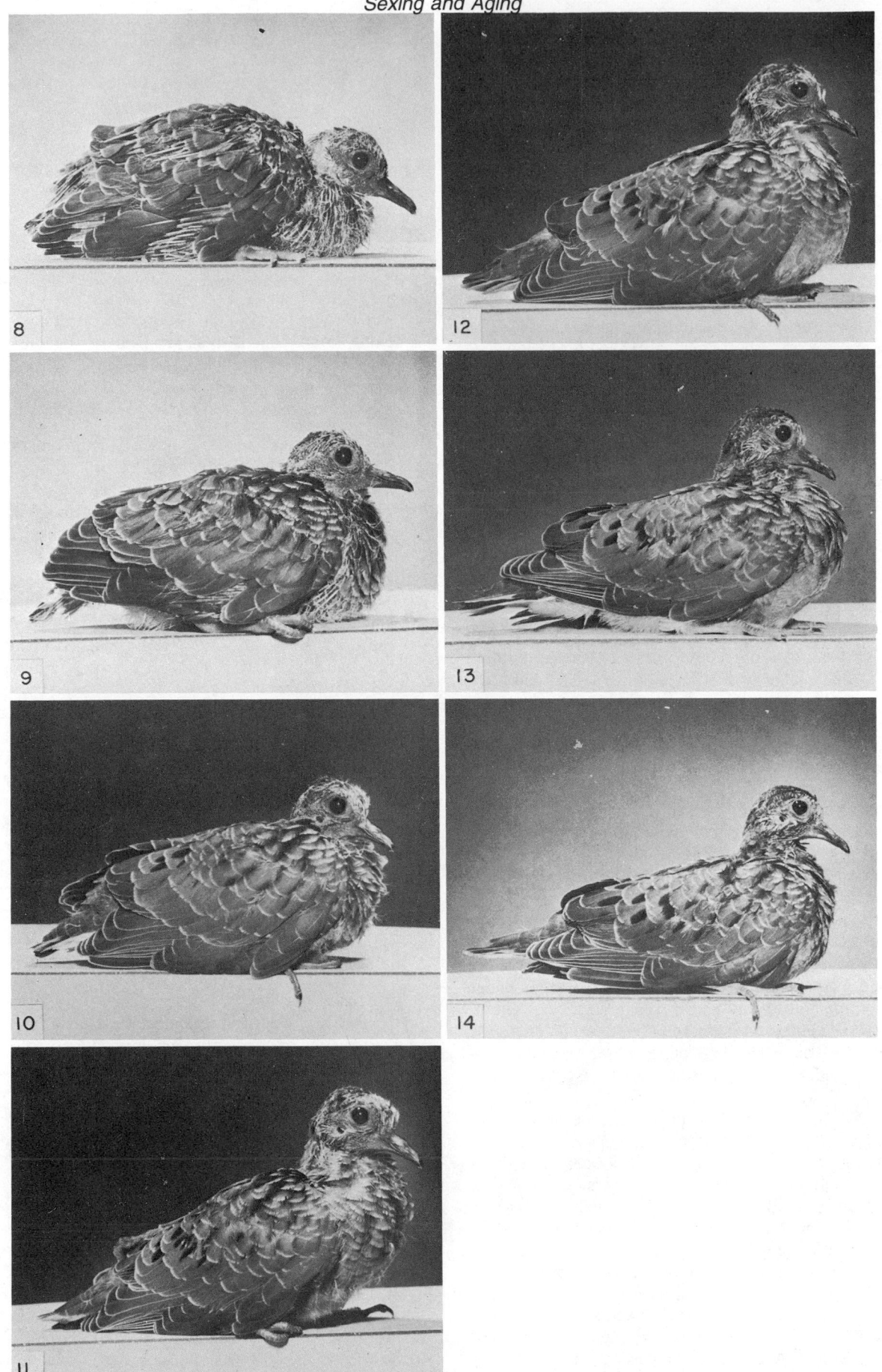

Fig. 1. Daily growth stages of nestling mourning doves (from Hanson and Kossack 1963). Ages in days are indicated by numerals.

cells, characteristic of tissue obtained from developing feathers. Cells are cultured, stained, and examined to identify the sex of the bird based upon characteristics of the sex chromosomes.

Biochemical analysis of collagen from bird tendons shows promise for aging birds. In metabolically inert proteins of living systems, the originally produced L-amino acid racemizes to the D-amino acid over time. Hunter (1989) measured D/L aspartic acid ratios in tendon collagen of brown pelicans and eastern bluebirds. D/L ratios increased significantly with age in both species.

Subcellular examinations for sex or age typically require sophisticated equipment and techniques to be successful. They offer significant promise, particularly for species that are strongly monomorphic and for which noninvasive techniques are necessary.

Embryonic Development

The degree of development of bird embryos prior to hatching varies among taxonomic groups. The young of doves and pigeons, raptors, and many songbirds are weakly feathered, blind, and helpless at hatching, defined by the term *altricial.* Gallinaceous birds, waterfowl, shorebirds, and cranes are covered with down at hatching, have open eyes, and have neurological systems mature enough to permit strong locomotion. These young are described as *precocial.*

Incubation period is typically shorter for altricial young. Stages in the embryonic development of the mourning dove were presented by Muller et al. (1984). Total incubation period for mourning doves is 14 days. Northern bobwhites at 14–15 days of a 23-day incubation period and wild turkeys at 18 days of a 26-day incubation period are near the same stage of embryonic development reached by mourning doves at the time of hatching. Roseberry and Klimstra (1965) and Stoll and Clay (1975) illustrated the daily changes in embryos of bobwhites and wild turkeys, respectively. Precocial young are prepared to leave the nest within a few hours after hatching.

Postnatal Development

Altricial young remain in the nest until fledging. This is a time of comparatively rapid neuromuscular development and feather growth. Mourning dove nestlings fledge at approximately 14 days, at which time they resemble adults superficially and are soon to become independent. Hanson and Kossack (1963) provided an excellent photographic guide to aging nestling mourning doves (Fig. 1).

Precocial young of some species can be aged in the field and in hand by the pattern of replacement of down with contour feathers. This pattern has been well documented for young waterfowl. The general sequence for dabbling ducks is illustrated in Fig. 2 (Bellrose 1980:27, from Gollop and Marshall 1954). The specific length of time required to complete each stage varies among species and with latitude and other environmental variables. In addition to plumage characters, size of young relative to the accompanying adult and the skill and strength of flight are useful in estimating approximate age. Williams and Austin (1988) provided a detailed description of age characteristics of turkey poults; similar descriptions have been published for northern bobwhites (Stoddard 1931), ruffed

grouse (Bump et al. 1947), and blue grouse (Smith and Buss 1963).

Waterfowl

GENERAL CHARACTERISTICS

The North American waterfowl fauna comprises about 45 species of swans, geese, and ducks, some of which have two or more recognized races. An additional five or more European or Asian species rarely or occasionally visit the continent. The mute swan is an exotic species that has established feral populations in a few locations in the United States.

Waterfowl pass through a sequence of plumages from hatching to adulthood, then most ducks (but not swans, geese, and whistling-ducks) cycle through two distinct plumages annually. Bellrose's (1980) thorough discussion of these plumages is summarized as follows:

Natal—Newly hatched waterfowl are completely covered with downy feathers, the follicles of which will later produce the definitive contour feather.

Juvenal—In 2.5–16 weeks, the natal down feathers are replaced by juvenal contour and flight feathers. Juvenal plumage is complete and flight is attained as soon as 6 weeks post-hatching in teals and up to 14–16 weeks in swans. Downy tail feathers of the natal plumage are attached to the ends of the juvenal tail feathers (Fig. 3). These tips break off, creating the notched tips that are useful as a criterion for aging waterfowl during autumn and early winter.

Immature—This short-lived plumage, difficult to detect, closely resembles the juvenal plumage, but males of some species may clearly show plumage characters intermediate between juvenal and adult. Wing feathers are not shed, but the notched juvenal tail feathers are replaced with pointed or rounded "adult" feathers, limiting the time this character can be used for aging.

Adult—Geese, swans, and whistling-ducks have only one plumage per year. It is their *definitive* plumage. Ducks have at least two body plumages annually that may differ markedly in color by season and by sex (oldsquaws have three). The *nuptial* (also called *breeding* or *alternate*) plumage is the definitive plumage for both sexes of ducks. Most ducks display this plumage for most of the annual cycle. Among males, the *eclipse* (also called *nonbreeding* or *basic*) plumage immediately follows the breeding season. It is markedly different from the nuptial plumage of sexually dimorphic ducks, the males resembling females during this period. Females also undergo a post-nuptial (eclipse) molt, but color differences between the breeding and nonbreeding plumage are not pronounced.

Sex Determination

In nuptial plumage, all North American ducks except the black duck, mottled duck, Florida duck, and New Mexican duck are strongly to moderately sexually dimorphic. In post-nuptial plumage and in immature plumages, males of sexually dimorphic species closely resemble females. Consequently, sex of most ducks can be readily distinguished by plumage characters for birds in hand and, usually, in the field. We refer the reader to Bellrose (1980)

Adult Dabbling Duck

Class I A, 1-7 days of age

Class I B, 8-13 days of age

Class I C, 14-18 days of age

Class II A, 19-27 days of age

Class II B, 28-36 days of age

Class II C, 37-42 days of age

Class III, 43-55 days of age, flying at 56-60 days

Fig. 2. Plumage development of young waterfowl (modified from Bellrose 1980:27, after Gollop and Marshall 1954).

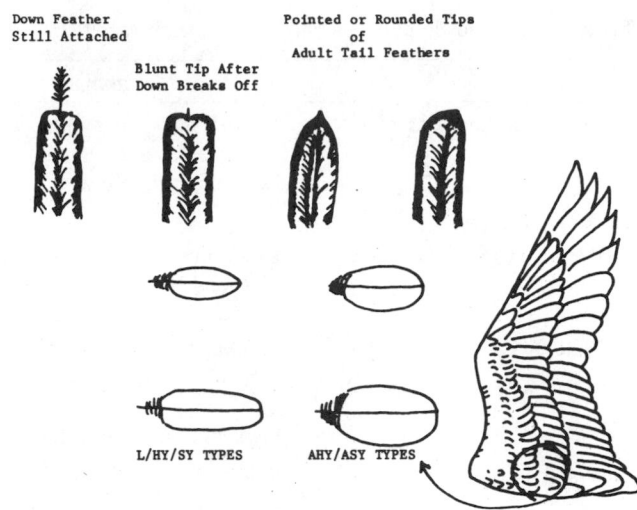

Down Feather Still Attached

Blunt Tip After Down Breaks Off

Pointed or Rounded Tips of Adult Tail Feathers

L/HY/SY TYPES

AHY/ASY TYPES

Fig. 3. Tail feather and wing feather criteria for distinguishing among age classes of North American ducks (from U.S. Fish and Wildlife Service and Canadian Wildlife Service 1977).

or to any one of several field guides for sex characteristics of individual duck species.

Plumages of all swans, geese, and whistling-ducks are monomorphic. Sex generally should be verified by cloacal examination (Fig. 4). Color of the bill and feet can be used to sex living adult black ducks, but colors of all fleshy parts fade after death and are not reliable for sexing dead birds.

Age Determination

Plumage characteristics can be used to place waterfowl into hatching year (HY) or after hatching year (AHY) age classes (Fig. 3). Species exhibiting subadult plumage in their second year may be further classified into an "after second year" age group (ASY). Tail feathers of HY birds typically are notched or have downy plumes attached to the tips of the shafts; comparable feathers of AHY birds are rounded or pointed, lacking notches or downy tips (Fig. 3). The length of time the notched tail feathers are retained varies among species, and even within species, as documented for Canada geese by Hanson (1962); this character is usually valid until late autumn or winter. If only a wing is available, age may be discerned by shape of the greater and middle tertial coverts. HY birds, and sometimes second year (SY) immatures, display coverts that are narrow and/or pointed (Fig. 3). AHY and ASY birds have broad coverts with rounded tips. Waterfowl that require more than 1 year to attain sexual maturity frequently display entire or partial juvenal plumage during

their first year of life. For some late-maturing species (e.g., common eider), adult breeding plumage may not be attained for 2 or more years.

Depth of the bursa of Fabricius decreases with advancing age and closes at sexual maturity (Fig. 4). The bursa can be measured through the cloacal aperture with a probe. For species with delayed sexual maturity (e.g., Canada geese), bursal depth measurements in combination with other characteristics can permit separation of yearlings from older geese (Hanson 1962). Among females, a membrane occludes the opening of the oviduct in immatures. Among males, the penis enlarges and darkens with

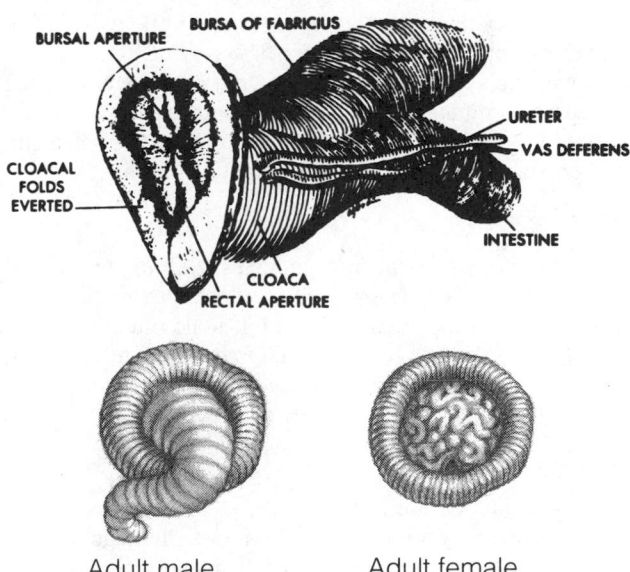

BURSAL APERTURE

BURSA OF FABRICIUS

URETER

VAS DEFERENS

CLOACAL FOLDS EVERTED

INTESTINE

CLOACA RECTAL APERTURE

Adult male

Adult female

The cloaca of a juvenile goose will appear much like that of an adult female. A male will have a small, dark, fleshy projection in the 8 o'clock position.

Fig. 4. Cloacal structures useful for aging and sexing waterfowl (top, from Godin 1960; bottom, from Moser et al. n.d.).

Table 1. Diagnostic plumage and appendage characteristics of immature and adult geese (after Bellrose 1980).

Species	Immature[a]	Adult
White-fronted goose	Grayish-brown body plumage, lacking white face patch and blotches on belly Legs and bill yellow	White face patch, black and brown blotches on belly Legs orange, bill pink
Lesser snow goose		
Blue phase	Drab, brownish-gray head and body plumage Legs and bill grayish brown	Slate gray body plumage, white head plumage Legs rose-red, bill pink
White phase	Head and dorsal plumage sooty gray, lower plumage mostly white Black wing tips Legs and bill grayish-brown	Plumage completely white, except black wing tips Legs rose-red, bill pink
Greater snow goose	Similar to white phase of lesser snow goose	Similar to white phase of lesser snow goose
Ross' goose	Body plumage pale gray Black wing tips	Body plumage snowy white Black wing tips
Emperor goose	Head and neck black-brown, becoming white, flecked with dark feathers in late Oct. Legs and bill black	Head and upper neck white Legs yellow, bill pink
Canada goose	Plumages alike in all age classes	Plumages alike in all age classes
Brant		
Atlantic race	No white on neck until midwinter Greater and middle wing coverts tipped with white	White crescent each side of neck Greater and middle wing coverts uniformly dark brown
Black race	Body plumage entirely dark, except white undertail coverts Light gray edging of wing coverts	Sides and flanks strongly barred gray and white Wing coverts uniformly dark

[a]Notched tail feathers characterize immatures of all species, but for varying periods of time within and among species.

age. Size and color of this organ can be used to age waterfowl up to sexual maturity. For early maturing species, such as many ducks, this criterion may not be useful after November or December of their first year of life. For species with delayed sexual maturity, the penis may be used to separate immatures, yearlings, and other adults (Hanson 1962). The cloacal sphincter muscle increases in size and darkens in color with increasing age in both sexes.

SWANS

Plumage is identical in both sexes. Sex of tundra and trumpeter swans is determined by cloacal examination in living or fresh specimens. The adult male mute swan has a fleshy knob on the forehead that is more prominent than the knob on the female. Plumage of immature swans is dull mouse-gray, adult plumage is snowy white.

GEESE

Plumage is identical in both sexes of all species. Sex is determined by cloacal examination. Plumage of immature geese may be similar to or distinctly different from that of adults (Table 1).

WHISTLING-DUCKS

Plumage is similar in males and females, but coloration of adult females is duller than that of males in black-bellied and fulvous whistling-ducks. Plumage of immature whistling-ducks is duller than plumage of adults. Juvenile

wing pattern is retained until after first breeding season (Bellrose 1980).

OTHER DUCKS

The strong to moderate sexual dimorphism of the adult body plumage of most species permits ready identification of the sex of birds in hand and in the field. For birds in hand, the presence or absence of a penis verifies sex for birds of questionable plumage (some juveniles, eclipse plumage). The U.S. Fish and Wildlife Service and Canadian Wildlife Service (1977) provided a useful key to sex and age characters of several important duck species. An example from the detailed manual, American green-winged teal, is presented in Fig. 5.

Particularly significant to North American waterfowl management is the annual analysis of sex and age ratios based upon samples of duck wings contributed by waterfowl hunters. Characters of sex and age based entirely upon wing plumage were described by Carney (1992). An example from this report is provided in Fig. 6.

Information provided by these two sources is too voluminous to reprise in this text. Professionals actively engaged in waterfowl management or bird banding may obtain the complete references from the U.S. Fish and Wildlife Service, Office of Migratory Bird Management, Laurel, MD 20811. Bellrose (1980) provided descriptions and color photos with helpful clues to sex and age characters of all North American waterfowl.

KEY TO AGE AND SEX: *Caution* locals can be sexed only by cloacal examination unless tertials are present.

1A   Stripe on most distal tertial black and sharply delineated from basic feather colour; penis present . . . . Male (see2)

1B   Stripe on most distal tertial blackish to brownish, grading into basic feather colour, penis absent . . . . Female (see 3)

      2A(1)   Greater tertial coverts long and narrow with fine light edging, dull and faded . . . . HY/SY

      2B   Greater tertial converts tapering to blunt point, sometimes with a narrow buffy edging, uniform gray . . . . AHY/ASY

3A(1)   Tertials with frayed tips; tertial coverts narrow; primary coverts heavily light-edged . . . . HY/SY

3B   Tertials with unfrayed tips; tertial coverts broadly rounded; primary coverts unedged to faintly light-edged . . . . AHY/ASY

SIMILAR SPECIES: *Blue-winged Teal* have bright blue wing patches; *Cinnamon* are reddish in colour.

MOLTS: Post-juvenal partial (except some wing feathers), Sep.-Dec.; pre-nuptial complete, Sep.-Mar.; post-nuptial partial (body, scapulars only), Jun.-Aug.

INCUBATION: 21-23 days. FLYING YOUNG: 35-44 days. BANDING: ? days.

REFERENCES: Carney 1964. FWS. SSR. No. 82.

USUALLY ACCEPTABLE AGE-SEX CODES BY MONTH

Age-Sex	JAN	FEB	MAR	APR	MAY	JUN	JUL	AUG	SEP	OCT	NOV	DEC
L-U/M/F												
HYU/M/F												
SY-M/F												
AHY-M/F												
ASY-M/F												

American Green-winged Teal codes.

1/   Species treated in this report: mallard, black duck, American wigeon, green-winged teal, blue-winged teal, cinnamon teal, northern shoveller, pintail, wooduck, redhead, canvasback, greater scaup, common goldeneye, Barrow's goldeneye, and bufflehead.

2/   Varying amounts of information on plumage characters for aging of other species is provided in narrative by Bellrose (1980).

Fig. 5. Example of a key to sex and age criteria for a species of duck (green-winged teal) (U.S. Fish and Wildlife Service and Canadian Wildlife Service 1977).

Gallinaceous Birds

GENERAL CHARACTERISTICS

Sixteen native game birds and three introduced exotics comprise the gallinaceous bird fauna of North America. Native species include the wild turkey, six species of quail, six of grouse, and three of ptarmigan. Ring-necked pheasants, gray partridge, and chukar are well-established exotic species.

Definitive plumages range from strongly sexually dimorphic (e.g., ring-necked pheasant) to monomorphic (e.g., mountain quail, chukar). Secondary sexual characteristics, including plumage and soft parts, may become pronounced during the breeding season for some species (sage grouse) or not change for others (all species of quail).

Most gallinaceous birds can be identified as juveniles or adults by plumage characteristics. Primary flight feathers are molted sequentially, beginning with the proximal feather, P1, and progressing distally in a fairly regular time pattern. Typically, primaries 9 and 10 will be retained until after the first breeding season. Consequently, these two feathers might be worn, duller in color, and more pointed in juveniles than in adults. Among juveniles of many species, the pattern of replacement of primary feathers is sufficiently consistent to permit aging individuals in days or weeks up to the time primary 8 is completely grown. Primary coverts of most quail are retained through the first breeding season; shape and color of these

SURF SCOTER

All surf scoter wings are dark and unpatterned on both upper and under surfaces. Only adult males are black. Wings of all other sex-age categories are dark brown. Among these, adult females can be identified by their broadly rounded tertials and greater coverts over both secondaries and terials. On immature birds, tertials are pointed and usually frayed and faded at their tips. Greater coverts over both secondaries and tertials are quite narrow and have frayed and faded tips.

Wing Character	Male		Female	
	Adult	Immature	Immature	Adult
Primaries	Outermost primary similar to and as long as or longer than the adjacent primary			
	Outer webs black	Outer webs dark blackish brown		
Tertials	Shiny black and bluntly pointed; approximately 20 mm. longer than most secondaries	Dark brown and pointed; may be faded at their tips		Very dark blackish brown; tips bluntly pointed; usually less than 20 mm. longer than most secondaries
Tertial	shiny black	Dark brown; noticeably narrower than those of adults; often faded at their tips		Very dark blackish brown; smoothly rounded tips
Greater, middle, and lesser coverts	Entirely black; appear smooth	Dark brown; most greater coverts are faded at their tips; they often appear rough		Very dark blackish brown; some are slightly faded at their tips; all appear smooth

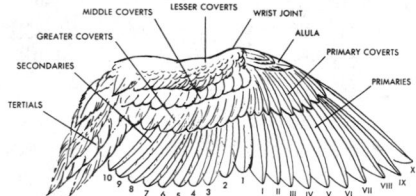

Fig. 6. Key for aging surf scoter wings (text based on Carney 1992).

feathers are excellent diagnostic characters of age (Petrides 1942). Typically, coverts of juveniles are slightly more pointed, often with lighter colored tips, and duller in overall color than coverts of adults (Fig. 7). The bursa of Fabricius is also a useful guide to age, but complete closure of the bursa might not occur in some species at the time sexual maturity is reached.

WILD TURKEY

Four races of wild turkeys occupy North America north of Mexico (Williams and Austin 1988). General characters of sex and age apply among these races, though the timing of specific molts and the rate of development of plumage and other characters might differ among populations (Healy and Nenno 1980, Williams and Austin 1988).

Differentiation between the sexes is apparent 10–14 weeks post-hatching (Williams and Austin 1988:79). In males caruncles of the neck begin to enlarge, and the head

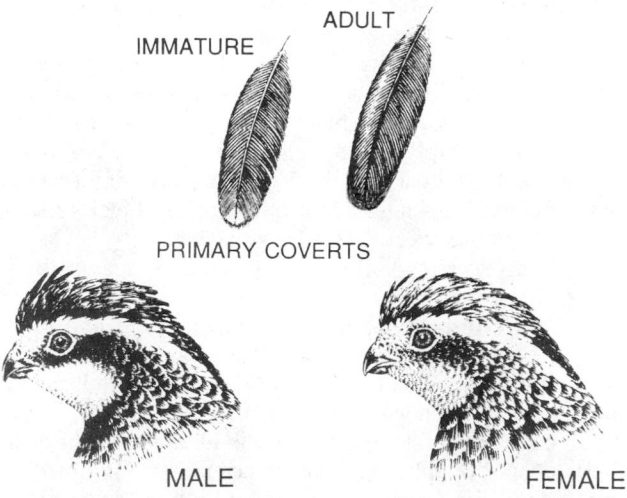

Fig. 7. Sex and age characters of northern bobwhites (Dimmick 1992).

and neck show fewer feathers than in females. The skin on the sides of the head is pink in males at 11–13 weeks; females of all ages lack pink skin on the sides of the head. The beard of the young male emerges from the skin as a cornified epidermal protuberance at 16–20 weeks. Among mature birds, the head, throat, and neck of males are bare to nearly bare, and the skin is predominantly reddish. The head, throat, and neck of females are moderately feathered, and the skin is gray to grayish-blue. The beard of males in their second autumn may reach >12 cm in length; the beard displayed by an occasional female seldom exceeds 7.6 cm (Edminster 1954:61–62). Wallin (1982) sexed juvenile, autumn-harvested wild turkeys in Vermont using the length of primary 10 measured from the rotator muscle to the tip. All birds with 10th primaries ≤22.9 cm were identified as females, and all birds with primaries >22.9 cm were classified males. Error rate was 1.8%, and about equal numbers of each sex were misclassified. Breast feathers of males are black-tipped; those of females are buff-tipped.

Three plumage characteristics are useful for distinguishing between juvenile and older wild turkeys during autumn and winter. The most pronounced character is the extended central three pairs of tail feathers of first-year birds versus the uniform length of all tail feathers of adults (Fig. 8). Both sexes display this trait, though it is more obvious in males. The greater upper secondary covert patch is narrower in first-year birds and is also duller in color (Fig. 8). This character is detectable in live turkeys at a distance (Williams and Austin 1988). Juvenal primary feathers 9 and 10 are typically retained into the first winter, though in the Florida race P9 may be molted early, and more than 5% will molt all 10 primaries (Williams and Austin 1970). Juvenal primaries are pointed, lack barring in the distal portion, and may be ragged and dull (Fig. 8). Spur length of gobblers can be a reliable indicator for distinguishing among year classes. Kelly (1975) reported that spur length was more highly correlated with age than any other single variable. Body weight was lighter and beard length was significantly shorter among first-year birds than older birds, but neither of these criteria permitted aging birds to year class after the first year.

NORTHERN BOBWHITE

Females have buffy chins, upper throats, and eyestripes (Fig. 7). These markings are white on males. Feathers of the middle wing coverts of females have wide, dull-gray bands, lacking distinct contrast (Thomas 1969). Middle wing coverts of males display fine, black, sharply pointed undulations, sharply contrasting with adjacent colors on the feathers. The base of the lower mandible of females is yellow, whereas in males it is uniformly black (distinguishable at 6–8 weeks) (Loveless 1958).

The upper greater primary coverts of immature bobwhites have buffy tips and are dull brown and tapered (Fig. 7). Corresponding feathers of adults are uniformly gray or gray-brown, shiny, and broadly rounded. The outer two primaries (P9 and P10) are pointed and dull brown in immatures, rounded and grayish in adults.

An estimate of age in days can be obtained for birds in the process of replacing juvenal primaries. This method relies upon the replacement and growth of primaries 1 through 8. It is valid to about 150 days post-hatching,

when P8 has been replaced and is fully grown (Petrides and Nestler 1952).

SCALED QUAIL

Sex of scaled quail is difficult to distinguish for birds in hand and virtually impossible for birds observed in the field. The most striking and consistent differences occur in plumage on the head and throat (Wallmo 1956). The plumage of the side of the face of females is streaked and dirty gray in color due to longitudinal black streaks on background color of gray or grayish-white. The side of the face of males is uniformly pearl gray except for a brownish ear patch. The throat of females is streaked; in males it is clear white behind the mandible, blending into a yellowish or buffy wash. These characters become evident in juveniles at about 17 weeks.

Immatures are characterized by primary coverts that are tipped, edged, or mottled with white. In adults, these are all uniformly gray. This character separates first-year birds up to about 1 year of life (Wallmo 1956).

GAMBEL'S QUAIL, CALIFORNIA QUAIL

Females have dark brown crests and lack black throats. Males have black crests and black throats.

Immatures of both species have mostly buff-tipped and pointed greater upper primary coverts, adults have uniformly gray, rounded coverts. The outer two primaries (P9 and P10) are more pointed and frayed in immatures, rounded in adults.

MOUNTAIN QUAIL

Females have shorter and browner plumes than males (Johnsgard 1975). Except for a population in Monterrey County, California, however, separation of sexes by plume length is sufficient for this character to be diagnostic of sex (Brennan and Block 1985). Brown color of the back extends to the top of the head of females; the hind neck of males is grayish-blue.

Immatures have buff-tipped primary coverts and pointed, frayed, outer primaries (P9 and P10). Adult primary coverts are uniformly gray; P9 and P10 are more rounded, not noticeably different from P1–P8.

HARLEQUIN QUAIL

Sexes are markedly different in plumage but similar in size. The head and neck of females are mottled brown and buff with whitish chin (Leopold 1959). The face and throat of males are boldly marked with a black and white pattern. Male head feathers are elongated, forming a broad, tan hood streaked with dark hues.

Greater upper primary coverts of immatures are edged with buff or barred near the base with buff. These feathers on adults are spotted with white (males) or barred with wide white markings (females) (Johnsgard 1973).

RING-NECKED PHEASANT

Adult and older juvenile pheasants are strongly sexually dimorphic. Distinguishing sex is relatively simple at 8 weeks or older; males exhibit brightly colored plumage and females present mottled shades of brown and buff coloration. However, the widespread popularity of this game bird, its extensive production in private and public game farms, and a pattern of hunting regulations prohib-

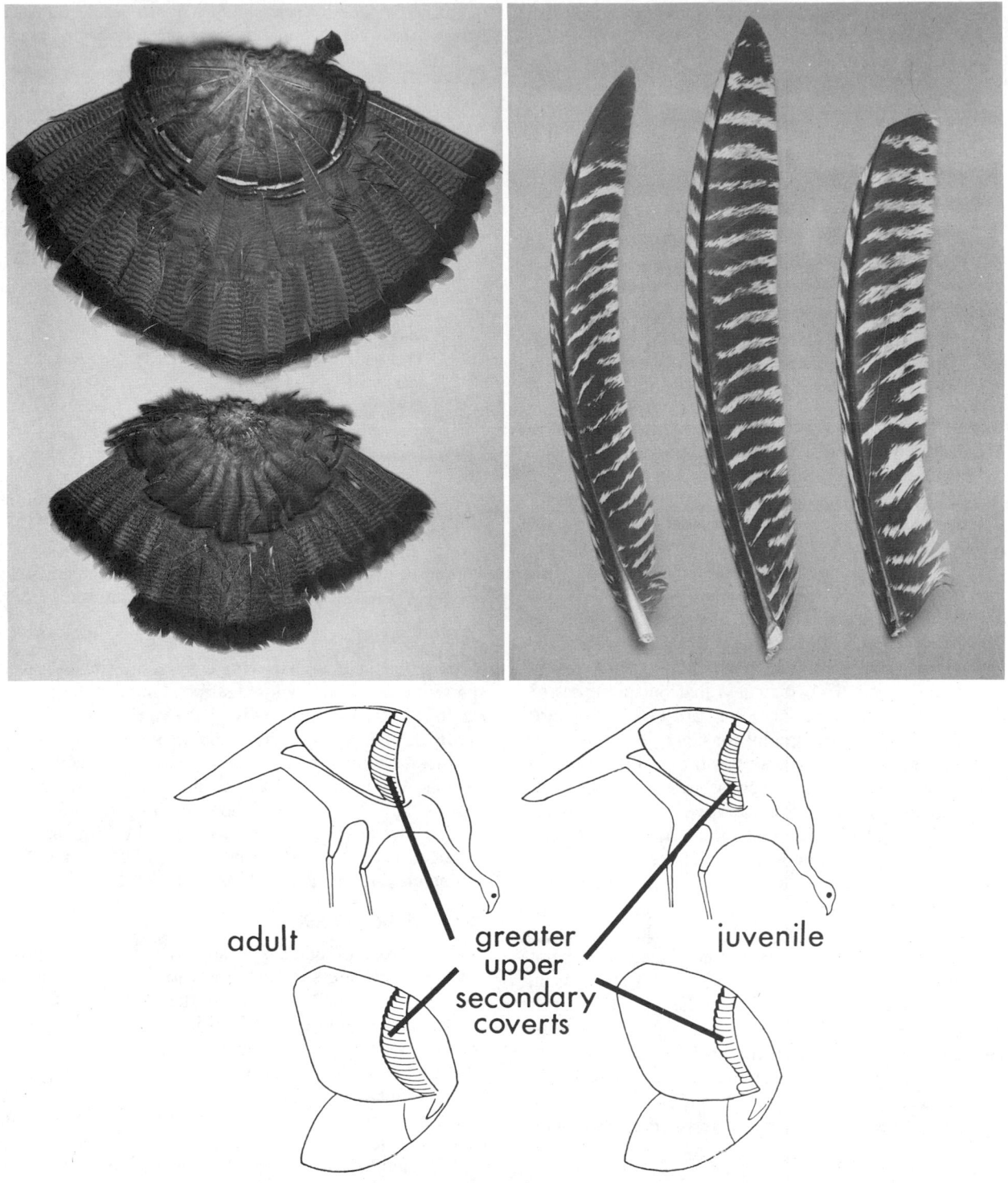

Fig. 8. Diagnostic plumage characteristics of adult and juvenile wild turkeys (based on Williams 1961, original report by Petrides 1942). Upper left: tail fans of adult (top) and juvenile. Upper right: outer primaries of juvenile (left) and adult (center and right). Blunt tip of right feather caused by dragging on ground during strut. Bottom: shape of secondary covert patch on folded wing.

iting or restricting the take of females have combined to generate research on a variety of criteria for determining age and sex of this species. Sex of day-old chicks can be determined quickly and accurately by the presence in males of an infantile wattle—a small flap of unfeathered papillary tissue just below the eye, partially hidden by natal down (Fig. 9, Woehler and Gates 1970). This tech-

nique produced 90% accuracy among males and 98% among females at 24–36 hours post-hatching.

Primary feathers are useful for sexing birds when only wings are available (Linder et al. 1971). Primaries from females typically show light-colored bars that meet the rachis at right angles along its entire length. Males typically exhibit no barring at the tips of primaries. Bars on

Fig. 9.   Heads of day-old pheasant chicks showing regions of maximum wattle development. Male chick on left, female on right (from Woehler and Gates 1970).

male primaries meet the rachis at a sharp angle, and the pattern is often diffuse. Accuracy of this technique was greater than 90% for all sex-age classes except wild males that had not completed the post-juvenal molt (63%). Dressed carcasses can be sexed by the presence of a spur on the leg of males and by plumage color on the head, providing these body parts are retained. When they are removed, the larger body size of males enables sexing the carcass with measurements of the breast. Oates et al. (1985) and Rodgers (1985) provided breast dimensions and methods for determining sex of dressed pheasants.

Depth of the bursa of Fabricius is a reliable method for separating juvenile from adult pheasants (Wishart 1969). Larson and Taber (1980) indicated that bursal depths of males ≤8 mm denote adults. Johnsgard (1975:106) also used 8 mm as a separation point for males, but noted that depths of adult female bursa were ≤6 mm. In contrast to most North American gallinaceous birds, pheasant wing feathers do not provide readily observed qualitative clues to age. Wishart (1969) separated adults from juveniles using combined measurements of shaft diameter and length of primary 1. His technique was useful in autumn and spring for pen-reared and wild Alberta pheasants. Greenberg et al. (1972) reported that P1 shaft diameter alone yielded a reliable and relatively simple separation point for Illinois pheasants. When adjusted for sex and season, P1 shaft diameter provided 90–98% reliability. Etter et al. (1970) provided criteria based on length of P10 for estimating the age in weeks of juvenile pheasants. Spur length (Stokes 1957), qualitative spur characters (Gates 1966), and eye-lens weight (Dahlgren et al. 1965) are not reliable techniques for separating juveniles from adult pheasants.

CHUKAR

Male and female chukars cannot be differentiated by qualitative plumage and structural characteristics. Combinations of various wing-feather measurements are necessary for identifying the sex of chukars (Table 2) (Weaver and Haskell 1968). Christensen (1970), however, noted significant bias favoring females when results obtained by using Weaver and Haskell's wing key were compared with results from internal sexing.

Juveniles less than about 14 weeks old possess mottled secondaries, whereas secondaries of older juveniles and adults lack mottling. Primary covert 9 measures <29 mm among juveniles throughout the first winter. This charac-

Table 2.   A key for determining age and sex of chukar partridge from wings, from mid-September through December (from Weaver and Haskell 1968).

1a.	Mottled secondaries absent	2
1b.	Mottled secondaries present	juvenile 5
2a.	Neither primary 9 nor 10 in stage of molt	3
2b.	Either 9 or 10 or both in stage of molt	adult 8
3a.	Upper primary covert 9 is <29 mm	4
3b.	Upper primary covert 9 is ≥29 mm	adult 8
4a.	Outer two primaries pointed at tips, only slightly faded, showing little wear	juvenile 5
4b.	Outer two primaries faded, showing wear	adult 8
5a.	Primary 3 is fully grown, is at least 4 mm longer than primary 2	6
5b.	Primary 3 is in stage of molt, not fully grown	7
6a.	Primary 3 is <135 mm	juvenile female
6b.	Primary 3 is ≥135 mm	juvenile male
7a.	Primary 1 is ≤119 mm	juvenile female
7b.	Primary 1 is >119 mm	juvenile male
8a.	Primary 3 is ≤136 mm	adult female
8b.	Primary 3 is >136 mm	adult male

teristic and pointed primaries 9 and 10 are reliable indicators of juveniles beyond the post-juvenal molt.

GRAY PARTRIDGE

Scapulars and median wing coverts of females typically present a wide, buff-colored stripe along the shaft and two to four buffy crossbars (Fig. 10) (McCabe and Hawkins 1946). Outer edges of the scapulars show vermiculations. Comparable male feathers lack crossbars, have a narrow, median, longitudinal stripe, and have vermiculations across the entire width of the scapulars.

The outer two primaries are pointed on immatures but rounded on adults. Also, the covert of P9 is pointed on immatures, rounded on adults (Petrides 1942).

RUFFED GROUSE

The sexes of adult and somatically mature juvenile ruffed grouse are similar but not identical in appearance. Females are smaller, have shorter ruffs and tails than males, and exhibit qualitative differences in markings on certain portions of their plumage. The species is widely distributed in North America and has many subspecies; it varies clinally in some of the qualitative measurements diagnostic of sex, and the reliability of qualitative plumage characteristics appears to differ among populations.

The number of whitish dots on the terminal ends of rump feathers is a reliable sex criterion for birds 13 weeks old or older; females possess only one dot, males two or three (Fig. 11) (Roussel and Ouellet 1975). In Quebec, only one female of 366 grouse sexed by this method was misclassified. Servello and Kirkpatrick (1986) correctly classified each of 62 southeastern ruffed grouse, and Kalla (1991) reported a 2.6% error rate for 235 Tennessee birds. Length of the plucked central tail feather is a widely used indicator of sex. Central rectrices <~15 cm characterize females, longer feathers denote males (Fig. 12) (Hale et al. 1954). More recent studies indicated that the separation point varies with geographic region and with age of the individual, more southerly populations tending toward

Fig. 10. Scapular feathers and wings from gray partridge. Note central stripe in feather from male on right and barring on feather from female (after McCabe and Hawkins 1946).

greater lengths (Uhlig 1953, Davis 1969, Servello and Kirkpatrick 1986). Using different separation points for juveniles and adults increases reliability. In Tennessee a separation point of 16.5 cm for adults and 15.5 cm for juveniles yielded 2.4% error for each group; a combined separation point of 16.0 cm yielded an error rate of 6.0% (n = 235) (Kalla 1991). Coloration of the eye patch and completeness of the tail band are somewhat ambiguous characters. Their use for adults typically results in an unacceptable number of unclassifiable birds and a high rate of misclassifications (Kalla 1991). Coloration of the eye patch permits distinction of the sexes from about 8 to 9 weeks until the spotted rump feathers appear. Females lack color in the eye patch, males display vivid or moderate reddish-orange color. Palmer (1959) reported 95% accuracy on live immature birds with this method.

Criteria for aging ruffed grouse are less reliable than sexing criteria, particularly beyond midwinter. Presence of a bursa is the most dependable indicator for immatures but is useful only until about January (Kalla 1991). Sharply pointed tips of primaries 9 and 10 are indicative of immatures (Fig. 13) (Hale et al. 1954), but the character declines in reliability as the season progresses (Kalla 1991). The usefulness of sheathing on the base of P8 and its absence on P9 and P10 for identifying immatures likewise decreases in reliability in late winter (Kalla 1991). The diameter of the calamus of P9 is smaller in immatures of both sexes and is a useful criterion of age among sexed birds beyond midwinter (Davis 1969). The ratio of P9:P8 calamus diameters provides some increase in reliability, immatures having a lower ratio than adults (Rodgers 1979).

BLUE GROUSE

Sexes are differentiated by white feathers tipped with bluish-black around the cervical sacs of males; among females, feathers of the cervical region are barred grayish-brown (Caswell 1954). This character is diagnostic as early as 6 weeks and often can be observed in the field

as well as on birds in hand. Wings of female blue grouse present a more mottled brown appearance than do male wings (Fig. 14) (Mussehl and Leik 1963). Female marginal coverts at the base of the alula have numerous blotches of brown mottling; male marginal coverts are gray and less mottled. This character is evident on adults and on juveniles 10 weeks or older (Hoffman 1985). Coloration and pattern of barring of upper tail coverts permit sexing birds as young as 6 weeks (Nietfeld and Zwickel 1983). Females show black to blackish-brown coverts with bold cinnamon or buffy brown crossbars; male coverts are black with gray flecking and have whitish-gray, narrow bars. Fewer than 2% of juveniles of the sooty race of blue grouse were misclassified or not classifiable by tail covert markings.

Primaries 9 and 10 are pointed on juveniles and rounded on adults. Hoffman (1985) provided tables for estimating age of juvenile blue grouse in weeks based on stage of development of post-juvenal primary feathers.

SPRUCE GROUSE

Spruce grouse exhibit marked plumage differences among races, but all races are strongly sexually dimorphic. Markings on breast feathers are diagnostic from about 5 to 6 weeks post-hatching. Breast feathers of females are tipped with white or buffy brown and have one to three buffy brown bars on a black background (Ellison 1968). Male breast feathers are black, tipped with 1–4 mm of white. Chin and cheek feathers of females are barred with brown, those of males are black. A white eye stripe and white cheek stripe are more pronounced on males. The rectrices of females are black with vermiculated brown barring often extending from base to tip; male rectrices are black, and brown flecking is limited mostly to the basal two-thirds of the feather (Zwickel and Martinsen 1967).

Juvenile spruce grouse retain a bursa until at least December but have lost it by April (Ellison 1968). The pointed tips of P9 and P10 also characterize juveniles; how-

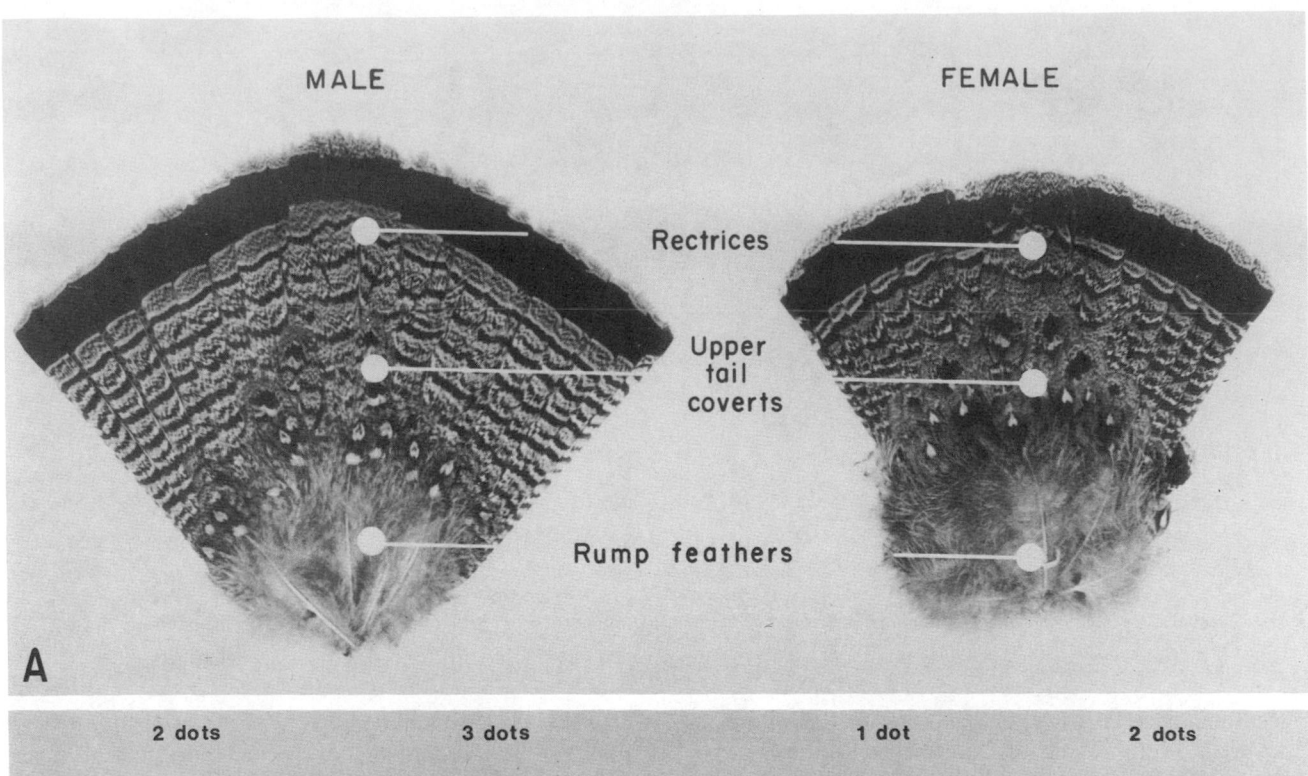

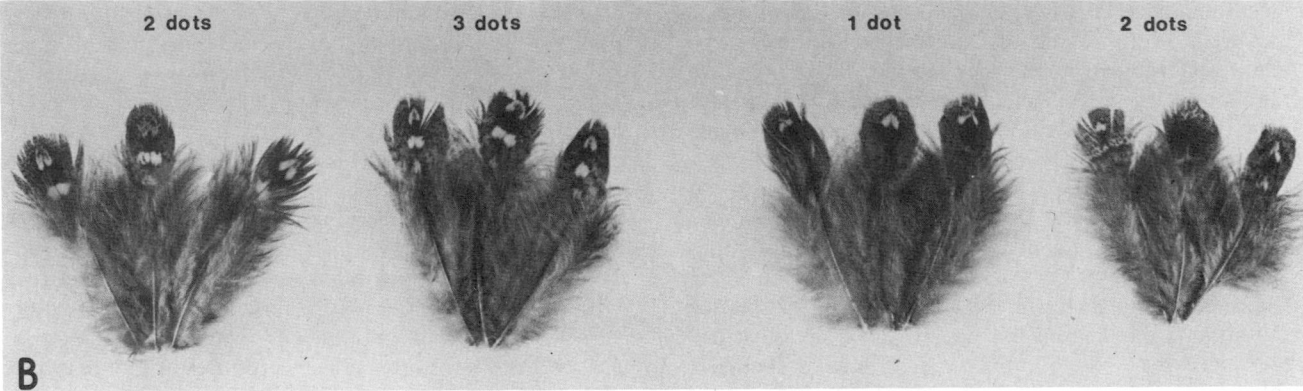

Fig. 11. Markings on tail feathers of male and female ruffed grouse. A. Location of rump feathers. B. Dot configuration on individual rump feathers (Roussel and Ouellet 1975).

ever, this character is subjective and regarded as unreliable by some researchers. Zwickel and Martinsen (1967) separated juvenile from adult Franklin's spruce grouse by the narrower, light-colored tip stripe of the upper tail coverts of both sexes of juveniles. Ellison (1968), however, could not separate age classes of Alaskan spruce grouse by this criterion. Using shaft diameters or length of various primaries might permit separation of age classes, including distinguishing between older yearlings and adults. McKinnon (1983) separated adult from yearling Franklin's spruce grouse in southwestern Alberta using the calamus diameter of P9. This was a highly reliable criterion, but specific separation points may vary with geographic region. Szuba et al. (1987) reported that shaft diameter of P1 reliably separated age classes of Hudsonian spruce grouse in Ontario in all seasons except summer. McCourt and Keppie (1975) and Towers (1988) developed growth curves for specific primary feathers to age juvenile spruce grouse from Alberta, and New Brunswick and Ontario, respectively.

SAGE GROUSE

Adult male sage grouse are nearly twice as large as adult females. In nuptial plumage the sexes are markedly different. The breeding cock presents a black chin, narrow-white throat band, and white breast (Dalke et al. 1963). The female shows no black and white patterns but has gray feathering on the throat, neck, and breast and a light gray chin. During summer, chin, neck, and throat feathering of both sexes is similar. By September-October, some of the black chin and throat feathers are showing on adult and juvenile cocks. Sexes are also differentiated by the typical black and white pattern on the longest undertail coverts. The tips of male coverts are white and have no other white except the rachis; female coverts have white markings in other vaned areas (Dalke et al. 1963). Coverts exhibit diagnostic markings at about 12 weeks. Minor wing coverts of males are dark, occasionally with some white in the rachis. Female coverts show more white, often causing the feathers to appear barred.

The two outer primaries of immature sage grouse are

Fig. 12. Central tail feathers of ruffed grouse, showing sex characters of length and subterminal tail band pattern (based on Hale et al. 1954). Male on left.

birds as young as 12 weeks. Crown feathers of females are cross-barred with alternating light and dark bands; male feathers are dark with a buff-colored edge (Henderson et al. 1967). Crown feathers are less useful than tail characters for sexing pinnated grouse.

Immature birds have conspicuous spotting on the anterior portions of P9 and P10 all the way to the tips; in adults spotting does not extend to the tips (Campbell 1972). P9 and P10 of immatures are worn, faded, and pointed, contrasted with feathers of adults. Copelin (1963) observed white coloration in the distal portion of the shaft of the outer primary wing coverts of immatures, but not adults. Baker (1953) presented a sequence of photos and descriptions of juvenile greater prairie chickens by 1-week intervals.

SHARP-TAILED GROUSE

The sexes are similar in size and general appearance but can be distinguished in hand by tail feather and crown feather markings. Crown feather markings (7.0% error) are more reliable than tail feather markings (13.0%) (Henderson et al. 1967). Crown feathers of females show alternating dark and buff-colored crossbars; male feathers are dark with buff edges. Central tail feathers of females are cross-barred, whereas male feathers are longitudinally striped along most of their length.

Immatures are distinguished by having P9 and P10 more pointed than those of adults. P9 and P10 will also have frayed, worn tips contrasted with even-edged, less-worn tips of adult outer primaries (Hillman and Jackson 1973).

PTARMIGAN

North American ptarmigan exhibit distinctly different plumages in summer and winter. The sexes are similar in body size but show plumage differences that are more noticeable in summer than in winter. Adult females of willow and rock ptarmigan lack the conspicuous red ''eyebrows'' of adult males, but both sexes of white-tailed ptarmigan exhibit eyecombs (Johnsgard 1973). Females of all three species in summer plumage are more heavily barred on the breast and flanks than are males. Female willow ptarmigan have shorter wings and tails than males, and they have brown pigment rather than black on the rectrices and central pair of upper tail coverts (Bergerud et al. 1963). These characters are diagnostic in all seasons.

Immature willow ptarmigan show more dark pigmentation on P9 than on P8. Adults have similar amounts on P8 and P9 or more on P8 (Bergerud et al. 1963). Immatures also have a greater amount of gloss on P8 than on P9 or P10, whereas adults show similar amounts on all three feathers. This characteristic was about 98% accurate for all sex and age groups of Alaskan and Scottish rock ptarmigan (Weeden and Watson 1967). Shape of the outer primaries is not a good indicator of age. Among white-tailed ptarmigan, immatures display black pigmentation on P9 and/or P10 and on the outer primary covert. Adults lack pigmentation in these areas (Braun and Rogers 1967 *in* Johnsgard 1973:242). Criteria for aging juveniles in days post-hatching were described for willow ptarmigan in Newfoundland (Bergerud et al. 1963), white-tailed ptarmigan in Colorado (Giesen and Braun 1979), and red grouse in Scotland (Parr 1975).

pointed and frayed, contrasted with the rounded tips of the same adult feathers (Eng 1955). Lengths of wing and specific primaries are shorter in females than in males. Crunden (1963) provided division points for separating juveniles from adults that gave a high degree of reliability. However, primary numbers assigned by Crunden (1963) (and several others reporting on sage grouse) were the reverse of numbers typically assigned, i.e., Crunden's P1 is usually designated P10.

PINNATED GROUSE

The sexes of all races of pinnated grouse are generally similar in external appearance, but tail and crown plumage provide reliable characters for distinguishing between sexes. Tail feathers of hens (adults and juveniles) are entirely or partially barred; tails of males are black or lightly barred (Fig. 15) (Copelin 1963). This technique is highly reliable for adults and juveniles with fully developed remiges (tail feathers). The undertail coverts of hens are barred, those of males are black with a round white spot on the end (Copelin 1963). This character enables sexing

Fig. 13. Typical completely molted adult (right) and juvenile ruffed grouse wings, showing age difference in contours of outer two primaries (based on Hale et al. 1954).

Shorebirds

GENERAL CHARACTERISTICS

Two native game birds, American woodcock and common snipe, represent this large group of birds in North America. Neither species exhibits pronounced sexual dimorphism in plumage, though woodcock females are noticeably larger than males. Juveniles attain adult size and general plumage characteristics within about 4 weeks after hatching (Fogarty et al. 1977, Owen et al. 1977).

AMERICAN WOODCOCK

Female woodcock are heavier than males, weight ranging from 160 g to 240 g for females and from 125 g to 190 g for males (Owen and Krohn 1973). Significant overlap in body weights, however, limits their utility for distinguishing between sexes. Beak length, combined width of the outer three primary feathers, and wing length are reliable indicators of sex used independently or in combination. Beak lengths >72 mm characterize females, and lengths <64 mm indicate males (Fig. 16). However, 17% of woodcock could not be sexed by this criterion (Mendall and Aldous 1943). Combined width of the outer three primaries of females (measured 2 cm from the tip) is ≥12.6 mm; for males it is ≤12.4 mm (Blankenship 1957:89). This technique is highly reliable when all three outer primaries are present. With some practice, a technician can sex woodcock correctly by inspection without measurement, because male primaries are noticeably nar-

rower than those of females. Artmann and Shroeder (1976) refined the use of total wing length by measuring from the tip of primary 6 or 7 to the notch at the bend of the wing. Wings measuring ≥134 mm were from females, ≤133 mm were from males in 99.7% of 700 wings.

Depending on the time of year, two or three age classes can be recognized: immatures (flying young); subadults (birds hatched in the preceding calendar year that have retained juvenal secondaries); and adults (birds hatched earlier than the preceding year) (Martin 1964). Proximal secondary feathers of immatures have light tips and well-defined, dark, subterminal bars (Fig. 16). Subadults retain these secondaries but can be distinguished from immatures during April-September by the greater amount of wear on their primaries and by the occurrence of primary and secondary feather molt that begins about July. Adults exhibit secondaries lacking the contrasting, light-colored band around the tip and the well-defined, dark, subterminal bar.

COMMON SNIPE

Sexes cannot be distinguished by plumage or cloacal characteristics (U.S. Fish and Wildlife Service and Canadian Wildlife Service 1977). Immatures can be separated from adults by lesser and median secondary coverts, at least in September and early October (Dwyer and Dobell 1979). Immatures show a faint, black tip on some of these coverts; adults show a wide, dark, terminal shaft line on these feathers.

Fig. 14. Adult blue grouse wings, showing the difference in the amount of mottling on the female (left) and male (right) (based on Mussehl and Leik 1963).

Doves and Pigeons

GENERAL CHARACTERISTICS

Definitive plumages of doves and pigeons are monomorphic. Distinguishing sex and age is not practical except for birds in hand, although courting behavior during breeding season (e.g., cooing and puffing of throat by males) can permit a careful observer to identify the sex of some birds. Slight but recognizable differences in color of body plumage denote sex of somatically mature birds. Wing plumage characters separate juveniles from adults until the post-juvenal wing molt is completed.

MOURNING DOVE

The crown and nape of the head region of females are brown or brownish-gray; on males this region is blue or blue-gray (Reeves et al. 1968). The breast and throat area of females is tan; it is washed with a pink or rosy hue on males. Cloacal examination will expose the oviduct opening of females or the genital papillae of males. This technique is highly accurate, but time consuming.

Correctly delineating the age structure of a mourning dove population is made difficult because the nesting season is long, and full adult plumage is reached in only 4.5–5 months (Reeves et al. 1968). Immatures are identified by the presence of at least one white or buffy-tipped primary covert. If all coverts are uniformly gray, but the 9th or 10th primary has a smooth, whitish edge, the bird is immature. If primary coverts are all gray and primary 9 or 10 has frayed, worn edges, the bird is an adult. If the bird has completely replaced all primaries, it usually cannot be aged.

BAND-TAILED PIGEON

Color of the breast and crown of band-tailed pigeons is an indicator of sex (White and Braun 1978). Feathers of these regions are dull brown to gray on females, purplish to vinaceous on males. Post-juvenal molt replaces at least some of the breast feathers indicative of sex as early as 45 days. Consequently, band-tailed pigeons can be sexed at an early age; 96% of immatures examined at ≤80 days were classified correctly by plumage characters (White and Braun 1978).

Immature band-tailed pigeons can be distinguished from adults to about 340 days on the basis of plumage characters, particularly the presence of juvenal primaries, secondaries, and secondary coverts (White and Braun 1978). Primary wing feathers of immatures exhibit white or buffy edging (Silovsky et al. 1968); adult primaries lack this edging, and the outer two primaries show wear at the tips. Primary characters are dependable through about October. Secondaries 6 and 7 are particularly important for separating older juveniles from adults, for they are the last juvenal feathers molted (340 days) (White and Braun 1978). Adults show wear on the tips and leading edges of unmolted secondaries, immature pigeons do not (Silovsky et al. 1968).

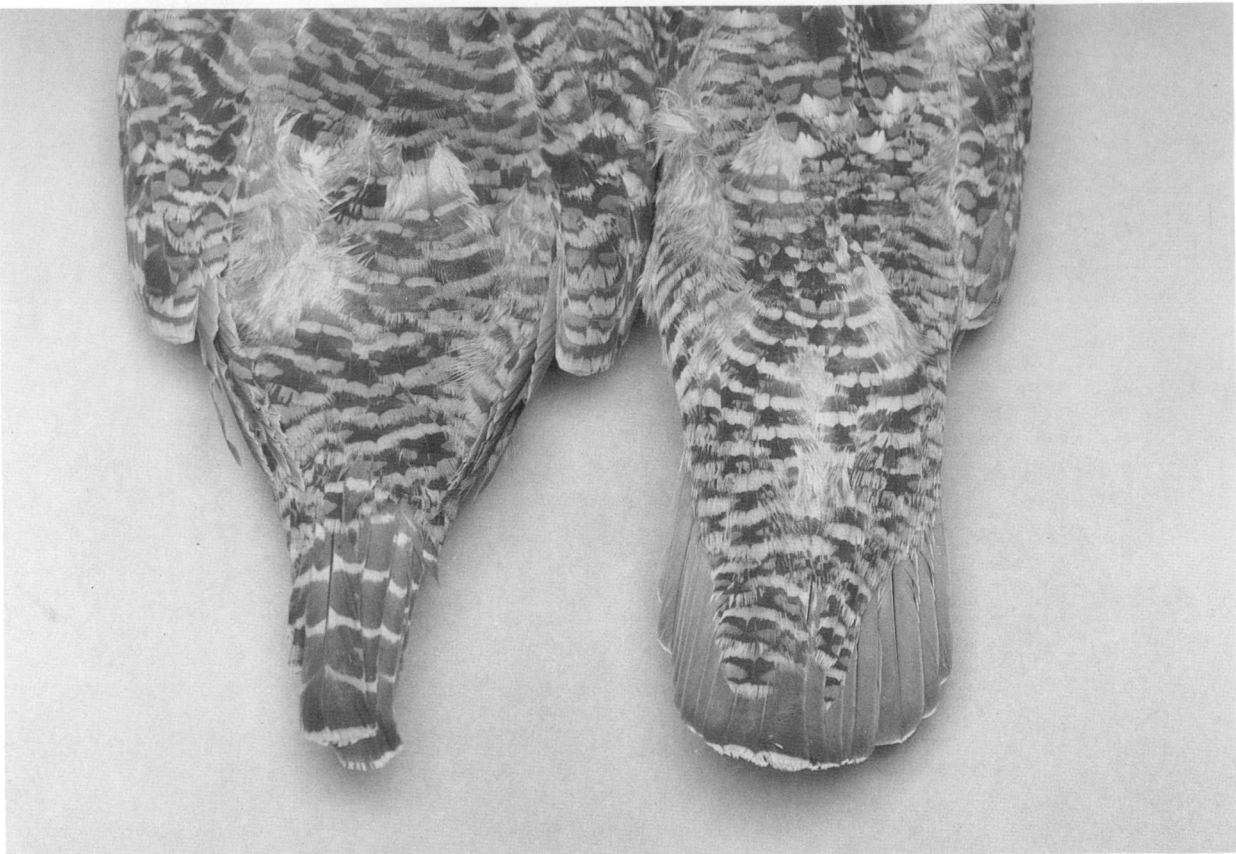

Fig. 15. Marking patterns of rectrices of pinnated grouse used to ascertain sex. Female (left) has bars across rectrices (upper tail coverts removed), male rectrices lack bars. Based on Copelin (1963).

WHITE-WINGED DOVE

Sexes are almost indistinguishable in the field, but the male is somewhat larger and more colorful (Cottam and Trefethen 1968). On the adult female, plumage of the crown, nape, and hindneck is brown, duller than on the male. These characteristics are similar to, but much less conspicuous than, those on mourning doves. Primary coverts of juveniles have pale tips, and the edges of the juvenal primaries are whitish or buffy as on mourning doves.

Cranes

Definitive plumages of North American cranes are strongly monomorphic. Juveniles are recognizable by plumage characters until the post-juvenal molt is complete. In sandhill cranes, this transpires as early as October of their first year post-hatching (Walkinshaw 1949:20).

SANDHILL CRANE

Sexes are not distinguishable on the basis of plumage. Cloacal examination for the presence of cloacal papillae (males) permitted the correct identification of only 66% of post-juvenile sandhill cranes (Tacha and Lewis 1978). The sex of juveniles could not be determined by this method. Reliable sex ratios are determined by inspecting gonads of dissected birds and by chromosome analysis of blood feathers (Van Turner and Valentine 1983).

Juvenile sandhill cranes typically display brownish body plumage versus gray body plumage of adults, but body plumage is sometimes quite similar between age classes because the plumage of adults is stained rusty brown (Lewis 1979). Consequently, definite identification of juveniles is based on presence or absence of feathering on the head. "Young cranes acquire adult body plumage and head characteristics during fall to spring. In full juvenal plumage there are tawny feathers on the crown, occiput, and nape, and short pale-gray feathers on the forehead. . . . In full adult plumage, cranes have pallid or pale mouse-gray feathers on the occiput and nape, and red papillose skin, covered with short black bristles, above the orbits and extending across the crown, forehead, and lores to the upper culmen" (Lewis 1979:212).

WHOOPING CRANE

Sexes cannot be distinguished by plumage characters (Walkinshaw 1973). Juvenile whooping cranes are predominantly white but show many brownish or buffy-tipped feathers. The head is feathered, and color ranges from dark russet brown to yellowish or pinkish buff. The adults are white except for black wing tips. The forehead, crown, and occiput of adults are bare and have warty or granulated reddish skin. The transition to adult plumage transpires gradually from October to May.

Rails

Sexes of most rails cannot be distinguished by plumage characteristics. On female soras, the black face patch is duller and more restricted than on males, the mantle is more spotted with white, and colors are generally less intense (Odom 1977).

Techniques for separating immature from adult rails by plumage characters are poorly documented. Adams and Quay (1958) assessed the age of collected clapper rails from the presence (juvenile) or absence of a bursa. Immatures up to about 10 weeks were identified by combinations of plumage and soft-part characters, but at 10 weeks they were indistinguishable from adults. Immature soras lack the black throat patch of adults (Peterson 1980).

Gallinules and Moorhens

Gallinules and moorhens are sexually monomorphic on the basis of plumage characteristics and coloration of bill, feet, and legs (Peterson 1980). Immatures are brownish (purple gallinule) or grayish (common moorhen) and have white feathers in the throat region. Bills of both species lack the bright red and yellow colors of adults. Some evidence of immaturity, such as white feathers in the throat, may persist until spring (Holliman 1977).

American Coot

American coots exhibit plumage that is sexually monomorphic. Females are smaller than males, on average having shorter culmens, wings, and metatarsus-midtoe lengths (Fredrickson 1968). However, a high degree of overlap exists between the sexes for all of these measurements, and no single measurement or combination of measurements permits accurate designation of sex.

Immatures closely resemble adults but are paler in color and have duller bills (Peterson 1980). Depth of bursa, frequently a dependable measure of maturity for birds, does not appear to correlate well with age or sexual maturity in coots (Fredrickson 1968). However, Fredrickson (1968: 411) noted that in presumed young, the bursa wall was composed of "fatty-appearing material that was several millimeters thick. In adults the wall of the bursa was very thin and transparent."

Raptors

This large group includes eagles, hawks, owls, vultures, and their relatives. Distinguishing sex and age of living birds in the field or in hand ranges from a simple task for some species to extremely difficult for others.

Wing chord length is the accepted method for sexing raptors, but buteo hawks and owls cannot be sexed by this criterion (Dunne 1987). The wing chord length is measured from the carpal joint of the bent wing to the tip of the longest primary (Pyle et al. 1987). The ranges and cutoff points for males and females are given in the *North American Bird Banding Manual* (U.S. Fish and Wildlife Service and Canadian Wildlife Service 1977). Differences in plumage characteristics may be used to sex some raptor species. The adult male northern harrier is gray and the female is brown (Peterson 1980). The male American kestrel has pale blue-gray wings; the female has rusty wings (Bull and Farrand 1977).

Most species of hawks have a juvenal plumage that is markedly different from adult plumage (Dunne 1987). Juvenal plumage in the first autumn will be fresh. Feathers will show uniform and little wear. Fault lines cutting across the width of the feather will be uniformly situated on flight or tail feathers of juveniles. Eye color is related

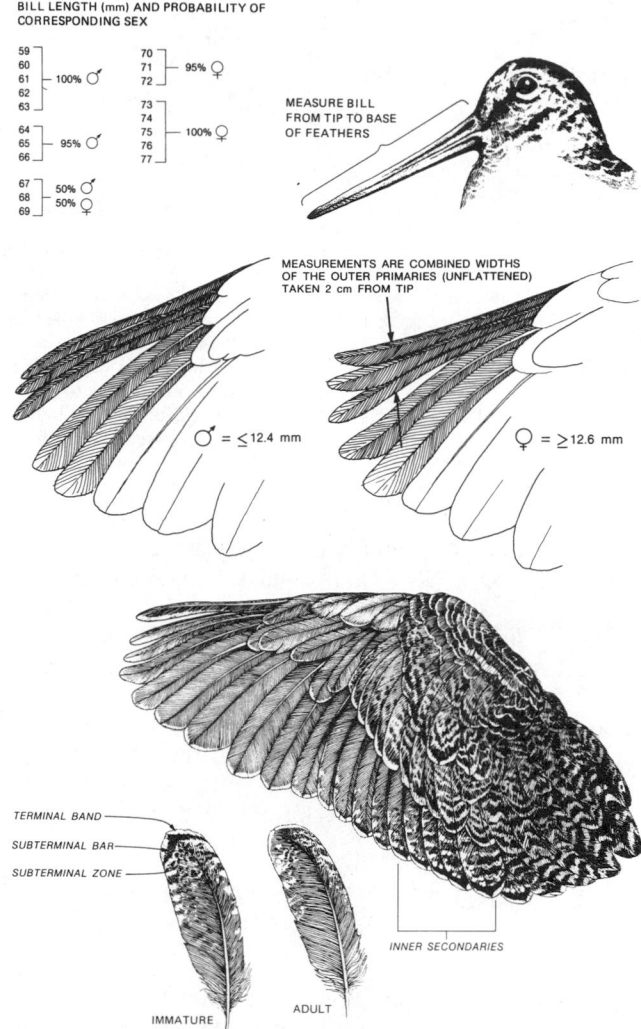

Fig. 16. Ascertaining sex and age in American woodcock. Bill lengths (top) and outer primary widths (center) distinguish sex. Bottom: patterns of inner secondary feathers distinguish age classes. Based on Roberts (1988), adapted from Liscinsky (n.d.) and Martin (1964).

to age in some species (e.g., juvenile accipiters have yellow eyes; adults have red, orange, or brown eyes) (Dunne 1987).

Owls show little or no sex-/or age-related plumage differences, though females may be larger than males. Immature saw-whet owls and boreal owls are less spotted on the forehead, have pronounced white or grayish eyebrows, and are more uniformly brownish-tan than adults (Peterson 1980).

Bald eagles are sexually monomorphic in plumage and not exclusively sexually dimorphic in size (Bortolotti 1984). The only reliable method for sexing is to examine the gonads. Four plumage classes are closely associated with age but do not represent unequivocal age classes: (1) young immature, (2) old immature, (3) subadult, and (4) adult. Increasing age is accompanied by increasing whiteness of the tail and head, culminating in adults with bright white heads and tails and dark brown body plumage.

Passerines

A detailed treatment of the several hundred additional species of North American birds is beyond the scope of

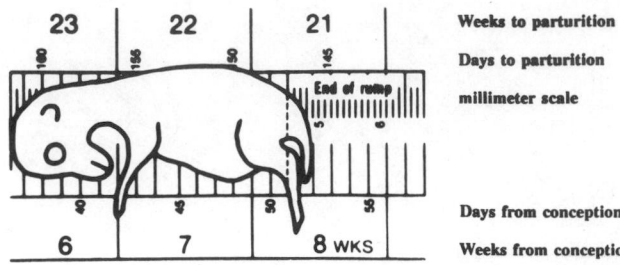

Fig. 17. Using an aging ruler to age a white-tailed deer embryo, back-date to conception, and predict date of parturition. A doe was killed on 15 December and the age of the fetus was 51 days. The Julian date of 15 December is 349. This number minus the fetal age in days is 298 (25 Oct). This is the date of conception. The number of days to parturition was 147. This number added to the Julian date of kill (349) is 496 (11 May). That is the date of parturition. Ruler based on data from Hamilton et al. (1985).

this chapter. Pyle et al. (1987) provided such a treatment for 276 species (28 families and subfamilies) of North American Passeriformes. The authors described and illustrated techniques for measuring wing, tail, bill, crown patch, and tarsus length. They also described in detail wing formulas, the art of skulling, and the use of molts and plumages for assigning age and sex to individuals in hand. General information on age and sex characteristics of other North American birds is often included in the species accounts in the several excellent field guides to North American birds.

SEX AND AGE CHARACTERISTICS OF MAMMALS

The ability to assess accurately the sex or age of a mammal depends upon whether the animal is observed at a distance, examined alive in hand, or examined as a carcass or any part thereof. Obviously, the ability to detect sex or age of free-ranging mammals has limitations. Sex ratios and general age groups can sometimes be derived from a distance, depending on season, habitat characteristics, behavior, and anatomical characteristics. With an animal ''in hand'' either dead or alive, capabilities are greatly enhanced and limitations may be more a matter of time and resources.

All that may be needed from some populations is the sex ratio and the proportion that is juveniles (nonbreeders) versus adults (breeders). However, knowing age-specific natality is important for many population studies. Thus, the necessity of determining year classes becomes important. A quick examination of tooth wear from a mandible likely will not suffice. For most mammals, collection of a tooth for cementum annuli analysis is the method of choice. This involves a more costly and time-consuming routine, but the results can add significantly to information about the population being studied.

Subcellular Sex Determination

Wildlife biologists may employ subcellular techniques or use primary and secondary physical characters to determine sex of mammals. Subcellular techniques are useful for determining sex of unborn young or when only a portion of a carcass is available.

Sex may be detected in stained cells of many mammals

(Moore 1966). A planoconvex, dark-staining mass against the inner surface of the nuclear membrane is associated with the second X chromosome in females. Usually surgery or necropsy is necessary to obtain the proper internal nerve or epithelial tissues necessary to display the evidence. Segelquist (1966) and Crispens and Doutt (1970) employed the technique to sex fetal deer. In some species appropriate cells can be obtained from the sheath of hair roots. Schmid (1967) and DeGraaf and Larson (1972) tested the technique on a wide range of mammals. Scrapings of the internal cheek lining are commonly used to obtain epithelial cell specimens from humans, but the samples are contaminated by bacteria and degenerating cells, making microscopic examination more difficult. The same problem could be expected in wild mammals. Another cell that is easily obtained and displays visible sexual dimorphism in some species is the polymorphonuclear neutrophil leucocyte, one of the white blood cells. Larson and Knapp (1971) employed this approach to sex beavers. Mittwoch (1963) described several sex differences in cells of mammals.

Until recently, subcellular indices of sex focused on characters associated with X chromatin. Hoekstra and Carr (1977) identified white-tailed deer sex through fluorescent Y chromatin techniques, using blood and lymphocytes. Because samples can be taken from dry blood on either permeable or impermeable surfaces, or from meat pieces or frozen samples, this is a useful forensic technique.

Embryonic Development

The degree of fetal development is indicative of the age in days of prenatal mammals (e.g., Bookhout 1964). With this knowledge, birth dates can be established for a population; this in turn can assist in developing management guidelines or regulations such as the timing and length of hunting seasons as well as correlating environmental events with breeding or parturition success. Crown-rump, forehead-rump, or limb/foot measurements are typically used for assessing prenatal age.

A method for measuring the crown-rump dimension on white-tailed deer was illustrated by Hamilton et al. (1985) (Fig. 17). The stages of fetal development in white-tailed deer and mule deer were described by Armstrong (1950) and Hudson and Browman (1959) (Table 3). Salwasser and Holl (1979) reported hind-foot length to be the best parameter for aging late-term mule deer in California. Ozoga and Verme (1985) successfully determined fetal age of live white-tailed deer with a portable x-ray unit. Development and measurements can be expected to vary among populations and among individuals at different nutritional levels. Even with potential variation, current techniques are rather accurate. However, a technique for aging fetuses has not been reported for most mammalian species.

Postnatal Development

During the period of growth from birth to sexual maturity skull sutures and epiphyseal cartilage in long bones are universally present in mammals (Figs. 18, 19). In addition, sexually immature mammals exhibit an array of characteristics measurably different from older, sexually mature individuals. Body size and weight, pelt size, pel-

Table 3. Stages in fetal development (days of gestation) in white-tailed deer (from New York) (Armstrong 1950) and mule deer (from Montana) (Hudson and Browman 1959), showing physical characteristics, crown–rump,[a] hind foot,[b] and hind leg[c] measurements (in millimeters) (from Larson and Taber 1980).

Age in days	White-tailed deer	Mule deer
37–40	Eyelids absent; no vibrissae follicles noticeable. C-R 17.1–27.0	
41–44	Vibrissae follicles present above eye, under eye, on muzzle, on cheeks. C-R 27.7–29.6	
45–52	Eyelids absent; mouth open. 48 days: C-R 37.8	48 days: vibrissae follicles just present above eye. C-R 32.4
53–60	Eyelids formed; mouth closed. 60 days: C-R 62.7; HL 20.7	57 days: eyelids cover eye. C-R 59.2; HF 20.5
61–65	Fetus loses fishhook shape; angle formed between long axis of body and straight line drawn from muzzle through ear ≥90	61 days: vibrissae follicles visible on chest and upper forelimbs. C-R 74.3; HF 29.0
66–68	Preorbital fold at anterior median side of eye. 66 days: C-R 83; HL 30.5	68 days: vibrissae follicles appear on abdomen and trunk. C-R 94.7; HF 37.2
73–75	75 days: C-R 113; HL 43.6	73 days: brown pigment on nose between nostrils and down to lip. C-R 110.7; HF 45.7
76–85	Gray pigment appears on top of nose	
86–90	Vibrissae broken through skin over eye, on muzzle, on cheeks; black pigment on dorsal area of nose; brown pigment on anterior surface of nose. 90 days: C-R 167.3; HL 69.8	86 days: C-R 155; HF 73.9 89 days: C-R 164; HF 77.0
91–95	Brown pigment on surface of lower lip; black pigment along closing surface of eyelids; metatarsal gland appears on oblong white spot on tarsus	
96–105	Hooves black-pigmented. 98 days: C-R 197.2; HL 98.8	
106–110	Nostrils open. 107 days: C-R 224.2; HF 114.5. 110 days: C-R 233.2; HL 124	
111–120	Incomplete hair covering; light pigment spots on trunk; eyelashes grown; dark spots (both sexes) at position of antlerbud. 115 days: C-R 252; HL 127	111 days: nostrils open; legs and hooves brownish-black; hair on muzzle. C-R 232; HF 127. 117 days: hair appears on legs. C-R 252; HF 143
121–132	Hair present on anterior and posterior surfaces of skin covering proximal ends of femur and humerus; tarsal glands present	
133–150	Appearance of hair on legs; row of stiff bristles around top of hoof. 135 days: C-R 318.5, HL 192	137 days: hair surrounds metatarsal and covers tarsal gland. C-R 311; HF 184. 144 days: short hair present just above hooves; teeth still covered with membrane. C-R 327; HF 193
151–180	Hair covering as in adult; incisors still covered by membrane, black pigment covers nose; metatarsal gland with complete hair covering. 159 days: C-R 396; HL 251	161 days: tips of incisors and canines exposed. C-R 397; HF 262. 174 days: hair covering as in newborn fawn. C-R 443; HF 278
181–200	Incisors erupted; tarsal gland with complete hair covering. 181 days: C-R 445; HL 304. 192 days: C-R 459; HL 314	

[a]C-R = crown–rump or forehead–rump measurement.

[b]HF = measurement of hind feet from the tip of the hoof to the angle of hock.

[c]HL = measurement of the hind leg from the tip of the hoof to the tubercle of the tibio-fibula.

age differences, eye-lens weights, changes in dentition, teat color and size, and testicular characteristics can be used with varying degrees of success for aging most mammals. These techniques normally permit separating a population into two to four general age classes. Because young animals often comprise the largest segment of the harvest, identifying this age group quickly and easily reduces the time required to establish the age structure (Johnston et al. 1987).

Determining the sex of sexually immature individuals can be difficult, particularly in rodents. In many species the distance between the urogenital opening and the anus is a good criterion; that distance for females is one-half or less than for males of the same species.

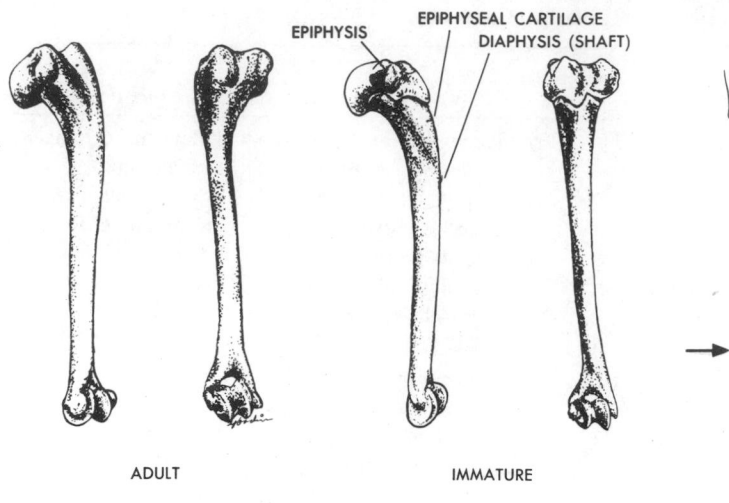

EPIPHYSIS   EPIPHYSEAL CARTILAGE
DIAPHYSIS (SHAFT)

ADULT   IMMATURE

Fig. 18. Lateral and posterior view of humerus of cottontail. Note epiphysis and diaphysis in the immature and its absence in the adult (from Hale 1949, after Godin 1960).

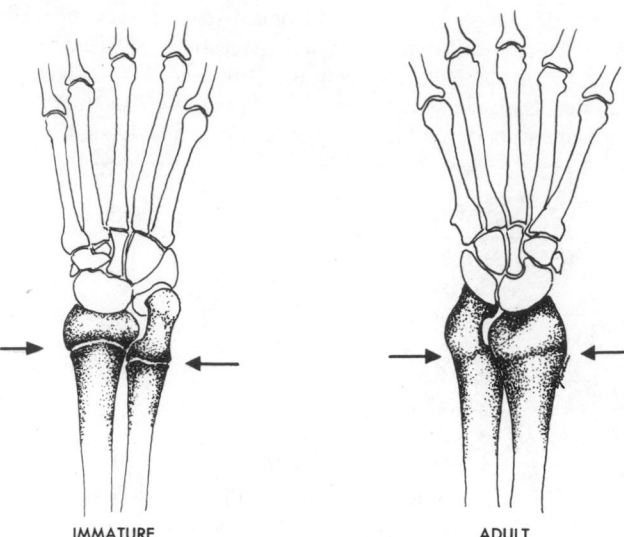

IMMATURE   ADULT

Fig. 19. Drawing made from x-ray photograph of radii and ulnae of raccoon showing open (immature) and closed (adult) epiphyses (after Sanderson 1961, from Godin 1960).

Eye-Lens Weight

The crystalline eye lens of vertebrates grows throughout the life of the animal and is the only organ that does not shed cells (Bloemendal 1977). These special features make it an indicator of age in many mammal species. This technique requires special preservation, drying, and weighing of the lens from freshly killed specimens only; freezing adversely affects the lens. The degree of error typically is greater with lenses of small animals. The lens in mammals grows measurably after sexual maturity, and variation in lens weights in mammals is high among adults of the same species. Eye-lens weights probably are most useful for separating juveniles from adults and are not practical indicators of year class among adults. Friend (1967) presented a complete review of the technique.

Another characteristic of the eye lens may be a more important means for determining age. Tyrosine, an insoluble protein, accumulates in the lens throughout life and affords an accurate method for estimating the age of small mammals (Dapson and Irland 1972, Otero and Dapson 1972, Birney et al. 1975). Ludwig and Dapson (1977) showed it to be effective also for aging white-tailed deer. Freezing affects the lens adversely and, because processing large numbers of fresh specimens is impractical, the technique has certain limitations. If fresh specimens are available, it is superior to tooth-wear procedures and is less costly and time consuming than counting cementum annuli.

Cementum Annuli

Layers in the cementum of teeth and in the periosteal zone of bones can be accurate indicators of year class (Klevezal' and Kleinenberg 1967). Cementum is deposited on the roots of teeth each year in bands so that the bands close to the dentine are from the earlier years and the layers of the current year lie on the exterior of the root. Age of the individual is determined by counting the annual cementum growth layers. Klevezal' and Mina (1973) reported that the pattern of layers is not affected by sex or a change of physiological state associated with rut or pregnancy. Nor could it be attributed to specific

conditions of the year when the layers are formed. Variability of the pattern is least in populations in a continental climate and is greatest among those in slightly continental climate and sea climates. Jacobson and Reiner (1989) observed that tooth eruption and wear were more accurate methods of aging Mississippi white-tailed deer than cementum annuli counts for age classes <3.5 years. But cementum annuli occur in virtually all mammals, and the technique is likely to be effective for any mammal if the equipment and skills to expose the layers are available.

In a few species (e.g., beavers), the teeth and layers are sufficiently large and distinct so that simply grinding and polishing a sagittal section will show the layers well under a dissecting microscope (Van Nostrand and Stephenson 1964). For most species and all small teeth, it is necessary to decalcify the tooth, cut thin histological sections with either a microtome or cryostat (Child 1973), stain, and read the layers under higher magnification. All teeth have layers, but the tooth of choice varies among species and collecting conditions. Some teeth, such as incisors and premolars, are easier to extract and may even be removed from live animals without adverse effect. In trophy specimens, the tooth extracted may be influenced by the requirements of the taxidermist. Cementum aging is considered to be more accurate than tooth wear for aging older mammals. Among 120 known-age, land-mammal teeth (12 species) processed at Matson's Laboratory (P.O. Box 308, Milltown, MT 59851), exact agreement occurred between known age and cementum age in 94 individuals; 21 were within 1 year of known age, and 5 were incorrect by more than 1 year (G. Matson, pers. commun.).

Cementum aging is unnecessary for age classes of young mammals that can be easily identified by characteristics such as the thin root walls and wide-open root tips of some furbearer canines or the presence of deciduous teeth (Johnston et al. 1987). Radiography (x-ray) is a low-cost method of identifying canine teeth of juvenile furbearers when the root tips close early in life. Juvenile furbearers taken during the earlier part of harvest season have pulp cavities that are so much larger than those of adults that measurement is not needed to identify them. Later in the year, the distinction between juveniles and yearlings is more difficult and less accurate. A sheet that describes techniques for collecting, processing, and aging is available from Matson's Laboratory, P. O. Box 308, 8140 Flagler Road, Milltown, MT 59851.

COLLECTING THE TOOTH

The teeth selected should be those that are standards for cementum aging. For all ungulates, the standard is the central incisor, I1. For most carnivores, it is the canine. The standard for bears and wolves is premolar 1; for cougars it is upper premolar 2. Marten are aged by canines, premolar 3, or premolar 4 (for marten, different laboratories and technicians have their own preferences). Incisor 4 may be removed for aging live ungulates, as may the upper canine for aging elk.

If a nonstandard tooth type is selected for cementum aging, the type must be identified, because the differences in eruption time require different interpretations of cementum growth layers. Errors of at least 1 year can result when an unidentified, nonstandard tooth type is substituted.

Care should be exercised during tooth removal to prevent breaking off the root tip. A 1-cm portion of the root that includes the extreme root tip is most important for accurate age determination. Bears are exceptional in that accurate age estimates are often possible from broken teeth because of a thickened cementum layer near the gum line. A dental elevator, available in different sizes from veterinarian supply sources, is useful for removing teeth from their sockets. Teeth of freshly killed ungulates are not difficult to remove with a pulling, twisting motion after either side of the gum tissue is cut as far down as possible. Canine teeth from carnivores are difficult to remove and must be heated to 60–80 C for up to 12 hours before they can be pulled. Autoclaving damages or destroys dental tissues and should not be done. Contaminating the teeth with dirt should be avoided. Particles of sand or dirt do not soften during treatment with acid and will dull the delicate sectioning knife.

Teeth should be stored in paper envelopes. Plastic containers seal in moisture that permits spoilage. If teeth are to be kept for several months before being processed for cementum aging, they should be stored frozen. Paper and cardboard containers should be used to prevent moisture damage after thawing. The exact effects of long, unprotected storage upon the physical properties and biological stain reactivity are unknown. Freezing is recommended as the safest precaution. Chemical preservatives can have deleterious effects upon reactivity to biological stains, and their use is not recommended.

Dirt can be removed from dried teeth by heating the tooth in water at 60–80 C for 10–20 minutes and carefully pulling away the softened, dirt-contaminated periodontal tissue. Direct scraping on the cementum with any sharp instrument will remove the last-formed cementum layers and cause error in the age determination.

Self-mailers for hunters of game animals are a successful means for obtaining large samples of teeth. The hunter removes a tooth at the time of kill and mails it in an envelope supplied by the game management agency. Both U.S. and Canadian postal authorities have expressed concern about the practice, because of machinery damage from handling the hard, unpadded teeth and fears about disease spread by blood contamination. Careful attention should be given to a tooth envelope design that will be approved by local and regional postal personnel. Recent designs have incorporated the use of colored materials to permit the envelope to be spotted for hand cancelling and an explanatory note to postal workers stating that the contents are dried animal teeth for laboratory analysis and are nonperishable, nonhazardous, and nonetiologic (not disease carrying).

LABORATORY PROCESSING

The cementum of most ungulates is thickened on the anterior and posterior (rostral and caudal) portions of the root tip. Laboratory processing should maximize the exposure of cementum from these areas.

Tooth roots should be sectioned longitudinally, in the mid-sagittal plane. Longitudinal tooth root sections have the following advantages over cross sections: (1) a larger area of cementum is available for study, and (2) greater opportunity exists to study the extent of the first dark growth layer to determine its relationship to the dentine-cementum junction and characteristics at the extreme root tip. Cross sections have the following disadvantages: (1) section characteristics at fixed distances above the root tip vary according to age, and (2) sections are more likely to curl and loosen from slides during staining.

AGING THE TOOTH

A separate aging method is required for each species because of differences in cementum growth-layer characteristics. A fixed set of rules for interpreting the cementum must be written for each species and supplemented by photographs and explanatory diagrams. Particular attention must be given to the characteristics and age designation of the first dark cementum growth layer, the characteristics of subsequent growth layers, the presence of nonannual dark cementum, and the presence of a "juvenile" dark layer produced before the age of 1 year.

The status of cementum aging among a variety of wild mammals is summarized in Table 4. A schematic of a tooth section is depicted in Fig. 20, and various cementum annuli in black bears and their interpreted ages are illustrated in Fig. 21.

Even-Toed Ungulates

GENERAL CHARACTERISTICS

The order Artiodactyla comprises 12 species in North America, representing five families (Suidae, Tayassuidae, Cervidae, Antilocapridae, and Bovidae). Additional introduced exotic species in these families currently occur outside confinement and are of increasing interest to biologists and managers.

Sex Determination

The most noteworthy characteristic of this order is the presence of horns or antlers in adults of three of the five

Table 4. Cementum aging status of some common North American mammals (data in this table provided by G. Matson Laboratory, Milltown, MT 59851, unpubl. data, 1990).

Species	Standard tooth	Substitute tooth	Identification of juvenile[a]	Cementum pattern[b]	Experience[c]	Accuracy[d]	Disagreement among experienced technicians[e]
White-tailed deer	I1	I2–I4, PM	M	V	4	3	
Mule deer	I1	I2–I4, PM	M	V	4	3	
Black-tailed deer	I1		M	V	4	3	
Elk	I1	I2–I4, Up. C	M	D	4	4	
Moose	I1	I2–I4	M	V	3	3	
Caribou	I1	I2–I4, PM	M	V	3	3	
Pronghorn	I1		M	V	2	3	
Goat	I1	I2–I4	M	D	4	4	
Sheep	I1	I2–I4	M	V	2	3	
Bobcat	C		O, R	V	4	3	
Lynx	C		O, R	D	2	3	
Cougar	PM2			I	2, D	2	
All fox	C		R	D	3	3	
Coyote	C	PM1	O, R	V	3	3	
Wolf	PM1	C	O, R	I	1, D	2	
River otter	C		O, R	C	3	2	
Mink	C		R	C	2	2	
Fisher	C		R	D	2	3	
Marten	C		R	C	3, D	2	T, M
Badger	C		R	C	1	2	
Wolverine	C		R	C	1	2	
Black bear	PM1	All		C	4	3	M
Grizzly bear	PM1	All		C	3	3	M
Raccoon	C	I1	O, R	D	3	3	

[a]M = morphological, O = open root tip, R = radiography.
[b]V = variable, D = distinct, I = indistinct, C = complex.
[c]4 = great, 1 = small, D = aging method under special development.
[d]4 = accurate, 1 = approximate.
[e]T = tooth type, M = method.

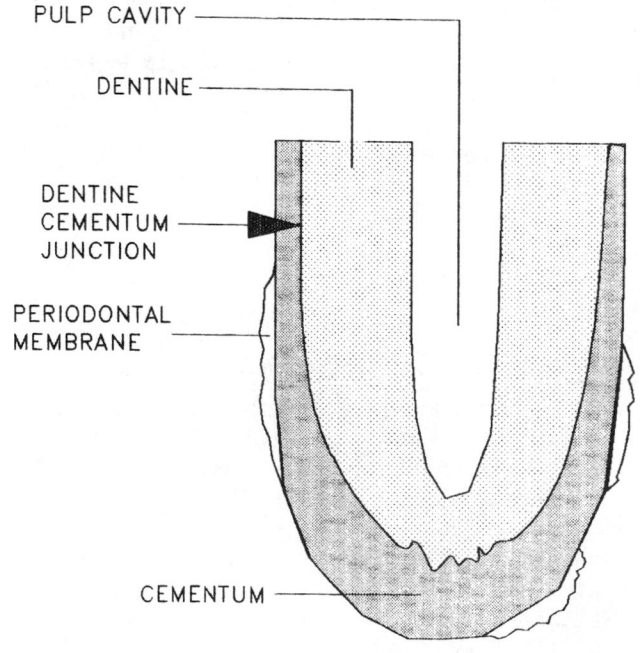

Fig. 20. Schematic of a mammalian tooth section (from G. Matson, unpubl. data).

represented families (Cervidae, Bovidae, Antilocapridae). With the exception of caribou and rare anomalous conditions, only males have antlers in the family Cervidae. Both sexes of the family Bovidae have horns, but males of the five represented species (bison, mountain goat, muskox, mountain sheep, and Dall sheep) have more pronounced horn development than do females. Antilocapridae is represented by a single species, the pronghorn. Male pronghorns have horns and the females may or may not.

Age Determination

Most of the even-toed ungulates can be aged by tooth replacement or wear techniques and by counting cementum annuli. Because of the ready availability of mandibles for inspection, much effort has gone into developing aging techniques that are useful in the field for classifying these species into yearly age classes. Tooth replacement techniques usually are limited to estimating year class or the age of juveniles. Tooth-wear criteria permit placing individuals into several year classes. Diet and soil conditions contribute to significant geographic variation in the degree of wear of teeth for any species.

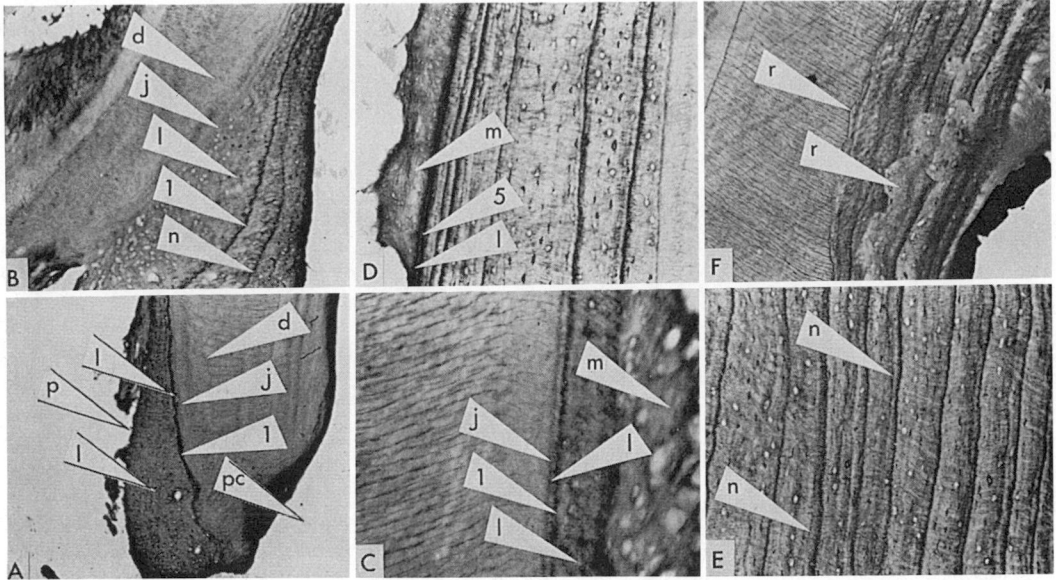

Fig. 21. Annulations in cementum of black bear. No ages are known. 1,2, ... = annual layers of dark cementum, numeral indicates year; d = dentine; j = dentine-cementum junction; l = light cementum; m = periodontal membrane; n = nonannual layer of dark cementum; p = peripheral margin of tooth root; pc = pulp cavity; r = cementum resorbed, filled in by new cementum deposition. A: PM1 collected September; age 1A; root tip; 35×. The highly developed root rules out juvenile age; the single, prominent annual dark band near the dentine does not extend around the root tip. B: PM1 collected September; age 1A; root tip; 35×. The first annual band is broadly separated from the dentine (compare with A). C: same tooth section as A; 3 mm above root tip; 150×. A thin layer of light cementum separates the 1-year band from the junction. D: PM1 collected May; age 5A; 3 mm above root tip; 95×. The last-formed annual band is just visible at the periphery of the section. E: PM1 collected September; age 11A; 6 mm above root tip; 95×. Nonannual dark bands recur regularly, year after year. Such recurrence is usually in older bear teeth with complex annual bands. F: PM1 collected August; age 7A; near root tip; 95×. Resorption removed not only cementum, but also dentine. Deposition of cementum subsequently filled in the areas of tissue loss. Annual bands removed in this area of the section were unaffected in other areas.

WHITE-TAILED DEER

Skeletal remains of males (>2 years old) may be distinguished from those of females by the presence of tuberosities where the ligaments supporting the penis attach (Taber 1956) (Fig. 22). Sexual dimorphism of the iliopectineal eminence of the pelvic girdle was useful for animals as young as 1 year old (Edwards et al. 1982). The pattern of tooth replacement defines individuals as fawns or yearlings (Table 5, Fig. 23, Severinghaus 1949). Tooth eruption is complete at about 21 months. Deer >1.5 years are aged by the relative amount of tooth wear (Severinghaus 1949, Hesselton and Hesselton 1982). However, tooth-wear characteristics are too varied for accurate year-class determination among older animals (Gilbert and Stolt 1970). Known-age mandibles from deer representative of all habitat types should be used for comparison (Hesselton and Hesselton 1982). When used properly, this technique is valid for younger age classes and more general categorizations of age. Year class is most accurately determined from cementum annuli of incisor (Lockard 1972) or molar teeth (McCullough and Beier 1986). However, Jacobson and Reiner (1989) reported that tooth eruption and wear were more accurate than cementum annuli counts for age classes <3.5 years.

MULE AND BLACK-TAILED DEER

McCullough (1965) reported that the hooves of adult black-tailed deer are sexually dimorphic to the extent that tracks of adult males and larger yearling males have a larger arc width and can be recognized with certainty. For animals up to 33 months (Rees et al. 1966), the eruption pattern of mandibular teeth is useful (Table 5, Fig. 23). Connolly et al. (1969a) and Erickson et al. (1970) observed that after 24–28 months, tooth-wear, eye-lens, and molar tooth-ratio techniques were inadequate. However, when reference sets of sex-specific mandibles are available, the accuracy of the tooth-wear technique is acceptable (Thomas and Bandy 1975), and counts of cementum annuli from incisor sections are accurate for all ages (Thomas and Bandy 1973).

ELK

Only male elk have upper canines. Morphological characteristics also permit placing animals in age classes of calves, yearlings, and 2 years and older (Greer and Yeager 1967). Teeth erupt sequentially (Table 6), and complete, permanent dentition is present at 3 years of age (Peek 1982). Cementum annuli in first incisors of elk were accurate indicators of year class, but only 50% could be aged to year correctly by tooth wear (Keiss 1969).

MOOSE

Adult antlerless moose and most calves can be sexed from the air by the presence (females) or absence of a white vulval patch (Roussel 1975). Cementum annuli of incisors are valid indicators of year class (Gasaway et al. 1978, Haagenrud 1978).

CARIBOU

Miller (1982) listed 10 criteria for determining sex and age of caribou during aerial and ground surveys. Both males and females possess antlers, but antlers of males

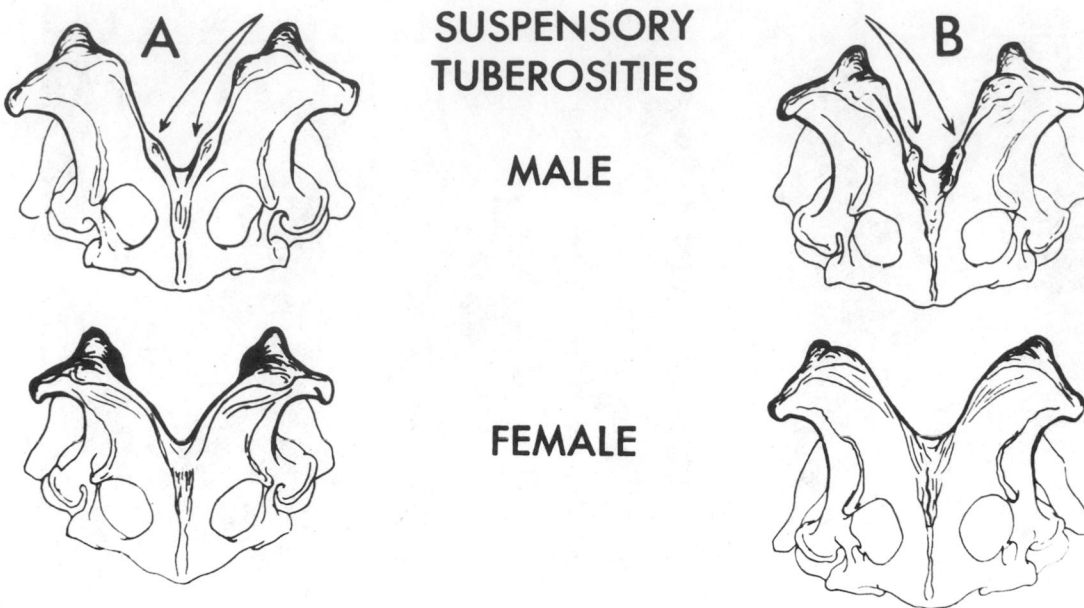

SUSPENSORY
TUBEROSITIES

MALE

FEMALE

Fig. 22. Pelvic girdle of the white-tailed (A) and black-tailed (B) deer, viewed from the rear, showing the suspensory tuberosities for the attachment of the penis ligments in the male and their absence in the female (after Taber 1956).

are larger and more ornate than those of females. Sex also can be determined by comparing mandible lengths (Bergerud 1964, Miller and McClure 1973). The eruption patterns of molars and permanent premolars were used to age Newfoundland caribou up to age 27.5 months (Bergerud 1970) and barren ground caribou up to 2 years (Miller 1974*b*). An accurate chart was developed depicting the tooth eruption pattern for the species (Table 7, Miller 1974*a*). Most caribou have a full set of mandibular teeth by the 29th month (Miller 1982). Relative tooth wear can be supplemented with linear tooth measurements to improve accuracy (Miller and McClure 1973, Miller 1974*a*).

The cementum annuli technique is valid for barren ground caribou. The second or third incisor can be removed from living animals for aging (Bergerud and Russell 1966).

MUSKOX

Sex and age of muskox are best determined by the appearance of horns, because most animals must be classified while alive. Calves are readily identified by their size. Yearlings are small and have small, straight horn projections; the length of the horn sheath is about 100 mm in males and about 66 mm in females.

Table 5. Tooth eruption in the New York white-tailed deer (after Severinghaus 1949) and lower jaw of mule deer (from Cowan 1936 and Taber and Dasmann 1958). D = milk or deciduous tooth; P = permanent tooth; parentheses indicate that tooth is in process of erupting.

	Incisors			Canine	Premolars				Molars		
Age	1	2	3	1	2	3	4	1	2	3
					White-tailed deer						
1 to 3 weeks	(D)	(D)	(D)	(D)	(D)	(D)	(D)			
2 to 3 months	D	D	D	D	D	D	D	(P)		
6 months	P	D	D	D	D	D	D	(P)		
12 months	P	P	P	P	D	D	(P)	P	(P)	
18 months	P	P	P	P	P	(P)	P	P	P	P
24 months	P	P	P	P	P	P	P	P	P	P
					Mule deer						
1 to 3 weeks	D	D	D	D	D	D	D			
2 to 3 months	D	D	D	D	D	D	D	(P)		
6 months	D	D	D	D	D	D	D	(P)	(P)	
12 months	P[a]	DP	D	D	D	D	D	P	(P)	
18 months	P	P	P	D	D	D	D	P	P	(P)
24 months	P	P	P	P	(P)	(P)	(P)	P	P	(P)
30 months	P	P	P	P	P	P	P	P	P	P

[a]Replacement and eruption are taking place at this time.

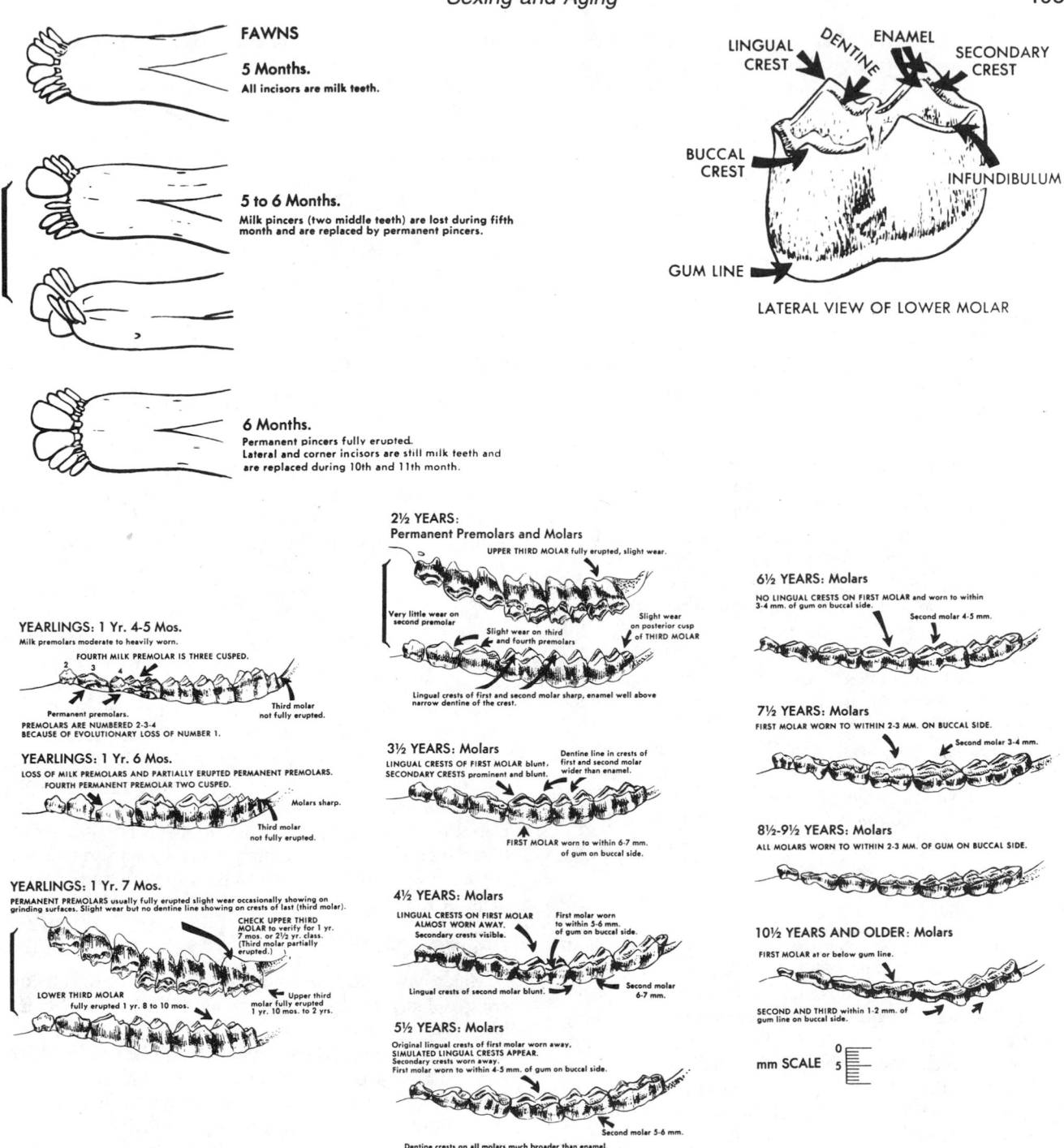

Fig. 23. Sequence of eruption and wear of teeth of white-tailed deer in New York (from Larson and Taber 1980), and nomenclature of the ungulate molariform tooth (Godin 1960).

Immature animals include bulls of 2.5–5.5 years and cows of 2.5–3.5 years. Sex is difficult to determine in 2.5-year-old animals, although the horns of bulls are whiter and project more nearly straight from the head than those of females. During the fourth year, the basal depressions of the horns of females reach their maximum development, almost touching the jaw, and the apical portions turn upward and out; they are then considered adults. Bulls are considered adults from their sixth year, which is marked by the growth of their horns completely over the forehead (Tener 1954).

Tooth emergence can be used for aging to 6 years (Tener 1965). Cementum annuli counts have been used (Parker et al. 1975). Use of a fluoroscope may be better than staining for annuli counts (Hinman 1979).

BISON

Sexual dimorphism is apparent among adult bison, but females generally resemble males in color, body configuration, and presence of permanent horns (Reynolds et al. 1982). The horns of the female are more slender and curve inward more than those of the males. The horn cores and

Table 6. Tooth eruption in the Rocky Mountain elk (from Quimby and Gaab 1952). D = milk or deciduous tooth; P = permanent tooth; parentheses indicate that tooth is in process of erupting.

Age (yr)	Incisors			Canine	Premolars			Molars		
	1	2	3	1	2	3	4	1	2	3
0.5	D	D	D	D	D	D	D	(P)		
1.5	P	DP	D	D	D	D	D	P	P	
2.5	P	P	P	P	D	D	D			
					(P)	(P)	(P)	P	P	P
					P	P	P			
3.5	P	P	P	P	P	P	P	P	P	

burrs are more pronounced in males than in females (Skinner and Kaisen 1947). Duffield (1973) identified sexual differences in several postcranial skeletal measurements.

Only gross age classification is possible for free-roaming bison as determined by conformation and horn development (Reynolds et al. 1982). Four age classes of females and five of males were derived on the basis of horn growth (Fuller 1959). Counting annual growth rings on horns was not a useful aging criterion. The degree of fusion of cranial sutures can be used to place skulls of males into two age classes (Shackleton et al. 1975). A table for calculating age based on closure of the epiphyses was developed by Duffield (1973). Eye-lens weight is not a useful indicator of age due to within-age-class variation (Novakowski 1965). Skinner and Kaisen (1947) recognized six general age categories, and Frison and Reher (1970) established seven based on tooth replacement and wear (Table 8). Five yearly age classes, based on tooth eruption and wear, were identified by Fuller (1959), but only the cementum annuli technique is considered valid for year-class determination (Novakowski 1965).

WILD SHEEP

The most intensively studied wild sheep is the bighorn sheep, but sex criteria probably apply equally well to Dall sheep. At a distance, lambs can be distinguished but not sexed. The horns of males are noticeably larger in sheep older than lambs (Lawson and Johnson 1982). Yearlings, especially rams, cannot be clearly distinguished from adult ewes unless they can be closely examined for the presence of a scrotum. Two-year-old and older rams are easily distinguished by their large horns. The younger adult rams (<¾ curl) may be separated from older ones (>¾ curl) (Jones et al. 1954). The ram's horn becomes much larger at the base and develops a curl with age, whereas horns on ewes retain the lamb shape (Lawson and Johnson 1982).

Horn segment counts also are valid on bighorns (Geist 1966). Sheep, like other North American Bovidae, are slower to achieve adult dentition than members of the deer family, 4 years being required for completion of tooth replacement and eruption (Table 9) (Deming 1952, Hemming 1969, Lawson and Johnson 1982). After 4 years, the length : width ratios and wear of teeth of males can be used to estimate age (Lawson and Johnson 1982). Cementum annuli are valid criteria of year class in Rocky Mountain bighorn, Nelson's bighorn, Peninsular bighorn, and Dall and Stone sheep (Turner 1977).

MOUNTAIN GOAT

Sex of all ages is indicated by urination posture. Males stand or stretch and females squat during urination. The scrotum of the yearling and older male is visible during the summer. Horns of males are thicker at the base than those of females and, when viewed frontally, show less space between the horn bases. However, using horn characteristics is a poor method for delineating sex in the field unless the observer is experienced and close to the goat (Wigal and Coggins 1982). Yearlings and older females show a black vulval patch under the tail.

Kids (animals born during the current summer) have horns barely visible to less than one-half ear length by autumn. Yearlings have horns in early summer less than ear length and to about ear length in the autumn. Two-year-olds and adults have horns longer than ears. Adults have faces larger and more angular than do 2-year-olds, but this distinction is difficult to discern by late summer (L. Nichols, Fed. Aid Proj. W-17-9 and W-17-10, Alaska Dep. Fish Game, Juneau, 1978). Aging is possible by the intermittent growth of the horn, causing the formation of annual rings (Fig. 24), and by the eruption and replacement of the teeth (Table 10).

PRONGHORN

Males have horns (technically, a deciduous sheath covering a large cone); females may or may not. Female horns average only 42 mm long (O'Gara 1968) and have no prongs or only rudimentary ones (O'Gara 1969). Male horns start to grow at 2 months of age, whereas those of females begin during their second year. In general, animals with horns longer than the ears are adult males. The adult male is marked with a black mask covering the face up to the horns; the female has a black nose, but only a faint shadow of dark hair extends upward on her face (Einarsen 1948). The sequence of tooth eruption and replacement is presented in Table 11. Year class is determined by cementum annuli in the first permanent incisor (McCutchen 1969). Histological sectioning of teeth to expose annuli is superior to cut and grinding procedures (Kerwin and Mitchell 1971). Techniques based on wear are not valid.

COLLARED PECCARY

No external sexual dimorphism is obvious except for genitals. However, suspensory tuberosities on the pelvic girdle are prominent in males and absent in females (Lochmiller et al. 1984).

Table 7. Eruption of teeth of Kaminuriak caribou according to age during first 29 months of life (from Miller 1974*b*). Numerical values are percentages of frequency of occurrence. When no values are given, occurrence is 100%. Incisors and premolars = D E P; molars = A E P.

Age (months)	Mandibular incisiform and molariform tooth rows									
	i1	i2	i3	c1	p2	p3	p4	m1	m2	m3
0	Dᵃ	D	D	D	D / Eᵇ	D / E	D / E	Aᶜ	A	A
1	D	D	D	D	D	D	D	A	A	A
3	D	D	D	D	D	D	D	Pᵈ / E	A	A
5	D	D	D	D	D	D	D	P / E	A	A
10	D 46 / E 21 / P 33	D	D	D	D	D	D	P	A 62 / E 38 / P 0	A
12	D 14 / E 3 / P 83	D 41 / E 21 / P 38	D 52 / E 22 / P 26	D 59 / E 22 / P 19	D	D	D	P	A 7 / E 77 / P 16	A
13	P	D 58 / E 0 / P 42	D 67 / E 0 / P 33	D 0 / E 8 / P 92	D	D	D	P	A 40 / E 33 / P 27	A
15	P	P	P	P	D	D	D	P	P	A 92 / E 8 / P 0
17	P	P	P	P	D	D	D	P	P	A 65 / E 35 / P 0
22	P	P	P	P	D 74 / E 21 / P 5	D 86 / E 12 / P 2	D 95 / E 3 / P 2	P	P	A 10 / E 89 / P 1
24	P	P	P	P	D 34 / E 20 / P 46	D 28 / E 23 / P 49	D 50 / E 15 / P 35	P	P	A 0 / E 47 / P 53
25	P	P	P	P	D 17 / E 11 / P 72	D 17 / E 6 / P 77	D 39 / E 3 / P 58	P	P	A 0 / E 39 / P 61
27	P	P	P	P	D 13 / E 7 / P 80	D 10 / E 3 / P 87	D 19 / E 0 / P 81	P	P	A 0 / E 6 / P 94
29	P	P	P	P	P	P	P	P	P	P

ᵃD = milk tooth.
ᵇE = erupting tooth. An erupting tooth has a stained portion but has not migrated to its position of permanent orientation.
ᶜA = absent tooth (permanent tooth not yet erupted).
ᵈP = permanent tooth.

Table 8. Tooth eruption and replacement in the lower jaw of the bison (after Hogben *in* Larson and Taber 1980). D = milk or deciduous tooth; P = permanent tooth; parentheses indicate that tooth is in process of erupting.

Age (years)	Incisors 1	2	3	Canine 1	Premolars 2	3	4	Molars 1	2	3
1	D	D	D	D	D	D	D	P	(P)	
2	P	D			P					
3	P	P	D	D	(P)	(P)	D	P	P	P
					P	P	(P)			
4	P	P	P	D	P	P	P	P	P	P
5	P	P	P	(P)	P	P	P	P	P	P

Kirkpatrick and Sowls (1962) described a technique based on tooth-replacement patterns that places animals up to 21.5 months in six age classes (Table 12). Eye-lens weight was of limited value for aging adults (Richardson 1966). Pelvic sutures are barely visible in 12-month-old animals (Lochmiller et al. 1984).

Terrestrial Carnivores

In North America the order Carnivora is represented by five families and more than 30 species. Obtaining sex and age data from some of these species is difficult because of their inherent or human-caused scarcity. Sample sizes often are small, consequently the need is great for accurate age estimates of those sampled. Many male carnivores possess bacula and can be palpated to ascertain sex or age (Petrides 1950, Newby and Hawley 1954, Thompson 1958). The presence of a penal scar or opening on a removed pelt is indicative of males in these furbearers. Males are generally larger than females within a population, but a large female in an older age class can be confused with a small male in a young age class. In addition, significant geographic variation exists among populations of a species, often due to wide differences in nutrition. Detailed information on sexing and aging of individual species is available in Novak et al. (1987) and Chapman and Feldhammer (1982). Johnston et al. (1987) provided a thorough overview of aging furbearers in North America.

WOLF

Only urination posture and behavior indicate sexes from a distance (Carbyn 1987). Sexing is accomplished by examining nipples and penal scar/opening; this can be done on live wolves, carcasses, and pelts.

Pups can be separated from adults by size for only the first 6–8 months (Carbyn 1987). Deciduous teeth are replaced at 16–26 weeks (Schonberner 1965); deciduous teeth are shorter and appear less massive than permanent teeth. Canines <21 mm in length distinguish pups from adults (Van Ballenberghe and Mech 1975). The complete fusion of the epiphyses to the diaphyses of the radius and ulna occurs at 12–14 months (Rausch 1967). Wolves are fully grown at 18 months of age (Young and Goldman 1944). The first observable cementum annuli appear at 18–22 months (Goodwin and Ballard 1985), and aging by counting annuli is as useful in this species as in other carnivores.

COYOTE

Coyotes or coyote pelts can be sexed by the presence or absence of nipples or a penal opening (Voigt and Berg 1987). The sagittal crest on the skull of males shows greater development than that of females (Bekoff 1982), but skull morphology is less reliable than other criteria (Gier 1968).

Permanent canines erupt at 4–5 months. The root canal of the canine tooth closes at approximately 8–12 months.

Table 9. Wild sheep permanent-tooth eruption patterns (from Chapman and Feldhamer 1982). Numbers are months.[a]

Tooth	Dall	Rocky Mountain bighorn	Desert bighorn
M_1	1–4	1–4	6
M_2	8–13	8–13	16
I_1	13–16	13–16	12
I_2	25–28	25–28	24
P_2	27–32	25	24
P_3	25–30	25	24
P_4	25–30	25–30	24
M_3	22–40	22–40	30
I_3	33–36	33–36	36
C	45–48	45–48	48

[a]Compiled from Deming (1952) and Hemming (1969).

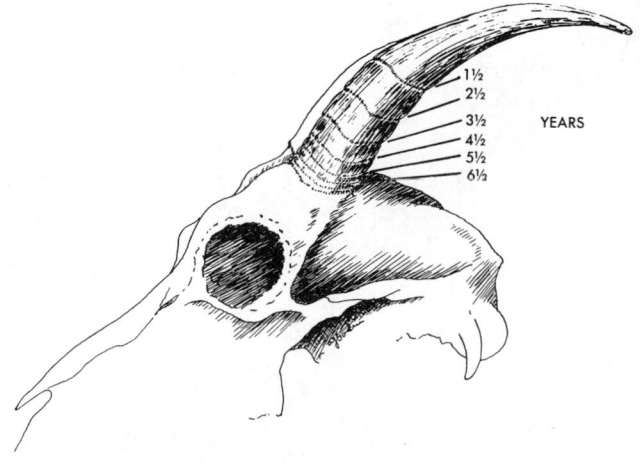

Fig. 24. Annual rings on the horn of the mountain goat (after Brandborg 1955).

Table 10. Tooth eruption and replacement in the lower jaw of the mountain goat (after Brandborg 1955). D = milk or deciduous tooth; P = permanent tooth; parentheses indicate that tooth is in process of erupting.

Age	Incisors			Canine	Premolars			Molars		
	1	2	3	1	2	3	4	1	2	3
1 week	(D)	(D)	(D)		(D)	(D)	(D)			
6 months	D	D	D	D	D	D	D	(P)		
10 months	D	D	D	D	D	D	D	(P)	(P)	
15–16 months	(P)	D	D	D	D	D	D	P	(P)	(P)
23 months	P	D	D	D	D	D	D	P	P	(P)
26–29 months	P	(P)	D	D	(P)	(P)	(P)	P	P	(P)
38–40 months	P	P	(P)	D	P	P	P	P	P	P
48 months	P	P	P	(P)	P	P	P	P	P	P

Radiographs reveal an open root canal for juveniles and a closed or partially closed canal for adults (Voigt and Berg 1987). The first dark annulus in the cementum forms at 20 months (Linhart and Knowlton 1967). Young coyotes (0–8 months) can be aged by total body weight with regression equations presented by Bekoff (1982) and Barnum et al. (1979).

Cementum annuli of tooth sections yield an accurate year-class age determination (Linhart and Knowlton 1967, Nellis et al. 1978, Bowen 1982). Timing of the deposition of cementum annuli may vary considerably among geographic regions (Allen and Kohn 1976). Roberts (1978) recommended the lower canine as the tooth of choice for examining cementum layers because the number of annuli varied among different kinds of teeth. Premolars from live animals can be removed to count cementum annuli (Voigt and Berg 1987), but canines removed from carcasses are more accurate (Roberts 1978). Nellis et al. (1978) described a technique for estimating age up to 50 months based on progressive closure of the canine socket.

FOX (RED AND GRAY)

Pelt or whole-body examination can be used to sex foxes by the presence of penal scar or opening, or nipples. Sexes cannot be differentiated by remote observation in the field (Fritzell 1987). Though males tend to be slightly heavier, size or color is not discernibly different. Palpating can detect the baculum in males.

Geiger et al. (1977) and Harris (1978) reviewed the various techniques for aging red foxes; these include eye-lens weight, baculum and body measurements, cranial suture closure, skull measurements, and epiphysial closure. A closed root canal of a canine tooth denotes a red fox older than 1 year. Accuracy of cementum annuli counts decreases as the number of annuli increases. Premolars can be extracted from live foxes and used for aging as described by Allen (1974), Johnston and Watt (1981), and Johnston et. al. (1987). Voight (1987) used annuli from longitudinal sections of root to show age up to 4–5 years old.

Gray fox juveniles can be distinguished from adults with varying degrees of accuracy by the degree of ossification of the radius and ulna (Sullivan and Haugen 1956), tooth wear (Root and Payne 1984), pelage and body weights (Wood 1958, Lord 1961a), eye-lens weight (Wood 1958, Lord 1961b, Nicholson and Hill 1981, Root and Payne 1984), width of the canine pulp cavity, size of the apical foramen of the canine (Tumlison and McDaniel 1984), and height of the enamel line (Root and Payne 1984). The baculum is shorter and lighter in weight among subadults (Tables 13, 14).

BEAR (BLACK, BROWN, AND POLAR)

Size alone generally cannot be used to determine sex of bears. Although adult males may be twice the weight of adult females, weights of one age class may overlap those of females of an older age class (Pearson 1975, Alt 1980, Craighead and Mitchell 1982). A combination of skull, dental, and body measurements generally will separate males and females, females having smaller measurements. Black bear canines are sexually dimorphic; maximum root width and thickness of lower canines are the best criteria of sex (Fig. 25) (Sauer 1966). Gordon and Morejohn (1975) distinguished sexes using combinations of length of the mandibular canine alveolus and width of the second mandibular molar.

Table 11. Tooth eruption and replacement in the lower jaw of the pronghorn (after Dow 1952, Dow and Wright 1962). D = milk or deciduous tooth; P = permanent tooth; parentheses indicate that tooth is in process of erupting.

Age	Incisors			Canine	Premolars			Molars		
	1	2	3	1	2	3	4	1	2	3
Birth	D						D			
6 weeks	D	D	D	D	D	D	D	P		
15–17 months	P	D	D	D	D	D	D	P	P	P
27–29 months	P	P	D	D	P	P	P	P	P	P
39–41 months	P	P	P?	(P)	P[a]	P[a]	P[a]	P	P	P

[a]Total of 24 infundibula.

Table 12. Tooth eruption in the lower jaw of the collared peccary (after Kirkpatrick and Sowls 1962). D = milk or deciduous tooth; P = permanent tooth; parentheses indicate that tooth is in the process of erupting.

Age (months)	Incisors			Canine	Premolars			Molars		
	1	2	3	1	1	2	3	1	2	3
2–6	D	D	D	D	D	D	D	D	D	D
7–10	D	D	D	D	D	D	D	P	D	D
11–12	D	D	D	P	D	D	D	P	D	D
13–18	D	D	D	P	D	D	D	P	P	D
19–21.5	D	D	D	P	D	D	D	P	P	(P)
>21.5	D	P	P	P	P	P	P	P	P	P

Canine teeth of black bears do not erupt fully until the bears are 14–16 months old (Marks and Erickson 1966). The degree of closure of the root canal can be used to assign ages to female black bears up to 3 years and males to 4 years (Sauer et al. 1966, Poelker and Hartwell 1973).

A complete, permanent dentition is acquired by brown bears by 2 years of age (Couturier 1954). Adult-juvenile age categories of brown bears can be derived by baculum weight (Pearson 1975) or degree of closure of the canine root tip canal (Rausch 1969). Age determination for polar bears is most reliable when a combination of age indicators is used (Hensel and Sorensen 1980); this combination includes general body size, reproductive status, tooth wear and replacement, and cementum annuli (Kolenosky 1987). Premolars are used for cementum annuli counts primarily because they can be easily extracted from live bears or carcasses without harm or mutilation of the animal (Kolenosky and Strathearn 1987). Calcified tooth sections examined under ultraviolet light exhibit zones of autofluorescence that can be counted to ascertain age (Johnston and Watt 1981).

Autumn annuli and double annuli within a year can present problems. In polar bears the accuracy of cementum annuli counts often is complicated by uneven or double layers, especially in older bears (Kolenosky 1987). Most investigators overestimate the ages of younger bears and underestimate the age of older bears (Hensel and Sorensen 1980).

RACCOON

Although males are slightly larger than females, much overlap occurs among age classes of this species; size alone is not a good criterion of sex or age. Testes in males are always descended and the baculum is easily palpated. Scars or nipples can often be located on pelts or carcasses.

Closure of the epiphyses, baculum size and shape, and eye-lens weights separate raccoons into two age classes—

juveniles and adults (Sanderson 1961). The baculum of a juvenile is relatively small, straight, and porous at the base, has a cartilaginous tip, weighs <1.2 g, and is <90 mm long (Kaufmann 1982). The gradual disappearance of cranial sutures is useful up to 122 months old (Junge and Hoffmeister 1980). Tooth eruption is useful up to 110 days (Montgomery 1964) (see Table 15). Cementum annuli can effectively separate raccoons into four age classes (Grau et al. 1970, Johnson 1970).

MINK

Sex of skins may be determined by presence or absence of the penis scar (Table 16) or the presence of nipples (Petrides 1950). Testes are permanently descended, so sexing a live mink is simple (Eagle and Whitman 1987). Condyle-premaxillae measurements reveal sex as well (Birney and Fleharty 1966).

Tooth replacement and wear are helpful up to 3 months of age (Aulerich and Swindler 1968). Age class can be estimated by the morphology of the baculum in males (heavier, more massive, and longer, rougher proximal end in adult) and by the status of the lateral supra-sesamoid tubercle (Fig. 26) and the fugal-squamosal suture in females (Tables 17, 18) (Lechleitner 1954, Greer 1957). Birney and Fleharty (1968) reported best results from using the status of the tubercle and the texture of the femur in the region of the epiphyseal closure (porous in juve-

Table 13. Measurements (early winter) of gray fox bacula in central Ohio (after Petrides 1950).

Age class (according to baculum shape)	N	Length (mm)	Weight (mg)
Subadult	5	51 ± 1.7	280 ± 62
Adult	5	57 ± 2.6	528 ± 100

Table 14. Age characters for red and gray foxes (from Larson and Taber 1980). Characters for young are italicized.

Anatomical structure	Characters
Baculum[a]	Larger, heavier, with enlarged and roughened basal area; *smaller, lighter without large, rough basal area.*
Teats[a]	More than 2 mm diam., dark, obvious to touch in dry pelts; *<1 mm diam., light-colored, scarcely raised* (separates those that have bred from others
Epiphyses of radius and ulna[b]	No cartilage plate at distal end of (by x-ray) radius and ulna; *cartilage plate at distal end of radius and ulna (at distal epiphyseal gap)* separates young up to 8–9 months from adults

[a]Petrides 1950.
[b]Sullivan and Haugen 1956.

WIDTH (mm)   THICKNESS (mm)

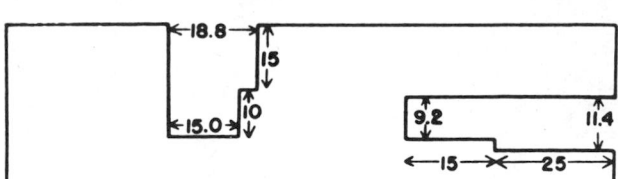

Fig. 25. The gauge devised for sexing black bears by the size of the lower canine tooth. It indicates sex as follows: female, tooth fits into smaller part of slot for either width or thickness (but not necessarily both); male, tooth does not fit into larger part of slot for either width or thickness (but not necessarily both). If tooth fits into the large part of both slots, but into the smaller part of neither, it must be measured as a summation of the width and thickness made as described in Sauer (1966).

Table 15. Mean ages of raccoons at eruption of deciduous and permanent teeth (from Montgomery 1964).

Tooth	Upper		Lower	
	Mean age (days)	SE	Mean age (days)	SE
Deciduous				
First incisor	34.0	2.8	28.5	2.2[a]
Second incisor	25.4	2.2[a]	37.3	2.1
Third incisor	26.2	1.2	33.0	2.4
Canine	29.3	0.9	29.3	0.8
First premolar	64.5	1.9	60.7	1.8
Second premolar	46.2	1.0	43.4	1.6
Third premolar	49.2	1.2	48.4	1.2
Fourth premolar	48.7	1.0	48.7	1.0
Permanent				
First incisor	65.6	1.4	65.9	1.0
Second incisor	73.3	1.3	72.6	1.2
Third incisor	96.6	1.7	85.5	1.9
Canine	111.7	3.9	105.6	3.6
First molar	81.0	1.2	78.1	1.5

[a]The following percentages of these teeth either did not erupt or erupted and were shed before examination of the animals; upper first incisor and lower second incisor, 66.7; lower first incisor, 16.7; upper second incisor, 5.5; lower third incisor, 33.3.

niles, smooth in adults). However, G. P. Dellinger (unpubl. data) reported 13.5–18.3% overlap when the tubercle was used as an age indicator in Missouri, indicating that distinct variations may occur among different populations.

Birney and Fleharty (1968) observed nasal sutures in juveniles of both sexes and an absence of bony deposits on the ischia of juvenile males. Adults had heavily worn teeth. Eye-lens weights separated adults from juveniles but not without overlap. Stained sections of teeth and mandibles were useful for aging (Klevezal' and Kleinenberg 1969).

PINE MARTEN

The presence or absence of the baculum (determined by palpation) or vulva, or the larger overall size and broader head of the male are criteria suggested for sexing live pine martens (Newby and Hawley 1954). The sagittal crest also is more pronounced in males. Most skull measurements separate males from females (Brown 1983), but variation may be too great for separating age groups (Strickland et al. 1982). The presence of a preputial orifice on a pelt is definitive for males (Strickland and Douglas 1987). Length, width, and thickness of the canine tooth separate males from females (Brown 1983).

Permanent dentition is complete by 18 weeks (Brassard and Bernard 1939); tooth wear and replacement are not useful for age determination during the trapping season (Strickland and Douglas 1987). Radiographs of the pulp cavity of canines separate adults from juveniles (Berg and Kuehn 1980, Dix and Strickland 1986). Fusion of the distal femoral epiphysis is not a reliable indicator of age in winter-trapped animals, because fusion is a function of size and possibly sex (Dagg et al. 1975). Formation of the suprafabellar tubercle on the femur separates adults from juveniles (Leach et al. 1982). A baculum weighing <0.1 g denotes a juvenile male (Marshall 1951, Brown 1983). Counts of cementum annuli are valid for aging marten (Strickland et al. 1982, Archibald and Jessup 1984).

RIVER OTTER

River otters cannot be sexed or aged reliably on the basis of tracks or visual observations in the wild. Adults in hand may be sexed by the relative positions of the anus and urogenital opening (Fig. 27). Considerable size overlap exists between yearlings and adults (Melquist and Hornocker 1983). Radiographs of teeth of juvenile river otters exhibit open apical foramina; pulp cavities constitute more than one-half the tooth width (Melquist and Dronkert 1987). Closure of the epiphyses of the long bones is useful in grouping river otters into juvenile, yearling, and adult age classes (Hamilton and Eadie 1964). Eye-lens weight also can be useful (Lauhachinda 1978). Other age criteria include characteristics of the baculum, development of the testes, tooth eruption patterns, body size, and skull characteristics (Toweill and Tabor 1982). The most reliable and useful is the number of cementum annuli. The initial band is deposited in spring or summer at 1 year of age (Tabor 1974). Canine teeth present reliable annuli counts (Stephenson 1977).

WOLVERINE

Genitalia (scars and holes) for both sexes and nipples on females are apparent on live animals, carcasses, and pelts (Hash 1987). Females average 30% less in weight than males (Hall 1981). The condylobasal length measurement of the skull separates males from females and overlaps only slightly (Magoun 1985).

Reproductive organs, long bones, and cranial sutures separate young-of-the-year from adults only if the young are <10–11 months old (Rausch and Pearson 1972). Eye-lens weight is unsuitable for aging. Four age categories were derived by Whitman et al. (1986) based on body weight, overall condition of teeth, general physical condition of teeth, and physiological signs of aging in immobilized live animals and carcasses. Cementum annuli counts are best for determining year class beyond 1 year (Rausch and Pearson 1972).

Table 16. Sex and age characters in opossum, raccoon, mink, badger, skunks, and long-tailed weasel.

Species	Criteria for distinguishing sex		Criteria for distinguishing age	
	Female	Male	Immature	Mature
Opossum	*Secondary sex characters (reliable from 17 days)*		*Female pouch*[a]	
	Pouch outline and nipple rudiments (Reynolds 1945)	Scrotum	Pouch white, shallow, or practically absent. On pelts, pouch not flabby, fatty, dark, or prone to tear (Petrides 1949)	Pouch rusty inside and border, teats dried (winter), 3 mm diam. On pelts, pouch flabby, fatty, dark, and prone to tear
	Distance from external urinary opening to anus		*Baculum*	
	<26 mm	>26 mm	Distal end cartilaginous, basal end porous	Plate or knob at distal end(s), bone of basal end not porous (Dellinger 1954)
Raccoon	*Urinary papilla*		*Uterine horns*[b]	
	No penis bone (Stuewer 1943)	Penis bone felt by palpation	Translucent, 1–3 mm diam., w/o placental scars (Sanderson 1950)	Opaque, 4–7 mm diam., with placental scars (Sanderson 1950)
	Cased skin (if not too fat)			
	No rough area, teats large (Sanderson 1950)	Roughened area near middle of belly (site of preputial orifice)		
Mink	Penis scar absent	Penis scar present	*Teats (of female)*	
			Scarcely raised, <1 mm diam. (Petrides 1950)	Dark, raised, >1 mm diam.
Badger	Same as mink		*Teats (of female)*	
			1.5 mm diam. and 1 mm long in dried skins (Petrides 1950). Also see text	4–6 mm diam., 4–10 mm long even in dried skins
Skunk, striped and spotted	Same as mink		*Teats (of female)*	
			<1 mm, usually flesh-covered (Petrides 1950). Also see text	At least 2 mm diam. and 2.5 mm long, usually dark
Long-tailed weasel	Same as mink		*Teats (of female Jul–Oct)*	
			Not visible (Wright 1948)	Enlarged

[a] Distinguishes those that have bred from those that have not.
[b] Distinguishes those that have had litters from those that have not.

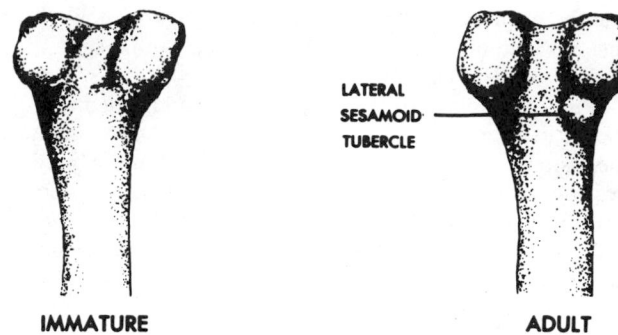

LATERAL
SESAMOID
TUBERCLE

IMMATURE   ADULT

Fig. 26. Femurs of mink showing lateral supra-sesamoid tubercle of adult and its absence in the immature (from Lechleitner 1954, by Godin 1960).

FISHER

Although males are nearly twice as large as females, the sizes of their feet are not so different; thus track size is not a useful criterion for sexing fishers (Coulter 1966, Johnson 1984). Sex is ascertained by examining external genitalia or nipples on live animals or pelts. The maximum root width of lower canines is indicative of sex; widths >5.64 mm are of males (Parsons et al. 1978). Canine length also is useful (Kuehn and Berg 1981, Jenks et al. 1984, Dix and Strickland 1986). Several bone measurements reveal the sex (Leach 1977, Leach and de Kleer 1978), including zygomatic width, skull length, and skull weight (Strickland 1978).

Adults can be separated from juveniles by the presence of the suprafabellar tubercle on the adult femur (Leach et al. 1982). Radiographs of the pulp cavity in the canines also are useful for separating adults from juveniles (Kuehn and Berg 1981, Dix and Strickland 1986). Adult males have prominent sagittal crests but they may not always be indicative of age (Douglas and Strickland 1987). Permanent dentition is complete at 7 months. Epiphyses of long bones and cranial sutures also fuse early in some animals. These techniques are not considered reliable for aging

Table 18. Use of femur tubercle (present on at least one side in adults, absent on both sides in juveniles) and jugal-squamosal suture (absent on at least one side in adults, present on both sides in juveniles) as aging criteria in combination for mink (after Greer 1957).

Age class	Total number examined	Number (%) with character
Juveniles	495	468 (95)
Adults	388	375 (97)

fishers taken during the trapping season (Dagg et al. 1975).

Counting cementum annuli is the only technique available for aging adults (Douglas and Strickland 1987). This technique can be used with live animals by extracting the first premolar, which is vestigial.

AMERICAN BADGER

Although some body and skull measurements are sexually dimorphic (Messick and Hornocker 1981), they have limited use because of variations in age, geography, and nutritional status (Messick 1987). Sex of live animals or pelts is easily detected by the testes/penis or penis scar of males, or the vulva or teats of females (Petrides 1950).

Several techniques have been used to separate juveniles from adults, including teat size (Petrides 1950), closure of skull sutures, development of a prominent midsagittal ridge (Messick 1987), dried eye-lens weight (Wright 1969), and baculum length, weight, and appearance (Petrides 1950, Wright 1969, Lindzey 1971, Messick and Hornocker 1981). Cementum annuli are the only reliable indicator of adult year classes (Lindzey 1971, Crowe and Strickland 1975, Todd 1980, Messick and Hornocker 1981).

STRIPED AND SPOTTED SKUNK

Maximum root thickness, maximum root width, and minimum and maximum root length of lower canines of

Table 17. Criteria of age based on the baculum of some mustelids (the term "head" means the basal, or proximal, enlargement) (from Larson and Taber 1980).

Species	Characteristics of baculum in young[a]	Characters of baculum in old[a]	Authority
Mink	No ridge, head not always morphologically distinct. Wt. 172 ± 34.2 mg (1 SD)	Distinct ridge on baculum at head, which is distinct. Wt. 398 ± 97.0 mg (1 SD)	Lechleitner 1954
Long-tailed weasel	Head scarcely larger than shaft. Wt. 14–29 mg	Head greatly expanded. Wt. 53–101 mg	Wright 1947
Striped skunk	Head not enlarged, shaft irregularly curved.	Head enlarged, shaft more straight.	Petrides 1950
Badger	Short, lightweight, only shallow grooves, no protuberances, head only slightly enlarged and never ridged.	Long, heavy, prominent grooves and protuberances, head much enlarged and often sharply ridged.	Petrides 1950
Pine marten	Wt. <100 mg	Wt. >100 mg	Marshall 1951
Wolverine	Wt. 653–1458 mg (avg. 1134 mg)	Wt. 1780–2940 mg (avg. 2338 mg)	Wright and Rausch 1955

[a]Young means young-of-the-year in the winter; old means older.

striped skunks adequately separate males and females (Fuller et al. 1984). Measurements are greater in males than females and overlap little. Several aging techniques are inadequate for skunks except for general age categories. Ossification of the epiphyses, baculum, cranial sutures, placental scars, tooth wear, eye-lens weight, and general appearance are unreliable (Allen 1939, Petrides 1950, Mead 1967, Upham 1967, Verts 1967, Bailey 1971, Bjorge 1977, Leach et al. 1982). Counts of cementum annuli are accurate (Nicholson and Hill 1981).

FELIDS

The male genitalia of felids are less obvious than those of other carnivores (Rolley 1987); untrained personnel often misidentify the sex of bobcats (McCord and Cardoza 1982). Nevertheless, sex of live bobcats can be determined by palpation of the genitals. However, accurate sex classification of skinned carcasses of bobcats may require internal examination (Rolley 1987). Sex also can be determined by measuring the maximum cross-sectional area of the lower canines (Friedrich et al. 1983). No totally reliable methods are known for rapidly sexing live mountain lions (Lindzey 1987). However, various cranial and skeletal measurements are of some value if primary sex characteristics are absent from a carcass.

Tooth replacement is a useful aging technique for bobcats and lynx; this technique can be applied to bobcats up to 240 days (Crowe 1975) and is believed similarly applicable to lynx (McCord and Cardoza 1982). Permanent teeth are acquired during the first winter. The foramen of the canine tooth closes at 13–18 months in both bobcat and lynx (Saunders 1961, 1964, Crowe 1972). However, Johnson et al. (1981) cautioned that kittens from late litters may have open root canals during early months of their second winter. This could cause errors in age classification. Counts of cementum annuli are used for aging older animals. The first cementum layer apparently is deposited during the animal's second winter (Crowe 1972, Nellis et al. 1972, Stewart 1973), so age is determined by adding 1 to the count of the annuli (Nava 1970). Mountain lions can be separated into three age classes on the basis of weight, pelage characteristics, tooth eruption and wear, and tissue changes indicating breeding by females (Lindzey 1987). However, several physiological and morphological variables proved insufficient for more specific age classification (Currier 1979). Counts of cementum annuli are not useful for aging mountain lions (Lindzey 1987).

Other Selected Mammals

PINNIPEDIA

In his review of 18 species, Laws (1962) showed that teeth of pinnipedia generally have external ridges, dentine layers, or cementum layers (or combinations of these) that are accurate for determining age. Year-class determination of northern fur seals is based on ridges on the external surface of whole upper canine teeth and on annuli in sectioned teeth for females aged 3–7 years. Whole or sectioned teeth are equally good for males aged 2–5 years (Anas 1970). Lens weights are useful only through age 2 years and only if sex is known (Bauer et al. 1964). Northern fur seals and their teeth are markedly sexually dimorphic in gross size. Northern sea lion pups up to 15 months can be aged by general size (to the nearest 6 weeks). The pattern of tooth eruption is also a guide to this age group (Spalding 1966).

Pinnipeds generally display annuli in the dentine deposits within the pulp cavity of canines and cementum deposits on the apex of the root. Both of these may be examined in longitudinal sections of the teeth (Kenyon and Fiscus 1963). In some genera (*Callorhinus, Eumetopias, Zalophus, Odobenus,* and *Mirounga,* at least), annuli are visible as ridges on the outer surfaces of the canines (Scheffer 1950, Laws 1953, Mansfield 1958, Kenyon and Fiscus 1963).

RABBITS AND HARES

When the live cottontail is relaxed, the penis of the male and the clitoris of the female (which somewhat resembles the penis) are withdrawn into the body. These organs can be erected by applying downward pressure with a thumb and forefinger placed in front of and behind the genital region, respectively.

The penis is a cylindrical organ whose basal sheaths unfold on erection in the manner of a small telescope. A tiny terminal opening is apparent on close inspection.

The clitoris of the female is nearly as large as the penis of a young male and might be mistaken for it. However, it differs in being flattened posteriorly and in having no terminal opening. The vaginal opening is located between the base of the clitoris and the anus but is not always visible. In young females it is covered by the vaginal membrane.

Small rabbits are difficult to sex. In a young male, the distal portion of the penis may "open up" along the midventral line to somewhat resemble the clitoris. However, this flattening of the terminal region of the penis does not extend to the base of the organ as it does in the clitoris. Fox and Crary (1972) developed and illustrated a technique useful for rabbits as young as 27 days based on examination of the urogenital papilla.

Sex and age determination for hares is similar to that for cottontails (Lechleitner 1957). Young-of-the-year may be separated from older animals on the basis of length of hind foot and dry weight of the eye lens. The epiphyseal closure of the humerus allows separation of two groups of animals <10 months of age (Bothma et al. 1972, Table 19). Conception dates of prenatal animals can be estimated by use of a photographic key developed by Rongstad (1969). Freezing eye lenses dramatically reduces their weights (Pelton 1970). Weights vary so much among adults that this technique can separate only young-of-the-year from adults (Rongstad 1966). Sullins et al. (1976) showed that periosteal layers are absent in mandibles of young-of-the-year eastern and Nuttall's cottontails and present in those that were 1 year older. They suggested that annuli are useful for separating year classes of cottontails >1 year, but this technique has no particular advantage over lens weight for separating adults from young-of-the-year.

Eye-lens weight is superior to epiphyseal closure up to 140 days in black-tailed jack rabbits (Connolly et al. 1969*b*). Tiemeier and Plenert (1964) demonstrated that lens weight increased to at least 680 days, but beyond 200 days their sample size was not sufficient to assess the amount of variability among individuals of the same age.

Table 19. Sex and age criteria for rabbits, hares, and muskrats.

Species	Criteria for distinguishing sex		Criteria for distinguishing age	
	Female	Male	Immature	Mature
Cottontail and hares	See text and Lechleitner (1957)		Epiphyseal line or groove present on humerus (up to 9 months) (Hale 1949)	Epiphyseal line or groove on humerus absent (Hale 1949)
Muskrat	Nipples present on pelt (Schofield 1955)	Nipples absent on pelt (Schofield 1955)	*Pelt primeness pattern*	
	Urethral papilla (fresh or live animals)		Regular, longitudinal arrangement in subadult, lyre-shaped unprime area on dorsal side in juveniles (Applegate and Predmore 1947)	Irregular, spotted, or mottled (Applegate and Predmore 1947)
	Penis absent (Baumgartner and Bellrose 1943)	Penis present (Baumgartner and Bellrose 1943)	*Penis*	
			<5.15 mm diam., lighter red, with knob-shaped tip (Schofield 1955, Baumgartner and Bellrose 1943)	>5.15 mm diam., dark, with blunt rounded tip (Schofield 1955, Baumgartner and Bellrose 1943)
			Testis length in autumn and early winter	
			<11.65 mm (Schofield 1955)	>11.65 mm (Schofield 1955)
			Vaginal orifice	
			Closed by thick membrane (Schofield 1955)	Membrane thin or missing (Schofield 1955)
			Placental scars	
			Absent (Schofield 1955)	Present (if no barren adults) (Schofield 1955)

These authors placed jack rabbits into three age classes on the basis of condition of the proximal epiphyseal groove of the humerus: I (0–5 months), definite groove; II (5–14 months), definite line; III (>14 months), no line. Ear and hind-foot length reached full size within age class I, but age classes II and III probably could be distinguished by lens weight.

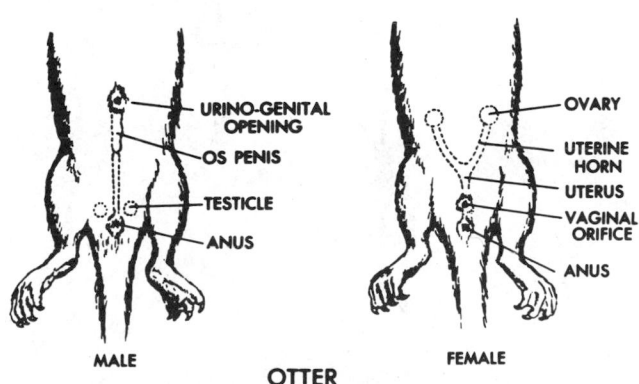

Fig. 27. Normal position of genitalia in male and female river otters. The presence of a baculum is readily determined in the male river otter by palpation in the live or dead animal (sketch adapted from Thompson 1958).

MUSKRAT

Sex may be determined by the presence or absence of the penis (Table 19) if the urinary papilla is grasped between forefinger and thumb and stripped posteriorly. The penis, if present, will be either felt or exposed. In very young rats not fully furred, the presence or absence of visible nipples will also indicate sex (Fig. 28).

The molt pattern on the inside of the skin is a good indicator of age (Fig. 29) until the molt is completed in February or March. Thereafter the most reliable way to distinguish juveniles (almost 1 year old) from adults is by the appearance of the first upper molar. Sather (1954) described the juvenile first upper molar as having fluting that runs deep into the alveolar socket, so that the end is not visible even in the cleaned skull. The adult, in contrast, has fluting that extends only part way along the tooth so that the end of the fluting is visible in the cleaned skull. In addition, the anterior face of the adult tooth is discernibly humped, whereas that of the juvenile is straight. Sather suggested that in the freshly killed animal it may be necessary to cut the gum away to see these characters clearly.

Olsen (1959) further refined this technique by distinguishing (in muskrats trapped in March and April) among three age classes as follows (description applies to the upper right molar): (1) highly developed roots and end of

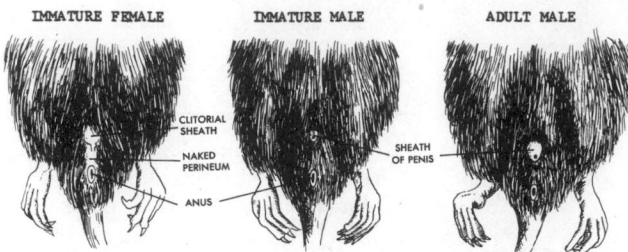

Fig. 28. Sex criteria as shown in genitalia in muskrats. Left: immature female with naked perineum; center: immature male, note size difference of penis sheath; right: adult male (after Dozier 1942, from Godin 1960).

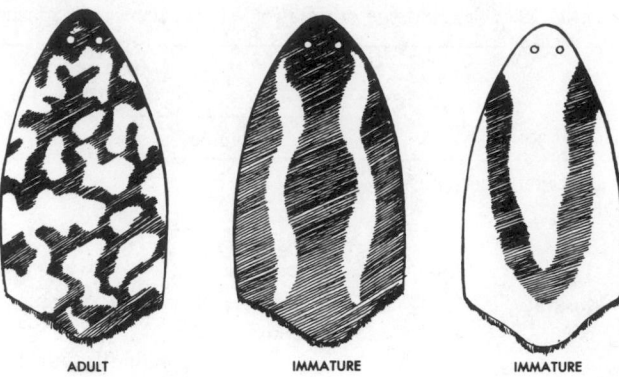

Fig. 29. The appearance of stretched muskrat pelts (skin side) showing dark and light color patterns related to age class. White areas denote primed section of hide; shaded areas are unprime (after Dozier 1942, from Godin 1960).

fluting extruded well below the bone line = adults; (2) moderate root development and end of the fluting just barely or not quite emerged from the bone line = subadults of about 10 months of age; and (3) little or no root development and fluting ending deep in the alveolar socket = juveniles averaging about 7 months of age. Doude Van Trootswijk (1976) developed a formula for estimating age in months of muskrats 1 month of age or older, based on the crown length and total length of the first molar:

$$\text{age (months)} = \frac{\left[\dfrac{100 - \text{Crown Length M}^1}{\text{Total Length M}^1} \times 100 \pm 1.98 \right]}{3.97} + 1.$$

She reported that the formula becomes less accurate as animals approach the age of 2 years. Vincent and Quere (1972) constructed a curve to estimate age, up to 36 months, based on eye-lens weights.

The zygomatic breadth was used to separate age classes of freshly skinned muskrats; subadults measured <4.16 mm and adults more than this (Alexander 1951). However, this measurement decreased as the skull dried, declining by 0.5 mm over the first 5 days. This shrinkage totalled 0.7 mm by the end of the first year. Summer humidity (70–80%) caused an increase of 0.3 mm in this measurement (Alexander 1960). Ossification of the baculum is a reliable age criterion for Missouri muskrats (Elder and Shanks 1962, Fig. 30).

BEAVER

Live adult beavers can be sexed, but only with experience, by palpation for the testes and baculum. The presence or absence of testes may be determined as follows. Place the beaver in a normal standing position, with head covered. Place one hand so that it lies lateral to the pubic symphysis with the fingertips anterior to the pubis and resting on the soft abdomen. Press lightly and draw the hand posteriorly. If the animal is a male, the testis can be felt as it slips anteriorly under the fingertips. If no testis is felt, a check may be made by palpation for the baculum. This is done by placing the thumb and forefinger immediately posterior to the pubic symphysis and passing them back toward the vent between the castor glands. Care must be taken not to misinterpret concretions in the castors. Another difficulty lies in the variability of the position of the penis. It may be at one side and in close proximity to the castors in old males; in

young males it is always in a median position (Osborn 1955). A. H. Kennedy (The sexing of beaver, Ont. Dep. Lands For., Fish Wildl. Div. Mimeogr., 1952) recommended palpating by inserting the index finger into the cloaca and urogenital orifice. The finger is passed ante-

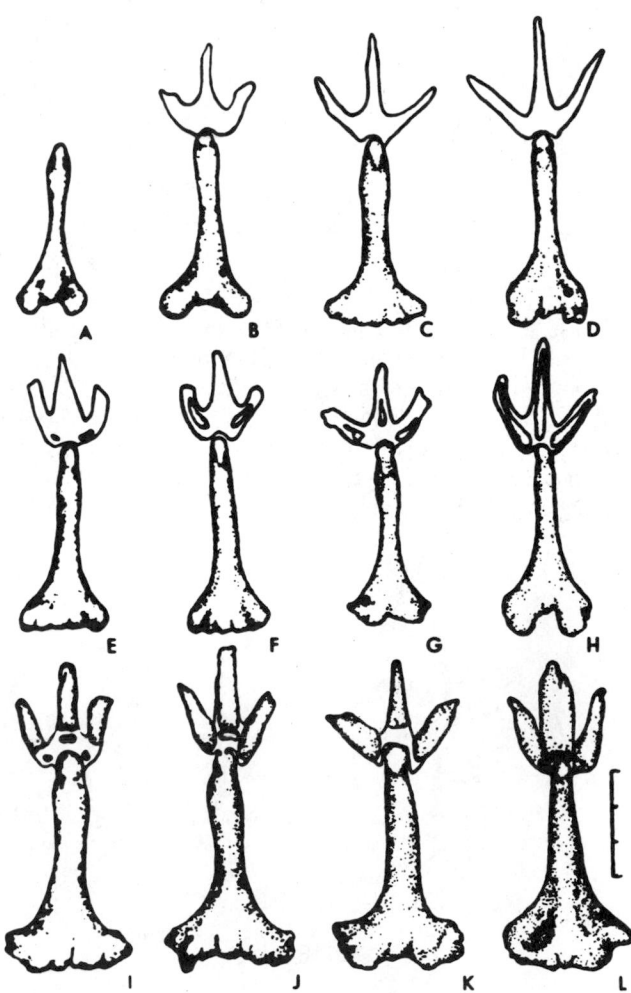

Fig. 30. Age changes in the baculum of the muskrat. A–D = about 4–8 months old; E–H = about 8–15 months old; I–L = >15 months old (adult). Scale in millimeters (from Elder and Shanks 1962).

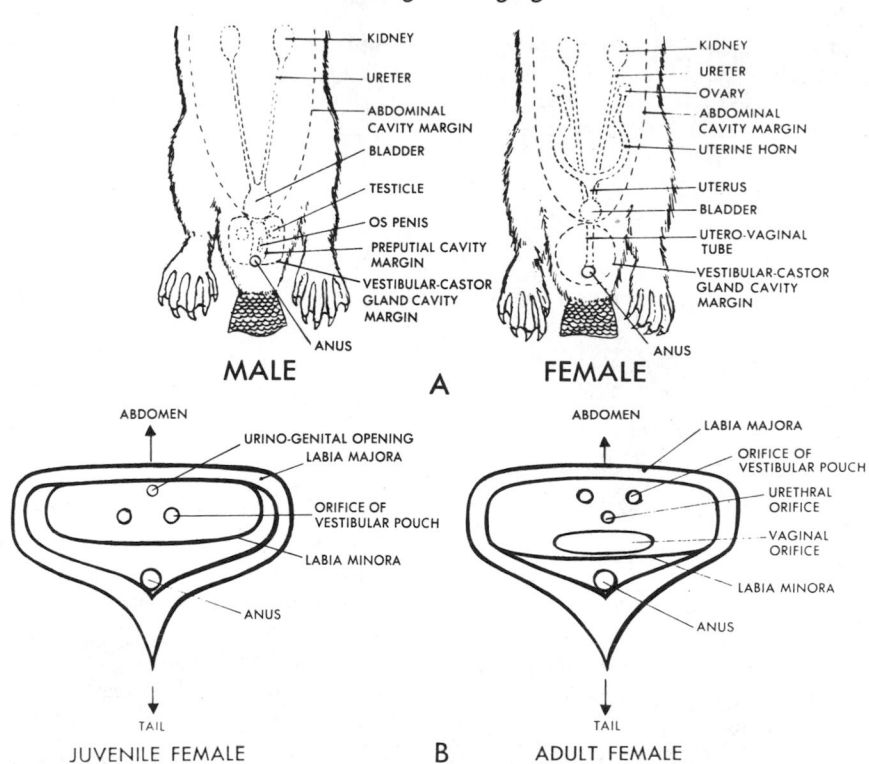

Fig. 31. Some age and sex characteristics of beavers. A. Schematic representation of penis and testis in the vestibular castor cavity of male (left) and normal position of the uterus of the female (right). Dissection is required to identify these organs. B. Diagrammatic representation of the anal-urogenital opening when stretched laterally as by the forefingers. This procedure can be performed on live or dead beavers (sketches by G. J. Knudson *in* Thompson 1958).

riorly into the vestibule or cavity that exists between the castor glands. The finger, moved from side to side about 2.5 cm from the external opening, will detect the penis if the animal is male. This method was first described by Bradt (1938). The appearance and location of beaver genitalia are illustrated in Fig. 31. Cementum annuli counts from ground and polished teeth are the most accurate indicator of year class (Van Nostrand and Stephenson 1964, Larson and Van Nostrand 1968). Buckley and Libby (1955) provided detailed descriptions of skull characteristics useful for aging beaver.

SQUIRREL (GRAY, FOX, AND RED)

Squirrels are born hairless and with the eyes closed; their subsequent postnatal development to 6 weeks is as shown in Table 20. Squirrels are sexed by examination of the external genitalia.

Several techniques have been developed to permit aging squirrels as immatures or adults. Body weights have been used with the reservation that weights are dependent on many external conditions independent of age (Table 21). Gray squirrels can be aged by cementum annuli (Fogl and Mosby 1978).

The degree of development of the external genitalia (Table 19) also has been used, but this criterion has limitations because fully adult males can be mistaken for immature males during the sexually quiescent period. Under favorable conditions, immature females will breed and appear to be >1 year old.

A normal November sample will contain spring-born juveniles (8–9 months of age), summer-born juveniles (3–4 months), and adults (>1 year). The three classes can be separated by the characters of the tail pelage, as viewed from below: young juvenile—two, sometimes three, dark lines run through the reddish-brown primary hairs of the tail, lower or proximal one-third of tail naked beneath; older juvenile (subadult)—dark lines as in juvenile, lower one-third of tail covered with short, appressed hairs; adult—tailbone obscured by appressed secondary hairs that radiate over and partially cover the long primary hairs of the tail. The lines or bars so prominent in young-of-the-year have become weakened in color intensity and are diffuse in the tails of adults (Fig. 32) (Sharp 1958). Pelage characteristics that reveal age are particularly advantageous in that live animals can be aged and then released to give known-age individuals in the population. Tail collagen strength has been used to estimate age in Belding's ground squirrels and may be applicable to other squirrels (Sherman et al. 1985).

Another readily applied field technique for aging gray squirrels that is based upon pelage characteristics was developed by Barrier and Barkalow (1967). In winter pelage, individuals can be identified as: summer juvenile—pelage in rump region when separated and laid flat by the thumbs reveals no yellow prebasal band in the black underfur, and most banded guard hairs are black-tipped; spring juvenile—the yellow prebasal band in the black underfur is absent or indistinct, and all banded guard hairs are white-tipped; adult—a distinct, yellow prebasal band is present in the black underfur, and all banded guard hairs are white-tipped.

One skeletal method separating juvenile from adult gray squirrels is the extent of the distal epiphyses of the radius

Table 20. Developmental characters of young fox and gray squirrels (from Larson and Taber 1980).

Age	Fox squirrel[a]	Gray squirrel[b]
Newborn	Wt. 14.2 g	
1 week	Wt. 28.3 g; first hair appears on back of head and shoulders	
2 weeks		Emerging hair darkens dorsal surface
3 weeks	Wt. 56.6 g; covered with dark hair about 1 mm long; hair beginning to turn brown on tail and around eyes and mouth; lower incisors appear and ears open	Ears open; lower incisors erupted
4 weeks		Silver hair on tail about 2 mm long; upper incisors erupting
5 weeks	Wt. 70.8–85 g; eyes open; hair appearing under tail, the last part to become furred	At least one eye open
6 weeks		Underside of tail covered with hair

[a]After Allen (1943).
[b]After Uhlig (1955).

Table 21. Characteristics for distinguishing between adult and juvenile fox and gray squirrels in October and November (from Uhlig 1956).

Adult	Juvenile
Male	
Ventral and posterior end of scrotum darkened and generally free of hair.	Posterior end of scrotum with smooth skin, brown to black, and possibly free of hair. Summer juveniles with scrotum covered with hair, small testes, may be difficult to detect. Spring males may sometimes be mistaken for adults.
Female	
Mammary glands large and noticeable, not hidden by hair; teats are black-tipped in fox squirrels. On gray squirrels the black spots may be absent or about the size of a pin point.	Teats inconspicuous, more or less hidden by the hair. Spring females occasionally have young before the hunting season and are classified as adults.
Male and female	
Tail rectangular, block-shaped, sides parallel or nearly so. Unless in an emaciated condition, adults weigh >396.2 g.	Tail pointed, triangular, sides not parallel. Spring juveniles will weigh >396.2 g. Summer juveniles will be <396.2 g. A rough index to age of summer juveniles is: 8 weeks—141.5 g; 10 weeks—198.1 g; 14 weeks—311.3 g; 16 weeks—367.9 g; 18 weeks—396.2 g

and ulna, which close with maturation of these long bones. X-ray reveals that the epiphyses remain open through the 18th week of life and that an epiphyseal line may still be detected until the 12th month. Thereafter, the epiphyseal line is absent (Petrides 1951, Carson 1961). Gray squirrels may be separated into young-of-the-year and adults on the basis of eye-lens weight (Fisher and Perry 1970).

The weight of the eye lens was used to distinguish age classes in fox squirrels shot in October and November in Michigan. The age classes and their respective lens weights were: summer born—to about 28 mg; spring born—29–39 mg; older—>39 mg, with some evidence of different year classes but with considerable overlap among them (Beale 1962). The lens-weight technique can be used without known-age weight or curves by plotting lens weight by frequency and noting natural breaks in the frequency distribution, provided a substantial sample is processed. This method would allow separation of summer young, spring young, and adults.

McCloskey (1977) tested various aging techniques on the same fox squirrel specimens and concluded that x-ray of the epiphyseal line in the forefoot was the most accurate indicator of age class. Among the field techniques, coloration and appearance of nipples of females, tail pelage (both sexes), and scrotal pigmentation of males were best.

Nellis (1969) observed cranial measurements overlapped too much to separate sexes of red squirrels. Lemnell (1974) reported cementum annuli in premolar and molar roots of red squirrels to be the most accurate aging

method, but he suggested that a more rapid method is to first sort out the juveniles on the basis of ephiphyseal closure or eye-lens techniques.

Although several aging techniques are applicable to squirrels, many are subject to external factors not related to age, such as range condition, health, and the experience and ability of the investigator. Therefore, one should always employ as many of the techniques as possible to enhance the accuracy of aging.

WOODCHUCK

Woodchucks in spring may be placed into young, yearling, and adult age classes based on the following criteria (Davis 1964):

Young: weigh 300–450 g about 15 May and gain about 19 g/day from June through September; do not begin molt until early July, pelage remains shorter and finer later in season than that of older animals, up to September; incisors narrow and pointed; mean eye-lens weight 12.32 mg (SD 2.8).

Yearling (young of previous year): in March and April size, head shape, and incisors like young; from February through April testes white (although some yearlings

Table 22. Age classes and approximate ages in months for *Didelphis* spp. based on eruption sequence of molars and replacement of deciduous (d) third (last) premolar. Parentheses around tooth numbers indicate erupting teeth.

Pre-molar (third)	Molars 1	2	3	4	Tyndale-Biscoe and Mackenzie[a] (1976)	Gardner[b] (1973)	Lowrance[c] (1949)	Gilmore[a] (1943)	Petrides[c] (1949)	VanDruff[c] (1971)
d3/d3	0/(1)	0/0	0/0	0/0	1[d]	immature			80 days +	
d3/d3	(1)/1	0/0	0/0	0/0						
d3/d3	1/1	0/(2)	0/0	0/0	2	1		juvenile 6–8	4	
d3/d3	1/1	(2)/2	0/(3)	0/0						4–6
d3/(3)	1/1	2/2	(3)/3	0/(4)	3	2	1		5–8.5	5–7
(3)/3	1/1	2/2	3/3	0/4	4	3	2	subadult	7–11	7–8
3/3	1/1	2/2	3/3	(4)/4		4	3			9–10
3/3	1/1	2/2	3/3	4/4	5	5	4	adult 10+	10+	10+
Wear on M$^{1-2}$					6	6				
Wear on all molars					7					

[a] *D. marsupialis* and *D. albiventris*.
[b] *D. marsupialis* and *D. virginiana*.
[c] *D. virginiana*.
[d] See Tyndale-Biscoe and Mackenzie (1976:252).

have pigmented testes in March and April); mean lens weight 21.78 mg (SD 1.7).

Adult: incisor broader, with worn points and darkly stained; testes light to dark brown; mean lens weight 28.53 mg (SD 4.5).

OPOSSUM

Sex of adult opossums can be determined by canine size; male canines are longer and heavier than female canines (Gardner 1982). A scrotum on males or a pouch on females is clear evidence of the sex on live animals, carcasses, and pelts (Table 16, Gardner 1982). Aging information is summarized in Tables 16 and 22.

SMALL MAMMALS

Eye-lens weights have been used extensively to attempt to age small mammals, but Dapson and Irland (1972) pointed out that because of the small size of the lens, the relative magnitude of error is high. They developed a technique based on the insoluble-protein (tyrosine) con-

tent of the lens, which is accurate up to at least 750 days. Birney et al. (1975) reported that for cotton rats lens weights are acceptable up to 130 days, but the insoluble-protein technique is necessary for older animals. Gourley and Jannett (1975) reported that if elaborately careful methods were used, lens weights were good in pine and montane voles up to 112 weeks. Tooth eruption is a good guide, up to 1 year, in arctic ground squirrels (Mitchell and Carsen 1967). Adhesion lines in the periosteal layer of lower jaws have been used in pika (Millar and Zwickel 1972) and cementum annuli in Uinta ground squirrels (Montgomery et al. 1971) and California ground squirrels (Adams and Watkins 1967). Apparently, cementum annuli and insoluble-protein content of the eye lens generally will be the most accurate techniques for most small mammals.

Beg and Hoffman (1977) used the pattern of maxillary tooth eruption and amount of wear on molariform teeth to determine age classes of the red-tailed chipmunk. Three age classes were established for wild juveniles between

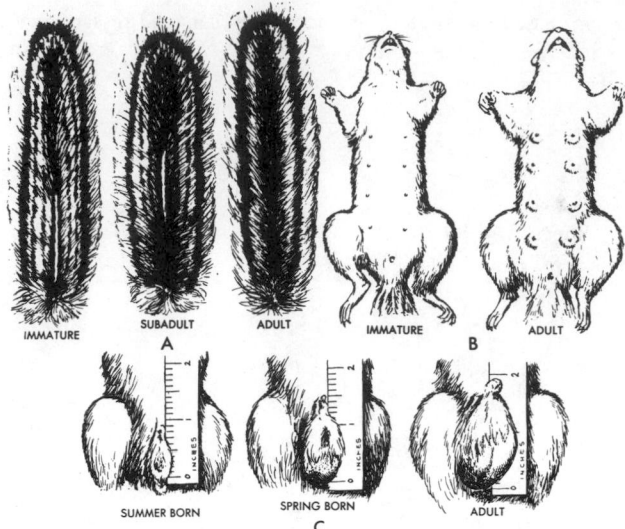

IMMATURE   SUBADULT   ADULT   IMMATURE   ADULT
                A   B

SUMMER BORN   SPRING BORN   ADULT
                    C

Fig. 32. Sex and age criteria for squirrels. A. Age may be determined by examination of the ventral surface of the tail. Left: juvenile, the shorter secondary hairs are absent on the lower side of the tailbone. Center: subadult, short appressed hairs are present on the lower third of the tailbone (after Sharp 1958). B. Mastology of the female squirrel. Left: juvenile, with nipples minute and barely discernible. Right: lactating adult, nipples pigmented black with most of hair worn off. C. Scrotal measurements of male squirrels. Left: summer born, the testes are abdominal and the skin is just beginning to pigment. Center: spring born, the testes are large and the scrotum is pigmented but heavily furred. Right: adult has shed most of the fur from the scrotum (after Allen 1943, from Godin 1960).

39 and 79 days of age and five age classes for adults between 10 and 64+ months of age.

Sex of small rodent (voles, lemmings, deer mice) skeletons found in raptor pellets, e.g., can be ascertained from the shape of the pelvic girdle (the three bones—ilium, ischium, and pubis constitute the innominate; the two innominates constitute the pelvic girdle). The pubic arm of the pelvis tends to be narrower and longer in proportion to the ischial arm in the male (Dunmire 1955).

LITERATURE CITED

ADAMS, D. A., AND T. L. QUAY. 1958. Ecology of the clapper rail in southeastern North Carolina. J. Wildl. Manage. 22:149–156.

ADAMS, L., AND S. G. WATKINS. 1967. Annuli in tooth cementum indicate age in California ground squirrels. J. Wildl. Manage. 31: 836–839.

ALEXANDER, M. M. 1951. The aging of muskrats on the Montezuma National Wildife Refuge. J. Wildl. Manage. 15:175–186.

———. 1960. Shrinkage of muskrat skulls in relation to aging. J. Wildl. Manage. 24:326–329.

ALLEN, D. L. 1939. Winter habits of Michigan skunks. J. Wildl. Manage. 3:212–228.

———. 1943. Michigan fox squirrel management. Mich. Dep. Conserv. Game Div. Publ. 100. 404pp.

ALLEN, S. H. 1974. Modified techniques for aging red fox using canine teeth. J. Wildl. Manage. 38:152–154.

———, AND S. C. KOHN. 1976. Assignment of age-classes in coyotes from canine cementum annuli. J. Wildl. Manage. 40:796–797.

ALT, G. L. 1980. Rate of growth and size of Pennsylvania black bears. Pa. Game News 51:7–17.

ANAS, R. E. 1970. Accuracy in assigning ages to fur seals. J. Wildl. Manage. 34:844–852.

APPLEGATE, V. C., AND H. G. PREDMORE, JR. 1947. Age classes and patterns of primeness in a fall collection of muskrat pelts. J. Wildl. Manage. 11:324–330.

ARCHIBALD, W. R., AND R. H. JESSUP. 1984. Population dynamics of the pine marten (*Martes americana*) in the Yukon Territory. Pages 81–97 *in* R. Olson, R. Hastings, and F. Geddes, eds. Northern ecology and resource management: memorial essays honoring Don Gill. Univ. Alberta Press, Edmonton.

ARMSTRONG, R. A. 1950. Fetal development of northern white-tailed deer (*Odocoileus virginianus borealis* Miller). Am. Midl. Nat. 43: 650–666.

ARTMANN, J. W., AND L. D. SCHROEDER. 1976. A technique for sexing woodcock by wing measurement. J. Wildl. Manage. 40:572–574.

AULERICH, R. J., AND D. R. SWINDLER. 1968. The dentition of mink (*Mustela vison*). J. Mammal. 49:488–494.

BAILEY, T. N. 1971. Biology of striped skunks on a southwestern Lake Erie marsh. Am. Midl. Nat. 85:196–207.

BAKER, M. F. 1953. Prairie chickens of Kansas. Univ. Kansas Mus. Nat. Hist. Misc. Publ. 5. 68pp.

BARNUM, D. A., J. S. GREEN, J. T. FLINDERS, AND N. L. GATES. 1979. Nutritional levels and growth rates of handreared coyote pups. J. Mammal. 60:820–823.

BARRIER, M. J., AND F. S. BARKALOW, JR. 1967. A rapid technique for aging gray squirrels in winter pelage. J. Wildl. Manage. 31: 715–719.

BAUER, R. D., A. M. JOHNSON, AND V. B. SCHEFFER. 1964. Eye lens weight and age in the fur seal. J. Wildl. Manage. 28:374–376.

BAUMGARTNER, L. L., AND F. C. BELLROSE, JR. 1943. Determination of sex and age in muskrats. J. Wildl. Manage. 7:77–81.

BEALE, D. M. 1962. Growth of the eye lens in relation to age in fox squirrels. J. Wildl. Manage. 26:208–211.

BEG, M. A., AND R. S. HOFFMAN. 1977. Age determination in the red-tailed chipmunk, *Eutamias ruficaudus*. Murrelet 58:26–36.

BEKOFF, M. 1982. Coyote. Pages 447–459 *in* J. A. Chapman and G. A. Feldhamer, eds. Wild mammals of North America. The Johns Hopkins Univ. Press, Baltimore, Md.

BELLROSE, F. C. 1980. Ducks, geese & swans of North America. Stackpole Books, Harrisburg, Pa. 540pp.

BERG, W. E., AND D. W. KUEHN. 1980. Radiographs as a carnivore aging technique. Abstr. Midwest Fish Wildl. Conf. 42:51.

BERGERUD, A. T. 1964. Relationship of mandible length to sex in Newfoundland caribou. J. Wildl. Manage. 28:54–56.

———. 1970. Eruption of permanent premolars and molars for Newfoundland caribou. J. Wildl. Manage. 34:962–963.

———, S. S. PETERS, AND R. MCGRATH. 1963. Determining sex and age of willow ptarmigan in Newfoundland. J. Wildl. Manage. 27: 700–711.

———, AND H. L. RUSSELL. 1966. Extraction of incisors of Newfoundland caribou. J. Wildl. Manage. 30:842–843.

BIRNEY, E. C., AND E. D. FLEHARTY. 1966. Age and sex comparisons of wild mink. Trans. Kansas Acad. Sci. 69:139–145.

———, AND ———. 1968. Comparative success in the application of aging techniques to a population of winter-trapped mink. Southwest. Nat. 13:275–282.

———, R. JENNESS, AND D. D. BAIRD. 1975. Eye lens proteins as criteria of age in cotton rats. J. Wildl. Manage. 39:718–728.

BJORGE, R. R. 1977. Population dynamics, denning, and movements of striped skunks in central Alberta. M.S. Thesis, Univ. Alberta, Edmonton. 96pp.

BLANKENSHIP, L. H. 1957. Investigations of the American woodcock in Michigan. Mich. Dep. Conserv. Rep. 2123. 217pp.

BLOEMENDAL, H. 1977. The vertebrate eye lens. Science 197:127–138.

BOOKHOUT, T. A. 1964. Prenatal development of snowshoe hares. J. Wildl. Manage. 28:338–345.

BORTOLOTTI, G. R. 1984. Sexual size dimorphism and age-related size variation in bald eagles. J. Wildl. Manage. 48:72–81.

BOTHMA, J. DU.P., J. G. TEER, AND C. E. GATES. 1972. Growth and age determination of the cottontail in south Texas. J. Wildl. Manage. 36:1209–1221.

BOWEN, W. O. 1982. Determining age of coyotes, *Canis latrans,* by tooth sections and tooth wear patterns. Can. Field-Nat. 96:339–341.

BRADT, G. W. 1938. A study of beaver colonies in Michigan. J. Mammal. 19:139–162.

BRANDBORG, S. M. 1955. Life history and management of the mountain goat in Idaho. Idaho Dep. Fish Game Wildl. Bull. 2. 142pp.

BRASSARD, J. S., AND R. BERNARD. 1939. Observations on breeding and development of martens, *Martes a. americana* (Ken). Can. Field-Nat. 53:15–21.

BRAUN, C. E., AND G. E. ROGERS. 1967. Determination of age and sex of the southern white-tailed ptarmigan. Colo. Game, Fish, Parks Dep. Game Inf. Leafl. 54. unnumb.

BRENNAN, L. A., AND W. M. BLOCK. 1985. Sex determination of mountain quail reconsidered. J. Wildl. Manage. 49:475–476.

BROWN, M. W. 1983. A morphometric analysis of sexual and age variation in the American marten (*Martes americana*). M.S. Thesis, Univ. Toronto, Toronto, Ont. 190pp.

BUCKLEY, J. L., AND W. L. LIBBY. 1955. Growth rates and age determination in Alaskan beaver. Trans. North Am. Wildl. Conf. 20: 495–507.

BULL, J., AND J. FARRAND, JR. 1977. The Audubon Society field guide to North American birds—eastern region. Alfred A. Knopf, New York, N.Y. 784pp.

BUMP, G., R. W. DARROW, F. C. EDMINSTER, AND W. F. CRISSEY. 1947. The ruffed grouse: life history, propagation, management. New York State Conserv. Dep., Albany. 915pp.

CAMPBELL, H. 1972. A population study of lesser prairie chickens in New Mexico. J. Wildl. Manage. 36:689–699.

CARBYN, L. N. 1987. Gray wolf and red wolf. Pages 358–376 *in* M. Novak, J. A. Baker, M. E. Obbard, and B. Malloch, eds. Wild furbearer management and conservation in North America. Minist. Nat. Resour., Toronto, Ont.

CARNEY, S. M. 1992. Species, age and sex identification of ducks using wing plumage. U.S. Fish Wildl. Serv., Washington, D.C. 144pp.

CARSON, J. D. 1961. Epiphyseal cartilage as an age indicator in fox and gray squirrels. J. Wildl. Manage. 25:90–93.

CASWELL, E. B. 1954. A method for sexing blue grouse. J. Wildl. Manage. 18:139.

CHAPMAN, J. A., AND G. A. FELDHAMER, EDITORS. 1982. Wild mammals of North America. The Johns Hopkins Univ. Press, Baltimore, Md. 1147pp.

CHILD, K. N. 1973. The cryostat: a tool for the big game biologist. Can. J. Zool. 51:663–664.

CHRISTENSEN, G. C. 1970. The chukar partridge: its introduction, life history, and management. Nev. Dep. Fish Game Biol. Bull. 4. 82pp.

CONNOLLY, G. E., M. L. DUDZIŃSKI, AND W. M. LONGHURST. 1969a. An improved age-lens weight regression for black-tailed deer and mule deer. J. Wildl. Manage. 33:701–704.

———, ———, AND ———. 1969b. The eye lens as an indicator of age in the black-tailed jack rabbit. J. Wildl. Manage. 33:159–164.

COPELIN, F. F. 1963. The lesser prairie chicken in Oklahoma. Okla. Wildl. Conserv. Dep. Tech. Bull. 6. 58pp.

COTTAM, C., AND J. B. TREFETHEN, EDITORS. 1968. Whitewings: the life history, status and management of the white-winged dove. D. Van Nostrand Company, Inc., Princeton, N.J. 348pp.

COULTER, M. W. 1966. Ecology and management of fishers in Maine. Ph.D. Thesis, Syracuse Univ., Syracuse, N.Y. 196pp.

COUTURIER, M. A. J. 1954. L'ours brun, *Ursus arctos*. Dr. M. Couturier, Grenoble, France. 906pp.

COWAN, I. McT. 1936. Distribution and variation in deer (genus *Odocoileus*) of the Pacific coastal region of North America. Calif. Fish Game 22:155–246.

CRAIGHEAD, J. J., AND J. A. MITCHELL. 1982. Grizzly bear. Pages 515–556 *in* J. A. Chapman and G. A. Feldhamer, eds. Wild mammals of North America. The Johns Hopkins Univ. Press, Baltimore, Md.

CRISPENS, C. G., JR., AND J. K. DOUTT. 1970. Studies of the sex chromatin in the white-tailed deer. J. Wildl. Manage. 34:642–644.

CROWE, D. M. 1972. The presence of annuli in bobcat tooth cementum layers. J. Wildl. Manage. 36:1330–1332.

———. 1975. Aspects of aging, growth, and reproduction of bobcats from Wyoming. J. Mammal. 56:177–198.

———, AND M. D. STRICKLAND. 1975. Population structures of some mammalian predators in southeastern Wyoming. J. Wildl. Manage. 39:449–450.

CRUNDEN, C. W. 1963. Age and sex of sage grouse from wings. J. Wildl. Manage. 27:846–849.

CURRIER, M. J. P. 1979. An age estimation technique and some normal blood values for mountain lions (*Felis concolor*). Ph.D. Thesis, Colorado State Univ., Ft. Collins. 81pp.

DAGG, A. I., D. LEACH, AND G. SUMNER-SMITH. 1975. Fusion of the distal femoral epiphysis in male and female marten and fisher. Can. J. Zool. 53:1514–1518.

DAHLGREN, R. B., C. M. TWEDT, AND C. G. TRAUTMAN. 1965. Lens weight of ring-necked pheasants. J. Wildl. Manage. 29:212–214.

DALKE, P. D., D. B. PYRAH, D. C. STANTON, J. E. CRAWFORD, AND E. F. SCHLATTERER. 1963. Ecology, productivity, and management of sage grouse in Idaho. J. Wildl. Manage. 27:811–841.

DAPSON, R. W., AND J. M. IRLAND. 1972. An accurate method of determining age in small mammals. J. Mammal. 53:100–106.

DAVIS, D. E. 1964. Evaluation of characters for determining age of woodchucks. J. Wildl. Manage. 28:9–15.

DAVIS, J. A. 1969. Aging and sexing criteria for Ohio ruffed grouse. J. Wildl. Manage. 33:628–636.

DeGRAAF, R. M., AND J. S. LARSON. 1972. A technique for the observation of sex chromatin in hair roots. J. Mammal. 53:368–371.

DELLINGER, G. P. 1954. Breeding season, productivity, and population trends of raccoon in Missouri. M.A. Thesis, Univ. Missouri, Columbia. 86pp.

DEMING, O. V. 1952. Tooth development of the Nelson bighorn sheep. Calif. Fish Game 38:523–529.

DIMMICK, R. W. 1992. Northern bobwhite (*Colinus virginianus*): Section 4.1.3. U.S. Army Corps of Engineers Wildlife Resources Management Manual. Tech. Rep. EL-92-18, U.S. Army Eng. Waterways Exp. Stn., Vicksburg, Miss. 74pp.

DIX, L. M., AND M. A. STRICKLAND. 1986. Use of tooth radiographs to classify martens by sex and age. Wildl. Soc. Bull. 14:275–279.

DOUDE VAN TROOSTWIJK, W. J. 1976. Age determination in muskrats, *Ondatra zibethicus* (L.) in the Netherlands. Lutra 18:33–43.

DOUGLAS, C. W., AND M. A. STRICKLAND. 1987. Fisher. Pages 511–529 *in* M. Novak, J. A. Baker, M. E. Obbard, and B. Malloch, eds. Wild furbearer management and conservation in North America. Minist. Nat. Resour., Toronto, Ont.

DOW, S. A. 1952. Antelope ageing studies in Montana. Proc. Western Assoc. State Game and Fish Comm. 32:220–224.

———, AND P. L. WRIGHT. 1962. Changes in mandibular dentition associated with age in pronghorn antelope. J. Wildl. Manage. 26: 1–18.

DOZIER, H. L. 1942. Identification of sex in live muskrats. J. Wildl. Manage. 6:292–293.

DUFFIELD, L. F. 1973. Aging and sexing the post-cranial skeleton of bison. Plains Anthropol. 18:132–139.

DUNMIRE, W. W. 1955. Sex dimorphism in the pelvis of rodents. J. Mammal. 36:356–361.

DUNNE, P. 1987. Introduction to raptor identification, aging and sexing techniques. Pages 13–21 *in* B. A. Giron Pendleton, B. A. Millsap, K. W. Cline, and D. M. Bird, eds. Raptor management techniques manual. Natl. Wildl. Fed., Washington, D.C.

DWYER, T. J., AND J. V. DOBELL. 1979. External determination of age of common snipe. J. Wildl. Manage. 43:754–756.

EAGLE, T. C., AND J. S. WHITMAN. 1987. Mink. Pages 615–624 *in* M. Novak, J. A. Baker, M. E. Obbard, and B. Malloch, eds. Wild furbearer management and conservation in North America. Minist. Nat. Resour., Toronto, Ont.

EDMINSTER, F. C. 1954. American game birds of field and forest. Charles Scribner's Sons, New York, N.Y. 490pp.

EDWARDS, J. K., R. L. MARCHINTON, AND G. F. SMITH. 1982. Pelvic girdle criteria for sex determination of white-tailed deer. J. Wildl. Manage. 46:544–547.

EINARSEN, A. S. 1948. The pronghorn antelope and its management. Wildl. Manage. Inst., Washington, D.C. 238pp.

ELDER, W. H., AND C. E. SHANKS. 1962. Age changes in tooth wear and morphology of the baculum in muskrats. J. Mammal. 43:144–150.

ELLISON, L. N. 1968. Sexing and aging Alaskan spruce grouse by plumage. J. Wildl. Manage. 32:12–16.

ENG, R. L. 1955. A method for obtaining sage grouse age and sex ratios from wings. J. Wildl. Manage. 19:267–272.

ERICKSON, J. A., A. E. ANDERSON, D. E. MEDIN, AND D. C. BOWDEN. 1970. Estimating ages of mule deer—an evaluation of technique accuracy. J. Wildl. Manage. 34:523–531.

ETTER, S. L., J. E. WARNOCK, AND G. B. JOSELYN. 1970. Modified wing molt criteria for estimating the ages of wild juvenile pheasants. J. Wildl. Manage. 34:620–626.

FISHER, E. W., AND A. E. PERRY. 1970. Estimating ages of gray squirrels by lens-weights. J. Wildl. Manage. 34:825–828.

FOGARTY, M. J., K. A. ARNOLD, L. McKIBBEN, L. B. POSPICHAL, AND R. J. TULLY. 1977. Common snipe. Pages 189–209 *in* G. C. Sanderson, ed. Management of migratory shore and upland game birds in North America. Int. Assoc. Fish Wildl. Agencies, Washington, D.C.

FOGL, J. G., AND H. S. MOSBY. 1978. Aging gray squirrels by cementum annuli in razor-sectioned teeth. J. Wildl. Manage. 42:444–448.

FOX, R. R., AND D. D. CRARY. 1972. A simple technique for the sexing of newborn rabbits. Lab. Anim. Sci. 22:556–558.

FREDRICKSON, L. H. 1968. Measurements of coots related to sex and age. J. Wildl. Manage. 32:409–411.

FRIEDRICH, P. D., G. E. BURGOYNE, T. M. COOLEY, AND S. M. SCHMITT. 1983. Use of lower canine tooth for determining the sex of bobcats in Michigan. Mich. Dep. Nat. Resour. Wildl. Div. Rep. 2960. 5pp.

FRIEND, M. 1967. A review of research concerning eye-lens weight as a criterion of age in animals. N.Y. Fish Game J. 14:152–165.

FRISON, G. C., AND C. A. REHER. 1970. Age determination of buffalo by teeth eruption and wear. Plains Anthropol. 15:46–50.

FRITZELL, E. K. 1987. Gray fox and island gray fox. Pages 408–421 in M. Novak, J. A. Baker, M. E. Obbard, and B. Malloch, eds. Wild furbearer management and conservation in North America. Minist. Nat. Resour., Toronto, Ont.

FULLER, T. K., D. P. HOBSON, J. R. GUNSON, D. B. SCHOWALTER, AND D. HEISEY. 1984. Sexual dimorphism in mandibular canines of striped skunks. J. Wildl. Manage. 48:1444–1446.

FULLER, W. A. 1959. The horns and teeth as indicators of age in bison. J. Wildl. Manage. 23:342–344.

GARDNER, A. L. 1973. The systematics of the genus Didelphis (Marsupialia: Didelphidae) in North and middle America. Mus. Texas Tech Univ. Spec. Publ. 4. 81pp.

———. 1982. Virginia opossum. Pages 3–36 in J. A. Chapman and G. A. Feldhamer, eds. Wild mammals of North America. The Johns Hopkins Univ. Press, Baltimore, Md.

GASAWAY, W. C., D. B. HARKNESS, AND R. A. RAUSCH. 1978. Accuracy of moose age determinations from incisor cementum layers. J. Wildl. Manage. 42:558–563.

GATES, J. M. 1966. Validity of spur appearance as an age criterion in the pheasant. J. Wildl. Manage. 30:81–85.

GEIGER, G., J. BROMEL, AND K. H. HABERMEHL. 1977. Concordance of various methods of determining the age of the red fox (*Vulpes vulpes* L., 1758). Z. Jagdwiss. 23:57–64.

GEIST, V. 1966. Validity of horn segment counts in aging bighorn sheep. J. Wildl. Manage. 30:634–635.

GIER, H. T. 1968. Coyotes in Kansas. Kans. Agric. Exp. Stn. Bull. 393. 118pp.

GIESEN, K. M., AND C. E. BRAUN. 1979. A technique for age determination of juvenile white-tailed ptarmigan. J. Wildl. Manage. 43:508–511.

GILBERT, F. F., AND S. L. STOLT. 1970. Variability in aging Maine white-tailed deer by tooth-wear characteristics. J. Wildl. Manage. 34:532–535.

GILMORE, R. M. 1943. Mammalogy in an epidemiological study of jungle yellow fever in Brazil. J. Mammal. 24:144–162.

GODIN, A. J. 1960. A compilation of diagnostic characteristics used in aging and sexing game birds and mammals. M.S. Thesis, Univ. Massachusetts, Amherst. 160pp.

GOLLOP, J. B., AND W. H. MARSHALL. 1954. A guide for aging duck broods in the field. Miss. Flyway Counc. Tech. Sect. Rep. 14pp.

GOODWIN, E. A., AND W. B. BALLARD. 1985. Use of tooth cementum for age determination of gray wolves. J. Wildl. Manage. 49:313–316.

GORDON, K. R., AND G. V. MOREJOHN. 1975. Sexing black bear skulls using lower canine and lower molar measurement. J. Wildl. Manage. 39:40–44.

GOURLEY, R. S., AND F. J. JANNETT, JR. 1975. Pine and montane vole age estimates from eye lens weights. J. Wildl. Manage. 39:550–556.

GRAU, G. A., G. C. SANDERSON, AND J. P. ROGERS. 1970. Age determination of raccoons. J. Wildl. Manage. 34:364–372.

GREENBERG, R. E., S. L. ETTER, AND W. L. ANDERSON. 1972. Evaluation of proximal primary feather criteria for aging wild pheasants. J. Wildl. Manage. 36:700–705.

GREER, K. R. 1957. Some osteological characters of known-age ranch minks. J. Mammal. 38:319–330.

———, AND H. W. YEAGER. 1967. Sex and age indications from upper canine teeth of elk (wapiti). J. Wildl. Manage. 31:408–417.

HAAGENRUD, H. 1978. Layers in secondary dentine of incisors as age criteria in moose (*Alces alces*). J. Mammal. 59:857–858.

HALE, J. B. 1949. Aging cottontail rabbits by bone growth. J. Wildl. Manage. 13:216–225.

———, R. F. WENDT, AND G. C. HALAZON. 1954. Sex and age criteria for Wisconsin ruffed grouse. Wis. Conserv. Dep. Tech. Wildl. Bull. 9. 24pp.

HALL, E. R. 1981. The mammals of North America. Vol. II. John Wiley & Sons, New York, N.Y. 1181pp.

HAMILTON, R. J., T. L. IVEY, AND M. L. TOBIN. 1985. Aging fetal

white-tailed deer. Proc. Annu. Conf. Southeast. Assoc. Fish Wildl. Agencies 39:389–394.

HAMILTON, W. J., JR., AND W. R. EADIE. 1964. Reproduction in the river otter, *Lutra canadensis*. J. Mammal. 45:242–252.

HANSON, H. C. 1962. Characters of age, sex, and sexual maturity in Canada geese. Ill. Nat. Hist. Surv. Biol. Notes 49. 15pp.

———, AND C. W. KOSSACK. 1963. The mourning dove in Illinois. Ill. Dep. Conserv. Tech. Bull. 2. 133pp.

HASH, H. S. 1987. Wolverine. Pages 575–585 in M. Novak, J. A. Baker, M. E. Obbard, and B. Malloch, eds. Wild furbearer management and conservation in North America. Minist. Nat. Resour., Toronto, Ont.

HEALY, W. M., AND E. S. NENNO. 1980. Growth parameters and sex and age criteria for juvenile eastern wild turkeys. Natl. Wild Turkey Symp. 4:168–185.

HEMMING, J. E. 1969. Cemental deposition, tooth succession, and horn development as criteria of age in Dall sheep. J. Wildl. Manage. 33:552–558.

HENDERSON, F. R., F. W. BROOKS, R. E. WOOD, AND R. B. DAHLGREN. 1967. Sexing of prairie grouse by crown feather patterns. J. Wildl. Manage. 31:764–769.

HENSEL, R. J., AND F. E. SORENSEN, JR. 1980. Age determination of live polar bears. Int. Conf. Bear Res. Manage. 4:93–100.

HESSELTON, W. T., AND R. M. HESSELTON. 1982. White-tailed deer. Pages 878–901 in J. A. Chapman and G. A. Feldhamer, eds. Wild mammals of North America. The Johns Hopkins Univ. Press, Baltimore, Md.

HILLMAN, C. N., AND W. W. JACKSON. 1973. The sharp-tailed grouse in South Dakota. S.D. Dep. Game, Fish, Parks Tech. Bull. 3. 62pp.

HINMAN, R. A., EDITOR. 1979. Annual report of survey inventory activities. Part 4: Sheep, mountain goat, bison, musk-oxen, marine mammals. Vol. 9. Alas. Dep. Fish Game, Juneau. 123pp.

HOEKSTRA, T. W., AND P. G. CARR. 1977. Sex determination in white-tailed deer tissues. Pages 212–232 in Proc. forensic science: a tool for modern fish and wildlife science. Alberta Recreation, Parks, Wildl., Fish Wildl. Div., Calgary.

HOFFMAN, R. W. 1985. Blue grouse wing analyses: methodology and population inferences. Colo. Div. Wildl. Spec. Rep. 60. 21pp.

HOLLIMAN, D. C. 1977. Purple gallinule. Pages 105–109 in G. C. Sanderson, ed. Management of migratory shore and upland game birds in North America. Int. Assoc. Fish Wildl. Agencies, Washington, D.C.

HUDSON, P., AND L. G. BROWMAN. 1959. Embryonic and fetal development of the mule deer. J. Wildl. Manage. 23:295–304.

HUNTER, S. A. 1989. Aspartic acid racemization in tendons as an indication of age in three avian species. Ph.D. Thesis, Southern Illinois Univ., Carbondale. 54pp.

JACOBSON, H. A., AND R. J. REINER. 1989. Estimating age of white-tailed deer: tooth wear versus cementum annuli. Proc. Annu. Conf. Southeast. Assoc. Fish Wildl. Agencies 43:286–291.

JENKS, J. A., R. T. BOWER, AND A. G. CLARK. 1984. Sex and age-class determination for fisher using radiographs of canine teeth. J. Wildl. Manage. 48:626–628.

JOHNSGARD, P. A. 1973. Grouse and quails of North America. Univ. Nebraska Press, Lincoln. 553pp.

———. 1975. North American game birds of upland and shoreline. Univ. Nebraska Press, Lincoln. 183pp.

JOHNSON, A. S. 1970. Biology of the raccoon (*Procyon lotor varius* Nelson and Goldman) in Alabama. Auburn Univ. Agric. Exp. Stn. Bull. 402. 148pp.

JOHNSON, N. F., B. A. BROWN, AND J. C. BOSOMWORTH. 1981. Age and sex characteristics of bobcat canines and their use in population assessment. Wildl. Soc. Bull. 9:203–206.

JOHNSON, S. A. 1984. Home range, movements, and habitat use of fishers in Wisconsin. M.S. Thesis, Univ. Wisconsin, Stevens Point. 78pp.

JOHNSTON, D. H., ET AL. 1987. Aging furbearers using tooth structure and biomarkers. Pages 228–243 in M. Novak, J. A. Baker, M. E. Obbard, and B. Malloch, eds. Wild furbearer management and conservation in North America. Minist. Nat. Resour., Toronto, Ont.

———, AND I. D. WATT. 1981. A rapid method for sectioning undecalcified carnivore teeth for aging. Pages 407–422 in J. A. Chapman and D. Pursley, eds. Proc. Worldwide Furbearer Conf., Frostburg, Md.

JONES, F. L., G. FLITTNER, AND R. GARD. 1954. Report on a survey of bighorn sheep and other game in the Santa Rosa Mountains,

Riverside County (California). Calif. Dep. Fish Game, Sacramento. 26pp. (mimeogr.)

JUNGE, R., AND D. F. HOFFMEISTER. 1980. Age determination in raccoons from cranial suture obliteration. J. Wildl. Manage. 44:725–729.

KALLA, P. I. 1991. Studies on the biology of ruffed grouse in the southern Appalachian mountains. Ph.D. Thesis, Univ. Tennessee, Knoxville. 101pp.

KAUFMANN, J. H. 1982. Raccoon and allies. Pages 567–585 *in* J. A. Chapman and G. A. Feldhamer, eds. Wild mammals of North America. The Johns Hopkins Univ. Press, Baltimore, Md.

KEISS, R. E. 1969. Comparison of eruption-wear patterns and cementum annuli as age criteria in elk. J. Wildl. Manage. 33:175–180.

KELLY, G. 1975. Indexes for aging eastern wild turkeys. Proc. Natl. Wild Turkey Symp. 3:205–209.

KENYON, K. W., AND C. H. FISCUS. 1963. Age determination in the Hawaiian monk seal. J. Mammal. 44:280–282.

KERWIN, M. L., AND G. J. MITCHELL. 1971. The validity of the wear-age technique for Alberta pronghorns. J. Wildl. Manage. 35:743–747.

KIRKPATRICK, R. D., AND L. K. SOWLS. 1962. Age determination of the collared peccary by the tooth-replacement pattern. J. Wildl. Manage. 26:214–217.

KLEVEZAL', G. A., AND S. E. KLEINENBERG. 1967. Age determination of mammals from annual layers in teeth and bones. USSR Acad. Sci., Severtsov Inst. Anim. Morphol. Clearinghouse Fed. Sci. Tech. Inf. U.S. Dep. Commer., Springfield, Va. 128pp. (Translated from Russian)

———, AND M. V. MINA. 1973. Factors determining the pattern of annual layers in dental tissue and bones of mammals. Zh. Obshch. Biol. 34:594–604.

KOLENOSKY, G. B. 1987. Polar bear. Pages 475–485 *in* M. Novak, J. A. Baker, M. E. Obbard, and B. Malloch, eds. Wild furbearer management and conservation in North America. Minist. Nat. Resour., Toronto, Ont.

———, AND S. M. STRATHEARN. 1987. Black bear. Pages 442–455 *in* M. Novak, J. A. Baker, M. E. Obbard, and B. Malloch, eds. Wild furbearer management and conservation in North America. Minist. Nat. Resour., Toronto, Ont.

KUEHN, D. W., AND W. E. BERG. 1981. Use of radiographs to identify age-classes of fisher. J. Wildl. Manage. 45:1009–1010.

LARSON, J. S., AND S. J. KNAPP. 1971. Sexual dimorphism in beaver neutrophils. J. Mammal. 52:212–215.

———, AND R. D. TABER. 1980. Criteria of sex and age. Pages 143–202 *in* S. D. Schemnitz, ed. Wildlife techniques manual. Fourth ed. The Wildl. Soc., Washington, D.C.

———, AND F. C. VAN NOSTRAND. 1968. An evaluation of beaver aging techniques. J. Wildl. Manage. 32:99–103.

LAUHACHINDA, V. 1978. Life history of the river otter in Alabama with emphasis on food habits. Ph.D. Thesis, Auburn Univ., Auburn, Ala. 169pp.

LAWS, R. M. 1953. A new method of age determination in mammals with special reference to the elephant seal (*Mirounga leonina* Linnaeus). Falk. Isl. Depend. Surv. Sci. Rep. 2:1–12.

———. 1962. Age determination of pinnipeds with special reference to growth layers in the teeth. Z. Saeugetierkd. 27:129–146.

LAWSON, B., AND R. JOHNSON. 1982. Mountain sheep. Pages 1036–1055 *in* J. A. Chapman and G. A. Feldhamer, eds. Wild mammals of North America. The Johns Hopkins Univ. Press, Baltimore, Md.

LEACH, D. 1977. The descriptive and comparative postcranial osteology of marten (*Martes americana* Terton) and fisher. Can. J. Zool. 55:199–214.

———, AND U. S. DE KLEER. 1978. The descriptive and comparative postcranial osteology of marten (*Martes americana* Terton) and fisher (*Martes pennanti* Erxleben): the axial skeleton. Can. J. Zool. 56:1180–1191.

———, B. K. HALL, AND A. I. DAGG. 1982. Aging marten and fisher by development of the suprafabellar tubercle. J. Wildl. Manage. 46:246–247.

LECHLEITNER, R. R. 1954. Age criteria in mink (*Mustela vison*). J. Mammal. 35:496–503.

———. 1957. The black-tailed jackrabbit on Grey Lodge Refuge, California. Ph.D. Thesis, Univ. California, Berkeley. 179pp.

LEMNELL, P. A. 1974. Age determination in red squirrels, (*Sciurus vulgaris*). Trans. Int. Congr. Game Biol. 11:573–580.

LEOPOLD, A. S. 1959. Wildlife of Mexico. Univ. California Press, Berkeley. 568pp.

LEWIS, J. C. 1979. Field identification of juvenile sandhill cranes. J. Wildl. Manage. 43:211–214.

LINDER, R. L., R. B. DAHLGREN, AND C. R. ELLIOTT. 1971. Primary feather pattern as a sex criterion in the pheasant. J. Wildl. Manage. 35:840–843.

LINDZEY, F. 1987. Mountain lion. Pages 658–668 *in* M. Novak, J. A. Baker, M. E. Obbard, and B. Malloch, eds. Wild furbearer management and conservation in North America. Minist. Nat. Resour., Toronto, Ont.

LINDZEY, F. G. 1971. Ecology of badgers in Carlew Valley, Utah and Idaho, with emphasis on movement and activity patterns. M.S. Thesis, Utah State Univ., Logan. 50pp.

LINHART, S. B., AND F. F. KNOWLTON. 1967. Determining age of coyotes by tooth cementum layers. J. Wildl. Manage. 31:362–365.

LISCINSKY, S. A. n.d. The American woodcock in Pennsylvania. Pa. Game Comm., Harrisburg. 32pp.

LOCHMILLER, R. L., E. C. HELLGREN, AND W. E. GRANT. 1984. Sex and age characteristics of the pelvic girdle in the collared peccary. J. Wildl. Manage. 48:639–641.

LOCKARD, G. R. 1972. Further studies of dental annuli for aging white-tailed deer. J. Wildl. Manage. 36:46–55.

LORD, R. D., JR. 1961a. A population study of the gray fox. Am. Midl. Nat. 66:87–109.

———. 1961b. The lens as an indicator of age in the gray fox. J. Mammal. 42:109–111.

LOVELESS, C. M. 1958. The mobility and composition of bobwhite quail populations in south Florida: with notes on the post-nuptial and post-juvenal molts. Fla. Game Freshwater Fish Comm. Tech. Bull. 4. 64pp.

LOWRANCE, E. W. 1949. Variability and growth of the opossum skeleton. J. Morphol. 85:569–593.

LUDWIG, J. R., AND R. W. DAPSON. 1977. Use of insoluble lens proteins to estimate age in white-tailed deer. J. Wildl. Manage. 41:327–329.

MAGOUN, A. J. 1985. Population characteristics, ecology, and management of wolverines in northwestern Alaska. Ph.D. Thesis, Univ. Alaska, Fairbanks. 211pp.

MANSFIELD, A. W. 1958. The biology of the Atlantic walrus *Odobenus rosmarus rosmarus* (Linnaeus) in the eastern Canadian Arctic. Fish. Res. Board Can. Manuscript Rep. Ser. 653. 146pp.

MARKS, S. A., AND A. W. ERICKSON. 1966. Age determination in the black bear. J. Wildl. Manage. 30:389–410.

MARSHALL, W. H. 1951. An age determination method for the pine marten. J. Wildl. Manage. 15:276–283.

MARTIN, F. W. 1964. Woodcock age and sex determination from wings. J. Wildl. Manage. 28:287–293.

MCCABE, R. A., AND A. S. HAWKINS. 1946. The Hungarian partridge in Wisconsin. Am. Midl. Nat. 36:1–75.

MCCLOSKEY, R. J. 1977. Accuracy of criteria used to determine age of fox squirrels. Proc. Ia. Acad. Sci. 84:32–34.

MCCORD, C. M., AND J. E. CARDOZA. 1982. Bobcat and lynx. Pages 728–766 *in* J. A. Chapman and G. A. Feldhamer, eds. Wild mammals of North America. The Johns Hopkins Univ. Press, Baltimore, Md.

MCCOURT, K. H., AND D. M. KEPPIE. 1975. Age determination of juvenile spruce grouse. J. Wildl. Manage. 39:790–794.

MCCULLOUGH, D. R. 1965. Sex characteristics of black-tailed deer hooves. J. Wildl. Manage. 29:210–212.

———, AND P. BEIER. 1986. Upper vs. lower molars for cementum annuli age determination of deer. J. Wildl. Manage. 50:705–706.

MCCUTCHEN, H. E. 1969. Age determination of pronghorns by the incisor cementum. J. Wildl. Manage. 33:172–175.

MCKINNON, D. T. 1983. Age separation of yearling and adult Franklin's spruce grouse. J. Wildl. Manage. 47:533–535.

MEAD, R. A. 1967. Age determination in the spotted skunk. J. Mammal. 48:606–616.

MELQUIST, W. E., AND A. E. DRONKERT. 1987. River otter. Pages 627–641 *in* M. Novak, J. A. Baker, M. E. Obbard, and B. Malloch, eds. Wild furbearer management and conservation in North America. Minist. Nat. Resour., Toronto, Ont.

———, AND M. G. HORNOCKER. 1983. Ecology of river otters in west central Idaho. Wildl. Monogr. 83. 60pp.

MENDALL, H. L., AND C. M. ALDOUS. 1943. The ecology and management of the American woodcock. Maine Coop. Wildl. Res. Unit, Orono. 201pp.

MESSICK, J. P. 1987. North American badger. Pages 587–597 *in* M. Novak, J. A. Baker, M. E. Obbard, and B. Malloch, eds. Wild furbearer management and conservation in North America. Minist. Nat. Resour., Toronto, Ont.

————, AND M. G. HORNOCKER. 1981. Ecology of the badger in southwestern Idaho. Wildl. Monogr. 76. 53pp.

MILLAR, J. S., AND F. C. ZWICKEL. 1972. Determination of age, age structure, and mortality of the pika, *Ochotona princeps* Richardson. Can. J. Zool. 50:229–232.

MILLER, F. L. 1974*a*. Age determination of caribou by annulations in dental cementum. J. Wildl. Manage. 38:47–53.

————. 1974*b*. Biology of the Kaminuriak population of barren ground caribou. Part II: Dentition as an indicator of sex and age; composition and socialization of the population. Can. Wildl. Serv. Rep. Ser. 31. 88pp.

————. 1982. Caribou. Pages 923–959 *in* J. A. Chapman and G. A. Feldhamer, eds. Wild mammals of North America. The Johns Hopkins Univ. Press, Baltimore, Md.

————, AND R. L. McCLURE. 1973. Determining age and sex of barren ground caribou from dental variables. Trans. Northeast. Sec. The Wildl. Soc. 30:79–100.

MITCHELL, O. G., AND R. A. CARSEN. 1967. Tooth eruption in the Arctic ground squirrel. J. Mammal. 48:472–474.

MITTWOCH, V. 1963. Sex differences in cells. Sci. Am. 209:54–62.

MONTGOMERY, G. G. 1964. Tooth eruption in preweaned raccoons. J. Wildl. Manage. 28:582–584.

MONTGOMERY, S. J., D. F. BALPH, AND D. M. BALPH. 1971. Age determination of Uinta ground squirrels by teeth annuli. Southwest. Nat. 15:400–402.

MOORE, K. L., EDITOR. 1966. The sex chromatin. W.B. Saunders Co., Philadelphia, Pa. 474pp.

MOSER, T. J., S. R. CRAVEN, AND B. K. MILLER. n.d. Canada geese in the Mississippi Flyway: a guide for goose hunters and goose watchers. U.S. Fish Wildl. Serv. Off. Ext. Publ., Washington, D.C. 24pp.

MULLER, L. I., T. T. BUERGER, AND R. E. MIRARCHI. 1984. Guide for age determination of mourning dove embryos. Ala. Agric. Exp. Stn. Circ. 272. 11pp.

MUSSEHL, T. W., AND T. H. LEIK. 1963. Sexing wings of adult blue grouse. J. Wildl. Manage. 27:102–106.

NAVA, J. A. 1970. The reproductive biology of the Alaska lynx (*Lynx canadensis*). M.S. Thesis, Univ. Alaska, Fairbanks. 141pp.

NELLIS, C. H. 1969. Sex and age variation in red squirrel skulls from Missoula County, Montana. Can. Field-Nat. 83:324–330.

————, S. P. WETMORE, AND L. B. KEITH. 1972. Lynx-prey interactions in central Alberta. J. Wildl. Manage. 36:320–329.

————, ————, AND ————. 1978. Age-related characteristics of coyote canines. J. Wildl. Manage. 42:680–683.

NEWBY, F. E., AND V. D. HAWLEY. 1954. Progress on a marten live-trapping study. Trans. North. Am. Wildl. Conf. 19:452–462.

NICHOLSON, W. S., AND E. P. HILL. 1981. A comparison of tooth wear, lens weight, and cementum annuli as indices of age in the gray fox. Pages 355–367 *in* J. A. Chapman and D. Pursley, eds. Worldwide Furbearer Conf., Frostburg, Md.

NIETFIELD, M. T., AND F. C. ZWICKEL. 1983. Classification of sex in young blue grouse. J. Wildl. Manage. 47:1147–1151.

NOVAK, M., J. A. BAKER, M. E. OBBARD, AND B. MALLOCH, EDITORS. 1987. Wild furbearer management and conservation in North America. Minist. Nat. Resour., Toronto, Ont. 1150pp.

NOVAKOWSKI, N. S. 1965. Cemental deposition as an age criterion in bison, and the relation of incisor wear, eye lens weight, and dressed bison carcass weight to age. Can. J. Zool. 43:173–178.

OATES, D. W., G. I. HOILIEN, AND R. M. LAWLER. 1985. Sex identification of field-dressed ring-necked pheasants. Wildl. Soc. Bull. 13:64–67.

ODOM, R. R. 1977. Sora. Pages 57–65 *in* G. C. Sanderson, ed. Management of migratory shore and upland game birds in North America. Int. Assoc. Fish Wildl. Agencies, Washington, D.C.

O'GARA, B. W. 1968. A study of the reproductive cycle of the female pronghorn (*Antilocapra americana* Ord.). Ph.D. Thesis, Univ. Montana, Missoula. 161pp.

————. 1969. Horn casting by female pronghorns. J. Mammal. 50:373–375.

OLSEN, P. F. 1959. Dental patterns as age indicators in muskrats. J. Wildl. Manage. 23:228–231.

OSBORN, D. J. 1955. Techniques of sexing beaver, *Castor canadensis*. J. Mammal. 36:141–142.

OTERO, J. G., AND R. W. DAPSON. 1972. Procedures in the biochemical estimation of age in vertebrates. Res. Popul. Ecol. (Kyoto) 13:152–160.

OWEN, R. B., JR., ET AL. 1977. American woodcock. Pages 149–186 *in* G. C. Sanderson, ed. Mangement of migratory shore and upland game birds in North America. Int. Assoc. Fish Wildl. Agencies, Washington, D.C.

————, AND W. B. KROHN. 1973. Molt patterns and weight changes of the American woodcock. Wilson Bull. 85:31–41.

OZOGA, J. J., AND L. J. VERME. 1985. Determining fetus age in live white-tailed does by x-ray. J. Wildl. Manage. 49:372–374.

PALMER, W. L. 1959. Sexing live-trapped juvenile ruffed grouse. J. Wildl. Manage. 23:111–112.

PARKER, G. R., D. C. THOMAS, E. BROUGHTON, AND D. R. GRAY. 1975. Crashes of muskox and Peary caribou populations in 1973–1974 in the Parry Islands, Arctic Canada. Can. Wildl. Serv. Prog. Notes 56. 10pp.

PARR, R. 1975. Aging red grouse chicks by primary molt and development. J. Wildl. Manage. 39:188–190.

PARSONS, G. R., M. K. BROWN, AND G. B. WILL. 1978. Determining the sex of fisher from the lower canine teeth. N.Y. Fish Game J. 25:42–44.

PEARSON, A. M. 1975. The northern interior grizzly bear *Ursus arctos* L. Can. Wildl. Serv. Rep. Ser. 34. 84pp.

PEEK, J. M. 1982. Elk. Pages 851–861 *in* J. A. Chapman and G. A. Feldhamer, eds. Wild mammals of North America. The Johns Hopkins Univ. Press, Baltimore, Md.

PELTON, M. R. 1970. Effects of freezing on weights of cottontail lenses. J. Wildl. Manage. 34:205–207.

PETERSON, R. T. 1980. A field guide to the birds. Houghton Mifflin Co., Boston, Mass. 384pp.

PETRIDES, G. A. 1942. Age determination in American gallinaceous game birds. Trans. North Am. Wildl. Conf. 7:308–328.

————. 1949. Sex and age determination in the opossum. J. Mammal. 30:364–378.

————. 1950. The determination of sex and age ratios in fur animals. Am. Midl. Nat. 43:355–382.

————. 1951. Notes on age determination in squirrels. J. Mammal. 32:111–112.

————, AND R. B. NESTLER. 1952. Further notes on age determination in juvenile bobwhite quails. J. Wildl. Manage. 16:109–110.

POELKER, R. J., AND H. D. HARTWELL. 1973. The black bear of Washington. Wash. State Game Dep. Biol. Bull. 14. 180pp.

PYLE, P., S. N. G. HOWELL, R. P. YUNICK, AND D. F. DESONTE. 1987. Identification guide to North American passerines. Slate Creek Press, Bolinas, Calif. 278pp.

QUIMBY, D. C., AND J. E. GAAB. 1952. Preliminary report on a study of elk dentition as a means of determining age classes. Proc. Western Assoc. State Game and Fish Comm. 32:225–227.

RAUSCH, R. A. 1967. Some aspects of the population ecology of wolves, Alaska. Am. Zool. 7:253–265.

————. 1969. Morphogenesis and age-related structure of permanent canine teeth in the brown bear, *Ursus arctos* L., in arctic Alaska. Z. Morphol. Tiere 66:167–188.

————, AND A. M. PEARSON. 1972. Notes on the wolverine in Alaska and the Yukon Territory. J. Wildl. Manage. 36:249–268.

REES, J. W., R. A. KAINER, AND R. W. DAVIS. 1966. Chronology of mineralization and eruption of mandibular teeth in mule deer. J. Wildl. Manage. 30:629–631.

REEVES, H. M., A. D. GEIS, AND F. C. KNIFFEN. 1968. Mourning dove capture and banding. U.S. Fish Wildl. Serv. Spec. Sci. Rep. Wildl. 117. 63pp.

REYNOLDS, H. C. 1945. Some aspects of the life history and ecology of the opossum in central Missouri. J. Mammal. 26:341–379.

REYNOLDS, H. W. R., R. D. GLAHOLT, AND A. W. L. HAWLEY. 1982. Bison. Pages 972–1007 *in* J. A. Chapman and G. A. Feldhamer, eds. Wild mammals of North America. The Johns Hopkins Univ. Press, Baltimore, Md.

RICHARDSON, G. L. 1966. Eye lens weight as an indicator of age in the collared peccary (*Pecari tajacu*). M.S. Thesis, Univ. Arizona, Tucson. 47pp.

ROBERTS, J. D. 1978. Variation in coyote age determination from annuli in different teeth. J. Wildl. Manage. 42:454–456.

ROBERTS, T. H. 1988. American woodcock (*Scolopax minor*). U.S. Army Corps Eng. Wildl. Resour. Manage. Manual, Tech. Rep. EL-88. 56pp.

RODGERS, R. D. 1979. Ratios of primary calamus diameters for determining age of ruffed grouse. Wildl. Soc. Bull. 7:125–127.

————. 1985. A field technique for identifying the sex of dressed pheasants. Wildl. Soc. Bull. 13:528–533.

ROLLEY, R. E. 1987. Bobcat. Pages 671–681 *in* M. Novak, J. A. Baker, M. E. Obbard, and B. Malloch, eds. Wild furbearer management and conservation in North America. Minist. Nat. Resour., Toronto, Ont.

RONGSTAD, O. J. 1966. A cottontail rabbit lens-growth curve from southern Wisconsin. J. Wildl. Manage. 30:114–121.

———. 1969. Gross prenatal development of cottontail rabbits. J. Wildl. Manage. 33:164–168.

ROOT, D. A., AND N. F. PAYNE. 1984. Evaluation of techniques for aging gray fox. J. Wildl. Manage. 48:926–933.

ROSEBERRY, J. L., AND W. D. KLIMSTRA. 1965. A guide to age determination of bobwhite quail embryos. Ill. Nat. Hist. Surv. Biol. Notes 55. 4pp.

ROUSSEL, Y. E. 1975. Aerial sexing of anterless moose by white vulval patch. J. Wildl. Manage. 39:450–451.

———, AND R. OUELLET. 1975. A new criterion for sexing Quebec ruffed grouse. J. Wildl. Manage. 39:443–445.

SALWASSER, H., AND S. A. HOLL. 1979. Estimating fetus age and breeding and fawning periods in the North Kings River deer herd. Calif. Fish Game 65:159–165.

SANDERSON, G. C. 1950. Methods of measuring productivity in raccoons. J. Wildl. Manage. 14:389–402.

———. 1961. Techniques for determining age of raccoons. Ill. Nat. Hist. Surv. Biol. Notes 45. 16pp.

SATHER, J. H. 1954. The dentition method of aging muskrats. Chicago Acad. Sci. Nat. Hist. Misc. Publ. 130. 3pp.

SAUER, P. R. 1966. Determining sex of black bears from the size of the lower canine tooth. N.Y. Fish Game J. 13:140–145.

———, S. FREE, AND S. BROWNE. 1966. Age determination in black bears from canine tooth sections. N.Y. Fish Game J. 13:125–139.

SAUNDERS, J. K. 1961. The biology of the Newfoundland lynx. Ph.D. Thesis, Cornell Univ., Ithaca, N.Y. 114pp.

———. 1964. Physical characteristics of the Newfoundland lynx. J. Mammal. 45:36–47.

SCHEFFER, V. B. 1950. Growth layers on the teeth of pinnipedia as an indicator of age. Science 112:309–311.

SCHMID, W. 1967. Sex chromatin in hair roots. Cytogenetics 6:342–349.

SCHOFIELD, R. D. 1955. Analysis of muskrat age determination methods and their application in Michigan. J. Wildl. Manage. 19:463–466.

SCHONBERNER, V. D. 1965. Beobachtungen zur fortpflanzungsbiologie de wolfes, *Canis lupus.* Z. Saeugetierkd. 30:171–178.

SEGELQUIST, C. A. 1966. Sexing white-tailed deer embryos by chromatin. J. Wildl. Manage. 30:414–417.

SERVELLO, F. A., AND R. L. KIRKPATRICK. 1986. Sexing ruffed grouse in the Southeast using feather criteria. Wildl. Soc. Bull. 14:280–282.

SEVERINGHAUS, C. W. 1949. Tooth development and wear as criteria of age in white-tailed deer. J. Wildl. Manage. 13:195–216.

SHACKLETON, D. M., L. V. HILLS, AND D. A. HUTTON. 1975. Aspects of variation in cranial characters of Plains bison (*Bison bison bison* Linnaeus) from Elk Island National Park, Alberta. J. Mammal. 56:871–887.

SHARP, W. M. 1958. Aging gray squirrels by use of tail-pelage characteristics. J. Wildl. Manage. 22:29–34.

SHERMAN, P. W., M. L. MORTON, L. M. HOOPES, J. BOCHANTIN, AND J. M. WATT. 1985. The use of tail collagen strength to estimate age in Belding's ground squirrels. J. Wildl. Manage. 49:874–879.

SILOVSKY, G. D., H. M. WIGHT, L. H. SISSON, T. L. FOX, AND S. W. HARRIS. 1968. Methods for determining age of band-tailed pigeons. J. Wildl. Manage. 32:421–424.

SKINNER, M. F., AND O. C. KAISEN. 1947. The fossil bison of Alaska and preliminary revision of the genus. Bull. Am. Mus. Nat. Hist. 89:123–256.

SMITH, N. D., AND I. O. BUSS. 1963. Age determination and plumage observations of blue grouse. J. Wildl. Manage. 27:566–578.

SPALDING, D. J. 1966. Eruption of permanent canine teeth in the northern sea lion. J. Mammal. 47:157–158.

STEPHENSON, A. J. 1977. Age determination and morphological variation of Ontario otters. Can. J. Zool. 55:1577–1583.

STEWART, R. R. 1973. Age determination, reproductive biology and food habits of Canada lynx, *Lynx canadensis* Kerr, in Ontario. M.S. Thesis, Univ. Guelph, Guelph, Ont. 61pp.

STODDARD, H. L. 1931. The bobwhite quail: its habits, preservation and increase. Charles Scribner's Sons, New York, N.Y. 559pp.

STOKES, A. W. 1957. Validity of spur length as an age criterion in pheasants. J. Wildl. Manage. 21:248–250.

STOLL, R. J., JR., AND D. CLAY. 1975. Guide to aging wild turkey embryos. Ohio Dep. Nat. Resour., Div. Wildl., Ohio Fish Wildl. Rep. 4. 19pp.

STRICKLAND, M. A. 1978. Fisher and marten study. Ont. Minst. Nat. Resour., Algonquin Reg. Prog. Rep. 5. 106pp.

———, AND C. W. DOUGLAS. 1987. Marten. Pages 531–546 *in* M. Novak, J. A. Baker, M. E. Obbard, and B. Malloch, eds. Wild furbearer management and conservation in North America. Minist. Nat. Resour., Toronto, Ont.

———, ———, M. NOVAK, AND N. P. HUNZIGER. 1982. Marten. Pages 599–612 *in* J. A. Chapman and G. A. Feldhamer, eds. Wild mammals of North America. The Johns Hopkins Univ. Press, Baltimore, Md.

STUEWER, F. W. 1943. Reproduction of raccoons in Michigan. J. Wildl. Manage. 7:60–73.

SULLINS, G. L., D. O. MCKAY, AND B. J. VERTS. 1976. Estimating ages of cottontails by periosteal zonations. Northwest Sci. 50:17–22.

SULLIVAN, E. G., AND A. O. HAUGEN. 1956. Age determination of foxes by x-ray of forefeet. J. Wildl. Manage. 20:210–212.

SZUBA, K. J., J. F. BENDELL, AND B. J. NAYLOR. 1987. Age determination of Hudsonian spruce grouse using primary feathers. Wildl. Soc. Bull. 15:539–543.

TABER, R. D. 1956. Characteristics of the pelvic girdle in relation to sex in black-tailed and white-tailed deer. Calif. Fish Game 42:15–21.

———, AND R. F. DASMANN. 1958. The black-tailed deer of the chaparral. Calif. Dep. Fish Game, Game Bull. 8. 163pp.

TABOR, J. E. 1974. Productivity, survival, and population status of river otter in western Oregon. M.S. Thesis, Oregon State Univ., Corvallis. 62pp.

TACHA, T. C., AND J. C. LEWIS. 1978. Sex determination of sandhill cranes by cloacal examination. Pages 81–83 *in* 1978 Crane Workshop. Int. Crane Found., Baraboo, Wis.

TENER, J. S. 1954. A preliminary study of the musk-oxen of Fosheim Peninsula, Ellesmere Island, N.W.T. Can. Wildl. Serv., Natl. Parks Branch, Wildl. Manage. Bull. Ser. 1, No. 9. 34pp.

———. 1965. Musk-oxen in Canada: a biological and taxonomic review. Can. Wildl. Serv. Monogr. 2. 166pp.

THOMAS, D. C., AND P. J. BANDY. 1973. Age determination of wild black-tailed deer from dental annulations. J. Wildl. Manage. 37:232–235.

———, AND ———. 1975. Accuracy of dental-wear age estimates of black-tailed deer. J. Wildl. Manage. 39:674–678.

THOMAS, K. P. 1969. Sex determination of bobwhites by wing criteria. J. Wildl. Manage. 33:215–216.

THOMPSON, D. R. 1958. Field techniques for sexing and aging game animals. Wis. Conserv. Dep. Spec. Wildl. Rep. 1. 44pp.

TIEMEIER, O. W., AND M. L. PLENERT. 1964. A comparison of three methods for determining the age of blacktailed jackrabbits. J. Mammal. 45:409–416.

TODD, M. 1980. Ecology of badger in southcentral Idaho, with additional notes on raptors. M.S. Thesis, Univ. Idaho, Moscow. 164pp.

TOWEILL, D. E., AND J. E. TABOR. 1982. River otter. Pages 688–703 *in* J. A. Chapman and G. A. Feldhamer, eds. Wild mammals of North America. The Johns Hopkins Univ. Press, Baltimore, Md.

TOWERS, J. 1988. Age determination of juvenile spruce grouse in eastern Canada. J. Wildl. Manage. 52:113–115.

TUMLISON, R., AND V. R. MCDANIEL. 1984. Gray fox age classification by canine tooth pulp cavity radiographs. J. Wildl. Manage. 48:228–230.

TURNER, J. C. 1977. Cemental annulations as an age criterion in North American sheep. J. Wildl. Manage. 41:211–217.

TYNDALE-BISCOE, C. H., AND R. B. MACKENZIE. 1976. Reproduction in *Didelphis marsupialis* and *D. albiventris* in Columbia. J. Mammal. 57:249–265.

UHLIG, H. G. 1953. Weights of ruffed grouse in West Virginia. J. Wildl. Manage. 17:391–392.

———. 1955. The determination of age of nestling and sub-adult gray squirrels in West Virginia. J. Wildl. Manage. 19:479–483.

———. 1956. The gray squirrel in West Virginia. W. Va. Conserv. Comm. Div. Game Manage. Bull. 3. 83pp.

UPHAM, L. L. 1967. Density, disperal, and dispersion of the striped skunk (*Mephitis mephitis*) in southeastern North Dakota. M.S. Thesis, North Dakota State Univ., Fargo. 63pp.

U.S. FISH AND WILDLIFE SERVICE AND CANADIAN WILDLIFE SERVICE. 1977. North American birdbanding manual. Vol. II. U.S. Dep. Inter., Washington, D.C. Var. pagin.

VAN BALLENBERGHE, V., AND L. D. MECH. 1975. Weights, growth, and survival of timber wolf pups in Minnesota. J. Mammal. 56:44–63.

VANDRUFF, L. W. 1971. The ecology of the raccoon and opossum, with emphasis on their role as waterfowl nest predators. Ph.D. Thesis, Cornell Univ., Ithaca, N.Y. 140pp.

VAN NOSTRAND, F. C., AND A. B. STEPHENSON. 1964. Age determination for beavers by tooth development. J. Wildl. Manage. 28:430–434.

VAN TURNER, P., AND M. VALENTINE. 1983. Cytological sex determination in cranes. Pages 571–574 *in* G. W. Archibald and R. F. Pasquitt, eds. Proc. 1983 Int. Crane Workshop, Baraboo, Wis.

VERTS, B. J. 1967. The biology of the striped skunk. Univ. Illinois Press, Urbana. 218pp.

VINCENT, J.-P., AND J.-P. QUERE. 1972. Quelques donnees sur la reproduction et sur la dynamique des populations de rat musque *Ondatra zibethica* L. dans le nord de la France. Ann. Zool. Ecol. Anim. 4:395–415.

VOIGT, D. R. 1987. Red fox. Pages 379–392 *in* M. Novak, J. A. Baker, M. E. Obbard, and B. Malloch, eds. Wild furbearer management and conservation in North America. Minist. Nat. Resour., Toronto, Ont.

———, AND W. E. BERG. 1987. Coyote. Pages 344–357 *in* M. Novak, J. A. Baker, M. E. Obbard, and B. Malloch, eds. Wild furbearer management and conservation in North America. Minist. Nat. Resour., Toronto, Ont.

WALKINSHAW, L. H. 1949. The sandhill cranes. Cranbrook Inst. Sci., Bloomfield Hills, Mich. 202pp.

———. 1973. Cranes of the world. Winchester Press, New York, N.Y. 370pp.

WALLIN, J. A. 1982. Sex determination of Vermont fall-harvested juvenile wild turkeys by the 10th primary. Wildl. Soc. Bull. 10:40–43.

WALLMO, O. C. 1956. Determination of sex and age of scaled quail. J. Wildl. Manage. 20:154–158.

WEAVER, H. R., AND W. L. HASKELL. 1968. Age and sex determination of the chukar partridge. J. Wildl. Manage. 32:46–50.

WEEDEN, R. B., AND A. WATSON. 1967. Determining the age of rock ptarmigan in Alaska and Scotland. J. Wildl. Manage. 31:825–826.

WHITE, J. A., AND C. E. BRAUN. 1978. Age and sex determination of juvenile band-tailed pigeons. J. Wildl. Manage. 42:564–569.

WHITMAN, J. S., W. B. BALLARD, AND C. L. GARDNER. 1986. Home range and habitat use by wolverines in southcentral Alaska. J. Wildl. Manage. 50:460–463.

WIGAL, R. A., AND V. L. COGGINS. 1982. Mountain goat. Pages 1008–1020 *in* J. A. Chapman and G. A. Feldhamer, eds. Wild mammals of North America. The Johns Hopkins Univ. Press, Baltimore, Md.

WILLIAMS, L. E., JR. 1961. Notes on wing molt in the yearling wild turkey. J. Wildl. Manage. 25:439–440.

———, AND D. H. AUSTIN. 1970. Complete post-juvenal (pre-basic) primary molt in Florida turkeys. J. Wildl. Manage. 34:231–233.

———, AND ———. 1988. Studies of the wild turkey in Florida. Fla. Game Freshwater Fish Comm. Tech. Bull. 10. 232pp.

WISHART, W. 1969. Age determination of pheasants by measurement of proximal primaries. J. Wildl. Manage. 33:714–717.

WOEHLER, E. E., AND J. M. GATES. 1970. An improved method of sexing ring-necked pheasant chicks. J. Wildl. Manage. 34:228–231.

WOOD, J. E. 1958. Age structure and productivity of a gray fox population. J. Mammal. 39:74–86.

WRIGHT, P. L. 1947. The sexual cycle of the male long-tailed weasel (*Mustela frenata*). J. Mammal. 28:343–352.

———. 1948. Breeding habits of captive long-tailed weasels. Am. Midl. Nat. 39:338–344.

———. 1969. The reproductive cycle of the male American badger (*Taxidea taxus*). J. Reprod. Fertil. Suppl. 6:435–445.

———, AND R. RAUSCH. 1955. Reproduction in the wolverine (*Gulo gulo*). J. Mammal. 36:346–355.

YOUNG, S. P., AND E. A. GOLDMAN. 1944. The wolves of North America. Am. Wildl. Inst., Washington, D.C. 636pp.

ZWICKEL, F. C., AND C. F. MARTINSEN. 1967. Determining age and sex of Franklin spruce grouse by tails alone. J. Wildl. Manage. 31:760–763.

9

ESTIMATING THE NUMBER OF ANIMALS IN WILDLIFE POPULATIONS

Richard A. Lancia, James D. Nichols, and Kenneth H. Pollock

INTRODUCTION

In 1938 Howard M. Wight devoted nine pages, which was an entire chapter of the first wildlife management techniques manual, to what he termed "census" methods (Wight 1938). As the length of the chapter you are reading attests, the volume of recent literature on this subject has grown tremendously, spanning a range from complex mathematical-statistical models to cookbook applications. Our intent in this chapter is to present an overview of the basic and most widely used population estimation techniques and to provide an entree to the relevant literature.

Several possible approaches could be taken in writing a chapter dealing with population estimation. For example, we could provide a detailed treatment focusing on the statistical models themselves and on the derivation of estimators based on these models. Although a chapter using this approach might be a valuable reference for quantitative biologists and biometricians, it would likely be of limited use to many field biologists and wildlife managers. Another approach would be to focus on the details of actually applying various population estimation techniques. Such an approach would include both field application (e.g., how to set out a trapping grid or conduct an aerial survey) and detailed instructions on how to use the resulting data with appropriate estimation equations. We are reluctant to attempt such an approach, however, because of the tremendous diversity of real-world field situations as defined by such factors as the animal being studied, habitat, and available resources, and because of our resultant inability to provide detailed instructions for all possible situations.

Instead, we believe providing the reader with the conceptual basis underlying the various estimation methods is more useful. Thus, we have tried to provide intuitive explanations for how the basic methods work. In doing so, we present the relevant estimation equations for most methods and provide citations of more detailed treatments covering both statistical considerations and field applications. We have chosen to present methods that are representative of classes of estimators, rather than address every available one. Our hope is that this chapter will provide the reader with enough background to make an informed decision about what general method(s) will likely perform well in any particular field situation. Readers with a more quantitative background may then be able to consult detailed references and tailor the selected method to suit their particular needs. The less quantitative readers should consult a biometrician, preferably one with experience in wildlife studies, for this "tailoring," with the hope that they will be able to do so with a basic understanding of the general method, thereby permitting useful interaction and discussion with the biometrician.

Why Estimate Population Size?

The goals of managing natural animal populations are frequently expressed in terms of population size. When dealing with rare or endangered species, for example, wildlife managers often try to increase population size, whereas for undesirable (pest) species they try to reduce population size. For harvested populations, population size is maintained at a desirable level, and a harvest is permitted. Population size is thus the currency by which the success of many management programs ultimately is judged.

Despite the central role of population size in wildlife management, the importance of this quantity sometimes has been oversold. For example, a manager interested in the status or health of a particular population might decide to estimate population size as the means for assessing status. However, a single estimate of population size at one point in space and time is usually of limited value, and it provides much less information about status than is commonly thought. Instead, additional estimates of population size, for example at the same site in different years or at different sites and habitats in the same year, would help place a particular estimate in proper perspective. Such information allows inferences about population status relative to previous years, or other areas or habitats. Sometimes we are interested in population "trend," a statistic reflecting average direction and magnitude of change over a specified period of time. Trend estimation is an important topic, and we recommend the volume by Sauer and Droege (1990) for a discussion of this problem.

Any biologist/manager considering expending the effort to estimate population size properly should carefully consider the need for an estimate, and then how the estimate will be used. We recommend that the biologist/manager simply ask the following question: "What will I do with the estimate once I get it?" For example, if a population estimate of 400 deer versus one of 700 deer is not likely to lead to different management responses, then devoting a lot of effort to achieve a very precise and accurate population estimate might not be necessary.

Despite these cautions, population size estimates have many important uses. We agree with Macnab (1983) and Sinclair (1991) that wildlife biologists/managers often fail to take advantage of management manipulations because these manipulations are not often conducted so that their success can be evaluated. Much can be learned simply by comparing population estimates before, and then after, such a manipulation. Naturally, inferences are much stronger if manipulations are replicated over time and space (Nichols 1991). Skalski and Robson (1992) provided an excellent discussion of experimental design using capture-recapture estimates of population size. Monitoring programs in which estimates of population size are obtained for the same area over time is critical, but is seldom a standard part of management activities. Comparative estimates of population size at one point in time for different areas or habitats also can be useful in drawing inferences about population status and sometimes about habitat preferences.

Finally, even when estimates of population size are useful, as when used in conjunction with experimental manipulations, they seldom provide all the answers. Properly designed experiments involving estimates of population size before and after a manipulation can lead to strong inferences about the ability of the manipulation to influence population size, but a more fundamental, mechanistic explanation of exactly how and why the population responds to the manipulation is frequently more important (Gavin 1989). All changes in population size result from the action of four fundamental demographic variables—mortality, reproduction, emigration, and immigration. If we are interested in detailed mechanistic explanations of population response to some manipulation, we will want to obtain estimates of these rates in addition to estimates of population size (see Chapter 16).

Definitions and Statistical Concepts

The following definitions are used throughout this chapter. These terms, defined in relation to population estimation to help the reader understand the material in our chapter, are based on Overton (1969), Caughley (1977), White et al. (1982), and Verner (1985).

- A *population* is a group of animals that occupy a certain area at a certain time as defined by the people interested in the group. A population could be the deer on Remington Farms, the grizzly bears in Yellowstone National Park, the northern spotted owls in the Pacific Northwest, or the continental mallard population. We emphasize that this is an operational definition for this chapter and that it is not adequate from an ecological or genetic perspective.

- *Abundance* or *population size* refers to the number of individual animals, e.g., 49 white-footed mice. It can be expressed as *relative abundance*, wherein populations are ranked according to population size, or as *absolute abundance*, wherein the number of individuals in the population is known or estimated. For example, the following statements concern relative abundance: area A has more mice than area B, or A has 25% more mice than B. However, the statement that area A has 50 mice is a statement of absolute abundance.

- *Population density* is the number of individuals per unit area, e.g., 1.2 squirrels/ha or 10 elephants/km^2. Note that both abundance and area are relevant to density; consequently, density is frequently difficult to estimate. For example, trapping grids commonly are used to estimate the abundance of small mammals, but density estimates are more difficult to obtain because the effective area (i.e., the area from which trapped animals are drawn) of the trapping grid must also be estimated. Abundance is often much easier to estimate than density and will be sufficient for many management decisions.

- *Relative density* refers to the ranking of populations by density. For example, area A has 40% more mice per hectare than area B.

- A *census* is a complete count of an entire population of animals. Examples are a complete count of a raft of snow geese or a count of the number of wood ducks flying to a roost.

- A *population estimate* is an approximation of the true population size based on some method of sampling animals, such as by capturing or counting them. Although it would be ideal if all population estimates were unbiased (see below), in practice most are characterized by some degree of bias. A *robust* population estimate is still valid even when some of the assumptions of the estimation procedure are violated to some degree. Note that estimates of parameters (see below) are denoted by a small caret or "hat" over the symbol for the parameter. For example, N represents the true population size, and $\hat{N}$ ("N hat") represents an estimate of population size calculated from sample data.

- The term *population closure* has two components: *demographic closure*, in which neither births (natality) nor deaths (mortality) occur between sample periods, and *geographic closure*, in which no movements into (immigration) or out of (emigration) a population take place between sample periods. Thus, neither the size of the population nor the individuals in the population change for a population that is demographically and geographically closed. Population closure is frequently a basic assumption for population estimation procedures that are based on repeated observations over time, such as those involving capture and marking or removing individuals. In practice, determining whether the closure assumption is met is difficult. For counts that are essentially snapshots in time, such as an aerial count of kangaroos on a transect, the closure assumption is not a concern.

- An *open population* is one that is not closed.
- A *population index* is a statistic that is related to population size (see Caughley 1977). Use of indices is often restricted to comparisons between populations on the same area over time or between different areas at the same time, because the exact relationship between the index and the true population frequently is not known.
- *Frequency of occurrence* is a count of something, such as traps or plots, that has a particular attribute. An example would be the number of sardine bait stations disturbed by bears or traps that caught an animal. Because this is an either/or situation, frequencies are commonly expressed as a proportion of the total possible count, and thus range from 0 to 1.
- An *expected value* of a population estimate is the average value if the estimation procedure were repeated many times under exactly the same conditions. The notation E(C) represents the expected value of random variable C.
- *Accuracy* is a measure of how close a population estimate is to the true population size.
- Accuracy is measured by *mean squared error* (*MSE*), which is the average of the squared deviations between the true population size and the population estimate repeated many times.
- *Bias* is the difference between the expected value of a population estimate and the true population size. If the expected value of a population estimator is equal to the true population size, then the estimator is *unbiased*.
- *Precision* is a measure of how close a population estimate is to its expected value.
- Precision is measured by *variance* (*VAR*), which is the average of the squared deviations between a population estimate repeated many times and its expected value. The relationship among MSE, VAR, and bias is MSE $= \text{VAR} + \text{bias}^2$.
- *Standard error* (*SE*) is the square root of the variance. It is also a measure of precision; one use of the SE is to calculate confidence intervals (see below).
- A *confidence interval* (*CI*) reflects the reliability of an estimate and is usually written as $P[a \leq \theta \leq b] = 1 - \alpha$, where a and b are the upper and lower bounds, θ is the parameter of interest, and $1 - \alpha$ is the CI. If the estimator is normally distributed, then the 95% (i.e., $1 - 0.05$) CI would be $\theta \pm 1.96 \, \text{SE}(\theta)$, where θ is the estimate and $\text{SE}(\theta)$ is the SE of the estimate. The CI implies that if the estimate were repeated many times, then 95% of the CIs corresponding to each estimate would include the true value of the parameter θ. For a more detailed discussion, refer to a basic statistics text such as Moore and McCabe (1989).
- A *model* is an abstraction and simplification of reality that contains relevant features needed to develop a population estimate. A model is constructed so that unknown quantities such as population size are expressed in terms of known quantities such as counts or captures of animals.
- A *parameter* is a constant, usually an unknown quantity characterizing a population, that is used in a model. For example, capture probability is usually an unknown parameter.
- A *statistic* is a quantity derived from observed values.

The following analogy of a rifle firing at a target (Overton 1969) should help illustrate the definitions of statistical terms presented above. The bull's-eye is analogous to the value of the parameter being estimated (Fig. 1). The process of aiming and firing under a particular set of conditions is analogous to the process of collecting data and calculating an estimate under a particular set of conditions. The location of the bullet strike is analogous to the value of an estimate.

Now, imagine a *very* large number of shots fired at the target. The average or mean point of impact is analogous to the *expected value* of the estimator. This is where the rifle is firing, on the average. The *precision* of the estimator is analogous to the spread of the group about the mean point of impact. The greater the spread, the less the precision. *Variance* is used to measure precision; the smaller the variance, the better the precision. *Accuracy*, measured as the MSE of the estimator, is analogous to the spread of the group about the bull's-eye. The smaller the MSE, the better the accuracy of the estimator. Last, the distance from the mean point of impact to the bull's-eye is analogous to the bias of the estimator. The bias reflects the degree of incorrect sighting of the rifle for the conditions during firing. Thus, an unbiased estimate would be analogous to a rifle that is correctly sighted-in.

In practice, population estimates need to be both precise and accurate to be useful. If an experiment could be replicated many times, perhaps precision would not be as great a concern because averaging the replicates would tend to counteract a lack of precision. However, a biologist is hardly ever afforded the luxury of being able to replicate a population estimate many times. In any event, accuracy is still crucial.

Other Sources of Information

Numerous technical papers and reviews on population estimation procedures are available. Some of the more significant and recent ones are identified below; other important papers are cited to support discussion of particular techniques. Seber's book, *The Estimation of Animal Abundance and Related Parameters* (Seber 1982) is a classic source for mathematical and statistical models of population estimation. Caughley (1977) devoted several excellent chapters in his book, *Analysis of Vertebrate Populations*, to population estimation techniques, as did Krebs in a more recent book, *Ecological Methodology* (Krebs 1989). Gates (1979) and Burnham et al. (1980) described line-transect methods, Otis et al. (1978) and White et al. (1982) reviewed capture-recapture and removal methods, and Pollock et al. (1990) detailed the design, analysis, and interpretation of capture-recapture studies. Ralph and Scott (1981) and Verner (1985) provided comprehensive reviews of procedures for estimating bird populations.

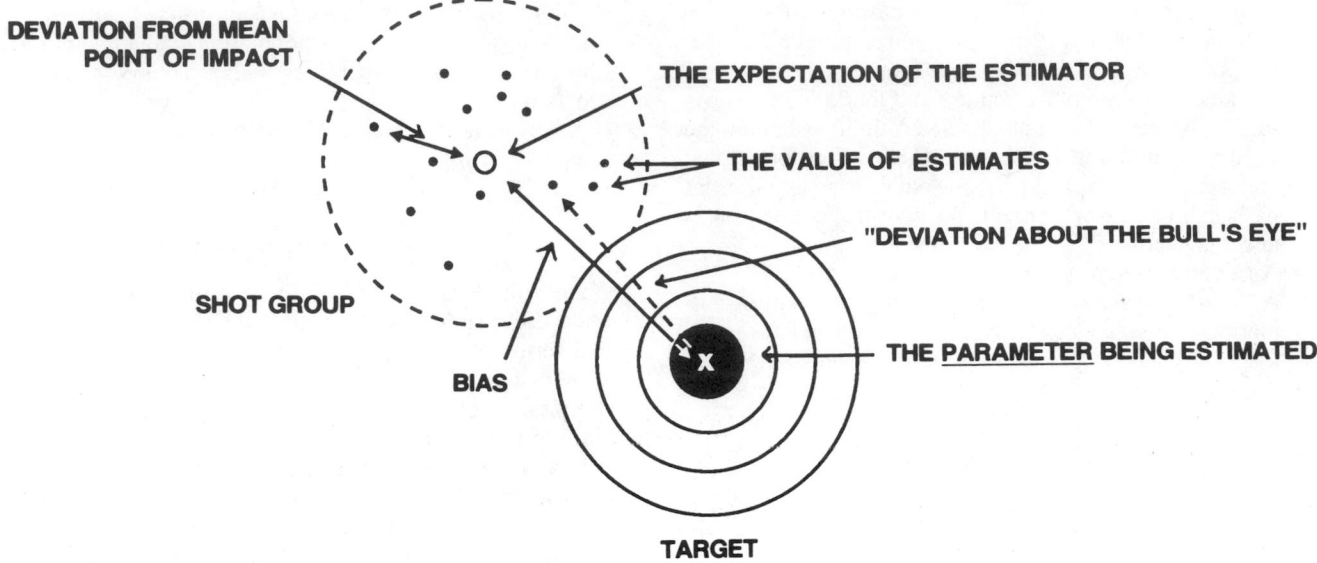

Fig. 1. The analogy between estimation and firing a rifle. A rifle with a tight shot group has high "precision." Adjusting the gun sight for elevation and windage will change the "bias" and "accuracy" but not "precision." The accuracy of a gunner is a measure of the distance of the shots from the center bull's-eye. If proper adjustment for elevation and windage has been made, the "bias" will be 0 (from Overton 1969:405).

Eberhardt et al. (1979) provided an overview of population estimation methods for marine mammals.

As a final introductory note, in this chapter we attempt to retain the notation that is traditional in the literature for particular methods. Consequently, the notation might not be consistent among all the methods we discuss. We will, however, define the notation as it is used in each method.

A CONCEPTUAL FRAMEWORK

In this chapter, we make an effort to present a unified view of population estimation. The methods we discuss form a diverse group of field applications and statistical estimation methods. It is far too easy to be caught up in the details of the various approaches and to be overwhelmed and confused by their diversity. Rather than simply present different methods as unrelated entries in a cookbook, we emphasize the features they share.

Two basic problems confront any biologist who wants to estimate population size—observability and sampling. All the methods we discuss represent solutions to one, or both, of these basic problems. Thus, all superficially diverse solutions can be expressed in a single general form.

Observability

The first problem is that most methods of surveying animals, such as direct observation or capture in traps, do not result in counts or captures of all animals present on an area. Instead, the probability (β) of seeing or catching an animal will generally be less than 1. We can write this relationship between a count of animals, denoted as the random variable C, and the true population size, N, as:

$$\mathrm{E}(C) = \beta N, \qquad (1)$$

where $\mathrm{E}(C)$ denotes the expected value of the count C. Thus, to translate a count resulting from any survey method into an estimate of population size, we must estimate the proportion of animals counted, β, and then divide our count by this estimate:

$$\hat{N} = \frac{C}{\hat{\beta}}. \qquad (2)$$

[handwritten annotation: C – count (seen); β – proportion of seen to total pop]

For example, if 20 birds are seen in a woodlot and the birds we see are only 25% of the total number of birds actually present, then $C = 20$, $\beta = 0.25$, and $\hat{N} = 20/0.25 = 80$. The bulk of the effort in developing population estimation methods for animal populations has involved ways of estimating β. We will refer to problems of this sort in general as observability concerns.

Sampling

The second basic problem is that time and money are nearly always limited, so a particular survey method cannot be applied to the entire area of interest. Therefore, sample area(s) must be selected that represent a fraction, α, of the total area of interest. Unlike β, this sampling fraction is frequently known with reasonable accuracy. If $\hat{N}'$ represents the estimated number of animals on a representative sample area, then the population size of the entire area, $\hat{N}$, can be estimated as:

$$\hat{N} = \frac{\hat{N}'}{\alpha}, \qquad (3)$$

[handwritten annotation: N̂′ – estimated # of animals on sample area; α – sample area]

where α again is the proportion of the total area to which $\hat{N}'$ pertains. For example, if 32 wildebeests were estimated on sample plots representing 10% of the total study area, then $\hat{N}' = 32$, $\alpha = 0.10$, and $\hat{N} = 32/0.10 = 320$, which is our estimate of the number of animals in the entire study area. Sampling designs can range from simple random sampling to more involved designs such as stratified random sampling or double sampling (see Chapter 2). We will refer to problems of this sort in general as *sampling* concerns.

Thus, observability (eqs. 1 and 2) and sampling (eq. 3) are two basic concerns involved with estimating populations of animals. Combining equations (2) and (3) yields the following general population size estimator:

$$\hat{N} = \frac{C}{\alpha\hat{\beta}}, \qquad (4)$$

where $\hat{N}$ is the estimate of population size, C is the count of animals, α is the proportion of the entire area that is surveyed, and $\hat{\beta}$ is the estimate of the proportion of the animals counted. Virtually all estimators of population size can be expressed in the form of equation (4), and we will point this out throughout the chapter.

If α is known and if there is no sampling correlation between C and $\hat{\beta}$, the variance of the estimator (eq. 4) is:

$$var(\hat{N}) = (N^2)\left\{\left(\frac{var(C)}{C^2}\right)(1-\alpha) + \left(\frac{var(\hat{\beta}^2)}{\beta^2}\right)\right\}, \qquad (5)$$

where var denotes sampling variance. The two additive terms within the brackets of equation (5) are the central components of variation contributing to the variance of $\hat{N}$. The component associated with var(β) depends on the method of estimating β, and it can be large. The component associated with var(C) depends on the variation in true abundance over the area of interest and the survey procedure. The latter component is smaller when animals are distributed uniformly and larger when animals are distributed in a patchy or clumped manner. This geographic component of variation is best estimated from replicate samples (subareas) within the total area. The $(1-\alpha)$ term is the finite population correction factor. The variance of $\hat{N}$ also becomes smaller as α becomes larger, i.e., approaches 1. This corresponds to the intuitive idea that var($\hat{N}$) decreases as the proportion of the total area sampled becomes large. The magnitude of β is also very important, with lower var($\hat{N}$) associated with larger β. Again, this corresponds with our intuition that the greater the proportion of the population counted, the less variable the population estimate. If α must be estimated, equation (5) must be modified to incorporate an additional variance component associated with estimation of α. Finally, if there is correlation (i.e., a measure of the joint variation of two variables) between C and β, an additional term (covariance) also must be included in equation (5) (Mood et al. 1974).

The above paragraphs provide a brief, but fundamental, framework for all estimators of population size discussed in this chapter. Although the focus of the chapter is on population size, we sometimes refer to the closely related quantity, population density. Density is simply the number of animals per unit area. If we estimate population size (N) at some specified location of total area, A, density is estimated as:

$$\hat{D} = \frac{\hat{N}}{A}. \qquad (6)$$

If A is known exactly, then var($\hat{D}$) can be estimated as:

$$var(\hat{D}) = \left[\frac{1}{A}\right]^2 var(\hat{N}), \qquad (7)$$

where var($\hat{N}$) is the estimated variance of $\hat{N}$. Sometimes, the exact area to which an estimate of population size applies is not known but must be estimated. For example, capture-recapture studies of small mammals on trapping grids can be used to estimate population size, but the area

to which such an estimate applies is often larger than the area covered by the grid. In any case, if A is not known but must be estimated, equation (7) must be modified to incorporate this additional source of variation in var($\hat{D}$).

The organization of our chapter, in accord with the conceptual framework described above, is present diagrammatically in Fig. 2. Indices are separated from all population estimation methods because they are a special case for which no population estimate is intended. The application of indices to make inferences about differences in population size principally involves *observability* concerns. We divided population estimation methods into two major categories depending on whether all individuals can be observed or whether only a portion of the individuals can be observed. In the former, *sampling* concerns predominate, and in the latter, *observability* concerns are also important. We have further subdivided these major categories into methods that involve capture of animals and those that involve only counting animals. For some readily observable species, a counting technique might be possible. For others that are not easily seen but can be captured, capture methods might be the only feasible alternative.

INDICES

Animal ecologists long have dealt with statistics that do not actually estimate animal abundance but are thought to be correlated with abundance. For example, the number of birds seen or heard singing in a woodlot, the number of small mammals caught in a grid of traps set in an old field, and the number of deer pellets counted in a plot of woods are all thought to provide some information on animal abundance. Few people would claim that these statistics are good estimators of abundance, but most believe that they are related to abundance in some manner. Caughley (1977:12) defined a density index as "any measurable correlative of density" and proceeded to describe several possible functional relationships between a density index and absolute density (see Eberhardt 1978).

Most uses of indices of either abundance or density involve comparisons between populations from the same location at different times or between populations from different locations at the same time. Thus, indices are used to indicate relative differences in abundance. Indices that are best suited to such uses are related to abundance as in the equation (which is the same as eq. 1) below:

$$E(C) = \beta N, \qquad (8)$$

where $E(C)$ denotes the expected value of the index (C), N denotes true abundance (the number of animals in some specified area), and β is a proportionality constant (i.e., an "observability" constant) relating C and N. Indices that are related to true abundance as defined by equation (8) are *constant-proportion indices*. They are, by far, the most useful in comparative population studies and thus will be the subject of most of our subsequent discussion. However, we will briefly discuss other indices at the end of this section.

Constant-Proportion Indices

The key to the reasonable use of any statistic thought to be a constant-proportion index is to ensure that equation (8) holds true for the intended comparison. For ex-

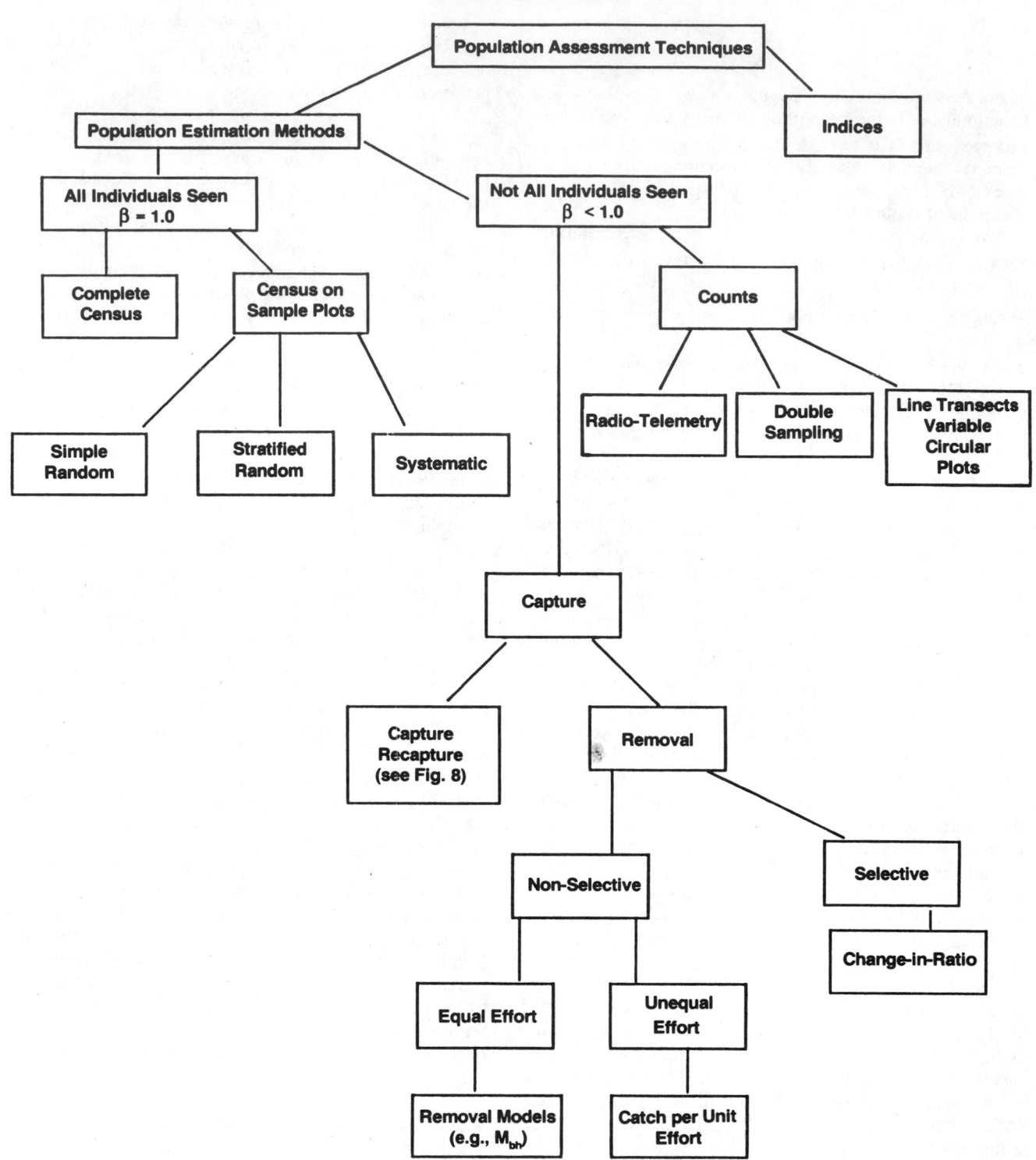

Fig. 2. The relationships among various population estimation techniques. Major groupings are based on whether all individuals can be seen (i.e., the probability of observing an animal is $\beta = 1.0$) or not seen ($\beta < 1.0$) and on whether animals can be counted or captured.

ample, assume that we are considering using the number of muskrats caught during 2 nights of trapping with 100 traps as a possible index to abundance in a marsh and that we are interested in estimating the rate of change in population size (i.e., the finite rate of increase λ, see Chapter 16) between sampling periods in 2 successive years, 1 and 2. We thus would like to estimate $\lambda = N_2/N_1$, using C_2/C_1, wherein the subscripts denote year. We can approximate the expected value of our estimator, $\mathrm{E}(C_2/C_1)$, as:

$$\frac{\mathrm{E}(C)_2}{\mathrm{E}(C)_1} = \frac{\beta_2 N_2}{\beta_1 N_1},\qquad(9)$$

where β_i is the capture probability or the proportion of the population caught in the sampling effort of year i. If the same proportion of the muskrat population is captured in the 2 different years of the study (i.e., if $\beta_1 = \beta_2 = \beta$), then the β's in equation (9) cancel, and C_2/C_1 is a reasonable (but still not unbiased [Barker and Sauer

1992]) estimator for the population change. However, if $\beta_1 \neq \beta_2$, we are not catching the same fraction of the population each year, and C is not a good index. In this latter instance, our estimate of $\hat{\lambda}$ is a function of the change in "capture probability" between years as well as the change in population size, i.e., capture probability and the change in population size are confounded. In this instance, the estimator should not be used to draw inferences about population change.

The question of whether some statistic, C, is a reasonable index of abundance for a specific comparative purpose thus involves a test of the hypothesis that the same proportion of the population is being sampled, or $\beta_i = \beta$ for all i's being compared, where i might denote year or geographic location. We can separate constant-proportion indices of abundance into two categories, depending on whether they are collected as part of a formal method for estimating population size.

INDICES ASSOCIATED WITH FORMAL POPULATION ESTIMATION METHODS

As pointed out above, all of the formal methods developed for estimating population size to be discussed later in this chapter (except change-in-ratio estimates) can be viewed in the context of equation (8). They all involve some observation-based statistic, C, and additional information that can be used to estimate the proportionality constant, β, relating C and population size, N. Estimation of N is then accomplished as in equation (10) (which is the same as eq. 2):

$$\hat{N} = \frac{C}{\hat{\beta}}. \qquad (10)$$

In capture-recapture studies, for example, C is the number of animals caught during a particular sampling period, and β is the average capture probability, or the proportion of the population represented by C. In this example, β is estimated from data on the capture histories of marked animals. Similarly, in removal studies in which constant effort is applied such as on a trapping grid, C is the total number of animals captured, and β again represents the capture probability. In this situation, β is estimated from information on the change in number of new individuals caught in successive sampling periods. In catch-per-unit-effort removal studies, C is essentially the catch of animals standardized for variable effort (i.e., catch-per-unit-effort) and β is a "catchability constant" estimated from the cumulative removals in relation to the numbers of animals caught each time period. In line-transect studies, C is the number of animals actually seen from the transect, and β can be thought of as the average probability of sighting an animal from the transect. Here, β is estimated from the distribution of right-angle distances of sighted animals along the transect line. Aerial surveys of animals often use a double-sampling approach in which complete ground counts are obtained for subsamples of the entire surveyed area. In this situation, C is the total aerial count. The ratio of the mean aerial count on the subsampled area to the mean ground count on this area then estimates β, the proportion of animals seen from the air.

Change-in-ratio (CIR) estimators fall into a unique class of population estimation techniques in which capture or observation probabilities are not estimated directly. Instead, observabilities of different classes of animals (for example, males and females) are expressed relative to one another as "relative observability." A common assumption is that the observabilities of the two classes of animals are equal, so they cancel out and do not appear in CIR equations. If the observabilities are not equal, a major assumption of the basic method is violated, but adjustments for unequal observability can be made in the equations or in the way the CIR estimate is implemented.

For most of these formal estimation methods, statistical tests of the hypothesis of constant β can be developed. For example, Skalski et al. (1983) presented both likelihood ratio and conditional contingency table tests of the null hypothesis of equal capture probabilities for two times or locations sampled with two-sample capture-recapture experiments. If the null hypothesis is not rejected (i.e., the capture probabilities are not different), total animals captured can be used to index abundance, and comparative tests based on this index can be made. A similar test for populations being compared with constant-effort removal sampling was provided by Skalski et al. (1984). For capture-recapture experiments on open populations, the null hypothesis of constant capture probability for a single population over time can be tested (Jolly 1982). If capture probabilities remain constant over time, it is reasonable to use number of animals caught at each sampling period to index population size for inferences involving temporal changes in abundance. However, if the null hypothesis of equal capture probabilities is rejected, number of animals caught will not provide a good index, because temporal changes will reflect changes in both abundance and capture probability.

As indicated earlier, aerial surveys of animals often use a double-sampling approach. Contingency tables can be used to test hypotheses about variation in these β_i over time and space. If the β_i are constant, inferences can be based directly on the aerial count data.

Why should the hypothesis of equal β_i be tested when indices are obtained in conjunction with a formal population estimation method? The biological hypothesis of interest can be tested with the actual population estimates themselves, $\hat{N}_i$ (e.g., from eq. 10), rather than the statistics, C_i, on which they are based. The reason for using the statistics when β_i's are equal is that the index statistics generally will have smaller variances than their associated population estimators (Eberhardt and Simmons 1987). This can be seen by examining equation (5) and recalling that the variance of $\hat{N}$ includes two main components, one associated with the statistic C and the other associated with the estimation of β. The variance of C alone, however, does not include variation associated with β.

An important consequence of the smaller relative variance of C than $\hat{N}$ is that hypothesis tests that use C will tend to have greater power than tests that use $\hat{N}$. Similarly, estimates of rate of change (when comparisons involve a single population at different points in time) and proportional abundance (when comparisons involve different populations at one point in time) tend to be more precise when based on C_i than when based on $\hat{N}_i$. For example, estimates of proportional abundance ($\hat{N}_2/\hat{N}_1$) based on the total number of animals captured in a two-sample, capture-recapture experiment (C_2/C_1) are from 2 to 20 times more efficient than estimates based on Lincoln-Petersen estimates ($\hat{N}_2/\hat{N}_1$) (Skalski et al. 1983). In constant-effort removal experiments, estimates of rate of change and proportional abundance based on total animals captured are more efficient than those based on the population size estimators of models M_h and M_{bh} of Otis et al. (1978) (see CAPTURE-RECAPTURE METHODS [p. 239], Skalski et al. 1984).

In view of the potential use of index statistics collected in conjunction with formal estimation methods, these methods serve two primary functions in studies comparing population size. First, data collected for population estimates provide the basis for testing the hypothesis of equal β_i among populations being compared. Second, if the hypothesis of equal β_i is rejected, the population estimates can be used to compare populations. The population estimation methods "correct" index statistics for unequal β_i, thus permitting reasonable tests of hypotheses about comparative population size.

The power of the test for equal β_i's is important to consider when use of index statistics to compare population sizes is contemplated. For example, if we fail to reject the hypothesis that $\beta_i = \beta$, but the power (i.e., the probability of rejecting the null hypothesis in the event that it is really false) of this test is low (say 0.4), we will not have much confidence that the β_i's are really equal. Given this uncertainty about the equality of the β_i's, we might decide that it is better to use population size estimates to compare population sizes, rather than index statistics. This approach is conservative in the sense that we sacrifice power in our test for differences in population size, but we are confident that our specified Type I error (the probability of falsely rejecting the hypothesis of no difference, when no difference actually exists) probably is correct. The decision of whether to use population estimates or index statistics in such situations will depend on the investigator's prior knowledge and intuition about β_i (i.e., is it likely that the β_i's really are equal) and on the relative seriousness of making a Type I or Type II error when we test for differences in population size. For a more detailed discussion of hypothesis testing and power, see a basic statistics text such as Moore and McCabe (1989).

INDICES NOT ASSOCIATED WITH FORMAL POPULATION ESTIMATION METHODS

Many statistics considered as potential indices to abundance are not obtained with a formal estimation model in mind. In such situations the ancillary data needed to estimate and test hypotheses about β_i generally are not obtained as part of the standard, data-collection process. Inferences about β_i require special efforts to estimate the true population size to which the index is thought to apply and then to "calibrate" the index to population size (Eberhardt and Simmons 1987). If estimates of population size can be obtained for several time periods or locations (depending on the intended comparative use of the index), a regression of population estimates on index values permits inferences about β. If the selected statistic is a true constant-proportion index, the relationship between the population estimates and index values will be linear, the intercept will be 0, and the slope of the regression will estimate β. Each ratio of the index statistic to the population size estimate, C_i/N_i, will also estimate β_i, and these

ratios often can be used to devise tests of the hypothesis that the β_i's are equal.

When the comparison of population estimates and potential index statistics leads to the conclusion that the β_i's are not constant, the selected statistic, C_i, does not meet the critical criterion of a constant-proportion index. However, measurable, exogenous variables, such as weather conditions or observer identity, may account for most of the variation in β_i. If so, multiple regression analysis can be used to model population size as a linear function of the index statistic, β_i, and relevant exogenous variables (Overton 1969). The other approach to dealing with relevant exogenous variables is to collect index statistic data under standard conditions. Such standardization might include restriction of data collection to periods with a specified range of weather conditions, to particular phenological periods or times of day such as early morning for birds, to particular observers, or to observers who have undergone a specific training program.

Many indices are based on actual counts of animals seen, heard, caught, or harvested. Three nationwide surveys coordinated by the U.S. Fish and Wildlife Service are based on counts of birds seen and heard at established stops on permanent, roadside routes: the mourning dove call-count survey (see Dolton 1993), the woodcock singing-ground survey (see Tautin 1982, Tautin et al. 1983), and the North American Breeding Bird Survey (Robbins et al. 1986). Detailed written instructions are provided to all observers in these surveys, and counts are "standardized" to the degree possible with respect to season, time of day, and weather. Also, records are kept of observer identity, and this variable is often used as a covariate in analyses of index data (Geissler and Sauer 1990). Despite these efforts to account for some variables via standardization and others by direct incorporation in analyses, the important question remains: do these count statistics reflect a constant proportion of the total populations being sampled?

Baskett et al. (1978) considered the utility of indices resulting from the mourning dove call-count survey and reviewed results of research on potential effects of dove pair status, position in the nesting cycle, time of day, population density, and weather. They concluded that all of these variables can influence call-count results, but that most "pose no major problems within limits of current call-count survey procedures" (Baskett et al. 1978:174). However, they singled out pair status as an important variable affecting cooing rate and noted that future studies should investigate how the proportion of mated males in the population varies over time and space. Baskett et al. (1978) also reviewed results of studies in which call-count data from local study areas were compared with other statistics thought to reflect numbers of breeding doves. They (p. 173) were "very doubtful that numbers of cooing males can be used to make reliable estimates of breeding densities, numbers of nests, or young doves produced within the radius of audibility of a call-count stop."

The woodcock singing-ground survey also involves counts of a specific segment of a woodcock population—singing males. Tautin et al. (1983) reviewed four studies of the relationship between singing male counts and local population size. The only study they believed yielded reasonable inferences was conducted by Dwyer et al. (1988)

at the Moosehorn National Wildlife Refuge, Maine, during 1976–80. Numbers of singing males counted varied little over the years of the study, but independent capture-recapture estimates of numbers of adult male woodcock obtained from the partially open model of Darroch (1959) showed evidence of a steady increase over the study period. Dwyer et al. (1988) concluded that the proportion of adult males in the population that had singing grounds varied over time (i.e., that β varied over time).

The Breeding Bird Survey differs from the mourning dove and woodcock surveys in not being restricted to a particular species. Sources of variation in the proportion of populations counted in this and similar surveys were discussed extensively in Ralph and Scott (1981). Sources of variation were organized into the broad categories of species variation, observer variation, and environmental influences, including season, time of day, habitat, and weather. Most of the contributions in Ralph and Scott (1981) acknowledged the likelihood that detection probability does vary (i.e., β varies), although few good tests of this hypothesis were available. Wilson and Bart (1985) investigated detection probabilities of house wrens and concluded that year-to-year phenological differences could result in substantial differences in the probability of a bird singing during a 3-minute listening period. Bart and Schoultz (1984) reported that the proportion of birds detected in surveys of singing birds decreased as true density increased, thereby violating the constant-proportion assumption.

Possible variation in detection probability, β, is important to interpreting results of the mourning dove, woodcock, and Breeding Bird surveys. Efforts to test the hypothesis of $\beta_i = \beta$ generally have led to the conclusion that detection probability varies over time and space. However, as in most areas of wildlife ecology, it is easy to find problems with existing methodology, but much more difficult to suggest preferable alternatives. Most workers who have studied these three surveys have expressed the hope (in some instances the opinion) that variation in detection probability may be small enough to permit reasonable use of index values to detect substantial changes in population size occurring over broad geographic areas (e.g., see Baskett et al. 1978, Dwyer et al. 1988).

In addition to extensive national surveys, counts of birds seen and heard are used to index local abundance of raptors, crowing male pheasants, drumming ruffed grouse, and whistling bobwhites (Bull 1981, Fuller and Mosher 1981). Spotlight counts of deer and other animals at night are sometimes used as indices. For example, night counts of alligators were used to index abundance at local and regional levels (Chabreck 1966, 1973, Taylor and Neal 1984). Woodward and Marion (1978) studied sources of variation in alligator night counts and identified water level, water temperature, and moonlight as factors influencing the proportion of animals seen. Roadside counts of cottontails at night and in early morning were used to monitor statewide population trends in Illinois (Preno and Labisky 1971). Sightings of red foxes by rural mail-carriers were used to index fox abundance in North Dakota (Allen and Sargeant 1975). Fox sightings were correlated with abundance estimates based on aerial surveys, leading Allen and Sargeant (1975) to conclude that

the index was a good one. Thus, statistics based on direct counts may sometimes provide reasonable indices to abundance. For most of these indices, however, the question of constancy of the proportion of the population seen or heard has not been adequately addressed.

The number of animals caught in trapping efforts is a commonly used population index in studies of small mammals (e.g., Dice 1941, Keller and Krebs 1970), and recommendations have been made for establishing standardized trap lines to obtain such statistics (Calhoun 1948). However, capture-recapture studies of small mammals that use model-based estimators typically show evidence of variation in capture probability over time (Nichols and Pollock 1983) and between species (Nichols 1986). The number of animals caught on traplines has also been used to index furbearer abundance (e.g., Wood 1959, Wood and Odum 1964). Recent capture-recapture studies of furbearers, however, have provided strong evidence of year-to-year variation in capture probability (Smith et al. 1984). We suspect that the number of animals caught generally will not represent a constant proportion of the population and that this statistic typically will be a poor index.

The number of animals harvested has been used to index the size of animal populations exposed to open hunting, fishing, or trapping seasons. For harvests to be a reasonable index, the proportion of the total population that is harvested (harvest rate) must be constant for areas or time periods being compared. Investigations of harvest rate, however, nearly always find evidence of variation over both time and space (e.g., Anderson 1975, Clark 1987), indicating that harvest generally does not provide a good population index.

In addition to counts of animals seen, heard, trapped, or harvested, many indirect indices are based on signs of animal presence and activity (see reviews in Scattergood 1954, Overton 1969). Counts of animal tracks crossing roads or trails have been used to index abundance, especially for deer (Tyson 1959, Connolly 1981). Tyson (1959) used track-count data in conjunction with ancillary data on deer movement patterns to develop an actual estimator of population size. He obtained independent estimates of deer numbers from drive counts and found good agreement between these and estimates based on track counts. However, he doubted that the relationship between tracks and deer would remain constant throughout the year. Downing et al. (1965) compared track counts with known numbers of deer in an enclosure, and results indicated variation in the relationship between tracks and deer number.

Fecal or pellet counts have been used as an abundance index of deer and other ungulates (reviewed in Neff 1968), rabbits (Cochran and Stains 1961), and even birds (Bull 1981). Ancillary information on rates of defecation and fecal decomposition can be used to estimate β, and hence to estimate population size from fecal counts (Eberhardt and Van Etten 1956, Neff 1968). Neff (1968) reviewed studies comparing pellet-count estimates of deer abundance with known numbers of deer. Data from two of these studies were adequate to address the assumption of constant β over time, and some temporal variation appeared to be present in both studies. More recently, Fuller (1991) found no relationship between pellet counts and aerial surveys that were corrected for observability bias

Estimating Numbers

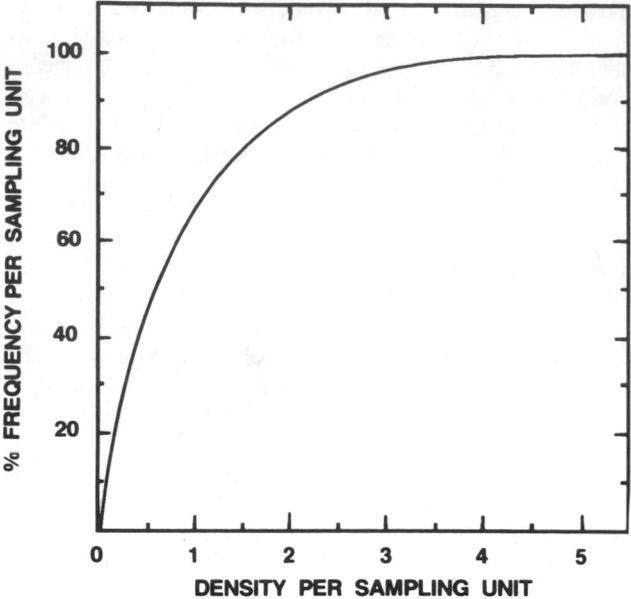

Fig. 3. The relationship between frequency and density when animals are randomly distributed (from Caughley 1977).

with radio-marked deer. In our opinion, indices of this sort are not likely to meet the constant proportion assumption.

Counts of conspicuous structures made by animals also are used as population indices. For example, indices have been based on counts of muskrat houses (Dozier et al. 1948), beaver lodges (Hay 1958), and nests of squirrels (Uhlig 1956), alligators (Chabreck 1966, Taylor and Neal 1984), and various bird species (e.g., Nettleship 1976, Bull 1981, Fuller and Mosher 1981). Independent estimates of beaver population size did not show a constant relationship to number of active lodges across colonies (Hay 1958), indicating that lodge counts do not provide an adequate index. Uhlig (1956) reported that the number of squirrels per leaf nest on more than 1,500 study plots varied little over 3 years, indicating that nest counts may provide a reasonable index to squirrel numbers in some situations. Efforts to relate counts of bird nests to independent estimates of bird abundance are few. However, variation in age structure of the population, proportion of breeding-age birds that nest, timing of the breeding season, and tendency to renest can easily result in substantial variation in β, the birds-per-nest ratio. When counts of animal structures are used to index animal abundance, a common assumption is that these structures are detected with probability 1.0, i.e., all existing structures are seen. A recent study of nest counts in white-winged dove colonies indicated that the proportion of nests seen by observers ranged from 0.93 to only 0.57, so that formal estimation methods may be necessary to properly estimate the number of structures in some situations (Nichols et al. 1986).

In summary, many different statistics not collected in conjunction with a formal population estimation method have been proposed as possible constant-proportion indices. Although the assumption of constant β is required for the reasonable use of such indices to compare population sizes, the data required to test this assumption are not routinely collected as a part of a standard data collection

process. Often the assumption of constant β is simply not tested. When special efforts have been made to test this assumption, they have typically identified sources of variation in β. Some sources of variation can be overcome by standardizing data collection procedures, but many others either cannot be handled in this manner or cannot even be identified. Thus, we recommend a large degree of caution and skepticism when these indices are used and interpreted. When population changes estimated from indices are used in population management, the assumption of constant β probably merits detailed study.

Frequency Indices

Although constant-proportion indices are by far the most useful, there may be situations when other indices have some utility, depending on whether something is known about the relationship between population size and the index. This represents a "Catch-22" because the only way to obtain knowledge of the relationship is to estimate population size and "calibrate" the index. However, the reason for considering use of an index in the first place may be because the cost and effort involved in estimating population size are prohibitive.

The only index we will discuss in this category is the frequency index. Frequency indices are based on the proportion of sample units that contain at least one of something such as an animal or sign of animal activity (Scattergood 1954, Caughley 1977, Seber 1982). Frequency indices generally are not related linearly to abundance or density as in equation (8). In the best situations, the relationship between index and true population size is positive, but nonlinear (Fig. 3). Comparative uses are thus restricted to ranking density or abundance, and, in general, neither rate of change nor proportional abundance can be estimated. However, Caughley (1977:22) noted that for frequencies less than about 0.2, the frequency-density relationship is nearly linear and may provide useful comparisons.

Frequency indices have been used in conjunction with several different sampling methods. For example, direct counts of animals on quadrats or other sampling units, numbers of animals caught in a specified number of traps, and counts of animal tracks or other sign on a specific number of sampling units have been used to compute frequency indices. The proportion of transects on which groups of howler monkeys were seen provided a useful frequency index (Subcommittee on Conservation of Natural Populations 1981). Wood (1959) concluded that the number of trapline stations catching at least one fox provided a reasonable frequency index to fox abundance (but see Smith et al. 1984). The proportion of established scent stations where animal activity is detected by the presence of tracks provides an often-used frequency index to abundance of many furbearer species (Wood 1959, Conner et al. 1983). Conner et al. (1983:151) compared scent-station indices to abundance estimates and concluded that the indices "accurately reflected trends in the population abundance of bobcats, raccoons, and gray foxes, but not of opossums."

Assumptions underlying the reasonable use of frequency indices are similar to those required for the more useful constant-proportion indices. The probability of catching, counting, or otherwise detecting an animal in sample units from two areas or time periods being compared should be similar. As Seber (1982) noted, the statistical distribution of animals over space should also be similar for areas or time periods being compared. For example, if direct counts are used to ascertain presence or absence on sample quadrats, a population with a highly clumped distribution will yield a lower frequency index (proportion of quadrats with at least one animal) than a population of similar density that exhibits a more uniform distribution over space. Finally, we note that although frequency data usually are treated as indices, certain distributional assumptions permit estimation of population size in some situations (see below).

ESTIMATES OF ABUNDANCE AND DENSITY
All Individuals Observed—Complete Counts

TOTAL COUNTS

Seldom, if ever, will it be possible to obtain a total count of animals over an entire area of interest. If the count is purported to be a total census of all the animals, inferential statistics such as variance and confidence interval are not needed because the entire population is counted; no sampling is employed. In such situations, both α and β in equation (4) equal 1, so $N = C$, and the count represents population size. However, an error-free census is unlikely, so "... a census datum should be accompanied by a critical evaluation of its accuracy and by an explicit statement of the constraints and definitions under which it was collected" (Overton 1969:419).

Drive counts (Morse 1943) of deer or other ungulates are sometimes purported to permit accurate population counts, but most workers using this method concede that an unknown number of animals likely is missed (e.g., Tyson 1959, McCullough 1979). Aerial photography can be used to obtain nearly complete counts of animals in certain special situations. For example, Haramis et al. (1985) used 35-mm photography from low-flying aircraft to count canvasbacks in flocks throughout Chesapeake Bay and coastal North Carolina. We give some brief descriptions of several complete count methods below, but we refer the reader to reviews by Scattergood (1954), Overton (1969), Eberhardt et al. (1979), Seber (1982), and Miller (1984).

Drive Counts

As the name implies, animals are driven by counters to census the total number of animals in a defined area. The technique is well suited to species such as deer and pheasants that inhabit relatively open habitat. Overton (1969) gave a brief overview of the technique. Drivers, spaced along a line, sweep across an area with well-defined boundaries. Additional observers may be situated along the boundaries to count animals that move into or out of the census area. The census is simply the sum of the number of animals moving out of the area ahead of the line of drivers, plus those moving from ahead of the observers through the line, minus any moving into the area ahead of the drivers.

McCullough (1979) used drive counts to census a fenced-in deer population on the George Reserve in Michigan. He compared drive counts with the "known" population that was reconstructed from the age of death of individuals in the population (see Population Reconstruction, next page). He concluded that at low populations

drive counts underestimated the true population, and at high populations they overestimated the true population. Errors could be as large as 20–30%. Thus, drive counts are probably best viewed as an index of population size. As with other indices, efforts to test hypotheses about possible variation in β should precede serious use of drive-count data for management purposes.

Total Mapping of Bird Territories

This approach is similar to spot mapping (described below) except that an effort is made to color band and follow all marked individuals to delineate their territories or home ranges. In most total mapping studies, however, population estimation is not the primary objective of the study. Verner (1985:266) believed that, when thoroughly executed, total mapping is probably the most accurate method of estimating population density of breeding birds, and "... total mapping should be used as a standard for evaluating the accuracy of other methods of estimating the densities of birds." Note that this method estimates only the population of birds holding territories, not floaters or transients.

Spot-Mapping or Territorial-Mapping Method

Spot mapping (Verner 1985) gives estimates of breeding bird population density. The technique is most suited to passerine birds that regularly sing or call within exclusive territories. Nonterritorial birds (floaters) are not surveyed by this technique.

Spot mapping involves plotting locations of individual birds on a gridded map during repeated visits to a study area. Then, clusters of locations, assumed to represent centers of activity of individual territories during the breeding season, are identified on the map. The total number of clusters in the study area equals the number of clusters completely inside the area plus the sum of fractional parts of clusters on the boundaries. The total number of birds is then estimated by multiplying the number of clusters by the mean number of birds per cluster, which is normally two (presence of a breeding pair is assumed).

Assumptions of the method (after Verner 1985) are (1) populations are constant, and birds remain within exclusive spaces or territories during the sampling period, (2) birds on territories produce cues frequently enough to permit repeated location on successive observational visits, (3) estimated proportions of territories along boundaries are accurate, (4) the estimated mean number of birds represented by each cluster is accurate, and (5) birds are correctly identified.

Meeting the assumptions of the method presents many problems. For example, interpreting the spatial arrangement of clusters for some species varies considerably among observers (Best 1975, O'Conner and Marchant 1981). Recently, Verner and Milne (1990) provided strong evidence that spot-mapping results should not be considered to be complete counts and that these results can vary substantially among observers and map analysts. Thus, at best, spot mapping yields an index; the variation in β among observers and analysts suggests caution in using this method.

Thermal Scanners

Remote sensing with thermal infrared (IR) scanners (3–5 and 8–14 μm) has been proposed as a technique to census animal populations. Parker and Driscoll (1972) suggested that thermal scanning for wildlife was feasible, but the appropriate equipment needed to be developed. More recently, Wyatt et al. (1980:401) concluded that ". . . a thermal scanner can successfully detect deer against a snow-covered background; however, such a system would exhibit large errors in detecting deer . . ." under certain conditions. They believed that thermal contrast, by itself, was not of value in censusing deer populations. Best and Fowler (1981), studying geese, also concluded that aerial thermography could be used as a census technique under suitable environmental conditions.

Multispectral Scanners

Wyatt et al. (1984) described a multispectral system using energy in the 0.7–1.1 μm region of the electromagnetic spectrum. This system, the result of intensive laboratory studies (Wyatt et al. 1985), used nearly 7,000 photodiode detectors and sensed spectral classification in four "colors"; it is still in the developmental stage, however (D. R. Anderson, pers. commun.)

The spatial resolution of remote sensing instruments relative to the size of the animals has been a major problem for wildlife census (L. L. Strong, pers. commun.). Recently, Strong et al. (1991:250), using a multispectral scanner to inventory winter concentrations of geese on a water background, suggested that "it was possible to inventory white geese and dark geese in mixed concentrations on water from visible and shortwave infrared reflectance measurements." They used an image-processing mathematical model, called a "mixture model," with an instantaneous-field-of-view scanner to estimate the proportions of animals and background within the view of the scanner.

Population Reconstruction

If all dead animals from a population can be located or otherwise known and if the year and age at death can be determined, the population can be "reconstructed" based on how long each individual was known to be in the population. Thus, population size can be determined for a given year in the past only after all the individuals alive in that year have died. Population reconstruction is distinct from life-table analysis (see Chapter 16), in which the relative numbers of animals dying (surviving) in different age classes are used to estimate survival rates, not population size. McCullough (1979) used population reconstruction to determine the size of the George Reserve deer herd. The method is completely accurate if all individuals can be accounted for, but in most field applications this is extremely unlikely.

Aerial Photography

Low-altitude photographs of flocks of birds (or other groups of animals) often are used as a census technique. The entire assemblage of animals is photographed and later counted to give a complete census. However, it is often difficult to determine if all individuals are "visible" to be photographed, and errors in counting undoubtedly are made. This approach is distinct from aerial surveys in which the population is sampled by counting animals seen on a transect or on quadrats as the observer flies over.

Haramis et al. (1985) conducted a photographic census of the wintering population of canvasbacks in Chesapeake

Bay and North Carolina. The authors believed the photographic survey represented ". . . nearly a complete census of open-water habitats in this region" (Haramis et al. 1985:449). A hand-held, 35-mm camera, equipped with a 35-mm, wide-angle lens and color slide film (Kodachrome-X, ASA 64) to facilitate distinguishing species and sexes, was used to photograph canvasback flocks from the window of a low-flying (<60 m altitude), fixed-wing airplane. A sequence of photos was often necessary to cover an entire flock clearly. Slides were projected on paper, and ducks were identified by species and sex.

TOTAL COUNTS ON SAMPLE PLOTS

Obtaining complete counts of animals on suitably sized (relative to the organism being considered) sample plots within some larger area of interest may be possible. In terms of equation (4), $\hat{\beta} = 1$ and thus $\hat{N} = C/\alpha$, so no component of variation is associated with estimating the proportion of animals seen on sample plots. Instead, we are concerned only with geographic (plot-to-plot) variation, which becomes relevant when we extrapolate from the plot counts to draw inferences about the larger area from which the samples are drawn.

The estimation methods used for these situations come directly from standard statistical sampling theory (Cochran 1977). Consider an example of simple random sampling in which the following notation (after Seber 1982) is used:

A = total area occupied by the population (assumed to be known),

N = total population size (the quantity we would like to estimate),

s = number of randomly selected sample plots on which counts are made,

a = area of each sample plot,

$S = A/a$ = total number of potential sample plots in A from which the s plots are selected,

x_i = number of animals counted on plot i,

$\bar{x} = \sum_{i=1}^{s} \dfrac{x_i}{s}$ = mean number of animals counted per sample plot,   and

$\widehat{\text{var}}(x_i) = \sum_{i=1}^{s} \dfrac{(x_i - \bar{x})^2}{(s - 1)}$ = the estimated sampling variance of the x_i.

As shown by Seber (1982) and Cochran (1977), total population size, N, can be estimated as:

$$\hat{N} = \frac{\left[\sum_{i=1}^{s} x_i\right]}{\left[\dfrac{s}{S}\right]} = \bar{x}S \qquad (11)$$

and the variance of $\hat{N}$ as:

$$\text{var}(\hat{N}) = S^2 \left(\frac{\text{var}(x_i)}{s}\right)\left(1 - \frac{s}{S}\right). \qquad (12)$$

Table 1. Estimates of the wintering mule deer population based on a stratified random sample of eight strata on the Uncompahgre Plateau, Colorado, 1977–79 (from Kufeld et al. 1980).

Parameter	Year	Stratum 1	2	3	4	5	6	7	8	Total
Area (km^2)	All	133	285	262	317	150	207	103	231	1,688
No. quadrats										
Total	All	206	440	405	489	232	319	159	357	2,607
Sample	All	12	35	23	38	13	25	9	38	193
Deer population	1977	189	289	581	1,361	428	2,386	724	5,440	11,401
estimates	1978	275	1,307	352	3,861	892	4,772	901	5,524	17,884
	1979	155	1,446	669	2,458	339	5,589	247	6,182	17,085
90% CL	1977									±2,205
	1978									±4,042
	1979									±2,951

Population size for the entire area is thus estimated (eq. 11) as the total number of animals counted on the sampled area divided by the proportion of the total area covered by the sample ($s/S = \alpha$). This expression thus fits into our conceptual framework for dealing with sampling and geographic variation (eq. 4). The population size estimator can also be written more simply as the product of the average number of animals per sample plot and the total number of potential plots in the area (eq. 11).

Density is easily estimated because the area of each plot (a) and the total area (A) are known, so

$$\hat{D} = \frac{\hat{N}}{A} \quad \text{and}$$

$$\text{var}(\hat{D}) = \frac{1}{A^2}\text{var}(\hat{N}).$$

In some situations, precision can be increased by stratifying the area sampled based on habitat or expected variation in density. In such situations we essentially use equation (11) to estimate population size and equation (12) to estimate its variance for each stratum, and then add stratum estimates (if the strata are all the same size) to obtain the overall $\hat{N}$ and $\widehat{\text{var}}(\hat{N})$ (see Seber 1982).

A recent example involves the estimation of American black duck numbers on the Atlantic coast from aerial transect surveys in 1983–86 (Conroy et al. 1988) by the U.S. Fish and Wildlife Service. The authors used a stratified, random sampling design, but their computational procedures were slightly different from those presented above because of special aspects of the survey. For example, the area sampled by each transect (analogous to plots in the above discussion) was not a constant but varied because some transects were longer than others. Although the above methods were used with the black duck survey, the authors did not believe they were obtaining complete counts on the transects (sample plots). Instead, they believed that they were likely seeing the same (unknown) proportion of the population ($\beta < 1$) each year. They thus recommended that their survey-based estimates be considered a constant-proportion index.

Kufeld et al. (1980) used total counts on sample plots (quadrats) and a stratified random sampling design to es-

timate the wintering population of mule deer on the Uncompahgre Plateau of Colorado. They divided the 1,688-km^2 study area into eight strata based on "educated guesses" of relative deer densities. Each stratum was further divided into 0.6475-km^2 (0.25 mi^2) quadrats. A sample of 7.4% of the quadrats was selected (random selection within strata) and quadrats were marked on the ground so they could be identified from a helicopter. The number of sampled quadrats in each stratum was proportional to both the area of the stratum and the educated guesses of relative deer densities in the stratum. Strata with higher densities received more samples because they were more likely to have greater variances.

The sample quadrats were "censused" by helicopter with three observers in 1977, 1978, and 1979. The results of the survey are given in Table 1. Although some problems were encountered the first year, the authors estimated population means to be within about 20% of the true value with 90% confidence. To us, this is a reasonable level of precision and accuracy for large-scale field studies. Kufeld et al. (1980) also determined that the stratified random design reduced the variance of the mean number of deer seen per quadrat by about one-third compared to a simple random design without stratification.

The major concern with this type of study is whether all the deer on each surveyed quadrat are actually seen. Usually this is probably not true; but, because of the relatively open habitat and the use of a noisy helicopter with three observers (as opposed to a fixed-wing aircraft), the authors believed few deer were missed. Furthermore, they cited a previous study where all the mule deer known to be in three fenced pastures in the same vegetation type as the Uncompahgre Plateau were counted by a helicopter survey. In any case, these estimates should be considered conservatively as minimum estimates or, perhaps, as a constant-proportion index.

All Individuals Not Observed—Counting Methods

Incomplete counts simply refer to counts of animals in the usual situation where we cannot assume that all animals present are observed. Instead, the observers count some fraction, β, of the total animals present, and this fraction must be estimated to translate incomplete counts into estimates of population size (as in eq. 2). Several approaches are available for estimating β in conjunction with animal counts. Some of these approaches were developed with ground surveys in mind, whereas others were developed for aerial surveys. Although the general estimation methods are applicable regardless of the observer's location (air or ground), our description of each method is based on the survey situation to which the method is most commonly applied. However, the reader should realize that extension of the described methods to other applications is straightforward.

DOUBLE SAMPLING

Sometimes complete counts of animals can be made on small subsamples of a larger area where incomplete counts are made. Then double sampling from formal sampling theory (Cochran 1977) can be applied (Box 1). In the standard wildlife application, the incomplete counts are made over an extensive area from the air (via fixed-wing airplane or helicopter), and the complete counts are made on the ground. For example, in many aerial surveys, the airplane flies along a predetermined flight path or transect line, and an observer counts animals seen in the strip extending from the transect line to some specified distance on either side of the line. Ground sampling then involves an intensive search of the surveyed strip on some smaller sample of the transects. Ideally, ground and air counts should be made simultaneously. If this is not possible, the two counts should be separated by as little time as possible to minimize the chance of movement changing the number of animals in the sampled area.

The natural estimator for β, the proportion of animals seen from the air, is simply the ratio of the mean aerial count ($\bar{y}$) to the mean ground count ($\bar{x}$) on the air-ground subsample

$$\hat{\beta} = \frac{\bar{y}}{\bar{x}}. \tag{13}$$

It follows that the total number of animals present on the transects surveyed from the air can be determined by using equation (2)

$$\hat{N} = \frac{C}{\hat{\beta}},$$

where $\hat{\beta}$ is from equation (13) and C is the total number of animals seen from the air. To estimate population size for the entire area, we can use equation (4)

$$\hat{N} = \frac{C}{\alpha\hat{\beta}}, \tag{14}$$

where β and C were defined above and α is the fraction of the total area sampled from the air. An estimator for the variance of $\hat{N}$ was presented by Jolly (1969a,b) and Pollock and Kendall (1987).

"Ground" counts have been used with aerial surveys to estimate sea otter numbers in California (Eberhardt et al. 1979). The extensive aerial survey of breeding waterfowl conducted each May by the U.S. Fish and Wildlife Service, the Canadian Wildlife Service, and various state and provincial wildlife agencies uses ground counts to es-

---

---

timate α (Martin et al. 1979). In our opinion, the most critical consideration in the practical application of double sampling is the accuracy of the ground count, because the method assumes that ground counts are entirely accurate. If all animals on subsampled areas are not seen on the ground, estimates of sighting probability will be too large, and estimates of population size will be biased low. Additionally, the timing of the ground and aerial counts must coincide so they reflect the same population of animals. Jolly (1969a, b) discussed some other considerations relevant to the design of aerial surveys that use ground counts.

MARKED SUBSAMPLE

Marked subpopulations can be used to estimate sighting probabilities. In this approach animals are marked individually so that at the time of the survey a known number of marked animals is in the area being surveyed. During the survey, marked and unmarked animals are counted.

Assume that we are dealing with an aerial survey and define the following notation:

N = total population size on the surveyed area,
n_1 = number of marked animals present on the area at the time of the aerial survey,
n_2 = number of animals (both marked and unmarked) seen during the aerial survey, and
m = number of marked animals seen during the aerial survey.

A natural estimator for the proportion of animals seen from the air, β, is:

$$\hat{\beta} = \frac{m}{n_1}, \tag{15}$$

and our estimator for population size based on equation (2) is thus:

$$\hat{N} = \frac{n_2}{\hat{\beta}} = \frac{[n_2 n_1]}{m}. \tag{16}$$

Box 2. Using a marked subpopulation to estimate observability and population size with the bias-adjusted Lincoln-Petersen estimator.

Rice and Harder (1977) used aerial survey in conjunction with a marked subsample to estimate white-tailed deer numbers on Plum Brook Station, a fenced, 2,176-ha NASA research facility near Sandusky, Ohio. Deer were captured during September 1974 to February 1975 and fitted with radio collars. Marked and unmarked animals were then counted during five helicopter flights in January–February 1975. The number of deer in an enclosed 122-ha test area within the facility was determined with an intensive (80-person) drive. Of the 155 deer counted in the drive, 10 were marked. The first of the five helicopter counts yielded 106 deer, 8 of which were marked. These statistics ($n_1 = 10$, $n_2 = 106$, $m = 8$) were used with the bias-adjusted Lincoln-Petersen estimator (eq. 49) to yield an estimate of 130 animals. This estimate is reasonably close to the known population size of 155. The average of the estimates from the five replicate counts was 159 ($\widehat{SE} = 32$), very close to the true value. Rice and Harder (1977) estimated the deer population of the entire facility to be 2,499 ($\widehat{SE} = 47$).

This estimator (eq. 16) is simply the unmodified Lincoln-Petersen estimator from capture-recapture. In practice we recommend use of the bias-adjusted modification of this estimator and the associated variance estimator provided by Chapman (1951) (see eq. 49, p. 240). An example is provided in Box 2.

Although this approach sounds very straightforward, the practical aspects of application to any particular situation require careful consideration. The nature of the mark, for example, is important. The mark must be visible from the air such that the aerial observer knows whether each animal that is seen is marked, and yet marked and unmarked individuals must have the same probability of being seen. Thus, the mark must be distinct and readily visible so that no marked animals are seen but recorded as unmarked, but not so obvious that it draws attention to marked animals, making them more visible than unmarked animals. Radiotelemetry can be used to determine the number of radio-marked animals in the surveyed area at the time of the survey (e.g., see Packard et al. 1985). A receiver in the airplane can be used to determine whether each animal seen from the air is marked. If radio-marked individuals cannot be positively identified, these animals will require some additional, unique mark visible from the air.

A major advantage of using radio-marked animals as the marked subpopulation is that determining the number of marked animals in the surveyed area at the time of the aerial survey is relatively easy (but see DeYoung et al. 1989). If other markers are used (e.g., identifica-

tion collars on deer or other ungulates), different considerations become important. If animals are marked and released just before (e.g., within a few days of) the air count, all released animals can be assumed to be available to be seen. It is not necessary to have individually identifiable animals from the air, and "batch" marks (e.g., collars with no alphanumeric identification code) will suffice. However, capturing and marking an adequate number of animals in a relatively short period of time is often difficult, necessitating introduction of marked animals to the population over an extended period (e.g., several weeks or even months). In these situations, marked animals could die or emigrate from the survey area before the aerial survey is conducted. Special efforts, such as using radiotelemetry, may be necessary to locate marked animals immediately before or after the aerial survey to establish which marked animals are known to have been present at the time of the aerial survey. Thus, n_1 in equations (15) and (16) is the number of marked animals known to be present from efforts to locate marked animals on the ground and not the total number of marked animals released. Similarly, m is the number of these n_1 animals seen from the air. Marked animals seen from the air, but not "known" to be present from earlier efforts on the ground, are not included in n_1. They are treated as unmarked in the aerial survey data so they are included in n_2 but not in m.

MULTIPLE OBSERVERS

Independent Observers

Two observers in the same airplane might be able to record the sighting location of individual animals on a map even if the animals are not individually marked. If the two observers do not communicate and if they sight animals independently of each other, the mapped sighting locations can be used with the Lincoln-Petersen estimator (the bias-adjusted version of equation 16 [see eq. 49]) to estimate total number of animals in the surveyed area (see Grier et al. 1981, Caughley and Grice 1982, Pollock and Kendall 1987). In this situation, n_1 of equation (16) is the total number of animals seen by one observer, n_2 is the number seen by the other observer, and m is the number seen by both observers. This method requires precise and detailed maps of sighting locations to ensure no ambiguity about whether one or both observers sighted a particular animal. Independence of sightings between the observers is also an important requirement that may be difficult to achieve. For example, the independence assumption will be violated if some activity of one observer, such as speaking into a tape-recorder microphone or writing on a map, alerts the other observer to the possibility of an animal nearby. Another assumption is that all animals have equal sighting probabilities for a particular observer, but they can differ between the two observers. If certain animals are much more visible than others, the resulting heterogeneity will produce negative bias in the Lincoln-Petersen estimator (see CAPTURE-RECAPTURE METHODS [p. 239]).

When mobile animals are considered, the mapping of locations must occur at the same time to ensure reasonable confidence of animal identification. However, this general

> **Box 3.   Using two independent observers with the bias-adjusted Lincoln-Petersen estimator to estimate the size of a population of an immobile object—alligator nests.**
>
> Two independent observers were used to estimate the number of alligator nests on Orange Lake, Florida, in the summer of 1986. The work was conducted by A. R. Woodward and M. Jennings of the Florida Game and Fresh Water Fish Commission and H. F. Percival of the Florida Cooperative Fish and Wildlife Research Unit. Two helicopter surveys with different observers were conducted on the lake during the early incubation period, and nest locations were mapped, permitting unambiguous identification. The numbers of nests seen by the first and second observers were $n_1 = 34$ and $n_2 = 37$, respectively, and the number of nests seen by both observers was $m = 20$. Use of these data in conjunction with the bias-adjusted Lincoln-Petersen estimator yielded $\hat{N} = 62$ (SE = 5.7) total nests.

approach also can be used with immobile objects such as muskrat lodges, bird nests, or crocodile nests because their location does not change, and the two observers need not be in the same aircraft at the same time (Box 3). Independence of the two observers could be ensured by making either two separate aerial surveys or an aerial survey and corresponding ground survey. Different observers are used for the two surveys, and locations are mapped for identification as before. The bias-adjusted version of equation (16) (see eq. 49) can be used, with n_1 and n_2 corresponding to the total number of objects counted by observers 1 and 2, respectively, and m corresponding to the number of objects seen by both observers. Henny et al. (1977) used this general approach to estimate numbers of osprey nests, Magnusson et al. (1978) used it to estimate the number of saltwater crocodile nests, and Estes and Jameson (1988) used it to estimate a sighting probability for sea otters.

Dependent Observers

Another estimation approach involves two observers in the same plane who communicate with each other. One is designated as the "primary" observer, the other as the "secondary" observer. Here, we are interested in the number of animals seen by the primary observer and the number of additional animals (i.e., in addition to those seen by the primary observer) seen by the secondary observer. Under the assumption of equal sighting abilities of the two observers, these two statistics permit estimation of population size under the two-sample removal model (Pollock and Kendall 1987). Heterogeneous sighting probabilities again produce negatively biased estimates of population size.

A similar approach that does not require the assumption of equal sighting abilities of the two observers was developed by Cook and Jacobson (1979). Under this design,

the primary observer records animals seen and the secondary observer records additional animals seen, as before. However, the two observers switch roles halfway through the survey. We assume that the sighting probability associated with each observer does not change with their role as primary or secondary observer. The estimators associated with this method were presented by Cook and Jacobson (1979) (also see Pollock and Kendall 1987).

Sighting Probability Models

All of the preceding methods for estimating population size from incomplete count data involve (either explicitly or implicitly) efforts to estimate sighting probability (β) during the survey. The incomplete counts themselves and the data to estimate sighting probabilities associated with those counts are obtained concurrently.

An alternative approach is to conduct experimental surveys designed to identify variables likely to influence sighting probability and then to use these variables to develop a model for predicting sighting probability (e.g., Caughley et al. 1976, Samuel et al. 1987). During subsequent surveys, variables in the models are measured and used to predict sighting probability. These sighting probability estimates are used with the incomplete counts from subsequent surveys to estimate population size.

An excellent example was provided by Samuel et al. (1987), who used data from radio-marked elk to develop a sighting probability model for use with aerial surveys in Idaho. Before each experimental survey flight, Samuel et al. (1987) located radio-marked individuals and measured variables that they thought affected observability. Then, they recorded whether each animal was seen during the subsequent survey. The resulting data were used with binary regression analysis to develop models for predicting sighting probability. They found that group size and percent vegetation cover were the most important determinants of sighting probability. The model based on these two variables was used to compute a sighting probability for each group of elk observed during operational surveys. The number of groups of a given size and in a given vegetational cover class was adjusted by dividing by the estimated sighting probability, as in equation (2). The total population estimate for the surveyed area was computed as the sum of these group estimates.

The appeal of a model-based approach is that the often costly process of estimating sighting probabilities is done only once during the initial experimental period of model development. After the model has been developed and satisfactorily tested, operational survey efforts require only recording information on the model variables; sighting probabilities are not estimated in subsequent surveys. A limitation, of course, is the possibility that the model works well only under the exact conditions when the model was developed.

LINE TRANSECTS

In line-transect methods, a transect or line of length, L, is set out randomly within some area to be sampled. Counts of all animals seen are then made by the observer traveling along this line. Sometimes a maximum observation distance, w (perpendicular to the line on each side), beyond which no animals are counted, is established. In other applications all animals are counted regardless of distance from the line (see Box 4).

Sometimes the term *line transect* or, more commonly, *strip transect* has been used to refer to transects of fixed half-width, w, for which counts within the strip are assumed to be complete (i.e., the observer sees all animals actually present within the strip). Such strip transects are simply a special case of complete counts on sample plots (see TOTAL COUNTS ON SAMPLE PLOTS [p. 226]), and the plot is a rectangular strip.

In the line-transect methods considered here, the counts are assumed to be incomplete. Thus, the proportion of animals present that are actually seen (β) must be estimated, and the actual counts must be corrected by these sighting or detection probabilities. Either perpendicular distance data or both sighting distance and sighting angle data are required to estimate sighting probabilities. These additional data are defined as follows: x_i = the perpendicular distance from the line to the detected animal i (or nest or center of a group of animals), r_i = distance from the observer to detected animal i at the moment of detection, and θ_i = the angle between the line of travel and the line of sight to detected animal i at the moment of detection (Fig. 4). Thus, the data resulting from any line-transect survey are the total number of animals detected, n, and either the corresponding perpendicular distances, x_i, or both the sighting distances, r_i, and angles, θ_i. Estimation proceeds with these data and the known transect length, L, and width, $2w$. Recall that w is the maximum observation distance, hence the width of the surveyed strip is $2w$.

The use of line-transect methodology requires four assumptions, listed here as suggested by Burnham et al. (1980:14) in order from most to least critical: (1) all animals located directly on the line are detected (i.e., detection probability on the line is 1); (2) animals are fixed at the location where they are initially sighted (i.e., they do not move before being sighted) and no animals are counted twice; (3) distances (and angles if recorded) are measured exactly; and (4) sightings are independent events (e.g., the flushing of one animal does not cause another to flush).

Perpendicular Distance Data

The basic idea underlying estimation from line-transect data, or similar "distance sampling" approaches, is that the probability of detecting an animal decreases with increasing distance from the line, i.e., increasing x. The distance data, x_i, are then used to estimate the specific shape of this function, $g(x)$, relating detection probability to distance from the transect. We can define the detection function, $g(x)$, as the conditional probability of observing an animal, given that it is located at distance x from the line, or mathematically:

$$g(x) = \text{Pr}\{\text{animal observed} \mid x\}.$$

To obtain an idea of what $g(x)$ looks like, we can plot a histogram of detections grouped by small distance intervals from the center of the transect to the maximum sighting distance. If our sample size is large and we detect a large number of animals, n, we can approximate the shape of $g(x)$ by drawing a smooth curve through the histogram

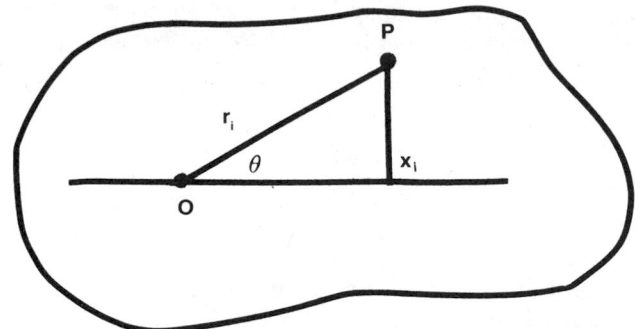

Fig. 4. A diagram of sighting distance (r_i), sighting angle (θ_i), and perpendicular distance (x_i) from the transect line to the sighted animal i. O is the position of the observer when an object is detected at point P (from Burnham et al. 1980:29).

(Fig. 5). In practice, sample sizes are often too small, so this procedure is not very satisfactory.

Even if we know $g(x)$, we still must estimate population size. This involves estimating an average detection prob-

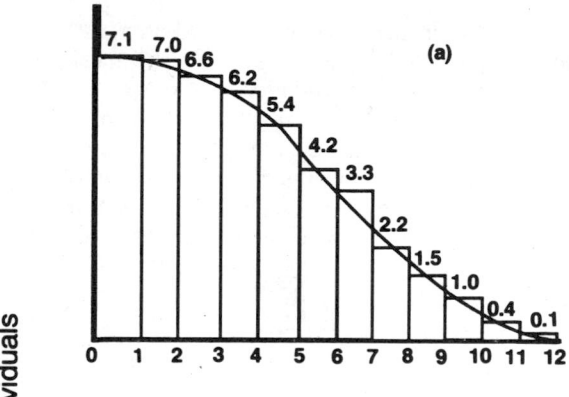

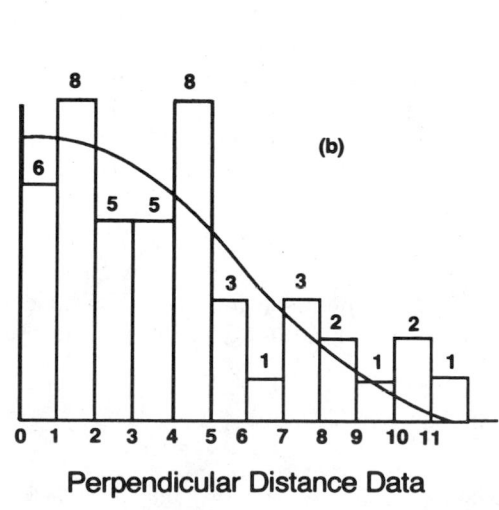

Perpendicular Distance Data

Fig. 5. (a) Expected histogram of perpendicular distance data (arbitrary units) if $g(x)$ has the shape shown for a sample size of $n = 45$ (from Burnham et al. 1980:16) and (b) an example of actual perpendicular distance data (from Burnham et al. 1980).

ability, P_w, which is equivalent to β in our general equation (2). Thus, the number of detected objects and P_w are used to estimate N:

$$\hat{N}_w = \frac{n}{\hat{P}_w}, \qquad (17)$$

where $\hat{N}_w$ is the estimated number of animals present in the strip defined by length L and width $2w$ (where w is the predetermined maximum sighting distance). Because of the random placement of the transect line, animals are equally likely to be located (not detected) at all distances between 0 and w. Thus, we can estimate the average detection probability, P_w, as:

$$\hat{P}_w = \frac{\left[\int_0^w g(x) \, dx \right]}{w}. \qquad (18)$$

Those who prefer not to think in terms of integrals of functions may prefer to think of $g(x)$ as a histogram of detection probabilities at different distances. Equation (18) is roughly equivalent to computing the average of the probabilities for different distance categories of the histogram.

As noted earlier, the observed distribution of detection distances is closely related to $g(x)$, but we still must find a way to use this observed distribution to estimate $\int_0^w g(x) \, dx$. Denote this integral as a, and denote the probability density function of the perpendicular distance data as $f(x)$. This function, $f(x)$, can be thought of as the underlying probability distribution from which the observed distance data were generated. It can be shown (Burnham and Anderson 1976) that $f(x)$ and $g(x)$ are related by:

$$f(x) = \frac{g(x)}{\int_0^w g(x) \, dx} = \frac{g(x)}{a}. \qquad (19)$$

As noted by Burnham et al. (1980), this equation shows that $f(x)$ is simply $g(x)$ scaled to integrate to 1 (and hence to be a probability distribution function).

The critical assumption permitting estimation from distance data is that all animals located directly on the line (distance = 0) are detected, i.e., $g(0) = 1$. This assumption permits us to use equation (19) and write $f(0) = 1/a$. Thus, if we can estimate $f(0)$ using the observed distances, x_i, then we can estimate a as:

$$\hat{a} = \frac{1}{\hat{f}(0)}. \qquad (20)$$

We can substitute equation (20) into equation (18) [remember $a = \int_0^w g(x) \, dx$] and obtain our estimate of P_w as:

$$\hat{P}_w = \frac{1}{w\hat{f}(0)}. \qquad (21)$$

The resulting estimate of population size in the strip defined by L and w is thus (from eqs. 17 and 21):

$$\hat{N}_w = nw\hat{f}(0). \qquad (22)$$

Note that our discussion has focused on the estimation

of population size, N_w, in the strip defined by L and w. We did this because the chapter deals with population size and because the estimator fits our general framework explained earlier. In contrast, in their development of line-transect estimators, Burnham et al. (1980) focused on the direct estimation of population density (population size/ unit area) rather than population size. Their treatment is thorough and understandable, and we recommend it highly for those who intend to use line-transect methodology. Direct estimation of density, rather than population size, permits the generalization of the described sampling situation to transects with no predetermined width.

For completeness, we will present the basic estimator for density, $\hat{D}$. We can express density in the surveyed strip defined by L and w as:

$$D = \frac{N_w}{2Lw}. \qquad (23)$$

Using our estimator for N_w in equation (22), we can estimate density as:

$$\hat{D} = \frac{n\hat{f}(0)}{2L} = \frac{n}{2L\hat{a}}. \qquad (24)$$

Note that a [recall that $a = \int_0^w g(x) \, dx$] sometimes is referred to as the "effective half-width" or "one-half the effective strip width" of the transect. This interpretation equates a with the half-width of a strip on which one would expect n animals to be located. The basis for this interpretation can be seen by examining equation (24) and recalling the definition of density as animals per unit area.

Variance estimation in line-transect work usually is expressed in terms of density. Burnham et al. (1980) showed that the estimator for the variance of estimated density, $\widehat{var}(\hat{D})$, can be written as:

$$\widehat{var}(\hat{D}) = \hat{D}^2 \left[\frac{\widehat{var}(n)}{n^2} + \frac{\widehat{var}(\hat{f}(0))}{\hat{f}(0)^2} \right], \qquad (25)$$

where $\widehat{var}$ denotes the estimated sampling variance. Thus, the two components of variation are the variance associated with the estimation of $f(0)$ and the sampling variance of n. The sampling variance of $f(0)$ is obtained directly from the estimation process used to estimate $f(0)$. However, the sampling variance of n is not easily estimated and depends on the unknown spatial distribution of animals in the surveyed area. For example, if animals are distributed randomly (according to a Poisson distribution), then $var(n) = n$, whereas aggregation usually results in $var(n)$ exceeding n (i.e., $var(n) > n$). A preferable alternative to separately estimating the two components of equation (25) is to estimate $var(\hat{D})$ directly from the $\hat{D}$ obtained on replicate transects (Burnham et al. 1980). If sample sizes are not adequate for this direct approach, a more complex (jackknife) approach to variance estimation also may be appropriate in some situations (Burnham et al. 1980).

The relationship between population size and density ($N = D2Lw$) leads to the following estimator for the sampling variance of population size estimates resulting from line-transect studies:

$$\widehat{var}(\hat{N}_w) = (2Lw)^2 \widehat{var}(\hat{D}), \qquad (26)$$

Box 4. Use of the line-transect method to estimate the density of duck nests.

Anderson and Pospahala (1970) used line-transect methods during spring-summer 1967 and 1968 to estimate the density of duck nests on Monte Vista National Wildlife Refuge, Colorado. Perpendicular distance data were obtained for 534 nests and were grouped into eight 1-foot intervals (Fig. 6). Burnham et al. (1980) used program TRANSECT to fit a Fourier series model to these data. The goodness-of-fit test indicated reasonable fit ($\chi^2 = 4.4$, 6 df, $P = 0.63$) for a one-term Fourier series. Resulting estimates were $\hat{a}_1 = 0.02269$ ($\widehat{SE} = 0.0076$) and $\hat{f}(0) = 0.1477$ ($\widehat{SE} = 0.0076$), yielding a density estimate of $\hat{D} = 50.2$ nests/km^2 ($\widehat{SE} = 3.3$).

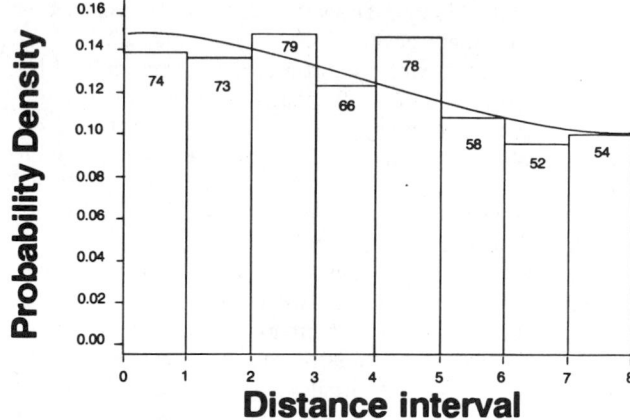

Fig. 6. A histogram of waterfowl nest data from line transects (from Anderson and Pospahala 1970).

where L and w are known constants and $\widehat{var}(\hat{D})$ is estimated as described above.

In the estimators for population size (eq. 22) and density (eq. 24), the only quantity to be estimated is f(0). The main problem in line-transect estimation involves developing an appropriate model for f(x) and then using this model to estimate f(0). Burnham et al. (1979, 1980) discussed the criteria to judge the performance of a particular model for f(x) and an associated estimator, f(0). They noted that an estimator should be based on a flexible model capable of fitting a variety of shapes and should perform well in the face of variation in detection probability for fixed distance, x (Burnham et al. 1979, 1980). They also proposed that the true detection function g(x) will likely have a "shoulder" near $x = 0$, i.e., there should be a region near the transect line where g(x) is 1, and suggested that a good estimator should also meet this "shape criterion." Finally, they emphasized estimator efficiency, noting that f(0) should have the smallest possible sampling variance (Burnham et al. 1979, 1980).

A variety of parametric and nonparametric estimation models has been proposed for use in line-transect estimation. Rather than review these models and estimators, we recommend the excellent reviews by Gates (1979) and Burnham et al. (1980). Two comprehensive computer programs, TRANSECT (Laake et al. 1979) and LINETRAN (Gates 1980), are available to compute estimates under several possible models. For example, TRANSECT uses five different estimators—Fourier series, exponential power series, exponential polynomial, negative exponential, and half-normal—to compute estimates. Burnham et al. (1980) studied these estimators extensively with computer simulation and concluded that the Fourier series estimator generally performed best according to their four criteria listed above (also see Quinn 1981).

The actual application of line-transect methods involves many decisions and considerations specific to a particular situation. For example, many animals exhibit gregarious behavior and tend to occur in groups. The density of groups can be estimated with distance measurements taken from the line to the geometric center of each ob-

served group (Burnham et al. 1980, Quinn 1981). The number of animals in each detected group also must be recorded to estimate density or population size. Drummer and McDonald (1987) and Otto and Pollock (1990) discussed models for use when detection probability for fixed distance, x, depends on group size. Other special considerations involve the possible existence of extreme values (sightings at extreme distances) or outliers. Burnham et al. (1980) recommended truncation of data at distances greater than some distance w*, beyond which observations seem likely to be outliers.

Another consideration involves the grouping of data. Accurate measurement of distances in the field may not be possible, so detections may be grouped by distance categories. Even when direct distance measurements are recorded, anomalous patterns may be apparent, such as few objects detected at very short distances, clumped detections at commonly rounded measurements, or a relatively large number of detections near the boundary distance, w. In these situations, data may be grouped into a histogram before analysis as a "smoothing" technique (Burnham et al. 1980).

Guidelines for designing and conducting a line-transect study were presented by Anderson et al. (1979) and Burnham et al. (1980), and we recommend these papers to those planning to use line transects in their studies. We also suggest that L be selected to provide a minimum of 40 animals detected, and preferably 60–80 (Anderson et al. 1979, Burnham et al. 1980).

Sighting Distance and Angle Data

We recommend using perpendicular distance data whenever possible, because methods that use sighting distance and angle data require additional assumptions about the detection process that are not required by methods that use perpendicular distance data. The relationship among x = perpendicular distance, r = sighting distance, and θ = sighting angle, is:

$$x = r[\sin(\theta)]. \tag{27}$$

If the ability to meet the assumptions required for analysis of sighting distance and angle data is questionable, the data can be transformed to perpendicular distance data with equation (27) and analyzed with the Fourier series estimator (Burnham et al. 1980).

Three estimators were incorporated in program TRAN-SECT and presented by Burnham et al. (1980) for use with sighting angle and distance data. All are based on the idea that sighting distance, r, is a random variable with some unspecified probability distribution. A critical assumption underlying all three estimators can be addressed by testing whether r_i and θ_i (where i indexes animal, $i = 1, 2, \ldots, n$) are independent. If a simple correlation analysis reveals an association between sighting distance and angle, the three proposed estimators should not be used, and the data should be transformed (eq. 27) for use with perpendicular distance estimators.

The three estimators are described below. The Hayne (1949) estimator is based on the idea that animals are sighted whenever the observer crosses the boundary of an imaginary "flushing circle" around an animal. The Hayne estimator requires the assumption that $\sin(\theta)$ is a uniform random variable, and hence that the average expected sighting angle is 32.7°. Tests of this assumption are computed in program TRANSECT (Laake et al. 1979, Burnham et al. 1980).

The generalized Hayne estimator of Burnham (1979) is again based on the idea of sighting occurring when the observer crosses the boundary of an imaginary flushing curve, but now the curve is envisioned as an ellipse, rather than specifically as a circle (the Hayne model is a special case of this more general model). An alternative estimator for the situation of an elliptic flushing curve was presented by de Vries (1979); also see Otten and de Vries (1984).

The third estimator is a modified Hayne estimator (Burnham and Anderson 1976, Burnham et al. 1980). It is not based on a specific flushing model but is more general and can even be applicable, for example, in situations in which detection depends on active searching rather than on a response by the animal.

VARIABLE CIRCULAR PLOTS

Reynolds et al. (1980) presented a field methodology for surveying birds in tall, structurally complex vegetation and rough terrain. They believed that observers traveling along line transects in such habitats tended to watch the path of travel so that their ability to detect birds was reduced. Consequently, they recommended establishing equally spaced observer stations positioned along a transect. The observer proceeds to a station (simply a point on the transect line) and allows a rest period of specified duration for equilibration of bird activity (Reynolds et al. 1980). The observer then detects birds (by both sight and sound) for a specified count period. All detected birds and their distances from the observer are recorded. Distances can be true measurements, or birds can be assigned to distance categories representing concentric bands around the station.

Estimating population density with variable circular plots is similar to line-transect sampling with perpendicular distance data because estimation is based on a detection function $g(x)$ that specifies detection probability as a function of distance from the observation point. However, instead of animals being located perpendicular to the line as in line-transect sampling, animals can be sighted, and corresponding measurements taken, in all possible directions from the point of observation. The density estimator for variable circular plot sampling has the form:

$$\hat{D} = \frac{n}{\pi\rho^2}, \tag{28}$$

where n is again the number of birds observed, and ρ, analogous to a in the perpendicular distance development, is the parameter to be estimated. The denominator of equation (28) is simply the area of a circle of radius ρ. The definition of density leads to the interpretation of ρ as the "effective radius of detection" (Ramsey and Scott 1979) or the distance from the observation point within which we would expect n birds to be located. Estimation of ρ can proceed by assuming a parametric form for $g(x)$ or by using a more general estimation approach, such as the Fourier series, in a manner similar to that used for perpendicular distance data.

Assumptions of the variable circular plot method are similar to those required for line-transect sampling data. However, the analog of the $g(0) = 1$ assumption, i.e., all animals on the transect line are seen, is that within some distance, r, from the observation point, all birds are detected (i.e., $g(x) = 1$ for all $x < r$). Ramsey and Scott (1979) labeled this distance, r, the "basal radius" and discussed several possible methods for estimating it. They noted that $\hat{r}$ could be used in the following density estimator

$$\hat{D} = \frac{n(0, \hat{r})}{\pi\hat{r}^2}, \tag{29}$$

where $n(0, \hat{r})$ denotes the number of birds seen in the circle of radius $\hat{r}$ centered on the observation point. Buckland (1987) developed binomial models for variable circular plot data in which birds are categorized simply according to whether they are within or beyond a single specified distance. Mountainspring and Scott (1985) and Scott et al. (1986) used the variable circular plot method to survey Hawaiian forest birds.

All Individuals Not Observed—Capture Methods

Capture methods, as the name implies, involve handling animals in some way. Most of these methods require the assumption of a closed population, but some open population models have been developed. Biologists usually resort to capture methods when animals are difficult to observe and count, but there is a reasonable chance of capturing them.

REMOVAL METHODS

Removal methods of population estimation are old and have been analyzed by numerous investigators over the years. These methods are attractive because removal data often can be collected by someone other than the investigator, such as hunters. Thus, the investigator may not have to actually capture animals to develop population estimates based on removals, which often makes these methods inexpensive to implement in the field.

Removal methods can be categorized according to whether the removals are "selective" (Fig. 2). If the proportions of "types" (e.g., sexes, age classes, species) of animals in the removals are substantially different from the proportions of the same types in the preremoval population, change-in-ratio (CIR) estimators can be used. If removals are not selective, either standard removal models or catch-per-unit-effort (C/E) models can be used to estimate population size. Standard removal models assume equal effort is expended on catching/removing animals at each sampling occasion, whereas C/E models can be applied when sampling effort varies among sampling periods but is known (or can be estimated). CIR and C/E estimators are covered in this section, but standard removal models are discussed later.

Single-Stage Change-in-Ratio

We now briefly describe the equations used to estimate population size with the CIR technique. These equations are essentially algebraic solutions to simultaneous equations relating population sizes of classes of animals before and after known removals. For more detail on derivation and a complete description of the technique see Seber (1982), Overton (1969), and Paulik and Robson (1969).

The basic CIR method assumes a closed population with two classes of animals, x-type and y-type. These could be male and female pheasants, antlered and antlerless deer, adults and juveniles, or even two different species. If the proportion of x- and y-type animals in the population changes due to the removal of a known number of animals (see Box 5), we can write the new (postremoval) proportion of x-type animals as:

$$P_2 = \frac{X_1 - R_x}{N_1 - R} = \frac{P_1 N_1 - R_x}{N_1 - R},$$

where

R_x = the number of x-types removed (known),
R_y = the number of y-types removed (known),
$R = R_x + R_y$ = the total number of animals removed (known),
X_1 = the number of x-type animals in the initial (preremoval) population,
Y_1 = the number of y-type animals in the initial (preremoval) population,
$P_1 = X_1/N_1$ = the proportion of x-type animals before the removal (where N_1 is the total population size before the removal), and
$P_2 = X_2/N_2$ = the proportion of x-type animals after the removal (where N_2 is the total population size after the removal).

Solving for N_1 yields the following estimator of total population size before the removal:

$$\hat{N}_1 = \frac{(R_x - RP_2)}{(P_1 - P_2)}. \tag{30}$$

Note that P_1 and P_2 are estimated by some sampling scheme, such as road counts of antlered and antlerless deer, as $\hat{P}_1$ and $\hat{P}_2$ and, together with the numbers removed, are substituted into equation (30) to estimate the preremoval population size. The number of x-type animals in the initial (preremoval) population is estimated by:

$$\hat{X}_1 = \hat{P}_1 \hat{N}_1 \tag{31}$$

If independent estimates of P_1 and P_2 are assumed,

variance estimates for $\hat{N}_1$ and $\hat{X}_1$ can be calculated (see Seber 1982) as:

$$\text{var}(\hat{N}) = \frac{[N_1^2 \text{var}(\hat{P}_1) + N_2^2 \text{var}(\hat{P}_2)]}{[P_1 - P_2]^2} \quad \text{and} \tag{32}$$

$$\text{var}(\hat{X}_1) = \frac{[N_1^2 P_2^2 \text{var}(\hat{P}_1) + N_2^2 P_1^2 \text{var}(\hat{P}_2)]}{[P_1 - P_2]^2}. \tag{33}$$

For random sampling with replacement, that is, the animals that are sighted to estimate the P ratios can be resighted on different occasions, the variance of the estimated P ratios ($\hat{P}_i = x_i/n_i$) is:

$$\text{var}(\hat{P}_i) = \frac{[\hat{P}_i(1 - \hat{P}_i)]}{n_i}, \tag{34}$$

where n_i is the total number of animals seen and x_1 is the total number of x-type animals seen while sampling to estimate the P ratios.

From the estimates of $\hat{N}_1$ and $\hat{X}_1$, and the removals, the following can be calculated:

$$\hat{Y}_1 = \hat{N}_1 - \hat{X}_1, \tag{35}$$

$$\hat{X}_2 = \hat{X}_1 - \hat{R}_x, \tag{36}$$

$$\hat{Y}_2 = \hat{Y}_1 - \hat{R}_y, \quad \text{and} \tag{37}$$

$$\hat{N}_2 = \hat{X}_2 + \hat{Y}_2 \quad \text{or} \quad \hat{N}_2 = \hat{N}_1 - \hat{R}. \tag{38}$$

The assumptions of the CIR method are:

(1) The observed proportions of x- and y-type animals are unbiasied estimates of the true proportions in the population. Expressed another way, the x- and y-type animals have an equal probability of being sampled, i.e., they are equally observable.

The CIR method can be applied when only a single type is removed, such as x-types being removed in an antlered-bucks-only hunt. In this special situation the population estimate of the type removed (bucks) is appropriate regardless of whether x- and y-types are equally sightable (Seber 1982). This relationship is exploited in the two-stage CIR explained below.

(2) The population is closed except for the removals. This assumption can best be met by keeping the removal period and the time between the two estimates of the P ratios as short as possible.

(3) The number of removals of x- and y-type animals is known. The method still can be applied if unknown removals can be estimated (Paulik and Robson 1969, Seber 1982).

(4) The proportion of the x-types in the harvest is different from that in the population. If the x-types are removed in the same proportion in which they occur in the population, the P ratios do not change from before to after the removal. Therefore $P_1 = P_2$, the denominator in equation (30) is 0, and the method fails. This can be a significant shortcoming in the technique if the management objective is to maintain a balanced ratio of x- to y-types in the population such as in quality deer herd management (Lancia et al. 1988). However, the two-stage CIR (see below) circumvents this problem.

Several authors investigated the effect of sample size

and variability in samples used to estimate the P ratios on the accuracy and precision of CIR estimates (Paulik and Robson 1969, Seber 1982, Pollock et al. 1985, Conner et al. 1986). The initial P ratio (P_1), the change in P ratios ($\Delta P = P_1 - P_2$), and the number of animals that are observed to estimate the P ratios all affect the accuracy and precision of CIR estimates. In general, CIR is most accurate when ΔP is large. Large removals of a single type, or removals of both types widely disproportionate to their representation in the population, will produce large ΔPs. If ΔP is small, the method is likely to produce excessively large or small population estimates or model failures (i.e., negative population estimates). Accuracy also improves as sample sizes used to estimate the P ratios increase. From computer simulations, J. W. Bishir (unpubl. data) suggested that removing 70–80% or more of x-types in a single-type removal from populations of from 50 to 1,000 animals, with a large number sighted to estimate the P ratios, yielded accurate CIR population estimates. Removals of this magnitude, for example, could occur where only antlered deer are removed.

The above analyses of accuracy and precision assume that the underlying assumptions of the method are met and that observations of the P ratios and the numbers of animals removed are binomial random variables. In practice the assumption of equal observability (assumption 1) is most difficult to meet (see Box 6). The relative observability of the two types of animals, λ, can be expressed mathematically as λ = observability of y-type animals/ observability of x-type animals. When both types are equally observable, $\lambda = 1$. Antlerless deer are usually more observable than are antlered bucks (Conner et al. 1986), and buck observability probably also varies with the age of the animal and other things. If λ is known or can be estimated, the P ratios can be adjusted to reflect λ, yielding unbiased CIR estimates (Conner et al. 1986). Alternatives are the two-stage CIR that relaxes the equal catchability assumption, single-type removals that are not affected by unequal observability, and brief removal periods that limit the time over which changes in observability could occur.

Two-Stage Change-in-Ratio

A brief description of the two-stage CIR method follows; for details see Pollock et al. (1985). Consider a closed population in which the relative observability of x- and y-type animals is not equal ($\lambda \neq 1.0$). If animals are removed in two separate, single-type hunts, then λ and the size of the x- and y-type portions of the population before and after the removals can be estimated. The observed proportions of x-type animals in the population at three times (t_1, t_2, t_3), separated by the two single-type removals, and the numbers removed are used to estimate population size (Pollock et al. 1985). A typical application would be removing antlered and antlerless deer in two separate, single-type (sex) hunts. The expected observed proportions (P) of antlered deer in the population before and after the removals can be written as:

Box 5.   Using the CIR method to estimate population size of antlered and antlerless deer.

Antlered (x-type) and antlerless (y-type) deer were observed during 54 prehunt and 52 posthunt road counts on Remington Farms, Maryland (Conner et al. 1986). One hundred and twenty x-type and 1,126 y-type animals were observed before 56 antlered deer (R_x) and 54 antlerless deer (R_y) were removed by hunters during a 1-week season. After the season, 43 x-type and 1,086 y-type deer were observed. Therefore,

$$\hat{P}_1 = x_1/n_1 = 120/1{,}246 = 0.0963$$

$$\hat{P}_2 = x_2/n_2 = 43/1{,}129 = 0.0381$$

$$\hat{N}_1 = (R_x - R\hat{P}_2)/(\hat{P}_1 - \hat{P}_2)$$

$$= \{56 - 110(0.0381)\}/(0.0963 - 0.0381)$$

$$= 51.809/0.0582$$

$$= 890 \ (\widehat{SE} = 149)$$

$$\hat{X}_1 = \hat{P}_1 \hat{N}_1 = 0.0963(890)$$

$$= 86 \ (\widehat{SE} = 14).$$

In this example both x and y types were removed. Because the y-type (antlerless) animals were probably more observable than the x-type animals, λ was probably >1.0, and the population estimates were likely to be biased high.

$$E(\hat{P}_1) = X_1/(X_1 + \lambda Y_1), \tag{39}$$

$$E(\hat{P}_2) = (X_1 - R_x)/(X_1 - R_x + \lambda Y_1), \quad \text{and} \tag{40}$$

$$E(\hat{P}_3) = (X_1 - R_x)/[X_1 - R_x + \lambda(Y_1 - R_y)]. \tag{41}$$

Solving equations (39), (40), and (41) for X_1, Y_1, and λ yields:

$$\hat{X}_1 = [R_x \hat{P}_1 (1 - \hat{P}_2)]/(\hat{P}_1 - \hat{P}_2), \tag{42}$$

$$\hat{Y}_1 = [R_y \hat{P}_3 (1 - \hat{P}_2)]/(P_3 - \hat{P}_2), \quad \text{and} \tag{43}$$

$$\hat{\lambda} = [R_x (1 - \hat{P}_1)(\hat{P}_3 - \hat{P}_2)]/[R_y \hat{P}_3 (\hat{P}_1 - \hat{P}_2)]. \tag{44}$$

Variance estimates for $\hat{X}_1$, $\hat{Y}_1$, and $\hat{\lambda}$ can be found in a recent examination of CIR methods by Udevitz (1989).

The assumptions of the two-stage CIR are the same as in the traditional CIR with one exception. In the traditional CIR the observability of the x- and y-types must be equal. In the two-stage CIR the observabilities need only to be constant from t_1 to t_3. In other words, λ must be constant from before the first to after the last removal. This assumption still could be difficult to meet, but it is much less severe than the assumption of equal observability that is required for the traditional CIR method. An additional advantage is that λ can also be estimated. An example is provided in Box 7.

Box 6. An assumption of the standard change-in-ratio method is that the two classes of animals are equally observable. The following illustrates that observabilities of two classes of deer can be different.

This is an example of estimating $\hat{\lambda}$ based on observations made by McCullough (1982) of the George Reserve deer herd. In November, 50 fawns/100 females and 25 males/100 females were observed during spotlight counts while 110 fawns/100 females and 60 males/100 females were known to be in the population. The y-type animals are antlerless deer, i.e., fawns and females, and the x-type animals are antlered deer, i.e., males. Therefore, the proportion of y-type animals in the population that was observed, $\hat{p}$, is

$$\hat{p} = y_{\text{obs}}/n_{\text{obs}} = (100 + 50)/(100 + 50 + 25)$$
$$= 150/175 = 0.857,$$

where

y_{obs} = the number of y-types observed,   and

n_{obs} = the total number of animals observed.

Then, the proportion of x-type animals in the population that was observed is

$$1 - \hat{p} = 1 - 0.857 = 0.143.$$

Similarly, the true proportions of x- and y-type animals are

$$p = y_{\text{true}}/n_{\text{true}} = (100 + 110)$$
$$\div (100 + 110 + 60)$$
$$= 210/270 = 0.778, \quad \text{and}$$
$$1 - p = 1 - 0.778 = 0.222.$$

Therefore, the observabilities of the y- and x-type animals are $\hat{p}/p$ and $(1 - \hat{p})/(1 - p)$. Thus,

$$\hat{\lambda} = \text{observability of } y\text{-type animals}$$
$$\div \text{ observability of } x\text{-type animals}$$
$$= (\hat{p}/p)/\{(1 - \hat{p})/(1 - p)\}$$
$$= (0.857/0.778)/(0.143/0.222)$$
$$= 1.10/0.64$$
$$= 1.72.$$

In this example the antlerless deer are more observable than the antlered deer (1.10 vs. 0.64) and $\hat{\lambda}$ (= 1.72) is greater than 1.0. Thus, a population estimate calculated from these observabilities likely would be biased high.

The accuracy and precision of the two-stage CIR was evaluated by Pollock et al. (1985). In general, larger changes in P ratios yield better estimates. Also, large sample sizes to estimate the P ratios increase the accuracy of the two-stage CIR estimates (Pollock et al. 1985). For extensions of the CIR method see Udevitz (1989).

Catch-per-Unit-Effort

Catch-per-unit-effort (C/E) estimators have been examined by many individuals including Leslie and Davis (1939), Chapman (1954), Ricker (1958), and Seber (1982). Overton (1969) provided a clear, simple description of the derivation of the basic C/E estimator.

Catch-per-unit-effort is based on the premise that as more and more animals are removed from a population, fewer are available to be "caught," and catch per unit of effort should decline. For example, fewer animals can be seen per hour or fewer can be harvested per hunter-day as more animals are removed. Eventually, if all the animals could be removed, the expected catch would be zero, and the total number of animals removed would be equivalent to the initial population size. Because it is generally not desirable (or possible) to remove all the individuals in a population, the C/E method estimates the cumulative catch (total animals removed) at which the expected catch-per-unit-effort is zero, which corresponds to the initial population size. An advantage of C/E is that population estimates can be derived from removals that are a part of a routine management activity such as hunter harvests (see Box 8).

The traditional C/E method involves developing a linear regression of catch-per-unit-effort on the cumulative total number of animals removed. Repeated observations by trap-nights, days, weeks, or other time unit of catch-per-unit-effort and cumulative number removed are used to derive the linear regression. Thus,

$$y_i = A + Bx_i,$$

where

y_i = observations of catch-per-unit-effort,
A = y-intercept estimated by the regression equation,
B = the slope of the regression, and
x_i = the observed cumulative removals.

Note that catchability, K, or the probability a given animal is "caught" on a given period (e.g., day, week) by a given observer (e.g., investigator, hunter), is equal to $-B$. The estimated preremoval population size, N, is the cumulative catch (population size) for which the expected catch-per-unit-effort is zero ($y = 0$), or the x-intercept. Thus,

$$N = A/-B \quad \text{or} \tag{45}$$
$$= A/K. \tag{46}$$

In terms of our general framework, the y-intercept (A) represents a "count" of the animals in the preremoval population (i.e., the initial catch standardized for variable effort), and the slope of the regression ($K = -B$) represents β, the "observability" proportion.

Although maximum likelihood and weighted least-squares estimators for the C/E method have been presented (e.g., Seber 1982, Pollock et al. 1984), no computer

Box 7.   When observabilities of classes of animals are different, the two-stage CIR can be used to estimate population size.

A 12-m bag seine was used to catch fish and estimate the ratios of two size classes in the population (Pollock 1985). The x types were >12.7 cm long and the y types were <12.7 cm long. So that the P ratios could be estimated, between the first and second samples 274 x-type fish (R_x) were "removed" by clipping the dorsal fin, and between the second and third samples 159 y-type fish (R_y) were "removed." The estimated P ratios were: $\hat{P}_1 = 0.3821$, $\hat{P}_2 = 0.3142$, and $\hat{P}_3 = 0.3914$. Therefore,

$$\hat{X}_1 = [R_x\hat{P}_1(1 - \hat{P}_2)]/(\hat{P}_1 - \hat{P}_2)$$

$$= [274(0.3821)(1 - 0.3142)]$$

$$\div (0.3821 - 0.3142)$$

$$= 71.800/0.0679$$

$$= 1,057$$

$$\hat{Y}_1 = [R_y\hat{P}_3(1 - \hat{P}_2)]/(\hat{P}_3 - \hat{P}_2)$$

$$= [159(0.3914)(1 - 0.3142)$$

$$\div (0.3914 - 0.3142)$$

$$= 42.679/0.0772$$

$$= 553$$

$$\hat{\lambda} = [R_x(1 - \hat{P}_1)(\hat{P}_3 - \hat{P}_2)]/[R_y\hat{P}_3(\hat{P}_1 - \hat{P}_2)]$$

$$= [274(1 - 0.3821)(0.3914 - 0.3142)]$$

$$\div [159(0.3914)(0.3821 - 0.3142)]$$

$$= 13.070/4.226$$

$$= 3.09$$

$$\hat{N}_1 = \hat{X}_1 + \hat{Y}_1$$

$$= 1,057 + 553$$

$$= 1,610.$$

The large value of $\hat{\lambda}$ ($= 3.09$) indicates that the fish were not equally "observable," and the two-stage CIR was a more appropriate method than the traditional CIR. (Note that we did not include standard errors because Udevitz [1989] showed that our original ones [Pollock et al. 1985] were incorrect. See Udevitz [1989] for further details.)

programs for these estimation procedures are currently available, and the traditional regression method typically has been used to implement the C/E method for closed populations. More work is required to develop C/E computer programs, analogous to program CAPTURE (see CAPTURE-RECAPTURE METHODS [p. 239]). Recently, Novak et al. (1991) used an open C/E model and maximum likelihood estimates (Dupont 1983) to estimate the size of a deer population.

The meanings of "removed" and "catch" require elaboration. Removed animals can be physically taken from the population by being killed or livetrapped and removed, or animals can be figuratively "removed" by being marked (e.g., see CAPTURE models M_b and M_{bh} [p. 242]). In the latter situation, observations of marked animals would be ignored in subsequent catches. Removals can be by any means; they do not have to correspond to the catch used in the estimator. For example, removals can be hunter kills and catch can be animals seen per day. All sources of removals, such as accidental road-kills, poached animals, or unretrieved kills, are included in the cumulative total removed. Animals "caught" during the catch-per-unit-effort can be shot, trapped, or sighted. They do not have to be physically taken or removed to be "caught."

Although the form of the C/E regression equation looks familiar, it is not a typical regression because y, the catch-per-unit-effort, and x, the cumulative removals, can depend on the same removals. This lack of independence makes calculation of variances and confidence intervals difficult. J. W. Bishir (unpubl. data) showed that the estimates of N do not follow a normal distribution and, therefore, standard variance equations are not appropriate.

The assumptions of the C/E method are similar to CIR:

(1) The population is closed (except for the removals). This assumption can best be met by keeping the removal period as short as possible. Note that some models (e.g., DuPont 1983) permit relaxation of this assumption.

(2) For each period (e.g., day, week), all individual animals have an equal probability (K) of being caught by a particular unit of effort, and K is constant over time. This is the equal catchability or equal observability assumption. A qualitative test of this assumption is to examine the trajectory of the plot of catch-per-unit-effort on the cumulative removals. If it is not linear, the equal catchability assumption is violated, and the technique should be abandoned (Caughley 1977). If the units of effort are constant, such as the same number of traps being set each night or the same number of hunters each day and the units of effort are trap-nights or hunter-days, respectively, the catch-per-unit-effort is simply the number of animals caught on successive occasions. In this situation the CAPTURE program models M_b and M_{bh} (p. 242), which are constant-effort models, can be used to estimate population size. Model M_{bh} relaxes the assumption of equal catchability and allows individual differences in catchability. The M_{bh} model is a generalized removal model that should be used to estimate population size when constant effort is employed.

(3) All the removals are known. The likelihood of vi-

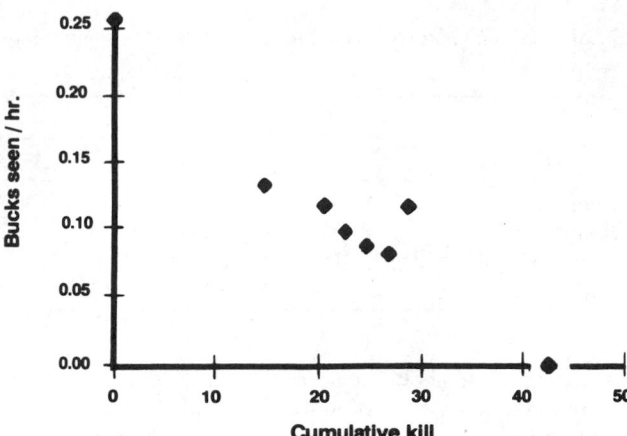

Box 8. **Removal methods often can be used in conjunction with harvests of hunted species as illustrated below.**

The following example is from J. W. Bishir (unpubl. data). The number of bucks in the Remington Farms deer herd was estimated from hunter diaries of the number of bucks seen per hour and check-station data of the number of bucks killed per day during a 1-week hunting season in 1983 (Fig. 7, Table 2). The total kill was 34 bucks, and the population estimate was 42.6 (SE = 4.0). Because a large proportion of the population was removed (about 80%), this estimate is probably reasonably accurate.

Fig. 7. Catch-per-unit-effort, with catch expressed as the number of bucks seen per hour by hunters and removals are harvested deer. The data are from Table 2.

olating this assumption can be minimized by keeping the removal period as short as possible and by searching the study area for unreported removals.

J. W. Bishir (unpubl. data) used computer simulation to examine the accuracy and precision of C/E estimates. The method can fail if the slope of the regression line (B) is positive, resulting in a negative population estimate (negative *x*-intercept) (Overton 1969). Similarly, excessively large estimates result if B is negative but very small. These occasional large values skew distributions of C/E estimates towards larger values. Consequently, the mean and variance are often unreliable indicators of central tendency and dispersion, respectively. Model failures and excessively large estimates are more likely to occur if a small proportion of the population is removed. Catch-per-unit-effort estimates are likely to be accurate and precise if >70–80% of the population is removed.

CAPTURE-RECAPTURE METHODS

Capture-recapture methods have a long history of use in ecology (Le Cren 1965), and now a large body of literature is available on the statistics of capture-recapture sampling models. Recent reviews were given by Cormack et al. (1979), Nichols et al. (1981), Pollock (1981*b*), Seber (1982), and Pollock et al. (1990). Skalski and Robson (1992) considered the design and analysis of capture-recapture methods for field studies and environmental impact assessment. This section closely follows the presentation in Pollock et al. (1990).

In capture-recapture studies the population is sampled two or more times, generally by livetrapping but sometimes by reobservation without actual recapture. Each time, every unmarked animal that is captured is uniquely marked, previously marked animals are recorded, and all animals are released into the population, so at the end of the study each animal has a complete capture history. Batch marks, wherein individual animals cannot be distinguished, do not provide individual capture histories and should be avoided, except perhaps for simple Lincoln-Petersen estimates (see below).

Capture-recapture models are classified into those suitable for closed or open populations. Because closed population models have fewer parameters, they are usually simpler than open models. Otis et al. (1978) and White et

al. (1982) presented excellent reviews, written for biologists, of closed models. A major disadvantage of closed population models is that their use is limited to short-term studies for which births (and immigration) and deaths (and emigration) are negligible and can be ignored. Open population models circumvent this limitation, but distinguishing deaths from emigration and births from immigration is difficult. However, as we shall see, the rigid separation between open and closed models is rather artificial. In many instances study designs that incorporate both types of models (Pollock 1982) can be used.

The typical capture-recapture study provides two distinct types of information: the recapture of marked animals and proportions of marked and unmarked animals captured at each sampling time. Data from the former are used to estimate "survival" rates, whereas both types of information are necessary to estimate population abundance and the number of "births." Note that survival rates include deaths and emigration and births include births and immigration.

In some studies estimation of survival rate is of primary concern, so only information on the recovery of marked animals is taken. One example is band recovery, when hunters return the bands or tags of animals they have killed. These models, discussed in Chapter 16, are closely related to the models based on live recaptures that are discussed below.

This section, organized according to whether one or more recapture occasions have occurred and whether populations are open, closed, or both (Fig. 8), presents an overview of capture-recapture models. For additional details, see Pollock et al. (1990).

Lincoln-Petersen

The Lincoln-Petersen model, the simplest capture-recapture method, dates back to Laplace, who, in 1786, used it to estimate the human population of France (Seber 1982). We present it to set the stage for the more complex models that follow. A detailed discussion of the method was given by Seber (1982).

A sample of n_1 animals is captured, marked, and released. Later, a second sample of n_2 animals is captured, some of which, m_2, are marked. Intuitively, the proportion

Table 2. Summary of daily hunter reports, Remington Farms, 1983 (E. C. Soutiere, Remington Farms, Chestertown, Md., pers. commun.).

No.	Day of the season						
	1	2	3	4	5	6	7
Hunters	33	35	24	16	21	19	18
Hours hunted	161.2	198.7	129.5	63	95.5	87.8	80.3
Bucks killed	15	6	2	2	2	2	5
Bucks seen	40	26	15	6	8	7	9

of marked animals in the second sample should be equivalent to the proportion of marked animals in the total population (capture probability is assumed to be independent of marking status) so that

$$m_2/n_2 = n_1/N, \qquad (47)$$

where N is the total population size. Rearranging terms yields the estimator

$$\hat{N} = n_1 n_2/m_2. \qquad (48)$$

A modified version with less bias was originally developed by Chapman (1951) as

$$\hat{N}_c = \left[\frac{(n_1 + 1)(n_2 + 1)}{(m_2 + 1)}\right] - 1. \qquad (49)$$

The variance of $\hat{N}_c$ is (Seber 1982:60):

$$\text{var}(\hat{N}_c) = \frac{(n_1 + 1)(n_2 + 1)(n_1 - m_2)(n_2 - m_2)}{(m_2 + 1)^2 (m_2 + 2)}.$$

The Lincoln-Petersen model is based on the following assumptions: (1) the population is closed, (2) all animals are equally likely to be captured in each sample, and (3) marks are not lost, gained, or overlooked. The first assumption usually can be met if the interval between samples is short. The second assumption, the equal catchability assumption, is relaxed in models presented later. The last assumption also can be met if an appropriate marking technique is used (Chapter 7). If marks are lost, m_2 would be too small and N too large, yielding a positively biased estimate. If tag loss is a serious problem, corrections based on a double marking scheme (Caughley 1977, Seber 1982) can be employed.

The assumption of equal catchability is unlikely to be true for many wildlife populations, because capture probabilities of each individual animal might be different or might vary in response to the traps. When the probability of capture is a property of individual animals and each animal has a unique capture probability, the situation is termed *heterogeneity*. Variation in capture probabilities among individuals could result from many factors such as sex, age, social status, or spatial distribution of animals and capture efforts. If heterogeneity is present, individuals with high capture probabilities tend to be captured in the first sample and recaptured in the second sample. This means m_2 is too large and hence $\hat{N}$ is too small. Thus, $\hat{N}$ is negatively biased when capture probabilities are heterogeneous.

Another possibility is that capture probabilities may depend on whether an animal has been captured previously. For example, marked animals may become "trap shy" and have a lower capture probability, or they may become "trap happy" and have a higher capture probability than unmarked (not previously captured) individuals. If animals are trap happy, m_2 will again be too large and $\hat{N}$ too small, whereas the converse is true for a population with trap-shy individuals. In summary, a *trap response* or *behavior response* in capture probabilities can result in population estimates that are either negatively biased (too low) due to trap-happy animals or positively biased (too high) due to trap-shy animals (see Box 9).

K-Sample Closed Population Models

Because closed population models do not permit unknown changes in the size of the population while the estimation procedure is being implemented, these methods are generally conducted over a relatively short period of time (5–10 days). Capture histories of every animal caught are needed for obtaining estimates with closed models that allow relaxation of the equal catchability assumption. Pollock (1974) considered these models, then Otis et al. (1978), White et al. (1982), and Pollock and Otto (1983) added improvements.

Closed population models differ in the way capture probabilities are modeled in that three basic sources of variation are considered: heterogeneity, trap response or behavior, and time. Heterogeneity and trap response were discussed under the Lincoln-Petersen method. Variability in capture probabilities in *time* means that capture probabilities are different for each day of trapping. For example, different weather conditions might change capture probabilities.

The following models have been developed to account for different sources of variation in capture probabilities:

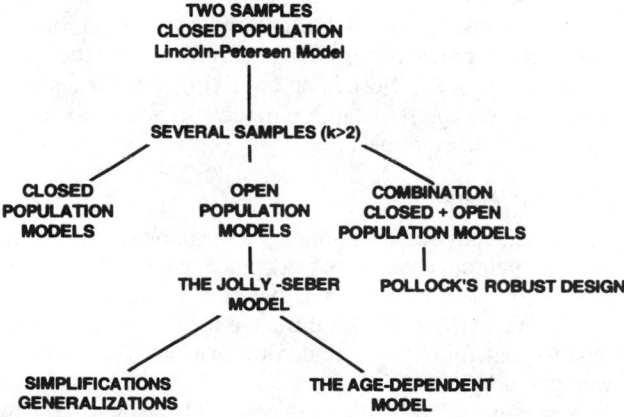

Fig. 8. The relationships among capture-recapture methods (from Pollock et al. 1990:Fig. 1.1).

Box 9. Examples of the Lincoln-Petersen mark-recapture method. In both examples different methods were used to initially "capture" the animals and then to "recapture" them. Using different methods helps reduce problems associated with heterogeneous capture probabilities and behavorial response.

In August 1974 a sample of 87 Nuttall's cottontail rabbits was livetrapped and marked with picric acid dye on their tails and hind legs (Skalski et al. 1983). About 1 month later rabbits were counted on a drive. Fourteen rabbits were seen, seven of which were marked. Chapman's estimate from equation (49) is

$$\hat{N}_c = \left[\frac{(n_1 + 1)(n_2 + 1)}{(m_2 + 1)}\right] - 1$$

$$= \left[\frac{(88)(15)}{8}\right] - 1$$

$$= 164$$

and the variance estimate is

$$\hat{\text{var}}(\hat{N}_c)$$

$$= \frac{(n_1 + 1)(n_2 + 1)(n_1 - m_2)(n_2 - m_2)}{(m_2 + 1)^2(m_2 + 2)}$$

$$= \frac{88(15)(80)(7)}{8^2(9)}$$

$$= 1{,}283.33.$$

An approximate 95% CI (normality for $\hat{N}_c$ is assumed) is

$$N_c \pm 1.965\sqrt{\hat{\text{var}}(\hat{N}_c)}$$

$$164 \pm 1.965\sqrt{1283.33}$$

$$164 \pm 70.$$

The approximate 95% CI would be from 94 to 234, a rather wide range. Precision would improve if more animals had been seen on the drive count. Because two different sampling techniques were used, concern for heterogeneity or behavioral response of the capture probabilities should have been minimal.

In northern Florida Conner et al. (1983) captured 48 raccoons in live traps. An unusual marking scheme was used; the animals were injected with a small amount of radioactive isotope. The "recapture" sample involved an intensive search of the study area for scats, and scats that were "marked" were detected with a special scintillation counter. This is an example of batch marking, which restricts analysis to the Lincoln-Petersen method only.

Conner et al. (1983) searched for scats for 5 weeks and treated each week separately, so they obtained five different Lincoln-Petersen estimates. With five estimates they could consider variation in the estimates during the study. The data estimates were as follows:

Week of collection	No. marked (n_1)	No. scats collected (n_2)	No. marked scats (m_2)	Lincoln-Petersen, $\hat{N}$
1	48	71	31	109.9
2	48	22	11	96.0
3	48	74	35	101.5
4	48	28	9	149.3
5	48	35	19	88.4

If we assume these five independent, normally distributed estimates of N, 109 ($\hat{SE}$ = 10.6) is the "best" point estimate (obtained as the arithmetic mean of the five weekly estimates of $\hat{N}$), with a 95% CI of 82–136.

---

equal catchability (M_0), heterogeneity (M_h), behavior (M_b), time (M_t), behavior-heterogeneity (M_{bh}), time-heterogeneity (M_{th}), time-behavior (M_{tb}), and time-behavior-heterogeneity (M_{tbh}). No estimator is available for M_{tbh}.

We recommend using the computer program, CAPTURE (Rexstad and Burnham, User's guide for interactive program CAPTURE, Colorado Coop. Fish Wildl. Unit, Ft. Collins, 1991), for analysis of data using M_0, M_h, M_b, M_t, M_{bh}, M_{th}, and M_{tb} models. To aid users, the CAPTURE program also provides a model selection procedure that chooses the model to best fit a particular data set (see Box 10).

M_0 The Equal Catchability Model.—This model is unlikely to be realistic because it assumes that every animal in the population has the same probability of capture (p) for each sampling period in the study. It is included mainly for pedagogic reasons to provide a basis for generalizations.

The M_0 model uses a Maximum Likelihood estimator (ML) of N (and an approximate SE) that is computed iteratively by CAPTURE. However, this estimator can be highly biased if the equal catchability assumption is violated. Otis et al. (1978) observed that $\hat{N}_0$ is reasonably *robust* to changes in capture probabilities over time, so that model M_0 can give reasonable estimates even if capture probabilities vary over time. When recaptures are infrequent, the M_0 model is usually chosen by the model selection procedure as the most appropriate estimator. Few recaptures provide only limited information, thereby preventing detection of different sources of variation in capture probability and, hence, use of other models.

M_h The Heterogeneity Model.—The heterogeneity model assumes each animal has a unique capture probability (p_j, where $j = 1, \ldots, N$ animals in the population) that remains constant over all the trapping occasions. Furthermore, the capture probabilities are assumed to be a random sample of all the individuals in the population. Under this model the number of animals captured 1, 2, 3, ..., k times contains all the information used for estimating N.

The M_h model was first considered by Burnham (1972) and later by Burnham and Overton (1978, 1979), who developed a jackknife estimation procedure. More recently, Chao (1988) proposed a moment estimator under the heterogeneity model. Based on computer simulation, her moment estimator was superior to the jackknife procedure when capture probabilities were small so that animals were recaptured only one or two times. Both methods are now implemented in program CAPTURE.

M_b The Trap Response Model.—Model M_b allows a change in capture probabilities caused by a response to trapping. The model assumes every unmarked (not previously captured) individual in the population has the same initial capture probability (p) for all trapping occasions, and every marked individual has the same probability of recapture (c) for all trapping occasions after its initial capture. Thus, all individuals in the population have the same initial capture probability, p, and all marked animals have the same recapture probability, c, but p and c are not equal.

The M_b model is essentially a constant effort, catch-per-unit-effort removal model (see Catch-per-Unit-Effort [p. 237]) with one important exception—the "removed" individuals need not be physically taken out of the population. Rather, marked individuals are ignored in subsequent samples, so only initial capture data are used to estimate the population. An assumption of the M_b model is that the initial capture probability for all animals is the same (equal catchability). The M_{bh} model explained below relaxes the equal catchability assumption.

The CAPTURE program uses an ML estimator of N rather than the traditional catch-per-unit-effort regression technique that is explained in the Catch-per-Unit-Effort section. In practice, the difference between the ML and regression estimates appears to be slight (J. W. Bishir, North Carolina State Univ., pers. commun.).

M_{bh} The Heterogeneity and Trap Response Model.—The M_{bh} model is based on the assumption that each animal has its own unique pair of potential capture probabilities, p_j and c_j ($j = 1, \ldots, N$ animals in the population), where p_j is the initial capture probability and c_j is the recapture probability. The capture probabilities are assumed to remain constant for all trapping occasions. This model was first considered by Pollock (1974). Subsequently, Otis et al. (1978) developed a "generalized removal method" and Pollock and Otto (1983) developed a jackknife method; both of these are now implemented in program CAPTURE.

M_t The Time Variation (Schnabel) Model.—The M_t model is based on the assumption that every individual in the population has the same capture probability for a given sampling occasion, but these probabilities can vary at each sampling time. Thus, the capture probabilities are p_i, $i = 1, 2, \ldots, k$ sampling occasions. This is the classic closed population model allowing temporal variation; it was first developed by Schnabel (1938).

The CAPTURE program uses an ML estimator to calculate N. Although Schnabel estimates are easy to compute manually, the CAPTURE program should be used because it can also help select the most appropriate model for a given data set. Model M_t estimates can be highly biased if capture probabilities are not equal for all individuals within a given time period. The effect is similar to the bias created by unequal catchability when model M_0 is used.

Other Time-Dependent Models: M_{th}, M_{tb}, M_{tbh}.—Conceptually, three of the models given earlier, M_h, M_b, M_{bh}, can be generalized to include a time factor as M_{th}, M_{tb}, M_{tbh}, but estimation procedures for these have not been published. However, recently M_{th} (Chao et al. 1992) and M_{tb} (Rexstad and Burnham, User's guide for interactive program CAPTURE, Colorado Coop. Fish Wildl. Unit, Ft. Collins, 1991) estimation procedures have been incorporated in the CAPTURE program (Rexstad and Burnham, User's guide for interactive program CAPTURE, Colorado Coop. Fish and Wildl. Unit, Ft. Collins, 1991).

The CAPTURE program includes a procedure to choose the best model among the models described above that is based on goodness-of-fit tests and tests between models. The model selection procedure should be used with caution, however, because the tests often have low power, especially for small populations (Chapman 1980, Menkens and Anderson 1988). In some situations a Lincoln-Petersen estimate based on pooling samples into an early versus late sample might be preferable to CAPTURE's model choice.

Biological information should be used, if possible, to reduce the number of reasonable models. For example, evidence may exist, based on the behavior of a particular species and the trapping method used, that trap response is unlikely, so models that allow trap response should not be considered. Further, if a model is selected for which there is no estimator in CAPTURE, another model should be chosen based on the investigator's experience and judgment.

Density Estimation

Although this chapter is devoted to estimation of population size, the closely related concept of density is important also. Here we describe two methods for estimating density from capture-recapture data for closed populations.

Nested Grids.—This method of estimating density was developed by Otis et al. (1978) for use in conjunction with the closed population capture-recapture models described above. Initially, it would seem to be a simple matter to estimate density from trapping-grid data. One simply estimates population size with one of the capture-recapture models described above and then divides by area as specified in equation (6). However, even though the area covered by the traps on a grid can be computed easily, it is not so simple to compute the area from which the captured animals were drawn, and thus the area to which the population estimate pertains. It has been traditional (e.g., Dice 1938) to think of a boundary strip of constant width extending beyond the traps along the periphery of the trapping grid, from which the animals trapped on the grid were drawn. Thus, the area used by the population of animals exposed to grid trapping is the sum of the areas of the grid itself and of the boundary strip.

Otis et al. (1978) noted that population size associated

Box 10. **An example of using the CAPTURE program to estimate the population size of meadow voles on a live-trapping grid.**

Nichols and Pollock set up a small mammal live-trapping grid in old-field habitat at the Patuxent Wildlife Research Center in Laurel, Md. The grid contained a 10×10 matrix of Fitch live traps (Rose 1973) at 7.6-m intervals. Hay and dried grass were placed in the nest box section of the traps, and whole corn was used as bait. Traps were set each evening and checked each morning for 5-day periods once each month from June 1981 through January 1982. Captured animals were marked with ear tags and released at the capture site. For more complete information on the trapping procedure see Nichols et al. (1984b).

Five consecutive trapping days from October 1981 are used as an example of the model section prodedure and population estimation from the CAPTURE program. Data are the capture histories of adult meadow voles; following is the output from the model selection procedure:

Model	M_0	M_h	M_b	M_{bh}	M_t	M_{th}	M_{tb}	M_{tbh}
Criterion	0.80	1.00	0.38	0.59	0.00	0.32	0.52	0.98

The highest value for the model selection procedure indicates the most appropriate model for these data. Thus, the high values for M_h and M_{tbh} suggest that they are better models than the others. The best is the jackknife estimator corresponding to M_h; there is no estimator for M_{tbh}.

The output for model M_h is presented in Table 3. The estimated average capture probability, $\hat{p} = 0.44$, is high so that the precision of the population estimate is good.

Table 3. Output from the heterogeneity model (M_h) of the CAPTURE program for meadow vole data collected at Patuxent Wildlife Research Center, Laurel, Maryland, 1981.

	Frequencies of capture[a]				
i	1	2	3	4	5
F(i)	29	15	15	16	27

Number of animals captured = 102[b]
Average $\hat{p}$ = 0.44
Interpolated population estimate = 139 (SE = 10.85)
Approximate 95% CI = 117 − 161

[a]Frequencies of capture are the number of animals captured one time, two times, three times, up to the number of capture occasions.
[b]Number of different individuals captured one or more times.

sively less biased as grid size increases, and the estimates from the different grids are used (in conjunction with generalized nonlinear least squares) to estimate true density and boundary-strip width.

Simulation studies by Wilson and Anderson (1985a) showed large, positive biases and large estimated variances associated with density estimates based on the nested grid approach. However, when capture probabilities are high and sample sizes are large, the nested grid approach can perform relatively well (see example of Jett and Nichols 1987).

Trapping Web.—This method of density estimation relies on captures of animals but is free of specific assumptions about sources of variation in capture probabilities (Anderson et al. 1983). The method is not based on underlying models of capture history data and thus differs conceptually from all the capture-recapture models discussed above. Distance sampling is the conceptual basis for estimation with the trapping web, which is analogous to formal transect models and the variable circular plots described previously.

With the line-transect and the variable circular plot, sighting or detection probabilities decrease with distance from the transect line or center of the circular plot. The trapping web is a configuration of traps designed to yield a similar gradient of detection (i.e., capture) probabilities. The web consists of some number (e.g., 16) of equally spaced lines of equal length radiating from a randomly chosen center point. Traps (e.g., 20/line) are then placed on the radial lines at fixed distances from the center point (the center of the web), with a constant distance separating successive traps on a particular line (Fig. 9). The design thus yields a series of concentric rings of traps. The ring nearest the web center has traps spaced close together, and the distance between traps of the same ring on adjacent radial lines increases with increasing distance of the ring from web center. Trap density, and hence the probability that an animal in a particular area will be caught, decreases with distance from the web center, producing a gradient in detection probabilities similar to that of the line-transect and variable circular plot sampling situations. Traps are run for several consecutive days (e.g., 5), and the concentric ring of initial capture is recorded for each animal. Because only initial capture data are used, the method can be used with removal data.

Rings, defined by the points halfway between adjacent

with a trapping grid can be written as a function of two known parameters—grid area and perimeter—and two unknown parameters—boundary-strip width and true animal density. A population estimate from a single grid does not permit estimation of the two unknown parameters, but if estimates of population size are available for several grids of different size, estimation is possible. A single trapping grid can be viewed as a series of nested grids. For example, assume that we have a 10×10 square trapping grid. If we omit captures from the outermost two squares, our data correspond to an 8×8 grid. Similarly, we can omit the outermost four and six squares of traps to yield 6×6 and 4×4 grids, respectively.

To estimate density, Otis et al. (1978) first used the closed population models described above to estimate population size in each nested grid within the overall trapping grid (four grids in our 10×10 example). They then wrote their "naive density" estimate (i.e., population size estimate divided by area covered by the grid) as a function of the two quantities of interest—true density and boundary-strip width. The naive density estimates are succes-

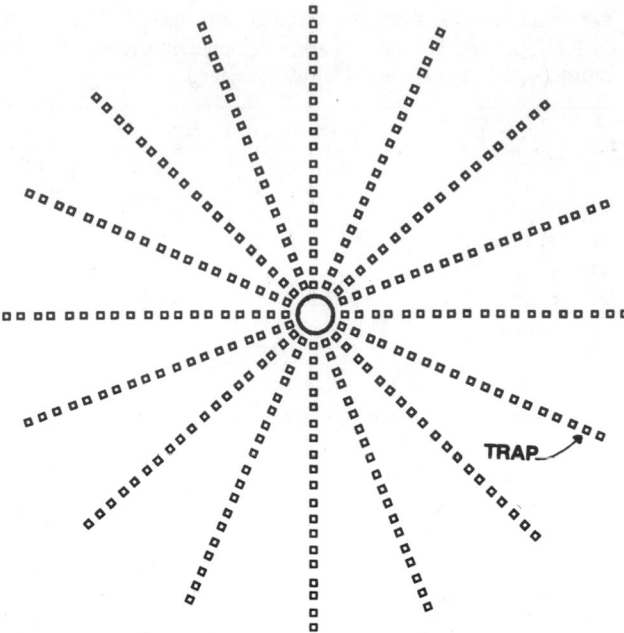

Fig. 9. An example of a trapping web with 16 radial lines and 20 traps/line (from Anderson et al. 1983:676).

traps on a radial line, are used to compute the area trapped by a particular ring of traps. The data on area trapped and the number of animals caught are available for each ring and form the basis for estimation of f(c), the probability density function of the area sampled. After f(c) is estimated, density is estimated in a manner similar to that used in line-transect estimation:

$$\hat{D} = M_{t+1}\hat{f}(0),$$

where M_{t+1} denotes the total number of animals captured during web operation. The Fourier series or other estimators from TRANSECT (Burnham et al. 1980) can be used to estimate density. The variance, i.e., VAR($\hat{D}$), can be estimated with TRANSECT output in conjunction with the estimator of Wilson and Anderson (1985*b*).

Four assumptions underlie estimating density with the trapping web: (1) all animals at the web center are caught with probability 1.0; (2) animal movements are "stable" (i.e., there are no preferential movements toward or away from the web center); (3) distances from the web center to each trap are measured accurately when the web is laid out; and (4) animal captures are independent events (this is required for variance estimation). Computer simulation studies led Wilson and Anderson (1985*b*) to conclude that the trapping web performed well. Examples were given by Anderson et al. (1983), Jett and Nichols (1987), and Parmenter et al. (1989); the last-named authors presented evidence that the web estimates performed well in field tests on populations of known density.

Open Population Models

The basic open population model that can be used to estimate population size, survival rates, and births is the Jolly-Seber model (Jolly 1965, Seber 1965). Cormack (1973) gave a brief, intuitive description of this model and its estimators, and Seber (1982) gave a detailed presentation. Computer programs that calculate Jolly-Seber es-

timates of population size include POPAN (Arnason and Baniuk 1980) and JOLLY (Pollock et al. 1990) (see Chapter 3).

The Jolly-Seber Model.—The Jolly-Seber model allows estimation of population size at each sampling time as well as "survival" rates and "births" between sampling occasions. As mentioned earlier, the mathematical complement (i.e., 1 − survival) of "survival rate" includes mortality and emigration, and "births" include immigration. The following discussion closely follows Seber (1982).

The following are assumptions of the Jolly-Seber model:

(1) Every animal present in the population at the time of the ith sample, where $i = 1, 2, 3, \ldots, k$ trapping occasions, has the same probability of capture, p_i. Thus, the Jolly-Seber model requires the assumption of equal catchability.

(2) Every marked animal present in the population immediately after the ith sample has the same probability of survival, ϕ_i, until the next sampling time, $i + 1$, where $i = 1, 2, 3, \ldots, k - 1$ trapping occasions. There are no survival estimates for the last two sampling occasions.

(3) Marks are not lost or overlooked.

(4) All samples are instantaneous, and each release is made immediately after the sample.

Assumptions 1 and 3 are required under the Lincoln-Petersen model. Only the marked animals are used to estimate survival rates, so an assumption that marked and unmarked animals have equal survival is not necessary. However, in practice the survival estimates are interpreted as though they apply to the entire population.

An intuitive discussion of parameter estimation in the Jolly-Seber model follows. Notation for the Jolly-Seber model is summarized in Table 4. Imagine that the number of marked animals in the population, M_i, just before the ith sample is known for all values of i from $i = 2, \ldots, k$, where k is the number of sampling occasions. No marked animals are present before the first sample, so $M_1 = 0$. Later, we will discuss how to estimate the M_i's, which are unknown in an open population because some mortality (and/or emigration) can occur.

An intuitive estimator, $\hat{N}_i$, of population size at time i, is the Lincoln-Petersen estimator discussed previously. Thus,

$$m_i/n_i = M_i/N_i, \qquad (50)$$

where m_i is the number of marked animals that are recaptured in the ith sample, and n_i is the total number of animals captured in the ith sample. Solving for N_i yields the estimator

$$\hat{N}_i = n_i\hat{M}_i/m_i. \qquad (51)$$

Estimates of N_i are defined only for $i = 2, \ldots, k - 1$ sampling times.

The survival rate is simply the ratio of the number of marked animals in the $i + 1$th sample to the number of marked animals in the ith sample. The number of marked animals in the ith sample is the number of marked animals in the population just prior to the ith sample, M_i, plus the unmarked animals, U_i, that are newly marked in the ith

Table 4. Notation for the Jolly-Seber model.

M_i = the number of marked animals in the population at the time of the ith sample ($i = 1, \ldots, k$; $M_1 = 0$).

N_i = the total number of animals in the population at the time of the ith sample ($i = 1, \ldots, k$).

B_i = the total number of new animals entering the population between the ith and $i + 1$th sample and still in the population at the time of the $i + 1$th sample ($i = 2, \ldots, k - 2$).

ϕ_i = the survival probability for all animals between the ith and $i + 1$th sample ($i = 1, \ldots, k - 2$).

p_i = the capture probability for all animals in the ith sample ($i = 2, \ldots, k$).

m_i = the number of marked animals captured in the ith sample ($i = 1, \ldots, k$; $m_1 = 0$).

u_i = the number of unmarked animals captured in the ith sample ($i = 1, \ldots, k$).

n_i = $m_i + u_i$, the total number of animals captured in the ith sample ($i = 1, \ldots, k$).

R_i = the number of the n_i that are released after the ith sample ($i = 1, \ldots, k - 1$). This may not include all the n_i because of losses during capture.

r_i = the number of R_i animals released at i that are captured again ($i = 1, \ldots, k - 1$).

z_i = the number of animals captured before i that are not captured at i, but are captured again later ($i = 2, \ldots, k - 1$).

---

sample. Because some animals may be captured but not released back into the population (due to trap mortality for example), the number of unmarked animals is expressed in a more general way as $U_i = R_i - m_i$, where R_i is the number of animals that are released. Thus, the number of marked animals in the ith sample is $M_i - m_i + R_i$. The number of animals still alive in the population just before the $i + 1$th sample is M_{i+1}. Therefore, the survival rate at sampling time i is estimated as:

$$\hat{\phi}_i = \hat{M}_{i+1}/(\hat{M}_i - m_i + R_i). \qquad (52)$$

Estimates of ϕ_i are defined only for $i = 1, \ldots, k - 2$ sampling times.

The number of "births" or recruitment in time interval i to $i + 1$ is estimated as the difference between the size of the population at time $i + 1$, which is N_{i+1}, and the expected number of survivors from i to $i + 1$, which is the product of the survival rate and the number of animals at time i or $\phi_i N_i$. The expected number of survivors can be written as $\phi_i(N_i - n_i + R_i)$ ($n_i = R_i$ if there are no trap mortalities). Thus, the number of births at sampling time i is estimated as:

$$\hat{B}_i = \hat{N}_{i+1} - \hat{\phi}_i(\hat{N}_i - n_i + R_i). \qquad (53)$$

Estimates of B are defined only for $i = 2, \ldots, k - 2$ sampling times.

The probability of capture, p_i, can be estimated as the proportion of marked animals alive at time i that are captured at time i or the proportion of the total (marked + unmarked) animals alive at time i that are captured at time i. Thus,

$$\hat{p}_i = m_i/\hat{M}_i = n_i/\hat{N}_i. \qquad (54)$$

Estimates of p_i are defined only for $i = 2, \ldots, k - 1$ sampling times.

Finally, M_i is unknown in an open population and must be estimated by equating the two ratios

$$z_i/(M_i - m_i) = r_i/R_i, \qquad (55)$$

which are the future recapture rates of two distinct groups of marked animals in the population at sampling period i. $M_i - m_i$ = the marked animals not captured at time i, and R_i is the animals captured at time i, marked (if they do not already have a mark) and released for possible recapture. Note that z_i and r_i are animals from the groups $M_i - m_i$ and R_i, respectively, that are captured again at least once. Thus, r_i is the number of animals recaptured later from the animals released at time i (i.e., R_i), and z_i is the number of animals captured before time i that are not recaptured at time i (i.e., they are members of $M_i - m_i$) but are recaptured again during a subsequent capture session. The estimator of M_i is

$$\hat{M}_i = m_i + R_i z_i/r_i. \qquad (56)$$

Estimates of M_i are defined only for $i = 2, \ldots, k - 1$ sampling times. Seber (1982) presented approximately unbiased versions of these estimators and gave equations for variances and covariances.

A concern when the Jolly-Seber model is used is whether the equal catchability assumption can be met. Heterogeneity of trap response in capture probabilities can significantly affect estimates of population size because the sample ratio, m_i/n_i, will not accurately reflect the population ratio, M_i/N_i. Heterogeneity and some forms of trap response also will affect survival estimates but to a lesser degree than population size estimates. This is because survival is estimated from ratios of only marked animals so that variations in capture probabilities tend to cancel out. When trap response is permanent, survival estimates are not affected at all (Nichols et al. 1984a).

The following is an example of using the Jolly-Seber capture-recapture method to estimate births, survival, and population size. In this example we analyze data from a 2-year study of gray squirrels in a mature oak woodland in Surrey, England, by A. Duboek. Squirrels were captured at approximately monthly intervals from November 1972 until September 1974. Multiple-capture traps, dispersed throughout the woodland, were baited with grain. Captured squirrels were uniquely marked by toe clipping.

Jolly-Seber estimates are given in Table 5. We present no estimates for September through November 1973 because we felt the small number of captures would give misleading estimates. Also, survival rate estimates that were >1.0 were recorded as 1.0, and some birth number estimates that were negative were recorded as 0.

The estimates in this example were precise because of high capture and survival probabilities. Once a squirrel was captured and marked, it tended to stay in the population and was recaptured often. Consequently, recaptures provided much information for the estimation procedures. However, the precision of the estimates varied considerably during the study due to changes in capture probabilities. Even when capture probabilities remained nearly constant, precision was highest during the middle of the study because the marked population gradually increased, and many sampling periods remained in the study to pro-

Table 5. Capture-recapture data (collected by A. Duboek) and Jolly-Seber estimates and approximate standard errors[a] for a gray squirrel population at Alice Holt Forest Research Station, Surrey, England, November 1972 to September 1974.

Date	n_i	m_i	R_i	r_i	z_i	$\hat{N}_i$	$\widehat{SE}$	$\hat{\phi}_i$	$\widehat{SE}$	$\hat{B}_i$	$\widehat{SE}$
Nov '72	46		46	43				0.94	0.03		
Dec	46	42	46	44	1	47.1	0.4	0.96	0.03	6.3	0.7
Jan '73	48	42	48	48	3	51.3	0.7	1.00	0.00	4.5	1.3
Feb	46	42	46	45	9	56.0	1.2	0.99	0.02	5.1	1.5
Mar	51	46	50	46	8	60.5	1.5	0.94	0.04	0.0	1.1
Apr	37	37	37	35	17	54.9	1.2	0.95	0.04	0.0	0.0
May	41	41	41	40	11	52.3	0.6	1.00	0.03	3.9	1.2
May–Jun	42	39	42	37	12	56.5	2.1	0.90	0.05	3.7	1.4
Jun	47	43	47	40	6	54.6	1.6	0.92	0.07	8.7	3.3
Jul	31	26	31	26	20	58.9	4.6	0.84	0.07	2.2	6.6
Aug	8	7	8	8	39	51.8	6.0	1.00	0.00	0.0	6.0
Sep	2	2	2	2	45						
Oct	1	0	1	1	47						
Nov	4	3	4	3	45						
Dec	9	8	9	8	40	58.3	9.2	0.93	0.12	1.0	6.6
Jan '74	19	17	18	17	31	55.3	4.3	0.98	0.07	13.1	8.2
Feb	19	14	19	18	34	66.4	8.1	1.00	0.07	6.8	10.1
Mar	27	20	27	24	32	74.5	7.9	0.93	0.07	0.0	6.3
Apr	36	36	36	32	20	58.4	2.1	0.99	0.07	18.2	4.2
May	45	34	44	33	18	76.0	6.1	1.00	0.17	33.9	8.9
Jul	74	46	73	15	5	110.3	18.1	0.21	0.05	0.0	2.2
Aug	22	20	22	2	0	21.9	0.0				
Sep	3	2	2								

[a] $\widehat{SE}(\hat{N}_i)$ includes only sampling variation or "error of estimation"; $\widehat{SE}(\hat{\phi}_i)$ and $\widehat{SE}(\hat{B}_i)$ were obtained using the full variance estimators of Jolly (1965).

vide recapture data. In other words, r_i must be large for highly precise estimates.

Many factors affected the accuracy of the estimates. Movement into and out of the population was believed to be negligible and therefore Duboek interpreted the estimates of survival and births to reflect only mortality and reproduction. Because of possible differences in capture and survival probabilities for the different sex and age classes in the population (heterogeneity) and because the animals were probably trap happy (trap response), capture probabilities likely were not equal among individuals as assumed in the model. Both of these departures tended to cause negatively biased estimates of population size and to a lesser extent underestimation of survival rates.

Young animals should have been joining the catchable population in April and May. This showed up as a high estimate of the number of births during these months in 1974 but not in 1973. Duboek predicted that 1973 would be a poor year for squirrel production. After a large number of squirrels joined the population in the spring of 1974, survival dropped substantially. Unfortunately, this occurred at the end of the study so the estimates probably were unreliable.

Extensions of the Jolly-Seber Model.—A variety of restricted versions of the Jolly-Seber model is available in the literature (Pollock et al. 1990). For populations in which births and immigration are negligible, a model that allows only for losses (deaths and emigration) may be useful (Darroch 1959). Darroch (1959) also proposed a model for which only additions (births and immigration) are allowed. Other restricted models that are useful are

the constant survival and capture rate models described by Jolly (1982) and developed further by Brownie et al. (1986).

Sometimes a species may have different identifiable age classes that can be modeled most realistically with age-dependent survival and capture rates. For these situations, models developed by Pollock (1981a) and Stokes (1984) may be useful, and computer program JOLLYAGE is available (Pollock et al. 1990). Another modification allows marking to have a short-term effect on survival and capture rates (Robson 1969, Pollock 1975, Brownie and Robson 1983). Finally, a cohort Jolly-Seber model was described by Buckland (1982) and Loery et al. (1987) for animals marked at a known age.

Combination of Open and Closed Models— Pollock's Robust Design

The distinction between open and closed populations is made to simplify the models used to estimate population parameters of interest. The simplifications are a result of assumptions that may or may not be met in field applications. Biologists must be aware of the assumptions that underlie these models and design studies that satisfy the assumptions as closely as possible.

Pollock (1982) was motivated by the desire to find a design for long-term studies that was robust to heterogeneity and/or trap response in capture probabilities. He proposed a design that combined open and closed population models to exploit the advantages and minimize the shortcomings of both (see Box 11). For example, some closed population models (e.g., CAPTURE models) relax the assumption of equal catchability, and the Jolly-Seber open population model allows for changes in the size of the population. Thus, an experimental design in which short-term studies are used to estimate population size with closed population models, combined with Jolly-Seber open population model estimates of survival and births in between the short-term studies, may be the most robust design for many long-term field applications.

Estimation of Population Parameters.—For the combined design consider a capture-recapture sampling study in which we have K primary sampling periods, such as seasons or years (Fig. 10). Within each of these we have n secondary sampling periods over a short time, such as n consecutive days of trapping. The combined design can be used to estimate population size for each of the secondary sampling periods ($N_1, N_2, \ldots, N_k$) if population size is assumed constant within each of these sampling periods (i.e., the population is closed over the secondary sampling periods within each primary period). Closed population models in CAPTURE should be used to estimate the Ns.

Survival rates, $\phi_1, \phi_2, \ldots, \phi_{k\ 2}$, can be estimated by "pooling" the secondary samples within each primary sample and using the Jolly-Seber model. Pooling the secondary samples simply means recording whether an animal is captured at least once during each primary sample. The number of individuals entering the population ("births") between the primary sampling periods, $B_1, B_2, \ldots, B_{k-2}$, can be estimated with the closed model N_i and the open model ϕ_i (see eq. 53).

Although the Jolly-Seber model could be used in the standard way on all the data as in the squirrel example presented above, not all population parameter estimates are defined. In the Jolly-Seber model, estimates of population size are not possible for the first (N_1) and last (N_k) primary sampling periods, and a birth estimate is not possible for the first (B_1) primary sampling period. However, all of these can be estimated with the combined design.

The approximate variances for survival rates and births were given in Pollock et al. (1990). The variance estimates for population size are obtained from CAPTURE.

SELECTION AND COMPARISON OF METHODS

Although selection of a method is the first task a biologist has to consider, we have left this topic until all the methods were presented. We believe that a biologist must be aware of the breadth of methods and the underlying assumptions before selecting a population estimation method.

As emphasized throughout this chapter, the selection of a method for estimating population size will involve observability and sampling considerations, i.e., the investigator will need to consider both β and α of equation (4). The first sampling consideration is simply whether the area of interest is so large that sampling is required. When sampling is necessary, the investigator must make decisions about the size of the sample plots (or transects). These decisions will depend largely on logistical considerations associated with the method selected for dealing with observability concerns. For example, larger plots would typically be used in aerial surveys with multiple observers than in line transects with ground observers.

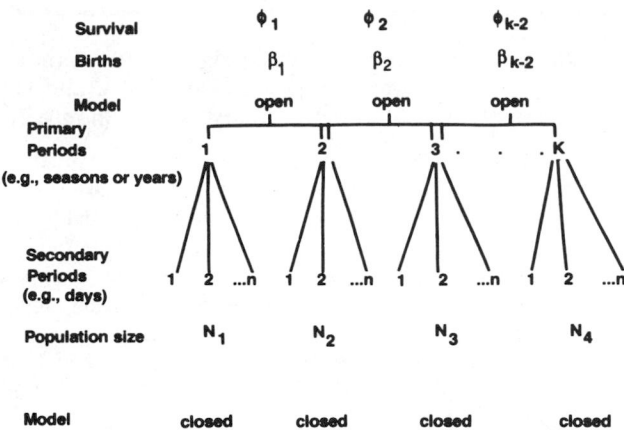

Fig. 10. The relationship between primary and secondary periods in Pollock's Robust Design.

The number of sample plots, and hence the sampling proportion, then must be decided. Here, as in all sampling problems, the investigator must balance the desirability of increased precision (the variance of the estimate decreases as α increases) with the costs of sampling. A sampling scheme also must be selected. If plot-to-plot variation in animal abundance is thought to be large, and if the investigator has (or can easily obtain) some a priori information about relative abundances over the entire study area, stratified random sampling is generally advisable. However, if there is no basis for stratification (i.e., no prior information about relative abundance is available, nor is an inexpensive way to obtain such information), simple random sampling is the most reasonable approach. In some situations, logistical considerations may dictate a systematic rather than a random sampling scheme. We have not considered sampling approaches in detail because that goes beyond the scope of this chapter; we recommend Cochran (1977) and Chapter 1 for further information. Jolly (1969a,b), Jolly and Watson (1979), Hankin (1984), and Hankin and Reeves (1988) discussed sampling issues for specific kinds of animal surveys, and these discussions are relevant to problems of estimating animal abundance.

Although the investigator interested in estimating animal abundance can obtain guidance on sampling considerations from traditional statistical work on sampling problems (e.g., Cochran 1977), some sampling issues are peculiar to animal estimation problems. In particular, issues involving the interplay between sampling and observability have received inadequate attention and should provide fertile ground for future statistical research. Examination of our general expression for the variance of a population estimate (eq. 5) shows that sampling, i.e., α, var(C), and observability, i.e., β, var($\hat{\beta}$), components are important determinants of precision. Because we always face resource limitations, the tradeoffs between these components must be considered. For example, we are faced with questions such as: Should I increase α at the expense of obtaining a smaller β with less precision?

One issue associated with this tradeoff involves double sampling—the approach of obtaining some count statistic, C, on all sample plots and estimating observability, β, on a subset of these sample plots. On a given plot, the esti-

Box 11. **Pollock's Robust Design, which combines open and closed population models, can be used to estimate population size, births, and survival. Parameters can be estimated for more time intervals with the robust design than with the Jolly-Seber method alone.**

Live-trapping data for meadow voles at Patuxent Wildlife Research Center, presented in an earlier example of the closed population models available in CAPTURE, are also used here as an example of the combined method (Nichols et al. 1984b). For each of six monthly (primary) sampling periods from June 1981 through December 1981, trapping was conducted for 5 consecutive days (secondary sampling periods).

The CAPTURE program was used to analyze the data (Table 6) from the secondary sample periods under assumption of a closed population. CAPTURE's model selection procedure indicated that heterogeneity in capture probabilities was present (Pollock et al. 1990), so model M_h was used to estimate population size for each month.

The Jolly-Seber model was used to estimate survival rates between the primary sample periods. The number of births was estimated with closed-model $\hat{N}_i$ and the Jolly-Seber $\hat{\phi}_i$ under the combined approach (Table 7). The Jolly-Seber model also was used in the traditional way to estimate population size, births, and survival rates, and the results were compared to those of the combined method (Table 7).

In this example, M_h estimates of average capture probability (i.e., the probability that an animal will be captured on any particular day or secondary sampling period) ranged from 0.35 to 0.56, resulting in reasonably precise estimates of population size. Jolly-Seber estimates of capture probability (i.e., the probability that an animal will be caught at least once during a 5-day primary period) are also very high (mean P = 0.91). Again, precise estimates of population size result. However, the Jolly-Seber estimates should be negatively biased in the presence of heterogeneity, but the bias should be relatively small when capture probabilities are high (Carothers 1973, Gilbert 1973). Indeed, all four Jolly-Seber estimates of population size are slightly smaller than the M_h estimates. Thus, even though the Jolly-Seber estimates of population size are more precise than the M_h estimates, we prefer the M_h estimates because they should be less biased. This is because M_h allows heterogeneous capture probabilities.

The Jolly-Seber survival rate estimates (Table 7) are robust to heterogeneity (Gilbert 1973) and are not biased by permanent trap response (Nichols et al. 1984a). Therefore, the survival estimates should be accurate. Because of the high capture probabilities, they are also precise.

Estimates of the number of births or recruitment between the primary sample periods are generally higher but less precise for the combined method than for the Jolly-Seber method. However, the combined method estimates B_1 but Jolly-Seber does not. Finally, the combined method estimates of births should be less biased than the Jolly-Seber estimates.

Table 6. Meadow vole capture frequency data for 5-day secondary sampling periods within each of six monthly primary sampling periods, Laurel, Maryland, 1981.

Primary period (month)		Secondary period (day)				
		1	2	3	4	5
1	Animals caught[a]	63	72	74	65	63
	Frequency[b]	20	15	21	21	28
2	Animals caught	66	81	82	0[c]	0[c]
	Frequency	35	37	40	0[c]	0[c]
3	Animals caught	53	54	46	47	43
	Frequency	37	23	16	13	12
4	Animals caught	60	62	61	52	68
	Frequency	29	15	15	16	27
5	Animals caught	60	67	65	56	64
	Frequency	19	19	19	17	26
6	Animals caught	87	89	79	85	64
	Frequency	40	28	32	28	20

[a]Number of animals caught on each day of trapping.
[b]Number of aimals caught on one, two, . . . , five occasions.
[c]A raccoon tipped over a large number of traps, so these days were omitted from the analysis.

mation of β is typically much more expensive in terms of time, effort, and money than is obtaining the count statistic. Therefore, an obvious question is: On how many sample plots must β be estimated? In many situations, the precision of our estimate of β on a sample plot will be a function of effort expended on its estimation, so another question involves how much effort to expend on each sample plot. Extremely useful considerations of expenditure of effort in double sampling problems for animal populations were presented by Jolly (1969a,b) and Jolly and Watson (1979) for aerial survey work, Hankin (1984) and Hankin and Reeves (1988) for stream fish surveys, and Skalski (1985a,b) for capture-recapture surveys (with a small mammal example). Despite the different specific examples considered by these authors, a common conclusion seemed to involve the greater importance of geographic variation among sample plots relative to variation associated with estimation of β within plots. Skalski (1985a,b) considered the development of cost functions as a means of exploring optimal sampling designs. We believe that this kind of work is extremely important and merits much more attention.

We noted in our introduction that most previous statistical work on estimating abundance has focused on the

Table 7. A comparison of population parameter estimates from the combined method and the Jolly-Seber model for meadow vole data, Laurel, Maryland, 1981.

Month	Population size				Survival rates		Births			
	Comb.[a]	$\widehat{SE}$	J-S	$\widehat{SE}$	J-S	$\widehat{SE}$	Comb.	$\widehat{SE}$	J-S	$\widehat{SE}$
1	123	5.2			0.88	0.021	39	8.7		
2	144	6.9	138	4.3	0.66	0.023	50	11.5	31	3.6
3	141	10.0	118	4.5	0.69	0.022	43	13.2	29	2.9
4	140	10.9	109	3.1	0.63	0.015	28	8.5	43	3.1
5	115	4.7	111	3.1						
6	189	10.3								

[a]Model M_h from CAPTURE was used to estimate population size. Trap losses were added to the M_h estimates.

estimation of β, and indeed most of this chapter has concerned methods of dealing with and estimating observability. The remainder of our discussion will not emphasize sampling issues but instead will focus on selection of a method of counting animals and estimating β.

When all animals can be observed, total counts (either on sample plots or on the entire study area) are recommended. It is our experience, however, that even the most visible animals in the most open habitat seldom can be observed with probability 1.0, and we strongly recommend that investigators intending to use total counts first test the hypothesis that $\beta = 1.0$.

In situations involving relatively large areas and animals that are readily observable from the air, aerial surveys provide an excellent means of counting animals and obtaining *C*. We discussed several methods for estimating β in conjunction with aerial surveys. We do not view any of these as an obvious "front-runner"; in different situations they all have limitations and advantages. When total ground counts are possible, the double sampling approach of estimating β from ground counts on a subset of sample plots should be useful. However, we emphasize that this method depends on the detection of all animals in ground counts, and this assumption should be tested. The use of a marked subpopulation also enables estimation of β for aerial surveys, although the requirements of this approach will prohibit its use in many situations. Determining the number of marked animals available to be seen during the survey can be difficult if standard marks are used, but it is relatively easy when individuals are marked with radio transmitters. A more difficult requirement is that all members of the marked subpopulation seen from the air can be identified as marked, but the mark cannot change the detection probability of a marked animal relative to that of an unmarked counterpart.

The use of independent observers in the aircraft requires that animal sightings be accurately mapped and that activity by one observer does not alert the other to a sighted animal. Although probably not useful for highly mobile animals or animals occurring in groups, this method is recommended for counts of stationary objects such as bird nests, alligator nests, or beaver lodges. Such structures can be mapped accurately, and observers can survey the study area on different flights, thereby assuring independence.

Methods involving the development of a sighting probability model show good promise for use with mobile animals. Although model development could be expensive, subsequent use of the model involves little work in addition to the actual counting. For this reason, we believe that this approach will be especially useful in operational management surveys. The greatest potential problem is that conditions affecting sighting probability when the model is developed differ from those affecting sighting probability when the model is applied.

When aerial surveys are not used, but when animals are readily visible from the ground, "distance sampling methods" (e.g., line transects and variable circular plots) provide a rigorous means of estimating density and population size. Line-transect methods should be useful for highly visible animals inhabiting relatively open habitat. The primary limitations are (1) the assumption that all animals on the line are seen and (2) the practicality of measuring or estimating perpendicular distances from the line to the sighted animals. In many situations the assumption of seeing all the animals on the line probably can be met easily. Measurement of distances sometimes will be practical in some situations, and even when it is not, observers with sufficient training can learn to estimate distances accurately. Thus, line transects are useful for estimating population size in many circumstances.

Variable circular plots are closely related to line transects and have been used extensively in bird population surveys. The main advantages of variable circular plots over line transects are increased safety and ease of use in rough terrain, decreased chance of missing animals close to the observer, and increased ability to define habitat variables associated with each observation station. The relative disadvantages are that less area is covered (and hence sample size is often reduced) and the opportunity to count animals more than once is greater. The variable circular plot was developed for use in structurally complex vegetation, but in these situations recording distances accurately may be difficult, especially when birds are identified only by song or call. However, the use of the method in extensive surveys of Hawaiian birds (Mountainspring and Scott 1985, Scott et al. 1986) provides a good demonstration of its utility.

When animals can be seen readily, observation-based methods are generally preferable. Managers and biologists generally have recognized this and have tended to use observation-based methods for estimating population size of birds and large mammals. However, other smaller and more secretive vertebrates and most invertebrates are not readily observed. For these animals, capture-recapture and removal methods frequently will be the most reasonable means of estimating population size.

When population size is the only parameter of interest, capture-recapture studies should be carried out over a short period of time (e.g., trapping small mammals on a grid for 5 consecutive days), thereby increasing the likelihood that the population is closed to gains and losses of animals over the study period. Closure is desirable because of the variety of available closed, capture-recapture models that permit various sources of variation in capture probability. If tests indicate failure of the closure assumption, partially and completely open models could be used. Studies involving several (more than four) different capture periods are desirable, because tests of assumptions about closure and variation in capture probability can be made. These tests then lead to selection of the most appropriate model to fit the data.

Historically, two-sample capture-recapture studies have been popular, relying on the Lincoln-Petersen estimator. However, data from two-sample studies (i.e., two trapping occasions) are not adequate to test the assumptions on which the method is based. If the recapture (or reobservation) method is different from and independent of the initial capture method, however, the two-sample capture-recapture method is recommended. Different capture and recapture methods were used in both of our examples of the Lincoln-Petersen estimator. In such situations, heterogeneous capture probabilities and trap response are not likely to present problems, and resulting estimates should be reasonable. Two-sample studies should be conducted over a relatively short time period to try to ensure at least partial population closure.

In many studies of animal population dynamics, the investigator will be interested in periodic estimates not only of population size but also of survival and recruitment rates. Open capture-recapture models can be used to estimate gains and losses to the population. For long-term, capture-recapture studies of animal population dynamics, we recommend the combined use of closed and open models (i.e., Pollock's Robust Design). This design enables estimation of population size with a variety of flexible, closed population models. Survival rate is then estimated with capture-recapture data for the primary sampling periods, and recruitment is estimated from these population size and survival estimates. This design offers great potential for studying long-term population dynamics.

For species heavily exploited by hunting, trapping, or other means of population reduction, removal methods might be useful for estimating population size. These methods require data on the numbers of animals removed and hence have application in controlled, local situations such as deer hunting on a management area or trapping a specific marsh, in which harvested animals are brought through check stations or otherwise tallied. Statistics reflecting "effort," for example the number of hunters per unit time, the number of trap-nights, or the number of animals seen per unit time, are also frequently available in these controlled situations, permitting the use of catch-effort models and, perhaps, the special-case, constant-effort removal models. Catch-effort models have been used extensively in fisheries applications, and we believe that they offer great potential for use in wildlife studies associated with closely monitored harvests.

If the exploitation in such controlled situations is likely to be selective for different types of animals (e.g., antlered-bucks-only deer hunting, size-selective alligator hunting), change-in-ratio methods could be useful. These methods require additional surveys to estimate the proportions of the different types of animals in the population. When the different types of animals are equally observable in these ancillary surveys, only two surveys are necessary—one before and one after the removal period. When observabilities of the different types are not equal, other approaches can be used such as single-type removals, the two-stage change-in-ratio, or direct estimation of observabilities and "correction" of the standard change-in-ratio technique. We doubt that the change-in-ratio methods will have the general applicability of catch-effort models, but they do seem to offer good potential in some special instances of controlled, selective exploitation.

In summary, it makes no sense to speak of a single, "best" method for estimating population size. If we believed that one method was universally preferable, this chapter would have been considerably shorter. Rather, virtually all the methods considered here are potentially useful in certain situations. Selection of the most reasonable method for a given situation will depend on the particular details associated with that situation. Such details, which must come from a wildlife biologist or manager, include information on the biology and habits of the animal species under study. In addition, because of the variety and complexity of methods available for estimating animal population size, it is becoming increasingly important to involve a statistician or quantitative population ecologist in the selection and application of a method.

Our aim has been to provide biologists and managers with a sufficient understanding of the concepts underlying population estimation methods to enable useful dialogue between them and a statistician. Such dialogue is essential, not only to method selection (and frequently tailoring the method to a particular set of circumstances), but also to study design. We hope that this collaboration will become more common and will lead to a better understanding of animal populations.

LITERATURE CITED

ALLEN, S. H., AND A. B. SARGEANT. 1975. A rural mail-carrier index of North Dakota red foxes. Wildl. Soc. Bull. 3:74–77.

ANDERSON, D. R. 1975. Population ecology of the mallard. V. Temporal and geographic estimates of survival, recovery and harvest rates. U.S. Fish Wildl. Serv. Resour. Publ. 125. 110pp.

———, K. P. BURNHAM, G. C. WHITE, AND D. L. OTIS. 1983. Density estimation of small-mammal populations using a trapping web and distance sampling methods. Ecology 64:674–680.

———, J. L. LAAKE, B. R. CRAIN, AND K. P. BURNHAM. 1979. Guidelines for line transect sampling of biological populations. J. Wildl. Manage. 43:70–78.

———, AND R. S. POSPAHALA. 1970. Correction of bias in belt transect studies of immotile objects. J. Wildl. Manage. 34:141–146.

ARNASON, A. N., AND L. BANIUK. 1980. A computer system for mark-recapture analysis of open populations. J. Wildl. Manage. 44:325–332.

BARKER, R. J., AND J. R. SAUER. 1992. Modelling population change from time series data. Pages 182–194 in D. R. McCullough and R. H. Barrett, eds. Wildlife 2001: populations. Elsevier Scientific Publ. Co., London, U.K.

BART, J., AND J. D. SCHOULTZ. 1984. Reliability of singing bird surveys: changes in observer efficiency with avian density. Auk 101:307–318.

BASKETT, T. S., M. J. ARMBRUSTER, AND M. W. SAYRE. 1978. Biological perspectives for the mourning dove call-count survey. Trans. North Am. Wildl. Nat. Resour. Conf. 43:163–180.

BEST, L. B. 1975. Interpretational errors in the "mapping method" as a census technique. Auk 92:452–460.

BEST, R. G., AND R. FOWLER. 1981. Infrared emissivity and radiant surface temperatures of Canada and snow geese. J. Wildl. Manage. 45:1026–1029.

BROWNIE, C., J. E. HINES, AND J. D. NICHOLS. 1986. Constant-parameter capture-recapture models. Biometrics 42:561–574.

——, AND D. S. ROBSON. 1983. Estimation of time-specific survival rates from tag-resighting samples: a generalization of the Jolly-Seber model. Biometrics 39:437–453.

BUCKLAND, S. T. 1982. A mark-recapture survival analysis. J. Anim. Ecol. 51:833–847.

——. 1987. On the variable circular plot method of estimating animal density. Biometrics 43:363–384.

BULL, E. L. 1981. Indirect estimates of abundance of birds. Pages 76–80 in C. J. Ralph and J. M. Scott, eds. Estimating the numbers of terrestrial birds. Stud. Avian Biol. 6.

BURNHAM, K. P. 1972. Estimation of population size in multiple capture-recapture studies when capture probabilities vary among animals. Ph.D. Thesis, Oregon State Univ., Corvallis. 168pp.

——. 1979. A parametric generalization of the Hayne estimator for line transect sampling. Biometrics 35:587–595.

——, AND D. R. ANDERSON. 1976. Mathematical models for non-parametric inferences from line transect data. Biometrics 32:325–336.

——, ——, AND J. L. LAAKE. 1979. Robust estimation from line transect data. J. Wildl. Manage. 43:992–996.

——, ——, AND ——. 1980. Estimation of density from line transect sampling of biological populations. Wildl. Monogr. 72. 202pp.

——, AND W. S. OVERTON. 1978. Estimation of the size of a closed population when capture probabilities vary among animals. Biometrika 65:625–633.

——, AND ——. 1979. Robust estimation of population size when capture probabilities vary among animals. Ecology 60:927–936.

CALHOUN, J. B. 1948. North American census of small mammals. John Hopkins Univ., Rodent Ecol. Program, Release 1. 9pp. (Mimeogr.)

CAROTHERS, A. D. 1973. The effects of unequal catchability on Jolly-Seber estimates. Biometrics 29:79–100.

CAUGHLEY, G. 1977. Analysis of vertebrate populations. John Wiley & Sons, New York, N.Y. 234pp.

——, AND D. GRICE. 1982. A correction factor for counting emus from the air and its application to counts in western Australia. Aust. Wildl. Res. 9:253–259.

——, R. SINCLAIR, AND D. SCOTT-KEMMIS. 1976. Experiments in aerial survey. J. Wildl. Manage. 40:290–300.

CHABRECK, R. H. 1966. Methods of determining the size and composition of alligator populations in Louisiana. Proc. Annu. Conf. Southeast. Assoc. Game and Fish Agencies 20:105–112.

——. 1973. Population status surveys of the American alligator in the southeastern United States. Proc. Second Meet. Crocodile Specialists, Suppl. Pap. 41:14–21.

CHAO, A. 1988. Estimating animal abundance with capture frequency data. J. Wildl. Manage. 52:295–300.

——, S. M. LEE, AND S. L. JENG. 1992. Estimating population size for capture-recapture data when capture probabilities vary by time and individual animal. Biometrics 48:201–216.

CHAPMAN, D. G. 1951. Some properties of the hypergeometric distribution with applications to zoological sample censuses. Univ. Calif. Publ. Stat. 1:131–160.

——. 1954. The estimation of biological populations. Ann. Math. Stat. 25:1–15.

——. 1980. Review of statistical inference from capture data on closed animal populations. Biometrics 36:362.

CLARK, W. R. 1987. Effects of harvest on annual survival of muskrats. J. Wildl. Manage. 51:265–272.

COCHRAN, G. A., AND H. J. STAINS. 1961. Deposition and decomposition of fecal pellets by cottontails. J. Wildl. Manage. 25:432–435.

COCHRAN, W. G. 1977. Sampling techniques. Third ed. John Wiley & Sons, New York, N.Y. 428pp.

CONNER, M. C., R. F. LABISKY, AND D. R. PROGULSKE, JR. 1983. Scent-station indices as measures of population abundance for bobcats, raccoons, gray foxes, and opossums. Wildl. Soc. Bull. 11:146–152.

——, R. A. LANCIA, AND K. H. POLLOCK. 1986. Precision of the change-in-ratio technique for deer population management. J. Wildl. Manage. 50:125–129.

CONNOLLY, G. E. 1981. Assessing populations. Pages 287–345 in O. C. Wallmo, ed. Mule and black-tailed deer of North America. Univ. Nebraska Press, Lincoln.

CONROY, M. J., J. R. GOLDSBERRY, J. E. HINES, AND D. B. STOTTS. 1988. Evaluation of aerial transect surveys for wintering American black ducks. J. Wildl. Manage. 52:694–703.

COOK, R. D., AND J. O. JACOBSON. 1979. A design for estimating visibility bias in aerial surveys. Biometrics 35:735–742.

CORMACK, R. M. 1973. Commonsense estimates from capture-recapture studies. Pages 225–234 in M. S. Bartlett and R. W. Hiorns, eds. The mathematical theory of the dynamics of biological populations. Academic Press, New York, N.Y.

——, G. P. PATIL, AND D. S. ROBSON. 1979. Sampling biological populations. Stat. Ecol. Ser., Vol. 5. Int. Coop. Publ. House, Fairland, Md.

DARROCH, J. N. 1959. The multiple-recapture census II: estimation when there is immigration or death. Biometrika 46:336–351.

DE VRIES, P. G. 1979. A generalization of the Hayne-type estimator as an application of line intersect sampling. Biometrics 35:743–748.

DEYOUNG, C. A., F. S. GUTHERY, S. L. BEASOM, S. P. COUGHLIN, AND J. R. HEFFELFINGER. 1989. Improving estimates of white-tailed deer abundance from helicopter surveys. Wildl. Soc. Bull. 17:275–279.

DICE, L. R. 1938. Some census methods for mammals. J. Wildl. Manage. 2:119–130.

——. 1941. Methods for estimating populations of mammals. J. Wildl. Manage. 5:398–407.

DOLTON, D. D. 1993. The call-count survey: historic development and current procedures. Pages 233–252 in T. S. Baskett, M. W. Sayre, R. E. Tomlinson, and R. E. Mirarchi, eds. Ecology and management of the mourning dove. Stackpole Books, Harrisburg, Pa.

DOWNING, R. L., W. H. MOORE, AND J. KIGHT. 1965. Comparison of deer census techniques applied to a known population in a Georgia enclosure. Proc. Annu. Conf. Southeast. Assoc. Fish Wildl. Agencies 19:26–30.

DOZIER, H. L., M. H. MARKLEY, AND L. M. LLEWELLYN. 1948. Muskrat investigations on the Blackwater National Wildlife Refuge, Maryland, 1941–1945. J. Wildl. Manage. 12:177–190.

DRUMMER, T. D., AND L. L. McDONALD. 1987. Size bias in line transect sampling. Biometrics 43:13–21.

DUPONT, W. D. 1983. A stochastic catch-effort method for estimating animal abundance. Biometrics 39:1021–1033.

DWYER, T. J., G. F. SEPIK, E. L. DERLETH, AND D. G. McAULEY. 1988. Demographic characteristics of a Maine woodcock population and effects of habitat management. U.S. Fish Wildl. Serv. Fish Wildl. Res. 4. 29pp.

EBERHARDT, L. L. 1978. Appraising variability in population studies. J. Wildl. Manage. 42:207–238.

——, D. G. CHAPMAN, AND J. R. GILBERT. 1979. A review of marine mammal census methods. Wildl. Monogr. 63. 46pp.

——, AND M. A. SIMMONS. 1987. Calibrating population indices by double sampling. J. Wildl. Manage. 51:665–675.

——, AND R. C. VAN ETTEN. 1956. Evaluation of the pellet group count as a deer census method. J. Wildl. Manage. 20:70–74.

ESTES, J. A., AND R. J. JAMESON. 1988. A double-survey estimate for sighting probability of sea otters in California. J. Wildl. Manage. 52:70–76.

FULLER, M. R., AND J. A. MOSHER. 1981. Methods of detecting and counting raptors: a review. Pages 235–246 in C. J. Ralph and J. M. Scott, eds. Estimating numbers of terrestrial birds. Stud. Avian Biol. 6.

FULLER, T. K. 1991. Do pellet counts index white-tailed deer numbers and population change? J. Wildl. Manage. 55:393–396.

GATES, C. E. 1979. Line transects and related issues. Pages 71–154 in R. M. McCormick, P. Patil, and D. S. Robson, eds. Sampling biological populations. Stat. Ecol. Ser., Vol. 5. Int. Coop. Publ. House, Fairland, Md.

——. 1980. Linetran, a general computer program for analyzing line-transect data. J. Wildl. Manage. 44:658–661.

GAVIN, T. A. 1989. What's wrong with the questions we ask in wildlife research. Wildl. Soc. Bull. 17:345–350.

GEISSLER, P. H., AND J. R. SAUER. 1990. Topics in route-regression analysis. Pages 54–57 in J. R. Sauer and S. Droege, eds. Survey designs and statistical methods for the estimation of avian population trends. U.S. Fish Wildl. Serv. Biol. Rep. 90(1).

GILBERT, R. O. 1973. Approximations of the bias in the Jolly-Seber capture-recapture model. Biometrics 29:501–526.

GRIER, J. W., J. M. GERRARD, G. D. HAMILTON, AND P. A. GRAY. 1981. Aerial-visibility bias and survey techniques for nesting bald eagles in northwestern Ontario. J. Wildl. Manage. 45:83–92.

HANKIN, D. G. 1984. Multistage sampling designs in fisheries research: application in small streams. Can. J. Fish. Aquat. Sci. 41:1575–1591.

———, AND G. H. REEVES. 1988. Estimating total fish abundance and total habitat area in small streams based on visual estimation methods. Can. J. Fish. Aquat. Sci. 45:834–844.

HARAMIS, G. M., J. R. GOLDSBERRY, D. G. MCAULEY, AND E. L. DERLETH. 1985. An aerial photographic census of Chesapeake Bay and North Carolina canvasbacks. J. Wildl. Manage. 49:449–454.

HAY, K. G. 1958. Beaver census methods in the Rocky Mountain region. J. Wildl. Manage. 22:395–402.

HAYNE, D. W. 1949. An examination of the strip census method for estimating animal populations. J. Wildl. Manage. 13:145–157.

HENNY, C. J., M. A. BYRD, J. A. JACOBS, P. D. MCLAIN, M. R. TODD, AND B. F. HALLA. 1977. Mid-Atlantic coast osprey population: present numbers, productivity, pollutant contamination, and status. J. Wildl. Manage. 41:254–265.

JETT, D. A., AND J. D. NICHOLS. 1987. A field comparison of nested grid and trapping web density estimators. J. Mammal. 68:888–892.

JOLLY, G. M. 1965. Explicit estimates from capture-recapture data with both death and immigration-stochastic model. Biometrika 52:225–247.

———. 1969a. Sampling methods for aerial censuses of wildlife populations. East Afr. Agric. For. J. (spec. issue) 34:46–49.

———. 1969b. The treatment of errors in aerial counts of wildlife populations. East Afr. Agric. For. J. (spec. issue) 34:50–56.

———. 1982. Mark-recapture models with parameters constant in time. Biometrics 38:301–321.

———, AND R. M. WATSON. 1979. Aerial sample survey methods in the quantitative assessment of ecological resources. Pages 203–216 in R. M. Cormack, G. P. Patil, and D. S. Robson, eds. Sampling biological populations. Stat. Ecol. Ser., Vol 5. Int. Coop. Publ. House, Fairland, Md.

KELLER, B. L., AND C. J. KREBS. 1970. Microtus population biology; III. Reproductive changes in fluctuating populations of *M. ochrogaster* and *M. pennsylvanicus* in southern Indiana, 1965–67. Ecol. Monogr. 40:263–294.

KREBS, C. J. 1989. Ecological methodology. Harper & Row Publ., New York, N.Y. 654pp.

KUFELD, R. C., J. H. OLTERMAN, AND D. C. BOWDEN. 1980. A helicopter quadrat census for mule deer on Uncompahgre Plateau, Colorado. J. Wildl. Manage. 44:632–639.

LAAKE, J. L., K. P. BURNHAM, AND D. R. ANDERSON. 1979. User's manual for program TRANSECT. Utah State Univ. Press, Logan. 26pp.

LANCIA, R. A., K. H. POLLOCK, J. W. BISHIR, AND M. C. CONNER. 1988. A white-tailed deer harvesting strategy. J. Wildl. Manage. 52:589–595.

LE CREN, E. D. 1965. A note on the history of mark-recapture population estimates. J. Anim. Ecol. 34:453–454.

LESLIE, P.H., AND D. H. S. DAVIS. 1939. An attempt to determine the absolute number of rats on a given area. J. Anim. Ecol. 8:94–113.

LOERY, G., K. H. POLLOCK, J. D. NICHOLS, AND J. E. HINES. 1987. Age-specificity of black-capped chickadee survival rates: analysis of capture-recapture data. Ecology 68:1038–1044.

MACNAB, J. 1983. Wildlife management as scientific experimentation. Wildl. Soc. Bull. 11:397–401.

MAGNUSSON, W. E., G. J. CAUGHLEY, AND G. C. GRIGG. 1978. A double-survey estimate of population size from incomplete counts. J. Wildl. Manage. 42:174–176.

MARTIN, F. W., R. S. POSPAHALA, AND J. D. NICHOLS. 1979. Assessment and population management of North American migratory birds. Pages 187–239 in G. P. Patil, J. Cairns, and W. E. Waters, eds. Environmental biomonitoring, assessment, prediction, and management—certain case studies and related quantitative issues. Stat. Ecol. Ser., Vol. 11. Int. Coop. Publ. House, Fairland, Md.

MCCULLOUGH, D. R. 1979. The George Reserve deer herd. Univ. Michigan Press, Ann Arbor. 271pp.

———. 1982. Evaluation of night spotlighting as a deer study technique. J. Wildl. Manage. 46:963–973.

MENKENS, G. E., JR., AND S. H. ANDERSON. 1988. Estimation of small-mammal population size. Ecology 69:1952–1959.

MILLER, S. A. 1984. Estimation of animal production numbers for national assessments and appraisals. U.S. For. Serv. Gen. Tech. Rep. RM-105. 23pp.

MOOD, A. M, F. A. GRAYBILL, AND D. C BOES. 1974. Introduction to the theory of statistics. Third ed. McGraw-Hill, New York, N.Y. 564pp.

MOORE, D. S., AND G. P. MCCABE. 1989. Introduction to the practice of statistics. W. H. Freeman & Co., New York, N.Y. 790pp.

MORSE, M. A. 1943. Technique for reducing man-power in the deer drive census. J. Wildl. Manage. 7:217–220.

MOUNTAINSPRING, S., AND J. M. SCOTT. 1985. Interspecific competition among Hawaiian forest birds. Ecol. Monogr. 55:219–239.

NEFF, D. J. 1968. The pellet-group count technique for big game trend, census, and distribution: a review. J. Wildl. Manage. 32:597–614.

NETTLESHIP, D. N. 1976. Census techniques for seabirds of arctic and eastern Canada. Can. Wildl. Serv. Occas. Pap. 25. 33pp.

NICHOLS, J. D. 1986. On the use of enumeration estimators for interspecific comparisons, with comments on a 'trappability' estimator. J. Mammal. 67:590–593.

———. 1991. Science, population ecology, and the management of the American black duck. J. Wildl. Manage. 55:790–799.

———, J. E. HINES, AND K. H. POLLOCK. 1984a. Effects of permanent trap response in capture probability on Jolly-Seber capture-recapture model estimates. J. Wildl. Manage. 48:289–294.

———, B. R. NOON, S. L. STOKES, AND J. E. HINES. 1981. Remarks on the use of mark-recapture methodology in estimating avian population size. Pages 121–136 in C. J. Ralph and J. M. Scott, eds. Estimating numbers of terrestrial birds. Stud. Avian Biol. 6.

———, AND K. H. POLLOCK. 1983. Estimation methodology in contemporary small mammal capture-recapture studies. J. Mammal. 64:253–260.

———, ———, AND J. E. HINES. 1984b. The use of a robust capture-recapture design in small mammal population studies: a field example with *Microtus pennsylvanicus*. Acta Theriol. 29,30:357–365.

———, R. E. TOMLINSON, AND G. WAGGERMAN. 1986. Estimating nest detection probabilities for white-winged dove nest transects in Tamaulipas, Mexico. Auk 103:825–828.

NOVAK, J. M., K. T. SCRIBNER, W. D. DUPONT, AND M. H. SMITH. 1991. Catch-effort estimation of white-tailed deer population size. J. Wildl. Manage. 55:31–38.

O'CONNER, J. R., AND J. H. MARCHANT. 1981. A field evaluation of some common birds census techniques. Br. Trust Ornithol. Rep., Nat. Conserv. Counc., Huntingdon, U.K.

OTIS, D. L., K. P. BURNHAM, G. C. WHITE, AND D. R. ANDERSON. 1978. Statistical inference from capture data on closed animal populations. Wildl. Monogr. 62. 35pp.

OTTEN, A., AND P. G. DE VRIES. 1984. On line-transect estimators for population density, based on elliptic flushing curves. Biometrics 40:1145–1150.

OTTO, M. C., AND K. P. POLLOCK. 1990. Size bias in line transect sampling: a field test. Biometrics 46:239–245.

OVERTON, W. S. 1969. Estimating the numbers of animals in wildlife populations. Pages 403–455 in R. H. Giles Jr., ed. Wildlife management techniques manual. Third ed. The Wildl. Soc., Washington, D.C.

PACKARD, J. M., R. C. SUMMERS, AND L. B. BARNES. 1985. Variation of visibility bias during aerial surveys of manatees. J. Wildl. Manage. 49:347–351.

PARKER, H. D., JR., AND R. S. DRISCOLL. 1972. An experiment in deer detection by thermal scanning. J. Range Manage. 25:480–481.

PARMENTER, R. P., J. A. MACMAHON, AND D. R. ANDERSON. 1989. Animal density estimation using a trapping web design: field validation experiments. Ecology 70:169–179.

PAULIK, G. J., AND D. S. ROBSON. 1969. Statistical calculations for change-in-ratio estimators of population parameters. J. Wildl. Manage. 33:1–27.

POLLOCK, K. H. 1974. The assumption of equal catchability of animals in tag-recapture experiments. Ph.D. Thesis, Cornell Univ., Ithaca, N.Y. 82pp.

———. 1975. A K-sample tag-recapture model allowing for unequal survival and catchability. Biometrika 62:577–583.

———. 1981a. Capture-recapture models allowing for age-dependent survival and capture rates. Biometrics 37:521–529.

———. 1981b. Capture-recapture models: a review of current methods, assumptions and experimental design. Pages 426–435 in C. J. Ralph and J. M. Scott, eds. Estimating numbers of terrestrial birds. Stud. Avian Biol. 6.

————. 1982. A capture-recapture design robust to unequal probability of capture. J. Wildl. Manage. 46:752–757.

————, J. E. HINES, AND J. D. NICHOLS. 1984. The use of auxiliary variables in capture-recapture and removal experiments. Biometrics 40:329–340.

————, AND W. L. KENDALL. 1987. Visibility bias in aerial surveys: a review of estimation procedures. J. Wildl. Manage. 51:502–510.

————, R. A. LANCIA, M. C. CONNER, AND B. L. WOOD. 1985. A new change-in-ratio procedure robust to unequal catchability of types of animal. Biometrics 41:653–662.

————, J. D. NICHOLS, C. BROWNIE, AND J. E. HINES. 1990. Statistical inference for capture-recapture experiments. Wildl. Monogr. 107. 97pp.

————, AND M. C. OTTO. 1983. Robust estimation of population size in closed animal populations from capture-recapture experiments. Biometrics 39:1035–1049.

PRENO, W. L., AND R. F. LABISKY. 1971. Abundance and harvest of doves, pheasants, bobwhites, squirrels, and cottontails in Illinois, 1956–69. Ill. Dep. Conserv., Springfield. 76pp.

QUINN, T. J., II. 1981. The effect of group size on line transect estimators of abundance. Pages 502–508 in C. J. Ralph and J. M. Scott, eds. Estimating numbers of terrestrial birds. Stud. Avian Biol. 6.

RALPH, C. J., AND J. M. SCOTT, EDITORS. 1981. Estimating numbers of terrestrial birds. Stud. Avian Biol. 6. 630pp.

RAMSEY, F. L., AND J. M. SCOTT. 1979. Estimating population densities from variable circular plot surveys. Pages 155–182 in R. M. Cormack, G. P. Patil, and D. S. Robson, eds. Sampling biological populations. Stat. Ecol. Ser., Vol 5. Int. Coop. Publ. House, Fairland, Md.

REYNOLDS, R. T., J. M. SCOTT, AND R. A. NUSSBAUM. 1980. A variable circular-plot method for estimating bird numbers. Condor 82:309–313.

RICE, W. R., AND J. D. HARDER. 1977. Application of multiple aerial sampling to a mark-recapture census of white-tailed deer. J. Wildl. Manage. 41:197–206.

RICKER, W. R. 1958. Handbook of computations for biological statistics of fish populations. Fish. Res. Board Can. Bull. 119. 300pp.

ROBBINS, C. S., D. BYSTRAK, AND P. H. GEISSLER. 1986. The breeding bird survey: its first fifteen years, 1965–1979. U.S. Fish Wildl. Serv. Resour. Publ. 157. 196pp.

ROBSON, D. S. 1969. Mark-recapture methods of population estimation. Pages 120–140 in N. L. Johnson, and H. Smith, Jr., eds. New developments in survey sampling. John Wiley & Sons, New York, N.Y.

ROSE, R. K. 1973. A small mammal live trap. Trans. Kansas Acad. Sci. 76:14–17.

SAMUEL, M. D., E. O. GARTON, M. W. SCHLEGEL, AND R. G. CARSON. 1987. Visibility bias during aerial surveys of elk in northcentral Idaho. J. Wildl. Manage. 51:622–630.

SAUER, J. R., AND S. DROEGE, EDITORS. 1990. Survey designs and statistical methods for the estimation of avian population trends. U.S. Fish Wildl. Serv. Biol. Rep. 90(1). 166pp.

SCATTERGOOD, L. W. 1954. Estimating fish and wildlife populations: a survey of methods. Pages 273–285 in O. Kempthorne, T. A. Bancroft, J. W. Gowen, and J. L. Lush, eds. Statistics and mathematics in biology. Iowa State College Press, Ames.

SCHNABEL, Z. E. 1938. The estimation of the total fish population of a lake. Am. Math. Mon. 45:348–352.

SCOTT, J. M., S. MOUNTAINSPRING, F. L. RAMSEY, AND C. B. KEPLER. 1986. Forest bird communities of the Hawaiian islands: their dynamics, ecology, and conservation. Stud. Avian Biol. 9. 431pp.

SEBER, G. A. F. 1965. A note on the multiple-recapture census. Biometrika 52:249–259.

————. 1982. The estimation of animal abundance and related parameters. Second ed. Macmillian Publ. Co., Inc., New York, N.Y. 653pp.

SINCLAIR, A. R. E. 1991. Science and the practice of wildlife management. J. Wildl. Manage. 55:767–773.

SKALSKI, J. R. 1985a. Construction of cost functions for tag-recapture research. Wildl. Soc. Bull. 13:273–283.

————. 1985b. Use of capture data to quantify change and test for effects on the abundance of wild populations. Ph.D. Thesis, Cornell Univ., Ithaca, N.Y. 404pp.

————, AND D. S. ROBSON. 1992. Techniques for wildlife investigations design and analysis of capture data. Academic Press, San Diego, Calif. 237pp.

————, ————, AND M. A. SIMMONS. 1983. Comparative census procedures using single mark-recapture methods. Ecology 64:752–760.

————, M. A. SIMMONS, AND D. S. ROBSON. 1984. The use of removal sampling in comparative censuses. Ecology 65:1006–1015.

SMITH, L. M., I. L. BRISBIN, JR., AND G. C. WHITE. 1984. An evaluation of total trapline captures as estimates of furbearer abundance. J. Wildl. Manage. 48:1452–1455.

STOKES, S. L. 1984. The Jolly-Seber method applied to age-stratified populations. J. Wildl. Manage. 48:1053–1059.

STRONG, L. L., D. S. GILMER, AND J. A. BRASS. 1991. Inventory of wintering geese with a multispectral scanner. J. Wildl. Manage. 55: 250–259.

SUBCOMMITTEE ON CONSERVATION OF NATURAL POPULATIONS. 1981. Techniques for the study of primate population ecology. Natl. Acad. Press, Washington, D.C. 233pp.

TAUTIN, J. 1982. Assessment of some important factors affecting the singing-ground survey. Pages 6–11 in T. J. Dwyer and G. L. Storm, eds. Woodcock ecology and management. U.S. Fish Wildl. Serv. Wildl. Res. Rep. 14.

————, P. H. GEISSLER, R. E. MUNRO, AND R. S. POSPAHALA. 1983. Monitoring the population status of American woodcock. Trans. North Am. Wildl. Nat. Resour. Conf. 48:376–388.

TAYLOR, D., AND W. NEAL. 1984. Management implications of size-class frequency distributions in Louisiana alligator populations. Wildl. Soc. Bull. 12:312–319.

TYSON, E. L. 1959. A deer drive vs. track census. Trans. North Am. Wildl. Nat. Resour. Conf. 24:457–464.

UDEVITZ, M. S. 1989. Change-in-ratio methods for estimating the size of closed populations. Ph.D. Thesis, North Carolina State Univ., Raleigh. 105pp.

UHLIG, H. G. 1956. The gray squirrel in West Virginia. W.Va. Conserv. Comm., Div. Game Manage. Bull. 3. 83pp.

VERNER, J. 1985. Assessment of counting techniques. Curr. Ornithol. 2:247–302.

————, AND K. A. MILNE. 1990. Analyst and observer variability in density estimates from spot mapping. Condor 92:313–325.

WHITE, G. C., D. R. ANDERSON, K. P. BURNHAM, AND D. L. OTIS. 1982. Capture-recapture and removal methods for sampling closed populations. Los Alamos Natl. Lab., LA-8787-NERP. 235pp.

WIGHT, H. M. 1938. Field and laboratory technic in wildlife management. Univ. Michigan Press, Ann Arbor. 107pp.

WILSON, D. M., AND J. BART. 1985. Reliability of singing bird surveys: effects of song phenology during the breeding season. Condor 87: 69–73.

WILSON, K. R., AND D. R. ANDERSON. 1985a. Evaluation of a density estimator based on a trapping web and distance sampling theory. Ecology 66:1185–1194.

————, AND ————. 1985b. Evaluation of a nested grid approach for estimating density. J. Wildl. Manage. 49:675–678.

WOOD, J. E. 1959. Relative estimates of fox population levels. J. Wildl. Manage. 23:53–63.

————, AND E. P. ODUM. 1964. A nine-year history of furbearer populations on the AEC Savannah River Plant area. J. Mammal. 45:540–551.

WOODWARD, A. R., AND W. R. MARION. 1978. An evaluation of factors affecting night-light counts of alligators. Proc. Annu. Conf. Southeast. Assoc. Fish Wildl. Agencies 32:291–302.

WYATT, C. L., D. R. ANDERSON, R. HARSHBARGER, AND M. TRIVEDI. 1984. Deer census using a multispectral linear array instrument. Proc. Int. Symp. on Remote Sensing of Environment 18:1475–1487.

————, M. TRIVEDI, AND D. R. ANDERSON. 1980. Statistical evaluation of remotely sensed thermal data for deer census. J. Wildl. Manage. 44:397–402.

————, ————, ————, AND M. C. PATE. 1985. Measurement techniques for spectral characterization for remote sensing. Photogrammetric Eng. Remote Sensing 51:245–251.

10

MEASURING VERTEBRATE USE OF TERRESTRIAL HABITATS AND FOODS

John A. Litvaitis, Kimberly Titus, and Eric M. Anderson

INTRODUCTION

The abundance of animals and distribution of their populations vary in space and time, often with the availability of the environmental components necessary for life. These *life requisites* include food, water, cover, and nesting or denning sites. Each species exploits a set of resources, so an understanding of the habitat and food use by a species is essential before any management efforts are initiated. Although many studies on these topics have been published for common game and furbearing species, we still know little about the life requisites of most terrestrial vertebrates. In addition, information about the habitats and foods used by a particular species may be needed for a specific region or time period. For example, wildlife biologists have become increasingly involved in assessing the effects of human activities such as highway construction, powerline development, and urbanization on environmental quality. These assessments often require identification of important habitat patches and food resources in the affected area. As a result, a biologist must collect site-specific information on patterns of habitat and food use. But how is such information obtained? What should be considered when a study is designed to identify habitat or food use? This chapter will provide an outline of the major techniques used to study these parameters and some of the problems likely to be encountered. We have intentionally taken different approaches in addressing aspects of habitat and food use. This was necessary to take full advantage of other sections of this book. Chapter 22, on sampling vegetation, outlines many of the techniques used to describe wildlife habitat and would make any summary in our chapter redundant.

Therefore, we have focused on the conceptual issues of investigating habitat use and selection. Much of this summary is also relevant to investigating food use. Chapter 12, on wildlife nutrition, provides additional background for understanding food use patterns. As a result, we have outlined the techniques used to investigate food use and availability.

Before any study of habitat or food use begins, an understanding of how the results will be used is essential. Is the objective of the study to describe habitat use patterns or a diet for the entire year or during one season that is considered critical? Is it to identify limiting factors or simply document use? Much of the wildlife research to date has been directed at addressing the descriptive questions of how, what, when, and where (Keppie 1990, Gavin 1991). These investigations have provided us with a detailed foundation on the natural history of many species. However, understanding *why* an animal occupies a specific habitat (for thermal cover, food abundance, or predator avoidance) or selects a particular forage grass (to maximize energy intake, obtain a specific nutrient, or minimize toxin intake) may reveal much more about the factors that limit a species than simply documenting patterns of use (Gavin 1991). Although it may seem obvious, taking the time to "think through a study" and articulate the specific question(s) being addressed is time well spent. One should be able to state concisely to someone else the research question that is being addressed. Results will be only as coherent as the initial conception of the problem (Green 1979). A thorough review of existing literature can help investigators develop an understanding of the variability in resource use patterns and avoid the common pitfall of collecting descriptive information simply because it has not been collected in their specific study areas (Hunter 1989). The final question to consider concerns the application of results. Are the conclusions of the study to be extrapolated from samples collected at one area and applied to other regions? Without consideration of spatial and temporal variations, any extrapolations may lead to incorrect conclusions.

Use, Selection, and Preference

The words *use, selection*, and *preference* have been applied widely and often interchangeably when information on patterns of resource exploitation is presented. *Use* simply indicates an association or consumption when habitat or food resources, respectively, are discussed. *Selection*, however, implies that an animal is choosing among alternative habitats or foods that are available to it. Use is selective if components are exploited disproportional to their availability (Johnson 1980). *Preference* is determined independent of availability; that is, an animal is allowed access to different resources on an equal basis. This information can be obtained only under unique conditions, such as enclosure experiments to determine habitat preference, or cafeteria experiments wherein captive animals are presented a variety of foods and allowed to choose among them. Because of the unique nature of preference experiments, we will focus our attention on developing an understanding of habitat and food selection.

Levels of Selection and Effects of Scale

Theories of habitat selection are well developed (e.g., Fretwell 1972, Rosenzweig 1981, Fagen 1988, Hobbs and Hanley 1990), but their application to field measurements is not always clear. Habitat selection can occur at a variety of levels or scales. The concept that an animal may select its habitat according to a hierarchical scheme was first offered by Hilden (1965, but also see Johnson 1980), and this concept has practical value. These scales include the **biogeographic** (e.g., the eastern deciduous forest), **home range** (e.g., mature hardwood or oak-hickory forest), and the finest scale, that of some **activity point**, such as a den, nest, or roost site within a home range. The factors that influence selection at each of these scales also vary. For example, climatic extremes may determine the geographic range of a species, whereas habitat structure may influence home-range size and shape, and competition with conspecifics may influence territory placement within a home range. The distribution of food and cover is probably most influential in determining local movements within a home range.

The choice of an appropriate scale of measurement, both in time and space, will directly influence the results and their interpretation (Wiens 1981, 1983, Karr 1983). Although thinking of scale in discrete levels (e.g., time = daily, seasonal, annual intervals; space = feeding site, home range, geographic range) is convenient, it is important to recognize that scales of measurement and environmental heterogeneity are continua (Karr 1983). Choosing the wrong scale of measurement may lead to the interpretation that a species is generalized or specialized in its use of available habitat, whereas another scale of measurement might lead to a very different interpretation. For example, Wiens (1989) observed that the biogeographic

range of Brewer's sparrows was associated with shrub-dominated habitats. However, at a regional scale (multiple study sites), the abundance of sparrows was negatively associated with shrub coverage, and at a local scale (single study area) shrub abundance and sparrow abundance were not related.

Habitats also can be characterized on a "macro" scale according to dominant biome or cover type (e.g., grassland, hardwood forest, wetland), or on a "micro" scale according to such features as stem density, litter depth, or canopy closure (Calder 1973). Wildlife biologists often restrict their studies to either of these scales; however, an examination of both scales may provide the greatest insight to animal-habitat relations (Morris 1984, Snyder and Best 1988). For example, the macro (forest-cover type) and microhabitat components (canopy closure and snow depth) of white-tailed deer wintering areas (yards) are important management considerations for this species in the northern portion of its range (Verme 1968). Macrohabitats in this example may describe the interspersion of food and cover used by deer. On the other hand, the microhabitat features may have a direct effect on thermoregulation (a factor influenced by variation in canopy closure) and the energy costs of travel or ability to escape predators (factors influenced by snow depth).

Many resource management agencies are now using such tools as Geographic Information Systems (GIS, see Chapter 21) for discerning land-cover/land-use associations of wildlife species. At this macro scale, habitat data may be acquired remotely, such as from maps (Mosby 1969), aerial photographs (Avery 1968), or satellite images (Short 1982). Unfortunately, the associations between animals and habitat attributes at this level of resolution often are very general. Applications of such information, therefore, should be restricted to broad areas and not specific sites. Such problems can be avoided best if information is gathered at a scale comparable to the scale at which the research or management question is being observed. An example provided by T. Nudds (pers. commun.) illustrates this point. Species richness of forest birds is known to vary with the size of woodlots, but many variables are needed to explain variation among sites within woodlots (e.g., canopy height, understory density, and snag abundance). Because the decline in species richness in small woodlots is a widespread, regional phenomenon, no amount of information about what covaries with species richness in a *single woodlot* is relevant to counter the decline of species richness. As a result, monitoring size *among woodlots* is the appropriate measure for understanding changes in regional populations.

Management Implications of Resource Selection

It is important to recognize the limits of the data when applications of research conclusions to manipulations of habitats or populations are considered. Use/availability or related analyses can be helpful in identifying patterns of habitat or food selection. However, biologists should not conclude **biological need** from such patterns. For example, suppose that a fictitious species (the blue-nosed yak) has demonstrated selective use of forests 40–80 years old; that is, yaks are most abundant or spend a disproportional amount of time in this habitat. Would we be correct to conclude that if all forests 40–80 years old were eliminated from the range of blue-nosed yaks this species would decline in abundance or go extinct? Probably not. Although we have demonstrated selection, we have not shown how the fitness (e.g., survival or reproductive success) of yaks varies with different amounts of the selectively used habitat. As a result, we cannot make a "biological leap of faith" and assume that if we increase the amount of the selected habitat (or food) we will have more yaks. Van Horne (1982) showed that population density and habitat quality (based on animal fitness) can be inversely correlated. Subordinant individuals (especially juveniles) may become locally abundant in "sink" habitats as a result of avoiding contact with dominant individuals that occupy the sites that have an abundance of food and cover ("source" habitats). As a result, survival and reproductive success in sink habitats are low. Therefore, if our objective is to determine the biological importance of a particular habitat, we should consider some type of manipulation experiment in which the amounts of the selected habitat (or food) are varied and fitness is monitored (Van Horne 1983). Although such studies are not always practical, they are essential to demonstrate such habitat associations and may be possible when applied to habitat management programs or large-scale habitat manipulations, such as impounding a river with a dam or logging a forest (see Macnab 1983, Sinclair 1991). Such experiments will require considerable planning because biologists usually do not have control over the manipulation. Additionally, some habitat or food-based questions may allow the researcher to compare a measure of "success" among used and less-used resources and evaluate the features that lead to "success" and presumably the basis of selection (e.g., waterfowl nest success and vegetation features that influence concealment).

MEASURING HABITAT USE, AVAILABILITY, AND SELECTION

It will be useful in this chapter to define habitat in its broadest context, including all abiotic and biotic features of the environment. Although other definitions of habitat exist (e.g., Karr 1981), we will consider habitat use as the association of an animal with these features. Selection and use of a particular area by an animal are the result of *proximate* and *ultimate* factors (Hilden 1965, Partridge 1978). *Proximate factors* are those features used as cues when an animal evaluates a site. They may include structural features such as understory cover, canopy height, or slope. The presence/absence of other animals that may act as competitors or predators also may influence habitat use. Animals may use such features as cues, but they may not be the same as the factors that have resulted in evolutionary associations between animals and habitat. *Ultimate factors* are those parameters that determine how successful an animal is within a particular habitat. An individual's abilities to reproduce, obtain food, and avoid predators are examples of ultimate factors that influence habitat selection. Except for the assessment of food availability, studies of habitat use usually involve some measure of the proximate factors.

The relationship between a habitat feature being measured and its biological link to the animal often is clear. For example, understory stem density frequently is used as an index of escape cover for small or medium-sized mammals, such as snowshoe hares (Litvaitis et al. 1985). However, in other instances the animal-habitat relationship may be less obvious, such as using the abundance of snags as an index of insect availability to pileated woodpeckers. In this situation, a structural feature is believed to be correlated with a proximate factor. Because it is much easier to inventory and manage snag abundance than insect abundance, one is tempted to use this association when investigating woodpecker-habitat relations. However, the relationship between snags and insects should be verified to ensure that subsequent research conclusions are reliable.

Techniques To Measure Habitat Use

Direct and indirect methods have been used to index wildlife habitat use (Table 1). Direct methods include observation, capture, and radiotelemetry, whereas indirect methods are dependent on some evidence of animal activity within an area or specific site (e.g., bed sites, browsed twigs, feces, nests, or tracks). These indices may be used to sample habitat use along systematic transects, such as a small-mammal trapping grid, or with other sampling designs appropriate to the animal of interest (Fig. 1). The basic assumption of these methods is that the index increases with the amount of time an animal spends at a site, or that population density increases as the index increases. However, several factors may influence this relationship, including observer bias, or the accumulation rates of indices may vary independent of time spent within a site. For example, Collins and Urness (1981) observed that defecation rates of mule deer varied with activity (feeding versus traveling). As a result, the distribution of pellet groups gave a biased impression of habitat use. Therefore, using more than one index to help avoid erroneous conclusions may be necessary.

Frequently, investigators have focused on examining the characteristics of a site that is used by an animal while it is feeding, resting, or rearing young (Stinnett and Klebenow 1986). Such activity points are detected by using the direct or indirect methods mentioned earlier. These sites are then compared to sites within a study area where use was not detected, or use may be partitioned into categories (e.g., never used, occasionally used, frequently used). Comparisons also may be made between portions of the home ranges of study animals (activity core versus outside the core). A "shotgun" approach is then used to sample a variety of features at each site or within portions of individual home ranges (James and Shugart 1970, Dueser and Shugart 1978, Fridell and Litvaitis 1991). The procedures for sampling used or unused sites are a compilation of many other techniques, most of which were adopted from plant ecology (Greig-Smith 1964, Mueller-Dombois and Ellenberg 1974, Bonham 1989), forestry (Avery and Burkhart 1983), and range management (Cook and Stubbendieck 1986) and are now widely applied to inventories of wildlife habitat (Hays et al. 1981; also see Chapter 22 for a review of many of these methods). The features selected to describe a site (e.g., litter depth, understory stem density, canopy closure) are assumed to represent or be highly correlated with the factors used by animals to evaluate a site and often include some measurement of food abundance, cover, and structural char-

Table l. Summary of direct and indirect methods used to evaluate habitat use by terrestrial vertebrates.

Method	Advantages	Concerns	Examples
Direct			
Observation	Sample large segment of population	Differential visibility among habitats	Biggins and Pitcher 1978, Stinnett and Klebenow 1986
	Distinguish activities within habitats	Restricted to diurnal periods	
	Inexpensive		
Capture	Can examine age/sex difference in habitat use	Differential vulnerability to capture among segments of population	Parren and Capen 1985
	Can be combined with mark–recapture	Bait may attract animals into habitats normally not used	
	Good in small or dense populations		
Radiotelemetry	Can examine age/sex differences	Accuracy may limit application in a patchy environment	See Chapter 15, Nams 1989
	Ability to follow annual patterns of known individuals	Sample size usually small	
	Can be used to obtain information on important habitat components (e.g., den sites, roost sites)	Expensive	
Indirect			
Track counts	Sample all segments of a population	Distance traveled within a habitat may not be correlated with time spent within habitat	Litvaitis et al. 1985, Thompson et al. 1989
	Sample a large area in a short time	Seasonal and regional limitations if relying on snow	
	Inexpensive		
Pellet counts	Sample all segments of a population	Defecation rates may vary with activity	Collins and Urness 1981, Orr and Dodds 1982
	Can provide information on seasonal habitat use if sample plots are cleared	Potential for differential decomposition rates among habitats	
	May be used to provide density estimate if deposition rate is known		
Browsed twigs counts	Sample all segments of a population	Obvious bias, restricted to sites where browse is available	
	May also provide information on food habits and relative position of population to carrying capacity		
Drey (squirrel tree nest) counts	Sample large areas, provide good information on relative differences in density of squirrels between habitats	Drey counts not easily converted to squirrel numbers	Wauters and Dhondt 1988

acteristics (Fig. 2). Because the actual habitat features influencing selection are not known, measuring several features is appropriate (e.g., Rice et al. 1984). However, as the number of features measured becomes large, the chances of detecting spurious relationships also increases. Therefore, the list of features to be sampled should be limited to those based on biological considerations for the relationships between animals and their habitat (Green 1979).

Techniques To Measure Habitat Availability

The delineation of available habitat will profoundly influence the interpretation of habitat selection and any subsequent management recommendations. Assessing availability of habitats from the point of view of an animal is not possible; therefore any effort to estimate availability is naturally beset with problems (Chesson 1978, Jaenike 1980, Johnson 1980). Biologists often choose administrative units such as a parks, forests, or refuges to arbitrarily represent the habitat available to study animals because of the obvious management ramifications. This can have significant drawbacks. Study animals may use a larger area, resulting in a biased interpretation of selection. On the other hand, the habitat considered to be available may be much larger than the

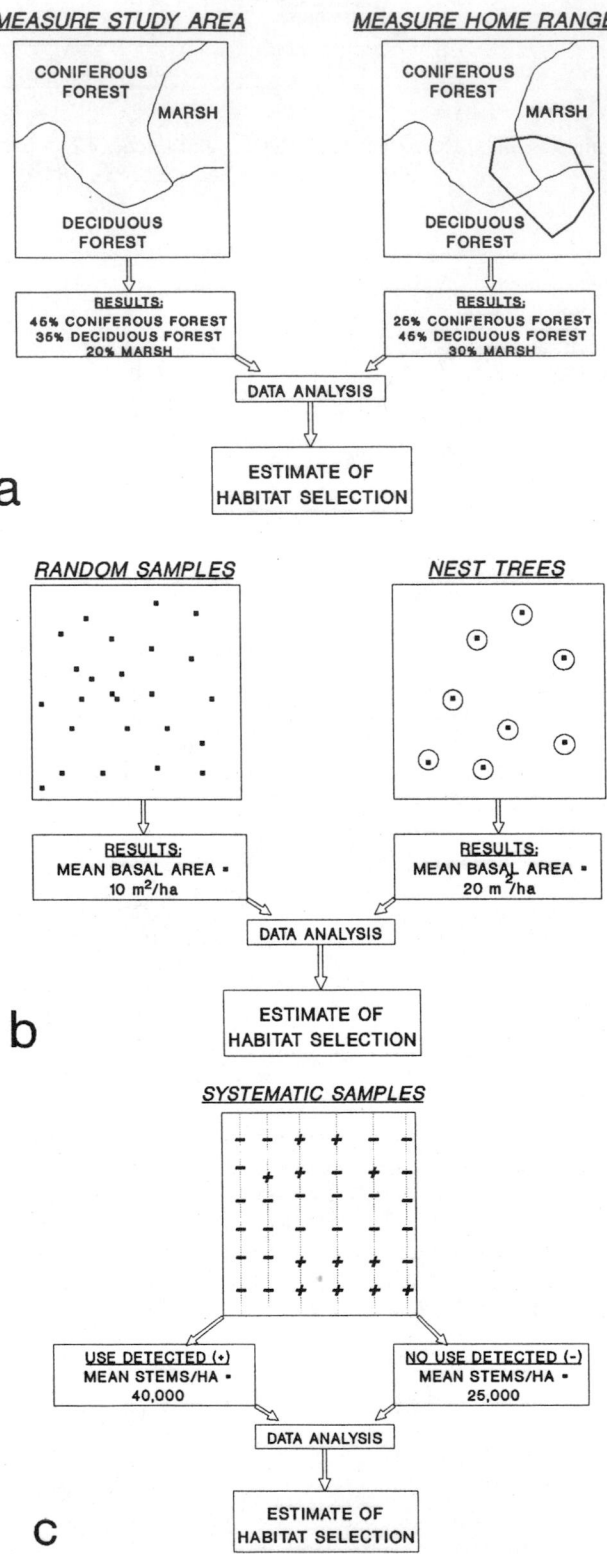

Fig. 1. Representative methods that can be used to examine habitat use patterns. Available habitat is inventoried and compared to the composition of an animal's home range (a), or random samples are compared to characteristics of sites where use has been detected, such as a nest or roost site (b), or systematic plots are established and features are compared between sites where use was detected (via captures, tracks, feces, e.g.) and sites where use did not occur (c).

area actually occupied by the study animals, and this also would produce biased results.

Knowing something of the habitat associations of the animal being studied is essential before study area boundaries are delineated. For example, including open fields as available habitat for forest-interior songbirds (e.g., ovenbird) probably would yield highly significant results in terms of their habitat selection. Yet these results would be trivial because we already know such areas are rarely used by these species. At the other end of the spectrum, delineating study area boundaries or habitat availability should not be so restrictive that potentially available habitats are eliminated.

The distribution and size of cover types within a study area also can influence our ability to detect selection patterns. Porter and Church (1987) illustrated this problem by comparing an area where cover types were regularly distributed to an area where they were clumped. In areas that had regular or random distributions of cover types, the delineation of the study area had little influence on the analysis of habitat use versus availability. However, if cover types were in an aggregated pattern, the delineation of study area boundaries substantially influenced the analysis of selection. The guidelines listed below may be helpful when study area boundaries are delineated, but each study is unique.

1. Size of the study area should be substantially larger than the home range of the study species.
2. Numbers of study animals, groups, or social units present on the study area should be, as far as possible, adequate for study.
3. An opportunity should exist for true replication of samples within the study area.
4. Study area boundaries should be chosen with consideration of the biology of the animal. Physical barriers such as rivers or mountain ranges might make better boundaries than an arbitrary (geopolitical) straight line on a map.

Availability of cover types within a study area often is measured directly from aerial photographs, maps, or satellite images. In these situations, we are dealing with known amounts that, although having measurement error associated with them, do not have sampling error associated with them. Biologists often have access to cover-type maps produced for multiple-use planning, such as timber type and plant association maps that are produced for national forests or private forest industry lands. Although the inventory may cover the area of interest, various approximations and measurement errors are part of these products (e.g., smallest forest stand inventoried is often >1 ha), and therefore an understanding of these limitations is required prior to any integration with wildlife data. In large study areas, however, the availabilities of cover types often are estimated by some random sampling procedure (Marcum and Loftsgaarden 1980). When available habitat is sampled by some type of random, systematic, stratified, or clustering procedure, the delineation of study area boundaries may be less important so long as the overall study area represents a random sample of the larger area of interest. For some administrative boundary, such as a national forest, the study area may be very large, and therefore the key issue is to randomly

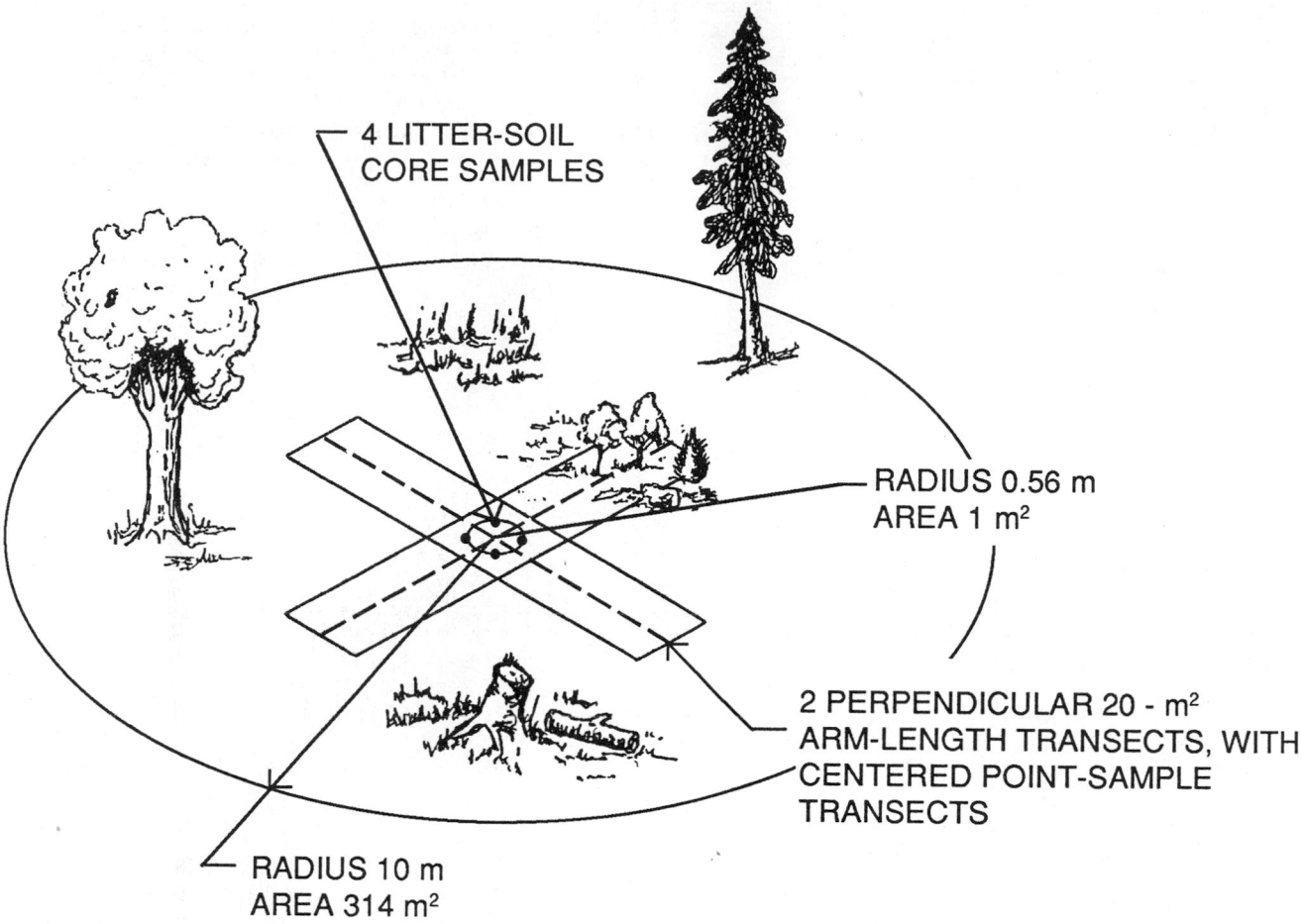

4 LITTER-SOIL
CORE SAMPLES

RADIUS 0.56 m
AREA 1 m²

2 PERPENDICULAR 20 - m²
ARM-LENGTH TRANSECTS, WITH
CENTERED POINT-SAMPLE
TRANSECTS

RADIUS 10 m
AREA 314 m²

Fig. 2. An example of nested plots used to sample ground litter, understory stem density, and overstory composition (modified from Dueser and Shugart 1978).

place the small samples throughout the study area. If availabilities are estimated and thus have a sampling error, subsequent comparisons with use should be analyzed differently than if the availabilities are known (Thomas and Taylor 1990, White and Garrott 1990).

Researchers also have used characteristics of the study animals to define the sampling area. For instance, if study animals are equipped with transmitters, an effective study area can be delineated by connecting the extreme locations of all marked animals. Miller and Litvaitis (1992) used this method to investigate habitat segregation between sexes of moose. This method should be applied with an understanding that it may impose a result on the study. Animals may be selecting habitats at a larger scale than recognized by the investigator when the boundary of available habitat is delineated, and selection patterns may not be detected in subsequent analyses. As a result, we may need to consider evaluating habitat availability and selection within several scales—for example, local cover types and regional landscapes (Steventon and Major 1982). Partitioning home ranges into discrete grid cells (e.g., 100 × 100 m) provides an alternative approach to identifying important habitat features (Witmer and de-Calesta 1983, Litvaitis et al. 1986, Nicholls and Fuller 1987). Cells are categorized according to the number of captures, observations, or other index of use by study an-

imals. Habitat features then are sampled within each cell or a subset of cells, and comparisons based on intensity of use, or used versus unused cells, can be made (Porter and Church 1987).

So far we have examined site-specific aspects of habitat. Not only are specific attributes of habitat (amount and size) important, but the juxtapositions among habitats and the variability between habitats may be influential in determining habitat suitability. One of the most important components of habitat structure is the spatial heterogeneity or patchiness. It integrates not only the absolute values of the vegetation or physiography, but also their variation in space (Wiens 1976). Many birds and mammals rely on more than one habitat for feeding, mating, nesting, or denning. A specific habitat may have an abundance of one resource, such as food, and not be used by an animal because it lacks or is distant from sites that provide another necessary resource, such as cover. Additionally, habitat patchiness may have ramifications on habitat suitability, such as influencing a predator's ability to stalk or ambush prey. Therefore, some measure of habitat variation is important to these species.

Leopold (1933) was among the first to note the importance of habitat heterogeneity for wildlife, so the concept is not new. Generally, habitat heterogeneity can be viewed both at a coarse-grained (e.g., between cover types) and

Table 2. Examples of methods for measuring habitat heterogeneity.

Method and equation	Definitions/procedure	Source	Comments
Simple statistics of dispersion			
Range: Range = $y_{max} - y_{min}$	y_{max} = largest item in sample y_{min} = smallest item in sample	Introductory statistics texts	Larger range correlates with more heterogeneity; highly sensitive to outliers.
Variance: $s^2 = \Sigma\,(y - \bar{y})^2/(n - 1)$	$\bar{y} = \Sigma\,y/n$ n = number of subsamples	Introductory statistics texts	Larger variance correlates to more heterogeneity; requires adequate subsamples to avoid being too sensitive to outliers.
Coefficient of variation: $CV = s/\bar{x}(100\%)$	n = number of subsamples s = subsample standard deviation $\bar{x}$ = subsample mean	Zar 1974	Expresses variability relative to the sample or subsample mean, allowing comparisons among different units of measurement; Roth (1976) termed this an index of heterogeneity using the point-centered quarter technique of Cottam and Curtis (1956).
Simple method for summing			
Total: $\displaystyle\sum_{i=1}^{n} y_i$	y_i = value of the ith sample		If all sampling is equal, larger sum equates to more heterogeneity.
Heterogeneity indices			
Wiens' heterogeneity index: $H = \Sigma\,(max - min)/N$	max = maximum value of attribute within sample unit min = minimum value of attribute within sample unit N = total number of sample points	Wiens 1974, Rotenberry and Wiens 1980	Designed to depict within-sample unit variation; assumes the same degree of difference has equal importance in different habitat situations.
Wiens' heterogeneity index: $HI = \Sigma\,(max - min)/\Sigma\,\bar{x}$	max = maximum value of attribute within sample unit min = minimum value of attribute within sample unit $\bar{x}$ = mean within the sample unit	Wiens 1974, Rotenberry and Wiens 1980	Corrects for the bias in the above index by weighting by the mean.
Interspersion: $I_s = \Sigma\,C_c/\Sigma\,C_T$	C_c = total number of cover-type changes surrounding a cover-type cell C_T = total possible number of cover-type changes	Heinen and Cross 1983	Cover types must be defined and mapped a priori.

Juxtaposition: No equation, stepwise procedure	1. identify all combinations of edge types
	2. assign numerical ratings
	   1 = "diagonal" edges
	   2 = "vertical" or "horizontal" edges
	3. assign relative weighting factor from 0 to 1 as index of quality
	4. multiply quality factor by numerical rating
	5. sum and divide by standardizing value
		Heinen and Cross 1983	

Table 2. Continued.

Method and equation	Definitions/procedure	Source	Comments
Spatial diversity index: $Sd_A = ([\sigma_A I_s] + [\alpha_A J_x])$ $\cdot (1_A)(2_A)(3_A)$	A = indicates a particular species σ_A = relative importance of interspersion α_A = relative importance of juxtaposition $\sigma_A + \alpha_A$ must = 1 $1_A, 2_A, 3_A$ are variable number of exclusion factors ranging from 0 to 1	Mead et al. 1981, Heinen and Cross 1983	Any number of exclusion factors may be used depending on the area and species being considered. An exclusion factor may have a positive or negative effect. For example, a negative exclusion factor might be a habitat disturbance within 1 km that reduces the suitability to 0, so if present a 0 is assigned.
Diversity index – shape of patch: $DI = TP/2\sqrt{A \cdot \pi}$	TP = total perimeter around the area plus any linear edge within the area A = area	Patton 1975, Thomas et al. 1979	Measures irregularity of a patch based on the area-to-perimeter ratio of a circle.
Inherent diversity index: Inherent $DI = TE_c/2\sqrt{A \cdot \pi}$	TE_c = total edge between plant communities in length found within or on the perimeter of the area under consideration A = area	Thomas et al. 1979	Report as percent by multiplying by 100. Inherent edges are site-related and occur when plant communities meet.
Induced diversity index: Induced $DI = TE_s/2\sqrt{A \cdot \pi}$	TE_s = total length of edges created between successional stages A = area	Thomas et al. 1979	Induced edges are produced by the land manager and result from all activities that alter the vegetation. Report as percent by multiplying by 100.
Total diversity index: Total $DI = TE_{c+s}/2\sqrt{A \cdot \pi}$	TE_{c+s} = total length of all inherent and induced edges A = area	Thomas et al. 1979	Report as percent by multiplying by 100.
Baxter-Wolfe interspersion index: Sum the number of changes in habitat types occurring along a transect		Baxter and Wolfe 1972	
Relief index: Sum the number of contour lines crossed by transects radiating from some activity point			
Land surface ruggedness index: Sum the number of dot-contour line intersections for a given area and map type		Beasom et al. 1983	Based on the assumption that ruggedness is a function of the total length of all contour lines traversing a given area.

a fine-grained (within cover types) scale. The choice of scale and the method for assessing heterogeneity or patchiness should always be organism-defined and not in terms of the perceptions of the investigator. Heterogeneity can be expressed in vertical and horizontal dimensions. The layering of vegetation in plant communities is a common way to express vertical heterogeneity. Techniques such as those that use a vertical density board have been used to describe it (De Vos and Mosby 1969, Nudds 1977, Noon 1981, Robbins et al. 1989). Other measures of habitat heterogeneity include some estimate of variance from grouped samples or subsamples. Biologists frequently apply the arithmetic mean of a number of habitat samples as an index of a particular site. The coefficient of variation (CV = standard deviation ÷ mean) is an index of within-site variation that is easy to obtain. Many methods have been developed for use at a variety of scales (Table 2).

Biologists working at the level of landscapes evaluate such parameters as habitat patch dispersion and corridor development (Forman and Godron 1986). However, the biological interpretation of such characteristics may be less intuitive than other habitat features and should be addressed before such information is collected.

Considerations for Study Design, Data Analysis, and Interpretation of Study Results

Because an objective of investigating habitat selection often is to extrapolate the results either spatially or temporally, it is necessary to have valid, replicate samples from which inferences can be drawn. Therefore, habitat samples need to be truly independent in space and time. Correctly distinguishing sample units in a study is essential, but this distinction is not always obvious (Hurlbert 1984). For example, in most studies of habitat use involving radiotelemetry, the sample unit is the individual animal and not the number of locations generated (these are subsamples). In such studies, it is tempting to pool all locations and perform a single evaluation of use versus habitat availability. This would be appropriate if each animal contributed only a few locations. Otherwise, pooled observations may lead to serious bias if one or two individuals contribute a large proportion of the total observations. Additionally, misusing the subsamples (as "pseudoreplicates") would mask any age or sex differences in habitat use (Thomas and Taylor 1990). As a result, the "averaged" habitat-use pattern may not be representative of any individuals in the sample population.

It is also important to consider the concept of independent sampling with respect to time. Again, a biologist can acquire large amounts of information on habitat use from a sample of radio-equipped animals in a relatively short period of time, such as recording the location of an animal every 15 minutes during a 24-hour sampling period. This sampling schedule would result in hundreds of locations in a relatively short period. However, in most instances the data would not be a collection of statistically independent observations. Such data are considered *autocorrelated* because the interval between locations was not long enough to enable the study animal to enter another habitat. As a result, the location of an animal at 1100 hours will influence the location of the same animal at 1115 hours. Therefore, it is important to develop a sampling schedule that ensures independence among observations (Swihart and Slade 1985).

After a data set is acquired, it can be analyzed with a variety of methods. Often, biologists are interested in comparing use to availability to identify selected habitats (Box 1). Alldredge and Ratti (1986, 1992) compared the performance of several techniques used to evaluate habitat selection. Although these authors indicated there is no clear choice or best method to evaluate use/availability data, the number of animals monitored and the number of habitats designated as available influence the Type I (rejecting the null hypothesis when it is true) and Type II (accepting the null hypothesis when it is not true) error rates. These results indicate the importance of first identifying available habitats and estimating the samples (animals) needed to properly evaluate selection.

Evidence of selection of one or more habitat feature does not necessarily mean that animals are actually selecting habitats. As we have already indicated, the observed pattern of selection may be a result of the limitations imposed by the estimation of available habitats. Furthermore, selection may be based on a habitat feature that was not sampled but that is correlated with a feature that was sampled and found to be selected for. In our example of snowshoe hare habitat, we indicated that understory stem density was an important feature because it influenced the vulnerability of hares to predators. However, if the benefits of understory stems were actually more relevant to thermal cover, then managing stem densities to reduce local predation rates may be unsuccessful unless the benefits of enhanced thermoregulation and reduced predation covary directly.

Wildlife habitat assessments usually are not true experiments because the biologist usually does not involve the use of some external factor (manipulation) on the experimental units (Hurlbert 1984, Diamond 1986). Many of these studies can be made more rigorous (Macnab 1983). For example, a biologist may be studying deer use of habitat patches, some of which have been clearcut and others that have not. The observed patterns of deer movement or density may be correlated with the distribution of clearcuts and we would therefore infer some association between deer and clearcuts (e.g., selection or avoidance). However, a better design would be to first monitor deer movements in areas where cuts are scheduled and then compare movements before and after cutting. If control sites were monitored simultaneously, more definitive conclusions could be drawn from the study and applied with greater confidence. More detailed presentations of the issues that concern study design were provided by Green (1979), Stewart-Oaten et al. (1986), and Chapter 1.

MEASURING FOOD USE, AVAILABILITY, AND SELECTION

The abundance and distribution of food resources are among the major environmental features that influence habitat selection. As a result, food acquisition or foraging can be considered as a demonstration of how an animal *actively* uses its habitat (Morrison et al. 1992). In the past 2 decades, a substantial body of research has developed in this area and is collectively called the optimal foraging theory. A central theme of this area of research is to understand why an animal decides to forage in a particular area. Two distinct approaches have developed to address this issue. The first considers that an animal selects among various food items or prey that are distributed in some fashion (e.g., clumped) throughout a generally suitable habitat. The second approach examines how animals discriminate among various patches of habitat that vary in productivity and suitability (Morrison et al. 1992). The latter approach also can be viewed as an evaluation of habitat selection. Because we have already addressed the major concepts associated with an investigation of habitat selection, we will not review the major findings of foraging research that address this issue (but we encourage readers to review recent publications on this topic, e.g., Stephens and Krebs 1986, Kamil et al. 1987, Stephens 1990). We will, however, discuss the pragmatic aspects of estimating food availability and use and some of the problems encountered when patterns of food selection are interpreted.

Studies of food habits have intrinsic value because they are important components of an animal's life history, and substantial information has been collected on the diets of many terrestrial vertebrates (e.g., Martin et al. 1961). Yet, variation exists within diets according to location, season, and age/sex of the animal. In addition, information on food use is an essential component of research efforts addressing such issues as the impact of predation on prey populations, extrinsic factors that influence reproductive success, and assessments of productivity of local habitats. As a result, biologists should be familiar with the techniques used to investigate wildlife food habits.

Techniques To Measure Food Use

Many techniques have been developed to investigate wildlife food use. Most of these methods fall into one of three broad categories: (1) observational, where animals

Box 1. Evaluating wildlife habitat preference.

Radiotelemetry and live captures are among the most frequently used methods to investigate wildlife habitat associations. The following is a data set that resulted from monitoring a transmitter-equipped black bear in a 100-km^2 study area in northern New Hampshire.

Cover type	Study area (%)	Home range (%)	Locations (%)
Aspen	40	50	35
Spruce-fir	25	10	5
Clearcut	15	30	40
Bog	20	10	20
Total:	100	100	100

For ease of presentation, assume that 100 locations were obtained on the bear. Now, we can approach an evaluation of habitat preference using the procedures described by Neu et al. (1974) and Byers et al. (1984). This method combines an initial chi-square comparison with confidence intervals on observed use. First, we must generate the expected number of locations by using the portion of the study area or home range, depending on the comparison we plan to make.

Cover type	Observed use	Expected use*
Aspen	35	0.40 × 100 = 40
Spruce-fir	5	0.25 × 100 = 25
Clearcut	40	0.15 × 100 = 15
Bog	20	0.20 × 100 = 20
Total:	100	100

*Based on the composition of the study area.

$$\chi^2 = \frac{(35-40)^2}{40} + \frac{(5-25)^2}{25} + \frac{(40-15)^2}{15}$$
$$+ \frac{(20-20)^2}{20} = 58.3$$

This value of chi-square with 3 df is highly significant. Therefore, we can conclude that the bear is not using the four cover types in proportion to their availability within the study area. Next, we examine the selection or avoidance of each cover type by constructing confidence intervals for the proportion of use in each type using the following formula:

$$\hat{p}_i - z_{\alpha/2k}\left[\frac{\hat{p}_i(1-\hat{p}_i)}{n}\right]^{\frac{1}{2}} < \hat{p}_i < \hat{p}_i + z_{\alpha/2k}\left[\frac{\hat{p}_i(1-\hat{p}_i)}{n}\right]^{\frac{1}{2}},$$

where $\hat{p}_i$ is the proportion of locations in habitat type i and $z_{\alpha/2k}$ is the upper standard normal variate corresponding to the probability tail of $\alpha/2k$. Because we are making multiple simultaneous comparisons, we should adjust the probability level (α) to compensate by dividing by the number of comparisons (or cover types examined = 4). In this example $z_{\alpha/2k} = 2.50$ because we are using $\alpha = 0.05$.

Cover type	Confidence interval on use
Aspen	0.231 < 0.350 < 0.469
Spruce-fir	0.000 < 0.050 < 0.104 −
Clearcut	0.278 < 0.400 < 0.522 +
Bog	0.100 < 0.200 < 0.300

By comparing the confidence interval on the observed use to the proportion of the study area in each cover type, we can conclude that the bear selected clearcuts and avoided spruce-fir stands. These results are based on comparisons of use to availability within the study area. If we based availability on the composition of the bear's home range, our results would have been different.

are observed ingesting food; (2) feeding site surveys, where the amount of vegetation removed by foraging animals is measured or estimated; and (3) post-ingestion samples, where the remains of food in gastrointestinal tracts, feces, or regurgitated pellets are identified.

OBSERVATIONAL

Direct observation has been a widely used technique for estimating food habits of large herbivores. Individual animals are watched through binoculars or a spotting scope as they graze or browse, and the type and frequency of plant species consumed are recorded. Observations are quantified either as bite-counts (number of bites of a particular plant species) or as feeding minutes (time spent foraging on a particular plant species). These values are then translated into relative occurrence in the diet by considering the number of bites or minutes of grazing on a particular species compared to the total for the entire observation period. Forage intake for an individual can be estimated by multiplying the bite counts for each species by the average mass per bite for those species. The estimated biomass consumed of each plant species can be divided by the total biomass of forage consumed to give a percentage of each species in the diet (Smith and Hubbard 1954). Obviously, this technique is simple, requires little equipment, and may be easily accomplished. Unfortunately, the accuracy of identifying different forage species at a distance is highly variable and depends on the distance to the animal, the familiarity of the observer with forage of the area, the complexity of the plant community, and the phenological development of individual plants (Holechek and Gross 1982). The technique also is limited to diurnal/crepuscular herbivores living in relatively open habitats.

Some of the concerns about accuracy of direct observations have been addressed by the use of tame, hand-reared animals. These individuals will feed with a researcher next to them, so misidentification of forage species is minimized. Gill et al. (1983) reported that bite-count analysis with tame animals gave accurate results that were repeatable among observers. Despite the apparent accuracy, there are serious concerns that tame animals may not reflect the food habits of their wild counterparts. Physiological condition, degree of hunger, topography, other animals present, and past grazing experience may influence forage selection by a tame animal (Wallmo and Neff 1970, Wallmo et al. 1973). Wallmo et al. (1973) concluded that bite-counts were a poor technique for quantifying the amount of forage consumed by large herbivores, but this technique is useful for identifying the principal food items in their diet. However, recent modifications of the lead-animal technique may provide new insight into food use and selection. Heim (1988) hand-reared several white-tailed deer fawns, equipped them with transmitters, and released them into the wild. After an acclimation period of several months, he was still able to walk beside the deer and record food availability and use.

Examination of feeding sites shortly after an herbivore has been observed feeding there also has been used to determine food habits. Recently grazed plants or twigs in the exact location in which an animal was observed are counted and totals are converted into percentages of materials grazed or browsed. This technique works best where the animal of interest has pawed through snow to eat plants. Even under the best of circumstances, this method ignores the "invisible use" problem of entire plants being grazed and leaving no residual indication of their presence (McInnis et al. 1983), as well as the difficulty in determining if all evidence of grazing or browsing was produced by the most recently feeding animal.

Carnivores generally do not lend themselves to the study of food habits by direct observation. Some large carnivores that are abundant and occur in open habitats, such as in the Serengeti, can be observed directly (Schaller 1972). Mech (1966) used this technique and feeding-site surveys to determine the age and sex composition of moose that were killed by wolves. Similarly, observations at raptor nests during the nesting period can provide information on the food items brought to the nest for the young (Errington 1932, Marti 1987, Bielefeldt et al. 1992). Advances in photographic equipment also have enabled researchers to place cameras at nest sites of even small passerines that may be difficult to observe otherwise.

FEEDING SITE/BROWSE SURVEYS

Feeding-site surveys were among the earliest approaches used to estimate diets of grazing animals and were developed originally for use on domestic livestock. These methods attempt to estimate the amount of vegetation removed by foraging animals from an area during a given time period (Edlefsen et al. 1960, Smith et al. 1962, Telfer 1969, Martin 1970, Cooperrider 1986). Compared to other techniques used for wild, free-ranging herbivores, feeding-site surveys provide only general dietary information.

Survey methods can be divided into difference and grazed-plant estimates. Difference estimates are obtained either by comparing single plots before and after grazing or by comparing grazed with ungrazed plots. The precision of estimates derived from *before* and *after* plots can be improved by selecting two similar plots, clipping all vegetation off one plot prior to grazing, and clipping the other plot at the end of the grazing season. Difference in the dried biomass between plots gives an approximation of the amount of forage removed by the grazers (Cook and Stubbendieck 1986).

Comparisons of grazed with ungrazed plots generally are accomplished with small, wire cages that exclude grazers. Paired plots with similar species composition and production are established prior to the grazing season, and one plot is randomly chosen to be caged. At the end of the grazing season, vegetation in both plots is clipped and the mass is recorded, or use is estimated by some other method. This technique should not be applied to estimates of use during the growing season because protected plants grow at different rates than grazed plants. Woody twig use also can be estimated by a similar technique. The biomass of twigs clipped in spring from plots subject to browsing is compared to the weight of twigs clipped from equal-sized fenced plots, the difference representing an estimate of use (Bobek et al. 1975). In general, difference methods are expensive, time consuming, and imprecise and do not detect small differences in use. They should not be considered unless use is expected to be >50% on the forage species (Cooperrider 1986).

Grazed-plant methods rely on estimating frequency of use, height/length conversions, or form-class/ocular estimates. These techniques are applicable to herbaceous plants as well as to browse species. Frequency methods require counting individual plants of each species in a given area and recording the percentage that has been grazed. Regression tables, specific for each plant species, then can be used to estimate percent use from the proportion of the plants that were grazed (Cook and Stubbendieck 1986). Aldous (1944) used circular plots to estimate the percentage cover for browse species within reach of browsers, and the ratio of browsed to unbrowsed stems on a subsample of twigs within the plot to provide an estimate of the amount consumed for each species. A similar "twig-count" technique involves counting all twigs of browse species within different height categories of sample plots to determine a use factor for each species (Passmore and Hepburn 1955). This technique does not work well when consumption is >60% and does not allow for between-species comparisons because of interspecific differences in twig weight (Jensen and Scotter 1977).

A labor-intensive, but more accurate, technique for assessing use has been widely used in the western United States (Nelson 1930, Smith and Urness 1962). Stems of potential browse species are tagged, and total length of annual twig growth is measured during autumn. To determine use, the observer returns during spring to measure the length of annual growth that remains. Although the technique does not directly measure biomass removed, the length of the twig removed is highly correlated with the amount of forage removed (Smith and Urness 1962).

POST-INGESTION SAMPLES

The most common technique for analyzing food habits of vertebrates involves sampling either during or after the digestive process. All post-mastication sampling requires identification of materials that may not be easily recognized. In herbivores, this often entails microscopic examination of vegetative fragments, many of which cannot be identified. Although many prey remains of carnivores can be identified macroscopically, some parts of smaller animals may be crushed into indistinguishable pieces. As a result, a portion of most carnivore and herbivore samples will be classified as unidentified.

Sample preservation techniques are similar for carnivores and herbivores. Stomach or intestinal contents can be stored with the least alteration of constituent materials by freezing. Under field conditions, immersion in 5% formalin for small to mid-sized species and in 10% formalin for larger species is effective, as is preservation with 70% alcohol. Soft-bodied invertebrates may disintegrate during storage, so it may be necessary to identify these prey at the time of collection. Feces or regurgitated pellets can be stored for later analysis by oven drying at 80–85 C for several hours to arrest microbial and insect degradation. Samples then can be stored in plastic or paper bags with a fumigant (naphthalene, paradichlorobenzene) to prevent insect infestation. In the field, fecal samples can be stored in an equal quantity of table salt until they are analyzed.

Carnivores

Predator diets generally are assessed by examining feces (scats), stomach or intestinal contents, and regurgitated pellets. Scats and regurgitated pellets offer the advantage of being available year-round and can be collected without harming or interfering with the target species. Even so, biases are involved with scat or pellet analyses of some predators that might require long-term data sets (e.g., Mattson et al. 1991), as might biases in the identification of prey remains (e.g., Mersmann et al. 1992). Feces of most species can be identified by size, conformation, composition, and odor (Murie 1974). Some carnivores produce feces are that often indistinguishable by appearance from those of other species. Under those circumstances, thin-layer chromatography can be used to identify bile acids that leave characteristic ''signatures'' for each species (Major et al. 1980). However, the cost and time required to conduct the analysis have limited its application.

Fecal material and pellets from predators require minimal preparation. These samples can be broken apart by hand or teased apart with forceps. However, consider autoclaving scats or wearing a gauze mask, because storage does not destroy tapeworm eggs or other endoparasites that may occur in carnivore feces. Feces also can be washed through a fine sieve or placed in a nylon bag that is washed in a household washing machine (gentle cycle) to separate hair, teeth, and other identifiable parts. If fur, feathers, and other nonbony materials are not crucial for identification, samples can be soaked in an 8% solution of NaOH for 12 hours and then rinsed over a small (18-mesh) sieve (Schueler 1972, Green et al. 1986). This technique dissolves nonbony components and leaves behind

Table 3. Diameter of sieve mesh and equivalent U.S. standard mesh.

Diameter (mm)	U.S. standard mesh
1.682	12
1.000	18
0.841	20
0.500	35
0.149	100
0.105	140
0.074	200

only scoured bones, teeth, insect exoskeletons, and pieces of reptile skin.

Gastrointestinal tracts that are collected in conjunction with legal hunting or trapping seasons also provide information on food use. However, these seasons usually span only a short portion of the year, limiting the application of this technique. Harvest samples also may be biased by foods used as bait. On the other hand, information about the sex, age, and body condition of the sampled animal, as well as volume of prey consumed, are advantages of this method.

Stomach contents from predators should be washed with hot water through a large soil sieve (12–20 mesh, Table 3). Care should be taken whenever alimentary tract samples are handled. Predators are host to endoparasites, viruses, and bacteria that may infect humans and can be transmitted through direct contact with samples or as an aerosol. Depending on the size of prey and the level of mastication, much of the material in a stomach of a carnivore may be identified with field guides to birds, mammals, and insects for comparison. Other items may require the use of bones, teeth, hair, feathers, or scale patterns to identify them. Reference materials, therefore, may include complete skeletons of vertebrates, samples of hair, feathers, scales of fish and reptiles, and exoskeletons of insects. A collection of dorsal guard hairs of the mammals likely to be encountered will be particularly useful because of the characteristic features of color banding, medullary pigment patterns, and morphology of cuticular scales (Adorjan and Kolenosky 1969). For preparation of reference slides, hairs are cleaned in ether (or similar solvent), dried, placed on a microscope slide with mounting medium (Permount), and sealed with a coverslip (DeBlase and Martin 1981). Methods for preparing permanent or temporary slides of cuticular scale patterns have been described (Williamson 1951, Spence [Wyo. Game and Fish Comm. P-R Proj. FW-3-R, 1963], Korschgen 1980, DeBlase and Martin 1981). These techniques rely on leaving a cast of the cuticular pattern on a soft, impressionable material such as heated plastic coverslips (DeBlase and Martin 1981), heated media that gel when cool (L. E. Spence, Wyo. Game and Fish Comm. P-R Proj. FW-3-R, 1963), a variety of resins (Korschgen 1980), or partially dried, clear fingernail polish. In addition to a reference collection of dorsal guard hairs, several regional and family-specific keys are available (Mathiak 1938, Nason 1948, Mayer 1952, Stains 1958, Day 1966, Moore et al. 1974). Broley (1950) and Day (1966) photographed some feathers, but a comprehensive guide to feathers of birds is currently unavailable. For invertebrate prey items, wet and dry reference specimens may be available at a local university or museum. Depending on the degree of digestion and initial mastication, invertebrate remains may not be identifiable without the use of a microscope. Reference and sample slides can be prepared as described by Hansson (1970), Krantz (1978), and DeBlase and Martin (1981).

Herbivores

Samples may be collected from various stages of digestion for use in identifying food habits of mammalian and avian herbivores. A common technique used is fecal pellet analysis, because it is nondestructive and large samples can be collected easily. This technique has been applied with equal success to waterfowl (Owen 1975) and upland birds (Eastman and Jenkins 1970), as well as to small and large-sized mammals. Field guides to fecal materials are available (Webb 1943, Murie 1974), although specialized techniques such as pH analysis may be necessary to distinguish feces of some sympatric herbivores (Howard 1967).

Contents of alimentary tracts generally are collected only from wild animals with large populations because they usually involve sacrificing the animal. Exceptions to this are esophageal and rumen fistulated animals. Fistulating an animal involves installing a permanent device in the digestive tract of living animals that allows samples to be taken of food items passing that point in the digestive process (Torell 1954, Short 1962, McManus 1981). Fistulation has been used extensively in defining the diets of domestic animals (Vavra et al. 1978) but rarely has been applied to wild ruminants (Rice 1970). Taming or hand-rearing fistulated animals is the only way a researcher can approach them to collect samples. But, as with using domesticated ''wild'' herbivores for bite-count data collection, the time and money required are generally prohibitive, and tame animals may not reflect the true food habits of their wild counterparts. Emetics, flushing tubes, and manual expression of the gullet also have been used, primarily on birds, to purge the upper portion of the digestive tract without harming the individual (Errington 1932, Vogtman 1945). Excellent reviews of the advantages and disadvantages of the various methods for determining the food habits of herbivores can be found in Medin (1970), Van Dyne et al. (1980), Holechek et al. (1982*b*), and McInnis et al. (1983).

Although some partially digested plant parts, such as seeds and fruits, can be identified macroscopically, most of the analysis of herbivorous materials relies on microhistological techniques to identify characteristic cells and structures of foods consumed. This technique is applicable to materials gathered anywhere in the digestive process. Numerous authors described techniques for preparing samples and reference slides for microhistological identification (Baumgartner and Martin 1939, Dusi 1949, Sparks and Malechek 1968, Hansson 1970, Voth and Black 1973, Meserve 1976, DeBlase and Martin 1981). Box 2 outlines the technique as summarized by Sparks and Malechek (1968) and Hansen et al. (1976). As with carnivores, a reference collection of potential food items is crucial. Microscope slides should be prepared from all potential food plants in the same manner as sample materials. In addition, a collection of local seeds and fruits also should be made for reference.

Generally, less than one-half of what appears on a typical slide will be identifiable plant fragments. The cellular characteristics used to identify plant fragments are those that survive the mastication and digestive process and generally are composed of epidermal tissue (Storr 1961). These include cuticle, stomata, cell walls, aperites, glands, trichomes, silica cells, druses, crystals, starch grains, and silica-suberose couples as well as general cellular configurations, size, and other structural characteristics. The relative proportions of plant species in each sample can be quantified by examining 20 microscope fields at 125× magnification for each of the five slides created for each diet sample (Holechek and Vavra 1981). The *percent density* of each food item can be estimated by dividing the number of plant fragments of a species by the total number of identifiable plant fragments in the sample. To avoid the tedious and time-consuming process of counting all identifiable fragments in a microscope field, Sparks and Malechek (1968) developed a conversion technique from frequency of occurrence to mean density of fragments per field. *Frequency of occurrence* can be calculated for each species by reporting the number of microscope fields that contained evidence of the species in 100 fields examined per sample. Percent occurrence can then be converted into density of identifiable fragments per microscope field either by a table provided by Fracker and Brischle (1944) or by the formula $F = 100 (1 - e_{-D})$, where F is frequency of occurrence and D is the average density of fragments per field examined (Fracker and Brischle 1944). The average density of fragments per field per species (F) then can be converted to relative density (RD) by dividing the density of discerned fragments for a species (F) by the sum of densities for all observed species (ΣF) and multiplying by 100. A simpler, alternate method for converting frequency to density was developed by Holechek and Gross (1982). The number of frequency observations of each species is divided by the total number of frequency observations for all species. This value is then multiplied by 100 to give the relative percentage by weight that each species represents in the diet. Sparks and Malechek (1968) demonstrated that for at least some grasses and forbs sampled at the esophagus, rumen, and stomach, relative density accurately reflected the dry biomass percentages consumed by individuals. However, Curtis and McIntosh (1950) cautioned that before frequency of occurrence can be converted to density, plant fragments must be distributed randomly on the slides, and the most common species should not occur in more than 86% of the microscope fields examined. Because the ratio of identifiable to nonidentifiable fragments changes during digestion and sample preparation (Havstad and Donart 1978, Holechek 1982), and because certain browse species have a low proportion of epidermal material in relation to their biomass (Westoby et al. 1976), correction factors may be developed to improve the approximation of diet composition (Dearden et al. 1975). Several researchers have recommended that hand compounded diets be used to test the assumption that the actual diet matches the diet estimated from microhistological analysis (Westoby et al. 1976, Vavra and Holechek 1980, Holechek et al. 1982*a*). Others have suggested that the differences are

---

Box 2. Preparing sample and reference material for microhistological investigations of herbivore diets.

Washing.—Samples (esophageal, rumen, stomach, fecal) should be washed initially in fine mesh nylon bags in a washing machine, or soaked and washed over an 18-mesh screen to remove solubles and small, unidentifiable plant fragments.

Homogenizing.—Microscope slide preparation requires that the materials be ground or macerated to approximately the same size fragments (≥18 mesh). This permits preparation of thinner slides, yields uniformity among reference samples and digested residues of unknown samples, and allows quantification of the various plant species occurring in an animal's diet regardless of differences in the degree of maceration exhibited by the digestive process in different herbivores. To accomplish this, dry the samples and grind them in a Wiley mill with an 18-mesh screen.

Known botanical materials for preparing reference slides can be freshly collected, dried, or preserved in another form. They should be dried and ground as described for the samples above.

Clearing.—Ground samples can be cleared by use of either household bleach or Hertwig's clearing solution (Baumgartner and Martin 1949). Bleach often will clear a sample but destroy identifying characteristics of lichens. After being cleared in bleach, samples should be thoroughly rinsed with hot water under a 200-mesh screen. Enough sample material should be placed on a microscope slide to provide an average of three identifiable plant fragments per microscope field viewed at 100×.

Hertwig's clearing solution should be used when bleach does not adequately clear samples. The solution is comprised of 270 g chloral hydrate crystals combined with 19 ml of 1 normal HCl, 150 ml distilled water, and 60 ml glycerine. Add two drops of the solution to the rinsed sample on the slide and heat over an alcohol lamp or hot plate until the solution evaporates, but does not burn.

Staining.—Although staining is not necessary for identification of most materials, it can be accomplished by various techniques described by Dusi (1949), Williams (1962), and Hansson (1970).

Mounting.—Samples are mounted in Hoyer's mounting medium (Baker and Wharton 1952) or any commercially available water-soluble mounting medium. Hoyer's medium can be made by combining, in sequence, 50 ml distilled water, 30 g photopurified gum arabic, 200 g chloral hydrate, and 20 ml glycerine. After adding one or two drops of the mounting medium to the sample, place a coverslip over the sample and heat until the sample is bubbling evenly. Cool the slide on a wet sponge until the large air bubbles disappear from under the coverslip and dry the slide in an oven at 45–60 C for 24–48 hours. If the slide is to be a permanent reference, the edges of the coverslip should be sealed with additional Hoyer's medium or another waterproof sealant. Generally, five slide mounts are prepared for each diet sample.

---

either too small to justify developing correction factors, or that correction factors do not consistently improve the estimation of diet composition, particularly when the diets contain a diversity of grasses, forbs, and browse (Hansson 1970, Gill et al. 1983).

Fecal analysis has been criticized for identifying fewer species than can be found in rumen samples. Generally, easily digested forbs are underestimated, and the less digestible items are overestimated (Anthony and Smith 1974, Vavra et al. 1978, Smith and Shandruk 1979, McInnis et al. 1983). Inaccuracy of the technique, particularly when applied to diets of "mixed" (grasses, forbs, and browse combined) feeders, has led some researchers to question its usefulness for animals other than grazers (Gill et al. 1983). Experience and training of the technician in identifying plant fragments often are cited as the most important sources of error in using the microhistological technique (Holechek et al. 1982a). Johnson and Pearson (1981) stressed that microscopic analysis of botanical compositions "is as much an art as a science." Several reference articles and books are available for identifying plant fragments (Howard and Samuel 1979), but becoming proficient with the technique requires much time and effort.

Crop contents of granivores are a much different sample because only limited digestion has occurred. Often the investigator can separate and identify most of the sample.

The volume of each seed type also can be estimated with a graduated cylinder or by the displacement of a known quantity of water in a burette (Inglis and Barstow 1960). Seeds and fruits can be identified by comparison to a local reference collection or use of reference books (Musil 1963).

Techniques To Measure Food Availability

Wildlife food abundance can be estimated for an area by measuring the annual production of herbaceous plants, woody stems, fruits, and seeds or by assessing the size and distribution of vertebrate and invertebrate populations. However, estimates of *abundance* may or may not have a direct relationship to food *availability*. Availability connotes that a resource is accessible and usable (Morrison et al. 1992). Access to food resources can vary substantially with such factors as weather, the presence of competitors or predators, and prey behavior. Among herbivores, snow accumulation can influence food availability by burying some plants and elevating animals to other plants or stems that were previously out of reach (Keith et al. 1984). Snow also can influence the availability (or vulnerability) of mammalian prey to carnivores (Halpin and Bissonette 1988, Fuller 1991). Likewise, some species may be abundant but not easily captured by a predator. This dynamic nature of food availability may require investigators to stratify a study into spatial or temporal units that are relatively similar with respect to access of food resources. Because the factors that influence access to food are probably unique to each study, we will not attempt to make recommendations on them and will restrict our discussion to methods of estimating forage abundance. Chapter 9 describes techniques to estimate animal (prey) populations.

GRASSES AND FORBS

Clipping and weighing dried samples of aboveground vegetation is the most accurate, but most time-consuming, technique for determining availability of herbaceous plants. More rapid techniques have been developed that use observers to estimate vegetative biomass from small quadrats (Pechanec and Pickford 1937, Shoop and McIlvain 1963). Robel et al. (1970) developed a technique that used visual obstruction of a 1-m-high pole to predict biomass of grassland vegetation. Range scientists also have developed two simple techniques to monitor and inventory rangelands that rely on ground coverage to estimate relative abundance of herbaceous plants. One method involves estimating coverage by species in small sample plots (Stewart and Hutchings 1936, Daubenmire 1958). To aid observers using this technique, Daubenmire (1958) suggested using classes of coverage (0–5%, 5–25%, 25–50%, 50–75%, 75–95%, 95–100%) for each species encountered. The second method (point step) involves sampling ground cover at specific points to estimate coverage by each plant species (Evans and Love 1957, Owensby 1973). Basically, this technique involves notching or marking a spot on the edge of an observer's boot. The observer then walks through an area and, at regular intervals, records the species of plant that is beneath the notch on the boot. This technique is particularly useful for determining coverage and species composition in large study areas. Thorough reviews of these techniques and others can be found in Cook and Stubbendieck (1986), and in Chapter 22 of this book.

BROWSE

Like herbaceous plants, annual browse production is most accurately measured by clipping, drying, and weighing twigs produced during the growing season (Harlow 1977). This technique, however, is too time consuming to be of practical value. As a result, fairly accurate predictive equations have been developed that relate measures of shrub size to forage production (Lyon 1968, Bobek and Bergstrom 1978). This approach does, however, require that unique predictive equations be developed for each species, often for each study site.

The twig-count method estimates available biomass of browse by determining the average weight of edible material in a single twig and multiplying that by the number of twigs available (Shafer 1963). A random sample of 100 previously browsed twigs is used to determine the average browsing diameter for a particular forage species. Average mass of a browsed twig is then calculated by determining the mass of 50 twigs that are clipped to the size of the average browsed twig. Densities of twigs are then estimated from counts on circular plots (Shafer 1963) or belt transects (Irwin and Peek 1979), and available browse biomass is calculated per unit area. Modifications of this technique include development of equations that use unbrowsed twig length or basal diameter to estimate twig mass (Basile and Hutchings 1966, Telfer 1969).

FRUITS AND SEEDS

Annual production of fruit and seeds can be extrapolated from complete counts, collections of production in sample areas, or use of seed traps. Fruits and nuts from individual low-growing herbs and shrubs can be counted and averaged per plant and then combined with density information on plants to estimate fruit or seed production per unit area. Similarly, hard mast crops, such as acorns, can be collected in funnel-type seed traps that sample a fixed area under the canopy (Gysel 1956). Although the traps usually prevent animals from taking mast once it has fallen from the tree, information on production can be biased if seeds are consumed before falling to the ground or if collected seeds have little value to wildlife due to insect damage (Gysel and Lyon 1980).

Considerations for Study Design, Data Analysis, and Interpretation of Results

Regardless of the technique used to determine food habits, the concern for sample size and sampling period is crucial. Sample sizes required for food-habits studies will depend on season, intraspecific variability in food habits, and the breadth of a species' diet. Although several studies specifically address the issue of sample size for various species (Davison 1940, Korschgen 1948, Anthony and Smith 1974, Hansen et al. 1976, Holechek and Vavra 1983), circumstances of a particular study are so variable that no generally agreed upon minimum sample sizes exist. Hanson and Graybill (1956) offered statistical guidelines for determining adequate sample sizes based on the variance in the most important prey item. Korschgen (1980) reviewed several techniques for determining sample size and generally concluded that a sample is large enough when additional samples offer no new or different information.

The time frame in which samples are collected also is crucial and depends on seasonal variation within the diet of the specific species. For example, feces collected in Texas during July–August indicated coyotes ate fruits exclusively, whereas scats collected during January reflected a diet of more than 90% mammalian prey (Andelt et al. 1987). The availability of different foods may change seasonally and may not be consistent from 1 year to the next because of fluctuations in prey populations, insect outbreaks, changes in mast and fruit production, changes in the palatability of species, or differential vulnerability of prey (e.g., Mattson et al. 1991). It is, therefore, often more meaningful to base food-habits analyses on biologically significant time periods (e.g., gestation, nesting, breeding) and not strictly on arbitrary chronologies.

Several techniques are used to quantify food habits of carnivores. Results of fecal and pellet analyses are generally presented as percent occurrence. Only the presence or absence of a species in a sample is recorded, regardless of the number of individuals of a given prey type that appear in each sample. Although this is a widely used technique to represent food habits of predators, it does not necessarily reflect the proportion by volume, weight, or energy that different food items contribute to the diet. Remains of a single mouse in a sample receive the same relative importance as remains of a deer. The problem is further compounded by differential digestibility of prey items. Some prey items are more disintegrated by the digestive process than others, hence the absolute proportions of residue in feces or pellets may not be representative of proportions actually ingested (Lockie 1959, Floyd et al. 1978, Dickman and Huang 1988). Attempts have been made to determine biomass of the prey items consumed based on fecal remains (Lockie 1959, Floyd et al. 1978). Correction factors have been developed for converting the weight of identifiable remains in scats to actual weights of foods eaten. The technique has not been used widely, however, despite the high correlation between weights of prey consumed and weights of identifiable remains (Floyd et al. 1978). The primary difficulty is identifying the portion of a scat composed of each prey type in scats that contain the remains of more than one prey species.

Results from stomach or crop samples often are reported as percent volume or, less frequently, percent dry mass. Percent volume is determined by adding wet prey items to a known volume of water, usually in a large graduated cylinder, and recording the increase in volume. Prey items also can be oven-dried and weighed to calculate a percent dry mass, although this is rarely done. Results of either technique can be aggregated by prey item across all stomachs/crops and reported or calculated as a percentage of the contents of each sample and averaged across all samples (Table 4). Aggregate percentage is calculated by summing the proportion of sample volume a particular food item represents in each animal and dividing it by the total number of animals in the sample. Aggregate volume is determined by summing the volume of a particular food item in all the animals and dividing it by the volume of all foods in all the animals. The aggregate volume method gives importance to the absolute volume of food consumed by *all* animals, whereas aggregate percentage gives equal importance to the percent composition of food items in *each* animal (Swanson and Bartonek 1970).

Once use and availability are sampled, several mathematical and statistical approaches can be used for estimating selection of food items (Cock 1978). Many of the methods used to analyze habitat selection can be applied to measuring food selection. Krebs (1989) reviewed some of the more common methods, including the forage ratio (Williams and Marshall 1938), Ivlev's (1961) electivity index, Murdoch's (1969) index, Manly's alpha (Manly et al. 1972), and Johnson's (1980) rank preference index. He recommended Manly's alpha and the rank preference index as the best and most easily understood measures of selection for most situations. Hobbs and Bowden (1982) also provided a method to generate confidence intervals on selection indices.

INDICES OF OVERLAP IN RESOURCE USE

Data collected from habitat or food-use studies of sympatric species often are used to evaluate niche overlap and then provide a foundation to speculate on the prevalence of interspecific competition (Reynolds and Meslow 1984, Thill and Martin 1986, Major and Sherburne 1987). The methods used to index niche overlap often rely on a comparison of similarity in resource-use patterns between two species. An index value is generated as an indicator of

Table 4. Sample results of crop analysis as calculated by aggregate percentage and aggregate volume.

	Volume (cc)			
Food item	Bird 1	Bird 2	Bird 3	Total
Corn	1.9	5.3	0.3	7.5
Soybeans	2.1	4.7	1.8	8.6
Weed seeds	1.2	1.1	0.6	2.9
Total	5.2	11.1	2.7	19.0

Aggregate percentage:

$$\frac{\Sigma P_i}{n} \quad P_i = \text{proportion food item i is of each crop}$$
$$n = \text{total number of crops}$$

Corn

$$\Sigma P_i = \frac{1.9}{5.2} + \frac{5.3}{11.1} + \frac{0.3}{2.7} = 0.954$$
$$\frac{\Sigma P_i}{n} = \frac{0.954}{3} = 0.318$$

Soybeans

$$\Sigma P_i = \frac{2.1}{5.2} + \frac{4.7}{11.1} + \frac{1.8}{2.7} = 1.494$$
$$\frac{\Sigma P_i}{n} = \frac{1.494}{3} = 0.498$$

Weed seeds

$$\Sigma P_i = \frac{1.2}{5.2} + \frac{1.1}{11.1} + \frac{0.6}{2.7} = 0.552$$
$$\frac{\Sigma P_i}{n} = \frac{0.552}{3} = 0.184$$

Aggregate volume:

$$\frac{\Sigma V_i}{V} \quad V_i = \text{volume of food item i in each crop}$$
$$V = \text{total volume of all food items in all crops}$$

Corn

$$\frac{\Sigma V_i}{V} = \frac{7.5}{19.0} = 0.395$$

Soybeans

$$\frac{\Sigma V_i}{V} = \frac{8.6}{19.0} = 0.453$$

Weed seeds

$$\frac{\Sigma V_i}{V} = \frac{2.9}{19.0} = 0.153$$

similarity, often ranging from 0 to 1, where 0 = no overlap and 1 = complete overlap in resource use. Brower and Zar (1984) and Krebs (1989) provided computer programs that can calculate several of these indices. Numerous authors have addressed the biases and assumptions of the different measures of niche overlap (Hurlbert 1978, Abrams 1980, Ricklefs and Lau 1980, Smith and Zaret 1982, Krebs 1989). There is no general agreement on which index is most appropriate; each measure has its advantages and disadvantages (Table 5). The bias associated with all indices increases as the number of resources increases, but decreases as the sample size in-

Table 5. Formulae and evaluation of eight measures of dietary overlap (modified from Krebs 1989).

Index (reference)	Formula[a]	Comments
Percent overlap (Schoener 1970)	$P_{jk} = \left[\sum (\text{minimum of } P_{ij}, P_{ik}) \right] 100$	Simple to calculate and interpret Underestimates true overlap Bias increases as the number of food items increases or sample size decreases
Spearman's Rank Correlation	$r_s = 1 - \dfrac{6 \sum d_i^2}{n^3 - n}$	
Pianka (Pianka 1974)	$O_{jk} = \dfrac{\sum P_{ij} P_{ik}}{\sqrt{\sum P_{ij}^2 \sum P_{ik}^2}}$	0 = no shared food items use, 1 = complete overlap Similar results as simplified by Morisita but slightly less precise
Morisita (Morisita 1959)	$C = \dfrac{2 \sum P_{ij} P_{ik}}{\sum\limits^{n} P_{ij}[(n_{ij} - 1)/(N_j - 1)] + \sum\limits^{n} P_{ik}[(n_{ik} - 1)/(N_k - 1)]}$	Requires counts of individuals consumed, not biomass or proportions
Simplified Morisita (Horn 1966)	$C_H = \dfrac{2 \sum P_{ij} P_{ik}}{\sum P_{ij}^2 + \sum P_{ik}^2}$	
Horn (Horn 1966)	$R_o = \dfrac{\sum (P_{ij} + P_{ik})\log(P_{ij} + P_{ik}) - \sum P_{ij}\log P_{ij} - \sum P_{ik}\log P_{ik}}{2 \log 2}$	
Hurlbert (Hurlbert 1978)	$L = \sum (P_{ij} P_{ik}/a_i)$	Value changes when totally unused food items are included 0 = no shared food items, 1 = both spp. use each food item in proportion to its availability, >1 = both spp. use certain food items more intensively than others and resource preference coincides

P_{ij} = proportion item i is of total foods used by species j; P_{ik} = proportion item i is of total foods used by species k; n = total number of food items; a_i = proportional availability of food item i; n_{ij} = number of individuals of food item i in samples of species j; n_{ik} = number of individuals of food item i in samples of species k; N_j = total number of individuals of each food item in sample of species j ($\sum n_{ij} = N_j$); N_k = total number of individuals of each food item in sample of species k ($\sum n_{ik} = N_k$); d = difference in rank of food item i between species j and k.

creases. To minimize bias, Smith and Zaret (1982) recommended using Morisita's (1959) measure. Unfortunately, that measure requires the number of individuals of each food item to be recorded so is primarily limited to crop/stomach studies of granivores. If the data are not in that form, Horn's (1966) index may offer the least biased alternative (Krebs 1989). Greene and Jaksić (1983) also pointed out the influence different taxonomic levels of prey identification can have on indices of food niche overlap. They demonstrated that analyses done on items identified only to order rather than species tend to simplify diets and, as a result, overestimate diet overlap. Analysis of mixed data (some identified to species level and some to order) will give similarly misleading results.

Historically, researchers have been tempted to equate overlap in resource use with competition. However, the relationship between niche overlap and competition is poorly understood. For example, does an overlap index of zero indicate that competition is absent, or that one species is consistently denied access to the array of resources used by a dominant species? Likewise, high indices of overlap may simply mean an abundance of the resources is available and is being used by both species. Interpretation of overlap indices should, therefore, be approached with caution (Holt 1987).

LITERATURE CITED

ABRAMS, P. 1980. Some comments on measuring niche overlap. Ecology 61:44–49.

ADORJAN, A. A., AND G. B. KOLENOSKY. 1969. A manual for the identification of hairs of selected Ontario mammals. Ont. Dep. Lands For. Res. Rep. (Wildl.) 90. 64pp.

ALDOUS, S. E. 1944. A deer browse survey method. J. Mammal. 25: 130–136.

ALLDREDGE, J. R., AND J. T. RATTI. 1986. Comparison of some statistical techniques for analysis of resource selection. J. Wildl. Manage. 50:157–165.

———, AND ———. 1992. Further comparison of some statistical techniques for analysis of resource selection. J. Wildl. Manage. 56: 1–9.

ANDELT, W. F., J. G. KIE, F. F. KNOWLTON, AND K. CARDWELL. 1987.

Variation in coyote diets associated with season and successional changes in vegetation. J. Wildl. Manage. 51:273–277.

ANTHONY, R. G., AND N. S. SMITH. 1974. Comparison of rumen and fecal analysis to describe deer diets. J. Wildl. Manage. 38:535–540.

AVERY, T. E. 1968. Interpretation of aerial photographs. Second ed. Burgess Publ., Minneapolis, Minn. 324pp.

———, AND H. E. BURKHART. 1983. Forest measurements. Third ed. McGraw-Hill, New York, N.Y. 331pp.

BAKER, E. W., AND G. W. WHARTON. 1952. An introduction to acarology. Macmillan Co., New York, N.Y. 465pp.

BASILE, J. V., AND S. S. HUTCHINGS. 1966. Twig diameter-length-weight relationships of bitterbrush. J. Range Manage. 19:34–38.

BAUMGARTNER, L. L., AND A. C. MARTIN. 1939. Plant histology as an aid in squirrel food-habit studies. J. Wildl. Manage. 3:266–268.

BAXTER, W. L., AND C. W. WOLFE. 1972. The interspersion index as a technique for evaluation of bobwhite quail habitat. Proc. Natl. Bobwhite Quail Symp. 1:158–165.

BEASOM, S. L., E. P. WIGGERS, AND J. R. GIARDINO. 1983. A technique for assessing land surface ruggedness. J. Wildl. Manage. 47:1163–1166.

BIELEFELDT, J., R. N. ROSENFIELD, AND J. M. PAPP. 1992. Unfounded assumptions about the diet of the Coopers' hawk. Condor 94:427–436.

BIGGINS, D. E., AND E. J. PITCHER. 1978. Comparative efficiencies of telemetry and visual techniques for studying ungulates, grouse, and raptors on energy development lands in southeastern Montana. Pecora 4:188–193.

BOBEK, B., AND R. BERGSTROM. 1978. A rapid method of browse biomass estimation in a forest habitat. J. Range Manage. 31:456–458.

———, S. BOROWSKI, AND R. DZIECIOLOWSKI. 1975. Browse supply in various forest ecosystems. Pol. Ecol. Stud. 1:17–32.

BONHAM, C. D. 1989. Measurements of terrestrial vegetation. John Wiley & Sons, New York, N.Y. 338pp.

BROLEY, J. 1950. Identifying nests of the Anatidae of the Canadian prairies. J. Wildl. Manage. 14:452–456.

BROWER, J. E., AND J. H. ZAR. 1984. Field and laboratory methods for general ecology. W.C. Brown Publ., Dubuque, Ia. 226pp.

BYERS, C. R., R. K. STEINHORST, AND P. R. KRAUSMAN. 1984. Clarification of a technique for analysis of utilization-availability data. J. Wildl. Manage. 48:1050–1053.

CALDER, W. A. 1973. Microhabitat selection during nesting of hummingbirds in the Rocky Mountains. Ecology 54:127–134.

CHESSON, J. 1978. Measuring preference in selective predation. Ecology 59:211–215.

COCK, M. J. W. 1978. The assessment of preference. J. Anim. Ecol. 47:805–816.

COLLINS, W. B., AND P. J. URNESS. 1981. Habitat preferences of mule deer as rated by pellet-group distributions. J. Wildl. Manage. 45:969–972.

COOK, C. W., AND J. STUBBENDIECK. 1986. Methods of measuring herbage and browse utilization. Pages 120–121 in C. W. Cook and J. Stubbendieck, eds. Range research: basic problems and techniques. Soc. Range Manage., Denver, Colo.

COOPERRIDER, A. Y. 1986. Food habits. Pages 699–710 in A. Y. Cooperrider, R. J. Boyd, and H. R. Stuart, eds. Inventory and monitoring of wildlife habitat. U.S. Dep. Inter. Bur. Land Manage. Serv. Cent., Denver, Colo.

COTTAM, G., AND J. T. CURTIS. 1956. The use of distance measures in phytosociological sampling. Ecology 37:451–460.

CURTIS, J. T., AND R. P. MCINTOSH. 1950. The interrelations of certain analytic and synthetic phytosociological characters. Ecology 31:434–455.

DAUBENMIRE, R. F. 1958. A canopy-coverage method of vegetational analysis. Northwest Sci. 53:43–64.

DAVISON, V. E. 1940. A field method of analyzing game bird foods. J. Wildl. Manage. 4:105–116.

DAY, M. G. 1966. Identification of hair and feather remains in the gut and feces of stoats and weasels. J. Zool. Proc. 148:201–217.

DEARDEN, B. L., R. E. PEGAU, AND R. M. HANSEN. 1975. Precision of microhistological estimates of ruminant food habits. J. Wildl. Manage. 39:402–407.

DEBLASE, A. F., AND R. E. MARTIN. 1981. A manual of mammalogy with keys to families of the world. W.C. Brown Publ., Dubuque, Ia. 436pp.

DE VOS, A., AND H. S. MOSBY. 1969. Habitat analysis and evaluation. Pages 135–172 in R. H. Giles, Jr., ed. Wildlife management techniques. Third ed. The Wildl. Soc., Washington, D.C.

DIAMOND, J. 1986. Overview: laboratory experiments, field experiments, and natural experiments. Pages 3–22 in J. Diamond and T. J. Case, eds. Community ecology. Harper & Row Publ., New York, N.Y.

DICKMAN, C. R., AND C. HUANG. 1988. The reliability of fecal analysis as a method for determining the diet of insectivorous mammals. J. Mammal. 69:108–113.

DUESER, R. D., AND H. H. SHUGART. 1978. Microhabitats in a forest-floor small-mammal fauna. Ecology 59:89–98.

DUSI, J. L. 1949. Methods for the determination of food habits by plant microtechniques and histology and their application to cottontail rabbit food habits. J. Wildl. Manage. 13:295–298.

EASTMAN, D. S., AND D. JENKINS. 1970. Comparative food habits of red grouse in northeast Scotland, using fecal analysis. J. Wildl. Manage. 34:612–620.

EDLEFSEN, J. L., C. W. COOK, AND J. T. BLAKE. 1960. Nutrient content of the diet as determined by hand-plucked and esophageal samples. J. Anim. Sci. 19:560–563.

ERRINGTON, P. L. 1932. Techniques of raptor food habits study. Condor 34:75–86.

EVANS, R. A., AND R. M. LOVE. 1957. The step point method of sampling—a practical tool in range research. J. Range Manage. 10:208–212.

FAGEN, R. 1988. Population effects of habitat change: a quantitative assessment. J. Wildl. Manage. 52:41–46.

FLOYD, T. J., L. D. MECH, AND P. A. JORDAN. 1978. Relating wolf scat content to prey consumed. J. Wildl. Manage. 42:528–532.

FORMAN, R. T. T., AND M. GODRON. 1986. Landscape ecology. John Wiley & Sons, New York, N.Y. 619pp.

FRACKER, S. B., AND H. A. BRISCHLE. 1944. Measuring the local distribution of Ribes. Ecology 25:283–303.

FRETWELL, S. D. 1972. Populations in a seasonal environment. Princeton Univ. Press, Princeton, N.J. 217pp.

FRIDELL, R. A., AND J. A. LITVAITIS. 1991. Influence of resource distribution and abundance on home-range characteristics of southern flying squirrels. Can. J. Zool. 69:2589–2593.

FULLER, T. K. 1991. Effect of snow depth on wolf activity and prey selection in north central Minnesota. Can J. Zool. 69:283–287.

GAVIN, T. A. 1991. Why ask "why": the importance of evolutionary biology in wildlife science. J. Wildl. Manage. 55:760–766.

GILL, R. B., L. H. CARPENTER, R. M. BARTMANN, D. L. BAKER, AND G. G. SCHOONVELD. 1983. Fecal analysis to estimate mule deer diets. J. Wildl. Manage. 47:902–915.

GREEN, G. A., G. W. WITMER, AND D. S. DECALESTA. 1986. NaOH preparation of mammalian predator scats for dietary analysis. J. Mammal. 67:742.

GREEN, R. H. 1979. Sampling design and statistical methods for environmental biologists. John Wiley & Sons, New York, N.Y. 257pp.

GREENE, H. W., AND F. M. JAKSIĆ. 1983. Food-niche relationships among sympatric predators: effects of level of prey identification. Oikos 40:151–154.

GREIG-SMITH, P. 1964. Quantitative plant ecology. Butterworth, London, U.K. 256pp.

GYSEL, L. W. 1956. Measurement of acorn crops. For. Sci. 2:305–313.

———, AND L. J. LYON. 1980. Habitat analysis and evaluation. Pages 305–327 in S. D. Schemnitz, ed. Wildlife management techniques manual. Fourth ed. The Wildl. Soc., Washington, D.C.

HALPIN, M. A., AND J. A. BISSONETTE. 1988. Influence of snow depth on prey availability and habitat use by red fox. Can. J. Zool. 66:587–592.

HANSEN, R. M., T. M. FOPPE, M. B. GILBERT, R. C. CLARK, AND H. W. REYNOLDS. 1976. The microhistological analyses of feces as an estimator of herbivore diet. Range Sci. Composition Anal. Lab., Colorado State Univ., Ft. Collins. 6pp.

HANSON, W. R., AND F. GRAYBILL. 1956. Sample size in food-habits analyses. J. Wildl. Manage. 20:64–68.

HANSSON, L. 1970. Methods of morphological diet micro-analysis in rodents. Oikos 21:255–266.

HARLOW, R. F. 1977. A technique for surveying deer forage in the southeast. Wildl. Soc. Bull. 5:185–191.

HAVSTAD, K. M., AND G. B. DONART. 1978. The microhistological technique: testing two central assumptions in south central New Mexico. J. Range Manage. 31:469–470.

HAYS, R. L., C. SUMMERS, AND W. SEITZ. 1981. Estimating wildlife habitat variables. U.S. Fish Wildl. Serv. FWS/OBS-81-47. 111pp.

HEIM, S. J. 1988. Late winter and spring food habits of tame free-ranging white-tailed deer in southern New Hampshire. M.S. Thesis, Univ. New Hampshire, Durham. 51pp.

HEINEN, J., AND G. H. CROSS. 1983. An approach to measure interspersion, juxtaposition, and spatial diversity from cover-type maps. Wildl. Soc. Bull. 11:232–237.

HILDEN, 0. 1965. Habitat selection in birds: a review. Ann. Zool. Fenn. 2:53–75.

HOBBS, N. T., AND D. C. BOWDEN. 1982. Confidence intervals for food preference indices. J. Wildl. Manage. 46:505–507.

———, AND T. A. HANLEY. 1990. Habitat evaluation: do use/availability data reflect carrying capacity? J. Wildl. Manage. 54: 515–522.

HOLECHEK, J. L. 1982. Sample preparation techniques for microhistological analysis. J. Range Manage. 35:267–268.

———, AND B. D. GROSS. 1982. Evaluation of different calculation procedures for microhistological analysis. J. Range Manage. 35: 721–723.

———, AND M. VAVRA. 1981. The effect of slide and frequency observation numbers on the precision of microhistological analysis. J. Range Manage. 34:337–338.

———, AND ———. 1983. Fistula sample numbers required to determine cattle diets on forest and grassland ranges. J. Range Manage. 36:323–326.

———, ———, S. MADY DABO, AND T. STEPHENSON. 1982a. Effects of sample preparation, growth stage, and observer on microhistological analysis of herbivore diets. J. Wildl. Manage. 46:502–505.

———, ———, AND R. D. PIEPER. 1982b. Botanical composition determination of range herbivore diets: a review. J. Range Manage. 35:309–315.

HOLT, R. D. 1987. On the relation between niche overlap and competition: the effect of incommensurable niche dimensions. Oikos 48:110–114.

HORN, H. S. 1966. Measurement of "overlap" in comparative ecological studies. Am. Nat. 100:419–424.

HOWARD, G. S., AND M. J. SAMUEL. 1979. Atlas of epidermal plant species fragments ingested by grazing animals. U.S. Dep. Agric. Tech. Bull. 1582. 143pp.

HOWARD, V. W., JR. 1967. Identifying fecal groups by pH analysis. J. Wildl. Manage. 31:190–191.

HUNTER, M. L., JR. 1989. Aardvarks and Arcadia: two principles of wildlife research. Wildl. Soc. Bull. 17:350–351.

HURLBERT, S. H. 1978. The measurement of niche overlap and some relatives. Ecology 59:67–77.

———. 1984. Pseudoreplication and the design of ecological field experiments. Ecol. Mongr. 54:187–211.

INGLIS, J. M., AND C. J. BARSTOW. 1960. A device for measuring the volume of seeds. J. Wildl. Manage. 24:221–222.

IRWIN, L. L., AND J. M. PEEK. 1979. Shrub production and biomass trends following five logging treatments in the cedar-hemlock zone of northern Idaho. For. Sci. 25:415–426.

IVLEV, V. S. 1961. Experimental ecology of the feeding of fishes. Yale Univ. Press, New Haven, Conn. 302pp.

JAENIKE, J. 1980. A relativistic measure of variation in preference. Ecology 61:990–991.

JAMES, F. C., AND H. H. SHUGART, JR. 1970. A quantitative method of habitat description. Audubon Field Notes 24:727–736.

JENSEN, C. H., AND G. W. SCOTTER. 1977. A comparison of twig-length and browsed-twig methods of determining browse utilization. J. Range Manage. 30:64–67.

JOHNSON, D. H. 1980. The comparison of usage and availability measurements for evaluating resource preference. Ecology 61:65–71.

JOHNSON, M. K., AND H. A. PEARSON. 1981. Esophageal, fecal, and exclosure estimates of cattle diets on a longleaf pine-bluestem range. J. Range Manage. 34:232–235.

KAMIL, A. C., J. R. KREBS, AND H. R. PULLIAM, EDITORS. 1987. Foraging behavior. Plenum Press, New York, N.Y. 686pp.

KARR, J. R. 1981. Rationale and techniques for sampling avian habitats: introduction. Pages 26–28 in D. E. Capen, ed. The use of multivariate statistics in studies of wildlife habitat. U.S. For. Serv. Gen Tech. Rep. RM-87.

———. 1983. Commentary. Pages 403–410 in A. H. Brush and G. A. Clark, Jr., eds. Perspectives in ornithology. Cambridge Univ. Press, Cambridge, U.K.

KEITH, L. B., J. R. CARY, O. J. RONGSTAD, AND M. C. BRITTINGHAM. 1984. Demography and ecology of a declining snowshoe hare population. Wildl. Monogr. 90. 43pp.

KEPPIE, D. M. 1990. To improve graduate student research in wildlife education. Wildl. Soc. Bull. 18:453–458.

KORSCHGEN, L. J. 1948. Late-fall and early-winter food habits of bobwhite quail in Missouri. J. Wildl. Manage. 12:46–57.

———. 1980. Procedures for food-habits analyses. Pages 113–127 in S. D. Schemnitz, ed. Wildlife management techniques manual. Fourth ed. The Wildl. Soc., Washington, D.C.

KRANTZ, G. W. 1978. Collection, rearing, and preparation for study. Pages 77–98 in G.W. Krantz, ed. A manual of acarology. Oregon State Univ., Corvallis.

KREBS, C. J. 1989. Ecological methodology. Harper & Row, New York, N.Y. 654pp.

LEOPOLD, A. 1933. Game management. Charles Scribner's Sons, New York, N.Y. 481pp.

LITVAITIS, J. A., J. A. SHERBURNE, AND J. A. BISSONETTE. 1985. A comparison of methods used to examine snowshoe hare habitat use. J. Wildl. Manage. 49:693–695.

———, ———, AND ———. 1986. Bobcat habitat use and home range size in relation to prey density. J. Wildl. Manage. 50:110–117.

LOCKIE, J. D. 1959. The estimation of the food of foxes. J. Wildl. Manage. 23:224–227.

LYON, L. J. 1968. Estimating twig production of serviceberry from crown volumes. J. Wildl. Manage. 32:115–119.

MACNAB, J. 1983. Wildlife management as scientific experimentation. Wildl. Soc. Bull. 11:397–401.

MAJOR, J. T., AND J. A. SHERBURNE. 1987. Interspecific relationships of coyotes, bobcats, and red foxes in western Maine. J. Wildl. Manage. 51:606–616.

MAJOR, M., M. K. JOHNSON, W. S. DAVIS, AND T. F. KELLOGG. 1980. Identifying scats by recovery of bile acids. J. Wildl. Manage. 44: 290–293.

MANLY, B. F. J., P. MILLER, AND L. M. COOK. 1972. Analysis of a selective predation experiment. Am. Nat. 106:719–736.

MARCUM, C. L., AND D. O. LOFTSGAARDEN. 1980. A nonmapping technique for studying habitat preferences. J. Wildl. Manage. 44: 963–968.

MARTI, C. D. 1987. Raptor food habits studies. Pages 67–80 in B. A. Giron Pendleton, B. A. Milsap, K. W. Cline, and D. M. Bird, eds. Raptor management techniques manual. Natl. Wildl. Fed., Washington, D.C.

MARTIN, A. C., H. S. ZIM, AND A. L. NELSON. 1961. American wildlife & plants: a guide to wildlife food habits. Dover Publ., New York, N.Y. 500pp.

MARTIN, S. C. 1970. Relating vegetation measures to forage consumed by animals. Pages 93–100 in Range and wildlife habitat evaluation—a research symposium. U.S. For. Serv. Misc. Pub. 1147.

MATHIAK, H. A. 1938. A key to hairs of the mammals of southern Michigan. J. Wildl. Manage. 2:251–268.

MATTSON, D. J., B. M. BLANCHARD, AND R. R. KNIGHT. 1991. Food habits of Yellowstone grizzly bears, 1977-1987. Can. J. Zool. 69: 1619–1629.

MAYER, W. V. 1952. The hair of California mammals with keys to the dorsal guard hairs of California mammals. Am. Midl. Nat. 38:480–512.

McINNIS, M. L., M. VARVA, AND W. C. KRUEGER. 1983. A comparison of four methods used to determine the diets of large herbivores. J. Range Manage. 36:302–307.

McMANUS, W. R. 1981. Oesophageal fistulation technique as an aid to diet evaluation of the grazing ruminant. Pages 249–260 in J. L. Wheeler and R. D. Mochrie, eds. Forage evaluation: concepts and techniques. Am. Forage Grassland Counc., Lexington, Ky.

MEAD, R. A., T. L. SHARIK, S. P. PRISELY, AND J. T. HEINEN. 1981. A computerized spatial analysis system for assessing wildlife habitat from vegetation maps. Can. J. Remote Sensing 7:34–40.

MECH, L. D. 1966. The wolves of Isle Royale. U.S. Natl. Park Serv. Fauna Ser. 7. 210pp.

MEDIN, D. E. 1970. Stomach content analyses: collections from wild herbivores and birds. Pages 133–145 in Range and wildlife habitat evaluation—a research symposium. U.S. For. Serv. Misc. Publ. 1147.

MERSMANN, T. J., D. A. BUEHLER, J. D. FRASER, AND J. K. D. SEEGAR. 1992. Assessing bias in studies of bald eagle food habits. J. Wildl. Manage. 56:73–78.

MESERVE, P. L. 1976. Food relationships of a rodent fauna in a California coastal sage scrub community. J. Mammal. 57:300–319.

MILLER, B. K., AND J. A. LITVAITIS. 1992. Habitat segregation by moose in a boreal forest ecotone. Acta Theriol. 37:41–50.

MOORE, T. D., L. E. SPENCE, C. E. DUGNOLLE, AND W. G. HEPWORTH. 1974. Identification of the dorsal guard hairs of some mammals of Wyoming. Wyo. Game and Fish Dep. Bull. 14. 177pp.

MORISITA, M. 1959. Measuring of interspecific association and similarity between communities. Mem. Fac. Sci. Kyushu Univ. Ser. E (Biol.) 3:65–80.

MORRIS, D. W. 1984. Patterns and scale of habitat use in two temperate-zone, small mammal faunas. Can. J. Zool. 62:1540–1547.

MORRISON, M. L., B. G. MARCOT, AND R. W. MANNAN. 1992. Wildlife-habitat relationships. Univ. Wisconsin Press, Madison. 343pp.

MOSBY, H. S. 1969. Reconnaissance mapping and map use. Pages 119–134 in R. H. Giles, Jr., ed. Wildlife management techniques. Third ed. The Wildl. Soc., Washington, D.C.

MUELLER-DOMBOIS, D., AND H. ELLENBERG. 1974. Aims and methods of vegetation ecology. John Wiley & Sons, New York, N.Y. 547pp.

MURDOCH, W. W. 1969. Switching in general predators: experiments on predator specificity and stability of prey populations. Ecol. Monogr. 39:335–354.

MURIE, O. J. 1974. Animal tracks. Houghton Mifflin, Boston, Mass. 375pp.

MUSIL, A. F. 1963. Identification of crop and weed seeds. U.S. Dep. Agric. Handb. 219. 171pp.

NAMS, V. O. 1989. Effects of radiotelemetry error on sample size and bias when testing for habitat selection. Can. J. Zool. 67:1631–1636.

NASON, E. S. 1948. Morphology of hair of eastern North American bats. Am. Midl. Nat. 39:345–361.

NELSON, E. W. 1930. Methods of studying shrubby plants in relation to grazing. Ecology 11:764–769.

NEU, C. W., C. R. BYERS, AND J. M. PEEK. 1974. A technique for analysis of utilization-availability data. J. Wildl. Manage. 38:541–545.

NICHOLLS, T. H., AND M. R. FULLER. 1987. Territorial aspects of barred owl home range and behavior in Minnesota. Pages 121–128 in R. W. Nero, R. J Clark, R. J. Knapton, and R. H. Hamre, eds. Biology and conservation of northern forest owls. U.S. For. Serv. Gen. Tech. Rep. RM-142.

NOON, B. R. 1981. Techniques for sampling avian habitats. Pages 42–52 in D. E. Capen, ed. The use of multivariate statistics in studies of wildlife habitat. U.S. For. Serv. Gen Tech. Rep. RM-87.

NUDDS, T. D. 1977. Quantifying the vegetative structure of wildlife cover. Wildl. Soc. Bull. 5:113–117.

ORR, C. D., AND D. G. DODDS. 1982. Snowshoe hare habitat preference in Nova Scotia spruce-fir forests. Wildl. Soc. Bull. 10:147–150.

OWEN, M. 1975. An assessment of fecal analysis technique in waterfowl feeding studies. J. Wildl. Manage. 39:271–279.

OWENSBY, C. E. 1973. Modified step-point system for botanical composition and basal cover estimates. J. Range Manage. 26:302–303.

PARREN, S. G., AND D. E. CAPEN. 1985. Local distribution and coexistence of two species of Peromyscus in Vermont. J. Mammal. 66:36–44.

PARTRIDGE, L. 1978. Habitat selection. Pages 351–376 in J. R. Krebs and N. B. Davies, eds. Behavioural ecology: an evolutionary approach. Sinaurer Assoc., Sunderland, Mass.

PASSMORE, R. C., AND R. L. HEPBURN. 1955. A method for appraisal of winter range of deer. Ont. Dep. Lands For. Res. Rep. 29. 7pp.

PATTON, D. R. 1975. A diversity index for quantifying habitat "edge." Wildl. Soc. Bull. 3:171–173.

PECHANEC, J. F., AND G. D. PICKFORD. 1937. A weight-estimate method for determination of range or pasture production. J. Am. Soc. Agron. 29:894–904.

PIANKA, E. R. 1974. Niche overlap and diffuse competition. Proc. Natl. Acad. Sci. USA 71:2141–2145.

PORTER, W. F., AND K. E. CHURCH. 1987. Effects of environmental pattern on habitat preference analysis. J. Wildl. Manage. 51:681–685.

REYNOLDS, R. T., AND E. C. MESLOW. 1984. Partitioning food and niche characteristics of coexisting Accipiter during breeding. Auk 101:761–779.

RICE, J., B. W. ANDERSON, AND R. D. OHMART. 1984. Comparison of the importance of different habitat attributes to avian community organization. J. Wildl. Manage. 48:895–911.

RICE, R. W. 1970. Stomach content analyses: a comparison of the rumen vs. esophageal techniques. Pages 127–132 in Range and wildlife habitat evaluation—a research symposium. U.S. For. Serv. Misc. Pub. 1147.

RICKLEFS, R. E., AND M. LAU. 1980. Bias and dispersion of overlap indices: results of some Monte Carlo simulations. Ecology 61:1019–1024.

ROBBINS, C. S., D. K. DAWSON, AND B. A. DOWELL. 1989. Habitat area requirements of breeding forest birds of the Middle Atlantic states. Wildl. Monogr. 103. 34pp.

ROBEL, R. J., J. N. BRIGGS, A. D. DAYTON, AND L. C. HULBERT. 1970. Relationships between visual obstruction measurements and weight of grassland vegetation. J. Range. Manage. 23:295–297.

ROSENZWEIG, M. L. 1981. A theory of habitat selection. Ecology 62:327–335.

ROTENBERRY, J. T., AND J. A. WIENS. 1980. Habitat structure, patchiness, and avian communities in North American steppe vegetation: a multivariate analysis. Ecology 61:1228–1250.

ROTH, R. R. 1976. Spatial heterogeneity and birds species diversity. Ecology 57:773–782.

SCHALLER, G. B. 1972. The Serengeti lion: a study of predator-prey relations. Univ. Chicago Press, Chicago, Ill. 480pp.

SCHOENER, T. W. 1970. Nonsynchronous spatial overlap of lizards in patchy habitats. Ecology 51:408–418.

SCHUELER, F. W. 1972. A new method of preparing owl pellets: boiling in NaOH. Bird Banding 43:142.

SHAFER, E. L. 1963. The twig-count method for measuring hardwood deer browse. J. Wildl. Manage. 27:428–437.

SHOOP, M. C., AND E. H. MCILVAIN. 1963. The micro-unit forage inventory unit. J. Range Manage. 16:172–179.

SHORT, H. L. 1962. The use of a rumen fistula in a white-tailed deer. J. Wildl. Manage. 26:341–342.

SHORT, N. M. 1982. The Landsat tutorial workbook. NASA Ref. Publ. 1078. Natl. Aeronaut. and Space Adm., Washington, D.C. 553pp.

SINCLAIR, A. R. E. 1991. Science and the practice of wildlife management. J. Wildl. Manage. 55:767–773.

SMITH, A. D., AND R. L. HUBBARD. 1954. Preference ratings for winter deer forages from northern Utah ranges based on browsing time and forage consumed. J. Range Manage. 7:262–265.

————, AND J. L. SHANDRUK. 1979. Comparison of fecal, rumen, and utilization methods for ascertaining pronghorn diets. J. Range Manage. 32:275–279.

————, AND P. J. URNESS. 1962. Analysis of the twig-length method of determining utilization of browse. Utah State Dep. Fish Game Publ. 69–9. 35pp.

SMITH, D. R., P. O. CURRIE, J. V. BASILE, AND N. C. FRISCHKNECHT. 1962. Methods of measuring forage utilization and differentiating use by different classes of animals. Pages 93–98 in Range research methods. U.S. For. Serv. Misc. Publ. 940.

SMITH, E. P., AND T. M. ZARET. 1982. Bias in estimating niche overlap. Ecology 63:1248–1253.

SNYDER, E. J., AND L. B. BEST. 1988. Dynamics of habitat use by small mammals in prairie communities. Am. Midl. Nat. 119:128–136.

SPARKS, D. R., AND J. C. MALECHEK. 1968. Estimating percentage dry weight in diets using a microscopic technique. J. Range Manage. 21:264–265.

STAINS, H. J. 1958. Field key to guard hair of middle western furbearers. J. Wildl. Manage. 22:95–97.

STEPHENS, D. W. 1990. Foraging theory: up, down, and sideways. Stud. Avian Biol. 13:444–454.

————, AND J. R. KREBS. 1986. Foraging theory. Princeton Univ. Press, Princeton, N.J. 247pp.

STEVENTON, J. D., AND J. T. MAJOR. 1982. Marten use of habitat in a commercially clear-cut forest. J. Wildl. Manage. 46:175–182.

STEWART, G., AND S. S. HUTCHINGS. 1936. The point-observation-plot (square-foot-density) method of vegetation survey. J. Am. Soc. Agron. 28:714–722.

STEWART-OATEN, A., W. W. MURDOCH, AND K. R. PARKER. 1986. Environmental impact assessment: "pseudoreplication" in time? Ecology 67:929–940.

STINNETT, D. P., AND D. A. KLEBENOW. 1986. Habitat use of irrigated lands by California quail in Nevada. J. Wildl. Manage. 50:368–372.

STORR, G. M. 1961. Microscopic analysis of faeces: a technique for ascertaining the diet of herbivorous mammals. Aust. J. Biol. 14:157–164.

SWANSON, G. A., AND J. C. BARTONEK. 1970. Bias associated with food analysis in gizzards of blue-winged teal. J. Wildl. Manage. 34:739–746.

SWIHART, R. K., AND N. A. SLADE. 1985. Testing for independence of observations in animal movements. Ecology 66:1176–1184.

TELFER, E. S. 1969. Twig weight-diameter relationships for browse species. J. Wildl. Manage. 33:917–921.

THILL, R. E., AND A. MARTIN, JR. 1986. Deer and cattle diet overlap on Louisiana pine-bluestem range. J. Wildl. Manage. 50:707–713.

THOMAS, D. L., AND E. J. TAYLOR. 1990. Study designs and tests for comparing resource use and availability. J. Wildl. Manage. 54:322–330.

THOMAS, J. W., C. MASER, AND J. E. RODEIK. 1979. Edges. Pages 48–59 in J. W. Thomas, ed. Wildlife habitat in managed forests—the Blue Mountains of Oregon and Washington. U.S. For. Serv. Agric. Handb. 533.

THOMPSON, I. D., I. J. DAVIDSON, S. O'DONNELL, AND F. BRAZEAU. 1989. Use of track transects to measure the relative occurrence of some boreal mammals in uncut and regenerating stands. Can. J. Zool. 67:1816–1823.

TORELL, D. T. 1954. An esophageal fistula for animal nutrition studies. J. Anim. Sci. 13:878–882.

VAN DYNE, G. M., N. R. BROCKINGTON, Z. SZOCS, J. DUEK, AND C. A. RIBIC. 1980. Large herbivore subsystem. Pages 269–537 in A. I. Breymeyer and G. M. Van Dyne, eds. Grasslands, systems analysis and management. Cambridge Univ. Press, Cambridge, Mass.

VAN HORNE, B. 1982. Niches of adult and juvenile deer mice (Peromyscus maniculatus) in seral stages of coniferous forest. Ecology 63:992–1003.

———. 1983. Density as a misleading indicator of habitat quality. J. Wildl. Manage. 47:893–901.

VAVRA, M., AND J. L. HOLECHEK. 1980. Factors influencing microhistological analyses of herbivore diets. J. Range. Manage. 33:371–374.

———, R. W. RICE, AND R. M. HANSEN. 1978. A comparison of esophageal fistula and fecal material to determine steer diets. J. Range Manage. 31:11–13.

VERME, L. J. 1968. An index of winter weather severity for northern deer. J. Wildl. Manage. 32:566–574.

VOGTMAN, D. B. 1945. Flushing tube for determining food of game birds. J. Wildl. Manage. 9:255–257.

VOTH, E. H., AND H. C. BLACK. 1973. A histologic technique for determining feeding habits of small herbivores. J. Wildl. Manage. 37:223–231.

WALLMO, O. C., R. B. GILL, L. H. CARPENTER, AND D. W. REICHERT. 1973. Accuracy of field estimates of deer food habits. J. Wildl. Manage. 37:556–562.

———, AND D. J. NEFF. 1970. Direct observation of tamed deer to measure their consumption of natural forage. Pages 105–109 in Range and wildlife habitat evaluation—a research symposium. U.S. For. Serv. Misc. Pub. 1147.

WAUTERS, L. A., AND A. A. DHONDT. 1988. The use of red squirrel (Sciurus vulgaris) dreys to estimate population density. J. Zool. (Lond.) 214:179–187.

WEBB, J. 1943. Identification of rodents and rabbits by their fecal pellets. Trans. Kansas Acad. Sci. 43:479–481.

WESTOBY, M., G. R. ROST, AND J. A. WEIS. 1976. Problems with estimating herbivore diets by microscopically identifying plant fragments from stomachs. J. Mammal. 57:167–172.

WHITE, G. C., AND R. A. GARROTT. 1990. Analysis of wildlife radio-tracking data. Academic Press, San Diego, Calif. 383pp.

WIENS, J. A. 1974. Habitat heterogeneity and avian community structure in North American grasslands. Am. Midl. Nat. 91:195–213.

———. 1976. Population responses to patchy environments. Ann. Rev. Ecol. Syst. 7:81–120.

———. 1981. Scale problems in avian censusing. Stud. Avian Biol. 6:513–521.

———. 1983. Avian community ecology: an iconclastic view. Pages 355–403 in A. H. Brush and G. A. Clark, Jr., eds. Prespectives in ornithology. Cambridge Univ. Press, Cambridge, U.K.

———. 1989. The ecology of bird communities. Vol. 2. Cambridge Univ. Press, New York, N.Y. 316pp.

WILLIAMS, C. S., AND W. H. MARSHALL. 1938. Duck nesting studies, Bear River Migratory Bird Refuge, Utah, 1937. J. Wildl. Manage. 2:29–48.

WILLIAMS, O. 1962. A technique for studying microtine food habits. J. Mammal. 43:365–368.

WILLIAMSON, V. H. H. 1951. Determination of hairs by impressions. J. Mammal. 32:80–84.

WITMER, G. W., AND D. S. DECALESTA. 1983. Habitat use by female Roosevelt elk in the Oregon coast range. J. Wildl. Manage. 47:933–939.

ZAR, J. H. 1974. Biostatistical analysis. Prentice-Hall, Inc., Englewood Cliffs, N.J. 620pp.

11

PHYSIOLOGICAL METHODS IN WILDLIFE RESEARCH

John D. Harder and Roy L. Kirkpatrick

INTRODUCTION

The central role of nutrition in recruitment and loss in animal populations is widely recognized. Emigration, immigration, and mortality often are influenced directly by the abundance and distribution of food. Perhaps most thoroughly documented is the relationship between nutritional status and reproductive rate and, consequently, natality. During the last 3 decades, many studies of domestic and wild species have revealed graded responses to variation in nutrition in a variety of reproductive characteristics including age at puberty, ovulation rate, and neonatal survival (Verme 1969, Kirkpatrick 1988, Bronson 1989). Consequently, indices of nutritional condition such as body weight, fat reserves, and measures of reproductive rate, e.g., clutch size, are included routinely in wildlife investigations. The resulting information not only provides a sound biological framework for management decisions but also contributes to basic comparative biology. The value of detailed laboratory studies on domestic species is enhanced when comparable data are available from related species studied in nature.

Wildlife biologists now have statistically valid, cost-effective methods for estimation of population size. Unfortunately, wide confidence intervals on these estimates, large sample sizes notwithstanding, can frustrate attempts to detect the effects of population or habitat management. Measures of nutritional condition and reproductive rate offer additional information that potentially can indicate changes in environmental quality and even predict changes in population size. In this way, physiological indices enhance both the reliability and value of population estimators.

The goal of this chapter is to review procedures for assessing the nutritional status and reproductive capacity of wildlife populations, particularly those of birds and mammals. We conclude with a section on physiological indicators of stress. A wide array of nutritional indices is evaluated, ranging from body weight and fat analyses to biochemical measurements in blood and urine. The events in the reproductive cycles of birds and mammals form an outline for consideration of the techniques and procedures currently used in the study of reproduction. Detailed instructions for some of the most commonly used methods are provided; for others the reader is referred to primary sources. Our primary intention is to increase awareness of and appreciation for a full spectrum of physiological techniques and their potential application to the study and conservation of wildlife populations.

INDICES OF NUTRITIONAL CONDITION

That wildlife population sizes are a function of habitat adequacy is commonly accepted in wildlife management. In many instances, habitats appear to be important primarily from the standpoint of nutrition. As wildlife management becomes more intense, good measures or indices of nutritional status will be needed to evaluate the adequacy of habitats to support a given number of wild animals.

The use of nutritional indices to assess the well-being of wildlife has increased substantially during the last 10–20 years. Owen and Cook (1977) defined nutritional condition as the ability of an animal to cope with its present and future needs. In extending this line of thought, Grubb (In Press) defined *nutrition* as the "rate of ingestion of assimilable energy and nutrients" and *nutritional condition* as the "state of body components controlled by nutrition and which, in turn, influence an animal's fitness." In reality, most indices of nutritional status are measures of fat or energy stores in the body, although body protein and calcium stores also have been assessed (Ankney and

MacInnes 1978). Fats or lipids are the primary mode of storing energy in vertebrates and play an important role in reproduction (particularly in egg laying and lactation), migration, hibernation, and thermoregulation. However, carrying large amounts of fat confers both advantages and disadvantages. Large fat depots may restrict locomotion and thus may be detrimental to both predator and prey. As a result, wild animals appear to walk a thin line between carrying enough fat to assure meeting energy needs and carrying so much that it is burdensome (Rogers 1987). Most, apparently, have adapted fat deposition and use on a seasonal basis so that deposits are laid down only to meet predictable future needs. Much of the early work on condition indices focused on the family Cervidae, but recently smaller mammals, small birds, and waterfowl have been studied as well. Indices presently in use vary considerably in form and accuracy. Some can be obtained from animals only after death (e.g., femur fat, kidney fat, or gizzard fat), whereas others can be obtained only from living animals or immediately after death, e.g., blood samples. Still others can be obtained from live or dead animals (antler measurements, body measurements, and weights). Desirable characteristics of a good index of nutritional status were described by Riney (1955) and LeResche et al. (1974). These can be summarized as follows:

(1) The index should be sensitive to slight changes in nutritional status.
(2) It should be specific in its indications, e.g., capable of indicating protein or energy reserves or mineral balance.
(3) It should involve collection of tissues or measurements easily obtained from live or dead animals by relatively unskilled personnel.
(4) It should measure condition of different age groups and sexes at different times of the year and be little affected by the stress of collection.
(5) It should be objective and reproducible.

Many of the indices of nutritional status focus on one aspect, digestible energy intake, and are based on measurements of the various fat depots of the body. Ideally, each of these reflects whole body fat that accumulates or is depleted in individuals over a period of time, perhaps a matter of weeks or months in large species. In this section, we first review methods of measuring whole body fat, then review indices based on specific depots. These are followed by reviews of biochemical measurements in blood, which are often used to monitor short-term physiological events, e.g., hormone dynamics, or to detect qualitative changes in nutritional status, primarily protein metabolism.

Mammals

Riney (1955) described the order of fat catabolism in mammals on a declining nutritional plane as follows: (1) subcutaneous fat over the rump and saddle disappears; (2) abdominal cavity fat is used; and (3) bone marrow fat stores decline. As fat stores are replenished, the opposite order is followed. Of course, deposition and use of various fat deposits are not so simple, and there is considerable overlap as various deposits are used simultaneously. For example, mobilization of bone marrow fat begins before abdominal fat is exhausted (Ransom 1965).

WHOLE BODY FAT

Whole body fat has been used as an index of nutritional status primarily in small mammals (Fleharty et al. 1973, Cengel et al. 1978). Differences among sexes, seasons, and areas have been demonstrated in small mammals and between various stages of migration in birds. Whole body fat also has been determined in at least three studies of deer and pronghorns (Finger et al. 1981, Torbit et al. 1985, Depperschmidt et al. 1987). The most obvious and accurate method for determining fat stores of animals is by ether extraction of dried, homogenized carcasses. Although numerous references indicate fat extraction is expensive and time consuming, in reality it is neither when compared to the usual expense and effort of obtaining animals for extraction. A simple method for fat extraction was given by Williams (1984), and an adaptation of this method for wild animals is presented in Box 1. Fat content of the body traditionally has been expressed as a percentage (grams fat/grams dry body mass $\times$ 100). However, Johnson et al. (1985) presented good arguments for expressing fat content as a lipid index, i.e., grams fat/grams fat-free body mass $\times$ 100. In essence, both of the above expressions are ratios in which numerator and denominator may change; however, in the first expression (grams fat/grams dry body weight $\times$ 100), the denominator automatically changes every time the numerator does because the latter is a part of the former. Fat-free body mass, in contrast, is relatively more stable and is the preferred scaling factor for relating fat stores to structural size. If animals do not vary much in structural size (such as when animals of single sex and age are being compared), grams of fat may be the preferred method of expression.

SKELETAL MEASUREMENTS AND WEIGHTS

Various skeletal measurements, body weights, and their associated ratios have been used as indices of growth, condition, and nutritional status. Body weight alone can be used as a condition index in certain situations. For example, state game agencies often use dressed carcass weights of 1.5-year-old male deer taken during the autumn hunting season as a measure of deer condition and habitat adequacy. In this instance, body weight is a function of fat content and structural size of the animal, both of which are of interest. By monitoring these weight trends over several years, biologists can determine if deer herd condition is declining or remaining stable. This is an example of a measure that is not specific in its indications (body weight reflects both structural size and degree of fatness) but is useful from a management viewpoint. We should note, however, that body weights of all male deer decline markedly during and after the rut due to increased activity and voluntary food restriction (Warren et al. 1981). Therefore, eviscerated body weights of males measured at various times in late autumn or winter are of limited value. Anderson et al. (1972) concluded that eviscerated carcass weight (here, whole body weight minus all fat within the body cavity and all viscera except the esophagus and trachea) was a good index of condition in female mule deer but not in males.

Box 1. Soxhlet procedure for estimating total body fat.

1. Collect and freeze carcasses of animals to be extracted and analyzed. Double-bag in sealed plastic bags if percent wet or dry weight is needed.
2. Remove the head and feet from birds and pluck or shear the feathers. This is optional, but ground feathers can cause problems in obtaining homogeneous samples.
3. Remove the gastrointestinal (GI) tract, slit it to remove contents, and return the GI tract and all mesenteric fat to the body cavity. Alternatively, remove the GI tract intact, strip the attached fat from it, and return the fat to the body cavity for analysis with the rest of the carcass. In this way the stomach or crop and gizzard and their contents can be frozen separately or placed in a preservative for future analysis, e.g., in feeding habit studies. The description of methods must state clearly what was removed from the carcass prior to homogenization so others can compare data from various studies accurately.
4. Grind or mince carcass. A hamburger-type meat grinder is used for larger carcasses, a commercial-grade food blender for smaller ones. If a meat grinder is not available, larger carcasses (ducks, geese) can be cut into smaller pieces and homogenized either before or after drying the sample.
5. Freeze-dry or oven-dry the sample. If body parts are dried before homogenization, a second drying should be done to ensure a stable dry weight. Kerr et al. (1982) detected no difference in fat levels of samples that were freeze-dried or oven-dried up to temperatures of 120 C.
6. An aliquot (2–20 g) or sample of the homogenate is next wrapped securely in tared filter paper and tied with a string (also tared) for fat extraction in a Soxhlet apparatus.
7. Eight to 10 packets can be extracted per Soxhlet unit, which permits extraction of 48–60 packets in a six-unit system.
8. Packets are extracted for 24–48 hours (depending on heat level and, therefore, the number of flushes with ether per 24 hours). The packets then are removed and left in an exhaust hood for 2 hours to allow the ether to evaporate, dried in an oven for 12 hours, allowed to cool, and weighed.
9. The difference in the weight of the original sample (minus filter paper and string weight) and the weight of the extracted sample (minus filter paper and string weight) is the weight of the fat in the original sample.
10. To calculate the lipid index, divide the weight of fat by the weight of the extracted sample (residue) and multiply by 100.
11. To calculate percent fat, divide the weight of the fat by the weight of the original sample and multiply by 100.

Riney (1955) reported high correlations between body weight and heart girth. However, he concluded from his work and from a review of the literature that both of these measures were only gross indicators of total fat reserves. Bandy et al. (1956) predicted body weights from heart girth measurements and hind foot length and compared the ratio of the two to estimate recent nutritional status of Columbian black-tailed deer. They reasoned that hind foot length, once attained, is not affected as much as is heart girth by nutritional levels. If the ratio, body weight estimated from heart girth to body weight estimated from hind foot length, was <1, the animal was supposedly in poor condition; if >1, it was in very good condition. Klein (1964) used femur : hind foot ratio to compare long-term nutritional status of two populations of Sitka black-tailed deer. This index is based on the fact that the growth of the metatarsals (which compose most of the length of the hind foot) is relatively more complete at birth than is the femur. Thus, the ratio of the two in an adult deer can indicate the amount of skeletal growth occurring over the lifetime of an animal and hence the relative long-term nutritional regime. A low femur : hind foot ratio would be indicative of poor nutrition, whereas a high ratio would be expected in animals on better diets. McEwan and Wood (1966), comparing captive and wild-reared caribou, suggested that body weight : hind foot ratio was a good index of growth rate and long-term nutritional status in this species.

Bailey (1968) calculated the following formula for computing an index to physical condition in cottontail rabbits in Illinois based on weight and length relationships:

$$CI = (W - 16)L^3,$$

where

CI = condition index,
W = weight in grams, and
L = length in decimeters.

Any animal having a CI >5.48 in his study was heavier than average for its length class.

Antler beam diameters often are used as indices to nutritional status in cervids (Severinghaus et al. 1950, Riney 1955). Rasmussen (1985) reviewed the use of antler measurements as an index of deer physical condition and range quality and concluded (p. 112) that measurements of "antler size for yearling males is a reliable measure of physical condition and that annually monitoring trends in antler beam diameter of such deer appears to be the most practical means of assessing the health and vitality of deer populations." He also believed that data on number of antler points and main antler beam length provided useful, supplemental information. He found fairly high correlations ($r = 0.72–0.88$) between antler beam diameter (usually taken 2.54 cm above the burr) and whole body weight. Rasmussen's evaluation is probably correct for

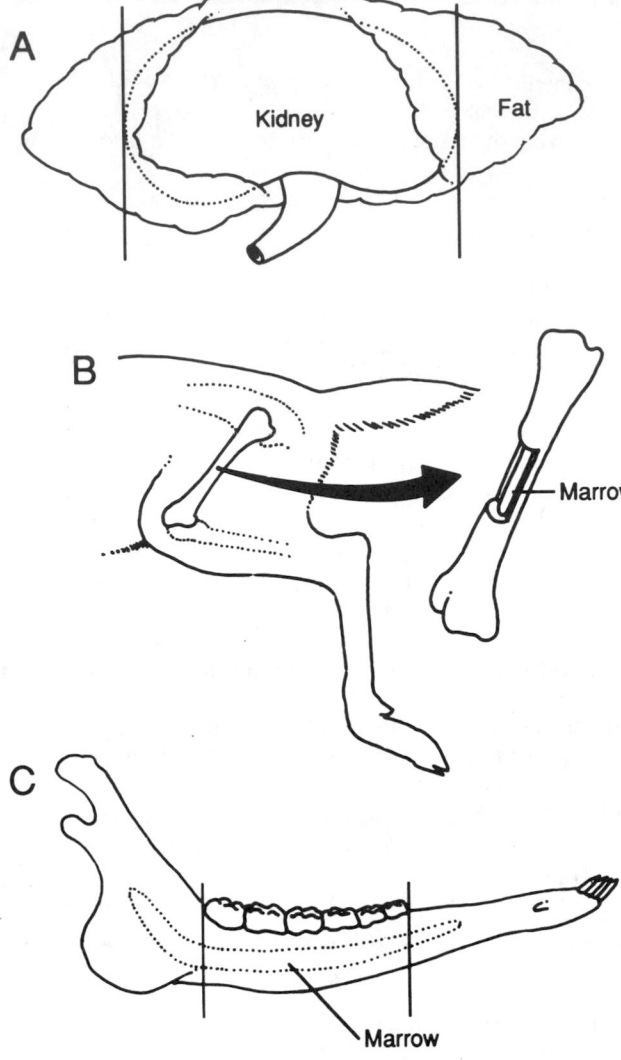

Fig. 1. The amount of fat surrounding the kidney (A) provides an index to visceral fat (from Riney 1955). The amount fat in the marrow of mammalian long bones (B) (e.g., femur from Cheatum 1949*a*) or the mandible (C) (from Nichols and Pelton 1974) provides a measure of energy in this depot of last resort. Fat (in A) or bone (in C) lateral to the vertical lines is trimmed and discarded.

management purposes. However, Ullrey (1982) presented evidence that protein and energy supplies in the month prior to the start of antler development are particularly important in determining antler size.

KIDNEY FAT INDEX

Kidney fat often is measured as an indicator of abdominal fat reserves. The kidney fat index (KFI), as developed for red deer by Riney (1955), is obtained by removing the kidney and its surrounding (perirenal) fat from the abdominal cavity. The fat is cut exactly at both ends of the kidney, perpendicular to the main kidney axis (Fig. 1A); tissue that does not remain affixed to the kidney is discarded. The ratio of the weight of the remaining fat to the weight of the kidney × 100 is the kidney fat index. Riney (1955) believed that KFI measured nutritional status in all seasons, enabled different-sized red deer to be compared on a uniform basis, and permitted valid measurements over a wide range of environmental conditions, a conclusion

also reached by Flux (1971), who studied red deer and hares. However, Batcheler and Clarke (1970) and Dauphiné (1975) reported that kidney weights of red deer and caribou fluctuated seasonally, and thus KFI distorts the true seasonal changes in fatness by displacing the measure of maximum fatness from late autumn–early winter towards midwinter. Dauphiné (1975) found little justification for using KFI in place of perirenal fat (defined in this case as the fat used for obtaining KFI) in caribou. He recommended grouping animals by age to correct for gross differences in body size in caribou without risk of introducing a seasonal variable. Van Vuren and Coblentz (1985) did not discount the use of KFI in feral sheep but recommended standardizations to control for effects of season and age. In the only study we found that attempted to predict whole body fat from KFI (white-tailed deer, Finger et al. 1981), KFI explained 75% of the variance (i.e., $r^2 = 0.75$) in percent body fat, indicating that KFI is a good predictor of total fat stores.

Ransom (1965) suggested using a combination of the kidney fat index and femur marrow fat to estimate nutritional status over a wide range of conditions in white-tailed deer. His data indicated that it is best to use KFI down to a value of 30 (perirenal fat 30% of kidney weight) and then shift to femur marrow fat, because marrow fat seemed to begin to decline at this level of KFI. Connolly (1981) suggested shifting at a KFI value of 20 for mule deer; Kie et al. (1983) suggested a shift at a KFI of 15 for white-tailed deer in southern Texas.

In summary, the kidney fat store seems to be a good indicator of nutritional status in many mammals, particularly ungulates (Smith 1970) and lagomorphs (Flux 1971, Jacobson et al. 1978*a*). However, it is not applicable to all species, one example being the brush-tailed possum in which perirenal fat does not form a discrete mass (Bamford 1970). Also, KFI probably is not appropriate for evaluating seasonal changes in fat stores in some species because of the seasonal changes in kidney weights. Thus, the choice of a particular measurement or ratio (perirenal fat [Dauphiné 1975], total fat associated with the kidney, or KFI) should depend upon the species being studied and the environmental variables of interest.

MARROW FAT

One of the most widely known and used indices of nutritional status in large mammals is the fat level in bone marrow. This technique was first described by Cheatum (1949*a*) and was used to determine if dead white-tailed deer found in the spring in the northern U.S. succumbed to starvation. Its use has been extended to most of the Cervidae as well as several other mammalian taxa under a variety of conditions. Because marrow fat is believed to be the last fat source depleted in a poorly nourished animal, low bone marrow fat is indicative of poor nutrition over a relatively long period of time. In fact, Mech and DelGiudice (1985) stated that marrow fat is a one-way index and that any loss of this fat reserve is indicative of poor condition. They noted that researchers and managers often assume, erroneously, that animals are in good or excellent condition if marrow fat is close to 100%.

Traditionally, marrow in the middle third of the femur is examined (Fig. 1B), and the percentage of fat is determined chemically. Although useful at one time, visual rating schemes are extremely subjective and have been largely replaced by more quantitative procedures, such as the oven-drying method described by Neiland (1970) and the reagent dry technique of Verme and Holland (1973). Neiland (1970) pointed out that the nonfat residue in bone marrow was relatively insignificant compared to the proportions of the other two components, water and fat. He obtained wet weight of a marrow sample and dried it to constant weight in an oven at 60–65 C. The dry weight divided by the wet weight gave a reliable estimate of percent fat. If maximum accuracy is desired, Neiland suggested determining nonfat residue levels corresponding to a given dry weight and subtracting this value from that dry-weight value to give the corrected percentage of fat.

Verme and Holland (1973) reported an alternative method for obtaining dried weight of the marrow that does not require an oven. In their method, a 2-g to 3-g plug of marrow from the middle third of the femur is weighed, macerated to a puttylike consistency, and mixed with 10 ml of a 2:1 solution of chloroform and methanol (Bloor's reagent). The mixture is set aside near a low heat source until the chemicals and water contained in the marrow evaporate. The marrow fat dissolves in the chloroform, and the water dissolves in the methanol. This procedure permits rapid evaporation of the water without spoilage of the marrow sample. The dry weight obtained contains both the fat and nonfat residues. If greater accuracy is desired, a method for correction of the value for nonfat residue similar to that described by Neiland (1970) could be developed.

The amount of marrow fat in the mandibular cavity (Fig. 1C) also has been used as an index of the nutritional status of white-tailed deer (Nichols and Pelton 1974) and moose (Cederlund et al. 1986). The procedures are similar to those used to measure femur marrow fat, but use of mandibles is more convenient because the mandible is routinely collected for aging purposes. Nichols and Pelton (1974) reported that mandibular cavity fat in deer separates into more distinguishable condition classes than does fat from the femur marrow tissues. Fat levels in bone marrow often are expressed on a wet-weight basis, but apparently much less variability, in relation to size of the mean, is found in mandibular fat determinations when they are expressed on a dry-weight basis. The source of high variation in the wet-weight samples is unknown, but a major portion probably arises because of dehydration of some samples before or during initial weighing. If sample values are expressed on a wet-weight basis, extreme caution should be taken to minimize water loss of marrow by sealing samples in small, airtight containers immediately upon collection and obtaining wet weights as rapidly as possible after opening the containers.

Although marrow fat levels have been used primarily in large ruminants, they are equally useful for evaluating condition in smaller mammals. Jacobson et al. (1978b) reported differences in percentage fat in femurs and tibias of wild cottontails as affected by season and sex. Warren and Kirkpatrick (1978) also reported a close relationship between percentage fat in these bones (expressed on a dry-weight basis) and known nutritional intake in cottontails. Bamford (1970) concluded that femur marrow fat (expressed on a percentage dry-weight basis) was a good

indicator of total fat reserves in the brush-tailed possum when fat reserves were low.

OTHER FAT INDICES

Several other fat stores have been measured in mammals, but most lack either objectivity or practicality. Bear (1971) reported that a ratio of total visceral fat (including mesenteric and kidney fat) to the eviscerated carcass weight showed similar seasonal trends to the KFI in pronghorns. A mesogastric fat index (mesogastric fat weight divided by body length × 2.8) was the best and most objective measure of fat reserves in brush-tailed possums (Bamford 1970). Anderson et al. (1972) recommended carcass density (mg/ml) as a good index of nutritional status in mule deer.

BLOOD AND URINE CHARACTERISTICS

Numerous blood characteristics have been investigated and used as indices of the current nutritional status of wild animals. These were reviewed in some detail by LeResche et al. (1974), Hanks (1981), and Franzmann (1985, Table 1). Only those that currently show the most promise for applicability will be described here.

Blood urea nitrogen (BUN) (LeResche et al. 1974) is undoubtedly one of the most widely used indices of nutritional status. BUN is relatively unaffected by the stress of handling or immobilization with drugs in Cervidae (Seal et al. 1972, Wesson et al. 1979b) and is a good indicator of protein nutrition, being directly related to protein ingested so long as energy intake is constant and above maintenance level (Kirkpatrick et al. 1975). However, care must be exercised in interpreting these data. For example, high dietary energy levels depress BUN in white-tailed deer (Kirkpatrick et al. 1975) and domestic steers (Preston et al. 1961). This has been attributed to a more efficient use of protein by rumen microbes and a subsequent reduction in ammonia production and urea formation when energy intake is high. Also, when energy levels drop below maintenance levels, BUN can rise as a result of tissue catabolism. Thus, an index of energy intake should accompany BUN analyses to assure proper interpretation of BUN data.

Blood indices of energy nutrition were thoroughly reviewed by Franzmann (1985:247) for herbivores; he concluded that there is no "overwhelming choice of a blood parameter" for assessing energy intake, and we agree. Serum cholesterol, nonesterified fatty acids (NEFA), and ketones have been investigated thoroughly, and all have shown promise as energy indices in individual studies (Vogelsang 1977, Seal and Hoskinson 1978, Seal et al. 1978a,b, Card et al. 1985, DelGiudice et al. 1987a), but none has proven to be consistent enough to be used reliably across a range of energy intake levels (Warren et al. 1981, 1982, Card et al. 1985). Other blood characteristics, such as packed cell volume, various measures of hemoglobin, electrolytes, enzymes, hormones, and amino acids, have been proposed as indices of nutritional status (Franzmann 1985, Table 1) but have been inconsistent in their indications or require further verification in controlled studies.

BUN and cholesterol also have been tested as nutritional indices in cottontails. Warren and Kirkpatrick (1978) reported no difference in cholesterol levels in cot-

Table 1. Response of blood parameters to nutritional boundary conditions in large mammalian herbivores as tested in controlled studies (abridged version of Table 3, Franzmann 1985, with permission). Response symbols: + = positive correlation of blood parameter to the nutritional condition; − = negative correlation; 0 = no correlation; * = significant change.

Blood parameters	Protein intake		Energy intake			Protein and energy		Starvation		Season (* indicates fluctuation)			Condition	
	Bison	White-tailed deer	Bison	White-tailed deer	Black-tailed deer	Bison	White-tailed deer	White-tailed deer	Moose	Bison	White-tailed deer	Cari-bou	Impala	Moose
Hematology														
Red blood cells (RBC)	+	+		+			0				*			
Hemoglobin (Hb)	+	+0	+	0	0	+0	0		*	*	*			+
Packed cell volume	+	0	+	0	0	−0	0		*	*0	*		0	+
Mean corpuscular Hb	+	+	0	+		0	+		*	*	0			
Mean corpuscular volume	+	+	0	0		0	0			*	0			
Mean corpuscular Hb concentration	0	0	0	0		0	+			0				
White blood cells (WBC)	+	0	0	0		+−0	0			*0	0			
WBC differential						+								
Sedimentation rate					0									
Nonprotein nitrogen														
Urea nitrogen	+	+	+	−0	0	+0	+0	+	*	*	*		0	0
Creatinine						+−				0				0
Bilirubin	+			0					*					
Urea/creatinine ratio	+	+												
Proteins														
Total protein	−	0	+−	0	0	0	0	−	*	*0	*		0	+
Albumin	0	0	0	0	0	0	0		*	*	*		0	+
Globulin					0				*	*				0
Alpha globulin	+	0	+			0			*	0				0
Beta globulin	+	0	0			+		*	*	*				+
Gamma globulin	0		−		0							0		
Fibrinogen	0	0		0	−	0	0			0	*			
Sialic acid							+							
Lipids and fatty acids														
Cholesterol	+	+	−	0		+	+		*	*	*		0	0
Triglyceride		0		0			0				0			
Nonesterified fatty acid	0	0		−0										
Ketones	+	+		0										
Carbohydrates														
Glucose	−	0	+	0	−	+0	+0		*	*	0	*		+

Table 1. Continued.

Blood parameters	Protein intake		Energy intake		Protein and energy		Starvation		Season (* indicates fluctuation)			Condition	
	White-tailed deer	Bison	White-tailed deer	Black-tailed deer	Bison	White-tailed deer	White-tailed deer	Moose	Bison	White-tailed deer	Caribou	Impala	Moose
Electrolytes, minerals and trace elements													
Sodium	+		+		0	0			*	0			
Potassium	0		0		0	0				*			
Chloride	+		+		0	+		*	0	0			
Calcium					0	0		*	0	*		+	+
Phosphorous	0	+	–		0	0		0					
Magnesium					0	0		*	0				0
Enzymes													
Alkaline phosphatase	0	+	+		0	0		0	*	0			0
Glutamine oxalycetic transaminase	0	+			+0	0			0				
Glutamine pyruvic transaminase				0				*		*			
Creatinine kinase	0	+	++		+	0			*0				0
Lactic dehydrogenase	0					0	0		0				
Hormones													
Triiodothyronine	0		+	0	0	+0	–		0	*	0		
Thyroxine	0					+0	–						
Growth hormone					0	0				0	*		
Cortisol	0		++			+0							
Progestin	0		+			+							
Insulin	0		+			0							
Amino acids													
Isoleucine	0		+			0							
Leucine	0		+			0							
Phenylalanine	0		+			0							
Histidine	0		+			0							
Threonine	0		+			0							
Glycine	+		+			0							
Valine	0		+			+							
Citrulline	0		0			+							
Taurine	0		0			+							
Glutamic acid	0		0			+							
Aspartic acid	0		0			+							
Glutamine/asparagine	0		0										
Amino acid nitrogen							+				*		

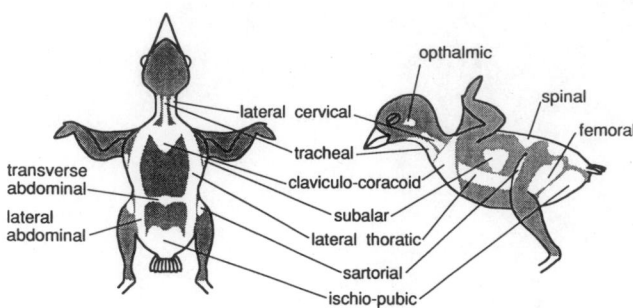

Fig. 2. Distribution of subcutaneous fat in the white-crowned sparrow (copied with permission from King and Farner 1965).

tontails maintained on two nutritional levels. BUN, however, was higher in animals that were on restricted diets and were losing weight, indicating protein catabolism.

BUN appears to be affected by stressors more so in cottontails than in ruminants. Jacobson et al. (1978a) observed significantly higher BUN levels in rabbits confined in box traps than in those obtained by shooting. Jacobson et al. (1978a), however, also demonstrated distinct seasonal differences in BUNs of shot cottontails, which were believed to be related to differences either in parasite load or nutritional intake.

A word of caution is in order for those contemplating the use of blood biochemistry in nutritional studies. Many blood characteristics have daily rhythms, which are not well understood, and most can be influenced by stressors, including those involved with sample collection (e.g., trapping and restraint) (Wesson et al. 1979a,b,c). Hence, it is extremely important that all blood samples be taken and handled as uniformly as possible. This includes making collections at approximately the same time of day and collecting and storing samples on ice as quickly as possible after an animal is shot or restrained.

Metabolites in urine have been investigated recently as indicators of condition in cervids and wolves (Warren et al. 1981, 1982, DelGiudice et al. 1987b, 1989, Mech et al. 1987). Because urine volume, and hence concentrations of urinary metabolites, varies over time and with type and amount of food ingested, metabolites are commonly expressed as a ratio with urinary creatinine. Creatinine is an end product of muscle metabolism that can be used to standardize concentrations of metabolites because it is excreted at a relatively constant rate (Bovee 1984). Mech et al. (1987) and DelGiudice et al. (1989) used this index in the field to estimate condition of wolves and deer, respectively, by collecting "snow urine." The further dilution of urine by snow has no effect because the metabolite of interest and its denominator (creatinine) are diluted to the same extent by snow water.

DelGiudice et al. (1989) compared deer in four winter yards in Minnesota by measuring sequential (at 2-week intervals) changes in urea nitrogen, potassium, sodium, calcium, and phosphorus in snow urine and expressing them as a ratio with urinary creatinine. Snow urine samples were collected within 72 hours of a recent snowfall. They concluded that differences in metabolite:creatinine ratios in snow urine, if taken every other week during winter, resulted in a correct assessment of deer condition in the four yards. Mech et al. (1987) presented evidence

from captive wolves that metabolites in snow urine accurately assessed condition in this species as well.

The snow urine approach to assessment of nutritional condition is not without problems of interpretation. Identifying the individual or even the sex or age of the animal being sampled seldom is possible, and this can introduce large amounts of variation into the data. For example, fawns in northern deeryards would be expected to decline in condition more rapidly over winter than adults. If snow urine samples from one deeryard were heavily dominated by adults, one might conclude that the condition of the deer or their habitats was different when in fact it might not be. Also, urinary urea nitrogen, like BUN, first declines and then increases when the animal begins metabolizing its protein stores, and accurately identifying just where the animal may be on this U-shaped curve is often difficult.

Several other measurements have been used or suggested as indices of nutritional status and habitat conditions in wild mammals. The Southeastern Cooperative Wildlife Disease Study at the University of Georgia developed an abomasal parasite index for deer (Eve and Kellogg 1977) that is used widely in the southeastern United States. The basis for this index is that abomasal parasite counts increase as deer populations approach carrying capacity. Ozoga and Verme (1978) presented evidence indicating that thymus gland weights also might be useful as an index of nutritional status of white-tailed deer.

Birds

The pattern of fat deposition and metabolism in birds is somewhat less clear and shows more interspecific variation than in mammals. Blem (1976) reported that subcutaneous fat (Fig. 2) is deposited first, followed by abdominal fat, a conclusion that seems to be based largely on field notes of McCabe (1943). This is at variance with King's (1967) observations of white-crowned sparrows, which indicated that fat deposition begins earlier in mesenteric adipose tissue than in subcutaneous tissue and that abdominal fat deposits are about two times larger than subcutaneous deposits. Woodall (1978) concluded that subcutaneous and internal fat in red-billed teals were laid down in equal amounts. However, Whyte and Bolen (1984) showed that subcutaneous fat made up 59–67% of total body fat in mallards. Raveling (1979:246) observed an analogous pattern in declining fat deposits in incubating Canada geese and concluded that "abdominal and subcutaneous fat were used simultaneously, but subcutaneous deposits were the last to be depleted." As with mammals, the best indication of fat stores in birds is the measurement of whole body fat through extraction in a Soxhlet apparatus (Box 1). However, many of the physical measurements described in the following sections provide an index to whole body fat.

BODY WEIGHT AND STRUCTURAL MEASUREMENTS

In live birds, body weight alone can be used as a condition index or as a predictor of total body fat, but it usually explains only 40–60% of the variance (i.e., $r^2 = 0.40$–0.60) in body fat (Bailey 1979, Whyte and Bolen 1984, Johnson et al. 1985). Combining body weight with a measure of structural size usually improves considerably the correlation with grams fat or grams fat/grams fat-free dry weight. Wing length has been used most often as a scaler for structural size (Owen and Cook 1977, Whyte and Bolen 1984, Johnson et al. 1985), but keel length, bill length, tarsus length, culmen length, total body length, and various combinations of these have been used on live waterfowl (Bailey 1979, Chappell and Titman 1983, Hohman and Taylor 1986). Servello and Kirkpatrick (1987, 1988) concluded that body weight was not a good or consistent predictor of body fat in ruffed grouse in that body weight alone explained 4–55% of the variance in body fat. Thomas et al. (1975) also observed no simple relationship between body weight and body fat in ruffed grouse. Owen (1981) placed free-ranging barnacle geese (observed at distances up to 100 m) into one of four fat reserve classes based on the shape of their abdominal profiles.

DISCRETE FAT DEPOTS

Various discrete fat deposits have been used as condition indices or to estimate total body fat, primarily in waterfowl. The most common of these are abdominal or omental fat and subcutaneous fat (usually wet skin weight) (Hohman and Taylor 1986). Delineation of these fat masses is not always easy. Abdominal fat has been described as "a discrete mass of fat in the abdominal cavity immediately posterior to the gizzard" (Thomas et al. 1983:1115) and as fat in the body cavity under subcutaneous fat and around the pubic bones (Bailey 1979). Omental fat was described by Whyte and Bolen (1984) as fat on the gizzard and lower omentum and by Woodall (1978) as the fat in the omentum and surrounding the gizzard and cloaca (but excluding fat surrounding the intestines).

Abdominal or omental fat is a good predictor of total body fat, usually explaining 80–95% of the variance in total body fat. However, wet skin weight (with adhering fat), alone, is one of the best predictors, usually explaining about 90% of the variance in total body fat.

Visceral fat, described by Bailey (1979) as fat around the heart, ventriculus, and mesenteries of the small intestine and by Whyte and Bolen (1984) as fat stripped from the intestines, usually explains only about 70% of the variance in total body fat. Ankney and MacInnes (1978) used the combined wet weight of subcutaneous, mesenteric, and abdominal fat as a fat reserve index in lesser snow geese. Various combinations of these three fat depots often explain >90% of the variance in total body fat.

Gizzard fat (g) and a gizzard fat index (GFI), calculated similarly to KFI for mammals, have been reported to be suitable indices of body fat in ring-necked pheasants (Dowell and Warren 1982), ruffed grouse (Servello and Kirkpatrick 1987), and northern bobwhites (Koerth and Guthery 1988). Gizzard fat predicted the lipid index in ruffed grouse with an r^2 of 0.90.

BONE MARROW FAT

Bone marrow fat received almost no attention in birds until relatively recently. Hutchinson and Owen (1984) examined bones of 23 species of waterfowl and observed marrow lipid in all bones examined except in the humeri of nine species, which were pneumatized (air sacs) and devoid of both marrow and lipid. They recommended the ulna as first choice as an index to starvation and presented a graph showing that percent ulna fat declines rapidly when total body fat drops below 20%.

PERCENT WATER

Little water apparently is required in the storage of fat (Odum et al. 1964), and several research studies have noted a high negative correlation (≥ 0.95) between body water and body fat (Bailey 1979, Johnson et al. 1985). Johnson et al. (1985) evaluated three techniques based on body water content for estimating either grams fat or grams fat/grams fat-free dry weight and observed that all three methods gave values of $r^2 > 0.80$. In fact, the methods of Child and Marshall (1970) and Campbell and Leatherland (1980), which are based on the nearly constant relationship of water content and fat-free weight, gave values of $r^2 > 0.96$.

Methods for estimating body water and, hence, total body fat in live animals are available but are time consuming and complex or have not been tested adequately on different species. A technique based on turnover of radiolabeled (tritiated) water (^3H_2O) has been available for some time, but it has seen only limited use in wildlife studies (Crum et al. 1985, Torbit et al. 1985). This method involves injecting a known amount of ^3H_2O into an organism, allowing it to equilibrate with normal body water, taking a blood sample, and determining the dilution factor, hence total body water. This technique normally overestimates body water slightly, but if used to estimate fat on live animals it probably would improve prediction measurably over currently used techniques, i.e., body weight or relationships of weight to structural size.

ELECTRICAL CONDUCTIVITY AND ULTRASOUND

Electrical conductivity techniques for estimating fat are based primarily on the percent H_2O in the body, which is positively correlated with percent fat-free dry weight and, hence, negatively correlated with fat content. Consequently, one can estimate fat content using measures of conductivity or impedance. The most promising of these methods employs an instrument originally designed to estimate fat levels in ground meat (Walsberg 1988). The animal is placed in a plastic cylinder, which is inserted into the center of an electric coil. The presence of the animal body alters the electromagnetic inductance of the coil, which is measured by changes in the phase relation of voltage and current in an electrical signal passed through the coil (Walsberg 1988). Conductivity of fat tissue is only about 4.5% of that in lean tissue, body fluids, and bone. Therefore, the primary determinant of inductance change is lean body mass. Comparison of total body mass with lean body mass, as determined from conductivity, gives an estimate of fat content. Instruments for measuring conductivity of live animals are sold by EmScan, Inc., Springfield, Ill., as the SA-3000 Multi-Detector Small Animal Body Composition Analysis System. Seven detection chambers for animals of different sizes are currently available and range in size from 30 to 203 mm inside diameter,

High r^2s between predicted and actual lean body masses have been obtained with conductivity measurements on several species of small birds (40–170 g) and mammals

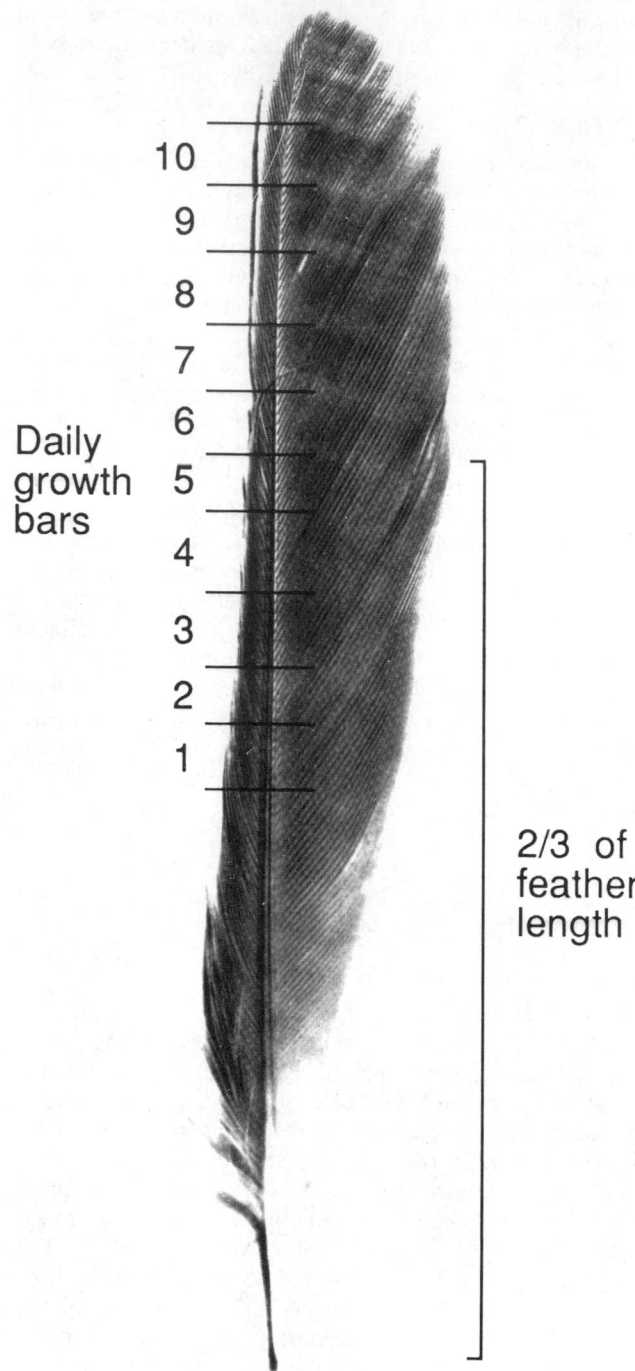

Daily growth bars

10
9
8
7
6
5
4
3
2
1

2/3 of feather length

Fig. 3. The outermost right rectrix of a male white-crowned sparrow. Growth bars appear as pairs of light and dark bands, roughly perpendicular to the rachis. The distance between the distal edge of one dark band and the distal edge of the next represents 24 hours of growth. Daily feather growth is estimated from the average width of 10 bars centered (5 proximal and 5 distal) on a standard point, which is located two-thirds of the feather's length from the proximal (calamus) end (Grubb 1989) (photograph by T. Grubb).

(40–600 g) (Walsberg 1988). The r^2 for prediction of fat will be much lower in most instances because the absolute error (in grams of tissue) is identical for prediction of both lean body mass and fat mass (Morton et al. 1991). Because fat usually makes up a smaller proportion of total body mass than does lean mass, absolute error in the estimation of fat mass is proportionately greater than ab-

solute error in estimation of lean mass, resulting in a less accurate prediction of fat mass. Reported r^2, therefore, should be for prediction of fat mass and not for lean mass (see Morton et al. 1991), as in some of the early research on this subject. Castro et al. (1990) further tested the total body electrical conductivity method and concluded that it accurately estimated lean mass within as well as between bird species. They further showed that the estimation of lean mass was not affected by the presence of metal leg bands on the birds and that dead birds gave significantly different readings from those of live birds.

Bioelectrical impedance has been used to estimate body fat levels in humans (Cohn 1985, Lukaski et al. 1986), but its accuracy has been questioned. Electrical impedance is a measure of the hindrance of a small electrical current passed through the body; impedance is negatively correlated with the water content of the body and, therefore, positively correlated with fat content. An instrument for measuring body impedance, available from RJL Systems, Inc., apparently used impedance successfully to estimate fat levels of black and grizzly bears (S. D. Farley and C. T. Robbins, unpubl. data).

Baldassarre et al. (1980) used ultrasonic sound to estimate fat depots in plucked mallards. They used a commercially available ultrasonic flow detector (Model USL-31, Krautkramer-Branson Co., Stratford, CT 06497) and could explain 58%, 65%, and 59% of the variation in total, subcutaneous, and omental fat, respectively.

PTILOCHRONOLOGY

Ptilochronology is a new and promising approach to studying day-to-day and week-to-week variation in the nutritional status of birds. Ptilochronology is the study of growth rates of feathers by the measurement of growth bars or cross-banding on feathers (Michener and Michener 1938) that indicate 24-hour periods of growth (Grubb 1989). This pattern (Fig. 3) apparently reflects variation in available energy and nutrients on a 24-hour cycle, analogous to cementum annuli in mammalian teeth.

The method involves plucking the outermost rectrix and releasing the bird to its habitat while the replacement feather is grown, a matter of several (5–6) weeks. In this way the growth bars on the newly grown feather can be dated, providing a daily record of the nutritional status of the bird (Grubb 1989). Measurements are standardized by comparing the mean width of 10 growth bars at a standard position on the first feather with the same bands (by location) in the regrown feather (Fig. 3).

Ptilochronology has not been widely tested in the field, but the data published thus far confirm its potential as a unique index of nutritional status. Growth bars were narrower when food was lacking (Grubb 1991) and wider when birds had supplemental food (Grubb and Cimprich 1990, Waite 1990). Ptilochronology revealed the parental costs of rearing young starlings to be a function of brood size (White et al. 1991).

PROTEIN INDEX

Relatively few workers have tried to estimate or develop an index of protein reserves in birds. However, interest in the fate of protein reserves in reproduction in waterfowl has stimulated research in this area. The most accurate method for determining protein is to perform a

Kjeldahl analysis (for N) on the fat-free residue obtained after fat is extracted from a sample. Alternatively, one can ash this residue in a muffle furnace and determine ash (as the residue left after combustion) directly and protein by the difference. Ankney and MacInnes (1978) used total dry weight of sternal muscles, leg muscles, and gizzard as a protein reserve index in lesser snow geese. Raveling (1979) believed that the non-ether extractable residue of cackling geese was a reasonable estimate of their protein content. Hohman and Taylor (1986) used linear regression to estimate ash-free, lean dry weight from a combination of eviscerated carcass weights, breast muscle weights, and bill measurements. Other workers have used breast muscles or gizzard weights alone as indices of protein reserves.

BLOOD CHARACTERISTICS

Blood characteristics have not been used as physiological indices as frequently in birds as in mammals, and most of the work has centered on determining the extent of nutritional stress during egg laying and incubation. Harris (1970) reported slight increases in free fatty acids (FFA, another term for NEFA) in blue-winged teal hens between egg laying and incubation. FFA first increased during incubation and then declined. Plasma glucose and nonprotein nitrogen (NPN) increased between prelaying and incubation, indicating a depletion of fat stores and a consequent increase in gluconeogenesis (Harris 1970). Korschgen (1977) reported decreases in hematocrit, plasma proteins, and NPN and an increase in FFA in female eiders during incubation. He believed that eiders were using fat as energy but that they had not totally depleted fat stores at the time of sampling. Uric acid and BUN increased in wintering green-winged teals with increasing wind velocity and relative humidity (Bennett and Bolen 1978).

Triglycerides increased eight-fold at the onset of laying in northern bobwhites (McRae and Dimmick 1982). Serum albumin and globulin also increased. Triglycerides are used in yolk formation, and serum albumin is involved in transport of FFA and calcium to the shell gland.

ASSESSMENT OF REPRODUCTIVE PATTERNS AND PERFORMANCE

Estimation of natality, i.e., number of young produced per unit population per unit time, is a fundamental requirement for understanding the dynamics of a wild population. In some instances, this information can be obtained indirectly through mark-recapture procedures. However, estimation of the number of young entering a population in this manner often is difficult, because mortality of newborn young is high and the younger age groups are difficult to trap for marking. Consequently, it is necessary to measure clutch sizes or examine ovaries and reproductive tracts collected during the breeding season to measure a reproductive characteristic, such as the average number of preovulatory follicles or corpora lutea (for ovulation rate) or fetuses per female. Data on reproductive rate and sex and age structure from other sources then can be used to calculate the number of young born or hatched per unit population, i.e., gross natality.

Reproductive rates can be measured at every point in the cycle, beginning with courtship behavior and ending with fledging or weaning and dispersal of offspring. The value of estimates made at different points varies with the species or taxon and, of course, with the goals of the investigation. For example, observations of singing males or nests in a given area establish or identify breeding habitat for birds, whereas counts of placental scars in squirrels at necropsy provide a size estimate of previous litters.

Knowledge of the various measurements of reproductive performance that can be made throughout the reproductive cycle of a given species provides the investigator with more options for meeting the specific objectives of the study. Such knowledge also allows the investigator to consider techniques that might be applied at different times of the year and thereby to achieve more efficient use of fiscal resources and biological material. Accordingly, the sequential events in the reproductive cycle from breeding to fledging or weaning form an outline in this section for review of techniques and measurements used in the study of reproduction.

Reproductive Characteristics in Male Birds and Mammals

TESTIS SIZE

The gonads and reproductive tracts of male and female birds remain in a completely regressed state during the nonbreeding season. The testis of a mature male white-crowned sparrow will grow from <10 mg to >600 mg during the height of the breeding season (Wingfield and Farner 1980); similar increases at the onset of the breeding season are seen in other avian taxa. Seasonal growth and recrudescence of the mammalian testis and accessory glands are much less pronounced and variable across taxa than in birds. In some, such as tammar wallaby and brush-tailed possum, testis weight changes little throughout the year, although prostate weight and testosterone levels show marked seasonal variation (Gilmore 1969, Inns 1982). Most seasonally breeding mammals, however, show noticeable to marked increases in testis size and in androgen levels (Mirarchi et al. 1977b). Weight and volume (by water displacement) are obtained at necropsy, but linear measurements of scrotal testes on live males also provide a valid index of testicular volume and reproductive status. For example, testis volume in deer (measured on both live and dead animals) increased from a low of 50 cc in June to >150 cc in November (breeding season) during the same time that plasma testosterone increased from basal levels to >3 ng/ml (McMillin et al. 1974).

SPERMATOZOA COUNTS

The presence of spermatozoa in the testes or epididymides provides a qualitative expression of reproductive status and is particularly useful for determining the age at puberty or seasonal differences in male reproductive activity (Mirarchi et al. 1977a), social stress (Sullivan and Scanlon 1976), and exposure to environmental contaminants (Sanders and Kirkpatrick 1975).

An estimate of sperm density or total spermatozoa per testis or epididymis may be obtained by homogenizing a known mass of sliced tissue in a blender or tissue homogenizer with culture medium, such as Hank's solution, or physiological saline. Triton X-100 (J. T. Baker Chemical Co., Phillipsburg, NJ 08865) can be added (0.01–

0.05% by volume) to prevent foaming in the blender (Amann and Lambiase 1969, Sullivan and Scanlon 1976). Alternatively, testes or sections of epididymides of small species can be minced thoroughly with scissors and rinsed repeatedly with a known volume of culture medium to collect spermatozoa. An aliquot of homogenate or rinse then is removed and added to both chambers of a standard hemocytometer. Sperm density relative to tissue mass then can be calculated with the appropriate dilution factors (Box 2). A simple test for the presence of motile spermatozoa involves cutting the tail of the epididymis from a freshly killed specimen and making a sperm smear on a slide for microscopic examination (Kibbe and Kirkpatrick 1971).

TESTICULAR BIOPSY

Counts of spermatozoa usually are made on tissues or fluid collected at necropsy or in ejaculates. However, needle-biopsy procedures offer an opportunity to obtain testicular tissue and spermatogenic cells from living animals. The animal is anesthetized and the surface of the scrotum is thoroughly cleansed and disinfected. A testis is punctured with a 19- to 22-gauge needle, using care to avoid the epididymis (Sundqvist et al. 1986). Slight pressure

---

Box 2.   Procedure for estimating the density of spermatozoa and other types of cells with a hemacytometer.

A hemacytometer is a thick, glass microscope slide with two identical counting chambers of known volume. It was originally developed to provide standard conditions for human blood-cell counts, but it has found application in a variety of cytological work, including estimation of sperm density for assessment of male fertility. The highly polished floor of each counting chamber is gridded with fine lines spaced at exact intervals as shown below.

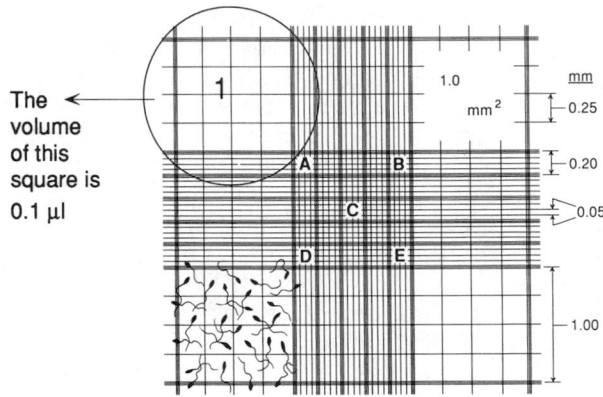

Ridges on the sides of the hemacytometer hold the coverslip exactly 0.1 mm above the floor of the counting chamber. Consequently, cells counted within a given section of the chamber are related to volume, and cell density can be calculated. Counts can be made on fresh, frozen, or fixed material, but fresh material provides the option of rating spermatozoa for motility.

In reality, a hemacytometer simply provides a convenient system of grids that can be readily applied to a variety of microscopic sampling problems.

Procedure

1. Spermatozoa may be obtained from a variety of sources: ejaculate, homogenized testis, or minced epididymis. If spermatozoa are to be evaluated for motility, semen or tissue homogenate should be diluted with an appropriate culture medium, such as Hank's solution (rather than common physiological saline). Both the medium and hemacytometer should be held on a slide warmer at normal scrotal temperature, e.g., 35 C.

2. The source material is diluted with a known volume of medium or saline, and a drop of the suspension is added to both chambers of the hemacytometer.

3. After the spermatozoa have settled for 5 minutes, count the number of spermatozoa in sections A, B, C, D, and E of the center grid and multiply the total by 5 to obtain an estimate of the total number in the center grid (1 mm or 0.01 cm). Alternatively, select one of the four corner grids (comprising 16 squares) at random and count all spermatozoa with heads inside the perimeter lines. If sperm density is high (>40 cells/square), count within a randomly chosen row of four squares and multiply the total by four to obtain an estimate of the total in all 16 squares.

4. This procedure should be repeated in a grid on the other counting chamber. In this way, a mean and standard deviation can be calculated and an estimate of counting precision can be obtained.

5. The coverslip is held 0.1 mm above the floor of the counting chamber, which in this example creates a chamber volume of 1×10^{-4} cc or 0.1 μl.

6. The concentration or density of sperm in the source material is calculated with the appropriate dilution factors. For example, a 150-mg epididymis is minced and rinsed with 600 μl of saline, and a sample of the resulting cell suspension is placed in a hemacytometer. If 500 spermatozoa are counted in a 1.0-mm^2 section (0.1 μl) of the hemacytometer, sperm density or concentration (C, i.e., number of spermatozoa per mg of epididymis) would be calculated as follows: C = N $\times$ D/0.1 $\div$ S, where N = the number of spermatozoa counted in 0.1 μl of the hemacytometer, D = the volume (μl) of saline used to rinse or dilute the sample, and S = the volume or weight of the original sample. C = 500 $\times$ 600/0.1 $\div$ 150 = 20,000 spermatozoa/mg of epididymis.

applied to the testis or negative pressure in an attached syringe will ensure aspiration of tissue into the needle. The needle is withdrawn from the testis, and the material contained in the needle is spread onto a microscope slide, dried, and stained prior to examination of the cells. The biopsy smears are scored as follows: low (1–2), when only sertoli cells or spermatogonia are present, to high (9–10), when large numbers of mature spermatids are counted. This procedure, used in diagnosing human infertility, has been applied to ranch mink, in which up to 20% of the males are infertile. Mink with biopsy scores <7 are considered infertile (Sundqvist et al. 1986). Stages of spermatogenesis were described by van Tienhoven (1983), and methods for quantifying mammalian spermatogenesis may be found in Berndtson (1977).

Reproductive Characteristics in Female Birds

OVARIAN ACTIVITY

Onset of the breeding season is heralded by an increase in the size of the ovary, stemming from an increase in the number and size of follicles before ovulation. Counts of large, yolk-filled, preovulatory follicles, i.e., tertiary follicles, have been used as measures of reproductive activity in band-tailed pigeons (March and Sadlier 1970) and mourning doves (Guynn and Scanlon 1973). Ankney and MacInnes (1978) were able to distinguish a group of large (>20 mm diameter), highly vascularized preovulatory follicles from smaller (<10 mm) ones (Fig. 4) and thereby estimate potential clutch size in a sample of snow geese shot as they arrived on the nesting ground. Because these estimates of ovulation rate are collected at necropsy, they can be analyzed with reference to other measurements such as carcass weight, nutrient reserves, and hormone levels.

POSTOVULATORY FOLLICLES

After ovulation, the collapsed wall of the avian follicle does not form a corpus luteum (as in mammals) or any functional counterpart. Instead, postovulatory follicles (POFs) regress, the process being completed in many avian species within a month after ovulation (Payne 1973). However, in some species, such as the ring-necked pheasant, POFs persist for months as small (1–2 mm), pigmented (reddish-brown) structures that can be viewed macroscopically (Fig. 4). Kabat et al. (1948) reported a high correlation between the number of eggs laid and the number of POF counted in captive pheasants killed up to 100 days after ovulation, thus providing a potential method for estimating clutch sizes in summer from birds killed the following autumn. In contrast, Hannon (1981) determined that POFs in blue grouse could not be counted macroscopically more than 25 days after ovulation, although they served to distinguish laying from nonlaying hens shot during the hunting season.

LAPAROTOMY

Observation of the avian ovary need not be limited to necropsies. Laparotomy is a widely used and accepted procedure for identifying sex of monomorphic birds that are captured live in mist nets and other types of traps (Risser 1971). An incision is made on the left side to expose the left testis or ovary, the latter being distin-

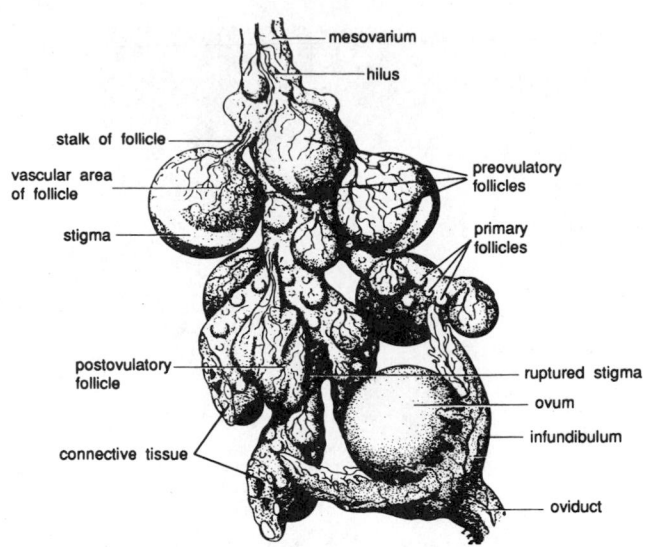

Fig. 4. Avian ovary showing a presumptive clutch (group of preovulatory follicles) within a hierarchy of follicles (copied with permission from Nelsen 1953). Also shown are a postovulatory (collapsed) follicle with a ruptured stigma and an ovum entering the infundibulum of the oviduct.

guished by the presence of follicles (Bailey 1953). The incision is small, and the procedure can be accomplished in the field without anesthetizing the bird (Wingfield and Farner 1976). Although an accurate classification or count of preovulatory follicles is not feasible under such conditions, laparotomy can provide useful information on the stage of follicular development and the proportion of birds nearing the egg-laying stage. If repeated observation of gonads and other organs in the same animal is required, the use of a fiber-optic scope (4–8 mm diam) is recommended (see LAPAROTOMY, LAPAROSCOPY, AND ULTRASONOGRAPHY [p. 291]).

CLUTCH SIZE

Estimates of clutch size are by far the most popular and, usually, the most practical method of estimating ovulation rate in birds. Fully active birds lay up to one egg per day. Release of the egg from the ovary, i.e., ovulation, precedes laying or deposition of the egg by approximately 26 hours. Thus, close observation of a laying bird provides immediate, real-time data on ovulation rate and an opportunity to study the temporal relationships among courtship behavior, copulation, ovulation, and associated hormone changes.

Nests found early in the incubation period provide not only a good estimate of the number of eggs produced but also a basis for estimating egg loss and hatching success. Furthermore, sampling procedures normally provide estimates of nest density, which in turn can be used to calculate the size of the breeding population of females and males in monogamous species. In this way, all components necessary for calculation of gross natality are available in a situation that seldom can be duplicated in mammalian studies.

EGG QUALITY

Because the avian egg contains all the nutrients and energy required by the young from conception through

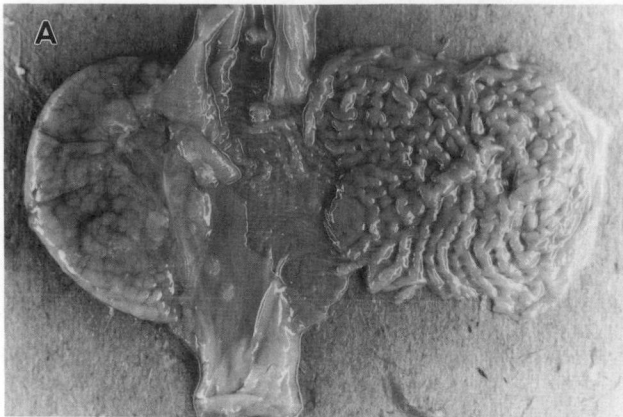

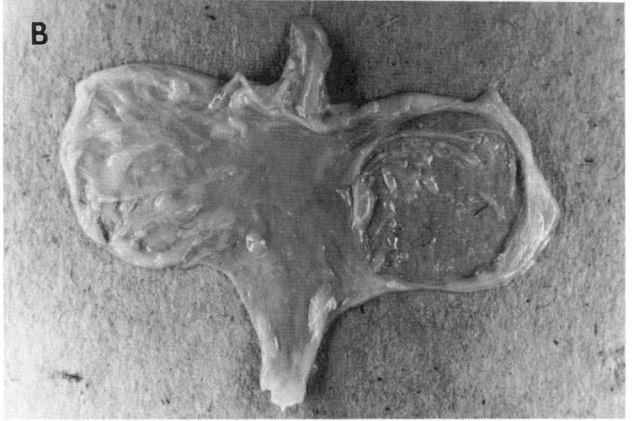

Fig. 5. An active (A) and inactive (B) crop gland from the mourning dove (photographs by R. Mirarchi).

the immediate posthatching period, egg size and quality are major factors determining the hatchability and survival of avian young. Ankney (1980) stimulated interest in this relationship by demonstrating a positive effect of increased egg size (including yolk and albumen content) on survivability of snow goose goslings. Beckerton and Middleton (1982) demonstrated that increased protein content (varying from 7.6% to 20.1%) in isocaloric diets was associated with linear increases ($P < 0.025$) in a series of nine reproductive parameters ranging from clutch size and egg weight to chick survival. Vangilder and Peterle (1981) observed a reduction in eggshell thickness and the proportion of yolk contained in eggs laid by mallards fed crude oil or DDE.

CROP GLAND OBSERVATIONS

Crop gland development is an index of reproductive activity that can be examined at necropsy. Doves, pigeons, and other members of the family Columbidae feed their young on a curd-like material (''pigeon milk'') produced by desquamation of epithelial cells lining the crop (Levi 1969). Development of the crop glands of pigeons and doves is apparent in both sexes by macroscopic appearance and increased weight (Guynn and Scanlon 1973) between the 9th day of incubation and the 14th day posthatching (Levi 1969:267–268, Mirarchi and Scanlon 1980:212) (Fig. 5). March and Sadlier (1970) described macroscopic changes in the appearance of the crop gland of band-tailed pigeons that distinguished inactive, growing, active, and declining crop glands. Because mourning doves breed throughout much of the year, crop glands from birds shot in early autumn hunting seasons can indicate the proportion of birds that are incubating or rearing young. However, crop gland regression may be prolonged in late summer, particularly in males (Books-Blenden et al. 1984), producing an overestimation of the proportion of nesting doves in the autumn harvest.

Reproductive Characteristics in Female Mammals

DETECTION OF ESTRUS

Beginning in the mid-1970s, wildlife biologists realized the importance of captive breeding in wildlife conservation. Not only have several endangered species programs had to rely upon captive breeding when prospects for natural reproduction in the wild were diminished (e.g., California condors and black-footed ferrets), but zoos also have entered the conservation arena and begun to employ more advanced techniques such as artificial insemination and embryo transfer in their programs. Estrus is the behavioral state of sexual receptivity associated with elevated estradiol and immediately preceding or coincident with ovulation. The ability to predict or at least detect the time of estrus is essential to efficient management of safe and compatible animal pairing in captive breeding programs.

Also, artificial insemination requires placement of an adequate number of viable sperm into the vagina within a relatively narrow time ''window'' of 6–24 hours of ovulation. This can be accomplished by monitoring follicular development around estrus by laparoscopy or ultrasonography (Ginther 1990).

The approach used by the livestock industry to detect estrus can be applied readily to many larger mammals. Females normally undergo a change in behavior that can be observed directly, i.e., greater tolerance or solicitation of male advances or greater physical activity in general. Indirect evidence of mating can be seen from roughened dorsal pelage or marks left by the brisket of the male that has been smeared with black grease. Ozoga and Verme (1975) detected a sharp increase in nocturnal activity of estrous does as evidenced by their frequency of passage through a microswitched gate installed in breeding pens.

Hormonal changes that occur throughout the estrous cycle induce changes in the relative proportion of leukocytes and epithelial cells found in the vaginae of rodents and certain other mammals (Fig. 6). In particular, a rise in circulating estradiol that occurs at estrus stimulates a rapid proliferation and sloughing of epithelial cells into the lumen of the vagina, such that a smear or aspirate of the vagina reveals a heavy concentration of cornified epithelial cells characteristic of estrus. Because each stage of the cycle is characterized by a particular vaginal cytology, estrous cycles can be monitored, and time of estrus can be predicted by this method (Box 3). The technique was first described for domestic rodents (guinea pigs, rats, and mice, Zarrow et al. 1964), and it has had its greatest application in this group, including some wild species such as deer mice (Clark 1936), pine voles (Kirkpatrick and Valentine 1970), and beavers (Doboszynska 1976). However, vaginal cytology also has been used successfully to

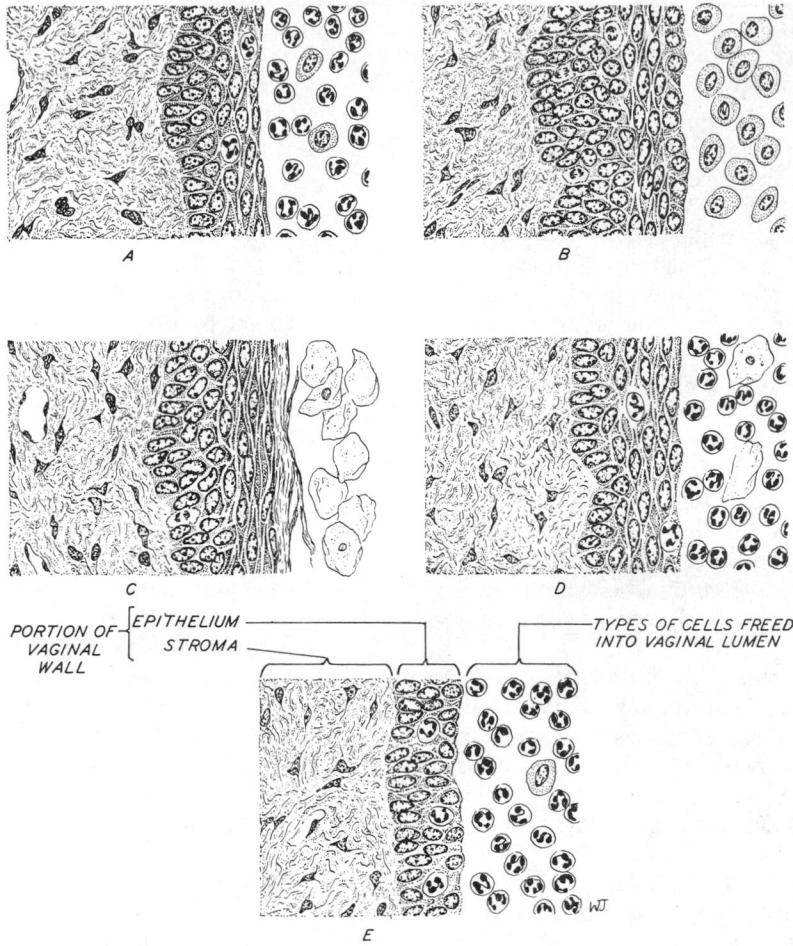

PORTION OF VAGINAL WALL { EPITHELIUM — STROMA — } TYPES OF CELLS FREED INTO VAGINAL LUMEN

Fig. 6. Sections through the vaginal wall of the rat illustrating changes in the proportion of epithelial cells and leukocytes found in smears taken from the vaginal lumen during each stage of the estrous cycle. (A) diestrus; (B) proestrus; (C) estrus; (D) metestrus; (E) adult animal that had been ovariectomized for 6 months (copied with permission from Turner and Bagnara 1976).

monitor estrous cycles in dogs and cats (Stabenfeldt and Shille 1977), coyotes (Kennelly and Johns 1976), and Virginia opossums (Jurgelski and Porter 1974). Trials on other species undoubtedly would extend this list.

OVULATION AND OVARIAN ANALYSIS

Next to neonatal survival, ovulation rate (number of ova shed per estrous cycle per female) is probably the single, most important reproductive characteristic in mammals. It is highly sensitive to nutritional condition, age, and, in some species (e.g., wolves), social status of the individual. In species with high fertilization and in utero survival rates, estimation of the ovulation rate is a reasonable (sometimes the only) alternative to direct estimation of litter size.

The eggs of eutherian mammals are small (70–120 μm) (Austin 1982) and retained within the female reproductive tract; therefore, they cannot be counted directly as is so conveniently accomplished with avian clutches. Direct enumeration of the number of eggs released in a given estrous cycle requires flushing eggs from the oviduct or uterus with saline solution and microscopic examination of the resulting fluid, a technique used routinely in embryology and reproductive physiology. Although this procedure is not difficult, it is time consuming, somewhat

tedious, and impractical in most wildlife investigations. Instead, ovaries are examined for follicular development or evidence of ovulation by the presence a corpus luteum (plural = corpora lutea) and related scar structures.

OVARIAN FUNCTION

Ovarian analysis requires a basic understanding of ovarian anatomy and physiology with respect to the events preceding and following ovulation. The mammalian ovary contains a lifelong supply of oogonia at birth, nearly all contained within small, primordial follicles (Fig. 7). With each estrous cycle, a small fraction of these follicles is recruited into a pool of more rapidly developing follicles, each of which grows larger and eventually develops a fluid-filled cavity or antrum. Granulosa and thecal cells comprising the walls of these growing follicles secrete increasing amounts of estrogen, which stimulates estrus and, indirectly, ovulation. Some of the antral (Graafian) follicles reach precisely the appropriate stage of preovulatory development to respond to the LH surge at estrus and ovulate. Most follicles, however, do not reach the preovulatory state, but instead undergo atresia, a degenerative process leading to disassociation of the granulosa layers of the follicle and death of the oocyte. In this regard, the ovulation rate in any given estrous cycle can be thought

Box 3. Characteristic vaginal cytology for each stage of the estrous cycle and procedure for collection (vaginal smear or lavage) and staining of cells from the vagina.

The estrous cycle is a sequence of changes in ovarian activity and physiology of the reproductive tract punctuated with recurring periods of sexual receptivity (estrus) and ovulation. Although estrous cycles of different species vary considerably in length (from a few days to several weeks) and in the timing of cytological changes around estrus, the following description of the 4- to 5-day estrous cycle of the laboratory rat (adapted from Turner and Bagnara 1976) is reasonably representative of the four stages in other species that exhibit cyclic vaginal cytology.

Diestrus.—This is a relatively long stage (60–70 hr) that extends into pregnancy if fertilization occurs or into anestrus during the nonbreeding season of many wild mammals. Corpora lutea begin to regress, and progesterone levels decline late in this stage. Leukocytes migrate through the thin vaginal mucosa and appear as the predominant cell type in vaginal smears (Fig. 6A).

Proestrus.—This stage, which precedes estrus and lasts for 17–21 hours, is also known as the follicular phase in species with longer estrous cycles. It is characterized by growth of preovulatory follicles, elevated estrogen levels, and swelling of the uteri. Nucleated epithelial cells dominate the vaginal smears collected at this time (Fig. 6B).

Estrus.—Sexual receptivity is high and limited to this period, which lasts for 9–15 hours. Ovulation occurs during or immediately after estrus. The vaginal epithelium proliferates rapidly, causing the upper layers to exfoliate into the vaginal lumen. Vaginal smears taken at this time are dominated by cornified (wrinkled) epithelial cells; few leukocytes are present (Fig. 6C).

Metestrus.—This stage, which lasts for 10–14 hours in the rat, begins with formation of CL following ovulation and is characterized by elevated progesterone levels. Metestrus and diestrus generally are known as the luteal phase in species with longer estrous cycles. Large numbers of leukocytes invade the vaginal lumen and often appear clumped around a few cornified epithelial cells in the vaginal smear (Fig. 6D).

The Vaginal Smear Procedure

1. Appropriate animal restraint varies with size and behavior of the animal under study, but many species, ranging from mice to opossums in size, can be handled by grasping the tail. The animal is allowed to stand on the top of a bench or a cage while the tail is raised to expose the vaginal orifice. With this approach, the animal's struggling is reduced and focused on escape, which directs the head and teeth away from the handler.

2a. Cells are collected from the vaginal lumen with a cotton swab that is moistened with physiological saline, inserted into the vagina, rotated, and removed. Cells are transferred by rolling the tip of the swab over the surface of a microscope slide.

2b. Alternatively, cells can be collected by vaginal lavage, the recommended approach for mouse-sized animals. With this procedure, the tip of a fire-polished Pasteur pipet or medicine dropper containing a drop of physiological saline is inserted a few millimeters into the vagina. The saline solution is aspirated several times to rinse the vaginal lumen and collect cells. A drop of the aspirated cell suspension is then placed on a microscope slide to dry.

3. The dried vaginal smear may be fixed by immersing the slide in methanol and allowing the smear to dry once more.

4. The smear then is stained by immersion in a methylene blue solution for 10–15 minutes. After the smear is rinsed with distilled water and dried, it is ready for microscopic examination.

Reliable monitoring of estrous cycles by vaginal cytology requires that observers be able to recognize three types of cells present in the smears:

Polymorphonuclear Leukocyte.—A small cell (less than one-half the size of epithelial cells) with a large, lobed nucleus and little visible cytoplasm. This cell is represented by small, dark-staining, C-shaped nuclei in vaginal smears.

Nucleated Epithelial Cell.—This large, rounded cell with a prominent nucleus (basal cell) appears in greatest numbers during proestrus, but it can be found in smears collected at all stages of the cycle.

Cornified Epithelial Cell.—This is a large, squamous cell with a wrinkled, "potato chip" appearance. The nucleus is degenerate and often not visible, even in stained preparations. It is present in such large numbers at estrus that a vaginal lavage takes on a milky appearance.

---

of as being determined by a dynamic balance of follicular development and atresia.

Ovulation results in rupture of the follicle (Fig. 7) that leaves a corpus hemorrhagicum (the ovulation point or blood spot) and initiates immediate luteinization of thecal and granulosa cells, i.e., they grow and sequester lipids. This process results in the filling of the cavity and formation of the corpus luteum (CL) (Fig. 7), a transient endocrine gland that secretes progesterone during the luteal phase of the estrous cycle.

If fertilization or pregnancy fails, the CL regresses and a new estrous cycle begins with growth of preovulatory follicles. If conception occurs and embryos implant in the uterus, the CL persists and secretes large amounts of progesterone throughout a substantial portion of the gestation period. A CL of pregnancy becomes large, often occupying much of the volume of the enlarged ovary (Fig. 7). Prior to or coincident with parturition, the CL regresses (i.e., degenerates), decreases in size, and ceases progesterone secretion. In some species, particularly those with long gestations (>3 months) and large CL of pregnancy, luteal regression is a prolonged process leading to the formation of pigmented scars known as corpora rubra, if red, or corpora albicantia if white (Fig. 7).

LAPAROTOMY, LAPAROSCOPY, AND ULTRASONOGRAPHY

Ovarian analysis traditionally has been based on material collected during necropsy of carcasses obtained at hunter check stations or on other sources, such as roadkills. However, physiological studies and reproductive manipulations such as artificial insemination require repeated observations of ovarian activity in live animals. Ovaries and reproductive tracts can be observed through laparotomy. Follis et al. (1972) described this procedure as performed on elk. Properly done, the procedure involves minimal risk. Only one mortality was observed in 51 mule deer that were observed in captivity or radio-monitored in the field after laparotomy (Zwank 1981). As with any surgery, laparotomy should be performed only by individuals with training in appropriate aseptic surgical technique and in accordance with governmental regulations.

With the application of fiber optics to surgical instruments, internal examination of live animals has become safer and more convenient through a technique known as laparoscopy. The abdominal wall is punctured with a large needle cannula through which a fiber-optic scope (4–8 mm diam) is inserted. Organs can be manipulated with a probe inserted through a second cannula. With this technique, ovarian follicles, CL, and, sometimes, uterine swellings can be observed. The animal must be anesthetized, but incisions per se are not made. Therefore, surgical trauma and risk of infection are minimized. Although laparoscopy is used extensively in medicine and reproductive physiology, applications in wildlife research have been limited. Nelson and Woolf (1983) used a portable generator and laparoscope to observe ovaries of white-tailed deer captured or immobilized in the field. No mortalities or complications related to this procedure were observed in 20 radio-collared deer monitored after the operation. Laparotomy and laparoscopy require special equipment and training and therefore are recommended only in situations where (1) multiple observations on the same animals are required, as in pen-based experiments, (2) the animal under study is rare or endangered, or (3) the animals are very expensive and little is known of their basic physiology, e.g., zoo animals.

Echoes of high-frequency sound (3–8 mHz), processed by real-time, computerized video displays, can be used to reveal internal morphology (Fig. 8). This technique, known as ultrasound or ultrasonography and used routinely in human medicine, is recognized as a practical and reliable approach to monitoring estrous cycles and gestation in domestic livestock and several wild species such as dolphins (Williamson et al. 1990), llamas (Adams et al. 1989), red deer, rhinos, giraffe, and guar (Ad-

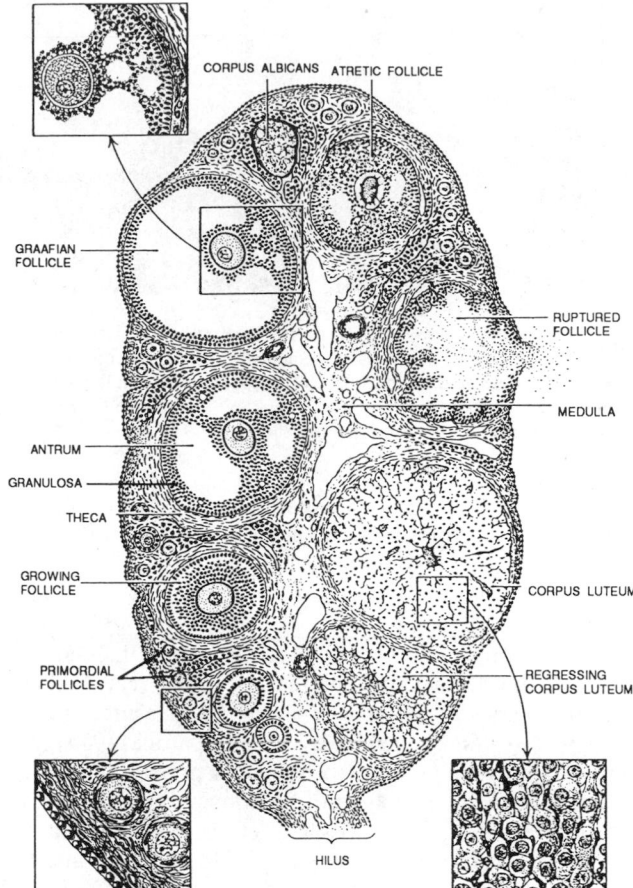

Fig. 7. Drawing of the mammalian ovary illustrating (in clockwise progression from lower left) follicular development (from primordial to Graafian follicle), ovulation, development of a corpus luteum (CL), and regression of the CL. The ovum is released with follicular fluid (shown escaping) from the ruptured follicle (copied with permission from Turner and Bagnara 1976).

ams et al. 1991). It is particularly effective in visualizing cyst-like structures such as ovarian follicles, because the liquid phase absorbs ultrasound and appears black in contrast to tissues that emit strong ultrasound echoes. A real-time ultrasonograph with transducer was used to monitor follicular development in mares (Palmer and Driancourt 1980, Ginther 1990). Pierson and Ginther (1987) achieved remarkable resolution in cattle (Fig. 8); follicles ≥2 mm in diameter were measured and counted, and growth of individual follicles (5–15 mm diam) was monitored with this technique in cows (Sirois and Fortune 1988). Ultrasonography with this level of reliability appears to have considerable potential for application in ovarian analysis of zoo and wildlife species, e.g., larger carnivores and ungulates. With this technique many aspects of ovarian activity, e.g., impending ovulation, can be monitored in living animals without the use of invasive surgical procedures.

ANALYSIS OF FOLLICLES

Qualitative assessment of the reproductive status of female mammals can be made through macroscopic examination of ovaries. However, in contrast to the situation in some birds, the ovulation rate cannot be estimated accu-

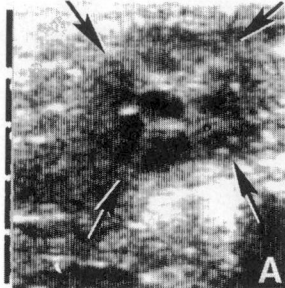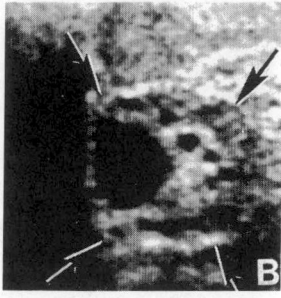

Fig. 8. Ultrasound images of ovaries in cattle; arrows mark the periphery of the ovary. (A) Three small follicles (5–7 mm in diameter) are visible as dark, round objects. (B) Several small (2–3 mm) follicles are visible to the right of the large (12 mm) follicle on this ovary (copied with permission from Pierson and Ginther 1988).

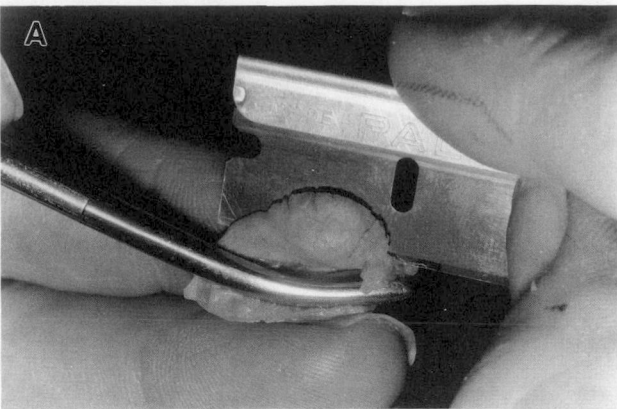

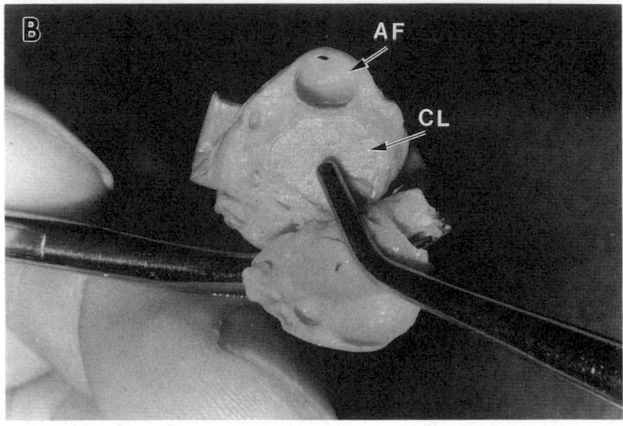

Fig. 9. (A) Procedure for slicing a fixed ovary with a razor blade and (B) a view of the sliced ovary showing an antral follicle (AF) and a CL (photos by D. Dennis).

rately or predicted from a count of preovulatory follicles. Many of these follicles become atretic and do not release ova. Furthermore, some large follicles and CL lie hidden below the surface of the ovary and can be seen only after the ovary is sliced at necropsy.

Counts of unruptured follicles can reveal the status of animals of different ages and seasons of the year. Follicles on the ovaries of larger species can be measured grossly and placed into diameter classes (Kirkpatrick 1974). Microscopic examination of histological preparations is required for ovarian analysis of most mammals (at least up to the size of squirrels) (Cowles et al. 1977), and it can be used to advantage in studies of larger mammals. For example, histology of white-tailed deer ovaries collected from August through November revealed no increase in the number of large (>3 mm) antral follicles, but the proportion of these follicles that were atretic decreased from 80% in September to 47% in early November, just before the peak date of conception (Harder and Moorhead 1980).

ENUMERATION OF CORPORA LUTEA

Although follicle counts will not predict ovulation rate in mammals, the ovary develops an unambiguous sign of ovulation, the CL, which in many species grows to occupy most of the volume of the ovary. In fact, in large mammals such as domestic cattle, CL can be palpated on the ovary through the overlying wall of the rectum with a gloved hand. This procedure, routinely used in the dairy and beef industry, has limited utility in wildlife studies because of size limitations, although it has been used with wapiti (Greer and Hawkins 1967).

CL can be counted, and ovulation rate estimated, in medium to large mammals (cottontails and larger) through gross (or with the aid of a dissecting microscope) examination of sliced ovaries obtained at necropsy (Fig. 9B). This approach has been applied widely to many species, including beavers (Provost 1962), moose (Simkin 1965, Hawley et al. 1982), badgers (Wright 1966), red foxes (Oleyar and McGinnes 1974), and cottontails (Zepp and Kirkpatrick 1976). Accessory CL (unovulated, luteinized follicles) and other structures unrelated to ovulation sometimes can be distinguished by size and appearance.

In general, CL counts provide an accurate measure of ovulation rate but only an index of the number of young in utero. This is because CL form during the normal course of each estrous cycle whether or not conception

occurs, and even though most animals collected from the wild with active CL will be pregnant, each ovulated follicle in a pregnant animal forms a CL whether or not the egg from each follicle is fertilized and develops. Thus, if the fertilization rate in a given species is low or if embryonic or fetal losses are high, CL counts will overestimate the number of young produced. For example, the Virginia opossum has a high ovulation rate (30 CL/ovary/cycle, Fleming and Harder 1983) but gives birth to only 10–20 young and weans 6–8. Fertilization rates and in utero survival vary considerably among species and must be determined for each (Brambell 1948), but in some species, such as white-tailed deer, they are high and ovulation rate is closely correlated with litter size.

DEER OVARIAN ANALYSIS

All CL leave scar tissue in the ovary as they regress. In most mammals they are visible only in microscopic examination as whitish bodies (corpora albicantia [CA]) of connective tissue. However, the large, long-lived CL of pregnancy in cervids regress slowly after parturition and are grossly visible for at least 8 months as pigmented CA (Cheatum 1949b, Box 4). Deer hunter check stations traditionally have provided biologists with the best opportunity for collection of a large number of deer ovaries. Unfortunately, hunting seasons often coincide with the breeding season (November, in many regions, into Janu-

Box 4.   Preservation of tissues and gross ovarian analysis.

Preservation of Tissues at Necropsy

Postmortem changes are slowed considerably at low temperatures (e.g., 4 C). Organs can be frozen for subsequent gross examination, but the ice crystals that form in the cytoplasm ruin the cells for microscopic study. If histology is planned, organs and tissues must be placed in a fixative solution at necropsy.

Fixatives are used prior to histology to (1) prevent putrification, (2) coagulate protein, and (3) protect the tissue against shrinkage and distortion in subsequent procedures. Buffered 10% formalin (1:10 dilution of 40% formaldehyde) is used widely, but it is only one of many options. A histology manual, e.g. Humason (1979), should be consulted for specific recommendations. The volume of fixative should exceed that of the tissue by 5 or 10 times to avoid excessive dilution of the fixative by water from the tissue.

Gross Examination of Deer Ovaries (after Cheatum 1949*b*)

1. The ovaries are removed by cutting the mesovarium, the mesentery that suspends the ovary in the body cavity near the ostium of the oviduct. Ovaries are more easily manipulated if some mesovarium remains with the ovaries, and left and right ovaries can be identified later if extra mesovarium is routinely left on the ovary from one side.

2. After ovaries have been in a fixative such as formalin for 36 hours, they will harden sufficiently to withstand slicing. Each is removed from the fixative and rinsed thoroughly in tap water. Latex gloves should be worn to reduce damage to skin on the hands.

3. An ovary can be secured by grasping the mesovarium close to the ovary with curved forceps. It is then sliced along the long axis with a scalpel or razor blade, cutting toward the mesovarium and forceps (Fig. 9A). With practice, horizontal slices of about 2-mm thickness can be cut, stopping just before the mesovarium is reached. In this way, the sliced ovary will stay together like pages of a book, ready for thorough examination. Hawley (1982) described a razor-blade device used to slice moose ovaries into uniform 1.5-mm sections.

4. Ovaries collected during the breeding season will contain several follicles of mixed size and, perhaps, recently ovulated follicles or new CL with ovulation points still evident. New CL of pregnancy grow to near full size (7-mm diam) within the first 2 or 3 weeks after ovulation and eventually occupy most of the ovarian volume of the pregnant deer. As such, they can be "followed" through several slices from one side of the ovary to the other. The sliced surface is solid, cheesy in texture, and creamy white in color. The color varies from yellowish to gray in other species.

5. Far less evident are the small corpora albicantia (CA), pigmented scars of the regressing corpora lutea of the previous pregnancy. Each slice of ovary must be examined carefully on both sides for these small (1–3-mm diam), rust-colored structures that often are compressed into triangular or crescent shapes by surrounding follicles and growing CL. Color is the primary distinguishing characteristic, but this too can vary from dark yellow to deep brownish orange.

6. Ovaries to be saved for further macroscopic or microscopic examination should be stored in 70% ethanol to prevent excessive hardening.

---

ary in southern areas such as Texas [Barron and Harwell 1973]), i.e., before all does have had an opportunity to ovulate and before fetuses are visible in utero. Therefore, considerable attention has focused on CA as a basis for estimating the number of fetuses carried to term in the previous pregnancy (preceding spring).

The use of CA in estimation of average litter size in previous pregnancies depends on knowledge of three variables: (1) the fertilization rate, (2) in utero survival, and (3) longevity of the CA. Fortunately, fertilization rate in deer is high and remarkably constant. A ratio of fetuses per CL (all does included) in midgestation in white-tailed deer is 86–87% (Roseberry and Klimstra 1970, Woolf and Harder 1979). The percentage is even higher (95–98%) when calculations are based on pregnant deer only (Ransom 1967, Harder and Peterle 1974). Thus, if CA remained visible for only 6–12 months after birth, they would provide an ideal substitute for uterine analysis of pregnant deer. Unfortunately, these regressing CL of pregnancy remain grossly visible for more than a year (Golley 1957, Trauger and Haugen 1965, Mansell 1971). CA accumulate (and perhaps even fragment) in older does so

that their inclusion in calculations leads to overestimation of ovulation and fetal rates. Determining the age of CA through histological examination (Mansell 1971) would largely eliminate biases in the estimation of previous litter sizes, but this seldom has been done.

Potential errors associated with CA data, particularly ambiguities concerning the number of reproductive seasons represented, have led to a reduction in the use of this technique for assessing reproductive rate in deer. This is unfortunate, because the method has great utility and practicality in providing an accurate estimate of the percentage of fawns breeding in a population. This demographic parameter is tremendously important, because the 6-month age class is by far the largest in any population and, therefore, one that has great potential in many parts of the range for impact on natality, population growth, and sustainable yield. The proportion of fawns that are pregnant in a population varies considerably, from 0 (Woolf and Harder 1979) to 77–82% (Nixon 1971, Haugen 1975). With few exceptions, the maximum ovulation rate for fawns that reach puberty and ovulate is one per female. Thus, yearlings that are killed during the hunting season

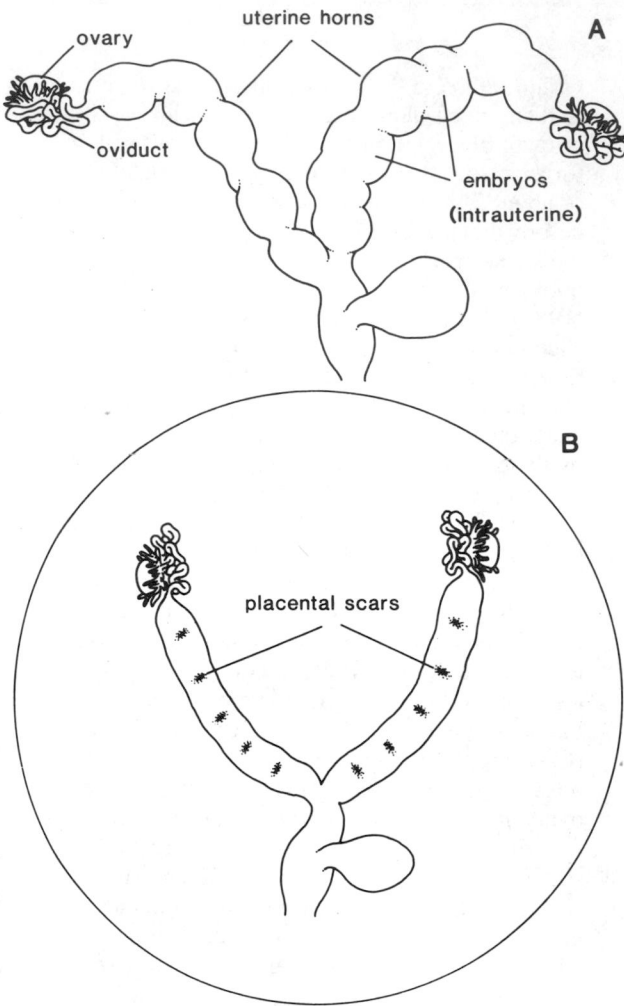

Fig. 10. (A) Uterine swellings in a pregnant white-footed mouse, and (B) uterine horns compressed between the lid and inverted base of a Petri dish to reveal placental scars (drawings by D. Dennis).

or from other causes (e.g., collision with automobiles) during the months of July through February will yield reliable CA data because, for all practical purposes, only two possibilities exist: zero or one CA per doe.

UTERINE ANALYSIS AND FETAL COUNTS

Enumeration of embryos or fetuses in utero has been a popular and convenient measure of the reproductive performance of a population. It is popular because the number in utero, especially during the third trimester of gestation, is often a reliable indicator of the number of young that will be born, i.e., litter size. After placental attachment and pregnancy are firmly established (after the first trimester of gestation), in utero mortality and resorption are reduced substantially in most species. Uterine examination for embryos or fetuses is convenient, particularly in later stages of gestation, because fetuses are grossly visible and can be counted reliably by individuals with minimal training, including hunters or law enforcement personnel at road-kills. The entire uterus with fetuses may be collected and frozen or fixed for later study, or it can be inspected on the spot (Fig. 10A). Crown-rump length measurements can be used to age deer (Armstrong 1950)

and coyote (Kennelly and Johns 1976) fetuses and, by backdating, to estimate breeding dates. This technique is detailed in Chapter 8. Also, a fetus scale for white-tailed deer that converts fetal size to conception date is available from Forestry Suppliers, Inc. (Jackson, MS 39384). The primary sex ratio for a population also can be estimated from examination of fetuses, a parameter of practical value in population models and of considerable interest among theoretical and experimental ecologists (Trivers and Willard 1973, Austad and Sunquist 1986, Gosling 1986).

Fetal counts are most often done at necropsy, but in many instances this information must be obtained from living animals, such as with investigations on rare and endangered species or zoo specimens. Uterine swellings, indicative of fetuses, and CL of pregnancy can be counted in living animals by laparotomy. This approach has been used for several species, including cottontails (Murphy et al. 1973), mule deer (Zwank 1981), and elk (Follis et al. 1972). Transabdominal ultrasonography has been used to diagnose pregnancy in bighorn sheep (Harper and Cohen (1985), fallow deer (Mulley et al. 1987), and bottle-nosed dolphins (Williamson et al. 1990). Several noninvasive methods, based on hormonal and biochemical changes in blood, are available for diagnosing pregnancy and are discussed later.

PLACENTAL SCARS

Placental scars are pigmented areas of uterine tissue marking sites of previous placental attachment (Fig. 10B). Their formation, described by Deno (1937) and Martin et al. (1976), is limited to mammalian taxa with deciduous placentae (Wydoski and Davis 1961). In these species there is an erosion of the uterine endometrium by the embryonic membranes and an interdigitation of uterine and chorionic tissue such that endometrial tissue is torn away when the placenta is expelled at birth, i.e., the placenta is deciduous (Vaughan 1986). As the new uterine endometrium grows over this wound, stagnant pools of blood become trapped, and the hemoglobin in the red blood cells is then degraded to hemosiderin (an iron-containing pigment) by macrophages. The entrapped hemosiderin remains visible as a placental scar for varying lengths of time, depending on the species.

Species that develop prominent placental scars belong primarily to the orders Insectivora, Chiroptera, Lagomorpha, Rodentia, and Carnivora, although they have also been described in elephants (Laws 1967). Litter size has been estimated from placental scars in a variety of carnivores including brown bears (Hensel et al. 1969), raccoons (Sanderson 1950), badgers (Wright 1966) and gray foxes (Oleyar and McGinnes 1974). Placental scars are most useful in mammals that have only one or two litters per year, such as beavers (Henry and Bookhout 1969) or gray squirrels (Nixon et al. 1975). The smaller rodents, which have several litters per season, often have uteri with two or more "sets" of scars that differ in size and opacity. These sets represent different litters or pregnancies (Rolan and Gier 1967, Martin et al. 1976) and should be enumerated separately.

In many species, placental scars can be seen easily in fresh, thawed, or preserved tissue without special treatment. They stand out as darkened spots or bands in the uterine horns. If females are collected soon after parturition, the tissue around the old implantation sites will be swollen. With increasing time after parturition, however, the scars fade and additional steps must be taken to clearly visualize the scars. Perhaps the simplest first step, at least with small mammals, is to compress the uterine horns between two microscope slides or between the nested lid and base of a petri dish (Fig. 10B). The scars then can be viewed on a dissecting scope (10 ×), and with backlighting it is possible to distinguish and count scars in sets, based on similarities in size and opacity. With larger species, it usually is necessary to expose the endometrium of the uterus by cutting longitudinally along the length of each horn with scissors. The scars then appear as darkened bands or discs in the uterine lumen.

In some species, placental scars are indistinct, and special clearing or staining procedures are required to make them more visible. The two most common are the clearing technique (Orsini 1962) and the Prussian Blue reaction (Humason 1979). Because these procedures are complex, the reader is referred to the original publications for details.

As with most indices, placental scars provide only an approximation of litter sizes. They represent, however, a relatively close estimate for most species in which their reliability has been verified. One source of error arises from the fact that resorbed fetuses also leave placental scars, which can be indistinguishable from those left by fetuses developing to term (Conaway 1955).

LACTATION

The transition from late gestation to lactation is a critical period in the reproductive cycle of mammals. Mammary glands of lactating females grow and fill with milk, which can be expressed from the teats of all but the smallest species. If milk cannot be expressed from a female at necropsy, the mammary gland should be sliced open and inspected for pools of milk within the tissue. The nipples of lactating females become swollen and pinkish, and often the fur immediately surrounding the nipple is thin or absent. These and other signs of lactation are only indirect signs that a female is nursing or has recently weaned her young. They should be verified by each investigator, preferably through measurements of females known to be nursing young. Sauer and Severinghaus (1977) compared, in autumn, the teat length of yearling does with milk in their udders (recently suckling young) to that of yearling does without. They concluded that any yearling with one of four teats <10 mm was without young and those with one >15 mm was nursing or had recently weaned a fawn. Those with teats of intermediate size were placed in the uncertain category. This index needs additional testing and verification, but it is particularly attractive because it reflects net natality for the largest age class in most populations and because data can be collected from field-dressed carcasses brought to hunter check stations.

Given the reality of substantial neonatal mortality in many wildlife populations, an estimate of the proportion of females in a population that are lactating might be more relevant to reproductive success than are parameters measured earlier in the reproductive cycle, e.g., ovulation rate. Coupled with estimates of average litter sizes, lactation indices could, in some situations, substantially improve

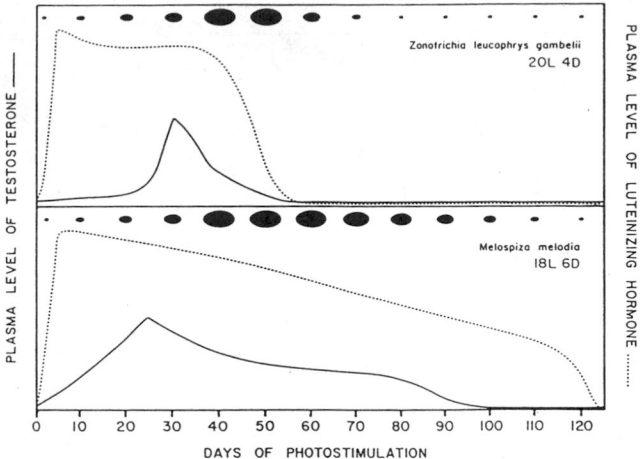

Fig. 11. Plasma levels of luteinizing hormone (LH), testosterone, and relative testis size (solid ovals) in photostimulated male Gambel's sparrows and song sparrows (copied with permission from Wingfield and Moore 1987).

estimates of net natality, i.e., the number of young weaned per female. At the very least, inspection of mammary glands and nipples or teats for evidence of recent activity is a convenient, noninvasive method for monitoring reproductive activity in a population.

REPRODUCTIVE ENDOCRINOLOGY
Hormone Profiles

Each event or stage in the reproductive cycle from courtship and fertilization to egg incubation or lactation is controlled by a sequence of hormonal signals. These are best visualized as plots of blood hormone concentrations (measured in plasma or serum) over time, known as profiles (Figs. 11–13). Many hormones are involved in the control of male and female reproductive processes. However, much has been learned of these processes in wildlife species through the study of a relatively few pituitary hormones and gonadal steroids, providing the biologist with new techniques and alternatives for the study of wildlife reproduction (Table 2). The literature on wildlife reproductive endocrinology has expanded rapidly since the mid-1970s to the point that for some species, such as the tammar wallaby, our understanding of hormonal mechanisms controlling ovarian function, seasonal breeding, and parturition rivals that for more well-established laboratory and domestic species. The extent to which wildlife reproductive endocrinology can advance ecology, systematics, and the frontiers of reproductive biology is beautifully illustrated in a recent book on the reproductive physiology of marsupials by Tyndale-Biscoe and Renfree (1987).

MALE BIRDS AND MAMMALS

The marked growth of the avian testis during the onset of the breeding season is associated with rapid increases in secretion of follicle stimulating hormone (FSH) and luteinizing hormone (LH) from the anterior pituitary gland and testosterone from the testes (Fig. 11, Table 2). Modern techniques permit measurements of hormones in small blood samples (100–500 µl), volumes that can be obtained from small wild birds. This has facilitated diverse experiments in field endocrinology (Wingfield and Moore

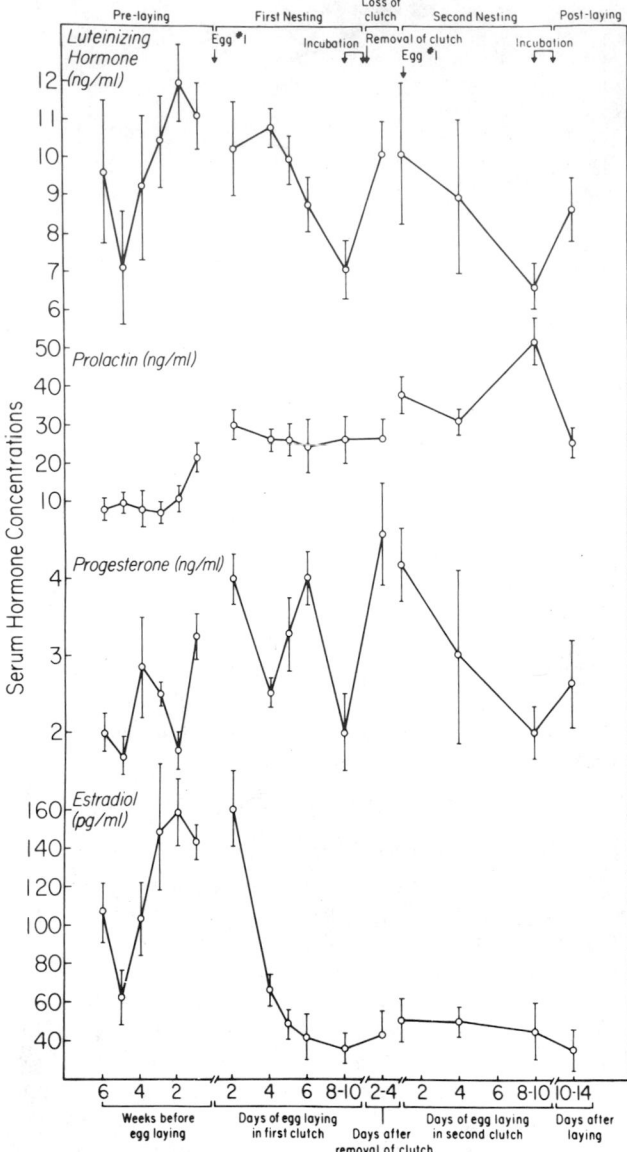

Fig. 12. Serum levels of luteinizing hormone (LH), prolactin, estradiol, and progesterone during the laying cycle of canvasbacks (copied with permission from Bluhm et al. 1983).

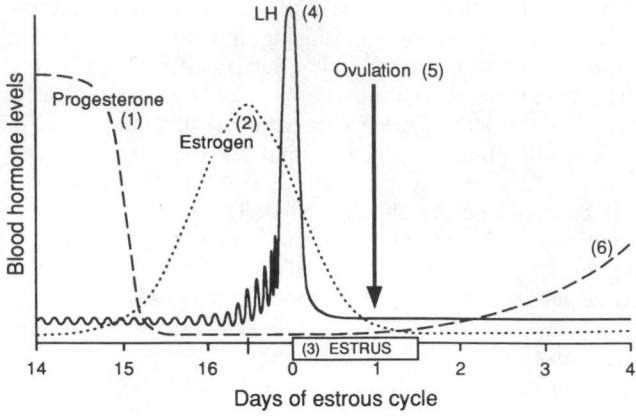

Fig. 13. Temporal relationships of estrus and ovulation to circulating levels of progesterone (P), estrogen (E), and LH during the estrous cycle of the ewe, which has preovulatory hormone dynamics typical of many mammals. A decline in P (1) is followed by a rise in E (2) which stimulates estrus (3) and the LH surge (4). The LH surge stimulates ovulation (5) and formation of the CL, which is accompanied by rising levels of P (6) (from Short 1972) (reprinted with permission from Cambridge University Press).

or reproductively active males, even in the absence of detailed behavioral observations.

FEMALE BIRDS

The sequence of hormonal signals responsible for control of the circadian cycle of ovulation and egg laying is complex, but a major component in the domestic hen is a positive feedback relationship between progesterone from the ovarian follicle and LH release from the anterior pituitary, which stimulates ovulation (Johnson 1986). Prior to ovulation, elevated estrogen also induces courtship behavior (Fig. 12, Table 2) and stimulates mobilization of yolk precursors from the liver and their deposition in the egg while it is in the follicle (van Tienhoven 1983). Incubation behavior and brood patch formation are stimulated by high prolactin levels in association with estradiol and progesterone (Fig. 12). Prolactin from the anterior pituitary also stimulates production of "pigeon milk" in the crops of doves and pigeons.

Measurements of circulating hormones are useful in studies of toxic substances and in investigations of low reproductive performance of birds. Progesterone levels peaked earlier relative to oviposition in mourning doves treated with dietary polychlorinated biphenyls than in control birds (Koval et al. 1987). Nonlaying canvasbacks had lower serum concentrations of prolactin and LH than did laying ducks, and progesterone levels in laying ducks increased during the breeding season, whereas those in nonlaying birds declined (Bluhm et al. 1983).

FEMALE MAMMALS

The estrous cycle is a sequence of interrelated physiological events in the hypothalamus, anterior pituitary, ovary, and reproductive tract marked by a period of recurring sexual receptivity (estrus) and ovulation. Preovulatory follicles secrete increasing amounts of estradiol that stimulate estrus and the release of a large amount of LH from the anterior pituitary. This LH surge, in turn, stimulates ovulation and formation of the CL (Fig. 13). This general

1987) that have revealed, for example, significant increases in circulating levels of testosterone in male song sparrows actively defending territories compared to those with stable (uncontested) territorial boundaries (Wingfield 1985).

In mammals, circulating levels of testosterone and other androgens also increase substantially during the breeding season (Table 2), a requisite condition for full spermatogenic activity and breeding behavior. Circulating testosterone levels also are high during musth (an aggressive behavioral state) in Asian elephants (Cooper et al. 1990). In some species, such as the tammar wallaby and domestic cattle, full seasonal elevation of testosterone appears to be dependent on direct association with females in breeding condition (Kantongol et al. 1971, Catling and Sutherland 1980). If this occurs in other polygynous species, it would provide a potential method for identifying the dominant

Table 2. Examples of hormone data applicable to the assessment of reproductive activity in selected birds and mammals. All data, except testosterone, are from females. Hormone concentrations are pg/ml serum or plasma for estrogens and ng/ml serum or plasma for all others, except steroid conjugates in urine which are expressed in μg, ng, or pg/mg of creatinine. Two concentrations indicate the approximate lows and highs associated with a given reproductive event.

Species	Hormone	Change in concentration	Reproductive event	Reference
Turkey (domestic)	Prolactin	90–709	Elevated during incubation, low in laying and brooding hens	Wentworth et al. 1983
Turkey (domestic)	Growth hormone	7–31	Increases from incubation to high levels during brooding	Wentworth et al. 1983
Turkey (wild)	Cholesterol	138–191	Increases during breeding season	Martin et al. 1981
White-crowned sparrow	Estradiol	35–400	Increases with courtship and egg laying	Wingfield and Farner 1980
Canvasback	Progesterone	2–4	Low levels 4 weeks prior to laying, highest at laying	Bluhm et al. 1983
Japanese quail	Testosterone	0–5	Increases with day length and onset of breeding season	Follet and Maung 1978
White-crowned sparrow	Testosterone	1–4	Increases from migration to courtship and egg laying	Wingfield and Farner 1979
Woodchuck	Testosterone	0–3	Increases during breeding season	Baldwin et al. 1985
White-tailed deer	Testosterone	0–3	Increases with hardened antlers and rut	McMillin et al. 1974
Little brown bat	Progesterone	7–136	Increased from early to late pregnancy	Buchanan and Young-lai 1986
Woodchuck	Progesterone	0–60	Progesterone higher in postpartum than pregnant females	Concannon et al. 1983
White-tailed deer	Progesterone	0–6	Rises with CL formation and in early pregnancy	Plotka et al. 1977, Harder and Moorhead 1980
White-tailed deer	Estrogens	119–295	Increases near parturition	Harder and Woolf 1976
Rhinoceros, Indian	Estrone sulfate	47–1	High at estrus and drops rapidly at ovulation	Kasman et al. 1986
Okapi (Giraffidae)	Pregnanediol-3-glucuronide	1–24	Rises with luteal phase of cycle (active CL)	Loskutoff et al. 1982
Killer whale	Estrone conjugates	0–35	Increases prior to presumptive ovulation	Walker et al. 1988
Killer whale	Pregnanediol-3-glucuronide	0–100	Elevated during pregnancy	Walker et al. 1988

pattern appears to hold true for many mammals as evidenced from data from such widely divergent taxa as wallabies (Harder et al. 1985), rats (Nequin et al. 1979), deer (Plotka et al. 1980), and sheep (Hauger et al. 1977). The periodic rise in estrogen around the time of estrus confirms normal ovarian activity and is potentially valuable as a predictor of the time of ovulation in mammals that ovulate spontaneously. Such knowledge is essential for artificial insemination and embryo transfer, techniques that are being used with increasing frequency by zoos, endangered species programs, and modern game farms. Jacobson et al. (1989) achieved conception in 75% of 53 trials of artificial insemination in white-tailed deer.

Although valuable in assessing ovarian activity, changes in estrogen levels prior to estrus and ovulation often are small, of short duration, and difficult to detect. By contrast, the events following ovulation, namely CL formation and the luteal phase of the estrous cycle or gestation, are of relatively long duration (several days to months, in larger species) and are characterized by elevated levels of circulating progesterone (secreted by the CL and/or placenta) (Fig. 13). Therefore, useful progesterone profiles can be obtained even with the relatively low blood-sampling frequencies that are typical of many wildlife studies. In fact, if breeding is highly synchronized, much can be learned from single samples taken from animals soon after they are shot (Wesson et al. 1979b). In white-tailed deer, this approach revealed a shortened, nonfertile cycle that preceded the first estrus and normal luteal cycle during the onset of the breeding season (Harder and Moorhead 1980). Progesterone profiles are considered the only reliable means of monitoring the 15-week estrous cycles of captive Asian and African elephants (Plotka et al. 1988). In fact, until a validated, sensitive, and accurate progesterone radioimmunoassay (RIA) was applied to a long series of serum samples from the same cows (Hess et al. 1983), the estrous cycle of the Asian elephant was believed to be only 3 weeks long!

The practical value of progesterone data for detecting pregnancy in large mammals has been recognized widely. Blood can be collected in the field from trapped or tranquilized animals (see below for cautions) and assayed later, whereas other techniques, e.g., laparotomy or ultrasonography, require transport of the animal or equipment. For example, plasma progesterone concentrations in pregnant white-tailed deer are generally >2 ng/ml, whereas those of nonpregnant deer are <1 ng/ml (Abler et al. 1976). The potential errors associated with this generalization (Plotka et al. 1983) notwithstanding, properly validated progesterone assays have permitted accurate (<2% error) diagnosis of pregnancy in white-tailed deer and mule deer (Wood et al. 1986). Gadsby et al. (1972) reported higher progesterone levels in domestic ewes carrying two fetuses than in ewes with a single fetus. Vogelsang (1977) reported a similar relationship in white-tailed deer, but this has not been confirmed by other studies or clearly documented in other wildlife species.

The uterus and placenta secrete numerous nutritive and regulatory proteins during gestation, some of them unique to gestation and therefore useful in identifying pregnant animals. Wood et al. (1986) used a qualitative test for pregnancy-specific protein B (bovine) to identify (with 4% error) pregnant mule deer and white-tailed deer. Similar results were reported with this pregnancy test in mountain goats (Houston et al. 1986) and muskoxen (Rowell et al. 1989).

Hormone Metabolites in Urine and Feces

For many wild species, even pen-reared individuals, stress associated with restraint and venipunctures often precludes collection of blood samples. Fortunately, another approach is available. Steroids (e.g., progesterone) are metabolized in the liver and excreted as conjugates (primarily sulfates and glucuronides), and during the last decade considerable progress has been made in monitoring gestation and estrous cycles of diverse mammals such as rhinoceroses, primates, and pandas through assay of hormone metabolites in their urine and feces (for reviews see Loskutoff et al. 1983 and Safar-Hermann et al. 1987). In some species, such as the pig-tailed macaque, concentrations of estrogens and progestins in the feces can be used to distinguish the follicular, preovulatory, and luteal phases of the cycle as well as pregnancy (Wasser et al. 1988). Kirkpatrick et al. (1988) described detailed validation experiments and procedures for estimation of estrone sulfate concentrations in soil soaked with urine by free-roaming feral horses. Twelve of 15 mares with estrone sulfate >1.0 μg/mg creatinine later produced foals, whereas none of the mares with lower concentrations foaled.

The foregoing summaries of hormone dynamics in avian and mammalian reproductive cycles are intentionally brief and simplified. They are offered as an introduction to the major hormonal signals underlying the reproductive events discussed in previous sections and as background for interpretation of endocrinological techniques and results applied to wildlife research and management. For more thorough treatments of these topics, the reader is referred to van Tienhoven (1983), Austin and Short (1984), Johnson (1986), and Knobil and Neill (1988).

Collection of Samples and Hormone Measurements

BLOOD SAMPLING PROCEDURES

Hormones can be measured in all tissue types with the application of appropriate tissue homogenization and hormone extraction techniques, but blood is by far the type most commonly assayed. The goal in any sampling procedure should be to collect an adequate volume of blood quickly and efficiently while minimizing stress to the animal. Reaching this goal often requires the administration of a sedative or anesthetic, although this step might be omitted for domesticated species or conditioned animals.

The effect of restraint or anesthetics on hormone levels should be investigated, particularly if progesterone or adrenal hormones are to be measured. For example, immobilization with succinylcholine chloride elevated circulating progesterone levels in white-tailed deer (Wesson et al. 1979c), and certain anesthetics depressed serum concentrations levels of this steroid (Plotka et al. 1983). Also, the stress of prolonged (15–45 min) restraint can induce significant release of corticosteroids, including progesterone (Plotka et al. 1983).

The effects of restraint or anesthesia on blood hormone levels can be investigated by collecting blood samples from quiet animals fitted with an indwelling catheter. Periodic samples will reveal the natural diel pattern of circulating levels, including episodic changes that can be distinguished from those that might be related to the stress of handling or anesthesia in the same animal. An alternative (and less instructive) approach is to collect blood samples immediately upon restraint of the animal. The animal then can be anesthetized or restrained with standard procedures for collection of blood samples at intervals over a time period that would be in excess of the maximum required for any routine collection. If the blood sampling procedure alters secretion of the hormone(s) under study, blood concentrations will change in successive samples relative to that in the initial sample. Such experiments will determine the need to standardize the timing of blood collection, i.e., time from restraint or anesthesia.

Peripheral blood (i.e., from the heart or any vein not directly draining the endocrine gland under study) is usually collected from the jugular or a prominent leg or tail vein. The orbital plexus in the corner of the eye may be used to obtain small volumes from mice. The brachial (wing) vein often is used on birds, although Arora (1979) concluded that the jugular vein was the best source of blood from Japanese quail. Blood collection from living animals should be performed only by fully trained personnel. This is particularly important when blood is obtained by cardiac puncture, in which case the animal should be anesthetized or sedated to alleviate pain and prevent its movement while the needle is in the heart.

If plasma is to be separated from blood, the needle and syringe first are rinsed with a sterile heparin solution or other anticoagulant. Syringes or tubes containing blood should be placed in crushed ice to cool before the plasma is separated by centrifugation. Alternatively, if serum is to be obtained, blood is allowed to clot for 2–3 hours at room temperature or overnight in a refrigerator. The serum then can be poured carefully from the tube or removed by pipette; any remaining cellular material is separated from the serum by centrifugation.

Hormone concentrations can be measured in either plasma or serum, but plasma sometimes is preferred because the blood can be chilled and centrifuged immediately after collection. This is important if the hormone under study is temperature sensitive or subject to degradation when it is in contact with blood cells before separation of serum or plasma (Vahdat et al. 1984), as is progesterone in cattle (Vahdat et al. 1984) and muskoxen (Rowell and Flood 1987). In any event, blood destined for hormone assays should be chilled immediately after collection and centrifuged as quickly as possible with a uniform time interval between collection and centrifugation for all samples (Wiseman et al. 1982). Blood and other biological specimens also should be protected from direct sunlight to avoid possible photo-oxidation of compounds under study. Plasma and serum should be stored frozen at -15 C or lower and assayed as soon as practical to avoid degradation of hormones, although steroid hormone concentrations do not appear to change in plasma and serum stored frozen for periods of 3–8 years.

For collection of urine, midstream samples are preferred, and many species can be trained for this procedure, even killer whales (Table 2). Individualized urine or feces (from observed animals) also can be collected from the floor of pens, small enclosures, or even open range (Kirkpatrick et al. 1988). Urine and feces may be stored at -15 C or lower, but if fecal samples are contaminated with urine, they probably should be stored in ethanol to eliminate conversion of urinary conjugates (Wasser et al. 1988). Lyophilization of samples to adjust for water content improved the estimates of estradiol and progestin concentrations in macaque feces (Wasser et al. 1988).

Review of Figs. 11–13 should make clear the importance of timing in collection of blood and other tissue samples for studies of reproductive physiology. Hormones often are secreted episodically and some, such as testosterone, may show a clear, diel pattern. Thus, blood samples should be collected with sufficient frequency (e.g., every 4–6 hr) to detect such changes or at the same time of day (in studies employing low sampling frequency) to reduce variance in the estimated values. In most field studies, the exact stage of egg laying or estrous cycle is not known, and blood samples must be grouped to represent broad categories of reproductive activity, e.g., courtship, incubation, pregnancy, or lactation. The problem of temporal specificity in blood sampling is particularly complex relative to the ovulatory cycle of birds, because many changes occur within a 24- to 30-hour period.

RADIOIMMUNOASSAY

Because radioimmunoassay (RIA) is highly specific and sensitive, reliable estimates of hormone concentrations can be obtained from small plasma samples, typically 0.1–0.5 ml; sometimes as little as 0.02 ml is adequate. Not surprisingly, RIA not only has revolutionized the study of reproductive physiology since its first wide application to the field in the early 1970s, but it also has found ready application in field-oriented disciplines such as wildlife biology and has legitimized the notion of true field endocrinology (Wingfield and Moore 1987).

RIA employs three key reagents: (1) an antiserum that will selectively bind the hormone under study, (2) a radiolabeled form of the hormone, e.g., ^{3}H-progesterone, and (3) known, standard concentrations of the unlabeled hormone (e.g., progesterone). The antiserum is diluted to the point that it will bind only about 50% of a fixed amount of labeled hormone so that the addition of unlabeled hormone (from standard solutions or a sample) to the same tubes will displace some of the labeled hormone from the antibody in a dose-related manner. Unbound steroid is adsorbed on charcoal, and the radiolabeled hormone bound to the antibody is counted in a liquid scintillation spectrometer. The technique is highly sensitive and capable of measuring 2–5 pg/ml plasma (a picogram [pg] $= 1 \times 10^{-12}$ g and a nanogram [ng] $= 1 \times 10^{-9}$ g). A more thorough explanation of the principles underlying RIA and their application to the measurement of a variety of hormones can be found in Jaffe and Behrman (1974).

A word of caution is in order regarding the use of RIA in wildlife research. Underlying seemingly straightforward procedures are complex interactions that must be controlled and standardized carefully. Anyone contemplating the use of RIA should first train in a laboratory that specializes in routine application of the procedure. Most importantly, each laboratory must establish and validate procedures for each hormone in each species under study. Results of validation experiments, designed to demonstrate accuracy, precision, and quality control (Abraham et al. 1977, Jeffcoate 1981) should be published and, in fact, are required in manuscripts submitted to many endocrine journals (see Notice to Contributors, *Journal of Reproduction and Fertility,* 1991, 93[1]:ii–iv).

Reagents and procedures set forth in commercial RIA "kits" most often have been established only for human, rat, or monkey plasma and should not be used for other species or tissues unless validated. Improper application of such kits to biological material from other species can lead to highly inaccurate and misleading results. Common problems include variable hormone extraction efficiency and nonspecific interference of hormone-antibody binding. For example, the high lipid content of blood of laying hens can interfere with steroid extraction as well as the binding and charcoal separation phases of RIA. Also, gonadotropic hormones such as LH exhibit species-specific molecular structures that complicate the validation of assays based on antibodies raised to the gonadotropin of another species. For example, anti-LH raised against sheep LH has been used to measure LH in a wide range of mammalian species, but such heterologous assays require rigorous validation and usually are less sensitive than when the anti-sheep LH is used to measure LH in samples from sheep.

PHYSIOLOGICAL RESPONSES TO STRESS

More than 50 years ago, Selye's (1936) landmark paper on the symptoms of stress on human health was published. Stress, according to Selye (1976), is the nonspecific response of the body to any demand, or, in other words, nonspecific changes in an organism caused by an emotional or physical disturbance. A stressor is any stimulus that elicits stress, e.g., pain, fear, cold, blood loss, environmental contaminants, pathogenic microbes, or social tension.

Prolonged or chronic stress leads to activation of the hypothalamic-pituitary-adrenal (HPA) axis, and elevated blood levels of corticotropin-releasing factor (CRF) stimulate the release of adrenocorticotropic hormone (ACTH), which stimulates the secretion of steroids by the adrenal cortex. These include those that regulate glucose metabolism (i.e., glucocorticoids), primarily cortisol and corticosterone (Asterita 1985). This response is adaptive in that it allows individuals to maintain daily activities in the presence of a stressor. However, prolonged activation of the HPA axis also is associated with pathological conditions such as gastrointestinal ulcers (Moberg 1985) and reduced reproductive performance. CRF, ACTH, β-endorphin, and corticosteroids appear to play key roles in modulating the effects of stress on reproductive functions; a current understanding of the hormonal mechanisms involved in this modulation was summarized by Rivier and Rivest (1991).

Christian (1950) and colleagues were quick to recognize the potential relevance for population ecology of Selye's (1946) observations. They postulated that elevated corticosteroid levels interfered with hormonal control of reproduction and induced mortality in high-density populations (Christian 1963, Christian and Davis 1964). This chapter cannot do justice to the numerous studies that have addressed this controversial hypothesis, but, in summary, emigration and other factors such as nutrition appear to prevent most natural populations from attaining densities sufficiently high to evoke a pathological hormonal response that would be regulatory at the population level. Experiments with high-density snowshoe hare (Windberg and Keith 1976, Vaughan and Keith 1981) and deer populations (Seal et al. 1983) support this conclusion. By contrast, studies that have focused on segments of a population (e.g., sex or social status) have produced the clearest evidence for endocrine responses to social stress (Carrick 1963, McDonald et al. 1981, Sapolsky 1987), which suggests that more attention should be given to stress as a factor in the natural structure and functioning of populations. With increasing emphasis on work with endangered species, biologists also need to consider the impact of acute stress on trapped or immobilized animals and chronic stress on animals held captive for controlled research.

Measures of physiological response of stress are numerous, but physiological responses to acute stress should be measured under controlled laboratory conditions (Carruthers and Path 1983). The techniques described below are those most often used in field or pen studies involving prolonged stress.

Adrenal Weights

The most consistent and well-documented response to prolonged stress is an increase in ACTH and the resultant stimulation of corticosteroid secretion. Increased secretory activity of the adrenal cortex is accompanied by an increase in size of the gland such that adrenal weight generally has been accepted as a valid index of adrenocortical activity (Christian 1963, Bronson and Eleftheriou 1964, Adams and Hane 1972). The adrenal gland is actually two glands in one, being made up of chromaffin tissue (neural tissue that secretes epinephrine and norepinephrine) and steroidogenic tissue. In mammals, chromaffin tissue is localized in the medulla and steroidogenic tissue in the cortex, but in the adrenals of birds and other vertebrates, chromaffin tissue is scattered in pockets throughout the steroidogenic tissue. The increased adrenal weight that occurs after prolonged exposure to a stressor stems from an enlargement of the cortex or steroidogenic tissue; the contribution of the medulla or chromaffin tissue to adrenal weight change is negligible (Christian 1963).

Adrenal weights (usually paired) can be obtained readily from glands collected at necropsy in laboratory and field studies. Fresh or fixed (10% formalin) glands can be weighed (after removal of adhering tissue), but fixed material is preferred for small animals because of problems in cleaning and rapid dehydration of their small adrenals. Also, fixed material can be saved for later histological analysis.

Sample data should be grouped by age, sex, reproductive condition, and social status of the study animals, but adrenal weights should not be adjusted for body size. Transformation (e.g., mg adrenal weight/100 mg body weight) tends to overcorrect for lighter animals and undercorrect for heavier animals. It is particularly unwise to make any adjustment if the independent variable (body weight or length in the above example) is affected by the experimental treatment (Steel and Torrie 1960). For example, if adrenal weights are adjusted for body weight, either by covariance analysis or by a simple adrenal weight/100 g body weight ratio, and body weights are affected by experimental treatment, one can show a difference in relative or adjusted adrenal weights when there was no change in adrenal weight and activity but, rather, a change only in body weight. A statistician should be consulted before any morphometric data for body weight or size are adjusted.

ACTH and Corticosteroid Levels in Blood

Circulating levels of ACTH and corticosteroids such as cortisol or corticosterone provide a direct measure of the endocrine response to stress. With RIA, hormone concentrations in relatively small (0.05–0.3 ml) plasma samples can be measured. Thus, even small animals can be sampled and returned to the experimental group or population, an obvious advantage over procedures that require removal of glands at necropsy. Cortisol is most often measured to quantify the stress response, but in rodents corticosterone is more instructive in this regard. Blood samples must be taken quickly after restraint or corticosteroids will reflect the stress due to handling rather than that due to natural factors in the population. Alternatively, increases in levels of corticosteroids in blood collected at intervals over a standard time (e.g., 30 min) following capture might be used as an index to an individual's vulnerability or resistance to stress. For example, plasma levels of corticosterone increase rapidly following capture of female and immature white-throated sparrows but not of adult males (Schwabl et al. 1988).

CONCLUSIONS AND RECOMMENDATIONS

Physiological measurements have considerable potential for detecting deviations from the "normal" in wildlife populations, but interpretation of the data sometimes is compromised by an incomplete knowledge of the normal value for a given parameter in a given season or geographic area and by inadequate sample sizes. In addition, many physiological parameters are measured only as components of specific, often short-term, research projects wherein values from the control group and published literature are used for reference. Much has been learned from such efforts, but limitations in the temporal and geographic distribution of samples in a typical 2- to 5-year research project can limit the generality of results and conclusions.

Coupling of research projects with long-term, perhaps follow-up, sampling of the same physiological indices over wide geographic areas would enhance the reliability of conclusions reached and improve their potential application to management programs. Whenever feasible, consideration should be given to establishing programs for collection of data on key species throughout the year. For example, many state game agencies collect data on deer at hunter check stations (Harder 1980), but often little is known about the status of animals at other times of the year. Data can be obtained from a variety of sources including check stations, road-kills, mark-recapture experiments, and pen studies, e.g., those of Verme and Ullrey (1984). With advance planning, several measurements can be made on the same animal, thereby achieving economy of effort. Also, commitments by institutions and agencies to long-term data collection eventually would produce invaluable databases for rare or infrequently encountered species that might not be studied otherwise.

LITERATURE CITED

ABLER, W. A., D. E. BUCKLAND, R. L. KIRKPATRICK, AND P. F. SCANLON. 1976. Plasma progestins and puberty in fawns as influenced by energy and protein. J. Wildl. Manage. 40:442–446.

ABRAHAM, G. E., F. S. MANLIMOS, AND R. GAZARA. 1977. Radioimmunoassay of steroids. Pages 591–999 in G. E. Abraham, ed. Handbook of radioimmunoassay. Marcel Dekker, Inc., New York, N.Y.

ADAMS, G. P., P. G. GRIFFIN, AND O. J. GINTHER. 1989. In situ morphologic dynamics of ovaries, uterus, and cervix in llamas. Biol. Reprod. 41:551–558.

———, E. D. PLOTKA, C. S. ASA, AND O. J. GINTHER. 1991. Feasibility of characterizing reproductive events in large nondomestic species by transrectal ultrasonic imaging. Zoo Biol. 10:247–259.

ADAMS, L., AND S. HANE. 1972. Adrenal gland size as an index of adrenocortical secretion rate in the California ground squirrel. J. Wildl. Dis. 8:19–23.

AMANN, R. P., AND J. T. LAMBIASE, JR. 1969. The male rabbit. III. Determination of daily sperm production by means of testicular homogenates. J. Anim. Sci. 28:369–374.

ANDERSON, A. E., D. E. MEDIN, AND K. C. BOWDEN. 1972. Indices of carcass fat in a Colorado mule deer population. J. Wildl. Manage. 36:579–594.

ANKNEY, C. D. 1980. Egg weight, survival, and growth of lesser snow goose goslings. J. Wildl. Manage. 44:174-182.

———, AND C. D. MACINNES. 1978. Nutrient reserves and reproductive performance of female lesser snow geese. Auk 95:459-471.

ARMSTRONG, R. A. 1950. Fetal development of the northern white-tailed deer (*Odocoileus virginianus borealis* Miller). Am. Midl. Nat. 43:650–666.

ARORA, K. L. 1979. Blood sampling and intravenous injections in Japanese quail (*Coturnix coturnix japonica*). Lab. Anim. Sci. 29:114–118.

ASTERITA, M. F. 1985. The physiology of stress. Human Sci. Press, Inc., New York, N.Y. 264pp.

AUSTAD, S. N., AND M. E. SUNQUIST. 1986. Sex-ratio manipulation in the common opossum. Nature 324:58–60.

AUSTIN, C. R. 1982. The egg. Pages 46–62 in C. R. Austin and R. V. Short, eds. Germ cells and fertilization. Second ed. Cambridge Univ. Press, New York, N.Y.

———, AND R. V. SHORT. 1984. Hormonal control of reproduction. Second ed. Cambridge Univ. Press, New York, N.Y. 244pp.

BAILEY, J. A. 1968. A weight-length relationship for evaluating physical condition of cottontails. J. Wildl. Manage. 32:835–841.

BAILEY, R. E. 1953. Surgery for sexing and observing gonad condition in birds. Auk 70:497–499.

BAILEY, R. O. 1979. Methods of estimating total lipid content in the redhead duck (*Aythya americana*) and an evaluation of condition indices. Can. J. Zool. 57:1830–1833.

BALDASSARE, G. A., R. J. WHYTE, AND E. G. BOLEN. 1980. Use of ultrasonic sound to estimate body fat depots in the mallard. Prairie Nat. 12:79–86.

BALDWIN, B. H., B. C. TENNANT, T. J. REIMERS, R. G. COWAN, AND P. W. CONCANNON. 1985. Circannual changes in serum testosterone concentrations of adult and yearling woodchucks (*Marmota monax*). Biol. Reprod. 32:804–812.

BAMFORD, J. 1970. Estimating fat reserves in the brush-tailed possum, *Trichosurus vulpecula* Kerr (Marsupialia: Phalangeridae). Aust. J. Zool. 18:415–425.

BANDY, P. J., I. McT. COWAN, W. D. KITTS, AND A. J. WOOD. 1956. A method for the assessment of the nutritional status of wild ungulates. Can. J. Zool. 34:48–52.

BARRON, J. C., AND W. F. HARWELL. 1973. Fertilization rates of south Texas deer. J. Wildl. Manage. 37:179–182.

BATCHELER, C. L., AND C. M. H. CLARKE. 1970. Note on kidney weights and the kidney fat index. N.Z. J. Sci. 13:663–668.

BEAR, G. D. 1971. Seasonal trends in fat levels of pronghorns, *Antilocapra americana*, in Colorado. J. Mammal. 52:583–589.

BECKERTON, P. R., AND A. L. A. MIDDLETON. 1982. Effects of dietary protein levels on ruffed grouse reproduction. J. Wildl. Manage. 46:569–579.

BENNETT, J. W., AND E. G. BOLEN. 1978. Stress response in wintering green-winged teal. J. Wildl. Manage. 42:81–86.

BERNDTSON, W. E. 1977. Methods for quantifying mammalian spermatogenesis: a review. J. Anim. Sci. 44:818–833.

BLEM, C. R. 1976. Patterns of lipid storage and utilization in birds. Am. Zool. 16:671–684.

BLUHM, C. K., R. E. PHILLIPS, AND W. H. BURKE. 1983. Serum levels of luteinizing hormone (LH), prolactin, estradiol, and progesterone in laying and nonlaying canvasback ducks (*Aythya valisineria*). Gen. Comp. Endocrinol. 52:1–16.

BOOKS-BLENDEN, P., T. S. BASKETT, AND M. W. SAYRE. 1984. Crop gland activity vs. nesting records for assessing September nesting of mourning doves. Wildl. Soc. Bull. 12:376–381.

BOVEE, K. C. 1984. Clinical and laboratory evaluation of renal function. Pages 219–233 in K. C. Bovee, ed. Canine nephrology. Harwal Publ. Co., Media, Pa.

BRAMBELL, F. W. R. 1948. Prenatal mortality in mammals. Biol. Rev. Camb. Philos. Soc. 23:370–405.

BRONSON, F. H. 1989. Mammalian reproductive biology. Univ. Chicago Press, Chicago, Ill. 324pp.

———, AND B. E. ELEFTHERIOU. 1964. Chronic physiological effects of fighting in mice. Gen. Comp. Endocrinol. 4:9–14.

BUCHANAN, G. D., AND E. V. YOUNGLAI. 1986. Plasma progesterone levels during pregnancy in the little brown bat *Myotis lucifugus* (Vespertilionidae). Biol. Reprod. 34:878–884.

CAMPBELL, R. R., AND J. F. LEATHERLAND. 1980. Estimating body protein and fat from water content in lesser snow geese. J. Wildl. Manage. 44:438–446.

CARD, W. C., R. L. KIRKPATRICK, K. E., WEBB, JR., AND P. F. SCANLON. 1985. Nutritional influences on NEFA, cholesterol, and ketones in white-tailed deer. J. Wildl. Manage. 49:380–385.

CARRICK, R. 1963. Ecological significance of territory in the Australian magpie, *Gymnorhina tibicen*. Int. Ornithol. Congr. 13:740–753.

CARRUTHERS, M., AND M. R. C. PATH. 1983. Instrumental stress tests. Pages 331–362 in H. Selye, ed. Selye's guide to stress research. Second ed. Scientific and Academic Editions, New York, N.Y.

CASTRO, G., B. A. WUNDER, AND F. L. KNOPF. 1990. Total body electrical conductivity (TOBEC) to estimate total body fat of free-living birds. Condor 92:496–499.

CATLING, P. C., AND R. L. SUTHERLAND. 1980. Effect of gonadectomy, season, and the presence of females on plasma testosterone, luteinizing hormone, and follicle stimulating hormone levels in male tammar wallabies (*Macropus eugenii*). J. Endocrinol. 86:25–33.

CEDERLUND, G. N., R. J. BERGSTRÖM, F. V. STÅLFELT, AND K. DANELL. 1986. Variability in mandible marrow fat in 3 moose populations in Sweden. J. Wildl. Manage. 50:719–726.

CENGEL, D. J., J. E. ESTEP, AND R. L. KIRKPATRICK. 1978. Pine vole reproduction in relation to food habits and body fat. J. Wildl. Manage. 42:822–833.

CHAPPELL, W. A., AND R. D. TITMAN. 1983. Estimating reserve lipids in greater scaup (*Aythya marila*) and lesser scaup (*A. affinis*). Can. J. Zool. 61:35–38.

CHEATUM, E. L. 1949a. Bone marrow as an index of malnutrition in deer. N.Y. State Conserv. 3:19–22.

———. 1949b. The use of corpora lutea for determining ovulation incidence and variations in fertility of white-tailed deer. Cornell Vet. 39:282–291.

CHILD, G. I., AND S. G. MARSHALL. 1970. A method of estimating carcass fat and fat-free weights in migrant birds from water content of specimens. Condor 72:116–119.

CHRISTIAN, J. J. 1950. The adreno-pituitary system and population cycles in mammals. J. Mammal. 31:247–259.

———. 1963. Endocrine adaptive mechanisms and the physiologic regulation of population growth. Pages 189–353 in W. V. Mayer and R. C. Van Gelder, eds. Physiological mammalogy. Vol. I: Mammalian populations. Academic Press, New York, N.Y.

———, AND D. E. DAVIS. 1964. Endocrines, behavior, and population: social and endocrine factors are integrated in the regulation of growth of mammalian populations. Science 146:1150–1560.

CLARK, F. H. 1936. The estrous cycle of the deer-mouse, *Peromyscus maniculatus*. Univ. Michigan Contrib. Lab. Vertebr. Genet. 1:1–7.

COHN, S. T. 1985. How valid are bioelectric impedance measurements in body composition studies. Am. J. Clin. Nutr. 42:889–890.

CONAWAY, C. H. 1955. Embryo resorption and placental scar formation in the rat. J. Mammal. 36:516–532.

CONCANNON, P., B. BALDWIN, J. LAWLESS, W. HORNBUCKLE, AND B. TENNANT. 1983. Corpora lutea of pregnancy and elevated serum progesterone during pregnancy and postpartum anestrus in woodchucks (*Marmota monax*). Biol. Reprod. 29:1128–1134.

CONNOLLY, G. E. 1981. Assessing populations. Pages 287–345 in O. C. Wallmo, ed. Mule and black-tailed deer of North America. Univ. Nebraska Press, Lincoln.

COOPER, K. A., ET AL. 1990. Serum testosterone and musth in captive male African and Asian elephants. Zoo Biol. 9:297–306.

COWLES, C. J., R. L. KIRKPATRICK, AND J. O. NEWELL. 1977. Ovarian follicular changes in gray squirrels as affected by season, age and reproductive state. J. Mammal. 58:67–73.

CRUM, B. G., J. B. WILLIAMS, AND K. A. NAGY. 1985. Can tritiated water-dilution space accurately predict total body water in chukar partridges. J. Appl. Physiol. 59:1383–1388.

DAUPHINÉ, T. C., JR. 1975. Kidney weight fluctuations affecting the kidney fat index in caribou. J. Wildl. Manage. 39:379–386.

DELGIUDICE, G. D., L. D. MECH, AND U. S. SEAL. 1989. Physiological assessment of deer populations by analysis of urine in snow. J. Wildl. Manage. 53:284–291.

———, ———, ———, AND P. D. KARNS. 1987a. Effects of winter fasting and refeeding on white-tailed deer blood profiles. J. Wildl. Manage. 51:865–873.

———, U. S. SEAL, AND L. D. MECH. 1987b. Effects of feeding and fasting on wolf blood and urine characteristics. J. Wildl. Manage. 51:1–10.

DENO, R. A. 1937. Uterine macrophages in the mouse and their relation to involution. Am. J. Anat. 60:433–471.

DEPPERSCHMIDT, J. D., S. C. TORBIT, A. W. ALLDREDGE, AND R. D. DEBLINGER. 1987. Body condition indices for starved pronghorns. J. Wildl. Manage. 51:675–678.

DOBOSZYNSKA, T. 1976. A method for collecting and staining vaginal smears from the beaver. Acta Theriol. 21,22:299–306.

DOWELL, J. H., AND R. J. WARREN. 1982. Variations in nutritional indices of Texas ring-necked pheasants. Proc. Annu. Conf. Southeast. Assoc. Fish Wildl. Agencies 36:463–472.

EVE, J. H., AND F. E. KELLOGG. 1977. Management implications of abomasal parasites in southeastern white-tailed deer. J. Wildl. Manage. 41:169–177.

FINGER, S. E., I. L. BRISBIN, JR., M. H. SMITH, AND D. F. URBSTON. 1981. Kidney fat as a predictor of body condition in white-tailed deer. J. Wildl. Manage. 45:964–968.

FLEHARTY, E. D., M. E. KRAUSE, AND D. P. STINNETT. 1973. Body composition, energy content and lipid cycles of four species of rodents. J. Mammal. 54:426–438.

FLEMING, M. W., AND J. D. HARDER. 1983. Luteal and follicular populations in the ovary of the opossum (*Didelphis virginiana*) after ovulation. J. Reprod. Fert. 67:29–34.

FLUX, J. E. C. 1971. Validity of the kidney fat index for estimating the condition of hares: a discussion. N.Z. J. Sci. 14:238–244.

FOLLETT, B. K., AND S. L. MAUNG. 1978. Rate of testicular maturation, in relation to gonadotrophin and testosterone levels, in quail exposed to various artificial photoperiods and to natural daylengths. J. Endocrinol. 78:267–280.

FOLLIS, T. B., W. C. FOOTE, AND J. J. SPILLETT. 1972. Observation of genitalia in elk by laparotomy. J. Wildl. Manage. 36:171–173.

FRANZMANN, A. W. 1985. Assessment of nutritional status. Pages 239–260 in R. J. Hudson and R. G. White, eds. Bioenergetics of wild herbivores. CRC Press, Inc., Boca Raton, Fla.

GADSBY, J. E., R. B. HEAP, D. G. POWELL, AND D. E. WALTERS. 1972. Diagnosis of pregnancy and of the number of foetuses in sheep from plasma progesterone concentrations. Vet. Res. 90:339–342.

GILMORE, D. P. 1969. Seasonal reproductive periodicity in the male Australian brush-tailed possum. J. Zool. 157:75–98.

GINTHER, O. J. 1990. Folliculogenesis during the transitional period and early ovulatory season in mares. J. Reprod. Fert. 90:311–320.

GOLLEY, F. B. 1957. An appraisal of ovarian analyses in determining reproductive performance of black-tailed deer. J. Wildl. Manage. 21:62–65.

GOSLING, L. M. 1986. Biased sex ratios in stressed animals. Am. Nat. 127:893–896.

GREER, K. R., AND W. W. HAWKINS, JR. 1967. Determining pregnancy in elk by rectal palpation. J. Wildl. Manage. 31:145–149.

GRUBB, T. C. 1989. Ptilochronology: feather growth bars as indicators of nutritional status. Auk 106:314–320.

———. 1991. A deficient diet narrows growth bars on induced feathers. Auk 108:725–727.

———. 1995. On induced anabolism, induced caching and induced construction as unambiguous indices of nutritional condition. Proc. West. Found. Vertebr. Zool. 6:258–263.

———, AND D. A. CIMPRICH. 1990. Supplementary food improves the nutritional condition of wintering woodland birds: evidence from ptilochronology. Ornis Scand. 21:277–281.

GUYNN, D. E., AND P. F. SCANLON. 1973. Crop-gland activity in mourning doves during hunting seasons in Virginia. Proc. Annu. Conf. Southeast. Assoc. Game and Fish Comm. 27:36–42.

HANKS, J. 1981. Characterization of population condition. Pages 47–74 in C. W. Fowler and T. D. Smith, eds. Dynamics of large mammal populations. John Wiley & Sons, New York, N.Y.

HANNON, S. J. 1981. Postovulatory follicles as indicators of egg production in blue grouse. J. Wildl. Manage. 45:1045–1047.

HARDER, J. D. 1980. Reproduction of white-tailed deer in the north central United States. Pages 23–35 in R. L. Hine and S. Nehls, eds. White-tailed deer population management in the north central states. North-Cent. Sect., The Wildl. Soc., Urbana, Ill.

———, L. A. HINDS, C. A. HORN, AND C. H. TYNDALE-BISCOE. 1985. Effects of removal in late pregnancy of the corpus luteum, Graafian follicle or ovaries on plasma progesterone, oestradiol, LH, parturition and post-partum oestrus in the tammar wallaby, *Macropus eugenii*. J. Reprod. Fert. 75:449–459.

———, AND D. L. MOORHEAD. 1980. Development of corpora lutea and plasma progesterone levels associated with the onset of the breeding season in white-tailed deer (*Odocoileus virginianus*). Biol. Reprod. 22:185–191.

———, AND T. J. PETERLE. 1974. Effect of diethylstilbestrol on reproductive performance of white-tailed deer. J. Wildl. Manage. 38:183–196.

———, AND A. WOOLF. 1976. Changes in plasma levels of oestrone and oestradiol during pregnancy and parturition in white-tailed deer. J. Reprod. Fert. 47:161–163.

HARPER, W. L., AND R. D. H. COHEN. 1985. Accuracy of Doppler ultrasound in diagnosing pregnancy in bighorn sheep. J. Wildl. Manage. 49:793–796.

HARRIS, L. E. 1970. Nutrition research techniques for domestic and wild animals. Vol. I. An international record system and procedures for analyzing samples. Utah State Univ., Logan. 233pp.

HAUGEN, A. O. 1975. Reproductive performance of white-tailed deer in Iowa. J. Mammal. 56:151–159.

HAUGER, R. L., F. J. KARSCH, AND D. L. FOSTER. 1977. A new concept of the control of the estrous cycle of the ewe based on the temporal relationships between luteinizing hormone, estradiol and progesterone in peripheral serum and evidence that progesterone inhibits tonic LH secretion. Endocrinology 101:807–817.

HAWLEY, A. W. L., S. SYLVÉN, AND M. WILHELMSON. 1982. A simple device for sectioning ovaries. J. Wildl. Manage. 46:247–249.

HENRY, D. B., AND T. A. BOOKHOUT. 1969. Productivity of beavers in northeastern Ohio. J. Wildl. Manage. 33:927–932.

HENSEL, R. J., W. A. TROYER, AND A. W. ERICKSON. 1969. Reproduction in the female brown bear. J. Wildl. Manage. 33:357–365.

HESS, D. L., A. M. SCHMIDT, AND M. J. SCHMIDT. 1983. Reproductive cycle of the Asian elephant (*Elephus maximus*) in captivity. Biol. Reprod. 28:767–773.

HOHMAN, W. L., AND T. S. TAYLOR. 1986. Indices of fat and protein for ring-necked ducks. J. Wildl. Manage. 50:209–211.

HOUSTON, D. B., C. T. ROBBINS, C. A. RUDER, AND R. G. SASSER. 1986. Pregnancy detection in mountain goats by assay for pregnancy-specific protein B. J. Wildl. Manage. 50:740–742.

HUMASON, G. L. 1979. Animal tissue techniques. Third ed. W. H. Freeman and Co., San Francisco, Calif. 641pp.

HUTCHINSON, A. E., AND R. B. OWEN. 1984. Bone marrow fat in waterfowl. J. Wildl. Manage. 48:585–591.

INNS, R. W. 1982. Seasonal changes in the accessory reproductive system and plasma testosterone levels of the male tammar wallaby, *Macropus eugenii*, in the wild. J. Reprod. Fertil. 66:675–680.

JACOBSON, H. A., H. J. BEARDEN, AND D. B. WHITEHOUSE. 1989. Artificial insemination trials with white-tailed deer. J. Wildl. Manage. 53:224–227.

———, R. L. KIRKPATRICK, H. E. BURKHART, AND J. E. DAVIS. 1978a. Hematologic comparisons of shot and live-trapped cottontail rabbits. J. Wildl. Dis. 14:82–88.

———, ———, AND B. S. MCGINNES. 1978b. Disease and physiologic characteristics of two cottontail populations in Virginia. Wildl. Monogr. 60. 53pp.

JAFFE, B. M., AND H. R. BEHRMAN. 1974. Methods of hormone radioimmunoassay. Academic Press, New York, N.Y. 520pp.

JEFFCOATE, S. L. 1981. Efficiency and effectiveness in the endocrine laboratory. Academic Press, New York, N.Y. 223pp.

JOHNSON, A. L. 1986. Reproduction in the female. Pages 403–431 in P. D. Sturkie, ed. Avian physiology. Fourth ed. Springer-Verlag, New York, N.Y.

JOHNSON, D. H., G. L. KRAPU, K. J. REINECKE, AND D. G. JORDE. 1985. An evaluation of condition indices for birds. J. Wildl. Manage. 49:569–575.

JURGELSKI, W., JR., AND M. E. PORTER. 1974. The opossum (*Didelphis virginiana*) as a biomedical model. III. Breeding in captivity: methods. Lab. Anim. Sci. 24:412–425.

KABAT, C., I. O. BUSS, AND R. K. MEYER. 1948. The use of ovulated follicles in determining eggs laid by the ring-necked pheasant. J. Wildl. Manage. 12:399–416.

KANTONGOL, C. B., F. NAFTOLIN, AND R. V. SHORT. 1971. Relationship between blood levels of luteinizing hormone and testosterone in bulls, and the effects of sexual stimulation. J. Endocrinol. 50:457–456.

KASMAN, L. H., E. C. RAMSAY, AND B. L. LASLEY. 1986. Urinary steroid evaluations to monitor ovarian function in exotic ungulates: III. Estrone sulfate and pregnanediol-3-glucuronide excretion in the Indian rhinoceros (*Rhinoceros unicornis*). Zoo Biol. 5:355–361.

KENNELLY, J. J., AND B. E. JOHNS. 1976. The estrous cycle of coyotes. J. Wildl. Manage. 40:272–277.

KERR, D. C., C. D. ANKNEY, AND J. S. MILLAR. 1982. The effect of drying temperature on extraction of petroleum ether soluble fats of small birds and mammals. Can. J. Zool. 60:470–472.

KIBBE, D. P., AND R. L. KIRKPATRICK. 1971. Systematic evaluation of late summer breeding in juvenile cottontails, *Sylvilagus floridanus*. J. Mammal. 52:465–467.

KIE, J. G., M. WHITE, AND D. L. DRAWE. 1983. Condition parameters of white-tailed deer in Texas. J. Wildl. Manage. 47:583–594.

KING, J. R. 1967. Adipose tissue composition in experimentally induced fat deposition in the white-crowned sparrow. Comp. Biochem. Physiol. 21:393–404.

———, AND D. S. FARNER. 1965. Fat deposition in migratory birds. N.Y. Acad. Sci. 131:422–445.

KIRKPATRICK, J. F., L. H. KASMAN, B. L. LASLEY, AND J. W. TURNER, JR. 1988. Pregnancy determination in uncaptured feral horses. J. Wildl. Manage. 52:305–308.

KIRKPATRICK, R. L. 1974. Ovarian follicular and related characteristics of white-tailed deer as influenced by season and age in the Southeast. Proc. Annu. Conf. Southeast. Assoc. Game and Fish Comm. 28:587–594.

———. 1988. Comparative influences of nutrition on reproduction and survival of wild birds and mammals—an overview. Caesar Kleberg Wildl. Res. Inst., Kingsville, Tex. 57pp.

———, D. E. BUCKLAND, W. A. ABLER, P. F. SCANLON, J. B. WHELAN, AND H. E. BURKHART. 1975. Energy and protein influences on blood urea nitrogen of white-tailed deer fawns. J. Wildl. Manage. 39:692–698.

———, AND G. L. VALENTINE. 1970. Reproduction in captive pine voles, *Microtus pinetorum*. J. Mammal. 51:779–785.

KLEIN, D. R. 1964. Range-related differences in growth of deer reflected in skeletal ratios. J. Mammal. 45:226–235.

KNOBIL, E., AND J. D. NEILL. 1988. The physiology of reproduction. Raven Press, New York, N.Y. 2414pp.

KOERTH, N. E., AND F. S. GUTHERY. 1988. Reliability of body fat indices for northern bobwhite populations. J. Wildl. Manage. 52:150–152.

KORSCHGEN, C. E. 1977. Breeding stress of female eiders in Maine. J. Wildl. Manage. 41:360–373.

KOVAL, P. J., T. J. PETERLE, AND J. D. HARDER. 1987. Effects of polychlorinated biphenyls on mourning dove reproduction and circulating progesterone levels. Bull. Environ. Contam. Toxicol. 39:663–670.

LAWS, R. M. 1967. Occurrence of placental scars in the uterus of the African elephant (*Loxodonta africana*). J. Reprod. Fert. 14:445–449.

LERESCHE, R. E., U. S. SEAL, P. D. KARNS, AND A. W. FRANZMANN. 1974. A review of blood chemistry of moose and other cervidae with emphasis on nutritional assessment. Nat. Can. 101:263–290.

LEVI, W. M. 1969. The pigeon. Levi Publ. Co. Inc., Sumter, S.C. 667pp.

LOSKUTOFF, N. M., J. E. OTT, AND B. L. LASLEY. 1982. Urinary steroid evaluations to monitor ovarian function in exotic ungulates: I. Pregnanediol-3-glucuronide immunoreactivity in the okapi (*Okapia johnston*). Zoo Biol. 1:45–53.

———, ———, AND ———. 1983. Strategies for assessing ovarian function in exotic species. J. Zoo Anim. Med. 14:3–12.

LUKASKI, H. C., W. W. BOLONCHUK, C. B. HALL, AND W. A. SIDERS. 1986. Validation of tetrapolar bioelectrical impedance method to assess human body composition. J. Appl. Physiol. 60:1327–1332.

MANSELL, W. D. 1971. Accessory corpora lutea in ovaries of white-tailed deer. J. Wildl. Manage. 35:369–374.

MARCH, G. L., AND R. M. F. S. SADLIER. 1970. Studies on the band-tailed pigeon (*Columba fasciata*) in British Columbia. 1. Seasonal changes in gonadal development and crop gland activity. Can. J. Zool. 48:1353–1357.

MARTIN, K. H., R. A. STEHN, AND M. E. RICHMOND. 1976. Reliability of placental scar counts in the prairie vole. J. Wildl. Manage. 40:264–271.

MARTIN, R. M., M. E. LISANO, AND J. E. KENNAMER. 1981. Plasma estrogens, total protein, and cholesterol in the female eastern wild turkey. J. Wildl. Manage. 45:798–802.

MCCABE, T. T. 1943. An aspect of a collector's technique. Auk 60:550–558.

MCDONALD, I. R., A. K. LEE, A. J. BRADLEY, AND K. A. THAN. 1981. Endocrine changes in dasyurid marsupials with differing mortality patterns. Gen. Comp. Endocrinol. 44:292–301.

MCEWAN, E. H., AND A. J. WOOD. 1966. Growth and development of the barren ground caribou. 1. Heart girth, hind foot length, and body weight relationships. Can. J. Zool. 44:401–411.

MCMILLIN, J. M., U. S. SEAL, K. D. KEENLYNE, A. W. ERICKSON, AND J. E. JONES. 1974. Annual testosterone rhythm in the adult white-tailed deer (*Odocoileus virginianus borealis*). Endocrinology 94:1034–1040.

MCRAE, W. A., AND R. W. DIMMICK. 1982. Body fat and blood-serum values of breeding wild bobwhites. J. Wildl. Manage. 46:268–271.

MECH, L. D., AND G. D. DELGIUDICE. 1985. Limitations of the marrow-fat technique as an indicator of body condition. Wildl. Soc. Bull. 13:204–206.

———, U. S. SEAL, AND G. D. DELGIUDICE. 1987. Use of urine in snow to indicate condition of wolves. J. Wildl. Manage. 51:10–13.

MICHENER, H., AND J. MICHENER. 1938. Bars in flight feathers. Condor 40:149–160.

MIRARCHI, R. E., AND P. F. SCANLON. 1980. Duration of mourning dove crop gland activity during the nesting cycle. J. Wildl. Manage. 44:209–213.

———, ———, AND R. L. KIRKPATRICK. 1977a. Annual changes in spermatozoan production and associated organs of white-tailed deer. J. Wildl. Manage. 41:92–99.

———, ———, ———, AND C. B. SCHRECK. 1977b. Androgen levels

and antler development in captive and wild white-tailed deer. J. Wildl. Manage. 41:178–183.

MOBERG, G. P. 1985. Biological response to stress: key to assessment of animal well-being? Pages 27–49 *in* G. P. Moberg, ed. Animal stress. First ed. Waverly Press, Inc., Baltimore, Md.

MORTON, J. M., R. L. KIRKPATRICK, AND E. P. SMITH. 1991. Comments on estimating total body lipids from measures of lean mass. Condor 93:463–465.

MULLEY, R. C., A. W. ENGLISH, R. J. RAWLINSON, AND R. S. CHAPPLE. 1987. Pregnancy diagnosis of fallow deer by ultrasonography. Aust. Vet. J. 64:257–258.

MURPHY, W. F., JR., P. F. SCANLON, AND R. L. KIRKPATRICK. 1973. Examination of ovaries in living cottontail rabbits by laparotomy. Proc. Annu. Conf. Southeast. Assoc. Game and Fish Comm. 27:343–344.

NEILAND, K. A. 1970. Weight of dried marrow as indicator of fat in caribou femurs. J. Wildl. Manage. 34:904–907.

NELSEN, O. E. 1953. Comparative embryology of the vertebrates. Mc-Graw-Hill Book Co., New York, N.Y. 982pp.

NELSON, T. A., AND A. WOOLF. 1983. Field laparoscopy of female white-tailed deer. J. Wildl. Manage. 47:1213–1216.

NEQUIN, L. G., J. ALVAREZ, AND N. B. SCHWARTZ. 1979. Measurement of serum steroid and gonadotropin levels and uterine and ovarian variables throughout 4 day and 5 day estrous cycles in the rat. Biol. Reprod. 20:659–670.

NICHOLS, R. G., AND M. R. PELTON. 1974. Fat in the mandibular cavity as an indicator of condition in deer. Proc. Annu. Conf. Southeast. Assoc. Game and Fish Comm. 28:540–548.

NIXON, C. M. 1971. Productivity of white-tailed deer in Ohio. Ohio J. Sci. 71:217–225.

———, M. W. MCCLAIN, AND R. W. DONOHOE. 1975. Effects of hunting and mast crops on a squirrel population. J. Wildl. Manage. 39:1–25.

ODUM, E. P., D. T. ROGERS, AND D. L. HICKS. 1964. Homeostasis of the nonfat components of migrating birds. Science 143:1037–1039.

OLEYAR, C. M., AND B. S. MCGINNES. 1974. Field evaluation of diethylstilbestrol for suppressing reproduction in foxes. J. Wildl. Manage. 38:101–106.

ORSINI, M. W. 1962. Technique of preparation, study and photography of benzyl-benzoate cleared material for embryological studies. J. Reprod. Fert. 3:283–287.

OWEN, M. 1981. Abdominal profile—a condition index for wild geese in the field. J. Wildl. Manage. 45:227–230.

———, AND W. A. COOK. 1977. Variations in body weight, wing length and condition of mallard *Anas platyrhynchos platyrhynchos* and their relationship to environmental changes. J. Zool. (Lond.) 183:377–395.

OZOGA, J. J., AND L. J. VERME. 1975. Activity patterns of white-tailed deer during estrus. J. Wildl. Manage. 39:679–683.

———, AND ———. 1978. The thymus gland as a nutritional status indicator in deer. J. Wildl. Manage. 42:791–798.

PALMER, E, AND M. A. DRIANCOURT. 1980. Use of ultrasonic echography in equine gynecology. Theriogenology 13:204–211.

PAYNE, R. B. 1973. Individual laying histories and the clutch size and numbers of eggs of parasitic cuckoos. Condor 75:414–438.

PIERSON, R. A., AND O. J. GINTHER. 1987. Reliability of diagnostic ultrasonography for identification and measurement of follicles and detecting the corpus luteum in heifers. Theriogenology 28:929–936.

———, AND ———. 1988. Ultrasonic imaging of the ovaries and uterus in cattle. Theriogenology 29:21–37.

PLOTKA, E. D., U. S. SEAL, G. C. SCHMOLLER, P. D. KARNS, AND K. D. KEENLYNE. 1977. Reproductive steroids in the white-tailed deer (*Odocoileus virginianus borealis*). I. Seasonal changes in the female. Biol. Reprod. 16:340–343.

———, ———, L. J. VERME, AND J. J. OZOGA. 1980. Reproductive steroids in deer. III. Luteinizing hormone, estradiol and progesterone around estrus. Biol. Reprod. 22:576–581.

———, ———, ———, AND ———. 1983. The adrenal gland in white-tailed deer: a significant source of progesterone. J. Wildl. Manage. 47:38–44.

———, ET AL. 1988. Ovarian function in the elephant: luteinizing hormone and progesterone cycles in African and Asian elephants. Biol. Reprod. 38:309–314.

PRESTON, R. L., L. H. BREUER, AND G. B. THOMPSON. 1961. Blood urea in cattle as affected by energy, protein and stilbestrol. J. Anim. Sci. 20:977. (Abstr.)

PROVOST, E. E. 1962. Morphological characteristics of the beaver ovary. J. Wildl. Manage. 26:272–278.

RANSOM, A. B. 1965. Kidney and marrow fat as indicators of white-tailed deer condition. J. Wildl. Manage. 29:397–398.

———. 1967. Reproductive biology of white-tailed deer in Manitoba. J. Wildl. Manage. 31:114–123.

RASMUSSEN, G. P. 1985. Antler measurements as an index to physical condition and range quality with respect to white-tailed deer. N.Y. Fish Game J. 32:97–113.

RAVELING, D. G. 1979. The annual cycle of body composition of Canada geese with special reference to control of reproduction. Auk 96:234–252.

RINEY, T. 1955. Evaluating condition of free ranging red deer (*Cervus elaphus*), with special reference to New Zealand. N.Z. J. Sci. Technol. Sect. B 36:429–463.

RISSER, A. C., JR. 1971. A technique for performing laparotomy on small birds. Condor 73:376–379.

RIVIER, C., AND S. RIVEST. 1991. Effects of stress on the activity of the hypothalamic-pituitary-gonadal axis: peripheral and central mechanisms. Biol. Reprod. 45:523–532.

ROGERS, C. M. 1987. Predation risk and fasting capacity: do wintering birds maintain optimal body mass? Ecology 68:1051–1061.

ROLAN, R. G., AND H. T. GIER. 1967. Correlation of embryo and placental scar counts of *Peromyscus maniculatus* and *Microtus ochrogaster*. J. Mammal. 48:317–319.

ROSEBERRY, J. L., AND W. D. KLIMSTRA. 1970. Productivity of white-tailed deer on Crab Orchard National Wildlife Refuge. J. Wildl. Manage. 34:23–28.

ROWELL, J., AND P. F. FLOOD. 1987. Changes in muskox blood progesterone concentration between collection and centrifugation. J. Wildl. Manage. 51:901–903.

———, ———, C. A. RUDER, AND R. G. SASSER. 1989. Pregnancy-specific protein in the plasma of captive muskoxen. J. Wildl. Manage. 53:899–901.

SAFAR-HERMANN, N., M. N. ISMAIL, H. S. CHOI, E. MOSTL, AND E. BAMBERG. 1987. Pregnancy diagnosis in zoo animals by estrogen determination in feces. Zoo Biol. 6:189–193.

SANDERS, O. T., AND R. L. KIRKPATRICK. 1975. Effects of a polychlorinated biphenyl (PCB) on sleeping times, plasma corticosteroids, and testicular activity of white-footed mice. Environ. Physiol. Biochem. 5:308–313.

SANDERSON, G. C. 1950. Methods of measuring productivity in raccoons. J. Wildl. Manage. 14:389–402.

SAPOLSKY, R. M. 1987. Stress, social status, and reproductive physiology in free-living baboons. Pages 291–322 *in* D. Crews, ed. Psychobiology of reproductive behavior. First ed. Prentice-Hall, Inc., Englewood Cliffs, N.J.

SAUER, P. R., AND C. W. SEVERINGHAUS. 1977. Determination and application of fawn reproductive rates from yearling teat length. Trans. Northeast. Sect., The Wildl. Soc. 33:133–144.

SCHWABL, H., M. RAMENOFSKY, I. SCHWABL-BENZINGER, D. S. FARNER, AND J. C. WINGFIELD. 1988. Social status, circulating levels of hormones, and competition for food in winter flocks of the white-throated sparrow. Behaviour 107:107–121.

SEAL, U. S., AND R. L. HOSKINSON. 1978. Metabolic indicators of habitat condition and capture stress in pronghorns. J. Wildl. Manage. 42:755–763.

———, M. E. NELSON, L. D. MECH, AND R. L. HOSKINSON. 1978a. Metabolic indicators of habitat differences in four Minnesota deer populations. J. Wildl. Manage. 42:746–754.

———, ———, ———, AND E. D. PLOTKA. 1983. Metabolic and endocrine responses of white-tailed deer to increasing population density. J. Wildl. Manage. 47:451–462.

———, J. J. OZOGA, A. W. ERICKSON, AND L. J. VERME. 1972. Effects of immobilization on blood analyses of white-tailed deer. J. Wildl. Manage. 36:1034–1040.

———, L. J. VERME, AND J. J. OZOGA. 1978b. Dietary protein and energy effects on deer fawn metabolic patterns. J. Wildl. Manage. 42:776–790.

SELYE, H. 1936. A syndrome produced by diverse nocuous agents. Nature 138:32–34.

———. 1946. The general adaptation syndrome and the diseases of adaptation. J. Clin. Endocrinol. 6:117–230.

———. 1976. Stress in health and disease. Butterworths, Boston, Mass. 1256pp.

SERVELLO, F. A., AND R. L. KIRKPATRICK. 1987. Fat indices for ruffed grouse. J. Wildl. Manage. 51:173–177.

————, AND ————. 1988. Nutrition and condition of ruffed grouse during the breeding season in southwestern Virginia. Condor 90: 836–842.

SEVERINGHAUS, C. W., H. F. MAGUIRE, R. A. COOKINGHAM, AND J. E. TANCK. 1950. Variations by age class in the antler beam diameters of white-tailed deer related to range conditions. Trans. North Am. Wildl. Conf. 15:551–570.

SHORT, R. V. 1972. The role of hormones in sex cycles, Book 3. Pages 42–72 *in* C. R. Austin and R. V. Short, eds. Hormones in reproduction. Cambridge Univ. Press, New York, N.Y.

SIMKIN, D. W. 1965. Reproduction and productivity of moose in northwestern Ontario. J. Wildl. Manage. 29:740–750.

SIROIS, J., AND J. E. FORTUNE. 1988. Ovarian follicular dynamics during the estrous cycle in heifers monitored by real-time ultrasonography. Biol. Reprod. 39:308–317.

SMITH, N. S. 1970. Appraisal of condition estimation methods for East African ungulates. E. Afr. Wildl. J. 8:123–129.

STABENFELDT, G. H., AND V. M. SHILLE. 1977. Reproduction in the dog and cat. Pages 499–527 *in* H. H. Cole and P. T. Cupps, eds. Reproduction in domestic animals. Third ed. Academic Press, New York, N.Y.

STEEL, R. G. D., AND J. H. TORRIE. 1960. Principles and procedures of statistics. McGraw-Hill Book Co., Inc., New York, N.Y. 481pp.

SULLIVAN, J. A., AND P. F. SCANLON. 1976. Effects of grouping and fighting on the reproductive tracts of male white-footed mice (*Peromyscus leucopus*). Res. Popul. Ecol. 17:164–175.

SUNDQVIST, C., A. LUKOLA, AND M. PARVINEN. 1986. Testicular aspiration biopsy in evaluation of fertility of mink. J. Reprod. Fert. 77:531–535.

THOMAS, V. G., H. G. LUMSDEN, AND D. H. PRICE. 1975. Aspects of winter metabolism of ruffed grouse (*Bonasa umbellus*) with special reference to energy reserves. Can. J. Zool. 53:434–440.

————, S. H. MAINGUY, AND J. P. PREVETT. 1983. Predicting fat content of geese from abdominal fat weight. J. Wildl. Manage. 47: 1115–1119.

TORBIT, S. C., L. H. CARPENTER, A. W. ALLDREDGE, AND D. M. SWIFT. 1985. Mule deer body composition—a comparison of methods. J. Wildl. Manage. 49:86–91.

TRAUGER, D. L., AND A. O. HAUGEN. 1965. Corpora lutea variations of white-tailed deer. J. Wildl. Manage. 29:487–492.

TRIVERS, R. L., AND D. E. WILLARD. 1973. Natural selection of parental ability to vary the sex ratio of offspring. Science 179:90–92.

TURNER, C. D., AND J. T. BAGNARA. 1976. General endocrinology. W.B. Saunders Co., Philadelphia, Pa. 596pp.

TYNDALE-BISCOE, C. H., AND M. B. RENFREE. 1987. Reproductive physiology of marsupials. Cambridge Univ. Press, New York, N.Y. 476pp.

ULLREY, D. E. 1982. Nutrition and antler development in white-tailed deer. Pages 49–59 *in* R. D. Brown, ed. Antler development in Cervidae. Caesar Kleberg Wildl. Res. Inst., Kingsville, Tex.

VAHDAT, F., B. E. SEGUIN, H. L. WHITMORE, AND S. D. JOHNSTON. 1984. Role of blood cells in degradation of progesterone in bovine blood. Am. J. Vet. Res. 45:240–243.

VANGILDER, L. D., AND T. J. PETERLE. 1981. South Louisiana crude oil or DDE in the diet of mallard hens: effects on egg quality. Bull. Environ. Contam. Toxicol. 26:328–336.

VAN TIENHOVEN, A. V. 1983. Reproductive physiology of vertebrates. Second ed. Cornell Univ. Press, Ithaca, N.Y. 491pp.

VAN VUREN, D., AND B. E. COBLENTZ. 1985. Kidney weight variation and the kidney fat index: an evaluation. J. Wildl. Manage. 49:177–179.

VAUGHAN, M. R., AND L. B. KEITH. 1981. Demographic response of experimental snowshoe hare populations to overwinter food shortage. J. Wildl. Manage. 45:354–380.

VAUGHAN, T. A. 1986. Mammalogy. Third ed. Saunders Coll. Publ., New York, N.Y. 576pp.

VERME, L. J. 1969. Reproductive patterns of white-tailed deer related to nutritional plane. J. Wildl. Manage. 33:881–887.

————, AND J. C. HOLLAND. 1973. Reagent-dry assay of marrow fat in white-tailed deer. J. Wildl. Manage. 37:103–105.

————, AND D. E. ULLREY. 1984. Physiology and nutrition. Pages 91–128 *in* L. K. Halls, ed. White-tailed deer: ecology and management. Stackpole Books, Harrisburg, Pa.

VOGELSANG, R. W. 1977. Blood urea nitrogen, serum cholesterol and progestins as affected by nutritional intake, pregnancy and the estrous cycle in white-tailed deer. M.S. Thesis, Virginia Polytechnic Inst. State Univ., Blacksburg. 109pp.

WAITE, T. A. 1990. Effects of caching supplemental food on induced feather regeneration in wintering gray jays (*Perisoreus canadensis*): a ptilochronology study. Ornis Scand. 21:122–128.

WALKER, L. A., ET AL. 1988. Urinary concentrations of ovarian steroid hormone metabolites and bioactive follicle-stimulating hormone in killer whales (*Orcinus orchus*) during ovarian cycles and pregnancy. Biol. Reprod. 39:1013–1020.

WALSBERG, G. E. 1988. Evaluation of a nondestructive method for determining fat stores in small birds and mammals. Physiol. Zool. 61:153–159.

WARREN, R. J., AND R. L. KIRKPATRICK. 1978. Indices of nutritional status in cottontail rabbits fed controlled diets. J. Wildl. Manage. 42:154–158.

————, A. OELSCHLAEGER, P. F. SCANLON, AND F. C. GWAZDAUSKAS. 1981. Dietary and seasonal influences on nutritional indices of adult male white-tailed deer. J. Wildl. Manage. 45:926–936.

————, ————, ————, K. E. WEBB, JR., AND J. B. WHELAN. 1982. Energy, protein, and seasonal influence on white-tailed deer fawn nutritional indices. J. Wildl. Manage. 46:302–312.

WASSER, S. K., L. RISLER, AND R. A. STEINER. 1988. Excreted steroids in primate feces over the menstrual cycle and pregnancy. Biol. Reprod. 39:862–872.

WENTWORTH, B. C., J. A. PROUDMAN, H. OPEL, M. J. WINELAND, N. G. ZIMMERMANN, AND A. LAPP. 1983. Endocrine changes in the incubating and brooding turkey hen. Biol. Reprod. 29:87–92.

WESSON, J. A., III, P. F. SCANLON, R. L. KIRKPATRICK, AND H. S. MOSBY. 1979a. Influence of chemical immobilization and physical restraint on packed cell volume, total protein, glucose, and blood urea nitrogen in blood of white-tailed deer. Can. J. Zool. 57:756–767.

————, ————, ————, AND ————. 1979b. Influence of time of blood sampling after death on blood measurements of the white-tailed deer. Can. J. Zool. 57:777–780.

————, ————, ————, ————, AND R. L. BUTCHER. 1979c. Influence of chemical immobilization and physical restraint on steroid hormone levels in blood of white-tailed deer. Can. J. Zool. 57:768–776.

WHITE, D. W., E. D. KENNEDY, AND P. C. STOUFFER. 1991. Feather regrowth in female European starlings rearing broods of different sizes. Auk 108:889–895.

WHYTE, R. J., AND E. G. BOLEN. 1984. Variation in winter fat depots and condition indices of mallards. J. Wildl. Manage. 48:1370–1373.

WILLIAMS, S., EDITOR. 1984. Official methods of analysis. Fourteenth ed. Assoc. Off. Anal. Chem., Washington, D.C. 1141pp.

WILLIAMSON, P., N. J. GALES, AND S. LISTER. 1990. Use of real-time B-mode ultrasound for pregnancy diagnosis and measurement of fetal growth rate in captive bottlenose dolphins (*Tursiops truncatus*). J. Reprod. Fert. 88:543–548.

WINDBERG, L. A., AND L. B. KEITH. 1976. Snowshoe hare population response to artificial high densities. J. Mammal. 57:523–553.

WINGFIELD, J. C. 1985. Short-term changes in plasma levels of hormones during establishment and defense of a breeding territory in male song sparrows, *Melospiza melodia*. Horm. Behav. 19:174–187.

————, AND D. S. FARNER. 1976. Avian endocrinology—field investigations and methods. Condor 78:570–573.

————, AND ————. 1979. Some endocrine correlates of renesting after loss of clutch or brood in the white-crowned sparrow, *Zonotrichia leucophrys gambelii*. Gen. Comp. Endocrinol. 38:322–331.

————, AND ————. 1980. Control of seasonal reproduction in temperate-zone birds. Pages 62–101 *in* R. J. Reiter and B. K. Follet, eds. Progress in reproductive biology. Vol. 5. S. Karger, Basel, Switzerland.

————, AND M. C. MOORE. 1987. Hormonal, social, and environmental factors in the reproductive biology of free-living male birds. Pages 148–175 *in* D. Crews, ed. Psychobiology of reproductive behavior. First ed. Prentice-Hall,Inc., Englewood Cliffs, N.J.

WISEMAN, B. S., D. L. VINCENT, P. J. THOMFORD, N. S. SCHEFFRAHN, G. F. SARGENT, AND D. J. KESLER. 1982. Changes in porcine, ovine, bovine and equine blood progesterone concentrations between collection and centrifugation. Anim. Reprod. Sci. 5:157–165.

WOOD, A. K., R. E. SHORT, A. E. DARLING, G. L. DUSEK, R. G. SASSER, AND C. A. RUDER. 1986. Serum assays for detecting pregnancy in mule and white-tailed deer. J. Wildl. Manage. 50:684–687.

WOODALL, P. F. 1978. Omental fat: a condition index for redbilled teal. J. Wildl. Manage. 42:188–190.

WOOLF, A., AND J. D. HARDER. 1979. Population dynamics of a captive white-tailed deer herd with emphasis on reproduction and mortality. Wildl. Monogr. 67. 53pp.

WRIGHT, P. L. 1966. Observations on the reproductive cycle of the American badger (*Taxidea taxus*). Pages 27–45 *in* I. W. Rowlands, ed. Comparative biology of reproduction in mammals. Symp. Zool. Soc. Lond. 15.

WYDOSKI, R. S., AND D. E. DAVIS. 1961. The occurrence of placental scars in mammals. Proc. Penn. Acad. Sci. 35:197–204.

ZARROW, M. X., J. M. YOCHIM, AND J. L. McCARTHY. 1964. Experimental endocrinology: a sourcebook of basic techniques. Academic Press, New York, N.Y. 519pp.

ZEPP, R. L., JR., AND R. L. KIRKPATRICK. 1976. Reproduction in cottontails fed diets containing a PCB. J. Wildl. Manage. 40:491–495.

ZWANK, P. J. 1981. Effects of field laparotomy on survival and reproduction of mule deer. J. Wildl. Manage. 45:972–975.

12

TECHNIQUES FOR WILDLIFE NUTRITIONAL ANALYSES

Jonathan B. Haufler and Frederick A. Servello

INTRODUCTION

Food contains the organic and inorganic substances that are the sources of energy and nutrients for life processes. Wild species vary considerably in their selection of food, because food availability and quality, animal digestive and behavioral adaptations, and plant and animal defensive strategies produce a large matrix of consumer-prey possibilities. This chapter discusses techniques available to wildlife biologists for analysis and evaluation of wildlife foods and nutrition. Feeds, prepared diets for animals, are not addressed in this chapter. Ration preparation for captive or zoo animals is a complicated task and is beyond the focus of this chapter. Feeding of free-ranging wildlife can result in an artificial dependence of wildlife on humans, and the loss of natural ties of an animal to its habitat. The more we understand about the foraging strategies of wild species, the more feeding becomes an unacceptable practice of wildlife management.

The nutrition techniques discussed in this chapter usually are applicable to all wild species. Where this is not true, appropriate uses are noted. The chapter also differentiates between techniques applicable in a field setting and those that typically require a laboratory or penned animal setting.

A basic understanding by the reader of nutrients and animal digestive strategies is assumed. Texts (Maynard et al. 1979, Church and Pond 1988) describing nutrients and nutritional concepts in animal nutrition provide good background material. Robbins (1983) discussed many concepts of wildlife nutrition, and chapters reviewing various aspects of wildlife nutrition were written by Mautz (1978) and Schwartz and Hobbs (1985).

CHEMICAL COMPOSITION OF FOODS

The nutritional quality of a food is determined by the nutrients contained in the food and an animal's ability to digest or utilize these nutrients. The chemical composition of a food (Fig. 1) varies with such factors as the type of food (plant or animal), species, season, age, or site effects. Many studies over the last 20 years have determined the basic chemical constituents in common wildlife foods of either plant or vertebrate prey species. Foods of insectivores and the composition and role of plant defensive chemicals have not been as widely researched and are worthy of some mention.

Insectivores consume diets that are moderately digestible (60–90%) (Hawkins and Jewell 1962, Pernetta 1977, Nagy et al. 1978, Balakrishnan and Alexander 1979, Allen 1989) and have varying fat, ash, and element levels (Allen 1989, Allen and Oftedal 1989). Total nitrogen levels are fairly constant, although not all of this nitrogen may be available (Allen 1989). Additional aspects concerning the nutrition of insectivory need further research.

A rapidly expanding area of research in wildlife nutrition is the effects of plant defensive chemicals on mammalian and avian herbivores. Several classes of plant chemicals, also called secondary plant metabolites or allelochemicals, appear to serve a role in defense against herbivory (Levin 1976, Rhoades and Cates 1976, Rosenthal and Janzen 1979), but phenolics and terpenoids have received the most attention in wildlife investigations. Alkaloids have received less study as they apply to wildlife

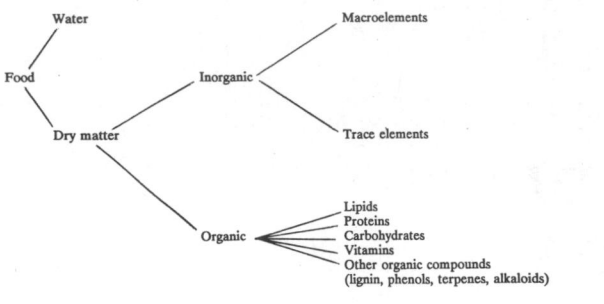

Fig. 1. Basic chemical composition of wildlife foods.

but should also be mentioned because of their widespread distribution in the plant kingdom and their well-documented pharmacological effects on animals (Robinson 1979, Fowler 1983). For these and other secondary plant metabolites, Rosenthal and Janzen (1979) provided a review of chemistry, distribution, methods of determination, and effects on herbivores.

Phenolics are often broadly categorized functionally and structurally as tannin or non-tannin. Tannins are large compounds (500–3,000 molecular weight) that precipitate protein from an aqueous medium (Martin and Martin 1982). The non-tannin phenolics primarily include smaller flavonoids, which do not precipitate protein (Peri and Pompei 1971, Harborne 1979). Phenolics are widespread and occur in all classes of vascular plants (Swain 1979); they are found in 17% of nonwoody annuals, 14% of herbaceous perennials, 79% of deciduous woody plants, and 87% of evergreen woody plants (Rhoades and Cates 1976). Phenolic concentrations vary seasonally and by plant part (Feeny and Bostock 1968, Bryant 1981, Palo et al. 1985, Van Horne et al. 1988), which should be a major sampling consideration for vegetation. Tannins reduce food intake, deter feeding, decrease protein digestibility, increase metabolic rates, and occasionally are toxic to wild herbivores (Buchsbaum et al. 1984, Lindroth and Batzli 1984, Smallwood and Peters 1986, Robbins et al. 1987*a,b,* Thomas et al. 1988). Non-tannin phenolics also can reduce food intake (Lindroth and Batzli 1984, Robbins et al. 1987*a*).

Terpenoids are a large class of biological compounds that includes the sesquiterpene lactones, volatile terpenes, and higher terpenes (Mabry and Gill 1979). Plant substances commonly referred to as essential oils, volatile oils, and resins contain terpene components (Nagy et al. 1964, Schwartz et al. 1980, Fowler 1983). Wildlife studies have focused on the negative effects of terpenes, primarily in Douglas-fir, juniper, and sagebrush, on food selection and microbial digestion in ruminants (Oh et al. 1967, Nagy and Tengerdy 1968, Radwan 1972, Schwartz et al. 1980, Cluff et al. 1982, Risenhoover et al. 1985, Personius et al. 1987).

Alkaloids are heterocyclic nitrogen compounds (Robinson 1979) that are present in 15–20% of vascular plants (Levin 1976). Their defensive properties are largely a result of their toxicities (Fowler 1983).

CHEMICAL AND ENERGETICS ANALYSES OF FOODS
Sample Collection and Preparation

Herbivore foraging behavior and nutrition are highly influenced by variation in the chemical composition of plants. Therefore, it is important to follow basic sampling considerations. Foods collected for chemical analyses or feeding trials should be similar to those selected by the animal.

Food samples should be collected from the same seasonal habitat as that used by the herbivore. The selection of species to be investigated should be based upon a knowledge of the foods eaten by the wildlife species of interest. As much as possible, information on foraging behavior should be incorporated into the sampling plan to simulate selection by the animal. For example, Regelin et al. (1974), Schwartz et al. (1977), and Hobbs et al. (1983) attempted to duplicate food selection of ungulates by observing tame animals as they fed and collecting samples for nutritional analyses by simultaneous hand-picking of the food species.

Methods used to collect, store, and prepare plant material for chemical analyses or feeding trials can significantly impact results. Physical damage to leaves can affect levels of phenolics (Swain 1979), because tannin phenolics are sequestered in vacuoles in the plant and will form complexes with plant proteins when leaves are crushed (McLeod 1974, Swain 1979). Damaged leaves containing high concentrations of phenolics sometimes will develop a brown or black coloration, indicating that the oxidation of phenolics has occurred (Ribéreau-Gayon 1972). Crushing leaves also can cause the release of volatile terpenes (Mabry and Gill 1979).

Samples can be washed after collection but usually are not. Herbivores do not wash plants prior to consuming them, so substances adhering to the plant will be ingested. Also, some nutrients can be leached from a plant by washing (Tukey 1966). However, in certain experimental designs such as comparing treatment effects on plant nutrition at different times, differing rainfall patterns could influence the amount of dust or soil adhering to plants and influence element analyses or ash content. In these situations, washing of samples may be recommended.

In general, collected plant samples should always be kept cool (in *some* situations frozen) after collection and analyzed as soon as possible. Even after plants are picked, dry matter losses of sugars from respiration and enzymatic conversions of sugars to starches can occur (Smith 1973). These losses can be reduced by cold storage. Volatile terpene loss is reduced in collected samples by freezing with dry ice or liquid nitrogen (Schwartz et al. 1980, Welch and McArthur 1981).

Freezing fresh plant material is a common method of sample storage, but it can cause problems with phenolic and forage fiber analyses. Thawing samples can produce black coloration in leaves high in phenolics, causing an apparent loss of phenolics (Mould and Robbins 1981*a,* Servello et al. 1987). Reductions in measured phenolic content also result in underestimates of neutral detergent solubles (detergent analysis) and, therefore, overestimates of neutral detergent fiber (Mould and Robbins 1981*a,* Servello et al. 1987).

If sample drying is required, it should be completed as soon as possible, because chemical changes are less likely at low moisture levels. However, the drying method used can alter chemical composition. Oven-drying, even at relatively low temperatures (40–60 C), can substantially decrease phenolic levels (Julkunen-Tiitto 1985, Servello et al. 1987, Nastis and Malechek 1988), and frozen storage before oven-drying reduces phenolic levels more than oven-drying fresh material (Servello et al. 1987). Air-drying or oven-drying at room temperature (20–25 C) reduces phenolic levels of leaves, but less than at higher oven-drying temperatures (Servello et al. 1987, Nastis and Malechek 1988). Smith (1973) reported that oven-drying leaves below 50 C provides time for dry matter losses of nonstructural carbohydrates by respiration and enzymatic conversion, and drying above 80 C can result in thermochemical degradation. Drying above 50 C can cause enzymatic browning, resulting in artifact lignin in the detergent analysis (Van Soest 1965*b*).

Lyophilizing (freeze-drying), in contrast to other drying methods, results in greater phenolic, neutral detergent-soluble, and in vitro digestibility values for leaves high in phenolics (Servello et al. 1987, Nastis and Malechek 1988) and therefore is probably the mildest drying treatment. Frozen leaf samples can be transferred directly to a freeze-dryer, thus reducing changes associated with thawing described above. Smith (1983) discussed the feasibility of using microwave ovens for drying plant samples, but did not consider the effect of this method on chemical composition of the samples other than noting that charring of samples could occur.

In summary, there is not one best drying method for plant samples. We recommend lyophilizing whenever possible, particularly when plant samples have been stored frozen. If oven-drying is the only option, we suggest drying at 40 C to minimize chemical alteration. Working with fresh plant material is usually awkward and difficult, and we do not recommend it unless required (e.g., for terpenes). Most importantly, we suggest that researchers carefully review the potential effects of drying methods on the plant species and chemical constituents being studied and select the most appropriate drying method. We recommend lyophilizing for all animal tissues.

Terpenes must be extracted from fresh or fresh-frozen plant material, because some terpene compounds are highly volatile (Mabry and Gill 1979, Personius et al. 1987). Extracts of fresh material can be used for phenolic determinations as well (Mould and Robbins 1981*a*).

Most samples must be ground prior to chemical analysis. This step allows a more homogeneous sample to be analyzed and ensures complete contact of the sample with the chemicals. Generally, samples should be ground to pass a 0.5-mm or 1.0-mm sieve.

Chemical Analyses

Determination of the chemical composition of a food allows it to be evaluated in terms of its digestibility by an animal or its adequacy in supplying required nutrients to an animal. Unfortunately, most chemical analyses for nutritional purposes measure only groups of chemically similar compounds. A concern with this approach is that all compounds in a fraction may not react in the same way in the digestive tract of an animal, which can lead to problems when these chemical measures are compared among foods. In addition, most chemical analysis systems have some analytical flaws in the division of chemical fractions. Even with these problems, chemical analyses allow information on the composition of foods to be ob-

---

Box 1. **Steps for conducting proximate analysis (from Church and Pond 1988).**[a]

Step 1. Dry and grind sample.

Step 2. Determine true percent dry matter (dry in 100 C oven). Use dry-matter percentage correction for all subsequent steps.

Step 3. Conduct Kjelkahl digestion and analyze for nitrogen content. Calculate crude protein by multiplying nitrogen percentage times 6.25.

Step 4. On a separate sample, conduct ether extract procedure to determine crude fat content.

Step 5. Using fat-free residue from Step 4, use acid-base treatment to determine crude fiber component, which also includes ash.

Step 6. Burn residue from Step 5 in 500–600 C muffle furnace to determine ash content.

Step 7. Subtract crude protein, crude fat, crude fiber, and ash percentages from 100% to calculate nitrogen-free extract (NFE).

---

[a]Although this complete method is no longer recommended (particularly steps 5 and 7), steps 2, 3, 4, and 6 are still routinely used.

---

tained in a standardized and efficient manner with relatively small amounts of food in replicated samples.

Proximate analysis, or the Weende Method, was developed in the 1800s (Box 1) and has been used extensively for wildlife studies. However, the method has several problems with some of its analyses, as discussed below, which has led to the development of other analyses for certain food components. Although parts of the proximate analysis procedure are no longer recommended, we mention it because some of the procedures are still used for analysis of some chemical constituents, and because of the large body of literature reporting results from their use. Proximate analysis separates food into six chemical groupings: water, ether extract, crude fiber, nitrogen-free extract (NFE), crude protein, and ash. Crampton and Harris (1969) provided a good description of proximate analysis.

Water content of foods is determined primarily so that other constituents can be evaluated on a dry-matter basis, because the water content of foods varies considerably. The true percent dry matter can be determined from samples that have been initially dried and ground, then oven-dried at 100 C for 24 hours. The weight loss resulting from sample drying and the true dry-matter determination is the water content of the sample. Further analyses of dried samples are adjusted to a true dry-matter basis by the percent dry matter.

Ether extract, or crude fat, is determined by extracting the portion of a food sample that can be dissolved in anhydrous ethyl ether (Horwitz 1975). The problem with ether extract is that it dissolves more than the fats and oils it is designed to measure, including fat-soluble vitamins, chlorophyll, alkalis, resins, waxes, and volatile oils. Crude

Table 1. Determination of chemical composition of dried forages with proximate analysis and detergent analysis systems.

	Proximate analysis	Detergent analysis
Nitrogenous compounds		
Soluble proteins	Crude protein	NDS[a]
Cell-wall nitrogen	Crude protein	NDF[b]
Lignified nitrogen	Crude protein	ADF[c]
Carbohydrates		
Sugars, starches	NFE[d]	NDS
Hemicellulose	NFE	NDF
Cellulose	Crude fiber	ADF
Lipids	Crude fat	NDS
Vitamins		
Water-soluble	NFE	NDS
Fat-soluble	Crude fat	NDS
Ash	Ash	Ash
Lignin		
Alkali-soluble	NFE	ADL[e]
Insoluble lignin	Crude fiber	ADL
Secondary plant compounds	Crude fat or NFE	NDS

[a]Neutral detergent solubles.
[b]Neutral detergent fiber.
[c]Acid detergent fiber.
[d]Nitrogen-free extract.
[e]Acid detergent lignin.

fat is considered to be the readily digestible, high-energy component of the food. However, additional compounds are removed in the extraction, some of which can have inhibitory effects on digestion, so the energy attributed to crude fat must be viewed with caution.

Crude protein is estimated by measuring the nitrogen content of a food. The commonly used Kjeldahl procedure for measuring nitrogen involves digesting the sample in H_2SO_4, neutralizing with NaOH, distilling the resulting ammonium, and titrating with acid (Horwitz 1975, Church and Pond 1988). The nitrogen percentage is then multiplied by 6.25 to calculate the crude protein content, because proteins, on the average, contain 16% nitrogen ($100/16 = 6.25$). Problems with crude protein estimation arise in the conversion of nitrogen to crude protein because not all proteins contain 16% nitrogen, and nonprotein nitrogen (NPN) compounds will be present in varying amounts. The NPN fraction is usually small; however, crude protein overestimates true protein by 22–52% in some plants (Sedinger 1984). Crude protein does not provide information on protein quality (i.e., amino acid composition), which can be important to nonruminants. Also, tannins in plants can bind with protein, making some of the measured crude protein unavailable to the animal (Robbins et al. 1987b).

Ash is determined by burning a sample in a muffle furnace (500–600 C) and weighing the residue (Horwitz 1975). Washing a sample after collection in the field can significantly change measured ash content due to soils or

dust adhering to the plant. Determining ash content does not provide any information on the elements contained in the ash.

Crude fiber, theoretically the undigestible fiber, is determined by using the residual sample after ether extraction (Church and Pond 1988), treating it with dilute acid and base, and burning the sample to account for ash content (Horwitz 1975). Much of the lignin in the sample is dissolved in the process, as is some of the hemicellulose, so crude fiber estimates may be misleading. In addition, most animals can digest at least some of the crude fiber in a food (Crampton and Harris 1969), depending largely on the amount of lignin in the fiber fraction.

Lastly, NFE is determined by subtracting the percentages of water, crude protein, crude fat, ash, and crude fiber from 100%. NFE is considered the readily digestible carbohydrate in the food. Because NFE is determined by difference, all of the errors discussed for the other fractions are combined in the NFE estimate.

Detergent analysis (Van Soest 1963a,b, 1965a, 1967, 1982, Goering and Van Soest 1970) is a commonly used method of separating the fiber components of a food. This method is commonly used in place of the crude fiber and NFE fractions of proximate analysis; it is theoretically based on an initial chemical separation of a food sample into a nearly completely digestible-cell-contents fraction (Jones and Wilson 1987) and a variably digestible-cell-wall fraction. This division has some error, however, so these fractions are more accurately referred to as neutral detergent solubles (cell contents) and neutral detergent fiber (cell walls) (Mould and Robbins 1981a). Subsequent steps in the analysis divide the cell-wall fraction into cellulose, hemicellulose, lignin, and ash. Mould and Robbins (1981b) modified the original method with a recommendation that Na_2SO_3 not be used in the neutral detergent because it dissolved some lignin. Secondary plant compounds, if they occur, will be separated into the neutral detergent-soluble fraction and may significantly reduce neutral detergent digestibility (Mould and Robbins 1982, Robbins et al. 1987a). A comparison of the division of chemical compounds by proximate analysis and detergent analysis is given in Table 1.

Additional methods are available for further characterizing the carbohydrates in a food, especially the hemicelluloses. Van Soest (1982) presented a good overview of some of these methods.

Element contents of foods also may be of interest. The most commonly used method of analysis involves atomic absorption spectrophotometry, which can be used for many macro-/and microelements (Dahlquist and Knoll 1978, DeBolt 1980).

Phenolics are extracted from fresh or dried (see cautions under collection and storage of plant material) plant material with a polar solvent, usually methanol, acetone, ethanol, or ethyl acetate in an aqueous mixture (Swain 1979). Martin and Martin (1982) and Robbins et al. (1987a) used boiling aqueous methanol extractions. Total phenolics are commonly assayed colorimetrically in plant extracts by the Folin-Denis procedure (Burns 1963), but the Folin-Ciocalteu procedure is an improvement over the former method (Singleton and Rossi 1965). Total phenolic information alone has limited value for describing nutritional quality because of poor correlations between total phenolics and levels of protein-precipitating phenolics (Martin and Martin 1982). However, total phenolic assays provide a measure of a nutritionally useless fraction of plants that can constitute a substantial portion of plant dry matter (Mould and Robbins 1982, Servello et al. 1987).

Tannin assays are still a subject of much research (Mole and Waterman 1987, Wisdom et al. 1987), but assay methods are not standardized. The total tannin fraction in extracts cannot be measured directly. Vanillan-HCl (Price et al. 1978) and chloroform-HCl (Walton et al. 1983) methods are used to assay condensed tannins in domestic grains. Protein-binding methods have received recent study and appear to provide the most biologically meaningful measure of tannin content because they mimic the expected effect of tannins on digestion. The most commonly used or recently developed protein-binding methods are hemoglobin (Bate-Smith 1973), bovine serum albumin (BSA) (Hagerman and Butler 1978, Martin and Martin 1982), ribulose-1,5 biphosphate carboxylase oxygenase (RuBPc) (Martin and Martin 1983), and dye-labelled BSA (Asquith and Butler 1985). Hagerman (1987) recently developed a relatively simple and inexpensive BSA precipitation method that is based on the diffusion of tannin in a protein-containing agar slab. These methods do not give identical results because they are not a measure of tannin content, and different amounts of protein may be precipitated by different types of tannins (Martin and Martin 1982); however, the relative ranking of tannin content should be similar.

Terpenes usually are extracted by steam distillation of fresh plant material and collection in ether (Radwan 1972, Risenhoover et al. 1985). Schwartz et al. (1980) fractionated oils into three groups (monoterpenes, oxygenated monoterpenes, and sesquiterpenes) by distilling at three temperature ranges. Quantification of specific fractions is done chromatographically (Mabry and Gill 1979, Schwartz et al. 1980, Welch and McArthur 1981).

Two qualitative tests are commonly used to determine the presence of alkaloids; one uses Dragendorff's reagent and the other uses Meyer's reagent (Robinson 1979). Identification and quantification of alkaloids involve complex chemistry beyond the scope of this chapter.

Energy Content

Energy content of food samples is measured by bomb calorimetry. Bomb calorimeters are either adiabatic or nonadiabatic, although most laboratories today use the adiabatic type. In adiabatic calorimetry, food samples are combusted in oxygen and the released heat is measured in a surrounding waterbath. A good source of information on bomb calorimetry can be found in the instruction book for the Parr adiabatic calorimeter (Parr Instrument Co., Moline, IL 61265). Gessaman (1987) also provided a good description of bomb calorimetry and sampling considerations. Energy content of foods ranges from about 4 kcal/g for foods high in carbohydrates to about 9 kcal/g for foods high in fats (Church and Pond 1988:144). Bomb calorimetry produces accurate estimates of total energy content commonly called gross energy. However, the percentage of the energy available to an animal varies considerably, depending on the chemical composition of the sample and its digestibility in different species. Available energy must be determined by digestion trials. Trials can be in vitro digestion trials or actual feeding trials.

FOOD DIGESTIBILITY MEASUREMENTS
In Vitro Digestion

Chemical analyses, although providing information on the chemical composition of foods, do not provide information on actual digestion or nutrient supply to an animal. Feeding trial methods provide specific measures of food utilization by an animal but require the maintenance of experimental animals, large amounts of test foods for the trial, and a major commitment of time and money. In vitro digestion methods are designed to simulate the ruminant digestion process in the laboratory and require using only small amounts of the test foods as in chemical analyses. Thus, in vitro methods are a more efficient means of determining digestibility than are feeding trials, particularly when information is needed on large numbers of foods.

Various in vitro digestion procedures have been developed (Pearson 1970), but the method developed by Tilley and Terry (1963) has become the most commonly used. This method involves inoculating a sample of food with rumen fluid and a buffer solution designed to simulate saliva and maintaining these in a hot water bath for 48 hours. In the second stage, the digested contents are treated with pepsin and mild acid. Modifications of this method include the addition of a phosphate-carbonate buffer to reduce foaming and a smaller ratio of rumen fluid to buffer solution (Campa et al. 1984), use of McDougall's solution to allow direct acidification and addition of pepsin (Van Soest 1982), or substituting the second stage of pepsin and mild acid with a neutral detergent extraction (Van Soest 1982).

In vitro digestion has been compared with in vivo digestion trials (with live animals) in several studies and has produced varying results. Robbins (1983) discussed some of these comparisons.

The source of rumen fluid can significantly influence in vitro digestibilities. Rumen fluid collected from wild deer produced significantly different digestibility estimates compared to rumen fluid collected from captive deer on a pelleted ration or from a fistulated cow (Campa et al. 1984). Jenks and Leslie (1988) reported lower in vitro digestibilities with cow inoculum than with deer inoculum. Thus, although in vitro techniques with captive animals of the same or different species may be used to provide relative comparisons of food samples, extrapolating digestibilities to animals in the wild is not recommended. In addition, Clary et al. (1988) reported considerable intraspecific variation in the ability of rumen inocula to digest forages. They suggested that multiple rumen fluid donors are important, as is the use of standard reference forages. Palmer and Cowan (1980) also suggested the use of standard reference forages for in vitro digestion trials.

Alternative methods of estimating digestion of foods have been proposed in which various combinations of pepsin, fungal cellulose enzymes, or other enzymes are used (Clark and Beard 1977, Goto and Minson 1977, McLeod and Minson 1978, 1982, Choo et al. 1981, Clarke et al. 1982, Dowman and Collins 1982, Pace et al. 1984, Barnes 1988). The accuracy of these methods in estimat-

Box 2. **Equations for calculating apparent dry matter digestibility (ADDM), apparent digestible energy (ADE), apparent digestibility of a specific nutrient, and apparent metabolizable energy (AME) from digestion or metabolism trial data.**

$$\text{ADDM (\%)} = \frac{\text{Food intake} - \text{Fecal dry matter}}{\text{Food intake}} \times 100$$

$$\text{ADE (\%)} = \frac{(\text{intake}_g \times \text{GE food}) - (\text{feces}_g \times \text{GE feces})}{(\text{intake}_g \times \text{GE food})} \times 100$$

$$\text{GE} = \text{gross energy}$$

$$g = \text{grams}$$

Apparent Digestibility (%) of Nutrient A

$$= \frac{(\text{Food intake} \times \% \text{ A in food}) - (\text{Fecal dry matter} \times \% \text{ A in feces})}{(\text{Food intake} \times \% \text{ A in food})} \times 100$$

$$\text{AME (\%)} = \frac{\text{E intake} - (\text{fecal E} + \text{urinary E} + \text{gaseous E}^a)}{\text{E intake}} \times 100$$

$$\text{E} = \text{energy.}$$

---

[a] In nonruminants, gaseous E is small and ignored.

---

ing true digestibility has been questioned (Barnes 1988), and additional work on comparisons of techniques is needed.

Another technique, the nylon bag method, involves placing a food sample in a nylon bag suspended through a fistula into the rumen of a live animal (Johnson 1966). This permits actual rumen action on the food. The diet of the fistulated animal must be considered in interpreting results, and some errors may result, because small, undigested particles of food may either enter the nylon bag from the rumen or exit the nylon bag with washing and be incorrectly attributed to digestion. Person et al. (1980) compared nylon bag analysis and in vitro analysis of reindeer and caribou forages and reported varying results with different types of forages.

Digestion and Metabolism Trials

Digestion and metabolism trials are used to determine the digestibility and metabolism of foods and food constituents. This section will review the basics of digestion and metabolism trials. However, these general methods sometimes must be modified for individual species. The following list of species for which digestibility studies have been made can serve as the starting point for designing digestion trials for particular species: deer and elk (Baker and Hansen 1985), reindeer (White et al. 1984), moose (Hjeljord et al. 1982), mountain sheep (Baker and Hobbs 1987), howler monkey (Nagy and Milton 1979), grey kangaroo (Kempton et al. 1976), collared peccary (Carl and Brown 1985), snowshoe hare (Holter et al. 1974), black-tailed prairie dog (Hansen and Cavender 1973), fox squirrel (Havera and Smith 1979), voles (Batzli and Cole 1979), badger (Harlow 1981), red fox (Litvaitis and Mautz 1976), coyote (Litvaitis and Mautz 1980), bobcat (Johnson and Aldred 1982), fisher (Davison et al. 1978), mink (Farrell and Wood 1968), weasel (Moors

1977), pinnipeds (Helm 1984), bald eagle (Stalmaster and Gessaman 1982), lesser snow goose (Burton et al. 1979), northern bobwhite (Case and Robel 1974), Egyptian goose (Halse 1984), northern pintail, gadwall, northern shoveler, mallard (Sugden 1971, Miller 1984), king penguin (Adams 1984), ruffed grouse (Servello et al. 1987), wild turkey (Billingsley and Arner 1970), Canada goose, Atlantic brant (Buchsbaum et al. 1986), sharp-tailed grouse (Evans and Dietz 1974), spruce grouse (Pendergast and Boag 1971), willow grouse (West 1968), rock ptarmigan (Gasaway et al. 1976), mourning dove (Shuman et al. 1988), and small songbirds (Willson and Harmeson 1973, Holthuijzen and Adkisson 1984).

TOTAL COLLECTION METHOD

The total collection method is the standard method of measuring the digestive capability of animals and the digestibility and metabolism of energy and nutrients in foods. Individual animals are fed weighed amounts of a test food over a number of days. The excreta resulting from the test food are collected and weighed. The apparent dry matter digestibility (ADDM) is the percentage of dry matter not excreted in the feces (Box 2). The apparent digestibility of a specific nutrient or food component (e.g., fiber) can be determined by measuring the percentage of the nutrient in the food and feces and calculating the percentage of the nutrient eaten that was digested (Box 2).

Similarly, percent apparent digestible energy (ADE) can be determined by measuring the gross energy (GE) in food and fecal samples with a bomb calorimeter and converting dry-matter measurements to energy equivalents (Box 2). Apparent metabolizable energy (AME) is determined by also measuring and subtracting urinary and gaseous energy losses (Box 2). Methane (gas) production is low in nonruminants (Robbins 1983) and therefore ignored in those species. AME is routinely reported for birds, because feces and urine mix in the cloaca of birds.

Digestibility values for dry matter, energy, and some nutrients are reported as "apparent" because some fecal and urinary dry matter does not originate from the test food. These endogenous losses include digestive enzymes, gastrointestinal epithelial cells, microbes, and excreted end products of metabolism (Maynard et al. 1979). "True" digestibility or metabolism values can be determined by measuring the amount of endogenous dry matter, energy, or nutrients produced and correcting apparent values or with more complex techniques to estimate true values directly (see NUTRITIONAL REQUIREMENTS or true ME method described below). Low food or nutrient intake increases the difference between apparent and true values because endogenous losses are relatively constant and comprise a larger fraction of the excreta at low intake levels (Sibbald 1975, Robbins 1983). However, for evaluating diets, the practical significance of "true" values is questionable because endogenous losses represent actual losses that must be replaced by the diet (Maynard et al. 1979).

The ability of the animal to extract or use energy in the test food is measured as percent digestible or metabolizable energy, respectively. However, the energetic value of food should be reported as the amount of DE or ME (kcal) per gram of food dry matter (kcal/g) rather than as a percentage, because foods with identical percent DE or percent ME values can differ in gross energy content.

Metabolizable energy values for foods (kcal/g) are often calculated as nitrogen-corrected ME. Individual animals losing or gaining differing amounts of body tissue during a feeding trial will vary in the amount of endogenous urinary nitrogen in their excreta. ME estimates are standardized as follows: deviations from nitrogen balance (gains or losses of nitrogen) are corrected by adding or subtracting the energy content of urinary nitrogen; an equivalent of 8.22 kcal/g is used for birds (Scott et al. 1982:537), and 7.45 kcal/g is used for mammals (Maynard et al. 1979:196) for each gram of nitrogen retained or lost. In practice, though, nitrogen-corrected ME values differ little from uncorrected values (e.g., Burton et al. 1979, Beckerton and Middleton 1982, Scott et al. 1982).

Procedures

Animals must be acclimated to the test diet prior to the fecal collection period. For monogastric species, including herbivorous and carnivorous birds and mammals, 3- to 5-day acclimation and 3- to 5-day collection periods are most commonly used (e.g., Short 1976, Robel et al. 1979). Seven- to 10-day acclimation periods and 7-day fecal collection periods are most common for ruminants. Mothershead et al. (1972) reported that a 10-day collection period was the most accurate for white-tailed deer, but 7 days was not much of a disadvantage.

Metabolism cages designed for collection of spilled food, feces, and urine are available in a variety of sizes. Cowan et al. (1969) described a cage for deer. HCl or H_2SO_4 should be added to urine collection bottles to maintain solution acidity to prevent loss of ammonia.

Maintaining a regular and adequate food intake is important for accurate results. Animals should be fed at the same time each day. Test diets are often fed ad libitum, but constant intake can be assured and selective feeding can be avoided by feeding at a slight reduction (e.g., 90% ad libitum). Chopping or pelleting plant material or homogenizing animal tissue often is used to prevent selective feeding.

Whether birds need access to grit during metabolism trials is not clear. Robel and Bisset (1979) reported that supplemental grit did not change ME values for seeds and a commercial feed. However, McIntosh et al. (1962) reported that access to grit improved ME values for chickens in some instances. Providing grit during a trial adds the problem of having to separate excreted grit from fecal matter.

Other feeding considerations may be necessary for carnivores that adapt to periods of prey scarcity. Harlow (1981) reported that badgers that had been fasted before the collection period had a slower food passage rate and metabolized 11% more of the energy in food than badgers fed on a daily basis prior to the trial.

Experimental Animals

Experimental animals must be acclimated or trained for handling and confinement in metabolism cages. Wild-captured animals of some small species often are used for feeding trials, but many species or individual animals do not adapt to confinement. For this reason, most experimental animals are reared in captivity. However, even captive-raised animals must be acclimated to cage confinement. Mautz (1971b) observed that food intake decreased with some confined, captive-raised deer, which required 9–12 days to return to preconfinement levels.

Feeds used to raise and maintain animals for digestion studies may influence results. The digestive tract (gizzard, intestines, ceca) of herbivorous birds increased in size with increasing dietary fiber levels (Moss 1972, Miller 1975, Halse 1984). Therefore, maintenance diets should be similar to natural diets in fiber and energy levels. Also, feeding whole forages (e.g., leaves, twigs) in trials to birds maintained on commercial feeds may be a problem because of insufficient gizzard development, but this has not been studied.

Mothershead et al. (1972) suggested that five deer per test diet were adequate for most digestion trials. Three to six animals per trial are commonly used for most species. If the animals are healthy and eating normally, standard errors for DDM, DE, and ME estimates are usually relatively small with these sample sizes.

DIFFERENCE METHOD

Some forages are unpalatable or too low in nutritional quality to be fed alone in a total collection digestion trial. Such forages can be mixed with a highly palatable basal diet (usually a commercial feed) and fed in a total collection trial. By conducting a similar trial for the basal diet alone with the same animals (usually prior to the experimental trial), one can calculate the digestibility of the test forage:

ADDM (%) = [ADDM of test diet

− (proportion of basal diet in test diet

× ADDM basal diet)]

÷ (% forage in test diet).

It is assumed that inclusion of the basal diet has no effect on the digestibility of the test forage. This method has been used for moose (Schwartz et al. 1988*b*), deer (Robbins et al. 1975), ruffed grouse (Hill et al. 1968), and pine voles (Servello et al. 1984) and is the standard method for testing individual forages for poultry (Scott et al. 1982: 536).

A related method is the use of multiple test foods in mixed diets. Digestibilities of individual foods are calculated by solving simultaneous equations (Pekins and Mautz 1988). Several test foods are all included in various combinations in diets fed in total collection trials. The diet digestibilities and proportions of the forages in the diets are then used to solve for digestibilities of individual foods (Pekins and Mautz 1988).

TRUE METABOLIZABLE ENERGY METHOD FOR BIRDS

Sibbald (1976, 1979) developed a method for determining the true ME (TME) of foods for poultry. The TME method is relatively fast and requires less of the test food than a conventional feeding trial and has been proposed for use on wild avian species (Miller and Reinecke 1984). Several modifications have been tested, but the general procedure for chickens is to first deprive the birds of food for 24 hours. Species with high metabolic rates may need to be kept on a maintenance diet prior to the test (Hoffman and Bookhout 1985). An experimental bird is then force-fed a known quantity of test food at the start of a 48-hour excreta collection period. A control bird of similar weight continues to be fasted during the 48-hour collection period. The excreta of the control bird is collected and weighed to estimate metabolic and endogenous dry matter and energy losses. TME is calculated by subtracting the endogenous and metabolic energy losses of the control bird (C) from the fecal and urinary energy of the fed bird (F):

$$\text{TME}(\%) = [\{E \text{ intake}_{(F)} - ([\text{fecal} + \text{urinary } E_{(F)}]$$
$$- [\text{fecal} + \text{urinary } E_{(C)}])\}$$
$$\div E \text{ intake}_{(F)}] \times 100.$$

The TME method has been used with ruffed grouse (Norman 1980), mallards, northern pintails (Hoffman and Bookhout 1985), and black ducks (Jorde and Owen 1988), but it has not been compared with standard digestion trials for any wild species. Wild avian herbivores have more complex digestive systems than poultry and feed on more complex foods than commercial feeds; thus the TME method should be used with caution. For example, length of the collection period influences TME results differently for different foods (Chami et al. 1980), which may be particularly troublesome with wild avian herbivores with complex digestive systems (e.g., large ceca and gizzards).

Indicator Methods

Indicator methods that do not require the laborious accounting of intake and excretion are sometimes used to determine digestibility. Either a naturally occurring indigestible indicator substance or one added to the test diet is measured in a sample of the food and feces, and apparent digestibility is calculated for dry matter, energy, or any food component:

Apparent digestibility (%)

$$= 1 - \left[\left(\frac{\% \text{ indicator in food}}{\% \text{ indicator in feces}} \right) \times \left(\frac{\% \text{ A or energy content in feces}}{\% \text{ A or energy content in food}} \right) \right]$$

where A is a particular forage component, e.g., fiber, N. This method assumes that the indicator is indigestible or is not changed in the digestive tract and that it mixes and moves uniformly with the food. Naturally occurring indicators that have been used include lignin (Buchsbaum et al. 1986), cellulose (Inman 1973), magnesium (Moss 1973), and ash (Johnson and Groepper 1970). However, cellulose and lignin are digested or changed in form to some extent (Inman 1973, Thonney et al. 1979, Servello et al. 1983), and magnesium and ash require that the animal be in mineral balance. Radioactive chromic chloride and Cr-EDTA are indicators added to the food that generally give results similar to the total collection method (Mautz 1971*a*, Gasaway et al. 1976, Han et al. 1976).

The importance of rates of digestion or food passage as a significant factor limiting food intake and nutrition of some species (Demment and Van Soest 1985) has resulted in extensive study of food passage rates. Several indicators or markers, including chromic oxide, Cr-EDTA, Ce-144, polyethylene glycol, barium, and ytterbium, are used to study food passage, digesta flow, and digestive efficiency of mammals and birds (Gasaway et al. 1975, 1976, Björnhag and Sperber 1977, Warner 1981, Baker and Hobbs 1987).

NUTRITIONAL REQUIREMENTS

An important but complex endeavor in wildlife nutrition research is the determination of energy and nutrient requirements for individual species. Requirements vary with life functions (maintenance, growth, reproduction) and season and are influenced by physiological adaptations. Nutrient or energy requirements for diets also are sometimes interactive. For example, food intake is dependent to a certain extent on the energy level of the diet (Ammann et al. 1973, Batzli and Cole 1979, Scott et al. 1982). Also, the protein-to-energy ratio is as important as the protein content of the diet in determining performance in poultry (Scott et al. 1982). Requirements also may be influenced significantly by underlying physiological adaptations. For example, white-tailed deer voluntarily reduce food and energy intake in winter (Silver et al. 1969, Thompson et al. 1973).

Energy requirements of captive animals can be determined by direct measures of energy expenditure or with feeding trials. Nutrient requirements usually are determined through specialized feeding trials with captive animals. Optimal nutrient or energy requirements have been quantified for few species, and even less is known about suboptimal tolerances. Because diets of suboptimal quality are probably common in wild populations, understanding the effects of nutrient intake over a range of suboptimal levels is important.

Metabolism Techniques

Energy expended by an animal can be for different purposes. A certain amount of energy is needed to maintain basic life processes and cellular activity of an animal, called the basal metabolic rate (BMR). Homeotherms have additional energy demands to maintain their body temperature, or thermoregulatory energy demands. Numerous other activities necessary for survival of the animal, such as feeding, predator avoidance, social interactions, growth, and migration, also require energy expenditure. Active metabolic rates of animals are difficult to determine, especially for free-ranging animals, as are total energy requirements on a year-round basis.

Feeding trials involve placing an animal on diets of differing levels of digestible or metabolizable energy intake and determining the energy level necessary to maintain body weight. The most common approach is to vary the amount of food offered to captive animals and to plot weight change versus energy intake. Regression analysis is used, and the point on the regression line where body weight change equals zero is taken as the energy requirement for maintenance. Variations on this method include altering the amount of food given to each animal until weight stabilizes (Keiver et al. 1984) or taking as an estimate the energy intake level for a time period when weight change was stable (<1–2%) (Case and Robel 1974, Williams and Kendeigh 1982). These measures are commonly referred to as existence energy or metabolizable energy requirements. Body composition must be considered, though it is difficult to assess without slaughtering the animals, because fat reserves can be replaced by protein and water, and energy is supplied to the animal without measurable weight loss (Hudson and Christopherson 1985). Feeding trials provide a measure of maintenance energy for captive animals. This is greater than basal metabolism because the animal is feeding and conducting some activities, but it will generally be less than the active metabolic rate of a free-ranging animal (Robbins 1983).

Maintenance energy requirements have been determined through feeding trials for many captive wildlife species, such as white-tailed deer (Ullrey et al. 1970, Thompson et al. 1973, Verme and Ozoga 1980), mule deer (Robinette et al. 1973, Baker et al. 1979), barren ground caribou (McEwan 1970), moose (Schwartz et al. 1988a), collared peccary (Zervanos and Day 1977), raccoon (Teubner and Barrett 1983), bald eagle (Stalmaster and Gessaman 1982), Canada goose (Williams and Kendeigh 1982), blue jay (Clemans 1974), pine vole (Lochmiller et al. 1983), snowshoe hare (Holter et al. 1974), red fox (Vogtsberger and Barrett 1973), gray seal (Ronald et al. 1984), black-bellied whistling-duck (Cain 1976), common barn-owl (Wallick and Barrett 1976), and white ibis (Kushlan 1977).

Energy expenditure can be determined directly through heat production by an animal. Although this method is technologically feasible, it has not been widely used for wildlife species because indirect measures are easier and more economical (Mautz 1978). Indirect measures of energy metabolism of an animal can be made with several techniques. An accurate and widely used method is respiratory gas exchange, which measures the ratio of CO_2 production to O_2 consumption, or the respiratory quotient (RQ) of an animal. This method has been used in closed respiration chambers to determine the basal or fasting metabolic rate of many species. Basal metabolism is the

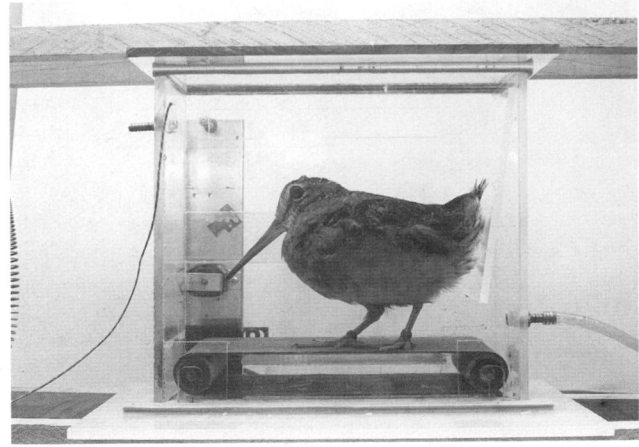

Fig. 2. Energy expenditure measured for woodcock walking on a treadmill in a respiration chamber (photo by Vander Haegen).

heat production of an animal in a post-absorptive state at rest in a thermoneutral environment. A post-absorptive state assumes an empty digestive tract in the animal. Fasting metabolic rate is similar but does not assume the animal is in a post-absorptive state, because this is difficult to achieve in many species, especially ruminants. If temperature control is also possible in the chamber, information on thermoregulatory energy can be determined. A detailed description of the RQ method was provided by Gessaman (1987). Modifications to this approach have involved developing face masks or tracheal fistulas (Mautz 1978) to determine energy expenditure of animals in various activities (Corts and Lindzey 1984, Parker et al. 1984, Wickstrom et al. 1984) or increasing activity of animals in a respiration chamber (Fig. 2). Numerous studies have determined metabolic rates of animals through respiratory gas exchange methods. Examples of the breadth of species studied include deer mouse, brush mouse (Mazen and Rudd 1980), armadillo (McNab 1980), pine marten (Worthen and Kilgore 1981), mallard (Smith and Prince 1973), coyote, kit fox (Golightly and Ohmart 1983), zebra finch (Vleck 1981), white-crowned sparrow (DeJong 1976, Maxwell and King 1976), ring-billed gull (Dawson et al. 1976), long-eared owl, short-eared owl, saw-whet owl (Graber 1962), red-tailed hawk, great horned owl (Pakpahan et al. 1989), wolf (Okarma and Koteja 1987), wild turkey (Gray and Prince 1988), ruffed grouse (Thompson and Fritzell 1988a), bobcat (Mautz and Pekins 1989), mule deer (Kautz et al. 1982, Parker et al. 1984), white-tailed deer (Silver et al. 1959, 1969, Mautz and Fair 1980), elk (Gates and Hudson 1979, Robbins et al. 1979, Parker et al. 1984), pronghorn (Wesley et al. 1973), moose (Regelin et al. 1985), caribou (Luick and White 1986), cougar (Corts and Lindzey 1984), and California ground squirrel (Schitoskey and Woodmansee 1978).

A problem with respiratory exchange methods is the difficulty of collecting gases expelled from an animal in a free-ranging state. Attempts have been made to determine energy expenditure by measuring heart rate (Holter et al. 1976, Wooley and Owen 1978, Mautz and Fair 1980, Kautz et al. 1981, Freddy 1984, Fancy and White 1985). Gessaman (1980) reported that heart rate satisfac-

Fig. 3. Grouse mount used for determination of relative energy expenditure by grouse in different thermal covers (photo by F. R. Thompson and E. K. Fritzell).

torily measured energy expenditure of some American kestrels he examined, but not of others. Seasonal variations also were a problem. Holter et al. (1976) observed that heart rate accounted for 78% of the variation observed in metabolic rates, whereas Mautz and Fair (1980) reported that it accounted for only 36% of the variance in energy expenditures. Thus, heart rate can be monitored remotely in free-ranging animals through telemetry, but variability in its results restricts its application.

A final method of determining energy requirements of animals is through the use of stuffed mounts of a species with implanted heaters (Fig. 3). The amount of heat required to maintain the mount at a certain temperature can be measured in different environmental settings or in different types of thermal cover. The heat required can be calibrated to the energy expenditure of a live animal in a laboratory thermal chamber by respiratory exchange. This method has been used for several wildlife species (Heller 1972, Thorkelson and Maxwell 1974, Chappell 1980, Bakken et al. 1983, Thompson and Fritzell 1988*b*).

Nutrient Requirements

Nutrient requirements are determined primarily through feeding trials. Most work has looked at protein or nitrogen requirements. The nitrogen requirement for maintenance is the level of digestible nitrogen intake that produces tissue nitrogen balance (TNB). TNB (also called nitrogen retention) equals 0 when nitrogen intake is equal to endogenous nitrogen excretion and nitrogen assimilation for normal tissue replacement (e.g., hair replacement). Nitrogen is lost from the body as either endogenous urinary nitrogen (EUN) or metabolic fecal nitrogen (MFN). EUN is the excreted nitrogen resulting from normal metabolism and is a constant proportion of metabolic weight (Mould and Robbins 1981*b*). MFN consists of microbes, digestive enzymes, mucus, and gastrointestinal epithelial cells accumulated during digestion and is proportional to feed intake (Mould and Robbins 1981*b*). Except during periods of substantial adult tissue growth (e.g., molt), nitrogen costs for adult growth are small and generally ignored (Maynard et al. 1979). MFN and EUN can be measured in mammals, and their sum is a minimal estimate of maintenance nitrogen requirements (Mould and Robbins

1981*b*). Separate estimates of MFN and EUN cannot be readily obtained for birds because fecal matter and uric acid mix in the cloaca.

A commonly used alternative approach for estimating nutrient requirements is to relate nitrogen intake to nitrogen balance (Holter et al. 1979, Carl and Brown 1985, Priebe and Brown 1987). Both estimates are derived from the same experimental designs and data. Experimental diets containing varying levels of nitrogen content are fed to captive animals to produce varying levels of nitrogen intake. Experimental diets must contain sufficient energy to maintain energy balance, otherwise part of urinary nitrogen may result from tissue catabolism and inflate estimated nitrogen requirements (Maynard et al. 1979, Carl and Brown 1985). Protein requirements have been assessed for several wildlife species including eastern cottontail (Snyder et al. 1976), white-tailed deer (Holter et al. 1979), elk (Mould and Robbins 1981*b*), moose (Schwartz et al. 1987), snowshoe hare (Holter et al. 1974), collared peccary (Carl and Brown 1985), and ruffed grouse (Beckerton and Middleton 1983).

Balance Trial

In a balance trial, the gain or loss of nitrogen or a macroelement (e.g., Na, K, Ca, Mg, P) by the animal is determined as a measure of diet quality. Balance trials are similar to metabolism trials in that food intake and feces and urine production are measured during a collection period. The test food is provided ad libitum. After the element of interest in the food, feces, and urine is measured, animal performance is calculated as grams of the element lost or gained per day, expressed on a metabolic weight basis (Havera and Smith 1979, Beckerton and Middleton 1983).

NUTRITIONAL AND ENERGETICS TECHNIQUES FOR WILD POPULATIONS

Ultimately the goal of wildlife nutrition research is to understand the nutritional ecology of wildlife populations in natural conditions. Several methods have been used to estimate food intake, diet digestibility, and energy expenditure of free-ranging animals and populations. Sometimes these methods have been used to avoid assumptions, inadequacies, or difficulties of captive animal methods. For example, feeding trial data have limited value for estimating energy requirements of free-ranging animals. Also, analyzing or experimenting with all combinations of individual foods in the varied diets of some species can be impractical. Therefore, techniques that can provide information on the natural diets selected by the animal are of value. The following is a summary of methods that have been used to estimate diet quality, food intake, and energy expenditures in free-ranging wildlife.

Combining Food Habits and Diet Quality Data

A simple and commonly used method for estimating the quality of natural diets is to mathematically combine food habits data (percentages of foods in diets) with digestibility or nutrient data from nutritional analyses and feeding trials (e.g., Schwartz et al. 1977, Hobbs et al. 1982, Leslie and Starkey 1985). This method assumes that hand-collected forages for analyses are representative of forages selected by the animal.

Table 2. Equations for predicting the percent apparent digestible dry matter (DDM), digestible energy (DE), or metabolizable energy in diets from chemical analyses of foods.

Species	Foods analyzed	Equation[a]	Source
Elk	Forages, diets[b]	$DDM = 1.11 NDS - 21.88$ $+ NDF \dfrac{(176.92 - 40.50 \, \text{Log} \, e^x)}{100}$	Mould and Robbins 1982
White-tailed and mule deer	Forages, diets	$DDM = [0.9231e^{-0.0451x} - 0.03z](NDF)$ $+ [(-16.03 + 1.02 NDS) - 2.8P]$ where $P = -0.01 + (11.82 \, \text{BSA precipitation})$	Robbins et al. 1987*a, b*
Ruffed grouse	Diets, foods or crop contents	$ME = 0.87(NDS - \text{total phenolics})$ $+ 0.18(\% \text{ acorn meat}) - 5.76$	Servello et al. 1987
Pine vole	Diets, foods	$DDM = 1.18 NDS - 19.42$ $DE = 1.12 NDS - 14.31$	Servello et al. 1983
	Stomach contents	$DDM = 1.14 AFNDS - 14.89$ $DE = 1.07 AFNDS - 8.50$	Servello et al. 1983
Meadow vole	Diets, foods	$DDM = 1.09 NDS - 11.12$ $DE = 1.09 NDS - 11.84$	MacPherson et al. 1985
	Stomach contents	$DDM = 1.08 AFNDS - 1.3$ $DE = 1.07 AFNDS - 1.6$	MacPherson et al. 1985

[a] Chemical composition abbreviations: NDS = neutral detergent solubles; NDF = neutral detergent fiber; AFNDS = acid-insoluble, ash-free neutral detergent solubles; x = lignin and cutin content (7%) of the neutral detergent fiber; z = biogenic silica content (7%) of grasses; P = reduction in protein digestion; BSA = bovine serum albumin.

[b] Forages, diets low in phenolic content.

Equations for predicting the digestible or metabolizable energy content of forages from chemical composition (Table 2) are available for white-tailed deer and elk (Mould and Robbins 1982, Robbins et al. 1987*a*), ruffed grouse (Servello et al. 1987), and voles (Servello et al. 1983, MacPherson et al. 1985). Because only a small amount of forage is needed for chemical analyses (compared to that needed for feeding trials), estimates of DE and ME values can be obtained for a large number of forages and for specific plant parts in all seasons or under specific environmental conditions to obtain more refined estimates of dietary energy for these species.

Analyses of Stomach and Crop Contents

Servello et al. (1984) and MacPherson et al. (1988) measured the digestible energy in the diets of wild voles using equations to predict digestible energy from a chemical analysis of stomach contents. Dietary ME for ruffed grouse can be predicted similarly from chemical analyses of crop contents (Servello and Kirkpatrick 1987). Similar efforts were made to measure nutrient content in deer rumen contents (Kirkpatrick et al. 1969). These methods eliminate the bias between hand-picked samples and those selected by the animal. However, there is an assumption that the plant material is not substantially altered in the animal before collection. This assumption does not appear to be a significant problem with voles (Servello et al. 1983) and ruffed grouse (Servello and Kirkpatrick 1987) but is likely a problem with ruminants.

Indicator Techniques

Variations of the indicator techniques described for feeding trials have been applied to wild populations to determine the digestibility of natural diets. Lignin, ash, and indigestible cell-wall concentrations measured in the stomach contents and feces from the colon of collected individuals have been used to calculate diet digestibility in small rodents (Johnson and Groepper 1970), pine voles (Noffsinger 1976), and rabbits (Wallage-Drees and Deinum 1986). Slightly different approaches have been reported for geese (Buchsbaum et al. 1986), red grouse (Moss 1977), and rock ptarmigan (Moss 1973). Buchsbaum et al. (1986) and Moss (1973, 1977) measured lignin and magnesium, respectively, in hand-picked samples to simulate the diet and in fecal samples collected on the site from wild or free-ranging captive animals to calculate diet digestibility. As discussed with feeding trials, a major problem with indicator techniques is the apparent digestibility of some indicators.

Indices of Diet Quality

Fecal nitrogen concentration has been proposed as an index of diet quality (Kie and Burton 1984, Leslie and Starkey 1985); however, there is considerable disagreement on its usefulness (Hobbs 1987, Leslie and Starkey 1987). Robbins et al. (1987*b*) reported that high dietary tannin levels increased fecal nitrogen, which can lead to inaccurate conclusions. DAPA, 2,6 diaminopinelic acid, which is found in rumen bacteria, is another proposed fecal index. It is hypothesized that diet quality changes that alter rumen bacterial numbers will result in correlated changes in DAPA concentrations in ruminant feces (Kie and Burton 1984).

Methods To Estimate Food Intake

Esophageal fistulation frequently is used to measure food intake for grazing ungulates (Holleman et al. 1979, Wickstrom et al. 1984). A more complex method used by Renecker and Hudson (1985) involved clipping plant samples to simulate the diet of moose observed in an enclosure and determining diet digestibility by the nylon-bag technique with fistulated moose. After total fecal collections from individual moose feeding in the enclosure were made for 24 hours, daily food intake was back-calculated.

The bite-count method has been used frequently with captive ungulates to estimate food intake (Collins et al. 1978, Bengtson 1983, Wickstrom et al. 1984). For this method, estimates are obtained for bite rate, simulated bite weight, and total foraging time to calculate intake. Alldredge et al. (1974) and Holleman et al. (1979) reported on the use of the natural fallout of radiocesium (cesium-137) to estimate intake for ungulates.

Estimating Total Daily Energy Expenditure

The time-energy budget (TEB) method and the doubly-labeled water method arc the two most commonly used techniques for estimating total daily energy expenditures of free-ranging animals. As its name implies, the TEB method has two parts: an accounting is made of time spent in major activities or behaviors (e.g., foraging, resting) by the animal, and then the activity data are converted to energetic equivalents from estimates of energy costs for each activity determined in laboratory or controlled studies (Weathers et al. 1984). This method is most commonly used with birds because of the relative ease of collecting activity data (e.g., Ashkenazie and Safriel 1979, Stalmaster and Gessaman 1984, Morton et al. 1989). The doubly-labeled water method involves injection (labeling) of oxygen (oxygen-18) and hydrogen (tritium or deuterium) isotopes into an animal prior to its release and calculating the rate of CO_2 production, which can be equated to metabolic rate, from the relative turnover rates of the isotopes measured upon recapture of the animal (Nagy 1980, Williams and Nagy 1984, Kam et al. 1987). This measure has been termed field metabolic rate (Nagy 1987). Its use has been limited in the past to smaller wildlife (Ricklefs and Williams 1984, Bryant et al. 1985, Williams and Nagy 1985, Tatner and Bryant 1986, Williams and Prints 1986, Gabrielsen et al. 1987) because of the high costs of working with isotopes (Nagy 1989). However, better analytical methods now make possible its use on larger species.

FEEDING STRATEGIES

One of the most fascinating areas of wildlife nutrition is the determination of wildlife feeding or foraging strategies. Each species has evolved its own feeding niche based on its size and corresponding metabolic rate, digestive tract anatomy and physiology (i.e., ruminant, cecal fermentor, simple digestive tract), the ratio of the volume of the digestive tract to the size of the animal (Hanley 1982, Demment and Van Soest 1985), and specialized feeding structures (i.e., modified mouth, appendages, or body form). These factors determine the specific foods that are most suitable for the animal. This goes beyond simple food-habits determination in that it addresses the reasons why certain foods are important to the species. It not only supplies information on the habitat requirements of the species, but it often supplies important information on interspecific relationships and has revealed some interesting occurrences of commensalism.

Although some research has addressed feeding strategies of carnivores, such as the work on coyotes reported by MacCracken and Hansen (1987), much of the study of feeding strategies has been directed toward herbivores. Work on African herbivores revealed interesting species interactions, including sequential grazing of areas by different species (Vesey-Fitzgerald 1960, Gwynne and Bell 1968, Bell 1971, Jarman 1974). More recent work in North America has revealed foraging strategies and interrelationships of many ungulate species (Schwartz et al. 1977, Hanley and Hanley 1982, Hobbs et al. 1983, Krueger 1986, Baker and Hobbs 1987, Jenkins and Wright 1988). Feeding strategies also may influence intraspecific habitat selection by ungulates (Main and Coblentz 1990). Chivers et al. (1984) discussed food acquisition and processing by primates. Considerably more work is needed on feeding strategies of most species to better understand their feeding relationships and requirements.

MANAGEMENT IMPLICATIONS AND FUTURE DIRECTIONS

Using the techniques described in this chapter, one can conduct nutritional analyses for many purposes. One purpose may be to assess the ability of an area to meet the nutritional requirements of one or more species of wildlife. They also allow for evaluation of the contribution of different vegetation types to the nutritional status of a selected species. For example, the quantity of a food available in a type must be evaluated relative to its nutritional quality within that type, because factors such as shading by overstory vegetation may influence tannin levels, and thus protein availability or digestibility of the forage species (Robbins et al. 1987a,b). If nutritional problems are found, methods can be applied to improve the nutritional quality of the area. These can include a variety of habitat management techniques including timber harvesting or mechanical treatments; burning to alter vegetation and release nutrients to the soil; fertilizing with chemicals, manure, or selected sludges; planting or seeding with species of higher nutritional quality; irrigating; and manipulating grazing pressure. Chapters 25–28 discuss many of the techniques available relative to these manipulations.

As we learn more about wildlife nutrition, the more we understand some of the complexities in assessing a species' food requirements and supply. Additional research is needed to add to our understanding in many areas, including the role that plant chemical defenses play in influencing foraging strategies, the optimal and suboptimal supply of nutrients to many species, the reproductive response of animals to improvements in the supply of selected nutrients, and the effects of habitat manipulations on the nutritional quality of most species.

Wildlife nutrition has expanded considerably in its knowledge base and methodologies over the last 10 years. Continued expansion of the field is important to our thorough understanding of one of the most basic components of wildlife habitat, the supply of food to an animal.

LITERATURE CITED

ADAMS, N. J. 1984. Utilization efficiency of a squid diet by adult king penguins (*Aptendytes patagonicus*). Auk 101:884–886.

ALLDREDGE, A. W., J. F. LIPSCOMB, AND F. W. WHICKER. 1974. Forage intake rates of mule deer estimated with fallout cesium-137. J. Wildl. Manage. 38:508–516.

ALLEN, M. E. 1989. Nutritional aspects of insectivory. Ph.D. Thesis, Michigan State Univ., East Lansing. 205pp.

————, AND O. T. OFTEDAL. 1989. Dietary manipulation of the calcium content of feed crickets. J. Zoo Wildl. Med. 20:26–33.

AMMANN, A. P., R. L. COWAN, C. L. MOTHERSHEAD, AND B. R. BAUMGARDT. 1973. Dry matter and energy intake in relation to digestibility in white-tailed deer. J. Wildl. Manage. 37:195–201.

ASHKENAZIE, S., AND U. N. SAFRIEL. 1979. Time-energy budget of the semipalmated sandpiper *Calidris pusilla* at Barrow, Alaska. Ecology 60:783–799.

ASQUITH, T. N., AND L. G. BUTLER. 1985. Use of dye-labeled protein as spectrophotometric assay for protein precipitants such as tannin. J. Chem. Ecol. 11:1535–1544.

BAKER, D. L., AND D. R. HANSEN. 1985. Comparative digestion of grass in mule deer and elk. J. Wildl. Manage. 49:77–79.

————, AND N. T. HOBBS. 1987. Strategies of digestion: digestive efficiency and retention time of forage diets in montane ungulates. Can. J. Zool. 65:1978–1984.

————, D. E. JOHNSON, L. H. CARPENTER, O. C. WALLMO, AND R. B. GILL. 1979. Energy requirements of mule deer fawns in winter. J. Wildl. Manage. 43:162–169.

BAKKEN, G. S., D. J. ERSKINE, AND W. R. SANTEE. 1983. Construction and operation of heated taxidermic mounts used to measure standard operative temperature. Ecology 64:1658–1662.

BALAKRISHNAN, M., AND K. M. ALEXANDER. 1979. A study on aspects of feeding and food utilization of the Indian musk shrew, *Suncus murinus viridescens* (Blyth). Physiol. Behav. 22:423–428.

BARNES, T. G. 1988. Digestion dynamics in white-tailed deer. Ph.D. Thesis, Texas A&M Univ., College Station. 153pp.

BATE-SMITH, E. C. 1973. Haemanalysis of tannins: the concept of relative astringency. Phytochemistry 12:907–912.

BATZLI, G. O., AND F. R. COLE. 1979. Nutritional ecology of microtine rodents: digestibility of forage. J. Mammal. 60:740–750.

BECKERTON, P. R., AND A. L. A. MIDDLETON. 1982. Effects of dietary protein levels on ruffed grouse reproduction. J. Wildl. Manage. 46:569–579.

————, AND ————. 1983. Effects of dietary protein levels on body weight, food consumption, and nitrogen balance in ruffed grouse. Condor 85:53–60.

BELL, R. H. V. 1971. A grazing ecosystem in the Serengeti. Sci. Am. 225:86–93.

BENGTSON, J. L. 1983. Estimating food consumption of free-ranging manatees in Florida. J. Wildl. Manage. 47:1186–1192.

BILLINGSLEY, B. B., JR., AND D. H. ARNER. 1970. The nutritive value and digestibility of some winter foods of the eastern wild turkey. J. Wildl. Manage. 34:176–182.

BJÖRNHAG, G., AND I. SPERBER. 1977. Transport of various food components through the digestive tract of turkeys, geese, and guinea fowl. Swed. J. Agric. Res. 7:57–66.

BRYANT, D. M., C. J. HALLS, AND R. PRYS-JONES. 1985. Energy expenditure by free-living dippers (*Cinclus cinclus*) in winter. Condor 87:177–186.

BRYANT, J. P. 1981. Phytochemical deterrence of snowshoe hare browsing by adventitious shoots of four Alaskan trees. Science 213:889–890.

BUCHSBAUM, R., I. VALIELA, AND T. SWAIN. 1984. The role of phenolic compounds and other plant constituents in feeding by Canada geese in a coastal marsh. Oecologia 63:343–349.

————, J. WILSON, AND I. VALIELA. 1986. Digestibility of plant constituents by Canada geese and Atlantic brant. Ecology 67:386–393.

BURNS, R. E. 1963. Methods of tannin analysis for forage crop evaluation. Univ. Georgia Tech. Bull. N.S. 32. 14pp.

BURTON, B. A., R. J. HUDSON, AND D. D. BRAGG. 1979. Efficiency of utilization of bulrush rhizomes by lesser snow geese. J. Wildl. Manage. 43:728–735.

CAIN, B. W. 1976. Energetics of growth for black-bellied tree ducks. Condor 78:124–128.

CAMPA, H., III, D. K. WOODYARD, AND J. B. HAUFLER. 1984. Reliability of captive deer and cow in vitro digestion values in predicting wild deer digestion levels. J. Range Manage. 37:468–470.

CARL, G. R., AND R. D. BROWN. 1985. Protein requirement of adult collared peccaries. J. Wildl. Manage. 49:351–355.

CASE, R. M., AND R. J. ROBEL. 1974. Bioenergetics of the bobwhite. J. Wildl. Manage. 38:638–652.

CHAMI, D. B., P. VOHRA, AND F. H. KRATZER. 1980. Evaluation of a method for determination of true metabolizable energy of feed ingredients. Poult. Sci. 59:569–571.

CHAPPELL, M. A. 1980. Thermal energetics and thermoregulatory costs of small arctic mammals. J. Mammal. 61:278–291.

CHIVERS, D. I., P. ANDREWS, H. PREUSCHOFT, A. BILSBOROUGH, AND B. A. WOOD. 1984. Food acquisition and processing in primates: concluding discussion. Pages 545–556 *in* D. I. Chivers, B. A. Wood, and A. Bilsborough, eds. Food acquistion and processing in primates. Plenum Press, New York, N.Y.

CHOO, G. M., P. G. WATERMAN, D. B. MCKEY, AND J. S. GARTLAN. 1981. A simple enzyme assay for dry matter digestibility and its value in studying food selection by generalist herbivores. Oecologia 49:170–178.

CHURCH, D. C., AND W. G. POND. 1988. Basic animal nutrition and feeding. Third ed. John Wiley & Sons, New York, N.Y. 472pp.

CLARK, J., AND J. BEARD. 1977. Prediction of the digestibility of ruminant feeds from their solubility in enzyme solutions. Anim. Feed Sci. Technol. 2:153–159.

CLARKE, T., P. C. FLINN, AND A. A. MCGOWAN. 1982. Low-cost pepsin cellulose assays for prediction of digestibility of herbage. Grass Forage Sci. 37:147–150.

CLARY, W. P., B. L. WELCH, AND G. D. BOOTH. 1988. In vitro digestion experiments: importance of variation between inocula donors. J. Wildl. Manage. 52:358–361.

CLEMANS, R. J. 1974. The bioenergetics of the blue jay in central Illinois. Condor 76:358–360.

CLUFF, L. K., B. L. WELCH, J. C. PEDERSON, AND J. D. BROTHERSON. 1982. Concentration of monoterpenoids in the rumen ingesta of wild mule deer. J. Range Manage. 35:192–194.

COLLINS, W. B., P. J. URNESS, AND D. D. AUSTIN. 1978. Elk diets and activities on different lodgepole pine habitat segments. J. Wildl. Manage. 42:799–810.

CORTS, K. E., AND F. G. LINDZEY. 1984. Basal metabolism and energetic cost of walking in cougars. J. Wildl. Manage. 48:1456–1458.

COWAN, R. L., E. W. HARTSOOK, J. B. WHELAN, T. A. LONG, AND R. S. WETZEL. 1969. A cage for metabolism and radioisotope studies with deer. J. Wildl. Manage. 33:204–208.

CRAMPTON, E. W., AND L. E. HARRIS. 1969. Applied animal nutrition. Second ed. W. H. Freeman and Co., San Francisco, Calif. 753pp.

DAHLQUIST, R. L., AND J. W. KNOLL. 1978. Inductively coupled plasma-atomic emission spectrometry: analysis of biological materials and soils for major trace and ultra-trace elements. Appl. Spectrosc. 32:1–30.

DAVISON, R. P., W. W. MAUTZ, H. H. HAYES, AND J. B. HOLTER. 1978. The efficiency of food utilization and energy requirements of captive female fishers. J. Wildl. Manage. 42:811–821.

DAWSON, W. R., A. F. BENNETT, AND J. W. HUDSON. 1976. Metabolism and thermoregulation in hatchling ring-billed gulls. Condor 78:49–60.

DEBOLT, D. C. 1980. Multielement emission spectroscopic analysis of plant tissue using DC argon plasma source. J. Assoc. Off. Agric. Chem. 63:802–805.

DEJONG, A. A. 1976. The influence of simulated solar radiation on the metabolic rate of white-crowned sparrows. Condor 78:174–179.

DEMMENT, M. W., AND P. J. VAN SOEST. 1985. A nutritional explanation for body-size patterns of ruminant and nonruminant herbivores. Am. Nat. 125:641–672.

DOWMAN, M. G., AND F. C. COLLINS. 1982. The use of enzymes to predict the digestibility of animal feeds. J. Sci. Food Agric. 33:689–696.

EVANS, K. E., AND D. R. DIETZ. 1974. Nutritional energetics of sharp-tailed grouse during winter. J. Wildl. Manage. 38:622–629.

FANCY, S. G., AND R. G. WHITE. 1985. Energy expenditures by caribou while cratering in snow. J. Wildl. Manage. 49:987–993.

FARRELL, D. J., AND A. J. WOOD. 1968. The nutrition of the female mink (*Mustela vison*). II. The energy requirement for maintenance. Can. J. Zool. 46:47–52.

FEENY, P. P., AND H. BOSTOCK. 1968. Seasonal changes in the tannin content of oak leaves. Phytochemistry 7:871–880.

FOWLER, M. E. 1983. Plant poisoning in free-living wild animals: a review. J. Wildl. Dis. 19:34–43.

FREDDY, D. J. 1984. Heart rates for activities of mule deer at pasture. J. Wildl. Manage. 48:962–969.

GABRIELSEN, G. W., F. MEHLUM, AND K. A. NAGY. 1987. Daily energy expenditure and energy utilization of free-ranging black-legged kittiwakes. Condor 89:126–132.

GASAWAY, W. C., D. F. HOLLEMAN, AND R. G. WHITE. 1975. Flow of digesta in the intestine and cecum of the rock ptarmigan. Condor 77:467–474.

————, R. G. WHITE, AND D. F. HOLLEMAN. 1976. Digestion of dry matter and absorption of water in the intestine and cecum of rock ptarmigan. Condor 78:77–84.

GATES, C. C., AND R. J. HUDSON. 1979. Effects of posture and activity on metabolic responses of wapiti to cold. J. Wildl. Manage. 43: 564–567.

GESSAMAN, J. A. 1980. An evaluation of heart rate as an indirect measure of daily energy metabolism of the American kestrel. Comp. Biochem. Physiol. 65(A):273–289.

———. 1987. Energetics. Pages 289–320 in B. A. Giron Pendleton, B. A. Millsap, K. W. Clire, and D. M. Bird, eds. Raptor management techniques manual. Natl. Wildl. Fed., Washington, D. C.

GOERING, H. K., AND P. J. VAN SOEST. 1970. Forage fiber analyses (apparatus, reagents, procedures, and some applications). U.S. Dep. Agric. Agric. Handb. 379. 20pp.

GOLIGHTLY, R. T., JR., AND R. D. OHMART. 1983. Metabolism and body temperature of two desert canids: coyotes and kit foxes. J. Mammal. 64:624–635.

GOTO, I., AND D. J. MINSON. 1977. Prediction of the dry matter digestibility of tropical grasses using a pepsin-cellulase assay. Anim. Feed Sci. Tech. 2:245–253.

GRABER, R. R. 1962. Food and oxygen consumption in three species of owls (Strigidae). Condor 64:473–487.

GRAY, B. T., AND H. H. PRINCE. 1988. Basal metabolism and energetic cost of thermoregulation in wild turkeys. J. Wildl. Manage. 52:133–137.

GWYNNE, M. D., AND R. H. V. BELL. 1968. Selection of vegetation components by grazing ungulates in the Serengeti National Park. Nature 220:390–393.

HAGERMAN, A. E. 1987. Radical diffusion method for determining tannin in plant extracts. J. Chem. Ecol. 13:437–449.

———, AND L. G. BUTLER. 1978. Protein precipitation method for the quantitative determination of tannins. J. Agric. Food Chem. 26: 809–812.

HALSE, S. A. 1984. Diet, body condition, and gut size of Egyptian geese. J. Wildl. Manage. 48:569–573.

HAN, I. K., H. W. HOCHSTETLER, AND M. L. SCOTT. 1976. Metabolizable energy values of some poultry feeds determined by various methods and their estimation using metabolizability of the dry matter. Poult. Sci. 55:1335–1342.

HANLEY, T. A. 1982. The nutritional basis for food selection by ungulates. J. Range Manage. 35:146–151.

———, AND K. A. HANLEY. 1982. Food resource partitioning by sympatric ungulates on Great Basin rangeland. J. Range Manage. 35: 152–158.

HANSEN, R. M., AND B. R. CAVENDER. 1973. Food intake and digestion by black tailed prairie dogs under laboratory conditions. Acta Theriol. 18:191–200.

HARBORNE, J. B. 1979. Flavonoid pigments. Pages 619–655 in G. A. Rosenthal and D. H. Janzen, eds. Herbivores: their interaction with secondary plant metabolites. Academic Press, New York, N.Y.

HARLOW, H. J. 1981. Effects of fasting on rate of food passage assimilation efficiency in badgers. J. Mammal. 62:173–177.

HAVERA, S. P., AND K. E. SMITH. 1979. A nutritional comparison of selected fox squirrel foods. J. Wildl. Manage. 43:691–704.

HAWKINS, A. E., AND P. A. JEWELL. 1962. Food consumption and energy requirements of captive British shrews and the mole. Proc. Zool. Soc. London 138:137–155.

HELLER, H. C. 1972. Measurements of convective and radiative heat transfer in small mammals. J. Mammal. 53:289–295.

HELM, R. C. 1984. Rate of digestion in three species of pinnipeds. Can. J. Zool. 62:1751–1756.

HILL, D. C., E. V. EVANS, AND H. G. LUMSDEN. 1968. Metabolizable energy of aspen flower buds for captive ruffed grouse. J. Wildl. Manage. 32:854–858.

HJELJORD, O., F. SUNDSTOL, AND H. HAAGENRUD. 1982. The nutritional value of browse to moose. J. Wildl. Manage. 46:333–343.

HOBBS, N. T. 1987. Fecal indices to dietary quality: a critique. J. Wildl. Manage. 51:317–320.

———, D. L. BAKER, J. E. ELLIS, D. M. SWIFT, AND R. A. GREEN. 1982. Energy- and nitrogen-based estimates of elk winter-range carrying capacity. J. Wildl. Manage. 46:12–21.

———, ———, AND R. B. GILL. 1983. Comparative nutritional ecology of montane ungulates during winter. J. Wildl. Manage. 47:1–16.

HOFFMAN, R. D., AND T. A. BOOKHOUT. 1985. Metabolizable energy of seeds consumed by ducks in Lake Erie marshes. Trans. North Am. Wildl. Nat. Resour. Conf. 50:557–565.

HOLLEMAN, D. F., J. R. LUICK, AND R. G. WHITE. 1979. Lichen intake estimates for reindeer and caribou during winter. J. Wildl. Manage. 43:192–201.

HOLTER, J. B., H. H. HAYES, AND S. H. SMITH. 1979. Protein requirement of yearling white-tailed deer. J. Wildl. Manage. 43:872–879.

———, G. TYLER, AND T. WALSKI. 1974. Nutrition of the snowshoe hare (Lepus americanus). Can. J. Zool. 52:1553–1558.

———, W. E. URBAN, JR., H. H. HAYES, AND H. SILVER. 1976. Predicting metabolic rate from telemetered heart rate in white-tailed deer. J. Wildl. Manage. 40:626–629.

HOLTHUIJZEN, A. M. A., AND C. S. ADKISSON. 1984. Passage rate, energetics, and utilization efficiency of the cedar waxwing. Wilson Bull. 96:680–684.

HORWITZ, W., EDITOR. 1975. Official methods of analysis of the Association of Official Analytical Chemists. 12th ed. Assoc. Off. Anal. Chem., Washington, D.C. 1094pp.

HUDSON, R. J., AND R. J. CHRISTOPHERSON. 1985. Maintenance metabolism. Pages 121–142 in R. J. Hudson and R. G. White, eds. Bioenergetics of wild herbivores. CRC Press, Inc., Boca Raton, Fla.

INMAN, D. L. 1973. Cellulose digestion in ruffed grouse, chukar partridge and bobwhite quail. J. Wildl. Manage. 37:114–121.

JARMAN, P. J. 1974. The social organisation of antelope in relation to their ecology. Behaviour 48:215–267.

JENKINS, K. J., AND R. G. WRIGHT. 1988. Resource partitioning and competition among cervids in the northern Rocky Mountains. J. Appl. Ecol. 25:11–24.

JENKS, J. A., AND D. M. LESLIE, JR. 1988. Effect of lichen and in vitro methodology on digestibility of winter deer diets in Maine. Can. Field-Nat. 102:216–220.

JOHNSON, D. R., AND K. L. GROEPPER. 1970. Bioenergetics of north plains rodents. Am. Midl. Nat. 84:537–548.

JOHNSON, M. K., AND D. R. ALDRED. 1982. Mammalian prey digestibility by bobcats. J. Wildl. Manage. 46:530.

JOHNSON, R. R. 1966. Techniques and procedures for in vitro and in vivo rumen studies. J. Anim. Sci. 25:855–875.

JONES, D. I. H., AND A. D. WILSON. 1987. Nutritive quality of forage. Pages 65–89 in J. B. Hacker and J. H. Ternouth, eds. The nutrition of herbivores. Academic Press, New York, N.Y.

JORDE, D. G., AND R. B. OWEN, JR. 1988. Efficiency of nutrient use by American black ducks wintering in Maine. J. Wildl. Manage. 52:209–214.

JULKUNEN-TIITTO, R. 1985. Phenolic constituents in the leaves of northern willows: methods for the analysis of certain phenolics. J. Agric. Food Chem. 33:213–217.

KAM, M., A. A. DEGEN, AND K. A. NAGY. 1987. Seasonal energy, water, and food consumption of negev chukars and sand partridges. Ecology 68:1029–1037.

KAUTZ, M. A., W. W. MAUTZ, AND L. H. CARPENTER. 1981. Heart rate as a predictor of energy expenditure of mule deer. J. Wildl. Manage. 45:715–720.

———, G. M. VAN DYNE, L. H. CARPENTER, AND W. W. MAUTZ. 1982. Energy cost for activities of mule deer fawns. J. Wildl. Manage. 46:704–710.

KEIVER, K. M., K. RONALD, AND F. W. H. BEAMISH. 1984. Metabolizable energy requirements for maintenance and faecal and urinary losses of juvenile harp seals (Phoca groenlandica). Can. J. Zool. 62:769–776.

KEMPTON, T. J., R. M. MURRAY, AND R. A. LENG. 1976. Methane production and digestibility measurements in the grey kangaroo and sheep. Aust. J. Biol. Sci. 29:209–214.

KIE, J. G., AND T. S. BURTON. 1984. Dietary quality, fecal nitrogen, and 2, 6 diaminopimelic acid in black-tailed deer in northern California. U.S. For. Serv. Res. Note PSW-364. 3pp.

KIRKPATRICK, R. L., J. P. FONTENOT, AND R. F. HARLOW. 1969. Seasonal changes in rumen chemical components as related to forages consumed by white-tailed deer of the Southeast. Trans. North Am. Wildl. Nat. Resour. Conf. 34:229–238.

KRUEGER, K. 1986. Feeding relationships among bison, pronghorn, and prairie dogs: an experimental analysis. Ecology 67:760–770.

KUSHLAN, J. A. 1977. Growth energetics of the white ibis. Condor 79:31–36.

LESLIE, D. M., JR., AND E. E. STARKEY. 1985. Fecal indices to dietary quality of cervids in old-growth forests. J. Wildl. Manage. 49:142–146.

———, AND ———. 1987. Fecal indices to dietary quality: a reply. J. Wildl. Manage. 51:321–325.

LEVIN, D. A. 1976. The chemical defenses of plants to pathogens and herbivores. Annu. Rev. Ecol. Syst. 7:121–159.

LINDROTH, R. L., AND G. O. BATZLI. 1984. Plant phenolics as chemical defenses: effects of natural phenolics on survival and growth of prairie voles (*Microtus ochrogaster*). J. Chem. Ecol. 10:229–244.

LITVAITIS, J. A., AND W. W. MAUTZ. 1976. Energy utilization of three diets fed to captive red fox. J. Wildl. Manage. 40:365–368.

———, AND ———. 1980. Food and energy use by captive coyotes. J. Wildl. Manage. 44:56–61.

LOCHMILLER, R. L., J. B. WHELAN, AND R. L. KIRKPATRICK. 1983. Seasonal energy requirements of adult pine voles, *Microtus pinetorum*. J. Mammal. 64:345–350.

LUICK, B. R., AND R. G. WHITE. 1986. Oxygen consumption for locomotion by caribou calves. J. Wildl. Manage. 50:148–152.

MABRY, T. J., AND J. E. GILL. 1979. Sesquiterpene lactones and other terpenoids. Pages 502–537 *in* G. A. Rosenthal and D. H. Janzen, eds. Herbivores: their interaction with secondary plant metabolites. Academic Press, New York, N.Y.

MACCRACKEN, J. G., AND R. M. HANSEN. 1987. Coyote feeding strategies in southeastern Idaho: optimal foraging by an opportunistic predator? J. Wildl. Manage. 51:278–285.

MACPHERSON, S. L., F. A. SERVELLO, AND R. L. KIRKPATRICK. 1985. A method of estimating diet digestibility in wild meadow voles. Can. J. Zool. 63:1020–1022.

———, ———, AND ———. 1988. Seasonal variation in diet digestibility of pine voles. Can. J. Zool. 66:1484–1487.

MAIN, M. B., AND B. E. COBLENTZ. 1990. Sexual segregation among ungulates: a critique. Wildl. Soc. Bull. 18:204–210.

MARTIN, J. S., AND M. M. MARTIN. 1982. Tannin assays in ecological studies: lack of correlation between phenolics, proanthocyanidins and protein-precipitating constituents in mature foliage of six oak species. Oecologia 54:205–211.

———, AND ———. 1983. Tannin assays in ecological studies: precipitation of ribulose-1,5 biphosphate carboxylase/oxygenase by tannic acid, quebracho, and oak foliage extracts. J. Chem. Ecol. 9:285–294.

MAUTZ, W. W. 1971a. Comparison of the $^{51}CrCl_3$ ratio and total collection techniques in digestibility studies with a wild ruminant, the white-tailed deer. J. Anim. Sci. 32:999–1002.

———. 1971b. Confinement effects on dry-matter digestibility coefficients displayed by deer. J. Wildl. Manage. 35:366–368.

———. 1978. Nutrition and carrying capacity. Pages 321–348 *in* J. L. Schmidt and D. L. Gilbert, eds. Big game of North America: ecology and management. Stackpole Books, Harrisburg, Pa.

———, AND J. FAIR. 1980. Energy expenditure and heart rate for activities of white-tailed deer. J. Wildl. Manage. 44:333–342.

———, AND P. J. PEKINS. 1989. Metabolic rate of bobcats as influenced by seasonal temperatures. J. Wildl. Manage. 53:202–205.

MAXWELL, C. S., AND J. R. KING. 1976. The oxygen consumption of the mountain white-crowned sparrow (*Zonotrichia leucophrys oriantha*) in relation to air temperature. Condor 78:569–570.

MAYNARD, L. A., J. K. LOOSLI, H. F. HIRTZ, AND R. G. WARNER. 1979. Animal nutrition. Seventh ed. McGraw-Hill Book Co., New York, N.Y. 602pp.

MAZEN, W. S., AND R. L. RUDD. 1980. Comparative energetics in two sympatric species of *Peromyscus*. J. Mammal. 61:573–574.

MCEWAN, E. H. 1970. Energy metabolism of barren ground caribou (*Rangifer tarandus*). Can. J. Zool. 48:391–392.

MCINTOSH, J. I., S. J. SLINGER, I. R. SIBBALD, AND G. C. ASHTON. 1962. Factors affecting the metabolizable energy content of poultry feeds. 7. The effects of grinding, pelleting, and grit feeding on the availability of the energy of wheat, corn, oats, and barley. 8. A study of the effects of dietary balance. Poult. Sci. 41:445–456.

MCLEOD, M. N. 1974. Plant tannins—their role in forage quality. Nutr. Abstr. Rev. 11:803–815.

———, AND D. J. MINSON. 1978. The accuracy of the pepsin-cellulase technique for estimating the dry matter digestibility in vivo of grasses and legumes. Anim. Feed Sci. Tech. 3:277–287.

———, AND ———. 1982. Accuracy of predicting digestibility by the cellulase technique: the effect of pretreatment of forage samples with neutral detergent or acid pepsin. Anim. Feed Sci. Tech. 7:83–92.

MCNAB, B. K. 1980. Energetics and the limits to a temperate distribution in armadillos. J. Mammal. 61:606–627.

MILLER, M. R. 1975. Gut morphology of mallards in relation to diet quality. J. Wildl. Manage. 39:168–173.

———. 1984. Comparative ability of northern pintails, gadwalls, and northern shovelers to metabolize foods. J. Wildl. Manage. 48:362–370.

———, AND K. J. REINECKE. 1984. Proper expression of metabolizable energy in avian energetics. Condor 86:396–400.

MOLE, S., AND P. G. WATERMAN. 1987. A critical analysis of techniques for measuring tannins in ecological studies. I. Techniques for chemically defining tannins. Oecologia 72:137–147.

MOORS, P. J. 1977. Studies of the metabolism, food consumption and assimilation efficiency of a small carnivore, the weasel (*Mustela nivalis* L.). Oecologia 27:185–202.

MORTON, J. M., A. C. FOWLER, AND R. L. KIRKPATRICK. 1989. Time and energy budgets of American black ducks in winter. J. Wildl. Manage. 53:401–410.

MOSS, R. 1972. Effects of captivity on gut lengths in red grouse. J. Wildl. Manage. 36:99–104.

———. 1973. The digestion and intake of winter foods by wild ptarmigan in Alaska. Condor 75:293–300.

———. 1977. The digestion of heather by red grouse during the spring. Condor 79:471–477.

MOTHERSHEAD, C. L., R. L. COWAN, AND A. P. AMMANN. 1972. Variations in determinations of digestive capacity of the white-tailed deer. J. Wildl. Manage. 36:1052–1060.

MOULD, E. D., AND C. T. ROBBINS. 1981a. Evaluation of detergent analysis in estimating nutritional value of browse. J. Wildl. Manage. 45:937–947.

———, AND ———. 1981b. Nitrogen metabolism in elk. J. Wildl. Manage. 45:323–334.

———, AND ———. 1982. Digestive capabilities in elk compared to white-tailed deer. J. Wildl. Manage. 46:22–29.

NAGY, J. G., H. W. STEINHOFF, AND G. M. WARD. 1964. Effects of essential oils of sagebrush on deer rumen microbial function. J. Wildl. Manage. 28:785–790.

———, AND R. P. TENGERDY. 1968. Antibacterial action of essential oils of *Artemisia* as an ecological factor. II. Antibacterial action of the oils of *Artemisia tridentata* (big sagebrush) on bacteria from the rumen of mule deer. Appl. Microbiol. 16:441–444.

NAGY, K. A. 1980. CO_2 production in animals: analysis of potential errors in the doubly labeled water method. Am. J. Physiol. 238: R466–R473.

———. 1987. Field metabolic rate and food requirement scaling in mammals and birds. Ecol. Monogr. 57:111–128.

———. 1989. Field bioenergetics: accuracy of models and methods. Physiol. Zool. 62:237–252.

———, AND K. MILTON. 1979. Energy metabolism and food consumption by wild howler monkeys (*Alouatta palliata*). Ecology 60: 475–480.

———, R. S. SEYMOUR, A. K. LEE, AND R. BRAITHWAITE. 1978. Energy and water budgets in free-living *Antechinus stuartii* (Marsupialia: Dasyuridae). J. Mammal. 59:60–68.

NASTIS, A. S., AND J. C. MALECHEK. 1988. Estimating digestibility of oak browse diets for goats by in vitro techniques. J. Range Manage. 41:255–258.

NOFFSINGER, R. E. 1976. Seasonal variation in the natality, mortality, and nutrition of the pine vole in two orchard types. M.S. Thesis, Virginia Polytechnic Inst. State Univ., Blacksburg. 128pp.

NORMAN, G. W. 1980. Nutritional ecology of ruffed grouse in southwest Virginia. M.S. Thesis, Virginia Polytechnic Inst. State Univ., Blacksburg. 134pp.

OH, H. K., T. SAKAI, M. B. JONES, AND W. M. LONGHURST. 1967. Effect of various essential oils isolated from Douglas fir needles upon sheep and deer rumen microbial activity. Appl. Microbiol. 15: 777–784.

OKARMA, H., AND P. KOTEJA. 1987. Basal metabolic rate in the gray wolf in Poland. J. Wildl. Manage. 51:800–801.

PACE, V., M. T. BARGE, D. SETTINERI, AND F. MALOSSINI. 1984. Comparison of forage digestibility in vitro with enzymic solubility. Anim. Feed Sci. Tech. 11:125–136.

PAKPAHAN, A. M., J. B. HAUFLER, AND H. H. PRINCE. 1989. Metabolic rates of red-tailed hawks and great horned owls. Condor 91:1000–1002.

PALMER, W. L., AND R. L. COWAN. 1980. Estimating digestibility of deer foods by an in vitro technique. J. Wildl. Manage. 44:469–472.

PALO, R. T., K. SUNNERHEIM, AND O. THEANDER. 1985. Seasonal variation of phenols, crude protein and cell wall content of birch (*Betula pendula* Roth.) in relation to ruminant in vitro digestibility. Oecologia 65:314–318.

PARKER, K. L., C. T. ROBBINS, AND T. A. HANLEY. 1984. Energy expenditures for locomotion by mule deer and elk. J. Wildl. Manage. 48:474–488.

PEARSON, H. A. 1970. Digestibility trials: *in vitro* techniques. Pages 85–92 *in* Range and wildlife habitat evaluation—a research symposium. U.S. For. Serv. Misc. Publ. 1147.

PEKINS, P. J., AND W. W. MAUTZ. 1988. Digestibility and nutritional value of autumn diets of deer. J. Wildl. Manage. 52:328–332.

PENDERGAST, B. A., AND D. A. BOAG. 1971. Nutritional aspects of the diet of spruce grouse in central Alberta. Condor 73:437–443.

PERI, C., AND C. POMPEI. 1971. Estimation of different phenolic groups in vegetable extracts. Phytochemistry 10:2187–2189.

PERNETTA, J. C. 1977. Anatomical and behavioural specialisations of shrews in relation to their diet. Can. J. Zool. 55:1442–1453.

PERSON, S. J., R. E. PEGAU, R. G. WHITE, AND J. R. LUICK. 1980. In vitro and nylon-bag digestibilities of reindeer and caribou forages. J. Wildl. Manage. 44:613–622.

PERSONIUS, T. L., C. L. WAMBOLT, J. R. STEPHENS, AND R. G. KELSEY. 1987. Crude terpenoid influence on mule deer preference for sagebrush. J. Range Manage. 40:84–88.

PRICE, M. L., S. V. SCOYOC, AND L. G. BUTLER. 1978. A critical evaluation of the vanillin reaction as an assay for tannin in sorghum grain. J. Agric. Food Chem. 26:1214–1218.

PRIEBE, J. C., AND R. D. BROWN. 1987. Protein requirements of subadult nilgai antelope. Comp. Biochem. Physiol. 88A:495–501.

RADWAN, M. A. 1972. Differences between Douglas-fir genotypes in relation to browsing by black-tailed deer. Can. J. For. Res. 2:250–255.

REGELIN, W. L., C. C. SCHWARTZ, AND A. W. FRANZMANN. 1985. Seasonal energy metabolism of adult moose. J. Wildl. Manage. 49:388–393.

———, O. C. WALLMO, J. NAGY, AND D. R. DIETZ. 1974. Effect of logging on forage values for deer in Colorado. J. For. 72:282–285.

RENECKER, L. A., AND R. J. HUDSON. 1985. Estimation of dry matter intake of free-ranging moose. J. Wildl. Manage. 49:785–792.

RHOADES, D. F., AND R. G. CATES. 1976. Toward a general theory of plant antiherbivore chemistry. Pages 168–213 *in* J. W. Wallace and R. L. Mansell, eds. Biochemical interaction between plants and animals. Rec. Adv. Phytochem. 10.

RIBÉREAU-GAYON, P. 1972. Plant phenolics. Oliver and Boyd, Ltd., Edinburgh, Scotland. 254pp.

RICKLEFS, R. E., AND J. B. WILLIAMS. 1984. Daily energy expenditure and water-turnover rate of adult European starlings (*Sturnus vulgaris*) during the nesting cycle. Auk 101:707–716.

RISENHOOVER, K. L., L. A. RENECKER, AND L. E. MORGANTINI. 1985. Effects of secondary metabolites from balsam poplar and paper birch on cellulose digestion. J. Range Manage. 38:370–372.

ROBBINS, C. T. 1983. Wildlife feeding and nutrition. Academic Press, New York, N.Y. 343pp.

———, Y. COHEN, AND B. B. DAVIT. 1979. Energy expenditure by elk calves. J. Wildl. Manage. 43:445–453.

———, T. A. HANLEY, A. E. HAGERMAN, O. HJELJORD, D. L. BAKER, C. C. SCHWARTZ, AND W. W. MAUTZ. 1987a. Role of tannins in defending plants against ruminants: reduction in protein availability. Ecology 68:98–107.

———, S. MOLE, A. E. HAGERMAN, AND T. A. HANLEY. 1987b. Role of tannins in defending plants against ruminants: reduction in dry matter digestion? Ecology 68:1606–1615.

———, P. J. VAN SOEST, W. W. MAUTZ, AND A. N. MOEN. 1975. Feed analyses and digestion with reference to white-tailed deer. J. Wildl. Manage. 39:67–79.

ROBEL, R. J., AND A. R. BISSET. 1979. Effects of supplemental grit on metabolic efficiency of bobwhites. Wildl. Soc. Bull. 7:178–181.

———, ———, T. M. CLEMENT, JR., AND A. D. DAYTON. 1979. Metabolizable energy of important foods of bobwhites in Kansas. J. Wildl. Manage. 43:982–987.

ROBINETTE, W. L., C. H. BAER, R. E. PILLMORE, AND C. E. KNITTLE. 1973. Effects of nutritional change on captive mule deer. J. Wildl. Manage. 37:312–326.

ROBINSON, T. 1979. The evolutionary ecology of alkaloids. Pages 413–448 *in* G. A. Rosenthal and D. H. Janzen, eds. Herbivores: their interaction with secondary plant metabolites. Academic Press, New York, N.Y.

RONALD, K., K. M. KEIVER, F. W. H. BEAMISH, AND R. FRANK. 1984. Energetic requirements for maintenance and faecal and urinary losses of the grey seal (*Halichoerus grypus*). Can. J. Zool. 62:1101–1105.

ROSENTHAL, G. A., AND D. H. JANZEN. 1979. Herbivores: their inter-action with secondary plant metabolites. Academic Press, New York, N.Y. 718pp.

SCHITOSKEY, F., JR., AND S. R. WOODMANSEE. 1978. Energy requirements and diet of the California ground squirrel. J. Wildl. Manage. 42:373–382.

SCHWARTZ, C. C., AND N. T. HOBBS. 1985. Forage and range evaluation. Pages 25–51 *in* R. J. Hudson and R. G. White, eds. Bioenergetics of wild herbivores. CRC Press, Inc., Boca Raton, Fla.

———, M. E. HUBBERT, AND A. W. FRANZMANN. 1988a. Energy requirements of adult moose for winter maintenance. J. Wildl. Manage. 52:26–33.

———, J. G. NAGY, AND W. L. REGELIN. 1980. Juniper oil yield, terpenoid concentration, and antimicrobial effects on deer. J. Wildl. Manage. 44:107–113.

———, ———, AND R. W. RICE. 1977. Pronghorn dietary quality relative to forage availability and other ruminants in Colorado. J. Wildl. Manage. 41:161–168.

———, W. L. REGELIN, AND A. W. FRANZMANN. 1987. Protein digestion in moose. J. Wildl. Manage. 51:352–357.

———, ———, AND ———. 1988b. Estimates of digestibility of birch, willow, and aspen mixtures in moose. J. Wildl. Manage. 52:33–37.

SCOTT, M. L., M. C. NESHEIM, AND R. J. YOUNG. 1982. Nutrition of the chicken. Third ed. M. L. Scott and Associates, Ithaca, N.Y. 562pp.

SEDINGER, J. S. 1984. Protein and amino acid composition of tundra vegetation in relation to nutritional requirements of geese. J. Wildl. Manage. 48:1128–1136.

SERVELLO, F. A., AND R. L. KIRKPATRICK. 1987. Regional variation in the nutritional ecology of ruffed grouse. J. Wildl. Manage. 51:749–770.

———, ———, AND K. E. WEBB, JR. 1987. Predicting the metabolizable energy in the diet of ruffed grouse. J. Wildl. Manage. 51:560–567.

———, ———, ———, AND A. R. TIPTON. 1984. Pine vole diet quality in relation to apple tree root damage. J. Wildl. Manage. 48:450–455.

———, K. E. WEBB, JR., AND R. L. KIRKPATRICK. 1983. Estimation of the digestibility of diets of small mammals in natural habitats. J. Mammal. 64:603–609.

SHORT, H. L. 1976. Composition and squirrel use of acorns of black and white oak groups. J. Wildl. Manage. 40:479–483.

SHUMAN, T. W., R. J. ROBEL, A. D. DAYTON, AND J. L. ZIMMERMAN. 1988. Apparent metabolizable energy content of foods used by mourning doves. J. Wildl. Manage. 52:481–483.

SIBBALD, I. R. 1975. The effect of level of feed intake on metabolizable energy values measured with adult roosters. Poult. Sci. 54:1990–1997.

———. 1976. A bioassay for true metabolizable energy in feedingstuffs. Poult. Sci. 55:303–308.

———. 1979. A bioassay for available amino acids and true metabolizable energy in feedingstuffs. Poult. Sci. 58:668–673.

SILVER, H., N. F. COLOVOS, AND H. H. HAYES. 1959. Basal metabolism of white-tailed deer—a pilot study. J. Wildl. Manage. 23:434–438.

———, ———, J. B. HOLTER, AND H. H. HAYES. 1969. Fasting metabolism of white-tailed deer. J. Wildl. Manage. 33:490–498.

SINGLETON, V. L., AND J. A. ROSSI, JR. 1965. Colorimetry of total phenolics with phosphomolybdic-phosphotungstic acid reagents. Am. J. Enol. Vitic. 16:144–158.

SMALLWOOD, P. D., AND W. D. PETERS. 1986. Grey squirrel food preferences: the effects of tannin and fat concentration. Ecology 67:168–174.

SMITH, D. 1973. Influence of drying and storage conditions on nonstructural carbohydrate analysis of herbage tissue—a review. J. Br. Grassl. Soc. 28:129–134.

SMITH, K. G., AND H. H. PRINCE. 1973. The fasting metabolism of subadult mallards acclimatized to low ambient temperatures. Condor 75:330–335.

SMITH, M. C. 1983. The feasibility of microwave ovens for drying plant samples. J. Range Manage. 36:676–677.

SNYDER, W. I., M. E. RICHMOND, AND W. G. POND. 1976. Protein nutrition of juvenile cottontails. J. Wildl. Manage. 40:484–490.

STALMASTER, M. V., AND J. A. GESSAMAN. 1982. Food consumption and energy requirements of captive bald eagles. J. Wildl. Manage. 46:646–654.

———, AND ———. 1984. Ecological energetics and foraging behavior of overwintering bald eagles. Ecol. Monogr. 54:407–428.

SUGDEN, L. G. 1971. Metabolizable energy of small grains for mallards. J. Wildl. Manage. 35:781–785.

SWAIN, T. 1979. Tannins and lignins. Pages 657–682 *in* G. A. Rosenthal and D. H. Janzen, eds. Herbivores: their interaction with secondary plant metabolites. Academic Press, New York, N.Y.

TATNER, P., AND D. M. BRYANT. 1986. Flight cost of a small passerine measured using doubly labeled water: implications for energetics studies. Auk 103:169–180.

TEUBNER, V. A., AND G. W. BARRETT. 1983. Bioenergetics of captive raccoons. J. Wildl. Manage. 47:272–274.

THOMAS, D. W., C. SAMSON, AND J. M. BERGERON. 1988. Metabolic costs associated with ingestion of plant phenolics by *Microtus pennsylvanicus*. J. Mammal. 69:512–515.

THOMPSON, C. B., J. B. HOLTER, H. H. HAYES, H. SILVER, AND W. E. URBAN, JR. 1973. Nutrition of white-tailed deer. I. Energy requirements of fawns. J. Wildl. Manage. 37:301–311.

THOMPSON, F. R., III, AND E. K. FRITZELL. 1988*a*. Ruffed grouse metabolic rate and temperature cycles. J. Wildl. Manage. 52:450–453.

———, AND ———. 1988*b*. Ruffed grouse winter roost site preference and influence on energy demands. J. Wildl. Manage. 52:454–460.

THONNEY, M. L., D. J. DUHAIME, P. W. MOE, AND J. T. REID. 1979. Acid insoluble ash and permanganate lignin as indicators to determine digestibility of cattle rations. J. Anim. Sci. 49:1112–1116.

THORKELSON, J., AND R. K. MAXWELL. 1974. Design and testing of a heat transfer model of a raccoon (*Procyon lotor*) in a closed tree den. Ecology 55:29–39.

TILLEY, J. M. A., AND R. A. TERRY. 1963. A two stage technique for the *in vitro* digestion of forage crops. J. Br. Grassl. Soc. 18:104–111.

TUKEY, H. B., JR. 1966. Leaching of metabolites from above-ground plant parts and its implications. Bull. Torrey Bot. Club 93:385–401.

ULLREY, D. E., W. G. YOUATT, H. E. JOHNSON, L. D. FAY, B. L. SCHOEPKE, AND W. T. MAGEE. 1970. Digestible and metabolizable energy requirements for winter maintenance of Michigan white-tailed does. J. Wildl. Manage. 34:863–869.

VAN HORNE, B., T. A. HANLEY, R. G. CATES, J. D. MCKENDRICK, AND J. D. HORNER. 1988. Influence of seral stage and season on leaf chemistry of southeastern Alaska deer forage. Can. J. For. Res. 18:90–99.

VAN SOEST, P. J. 1963*a*. Use of detergents in the analysis of fibrous feeds. I. Preparation of fiber residues of low nitrogen content. J. Assoc. Off. Agric. Chem. 46:825–829.

———. 1963*b*. Use of detergents in the analysis of fibrous feeds. II. A rapid method for the determination of fiber and lignin. J. Assoc. Off. Agric. Chem. 46:829–835.

———. 1965*a*. Nonnutritive residues: a system of analysis for the replacement of crude fiber. J. Assoc. Off. Agric. Chem. 49:546–551.

———. 1965*b*. Use of detergents in analysis of fibrous feeds. III. Study of effects of heating and drying on yield in fiber and lignin in forages. J. Assoc. Off. Agric. Chem. 48:785–790.

———. 1967. Development of a comprehensive system of feed analyses and its application to forages. J. Anim. Sci. 26:119–128.

———. 1982. Nutritional ecology of the ruminant. O and B Books, Inc., Corvallis, Oreg. 374pp.

VERME, L. J., AND J. J. OZOGA. 1980. Effects of diet on growth and lipogenesis in deer fawns. J. Wildl. Manage. 44:315–324.

VESEY-FITZGERALD, D. F. 1960. Grazing succession among East African game animals. J. Mammal. 41:161–172.

VLECK, C. M. 1981. Energetic cost of incubation in the zebra finch. Condor 83:229–237.

VOGTSBERGER, L. M., AND G. W. BARRETT. 1973. Bioenergetics of captive red foxes. J. Wildl. Manage. 37:495–500.

WALLAGE-DREES, J. M., AND B. DEINUM. 1986. Quality of the diet selected by wild rabbits (*Oryctolagus cuniculus* (L.)) in autumn and winter. Netherlands J. Zool. 36:438–448.

WALLICK, L. G., AND G. W. BARRETT. 1976. Bioenergetics and prey selection of captive barn owls. Condor 78:139–141.

WALTON, M. F., F. A. HASKINS, AND H. J. GORZ. 1983. False positive results in the vanillin-HC1 assay of tannins in sorghum forage. Crop Sci. 23:197–200.

WARNER, A. C. I. 1981. Rate of passage of digesta through the gut of mammals and birds. Nutr. Abstr. Ser. B. 51:789–819.

WEATHERS, W. W., W. A. BUTTEMER, A. M. HAYWORTH, AND K. A. NAGY. 1984. An evaluation of time-budget estimates of daily energy expenditure in birds. Auk 101:459–472.

WELCH, B. L., AND E. D. MCARTHUR. 1981. Variation of monoterpenoid content among subspecies and accessions of *Artemisia tridentata* grown in a uniform garden. J. Range Manage. 34:380–384.

WESLEY, D. E., K. L. KNOX, AND J. G. NAGY. 1973. Energy metabolism of pronghorn antelopes. J. Wildl. Manage. 37:563–573.

WEST, G. C. 1968. Bioenergetics of captive willow ptarmigan under natural conditions. Ecology 49:1035–1045.

WHITE, R. G., E. JACOBSEN, AND H. STAALAND. 1984. Secretion and absorption of nutrients in the alimentary tract of reindeer fed lichens or concentrates during the winter. Can. J. Zool. 62:2364–2376.

WICKSTROM, M. L., C. T. ROBBINS, T. A. HANLEY, D. E. SPALINGER, AND S. M. PARISH. 1984. Food intake and foraging energetics of elk and mule deer. J. Wildl. Manage. 48:1285–1301.

WILLIAMS, J. B., AND K. A. NAGY. 1984. Daily energy expenditure of savannah sparrows: comparison of time-energy budget and doubly-labeled water estimates. Auk 101:221–229.

———, AND ———. 1985. Daily energy expenditure by female savannah sparrows feeding nestlings. Auk 102:187–190.

———, AND A. PRINTS. 1986. Energetics of growth in nestling savannah sparrows: a comparison of doubly labeled water and laboratory estimates. Condor 88:74–83.

WILLIAMS, J. E., AND S. C. KENDEIGH. 1982. Energetics of the Canada goose. J. Wildl. Manage. 46:588–600.

WILLSON, M. F., AND J. C. HARMESON. 1973. Seed preferences and digestive efficiency of cardinals and song sparrows. Condor 75:225–234.

WISDOM, C. S., A. GONZALEZ-COLOMA, AND P. W. RUNDEL. 1987. Ecological tannin assays. Evaluation of proanthocyanidins, protein binding assays and protein precipitation potential. Oecologia 72:395–401.

WOOLEY, J. B., JR., AND R. B. OWEN, JR. 1978. Energy costs of activity and daily energy expenditure in the black duck. J. Wildl. Manage. 42:739–745.

WORTHEN, G. L., AND D. L. KILGORE, JR. 1981. Metabolic rate of pine marten in relation to air temperature. J. Mammal. 62:624–628.

ZERVANOS, S. M., AND G. I. DAY. 1977. Water and energy requirements of captive and free-living collared peccaries. J. Wildl. Manage. 41:527–532.

13

EVALUATION OF CAUSES OF WILDLIFE MORTALITY

Thomas J. Roffe, Milton Friend, and Louis N. Locke

INTRODUCTION

Timely and accurate diagnosis of wildlife mortality is a critical first step in prevention and mitigation against additional losses. Field biologists have an important role in this process, because often they are the first to observe dead and dying wildlife in free-ranging populations. The purpose of this chapter is to provide field biologists with guidance regarding the investigation and documentation of wildlife mortality, including field necropsy, and the collection and preservation of samples taken from animal carcasses. The descriptions of sample collection and preservation described in this chapter are those made under ideal conditions, and the authors realize that in most situations ideal conditions cannot be met nor can biologists in the field be expected to use the ideal tools, containers, and preservatives. The authors frequently offer improvised solutions that they have used in the field. In almost every instance, however, field samples collected under less than ideal conditions are *better* than none. Determination of the specific causes of mortality is an area of specialized study beyond the scope or purpose of this chapter.

The quality of field observations, and the appropriateness and quality of specimens received for examination, directly enhance or complicate the ability of disease specialists to provide timely, conclusive, cause-of-death evaluations. Complete and properly recorded data gathered on-site by field biologists with good powers of observation and logic, and the proper selection, preservation, packaging, and shipment of specimens, are critical in reaching an accurate and timely diagnosis. Good field observations and proper selection and handling of specimens are important regardless of whether the cause of death is a toxicant, an infectious agent, a mechanical process, or some other factor. Many situations will be encountered where diagnostic assistance is not readily available in the field. Even for those situations, retrospective evaluations by field biologists can be of considerable value provided the observations and recording of findings associated with the event and field necropsies are of high quality. Also, long-term sample preservation for diagnostic purposes is possible for some types of specimens.

The most efficient and effective process for resolving unexplained wildlife mortality involves field biologists and disease specialists functioning as an integrated team. Establishment of communications and understanding regarding the strengths and limitations of each party, and preparation before being confronted with a mortality event, are invaluable and strongly encouraged. Individuals having sufficient interest to read this chapter should stop at this point and ask themselves, "If I encounter unexplained wildlife mortality, do I know where I will seek assistance in obtaining a diagnosis?" A yes response to this question provides a focus for communications and actions needed for preplanning, including obtaining required training, and beneficial guidance regarding field response.

It is also important to recognize and be constantly aware that humans can be affected by many of the same agents causing illness and death in wildlife. Diseases transmitted from animals to humans are called zoonotic diseases. Personal safety should never be compromised during the investigation of wildlife mortality. The handling of sick and dead wildlife and the dissection of wildlife carcasses have **inherent personal risks.** These risks should be addressed through proper technique, protective clothing, and appropriate immunization as determined by qualified medical personnel. As a minimum, clinically ill wildlife and wildlife dead from unexplained causes should **never** be handled with **bare** hands. Handling dead animals always should be done with protective gloves or other barriers impervious to blood and other body fluids.

Field biologists should have a general familiarity with the diseases of wildlife, especially the zoonotic diseases. They should have a more detailed knowledge of those diseases that occur in the wildlife species with which they are working and in their geographic region, and the means by which those diseases are transmitted. One of the most important concepts wildlife biologists need to understand is how diseases are transmitted from animal to animal and, if a zoonotic disease, how it is transmitted to humans.

In general, mammals represent a higher potential risk of zoonotic disease for humans than do birds. However, this does not diminish the need for proper precautions when bird carcasses are handled. Moreover, ectoparasites from wildlife also can transmit disease to humans. For example, tularemia can be acquired from infected ticks from mammals and birds (Olsen 1975, Bell 1980). Obvious exposures of persons to potentially harmful diseases of wildlife, such as being bitten by wildlife that are reservoirs of rabies, or unexplained personal illness after working with wildlife should be dealt with promptly by medical personnel. The attending physician should be made aware of all contacts with wildlife and the nature of those contacts. Detailed and complete information should be provided, because the incubation periods for some diseases can involve weeks to months between exposure and onset of symptoms.

FIELD OBSERVATIONS

Wildlife mortality is the end-product of a sequence of events involving the host animal, its environment, and the lethal or causative agent. Observations made in the field by an investigating biologist can provide important information needed in focusing the laboratory's diagnostic efforts in attempting to determine the cause of the losses. For example, observations of environmental conditions may be consistent with profiles for specific disease or eliminate certain diseases from consideration. The distribution pattern of carcasses within the area can reflect acute versus chronic causes of death, point-source versus generalized types of exposure, and association with other species. The presence or absence of field signs such as whether sick animals as well as dead ones are present, findings such as specifically colored feces, and the behavior/appearance of sick animals all have relevance to identify potential causes of the problem. The following data are sought by the National Wildlife Health Center and other diagnostic laboratories in evaluating causes of wildlife mortality and serve to further illustrate the value of specific types of field observations.

Environment

A diverse range of environmental changes can lead to wildlife mortality either directly, e.g., exposure to extremes of freezing weather, or indirectly through increased exposure of wildlife to infectious agents (e.g., bacteria, viruses, fungi), insect pests, or toxicants. Biologists investigating wildlife losses need to document such environmental changes. For example, were pesticides applied to control insects? Did severe weather such as hail, lightning, precipitous drops or increases in temperature, heavy snows, or freezing rain occur? Was there a major eruption of arthropod populations such as ticks or mosquitoes? Was sewage/wastewater discharged into the wetland? Did an algal bloom occur recently? Did major changes in cropping practices alter the quantity/type of food available? Have water management regimes been altered or the source of water changed; has flooding, drawdown, or drought occurred? Have any wildlife been released on the area, or have new industries or domestic animal populations been established in close proximity to the site? Are unusual concentrations of wildlife on the area either in terms of populations numbers or species composition?

All of the above conditions have been associated with wildlife mortality, but all could also occur without causing any wildlife mortality. Therefore, this type of information should be gathered as objectively as possible rather than from a perspective of trying to incriminate a known change as the cause. Sound observations provide important perspectives to guide investigations. *Premature judgements* can result in failure to record other important observations and cause investigators to pursue the wrong factors.

Onset of Problems

The date of onset has direct value in establishing linkages with events recorded under evaluation of the environment. Carcass condition relative to stage of decomposition and amount of scavenging can help establish a time of onset. Ambient temperatures affect the rate of carcass decomposition and should be considered and recorded. The presence of on-site activities that could have resulted in the detection of sick or dead animals if they were present also must be considered relative to the date sick or dead animals first were observed.

Species Affected

One of the most important field actions that can be taken is to record the species present and those that are being affected. The narrow host range for some disease agents can focus attention on those agents or eliminate them from consideration on the basis of species mortality. For example, duck plague kills only Anseriformes (Leibovitz 1967), avian cholera has a broad host range in birds (Rosen 1971), and tularemia can kill a broad spectrum of mammals and birds (Bell 1980). To the extent possible, morbidity/mortality estimates also should reflect the percentage of species present that are affected. For example, "10% of an estimated population of 1,000 muskrats died, but none of the estimated 100 raccoons on the area were affected" would be an important observation.

Age

Some disease agents kill only young animals because of age-related resistance in older animals; others kill animals of all ages. Therefore, observations regarding population age structure and age structure imbalances of affected animals are important. Disease transmission also may be influenced by behavior of animals of different ages.

Sex

Differential mortality between sexes may occur as the result of actual differences in the relative susceptibility of each sex or, more commonly, as the result of differences in behavior patterns. Severe losses of female eiders during the nesting season have been caused by outbreaks of avian cholera on their island nesting grounds (Korschgen et al. 1978); the males suffer little or no losses, because they have left the nesting grounds and are rafting at sea. In contrast, the segregation of molting male mallards has resulted in a disproportionate mortality among these birds from botulism (Nat. Wildl. Health Cent., unpubl. data).

Number Sick or Dead

The number of sick or dead animals found at a die-off site may vary not only with the actual numbers of animals present, but also with the rapidity with which a mortality factor can cause illness or death (its acuteness). Agents such as the extremely toxic organophosphate pesticides are apt to cause a die-off with numbers of dead birds but few, if any, observed "sick" birds. In contrast, botulism outbreaks, caused by a generally much slower acting toxin, tend to be characterized by large numbers of both dead and sick birds. Sick birds will hide in dense cover, thus resulting in low estimates of the numbers actually involved. Field observations involving a large number of freshly dead carcasses and few or no sick animals eliminate consideration of more chronic forms of mortality and diseases with periods of visible illness prior to death. Follow-up observations of changes in observed ratios provide additional important information for guiding mortality evaluations. The effectiveness of predators in removing sick animals must also be taken into consideration. Therefore, the types of predators and scavenger species present, their relative numbers, and whether any major influx of those species has occurred as a result of the mortality event should be noted. Any evidence that suggests that predators or scavengers are becoming sick and dying after feeding on the carcasses should be recorded.

Clinical Signs

The physical appearance of sick animals should be described in as much detail as practical (e.g., "haircoat of affected animals is unkempt and patches of hair are missing, especially around the neck and ears"), and clinical signs of disease should be recorded. Abnormal behavior patterns and selection of habitat and aberrant posture also should be noted. For example, the animal may exhibit tumors, difficulty in walking, wing droop, and inability to hold the head erect; may not be reactive to external stimuli that normally would result in aggression or flight; or may be located in an unusual geographic location or in a habitat type it normally does not use. Each of these observations provides guidance, in combination with other factors, for focusing evaluations. Photographs (including video) can be especially useful if sufficient detail is provided to faithfully record the signs observed. Audiovisual equipment has the advantage of being able to record distress sounds made by the animal and any aberrant changes in the animal's call. Still photography can be combined with audio recordings on a tape when video equipment is not available. The importance of clinical signs is such that a *negative notation* should always be made when none are observed within the population being investigated.

Population at Risk

The local populations of those species involved in the losses and those using the die-off site should be estimated. This information has value in assessing probable causes due to density-dependent diseases versus diseases in which numbers of individuals present are not a major factor. Additionally, this information provides insight regarding mortality involving interspecies linkages.

Population Movement

Recent changes in the composition and size of animal populations on the area should be noted. Knowledge of animal movements related to population changes should be reported. This information can be of value in linking the mortality event at the location being investigated with events at other locations. Perhaps of greater importance are daily movement patterns of animals between the problem site and surrounding locations. Animals may die on an area as a result of exposures elsewhere. Movements that place animals in contact with other species such as deer with livestock or waterfowl with poultry or local zoo flocks should be noted. This information can broaden perspectives regarding possible causes of the mortality being observed.

Specific Features of the Problem Area

The full spectrum of potential causes of mortality must be considered initially by the diagnostician, and selective judgements must be made to focus on productive areas for pursuit. Descriptions of the problem area by field biologists and identification of any special features involved can be of great value in guiding that focus. Observations that the dead animals are in agricultural fields, in a small marsh with shallow water, in a deep-water area of a lake, or in the interior of a dense forest immediately bring specific types of diseases to mind and exclude others. Field observations regarding the type of habitat in which affected animals are found should be expanded to indicate whether this habitat is used as feeding sites, roosting sites, or loafing areas and any other specifics that may link the species to the site. Terrain characteristics such as steep slopes; physical features such as power transmission lines; and activities taking place on the area, such as predator control involving deployment of specific poisons or insect control by spraying once a month with compound X, also should be recorded.

Items listed in Table 1 should be used as a checklist to develop a complete and thorough field history of the event. Each factor should be addressed, and a negative notation should be recorded where no observation is appropriate. Use of photography can greatly enhance the value of information gathered. A copy of the field history (including photographic records) should be submitted with carcasses/specimens sent to the diagnostic laboratory. Findings from the diagnostic laboratory should become part of the biologist's data file associated with the mortality event.

Table 1. Data to be collected as a field history of wildlife mortality events.

1. Observer's name, address, and telephone number.
2. Observer's position, title, and agency of employment.
3. Dates observations were made.
4. Specific location of mortality event (state or province, county, closest town, and specific location that can be located on a map by someone not familiar with the area).
5. Environmental factors (record such conditions as storms, precipitation, temperature changes, or other changes that may contribute to stress).
6. Disease onset (your best estimate of when the outbreak started).
7. Species affected (the diversity of species affected may provide clues to the disease involved).
8. Age/sex (any selective mortality related to age and sex).
9. Morbidity/mortality (ratio of sick animals to dead animals).
10. Known number of dead (actual number picked up).
11. Estimated number of dead (consider removal by scavengers or other means).
12. Clinical signs (any unusual behavior and physical appearance).
13. Population at risk (number of animals in the area that could be exposed to the disease).
14. Population movement (recent changes in the number of animals on the area and their source or destination, if known).
15. Problem area description (land use, habitat types, and other distinctive features).
16. Other comments.
17. Types and number of specimens collected and dates of collection.
18. Types of photographs and other supplemental information.
19. Date specimens were submitted for evaluation.
20. Identification of person(s) to whom specimens were sent (name, organization, address, and telephone number).
21. Methods of specimen preservation and shipment.
22. Shipment data (carrier and identifying numbers for tracking, such as airbill number).

SPECIMEN COLLECTION AND PRESERVATION

Specimens collected for cause-of-death evaluations must be handled properly to avoid complicating the diagnosis by contamination, by creating artifacts from preservation methods, by lost labels, and by other factors. Improperly collected and improperly preserved specimens are of little value and result in wasted collection activities and diagnostic evaluations. Specimens intended for law enforcement investigations must comply with formal rules of evidence that require documentation of a chain-of-custody from the time of collection through all aspects of specimen processing. Time constraints relative to the admissability of evidence in court also apply. These facts emphasize the need for preplanning and knowing what to do and how to do it in dealing with assessments of wildlife mortality. The following is provided as general guidance. Individual innovation often can overcome specific problems in specimen collection and preservation, but basic principles still must be satisfied.

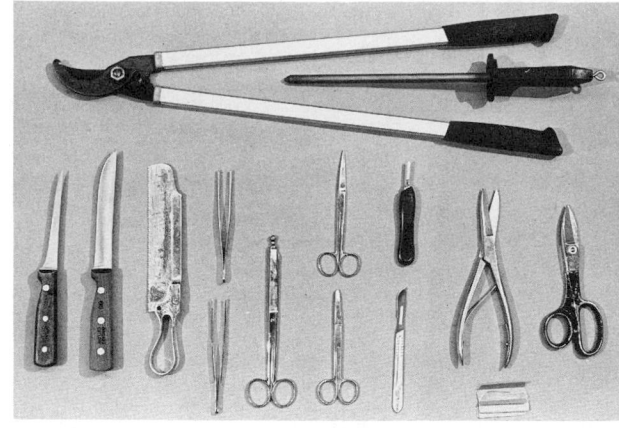

Fig. 1. Basic field necropsy kit.

Basic Supplies

Anyone planning to investigate causes of mortality among wildlife species should assemble a field kit that includes all the instruments and materials needed to perform postmortem examinations, to collect appropriate specimens for laboratory study later, and to store the specimens for transport to the laboratory. The instruments needed in a field kit (Fig. 1) should include, as a minimum, scalpel handle, scalpel blades, skinning knife, straight surgical scissors, forceps, bone saw, poultry shears, and pruning shears. Items that should be included in a field necropsy kit are listed in Table 2.

Specimen Selection

Submission of fresh, entire, *unopened* carcasses to a diagnostic laboratory for examination almost always will result in having the best specimens available for obtaining a correct diagnosis. Obviously, if the biologist encounters a dead sperm whale or is doing fieldwork in the Arctic National Wildlife Refuge in Alaska, sending the carcass to a laboratory may be impossible, but in many other instances specimens can be shipped to a laboratory in a number of ways. If the die-off site is located far from a diagnostic laboratory, a series of relays can be established to transport fresh carcasses to the laboratory; e.g., during the 1973 outbreak of duck plague at Lake Andes National Wildlife Refuge in South Dakota, fresh waterfowl carcasses were relayed by a series of vehicles driven by state game department personnel to the diagnostic laboratory at Brookings, South Dakota, a distance of >250 km.

Carcasses can be frozen and submitted to the laboratory in that condition if rapid submission of fresh specimens is not possible. Suitable freezers are almost always available in wildlife agency offices or in nearby towns. Although freezing and the subsequent thawing for examination will produce artifacts that can limit the usefulness of microscopic evaluation (histopathological studies) of the specimen's tissues, freezing tends to preserve many of the infectious agents, prevent loss of toxicants from tissues and the alimentary canal, and preserve some parasites. Thus, a fresh, intact, chilled carcass is the *best* specimen, but a frozen carcass is *very* valuable for diagnostic examinations.

Biologists always should contact personnel at the di-

Table 2. Desirable components of a field necropsy kit. Asterisks denote essential items.

No.	Item needed
1	Hemostat
*3	Scalpel handles
*1	Large forceps
*2	Large scissors
*2	Small scissors
*3	Toothed forceps
*2	Plain forceps
*1	Shears
1	Bone snips
*12 dz	Scalpel blades
*150	Small whirl-pack bags
150	Large whirl-pack bags
*50	Small (duck) plastic bags
*25	Medium (goose) plastic bags
*10	Large plastic bags
*100	Ties for plastic bags
100	18G 1½ needles
100	20G 1½ needles
100	21G 1½ needles
100	22G 1 needles
50	5-cc syringe without needle
50	10-cc syringe without needle
*1	Plastic bottle 40% formaldehyde
*1	Plastic undiluted Roccal®
1	Plastic bottle methanol
1 gr	Microscopic slides
1 bx	Coverslips
*3	Waterproof markers
*3	Pens
*3	Pencils
*100	Necropsy sheets
1	Cipboard
2	Dissecting pans
*1	Stainless steel pan w/cover
100	5-ml serum monovettes
	Zip-loc bags
	Paper towels
	Face masks
	Instrument tray with cover
*2	Bottles 10% formalin
	Specimen bottles
	Dry ice shipping labels & instructions
1 gr	Addressograph labels
*6 pr	Nondisposable gloves
*6 pr	Disposable gloves (large)
1 pr	Heavy-duty gloves
*1	Ruler
25	Tongue depressors
1 bx	Parafilm
*2 rolls	Strapping tape
*2 rolls	Masking tape
1 roll	Clear tape
1 bx	Disposable capillary pipettes
6 m	0.64-cm rope
25	Transport media—virology
50	Transport media—bacteriology
3	Pipette bulbs
*1 roll	Aluminum foil

Table 2. Continued.

No.	Item needed
1	Wire cage box
1 bx	Kimwipes (small)
1 gr	Ringing sticks
2	Test tube racks
100	Centrifuge tubes
1 roll	Cotton
1 roll	Gauze
2	Slide boxes
144	1-dram vials w/screw caps
1 roll	Flagging ribbon
100	Swabs
1	Hatchet
1	Alcohol lamp
1	Lighter
1	Bottle fuel for lamp
1	Boot brush
2 rolls	Label tape
*1 roll	Footguards
	Transport tubes
*	Blue ice
	Bar magnet
	Thermometer
5	35-cc syringes
*1	Straight butcher knife
*1	Curved butcher knife
1	Bone saw

agnostic laboratory, follow their instructions as closely as possible, and continue to work *closely* with the laboratory personnel throughout the investigation. Carcass selection should be representative of the species affected, including sex and age. If sick and dead animals are present, some of both should be collected. In some instances unaffected control animals may be required for comparative purposes, e.g., situations where comparative brain cholinesterase levels are needed for assessment of exposure to contaminants. Unaffected animals needed as controls for environmental contaminant studies often must be collected at sites some distance from the site of the die-off to preclude obtaining abnormal results from animals suffering from a nonlethal exposure to the suspected chemicals. Established humane methods for euthanasia must be used when sick animals are collected (American Veterinary Medical Association Panel on Euthanasia 1993), and the method of euthanasia must *not* compromise the diagnostic tests. For example, if an animal is suspected of having rabies, its brain is a primary tissue for demonstrating the rabies virus and should be kept intact without damage to the protecting encasement of the skull. The method of euthanasia *must* be recorded on all specimen tags and within the field history.

Concurrent and multiple causes of mortality are common in wildlife die-offs and are a primary reason for assuring that representative carcasses are examined. Collection of sick animals to obtain fresh carcasses may bias the sample for factors that do not kill quickly, whereas selection of only dead carcasses may bias the sample for acute causes of mortality. Sample bias was the cause of a discrepancy noted in snow geese submitted by two different groups of biologists investigating the same mortality event. Lead poisoning was diagnosed in all sick birds collected, but avian cholera was diagnosed in those found dead (M. Friend, unpubl. data).

If a variety of species is affected, a minimum of five carcasses of each species should be examined if practical. Several carcasses often are required to view the full range of pathology associated with a disease, and all animals may not have died from the same cause. Procedures regarding carcass collection and selection for examination are the same if carcasses are to be sent to a diagnostic laboratory or if field necropsies are to be conducted.

Postmortem decomposition rapidly reduces the value of most carcasses for cause-of-death evaluations as ambient temperatures increase above freezing. Carcasses that have been scavenged may be of limited value, even if freshly dead, because of damage to tissues and organs and contamination by bacteria that gain entry through openings created in the carcass.

However, for some diseases examination of moderately decomposed carcasses and carcass remains can provide valuable information. Prominent lesions such as cutaneous forms of pox, the presence of tumors and fibromas, proventricular impactions and associated gizzards containing lead shot, physical evidence of gunshot or entanglement with nets and plastic products, and fractures associated with blunt trauma are examples of persistent evidence associated with wildlife mortality. The sterility of brain tissue encased in an intact skull provides an extended period for isolation of microbes that may have caused death and as a tissue for laboratory assays in carcasses. *Pasteurella multocida,* the causative agent of avian cholera, can be isolated from avian bone marrow for at least a month after death under arctic summer conditions. Successful isolations have been made from carcass remains that consisted of little more than skin, bones, and feathers (Wobeser 1981).

More than carcasses and tissues from those carcasses may be needed to reach a diagnosis. Environmental samples such as water, soil, vegetation, and various food items, from insects to grain, may be needed, depending on the circumstances involved. Judgement of what to collect is based on field observations. A concentration of dead animals in and around croplands where animals are feeding would suggest a toxicant such as a pesticide, plant toxins or perhaps toxins, such as aflatoxins, produced on moldy grain or moldy peanuts. Other circumstances may suggest a toxic substance in water and the need for water samples. Field necropsies with findings of no detectable abnormalities, moderate to good body condition (no wasting of musculature and presence of some visceral fat), and a consistent type of food item in stomach contents strongly suggest the need for chemical analyses of samples of stomach contents and additional samples of that food item collected from the immediate environment.

Sample Collection

Once a decision has been made regarding the selection of samples, they must be properly handled from the point of initial processing (collection) through completion of assays and evaluation. The first step in this process is to ensure *personal* safety.

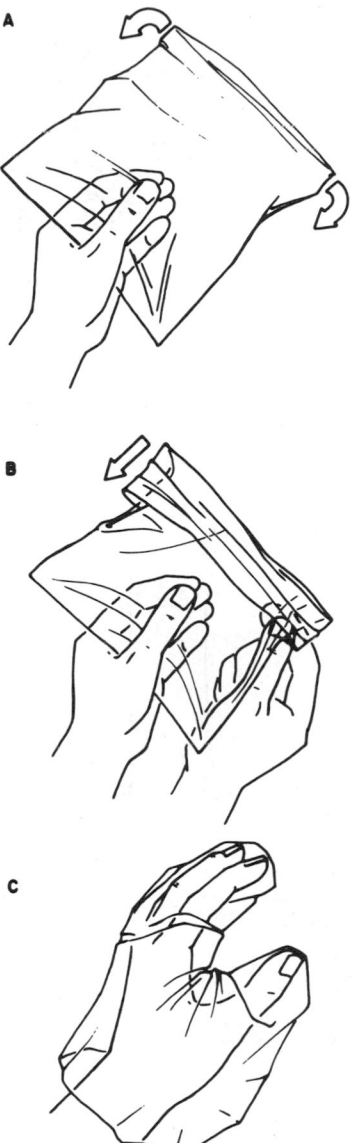

Fig. 2. Use of a plastic bag to protect hands (from Friend 1987).

Reasonable precautions should be taken to prevent direct skin contact with carcasses being collected or necropsied and with the tissues from these carcasses. The use of coveralls or other clothing that can be worn over outer clothing and then removed upon the completion of specimen collection, processing, and handling is highly recommended. Rubber footwear or boots that can withstand disinfection solutions such as chlorine bleach should be worn. Hands always should be covered with rubber or plastic gloves. If gloves are not available or become torn, plastic bags may be substituted (Fig. 2). After completion of any work with sick or dead animals, biologists should thoroughly wash with soap and water. Simple washing is a highly effective means of protection.

After personal protection has been provided, securely tag/label the carcass/sample and assign it a field identification number. *All* samples originating from a single carcass, including the carcass itself, should have a *common* field identification number. That number also should be

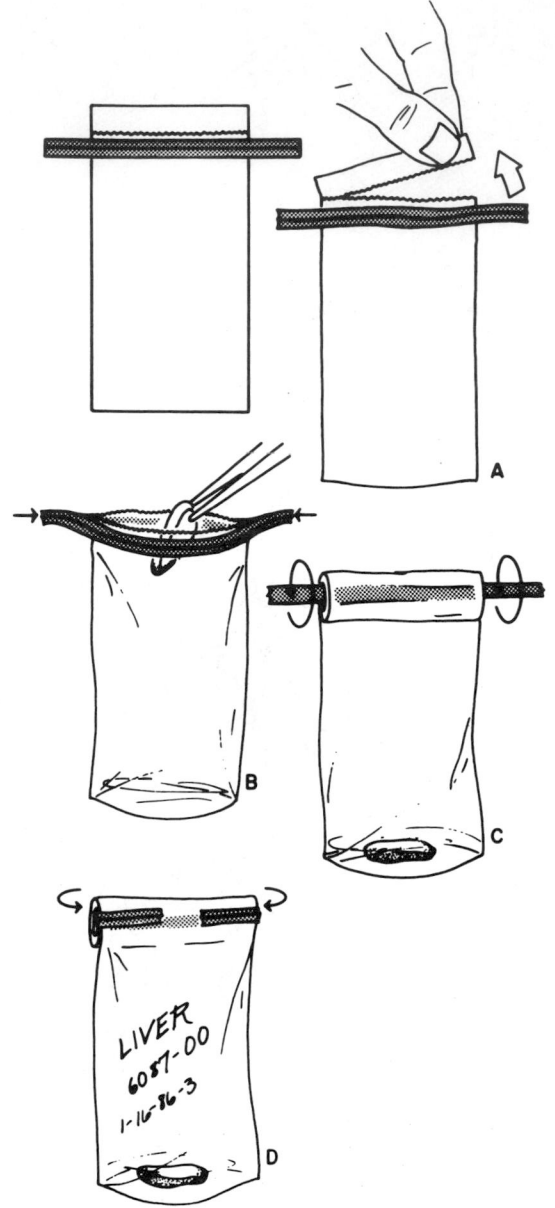

Fig. 3. Use of Whirl-Pak bag for specimen collection (from Friend 1987).

included in the information provided as part of the field history. Tags used on carcasses should be of durable material and fastened by wire to a leg. If such tags are not available, use an index card or piece of paper, and after recording appropriate information on this "tag," seal it within a plastic bag and tape or tie it to the animal's leg. Use soft pencil or indelible marker, never ballpoint pen, to record information on the tag. For carcasses being sent to others for processing, tag information should include name, address, and telephone number of the submitter, collection site (state or province and specific locality), date, field identification number, species, whether the animal was found dead or euthanized, and the method of euthanasia if one was used. The reverse side of the tag should include a brief summary of clinical signs or abnormal behavior the animal exhibited.

The carcass/specimen field identification number, date

of collection, and specimen identification should appear on all specimen containers. In most instances this should be written directly on the container with a fine-point indelible marker. We recommend the use of Whirl-Pak® (Nasco, 901 Janesville Ave., Fort Atkinson, WI 53538-0901) bags as primary containers for most small specimens. These bags have a sterile interior, are lightweight and easy to carry in the field, seal well, are of durable material, allow bag contents to be seen easily, and have a surface that is easy to write on with an indelible marker and that maintains the markings well (Fig. 3).

Tissue Selection and Preservation

GENERAL

Table 3 provides a general summary regarding specimen selection and preservation when field necropsies are undertaken because circumstances prevent intact carcasses from being sent to a diagnostic laboratory. Costs associated with specimen processing are high. Therefore, many laboratories do not accept tissues for assays unless they have been collected by properly trained personnel under conditions that minimize sample contamination during collection and preserve the integrity of the sample during the time between collection and the samples reaching the laboratory.

Two basic procedures must be followed when field necropsies are conducted: one is to minimize contamination of tissues to be examined for microbial pathogens; the second is to preserve collected tissues in a manner that will maintain the viability of primary disease agents present so they can be identified when the tissues reach the diagnostic laboratory. A third objective is to prevent cross-contamination among samples when collections are made from multiple carcasses. These considerations clearly establish that tissue collections made in the field are a poor alternative for the identification of many diseases when options exist between conducting field necropsies and shipping of intact carcasses to a diagnostic laboratory. *Specimen collection should be limited to those tissues and materials that can be preserved adequately under the circumstances at hand when field necropsies must be undertaken.*

PHOTOGRAPHY

To the extent possible, abnormal-appearing tissues seen at necropsy should be photographed. Several considerations must be addressed when photography is used. Perhaps the two most important are recording the abnormality on film with sufficient image size for identification of what is being viewed and displaying sufficient depth of field to provide for clarity of the image.

At close focusing distances, depth-of-field (that portion of the scene that is in clear focus) is very limited. Maximum depth-of-field is obtained by using the smallest aperture setting possible (largest f-stop number, such as f16 or f22). However, small aperture settings result in minimum light reaching the film and reduce the shutter speed that can be used. Photographs taken without a tripod or other secure support generally will be blurred because of camera movement caused by the person holding the camera at shutter speeds below $1/60$th or $1/30$th of a second when a 50-mm lens is used. Higher magnification lenses also magnify slight camera movement and correspondingly add to this problem. High-speed film (e.g., those with an ASA rating of ≥1000) is readily available and can be used to facilitate the use of small lens openings at adequate shutter speeds. The ASA rating of some lower-speed films can be extended by exposing them at a higher ASA rating and then having the film processed at that rating. Because of quality deterioration and increased processing costs, this alternative should be reserved for emergency situations. Also, prior knowledge of how far the ASA rating can be extended is required.

The limited depth-of-field at close focusing ranges requires that the camera be maintained in a position that is parallel to the subject matter being photographed. Take several exposures that vary in distance from the subject. Illustrate areas of interest and the relationship of that subject matter to a larger perspective of the subject being photographed. For example, one photograph might be a full image of an abnormal growth and another show the location of the growth on an identifiable part of the body. Photographs should receive the same field identification number as other specimens collected from the carcass and should be noted on the field history form as having been taken.

MICROSCOPIC EXAMINATION (HISTOLOGY)

Tissues can be collected and easily preserved for microscopic examination by the use of 10% buffered formalin solution. This solution is a dilution of 10 parts of 100% formalin, which is 36–38% formaldehyde solution in water, with 90 parts water. After mixing throughly, add 4 g of sodium phosphate monobasic and 6.5 g of sodium phosphate anhydrous per 1,000 ml of solution and remix. A small amount of concentrated formalin can be carried in the field and diluted when one is working in remote areas under conditions that preclude carrying sufficient volumes of prepared solution. Wide-mouth, laboratory-quality plastic bottles with threaded caps made by Nalgene®, Inc., (Nalge Co., subsidiary of Syborn Corp., P.O. Box 20365, Rochester, NY 14602) are recommended for initial preservation of tissues. Premarking these containers with concentrate and 100% volume lines (leaving about 10% space at the top for displacement by tissues) facilitates preparation of the solution in the field when concentrated formalin is to be diluted on-site during field collections and estimation of maximum volume of tissue to be placed in the container. Concentrated formalin can be taken into the field in one container and transferred to other containers and diluted to achieve a larger volume of 10% formalin for specimen preservation. Do **not** add buffers to concentrated formalin. However, impurities in the available water source (snow, ice, free-water) and lack of buffering salts may compromise specimen quality by causing artifacts to appear that will be difficult to separate from pathological changes in tissues due to the disease process being investigated. Formalin is hazardous to human health and should be handled in an appropriate manner.

When tissues are collected for microscopic examination, five basic rules must be followed: (1) the piece of tissue collected (except for fluids such as blood) should include a portion of the abnormality to be examined and adjacent normal appearing tissue, (2) the piece of tissue should be no larger than 10–20 mm long × 4–6 mm thick (smaller pieces will fix more readily but may curl or may be lost), (3) the tissue should be cut with a sharp blade (knife, scalpel blade, razor blade), (4) the tissue should be placed in a volume of 10% buffered formalin solution equal to at least 10 times the tissue volume for adequate fixation to occur, and (5) the formalin solution should be changed after 2–3 days (longer period for larger amounts of tissue).

Formalin fixation is a relatively slow process that begins at the tissue surface and permeates inward. The purpose of fixation is to stop autolytic processes that break down tissue structure. The thicker the tissue placed in formalin, the more time that is required for formalin penetration to occur and the more degradation of cell structure that occurs. The purpose of microscopic examinations is to detect changes in the tissues caused by pathological processes. This evaluation is confounded by postmortem change and those resulting from such processes as freezing and thawing of tissues. Therefore, the best results will be obtained from properly collected and preserved tissues from freshly dead animals. Also, the permeability of formalin through tissues is higher at elevated temperatures, as is the decomposition of tissues. Warming by placing containers of the solution in internal garment pockets in cold climates and cooling in hot climates by whatever means are available (placing containers in shade or waters below ambient temperatures) can be of value in balancing formalin penetration and tissue decomposition rates. Tissues placed in formalin must be protected from freezing.

Care must be taken not to crush the tissue being collected for formalin fixation. Mechanical compaction of tissue structure when the tissue is held for cutting, and the cut itself if the blade edge is not sharp, can greatly reduce the value of specimens collected. Therefore, cut through the tissue with a clean, single motion if possible, rather than with a "sawing" motion. Also, if the tissue is to be held with forceps, the gripping point should not be located in close proximity to the tissue that is to be cut.

Cross-contamination is not a problem when tissues are fixed in formalin for microscopic examination. Therefore, multiple tissues can be placed in the same container, but tissue from different animals should not be mixed or individual identity will be lost. If possible, tissues that float (e.g., lung) should be wrapped loosely in a permeable wrapping (e.g., cheesecloth or a piece of a handkerchief) that can be weighted with a small stone or some other object to keep the tissue submerged in the solution. After initial fixation the formalin should be changed and the volume reduced to just enough to keep the tissues constantly wet.

Nalgene plastic bottles can be reused. Therefore, waterproof tape with strong bonding qualities should be placed on these bottles for recording specimen information. After initial fixation of specimens, their transfer to Whirl-Pak plastic bags reduces space and weight of specimen storage. Special care must be taken to properly label original specimen containers and to accurately transfer this information to the Whirl-Paks when specimen transfer is made. Data should be recorded directly on the bags with indelible markers. Depending on their size, the now-emptied, wide-mouth plastic bottles can be used as containers for the Whirl-Paks with fixed tissues and serve as leak-

Table 3. Sample selection and preservation from field necropsy when entire carcass cannot be submitted and circumstances/necropsy findings suggest specific causes may be involved.

Sample	Projected tests	Method of preservation[a]	Comments
When microbial infections are suspected			
Observed lesions	Microbiology	Frozen	Lesions (abnormal-appearing tissue): a portion of each lesion should be saved frozen and fixed.
Heart	Bacteriology	Frozen	Entire heart from birds and small mammals; selected portions from larger mammals.
Liver	Bacteriology	Frozen	Entire lobe from birds and small mammals; several pieces up to 2 cm^2 or larger in larger mammals.
Blood/serum	Bacteriology/virology	Frozen	Serum also useful for serology.
Spleen	Bacteriology/virology	Frozen	Entire spleen from birds and small mammals; selected portions from larger mammals. Fix remainder.
Intestine (small fragment)	Bacteriology/virology	Frozen	Segments from middle or distal (ileum) of the small intestine.
Brain	Bacteriology/virology	Frozen	If animal exhibited abnormal behavior, save entire head; submit intact head to laboratory for removal of brain by laboratory personnel.
When toxicants are suspected			
Lesions	As appropriate	Frozen	Lesions (abnormal-appearing tissue): a portion of each lesion should be saved frozen. Fixed tissue important.
Liver	Heavy metals (Pb, Tl)	Frozen	Entire liver from birds and small mammals; selected portions from larger mammals. Fixed tissue important.
Kidney	Heavy metals (Pb, Hg, Tl, Fe, Cd, Cr)	Frozen	Entire kidneys from birds and small mammals; selected portions from larger mammals. Fixed tissue important.
Stomach contents	Organophosphates, carbamates, plant poisons, strychnine, cyanide, mycotoxins	Frozen	Save entire contents. Samples to be checked for cyanide or H_2S must be placed in airtight container to prevent loss of these toxic gasses into the air.
Brain	Brain cholinesterase, organochloride residues, organomercuric compounds	Frozen	If brain is removed for chemical analysis, the brain must be wrapped in clean aluminum foil then placed inside a chemically clean glass bottle. Fixed tissue important.
Blood	Lead, cyanide, H_2S, nitrites	Frozen	Samples to be checked for cyanide or H_2S must be placed in air-tight container to prevent loss of these toxic gasses into the air.
Lungs	H_2S, cyanide	Frozen	Samples to be checked for cyanide or H_2S must be placed in air-tight container to prevent loss of these toxic gasses into the air.
For microscopic study			
Lesions	Specimen is fixed, sectioned, and stained for microscopic study	10% formalin	Lesions (abnormal-appearing tissue): a portion of each lesion should be saved frozen.
Liver	Specimen is fixed, sectioned, and stained for microscopic study	10% formalin	Specimen portions should not exceed 6 mm in thickness.
Kidney	Specimen is fixed, sectioned, and stained for microscopic study	10% formalin	Specimen portions should not exceed 6 mm in thickness.

Table 3. Continued.

Sample	Projected tests	Method of preservation[a]	Comments
Gonads	Specimen is fixed, sectioned, and stained for microscopic study	10% formalin or Bouin's[b]	Specimen portions should not exceed 6 mm in thickness.
Intestinal tract	Specimen is fixed, sectioned, and stained for microscopic study	10% formalin or Bouin's	Snippet of stomach at the ileocecal junction, piece of duodenum (near the pancreas), and colon.
Brain, nervous tissues, eyes	Formalin-fixed material will be sectioned and stained.	10% formalin	Divide brain in half (sagittal); place one half in formalin and save the other half frozen.
Impression smear	Can be made by touching glass slide to cut surface of any organ	Air-dry	Air-dried slide can be used for many laboratory tests.
Heart, lung, skeletal muscle, lymph nodes, spleen, thymus	Specimen is fixed, sectioned, and stained for microscopic study	10% formalin	Specimen portions should not exceed 6 mm in thickness.

[a]Chilled specimens are preferred; however, if specimen is not delivered to laboratory within 48 hours (maximum), it should be frozen.
[b]Bouin's fixation: picric acid, saturated aqueous, 75.0 ml; concentrated formalin, 25.0 ml; glacial acetic acid, 5.0 ml.

proof, nonbreakable secondary containers to be used in packing specimens for shipment.

PRESERVATION OF PARASITES

Most parasites encountered during field necropsy procedures also can be preserved adequately for later identification provided basic preservation media are available (Table 3). A good general preservative consists of 70% alcohol with 5% glycerin. To prepare this solution, dilute 70 parts of pure alcohol with 30 parts of water. Mix well and then add 5 parts of glycerin (obtainable from drugstores and biomedical supply houses) to 95 parts of the 70% alcohol mixture. Ethyl or grain alcohol is *recommended* as the alcohol component, but good-quality isopropyl alcohol (rubbing alcohol) is also suitable and easier to obtain (retail stores). Methyl or wood alcohol can be used if this is the only source of alcohol available, but it is significantly inferior to the other alcohols for parasite preservation.

Ectoparasites such as ticks, fleas, lice, and mites can be placed directly into 70% ethyl alcohol or alcohol/glycerin solution. These parasites also can be kept alive in tightly closed containers for several days or longer, depending on the species. Bots and warbles (larval stages of insects) can be handled in the same manner.

Some internal parasites require special handling to provide specimens for species identification. Nematodes (''roundworms'') such as many of the parasites present in the gastrointestinal tract, lungs, and heart should be collected alive and placed in a hot (but not boiling) alcohol/glycerin solution with the above concentrations. The solution can be heated in the field with a cigarette lighter provided it is placed in an open container that will withstand this level of heat. Placing the worms in this heated solution straightens them, and the glycerin keeps them pliable so they can be examined for identification. Also, they can be fixed in hot alcohol and transferred later to glycerin alcohol.

Tapeworms, in contrast, are flatworms of varying length and width. These also should be collected alive but placed in cold water until they straighten and stop moving. The best preservative for tapeworms is AFA (alcohol-formalin-acetic acid). Its formula is: formalin, 6 parts; ethyl alcohol, 50 parts; glacial acetic acid, 4 parts; and distilled water, 40 parts. After fixing in AFA, the tapeworms are stored in 70% ethyl alcohol. Care must be taken not to separate the head (scolex) from the remainder of the tapeworm during collection. If the tapeworms are imbedded into the intestinal wall, they should be collected by including a small amount of intestinal tissue to which the worms are attached to assure that scolex characteristics are available for species identification. Flukes such as those found in the liver and bile ducts should be preserved by the same procedure as described for tapeworms.

Volume of preservative relative to that of parasites is not critical but should be at least 2:1. Whirl-Paks are good containers for parasites and their alcohol/glycerine preservative but should not be used for hot solutions. All containers should be labeled adequately regarding collection data and also should contain a notation of where in/on the body the parasites were found.

BLOOD SMEARS

Blood smears are another type of sample that can be preserved easily under field conditions and one that can be useful in reaching diagnostic conclusions. Blood smears can provide evidence of infection with protozoan parasites such as those that cause avian malaria or identify active infection of the circulating blood with gram-negative and gram-positive bacteria, and they help provide evidence of nonspecific disease processes on the basis of the types and composition of the cells present. Anyone can learn to make good-quality blood smears. However, this technique should be mastered before field collections are attempted. The qualities sought are an ultrathin blood film that is evenly distributed over the slide. Figure 4 illustrates the procedure to follow (Kolmer et al. 1951).

Blood smears must be made from blood that has not yet begun clotting. The small quantity required (one drop or less per slide) can be obtained easily from the heart

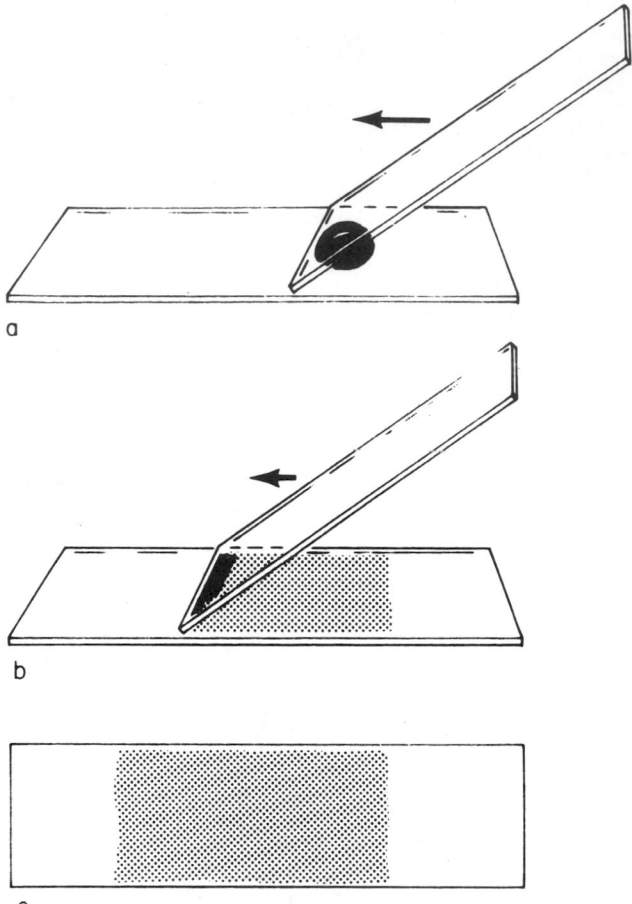

a

b

c

Fig. 4. Preparation of blood smear for later microscopic examination (from Friend 1987).

and major vessels of the vascular system in *recently* dead animals. Slides should be air-dried and then labeled with collection information that can be referred to the field history sheet. A minimum of two or three slides should be made per animal examined; the use of a glass-marking pencil (diamond point) eliminates the problem of obliterating specimen information when the slides are stained at the laboratory. Slides with frosted ends for marking with a pencil can be used. Once air-dried, the slides should be placed in dust-proof containers (such as sealed slide boxes or Whirl-Paks) until they can be sent to a laboratory for evaluation. This should be done as soon as possible because of eventual deterioration of the blood film without fixation and staining. The side of the slide on which the blood film has been made must be protected from contact with other slides to avoid scratching of the film or the sticking together of the slides because of inadequate drying or other conditions. Slides of blood films should be stored in as cool (nonrefrigerated) and dry an environment as possible to minimize fungal growth and blood-film deterioration.

BLOOD SAMPLES

Collecting usable blood samples from recently dead birds and mammals is sometimes possible. The major vessels leading to the heart, such as the posterior vena cava (between the heart and liver), and the heart itself generally

contain a substantial quantity of blood. Blood can be collected by syringe and needle or by careful incision of vessels or the heart to provide an opening for flow induced by palpation of the surrounding tissue. Blood collected during field necropsies can be used later for detection of toxicants, e.g., botulism, blood chemistry, and serologic tests. Gaining information on the animal's health from blood-chemistry values requires specialized anticoagulants to be used and blood samples that reach the laboratory for processing within a few hours. In addition, blood for future evaluation regarding exposure to disease agents that produce measurable antibodies can be obtained from the serum fraction of the blood or, for a few diseases, by immersing special, small paper discs into blood or serum and then air-drying the discs and maintaining them without further preservation until they can be submitted for analysis (Karstad et al. 1957). Discs can be obtained from Carl Schleicher and Schuell Co. (Keene, NH 03431).

Serum can be maintained for 1 to 2 weeks, sometimes longer, under field conditions if the temperatures remain at ≤4 C (serum samples can be kept in a refrigerator for 2 or 3 weeks or more if the serum is sterile). Freezing is the preferred method for preservation. In warm weather, samples will degrade within 1 to 2 days or less if contaminated, and they should **not** be collected if they cannot be kept cool. Contamination by bacteria is a major problem in warm weather because serum is a growth enhancement media for many types of bacteria, thus sterile collection techniques are helpful.

The following procedures are useful for collecting serum under field conditions when syringes, needles, collection tubes, and other materials normally used are not available.

1. Collect unclotted blood in a clean container (tube, Whirl-Pak bag) that is filled to no more than three-fourths capacity. The more cylindrical the collection vessel the better.
2. Place the container on an inclined angle of about 45° while the blood is clotting. The blood should be maintained at about 15–21 C for the clotting period (about 1–2 hr) if possible.
3. Once the blood has clotted, carefully separate the clot from its adhesion with the walls of the collection container by inserting sticks made for this purpose, a thin and clean knife blade, a smooth, thin twig (debarked), or similar object between the clot and the wall of the collection container. Then pull the object being used carefully and completely around the entire collection container, placing the pressure towards the wall of the container and avoiding entry into the clot or agitation of the sample.
4. Cool the sample while maintaining an inclined angle during storage, preferably by placing the serum in a refrigerator. Refrigerated temperatures are thought to aid the shrinkage of the clot and release of the serum fraction of the blood. Freezing must be avoided because it will result in rupture of the red blood cells and destroy much of the value of the sample.
5. The following day, carefully decant into a clean container the serum (clear yellow fluid) that has separated from the clot. Label the sample and maintain it at as

cool a temperature as possible (freezing is preferred) until the samples can be sent to a laboratory for analysis.

Specimen Preservation Under Temperature Extremes

Temperature extremes provide special challenges for specimen preservation under field conditions. Temperatures below freezing will cause many plastic containers to become brittle, adhesives to fail, and inks in marking pens to freeze. Therefore, special consideration must be given to the type and quality of materials taken into the field for specimen collections under winter and arctic conditions. If collection materials are satisfactory for cold temperatures, an additional problem is preventing freezing of samples that need to be protected from freezing. Insulated chests and battery-operated refrigerators may not be available or cannot be taken into the field because of logistical considerations. Enclosures such as "miniature igloos" that protect specimen containers from the increased cooling of the wind can be built and warmed by placing hand warmers in them when specimens must be preserved for extended periods under field conditions. Specimen containers also can be placed within inner layers of clothing where body heat and protection from wind prevent freezing.

High ambient temperatures generally destroy tissues and quickly reduce the biological value of specimens to the point that processing is not worthwhile. Freezing is advantageous for some tissue samples preserved for microbiological, toxicological, and serological assays. Liquid nitrogen tanks are available in various sizes and provide a means for quick-freezing and preservation of those samples that can be frozen. Dry ice is another means for freezing specimens that, in combination with acetone, provides a quick-freezing method. Acetone is poured into a wide-mouth container capable of withstanding subzero temperatures. A piece of dry ice carefully added to the acetone will supercool the liquid. Tissues to be frozen are placed in a collection container, such as a Whirl-Pak bag, capable of withstanding below-zero temperature, sealed, and labeled. The collection container is then carefully lowered into the supercooled container of acetone. Immediate freezing takes place. The specimen collection container then is transferred to a storage container that will keep the specimen frozen. This may be an insulated chest containing refrigerants, the container in which dry ice is being stored, or other suitable storage unit.

Skin contact with liquid nitrogen or with dry ice results in severe burns. Therefore, gloves designed for handling this material or other protective devices should be used. Eye protection should be worn to guard against splatter when dry ice is placed in acetone. Eye protection also is recommended to guard against shattering of glass specimen tubes or other containers that may not withstand the cold temperatures of dry ice or liquid nitrogen. Also, dry ice containers must be permeable to prevent explosion due to gas buildup as the dry ice reverts to a gaseous state (CO_2). Acetone is a carcinogen and also must be handled with care.

Chemical ice packs are preferred over wet ice for specimen preservation. The ice packs can be frozen and tightly packed in small styrofoam and other insulated containers carried into the field. Once specimen collection begins, half the ice packs can be removed, leaving the remaining space to be filled by specimens that become chilled, but not frozen, for transport from the field. This method is adequate for most situations involving 1–2 days. The insulated container should be protected from the sun as much as possible. Storage of the container below ground level or in a spring or other water source with temperatures less than air temperatures can be helpful. The container should be placed in plastic bags that protect against soil or water entering the storage compartment or be of a type that is impervious to those substances and that can be sealed to prevent intrusion.

When circumstances allow wet ice to be taken into the field for specimen preservation, the ice should be placed in clean plastic bags and the entire volume of the container it is placed within should be filled. The ice container should be well insulated and leakproof, or placed in a secondary, larger container that is well insulated. Whenever possible, the inside of the larger container should be well cooled before the ice is placed in it. For example, a picnic-type insulated chest might be kept in a walk-in freezer with the lid open for several hours before it is filled with supplies, including the container with ice, sealed, and taken into the field.

SPECIMEN SHIPMENT

Efforts expended in field collection of specimens can be rendered useless by improper shipment of specimens. Whenever possible, guidance for specimen collection, preservation, and shipment should be obtained from the persons who will process the specimens. Guidance received will generally address five basic components of proper specimen shipment: (1) prevent cross-contamination from specimen to specimen, (2) prevent specimen decomposition, (3) prevent leakage of fluids, (4) preserve individual specimen identity, and (5) assure proper labeling of the package (Franson 1987). Basic supplies for specimen shipment are shown in Fig. 5.

Tightly sealed, heavyweight plastic bags are a simple and effective means for separating individual specimens and preventing cross-contamination between specimens. These bags should be strong enough to resist being punctured by materials contained within them and from abrasions resulting from contact with other containers being shipped in the same package. Thoroughly cooled rather than frozen specimens generally are preferred if the specimens are to reach the processing laboratory within 24–36 hours. Sealed chemical coolant units such as blue ice packs are preferable to wet ice, because leakage is not a problem if the coolant melts. Similarly, plastic jugs, plastic soda bottles, and waxed cardboard juice and milk containers filled with water, sealed, and frozen are preferred over ice cubes and block ice as a coolant. The lids on these containers must be taped closed to prevent their being jarred open during shipment.

Frozen specimens generally are preferred if an extended period of time is required for the specimens to reach the processing laboratory and when high ambient temperatures precludes adequate cooling during transit. Dry ice is preferred to keep frozen materials frozen. However, because of the ultra-low temperature of dry ice, chilled specimens placed in the same shipping container also will

Specimen Shipment

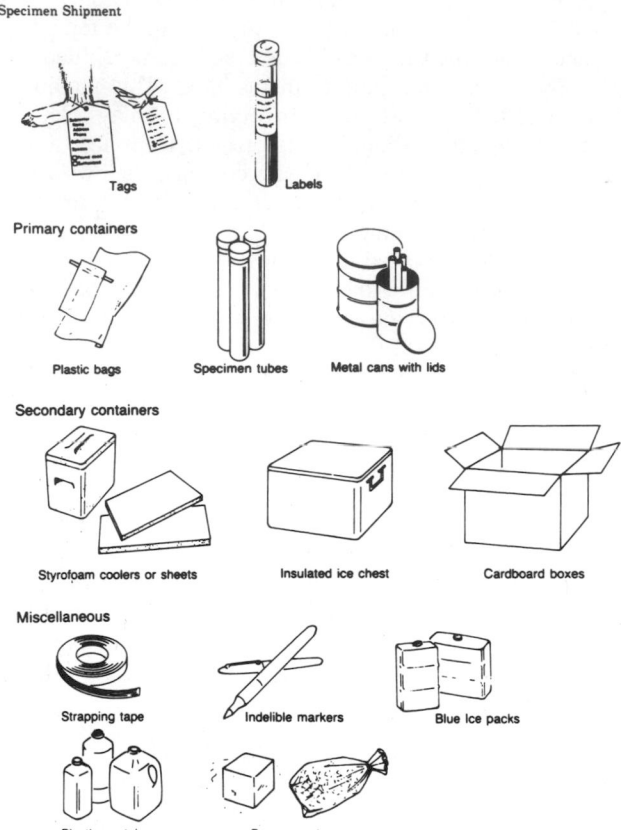

Fig. 5. Basic supplies used for shipment of specimens (from Friend 1987).

freeze. Use of dry ice in shipment is regulated and must be cleared with the carrier before shipping to avoid rejection of the shipment because of improper packaging, exceeding allowed limits, or other factors. The container must be permeable or properly vented to prevent the carbon dioxide gas given off by dry ice from building up so much pressure that the container is destroyed. Also, this gas can destroy some disease agents. This is of more concern when tissues rather than whole carcasses are shipped. Regardless of whether samples are shipped cooled or frozen, specimens within the shipping container should be protected from direct contact with the coolants or freezing material used.

Delays in transit and other factors may result in coolants melting, frozen materials thawing, and containers with liquid samples (e.g., blood) becoming broken during shipment. Packaging must prevent leakage of fluids to the outside of the shipment container. Absorbent materials should be placed throughout the shipping container to capture fluids that escape individual components of the shipment.

Styrofoam coolers shipped in cardboard boxes are inexpensive, high-quality shipping containers. Styrofoam at least 2.54 cm thick should be used when possible, and coolers with straight sides are preferred. Those that are wider at the top than bottom are more likely to break during transit. Filling the space between the outside of the styrofoam cooler and the cardboard box with crumpled newspaper or other shock-absorbent packing materials is helpful in preventing cooler breakage. The cardboard box

protects the styrofoam cooler from being crushed and serves as containment for the entire package. The strength of the box should be consistent with the weight of the package. Sheets of styrofoam insulation can be cut to fit inside cardboard boxes if ready-made coolers are not available (Franson 1987).

The following guidance (Franson 1987) for submission of specimens is provided as basic procedures. Instructions from other laboratories may be different.

1. Double-bag the carcasses (Fig. 6a) and place them in a styrofoam cooler lined with a plastic bag. When both frozen and fresh whole carcasses are submitted in the same container, the frozen carcasses can be used as a refrigerant to keep the fresh carcasses chilled by interspersing individually bagged frozen carcasses among the individually bagged fresh carcasses or by placing the fresh carcasses between two layers of frozen carcasses (Fig. 6b). Blood tubes and other breakable containers of uniform size can be protected by packing them in a common plastic bag that is sealed within a coffee can (Fig. 6c). Any space around the specimen containers in the coffee can (side and top) should be packed with paper or some other absorbent material to prevent jarring that could cause breakage and to collect fluids if breakage does occur. Then, the coffee can is sealed in a plastic bag before it is placed in the styrofoam cooler.

2. If blue ice packs are used, they should be interspersed among specimens; other types of coolants should be distributed in locations within the styrofoam container to provide maximum cooling for all contents and to keep everything frozen when dry ice is used (Fig. 6d). All empty space in the styrofoam cooler should be filled with newspaper to prevent materials from being tossed around during transit. The insulating properties of newspaper also will help maintain cool temperatures within the package, and its absorbent qualities will help prevent leakage of fluids to the outside of the box or container.

3. The plastic bag lining the cooler should be closed and the lid should be sealed with strapping tape (Fig. 6e). The specimen data sheet and history, contained in an envelope in a waterproof plastic bag, should be taped to the top of the cooler.

4. The styrofoam cooler should be enclosed in a cardboard box and the contents should be secured with strapping tape (Fig. 6f).

In the United States, packaging and labeling of specimens must conform to the Federal Shipping Regulations for Packaging and Labeling. Under Fish and Wildlife Regulations (50 Code of Federal Regulations [CFR] 14), containers with wildlife specimens must bear the name and address of the shipper and consignee, and an accurate statement of the contents (by species and number of each species) must be marked conspicuously on the outside of the container. An alternative to this requirement is to conspicuously mark the outside of each package or container with the word "wildlife" or the common names of the species contained in the package. In addition, an invoice or packing list that includes the name and address of the consignee and shipper, and which accurately states the types and quantities of each species contained in the ship-

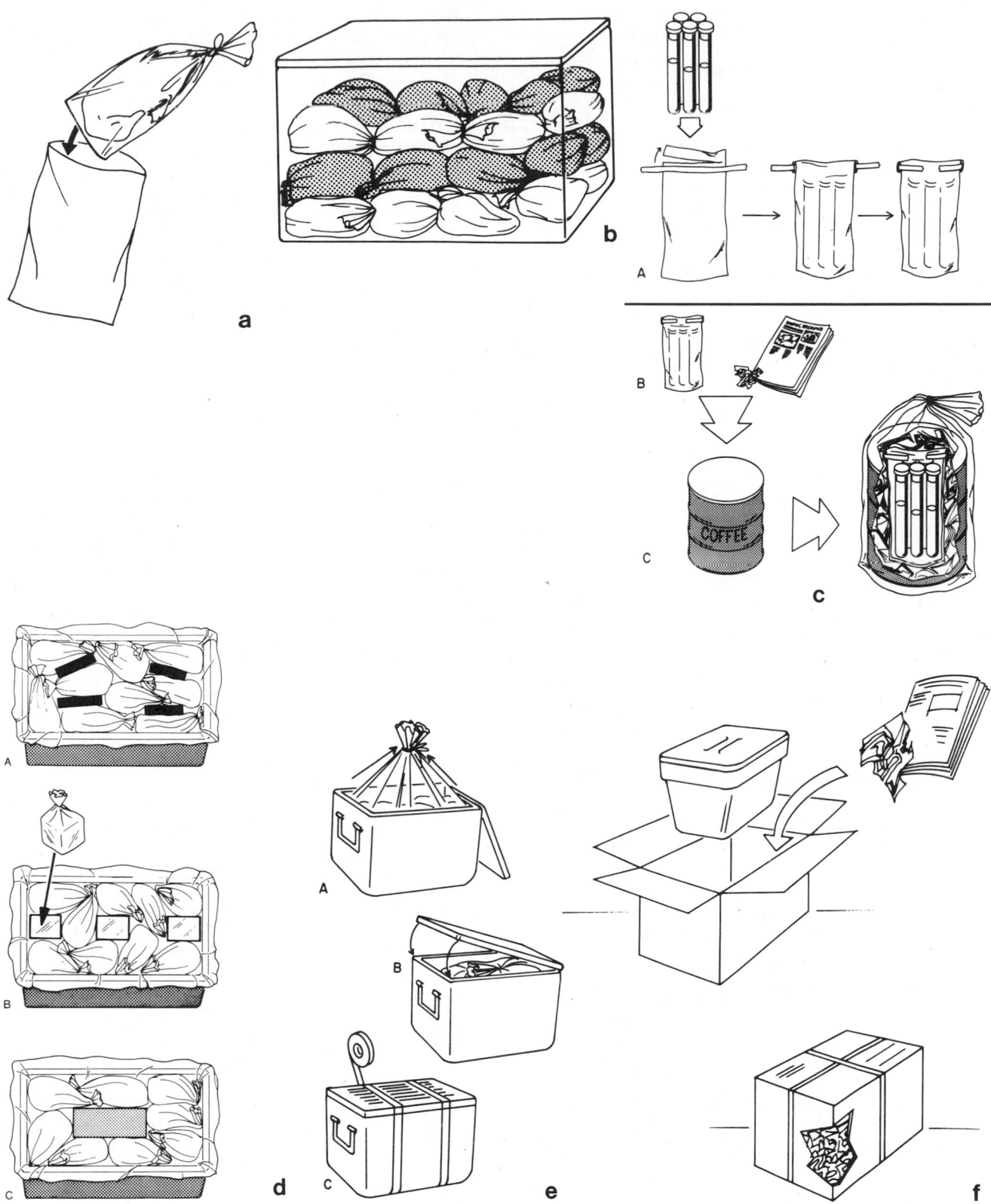

Fig. 6. a. Double-bagging of specimens. Individual samples should be double-bagged to prevent leakage of fluids and cross-contamination of specimens. b. Frozen specimens (white bags) can be used to keep fresh, unfrozen specimen (dark bags) chilled during shipment. c. Packing sequence of blood tubes. (A) blood tubes placed within Whirl-Pak or other plastic bag; (B) bag is closed and placed within can or hard-plastic container; container is then packed with newspaper or other absorbent material; (C) closed can is then placed in a plastic bag and sealed. d. Packing specimens for shipment when (A) blue ice, (B) wet ice, or (C) dry ice is used as a coolant. e. Closure of specimen container: secure large, plastic bag containing bagged specimens by tying; close cooler lid and secure container with strapping tape. Attach specimen data sheet and history to outside of cooler and place sealed cooler in large cardboard box. f. Place sealed cooler in large cardboard box packed with newspapers or other packing material (from Friend 1987).

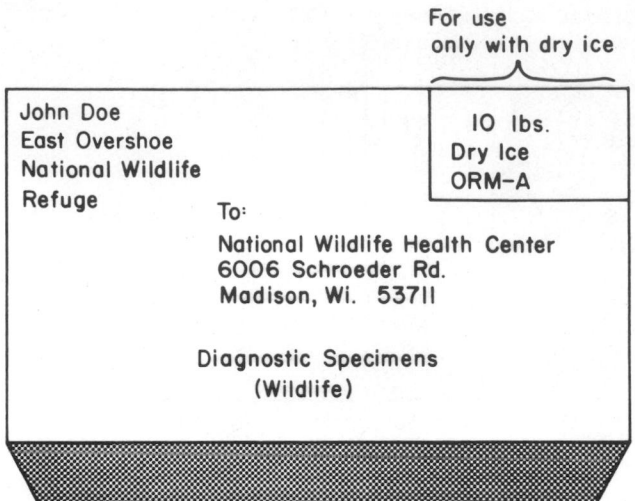

Fig. 7. Example of required shipping labels that must be affixed to outside of cardboard box with and without dry ice as coolant (from Friend 1987).

ment, should be secured to the outside of one container in the shipment.

In addition to Fish and Wildlife Service regulations, the interstate shipment of diagnostic specimens is subject to applicable packaging, labeling, and shipping requirements for etiologic agents (42 CFR Part 72). These regulations do not require identifying diagnostic specimens as etiologic agents when the disease agent is not known or only suspected. It is also helpful to prominently label the package with the words "DIAGNOSTIC SPECIMENS." Packaging requirements under 42 CFR Part 72 are met by following the preceding recommendations 1 through 4 for enclosing specimens within two containers before they are enclosed with the package.

Hazardous Material Regulations of the U.S. Department of Transportation apply whenever dry ice is contained in the shipping container (49 CFR Parts 172-173). The amount of dry ice contained in the package must be written clearly in the upper right-hand corner on the outside of the container. The letters "ORM-A" must appear under the amount of dry ice, and this information must be written in a rectangular area marked on the outside of the package. The words "DIAGNOSTIC SPECIMENS" must appear under that designation. Properly labeled containers for use with and without dry ice are illustrated in Fig. 7. Labeling should be done with permanent markers.

FIELD NECROPSY PROCEDURES

A postmortem examination, or necropsy, is a standardized procedure for the methodical, anatomical evaluation of an animal after death. The usual purpose is to determine the cause of death. Other purposes include studies of comparative anatomy, monitoring of population health, and collection of samples. The following provides guidance to assist field biologists in conducting meaningful field necropsies of wildlife when carcass transport to a diagnostic laboratory is not possible or feasible. The field necropsy is one of several valuable tools for evaluating causes of wildlife mortality but is seldom an end in itself. Evaluation of abnormal findings at necropsy requires abil-

ity to differentiate abnormal from normal and properly interpret the causes involved. Microscopic examination of abnormal tissues and appropriate laboratory tests involving bacteriology, virology, parasitology, toxicology, and serology usually are required for a diagnosis.

Wildlife biologists contemplating conducting frequent necropsies in the field or at facilities where they are employed should obtain formal instruction by working with appropriately trained individuals and taking a formal course (with laboratory) in animal pathology. Also, much can be learned by conducting necropsies on road-kills and other specimens to develop necropsy skills and enhance recognition of normal versus diseased tissue color, texture, and appearance; organ size; and postmortem changes. Use of road-kills for this purpose should be restricted to birds and large mammals, because road-killed carnivores and other small mammals in some geographic areas often have greater probability of being infected with rabies or other diseases transmissible to humans.

Wildlife disease specialists are employed by some state conservation agencies, the U.S. Fish and Wildlife Service, and many universities and state veterinary diagnostic laboratories. These individuals are valuable sources of information and, in some instances, sources for obtaining training. Assistance (technical advice and specimen processing) also is often obtainable from Department of Agriculture (state and federal) diagnostic facilities, schools of veterinary medicine, and veterinary science departments. The largest wildlife disease programs in the United States are the National Biological Service National Wildlife Health Center, 6006 Schroeder Road, Madison, WI 53711, and the Southeastern Cooperative Wildlife Disease Study, College of Veterinary Medicine, University of Georgia, Athens, GA 30602. Both programs give wildlife disease workshops and are good sources of guidance for technical assistance.

General Guidelines

Postmortem examinations should be done in a standardized manner to facilitate observations and avoid compromising sample collection and integrity. Four basic guidelines must be kept in mind. If followed, these considerations will maximize the value of findings and provide greater return from the investments of time and effort.

1. *Be methodical, thorough, and complete.* More is lost by not looking than by not knowing. Most people will recognize an abnormality even if they do not know what it is. However, abnormalities present will not be detected in organs and tissues that are not examined. There is no one right way to do a necropsy, but there are many wrong ways. One should develop and use a systematic method and avoid being sidetracked by examining only that which appears abnormal at first glance. A good necropsy technique ensures examination of all body systems, causes minimal disturbance of organs and tissues before they are visually examined, and allows organs and tissues to be examined in place, in proper anatomical orientation, and in relation to other organs.

2. *Describe and record what you see.* Be specific in terms of color (e.g., bright red, mottled red/black), texture

(waxy), size (10 mm in diameter), shape (e.g., tear-drop), consistency (e.g., semisolid, liquid, creamy), presence of exudates (yellow pus), and any other distinguishing features ("a 4-cm nodule with a surface formed by adjoining flat polygons"). The purpose of the descriptors and measurements is to allow other individuals to clearly visualize what you observed. Good descriptions by individuals not trained in pathology can greatly assist disease specialists reach valid conclusions for diseases that have characteristic lesions. For example, describing the lungs as red/purple and firm over the lower half suggests pneumonia in a deer, and an enlarged spleen with dark red/black color and soft, gelatinous consistency might suggest anthrax. Photographs are an excellent way to complement written descriptions.

3. *Maintain complete and accurate records.* All pertinent information should be recorded relative to the animal being necropsied. Record-keeping should follow the same systematic process as the necropsy and should be done while observations are fresh and omissions can be corrected. Tape recorders are effective for recording data, but the taping should be checked periodically to guard against mechanical malfunction. Having a second person take notes during the necropsy is also effective when assistance is available. Complete, accurate observations provide a database for the development of inferences regarding specific diseases. Similar findings associated with multiple necropsies can provide a basis for retrospective diagnosis by wildlife disease specialists and pathologists. All components of the case (records, samples, photographs) should have the same unique case number.

4. *Guard against contamination of samples.* Field necropsies are not done in a sterile environment and occasionally are complicated by physical and environmental factors that may limit the procedure to a process for collection of observational data (including photographic records) rather than tissues for further analyses. High sensitivity must be maintained for potential sources of sample contamination, and common sense must prevail in the selection of locations in which to conduct the necropsy when the animal is small enough to be moved and terrain allows carcass movement. To the extent possible, the necropsy should be done in a well-lit area on dry land in a location protected from wind and blowing debris. If needed, a wind break can be made from a poncho, other clothing, large plastic bags, or other materials. Tissues being collected for environmental-contaminant analysis should not come in contact with the environment (soil, water, vegetation) of the field necropsy site.

Biosafety is an additional general principle that deserves great attention. *Protect yourself, others, and the environment from any pathogens that may be present.* As mentioned earlier, many diseases of animals are *transmissible* to humans. Not only must personal care be taken to protect oneself, but proper disposal of the carcass and subsequent decontamination treatment of the necropsy site also must be considered. The postmortem procedure will soil the immediate environment with body fluids that may contain infectious agents. Some agents such as those causing tuberculosis and anthrax have an environmental longevity that is measured in years. Therefore, care should be taken to minimize contamination of the field necropsy site.

To the extent possible, field necropsy activities should be done on plastic sheets. Carcass disposal by on-site incineration is preferred. If carcasses are buried, depth should be sufficient to prevent scavengers from digging up the carcass and spreading infectious material. Potential water-table contamination must be considered and prevented. Carcasses that must be transported to a distant site for disposal should be enclosed in plastic bags that have been disinfected on the outer surfaces. Mechanical transmission of disease agents from the field necropsy site to other areas can occur by contaminated clothing and body surfaces. Therefore, whenever possible, field necropsy work should be followed by cleaning and disinfecting of protective clothing, a shower, and a change of clothing; further, working with other susceptible animals during at least the same day should be avoided.

When field activities are specifically scheduled for necropsy work, the following precautions should be taken. Vaccination of personnel for high-risk diseases prevalent in an area should be discussed with medical personnel, and, if recommended, a vaccination schedule should be planned for all involved personnel. Well in advance of necropsy work, protective outer clothing that can be removed or disinfected should be worn. Coveralls, boots, and gloves are standard body coverings. Masks are desirable when one works in bird roosts and bat caves containing a large amount of guano and in other situations where diseases easily transmitted by aerosol may be present. An effective cleaning agent (e.g., soap) should be taken into the field to clean instruments, clothing, and the work site. After these are cleaned, a 10% solution of household bleach can be used as a disinfectant. Instruments, gloves, and working surfaces should be disinfected and then rinsed to clean off the disinfectant between necropsies to prevent cross-contamination and potential erroneous results from tissue assays. Boots and the outer surfaces of specimen containers should be disinfected before one leaves the site, and outer work clothes should be removed and placed in plastic bags that are disinfected after they have been sealed. Using propane torches, igniting instruments dipped in alcohol, and dipping instruments in formalin are means of sterilizing instruments between specimens. Upon completion of necropsy or after handling any sick or dead animals, a thorough washing with soap and water is an excellent and essential protection against possible infection.

Equipment

The size and species of animal dictate the type of equipment needed to conduct a necropsy. Personal preferences influence the styles of equipment selected, such as straight or curved knife blades. Table 2 provides an extensive list that will accommodate most situations.

Basic Principles of Necropsy

Although the anatomic procedure varies by species, all necropsies should follow the same general pattern. The *first step* is to consider the history of what is known about the circumstances that may have led to the animal's death. The field site in the immediate vicinity of the carcass should be observed carefully, and any pertinent observations should be recorded just as one might do at a human crime scene. This information may provide important clues regarding the cause of death and should be the first information recorded. The second phase of the process is to carefully conduct an external examination of the carcass, paying particular attention to body openings and recording a description of any discharges from those sites. Relative amount, color, and consistency of any material discharged should be recorded. Appendages should be examined thoroughly by hand, through touch and manipulation, to determine whether fractures or joint problems are evident. The skin, scales, fur, or feathers should be examined for burn marks, broken feathers, soiling, aberrant hair coat, and wounds. All surfaces of the animal should be examined and observations should be recorded. If no abnormalities are found, this should also be recorded to dispel questions at a later time of whether certain conditions were present but not looked for. The internal examination is the third phase of the process and varies with the species involved. General procedures follow.

Mammal Necropsy

Once the external examination is completed, the animal is positioned for postmortem examination. Because of the diversity of animal types, positioning the animal depends on the species being examined. In most instances, the animal should be laid on its side. Carnivores can be positioned on either side, generally with the animal's left side down. Ruminants should be necropsied with the left side down because this places the rumen on the down side and allows the examination of abdominal viscera. Equines should be placed with the right side down because of the anatomical structure and frequent displacements of the large colon. The following method is general for mammals and includes a description of the technique for handling the ruminant gastrointestinal tract.

Place the animal left side down (legs toward examiner, head pointing to the right). Cut the right fore- and hind legs and reflect them over the back (Fig. 8). Cut the hind limb through the skin, close to where the leg and body join, through the muscle and into the hip joint. Note subcutaneous tissue, muscle, joint (surfaces and fluid), and superficial lymph nodes. Similarly reflect the forelimb by a cut between the body and limb. There is no bony joint to cut on the forelimb.

Make a midline incision through the skin from the pelvis to the tip of the lower jaw, just to the right (off midline) of the penis or mammary gland (leaving these structures on the down side). Remove the skin from the right side of the carcass up to the head by separating it from the musculature with a knife. If this is done properly, the skin will come off easily. Free the skin along the bottom of the jaw. One technique for the head is to puncture through the skin with a knife as far dorsally as possible, then cut directly along the surface of the face (through the ear canal).

Reflect a small amount of skin on the down (left) side from the midline incision along the length of the carcass. This helps prevent contamination with hair. The entire right side of the carcass is now skinned and the legs are reflected out of the way.

INITIAL OPENING

Remove the right half of the jaw by first cutting the musculature on the outside and tongue sides of the jaw. Use pruning shears to separate the jaw at the chin. The right mandible is removed by twisting backward and upward (towards the right front leg), providing complete access to the oral cavity. Examine the tongue, all surfaces of the teeth, the roof of the mouth, and tonsillar area. Firmly grip the tongue and pull it towards you and locate the hyoid bones on either side of the larynx. Maintain tension on the tongue and cut through these bones (either at the joint or with a bone cutter) and continue cutting to the trachea (windpipe) and esophagus until the thoracic inlet is reached. The lymph nodes at the very back of the throat, where ruminants often develop abscesses, and the internal parts of the oropharynx, where bot larvae (parasitic fly larvae) often are found, can be examined. Other structures such as salivary glands can be examined easily.

Move to the abdomen and palpate the highest point of the right side behind the last rib. Make a small incision into the abdominal cavity (Fig. 9). A preliminary assessment of the abdomen can reveal whether excessive quantities of fluid are present; if so, samples of it can be taken with a syringe at this time. Extend the incision upward (toward the backbone), then back toward the tail, and finally down toward the penis or mammary gland. Both ends of the incision now can be extended ventrally to provide an adequate view of the abdominal viscera while a flap of abdominal wall contains fluids and viscera.

Reaching under the last rib, locate and cut the diaphragm away from the rib cage. Note the presence or absence of negative pressure (negative pressure is expected in the chest cavity). Cut the ribs close to the backbone and sternum to remove the right side of the chest cavity (Fig. 10). Before discarding the rib cage, examine its internal surface. Most internal organs (aside from central nervous system) now are exposed and should be examined visually before any additional handling is done. Organs should be observed for abnormalities in placement, particularly twists of the gastrointestinal tract (GI) or herniation of the GI tract into the thoracic cavity. This is also a good time to make a preliminary assessment of possible samples to be collected for microbiology.

Removal of organs follows the initial opening, and samples for later microbiological studies or chemical analysis should be taken at this time to avoid contamination. Many pathologists prefer to remove all internal organs before examining individual organs. However, for the sake of chapter organization, descriptions of internal organ removal are followed immediately by a discussion of examination of that particular organ. Either method can be used.

THORACIC ORGANS

Grasp the trachea/esophagus and, placing tension on the thoracic organs, remove the organs by cutting along the backbone, the sternum, and the surface of the diaphragm. These cuts require cutting across the main blood supplies to the heart and generally result in blood partially filling the thoracic cavity; to prevent this, you may wish to tie

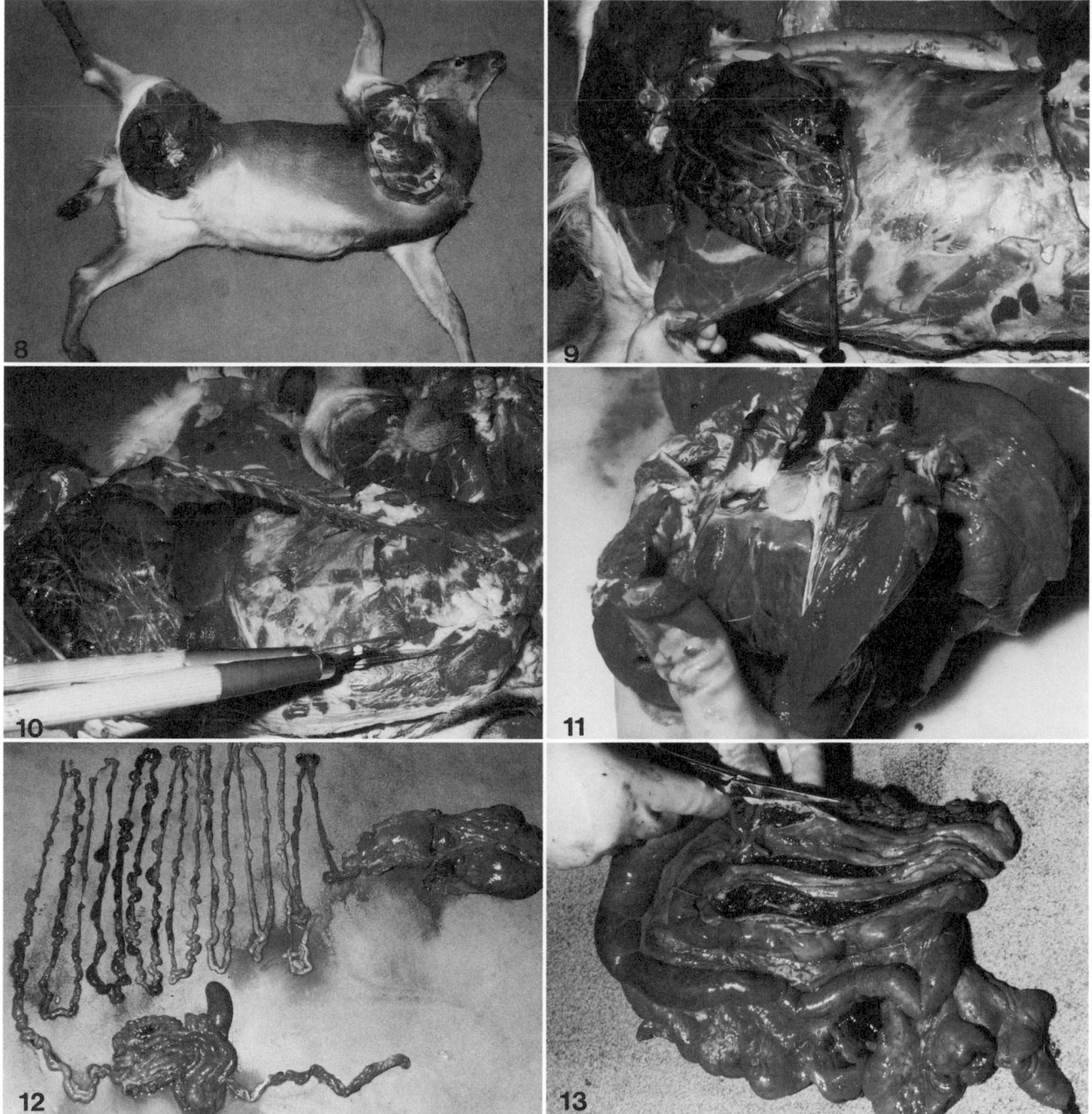

Fig. 8. Necropsy of deer. Attachments of right legs have been cut and the legs have been reflected dorsally. *Fig. 9.* Small incision made into abdominal cavity exposes abdominal viscera. A flap is made by extending both ends of incision ventrally. Flap serves to hold abdominal viscera in place and can serve to retain any excessive fluid present in abdominal cavity. *Fig. 10.* Incision is extended anteriorly and rib cage is removed with pruning shears. *Fig. 11.* Examine internal surface (endocardium) of the heart; examine valves and their attachments. *Fig. 12.* Intestinal tract is removed from carcass and laid out for examination. *Fig. 13.* Intestinal tract should be opened with scissors.

off the aorta and caudal vena cava before severing them. The esophagus also is cut at the level of the diaphragm in the thoracic cavity. If the stomach or rumen is full, consider tying off the esophagus before severing it to prevent stomach contents from filling the thoracic cavity.

Open the esophagus lengthwise with scissors and examine the mucosa (internal lining). Examine the lungs visually. *Gently* feel the lung for abnormalities. Lungs are the one organ for which more can be determined by touch than by sight. Postmortem discolorations of lung tissue are common. Locate the thyroids along the trachea just behind the larynx and examine them before proceeding with the respiratory system. Open larynx, trachea, and bronchi with scissors or knife through to the smallest bronchi possible. Lungworms may be present in airways. Cut cross sections of lung, especially noting any exudate. Samples for microscopic study should be taken before too much manipulation has occurred and preferably from areas that have not been handled. Because of the nonuniformity of lungs and lung lesions, multiple samples should be taken from different segments of the lungs.

Examine the membrane-like sac enclosing the heart (pericardium) and note the amount of fluid in the sac. Samples of excessive pericardial fluid should be saved for further laboratory studies. Open the pericardium and examine the external surface (epicardium) of the heart. The heart is opened in the order of blood flow. Starting with right atrium and ventricle, follow through to the pulmonary outflow via the pulmonary valve. Examine the internal surface (endocardium) including the valves and their attachments (Fig. 11). Turn the heart over and open the left side from atrium through the ventricle and out the aorta. This method ensures that all structures are examined before they are destroyed by subsequent cuts. Samples saved for microscopic study should include papillary muscle and septum.

LIVER

Remove the liver by cutting its attachments to the diaphragm, GI tract, and kidneys. Care must be made not to cut the intestine. Before cutting the bile duct, note its entry into the small intestine. In species with a gallbladder (deer and several other species do not have one), gentle pressure on the bladder will distend the bile duct, confirming bile flow into the intestine is not obstructed.

Carefully examine both liver surfaces and all lobes in those species with a multilobulated liver. In species with a gallbladder, open and evert the bladder, noting the consistency of bile and the presence of any gallstones. Slice the liver as one would slice bread except avoid having the cut separate the slices from the organ itself. Examine the cut surfaces and collect samples as appropriate. Large flukes are a common finding within the liver tissue of white-tailed deer and other members of the Cervidae in some regions. Note the color and texture of the liver. Tough livers may indicate normal liver tissue has been replaced with fibrous connective tissue.

GASTROINTESTINAL (GI) TRACT

The GI tract is examined to a greater or lesser extent depending on completeness of the necropsy and purpose of the investigator. A complete intestinal exam is time consuming, but it may be rewarding because some diseases/parasites affect specific segments of the intestine. Complete examination may be necessary for the enumeration of parasite loads. Alternatively, the GI tract may be spot-checked along its length; when this method is used, the entire GI tract should be examined from its external surface (serosa), and abnormally appearing areas should be cut open and examined. Complete examinations are best accomplished by separating intact the GI tract and stretching it out in normal configuration. This method organizes and greatly speeds up the internal examination, allows the examiner to know at any time where in the GI tract an observation is made, and keeps the entire organ system together (Fig. 12). Total area affected relative to GI length can be determined at a glance.

Regardless of whether a complete or spot-check examination of the intestine is conducted, the first step is to examine the adjacent mesenteric lymph nodes prior to removing the GI tract. These are located in the mesenteries attaching the small intestine to the dorsal abdominal cavity. Enlargement, discoloration, and excessive wetness upon cutting through this tissue are all indications that the intestine should be examined closely. If the intestines are to be strung out, grasp *any* loop of small intestine and cut the mesenteric attachments forward and backward as far as possible. This is the only step that is different between a complete and spot-check exam.

Locate the last part of the colon as it disappears into the pelvic cavity and cut across it. Rarely is it necessary to tie off the colon, because feces can be manipulated away from the site of the cut. Cut the mesentery attaching the colon. The entire GI tract now is held only by the root of the mesentery along the backbone and the attachments of the esophagus to the diaphragm. Remove the GI tract by cutting the mesentery at its root along the backbone between the kidneys and GI tract. In carnivores, other small mammals, or small ruminants, a direct cutting of the root can be done. In most ruminants, particularly large animals, removal of the GI tract is facilitated by placing the GI tract outside the body, *above* the animal, and cutting as the weight of the rumen places traction on the mesenteric root. The adrenals are located in front of the kidneys and towards the midline. If the GI tract is removed carefully, these organs will not be damaged. Remove the spleen.

The four stomachs of ruminants are opened lengthwise; the quantity and consistency of feed should be especially noted. The mucosal linings and functions of each compartment differ. If desired, the abomasum (glandular stomach) can be opened separately and washed into a bucket to enumerate parasites. Counts of abomasal parasites frequently are used to provide an index of herd health (particularly white-tailed deer in southeastern United States) (Davidson et al. 1981).

Examine the intestine by opening it lengthwise with scissors (Fig. 13). Characteristics of the intestinal contents (color, consistency, odor, quantity) are important and should be recorded. Microbiological samples are best taken by tying off a loop of unopened intestine and preserving the loop with its contents by refrigeration or freezing if necessary. Preservation by freezing is especially valuable if toxic chemicals are believed to be involved in the animal's death. Note and inspect the pancreas located adjacent to the duodenum (initial segment of the small intestine). The spiral colon (tightly coiled structure unique to ruminant colons) can be examined easily without uncoiling it. Conclude the GI examination by opening the remaining colon.

SPLEEN

After examining both surfaces, cut the spleen at intervals and examine the cut surfaces. Spleen is often saved as a routine tissue for later virological workup. Any changes in the color or consistency of the spleen should be recorded.

ADRENALS

These are located between the anterior ends of both kidneys. Remove the adrenals and inspect by cutting. Note the condition of the adrenal cortex and the adrenal medulla.

UROGENITAL SYSTEM

A complete examination requires exposure of the pelvic canal. This is accomplished by removing the hemipelvis closest to the examiner (the "up" side). The hemipelvis can be removed by cutting through the bones with pruning shears. Cutting soft tissue attachments and removing the severed bone will allow in-place visualization of the entire urogenital system.

In males, remove and examine the testicles individually. Then cut the testicles into transverse sections and examine the cut surfaces. Note any exuding fluids. The remainder of the urogenital system is removed en masse. The penis is cut from its attachments along the abdominal surface and extended as far back as possible. The kidneys are cut free of their attachments, including cutting through the renal artery but *not* the ureter. Gentle traction on the kidney with some cutting of attachments will expose the ureters and their attachment to the urinary bladder.

In females the ovaries are freed from their attachments to the body wall. All structures (ureters and urinary bladder, descending colon, and uterus in females) are gently retracted, and soft tissue attachments are cut back to the anus. Skin around the anus is cut and the urogenital tract is removed intact. The entire system can then be laid out in its normal configuration, allowing proper orientation during examination.

The kidneys are examined by cutting them longitudinally. Note the three well-delineated areas of the cortex, medulla, and pelvis. Samples for histology should be taken to incorporate all three areas. Examine the ureters externally. A urine sample can be taken by needle and syringe prior to opening the bladder. The bladder can be snipped open and everted as it is opened lengthwise. Continue the incision through the urethra, noting the prostate gland (males only, in some species). Note characteristics of the urine. Ovaries can be fixed whole or sectioned. The uterus and vagina are examined by opening lengthwise, as is the small piece of attached descending colon.

REMAINDER OF CARCASS

In ruminants, the carotid artery from the neck to the base of the head where the artery branches should be opened and examined carefully. This is the primary site for the arterial worm *Elaeophora schneideri*, a fairly common parasite of black-tailed deer and mule deer (Herman 1945) in which it usually causes no problems; however, it is a dangerous parasite to elk (Hibler and Adcock 1971) and white-tailed deer (Couvillion et al. 1985, 1986).

Disarticulate the head at its attachment to the spinal cord, remove the skin covering the top of the skull, and open the skull by any method that does not damage the brain and the membrane sheath covering it (meninges). In the field, the easiest and most convenient method, without special equipment, is to use a small saw to cut away the complete top of the skull, including the opening in the posterior portion of the skull where the spinal cord exits (foramen magnum). Examine the meninges. Remove the brain by cutting the ventral attachments with scissors as the head is held vertically, nose up. Allow the weight of the brain to pull the brain free as the attachments are cut. The brain is examined superficially, and either the whole brain or a longitudinal half brain can be fixed for histological exam.

If possible, the spinal cord should be examined. This is a time-consuming operation and often is done only in special circumstances. The cord is exposed by removal of the top portions of the vertebrae. This is accomplished by cutting the vertebral arches with bone shears, hatchet, or saw after most of the muscles around the vertebrae are removed.

Examine the carpal and stifle joints. First remove the skin surrounding these joints by making a longitudinal incision and reflecting the covering skin. Open the joint by cutting across it. Note the articular surfaces and the quantity, consistency, and color of joint fluid.

Skin the left side of the carcass, examining subcutaneous surfaces and musculature. Several incisions into the large muscle masses of the hind limb are useful to detect abnormalities in the muscle. The mammary gland in females should be palpated and inspected by sectioning. Abnormal-appearing fluids and tissue should be collected for microbiological assays.

SPECIAL CONSIDERATIONS

The techniques described above are applicable to ruminants and may need to be adapted due to unique anatomical features of some species and conditions encountered in the field. The following comments provide guidance to facilitate necropsy for some species groups and evaluation of observations made.

SMALL MAMMALS

Small mammals generally can be necropsied best while they are lying on their backs and attached by pins (or something similar) to the working surface. Exposure of the thoracic viscera is accomplished by cutting the ribs on both sides and removing most of the rib cage and breast bone (sternum). After in-place examination and removal of thoracic organs, the abdominal viscera (with the exception of urogenital tract) are removed en masse. The pelvic cavity can be examined best by cutting both sides of the pubis and ischium and removing the ventral portion of the pelvis. The urogenital tract is removed as described above.

CETACEA

All marine mammals have extensive modifications of limbs and external shape as adaptations to the marine environment. Generally these modifications do not affect postmortem examination. Cetacea are examined best in left lateral position, although the larger cetacea (whales) are examined in any manner possible. These larger species require specialized equipment such as large fleshing knives, winches, chainsaws, and axes. The investigator may need to walk into the body cavity to collect specimens.

The subcutaneous fat layer (blubber) is thick in cetacea and contains a dense collagen matrix. This means knives and other cutting instruments will require *frequent* sharpening. Because of this blubber layer, the skin and blubber are best peeled off the carcass in contrast to the routine skinning of other mammal carcasses.

Anatomically the abdominal cavity is relatively small, which may cause some problems with organ observation. The digestive system is functionally compatible with that of monogastric mammals; however, the system differs structurally. Cetacea have three "stomachs." The initial portion is divided into glandular and nonglandular portions similar to that of horses. The second stomach is merely a connecting tube between the first and third. The third stomach is the pyloric region and contains primarily mucus-secreting glands. In addition, the duodenum has a dilatation often mistaken for a "fourth" stomach. The intestine is extra long, posteriorly displaced, and compressed in the small abdominal space. The liver is large; cetacea lack gallbladders.

The upper respiratory system is radically modified and terminates in external nares on top of the head. In toothed whales (Odontoceti), a single naris or blow hole is present, whereas baleen whales (Mysticeti) have twin blow holes. The larynx is an elongated tubular structure oriented in a vertical position and ventral to the entrance of the blow hole. The esophageal openings are located on both sides of the larynx. The short, wide trachea has heavy cartilaginous rings and branches a short distance beyond the larynx. The lungs contain cartilaginous rings to the terminal airways and are covered with a thick pleura. Although inflammation of the pleura (pleuritis) is common in cetacea, the thick pleural lining often is mistaken for pleuritis.

The kidneys are extensively lobulated into individual segments. Each segment acts as an independent renal unit with cortex, medulla, and papilla. The urinary bladder is small and muscular.

Cetacea contain proportionally more blood than other mammals, in some species upwards of 25% body weight. This results in the apparent congestion of many organs at necropsy. This congestion should not be confused, however, with the normal dark red/purple color of marine mammal muscle that is due to extensive myoglobin storage. Extensive modifications of the vascular system have resulted in the development of rete systems. Retes are interwoven, anastomosing beds of blood vessels that contain variable amounts of blood and serve as a mechanism for blood redistribution during diving or as countercurrent heat exchange units. The retes of the thoracic aorta and sublumbar region (deep to the kidneys) are most noticeable. Other anatomic modifications of cetacea include a relatively small, dark, round spleen; intra-abdominal testes; and large adrenals.

PINNIPEDIA

Pinnipeds are not nearly as modified internally as cetacea. Necropsies are best conducted with the animal lying on its back. During necropsy the ribs should be cut on both sides of the sternum widely enough to expose the thoracic cavity. Blubber layers vary in thickness, depending on species; however, the substance is similar to subcutaneous fat of other species and lacks the dense collagen of cetacea. Internal anatomy is similar to that of other carnivorous animals.

The stomach is single chambered and the intestines are very long. The respiratory system is modified only in the extent of the cartilage in airways in some species and the early branched (i.e., short) trachea in eared seals (Otariidae).

Pinnipeds have a variety of minor modifications to handle pressure and volume changes in blood during diving. For example, in earless or true seals (Phocidae), the dilatation of the aorta at the aortic arch can be mistaken for an aneurysm. The caudal vena cava and hepatic sinus often are dilated and engorged with blood, particularly in elephant seals. This blood engorgement ("diving response"), inappropriately stimulated, has been implicated in the death of northern elephant seals kept in captivity.

SEA OTTERS

Although unique in their fur and limb adaptations for the marine environment, sea otters can be handled identically to other mammals except necropsies often are done with the otter lying on its back. Internal anatomy is similar to that of other carnivores.

Bird Necropsy

Avian anatomy is unique, and necropsies are conducted quite differently from those for mammals. External examination is similar, but wings should be spread to examine feathering. After external examination, birds are placed on their back with the head pointed away from the examiner or the head pointed to the left (right-handed individual) as personal preference dictates. A soapy solution applied to the ventral feathers helps keep them out of the necropsy field and from flying about the necropsy area. An incision, just off the midline, is made through the skin (Fig. 14). The incision need not be large, extending only 5–6 cm in a mallard-size bird to 8–10 cm in a large, goose-size bird. *Blunt* dissection (primarily thumb and fingers) is used to easily remove the skin from the side of the carcass on which the incision is made, including the area around the thighs and lower legs. A scalpel can be used to cut the skin from the keel, and the other half of the carcass can be skinned by blunt dissection. At this time carcass condition can be assessed by examining the amount and color of subcutaneous fat and the degree of wasting, if any, of the major breast muscles (pectorals). The skin incision should be extended with scissors along the neck to the bill, exposing the structures of the neck (Fig. 15).

INITIAL OPENING

A scalpel is used to incise the breast muscle along the sternum of the carcass. Cut the muscle tissue from the thoracic inlet to the abdomen by cutting *parallel* to the keel and deep to the underlying bone. Grasp the muscle by placing the fingers and thumb in the cut surfaces, slightly elevate the sternum, and carefully cut the attachment of the abdominal muscles to the sternum. This cut is *perpendicular* to the keel. Continuing to hold the muscles, insert poultry shears to cut the sternum along the same cuts already made in the breast muscle. The tenden-

→

Fig. 14. A ventral incision is made just lateral to sternal keel and skin is removed by blunt dissection. *Fig. 15.* Ventral skin has been removed to expose ventral abdominal wall, pectoral muscles, and structures of the neck. *Fig. 16.* Sternum and ventral abdominal musculature have been removed and internal organs are visible. Note condition of heart and liver. *Fig. 17.* Note the occurrence of or the lack of fat along the outer surfaces of the coronary arteries of the heart. *Fig. 18.* An excised liver cut into transverse sections for examination. *Fig. 19.* Method of opening a gizzard. Gizzard is held in one hand; with scissors cut through the gizzard along its lumen. *Fig. 20.* An opened gizzard. Note color of its internal koilin layer and check for presence of ingested shot.

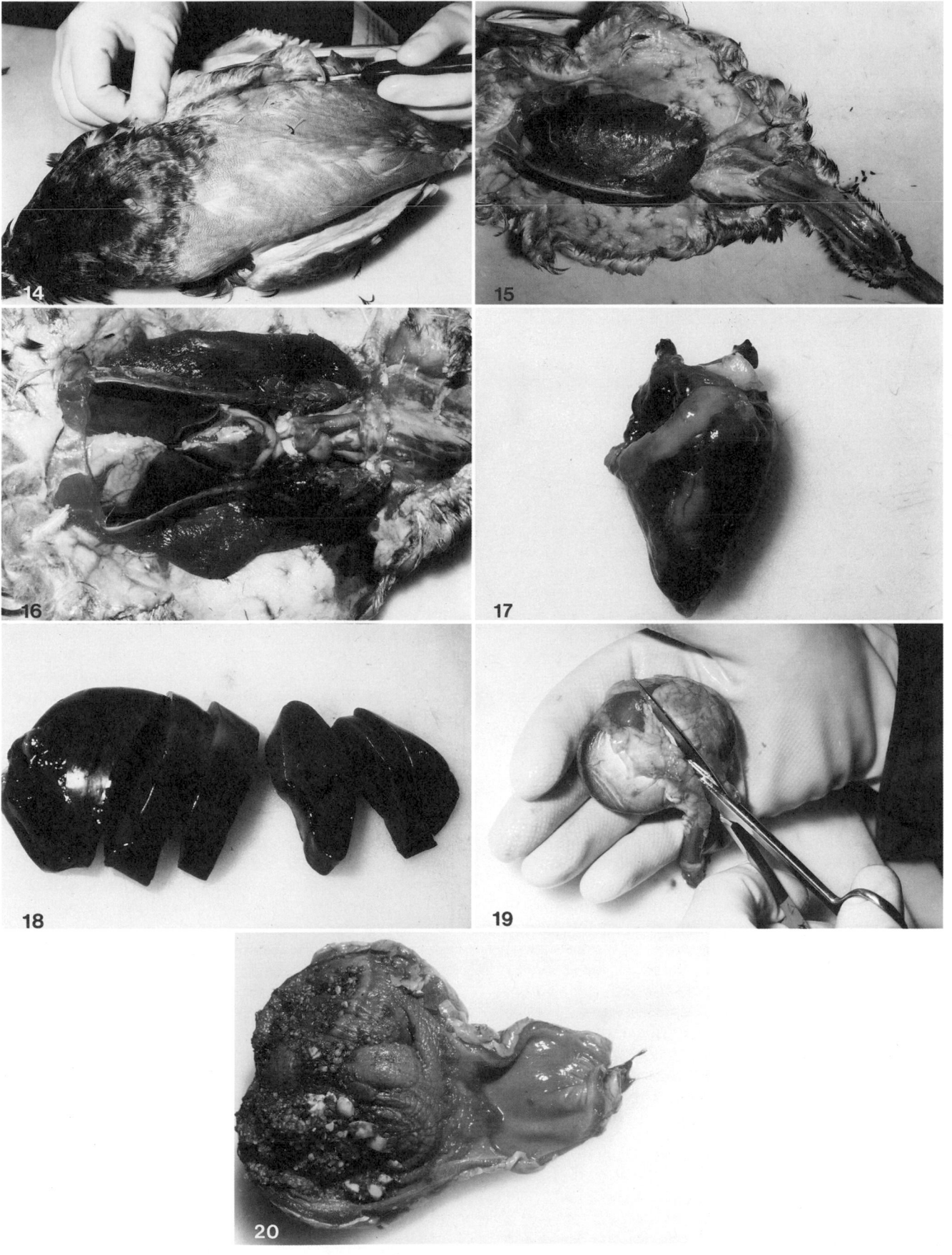

cy is to cut too far toward the backbone, thus cutting into the wing or having difficulty removing the sternum. If this procedure is done properly, the bones cut easily and two distinct, large bones are cut on each side near the thoracic inlet. Maintain traction on the sternum by holding the muscles and separate the sternum from the underlying viscera by gently using a scalpel. Birds have air sacs rather than a diaphragm. The thoracic air sacs should be examined during and after removal of overlying tissues.

Using small forceps, grasp the abdominal musculature where it was cut from the sternum and remove the muscle with scissors to expose the abdominal area. The internal organs are now visible in place (Fig. 16). Varying amounts of fat will overlie the abdominal viscera depending on time of year, age of bird, and other factors. Before proceeding further, inspect the pericardium and air sacs closely. The air sacs should be transparent and thin. Note the presence of any growths or thickened appearance of this tissue and collect samples of abnormalities for bacteriology or mycology. As with mammals, many investigators prefer to remove all internal organs before proceeding with the examination of individual organs. For purposes of this chapter, removal and examination of organs are described together.

Heart

Because of its anatomical location in the bird relative to other organs, the heart is best removed first and independently of other thoracic organs. Gentle traction at the apex of the pericardium while the large blood vessels are sectioned as far from the base of the heart as possible will remove the heart. The pericardium should be examined on both surfaces, and any excessive fluid in the pericardium should be recorded. Note whether deposits of fat are present along the outer surfaces of the coronary arteries (Fig. 17). Some fat should be present at this site, and its absence could reflect a chronic debilitating infection, chronic lead poisoning, or organochlorine poisoning.

The avian heart has four chambers. The right ventricular wall is very thin and the right ventricular space forms a thin crescent around half of the thick muscular wall of the left ventricle. The left ventricular space is roughly a tapered cylinder. After inspection of the epicardium, coronary fat, and general heart shape, make a cross section of the heart through the middle of the ventricles. The sectioned surface permits examination of the relative size and shape of the ventricles and a view of the myocardium of the left and right ventricular free walls and the septum. A second slice a few millimeters towards the base will produce a disc-like section containing all these features for histopathology. Using scissors, cut the ventricular walls toward the base into the atria, pulmonary vessels, and aorta. This provides the opportunity to examine the endocardium, valves, and inner lining of attached blood vessels. Blood collected from the heart is useful for testing for avian botulism and should be frozen as soon after collection as possible and kept frozen until tested.

LIVER

Remove the liver by gently cutting attachments to the gizzard and intestine. This is best accomplished with sharp scissors. Careful dissection where the gallbladder is adjacent to the duodenum will prevent cutting the gall-

bladder and the intestine. Examine the liver and note its color and consistency, and whether it appears enlarged or swollen. Examine the surface of the liver and note any spots or other lesions (abnormalities) present. Make several slices into the tissue and examine the cut surfaces (Fig. 18). The liver is an important tissue for laboratory assays for diseases caused by bacteria, viruses, and some toxicants. Measure the gallbladder, note its color, then open it and note the amount and consistency of the bile.

GASTROINTESTINAL (GI) TRACT

Cut the attachments of the gizzard (e.g., waterfowl) or muscular stomach (e.g., raptors and fish-eating birds) to the abdominal wall. Grasp the intestines and gizzard, cutting the air sac and mesenteric attachments to these structures with scissors. Maintain traction on the GI tract, and cut through the esophagus just above the proventriculus or stomach. Gentle traction towards the legs will remove the entire GI tract, with the exception of the distal colon, cloaca, and spleen. After examining the attachments of the urogenital system to the cloaca, apply traction to the GI tract and use a sharp scalpel to cut the remaining attachments around the cloaca. If this procedure is done properly, the colon, cloaca, and bursa of Fabricius will remain intact and the entire GI tract, including some skin and feathers around the cloaca, will be removed. Gentle traction and some minor cutting of the mesentery are used to straighten the intestine with the exception of the intestinal loop containing the pancreas.

Free the proventriculus and gizzard from the intestinal tract and open them with scissors. Examine the anterior (esophageal-proventricular) and the posterior (proventricular-ventricular) ends of the proventriculus for abnormalities; in some species of waterfowl, parasites can produce ulcers at these sites. Examine the walls of the proventriculus for parasite-induced changes. Open the gizzard by cutting along its lumen (Fig. 19); examine any gizzard contents, noting their odor, color, consistency, and nature (Fig. 20). If poisons such as organophosphates, carbamates, or strychnine are believed to have been involved, the gizzard contents should be saved intact and immediately frozen for later chemical analysis. Otherwise, the contents should be placed in a pan and examined for ingested lead shot, steel shot, or other foreign bodies. The color and condition of the "keratinized" or "horny" layer of the gizzard should be recorded; e.g., in lead-poisoned waterfowl the gizzard lining often is stained dark green by regurgitated bile. Excessive buildup of the "horny" layer generally indicates that the gizzard is not functioning properly as a grinding organ. Gizzard muscles should be cut through and examined, and any abnormal coloring should be recorded (lack of vitamin E often will cause a pale streaking of the gizzard muscles). Open the intestinal tract lengthwise; examine it for parasitic worms, color and consistency of items in its lumen, and abnormalities (e.g., ulcers, hemorrhage). Lymphoid tissues (source of antibodies) are sometimes visible along the intestinal tract, lying just beneath the mucosa. These lymphoid tissues are often disc-like, occasionally ring-like, and scattered along the intestine; they should be examined carefully for hemorrhages, ulcers, or other abnormal changes. The bursa of Fabricius, located near the cloaca, is a major site of immune cells in young birds and should be opened and inspected. Finally, the cloaca is opened lengthwise so its walls can be examined.

SPLEEN

Spleen shape varies among the various orders of birds, ranging from almost spherical (psittacines) to oval, to almost triangular (anseriforms), to long, narrow, and elliptical (passerines and columbiforms). The spleen is best examined before visceral tissues and organs are disrupted because its small size in some birds and varying shape in different species can make this organ difficult to locate once the viscera have been disturbed. This organ is important for virus isolation attempts, and abnormally large spleens often indicate the presence of infectious diseases such as ornithosis or the parasitic disease known as leucocytozoan infection.

UROGENITAL SYSTEM

The ovaries and testes of birds undergo marked variation in sizes in conjunction with reproductive cycles. For example, the testis of an adult mallard in the nonbreeding/post-breeding phase will be 8–15 mm long × 2–3 mm in diameter. During the height of the breeding cycle, the testis will be 40–50 mm long × 30–35 mm in diameter. In many species of birds, including the northern bobwhite, the testis of an inactive, nonbreeding adult bird is black and measures 4–6 mm × 2–3 mm; however, in the breeding male the testis is pale cream-colored and is two to three times larger in size. This pronounced change in color reflects the fact that the number of melanocytes (black pigment-bearing cells) in the outer layer of the testis is fixed, and as the testis enlarges it becomes lighter, grayer, and finally dull cream-white. Similarly, the ovary and the oviduct in the female undergo marked enlargement and regression in conjunction with the breeding cycle.

Ovary and testis should be measured before they are removed from the carcass, and both usually are examined only externally unless abnormalities are seen. During the breeding season, the testes should be cut through to observe the presence or absence of seminal fluid. The oviduct of laying females should be opened longitudinally so the inner layer can be examined. Occasionally a ruptured yolk or egg is present within the oviduct, and this can be an occasional cause of death among laying hens, mallard ducks, and shorebirds, and probably other species of birds. Infections of the oviduct are observed frequently.

The adrenals are located at the anterior end of the kidneys near the attachment of the testes or ovary and should be examined at this time. Adrenals tend to be well developed in chicks and ducklings and appear large relative to the size of the animal.

The kidneys lie within a bony recess of the pelvic structure, and in healthy birds they usually are bordered by deposits of fat. Normal kidneys usually are deep red-purple to purple and the renal network of tubules is not identifiable without a microscope. Kidneys should be examined first while in place, then they should be removed and cut through transversely in several sites. These kidney cross sections should be examined carefully for the presence of small white to yellow spots that indicate accumulation of urates in the tubules. Occasionally, the kidneys are pale pinkish-white with red spots; cutting through the kidneys will reveal that the kidney tubules are packed with urates. In these latter instances, usually the ureters also are packed with urates.

Even kidneys appearing normal to the eye ("grossly normal") are extremely valuable specimens for later chemical analytical studies, particularly if one of the heavy metals is suspected as a possible cause of death. In this situation, some kidney should be saved for analysis.

RESPIRATORY SYSTEM

Cut the bill or beak with scissors or poultry shears to expose the internal nares and the oral cavity, extending the incision down the trachea to the bronchi and lungs. The adjacent esophagus usually is opened at this time. Examine the posterior portion of the oral cavity for abnormalities (e.g., the large "cheesy" [caseous] mass often seen in fatal cases of trichomoniasis in mourning doves) and check the tongue and surrounding floor of the mouth for lesions. The nasal passages can be opened by scissors or (for bigger birds) poultry shears by cutting into the external nares and extending the cut toward the brain case. This incision will expose the nasal turbinates. Examine these carefully for nasal mites, parasitic nematodes, and, in waterfowl, for nasal leeches.

Remove the lungs by blunt dissection (the back end of the scalpel handle works well) along the rib cage and cutting of attachments to the backbone. Then cut transversely through the excised lungs.

NASAL GLANDS

Incise the skin over the skull and bluntly dissect the skin from the head by pulling it to either side. The nasal glands will be recognized as light-brownish-purple to purple crescents of glandular tissue located in the eye socket above each eye. These nasal glands are most highly developed in marine birds (e.g., albatrosses, petrels, pelicans, gulls) in which the gland serves a major role in the excretion of electrolytes (Na^+). Poorly developed glands in birds using brackish or coastal marshes, or in marine birds, should be examined microscopically by a pathologist. Subclinical poisoning by certain organophosphate pesticides will significantly interfere or block the functioning of these glands.

NERVOUS SYSTEM

In most small birds the top of the exposed skull can be removed easily with the pointed end of scissors or poultry shears. Large birds (e.g., golden eagles) may require the use of a reciprocating bone saw. Alternatively, small bone forceps can be used to chip away the skull. Examine the outer surface of the brain prior to its removal. Turn the head to a vertical position (bill or beak up); use scissors to free the attachments of the brain to the bottom of the braincase and the brain will fall gently to the table.

Decisions on how to handle the brain after its removal from the skull are based on how good a history of the die-off the biologist has obtained prior to beginning the necropsy. If the history suggests that organochlorines (such as DDT, dieldrin, or PCBs), organophosphates, or carbamates might have been involved, the entire brain should be saved and frozen for later chemical analyses. If a parasitic or infectious disease is suspected, half the brain should be fixed in 10% formalin and half frozen for microbiological, parasitological, or virological studies. Outbreaks of salmonellosis at bird feeders involving house sparrows, pine siskins, and American tree sparrows often include individual birds exhibiting peculiar neurological behavior (e.g., tumbling, inability to right itself, variable amount of "paralysis"). Microscopic examination and bacteriological tests of the brain often will reveal a large "abscess-like" structure in the brain of such afflicted birds.

If the history is incomplete, it is usually best to divide the brain sagittally into left and right halves, then freeze one half and fix the other in formalin as discussed above. The spinal cord can be removed by cutting the dorsal vertebral arches from a skinned bird with bone forceps.

The pelvic nerves (located off the spinal cord deep to the kidneys, radiating towards the legs), ischiatic nerve (located along the caudal side of the femur), and vagus (located in the neck along the jugular vein) should be identified and examined individually. The brachial nerves, large nerves in the axillary area, should be saved for microscopic study if it is believed that mercury poisoning might be involved.

SKELETAL SYSTEM

Mineralization of bone can be assessed subjectively by bending and breaking a metatarsal bone. A distinct crack should be heard on breaking. A metatarsal bone that bends without breaking suggests poor mineralization. The knee joints and hock joints should be examined, and if excessive fluid is found within the joint, some of the fluid should be saved for microbiological studies. Brown pelicans dying of erysipelas infection in California had swollen, fluid-filled leg joints (Nat. Wildl. Health Res. Cent., unpubl. data).

LITERATURE CITED

AMERICAN VETERINARY MEDICAL ASSOCIATION PANEL ON EUTHANASIA. 1993. Report of the American Veterinary Medical Association Panel on Euthanasia. J. Am. Vet. Med. Assoc. 202:229–249.

BELL, J. F. 1980. Tularemia. Pages 161–193 in J. H. Steele, ed. Handbook series in zoonoses. Section A, Vol. II. CRC Press, Boca Raton, Fla.

COUVILLION, C. E., W. R. DAVIDSON, AND V. F. NETTLES. 1985. Distribution of *Elaeophora schneideri* in white-tailed deer in the southeastern United States, 1962-1983. J. Wildl. Dis. 21:451–453.

——, V. F. NETTLES, C. A. RAWLINGS, AND R. L. JOYNER. 1986. Elaeophorosis in white-tailed deer: pathology of the natural disease and its relation to oral food impactions. J. Wildl. Dis. 22:214–223.

DAVIDSON, W. R., F. A. HAYES, V. F. NETTLES, AND F. E. KELLOGG. 1981. Diseases and parasites of white-tailed deer. Tall Timbers Res. Stn. Misc. Publ. 7. 458pp.

FRANSON, J. C. 1987. Speciment shipment. Pages 13–20 in M. Friend, ed. Field guide to wildlife diseases. U.S. Fish Wildl. Serv. Resour. Publ. 167.

FRIEND, M., EDITOR. 1987. Field guide to wildlife diseases. U.S. Fish Wildl. Serv. Resour. Publ. 167. 225pp.

HERMAN, C. M. 1945. Some worm parasites of deer in California. Calif. Fish Game 31:201–208.

HIBLER, C. P., AND J. L. ADCOCK. 1971. Elaeophorosis. Pages 263–278 in J. W. Davis and R. Anderson, eds. Parasitic diseases of wild mammals. Iowa State Univ. Press, Ames.

KARSTAD, L., J. SPALATIN, AND R. P. HANSON. 1957. Application of the paper disc technique to the collection of whole blood and serum samples. J. Infect. Dis. 101:295–299.

KOLMER, J. A., E. H. SPAULDING, AND H. W. ROBINSON. 1951. Approved laboratory technic. Fifth ed. Appleton-Century-Crofts, Inc., New York, N.Y. 1180pp.

KORSCHGEN, C. E., H. C. GIBBS, AND H. L. MENDALL. 1978. Avian cholera in eider ducks in Maine. J. Wildl. Dis. 14:254–258.

LEIBOVITZ, L. 1969. The comparative pathology of duck plague in wild Anseriformes. J. Wildl. Manage. 33:294–303.

OLSEN, P. F. 1975. Tularemia. Pages 191–223 in W. T. Hubbert, W. F. McCulloch, and P. R. Schnurrenberger, eds. Diseases transmitted from animals to man. Sixth ed. Charles C Thomas, Springfield, Ill.

ROSEN, M. N. 1971. Avian cholera. Pages 59–74 in J. W. Davis, R. C. Anderson, L. Karstad, and D. O. Trainer, eds. Infectious and parasitic diseases of wild birds. Iowa State Univ. Press, Ames.

WOBESER, G. 1981. Diseases of wild waterfowl. Plenum Press, New York, N.Y. 300pp.

Appendix I. Supplemental Reading
General

DAVIS, J. W., AND R. C. ANDERSON, EDITORS. 1971. Parasitic diseases of wild mammals. Iowa State Univ. Press, Ames. 364pp.

——, ——, L. H. KARSTAD, AND D. O. TRAINER, EDITORS. 1971. Infectious and parasitic diseases of wild birds. Iowa State Univ. Press, Ames. 344pp.

——, L. H. KARSTAD, AND D. O. TRAINER, EDITORS. 1981. Infectious diseases of wild mammals. Second ed. Iowa State Univ. Press, Ames. 446pp.

EKLUND, M. W., AND V. R. DOWELL, JR., EDITORS. 1987. Avian botulism—an international perspective. Charles C Thomas, Springfield, Ill. 405pp.

HOFF, G. L., AND J. W. DAVIS, EDITORS. 1982. Noninfectious diseases of wildlife. Iowa State Univ. Press, Ames. 174pp.

HUBBERT, W. T., W. F. McCULLOCH, AND P. R. SCHNURRENBERGER, EDITORS. 1975. Diseases transmitted from animals to man. Sixth ed. Charles C Thomas, Springfield, Ill. 1206pp.

JENSEN, W. I., AND C. S. WILLIAMS. 1964. Botulism and fowl cholera. Pages 333–341 in J. P. Linduska and A. L. Nelson, eds. Waterfowl tomorrow. U.S. Dep. Inter., Washington, D.C. 770pp.

PAGE, L. A., EDITOR. 1976. Wildlife diseases—proceedings of third international wildlife disease conference, Munich, 1975. Plenum Press, New York, N.Y. 686pp.

By Region
Alaska

DIETERICH, R. A. 1981. Alaskan wildlife diseases. Univ. Alaska Inst. Arctic Biol., Fairbanks. 524pp.

Rocky Mountains

ADRIAN, W. J., EDITOR. 1981. Manual of the common wildlife diseases in Colorado. Colo. Div. Wildl., Ft. Collins. 139pp.

——, EDITOR. 1992. Wildlife forensic field manual. Colo. Div. Wildl., Ft. Collins. 179pp.

THORNE, E. T., EDITOR. 1981. Diseases of wildlife in Wyoming. Wyo. Game and Fish Dep., Cheyenne. 353pp.

Southeastern U.S.

DAVIDSON, W. R., AND V. F. NETTLES. 1988. Field manual of wildlife diseases in the southeastern United States. Southeast. Coop. Wildl. Dis. Stud., Athens, Ga. 309pp.

FORRESTER, D. J. 1992. Parasites and diseases of wild mammals in Florida. Univ. Florida Press, Gainesville. 459pp.

Canada

FYVIE, A. 1964. Manual of common parasites, diseases and anomalies of wildlife in Ontario. Ont. Dep. Lands For., Maple. 100pp.

14

SAMPLING INVERTEBRATES IN AQUATIC AND TERRESTRIAL HABITATS

Henry R. Murkin, Dale A. Wrubleski, and Frederic A. Reid

INTRODUCTION

Advances in our understanding of the importance of invertebrates to many fish and wildlife species have led to increased need for information on this diverse group of organisms. The function of invertebrates as an essential food resource has been well documented (e.g., Waters 1969, Scott and Crossman 1973, Johnsgard 1981, Murkin and Batt 1987). Their roles in detrital processing and nutrient cycling (Edwards et al. 1970, Petersen and Luxton 1982, Merritt et al. 1984a), as pests and disease vectors (Merritt and Newson 1978, Wobeser 1981), and in regulation of plant community dynamics (Schowalter et al. 1986, Crawley 1989) also have attracted growing attention.

Practical and efficient research on invertebrates requires a basic understanding of the principles involved in the design of any biological study and of factors that are specific to studies dealing with invertebrates. Invertebrates pose special sampling problems because of their small size, different life stages, mobility, high reproductive rates, patchy spatial distributions, and the diversity of habitats that they occupy (Waters and Resh 1979).

Poorly designed programs for sampling invertebrates can result in wasted time and resources. Commonly discarded during the inventory of storerooms of fish and wildlife agencies and university departments are old, un-

sorted invertebrate samples collected as part of various habitat studies. This usually arises from a lack of understanding regarding the effort required to process the samples. In other instances, many hours are spent processing invertebrate samples only to find that the resulting data cannot be used in statistical tests and comparisons.

The objectives of this chapter are to provide general guidelines and background information to assist in the design and implementation of invertebrate studies in wildlife research programs. We will focus primarily on monitoring invertebrate abundance or biomass. For more specialized techniques such as mark and recapture, nearest neighbor, sequential, or removal sampling, see Southwood (1978); for determination of secondary productivity see Petrusewicz and Macfadyen (1970), Phillipson (1971), and Downing and Rigler (1984).

SETTING OBJECTIVES OF A SAMPLING PROGRAM

The initial consideration and probably the most important step in any research or other wildlife investigation is clearly stating the study objectives (Fig. 1, Elliott 1977, and see Chapter 1). These may range from general to detailed; however, they should be clearly identified, because they will guide every step in the design and implementation of the study (Green 1979). If the objective is simply

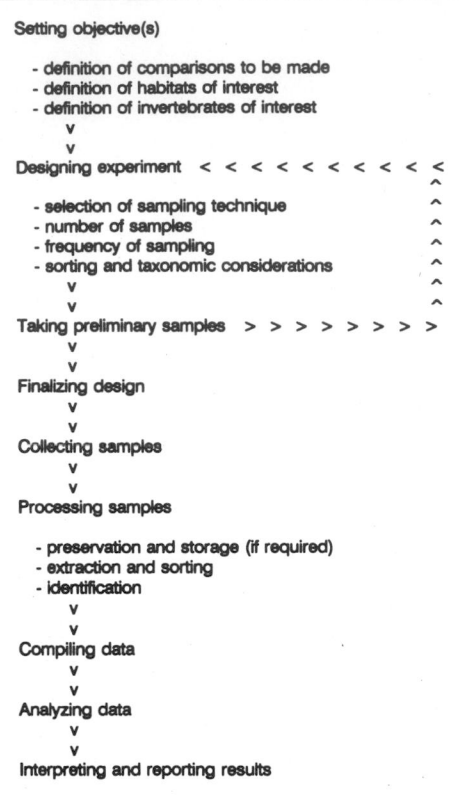

```
Setting objective(s)

  - definition of comparisons to be made
  - definition of habitats of interest
  - definition of invertebrates of interest
       v
       v
Designing experiment  < < < < < < < < < <
                                         ^
  - selection of sampling technique       ^
  - number of samples                     ^
  - frequency of sampling                 ^
  - sorting and taxonomic considerations  ^
       v                                  ^
       v                                  ^
Taking preliminary samples  > > > > > > > >
       v
       v
Finalizing design
       v
Collecting samples
       v
       v
Processing samples

  - preservation and storage (if required)
  - extraction and sorting
  - identification
       v
       v
Compiling data
       v
       v
Analyzing data
       v
Interpreting and reporting results
```

Fig. 1.   Chronology of steps in an invertebrate sampling program.

to determine whether invertebrates are present within a particular habitat, nothing more is required than a walk or paddle through the area of interest or a few qualitative samples. As the objective becomes more complicated, so will the design of the sampling program. For example, assessing the response of an invertebrate community to an experimental manipulation within a particular habitat or series of habitats over a number of years is a complex task. The study objectives also will determine the type and precision of data required, number of samples, sample type, size of individual samples, where to sample, frequency and duration of sampling, sample processing requirements, and identification needs.

The initial step is to state the study objective as a question in common-sense terms (Green 1979). For example, does fire affect invertebrate densities in tallgrass prairie? Then rephrase the question as a statement: fire affects invertebrate densities in tallgrass prairie. The precision of the statement is increased by including information about the actual factors or variables of interest. Are all invertebrates of interest or only a particular species or group of species? Which habitats are to be considered, tallgrass prairie in general or a particular subhabitat within the prairie? For example, the above statement could be modified to reflect specific interests as follows: fire reduces grasshopper abundance in tallgrass prairie adjacent to agricultural land. Next, formulate the null hypothesis: fire does not reduce grasshopper abundance in tallgrass prairie adjacent to agricultural land. By carefully defining and re-

fining the study objectives, one will be better prepared to design a study to address specific needs.

It is important to consider the resources available in defining objectives. Setting objectives that require an unrealistic amount of time is a common trap. Setting realistic objectives involves knowledge of the effort needed to carry out the proposed study design. This knowledge can be gained by consultation with individuals familiar with invertebrate sampling and by preliminary sampling.

DESIGNING A SAMPLING PROGRAM
Selection of Sample Sites

The study objectives will determine location of the sample sites. Selection of sample sites must be either completely random or randomly located within strata of relatively homogeneous habitat types (Elliott 1977, and see Chapter 1). Systematic sampling based on random placement of the initial site is also appropriate. Placing sample sites in typical or representative sites is not random sampling.

Prior knowledge of invertebrate distribution and abundance may permit stratified sampling within the study area. If the area of interest contains several habitats, samples should be stratified by habitat type (Green 1979). An important component of the variability between samples will be the inherent difference between sampling sites. For example, suppose the objective is to compare benthic invertebrate abundances between a wetland treated with insecticide and an untreated control. If the study areas consist of a combination of open water interspersed with dense cattail stands, establishing random sample sites across the study areas without regard for the different habitat types present would be inefficient. The error variation within each wetland would be inflated by differences in abundance of benthic invertebrates between open-water habitats and cattail stands (Murkin and Kadlec 1986, Wrubleski and Rosenberg 1990), and therefore the ratio of error among (between treated and control wetlands) to error within would be reduced. If information is not available from other studies in similar habitats, preliminary sampling will provide some initial indication of differences in abundance of macroinvertebrates among habitat types present in the study areas. Because distribution and abundance of invertebrates are affected by habitat structure and complexity, stratifying sampling by habitat type even without prior information would be appropriate. If analysis indicates no difference between habitats, the data can be pooled for the final analyses.

Timing and Frequency of Sampling

The timing and frequency of sampling will be determined by the study objectives, the invertebrates of interest, and the field logistics involved (Fig. 1). Many invertebrates show marked seasonal changes in abundance and habitat use, which necessitates sampling at different times throughout the year. Frequency of sampling depends on the sampling device chosen and life-cycle duration of the invertebrates. Some sampling devices sample continuously and can operate over an entire sampling season (e.g., emergence traps, pitfall traps), whereas other devices require regular visits to collect samples (e.g., grabs, sweep nets). Invertebrates with short life cycles will require frequent sampling or they could be missed entirely. A sampling regime must have a sampling frequency that is shorter than the life cycle of the invertebrate group of interest. Weekly or biweekly sampling periods are common in studies of invertebrate communities.

Diel activity patterns of many invertebrate species vary greatly (e.g., Elliott 1970, Costa and Crossley 1991), so timing of sampling during the day also requires consideration. Weather has a marked effect on diel activity and microhabitat selection. Many terrestrial invertebrates move to the upper, more exposed parts of vegetation during warm weather and down closer to the soil during cold, damp periods. This behavior influences the efficiency of many sampling devices, such as terrestrial sweep-net samples (Hughes 1955, Saugstad et al. 1967). Sampling should be postponed during inclement weather (Strickland 1961).

Selecting a Sampling Device

A variety of samplers is available for invertebrates (e.g., Martin 1977, Southwood 1978, Downing and Rigler 1984, Merritt et al. 1984b). The one selected will depend on the invertebrate group of interest, its life-cycle characteristics, the habitat or habitats to be sampled, and the precision required (Fig. 1, Resh 1979). No single device samples all invertebrate groups or habitats efficiently. A combination of samplers usually is required to sample the entire invertebrate community or to follow a single invertebrate species over its life cycle (Malley and Reynolds 1979). In addition, most samplers have associated biases that must be considered before use (Southwood 1978, Resh 1979). For example, one invertebrate species may be more mobile than another and able to avoid a particular sampler. As a result, the more mobile species would not be as well represented in the sample.

When a sampling device is selected, it is important to consider whether absolute (i.e., number and biomass per unit area or volume of habitat) or relative (index) measures of abundance are required (Spence 1980, Murkin et al. 1983). Devices that sample a known area or volume can be used to develop absolute abundance estimates that then can be converted to more common units such as number of individuals per square meter. This type of data is required for comparisons with other studies. If comparisons between treatments or changes over time are of interest, a simple relative measure of abundance may be sufficient (Eberhardt 1978, Spence 1980). The decision is important because many sampling techniques cannot be used to produce absolute abundance estimates. Moreover, relative indices often are much easier to obtain than absolute values. For example, determining the number of beetles per unit area of habitat requires considerable time and effort. The beetles must be collected and separated from the vegetation and litter within the sampling unit. In contrast, a simple pitfall trap (see Selecting a Sampling Device [p. 32]) provides a relatively clean sample of beetles. However, because the actual area sampled by the trap is unknown, the resulting data can be considered only an index to beetle abundance. The tradeoff between reduced processing time and the use of an index must be considered by the researcher on the basis of the objectives of the study.

The sampling technique chosen must provide samples of consistent quality (Elliott 1977, Resh 1979, and see Chapter 1). This is a particularly important feature of operator-controlled sampling devices such as sweep nets or benthic grab samplers. These types of samplers have the potential for a great deal of variability that will lead to increased error variation (among samples within treatments) and, as discussed earlier, less chance of detecting among-treatment differences. For example, one individual may take longer or shorter strokes than other individuals in the sampling crew when sweep nets are used to obtain samples of insects from stands of grass. The amount of pressure applied to a pole-mounted benthic grab sampler will affect the depth to which it penetrates the substrate. These problems are reduced in studies in which all sampling is done by the same individual (Elliott 1977), as is often the situation in wildlife graduate studies. However, in long-term monitoring programs, several field technicians usually are involved within any single field season or over the years, thereby increasing the likelihood for operator variability (Murkin et al. 1983). Sampling devices that do not depend on operator-controlled motion (e.g., pitfall traps) reduce this problem.

Another important consideration when a sampling device is selected is the amount of sample processing required. Processing is by far the most time-consuming and tedious part of invertebrate research (Karlsson et al. 1976). Samplers that provide relatively clean samples, like the beetle pitfall trap above, are usually preferable to devices producing samples in which the invertebrates must be removed from large amounts of extraneous material. Once again the study objectives are important. For example, in a study of wetland invertebrates, should emergence traps (relatively clean samples) or benthic corers (samples require substantial processing) be used? If the objective of the study is to determine potential food items for bottom-feeding waterfowl, samples should be taken in the microhabitats where the birds are actually foraging, and this would require bottom core samples. However, if the objective of the study is to determine the impact of an herbicide or some other treatment in wetland habitats, selecting emerging insects to indicate treatment effects may be appropriate.

When the study objectives require samples that generate a considerable amount of processing, it may be possible, and in some cases advantageous, to reduce the size of individual samples (Downing 1979). For example, the number of earthworms per unit volume of soil recovered by hand sorting decreased with increasing size of the soil sample, indicating increased inefficiency with larger samples (Zicsi 1958). The actual size of the sample will be based on the density of invertebrates (Downing 1979, Morin 1985). For higher densities, smaller samples are normally sufficient. Preliminary sampling will provide necessary information on densities. It also will provide information on invertebrate distribution that may help to further reduce sample size. For example, if preliminary sampling indicates that most invertebrates in soil samples are within 5 cm of the surface, sampling below this depth is unnecessary. A cautionary note is that the smaller the sample collected, the more likely that rare species may be missed (Paterson and Fernando 1971). Selection of an extraction technique that speeds up separation of invertebrates from the extraneous material also should be con-

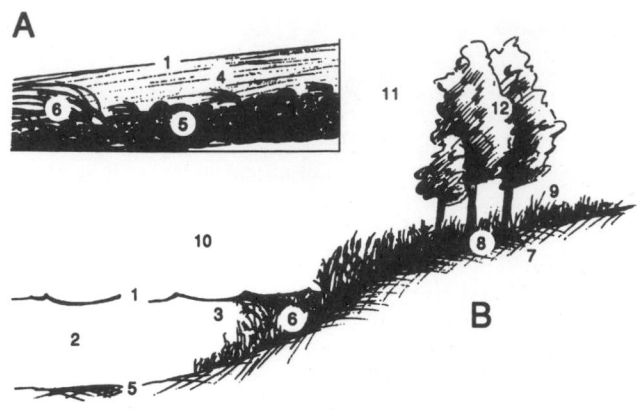

Fig. 2. Invertebrates and their habitats for which sampling methods are described. A. A lotic habitat. B. Lentic and terrestrial habitats. 1 = neuston; 2 = zooplankton; 3 = nekton; 4 = drift; 5 = benthos; 6 = epiphytic invertebrates; 7 = soil; 8 = litter; 9 = herbaceous vegetation; 10 = emerging insects; 11 = aerial; 12 = arboreal.

sidered, and it may reduce sorting times greatly (see PROCESSING SAMPLES [p. 360]).

Another point to consider is the size of the invertebrates under investigation. Many samplers, such as sweep nets, drift nets, or box-type samplers, use mesh or netting in their construction. In others, the contents of the sampler are passed through a sieve or series of sieves to separate the invertebrates from extraneous material. Small mesh sizes capture invertebrates that are smaller than may be required and increase sorting time because a great deal of extraneous material is retained. In contrast, large mesh sizes result in the loss of many small invertebrates (e.g., Zelt and Clifford 1972). The study objectives will determine the mesh size to be used in these types of samples.

Below we describe a variety of common devices for sampling invertebrates. We provide a short description and references for each, as well as alternative techniques and important review articles that may be helpful in the selection of appropriate sampling procedures.

SAMPLING AQUATIC INVERTEBRATES
General Considerations

A wide array of sampling devices is available for aquatic invertebrates (e.g., see Elliott and Tullett 1978, 1983, Rosenberg 1978, Downing and Rigler 1984, Merritt et al. 1984*b*, Klemm et al. 1990, Resh et al. 1990). When a sampling device is selected, attention must be paid to the life-history characteristics of the invertebrates involved and the habitats occupied (Malley and Reynolds 1979, Resh 1979). Benthic invertebrates (benthos) live within or on bottom substrates (e.g., sediments) and submersed surfaces (e.g., aquatic vegetation, moss, algae, wood, or rocks) (Fig. 2). Within the water column, invertebrates can be free-floating (zooplankton) or free-swimming (nekton), or in running waters they may drift with the current. Neustonic invertebrates live on the water surface, and some aquatic insect species emerge from the water surface as adults and become aerial (Fig. 2).

Bottom Substrates
STANDING WATERS

Benthos abundance in the bottom sediments normally is expressed as number of individuals or mass per unit area of bottom surface (e.g., number per square meter). In areas of standing water (lentic habitats), a sample of sediments, from a known surface area, is collected and the invertebrates are sorted and counted.

A variety of grab-type samplers has been devised for sampling benthic invertebrates in particular habitats or substrates (e.g., Flannagan 1970, Elliott and Drake 1981, Downing 1984). Probably the most common grab for sampling benthos in soft, unconsolidated sediments is the Ekman (Birge-Ekman) grab (Fig. 3A) or modifications of it (Blomqvist 1990). It is basically an open-ended box with spring-loaded jaws on the open end. The box is lowered into the sediments and then a trigger mechanism trips the jaws to enclose the sample. It is returned to the surface, the jaws are reopened, and the sample is removed. In shallow waters the grab can be mounted on the end of a pole, and a connecting rod within the handle is used to trip the jaws. In deeper water the grab is suspended from a rope or cable. Its weight (additional weights often are added) causes it to penetrate into the sediments. A sender weight is then attached to the main cable and dropped to trip the jaw mechanism. These samplers are readily available through scientific suppliers, sample a known area of sediment (area of box opening), are relatively simple to use, and can be used in a variety of water depths. However, they have several disadvantages (Downing 1984, Blomqvist 1990). The depth to which they penetrate can vary, depending on the composition of the substrate. Obstructions (e.g., rocks or roots) within the sediments can cause incomplete closure of the jaws and loss of the sample upon retrieval. They also tend to collect large samples that require considerable effort to process (Paterson and Fernando 1971, Karlsson et al. 1976).

Another common sampling device for benthic invertebrates is the core sampler (Fig. 3B–D). It consists of a tube that penetrates into the sediments to a known depth and then is removed with the sample held within the tube. The tube is constructed of metal or plastic depending on the composition of the sediments (Karlsson et al. 1976, Swanson 1978*c*, 1983). The lower end of the tube can be sharpened or fitted with a cutting blade to facilitate penetration into the sediments, particularly if roots are present (Swanson 1983). The top of the tube is fitted with a stopper that is left open as the tube enters the sediment. After the corer has penetrated to the desired depth in the sediments, the stopper is closed to create a vacuum that holds the sample in place while the tube is removed.

The sediment sample can be removed from the tube by use of a small hand pump (a rubber bulb with a one-way valve works well) attached to a one-holed stopper of the same diameter as the stopper on the corer (Swanson 1983). The regular stopper is replaced with the hand pump stopper and air is pumped into the top of the corer, forcing the sample out of the tube. A simple plunger device also can be used to remove the sample from the tube.

The corer can be hand-operated or mounted on a pole (Fig. 3B) for use in shallow water (Gale 1971, Swanson 1978*a*, 1983). It also can be suspended from a cable for sampling in deeper water (Fig. 3C); a sender device is used to close the stopper (Brinkhurst 1974). Multiple core samplers (Fig. 3D) are useful for obtaining replicate samples (Euliss et al. 1992), especially in deeper waters where considerable effort is required to obtain a sample (Flan-

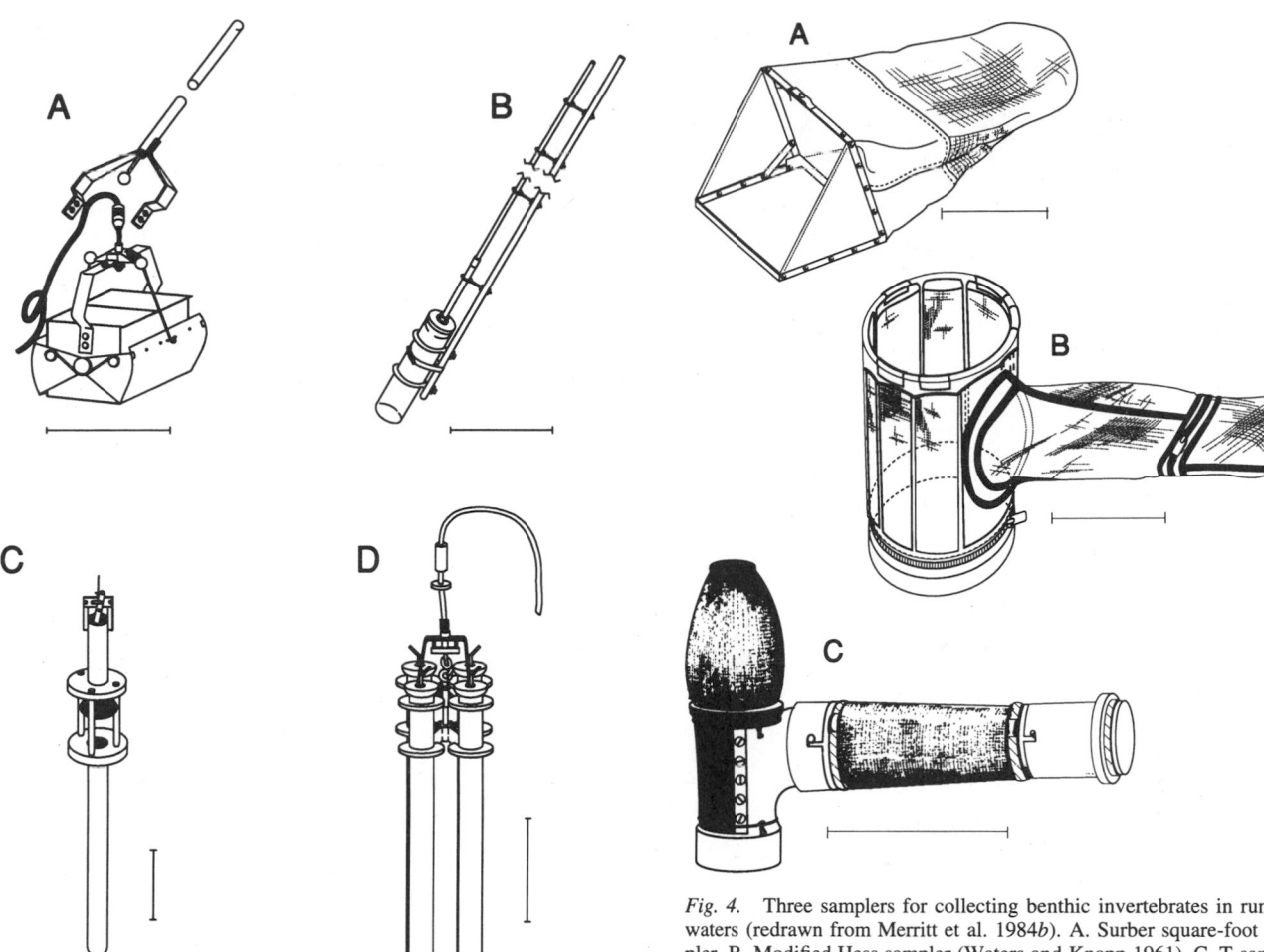

Fig. 3. Four samplers for collecting benthic invertebrates in lentic bottom sediments (redrawn from Merritt et al. 1984*b*). A. Ekman grab. B. Pole-mounted corer. C. Cable-mounted corer (Brinkhurst 1974). D. Multiple corer (Hamilton et al. 1970). Scale bar = 25 cm.

Fig. 4. Three samplers for collecting benthic invertebrates in running waters (redrawn from Merritt et al. 1984*b*). A. Surber square-foot sampler. B. Modified Hess sampler (Waters and Knapp 1961). C. T-sampler (Mackie and Bailey 1981). Scale bar = 25 cm.

nagan 1970, Hamilton et al. 1970, Milbrink and Wiederholm 1973). Corers do not function well in loose, unconsolidated sediments, sand, or gravel (Flannagan 1970).

An advantage of the core sampler is that tubes of different diameters can be used to suit invertebrate abundance. Corers used for benthic studies range in size from 3 cm^2 to 855 cm^2 (Downing 1984), but 10–40 cm^2 is probably the most common range of sizes. The higher the abundance of invertebrates, the smaller the diameter of tube that should be used (Downing 1979). A small-diameter tube also reduces the size of the sample and subsequent processing time. In addition, the corer can be modified to sample to a desired depth by mounting a bracket on the side of the tube to prevent it from sinking beyond the depth required. Most benthic invertebrates live within the upper 5–10 cm of the sediments (Mundie 1957, Downing 1984 and references therein), but this is dependent upon the type of sediments, time of the year, and invertebrates of interest (e.g., Nalepa and Robertson 1981).

RUNNING WATERS

Qualitative samples of benthic invertebrates can be obtained in shallow streams by kick sampling (Frost et al.

1971, Mackey et al. 1984, Storey et al. 1991). The investigator holds a long-handled pond net against the stream bottom and disturbs the substrate upstream of the net by kicking for a standardized period of time. The invertebrates are dislodged and carried into the net by the current. This technique is easy to use and relatively inexpensive, but it is not quantitative because the exact area from which the invertebrates are collected is unknown. Operator variability and the types of substrates sampled also will affect kick sampling efficiency (Pollard 1981, Mackey et al. 1984, Storey et al. 1991).

Quantitative sampling of stream (lotic) substrates often is done with the Surber sampler (Fig. 4A). The sampler is placed over an area of substrate with the net trailing downstream. The substrate within the frame of the sampler is disturbed, dislodging the invertebrates, which are swept into the collecting net by the current. Rocks and all other substrate materials are then scrubbed thoroughly within the sampler to remove tightly adhering invertebrates (Lavery and Costa 1972). Collection bottles can be added to the ends of the nets to simplify sample removal (e.g., Lane 1974). Problems with specimen loss because of outwash, and the limited water depth in which the Surber sampler can be used (e.g., Kroger 1972, Resh 1979), have contributed to the development of other samplers such as the Hess (Fig 4B) (Waters and Knapp 1961) and T-samplers (Fig. 4C) (Mackie and Bailey 1981, English 1987).

A **B**

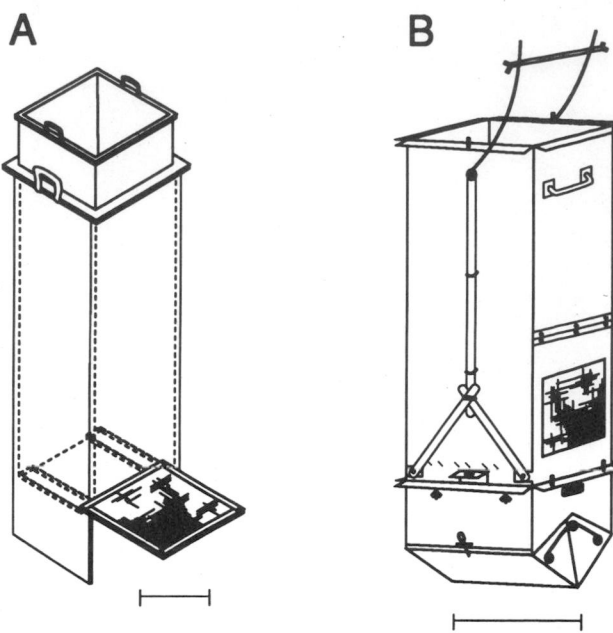

Fig. 5. Two box-type samplers for collecting aquatic vegetation and associated invertebrates. A. Gerking (1957) sampler (redrawn from Merritt et al. 1984*b*). B. Gates et al. (1987) sampler. Scale bar = 25 cm.

Lotic samplers require the current to carry invertebrates into the collection nets. In slow-moving streams or rivers, these samplers may not function properly. Coarse substrates in these habitats often preclude the use of grab or core samplers (Elliott and Drake 1981), therefore other means must be used to obtain benthic samples. Air-lift or pump-type samplers have been used in these situations (Drake and Elliott 1982, Boulton 1985, Brown et al. 1987).

Submersed Surfaces

AQUATIC VEGETATION

In standing water (lentic) habitats, common submersed surfaces for invertebrate colonization include aquatic plants. Invertebrates living on aquatic vegetation have been called macroperiphytonic fauna, phytophilous fauna, phytomacrobenthos, phytomacrofauna, and epiphytic invertebrates; the last two terms are the most commonly used, however. Sampling these habitats quantitatively is difficult, and resulting samples can require much time to process (Downing and Cyr 1985). An important initial consideration is whether invertebrate abundance is to be expressed per unit area of plant surface or per unit area of pond or lake bottom. Determining the surface area of submersed vegetation, especially those species with finely dissected leaves, can be laborious (see Harrod and Hall 1962, Cattaneo and Carignan 1983, and Brown and Manny 1985 for some methods). This will be necessary if the objectives of the study are to compare invertebrate populations on several species of aquatic plants. In most studies, however, invertebrate abundances are expressed simply as number or biomass per square meter of bottom area.

The simplest method of sampling epiphytic invertebrates is with a sweep net, but this method is considered to be too unreliable and is not recommended (Downing 1984). A variety of box or trap-type samplers (Fig. 5) is available (see review in Downing 1984, Downing and Cyr 1985, Gates et al. 1987). These devices enclose a known volume of water with the aquatic vegetation and associated invertebrates. The plants are cut off at or near the sediments and retrieved. Such samplers are cumbersome to operate, they are restricted to shallow water, and they usually generate large samples. Quadrat sampling by SCUBA is often more accurate than these kinds of samplers (Downing and Cyr 1985). This technique involves the gentle clipping of all plants from a known area and placing them in plastic or mesh bags.

Menzie (1980) and Downing (1986) described a two-stage approach to determining the number or biomass of epiphytic invertebrates. The first stage involves the collection of aquatic vegetation with its associated invertebrates. Much smaller samplers than described above can be used (Menzie 1980, Downing 1986). Number or biomass of invertebrates per unit of plant dry weight are determined and then used to develop a regression equation between macrophyte biomass and invertebrate abundances. In the second stage, quadrat samples are used to estimate macrophyte biomass per unit of bottom area. Multiplying the macrophyte densities by the invertebrate abundances per unit of macrophyte will yield an estimate of invertebrate abundance per unit of bottom area. This method is considered easier and more accurate than the box, trap, or quadrat methods described previously (Downing 1986).

GENERAL SUBMERSED SURFACES

Artificial substrates can be used to sample benthic invertebrates in lentic and lotic habitats. Instead of collecting a unit of natural habitat and its associated invertebrates, an artificial substrate designed to mimic invertebrate habitat (Rosenberg and Resh 1982) can be introduced into the habitat of interest. The substrate is colonized by invertebrates and is then retrieved after a predetermined period of time. Artificial substrates are used most commonly in situations where conventional sampling is difficult. Advantages include known surface areas, ease of use in a variety of habitats, and minimal sampling disturbance.

Two basic categories of artificial substrates are used: those made of natural habitat materials (called representative artificial substrates or RAS), and those that offer a standardized habitat (called standardized artificial substrates or SAS) that differs from the natural substrate (Rosenberg and Resh 1982).

Wire baskets filled with rocks (Fig. 6) are common RAS used in stream and river habitats. The rocks are first cleaned, their total surface area is determined (see McCreadie and Colbo 1991 for methods), and they are placed in a wire basket and submersed in the habitat under investigation. After a period of time (usually 2 weeks or more), the basket is retrieved, rocks are removed, and invertebrates are collected from the rocks. The same basket and rocks with their known surface area then can be reused. Leaf packets, pieces of wood, and other natural substrates also may be used as artificial substrates (e.g., Petersen and Cummins 1974, Voshell and Simmons 1977, Flannagan and Rosenberg 1982).

Multiplate samplers are commonly used SAS. The Hester-Dendy type of sampler (Fig. 6) is a stack of tempered hardboard (e.g., Masonite) pieces separated by spacers and held together by a central bolt. After retrieval of the stack, the central bolt is removed and the sampler is disassembled so the individual pieces can be searched for invertebrates. Some Chironomidae (Diptera) can use hardboard as a food source (Ferrington and Christiansen 1985). Other SAS samplers include baskets filled with spheres (e.g., porcelain balls), conservation webbing, artificial vegetation, and plastic strips (Flannagan and Rosenberg 1982).

When artificial substrates are used, an assumption is made that the invertebrate communities that develop on these substrates are similar to those found on natural substrates. This is not always true (e.g., Peckarsky 1984). In addition, the period between placement and retrieval must be sufficient to allow the invertebrates enough time to colonize the new substrate. Prior information and preliminary sampling will help in this regard. Rosenberg and Resh (1982) provided an in-depth review of the advantages and disadvantages of using artificial substrates.

Water Column

STANDING WATER

The general objective of sampling nekton and zooplankton in the water column (Fig. 2) is to count or remove the invertebrates from a known volume of water. Various tube and box-type samplers are used in shallow, lentic habitats. For example, Swanson (1978*b*) used a plastic, graduated cylinder with the lower end removed (Fig. 7A). The open cylinder is lowered through the water column, and the sample is secured by placing a stopper in the lower end of the cylinder at the desired depth. Other designs include tubes with closing door mechanisms (Gilbert and Ruber 1986). After collection, the invertebrates are removed by pouring the contents of the tube through a sieve. Bendell and McNicol (1987) used a dip net to capture enclosed invertebrates from a large tube sampler. Tube diameters can be varied depending on the abundance of the invertebrates present; smaller diameters are suitable for higher densities. Because the inside diameter of the tube and the depth of the water column sampled are known, the volume of water sampled can be calculated. This information can be used to determine the number or biomass of invertebrates per unit volume of water.

Commonly used in shallow lentic habitats are open-ended box-type samplers with a sliding door on the lower end (Fig. 7B). The box is lowered through the water column with the door open. At the desired depth the door is closed. The door and sides are fitted with screens to let the water drain as the sampler is removed from the water. The invertebrates are removed from the bottom screen. Box-type samplers designed for collecting epiphytic invertebrates (e.g., Gerking 1957, Gates et al. 1987) also can be used to sample invertebrates in the water column.

Probably the most common sampling device for invertebrates in shallow standing water is the sweep net (Fig. 7C, Whitman 1974, Kaminski and Murkin 1981, Murkin et al. 1983, Bendell and McNicol 1987). Unfortunately the ways in which sweep nets are used vary greatly, thereby making comparison of results difficult. Sweep nets are used most often to provide a relative index to invertebrate

Fig. 6. Two commonly used artificial substrates (redrawn from Merritt et al. 1984*b*). Top, rock-filled wire basket. Bottom, multiplate sampler. Scale bar = 10 cm.

abundance. The net is swept through the water column (often in a figure-eight pattern) for a set period of time or distance (e.g., Whitman 1974, Bendell and McNicol 1987). Because the actual volume of water sampled is unknown, absolute densities cannot be determined.

A modified sweep net can be used in shallow aquatic habitats to provide quantitative estimates of invertebrate abundances within the water column (Voigts 1976, Kaminski and Murkin 1981, Murkin et al. 1983). The net frame is bent at a 45° angle so the handle projects out of the water when the net frame lies flat on the bottom. To sample, one lowers the net frame with the net folded and places it flat on the bottom. It then is moved forward away

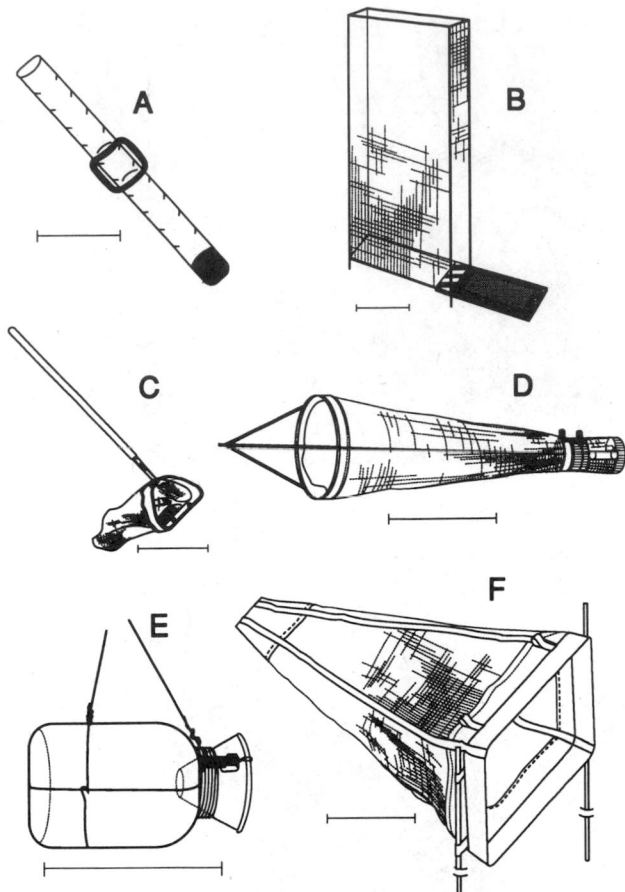

Fig. 7. Samplers for nekton, zooplankton, and drifting invertebrates. A. Swanson (1978*b*) water column sampler. B. Kaminski and Murkin (1981) water column sampler. C. D-frame sweep net (redrawn from Merritt et al. 1984*b*). D. Plankton tow net (redrawn from Merritt et al. 1984*b*). E. Aquatic activity trap (after Murkin et al. 1983). F. Drift net (redrawn from Merritt et al. 1984*b*). Scale bar = 25 cm.

from the area disturbed by the net's descent and lifted straight up through the water column to the surface. The water volume sampled can be calculated from the area of the net frame and the water depth. Kaminski and Murkin (1981) reported no differences between sweep net samples collected in this manner and a modified Gerking device (Fig. 7B). This technique is most suitable in areas with little or no vegetation. The size of the net frame will be determined by the density of invertebrates of interest, and the shape of the net frame will depend partly on the habitat sampled. Rectangular net frames are recommended for open-water sites, but any shape can be used.

In deeper habitats, a plankton net (Fig. 7D) can be used. The net can be towed behind a boat for a known distance, and the volume sampled is determined from the area of the net opening and the distance travelled. A vertical sample of the water column is obtained by lowering the apparatus to the bottom and hauling it straight up to the surface. The volume sampled then can be calculated from the area of the net opening and water depth. George and Owen (1978) described a long, flexible tube sampler that can be used in deep water, much like the tube samplers for shallow water described above. Lasenby and Sherman (1991) described a bottom-closing drop net that samples the entire water column. Box-type samplers, such as the

Juday and Schindler-Patalas traps (Schindler 1969), as well as various bottle samplers (e.g., Van Dorn sampler), can be used to sample invertebrates at discrete depths. An extensive review of zooplankton sampling methods was provided by de Bernardi (1984).

All of the above techniques require operator-controlled motion of the sampler. All have problems of varying degrees if a substantial amount of vegetation is in the water column. Also, mobile invertebrates may be able to avoid these devices. Activity traps are stationary samplers designed to overcome many of these problems. They are not susceptible to operator bias, they produce samples relatively free of extraneous material, and they sample over long periods, thus incorporating invertebrate diel activity patterns (Murkin et al. 1983, Hilsenhoff and Tracy 1985). An example is the funnel trap (Fig. 7E) that is suspended in the water column to sample over a 24-hour period (Whitman 1974, Swanson 1978*c,* Murkin et al. 1983). Orientation of the trap in the water column will be determined by study objectives (e.g., Swanson 1978*c,* Aiken and Roughley 1985, Hilsenhoff and Tracy 1985). As with most techniques, activity traps have several disadvantages (Murkin et al. 1983). The samples obtained provide only relative abundances (i.e., the actual area or volume of water sampled is unknown). The data may be biased with respect to certain species of predaceous invertebrates that may be attracted to the traps. Fish may enter the bottles and affect the numbers of invertebrates collected. Finally, activity traps require one more trip to the sampling site, a trip to set it, and another to retrieve it.

RUNNING WATER

Invertebrate drift in streams and rivers is often an important consideration in fisheries studies (e.g., Waters 1969, Jenkins et al. 1970, Elliott 1973). In slowly moving water, the techniques for sampling invertebrates in standing water described above may be suitable; however, as current speed increases, different techniques must be used. The usual procedure is to use a stationary sampler such as a drift net (Fig. 7F) and determine the volume of water flowing through it over a period of time (Waters 1969, Elliott 1970). In shallow running water, a single net that extends from the substrate to the water surface can be used. In deeper water, a series of stacked nets may be required. The study objectives and preliminary sampling will determine whether the entire water column must be sampled. Using the current speed, the time elapsed, and the area of the net opening will permit the number or biomass of invertebrates per unit volume of water to be determined. Samples usually are collected over a 24-hour period to incorporate invertebrate diel drift patterns (Elliott 1977, Brittain and Eikeland 1988); however, the net may have to be emptied at shorter intervals depending on the objectives of the study, the flow rate of the stream, the abundance of invertebrates, and the amount of litter in the flowing water. The litter caught by the nets eventually will block water flow (Resh 1979), which will affect the sampling efficiency of the net, increase net drag, and possibly cause the net to be pulled out by the current.

Mesh size used in a drift net can significantly affect the results obtained (Slack et al. 1991). Large mesh sizes (e.g., >500 μm) are often used to prevent net clogging, but they underestimate drift by small invertebrates. These organisms can be sampled with nested nets (Slack et al. 1991) or pump sampling (Armitage 1978, Williams 1985). Elliott (1970), Allan and Russek (1985), and Brittain and Eikeland (1988) reviewed drift-sampling techniques.

Water Surface

NEUSTON

Neuston (sometimes referred to as pleuston or epineuston) include such groups as Gerridae (water striders), Collembola (springtails), and some spiders and adult Diptera (Fig. 2). The information of interest for this group of invertebrates is the number of individuals or biomass per unit area of water surface. For small-sized individuals, this will simply require placing a floating quadrat of known area on the water surface and counting or collecting the invertebrates within it (Spence 1980). Sampling larger, less numerous invertebrates requires designating a larger area of water surface and counting or collecting the individuals of interest. In a comparison of sweep net and quadrat counts, Spence (1980) generated regression equations that enabled him to use sweep net collections to estimate absolute population densities for water striders. Floating sticky traps and pan traps also can be used to obtain relative estimates of neustonic invertebrate abundance (Deonier 1972).

EMERGING INSECTS

Many insects spend the early stages of their life cycles in the aquatic environment and then emerge into the terrestrial environment as adults for mating and dispersal (Fig. 2). Numbers and biomass of emerging insects represent the cumulative effects of growth and mortality within the aquatic habitat.

Insects that have emerged can be sampled by sweep netting the surrounding vegetation or by using light traps. However, the individuals collected by these techniques cannot be related to a particular unit of habitat. Quantitative estimates of emergence can be obtained by intercepting insects in emergence traps either on their way to the water surface or after they have emerged (Fig. 8). The diversity of habitats occupied by aquatic, immature insects is great, so no single emergence trap is suitable for all habitats. All emergence trap designs should protect the insects collected from wind, waves, predation, extremes of temperature, and deterioration after death. In addition, a trap should let as much light through as possible so that insects approaching the surface will not avoid the trap (Daniel et al. 1985). Davies (1984) provided an excellent review of sampling aquatic insects upon emergence. Two basic designs of emergence traps are used in standing-water (lentic) habitats: floating and submersed. Floating traps (Fig. 8A) enclose an area of the water surface (usually 0.25–0.5 m^2) inside a tent or cage supported by a frame of wood or metal on a floating base. The traps are tied to anchors or stakes depending on the water depth. Clear plastic and light-colored mesh are used in construction to keep the traps as transparent as possible. Clear plastic coverings or aprons made of polyethylene film can be used to protect the trap and sample from wind and rain. Some mesh covering is required to allow ventilation within the trap to prevent condensation, which will trap smaller insects. Mesh fabrics used should have maximum open-

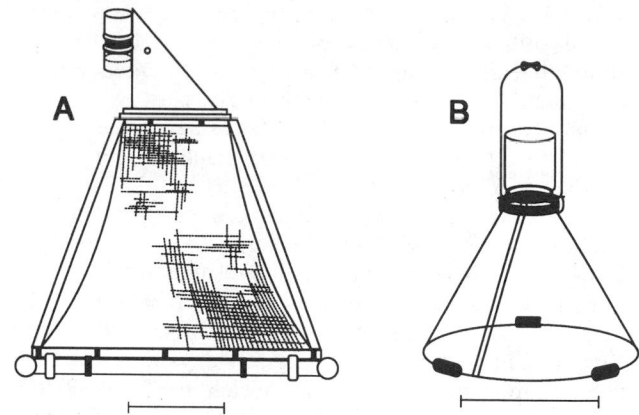

Fig. 8. Emergence traps. A. Floating trap (Wrubleski 1984). B. Submersed trap (redrawn from Davies 1984). Scale bar = 25 cm.

ings of 250 μm to retain the smallest insects (Davies 1984). Insects are collected by lifting the traps with the base covered and removing the insects manually. Aspiration of insects can lead to allergic reactions and is not recommended. Small, battery-powered vacuum devices, such as a modified Dustbuster (Marshall 1982), can be used instead. Some trap designs have a recloseable opening (sleeve, zipper, or Velcro) in the mesh to permit removal of the insects (Davies 1984).

Because manual removal of the insects is a time-consuming and labor-intensive process, many floating-trap designs incorporate a collecting jar containing preservative that collects and preserves the emerging insects (e.g., Fig 8A). These traps can be left for long periods (about 1 week) between visits and can be serviced quickly. Other emergence traps use removable plates coated with a sticky material (e.g., Tanglefoot, Boltac) to capture the insects (Mason and Sublette 1971, Street and Titmus 1979), but they are messy to use. The sticky material must be dissolved to remove the insects (Murphy 1985).

Several factors must be considered when floating emergence traps are used. If the area is susceptible to wave action, the trap must be strong enough to withstand the waves and protect the sample. The larger and more sturdy the trap, the more difficult it is to handle and make less obvious to emerging insects. In areas with emergent vegetation, the trap must fit over the vegetation. In shallow habitats, traps must be moved periodically so they do not affect the environment beneath the trap and, as a result, insect emergence. This can be solved by rotating the trap around a central anchor pole. The trap floats should not be constructed of materials that are susceptible to insect burrowing (Wrubleski and Rosenberg 1984). It is also important to regularly clean all parts of the trap in contact with water, because insects will colonize these parts and add to the emergence collected. PVC pipe with 90° elbow corners is relatively lightweight, durable, easily cleaned, and not susceptible to burrowing.

Emergence traps submersed within the water column (Fig. 8B) are less susceptible to weather and the condensation problems associated with floating traps. Insect emergence is restricted to the small air space inside the removable bottle at the top of the trap. The sample is collected simply by removing and capping the sample bottle while it is held below the water surface. A new bottle with a suitable air pocket is then attached to the trap. Although earlier models of submersed emergence traps were made from a variety of materials, plastics are now commonly used (Davies 1984). The size of the trap is limited to 0.01–0.25 m^2 because of handling constraints and the capacity of the sample bottle to hold insects. Disadvantages of submersed emergence traps include: (1) a requirement for water without emergent vegetation, (2) a build-up of algae on the trap, (3) some insects (e.g., Ephemeroptera) cannot complete development within the small air space of the collection jar (Davies 1984), and (4) a need to visit the traps every few days to remove the insects before they decompose. Welch et al. (1988) developed a submersed emergence trap that contains a preservative and can be left for periods of up to 18 days between visits.

Many of the emergence traps used in lentic habitats also can be used in running waters (lotic) if current speeds are not too great. A primary consideration for sampling emergence in running waters is knowing the habitats from which the insects originate. Floating traps (e.g., LeSage and Harrison 1979) will capture insects emerging from the substrate beneath the trap and from the drift. Traps that enclose an area of the substrate (e.g., Hamilton 1969) will capture only those insects originating from beneath the trap. Mundie (1964) described an emergence trap designed to sample emergence from the drift. In fast running water or streams subject to fluctuating water levels, floating traps are recommended.

SAMPLING TERRESTRIAL INVERTEBRATES
General Considerations

A variety of sampling techniques is available to sample terrestrial invertebrates (e.g., Petrusewicz and Macfadyen 1970, Martin 1977, Southwood 1978). Many methods and sampling strategies have been developed and refined as part of crop or forest pest management programs (e.g., Strickland 1961, Waters and Resh 1979). The following describes some common sampling devices used in each of the terrestrial habitats shown in Fig. 2.

Soil, Litter, and Soil Surface

SOIL

Invertebrates perform many important functions in the soil (e.g., Crossley 1977, Petersen and Luxton 1982, Spence 1985, Edwards et al. 1988). In habitats such as tallgrass prairie, biomass of invertebrates below ground can be 2–10 times greater than biomass above ground (Seastedt 1984). Soil invertebrates are also an important food resource for many wildlife species (e.g., Bengtson et al. 1976).

Soil samples usually are taken with a simple core sampler when absolute density estimates are required. More elaborate designs such as a split corer (O'Connor 1957) are available if vertical distribution of the invertebrates within the soil is of interest. Most soil invertebrates are found within the top 5–10 cm (Petersen and Luxton 1982), and some may require special sampling techniques. For example, soil expellants such as dilute formaldehyde or formalin have been used in studies of earthworm populations (Raw 1959, Satchell 1971, Bengtson et al. 1976).

Thorough reviews of soil sampling techniques were provided by Macfadyen (1962), Phillipson (1971), Southwood (1978), and Edwards (1991).

Most sorting procedures for soil samples involve live extraction (see Live Extraction Techniques [p. 39]) because manual sorting is tedious and labor intensive and usually misses small, cryptic species (Petersen and Luxton 1982). The soil sample must be treated with care to ensure survival of the invertebrates. Sample compaction during collection may affect the invertebrates and reduce extraction efficiencies. Samples must be kept cool during transport and storage prior to extraction. If samples are not to be extracted immediately, they can be safely stored at 5 C for up to a week (Edwards and Fletcher 1971). Core samples also should be kept intact during collection, transport, and storage to enhance extraction efficiency (Macfadyen 1962).

LITTER

Litter is the material on the soil surface composed mainly of dead plants and their shed organs (Medwecka-Kornas 1971). Invertebrates within litter can be collected with the same core samplers used to obtain soil samples (e.g., Seastedt and Crossley 1981). Densities normally are expressed as number or biomass per unit weight of litter. A more common technique is the use of litter bags (Crossley and Hoglund 1962, Wiegert 1974, Schowalter and Sabin 1991). Known amounts of litter are put in preweighed mesh bags and placed in the appropriate habitat. After set periods of time, the bags are retrieved and invertebrates are extracted from the samples. Live invertebrates in litter samples can be extracted with the Berlese-Tullgren funnel methods described below.

SOIL SURFACE

Beetles (Coleoptera) and spiders (Arachnida) are important components of the diverse invertebrate community living on the soil surface. Quadrat sampling with a steel frame or other similar device can be used to determine number or biomass of soil-surface invertebrates. The quadrat is placed quickly on the surface before mobile invertebrates are able to escape, the area within is thoroughly inspected, and all invertebrates are collected. This method provides a good measure of absolute abundance of surface-dwelling invertebrates, but it is extremely time consuming and labor intensive and can be destructive to the habitat.

The use of pitfall traps is an alternative to quadrat sampling (Fig. 9). These traps consist of a container sunk in the ground with the open end flush with the soil surface. Invertebrates (particularly spiders and beetles) fall into the traps and are unable to escape. Traps can be modified in many ways, including the use of various types of containers, rain roofs, double-container designs for ease of emptying, drain holes, ramps, aprons, barriers, and preservatives (e.g., Luff 1975, Morrill 1975, Houseweart et al. 1979, Durkis and Reeves 1982, Bostanian et al. 1983, Epstein and Kulman 1984, Waage 1985). Pitfall traps are inexpensive, require little labor to set and operate, provide a clean sample, are independent of operator error, and function continuously over a 24-hour period. Traps also can be baited for the capture of specific invertebrate groups (Newton and Peck 1975, Hunt and Raffa 1989).

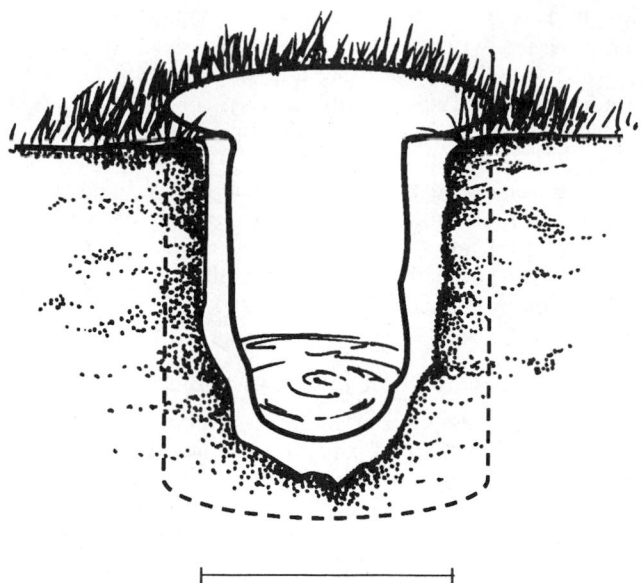

Fig. 9. Pitfall trap for sampling soil-surface invertebrates. Scale bar = 10 cm.

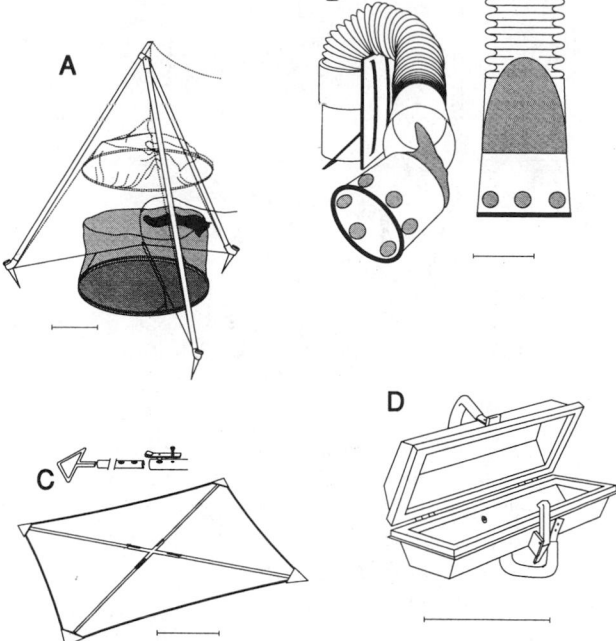

Fig. 10. Sampling devices for invertebrates on herbaceous vegetation and trees. A. Turnbull and Nicholls (1966) drop trap. B. Backpack version of the D-vac (Dietrick 1961). C. Beating sheet (redrawn from Martin 1977). D. Dempster (1961) box trap. Scale bar = 25 cm.

Considerable debate continues regarding the use of pitfall traps for providing accurate population estimates of soil-dwelling invertebrates (Greenslade 1964, Gist and Crossley 1973, Southwood 1978, Baars 1979, Desender and Maelfait 1986). These traps are essentially activity traps and, as such, preferentially collect active fauna. Activity can be affected by weather, habitat, and life cycle of the invertebrates (Greenslade 1964, Chiverton 1984). Some invertebrates avoid these traps (Halsall and Wratten 1988) or are able to escape from them. Killing agents and preservatives reduce escapes but may affect the attractiveness or repulsiveness of the traps to some groups (Luff 1968, Greenslade and Greenslade 1971, but also see Waage 1985). Ethylene glycol (antifreeze) is a commonly used preservative, but it should be used with caution because it is a major cause of poisoning in animals (Hall 1991). Although pitfall traps have limitations, they are still a useful means of sampling soil-surface invertebrates.

Herbaceous Vegetation

Terrestrial vegetation is a heterogeneous habitat that changes over the growing season, making accurate sampling difficult (Southwood 1978). Many sampling devices are available, but most provide only a relative measure of invertebrate abundance. Specialized methods have been developed for specific invertebrates or parts of plants (i.e., flowers, leaves, stems, or branches).

Quadrat or whole-plant samples provide the most accurate estimates of invertebrate numbers or biomass on plants; however, these methods are extremely time consuming and tedious. Part or all of a plant is isolated, usually within a tube, box, or net (e.g., Turnbull 1966, Onsager 1977). All invertebrates within the enclosure then are captured. A chemical agent can be used to kill or stun the invertebrates (Schotzko and O'Keeffe 1986, 1989). A variety of vacuum devices has been described to simplify the collection of invertebrates from within the enclosures (Johnson et al. 1957, Turnbull 1966, Marshall 1982).

Placing enclosures without disturbing the vegetation and the invertebrates is difficult. Disturbance can be reduced by a drop trap (Fig. 10A) (Turnbull and Nicholls 1966), which is placed over the vegetation 24 hours prior to sampling. This allows time for the invertebrates to redistribute themselves after the initial disturbance. A spring-operated device, triggered some distance from the trap, drops a screened enclosure onto the vegetation. This method works best in grass <20 cm tall (Turnbull and Nicholls 1966).

The D-vac sampler (Fig. 10B) combines quadrat sampling and a vacuum collector into a single apparatus (Dietrick et al. 1959, Dietrick 1961). The collection head isolates a unit of vegetation and the vacuum collects the enclosed invertebrates, which are retained in a cloth bag in the collection head. Air speed in the collection head must be around 100 km/h if high rates of extraction are to be obtained (Southwood 1978). Ellington et al. (1984) developed a large, self-propelled vacuum-sampling device.

Sweep nets are probably the most commonly used sampling device for collecting invertebrates on vegetation (Southwood 1978). They are easy to use and provide a large amount of material with little effort. A strong net bag, frame, and handle must be used to withstand continual sweeping through the vegetation. A common procedure is to take a predetermined number of sweeps through the vegetation while one walks along a transect in the habitat of interest. Sweep-net efficiency varies with the structure of the plant community being sampled and weather conditions. As discussed earlier, weather affects invertebrate activity and position on the plants (Hughes 1955, Saugstad et al. 1967, Cherry et al. 1977). Further, no two operators are likely to collect sweep samples in

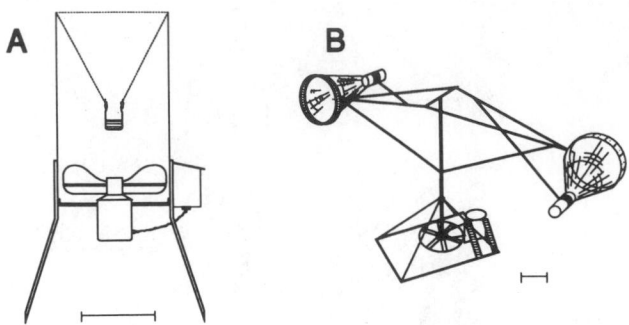

Fig. 11. Sampling devices for flying insects. A. Suction trap (Johnson and Taylor 1955; redrawn from Southwood 1978). B. Whirligig or rotary sweep net (after Juillet 1963). Scale bar = 25 cm.

the same manner, thereby introducing additional variation in the samples collected. The data provided usually are considered an index to invertebrate abundance. Tonkyn (1980) provided a formula that permits sweep-net data to be reported as number caught per unit volume of vegetation sampled.

Trees

Sampling in trees is difficult because of accessibility. Many different techniques can be used, but most provide only relative measures of invertebrate abundance. Beating is a method commonly used in forest insect surveys (Harris et al. 1972). The branches of a tree are struck with a pole or bar, dislodging the invertebrates, which fall onto a canvas sheet (Fig. 10C) or into an aluminum funnel held under the tree (Bostanian and Herne 1980, Herms et al. 1990). This technique is easy to use, rapid, and inexpensive, but the proportion of invertebrates dislodged is unknown. Canopy fogging or spraying with a knock-down pesticide operates on a similar principle (e.g., Martin 1966, Southwood et al. 1982). The D-vac sampler also can be used to provide a relative measure of invertebrate populations on trees (Dietrick 1961, Herms et al. 1990).

Invertebrates on trees can be sampled quantitatively by isolating branches and removing the invertebrates (e.g., Gibb and Betts 1963). Box-type traps (Fig. 10D, Dempster 1961) or nets with closures (Schowalter et al. 1981, Blanton 1990, Costa and Crossley 1991) have been used to trap arboreal invertebrates. After the invertebrates are trapped, they are stunned with CO_2 or a fumigant and collected. The leaves and branches also are collected so invertebrate abundances per weight of plant material can be determined. Waters and Resh (1979) reviewed sampling strategies for forest pest insects.

Aerial

The most reliable quantitative estimates of aerial insect abundances are provided by suction traps (Fig. 11A, Southwood 1978). These traps draw in a measured volume of air, allowing estimates of insect density to be made (Johnson 1950a,b, Taylor 1951, 1962, Johnson and Taylor 1955). Wind speed is an important variable affecting the functioning of these traps (Southwood 1978). A simple sweep net also can be used to capture flying insects. If the volume of air swept is known, number or biomass per volume can be estimated. Sweep nets also have been incorporated into stationary and rotary or whirl-

igig traps (Fig. 11B, Johnson 1950b, Nicholls 1960, Juillet 1963, Graham 1969).

Malaise and other tent-like traps (Townes 1962, Juillet 1963, Graham 1969), flight intercept or window traps (Chapman and Kinghorn 1955, Jónsson et al. 1986), sticky traps (Juillet 1963, Williams 1973), and light traps provide relative estimates of aerial insect abundance. Many types of baited or attractant-type traps have been developed for monitoring populations of biting flies such as mosquitoes, black flies, and horseflies (e.g., Graham 1969, Adkins et al. 1972, Service 1976, 1987, Southwood 1978).

PROCESSING SAMPLES

Sample processing is often the most labor-intensive and time-consuming part of any invertebrate sampling program. Therefore, handling of samples after collection should be an important consideration in the selection of a sampling technique. For example, if two sampling techniques produce similar information, it may be wise to choose the one that produces the lesser amount of extraneous material. Better still, if the study objectives require only a count of the invertebrates within the sample, sorting may not be necessary.

Sample processing involves several distinct steps that vary with each study and sample type. It is important to consider initially whether the sample can be sorted in the field immediately after collection to avoid having to package and transport it to the laboratory. Invertebrates in clean samples, or conspicuous invertebrates, can be counted or removed in the field, and the sample then is discarded. For example, invertebrates collected in pitfall traps can be counted quickly in the container and discarded. This also avoids the time and effort to transfer the sample to another container and the potential loss of invertebrates from handling.

Most sampling techniques require that the samples be taken to the laboratory for processing. Depending on the type of sample, it may be possible to sort and count the samples while the invertebrates are still alive. A moving invertebrate is easier to sort from the surrounding extraneous material than a dead one. Irritants such as dilute acids or alcohol can be added to aquatic samples to ensure invertebrate movement. Live sorting may miss small, cryptic invertebrates (Lackey and May 1971), therefore, depending on the study objectives, preserving the residual material for later examination may be required. If the invertebrates cannot be removed soon after collection, they can be extracted later by methods described below, or the samples can be preserved for later processing.

Live-Extraction Techniques

Techniques that utilize the behavior and mobility of the invertebrates to separate them from the extraneous material are referred to as behavioral or dynamic extractions. Heat, moisture, chemical, or electrical stimuli can be used to drive the invertebrates from the sample. A chief advantage of these methods is that once they are set up and running, they can be left unattended, and many samples can be processed at one time. Extraction efficiencies vary with each invertebrate group, the condition of the sample, and the extraction method used (Edwards and Fletcher 1971, Edwards 1991). Efficiency should be determined as part of any invertebrate sampling program that uses live-extraction techniques (Petersen and Luxton 1982).

The extraction method selected will depend upon completeness of extraction required, extraneous material in the sample, sources of stimuli (e.g., heat) available, and invertebrates under investigation (Macfadyen 1962, Edwards 1991). No single extraction method works well with all materials and for all invertebrate groups. Live extraction is unsuitable for nonmobile life stages (i.e., eggs and pupae).

The most common device for the extraction of invertebrates from soil, litter, and vegetation samples is the Berlese-Tullgren funnel (Fig. 12). Samples are placed on a screen situated partway down a large funnel. A heat source at the top of the funnel forces the invertebrates to move out of the sample and down through the screen into a collection jar. Many modifications have been developed, including models with humidity gradients, light sources below the funnel for positively phototactic animals, traps to prevent litter and soil from falling into the collection jars, and collapsible models for use in the field (Dietrick et al. 1959, Macfadyen 1962, Edwards and Fletcher 1971, Southwood 1978). Edwards and Fletcher (1971) discussed the efficiencies of a variety of dynamic extraction devices for soil invertebrates. An aquatic version of the Berlese-Tullgren funnel was developed to extract invertebrates from aquatic vegetation and litter (Fairchild et al. 1987). Electricity also can be used to extract aquatic invertebrates from detritus (Fahy 1972).

Sample Preservation

If the samples cannot be processed soon after collection, they must be preserved to prevent decomposition. The type of preservation used will depend on the invertebrate group, type of sample, and type of information required. Sample preservation will affect biomass estimates. Freezing and liquid preservation, the two most common means of preserving samples, result in invertebrate weight loss, which cannot be standardized readily (Howmiller 1972, Stanford 1973, Donald and Paterson 1977, Leuven et al. 1985, Salonen and Sarvala 1985, Nolte 1990).

Freezing is used most often for large invertebrates with little extraneous material (e.g., sweep-net or light-trap samples). Freezing aquatic samples can result in the disintegration of small, soft-bodied invertebrates (e.g., oligochaetes), but cryoprotectants can be used to minimize this problem (Salonen and Sarvala 1985).

The most common liquid preservatives are 70% ethyl alcohol or 4% formaldehyde (= 10% formalin). Formaldehyde has the disadvantage of an unpleasant smell and is poisonous and carcinogenic (Leuven et al. 1985). The addition of rose bengal with liquid preservation will stain invertebrates a bright red color, increasing efficiency and reducing sorting time (Mason and Yevich 1967, Lackey and May 1971, Williams and Williams 1974).

Sample Sieving

Prior to whole-sample preservation, sieving the sample to reduce its volume may be useful; this procedure frequently is used with benthic samples. A smaller sample is easier to preserve and store and, as noted previously, smaller samples require less sorting time. The size of

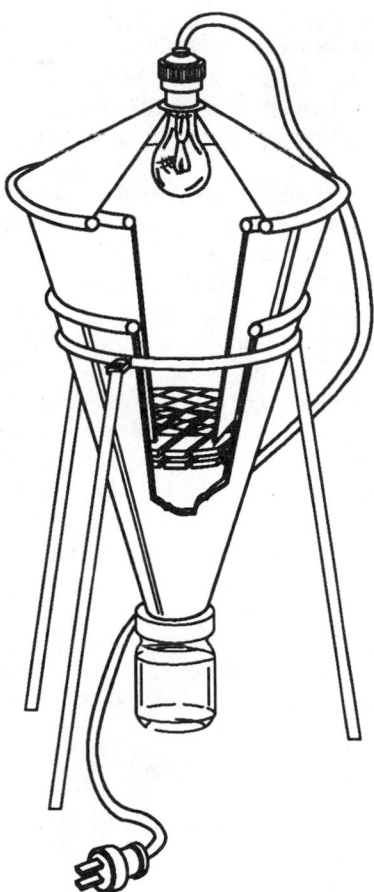

Fig. 12. Berlese-Tullgren funnel for the extraction of soil-dwelling invertebrates (redrawn from Martin 1977).

mesh used during sieving is critical because invertebrates smaller than the minimum mesh size may be lost from the sample (Jónasson 1955, 1958, Barber and Kevern 1974, Storey and Pinder 1985). This loss of small individuals will affect density estimates, but it will have a lesser effect on biomass estimates (Zelt and Clifford 1972, Barber and Kevern 1974).

The sample normally is placed on the screen and washed with a gentle stream of water. During this procedure large pieces of extraneous material (e.g., rocks and roots) can be removed as well. Although the screening procedure does not remove all extra material, it will greatly reduce the amount of material to be preserved and sorted. Samples also can be sieved in the field before they are taken to the laboratory (Euliss and Swanson 1989, Mason 1991).

Substrate cores may contain large amounts of clay, especially if taken in floodplain alluvium. This clay will tend to bind the core together and make sieving difficult. Adding a small amount of calcium carbonate to the sample during sieving will dissolve the colloids and the sample can be sieved more efficiently.

Subsampling

Before the sample is sorted, the decision of whether to subsample must be made. Subsampling involves counting the invertebrates from a known portion of the sample and then extrapolating to the entire sample. Subsampling devices and procedures have been described for zooplankton (e.g., Sell and Evans 1982, George et al. 1984), benthic invertebrate (e.g., Hickley 1975, Reger et al. 1982, Wrona et al. 1982, Sebastien et al. 1988, Meyer 1990, Mason 1991), and light-trap (Van Ark and Pretorius 1970) samples.

Subsampling generally involves the thorough mixing of the sample before a known amount (by volume or weight) is removed for sorting. Sieving the entire sample to remove large extraneous materials may be necessary to ensure a homogeneous sample. Counts or biomass estimates are then made on the subsample. Reger et al. (1982) suggested that a sample be divided into several subsamples (eight in their example). Subsamples then are chosen at random, sorted, and counted until a minimum of 50 individuals of the taxa of interest is obtained. Others have suggested that at least 100 individuals of each invertebrate group/species must be counted to ensure reasonably accurate estimation of abundances (Hickley 1975, Elliott 1977). Rare species require special consideration in the selection of subsampling techniques (Sebastien et al. 1988, Meyer 1990). The accuracy of a subsampling procedure should always be checked prior to its use in any study (see Venrick 1971, Elliott 1977).

Sorting

The next step in sample processing involves removing the invertebrates from the extraneous material. This can be time consuming for large samples and often is the point of failure of many invertebrate studies. Sorting can range from simply picking through the sample and counting or removing the invertebrates to using various mechanical or washing techniques. Picking through the sample is the most tedious procedure, but for some sample types and invertebrate groups it cannot be avoided. Moreover, invertebrates that bind tightly to or burrow within the substrate or vegetation likely require individual attention. If some cleaning of the sample is required, the first step is to wash out some of the finer material. This will involve the further use of a sieve or series of sieves as described above.

Flotation techniques, which involve submersing the sample in a solution with a specific gravity higher than that of the invertebrates, are especially useful for samples that have been preserved (Edwards and Fletcher 1971, Pask and Costa 1971). The invertebrates float to the surface where they can be removed easily. Large amounts of vegetation (e.g., peat) also may float (Edwards and Fletcher 1971), making sorting more difficult. Invertebrates that live in shells or soil tubes may not be suitable for flotation techniques (Flannagan 1973). Flotation solutions used include sugar (Anderson 1959, Lackey and May 1971, Pask and Costa 1971, Flannagan 1973), sodium chloride (Dondale et al. 1971, Edwards and Fletcher 1971), magnesium sulfate (Hale 1964, Lawson and Merritt 1979), benzene (Karlsson et al. 1976), and kerosene (Barmuta 1984). Flotation has been incorporated with other techniques such as wet sieving and mechanized separators (e.g., Edwards et al. 1970, Lawson and Merritt 1979). Once again, the efficiency of flotation methods should be determined before they are used extensively.

Biomass

Biomass is a quantitative estimate of total mass of organisms measured as fresh, dry, or ash-free dry weight (AFDW) or energy (calories). It is often referred to as standing stock or standing crop. Biomass is an important consideration in quantifying invertebrate abundance, calculating secondary production, and assessing matter or energy flow among trophic levels.

Dry weight is a more accurate measurement than fresh weight because of variations in moisture content of live invertebrates. Dry weight is determined by drying the invertebrates at a predetermined temperature to constant weight. The temperature used should be low enough to prevent the loss of volatile materials such as lipids. A recommended temperature is 60 C (Southwood 1978). As discussed earlier, directly determining dry weight from invertebrates that have been preserved is impossible.

Biomass can be calculated from the relationship between body size (length or volume) and weight. Dumont et al. (1975), Rogers et al. (1976, 1977), Smock (1980), Meyer (1989), and Nolte (1990) provided regression equations for length-weight relationships of a variety of invertebrates. Ciborowski (1983) provided a means of determining wet and dry weights of invertebrates from their volume. Taxon-specific regressions will provide more accurate data (Rogers et al. 1976, Wenzel et al. 1990) and should be considered when possible. McCauley (1984) reviewed methods of estimating dry weight for zooplankton.

Invertebrate mass also can be expressed as AFDW, a measurement that is equivalent to organic weight. Ash content of invertebrates is determined by burning off the organic matter in a muffle furnace at 550 C for approximately 3 hours. AFDW then is calculated by subtracting the weight of the ash from the initial dry weight (see Chapter 12). AFDW should be considered when movement of organic matter in ecosystems is under investigation. Expressing biomass as AFDW makes comparisons among taxa more accurate, because variations in ash content of invertebrates (e.g., calcareous exoskeletons) are removed.

Energy content of invertebrates is expressed as calories per gram AFDW and is important in the determination of energy transfer among trophic levels (Paine 1971). Bomb calorimetry, which measures the amount of energy (heat) liberated when material is burned, is the most common method of determining energy content of invertebrates (Southwood 1978). Energy values for a variety of invertebrates were provided by Wissing and Hasler (1968, 1971), Schindler et al. (1971), Thayer et al. (1973), Driver et al. (1974), McCauley and Tsumura (1974), and Gardner et al. (1985). Energy values vary with location, season, and developmental stage and are dependent on lipid reserves (Wissing and Hasler 1971, Gardner et al. 1985). This variation should be considered when energy values are taken from the published literature.

Identification

The taxonomic level to which invertebrates are identified depends on the objectives of the study, the time and effort that can be expended, and the expertise available. Species are the basic biological units, and it is at this level that changes in invertebrate communities are best docu-

Table 1. Sources of general keys and other information for the identification of invertebrates.

Reference	Group	Life stage identified	Level of identification
Bland and Jaques (1978)	Insects	Adults	Family (with notes on common species)
Borror et al. (1989)	Insects	Adults	Family and subfamily (with notes for each, including important species)
Borror and White (1970)	Insects	Adults and immatures	Family (field guide, no keys)
Burch (1989)	Snails		Species (with notes on distributions)
Kaston (1978)	Spiders	Adults	Family (with notes)
Kevan and Scudder (1989)	Millipedes, centipedes	Adults and immatures	Family
Klemm (1985)	Annelids		Species (with notes on distributions)
Lehmkuhl (1979)	Insects	Adults and immatures	Family (with notes on each)
Merritt and Cummins (1984)	Insects	Adults and immatures	Family and some to genus
Peckarsky et al. (1990)	Aquatic	Adults and/or immatures	Genus for most and some to species
Pennak (1989)	Protozoa to mollusca	Adults and immatures	Family and genus (with some keys to species)
Stehr (1987, 1990)	Insects	Immatures	Family and subfamily (with notes on each)
Thorp and Covich (1991)	Aquatic invertebrates	Adults and immatures	Family and many to genera

mented and explained (Resh and Unzicker 1975, Rosenberg et al. 1986, Danks 1988). However, species-level keys are unavailable for many invertebrate groups, and the time and effort required for such identification are beyond the resources of most studies. Many invertebrate groups have large numbers of species and include small individuals that necessitate specialized techniques for identification. Identification to order or family is generally the most widely used in wildlife studies. Table 1 lists references that provide general keys to invertebrate identification.

An alternative to a taxonomic classification is one based on a resource utilization, often termed functional feeding group or guild (Cummins 1973, Cummins and Klug 1979, Bahr 1982, Hawkins and MacMahon 1989). This system, which classifies each invertebrate by its feeding mechanism, has been used to evaluate ecological roles and resource utilization, especially in aquatic invertebrate studies (Cummins 1973, Wiggins and Mackey 1978, Cummins and Klug 1979, Merritt and Cummins 1984).

STATISTICAL CONSIDERATIONS

Statistical considerations are a critical step in the development of any study design. In addition to the basic considerations of any research program (see Chapter 1), the clumped spatial distributions and high variability of invertebrate populations require special attention. The following authors discuss the statistical analysis of invertebrate data and the design of invertebrate studies and should be required reading for anyone developing an invertebrate sampling program: Elliott (1977), Green (1979), Underwood (1981), Fowler and Witter (1982), Allan (1984), and Prepas (1984).

Chapter 1 emphasizes the importance of a pilot study in developing a research project. Preliminary sampling is

critical in invertebrate sampling programs. An understanding of the distribution and variability in the study populations is necessary for the determination of sample site location, individual sample size, and number of samples taken per experimental unit.

Perhaps the most commonly asked statistical question in planning invertebrate studies is the number of samples to take. This number should be based on knowledge of the existing invertebrate populations as revealed by preliminary sampling. Invertebrate populations usually have a clumped (patchy, contagious) spatial distribution (Downing 1979, Resh 1979, Kuno 1991) that produces high variation (i.e., variance > mean), and small sample numbers are often insufficient to develop a reasonable estimate of population size. The solution is to take a large number of samples (>50); however, the time and effort required to collect, process, and identify this number of samples are usually unattainable. Selection of sample numbers should balance desired precision with the ability to process those samples. Preliminary sampling provides an estimate of sample variance that can be used together with a desired level of precision to calculate an acceptable sample number (e.g., Downing 1979, Fowler and Witter 1982, Allan 1984, Chew 1984).

Parametric statistical procedures assume normal distributions and equality of variances among samples. The presence of non-normal distributions (usually caused by patchy spatial distributions) can be tested by such methods as the Shapiro-Wilk (W) statistic (Neter et al. 1990). When non-normal distributions are indicated, two statistical options are available—nonparametric procedures or parametric tests on transformed data (see Chapter 2 for a detailed account of data analysis). Nonparametric methods are simple to apply, not subject to the assumptions of normality and equality of variances, and for simple experimental designs have nearly the power of parametric equivalents (Allan 1984). However, for more complicated designs or those for which the researcher wants to use the full power of parametric procedures, using standard parametric tests is appropriate once suitable data transformations have been made (Green 1979).

Many investigators simply transform data with standard textbook techniques and assume distributional normality, but the transformed data set also should be tested for normality and equality of variances. The most common transformations for invertebrate data are (1) log x, (2) log (x + 1) when many zeros are present in the data, (3) square root, and (4) fourth root (Elliott 1977, Downing 1979). Taylor's Power Law (Taylor 1961, Downing 1979) can be used to develop an exact transformation. In any case, transformation of data should be described clearly in all reports and publications. When data are presented in graphic or tabular form, the means and variance of the transformed data should be used because these are the data on which the statistical tests and comparisons were made.

Acknowledgments.—We thank N. H. Euliss, Jr., L. C. M. Ross, and D. M. Rosenberg for reviewing earlier drafts of this manuscript. T. Gregg and A. Guzzi assisted in the production of the figures. This is paper number 4 of the Institute for Wetland and Waterfowl Research.

LITERATURE CITED

ADKINS, T. R., JR., W. B. EZELL, JR., D. C. SHEPPARD, AND M. M. ASKEY, JR. 1972. A modified canopy trap for collecting Tabanidae. J. Med. Entomol. 9:183–185.

AIKEN, R. B., AND R. E. ROUGHLEY. 1985. An effective trapping and marking method for aquatic beetles. Proc. Acad. Nat. Sci. Philadelphia 137:5–7.

ALLAN, J. D. 1984. Hypothesis testing in ecological studies of aquatic insects. Pages 484–507 *in* V. H. Resh and D. M. Rosenberg, eds. The ecology of aquatic insects. Praeger Publ., New York, N.Y.

———, AND E. RUSSEK. 1985. The quantification of stream drift. Can. J. Fish. Aquat. Sci. 42:210–215.

ANDERSON, R. O. 1959. A modified flotation technique for sorting bottom fauna samples. Limnol. Oceanogr. 4:223–225.

ARMITAGE, P. D. 1978. Catches of invertebrate drift by pump and net. Hydrobiologia 60:229–233.

BAARS, M. A. 1979. Catches in pitfall traps in relation to mean densities of carabid beetles. Oecologia 41:25–46.

BAHR, L. M., JR. 1982. Functional taxonomy: an immodest proposal. Ecol. Model. 15:211–233.

BARBER, W. E., AND N. R. KEVERN. 1974. Seasonal variation of sieving efficiency in a lotic habitat. Freshwater Biol. 4:293–300.

BARMUTA, L. A. 1984. A method for separating benthic arthropods from detritus. Hydrobiologia 112:105–107.

BENDELL, B. E., AND D. K. MCNICOL. 1987. Estimation of nektonic insect populations. Freshwater Biol. 18:105–108.

BENGTSON, S.-A., A. NILSSON, S. NORDSTRÖM, AND S. RUNDGREN. 1976. Effect of bird predation on lumbricid populations. Oikos 27:9–12.

BLAND, R. G., AND H. E. JAQUES. 1978. How to know the insects. Third ed. William C. Brown Co. Publ., Dubuque, Ia. 409pp.

BLANTON, C. M. 1990. Canopy arthropod sampling: a comparison of collapsible bag and fogging methods. J. Agric. Entomol. 7:41–50.

BLOMQVIST, S. 1990. Sampling performance of Ekman grabs—*in situ* observations and design improvements. Hydrobiologia 206:245–254.

BORROR, D. J., C. A. TRIPLEHORN, AND N. F. JOHNSON. 1989. An introduction to the study of insects. Sixth ed. Saunders Coll. Publ., Philadelphia, Pa. 875pp.

———, AND R. E. WHITE. 1970. A field guide to the insects of America north of Mexico. Houghton Mifflin Co., Boston, Mass. 404pp.

BOSTANIAN, N. J., G. BOIVIN, AND H. GOULET. 1983. Ramp pitfall trap. J. Econ. Entomol. 76:1473–1475.

———, AND D. H. C. HERNE. 1980. A rapid method of collecting arthropods from deciduous fruit trees. J. Econ. Entomol. 73:832–833.

BOULTON, A. J. 1985. A sampling device that quantitatively collects benthos in flowing or standing waters. Hydrobiologia 127:31–39.

BRINKHURST, R. O. 1974. The benthos of lakes. St. Martin's Press, New York, N.Y. 190pp.

BRITTAIN, J. E., AND T. J. EIKELAND. 1988. Invertebrate drift—a review. Hydrobiologia 166:77–93.

BROWN, A. V., M. D. SCHRAM, AND P. P. BRUSSOCK. 1987. A vacuum benthos sampler suitable for diverse habitats. Hydrobiologia 153:241–247.

BROWN, C. L., AND B. A. MANNY. 1985. Comparison of methods for measuring surface area of submersed aquatic macrophytes. J. Freshwater Ecol. 3:61–68.

BURCH, J. B. 1989. North American freshwater snails. Malacological Publ., Hamburg, Mich. 365pp.

CATTANEO, A., AND R. CARIGNAN. 1983. A colorimetric method for measuring the surface area of aquatic plants. Aquat. Bot. 17:291–294.

CHAPMAN, J. A., AND J. M. KINGHORN. 1955. Window flight traps for insects. Can. Entomol. 87:46–47.

CHERRY, R. H., K. A. WOOD, AND W. G. RUESINK. 1977. Emergence trap and sweep net sampling for adults of the potato leafhopper from alfalfa. J. Econ. Entomol. 70:279–282.

CHEW, V. 1984. Number of replicates in experimental research. Southwest. Entomol. Suppl. 6:2–9.

CHIVERTON, P. A. 1984. Pitfall-trap catches of the carabid beetle *Pterostichus melanarius,* in relation to gut contents and prey densities, in insecticide treated and untreated spring barley. Entomol. Exp. Appl. 36:23–30.

CIBOROWSKI, J. J. H. 1983. A simple volumetric instrument to estimate biomass of fluid-preserved invertebrates. Can. Entomol. 115:427–430.

COSTA, J. T., III, AND D. A. CROSSLEY, JR. 1991. Diel patterns of canopy arthropods associated with three tree species. Environ. Entomol. 20:1542–1548.

CRAWLEY, M. J. 1989. Insect herbivores and plant population dynamics. Annu. Rev. Entomol. 34:531–564.

CROSSLEY, D. A., JR. 1977. The roles of terrestrial saprophagous arthropods in forest soils: current status of concepts. Pages 49–56 *in* W. J. Mattson, ed. The role of arthropods in forest ecosystems. Springer-Verlag, New York, N.Y.

———, AND M. P. HOGLUND. 1962. A litter-bag method for the study of microarthropods inhabiting leaf litter. Ecology 43:571–573.

CUMMINS, K. W. 1973. Trophic relations of aquatic insects. Annu. Rev. Entomol. 18:183–206.

———, AND M. J. KLUG. 1979. Feeding ecology of stream invertebrates. Annu. Rev. Ecol. Syst. 10:147–172.

DANIEL, P. M., K. LYNK, AND M. W. BOESEL. 1985. A comparison of clear and opaque funnel traps for emerging insects in a southwestern Ohio pond. Ohio J. Sci. 85:199–202.

DANKS, H. V. 1988. Systematics in support of entomology. Annu. Rev. Entomol. 33:271–296.

DAVIES, I. J. 1984. Sampling aquatic insect emergence. Pages 161–227 *in* J. A. Downing and F. H. Rigler, eds. A manual on methods for the assessment of secondary productivity in fresh waters. Second ed. IBP Handb. 17. Blackwell Sci. Publ., Oxford, U.K.

DE BERNARDI, R. 1984. Methods for the estimation of zooplankton abundance. Pages 59–86 *in* J. A. Downing and F. H. Rigler, eds. A manual on methods for the assessment of secondary productivity in fresh waters. Second ed. IBP Handb. 17. Blackwell Sci. Publ., Oxford, U.K.

DEMPSTER, J. P. 1961. A sampler for estimating populations of active insects upon vegetation. J. Anim. Ecol. 30:425–427.

DEONIER, D. L. 1972. A floating adhesive trap for neustonic insects. Ann. Entomol. Soc. Am. 65:269–270.

DESENDER, K., AND J.-P. MAELFAIT. 1986. Pitfall trapping within enclosures: a method for estimating the relationship between the abundances of coexisting carabid species (Coleoptera: Carabidae). Holarct. Ecol. 9:245–250.

DIETRICK, E. J. 1961. An improved backpack motor fan for suction sampling of insect populations. J. Econ. Entomol. 54:394–395.

———, E. I. SCHLINGER, AND R. VAN DEN BOSCH. 1959. A new method for sampling arthropods using a suction collecting machine and modified Berlese funnel separator. J. Econ. Entomol. 52:1085–1091.

DONALD, G. L., AND C. G. PATERSON. 1977. Effect of preservation on wet weight biomass of chironomid larvae. Hydrobiologia 53: 75–80.

DONDALE, C. D., C. F. NICHOLLS, J. H. REDNER, R. B. SEMPLE, AND A. L. TURNBULL. 1971. An improved Berlese-Tullgren funnel and a flotation separator for extracting grassland arthropods. Can. Entomol. 103:1549–1552.

DOWNING, J. A. 1979. Aggregation, transformation, and the design of benthos sampling programs. J. Fish. Res. Board Can. 36:1454–1463.

———. 1984. Sampling the benthos of standing waters. Pages 87–130 in J. A. Downing and F. H. Rigler, eds. A manual on methods for the assessment of secondary productivity in fresh waters. Second ed. IBP Handb. 17. Blackwell Sci. Publ., Oxford, U.K.

———. 1986. A regression technique for the estimation of epiphytic invertebrate populations. Freshwater Biol. 16:161–173.

———, AND H. CYR. 1985. Quantitative estimation of epiphytic invertebrate populations. Can. J. Fish. Aquat. Sci. 42:1570–1579.

———, AND F. H. RIGLER, EDITORS. 1984. A manual on methods for the assessment of secondary productivity in fresh waters. Second ed. IBP Handb. 17. Blackwell Sci. Publ., Oxford, U.K. 501pp.

DRAKE, C. M., AND J. M. ELLIOTT. 1982. A comparative study of three air-lift samplers used for sampling benthic macro-invertebrates in rivers. Freshwater Biol. 12:511–533.

DRIVER, E. A., L. G. SUGDEN, AND R. J. KOVACH. 1974. Calorific, chemical and physical values of potential duck foods. Freshwater Biol. 4:281–292.

DUMONT, H. J., I. VAN DE VELDE, AND S. DUMONT. 1975. The dry weight estimate of biomass in a selection of Cladocera, Copepoda and Rotifera from the plankton, periphyton and benthos of continental waters. Oecologia 19:75–97.

DURKIS, T. J., AND R. M. REEVES. 1982. Barriers increase efficiency of pitfall traps. Entomol. News 93:8–11.

EBERHARDT, L. L. 1978. Appraising variability in population studies. J. Wildl. Manage. 42:207–238.

EDWARDS, C. A. 1991. The assessment of populations of soil-inhabiting invertebrates. Agric. Ecosyst. Environ. 34:145–176.

———, AND K. E. FLETCHER. 1971. A comparison of extraction methods for terrestrial arthropods. Pages 150–185 in J. Phillipson, ed. Methods of study in quantitative soil ecology: population, production and energy flow. IBP Handb. 18. Blackwell Sci. Publ., Oxford, U.K.

———, D. E. REICHLE, AND D. A. CROSSLEY, JR. 1970. The role of soil invertebrates in turnover of organic matter and nutrients. Pages 147–172 in D. E. Reichle, ed. Analysis of temperate forest ecosystems. Springer-Verlag, New York, N.Y.

———, B. R. STINNER, D. STINNER, AND S. RABATIN, EDITORS. 1988. Biological interactions in soil. Agric. Ecosyst. Environ. 24:1–377.

———, A. E. WHITING, AND G. W. HEATH. 1970. A mechanized washing method for separation of invertebrates from soil. Pedobiologia 10:141–148.

ELLINGTON, J. J., K. KISER, M. CARDENAS, J. DUTTLE, AND Y. LOPEZ. 1984. The insectavac—a high clearance, high volume arthropod vacuuming platform for agricultural ecosystems. Environ. Entomol. 13:259–265.

ELLIOTT, J. M. 1970. Methods of sampling invertebrate drift in running water. Ann. Limnol. 6:133–159.

———. 1973. The food of brown and rainbow trout (*Salmo trutta* and *S. gairdneri*) in relation to the abundance of drifting invertebrates in a mountain stream. Oecologia 12:329–347.

———. 1977. Some methods for the statistical analysis of samples of benthic invertebrates. Second ed. Freshwater Biol. Assoc. Sci. Publ. 25. 156pp.

———, AND C. M. DRAKE. 1981. A comparative study of seven grabs used for sampling benthic macroinvertebrates in rivers. Freshwater Biol. 11:99–120.

———, AND P. A. TULLETT. 1978. A bibliography of samplers for benthic invertebrates. Freshwater Biol. Assoc. Occas. Publ. 4. 61pp.

———, AND ———. 1983. A supplement to a bibliography of samplers for benthic invertebrates. Freshwater Biol. Assoc. Occas. Publ. 20. 27pp.

ENGLISH, W. R. 1987. Three inexpensive aquatic invertebrate samplers for the benthos, drift and emergent fauna. Entomol. News 98:171–179.

EPSTEIN, M. E., AND H. M. KULMAN. 1984. Effects of aprons on

pitfall trap catches of carabid beetles in forest and fields. Great Lakes Entomol. 17:215–221.

EULISS, N. H., JR., AND G. A. SWANSON. 1989. Improved self-cleaning screen for processing benthic samples. Calif. Fish Game 75: 124–128.

———, ———, AND J. MACKAY. 1992. Multiple tube sampler for benthic and pelagic invertebrates in shallow wetlands. J. Wildl. Manage. 56:186–191.

FAHY, E. 1972. An automatic separator for the removal of aquatic insects from detritus. J. Appl. Ecol. 9:655–658.

FAIRCHILD, W. L., M. C. A. O'NEILL, AND D. M. ROSENBERG. 1987. Quantitative evaluation of the behavioral extraction of aquatic invertebrates from samples of sphagnum moss. J. North Am. Benthol. Soc. 6:281–287.

FERRINGTON, L. C., JR., AND C. CHRISTIANSEN. 1985. Statistical and biological significance of *Stenochironomus* larvae on multiplate artificial substrate samplers with Masonite® discs. J. Kans. Entomol. Soc. 58:724–726.

FLANNAGAN, J. F. 1970. Efficiencies of various grabs and corers in sampling freshwater benthos. J. Fish. Res. Board Can. 27:1691–1700.

———. 1973. Sorting benthos using floatation media. Fish. Res. Board Can. Tech. Rep. 354. 14pp.

———, AND D. M. ROSENBERG. 1982. Types of artificial substrates used for sampling freshwater benthic macroinvertebrates. Pages 237–266 in J. Cairns, Jr., ed. Artificial substrates. Ann Arbor Sci. Publ., Ann Arbor, Mich.

FOWLER, G. W., AND J. A. WITTER. 1982. Accuracy and precision of insect density and impact estimates. Great Lakes Entomol. 15: 103–117.

FROST, S., A. HUNI, AND W. E. KERSHAW. 1971. Evaluation of a kicking technique for sampling stream bottom fauna. Can. J. Zool. 49:167–173.

GALE, W. F. 1971. Shallow-water core sampler. Prog. Fish-Cult. 33: 238–239.

GARDNER, W. S., T. F. NALEPA, W. A. FREZ, E. A. CICHOCKI, AND P. F. LANDRUM. 1985. Seasonal patterns in lipid content of Lake Michigan macroinvertebrates. Can. J. Fish. Aquat. Sci. 42:1827–1832.

GATES, T. E., D. J. BAIRD, F. J. WRONA, AND R. W. DAVIES. 1987. A device for sampling macroinvertebrates in weed ponds. J. North Am. Benthol. Soc. 6:133–139.

GEORGE, D. G., M. A. HURLEY, AND B. WINSTANLEY. 1984. A simple plankton splitter with a note on its reduced subsampling variance. Limnol. Oceanogr. 29:429–433.

———, AND G. H. OWEN. 1978. A new tube sampler for crustacean zooplankton. Limnol. Oceanogr. 23:563-566.

GERKING, S. D. 1957. A method for sampling the littoral macrofauna and its application. Ecology 38:219–226.

GIBB, J. A., AND M. M. BETTS. 1963. Food and food supply of nestling tits (Paridae) in Breckland Pine. J. Anim. Ecol. 32:489–533.

GILBERT, A. T., AND E. RUBER. 1986. A water column sampler for invertebrates in salt-marsh tidal pools. Estuaries 9:380–381.

GIST, C. S., AND D. A. CROSSLEY, JR. 1973. A method of quantifying pitfall trapping. Environ. Entomol. 2:951–952.

GRAHAM, P. 1969. A comparison of sampling methods for adult mosquito populations in central Alberta, Canada. Quaest. Entomol. 5:217–261.

GREEN, R. H. 1979. Sampling design and statistical methods for environmental biologists. John Wiley & Sons, New York, N.Y. 257pp.

GREENSLADE, P., AND P. J. M. GREENSLADE. 1971. The use of baits and preservatives in pitfall traps. J. Aust. Entomol. Soc. 10:253–260.

GREENSLADE, P. J. N. 1964. Pitfall trapping as a method for studying populations of Carabidae (Coleoptera). J. Anim. Ecol. 33:301–310.

HALE, W. G. 1964. A flotation method for extracting Collembola from organic soils. J. Anim. Ecol. 33:363–369.

HALL, D. W. 1991. The environmental hazard of ethylene glycol in insect pit-fall traps. Coleopt. Bull. 45:193–194.

HALSALL, N. B., AND S. D. WRATTEN. 1988. The efficiency of pitfall trapping for polyphagous predatory Carabidae. Ecol. Entomol. 13: 293–299.

HAMILTON, A. L. 1969. A new type of emergence trap for collecting stream insects. J. Fish. Res. Board Can. 26:1685–1689.

———, W. BURTON, AND J. F. FLANNAGAN. 1970. A multiple corer

for sampling profundal benthos. J. Fish. Res. Board Can. 27: 1867–1869.

HARRIS, J. W. E., D. G. COLLIS, AND K. M. MAGAR. 1972. Evaluation of the tree-beating method for sampling defoliating forest insects. Can. Entomol. 104:723–729.

HARROD, J. J., AND R. E. HALL. 1962. A method for determining the surface areas of various aquatic plants. Hydrobiologia 20:173–178.

HAWKINS, C. P., AND J. A. MACMAHON. 1989. Guilds: the multiple meaning of a concept. Annu. Rev. Entomol. 34:423–451.

HERMS, D. A., D. G. NIELSEN, AND T. D. SYDNOR. 1990. Comparison of two methods for sampling arboreal insect populations. J. Econ. Entomol. 83:869–874.

HICKLEY, P. 1975. An apparatus for subdividing benthos samples. Oikos 26:92–96.

HILSENHOFF, W. L., AND B. H. TRACY. 1985. Techniques for collecting water beetles from lentic habitats. Proc. Acad. Nat. Sci. Philadelphia 137:8–11.

HOUSEWEART, M. W., D. T. JENNINGS, AND J. C. REA. 1979. Large capacity pitfall trap. Entomol. News 90:51–54.

HOWMILLER, R. P. 1972. Effects of preservatives on weights of some common macrobenthic invertebrates. Trans. Am. Fish. Soc. 101: 743–746.

HUGHES, R. D. 1955. The influence of the prevailing weather on the numbers of *Meromyza variegata* Meigen (Diptera, Chloropidae) caught with a sweepnet. J. Anim. Ecol. 24:324–335.

HUNT, D. W. A., AND K. F. RAFFA. 1989. Attraction of *Hylobius radicis* and *Pachylobius picivorus* (Coleoptera: Curculionidae) to ethanol and turpentine in pitfall traps. Environ. Entomol. 18:351–355.

JENKINS, T. M., JR., C. R. FELDMETH, AND G. V. ELLIOTT. 1970. Feeding of rainbow trout (*Salmo gairdneri*) in relation to abundance of drifting invertebrates in a mountain stream. J. Fish. Res. Board Can. 27:2356–2361.

JOHNSGARD, P. A. 1981. The plovers, sandpipers, and snipes of the world. Univ. Nebraska Press, Lincoln. 493pp.

JOHNSON, C. G. 1950a. A suction trap for small airborne insects which automatically segregates the catch into successive hourly samples. Ann. Appl. Biol. 37:80–91.

———. 1950b. The comparison of suction trap, sticky trap and townet for the quantitative sampling of small airborne insects. Ann. Appl. Biol. 37:268–285.

———. 1958. The mesh factor in sieving techniques. Verh. Internat. Ver. Limnol. 13:860–866.

———, T. R. E. SOUTHWOOD, AND H. M. ENTWISTLE. 1957. A new method of extracting arthropods and molluscs from grassland and herbage with a suction apparatus. Bull. Entomol. Res. 48:211–218.

———, AND L. R. TAYLOR. 1955. The development of large suction traps for airborne insects. Ann. Appl. Biol. 43:51–62.

JÓNASSON, P. M. 1955. The efficiency of sieving techniques for sampling freshwater bottom fauna. Oikos 6:183–207.

JÓNSSON, E., A. GARDARSSON, AND G. GÍSLASON. 1986. A new window trap used in the assessment of the flight periods of Chironomidae and Simuliidae (Diptera). Freshwater Biol. 16:711–719.

JUILLET, J. A. 1963. A comparison of four types of traps used for capturing flying insects. Can. J. Zool. 41:219–223.

KAMINSKI, R. M., AND H. R. MURKIN. 1981. Evaluation of two devices for sampling nektonic invertebrates. J. Wildl. Manage. 45: 493–496.

KARLSSON, M., T. BOHLIN, AND J. STENSON. 1976. Core sampling and flotation: two methods to reduce costs of a chironomid population study. Oikos 27:336–338.

KASTON, B. J. 1978. How to know the spiders. William C. Brown Co. Publ., Dubuque, Ia. 272pp.

KEVAN, D. K. McE., AND G. G. E. SCUDDER. 1989. Illustrated keys to the families of terrestrial arthropods of Canada. 1. Myriapods (Millipedes, Centipedes, etc). Biol. Surv. Can., Taxonomic Ser. 1. Ottawa, Ont. 88pp.

KLEMM, D. J., EDITOR. 1985. A guide to the freshwater Annelida (Polychaeta, Naidid and Tubificid Oligochaeta, and Hirudinea) of North America. Kendall/Hunt Publ. Co., Dubuque, Ia. 198pp.

———, P. A. LEWIS, F. FULK, AND J. M. LAZORCHAK. 1990. Macroinvertebrate field and laboratory methods for evaluating the biological integrity of surface waters. U.S. Environ. Prot. Agency EPA/600/4-90/030. 256pp.

KROGER, R. L. 1972. Underestimation of standing crop by the Surber sampler. Limnol. Oceanogr. 17:475–478.

KUNO, E. 1991. Sampling and analysis of insect populations. Annu. Rev. Entomol. 36:285–304.

LACKEY, R. T., AND B. E. MAY. 1971. Use of sugar flotation and dye to sort benthic samples. Trans. Am. Fish. Soc. 100:794–797.

LANE, E. D. 1974. An improved method of Surber sampling for bottom and drift fauna in small streams. Prog. Fish-Cult. 36:20–22.

LASENBY, D. C., AND R. K. SHERMAN. 1991. Design and evaluation of a bottom-closing net used to capture mysids and other suprabenthic fauna. Can. J. Zool. 69:783–786.

LAVERY, M. A., AND R. R. COSTA. 1972. Reliability of the Surber sampler in estimating *Parargyractis fulicalis* (Clemens) (Lepidoptera: Pyralidae) populations. Can. J. Zool. 50:1335–1336.

LAWSON, D. L., AND R. W. MERRITT. 1979. A modified Ladell apparatus for the extraction of wetland macroinvertebrates. Can. Entomol. 111:1389–1393.

LEHMKUHL, D. M. 1979. How to know the aquatic insects. William C. Brown Co. Publ., Dubuque, Ia. 168pp.

LESAGE, L., AND A. D. HARRISON. 1979. Improved traps and techniques for the study of emerging aquatic insects. Entomol. News 90:65–78.

LEUVEN, R. S. E. W., T. C. M. BROCK, AND H. A. M. VAN DRUTEN. 1985. Effects of preservation on dry- and ash-free dry weight biomass of some common aquatic macro-invertebrates. Hydrobiologia 127:151–159.

LUFF, M. L. 1968. Some effects of formalin on the numbers of Coleoptera caught in pitfall traps. Entomol. Mon. Mag. 104:115–116.

———. 1975. Some features influencing the efficiency of pitfall traps. Oecologia 19:345–357.

MACFADYEN, A. 1962. Soil arthropod sampling. Adv. Ecol. Res. 1:1–34.

MACKEY, A. P., D. A. COOLING, AND A. D. BERRIE. 1984. An evaluation of sampling strategies for qualitative surveys of macroinvertebrates in rivers, using pond nets. J. Appl. Ecol. 21:515–534.

MACKIE, G. L., AND R. C. BAILEY. 1981. An inexpensive stream bottom sampler. J. Freshwater Ecol. 1:61–69.

MALLEY, D. F., AND J. B. REYNOLDS. 1979. Sampling strategies and life history of non-insectan freshwater invertebrates. J. Fish. Res. Board Can. 36:311–318.

MARSHALL, S. A. 1982. Techniques for collecting and handling small Diptera. Proc. Entomol. Soc. Ont. 113:73–74.

MARTIN, J. E. H. 1977. Collecting, preparing, and preserving insects, mites, and spiders. The insects and arachnids of Canada, Part 1. Agric. Can. Publ. 1643. 182pp.

MARTIN, J. L. 1966. The insect ecology of red pine plantations in central Ontario. IV. The crown fauna. Can. Entomol. 98:10–27.

MASON, W. T., JR. 1991. Sieve sample splitter for benthic invertebrates. J. Freshwater Ecol. 6:445–449.

———, AND J. E. SUBLETTE. 1971. Collecting Ohio River basin Chironomidae (Diptera) with a floating sticky trap. Can. Entomol. 103:397–404.

———, AND P. P. YEVICH. 1967. The use of phloxine B and rose bengal stains to facilitate sorting of benthic samples. Trans. Am. Microsc. Soc. 86:221–223.

MCCAULEY, E. 1984. The estimation of the abundance and biomass of zooplankton in samples. Pages 228–265 in J. A. Downing and F. H. Rigler, eds. A manual on methods for the assessment of secondary productivity in fresh waters. Second ed. IBP Handb. 17. Blackwell Sci. Publ., Oxford, U.K.

MCCAULEY, V. J. E., AND K. TSUMURA. 1974. Calorific values of Chironomidae (Diptera). Can. J. Zool. 52:581–586.

MCCREADIE, J. W., AND M. H. COLBO. 1991. A critical examination of four methods of estimating the surface area of stone substrate from streams in relation to sampling Simuliidae (Diptera). Hydrobiologia 220:205–210.

MEDWECKA-KORNAS, A. 1971. Plant litter. Pages 24–33 in J. Phillipson, ed. Methods of study in quantitative soil ecology: population, production and energy flow. IBP Handb. 18. Blackwell Sci. Publ., Oxford, U.K.

MENZIE, C. A. 1980. The chironomid (Insecta: Diptera) and other fauna of a *Myriophyllum spicatum* L. plant bed in the lower Hudson River. Estuaries 3:38–54.

MERRITT, R. W., AND K. W. CUMMINS, EDITORS. 1984. An introduction to the aquatic insects of North America. Second ed. Kendall/Hunt Publ. Co., Dubuque, Ia. 722pp.

———, ———, AND T. M. BURTON. 1984a. The role of aquatic insects in the processing and cycling of nutrients. Pages 134–163 *in* V. H. Resh and D. M. Rosenberg, eds. The ecology of aquatic insects. Praeger Publ., New York, N.Y.

———, ———, AND V. H. RESH. 1984b. Collecting, sampling, and rearing methods for aquatic insects. Pages 11–26 *in* R. W. Merritt and K. W. Cummins, eds. An introduction to the aquatic insects of North America. Second ed. Kendall/Hunt Publ. Co., Dubuque, Ia.

———, AND H. D. NEWSON. 1978. Ecology and management of arthropod populations in recreational lands. Pages 125–162 *in* G. W. Frankie and C. S. Koehler, eds. Perspectives in urban entomology. Academic Press, Inc., New York, N.Y.

MEYER, E. 1989. The relationship between body length parameters and dry mass in running water invertebrates. Arch. Hydrobiol. 117:191–203.

———. 1990. A simple subsampling device for macroinvertebrates with general remarks on the processing of stream benthos samples. Arch. Hydrobiol. 117:309–318.

MILBRINK, G., AND T. WIEDERHOLM. 1973. Sampling efficiency of four types of mud bottom samplers. Oikos 24:479–482.

MORIN, A. 1985. Variability of density estimates and the optimization of sampling programs for stream benthos. Can. J. Fish. Aquat. Sci. 42:1530–1534.

MORRILL, W. L. 1975. Plastic pitfall trap. Environ. Entomol. 4:596.

MUNDIE, J. H. 1957. The ecology of Chironomidae in storage reservoirs. Trans. R. Entomol. Soc. Lond. 109:149–232.

———. 1964. A sampler for catching emerging insects and drifting materials in streams. Limnol. Oceanogr. 9:456–459.

MURKIN, H. R., P. G. ABBOTT, AND J. A. KADLEC. 1983. A comparison of activity traps and sweep nets for sampling nektonic invertebrates in wetlands. Freshwater Invertebr. Biol. 2:99–106.

———, AND B. D. J. BATT. 1987. The interactions of vertebrates and invertebrates in peatlands and marshes. Pages 15–30 *in* D. M. Rosenberg and H. V. Danks, eds. Aquatic insects of peatlands and marshes. Mem. Entomol. Soc. Can. 140.

———, AND J. A. KADLEC. 1986. Responses by benthic macroinvertebrates to prolonged flooding of marsh habitat. Can. J. Zool. 64:65–72.

MURPHY, W. L. 1985. Procedure for the removal of insect specimens from sticky-trap material. Ann. Entomol. Soc. Am. 78:881.

NALEPA, T. F., AND A. ROBERTSON. 1981. Vertical distribution of the zoobenthos in southeastern Lake Michigan with evidence of seasonal variation. Freshwater Biol. 11:87–96.

NETER, J. W., W. WASSERMAN, AND M. H. KUTNER. 1990. Applied linear statistical models: regression, analysis of variance, and experimental designs. Third ed. R.D. Irwin, Homewood, Ill. 1181pp.

NEWTON, A., AND S. B. PECK. 1975. Baited pitfall traps for beetles. Coleopt. Bull. 29:45–46.

NICHOLLS, C. F. 1960. A portable mechanical insect trap. Can. Entomol. 92:48–51.

NOLTE, U. 1990. Chironomid biomass determination from larval shape. Freshwater Biol. 24:443–451.

O'CONNOR, F. B. 1957. An ecological study of the enchytraeid worm population of a coniferous forest soil. Oikos 8:161–199.

ONSAGER, J. A. 1977. Comparison of five methods for estimating density of rangeland grasshoppers. J. Econ. Entomol. 70:187–190.

PAINE, R. T. 1971. The measurement and application of the calorie to ecological problems. Annu. Rev. Ecol. Syst. 2:145–164.

PASK, W. M., AND R. R. COSTA. 1971. Efficiency of sucrose flotation in recovering insect larvae from benthic stream samples. Can. Entomol. 103:1649–1652.

PATERSON, C. G., AND C. H. FERNANDO. 1971. A comparison of a simple corer and an Ekman grab for sampling shallow-water benthos. J. Fish. Res. Board Can. 28:365–368.

PECKARSKY, B. L. 1984. Sampling the stream benthos. Pages 131–160 *in* J. A. Downing and F. H. Rigler, eds. A manual on methods for the assessment of secondary productivity in fresh waters. Second ed. IBP Handb. 17. Blackwell Sci. Publ., Oxford, U.K.

———, P. R. FRAISSINET, M. A. PENTON, AND D. J. CONKLIN, JR. 1990. Freshwater macroinvertebrates of northeastern North America. Cornell Univ. Press, Ithaca, N.Y. 442pp.

PENNAK, R. W. 1989. Fresh-water invertebrates of the United States. Protozoa to Mollusca. Third ed. John Wiley & Sons, New York, N.Y. 628pp.

PETERSEN, H., AND M. LUXTON. 1982. A comparative analysis of soil fauna populations and their role in decomposition processes. Oikos 39:287–388.

PETERSEN, R. C., AND K. W. CUMMINS. 1974. Leaf processing in a woodland stream. Freshwater Biol. 4:343–368.

PETRUSEWICZ, K., AND A. MACFADYEN. 1970. Productivity of terrestrial animals: principles and methods. IBP Handb. 13. Blackwell Sci. Publ., Oxford, U.K. 190pp.

PHILLIPSON, J., EDITOR. 1971. Methods of study in quantitative soil ecology: population, production and energy flow. IBP Handb. 18. Blackwell Sci. Publ., Oxford, U.K. 297pp.

POLLARD, J. E. 1981. Investigator differences associated with a kicking method for sampling macroinvertebrates. J. Freshwater Ecol. 1:215–224.

PREPAS, E. E. 1984. Some statistical methods for the design of experiments and analysis of samples. Pages 266–335 *in* J. A. Downing and F. H. Rigler, eds. A manual on methods for the assessment of secondary productivity in fresh waters. Second ed. IBP Handb. 17. Blackwell Sci. Publ., Oxford, U.K.

RAW, F. 1959. Estimating earthworm populations by using formalin. Nature (Lond.) 184:1661–1662.

REGER, S. J., C. F. BROTHERSEN, T. G. OSBORN, AND W. T. HELM. 1982. Rapid and effective processing of macroinvertebrate samples. J. Freshwater Ecol. 1:451–465.

RESH, V. H. 1979. Sampling variability and life history features: basic considerations in the design of aquatic insect studies. J. Fish. Res. Board Can. 36:290–311.

———, J. W. FEMINELLA, AND E. P. McELRAVY. 1990. Sampling aquatic insects. Videotape. Off. Media Serv., Univ. Calif., Berkeley.

———, AND J. D. UNZICKER. 1975. Water quality monitoring and aquatic organisms: the importance of species identification. J. Water Pollut. Control Fed. 47:9–19.

ROGERS, L. E., R. L. BUSCHBOM, AND C. R. WATSON. 1977. Length-weight relationships for shrub-steppe invertebrates. Ann. Entomol. Soc. Am. 70:51–53.

———, W. T. HINDS, AND R. L. BUSCHBOM. 1976. A general weight vs. length relationship for insects. Ann. Entomol. Soc. Am. 69:387–389.

ROSENBERG, D. M. 1978. Practical sampling of freshwater macro-zoobenthos: a bibliography of useful texts, reviews, and recent papers. Fish. Mar. Serv. Tech. Rep. 790. Fish. Environ. Can., Ottawa, Ont. 15pp.

———, H. V. DANKS, AND D. M. LEHMKUHL. 1986. Importance of insects in environmental impact assessment. Environ. Manage. 10:773–783.

———, AND V. H. RESH. 1982. The use of artificial substrates in the study of freshwater benthic macroinvertebrates. Pages 175–235 *in* J. Cairns, Jr., ed. Artificial substrates. Ann Arbor Sci. Publ., Ann Arbor, Mich.

SALONEN, K., AND J. SARVALA. 1985. Combination of freezing and aldehyde fixation. A superior preservation method for biomass determination of aquatic invertebrates. Arch. Hydrobiol. 103:217–230.

SATCHELL, J. E. 1971. Earthworms. Pages 107–127 *in* J. Phillipson, ed. Methods of study in quantitative soil ecology: population, production and energy flow. IBP Handb. 18. Blackwell Sci. Publ., Oxford, U.K.

SAUGSTAD, E. S., R. A. BRAM, AND W. E. NYQUIST. 1967. Factors influencing sweep-net sampling of alfalfa. J. Econ. Entomol. 60:421–426.

SCHINDLER, D. W. 1969. Two useful devices for vertical plankton and water sampling. J. Fish. Res. Board Can. 26:1948–1955.

———, A. S. CLARK, AND J. R. GRAY. 1971. Seasonal calorific values of freshwater zooplankton, as determined with a Phillipson bomb calorimeter modified for small samples. J. Fish. Res. Board Can. 28:559–564.

SCHOTZKO, D. J., AND L. E. O'KEEFFE. 1986. Comparison of sweep-net, D-vac, and absolute sampling for *Lygus hesperus* (Heteroptera: Miridae) in lentils. J. Econ. Entomol. 79:224–228.

———, AND ———. 1989. Comparison of sweep net, D-vac, and absolute sampling, and diel variation of sweep net sampling estimates in lentils for pea aphid (Homoptera: Aphididae), nabids (Hemiptera: Nabidae), lady beetles (Coleoptera: Coccinellidae), and lacewings (Neuroptera: Chrysopidae). J. Econ. Entomol. 82:491–506.

SCHOWALTER, T. D., W. W. HARGROVE, AND D. A. CROSSLEY, JR.

1986. Herbivory in forested ecosystems. Annu. Rev. Entomol. 31:177–196.

———, AND T. E. SABIN. 1991. Litter microarthropod responses to canopy herbivory, season and decomposition in litterbags in a regenerating conifer ecosystem in western Oregon. Biol. Fert. Soils 11:93–96.

———, J. W. WEBB, AND D. A. CROSSLEY, JR. 1981. Community structure and nutrient content of canopy arthropods in clearcut and uncut forest ecosystems. Ecology 62:1010–1019.

SCOTT, W. B., AND E. J. CROSSMAN. 1973. Freshwater fishes of Canada. Fish. Res. Board Can. Bull. 184. 966pp.

SEASTEDT, T. R. 1984. Belowground macroarthropods of annually burned and unburned tallgrass prairie. Am. Midl. Nat. 111:405–408.

———, AND D. A. CROSSLEY, JR. 1981. Microarthropod response following cable logging and clear-cutting in the southern Appalachians. Ecology 62:126–135.

SEBASTIEN, R. J., D. M. ROSENBERG, AND A. P. WIENS. 1988. A method for subsampling unsorted benthic macroinvertebrates by weight. Hydrobiologia 157:69–75.

SELL, D. W., AND M. S. EVANS. 1982. A statistical analysis of subsampling and an evaluation of the Folsom plankton splitter. Hydrobiologia 94:223–230.

SERVICE, M. W. 1976. Mosquito ecology: field sampling methods. John Wiley & Sons, New York, N.Y. 583pp.

———. 1987. Monitoring adult simuliid populations. Pages 187–200 in K. C. Kim and R. W. Merritt, eds. Black flies: ecology, population management, and annotated world list. Pennsylvania State Univ. Press, University Park.

SLACK, K. V., L. J. TILLEY, AND S. S. KENNELLY. 1991. Mesh-size effects on drift sample composition as determined with a triple net sampler. Hydrobiologia 209:215–226.

SMOCK, L. A. 1980. Relationships between body size and biomass of aquatic insects. Freshwater Biol. 10:375–383.

SOUTHWOOD, T. R. E. 1978. Ecological methods with particular reference to the study of insect populations. Second ed. Chapman and Hall, London, U.K. 524pp.

———, V. C. MORAN, AND C. E. J. KENNEDY. 1982. The assessment of arboreal insect fauna: comparisons of knockdown sampling and faunal lists. Ecol. Entomol. 7:331–340.

SPENCE, J. R. 1980. Density estimation for water striders (Heteroptera: Gerridae). Freshwater Biol. 10:563–570.

———, EDITOR. 1985. Faunal influences on soil structure. Quaest. Entomol. 21:371–694.

STANFORD, J. A. 1973. A centrifuge method for determining live weights of aquatic insect larvae, with a note on weight loss in preservative. Ecology 54:449–451.

STEHR, F. W., EDITOR. 1987. Immature insects. Vol. 1. Kendall/Hunt Publ. Co., Dubuque, Ia. 754pp.

———, EDITOR. 1990. Immature insects. Vol. 2. Kendall/Hunt Publ. Co., Dubuque, Ia. 975pp.

STOREY, A. W., D. H. D. EDWARD, AND P. GAZEY. 1991. Surber and kick sampling: a comparison for the assessment of macroinvertebrate community structure in streams of south-western Australia. Hydrobiologia 211:111–121.

———, AND L. C. V. PINDER. 1985. Mesh-size efficiency of sampling of larval Chironomidae. Hydrobiologia 124:193–197.

STREET, M., AND G. TITMUS. 1979. The colonisation of experimental ponds by Chironomidae (Diptera). Aquat. Insects 1:233–244.

STRICKLAND, A. H. 1961. Sampling crop pests and their hosts. Annu. Rev. Entomol. 6:201–220.

SWANSON, G. A. 1978a. A simple lightweight core sampler for quantitating waterfowl foods. J. Wildl. Manage. 42:426–428.

———. 1978b. A water column sampler for invertebrates in shallow wetlands. J. Wildl. Manage. 42:670–672.

———. 1978c. Funnel trap for collecting littoral aquatic invertebrates. Prog. Fish-Cult. 40:73.

———. 1983. Benthic sampling for waterfowl foods in emergent vegetation. J. Wildl. Manage. 47:821–823.

TAYLOR, L. R. 1951. An improved suction trap for insects. Ann. Appl. Biol. 38:582–591.

———. 1961. Aggregation, variance and the mean. Nature 189:732–735.

———. 1962. The absolute efficiency of insect suction traps. Ann. Appl. Biol. 50:405–421.

THAYER, G. W., W. E. SCHAAF, J. W. ANGELOVIC, AND M. W.

LACROIX. 1973. Caloric measurements of some estuarine organisms. Fish. Bull. 71:289–296.

THORP, J. H., AND A. P. COVICH, EDITORS. 1991. Ecology and classification of North American freshwater invertebrates. Academic Press, Inc., San Diego, Calif. 911pp.

TONKYN, D. W. 1980. The formula for the volume sampled by a sweep net. Ann. Entomol. Soc. Am. 73:452–454.

TOWNES, H. 1962. Design for a Malaise trap. Proc. Entomol. Soc. Wash. 64:253–262.

TURNBULL, A. L. 1966. A population of spiders and their potential prey in an overgrazed pasture in eastern Ontario. Can. J. Zool. 44:557–583.

———, AND C. F. NICHOLLS. 1966. A "quick trap" for area sampling of arthropods in grassland communities. J. Econ. Entomol. 59:1100–1104.

UNDERWOOD, A. J. 1981. Techniques of analysis of variance in experimental marine biology and ecology. Oceanogr. Mar. Biol. Annu. Rev. 19:513–605.

VAN ARK, H., AND L. M. PRETORIUS. 1970. Subsampling of large light trap catches of insects. Phytophylactica 3:29–32.

VENRICK, E. L. 1971. The statistics of subsampling. Limnol. Oceanogr. 16:811–818.

VOIGTS, D. K. 1976. Aquatic invertebrate abundance in relation to changing marsh vegetation. Am. Midl. Nat. 95:313–322.

VOSHELL, J. R., JR., AND G. M. SIMMONS, JR. 1977. An evaluation of artificial substrates for sampling macrobenthos in reservoirs. Hydrobiologia 53:257–269.

WAAGE, B. E. 1985. Trapping efficiency of carabid beetles in glass and plastic pitfall traps containing different solutions. Fauna Norv. Ser. B. 32:33–36.

WATERS, T. F. 1969. Invertebrate drift-ecology and significance to stream fishes. Pages 121–134 in T. G. Northcote, ed. Symposium on salmon and trout in streams. Inst. Fish., Univ. British Columbia, Vancouver.

———, AND R. J. KNAPP. 1961. An improved stream bottom fauna sampler. Trans. Am. Fish. Soc. 90:225–226.

WATERS, W. E., AND V. H. RESH. 1979. Ecological and statistical features of sampling insect populations in forest and aquatic environments. Pages 569–617 in G. P. Patil and M. Rosenzweig, eds. Contemporary quantitative ecology and related econometrics. Int. Coop. Publ. House, Fairland, Md.

WELCH, H. E., J. K. JORGENSON, AND M. F. CURTIS. 1988. Measuring abundance of emerging Chironomidae (Diptera): experiments on trap size and design, set duration, and transparency. Can. J. Fish. Aquat. Sci. 45:738–741.

WENZEL, F., E. MEYER, AND J. SCHWOERBEL. 1990. Morphometry and biomass determination of dominant mayfly larvae (Ephemeroptera) in running waters. Arch. Hydrobiol. 118:31–46.

WHITMAN, W. R. 1974. The response of macro-invertebrates to experimental marsh management. Ph.D. Thesis, Univ. Maine, Orono. 103pp.

WIEGERT, R. G. 1974. Litterbag studies of microarthropod populations in three South Carolina old fields. Ecology 55:94–102.

WIGGINS, G. B., AND R. J. MACKAY. 1978. Some relationships between systematics and trophic ecology in Nearctic aquatic insects, with special reference to Trichoptera. Ecology 59:1211–1220.

WILLIAMS, C. J. 1985. A comparison of net and pump sampling methods in the study of chironomid larval drift. Hydrobiologia 124:243–250.

WILLIAMS, D. D., AND N. E. WILLIAMS. 1974. A counterstaining technique for use in sorting benthic samples. Limnol. Oceanogr. 19:152–154.

WILLIAMS, D. F. 1973. Sticky traps for sampling populations of *Stomoxys calcitrans*. J. Econ. Entomol. 66:1279–1280.

WISSING, T. E, AND A. D. HASLER. 1968. Calorific values of some invertebrates in Lake Mendota, Wisconsin. J. Fish. Res. Board Can. 25:2515–2518.

———, AND ———. 1971. Intraseasonal change in caloric content of some freshwater invertebrates. Ecology 52:371–373.

WOBESER, G. A. 1981. Diseases of wild waterfowl. Plenum Press, New York, N.Y. 300pp.

WRONA, F. J., J. M. CULP, AND R. W. DAVIES. 1982. Macroinvertebrate subsampling: a simplified apparatus and approach. Can. J. Fish. Aquat. Sci. 39:1051–1054.

WRUBLESKI, D. A. 1984. Species composition, emergence phenologies, and relative abundances of Chironomidae (Diptera) from the

Delta Marsh, Manitoba, Canada. M.S. Thesis, Univ. Manitoba, Winnipeg. 115pp.

————, AND D. M. ROSENBERG. 1984. Overestimates of Chironomidae (Diptera) abundance from emergence traps with polystyrene floats. Am. Midl. Nat. 111:195–197.

————, AND ————. 1990. The Chironomidae (Diptera) of Bone Pile Pond, Delta Marsh, Manitoba, Canada. Wetlands 10:243–275.

ZELT, K. A., AND H. F. CLIFFORD. 1972. Assessment of two mesh sizes for interpreting life cycles, standing crop, and percentage composition of stream insects. Freshwater Biol. 2:259–269.

ZICSI, A. 1958. Determination of number and size of sampling units for estimating lumbricid populations of arable soils. Pages 68–71 *in* P. W. Murphy, ed. Progress in soil ecology. Butterworths, London, U.K.

15

WILDLIFE RADIOTELEMETRY

Michael D. Samuel and Mark R. Fuller

INTRODUCTION

The development of radiotelemetry techniques has tremendously influenced the direction of wildlife research. This "high-tech" approach has provided increased opportunity to examine detailed ecological and management questions related to movement, behavior, habitat use, survival, productivity, and many other related questions of interest for individual animals. Radiotelemetry has enhanced the ability of wildlife ecologists to locate animals at will, make observations of their habits, detect and determine the proximate cause of mortality, and record changes in physiological processes occurring within the animals. Perhaps, in many respects, these numerous opportunities to acquire detailed and unique knowledge of animal ecology have been a mixed blessing. Many research studies have used the attachment of radio transmitters and collection of telemetry data as a substitute for the innovative application of radiotelemetry and other techniques to answer research questions. Such studies have few or poorly defined research objectives, lack clear definition of experimental units, and are typically characterized by post-facto (or ad hoc) interpretation of the data in hand (Lance and Watson 1980). This approach has generated and undoubtedly will continue to generate interesting research questions and insights into animal ecology. However, recognition of radiotelemetry as a research technique within the framework of the scientific method will provide more rapid and reliable advances in biological understanding.

Many telemetry users have concentrated their efforts on particular aspects of telemetry studies. Some have stressed sophisticated equipment and field protocols but have used poor analytical procedures or hoped that someone else could make sense of the data when the study was finished. Others have used sophisticated analytical methods, but ignored sampling and design considerations. All four aspects of a telemetry study (design, equipment, field methods, and data analysis) are critical to the successful completion of a research study. Each phase must be given appropriate consideration to ensure that the entire project does not fail. Researchers should recognize that even the most sophisticated analysis cannot salvage a poorly designed study, nor can a complex design overcome unsuitable or unreliable equipment and field procedures.

This chapter identifies and briefly describes some of the basic considerations in designing telemetry studies, selection of radiotelemetry equipment, and uses of telemetry equipment under field conditions and discusses appropriate analyses of telemetry data. The equipment used for telemetry studies and the methods available to analyze telemetry data are evolving rapidly in diversity and complexity. Researchers and managers who need to investigate the latest techniques must pay close attention to the pertinent literature and establish a network of colleagues to exchange advice and results. Equipment limitation, especially related to transmission range and life-span, probably will continue to pose the most severe restrictions on telemetry applications. Within these limitations, the electronics have progressed enough to accommodate most research needs provided the necessary money and expertise are available. Similar advances also have been made in the development of computer software to analyze and interpret telemetry data.

STUDY DESIGN
Initial Considerations
STUDY OBJECTIVES

Radiotelemetry should be viewed as a technique that can increase the efficiency of collecting information. Establish biological objectives, then determine if radiotelemetry might be a useful method for achieving those objectives. Researchers should remember that the most frequent application of telemetry is to study differences among individual animals (or closely associated social groups). Ecological understanding of how, when, and why such differences occur are likely to be the focus of well-designed studies using radiotelemetry.

Study objectives usually dictate the type of equipment needed, data-collection procedures, costs, and data analyses. Some project objectives often are not completely compatible with the requirements of other objectives. Hence, compromise among different objectives might be a necessity. A precise delineation of study objectives and their relative priorities is the first step in achieving the appropriate study design and successful results. Because telemetry studies tend to be expensive in personnel and equipment cost, they also tend to be multipurpose so that many questions can be addressed from a single data set. Therefore, careful planning is essential to avoid wasting money and effort on ill-conceived radiotelemetry studies. Successful radiotelemetry studies do not happen by accident. Typically, the best studies are a result of careful design, compromise between competing objectives, and realistic assessment of the technical capabilities (Sargeant 1980).

ALTERNATIVE METHODS

A thorough assessment of any research plan should involve a consideration of the alternative methods suitable to accomplish the research goals. Such an assessment must evaluate a variety of factors, including probability of successfully meeting the research objectives, project costs, cost per unit of data, time available, methodological assumptions and limitations, and availability of necessary personnel or equipment. Because of the diversity of applications, radiotelemetry is likely to be a method commonly used in many research projects; however, researchers should always consider alternative tools. Can the necessary data be obtained simply by observing individual animals (either marked or unmarked)? Can the objectives be achieved by experimental methods (e.g., enclosure or exclosure studies)? Because attachment of a radio transmitter may influence the behavior or even survival of some animals, alternative methods may provide more reliable information.

The evaluation of alternative methods requires consideration of the number of animals required, sampling frequency, behavioral characteristics of the animals to be studied, necessity of identifying individual animals, potential movement of animals, and other logistic considerations. Kenward (1987) discussed several alternative techniques for obtaining appropriate data. He suggested that consideration be given to visual markers (Stonehouse 1978), including Beta lights (Buchler 1976, Wolcott 1980, Mullican 1988), radioactive tags (Bailey et al. 1973, Linn 1978), and radioisotope marking (Kruuk et al. 1980, Jenkins 1980). In addition, potential biases and reliability of the data collected as a result of the particular method should be seriously evaluated. For example, the inability to locate accurately a radioed animal in relation to habitat characteristics will negate the use of radiotelemetry in habitat analysis.

COSTS

The costs associated with any research will depend largely on the specifics of the study. Radiotelemetry novices should contact experienced telemetry researchers to determine logistical requirements and estimate necessary costs from similar studies. The development and formal review of project proposals that include estimated expenses are typically a valuable exercise.

The typical costs associated with radiotelemetry projects are the purchase of radio equipment (transmitters, antennas, and receivers), salaries and expenses for field personnel, and costs for transportation (e.g., vehicles and aircraft). Personnel expenses will be determined partially by study design, including the number of radio-marked animals, the frequency of obtaining locations, and the difficulty in obtaining location data. Researchers should consider the costs and personnel associated with computer data entry, computer processing and analysis, report preparation, and publication. This phase of the project typically is given the least consideration in planning research projects. When a substantial amount of telemetry data has been collected, this final phase can require considerable time and expense.

Cost undoubtedly will influence study design because it limits how many animals can be studied, how often animals will be located, and many other logistic concerns. Frequently, a good study design will be too costly to achieve, but the biologist will continue anyway in hopes of gaining at least some useful information (Pollock 1987). Although these efforts might not be a waste of time, researchers should be aware of the limitations of their studies and attempt to improve the design where possible.

Research Design

Research design is concerned with planning a study to obtain the maximum amount of information from the available resources. An important step in research design is the determination of the appropriate experimental unit for data collection and analysis. This is not a trivial undertaking, because different levels or sizes of experimental units can be used at a particular phase of a research project. For example, previous studies on animal survival have analyzed the animal-days of life, others have analyzed the number of animals alive or dead during sequential time intervals, and still others have been concerned simply with the length of time each animal lives. Such differences in experimental units generate a degree of ambiguity and confusion related to sample size and statistical conclusions. Readers should refer to Ratti and Garton (Chapter 1) for a detailed discussion of research and experimental design methods.

Telemetry studies typically involve two levels of experimental units—the individual animals and the measurement of each animal's response over time. For instance, one bird followed for 100 days, 100 birds followed for 1 day, or five birds followed for 20 days can provide a similar number of animal exposure days, but very different conclusions about survival rates. Monitoring the survival of only a single animal is obviously a poor design because it fails to provide replication of animals when different populations are compared. Conversely, monitoring 100 animals for only 1 day can be impractical logistically and produce a poor measurement of the survival time for each animal. The intermediate strategy of monitoring five animals for 20 days provides an improved design with replication of animals and days of monitoring.

The concept of levels of experimental units can be illustrated further by monitoring the home range and habitat use by random telemetry locations of hunted and unhunted populations of deer. A researcher can use the set of locations for each animal to estimate an individual home range (first level experimental unit) for each of the hunted and unhunted animals. Mean home-range size then can be compared for statistically significant differences between populations. In a second analysis, a researcher can estimate proportional habitat use from the set of locations (second level experimental unit) for each animal and test for habitat selection by each animal. The estimate of individual habitat use applies to the collection of all possible locations of each individual. It is a valid measure for each individual monitored, but not for either population of deer. Valid measures of habitat use for the hunted or unhunted population should consider that an individual animal (first level) is the appropriate experimental unit, rather than each animal's location (second level).

These apparent conflicts in study design can be avoided by careful definition of the appropriate experimental units and an understanding that a hierarchy of experimental units is possible. Each level of experimental unit then is associated with different levels of research questions within the study design. Different data analysis and research conclusions then can be based on a particular selection of experimental units from different population levels in the research design. Unfortunately, increasing replication (and therefore precision) at one level of the design usually reduces the replication at a different level. Frequently such trade-offs can lead to a research design that is inefficient. Researchers should use care to ensure that study objectives correspond with the appropriate number of experimental units, so that objectives at the desired level of inference (research population) can be answered.

Telemetry Sampling

A unique aspect of telemetry studies is that animal locations and other response measures involve a spatial component and a time component. This time component has long been recognized in sampling the behavior of animals (Altmann 1974), but the importance of a time component only recently has been considered in spatial telemetry research (Dunn and Gipson 1977, Swihart and Slade 1985a, Reynolds and Laundre 1990). A variety of sampling designs that recognize the importance of sampling animals through time is available for use in telemetry studies (Cochran 1977, Scheaffer et al. 1986). Designs including simple random sampling, stratified random sampling, systematic sampling, and cluster sampling (Chapter 1) are likely to be applicable to radiotelemetry studies.

Radiotelemetry involves the study of individual animals and therefore is ideally suited to comparisons among individuals. However, if the research goal is to compare population parameters, attention must be given to the sampling design and sample size. A primary objective in selecting a sampling design involves the organization of experimental units (animals) into homogeneous groups (e.g., age, sex, reproductive status). This grouping reduces variability associated with ancillary factors and provides a more powerful test of differences between sample populations. Effective use of sampling designs will reduce the number of experimental units required, thus reducing cost and increasing research efficiency.

Telemetry involves resampling animals through time. Thus, a secondary criterion in selecting a sampling design is related to the frequency of sampling individuals. This procedure also must consider logistical efficiency and cost per unit of information collected. Random sampling usually is preferred in many telemetry studies, because the resulting data are statistically independent and simpler to analyze. However, when animals are located to obtain several measurements (e.g., location, habitat use, and behavior), frequently some measurements will be serially correlated (autocorrelated). Measurements that change rapidly, such as behavior, can be sampled more frequently than location or habitat use to obtain independent samples. Serially correlated samples should be analyzed with appropriate statistical methods. Reynolds and Laundre (1990) located coyotes and pronghorns at 30-minute intervals during 24-hour tracking sessions. The minimum time before sequential locations became statistically independent was 4 hours for pronghorns and 6 hours for coyotes. Carey et al. (1989) determined that spotted owl locations were statistically independent when obtained at 3- to 5-day intervals. The effect of using only independent relocations reduced their data set by 60–70%. As an alternative, the measurements can be subsampled to remove serial correlation. The latter approach is not always readily accepted, because some data will be disregarded in this analysis. Researchers are urged to consider the problems of sampling frequency in the design of a research project to avoid difficult decisions during the analysis.

In many telemetry studies, the logistics required to initially locate an animal may be far greater than obtaining sequential data once the animal is located. In this situation, cluster sampling may provide a reasonable scheme for collecting information about an animal. In cluster sampling, an animal is located at a random time and data are collected for a fixed period to provide information on the animal's activities. This method of collecting a burst (cluster) of information on one animal may be more efficient than attempting to obtain only independent locations (Reynolds and Laundre 1990). Andersen and Rongstad (1989) used bursts of relocation data to calculate the home ranges of hawks. Samuel and Garton (1987) reported a method for analyzing behavioral use patterns based on bursts of observation data.

A final sampling design that has been used frequently in radiotelemetry studies is called systematic sampling, in which data on individual animals are collected at fixed time intervals (e.g., daily). This method is used commonly to monitor the daily survival of each study animal. However, this approach also has been used, in conjunction with automatic recording systems, to record presence/absence of animals on a study area (Williams and Williams 1970), animal activity (Cooper and Charles-Dominique 1985), and home-range patterns (Heezen and Tester 1967).

Alternative sampling designs such as stratified, proportional to size, and other procedures (Chapter 1) also have been devised for sampling. However, when these procedures are used, the data should be analyzed according to the appropriate sampling scheme. This approach will require the assistance of a statistician in study design and data analysis. Because many of the methods commonly used to analyze telemetry data assume a random sample, modifying these methods or developing new methods could be necessary if telemetry data cannot be gathered by a random sampling design.

Generalizations on the frequency of sampling are difficult to provide and depend on the study objectives, the variables being measured, and the tradeoffs between more data on a few animals versus fewer data on more animals. Collecting samples so frequently that they have high serial correlations will provide a limited amount of additional new information over that achieved with fewer samples. A priori knowledge of an animal's behavior and movement patterns can assist in selecting a sampling scheme. Pilot studies (Smith et al. 1981, Carey et al. 1989) can be used to provide adequate estimates of experimental variation and independence. Reynolds and Laundre (1990) provided an example of the problems involved in determining a sampling frequency to simultaneously obtain movement and home-range estimates with radiotelemetry data.

Collection of considerable data on a few individuals at the expense of sampling more individuals should not be overvalued. It is important to develop a balance between the accuracy of measuring a characteristic for each individual and the number of individuals to be monitored (Alldredge and Ratti 1986). A general tendency is to intensively monitor a few animals. These results are interesting and provide potentially detailed understanding about these specific individuals. However, because few individuals are monitored, little information will be gained about the population of animals. Investigators should acknowledge the limitations of their results and avoid far-reaching statements from limited data sets.

Determination of the amount of replication required for each study also can be a difficult procedure. Typically, this determination requires an estimate of the variation associated with experimental units, specification of the level of statistical certainty (Type I and Type II errors), and a hypothetical difference between the sample populations. Investigators should refer to texts by Zar (1984) and Cohen (1977), or to Ratti and Garton (Chapter 1), and consult a statistician to determine sample size requirements and Type II error rates for particular statistical tests and research studies.

TELEMETRY EQUIPMENT

Wildlife telemetry equipment, like that described here and in the literature, is available from commercial manufacturers (Appendix I); however, designs frequently are modified to take advantage of new components and improved assembly techniques. These developments mean biologists must rely on the recent experience of others, the equipment specifications, and the results of a pilot study to determine the adequacy of equipment for meeting their requirements. We encourage biologists to discuss their requirements with manufacturers during the planning stages of a telemetry study. Manufacturers should assist biologists to modify, repair, or replace equipment when necessary. Smith and Amlaner (1989) defined many terms, briefly discussed concepts, and described basic electronic circuitry used for wildlife radiotelemetry. We provide the biologist only an introduction to the equipment and terminology associated with wildlife radiotelemetry.

Transmitting Systems

The radio transmitter often is the first piece of equipment that biologists consider for wildlife tracking or biotelemetry. The signal a transmitter can send (e.g., "beeps" for direction finding, calibrated pulses for activity or temperature sensing), the size and mass of the transmitter package, the strength of the radio signal, and duration of operation are the critical characteristics that determine if radiotelemetry is a potentially useful tool. Complete transmitter packages (or radio-tags) comprise electrical circuitry, a power source, a transmitting antenna, encapsulation (often called potting), and material for attaching the package to an animal (Fig. 1). These transmitter components should be selected to meet research objectives and to ensure that radio-marking does not adversely affect the animal. Kenward (1987) listed components for construction of several common wildlife transmitters and provided circuit diagrams and step-by-step instructions for radio-tag assembly.

BASIC TRANSMITTER CIRCUITS

Radio transmitters are designed to minimize spurious emissions that waste power and clutter the frequency spectrum. Quality components and careful construction are required to control the length of the transmitted pulses (pulse width) and the time between pulses (pulse interval). Pulsing, rather than continuous transmission, conserves power and is useful for radio-tracking and data transmission. Limitations in receiver technology and human perception dictate that pulse width be 20–35 msec and that pulse rate be 45–80 pulses/minute. Longer pulses are easier to distinguish during noisy reception, and determination of the direction from which rapid pulses are coming is easier. Experienced trackers can use the lower end of these ranges and thus conserve battery energy.

Animals usually are marked with radio transmitters that send a signal on a unique radio frequency (Box 1). The fundamental radio signal for a wildlife transmitter is generated by a stable quartz crystal oscillator designed to operate at a specific frequency. Additional circuitry is added to multiply this fundamental signal to the desired transmission frequency (e.g., 75 MHz $\times$ 2 = 150 MHz). However, quartz crystals have a prescribed tolerance within a defined temperature range, which commonly causes frequency drift of about ± 1–2 KHz in a completed transmitter. Because of these inherent transmitter variations, biologists must specify adequate spacing between transmitter frequencies to ensure that misidentification does not occur. The amount of spacing required will depend on the characteristics of both the transmitters and the receiver, but 10 KHz usually is recommended.

Basic transmitter circuits can be wired as discrete components on either a printed circuit board or soldered together in point-to-point assemblies. Some transmitters incorporate integrated circuits for improved signal pro-

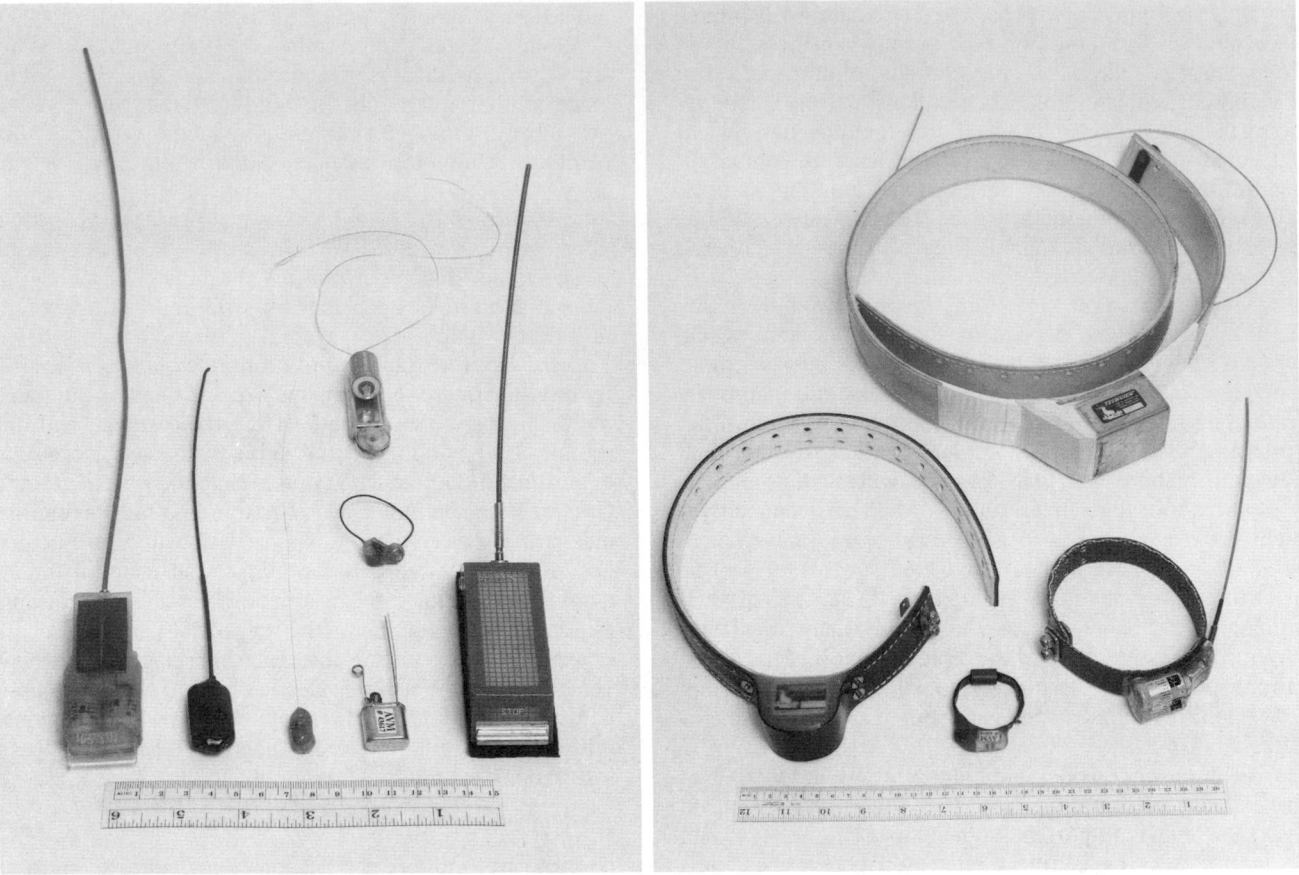

Fig. 1. Wildlife radio transmitters. From left to right: solar power and NiCad batteries (55 g) for large birds; lithium battery power (12 g) for medium birds; mercury battery power (2.3 g) for small birds; top—implant transmitter components and leads for sensing ECG (<50 g, when potted); middle—fixed-loop antenna collar (2.8 g) for small mammals; bottom—case in which transmitter components are hermetically sealed; platform transmitter terminal (PTT) to operate with Argos satellite system (85 g); four sizes of adjustable collars for radio-marking mammals, clockwise: 13 g, 107 g, 768 g, and 268 g.

cessing and digital logic. For example, integrated circuits (e.g., complementary symmetrical metal oxide semiconductors or CMOS) are used for pulse control and switching applications. Some transmitters use a secondary quartz clock for timing and modulation. Timing devices in the circuit also can be used to conserve power by shutting off transmission at a certain time of day, or by delaying the onset of transmission for some time after the animal is radio-marked (e.g., until dispersal or onset of migration). Normally, transmitters are turned on by soldering leads (which then must be potted) or by removing an external magnet to close a reed switch and power the circuit.

Wildlife radio transmitters usually are classified as one-stage or two-stage (some three-stage are built), depending on the number of transistors used in their construction. With current technology, a transmitter can weigh as little as 0.5 g (excluding battery). The total mass, with battery and potting, can be 0.8–1.2 g for a unit that will transmit for 20–30 days. However, one-stage transmitters usually operate on one battery (1.35–1.5 V) and radiate ≤ -10 dBm (Box 2). The "second stage" of a two-stage transmitter increases radiated power by draining more current from the battery, which reduces battery life. A two-stage transmitter operating at the same voltage and antenna length should radiate about

10 times more signal power than a one-stage transmitter. The signal power of two-stage transmitters can be enhanced by increasing battery voltage (3.5–15 V); however, this increases transmitter mass and design complexity. Kenward (1987:24–25) recommended two-stage transmitters for wide-ranging animals, because signal power might be needed to increase reception range or to overcome noise, such as for automatic receiving and data recording.

The power of a transmitted signal and the operational life of the transmitter usually are the major concerns of the biologist. Unfortunately, the batteries that provide power and operational life are also the bulkiest and heaviest part in most transmitter packages, and most animals are limited in the size and mass of a transmitter they can carry without incurring energetic or behavioral costs. Thus, specifications for signal strength and operational life also must consider the animal's welfare. Tuning of the circuitry, voltage of the power source, and characteristics of the transmitting antenna also influence effective radiated power (ERP), which is commonly measured in dBm at the transmitting antenna. Biologists will find differences among transmitters with the same specifications because manufacturers use different components and circuit designs and because the matching and tuning of components vary.

Box 1.   Description and measurement of radio signals.

Radio frequencies are measured in cycles per second, a Hertz (Hz); thus 100 Hz = 100 cycles per second. Radio signals occur in a range from thousands to billions of cycles per second with quantities abbreviated by the conventional metric prefixes:

G = giga = 1 billion (10^9)
M = mega = 1 million (10^6)
K = milo = 1 thousand (10^3)
m = milli = 1 thousandth (10^{-3})
U = micro = 1 millionth (10^{-6})
N = nano = 1 billionth (10^{-9})
P = pico = 1 trillionth (10^{-12})

Therefore, 164.000 megaHertz (MHz) is 164,000,000 cycles per second.

Radio signals travel in the form of sine waves, and a cycle is a wavelength. A high-frequency radio signal has a short wavelength and has more cycles per second (e.g., 216 MHz) than a low-frequency signal (e.g., 40 MHz).

Wavelength (λ) usually is expressed in meters and can be determined by the formula:

$$\lambda = \frac{300}{\text{frequency (MHz)}}.$$

The value 300 is the speed ($\sim 300 \times 10^6$ M s^{-1}) at which electromagnetic waves (e.g., radio, radar, microwaves, infrared, x-ray) travel. The wavelength of a radio frequency is an important characteristic because it dictates antenna size.

Radio waves are categorized into spectral bands such as long waves, short waves, Very High Frequency (VHF), and Ultra High Frequency (UHF), each containing a range of radio frequencies (VHF = 30–300 MHz). However, the term band or bandwidth also is used to describe a specific range of frequencies (e.g., a 2-MHz band of 164–166 MHz).

Box 2.   Decibels—measurement of transmitter power.

A decibel (dB) is a measure of power, equal to 10 times the base 10 logarithm of the power in milliwatts of electricity; dBm = 10 log P/1 mw.

Power (milliwatts)	dBm
1000 (1 watt)	+30
100	+20
10	+10
4	+6
2	+3
1	0
0.1	−10
0.01	−20
0.001	−30
...	...

The decibel also is used for expressing transmission gains, losses, and levels for antennas. In this case, the measure can be relative to a theoretical isotropic point source, dBi, or a λ/2 dipole antenna, dBd (dBd = dBi + 2.14 dB). A well-tuned, three-element Yagi antenna has a gain of about 7 dB over a λ/2 dipole.

TRANSMITTING ANTENNAS

Transmitting antennas often are the weakest link in a wildlife radiotelemetry system because they are inherently inefficient and structurally weak. The point at which the antenna exits the transmitter frequently is a place where moisture enters the circuitry and where whip antennas are likely to break. Moisture in the circuitry or antenna damage often causes transmitter failure before the battery power is depleted. A damaged antenna destroys the performance characteristics of the transmitter; therefore, biologists should consider carefully how the antenna is incorporated into the transmitter package for each species. Ideally, the transmitting antenna should be one-quarter wavelength long (see below), perpendicular to the earth's surface, and attached so it is not against the animal's body. However, antennas are always a compromise because an animal's size usually precludes it from carrying an antenna of the ideal length. Furthermore, an animal's behavior toward a long antenna (e.g., biting a protruding antenna or breaking it by rolling on it or rubbing it against objects) can prohibit use of optimal length or placement.

Loop antennas are made of a brass or copper band, or a coated wire, and incorporated into a collar attachment for animals the size of a raccoon or smaller (Fig. 1). Resonant tuning is critical to effectively radiate a signal and is best accomplished when the transmitter is attached to the animal. If the manufacturer tunes the loop before attachment, the biologist must not change the diameter of the loop.

Whip antennas are less sensitive than loops and can be tuned with the transmitter off the animal. Whip antennas are more efficient signal radiators than loops, especially when they are cut to optimal lengths relative to the transmitter frequency and suitability for the animal (Amlaner 1980). Quarter wavelengths (λ/4) often are recommended and can be calculated in centimeters by: 7,500 cm/frequency in MHz. Shorter antennas will greatly reduce the ERP of the transmitter (Fig. 2). Kenward (1987: 32) recommended whip antennas of at least λ/8, in conjunction with a ground wire two-thirds as long as and perpendicular or opposite in direction to the main whip. Whip antennas often are positioned against or near the animal's body to reduce aerodynamic drag and the antenna's bothering the animal or snagging on objects. Cochran (1980) noted that positioning antennas close to an animal's body can result in a 20-dB loss, and additional loss will occur if the antenna is within λ/2 of the ground (about 1 m at 150 MHz). Consequently the type, length, and

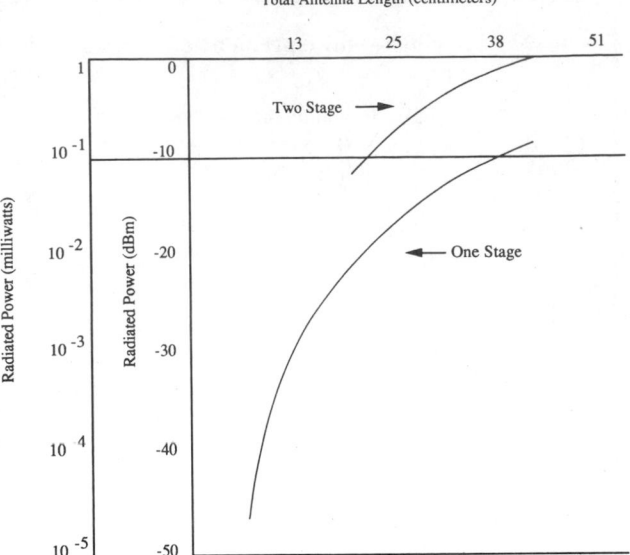

Fig. 2. Relationships between transmitter antenna length and effective radiated power (modified from graph by A. L. Kolz, U.S. Dep. Agric., Denver Wildl. Res. Cent.).

position of the transmitting antenna are important factors in determining the ERP of a transmitter.

Whip antennas usually are made of strong materials including guitar string, single or multiple strands of nickel stainless steel, twisted multistrand dental (orthodontics) wire, fishing line leader, and multistrand stainless steel cable. Often the wires are covered with tough plastic coating or shrink tubing that is heated to mold tightly to the surface of wires and solder junctions. Different lengths of shrink tubing can be layered to reinforce an antenna at the point it enters the transmitter package. Alternatively, a metal spring or cone of silicon sealant at the antenna base can alter the bending radius and thereby decrease wire fatigue. The distal end of the tubing over multistranded wire should be sealed to retard the wicking of moisture along the antenna. Shrink tubing and sealant also help to retard moisture penetration of the package.

POWER SOURCES

In the context of wildlife telemetry, "power" refers to signal strength (dBm) or to the energy sources (batteries and solar cells) for a transmitter. A variety of battery types and solar cell circuits is available to meet the requirements for signal strength and operational life in conjunction with size, mass, and attachment methods. Biologists will encounter many terms that are associated with these power-source options. We briefly introduce readers to some of these terms and recommend Smith and Amlaner (1989) and Kenward (1987) for additional information about transmitter batteries.

Chemical (lithium, mercury, zinc, silver oxide) reactions create electrical current in batteries. The loss of energy (current drain) from a battery running a transmitter is measured in amperes (amps, A), and the energy capacity of a battery is given in milliamp-hours (mA-hr). The operational life can be estimated by dividing the battery capacity (mA-hr) by the transmitter current drain. This provides a reasonable estimate for lithium batteries, but at

least 3 months for each year of estimated life must be subtracted for other types of batteries.

Battery voltage provides the force that causes movement of electrons, which produces electrical current. Batteries commonly used in wildlife telemetry have 1.35–3.6 volts (V) and can be wired in series to increase the power (1.5 V + 1.5 V = 3.0 V) or in parallel to increase the capacity. Low temperatures reduce a battery's capacity and its ability to deliver current. The power available from most batteries used in wildlife transmitters remains steady until a few days before it is depleted, then the signal pulse rate often increases.

Lithium batteries (2.9–3.9 V) have twice the energy: weight ratio as mercury batteries. Stored lithium cells lose only about 10% of their capacity over 5 years at room temperature. Lithium cells require careful handling because they can be short-circuited and explode. Mercury batteries (1.35–1.4 V) have a short "shelf-life" and therefore should not be stored more than 3 months, except in a freezer, which will extend their life. The voltage of mercury batteries drops near 0 C. They have about the same energy : volume ratio of lithium batteries. Silver oxide batteries (1.55 V) have about the same energy : weight ratio as mercury batteries, but they also have a short shelf-life and are affected by cold. Zinc air batteries (1.45 V) have two times the energy density of mercury or silver oxide cells. They can be used only in situations where moisture and dirt do not enter the vent hole and preclude air from reaching the zinc where it produces the chemical reaction.

Photovoltaic solar cells are an alternative source of power for some wildlife telemetry applications. Solar cells can be used with a capacitor to operate transmitters, even in low light conditions, but not in the dark. Special mounts may be required to keep the solar cells clear of obstructions (e.g., Snyder et al. 1989). A 2-stage transmitter, with solar cells and capacitor, can weigh about 8 g and will operate indefinitely if the transmitter is not damaged. Solar cells also can be combined with rechargeable nickel-cadmium (NiCad) batteries (1.35 V). Generally, sunlight of 4–5 hours is needed to charge the NiCad batteries, but charging capacity is limited in hot or cold temperatures. Depending on the number of solar cells, NiCads have a recharge time : operation time ratio of anywhere from 2:12 to 2:50. For wildlife transmitters, which seldom have a complete discharge, extra time is required to re-establish the recharge:operation ratio. Furthermore, irregular discharge and recharge can cause reduced capacity and ultimately failure of the NiCads. Another combination uses solar cells and a capacitor to power the transmitter during the day, then the circuitry switches to a lithium cell for operation in the dark. Upon depletion of the lithium battery, the transmitter will function only in light.

POTTING

Encapsulation ("potting") of the transmitter circuitry is designed to protect it against moisture, to keep animals from damaging components, and to prevent impact damage during running, flying, and other activities. Moisture will short-circuit or corrode transmitters quickly, causing them to malfunction or fail. A way to ensure moisture will not reach most components is to place them in a hermetically sealed metal canister (Fig. 1). Hermetically sealed transmitters weigh more and cost more than the same design potted in other material.

Transmitters usually are potted in epoxy resins, which come in a variety of chemical combinations with different rates of water penetration, but none is waterproof. Some, such as 3M Scotchcast electrical resins, are intended for use with electronics. Kenward (1987:80) suggested Rapid Araldite for potting, which can be mixed to dry either a bit harder or more flexible, depending on the proportions of hardener to adhesive. Dental acrylic is a hard, tough material that also is used for potting transmitters. Biologists should describe their field conditions to the transmitter manufacturers, especially if saltwater will be encountered, so the most appropriate potting can be applied.

The potting material is applied by "painting" it on very small transmitters or dipping small to medium units in the epoxy repeatedly for thicker coatings. If transmitters are designed to last more than a month, or they are exposed to the attention of the animal, they might require several coatings for adequate protection. Large, long-lived transmitters like those in deer or bear collars usually are cast in a form filled with potting. The potting should be free of air bubbles where moisture can condense and spread toward the components. Biologists should inspect radio tags for cracks in the potting and for "gaps," especially where the antenna or leads enter the potting. Cracks can develop during or after potting, and epoxy might not adhere in some places. Remember to check for a seal at the distal end of a whip antenna and to ensure the base of the whip, where it enters the potting, is sealed and reinforced.

Most potting methods should protect transmitters in terrestrial applications for at least a year, but transmitters to be implanted in an animal require special treatment. Some epoxies, and dental acrylic, are designed to prevent rapid penetration of body fluids and adverse reaction from tissues. However, beeswax has one of the best moisture barriers and has been used as potting since the inception of wildlife radiotelemetry. For implants, beeswax can be adequate by itself (in external units it usually is coated with a harder material). Some special wax formulations, such as Elvax, have better properties relative to melting point, flexibility, and physiological response. Beeswax and wax-based formulations can be cold sterilized in zephirran chloride or chlorhexidine diacetate.

SENSORS AND SPECIAL-PURPOSE TRANSMITTERS

Biologists wishing to use sensors must buy transmitters in which the sensors, or leads from sensors, are built into the circuitry before potting. These transmitters usually require custom circuitry or package design to meet specific project requirements. Sensors have been developed to measure dozens of physical and physiological parameters used in wildlife telemetry (Table 1). Sensor calibration, surgical implantation methods, and sensor site selection (e.g., Diehl and Helb 1986) may require special planning. Implanted transmitters often are used with physiological sensors, and this telemetry requires consideration of many factors (Stohr 1989, see Receiving Systems [p. 379] and Transmitter Attachment Methods [p. 386]).

Sensors

The measurement of movement and temperature is the most common use of sensors in wildlife telemetry. A mercury activity switch or a thermistor (heat-sensitive semiconductor) can be included in the circuitry of one- or two-stage transmitters (Kenward 1987:61) with little increase in cost, mass, or size of the transmitter. A mercury switch (~2.0 g) is a tube in which a ball of mercury can roll back and forth, causing intermittent switch closures and thus increasing or decreasing the transmission pulse rate. In addition to sensing activity (Kunkel et al. 1992), mercury switches have been used for detecting certain behaviors, prolonged inactivity (e.g., hibernation), or mortality (Kenward 1987). With a CMOS circuit and a mercury switch, a specified pulse rate (e.g., 80 pulses/min) can be transmitted after a prescribed period (e.g., 24 hr) of inactivity and thus might indicate mortality. A decrease in body temperature sensed with an implant or thermistor next to the body also is used as an indication of mortality (Lotimer 1980). However, any sensor that increases pulse rate, and thus power consumption, shortens transmitter operation life. In some circumstances the frequency fluctuations caused by antenna movement can be an indication of activity (motion) or certain behaviors (Cederlund and Lemnell 1980, Holthuijzen et al. 1985, Kenward 1987).

Activity or temperature data also can be transmitted by pulse interval modulation (PIM) coding, a change in pulse rate calibrated to a change in the sensor (Anderka 1980). Logic circuits can monitor 300- to 2,000-msec intervals, corresponding to a sensitivity of ≤ 0.1 C. A pulse-interval timer is useful to decode fast pulse rates. A CMOS component also can be used to time transmission at fixed intervals or to store data in a transmitter memory chip (Mohus 1987) for subsequent data analysis (Strikwerda et al. 1985, Fancy et al. 1988, Cupal and Weeks 1989). It is possible to incorporate more than one sensor and circuitry so that several types of data can be sent from a transmitter (Lotimer 1980).

Coded Transmitters

In addition to using coding to convey data from a sensor (Cupal and Weeks 1989), a coded signal can uniquely identify a transmitter. Lotimer (1980) used pulse-interval modulation to distinguish a few transmitters operating on the same frequency. Digital coding can be used to uniquely identify many (theoretically up to 65,000) transmitters operating on the same frequency (Anderka 1984, Howey et al. 1989) and monitor presence-absence of radio-tagged animals for up to 1 year. This configuration cannot be used for conventional direction finding, but can indicate general direction by switching among four-element antennas pointing in different directions (e.g., east, north, west, and south).

Egg Telemetry

Data on temperature, relative humidity, incubation, and relative egg position in the nest have been transmitted from sensors built into artificial bird eggs (Howey et al. 1977, 1987, Schwartz et al. 1977). Artificial eggs also have been equipped with a circular mercury switch (Boone and Mesecar 1989).

Table 1. Sensor types and applications for wildlife telemetry.

Sensor	Application	Animal	Reference
Pressure transducer	Air pressure	Birds	Bögel and Bruchard 1992
Activity (mercury switch)	Behavior	Birds	Kenward et al. 1982
	Daily pattern	Waterfowl	Swanson et al. 1976
		Black-tailed deer	Gillingham and Bunnel 1985
		Wolves, deer	Kunkel et al. 1992
		Mongoose	Palomares and Delibes 1991
Temperature	Deep body temp.	Small birds	Reinertsen 1982
	Temp. cycles	Small mammals	Vogt et al. 1983
	Temp., activity patterns	Small mammals	Osgood 1980
	Thermoregulation	Wombats	Brown and Taylor 1984
	Detecting flight	Woodcock	Kenward et al. 1982
	Body temp.	Turtles	Brown et al. 1990
Heart rate	Heart rate, temp.	Domestic fowl	Duncan and Filshie 1980
		Sandhill cranes	Klugman and Fuller 1990
	Heart rate, harassment	Bighorn sheep	MacArthur et al. 1979
		Herring gull	Ball and Amlaner 1980
	Heart rate, behavior	Blackbird	Diehl and Helb 1986
	Heart rate, temp.	Ghost crabs	Wolcott 1980
	Heart rate, respiration rate	Waterfowl	Woakes and Butler 1989
Blood flow	Blood flow	Animals ≥100 g	Smith and Barnes 1989
Electrocardiogram	ECG, heart rate	Captive small mammals	Stohr 1989
	ECG, temperature	Animals ≥50 g	Smith and Moore 1989
Air flow	Respiration	Waterfowl	Woakes and Butler 1989
Neural activity	Electroencephalogram	Northern bobwhite	Schmidt et al. 1989
Muscle activity	Gastric motility	Barred owl	Kuechle et al. 1987
	Chewing	Reindeer	Kokjer and White 1986
Water	Salinity	Polar bear	Garner et al. 1989
	Urine	Primate	Charles-Dominique 1977
Light	Light	Rabbit	Althoff et al. 1989
Sound	Vocalizations	Monkey	Gautier 1980
	Vocalizations, sounds	Porcupine	Alkon and Cohen 1986
Eggs	Temp., humidity, light, position	Waterfowl	Howey et al. 1977

Transmitters for Capture Assistance

Radiotelemetry can be used to trigger a release device on traps and to monitor the status of a trap (Hayes 1982, Nolan et al. 1984). Transmitters can be attached to prey that might be taken to a den, nest, or nest site. Also, transmitters can be dropped at areas of interest (e.g., nest) during aerial surveys, then subsequently located quickly on the ground (e.g., Nicholls et al. 1981). Transmitters are available for use with anesthetic darts, enabling biologists to track animals until the drugs take effect (Lovett and Hill 1977). Radiotelemetry also can be used to activate an anesthetic dart placed at a bird's nest (Wilson and Wilson 1989) or incorporated in the radio-transmitter collar of large mammals (Delgiudice et al. 1990, Mech et al. 1990).

Telemetry via Satellite

Platform Transmitter Terminals (PTTs) work with the Argos-Tiros satellite system (Fancy et al. 1988) and incorporate many of the features presented above. However, PTTs operate on an Ultra High Frequency of 401.650 MHz, transmitting identification and data from up to eight sensors. The signals are digitally encoded on a pulse width of ~0.33 second and a pulse interval of every 50–90 seconds. The frequency of the signal must be very stable (<2 Hz drift), and the radiated power must be relatively high (~0.25–2.0 W). Consequently, the circuitry is more complex than conventional VHF wildlife transmitters, and PTTs cost ~$2,000–$3,000 each. The signal is received by the satellites, then transmitted to processing equipment on the ground. Locations are estimated by the Doppler principle. Location estimates vary from ±150 m to many kilometers, depending on animal behavior, environmental variables, and data-processing options (Keating et al. 1991).

Wildlife telemetry via satellites can be cost effective for wide-ranging species (Craighead and Craighead 1987) and in remote areas. Advances in electronics (Fuller et al. 1984, Strikwerda et al. 1985) and data processing and analysis (Harris et al. 1990b) continue to make it more applicable to a wider variety of species (Strikwerda et al. 1986, Tanaka et al. 1989, Jouventin and Weimerskirch 1990, Harris et al. 1990b) and objectives.

Receiving Systems

Receiving systems comprise radio receivers, receiving antennas, cables to connect the antenna to the receiver, accessories (e.g., headphones, chargers), counters and decoders, and recording devices. Receiving systems are electronically more complex than transmitters, and manufacturers usually offer few choices of components and options for receivers. Without a receiver, transmitter signals cannot be detected. Equipment failures are not uncommon, therefore biologists always should have a back-up receiver.

RADIO RECEIVERS

Biologists should purchase radio receivers designed specifically for wildlife radiotelemetry or biotelemetry. The receiver must amplify the relatively weak radio signals from wildlife transmitters and reject the stronger signals from other sources on slightly different frequencies (Smith and Amlaner 1989). A tuned radio frequency amplifier, oscillator, and other receiver components are necessary to process the transmitter signal on each frequency, convert the processed signal to an audio tone, and produce other signals for further processing by demodulators, decoders, or pulse counters. Some physiological data from transmitters operated at relatively high power, or over comparatively short distances, can be received with a conventional AM or FM radio (e.g., Smith and Barnes 1989); however, these receivers are not sensitive enough for most wildlife telemetry. Radio-tracking and biotelemetry of free-ranging wildlife usually require a receiver with a sensitivity of -140 to -150 dBm. Most wildlife receivers of this sensitivity are crystal controlled or "crystal synthesized."

Usually, a "noise blocker" is incorporated in the receiver circuitry to reduce interference from common sources, such as auto or aircraft engines. The smaller, simpler receivers (Fig. 3) can be used with about 5–20 transmitter frequencies. Transmitter frequency stability should be $< \pm 0.1$ KHz, and frequency drift over the temperature operating range (-40 to $+70$ C for manual receivers, -10 to $+50$ C for programmable receivers) should be $< \pm 0.5$ KHz. More complex (and more expensive) receivers cover from 1 to 6 MHz bandwidths. Dials or buttons typically are used to select the frequency (channel) to be received. Some receivers have a "fine-tune knob" to achieve the finest resolution (e.g., 1 KHz), and other receivers have digital tuning. Some include a "sweep" option that automatically searches a range (≤ 10 KHz) around the selected frequency. The frequency to which the receiver is tuned usually is displayed on or around the tuning dials and often is given in an LCD display.

Wildlife radio receivers include a control to increase the gain (sensitivity), and some receivers have an automatic gain control (AGC). With AGC, the receiver output is maintained within a range (e.g., 40 dB), even as the signal strength increases or decreases. Commonly, signal strength is indicated by a meter, but some programmable receivers display the signal strength on an LCD. The volume control does not affect the gain, it simply makes the audio signal (including background noise) louder. Receivers should have a built-in speaker and a jack for connecting earphones. Earphones are useful in a noisy environment because external noise is reduced, and the listener can adjust gain and volume to hear weak signals more easily and discern their peak amplitude. Adjusting the sound is important, because operators can suffer hearing damage from radio-tracking. Receivers can have a bayonet (BNC) or threaded (PL259) connector for attaching the coaxial cable from the antenna. Many receivers also include a jack for connecting a recorder or computer (e.g., via RS232 cable).

Wildlife receivers should have low power consumption (e.g., ~40 mA depending on gain setting). They are powered by nonrechargeable (8–10 hr of operation) or rechargeable batteries (5–8 hr), and most receivers have a meter to indicate the status of the supply voltage. Biologists should have extra batteries or battery packs, or a transformer for recharging NiCads, and a back-up receiver with fresh batteries. Most wildlife receivers can be operated or charged with a converter for use with a car battery (12 V).

A programmable, automatic frequency scanning capability is an option for some receivers (Fig. 3). Depending on the type of scanning receiver, the period during which each frequency can be sampled ("listened" for) varies from 0.5 second to 10 minutes. The scanning function can be interrupted manually when a radio signal is detected. Programmable receivers are useful when many transmitters are in the area of reception, when a large area is covered in which many signals might be received, when an area is searched quickly, such as from aircraft, or when the receiver can be left unattended and its output is recorded automatically.

Programmable receivers are larger, more complex, and potentially more delicate than manual models, but all receiving equipment is vulnerable to damage from dirt, moisture, and shock. Manufacturers can advise investigators about the vulnerability of various pieces of equipment and the need for protection with plastic bags, umbrellas, or special carrying cases. Portability of receivers becomes an important factor in dense vegetation, or at the end of a long monitoring session when one is juggling a hand-held antenna, a compass, and a notebook and pen, and when neck and shoulder muscles are sore from a day of supporting the weight of equipment. Biologists should consider carefully the size of equipment and accessories such as protective cases and carrying straps.

RECORDERS, COUNTERS, AND DECODERS

Humans are by far the most common recorders and decoders of signals from a receiver. The human ear and brain are more sensitive listeners than mechanical processors, and they are capable of much more complex programming, including feedback and response. However, humans are prone to fatigue and they are notoriously difficult to standardize and expensive to obtain and maintain. Therefore, some attempts have been made to mechanize the processing of radiotelemetry signals and data. Most of the equipment available from manufacturers measures the intervals between pulses or records changes in signal amplitude, marks the presence or absence of a signal, or decodes a signal and records an analog output. Additional processing capabilities are possible with custom-made systems, often developed in university or public research

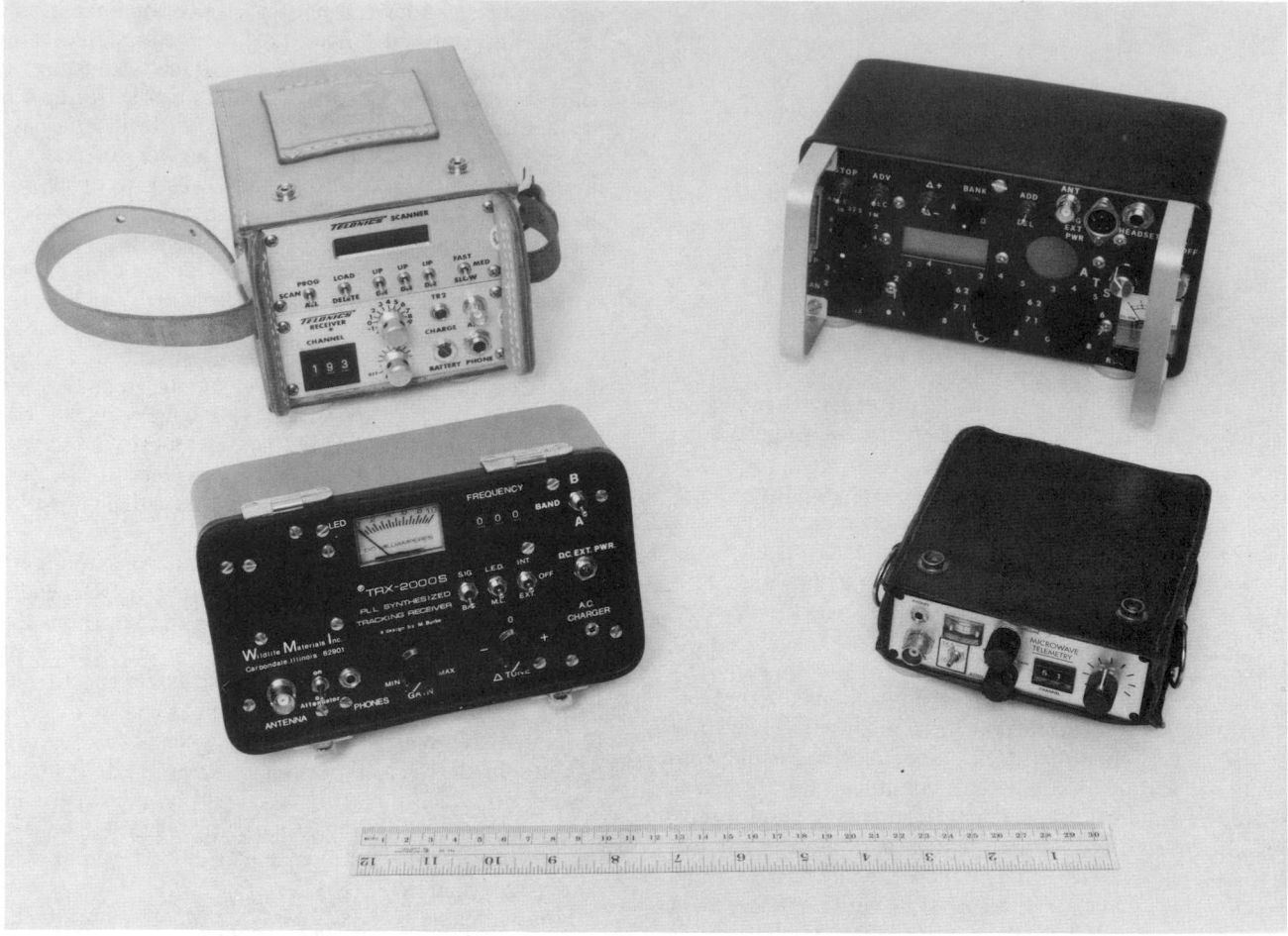

Fig. 3. Wildlife radio receivers. Top left: manual receiver for 180 or 360 frequencies at 10-KHz separation (lower unit, 860 g) with detachable automatic scanning module for 400 frequencies (upper unit, 580 g). Top right: automatic scanning receiver with computer interface (2.3 kg). Lower left: manual receiver for 200 frequencies (900 g). Lower right: manual receiver for 200 frequencies (550 g).

laboratories. Few systems have been developed for automatic radio-tracking.

Recorders

The simplest automatic recording system involves connecting a paper chart (often thermosensitive) recorder to a receiver and recording the presence or absence of a radio signal within range of the receiving antenna (Licht et al. 1989). One- and two-channel chart recorders can be combined with a scanning receiver, but they are limited in the number of frequencies that can be monitored (e.g., Kenward 1987:156). Biologists must consider how often they want a signal recorded and how often they can change the chart paper. Chart speed can vary with input voltage, which is an important factor if the chart recorder is used to time presence-absence or activity events reflected by amplitude changes in the radio signal (Cederlund and Lemnell 1980, Widen 1982). A timing device can be added to strip-chart recorders to provide accurate timing marks on the chart (Gillingham and Parker 1992). Strip charts also can record physiological data including muscular contractions (Kuechle et al. 1987), light and temperature (Althoff et al. 1989), ECG and heart rate (Stohr 1989), and EEG (Schmidt et al. 1989). Deep-cycle bat-

teries designed to be discharged and recharged many times are important equipment for use in the field to power recording and decoding devices.

Signals also can be tape recorded (Kenward 1987). Tape recorders can run continuously, be timed to sample data periodically (Macdonald and Amlaner 1980), or accumulate data over a long duration (Schober et al. 1989, Stanner and Farhi 1989). Tape recorders can be used as the primary recording mechanism (Diehl and Helb 1986, Smith and Aitken 1989), as an alternative to a chart recorder (Stohr 1989), or as permanent storage after some processing (Schober and Oehry 1987).

Computers, from microprocessors to personal computers, can serve as data recording and storage devices as well as processors of the received signal (Janeau et al. 1987, Kuechle et al. 1989). Howey et al. (1987) programmed a computer to run a scanning receiver that sequentially monitored numerous frequencies for heart rate and body temperature data. The received signals were screened for error, pulses were counted, and mean values were periodically transferred to a diskette where they were saved for storage and retrieval. Schober et al. (1989) described computer software for extensive processing of incoming radio signals for decoding and storage.

Counters and Decoders

All data (except presence/absence) recorded on charts, tapes, and computers require some type of decoding. Decoders can count pulses or measure pulse intervals and pulse widths, and convert them to voltage, using a digital counter, with a digital-to-analog option for output (Kuechle et al. 1987). Schmidt et al. (1989) used an analog-to-digital converter to process EEG signals. Althoff et al. (1989) and Stohr (1989) described other customized decoders, and Strikwerda et al. (1985), Cupal and Weeks (1989), and Kunkel et al. (1992) described a microprocessor-controlled decoder built into the transmitter. These systems deal primarily with detecting, processing, and recording physiological signals.

"AUTOMATIC" RADIO-TRACKING

An automatic receiving system processes and stores information without a human operator to dial-in transmitter frequencies or estimate the direction from which the signal has been transmitted. Many automatic direction-finding or location-estimating systems are available for navigation during radio-tracking, including LORAN-C (Patric et al. 1988, Fuller et al. 1989) and the Global Positioning Satellite system. However, these systems currently require transmitter batteries that are too large for most animals to carry. Automatic wildlife tracking systems usually are costly and limited to a small area where transmitter signals can be received. The first, and longest, operating automatic wildlife tracking system (almost 20 years) included two mechanically rotated, stacked Yagi antenna arrays on towers (20 and 30 m in height) about 1 km apart (Cochran et al. 1965). Coverage with this system ranged from several hundred meters to tens of kilometers, depending on transmitter signal strength and position. A similar system operated for several years in France (Deat et al. 1980). Recently, a microprocessor-controlled receiver and automatic, mechanically controlled rotating antennas have been used in the European Alps to obtain bearings about every 2 minutes with an accuracy of ±7° for a stationary transmitter and ±12° for a moving transmitter (Bögel 1991). The large Yagi receiving antennas and mechanical rotating devices in these systems require large support towers and power sources that limit their portability and make them costly. Single or stacked dipole antennas spread around a study area (e.g., Kenward 1987), or electronic switching among arrays of dipoles on masts (Burchard 1989a), hold promise for automatic tracking. Development of better receivers could make automation more practical (Kolz and Castles 1983).

Several other approaches to automatic radio-tracking currently have limited application. A grid of wires placed just above the surface of small areas such as a field, pasture, or pond can locate small mammals (Chute et al. 1974, Zinnel and Tester 1984) or aquatic organisms (Cunningham et al. 1983). An approach involving measurement of the time of arrival of a radio signal at different receiving stations has been used for large terrestrial mammals (Lemnell et al. 1983) and in aquatic environments (O'Dor et al. 1989). Lastly, there is ongoing development of wildlife radiotelemetry via the Argos satellite system to obtain location estimates and sensor data from transmitters anywhere around the earth (Strikwerda et al. 1986,

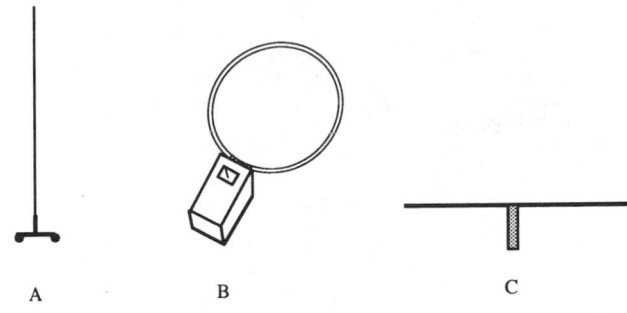

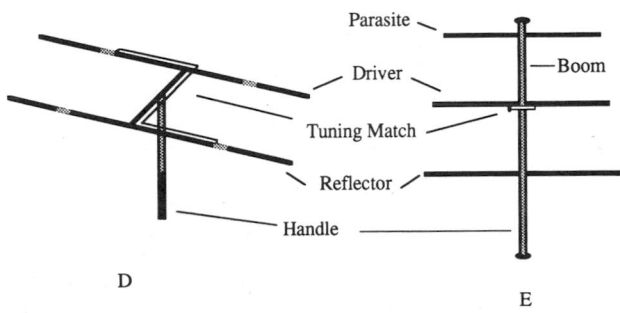

Fig. 4. Receiving antennas: (A) omnidirectional whip, (B) loop (attached to hand-held receiver), (C) dipole, (D) Adcock (H-type), (E) three-element Yagi.

Fancy et al. 1988). Biologists should consult the recent literature and ask manufacturers about advances in automatic wildlife tracking.

RECEIVING ANTENNAS

The receiving antenna, like the transmitting antenna, is a critical component in a radiotelemetry system. Transmitting antennas send the radio signals into the air, and receiving antennas intercept these signals. Biologists frequently underestimate the importance of antennas and the characteristics that influence their effective use. A thorough introduction to antenna fundamentals was given in the American Radio Relay League Antenna Book (ARRL 1988:Chapter 2), and details relevant to wildlife radiotelemetry antennas were given by Amlaner (1980) and Kenward (1987). Hundreds of designs for receiving antennas (ARRL 1988) exist, but only a few are used commonly for wildlife telemetry, mainly because portable antennas often are required so trackers can stay within range of mobile animals. We describe five common, basic antenna designs (Fig. 4) and discuss a few specialty antennas. However, before describing the antenna types, we review some basic characteristics of receiving antennas.

Characteristics of Receiving Antennas

The receiving antenna has signal-gathering capacity (gain) that contributes power to the receiving process. The gain of a receiving antenna is measured in decibels (Box 2). Receiving antennas generate a three-dimensional pattern based upon their physical configuration (Fig. 5). Depending on the antenna geometry, signals can be received more efficiently in certain directions relative to the anten-

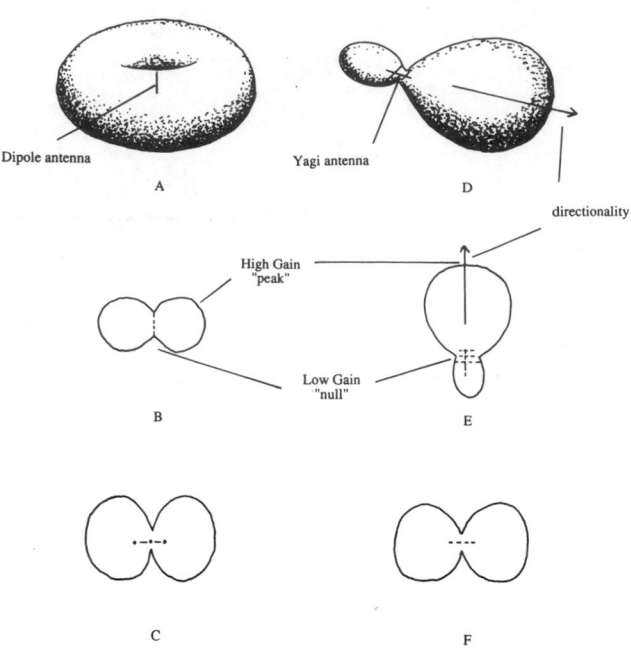

Fig. 5. Reception patterns of receiving antennas: (A) three-dimensional pattern of a vertical whip dipole, (B) cross section of a vertical whip dipole, (C) cross section of an Adcock ("H") antenna, (D) three-dimensional pattern of a three-element Yagi antenna, (E) the horizontal cross section pattern of a three-element Yagi, and (F) a loop antenna. Direction from the Yagi to the source of the radio signal is most readily detected when the area of high gain is pointing toward the transmitter (modified from Kenward 1987; used with permission from Academic Press).

na orientation. Thus, the peak intensity of a radio signal will be detected by a receiving antenna when the lobe(s) of the pattern with the greater gain is directed toward the signal. In addition to this "peak" signal, there will be a "null" with comparatively weak or no reception between the lobes (Fig. 5). The actual pattern of reception will most resemble the theoretical pattern when the antenna is deployed at distances $>\frac{1}{2}\lambda$ from anything that might affect its properties. Conductors such as wires, a car roof, moisture, or a person's body can distort patterns, reduce directionality, and lower gain. If biologists do not correct for or eliminate these factors, variability is introduced into the accuracy of the receiving process (see Radio Wave Propagation [p. 389]).

Most receiving antennas consist of a boom, metal elements, a tuning device, and connecters to a transmission line for conducting power from the antenna to the receiver. The basic factor dictating antenna design and size is the radio frequency wavelength to be received. Lower radio frequencies (e.g., 40 MHz) have longer wavelengths, and thus longer antenna element(s) are required for signal reception. More directional gain can be achieved by using longer elements or more elements, but at the expense of portability.

Coaxial Cables and Preamplifiers

The radio frequency power from the antenna is transferred to the receiver by a transmission line (ARRL 1988), which typically is coaxial cable. This cable provides electromagnetic shielding against extraneous noise, and its impedance (resistance value) matches the antenna to the re-

ceiver. The inner conductor wire of the cable is enclosed in dialectic insulation (polyethylene or Teflon), surrounded by braided copper conductor, and covered by a vinyl jacket. Breaks, nicks, abrasions, or kinks in the coaxial cable disrupt its characteristic impedance, causing a loss of the transmitted power between the antenna and the receiver. Coaxial cable is attached to the antenna and receiver by screw-together UHF connectors or BNC bayonet-mount connectors. Breakage most commonly occurs at the cable and connector interface. ARRL (1988:Chapter 24) suggested using RG8 or RG11 (75 Ω resistance) coaxial cable for lengths $\geq$15 m and avoiding the smaller diameter RG58U and RG59U (50 Ω resistance) with any VHF frequencies. Before going into the field, biologists should acquire extra cable and connectors and practice joining them together.

Power loss along a transmission line (decibels per unit length) increases logarithmically with the cable length and varies with operating frequency (ARRL 1988). For example, from 138 to 174 MHz, there is a 2-dB to 3-dB attenuation of signal strength along a 10-m cable, depending on the type of cable. Preamplifiers can be placed between the antenna and the transmission line if power loss interferes with reception (Kenward 1987, Howey et al. 1989). Biologists and equipment suppliers should discuss the distances among and the surroundings of the components of the receiving system to ensure maximum signal reception.

Additional electrical properties make it important to carefully maintain a fully functional, specific combination of antenna, coaxial cable, and receiver. Maximum performance from the receiving antenna is obtained when the receiver is matched to the impedance at the point of transmission line connection with the antenna. Under optimum conditions, half the power from the antenna is delivered to the receiver (half is re-radiated). If components are mismatched, less power reaches the receiver (ARRL 1988) and reception range is reduced. Equipment manufacturers should provide the best combinations of components and tune them for matching impedance.

Types of Receiving Antennas

Dipole antennas are the standard to which the gain of other antennas is compared. The dipole usually is not used for direction finding, but rather for presence/absence of a signal or recording coded data (e.g., physiological). However, Parish (1980) described tilting the dipole element about 15° from horizontal to obtain a narrower null on the side of the element toward the animal, thereby indicating direction to the signal. Also, he described construction of a dipole that collapses (20 cm × 4 cm) to facilitate transport.

The element of a *loop antenna* is configured in a circle or diamond. The main advantage of loop design is that the dimensions can be relatively small compared to other directional antennas for the same frequency, but at the expense of gain. A loop antenna for frequencies in the 30- to 40-MHz range can be hand-held. These loops have a bearing accuracy of about 5°, with nulls in two directions (Amlaner 1980). Portable loop antennas also can be constructed for higher frequencies (e.g., 150 MHz) at small diameters (20 cm), which are useful for short range (e.g., <20 m) reception (Kenward 1987:17).

Omni-directional antennas also are called "whip" antennas. Their reception pattern is uniform through 360°, with a gain of 0–3 dBi depending on grounding methods and other circuitry options. Omni-directional antennas are used commonly to detect presence of signals in relatively confined areas from fixed or mobile receiving sites. Whip antennas are adapted easily for magnetic or bolt-on attachment to vehicles and aircraft.

The *Yagi antenna* is a common wildlife receiving antenna, providing desirable gain and directionality. It comprises two or more linear dipole elements including a driver, parasite director(s) in front of the driver, and reflector element(s) behind the driver, all of which are mounted on a boom (Fig. 4E). The dimensions of a Yagi are critical; the lengths of the elements and their spacing depend on the radio frequency to be received. Element lengths are ±5% of 0.5 λ (the reflector is longest, the driver intermediate, and the directors shortest) and they are spaced 0.1–0.2 λ apart (Amlaner 1980). A four-element Yagi for 164 MHz will be about 100 cm × 113 cm, a five-element Yagi for 216 MHz about 68 cm × 120 cm.

Tuning of a Yagi is accomplished with the gamma or balun circuit, which resembles a trombone slide, on the driver element. Impedance matching in the field is possible by changing the length of the slide until maximum signal strength is obtained from a source greater than 20 λ from the antenna (Amlaner 1980). A Yagi antenna has more gain in front of the antenna than behind it, which creates a reception pattern that facilitates determining the direction from which the signal is arriving (Fig. 5). A three-element Yagi provides a bearing accuracy of about ±5°. Increasing the number of elements provides more gain and directionality (Fig. 6). Thus, a 12-element Yagi has more power than a 3-element Yagi, and provides about ±3.0° bearing accuracy (Kenward 1987:18–19). Additional elements increase the boom length and mass, making maneuvering the antenna more difficult. Usually, Yagi antennas with more than five elements are restricted to use on a fixed mast or are mounted securely on a vehicle (see below). Small folding or collapsible Yagis, which are less sensitive, are available from some manufacturers.

Two Yagi antennas can be mounted about 1 λ or λ/4 apart (Voigt and Lotimer 1981) and electrically connected by joining their coaxial cables between the antennas and the receiver. Signals from the two Yagi antennas can be added or subtracted to increase gain and give a directional error of about ±1° or switched to create a narrow "beam width" null with ±0.5° of error (Amlaner 1980, Anderka 1987). Such precision is highly desirable, but these "null-peak" systems require careful matching and cable phasing. If elements are bent or jarred out of place, or if coaxial cable from one antenna is damaged, the peak or null will not be accurate. The null-peak system must be securely mounted and checked frequently for accuracy.

The *Adcock antenna* (or *H antenna*) has a driver (in front) and reflector, each λ/2 in length. For any given frequency, the Adcock has smaller dimensions than a Yagi and thus is more portable. Furthermore, some manufacturers use folding or screw-together elements so the Adcock can be disassembled for convenient transport. Livezey (1988) described polyvinylchloride element covers to protect the antenna during use without affecting its

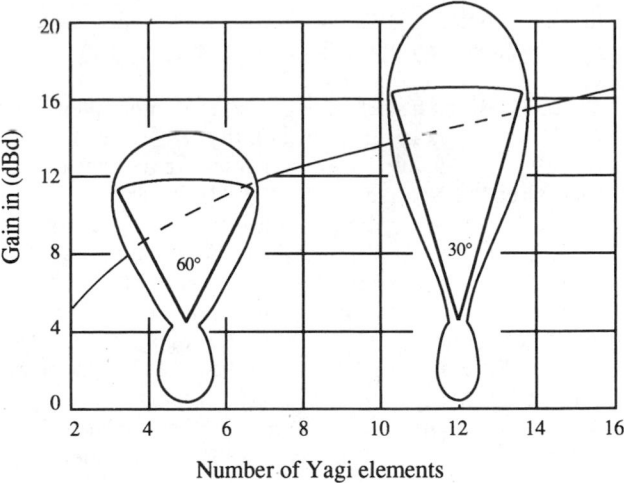

Fig. 6. The gain and directionality of a Yagi antenna in relation to the number of parasite elements. Reception range is doubled for each increase of 6 dBd. The reception patterns of 3- and 12-element Yagi antennas illustrate how the increase in gain is accompanied by a narrower (30° vs. 60°) field of reception. The field is defined by reception in which the signal gain decreases by 3 dBd on each side of the peak (maximum) signal. The peak signal is obtained by pointing a well-tuned Yagi, held at least 2λ from any surface, directly at the signal source. In practice, the reception area usually is wider than this field (see Fig. 5) (modified from Kenward 1987; used with permission from Academic Press).

performance. The maximum gain of an Adcock is less than that for a three-element Yagi (Amlaner 1980). The reception pattern of an Adcock has a narrow null, allowing the signal location to be estimated within about 2–3°. However, two nulls occur at 180° from each other, so the actual direction from which the signal comes must be determined by triangulation or prior knowledge of the animal's general location (Kenward 1987:17). The Adcock antenna, or a similar design, is often called an H antenna.

Mounting Devices for Antennas

Commonly, receiving antennas are hand-held because biologists move about frequently to stay within reception range of a radio-tagged animal and to maximize signal strength and clarity by positioning themselves relative to the transmitter. Many factors affect radio-wave propagation (ARRL 1988: Chapter 23, see also Radio-Tracking [p. 58]), and, generally, holding the antenna above one's head will maximize reception. Standing on a high point relative to the terrain also is useful. Antennas should not be held by the elements or by the boom between elements. A person's body, hand, and arms can alter the reception pattern of antennas.

Receiving antennas can be mounted on masts or towers to increase their height and enhance radio-signal reception. Biologists should consult equipment manufacturers about mounting materials and limits on the proximity of the antenna to materials that interfere with reception. The effects of those materials will vary with antenna design and radio frequency. Many types of poles, including telescoping or crank-up masts, can be secured by guy wires or tripods, or placed in a hole or cement foundation, depending on the height and strength required to support the antenna(s) (ARRL 1988: Chapter 22). Many mounting devices can be obtained from suppliers of radio communications equipment. O'Connor et al. (1987) described a typical fixed-site mount, 5.5 m in height, that can be rotated.

Several designs have been devised to mount antenna(s) on cars and trucks for greater mobility. However, the antenna must be elevated to avoid interference caused by the metal on the vehicle. Bray et al. (1975) and Kolz and Johnson (1975) described mounts for car-top carriers. Hegdal and Gatz (1987) placed the mast through a hole in the roof so the antenna could be rotated from inside the vehicle. The design by Cederlund and Lemnell (1980) also allowed rotation and elevation through a hole in the roof (see also Kenward 1987). Sturdy mast systems were described by Medina and Smith (1986).

Mounting antennas on aircraft is possible from the standpoint of electronic function, but safe attachment is a major consideration (Gilmer et al. 1981). Mounting methods must be adapted to different types and models of aircraft (e.g., high wing, low wing, helicopter). Gilmer et al. (1981) described a removable cuff that wraps around a wing strut and a mount for a helicopter skid that have been used in the United States. Voigt and Lotimer (1981) described a cuff mount used in Canada. Inglis (1981) also described mounting Yagi antennas on aircraft. In Australia, Whitehouse and Steven (1977) used a design with the elements installed permanently in the airplane wings. Multiple-element antennas also have been mounted on a mast that extends through a hole in the belly of aircraft (Judd and Knight 1977, LeCount and Carrel [Ariz. Game and Fish Dep., Fed. Aid Wildl. Restor. Proj. W-78-R-20, 1980]). Whip antennas, which are used commonly for communication systems in aircraft, can be attached to most aircraft.

In most countries the addition of any object to the outside of aircraft must be approved. Approval is not universal and usually is dependent on the ownership of the aircraft (e.g., government, private, commercial) and the concerns of individual mechanics and inspectors. Biologists always should ensure that the pilot is familiar with the mount, the antenna(s), and their effect on aircraft performance. Biologists should consult national authorities or the International Association of Natural Resources Pilots for information about successful methods. In the United States, biologists should contact the local office of the Federal Aviation Administration, which has district offices throughout the country.

Specialized Receiving Systems

Many innovations have been developed to facilitate radio-tracking. Burchard (1989b) designed a two-element directional antenna for close-range reception in areas of heavy vegetation. Cederlund and Lemnell (1980) mounted an electronic compass just under the antenna at the top of the mast. A line from the compass to the base of the mount allowed a relatively precise and accurate bearing to be read inside the vehicle. Smith and Trevor-Deutsch (1980) developed a system in which motorized rotors (operated by remote control) turned a null-peak Yagi array on top of masts (also see Spencer et al. 1987).

Automatic tracking systems also have been developed that are based on the time of signal arrival at different receiving sites. Lemnell et al. (1983) estimated locations (± 40 m) of ≤ 60 large mammals in an area of about 3,000 ha by sending a signal to the transmitter (250–400 g including a built-in receiver), which then radiated an 8-W to 20-W radio signal to three ground stations. Yerbury (1980) used the same concept to locate (± 50 m at 5 km) crocodiles marked with 10-mW to 20-mW, 140-g transmitters. These systems, one of which is available commercially (see Appendix I), have limited range and require a minimum investment of about \$50,000 to install the receiving stations. Even greater automation has been achieved with fixed-tower conventional triangulation methods (see "Automatic" Radio-Tracking [p. 31]). Like Kenward (1987), we warn biologists that semiautomated and automated systems are largely experimental, and application likely would require considerable investment of time and money in comparison to antennas and receivers directly operated by field personnel.

FIELD PROCEDURES
Radio Frequency Selection

The radio frequencies available for use on free-ranging animals are regulated in most countries, and they usually are restricted to a limited spectrum at a low power of transmission (e.g., 1.0–10 mW). Adherence to frequency allocations is necessary for legal use and for avoiding frequency interference with other radiotelemetry, including commercial radio and television, data transmissions, and communication such as telephone, ship-to-shore, or citizens band radio. The radio frequencies allocated to wildlife uses vary among countries; for example, in the United Kingdom biologists can use 104.6–105.0 MHz and 173.70–174.00 MHz, and in Finland the designated frequency is 230 MHz (Kenward 1987). In the United States, frequency allocations and technical standards are regulated by the Federal Communications Commission (FCC) for persons who operate under state or local governments or in the private sector, and by the National Telecommunications and Information Administration (NTIA) for federal government projects. These two regulatory groups have allocated the spectral bands of 40.66–40.70 and 216–220 MHz for wildlife telemetry. Also, about 18 specific frequencies in the 30-MHz band and 30 frequencies in the 164–167 band have been assigned for exclusive use to the U.S. Department of Interior for wildlife telemetry. Some state governments are licensed to use frequencies in the 150-MHz area. All users must obtain a license to use any of the wildlife telemetry frequencies. Biologists may apply for temporary use of other frequencies under an experimental designation, but this can be a prolonged process and few frequencies are available. Applications for licenses can be obtained from the FCC if the transmitter owner is non-federal. Federal personnel should inquire about NTIA licenses thorough their agency's communications coordinator. Equipment manufacturers should provide applicants with the technical data required to complete the application. Kolz (1983) reviewed the regulations that are pertinent to wildlife telemetry in the United States.

All wildlife telemetry in the United States is considered a secondary radio service and is afforded no protection from interference by other authorized use, most of which occurs at higher power levels than wildlife transmissions. Before ordering transmitters, biologists should visit the study area and use their planned receiving techniques and equipment at representative sites to listen on the frequencies they anticipate using. Persons listening for interference should not stand next to running motors, power lines, electrical transformers, and broadcast antennas, which are sources of interference. Urban areas, airports, and industrial complexes are often general sources of interference. Ordering transmitters at frequencies on which there is unacceptable interference should be avoided, but biologists should be aware that some noise (e.g., hiss, occasional tones) or other interference occurs under most circumstances.

Increasingly, wildlife studies by other biologists are becoming sources of interference because of more applications of wildlife radiotelemetry, larger samples of radio-tagged animals per study, unpredictable movements of some animals, and simultaneous, proximate studies. Land managers, state wildlife agencies, bird-banding laboratories, and other biologists can provide names of investigators they know are using telemetry. Coordination of frequency use among biologists is necessary to avoid misinterpretation of data and time wasted from tracking the wrong animal. Also, coordination among projects is beneficial when equipment, monitoring tasks, and even study animals can be shared. When too few frequencies are available to be apportioned among the necessary sample of animals, investigators can consider coded transmitters that send a unique identification on a single frequency (see Coded Transmitters [p. 377]) or apply for a license for experimental frequencies.

Equipment Acquisition and Testing

Numerous companies in North America and Europe specialize in equipment for wildlife telemetry (Appendix I). Many companies have experience with different species, geographic areas, and research objectives and can provide valuable advice about appropriate equipment. We suggest that biologists obtain the manufacturers' catalogues and correspond with the companies after identifying the study objective(s), reading relevant literature, and contacting researchers with experience in the application of interest. Information about the study species' habits, the topography and vegetation in which it occurs, and especially about previous radio-tagging is useful for selecting equipment. When ordering, the buyer should state the specifications for the equipment. For example, when ordering transmitters specify operating frequency, including separation between channels and allowable frequency drift, pulse rate, pulse width, radiated power, operating temperature range, required operation life, allowable size, mass, antenna length, and method of attachment to the animal.

The cost of telemetry equipment is expensive relative to some other methods. Most transmitters cost \$100–\$300 and receivers \$800–\$4,000 (U.S. dollars). This might tempt investigators to construct, rather than purchase, equipment. However, we urge careful consideration of the materials and expertise required to construct quality telemetry equipment. Expensive electronic tools and test instruments are required to build telemetry equipment. Time is required to determine what components are needed, order the components, and assemble them. Quartz crystals, batteries, and other circuitry components are subject to

availability. Considerable time also might be devoted to learning the "art" of building telemetry equipment. Kenward (1987) provided lists of materials and step-by-step directions for construction of numerous types of transmitters. Proceedings of biotelemetry conferences and workshops, and occasionally journals, contain papers describing new circuitry and construction techniques (e.g., Amlaner and Macdonald 1980, Amlaner 1989). Generally, we recommend purchasing equipment from established suppliers of wildlife telemetry products.

Scheduling equipment acquisition should include time for assembly and testing. Manufacturers do not always have a large stock of items on hand, and a large order can require procurement and construction that requires months to complete. Also, lead time frequently is needed if biologists' equipment specifications require manufacturers to modify or design and test circuitry. Thus, inquiring about equipment 12 months before the study is not unreasonable. To preclude "downtime," equipment acquisition should include some backup parts for the system, particularly for field studies where damage or loss is possible.

Beyond simply connecting parts of the receiving system (e.g., antenna–coaxial cable–receiver) and turning on transmitters and receivers to check function, each transmitter should be tuned on each receiver. Often a nominal transmitter frequency, such as 216.123 MHz, will be received at a different place on the frequency scale on each receiver, such as at 216.120 or 216.129 MHz. If more than four or five transmitters and two receivers are to be used, it is convenient to record the reception point for each transmitter on each receiver and to tape the list to the receiver or carrying case. Most receivers emit an audible tone when a signal is received, and this tone varies as the receiver is tuned slightly up or down the scale. Different persons often prefer slightly different tones and might select slightly different frequency settings to monitor the same transmitter signal. Confusion is avoided if these differences are identified before the study is initiated. Also, determine if the capacity of rechargeable batteries is adequate, especially for used equipment; if primary batteries are used, a supply of replacements should be ordered and properly stored (usually refrigerated). Some receiving-recording systems require adjustments after shipping or storage, or when they are operated in different temperature environments from one study to the next. Sensors in transmitters (e.g., temperature) might require calibration before the study begins.

Manufacturers' specifications usually are based on measurements obtained in the factory or laboratory. The performance of equipment can be altered dramatically by factors such as animal species, attachment method, topography, vegetation, and climate. Manufacturers can provide an indication of how performance might be affected by these factors; however, biologists should conduct their own tests to evaluate the reliability of specifications and of packaging (i.e., check for leakage), and the effects of temperature on general performance. Equipment testing should include operation in the field environment, and, when possible, trials with transmitters on the study species or a surrogate. Some transmitters (up to 20%, Cochran 1980) will fail before the predicted operation life is achieved. When a minimum sample is critical, it might be useful to estimate average operational life and failure rates from a sample of transmitters acquired for the study.

Transmitter Attachment Methods

The continued development and refinement of transmitter attachment methods provide an indication of the uncertainties of radio-marking, the diversity of animal shapes, and the differences in animal behavior (Cochran 1980). These variations limit the generalizations we can make about attachment methods. Therefore, biologists should not rely only on our brief descriptions of common methods, but must read the relevant papers, correspond with persons who have used the methods, and, above all, gain experience by visiting studies or practicing on captive animals. Fitting the transmitter on the animal is a critical factor, and transmitter attachment is "likely to remain an art" (Cochran 1980:515). Some attachment procedures require 30–60 minutes and considerable manipulation of the animal. Most species require restraint, sedation, or anesthesia while being radio-tagged. The care given to animals during capture, handling, and radio-marking is important for a successful study (Hill and Talent 1990, see also Chapter 6).

The special shapes and encapsulation required for implants, the attachment harness material, bolts, materials, and shapes for collars are integral parts of a wildlife radio transmitter. After biologists have determined that a species can be radio-marked, they must include specifications for the attachment method along with the electronics specifications in their order for transmitters. Ultimately the attachment device is as important as the power output or pulse interval. As Cochran (1980) noted, biologists should work with equipment manufacturers to explore and exploit trade-offs, for a variety of options, to fulfill research needs.

COLLARS

Incorporating a transmitter in a collar is a common attachment method for mammals (Pouliquen et al. 1990) and several species of birds. Collars should be shaped carefully to fit the animal's neck contours and should be wide and smooth enough to evenly distribute the transmitter mass to avoid cutting or severe chafing. Joints or other irregularities in the transmitter should be positioned on the sides of the collar. The collar material must be durable, yet flexible in response to neck, shoulder, or chest movement. The mass of the transmitter and collar material should be positioned to keep it from interfering with the animal's natural movements (e.g., Garcelon 1977) and to prevent swaying or flapping of the unit as the animal moves.

Collars for small mammals (neck circumference ≤18 cm) often are designed with a fixed-length loop antenna (Anderka 1987). Usually these antennas are embedded in the collar material, consequently the circumference cannot be adjusted. Biologists might have to prepare several collar sizes to accommodate variability among animals. Such loop antennas often must be soldered closed, bolted together, or adjusted and covered with heat-shrink tubing after being placed on the animal (Kenward 1987, Trout and Sunderland 1988). A stiff piece of paper placed between the collar and the animal's neck will protect the

animal from the heat and from snagging its hair during the attachment procedure.

Collars for medium and large mammals usually have a whip antenna running between layers of collar material. The antenna can be extended upward through the outer layer to improve signal transmission, provided the animal's habits will not cause it to be broken. The collar then is secured by bolts or rivets (Kenward 1987) that can be adjusted to fit individuals. Some mammal collars have been designed to accommodate the growth of young animals or a temporary neck expansion in rutting animals (Jullien et al. 1990). Foam rubber inserts and sewn pleats, which tear apart with expansion, have been used successfully on pronghorns (Beale and Smith 1973), mountain lions (Garcelon 1977), black bears (Strathearn et al. 1984), and bobcats (Jackson et al. 1985). Break-away collars that rely on the decomposition of rubber or wire (sea otters, Loughlin 1980) or cotton fabric (black bear, Hellgren et al. 1988) have been designed to detach from the animal after a period of time. However, biologists should not assume that these expansion and break-away designs always work reliably under varying environmental conditions (e.g., dampness, sunlight, and temperature) or on different species.

Collar necklaces for birds can be flexible, vinyl-coated fabric, and the transmitter is attached by sewn thread and adhesive. A hole shaped to the bird's neck is cut in the fabric, and the collar is slipped over the bird's head and worked into the neck and breast feathers. This attachment is used most commonly on grouse (Amstrup 1980), ring-necked pheasants (Marcstrom et al. 1989), quail (Shields and Mueller 1983), and waterfowl (Montgomery 1985, but see Sorenson 1989). Some large waterfowl have been radio-marked by gluing transmitters to plastic neck bands (G. Bartelt, pers. commun.).

HARNESSES

Transmitters secured by a harness are used on some mammals with head and neck shapes that will not retain a collar, and when it is necessary to use a large package on birds. When fitting the harness, biologists must accommodate for growth and other changes in body size and shape. Harnesses have been used on badgers (Cheeseman and Mallinson 1980), marine mammals (Broekhuizen et al. 1980, Jennings and Gandy 1980, Kolz et al. 1980), and sea turtles (Ireland 1980).

Harnesses for birds include numerous designs. Usually, the transmitter and battery are encapsulated in one package held on the center of the bird's back with the antenna trailing down the back and tail (Dunstan 1972, Dwyer 1972). A single body loop of harness material extending around the body, behind the wings but in front of the legs (Cochran 1972), has been used for short-term marking on small raptors (Dunstan 1972), woodcock (Coon et al. 1976), and mourning doves (Perry et al. 1981), usually in conjunction with some adhesive between the transmitter and feathers. In a few instances, the transmitter has been placed on the breast (Nicholls and Warner 1968, Siegfried et al. 1977), or the battery has been placed on the breast and the transmitter on the bird's back (Dumke and Pils 1973). Care in fitting is required to avoid interfering with wing movement, and a strap across the breast (Nesbitt et al. 1982) can ensure the transmitter does not shift forward or backward.

Double-loop harnesses are used to attach large transmitters to birds. Dwyer's (1972) design is applied most often to ducks and uses a continuous plastic-coated wire for the neck and body loops. The loops can be adjusted to each bird's body size so the transmitter remains free to move a bit after attachment, thus adjusting to the bird's posture. Another double-loop harness involves joining the neck loop and the body loop over the sternum. This design uses flexible materials such as woven Teflon ribbon (Dunstan 1972), elastic (Green 1985), or rubber or plastic tubing (Nesbitt et al. 1982) to hold the transmitter in place on birds that perch in an oblique or upright posture. The loops must be positioned to avoid impinging on the leading or trailing edges of the wings where they join the body. The strands extending from each corner of the transmitter are brought together at a point over the midline of the breast or, for birds with relatively long bodies, at the posterior and anterior ends of the sternum. In the latter configuration, the neck loop and body loop are linked by a strand along the sternum (Melvin and Temple 1987). The surest way to join the ends of the strands is to stitch them together. Details for fitting and securing a Teflon ribbon double-loop harness were given by Snyder et al. (1989). Wing loops and loops around legs also have been used to attach backpack transmitters (Nesbitt 1976, Rappole and Tipton 1991). Again, care in fitting is crucial to avoid interfering with the bird's natural movement.

Boshoff et al. (1984) designed a weak link into a harness to permit the neck loop to break. Karl and Clout (1987) used cotton threads to join the neck and body loop strands. The cotton will break should the harness or transmitter become snagged on an object, or the thread will decompose with time. Elastic harness material loosens with time and will permit the transmitter to fall off (Amlaner et al. 1978, Hirons and Owen 1982). As with expandable and breakaway collars, the weathering of the elastic harness and decomposition of the cotton are quite variable. The transmitter might detach before the study is completed, and a loosened harness will not fit properly.

ADHESIVES

Gluing or taping a transmitter to an animal provides an alternative to encumbering it with the weight and potential physical interference of a collar or harness. A variety of adhesives has been tried; however, caution is required because some adhesives are irritants to tissues. Generally, biologists can expect adhesives to be successful only with comparatively small, light transmitters and for relatively short periods (e.g., 2–24 days). Transmitters have been glued to bats (Stebbings 1982) and bears (Anderka 1987), but adhesives are used most commonly on birds (Johnson et al. 1991). Various methods have been used to glue transmitters to birds (Jackson et al. 1977, Raim 1978, Harrison and Stoneburner 1981, Perry et al. 1981, O'Connor et al. 1987, Sykes et al. 1990). Fiberglass resin (Wanless et al. 1988), Velcro fabric (Heath 1987), Tesa tape (Wilson and Wilson 1989), and automotive hose clamps (Kooyman et al. 1982) have been used to attach transmitters to seabirds.

TAIL MOUNTS

Small transmitters, generally ≤2% of body mass, can be mounted at the base of a bird's tail feathers (Giroux et al. 1990). When attached ventrally, the package is out of sight and kept warm and dry among the under-tail coverts. The feather(s) can be prepared by trimming the barbs from the base of the shaft and wiping it (them) clean with alcohol. Attachment is facilitated if the transmitter has a groove in which the shaft will lie. Usually, the whip antenna is tied and glued at one or two points along the shaft distally from the base of the tail. When attaching the transmitter, biologists must be careful not to bend the rectrices; excessive transmitter mass or manipulation causes the follicle to "release" the feather. Biologists have used a variety of methods, including fiber packing tape (Fuller and Tester 1973), plastic cable (Wanless et al. 1989), glue (Fitzner and Fitzner 1977, Kenward 1978, Pennycuick et al. 1990), and clips (Bray and Corner 1972, Kenward 1987) to attach transmitters to birds' tail feathers.

IMPLANTS

Radio transmitters can be surgically implanted, thus eliminating the external object that might disrupt the animal's normal behavior or impede its movements. However, implant procedures usually are long and complex and often require small, low-powered transmitters with ≥50% reduction in reception range. Veterinary medical techniques are required, including aseptic or antiseptic conditions, sedation or anesthesia, antibiotic treatment, and monitoring of postsurgery recovery. In addition, it might be required that a veterinarian conduct the surgery (e.g., Canada), or that a license be obtained to perform the implant procedure (e.g., United Kingdom). Implant procedures have been developed for relatively few species. Therefore, experimentation with captive animals often will be needed to assure successful methods.

Many of the initial radiotelemetry implants in mammals were conducted to obtain physiological data. These studies, including a standard intraperitoneal implant procedure, were reviewed by Folk and Folk (1980). The placement of the transmitter for intraperitoneal implants is critical to ensure that the transmitter does not interfere with body functions (Smith 1980, Williams and Siniff 1983). Mammals that have been radio-tracked with implants include river otters (Melquist and Hornocker 1979, Davis et al. 1984), ground squirrels and mink (Eagle et al. 1984), beavers (Guynn et al. 1987), yellow-bellied marmots (Van Vuren 1989), black bears (Jessup and Koch 1984), sea otters (Garshelis and Siniff 1983, Ralls et al. 1989), brown bears (Philo et al. 1981), and lions (McKenzie et al. 1990).

Physiological monitoring was also the objective of most implant methods with birds (Woakes and Butler 1975, Klugman and Fuller 1990), but only recently has radio-tracking been attempted. Korschgen et al. (1984) successfully implanted transmitters in the abdominal cavity of six species of waterfowl. Reception ranged from 0.4 km to 1.6 km from the ground and ≤2.4 km from an aircraft. Olsen et al. (1992) radio-tracked implanted canvasback ducks from aircraft at ranges of 5–11 km. Implanted transmitters also have been used for tracking reptiles (Weatherhead and Anderka 1984, Lutterschmidt and Reinert 1990) and amphibians (Stouffer et al. 1983, Smits 1984).

MISCELLANEOUS ATTACHMENT METHODS

Several other methods of attachment of radiotransmitters have received incidental use. Ear tags provide alternative attachment to a collar on some mammals (Servheen et al. 1981, Garrott et al. 1985). Swanson et al. (1976) mounted transmitters with activity sensors on the nasal saddles of ducks to monitor feeding behavior. Perry (1981) used cyanoacrylic glue to hold a transmitter on the top of the bill of canvasback ducks, a species adversely affected by backpack transmitters. However, attachment to the bill limited transmitter size. Sutures (Martin and Bider 1978, Mauser and Jarvis 1991) or glue and sutures (Wheeler 1991) have been used to attach backpack transmitters on birds. Melvin et al. (1983) used plastic leg bands as the base for battery- and solar-powered transmitters, and Kenward (1985) described a leg-band transmitter for raptors. Feather growth on nestlings or fledglings can preclude use of harnesses or tail mounts, and a leg mount provides an alternative means of transmitter attachment. However, the proximity of the antenna to the ground can reduce reception range, and the antenna is more likely to be broken on a leg mount.

Effects of Radio-Marking Animals

The capture and restraint of animals constitute an interruption to their normal activity, and attaching an object to them is associated with a variety of changes in behavior and other aspects of their life history (Marks and Marks 1987, Vaughan and Morgan 1992). We agree with Cochran (1972) that radio-marking will have some effects on wildlife; however, we cannot generalize easily about its effects. For example, some studies detect no detrimental effects on reproduction of radio-tagged birds (Kalas et al. 1989, Sodhi et al. 1991, Taylor 1991), whereas other results reveal an adverse effect (Massey et al. 1988, Paton et al. 1991, Foster et al. 1992). Observation and study of radio-tagged animals reveal different effects among species, times of year, and ages of animals. Also, animals respond differently to different size transmitters (Burger et al. 1991), to various attachment methods, and to how the transmitter and attachment conform to animals. The effects of radio-marking can be short- or long-term, overt or subtle (Brigham 1989). The following descriptions and associated literature on the diversity and magnitude of effects should guide the biologist in determining feasibility of radio-marking, in knowing how and what to look for after a radio-tagged animal is released, and in understanding the effects of radio-marking on study objectives.

Some animal responses to capture and radio-marking are immediate, but often they do not persist. For example, the animal does not return to the capture area for several days. Also, biologists commonly report a temporary change in an animal's activity pattern as a result of attention toward the transmitter, antenna, or harness (Hooge 1991). Kenward (1982) observed short-term responses in grey squirrels, but over a longer term, the radio-tagged squirrels had no differences in pregnancy rates or in body mass compared to squirrels that were not radio-tagged. Siegfried et al. (1977) documented increased preening by waterfowl for a day to a week after transmitter attachment

with a harness. Some responses, even of short duration, can be detrimental due to the timing of the radio-marking. Female woodcock, marked when their chicks were ≤2 days old, abandoned their broods (Horton and Causey 1984); however, spruce grouse were marked with no apparent effect, provided their clutch was halfway through incubation (Herzog 1979). Erikstad (1979) believed that transmitters attracted more predators to female willow grouse, which lost more chicks than unmarked females.

Long-term effects of radio-marking include chafing and feather loss that results in reduced insulation for waterfowl (Greenwood and Sargent 1973). After being marked with a harness and backpack, canvasback ducks spent an inordinate amount of time on shore, picking at the transmitter, which eventually resulted in significant weight loss (Perry 1981). Even after careful preparation, radio-marking can produce unexpected results. Despite encouraging experimental results with surrogate yellow-throated warblers, 1 year after radio-marking, significantly more banded Kirtland's warblers without transmitters returned to the breeding area than did radio-marked birds (Sykes et al. 1990, C. Kepler and P. Sykes, pers. commun.).

Some effects of radio-marking are more difficult to detect or are infrequent. Jackson et al. (1977) observed that the antennas used on red-cockaded woodpeckers occasionally became snagged in the cracks of tree bark; however, this problem was solved with the use of a more flexible antenna (Nesbitt et al. 1982). Webster and Brooks (1980) reported that radio-marked meadow voles decreased their feeding activity, which eventually decreased survival. Clute and Ozoga (1983) observed that expandable collars on white-tailed deer fawns accumulated heavy ice during a cold period. Occasionally, effects are manifested in only one segment of the population. Survival of radio-marked ring-necked pheasants decreased with increasing transmitter mass, but only for those birds with relatively low body mass at the time of radio-marking (Johnson and Berner 1980).

Experimental methods have been useful for evaluating the effects of radio-tagging. A series of studies including work with captive birds, field experiments, and observations was necessary to determine the effects of radio-marking red grouse (Boag 1972, Boag et al. 1973, Lance and Watson 1977, 1980). Herzog (1979) compared the maintenance and reproductive behaviors, movements, and nest success of spruce grouse that were radio-marked to those that were only leg-banded to evaluate the potential limitations of radiotelemetry. Similarly, Kenward (1978) compared the weight of northern goshawks that were radio-marked to those banded only. Amlaner et al. (1978) observed differential incubation behavior and clutch survival for nesting herring gulls marked with transmitters of different masses.

Different transmitter attachment methods also have been evaluated in field experiments. Marcstrom et al. (1989) reported higher survival for ring-necked pheasants wearing a necklace attachment than for those with backpack harnesses. Garrott et al. (1985) observed comparable survival of mule deer fawns wearing transmitter collars or ear tags. Such experiments have been informative, and much more can be learned from comparisons of attachment methods. However, all experimental approaches require careful design, consideration of statistical power, and sample size (White and Garrott 1990).

The use of radio transmitters also has implications for an animal's energy budget. Captive bald eagles, harnessed with backpack transmitters at 0 C, had higher metabolism than when they were not wearing the package. However, barred owls showed no change in energy metabolism with transmitters weighing 2%, 5%, or 10% of their body mass, at 20 C, 0 C, or −20 C (Gessaman et al. 1991). Wooley and Owen (1978) and Sedinger et al. (1990) reported no differences in metabolism of captive radio-marked and unmarked waterfowl. Gessaman and Nagy (1988) flew homing pigeons 90 km and 320 km without markers, with a harness only, and with transmitters weighing 2.5% and 5.0% of the birds' mass, respectively. The harness slowed birds by 15% on the 90-km flight, and transmitters further increased flight times. When carrying the 5.0% transmitter load over 320 km, the birds produced 85–100% more CO_2. Flight duration and loss of body water were affected by radio-marking of tippler pigeons (Gessaman et al. 1991). Pennycuick et al. (1990), using doubly labeled water to estimate energy metabolism, reported that radio-marked (2.0% body mass) white-tailed tropicbirds expended significantly more energy during foraging trips than did unmarked birds. A comparison of the duration of foraging flights and mass of food delivered revealed no difference between marked and unmarked birds. Their results suggested that feeding rates are not a sufficient measure of telemetry effects.

Different types of flight, such as long distance-migration, short, rapid flight, or slow-speed, maneuvering flight, are affected differently by transmitter mass and aerodynamic drag. Aldridge and Brigham (1988) suggested that for bats weighing <70 g, biologists could use a transmitter weighing 5% of the body mass without serious effects on maneuverability. However, for bats >70 g, Caccamise and Hedin's (1985) formula should be used to estimate allowable transmitter mass. Larger flying animals require greater muscle power in proportion to their mass (Pennycuick 1975). Thus, a package weighing 3% of body mass will require a greater proportion of the available muscle power for a goose than will be required by a robin. Studies like these reveal the inadequacy of a "rule-of-thumb" guide for determining the mass of a transmitter that an animal can carry.

Biomechanical models can be used to estimate the effects of transmitter mass and drag on various aerodynamic factors (Pennycuick 1975). Biologists can use computer software (Pennycuick 1989), with data about the drag of the bodies of birds, their mass, and wingspan, to estimate the effect of a transmitter on flight parameters. For example, a transmitter can reduce the maximum flight range 15–34%, depending on the amount of fat the bird has for fuel (Pennycuick and Fuller 1987). This biomechanical approach also has been used to demonstrate the importance of streamlining the shape of transmitters that extend above the body contour (Obrecht et al. 1988). Transmitter drag will have an inconsequential effect on a bird making short flights, such as to carry food to its young, but the additional mass of the package will reduce the power available for lifting (Pennycuick et al. 1989). Transmitter drag also is a factor that can change swimming speeds in aquatic birds significantly (Wilson and Wilson 1989). By using a few measurements or estimates in biomechanical models, biologists can better determine how radio-marking likely will affect the animal and the results of their study.

Radio-Tracking

Radio-tracking techniques are used when animal movement patterns and activity patterns limit the use of conventional methods such as direct observation or mark-recapture (or resight). However, radiotelemetry is not a panacea, because, for example, animal movement and errors in determining radio-signal bearings influence the accuracy with which the biologist estimates a location. Cochran (1972, 1980), L. Kolz (pers. commun.), and others emphasized the need for a basic understanding of radio-wave propagation and equipment function if biotelemetry techniques are to be used effectively in the field environment. This section deals with field procedures for locating radio-marked animals. Biologists should be familiar with this material so they can evaluate the utility of radio-tracking to accomplish their research objectives and design an effective study. We do not provide information about recording physiological or presence/absence data; however, the material about radio-signal propagation is relevant for most biotelemetry.

RADIO-WAVE PROPAGATION AND RECEPTION RANGE

Radio signals are electromagnetic waves with properties similar to those of light waves (ARRL 1988). However, the term "line-of-sight," which often is used in wildlife telemetry, is not the same as visual line-of-sight. Radio waves go through some objects that block our vision, and they can be affected by unseen factors (Cochran 1980). Radio waves are polarized vertically (perpendicular) or horizontally (parallel) to the earth's surface. The initial polarization of a radio wave is determined by the orientation of the transmitting antenna. Thus a whip antenna on a perched bird's back, vertical to the earth's surface, transmits vertically polarized radio waves. Radio waves tend to remain vertical over flat terrain and water, but dense vegetation can cause horizontal polarization. Above 100 MHz, many types of obstructions cause horizontal polarization (ARRL 1988). Polarization is not highly predictable in environments with many obstructions. In this situation, biologists need to turn the antenna to orient the elements through the horizontal to vertical planes to obtain the maximum gain and directionality of the receiving antenna (Cochran 1980).

The initial strength of a radio signal is determined by the effective radiated power from the transmitting antenna. As a radio wave leaves a transmitting antenna, it spreads over a large area and becomes weaker at a rate proportional to the square of the distance. In practice, the signal strength usually weakens at a much faster rate than this theoretical relationship, because signal propagation is affected immediately by the animal's body (a conductor) and many other factors. The earth (i.e., soil, rocks) is also a conductor that can dramatically influence transmitted radio waves. Cochran (1980) pointed out that reception range can be quadrupled by increasing the power of a transmitter 16× when the signal makes a path ≥20° above the earth's surface. If the radio waves are lower, thus intercepting the surface, power would have to be increased 100× to achieve a four-fold range increase. He noted that if the radio waves pass through a dense forest, a 1,000× increase in power might be required. Elevation of the transmitting and receiving antennas is the most important factor affecting reception range (Anderka 1987). If the antennas are 2λ (~4 m at 150 MHz) above obstructions, ranges of 15 km can be achieved, but as the transmitter approaches the earth, range is reduced to 1–3 km. Raising only the receiving antenna will offset this partially.

Radio waves lose power rapidly when they encounter other conductors such as a body, wires (e.g., telephone lines, fence), or metal structures. These obstructions, as well as the terrain and vegetation, also cause radio waves to be reflected (bounce) and diffracted (spread). When prairie vegetation began to grow, the reception range to radio-marked skunks decreased by 50% (Sargeant 1980). Certain material, especially metals, can block radio signals; however, some radio waves pass through most types of obstruction or are diffracted around obstructions, creating a "shadow" of weak signal. Signal loss occurs when radio waves are reflected because they travel a longer distance than when they are not obstructed. Reflected signals create another problem for radio-tracking, because the directional receiving antenna reveals only the path of the strongest wave, not the direction to the transmitter. Signal bounce, polarization, and power attenuation vary with the multitude of environmental conditions encountered during fieldwork and contribute error to the radio-tracking process.

ERROR

Determination of animal locations is affected not only by variability in radio-wave propagation, but by animal movement and by variability in equipment performance and operation. Consequently, radio-tracking usually provides only an estimate of the animal's actual location. Biologists should determine the location accuracy and precision required for their study objectives, then develop field procedures and quality control throughout the fieldwork that will ensure that accuracy and precision are obtained. Radio-tracking results should include estimates of error that can be used to interpret the results (Pyke and O'Connor 1990). Description of a continental migration route might require location data accurate within a few kilometers, but the accuracy of an animal's location during the breeding season might be useful only if location can be limited to a few meters. Determination of habitat selection or a study of the microclimate effects on energy budgets could require location estimates with a resolution within a few meters. White and Garrott (1990) presented comprehensive discussions about methods for analyzing errors and their effects on radio-tracking. Below we provide introductory material about sources of error and ways to estimate, reduce, and account for it.

Sources of Error

One major source of error is not knowing the actual location of the receiver site. Error can result from simply misjudging the receiver site location, from inaccurate maps, from inappropriate map (or air photo) scale, or from plotting mistakes. Even if the biologist "homes" (see HOMING [p. 392]) to make visual contact with the radio-

marked animal, there can be error in determining the location. The width of a pencil mark used to plot a location on a 1:24,000 scale map (i.e., U.S. Geological Survey 7.5' topographic map) can cover 5–10 m. This amount of error adds to that which is inherent in most direction finding. A 1° error in the compass bearing to a true location causes 17.5 m of linear error/km of distance from the receiving site. Using the Global Positioning Satellite system, field-workers can routinely determine their location to about 15–20 m. Aeronautical navigation systems provide locations for aerial telemetry with accuracy from 100 m to >11 km (LORAN-C, Patric et al. 1988) and about 1–3 km (DME-VOR, Fuller et al. 1989). The use of surveyed points and small-scale maps can reduce the component of error associated with receiver locations. However, as Cochran (1980) noted, consistent accuracy of ±1° requires better mapping accuracy and more electrical and physical stability of equipment than is commonly used in wildlife biotelemetry.

Radio-tracking equipment contributes to error as do equipment operators. Receiving antenna systems provide from ±0.5° to ~7° accuracy depending on design (Kenward 1987:15–21, Macdonald and Amlaner 1980; see also Receiving Systems [p. 379]). Equipment must be in good repair, especially coaxial cables, antenna elements, and alignment of null-peak arrays. Compass rosettes should be oriented carefully and checked periodically (White and Garrott 1990). Correction is necessary for magnetic declination from true north. Pace (1988) observed that incorrect orientation of the vehicle-mounted mobile antenna systems contributed as much or more to location error as the process of signal detection. The use of beacon transmitters at known locations (Kufeld et al. 1987) or careful vehicle orientation procedures (Hutton et al. 1976) can reduce this problem. Springer (1979), Lee et al. (1985), and Kufeld et al. (1987) reported no differences among operators' abilities to ascertain signal direction with mast-supported Yagi antennas, but Hoskinson (1976) detected differences among pilots during aerial tracking. Mills and Knowlton (1989) reported significant increases in accuracy when four operators knew they were being tested. All operators should be trained to use the equipment properly and to recognize those situations in which they might reduce error with careful technique.

It is important to recognize field conditions that can influence radio-wave propagation and increase errors associated with direction finding. Kufeld et al. (1987) judged signal quality and accuracy to be diminished with increasing height of surrounding terrain. Garrott et al. (1986) reported excessive error in 52% of the bearings taken when ridges blocked the path toward the transmitter. Hupp and Ratti (1983) recorded greater error in rugged terrain and in rolling forested areas than across flat, open country. Chu et al. (1989) observed that forest vegetation significantly affected error and recommended not using bearings when their intersection angles are <20° or >160°, or when the summed distance between the two receivers and the transmitter is >2 km. Other environmental factors such as wind blowing tree limbs (Hupp and Ratti 1983), animal movement (Kenward 1987, White and Garrott 1990), and movement of the transmitter antenna (causing a modulated signal) can contribute to error, but these factors are unpredictable (Lee et al. 1985) and difficult to detect or quantify during radio-tracking. Therefore, it is necessary to assess error under conditions similar to the proposed field study before the study is initiated.

Error Assessment

A pilot study of errors associated with field procedures provides many benefits: (1) operators become familiar with equipment; (2) equipment can be checked for proper function; (3) the study area can be mapped for regions of unusual radio wave attenuation, signal bounce, polarization, or environmental noise and radio interference; (4) field protocols can be established; (5) the estimates of error can be evaluated to determine if telemetry data will be accurate enough to meet study objectives; and (6) data management and analyses can be undertaken with the information gathered in the pilot study. White and Garrott (1990) devoted a chapter to designing and testing radio-tracking systems, so we will describe only briefly the basic steps in assessing error.

Initially, one must become familiar with the function of the receiving equipment by operating it with the type of transmitter to be used in the error assessment and study. The range of each transmitter should be established (Sargeant 1980, Mech 1983) because ranges will vary, despite transmitters having been ordered with the same specifications. Some transmitters might be deployed more effectively on animals in certain parts of the study area, depending on reception characteristics or interference patterns. First, transmitter range testing should be accomplished in a flat, open area. If possible, the antenna orientation on radio-marked animals should be simulated and a plastic bottle of saline solution should be used to mimic conductivity of the animal's body. These test results will be an indication of maximum range and permit comparison among transmitters. Next, transmitters should be placed at known locations that are randomly selected or are representative of the environment in the study area. Also, if the animals are known to use burrows, cavities, wetlands, field-forest edges, or other habitats that affect radio-wave propagation, some transmitters should be placed in those locations.

The mobility and dispersion of the study animals will dictate the specific radio-tracking methods (see HOMING, AERIAL TRACKING, and TRIANGULATION, below) and thus the deployment of receiving systems. Whenever possible, receiving sites should be selected that are readily accessible without disturbing the study animals and that provide elevation of the antenna above obstructions. Sites near wires, electrical lines, motors, or sources of loud sounds should be avoided. Usually, two or more of these receiving sites will be used simultaneously or in rapid sequence to obtain bearings for location estimation by triangulation. White and Garrott (1990) discussed strategies for positioning receiving sites relative to the study area. Their examples, which we summarize here, assume no interference by reflection, unusual attenuation, or other complications. If only two receiving sites are possible, they should be just off the study area, along one side. This arrangement ensures that no bearings for animals on the study area will intersect on or near the "baseline" between receiving sites. Bearings that intersect near the baseline produce relatively large error areas. Generally,

three or more equally spaced receiving sites should be located near the sides, but away from corners of the study area. White and Garrott (1990:94–110) provided the theoretically best locations for two to six antenna towers on square and rectangular study areas for two strategies: (1) to minimize the average error area for location estimates, or (2) to minimize the maximum error area for location estimates. They also presented methods for determining the number of receiver sites needed to achieve a certain level of precision.

Results from the pilot study will indicate limitations in reception range and can be the basis for estimating location error. During pilot studies, operators should not know the location of the transmitters. Bearings to the transmitters can be taken in random order. Multiple bearings to each transmitter should be acquired, assuring independence of replicates (Springer 1979, Lee et al. 1985, Garrott et al. 1986). These bearings can be the basis for location estimates from triangulation (see TRIANGULATION [p. 393]). We suggest that the standard deviation of the bearing errors be calculated for each receiver site and transmitter pair, and that error areas be estimated for every combination of two or more receiver sites to a transmitter site. Analyzing all combinations of signals received during the pilot study provides clues about those combinations that are likely to produce unreliable results during fieldwork. This information can be used for designing sampling strategies and field protocols. If, after analyses of the pilot study data, the accuracy is inadequate, it might be necessary to omit some bearings that intersect at certain angles or distances from the receiving sites (Heezen and Tester 1967, Dodge and Steiner 1986, Chu et al. 1989), increase the number of receiving sites to cover the area, relocate sites to avoid obstructions (White and Garrott 1990), increase the number of bearings used for each location estimate (see TRIANGULATION [p. 393]), or improve precision of the receiving or transmitting equipment.

Similar pilot study procedures should be used to identify receiving sites for hand-held antennas or antennas elevated on a mast that is placed on the ground or on a vehicle. After likely receiver sites are located, they must be marked on the base map, and a mast, antenna, coaxial cables, and compass must be installed if towers are used. A method must be devised for orienting the compass rose, or vehicle, or for checking the function of a hand-held compass. Methods for testing compass rose orientation were described by White and Garrott (1990).

DIRECTION FINDING

Detailed field procedures for direction finding and tracking, including suggestions for numerous special situations (e.g., short range and three-dimensional), were given by Mech (1983) and Kenward (1987). Before direction finding begins, biologists should select a coordinate system for recording each animal's location and they must have maps or aerial photos of the appropriate scale. We suggest using the Universal Transverse Mercator (UTM) system, which is explained in White and Garrott (1990). Dodge and Steiner (1986) prepared a program to use UTM for radio-tracking, and Dodge et al. (1986) wrote a program to convert latitude-longitude coordinates (e.g., from LORAN-C or GPS navigations) to UTM. The methods, examples, and programs in White and Garrott (1990) are based on UTM coordinates.

To begin the process of direction finding:

(1) Select a receiving site within the expected reception range of the transmitter. An ideal receiving site should be on an open high site, with the antenna 1.5–2 λ (3–4 m for 150 MHz) above ground (Kenward 1987:115–116, Anderka 1987). Alternatively, raise the antenna as high as possible and away from obstructions.

(2) Set the receiver gain and volume controls about halfway through their range. Headphones are a useful aid for perceiving subtleties of signal variation.

(3) Turn the receiving antenna 360° while listening for the signal.

(4) If no signal is heard, rotate the antenna from horizontal to vertical polarizations, and increase the gain. If no signal is heard, move closer to the expected animal location and return to step 1.

(5) When a signal is heard, adjust the gain and volume to detect the peak signal strength and nulls while turning the antenna. Too much gain will cause a broad peak and a relatively strong signal on the back side of a Yagi antenna pattern.

(6) With knowledge of the polar diagram reception pattern for the antenna (Fig. 5), point the antenna toward the direction of the strongest signal, then reduce the gain until the signal is barely audible.

(7) Direct the antenna to each side of the peak at the point where the peak signal fades, or to where the null begins, and note landmarks to which bearings can be taken (use hand-held compasses away from metal equipment). Sometimes "peaks" are heard through a large arc (versus a few degrees) and can be difficult to measure. Nulls are usually narrower than peaks, and detecting the difference between the signal sound and no signal is easier than among the gradients of signal strengths (Cochran 1980). Therefore, to estimate the bearing to a peak signal, most authors suggest using the mean of the nulls on either side of the peak (Springer 1979, Macdonald and Amlaner 1980, Kenward 1987) or the mean of the null of a null-peak system (Hupp and Ratti 1983, Lee et al. 1985, Garrott et al. 1986, Kufeld et al. 1987). Adcock (H), dipole, and loop antenna patterns have two peaks of similar strength (Fig. 5). To determine signal direction, the operator can hold the antenna in front of the body, turn 360°, and listen for a weaker signal when the operator is between the transmitter and the antenna (Kenward 1987:118–123).

(8) When the edges of the peak, or null, have been located, take compass bearings along the edges and divide in half the degrees between these bearings to obtain an estimate of the direction from which the signal is coming.

If no signal is received after step 4, check again to ensure that the equipment is working. Searches for undetected transmitters should begin at places and times at which the animal is likely to be detected (e.g., den, roost, nest). Maximize the elevation of the receiving antenna. Search the study area methodically at minimum reception range intervals, but consider how depletion of the transmitter battery, animal dispersal, or mortality (and thus transmitter antenna position) will affect signal strength and radio-wave propagation. Aerial tracking (see below) frequently is an effective strategy for finding "missing" animals.

If direction finding is to be used for homing to the transmitter or as a basis for estimating locations by triangulation, determining the variability associated with direction finding will be useful. White and Garrott (1990) provided a description of the relationships among bearings, angles, and map coordinates and procedures for measuring the accuracy of bearings to transmitters in known locations. These topics apply to hand-held, tower, and vehicle-mounted receiving antennas.

HOMING

Homing is a method by which the operator uses antenna directionality and signal strength information to move toward the transmitter, ultimately to make visual contact with the animal or to find the transmitter (Mech 1983). Following step 8 of DIRECTION FINDING, the operator should:

(9) Move toward the signal source while listening to the signal and moving the antenna back and forth in an arc toward the signal.

(10) While moving, periodically reduce the gain and confirm signal direction because the signal should become stronger as the transmitter is approached. If direction is uncertain, turn the antenna 360° to reestablish direction toward the signal peak.

(11) If the signal is lost or the direction is ambiguous, return to the last position where a directional signal was heard or walk a few meters in four directions 90° apart. Moving off the original path can help detect signal reflection and provide another bearing, thus supplementing homing with a component of triangulation (see Kenward 1987, Fig. 6.6).

(12) If a signal is detected when gain control is at the minimum setting, the transmitter should be relatively close and the operator might have to move only a few meters to lose the signal, indicating it has been passed. Move back and forth, left and right, to "box" in or circle the transmitter.

(13) A relatively strong signal at close range can "swamp" the receiver's gain control. It might be necessary to use a loop antenna (Cochran 1980, Hegdal and Colvin 1986), or only the coaxial cable or a short piece of wire (e.g., paper clip), as the antenna while repeating step 12.

AERIAL TRACKING

Radio-tracking from aircraft usually is a special case of homing, used to cover large areas with many transmitters. However, aerial tracking requires that many arrangements be made prior to the study. For example, authorized antenna attachment, including running uncrimped coaxial cable into the cockpit, and testing of equipment, reception range, and determination of location error are necessary. Each flight plan, tracking strategy, and objective should be discussed with the pilot (see Gilmer et al. 1981). Transmitters with a reception range of 3–4 km to receiving antennas near the ground can be received from 35 km (Hegdal and Colvin 1986) to 100 km (Anderka 1987) with antenna on aircraft flying up to 3 km above ground level (AGL). However, the maximum useful altitude can be limited by interference or signal strength (Cochran 1980). The number of animals that can be monitored during searches for presence-absence determination is limited by the aircraft speed, minimum reception range, transmitter pulse rate and width, and dwell time on a given frequency (Gilmer et al. 1981, Kenward 1987). We encourage biologists to carefully plan for these factors and to practice the methods with the pilot and aircraft before the study.

Methods for aerial radio-tracking were described in detail by Gilmer et al. (1981), Mech (1983), and Kenward (1987). The radiotelemetry equipment for aerial tracking is basically the same as for other wildlife telemetry and was described in preceding sections. Signal reception usually is accomplished with two directional antennas mounted under the wings (or on helicopter struts) and facing to the side and downward. Coaxial cable, through a window or port, leads to a switch box that allows the operator to select between the left or right antennas, or listen to both. Aircraft navigation can be integrated to estimate aircraft and animal location. The basic steps for estimating locations from the air follow:

(1) Test equipment function (directional antenna on each wing, coaxial cable, switch box, scanning receiver, headphones, and options, e.g., laptop computer, LORAN-C interface) with beacon transmitter during departure from the airport.

(2) Fly to the point of last known location or area where a radio-marked animal is expected. The use of aeronautical navigation and programs for a laptop computer can save time in transit among destinations (Dodge et al. 1986).

(3) Listen for signals at each antenna as the plane approaches the target area and circle the area from about 300 m to 3,000 m AGL until a signal is heard. If no signal is detected by the time the plane reaches 3,000 m AGL, go to step 13 to search a larger area.

(4) Switch reception between antennas to determine direction to the strongest signal. Reduce the gain if necessary.

(5) When the plane is flying toward the signal, there will often be a null or weak signal.

(6) As the plane approaches the transmitter, the signal will become stronger, often slightly to one side or the other. Reduce gain if necessary while approaching the transmitter.

(7) Circle (~3 km diameter) to the strong signal side while continuing to switch between antennas.

(8) If the stronger signal is outside the circle, turn toward the signal and return to step 6.

(9) If the plane passes the transmitter, a dramatic change from a strong to a weak signal should be detectable. Mech (1983) noted that the gain control at its lowest setting might not permit auditory detection of strong signal amplitude changes, but a signal meter might indicate these changes. If the plane passes over the transmitter, turn 180°, pass over it again, and repeat step 7.

(10) If the strongest signal remains in the circle, begin circling lower and tighter (~1.6 km diam) and reduce the gain. Continue the switch between antennas to confirm the side on which the stronger signal occurs.

(11) If the stronger signal switches to the opposite side while the plane is making a tight circle, begin circling in that direction, as in step 10. The transmitter location is within the smallest diameter circle the plane can safely circumscribe.

(12) It might be possible to delineate a smaller error area around the transmitter by flying across the small circle at right angles and determining the quadrant from which the strongest signal comes (Mech 1983).

(13) If no signal is detected where it is expected, begin a large area search by flying transects spaced at $\leq 2\times$ the minimum reception range (at a given AGL). In flat, open terrain, transects can be flown by circling outward from the initial search area, by starting on one side of the study area and flying parallel routes back and forth, or by searching likely paths of dispersal or migration (Kenward 1987). A higher altitude might be required over forested or mountainous habitat.

Few data are available from tests of the precision and accuracy of aerial radio-tracking, but ± 100–200 m is probably the best commonly achieved accuracy. Hoskinson (1976) flew unusually low (15–30 m AGL) and slowly (95–115 km/h) and circled for 5 minutes to obtain minimum errors from 7 m to 40 m and maximum errors from 40 m to 70 m with different pilots. The antenna-mounting method (Gilmer et al. 1981) and antenna symmetry (Cochran 1980) also can influence accuracy. Using a four-element Yagi configuration, Fuller et al. (1989) were able to circle radio-marked seabirds within a 1-km-diameter area and locate the aircraft with ± 1–3 km accuracy using DME and VOR aeronautical navigation.

TRIANGULATION

The location of a radio-marked animal can be estimated by using direction-finding methods to take bearings from two receiving sites. The point at which the bearings cross provides a simple estimate of the animal's location. An error area, or polygon, associated with the estimate is based on the standard deviation(s) of the bearings (Heezen and Tester 1967, Springer 1979, Hupp and Ratti 1983). White and Garrott (1990) presented the steps needed to convert bearings to an x-y coordinate system (e.g., UTM) and the methods to calculate the confidence area of the error polygon. The standard deviation usually is based on a pilot study involving multiple measurements of bearings from receiving sites to transmitters at known locations (White and Garrott 1990). If a fixed number of permanent receiving sites is to be used during the study, different standard deviations could be estimated for various combinations of receiver-transmitter pairs. This approach to estimating animal location and triangulation error is probably adequate for small study areas (in relation to signal strength and distances between radio-marked animals and receivers) with environmental conditions that do not contribute to large variability of radio-wave propagation.

Many wildlife radio-tracking studies, however, involve relatively wide-ranging, fast-moving animals in heterogeneous environments. A relatively small sample of receiving sites and of potential animal locations is unlikely to be representative of the diversity of radio-tracking conditions, such as animal behavior, weather, seasons, or differences in terrain, that affect radio-wave propagation and triangulation procedures. In these conditions we recommend obtaining at least three bearings for each location estimate because, from the calculations of Lenth (1981), each animal location will have an associated error ellipse (Chu et al. 1989, Nams and Boutin 1991). Consequently, the biologist has much more information about the accuracy and precision of the location data. The Lenth approach does accommodate the use of at least two bearings as a basis for the location estimate. However, an advantage of using at least three bearings is in detecting poor-quality locations and reducing the average size of the confidence ellipse by four to six times (White and Garrott 1990). White and Garrott (1984) provided a computer program for using Lenth's (1981) method. To date the method has been applied mostly to the study of telemetry field procedures and analytical methods (White 1985, Garrott et al. 1986, Saltz and White 1990, Samuel and Kenow 1992).

In practice, many attempts at direction finding can result in bearings that are not useful because of signal bounce, attenuation, poor relative angle, and other factors (Garrott et al. 1986, Chu et al. 1989). If at least four bearings are taken, "bad" signals (outliers) can be discarded and a location estimate still can be obtained. White and Garrott (1990) presented a method (including SAS programs) for estimating the number of useful bearings as a function of the probability of obtaining a bad signal and the number of receiving sites. To take full advantage of these possibilities, biologists need adequate field effort to obtain the bearings and a "real time" method for assessing the usefulness of the bearings.

Triangulation generally is used to obtain estimated locations for animals that can move rapidly and that use large areas. Therefore, simultaneous bearings are most desirable (White and Garrott 1990), and two-way communication is usually essential. However, the number of personnel can be limited for taking simultaneous bearings. We suggest a simple procedure with which each of three biologists can obtain two bearings in about 2 minutes. After obtaining the first bearing from a receiving point, the operator should move 1–2 m and repeat the direction-finding process of listening for peak signal strength and nulls throughout all 360°. Because many outliers are the result of signal reflection (bounce), moving just a few paces usually will shift the operator out of the path of a reflected signal (e.g., Kenward 1987). It is important that each operator scan all 360° during both efforts at direction finding; the first effort must not influence the determination of the signal direction during the second effort. A team of field-workers, coordinating simultaneous direction finding by two-way radio communication (or a preplanned sampling schedule), thus can double the number of bearings that can be sampled for each location estimate. One field-worker can carry a portable computer and enter bearing data at the end of each bout of direction finding. Using programs such as those of White and Garrott (1984, 1990) or Dodge and Steiner (1986), or LOCATE II (see Pacer in Appendix I), the biologist can evaluate the results in the field by applying predetermined censor criteria (White and Garrott 1990) to the sample and discarding bad bearings and duplicate bearings. If need be, additional bearings, perhaps from different receiving sites, can be taken to obtain a useful sample for location estimate.

ANALYSIS OF TELEMETRY DATA
Movement and Migration

Since its inception, radiotelemetry has been used to study the movement patterns of animals. Gradually, these studies have evolved from descriptive investigations to research aimed at specific aspects of animal space use patterns, habitat use patterns, survival studies, and behavioral studies. Although analysis of data pertaining to these specific aspects of animal movement is described later in this chapter, there is a continual need to display and evaluate animal movement data. A variety of animal movement patterns is usually of ecological interest, including the presence or absence of animals at a particular site, the daily movement of individuals, seasonal movements between winter and summer areas, and dispersal movements from the natal area (Sanderson 1966).

The simple recording of the presence or absence of individual animals at key areas such as den or nest sites can provide crude yet valuable data on activity patterns (Parish and Kruuk 1982). These data typically can be recorded automatically to facilitate continuous collection of data for extended time periods.

Perhaps the most common use of movement data from a biotelemetry study is calculation of a movement rate. This rate usually is calculated as the distance moved between two consecutive points divided by the elapsed time between locations. The resulting rates are used as a relative index to compare with other factors such as age, sex, or season (Laundre et al. 1987). However, because movements are influenced by a variety of factors (Sanderson 1966), choice of how often to measure rate of movement will be critical to the outcome. Laundre et al. (1987) illustrated some of the difficulties by comparing the 24-hour movement rates to movement determined from more frequent observations. They observed little consistent correlation between rates calculated from daily locations and the movement rates determined from frequent locations of four different species (two carnivores, one ungulate, and one bird). Small and Rusch (1989) concluded that movements between daily locations were so variable that "smoothing" the rate of movement over a 5-day period was necessary to determine the underlying pattern. Because animals seldom travel in a straight line for a lengthy period of time (except for long-distance movements such as dispersal or migration), the frequency of locations used to determine movements is critical. Researchers must ensure that movement rates are estimated at intervals that are biologically appropriate.

Migratory movement patterns of many bird species and some mammals occur over extensive distances and thus are difficult to document even with radiotelemetry techniques. However, the recent application of satellite telemetry methods (Strikwerda et al. 1986) has facilitated the tracking of animals during long-distance migrations. These studies can be a valuable supplement to information obtained from band recoveries, because the actual route followed by individual animals and the length of time spent in each stopover area can be documented. In many instances the analysis applied to these data will be a simple display of the geographic locations of each animal location and time of observation (Strikwerda et al. 1986).

The study of juvenile dispersal can provide critical information in the understanding of population structure, gene flow, social ecology, and population regulation. Storm et al. (1976) summarized several factors of possible interest in the study of dispersal, including (1) the time of dispersal and factors influencing onset, (2) proportion of the population dispersing, (3) dispersal distance, and (4) dispersal direction. The rate of dispersal also can be useful in identifying the initiation and termination of dispersal movements (Small and Rusch 1989).

Analysis of movement data will depend on the research objectives, frequency of recording animal locations, and geographic scale of the movement pattern. A simple first step in any analysis is a plot of the locations for individual animals. White and Garrott (1990) discussed several approaches to the display of movement data by plotting all points, sequential movements, and three-dimensional animation of movement. Analysis of data on distance and direction of movement must consider that the data are circular (i.e., compass bearing), with no absolute zero and arbitrary high and low values. Specialized statistical procedures that use circular distributions are necessary to analyze directional movement data (see Chapter 2). Zar (1984) described several procedures for descriptive statistics and hypotheses tests from circular distributions. White and Garrott (1990) provided several examples of the analysis of circular data. Additional references include Mardia (1972) and Batschelet (1981). Applications of circular statistics to animal movement data can be found in research on the dispersal of ruffed grouse (Small and Rusch 1989) and the distribution of band recoveries from waterfowl (Nichols and Haramis 1980, Diefenbach et al. 1988).

Space Use Patterns

The description of an animal's space use pattern traditionally has been called its home range. An animal's home range is believed to contain many of the essential requirements such as food, cover, and water. In contrast, territory usually is defined as a defended part of the home range (Burt 1943). Although home ranges of different individuals can overlap, territories generally do not. Home-range size has been correlated with several ecological factors that indicate the concept has importance in explaining biological patterns. Home-range sizes have been correlated with the body size and foraging strategies of animals (Schoener 1968, Harestad and Bunnell 1979). Home-range size also can depend on the amount of food resources (Brown 1964), their distribution (Ford 1983), preferred habitat (Gese et al. 1988), population density (Cooper 1978), or risk of predation (Covich 1976).

Several methods have been developed to estimate an animal's home range from telemetry locations. The primary application of these methods has been the identification of a home-range boundary and corresponding area. Methods also have been proposed to identify some of the internal aspects (Adams and Davis 1967) of home-range use. The center of activity has been used to describe the single, most important point in the animal's home range.

Lair (1987) observed that the harmonic mean provided a better estimate for a single activity center than the arithmetic mean or the median. Aggregation of animal locations also suggests some constraints or disproportionate use of favorable parts of the home range. Core areas containing the principal use sites, refuges, and most dependable food sources (Kaufmann 1962) have been proposed to identify these critical areas within a home range (Samuel et al. 1985a, Samuel and Green 1988, Harris et al. 1990a). Other aspects of internal home-range use such as behavioral patterns (Braun 1985, Samuel and Garton 1987), activity areas (Don and Rennolls 1983, Morrison and Caccamise 1985, Caccamise and Morrison 1986), and overlap with conspecifics (Harris et al. 1990a).

Many animals exhibit differential use of areas that are within their home range. These differential use patterns may be exaggerated in some species and result in atypical space use patterns. Melquist and Hornocker (1983) reported that river otters traveled water drainages and shorelines in Idaho. The shape and size of their home ranges were strongly dictated by drainage patterns. Taylor (1978) reported that rats exhibited a linear home-range pattern along hedgerows in agricultural habitats. Animals that follow such linear use patterns are difficult to analyze with many of the currently available home-range methods. For these situations researchers must determine if the animal's space use pattern can be represented appropriately by the available methods. Melquist and Hornocker (1983) concluded that home ranges of river otters in their study were best estimated by a linear representation of stream or shoreline used.

Telemetry data can contain several errors related to animal location, data entry mistakes, or other recording errors. Data should be reviewed for such errors prior to home-range analyses (or any other type of analysis). Researchers also must determine how to handle unusual movements by animals that are outside their "normal" home range. Such excursions usually are not considered to be part of the animal's usual home range (Burt 1943). These movements have important biological consequences, but typically little information is available to explain the reason for the movement. Regardless of the reason, these movements can have a substantial effect on the estimated home-range area (Schoener 1981, Samuel and Garton 1985). Three approaches have been developed to reduce the effect of excursions on home-range estimates. First, subjective evaluation has been employed by researchers to identify unusual movements. This approach is subject to researcher biases and might not provide a repeatable home-range estimate among different researchers. A second method is to include only a portion (e.g., 95%) of the animal's locations or utilization distribution when the home-range area is determined. A third method (Koeppl and Hoffmann 1985, Samuel and Garton 1985) attempts to reduce the influence of locations that are a considerable distance from the arithmetic center. This method is repeatable but assumes an underlying bivariate normal distribution of the animal's locations. However, the procedure can be helpful in identifying both animal excursions and some types of recording errors in telemetry data (Samuel and Garton 1985).

The definition of a home-range area that includes a portion of the animal locations or of the utilization distribu-

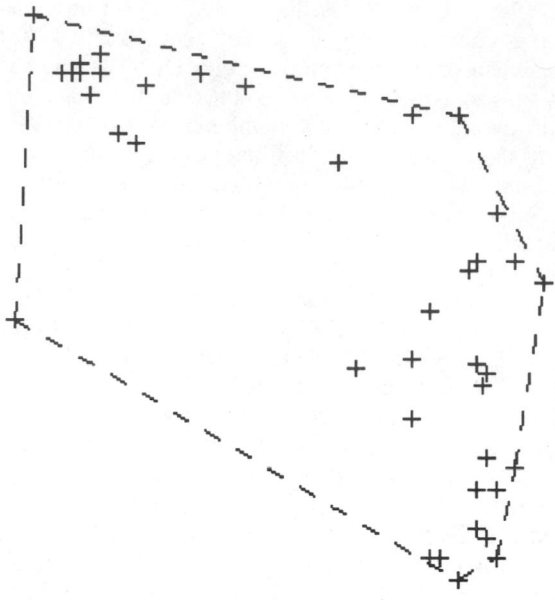

Fig. 7. Boreal owl home range (after Hayward 1989) represented by the convex polygon method.

tion appears somewhat arbitrary. Previous studies commonly have used 95% (Jennrich and Turner 1969, Van Winkle 1975) of the distribution for the bivariate normal method as a home-range boundary. However, other criteria can result in dramatic changes in home-range estimates (Schoener 1981) and the precision of these estimates (Anderson 1982). The principal difficulty in any of these choices is to obtain a home-range estimate that has a sound biological basis (White and Garrott 1990). This estimate will depend upon the research objectives and biological questions of the particular study.

Researchers also should be aware of the time-sampling nature of the collection of telemetry data. The home-range method selected cannot compensate for a poorly designed sampling scheme for collecting animal locations. This sampling scheme should correspond with the assumptions required by the method used to calculate home-range size. When the study objectives are to estimate the animal's space use pattern, particular attention should be given to sampling schemes that capture the actual activity patterns. Consideration should be given to the diurnal aspects of the animal's activity patterns, because many animals will be more active during specific times of the day. Other animals have seasonal patterns of movement that correspond to annual biological cycles. Casement displays (Geissler and Fuller 1985) can be helpful in determining diurnal and seasonal shifts in use patterns or changes in activity patterns for large data sets. All the methods assume stability of the home range (Worton 1987), so efforts should be made to reduce the chances that home-range shifts will occur during the study. Large shifts in home-range use can be detected by sequential plots of the locational data, but small changes will be nearly impossible to detect. Comparison of home ranges among different animals and studies should ensure that estimates are comparable in timing and intensity of telemetry data in relation to biological activities and changes in use patterns.

Finally, both the biological and methodological assumptions appropriate for each estimation technique should be considered prior to both collection and analysis of telemetry data. When possible, these assumptions should be tested to determine if that analytical method is appropriate. Unfortunately, few tests are available to validate present methods. In addition, many of the available tests typically indicate that the assumptions of the methods usually are violated. In some instances alternative methods might be more appropriate or the telemetry data can be manipulated to satisfy the assumptions. The principal statistical tests available enable (1) testing the independence of sequential observations (Swihart and Slade 1985*b*), (2) testing for a bivariate normal distribution (Smith 1983, Samuel and Garton 1985), and (3) testing for a uniform distribution (Samuel and Garton 1985). A brief description of the commonly used home-range methods, their biological assumptions, and statistical assumptions follows.

MINIMUM CONVEX POLYGON

The simplest and most commonly used method for determining home-range area is the minimum area polygon (Mohr and Stumpf 1966, Jennrich and Turner 1969). A convex polygon is constructed by connecting the outer locations obtained for an animal (Fig. 7). The polygon is easy to calculate and provides a boundary to the area in which the animal actually was observed. Home-range estimates from the minimum convex polygon method are useful for comparison to other studies; however, this method has substantial biological and statistical disadvantages. The estimated home-range area increases as the number of animal locations collected increases (Jennrich and Turner 1969, Anderson 1982, Bekoff and Mech 1984). Therefore, animals with home ranges based on different numbers of data points might not be comparable. Area estimates can be either too large because some unused areas are included inside the polygon boundary or too small because animals tend to use areas beyond those of the farthest sampled locations. The addition of boundary strips to increase the home-range area estimate has been proposed (Sanderson 1966). Alternatively, the elimination of a portion of the outermost points has been suggested to reduce home-range area (Kenward 1987, White and Garrott 1990).

Calculation of the minimum convex polygon involves only the outer locations of the distribution. Thus, the internal distribution of animal locations is ignored in determining the home-range area. When the animal's distribution of locations follows a uniform pattern (all areas within the home range used equally), the minimum convex polygon can provide a more reliable estimate than other methods. Many studies that calculate the interaction between animals based on home range overlap from the minimum convex polygon method implicitly assume that the use patterns follow a uniform distribution (Macdonald et al. 1980, Samuel and Garton 1985).

NORMAL DISTRIBUTION

The bivariate normal home-range model was proposed by Calhoun and Casby (1958) and extended by Jennrich and Turner (1969). This model has received common application as an alternative to the minimum convex poly-

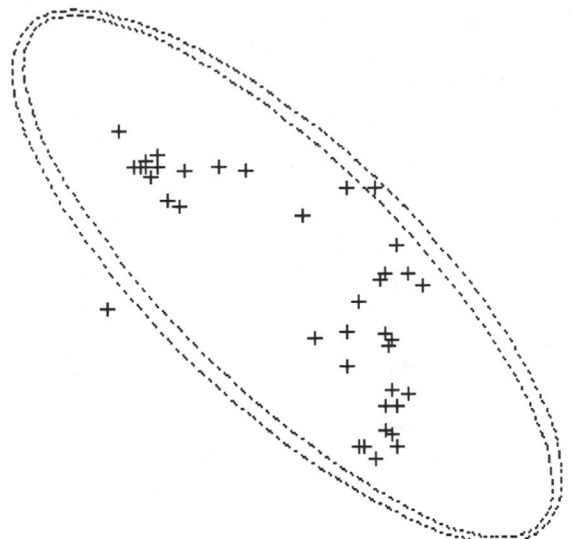

Fig. 8. Boreal owl home range represented by the 95% bivariate normal (larger ellipse) and 95% weighted-bivariate normal methods.

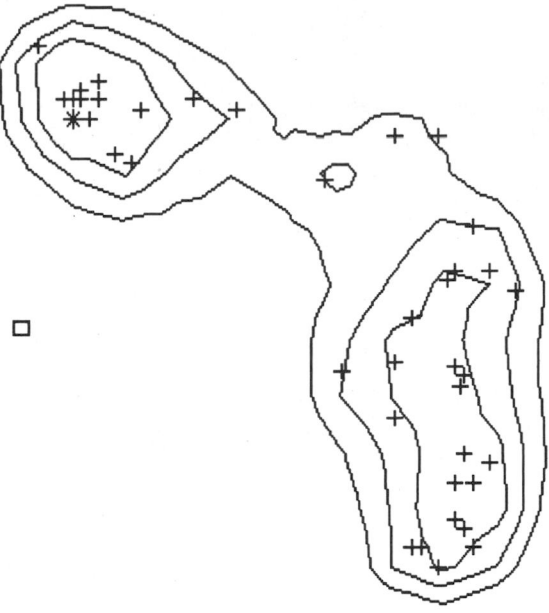

Fig. 9. Boreal owl home range represented by the 95%, 75%, and 50% contours from harmonic mean method. Animal locations (+), harmonic center of activity (*), and one location identified as an outlier (□) are also shown.

gon method. Calculation of home-range size by this method assumes an animal's activity is concentrated in the central area of the home range, and the probability of the animal occurring in an area decreases (according to a normal distribution) with increasing distance from the center of activity (Metzgar 1973). The home-range shape is always elliptical (Fig. 8) and has only a single center of activity that occurs at the arithmetic center of the distribution. However, the calculated center of activity may not be within the area most intensively used by the animal (Dixon and Chapman 1980). Schoener (1981) concluded that the bivariate normal use pattern might be reasonable for some types of animals (sit-and-wait predators and central place foragers) where use declines from the center of activity in a homogeneous environment. Extreme locations in the data set can pose significant problems in estimating the home-range area, because the distance between each location and the arithmetic center is squared in the bivariate normal calculation (Dixon and Chapman 1980). Samuel and Garton (1985) and Koeppl and Hoffmann (1985) proposed a weighting procedure to reduce the influence of these outliers on the home-range estimate (Fig. 8).

Statistical assumptions for the bivariate normal method require that animal locations are independent. Failure to achieve independent locations usually results when locations are collected too frequently. This failure will result in underestimates of the home-range size (Swihart and Slade 1985a). Swihart and Slade (1985b) proposed a procedure to validate this requirement and to adjust the location data set so that independence can be achieved. Dunn and Gipson (1977) developed a modification to the bivariate normal method that allows bursts of correlated animal locations. Goodness-of-fit tests were proposed by Smith (1983) and Samuel and Garton (1985) to verify that the location data meet the bivariate normal assumption. If the statistical assumptions of the bivariate normal method can be validated, this method will produce more reliable area estimates than other available methods. This method

also can provide an estimate of the variance associated with home range. This estimate is useful in determining the number of independent locations required to achieve a specified level of reliability for each home-range estimate. A further advantage over the minimum convex polygon method is that bivariate normal home range is independent of the number of animal locations. Therefore, unbiased estimates can be achieved with relatively small sample size; however, such estimates can have poor precision and are not recommended.

Don and Rennolls (1983) proposed an extension of the bivariate normal model to incorporate multiple activity centers. This method might be suitable when the activity centers (attraction points) are known. The utilization distribution proposed by Don and Rennolls represents a mixture of circular normal distributions centered at each attraction point. This model is identical to that proposed by Calhoun and Casby (1958) when only one attraction point exists (Worton 1987).

HARMONIC MEAN

The harmonic mean was proposed by Dixon and Chapman (1980) as an alternative home-range method based on the travel distance between a lattice (grid) of map coordinates and the animal locations. These travel distances are used to estimate activity contours that describe the animal's use pattern within the home range (Fig. 9). This method does not assume that the animal follows a specific pattern of utilization distribution, but allows the distribution to be estimated from the animal's actual use pattern. Home-range shapes can be irregular, and multiple disjoint activity centers (Samuel et al. 1985a) also can be described. Samuel and Garton (1987) suggested a modification to the harmonic mean method that can be used to incorporate the time an animal spent at a particular loca-

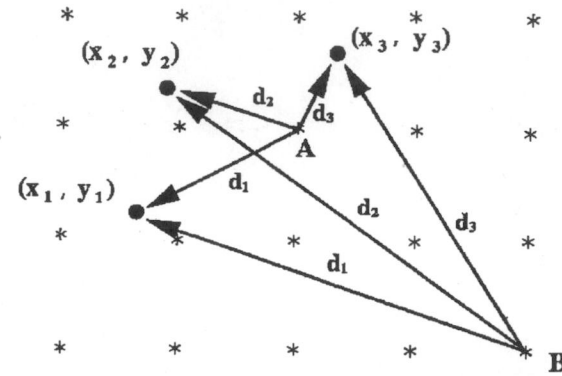

Fig. 10. The harmonic mean distances ($1/d_1$, $1/d_2$, $1/d_3$) are smaller at grid node A than at B. The area around grid node A will be included in a lower probability contour than will the area around node B.

tion during radio-tracking or to analyze behavioral activities within the home range.

The harmonic mean method uses the distances between all animal locations and all intersections of a grid lattice superimposed on the home-range area (Fig. 10). This method implicitly assumes that the distance between the grid points and each location is an indicator of potential biological activity. Therefore, the method can provide poor estimates of home range for animals with linear use patterns or traditional travel corridors. Spencer and Barrett (1984) showed that animal locations at the same coordinate as the grid point will cause a discontinuity in the harmonic mean calculation. Samuel et al. (1985b) attempted to correct this problem by relocating all observations to a mean distance from the nearest grid point. This procedure also accounts for changes in grid size and map scale associated with different animal species. Samuel et al. (1985b) also reported that harmonic mean estimates are dependent on grid cell size. They proposed an ad hoc algorithm to adjust grid cell size depending upon the number of animal locations and their density within the home-range boundary. Jaremovic and Croft (1987) proposed a grid system based on equilateral triangles, rather than squares, to estimate the harmonic home-range isopleths.

Because the harmonic mean method is based on average travel distances among animal locations and the lattice of grid points, it is not strictly required that sequential locations are independent. However, representation of an animal's space use pattern requires that locations be selected randomly (equal probability for all locations) through time. Use of nonrandom locations can occur when animals are sampled with different intensities during the study. This practice can result in a misrepresentation of the animal's use pattern that reflects the sampling effort rather than the actual use pattern (Samuel et al. 1985a).

FOURIER METHOD

Anderson (1982) proposed the Fourier transform method for representing an animal's utilization distribution. This approach is a two-dimensional extension of the Fourier method used in line transect estimation (Burnham et al. 1980). This method "smoothes" the observed histogram of animal locations by including the appropriate number of high- and low-frequency sine and cosine components in the Fourier transform series. Inclusion or exclusion of components is based on the objective criterion established by Tarter and Kronmal (1970). In addition, objective rules have been developed for determining the best grid cell size to use in calculating the Fourier smoothing functions. However, calculation of the appropriate frequency components and grid cell size requires that animal locations be independent.

Anderson (1982) observed that the Fourier method had difficulty estimating areas at the edge of the home range, primarily because few locations are usually available for these areas. Worton (1987) also noted that the method could produce regions near the home-range boundary where the estimated utilization distribution was negative. Thus, the Fourier method provided highly variable estimates for large percentages (90–95%) of the home range. Fourier home-range estimates near the activity centers (e.g., 50% of the utilization distribution) produced unbiased results (Anderson 1982).

The Fourier transform method provides an objective generalization of previously proposed grid cell methods (Siniff and Tester 1965, Voigt and Tinline 1980). These previous methods required an arbitrary specification of the cell influence on contiguous cells with influence rules typically defined in terms of chess moves: queen's (both diagonal and horizontal-vertical), rook's (only horizontal-vertical), and bishop's (only diagonal) rules (Worton 1987). However, the Fourier method overcomes the greatest problem with the grid cell method—determination of the appropriate grid cell size.

COMPARISON OF HOME-RANGE ESTIMATORS

Boulanger and White (1990) compared four home-range estimators using computer-simulated data from four different distribution patterns. They observed that all the estimators showed some bias in their estimates of the simulated distributions. In addition, the degree of bias was influenced by sample size, except for the bivariate normal estimates from a normal distribution. This method was substantially biased for all other simulated distributions. The minimum convex polygon provided poor performance in estimating the home range for all the distribution patterns (Boulanger and White 1990). The Fourier method also produced strongly biased estimates that depended on sample size and type of distribution pattern. Overall, Boulanger and White (1990) concluded that the harmonic mean method produced the least biased estimates; however, this method also had lower precision than the other methods. Larger sample sizes increased the precision for most methods, except the minimum convex polygon estimate for the normal distribution patterns.

Despite Boulanger and White's (1990) simulations, blanket recommendations on the use of home-range estimators cannot be given easily. Harris et al. (1990a) provided a helpful conceptual review and comparison of the common home-range methods, and Kenward (1992) compared the characteristics of different methods in meeting research objectives. Each method has its unique set of faults and limitations and can provide different interpretations of home-range use patterns (Fig. 11). However, because home-range methods will continue to be used, several criteria can be suggested for appropriate application of the available methods:

(1) The home-range method selected should not assume an underlying statistical distribution unless that distribution is biologically reasonable. If possible, the distribution of animal locations should be tested to ensure that this assumption is appropriate. A method should not be selected simply because it is easy to compute. This rationale is unacceptable, given the availability of computers and appropriate software.

(2) Methods should be selected that contribute to addressing the project's research objectives (Kenward 1992). Research studies that are designed to investigate the ecological aspects of an animal's space use pattern should avoid methods that simply determine a home-range boundary.

(3) Assumptions pertaining to sampling design and frequency of locations for the method should be either followed or tested to determine the effect of violation of the assumption. The number of locations required to achieve the desired level of precision or description of an animal's home range must be considered. Most methods will require at least 50 independent locations to achieve reasonable tests for underlying statistical distributions or to identify core areas. More typically, 100 or more locations will be required to achieve a reliable estimate of home range. One approach to determining the number of required locations is to plot the home-range area versus sequentially collected locations. Additional locations are collected until the home-range size reaches an asymptote (Harris et al. 1990a).

(4) Finally, investigators should consider whether home-range methods are appropriate for testing the research hypothesis. In some instances, research hypotheses can be tested without relying on home-range methods (White and Garrott 1990). Frequently, home-range estimates appear to be the only product from a telemetry study because no hypotheses were designed or tested (White and Garrott 1990).

Habitat Use Patterns

One of the principal applications of radiotelemetry methods has been the measurement of habitat use by animals. Habitat analyses have been an important component of wildlife management because habitat provides food, cover, and other factors essential for the population to survive. Radiotelemetry allows the frequent location of animals while minimizing disturbances that can affect their use pattern. Many species are elusive, so visual observations of their habitat use would be impossible. Some animals also use a variety of vegetation types that causes differential visibility. Visual location in these circumstances will produce biased use patterns that reflect the combination of habitat use and visibility differences. Indirect methods such as tracking or measurement of animal sign (e.g., pellet groups) also have been used to assess habitat use patterns. However, these indirect methods have limited application because of bias (Loft and Kie 1988) and because they typically cannot identify patterns associated with individual animals. The application of radiotelemetry to habitat studies requires that telemetry lo-

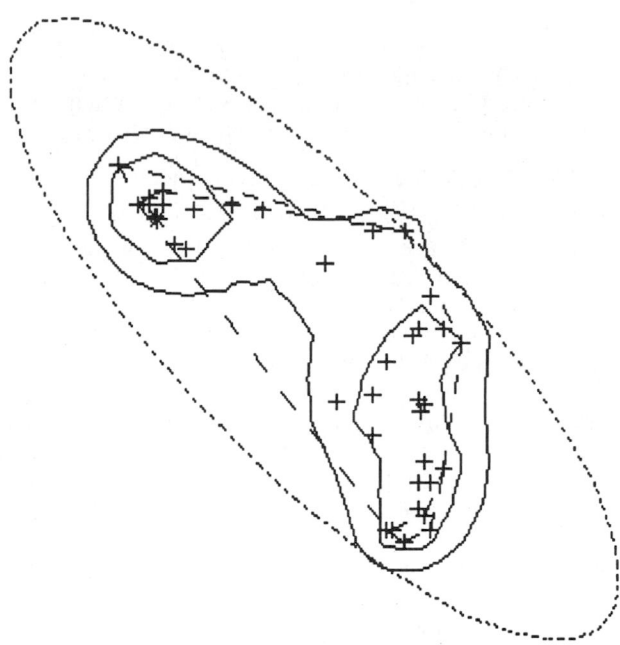

Fig. 11. Boreal owl home range (Hayward 1989) calculated with a 95% harmonic mean and core areas (———), convex polygon (– – –), and 95% bivariate normal (- - - -) methods. Animal locations (+) and harmonic center of activity (*) are shown.

cations are accurate enough to correctly distinguish which habitat the animal is using. Such an assessment will depend upon study objectives, accuracy of the telemetry locations, and patchiness of the habitat. When accuracy is poor, it may falsely appear that animals are simply using the available habitats at random.

Planning research studies that analyze habitat selection data requires decisions on two critical components that influence which analytical procedures will be used and interpretation of results. First, because of the hierarchical nature of habitat selection (Johnson 1980), the concept of selection order must be considered carefully. Research objectives should clearly state the order of selection to be investigated during the study. First-order selection considers the selection of the physical or geographical ranges of a species. Most studies involving radiotelemetry will focus on second-order or third-order habitat selection. Second-order habitat selection evaluates the selection of a home range within the study area or geographical range of the species (Johnson 1980, Thomas and Taylor 1990). Third-order selection evaluates the importance of habitat components within the animal's home range (Thomas and Taylor 1990). The importance of hierarchical selection has been conceptually recognized but should be explicitly determined as part of the research design, because the selection order implicitly identifies the population of interest, the experimental units, and the level of management interpretation (Thomas and Taylor 1990). Second-order research designs will emphasize selection within the study area. Third-order designs emphasize selection within the home range. The last-named design does not address the question of why a home range was selected (Thomas and Taylor 1990). A second critical element of the study design is the determination of how the use and availability data will be collected. The variety of proposed approaches primarily reflects data reliability, quantity of data recorded at each animal location, and the assumption about data independence among animals and animal locations. These approaches play an important role in determining which methods can be used to analyze the data collected.

DETERMINATION OF AVAILABILITY

Availability represents the amount of area of each habitat type that is available for use by the population or an individual animal. Unfortunately, the biologist's definition of availability might not correspond with what the animal perceives to be available. Thus, the determination of the area for each habitat may not provide an adequate means to measure habitat availability (Johnson 1980, White and Garrott 1990). In addition, definition of the appropriate study area can be problematic and influence availability (Porter and Church 1987). Recommendations on how to avoid such problems are limited (Johnson 1980, Porter and Church 1987) and probably depend on the species investigated. The occurrence of common habitat types that are rarely used may indicate that this habitat is not biologically suited for the species. Inclusion or exclusion of this habitat may have a substantial influence on habitat selection analyses (Johnson 1980, Thomas and Taylor 1990). Thomas and Taylor (1990) recommended that analyses be presented including and excluding these habitats.

The simplest method for determining availability is to measure the areas of each habitat type present. This measurement can be accomplished with a habitat map and planimeter. A second method is to cut the map into pieces representing each habitat type and weigh each type to obtain the proportional distribution (White and Garrott 1990). This distribution can be converted to habitat areas by determining the total area of the original map. Geographic Information Systems (GIS) or other computer software routines can be used to determine the habitat type areas from appropriately coded habitat maps (White and Garrott 1990). These methods of determining habitat availability assume that the areas can be measured exactly (without statistical error). This assumption may not be valid when habitats occur in small patches that are difficult to measure accurately. Difficulties also may arise when continuous measurements (e.g., slope, aspect, and elevation) of habitat use or availability data are collected, unless they can be divided into categorical groups and accurately measured.

When habitat availability cannot be measured easily, it can be estimated by several different approaches (Box 3). One method is to estimate the ranks of relative availability of the different habitat types (Johnson 1980). This method may be especially advantageous when only "quick and dirty" estimates of availability are required. A second method is to draw a random sample of locations to obtain an estimate of proportional availability for each habitat (Marcum and Loftsgaarden 1980). This method allows the simple determination of availability from a habitat map. The area in each habitat type can be estimated from its proportional distribution and the size of the study area. Because the random points provide an estimate of the proportional distribution (rather than an exact distribution), this method has sampling error that must be treated in the statistical analysis of habitat selection. These random points also can be used to record additional data beyond

---

Box 3. Comparison of several methods for determining habitat availability.

Habitat measurements based on a study area of 70 ha with four habitat types (A–D). Random locations ($n = 50$) were used to estimate the proportional distribution and area of each habitat type. Habitat ranks were determined by ordering the habitats by number of random locations.

	Habitat type			
	A	B	C	D
Measured area (ha)	32	6	12	20
No. random locations	26	6	7	11
Percent of locations (SE)[a]	52 (7.1)	12 (4.6)	14 (4.9)	22 (5.9)
Area (ha) estimated from random locations	36.4	8.4	9.8	15.4
Rank of habitats based on random locations	1	4	3	2

[a] $SE = [p(1 - p)/n]^{1/2}$.

---

what habitat type is available. Information on slope, aspect, elevation, soil conditions, distance to important resources, and other variables also can be obtained for each random point or grid (Clark et al. 1993). This collection of measurements forms a multivariate vector of data uniquely associated with each random point. In addition, field data can be obtained from the random locations (or proximity) to supplement the data available on maps or through GIS. This approach also requires that the same variables are collected for availability data and animal locations to make appropriate comparisons.

DETERMINATION OF USE

Methods to measure animal use patterns usually parallel those applied to availability data. However, no attempt is made to determine exactly the proportional distribution of habitat used by each animal, as can be achieved for availability. This goal would require knowledge of the exact amount of time an animal spent in each habitat. Sampling the animal's locations would be logistically infeasible and entirely too expensive, and it would negate the apparent advantage of radiotelemetry. Consequently, animal locations usually are treated in a manner similar to the random points used to determine availability. Animal locations can be used to determine proportional use by habitat type or to collect multivariate vectors of use data that correspond to that measured for availability. If analytical methods based on ranks are used, the data for proportional use can be converted easily to rank scores (see Johnson 1980).

Most methods for analyzing use data assume that animal locations are random and independent. Therefore, methods used to sample animal locations should be unbiased and serially independent. These assumptions imply that an animal is not more easily located in any habitat and that its use of one habitat during a single location is independent of its habitat use for the subsequent location.

Locations of different animals may not be independent if the animals aggregate or avoid other individuals. Statistical methods to evaluate these assumptions are lacking. Some investigators have employed ad hoc procedures to determine the appropriate sampling time between sequential locations. These procedures can be based on biological knowledge of movement rates (Carey et al. 1989) or actual observations of the distance moved between locations (Porter and Church 1987).

Accurate determination of habitat use requires that each animal location be classified correctly as to the true habitat used by the animal. Error in classification of habitat use caused by inaccurate telemetry locations will reduce the power of statistical tests for habitat selection (White and Garrott 1986, Nams 1989). Samuel and Kenow (1992) proposed a method to account for telemetry error for habitat selection tests by using the error distribution of each animal location to assess habitat misclassification.

METHODS FOR ANALYSIS OF HABITAT SELECTION

The Neu et al. (1974) method uses a chi-square goodness-of-fit analysis to determine whether observations of habitat use follow the same pattern of occurrence found for habitat availability. A statistically significant chi-square test indicates a difference in use and availability patterns. These differences can be investigated further by simultaneous Bonferroni confidence intervals on the difference in percent availability and percent use (Byers et al. 1984). The goodness-of-fit test assumes that habitat availability is measured without error and that all observations of habitat use are independent. This test has been applied to habitat use by combining locations from a collection of animals (Neu et al. 1974, Byers et al. 1984, Jenkins and Starkey 1984). The approach allows comparison of use versus availability for each habitat type across all animals. However, this comparison might not be particularly meaningful if individual animals have different habitat use or habitat availability patterns (Thomas and Taylor 1990, White and Garrott 1990). This approach also treats telemetry locations rather than individual animals as the experimental unit and lacks a clear interpretation in terms of hierarchical selection. In addition, the pooling of habitat use and availability data across animals prevents a detailed analysis of individual selection strategies. White and Garrott (1990) recommended combining chi-square test statistics from each animal, rather than pooling over animals, and then analyzing with a chi-square test. An alternative application of the Neu et al. (1974) method based on a comparison of habitat selection within the animal's home-range was used by Gese et al. (1988). Selection of a particular habitat type for a collection of animals was determined by combining the Bonferroni probabilities for each animal into a single overall test (Sokal and Rohlf 1969).

Marcum and Loftsgaarden (1980) proposed an alternative chi-square homogeneity analysis to test for habitat selection when habitat availability was sampled (rather than exactly known). Random locations are sampled, typically from a habitat map, to estimate availability. This test accounts for the additional statistical variability associated with estimating the proportions of available habitat. The method requires the same assumptions as the Neu et al. (1974) method, plus the requirement that random samples of habitat availability are independent. This method has been used to test for hierarchical habitat selection in the same manner as the Neu et al. (1974) method. When the number of random locations is very large (theoretically infinite), the error associated with estimating availability will be negligible. Then the Neu et al. (1974) and Marcum and Loftsgaarden (1980) methods provide identical results. Thus, the Neu et al. (1974) method can be considered to be the infinite (or large sample) version of the Marcum and Loftsgaarden (1980) method.

The Friedman (1937) test also can be used to check for differences in percent availability and percent use of each habitat type. The differences between availability and use for each habitat are ranked for each animal, and the ranks are used to compute a test statistic for the hypothesis that the ranks are the same for all habitats (Alldredge and Ratti 1986). Computational procedures for this test are provided in Zar (1984); habitats represent the "treatments" and animals represent the "blocks." When the Friedman test statistic rejects the hypothesis of equal ranks for all habitats, further investigation of habitat differences is required. Alldredge and Ratti (1986) chose the Fisher's least significant difference (LSD) procedure to compare all pairs of habitats for equal ranks. Zar (1984) provided an alternative multiple comparison procedure and suggestions for handling missing data (e.g., one or more habitat types are not available to all animals).

The Quade (1979) method (Alldredge and Ratti 1986) is a two-way analysis of variance procedure that tests the same hypotheses as the Friedman method. The Quade method assumes independence of ranks among animals, habitat types, and habitats for each animal. In contrast, the Friedman method allows for dependence of habitat ranks for each animal. This latter assumption seems conceptually more realistic for habitat use studies based on percentage of habitats available and used.

Johnson's (1980) method compares the ranks of utilized habitat with the ranks of available habitat. Thus, the habitat use pattern is ranked by importance of use separately from that for habitat availability. The differences between use rank and availability rank for each habitat are averaged across animals to obtain an estimate of the relative selection for that habitat (Alldredge and Ratti 1986). The size of the average differences provides an index to the order of habitat preference, large differences indicating a higher relative preference. Hotelling's T^2-statistic is used to test that the relative selections for all habitats are equal. Significant differences detected in this hypothesis are tested with the Waller-Duncan multiple comparison procedure for each pair of habitats (Johnson 1980). This method does not require actual estimates of the percentage of use or availability for each habitat; only the relative importance (rank) is required.

Multivariate methods also have been used extensively to evaluate resource selection, differences in resource use among populations, or differences in resource use among species. Many of the applicable multivariate methods were reviewed in Capen (1981). Factor analysis, principal components analysis (PCA), and similar methods have received considerable application (especially in bird habitat studies) for comparing the resource use patterns among several species. This approach attempts to reduce the relatively large number of correlated variables to a smaller set of independent factors that retain most of the information contained in the original variables. The methods have been useful in describing resource use patterns and provide a close correspondence with multidimensional niche theory. Several authors (Karr and Martin 1981, Stauffer et al. 1985, Rexstad et al. 1988) cautioned that these methods can provide statistically significant results when applied to random data. Therefore, objective interpretation of the results is critical.

Multivariate analysis of variance procedures have been used to test for significant differences in resource use patterns between populations or between sites used by animals and randomly selected sites that represent availability. Clark et al. (1993) proposed a multivariate Mahalanobis distance statistic to model the similarity between multivariate habitats used by a group of radio-marked animals and available habitat characteristics. In their approach, the Mahalanobis distance for each grid of available habitat was converted to a statistical *P*-value indicating the similarity (high *P*-values) or dissimilarity (low *P*-values) between the available habitat and that used by the animals. Other multivariate methods commonly applied to resource selection problems include discriminant function analysis (DFA) and logistic regression. These latter methods should be used cautiously, because habitat sites identified by the presence or absence of a species (or individual) usually are not well defined (Williams 1981), and classification results may be strongly influenced by sample size (Williams et al. 1990). The classification of ''absent'' can result for at least three reasons (Johnson 1981): (1) habitat at the site is unsuitable; (2) the site is not occupied for other biological reasons (i.e., low population or competition); or (3) the site is incorrectly classified as unsuitable because the sampling procedure failed. Multivariate habitat selection hypotheses also can be analyzed for data collected by categorical, rather than continuous, resource variables (Heisey 1985). Heisey's (1985) method uses log-linear model analysis to evaluate the ratio of habitat use and availability based on Manly's selectivity index (Manly et al. 1972, Manly 1974). This approach assumes that availabilities are known (c.f. Neu et al. 1974) and can be used to test for selection differences between individuals or for covariates (e.g., sex and age) associated with individual animals, or pooled across animals when differences are not significant.

COMPARISON AMONG METHODS

Selection of a method for analyzing habitat data will depend partly on the type of data collected, its reliability, how observations are weighed, assumptions required for the test, and the hypothesis to be tested. If habitat availability values are known exactly, the univariate method proposed by Neu et al. (1974) may be advantageous. Although Alldredge and Ratti (1986) made no strong recommendations, the Neu et al. (1974) procedure performed well in comparison to other methods. Heisey (1985) offered a multivariate approach for habitat analysis when availabilities were known exactly. Despite the similarity in assumptions and data requirements for the Neu et al. (1974) and Heisey (1985) methods, they are based on different tests of habitat use and availability. Neu et al. (1974) tested the differences between use and availability,

but Heisey's (1985) test is based on the ratio of use to availability. If habitat availability data are estimated from random locations, the Marcum and Loftsgaarden (1980) procedure should be considered. When only approximate estimates of the amount of habitat available or used have been obtained, procedures based on ranks (Johnson 1980, Alldredge and Ratti 1986) are most appropriate.

Multivariate methods are useful in analyzing more general questions related to resource selection. However, such analyses should be undertaken cautiously, given the apparent propensity to find statistically significant results that have limited biological meaning. These techniques can be especially useful in exploratory research to illuminate potentially meaningful ecological relations, which then can be validated by careful a priori design. Researchers should be cautioned that multivariate vectors of resource use are assumed to be measured without error. Thus, errors associated with the telemetry location (White and Garrott 1986) can cause substantial errors in the measurement of multivariate data. The magnitude and importance of such telemetry errors on multivariate analyses have not been quantified.

SAMPLING RECOMMENDATIONS

Specific recommendations on sample sizes (both the number of animals and the number of locations per animal) are difficult to make because of the individual goals and circumstances of each study. Study designs that emphasize few animals with many locations for each animal can provide detailed knowledge for these particular animals, but they have limited application to a larger population. Such designs might be more applicable when the research population is small (e.g., endangered species), if the study objectives are focused on a few individuals, or if the study is designed to collect preliminary information on a relatively unknown population. Designs that emphasize the opposite extreme (many individuals with few locations per individual) also should be avoided. This design will not provide sufficient information to assess the habitat selection patterns of individual animals and how these patterns might vary among animals.

Alldredge and Ratti (1986) compared habitat selection analyses for different numbers of animals, locations per animal, and number of habitat types available. They observed that several methods work well when the number of habitats is small, >20 animals are used, and 50 locations/animal are obtained. In general, as the number of animals or the number of observations per animal increases, the test is more powerful in detecting differences between habitat use and availability. They recommended against having few observations (e.g., 15) on few animals (<20), because this design can easily result in the conclusion that no habitat selection is occurring. Alldredge and Ratti (1986) consequently concluded that Type I and Type II error rates should be considered in selecting sample sizes when habitat selection studies are planned. Type II error rates for the Neu et al. (1974) and Marcum and Loftsgaarden (1980) methods can be calculated for planning sample sizes or evaluating habitat selection experiments by use of the methods in Cohen (1977), Zar (1984), and White and Garrott (1990). Thomas and Taylor (1990) recommended using a multinomial procedure developed by Thompson (1987) for determining the sample sizes required to estimate availability.

Several other considerations are of interest when the sample sizes required for a research project are determined. First, many researchers use the same number of random points as animal locations when estimating the proportional availability of habitat types. However, random points for estimating habitat availability are frequently less costly to obtain than animal locations, so design efficiency can be improved by increasing the number of random points. For the Marcum and Loftsgaarden (1980) method, increasing the number of random points reduces the variation associated with estimated availability of each habitat type. Second, errors in the location of an animal can cause a reduction in the power of tests for habitat selection (White and Garrott 1986). White and Garrott (1986) recommended that sample size be increased to compensate for reduced power. However, these classification errors also will introduce bias into the calculation of habitat use (Nams 1989, Samuel and Kenow 1992). Consequently, methods to reduce this bias also are necessary if habitat studies are to reflect animal use patterns accurately. Nams (1989) and Samuel and Kenow (1992) proposed methods for incorporating this misclassification error into the determination of habitat use.

Population Density Estimates

Radiotelemetry methods typically provide little direct information for determining animal population density. However, they can provide a useful method of validating the assumptions or developing correction factors for other population estimation procedures. Seber (1982) described a method of simultaneously estimating the population density and animal home-range size. This method has been used primarily with trapping data. Cooper (1978) concluded that the theoretical relationship between home range and density depended upon average home-range size and amount of overlap among animals. Little application of this theoretical result has occurred to make such approaches practical. However, Fuller and Snow (1988) described an approach for estimating the density of wolves over a large area by repeatedly locating radio-marked individuals in winter to identify packs, delineate territories, and count pack members (Mech 1982). Their method requires that all packs are identified and all individuals within packs are counted. Even under these ideal circumstances the number of wolves will be underestimated if lone wolves are not considered.

White and Garrott (1990) indicated that radios can be used in capture-recapture estimation to verify that marked animals are in the population prior to the recapture period. Individual Lincoln-Petersen estimates can use the radio-marked animals as the marked sample to determine the population size (White and Garrott 1990). This approach can be particularly useful for elusive species whose visual markers are not spotted readily (Kenward 1987). Such mark-resighting studies for closed populations have been analyzed by Lincoln-Petersen methods (Kenward 1987). Radiotelemetry also can be useful in trapping grid studies where animals live in high densities. In this situation, radios are used to determine the effective area covered by the trapping grid or to estimate the proportion of a radio-marked animal's time within the grid (Kenward 1987).

Aerial survey methods for estimating populations have benefitted substantially from radiotelemetry to assess the importance of visibility bias (Caughley 1974). Floyd et al. (1979) used radiotelemetry to determine the proportion of deer seen in different habitats during aerial surveys. Biggins and Jackson (1984) used a similar method to evaluate the importance of habitat, snow conditions, and other environmental factors on deer visibility. Gasaway et al. (1985) observed that moose visibility was influenced by animal behavior (bedded versus standing). Samuel et al. (1987) used radiotelemetry to develop a visibility model for elk based on group size and vegetation cover. Methods for appropriately using these correction factors in population estimates were presented by Steinhorst and Samuel (1989).

Survival Rates

Estimation of survival rates has become an important component in the assessment of natural factors (Fowler 1981, Gavin et al. 1984, Keith et al. 1984) and management factors (Anderson and Burnham 1976, Burnham and Anderson 1984, Nichols et al. 1984) regulating animal populations. Traditional methods of band recovery analysis are often inadequate for evaluating survival within a portion of the year or when it is difficult to band large numbers of animals and ensure high recovery rates of these banded animals. In addition, band recovery methods rely largely on bands reported by hunters and therefore are of limited value for nonhunted species. Band recovery methods also are of little use in determining causes of nonhunting mortality or for estimating survival during the nonhunting period. Alternatively, radiotelemetry studies can be extremely useful in identifying causes of mortality, rates of survival (1.0 − mortality), and factors that influence survival. Radiotelemetry also can be used to locate animals soon after death to determine the cause of mortality. In this context, radiotelemetry has been used to study mortality factors for snowshoe hares (Brand et al. 1975) and pheasants (Dumke and Pils 1973) and to determine the influences of predators on young ungulates (Schlegel 1976, Franzmann et al. 1980, Barrett 1984). Perhaps the more common use of radiotelemetry in survival analysis has been to calculate survival and to determine which factors (e.g., age and sex) might account for differences in survival rates.

Statistical techniques available to estimate survival rates from telemetry data range from a simple comparison of the proportion of animals that die during an interval of time to sophisticated methods that allow comparison of actual survival curves or test for the effect of time-specific covariates on survival. Choice of the most appropriate method depends on specific research objectives, when animals are radio-marked, whether animals lose their radios or leave the study area, and whether mortality rates are constant during the study period. Regardless of which method is selected, the following assumptions (Bunck 1987) should be met to assure correct survival estimates:

(1) The radio-marked sample must be representative of the population to be studied. Efforts should be made to obtain a random sample of the population.

(2) Radio-marked animals represent independent samples. Animals that are closely associated (e.g., nest-

lings) may be subjected to similar mortality factors and thus provide less information about mortality rates than data obtained from truly independent individuals.

(3) Radio-marking should not influence survival. Animals with transmitters should provide an unbiased estimate of survival rate for the population. These first three assumptions also are required of survival estimates from band recovery and mark-recapture studies (Jolly 1965, Seber 1965, Pollock 1981, Brownie et al. 1985).

(4) When the fate of an animal is unknown (commonly called censoring), the known survival time is assumed to be independent of the animal's actual fate. Censoring can occur because of radio failure, movement out of the study area, or termination of the study before all radio-marked animals die. This assumption implies that censored animals have the same probability of death as other radio-marked animals. The random-censoring assumption might be violated if a predator destroys the radio when killing the animal or because emigrating animals are stronger than those remaining (Pollock et al. 1989*b*).

(5) The exact time of death is known. This assumption can be relaxed, however, without substantially affecting the survival estimates (Johnson 1979, Bart and Robson 1982, Heisey and Fuller 1985). Determination of the time of mortality within 1–2 days of the actual time will be sufficient for most wildlife studies that estimate survival rates over a several-month period.

Two general types of statistical methods are available to estimate survival rates. The distinction between these two methods is based on estimating the survival rate for a specific time interval versus estimating a continuous survival curve. Interval survival estimates assume that mortalities occur at a constant rate during the study period, and thus the exact times of mortality are not of primary concern. However, the number of days each animal survives within the time interval is required for many interval survival methods. If the study period has a clearly defined origin (i.e., all animals are radio-marked at the beginning of the interval), the fate of all animals is known, and the survival rate is constant, interval survival methods should be adequate for most research objectives. Typically, some period of capturing and marking animals will occur before the study begins. Ideally, this period should be small relative to the interval over which survival is estimated. Alternatively, continuous survival methods will be more useful for those situations in which animals are captured and radio-marked over an extended time period during the beginning or throughout the length of the study, when censoring occurs, when mortality rates are not constant, or when more complex hypotheses related to the influence of time specific factors on survival are evaluated.

ANALYSIS OF INTERVAL SURVIVAL DATA

Hypothesis tests about the proportion of animals alive or dead at the end of the time interval (White and Garrott 1990) provide a simple test for differences in survival rates between two populations for a single time period. A chi-square test can be performed to test for equal probability of survival for the two populations (Box 4). If more

---

Box 4. Chi-square test for equal survival.

Sixty adult and 40 young were radio-tagged at the beginning of a hypothetical study. Survival rates were 83% for adults and 50% for young.

	Alive	Dead
Adult	50	10
Young	20	20

A chi-square test indicated a significantly higher ($\chi^2 = 12.7$, df $= 1$, $P < 0.01$) survival for adults during the study.

---

than two populations are being evaluated or if the influence of more than one factor (e.g., age and sex) on survival is being evaluated, methods such as logistic regression (Cox 1970, Lee 1980, White and Garrott 1990) will be useful in analyzing survival probabilities. In such an analysis, the survival data are treated as the dichotomous dependent variable—each animal is either alive or dead at the end of the interval. Logistic regression then can be used to evaluate explanatory variables (covariates) that possibly are associated with changes in the proportion surviving. Logistic regression is advantageous because categorical (e.g., age, sex, or population) and continuous variables (e.g., body weight or body condition) can be evaluated. The logistic regression method was applied in an analysis of banded and recovered ducks by Haramis et al. (1986) and Hepp et al. (1986).

When survival for more than one time period (i.e., several years) is considered or when new animals are radio-marked at specific intervals during each time period, a more complex analysis resembling band recovery methods might be appropriate. Such analyses can be conducted with the numerical techniques found in the computer program SURVIV (White 1983). White et al. (1987) used this approach to estimate annual survival rates of deer for a 3- to 4-year period on two different study areas. However, researchers should be cautious in using SURVIV unless they are familiar with construction of the algebraic expressions required to estimate parameters in band recovery models (Brownie et al. 1985).

Statistical analysis based on the probability of surviving a long time interval can be an inefficient use of the data collected from radiotelemetry studies, because only the status (alive or dead) of the animal at the end of the interval is recorded. The objective of many studies that use radiotelemetry will be to determine survival rates for a study period <1 year. These studies might wish to investigate survival rates during several time intervals within the study period or to investigate the relative effects of different causes of mortality on survival. More efficient statistical methods for the analysis of radiotelemetry data use the number of radio-transmitter or exposure days to calculate survival rates. The minimum exposure time for calculating survival usually coincides closely with the frequency of locating animals and actual determination of

Box 5. Calculation of daily survival rates.

The daily survival rate for the ith time period of a radiotelemetry study may be calculated by:

$$s_i = \frac{x_i - d_i}{x_i},$$

where x_i is the number of transmitter days and d_i is the number of deaths in interval i.

Variance of the daily survival rate for any interval (s_i) is computed by:

$$\text{var}(s_i) = \frac{s_i(1 - s_i)}{x_i}.$$

The daily mortality due to a specific cause j is:

$$m_{ij} = \frac{d_{ij}}{x_i},$$

where d_{ij} is the number of deaths in interval i due to source j.

Box 6. Survival times.

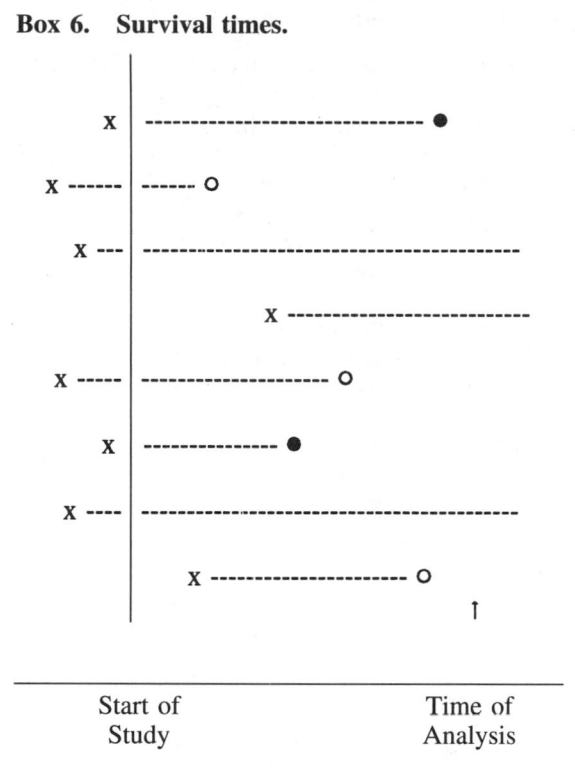

Start of Study	Time of Analysis

Illustration of marking (x) and survival times for a radiotelemetry study. Some animals are still alive at the end of the study, so the survival time for these animals is a minimum estimate (right censored). Some animals may be censored (○) during the study because their radios are lost or fail or they leave the study area. Some animals die during the study (●), and these have a known survival time. Some animals are marked after the study is initiated.

mortality during radiotelemetry studies. Statistical methods to analyze these data are based upon the methods proposed for analysis of nest success (Mayfield 1961, 1975). These methods assume that survival rates are constant throughout the study and that each time unit (i.e., day) is independent of the next time unit. Estimation of the daily survival rate then is calculated from the number of deaths and the number of transmitter days during the study interval (Box 5). This approach first was applied to a radiotelemetry study of cottontail rabbits by Trent and Rongstad (1974). Later, the method was generalized by Heisey and Fuller (1985) to determine cause-specific survival rates and to accommodate several time intervals within the study period where the survival rates are most likely to be constant.

ANALYSIS OF CONTINUOUS SURVIVAL DATA

The second major category of statistical techniques for analyzing telemetry data considers the survival time for each individual animal as a continuous measure. These techniques are mathematically complex, have been developed to analyze the results from medical studies with human subjects, and are relatively new to wildlife science. However, these methods provide a powerful set of tools for survival analysis. The methods already have been used by several wildlife researchers, and their application in wildlife research undoubtedly will increase. Researchers interested in continuous survival analysis should consult the text by Lee (1980) for an introduction to the methodology.

Continuous survival methods have several characteristics that might be unfamiliar to researchers conducting wildlife radiotelemetry studies. First, most methods use a separate origin for each subject and monitor the time until death or the termination of the study (Box 6). For wildlife studies, this approach implies that survival is primarily influenced by the length of time an animal survives after it was marked. However, seasonal influences can play an important role in survival of animal populations. This limitation can be minimized either by marking all animals at approximately the same time or by considering the origin for each animal to begin after all animals are marked. A second feature of continuous survival methods is the ability to calculate survival curves by using censored subjects. Animals alive at the end of the study are called right censored (Box 6) because their survival time is a minimum (censored) estimate of the actual time to mortality. Animals also can be added to the study at different times (staggered entry), but the survival period begins prior to the entry of all animals. Additionally, cause-specific survival rates can be estimated by assuming that deaths from other causes represent censored survival times (Kalbfleisch and Prentice 1980, Cox and Oakes 1984, Heisey and Fuller 1985). However, strong assumptions related to independence among mortality sources are required for this approach.

Continuous survival methods include parametric and nonparametric techniques to estimate the survival distribution. These techniques usually do not require the assumption of constant survival. They can be advantageous

Box 7.   Exponential survival model.

The exponential survival model is described by a single parameter:

$$S(t) = e^{-t\lambda},$$

where λ is the instantaneous probability of mortality (hazard function) for each time interval (t). This model is specified by a constant value for λ, indicating a constant hazard function and constant survival rate.

Under the exponential survival model, $\ln[S(t)]$ is linearly decreasing in time with slope λ. For a hazard rate of $\lambda = 0.0019$, the annual survival rate is $S(365) = e^{-(365*0.0019)} = 0.5$. Conversely, the hazard rate can be calculated for an estimated annual survival rate $\{-\ln[S(365)/365] = \lambda\}$. The mean survival time (M) also can be estimated by $M = 1/\lambda$ or $M \approx 526$ days.

for wildlife studies because constant survival rates might not be realistic, and display of the continuous survival distribution can provide additional biological insight about the timing of mortality.

Several parametric methods have been used in survival analysis, primarily in medical research. These methods are based on specific statistical distributions and estimated parameters (hence parametric) that describe the observed survival distribution. In a similar fashion a normal distribution is described by its parameters, the mean (μ) and variance (σ^2). The simplest and most important survival distribution is the exponential. The exponential distribution is characterized by a single parameter (λ), the constant rate of mortality (Box 7). This model implies that the instantaneous rate of mortality is constant and independent of previous events or time since marking (Fig. 12A). This constant survival assumption also is required for the interval survival methods described previously.

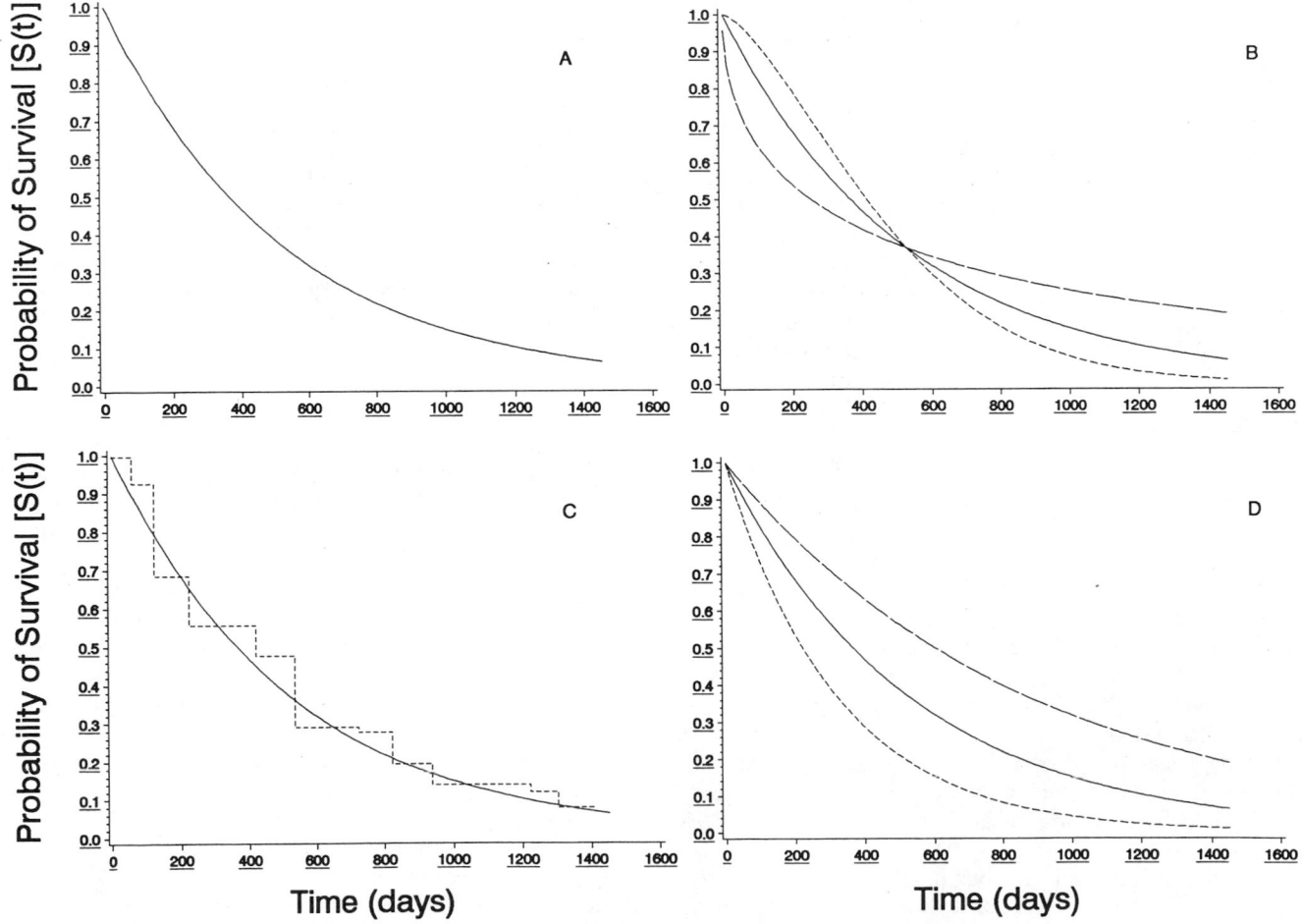

Fig. 12. Illustration of survival curves for the exponential (A), Weibull (B), Kaplan-Meier (C), and proportional hazards (D) models. In all figures (———) is the exponential survival of $S(365) = 0.5$ ($\lambda = 0.0019$). The Weibull model (B) with constant ($\alpha = 526.3 = 1/\lambda$, $\gamma = 1.0$) hazard (———), increasing ($\alpha = 526.3 = 1/\lambda$, $\gamma = 1.5$) hazard (- - - -), and decreasing ($\alpha = 526.3 = 1/\lambda$, $\gamma = 0.5$) hazard (– – –) survival curves. The Kaplan-Meier (C) survival curve (- - - -) compared to the exponential model (———) with the same survival rates. The Cox proportional hazards model (D) with the hazard [$h(t) = h_0(t)\exp(\beta_1 x_1)$], with a constant hazard $h_0(t) = \lambda t$, $\beta_1 = 0.5$, and x_1 the coded (adults = -1, juveniles = $+1$) effect of age on survival. $x_1 = 0$ is the exponential model (———), $x_1 = -1$ is the survival curve for adults (– – –), and $x_1 = +1$ is the survival curve for juveniles (- - - -).

Box 8.   Weibull survival model.

The Weibull model is based on two parameters:

$$S(t) = e^{-(t/\alpha)^\gamma},$$

where α and γ are considered the scale and shape parameters, respectively. When the shape parameter (γ) is 1.0, then $S(t)$ is the exponential model where $\lambda = 1/\alpha$.

Samuel et al. (1990) reported that neckbands on Canada geese followed a Weibull model with $\alpha = 1{,}406$ and $\gamma = 1.33$. They observed that γ was significantly >1.0, indicating that neckbands were not lost at a constant rate, rather the loss rate was lower soon after neckbands were applied and increased with time. From their model, the 1- and 2-year retention rates can be estimated:

$$S(365) = e^{-(365/\alpha)^\gamma} = 0.847$$

and

$$S(730) = e^{-(730/\alpha)^\gamma} = 0.658.$$

Box 9.   Kaplan-Meier survival rate.

The Kaplan-Meier estimate of the survival function $[S(t)]$ is the probability of an animal surviving t time units since the beginning of the study. This estimate is calculated for each of the time points when a death occurs. The formula is given by:

$$S(t_j) = \Pi \, (1 - d_j/r_j),$$

where d_j is the number of deaths and r_j is the number of radio-marked animals at risk at time t_j. The survival function $[S(t)]$ must be calculated by considering the product of all previous times when a death occurred. For example, $S(t_1) = (1 - d_1/r_1)$ and $S(t_2) = (1 - d_1/r_1)(1 - d_2/r_2)$.

Variance of the survival function can be obtained for an arbitrary time point (t) by:

$$\text{var}[S(t)] = [S(t)]^2 \sum \frac{d_j}{r_j(r_j - d_j)},$$

where the summation is over all death times which are $<$ time t.

The Weibull model (Lee 1980, Cox and Oakes 1984) provides a more flexible version of the exponential model, because the Weibull model does not assume a constant rate of instantaneous mortality. It is characterized by two parameters; one determines the general shape and the second determines the rate of the survival curve (Box 8). When the shape parameter (γ) is 1.0, the instantaneous mortality rate is constant; this simplification of the Weibull model is the exponential case. Thus, the Weibull model can be used to test for a constant rate of instantaneous mortality ($\gamma = 1.0$). When the shape parameter is >1.0, the instantaneous mortality rate increases and when the parameter is <1.0, it decreases with time (Fig. 12B). Increasing or decreasing mortality rates can indicate that the factors causing mortality are changing. When the mortality rate is increasing with time, the risk of mortality is higher later in the study, and conversely for decreasing mortality rates. Samuel et al. (1990) used the Weibull model to estimate the survival distribution of neckbands on Canada geese. Other parametric distributions such as the lognormal or gamma (Lee 1980, Pyke and Thompson 1986) also can be useful.

Graphic methods that compare the parametric survival curve to the actual survival data (Kaplan-Meier method) typically are used to determine which parametric model provides the best fit (Lee 1980, Pyke and Thompson 1986). Choice of the appropriate parametric model also has been based upon whether the mortality rates are believed to be constant, increasing, decreasing, or alternating over the study period. Statistical tests to determine the influence of covariates on survival rates can be conducted once the appropriate parametric model is identified. When parametric methods are applicable, they can provide a concise model for describing the survival distribution, determining whether mortality rates change with time, and evaluating the effect of explanatory variables on survival. Parametric methods also can be useful for validation of

the constant survival assumption required for interval survival methods.

Two nonparametric methods have received use in wildlife research: the Kaplan-Meier estimator (Kaplan and Meier 1958) and its generalization, the Cox proportional hazards model (Cox 1972). These nonparametric methods do not require the specification of a statistical distribution for survival time. The Kaplan-Meier or product limit estimator determines the probability of an animal surviving t time periods (usually days) since the beginning of the study. The survival probability is calculated sequentially at each time period that an animal dies (Box 9). The survival rate between each mortality is calculated from the number of radio-marked animals at risk and the number of deaths. Survival curves are calculated as the product of previous interval survival rates. A drop in the survival curve occurs at each death and produces a stair-shaped estimate of survival (Fig. 12C). Censoring is accomplished by removing censored animals from the number of animals at risk. The Kaplan-Meier method was used to estimate survival rates of wild turkeys (Kurzejeski et al. 1987), wild and game-farm pheasants (Krauss et al. 1987), mountain lions (Lindzey et al. 1988), and reintroduced ruffed grouse (Kurzejeski and Root 1988). A modification of the Kaplan-Meier method was developed by Pollock et al. (1989a) to allow new animals to be added (staggered entry) after the study has been initiated.

Nonparametric tests for significant differences between two survival curves usually are made with log rank or other appropriate statistical procedures (Kalbfleisch and Prentice 1980, Lee 1980, Cox and Oakes 1984). In essence, these log rank tests compare the actual number of deaths in one population to the expected number of deaths calculated by assuming both populations have equivalent survival curves. Large deviations between the actual and expected deaths increase the test statistic and decrease the probability that both populations have identical survival curves. Pollock et al. (1989a) provided a simple example of the calculation of a log rank test.

The proportional hazard method is another nonparametric statistical technique for investigating the relationship between covariates and survival time. Cox (1972) suggested that the instantaneous survival rate (hazard function) could be used to assess the relationship between the distribution of survival time t and the covariates. The underlying survival distribution, when all covariates are ignored, provides the basis for testing the effects of independent variables on survival (Box 10). This method assumes that the survival rates for different covariate values are proportional to this base survival distribution (Fig. 12D). This "proportional hazards" assumption implies that the relative risk of mortality associated with a significant covariate remains constant through time. This assumption is most likely to be satisfied when a single basic cause of mortality is acting continuously throughout the study period. If several mortality factors are important, they might have differing effects on portions of the radiomarked population. Similarly, problems might occur when primary mortality factors change during the study. Graphic procedures that compare the hazard functions typically are used to test the proportional hazards assumption (Kalbfleisch and Prentice 1980). Nonproportional hazard functions will be indicated by hazard curves that are not parallel (Lagakos 1982). The proportional hazards model requires that survival time for each animal begins at the date of marking or a common date when all animals have been marked. A version of the proportional hazards model that allows for time-dependent covariates is also possible (Lee 1980). The Cox proportional hazards model was used by Sievert and Keith (1985) to evaluate the effects of several covariates on survival of snowshoe hares and by White et al. (1987) to evaluate the influence of body size on fawn survival.

COMPARISON AMONG SURVIVAL METHODS

Analysis of survival data is presently in a transitional phase between the methods traditionally used in wildlife ecology (Mayfield 1961, Trent and Rongstad 1974) and the methods traditionally used in human medicine (Kaplan and Meier 1958, Cox 1972, Lee 1980). Most telemetry researchers are familiar with the traditional wildlife methods. However, these methods are limited because they require assumptions that can be unrealistic (e.g., constant survival), they are difficult to use for testing the importance of covariates on survival, and they usually assume that each radio-day is independent. Conversely, methods used in medicine are mathematically difficult to understand, they usually require a computer program to perform calculations, and they have assumptions that might not be applicable to wildlife studies (e.g., proportional hazards). Eventually, new procedures that combine the best features from both approaches will be developed. Firm conclusions about which methods are best are not possible presently but depend on the unique aspects of each research project. Guidelines to the relative merits of the alternative methods are presented below.

Interval survival methods that compare the probability of survival among different populations with chi-square analysis (or the more generalized approaches such as logistic regression or program SURVIV) can be especially

---

Box 10.   Cox proportional hazard model.

The Cox proportional hazard model is based on the relationship of

$$S(t) = e^{-h(t)},$$

where $h(t)$ is the hazard rate (instantaneous mortality rate) at time t. Cox (1972) suggested that this hazard function be evaluated by rewriting it as

$$h(t) = h_0(t)\exp(\Sigma\beta_j x_j),$$

where $h_0(t)$ is the hazard function for the underlying survival distribution when all predictor variables (x_j) are ignored. The principal interest is to determine if any of the predictor variables (e.g., body weight, age, sex) has a significant influence on survival. Multiple regression methods then can be used to evaluate the effect of the predictor variables (x_j's) on the hazard function by testing whether the β_j's are significantly different from 0.0. This method assumes that the mortality rate associated with each predictor variable is a constant proportion (or multiple) of the underlying hazard function [$h_0(t)$].

Sievert and Keith (1985:Table 3) reported that six covariates (x_j's) significantly influenced the survival of snowshoe hares released in Wisconsin. They used regression coefficients (β_j's) to quantify the effects of each covariate on 30-day survival rates. For example, they observed that hares with small body size had a 25% lower survival rate than hares with large body size.

---

useful for longer-term studies (≥ 1 year). For studies in which reliably estimating the time of death (relative to study length) is difficult, interval survival methods can provide the only practical approach to survival analysis. These methods also may be useful in making a preliminary evaluation of survival data. However, for most telemetry studies, alternative methods will be more efficient because they make use of the actual time an animal survived. In addition, these more complex methods allow some flexibility to test the assumption of constant survival.

The methods based on time intervals (Mayfield 1961, Trent and Rongstad 1974, Bart and Robson 1982, Heisey and Fuller 1985) provide a powerful approach to analyzing survival data when mortality rates are constant for each interval. These methods can be useful in identifying cause-specific mortality rates (Heisey and Fuller 1985) or comparing survival rates among groups of animals with different characteristics (e.g., age and sex classes). Survival might not be constant during the entire study interval, so the method proposed by Heisey and Fuller (1985) should be the most useful in analyzing survival data. The primary limitation of this approach is the difficulty in evaluating the effect of covariates on survival time.

Survival curves produced by the Kaplan-Meier method provide the actual distribution of survival during the study period. In addition, this method has considerable flexibil-

> **Box 11. Estimation of sample size for survival studies.**
>
> The number of animals (n) required in each population for a telemetry study can be estimated by:
>
> $$\sqrt{2n} = \frac{(\lambda_1 + \lambda_2)(Z_{\alpha/2} + Z_\beta)}{|\lambda_1 - \lambda_2|},$$
>
> where λ_1 and λ_2 are the exponential hazard rates for each population. Any value of λ can be calculated from the annual survival rate S(365) by:
>
> $$S(365) = e^{-365\lambda} \quad \text{or}$$
>
> $$\lambda = -\ln[S(365)]/365.$$
>
> For example, if $S_1(365) = 0.5$ and $S_2(365) = 0.4$, then $\lambda_1 = 0.0019$ and $\lambda_2 = 0.0025$. With $\alpha = 0.05$ and $\beta = 0.10$, we find $Z_{\alpha/2} = 1.96$ and $Z_\beta = 1.282$.
>
> $$\sqrt{2n} = \frac{(0.0019 + 0.0025)(1.96 + 1.282)}{|0.0019 - 0.0025|}.$$
>
> Thus, $n \approx 283$ animals should be radioed in each population.

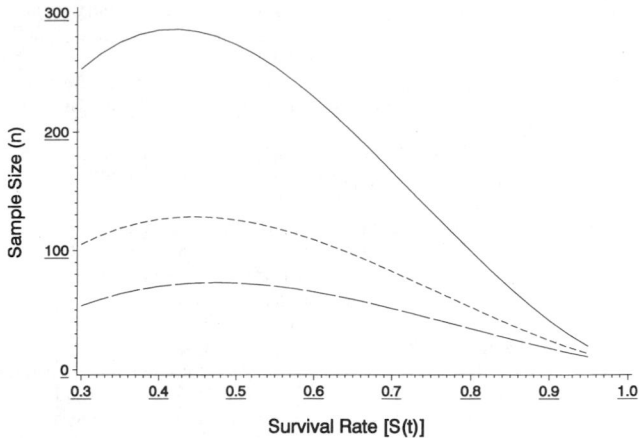

Fig. 13. Number of radioed animals required in each population (n) to show a statistically significant difference ($\alpha = 0.05$, $\beta = 0.10$) between populations. Sample sizes calculated for a 0.10 (——), 0.15 (- - - -), or 0.20 (– – –) difference in annual survival rates [$S_1(365) - S_2(365)$]. Sample size can be estimated for a given $S_1(365)$ and the curve representing the appropriate difference. Exact calculations for other values can be obtained by use of Box 11.

ity in respect to the practical aspects of conducting a wildlife research study. The major limitation of this method is the difficulty in testing biological hypotheses regarding the factors related to survival rates and cause-specific mortality. A useful point to begin a survival analysis is the Kaplan-Meier survival curve. This method provides a reference for determining which time intervals have a relatively constant survival rate and for continuing analyses with the interval survival methods. The use of continuous survival methods such as a parametric Weibull or other method to test the assumption of constant survival rate for the entire study or smaller time intervals also should be considered. If the survival data show a close correspondence with a parametric method, such an approach can be useful in determining the overall shape of the survival curve and testing the effect of important covariates on survival. If nonparametric methods are desired, the Cox model can be used, provided the proportional hazards and other assumptions can be validated. When animals are continually added to the study, the staggered-entry Kaplan-Meier method might be the only appropriate analysis.

SAMPLE SIZE RECOMMENDATIONS

Recommendations on the number of animals to radio-track will depend on the unique aspects of each study. Important factors to consider are the expected survival rates, the differences in survival rates among populations, and the desired statistical error rates in making the comparison. In some situations, a pilot study will be necessary to provide an estimate of the survival rates and variability. Sample size for simple comparisons, such as the chi-square test for difference in survival probability, can be calculated from the statistical formulas for differences between two proportions (e.g., Zar 1984). The text by Cohen (1977) is useful for calculating sample size for more com-

plex chi-square or logistic regression analyses of survival probability. Heisey and Fuller (1985) reported that variability in survival rates for discrete survival methods (e.g., Trent and Rongstad 1974, Bart and Robson 1982) decreased in direct relation to the change in number of animals radio-marked. They observed that as sample size doubled, variance was halved. Pollock et al. (1989a) indicated that at least 40–50 marked animals were required at all times to produce results with good precision from a study on northern bobwhites. They recommended that biologists be prepared to introduce additional radio-tagged animals prior to periods of high mortality. Alternatively, simulations can be performed to determine the number of radio-marked animals required for a particular study. In the absence of other information, a preliminary guideline can be calculated (Box 11) or estimated (Fig. 13) with an approximation presented by Lachin (1981) and the expected annual survival rates for each population.

Acknowledgments.—Fuller was introduced to the concepts and use of wildlife telemetry by J. R. Tester, V. B. Kuechle, R. A. Reichle, R. J. Schuster, and R. Huempfner. He also appreciates the many instructive discussions he has had with A. L. Kolz, P. W. Howey, T. K. Fuller, R. Kenward, and C. J. Amlaner, Jr. W. S. Seegar has encouraged and provided many opportunities for development and application of telemetry techniques. M. D. Samuel appreciates the advice and discussion on analysis of telemetry data with J. R. Cary, E. O. Garton, K. H. Pollock, and G. C. White. C. Hagen and J. Armstrong helped prepare portions of the manuscript. K. McDaniel provided line drawings, and M. Uehling provided photography. A. L. Kolz, M. R. Vaughan, S. S. Klugman, R. A. Garrott, R. J. Small, and G. L. Storm provided many useful comments about the draft manuscript.

LITERATURE CITED

ARRL. 1988. The ARRL antenna handbook. Am. Radio Relay League, Newington, Conn. Var. pagin.

ADAMS, L., AND S. D. DAVIS. 1967. The internal anatomy of home range. J. Mammal. 48:529–536.

ALDRIDGE, H. D. J. N., AND R. M. BRIGHAM. 1988. Load carrying and maneuverability in an insectivorous bat: a test of the 5% "rule" of radio-telemetry. J. Mammal. 69:379–382.

ALKON, P. U., AND A. COHEN. 1986. Acoustical biotelemetry for wildlife research: a preliminary test and prospects. Wildl. Soc. Bull. 14: 193–196.

ALLDREDGE, J. R., AND J. T. RATTI. 1986. Comparison of some statistical techniques for analysis of resource selection. J. Wildl. Manage. 50:157–165.

ALTHOFF, D. P., G. L. STORM, T. W. COLLINS, AND V. B. KUECHLE. 1989. Remote sensing system for monitoring animal activity, temperature, and light. Pages 116–124 in C. J. Amlaner, Jr., ed. Biotelemetry X. Univ. Arkansas Press, Fayetteville.

ALTMANN, J. 1974. Observational study of behavior: sampling methods. Behaviour 49:227–267.

AMLANER, C. J., JR. 1980. Design of antennas for use in radiotelemetry. Pages 251–261 in C. J. Amlaner, Jr., and D. W. Macdonald, eds. A handbook on biotelemetry and radio tracking. Pergamon Press, Oxford, U.K.

———, EDITOR. 1989. Biotelemetry X. Univ. Arkansas Press, Fayetteville. 733pp.

———, AND D. W. MACDONALD, EDITORS. 1980. A handbook on biotelemetry and radio tracking. Pergamon Press, Oxford, U.K. 804pp.

———, R. M. SIBLEY, AND R. H. MCCLEERY. 1978. The effects of telemetry transmitter weight on breeding success in herring gulls. Biotelem. Patient Monit. 5:154–163.

AMSTRUP, S. C. 1980. A radio-collar for game birds. J. Wildl. Manage. 44:214–217.

ANDERKA, F. W. 1980. Modulators for miniature tracking transmitters. Pages 181–184 in C. J. Amlaner, Jr., and D. W. Macdonald, eds. A handbook on biotelemetry and radio tracking. Pergamon Press, Oxford, U.K.

———. 1984. Digital coding of wildlife transmitters. Pages 405–408 in H. P. Kimmich and H. J. Klewe, eds. Biotelemetry VIII. Kimmich/Klewe, Nijmegan, Netherlands.

———. 1987. Radiotelemetry techniques for furbearers. Pages 216–227 in M. Novak, J. A. Baker, M. E. Obbard, and B. Malloch, eds. Wild furbearer management and conservation in North America. Ont. Minist. Nat. Resour., Toronto.

ANDERSEN, D. E., AND O. J. RONGSTAD. 1989. Home-range estimates of red-tailed hawks based on random and systematic relocations. J. Wildl. Manage. 53:802–807.

ANDERSON, D. J. 1982. The home range: a new nonparametric estimation technique. Ecology 63:103–112.

ANDERSON, D. R., AND K. P. BURNHAM. 1976. Population ecology of the mallard: VI. The effect of exploitation on survival. U.S. Fish Wildl. Serv. Resour. Publ. 128. 66pp.

BAILEY, G. N. A., I. J. LINN, AND P. J. WALKER. 1973. Radioactive marking of small mammals. Mammal Rev. 3:11–23.

BALL, N. J., AND C. J. AMLANER, JR. 1980. Changing heart rate of herring gulls when approached by humans. Pages 589–594 in C. J. Amlaner, Jr., and D. W. Macdonald, eds. A handbook on biotelemetry and radio tracking. Pergamon Press, Oxford, U.K.

BARRETT, M. W. 1984. Movements, habitat use, and predation on pronghorn fawns in Alberta. J. Wildl. Manage. 48:542–550.

BART, J., AND D. S. ROBSON. 1982. Estimating survivorship when the subjects are visited periodically. Ecology 63:1078–1090.

BATSCHELET, E. 1981. Circular statistics in biology. Academic Press, New York, N.Y. 371pp.

BEALE, D. M., AND A. D. SMITH. 1973. Mortality of pronghorn antelope fawns in western Utah. J. Wildl. Manage. 37:343–352.

BEKOFF, M., AND L. D. MECH. 1984. Simulation analyses of space use: home range estimates, variability, and sample size. Behav. Res. Methods Instrum. 16:32–37.

BIGGINS, D. E., AND M. R. JACKSON. 1984. Biases in aerial surveys of mule deer. Thorne Ecol. Inst. Tech. Publ. 14:60–65.

BOAG, D. A. 1972. Effect of radio packages on behavior of captive red grouse. J. Wildl. Manage. 36:511–518.

———, A. WATSON, AND R. PARR. 1973. Radio-marking versus back-tabbing red grouse. J. Wildl. Manage. 37:410–412.

BÖGEL, R. 1991. Automatic radio tracking. Pages 115–124 in Acta du colloque international: Suivi des vertebres terrestres par radiotelemetrie. Principante de Monaco, 12–13 December 1988. Parc National Mercantour, Nice, France.

———, AND D. BURCHARD. 1992. An air pressure transducer for tele-metering flight altitude of birds. Pages 100–106 in I. G. Priede and S. M. Swift, eds. Proc. 4th Eur. Conf. Wildl. Telem. Ellis Horwood, Chichester, U.K.

BOONE, R. B., AND R. S. MESECAR, III. 1989. Telemetric egg for use in egg-turning studies. J. Field Ornithol. 60:315–322.

BOSHOFF, A. F., A. S. ROBERTSON, AND P. M. NORTON. 1984. A radio-tracking study of an adult Cape griffon vulture Gyps coprotheres in the Southwestern Cape Province. S. Afr. J. Wildl. Res. 14:73–78.

BOULANGER, J. G., AND G. C. WHITE. 1990. A comparison of home-range estimators using Monte Carlo simulation. J. Wildl. Manage. 54:310–315.

BRAND, C. J., R. H. VOWLES, AND L. B. KEITH. 1975. Snowshoe hare mortality monitored by telemetry. J. Wildl. Manage. 39:741–747.

BRAUN, S. E. 1985. Home range and activity patterns of the giant kangaroo rat, *Dipodomys ingens*. J. Mammal. 66:1–12.

BRAY, O. E., AND G. W. CORNER. 1972. A tail clip for attaching transmitters to birds. J. Wildl. Manage. 36:640–642.

———, K. H. LARSEN, AND D. F. MOTT. 1975. Winter movements and activities of radio-equipped starlings. J. Wildl. Manage. 39: 795–801.

BRIGHAM, R. M. 1989. Effects of radio-transmitters on the foraging behavior of barn swallows. Wilson Bull. 101:505–506.

BROEKHUIZEN, S., C. A. VAN'T HOFF, M. B. JANSEN, AND F. J. J. NIEWOLD. 1980. Application of radio tracking in wildlife research in the Netherlands. Pages 65–84 in C. J. Amlaner, Jr., and D. W. Macdonald, eds. A handbook on biotelemetry and radio tracking. Pergamon Press, Oxford, U.K.

BROWN, G. D., AND L. S. TAYLOR. 1984. Radio-telemetry transmitters for use in studies of the thermoregulation of unrestrained common wombats, (Vombatus ursinus). Aust. Wildl. Res. 11:289–298.

BROWN, G. P., R. J. BROOKS, AND J. A. LAYFIELD. 1990. Radiotelemetry of body temperatures of free-ranging snapping turtles (Chelydra serpentina) during summer. Can. J. Zool. 68:1659–1663.

BROWN, J. L. 1964. The evolution of diversity in avian territorial systems. Wilson Bull. 76:160–169.

BROWNIE, C., D. R. ANDERSON, K. P. BURNHAM, AND D. S. ROBSON. 1985. Statistical inference from band recovery data—a handbook. U.S. Fish Wildl. Serv. Resour. Publ. 131. 305pp.

BUCHLER, E. R. 1976. Chemiluminescent tag for tracking bats and other small nocturnal animals. J. Mammal. 57:173–176.

BUNCK, C. M. 1987. Analysis of survival data from telemetry projects. J. Raptor Res. 21:132–134.

BURCHARD, D. 1989a. Direction finding in wildlife research by Doppler effect. Pages 169–177 in C. J. Amlaner, Jr., ed. Biotelemetry X. Univ. Arkansas Press, Fayetteville.

———. 1989b. Towards higher frequencies in outdoor applications. Pages 57–65 in C. J. Amlaner, Jr., ed. Biotelemetry X. Univ. Arkansas Press, Fayetteville.

BURGER, L. W., JR., M. R. RYAN, D. P. JONES, AND A. P. WYWIALOWSKI. 1991. Radio transmitters bias estimation of movements and survival. J. Wildl. Manage. 55:693–697.

BURNHAM, K. P., AND D. R. ANDERSON. 1984. Tests of compensatory vs. additive hypotheses of mortality in mallards. Ecology 65:105–112.

———, ———, AND J. L. LAAKE. 1980. Estimation of density from line transect sampling of biological populations. Wildl. Monogr. 72. 202pp.

BURT, W. H. 1943. Territoriality and home range concepts as applied to mammals. J. Mammal. 24:346–352.

BYERS, C. R., R. K. STEINHORST, AND P. R. KRAUSMAN. 1984. Clarification of a technique for analysis of utilization-availability data. J. Wildl. Manage. 48:1050–1053.

CACCAMISE, D. F., AND R. S. HEDIN. 1985. An aerodynamic basis for selecting transmitter loads in birds. Wilson Bull. 97:306–318.

———, AND D. W. MORRISON. 1986. Avian communal roosting: implications of diurnal activity centers. Am. Nat. 128:191–198.

CALHOUN, J. B., AND J. U. CASBY. 1958. Calculation of home range and density of small mammals. U.S. Public Health Serv. Public Health Monogr. 55. 24pp.

CAPEN, D. E., EDITOR. 1981. The use of multivariate statistics in studies of wildlife habitat. U.S. For. Serv. Gen. Tech. Rep. 87. 249pp.

CAREY, A. B., S. P. HORTON, AND J. A. REID. 1989. Optimal sampling for radiotelemetry studies of spotted owl habitat and home range. U.S. For. Serv. Resour. Pap. PNW-RP-416. 17pp.

CAUGHLEY, G. 1974. Bias in aerial survey. J. Wildl. Manage. 38:921–933.

CEDERLUND, G., AND P. A. LEMNELL. 1980. A simplified technique for mobile radio tracking. Pages 319–322 *in* C. J. Amlaner, Jr., and D. W. Macdonald, eds. A handbook on biotelemetry and radio tracking. Pergamon Press, Oxford, U. K.

CHARLES-DOMINIQUE, P. 1977. Urine marking and territoriality in *Galago alleni* (Waterhouse, 1837-Lorisoidea, Primates): a field study by radio-telemetry. Z. Tierpsychol. 43:113–138.

CHEESEMAN, C. L., AND P. J. MALLINSON. 1980. Radio tracking in the study of bovine tuberculosis in badgers. Pages 649–656 *in* C. J. Amlaner, Jr., and D. W. Macdonald, eds. A handbook on biotelemetry and radio tracking. Pergamon Press, Oxford, U.K.

CHU, D. S., B. A. HOOVER, M. R. FULLER, AND P. H. GEISSLER. 1989. Telemetry location error in forested habitat. Pages 188–194 *in* C. J. Amlaner, ed. Biotelemetry X. Univ. Arkansas Press, Fayetteville.

CHUTE, F. S., W. A. FULLER, P. R. J. HARDING, AND T. B. HERMAN. 1974. Radio tracking of small mammals using a grid of overhead wire antennas. Can. J. Zool. 52:1481–1488.

CLARK, J. D., J. E. DUNN, AND K. G. SMITH. 1993. A multivariate model of female black bear habitat use for a geographic information system. J. Wildl. Manage. 57:519–526.

CLUTE, R. K., AND J. J. OZOGA. 1983. Icing of transmitter collars on white-tailed deer fawns. Wildl. Soc. Bull. 11:70–71.

COCHRAN, W. G. 1977. Sampling techniques. Third ed. John Wiley & Sons, New York, N.Y. 428pp.

COCHRAN, W. W. 1972. Long-distance tracking of birds. Pages 39–59 *in* S. R. Galler, K. Schmidt-Koening, G. J. Jacobs, and R. E. Belleville, eds. Animal orientation and navigation. NASA SP-262. U.S. Gov. Printing Off., Washington, D.C.

———. 1980. Wildlife telemetry. Pages 507–520 *in* S. D. Schemnitz, ed. Wildlife management techniques manual. Fourth ed., rev. The Wildl. Soc., Washington, D.C.

———, D. W. WARNER, J. R. TESTER, AND V. B. KUECHLE. 1965. Automatic radio tracking system for monitoring animal movements. BioScience 15:98–100.

COHEN, J. 1977. Statistical power analysis for the behavioral sciences. Academic Press, New York, N.Y. 474pp.

COON, R. A., P. D. CALDWELL, AND G. L. STORM. 1976. Some characteristics of fall migration of female woodcock. J. Wildl. Manage. 40:91–95.

COOPER, H. M., AND P. CHARLES-DOMINIQUE. 1985. A microcomputer data acquisition-telemetry system: a study of activity in the bat. J. Wildl. Manage. 49:850–854.

COOPER, W. E., JR. 1978. Home range size and population dynamics. J. Theor. Biol. 75:327–337.

COVICH, A. P. 1976. Analyzing shapes of foraging areas: some ecological and economic theories. Ann. Rev. Ecol. Syst. 7:235–257.

COX, D. R. 1970. The analysis of binary data. Chapman and Hall, New York, N.Y. 142pp.

———. 1972. Regression models and life tables (with discussion). J. R. Stat. Soc. Ser. B 34:187–220.

———, AND D. OAKES. 1984. Analysis of survival data. Chapman and Hall, New York, N.Y. 201pp.

CRAIGHEAD, J. J., AND D. J. CRAIGHEAD, JR. 1987. Tracking caribou using satellite telemetry. Natl. Geogr. Res. 3:462–479.

CUNNINGHAM, C. R., J. F. CRAIG, AND W. C. MACKAY. 1983. Some experiences with an automatic grid antenna radio system for tracking freshwater fish. Int. Wildl. Biotelem. Conf. 4:135–149.

CUPAL, J. J., AND R. W. WEEKS. 1989. Digital encoding for the telemetering of biological data. Pages 39–50 *in* C. J. Amlaner, Jr., ed. Biotelemetry X. Univ. Arkansas Press, Fayetteville.

DAVIS, J. R., A. F. VON RECUM, D. D. SMITH, AND D. C. GUYNN, JR. 1984. Implantable telemetry in beaver. Wildl. Soc. Bull. 12:322–324.

DEAT, A., C. MAUGET, R. MAUGET, D. MAUREL, AND A. SEMPERE. 1980. The automatic, continuous and fixed radio tracking system of the Chize Forest: theoretical and practical analysis. Pages 439–451 *in* C. J. Amlaner, Jr., and D. W. Macdonald, eds. A handbook on biotelemetry and radio tracking. Pergamon Press, Oxford, U.K.

DELGIUDICE, G. D., K. E. KUNKEL, L. D. MECH, AND U. S. SEAL. 1990. Minimizing capture related stress on white-tailed deer with a capture collar. J. Wildl. Manage. 54:299–303.

DIEFENBACH, D. R., J. D. NICHOLS, AND J. E. HINES. 1988. Distribution patterns of American black duck and mallard winter band recoveries. J. Wildl. Manage. 52:704–710.

DIEHL, P., AND H. W. HELB. 1986. Radiotelemetric monitoring of heart-rate responses to song playback in blackbirds (*Turdus merula*). Behav. Ecol. Sociobiol. 18:213–219.

DIXON, K. R., AND J. A. CHAPMAN. 1980. Harmonic mean measure of animal activity areas. Ecology 61:1040–1044.

DODGE, W. E., AND A. J. STEINER. 1986. XYLOG: a computer program for field processing locations of radio-tagged wildlife. U.S. Fish Wildl. Serv. Tech. Rep. 4. 22pp.

———, D. S. WILKIE, AND A. J. STEINER. 1986. UTMEL: a laptop computer program for location of telemetry "finds" using LORAN-C. Massachusetts Coop. Fish Wildl. Res. Unit, Amherst. 21pp.

DON, B. A. C., AND K. RENNOLLS. 1983. A home range model incorporating biological attraction points. J. Anim. Ecol. 52:69–81.

DUMKE, R. T., AND C. M. PILS. 1973. Mortality of radio-tagged pheasants on the Waterloo Wildlife Area. Wisconsin Dep. Nat. Resour. Tech. Bull. 72. 53pp.

DUNCAN, I. J. H., AND J. H. FILSHIE. 1980. The use of radio telemetry devices to measure temperature and heart rate in domestic fowl. Pages 579–588 *in* C. J. Amlaner, Jr., and D. W. Macdonald, eds. A handbook on biotelemetry and radio tracking. Pergamon Press, Oxford, U.K.

DUNN, J. E., AND P. S. GIPSON. 1977. Analysis of radio telemetry data in studies of home range. Biometrics 33:85–101.

DUNSTAN, T. C. 1972. A harness for radio–tagging raptorial birds. Int. Bird Banding News 44:4–8.

DWYER, T. J. 1972. An adjustable radio-package for ducks. Bird Banding 43:282–284.

EAGLE, T. C., J. CHOROMANSKI-NORRIS, AND V. B. KUECHLE. 1984. Implanting radio transmitters in mink and Franklin's ground squirrels. Wildl. Soc. Bull. 12:180–184.

ERIKSTAD, K. E. 1979. Effects of radio packages on reproductive success of willow grouse. J. Wildl. Manage. 43:170–175.

FANCY, S. G., ET AL. 1988. Satellite telemetry: a new tool for wildlife research and management. U.S. Fish Wildl. Serv. Resour. Publ. 172. 54pp.

FITZNER, R. E., AND J. N. FITZNER. 1977. A hot melt glue technique for attaching radiotransmitter tail packages to raptorial birds. North Am. Bird Bander 2:56–57.

FLOYD, T. J., L. D. MECH, AND M. E. NELSON. 1979. An improved method of censusing deer in deciduous-coniferous forests. J. Wildl. Manage. 43:258–261.

FOLK, G. E., JR., AND M. A. FOLK. 1980. Physiology of large mammals by implanted radio capsules. Pages 33–43 *in* C. J. Amlaner, Jr., and D. W. Macdonald, eds. A handbook on biotelemetry and radio tracking. Pergamon Press, Oxford, U.K.

FORD, R. G. 1983. Home range in a patchy environment: optimal foraging predictions. Am. Zool. 23:315–326.

FOSTER, C. D., ET AL. 1992. Survival and reproduction of radio-marked adult spotted owls. J. Wildl. Manage. 56:91–95.

FOWLER, C. W. 1981. Density dependence as related to life history strategy. Ecology 62:602–610.

FRANZMANN, A. W., C. C. SCHWARTZ, AND R. O. PETERSON. 1980. Moose calf mortality in summer on the Kenai Peninsula, Alaska. J. Wildl. Manage. 44:764–768.

FRIEDMAN, M. 1937. The use of ranks to avoid the assumption of normality implicit in the analysis of variance. J. Am. Stat. Assoc. 32: 675–701.

FULLER, M. R., ET AL. 1984. Feasibility of bird-borne transmitter for tracking via satellite. Pages 375–378 *in* H. P. Kimmich and H. J. Klewe, eds. Biotelemetry VIII. Kimmich/Klewe, Nijmegan, Netherlands.

———, H. H. OBRECHT, III, C. J. PENNYCUICK, AND F. C. SCHAFFNER. 1989. Aerial tracking of white-tailed tropicbirds over the Caribbean Sea. Pages 133–138 *in* C. J. Amlaner, Jr., ed. Biotelemetry X. Univ. Arkansas Press, Fayetteville.

———, AND J. R. TESTER. 1973. An automated radio tracking system for biotelemetry. Raptor Res. 7:105–106.

FULLER, T. K., AND W. J. SNOW. 1988. Estimating winter wolf densities using radiotelemetry data. Wildl. Soc. Bull. 16:367–370.

GARCELON, D. K. 1977. An expandable drop-off transmitter collar for young mountain lions. Calif. Fish Game 63:185–189.

GARNER, G. W., S. C. AMSTRUP, D. C. DOUGLAS, AND C. L. GARDNER. 1989. Performance and utility of satellite telemetry during field studies of free-ranging polar bears in Alaska. Pages 66 76 *in* C. J. Amlaner, Jr., ed. Biotelemetry X. Univ. Arkansas Press, Fayetteville.

GARROTT, R. A., R. M. BARTMANN, AND G. C. WHITE. 1985. Comparison of radio-transmitter packages relative to deer fawn mortality. J. Wildl. Manage. 49:758–759.

————, G. C. White, R. M. Bartmann, and D. L. Weybright. 1986. Reflected signal bias in biotelemetry triangulation systems. J. Wildl. Manage. 50:747–752.

Garshelis, D. L., and D. B. Siniff. 1983. Evaluation of radio-transmitter attachments for sea otters. Wildl. Soc. Bull. 11:378–383.

Gasaway, W. C., S. D. Dubois, and S. J. Harbo. 1985. Biases in aerial transect surveys of moose during May and June. J. Wildl. Manage. 49:777–784.

Gautier, J. P. 1980. Biotelemetry of the vocalizations of a group of monkeys. Pages 535–544 in C. J. Amlaner, Jr., and D. W. Macdonald, eds. A handbook on biotelemetry and radio tracking. Pergamon Press, Oxford, U.K.

Gavin, T. A., L. H. Suring, P. A. Vohs, Jr., and E. C. Meslow. 1984. Population characteristics, spatial organization, and natural mortality in the Columbian white-tailed deer. Wildl. Monogr. 91. 41pp.

Geissler, P. H., and M. R. Fuller. 1985. Detecting and displaying the structure of an animal's home range. Am. Stat. Assoc., Stat. Comput. Sect. Proc. 1985:378–383.

Gese, E. M., O. J. Rongstad, and W. R. Mytton. 1988. Home range and habitat use of coyotes in southeastern Colorado. J. Wildl. Manage. 52:640–646.

Gessaman, J. A., M. R. Fuller, P. J. Pekings, and G. E. Duke. 1991. Resting metabolic rate of golden eagles, bald eagles, and barred owls with a tracking transmitter or an equivalent load. Wilson Bull. 103:261–265.

————, and K. A. Nagy. 1988. Transmitter loads affect the flight speed and metabolism of homing pigeons. Condor 90:662–668.

Gillingham, M. P., and F. L. Bunnell. 1985. Reliability of motion-sensitive radio collars for estimating activity of black-tailed deer. J. Wildl. Manage. 49:951–958.

————, and K. L. Parker. 1992. Simple timing device increases reliability of recording telemetric activity data. J. Wildl. Manage. 56:191–196.

Gilmer, D. S., L. M. Cowardin, R. L. Duval, L. M. Mechlin, C. W. Schaiffer, and V. B. Kuechle. 1981. Procedures for the use of aircraft in wildlife biotelemetry studies. U.S. Fish Wildl. Serv. Resour. Publ. 140. 19pp.

Giroux, J. F., D. V. Bell, S. Percival, and R. W. Summers. 1990. Tail-mounted radio transmitters for waterfowl. J. Field Ornithol. 61:303–309.

Green, P. 1985. Some results from the use of a long life radio transmitter package on corvids. Ringing Migr. 6:45–51.

Greenwood, R. J., and A. B. Sargeant. 1973. Influence of radio packs on captive mallards and blue-winged teal. J. Wildl. Manage. 37:3–9.

Guynn, D. C., Jr., J. R. Davis, and A. F. Von Recum. 1987. Pathological potential of intraperitoneal transmitter implants in beavers. J. Wildl. Manage. 51:605–606.

Haramis, G. M., J. D. Nichols, K. H. Pollock, and J. E. Hines. 1986. The relationship between body mass and survival of wintering canvasbacks. Auk 103:506–514.

Harestad, A. S., and F. L. Bunnell. 1979. Home range and body weight—a reevaluation. Ecology 60:389–402.

Harris, R. B., et al. 1990b. Tracking wildlife by satellite: current systems and performance. U.S. Fish Wildl. Serv. Fish Wildl. Tech. Rep. 30. 52pp.

Harris, S., W. J. Cresswell, P. G. Forde, W. J. Trewhella, T. Woollard, and S. Wray. 1990a. Home-range analysis using radio-tracking data—a review of problems and techniques particularly as applied to the study of mammals. Mammal. Rev. 20:97–123.

Harrison, C. S., and D. Stoneburner. 1981. Radiotelemetry of the brown noddy (*Anous stolidus*) of Manana Island (Oahu) Hawaii. Pac. Seabird Group Bull. 6:45.

Hayes, R. W. 1982. A telemetry device to monitor big game traps. J. Wildl. Manage. 46:551–553.

Hayward, G. D. 1989. Habitat use and population biology of boreal owls in the northern Rocky mountains, USA. Ph.D. Thesis, Univ. Idaho, Moscow. 113pp.

Heath, R. G. M. 1987. A method for attaching transmitters to penguins. J. Wildl. Manage. 51:399–401.

Heezen, K. L., and J. R. Tester. 1967. Evaluation of radio-tracking by triangulation with special reference to deer movements. J. Wildl. Manage. 31:124–141.

Hegdal, P. L., and B. A. Colvin. 1986. Radiotelemetry. Pages 679–698 in A. Y. Cooperrider, R. J. Boyd, and H. R. Stuart, eds. Inventory and monitoring of wildlife habitat. U.S. Bur. Land Manage. Serv. Cent., Denver, Colo.

————, and T. A. Gatz. 1987. Technology of radiotracking for various birds and mammals. Pages 204–206 in PECORA IV: a symposium on application of remote sensing data to wildlife management. Natl. Wildl. Fed. Sci. Tech. Ser. 3.

Heisey, D. M. 1985. Analyzing selection experiments with log-linear models. Ecology 66:1744–1748.

————, and T. K. Fuller. 1985. Evaluation of survival and cause-specific mortality rates using telemetry data. J. Wildl. Manage. 49:668–674.

Hellgren, E. C., D. W. Carney, N. P. Garner, and M. R. Vaughan. 1988. Use of breakaway cotton spacers on radio collars. Wildl. Soc. Bull. 16:216–218.

Hepp, G. R., R. J. Blohm, R. E. Reynolds, J. E. Hines, and J. D. Nichols. 1986. Physiological condition of autumn-banded mallards and its relationship to hunting vulnerability. J. Wildl. Manage. 50:177–183.

Herzog, P. W. 1979. Effects of radio-marking on behavior, movements, and survival of spruce grouse. J. Wildl. Manage. 43:316–323.

Hill, L. A., and L. G. Talent. 1990. Effects of capture, handling, banding, and radio-marking on breeding least terns and snowy plovers. J. Field Ornithol. 61:310–319.

Hirons, G. J. M., and R. B. Owen. 1982. Radio tagging as an aid to the study of woodcock. Pages 139–152 in C. L. Cheeseman and R. B. Mitson, eds. Telemetric studies of vertebrates. Academic Press, London, U.K.

Holthuijzen, A. M. A., L. Oosterhuis, and M. R. Fuller. 1985. Habitat use by migrating sharp-shinned hawks at Cape May Point, New Jersey. ICBP Tech. Publ. 5:317–327.

Hooge, P. N. 1991. The effects of radio weight and harnesses on time budgets and movements of acorn woodpeckers. J. Field Ornithol. 62:230–238.

Horton, G. I., and M. K. Causey. 1984. Brood abandonment by radio-tagged American woodcock hens. J. Wildl. Manage. 48:606–607.

Hoskinson, R. L. 1976. The effect of different pilots on aerial telemetry error. J. Wildl. Manage. 40:137–139.

Howey, P. W., R. G. Board, and J. Kear. 1977. A pulse-position-modulated multichannel radio telemetry system for the study of avian nest microclimate. Biotelemetry 4:169–180.

————, et al. 1987. A system for acquiring physiological and environmental telemetry data. Pages 347–350 in H. P. Kimmich and M. R. Neuman, eds. Biotelemetry IX. Doring-Druck, Braunschweig, Germany.

————, W. S. Seegar, M. R. Fuller, and K. Titus. 1989. A coded tracking telemetry system. Pages 103–107 in C. J. Amlaner, Jr., ed. Biotelemetry X. Univ. Arkansas Press, Fayetteville.

Hupp, J. W., and J. T. Ratti. 1983. A test of radiotelemetry triangulation accuracy in heterogeneous environments. Proc. Int. Wildl. Biotelem. Conf. 4:31–46.

Hutton, T. A., R. E. Hatfield, and C. C. Watt. 1976. A method for orienting a mobile radiotracking unit. J. Wildl. Manage. 40:192–193.

Inglis, J. M. 1981. The forward-null twin-Yagi antennal array for aerial radiotracking. Wildl. Soc. Bull. 9:222–225.

Ireland, L. C. 1980. Homing behavior of juvenile green turtles, *Chelonia mydas*. Pages 761–764 in C. J. Amlaner, Jr., and D. W. Macdonald, eds. A handbook on biotelemetry and radio tracking. Pergamon Press, Oxford, U.K.

Jackson, D. H., L. S. Jackson, and W. K. Seitz. 1985. An expandable drop-off transmitter harness for young bobcats. J. Wildl. Manage. 49:46–49.

Jackson, J. A., B. J. Schardien, and G. W. Robinson. 1977. A problem associated with the use of radio transmitters on tree surface foraging birds. Int. Bird Banding News 49:50–53.

Janeau, G., F. Spitz, E. Lecrivain, M. Dardaillon, and C. Kowalski. 1987. An automatic biotelemetry system for free ranging animals. Acta Oecol. Oecol. Appl. 8:333–341.

Jaremovic, R. V., and D. B. Croft. 1987. Comparison of techniques to determine eastern grey kangaroo home range. J. Wildl. Manage. 51:921–930.

Jenkins, D. 1980. Ecology of otters in northern Scotland. I. Otter (*Lutra lutra*) breeding and dispersion in mid-Deeside, Aberdeenshire in 1974–79. J. Anim. Ecol. 49:713–735.

Jenkins, K. J., and E. E. Starkey. 1984. Habitat use by Roosevelt elk in unmanaged forests of the Hoh Valley, Washington. J. Wildl. Manage. 48:642–646.

JENNINGS, J. G., AND W. F. GANDY. 1980. Tracking pelagic dolphins by satellite. Pages 753–755 *in* C. J. Amlaner, Jr., and D. W. Macdonald, eds. A handbook on biotelemetry and radio tracking. Pergamon Press, Oxford, U.K.

JENNRICH, R. I., AND F. B. TURNER. 1969. Measurement of non-circular home range. J. Theor. Biol. 22:227–237.

JESSUP, D. A., AND D. B. KOCH. 1984. Surgical implantation of a radiotelemetry device in wild black bears, *Ursus americanus*. Calif. Fish Game 70:163–166.

JOHNSON, D. H. 1979. Estimating nest success: the Mayfield method and an alternative. Auk 96:651–661.

———. 1980. The comparison of usage and availability measurements for evaluating resource preference. Ecology 61:65–71.

———. 1981. The use and misuse of statistics in wildlife habitat studies. Pages 11–19 *in* D. E. Capen, ed. The use of multivariate statistics in studies of wildlife habitat. U.S. For. Serv. Gen. Tech. Rep. 87.

JOHNSON, G. D., J. L. PEBWORTH, AND H. O. KRUEGER. 1991. Retention of transmitters attached to passerines using a glue-on technique. J. Field Ornithol. 62:486–491.

JOHNSON, R. N., AND A. H. BERNER. 1980. Effects of radio transmitters on released cock pheasants. J. Wildl. Manage. 44:686–689.

JOLLY, G. M. 1965. Explicit estimates from capture-recapture data with both death and immigration—stochastic model. Biometrika 52:225–247.

JOUVENTIN, P., AND H. WEIMERSKIRCH. 1990. Satellite tracking of wandering albatrosses. Nature 343:746–748.

JUDD, S. L., AND R. R. KNIGHT. 1977. Determination of grizzly bear movement patterns using biotelemetry. Proc. Int. Conf. Wildl. Biotelem. 1:93–100.

JULLIEN, J. M., J. VASSANT, AND S. BRANDT. 1990. An extensible transmitter collar designed for wild boar (*Sus scrofa scrofa*): study of neck size development in the species. Gibier Faune Sauvage 7:377–387.

KALAS, J. A., L. LAFOLD, AND P. FISKE. 1989. Effects of radio packages on great snipe during breeding. J. Wildl. Manage. 53:1155–1158.

KALBFLEISCH, J. D., AND R. L. PRENTICE. 1980. The statistical analysis of failure data. John Wiley & Sons, New York, N.Y. 321pp.

KAPLAN, E. L., AND P. MEIER. 1958. Nonparametric estimation from incomplete observations. J. Am. Stat. Assoc. 53:457–481.

KARL, B. J., AND M. N. CLOUT. 1987. An improved radio transmitter harness with a weak link to prevent snagging. J. Field Ornithol. 58:73–77.

KARR, J. R., AND T. E. MARTIN. 1981. Random numbers and principal components: further searches for the unicorn. Pages 20–24 *in* D. E. Capen, ed. The use of multivariate statistics in studies of wildlife habitat. U.S. For. Serv. Gen. Tech. Rep. 87.

KAUFMANN, J. H. 1962. Ecology and social behavior of the coati, *Nasua nirica* on Barro Colorado Island Panama. Univ. Calif. Publ. Zool. 60:95–222.

KEATING, K. A., W. G. BREWSTER, AND C. H. KEY. 1991. Satellite telemetry: performance of animal-tracking systems. J. Wildl. Manage. 55:160–171.

KEITH, L. B., J. R. CARY, O. J. RONGSTAD, AND M. C. BRITTINGHAM. 1984. Demography and ecology of a declining snowshoe hare population. Wildl. Monogr. 90. 43pp.

KENWARD, R. E. 1978. Radio transmitters tail-mounted on hawks. Ornis Scand. 9:220–223.

———. 1982. Techniques for monitoring the behaviour of grey squirrels by radio. Pages 175–196 *in* C. L. Cheeseman and R. G. Mitson, eds. Telemetric studies of vertebrates. Academic Press, London, U.K.

———. 1985. Raptor radio-tracking and telemetry. ICBP Tech. Publ. 5:409–420.

———. 1987. Wildlife radio tagging. Academic Press, London, U.K. 222pp.

———, G. J. M. HIRONS, AND F. ZIESEMER. 1982. Devices for telemetering the behaviour of the free-living birds. Pages 129–137 *in* C. L. Cheeseman and R. G. Mitson, eds. Telemetric studies of vertebrates. Academic Press, London, U.K.

KLUGMAN, S. S., AND M. R. FULLER. 1990. Effects of implanted transmitters on captive Florida sandhill cranes. Wildl. Soc. Bull. 18:394–399.

KOEPPL, J. W., AND R. S. HOFFMANN. 1985. Robust statistics for spatial analysis: the bivariate normal home range model applied to syntopic populations of two species of ground squirrels. Univ. Kans. Mus. Nat. Hist. Occas. Pap. 116. 18pp.

KOKJER, K. J., AND R. G. WHITE. 1986. A simple telemetry system for monitoring chewing activity of reindeer. J. Wildl. Manage. 50:737–740.

KOLZ, A. L. 1983. Radio frequency assignments for wildlife telemetry: a review of the regulations. Wildl. Soc. Bull. 11:56–59.

———, AND M. P. CASTLES. 1983. The development of correlation receivers for wildlife tracking. Proc. Int. Wildl. Biotelem. Conf. 4:112–134.

———, AND R. E. JOHNSON. 1975. An elevating mechanism for mobile receiving antennas. J. Wildl. Manage. 39:819–820.

———, J. W. LENTFER, AND H. G. FALLEK. 1980. Satellite radio tracking of polar bears instrumented in Alaska. Pages 743–752 *in* C. J. Amlaner, Jr., and D. W. Macdonald, eds. A handbook on biotelemetry and radio tracking. Pergamon Press, Oxford, U.K.

KOOYMAN, G. L., R. W. DAVIS, J. P. CROXALL, AND D. P. COSTA. 1982. Diving depths and energy requirments of king penguins. Science 217:726–727.

KORSCHGEN, C. E., S. J. MAXSON, AND V. B. KUECHLE. 1984. Evaluation of implanted radio transmitters in ducks. J. Wildl. Manage. 48:982–987.

KRAUSS, G. D., H. B. GRAVES, AND S. M. ZERVANOS. 1987. Survival of wild and game-farm cock pheasants released in Pennsylvania. J. Wildl. Manage. 51:555–559.

KRUUK, H., M. GORMAN, AND T. PARISH. 1980. The use of 65 Zn for estimating populations of carnivores. Oikos 34:206–208.

KUECHLE, V. B., M. R. FULLER, R. A. REICHLE, R. J. SCHUSTER, AND G. E. DUKE. 1987. Telemetry of gastric motility data from owls. Pages 363–366 *in* H. P. Kimmich and M. R. Neuman, eds. Biotelemetry IX. Doring-Druck, Braunschweig, Germany.

———, J. M. HAYNES, AND R. A. REICHLE. 1989. Use of small computers as telemetry data collectors. Pages 695–699 *in* C. J. Amlaner, Jr., ed. Biotelemetry X. Univ. Arkansas Press, Fayetteville.

KUFELD, R. C., D. C. BOWDEN, AND J. M. SIPEREK, JR. 1987. Evaluation of a telemetry system for measuring habitat usage in mountainous terrain. Northwest Sci. 61:249–256.

KUNKEL, K. E., R. C. CHAPMAN, L. D. MECH, AND E. M. GESE. 1992. Testing the "wildlink" activity system on wolves and white-tailed deer. Can. J. Zool. 69:2466–2469.

KURZEJESKI, E. W., AND B. G. ROOT. 1988. Survival of reintroduced ruffed grouse in north Missouri. J. Wildl. Manage. 52:248–252.

———, L. D. VANGILDER, AND J. B. LEWIS. 1987. Survival of wild turkey hens in north Missouri. J. Wildl. Manage. 51:188–193.

LACHIN, J. M. 1981. Introduction to sample size determination and power analysis for clinical trials. Controlled Clin. Trials 2:93–113.

LAGAKOS, S. W. 1982. Inference in survival analysis: nonparametric tests to compare survival distributions. Pages 340–364 *in* V. Mike and K. E. Stanley, eds. Statistics in medical research. John Wiley & Sons, New York, N.Y.

LAIR, H. 1987. Estimating the location of the focal center in red squirrel home ranges. Ecology 68:1092–1101.

LANCE, A. N., AND A. WATSON. 1977. Further tests of radio-marking on red grouse. J. Wildl. Manage. 41:579–582.

———, AND ———. 1980. A comment on the use of radio tracking in ecological research. Pages 355–359 *in* C. J. Amlaner, Jr., and D. W. Macdonald, eds. A handbook on biotelemetry and radio tracking. Pergamon Press, Oxford, U.K.

LAUNDRE, J. W., T. D. REYNOLDS, S. T. KNICK, AND I. J. BALL. 1987. Accuracy of daily point relocations in assessing real movement of radio-marked animals. J. Wildl. Manage. 51:937–940.

LEE, E. T. 1980. Statistical methods for survival data analysis. Lifetime Learning Publ., Belmont, Calif. 557pp.

LEE, J. E., G. C. WHITE, R. A. GARROTT, R. M. BARTMANN, AND A. W. ALLDREDGE. 1985. Accessing accuracy of a radiotelemetry system for estimating animal locations. J. Wildl. Manage. 49:658–663.

LEMNELL, P. A., G. JOHNSSON, H. HELMERSSON, O. HOLMSTRAND, AND L. NORLING. 1983. An automatic radio-telemetry system for position determination and data acquisition. Proc. Int. Wildl. Biotelem. Conf. 4:76–93.

LENTH, R. V. 1981. On finding the source of a signal. Technometrics 23:149–154.

LICHT, D. S., D. G. MCAULEY, J. R. LONGCORE, AND G. F. SEPIK. 1989. An improved method to monitor nest attentiveness using radiotelemetry. J. Field Ornithol. 60:251–258.

LINDZEY, F. G., B. B. ACKERMAN, D. BARNHURST, AND T. P. HEMKER. 1988. Survival rates of mountain lions in southern Utah. J. Wildl. Manage. 52:664–667.

LINN, I. J. 1978. Radioactive techniques for small mammal marking. Pages 177–191 *in* B. Stonehouse, ed. Animal marking: recognition marking of animals in research. Macmillan, London, U.K.

LIVEZEY, K. B. 1988. Protective frame for a 2-element hand-held Yagi antenna. J. Wildl. Manage. 52:565–567.

LOFT, E. R., AND J. G. KIE. 1988. Comparison of pellet-group and radio triangulation methods for assessing deer habitat use. J. Wildl. Manage. 52:524–527.

LOTIMER, J. S. 1980. A versatile coded wildlife transmitter. Pages 185–191 *in* C. J. Amlaner, Jr., and D. W. Macdonald, eds. A handbook on biotelemetry and radio tracking. Pergamon Press, Oxford, U.K.

LOUGHLIN, T. R. 1980. Radio telemetric determination of the 24-hour feeding activities of sea otters, *Enhydra lutris*. Pages 717–724 *in* C. J. Amlaner, Jr., and D. W. Macdonald, eds. A handbook on biotelemetry and radio tracking. Pergamon Press, Oxford, U.K.

LOVETT, J. W., AND E. P. HILL. 1977. A transmitter syringe for recovery of immobilized deer. J. Wildl. Manage. 41:313–315.

LUTTERSCHMIDT, W. I., AND H. K. REINERT. 1990. The effect of ingested transmitters upon the temperature preference of the northern water snake, *Nerodia s. sipedon*. Herpetologica 46:39–42.

MACARTHUR, R. A., R. H. JOHNSTON, AND V. GEIST. 1979. Factors influencing heart rate in free-ranging bighorn sheep: a physiological approach to the study of wildlife harassment. Can. J. Zool. 57:2010–2021.

MACDONALD, D. W., AND C. J. AMLANER, JR. 1980. A practical guide to radio tracking. Pages 143–159 *in* C. J. Amlaner, Jr., and D. W. Macdonald, eds. A handbook on biotelemetry and radio tracking. Pergamon Press, Oxford, U.K.

———, F. G. BALL, AND N. G. HOUGH. 1980. The evaluation of home range size and configuration using radio tracking data. Pages 405–424 *in* C. J. Amlaner, Jr., and D. W. Macdonald, eds. A handbook on biotelemetry and radio tracking. Pergamon Press, Oxford, U.K.

MANLY, B. F. J. 1974. A model for certain types of selection experiments. Biometrics 30:281–294.

———, P. MILLER, AND L. M. COOK. 1972. Analysis of a selective predation experiment. Am. Nat. 106:719–736.

MARCSTROM, V., R. E. KENWARD, AND M. KARLBOM. 1989. Survival of ring-necked pheasants with backpacks, necklaces, and leg bands. J. Wildl. Manage. 53:808–810.

MARCUM, C. L., AND D. O. LOFTSGAARDEN. 1980. A nonmapping technique for studying habitat preferences. J. Wildl. Manage. 44:963–968.

MARDIA, K. V. 1972. Statistics of directional data. Academic Press, New York, N.Y. 357pp.

MARKS, J. S., AND V. S. MARKS. 1987. Influence of radio collars on survival of sharp-tailed grouse. J. Wildl. Manage. 51:468–471.

MARTIN, M. L., AND J. R. BIDER. 1978. A transmitter attachment for blackbirds. J. Wildl. Manage. 54:62–66.

MASSEY, B. W., K. KEANE, AND C. BORDMAN. 1988. Adverse effects of radiotransmitters on the behavior of nesting least terns. Condor 90:945–947.

MAUSER, D. M., AND R. L. JARVIS. 1991. Attaching radio transmitters to 1-day-old mallard ducklings. J. Wildl. Manage. 55:488–491.

MAYFIELD, H. 1961. Nesting success calculated from exposure. Wilson Bull. 73:255–261.

———. 1975. Suggestions for calculating nest success. Wilson Bull. 87:456–466.

MCKENZIE, A. A., D. G. A. MELTZER, P. G. LE ROUX, AND R. A. GOSS. 1990. Use of implantable radio transmitters in large African carnivores. S. Afr. J. Wildl. Res. 20:33–35.

MECH, L. D. 1982. Wolves (radio-telemetry). Pages 227–228 *in* D. E. Davis, ed. Handbook of census methods for terrestrial vertebrates. CRC Press, Boca Raton, Fla.

———. 1983. Handbook of animal radio-tracking. Univ. Minnesota Press, Minneapolis. 107pp.

———, K. E. KUNKEL, R. C. CHAPMAN, AND T. J. KREEGER. 1990. Field testing of commercially manufactured capture collars on white-tailed deer. J. Wildl. Manage. 54:297–299.

MEDINA, A. L., AND H. D. SMITH. 1986. Designs for an antenna boom and masts for telemetry applications. Wildl. Soc. Bull. 14:291–297.

MELQUIST, W. E., AND M. G. HORNOCKER. 1979. Development and use of a telemetry technique for studying river otter. Proc. Int. Conf. Wildl. Biotelem. 2:104–114.

———, AND ———. 1983. Ecology of river otters in west central Idaho. Wildl. Monogr. 83. 60pp.

MELVIN, S. M., R. C. DREWIEN, S. A. TEMPLE, AND E. G. BIZEAU. 1983. Leg-band attachment of radio transmitters for large birds. Wildl. Soc. Bull. 11:282–285.

———, AND S. A. TEMPLE. 1987. Radio telemetry techniques for international crane studies. Pages 481–492 *in* G. W. Archibald and R. F. Pasquier, eds. Proc. 1983 Int. Crane Workshop. Int. Crane Found., Baraboo, Wis.

METZGAR, L. H. 1973. Home range shape and activity in *Peromyscus leucopus*. J. Mammal. 54:383–390.

MILLS, L. S., AND F. F. KNOWLTON. 1989. Observer performance in known and blind radio-telemetry accuracy tests. J. Wildl. Manage. 53:340–342.

MOHR, C. O., AND W. A. STUMPF. 1966. Comparison of methods for calculating areas of animal activity. J. Wildl. Manage. 30:293–304.

MOHUS, I. 1987. A storing telemetry-transmitter for recording bird activity. Ornis Scand. 18:227–230.

MONTGOMERY, J. 1985. A collar radio-transmitter attachment for wood ducks and other avian species. Proc. Int. Conf. Wildl. Biotelem. 5:19–27.

MORRISON, D. W., AND D. F. CACCAMISE. 1985. Ephemeral roosts and stable patches? A radiotelemetry study of communally roosting starlings. Auk 102:793–804.

MULLICAN, T. R. 1988. Radio telemetry and fluorescent pigments: a comparison of techniques. J. Wildl. Manage. 52:627–631.

NAMS, V. O. 1989. Effects of radiotelemetry error on sample size and bias when testing for habitat selection. Can. J. Zool. 67:1631–1636.

———, AND S. BOUTIN. 1991. What is wrong with error polygons? J. Wildl. Manage. 55:172–175.

NEU, C. W., C. R. BYERS, AND J. M. PEEK. 1974. A technique for analysis of utilization-availability data. J. Wildl. Manage. 38:541–545.

NESBITT, S. A. 1976. Use of radio telemetry techniques on Florida sandhill cranes. Pages 299–303 *in* J. C. Lewis, ed. Proc. Int. Crane Workshop. Oklahoma State Univ., Stillwater.

———, B. A. HARRIS, R. W. REPENNING, AND C. B. BROWNSMITH. 1982. Notes on red-cockaded woodpecker study techniques. Wildl. Soc. Bull. 10:160–163.

NICHOLS, J. D., M. J. CONROY, D. R. ANDERSON, AND K. P. BURNHAM. 1984. Compensatory mortality in waterfowl populations: a review of the evidence and implications for research and management. Trans. North Am. Wildl. Nat. Resour. Conf. 49:535–554.

———, AND G. M. HARAMIS. 1980. Sex-specific differences in winter distribution patterns of canvasbacks. Condor 92:406–418.

NICHOLLS, T. H., M. E. OSTRY, AND M. R. FULLER. 1981. Marking ground targets with radio transmitters dropped from aircraft. U.S. For. Serv. Res. Note NC-274. 4pp.

———, AND D. W. WARNER. 1968. A harness for attaching radio transmitters to large owls. Bird Banding 39:209–214.

NOLAN, J. W., R. H. RUSSELL, AND F. ANDERKA. 1984. Transmitters for monitoring Aldrich snares set for grizzly bears. J. Wildl. Manage. 48:942–945.

OBRECHT, H. H., III, C. J. PENNYCUICK, AND M. R. FULLER. 1988. Wind tunnel experiments to assess the effect of back-mounted radio transmitters on bird body drag. J. Exp. Biol. 135:265–273.

O'CONNOR, P. J., G. H. PYKE, AND H. SPENCER. 1987. Radio-tracking honeyeater movements. Emu 87:249–252.

O'DOR, R. K., D. M. WEBBER, AND F. M. VOEGELI. 1989. A multiple buoy acoustic-radio telemetry system for automated positioning and telemetry of physical and physiological data. Pages 444–452 *in* C. J. Amlaner, Jr., ed. Biotelemetry X. Univ. Arkansas Press, Fayetteville.

OLSEN, G. H., F. J. DEIN, G. M. HARAMIS, AND D. G. JORDE. 1992. Implanting radio transmitters in wintering canvasbacks. J. Wildl. Manage. 56:323–326.

OSGOOD, D. W. 1980. Temperature sensitive telemetry applied to studies of small mammal activity patterns. Pages 525–528 *in* C. J. Amlaner, Jr., and D. W. Macdonald, eds. A handbook on biotelemetry and radio tracking. Pergamon Press, Oxford, U.K.

PACE, R. M. 1988. Measurement error models for common wildlife radio-tracking systems. Minnesota Dep. Nat. Resour. Rep. 5. 19pp.

PALOMARES, F., AND M. DELIBES. 1991. Assessing three methods to estimate daily activity patterns in radio-tracked mongooses. J. Wildl. Manage. 55:698–700.

PARISH, T. A. 1980. A collapsible dipole antenna for radio tracking on 102 MHz. Pages 263–268 *in* C. J. Amlaner, Jr., and D. W. Macdonald, eds. A handbook on biotelemetry and radio tracking. Pergamon Press, Oxford, U.K.

———, AND H. KRUUK. 1982. The uses of radio tracking combined with other techniques in studies of badger ecology in Scotland. Pages 291–299 *in* C. L. Cheeseman and R. B. Mitson, eds. Telemetric studies of vertebrates. Academic Press, London, U.K.

PATON, W. C., C. J. ZABEL, D. L. NEAL, G. N. STEGER, N. G. TILGHMAN, AND B. R. NOON. 1991. Effects of radio tags on spotted owls. J. Wildl. Manage. 55:617–622.

PATRIC, E. F., T. P. HUSBAND, C. G. McKIEL, AND W. M. SULLIVAN. 1988. Potential of LORAN-C for wildlife research along coastal landscapes. J. Wildl. Manage. 52:162–164.

PENNYCUICK, C. J. 1975. Mechanics of flight. Pages 1–75 *in* D. S. Farner and J. R. King, eds. Avian biology. Vol. 5. Academic Press, New York, N.Y.

———. 1989. Bird flight performance. Oxford Univ. Press, Oxford, U.K. 153pp.

———, AND M. R. FULLER. 1987. Considerations of effects of radio-transmitters on bird flight. Pages 327–330 *in* H. P. Kimmich and M. R. Neuman, eds. Biotelemetry IX. Doring-Druck, Braunschweig, Germany.

———, ———, AND L. McALLISTER. 1989. Climbing performance of Harris' hawks (*Parabuteo unicinctus*) with added load: implications for muscle mechanics and for radiotracking. J. Exp. Biol. 142:17–29.

———, F. C. SCHAFFNER, M. R. FULLER, H. H. OBRECHT, III, AND L. STERNBERG. 1990. Foraging flights of the white-tailed tropicbird (*Phaethon lepturus*): radiotracking and doubly-labelled water. Colonial Waterbirds 13:96–102.

PERRY, M. C. 1981. Abnormal behavior of canvasbacks equipped with radio transmitters. J. Wildl. Manage. 45:786–789.

———, G. H. HAAS, AND J. W. CARPENTER. 1981. Radio transmitters for mourning doves: a comparison of attachment techniques. J. Wildl. Manage. 45:524–527.

PHILO, L. M., E. H. FOLLMANN, AND H. V. REYNOLDS. 1981. Field surgical techniques for implanting temperature-sensitive radio transmitters in grizzly bears. J. Wildl. Manage. 45:772–775.

POLLOCK, K. H. 1981. Capture-recapture models allowing for age-dependent survival and capture rates. Biometrics 37:521–529.

———. 1987. Experimental design of telemetry projects. J. Raptor Res. 21:129–131.

———, S. R. WINTERSTEIN, C. M. BUNCK, AND P. D. CURTIS. 1989*a*. Survival analysis in telemetry studies: the staggered entry design. J. Wildl. Manage. 53:7–15.

———, ———, AND M. J. CONROY. 1989*b*. Estimation and analysis of survival distributions for radio-tagged animals. Biometrics 45:99–109.

PORTER, W. F., AND K. E. CHURCH. 1987. Effects of environmental pattern on habitat preference analysis. J. Wildl. Manage. 51:681–685.

POULIQUEN, O., M. LEISHMAN, AND T. D. REDHEAD. 1990. Effects of radio collars on wild mice, *Mus domesticus*. Can. J. Zool. 63:1607–1609.

PYKE, D. A., AND J. N. THOMPSON. 1986. Statistical analysis of survival and removal rate experiments. Ecology 67:240–245.

PYKE, G. H., AND P. J. O'CONNOR. 1990. The accuracy of a radio tracking system for monitoring honeyeater movements. Aust. Wildl. Res. 17:501–509.

QUADE, D. 1979. Using weighted rankings in the analysis of complete blocks with additive block effects. J. Am. Stat. Assoc. 74:680–683.

RAIM, A. 1978. A radio transmitter attachment for small passerine birds. Bird Banding 49:326–332.

RALLS, K., D. B. SINIFF, T. D. WILLIAMS, AND V. B. KUECHLE. 1989. An intraperitoneal radio transmitter for sea otters. Mar. Mamm. Sci. 5:376–381.

RAPPOLE, J. H., AND A. R. TIPTON. 1991. New harness design for attachment of radio transmitters to small passerines. J. Field Ornithol. 62:335–337.

REINERTSEN, R. E. 1982. Radio telemetry measurements of deep body temperature of small birds. Ornis Scand. 13:11–16.

REXSTAD, E. A., D. D. MILLER, C. H. FLATHER, E. M. ANDERSON, J. W. HUPP, AND D. R. ANDERSON. 1988. Questionable multivariate statistical inference in wildlife habitat and community studies. J. Wildl. Manage. 52:794–798.

REYNOLDS, T. D., AND J. W. LAUNDRE. 1990. Time intervals for estimating pronghorn and coyote home ranges and daily movements. J. Wildl. Manage. 54:316–322.

SALTZ, D., AND G. C. WHITE. 1990. Comparison of different measures of the error in simulated radio-telemetry locations. J. Wildl. Manage. 54:169–174.

SAMUEL, M. D., AND E. O. GARTON. 1985. Home range: a weighted normal estimate and tests of underlying assumptions. J. Wildl. Manage. 49:513–519.

———, AND ———. 1987. Incorporating activity time in harmonic home range analysis. J. Wildl. Manage. 51:254–257.

———, ———, M. W. SCHLEGEL, AND R. G. CARSON. 1987. Visibility bias during aerial surveys of elk in northcentral Idaho. J. Wildl. Manage. 51:622–630.

———, AND R. E. GREEN. 1988. A revised test procedure for identifying core areas within the home range. J. Anim. Ecol. 57:1067–1068.

———, AND K. P. KENOW. 1992. Evaluating habitat selection with biotelemetry triangulation error. J. Wildl. Manage. 56:725–734.

———, D. J. PIERCE, AND E. O. GARTON. 1985*a*. Identifying areas of concentrated use within the home range. J. Anim. Ecol. 54:711–719.

———, ———, ———, L. J. NELSON, AND K. R. DIXON. 1985*b*. User's manual for program home range. Univ. Idaho For., Wildl. Range Exp. Stn., Moscow.

———, N. T. WEISS, D. H. RUSCH, S. R. CRAVEN, R. E. TROST, AND F. D. CASWELL. 1990. Neck-band retention for Canada geese in the Mississippi Flyway. J. Wildl. Manage. 54:612–621.

SANDERSON, G. C. 1966. The study of mammal movements—a review. J. Wildl. Manage. 30:215–235.

SARGEANT, A. B. 1980. Approaches, field considerations and problems associated with radio tracking carnivores. Pages 57–63 *in* C. J. Amlaner, Jr., and D. W. Macdonald, eds. A handbook on biotelemetry and radio tracking. Pergamon Press, Oxford, U.K.

SCHEAFFER, R. L., W. MENDENHALL, AND L. OTT. 1986. Elementary survey sampling. Third ed. Duxbury Press, Boston, Mass. 324pp.

SCHLEGEL, M. W. 1976. Factors affecting calf elk survival in north central Idaho: a progress report. Proc. Annu. Conf. West. Assoc. State Game and Fish Comm. 56:342–355.

SCHMIDT, D. F., J. P. SHAFFERY, N. J. BALL, D. LOENNEKE, AND C. J. AMLANER, JR. 1989. Electrophysiological sleep characteristics in bobwhite quail. Pages 339–344 *in* C. J. Amlaner, Jr., ed. Biotelemetry X. Univ. Arkansas Press, Fayetteville.

SCHOBER, F., W. M. BUGNAR, AND J. WAGNER. 1989. A software package for acquisition and evaluation of biotelemetric data from domestic and wild animals. Pages 700–708 *in* C. J. Amlaner, Jr., ed. Biotelemetry X. Univ. Arkansas Press, Fayetteville.

———, AND B. OEHRY. 1987. Automatic RF receiving system for carrier frequency pulses. Pages 351–354 *in* H. P. Kimmich and M. R. Neuman, eds. Biotelemetry IX. Doring-Druck, Braunschweig, Germany.

SCHOENER, T. W. 1968. Sizes of feeding territories among birds. Ecology 49:123–141.

———. 1981. An empirically based estimate of home range. Theor. Popul. Biol. 20:281–325.

SCHWARTZ, A., J. D. WEAVER, N. R. SCOTT, AND T. J. CADE. 1977. Measuring the temperature of eggs during incubation under captive falcons. J. Wildl. Manage. 41:12–17.

SEBER, G. A. F. 1965. A note on the multiple-recapture census. Biometrika 52:249–259.

———. 1982. The estimation of animal abundance and related parameters. Second ed. Oxford Univ. Press, New York, N.Y. 654pp.

SEDINGER, J. S., R. G. WHITE, AND W. E. HAUER. 1990. Effects of carrying radio transmitters on energy expenditure of Pacific black brant. J. Wildl. Manage. 54:42–45.

SERVHEEN, C., T. T. THIER, C. J. JONKEL, AND D. BEATY. 1981. An ear-mounted transmitter for bears. Wildl. Soc. Bull. 9:56–57.

SHIELDS, L. J., AND G. S. MUELLER. 1983. An alternative radio-transmitter attachment technique for small birds. Proc. Int. Conf. Wildl. Biotelem. 4:57–62.

SIEGFRIED, W. R., P. G. H. FROST, I. J. BALL, AND D. F. McKINNEY. 1977. Effects of radio packages on African black ducks. S. Afr. J. Wildl. Res. 7:37–40.

SIEVERT, P. R., AND L. B. KEITH. 1985. Survival of snowshoe hares at a geographic range boundary. J. Wildl. Manage. 49:854–866.

SINIFF, D. B., AND J. R. TESTER. 1965. Computer analysis of animal movement data obtained by telemetry. BioScience 15:104–108.

SMALL, R. J., AND D. R. RUSCH. 1989. The natal dispersal of ruffed grouse. Auk 106:72–79.

SMITH, E. N., AND E. G. AITKEN. 1989. Low power skin and muscle blood flow photo plethysmography biotelemetry system. Pages 325–331 in C. J. Amlaner, Jr., ed. Biotelemetry X. Univ. Arkansas Press, Fayetteville.

———, AND C. J. AMLANER, JR. 1989. Biotelemetry workshop. Pages 462–477 in C. J. Amlaner, Jr., ed. Biotelemetry X. Univ. Arkansas Press, Fayetteville.

———, AND G. L. BARNES. 1989. Miniature low power blood flow photo plethysmography biotelemetry system. Pages 125–130 in C. J. Amlaner, Jr., ed. Biotelemetry X. Univ. Arkansas Press, Fayetteville.

———, AND S. E. MOORE. 1989. Inexpensive magnetically switched temperature and egg biotelemetry system. Pages 552–557 in C. J. Amlaner, Jr., ed. Biotelemetry X. Univ. Arkansas Press, Fayetteville.

SMITH, G. J., J. R. CARY, AND O. J. RONGSTAD. 1981. Sampling strategies for radio-tracking coyotes. Wildl. Soc. Bull. 9:88–93.

SMITH, H. R. 1980. Growth, reproduction and survival in *Peromyscus leucopus* carrying intraperitoneally implanted transmitters. Pages 367–374 in C. J. Amlaner, Jr., and D. W. Macdonald, eds. A handbook on biotelemetry and radio tracking. Pergamon Press, Oxford, U.K.

SMITH, R. M., AND B. TREVOR-DEUTSCH. 1980. A practical, remotely-controlled, portable radio telemetry receiving apparatus. Pages 269–273 in C. J. Amlaner, Jr., and D. W. Macdonald, eds. A handbook on biotelemetry and radio tracking. Pergamon Press, Oxford, U.K.

SMITH, W. P. 1983. A bivariate normal test of elliptical home-range models: biological implications and recommendations. J. Wildl. Manage. 47:613–619.

SMITS, A. W. 1984. Activity patterns and thermal biology of the toad *Bufo boreas halophilus*. Copeia 1984:689–696.

SNYDER, N. F. R., S. R. BEISSINGER, AND M. R. FULLER. 1989. Solar radio-transmitters on snail kites in Florida. J. Field Ornithol. 60:171–177.

SODHI, N. S., I. G. WARKENTIN, P. C. JAMES, AND L. W. OLIPHANT. 1991. Effects of radiotagging on breeding merlins. J. Wildl. Manage. 55:613–616.

SOKAL, R. R., AND F. J. ROHLF. 1969. Biometry. W. H. Freeman Co., San Francisco, Calif. 776pp.

SORENSON, M. D. 1989. Effects of neck collar radios on female redheads. J. Field Ornithol. 60:523–528.

SPENCER, H. J., G. LUCAS, AND P. J. O'CONNOR. 1987. A remotely switched passive null-peak network for animal tracking and radio direction finding. Aust. Wildl. Res. 14:311–317.

SPENCER, W. D., AND R. H. BARRETT. 1984. An evaluation of the harmonic mean measure for defining carnivore activity areas. Acta Zool. Fenn. 171:255–259.

SPRINGER, J. T. 1979. Some sources of bias and sampling error in radio triangulation. J. Wildl. Manage. 43:926–935.

STANNER, M., AND E. FARHI. 1989. Computerized radio-telemetric system for monitoring free ranging snakes. Isr. J. Zool. 35:177–186.

STAUFFER, D. F., E. O. GARTON, AND R. K. STEINHORST. 1985. A comparison of principal components from real and random data. Ecology 66:1693–1698.

STEBBINGS, R. E. 1982. Radio tracking greater horseshoe bats with preliminary observations on flight patterns. Pages 161–173 in C. L. Cheeseman and R. B. Mitson, eds. Telemetric studies of vertebrates. Academic Press, London, U.K.

STEINHORST, R. K., AND M. D. SAMUEL. 1989. Sightability adjustment methods for aerial surveys of wildlife populations. Biometrics 45:415–425.

STOHR, W. 1989. Long term heart rate telemetry in small mammals. Pages 352–375 in C. J. Amlaner, Jr., ed. Biotelemetry X. Univ. Arkansas Press, Fayetteville.

STONEHOUSE, B., EDITOR. 1978. Animal marking: recognition marking of animals in research. Macmillan, London, U.K. 257pp.

STORM, G. L., R. D. ANDREWS, R. L. PHILLIPS, R. A. BISHOP, D. B. SINIFF, AND J. R. TESTER. 1976. Morphology, reproduction, dispersal, and mortality of midwestern red fox populations. Wildl. Monogr. 49. 82pp.

STOUFFER, R. H., JR., J. E. GATES, C. H. HOCUTT, AND J. R. STAUFFER, JR. 1983. Surgical implantation of a transmitter package for radiotracking endangered hellbenders. Wildl. Soc. Bull. 11:384–386.

STRATHEARN, S. M., J. S. LOTIMER, G. B. KOLENOSKY, AND W. M. LINTACK. 1984. An expanding break-away radio collar for black bear. J. Wildl. Manage. 48:939–942.

STRIKWERDA, T. E., H. D. BLACK, N. LEVANON, AND P. W. HOWEY. 1985. The bird-borne transmitter. Johns Hopkins Appl. Physics Lab. Tech. Digest 6:60–67.

———, M. R. FULLER, W. S. SEEGAR, P. W. HOWEY, AND H. D. BLACK. 1986. Bird-borne satellite transmitter and location program. Johns Hopkins Appl. Physics Lab. Tech. Digest 7:203–208.

SWANSON, G. A., V. B. KUECHLE, AND A. B. SARGEANT. 1976. A telemetry technique for monitoring diel waterfowl activity. J. Wildl. Manage. 40:187–190.

SWIHART, R. K., AND N. A. SLADE. 1985a. Influence of sampling interval on estimates of home-range size. J. Wildl. Manage. 49:1019–1025.

———, AND ———. 1985b. Testing for independence of observations in animal movements. Ecology 66:1176–1184.

SYKES, P. W., JR., J. W. CARPENTER, S. HOLZMAN, AND P. H. GEISSLER. 1990. Evaluation of three miniature radio transmitter attachment methods for small passerines. Wildl. Soc. Bull. 18:41–48.

TANAKA, S., N. KATO, K. TAKAO, AND M. SOMA. 1989. Tracking of bottlenose dolphins using satellite in Japan. Pages 411–416 in C. J. Amlaner, Jr., ed. Biotelemetry X. Univ. Arkansas Press, Fayetteville.

TARTER, M. E., AND R. A. KRONMAL. 1970. On multivariate density estimates based on orthogonal expansions. Ann. Math. Stat. 41:718–722.

TAYLOR, I. R. 1991. Effects of nest inspections and radiotagging on barn owl breeding success. J. Wildl. Manage. 55:312–315.

TAYLOR, K. D. 1978. Range of movement and activity of common rats (*Rattus norvegicus*) on agricultural land. J. Appl. Ecol. 15:663–677.

THOMAS, D. L., AND E. J. TAYLOR. 1990. Study designs and tests for comparing resource use and availability. J. Wildl. Manage. 54:322–330.

THOMPSON, S. K. 1987. Sample size for estimating multinomial proportions. Am. Stat. 41:42–46.

TRENT, T. T., AND O. J. RONGSTAD. 1974. Home range and survival of cottontail rabbits in southwestern Wisconsin. J. Wildl. Manage. 38:459–472.

TROUT, R. C., AND J. C. SUNDERLAND. 1988. A radio transmitter package for the wild rabbit (*Oryctolagus cuniculus*). J. Zool. (Lond.) 215:377–379.

VAN VUREN, D. 1989. Effects of intraperitoneal transmitter implants on yellow-bellied marmots. J. Wildl. Manage. 53:320–323.

VAN WINKLE, W. 1975. Comparison of several probabilistic home-range models. J. Wildl. Manage. 39:118–123.

VAUGHAN, M. R., AND J. T. MORGAN. 1992. Effect of radio transmitter packages on wild turkey roosting behavior. Proc. Int. Eur. Conf. Wildl. Telem. 4:628–632.

VOGT, F. D., G. R. LYNCH, AND S. SMITH. 1983. Radiotelemetric assessment of diel cycles in euthermic body temperature and torpor in a free-ranging small mammal inhabiting man-made nest sites. Oecologia 60:313–315.

VOIGT, D. R., AND J. S. LOTIMER. 1981. Radio tracking terrestrial furbearers: system design, procedures, and data collection. Pages 1151–1188 in J.A. Chapman and D. Pursley, eds. Worldwide Furbearer Conf., Frostburg, Md.

———, AND R. R. TINLINE. 1980. Strategies for analyzing radio tracking data. Pages 387–404 in C. J. Amlaner, Jr., and D. W. Macdonald, eds. A handbook on biotelemetry and radio tracking. Pergamon Press, Oxford, U.K.

WANLESS, S., M. P. HARRIS, AND J. A. MORRIS. 1988. The effect of radio transmitters on the behavior of common murres and razorbills during chick rearing. Condor 90:816–823.

———, ———, AND ———. 1989. Behavior of alcids with tail-mounted radio transmitters. Colonial Waterbirds 12:158–163.

WEATHERHEAD, P. J., AND F. W. ANDERKA. 1984. An improved radio transmitter and implantation technique for snakes. J. Herpetol. 18:264–269.

WEBSTER, A. B., AND R. J. BROOKS. 1980. Effects of radiotransmitters on the meadow vole, *Microtus pennsylvanicus*. Can. J. Zool. 58:997–1001.

WHEELER, W. E. 1991. Suture and glue attachment of radio transmitters on ducks. J. Field Ornithol. 62:271–278.

WHITE, G. C. 1983. Numerical estimation of survival rates from band-recovery and biotelemetry data. J. Wildl. Manage. 47:716–728.

———. 1985. Optimal locations of towers for triangulation studies using biotelemetry. J. Wildl. Manage. 49:190–196.

———, AND R. A. GARROTT. 1984. Portable computer system for field processing biotelemetry triangulation data. Colo. Div. Wildl. Game Inf. Leafl. 110. 4pp.

————, AND ————. 1986. Effects of biotelemetry triangulation error on detecting habitat selection. J. Wildl. Manage. 50:509–513.

————, AND ————. 1990. Analysis of wildlife radio-tracking data. Academic Press, Inc., San Diego, Calif. 383pp.

————, ————, R. M. BARTMANN, L. H. CARPENTER, AND A. W. ALLDREDGE. 1987. Survival of mule deer in northwest Colorado. J. Wildl. Manage. 51:852–859.

WHITEHOUSE, S., AND D. STEVEN. 1977. A technique for aerial radio tracking. J. Wildl. Manage. 41:771–775.

WIDEN, P. 1982. Radio monitoring the activity of goshawks. Pages 153–160 in C. L. Cheeseman and R. B. Mitson, eds. Telemetric studies of vertebrates. Academic Press, London, U.K.

WILLIAMS, B. K. 1981. Discriminant analysis in wildlife research: theory and applications. Pages 59–71 in D. E. Capen, ed. The use of multivariate statistics in studies of wildlife habitat. U.S. For. Serv. Gen. Tech. Rep. RM-87.

————, K. TITUS, AND J. E. HINES. 1990. Stability and bias of classification rates in biological applications of discriminant analysis. J. Wildl. Manage. 54:331–341.

WILLIAMS, T. C., AND J. M. WILLIAMS. 1970. Radio tracking of homing and feeding flights of a neotropical bat, *Phyllostomus hastatus*. Anim. Behav. 18:302–309.

WILLIAMS, T. D., AND D. B. SINIFF. 1983. Surgical implantation of radiotelemetry devices in the sea otter. J. Am. Vet. Med. Assoc. 183:1290–1291.

WILSON, R. P., AND M. T. J. WILSON. 1989. Tape: a package-attachment technique for penguins. Wildl. Soc. Bull. 17:77–79.

WOAKES, A. J., AND P. J. BUTLER. 1975. An implantable transmitter for monitoring heart rate and respiratory frequency in diving ducks. Biotelemetry 2:153–160.

————, AND ————. 1989. Wildlife studies in the laboratory. Pages 317–324 in C. J. Amlaner, Jr., ed. Biotelemetry X. Univ. Arkansas Press, Fayetteville.

WOLCOTT, T. G. 1980. Optical and radio optical techniques for tracking nocturnal animals. Pages 333–338 in C. J. Amlaner, Jr., and D. W. Macdonald, eds. A handbook on biotelemetry and radio tracking. Pergamon Press, Oxford, U.K.

WOOLEY, J. B., JR., AND R. B. OWEN, JR. 1978. Energy costs of activity and daily energy expenditure in the black duck. J. Wildl. Manage. 42:739–745.

WORTON, B. J. 1987. A review of models of home range for animal movement. Ecol. Model. 38:277–298.

YERBURY, M. J. 1980. Long range tracking of *Crocodylus porosus* in Arnhem Land, Northern Australia. Pages 765–776 in C. J. Amlaner, Jr., and D. W. Macdonald, eds. A handbook on biotelemetry and radio tracking. Pergamon Press, Oxford, U.K.

ZAR, J. H. 1984. Biostatistical analysis. Second ed. Prentice-Hall, Inc., Englewood Cliffs, N.J. 718pp.

ZINNEL, K. C., AND J. R. TESTER. 1984. Non-intrusive monitoring of plains pocket gophers. Bull. Ecol. Soc. Am. 65:166.

Appendix I. Suppliers of biotelemetry materials for wildlife telemetry.[a]

Advanced Telemetry Systems, Inc.
470 First Ave. South
Box 398
Isanti, MN 55040
(612)444-9267, FAX (612)444-9384

AF Antronics, Inc.
1906 Federal Dr.
Urbana, IL 61801
(217)328-0800
(receiving antennas)

[a]Suppliers provide common equipment such as transmitters, receivers, and antennas, unless indicated otherwise in parentheses.

Austec Electronics, Ltd.
17310 107th Ave.
Edmonton, Alberta,
Canada T5S 1E9
(403)486-0511, FAX (403)489-3697

AVM Instrument Co., Ltd.
2356 Research Dr.
Livermore, CA 94550
(510)449-2286, FAX (510)449-3980

Bally Ribbon Mills
23 N. 7th St.
Bally, PA 19503
(610)845-2211, FAX (610)845-8013
(Teflon ribbon harness material)

B & R Ingenieorgesellschaft mbH
Johann-Schill-Str.22
77806 March-Buchheim, Germany
7665-3885, FAX 761-123794

Biotrack
52 Furzebrook Road
Wareham, Dorset BH20 5AXJ
United Kingdom
(1929) 552 9922, FAX (1929)554 948
(Field equipment, analyses software)

Custom Electronics of Urbana, Inc.
2009 Silver Ct. West
Urbana, IL 61801
(217)344-3460, FAX (217)344-3460
(receivers and antennas)

Custom Telemetry & Consulting
1050 Industrial Drive
Watkinsville, GA 30677
(706)769-4024, FAX (706)769-4026

Detlef Burchard, Dipl.-Ing.
Box 14426
Riverside Dr. No. 45
Nairobi, Kenya
442371, FAX 442371

Hi-Tech Services
9 Devon Place
Camillus, NY 13031
(315)487-2484
(transmitters)

Holohil Systems Ltd.
3387 Stonecrest Rd.
Woodlawn, Ontario
Canada K0A 3M0
(613)832-3649, FAX (613)832-2728

L.L. Electronics
P.O. Box 420
Mahomet, IL 61853
(217)586-5327, (800)553-5328
FAX (217)586-5733

Lotek Engineering Inc.
115 Pony Dr.
Newmarket, Ontario
Canada L3Y 7B5
(905)836-6680, FAX (905)836-6455

Merlin Systems, Inc.
445 W. Ustick Rd.
Meridian, ID 83642
(208)884-3308, FAX (208)888-9528
(innovative telemetry equipment)

Microwave Telemetry Inc.
10280 Old Columbia Rd.
Suite 216
Columbia, MD 21046
(410)290-8672, FAX (410)290-8847
(general and for satellite system)

Mini-Mitter Co., Inc.
P.O. Box 3386
Sunriver, OR 97707
(503)593-8639, FAX (503)593-5604
(physiological telemetry)

Pacer
P.O. Box 1767
Dept. Biology—Agriculture College
Truro, Nova Scotia
Canada B2N 5Z5
(902)893-6607, FAX (902)895-4547
(LOCATE II, location estimate software)

Service Argos
18 Avenue Edouard Belin
31055 Toulouse Cedex, France
61-39 4700
or
1801 McCormick Dr., Suite 10
Landover, MD 20785
(301)925-4411, FAX (301)925-8995
or
4210 198th S.W., Suite 202
Lynnwood, WA 98036
(206)672-4699, FAX (206)672-8926
(satellite system)

Smith-Root, Inc.
14014 Northeast Salmon Cr. Ave.
Vancouver, WA 98686
(360)573-0202, FAX (360)286-1931
(receivers, aquatic)

Telemetry Systems, Inc.
P.O. Box 187
Mequon, WI 53092
(414)241-8335, FAX (414)241-8905

Televilt International AB
Box 53
S-71122 Lindesberg
Sweden
58117195, FAX 58117196

Telonics, Inc.
932 E. Impala Ave.
Mesa, AZ 85204-6699
(602)892-4444, FAX (602)892-9139
(general and for satellite system)

Toyo Communication Equipment Co., Ltd.
12-32, Konan 2-chome
Minato-ku, Tokyo 108, Japan
3-5462-9600, FAX 3-5462-9625
or
617 E. Golf Rd., Suite 112
Arlington Heights, IL 60005
(708)593-8780, FAX (708)593-5678
or
Bellenhöhe 5
4020 Mettman, Germany
02104-1-2009, FAX 02104-1-5546
(transmitters for satellite system)

Vemco
3895 Shad Bay, RR #4
Armdale, Nova Scotia
Canada B3L 4J4
(902)852-3047, FAX (902)852-4000
(aquatic)

Wildlife Computers
16150 NE 85th St. #226
Redmond, WA 98052
(206)881-3048, FAX (206)881-3405
(time-depth recorder, satellite link, software)

Wildlife Materials Inc.
1031 Autumn Ridge Rd.
Carbondale, IL 62901
(618)549-6330, FAX (618)457-3340

16

POPULATION ANALYSIS

Douglas H. Johnson

INTRODUCTION

Population analysis involves the study of population dynamics: the changes that occur over time and the causes of those changes. The population could be the number of aphids on a plant, the number of white-tailed deer in a woodlot, or the number of snow geese in North America. The analysis deals with *why* that many individuals are present—not more, not fewer—and what governs the number. What prevents the population from growing infinitely large? What keeps it from becoming extinct?

These questions are of obvious interest to wildlife managers and scientists. For species that are economic pests, ways of reducing numbers are sought. For game species, managers desire to maintain populations at levels that provide surpluses for harvest. For threatened species, the goal is to increase their numbers to avoid extinction. Meeting any of these objectives requires an understanding of the species' population dynamics as the first priority.

A useful definition of a *population* is a group of organisms of the same species living in a particular space at a particular time (Krebs 1985). For a population, it makes sense to speak of a birth rate, a death rate, a sex ratio, and an age structure (Cole 1957), concepts that lack meaning for lower levels of biological organization (e.g., individuals) or higher levels (communities or ecosystems).

The subject of population dynamics includes the number of individuals in a population and the factors that affect population size: (1) the survival of those individuals, (2) their reproduction, and (3) their movements into and out of a population (immigration and emigration). Population size is discussed in Chapter 9; the present chapter focuses on the variables that affect it, although immigration and emigration receive scant attention. If all factors were known for a population, understanding its dynamics would be straightforward. This happy state is never achieved, however, and the biologist is forced to make decisions based on an incomplete understanding.

Population analysis often requires modeling of some sort, especially to bridge gaps in knowledge. A *model* is some abstraction of a real system that enables us to think more clearly about the real one. A model may be a complex mathematical beast, incorporating thousands of variables and equations and requiring hours of mainframe computer time to analyze. Or it may be a simple heuristic concept, such that deer produce more fawns when browse is ample than when it is scarce. The kind of model appropriate to a scientific or management application depends on the objectives of the model. This chapter presents a variety of models, from simple, involving only a single parameter, to complex, involving numerous rates and relationships. Complexity increases with the realism. It will be clear, however, that even simple models can be difficult to fathom fully and may offer unanticipated insight into the dynamics of a population. Also, a variety of mathematical tools can be employed to understand them. The more complex models, contrary to intuition, may actually be easier to construct than simpler ones. But reaching a full understanding of a complex model is difficult, and computer simulation is often needed to apply the model.

In addition to varying in complexity, models of populations may vary in other attributes. In *discrete-time models*, events, such as births, occur only at certain times, such as a short breeding season within a year. In *continuous-time models*, events can occur throughout time. Or, as Starfield and Bleloch (1986) said, time jumps in a discrete model; time flows in a continuous one. Another distinction is whether random components are included in a

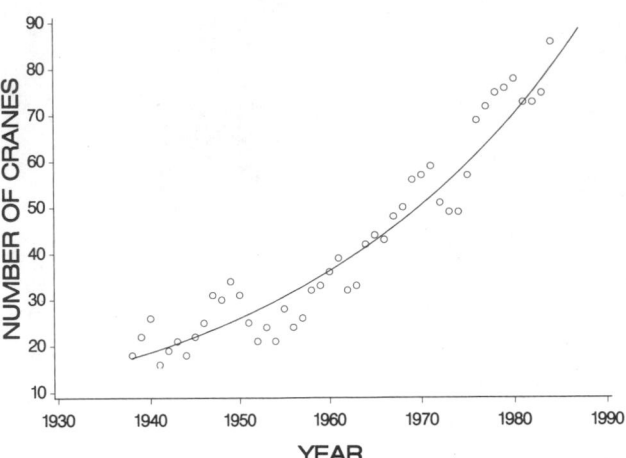

Fig. 1. Counts of wintering whooping cranes, 1938–84, fitted to exponential curve.

model. In a *deterministic model*, parameter values are fixed, and the result from the model depends only on the values of the input variables. In a *stochastic model*, certain parameters vary randomly; their statistical distributions rather than exact values are specified. If the variation of the system is important, stochastic models are usually more suitable than deterministic ones.

This chapter addresses, in addition to modeling population dynamics, the estimation of important parameters. Modeling logically precedes estimation, because modeling can tell us which variables are important. Conversely, the availability of usable estimates often dictates the kind of model that can be constructed, so modeling and estimation go hand in hand.

Examples are based on several species, but disproportionate attention is given to a few, especially the mallard and white-tailed deer. This emphasis reflects my own experience (mallards) and the amount of work that has been done on both species. It is worth recalling Durward Allen's remark that "numbers phenomena tend to be universal. They change only in detail as we shift from fish to fur to fowl" (Allen 1962:36).

A single chapter can only touch lightly on the diversity of techniques used in population analysis. For further reading, two books stand out. Seber (1982) provided a near-encyclopedic coverage of methods used to estimate the number of wild animals as well as their survival and related parameters; one update has appeared (Seber 1986) and another is scheduled. Caughley (1977) described in engaging prose his practical views on population analysis.

POPULATIONS WITH UNIMPEDED GROWTH

Model

The simplest model assumes that the number of animals in a population goes up (or down) by a constant ratio, say λ (lambda), with each unit of time (which we will assume is a year). That is, at time t the population size is λ times its value at time $t - 1$:

$$N_t = \lambda N_{t-1}.$$

The population is increasing if $\lambda > 1$, is constant if $\lambda = 1$, and is declining if $\lambda < 1$. Sometimes λ is called the *finite rate of population increase*. This formulation is

geared toward organisms that reproduce during a short breeding season (discrete growth; *birth-pulse fertility* of Caughley 1977). If N_0 is the population size at some initial year, then repeating the above equation t times gives

$$N_t = \lambda^t N_0. \qquad (1)$$

Consider the wild population of whooping cranes, which has been monitored on its wintering ground since 1938. Virtually the entire natural population congregates on an area around Aransas National Wildlife Refuge in Texas. Figure 1 illustrates the actual counts during 1938–84 versus the values fitted to equation 1, where the year t has been recoded so that 1938 is year 0 and $\hat{\lambda} = 1.0338$ is an estimate of λ. (The hat symbol denotes an estimator of a parameter.) Thus, the crane population was growing roughly like a bank deposit, with an interest rate of 3.38%, compounded annually. More detailed analyses of this population were presented by Binkley and Miller (1980, 1988), Boyce and Miller (1985), Boyce (1987), and Nedelman et al. (1987).

An alternative expression for population growth has some advantages. If we replace λ by e^r, then $\lambda^t = e^{rt}$. Here e is the base of natural logarithms, and r is termed the *instantaneous rate of increase*. Among the advantages of the exponential formulation (Caughley 1977:52) is the ability to convert easily between time units; e.g., if the growth rate of a population per year is 0.10, then the growth rate per day is 0.10/365. Also, the time required for a population to double is (log 2)/r = 0.69315/r. (Note: throughout this chapter natural logarithms are used.) This formulation is particularly appropriate for continuously growing populations (*birth-flow fertility* of Caughley 1977), but it works as long as the times (t) when the population is counted represent comparable times in the life cycle, such as the beginning of the breeding season. Then r is log (λ), and equation 1 becomes

$$N_t = N_0 e^{rt}. \qquad (2)$$

For the whooping crane population, we get $\hat{r} = $ log $(\hat{\lambda})$ = log (1.0338) = 0.03325, on an annual basis. At this rate, the population is growing 3.325% per year and will double every 0.69315/0.03325 = 20.8 years, if growth continues at this rate.

This model is termed the *exponential growth model* and may be realistic when growth is unhindered (i.e., resources are ample and competition is not a factor). Such situations often occur when a species initially invades an optimal habitat or, as with the cranes, when a population rebounds from near extinction and the habitat is adequate. It can also be useful for short-term forecasts (Eberhardt 1987). This approach is deterministic; that is, no allowance is made for variation caused by randomness or by variables not included in the model. It can be made stochastic (incorporating random events) by considering chance variations in births and deaths (Pielou 1969).

Estimation

ESTIMATING r FROM THE COUNTS

The simplest way to estimate the growth rate r in equation 2 from population counts is by taking logarithms of both sides, giving:

Table 1. Counts of wintering whooping cranes, Aransas National Wildlife Refuge, 1938–84 (from Boyce 1987).

Year	Adults	Young	Year	Adults	Young
1938	14	4	1962	32	0
1939	15	7	1963	26	7
1940	21	5	1964	32	10
1941	14	2	1965	36	8
1942	15	4	1966	38	5
1943	16	5	1967	39	9
1944	15	3	1968	44	6
1945	18	4	1969	48	8
1946	22	3	1970	51	6
1947	25	6	1971	54	5
1948	27	3	1972	46	5
1949	30	4	1973	47	2
1950	26	5	1974	47	2
1951	20	5	1975	49	8
1952	19	2	1976	57	12
1953	21	3	1977	62	10
1954	21	0	1978	68	7
1955	20	8	1979	70	6
1956	22	2	1980	72	6
1957	22	4	1981	71	2
1958	23	9	1982	67	6
1959	31	2	1983	68	7
1960	30	6	1984	71	15
1961	34	5			

$$\log(N_t) = \log(N_0) + rt. \qquad (3)$$

Linear regression of $\log(N_t)$ on t for a series of years provides estimates of the regression coefficient (the slope, equal to r) and the intercept. These values can be transformed by exponentiating to give us estimates of λ and N_0. As an example, consider the whooping crane counts during 1938–84 (Table 1). A linear regression of the logs of the counts (adults plus young) against year (recoded so that 1938 = 0) provides a slope of $\hat{r} = 0.03325$ (the estimated SE is 0.00168) and an intercept of 2.858 (SE = 0.045). Thus $\hat{N}_0 = e^{2.858} = 17.4$ (SE = 0.783 by the delta method, a procedure for obtaining estimated SEs of functions of random variables; see, e.g., Seber [1982:7]). These estimates were used to graph the curve in Fig. 1.

An alternative to linear regression on transformed variables is to use nonlinear regression directly on equation 2. Nonlinear regression is an iterative procedure, which computers generally do better than we do. As an example, applying the SAS™ (SAS Institute, Inc. 1987) procedure NLIN to the whooping crane data gives $\hat{r} = 0.0350$ (SE = 0.00155) and $\hat{N}_0 = 16.7$ (SE = 0.927). These estimates differ from the previous ones because the analytic methods are based on different assumptions. Errors in estimated population sizes are likely to increase with true population size, so the assumption of constant error variance, used in ordinary least squares regression, is more likely to be met with the linearized form of the model represented by equation 3 than by equation 2. For this reason, the linear approach usually is preferred.

ESTIMATING r FROM THE CHANGES IN POPULATION

The form of the exponential growth model lends itself to another method of estimating r. Consider the ratio of population sizes in successive years. From equation 2, this is

$$\frac{N_t}{N_{t-1}} = \frac{N_0 e^{rt}}{N_0 e^{r(t-1)}} = e^r.$$

Thus, the logarithms of the average of these ratios can be used to estimate r. For the whooping crane example, counts for the years 1938–84 provided 46 ratios N_t/N_{t-1}, which averaged 1.0450 (SE = 0.0212). The logarithm of this average gives the estimate $\hat{r} = 0.0440$ (SE = 0.0202).

In a comparison of the three estimates of whooping crane population growth (not shown here), the fit provided by the linearized model (eq. 3) was best, that of the nonlinear fit (eq. 2) was next best, and that of the ratios in successive years was worst. Eberhardt (1987) discussed other estimation methods for this model, including ratio estimators with various weights. He also considered variance estimators. For a spirited discussion of the estimation of population growth of a white-tailed deer herd, see McCullough (1982, 1983) and Van Ballenberghe (1983).

POPULATIONS WITH DENSITY-DEPENDENT GROWTH
Model

A CONTINUOUS-TIME FORMULATION

Consider now the number of bison on the National Bison Range during 1909–22, when no harvesting occurred (Table 2). Fitting the linearized model (eq. 3) to the first 10 years of data gives $\hat{N}_0 = 53.45$ (SE = 1.53) and $\hat{r} = 0.216$ (SE = 0.00535). The observed number of bison at the end of each of those years fits the exponential curve nicely (Fig. 2). Projections for the years 1919–22, however, are consistently higher than actual numbers (Fig. 2). It is conceivable that the slowdown in population growth can be attributed to a density-dependent response. In fact, the proportional annual change in population ($\hat{R}$) is negatively correlated with population size ($r = -0.77$, $P = 0.001$; here r denotes the correlation coefficient, not the population growth rate, a distinction between usages that should be clear from the context) (Fig. 3). A caution: relating population change to population size is not as straightforward as it appears; see the subsequent section on detecting density-dependence.

No population can continue to grow indefinitely at a constant rate. More likely, growth will slow down as the population becomes large and some limiting factor comes into play. Density-dependence is likely to operate. How can we include density-dependence in the model to make it more realistic and useful? In the model of equation 2, the population growth rate per animal,

$$\frac{1}{N_t} \frac{dN_t}{dt} = r,$$

is constant, regardless of population size. One way to make it depend on population size is to multiply it by a factor that has negligible effect when the population is small, but that reduces the growth rate to zero as the pop-

Table 2. Counts of bison on the National Bison Range, 1909–22 (from Fredin 1984).

Year	Number at start of year	Young born	Deaths	Number at end of year
1909	37	11	0	48
1910	51[a]	19	0	70
1911	70	16	1	85
1912	85	19	0	104
1913	104	26	0	130
1914	130	34	0	164
1915	164	32	2	194
1916	194	47	1	240
1917	240	56	1	295
1918	295	73	1	367
1919	367	58	5	420
1920	420	68	9	479
1921	479	82	7	554
1922	554	85	4	635

[a]Three animals added to existing herd.

ulation approaches some limit, K (which we might call the carrying capacity). The term $(K - N)/K$ does just that. (This is the simplest way; there are many others, as reviewed by May 1973.) It is nearly 1 when N is small, and converges to 0 as N approaches K. If we use this factor, then the per capita growth rate becomes

$$\frac{1}{N_t} \frac{dN_t}{dt} = r_m \frac{K - N_t}{K}, \qquad (4)$$

where r_m, the maximum rate of growth, replaces r. Equivalently,

$$\frac{dN_t}{dt} = r_m N_t \frac{K - N_t}{K},$$

which can be interpreted (Krebs 1985:213) as rate of increase of population per unit time, dN_t/dt, equals maximum rate of population growth per capita, r_m, times population size, N_t, times unused opportunity for growth ($K - N_t)/K$. Note that the factor modifying growth rate de-

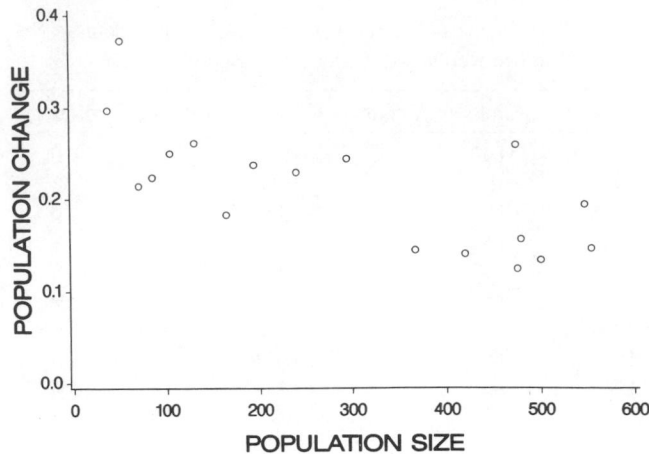

Fig. 3. Proportional change in bison population, $(N_{t+1} - N_t)/N_t$, on National Bison Range versus population size, N_t.

pends on both the population size (through N) and the environment (through K).

The equation has the solution

$$N_t = \frac{K}{1 + e^{a - r_m t}}, \qquad (5)$$

which is known as the *logistic equation*. K is the asymptote or carrying capacity, and r_m is the maximum rate of population growth, the rate that would result if the population were free of constraints caused by the density of the population. The value of the parameter a depends on the time origin used; it measures the size of the population at time 0 relative to the asymptotic size. Setting $t = 0$ in equation 5, we find $a = \log[(K - N_0)/N_0]$.

Fitted to the bison data, logistic growth assumes the S-shaped curve shown in Fig. 4. When the population is small, numbers increase rapidly, like an exponential curve. Growth then diminishes as the population approaches K (which is nearly 1,200), where it would level off.

The logistic equation has a rich history (Hutchinson 1978, Kingsland 1985) and is a mathematically convenient model that describes the growth of a variety of populations. It is not, however, any sort of *law* of population

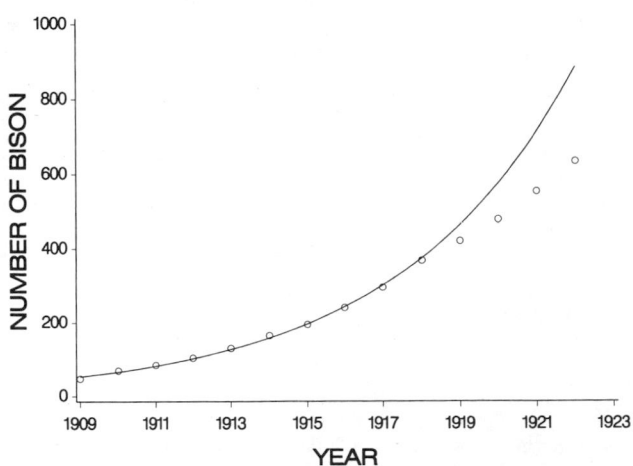

Fig. 2. Counts of bison on National Bison Range, 1909–18, fitted to exponential curve, also 1919–22 data and projections.

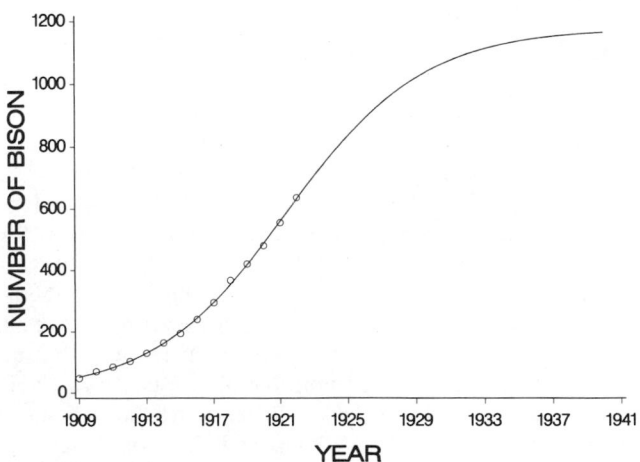

Fig. 4. Bison counts, 1909–22, fitted to logistic curve.

growth. Among the assumptions (e.g., Pielou 1969, Poole 1974, Krebs 1985) are: (1) all individuals, regardless of age, sex, or genotype, are equivalent with respect to survival, reproduction, and susceptibility to crowding; (2) the carrying capacity (K) is constant; (3) the growth rate of the population responds instantaneously to the population size; and (4) the effect of population size on growth rate is linear. Each of these restrictive assumptions can be relaxed, but with a loss of mathematical tractability. I later cover models in which assumptions 1 and 2 are eased (see Box 1). Assumption 3 can be modified either with discrete-time formulations (discussed in next section) or by introducing time lags into the continuous-time model (e.g., May 1973, Krebs 1985). One extension that overcomes assumption 4 is the generalized logistic equation (Gilpin and Ayala 1973, Eberhardt 1987). Pielou (1969) and others considered stochastic versions of the logistic model.

A DISCRETE-TIME FORMULATION

The logistic formulation above specifically applies to continuously reproducing organisms, although it suffices for populations with discrete breeding seasons if population size is measured at the same time each year, as with the bison. Otherwise, for animals with discrete breeding seasons, the discrete counterpart of equation 4 is

$$\frac{N_{t+1} - N_t}{N_t} = r_m \left(\frac{K - N_t}{K} \right),$$

with solution

$$N_{t+1} = N_t + r_m \left(1 - \frac{N_t}{K} \right) N_t. \qquad (6)$$

This discrete version implicitly has a time delay; the population growth rate at time $t + 1$ depends on the population size at time t. In contrast, equation 4 assumes the rate of population change responds instantaneously to changes in the size of the population. Because of the time delay in the discrete version, the behavior of the modeled population depends strikingly on the values of the parameters. The population can approach the asymptote smoothly, approach it in an oscillatory manner, cycle indefinitely, or fluctuate chaotically, depending on the value of r_m (May 1974, May and Oster 1976). That model (eq. 6), along with the realization that a simple deterministic mechanism could produce such a striking array of random-appearing behavior, was one of the early discoveries of what has now become the study of chaos.

Estimation

Nonlinear least squares can be used directly on equation 5 to estimate the parameters of the logistic equation. When the bison data for 1909–22 were fitted, the program NLIN (SAS Institute, Inc. 1987) gave estimates $\hat{K} = 1,172$ (SE = 77.4) for the asymptote, $\hat{r}_m = 0.2479$ (SE = 0.0078) for the rate parameter, and $\hat{a} = 3.069$ (SE = 0.046) for the origin parameter.

An alternative is to use the discrete form of the logistic model. From equation 6, we have

$$N_{t+1} = N_t + r_m(1 - N_t/K)N_t$$
$$= N_t(1 + r_m) + N_t^2(-r_m/K),$$

Box 1. The behavior of small populations—the Allee effect.

We have assumed density dependence will increase mortality rates or decrease birth rates or both as populations become large. Conversely, as populations grow small, mortality rates should decline and birth rates increase, according to the model. In reality, small populations may not enjoy such favorable demographic rates. Birth rates, especially, may decline rather than increase as populations dwindle. This may happen because finding mates is difficult when the population is small, or because breeding requires social stimulation. Another possibility is illustrated by colonial-nesting birds, in which larger colonies provide greater protection from predators and greater reproductive success (e.g., Birkhead 1977). This phenomenon of increased mortality rates or decreased birth rates at low population levels is known as the ''Allee effect,'' after W. C. Allee, who documented numerous situations in which it was manifested (Allee 1931).

so we can perform a regression of N_{t+1} on N_t and N_t^2, excluding an intercept term. The coefficient of N_t will be an estimate of $(1 + r_m)$ and the coefficient of N_t^2 will estimate $-r_m/K$. For the bison example, we obtain $\widehat{(1 + r_m)} = 1.2669$ (SE = 0.0266), so $\hat{r}_m = 0.2669$ (SE = 0.0266). Also, the estimate of $-r_m/K$ is -0.000238 (SE = 0.000061), so $\hat{K} = 0.2669/0.000238 = 1,121.43$ (SE = 183.24).

One statistical difficulty with this regression approach is the assumption that the explanatory variables, in this example N_t and N_t^2, are measured without error (Walters 1986). This is not a problem in our example, because we believe the bison counts are completely accurate, but the problem arises in most situations. We can illustrate the effect of measurement errors by reanalyzing the bison data, except that we include a small multiplicative error (each count is multiplied by e^z, where z is a normal random deviate with mean 0 and standard deviation [SD] 0.1), giving the values in Table 3. Results from this analysis give $\hat{r}_m = 0.4295$ (SE = 0.1138) and $\hat{K} = 594.88$ (SE = 72.43), values far different from estimates obtained using values measured without error (0.2669 and 1,121.43).

Of the two estimation techniques applied to the bison data, the nonlinear regression applied to equation 5 gave a better fit than regression of N_{t+1} on N_t and N_t^2. That superiority may not hold in general.

Detecting Density Dependence— Some Dangers

Discovering density dependence in a series of counts of a population is less straightforward than it might appear. First, population size and change in population tend to be negatively correlated, even if the change occurs independently of population size (e.g., Maelzer 1970, St.

Table 3. Actual counts of bison on the National Bison Range, 1909–22, and counts with multiplicative error.

Year	Actual count	Count with error
1909	48	48
1910	70	64
1911	85	84
1912	104	100
1913	130	144
1914	164	178
1915	194	199
1916	240	227
1917	295	292
1918	367	389
1919	420	387
1920	479	496
1921	554	598
1922	635	553

Table 4. Example illustrating the appearance of density dependence from annual counts of a population that varies randomly from year to year. N_t is actual population size in year t, O_t is observed population size, $\Delta_t = \log(N_t/N_{t-1})$ is actual change in population size, and $\Delta_t = \log(O_t/O_{t-1})$ is observed change in population size.

Year (t)	$\log(N_t)$	Δ_t	$\log(O_t)$	Δ_t
0	6.91		6.76	
1	7.15	0.24	6.95	0.19
2	7.18	0.03	7.58	0.63
3	7.12	−0.06	7.03	−0.55
4	7.10	−0.02	7.19	0.16
5	7.03	−0.07	7.23	0.04
6	7.06	0.03	7.12	−0.11
7	6.96	−0.10	6.94	−0.18
8	6.87	−0.09	6.89	−0.05
9	7.02	0.15	6.77	−0.12
10	7.12	0.10	7.22	0.45

Amant 1970). Second, any uncertainty in estimating population size tends to add to the appearance of density dependence.

Consider an example (Table 4) in which we started with $N_0 = 1,000$ animals; $\log(N_0) = \log(1,000) = 6.91$. The population in each successive year was generated by adding a random number to the logarithm of the previous population:

$$\log(N_{t+1}) = \log(N_t) + z,$$

where z is a random deviate with mean 0 and SD 0.1. Although z was generated independently of N_t, a negative correlation between the two variables was induced; in the example shown in Table 4 we have $r = -0.62$ ($P = 0.055$). The reason for this surprising result is that, even in an irregular sequence of numbers, an unusually high value tends to be followed by a decrease (if it were more likely followed by an increase, then it would no longer be an unusually high value), and vice versa (St. Amant 1970). Thus, a negative correlation between population change and previous population size cannot be construed as evidence for density dependence.

If the counts had been made subject to error, the situation is even worse. The appearance of density dependence increases, as the following illustrates. Suppose the population was underestimated in a particular year; this error will make the observed population size in that year more likely to be small than large. Also, it will make the change in observed population size larger than it should be, unless the population is underestimated again the following year. Thus, a smaller-than-expected population size will be associated with a larger-than-expected population change, and a negative correlation will be induced. Consider again Table 4, except that now the counts were measured with error rather than exactly (call the observed counts O_t):

$$\log(O_t) = \log(N_t) + y,$$

where y is another random deviate, normally distributed with mean 0 and SD 0.2. The correlation between observed population change and observed population size is stronger ($r = -0.73$, $P = 0.017$) than the correlation between true values.

From this we conclude that density dependence should not be inferred from regression analysis on counts of populations, even if they are measured exactly. The same problem arises when a regression of $\log(N_{t+1})$ on $\log(N_t)$ is performed; regression coefficients <1 are expected even if there is no density dependence (Maelzer 1970). For additional cautions see Eberhardt (1970), Slade (1977), Solow (1990), and especially Pollard et al. (1987). The converse problem also arises: Gaston and Lawton (1987) observed that methods for detecting density dependence from census data consistently failed to do so, even for populations known (from independent evidence) to be subject to density-dependent processes.

BIRTH AND DEATH MODELS
Models

We should recognize that population growth is the net result of births and deaths (ignoring emigration and immigration). In many situations we can profitably analyze the two processes separately, because they may be affected by different environmental variables. For example, the counts of bison can be divided into young-of-the-year and adult age classes (Table 2). The count of young-of-the-year can be considered the final outcome of the birth process, birth here including not only parturition, but also survival until autumn. The number of deaths of adults was also recorded.

Estimates of annual birth and death rates (defined per individual in the population at the start of the year) can be obtained from Table 2. Birth rates were smaller when the population was larger (Fig. 5), and death rates increased with population size (Fig. 6). If birth and death rates are similar functions of population size, as appears to be true for the bison data, we can work with their difference rather than with individual components. Only one or the other of birth and death rates may be density dependent, or the nature of the relationship may differ for the two processes. Then we should treat the two processes

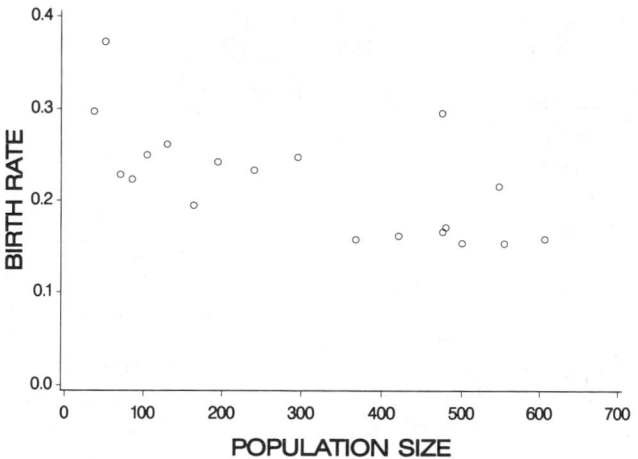

Fig. 5. Bison birth rates versus population size at start of year.

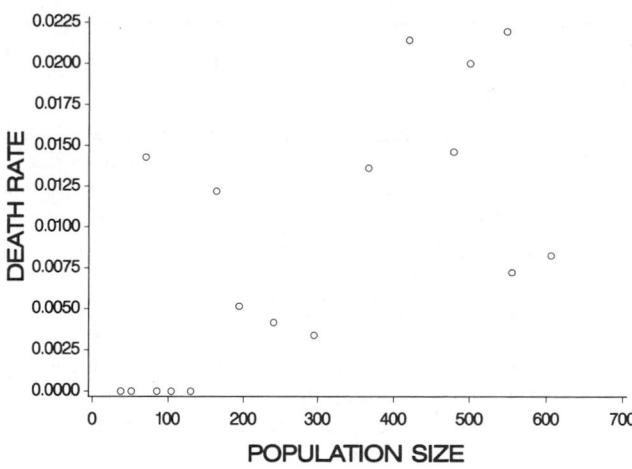

Fig. 6. Bison death rates versus population size at start of year.

separately. Suppose, for example, that instantaneous birth rates (logarithms of finite birth rates) varied with density, say

$$b = b_0 + b_1 N,$$

but that instantaneous death rates were independent of density,

$$d = d_0.$$

Then the population rate of increase $r = b - d$ is

$$r = (b_0 - d_0) + b_1 N$$

and depends on the population size N. That is, population growth is density dependent, but an analysis of the population size alone would not indicate which of the two processes, birth or death, was density dependent. Knowledge of specific relations is useful in understanding and managing a population (see Box 2).

Pielou (1969), among others, provided an introduction to stochastic birth and death processes. De Angelis (1976) applied these models to a population of Canada geese.

Estimating Birth Rates

The *fertility* of a population is the number of live births produced over some period of time, generally a year. Because it usually suffices to study the female segment of a population, fertility is often expressed as young females produced per female in the population. A related parameter is *fecundity*, the potential level of reproductive performance of a population, which is ordinarily much greater than the realized reproduction (fertility). (The terms fertility and fecundity are not always used consistently.) To calculate fertility, we need to know average litter size, average number of litters produced per time interval (year), and the sex ratio at birth (Caughley 1977).

The estimation of fertility rate has received only scattered attention. Typically, rates are based on different criteria for different species groups. For mammals, number of live births is an appropriate measure. For fish, reptiles, and birds, the number of eggs laid or hatched is often used. Because observing newborn young of many species is ordinarily difficult, fertility is often assessed by the number of young produced that attain a particular life

stage or size. In fishery work, for example, production to catchable size is the key measure. Likewise, in waterfowl studies some measures of reproduction involve counts of ducklings nearly ready to fledge, or as members of the autumn (hunted) population. For populations with synchronized, seasonal breeding (birth-pulse fertility), we calculate the number of births for females in a given age

---

Box 2.   How does density dependence work?

The exact role of density dependence in population regulation has long been a source of controversy (e.g., Krebs 1985). Density may influence survival or reproduction rates only at extreme densities (e.g., Strong 1986), exact values of which depend on the quantity and quality of the habitat available.

Evidence suggesting a relation between population density and fertility was given by Knowlton (1972), who compared the average number of uterine swellings per female coyote, an index to fertility, with the intensity of efforts used to control coyotes in seven counties in southern Texas. Although sample sizes were limited and the levels of coyote control were not randomly assigned to counties, an effect of population density on this index of fertility is suggested.

Control effort	County	Sample size	Average number of uterine swellings per female	
Intensive	Zavala	8	8.9	
	Dimmit	12	6.4	7.2
	Uvalde	10	6.2	
Moderate				
	Jim Wells	21	5.3	
	Hildago	11	3.7	4.5
Light	Jim Hogg	17	4.2	
	Duval	11	2.8	3.5

Box 3:   A common bias in estimating nest success.

Among birds, one of the most important factors determining the size of a population is the percentage of nests from which young are successfully fledged. Fortunately, it is a parameter that managers often can influence by manipulating habitat or predation.

Many studies and monitoring programs of nest success fell victim to a serious bias. Biologists reported nest success rate to be the percentage of the successful nests among the sample of nests they found. This intuitively reasonable procedure is acceptable if all nests can be found at initiation, or if destroyed nests are as likely to be found as successful ones. Many species of birds, however, are secretive about their nesting, and biologists are likely to find mostly nests tended by an adult. Once destroyed, a nest will be abandoned by the adults and may be difficult to detect by usual nest-searching methods (e.g., Klett et al. 1986). A successful nest, in contrast, will be tended by an adult from initiation until the young leave. For that reason, it will more likely be found. This disparity in chances of detecting failed and successful nests introduces a major bias into the usual nest success rate. Mayfield (1961) was among the first to recognize this problem, and he proposed a solution: he recommended computing a *daily mortality rate* for nests, based on the number of nests destroyed, divided by the total number of days nests were under observation. Subtracting this value from 1.0 gives a daily survival rate, which, when raised to a power equal to the number of days needed for a nest to proceed from initiation to success, gives a much better estimate of the true nest success rate. For example, suppose six nests were under observation for 13, 7, 9, 2, 12, and 7 days, during which two of the nests failed. The total time under observation is 50 days, so the daily mortality rate is $2/50 = 0.04$. The daily survival rate is $1.0 - 0.04 = 0.96$. If 20 days are required for a nest to succeed, then we estimate the proportion of successful nests as $0.96^{20} = 0.44$. This value contrasts with the apparent estimate, four of the six nests succeeding, or 0.67. A statistical model and standard errors for the Mayfield procedure were given by Johnson (1979). Johnson and Shaffer (1990) outlined situations in which the Mayfield method performed better than the apparent method.

tio techniques may be applicable in some limited circumstances (e.g., Hanson 1963, Seber 1982).

AGE RATIOS BASED ON DIRECT COUNTS

For the cranes or bison, we could count exactly the young produced (and surviving until time of census). More generally, biologists often relate the number of young seen to the number of adults seen and obtain an index to fertility. For example, ratios of fawns to does may be used for white-tailed deer, or number of placental scars in harvested squirrels, or number of successful nests or broods for birds or American alligators. For hunted species, the age ratio in the harvest is an index to recruitment, but it must be adjusted for differential vulnerability of age classes (e.g., Martin et al. 1979). Various errors can creep into such calculations of fertility. For example, does without fawns may be less conspicuous than those with fawns, or squirrels that had borne young may be more likely to be shot (and fertility thus measured) than those that did not, or successful bird nests may be more likely found than unsuccessful ones (Mayfield 1961).

More fundamentally, all the components of reproduction should be considered to gain a full understanding of the process. These include age at which animals first breed, incidence of nonbreeding among adults of breeding age, number of breeding cycles per year, size of clutch or litter, and survival to adult stage. For management purposes, a consistent index to reproduction, along with its standard error (SE), will often suffice.

MARK-RECAPTURE METHODS

Mark-recapture methods are described in Chapter 9 for *closed populations* (which do not change during the period of interest) and *open populations* (which allow births, deaths, and migration). The primary method for open populations involves the Jolly-Seber model. This model yields estimates not only of population size, but also of the number of individuals added to the population between trapping occasions (which includes births and immigrants) and the number removed from the population (which includes deaths and emigrants). If one can safely assume that no immigration has occurred, then the estimated number of additions to the population is a measure of births, although SEs are usually large. Chapter 9 provides formulas for the method and illustrates it with an example (Table 4 of Chapter 9).

INDIRECT MEASURES

Often, reproductive success of a population is evaluated in terms of some component of reproduction. For example, clutch size and nest success (see Box 3) are commonly used as measures in bird studies. Among mammals, characteristics of female reproductive tracts may be used (e.g., Kirkpatrick 1980). Such measures might be perfectly adequate indices to reproduction, but they are only part of the picture. Other factors must be considered; by focusing on only one or two of the components of reproduction, we are implicitly assuming that the others vary only slightly, if at all.

Consider how duck productivity might be monitored on a refuge, for example. Counts of breeding pairs are used to determine the size of the population, and studies of nests give an estimate of nest success rate (Cowardin and

class. For more-or-less continuous breeders (birth-flow fertility), the number of births for females in a specified age *interval* is appropriate.

Estimation of birth rate, especially by age class, is difficult. Three general approaches are discussed here. The first uses age ratios based on direct counts. The second employs mark-recapture methods. The third involves a potpourri of indirect measures. In addition, change-in-ra-

Blohm 1992). These two variables may be the key components in reproduction, but other variables could play a major, or even dominant, role (Johnson et al. 1992). Some members of the population might not breed, for example. Some might renest after a nest failure while others give up. Clutch size can differ, and individual eggs may be depredated from nests or not hatch for other reasons. Finally, of the ducklings that hatch from a successful nest, the fraction that ultimately fledge varies from 0 to 100%. Hence, reliance on only a subset of the components can cause misleading conclusions.

Estimating Mortality Rates

Five basic approaches are used for estimating a mortality rate (or survival rate, equal to 1 − mortality rate); each requires a different kind of data. Four of the approaches are discussed here; the fifth, based on age-structure data, will be presented under the age-dependent models.

OBSERVED MORTALITY

In some studies mortality can be observed directly. For captive or other closely watched populations, mortality rates can be calculated without difficulty. In other studies, markers attached to animals allow biologists to observe a subset of a population. Radiotelemetry especially affords an opportunity to monitor animals closely and record instances of mortality. Chapter 15 describes several methods of estimating mortality rates from telemetered animals. The same methods can be used for animals marked in other ways, as long as markers are retained and marked individuals can be found readily. Particularly troublesome are instances in which an animal's signal or marker cannot be located, so the observer is not sure if the transmitter failed, the animal (along with its transmitter) was destroyed, or the animal left the study area. Also, radio packages (Chapter 15) or other markers (e.g., Brodsky 1988, Kinkel 1989) may influence behavior and survival, and telemetry studies usually have been limited by small samples of animals and relatively short durations.

RATIOS OF POPULATION SIZES OR INDICES

If no migration into or out of a population occurs, then the mortality between times t and $t + 1$ is the population size at time t minus the number of those that still remain at time $t + 1$. If those survivors can be distinguished from the young that were added to the population, survival can be computed directly from the population sizes. Consider the whooping crane example. The adults counted in one winter represent the survivors of the total population (adults plus young) in the previous winter. From ratios of these counts, survival rates can be computed. In 1938, for instance, 18 birds were alive (14 adults and 4 young; Table 1). Of these, 15 were still alive in 1939, giving a survival rate of 15/18 = 0.83 (SE = 0.09, as a binomial variate). Survival rates for other years can be determined similarly.

To have exact counts from which survival rates can be calculated, such as we have for the cranes, is unusual. Often, however, indices to population size are available; if these faithfully represent a constant proportion of the population, they can be used equally as well. Consider as an example results from a banding study of female mal-

Table 5. Recoveries of female mallards banded as adults in Minnesota, 1968–70 (from Johnson 1974).

Year	Number banded	Number of recoveries in		
		1968	1969	1970
1968	338	16	9	5
1969	67		6	5
1970	93			12

lards in Minnesota (Table 5). In 1968, 338 adult females were banded. Assume that hunters took equal proportions of the banded populations in the 1968 and 1969 hunting seasons (a conclusion supported by a more rigorous analysis in Johnson 1974: Table 3). Then the 16 kills in 1968 represent the same fraction of the 1968 population that the 9 kills in 1969 represent of the 1969 population. From this, we can estimate the survival between hunting seasons to be 9/16 = 0.56, albeit with a large SE (SE = 0.23, as a ratio of two multinomial variates). This example is used only to illustrate how survival can be estimated from the ratio of population indices; better methods of analyzing banding data will be mentioned shortly.

Survival also can be estimated if indices do not represent constant fractions of the population, but are known to be in certain proportions. Such methods are based on *catch-effort models*. Suppose it is somehow known that recovery rates of female mallards banded in Minnesota varied during 1968–70 in the proportions 0.058, 0.056, and 0.100 (Johnson 1974: Table 3). The numbers of birds banded in 1968 and recovered in 1968, 1969, and 1970 were 16, 9, and 5, respectively. The ratios 16/0.058, 9/0.056, and 5/0.100 should then represent the same proportion of the population in each of the 3 years. These ratios, 275.86, 160.71, and 50.00, suggest that survival from 1968 to 1969 was 160.71/275.86 = 0.58 (minimum SE = 0.24, assuming that effort is known exactly) and survival from 1969 to 1970 was 50.00/160.71 = 0.31 (minimum SE = 0.17). Catch-effort models are most often applied to fisheries problems, in which fishing effort is well known, or to populations of small mammals, where trapping effort can be determined. The method is discussed in Chapter 9 relative to estimating population size, and Seber (1982, 1986) provided a full treatment in general.

A similar technique produces estimates of the survival of young animals from ratios of sizes of litters or broods at different ages. For example, Stoudt (1971) computed mortality of canvasback ducklings between young (Class I) and older (Class II) stages to be 1.2 ducklings per brood, based on average brood sizes in those two classes. The fact that some litters or broods may have been lost completely must be accounted for, however. Further, among young waterfowl, broods commonly split into two or more groups, or two or more broods combine into a larger aggregation. These processes can bias estimates of survival rate from brood counts.

CHANGE-IN-RATIO METHODS

The change-in-ratio technique, usually applied to estimating population size, can be used in some situations to estimate the rate of mortality from exploitation (Paulik

Table 6. Mark-recapture statistics[a] for a population of meadow voles trapped in Maryland in 1981 (Pollock et al. 1990:29).

Period	Dates	n_i	m_i	R_i	r_i	z_i	$\hat{S}_i$	SE
1	27 Jun–1 Jul	108	0	105	87	0	0.88	0.039
2	1 Aug–5 Aug	127	84	121	76	5	0.66	0.048
3	29 Aug–2 Sep	102	873	101	68	8	0.69	0.049
4	3 Oct–7 Oct	103	73	102	63	3	0.63	0.049
5	31 Oct–4 Nov	102	61	100	84	5		
6	4 Dec–8 Dec	149	89	148				

[a]For the ith occasion, n_i animals are captured, of which m_i had been marked already; R_i is the number of the n_i animals that are released after the ith sampling occasion; r_i is the number of the R_i animals released at i that are sometime captured again; z_i is the number of animals that were captured before i, not captured at i, but captured again later; and $\hat{S}_i$ is the estimated survival rate.

and Robson 1969, Seber 1982). To do so requires two distinguishable types of animals (male and female, say, or young and adult) and estimates of the fraction of each type in the population before harvest, in the harvest, and in the population after harvest. Assumptions required to give good estimates are stringent, however, and should be considered carefully before the method is adopted (Downing 1980).

MARK-RECAPTURE METHODS

Consider a study involving J occasions on which animals are captured, marked, and returned to the population. Suppose that all animals are alike in having the same chance of being captured on a particular occasion, call this probability c_i for the ith occasion, and in having the same probability of surviving from occasion i to occasion $i + 1$, say S_i. Define N_i to be the number of animals in the population on occasion i. Suppose that M_i of these had been marked previously. On the ith occasion, n_i animals are captured, of which m_i had been marked already and the remaining u_i had not been marked previously. From these values we can estimate the population size on occasions 2 through $J - 1$, as well as the number of combined births and immigrants (B_i) between occasions i and $i + 1$ for $i = 2$ through $J - 2$. Of special concern here, survival rates S_i, $i = 1$ through $J - 2$, can be estimated. These are:

$$\hat{S}_i = \frac{\hat{M}_{i+1}}{\hat{M}_i - m_i + R_i},$$

where

$$\hat{M}_i = m_i + R_i z_i / r_i,$$

and R_i is the number of the n_i animals that are released after the ith sampling occasion (normally this will be n_i minus any losses during capture), r_i is the number of the R_i animals released at i that are sometime captured again, and z_i is the number of animals that were captured before i, not captured at i, but captured again later. Estimated SEs of survival rates are available (e.g., Seber 1982, Pollock et al. 1990). Estimates of N_i, B_i, and c_i are given in Chapter 9. A readable justification for this, the Jolly-Seber model, was given by Cormack (1973). A computer program (JOLLY) to perform necessary calculations is available, as is one (JOLLYAGE) for populations structured by age (see Chapter 3). A variety of alternative models was described by Seber (1982). In a valuable review, Pol-

lock et al. (1990) discussed mark-recapture methods and developed some new models.

As an example, consider the data in Table 6, derived from a mark-recapture study of meadow voles. Animals were trapped on six occasions from June through December. From these recapture statistics, we can calculate estimates of survival rate from one occasion to the next (for the first four occasions). We get, for survival from occasion 1 to occasion 2,

$$\hat{M}_1 = m_1 + R_1 z_1 / r_1$$
$$= 0 + 105 \times 0/87$$
$$= 0$$

$$\hat{M}_2 = m_2 + R_2 z_2 / r_2$$
$$= 84 + 121 \times 5/76$$
$$= 91.96$$

so

$$\hat{S}_1 = \frac{\hat{M}_2}{\hat{M}_1 - m_1 + R_1}$$
$$= \frac{91.96}{0 - 0 + 105}$$
$$= 0.88,$$

and likewise for the remaining values. Pollock et al. (1990) also presented estimates of population size and the number of births.

It should be emphasized that births include all animals added to a population, whether by actual birth or by immigration. Also, the survival rate reflects not only actual survival, but also permanent emigration from the study area. The method can be used with different kinds of capturing on different occasions. Of particular interest is marking animals on the first occasion and using resightings of marked animals on subsequent occasions.

METHODS BASED ON RETURNS OF BIRD BANDS

Many models have been developed for use with data from banding programs for game birds. In those programs, large numbers of birds are captured each year and banded with individually identifiable bands. Hunters who recover a banded bird are encouraged to report the identification number. The situation is a mark-recapture study

Table 7A. Bandings and recoveries of wood ducks in 1964 and 1965 (from Brownie et al. 1985:22).

Year	Number banded	Number of recoveries in 1964	Number of recoveries in 1965
1964	1,603	127	44
1965	1,595		62

Table 7B. Expected numbers of bandings and recoveries of wood ducks in 1964 and 1965.[a]

Year	Number banded	Expected number of recoveries in 1964	Expected number of recoveries in 1965
1964	N_1	$N_1 c_1$	$N_1 S_1 c_2$
1965	N_2		$N_2 c_2$

[a] c_i is the recovery rate in the ith hunting season, and S_i is the probability that a bird survives from the beginning of the ith hunting season to the beginning of the next.

with many marking occasions (typically one per year for a series of years), but for an individual bird only a single recapture is possible. Consider a simple example. Suppose that wood ducks are banded for 2 years, just prior to the hunting season of each year (Table 7A). Define c_1 to be the recovery rate, the probability that a bird is shot and its band is reported during the first year. Similarly, c_2 is the probability that a bird, alive at the beginning of the hunting season in year 2, is shot and its band is reported. Let S_1 be the probability that a bird survives from the beginning of the first hunting season to the beginning of the second; this is the survival rate we wish to estimate. If 1,603 birds are banded in year 1, we expect 1,603 × c_1 to be shot and reported the first year (Table 7B); the actual number was 127. From this we determine an estimate of c_1: $\hat{c}_1 = 127/1{,}603 = 0.0792$. Of the 1,595 birds banded in the second year, we expect 1,595 × c_2 to be shot and reported, and the actual number was 62. From this we get $\hat{c}_2 = 62/1{,}595 = 0.0389$. Of the 1,603 banded the first year, we expect 1,603 × S_1 to survive to the beginning of the hunting season in the second year, and a fraction c_2 of them to be shot and reported. The actual number was 44. Thus, $44 = 1{,}603 \times S_1 \times 0.0389$, or $\hat{S}_1 = 0.7056$.

This procedure of equating observed to expected values applies only when the number of parameters to be estimated equals the number of equations, but it does illustrate the principle behind the construction of modern banding models. More generally, consider a recovery table based on 3 years of bandings and 5 years of recoveries (Table 8). N_i represents the number of birds banded in year i ($i = 1, 2, 3$). Of these, R_{ij} are recovered and reported in year j ($j = i, i + 1, \ldots, 5$). Also shown are summary statistics: row totals (R_i), column totals (C_j), and block totals (T_i), all of which are used in the formulas for the estimators.

Suppose that only adult birds are banded and released in the program. It may be reasonable to assume that recovery rates and survival rates vary annually but do not depend on the year when the bird originally was banded. This is Seber's (1970) model, which is termed Model 1 in the handbook by Brownie et al. (1985:15). Under this formulation, the expected numbers of recoveries, by year of banding (i) and year of recovery (j), are shown in Table 9. From the summary statistics we can obtain estimators of recovery rates c_i ($i = 1, 2, 3$) and survival rates S_i ($i = 1, 2$) as follows:

$$\hat{c}_i = \frac{R_i C_i}{N_i T_i}$$

and

$$\hat{S}_i = \frac{R_i}{N_i} \frac{(T_i - C_i)}{T_i} \frac{N_{i+1} + 1}{R_{i+1} + 1}.$$

Note that recovery rates can be estimated for each of the 3 years banding was conducted, but survival rates can be determined for only the first 2 years.

We illustrate this procedure with an example from Brownie et al. (1985:14) involving wood ducks (Table 10). From the summary statistics, we get

$$\hat{c}_1 = \frac{R_1 C_1}{N_1 T_1} = \frac{265 \times 127}{1{,}603 \times 265} = 0.0792,$$

$$\hat{c}_2 = \frac{R_2 C_2}{N_2 T_2} = \frac{210 \times 106}{1{,}595 \times 348} = 0.0401,$$

$$\hat{c}_3 = \frac{R_3 C_3}{N_3 T_3} = \frac{167 \times 195}{1{,}157 \times 409} = 0.0688,$$

and

$$\hat{S}_1 = \frac{R_1}{N_1} \frac{(T_1 - C_1)}{T_1} \frac{N_2 + 1}{R_2 + 1}$$

$$= \frac{265}{1{,}603} \frac{(265 - 127)}{265} \frac{1{,}596}{211}$$

$$= 0.6512$$

$$\hat{S}_2 = \frac{R_2}{N_2} \frac{(T_2 - C_2)}{T_2} \frac{N_3 + 1}{R_3 + 1}$$

$$= \frac{210}{1{,}595} \frac{(348 - 106)}{348} \frac{1{,}158}{168}$$

$$= 0.6311.$$

Estimated SEs are also available:

$$\text{SE}^2(\hat{c}_i) = (\hat{c}_i)^2 \left(\frac{1}{R_i} - \frac{1}{N_i} + \frac{1}{C_i} - \frac{1}{T_i} \right)$$

$$\text{SE}^2(\hat{S}_i) = (\hat{S}_i)^2 \left(\frac{1}{R_i} - \frac{1}{N_i} + \frac{1}{R_{i+1}} - \frac{1}{N_{i+1}} \right.$$

$$\left. + \frac{1}{T_{i+1} - R_{i+1}} - \frac{1}{T_i} \right),$$

which for the example give $\text{SE}(\hat{c}_1) = 0.00674$, $\text{SE}(\hat{c}_2) = 0.00415$, $\text{SE}(\hat{c}_3) = 0.00608$, $\text{SE}(\hat{S}_1) = 0.0675$, $\text{SE}(\hat{S}_2) = 0.0647$. An approximate 95% CI is the sample value minus and plus 1.96 times the SE. For the first-year recovery rate, for example, we get $0.0792 - 1.96 \times 0.00674 =$

Table 8. Table of recoveries for a 3-year banding program with 5 years of recoveries, and associated summary statistics.[a]

Year banded	Number banded	Year of recovery					Row total
		1	2	3	4	5	
1	N_1	$T_1 \vert R_{11}$	$\vert R_{12}$	$\vert R_{13}$	$\vert R_{14}$	$\vert R_{15}$	$R_1 = T_1$
2	N_2		$T_2 \vert R_{22}$	$\vert R_{23}$	$\vert R_{24}$	$\vert R_{25}$	R_2
3	N_3			$T_3 \vert R_{33}$	$T_4 \vert R_{34}$	$\vert R_{35}$	R_3
Column total		C_1	C_2	C_3	C_4	$C_5 = T_5$	

[a] Of the N_i birds banded in year i, R_{ij} are recovered and reported in year j; R_i are row totals, C_j are column totals, and T_l are block totals.

0.0660 as a lower limit, and $0.0792 + 1.96 \times 0.00674 = 0.0924$ as an upper limit. Hence, a 95% CI for c_1 is (0.0660, 0.0924). In addition to examining SEs, one should look at how well the model fits the data. Brownie et al. (1985) described goodness-of-fit tests, and the program ESTIMATE carries out these along with the estimation procedure.

More restrictive models might fit a particular data set adequately, in which case the relevant parameters may be estimated more precisely. A likely candidate is the model in which survival rates are assumed to be the same each year, but recovery rates vary. This model (Model 2 of Brownie et al. 1985:20) often fits data sets well, perhaps because true survival rates do not vary much, and the ability of actual banding data to detect those differences is weak. Another restriction is to assume that survival rates may vary annually, but recovery rates do not. This is Model 3 of Brownie et al. (1985:24) and is unlikely to be useful in most situations. The most restrictive model assumes that survival and recovery rates are the same each year. This model (Model 0 of Brownie et al. 1985:30) might fit small data sets but is likely to be true only in rare circumstances.

Often young birds and adult birds are banded in the same program. One cannot safely assume that birds of the two age groups have the same survival and recovery rates, so they must be treated differently. Yet, if the young birds survive long enough, they become adults, subject to adult survival and recovery patterns. Several useful models have been developed for this situation. One of the most general is that of Brownie and Robson (1976), termed Model H_1 by Brownie et al. (1985:59). As before, survival rates and recovery rates are assumed to vary by year, and young birds have different survival rates and recovery rates for their first year only. Estimators are presented in Brownie et al. (1985:60) and are calculated by the program BROWNIE.

Data illustrating an example are given in Table 11. Estimates are in Table 12.

As before, more restrictive models can give more precise estimates if they fit the data adequately. Another reasonable model allows survival rates for young birds and adults to vary from year to year but assumes that rates for the two age groups fluctuate in parallel. This model, proposed by Johnson (1974), has no closed-form solution and is not included in the program BROWNIE, but it can be fitted with program SURVIV (Chapter 3) or with general maximum likelihood programs. In addition, one can fit models that allow survival rates for the two age classes to vary from year to year in parallel, but with recovery rates varying independently, or vice versa.

Other restrictions include assuming that survival rates stay the same from year to year (Model H_{02} of Brownie et al. 1985:64), or that survival and recovery rates are constant (Model H_{01} of Brownie et al. 1985:69). Program BROWNIE estimates relevant parameters and SEs and carries out goodness-of-fit tests. Further, the procedure can be generalized to three age classes if birds can be distinguished by age class and some members from each age class are banded. This situation may pertain to geese, for example.

Two thoughts must be kept in mind when a banding program to estimate survival is planned. First, adults must

Table 9. Expected numbers of band recoveries under Model 1 of Brownie et al. (1985) for a banding study with 3 years of banding and 5 years of recoveries.[a]

Year banded	Number banded	Year of recovery				
		1	2	3	4	5
1	N_1	$N_1 f_1$	$N_1 S_1 f_2$	$N_1 S_1 S_2 f_3$	$N_1 S_1 S_2 S_3 f_4$	$N_1 S_1 S_2 S_3 S_4 f_5$
2	N_2		$N_2 f_2$	$N_2 S_2 f_3$	$N_2 S_2 S_3 f_4$	$N_2 S_2 S_3 S_4 f_5$
3	N_3			$N_3 f_3$	$N_3 S_3 f_4$	$N_3 S_3 S_4 f_5$

[a] f_i is the recovery rate in the ith hunting season, and S_i is the probability that a bird survives from the beginning of the ith hunting season to the beginning of the next.

Table 10. Banding and recovery data for male wood ducks (from Brownie et al. 1985:22).[a]

Year banded (i)	Number banded	Year of recovery (j)					R_i
		1964 1	1965 2	1966 3	1967 4	1968 5	
1964	1,603	127	44	37	40	17	265
1965	1,595		62	76	44	28	210
1966	1,157			82	61	24	167
		$C_j = 127$	106	195	145	69	
		$T_j = 265$	348	409	214	69	

[a] R_i are row totals, C_j are column totals, and T_j are block totals, of the number of recoveries.

Table 11. Data from a study of young and adult male mallards banded preseason in the San Luis Valley, Colorado, 1963–71.

Year banded	Number banded	Year of recovery								
		1963	1964	1965	1966	1967	1968	1969	1970	1971
					Banded as adults					
1963	231	10	13	6	1	1	3	1	2	0
1964	649		58	21	16	15	13	6	1	1
1965	885			54	39	23	18	11	10	6
1966	590				44	21	22	9	9	3
1967	943					55	39	23	11	12
1968	1,077						66	46	29	18
1969	1,250							101	59	30
1970	938								97	22
1971	312									21
					Banded as young					
1963	962	83	35	18	16	6	8	5	3	1
1964	702		103	21	13	11	8	6	6	0
1965	1,132			82	36	26	24	15	18	4
1966	1,201				153	39	22	21	16	8
1967	1,199					109	38	31	15	1
1968	1,155						113	64	29	22
1969	1,131							124	45	22
1970	906								95	25
1971	353									38

be included. If only young birds are banded, little can be estimated from the resulting recovery data unless some dubious assumptions are made (e.g., Burnham and Anderson 1979, Anderson et al. 1981). Second, sample size must be large to obtain meaningful estimates. The program BAND2 (Wilson et al. 1989) determines required sample sizes for various models. That program should be employed, and the handbook by Brownie et al. (1985)

should be carefully reviewed before a banding program is instituted.

POPULATIONS WITH AGE-DEPENDENT BIRTH AND DEATH RATES
Models
FERTILITY TABLES

Fertility and mortality are known to vary by age for many species, and considerable effort has gone into developing models with age-dependent birth and death rates. Consider an age-structured population with a maximum of I age classes recorded, say, in years. Suppose females of age x produce an average of m_x young females per year. A table giving the number of female offspring per year per female aged x is called a *fertility table*. An example for a population of white-tailed deer in central Michigan is included in Table 13. Notice that average fertility rates vary with age; they are 0 for young of the year, nearly 0

Table 12. Estimates of survival rates and recovery rates for young and adult birds, based on data in Table 11 fitted to Model H_1. Estimated SEs are in parentheses.

Year	Survival rate		Recovery rate	
	Adult	Young	Adult	Young
1963	0.576 (0.113)	0.471 (0.059)	0.0433 (0.0134)	0.0863 (0.0091)
1964	0.636 (0.076)	0.506 (0.070)	0.0856 (0.0092)	0.1467 (0.0134)
1965	0.666 (0.079)	0.589 (0.072)	0.0590 (0.0061)	0.0724 (0.0077)
1966	0.805 (0.098)	0.591 (0.072)	0.0628 (0.0067)	0.1274 (0.0096)
1967	0.650 (0.072)	0.478 (0.061)	0.0520 (0.0050)	0.0909 (0.0083)
1968	0.552 (0.058)	0.652 (0.072)	0.0633 (0.0055)	0.0978 (0.0087)
1969	0.572 (0.066)	0.464 (0.068)	0.0789 (0.0061)	0.1096 (0.0093)
1970	0.542 (0.129)	0.393 (0.113)	0.0888 (0.0080)	0.1049 (0.0102)
1971			0.0673 (0.0142)	0.1076 (0.0165)

Table 13. Survival and reproduction data for white-tailed deer in central Michigan (from Eberhardt 1969).

Age (x)	Survival rate (s_x)	Fertility rate (m_x)
0	0.58	0
1	0.70	0.047
2	0.70	0.503
3	0.70	0.663
4	0.70	0.733
5	0.70	0.743
6	0.70	0.771
>6	0.70	0.644

Table 14. Mortality schedule based on known deaths of 42 gray squirrels born in 1954 (from Downing 1980:256).

Age (years) (x)	Number in population (n_x)	Number of deaths (d_x)	Mortality rate (q_x)	Survival rate (s_x)
0–1	42	22	22/42 = 0.52	20/42 = 0.48
1–2	20	10	10/20 = 0.50	10/20 = 0.50
2–3	10	7	7/10 = 0.70	3/10 = 0.30
3–4	3	2	2/3 = 0.67	1/3 = 0.33
4–5	1	1	1/1 = 1.00	0/1 = 0

for yearlings, and increase with age up to age 6, after which they decline.

LIFE TABLES

Analogous to the fertility table is the mortality schedule, which describes the pattern of deaths by age class. Define the probability of a female surviving from the beginning of age class x to the beginning of age class $x + 1$ to be s_x. The survival data for the central Michigan deer population (Eberhardt 1969) are included in Table 13, along with fertility rates. The survival rates of age classes 1 and above did not differ significantly from one another, and the average rate, 0.70, was used to construct the table.

Consider a cohort of animals (a group born at about the same time) that begins with, say, 1,000 individuals (at age 0). Then there will be $1,000 \times s_0$ individuals the next year (at age 1), $1,000 \times s_0 \times s_1$ members the following year (at age 2), and so forth. The number of individuals surviving from birth to age class x will be termed n_x:

$$n_x = 1,000 s_0 s_1 \cdots s_{x-1}.$$

Often the mortality rate, rather than the survival rate, is expressed: $q_x = 1 - s_x$.

A life table (Table 14) gives these and other relevant values; it is basically a summary of the survivorship of a population. It can also be used to calculate or estimate mortality rates, by age, under certain assumptions. Life tables were developed for human populations, especially for insurance applications, but they have also been applied to wildlife populations. Human life tables generally involve large numbers of individuals for which exact times of death can be ascertained, whereas information for wild animals is typically incomplete. For most animal populations, information is based on a sample, thus the life table provides *estimates* of relevant parameters that are less exact than values for humans.

For many animals, survival and fertility rates differ more sharply by size or life stage than by age. Some life table methods can be used with size classes or stages. Lefkovitch (1965) developed population projection methods for such situations. See Usher (1972), Kirkpatrick (1984), Sauer and Slade (1987), and Caswell (1989) for further details and some applications.

A life table consists of several of the following six basic columns:

x—age, measured in years or some other convenient unit. This may actually be an interval, $[x, x + 1)$.

Box 4.　Variation in mortality rate by age.

Mortality rates of vertebrate populations normally vary by age. Typical patterns involve high mortality of young animals, lower mortality of animals in their prime, and increasing mortality with advancing age. Deviations from this pattern can occur, especially if reproduction imposes an added mortality risk. For the analysis of a population, the difference in mortality between young animals and prime-age animals is usually important. The difference between prime years and older years may be less important, especially in exploited populations in which few animals reach advanced ages.

---

n_x—the number of individuals surviving to the beginning of age x from an initial cohort of n_0 members.

d_x—the number of deaths in the age class $[x, x + 1)$.

$$d_x = n_x - n_{x+1}.$$

q_x—the mortality rate at age x.

$$q_x = d_x/n_x.$$

s_x—the survival rate at age x.

$$s_x = 1 - q_x.$$

l_x—the cumulative survival rate from birth until age x

$$l_x = s_0 \times s_1 \times \ldots \times s_{x-1} = n_x/n_0.$$

Notice that the definition of survival rates in the mortality table pertains to the period from the beginning of one age class to the beginning of the next. The fertility table describes reproduction per female in an age class. To use the survival rates and reproductive rates in combination, one must define the age classes similarly in the two tables. That is, if reproduction is determined by the number of young produced and surviving to autumn, the survival of adults should be assessed from autumn to autumn.

In Table 14 (here and elsewhere examples are given only to illustrate the method; sample sizes are too small to draw reliable conclusions) note that d_x can be computed from values of n_x by subtraction, and n_x can be determined by adding up entries in the d_x column from the bottom. Also, q_x is based on d_x and n_x, and conversely the table of n_x for $x > 0$ can be constructed from q_x values. Thus there is only one independent column, and all of the others can be calculated from the entries in any one column. Depending on the kinds of data available and the assumptions that can realistically be made, different types of life tables can be constructed.

Graphs of cohort size or cumulative survivorship (on a logarithmic scale) against age often approximate one of three characteristic shapes (Fig. 7), but possibly with a downward jag reflecting lower survival of newborns (Pearl 1928). Type I survivorship curves have low mortality early in life but higher rates among older individuals. Female elk in the northern Yellowstone herd exemplify this pattern, with the exception of a depressed

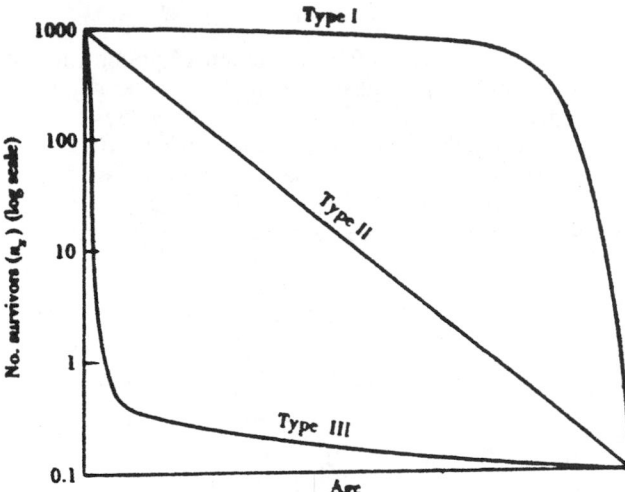

Fig. 7. Three characteristic survivorship curves (from Krebs 1985: 178).

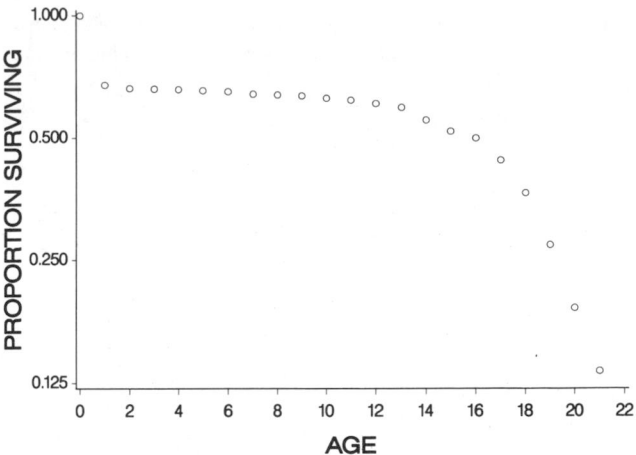

Fig. 8. Survivorship curve of female elk in the Northern Yellowstone herd (Houston 1982:55).

survival rate of the very young (Fig. 8; also see Box 4 concerning vertebrates in general). Type II survivorship curves have mortality rates roughly constant with age, leading to a straight-line relation on a log scale. Adult songbirds are suggested to have patterns such as this. The Type III survivorship curve involves high mortality among young and decreasing mortality as individuals age. Many invertebrates and fish display Type III survivorship; they are vulnerable when they are young and small, but age and growth impart greater security. Siler (1979) and Eberhardt (1985) discussed how survivorship functions might be partitioned into functions representing three stages of life: early life, maturity, and senescence.

THE STABLE AGE DISTRIBUTION

The *age distribution* of a population is the numbers of individuals of each age class in the population at a particular time. If age-dependent survival and fertility rates remain constant for a fairly long period of time, the proportion of animals in each age class will stabilize. This is true even if the population itself is not constant in size; that is, a population can be expanding or declining and still have constant proportions in each age class. The resulting fractions make up what is termed the *stable age distribution*. The fraction of the population in age class x will equal C_x:

$$C_x = \frac{e^{-rx} l_x}{\sum_i e^{-ri} l_i}, \qquad (7)$$

where r is the growth rate of the population once it attains a stable age distribution.

To see this, suppose the population were constant in size, say at N members. Then the stable age distribution at any time t would contain members in each age class proportional to the survivorship, i.e., $N \times l_0$ of age class 0, $N \times l_1$ of age class 1, and so on. But if the population has been changing at an annual rate λ, the number of members in age class x at time t would be the number born x years earlier ($N_{t-x} l_0$) times the survivorship of those members (l_x/l_0). And because of the population

growth, $N_t = N_{t-x} e^{\lambda x}$. Thus, in year t the fraction of the population in age class x will be $N_{t-x} l_0 \times l_x/l_0 = N_t e^{\lambda x} l_x$, which is the numerator of equation 7. The denominator is the sum of such values over all ages, which simply scales the numbers so that they total 1.0.

Alternatively, the size of the age class x relative to that of newborn is (Caughley 1977:114):

$$\frac{C_x}{C_0} = e^{-rx} l_x.$$

The value of r can be determined from the age-dependent survival and fertility rates according to the following equation:

$$1 = \sum_x e^{-rx} l_x m_x. \qquad (8)$$

This is the discrete version of what is termed *Lotka's equation* (sometimes called Euler's equation); see, e.g., Mertz (1970), Wilson and Bossert (1971), or Caughley (1977) for a derivation. Remember that it requires that survival and reproduction schedules remain constant for a long period of time, often an unlikely presumption. It is strictly appropriate only for a birth-pulse population in which births occur instantaneously and the age structure is observed at the same time (Michod and Anderson 1980).

Equation 8 can be solved for r from a schedule of age-dependent cumulative survival rates (l_x) and fertility rates (m_x), as will be shown later. For the white-tailed deer data (Table 13), the estimated value of r turns out to be $\hat{r} = -0.026$. This suggests, because $e^{-0.026} = 1 - 0.0257$, that the deer population was declining (because the rate is negative) at about 2.57% each year.

With this value of r, we can now determine the asymptotic age structure of the deer population, from equation 7. The number in age class x is proportional to $e^{-rx} l_x$, which gives the C_x values in Table 15. This distribution can be compared to the actual age distribution, if known, to test whether the underlying assumptions are met.

PROJECTING THE POPULATION: LESLIE MATRICES

If a population has attained a stable age distribution, and if we were fortunate enough to know the age-depen-

Table 15. Stable age distribution calculated from survival and reproduction data for white-tailed deer in central Michigan (from Eberhardt 1969).

Age (x)	Survival to age (l_x)	Fraction in age (C_x)
0	1.0000	0.3214
1	0.5800	0.1913
2	0.4060	0.1375
3	0.2842	0.0988
4	0.1989	0.0709
5	0.1392	0.0510
6	0.0974	0.0366
7	0.0682	0.0263
8	0.0478	0.0189
9	0.0334	0.0136
>9	$0.58(0.70)^{x-1}$	0.0337

dent survival and fertility rates, we could learn much about the population from studying those rates. Consider an age-structured population (with M age classes) that breeds seasonally and has survival and fertility rates that vary with age, but not annually. Suppose the population is censused for several years at the same time each year, say immediately after the birth season. Let $n_{x,t}$ be the number of individuals of age x in year t. The number of 1-year-olds in year $t + 1$ ($n_{1,t+1}$) will be the number that were born in year t ($n_{0,t}$) times the survival rate of 0-year-olds (s_0):

$$n_{1,t+1} = s_0 n_{0,t}.$$

Similarly, the number of 2-year-olds in year $t + 1$ equals the number of 1-year-olds in the previous year times their survival rate:

$$n_{2,t+1} = s_1 n_{1,t},$$

or, in general,

$$n_{i+1,t+1} = s_i n_{i,t}. \tag{9}$$

Consider next the number of births, which can be allocated according to the different age classes that reproduce. The number of 0-year-olds (births) in year $t + 1$ ($n_{0,t+1}$) represents the number of 1-year-olds in that year ($n_{1,t+1}$) times the fertility rate of 1-year-olds (m_1), plus the number of 2-year-olds in that year ($\widehat{n_{2,t+1}}$) times their fertility rate, and so forth. That is,

$$n_{0,t+1} = m_1 n_{1,t+1} + m_2 n_{2,t+1} + \ldots + m_M n_{M,t+1}.$$

We want to express this number in terms of the population in the previous year so, from equation 9, we get

$$n_{0,t+1} = m_1(s_0 n_{0,t}) + m_2(s_1 n_{1,t}) + \ldots$$
$$+ m_M(s_{M-1} n_{M-1,t})$$
$$= (m_1 s_0) n_{0,t} + (m_2 s_1) n_{1,t} + \ldots$$
$$+ (m_M s_{M-1}) n_{M-1,t}$$
$$= g_0 n_{0,t} + g_1 n_{1,t} + \ldots + g_{M-1} n_{M-1,t}, \tag{10}$$

where

$$g_i = m_{i+1} s_i, \quad \text{for } i = 0, \ldots, M - 1. \tag{11}$$

Equation 10 expresses the production of young (number of 0-year-olds) as a linear combination of the number in each age class in the previous year. The values of g_i indicate the number of young that are produced per individual aged i in year t and that will be alive in year $t + 1$. We can combine equation 9 for $i = 0, \ldots, M - 1$ and equation 10 into a single equation involving matrices:

$$
\begin{vmatrix}
n_0 \\
n_1 \\
n_2 \\
\cdot \\
\cdot \\
\cdot \\
n_M
\end{vmatrix}_{t+1}
=
\begin{vmatrix}
g_0 & g_1 & g_2 & \cdots & g_{M-1} & g_M \\
s_0 & 0 & 0 & \cdots & 0 & 0 \\
0 & s_1 & 0 & \cdots & 0 & 0 \\
\cdot & \cdot & \cdot & \cdots & \cdot & \cdot \\
\cdot & \cdot & \cdot & \cdots & \cdot & \cdot \\
\cdot & \cdot & \cdot & \cdots & \cdot & \cdot \\
0 & 0 & 0 & \cdots & 0 & s_{M-1}
\end{vmatrix}
\begin{vmatrix}
n_0 \\
n_1 \\
n_2 \\
\cdot \\
\cdot \\
\cdot \\
n_M
\end{vmatrix}_t
$$

Or, in matrix notation, $\mathbf{n}_{t+1} = L \times \mathbf{n}_t$. L is called the *population projection matrix*, or *Leslie matrix*. The term in any particular row and column can be thought of as the contribution of an individual in the age class represented by that *column* in year t to the age class represented by that *row* in year $t + 1$ (Jenkins 1988). Remember that we assumed survival and fertility rates were the same each year. This formulation was developed by Bernadelli (1941), Lewis (1942), and Leslie (1945, 1948). Van Groenendael et al. (1988) reviewed the method and applications and Caswell (1989) gave an excellent overview of the technique. It allows several interesting interpretations. For example, we can project the population from one year to the next or, by repeating the process, k years into future:

$$\mathbf{n}_{t+1} = L \times \mathbf{n}_t.$$

So

$$\mathbf{n}_{t+2} = L \times \mathbf{n}_{t+1} = L \times L \times \mathbf{n}_t,$$

and in general

$$\mathbf{n}_{t+k} = L^k \mathbf{n}_t.$$

We can also derive useful properties mathematically from this formulation, including the stable age distribution and population rate of change; see, e.g., Leslie (1945) and Pielou (1969).

The population projection matrix approach has been misused occasionally, generally by using fertility data as g_i values; Wethey (1985) and Jenkins (1988, 1989) presented examples. The parameter g_i is an odd one, incorporating both fertility and survival and measuring the fertility of a cohort aged $i + 1$ times the survival rate from age i to age $i + 1$ (eq. 11).

I described this formulation with censuses occurring immediately following the birth season. If counts are conducted at another time, the definition of g_i must be changed to incorporate survival from birth until the time of the census (e.g., Michod and Anderson 1980). In practice, the estimation of g_i is difficult at best (Taylor and Carley 1988).

AGE- AND DENSITY-DEPENDENT MODELS

Birth and death models in which the rates depend on both age and density can be constructed (Leslie 1948, 1959, Williamson 1959, Cooke and Leon 1976, Caswell

1989), but they do not have ready mathematical solutions and their properties are not well understood. Pennycuick et al. (1968) developed a computer program that allowed elements of a Leslie matrix to be density-dependent and also to have time lags. Little is known about how density dependence actually operates, however. More likely to be useful are models that partition birth and death rates into meaningful components. These components can be related to age, density, or environmental factors, as appropriate. Such models are covered later in the chapter.

Estimating Age-Dependent Death Rates

Earlier I covered ways of estimating mortality rates not specifically related to age. I now turn to the common problem of estimating mortality rates by age. A variety of methods is available, based either on following a cohort of animals or on examining the age distribution of a population. The appropriateness of estimators depends on the assumptions they require and how they are met by the population under study, and on how the data are collected.

ESTIMATING SURVIVAL BY FOLLOWING A COHORT

Knowing All Deaths

Suppose we knew the complete death history of a cohort of 42 squirrels born in 1954 (the d_x column of Table 14). Specifically, we knew that 22 of them died during their first year, 10 died during their second year, 7 died during their third year, 2 died during their fourth year, and 1 died in its fifth year. From this information, we can calculate exact mortality rates by age. If q_i is the probability of dying during year i (which also is during age i), then $q_0 = 22/42 = 0.52$, because 22 of the 42 squirrels died before their first birthday. Likewise, $q_1 = 10/20 = 0.50$, since 10 of the 20 survivors from the first year died during the second. Similarly, $q_2 = 7/10 = 0.70$, $q_3 = 2/3 = 0.67$, and $q_4 = 1/1 = 1.00$. Thus, age-dependent mortality rates can be computed exactly for this known population.

If the 42 animals can be considered a random sample from some larger population, statistical estimation is possible. Since q_0 is the mortality rate of animals in their first year of life, the number of animals expected to die by the beginning of the second year is $42q_0$. That number was actually 22, hence $\hat{q}_0 = 22/42 = 0.52$. Also, the number of animals alive at the beginning of each year is known at that time, and if we assume that individual animals live or die independently of one another, the number dying during year i can be treated as a binomial variate, with n_i representing the number of animals alive at the beginning of the year and rate = q_i. From this, the SE of $\hat{q}_i$ is estimated by

$$\sqrt{\hat{q}_i(1 - \hat{q}_i)/N_i}.$$

In our example, $SE(\hat{q}_0) = \sqrt{0.52 \times 0.48/42} = 0.077$. Similarly, $SE(\hat{q}_1) = 0.112$, $SE(\hat{q}_2) = 0.145$, $SE(\hat{q}_3) = 0.272$, $SE(\hat{q}_4) = 0$.

Knowing All the Living

Suppose that instead of knowing the age at which the individual squirrels died, we had censused the cohort at the beginning of each year. This information forms the basis of the n_x column of Table 14. There were 42 at the

beginning of the year 0 and 20 at the beginning of year 1. Thus, the survival rate during that year was 20/42 = 0.48, and the mortality rate was $1 - 0.48 = 0.52$. Mortality rates for the other years also coincide with those we determined just above from the information on the age of death. Likewise, if the sample of animals is representative of a larger population, we can treat the process as binomial and calculate the same estimates of SEs as above.

Life tables based on information from following a specific cohort are termed *cohort*, or sometimes *dynamic* or *age-specific, life tables*. Unfortunately, only for very closely monitored or captive populations do we face situations with such ideal knowledge, either of the ages at death or exactly how the size of a particular cohort changes over time.

Following More Than One Cohort

If more than one cohort is followed, an age-specific table can be generated for each of them. Then survival rates can be estimated that vary both by age and by year, although limited sample sizes usually preclude accurate measurements. Alternatively, estimates can be pooled across years to get age-dependent estimates (e.g., Downing 1980: Table 15.6) or pooled across ages to get year-dependent estimates. Which pooling is more appropriate depends on whether survival rates vary more by age or by year. Loery et al. (1987) presented an example of estimating survival rates by age and year for black-capped chickadees, based on a long-term mark-recapture study.

ESTIMATING SURVIVAL FROM AGE DISTRIBUTIONS

Suppose that we do not have complete information from following one or more cohorts through time, but we have the age composition of a sample of animals from the population at a particular time. That sample must accurately reflect either the *dying* members or the *living* members. We also require the population to have achieved a stable age distribution and to be constant in size (although the method can be adapted if the population is increasing or decreasing at a known rate). These are stringent assumptions that must be carefully regarded, and the methods that follow work better in theory than in practice.

The Age Distribution of the Living

Suppose that we have the age distribution of a sample from the living members of the population at a particular time in year t. The number of individuals of age x in year t ($n_{x,t}$) is the number that were aged $x - 1$ in year $t - 1$ ($n_{x-1,t-1}$) times the survival rate for those animals ($s_{x-1,t-1}$):

$$n_{x,t} = n_{x-1,t-1}s_{x-1,t-1},$$

from which we could estimate $s_{x-1,t-1}$:

$$\hat{s}_{x-1,t-1} = n_{x,t}/n_{x-1,t-1}.$$

We can do this if we have accurate age distribution data for successive years; without these, we cannot. By assuming the population is stationary, however, we can get estimates from a sample in a single year. Stationarity implies that survival rates (and fertility rates) are constant from year to year ($s_{x,t}$ is independent of t) and that the population size and age structure are the same from year

Table 16. Ages of male white-tailed deer prior to 1956 hunting season on George Reserve, Michigan (from McCullough 1979: 36).

Age (years)	Number in population
0	40
1	23
2	6
3	4
4	1
>4	0

Table 17. Life table based on age distribution of male white-tailed deer alive in 1956 (Table 16).

Age (years)	n_x	$\hat{d}_x$	$\hat{s}_x$	$SE(\hat{s}_x)$
0	40	17	0.575	0.150
1	23	17	0.261	0.120
2	6	2	0.667	0.430
3	4	3	0.250	0.280
4	1	1	0.000	0
>4	0	—	—	—

to year. That is, $n_{x,t}$ is independent of t (note: this is a *critical assumption*). Hence we have

$$\hat{s}_{x-1} = n_x/n_{x-1}.$$

Chapman and Robson (1960) recommended adding 1 to the denominator to reduce bias. Life tables formed this way are called *time-specific life tables* and represent a cross section of ages at a specific time.

A statistical model for these data can be developed as follows. In a sample of n_* animals in a particular year, the number of individuals aged x can be considered a multinomial variate. The probability ϕ_x that an individual will be in age class x is proportional to $n_0 s_0 s_1 ... s_{x-1}$. The proportion depends on the sampling intensity. These probabilities are estimated by $\hat{\phi}_x = n_x/n_*$. Also, due to the multinomial nature of the data,

$$E(\hat{\phi}_x) = \phi_x,$$

$$Var(\hat{\phi}_x) = \phi_x(1 - \phi_x)/n_*,$$

and the covariance between two survival rates is

$$Cov(\hat{\phi}_x, \hat{\phi}_y) = \phi_x\phi_y/n_*.$$

Survival rates are estimated by the ratio of successive $\hat{\phi}_x$ values:

$$\hat{s}_x = \hat{\phi}_{x+1}/\hat{\phi}_x,$$

with SE estimated from

$$SE^2(\hat{s}_x) = \frac{\hat{\phi}_{x+1}(\hat{\phi}_x + \hat{\phi}_{x+1})}{n_*\hat{\phi}_x{}^3}$$

$$= \frac{\hat{s}_x(1 + \hat{s}_x)}{n_*\hat{\phi}_x}.$$

The mortality rate, $\hat{q}_x$, will have the same SE as the survival rate $\hat{s}_x$.

We illustrate the procedure with the age distribution of male white-tailed deer on the George Reserve in Michigan, just before the 1956 hunting season (Table 16). These values can be used as an n_x column to generate a life table with estimated survival rate (Table 17). These estimates appear unrealistic, especially the higher survival rate for young animals than for older individuals. These aberrancies might in part reflect large SEs due to the small sample size, but more likely result from the population not being

stationary because of year-to-year variation in reproduction (McCullough 1979).

The Age Distribution of the Dying

Consider the example in Table 18 representing the ages of white-tailed deer found dead in surveys of carcasses. Thus the data reflect the age distribution of dying members of the population. Suppose the studied population is stationary. Then the age distribution can be used as the d_x column of a life table (Table 19). From these values we can estimate the n_x column, simply by adding the d_x entries from the bottom up. The ratio of d_x to $\hat{n}_x$ gives an estimate of q_x, the age-dependent mortality rate (Table 19). Note that the mortality rate for the last age class will always be 1.0.

Is the Sample of the Living or the Dying?

Surprisingly, the age composition of dead animals may not provide a suitable estimate of the age structure of animals dying (Caughley 1966). For example, if animals are shot unselectively with respect to age, the resulting sample will reflect the population *alive* at the beginning of the collection period and will not reflect the age structure of all animals that died (unless, of course, the shooting was the only mortality source). Such data would most appropriately be used in the n_x column of a life table. In contrast, if *all* of the animals that died during a year were recovered (like Table 18), resulting data would truly reflect the mortality and could be used in the d_x column.

The three main kinds of data that can be used to construct a time-specific life table are (Seber 1982:401–402): (1) number of animals of each age for a representative

Table 18. Ages of female white-tailed deer found dead in extensive mortality surveys (from Eberhardt 1969:488).

Age (years)	Number found dead
0–1	106
1–2	18
2–3	14
3–4	18
4–5	9
5–6	5
6–7	6
7–8	8
8–9	4
9–10	2
>10	8

sample of live animals, used as the n_x column; (2) number of animals of each age at death for a representative sample of animals killed by an agent independent of age (nonselective collection, natural catastrophe), also used as the n_x column; or (3) number of animals of each age at death for a representative sample of carcasses, used as the d_x column. Biased estimates can arise if younger age classes are less vulnerable to sampling, possibly because when they are alive, they are less detectable, and when they die their softer bones do not persist as long. Survival estimates for older age classes are unaffected by this bias (Caughley 1966, Seber 1982).

Is the Population Stationary?

We indicated that age-distribution data can be used to estimate survival if the population has a stable age distribution and is constant in size (i.e., the population is stationary). The method can be modified if the population is increasing or decreasing at a *known* rate (Caughley 1977, Eberhardt 1988). Knowing this requires independent information, such as estimates of population trend. The requirement of stable age distributions remains. Survival rates also can be estimated from data for a stable age distribution if appropriate fertility rates are available (Michod and Anderson 1980).

One cannot examine a single age distribution and determine whether the population is stationary (Caughley 1966, Seber 1982; but see Tait and Bunnell 1980 for a possible exception if the age at death is known for a large number of animals). A series of age distributions at different times may be used to determine stationarity.

It is tempting to assume a population is stationary, then to estimate survival rates from age-structure data, and then conclude from those results that the population is stable. Despite warnings to the contrary about the circularity of this argument (Caughley and Birch 1971), the practice persists (noted by Lancia and Bishir 1985 and Jenkins 1989).

Knowing Some of the Living and Some of the Dying

Opportunities to follow cohorts for long periods of time are rare, which precludes use of cohort analysis, and the critical assumption of a stable age distribution is so difficult to meet that it makes the use of age distributions suspect. Fryxell (1986) proposed an alternative method that overcomes these problems, but it requires three kinds of information about a population. He showed that age-dependent mortality rates can be obtained if one has estimates of (1) the age distribution of animals that die during a year, (2) the age distribution of animals alive at the beginning or end of the year, and (3) the mortality rate of the population as a whole. This method eliminates the need for the population to have settled into a stable age distribution. Indeed, it could be used when populations are fluctuating, periods of great interest to wildlife ecologists and managers. Unfortunately, obtaining the needed data is not easy.

Pooling Ages for Survival Estimation

Because of variation caused by small samples, smoothing either the observed age frequencies or the resulting

Table 19. Life table based on age distribution of female white-tailed deer found dead (Table 18).

Age (years)	$\hat{n}_x$	d_x	$\hat{q}_x$
0–1	198	106	0.535
1–2	92	18	0.196
2–3	74	14	0.189
3–4	60	18	0.300
4–5	42	9	0.214
5–6	33	5	0.152
6–7	28	6	0.214
7–8	22	8	0.364
8–9	14	4	0.286
9–10	10	2	0.200
>10	8	8	1.000

estimators is often necessary; Caughley (1977) illustrated the former, and I mention the latter below.

If one believes that mortality is constant for ages in a specified interval, pooled estimates of the rate can be obtained. Eberhardt (1985) noted that pooled mortality estimates are biased high if older animals survive at a lower rate than animals of prime age. For example, it seems reasonable that the mortality rate of the deer of Table 19 is roughly constant for individuals >1 year old. A pooled estimator of that adult mortality rate is

$$\frac{d_1 + d_2 + \ldots}{n_1 + n_2 + \ldots} = \frac{(n_1 - n_2) + (n_2 - n_3) + \ldots}{n_1 + n_2 + \ldots}$$

$$= \frac{n_1}{\sum_{j \geqslant 1} n_j},$$

which, in the present example, is

$$\frac{18 + 14 + \ldots + 8}{92 + 74 + \ldots + 8} = 92/383 = 0.240.$$

Average, rather than pooled, estimators also can be formed (Seber 1982). Better yet, the unbiased estimator of adult survival rate with smallest variance (Chapman and Robson 1960, Robson and Chapman 1961) is

$$\hat{s}_{CR} = T/(n + T - 1),$$

where

$$n = \sum_{j \geqslant 1} n_j$$

and

$$T = \sum_{j \geqslant 1} j n_j.$$

Its SE can be estimated from

$$\mathrm{SE}^2(\hat{s}_{CR}) = \hat{s}_{CR}\left(\hat{s}_{CR} - \frac{T-1}{n+T-2}\right).$$

In the example (Table 19), we have n = 92 + 74 + ... + 8 = 383, T = 1 × 92 + 2 × 74 + 3 × 60 + ... + 11 × 8 = 1,365, so $\hat{s}_{CR}$ = 1,365/(383 + 1,365 − 1) = 0.78134, and $\hat{q}$ = 1 − $\hat{s}$ = 0.219. Its SE is determined from

$$SE^2(\hat{q}) = SE^2(\hat{s}_{CR})$$

$$= 0.78134\left(0.78134 - \frac{1{,}365}{383 + 1{,}365 - 2}\right)$$

$$= 0.0000983,$$

so $SE = \sqrt{0.0000983} = 0.0099$.

Other Comments on Life Tables

Because age composition is typically determined from samples, rather than from entire populations, the entries in the life table are estimates, subject to sampling variation. Caughley (1977:95) suggested that life tables based on fewer than 150 age determinations were unlikely to be accurate enough for any purpose. Polacheck (1985) concluded from simulation that analyses based on even larger samples often provided misleading estimates of survival rate.

McCullough (1979:221) analyzed one of the best available sets of data on age structures (of white-tailed deer) and concluded: "Although numerous attempts have been made to apply life-table methods to the analysis of kill data, . . . most of these methods have not proven to be useful at the practical level." The major problem he identified was meeting the assumption of a stable age distribution, what with variable environmental factors having differential effects on different age classes. He suggested that time-specific life tables, although clearly not meeting the assumptions necessary to estimate survival rates, are valuable to the manager of exploited populations because they show the existence of strong year classes, indicative of good reproduction in a particular year. Seber (1982) cautioned that life tables may give an overall picture of a population, but they have limited accuracy and should be supported by other methods of estimation. Jenkins (1989) bemoaned the limited value of age-distribution data in light of the ease with which they can be obtained, especially by wildlife management agencies that monitor harvests.

Methods based on age-structure data have received much attention in the past, probably more than they merit considering their deficiencies. This chapter treats them in some detail, not to promote their use but to clarify the assumptions they require.

Estimating Population Growth From Birth and Death Rates

Knowing the birth and death rates of a population would appear to allow us to determine whether the population was increasing, holding steady, or declining. We can indeed calculate the growth rate of the population from such information. We discuss two approaches, a simplified one applicable when birth and death rates are not segregated into many age classes, and the Lotka equation for when they are. Also, the population projection matrix can be used to determine the growth rate of a population (Leslie 1945, Pielou 1969).

A DIRECT METHOD

Consider, following Martin et al. (1979), the female segment of the North American population of mallards. For the 1961–74 period, the average survival rate for adult

females was 0.555 and the average survival rate for immature females was 0.563. Survival was estimated between anniversary dates of 1 September in successive years. Suppose that the recruitment rate, measured as young females per adult female on this anniversary date, averaged 1.03. From these survival and fertility rates we can estimate the average annual change in the population of female mallards as follows. The number of adult females on 1 September of year $t + 1$ represents the adults from the previous year that survived, plus the immatures that survived. That is,

$$A_{t+1} = A_t(0.555) + Y_t(0.563),$$

where A_t is the number of adult females in year t and Y_t is the number of young females in year t. We also have from the recruitment rate

$$Y_t = A_t(1.03),$$

so

$$A_{t+1} = A_t(0.555) + A_t(1.03)(0.563),$$

or

$$A_{t+1} = A_t(1.135).$$

From this we can conclude one of three things: (1) the female segment of the mallard population was growing at a rate of 13.5%/year ($\lambda = 1.135$), (2) the estimates of survival or recruitment, or both, are wrong, or (3) the model is incorrect. Evidence from annual surveys of mallards during 1961–74 led Martin et al. (1979) to reject the first option, and the simplicity of the model argues against the third, so those authors concluded that certain estimates of survival or recruitment were the problem. In fact, they used this approach as a check on the consistency of their parameter estimates.

FROM LOTKA'S EQUATION

If age-dependent schedules of survival (l_x) and fertility (m_x) are available, the growth rate implied by those schedules can be computed from Lotka's formula (eq. 8). It is an iterative procedure. (Alternatively, Caughley [1977] presented a short FORTRAN computer program to perform the calculations.) We illustrate with the white-tailed deer data (Tables 13, 15). Let's try a value of $r = 0$ for growth rate, indicative of a steady population. Plugging $r = 0$ and values of l_x from Table 15 and m_x from Table 13 into the right-hand-side of equation 8, we get (since $e^0 = 1$)

$$1.000 \times 0 + 0.5800 \times 0.047 + 0.4060 \times 0.503 + \ldots,$$

all of which add up to 0.89. This value is less than 1, which means $r = 0$ is too high. Let's try $r = -0.10$. Using that value in equation 8 gives 1.44, far too large. The value of the sum we want is about 20% of the distance between 0.89 and 1.44, so we can try a value of r 20% of the way between 0 and -0.10, that is, $r = -0.02$. A value of $r = -0.02$ gives 0.97, still too small, but $r = -0.026$ results in a sum of 0.9995, close enough to stop the iteration. Since $e^{-0.026} = 1 - 0.0257$, this value suggests that the deer population was declining at about 2.57% each year.

This approach could also be used with the mallard data. Age-dependent survival and fertility rates give:

$$l_0 = 1$$

$$l_x = 0.563(0.555)^{x-1} \qquad x > 0$$

and

$$m_0 = 0$$

$$m_x = 1.03 \qquad x > 0.$$

Using these values with $r = \log \lambda = \log (1.135) = 0.1266$ in the right-hand side of equation 8 yields 0.99987, negligibly different from 1.0.

Another useful statistic is the *net reproductive rate*, the average number of young produced by an individual during its lifetime:

$$R_0 = \Sigma \, l_x m_x.$$

Values of $R_0 < 1$ indicate that members of the population are not replacing themselves, that is, the population is declining. Conversely, $R_0 > 1$ denotes an increasing population, and $R_0 = 1$ indicates a stable population.

For the deer population, we get $R_0 = 0.89$, again indicating a declining population. For the mallard example, we have

$$
\begin{aligned}
R_0 &= 1 \times 0 + 0.563 \times 1.03 \\
&\quad + 0.563 \times 0.555 \times 1.03 + \ldots \\
&= 0 + 0.580 + 0.322 + 0.179 + 0.099 \\
&\quad + 0.055 + 0.031 + 0.017 + 0.009 + 0.005 \\
&= 1.300,
\end{aligned}
$$

adding through age class 10 (notice how terms become small for older age classes, indicating that the few very old individuals have little effect on the size of the population). As before, this value suggests an increasing population.

INTERACTING SPECIES
Competition Models

Let us briefly consider the populations not of a single species but of two species that interact. Here we assume they compete for some resource, such as food. If that resource is limited, the habitat will support fewer of species 1 when species 2 is common than when species 2 is rare. Let α ($\alpha > 0$) be the relative impact of an individual of species 2 on the population growth rate of species 1. That is, N_2 individuals of species 2 have the same effect on species 1 as do N_1 individuals of species 1, or $N_1 = \alpha N_2$ in terms of effect on species 1. Then, if the logistic model of equation 5 is generalized, and the sizes of the two populations at time t are written as $N_1(t)$ and $N_2(t)$, the per capita growth rate of population 1 is modified, not just by

$$\frac{K_1 - N_1(t)}{K_1},$$

but by

$$\frac{K_1 - N_1(t) - \alpha N_2(t)}{K_1},$$

where $K_1 - \alpha N_2(t)$ can be considered the carrying capacity for species 1, as reduced by the presence of N_2 animals of species 2. From this we have

$$\frac{1}{N_1(t)} \frac{dN_1(t)}{dt} = r_1 \frac{K_1 - N_1(t) - \alpha N_2(t)}{K_1},$$

where r_1 and K_1 are the parameters of logistic growth for species 1 in the absence of species 2. Analogously, if β ($\beta > 0$) is the relative effect of an individual of species 1 on the population growth rate of species 2, then we get

$$\frac{1}{N_2(t)} \frac{dN_2(t)}{dt} = r_2 \frac{K_2 - N_2(t) - \beta N_1(t)}{K_2},$$

where r_2 and K_2 are defined correspondingly. (Many of the results can be obtained without recourse to the logistic formulation [Maynard Smith 1974], but it is a convenience.) The parameters α and β are termed the *competition coefficients* of the model, which was developed in the 1920s by Lotka (1925) and Volterra (1926). This system has been the basis of much theoretical work in competition (see, e.g., Levins 1968, MacArthur 1968, 1972, Vandermeer 1972, Pianka 1974, Berryman 1981) but has received little use in wildlife studies, in part because of the inherent difficulty of estimating the relevant parameters. Some mathematical results follow from the values of K_1, K_2, α, and β. Only if $K_1/\alpha > K_2$ and $K_2/\beta > K_1$ is it possible for the two species to coexist. Basically, this says that coexistence is possible only if the growth rate of each species is inhibited more by a member of its own species than by an individual of the other species. Their own density-dependent controls must cause growth to stop before they eliminate the competitor. One way this can happen is if the two species do not overlap completely in their resource use. Although results such as these are useful theoretically, most actual populations probably do not exhibit such pure behavior of this simple competition model, but are affected by a variety of other phenomena. For example, patchiness in resources reduces competition by favoring the first species in one kind of habitat and the second species in another kind. Also, because the environment changes with time, the relative competitive abilities of the species may vary.

Predator-Prey Models

A second kind of interaction involves predation. Predator-prey models in various forms have seen extensive use in wildlife studies. Suppose that species 1 serves as a prey for species 2, and that the population growth rate of species 1 is inhibited in direct proportion to the number of predators. Then

$$\frac{1}{N_1(t)} \frac{dN_1(t)}{dt} = r_1 - \gamma N_2(t),$$

where the *predation coefficient*, γ, indicates the removal rate of prey per predator. This model includes no inhibitory effects of the population of species 1; that is, in the absence of predators, the prey population would grow exponentially. Also, each predator consumes a number of prey proportional to the abundance of the prey. For predators (species 2), the per capita population growth rate is assumed to be

$$\frac{1}{N_2(t)} \frac{dN_2(t)}{dt} = \delta N_1(t) - d_2.$$

Here d_2 is the death rate of predators, which is assumed to be independent of the population of prey. The coefficient δ represents the *conversion rate* of prey to predators. Like the competition model, this one was developed by Lotka (1925) and Volterra (1926).

The model assumes (e.g., Ricklefs 1979): (1) exponential growth by the prey species in the absence of predators; that is, numbers of prey are limited only by predation; (2) exponential decay by predators in the absence of prey—mortality is independent of the density of predators; and (3) the rate at which prey are consumed is directly proportional to the product of the two species' densities (which can be construed as the chance of encountering one another if movements are random). The first two assumptions mean that the population growth of each species is controlled by the other species; these assumptions can be relaxed by including a logistic-type, self-inhibitory restraint on the population growth rates. The third assumption can be replaced by any of a variety of choices; see, for example, May (1973:81–84) or Maynard Smith (1974:25–33).

This model can be analyzed mathematically, under the assumption that there is no random variation. Depending on the values of the parameters, two outcomes of the model are feasible: either the populations of both species reach an equilibrium point and remain there, or populations of both species oscillate over time, and increases in the predator species lag behind increases in the prey species. Most investigations of actual populations of predators and prey involved invertebrate species in controlled laboratory situations; Tanner (1975) offered an exception that dealt with vertebrates. He concluded that vertebrate predator-prey systems were stable only if the prey species limited its own population or if it had lower (intrinsic) growth rate than the predator species. Populations of snowshoe hares and lynx, which have approximately equal growth rates, oscillate in a cyclic fashion. Caughley and Krebs (1983) provided a more general view of this issue.

Powell (1979) applied predator-prey modeling to a community involving the fisher and its primary prey, the porcupine. He prudently examined five variations and extensions of the basic predator-prey model, so that the conclusions he drew would be less susceptible to assumptions underlying any single model. He also considered the effects of two alternative prey species. Although space does not permit a detailed treatment of the models here, Powell's results suggested that the community was stable, but that only small increases in fisher mortality could cause local extinction of that predator.

These general predator-prey models, like other models, are unrealistically simplistic. Nonetheless, they offer useful insight into the general behavior of predator-prey systems, lead to more realistic models, which we will discuss next, and form the foundation for managing populations for optimal yield. This latter situation, in which the predator is human, is discussed in Chapter 17.

MODELS WITH COMPONENTS OF SURVIVAL OR BIRTH

The models presented thus far are, appearances to the contrary, rather simple—simple in the sense that they really depend only on time. Given the features of a model and the current status of a population, we can predict exactly what will happen at any future time (if the model were actually correct). This is extremely unrealistic, of course, but such simple models have nonetheless proven useful. Their major advantage lies in the way they can be treated mathematically. We now turn to models that are more complex but often, surprisingly, simpler to construct and analyze. The trade-off here is that we gain realism and complexity, but lose the ability to analyze the model mathematically; a computer is typically required. For that reason, most such models are simulation models. Some of the most useful population analyses today are based on simulation models.

We realize that a population goes up or down during a year depending on its annual survival rate and its birth rate for the year. The annual survival rate is an overall measure, encompassing the risks a population faces, which may vary season to season, or day to day, among individuals in the population, and from place to place. Examining survival rate in closer detail, such as by parts of the year, is often worthwhile. Likewise, fertility rates incorporate a multiplicity of components, which may be treated individually. For example, the measure of hatchings for our whooping crane example is the number of young recorded in the winter population. As such, it reflects the number of adult birds that are paired, the proportion of those that successfully lay eggs, the clutch size, the proportion of eggs that hatch, survival of young until fledging, and survival from fledging until the winter survey.

By dividing survival and fertility rates into finer components, we gain several advantages. First, we can consider environmental and other factors, beyond the density of conspecifics, that influence individual components. For example, the clutch size of mallards depends primarily on age of the female and date the clutch is initiated, so it can be modeled as a function of those factors. The success of the nest, however, depends mostly on predator numbers and nesting habitat, so those features can be used to model nest success. The second potential advantage is that we often can obtain better estimates of these individual components and the factors that influence them. For example, clutch size can be studied either passively or experimentally, and studies that manipulate clutch size might give good insight into that parameter but poor information about nest success. A third advantage is that we may gain a clearer understanding of the relationships involved in each component by their separate study. This is especially important for management applications, in which one or more components may be altered; the effect on the entire system needs to be understood or the population response may not be the desired one.

We illustrate this procedure with a model of the production of mallards in the prairie pothole region (Johnson et al. 1992). Only females are considered, and two age classes of breeding females, yearling and older, are identified. Let F_i ($i = 1,2$) be the number of yearling females and older females, respectively, in the breeding population, and F_0 be the number produced. F_0 can be apportioned according to age class of the adult:

$$F_0 = F_1 R_1 + F_2 R_2,$$

where R_i is the production rate for females in age class i. This value can be further decomposed according to nesting attempts, giving

$$R_i = D_i(Q_{i1} + Q_{i2} + Q_{i3} + Q_{i4} + Q_{i5}),$$

where D_i is the proportion of females of age class i that attempt to breed, and Q_{ij} is the production from the jth nesting attempt of females in age class i. This allows a maximum of five nesting efforts in a breeding season.

The production from a particular nesting attempt itself involves several factors, and can be expressed as the product:

$$Q_{ij} = A_{ij}C_{ij}HEB,$$

where A_{ij} is the probability that a female aged i will make the jth nesting attempt in a breeding season, C_{ij} is the average clutch size of the jth nesting attempt by a female of age i, H is the nest success rate, E is the survival rate of eggs in successful nests, and B is the survival rate of young.

Most parameters are indexed by age of the female and nesting attempt, because age and attempt are known to influence them. Effects of age and attempt on nest success rate (H), survival rate as eggs (E), and survival rate as young (B) have not been demonstrated clearly. The rates of incidence of breeding (D_i) vary most strongly with wetland conditions. Nesting probabilities were formulated to be higher for older females than yearlings, to be higher when wetland conditions were good, to decline with nesting attempt, and to be lower when nest success is high (because then nests are more likely to be destroyed later, when the female is in poorer condition). Clutch size was modeled to decline with nesting attempt and to be one egg smaller for yearling females than for older females.

Nest success of mallards is highly variable and is a component amenable to management. It varies according to predator abundance and the condition of the nesting habitat. Egg survival is generally high, and in the model it does not vary as a function of any environmental variable. The survival of young after hatch is lower, however, and likely depends on predators, weather, food supplies for ducklings, and possibly disease; some of these factors may operate in a density-dependent manner.

Johnson et al. (1992) executed the model described above by allowing the parameters to vary about as widely as they seem to do in natural populations. They contrasted results from the model for mallards with results for other species; they concluded that recruitment of the mallard was most dependent on predation and wetland conditions. Similar models (e.g., Johnson et al. 1987, Cowardin et al. 1988) have been used to evaluate management options in terms of expected production of mallards anticipated by manipulating various of the parameters.

Among the numerous species of wildlife to which simulation models have been applied are grizzly bear (Knight and Eberhardt 1985), bobcat (Crowe 1975), moose (Crête et al. 1981), white-tailed deer (Walters and Gross 1972), beaver (Molini et al. 1981), mallard (Anderson 1975), northern bobwhite (Roseberry 1979), and screech owl (North 1985). Most of these modeling efforts were directed at assessing the effect of various harvest strategies.

Good references on the construction of models for wildlife management include texts by Grant (1986) and Star-field and Bleloch (1986). The latter book describes the development of various kinds of models and includes the false starts likely to be encountered by modelers, novice and experienced alike.

Building models is easy, perhaps too easy in this day of ready access to computing power. Evaluating them is more difficult. One should compare model results with real data, independent of information used to construct the model and preferably obtained by direct experimentation. If that test is not feasible, a comparison with other models, built on different assumptions, is worthwhile, as is a comparison to analytic solutions.

IMMIGRATION AND EMIGRATION

Dispersal in its various forms is a critical process that allows individuals to persist despite degradation of the habitat they currently occupy. Virtually all plant and animal species exhibit dispersal during at least one life stage. Caughley (1977) defined *dispersal* as the movement of an animal from its point of origin to the place where it reproduces. He distinguished it from other types of movements, namely local movement within a home range, and migration (back-and-forth movements between discrete locations). Although dispersal is important in population dynamics, it is difficult to detect and even more difficult to measure. A biologist carrying out a population analysis typically ignores dispersal, assumes it to be nonexistent, or blithely hopes that immigration and emigration cancel one another.

Several biologists (e.g., Pielou 1969, Poole 1974, Caughley 1977) discussed models of dispersal, but its estimation has received little attention. Most techniques for detecting or estimating dispersal rely on marking the animals and observing where they go or recapturing them.

The observation of marked animals has provided most of the evidence of dispersal, including direction, distance, time of occurrence, and length of time between sightings. Fortuitous records, such as a coyote being trapped a long distance from where it had been marked, are interesting and informative but tell little about the dispersal patterns of coyotes in general. For a more complete picture we need telemetry studies in which all radio-equipped animals can be followed.

Mark-recapture studies can provide estimates of losses or gains to the population between trapping occasions, although they ordinarily are used to estimate the size of a population. With certain designs, losses can be partitioned into deaths and emigration, and gains can be separately estimated as births and immigration (Jackson 1939, also see Krebs 1985 and Manly 1985). Also, Nichols and Pollock (1990) presented a procedure for separately estimating births and immigrants from a mark-recapture study involving primary periods of trapping (well separated in time) and secondary trapping periods (closely spaced in time). The closeness of secondary trapping periods allows the reasonable assumption that no gains or losses to the population occur. Zeng and Brown (1987) proposed a method for distinguishing emigration from death in mark-recapture studies, but it requires the recapture of all animals that are still alive and have not dispersed. By comparing estimated survival rates based on local mark-recapture studies (which incorporate probabilities both of surviving and of returning to the study area) with survival rates from banding studies (which incorporate only survival), one can estimate the return rate (e.g., Anderson and Sterling 1974, Hepp et al. 1987). This value is 1 minus the probability of permanent dispersal from an area.

Hestbeck et al. (1991) developed models for the resighting of individually marked Canada geese in the Atlantic Flyway. In 3 years nearly 29,000 geese were marked and 102,000 resightings were made. The models included survival and resighting probabilities, and the probability of movement from one region to another in successive winters. They observed that annual changes in movement probabilities corresponded to variation in the severity of winter. Also, a model incorporating memory and tradition better fit the data, indicating that the wintering location of a goose depended not only on where it spent the previous winter, but also where it had been 2 years before.

CONCLUSIONS

We have followed a progression of models of population growth, beginning with exponential growth, in which population size grows (or declines) steadily with time. Then we considered models in which the population size, relative to carrying capacity, inhibits further growth of population. Despite resting on assumptions known to be invalid, these models are often useful, especially for short-term prediction and management decision-making (e.g., Eberhardt 1987). Next we looked at models in which survival or fertility varied by age. We then considered situations in which growth also depended on numbers of another species, i.e., a competitor, a prey, or a predator. We then briefly looked at more general models, in which births and deaths can be made functions of a variety of variables, including environmental factors. These models usually require computer execution to simulate population behavior, but they are often more useful than the older, deterministic approaches.

Along the way, we discussed how various parameters involved in these models could be estimated. Simple deterministic models have received increased attention in the past, and have greater generality, so more can be said about estimating parameters associated with them than about parameters involved in the more complex stochastic models. Stochastic models also require an understanding of the variation associated with the variables and what factors cause the variation.

Given the diversity of approaches that can be taken (only a fraction of which have been touched on here), how is a wildlife ecologist to choose? The first question must be: what are the exact objectives of the analyses? An analysis should include those relationships and variables suspected of being most influential to the dynamics of the population under study. Reasons for incorporating parameters include (Caughley 1977): (1) ease of estimation, (2) extent to which they describe the important features of the population, (3) their generality, and (4) how directly they relate to population processes. Any model must sacrifice at least one of three desiderata: generality, realism, or precision (Levins 1966). The trade-off between simplicity and complexity is a difficult one; a simple model is tractable, but may overlook key processes, whereas a complex one may satisfy only its builder.

Wildlife management in essence is based on only two primary tools, manipulation of habitat and control of harvest. These activities are effective only if they influence in the desired manner the population dynamics of the target species. To evaluate their actions and know they are doing the right thing, managers must understand those dynamics.

Acknowledgments.—M. D. Schwartz provided the graphics. I am grateful to D. R. Anderson, J. D. Carlson, Jr., M. J. Conroy, L. L. Eberhardt, G. Caughley, J. D. Nichols, T. L. Shaffer, D. R. Smith, and especially B. S. Bowen for comments on various drafts of this chapter.

LITERATURE CITED

ALLEE, W. C. 1931. Animal aggregations: a study in general sociology. Univ. Chicago Press, Chicago, Ill. 431pp.

ALLEN, D. L. 1962. Our wildlife legacy. Rev. ed. Funk & Wagnalls Co., Inc., New York, N.Y. 422pp.

ANDERSON, D. R. 1975. Optimal exploitation strategies for an animal population in a Markovian environment: a theory and an example. Ecology 56:1281–1297.

———, AND R. T. STERLING. 1974. Population dynamics of molting pintail drakes banded in south-central Saskatchewan. J. Wildl. Manage. 38:266–274.

———, A. P. WYWIALOWSKI, AND K. P. BURNHAM. 1981. Tests of the assumptions underlying life table methods for estimating parameters from cohort data. Ecology 62:1121–1124.

BERNADELLI, H. 1941. Population waves. J. Burma Res. Soc. 31:1–18.

BERRYMAN, A. A. 1981. Population systems: a general introduction. Plenum Press, New York, N.Y. 222pp.

BINKLEY, C. S., AND R. S. MILLER. 1980. Survivorship of the whooping crane, *Grus americana*. Ecology 61:434–437.

———, and ———. 1988. Recovery of the whooping crane *Grus americana*. Biol. Conserv. 45:11–20.

BIRKHEAD, T. R. 1977. The effect of habitat and density on breeding success in the common guillemot (*Uria aalge*). J. Anim. Ecol. 46:751–764.

BOYCE, M. S. 1987. Time-series analysis and forecasting of the Aransas/Wood Buffalo whooping crane population. Pages 1–9 *in* J. C. Lewis, ed. Proc. 1985 Crane Workshop. Whooping Crane Maintenance Trust, Grand Island, Nebr.

———, AND R. S. MILLER. 1985. Ten-year periodicity in whooping crane census. Auk 102:658–660.

BRODSKY, L. M. 1988. Ornament size influences mating success in male rock ptarmigan. Anim. Behav. 36:662–667.

BROWNIE, C., D. R. ANDERSON, K. P. BURNHAM, AND D. S. ROBSON. 1985. Statistical inference from band recovery data—a handbook. Second ed. U.S. Fish Wildl. Serv. Resour. Publ. 156. 305pp.

———, AND D. S. ROBSON. 1976. Models allowing for age-dependent survival rates for band-return data. Biometrics 32:305–323.

BURNHAM, K. P., AND D. R. ANDERSON. 1979. The composite dynamic method as evidence for age-specific waterfowl mortality. J. Wildl. Manage. 43:356–366.

CASWELL, H. 1989. Matrix population models. Sinauer Assoc., Inc., Sunderland, Mass. 328pp.

CAUGHLEY, G. 1966. Mortality patterns in mammals. Ecology 47:906–918.

———. 1977. Analysis of vertebrate populations. John Wiley & Sons, New York, N.Y. 234pp.

———, AND L. C. BIRCH. 1971. Rate of increase. J. Wildl. Manage. 35:658–663.

———, AND C. J. KREBS. 1983. Are big mammals simply little mammals writ large? Oecologia 59:7–17.

CHAPMAN, D. G., AND D. S. ROBSON. 1960. The analysis of a catch curve. Biometrics 16:354–368.

COLE, L. C. 1957. Sketches of general and comparative demography. Cold Spring Harbor Symp., Quant. Biol. 22:1–15.

COOKE, D., AND J. A. LEON. 1976. Stability of population growth determined by 2 × 2 Leslie matrix with density-dependent elements. Biometrics 32:435–442.

CORMACK, R. M. 1973. Commonsense estimates from capture-recapture studies. Pages 225–234 *in* M. S. Bartlett and R. W. Hiorns, eds. The mathematical theory of the dynamics of biological populations. Academic Press, New York, N.Y.

COWARDIN, L. M., AND R. J. BLOHM. 1992. Breeding population inventories and measures of recruitment. Pages 423–445 in B. D. J. Batt et al., eds. Ecology and management of breeding waterfowl. Univ. Minnesota Press, Minneapolis.

———, D. H. JOHNSON, T. L. SHAFFER, AND D. W. SPARLING. 1988. Application of a simulation model to decisions in mallard management. U.S. Fish Wildl. Serv. Tech. Rep. 17. 28pp.

CRÊTE, M., R. J. TAYLOR, AND P. A. JORDAN. 1981. Optimization of moose harvest in southwestern Quebec. J. Wildl. Manage. 45:598–611.

CROWE, D. M. 1975. A model for exploited bobcat populations in Wyoming. J. Wildl. Manage. 39:408–415.

DE ANGELIS, D. L. 1976. Application of stochastic models to a wildlife population. Math. Biosci. 31:227–236.

DOWNING, R. L. 1980. Vital statistics of animal populations. Pages 247–267 in S. D. Schemnitz, ed. Wildlife management techniques manual. Fourth ed. The Wildl. Soc., Washington, D.C.

EBERHARDT, L. L. 1969. Population analysis. Pages 457–495 in R. H. Giles, Jr., ed. Wildlife management techniques. Third ed. The Wildl. Soc., Washington, D.C.

———. 1970. Correlation, regression, and density dependence. Ecology 51:306–310.

———. 1985. Assessing the dynamics of wild populations. J. Wildl. Manage. 49:997–1012.

———. 1987. Population projections from simple models. J. Appl. Ecol. 24:103–118.

———. 1988. Using age structure data from changing populations. J. Appl. Ecol. 25:373–378.

FREDIN, R. A. 1984. Levels of maximum net productivity in populations of large terrestrial mammals. Pages 381–387 in W. F. Perrin, R. L. Brownell, Jr., and D. P. DeMaster, eds. Reports of the International Whaling Commission, Special Issue 6. Cambridge, U.K.

FRYXELL, J. M. 1986. Age-specific mortality: an alternative approach. Ecology 67:1687–1692.

GASTON, K. J., AND J. H. LAWTON. 1987. A test of statistical techniques for detecting density dependence in sequential censuses of animal populations. Oecologia (Berl.) 74:404–410.

GILPIN, M. E., AND F. J. AYALA. 1973. Global models of growth and competition. Proc. Natl. Acad. Sci. U.S.A. 70:3590–3593.

GRANT, W. E. 1986. Systems analysis and simulation in wildlife and fisheries science. John Wiley & Sons, New York, N.Y. 338pp.

HANSON, W. R. 1963. Calculation of productivity, survival, and abundance of selected vertebrates from sex and age ratios. Wildl. Monogr. 9. 60pp.

HEPP, G. R., R. T. HOPPE, AND R. A. KENNAMER. 1987. Population parameters and philopatry of breeding female wood ducks. J. Wildl. Manage. 51:401–404.

HESTBECK, J. B., J. D. NICHOLS, AND R. A. MALECKI. 1991. Estimates of movement and site fidelity using mark-resight data of wintering Canada geese. Ecology 72:523–533.

HOUSTON, D. B. 1982. The northern Yellowstone elk: ecology and management. Macmillan Publ. Co., New York, N.Y. 474pp.

HUTCHINSON, G. E. 1978. An introduction to population ecology. Yale Univ. Press, New Haven, Conn. 260pp.

JACKSON, C. H. N. 1939. The analysis of an animal population. J. Anim. Ecol. 8:238–246.

JENKINS, S. H. 1988. Use and abuse of demographic models of population growth. Bull. Ecol. Soc. Am. 69:201–207.

———. 1989. Comments on an inappropriate population model for feral burros. J. Mammal. 70:667–670.

JOHNSON, D. H. 1974. Estimating survival rates from banding of adult and juvenile birds. J. Wildl. Manage. 38:290–297.

———. 1979. Estimating nest success: the Mayfield method and an alternative. Auk 96:651–661.

———, J. D. NICHOLS, AND M. D. SCHWARTZ. 1992. Population dynamics of breeding waterfowl. Pages 446–485 in B. D. J. Batt et al., eds. Ecology and management of breeding waterfowl. Univ. Minnesota Press, Minneapolis.

———, AND T. L. SHAFFER. 1990. Estimating nest success: when Mayfield wins. Auk 107:595–600.

———, D. W. SPARLING, AND L. M. COWARDIN. 1987. A model of the productivity of the mallard duck. Ecol. Modelling 38:257–275.

KINGSLAND, S. E. 1985. Modeling nature: episodes in the history of population ecology. Univ. Chicago Press, Chicago, Ill. 267pp.

KINKEL, L. K. 1989. Lasting effects of wing tags on ring-billed gulls. Auk 106:619–624.

KIRKPATRICK, M. 1984. Demographic models based on size, not age, for organisms with indeterminate growth. Ecology 65:1874–1884.

KIRKPATRICK, R. L. 1980. Physiological indices in wildlife management. Pages 99–112 in S. D. Schemnitz, ed. Wildlife management techniques manual. Fourth ed. The Wildl. Soc., Washington, D.C.

KLETT, A. T., H. F. DUEBBERT, C. A. FAANES, AND K. F. HIGGINS. 1986. Techniques for studying nest success of ducks in upland habitats in the prairie pothole region. U.S. Fish Wildl. Serv. Resour. Publ. 158. 24pp.

KNIGHT, R. R., AND L. L. EBERHARDT. 1985. Population dynamics of Yellowstone grizzly bears. Ecology 66:323–334.

KNOWLTON, F. F. 1972. Preliminary interpretations of coyote population mechanics with some management implications. J. Wildl. Manage. 36:369–382.

KREBS, C. J. 1985. Ecology: the experimental analysis of distribution and abundance. Third ed. Harper & Row Publ., New York, N.Y. 800pp.

LANCIA, R. A., AND J. W. BISHIR. 1985. Mortality rates of beaver in Newfoundland—a comment. J. Wildl. Manage. 49:879–881.

LEFKOVITCH, L. P. 1965. The study of population growth in organisms grouped by stages. Biometrics 21:1–18.

LESLIE, P. H. 1945. On the use of matrices in certain population mathematics. Biometrika 33:183–212.

———. 1948. Some further notes on the use of matrices in population mathematics. Biometrika 35:213–245.

———. 1959. The properties of a certain lag type of population growth and the influence of an external random factor on a number of such populations. Physiol. Zool. 32:151–159.

LEVINS, R. 1966. The strategy of model building in population biology. Am. Sci. 54:421–431.

———. 1968. Evolution in changing environments. Princeton Univ. Press, Princeton, N.J. 120pp.

LEWIS, E. G. 1942. On the generation and growth of a population. Sankhya 6:93–96.

LOERY, G., K. H. POLLOCK, J. D. NICHOLS, AND J. E. HINES. 1987. Age-specificity of black-capped chickadee survival rates: analysis of capture-recapture data. Ecology 68:1038–1044.

LOTKA, A. J. 1925. Elements of physical biology. Williams and Wilkins, Baltimore, Md. 460pp.

MACARTHUR, R. H. 1968. The theory of the niche. Pages 159–176 in R. C. Lewontin, ed. Population biology and evolution. Syracuse Univ. Press, Syracuse, N.Y.

———. 1972. Geographical ecology: patterns in the distribution of species. Harper & Row, New York, N.Y. 269pp.

MAELZER, D. A. 1970. The regression of log N_{n+1} on log N_n as a test of density dependence: an exercise with computer-constructed density-independent populations. Ecology 51:810–822.

MANLY, B. F. J. 1985. The statistics of natural selection on animal populations. Chapman and Hall, New York, N.Y. 484pp.

MARTIN, F. W., R. S. POSPAHALA, AND J. D. NICHOLS. 1979. Assessment and population management of North American migratory birds. Pages 187–239 in J. Cairns, Jr., G. P. Patil, and W. E. Waters, eds. Environmental biomonitoring, assessment, prediction, and management—certain case studies and related quantitative issues. Int. Coop. Publ. House, Fairland, Md.

MAY, R. M. 1973. Stability and complexity in model ecosystems. Princeton Univ. Press, Princeton, N.J. 235pp.

———. 1974. Biological populations with nonoverlapping generations: stable points, stable cycles, and chaos. Science 186:645–647.

———, AND G. F. OSTER. 1976. Bifurcations and dynamic complexity in simple ecological models. Am. Nat. 110:573–599.

MAYFIELD, H. 1961. Nesting success calculated from exposure. Wilson Bull. 73:255–261.

MAYNARD SMITH, J. 1974. Models in ecology. Cambridge Univ. Press, New York, N.Y. 146pp.

MCCULLOUGH, D. R. 1979. The George Reserve deer herd. Univ. Michigan Press, Ann Arbor. 271pp.

———. 1982. Population growth rate of the George Reserve deer herd. J. Wildl. Manage. 46:1079–1083.

———. 1983. Rate of increase of white-tailed deer on the George Reserve: a response. J. Wildl. Manage. 47:1248–1250.

MERTZ, D. B. 1970. Notes on methods used in life-history studies. Pages 4–17 in J. H. Connell, D. B. Mertz, and W. W. Murdoch, eds. Readings in ecology and ecological genetics. Harper and Row, New York, N.Y.

MICHOD, R. E., AND W. W. ANDERSON. 1980. On calculating demographic parameters from age frequency data. Ecology 61:265–269.

MOLINI, J. J., R. A. LANCIA, J. BISHIR, AND H. E. HODGDON. 1981. A stochastic model of beaver population growth. Pages 1215–1245 *in* J. A. Chapman and D. Pursley, eds. Vol. 3. Worldwide Furbearer Conf., Frostburg, Md.

NEDELMAN, J., J. A. THOMPSON, AND R. J. TAYLOR. 1987. The statistical demography of whooping cranes. Ecology 68:1401–1411.

NICHOLS, J. D., AND K. H. POLLOCK. 1990. Estimation of recruitment from immigration versus in situ reproduction using Pollock's robust design. Ecology 71:21–26.

NORTH, P. M. 1985. A computer modelling study of the population dynamics of the screech owl (*Otus asio*). Ecol. Modelling 30:105–143.

PAULIK, G. J., AND D. S. ROBSON. 1969. Statistical calculations for change-in-ratio estimators of population parameters. J. Wildl. Manage. 33:1–27.

PEARL, R. 1928. The rate of living. Alfred A. Knopf, Inc., New York, N.Y. 185pp.

PENNYCUICK, C. J., R. M. COMPTON, AND L. BECKINGHAM. 1968. A computer model for simulating the growth of a population, or of two interacting populations. J. Theor. Biol. 18:316–329.

PIANKA, E. R. 1974. Evolutionary ecology. Harper & Row Publ., New York, N.Y. 356pp.

PIELOU, E. C. 1969. An introduction to mathematical ecology. John Wiley & Sons, Inc., New York, N.Y. 286pp.

POLACHECK, T. 1985. The sampling distribution of age-specific survival estimates from an age distribution. J. Wildl. Manage. 49:180–184.

POLLARD, E., K. H. LAKHANI, AND P. ROTHERY. 1987. The detection of density-dependence from a series of annual censuses. Ecology 68:2046–2055.

POLLOCK, K. H., J. D. NICHOLS, C. BROWNIE, AND J. E. HINES. 1990. Statistical inference for capture-recapture experiments. Wildl. Monogr. 107. 97pp.

POOLE, R. W. 1974. An introduction to quantitative ecology. McGraw-Hill Inc., New York, N.Y. 532pp.

POWELL, R. A. 1979. Fishers, population models, and trapping. Wildl. Soc. Bull. 7:149–154.

RICKLEFS, R. E. 1979. Ecology. Second ed. Chiron Press, New York, N.Y. 966pp.

ROBSON, D. S., AND D. G. CHAPMAN. 1961. Catch curves and mortality rates. Trans. Am. Fish. Soc. 90:181–189.

ROSEBERRY, J. L. 1979. Bobwhite population responses to exploitation: real and simulated. J. Wildl. Manage. 43:285–305.

SAS INSTITUTE, INC. 1987. SAS/STAT® guide for personal computers. Version 6 ed. SAS Institute, Inc., Cary, N.C. 1028pp.

SAUER, J. R., AND N. A. SLADE. 1987. Size-based demography of vertebrates. Annu. Rev. Ecol. Syst. 18:71–90.

SEBER, G. A. F. 1970. Estimating time-specific survival and reporting rates for adult birds from band returns. Biometrika 57:313–318.

———. 1982. The estimation of animal abundance and related parameters. Second ed. Macmillan Publ. Co., Inc., New York, N.Y. 600pp.

———. 1986. A review of estimating animal abundance. Biometrics 42:267–292.

SILER, W. 1979. A competing-risk model for animal mortality. Ecology 60:750–757.

SLADE, N. A. 1977. Statistical detection of density dependence from a series of sequential censuses. Ecology 58:1094–1102.

SOLOW, A. R. 1990. Testing for density dependence: a cautionary note. Oecologia 83:47–49.

ST. AMANT, J. L. S. 1970. The detection of regulation in animal populations. Ecology 51:823–828.

STARFIELD, A. M., AND A. L. BLELOCH. 1986. Building models for conservation and wildlife management. Macmillan Publ. Co., New York, N.Y. 253pp.

STOUDT, J. H. 1971. Ecological factors affecting waterfowl production in the Saskatchewan parklands. U.S. Fish Wildl. Serv. Resour. Publ. 99. 58pp.

STRONG, D. R. 1986. Density vagueness: abiding the variance in the demography of real populations. Pages 257–268 *in* J. Diamond and T. J. Case, eds. Community ecology. Harper & Row, New York, N.Y.

TAIT, D. E. N., AND R. L. BUNNELL. 1980. Estimating rate of increase from age at death. J. Wildl. Manage. 44:296–299.

TANNER, J. T. 1975. The stability and the intrinsic growth rates of prey and predator populations. Ecology 56:855–867.

TAYLOR, M., AND J. S. CARLEY. 1988. Life table analysis of age structured populations in seasonal environments. J. Wildl. Manage. 52:366–373.

USHER, M. B. 1972. Developments in the Leslie matrix model. Pages 29–60 *in* J. N. R. Jeffers, ed. Mathematical models in ecology. Blackwell Sci. Publ., Oxford, U.K.

VAN BALLENBERGHE, V. 1983. Rate of increase of white-tailed deer on the George Reserve: a re-evaluation. J. Wildl. Manage. 47:1245–1247.

VANDERMEER, J. H. 1972. Niche theory. Annu. Rev. Ecol. Syst. 3:107–132.

VAN GROENENDAEL, J., H. DE KROON, AND H. CASWELL. 1988. Projection matrices in population biology. Trends Ecol. Evol. 3:264–269.

VOLTERRA, V. 1926. Fluctuations in the abundance of a species considered mathematically. Nature 118:558–560.

WALTERS, C. J. 1986. Adaptive management of renewable resources. Macmillan Publ. Co., New York, N.Y. 374pp.

———, AND J. E. GROSS. 1972. Development of big game management plans through simulation modeling. J. Wildl. Manage. 36:119–128.

WETHEY, D. S. 1985. Catastrophe, extinction, and species diversity: a rocky intertidal example. Ecology 66:445–456.

WILLIAMSON, M. H. 1959. Some extensions of the use of matrices in population theory. Bull. Math. Biophysics 21:13–17.

WILSON, E. O., AND W. H. BOSSERT. 1971. A primer of population biology. Sinauer Assoc. Inc., Sunderland, Mass. 192pp.

WILSON, K. R., J. D. NICHOLS, AND J. E. HINES. 1989. A computer program for sample size computations for banding studies. U.S. Fish Wildl. Serv. Fish Wildl. Tech. Rep. 23. 19pp.

ZENG, Z., AND J. H. BROWN. 1987. A method for distinguishing dispersal from death in mark-recapture studies. J. Mammal. 68:656–665.

17

HARVEST MANAGEMENT

M. Dale Strickland, Harold J. Harju, Keith R. McCaffery, Harvey W. Miller, Loren M. Smith, and Robert J. Stoll

INTRODUCTION

Sport harvest has been the method of choice for controlling populations of game animals by managers since modern wildlife management began with the "Doctrine of Wise Use" advanced by Theodore Roosevelt and Gifford Pinchot in 1910. In the early days of wildlife management, the primary goal of harvest management was to prevent overharvest (Leopold 1933). As populations of big game increased, the goal became the control of populations to maintain them within habitat carrying capacity. Caughley (1977) listed three problems with which population managers must deal:

(1) treatment of a small or declining population to raise its density;
(2) exploitation of a population to take from it a sustained yield; and
(3) treatment of a population that is too dense, or which has an unacceptably high rate of increase, to stabilize or to reduce its density.

He labelled these problems as conservation, sustained yield management, and control, respectively. Examples of these three approaches are contained in (a) the conservative management of endangered species, (b) hunting strategies in deer management, and (c) the reduction of deer to well below carrying capacity in localized areas to reduce crop damage. Another important component of wildlife management is the public's desire for more or less wildlife.

Sport harvest produces economic benefits resulting from license sales. Money from the sale of hunting licenses supports game species management programs as well as other programs such as endangered and threatened species, nongame, education, and habitat acquisition and management. Harvest programs result in commercial development in nearby communities, including support facilities such as motels, restaurants, service stations, grocery stores, and sporting goods stores. Often landowners benefit from income from hunters. Some businesses, including outfitters, taxidermists, and some meat processors, have an almost complete dependence on hunters.

Subsistence harvest was important historically, but presently it is restricted primarily to Native Americans and Alaskans. Although subsistence hunting is important only locally, particularly in financially depressed communities, its effect on management of individual populations can be significant.

Harvest can be important in behavior modification of wildlife. Selective harvest has been used in managing big game to reduce damage to private lands. It also has been used in management of predator populations such as mountain lions and black bears. One of the primary justifications for harvesting grizzly bears in the Northern Continental Divide Ecosystem in Montana is for behavior modification of surviving individuals and the taking of bears that otherwise would have been removed by nuisance control actions (Montana Department of Fish, Wildlife, and Parks, Final programmatic environmental impact statement—The grizzly bear in northwestern Montana, Mont. Dep. Fish, Wildl. Parks, Helena, 1986). However, for harvest to have the desired effect, the animal that is harvested should be exhibiting the undesirable behavior such as damaging crops or preying on domestic sheep. Sometimes behavior modification can have an undesirable effect. Hunter harvest of elk migrating from Yellowstone National Park and the adjacent Teton Wilderness in Wyoming apparently has altered their migration routes, contributing to poor elk hunting on the adjacent national forests (Boyce 1989).

Finally, harvest programs result in political benefits. These benefits include political support for management programs, wildlife management agencies, habitat protection, land use planning, and multiple use on public lands.

MANAGEMENT OF BIG-GAME HARVESTS

Virtually all big-game animals fall into two taxonomic orders, Carnivora (bears and cats) and Artiodactyla (pigs, bovines, pronghorns, and cervids). They are relatively large animals that often rank high in hunter/public visibility and interest.

Harvest management for big game is the art of melding the objectivity of wildlife science and the subjectivity of public desires for the attainment of a management goal. The management goal should have broad public acceptance and provide a qualitative statement of a general management direction. An example of such a goal for deer could be: ''To provide a deer population that provides maximum recreational opportunity within the context of minimizing conflicts with humans and damage to native vegetation.'' A well-founded management goal is most important because it is what drives the management program. Harvest management is but one technique for accomplishing this goal. Habitat management, law enforcement, and education are other commonly integrated techniques.

Harvest regulations for a given species of big game often vary considerably among states. This can be explained, at least in part, by the subjective aspects of the harvest management process. In spite of this variation, some techniques are common to the management process, including the development of a harvest objective, management strategy, establishment of hunting regulations, and evaluation. Some states include the above management techniques in a long-range, comprehensive planning process, whereas other states respond to specific management situations as they arise. Whether the management process is guided by a long-range process or an expedient response to today's problems, it should be focused on an overall management goal.

The thrust of the following section will be to provide a general description of the basic techniques used to meet big-game harvest management goals and objectives. This section does not address the management of all big-game species of North America, but draws examples from well-tested harvest management programs that can be adapted and applied to specific situations and other species.

Planning is an integral first step in development of a harvest management program that meets management goals and objectives. The planning process relative to wildlife management was described adequately by Crowe (Comprehensive planning for wildlife resources, Wyo. Game and Fish Dep., Cheyenne, 1983) and the U.S. Fish and Wildlife Service (1973). The planning process under ideal conditions includes:

(1) Determining the status of the resource/population (Where are you?).

(2) Defining the goals and objectives of the management program (Where do you want to be?).
(3) Establishing management strategies to attain objectives (How do you get there?).
(4) Determining how closely the applied management strategy achieved the resource/population objectives (Did you get there?).

The harvest management strategy is the logical process to develop a hunting regulation. It has three basic components: (1) inventorying or determining population abundance, (2) identifying population and recreational goals and objectives, and (3) developing harvest management regulations directed at meeting the population and recreational goals and objectives.

Inventory

The first step in the harvest management strategy is inventory. Inventory includes the identification of management units and the determination of population status within the management units.

MANAGEMENT UNITS

The establishment of management units is perhaps the most important step in setting up a good management system. They form the biological and managerial subunits necessary to attain the overall management goal. Management units may be biological (e.g., drainages and watersheds that reflect more or less discrete animal populations), political (e.g., counties), or a combination (e.g., a grouping of counties that closely match a biological unit).

Biological units typically are used in western states where land form and climate produce discrete populations of big game (Fig. 1). Most inventories of big game are designed to estimate population size, taking into account measures of reproductive and mortality rates. However, these parameters are meaningful only if ingress and egress are quantified, considered negligible, or, in some circumstances, offsetting (i.e., equal). The most practical units are those designed to eliminate most ingress and egress. Intensive studies of seasonal distribution and migration patterns may be necessary in establishing good units.

Political units are more typical of eastern states where wildlife populations are more evenly distributed, migration is rare, and habitat and population dynamics are similar. Under these biological circumstances, the human familiarity associated with political units makes them desirable from the standpoint of inventory, identification of population objectives, administration of hunting regulations, and enforcement. Lang and Wood (1976) described a method used in Pennsylvania white-tailed deer management in which counties containing deer with similar biological characteristics were combined. The overriding consideration in any management unit scheme is the significance of unit boundaries to reproductive and mortality parameters.

The management unit, whether biological, political, or both, should be large enough to encompass habitat to support a viable population, facilitate the long-term collection of data with the needed precision, allow for determination of population objectives, and be recognized easily by hunters. For white-tailed deer, Wisconsin employed road boundaries to designate 96 management units based large-

JONES CREEK
ELK HERD

POPULATION OBJECTIVE — 4,200

LEGEND

⊠ NATIONAL FOREST

▨ BUREAU OF LAND MANAGEMENT

▥ STATE LAND

☐ PRIVATE LAND

◪ CRUCIAL WINTER RANGE

S SUMMER RANGE

S/F SPRING/FALL RANGE

W WINTER RANGE

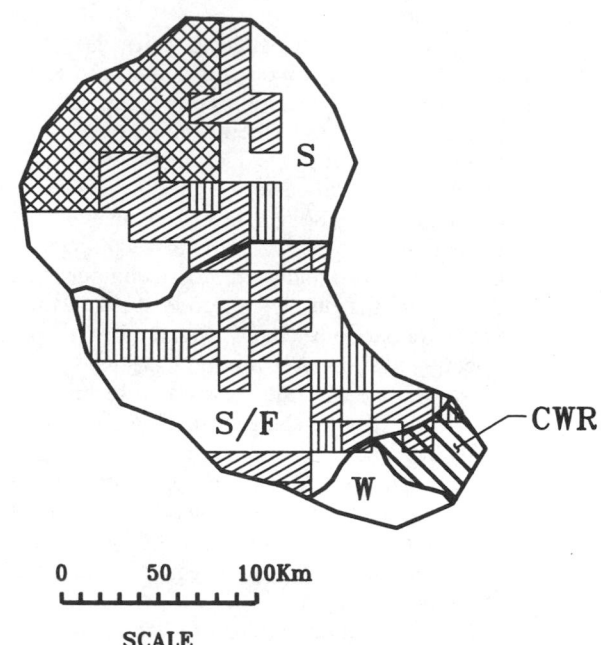

Fig. 1. Elk management unit in Wyoming showing seasonal range and land ownership.

ly on land use and averaging 1,500 km² (Creed et al. 1984), and Ohio used 88 county management units averaging 1,200 km² (Stoll and Mountz 1983). Practical constraints like tradition, historical data (e.g., harvest and hunting pressure information) related to animal populations, and agency budgets exert "real world" influence.

Landowners, hunters, and managers usually desire management units that are smaller than practical. When small units that do not support discrete populations are used, attainment of management objectives is difficult. Nevertheless, the concept of herd or population units is difficult to communicate to the public. And, landowners, hunters, and managers commonly desire more animals on public land, where access for hunting is free and relatively easy, and fewer animals on private land, where access for sportsmen may be difficult and expensive. Also, large numbers of ungulates on private land may compete for forage with domestic animals or consume valuable crops. Unfortunately, winter range often is the limiting resource for western big game and, more often than not, is on private land. In the eastern U.S., however, public land holdings often are small and fragmented and do not support discrete populations. If herd units are based on land ownership and not biological or managerial parameters, management objectives to increase animals on public land and decrease them on private land will be mutually exclusive. Normally, in this instance no objective is achieved and too few animals are on public land (angry hunters) and too many animals are on private lands (angry landowners).

Recently, the desire to privatize big-game management has increased as landowners realize the potential value of big game for fee hunting. Landowners wanting to capitalize on hunting often want fewer animals but more large bucks/bulls. Effective management of big game on landowner units is possible only with nonmigratory herds or where large land holdings provide year-round habitat. The latter situation is rare and makes the identification of logical management units all the more important when an agency tries to sell a management program to the public.

Some states have used a tiered unit system in which a management unit may contain one or more hunt units. The management unit contains a biologically discrete group of animals. At this level population parameters are measured and management objectives for population characteristics are established. Harvest objectives are established for the population and allocated to the hunt units based on animal distribution or crop damage complaints.

Management unit boundaries must be recognizable to the hunter. This is particularly important because harvest, the most important measure of human-caused mortality, generally is determined through surveys of hunters. Unit boundaries that are easily recognized on maps and in the field facilitate accurate harvest and hunting pressure determinations.

DETERMINING POPULATION STATUS

Once discrete management units are defined, the status of the population must be determined. The most powerful negotiating tool a manager has is the ability to provide the public with a thorough and objective analysis of data about wildlife populations. This analysis establishes the manager as an authority on the population and provides the public with information on which to base an opinion.

Procedures used to assess animal abundance include indexes, direct counts, and population reconstruction (see Chapter 9 for methodologies).

Generally, the techniques that agencies use to assess population status have evolved from compromise among management needs for precision, budget restrictions, and personnel availability. Assessment techniques vary widely among states. For example, Gladfelter (1980) listed 11 techniques used by midwestern states to assess white-tailed deer abundance. Some states rely heavily on one technique, but many use a variety of techniques to increase confidence. Most states employ techniques that: (1) administrators and the public can understand, (2) provide year-to-year comparisons (trends), and (3) can provide status projections to the next hunting season.

Even if not dictated by the assessment techniques employed, measures of reproductive performance and mortality usually are obtained because they are fundamental to population dynamics. Measures of reproduction and mortality provide data for status projections to next season and are required input data in population reconstruction techniques. These data are also of interest to sportsmen and other publics.

Estimates of reproductive performance can be obtained from fetal counts (e.g., Harder 1980), direct observations of young per adult female, and previous year's young per adult female in the harvest. In some western and northern states where climate (e.g., severe winters, drought) and other factors (e.g., disease outbreaks) may cause direct mortality or severely affect maternal condition because of poor habitat or lack of habitat accessibility, net reproductive performance must be estimated annually. In other areas where such factors are less variable, periodic estimates may be sufficient. Estimates of reproductive performance should measure as accurately as possible the current year's reproductive success and also make comparisons with previous years. The inventory should be designed to sample the entire reproductive segment of the population with standard, repeatable methodology. Thus, need for year-to-year comparisons makes sampling consistency extremely important.

Total mortality of adult males and females can be determined by sampling the sex and age composition of harvested animals if there is minimal differential selection or vulnerability among adults and sample sizes are sufficiently large and unbiased. The sampling effort should cover the entire geographic range and period of harvest. Mortality then can be separated into natural and human caused. Caution should be exercised in interpreting age composition of harvest data collected at check stations, roadside field checks, and voluntary reporting, which sample only a portion of the animals harvested.

Natural mortality is difficult to measure, although estimates are possible. Wisconsin estimates nonharvest mortality to range by unit from 10% to more than 40% of the total adult buck mortality, depending on harvest exploitation and climate (Wisconsin Department of Natural Resources, Management workbook for white-tailed deer, Wisconsin Dep. Nat. Resour., Madison, 1989). Research projects designed to estimate mortality under normal and extraordinary conditions are important but can be time consuming and expensive (e.g., Porath 1980). Subsequent recovery of radio-tagged animals provides the most precise measure of nonharvest mortalities but is prohibitively expensive to replicate on management units. Robinette et al. (1977) preferred belt- (strip-) transect surveys for estimating density of deer carcasses. These surveys habitually fail to detect 100% of the animals present in the strip (Anderson and Pospahala 1970). Line-transect surveys are the predominant tool for direct estimates of deer mortality (Connolly 1981). Theory and applications of line-transect sampling were reviewed comprehensively by Burnham et al. (1980).

For population inventory, the most important mortality parameters to measure when a major mortality event occurs are the timing, the number of animals or the portion of the population that dies, and the sex and age characteristics of dying animals. Although accurate estimates of each of these parameters may be impossible, it is important to estimate the effects of events such as severe winters and drought. As an example, Creed et al. (1984) described the use of a winter severity index to predict winter deer mortality, subsequent fawn production, and subsequent buck harvest.

Human-caused mortality usually is measured more easily than natural mortality. As with natural mortality, however, such losses from poaching (e.g., Beattie et al. 1980) and road-kills, and of unretrieved, wounded animals that later die, are difficult to quantify. Even so, estimating the magnitude of significant events is important. Some of these nonharvest mortalities, including road-kills and predation, may be fairly uniform or proportional from year to year, so annual measurements may be unimportant. In fact, many states use road-killed deer as an index to population trend (McCaffery 1973, Gladfelter 1980).

Legal harvest of big game is perhaps the most valuable human-caused mortality data available and the easiest to collect. Most states employ (1) some form of mandatory registration such as a report card or check stations, (2) a mail or telephone survey, or both, to estimate hunter harvest of big game (e.g., Aney 1974, Ryel 1980).

Mandatory registration requires hunters to register their animal(s) at a designated check station for official tagging, often within a designated time period. Mandatory registration provides a rapid means for obtaining an estimate of the legal harvest, location by management unit, and estimates of various population characteristics like recruitment, sex/age composition, and condition. This approach is a relatively inexpensive and traditional method for gathering harvest data (Aney 1974). Wisconsin spends $60,000–$100,000/year to register (i.e., physically tag) 125,000–300,000 animals but believes the public relations, public confidence, and precision of the information is worth the cost (F. P. Haberland, unpubl. data, Wisconsin Dep. Nat. Resour., Madison). Normally, mandatory registration does not provide estimates of local hunting pressure and success rates. Additionally, it provides a minimum count of harvest with no estimate of precision because of hunter noncompliance.

Potential disadvantages of the mandatory reporting system include a low return rate for mailed report cards, nonresponse and response biases, and difficulty in determining the precision of harvest estimates derived from partial reporting. The most accurate way to obtain hunter activity and harvest data is a total accounting. Anything less than a 100% compliance results in estimates based on a portion of the target population. The degree to which these data are required depends on the desired precision for a given species, in a given season, in a given area, and at a specified level of management.

Mail or telephone surveys of a sample of hunters provide a much slower means for obtaining an estimate of harvest by management unit, but they also provide estimates of hunting pressure (total hunters, days spent hunting) and success. Normally, such surveys do not provide reliable information on the age/sex composition and condition of animals in the harvest. The accuracy of questionnaire surveys is affected by various response/nonresponse biases that can be measured and taken into account when harvest is estimated. Typically, at least three biases are encountered in mail surveys that influence harvest estimates significantly: (1) sportsmen who do not hunt are less likely to respond to surveys, (2) unsuccessful hunters are less likely to respond than successful hunters, and (3) prestige bias causes respondents to inflate their success. These biases all tend to result in overestimates of harvest. The best method for harvest inventory, hunter success, and hunting pressure is a combination of mandatory registration and a well-designed hunter survey.

Harvest statistics also provide insight into changes in big-game abundance and can be used to verify the results of population reconstruction techniques. Statistics on hunting effort and success also can provide an index to population abundance. Buck/bull harvest trends, alone, often provide a valuable population index if a similar amount of hunting effort occurs each year. For example, in a declining population, harvesting an animal is generally more difficult, so hunter success (harvest per hunter or per days afield) and harvest should decrease. Conversely, an increasing population should result in greater harvest with a stable or increasing hunter success. These relationships generally hold, but factors such as weather, access to hunting units, and changing regulations can confuse the results and must be considered. Thus, there is virtue in seeking to employ a similar season framework (timing, length) from year to year.

The age and sex of the harvest may not be indicative of the actual population because these data often are affected by hunter behavior and weather (Downing 1981). Hunters permitted by regulation to take animals of any age or sex may select against young of the year, select for or against yearling males, select for larger adult males, and select females in the same proportion as they occur by age class in the population (Roseberry and Woolf 1988). However, in heavy cover such as in white-tailed deer habitat, or when seasons are short (e.g., <2 weeks in many whitetail regions) or hunting pressure is high, hunter selectivity may be less pronounced. In analyzing data for Pennsylvania deer herds, Lang and Wood (1976) assumed no hunter selection of one legal buck over another. This assumption probably is true also for hunters harvesting females, although there may be some selection for adult does with fawns when hunters are looking for a size comparison to avoid harvesting a fawn. The sex and age structure of the harvest and any inherent biases should be known, because harvest mortality can influence the population's subsequent growth rate and sex/age composition.

Computer modeling systems (e.g., Gross et al. 1973,

Moen et al. 1986) have produced population models that force managers to organize their data and discourage them from making management decisions based only on indicators (i.e., one or two data sets). Wildlife managers often use a favorite indicator to establish annual harvest regulations. These indicators can be meaningful but also can mislead the manager. Downing (1981) demonstrated that the number of deer of each sex in the harvest was an important parameter needed for management, whereas the often-used sex ratio of the harvest was not.

Eberhardt and Simmons (1987) pointed out that wildlife managers usually rely on an index of abundance supplemented by a few surveys. They suggested that, if indices are to be used, a great deal of effort should be expended to ensure that the indices accurately reflect changes in the population. They proposed the use of double sampling (see Chapters 1 and 2) as a way to verify indices.

Population modeling is a useful means for converting existing data and indices to estimates of abundance of big-game populations. A commonly used population model is ONEPOP, developed by Gross et al. (1973). ONEPOP and the various generations that followed (e.g., POP-II) are relatively simple deterministic models that do not include random impact variables. More sophisticated models exist but lack a key ingredient necessary to make them useful to managers; that is, managers and biologists need to be involved in model development (Eberhardt 1988) if they are to be expected to use them. ONEPOP models use commonly collected data on mortality and reproduction as either input variables or as known variables used in model verification. Gasson and Wollrab (1986) described the modeling process used for Wyoming pronghorn herds based on the POP-II model.

The usefulness of a model for making management decisions obviously varies with quality of input. Strickland (1982) described this process for an elk herd and demonstrated how observed postseason bull : cow ratios in the herd usually were lower than actual ratios, whereas observed preseason bull : cow ratios more accurately reflected the actual population characteristic. Good models have allowed better population estimates and management, and poor models have stimulated more and better data collection. Models also have allowed managers to simulate the effects of various harvest management options without actually applying them to populations.

The management of the Wyoming Range mule deer herd in western Wyoming offers an example of testing harvest options with a population model (Table 1). The winter of 1983–84 was severe, resulting in significant winter mortality in this herd and subsequent conservative hunter harvests. The conservative management implemented by the Department for the period 1985–89, although popular with sportsmen, resulted in a 1989 posthunt population estimate of 52,570 compared to a population objective of 38,000 (Wyoming Game and Fish Department, Annual big game herd unit reports—District I, Wyo. Game and Fish Dep., Cheyenne, 1984). The Department wished to reduce deer numbers to the objective of 38,000 by 1994. Various harvest strategies to accomplish this reduction were simulated with the POP-II population model (Fossil Creek Software 1986). The model predicted that a continuation of the conservative harvest strategy used in 1989 (Strategy A) would result in 58,860

Table 1. A computer simulation of three harvest strategies on the Wyoming Range mule deer herd in western Wyoming for the period 1979–94. Harvests from 1979 through 1989 in all three strategies are actual. Harvest strategy A represents a continuation of harvests at the 1989 level, strategy B represents a model-generated harvest for a 1990–94 period necessary to meet a winter population objective of 38,000, and strategy C represents the outcome from the actual harvest obtained in 1990 and a continuation of this harvest through 1994.

Year	Strategy A		Strategy B		Strategy C	
	Antlerless harvest	Winter pop.	Antlerless harvest	Winter pop.	Antlerless harvest	Winter pop.
1979	811	27,661	811	27,661	811	27,661
1980	1,195	33,164	1,195	33,164	1,195	33,164
1981	416	40,138	416	40,138	416	40,138
1982	929	35,499	929	35,499	929	35,499
1983	925	38,141	925	38,141	925	38,141
1984	266	23,499	266	23,499	266	23,499
1985	11	27,719	11	27,719	11	27,719
1986	0	31,646	0	31,646	0	31,646
1987	815	42,178	815	42,178	815	42,178
1988	1,078	56,755	1,078	56,755	1,078	56,755
1989	3,106	52,570	3,106	52,570	3,106	52,570
1990	3,106	58,860	6,092	49,671	3,634	57,184
1991	3,106	56,069	2,570	46,344	3,634	52,438
1992	3,106	55,596	4,427	43,837	3,634	49,732
1993	3,106	53,693	3,605	40,920	3,634	45,475
1994	3,106	51,268	3,450	38,000	3,634	40,550

wintering deer in 1990 and 51,268 deer by 1994. The model suggested a more liberal strategy (B), which called for a harvest of 6,092 antlerless and 7,970 antlered deer for 1990. Based on this analysis, the hunting season was liberalized by increasing season length and the number of antlerless permits. These changes resulted in a harvest in 1990 of 3,634 antlerless and 4,648 antlered deer. Although harvest was below desired levels, especially for antlerless deer, the 1990 hunting season reduced winter deer numbers to 57,184 in 1990, and a continuation of this harvest would reduce the population to just over 40,500 by 1994 (Strategy C).

Eberhardt (1988) challenged managers to become more sophisticated modelers by using experiments with proper controls. He also correctly pointed out that wildlife managers are handicapped by administrators and sportsmen who expect the manager's best immediate efforts be directed to management. It is incumbent on the research community to develop better models that are practical for managers to use and that have immediate application. As one example, Lancia et al. (1988) offered a harvesting strategy for white-tailed deer based on the development of a population model and a two-stage, change-in-ratio population estimation technique. McCullough et al. (1990) suggested an alternative strategy referred to as linked sex harvest strategy (LSHS) for use when accurate estimates of population size are absent. LSHS was proposed as a method to maximize harvest for a population at carrying capacity by manipulating the size of harvests of each sex in relationship to the other.

Establishing Population and Harvest Management Objectives

With inventories in place, the next step in the management strategy is establishing population goals (a qualita-

tive statement of management direction) and then population and harvest management objectives (quantitative targets) to meet those goals for each discrete management unit. Population objectives may be total number or density. Harvest management objectives (e.g., desired harvest level and recreational opportunities) are constrained by the population objective for each management unit. Establishing a well-defined population objective is essential to providing direction to the harvest management program as well as any habitat management programs. Without this direction, neither the manager nor the public knows where the program is headed, and harvest management decisions will lack the logic necessary for public acceptance.

ESTABLISHING POPULATION OBJECTIVES

The population objective for a given management unit should be expressed and should correspond to the measure(s) of animal abundance previously selected during the inventory process. This will enable the manager to assess population status relative to the population objective for each management unit. Generally, population objectives for a management unit will be determined as a function of biological (carrying capacity) and sociological (human tolerance) considerations.

Biological Carrying Capacity

It is generally accepted that big-game populations are K (carrying capacity) selected species; that is, a population will increase until it nears the biological carrying capacity (BCC) of its environment (McCullough 1979, 1984). This form of population growth is referred to as the logistic growth curve (Fig. 2). The optimum strategy for maximum harvest is to maintain the population near the midpoint (56% of K) of the growth curve. Sustained yield is lower at all other points on the growth curve.

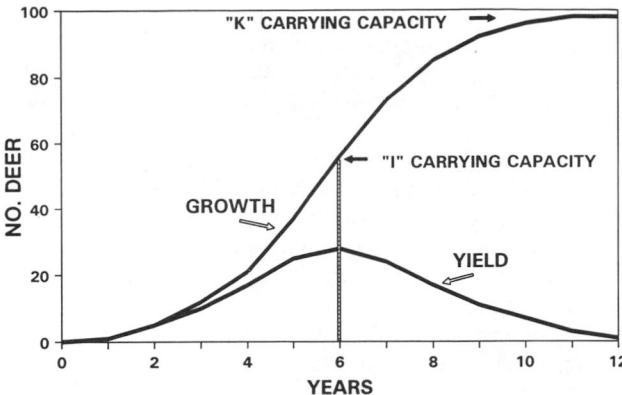

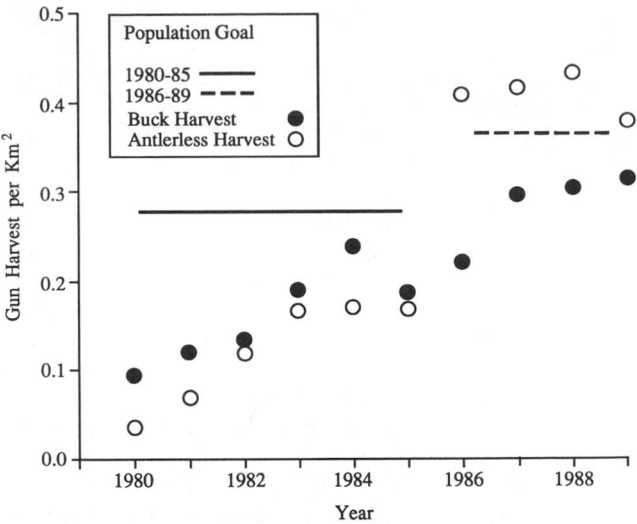

Fig. 2. Population growth and yield curves showing maximum biological carrying capacity (K) and point of maximum sustained yield (I). Growth curve (deer density) is from McCullough (1979) and yield curve is from Downing and Guynn (1983).

Fig. 3. Comparison of population index (solid circles) with goal (solid and dashed lines) for white-tailed deer in Pike County, Ohio. Goals for 1980–85 and 1986–90 update were derived from surveys of rural landowners (Stoll and Mountz 1983). Antlerless harvests (open circles) were increased as the index approached goals. Supplemental index information (not shown) included deer-vehicle accidents and population reconstruction estimates. Gun harvest estimates are from mandatory hunter registration.

It is not practical or possible to measure BCC annually. Annual BCC is influenced by several natural factors (especially weather) and human-caused influences and therefore is a moving target. Thus, BCC has been referred to as a "slippery shibboleth" (Macnab 1985). BCC can take on meaning for goal-setting only in a long-term context apart from the vagaries of short-term perturbations. In the long term, BCC is determined by two factors, climate and habitat. In forested environments neither factor changes dramatically in the short-term, and population objectives should reflect the average climate and habitat condition by management unit. In areas such as the Rocky Mountain West, where weather extremes result in a wide range in BCC, managing for the average may not be acceptable to the manager or the public. To do so means that much of the available habitat is underutilized most years. However, managing near maximum BCC means large-scale die-offs when severe weather (drought or cold) occurs.

McCullough (1979:93, 113) suggested estimating BCC by regressing a 10- to 20-year history of net recruitment (postharvest to preharvest) against winter population estimates. The estimate of BCC (K) by this method is where the regression line intersects with zero net increase. BCC for white-tailed deer also can be inferred from the incidence of fawn breeding (McCullough 1979:61, Downing and Guynn 1983), which is rare when herds exceed 60% of BCC. Yearling antler development on white-tailed deer also has been used as an index to BCC (Moen et al. 1986, D. R. Voigt [unpubl. rep., Ont. Min. Nat. Resour., Maple, 1989]). Vermont has tied its deer population objectives to measures of yearling antler beam diameter (Regan and Darling, Draft deer management plan for the state of Vermont, 1990–1995, Vt. Fish Wildl. Dep., Waterbury, 1989).

Precise estimates of BCC are rarely if ever available. Intuitive judgement and interpretation of long-term harvest data, population indexes, or physiological indexes may be necessary. Environmental factors (e.g., forage estimates) also can be used when goals are set. Maintenance of good vegetative condition, protection of rare plants, and soil stability should be management objectives. Obvious overuse of critical ranges demands that population numbers (objectives) be reduced to prevent further damage. A change in land use might remove habitat, which may require a reduction in the population objective. Conversely, habitat improvements (accelerated timber harvest, reduced cattle grazing) may permit an increase in the objective. Public preferences also change. Thus, it is important to monitor habitat condition and public preferences to adjust population objectives accordingly. For a more complete discussion of habitat inventory methods, the reader is referred to Chapter 22.

In agricultural ranges and urbanized areas, BCC may be of only academic interest. Normally, BCC in cropped areas greatly exceeds the tolerance of landowners and farmers. Similarly, in urbanized areas, deer-vehicle accidents, landscape damage, and fear of large animals limit the number tolerated by the public. Objective setting here focuses predominately on sociological concerns.

Sociological Carrying Capacity

Sociological carrying capacity (SCC), also called "cultural carrying capacity" (Ellingwood and Spignesi 1986), was defined by Decker and Purdy (1988) as the maximum wildlife population level in an area that is acceptable to people. As the population of a given species expands in a managed environment, it may conflict with human interest. Common examples are deer and agriculture, deer and motorists, and elk and ranchers. The manager's task is to identify these key human interest areas and determine the preference(s) for population abundance. This preference or maximum acceptable population level translates to the SCC for the management unit in question (Fig. 3). Population objectives imposed by SCC usually will be quite low relative to estimates of BCC. For example, in Wisconsin, BCC on some farmland is believed to exceed 40 deer/km² of deer range, but farmer tolerance (SCC) appears to be <12 deer/km^2 overwinter (McCaffery 1989).

Measuring public attitudes has been the subject of considerable recent interest by managers (Arthur and Wilson 1979, Decker and Purdy 1988). The key to measuring attitudes is the development of a survey instrument that is accurate and indicative of relative importance. Brown and Decker (1979) described such an approach for white-tailed deer in the intensively farmed Lake Plains region of New York. They surveyed farmer preferences with regard to deer abundance, related those preferences to actual inventory data, and identified a population objective (SCC) acceptable to farmers in the region. As pointed out by Decker and Purdy (1988), the SCC can be manipulated to some extent through such things as educational efforts.

The maximum objective based on BCC is a technical determination made by biologists. The public may choose a lower objective. Objectives based on SCC can be mostly a popular decision guided by technical advice.

Expressing Population Objectives

How population objectives are expressed will depend on how herd levels are monitored and on the manager's preference. Harvest data are the most basic population data. Trends in harvest of adult males may provide a simple index to herd status if hunting seasons and weather are similar from year to year. *Harvest objectives* are perhaps the simplest expression of an objective. Some northeastern states express objectives for white-tailed deer as a target buck take or adult buck harvest/km^2 (Dickenson 1986, Regan and Darling [Draft deer management plan for the state of Vermont, 1990–1995, Vt. Fish Wildl. Dep., Waterbury, 1989]). Adult bucks typically comprise 20–25% of the preseason herd in states with unlimited harvest of white-tailed bucks. If this proportion remains consistent, buck harvest can be an easily understood index to deer population trends. This type of objective can be easily adjusted up or down depending on antler development, farmer preference, or some other index.

States that employ computer models, field surveys, or herd reconstruction techniques to obtain estimates of abundance may choose to express objectives as prehunt or overwinter *population objectives*. As an example, Wyoming establishes objectives for wintering population size, total harvest, recreation days, harvest success rate, hunter effort, and area of occupied habitat (Wyoming Game and Fish Department, A strategic plan for the comprehensive management of wildlife in Wyoming 1990–1995, Wyo. Game and Fish Dep., Cheyenne, 1990). These population objectives per management unit are more meaningful to managers, but are more difficult to explain to the public because they require acceptance of population data and their relationship to harvest and recreation objectives.

Density objectives (e.g., animals/km^2 of range), such as used in Wisconsin (Creed et al. 1984), and occupied habitat objectives, such as used in Wyoming, are perhaps the most difficult to communicate to the public because they also include some assumptions about what constitutes suitable range. In forested units, there is usually good agreement on deer range estimates. However, in farmland units there is often debate on whether to define range in terms of permanent cover or also include certain crop fields. For white-tailed deer, Wisconsin has chosen permanent cover plus a perimeter area extending 100 m into agricultural fields (McCaffery, Wisconsin Dep. Nat. Resour. Final Rep., P-R Proj. W-141-R-23, 1988). This type of objective has several advantages. Density per unit of habitat permits comparisons among management units with widely different amounts of habitat. Density information also provides insight into unit carrying capacity and the threshold of human tolerance.

Managers or agencies with good population and habitat data may choose the luxury of employing any one or all of the above-mentioned expressions of objectives depending on their audience and the strength of their data.

Hunter Attitudes and Preferences

Population objectives, based either on BCC or SCC for the various management units, put limits on the abundance of the big-game species in question. Some hunters will want more animals than these limits allow. This may also be true of nonhunters, outfitters, conservation groups, and recreational landowners. Therefore, involving these groups early in the management process is imperative. A survey instrument designed to provide a representative assessment of their attitudes and preferences regarding abundance, recreational opportunity (e.g., season length, special seasons, hunter density, trophy hunting), and hunting pressure will provide important input when harvest management objectives are formulated. If this assessment indicates a desire for more animals than BCC or SCC, an educational effort is in order. Hunters and others need to be informed that wanting more animals than BCC is folly and that exceeding objectives based on BCC can be detrimental to the animals, the size of harvest, and the habitat. They also should know that exceeding SCC levels is at the expense (often economic) of the affected human interest group(s) and the management program if damage is compensated financially.

Special hunting seasons for archers and black powder enthusiasts are getting more attention as their popularity increases (Table 2). These seasons can be incorporated into the harvest management program for added opportunity. To better understand trends such as those represented by special interests, wildlife agencies often survey their constituency as well as other states. This information coupled with in-state hunter preference information enables the agency to respond to changing trends.

Gilbert (1971) observed that wildlife management is the interplay of wildlife populations, habitat, and people. How well the manager succeeds in this relationship will depend on the manager's ability to sell controversial ideas. The Institute for Participatory Planning (1981) called this process the development of substantial effective agreement on a course of action (SEACA). Effective agreement is reached when people affected by a management decision are willing to allow the decision to be implemented even if they do not totally agree. Anything less ultimately will block implementation and frustrate the management program. The best management program ever devised will fail if the public does not support it. The mastery of SEACA or similar techniques in public education is an absolute necessity in formulating and initiating management programs. This technique can be especially useful in facilitating conflict resolution involving hunters, nonhun-

Table 2. Special hunting seasons in the 20 states west of the Mississippi River and in Michigan, Minnesota, and Wisconsin in 1986.

Species	Number of states allowing special seasons			
	Muzzle-loader	Hand-gun	Ar-chery	Cross-bow
Pronghorn	3	0	12	2[a]
Bighorn sheep	1	0	3	1[a]
Deer	20	0	23	7[a]; 2[b]; 1[c]
Elk	7	0	10	1[a]
Moose	0	0	0	1[a]
Rocky Mountain goat	0	0	3	0

[a]Legal for handicapped during archery-only season.
[b]Legal during archery-only season.
[c]Legal during black powder season.

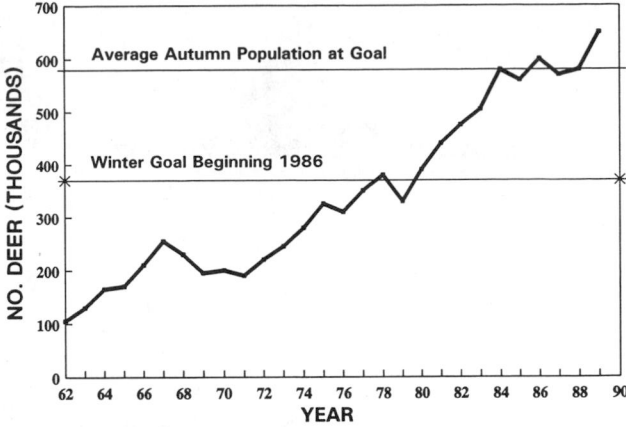

Fig. 4. Autumn population trends relative to the current goal in Wisconsin's farmland region. Goals (based on SCC) have been gradually increased during the past 30 years as landowner tolerance has been progressively tested. Recent populations have been maintained close to the latest goal by prescribed antlerless removals. Distance between winter goal and autumn population goal is based on average net recruitment when at goal. Normally, about 15% of the autumn population is removed annually by adult buck harvest. The remaining removal is by prescribed antlerless quota.

ters, protectionists, and the need for population management.

HARVEST MANAGEMENT GOALS AND OBJECTIVES

Management goals are implemented through the development of specific management objectives. Both objective (survey information) and subjective (tradition, empirical knowledge from past experience) information should be considered when the harvest management objective is formulated. Useful objectives should be quantifiable and measurable. This will enable the manager to evaluate progress toward meeting the objective. An example of a harvest management objective for a population below the desired population objective is: Provide a harvest of approximately n animals, which will permit 10% annual population growth while maintaining average hunting pressure at less than x hunters/km^2 and providing approximately y archery and z firearm recreation days.

In this particular example, the manager has identified and addressed in measurable terms (1) how the population objective will be met, (2) landowner concerns regarding hunting pressure, and (3) hunter preferences for recreational opportunity.

Objectives should be established for a period of several years and for discrete population units. Changing objectives based on year-to-year fluctuations in estimates of population characteristics and public attitude can cause dramatic fluctuation in management direction. Establishing objectives for subpopulations can result in conflicting management within the same population. Either of the above can result in loss of credibility with the public and difficulty in the attainment of objectives.

Within the constraints imposed by the population objectives, development of harvest management objectives tailored to hunter/public preferences can begin. A comparison of population status (inventory) with population objectives will guide the manager in determining harvest quotas (Fig. 3). Antlerless quotas are a function of the projected herd status and the population objective and are a technical determination that should not be delegated to public opinion. Management designed to maintain populations at desired levels, whether those levels are established by BCC or SCC, requires harvesting the net recruitment (Fig. 4). Information based on population change relative to past harvest levels, winter severity, population modeling, or population reconstruction techniques will aid the manager in setting harvest quotas. A long-term database (harvest history) for each management unit will aid the manager in setting harvest quotas. A history of harvest for each management unit is extemely important because harvest is often the most reliable data available on factors affecting population.

In addition to attainment of the population objective, the design of the hunt should address hunter preferences on how animals are harvested. These preferences are often expressed as recreational types (e.g., archery, firearm) and opportunities (number of hunting days). Other public preferences also can be addressed, for example landowner concerns about too many hunters. The degree to which these preferences can be addressed obviously will be constrained by the harvest quota needed to meet the population objective and how complex a regulation the manager and the agency are willing to live with.

Maximum Sustained Yield

Maximizing harvest from a K-selected (BCC) population is referred to as maximum sustained yield (MSY) management. MSY management requires the identification and maintenance of a population level below BCC, which provides maximum net recruitment and therefore maximum harvest. The MSY population level occurs near the midpoint (56% of K) of the logistic growth curve (Fig. 2). Above this level, reproduction and therefore net recruitment is reduced by density dependence (McCullough 1979).

In reality, MSY management seldom is realized. Annual carrying capacity is constantly changing, and data are often insufficient to determine the exact population and harvest levels. However, if goals are based on long-term

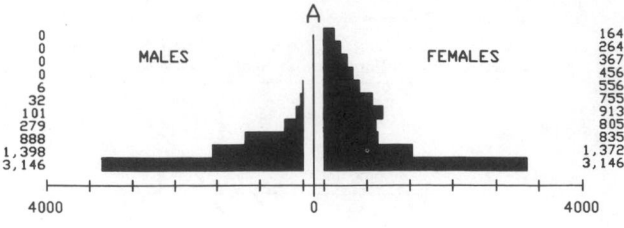

0
0
0
6
32
101
279
888
1,398
3,146

164
264
367
456
556
755
913
805
835
1,372
3,146

4000   0   4000

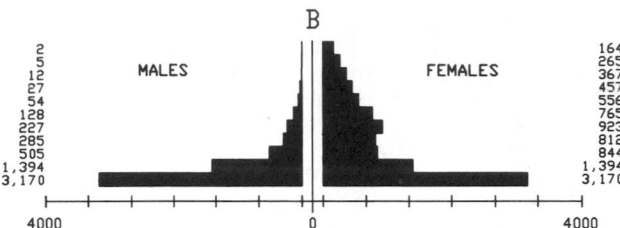

2
5
12
27
54
128
227
285
505
1,394
3,170

164
265
367
457
556
765
923
812
844
1,394
3,170

4000   0   4000

Fig. 5. Age structure of a population of mule deer managed so that only bucks with four-point or larger antlers are legal (A) and so that any buck is legal (B).

BCC, one can approximate MSY. Connolly (1981) suggested that to obtain maximum yield in deer herds, the manager should harvest primarily young animals and mature bucks, with a light culling of does. In practice, a liberal buck harvest with controlled but liberal harvest of females and young animals comes closest to MSY management for big game.

Trophy Management

In 1974, the Wyoming Game and Fish Department conducted a hunter attitude survey which indicated that less than 4% of deer, elk, and pronghorn hunters considered themselves trophy hunters. This obviously varies with species and geography. However, subsequent surveys of resident hunters in Wyoming indicated that most hunters simply wanted the opportunity to hunt. Nonetheless, an often vocal segment of the public is interested in the opportunity to harvest larger males and, thus, must be considered in harvest management objectives.

The best way to manage for older and larger animals is debatable. Connolly (1981) concluded that management for MSY was incompatible with trophy management. However, McCullough (1984) introduced a compelling argument that MSY management was the best way to manage for trophy deer. McCullough's hypothesis has been tested only experimentally on captive herds (Carpenter and Gill 1987).

Antler point restrictions often are used by managers as a method for improving buck quality. Carpenter and Gill (1987) pointed out that antler restrictions are a mixed blessing. They are popular with many hunters, guides, and outfitters, and they may reduce hunter density. They usually do not result in more bigger males and increased reproduction, however, and they always appear to increase the number of sublegal animals (too few antler points) shot and abandoned. However, in Colorado, Boyd and Lipscomb (1976) concluded that protecting yearling male

elk for 2 years increased total males in the population, decreased average harvest, increased illegal bull harvest, and resulted in fewer bulls of exceptional size. The trade-off for each increase of one young bull was 3.9 elk (yearling bulls, cows, and calves). They also projected a $3 million loss in Fish and Game revenue and a $32 million loss to the state economy. The change in the number of older mule deer bucks under two forms of management can be illustrated with computer models (Fig. 5). Under a four-point or larger restriction (only males with at least four points on one side may be legally taken) hunting pressure becomes focused on the older segment of the herd and the 6-years-and-older males in the population are virtually eliminated. However, under a strategy allowing the harvest of any male, bucks as old as 10 years remain in the population. Work with red deer by Clutton-Brock et al. (1987) suggests this decline in older males also may be attributed in part to higher mortality resulting from competition for food with females and increased numbers of yearling males that typically remain with females.

A popular ''large buck'' management strategy within some hunter circles is antlered-only hunting. For white-tailed deer, McCullough (1979) argued that this strategy was likely to have just the opposite effect; without a doe harvest, populations will increase to carrying capacity, resulting in poor antler development, relatively fewer quality males, and a longer time required for males to reach record antler size.

The Montana Department of Fish, Wildlife and Parks (1985) suggested that an effective public relations program may resolve many of the managers' problems associated with the big buck/bull controversy. Montana also correctly pointed out that a successful harvest management program should provide a diversity of opportunity and a freedom of choice for hunters. Carpenter and Gill (1987) suggested that wildlife agencies design their management after validly monitoring recreation demand. Most strategies to improve whitetail buck quality are too costly (in terms of lost hunting opportunity and reduced harvest) to implement except on very limited areas such as military reservations or parks not generally open to unrestricted public hunting (Langenau 1988).

Establishing Hunting Regulations

Management units have been inventoried and a population objective has been established for each management unit. The status of the population relative to the population objective for each management unit has been assessed. Overall hunt objectives (e.g., number of antlerless deer to harvest, recreational types and opportunities) have been determined that will best achieve the population objective. It now remains for the manager to develop a hunting regulation that will fulfill management objectives.

EVALUATION OF HUNTING REGULATION OPTIONS

The manager has five standard regulatory components that comprise a hunting regulation. They can be manipulated to form various regulatory options for achieving harvest management objectives. These components are: (1) timing of the season and opening date, (2) season length, (3) sex/age-specific harvest designation, (4) bag limit, and (5) legal hunting devices. Oftentimes, these regulatory components will have to be considered and established for a variety of season types like archery, muzzleloader, and firearms. Usually, the firearms season will provide the bulk of the hunting pressure and harvest and therefore will be the major means for accomplishing harvest management objectives. Past performance of the same or similar regulatory option in other management units or other states will guide the manager in selecting the regulatory options that are most likely to meet management objectives. If a computer model is used, various options can be tried on simulated populations. As an example, Fig. 5 illustrates the effect of two options for harvesting male mule deer on population age structure. Often, established traditions will limit the manager's options. Managers also must recognize that changing season structure from year to year may confound interpretation of harvest data and herd status and make attainment of population objectives more difficult.

Components of a Regulation and Their Effect

There is a paucity of literature on regulatory components and their evaluation. Generalizing how different season lengths, bag limits, and legal hunting devices will affect big-game populations is difficult. This difficulty is illustrated in Table 3 with Wyoming's experience in the management of several big-game species. Mohler and Toweill (1982) provided predictions of how elk hunting regulations can affect achieving various management objectives for that species. The Montana Department of Fish, Wildlife and Parks (1985) and the Colorado Division of Wildlife (Big game harvest regulation recommendations, Colo. Div. Wildl. DOW-M-S-2, 1974) also listed a variety of regulatory components and the expected effect on the most common big-game species including elk, mule deer, white-tailed deer, and pronghorns. Similarly, Denney (1978) listed big-game hunting season options from the most liberal to the most restrictive.

The effects of manipulating various regulatory components to change male : female ratios (e.g., antlered male only, any animal) or protect males (antler point restrictions) have been considered for objectives that include trophy concerns (e.g., Boyd and Lipscomb 1976, Carpenter and Gill 1987). Similarly, regulatory components can be manipulated to achieve different hunter success rates for obtaining the same harvest, thereby meeting varied public demands for recreational opportunities (success, recreation days) or agency funding needs (Table 4).

The relative effectiveness (range and accuracy) of various hunting devices, as well as hunter/public safety, also must be evaluated. Generally, modern firearms are the most effective, and because of higher hunter success rates, seasons are comparably short and more restrictive in the sex/age of the animals that may be hunted. In crowded regions of the East and Midwest, shotguns using slugs are preferred over the longer range, center-fire rifles because of safety concerns. Because hunter success is high, firearm seasons are usually the most popular. Thus, hunter overcrowding can be a concern from the standpoint of safety, landowner tolerance, and hunting aesthetics. Archery equipment is the shortest ranged and least accurate, but safest. As a consequence, hunter success is low, permitting seasons that are relatively liberal in length and sex/age designation; hunting pressure is light, well dis-

persed, and usually not a safety concern. Aesthetics usually are not a concern, although allegations of excessive wounding and unretrieved kills do occur. Muzzleloaders or primitive firearms are typically intermediate between modern firearms and archery equipment and fall more nearly in the category of shotguns with slugs. The use of special hunts may increase competition between special interests and increases the complexity of regulations.

Effect of External Factors

Factors often beyond the control of wildlife managers can profoundly affect performance of hunting regulations. Big game living in remote areas can withstand a much more liberal season than those living in an area with numerous roads. Private landowners often control harvest through their control of access. Landowners interested in commercializing hunting have even begun controlling the type of animal taken through charging differential access or trophy fees.

Land use practices like timber management can have a significant influence. Many big-game species depend on early successional stages of plant development for food, so creating more cuttings and forest openings can increase herd productivity. On the other hand, the concomitant loss of forest cover and increase in logging roads can improve hunter access and increase animal vulnerability.

Weather also can profoundly influence the performance of a regulation. Dry conditions can make hunting difficult in spite of the regulations. Too much snow can eliminate access to herds. A moderate amount of snow can concentrate animals and allow hunters to track animals, increasing their chance of success. Precipitation on key days of the hunt (opening day, first weekend, holidays) can seriously impair attainment of harvest objectives.

The manager also must consider constraints like hunting tradition, hunter unfamiliarity with new regulations, agency administrative, enforcement, and evaluation capabilities, and legalities. These practical constraints effectively narrow the range of options the manager has to consider.

Each set of regulations must be designed to meet management objectives, with full knowledge of the conditions in which they are to be applied. Identical regulations can have entirely different outcomes when applied to areas with a different topography, vegetation, land ownership, weather patterns, and population characteristics. Normally, a history of experience will be available to provide guidance in each locale.

REGULATORY PROCESS

From inception until becoming law, hunting regulations will run a gauntlet of in-house and public scrutiny (Fig. 6). This scrutiny places a premium on thorough analysis of data, including the evaluation of options discussed above. The public must be exposed to management objectives before seasons are set and effective agreement regarding objectives has been achieved. The regulatory process will work much better if discussion centers on the appropriate regulation to achieve objectives rather than the desirability of the objectives.

Public review and comment often occur at two levels. The recommended regulations are aired first at local meetings and then at county or district public hearings. Wild-

Table 3. Common components of hunting regulations for most big-game species and their expected effects in Wyoming.

Limitation	Effect
Timing of season	
Common opening date	Disperses hunters into surrounding areas and reduces harvest
Early season (Sep)	Adult bull elk most susceptible due to rutting behavior; increases hunters
	Adult buck pronghorns more susceptible due to rutting behavior; increases hunters
	Mule deer and white-tailed deer less susceptible due to dispersal and use of cover; decreases hunters
Firearms hunt (Oct)	Elk of both sexes less susceptible due to end of rut, dispersal, and use of cover
	Pronghorns less susceptible due to end of rut and increase in size of herds; decreases hunters
	Mule deer more susceptible due to concentration on autumn ranges
	White-tailed deer less susceptible due to dispersal and use of cover
Late hunt (Nov–Dec)	Elk of both sexes more susceptible as they become concentrated on autumn and winter ranges; increases hunters
	Pronghorns less susceptible due to size of herds; decreases hunters
	Mule deer bucks and white-tailed bucks most susceptible due to rut and concentration on autumn and winter range; increases hunters
Length of season	
Long season (>30 days)	Increases harvest of desirable animals such as males; increases hunters
Moderate season (15–30 days)	Allows some selection for desirable animals
Short season (<15 days)	Encourages the harvest of the first legal animal; decreases hunters
Timing of opening days	
Weekend opening	Concentrates hunters and increases harvest for all species
Weekday opening	Disperses hunters for all species; may or may not reduce harvest
Sex and age limitations	
Any animal	Elk: increases harvest of yearling males, females, and young; maintains or increases male : female ratios; maintains or reduces population
	Pronghorns: harvest is predominantly males; reduces male : female ratios; maintains or increases population
	Mule deer: increases harvest of young males, females, and young; maintains or increases male : female ratios; maintains or reduces population
	White-tailed deer: increases harvest of young males, females, and young; maintains or increases male : female ratios; maintains or reduces populations
Antlered only	Increases all populations and maintains or reduces male : female ratios
Female and young only	Reduces all populations; increases all male : female ratios
Antler restrictions (spikes excluded; three point or better required)	Elk: increases populations; increases young males; decreases older males; decreases hunters
	Mule deer: increases population; increases young males; decreases older males; decreases hunters
	White-tailed deer: increases population; increases young males; decreases older males; decreases hunters
Hunting devices	
Any legal bow or firearm	Maximizes hunter and harvest opportunity for all species
Primitive weapons (muzzleloading and archery)	Reduces hunters and harvest
Archery only	Lowest number of hunters and harvest
Limited entry hunt	Harvest remains the same depending on number of licenses, but success increases and hunter numbers decrease

Table 4. Hypothetical harvest strategy in which different hunter success rates are used to obtain the same harvest. Option A illustrates the recreation and revenue resulting from a strategy that emphasizes license sales by keeping hunter success relatively low. Options B and C illustrate the impact on recreation and revenue when hunter success is increased.

	Option A	Option B	Option C
Harvest	900	900	900
Hunters	6,920	3,460	1,800
Success (%)	13	26	50
Days per animal	30	17	10
Recreation days	27,000	15,300	9,000
License sales	422,300	211,150	109,854

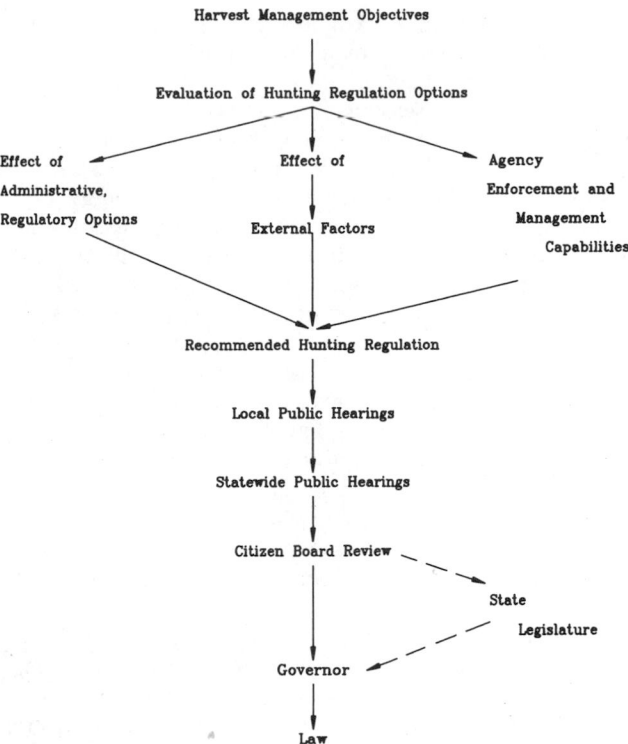

Fig. 6. Schematic of the general process used to establish big-game hunting regulations.

life managers are present to explain the regulations, answer questions, and record comments. After the local public hearings, the agency will review and, if necessary, revise the regulations before presenting them for final public review and comment at a statewide hearing.

Many state agencies have an official commission or citizens advisory board that has rule-making authority. The recommended hunting regulations, the agency's explanation for those regulations, and the results from the public hearings are presented to the board for review and decision. After a decision has been rendered by the board, the regulations may be promulgated directly into law (administrative code) or, in some instances, they may have to be submitted to the state's legislature for final review before becoming law.

EVALUATION OF HARVEST MANAGEMENT RESULTS

A great deal of effort has been expended on planning, inventory, identification of population objectives, establishment of harvest management objectives, and selection and approval of hunting regulations. Similar effort should be spent on an objective evaluation of management results. Analysis of results is the key to determining whether the management program performed as expected.

Management results include the harvest obtained, the recreation provided, public acceptance, and the economic benefits derived from implementation of the hunting regulation. The procedures for measuring these results were developed previously as part of the harvest management strategy. The evaluation phase of the program consists of a comparison of harvest management results against harvest management objectives.

An objective evaluation of management results is often one of the more neglected phases of the harvest management program. Yet, this phase offers many benefits: (1) it provides information on what succeeded and what did not, thereby enabling refinements to subsequent regulations, (2) it identifies needs for additional or new inventory or public survey data, (3) it identifies problem areas that may require special research, enforcement, management, or educational attention, and (4) it provides informational feedback to agency employees and the general public that promotes program understanding and acceptance. Wildlife managers are much more likely to win approval of hunt-

ing regulations that are based on careful analysis of results from a well-designed management program.

Summary

Harvest management for big game is the art of melding the objectivity of wildlife science and the subjectivity of public wants for the attainment of a management goal. Harvest management should be based on objectives established as a part of a comprehensive planned management program. The objective-setting process must be based on a thorough analysis of inventory information that includes sociological as well as biological data. Ultimately, a hunting regulation should emerge that

(1) addresses a set of fair and reasonable management objectives;
(2) is better than other obvious options;
(3) elicits voluntary compliance, is enforceable, and can be administered; and
(4) most everyone is willing to try.

Every hunting season should be evaluated thoroughly so the harvest management program remains dynamic and responsive to a changing management environment.

MANAGEMENT OF UPLAND GAME HARVESTS
Early Harvest Management (Early 1900s to 1950s)

"Management" of upland game harvests as practiced from the early 1900s to the 1950s consisted largely of harvest restrictions. Kinds of restrictions in use during that period included closing or shortening hunting seasons, creating refuges off-limits to hunting, limiting daily and seasonal bags, limiting kill to males, restricting weapon size and type, hunting by permit or lottery, and implementing shooting hours.

Closed hunting seasons or seasons of <2 weeks' duration were in effect over much of the historical range of the wild turkey, prairie-chicken, ruffed grouse, sage grouse, and northern bobwhite by the early 1930s. Leopold (1933:215) noted that ruffed grouse seasons had shrunk from 50 days to 10 days and bag limits from 25 to 7 in north-central United States between 1890 and 1930. The bobwhite season declined from 45 days to 20 days and the bag limit from 25 to 4 during the same period, and the prairie-chicken season was shortened from 45 days to 5 days. These restrictions, designed to reduce harvest without eliminating hunting, were combined with eliminating rifle hunting for most upland game birds, banning shotguns larger than 12 gauge, and restricting the number of shells allowed in shotguns. Use of nets and traps was outlawed. Shooting hours were implemented to prevent killing the wrong species or sex of upland game, and hunting was banned on some days. For species for which the sexes were easily distinguished (like wild turkeys and ring-necked pheasants [hereafter called pheasants]), only males were legal game, to eliminate adverse effects of hunting on reproduction. And, predator control was widely practiced, the theory being that killing predators reduced mortality of upland game, saving animals to reproduce and leaving more animals for the hunter.

Although these restrictions were effective and necessary, problems created by their use remain today. Restrictions like short or closed seasons, restrictive bag limits, shooting hours, half-day hunting, and sex limitations require concurrent public relations campaigns to sell the need for the management strategies. Once these strategies have been effective, or are perceived by the public to have been effective (even if they were not), they are nearly impossible to remove. Once they are eliminated, especially over public objections, the credibility of the management agency suffers. Despite considerable data to the contrary, predator control still is viewed as necessary by the public and even by some management agency personnel. Declines in numbers of upland game over a broad area, and sometimes in local areas, often are accompanied even now by demands for closed seasons and restrictive bag limits.

Modern Upland Game Harvest Management (1970s to Present)

Unfortunately, evaluation of the effects of various upland game management strategies has not kept pace with other wildlife research. Although learning how to sex and age nearly every kind of upland game animal is relatively easy, deciding what sort of hunting season to set is not. Many studies have been made of habitats selected and foods eaten by most upland game animals, but few good evaluations are available of the effects on populations of varying bag limits, harvest rates, and hunting season lengths. Those studies in existence often were done on small areas. Reasons are fairly evident. Management evaluations, sometimes on a statewide basis, rarely provide tightly controlled experiments that lend themselves to statistical analysis. Also, some individuals involved in wildlife research are not associated with a management agen-

cy, so they tend to think in theoretical, rather than applied, terms. Despite this, enough can be inferred from the basic biology of upland game and the existing literature to provide a framework for management. Lack of controlled studies or broad-scale management evaluations sometimes makes defense of upland game management policies within an agency or to the public much more difficult, however. Absence of "proof" of the effectiveness of an untested management strategy may mean the strategy can and will be resisted within an agency or by some portion of the interested publics. Results of studies from states other than the one in which the strategy is proposed sometimes are considered unacceptable by some publics, including agency personnel.

Modern upland game management is based upon several principles developed as upland game populations in many parts of North America recovered, some from near extinction. Among those are: (1) all upland game animals, except those with highly restricted distributions, annually produce harvestable surpluses of young; (2) hunting has seldom had adverse impacts on upland game populations; and (3) hunting pressure on upland game is subject to the "law of diminishing returns," and hunters either do not hunt or hunt less often when game is scarce or perceived to be. A corollary to this is, the more hunting pressure, the warier game becomes, and harvest declines as game becomes more difficult to approach.

Production of Large, Harvestable Surpluses

More than 50 years ago, Leopold (1933) noted the extremely large reproductive potential of upland game. Similar, more recent data were summarized by Johnsgard (1973), Sanderson (1977), and others. These data show that all upland game animals, from the prolific quail and partridge to the less prolific snipe and woodcock, produce sufficient young each year to support a harvest. Clutch sizes for grouse and quail average 5–16 eggs (Johnsgard 1973:68–70), those for snipe (Fogarty and Arnold 1977) and woodcock (Owen 1977) average four eggs, and litter sizes for cottontails, hares, and squirrels average 2–7. Except for squirrels, the mammals have from two to seven litters per year (Conaway et al. 1963, Keith et al. 1966, Nixon et al. 1974, Dolbeer and Clark, 1975, McKay and Verts 1978).

Obviously, animals with high reproductive rates must suffer high juvenile mortality rates, or the habitat would be overrun by the annual production. Despite high juvenile mortality, many young are available for harvest. For example, figures provided in Bump et al. (1947) showed that the average ruffed grouse hen lays 11.5 eggs. About 95% of each clutch hatches, and 61% of the nests are successful. Then, even if 62% of the young die, 2.5 young are produced by each successful hen each year. Fischer and Keith (1974) reported 2.0 juveniles per adult in an autumn ruffed grouse population. The same reasoning used above when applied to gray partridge data (Johnsgard 1973:88) demonstrated production of 3.5 young per successful hen. Data for mammals show similar but more dramatic results. Cottontails produce 9–15 young per female (Conaway et al. 1963, 1974, McKay and Verts 1978), squirrels 4–9 young per female (Nixon et al. 1974), and snowshoe hares 6–13 young per female (Keith et al. 1966, Dolbeer and Clark 1975).

Johnsgard (1973:89) summarized autumn and winter age ratios among grouse, quail, and partridge, which varied from 33% to 89.5% immature birds. All species but white-tailed ptarmigan had more than 50% immatures in the autumn-winter population. Johnsgard noted that a ratio of 50% immatures to 50% adults suggests a breeding season productivity of 100%. Rock ptarmigan, ruffed grouse, scaled and Gambel's quail, northern bobwhite, and gray and chukar partridge had ratios of 75% immatures to 25% adults, which indicates 300% productivity. This is high, indeed. This high proportion of immature animals in upland game populations has been apparent in reported harvests. If hunting is nonselective and hunters take animals in proportion to their availability in the population, much of the annual hunter harvest of all upland game should be immature (juvenile) animals. Proportions of juveniles reported in the harvest of various upland game animals vary from 41% to 87% for blue grouse (Hoffman, Colo. Div. Wildl. Job Prog. Rep., Fed. Aid Proj. W-37-R:265–317, 1980), cottontail and swamp rabbits (Martinson et al. 1961, Conaway et al. 1963, Trent and Rongstad 1974), fox squirrels (Allen 1943, Donohoe and Martinson 1961, Mosby 1969, Nixon et al. 1974), mourning doves (Dunks et al. 1982, Tomlinson et al. 1988), pheasants (Erickson et al. 1951, Baxter and Wolfe 1973), Montezuma quail (Brown 1979), ruffed grouse (Major and Olson 1980), sage grouse (Braun, Colo. Div. Wildl. Job Final Rep., Fed. Aid Proj. W-37-R:29–73, 1981), and sharp-tailed grouse (Hamerstrom et al., Wisconsin Dep. Nat. Resour. Job Compl. Rep. Proj. W-79-R-1:71–77, 1956).

Effects of Hunting

There is no question that uncontrolled hunting had devastating effects on upland game populations. Actions taken since the 1600s in North America, such as imposing and reducing bag limits, eliminating year-long hunting, restricting types of weapons used, and limiting shooting hours, were designed to reduce or eliminate those effects. Hunting of upland game, as practiced since the 1950s, has not been shown to have a measurable adverse effect upon upland game populations. Those animals taken by hunters are measurable, of course, and harvest estimates are made routinely by most state fish and wildlife agencies. Despite the use of check stations, telephone and mail surveys, mandatory reporting, and research, it has been difficult to show that an increase or decrease in harvest produces either an adverse or beneficial effect on most upland game populations. Rather, the literature suggests hunting mortality that is less than about 50% of the total mortality is nonadditive (compensatory) and is substituted for other natural mortality, or that factors such as immigration/emigration or density-dependent production are operating (Mosby and Handley 1943, Baumgartner 1944, Glading and Saarni 1944, Mosby and Overton 1950, Hickey 1955, Lobdell et al. 1972). This probably applies to most upland game, except spring hunting of male turkeys, which Lobdell et al. (1972) suggested is additive mortality. Apparently this is because spring hunting occurs late in the biological year, after all other mortality has been sustained. But, Lobdell et al. (1972) believed that removal of even 100% of the adult gobblers from a turkey population in spring could not happen quickly enough to inhibit breeding. Bergerud (1985) argued that all hunting mortality is additive. Perhaps the question is not whether game bird populations are affected, and, more particularly, whether they are so depressed they do not recover. Currently, there is not much literature to suggest this is a problem.

For many upland game animals, hunting has not been shown to increase mortality much above natural levels. Mosby (1969) observed 42% annual mortality of squirrels in an unhunted area, 38% in a hunted area. Mortality of 38% had no effect on average annual mortality or recruitment. Edminster (1937) showed that winter populations of ruffed grouse were the same on a refuge and in a hunted area. Palmer and Bennett (1963) saw similar 7-year declines in ruffed grouse numbers on hunted and unhunted areas. Gullion and Marshall (1968) reported that survival of male ruffed grouse was lower off refuges, so autumn hunting increased adult mortality in autumn. However, annual survival rates were the same in hunted and unhunted areas. In other ruffed grouse populations, autumn hunting had no measurable effect on population size the next spring (Dorney and Kabat 1960, Fischer and Keith 1974). Fischer and Keith (1974) shot males off territories in spring and saw no effect on brood production or autumn populations. Rose (1977) noted the mortality rate of rabbits was 84% when hunting was legal, 75% when it was not. This did not decrease rabbit populations the next autumn, apparently due to increased production or improved survival of young, dispersal, or some other undetected effect.

Under some circumstances, of course, a high level of harvest could produce additive mortality in an upland game population. Roseberry (1979) simulated effects of different harvest strategies on northern bobwhites and showed that harvests greater than 55% of the autumn bobwhite population would depress the following spring's population. Several authors (Allen 1943, Baumgartner 1944, Allen 1952, Nixon et al. 1974) concluded that fox squirrels in small woodlots could be overharvested, but squirrel populations in these areas were maintained by immigration from adjacent, unhunted areas. Where woodlots were small and widely separated, population recovery was slow because repopulation after removal of squirrels by hunting was slow. Where habitat is continuous, or nearly so, as in extensive forests, gray and fox squirrels are considered underhunted, that is, harvest is much less than annual recruitment to the population (Allen 1952, Uhlig [W. Va. Conserv. Comm. P-R Proj. 31-R, 1955], 1956, Mosby et al. 1977, Weaver and Mosby 1979). If all habitat had been subjected to the same level of harvest as the woodlots studied by Nixon et al. (1974), repopulation would have been impossible, and the population size would have been depressed.

Recently, biologists in Wisconsin (Small et al. 1991) working in fragmented habitats have proposed that ruffed grouse numbers might be depressed by high harvests. Yet, these birds are replaced each year by immigration from surrounding areas. Where that is not possible, as in severely fragmented habitats, populations might be extirpated locally, as undoubtedly has happended in some portions of the range of some game birds in the past.

The higher the harvest, the more likely it is to produce additive mortality. Theoretically, if hunting pressure was high enough, any population could be depressed. Fortunately, hunting pressure seldom, if ever, takes enough animals from an upland game population to depress the rapid reproductive rates of upland game animals or the density-dependent responses of fecundity and mortality (Shaw 1985). Nixon et al. (1975) suggested hunting pressure on public forests was not as important as squirrel density in determining total harvest. The magnitude of harvest has been examined for several upland game animals. The percentage of the population taken by hunting was <4% for white-winged doves (Brown et al. 1977), about 4% for blue grouse (Hoffman, Colo. Div. Wildl. Job Prog. Rep., Fed. Aid Proj. W-37-R:265–317, 1980), 5.5–12% for mourning doves (Dunks et al. 1982:1, Tomlinson et al. 1988:1), 7–11% for sage grouse (Braun and Beck 1985), 14% for band-tailed pigeons (Wight et al. 1967), 10–40% for turkeys (Lobdell et al. 1972, Weaver and Mosby 1979), 15–60% for squirrels (Fouch 1969, Mosby 1969, Nixon et al. 1974, 1975), and 45–70% for bobwhites (Roseberry 1979). Hickey (1955) suggested an upland game bird population can tolerate hunting mortality at least equal to one-half the natural mortality. Adult mortality rates for these species, which are lower than juvenile mortality rates, range from 40% to 80% (Lobdell et al. 1972, Johnsgard 1973, Hoffman [Colo. Div. Wildl. Job Prog. Rep., Fed. Aid Proj. W-37-R:265–317, 1980], Braun and Beck 1985), indicating harvest rates of at least 20–40% can be tolerated. Mosby (1969) reported that a 38% harvest of gray squirrels had no population effects. Peterle and Fouch (1969) believed a harvest of 60% of the fox squirrel population could be sustained. Nixon et al. (1974) said that a 60% harvest rate could not be sustained by resident fox squirrels in small woodlots without immigration from adjacent areas. Vance and Ellis (1972) reported that annual harvests of 70% were not excessive for bobwhites. However, Gullion and Marshall (1968) suggested that high harvest of ruffed grouse in late hunting seasons might depress grouse populations in Minnesota. Work in Wisconsin (Kubisiak 1984, DeStefano and Rusch 1986, Small et al. 1991) indicated hunting pressure in fragmented ruffed grouse habitats may depress ruffed grouse numbers. However, for now, immigration from nearby areas sustains grouse numbers. Overharvest might be a problem for turkey populations in some eastern states that experience heavy legal hunting, poaching, and habitat and weather problems. Harvest might come close to exceeding the acceptable harvest level rates in these locations, but probably not in most states.

Accessibility of habitat to the hunter has a positive influence on kill rates. As pointed out above, populations of squirrels in small woodlots on public lands that are easily hunted are more likely to be overharvested (Allen 1943, Allen 1952, Nixon et al. 1974), whereas squirrels in extensive forests, with large areas that are not hunted because they are distant from roads, are not (Uhlig [W. Va. Conserv. Comm. P-R Proj. 31-R, 1955], 1956, Mosby et al. 1977). Squirrels in small woodlots under private control, where hunter access is limited, are also less likely to be overharvested. Small public woodlots in the Midwest had four to five times the squirrel hunting pressure on national forests or private lands (Fouch 1969, Nixon et al. 1974). In Alberta, only 1% of marked ruffed grouse males >302 m from roads were shot, whereas 27% of males within 101 m of roads were shot (Fischer and Keith

1974). Ruffed grouse mortality was greater in areas with more roads. Small et al. (1991) noted that 60% of ruffed grouse on public hunting areas were harvested, whereas only 10% of grouse on private hunting areas were shot. Refuge areas of low hunting pressure from which animals can disperse into areas of high hunting pressure undoubtedly have had a great deal to do with the fact that upland game populations are seldom overharvested.

Effects of season length and timing on upland game have been debated for a long time. Typically, season length and timing are used to regulate harvest of upland game (Crawford 1982). Sometimes this affects the harvest, and sometimes it affects hunters more than it does game. A classic example of this is sage grouse hunting. For a long time, sage grouse hunting seasons in parts of the range, including Wyoming, opened in late August or early September. In the late 1970s, several states changed the timing of sage grouse seasons to mid-September. The idea behind this was to reduce hunting near water, which purportedly produced a higher harvest of females and young. Many biologists applied big-game logic to sage grouse and presumed a high harvest of hens and young meant too many were being harvested and populations were being adversely affected. The change in season timing was supposed to increase harvest of yearling and adult male sage grouse, because hunters would be hunting after broods broke up and young dispersed. Birds would be more dispersed, thus hunters would be more likely to encounter males. In Wyoming, harvest of male sage grouse, which was very low, increased only 10%, harvest of females and young stayed about the same, and hunters fond of a late August hunting season, when younger, tastier birds were easier to identify, were furious. More recently, sage grouse managers have reasoned that sage grouse populations are composed of mostly females and young (70–86%, Braun [Colo. Div. Wildl. Job Final Rep., Fed. Aid Proj. W-37-R:29–73, 1981]) at the time hunting seasons occur, so a harvest of mostly females and young is not necessarily bad. An autumn harvest of 70–91% females and young simply reflects the nonselectivity of hunters who are killing grouse roughly in proportion to their occurrence in the population.

Season length and timing can affect some species adversely in some areas. A season 88–98 days long that began in September depressed a squirrel population on a small woodlot (Nixon et al. 1974), whereas a season 22 days long in October did not depress a squirrel population even though the proportion of the population harvested was high (Peterle and Fouch 1969). Nixon et al. (1974) believed a delay in hunting until 15 September produced less impact because it reduced harvest of females. Mosby et al. (1977) said that squirrels were most vulnerable in early September, less so thereafter. Increased harvest required a hunting season opening before 1 October. The results mentioned earlier from Wisconsin (Kubisiak 1984, DeStefano and Rusch 1986, Small et al. 1991) that show depression of ruffed grouse numbers on public hunting areas seem to indicate that a much shorter hunting season, which does not occur when grouse are vulnerable, should be tried.

There is some disagreement in the literature about the effect of changing season lengths and bag/possession limits. Crawford (1982) reported that doubling the possession limit from two to four sage grouse increased harvest and that lengthening the hunting season more than doubled harvest. However, Braun (Colo. Div. Wildl. Job Final Rep., Fed. Aid Proj. W-37-R:29–73, 1981) and Braun and Beck (1985) concluded that increased bag limit and season length had no measurable effect on total harvest. Instead, sage grouse harvest was related more to environmental conditions prior to and during the hunting season. Increasing the length and bag limit of a restrictive season might increase harvest, but results in Wyoming agree with findings of Braun and Beck (1985). In 1981, the bag limit for sage grouse in southwestern Wyoming was reduced from three per day, six in possession, to two per day, four in possession. Season length was the same. Harvest declined by 45%. The next year, harvest increased 71% with the same season length and bag limit. Then, in 1985, the three/six bag/possession limit was restored, and harvest increased by 46%. In 1986 the bag/possession limit remained the same, but the season length was nearly doubled, to 16 days. Harvest declined by 8%. These results indicate factors other than changes in season lengths and bag limits are operative. The decline in harvest in 1986 corresponded with cold and foggy conditions on opening weekend. Birds were not concentrated in areas near water because all of the vegetation was moist, so they were scattered. Hunter success was low, and many hunters went home by noon. Much of the harvest occurs in the first weekend of the season, and this has a significant impact on total harvest.

For many upland game species, most of the harvest occurs in the first few days of the season. For example, 60–73% of sage grouse harvest in Colorado occurred on the first weekend of the hunting season (Braun and Beck 1985). Similar results were seen in Wyoming, as mentioned above. A longer hunting season for sage grouse in Colorado spread the harvest over a longer period and decreased harvest on the opening weekend, but 60% of the harvest was still taken during the first weekend. In Wyoming, we saw similar results. Overlapping the sage grouse season with the pronghorn season produced a second, but much smaller, peak of sage grouse harvest on the opening weekend of pronghorn season. In Colorado, from 27% to 44% of the blue grouse harvest is taken the first weekend, and 39–54% is taken on the first two weekends (Hoffman, Colo. Div. Wildl. Job Prog. Rep., Fed. Aid Proj. W-37-R:265–317, 1980). In Idaho, the popularity of the first few days of the pheasant season led to reduced bag and possession limits the first 5 days of the season and elimination of nonresident hunting during that period (Upland game species management plan, 1981–1985, Idaho Dep. Fish Game, Boise, 1981).

Generally, sex of upland game animals is indistinguishable in the field by the hunter, so seasons usually are set to take both sexes. Exceptions are pheasants and turkeys. In autumn turkey hunting seasons, there is probably little need for concern about harvest, even if either sex can be taken (Lobdell et al. 1972). A large body of opinion developed in the heyday of pheasant abundance suggested little detrimental effect of hunting hen pheasants (Madson 1962). As habitat has shrunk, however, some states have suggested that hens should not be hunted. This is not due to effects of hunting, but the additive combination of hunting, winter mortality, and nest destruction during al-

falfa harvest. Hunting and winter mortality together are not a problem, but when remaining hens are killed during alfalfa harvest, reproduction is severely depressed. Aside from this, separation of the harvest by sex is a concern only in spring turkey hunting. There, the mortality is additive, but is timed in such a way, or is of such low magnitude, that it is not a concern, as mentioned above. Many states allow bearded turkeys to be shot when sex restrictions are part of the hunting regulation, recognizing that some bearded hens will be shot, but that the proportion of bearded hens in the harvest will be small.

"Law of Diminishing Returns"

Hunters tend to avoid areas with low game numbers and to reduce hunting effort when game numbers are low or perceived to be low. Word of mouth (i.e., hunters telling other hunters what they have killed) is a factor encouraging or discouraging hunter effort. This appears particularly true of squirrel and cottontail harvest. If squirrels are scarce in September, hunter pressure declines rapidly, then all but disappears a few days after the opening day. If success is good, hunting pressure remains high (Mosby et al. 1977). Considerable evidence shows that positive or negative publicity about numbers of pheasants, grouse, quail, or rabbits can substantially influence harvest. When severe winters greatly impact upland game numbers, negative publicity about one species can carry over to another. For example, blue grouse harvest and hunters in Wyoming declined 38% and 32%, respectively, following the severe winter of 1983–84. However, blue grouse have not been shown to experience high winter mortality, and reproduction was good in 1984. The decrease in hunters seemed to be related to publicity concerning the adverse effects of the winter on other upland game, like cottontails and chukars.

Hunter numbers tend to be much greater earlier in the hunting season. Kubisiak (1984) and DeStefano and Rusch (1986) noted the much higher effort by hunters early in the Wisconsin ruffed grouse hunting season. Nixon et al. (1974) reported that 72.5% of squirrel hunters were in the field in the first month of the season, 23.1% in the second month, and only 4.4% in the third month. Early hunting pressure quickly removes animals that are less wary, thus more easily killed. As more upland game animals are harvested, the survivors become warier and more difficult to kill. Much more effort is needed to harvest game later in the season. This discourages many hunters, and they may give up. Nixon et al. (1974) noted that unsuccessful hunters spent 30–90 minutes less time per squirrel hunt than successful hunters. However, Nixon et al. (1974) also observed that hunting pressure often remained high on public hunting areas even when individual hunting success was low. Thus, high harvests occur each year because hunters who frequent these areas are not motivated to kill large numbers of game species—they seek enjoyment in the opportunity to hunt and to fraternize with their fellow hunters. Hamerstrom et al. (Wisconsin Dep. Nat. Resour. Job Compl. Rep., Proj. W-79-R-1:71–77, 1956) believed that the opening of the season in Wisconsin for sharp-tailed grouse in 1955 occurred after the birds had become wild and difficult to approach, which severely reduced the harvest. Anyone who has hunted ruffed grouse in the U.S. will readily agree that birds in the East, which are hunted harder, are much more wary and flush wilder, except early in the hunting season when many more younger, less wary birds are present. Western ruffed grouse are generally tamer, except where they receive more hunting pressure. This also applies to other game. Hoffman (Colo. Div. Wildl. Job Prog. Rep., Fed. Aid Proj. W-37-R:135–152, 1976) could not capture Colorado blue grouse with a noose pole as described by Harju (1974) in Wyoming because blue grouse on his study area were hunted more heavily and could not be approached as closely. Many sage grouse and sharp-tailed grouse hunters have experienced the frustration of having birds flush wildly, far out of gun range, as soon as they see hunters late in the season. Pheasants display similar behavior late in the hunting season, in addition to their irritating habit of running far ahead of the hunter before flushing.

Stocking of Game Birds

Because stocking of pen-reared birds is often a popular option in the minds of the public, it must be included in any discussion of upland game harvest. Stocking usually is done with pheasants and bobwhites. Much of the fondness for this practice among hunters stems from the success achieved when pheasants and other game birds were introduced to new areas. The fact that an empty niche was available to pheasants, and that the birds responded as most species would when introduced to space devoid of competitors, escapes the public. Since the pheasant was introduced to North America, small farms have become larger, fencerows have disappeared, and efficient farming has reduced cover and food. Much of the stocking of pheasants was done with birds trapped in the wild and relocated, not pen-reared. Pen-reared birds are unwary and do not seem to be able to forage as well as wild birds, and hens do not raise broods. Politicians and agency administrators, and many management personnel, have not shown much resistance to demands to stock pheasants and bobwhites partly due to fears of lost public support and partly due to threats of political reprisals.

All of the available evidence, largely collected on pheasants, indicates stocking is not an appropriate practice for harvest management. Captive-reared birds simply do not survive in very large numbers when they are placed into the wild, and the cost per bird returned to the bag is high. Pheasants have been stocked at various times of the year, and survival rates have been monitored. These attempts have included stocking hens just after hunting season, or in spring just before breeding season. Production by these birds was minimal. Besadny and Wagner (1963) reported that stocked hens that nested in Wisconsin produced only 0.2–0.4 cock/hen. In Illinois, 800 wild hens and 900 bird-farm hen pheasants were released in spring. Only the wild birds reproduced (Anderson 1964). This experiment, with similar results, has been repeated in most, perhaps all, states. Each was done because proponents of stocking refused to believe results from other areas. Of course, few pheasants reach old age, even in a wild population. Erickson et al. (1951) reported only 28% of pheasants in an unhunted population older than 1 year.

Band returns indicate stocked pheasants add little to the wild population or to the harvest. In Minnesota, only 3% of released pheasants were shot (Erickson et al. 1951).

Madson (1962:59) reported that only 9% of pen-reared pheasants released 30 days before the hunting season were shot, and only 2% of birds stocked the previous winter were harvested. Klimstra (1975) reported similar results with released, pen-reared bobwhites. Wisconsin experimented with a day-old chick program, in which pheasant chicks were given to sportsmen's clubs to raise and release (Besadny and Wagner 1963). This added 27% to the harvest in mediocre pheasant habitat, 38% to the harvest in "marginal" pheasant habitat, and only 5% to the harvest in the best pheasant habitat. The proportion of stocked birds in the harvest decreased as the season progressed. Generally, the closer to opening day pheasants are released, the greater the percentage harvested. Raising these birds is costly, so the more that are harvested, the less the cost per bird bagged. In Wyoming and other states, political pressure from sportsmen to maintain stocking of captive-reared pheasants has resulted in put-and-take hunting in which hunters are allowed on a public hunting area to hunt pheasants that are released the day before they are hunted. The hunt was created to lower the cost per pen-reared pheasant bagged by hunters. In Wyoming, the rate of harvest of these birds averages 66%. Rate of harvest of banded birds released in areas without a controlled hunt is 27–62% (mean = 40%). Lowest rate of return was in areas with the poorest habitat, apparently because the available cover would not hold many birds. Prior to the creation of this hunt, only 30–40% of stocked birds were shot. Klimstra (1975) reported harvest rates of 38.6%–60% of pen-reared bobwhites released during the hunting season. Cost per bird for pen-reared pheasants was $18–$20 in the 1950s in some states (Madson 1962:59), and it certainly is no lower 40 years later. In Wyoming, every pheasant released costs about $13, and the return is about $1 per pheasant harvested under even the most optimistic harvest estimates. And, concentration of so many hunters on public hunting areas has created lead-shot hot spots. The hunt is extremely popular with hunters, despite the cost, so it continues.

Current Trends in Upland Game Management

Accumulation of better data on the biology of upland game has resulted in several changes in upland game management since the restrictive regulations in effect at the middle of the century. Among these are longer hunting seasons, more liberal bag limits, and changes in emphasis of data collection for upland game.

Hunting seasons for upland game were lengthened first in the eastern and southern states, where big game was less abundant and there was a large human population with a tradition of upland game hunting. In western states, demand for upland game hunting was, for many years, of smaller magnitude and overshadowed by big-game hunting, so some upland game seasons were much more restrictive than necessary. Despite a considerable volume of literature to the contrary, it is not unusual today to see proposals to restrict bag limits, shorten hunting seasons, and close seasons after real or perceived declines in upland game populations. Table 5 lists season timing and season lengths for various upland game animals throughout North America. Obviously, hunting seasons for upland game have many "correct" times and lengths, and most hunting seasons are liberal. Along with liberal hunting seasons have come more liberal bag limits. These include two to five pheasants or grouse per day, 10–25 bobwhites per day, and 5–10 squirrels or rabbits per day.

In some states, dates for upland game bird seasons are set long before brood counts are made. Even if that were not so, brood counts have limited value in setting seasons. Because hunting has so little effect on upland game populations, the main value of brood counts is to predict the relative quality of hunting this year compared to last or to identify areas of habitat to protect. Many states have de-emphasized population data collection on upland game because the lack of harvest impact indicates little need for the data. Instead, emphasis has been placed on collection of harvest data, which can be used to evaluate what is happening to the population. Use of wing barrels (Hoffman and Braun 1975), into which hunters place wings of harvested birds, has increased in the West. Other methods include mandatory report cards and postcard or telephone surveys of a random sample of hunters.

Field checks of hunters long have been used in many states to collect management data. Generally, this is effective if done at a roadblock or other mandatory check station where a large sample of animals can be examined. The sample allows evaluation of the proportions of sexes and ages of the harvest of game birds. Some states that have a relatively low number of hunters and low harvest of upland game have proposed reducing the emphasis on wing collections to every other year or every third year.

Setting Hunting Seasons and Bag Limits

Hunting seasons for upland game are set at a variety of times in a variety of ways, depending upon the state or province involved. The hunting season may be set a year in advance, it may be set 6 months in advance, or it may be set 1 or 2 weeks before the season opens. In some places, the hunting seasons for certain species are standardized—they never change. Sometimes the season has been standardized by biologists, and sometimes it is set by the state legislature or a board or commission that oversees the agency. Often, the wildlife manager makes a recommendation for season timing, length, and bag limit, and that recommendation is approved or modified by agency administrators, a commission, or a board that oversees the agency, or by the legislature, as described earlier for big game.

Conversely, sometimes the timing of events dictates that hunting seasons be set before some data are available. Something as seemingly unimportant as the amount of time necessary to print the hunting regulations and distribute them to hunters and license agents may require setting hunting seasons before all the data have been collected. If a biologist has concerns about the status of a population of upland game and the impacts of hunting on that population, someone may need to insist on a greater level of data collection and analysis. Given the information presented earlier, however, concerning the lack of impact of hunting on populations of upland game, biologists need not be too concerned no matter when the hunting seasons are set. Hunting seasons should not be set for the convenience of the manager, but for the convenience and enjoyment of the hunting public, short of adversely affecting the target population.

Turkey hunting seasons in the United States illustrate

Table 5. Timing of upland game hunting seasons and season lengths in North America, 1989–90.

Species	Earliest opening date	Latest closing date	Most common dates	Season length (days) Short	Long	Average
Dove	Sep	Jan	Sep–Oct	30	122	78
Hare	Sep	Apr	Oct–Feb	4	365	163
Jack rabbit	—	—	Yearlong	32	365	197
Chukar	Sep	Feb	Oct–Dec	20	122	77
Gray partridge	Sep	Feb	Sep–Dec	45	144	77
Pheasant	Sep	Feb	Oct–Jan	14	153	61
Pigeon	Sep	Oct	Sep	30	30	30
Ptarmigan	Aug	Apr	Sep–Feb	76	263	167
Bobwhite	Oct	Mar	Nov–Feb	14	130	82
Mearns quail	Nov	Feb	Nov–Feb	80	80	80
Rabbit	Sep	Jun	Oct–Feb	6	365	138
Squirrel	May	Mar	Sep–Feb	31	365	125
Woodchuck	—	—	Yearlong	196	365	313
Turkey						
Spring	Mar	May	Apr–May	5	48	24
Autumn	Sep	Jan	Oct–Nov	4	88	29
Grouse						
All	Aug	May	Oct–Dec	82	288	115
Blue/spruce	Sep	Dec	Sep–Nov	20	106	64
Ruffed	Aug	May	Oct–Dec	53	137	94
Sage	Dec	Dec	Sep	1	86	19
Sharp-tailed	Sep	Dec	Oct–Nov	9	106	62
Prairie	Sep	Dec	Oct–Nov	1	85	31

the variety of factors considered when upland game hunting seasons are set (National Wild Turkey Federation 1986). Many of these factors are political rather than biological, but sometimes these political factors have more effect on the biologist's ability to manage animals than the biology of the species. Among the factors involved in setting spring and autumn turkey seasons are:

Spring hunting season	Autumn hunting season
Tradition	Tradition
Limit number of hunters to maintain hunt quality	Coincide with deer season
Coincide with peak of gobbling	Landowner tolerance
Landowner tolerance	Turkey numbers
Turkey numbers	Brood surveys
Season timed to coincide with hens beginning incubation	Access to turkey areas (weather)
Season ends before peak of hatch	Literature on turkeys
Access to turkey areas (weather)	Avoid deer season
Literature on turkeys	Autumn hunts might adversely affect the turkey population

In some areas, landowner tolerance of hunters and the ability to place hunters on private land is much more important than the number of animals available. If landowners in an area with predominantly private access object to a lengthy season and threaten to close access, the manager is faced with a dilemma. Sticking by one's principles can result in no hunting. In such a situation in Wyoming, a 10-day sage grouse season was approved in parts of two counties to prevent loss of access for all hunting on more than 40,469 ha controlled by two landowners who did not want a longer season. Areas without this problem had a 28-day season. The hunting season was shortened, and hunter opportunity was minimized but not eliminated. Each manager has to decide whether he or she has gone overboard in pleasing landowners at the expense of the public. The amount of consideration given landowner desires sometimes is based on the proportion of private land in the hunting area. Sometimes, education of the landowner or a good working relationship with the landowner is effective.

Similar factors have been part of season setting with other upland game animals. Although dove hunters in northern states complain that early, cool temperatures in September cause doves to leave, which severely reduces harvest, 1 September is the earliest date allowed for dove hunting. This is caused in part by the initial resistance to dove hunting seasons in some states.

Liberal blue/ruffed/spruce grouse seasons in the West are designed to allow big-game hunters to take these underhunted birds. These long seasons often have no weapon restrictions, so the birds can be taken by any weapon, thus encouraging harvest. Sage grouse seasons in some states have been set to coincide with pronghorn seasons, thus encouraging combination hunts. Pheasant seasons usually begin in October or November. Seasons this late ensure that young males are fully feathered and recognizable as cocks, for hens usually are not legal. In many places, late seasons also ensure that much corn is picked early in the season. After that, birds are more concentrated and easier to find by the much-reduced hunter numbers. Turkey seasons in autumn are in November or later, to make certain young turkeys are nearly full grown, thus more appealing to hunters. Some states have no autumn turkey season for fear of overharvest. Some states allow no Sunday hunting, and some do not allow the nonresident to hunt during the first 5 days of the season.

Although data are lacking, there has always been a feeling in some western states that sage grouse can be overhunted. Probably this stems from hunting of small, local flocks of birds, often associated with meadows. The fact that birds are present in those locations year after year despite the high annual harvest often escapes notice. As a result, sage grouse seasons are often short. Ruffed grouse are birds of denser, more timbered cover in the West, and data on these birds are much more difficult to collect. Yet, seasons for these birds may be 90 days long. There is no reason to believe sage grouse have less ability to become wild as they are hunted, yet the dichotomy persists. Idaho biologists believe sage grouse can be overhunted under some conditions (Upland game species management plan, 1981–1985, Idaho Dep. Fish Game, Boise, 1981), whereas Colorado biologists believe it is unlikely (Braun, Colo. Div. Wildl. Job Final Rep., Fed. Aid Proj. W-37-R:29–73, 1981).

Bag limits for upland game should not contribute to excessive harvest, and the "law of diminishing returns" probably prevents this from happening in most circumstances. The psychology of hunting suggests that bag limits should be low enough to be achievable by a significant number of hunters. A certain proportion of hunters is goal-oriented, interested in killing a limit of whatever animal is being hunted. Setting limits too high discourages these hunters and may result in their quitting the sport. Limits that are too high also can lead hunters to believe management is poor.

Most states have a possession limit that is two to four times the daily bag. The larger possession limit allows hunters to stay and hunt for several days. There is no evidence to suggest the larger possession limit increases harvest or affects populations. Some states impose a season limit, such as 50 birds/year. Such limits probably are not enforceable. Their intent apparently is to set some upper limit on the number of game animals taken. However, no data are available to show that people comply or that enough people achieve the season bag limit to affect game populations. In fact, the number of hunters who are so dedicated or skilled that they can reach the season bag limit is probably low. Unenforceable regulations should be avoided, because they can contribute to poor public relations, especially if they have no biological basis.

Shooting Preserves

In many parts of the United States, public access for hunting is limited by a lack of public land, and upland game numbers have declined as urban areas have increased in size. A combination of these two factors has stimulated interest in shooting preserves. These are private lands where upland game is raised and released for hunters to shoot. The area may be operated as a club, where each member shares in the cost of the preserve, or as a business, where the hunters pay for a certain number of hours of hunting or for each bird shot. A guide and a trained bird dog may or may not be provided. Most shooting preserves are licensed by state wildlife agencies. Hunting seasons usually start much earlier on shooting preserves and are longer.

On the positive side, shooting preserves provide hunting for people that otherwise might not hunt. They are likely to increase in importance as more of the nation becomes developed and access to private land continues to be restricted. On the negative side, these areas are fairly expensive and are more likely to be vulnerable to attacks from antihunting groups.

Summary

The available literature suggests hunting mortality of all upland game animals is compensatory; that factors such as immigration from refuge areas and density-dependent production operate in upland game populations; and that hunting does not significantly impact populations. Production of large, annual surpluses of young allows liberal harvest with little fear of population impacts in most areas. Seasons can be lengthy and bag limits generous with little concern for overharvest. The biology suggests nothing less. Political considerations dictated by tradition may suggest something else.

As with all wildlife, the quantity and quality of upland game habitat are more important in determining population size than any other factors. The vagaries of weather are more important than the number of hunters or the length of the hunting season, particularly for partridge and quail. Farming practices have much more impact on pheasants than does the hunting season. Management of upland game harvest is more affected by tradition, politics, and hunter or manager philosophy than anything else. Wildlife professionals should carefully analyze management strategies, their possible results, how to eliminate the strategy if necessary, and the cost in public relations and agency credibility if management strategies the public likes must be eliminated.

MANAGEMENT OF MIGRATORY BIRD HARVESTS

Background

Whereas most hunting in North America is regulated solely by the state, provincial, or territorial government having jurisdiction, hunting of migratory birds is regulated primarily by federal governments. Federal involvement is based upon bilateral treaties in 1916 between Great Britain (for Canada) and the United States and in 1936 between the United States and Mexico (Library of Congress 1974). Those treaties specify which migratory birds may be hunted and generally when and for how long they may be hunted. Those functionally responsible for compliance are the Minister of Environment in Canada, the Secretary of the Interior in the United States, and the Director General de Flora y Fauna Silvestre in Mexico.

Additionally, states, most provinces and territories, and even some local governments regulate hunting of migratory birds. In the United States, such regulations are subject to federal supremacy; accordingly, they may be the same or more restrictive but not more liberal than applicable federal regulation frameworks.

Regulations by federal governments and regulations by two or more levels of state, provincial, or regional government can have considerable impact upon the use of regulations in management of harvests. Purposes of this section are to review the authorities and processes for promulgation of regulations in the United States as they might affect that use and to illustrate some of that impact with examples of current use of regulations in harvest management. Although we discuss processes briefly for Canada, authorities and procedures are different enough there and in Mexico to make specific study advisable.

The first important authority for the United States government to be involved in regulation of hunting was the Lacey Act of 1900. That Act, still in effect, makes it a federal crime to transport across state lines wildlife taken in violation of laws of the state where taken. The Act was an outgrowth of the frustration of states unable to pursue poachers beyond their borders, often within habitats, for example, in the middle of a river.

The next effort to extend federal authority was fostered by growing concern over the plight of some migratory birds and the perceived effect of hunting in any one state on hunting in other states. In 1913, Congress passed legislation that declared that migratory birds were "... within the custody and protection ..." of the United States and that it was unlawful to shoot them except in accordance with regulations promulgated by the Secretary of Agriculture. This authority was soon challenged, and by 1915 two courts had declared the 1913 Act invalid because the Congress had not been delegated powers to regulate wildlife. Although the government appealed to the Supreme Court, the supporters of the Act chose to take another course before any decision was rendered.

Federal authority to regulate harvests of migratory birds was extended through a "Convention for the Protection of Migratory Birds" signed by representatives of the United States and Great Britain (for Canada) on 16 August 1916. The required enabling legislation in the United States, the Migratory Bird Treaty Act, was signed 3 July 1918 by President Woodrow Wilson. The Act, essentially unchanged except for amendments to incorporate the treaties with Mexico, Japan, and Russia, remains the key authority for the federal regulation of hunting migratory birds.

The Migratory Bird Treaty Act, although relatively short and succinct, is remarkably comprehensive. As an illustration, Section 703, in part, provides that "Unless and except as permitted by regulations . . . it shall be unlawful at anytime, by any means or in any manner, to pursue, hunt, take, capture, kill, . . . possess, offer for sale, sell, . . . purchase, . . . ship, export, import, . . . transport or cause to be transported, . . . any migratory bird, any part, nest, or eggs of any such bird, . . . included in the terms of the conventions." Further, Section 704, again in part, provides that "Subject to the provisions and . . . to carry out the purposes of the conventions, . . . the Secretary of the Interior is authorized and directed, . . . having due regard to the . . . abundance, . . . of such birds, to determine when, to what extent, if at all, . . . it is compatible with the terms of the conventions to allow hunting." The Act also accommodates regulation by states and territories, to wit Section 708: "Nothing in sections 703–711 . . . shall be construed to prevent . . . states and territories from making or enforcing laws or regulations not inconsistent with . . . said conventions or of said sections, . . . or laws or regulations which shall give further protection to migratory birds, their nests, and eggs." Thus, the authority of the federal government to regulate hunting of migratory birds is clear and, although the federal supremacy is unquestionable, the authority of states to regulate is clearly supported.

Annual Regulations

GENERAL

Federal regulations are considered either "basic" or "annual." "Basic" regulations include those that generally remain unchanged year after year, such as prohibiting more than three shells in a firearm or the use of live decoys. Although probably promulgated for some perceived effect on the ability of individuals to take migratory birds, basic regulations are not regularly used to make annual changes in the management of harvests. "Annual" regulations include especially season lengths and bag limits, and others, that are promulgated each year; these are adjusted in response to changes in the status of migratory birds. The annual promulgation of such regulations is consistent with the requirement for "... due regard of the abundance ..." of migratory birds (Martin and Carney 1977).

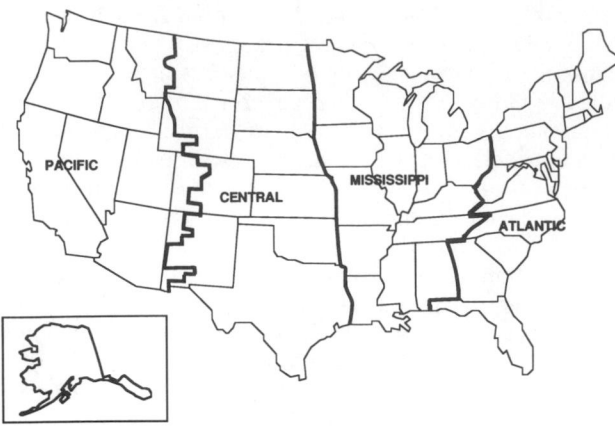

Fig. 7. The boundaries of the four waterfowl Flyway Councils in the United States.

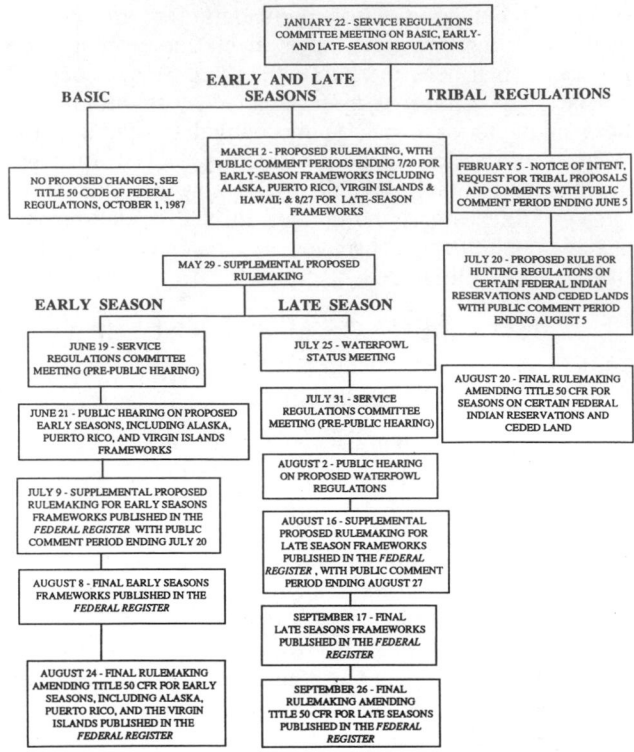

Fig. 8. Example of schedule of dates for migratory bird regulation meetings and publication deadlines for rules.

The processes of promulgating regulations have evolved substantially in the reflection of increased cooperation between federal and state governments, new laws and regulations prescribing government actions, and greatly increased public interest. An example of such cooperation is the organization of all states encompassing major migration routes into "Flyway Councils" for management of waterfowl (Fig. 7) and "management units" for specific populations or groups of migratory birds. The current processes are the subject of an Environmental Impact Statement (U.S. Department of Interior 1988), which is an update of a Final Environmental Statement (FES 75). These statements, in compliance with the National Environmental Policy Act, illustrate ancillary requirements that have greatly increased the complexities of the federal regulatory processes. Such requirements could be incentives to not change, i.e., to use regulations in the sense that any change may require analyses of potential impacts.

The annual regulatory processes involve a series of meetings and consultations beginning in late January or early February each year (Fig. 8). The process is designed to deal with early-season (September) and late-season (October and later) regulations separately. Early-season meetings concern most of the webless migratory game birds (e.g., woodcock and doves), sea ducks (e.g., scoters and eiders), and all migratory game birds in Alaska, the Virgin Islands, and Puerto Rico; late-season meetings involve most of the remaining waterfowl and American coot seasons.

The first meeting (e.g., 22 January, Fig. 8) concerns all migratory game species and is attended only by U.S. Fish and Wildlife Service (FWS) personnel. They examine annual hunting season frameworks, shooting hours, bag limits, and season lengths. Adjustments most commonly considered are in length of seasons and daily bag limits. At this meeting, regulation changes are considered and general harvest strategies are proposed but no final decisions are made. These recommendations are published in the *Federal Register* in early March. The four Flyway Councils (Fig. 7) then review these recommendations and make suggestions to the FWS at the annual North American Wildlife and Natural Resources Conference meeting in late March. Each Flyway Council has its own Technical Committee, made up of migratory-bird biologists from each state, which meets in February or early March. Recommendations from the Technical Committees are forwarded to the Councils prior to the North American meeting.

In February the U.S. Fish and Wildlife Service and representatives of some states hold waterfowl "wing bees"; here wings from ducks and tails from geese that have been mailed in by a selected sample of waterfowl hunters are identified by species, age, and sex, when possible. Some wing bee data are used at the March Technical Committee meetings, but most are not available for use until just before the late-season (July) meetings. The information is combined with harvest survey data and is used at the Waterfowl Status Meeting (see Late Season, Fig. 8).

Also, winter aerial surveys are conducted for ducks and geese. The goose information from this survey and the goose wing bee data are used by the Councils at the March meetings to make preliminary recommendations on future goose harvest. The duck population trend data from the mid-winter survey are poor for most species and are used in only very general terms at the Waterfowl Status Meeting. Exceptions to this may occur for species such as black ducks or for populations of ducks not monitored during the May breeding population survey.

After the end of the hunting season, band recovery information is available and can be used to index the harvest rate that occurred during the recent seasons. Harvest rate is the most reliable and useful indicator of the impact of regulations. Band recovery information also can be used to estimate survival rates and measure the distribution of

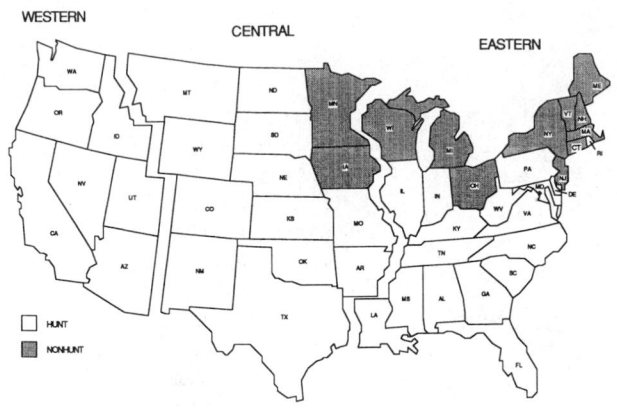

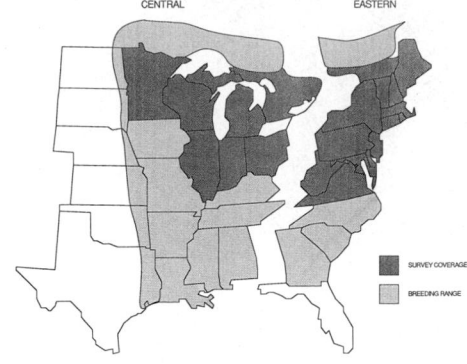

Fig. 10. Woodcock Management Units and call-count survey coverage.

Fig. 9. The boundaries of the three dove Management Units in the United States. Shaded states do not have hunting seasons for mourning doves.

the harvest from specific populations of ducks. This information is especially useful for mallards.

EARLY SEASONS

Meetings begin in June for the early-seasons regulations (Fig. 8). Call count and population trend data for mourning doves and woodcock are assembled in "status reports" and are considered at that meeting. State directors examine the status information and make recommendations for regulations through their Flyway Council. Call-count surveys for doves are conducted annually in the 48 conterminous states (Fig. 9), and >1,000 survey routes are conducted in late May (Dolton 1990). Whether these call counts are an accurate index to population size is not known, but they are used with the assumption that they reflect population change. A similar survey is conducted for white-winged doves in some of the western states. Woodcock also are surveyed with call-counts (Fig. 10). A woodcock wing-collection survey is used to monitor the harvest and age structure of the population.

An example of adjusting regulations in response to changes in call-count indices was the 1987 season for hunting mourning doves in the Western Management Unit (Fig. 9). Call-count surveys had indicated a significant downward population trend over the previous 21 years. Accordingly, both season length and bag limit were reduced from 70 days and 12 doves daily, 24 in possession (in 1986), to 30 days (45 in Arizona) and 10 doves daily, 20 in possession (in 1987).

Each year, sandhill cranes are aerially surveyed on staging areas during spring. In addition, every crane hunter must possess a federal crane hunting permit. One-half of the hunters are surveyed by questionnaire at the end of the season to determine the harvest. With this intensive sampling scheme, actual crane harvest can be monitored closely. These data and the aerial survey data then are used to establish crane hunting regulations.

Regulations recommended by the U.S. Fish and Wildlife Service for "early" hunting seasons that start in September are presented at a public hearing in late June and published for comments by about 20 July so that final rules can be available to hunters by 1 September. Important information on such things as that year's dove and

pigeon productivity rates simply are not available at the time regulations must be developed.

LATE SEASONS

In early May of each year a duck breeding population survey is initiated and continues through mid-June. The survey uses a stratified random, double-sampling scheme and is conducted by low-level aerial observations and ground "truthing" observations. The aerial portion of the survey allows extensive geographical coverage and results in a large sample; the ground survey samples a subset of the area to correct for observation bias. The survey, which has been conducted operationally since 1955, covers the habitat of most of the breeding waterfowl in North America with a series of transects (Fig. 11). In 1990, the survey was expanded through a pilot program into eastern Canada. These data can be used to compute breeding-population estimates and the number of wetlands available to breeding waterfowl in the survey area.

In July a second survey is conducted over most of the area that was surveyed in May to monitor duck production (Fig. 12). This survey provides an index to the number, age, and size of broods produced and the number of adults still on nesting territories. Wetlands also are counted and provide an index to the availability of brood habitat. Ground counts are not made in July to adjust for visibility bias.

Information from the May surveys is combined with that from the July surveys to provide an index of the number of young in autumn (recruitment) and an index to the fall flight of total ducks (FFI). The FFI for all ducks (excluding scoters, eiders, mergansers, and oldsquaws) is determined by the formula (Reynolds 1987:188):

$$FFI = BP(1 + P'),$$

where

BP = total duck breeding population in May, and
P' = $P[(LNI + BI)/BP]/[(LNI + BI)/BP]_{avg}$, where
P = constant base production rate,
LNI = late-nesting index, and
BI = brood index.

A separate fall flight index can be calculated for mallards ($FFI_{(m)}$) with information from the breeding population survey and a production rate estimate derived from har-

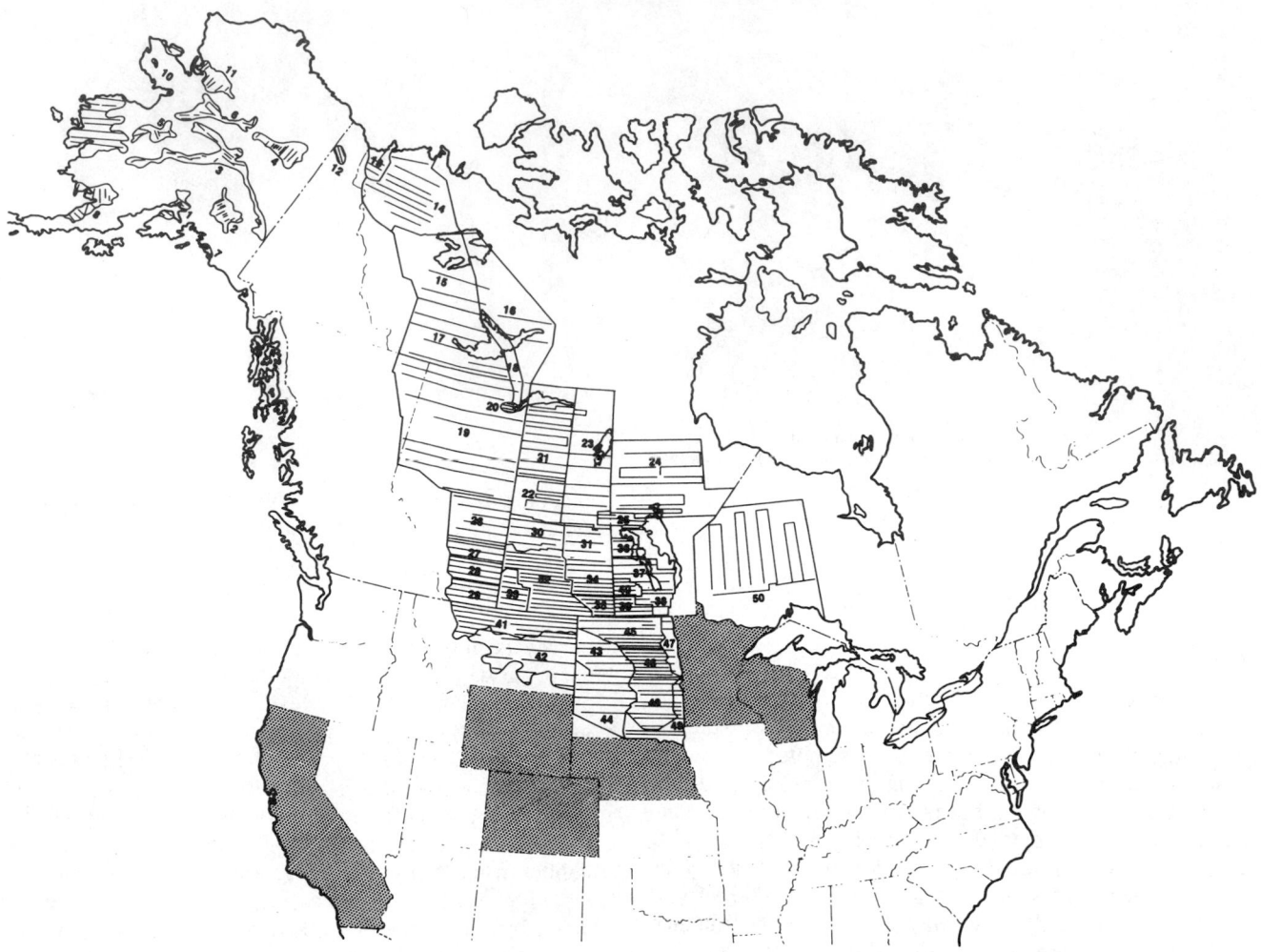

Fig. 11. Transects and strata for principal areas of waterfowl breeding population survey (source, U.S. Fish and Wildlife Service). Numbers simply identify strata.

vest data and banding studies. The formula is (Reynolds 1987:187–89):

$$FFI_{(m)} = N'_{TM} \ S'_m \ (1 + P) + N'_{TF} \ S'_F \ (1 + P),$$

where

N'_{TM} = spring population of adult males,
N'_{TF} = spring population of adult females,
S'_m = summer survival rate of males (15 May–15 Aug) = 0.90,
S'_F = summer survival rate of females (15 May–15 Aug) = 0.82, and
P = production rate (immature/adult in fall) = (y/a)/b,

where

y = the number of young mallards in the U.S. harvest,
a = the number of adult mallards in the U.S. harvest, and
b = the ratio of the band recovery rate of young mallards to the ratio of the band recovery rate of adult mallards.

Because appropriate harvest and band recovery data are not available until after the hunting season, the $FFI_{(m)}$ is not available prior to setting the season. It is useful for understanding variables that influence production and is used to develop models for predicting the mallard fall flight prior to the hunting season. Details of how the data are used in the formula and how the formula was derived are in Reynolds (1987). The fall flight forecasts are heavily relied upon to set the waterfowl seasons and bag limits.

The parts survey data from the February wing bee are tabulated in preliminary form in July. This survey provides estimates of age and sex composition of waterfowl harvested during the previous hunting season and, as noted above, is used with band recovery rates to estimate last year's P or to predict this year's P. In addition, a report is available in July for preliminary estimates of hunting activity, hunter success, and harvest by species for the previous year. Hunter activity and success are determined from a questionnaire sent to a sample of hunters who purchased duck stamps. This information is used with total hunter numbers derived from duck stamp sales to estimate waterfowl harvest. When it is combined with wing bee data, estimates are derived for species, age, and sex of the waterfowl harvest. This information is available by state, flyway, and total U.S. These data are used in conjunction with the fall flight estimates to provide population and harvest guidelines at the July Waterfowl Status Meeting (Fig. 8).

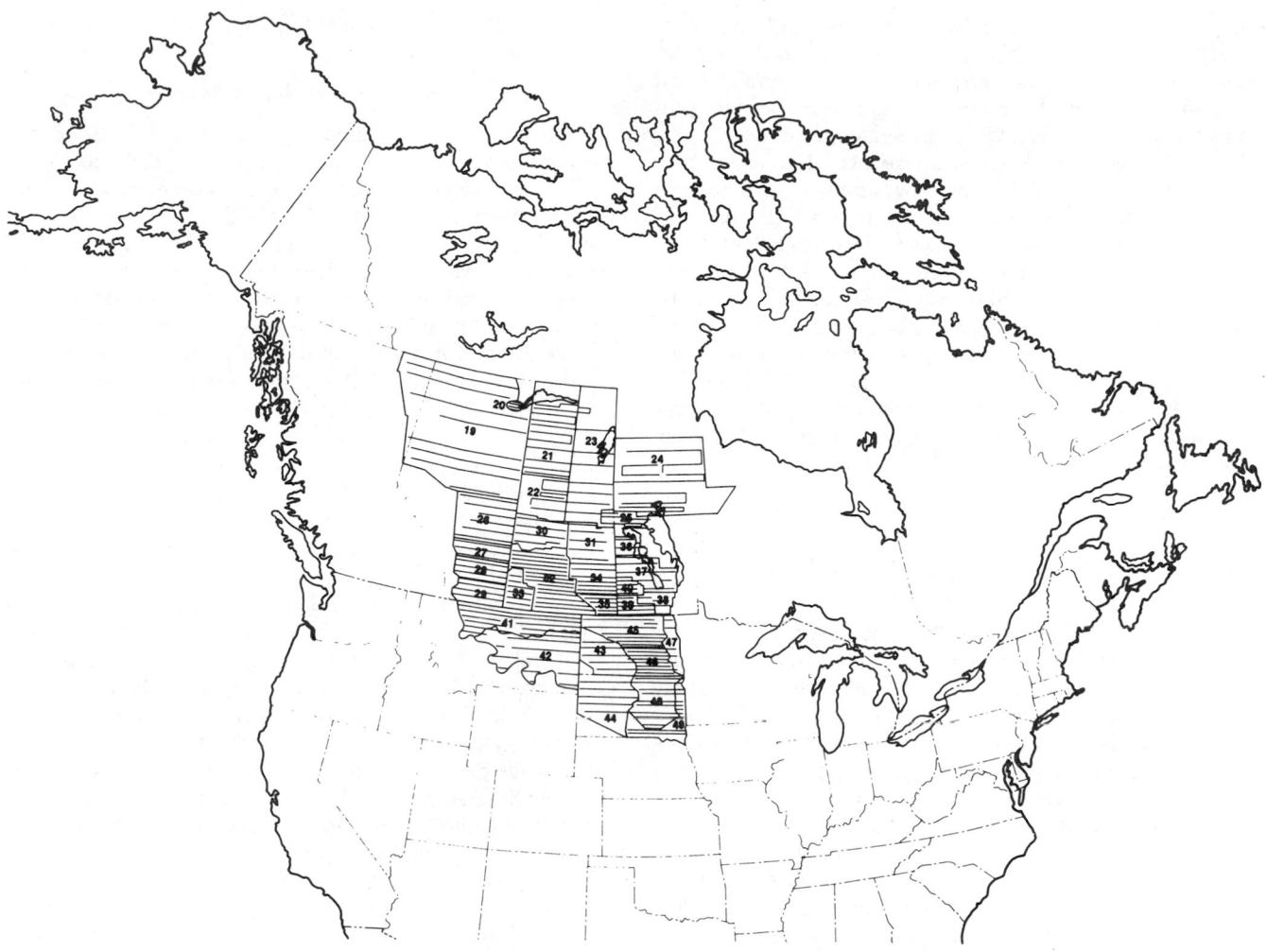

Fig. 12. Transects and strata for principal areas of July waterfowl production survey (source, U.S. Fish and Wildlife Service). Numbers simply identify strata.

At the end of July or the beginning of August, FWS personnel and flyway consultants meet to develop hunting regulation recommendations for waterfowl seasons that begin 1 October or later (Fig. 8). A public hearing is then held to allow public comment on the proposed hunting regulations. After the public hearing the Service Regulations Committee, with the flyway consultants, finalizes hunting season recommendations for publication in the *Federal Register*. Again the public is allowed a brief period (10–14 days) to comment in written form on the proposed regulations published in the *Federal Register*. The U.S. Fish and Wildlife Service then publishes the final hunting season frameworks in the *Federal Register* by mid-September. These then are adopted as law by the end of September. These schedules illustrate the time required to comply with rules, e.g., the Administrative Procedures Act, prescribing activities of the federal government.

An example of how the above data are used to adjust waterfowl regulations is the 1988–89 duck season. Fall-flight indices to autumn waterfowl populations (excluding scoters, eiders, oldsquaws, and mergansers) had decreased 4% from 1987 to 1988 and were 16% below the 1955–87 average. Breeding populations of many species were below population objectives, and the population of northern pintails was the lowest on record. Accordingly, season lengths were reduced 25% from 1987 hunting days (number of days varies among flyways), and bag limits were standardized at three ducks (with species restrictions, e.g., no more than one northern pintail) nationwide; this was a reduction from 1987 bag limits that varied from four to five (again with restrictions) among the flyways. Additionally, the earliest opening was changed from approximately 1 October to 8 October, latest closing from about 20 January to 8 January, and opening shooting hours from one-half hour before sunrise to sunrise. The U.S. Fish and Wildlife Service expected these restrictions to result in harvests at least 25% below the average of 1985–87 harvests (the Service reported that restrictions on hunting during the period 1985–87 already had reduced duck harvest by 25%).

Canada

In Canada, Parliament has jurisdiction over migratory birds because of the British North America Act of 1867, which gives legislative authority to Parliament for wildlife affected by treaty with other countries. Therefore, because of the Migratory Birds Convention Treaty of 1916 with the U.S., Parliament regulates the harvest of migratory birds (Boyd 1979). This system differs from the U.S. in that the U.S. Congress does not set U.S. migratory bird regulations. There have been disagreements among provincial governments over federal authority, but federal legislation prevails. However, provinces have been able to further restrict the harvest of migratory birds relative to federal regulation without conflict (Boyd 1979).

The Canadian Wildlife Service (Department of Environment), after consultation with provincial wildlife agencies, makes recommendations to Parliament for annual migratory bird hunting regulations. Canada does not have public hearings where individuals and groups can impact the regulatory process (Boyd 1979), nor does it have Flyway Councils. The Canadian regulations are adopted earlier in the year (e.g., May) so that most annual production data that are used in the U.S., beyond the earliest breeding population estimates, are not available for use in setting bag limits and seasons.

Effects of Hunting on Migratory Bird Populations

Most of the inferences that can be drawn as to the relationships between hunting and migratory bird populations come from studies of banded waterfowl, particularly mallards. On a nationwide scale, the mallard is the most important waterfowl species in the hunter bag and more mallards are banded than any other species. Therefore the best data for which to examine the potential impacts of hunting on populations exist for mallards.

Anderson and Burnham (1976) and Nichols et al. (1984) showed that the effect of hunting on mallard survival appears to be largely compensatory. That is, hunting mortality is compensated by reductions in other types of mortality, such as disease, and is not additive or causing the mallard population to decline. Most of their findings were based on adult mallards banded from the 1950s to the late 1970s throughout North America. The data were particularly strong for males (Burnham et al. 1984).

However, because of declines in continental waterfowl populations in the 1980s, some biologists were concerned that hunting may be detrimental to waterfowl numbers. Indeed, this is why regulations were used to attempt to restrict harvest by an additional 25% in 1987, as noted above. Smith and Reynolds (1992) suggested that reduced harvest rates, due to harvest restrictions, have resulted in an increased survival of mallards. Trost et al. (1987) reported that mallard and total duck harvest remained relatively constant between 1968–78 and 1979–84 but that hunter numbers declined during that period. Therefore, hunters were becoming more successful in hunting waterfowl, or the unsuccessful no longer were hunting. Mallard harvest in the lower Mississippi Flyway states has increased, and the flyway takes 45% of all mallards harvested in the U.S. (U.S. Department of Interior 1988). The U.S. Department of Interior (1988:49) hypothesized that this harvest "may be sustained in part because ducks are being concentrated in increasingly smaller areas of suitable wintering habitat and thus are more vulnerable to hunting." It should be obvious from the above discussion that the answer to the question of the effects of hunting on waterfowl populations remains speculative and is likely to vary on a case-by-case (species, time, area) situation.

A more detailed examination of this issue was provided by Nichols (1990).

Regulations and Harvest Management

The impacts of specific regulations on harvest of migratory birds are difficult to determine. Indeed, most studies have examined regulatory changes on harvest rate, not overall harvest. Harvest is closely tied to waterfowl population size, and therefore relating changes in hunting regulations to harvest is difficult (Nichols 1990). The regulations that are typically changed to impact harvest and harvest rate include bag limits, season lengths, and framework dates. If migratory bird populations change substantially, one or more of those three regulations also might be changed as a result. This makes it difficult to determine exactly how a particular regulation, such as season length, might affect subsequent harvest rate by hunters.

The effects of regulations on harvest rate of waterfowl have been examined in many different studies (e.g., Martin and Carney 1977, Martin et al. 1979, Rogers et al. 1979, Trost et al. 1987). Martin et al. (1979) examined how regulations in different years affected harvest rate of mallards. They simply classed certain years as having "liberal" or "restrictive" hunting regulations and did not separate out the individual components of the regulations such as bag limits or season lengths. They concluded that harvest rates of mallards in "restrictive" season years were significantly lower than during "liberal" regulation years. Similar results were presented by Rogers et al. (1979). It must be emphasized, however, that these regulations also varied with the fall flight forecast of waterfowl abundance for those years, and therefore population status of mallards is confounded with regulation changes. For example, Martin et al. (1979) also modeled mallard harvest rate as a function of hunting effort and mallard population size. Their model accounted for 97% of the variation in the annual mallard harvest, but they cautioned that many variables, such as season lengths and bag limits, and their impacts on hunter effort could not be separated individually in their analysis.

In an attempt to clarify some of the relationships between waterfowl harvest and hunting regulations, the U.S. Fish and Wildlife Service and Canada's Parliament stabilized waterfowl regulations for 5 years, 1980–85 (Brace et al. 1987). Regulations did not change during this period as waterfowl population levels changed. For example, in the Pacific Flyway seven mallards of either sex could be taken in the daily bag over the 5-year period. The relationships among mallard harvest and mallard population levels, hunter activity (numbers of hunters), mallard age ratio (index to productivity), and weather conditions were investigated by Trost et al. (1987) for the stabilized regulations period. They noted that mallard harvest was a function of numbers of hunters and seasonal hunter success. Mallard population size appeared to influence mallard harvest by affecting hunter success. Hunter effort and mallard population size were the two most important variables affecting mallard harvest. However, band recovery information showed that harvest rates remained constant or declined slightly during the stabilized regulations period (Trost 1987). R. E. Reynolds (pers. commun.) believes that recent banding studies conducted since 1985 demonstrate that restrictive regulations have been effective in reducing harvest rates to the lowest level on record for many species of ducks.

For some waterfowl populations that are geographically well defined, the effects of regulations on harvest are understood more easily. One of the most explicit uses of regulations in management of harvests is the season for hunting tundra swans. Hunting is by special permit only, and each permittee may take one swan per season. The number of permits (12,450 in 1988) authorized is based upon desired harvest rates, set forth in a cooperative management plan and adjusted for hunter success.

Harvests of several populations of geese also are managed relatively explicitly through "quotas," i.e., by desired or allowable harvests. For example, the 1988 season for hunting Canada geese in the "Southern Illinois Quota Zone" closed after 50 days or when 37,000 had been harvested, whichever occurred first, and the daily bag and possession limits were two and four, respectively. For another example, the 1988 season for hunting Canada geese in Arkansas could not exceed 2,400 geese or 16 days, whichever occurred first.

Summary

Migratory bird regulations are established each year following a well-defined set of meetings. At these meetings administrators and biologists try to develop migratory bird harvest guidelines based on sound population survey data when they are available. Along with these data, biologists examine the potential impacts of regulations on harvest and the impacts of harvest on waterfowl survival. An effective planning process should also include habitat in its considerations. After these factors are considered, regulations are presented to the public through an effective communications effort that takes public attitudes and desires into account.

LITERATURE CITED

ALLEN, D. L. 1943. Michigan fox squirrel management. Mich. Dep. Conserv. Game Div. Publ. 100. 404pp.

ALLEN, J. M. 1952. Gray and fox squirrel management in Indiana. Ind. Dep. Conserv. P-R Bull. 1. 112pp.

ANDERSON, D. R., AND K. P. BURNHAM. 1976. Population ecology of the mallard. VI. The effect of exploitation on survival. U.S. Fish Wildl. Serv. Resour. Publ. 125. 110pp.

———, AND R. S. POSPAHALA. 1970. Correction of bias in belt transects of immotile objects. J. Wildl. Manage. 34:141–146.

ANDERSON, W. L. 1964. Survival and reproduction of pheasants released in southern Illinois. J. Wildl. Manage. 28:254–264.

ANEY, W. W. 1974. Estimating fish and wildlife harvest, a survey of methods used. Proc. West. Assoc. Game and Fish Comm. 54:70–79.

ARTHUR, L. M., AND W. R. WILSON. 1979. Assessing the demand for wildlife resources: a first step. Wildl. Soc. Bull. 7:30–34.

BAUMGARTNER, F. M. 1944. Bobwhite quail populations on hunted vs. protected areas. J. Wildl. Manage. 8:259–260.

BAXTER, W. L., AND C. W. WOLFE. 1973. Life history and ecology of the ringnecked pheasant in Nebraska. Nebr. Game and Parks Comm., Lincoln. 58pp.

BEATTIE, K. H., C. J. COWLES, AND R. H. GILES, JR. 1980. Estimating illegal kill of deer. Pages 65–71 in R. L. Hine and S. Nehls, eds. White-tailed deer population management in the north central states. North-Cent. Sect., The Wildl. Soc., Urbana, Ill.

BERGERUD, A. T. 1985. The additive effect of hunting mortality on the natural mortality ranges of grouse. Pages 345–366 in S. L. Beasom and S. F. Roberson, eds. Game harvest management. Caesar Kleberg Wildl. Res. Inst., Kingsville, Tex.

BESADNY, C. D., AND F. H. WAGNER. 1963. An evaluation of pheasant

stocking through the day-old chick program in Wisconsin. Wis. Conserv. Dep. Tech. Bull. 28. 84pp.

BOYCE, M. S. 1989. The Jackson elk herd: intensive wildlife management in North America. Cambridge Univ. Press, Cambridge, Mass. 306pp.

BOYD, H. 1979. Federal roles in wildlife management in Canada. Trans. North Am. Wildl. Nat. Resour. Conf. 44:90–96.

BOYD, R. J., AND J. F. LIPSCOMB. 1976. An evaluation of yearling bull elk hunting restrictions in Colorado. Wildl. Soc. Bull. 4:3–10.

BRACE, R. K., R. S. POSPAHALA, AND R. L. JESSEN. 1987. Background and objectives on stabilized duck hunting regulations: Canadian and U.S. perspectives. Trans. North Am. Wildl. Nat. Resour. Conf. 52: 177–185.

BRAUN, C. E., AND T. BECK. 1985. Effects of changes in hunting regulations on sage grouse harvest and populations. Pages 335–343 in S. L. Beasom and S. F. Roberson, eds. Game harvest management. Caesar Kleberg Wildl. Res. Inst., Kingsville, Tex.

BROWN, D. E. 1979. Factors influencing reproductive success and population densities in Montezuma quail. J. Wildl. Manage. 43:522–526.

———, D. R. BLANKENSHIP, P. K. EVANS, W. H. KIEL, JR., G. L. WAGGERMAN, AND C. K. WINKLER. 1977. White-winged dove. Pages 247–272 in G. C. Sanderson, ed. Management of migratory shore and upland game birds in North America. Int. Assoc. Fish Wildl. Agencies, Washington, D.C.

BROWN, T. L., AND D. J. DECKER. 1979. Incorporating farmers' attitudes into management of white-tailed deer in New York. J. Wildl. Manage. 43:236–239.

BUMP, G., R. DARROW, F. EDMINSTER, AND W. CRISSEY. 1947. The ruffed grouse: life history, propagation, management. N.Y. State Conserv. Dep., Albany. 915pp.

BURNHAM, K. P., D. R. ANDERSON, AND J. L. LAAKE. 1980. Estimation of density from line transect sampling of biological populations. Wildl. Monogr. 72. 202pp.

———, G. C. WHITE, AND D. R. ANDERSON. 1984. Estimating the effect of hunting on annual survival rates of adult mallards. J. Wildl. Manage. 48:350–361.

CARPENTER, L. H., AND R. B. GILL. 1987. Antler point regulations: the good, the bad and the ugly. Proc. Annu. Conf. West. Assoc. Game and Fish Comm. 67:94–107.

CAUGHLEY, G. 1977. Analysis of vertebrate populations. John Wiley & Sons, New York, N.Y. 234pp.

CLUTTON-BROCK, T. H., G. R. IASON, AND F. E. GUINNESS. 1987. Sexual segregation and density-related changes in habitat use in male and female red deer (Cervus elaphus). J. Zool. (Lond.) 211:275–289.

CONAWAY, C. H., K. C. SADLER, AND D. H. HAZELWOOD. 1974. Geographic variation in litter size and onset of breeding in cottontails. J. Wildl. Manage. 38:473–481.

———, H. M. WIGHT, AND K. C. SADLER. 1963. Annual production by a cottontail population. J. Wildl. Manage. 27:171–175.

CONNOLLY, G. E. 1981. Limiting factors and population regulation. Pages 245–285 in O. C. Wallmo, ed. Mule and black-tailed deer of North America. Univ. Nebraska Press, Lincoln.

CRAWFORD, J. A. 1982. Factors affecting sage grouse harvest in Oregon. Wildl. Soc. Bull. 10:374–377.

CREED, W. A., F. HABERLAND, B. E. KOHN, AND K. R. MCCAFFERY. 1984. Harvest management: the Wisconsin experience. Pages 243–260 in L. K. Halls, ed. White-tailed deer: ecology and management. Stackpole Books, Harrisburg, Pa.

DECKER, D. J., AND K. G. PURDY. 1988. Toward a concept of wildlife acceptance capacity in wildlife management. Wildl. Soc. Bull. 16: 53–57.

DENNEY, R. N. 1978. Managing the harvest. Pages 395–408 in J. L. Schmidt and D. L. Gilbert, eds. Big game of North America: ecology and management. Stackpole Books, Harrisburg, Pa.

DESTEFANO, S., AND D. H. RUSCH. 1986. Harvest rates of ruffed grouse in northeastern Wisconsin. J. Wildl. Manage. 50:361–367.

DICKENSON, N. R. 1986. Testing selected harvest ratios for adult deer. N.Y. Fish Game J. 33:11–15.

DOLBEER, R. A., AND W. R. CLARK. 1975. Population ecology of snowshoe hares in the central Rocky Mountains. J. Wildl. Manage. 39: 535–549.

DOLTON, D. D. 1990. Mourning dove breeding population status, 1990. U.S. Fish Wildl. Serv., Washington, D.C. 12pp.

DONOHOE, R. W., AND R. K. MARTINSON. 1961. A preliminary report of age and sex ratios among Ohio squirrel populations. Ohio Div. Wildl. Rel. 75. 5pp.

DORNEY, R. S., AND C. KABAT. 1960. Relation of weather, parasitic disease and hunting to Wisconsin ruffed grouse populations. Wis. Conserv. Dep. Tech. Bull. 20. 66pp.

DOWNING, R. L. 1981. Deer harvest sex ratios: a symptom, a prescription, or what? Wildl. Soc. Bull. 9:8–13.

———, AND D. C. GUYNN, JR. 1983. A generalized sustained yield table for white-tailed deer. Pages 95–103 in Game harvest management. Caesar Kleberg Wildl. Res. Inst., Kingsville, Tex.

DUNKS, J. H., R. E. TOMLINSON, H. M. REEVES, D. D. DOLTON, C. E. BRAUN, AND T. P. ZAPATKA. 1982. Migration, harvest, and population dynamics of mourning doves banded in the Central Management Unit, 1967–77. U.S. Fish Wildl. Serv. Spec. Sci. Rep. Wildl. 249. 128pp.

EBERHARDT, L. C. 1988. Testing hypotheses about populations. J. Wildl. Manage. 52:50–56.

EBERHARDT, L. L., AND M. A. SIMMONS. 1987. Calibrating population indices by double sampling. J. Wildl. Manage. 51:665–675.

EDMINSTER, F. C. 1937. An analysis of the value of refuges for cyclic game species. J. Wildl. Manage. 1:37–41.

ELLINGWOOD, M. R., AND J. V. SPIGNESI. 1986. Management of an urban deer herd and the concept of cultural carrying capacity. Trans. Northeast Deer Tech. Comm., Vt. Fish Game Dep. 22:42–45.

ERICKSON, A. B., D. B. VESALL, C. E. CARLSON, AND C. T. ROLLINGS. 1951. Minnesota's most important game bird the pheasant. Flicker 23:23–49.

FISCHER, C. A., AND L. B. KEITH. 1974. Population responses of central Alberta ruffed grouse to hunting. J. Wildl. Manage. 38:585–600.

FOGARTY, M. J., AND K. A. ARNOLD. 1977. Common snipe. Pages 189–209 in G. C. Sanderson, ed. Management of migratory shore and upland game birds in North America. Int. Assoc. Fish Wildl. Agencies, Washington, D.C.

FOSSIL CREEK SOFTWARE. 1986. POP-II: system documentation version 6.0. Fossil Creek Software, Ft. Collins, Colo. 59pp.

FOUCH, W. R. 1969. Results of 3 years of early seasons at the Rose Lake Wildlife Research Area. Michigan Dep. Nat. Resour. Rep. 175. 4pp.

GASSON, W., AND L. WOLLRAB. 1986. Integrating population simulation modeling into a planned approach to pronghorn management. Proc. Pronghorn Antelope Workshop 12:86–98.

GILBERT, D. L. 1971. Natural resources and public relations. The Wildl. Soc., Washington, D.C. 320pp.

GLADFELTER, L. 1980. Deer population estimators in the Midwest farmland. Pages 5–11 in R. L. Hine and S. Nehls, eds. White-tailed deer population management in the north central states. North-Cent. Sect., The Wildl. Soc., Urbana, Ill.

GLADING, B., AND R. W. SAARNI. 1944. Effect of hunting on a valley quail population. Calif. Fish Game 30:71–79.

GROSS, J. E., J. E. ROELLE, AND G. L. WILLIAMS. 1973. Program ONE-POP and information processor: a systems modeling and communications project. Colorado Coop. Wildl. Res. Unit Program Rep., Ft. Collins. 327pp.

GULLION, G. W., AND W. H. MARSHALL. 1968. Survival of ruffed grouse in a boreal forest. Living Bird 7:117–167.

HARDER, J. D. 1980. Reproduction of white-tailed deer in the north-central United States. Pages 23–35 in R. L Hine and S. Nehls, eds. White-tailed deer population management in the north central states. North-Cent. Sect., The Wildl. Soc., Urbana, Ill.

HARJU, H. J. 1974. An analysis of some aspects of the ecology of dusky grouse. Ph.D. Thesis, Univ. Wyoming, Laramie. 142pp.

HICKEY, J. J. 1955. Some American population research on gallinaceous birds. Pages 326–396 in A. Wolfson, ed. Recent studies in avian biology. Univ. Illinois Press, Urbana.

HOFFMAN, R. W., AND C. E. BRAUN. 1975. A volunteer wing collection station. Colo. Div. Wildl. Game Inf. Leafl. 101. 3pp.

INSTITUTE FOR PARTICIPATORY PLANNING. 1981. Citizen participation handbook for public officials & other professionals servicing the public. Inst. Participatory Planning, Laramie, Wyo. 126pp.

JOHNSGARD, P. A. 1973. Grouse and quails of North America. Univ. Nebraska Press, Lincoln. 553pp.

KEITH, L. B., O. J. RONGSTAD, AND E. C. MESLOW. 1966. Regional differences in reproductive traits of the snowshoe hare. Can. J. Zool. 44:953–961.

KLIMSTRA, W. D. 1975. Harvest returns of pen-reared bobwhite quail. Trans. Ill. State Acad. Sci. 68:278–284.

KUBISIAK, J. F. 1984. The impact of hunting on ruffed grouse populations in the Sandhill wildlife area. Pages 151–168 in W. L. Robinson, ed. Ruffed grouse management: state of the art in the early 1980s. North-Cent. Sect., The Wildl. Soc., St. Louis, Mo.

LANCIA, R. A., K. H. POLLOCK, J. W. BISHIR, AND M. C. CONNER. 1988. A white-tailed deer harvesting strategy. J. Wildl. Manage. 52:589–595.

LANG, L. M., AND G. W. WOOD. 1976. Manipulation of the Pennsylvania deer herd. Wildl. Soc. Bull. 4:159–165.

LANGENAU, E. 1988. Managing Michigan herds for trophy bucks. Michigan Dep. Nat. Resour. Wildl. Div. Rep. 3080. 15pp.

LEOPOLD, A. 1933. Game management. Charles Scribner's Sons, New York, N.Y. 481pp.

LIBRARY OF CONGRESS. 1974. Treaties and other international agreements on fisheries, oceanographic resources, and wildlife to which the United States is party. U.S. Gov. Printing Off., Washington, D.C. 968pp.

LOBDELL, C. H., K. E. CASE, AND H. S. MOSBY. 1972. Evaluation of harvest strategies for a simulated wild turkey population. J. Wildl. Manage. 36:493–497.

MACNAB, J. 1985. Carrying capacity and related slippery shibboleths. Wildl. Soc. Bull. 13:403–410.

MADSON, J. 1962. The ring-necked pheasant. Conserv. Dep., Olin Mathieson Chem. Corp., East Alton, Ill. 104pp.

MAJOR, P. D., AND J. C. OLSON. 1980. Harvest statistics from Indiana's ruffed grouse hunting seasons. Wildl. Soc. Bull. 8:18–23.

MARTIN, E. M., AND S. M. CARNEY. 1977. Population ecology of the mallard. IV. A review of duck hunting regulations, activity and success, with special reference to the mallard. U.S. Fish Wildl. Serv. Resour. Publ. 130. 137pp.

MARTIN, F. W., R. S. POSPAHALA, AND J. D. NICHOLS. 1979. Assessment and population management of North American migratory birds. Pages 187–239 in J. Cairns, Jr., G. P. Patil, and W. E. Water, eds. Environmental and biomonitoring, assessment, prediction, and management—certain case studies and related quantitative issues. Statistical Ecology Vol. 11. Int. Coop. Publ. House, Fairland, Md.

MARTINSON, R. K., J. W. HOLTEN, AND G. K. BRAKHAGE. 1961. Age criteria and population dynamics of the swamp rabbit in Missouri. J. Wildl. Manage. 25:271–281.

MCCAFFERY, K. R. 1973. Road kills show trends in Wisconsin deer populations. J. Wildl. Manage. 37:212–216.

———. 1989. Deer population dynamics and management in Wisconsin. Proc. East. Wildl. Damage Control Conf. 4:155–161.

MCCULLOUGH, D. R. 1979. The George Reserve deer herd: population ecology of a K-selected species. Univ. Michigan Press, Ann Arbor. 271pp.

———. 1984. Lessons from the George Reserve, Michigan. Pages 211–242 in L.K. Halls, ed. White-tailed deer: ecology and management. Stackpole Books, Harrisburg, Pa.

———, D. S. PINE, D. L. WHITMORE, T. M. MANSFIELD, AND R. H. DECKER. 1990. Linked sex harvest strategy for big game management with a test case on black-tailed deer. Wildl. Monogr. 112. 41pp.

MCKAY, D. O., AND B. J. VERTS. 1978. Estimates of some attributes of a population of Nuttall's cottontails. J. Wildl. Manage. 42:159–168.

MOEN, A. N., C. W. SEVERINGHAUS, AND R. A. MOEN. 1986. Deer CAMP: computer-assisted management program operating manual and tutorial. CornerBrook Press, Lansing, N.Y. 170pp.

MOHLER, L. L., AND D. E. TOWEILL. 1982. Regulated elk populations and hunter harvests. Pages 561–597 in J. W. Thomas and D. E. Toweill, eds. Elk of North America. Stackpole Books, Harrisburg, Pa.

MONTANA DEPARTMENT OF FISH, WILDLIFE & PARKS. 1985. Antlered elk and deer management in Montana: past trends and current status. Montana Dep. Fish, Wildl. Parks, Helena. 68pp.

MOSBY, H. S. 1969. The influence of hunting on the population dynamics of a woodlot gray squirrel population. J. Wildl. Manage. 33:59–73.

———, AND C. O. HANDLEY. 1943. The wild turkey in Virginia: its status, life history, and management. Va. Commonw. Game Inland Fish., Richmond. 281pp.

———, R. L. KIRKPATRICK, AND J. O. NEWELL. 1977. Seasonal vulnerability of gray squirrels to hunting. J. Wildl. Manage. 41:284–289.

———, AND W. S. OVERTON. 1950. Fluctuations in the quail population on the Virginia Polytechnic Institute farms. Trans. North Am. Wildl. Conf. 15:347–355.

NATIONAL WILD TURKEY FEDERATION. 1986. Guide to the American wild turkey. Natl. Wild Turkey Fed., Edgefield, S.C. 189pp.

NICHOLS, J. D. 1990. Responses of North American duck populations to exploitation. Pages 488–515 *in* C. M. Perrins, J. D. Lebreton, and G. J. M. Hirons, eds. Bird population studies: their relevance to conservation and management. Oxford Univ. Press, Oxford, U.K.

———, M. J. CONROY, D. R. ANDERSON, AND K. P. BURNHAM. 1984. Compensatory mortality in waterfowl populations: a review of the evidence and implications for research and management. Trans. North Am. Wildl. Nat. Resour. Conf. 49:535–554.

NIXON, C. M., R. W. DONOHOE, AND T. NASH. 1974. Overharvest of fox squirrels from two woodlots in western Ohio. J. Wildl. Manage. 38:67–80.

———, M. W. MCCLAIN, AND R. W. DONOHOE. 1975. Effects of hunting and mast crops on a squirrel population. J. Wildl. Manage. 39:1–25.

OWEN, R. B. 1977. American woodcock. Pages 146–186 *in* G. C. Sanderson, ed. Management of migratory shore and upland game birds in North America. Int. Assoc. Fish Wildl. Agencies, Washington, D.C.

PALMER, W. L., AND C. L. BENNETT. 1963. Relation of season length to hunting harvest of ruffed grouse. J. Wildl. Manage. 27:634–639.

PETERLE, T. J., AND W. R. FOUCH. 1969. Exploitation of a fox squirrel population on a public shooting area. Mich. Dep. Conserv. Rep. 2251. 4pp.

PORATH, W. R. 1980. Fawn mortality estimates in farmland deer range. Pages 55–63 *in* R. L. Hine and S. Nehls, eds. White-tailed deer population management in the north central states. North-Cent. Sect., The Wildl. Soc., Urbana, Ill.

REYNOLDS, R. E. 1987. Breeding duck population, production and habitat surveys, 1979–85. Trans. North Am. Wildl. Nat. Resour. Conf. 52:186–205.

ROBINETTE, W. L., N. V. HANCOCK, AND D. A. JONES. 1977. The Oak Creek mule deer herd in Utah. Utah Div. Wildl. Resour. Publ. 77-15. 148pp.

ROGERS, J. P., J. D. NICHOLS, F. W. MARTIN, C. F. KIMBALL, AND R. S. POSPAHALA. 1979. An examination of harvest and survival rates of ducks in relation to hunting. Trans. North Am. Wildl. Nat. Resour. Conf. 44:114–126.

ROSE, G. B. 1977. Mortality rates of tagged adult cottontail rabbits. J. Wildl. Manage. 41:511–514.

ROSEBERRY, J. L. 1979. Bobwhite population responses to exploitation: real and simulated. J. Wildl. Manage. 43:285–305.

———, AND A. WOOLF. 1988. Evidence for and consequences of deer harvest data biases. Proc. Annu. Conf. Southeast. Assoc. Fish Wildl. Agencies 42:306–314.

RYEL, L. A. 1980. The legal deer kill—how it's measured. Pages 37–45 *in* R. L. Hine and S. Nehls, eds. White-tailed deer population

management in the north central states. Proc. North-Cent. Sect., The Wildl. Soc., Urbana, Ill.

SANDERSON, G. C. 1977. Management of migratory shore and upland game birds in North America. Int. Assoc. Fish Wildl. Agencies, Washington, D.C. 358pp.

SHAW, J. H. 1985. Introduction to wildlife management. McGraw-Hill, New York, N.Y. 316pp.

SMALL, R. J., J. C. HOLZWART, AND D. H. RUSCH. 1991. Predation and hunting mortality of ruffed grouse in central Wisconsin. J. Wildl. Manage. 55:512–520.

SMITH, G. W., AND R. E. REYNOLDS. 1992. Hunting and mallard survival, 1979–88. J. Wildl. Manage. 56:306–316.

STOLL, R. J., JR., AND G. L. MOUNTZ. 1983. Rural landowner attitudes toward deer and deer populations in Ohio. Ohio Dep. Nat. Resour., Div. Wildl. Fish Wildl. Rep. 10. 18pp.

STRICKLAND, M. D. 1982. Interpretation of post hunt sex ratios in Wyoming elk herds. Pages 129–141 *in* Proc. western states elk workshop, Flagstaff, Ariz.

TOMLINSON, R. E., D. D. DOLTON, H. M. REEVES, J. D. NICHOLS, AND L. A. MCKIBBEN. 1988. Migration, harvest, and population characteristics of mourning doves banded in the western management unit 1964–1977. U.S. Fish Wildl. Serv. Fish Wildl. Tech. Rep. I–IV. 101pp.

TRENT, T. T., AND O. J. RONGSTAD. 1974. Home range and survival of cottontail rabbits in southwestern Wisconsin. J. Wildl. Manage. 38:459–472.

TROST, R. E. 1987. Mallard survival and harvest rates: a reexamination of relationships. Trans. North Am. Wildl. Nat. Resour. Conf. 52:232–264.

———, D. E. SHARP, S. T. KELLY, AND F. D. CASWELL. 1987. Duck harvests and proximate factors influencing hunting activity and success during the period of stabilized regulations. Trans. North Am. Wildl. Nat. Resour. Conf. 52:216–232.

UHLIG, H. G. 1956. The gray squirrel in West Virginia. W.Va. Conserv. Comm. Div. Game Manage. Bull. 3. 83pp.

U.S. DEPARTMENT OF INTERIOR. 1988. Issuance of annual regulations permitting the sport hunting of migratory birds. U.S. Fish Wildl. Serv., Washington, D.C. 340pp.

U.S. FISH AND WILDLIFE SERVICE. 1973. Tactical planning in fish and wildlife management and research. U.S. Fish Wildl. Serv. Resour. Publ. 123. 19pp.

VANCE, D. R., AND J. A. ELLIS. 1972. Bobwhite populations and hunting on Illinois public hunting areas. Proc. Natl. Bobwhite Quail Symp. 1:165–174.

WEAVER, J. K., AND H. S. MOSBY. 1979. Influence of hunting regulations on Virginia wild turkey populations. J. Wildl. Manage. 43:128–135.

WIGHT, H. M., R. U. MACE, AND W. M. BATTERSON. 1967. Mortality estimates of an adult band-tailed pigeon population in Oregon. J. Wildl. Manage. 31:519–525.

18

IDENTIFICATION AND CONTROL OF WILDLIFE DAMAGE

Richard A. Dolbeer, Nicholas R. Holler, and Donald W. Hawthorne

INTRODUCTION

Wildlife management often is thought of in terms of protecting, enhancing, and nurturing wildlife populations and the habitat needed for their well-being. However, many species at one time or another require management actions to reduce conflicts with people or with other wildlife species. Examples include an airport manager modifying habitats to reduce gull activity near runways, a forester poisoning pocket gophers to increase tree seedling survival in a reforestation project, or a biologist trapping an abundant predator or competing species to enhance survival of an endangered species.

Wildlife damage control is an increasingly important part of the wildlife management profession because of expanding human populations and intensified land use practices. Concurrent with this growing need to reduce wildlife-people conflicts, public attitudes and environmental regulations are restricting use of some of the traditional tools of control such as poisons and traps. Agencies and individuals carrying out control programs are being scrutinized more carefully to ensure that their actions are justified, environmentally safe, and in the public interest. Thus, wildlife damage-control activities must be based on sound economic, ecological, and sociological principles and carried out as positive, necessary components of overall wildlife management programs.

Wildlife damage-control programs can be thought of as having four parts: (1) problem definition, (2) ecology of the problem species, (3) control methods application, and (4) evaluation of control. Problem definition refers to determining the species and numbers of animals causing the problem, the amount of loss or nature of the conflict, and other biological and social factors related to the problem. Ecology of the problem species refers to understanding the life history of the species, especially in relation to the conflict. Control methods application refers to taking the information gained from (1) and (2) to develop an appropriate management program to alleviate or reduce the conflict. Evaluation of control permits an assessment of the reduction in damage in relation to costs and impact of the control on target and nontarget populations. Increasingly, emphasis is being placed on integrated pest management whereby several control methods are used in combination and coordinated with other management practices being used at that time (Fig. 1).

This chapter focuses on techniques related to problem definition and methods application. Each major section on groups of wildlife species has three parts—one on assessment of damage; one on identification of damage by individual species; and one on control techniques, which is an elaboration of those listed under each of the species.

LEGAL REQUIREMENTS FOR CONTROL
Capturing or Killing Wildlife Species

Before action is taken to control wildlife damage, it is important to understand the laws covering the target wildlife species. The management of most wild mammals, reptiles, and amphibians in the United States and Canada is the responsibility of the individual states and provinces. The capture, possession, or killing of these vertebrates to achieve control of damage or nuisance situations is regulated by state or provincial laws. The main exception for

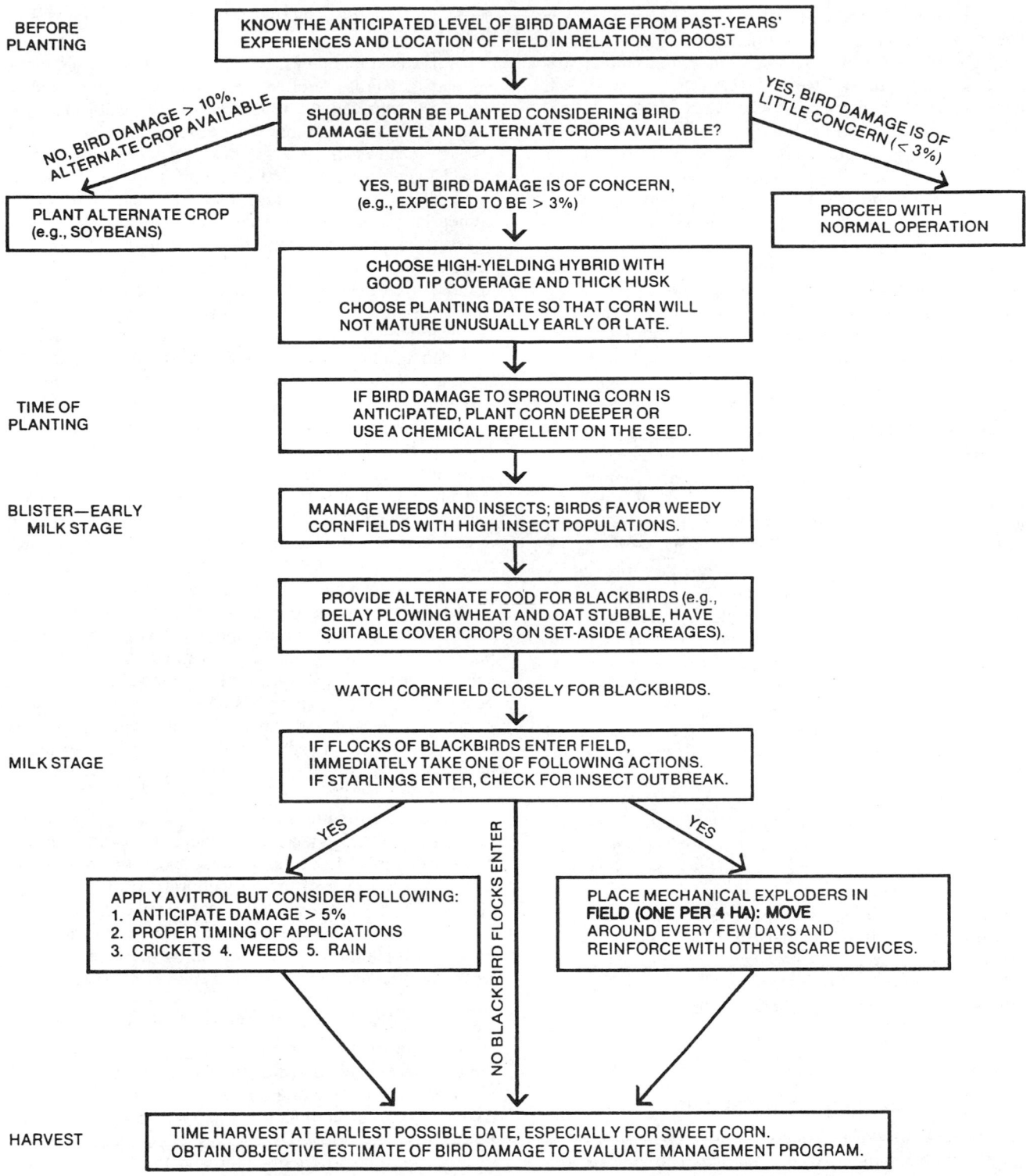

BEFORE PLANTING

KNOW THE ANTICIPATED LEVEL OF BIRD DAMAGE FROM PAST-YEARS' EXPERIENCES AND LOCATION OF FIELD IN RELATION TO ROOST

SHOULD CORN BE PLANTED CONSIDERING BIRD DAMAGE LEVEL AND ALTERNATE CROPS AVAILABLE?

NO, BIRD DAMAGE > 10%, ALTERNATE CROP AVAILABLE

YES, BIRD DAMAGE IS OF LITTLE CONCERN (< 3%)

YES, BUT BIRD DAMAGE IS OF CONCERN, (e.g., EXPECTED TO BE > 3%)

PLANT ALTERNATE CROP (e.g., SOYBEANS)

PROCEED WITH NORMAL OPERATION

CHOOSE HIGH-YIELDING HYBRID WITH GOOD TIP COVERAGE AND THICK HUSK

CHOOSE PLANTING DATE SO THAT CORN WILL NOT MATURE UNUSUALLY EARLY OR LATE.

TIME OF PLANTING

IF BIRD DAMAGE TO SPROUTING CORN IS ANTICIPATED, PLANT CORN DEEPER OR USE A CHEMICAL REPELLENT ON THE SEED.

BLISTER—EARLY MILK STAGE

MANAGE WEEDS AND INSECTS; BIRDS FAVOR WEEDY CORNFIELDS WITH HIGH INSECT POPULATIONS.

PROVIDE ALTERNATE FOOD FOR BLACKBIRDS (e.g., DELAY PLOWING WHEAT AND OAT STUBBLE, HAVE SUITABLE COVER CROPS ON SET-ASIDE ACREAGES).

WATCH CORNFIELD CLOSELY FOR BLACKBIRDS.

MILK STAGE

IF FLOCKS OF BLACKBIRDS ENTER FIELD, IMMEDIATELY TAKE ONE OF FOLLOWING ACTIONS. IF STARLINGS ENTER, CHECK FOR INSECT OUTBREAK.

YES

NO BLACKBIRD FLOCKS ENTER

YES

APPLY AVITROL BUT CONSIDER FOLLOWING:
1. ANTICIPATE DAMAGE > 5%
2. PROPER TIMING OF APPLICATIONS
3. CRICKETS 4. WEEDS 5. RAIN

PLACE MECHANICAL EXPLODERS IN FIELD (ONE PER 4 HA): MOVE AROUND EVERY FEW DAYS AND REINFORCE WITH OTHER SCARE DEVICES.

HARVEST

TIME HARVEST AT EARLIEST POSSIBLE DATE, ESPECIALLY FOR SWEET CORN. OBTAIN OBJECTIVE ESTIMATE OF BIRD DAMAGE TO EVALUATE MANAGEMENT PROGRAM.

Fig. 1. Schematic chart of integrated management program on farm to reduce blackbird damage to corn (from Dolbeer 1980).

mammals, reptiles, and amphibians in the United States regards endangered species that are regulated at the federal level by the Endangered Species Act of 1973, as amended.

Migratory birds, in contrast to these other vertebrates, are managed in North America at the federal level under the Migratory Bird Treaty Act of 1918, a treaty that has been amended several times and includes formal agreements with Canada, Mexico, Japan, and the Soviet Union (see Chapter 17). Federal regulations in the United States and Canada require that a depredation permit be obtained from the U.S. Fish and Wildlife Service and Canadian Wildlife Service, respectively, before any person may capture, kill, possess, or transport most migratory birds to control depredations. No federal permit is required merely to scare or herd depredating birds other than endangered or threatened species, or bald or golden eagles.

Introduced avian species in the United States such as

house sparrows, pigeons, starlings, and monk parakeets have no federal protection. Furthermore, a federal permit is not required to control yellow-headed, red-winged, tricolored, rusty, and Brewer's blackbirds, cowbirds, all grackles, crows, and magpies when they are found committing or about to commit depredations upon ornamental or shade trees, agricultural crops, livestock, or wildlife or when they are concentrated in such numbers and manner as to constitute a health hazard. However, federal provisions do not circumvent any state laws or regulations which may be more, but not less, restrictive.

In summary, anyone contemplating the capture or killing of a vertebrate species for damage control must first determine the state or provincial regulations for that species. For birds and endangered species, federal regulations also must be followed.

EPA Registration of Chemicals

The Federal Insecticide, Fungicide, and Rodenticide Act (FIFRA), as amended, requires all pesticides and other chemicals used in controlling or repelling organisms in the U.S. to be approved and registered by the Environmental Protection Agency (EPA). The registration process has become increasingly complex and costly, not only for new products being introduced but also for previously registered products being reviewed and re-evaluated (Hood 1978, Goldman 1988). Products federally registered for nationwide use under Section 3 of FIFRA may not be available for use in all states in the U.S., because many states have their own registration requirements that might be more restrictive. Some products have Section 24C registrations that are valid only for specific states that have localized problems. Occasionally, products are available temporarily in specific localities for emergency use under Section 18 provisions of FIFRA. Finally, many of the registered compounds, such as vertebrate toxicants, are classified as "restricted use" pesticides. These products can be used only by, or under the direct supervision of, a certified pesticide applicator. Each state has its own certification requirements. Thus, anyone contemplating use of chemicals in wildlife damage control must determine the status of and requirements for use of those chemicals in their particular locality. Jacobs (1994) provided a comprehensive list of registered chemicals for wildlife damage control.

BIRDS

Damage Assessment

Birds annually destroy many millions of dollars worth of agricultural crops in North America. The greatest loss appears to be from blackbirds feeding on ripening corn; a survey in 1981 indicated a loss of 272,154 metric tons worth $31 million in the United States (Besser and Brady 1986). Blackbird damage to sunflowers in the upper Great Plains states was estimated at $5 million in 1979 and $8 million in 1980 (Hothem et al. 1988). Damage by various bird species to fruit crops, peanuts, truck crops, and small grains also can be severe in localized areas (Besser 1986). Fish-eating birds can cause major losses at fish-rearing facilities. Economic losses from bird strikes to aircraft are perhaps more substantial than those in agriculture—at least $20 million annually each for U.S. commercial air

carriers (Steenblik 1983) and military aircraft (Merritt 1990).

Unlike most mammals, which are secretive when causing damage, birds are often highly visible and the damage is usually conspicuous. For these reasons, subjective estimates often overestimate losses as much as 10-fold (Weatherhead et al. 1982). Thus, objective estimates of bird damage to agricultural crops are important to accurately define the magnitude of the problem and to plan appropriate, cost-effective control actions (Dolbeer 1981).

To estimate losses to birds in agricultural crops, one must devise a sampling scheme to select the fields that are to be examined and then determine the plants or areas to be measured in the selected fields (Stickley et al. 1979). For example, to objectively estimate the amount of blackbird damage in a ripening corn or sunflower field, the estimator should examine at least 10 locations widely spaced in the field. If a field has 100 rows and is 300 m long, the estimator might walk staggered distances of 30 m along 10 randomly selected rows (e.g., 0–30 m in row 9, 31–60 m in row 20, and so on). In each 30-m length, the estimator should randomly select 10 plants and estimate the damage on each plant's ear or head. Bird damage to corn can be estimated by measuring the length of damage on the ear (DeGrazio et al. 1969) or by visually estimating the percent loss of kernels (Woronecki et al. 1980) and converting to yield loss per hectare. Fruit loss can be estimated by counting the numbers of undamaged, pecked, and removed fruits per sampled branch (Tobin and Dolbeer 1987). Sprouting rice removed by birds can be estimated by comparing plant density in exposed plots with that in adjacent plots with wire bird exclosures (Otis et al. 1983). The seeded surface area of sunflower heads destroyed by birds can be estimated with the aid of a clear plastic template (Dolbeer 1975).

Losses of agricultural crops to birds can be estimated indirectly through avian bioenergetics. By estimating the number of birds of the depredating species feeding in an area, the percentage of the agricultural crop in the birds' diet, the caloric value of the crop, and the daily caloric requirements of the birds, one can project the total biomass of crop removed by birds on a daily or seasonal basis (Weatherhead et al. 1982, White et al. 1985).

Species Damage Identification

Most bird damage occurs during daylight hours, and the best way to identify the species causing damage is by observation. Presence of a bird species in a crop receiving damage does not automatically prove the species guilty, however. For example, large, conspicuous flocks of common grackles in sprouting winter wheat fields were found, after careful observation and examination of stomach contents, to be eating corn residue from the previous crop. Smaller numbers of starlings were removing the germinating wheat seeds (Dolbeer et al. 1979). Below, the characteristics of damage for various groups of birds are described.

GULLS

Several gull species have adapted to existing in proximity to people, taking advantage of landfills for food. For example, the ring-billed gull population in the Great Lakes region has been increasing at about 10%/year since the early 1970s (Blokpoel and Tessier 1986). Gulls are the most serious bird threat to flight safety at airports (Solman 1981). They are increasingly causing nuisance problems in urban areas by begging for food, defacing property, contaminating municipal water supplies, and nesting on rooftops. In rural areas, gulls sometimes feed on fruit crops and at aquaculture facilities, eat duck eggs and kill ducklings, and compete with threatened bird species for nest sites.

Control Techniques.—Habitat manipulation, screening and wire grids, mechanical and chemical frightening agents, toxicants, shooting.

BLACKBIRDS AND STARLINGS

The term ''blackbird'' loosely refers to a group of about 10 species of North American birds, the most common of which are the red-winged blackbird, common grackle, and brown-headed cowbird. The starling, a European species introduced to North America in the late 1800s, superficially resembles native blackbirds and often associates with them. Together, blackbirds and starlings constitute the most abundant group of birds in North America, comprising a combined population of more than 1 billion (Dolbeer and Stehn 1983).

Blackbird damage to ripening corn, sunflowers, and rice can be serious (Dolbeer 1994). Much of this damage is done in late summer during the milk or dough stage of seed development. The seed contents of corn are removed, leaving the pericarp or outer coat on the cob (Fig. 2). Blackbird damage to sprouting rice in the spring can be serious in localized areas.

Starling depredations at feedlots in winter can cause substantial losses (Besser et al. 1968, Glahn et al. 1983). Although contamination of livestock feed by starling feces is often a concern of farmers, a study indicated this contamination did not interfere with food consumption or weight gain of cattle and pigs (Glahn and Stone 1984). Starlings can seriously damage fruit crops such as cherries and grapes.

Perhaps the greatest problem caused by blackbirds and starlings is their propensity to gather together in large, nocturnal roosting congregations, especially in winter. The noise, fecal accumulation, and general nuisance caused by millions of birds roosting together near human habitations can be significant (White et al. 1985). Roosting birds near airports can create a safety hazard for aircraft, and roost sites, if used for several years, can become focal points for the fungus that causes histoplasmosis, a respiratory disease in humans.

Control Techniques.—Habitat manipulation, cultural practices (e.g., resistant crop varieties), proofing and screening, mechanical and chemical frightening agents, repellents, toxicants, trapping, shooting, roost treatment with wetting agent (PA-14).

PIGEONS AND HOUSE SPARROWS

Pigeons and house sparrows are urban and farmyard birds whose droppings deface and deteriorate buildings. Around storage facilities they consume and contaminate grain. Pigeons and sparrows may carry and spread various diseases to people, primarily through their droppings (Weber 1979). Of particular concern, droppings that are allowed to accumulate over several years may harbor spores

Fig. 2. Damage to corn by blackbirds (top) and raccoons (bottom) can sometimes be confused. Blackbirds usually slit the husk and peck out the soft contents of kernels, leaving the pericarp. Raccoons and squirrels chew through the husk and bite off the kernels (photo, R. A. Dolbeer).

of the fungus that causes histoplasmosis. House sparrows can damage small grain crops, but this is normally of economic concern only around agricultural experiment stations with small but valuable research plots (Royall 1969). Sparrows build bulky grass nests in buildings, drain spouts, and other sites where they can cause fire hazards or other problems.

Control Techniques.—Screening and proofing, overhead wires, trapping, toxic and stupefying (alpha-chloralose) baits, shooting, and toxic perches.

CROWS, RAVENS, AND MAGPIES

Crows, ravens, and magpies are well-known predators of eggs and nestlings in other birds' nests. In certain situations, these species kill newborn lambs or other livestock by pecking their eyes (Larsen and Dietrich 1970). Magpies sometimes peck scabs on freshly branded cattle.

Crows occasionally damage agricultural crops such as sprouting and ripening corn, apples, and pecans. Most of this loss is localized and minor. Crow damage to apples can be distinguished from damage by smaller birds by the deep (up to 5 cm), triangular peck holes (Tobin et al. 1989). Roosting congregations of crows in trees in parks and cemeteries sometimes cause nuisance problems because of noise and feces.

Control Techniques.—Mechanical frightening devices, shooting, trapping, chemical frightening agents, toxicants.

HERONS, BITTERNS, AND CORMORANTS

These species sometimes concentrate at fish-rearing facilities and cause substantial losses (Salmon and Conte 1981). Salmon smolts released in rivers in the northeastern U.S. have sustained heavy depredation by cormorants. In recent years double-crested cormorants have caused serious losses at commercial fish ponds in the southern U.S. (Stickley and Andrews 1989). Nighttime observations are sometimes necessary to determine the depredating species, because herons and bitterns will feed at night.

Control Techniques.—Habitat modification, screening, overhead wires, frightening devices, shooting.

HAWKS AND OWLS

The raptors most often implicated in predation problems with livestock (primarily poultry and game-farm fowl) are goshawks, red-tailed hawks, and great horned owls (Hygnstrom and Craven 1994). Unlike mammalian predators, raptors usually kill only one bird per day. Raptor kills usually have bloody puncture wounds in the back and breast. Owls often remove the head. Raptors generally pluck birds, leaving piles of feathers. Plucked feathers that have small amounts of tissue clinging to their bases were pulled from a cold bird that probably died from other causes and was simply scavenged by the raptor. If the base of a plucked feather is smooth and clean, the bird was plucked soon after dying. Because raptors have large territories and are not numerous in any one area, the removal of one or two individuals generally will solve a problem.

Control Techniques.—Proofing and screening, habitat modifications, frightening devices, trapping and transplanting, shooting.

GOLDEN EAGLES

Golden eagles occasionally kill livestock, primarily lambs and kids on range. This predation can be locally severe in the sheep-producing areas from New Mexico through Montana (Phillips and Blom 1988).

Close examination is needed to identify an eagle kill. Eagles have three front toes opposing the hind toe, or hallux, on each foot. The front talons normally leave punctures about 2.5–5.0 cm apart in a straight line or small "V", and the wound from the hallux will be 10–15 cm from that of the middle toe. In contrast, mammalian predators almost always leave four punctures or bruises from the canine teeth. Talon punctures are usually deeper than tooth punctures, and tissue between the talon punctures is seldom crushed. If a puncture cannot be seen from the outside, skinning the carcass will reveal the pattern of talon or tooth marks. Often a young lamb is killed with a single puncture from the hallux in the top of the skull and punctures from the three opposing talons in the base of the skull or top of the neck (O'Gara 1978, 1994).

Control Techniques.—Modified herding techniques, mechanical frightening devices, trapping and transplanting, shooting.

WOODPECKERS

Woodpeckers at times cause damage to buildings with wood siding, especially cedar and redwood (Evans et al. 1983). The birds peck holes to locate insects, store acorns, or establish nest sites. They also damage utility poles. Sapsuckers attack trees to feed on the sap, bark tissues, and insects attracted to the sap. This feeding can sometimes kill the tree or degrade the quality of wood for commercial purposes (Ostry and Nicholls 1976). Woodpeckers occasionally annoy homeowners by knocking on metal rain gutters and stove pipes to proclaim their territories.

Control Techniques.—Exclusion, sticky repellents, live traps, snap traps, shooting, frightening devices.

DUCKS, GEESE, AND SANDHILL CRANES

Damage by ducks and cranes to swathed or maturing small-grain crops during the autumn harvest is a serious, localized problem in the northern Great Plains region (Knittle and Porter 1988). Damage occurs from direct consumption of grain and from trampling, which dislodges kernels from heads. Losses from trampling may be at least double the losses from consumption (Sugden and Goerzen 1979).

Canada and snow geese grazing on winter wheat and rye crops can reduce subsequent grain and vegetative yields (Kahl and Samson 1984, Conover 1988). Canada geese also can be a serious problem to sprouting soybeans in spring and in fields of standing corn in autumn. Canada geese have adapted to suburban environments in the past 20 years, creating nuisance problems around parks and golf courses through grazing and defecation (Conover and Chasko 1985).

Control Techniques.—Mechanical frightening devices, lure crops, hunting, trapping and transplanting, overhead wires, capture with drug (alpha-chloralose).

Control Techniques

MODIFICATIONS OF HABITAT AND CULTURAL PRACTICES

Habitat and cultural modifications can be implemented in many situations to make roosting, loafing, or feeding sites less attractive to birds. Although the initial investment of time and money may be high, these modifications often provide long-lasting relief. Thinning or pruning vegetation can cause roosting birds such as starlings to move, often increasing the commercial or aesthetic value of the trees at the same time (Good and Johnson 1978, Micacchion and Townsend 1983). Gull activity at airports can be reduced by eliminating standing water, allowing grass along runways to grow to 15 cm, and prohibiting landfills in close proximity. The U.S. Federal Aviation Administration's policy is that solid-waste disposal sites should not be located within 3 km of any runway used by turbine-powered aircraft (Harrison 1984).

The use of lure crops, where waterfowl or blackbirds are encouraged to feed, is sometimes cost-effective in reducing damage to nearby commercial fields of grain and sunflowers where bird-frightening programs are in place (Sugden 1976, Cummings et al. 1987). Bird-resistant cultivars of corn, sunflower, and sorghum have shown effectiveness in reducing damage. For example, cultivars of sweet corn with ears having long, thick husks difficult for blackbirds to penetrate have less damage than do cultivars with ears having short, thin husks (Dolbeer et al. 1988b). Planting crops so that they do not mature unusually early or late also can reduce damage by blackbirds (Bridgeland and Caslick 1983). Control of insects in cornfields can make those fields less attractive to blackbirds and reduce subsequent damage to the corn crop (Woronecki et al. 1981).

PROOFING AND SCREENING

Nylon or plastic netting is cost-effective in excluding birds from individual fruit trees or high-value crops such as blueberries or grapes (Fuller-Perrine and Tobin 1993) (Fig. 3). Netting or wire screening can be used to exclude birds from rafter areas of airport hangars, undersides of bridges, fish hatcheries, and vent openings of buildings. Ledges on buildings can be covered with slanting boards or other material placed at a 45° angle to prevent bird perching or nesting. Electrically charged wires installed on ledges and other sites can prevent birds from perching.

Parallel strands of monofilament lines or wires strung at 2.5- to 12-m intervals over ponds, landfills, and other structures can reduce gull activity (Blokpoel and Tessier 1984, McLaren et al. 1984). Monofilament lines at 30-cm to 60-cm intervals repelled house sparrows from feeding sites (Agüero et al. 1991). Gulls and house sparrows are reluctant to fly through these strands even though the spacing is larger than their wingspans. Overhead lines also have excluded birds from fish hatcheries. Recommended spacing between wires is 60 cm for mergansers and 30 cm for great blue herons (Salmon and Conte 1981). Heavy plastic (PVC) strips hung from open doorways will help exclude starlings and other birds from buildings (Johnson and Glahn 1994).

Fig. 3. Nylon netting can be a cost-effective means of eliminating bird damage from high-value crops such as in this vineyard on Long Island, New York (photo, M. E. Tobin).

FRIGHTENING

Mechanical Devices

Many devices are marketed, or homemade, to frighten birds. Birds usually habituate to such devices, no matter how effective they may be initially. Thus, two important rules are: (1) never rely solely on one type of device for frightening, and (2) vary the use of devices by altering the timing and location. Frightening devices are only as effective as the person deploying them.

Probably the most widely used frightening device is the propane cannon (Fig. 4), which produces a loud explosion at timed intervals. Several models are marketed, including ones with automatic timers and rotating barrels. To be effective in frightening birds from crops, at least one cannon should be used for each 2 ha and the cannons should be moved every few days. An occasional shotgun patrol to reinforce the exploders is important (Dolbeer 1980), using either live ammunition or shell crackers. Shell crackers, fired from a 12-gauge shotgun, shoot a projectile that explodes 50–75 m away. Other pyrotechnic devices for frightening birds include rockets and whistle bombs (Booth 1994).

Recorded alarm and distress calls of birds broadcast over a speaker system sometimes work well to frighten birds (Bomford and O'Brien 1990). Some airports have speakers mounted on vehicles from which personnel can broadcast these amplified calls for bird species frequently encountered during runway patrols. Shooting at birds with a shotgun often is used to reinforce the distress calls. These calls are commercially available for many bird species (Schmidt and Johnson 1983).

Ultrasonic devices emitting sounds with frequencies above the level of human hearing (20,000 Hz) are marketed for bird control in and around buildings. However, objective field tests have not demonstrated effectiveness of ultrasonic devices in repelling birds (Woronecki 1988). Most birds detect sounds in about the same range of frequencies as do humans.

Flags, helium-filled balloons with and without eyespots, and hawk-kites suspended from balloons or bamboo poles

Fig. 4.   Propane exploders are often used to frighten birds, especially blackbirds, from corn and other crops. For best results exploders should be elevated above the vegetation, moved around periodically, and occasionally supplemented with a shotgun patrol or other frightening device (photo, R. A. Dolbeer).

Fig. 5.   Mylar reflecting tape strung above the vegetation can reduce blackbird feeding activity in agricultural fields (photo, R. A. Dolbeer).

have been used with some success to repel birds from various agricultural fields (e.g., Conover 1984a). Mylar flags, 15 cm × 1.5 m in size, are used to keep geese from winter wheat, corn, and alfalfa. Ten flags per 4 ha are recommended (Heinrich and Craven 1990). Reflecting tape made of mylar, strung in parallel lines at 3-m to 7-m intervals, has reduced blackbird numbers in agricultural fields (Dolbeer et al. 1986) (Fig. 5).

Blackbird roosts containing up to several million birds can be moved by use of a combination of devices, particularly recorded distress calls, shell crackers, rockets, and propane cannons (Mott 1980). Strobe lights placed in the roost are also helpful. The operation should begin before sunset, when the first birds arrive, and end at dark. People with shotguns and shell crackers should be stationed on the perimeter of the roost to intercept flight lines as they enter the roost. Three to 5 nights of harassment may be required to achieve complete dispersal. If not done as a part of the disposal program, the habitat of the roost should be altered (e.g., tree thinning) after dispersal is achieved to discourage the roost from reforming.

Chemical Agents

Avitrol® is an EPA-registered frightening agent. The active ingredient, 4-aminopyridine, when ingested in small doses, causes the affected bird to emit distress calls while flying in erratic circles. The affected bird usually dies within 0.5 hour, but its initial behavior can act to frighten other birds away. Avitrol is registered for use on pigeons, gulls, house sparrows, starlings, and blackbirds around structures and nesting and roosting sites; for starlings in feedlots; for gulls at airports; and for blackbirds in corn and sunflower fields.

Avitrol-treated bait usually is diluted 1:10 or 1:99 with untreated bait so that only a portion of the birds feeding are affected. For use in standing corn and sunflowers, a 1:99 ratio of treated to untreated cracked corn bait is used. The bait is applied to about one-third of the field at a rate

of 3 kg/ha when birds first begin to feed in the field. Reapplication may be necessary at 5- to 10-day intervals, depending on rainfall, bird activity, and other factors (Dolbeer 1980).

Alpha-chloralose is a drug that can be mixed with corn or bread baits to immobilize and capture nuisance waterfowl and pigeons. Birds typically become immobilized 30 minutes to 1 hour after ingesting bait and fully recover 4–24 hours later (Woronecki et al. 1992). Alpha-chloralose is restricted by the U.S. Food and Drug Administration for use by U.S. Department of Agriculture biologists in the Animal Damage Control Program.

REPELLENTS

Birds have a poor sense of smell and taste in general, and repellents based on these senses usually are not effective. For example, naphthalene crystals, although registered as an odor repellent for starlings, pigeons, and house sparrows in indoor roosts, have not been effective in field trials (Dolbeer et al. 1988a). Taste repellents used as seed treatments to prevent consumption of germinating seeds are also of questionable value (Heisterberg 1983).

In contrast, chemicals that produce illness or adverse physiological response upon ingestion (i.e., conditioned aversion) appear to work well as bird repellents (Rogers 1974). Methiocarb, a carbamate insecticide, is a condition-aversive repellent that has been used as a seed treatment for corn (applied as a powder to the seed at planting) and as a spray treatment for ripening cherries and blueberries.

Several tactile repellents are available to prevent birds from roosting or perching on ledges and other structures. The materials must be placed on clean surfaces. Warm temperatures may cause them to run, and dust reduces their sticky properties (Williams and Corrigan 1994).

TRAPS

Starlings and certain blackbird species often can be captured in decoy traps. A decoy trap is a large (e.g., 6 × 6 × 1.8 m) poultry wire or net enclosure containing 5–20 decoy birds, food, water, and perches (Fig. 6). Birds enter the trap by folding their wings and dropping through an

Fig. 6. A typical blackbird and starling decoy trap showing elevated feed platform in center and gathering cage on far right. Birds fold their wings to enter the trap through a 0.6- × 1.2-m opening covered with 5- × 10-cm welded wire located directly above the feed platform (photo, R. A. Dolbeer).

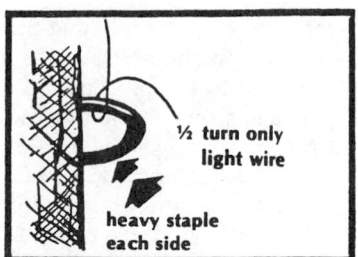

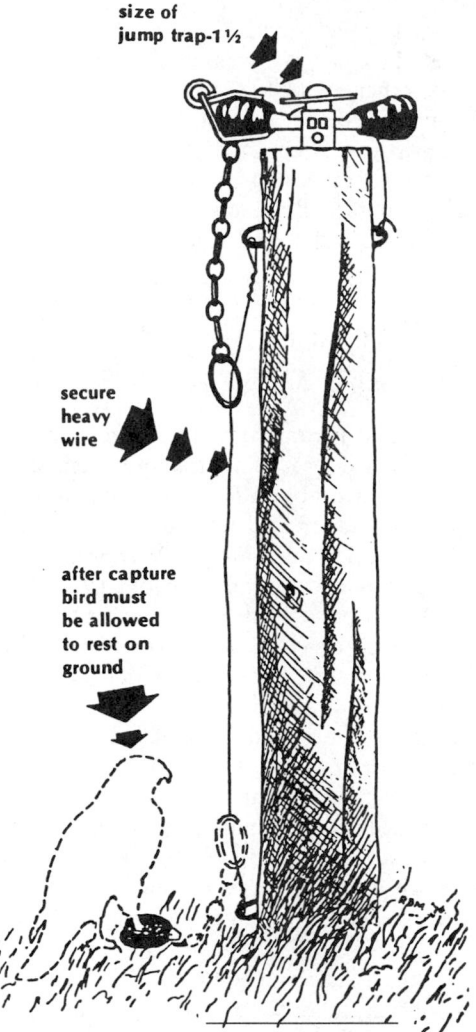

Fig. 7. Pole trap for capturing raptors (from U.S. Department of the Interior 1977).

opening (0.6 × 1.2 m) in the cage top covered with 5- × 10-cm welded wire to reach the food (cracked corn, millet) below. Decoy traps have been used to reduce local populations of starlings near cherry orchards (Bogatich 1967), to remove cowbirds from the nesting area of the endangered Kirtland's warbler (Kelly and DeCapita 1982), and to capture blackbirds for banding and research purposes. Pigeons and house sparrows can be captured in various walk-in or funnel traps (Fitzwater 1994, Williams and Corrigan 1994). Mist nets can be used to remove house sparrows around barns and small farm plots (Plesser et al. 1983).

Pole trapping is an effective method for capturing problem hawks and owls because of their preference to perch on tall, isolated poles. A #1½ steel trap with jaws padded with foam rubber or slit surgical tubing is recommended. The trap is placed on an isolated pole near where the killing is occurring. The trap must be rigged to slide down the pole so that the captured raptor can rest on the ground until it is removed for relocation (Fig. 7). The Swedish goshawk trap (Meng 1971) is also useful for capturing problem raptors. Golden eagles preying on livestock can be captured for transplanting with a net gun fired from a helicopter (O'Gara and Getz 1986).

SHOOTING

Shooting can be effective in reducing local populations of depredating birds if only a few birds are involved. Shooting has little effect on large numbers of birds other than the repelling value (Murton et al. 1974). This concept has been promoted in Wisconsin through a hunter referral program in which farmers allow goose hunters to shoot in agricultural fields sustaining chronic damage (Heinrich and Craven 1987).

The use of .22-caliber bird shot can be effective in removing a few starlings, house sparrows, or pigeons inside buildings, with minimal problems of ricochet or structural damage.

TOXICANTS

The use of toxic baits to kill pest birds without harming nontarget organisms requires patience and a thorough understanding of the habits and food preferences of the tar-

get species. Prebaiting for several days with untreated bait is critical, not only to enhance bait acceptance but to assess the amount of toxic bait to be used and possible nontarget hazards. Other nearby sources of preferred food should be restricted as much as possible during the prebait period. Strict control must be maintained over the toxic bait. Dead birds should be collected at least daily and buried.

DRC-1339 is a toxicant incorporated into poultry pellets and marketed as Starlicide Complete® for killing starlings at feedlots and poultry yards. DRC-1339, incorporated into bread baits, also is registered for killing certain gull species that compete with endangered or threatened bird species for nest sites. DRC-1339 affects the renal and circulatory systems, killing the bird 24–72 hours after ingestion. Strychnine-treated whole and cracked corn has been used to kill pigeons and house sparrows, respectively, in and around buildings. However, the EPA has greatly curtailed above-ground uses of strychnine in recent years.

The wetting agent PA-14 is registered by the U.S. Department of Agriculture (USDA) for killing blackbirds and starlings in upland roosts. The material is technically not a toxicant. When mixed with water and applied by aircraft or ground spray systems to roosting birds, the material allows water to penetrate the birds' feathers, cooling the birds so that they die of hypothermia when the air temperature is less than about 10 C (Stickley et al. 1986). The roost site must receive at least 1.3 cm of rain or sprayed water during the night of treatment.

Wick-type perches containing Endrin or Fenthion solution are registered for killing pigeons, sparrows, and starlings in buildings. The material is absorbed through the feet and skin.

UNGULATES

Damage Assessment

Ungulate damage to various agricultural, forestry, and ornamental crops caused by feeding, trampling, and antler rubbing is an increasing problem. Deer browsing in winter on buds of apple and other fruit trees can reduce yields the following year (Austin and Urness 1989) or adversely alter the growth pattern of tree limbs (Harder 1970). Similar browsing on nursery plants and in Christmas-tree plantations can reduce or eliminate their market values (Scott and Townsend 1985). Browsing of hardwood saplings and young fir trees in regenerating forests can reduce growth rates, misshape trees, and cause plantation failures (Crouch 1976, Tilghman 1989).

Damage to trees caused by antler rubbing can be severe (Scott and Townsend 1985). Small trees (16–25 mm in diameter at 15 cm above ground) with smooth bark such as green ash, plum, and cherry were preferred for antler rubbing by white-tailed deer in an Ohio nursery (Nielsen et al. 1982).

Objective estimates of economic loss from ungulate browsing and rubbing in orchards, nurseries, and reforestation projects are difficult to obtain. Losses in yield or tree value may accumulate for many years after damage occurs and vary with other stresses, including rodent damage, inflicted on the plants. In Ohio, growers reported average losses to deer in 1983 of $204/ha for orchards, $219/ha for Christmas-tree plantings, and $268/ha in nursery plantings (Scott and Townsend 1985). Losses apparently are in the millions of dollars annually in some U.S. states (Black et al. 1979, Craven 1983, Connelly et al. 1987).

Deer also feed on various agricultural crops, especially young soybean plants and ripening ears of corn. Hygnstrom and Craven (1988) estimated a mean loss of 2,680 kg of corn per ha for 51 unprotected cornfields in Wisconsin. Yield reductions in soybean fields are most severe when feeding occurs during the first week of sprouting (DeCalesta and Schwendeman 1978). Elk in some areas raid haystacks and cattle feedlots (Eadie 1954).

Species Damage Identification

Ungulates do not have upper incisors. Thus, twigs or plants nipped by these hoofed species do not show the neat, sharp-cut edge left by most rodents and lagomorphs, but instead show a rough, shredded edge and usually a square or ragged break. Pearce (1947) observed that deer in the Northeast seldom browse higher than 1.8 m from a standing position, but are able to reach to 2.5 m by rearing on their hind legs. Elk and moose browse to a height of about 3 m. Deer seldom browse on branches more than 2.5 cm in diameter. Moose and elk will gnaw the bark of aspen trees. When male ungulates rub the velvet from their antlers, the scarring is generally confined to the trunk area up to 1 m high (Pearce 1947).

Control Techniques

HABITAT MODIFICATIONS

Campbell (1974) reported that planting forbs preferred by deer and elk into areas with Douglas-fir seedlings reduced damage to the trees. Nielsen et al. (1982) recommended that trees not preferred by deer for antler rubbing (e.g., sweetgum, pin oaks) be planted in the remote areas of nurseries and highly preferred trees (e.g., plum, cherry) be planted in areas near human activity.

FENCING AND BARRIERS

Many different fence designs have been tested for excluding ungulates. The standard deer fence, a woven wire fence 2.4 m high and topped with barbed wire, is effective but also expensive, costing $4.10/m (Caslick and Decker 1979). Fence designs that use less material include the 1.5-m Penn State Vertical Electric Deer Fence consisting of five strands of high-tensile steel wire (Fig. 8). This design excluded deer in pen trials, whereas four other experimental designs did not (Palmer et al. 1985). Cost was about $0.72 to $1.00/m for materials. Single-strand electric wire fences, 0.6–1 m high and baited with peanut butter to entice deer to contact the wire with their muzzles, have shown effectiveness in reducing damage in orchards and cornfields. The peanut butter was either placed on aluminum foil flags at 10-m intervals or spread continuously on the wire (Porter 1983, Hygnstrom and Craven 1988). Benefit-to-cost ratios were favorable for these baited fences, which cost less than $0.50/m. Electric fences must be monitored routinely and kept clear of vegetation.

Individual seedling protectors made of photo-degradable plastics (e.g., Vexar tubes), as described in RODENTS AND OTHER SMALL MAMMALS [p. 483], are effective in reducing ungulate damage to young conifer trees (Campbell and Evans 1975, DeYoe and Schaap 1983). Individual saplings can be encircled with hardware cloth or chicken wire to prevent browsing or antler rubbing.

REPELLENTS

Numerous odor and taste repellents have been developed to reduce deer browsing on ornamental plants, fruit trees, and crops. High cost and variable effectiveness during the growing season generally make repellents impractical for use on low-value row crops such as corn (Hygnstrom and Craven 1988). Repellents are most effective on trees and shrubs during the dormant season, but results are inconsistent. Even under optimum conditions, some damage occurs.

Conover and Kania (1987) compared human hair, fermented egg solids (BGR®), and a blood meal-peppercorn mixture for reducing deer damage to young apple trees in winter. Trees had either a bag of hair or a bag of blood meal-peppercorns hung from them, or they were sprayed with BGR. All three repellents reduced browsing by about 50%, but whether the benefits would have exceeded costs is questionable.

Other repellents include bone tar oil (Magic Circle®), mothballs, capsaicin (Hot Sauce Animal Repellent®), tankage, the fungicide Thiram (marketed under several trade names), and ammonium soaps of high fatty acids (Hinder®). Results with these products have been mixed (e.g., McAninch et al. 1983, Palmer et al. 1983, Conover 1984*b,* 1987*a*), indicating that factors such as deer numbers, alternate food supply, target plant species, and weather can influence repellent effectiveness.

FRIGHTENING

Propane exploders, flashing lights, shell crackers, and other sonic devices deployed at night can provide temporary relief from ungulate damage. The proper deployment of these devices to maximize effectiveness is discussed in BIRDS [p. 476]. Ungulates adjust to these devices rather quickly, and they are generally not effective for an entire season.

SHOOTING AND TRAPPING

Effective use of the deer hunting season to reduce populations in areas of high damage is one of the best ways to control damage (Craven 1983). Some states also have special depredation permits that can be issued to a landowner to remove a specific number of deer at a problem site outside the normal hunting season if sufficient control cannot be achieved during the hunting season.

Deer can be captured with drop-door traps, rocket nets, or tranquilizer guns (Palmer et al. 1980). However, these methods of deer removal are usually at least twice as expensive as shooting. In addition, one then faces the problems of humanely holding the deer in captivity until they can be transported somewhere for release and finding suitable release sites. In areas such as arboretums, where shooting normally is prohibited, the use of a skilled marksman under permit is probably preferable to live capture (Ishmael and Rongstad 1984). Live capture and transplanting are generally the control option of last resort, mandated by safety or extreme public relations considerations.

RODENTS AND OTHER SMALL MAMMALS
Damage Assessment

Rodents and other small mammals often are not easily observed causing damage, and their damage frequently is difficult to measure and quantify. Nonetheless, assessments of damage that have been made indicate rodents and nonpredatory small mammals cause tremendous annual losses of food and fiber in the United States. Forest animal damage in Washington and Oregon was estimated to total $60 million annually to Douglas-fir and ponderosa pine, and the potential reduction in the total value of forest

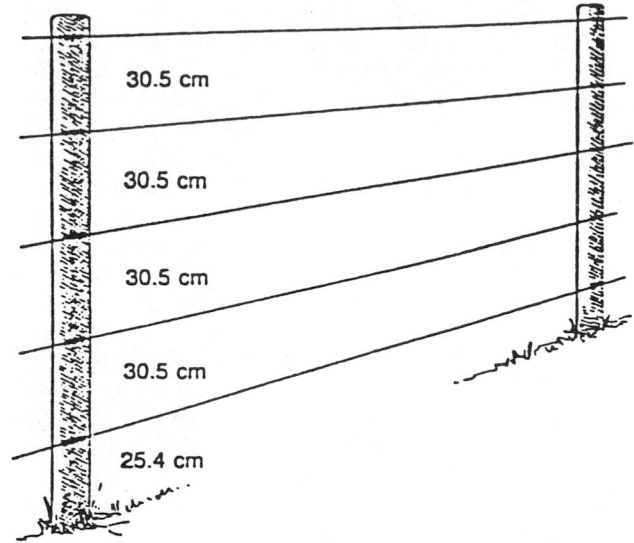

Fig. 8. Two-meter-high Penn State electrified deer fence (from Palmer et al. 1985).

resources was estimated to be $1.83 billion (Black et al. 1979, Brodie et al. 1979). Although these figures include losses attributable to ungulates, rodents and hares are responsible for much of the damage.

Miller (1987) surveyed forest managers and natural resource agencies in 16 southeastern states and estimated annual wildlife-caused losses, caused primarily by beavers, to be $11.2 million on 28.4 million ha. An additional $1.6 million was spent to control wildlife damage on this land. Arner and Dubose (1982) estimated that economic loss to beavers exceeded $4 billion over a 40-year period on 400,000 ha in the southeastern United States. Annual loss in Mississippi to nonimpounded timber was estimated to be $215 million over a period of at least 10 years (Bullock and Arner 1985).

Rats cause substantial losses to sugarcane. Lefebvre et al. (1978) estimated annual losses to be about $6 million ($235/ha) in one-third of the area producing sugarcane in Florida. Losses in Hawaii were reported to be in excess of $20 million per year (Seubert 1984). Ferguson (1980) estimated that in 1978 voles caused losses that approached $50 million to apple growers in the eastern United States. Losses of forage on rangelands to rodents, rabbits, and hares are also extensive; however, accurate estimates of the monetary losses are difficult to obtain because of the nature of the damage and the wide area over which it occurs (Marsh 1985*a*).

Pearson and Forshey (1978) compared yield of apple trees visibly damaged by voles to those not showing damage to determine the dollar losses in gross return per tree. Richmond et al. (1987) determined reductions in growth, yield, and fruit size of apple trees damaged by pine vole populations of known size maintained in enclosures around the trees.

An index of rodent damage to sugarcane was developed through sampling at harvest to determine the percentage of stalks damaged (Lefebvre et al. 1978). Clark and Young (1986) established transects in cornfields and noted rodent damage to individual seedlings over a 10-day period. Forage losses have been estimated by comparing production on areas with and without rodents (Turner 1969, Foster and Stubbendieck 1980, Luce et al. 1981). Sauer (1977) used exclusion cylinders to determine losses of forage to ground squirrels. Alsager (1977) described a method to determine forage production reductions from pocket gopher damage. These methods are useful in evaluating efficacy of control techniques. However, loss estimates must be converted to accurate assessments of dollar loss to enable cost/benefit evaluation of control programs. This conversion is difficult, given the vast areas involved and the variability in rodent populations.

In some situations (e.g., timber flooded by beavers, gopher damage to conifer seedlings, vole damage to apple trees), failure to initiate control may mean loss of the entire resource. Thus, potential loss in these situations is equal to the cost of replacement of the resource. In other situations, control may be necessitated irrespective of cost (e.g., bat colonies in homes).

These examples illustrate the complexity of damage situations and the need for better damage assessment methods, an area of high priority for future research. Lack of methods for determining damage levels has been a serious impediment to the development of cost-effective control strategies.

Species Damage Identification

Most wild mammals are secretive and not easily observed; many are nocturnal. Often the investigator must rely on various types of sign, such as tracks, trails, tooth marks, scats, or burrows, to identify the species doing the damage. Traps may be necessary to make a positive identification of small rodents; frequently, more than one species is involved.

Characteristics of the damage also may provide clues to the species involved. In orchards, for example, major stripping of roots usually is caused by pine voles, whereas damage at the root collar or on the trunk up to the extent of snow depth most often is caused by meadow voles. In sugarcane, various species of rats gnaw stalks so that they are hollowed out between the internodes but usually not completely severed. Rabbits, in contrast, usually gnaw through the stalks, leaving only the ring-shaped internodes. Damage to plants can be grouped generally as follows: root damage—pocket gophers and pine voles; trunk debarking—meadow voles, squirrels, porcupines, woodrats, rabbits, and mountain beavers; stem and branch cutting—beavers, rabbits, meadow voles, mountain beavers, pocket gophers, woodrats, squirrels, and porcupines; needle clipping—mice, squirrels, mountain beavers, porcupines, and rabbits; debudding—red squirrels and chipmunks. These characteristics can aid in identification of the species responsible, but positive identification should be made either by species-specific sign (e.g., tracks) or by capture of individuals.

ARMADILLOS

The armadillo has extended its range eastward and northward from Texas and now is found in all Gulf states and parts of New Mexico, Oklahoma, Kansas, Arkansas, and Missouri (Humphrey 1974). Armadillos feed primarily on invertebrates obtained by rooting in ground cover. When this takes place in lawns, golf courses, or gardens,

economic damage results. There is also concern about the impact of armadillos on forest-floor communities within their expanded range (Carr 1982). Armadillo burrows under orchard trees can cause root damage or excessive aeration (Marsh and Howard 1982). Nuisance problems result when armadillos burrow under structures. Armadillos carry the bacterium that causes leprosy in humans, but their importance in transmission of the disease to humans has not been determined (Davidson and Nettles 1988).

Control Techniques.—Exclusion fencing (25-cm-high poultry mesh), habitat manipulation (removal of cover), reduction of food through use of insecticides, live traps with wing fencing, conibear traps, leg-hold traps, shooting.

BATS

Bats, the only mammals capable of true flight, eat vast quantities of insects. Only a few of the 40 species of bats found in the United States and Canada cause problems, primarily when they form roosts or maternity colonies in human dwellings or structures. Those most commonly encountered in pest situations are: little brown bat, big brown bat, Brazilian free-tailed bat, pallid bat in the Southwest, and Yuma myotis in the West (Greenhall 1982, Frantz 1986). Species identification may be difficult but is important, because several bat species are endangered and protected by state and federal laws. Control operators unfamiliar with bat identification are urged to seek professional help from wildlife agencies or universities (Frantz 1986).

The presence of bats in a building usually is evidenced by noise (squeaking, scratching) and by the presence and distinctive, pungent odor of the accumulated fecal droppings and urine. Bat feces are readily identified from those of rodents by odor, insect content, and the ease with which they are crushed (Greenhall 1982).

Many people are fearful of bats and panic in their presence. Bats occasionally contract rabies, and, although few human deaths have resulted from bat-transmitted rabies (Greenhall 1982), contact with a rabid bat or a bite by a bat that escapes requires postexposure treatment of people and pets without current vaccinations (Frantz 1986). Where bat colonies are allowed to persist so that guano deposits accumulate, the fungus that causes histoplasmosis can develop. Bats roosting near airports may be hazardous to aircraft (Kincaid 1975).

Control Techniques.—Exclusion (including the use of valve devices that permit bats to leave but not return; this should be done after young reach flight stage), repellents, traps, artificial roosts, education to overcome phobias, toxicants (may increase risk of exposure to rabies and is not recommended in most situations).

BEAVERS

Beaver damage is easily identified by the distinctive, cone-shaped tree stumps resulting from their gnawing and often by the presence of their dams and lodges. The latter might not be present, however, in ponds or reservoirs or along swift mountain streams, where they burrow into banks. Usually, green sticks with the bark freshly peeled off may be found when beavers are active in an area.

Damage caused by beavers results from feeding behavior (tree cutting) and their efforts to control water levels (dam building). Tree cutting in certain situations results in selective elimination of preferred tree species, such as aspen and cottonwood, from the vicinity (Beier and Barrett 1987). Loss of timber and crops from flooding (Fig. 9) is of much greater importance, however, especially in the southeastern U.S. where beaver populations have increased dramatically as a result of a decline in trapping due to low pelt prices (Woodward 1983). Beavers often use sticks to plug road culverts or water-control structures in ponds and reservoirs. Additionally, beavers can cause extensive damage to levies and human-made dams by their burrowing.

Beavers are susceptible to infection by protozoan parasites (*Giardia* spp.) that can cause gastroenteritis and diarrhea in humans. Transmission to humans can be prevented by use of proper water treatment measures (Davidson and Nettles 1988).

Control Techniques.—Conibear traps, snares, leg-hold traps (#3 or larger), basket/suitcase-type live traps, shooting, explosives for dams, habitat manipulation, drain devices in dams or culverts.

CHIPMUNKS

Occasionally, chipmunks damage grain fields, garden seeds, flower bulbs, and plants through burrowing and feeding. They infrequently destroy eggs and nestling birds (Eadie 1954). They can establish residence in or under human dwellings. Chipmunks cause reforestation problems by consuming seeds, seedlings, and the terminal buds of older plants, and by caching seeds, often in large quantities (Marsh and Howard 1982). In parts of the western U.S., chipmunks are a potential reservoir for plague and are controlled in campgrounds (Marsh and Howard 1982). Chipmunks are easily observed due to their diurnal activity; their presence also can be determined by trapping.

Control Techniques.—Snap traps, live traps, toxic baits, repellents, shooting (.22-caliber with bird shot, shotgun, or air rifle), exclusion.

COTTON RATS

The hispid cotton rat, a common species in southern U.S. and Mexico, is the species of cotton rat most often causing damage. Two other species have localized occurrences in Arizona and New Mexico. They undergo major population fluctuations. Cotton rats are primarily vegetarian, but they also prey on eggs and young of ground-nesting birds (Hawthorne 1994). Most damage is a result of feeding in agricultural crops, especially melons and sugarcane. Cotton rats are active day and night and, when abundant, are observed often. Their presence also is indicated by well-developed runways through dense vegetation and the presence of grass cuttings 5–8 cm long placed in piles. Pale greenish-yellow droppings, about 9 mm long and 5 mm wide, sometimes are present in the runway. Cotton rat sign is similar to that of voles, but droppings, runways, and clippings of the cotton rat usually are larger (Hawthorne 1994). Cotton rats are often one of several rodent species causing damage in crops.

Control Techniques.—Habitat modification, toxic baits, snap traps.

Fig. 9. Damage to timber by beaver can be extensive in lowland areas where their activities result in permanent flooding of large areas (photo, F. Boyd, USDA/APHIS).

PEROMYSCUS (DEER MICE, WHITE-FOOTED MICE)

The genus *Peromyscus* is large, and one or more species are found in all parts of North America. These mice are nocturnal and active all year. *Peromyscus* populations may show large fluctuations. These mice are the most important seed predators in the Pacific Northwest, causing extensive damage in reforestation efforts (Sullivan 1978). Effects on reforestation have caused a shift to the use of hand-planted seedlings in many areas. *Peromyscus* also can cause significant losses to corn seedings in conservation tillage systems, but this damage may be offset by their consumption of harmful insects and weed seeds (Clark and Young 1986, Johnson 1986). *Peromyscus* invade homes where they eat stored food and damage upholstered furniture or other materials shredded for use in nest building. Trapping with snap or live traps is the best method to determine the species present.

Control Techniques.—Habitat modification, provision of alternative foods (Sullivan and Sullivan 1982), exclusion, snap traps, live traps, toxic baits, repellents.

GROUND SQUIRRELS

Ground squirrels, genus *Spermophilus,* are important pest species in north-central and western North America, causing serious losses of tree seeds and emergent seedlings. A careful search of an area showing damage will reveal opened seed hulls and caches. Ground squirrels can inflict serious damage to pastures, rangelands, grain fields, vegetable gardens, and fruit or nut crops. Their burrows can cause collapse of irrigation levees, increase erosion, and result in damage to farm machinery. Ground squirrels are an important predator of waterfowl eggs in the prairie pothole region (Sargeant and Arnold 1984). They carry several diseases transmissible to man, including plague; in plague-endemic areas, ground squirrel control should be combined with ectoparasite control (Marsh and Howard 1982).

Ground squirrels are diurnal and easily observed (Marsh 1985*a*). They hibernate and estivate and have ma-

Fig. 10. Meadow voles cause reduced apple production and sometimes loss of trees in orchards where they tunnel through snow and girdle trees by gnawing bark near the root collar and up the trunk as far as snow cover extends (photo by M. E. Tobin).

jor dietary shifts during the year (Marsh 1985a, 1986). Effective control strategies must consider these factors.

Control Techniques.—Habitat manipulation, toxic baits, live traps, leg-hold traps (#0–1½), snap traps, fumigants, exclusion, shooting.

KANGAROO RATS

Kangaroo rats are competitors of livestock on arid western rangelands (Marsh 1985a) when present in high populations, especially during drought. They also can retard recovery of overgrazed rangelands when cattle are removed (Howard 1994) and spread undesirable shrub species by caching of seeds (Reynolds and Glendening 1949, Marsh 1985a). Kangaroo rats cause significant damage to alfalfa and corn on irrigated sandy soils by consuming newly planted seeds and clipping off seedlings (Howard 1994). Sorghum, other grains, and garden crops also can be damaged in local areas.

Several species of kangaroo rats are endangered. Kangaroo rats are nocturnal, but their burrow systems, with above-ground mounds and interconnecting runways, are readily observed. Snap-trap surveys can identify the species present, provided the damage area is not within the range of one of the species listed as endangered.

Control Techniques.—Habitat manipulation, snap traps, live traps, toxic bait, exclusion from small areas, provision of alternate food.

MARMOTS

Marmots (woodchucks), like ground squirrels, can cause damage to many crops; forage production may be reduced markedly by marmot feeding and trampling (Marsh 1985a). They damage fruit trees and ornamental shrubs by gnawing or scratching woody vegetation (Bollengier 1994). Their burrows, often located along field edges, can cause damage to farm machinery and injure livestock; when located along irrigation ditches they can cause loss of water. In suburban areas burrows located under buildings or in landscaped areas cause problems (Marsh and Howard 1982). The presence of woodchucks is easily determined by direct observation of animals and burrows. During periods of forage growth, vegetation around burrows is noticeably shorter than in surrounding areas. Occupied burrows can be identified in spring by the presence of dirt pellets ranging from marble to fist size.

Control Techniques.—Fumigants, shooting, conibear traps, leg-hold traps (#1½–2), live traps.

VOLES

Voles (genus *Microtus*), also called meadow mice, field mice, and pine mice, cause extensive damage to forests, orchards, and ornamentals by gnawing bark and roots (Pearson and Forshey 1978, Byers 1984a, Pauls 1986, Sullivan et al. 1987, O'Brien 1994). Tree or shrub damage usually occurs under snow or dense vegetation; the bark is gnawed from small trees near the root collar and up the trunk as far as the snow extends. Voles gnaw through small trees or shoots up to about 6 mm in diameter (Fig. 10). Some species (e.g., pine vole) also cause extensive damage to root systems; this damage may not be detected until spring when it is reflected in the condition of new foliage. Voles also can damage field and garden crops; when vole populations are high, these losses can be catastrophic (Clark 1984, Marsh 1985a). Voles are carriers of bubonic plague and tularemia.

Vole populations are subject to large, rapid fluctuations. The presence of voles is determined most easily by searching for their runways and burrow systems. In orchards these can be found by pulling the grass and other debris from the bases of trees to expose the runways. Burrows of pine voles are usually subterranean. Gnawing on the trunks and roots of trees is usually less uniform than that of other rodents. Tooth marks can be at all angles, even on small branches, and may vary from light scratches to channels 3 mm wide, 2 mm deep, and 10 mm long. In hay crops, runways with numerous burrow openings, clipped vegetation, and feces can be detected in dense vegetation.

Control Techniques.—Screening, plastic mesh protectors for seedlings (Pauls 1986), habitat modification, toxic baits, snap traps (for small populations in local situations), provision of alternative foods (Sullivan and Sullivan 1988).

MOLES

Moles feed primarily on soil invertebrates, especially earthworms and grubs (beetle larvae). About 20% of their food is plant material, which may include garden vegetables and small grains (Silver and Moore 1941). Voles and mice also use the burrows of moles and can be responsible for some damage attributed to moles (Henderson 1994). Burrowing by moles may reduce production of forage crops by undermining and smothering vegetation and by exposing root systems to drying. Their surface burrows also can plug harvesting machinery and contaminate hay and silage (Wick and Landforce 1962). Moles can damage lawns and golf greens extensively through burrowing.

The presence of moles usually can be detected by the mounds of soil thrown up from extensive tunnels dug in search of food and by the raised soil of surface burrows. Mole mounds can be distinguished from those of pocket gophers by their more rounded contour and the lack of a burrow entrance or soil plug (Eadie 1954).

Control Techniques.—Harpoon, scissors, and choker traps; habitat manipulation (e.g., soil compaction); toxic bait; fumigants; repellents (thiram for protection of bulbs); insecticides (for removal of food source).

MOUNTAIN BEAVERS

Mountain beavers cause serious economic loss by burrowing through and feeding on garden vegetables, berry plants, and young trees. They use drainage ditches for burrow sites, and their burrows may undermine roadways.

Mountain beavers are a major factor limiting reforestation in the Pacific Northwest (Borrecco and Anderson 1980, Evans 1987a). Plantations are most susceptible to damage for 4 years after planting and when precommercially thinned at about 12–15 years (Evans 1987a). Mountain beavers clip seedlings and gnaw saplings and the stems and bark of larger trees.

Mountain beavers normally clip through seedlings at a 45° angle. On small seedlings this clipping may be difficult to distinguish from rabbit damage; however, rabbits seldom clip stems larger than 6 mm in diameter or 50 cm above ground level, whereas mountain beavers often cut stems larger than 13 mm in diameter and up to 3 m above ground (Lawrence et al. 1961). Mountain beavers leave branch stubs, cut at a 45° angle, protruding from the main stem. The bark of the main stem shows horizontal tooth marks and vertical claw marks (Packham 1970). Runways and burrows are present in or near the damaged area.

Control Techniques.—Conibear (#110) traps, live traps, leg-hold traps (#1½–2), toxic bait, plastic mesh tree protectors, habitat manipulation.

MUSKRATS

Muskrats most often cause problems where people have created or manipulated wetlands or where wetlands border agricultural crops. The most serious damage results from burrows in pond dams, levees, and irrigation canals. The burrow entrance is below water level and penetrates the embankment at an upward angle to allow for a room above the water level. Damage is increased when the water level rises and the burrow is extended higher to provide a dry chamber, thereby increasing chances of washouts and cave-ins. At times, muskrats cause severe damage to grain, such as rice, and to garden crops growing near water. Muskrats are primarily vegetarian, but they will feed on aquatic animals if vegetation is limited (Miller 1994).

Muskrats commonly construct cone-shaped houses projecting 0.5–1 m above the water surface. Muskrat presence is indicated by houses and burrow entrances. Underwater runs can be observed when the water is clear or after a winter drawdown of ponds or reservoirs (Miller 1994).

Control Techniques.—Conibear traps, leg-hold traps (#1–2), stovepipe traps, toxic baits, exclusion (specialized dam construction techniques [Miller 1994]), habitat manipulation.

NUTRIA

Nutria are semiaquatic, herbivorous mammals that feed on aquatic plants, roots, seeds, and crops grown close to waterways. The greatest losses from this introduced rodent are to sugarcane and rice, especially in fields adjacent to Gulf Coast marshes (LeBlanc 1994). Nutria may severely impede baldcypress regeneration (Conner and Toliver 1987). They also damage wooden structures and floating marinas.

Nutria presence is evidenced by tracks, droppings, and trails to and from the damage area. Nutria also may be observed in the damage area.

Control Techniques.—Habitat manipulation, toxic baits (most effective on floating bait stations [LeBlanc 1994]), leg-hold traps (#2), conibear traps (#210), shooting.

POCKET GOPHERS

Pocket gophers cause substantial damage to agricultural crops; lawns, rangeland, and tree plantings. Gophers feed primarily on the underground portions of plants and trees. Damage often is undetected until a tree shows aboveground signs of stress; by then damage is frequently lethal (Cummings and Marsh 1978). Pocket gophers also may damage plastic irrigation lines on agricultural lands as well as underground pipes and cables in other situations.

On rangeland, soil disturbance and mound building by pocket gophers result in increased plant diversity and a replacement of perennial by annual grasses (McDonough 1974, Foster and Stubbendieck 1980, Marsh 1985a). They can greatly reduce the carrying capacity of rangeland for livestock. Gophers can be a serious pest in alfalfa by feeding on the leaves, stems, and roots (Marsh 1985a). Gopher mounds can cause equipment breakage and increased wearing rate of haying machinery. Gopher tunnels result in water loss in irrigated areas (Case and Jasch 1994).

Pocket gophers are a major impediment to reforestation in the western U.S. (Crouch 1986). During winter they often forage above ground by tunneling in the snow. Coniferous trees have been debarked to a height of 3.5 m by pocket gophers working under the snow (Capp 1976). Gophers also fill some of the snow tunnels with soil, thus forming long, tubular ''casts'' that remain after the snow melts.

Pocket gopher presence is easily determined by fan-shaped soil mounds in contrast to the conical mounds of moles. Burrow entrances are usually plugged. Aboveground debarking injuries caused by pocket gophers show small tooth marks, differing from the distinct, broader grooves left by porcupines and the finely gnawed surface inflicted by meadow voles. Gophers sometimes pull saplings and vegetation into the burrow.

Control Techniques.—Toxic baits, lethal traps (Macabee, Victor, or California pocket gopher traps), fumigants, habitat modification (flood irrigation, crop rotation), seedling protection (plastic mesh), protective coverings for pipes and cables.

PORCUPINES

Porcupines are usually nocturnal and are active all year. During summer, they often feed on succulent plants, including garden and truck crops, in open meadows, in fields, and along the banks of streams and lakes. Greatest damage is caused in winter when porcupines feed on the inner bark of trees (Marsh and Howard 1982). Girdling in the upper bole of trees often results in dead tops (Evans 1987b). Basal girdling may occur on seedlings. Porcupines are attracted to anything containing perspiration salt: saddles, harnesses, belts, and tool handles.

Porcupine damage can be identified by broad incisor marks on exposed sapwood. Abundant oblong droppings about 2.5 cm long can be found under freshly damaged trees. Clipped twigs and tracks also may be found on snow. Top girdling in pines produces trees with a characteristic brushy crown.

Control Techniques.—Shooting, leg-hold traps (#1–3), proofing and screening (small areas or individual trees).

PRAIRIE DOGS

Prairie dogs were widespread on the Great Plains throughout the 1800s and reached peak numbers around 1900 after reduction of natural predators and establishment of cattle grazing. By 1921 the area occupied by prairie dogs was estimated to be 40 million ha. By 1971, following intensive control efforts, only 0.6 million ha were occupied. Populations have been expanding in recent years, commensurate with reduced control efforts (Fagerstone 1981).

Prairie dogs damage rangelands and pastures by clipping vegetation for food and nesting material and by clearing cover from the vicinity of burrows (Hygnstrom and Virchow 1994). This activity not only reduces available forage but can alter species composition of the vegetation in favor of forbs. Competition with cattle does not always exist, however, and in some situations beneficial effects of prairie dogs offset competition. Therefore, each conflict situation should be evaluated individually (Fagerstone 1981).

Crops planted near prairie dog colonies can receive serious damage from feeding and trampling. Also, damage to irrigation systems is common, and badgers digging for these rodents cause even greater damage. The burrows and mounds created by prairie dogs can increase soil erosion, cause drainage of irrigation water, and result in damage to farm implements. Prairie dogs also serve as a reservoir for bubonic plague (Hygnstrom and Virchow 1994).

Prairie dog colonies provide habitat for other species such as the endangered black-footed ferret. All lethal control should be preceded by a careful survey to ensure that ferrets are not present. The Utah prairie dog is a threatened species and should not be controlled.

Prairie dog colonies are easily identified by the conical mounds around burrow entrances and by the presence of the easily observed animals.

Control Techniques.—Toxic grain bait, fumigants, shooting, leg-hold traps (#0–2), conibear traps (#120), habitat modification (deferred grazing).

RABBITS AND HARES

Rabbits and hares can damage or completely destroy tree plantings, gardens, ornamentals, agricultural crops, and rehabilitated rangeland. In winter, they strip bark from and debud fruit trees, conifers, and other trees and shrubs (Craven 1994).

Rabbits are known vectors of tularemia, which is transmissible to humans, and they may carry larvated eggs of several ascarid roundworms that can produce disease if accidentally ingested (uncooked) by humans (Davidson and Nettles 1988).

Jack rabbits also damage orchards, gardens, ornamentals, and some agricultural crops, especially in areas adjacent to rangeland, and most frequently when natural vegetation is dry (Knight 1994). Jack rabbit populations show large fluctuations, and at times of high density, damage to rangeland vegetation and competition with livestock can be severe.

Trees clipped by rabbits and hares have a clean, oblique, knifelike cut on the stem. Rabbits and hares usually clip stems 6 mm in diameter or less at a height not more than 50 cm above the ground (Lawrence et al. 1961). Repeated clipping will deform seedlings. Rabbits and hares often can be observed at damage sites along with their tracks, trails, and droppings.

Control Techniques.—Habitat modification, fencing and proofing, repellents, live traps, body snares, shooting, toxic baits for jack rabbits in some localities.

TREE SQUIRRELS

Tree squirrels can be categorized into three groups: large tree squirrels (gray, fox, and tassel-eared), pine squirrels (red and Douglas'), and flying squirrels (northern and southern) (Jackson 1994). Squirrels eat plants and fruits, dig up newly planted bulbs and seeds, strip bark and leaves from trees and shrubs, invade homes, and consume bird eggs (Hadidian et al. 1987, Jackson 1994). They cause problems by shorting out transformers and gnawing on power and telephone lines (Marsh and Howard 1982, Hamilton et al. 1987).

Squirrels often can be observed at the damage site. Damage to conifer seeds is indicated by green, unopened cones scattered on the ground under mature trees and by the accumulated cone scales and "cores" at feeding stations. Bark stripping can be observed in trees, and bark fragments often are found on the ground, as are the tips of twigs and small branches.

Control Techniques.—Fencing and proofing, repellents, live traps, shooting, conibear traps, leg-hold traps (#0–1), toxicants.

WOODRATS

Woodrats, also called pack rats, brush rats, or trade rats, are attracted to food supplies left in buildings and will remove small objects such as spoons, forks, knives, and other items, sometimes leaving sticks or other objects "in trade." They often construct conspicuous stick houses in cabins, in abandoned vehicles, or in the upper branches of trees (Marsh and Howard 1982, Salmon and Gorenzel 1994). They will shred mattresses and upholstery.

Woodrats are agile climbers and consume fruits, seeds, and green foliage of herbaceous and woody plants (Lawrence et al. 1961). They strip and finely shred patches of bark from conifers and fruit trees to line nest chambers (Hooven 1959). They also will clip small branches. Their damage may be confused with that of tree squirrels and porcupines; however, woodrats leave a relatively smooth surface with a few scattered tooth marks and tend to litter the ground beneath the tree less than tree squirrels.

Several subspecies of woodrats are endangered. Local regulations should be checked before control efforts are undertaken.

Control Techniques.—Exclusion, repellents (mothballs have questionable efficacy), toxic baits, snap traps, live traps, shooting.

COMMENSAL RODENTS

The three species of commensal rodents (those that live primarily around human habitation) are Norway rats, roof (black) rats, and house mice. These omnivorous rodents consume millions of bushels of grain each year: in the field, on the farm, in the elevator, mill, store, and home, and in transit. They also waste many more millions of bushels by contamination. These rodents typically drop 25–150 pellets and void 10–20 cc of urine every 24 hours, and constantly shed fine hairs.

Rats cause extensive damage to sugarcane in Hawaii and Florida (Fig. 11), and roof rats are serious pests in Hawaiian macadamia nut plantations. These rodents will feed on poultry chicks and occasionally will attack adult poultry, wild birds, newborn pigs, lambs, and calves. Health departments annually report hundreds of human babies bitten by rats. Many viral and bacterial diseases are transmitted to humans by rodent feces and urine that contaminate food and water.

Gnawing by rodents causes considerable property damage. Fires are sometimes started when rats and mice gnaw the insulation of electric wiring. They also will use materials such as oily rags and matches for building nests, which can result in fires by spontaneous combustion. Extensive damage to foundations and concrete slabs sometimes results when Norway rats burrow under buildings. Burrows into dikes and outdoor embankments cause erosion.

Signs of commensal rodents are gnawing, droppings, tracks, burrows, and darkened or smeared areas along walls where they travel. Reviews of problems caused by these species and methods of control were provided by Meehan (1984), Jackson (1987), Baker et al. (1994), Marsh (1994), and Timm (1994).

Control Techniques.—Habitat modification, proofing and screening, snap traps, toxic baits (multiple dose and acute), tracking powder, fumigants.

Control Techniques

MODIFICATIONS OF HABITAT AND CULTURAL PRACTICES

All animals are dependent on food and shelter; therefore, elimination of one or both of these requirements will

Fig. 11. Rats damage sugarcane by gnawing internodal areas of stalks, creating canoe-shaped damage areas. This damage results in death of some stalks and loss of production in damaged stalks that survive (photo courtesy of Denver Wildl. Res. Cent., USDA/APHIS).

force them to move from the immediate area. This method of control, where practical, is often the most desirable and usually has the most permanent effect in stopping small mammal damage. One should recognize, however, that other species often are dependent upon the habitat being modified. Modifications of the habitat can result in greater adverse impacts to desirable nontarget species and natural communities than would careful use of a registered toxicant or other control tool. They also can create situations that result in other species becoming pests.

Many rodents and small mammals can be discouraged from using areas by removal of brush piles, weeds, old lumber piles, and other debris. Commensal rodent control can be greatly facilitated by removal of harborage, garbage, and refuse (Jackson 1987). Squirrel interference with power transformers may be reduced if vegetation near power poles is managed (Hamilton et al. 1987). Mountain beaver populations in cultivated areas may be decreased by removing surface shelters such as stumps, logs, and brush piles (Eadie 1954). High populations of round-tailed muskrats in Florida sugarcane are associated with trash remaining in fields after harvest (Steffen et al. 1981).

Control of pine voles with anticoagulant baits was enhanced in apple orchards cultivated two or three times a year (Byers 1976). Davis (1976) reported that pine vole damage in an apple orchard was reduced by mowing three times a year, clearing vegetation from under the trees, removing pruned branches, restricting the distribution of fertilizer, and, after harvest, inspecting and cleaning vulnerable parts of the orchard. Byers (1984b), however, observed that cultural controls (combinations of mowing, cultivation, and herbicide application) were much more expensive than the use of toxic baits and offered no advantages in vole control.

Provision of alternative foods will reduce conifer seed loss to mice in forest regeneration projects (Sullivan and Sullivan 1982) and also might be useful in reducing loss of corn seedlings in no-tillage fields (Johnson 1986). A study in British Columbia indicated that provision of alternate foods might reduce vole damage in apple orchards (Sullivan and Sullivan 1988). Pocket gopher infestations in logged areas can be reduced by prompt regeneration

Fig. 12. Mechanical devices, such as these T-culverts developed in New York, may be used to prevent beavers from stopping water flow through culverts (photo, K. Roblee, N.Y. State Div. Fish Wildl.).

and minimal site preparation. Selective cutting, when feasible, can be used in areas with a high potential for gopher infestations (Crouch 1986). The use of insecticides to reduce soil invertebrates can protect turf from armadillos and moles, but damage may increase initially because of increased food searching by animals already present (Henderson 1994).

Water levels behind beaver dams can be manipulated by installing a perforated pipe (Laramie 1978) or a three-log drain (Miller and Yarrow 1994) through the dam. Various mechanical methods have been developed to prevent beavers from stopping water flow through culverts (Roblee 1987) (Fig. 12). Muskrat damage to farm pond dams can be reduced by maintaining a 3:1 slope on the water side of the dam, a 2:1 slope on the outer face, and a top width ≥2.4 m (Miller 1994). The water level should be maintained at least 0.9 m below the top of the dam.

EXCLUSION

Exclusion involves the placement of barriers that prevent access by pest species into structures or areas, or their physical contact with specific objects. Proofing of structures is achieved most economically if it is considered prior to construction. Baker et al. (1994) provided detailed suggestions of ways to accomplish rodent-proof construction. Basically all openings or sites where rodents might create openings are protected with wire mesh, sheet metal, or concrete, providing long-term protection.

Exclusion is a necessary part of an effective program to remove bats from structures. Final closing of entrances to the structure should not be made until all young have reached the flight stage. At that time these openings can be closed with a valve device that permits bats to leave the structure but prohibits reentry (Greenhall 1982, Frantz 1986).

In small orchards, rabbit and rodent damage can be eliminated by wrapping trees with hardware cloth or burlap that is buried about 5 cm deep around the tree base. In England, wire netting and electrified netting fences have been effective in excluding rabbits from crop fields (McKillop and Wilson 1987). Fences made of 1.2- to 2.5-cm-mesh net wire 0.7–1 m high can protect small areas

against nonclimbing rodents and small mammals. Fences should be buried 15 cm deep with an "L" shape on the outside of the fence.

A 0.6-m-wide expandable metal band placed around tree trunks 2 m above the ground will keep squirrels out of isolated trees. Branches should be trimmed within 2 m of the ground or buildings. Steel-sheathed wire may be used on underground power and telephone lines to prevent pocket gopher gnawing. VEXAR® plastic seedling protectors or Remay sleeve protectors will protect conifer seedlings from pocket gophers, mountain beavers, and lagomorphs (Anthony et al. 1978, Evans 1987a). These plastic net-tubes, 76–90 cm tall and 5 cm in diameter, are placed over seedlings at planting. They allow branches to grow through the netting and provide protection for the terminal bud for about 3–5 years as it grows up through the tube. The protectors photodegrade.

FUMIGANTS

Fumigants produce gases that are lethal when inhaled; they are used to kill various burrowing mammals such as pocket gophers, commensal rodents, prairie dogs, ground squirrels, chipmunks, and woodchucks. When fumigants are used, all burrow openings should be closed after introduction of the pesticide. Gas cartridges are incendiary fumigants that produce carbon monoxide, causing death by suffocation (Dolbeer et al. 1991). Aluminum phosphide is a fumigant available in tablets or pellets that produces toxic phosphine gas when in contact with atmospheric moisture; this gas is flammable or explosive at some concentrations. Calcium cyanide is a fumigant that in the presence of moisture releases hydrocyanic acid (HCN), a colorless gas that is highly toxic by contact, ingestion, or inhalation. Calcium cyanide is extremely dangerous, requiring extra caution in its use. Amyl nitrite, an antidote, should always be immediately available when this fumigant is used. Some other registered fumigants are carbon disulfide, chloropicrin, magnesium phosphide, and methyl bromide. Jacobs (1994) provided information on specific fumigants.

TOXICANTS

Toxicants often are used to control damage by rodents and other small mammals. Efficacy of toxicant formulation and potential hazards to nontarget species must be considered when toxicants are used. Damage reduction is the goal of any control program, and this must be the final measure of efficacy. Efficacy of a control program sometimes can be increased by using several toxicants in combination or by periodically alternating those used; this can aid in overcoming developed resistance of the pest species to the primary toxicant (Marsh 1988a).

Hazards associated with the use of a toxicant are not necessarily related to the toxicity of the compound. They are associated more often with the use pattern. Hazards to nontarget wildlife can be reduced by properly selecting rodenticides, bait composition and formulation techniques (including bait color, size, shape, texture, and hardness), and bait delivery systems (Marsh 1985b).

Toxicants can best be discussed as anticoagulants and non-anticoagulants. Previously, anticoagulants were referred to as multidose or chronic toxicants and non-anticoagulants as single-dose or acute toxicants. New-generation anticoagulants, however, can be effective in a single feeding, and some new non-anticoagulants can be ingested by individuals of the target species over several days (Marsh 1988a).

Numerous rodenticide formulations are registered for use in commensal rodent control, around farm buildings, and in noncrop areas. Few rodenticides are registered for in-crop use, although such use may be necessary to achieve adequate control of damage (Lefebvre et al. 1985a). Development of registrations for in-crop use of rodenticides, particularly anticoagulants, is a high priority area for research.

Anticoagulants

Anticoagulant toxicants inhibit blood coagulation and result in internal bleeding leading to death (Meehan 1984). Early anticoagulants such as warfarin, pindone, diphacinone, and chlorphacinone generally require ingestion for 3–14 consecutive days to be effective. Bait shyness generally is not a problem because the animals do not associate ill effects with bait consumption. However, bait delivery procedures must consider the need for making toxicants available over a continuous period of days. Two of the newer anticoagulants, brodifacoum and bromadiolone, are highly toxic to rodents, and a single feeding on baits with an active ingredient concentration as low as 0.005% can produce death (Marsh 1988a). Certain rodents have developed resistance to some of the older anticoagulants.

Anticoagulants can be obtained in prepared baits or purchased as concentrates for mixing with fresh bait. Baits should be placed where the rodents feed, drink, or travel. For anticoagulants that require continuous exposure, bait stations, purchased from pesticide supply houses or constructed from wood or metal, are particularly useful in protecting the bait from weather and nontarget species. Old automobile tires and drainage tiles also have been used. Some baits are in packets that are gnawed open by rodents. Many anticoagulants are available in a paraffin-impregnated cereal bait for use in sewers or other damp locations.

Several anticoagulants are registered for use as a tracking powder; they are dusted into burrows and along runways where house mice or Norway rats travel. The animals lick the toxic dust from their feet and fur. Chlorophacinone (Rozol®) tracking powder is registered for bat control in dwellings in some states; however, the increased likelihood for human contact with dead or dying bats as a result of its use, with the potential for rabies exposure, should be considered (Greenhall and Frantz 1994).

Non-Anticoagulants

Rodenticides with different modes of action provide an obvious answer to anticoagulant resistance. Zinc phosphide, red squill, strychnine, and Compound 1080 (sodium monofluoroacetate) are non-anticoagulant toxicants used for many years. In recent years the use of strychnine and 1080 has been severely restricted through loss of EPA registrations. The need for safe, effective, non-anticoagulant rodenticides still exists. Several new compounds (cholecalciferol, bromethalin, and alpha-chlorohydrin) are now available (Marsh 1988a).

Zinc phosphide, one of the most commonly used single-dose rodenticides, is relatively safe to humans, and its use usually does not result in secondary poisoning of nontarget species. Efficacy is poor or inconsistent on some field rodents but often can be improved by prebaiting (Marsh 1988b). Zinc phosphide baits are prepared with sweet potatoes, carrots, or apples for nutria and muskrats; apples, cracked corn, or oats for voles and pocket gophers; oats for prairie dogs; and ground fish or meat for commensal rodents.

For muskrat and nutria control, bait can be placed on 1- × 1-m rafts constructed of marine plywood and anchored near the area of use. Prebaiting is necessary to assure success in nutria control (LeBlanc 1994).

Control methods for voles vary with the situation and species involved. Bait can be scattered along surface runways or placed in underground runways. In orchards, bait placed under boards or asphalt shingles inside the drip line of fruit trees takes advantage of the tendency for voles to burrow and nest under such objects. Tobin and Richmond (1987) described a bait station for voles made from polyvinylchloride (PVC) pipe.

A 2% zinc phosphide on steamrolled oats is used to control prairie dogs. After prebaiting with oats for 1–3 days, toxic bait is then scattered by hand around each burrow entrance (Tietjen 1976). A 2% zinc phosphide bait significantly reduced cotton rat (Holler and Decker 1989) but not roof rat populations (Lefebvre et al. 1985b) in Florida sugarcane fields. Strychnine- and 1080-treated grain also have been used to control various field rodents.

Red squill is an imported, relatively safe, plant-derived rodenticide that has shown only moderate effectiveness on Norway rats. A newer, more effective form has been developed and marketed in Europe (Marsh 1988a).

Cholecalciferol (vitamin D$_3$; marketed as Quintox® and Rampage®) is both a single- and multiple-feeding toxicant effective on commensal rodents (Marshall 1984). No secondary hazards have been associated with its use (Marsh 1988a). Bromethalin (marketed as Vengeance® and Assault®) is another new rodenticide effective on rats, including those resistant to warfarin (Marsh 1988a).

BURROW BUILDER

The burrow builder is a tractor-drawn mechanical tool that constructs an underground artificial burrow and places toxic grain baits therein for controlling pocket gophers (Fig. 13). During their underground travels, gophers intersect the artificial burrows, consume the toxic bait, and die underground. Artificial burrows are constructed 6–9 m apart, usually 20–30 cm deep. The proper depth to set the machine can be determined by locating and measuring the depth of gopher burrows by probing with a pointed instrument. Up to 40 ha of land can be treated in a day with this tool.

The trail builder is a variation of the burrow builder. The burrow is shallower and its diameter less than that constructed by the burrow builder. Zinc phosphide-treated grain typically is placed in the burrows to control vole damage in orchards or tree plantings (Anderson 1969).

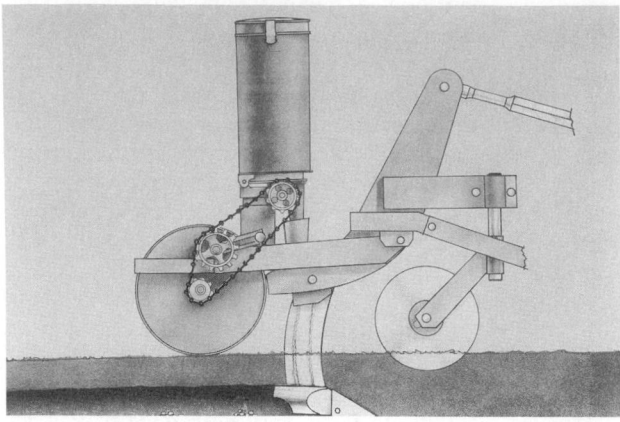

Fig. 13. Burrow builders are tractor-drawn mechanical devices used to create an artificial underground burrow and place toxic grain baits therein for controlling pocket gophers. Gophers intersect these burrows during their normal activity, feed on the bait, and die underground (sketch courtesy of Univ. California, Davis).

TRAPS

Live Traps

Live traps capture small mammals unharmed. They are excellent for use in residential areas or in situations where animals doing the damage may be transplanted to another location. These traps, in various shapes and sizes, can be homemade of wire or wood, or bought commercially. Some traps have doors at both ends, which allow animals to see through, therefore reducing reluctance to enter. Suggested baits include apple slices, sunflower seeds, peanut butter, and rolled oats.

The Bailey and Hancock live traps, used to capture beavers, are made of flexible mesh wire. When set, the Bailey trap resembles an open suitcase and the Hancock a half-open suitcase. When the triggering device is tripped the trap closes and the animal is caught between the two halves. These traps are best suited for use at entrance and exit routes of the lodge or in water travel lanes. The traps can be baited with an ear of corn or a fresh piece of aspen or other edible woody plant (Anderson 1969). They are used primarily to capture individual beavers for relocation; they are not efficient for intensive trapping efforts.

Leg-Hold Traps

Leg-hold traps, also called steel traps, are manufactured in several different sizes and are available with padded or unpadded jaws (Fig. 14). Their use is controversial; however, properly used they are effective and valuable. Some states prohibit their use, whereas others require that only traps with padded jaws be used. They are most extensively used for beaver, muskrat, and nutria control, but smaller sizes are used to capture tree and ground squirrels, rats, and woodchucks.

Traps can be set in travel lanes or near burrow openings without bait (blind sets), or they can be set adjacent to bait or various lures. Traps placed underwater for beavers and muskrats usually are set at burrow entrances or exit points from the water. Stakes or anchor material should be placed in the water in such a way that the trapped animals will seek deep water and drown, thus preventing them from twisting out. The Canadian Trappers Federa-

tion (no date) provides descriptions of various sets used for beavers and muskrats. Prairie dogs, ground squirrels, and mountain beavers can be caught by burying the traps near the burrows, using a pan cover, and covering the traps with soil. Scattered grain then is placed on the traps. Prebaiting may improve trapping success.

Body-Gripping Traps

Conibear traps are body-gripping traps chiefly used in water sets for muskrats, nutria, and beavers (Fig. 14). Manufactured in three sizes, they have the humane feature of killing quickly. This may also be a disadvantage, because any nontarget animal caught is killed as well. These traps have a pair of rectangular wires that close like scissors when released, killing the animal with a quick body blow. Conibear traps are lightweight and easy to use. They can be placed at the entrances of burrows and lodges and in dams, runs, and slides. The Canadian Trappers Federation (no date) also provides descriptions of sets for these traps. Care should be taken when large conibear traps are used because of the hazard to pets and children. A safety device is available that should be used when the large size is set. Some states prohibit their use in dry-land sets.

Somewhat similar body-gripping traps are available for moles and pocket gophers. For moles, the trap is placed in a section of the runway that has been pressed down. The trap is activated when the mole traveling the runway raises the depression, trips the trap, and is caught by the loops or scissors-like devices. The harpoon trap is used in a similar fashion, but instead of the mole being caught, it is speared by a spring-loaded harpoon.

Snap Traps

Advantages to using snap traps to control rats and mice include less danger to children or pets than with some chemicals, easy recovery of killed animals, and no contaminants. A snap trap's efficiency often can be increased by enlarging the bait pan with a heavy piece of cardboard or stiff screen wire. Obstacles such as boxes or boards can be used to funnel rodents to traps. Baits include peanut butter with uncooked oatmeal, a small piece of bacon or apple, or a raisin. These traps can be used outdoors to capture small field rodents when only a few animals are involved or to capture animals for identification or population indexing purposes.

SNARES

Beavers can be captured as effectively with snares as with conibear or leghold traps (Weaver et al. 1985). Snares cost and weigh less than traps and permit release of nontarget captives. Weaver et al. (1985) provided detailed instructions for their use. Snares are also effective in controlling small populations of rabbits. The animals must be traveling a well-defined trail or through a specific entrance such as a hole in a fence. Snares are made of a light wire or cable looped through a locking device, or a small nylon cord tied so it will tighten as the animals push against it. State game regulations should be checked before snares are used.

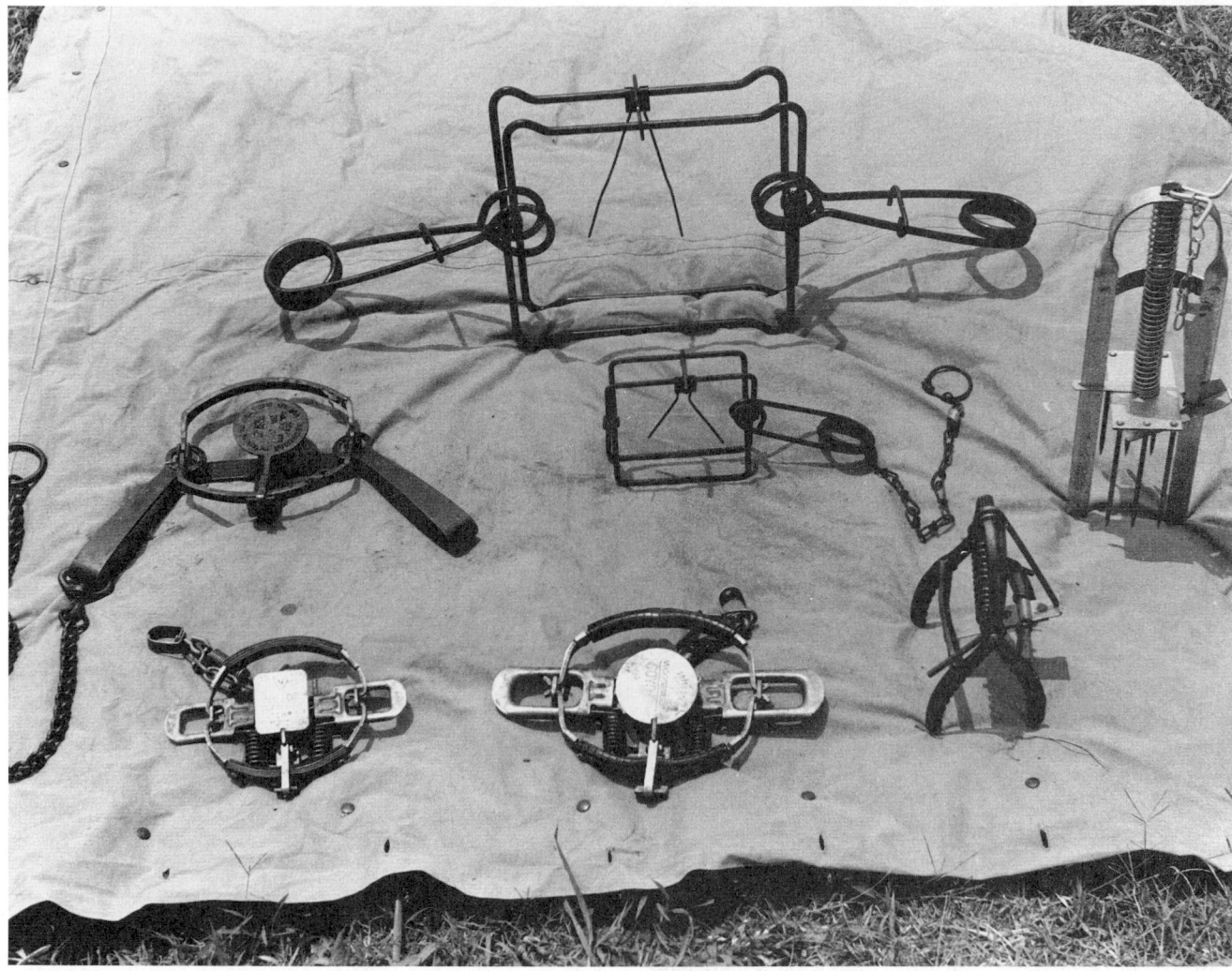

Fig. 14.   A variety of traps is used in the control of rodents and other small mammals. Shown here are: top—#330 conibear; middle row (left to right)—double-long spring leg-hold, #110 conibear, harpoon mole trap; bottom row (left to right)—#1½ Victor soft catch leg-hold, #3 Woodstream soft catch leg-hold, scissors mole trap (photo, F. Boyd, USDA/APHIS).

CHEMICAL REPELLENTS

Several compounds have been registered for use as small-mammal repellents (Jacobs 1994); however, definitive efficacy data for most of these are lacking. The use of some area repellents, such as naphthalene or para-dichlorobenzene, in structures often is limited because the vapors cannot be prevented from permeating areas occupied by people. Efficacy of repellents placed on plants or seeds is affected by availability of natural foods and ability to withstand weathering. Thiram, the most widely used taste repellent, can be applied to trees, tree seeds, seedlings, bulbs, and shrubs to protect them from various rodent species and moles. This compound cannot be used on plant parts eaten by humans or domestic animals. Fruit trees must be sprayed only in the dormant season. Thiram and methiocarb, when used as seed treatment, protected newly planted corn from rodent damage (Johnson et al. 1985, Holm et al. 1988).

SHOOTING

Shooting can be a selective method of eliminating individual pest mammals. Small-bore shotguns, rifles, and air guns can be used. Some animals such as beavers, muskrats, and nutria can be shot most effectively at night by using a spotlight with a red lens. Shooting is especially useful in controlling animals with low reproductive rates, such as porcupines. Local game codes should always be reviewed before shooting is used. Shooting at night, and in particular with a spotlight, is not legal in some states.

CARNIVORES AND OTHER MAMMALIAN PREDATORS
Damage Assessment

Mammalian predators have always been a concern to livestock producers. Wade (1982) estimated that the direct loss of sheep and goats to coyotes in the United States ranged from $75 million to $150 million annually. E. W. Pearson (unpubl. final rep., U.S. Fish Wildl. Serv., Denver Wildl. Res. Cent., 1986), using a summary of other studies and surveys, estimated the loss of sheep, lambs, and goats to predators, primarily coyotes, to be $68,160,000 in the 17 western states in 1984. Terrill (1988), using data from all 50 states, reported annual losses of sheep and lambs to coyotes and other predators ranged from $69 million to $83 million in 1985–87. Losses of poultry to predators, although not well documented, are also thought to be substantial.

Mammalian predators, especially red foxes, striped skunks, raccoons, and mink, seriously impact waterfowl nesting success in small wetland areas surrounded by agricultural lands. A study in North Dakota indicated nesting success of only 8% for mallards on such wetlands, one-half of what was needed to sustain the population (Cowardin et al. 1985). The red fox apparently is the most serious waterfowl predator because it is adept at catching nesting hens as well as destroying eggs (Sargeant et al. 1984).

Predation is rarely observed; therefore, accurate assessment of losses to specific predators often requires careful investigative work. The first action in determining the cause of death of an animal is to check for signs on the animal and around the kill site. Size and location of tooth marks often will indicate the species causing predation. Extensive bleeding usually is characteristic of predation. If external bleeding is not apparent, the hide can be removed from the carcass, particularly around the neck, throat, and head, and the area then is checked for tooth holes, subcutaneous hemorrhage, and tissue damage. Hemorrhage occurs only if skin and tissue damage occurs while the animal is alive. Animals that die from causes other than predation normally do not show external or subcutaneous bleeding, although bloody fluids may be lost from body openings (Bowns 1976). Animal losses are easiest to evaluate if examination is conducted when the carcass is fresh (Wade and Bowns 1982).

Animals may not always be killed by being bitten at the throat, but may be pulled down from the side or rear, often with blood on the sides, hind legs, and tail areas. Tails of calves may be chewed off, and the nose may have tooth marks or be completely chewed by the predator when the tongue is eaten (Bowns 1976).

Tracks and droppings alone are not proof of depredation or of the species responsible. They are evidence that a particular predator is in the area and, when combined with other characteristics of depredation, can help determine the species causing the problem.

Species Damage Identification

BADGERS

Badgers eat primarily rodents such as mice, prairie dogs, pocket gophers, and ground squirrels. They also will prey on rabbits, especially young. Badgers destroy nests of ground-nesting birds and occasionally kill small lambs and poultry, parts of which they sometimes bury in holes resembling their dens. Their dens in crop fields may slow harvesting or cause damage to machinery, and their digging can damage earthen dams or dikes (Lindzey 1994).

Badgers usually eat all of a prairie dog except the head and the fur along the back. This characteristic probably holds true for most larger rodents they eat; however, signs of digging near prey remains are the best evidence of badgers. Badger tracks often appear similar to coyote tracks, but on close examination they are distinctively pigeon-toed, and impressions from the long toenails are apparent in most situations.

Control Techniques.—Frightening devices, leg-hold traps (#3–4), shooting, snares.

BEARS

Black and grizzly bears prey on livestock. Black bears usually kill by biting the neck or by slapping the victim. Torn, mauled, and mutilated carcasses are characteristic of bear attacks. Often, the bear will eat the udders of female prey, possibly to obtain milk. The victim usually is opened ventrally and the heart and liver are consumed (Bowns and Wade 1980). The intestines often are spread out around the kill site, and the animal may be partially skinned while the carcass is fed upon. Smaller livestock such as sheep and goats may be consumed almost entirely, and only the rumen, skin, and large bones remain. Feces generally are found within the kill area, and a bedding site often is found nearby. Bears use their feet while feeding, so they do not slide the prey around as do coyotes. If the kill is made in the open, the carcass may be moved to a more secluded spot.

The grizzly has a feeding and killing pattern similar to that of the black bear. Murie (1948) observed that most cattle are killed by a bite through the back of the neck. A large prey often has claw marks on the flanks or hams. The prey's back is sometimes broken in front of the hips where the bear simply crushed it down. Young calves sometimes are bitten through the forehead.

The presence of bears has stampeded range sheep, resulting in death from suffocation or from falls over cliffs. A marauding bear searching for food also may play havoc with garbage cans, cabins, campsites, and apiaries (Maehr 1983).

Black bear damage to trees can be recognized by the large, vertical incisor and claw marks on the sapwood and ragged strips of hanging bark. Pole-size trees to small saw timber are preferred. Most bark damage occurs during May, June, and July (Packham 1970). After the bark is pulled away, bears scrape off the cambium layer of the tree with their incisor teeth, leaving vertical tooth marks (Murie 1954).

The bear track resembles that of a human but has distinctive claw marks. The little inside toes often leave no marks in dust or shallow mud, so the print appears to be four-toed (Murie 1954).

Control Techniques.—Hunting dogs, live trapping, foot snares, fencing, shooting, leg-hold traps (#5, 6, and 15) where legal.

BOBCATS AND LYNX

These related species occasionally prey on sheep, goats, deer, and pronghorns; however, they more commonly kill smaller animals such as porcupines, poultry, rabbits, rodents, birds, and house cats. Characteristically, bobcats kill adult deer by leaping on their back or shoulders, usually when the victim is lying down, and biting them on the trachea. The jugular vein may be punctured, but the victims usually die of suffocation and shock. Bowns (1976) reported that a lamb killed by a bobcat had hemorrhages produced by claws on both sides of the carcass, indicating the bobcat was holding the lamb with its claws while biting the neck. Small fawns, lambs, and other small prey often are killed by a bite through the top of the neck or head (Young 1958). The hindquarters of deer or sheep usually are preferred by bobcats, although the shoulder and neck region or the flank sometimes are eaten first. The rumen is often untouched. Poultry usually are killed by biting the head and neck (Young 1958); the heads usually are eaten. Both species reportedly prey on bird eggs.

Bobcat and lynx droppings are similar; in areas inhabited by both species, the tracks will help determine the responsible animal. The lynx has larger feet with much more hair, and the toes tend to spread more than they do on the more compact bobcat tracks.

Feline predators usually attempt to cover their kills with litter (Cook et al. 1971). Bobcats reach out 30–35 cm in scratching litter, compared to a 90-cm reach of a mountain lion (Young 1958). The distance between the canine teeth marks also will help distinguish a lion kill from that of a bobcat—3.8 cm vs. 1.9–2.5 cm, respectively (Wade and Bowns 1982).

Control Techniques.—Hunting dogs, snares, calling and shooting, leg-hold traps (#3–4), aircraft (under some specific circumstances), frightening.

COYOTES, WOLVES, AND DOGS

These predators prey on animals ranging from big game and livestock to rodents, wild birds, and poultry. Coyotes are the most common and most serious predator of livestock in the western United States (Wade and Bowns 1982) and are rapidly becoming a problem throughout the East.

Coyotes normally kill livestock with a bite in the throat, but they infrequently pull the animal down by attacking the side, hindquarters, and udder. The rumen and intestines may be removed and dragged away from the carcass. On small lambs, the upper canine teeth can penetrate the top of the neck or the skull. Calf predation by coyotes is most common when calves are young. Calves attacked, but not killed, exhibit wounds in the flank, hindquarter, or front shoulder; often their tails are chewed off near the top. Deer carcasses frequently are completely dismembered and eaten (Bowns 1976).

Complaints of pets being killed by coyotes have increased with urbanization (Howell 1982). Avocado producers using drip irrigation systems report that coyotes chew holes in plastic pipe and disrupt irrigation (Cummings 1973). Watermelons are damaged by coyotes biting a hole through the melons and eating out the center. This differs from raccoon damage to melons; raccoons make small holes in the melons and scoop out the pulp with their front paws. Coyotes also will damage other fruit crops.

Wolves prey on larger ungulates such as caribou, moose, elk, and cattle. Wolves usually bring down these animals by cutting or damaging the muscles and ligaments in the back legs or by seizing the victim in the flanks. Slash marks made by the canine teeth may be found on the rear legs and flanks. The downed animals usually are disembowelled.

Domestic dogs can be a serious problem to livestock, especially to sheep pastured near cities and suburbs. Dogs often attack the hindquarters, flanks, and head and rarely kill as effectively as coyotes (Green and Gipson 1994). Normally little flesh is consumed. They are likely to wound the animal in the neck and front shoulders; the

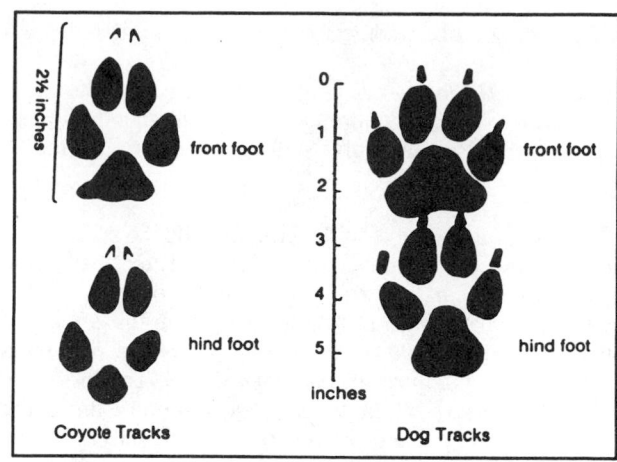

Fig. 15. Coyote and dog tracks are similar. Coyote tracks are more rectangular with the toes closer together, whereas dog tracks are more round with the toes spread apart (from Dorsett 1987).

ears often are badly torn. Attacking dogs often severely mutilate the victim (Bowns and Wade 1980).

Coyote and dog tracks are similar but distinguishable. Dog tracks are round with the toes spread apart. Toenail marks usually are visible on all toes (Dorsett 1987). Coyote tracks are more rectangular and the toes are closer together. If any toenail marks show, they are usually on the two middle toes (Fig. 15). Coyote tracks appear in a straight line, whereas those of a dog are staggered.

Control Techniques.—Fencing, herding, den hunting, calling and shooting, shooting from aircraft, guarding dogs, snares, M-44s, frightening, livestock protection collar, leg-hold traps (#3–4½).

FOXES

Gray and red foxes feed primarily on rabbits, hares, small rodents, poultry, birds, and insects. They also will consume fruits. The gray fox also eats fish, prey seldom eaten by the red fox. Gray and especially red foxes kill young livestock, although poultry is their more common domestic prey. Foxes usually attack the throat of lambs and birds but kill some prey by multiple bites to the neck and back (Wade and Bowns 1982). Normally, foxes taking fowl leave behind only a few drops of blood and feathers and carry the prey away from the kill location, often to a den. Eggs usually are opened enough to allow the contents to be licked out. The shells are left beside the nest and rarely are removed to the den, even though fox dens are noted for containing the remains of their prey, particularly the wings of birds.

Einarsen (1956) noted that the breast and legs of birds killed by foxes are eaten first and the other appendages are scattered about. The toes of the victims usually are drawn up in a curled position because of tendons pulled when the fox strips meat from the leg bone. Smaller bones are likely to be sheared off. The remains often are partially buried.

Foxes, like other wild canids, return to established denning areas year after year. They dig dens in wooded areas or open plains. Hollow logs also are used. Dens may be identified by the small, dog-like tracks or by fox hairs clinging to the entrance. The gray fox is the only fox that readily climbs trees, sometimes denning in a hollow cavity.

Control Techniques.—Dogs (hunting and guarding), leg-hold traps (#2–3) denning, calling and shooting, fencing, shooting from aircraft, M-44s, snares, frightening.

HOGS

Problems associated with feral or wild hogs have increased across the southern U.S. Rooting and wallowing by wild hogs can damage agricultural crops and timber and also damage farm ponds and irrigation dikes (Barrett and Birmingham 1994). Wild hogs also feed on young sheep and goats in certain parts of the U.S. The losses are difficult to determine at times because almost the entire carcass is either eaten or carried off, and the only evidence may be tracks and blood where feeding occurred (Wade and Bowns 1982).

Tracks of adult hogs resemble those made by a 90-kg calf. In soft ground dewclaws will show on adult hog tracks (Barrett and Birmingham 1994).

Control Techniques.—Live traps, snares, hunting dogs, shooting from aircraft.

MOUNTAIN LIONS

Often called cougar or puma, this large feline preys on deer, elk, and domestic stock, particularly horses, sheep, goats, and cattle. It also eats rodents and other small mammals when available. In one situation, according to Young (1933), a lone lion attacked a herd of ewes and killed 192 in 1 night. However, five to ten sheep killed in a single night is more typical (Shaw 1983).

Mountain lions, having relatively short, powerful jaws, kill with bites inflicted from above, often severing the vertebral column and breaking the neck. They also kill by biting through the skull (Bowns 1976). Lions usually feed first upon the front quarters and neck region of their prey. The stomach generally is untouched. The large leg bones may be crushed and ribs may be broken. Many times, after a lion has made a kill, the victim is dragged or carried into bushy areas and covered with litter. A lion might return to feed on a kill for 3 or 4 nights. They normally uncover the kill at each feeding and move it 10–25 m to recover it. After the last feeding the remains may be left uncovered, and a search of the area might reveal previous burial sites (Shaw 1983).

Adult lion tracks are approximately 10 cm in length and 11 cm in width; they have four well-defined impressions of the toes at the front, roughly in a semicircle. Lions have retractable claws; therefore, no claw prints will be evident. The untrained observer sometimes confuses large dog tracks with those of the lion; however, dog tracks normally show distinctive claw marks, are less round than lion tracks, and have distinctly different rear pad marks.

Control Techniques.—Dogs (guarding and hunting), snares, leg-hold traps (#4½ and 114).

OPOSSUMS

Opossums are omnivorous, eating fish, crustaceans, insects, mushrooms, fruits, vegetables, eggs, and carrion. They will raid poultry houses. The opossum usually kills one chicken at a time, often mauling the victim (Burkholder 1955). Eggs will be mashed and messy; the shells often are chewed into small pieces and left in the nest.

Opossums usually begin feeding on poultry at the cloacal opening. Young poultry or game birds are consumed entirely and only a few wet feathers remain.

Control Techniques.—Live traps, leg-hold traps (#1–1½), shooting, dogs, exclusion fencing.

RACCOONS

Raccoons eat mice, small birds, snakes, frogs, insects, crawfish, grass, berries, acorns, corn, melons—the list is almost endless. Garbage cans and dumps can be a major source of food in urban areas. Field crops or gardens near wooded areas may experience severe damage from raccoons. Ripening corn frequently is eaten and much is wasted (Conover 1987*b*) (Fig. 2). Raccoons raid nesting cavities of birds (Lacki et al. 1987). They occasionally kill small lambs, usually by chewing the nose.

Occasionally, raccoons enter poultry houses and take many birds in 1 night. The breast and crop can be torn and chewed, and the entrails sometimes are eaten. There may be bits of flesh near water. Eggs may be removed from poultry or game-bird nests and eaten away from the nest. Rearden (1951) reported that eggshells were located within 9 m of the nest.

The raccoon leaves a distinctive 5-toed track resembling a small human hand print. Tracks usually are paired, and the left hind foot is placed beside the right forefoot (Murie 1954). Raccoon and opossum tracks can be difficult to distinguish in soft sand where toes do not show.

Control Techniques.—Hunting dogs, live traps, leg-hold traps (#2–3), exclusion fencing, shooting.

SKUNKS

Insects, particularly grasshoppers, beetles, and crickets, make up a large portion of the skunk's diet. Skunks usually dig small, cone-shaped holes in lawns, golf courses, and meadows in search of beetle larvae. A common complaint of objectionable odor occurs when skunks take up residence under buildings. Skunks may depredate beehives.

Skunks kill few adult birds but are serious nest robbers (Einarsen 1956). Eggs usually are opened at one end; the edges are crushed as the skunk punches its nose into the hole to lick out the contents (Einarsen 1956, Davis 1959). The eggs may appear to have been hatched, except for the edges. When in a more advanced stage of incubation, eggs are likely to be chewed in small pieces. Eggs may be removed from the nest, but rarely more than 1 m away.

Most rabbit, chicken, and pheasant carcasses found at skunk dens are carrion that has been dragged to the den site (Crabb 1948). When skunks do kill poultry, they generally kill only one or two birds and maul them considerably. Crabb (1941) observed that spotted skunks help control rats and mice in grain-storage buildings. They kill these rodents by biting and chewing the head and foreparts; the carcasses are not eaten.

Inhabited dens can be recognized by fresh droppings near the mound or hole containing undigested insect parts. Hair and rub marks also may be present. Dens usually have a characteristic skunk odor, although it might not be strong.

Control Techniques.—Live traps, leg-hold traps (#1–1½), fumigants, shooting.

WEASELS AND MINK

Weasels and mink have similar feeding behaviors, killing prey by biting through the skull, upper neck, or jugular vein (Cahalane 1947). When they raid poultry houses at night, they often kill many birds, eating only the heads of the victims. Predation by rats usually differs in that portions of the body are eaten and carcasses are dragged into holes or concealed places.

Errington (1943) noted that mink, while eating large muskrats, make an opening at the back or side of the neck. As the mink eats away flesh, ribs, and pieces of the adjacent hide, the head and hindquarters are pulled out through the same hole and the animal is skinned. McCracken and Van Cleve (1947) noted similar feeding behavior by weasels eating small rodents.

Teer (1964) observed that blue-winged teal eggs destroyed by weasels were broken at the ends and had openings 15–20 mm in diameter. Close inspection of shell remains frequently will disclose finely chewed edges and tiny tooth marks (Rearden 1951).

Weasels den in the ground (e.g., in a mole or pocket gopher burrow), under a barn, in a pile of stored hay, or under rocks. Mink dig dens approximately 10 cm in diameter into banks. Mink also use muskrat burrows, holes in logs and stumps, and other natural shelters.

Control Techniques.—Leg-hold traps (#1–1½), conibear traps, fencing, barriers.

DOMESTIC CATS

Domestic cats rarely prey on anything larger than ducks, pheasants, rabbits, or quail. Einarsen (1956) noted the messy feeding behavior of these animals. Portions of their prey often are strewn over several square meters in open areas. The meaty portions of large birds are consumed entirely, and loose skin with feathers attached is left. Small birds generally are consumed and only the wings and scattered feathers remain. Cats usually leave tooth marks on every exposed bone of their prey. Nesting birds particularly are vulnerable to cat predation. In areas managed for game birds or waterfowl production, vagrant cat control is almost a necessity. Unlike their native cousins, domestic cats are observed readily in the daytime, although feral cats often are extremely wary.

Control Techniques.—Live traps, shooting, leg-hold traps (#1–1½).

Control Techniques

SHOOTING FROM AIRCRAFT

Various kinds of fixed-wing aircraft have been used to control wolves, coyotes, bobcats, and foxes, but the Piper Super Cub, with a 150-horsepower engine, is preferred. The pilot and gunner sit in tandem in this two-seat aircraft; this allows both occupants to see out both sides of the aircraft. Hunting is more effective on snow because target animals can be seen and tracked more easily. When the hunted animal is found, the pilot makes an approach over it at approximately 20 m of altitude, preferably into the wind. The ground speed of the aircraft is around 60–85 km/h at this point, but the airspeed should never be near the stalling speed of the aircraft. A 12-gauge semi-automatic shotgun is the most common weapon used, and number 4 buck-shot, BB, and number 2 shot are preferred.

Several modifications have been made to the Super Cub to increase safety and effectiveness. These modifications include a larger propeller called the Alaskan Super Prop and drooped wingtips to provide added power, stability, and maneuverability, particularly at higher altitudes. Larger balloon-type tires have been added to provide clearance for the longer propeller and to better utilize primitive runways for landings. The 160-horsepower engine is becoming more popular due to its added power and greater fuel efficiency (Vetterman 1985).

Rotary-wing aircraft (helicopters) have been used in recent years for predator control. The helicopter, with its ability to hover, can be used more effectively than fixed-wing aircraft in rougher and brushy terrain. In models with a plexiglass bubble cockpit, the visibility and consequent tracking ability are good.

The fixed-wing craft and helicopter sometimes are operated together. The helicopter is used for dispatching the animal, while the fixed-wing flies above the helicopter and maintains surveillance. This combination works well in areas with thick vegetation or where animals have been hunted heavily with the helicopter. These "chopper-wise" animals try to evade the hunters but can be spotted with the fixed-wing craft. Radio contact between the two aircraft is necessary.

Aerial hunting can be more efficient if one or more ground crews work with the aircraft. The ground crew induces coyotes to howl by using a horn, siren, voice, or recorded howl. When animals respond, the aircraft is directed into the area by two-way radio communication. Early morning and late afternoon are the most productive times for aerial hunting.

Federal law requires each state where aerial hunting is allowed to issue aerial hunting permits. Some states also require low-level flying waivers.

CALLING AND SHOOTING

Calling and shooting is a selective means to control coyotes, bobcats, and foxes. It has become a popular sport, and for some people it is not calling and shooting, but calling and photographing.

Several commercial calls are available, as are recorded calls. Open-reed predator or duck calls work well but require more practice. The call is blown to imitate the sound of a rabbit in distress. This sound either arouses the predator's curiosity or indicates an easy meal. Of course, some predators become wise to the call. Conversely, the call may be an effective method to attract a trap-wise animal.

Three factors must be kept in mind when calling is used: (1) ensure that the area being called to is upwind to prevent the predator from detecting the caller's scent before the animal comes into shooting range; (2) have a full view of the area being called so that the predator will be unable to approach unseen; and (3) avoid being seen by wearing camouflage clothing and hiding in vegetation.

The most effective times to call are early morning and late afternoon. The hunter can gain an added advantage by locating coyotes before beginning the call by inducing howls as previously described under aerial hunting. Calling at night and using a spotlight can be effective; however, local game laws should be checked.

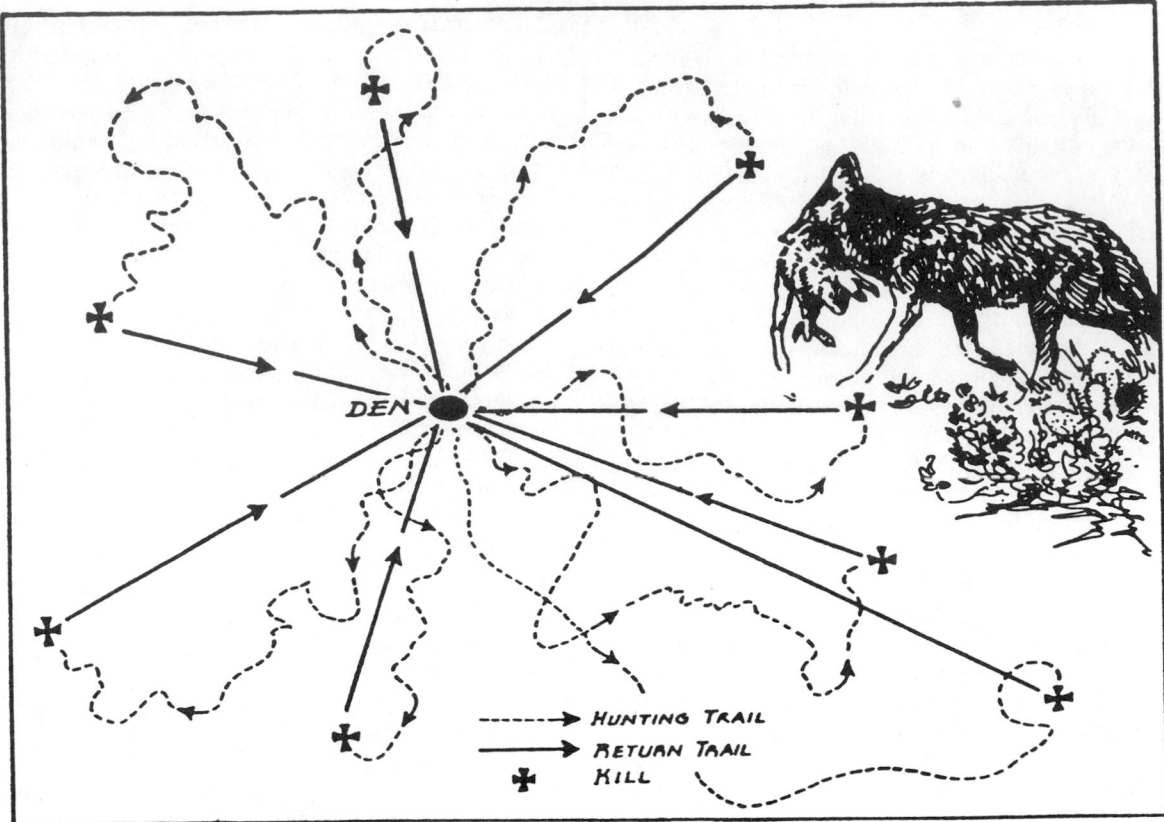

Fig. 16.   The tracks of a coyote that is hunting away from the den site will be meandering. When the coyote is returning to the den site, it will travel the straightest possible route (from Presnall 1950).

DENNING

In the spring, depredation on livestock and poultry by coyotes and foxes might indicate a nearby den that has increased the food requirement for support of the pups as well as the adults. Till and Knowlton (1983) reported that sheep losses to coyotes were greatly curtailed after removal of adults and pups or only pups of coyotes responsible for the losses. Removal of the entire litter usually will end the losses of livestock; however, this is dependent on the availability of other food sources and preferences.

Dens are located by tracking or observing the adults. Den hunting is based on the principle that adults tend to follow irregular routes while searching for prey, but once food is secured they return to the den in the most direct route possible (Fig. 16). The experienced observer can distinguish between these tracks.

An active den is evident by hairs around the entrance, fresh tracks, and, if the pups are large enough to come out of the den, matted and worn vegetation around the entrance. Fox dens usually have remains of prey brought in for food. This observation is not common at coyote dens.

Den hunting is difficult, time-consuming work, particularly on hard ground, in heavy cover, and during high winds. A good dog is a great help in locating dens. Some dogs are trained to return to the hunter when the adult predator tries to chase them out of the den area. This behavior usually will get the target animal within rifle range. A call blown to imitate a frightened or injured pup sometimes will bring adult coyotes within rifle range. Care should be taken while digging out dens because of

the possibility of cave-ins and ectoparasites. These hazards are eliminated if a gas cartridge is used in coyote dens.

Use of aircraft is a good method for locating coyote and fox dens. This is done during normal aerial hunting operations by looking for animal signs as well as the animals. Den signs include cleaned-out holes and worn vegetation (Vetterman 1985).

FRIGHTENING DEVICES

Devices such as lights, loud music or noise, scarecrows, plastic streamers, aluminum pie pans, and lanterns have been tried to frighten away predators. All of these devices can provide a temporary benefit in reducing damage or deterring predators. Changing the location or combination of techniques being used can prolong the frightening effect, but the effectiveness decreases when the predators become accustomed to the noise, lights, or objects.

Linhart (1984) reported that the use of warbling-type sirens and strobe lights in combination reduced lamb losses from coyotes by 44%. These battery-operated devices were activated in the evening by a photocell set on a schedule of 10-second bursts at 7- to 13-minute intervals. In a survey of North Dakota ranchers, Pfeifer and Goos (1982) reported the use of propane exploders delayed or prevented lamb losses to coyotes for a period of time to allow other control methods to be employed. The most important factors contributing to success were properly operating and maintaining the device, moving the device to different locations, and changing the firing intervals.

Fig. 17. The body snare, used primarily to capture coyotes, should be positioned directly under the fence with the top of the loop attached by a small, thin wire or a single strand of sewing thread. The attachment should release with the slightest pull (from Sims 1988).

HUNTING DOGS

Two types of dogs are used in predator control work. Dogs that hunt by sight, such as greyhounds, are kept in a box or cage until the predator is seen, then released to catch and kill the animal. This type of dog is effective only in relatively open terrain. The other type of dog is the trail hound, which follows an animal by its scent. Trail hounds hunt on bare ground; however, snow or heavy dew makes trailing easier. Hot, dry weather makes trailing difficult; therefore, early morning with dew is the most effective time. Several breeds such as bluetick, black and tan, Walker, and redbone, usually run in packs of two to five, are used.

Trained trail hounds are used to catch and "tree" raccoons, opossums, bobcats, bears, and mountain lions. Often these dogs are able to track the offending animal from a kill, thus making this control method highly selective. Local game codes should be checked before this type of control is used.

GUARDING DOGS

For centuries dogs have been used to work livestock, but only in the past 15 years has interest in guarding dogs caught the attention of the livestock industry. The three more common breeds used are the Great Pyrenees, komondor, and Akbash. All have been used effectively in fenced pasture situations, but Pyrenees have shown the most success on open-range flocks (Green et al. 1984). Mixed-breed dogs also have been used (Black and Green 1984). Guarding dogs must possess three behavioral traits: trustworthiness with sheep, attentiveness to the sheep, and aggressiveness to attack and chase potential predators (McGrew and Andelt 1986). The guarding dog puppy develops a bond with sheep by being placed with them at 6–8 weeks of age. For range operations, the ideal time to place a dog with the sheep is when the sheep are confined in a pasture or fenced area or after lambing when the main flock is being formed (Green and Woodruff 1983).

A critical factor in the success of guarding dogs is the handler's ability to train and use these animals, which requires patience and understanding. According to Green and Woodruff (1983), the most serious problem encountered by some producers is a disillusionment that the use of guarding dogs will be an immediate solution to their predator problem.

LIVESTOCK PROTECTION COLLAR

The livestock protection collar, also called the toxic collar, consists of several rubber pouches containing Compound 1080 attached with straps around the throat of a sheep or goat. Collars are designed to kill coyotes that puncture the pouches while attacking the throats of targeted livestock. The collar offers certain advantages over other methods of coyote control by specifically removing only those animals responsible for predation. It is particularly effective for coyotes that have become wary and avoid other control methods (Connolly and Burns 1990). The primary disadvantages of collars are the cost and labor of their application; the compartments being punctured by thorns, wire, or snags; and the EPA-required monitoring of the flock with collared animals (Wade 1985).

SNARES

Snares are made of varying lengths and sizes of wire or cable looped through a locking device that allows the snare to tighten. The two types of snares are body and foot. The body snare is used primarily on coyotes. This snare is set where the animals crawl under a fence, at a den entrance, or in some other narrow passageway. The device is looped so that the animal must put its head through the snare as it passes through the restricted area (Fig. 17). When the snare is felt around the neck, the animal normally will thrust forward and tighten the noose.

The foot snare is spring-activated. When the animal steps on the trigger the spring is released, lifting the noose and tightening it around the foot. This device has been used effectively to capture mountain lions and grizzly and black bears. The foot snare can be used in a bear pen or cubby set. This pen is just large enough to accommodate the bait, which usually is the carcass remains of an animal killed earlier by the predator. The pen can be built of brush or poles and has an open end where the snare is set. The pen and guide sticks will force the bear to step into the snare while trying to reach the bait. Bacus (1968) described a pipe snare set that consists of a 0.9-kg coffee can (or a similar length of 13-cm pipe) with a 2.5-cm slot cut down the side to accommodate the trigger. The can is buried, and the loop is laid loosely on the ground around the outside of it and covered with dirt. Bacon grease is melted into the can with a torch. A rock is placed on top of the can to prevent nontarget animals from tripping the snare. A bear can roll the rock off but, being unable to reach the bait in the bottom of the can with its mouth, will reach in with its front foot and spring the snare. Bears also can be caught with the foot snare in a trail set.

The foot snare also can be used to capture mountain lions. It should be set in a narrow trail known to be traveled by the target animal. Deer and livestock can be prevented from interfering with the snare with a pole or branch placed across the trail, directly over the set and about 0.9 m above the ground.

The selectivity of the foot snare may be improved by placing sticks under the trigger that break only under the weight of heavier animals. Foot snares have advantages over large bear traps in that they are lighter, easier to carry, and less dangerous to humans and nontarget animals.

TRAPS
Live Traps

Live traps, as discussed in RODENTS AND OTHER SMALL MAMMALS (p. 483), are available in various sizes to capture small predators as well as larger ones such as bears. Coyotes, foxes, and bobcats are difficult to live-trap because of their caution and reluctance to enter the confined area of the trap.

Canned dog or cat foods are effective baits to entice raccoons, opossums, skunks, and cats into live traps. Traps for skunks should be covered with a canvas or heavy cloth and provided a flap for the door. When a skunk is captured, the trapper can walk up to the trap on the covered side and drop the flap over the door. The skunk then can be transported to the release site. To release, the trapper should stand beside the trap and ease the flap and door open; the animal will flee and usually not look back.

Problem bears can be caught in a live trap made from steel culverts equipped with a trapdoor and trigger device. They normally are mounted on trailers to permit bears to be easily moved to other locations for release.

Leg-Hold Traps

Leg-hold or steel traps are manufactured in various sizes. The following trap sizes are recommended for the animals listed:

#0 and 1 for weasels and ground squirrels
#1 and 1½ for skunks, opossums, mink, feral cats, and muskrats
#2 and 3 for foxes, raccoons, small feral dogs, nutria, marmots, and mountain beavers
#3 and 4 for bobcats, coyotes, large feral dogs, badgers, and beavers
#4 and 4½ for wolves
#4½ and 114 for mountain lions

Success in trapping depends greatly on placing the trap where the predator regularly travels. A trap usually is set in the ground by digging a shallow trench the size of the trap (Fig. 18) and deep enough to allow the stake (or drag) and chain to be placed in the bottom of the trench and covered with soil. The trap is set firmly on top of this and should be about 11 mm below the soil surface. A canvas or cloth is placed over the pan and under the jaw to prevent soil from getting beneath the pan and preventing its release. The trap then is covered with soil and other material natural to the area surrounding the trap. The trap can be set unbaited in a trail being traveled by the target animal; this is called a "blind" or trail set. Traps also may be set off the trail and used with a lure. The lure set is more selective and is made more so by the type of lure used.

The dirt-hole set is effective for raccoons, foxes, and mink. The trap is set in the same manner as the baited set, but instead of placing the scent on the vegetation or ground, the lure is placed in a small hole, about 15 cm deep, dug on a slant behind the trap.

The bear trap is extremely large, powerful, and dangerous to humans, livestock, and pets. The bear foot snare is as effective and much safer to use, so bear traps are not recommended and are no longer legal in some states.

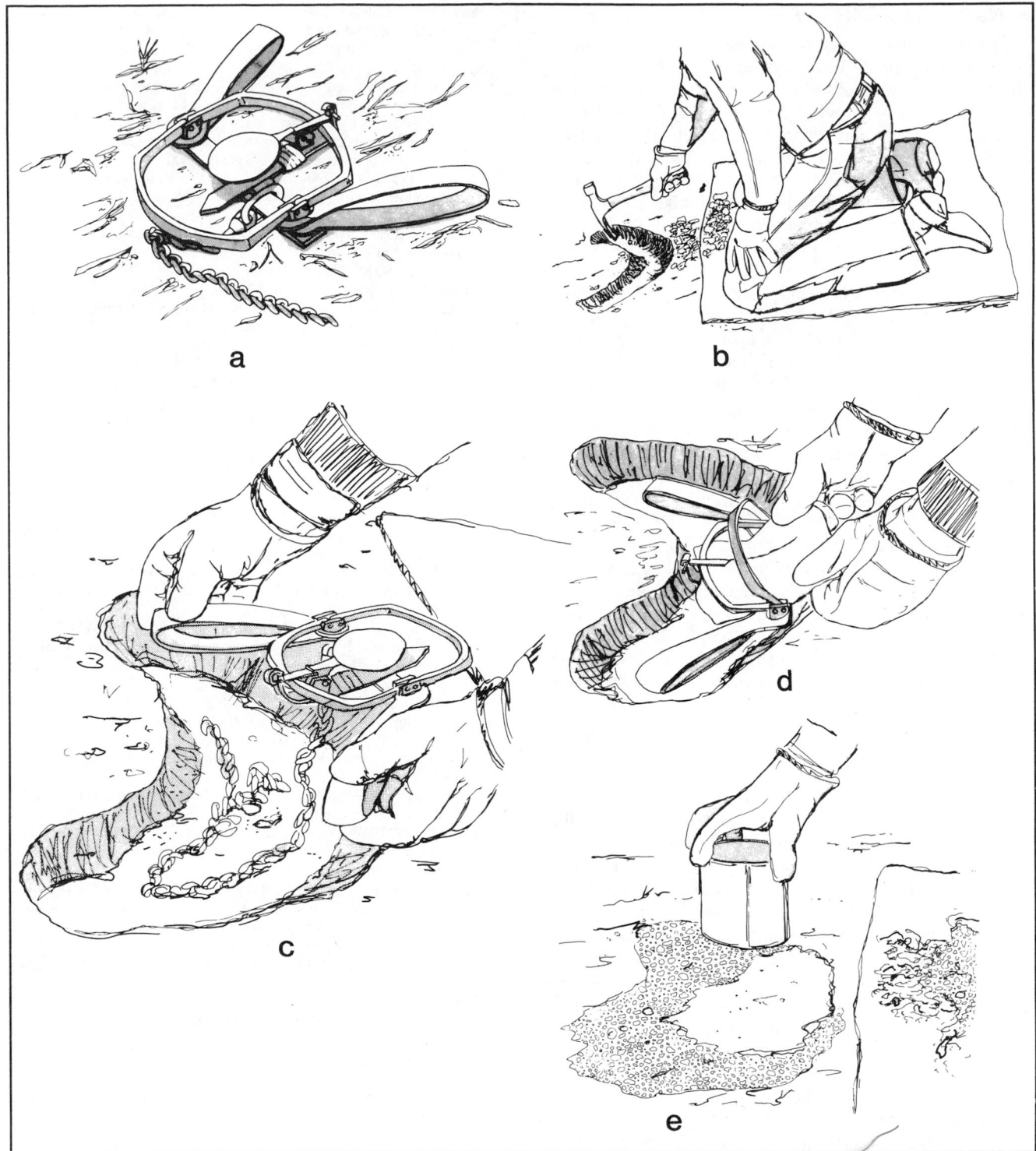

Fig. 18. a. A leg-hold trap is first laid on the ground to determine best location of hole. b. The hole should be about 11 cm deep and shaped to accommodate the trap. c. If a stake is used, it is driven into the bottom of the hole. If a drag is used, it is placed in the hole. The chain is then put into the hole and covered with soil until the hole is about 3 cm deep and packed to provide a firm foundation. d. The front jaw is raised and the pan cover is placed over the pan so soil cannot get under the pan. e. The trap is covered with finely sifted soil to a depth of 0.6–1.2 cm. A stick or whisk broom is used to touch up and make set appear as natural as possible (from Dorsett 1987).

The location of a trap set influences its selectivity. When placed beside a carcass, a trap can catch nontarget animals such as vultures, eagles, badgers, and other carrion-feeders. Nine meters away from the carcass normally is a safe distance to set traps to avoid nontarget animals. Weather also can affect the operation of traps. Frozen or wet ground can prevent a trap from springing.

Leg-hold traps must be checked often to prevent the lengthy restraint of captured animals. Most states have laws on the types of traps, baits and sets, and trap visitation schedule.

FENCING AND BARRIERS

Livestock, poultry, and crops can be protected from predation with properly placed fencing and barriers. Ordinary fencing will not keep most predators from gardens or poultry ranges. However, many of them can be excluded by adding a single wire strand electrified by a commercial fence charger, 20 cm out from the fence and 20 cm above the ground. Storer et al. (1938) reported success in keeping bears out of storehouses and other areas by the use of a specifically designed electric fence. An antipredator electric fence can provide some producers with a self-help method of effectively preventing coyote depredation of livestock (Nass and Theade 1988). One design is a fence 1.5 m high with 12 alternating ground and charged wires spaced 10–15 cm apart (Gates et al. 1978).

Skunks can be controlled around a poultry range by surrounding the range with a 0.9-m wire-netting fence set 0.6 m above ground and 0.3 m below the surface; a 15-cm length of the part below the surface is bent outwardly at right angles and buried 15 cm deep. Mink and weasels can be excluded from domestic animal quarters by covering all openings larger than 2.5 cm with metal or hardware cloth.

All holes in foundations of buildings should be closed or screened to prevent small predators such as skunks and opossums from living in or under them. If they have already established a home, all entrances except one should be closed. The soil should be loosened or flour should be sprinkled in front of the hole so a track can be detected. The area should be checked for tracks after dark, and if tracks indicate the animal has left the location, the opening should be sealed securely.

M-44

The M-44, registered by the EPA for the control of coyotes, foxes, and feral dogs, is a mechanical device that ejects sodium cyanide into the animal's mouth (Connolly 1988). The unit consists of a case holder wrapped with cloth, fur, wool, or steel wool; a plastic capsule or case that holds the cyanide; and a 7-cm ejector unit. The M-44 case is loaded with 12 grains (0.78 g) of sodium cyanide and an additive to reduce caking. A spring-loaded plunger ejects the cyanide. These components, when assembled, are encased in a tube driven into the ground. The cocked ejector with the case in the holder is screwed on top, placed into the tube, and baited. The bait usually is made from fetid meat, musks, and beaver castors. When an animal is attracted to the bait and tries to pick up the baited case holder with its teeth, the cyanide is ejected into its mouth. Dogs, skunks, raccoons, bears, and opossums sometimes are attracted to the bait used on M-44s; however, selectivity is enhanced by proper site and bait (scent) selections. The EPA and individual states have placed numerous restrictions on the use of M-44s.

LITERATURE CITED

AGÜERO, D. A., R. J. JOHNSON, AND K. M. ESKRIDGE. 1991. Monofilament lines repel house sparrows from feeding sites. Wildl. Soc. Bull. 19:416–422.

ALSAGER, D. E. 1977. Impact of pocket gophers (Thomomys talpoides) on the quantitative productivity of rangeland vegetation in southern Alberta: a damage assessment tool. Pages 47–57 in W. B. Jackson and R. E. Marsh, eds. Vertebrate pest control and management materials. Am. Soc. Test. Materials Spec. Publ. 625.

ANDERSON, T. E. 1969. Identifying, evaluating, and controlling wildlife damage. Pages 497–520 in R. H. Giles, ed. Wildlife management techniques. Third ed. The Wildl. Soc., Washington, D.C.

ANTHONY, R. M., V. G. BARNES, JR., AND J. EVANS. 1978. ''VEXAR'' plastic netting to reduce pocket gopher depredation of conifer seedlings. Proc. Vertebr. Pest Conf. 8:138–144.

ARNER, D. H., AND J. S. DUBOSE. 1982. The impact of the beaver on the environment and economics in the southeastern United States. Trans. Int. Congr. Game Biol. 14:241–247.

AUSTIN, D. D., AND P. J. URNESS. 1989. Evaluating production losses from mule deer depredation in apple orchards. Wildl. Soc. Bull. 17:161–165.

BACUS, L.C. 1968. The bear foot snare. U.S. Fish Wildl. Serv. Field Training Aid 2. 14pp.

BAKER, R. O., R. M. TIMM, AND G. R. BODMAN. 1994. Rodent-proof construction. Pages B137–B150 in S. E. Hygnstrom, R. M. Timm, and G. E. Larson, eds. Prevention and control of wildlife damage. Univ. Nebraska Coop. Ext. Serv., Lincoln.

BARRETT, R. H., AND G. H. BIRMINGHAM. 1994. Wild pigs. Pages D65–D70 in S. E. Hygnstrom, R. M. Timm, and G. E. Larson, eds. Prevention and control of wildlife damage. Univ. Nebraska Coop. Ext. Serv., Lincoln.

BEIER, P., AND R. H. BARRETT. 1987. Beaver habitat use and impact in Truckee River basin, California. J. Wildl. Manage. 51:794–799.

BESSER, J. F. 1986. A guide to aid growers in reducing bird damage to U.S. agricultural crops. Denver Wildl. Res. Cent. Bird Damage Res. Rep. 377. 91pp.

———, AND D. J. BRADY. 1986. Bird damage to ripening field corn increases in the United States from 1971 to 1981. U.S. Fish Wildl. Serv. Fish Wildl. Leafl. 7. 6pp.

———, J. W. DEGRAZIO, AND J. L. GUARINO. 1968. Costs of wintering starlings and red-winged blackbirds at feedlots. J. Wildl. Manage. 32:179–180.

BLACK, H. C., E. J. DIMOCK, II, J. EVANS, AND J. A. ROCHELLE. 1979. Animal damage to coniferous plantations in Oregon and Washington. Part I. A survey, 1963–75. Oregon State Univ. For. Res. Lab. Res. Bull. 25. 44pp.

BLACK, H. L., AND J. S. GREEN. 1984. Navajo use of mixed-breed dogs for management of predators. J. Range Manage. 38:11–15.

BLOKPOEL, H., AND G. D. TESSIER. 1984. Overhead wires and monofilament lines exclude ring-billed gulls from public places. Wildl. Soc. Bull. 12:55–58.

———, AND ———. 1986. The ring-billed gull in Ontario: a review of a new problem species. Can. Wildl. Serv. Occas. Pap. 57. 34pp.

BOGATICH, V. 1967. The use of live traps to remove starlings and protect agricultural products in the state of Washington. Proc. Vertebr. Pest Conf. 3:98–99.

BOLLENGIER, R. M., JR. 1994. Woodchucks. Pages B183–B187 in S. E. Hygnstrom, R. M. Timm, and G. E. Larson, eds. Prevention and control of wildlife damage. Univ. Nebraska Coop. Ext. Serv., Lincoln.

BOMFORD, M., AND P. H. O'BRIEN. 1990. Sonic deterrents in animal damage control: a review of device tests and effectiveness. Wildl. Soc. Bull. 18:411–422.

BOOTH, T. W. 1994. Bird dispersal techniques. Pages E19–E23 in S. E. Hygnstrom, R. M. Timm, and G. E. Larson, eds. Prevention and control of wildlife damage. Univ. Nebraska Coop. Ext. Serv., Lincoln.

BORRECCO, J. E., AND R. J. ANDERSON. 1980. Mountain beaver problems in the forests of California, Oregon, and Washington. Proc. Vertebr. Pest Conf. 9:135–142.

BOWNS, J. E. 1976. Field criteria for predator damage assessment. Utah Sci. 37(1):26–30.

———, AND D. A. WADE. 1980. Physical evidence of carnivore depredation. Tex. Agric. Ext. Serv., College Station (35-mm slide series and script).

BRIDGELAND, W. T., AND J. W. CASLICK. 1983. Relationships between cornfield characteristics and blackbird damage. J. Wildl. Manage. 47:824–829.

BRODIE, J. D., H. C. BLACK, E. J. DIMOCK, II, J. EVANS, C. KAO, AND J. A. ROCHELLE. 1979. Animal damage to coniferous plantations in Oregon and Washington—Part II. An economic evaluation. Oregon State Univ. For. Res. Lab. Res. Bull. 26. 22pp.

BULLOCK, J. F., AND D. H. ARNER. 1985. Beaver damage to nonimpounded timber in Mississippi. Southern J. Appl. For. 9:137–140.

BURKHOLDER, B. L. 1955. Control of small predators. U.S. Fish Wildl. Serv. Circ. 33. 8pp.

BYERS, R. E. 1976. Review of cultural and other control methods for reducing pine vole populations in apple orchards. Proc. Vertebr. Pest Conf. 7:242–243.

———. 1984a. Control and management of vertebrate pests in deciduous orchards of the eastern United States. Hort. Rev. 6:253–285.

———. 1984b. Economics of *Microtus* control in eastern U.S. orchards. Pages 297–302 *in* A. C. Dubock, ed. Organization and practice of vertebrate pest control. Imperial Chem. Industries PLC, Surrey, U.K.

CAHALANE, V. H. 1947. Mammals of North America. Macmillan Co., New York, N.Y. 682pp.

CAMPBELL, D.L. 1974. Establishing preferred browse to reduce damage to Douglas-fir seedlings by deer and elk. Pages 187–192 *in* H. C. Black, ed. Wildlife and forest management in the Pacific Northwest. Oregon State Univ., Corvallis.

———, AND J. EVANS. 1975. "Vexar" seedling protectors to reduce wildlife damage to Douglas fir. U.S. Fish Wildl. Serv. Leafl. 508. 11pp.

CANADIAN TRAPPER FEDERATION. No date. Canadian Trappers' Manual. Can. Trapper Fed., North Bay, Ont. Var. pagin.

CAPP, J. C. 1976. Increasing pocket gopher problems in reforestation. Proc. Vertebr. Pest Conf. 7:221–228.

CARR, A. 1982. Armadillo dilemma. Anim. Kingdom 85(5):40–43.

CASE, R. M., AND B. A. JASCH. 1994. Pocket gophers. Pages B17–B29 *in* S. E. Hygnstrom, R. M. Timm, and G. E. Larson, eds. Prevention and control of wildlife damage. Univ. Nebraska Coop. Ext. Serv., Lincoln.

CASLICK, J. W., AND D. J. DECKER. 1979. Economic feasibility of a deer-proof fence for apple orchards. Wildl. Soc. Bull. 7:173–175.

CLARK, J. 1984. Vole control in field crops. Proc. Vertebr. Pest Conf. 11:5–6.

CLARK, W. R., AND R. E. YOUNG. 1986. Crop damage by small mammals in no-till cornfields. J. Soil Water Conserv. 41:338–341.

CONNELLY, N. A., D. J. DECKER, AND S. WEAR. 1987. Public tolerance of deer in a suburban environment: implications for management and control. Proc. East. Wildl. Damage Control Conf. 3:207–218.

CONNER, W. H., AND J. R. TOLIVER. 1987. The problem of planting Louisiana swamplands when nutria (*Myocastor coypus*) are present. Proc. East. Wildl. Damage Control Conf. 3:42–49.

CONNOLLY, G. 1988. M-44 sodium cyanide ejectors in the animal damage control program, 1976–1986. Proc. Vertebr. Pest Conf. 13:220–225.

———, AND R. J. BURNS. 1990. Efficacy of compound 1080 livestock protection collars for killing coyotes that attack sheep. Proc. Vertebr. Conf. 14:269–276.

CONOVER, M. R. 1984a. Comparative effectiveness of Avitrol, exploders, and hawk-kites in reducing blackbird damage to corn. J. Wildl. Manage. 48:109–116.

———. 1984b. Effectiveness of repellents in reducing deer damage in nurseries. Wildl. Soc. Bull. 12:399–404.

———. 1987a. Comparison of two repellents for reducing deer damage to Japanese yews during winter. Wildl. Soc. Bull. 15:265–268.

———. 1987b. Reducing raccoon and bird damage to small corn plots. Wildl. Soc. Bull. 15:268–272.

———. 1988. Effect of grazing by Canada geese on the winter growth of rye. J. Wildl. Manage. 52:76–80.

———, AND G. G. CHASKO. 1985. Nuisance Canada goose problems in the eastern United States. Wildl. Soc. Bull. 13:228–233.

———, AND G. S. KANIA. 1987. Effectiveness of human hair, BGR, and a mixture of blood meal and peppercorns in reducing deer damage to young apple trees. Proc. East. Wildl. Damage Control Conf. 3:97–101.

COOK, R. S., M. WHITE, D. O. TRAINER, AND W. C. GLAZENER. 1971. Mortality of young white-tailed deer fawns in south Texas. J. Wildl. Manage. 35:47–56.

COWARDIN, L. M., D. S. GILMER, AND C. W. SHAIFFER. 1985. Mallard recruitment in the agricultural environment of North Dakota. Wildl. Monogr. 92. 37pp.

CRABB, W. D. 1941. Civets are rat killers. Iowa Farm Sci. Rep. 2(1):12–13.

———. 1948. The ecology and management of the prairie spotted skunk in Iowa. Ecol. Monogr. 18:201–232.

CRAVEN, S. R. 1983. New directions in deer damage management in Wisconsin. Proc. East. Wildl. Damage Control Conf. 1:65–67.

———. 1994. Cottontail rabbits. Pages D75–D80 *in* S. E. Hygnstrom, R. M. Timm, and G. E. Larson, eds. Prevention and control of wildlife damage. Univ. Nebraska Coop. Ext. Serv., Lincoln.

CROUCH, G. L. 1976. Deer and reforestation in the Pacific northwest. Proc. Vertebr. Pest Conf. 7:298–301.

———. 1986. Pocket gopher damage to conifers in western forests: a historical and current perspective on the problem and its control. Proc. Vertebr. Pest Conf. 12:196–198.

CUMMINGS, J. L., J. L. GUARINO, C. E. KNITTLE, AND W. C. ROYALL, JR. 1987. Decoy plantings for reducing blackbird damage to nearby commercial sunflower fields. Crop Prot. 6:56–60.

CUMMINGS, M. W. 1973. Rodents and drip irrigation. Proc. Drip Irrigation Semin. 4:25–30.

———, AND R. E. MARSH. 1978. Vertebrate pests of citrus. Pages 237–273 *in* W. E. Reuther, E. C. Calavan, and G. E. Garman, eds. The citrus industry. Vol. IV. Div. Agric. Sci., Univ. California, Davis.

DAVIDSON, W. R., AND V. F. NETTLES. 1988. Field manual of wildlife diseases in the southeastern United States. Southeast. Coop. Wildl. Dis. Stud., Univ. Georgia, Athens. 309pp.

DAVIS, D. E. 1976. Management of pine voles. Proc. Vertebr. Pest Conf. 7:270–275.

DAVIS, J. R. 1959. A preliminary progress report on nest predation as a limiting factor in wild turkey populations. Proc. Natl. Wild Turkey Manage. Symp. 1:138–145.

DeCALESTA, D. S., AND D. B. SCHWENDEMAN. 1978. Characterization of deer damage to soybean plants. Wildl. Soc. Bull. 6:250–253.

DeGRAZIO, J. W., J. F. BESSER, J. L. GUARINO, C. M. LOVELESS, AND J. L. OLDEMEYER. 1969. A method for appraising blackbird damage to corn. J. Wildl. Manage. 33:988–994.

DeYOE, D. R., AND W. SCHAAP. 1983. Comparison of 8 physical barriers used for protecting Douglas-fir seedlings from deer browse. Proc. East. Wildl. Damage Control Conf. 1:77–93.

DOLBEER, R. A. 1975. A comparison of two methods for estimating bird damage to sunflowers. J. Wildl. Manage. 39:802–806.

———. 1980. Blackbirds and corn in Ohio. U.S. Fish Wildl. Serv. Resour. Publ. 136. 18pp.

———. 1981. Cost-benefit determination of blackbird damage control for cornfields. Wildl. Soc. Bull. 9:44–51.

———. 1994. Blackbirds. Pages E25–E32 *in* S. E. Hygnstrom, R. M. Timm, and G. E. Larson, eds. Prevention and control of wildlife damage. Univ. Nebraska Coop. Ext. Serv., Lincoln.

———, G. E. BERNHARDT, T. W. SEAMANS, AND P. P. WORONECKI. 1991. Efficacy of two gas cartridge formulations in killing woodchucks in burrows. Wildl. Soc. Bull. 19:200–204.

———, M. A. LINK, AND P. P. WORONECKI. 1988a. Naphthalene shows no repellency for starlings. Wildl. Soc. Bull. 16:62–64.

———, AND R. A. STEHN. 1983. Population status of blackbirds and starlings in North America, 1966-81. Proc. East. Wildl. Damage Control Conf. 1:51–61.

———, A. R. STICKLEY, JR., AND P. P. WORONECKI. 1979. Starling (*Sturnus vulgaris*) damage to sprouting wheat in Tennessee and Kentucky, U.S.A. Prot. Ecol. 1:159–169.

———, P. P. WORONECKI, AND R. L. BRUGGERS. 1986. Reflecting tapes repel blackbirds from millet, sunflowers, and sweet corn. Wildl. Soc. Bull. 14:418–425.

———, ———, AND J. R. MASON. 1988b. Aviary and field evaluations of sweet corn resistance to damage by blackbirds. J. Am. Soc. Hort. Sci. 113:460–464.

DORSETT, J. 1987. Trapping coyotes. Tex. Anim. Damage Control Serv. Leafl. L-1908. 4pp.

EADIE, W. R. 1954. Animal control in field, farm and forest. Macmillan Co., New York, N.Y. 257pp.

EINARSEN, A. S. 1956. Determination of some predatory species by field signs. Oregon State Univ. Stud. Zool. Monogr. 10. 34pp.

ERRINGTON, P. L. 1943. An analysis of mink predation upon muskrat in north-central United States. Iowa State Coll. Agric. Exp. Stn. Res. Bull. 320:794–924.

EVANS, D., J. L. BYFORD, AND R. H. WAINBERG. 1983. A characterization of woodpecker damage to houses in east Tennessee. Proc. East. Wildl. Damage Control Conf. 1:325–330.

EVANS, J. 1987a. Mountain beaver damage and management. Pages 73–74 *in* D. M. Baumgartner, R. L. Mahoney, J. Evans, J. Caslick, and D. W. Brewer, co-chair. Animal damage management in Pacific Northwest forests. Coop. Ext. Serv., Washington State Univ., Pullman.

———. 1987b. The porcupine in the Pacific northwest. Pages 75–78 *in* D. M. Baumgartner, R. L. Mahoney, J. Evans, J. Caslick, and D. W. Brewer, co-chair. Animal damage management in Pacific Northwest forests. Coop. Ext. Serv., Washington State Univ., Pullman.

FAGERSTONE, K. A. 1981. A review of prairie dog diet and its variability among animals and colonies. Proc. Great Plains Wildl. Damage Control Workshop 5:178–184.

FERGUSON, W. L. 1980. Rodenticide use in apple orchards. Proc. East. Pine and Meadow Vole Symp. 4:2–8.

FITZWATER, W. D. 1994. House sparrows. Pages E101–E108 in S. E. Hygnstrom, R. M. Timm, and G. E. Larson, eds. Prevention and control of wildlife damage. Univ. Nebraska Coop. Ext. Serv., Lincoln.

FOSTER, M. A., AND J. STUBBENDIECK. 1980. Effects of the Plains pocket gopher (Geomys bursarius) on rangeland. J. Range Manage. 33:74–78.

FRANTZ, S. C. 1986. Batproofing structures with birdnetting checkvalves. Proc. Vertebr. Pest Conf. 12:260–268.

FULLER-PERRINE, L. D., AND M. E. TOBIN. 1993. A method for applying and removing bird-exclusion netting in commercial vineyards. Wildl. Soc. Bull. 21:47–51.

GATES, N. L., J. E. RICH, D. D. GODTEL, AND C. V. HULET. 1978. Development and evaluation of anti-coyote electric fencing. J. Range Manage. 31:151–153.

GLAHN, J. F., AND W. STONE. 1984. Effects of starling excrement in the food of cattle and pigs. Anim. Prod. 38:439–446.

———, D. J. TWEDT, AND D. L. OTIS. 1983. Estimating feed loss from starling use of livestock feed troughs. Wildl. Soc. Bull. 11:366–372.

GOLDMAN, D. S. 1988. Current and future EPA requirements concerning good laboratory practices relative to vertebrate pesticides. Proc. Vertebr. Pest Conf. 13:22–25.

GOOD, H. B., AND D. M. JOHNSON. 1978. Nonlethal blackbird roost control. Pest Control 46(9):14–18.

GREEN, J. S., AND P. S. GIPSON. 1994. Dogs (feral). Pages C77–C81 in S. E. Hygnstrom, R. M. Timm, and G. E. Larson, eds. Prevention and control of wildlife damage. Univ. Nebraska Coop. Ext. Serv., Lincoln.

———, AND R. A. WOODRUFF. 1983. Guarding dogs protect sheep from predators. U.S. Dep. Agric. Inf. Bull. 455. 27pp.

———, ———, AND R. HORMAN. 1984. Livestock guarding dogs and predator control. Rangelands 6(2):73–76.

GREENHALL, A. M. 1982. House bat management. U.S. Fish Wildl. Serv. Resour. Publ. 143. 33pp.

———, AND S. C. FRANTZ. 1994. Bats. Pages D5–D24 in S. E. Hygnstrom, R. M. Timm, and G. E. Larson, eds. Prevention and control of wildlife damage. Univ. Nebraska Coop. Ext. Serv., Lincoln.

HADIDIAN, J., D. MANSKI, V. FLYGER, C. COX, AND G. HODGE. 1987. Urban gray squirrel damage and population management: a case history. Proc. East. Wildl. Damage Control Conf. 3:219–227.

HAMILTON, J. C., R. J. JOHNSON, R. M. CASE, M. W. RILEY, AND W. W. STROUP. 1987. Fox squirrels cause power outages: an urban wildlife problem. Proc. East. Wildl. Damage Control Conf. 3:228.

HARDER, J. D. 1970. Evaluating winter deer use of orchards in western Colorado. Trans. North Am. Wildl. Nat. Resour. Conf. 35:35–47.

HARRISON, M. J. 1984. FAA policy regarding solid waste disposal facilities. Pages 213–218 in Proc. wildlife hazards to aircraft conference. U.S. Dep. Transp. Rep. DOT/FAA/AAS/84-1.

HAWTHORNE, D. W. 1994. Cotton rats. Pages B97–B99 in S. E. Hygnstrom, R. M. Timm, and G. E. Larson, eds. Prevention and control of wildlife damage. Univ. Nebraska Coop. Ext. Serv., Lincoln.

HEINRICH, J., AND S. CRAVEN. 1987. Distribution and impact of Canada goose crop damage in east-central Wisconsin. Proc. East. Wildl. Damage Control Conf. 3:18–19.

———, AND ———. 1990. Evaluation of three damage abatement techniques for Canada geese. Wildl. Soc. Bull. 18:405–410.

HEISTERBERG, J. F. 1983. Bird repellent seed corn treatment: efficacy evaluations and current registration status. Proc. East. Wildl. Damage Control Conf. 1:255–258.

HENDERSON, F. R. 1994. Moles. Pages D51–D58 in S. E. Hygnstrom, R. M. Timm, and G. E. Larson, eds. Prevention and control of wildlife damage. Univ. Nebraska Coop. Ext. Serv., Lincoln.

HOLLER, N. R., AND D. G. DECKER. 1989. Zinc phosphide rodenticide reduces cotton rat population in Florida sugarcane. Proc. East. Wildl. Damage Control Conf. 4:198–201.

HOLM, B. A., R. J. JOHNSON, D. D. JENSEN, AND W. W. STROUP. 1988. Responses of deer mice to methiocarb and thiram seed treatments. J. Wildl. Manage. 52:497–502.

HOOD, G. A. 1978. Vertebrate control chemicals: current status of registrations, rebuttable presumptions against registrations, and effects on users. Proc. Vertebr. Pest Conf. 8:170–176.

HOOVEN, E. F. 1959. Dusky-footed woodrat in young Douglas-fir. Oreg. For. Res. Cent. Res. Note 41. 24pp.

HOTHEM, R. L., R. W. DeHAVEN, AND S. D. FAIRAIZL. 1988. Bird damage to sunflower in North Dakota, South Dakota, and Minnesota, 1979–1981. U.S. Fish Wildl. Tech. Rep. 15. 11pp.

HOWARD, V. W., JR. 1994. Kangaroo rats. Pages B101–B104 in S. E. Hygnstrom, R. M. Timm, and G. E. Larson, eds. Prevention and control of wildlife damage. Univ. Nebraska Coop. Ext. Serv., Lincoln.

HOWELL, R. G. 1982. The urban coyote problem in Los Angeles County. Proc. Vertebr. Pest Conf. 10:55–61.

HUMPHREY, S. R. 1974. Zoogeography of the nine-banded armadillo (Dasypus novemcinctus) in the United States. BioScience 24:457–462.

HYGNSTROM, S. E., AND S. R. CRAVEN. 1988. Electric fences and commercial repellents for reducing deer damage in cornfields. Wildl. Soc. Bull. 16:291–296.

———, AND ———. 1994. Hawks and owls. Pages E53–E61 in S. E. Hygnstrom, R. M. Timm, and G. E. Larson, eds. Prevention and control of wildlife damage. Univ. Nebraska Coop. Ext. Serv., Lincoln.

———, AND D. R. VIRCHOW. 1994. Prairie dogs. Pages B85–B96 in S. E. Hygnstrom, R. M. Timm, and G. E. Larson, eds. Prevention and control of wildlife damage. Univ. Nebraska Coop. Ext. Serv., Lincoln.

ISHMAEL, W. E., AND O. J. RONGSTAD. 1984. Economics of an urban deer-removal program. Wildl. Soc. Bull. 12:394–398.

JACKSON, J. J. 1994. Tree squirrels. Pages B171–B175 in S. E. Hygnstrom, R. M. Timm, and G. E. Larson, eds. Prevention and control of wildlife damage. Univ. Nebraska Coop. Ext. Serv., Lincoln.

JACKSON, W. B. 1987. Current management strategies for commensal rodents. Pages 495–512 in H. H. Genoways, ed. Current mammalogy. Vol. 1. Plenum Press, New York, N.Y.

JACOBS, W. W. 1994. Registered vertebrate pesticides. Pages G1–G22 in S. E. Hygnstrom, R. M. Timm, and G. E. Larson, eds. Prevention and control of wildlife damage. Univ. Nebraska Coop. Ext. Serv., Lincoln.

JOHNSON, R. J. 1986. Wildlife damage in conservation tillage agriculture: a new challenge. Proc. Vertebr. Pest Conf. 12:127–132.

———, AND J. F. GLAHN. 1994. Starlings. Pages E109–E120 in S. E. Hygnstrom, R. M. Timm, and G. E. Larson, eds. Prevention and control of wildlife damage. Univ. Nebraska Coop. Ext. Serv., Lincoln.

———, A. E. KOEHLER, O. C. BURNSIDE, AND S. R. LOWRY. 1985. Response of thirteen-lined ground squirrels to repellents and implications for conservation tillage. Wildl. Soc. Bull. 13:317–324.

KAHL, R. B., AND F. B. SAMSON. 1984. Factors affecting yield of winter wheat grazed by geese. Wildl. Soc. Bull. 12:256–262.

KELLY, S. T., AND M. E. DeCAPITA. 1982. Cowbird control and its effect on Kirtland's warbler reproductive success. Wilson Bull. 94:363–365.

KINCAID, S. P. 1975. Bats, biology, and control. Proc. Great Plains Wildl. Damage Control Workshop 2:187–194.

KNIGHT, J. E. 1994. Jackrabbits. Pages D81–D85 in S. E. Hygnstrom, R. M. Timm, and G. E. Larson, eds. Prevention and control of wildlife damage. Univ. Nebraska Coop. Ext. Serv., Lincoln.

KNITTLE, C. E., AND R. D. PORTER. 1988. Waterfowl damage and control methods in ripening grain: an overview. U.S. Fish Wildl. Serv. Tech. Rep. 14. 17pp.

LACKI, M. J., S. P. GEORGE, AND P. J. VISCOSI. 1987. Evaluation of site variables affecting nest box use by wood ducks. Wildl. Soc. Bull. 15:196–200.

LARAMIE, H. A. 1978. Water level control in beaver ponds and culverts. N.H. Fish Game Dep., Concord. 5pp.

LARSEN, K. H., AND J. H. DIETRICH. 1970. Reduction of raven population on lambing grounds with DRC-1339. J. Wildl. Manage. 34:200–204.

LAWRENCE, W. H., N. B. KVERNO, AND H. D. HARTWELL. 1961. Guide to wildlife feeding injuries on conifers in the Pacific northwest. West. For. Conserv. Assoc., Portland, Oreg. 44pp.

LeBLANC, D. J. 1994. Nutria. Pages B71–B80 in S. E. Hygnstrom, R. M. Timm, and G. E. Larson, eds. Prevention and control of wildlife damage. Univ. Nebraska Coop. Ext. Serv., Lincoln.

LEFEBVRE, L. W., N. R. HOLLER, AND D. G. DECKER. 1985a. Comparative effectiveness of full-field and field-edge bait applications in delivering bait to roof rats in Florida sugarcane fields. J. Am. Soc. Sugar Cane Tech. 5:64–68.

———, ———, AND ———. 1985*b*. Efficacy of aerial application of a 2% zinc phosphide bait on roof rats in sugarcane. Wildl. Soc. Bull. 13:324–327.

———, C. R. INGRAM, AND M. C. YANG. 1978. Assessment of rat damage to Florida sugarcane in 1975. Proc. Am. Soc. Sugar Cane Tech. 7:75–80.

LINDZEY, F. C. 1994. Badgers. Pages C1–C3 *in* S. E. Hygnstrom, R. M. Timm, and G. E. Larson, eds. Prevention and control of wildlife damage. Univ. Nebraska Coop. Ext. Serv., Lincoln.

LINHART, S. B. 1984. Strobe light and siren devices for protecting fenced-pasture and range sheep from coyote predation. Proc. Vertebr. Pest Conf. 11:154–156.

LUCE, D. G., R. M. CASE, AND J. L. STUBBENDIECK. 1981. Damage to alfalfa fields by Plains pocket gophers. J. Wildl. Manage. 45:258–260.

MAEHR, D. S. 1983. Black bear depredation on bee yards in Florida. Proc. East. Wildl. Damage Control Conf. 1:133–135.

MARSH, R. E. 1985*a*. Competition of rodents and other small mammals with livestock in the United States. Pages 485–508 *in* S. M. Gaafar, W. E. Howard, and R. E. Marsh, eds. Parasites, pests and predators. Elsevier Sci. Publ. B. V., Amsterdam, The Netherlands.

———. 1985*b*. Techniques used in rodent control to safeguard non-target wildlife. Pages 47–55 *in* W. F. Laudenslayer, Jr., ed. Trans. West. Sect., The Wildl. Soc., Monterey, Calif.

———. 1986. Ground squirrel control strategies in Californian agriculture. Pages 261–276 *in* C. G. J. Richards and T. Y. Ku, eds. Control of mammal pests. Taylor and Francis, Inc., Philadelphia, Pa.

———. 1988*a*. Current (1987) and future rodenticides for commensal rodent control. Bull. Soc. Vector Ecol. 13:102–107.

———. 1988*b*. Relevant characteristics of zinc phosphide as a rodenticide. Proc. Great Plains Wildl. Damage Control Conf. 8:70–74.

———. 1994. Roof rats. Pages B125–B132 *in* S. E. Hygnstrom, R. M. Timm, and G. E. Larson, eds. Prevention and control of wildlife damage. Univ. Nebraska Coop. Ext. Serv., Lincoln.

———, AND W. E. HOWARD. 1982. Vertebrate pests. Pages 791–861 *in* A. Mallis, ed. Handbook of pest control. Sixth ed. Franzak and Foster Co., Cleveland, Oh.

MARSHALL, E. F. 1984. Cholecalciferol: a unique toxicant for rodent control. Proc. Vertebr. Pest Conf. 11:95–98.

MCANINCH, J. B., M. R. ELLINGWOOD, AND R. J. WINCHCOMBE. 1983. Deer damage control in New York agriculture. N.Y. State Dep. Agric. Markets Div., Plant Industry-ADC, Albany. 12pp.

MCCRACKEN, H., AND H. VAN CLEVE. 1947. Trapping: the craft and science of catching fur-bearing animals. Barnes Co., New York, N.Y. 196pp.

MCDONOUGH, W. T. 1974. Revegetation of gopher mounds on aspen range in Utah. Great Basin Nat. 34:267–275.

MCGREW, J. C., AND W. F. ANDELT. 1986. Livestock guarding dogs: a method for reducing livestock losses. Colorado State Univ., Coop. Ext. Serv. in Action 1.218. 4pp.

MCKILLOP, I. G., AND C. J. WILSON. 1987. Effectiveness of fences to exclude European rabbits from crops. Wildl. Soc. Bull. 15:394–401.

MCLAREN, M. A., R. E. HARRIS, AND W. J. RICHARDSON. 1984. Pages 241–251 *in* Proc. wildlife hazards to aircraft conference. U.S. Dep. Transp. Rep. DOT/FAA/AAS/84-1.

MEEHAN, A. P. 1984. Rats and mice: their biology and control. Rentokil Ltd., West Sussex, U.K. 383pp.

MENG, H. 1971. The Swedish goshawk trap. J. Wildl. Manage. 35:832–835.

MERRITT, R. L. 1990. Bird strikes to U.S. Air Force aircraft, 1988-89. Bird Strike Comm. Europe 20:511–518.

MICACCHION, M., AND T. W. TOWNSEND. 1983. Botanical characteristics of autumnal blackbird roosts in central Ohio. Oh. J. Sci. 83:131–135.

MILLER, J. E. 1987. Assessment of wildlife damage on southern forests. Pages 48–52 *in* J. G. Dickinson and D. E. Maughan, eds. Proc. management of southern forests for wildlife and fish. U.S. For. Serv. Gen. Tech. Rep. SO-65.

———. 1994. Muskrats. Pages B61–B69 *in* S. E. Hygnstrom, R. M. Timm, and G. E. Larson, eds. Prevention and control of wildlife damage. Univ. Nebraska Coop. Ext. Serv., Lincoln.

———, AND G. K. YARROW. 1994. Beaver. Pages B1–B11 *in* S. E. Hygnstrom, R. M. Timm, and G. E. Larson, eds. Prevention and

control of wildlife damage. Univ. Nebraska Coop. Ext. Serv., Lincoln.

MOTT, D. F. 1980. Dispersing blackbirds and starlings from objectionable roost sites. Proc. Vertebr. Pest Conf. 9:38–42.

MURIE, A. 1948. Cattle on grizzly bear range. J. Wildl. Manage. 12:57–72.

MURIE, O. J. 1954. A field guide to animal tracks. Houghton Mifflin Co., Boston, Mass. 374pp.

MURTON, R. K., N. J. WESTWOOD, AND A. J. ISAACSON. 1974. A study of wood-pigeon shooting: the exploitation of a natural animal population. J. Appl. Ecol. 11:61–81.

NASS, R. D., AND J. THEADE. 1988. Electric fences for reducing sheep losses to predators. J. Range Manage. 41:251–252.

NIELSEN, D. G., M. J. DUNLAP, AND K. V. MILLER. 1982. Pre-rut rubbing by white-tailed bucks: nursery damage, social role, and management options. Wildl. Soc. Bull. 10:341–348.

O'BRIEN, J. M. 1994. Voles. Pages B177–B182 *in* S. E. Hygnstrom, R. M. Timm, and G. E. Larson, eds. Prevention and control of wildlife damage. Univ. Nebraska Coop. Ext. Serv., Lincoln.

O'GARA, B. W. 1978. Sheep depredation by golden eagles in Montana. Proc. Vertebr. Pest Conf. 8:206–213.

———. 1994. Eagles. Pages E41–E48 *in* S. E. Hygnstrom, R. M. Timm, and G. E. Larson, eds. Prevention and control of wildlife damage. Univ. Nebraska Coop. Ext. Serv., Lincoln.

———, AND D. C. GETZ. 1986. Capturing golden eagles using a helicopter and net gun. Wildl. Soc. Bull. 14:400–402.

OSTRY, M. E., AND T. H. NICHOLLS. 1976. How to identify and control sapsucker injury on trees. North Cent. For. Exp. Stn., St. Paul, Minn. 6pp.

OTIS, D. L., N. R. HOLLER, P. W. LEFEBVRE, AND D. F. MOTT. 1983. Estimating bird damage to sprouting rice. Pages 76–89 *in* D. E. Kaukeinen, ed. Vertebrate pest control and management materials. Am. Soc. Test. Materials Spec. Tech. Rep. 817.

PACKHAM, C. J. 1970. Forest animal damage in California. U.S. Fish Wildl. Serv., Sacramento, Calif. 4pp.

PALMER, D. T., D. A. ANDREWS, R. O. WINTERS, AND J. W. FRANCIS. 1980. Removal techniques to control an enclosed deer herd. Wildl. Soc. Bull. 8:29–33.

PALMER, W. L., J. M. PAYNE, R. G. WINGARD, AND J. L. GEORGE. 1985. A practical fence to reduce deer damage. Wildl. Soc. Bull. 13:240–245.

———, R. G. WINGARD, AND J. L. GEORGE. 1983. Evaluation of white-tailed deer repellents. Wildl. Soc. Bull. 11:164–166.

PAULS, R. W. 1986. Protection with Vexar cylinders from damage by meadow voles of tree and shrub seedlings in northeastern Alberta. Proc. Vertebr. Pest Conf. 12:199–204.

PEARCE, J. 1947. Identifying injury by wildlife to trees and shrubs in northeastern forests. U.S. Fish Wildl. Serv. Res. Rep. 13. 29pp.

PEARSON, K., AND C. G. FORSHEY. 1978. Effects of pine vole damage on tree vigor and fruit yield in New York orchards. Hort. Sci. 13:56–57.

PFEIFER, W. K., AND M. W. GOOS. 1982. Guard dogs and gas exploders as coyote control tools in North Dakota. Proc. Vertebr. Pest Conf. 10:55–61.

PHILLIPS, R. L., AND F. S. BLOM. 1988. Distribution and magnitude of eagle/livestock conflicts in the western United States. Proc. Vertebr. Pest Conf. 13:241–244.

PLESSER, H., S. OMASI, AND Y. YOM-TOV. 1983. Mist nets as a means of eliminating bird damage to vineyards. Crop Prot. 2(4):503–506.

PORTER, W. F. 1983. A baited electric fence for controlling deer damage to orchard seedlings. Wildl. Soc. Bull. 11:325–327.

PRESNALL, C. C., EDITOR. 1950. Handbook for hunters of predatory animals. U.S. Dep. Inter., Washington, D.C. 67pp.

REARDEN, J. D. 1951. Identification of waterfowl nest predators. J. Wildl. Manage. 15:386–395.

REYNOLDS, H. G., AND G. E. GLENDENING. 1949. Merriam kangaroo rat: a factor in mesquite propagation on southern Arizona range lands. J. Range Manage. 2:193–197.

RICHMOND, M. E., C. G. FORSHEY, L. A. MAHAFFY, AND P. N. MILLER. 1987. Effects of differential pine vole populations on growth and yield of McIntosh apple trees. Proc. East. Wildl. Damage Control Conf. 3:296–304.

ROBLEE, K. J. 1987. The use of the T-culvert guard to protect road culverts from plugging damage by beavers. Proc. East. Wildl. Damage Control Conf. 3:25–33.

ROGERS, J. G., JR. 1974. Responses of caged red-winged blackbirds to two types of repellents. J. Wildl. Manage. 38:418–423.

ROYALL, W. C., JR. 1969. Trapping house sparrows to protect experimental grain crops. U.S. Fish Wildl. Serv. Leafl. 484. 4pp.

SALMON, T. P., AND F. S. CONTE. 1981. Control of bird damage at aquaculture facilities. Univ. California Coop. Ext. Wildl. Manage. Leafl. 475. 11pp.

———, AND W. P. GORENZEL. 1994. Woodrats. Pages B133–B136 *in* S. E. Hygnstrom, R. M. Timm, and G. E. Larson, eds. Prevention and control of wildlife damage. Univ. Nebraska Coop. Ext. Serv., Lincoln.

SARGEANT, A. B., S. H. ALLEN, AND R. T. EBERHARDT. 1984. Red fox predation on breeding ducks in midcontinent North America. Wildl. Monogr. 89. 41pp.

———, AND P. M. ARNOLD. 1984. Predator management for ducks on waterfowl production areas in the northern plains. Proc. Vertebr. Pest Conf. 11:161–167.

SAUER, W. C. 1977. Exclusion cylinders as a means of assessing losses of vegetation due to ground squirrel feeding. Pages 14–21 *in* W. B. Jackson and R. E. Marsh, eds. Vertebrate pest control and management materials. Am. Soc. Test. Materials Spec. Tech. Rep. 625.

SCHMIDT, R. H., AND R. J. JOHNSON. 1983. Bird dispersal recordings: an overview. Pages 43–65 *in* D. E. Kaukeinen, ed. Vertebrate pest control and management materials. Am. Soc. Test. Materials Spec. Tech. Rep. 817.

SCOTT, J. D., AND T. W. TOWNSEND. 1985. Characteristics of deer damage to commercial tree industries of Ohio. Wildl. Soc. Bull. 13:135–143.

SEUBERT, J. L. 1984. Research on nonpredatory mammal damage control by the U.S. Fish and Wildlife Service. Pages 553–571 *in* A. C. Dubbock, ed. Organization and practice of vertebrate pest control. Imperial Chem. Industries PLC, Surrey, U.K.

SHAW, H. G. 1983. Mountain lion field guide. Ariz. Game and Fish Dep. Spec. Rep. 9. 38pp.

SILVER, J., AND A. W. MOORE. 1941. Mole control. U.S. Fish Wildl. Serv. Conserv. Bull. 16. 17pp.

SIMS, B. 1988. Controlling coyotes with snares. Tex. Anim. Damage Control Serv. Leafl. L-1917. 4pp.

SOLMAN, V. E. F. 1981. Birds and aviation. Environ. Conserv. 8(1): 45–51.

STEENBLIK, J. W. 1983. Battling the birds. Air Line Pilot 52:18–23.

STEFFEN, D. E., N. R. HOLLER, L. W. LEFEBVRE, AND P. F. SCANLON. 1981. Factors affecting the occurrence and distribution of Florida water rats in sugarcane fields. Proc. Am. Soc. Sugar Cane Tech. 9:27–32.

STICKLEY, A. R., JR., AND K. J. ANDREWS. 1989. Survey of Mississippi catfish farmers on means, effort, and costs to repel fish-eating birds from ponds. Proc. East. Wildl. Damage Control Conf. 4:105–108.

———, D. L. OTIS, AND D. T. PALMER. 1979. Evaluation and results of a survey of blackbird and mammal damage to mature field corn over a large (three-state) area. Pages 169–177 *in* J. R. Beck, ed. Vertebrate pest control and management materials. Am. Soc. Test. Materials Spec. Tech. Publ. 680.

———, D. J. TWEDT, J. F. HEISTERBERG, D. F. MOTT, AND J. F. GLAHN. 1986. Surfactant spray system for controlling blackbirds and starlings in urban roosts. Wildl. Soc. Bull. 14:412–418.

STORER, T. I., G. H. VANSELL, AND B. D. MOSES. 1938. Protection of mountain apiaries from bears by use of electric fence. J. Wildl. Manage. 2:172–178.

SUGDEN, L. G. 1976. Waterfowl damage to Canadian grain. Can. Wildl. Serv. Occas. Pap. 24. 25pp.

———, AND D. W. GOERZEN. 1979. Preliminary measurements of grain wasted by field-feeding mallards. Can. Wildl. Serv. Prog. Notes 104. 5pp.

SULLIVAN, T. P. 1978. Biological control of conifer seed damage by the deer mouse (*Peromyscus maniculatus*). Proc. Vertebr. Pest Conf. 8:237–250.

———, J. A. KREBS, AND H. A. KLUGE. 1987. Survey of mammal damage to tree fruit orchards in the Okanagan Valley of British Columbia. Northwest Sci. 61:23–31.

———, AND D. S. SULLIVAN. 1982. The use of alternative foods to reduce lodgepole pine seed predation by small mammals. J. Appl. Ecol. 19:33–45.

———, AND ———. 1988. Influence of alternative foods on vole populations and damage in apple orchards. Wildl. Soc. Bull. 16:170–175.

TEER, J. G. 1964. Predation by long-tailed weasels on eggs of blue-winged teal. J. Wildl. Manage. 28:404–406.

TERRILL, C. E. 1988. Predator losses climb nationwide. Natl. Wool Grower 78(9):32–34.

TIETJEN, H. P. 1976. Zinc phosphide: its development as a control agent for black-tailed prairie dogs. U.S. Fish Wildl. Serv. Spec. Sci. Rep. Wildl. 195. 14pp.

TILGHMAN, N. G. 1989. Impacts of white-tailed deer on forest regeneration in northwestern Pennsylvania. J. Wildl. Manage. 53:524–532.

TILL, J. A., AND F. F. KNOWLTON. 1983. Efficacy of denning in alleviating coyote depredations upon domestic sheep. J. Wildl. Manage. 47:1018–1025.

TIMM, R. M. 1994. Norway rats. Pages B105–B120 *in* S. E. Hygnstrom, R. M. Timm, and G. E. Larson, eds. Prevention and control of wildlife damage. Univ. Nebraska Coop. Ext. Serv., Lincoln.

TOBIN, M. E., AND R. A. DOLBEER. 1987. Status of Mesurol as a bird repellent for cherries and other fruit crops. Proc. East. Wildl. Damage Control Conf. 3:149–158.

———, ———, AND P. P. WORONECKI. 1989. Bird damage to apples in the Mid-Hudson Valley of New York. Hort. Sci. 24:859.

———, AND M. E. RICHMOND. 1987. Bait stations for controlling voles in apple orchards. Proc. East. Wildl. Damage Control Conf. 3:287–295.

TURNER, G. T. 1969. Responses of mountain grassland vegetation to gopher control, reduced grazing, and herbicide. J. Range Manage. 22:377–383.

U.S. DEPARTMENT OF THE INTERIOR. 1977. Raptor control—protecting livestock from hawk and owl predation. U.S. Fish Wildl. Serv. A.D.C. Bull. 211. 77pp.

VETTERMAN, L. D. 1985. The use of fixed wing aircraft in predator control. Proc. Great Plains Wildl. Damage Control Workshop 7:177–180.

WADE, D. A. 1982. Impacts, incidence and control of predation on livestock in the United States with particular reference to predation of coyotes. Counc. Agric. Sci. Tech. Spec. Publ. 10. 20pp.

———. 1985. Applicator manual for Compound 1080. Tex. Agric. Ext. Serv. Bull. B-1509. 51pp.

———, AND J. E. BOWNS. 1982. Procedures for evaluating predation on livestock and wildlife. Tex. Agric. Ext. Serv. Bull. B-1429. 42pp.

WEATHERHEAD, P. J., S. TINKER, AND H. GREENWOOD. 1982. Indirect assessment of avian damage to agriculture. J. Appl. Ecol. 19:773–782.

WEAVER, K. M., D. H. ARNER, C. MASON, AND J. J. HARTLEY. 1985. A guide to using snares for beaver capture. Southern J. Appl. For. 9:141–146.

WEBER, W. J. 1979. Health hazards from pigeons, starlings and English sparrows. Thompson Publ., Fresno, Calif. 138pp.

WHITE, S. B., R. A. DOLBEER, AND T. A. BOOKHOUT. 1985. Ecology, bioenergetics, and agricultural impacts of a winter-roosting population of blackbirds and starlings. Wildl. Monogr. 93. 42pp.

WICK, W. Q., AND A. S. LANDFORCE. 1962. Mole and gopher control. Oregon State Univ. Coop. Ext. Bull. 804. 16pp.

WILLIAMS, D. E., AND R. M. CORRIGAN. 1994. Pigeons (rock doves). Pages E87–E96 *in* S. E. Hygnstrom, R. M. Timm, and G. E. Larson, eds. Prevention and control of wildlife damage. Univ. Nebraska Coop. Ext. Serv., Lincoln.

WOODWARD, D. K. 1983. Beaver management in the southeastern United States: a review and update. Proc. East. Wildl. Damage Control Conf. 1:163–165.

WORONECKI, P. P. 1988. Effect of ultrasonic, visual, and sonic devices on pigeon numbers in a vacant building. Proc. Vertebr. Pest Conf. 13:266–272.

———, R. A. DOLBEER, T. W. SEAMANS, AND W. R. LANCE. 1992. Alpha-chloralose efficacy in capturing nuisance waterfowl and pigeons and current status of FDA registration. Proc. Vertebr. Pest Conf. 15:72–78.

———, ———, AND R. A. STEHN. 1981. Response of blackbirds to Mesurol and Sevin applications on sweet corn. J. Wildl. Manage. 45:693–701.

———, R. A. STEHN, AND R. A. DOLBEER. 1980. Compensatory response of maturing corn kernels following simulated damage by birds. J. Appl. Ecol. 17:737–746.

YOUNG, S. P. 1933. Hints on mountain lion trapping. Bur. Biol. Surv. Leafl. 94. 8pp.

———. 1958. The bobcat of North America. Stackpole Co., Harrisburg, Pa., and Wildl. Manage. Inst., Washington, D.C. 193pp.

19

MANAGEMENT OF URBAN WILDLIFE

Larry W. VanDruff, Eric G. Bolen, and
Gary J. San Julian

INTRODUCTION
Urban Wildlife—The Resource

Although only about 4% of the earth's surface is occupied by our villages, suburbs, and cities, a majority of the human population now occupies such developed areas. And urbanization of the human population is predicted to continue in the foreseeable future. Abiotic, biotic, and cultural processes associated with urbanization produce a great diversity of habitats ranging from near natural to completely artificial in the urban ecosystem. Metropolitan areas contain a diverse remnant of native fauna, complemented by a large biomass of domestic, feral, and exotic species. Thus, most humans share the landscape and exist in daily association with urban wildlife. More importantly, the attitudes, experiences, and actions of urbanites determine the success or failure of traditional and future conservation and environmental programs.

Urban wildlife is the nondomestic animals—vertebrates and often conspicuous, attractive invertebrates—that exist in urban landscapes. Urban areas, in contrast to nonurban (usually exurban) and agricultural areas, are characterized by the presence of buildings, roads, artificial surfaces, and other constructed features.

Using the wildlife present in cities and suburbs, the student, researcher, planner, and natural resource manager can exploit a readily available resource. The justification for understanding, planning for, and managing urban wildlife is based on sound aesthetic, ecological, educational, and economic reasons. Wildlife managers, especially agency administrators, are beginning to realize the importance of programs that address wildlife that is in close contact with the tax-paying, policy-influencing public. This urban wildlife resource, heretofore largely neglected by professional ecologists, managers, and administrators (VanDruff 1979), now has emerged as an expanding and integral component of wildlife science.

The uniqueness of urban wildlife, in comparison to the regional or local fauna of a like-sized exurban area, can be succinctly characterized as follows:

a. The ecological dominance of humans is emphasized by their great density, modification of natural systems, and actions that constitute major ecological and environmental influences.

b. Large metropolitan areas contain a great diversity of wildlife habitats, ranging from nearly natural remnants of regional biomes to completely artificial habitats. The urban mosaic is characterized by small parcel size, diverse ownership, unexpected juxtaposition, intense human management, and rapidity of change. Locally, the habitats of urban wildlife, existing within fragmented landscapes, often are extensively altered (e.g., loss of soil cover or of the shrub layer); degraded by soil compaction, chemical influence (herbicides or insecticides), fill, or drainage; and isolated from similar or complementary habitats needed by mobile individuals and vagile species.

c. Exotic (non-native) wildlife (e.g., the Norway rat, European starling, house sparrow, and rock dove) often are abundant and conspicuous in the most densely developed areas.

d. Species richness generally is lower; in the more heavily developed areas, high numbers of a few species exist. Lacking from the urban fauna are species with large home ranges, those sensitive to human disturbance, those that have been extirpated by previous activities and land uses (e.g., logging and agriculture), and area-sensitive species incapable of using fragmented habitats.

e. The most common predators of vertebrates are free-roaming cats and dogs. The automobile and artificial elements (e.g., storm drains and channelized waterways, electrical hazards, and mowing and cultivation) are additional sources of mortality to wildlife.

f. Individual wild animals often lack wariness due to lack of native predators and human exploitation of the species, and to habituation to human presence, odors, and activities.

g. Humans may become familiar with, and relate to, specific individuals of a wildlife species from frequent and often close contact. Individual wild animals may become conditioned to respond to particular people or certain human activities, often for a reward of food.

h. In the urban system, humans are less aware and knowledgeable about other species. Birds are conspicuous, and most are enjoyed, whereas nocturnal mammals often are unnoticed and feared for their suspected ferocity or potential for disease transmission.

Working in conjunction with urban planners, developers, elected officials, environmental educators, citizens' groups, and others concerned for the natural environment, the wildlife biologist can manage habitat units and wildlife for the benefit of the citizenry. Similar to the management of wildlife resources on any large unit of land, the goals of urban wildlife programs may strive to:

(1) maintain viable populations of current species,
(2) increase the density of a species, a group (e.g., songbirds), or a guild (e.g., cavity nesters),
(3) decrease the density of a species, group, or guild,
(4) increase the diversity (species richness) of the faunal component of habitats or ecosystems,
(5) manage vegetation types (e.g., mast-producing species or the shrub stratum) or habitats, with secondary emphasis on wildlife,
(6) alter the location or activities of either desirable or undesirable wildlife,
(7) manipulate or regulate human contact or interaction with one or more species of wildlife,
(8) regulate ecosystem processes that affect wildlife (e.g., succession, predation, epizootics, biomagnification of toxics), and
(9) control the diversity, size, juxtaposition, and connectivity of wildlife habitat units and parcels of vegetation with aesthetic, economic, educational, ecological, or recreational values.

The urban setting with its humanized environments, anthropic ecosystems, and anthropogenic relationships has been characterized in the books of Kieran (1959), Bornkamm et al. (1982), Douglas (1983), and Spirn (1984), and in a compilation of abstracts by Dawe (1990). The best compilation of writings on urban wildlife appears in the proceedings of several national symposia (Noyes and Progulske 1974, Euler et al. 1975, Kirkpatrick 1978, Stenberg and Shaw 1986, Adams and Leedy 1987, 1991). Also, Thomas and DeGraaf (1975), Leedy (1979), Progulske and Leedy (1986), Adams (1988), and Robinson

and Bolen (1989) provided especially relevant information to serve as an introduction to urban wildlife. The needs and justification for urban wildlife management were amply presented in these sources. Brocke (1977) made a strong case for a more prominent role of professional wildlife biologists in the urban arena. A policy statement on urban wildlife by the professional society for wildlife conservationists attests to the significance of this rapidly expanding subject (The Wildlife Society 1992).

Current Issues in Urban Wildlife

The current issues in urban wildlife management, succinctly stated, fall into five broad categories:

—Habitat destruction, alteration, fragmentation, and isolation.
—Isolation of humans from the natural world.
—Lack of information, education, and awareness by the public.
—Inappropriate (positive or negative) responses by humans to wildlife.
—Critical need for a plan and continuing action.

This chapter discusses these issues and presents steps that address each condition.

HUMAN ATTITUDES, PREFERENCES, AND KNOWLEDGE

For success, urban wildlife management programs must have a sound foundation in human dimensions—that area of wildlife biology that addresses socioeconomic factors. Indeed, the basis for management priorities and activities relies on knowledge and characterization of the various public sectors, the current level of awareness of each, and their needs, desires, and commitment for wildlife and wildlife-oriented programs. Not surprisingly, surveys commonly focus on the knowledge and likes or dislikes of humans for various species or groups of wildlife. Intuitively, human preferences should play an important role in establishing priorities for public programs, hence we see the preoccupation of many researchers and managers with this subject. An attitudinal analysis is required because the reasons why wildlife is important to the public are often subtle and altruistic (Leuschner et al. 1989). This seems especially true in the complex world of urban life. Under some circumstances, the urban wildlife manager may choose to employ an acceptance capacity (Decker and Purdy 1988) for wildlife (i.e., the maximum or minimum level for a wildlife population that is acceptable to the resident human population) as a factor in planning local urban wildlife programs.

The role of nature and the human need for natural environments in the urban setting were presented by Kaplan (1984). Before selecting and promoting certain landscapes or wildlife habitats, managers should have an understanding of human acceptance of such elements and patterns. Human preferences for various landscapes or habitats have received limited attention. Ulrich (1986) reported that respondents often equated nature, vegetation, and naturalness. Physical qualities of vegetation may govern human response to the landscape more than species composition or the occurrence of natural processes within the system. Pudelkewicz (1981) studied residents in four neighborhoods in Columbia, Maryland; she asked residents for their preference for wildlife habitat next to their homes. Areas of mowed grass were rated lower than areas having the appearance of diverse wildlife habitat. She concluded that good-quality wildlife habitat could be incorporated into visually satisfying, open-space design through cooperation by urban planners and managers, landscape architects, and wildlife biologists. Wetlands retained for stormwater control and wildlife enhancement were considered valuable by residents of Columbia (Adams et al. 1984). Residents strongly agreed (94%) that fish and wildlife habitat was an important benefit of urban wetlands, and 75% of the homeowners believed that permanent bodies of water enhanced real-estate values. However, Talbot and Kaplan (1984), after interviewing 97 residents of low- and moderate-income, inner-city areas of Detroit, reported a fear of untouched and densely wooded areas. Concern for personal safety led residents to favor well-maintained areas incorporating built features. Similarly, others reported that urbanites show greater approval of undesigned remnant landscapes if those open-space units are located somewhere else in the city rather than across the street (Schauman et al. 1987). Understanding differences in preference for various natural landscapes among local socioeconomic and ethnic groups, as well as the effects of spatial relationships on human preferences, is requisite to understanding how urbanites may support habitat (i.e., natural areas) preservation.

Several studies have addressed the urban public's perceptions of, and preferences for, wildlife. Dagg (1970) surveyed 1,421 households in Waterloo, Ontario, and concluded that birds were almost universally liked, but providing birds with food did not significantly influence feeder's preferences. Chipmunks were the favorite mammal, liked by 86% of the respondents. Gilbert (1982) surveyed the knowledge and opinions of residents of Guelph, Ontario. He reported that avian species were tolerated and appreciated more than mammals, but many residents failed to recognize the role of habitat and vegetation in supporting wildlife species. The influence of urbanization and our resulting isolation from nature are emphasized by the fact that 89% of the respondents developed their perceptions of wildlife through television.

Brown et al. (1979) assessed the wildlife interests, needs, and attitudes of metropolitan New Yorkers. Although reptiles were never or seldom seen by 92% of the respondents, most respondents saw mammals weekly and reported daily sightings of birds. They indicated a preference for butterflies, robins, cardinals, sparrows, blue jays, squirrels, and hummingbirds around homes. Urbanites showed a distinct dislike around their home or in their neighborhood for pigeons, snakes, raccoons, foxes, and blackbirds-starlings. Dennis (1989) learned that observers preferred northern cardinals and black-capped chickadees first and second, respectively, among those birds that visit bird feeders in summer. The European starling and common grackle were the most disliked birds visiting the feeders. Witter et al. (1981) determined how urbanites value wildlife and pursue wildlife activities in Missouri. Using a professional polling firm to conduct telephone interviews, they reported that passive recreation, namely reading and watching television, was twice as popular as outdoor activities or sports. Nature-oriented activities with the most participants were watching nature programs on television, visiting a zoo or museum, going for a drive or a neighborhood walk, feeding or watching birds, and reading about nature. Importantly, "the nature-oriented opportunities most readily available to urbanites were those with the largest number of participants" (Witter et al. 1981:425).

Urban wildlife managers should be aware of differences in perceptions between genders. Kellert and Berry (1987) reported a significant difference between male and female attitudes, knowledge, and behavior regarding wildlife. Females valued individual animals as objects of affection, held anthropomorphic feelings toward animals, and were opposed to exploitation of, and dominance over, animals. Males were more willing to exploit animals, were more knowledgeable and less fearful of animals, and expressed greater concern about conservation of species and habitats. Kellert and Berry (1987:370) concluded that "gender is among the most important demographic factors in determining attitudes about animals in our society. Major efforts to broaden the scope and effectiveness of wildlife management should, thus, consider and understand the influence of gender."

Most surveys have focused on adults or households, but Schicker (1986) considered the perceptions and needs of the 52 million youth under age 18 living in urban areas. From direct contacts with these children, she emphasized the need for early hands-on contact with nature in residential wildlands that receive protection from development as a provision for the future. Kellert and Westervelt (1983) and Kellert (1984) summarized the trends of American children in their attitudes, knowledge, and behaviors toward wildlife and nature. Younger Americans, especially those in urban areas of 50,000–250,000 people, hold more wildlife protectionist views and show stronger emotional attachment to animals than do older persons. The marked difference in attitudes by age is attributed to a fundamental change that is occurring in America, as well as a change that occurs with personal maturation. Adams et al. (1987a) reported a disturbing lack of knowledge about wildlife among biology students in urban high schools in Texas. For example, 60% of the students incorrectly identified the opossum as a rat; 28% believed that the mink was extinct, 7% felt that it was abundant, and 25% thought that the mink never existed. Between 59% and 98% of the students did not know that urbanization promotes an increase of house mice, opossums, squirrels, bats, skunks, and raccoons.

Undeniably, urbanization has isolated humans from frequent contact with the natural world, resulting in a lower level of knowledge about wildlife compared with rural residents. Surprisingly, Burley and Martin (1986) observed no difference in homeowners' knowledge about wildlife between those living on open, rather sterile residential lots and those occupying nearby landscapes with numerous trees, ponds, and abundant wildlife. However, the small sample of homeowners in both categories, and the proximity of both groups to common resources, may have masked detection of true differences in attitude between the two survey groups.

In spite of their relative lack of knowledge of the natural history of wildlife around them, urbanites are aware of and value this natural resource. Using data from the *1980 National Survey of Hunting, Fishing, and Wildlife-*

Associated Recreation (U.S. Department of the Interior 1982), Shaw et al. (1985) analyzed the degree to which wildlife was enjoyed in residential areas. Because more than one-half of American adults annually engaged in wildlife-oriented activities near their homes, they concluded (p. 372) that "enjoyment of wildlife in residential environments could almost be termed a national pastime." In 1980, 62 million Americans fed wild birds—the most common wildlife-related activity—at an estimated annual expenditure of more than $500 million. Shaw et al. (1985) alerted wildlife managers and state wildlife agencies to focus greater attention on residential wildlife conservation issues and to use the urban wildlife resource to provide benefits to both consumptive (elsewhere) and nonconsumptive users near their homes. Cooperation of planners, engineers, developers, and architects can aid wildlife managers seeking to meet the needs of urban and suburban residents.

Attitudes, behaviors, and management preferences of urbanites toward nuisance occurrences have received some attention (Powell 1982, O'Donnell 1984, O'Donnell and VanDruff 1984). Understanding the human element in the management of nuisance urban/suburban deer (Shoesmith 1978, Kuser and Applegate 1985, Decker and Gavin 1985, 1987, Connelly et al. 1987, Decker 1987, Cypher et al. 1988), urban Canada geese (Conover and Chasko 1985, Conover 1987), and urban gray squirrels (Manski et al. 1981, Hadidian et al. 1987) is essential to the development and implementation of sound management plans for these species. Such can be said about most wildlife of the urban area.

Urban wildlife managers should base their programs on a solid foundation that focuses on human dimensions as much as on wildlife ecology. In addition to the above references, those on valuing wildlife (Decker and Goff 1987) and the newsletter of the Human Dimensions Working Group (James B. Armstrong, Auburn Univ.), a coalition of state, federal, and university investigators and program managers, will be helpful to persons seeking either useful techniques or the results of earlier studies.

Those collecting their own human-dimensions data on urbanites should consider use of the following survey instruments described in the cited studies:

1. Photo preference test by surveyor (Pudelkewicz 1981).
2. Questionnaire (written) (Dillman 1978) by mail (Brown et al. 1979).
3. Interview by telephone (Witter et al. 1981).
4. Interview on site (Kellert and Berry 1987).
5. Interview at household (Gilbert 1982).
6. Combination of photo and questionnaire (Schauman et al. 1987).
7. Combination of observation and interview (Hardin 1977).

LANDSCAPE ECOLOGY AND BIOLOGICAL DIVERSITY

Laurie (1979) presented the philosophic, ecologic, and humanistic context for landscape planning and management. Harris (1984) and Forman and Godron (1986) presented concepts of landscape ecology applicable to urban settings.

Regional landscape patterns that influence the occur-
rence, abundance, activities, or mortality of urban wildlife have received some attention in the published literature. For example, the effects of habitat fragmentation and isolation on remaining wildlife populations have been a prominent concern (Forman et al. 1976, Harris 1984, Lynch and Whigham 1984, Opdam et al. 1984). The role of wildlife reserves and corridors in the urban environment was summarized in an excellent review by Adams and Dove (1989). Aldrich and Coffin (1980) compared breeding bird populations during the transition of an area from forest to suburbs over a nearly 4-decade period. Anthony et al. (1990) assessed the land-use associations and population changes of raccoons during a rabies epizootic in Baltimore, Maryland. The study included nearly 1,500 capture and carcass records from the municipal animal shelter during a 4-year period.

The impacts of specific land uses and human activities associated with urbanization upon resident wildlife are poorly known. Undoubtedly, habitat fragmentation and isolation, reduction of various strata of vegetation, and influence of vehicular and pedestrian traffic across the regional landscape have an incremental and cumulative effect on the native wildlife community. Loss of wetlands and fragmentation of wetland habitats from encroaching development along regional river systems or coastal areas have severely reduced diversity and abundance of water-based invertebrate, fish, amphibian, and reptile communities.

Bird life has been affected by development and urbanization. Area-sensitive species of birds and minimum habitats required were listed by Wenger (1984). Batten (1972), Luniak (1983), Robbins (1984), Bezzel (1985), and DeGraaf (1987) noted responses in the avian community across urbanized landscapes that were similar in the U.S., Great Britain, and Europe. In summary, during its early stages, land development may result in a greater diversity of species as natural succession is selectively altered and new elements and habitat types are introduced. Ultimately, a reduction in the number of species is accompanied by a numerical dominance by the most abundant species, often exotics. Intensive human use of natural environments favors granivores, medium-sized omnivores, ground feeders, and sedentary species; fewer cavity and ground nesters, insectivorous migrants, and forest-interior species persist in the resultant avifauna.

Studies of urban mammals across diverse landscapes are not as common. Matthiae and Stearns (1981), Nilon (1986), VanDruff and Rowse (1986), and Nilon and VanDruff (1987) presented results for the United States. Studies of mammals in Europe generally concur with observations in North America that the larger species, those with larger home ranges, and specialized predators (e.g., native felids) are extirpated because of habitat fragmentation and restrictions on movements of individuals. Biodiversity and individual species of the small mammal community may reflect differences and management needs of urban greenspaces exhibiting various degrees of forest cover and past disturbance. Although fewer species of mammals than birds exist in urban systems, their impact (e.g., predation) or notoriety (e.g., white-tailed deer damage to ornamentals) may be greater. Under the extremes of urbanization, few native mammals exist and exotics (e.g., Norway rat and house mouse in North America) are prevalent.

Biological diversity—the variety of species, their ecological roles, and their genetic diversity—generally is reduced as the intensity of urbanization increases. Usually species with general habitat and resource requirements remain or colonize; these are the so-called generalists or broad-niched species and edge species. Biological diversity of urban areas, not only regionally but on the local level, is severely reduced by introduced species that prey on native species, compete for limited resources, and act as vectors for novel diseases and parasites to which native organisms can be particularly susceptible (Murphy 1988). Domestic pets have similar effects; house cats apparently are serious predators of urban birds and small mammals (Churcher and Lawton 1987). Soulé et al. (1988) demonstrated that removal of house cats by coyote predation resulted in greater avian species richness in urban San Diego County, California. For the most part, the relationships between native fauna and domestic or feral animals have not been determined for urban areas. The ecology of domestic and feral dogs and cats was addressed by Beck (1973, 1974), Childs (1986), Childs and Ross (1986), Calhoon and Haspel (1989), and Haspel and Calhoon (1989).

Consumptive uses of the wildlife resource have not been determined for urban areas. In particular, the contribution of urban animals to the annual game harvest of a region is not well known. Figley and VanDruff (1982) and Heusmann (1983) suggested that urban mallards are harvested at only 0.125–0.20 the rate of mallards banded in wild places. Suburban and urban-fringe populations of furbearers (e.g., opossum, raccoon, foxes) apparently undergo significant trapping harvests in some areas. Wildlife managers should consider the consumptive use of wildlife in developing local management alternatives.

WILDLIFE SURVEY AND RESEARCH TECHNIQUES
Previous Studies

Those charged with planning for and managing wildlife of local urban areas should assess the existing resources of their area. Bendell and Falls (1981) gave guidelines and practical suggestions for obtaining inventories, faunal lists, censuses, and relative and absolute abundance estimates for urban species. Adams and Dove (1989) suggested ways that land managers can increase and maintain biological diversity. The potential role of ecological restoration as a strategy for conserving biological diversity in an urbanized landscape deserves more attention (Jordan et al. 1988). A reexamination of the concept of edge effect (Reese and Ratti 1988) and habitat concepts (Harris and Kangas 1988) is especially germane to planning for biological diversity in urban environments. Excellent coverage of research and land management aspects of vertebrate-habitat relationships is presented in *Wildlife 2000*, the proceedings of a major international symposium on the topic (Verner et al. 1986). Proceedings of the conference *Wildlife 2001* focus on the conservation and management of wildlife populations (McCullough and Barrett 1992).

Field studies of urban wildlife are uncommon in the published literature in comparison to work on exurban wildlife populations, but a review of selected studies usually will reveal investigational techniques useful for several groups of urban wildlife (Table 1). In particular, amphibians and reptiles have received little attention. Destruction, fragmentation, isolation, and alteration of habitat, especially wetlands, have resulted in an extremely impoverished herptile fauna in most urbanized areas. Birds have been studied the most of any wildlife group. Although some investigators have studied single avian species (Howard 1974, Figley and VanDruff 1982, Wiley 1986, DeGraaf 1989), most field investigations have dealt with avian groups, especially the songbird community. Especially prominent during the 1970s and 1980s was the work conducted by the U.S. Forest Service Unit in Amherst, Massachusetts (Thomas and DeGraaf 1973, Thomas et al. 1977, Goldstein et al. 1983, 1986, DeGraaf 1986).

Avian studies have concentrated on elucidation of the habitat factors responsible for species richness, total abundance, or the density of particular species within the avian community. Most studies have focused on breeding birds, but Johnsen and VanDruff (1987) and Tilghman (1987b), among others, examined winter populations. Tzilkowski et al. (1986) reported on the relationship between street trees, including habitat features, and use by urban birds in the summer in northeastern United States.

Field techniques for the study of urban wildlife usually include variations on universally applied investigative methods. Due to the relative unwariness of animals in close contact with people, observations often are easy and reliable. For example, avian use of individual street-side trees can be quantified during an observation period (e.g., 10 min) (Tzilkowski et al. 1986), waterfowl on a pond can be censused by observation with relative ease (Adams et al. 1985), or interactions among conspecifics can be described and quantified (Gustafson and VanDruff 1990).

Frogs and toads have been censused by actual searches as well as by listening for calls after dusk, whereas reptiles have been counted on warm days by direct observation of individuals or shed skins (Dickman 1987). The location and activity of larger, more conspicuous urban species (e.g., deer or raptors) may be recorded from roadways.

To characterize variations in bird density or habitat selection, most investigators use a modification of the strip or belt transect that is popular for sampling bird populations. The investigator walks a street segment and records birds seen or heard within some estimated distance perpendicular to the line of travel. Such a method counts all birds within an imaginary belt, often 100×50 m. Habitat features such as strata and volume of natural and ornamental vegetation, characteristics of structures, and percentage of development (e.g., paving, buildings) should accompany the survey data. Socioeconomic characteristics of the human residents or the neighborhood for correlation with wildlife diversity and abundance may be obtained from the U.S. Bureau of Census and local metropolitan reports in addition to application of human dimensions methods reviewed earlier. Additional land use and large-scale features of the area obtained by aerial photogrammetry have been used to search for variables that best explain the variation observed in the diversity and abundance of the avian community.

Similar assessments of habitat appear in studies of urban mammalian communities in which mammal presence

Table 1. Selected field studies of urban fauna.

Group	Reference
Invertebrates	Faeth and Kane 1978, Luniak and Pisarski 1982, Arnold and Goins 1987
Vertebrates	Dickman 1987
Herptiles	Cook and Pinnock 1987
Amphibia	Campbell 1974, Schlauch 1978, Bascietto and Adams 1983, Dickman 1987
Reptilia	Dickman 1987
Aves	
General avifauna	Geis 1975, 1976a,b, 1980a, Thomas et al. 1977, Beissinger and Osborne 1982, Johnsen 1982, Luniak 1983, Goldstein et al. 1986, Gotfryd and Hansell 1986, Tzilkowski et al. 1986, Johnsen and VanDruff 1987, Tilghman 1987a,b, Cicero 1989, Mills et al. 1989
Exotics	Boudreau 1975, Weber 1979, Timm 1983
Waterbirds	Heusmann 1981, 1983, Figley and VanDruff 1982, Heusmann and Burrell 1984, Adams et al. 1985, Cooper 1987, 1991
Raptors	Thomsen 1971, Oliphant 1974, Spitzer and Poole 1980, Baker and Brooks 1981, Minor and Minor 1981, Runyan 1987, Wesemann and Rowe 1987, Gehlbach 1988, Gennaro 1988, Murphy et al. 1988, Ingraldi 1992, Plumpton and Lutz 1993
Songbirds	DeGraaf and Wentworth 1981, 1986, DeGraaf 1986, 1987
Other	Flycatcher—Wiley 1986 Robin—Howard 1974 Wood thrush—Roth 1987 Song sparrow—DeGraaf 1980 Killdeer—Ankney and Hopkins 1985 Crow—Knight et al. 1989 House finch/purple finch—Shedd 1990 Woodpecker—Moulton and Adams 1991 Chickadee–Brittingham and Temple 1992
Mammalia	
General	Matthiae and Stearns 1981, Dickman 1986, Nilon 1986, VanDruff and Rowse 1986, Dickman and Doncaster 1987, Nilon and VanDruff 1987
Opossums	Meier 1983
Bats	Geggie and Fenton 1985
Rodents	Gliwicz 1982
Tree squirrels	Williamson 1983, Gustafson and VanDruff 1990, Jodice and Humphrey 1992

Table 1. Continued.

Group	Reference
Carnivores	
General	Rosatte et al. 1991
Raccoons	Schinner and Cauley 1974, Hoffmann and Gottschang 1977, Slate 1985, Manski and Hadidian 1987, Bigler 1990, Hadidian et al. 1991, Feigley 1992
Foxes	Harris 1981, 1986, MacDonald and Newdick 1982, Harris and Rayner 1986, Doncaster et al. 1990
Coyotes	Shargo 1988, Atkinson and Shackleton 1991
Deer	Witham and Jones 1987, 1990, Cypher et al. 1988, Vogel 1989, Jones and Witham 1990
Other	
Cats	Childs 1986, Childs and Ross 1986, Churcher and Lawton 1987, Calhoon and Haspel 1989, Haspel and Calhoon 1989, Haspel and Calhoon 1993
Dogs	Beck 1973, 1974, Daniels and Bekoff 1989

and relative abundance were determined by hair-sampling tubes (Dickman 1987) or livetrapping (Matthiae and Stearns 1981, Nilon 1986, VanDruff and Rowse 1986, Nilon and VanDruff 1987). Livetrapping mammals in the urban environment presents challenges to the field crew. Because urban lands are parcelled into small units, much time must be spent seeking permission to work on private lots, and such efforts frequently require visits, often in the evening, with homeowners in the study area. Another challenge is the potential for vandalism and theft of live traps and other field equipment. Some biologists avoid this problem by concealing their collapsible traps in the daytime and setting the traps at dusk (e.g., VanDruff and Rowse 1986). Encounters with curious people—and sometimes investigating police!—also require additional time when urban wildlife studies are conducted. Examples of other field techniques reported in urban studies of mammals include time-area counts of gray squirrels (Williamson 1983), radiotelemetry of coyotes (Atkinson and Shackleton 1991) and raccoons (Hoffmann and Gottschang 1977), and use of various visual marking devices, such as colored neck collars (Vogel 1989).

Unique Research Opportunities

Wildlife living in close and daily association with humans offer unique opportunities for studying unconfined wild animals. Most individuals in urban wildlife populations lose a degree of wariness through habituation and conditioning, so field observation is accomplished more easily. Studies of gray squirrels in Lafayette Park, across Pennsylvania Avenue from the White House in Washington, D.C., are good examples (Manski et al. 1981, Hadidian et al. 1987). In such a setting, researchers compiled time and activity budgets for numerous squirrels and conducted other fieldwork, such as enticing squirrels from den cavities by whistling, that would be impossible with the greater wariness of squirrels in a rural setting. Gray squirrels also have been valuable, unwary subjects for studies of factors affecting gene frequency for the wild (gray) and melanic (black) morphs in an urban population (Tomsa 1987, Gustafson and VanDruff 1990) and of habitat utilization in an urban residential area (Howell 1982). Flyger (1974) discussed other aspects of the gray squirrel that support its merits as a research animal. Many other species of urban animals exhibit a lesser degree of wariness or more predictable activities, temporally or spatially, that permit research not possible with exurban populations of the same species.

Often the network of streets, highways, waterways, and other rights-of-way provides quick and ready access for researchers to livetrap or to obtain visual sightings or radiotelemetry locations of a resident wildlife population. Figley and VanDruff (1982) used residential streets and a network of canals to study mallards in a coastal New Jersey community. Similarly, Geis (1986) and Thomas et al. (1977), among others, used streets as sample units for bird censuses. The ready-made access to resident wildlife, their den sites, and their centers of activity, as well as features of the habitat, is a distinct benefit of working in the urban environment. Students of the biology and management of wildlife not only can conduct their science and art in close proximity to easily available support, but a large and curious public audience is eager to learn of the work and to be able to benefit from on-site or community educational programs.

Adams et al. (1987*b*) examined the research and educational activities at the university level that addressed urban wildlife. They reported that more than 30% of 95 North American colleges offering wildlife curricula were involved with research in urban wildlife ecology, management, planning, or education. Applied management was identified as the highest priority for additional research. Adams (1988) summarized recent research findings in urban wildlife; earlier, Progulske and Leedy (1986) outlined how biological and cross-disciplinary approaches might enhance additional research efforts. Most authors agree that the needs and opportunities for further study of the biological, ecological, and social aspects of urban wildlife are nearly limitless. Opportunities also exist in the urban environment for field-testing theoretical concepts of landscape ecology, population or community or ecosystem ecology, and conservation biology. Management of urban wildlife species, communities, or natural areas awaits creative application of traditional approaches. McPherson and Nilon (1987) provided an excellent example of the use of a Habitat Suitability Index (HSI) to identify habitat features and develop management suggestions for the gray squirrel in an urban cemetery.

A Suggested Approach

Because studying wildlife in the urban environment can present unique challenges, the following suggestions may aid those planning to conduct field studies. A careful review of the literature (see Table 1) will provide valuable suggestions from others. Use of tax maps may be necessary to determine ownership of many parcels, especially in undeveloped urban areas. Preparation of a high-quality

information leaflet is a vital first step to acquiring the confidence and approval of police, land managers, and residents before fieldwork begins. An aggressive, conspicuous promotion of the study, which spells out project personnel, duration, and goals, often will garner consent and support for study activities in the urban environment. A neighborhood meeting or door-to-door contact with residents is usually desirable. This may be vital, because fieldwork often necessitates work at night, stealth to avoid disturbing animals, and movements of crews across private and institutional properties. Well-trained, competent, and highly dependable field crews are essential to work in urban areas under the watchful eye of residents and the public. Animals should not be disturbed unless necessary to meet study objectives, and captured animals should be handled quickly and especially humanely. To avoid vandalism of field equipment (e.g., live traps) and distraction by curious residents, some investigators prefer to be as inconspicuous as possible at specific field sites. Residents may be valuable to report sightings of marked or injured study animals, but rarely do private individuals possess the time, skills, or dedication to collect reliable data over extended periods. An answering service or machine will permit receipt of incoming calls with a minimum commitment of time or inconvenience to field personnel.

PLANNING AND IMPLEMENTATION OF ENHANCEMENT PROGRAMS

Environmental awareness and concerns for maintaining the quality of life have prompted public demands for comprehensive planning and environmental regulation and management. Many federal and state agencies, regional planning boards, town/city environmental management councils, and community associations have responded with analysis, planning, and programs. Concerns about wildlife, especially threatened and endangered species, may emerge prominently in the deliberations of these groups. Whereas the National Environmental Policy Act (NEPA) of 1969 resulted in numerous federal initiatives, comprehensive and meaningful fish and wildlife programs have emanated slowly from Washington, D.C. The U.S. Fish and Wildlife Service assigned only one individual to urban wildlife research during the 1970s and 1980s. Although public interest in urban and nongame wildlife has increased steadily during the past 2 decades, the Fish and Wildlife Conservation Act of 1980 (the so-called ''Non-Game Bill'') has not yet received any federal funding appropriations. At best, federal leadership in efforts associated with urban wildlife is diffuse and inconspicuous (cf. Dunkle 1987).

European experiences in developing urban wildlife programs suggest useful models based on science incorporating land management and city planning, with the added consideration of community needs and the perceived need of the public (Emery 1986, Barker 1987). Incorporating a human ecological perspective into land-use planning may ensure strong interest in wildlife and increased feasibility of the plans (Jackson and Steiner 1985). Unfortunately, urban resource managers often must set priorities, design programs, and propose recommendations without adequate data on either the resource or relevant cultural factors. When necessity requires making fiscal and priority decisions without the benefit of adequate economic valuations, managers might follow the formal, rigorous decision strategy presented by Shafer and Davis (1989). Additional techniques for valuing wildlife were given by Decker and Goff (1987). The handbook by Andrews and Cranmer-Byng (1981), designed to narrow the gap between amateur naturalists and professional environmental scientists, contains practical suggestions and strategies useful to the understanding and management of natural areas.

The Role of Scale

Wildlife managers responsible for urban programs may be called upon to operate at many different levels, each with a different set of needs and priorities (Table 2). Employees of federal agencies and private groups, such as National Audubon Society or National Wildlife Federation, may need a regional—even national—perspective. Support and responsibility for interagency liasons necessitate a focus on a continental, national, or regional scale. The manager who must address the impacts of urbanization on large-scale ecosystems will want to use the tools of planning, design, and management with entire landscapes to meet the needs of both humans and wildlife (Rodiek and Bolen 1990).

At the state level, the scale permits and necessitates more attention to the needs and opportunities of individual cities. Working at the level of the city, the wildlife manager will increasingly value and utilize various constituencies from local sportsmen's clubs to preservationist groups. Goldstein et al. (1983) discussed the role of scale in planning for greenspace in residential developments. More attention can be directed to management of individual groups or species (e.g., hummingbirds, peregrines, white-tailed deer) or specific habitats. Restoration ecology plays a more important role as the focus becomes more site specific and the vegetation/fauna have been more severely disturbed by development.

Finally, working with urban residents who want to manage their lot—often <0.4 ha in size—requires more attention to on-site details. Often the scale and management prescriptions for urban wildlife require applications of landscape architecture, horticulture, and arboriculture on limited space. Nevertheless, due to the periodicity, secretiveness, mobility, and population characteristics of wildlife, the training and skills of a wildlife biologist are needed. We should prescribe management of urban wildlife at these different scales.

State and Regional Programs

State conservation and natural resource agencies began giving attention to urban wildlife in the 1970s. New York was the first state to assign a professional wildlife biologist to full-time duties for urban wildlife; that effort has continued with a principal emphasis on education (Matthews 1986). Likewise, Missouri, Kansas, Florida, and Arizona developed statewide programs dealing with urban wildlife (Shaw and Supplee 1987). State agencies attempting to increase their attention to urban wildlife would do well to adopt the guidelines prepared by the Urban Wildlife Committee of The Wildlife Society (Tylka et al. 1987). These succinct guidelines suggest program goals and objectives, program elements and job descriptions, and qualifications/training for urban wildlife biologists.

The suggested program elements and percentage of the program budget are: inventory and research, 5–25%; wildlife planning and management, 30–60%; public information, education, and extension services, 30–60%; and urban habitat acquisition, development, preservation, and conservation, 5–20%. The successes of a state conservation agency in the development of a habitat evaluation program, creation of a nongame division, and formal adoption of an urban wildlife policy were presented by Shaw and Supplee (1987). These experiences with the evolution of statewide urban wildlife programs will benefit other state agencies with similar needs. Lyons and Leedy (1984) summarized the status of urban wildlife programs among state and federal agencies in the mid-1980s.

Resource inventories often provide a database for subsequent programming. Aerial photography with ground truthing provided the basis for habitat mapping of the six major metropolitan areas of New York State (Matthews et al. 1988). Rapidly advancing technology in Geographic Information Systems (GIS) (see Chapter 21) provides a powerful tool for handling geographic and positional data (Hendrix et al. 1988). A GIS permits acquisition and ''stacking'' of numerous overlapping areal databases that portray many physiographic, biological, and cultural features. The GIS offers almost unlimited visual presentation of such data. Software programs, such as pMAP® and IDRISSI®, that can be run on a personal computer are especially useful for visualizing urban wildlife habitat relationships. Data files imported into graphics programs such as SURFER can provide three-dimensional depictions.

Regional planning that incorporates open-space and wildlife values has increased in importance as suburbanization and rural development increased during the 1980s. The integration of wildlife conservation and residential development is documented in the changes at Rancho Santa Margarita y las Flores (formerly >900 km^2) in coastal southern California (Froke 1986). The plan accommodates ecologically and economically integrated agriculture, wildland protection, and urban development on such tracts as Rancho Mission Viejo. Spanish, Mexican, and American settlement has progressively threatened and extirpated California condors, grizzlies, wolves, and jaguars, but mountain lions today are an integral part of the wildland/suburban edge at the Rancho. Public safety and sustainable mountain lion populations are major concerns as these large cats foray into residential and recreational areas at the natural/cultural ecotone. Conservation easements and creative land exchanges that provide economic benefits to all parties may greatly promote conservation of wide-ranging species (e.g., pronghorns) in the face of rapidly expanding urbanization (Andrews et al. 1986, Diehl and Barrett 1988). Those responsible for predicting, planning for, and managing regional effects on wildlife resulting from urbanization and other land-use changes should plan for wildlife before urban development takes place. Adams and Dove (1989) provided actual examples of the use of land transfers, performance zoning, conservation easements, real-estate transfer taxes, land bank arrangements, and outright purchases for ecological landscape planning and resource conservation. Hench et al. (1985), Sikorowski and Bissell (1986), and Shaw et al. (1986) should be consulted for models of successful plans.

At the intracounty level, wildlife conservation may be less encumbered legally, but the processes remain the same. The experiences of Boulder County, Colorado, may be instructive; a description of demographic, economic, and ecological resources, including a wildlife database, led to the designation of critical wildlife habitats. In Boulder County, critical habitat such as cottonwood river bottoms and montane willow-shrub wetlands represented sites that supported a high diversity of species. Corridors for elk migration also were considered critical. To implement a comprehensive plan, the planner and wildlife manager should use zoning regulations, subdivision provisions, and open-space purchases as tools while the plan receives public input, review, and hearings. The Colorado Division of Wildlife (CDW) attributes local control of land use to the success of the planning process (Bissell et al. 1987). CDW provides counties with traditional cartographic and faunal maps, "stacked" wildlife habitat composite maps, and worksheets. The worksheets provide the planner and the project's proponents with a cause-effect projection of the impacts, if any, of the proposed action. Worksheets also facilitate contact and discussion among CDW personnel, planners, and project proponents about potential impacts and alternative approaches.

Urban wildlife managers usually work in an arena of developed lands where competition for highly valued and scarce natural resources is intense. The public's perception of these resources and the response to proposed management goals may be heavily influenced by conflicts, perceived resource scarcity, and emotional and biased information. Economics-versus-amenity issues often influence the resolution of management goals for urban nature programs (Vining and Schroeder 1989).

Community Needs

Rarely have studies been conducted across the entire city or metropolitan area. Geis (1975, 1976*a,b,* 1980*a,* 1986) documented the effects on wildlife resulting from the development of rural and new town sites at Columbia, Maryland, and elsewhere. This work also described the effects of architectural and land-use designs on the bird communities that developed among neighborhoods. The findings and recommendations from these studies may apply to groups of wildlife besides songbirds in other cities:

(1) A much greater intensity of wildlife management can be justified in urban areas in comparison with efforts in rural sites.
(2) The amount of woody vegetation is the single, most important factor for promoting desirable populations of urban birds (i.e., diverse, native avifauna).
(3) Wildlife habitat is improved with ornamental plants, especially varieties that are small (e.g., shrubs), bear fruit, and have thorns.
(4) Natural succession, following curtailment of mowing, is the least expensive and easiest way to increase the variety of vegetation in urban areas.

Leedy et al. (1978) offered many useful guidelines for addressing wildlife habitat needs in the planning and de-

velopment of urbanizing areas. Bolen (1991) suggested that planners, researchers, and managers recognize and adopt analogs of familiar exurban habitat components within urban environments. Such examples of analogs as fencerows/ornamental shrubbery, natural wetlands/stormwater basins, continuous canopy/utility wire network (for gray squirrel movements) stimulate thoughts of many others. The experiences of others provide useful guides for those persons responsible for urban wildlife programs; anticipated professional responsibilities and tasks will include site rehabilitation (Johnson 1986, Cook and Pinnock 1987), wetland protection (Milligan and Raedeke 1986) or restoration (Kusler and Kentula 1990), critical and sensitive habitat preservation (Burns et al. 1986), impact assessment and mitigation (Postovit and Postovit 1987), stimulation of environmental contacts for children (Schicker 1987) and other user groups, identification of target publics (Schaefer 1987), and development and delivery of recreational (Hench et al. 1987) or public education programs (Houck 1987).

Although the sighting of an unusual species, or of wildlife in an unusual setting, may attract attention, long-range programs should be comprehensive and deal with self-sustaining, adequately sized populations in natural habitats. We suggest that urban wildlife research and enhancement programs place an emphasis on sustainable and ecologically functional wildlife populations (Conner 1988). Management of urban wildlife at the town level will employ some goals and techniques also applicable at the county and intracity level (see Table 2).

On-Site Management

Our treatment here cannot address all regional and site-specific conditions and situations. The wildlife manager or land steward should consult arboriculturists, horticulturists, and landscape architects for selection of plant materials suited for regional plant tolerance zones and local site conditions. The wildlife manager must determine the food and cover needs of the resident and potential wildlife community based on site history, present conditions, proximity of other populations as a source for natural colonization, severity of site disturbances, and other present and anticipated uses of the site. The following discussion will provide some suggestions and numerous sources of additional useful information.

Popular and semitechnical literature abounds on the subject of habitat enhancement, especially for wildlife living in backyards and on private lands. Many practical suggestions are found in Terres (1968), Leedy et al. (1978), Leedy and Adams (1984), Dennis (1985, 1989), Kress (1985), Henderson (1987), Stokes and Stokes (1987), and Tufts (1988).

An ongoing series of informational leaflets, which are amply researched and well presented for the homeowner, environmental educator, urban wildlife manager, or extension specialist, form the "Urban Wildlife Manager's Notebook" (available from the National Institute for Urban Wildlife, 10921 Trotting Ridge Way, Columbia, MD 21044). Titles such as *A Simple Backyard Pond, Housing for Nesting Birds, Natural Landscaping—Meadows, Brushpiles and Rock Piles,* and *Saving Snags for Urban Wildlife* indicate the breadth and application of the series

Table 2. Elements of an urban wildlife initiative according to scale.

Scale	Initiatives
Continental	Incorporation and integration of urbanized areas into programs for protection and enhancement of resources of national concern, such as wetlands, old-growth forests, unfragmented tracts, coastal dunes and estuaries, and threatened or endangered species (e.g., bald eagle and peregrine falcon). Planning and management strategies should be based on human ecology and on sound principles of landscape ecology, conservation biology, ecosystem ecology, and wildlife science. Although small in total area at this scale, urbanized areas may include critical habitats for species having impoverished or relict populations spread over large areas. Plans and management should be aimed at restoring or maintaining components and processes that result in a matrix of recognized biomes with urban islands. Educational programs must be directed at numerous sectors of the public.
Regional (multistate)	Attention should be given to individual land units with high value at the continental scale. Protecting major corridors, maintaining the integrity of major river systems of the continent, and addressing flyway concerns (e.g., protection of regional production, resting, and wintering areas for migrating waterbirds) merit special attention. Applying regional planning and wildlife management in a multidisciplinary approach at the landscape level will be the most productive.
State	In addition to adoption and cooperative enforcement of federal programs (e.g., NEPA or the Endangered Species Act), programs at the state level can be tailored to the wildlife resources, public attitudes and needs, and degree of urbanization at this level. Development and adoption of a statewide open-space plan that includes urban areas are first steps to regulating future urbanization. Educational efforts should be aimed at developing an informed and supportive electorate, as well as a strong identity of the state's natural resource agency with urban wildlife programs responsive to the needs of urbanites. Personnel and wildlife programs should be conspicuous to the public and centered in the major cities of the state. Creative funding mechanisms, yielding dependable allocations, will be necessary to significantly advance the urban wildlife program. Wildlife survey, research, management, and educational components should be integrated. See Tylka et al. (1987) for other specific suggestions.
County	Development and adoption of a comprehensive plan, incorporating wildlife values, for the regulation of development and other land uses, acquisition of conservation easements or purchase of high-value parcels, and accommodation of the needs of nearby urbanites will be a strong focus. Other values of greenspace systems, such as recreational (e.g., footpaths, cycleways, bridlepaths), economic (e.g., watershed protection), aesthetic (e.g., viewshed), and historical (e.g., canals), will help "sell" wildlife values; those responsible for wildlife should develop supportive liaisons with professionals and groups interested in these other values. Local land trusts can provide a focus for the concern and involvement of laypersons. At this scale, parcel size/shape and the protection of land units from loss or disturbance become major concerns, as do juxtaposition of units and continuity of corridors. Maintaining some large, forested tracts (or comparable "climax" communities of the region) should be given high priority; protection of habitats for sensitive species may necessitate severe restrictions on human usage.
City	Management efforts should begin with a survey of natural resources, community attitudes and conservation values, and agencies or individuals likely to provide support and seek results. Use of a multidisciplinary team and public input into the development of a comprehensive plan will provide long-term benefits. Succinctly, on a citywide level the wildlife manager should: a. Strive to counteract habitat loss, degradation, and fragmentation. b. Maintain larger units of wetland habitats and forested stands, and natural heritage or sensitive-species areas. c. Emphasize the unique cover types of larger parcels; otherwise strive for diversity of ample-sized habitat types and avoid monocultures. d. Protect units along lakeshores and those contiguous with dense cover (e.g., riparian fringe) or cover on sites adjacent to streams and rights-of-way to provide corridors and necklaces of open-space units. e. Use "planned neglect" and natural succession whenever possible, favor some tangled thickets with snags. f. Protect or create standing and flowing water with naturalistic borders. g. Manage the intensity of human usage of public open space, providing a range of areas from large urban wilds to downtown parks. h. Establish a "hot-line" for watchable wildlife events and provide telephone answers to urbanites' questions and need for assistance with urban ecology and wildlife initiatives.

Table 2. Continued.

Scale	Initiatives
Outskirts	Protection of remnant natural areas as community educational and ecological resources is an urgent need. Seek corporate support, and form a local conservation action group with a proprietary interest in high-value sites. Availability of areas merely for viewing wildlife in a natural setting is important. A nature center or informational pavilion, parking for vehicles, availability of public transportation, accommodation of outdoor recreation, and access and interpretation for those with disabilities should be included.
Residential	Management of the wildlife resource reaches its greatest intensity at this scale. Community needs for natural areas with watchable wildlife can be met with management of retention basins, ravines, lakefronts, institutional grounds, and cemeteries within the community. Diversity of habitat units from open to those vegetated with trees are needed. Increasing tree-crown volume and providing missing strata of vegetation, such as herbaceous ground cover and the shrub area, will benefit the avifauna and many other wildlife species. On institutional and residential lots, selection of vegetation may be governed by ornamental values, but judicious selection of particular species (or cultivars) of ground cover, flower garden plants, shrubs, and trees will benefit wildlife. Intensive management at minimal expense to the natural resource agency or other promoter (e.g., nature center or conservation society) is possible. Numerous urbanites experiencing wildlife annoyances and encountering diseased animals will demand attention from the wildlife manager. Educational efforts addressing nuisance problems are badly needed and are well received. Use of techniques suggested elsewhere in this chapter and in Chapter 18 will be necessary.
Central core	Largely the stronghold of unwanted and/or nuisance animals, the central business district and surrounding areas, as well as satellite shopping centers, sports complexes, and airports, hold promise, as well as challenges, for the wildlife manager. Emphasis usually will be upon reducing nuisances and protecting valued wildlife—often individuals—such as nesting raptors, endangered insects, or local amphibian populations. A citywide program of Watchable Wildlife, vigorously promoted through several media, will reach large numbers of urbanites of all age, racial, ethnic, and social groups using the city's center and its immediate surroundings. Nest boxes, signage, and protective surveillance by volunteers may be successful practices.
Small natural area or residential lot	Develop and follow a comprehensive plan for wildlife. Use local volunteers and donated materials on areas with public access. Save or plant natural, multilayered vegetation and permit ecological succession and senescence of plants. Select species and cultivars of plants with significant wildlife values. Leave dead trees and down wood. Add stone and brush piles. Leave borders or strips unmowed. Encourage briars, vines, and thorny vegetation. Restrict free-roaming pets, especially cats. Allow litter to accumulate and add mulch. Provide standing and moving water on the site if possible. Management reaches its greatest detail, intensity, and expense per unit of area at this scale. Site-specific objects and conditions, even microhabitats, become important.

to those seeking specific techniques for on-site management.

Managers and industrious urbanites often are limited solely by their own financial commitment and ingenuity in providing on-site amenities for desirable species of urban wildlife. Just as might be expected in other environments, the addition or maintenance of reliable sources of food, water, and secure shelter in urban settings should increase the abundance of wildlife. Native and ornamental herbaceous and woody vegetation provides the mainstay of foods and cover for many species of wildlife (Table 3). Bird feeding, an activity enjoyed by more than one-half of the households in the U.S., supplements natural foods for avifauna. Although the population effects of bird feeding generally are not known, bird feeders should be provisioned following the guidelines of Geis (1980*b*). He concluded that oil-type sunflower seeds and white proso millet were the best foods, whereas rice, hulled oats, wheat, milo, and peanut hearts were unattractive items. Leopold and Dedon (1983) demonstrated that bird feeders are essential to the mourning dove population of Berkeley, California, due largely to the absence of vacant lots for natural foraging.

Similarly, a diversity of vegetation, habitat strata, and cover types over larger areas usually will produce greater species richness. We agree with A. Geis (pers. commun.) that cultivated and ornamental vegetation is appropriate to create habitat that is both aesthetically pleasing and beneficial to wildlife in urban or suburban sites. The increased growth, vigor, disease resistance, ability to withstand soil compaction, and similar adaptations—as well as the production of showy flowers or persistent fruits—merit the selection of many ornamentals over native flora. Numerous plants are listed in the colorful presentation by Sharp (1977). DeGraaf and Witman (1979) presented the characteristics, including wildlife values, of trees and shrubs; for a resource guide to urban forestry, Moll and Ebenreck (1989) should be consulted. Henderson (1987) presented extensive lists of herbaceous and woody vegetation that include numerous native alternatives to exotics. Wenger (1984) included regional lists of food and cover plants useful for attracting wildlife to residential grounds (Table 4).

Table 3. The wildlife food (F), cover (C), and nesting (N) values of selected woody plants (from Wenger 1984:947–948).

Very high wildlife value			
Boxelder	F	Eastern redcedar	F
Black maple	F	Jack pine	C, N
Striped maple	F	Ponderosa pine	C, N
Red maple	F	Pitch pine	C, N
Sugar maple	F, C, N	Eastern white pine	F, C, N
Mountain maple	F	Pin cherry	F
Sweet birch	F	Black cherry	F
Yellow birch	F	Common chokecherry	F
River birch	F	White oak	F, N
Paper birch	F	Swamp white oak	F, N
Gray birch	F	Northern red oak	F, N
Common hackberry	F	Scarlet oak	F, N
Alternate-leaf dogwood	F, C, N	Shingle oak	F, N
Flowering dogwood	F, C, N	Bur oak	F
Blackjack oak	F	Chestnut oak	F
Chinkapin oak	F	Pin oak	F

High wildlife value			
Post oak	F	Black oak	F
Balsam fir	C, N	Bigtooth aspen	F, N
White fir	C, N	Quaking aspen	F, N
Hazel alder	C, N	Douglas-fir	F, C, N
Shadblow serviceberry	F	Flameleaf sumac	F
Allegheny serviceberry	F	Pussy willow	C, N
Devil's walkingstick	F, C	Black willow	C, N
Common persimmon	F	Nannyberry	F, C
American beech	F	Blackhaw	F, C

Moderate wildlife value			
Red mulberry	F	Common pricklyash	F, C
Engelmann spruce	F, C, N	American mountainash	F
Eastern cottonwood	F, N	Dotted hawthorn	F
Pignut hickory	F	Frosted hawthorn	F
Pecan	F	Eastern larch	N
Shagbark hickory	F	Prairie crabapple	F
Mockernut hickory	F	Black tupelo	N
Cockspur hawthorn	F, C, N	Eastern hemlock	C
Downy hawthorn	F, C, N	Smooth sumac	F
Glossy hawthorn	F, C, N	Staghorn sumac	F
Washington hawthorn	F		

Additionally, the following guidelines are offered:

Shrubs and Vines

- Establish in clumps with a volume, with herbaceous growth beneath, often to extend a woody border or reduce abruptness of a tree canopy.
- Permit natural succession of sumacs, dogwoods, multiflora rose, Virginia creeper, and grapes.
- If ornamentals are used, Boston ivy, currant, viburnums, barberry, bush honeysuckles, highbush cranberry, and junipers should be considered if they are suited to local conditions.

Trees

- Specimens often must withstand heat stress, water-related stress, compacted soil, physical injury, pollutants, and wastes.

- For most wildlife, the species is less important than size, structure, number, and location. Crown height and volume are particularly important. The larger the forested stand, the greater the species richness.
- Fruits, such as berries (mulberries) and mast (oaks and hickories), are especially attractive to songbirds and tree squirrels, respectively.
- Clumps of evergreens should be included to provide roosting and winter cover.
- Species to be considered include Norway, amur, hedge, and red maples; oaks; birches; ashes (female trees); ginkgo (male trees); white fir; Canadian hemlock; redcedar; spruces; pines; crabapples; serviceberries; dogwoods; hawthorns; cherries.
- Pin oak, American elm, and honeylocust are streetside trees preferred by birds in the Northeast.

Table 4. Food and cover plants useful for attracting wildlife to residential grounds (from Wenger 1984:954–956).

Flowers and grasses	Low shrubs and vines	Large shrubs	Small trees	Large trees
		Northeast		
Panicgrass	Blackberry	Autumn olive	Cherry	Beech
Sunflower	Spicebush	Dogwood	Crabapple	Birch
Timothy	Snowberry	Elderberry	Dogwood	Colorado spruce
Bristlegrass	Coralberry	Sumac	Hawthorn	Hemlock
Ragweed	Virginia creeper	Winterberry	Redcedar	Sugar maple
Knotweed	Greenbrier	Tartarian honeysuckle	Serviceberry	White oak
Pokeweed	Mapleleaf viburnum	Highbush blueberry	Mulberry	White pine
	Bittersweet	Multiflora rose		Blackgum
	Japanese honeysuckle	Firethorn		Red maple
		Highbush cranberry		Boxelder
		Northwest		
Filaree	Blackberry	Elderberry	Dogwood	California black oak
Sunflower	Oregon grape	Golden current	Hawthorn	Colorado spruce
Tarweed	Snowberry	Tartarian honeysuckle	Serviceberry	Douglas-fir
Timothy	Coralberry	Multiflora rose	Mountain ash	Lodgepole pine
Turkeymullein	Gooseberry	Firethorn	Thorn apple	Ponderosa pine
Bristlegrass	Buckthorn	Highbush cranberry	Squaw apple	Boxelder
Ragweed	Sagebrush	Russian olive		
Knotweed				
		Southeast		
Lespedeza	Bayberry	Dogwood	Cherry	Mountain ash
Panicgrass	Blackberry	Elderberry	Crabapple	Beech
Sunflower	Spicebush	Sumac	Dogwood	Hackberry
Bristlegrass	Virginia creeper	Tartarian honeysuckle	Hawthorn	Live oak
Ragweed	Greenbrier	Highbush blueberry	Holly	Loblolly pine
Knotweed	Mapleleaf viburnum	Multiflora rose	Palmetto	Pecan
Pokeweed	Japanese honeysuckle	Firethorn	Persimmon	Slash pine
		Arrowwood	Redcedar	Blackgum
			Serviceberry	Red maple
			Mulberry	Boxelder
		Southwest		
Filaree	Blackberry	Manzanita	Crabapple	Live oak
Sunflower	Juniper	Catclaw acacia	Sweet acacia	Pin oak
Turkeymullein	Pricklypear	Tartarian honeysuckle	Mesquite	Pinyon pine
Bristlegrass	Virginia creeper	Multiflora rose	Desert ironwood	Boxelder
Ragweed	Sagebrush	Firethorn	Mulberry	Saguaro (cactus)
Knotweed		Cholla (cactus)		

Specific Wildlife Attractants

- For hummingbirds, use trumpet vine, cardinal flower, bee balm, columbine, hollyhock, phlox, larkspur, and other species with red flowers.
- For cedar waxwings and mockingbirds, use roses (e.g., multiflora, hybrid tea) and European mountain ash.

The reduced availability of rotting trees, dead snags, and diverse vegetation in many urbanized areas makes nesting boxes and other structures important habitat-improvement techniques. Not only may such an effort reverse loss of nesting habitat, but the construction, erection, and monitoring of nesting structures heighten a participant's interest and knowledge of wildlife. Numer-ous guides to the construction of nesting structures are available—those by Ridlehuber and Teaford (1986), Teaford (1986), Henderson (1987), and Mitchell (1988) are particularly useful. Selection of durable wood, a design that permits water drainage and annual cleaning, and correct dimensions determine box longevity and use by wildlife. Variations on the basic bluebird box design can accommodate most cavity-dwelling vertebrates (Henderson 1984, Wenger 1984).

Habitat management need not be expensive. Natural succession often provides desirable habitats, widens ecotones, and yields a desirable intermixing of species, which thereby reduce monocultures. "Planned neglect" of lot corners, portions of parks, and segments of certain rights-

Table 5. Sensitivity of selected bird species to disturbance of vegetation, i.e., urbanization. Group I = relatively insensitive, i.e., species best adapted to built-up areas; Group II = benefitted by management for increased vegetation; Group III = species intolerant of loss of vegetation and undisturbed habitat (from Goldstein et al. 1986:382–383).

Group I	
Mourning dove	Wood thrush
Blue jay	European starling
Black-capped chickadee	Red-eyed vireo
Tufted titmouse	House sparrow
White-breasted nuthatch	Northern oriole
House wren	Common grackle
Northern mockingbird	Northern cardinal
Gray catbird	Chipping sparrow
American robin	Song sparrow

Group II	
Yellow-billed cuckoo	Black-and-white warbler
Northern flicker	Yellow warbler
Hairy woodpecker	Blackpoll warbler
Downy woodpecker	Common yellowthroat
Eastern kingbird	American redstart
Great crested flycatcher	Red-winged blackbird
	Brown-headed cowbird
Eastern phoebe	Scarlet tanager
Eastern wood-pewee	Rose-breasted grosbeak
Barn swallow	House finch
American crow	American goldfinch
Red-breasted nuthatch	Rufous-sided towhee
Brown thrasher	
Cedar waxwing	

Group III	
Ring-necked pheasant	Warbling vireo
Ruby-throated humming-bird	Blue-winged warbler
	Yellow-rumped warbler
Pileated woodpecker	Ovenbird
Least flycatcher	Eastern meadowlark
Tree swallow	Indigo bunting
Winter wren	Purple finch
Hermit thrush	Field sparrow
Swainson's thrush	Swamp sparrow
Blue-gray gnatcatcher	
Solitary vireo	

of-way is inexpensive. In fact, those responsible for the financial burden of extensive mowing or other maintenance programs may welcome such a natural source of relief. If human safety is not at risk, snags may be left standing to provide wildlife perching, foraging, and denning habitat.

Concomitant with efforts to increase the diversity and abundance of local wildlife should be an attempt to prevent a nuisance or pest situation (see Chapter 18). On-site, even communitywide, efforts at reducing mortality of urban wildlife, such as from deer-automobile or bird-window collisions, should occur (Shoesmith and Koonz 1977, Klem 1990). Farsighted and thorough wildlife management seeks a wholesome balance, maintained by

periodic reassessment and redirected efforts as necessary.

Special Concerns

SNAGS AND CAVITIES

Standing snags and large, over-mature trees offering excavation sites and cavities for wildlife are lacking in most urban habitats. Wildlife use dead or dying trees or dead limbs for denning and roosting sites, perches, and sources of food. Wenger (1984) listed snag sizes and numbers for key species of birds and mammals expected in the urban environment. The dearth of cavity-dwelling wildlife in many urban areas may be partially reversed by leaving standing snags and diseased specimens that pose no safety hazard. In addition to the provision of nest boxes described above, tree cavities can be artificially created. Working in an exurban forest, Gano and Mosher (1983) described how to make nesting cavities in living trees with a chain saw, wood chisel, hammer, and drill (such work should be undertaken only by an adult experienced in the safe use of these tools). Southern flying squirrels, white-footed mice, and white-breasted nuthatches responded to the artificial cavities excavated at the study area, demonstrating the usefulness of this technique for producing sites for secondary cavity-nesting birds and rodents. The researchers also girdled the trees to provide future snags for excavation by primary cavity nesters.

RESIDENTIAL AVIFAUNA

Urban wildlife biologists commonly direct their attention to the suburbs. Here much can be accomplished by working with relatively affluent homeowners who often are willing to spend money for water, food, and supplemental plantings. In this setting, birds provide a special attraction and generally are considered a desirable component of the urban landscape. DeGraaf (1986, 1987) and DeGraaf and Wentworth (1981, 1986) revealed interesting trends and provided useful guidelines for habitat enhancement for urban/suburban bird populations. Succinctly stated, their findings were:

(1) Ground, shrub, cavity, and tree-twig nesters are low in density or absent in urban areas.
(2) Wintering and breeding densities are higher in urban areas than in the suburbs, but total numbers of species are greater in the suburbs during both seasons.
(3) Woodlots and fields of the predevelopment landscape should be retained and patch size of woody vegetation maximized. Natural woodland is especially needed by breeding, migrant, and insectivorous birds.
(4) Maximizing crown volumes of shrubs and trees should increase the species richness of breeding birds. Shrub maturity is more important than numbers of shrubs.
(5) "The nearer a woodlot and open field, the smaller the lawn area, the more that weedy vegetation is permitted, and the lower the building density, the greater the species richness of the suburban bird community" (DeGraaf 1987:110).

Given the variability in occurrence, abundance, and habitat association of various birds detected in the urban environment, management efforts will yield variable results by species or groups. Goldstein et al. (1986) divided birds of residential areas in Amherst, Massachusetts, into three groups, each with variable responsiveness to management. Their results suggested that birds in Group I (Table 5) occupied a broad range of suburban habitats. These species are the best adapted to built-up areas. Birds in Group II were observed regularly and benefitted from management. They will occur in the suburban avifauna where an adequate volume of vegetation is provided. Birds in Group III were observed infrequently in built-up areas and were assigned a low priority for management. DeGraaf (1986) pointed out the need for a rich avifauna in the overall landscape from which to attract the greatest variety or selected species to smaller sites.

CORRIDORS

As the degree of urbanization increases, wildlife habitats become progressively degraded, fragmented, and isolated. The benefits of maintaining connectivity of habitats are several and worthy of the effort (Adams and Dove 1989). Conservation easements for maintaining naturalness along waterways, riparian habitats, utility rights-of-way, transportation corridors, and parkway belts augment these corridors. On a small scale, property boundaries may be left undeveloped, permitted to revert through natural succession, or re-created with restorative techniques. A network of continuous corridors and a necklace of similar habitat types across an urbanized landscape will provide mobile species with access to habitat for seasonal movements (e.g., migration of amphibians to water, postnatal dispersal of mammals) and for emigration from or repopulation of lower-quality habitats.

WATER

Water is an essential component of the habitat of many groups of wildlife, from fishes and amphibians to waterbirds, and water for drinking is required by most vertebrates. Any management that restores water quality, retains or restores bodies of water, including flowing water, reduces interruption of associated shoreline or riparian natural communities, or makes water for drinking available will benefit wildlife.

As mentioned earlier, corridor management can be combined with management of the larger lakes, rivers, and streams of urban areas. River conservation (Diamant et al. 1984) can be a grass-roots, citizens' effort, yielding numerous environmental benefits. Lacking connecting waterways, habitat patches >0.55 ha with permanent water may preserve the richness of reptiles and amphibians in a city (Dickman 1987). Nevertheless, herptile richness for a site increases with patch size and declines with distance from permanent water. In a community and residential setting, often associated with new developments, the design and management of retention ponds and detention basins used to control siltation or reduce the rate of stormwater flow can yield valuable wildlife habitat. The guidelines of Adams et al. (1986) should be followed (Table 6). Care must be exercised to avoid major water-quality problems when significant numbers of waterfowl are concentrated on urban ponds. Harris et al. (1981) documented the deleterious effects of waterfowl on water quality and the effectiveness of dredging as a pond renovation technique.

Table 6. Design guidelines to optimize the value of constructed urban stormwater control ponds and other wetland impoundments for wildlife (from Adams et al. 1986:258).

- Impoundements with gently sloping sides (on the order of 10:1) are preferable to impoundments with steep slopes. Gently sloping sides will encourage the establishment of marsh vegetation. Vegetation will provide food and cover for wildlife and help enhance water quality. Impoundments with gently sloping sides also are safer than steep-sided ponds for children who might enter the impoundments.
- Water depth should not exceed 61 cm for 25–50% of the water surface area, with approximately 50–75% having a depth not less than 1.1–1.2 m. A greater depth may be advisable for more northern areas subject to greater ice depths.
- An emergent vegetation/open-water ratio of about 50:50 should be maintained.
- For larger impoundments (≥2 ha), one or more small islands are recommended. The shape and position of islands should be designed to help direct water flow within the impoundment. Water flow around and between islands can help to oxygenate the water and prevent stagnation. Water quality can be enhanced by a flow-through system where water is continually flushed through the impoundment. Islands should be gently sloped, and the tops should be graded to provide good drainage. Appropriate vegetative cover should be established to prevent erosion and provide bird-nesting cover. Consideration should be given to including an overland flow area in the design of large impoundments.
- Impoundments should be designed with the capability to regulate water levels, including complete drainage, and with facilities for cleaning, if necessary.
- Locating permanent-water impoundments near existing wetlands generally will enhance the wildlife values of impoundments.

On-site sources of flowing water should be maintained on the surface or recreated as surface flowage when possible. On small sites, creative use of fountains and artificial streams and small ponds with associated vegetation and landscaping will provide a habitat attractive to people and beneficial to insects (e.g., dragonflies), amphibians, turtles, birds, and mammals.

NUISANCE SITUATIONS AND PROBLEM ANIMALS

A wealth of published literature addresses the control and management of nuisance wildlife (see Chapter 18). Techniques for nuisance reduction, especially those developed for the small-parcel, exurban landowner, often can be applied to suburban and urban residential lots as well as to larger parcels, such as city parks or institutional grounds.

Shoesmith (1978), Flyger et al. (1984), and San Julian (1984) discussed the diversity of urban wildlife problems, their specific nature, and some methods for resolving conflicts. Unlike the widespread and recurring wildlife damage to agricultural, nursery, or plantation crops, most wildlife damage experienced by suburban homeowners and most urbanites may be a one-time, or infrequent and isolated, event. Chronic problems are encountered more frequently in the central city, airports, sanitary landfill sites, or occasionally in yards with an abundance of horticultural plantings attractive to rodents or browsing cervids.

Scope

Most frequently, urban nuisance problems arise from, or are exacerbated by, one or more of the following conditions:

(1) Artificial feeding (deliberate or unwittingly).

Bird feeders and other venues for feeding wildlife, together with frequent availability of excess pet food and spilled household garbage, are major attractants to commensal and exotic species, as well as to native fauna. Animals that frequently become nuisances in these situations include rats, mice, gray squirrels, raccoons, opossums, house sparrows, starlings, pigeons, gulls, and mallards.

(2) Excess cover.

The presence of trash, scrap metal and lumber, discarded masonry, and tall, herbaceous growth around a residence or throughout a neighborhood, commercial area, or business district provides security to animals and promotes nuisance situations.

(3) Poorly designed, inappropriate, or deteriorating architecture.

Given some relief from natural predators, human exploitation, and extremes of weather in the urban environment, wildlife seek cover in, on, or beneath buildings and other anthropic structures. Structures ranging from streetlights, signs, and commercial buildings to private residences may serve as denning and nesting sites for unwanted wildlife. Structures that are deteriorating may be unusually attractive for some species, but the original design of many structures also may be appealing (e.g., rock doves roosting on flat window ledges).

(4) Poor environmental planning or design.

Among the most notorious results of poor planning is that of gulls frequenting airports located on landfills near coastal bays. These birds cause problems at water reservoirs, zoos, outdoor eateries, and public gathering sites, and on rooftops (Solmon et al. 1984). To preclude such problems from arising, town planners, developers, and architects should consult wildlife biologists. Case studies and other papers in Stenberg and Shaw (1986) will be useful when a plan is developed and implemented.

(5) Zoonoses, epizootics, and other mortalities.

The existence of more than 200 zoonoses in North America is sufficient justification for additional study and management efforts to reduce disease transmission in urban areas. Norway rats and feral pigeons are well-known vectors or reservoirs of diseases that threaten humans. An insidious disease of special concern is the visceral larval migrans (VLM) caused by the raccoon roundworm *Baylisascaris procyonis,* which is an ever-present threat whenever humans are in close association with raccoons (Kazacos 1985, Feigley 1992).

Rabies is a widely known and dreaded threat to human health. An outbreak of rabies, primarily in raccoons, in the mid-Atlantic states in the 1980s heightened awareness of the undesirable aspects of contacts between humans and wildlife. Rosatte et al. (1987) provided a model for addressing the danger of rabies, especially in densely populated urban areas such as Toronto, Ontario. Raccoons also serve as hosts for several species of ticks that are vectors of such human diseases as Rocky Mountain spotted fever, babesiosis, and the recently discovered and fast-spreading Lyme disease.

Locke (1974) and Karstad (1975) discussed the diseases and parasites of urban wildlife. Brittingham and Temple (1986) reported on bird mortality from diseases and unknown causes at bird feeders in Wisconsin. Platform feeders resulted in higher mortality at feeder sites than at sites without platform feeders.

(6) Insufficient knowledge, information, and assistance.

The isolation of urban-dwelling humans from natural systems and wild animals often results from limited past experience, uncertainty, and fear when a wild animal is encountered. Many urbanites view wildlife species as native pets and condition them to feeding areas near residences. Such close association with these animals (e.g., raccoons feeding on decks and porches) creates the potential for significant disease and injury problems. Urbanites usually have difficulty obtaining prompt response to their questions or pleas for help with a residential nuisance problem. Understaffed state and federal natural resource agencies cannot keep up with the increase in inquiries and demands for assistance from the urban public. Cooperative arrangements for meeting the public's need for assistance and for stronger identity of the agency or office responsible for nuisance and public health concerns are badly needed. Programs and publications of state cooperative extension offices should be directed toward the needs of urbanites. Brochures that address nuisance animals and reduction of problems should be widely available.

Prevention and Control

Undesirable situations and nuisance wildlife occurrences in urban areas deserve prompt attention to prevent erosion of the public's positive attitudes toward wildlife. The wildlife manager often must be creative in the selection and application of methodology. Generally, when dealing with urban wildlife nuisance situations, one should keep the following in mind:

(1) Small parcel sizes, concerns for humane treatment or a protective attitude toward wild animals, and municipal laws against discharge of firearms severely limit the application or effectiveness of population reduction with lethal methods. Many states prohibit the use of chemicals to control wildlife, although many states allow exceptions for agricultural uses.

(2) Concern for the safety of humans and pets prohibits the use of most lethal methods, but covert destruction of such commensals as Norway rats, house sparrows, starlings, and pigeons may be necessary. Extreme caution to protect the safety of people, pets, and nontarget wildlife must be exercised in all urban situations when lethal methods are used.

(3) Habitat modifications and management generally will be more acceptable and have longer lasting effects than destruction of the offending individuals or reduction of the abundance of the species.

(4) Generally a poor relationship exists between the degree of nuisance or damage and the population density of the offending species. Thus, a reduction in the abundance of the nuisance animals usually does not result in a commensurate reduction in the number or severity of problems.

(5) Public education may offer the most effective approach for dealing with nuisance situations. Ignorance, fear, or misinformation often results in exaggerated responses, including perceived ''problems.'' Increased human tolerance of the undesired animal and its activities may be the best ''solution'' to many urban nuisance problems. Furthermore, many nuisance problems are one-time or infrequent events that merit little more response from the wildlife biologist than telephone consultation with the resident or property manager.

(6) The goal should be to reduce the annoyance or damage to a level acceptable to the homeowner or urban public.

Controlling the Most Common Problems

Five species account for most of the nuisance wildlife complaints by urban/suburban homeowners. Urban wildlife biologists working in the geographic range of these species eventually will become involved in their management, either directly or indirectly (e.g., educational efforts). Reducing nuisances from these, and many other, urban wildlife species will require biological and social solutions. Continuing degradation of natural habitats and development of urban landscapes will increase problems during the foreseeable future. The following suggestions for management and nuisance reduction may be helpful.

GRAY SQUIRRELS

Although a delight to many people, this species frequently is cited as the number one pest in urban areas within its range. Because of their climbing and aerial abilities, gray squirrels are difficult to exclude from bird feeders or rooftops where tall shrubs, trees, or utility wires provide access. Popular articles usually recommend a variety of ingenious and often amusing devices for excluding squirrels from bird feeders, but squirrels denning in the attics or soffits of houses are more difficult to control. Power outages and fires, resulting from the activities of tree squirrels, represent serious problems in urban environments. Effective and long-lasting control of gray squirrels will require one or more of the following methods:

(1) Install baffles, counterweighted entrances, or other devices to prevent squirrels from climbing bird feeders. Conversely, feed birds exclusively with white proso millet, a highly preferred bird food that is of low palatability for squirrels (Geis 1980*b*).
(2) Attach metal flashing around isolated trees that produce mast or serve as activity centers for squirrels, or that provide access to buildings and rooftops. As an extreme, remove large-diameter, cavity-bearing, mast-producing trees.
(3) Install screening over windows, chimneys, and vents to exclude squirrels and other mammals and birds from entering houses.
(4) Keep buildings in good repair; use boards, metal flashing, and hardware cloth to repair holes and to exclude squirrels.
(5) Reduce damage to vegetation in urban parks by selecting proper plantings, reducing excess feeding by park visitors, reducing unnatural conditions such as monoculture in flower beds, and developing an interpretive program (Manski et al. 1981).

COTTONTAIL RABBITS

Rabbits are common in many urban environments and can cause problems in home gardens and yards throughout the year. In the spring, rabbits eat succulent garden crops such as beans and peas. When food supplies are scarce in the winter, rabbits gnaw bark or clip off tender plant stems. Rabbits nest in existing burrows, areas of heavy cover, and brushy fencerows, in addition to more manicured sites. Removal of nesting cover helps reduce the rabbit population; however, these structures also provide habitat for many other species of wildlife. Normally, simple exclusion structures should prevent most damage and enable the landowner to continue to enjoy viewing wildlife. Other methods can be employed where rabbits are causing severe economic damage or populations have risen to high levels:

(1) Small areas such as flower beds and gardens can be protected by a fence of woven wire 45–64 cm high. The bottom of the fence should be staked down or buried 15 cm in the ground.
(2) Temporary electric fences with strands at 10, 20, and 30 cm may be a worthwhile investment for a high-value cash crop.
(3) Small trees can be protected by placing hardware cloth cylinders around the tree or wrapping the tree with copper mesh; these devices should be 45–64 cm high.
(4) Live traps are a good option when only one or two animals are present and state and local laws allow trapping. Rabbits should be released a minimum of 8 km from the trap site (trap plans are given in Craven 1983).
(5) Repellents have varying degrees of success, and some must be reapplied after a rain. Only one product (Hinder®) is registered for use on food crops as a repellent. All chemicals should be applied according to label directions.
(6) Shooting may be an option if state and local regulations allow the taking of depredating animals and if local ordinances permit the discharge of a firearm in the area. Of course, personal and public safety must be the primary concern.
(7) No toxicant is registered for control of rabbits.

Rabbits in urban settings are more of a nuisance problem than a serious economic burden, and most homeowners enjoy watching these docile animals.

RACCOONS

Raccoons are perhaps the best adapted of the furbearers for urban living. Those individuals that are diseased, den in buildings, or become persistent visitors to trash cans or vegetable gardens present undesirable contacts with humans. The following actions should reduce the occurrence or probability of situations where raccoons become a nuisance:

- Stop deliberate or inadvertent (e.g., pet food, picnic supplies) feeding by residents, merchants, or park visitors.
- Use animal-proof garbage cans and covered trash dumpsters.
- Install raccoon-proof caps on chimneys.
- Deny access to buildings (undersides, attics) with screens, grillwork, louvers, and other barriers.
- Provide secure storm-sewer grates and covers.
- Attach metal flashing (climbing guards) on large, isolated trees, especially those with platforms or cavities suitable for denning.
- As an extreme measure, remove hedgerows, wooden fences, shrubbery, and other elements that provide secure travel lanes and escape cover near dwellings.

WHITE-TAILED AND MULE DEER

Dense deer populations increase opportunities for collisions with vehicles, excessive damage to gardens and other vegetation, and, for deer, traumas from encountering dogs, plate-glass windows, swimming pools, and other urban accoutrements. Unique situations in urban areas, such as deer at airports or in arboreta, often require special attention. Kuser and Applegate (1985:155) related a case history of efforts "to control deer populations effectively with a minimum of impacts offensive to significant proportions of our citizens" in rapidly urbanizing Princeton Township, New Jersey. Conover and Kania (1988) reported on the relative vulnerability of ornamental plants to deer browsing.

Deer confined by fencing around such areas as municipal watersheds, military and research installations, and exotic animal pens may reach high densities and cause severe habitat damage. If the herd cannot be reduced by hunters or sanctioned marksmen, other methods must be employed. Wemmer and Stüwe (1985) described the effectiveness of drives as a means of reducing a deer population confined in large enclosures. An analysis of the costs and effectiveness of various methods for removing deer from a 510-ha arboretum may be useful to others facing such management decisions (Ishmael and Rongstad 1984). Shooting deer over bait was the most efficient (13.5 hr/deer) of all the control methods tested, but for removing live deer, immobilization with a dart gun was the most efficient method (20.5 hr/deer). Total costs were $74 for each deer shot and killed and $412 per deer removed alive. Jones and Witham (1990) pointed out that post-release mortality of translocated deer may exceed 50% during the first year, an important factor to consider when the biological and social elements of management of urban deer herds are evaluated.

Porter (1991) compared the ecological effects, estimated costs, and probability of success of commonly suggested alternatives for management of high white-tailed deer densities in areas such as national parks in the East. Although not field-tested, selective removal of family units of deer may be an effective compromise solution where conflicts are most severe. Modeling suggests that a one-time removal of 12–20 individuals might affect reduction for 10–15 years on areas of ca. 400 ha. Such population control may operate because of apparent fidelity of does to seasonal ranges. Additionally, Porter pointed out the need for a clear distinction between goals and objectives in the value-driven arena of management.

CANADA GEESE

Management concerns for Canada geese now include local populations residing for some or all of the year in urban environments. Results of a survey of golf courses in the eastern United States showed, for example, that geese occurred on 42% of the fairways and, of these, 62% were considered nuisance flocks (Conover and Chasko 1985). The same survey indicated that the flocks in the mid-Atlantic states averaged 250 birds per golf course. Nationally, several hundred thousand Canada geese reside in urban settings, and large flocks occur in Seattle, Denver, Minneapolis, Chicago, Boston, and Nashville (Nelson and Oetting 1981).

To effect aversive conditioning, Conover (1985) sprayed turf with methiocarb, a nonlethal chemical that sickens geese grazing on the treated grass. The birds reduced their use of the sprayed areas by 71% for 8 weeks after the treatment. The cost, $110/ha, was acceptable to many golf course managers. However, until methiocarb gains the approval of the U.S. Environmental Protection Agency, biologists usually must face the expensive task of catching and removing nuisance geese. Moreover, some or all of the relocated birds may return to the original site. Adult females show especially strong tendencies to return, but most young birds that are relocated at least 32 km from the capture site do not return (Cooper 1987).

Current recommendations for relocating nuisance geese include (1) releasing the birds at least 800 km from the capture site, preferably to the south; (2) selecting release sites where hunting may regulate the goose population or sterilizing the birds where hunting is not feasible and where crop damage is unlikely; and (3) culling birds with diseases or abnormalities before they are released (U.S. Fish and Wildlife Service, Release 5-1, Adm. Man., Reg. 5, Newton Center, Mass., 1980).

More direct means of control include destroying the eggs of urban-nesting geese, but this measure may offend some segments of the public. More covert methods such as addling (shaking) eggs or replacing viable eggs with plastic ones may be effective and may be more acceptable to the public if most nests can be located and are accessible to field personnel. Indeed, control measures of any kind should be proposed to a governing civic body (e.g., city council) for local approval before they are implemented by public employees or on public property. Management decisions made in this manner gain a better chance of acceptance by the community at large and help avoid the unpleasant prospect of "outsiders," regardless of their authority, dictating local policy.

AVAILABLE RESOURCES

Those wishing to increase the coverage of urban wildlife in their personal or institutional library should consult Dove's (1985) comprehensive, categorical lists. Those wishing to obtain the most recent research and management findings should consult current issues of scientific journals such as *The Journal of Wildlife Management, Wildlife Society Bulletin, Landscape and Urban Planning,*

Transactions of the North American Wildlife and Natural Resources Conference, Environmental Management, Ecological Applications, Biological Conservation, and *Natural Areas Journal,* among others. The British quarterly, *Urban Wildlife News* (Nature Conservancy Council, Northminster House, Peterborough, U.K. PE1 1UA), summarizes local efforts for wildlife (plant and animal) restoration and preservation. Also, abstracts of studies and reports on numerous aspects of urban ecology, often local in scale, were compiled by Dawe (1990). Social, political, and economic aspects of cities are covered in the *Urban Affairs Quarterly, Urban Studies,* and *Journal of Urban Affairs.*

The staff of the National Institute for Urban Wildlife (formerly Urban Wildlife Research Center), at 10921 Trotting Ridge Way, Columbia, MD 21044, has produced or edited numerous publications of scientific interest. The urban wildlife literature contains numerous references to the publications of former or current staff members L. W. Adams, L. E. Dove, T. M. Franklin, and D. L. Leedy. The Institute's quarterly *Urban Wildlife News* and its supplement, *The Urban Wildlife Manager's Notebook,* provide current and useful information to subscribers.

Numerous state conservation agencies now have personnel responsible, at least part-time, for urban wildlife activities. State, regional, or national offices of National Audubon Society, The Nature Conservancy, National Wildlife Federation, and similar organizations may provide literature or point the way to valuable sources of additional information.

Educators, students, and employers will find guidelines for appropriate qualifications, training, and experience for urban wildlife biologists in Tylka et al. (1987). Recently, coursework in urban wildlife has been offered at Colorado State University (Ft. Collins) and State University of New York (Syracuse). Because resource managers, and especially urban managers, often face a broad array of decision-making opportunities, curricula designed to train competent personnel should include environmental values and ethics as well as fundamental courses in the biological sciences (Lemons 1989).

Wildlife managers addressing urban nuisance problems should begin with a survey of the general literature on the subject. A comprehensive analysis of the environmental, economic, aesthetic, and public-health issues for invertebrate and vertebrate pests was presented by the Committee on Urban Pest Management, Environmental Studies Board, Commission on Natural Resources of the National Research Council (National Research Council 1980). The NRC report included a thorough discussion of indoor invertebrate pests and associated human diseases. The ongoing series on Vertebrate Pest Control (Univ. of California) provides techniques and applications often appropriate to cities and suburbs. The Great Plains Wildlife Damage Control Workshops (e.g., Uresk et al. 1988) provide helpful suggestions for dealing with many species that may occur in an urban setting. Also useful are the proceedings of the Eastern Animal Damage Control Conferences (Decker 1984, Bromley 1985, Holler 1987, Craven 1989, Curtis et al. 1992). Perhaps the best single source of relevant information is the frequently updated handbook entitled *Prevention and Control of Wildlife Damage* edited by Timm (1983). Cooperative Extension offices (state or county) often can provide useful literature and suggestions. Species-specific fact sheets are available from the state offices of the USDA Office of Animal Damage Control. Again, one must follow federal, state, and local restrictions and guidelines, using caution with regard to pet and human safety, when selecting remedies for troublesome wildlife.

LITERATURE CITED

ADAMS, C. E., J. K. THOMAS, P. C. LIN, AND B. WEISER. 1987a. Urban high school students' knowledge of wildlife. Pages 83–86 *in* L. W. Adams and D. L. Leedy, eds. Integrating man and nature in the metropolitan environment. Natl. Inst. Urban Wildl., Columbia, Md.

ADAMS, L. W. 1988. Some recent advances in urban wildlife research and management. Proc. Southeast. Nongame Endangered Wildl. Symp. 3:213–224.

———, AND L. E. DOVE. 1989. Wildlife reserves and corridors in the urban environment. A guide to ecological landscape planning and resource conservation. Natl. Inst. Urban Wildl., Columbia, Md. 91pp.

———, ———, AND T. M. FRANKLIN. 1985. Mallard pair and brood use of urban stormwater-control impoundments. Wildl. Soc. Bull. 13:46–51.

———, ———, AND D. L. LEEDY. 1984. Public attitudes toward urban wetlands for stormwater control and wildlife enhancement. Wildl. Soc. Bull. 12:299–303.

———, T. M. FRANKLIN, L. E. DOVE, AND J. M. DUFFIELD. 1986. Design considerations for wildlife in urban stormwater management. Trans. North Am. Wildl. Nat. Resour. Conf. 51:249–259.

———, AND D. L. LEEDY, EDITORS. 1987. Integrating man and nature in the metropolitan environment. Natl. Inst. Urban Wildl., Columbia, Md. 249pp.

———, AND ———, EDITORS. 1991. Wildlife conservation in metropolitan environments. Natl. Inst. Urban Wildl., Columbia, Md. 264pp.

———, ———, AND W. C. McCOMB. 1987b. Urban wildlife research and education in North American colleges and universities. Wildl. Soc. Bull. 15:591–595.

ALDRICH, J. W., AND R. W. COFFIN. 1980. Breeding bird populations from forest to suburbia after thirty-seven years. Am. Birds 34:3–7.

ANDREWS, S. G., G. DICKENS, AND R. MILLER. 1986. Urbanization and pronghorn antelope. Pages 172–174 *in* K. Stenberg and W. W. Shaw, eds. Wildlife conservation and new residential developments. Univ. Arizona School Renewable Nat. Resour., Tucson.

ANDREWS, W. A., AND J. L. CRANMER-BYNG. 1981. Urban natural areas: ecology and preservation. Univ. Toronto Inst. Environ. Stud., Environ. Monogr. 2. 215pp.

ANKNEY, C. D., AND J. HOPKINS. 1985. Habitat selection by roof-nesting killdeer. J. Field Ornithol. 56:284–286.

ANTHONY, J. A., J. E. CHILDS, G. E. GLASS, G. W. KORCH, L. ROSS, AND J. K. GRIGOR. 1990. Land use associations and changes in population indices of urban raccoons during a rabies epizootic. J. Wildl. Dis. 26:170–179.

ARNOLD, R. A., AND A. E. GOINS. 1987. Habitat enhancement techniques for the El Segundo blue butterfly: an urban endangered species. Pages 173–181 *in* L. W. Adams and D. L. Leedy, eds. Integrating man and nature in the metropolitan environment. Natl. Inst. Urban Wildl., Columbia, Md.

ATKINSON, K. T., AND D. M. SHACKLETON. 1991. Coyote *Canis latrans* ecology in a rural-urban environment. Can. Field-Nat. 105:49–54.

BAKER, J. A., AND R. J. BROOKS. 1981. Raptor and vole populations at an airport. J. Wildl. Manage. 45:390–396.

BARKER, G. M. A. 1987. European approaches to urban wildlife programs. Pages 183–190 *in* L. W. Adams and D. L. Leedy, eds. Integrating man and nature in the metropolitan environment. Natl. Inst. Urban Wildl., Columbia, Md.

BASCIETTO, J. J., AND L. W. ADAMS. 1983. Frogs and toads of stormwater management basins in Columbia, Maryland. Bull. Md. Herpetol. Soc. 19:58–60.

BATTEN, L. A. 1972. Breeding bird species diversity in relation to increasing urbanisation. Bird Study 19:157–166.

BECK, A. M. 1973. The ecology of stray dogs: a study of free-ranging urban animals. York Press, Baltimore, Md. 98pp.

———. 1974. The ecology of urban dogs. Pages 57–59 *in* J. H. Noyes and D. R. Progulske, eds. A symposium on wildlife in an urbanizing environment. Univ. Massachusetts Coop. Ext. Serv., Amherst.

BEISSINGER, S. R., AND D. R. OSBORNE. 1982. Effects of urbanization on avian community organization. Condor 84:75–83.

BENDELL, J. F. S., AND J. B. FALLS. 1981. Wildlife. Pages 153–166 *in* W. A. Andrews and J. L. Cranmer-Byng, eds. Urban natural areas: ecology and preservation. Univ. Toronto Press, Toronto, Ont.

BEZZEL, E. 1985. Birdlife in intensively used rural and urban environments. Ornis Fenn. 62:90–95.

BIGLER, L. L. 1990. Selected zoonoses of a suburban raccoon (*Procyon lotor*) population located in Islip, (Suffolk County) N.Y. M.S. Thesis, State Univ. New York, Syracuse. 89pp.

BISSELL, S. J., K. DEMAREST, AND D. L. SCHRUPP. 1987. The use of zoning ordinances in the protection and development of wildlife habitat. Pages 37–42 *in* L. W. Adams and D. L. Leedy, eds. Integrating man and nature in the metropolitan environment. Natl. Inst. Urban Wildl., Columbia, Md.

BOLEN, E. G. 1991. Analogs: a concept for the research and management of urban wildlife. Landscape Urban Plann. 20:285–289.

BORNKAMM, R., J. A. LEE, AND M. R. D. SEAWARD, EDITORS. 1982. Urban ecology: the second European ecological symposium (Berlin, 8–12 September 1980). Blackwell Sci. Publ., Boston, Mass. 370pp.

BOUDREAU, G. W. 1975. How to win the war with pest birds. Wildl. Technol., Hollister, Calif. 174pp.

BRITTINGHAM, M. C., AND S. A. TEMPLE. 1986. A survey of avian mortality at winter feeders. Wildl. Soc. Bull. 14:445–450.

———, AND ———. 1992. Use of winter bird feeders by black-capped chickadees. J. Wildl. Manage. 56:103–110.

BROCKE, R. H. 1977. What future for wildlife management in an urbanizing society? Trans. Northeast. Fish Wildl. Conf. 34:71–79.

BROMLEY, P. T., EDITOR. 1985. Proc. Second Eastern Wildlife Damage Control Conf., Raleigh, N.C. 281pp.

BROWN, T. L., C. P. DAWSON, AND R. L. MILLER. 1979. Interests and attitudes of metropolitan New York residents about wildlife. Trans. North Am. Wildl. Nat. Resour. Conf. 44:289–297.

BURLEY, J. B., AND R. B. MARTIN. 1986. A study to determine variation in homeowner preference for naturalistic landscapes and tolerance of wildlife at a residential townhome community utilizing residential landscape composition as a preference indicator. Pages 56–65 *in* K. Stenberg and W. W. Shaw, eds. Wildlife conservation and new developments. Univ. Arizona School Renewable Nat. Resour., Tucson.

BURNS, J., K. STENBERG, AND W. W. SHAW. 1986. Critical and sensitive wildlife habitats in Tucson, Arizona. Pages 144–150 *in* K. Stenberg and W. W. Shaw, eds. Wildlife conservation and new developments. Univ. Arizona School Renewable Nat. Resour., Tucson.

CALHOON, R. E., AND C. HASPEL. 1989. Urban cat populations compared by season, subhabitat and supplemental feeding. J. Anim. Ecol. 58:321–328.

CAMPBELL, C. A. 1974. Survival of reptiles and amphibia in urban environments. Pages 61–66 *in* J. H. Noyes and D. R. Progulske, eds. A symposium on wildlife in an urbanizing environment. Univ. Massachusetts Coop. Ext. Serv., Amherst.

CHILDS, J. E. 1986. Size-dependent predation on rats (*Rattus norvegicus*) by house cats (*Felis catus*) in an urban setting. J. Mammal. 67:196–199.

———, AND L. ROSS. 1986. Urban cats: characteristics and estimation of mortality due to motor vehicles. Am. J. Vet. Res. 47:1643–1648.

CHURCHER, P. B., AND J. H. LAWTON. 1987. Predation by domestic cats in an English village. J. Zool. (Lond.) 212:439–455.

CICERO, C. 1989. Avian community structure in a large urban park: controls of local richness and diversity. Landscape Urban Plann. 17:221–240.

CONNELLY, N. A., D. J. DECKER, AND S. WEAR. 1987. Public tolerance of deer in a suburban environment: a case history. Proc. East. Wildl. Damage Control Conf. 3:207–218.

CONNER, R. N. 1988. Wildlife populations: minimally viable or ecologically functional? Wildl. Soc. Bull. 16:80–84.

CONOVER, M. R. 1985. Alleviating nuisance Canada goose problems through methiocarb-induced aversive conditioning. J. Wildl. Manage. 49:631–636.

———. 1987. The urban-suburban Canada goose: an example of short-sighted management? Proc. East. Wildl. Damage Control Conf. 3:346.

———, AND G. G. CHASKO. 1985. Nuisance Canada goose problems in the eastern United States. Wildl. Soc. Bull. 13:228–233.

———, AND G. S. KANIA. 1988. Browsing preference of white-tailed deer for different ornamental species. Wildl. Soc. Bull. 16:175–179.

COOK, R. P., AND C. A. PINNOCK. 1987. Recreating a herpetofaunal community at Gateway National Recreation Area, New York. Pages 151–154 *in* L. W. Adams and D. L. Leedy, eds. Integrating man and nature in the metropolitan environment. Natl. Inst. Urban Wildl., Columbia, Md.

COOPER, J. A. 1987. The effectiveness of translocation control of Minneapolis-St. Paul Canada goose populations. Pages 169–171 *in* L. W. Adams and D. L. Leedy, eds. Integrating man and nature in the metropolitan environment. Natl. Inst. Urban Wildl., Columbia, Md.

———. 1991. Canada goose management at the Minneapolis-St. Paul International Airport. Pages 175–183 *in* L. W. Adams and D. L. Leedy, eds. Wildlife conservation in metropolitan environments. Natl. Inst. Urban Wildl., Columbia, Md.

CRAVEN, S. R. 1983. Cottontail rabbits. Pages D69–D74 *in* R. M. Timm, ed. Prevention and control of wildlife damage. Great Plains Agric. Counc. and Nebraska Coop. Ext. Serv., Lincoln.

———, EDITOR. 1989. Proc. Fourth Eastern Wildlife Damage Control Conf., Madison, Wis. 258pp.

CURTIS, P. D., M. J. FARGIONE, AND J. E. CASLICK, EDITORS. 1992. Proc. Fifth Eastern Wildlife Damage Control Conf., Ithaca, N.Y. 225pp.

CYPHER, B. L., R. H. YAHNER, AND E. A. CYPHER. 1988. Seasonal food use by white-tailed deer at Valley Forge National Historical Park, Pennsylvania, USA. Environ. Manage. 12:237–242.

DAGG, A. I. 1970. Wildlife in an urban area. Nat. Can. 97:201–212.

DANIELS, T. J., AND M. BEKOFF. 1989. Population and social biology of free-ranging dogs, *Canis familiaris*. J. Mammal. 70:754–762.

DAWE, G. F. M. 1990. The urban environment—a sourcebook for the 1990s. Cent. Urban Ecol., Birmingham, U.K. 636pp.

DECKER, D. J., EDITOR. 1984. Proc. First Eastern Wildlife Damage Control Conf., Ithaca, N.Y. 379pp.

———. 1987. Management of suburban deer: an emerging controversy. Proc. East. Wildl. Damage Control Conf. 3:344–345.

———, AND T. A. GAVIN. 1985. Public tolerance of a suburban deer herd: implications for control. Proc. East. Wildl. Damage Control Conf. 2:192–204.

———, AND ———. 1987. Public attitudes toward a suburban deer herd. Wildl. Soc. Bull. 15:173–180.

———, AND G. R. GOFF, EDITORS. 1987. Valuing wildlife. Westview Press, Boulder, Colo. 424pp.

———, AND K. G. PURDY. 1988. Toward a concept of wildlife acceptance capacity in wildlife management. Wildl. Soc. Bull. 16:53–57.

DEGRAAF, R. M. 1986. Urban bird habitat relationships: application to landscape design. Trans. North Am. Wildl. Nat. Resour. Conf. 51:232–248.

———. 1987. Urban wildlife habitat research—application to landscape design. Pages 107–111 *in* L. W. Adams and D. L. Leedy, eds. Integrating man and nature in the metropolitan environment. Natl. Inst. Urban Wildl., Columbia, Md.

———. 1989. Territory sizes of song sparrows, *Melospiza melodia,* in rural and suburban habitats. Can. Field-Nat. 103:43–47.

———, AND J. M. WENTWORTH. 1981. Urban bird communities and habitats in New England. Trans. North Am. Wildl. Nat. Resour. Conf. 46:396–413.

———, AND ———. 1986. Avian guild structure and habitat associations in suburban bird communities. Urban Ecol. 9:399–412.

———, AND G. M. WITMAN. 1979. Trees, shrubs, and vines for attracting birds. A manual for the Northeast. Univ. Massachusetts Press, Amherst. 194pp.

DENNIS, J. V. 1985. The wildlife gardener. Alfred E. Knopf Publ., New York, N.Y. 293pp.

———. 1989. Summer bird feeding. Audubon Workshop, Northbrook, Ill. 136pp.

DIAMANT, R., J. G. EUGSTER, AND C. J. DUERKSEN. 1984. A citizen's guide to river conservation. The Conserv. Found., Washington, D.C. 113pp.

DICKMAN, C. R. 1986. A method for censusing small mammals in urban habitats. J. Zool. Ser. A (Lond.) 210:631–636.

———. 1987. Habitat fragmentation and vertebrate species richness in an urban environment. J. Appl. Ecol. 24:337–351.

———, AND C. P. DONCASTER. 1987. The ecology of small mammals in urban habitats. I. Populations in a patchy environment. J. Anim. Ecol. 56:629–640.

DIEHL, J., AND T. S. BARRETT. 1988. The conservation easement handbook: managing land conservation and historic preservation easement programs. Land Trust Exchange, Alexandria, Va. 269pp.

DILLMAN, D. A. 1978. Mail and telephone surveys: the total design method. John Wiley & Sons, New York, N.Y. 325pp.

DONCASTER, C. P., C. R. DICKMAN, AND D. W. MACDONALD. 1990. Feeding ecology of red foxes (*Vulpes vulpes*) in the city of Oxford, England. J. Mammal. 71:188–194.

DOUGLAS, I. 1983. The urban environment. Edward Arnold Publ., Baltimore, Md. 229pp.

DOVE, L. E. 1985. A guide to developing an urban wildlife library. Urban Wildlife Manager's Notebook 8. Natl. Inst. Urban Wildl., Columbia, Md. 11pp.

DUNKLE, F. H. 1987. Urban wildlife and the Fish and Wildlife Service: meeting a growing challenge. Pages 5–7 in L. W. Adams and D. L. Leedy, eds. Integrating man and nature in the metropolitan environment. Natl. Inst. Urban Wildl., Columbia, Md.

EMERY, M. 1986. Promoting nature in cities and towns: a practical guide. Croom Helm Ltd., London, U.K. 396pp.

EULER, D. L., F. GILBERT, AND G. MCKEATING, EDITORS. 1975. Proc. symp. wildlife in urban Canada. Univ. Guelph, Guelph, Ont. 134pp.

FAETH, S. H., AND T. C. KANE. 1978. Urban biogeography: city parks as islands for Diptera and Coleoptera. Oecologia 32:127–133.

FEIGLEY, H. P. 1992. The ecology of the raccoon in suburban Long Island, N.Y., and its relation to soil contamination with *Baylisascaris procyonis* ova. Ph.D. Thesis, State Univ. New York, Syracuse. 139pp.

FIGLEY, W. K., AND L. W. VANDRUFF. 1982. The ecology of urban mallards. Wildl. Monogr. 81. 40pp.

FLYGER, V. 1974. Tree squirrels in urbanizing environments. Pages 121–124 in J. H. Noyes and D. R. Progulske, eds. A symposium on wildlife in an urbanizing environment. Univ. Massachusetts Coop. Ext. Serv., Amherst.

———, D. L. LEEDY, AND T. M. FRANKLIN. 1984. Wildlife damage control in eastern cities and suburbs. Proc. East. Wildlife Damage Control Conf. 1:27–32.

FORMAN, R. T. T., A. E. GALLI, AND C. F. LECK. 1976. Forest size and avian diversity in New Jersey woodlots with some land-use implications. Oecologia 26:1–8.

———, AND M. GODRON. 1986. Landscape ecology. John Wiley & Sons, New York, N.Y. 619pp.

FROKE, J. B. 1986. Managing wildlife and development on the suburban/wildland edge in southern California. Pages 92–99 in K. Stenberg and W. W. Shaw, eds. Wildlife conservation and new developments. Univ. Arizona School Renewable Nat. Resour., Tucson.

GANO, R. D., JR., AND J. A. MOSHER. 1983. Artificial cavity construction—an alternative to nest boxes. Wildl. Soc. Bull. 11:74–76.

GEGGIE, J. F., AND M. B. FENTON. 1985. A comparison of foraging by *Eptesicus fuscus* (Chiroptera: Vespertilionidae) in urban and rural environments. Can. J. Zool. 63:263–266.

GEHLBACH, F. R. 1988. Population and environmental features that promote adaptation to urban ecosystems: the case of eastern screech-owls (*Otus asio*) in Texas. Proc. Int. Ornithol. Congr. 19:1809–1813.

GEIS, A. D. 1975. Urban planning and urban wildlife: a case study of a planned city near Washington, D.C. Pages 79–81 in D. L. Euler, F. Gilbert, and G. McKeating, eds. Proc. symp. wildlife in urban Canada. Univ. Guelph, Guelph, Ont.

———. 1976a. Bird populations in a new town. Atl. Nat. 31:141–146.

———. 1976b. Effects of building designs and quality on nuisance bird problems. Proc. Vertebr. Pest Conf. 7:51–53.

———. 1980a. Breeding and wintering bird populations at Cylburn and vicinity before the construction of Cold Spring Town. Atl. Nat. 33:5–8.

———. 1980b. Relative attractiveness of different foods at wild bird feeders. U.S. Fish Wildl. Serv. Spec. Sci. Rep. Wildl. 233. 11pp.

———. 1986. Wildlife habitat considerations in Columbia, Maryland and vicinity. Pages 97–99 in K. Stenberg and W. W. Shaw, eds. Wildlife conservation and new developments. Univ. Arizona School Renewable Nat. Resour., Tucson.

GENNARO, A. L. 1988. Breeding biology of an urban population of Mississippi kites in New Mexico. Pages 188–190 in R. L. Glinski, ed. Proc. Southwest raptor management symposium and workshop. Natl. Wildl. Fed. Sci. Tech. Ser. 11.

GILBERT, F. F. 1982. Public attitudes toward urban wildlife: a pilot study in Guelph, Ontario. Wildl. Soc. Bull. 10:245–253.

GOLDSTEIN, E. L., M. GROSS, AND R. M. DEGRAAF. 1983. Wildlife and greenspace planning in medium-scale residential developments. Urban Ecol. 7:201–214.

———, ———, AND ———. 1986. Breeding birds and vegetation: a quantitative assessment. Urban Ecol. 9:377–385.

GOTFRYD, A., AND R. I. C. HANSELL. 1986. Prediction of bird-community metrics in urban woodlots. Pages 321–326 in J. Verner, M. L. Morrison, and C. J. Ralph, eds. Wildlife 2000: modeling habitat relationships of terrestrial vertebrates. Univ. Wisconsin Press, Madison.

GUSTAFSON, E. J., AND L. W. VANDRUFF. 1990. Behavior of black and gray morphs of *Sciurus carolinensis* in an urban environment. Am. Midl. Nat. 123:186–192.

HADIDIAN, J., D. MANSKI, V. FLYGER, C. COX, AND G. HODGE. 1987. Urban gray squirrel damage and population management: a case history. Proc. East. Wildl. Damage Control Conf. 3:219–227.

———, AND S. RILEY. 1991. Daytime resting site selection in an urban raccoon population. Pages 39–45 in L. W. Adams and D. L. Leedy, eds. Wildlife conservation in metropolitan environments. Natl. Inst. Urban Wildl., Columbia, Md.

HARDIN, J. W. 1977. A study of human and waterfowl usage of urban ponds in the vicinity of Syracuse, N.Y. M.S. Thesis, State Univ. New York, Syracuse. 124pp.

HARRIS, H. J., JR., J. A. LADOWSKI, AND D. J. WORDEN. 1981. Water-quality problems and management of an urban waterfowl sanctuary. J. Wildl. Manage. 45:501–507.

HARRIS, L. D. 1984. The fragmented forest: island biogeography theory and the preservation of biotic diversity. Univ. Chicago Press, Chicago, Ill. 211pp.

———, AND P. KANGAS. 1988. Reconsideration of the habitat concept. Trans. North Am. Wildl. Nat. Resour. Conf. 53:137–144.

HARRIS, S. 1981. An estimation of the number of foxes (*Vulpes vulpes*) in the city of Bristol, and some possible factors affecting their distribution. J. Appl. Ecol. 18:455–465.

———. 1986. Urban foxes. Whittet Books, London, U.K. 128pp.

———, AND J. M. V. RAYNER. 1986. Urban fox (*Vulpes vulpes*) population estimates and habitat requirements in several British cities. J. Anim. Ecol. 55:575–591.

HASPEL, C., AND R. E. CALHOON. 1989. Home ranges of free-ranging cats (*Felis catus*) in Brooklyn, New York. Can. J. Zool. 67:178–181.

———, AND ———. 1993. Activity patterns of free-ranging cats in Brooklyn, New York. J. Mammal. 74:1–8.

HENCH, J. E., V. FLYGER, R. GIBBS, AND K. VAN NESS. 1985. Predicting the effects of land-use changes on wildlife. Trans. North Am. Wildl. Nat. Resour. Conf. 50:345–351.

———, K. V. NESS, AND R. GIBBS. 1987. Development of a natural resources planning and management process. Pages 29–35 in L. W. Adams and D. L. Leedy, eds. Integrating man and nature in the metropolitan environment. Natl. Inst. Urban Wildl., Columbia, Md.

HENDERSON, C. L. 1984. Woodworking for wildlife: homes for birds and mammals. Minnesota Dep. Nat. Resour., St. Paul. 48pp.

———. 1987. Landscaping for wildlife. Minnesota Dep. Nat. Resour., St. Paul. 145pp.

HENDRIX, W. G., J. G. FABOS, AND J. E. PRICE. 1988. An ecological approach to landscape planning using geographic information system technology. Landscape Urban Plann. 15:211–225.

HEUSMANN, H W. 1981. Movements and survival rates of park mallards. J. Field Ornithol. 52:214–221.

———. 1983. Mallards in the park—contribution to the harvest. Wildl. Soc. Bull. 11:169–171.

———, AND R. BURRELL. 1984. Park waterfowl populations in Massachusetts. J. Field Ornithol. 55:89–96.

HOFFMANN, C. O., AND J. L. GOTTSCHANG. 1977. Numbers, distribution, and movements of a raccoon population in a suburban residential community. J. Mammal. 58:623–636.

HOLLER, N. R., EDITOR. 1987. Proc. Third Eastern Wildlife Damage Control Conf., Gulf Shores, Ala. 362pp.

HOUCK, M. C. 1987. Urban wildlife habitat inventory: the Willamette River Greenway, Portland, Oregon. Pages 47–51 in L. W. Adams and D. L. Leedy, eds. Integrating man and nature in the metropolitan environment. Natl. Inst. Urban Wildl., Columbia, Md.

HOWARD, D. V. 1974. Urban robins: a population study. Pages 67–75 in J. H. Noyes and D. R. Progulske, eds. A symposium on wildlife in an urbanizing environment. Univ. Massachusetts Coop. Ext. Serv., Amherst.

HOWELL, R. R., JR. 1982. Habitat use in an urban residential area by the gray squirrel. M.S. Thesis, State Univ. New York, Syracuse. 42pp.

INGRALDI, M. F. 1992. The ecology of red-tailed hawks in an urban/suburban environment. M.S. Thesis, State Univ. New York, Syracuse. 78pp.

ISHMAEL, W. E., AND O. J. RONGSTAD. 1984. Economics of an urban deer-removal program. Wildl. Soc. Bull. 12:394–398.

JACKSON, J. B., AND F. R. STEINER. 1985. Human ecology for land-use planning. Urban Ecol. 9:177–194.

JODICE, P. G. R., AND S. R. HUMPHREY. 1992. Activity and diet of an urban population of Big Cypress fox squirrels. J. Wildl. Manage. 56:685–692.

JOHNSEN, A. M., III. 1982. Urban habitat use by house sparrows, rock doves, and starlings. M.S. Thesis, State Univ. New York, Syracuse. 75pp.

———, AND L. W. VANDRUFF. 1987. Summer and winter distribution of introduced bird species and native bird species richness within a complex urban environment. Pages 123–127 *in* L. W. Adams and D. L. Leedy, eds. Integrating man and nature in the metropolitan environment. Natl. Inst. Urban Wildl., Columbia, Md.

JOHNSON, C. W. 1986. The Ogden Nature Center: a case study in site rehabilitation to improve wildlife habitat with implications for new residential development. Pages 175–181 *in* K. Stenberg and W. W. Shaw, eds. Wildlife conservation and new developments. Univ. Arizona School Renewable Nat. Resour., Tucson.

JONES, J. M., AND J. H. WITHAM. 1990. Post-translocation survival and movements of metropolitan white-tailed deer. Wildl. Soc. Bull. 18:434–441.

JORDAN, W. R., III, R. L. PETERS, II, AND E. B. ALLEN. 1988. Ecological restoration as a strategy for conserving biological diversity. Environ. Manage. 12:55–72.

KAPLAN, R. 1984. Impact of urban nature: a theoretical analysis. Urban Ecol. 8:189–197.

KARSTAD, L. 1975. Disease problems of urban wildlife. Pages 69–78 *in* D. L. Euler, F. Gilbert, and G. McKeating, eds. Proc. symp. wildlife in urban Canada. Univ. Guelph, Guelph, Ont.

KAZACOS, K. R. 1985. Raccoon roundworms (*Baylisascaris procyonis*)—a cause of animal and human disease. Environ. Rev. 29:15–25.

KELLERT, S. R. 1984. Urban American perceptions of animals and the natural environment. Urban Ecol. 8:209–228.

———, AND J. K. BERRY. 1987. Attitudes, knowledge, and behaviors toward wildlife as affected by gender. Wildl. Soc. Bull. 15:363–371.

———, AND M. O. WESTERVELT. 1983. Children's attitudes, knowledge, and behaviors toward animals. Phase V of the study, "American attitudes, knowledge, and behaviors toward wildlife and natural habitats." U.S. Fish Wildl. Serv., Washington, D.C. 202pp.

KIERAN, J. 1959. A natural history of New York City. Houghton Mifflin Co., Boston, Mass. 428pp.

KIRKPATRICK, C. M., EDITOR. 1978. Wildlife and people. Proc. 1978 John S. Wright Forestry Conf., Purdue Univ., West Lafayette, Ind. 191pp.

KLEM, D., JR. 1990. Collisions between birds and windows: mortality and prevention. J. Field Ornithol. 61:120–128.

KNIGHT, R. L., D. J. GROUT, AND S. A. TEMPLE. 1989. Nest-defense behavior of the American crow in urban and rural areas. Condor 89:175–177.

KRESS, S. W. 1985. The Audubon guide to attracting birds. Charles Scribner's Sons, New York, N.Y. 377pp.

KUSER, J. E., AND J. E. APPLEGATE. 1985. Princeton Township: the history of a no-discharge ordinance's effect on deer and people. Trans. Northeast Sect., The Wildl. Soc. 41:150–155.

KUSLER, J. A., AND M. E. KENTULA, EDITORS. 1990. Wetland creation and restoration—the status of the science. Island Press, Covelo, Calif. 594pp.

LAURIE, I. C. 1979. Nature in cities. The natural environment in the design and development of urban green space. John Wiley & Sons, New York, N.Y. 428pp.

LEEDY, D. L. 1979. An annotated bibliography on planning and management for urban-suburban wildlife. U.S. Fish Wildl. Serv., Washington, D.C. 256pp.

———, AND L. W. ADAMS. 1984. A guide to urban wildlife management. Natl. Inst. Urban Wildl., Columbia, Md. 42pp.

———, R. M. MAESTRO, AND T. M. FRANKLIN. 1978. Planning for wildlife in cities and suburbs. U.S. Fish Wildl. Serv. Off. Biol. Serv., Washington, D.C. 64pp.

LEMONS, J. 1989. The need to integrate values into environmental curricula. Environ. Manage. 13:133–147.

LEOPOLD, A. S., AND M. F. DEDON. 1983. Resident mourning doves in Berkeley, California. J. Wildl. Manage. 47:780–789.

LEUSCHNER, W. A., V. P. RITCHIE, AND D. F. STAUFFER. 1989. Opinions on wildlife: responses of resource managers and wildlife users in the southeastern United States. Wildl. Soc. Bull. 17:24–29.

LOCKE, L. N. 1974. Diseases and parasites in urban wildlife. Pages 111–112 *in* J. H. Noyes and D. R. Progulske, eds. A symposium on wildlife in an urbanizing environment. Univ. Massachusetts Coop. Ext. Serv., Amherst.

LUNIAK, M. 1983. The avifauna of urban green areas in Poland and possibilities of managing it. Acta Ornithol. 19:3–61.

———, AND B. PISARSKI, EDITORS. 1982. Animals in urban environment. Proc. Symp. Institute Zoology of the Polish Academy of Sciences. Zaklad Narodowy im. Ossolinskich, Wroclaw, Poland. 175pp.

LYNCH, J. F., AND D. F. WHIGHAM. 1984. Effects of forest fragmentation on breeding bird communities in Maryland, USA. Biol. Conserv. 28:287–324.

LYONS, J. R., AND D. L. LEEDY. 1984. The status of urban wildlife programs. Trans. North Am. Wildl. Nat. Resour. Conf. 49:233–251.

MacDONALD, D. W., AND M. T. NEWDICK. 1982. The distribution and ecology of foxes, *Vulpes vulpes* (L.), in urban areas. Pages 123–135 *in* R. Bornkamm, J. A. Lee, and M. R. D. Seaward, eds. Urban ecology. Blackwell Sci. Publ., Oxford, U.K.

MANSKI, D. A., L. W. VANDRUFF, AND V. FLYGER. 1981. Activities of gray squirrels and people in a downtown Washington, D.C. park: management implications. Trans. North Am. Wildl. Nat. Resour. Conf. 46:439–454.

MATTHEWS, M. J. 1986. New York State's Urban Wildlife Program with emphasis on urban wildlife education. Pages 43–45 *in* K. Stenberg and W. W. Shaw, eds. Wildlife conservation and new developments. Univ. Arizona School Renewable Nat. Resour., Tucson.

———, S. O'CONNOR, AND R. S. COLE. 1988. Database for the New York State urban wildlife habitat inventory. Landscape Urban Plann. 15:23–37.

MATTHIAE, P. E., AND F. STEARNS. 1981. Mammals in forest islands in southeastern Wisconsin. Pages 55–66 *in* R. L. Burgess and D. M. Sharpe, eds. Forest island dynamics in man-dominated landscapes. Springer-Verlag, New York, N.Y.

McCULLOUGH, D. R., AND R. H. BARRETT, EDITORS. 1992. Wildlife 2001: populations. Elsevier Appl. Sci., New York, N.Y. 1163pp.

McPHERSON, E. G., AND C. NILON. 1987. A habitat suitability index model for gray squirrel in an urban cemetery. Landscape J. 6:21–30.

MEIER, K. E. 1983. Habitat use by opossums in an urban environment. M.S. Thesis, Oregon State Univ., Corvallis. 69pp.

MILLIGAN, D. A., AND K. J. RAEDEKE. 1986. Incorporation of a wetland into an urban residential development. Pages 162–171 *in* K. Stenberg and W. W. Shaw, eds. Wildlife conservation and new developments. Univ. Arizona School Renewable Nat. Resour., Tucson.

MILLS, G. S., J. B. DUNNING, JR., AND J. M. BATES. 1989. Effects of urbanization on breeding bird community structure in southwestern desert habitats. Condor 91:416–428.

MINOR, W. F., AND M. L. MINOR. 1981. Nesting of red-tailed hawks and great horned owls in central New York suburban areas. Kingbird 1981:68–76.

MITCHELL, W. A. 1988. Songbird nest boxes. Sect. 5.1.8, U.S. Army Corps Eng. Tech. Rep. EL-88-19. U.S. Army Waterways Exp. Stn., Vicksburg, Miss. 48pp.

MOLL, G., AND S. EBENRECK, EDITORS. 1989. Shading our cities: a resource guide for urban and community forests. Island Press, Washington, D.C. 333pp.

MOULTON, C. A., AND L. W. ADAMS. 1991. Effects of urbanization on foraging strategy of woodpeckers. Pages 67–73 *in* L. W. Adams and D. L. Leedy, eds. Wildlife conservation in metropolitan environments. Natl. Inst. Urban Wildl., Columbia, Md.

MURPHY, D. D. 1988. Challenges to biological diversity in urban areas. Pages 71–76 *in* E. O. Wilson, ed. BioDiversity. Natl. Acad. Press, Washington, D.C.

MURPHY, R. K., M. W. GRATSON, AND R. N. ROSENFIELD. 1988. Activity and habitat use by a breeding male Cooper's hawk in a suburban area. J. Raptor Res. 22:97–100.

NATIONAL RESEARCH COUNCIL. 1980. Urban pest management. A report of the Commission on Urban Pest Management, Committee on Natural Resources. Natl. Acad. Press, Washington, D.C. 273pp.

NELSON, H. K., AND R. B. OETTING. 1981. An overview of management of Canada geese and their recent urbanization. Int. Waterfowl Symp. 4:128–133.

NILON, C. H. 1986. Quantifying small mammal habitats along a gradient of urbanization. Ph.D. Thesis, State Univ. New York, Syracuse. 148pp.

———, AND L. W. VANDRUFF. 1987. Analysis of small mammal community data and applications to management of urban greenspaces. Pages 53–59 in L. W. Adams and D. L. Leedy, eds. Integrating man and nature in the metropolitan environment. Natl. Inst. Urban Wildl., Columbia, Md.

NOYES, J. H., AND D. R. PROGULSKE, EDITORS. 1974. A symposium on wildlife in an urbanizing environment. Univ. Massachusetts Coop. Ext. Serv., Amherst. 182pp.

O'DONNELL, M. H. 1984. Wildlife problems, human attitudes and response to wildlife in Syracuse, N.Y. metropolitan area. M.S. Thesis, State Univ. New York, Syracuse. 116pp.

———, AND L. W. VANDRUFF. 1984. Wildlife conflicts in an urban area: occurrence of problems and human attitudes toward wildlife. Proc. East. Wildl. Damage Control Conf. 1:315–323.

OLIPHANT, L. W. 1974. Merlins—the Saskatoon falcons. Blue Jay 32:140–147.

OPDAM, P., D. VAN DORP, AND C. J. F. TER BRAAK. 1984. The effect of isolation on the number of woodland birds in small woods in the Netherlands. J. Biogeography 11:473–478.

PLUMPTON, D. L., AND R. S. LUTZ. 1993. Influence of vehicular traffic on time budgets of nesting burrowing owls. J. Wildl. Manage. 57:612–616.

PORTER, W. F. 1991. White-tailed deer in eastern ecosystems: implications for management and research in national parks. U.S. Natl. Park Serv., Nat. Resour. Rep. 91/05. 57pp.

POSTOVIT, H. R., AND B. C. POSTOVIT. 1987. Impacts and mitigation techniques. Pages 183–213 in B. A. Giron Pendleton, B. A. Millsap, K. W. Cline, and D. M. Bird, eds. Raptor management techniques manual. Inst. Wildl. Res., Natl. Wildl. Fed., Washington, D.C.

POWELL, L. J. H. 1982. The occurrence and distribution of selected mammalian species and their interaction with residents in an urban area. M.S. Thesis, State Univ. New York, Syracuse. 113pp.

PROGULSKE, D. R., AND D. L. LEEDY. 1986. Urban wildlife management: the challenge at home. Trans. North Am. Wildl. Nat. Resour. Conf. 51:567–572.

PUDELKEWICZ, P. J. 1981. Visual response to urban wildlife habitat. Trans. North Am. Wildl. Nat. Resour. Conf. 46:381–389.

REESE, K. P., AND J. T. RATTI. 1988. Edge effect: a concept under scrutiny. Trans. North Am. Wildl. Nat. Resour. Conf. 53:127–136.

RIDLEHUBER, K. T., AND J. W. TEAFORD. 1986. Wood duck nest boxes. Sect. 5.1.2, U.S. Army Corps Eng. Tech. Rep. EL-86-12. U.S. Army Waterways Exp. Stn., Vicksburg, Miss. 21pp.

ROBBINS, C. S. 1984. Management to conserve forest ecosystems. Pages 101–107 in W. C. McComb, ed. Proc. workshop management of nongame species and ecological communities. Univ. Kentucky, Lexington.

ROBINSON, W. L., AND E. G. BOLEN. 1989. Wildlife ecology and management. Second ed. Macmillan Publ. Co., New York, N.Y. 574pp.

RODIEK, J., AND E. G. BOLEN, EDITORS. 1990. Wildlife and habitats in managed landscapes. Island Press, Washington, D.C. 250pp.

ROSATTE, R. C., P. M. KELLY-WARD, AND C. D. MACINNES. 1987. A strategy for controlling rabies in urban skunks and raccoons. Pages 161–167 in L. W. Adams and D. L. Leedy, eds. Integrating man and nature in the metropolitan environment. Natl. Inst. Urban Wildl., Columbia, Md.

———, M. J. POWER, AND C. D. MACINNES. 1991. Ecology of urban skunks, raccoons, and foxes in metropolitan Toronto. Pages 31–38 in L. W. Adams and D. L. Leedy, ed. Wildlife conservation in metropolitan environments. Natl. Inst. Urban Wildl., Columbia, Md.

———, ———, AND ———. 1992. Density, dispersion, movements and habitat of skunks (*Mephitis mephitis*) and raccoons (*Procyon lotor*) in metropolitan Toronto. Pages 932–944 in D. R. McCullough and R. H. Barrett, eds. Wildlife 2001: populations. Elsevier Appl. Sci., New York, N.Y.

ROTH, R. R. 1987. Assessment of habitat quality for wood thrush in a residential area. Pages 139–149 in L. W. Adams and D. L. Leedy, eds. Integrating man and nature in the metropolitan environment. Natl Inst. Urban Wildl., Columbia, Md.

RUNYAN, C. S. 1987. Location and density of nests of the red-tailed hawk, *Buteo jamaicensis*, in Richmond, British Columbia. Can. Field-Nat. 101:415–418.

SAN JULIAN, G. J. 1984. The need for urban animal control. Proc. East. Wildl. Damage Control Conf. 1:313–314.

SCHAEFER, J. M. 1987. Identifying and targeting urban publics. Pages 207–219 in L. W. Adams and D. L. Leedy, eds. Integrating man and nature in the metropolitan environment. Natl. Inst. Urban Wildl., Columbia, Md.

SCHAUMAN, S., S. PENLAND, AND M. FREEMAN. 1987. Public knowledge of and preferences for wildlife habitats in urban open spaces. Pages 113–118 in L. W. Adams and D. L. Leedy, eds. Integrating man and nature in the metropolitan environment. Natl. Inst. Urban Wildl., Columbia, Md.

SCHICKER, L. 1986. Children, wildlife, and residential developments. Pages 48–55 in K. Stenberg and W. W. Shaw, eds. Wildlife conservation and new developments. Univ. Arizona School Renewable Nat. Resour., Tucson.

———. 1987. Design criteria for children and wildlife in residential developments. Pages 99–105 in L. W. Adams and D. L. Leedy, eds. Integrating man and nature in the metropolitan environment. Natl. Inst. Urban Wildl., Columbia, Md.

SCHINNER, J. R., AND D. L. CAULEY. 1974. The ecology of urban raccoons in Cincinnati, Ohio. Pages 125–130 in J. H. Noyes and D. R. Progulske, eds. A symposium on wildlife in an urbanizing environment. Univ. Massachusetts Coop. Ext. Serv., Amherst.

SCHLAUCH, F. C. 1978. Urban geographical ecology of the amphibians and reptiles of Long Island. Pages 25–41 in C. M. Kirkpatrick, ed. Wildlife and people. Proc. 1978 John S. Wright Forestry Conf., Purdue Univ., West Lafayette, Ind.

SHAFER, E. L., AND J. B. DAVIS. 1989. Making decisions about environmental management when conventional economic analysis cannot be used. Environ. Manage. 13:189–197.

SHARGO, E. S. 1988. Home range, movements, and activity patterns of coyotes (*Canis latrans*) in Los Angeles suburbs. Ph.D. Thesis, Univ. California, Los Angeles. 113pp.

SHARP, W. C. 1977. Conservation plants for the Northeast. U.S. Soil Conserv. Serv., Washington, D.C. 40pp.

SHAW, W. W., J. M. BURNS, AND K. STENBERG. 1986. Wildlife habitats in Tucson: a strategy for conservation. Univ. Arizona School Renewable Nat. Resour., Tucson. 17pp.

———, W. R. MANGUN, AND J. R. LYONS. 1985. Residential enjoyment of wildlife resources by Americans. Leisure Sci. 7:361–375.

———, AND V. SUPPLEE. 1987. Wildlife conservation in rapidly expanding metropolitan areas: informational, institutional, and economic constraints and solutions. Pages 191–197 in L. W. Adams and D. L. Leedy, eds. Integrating man and nature in the metropolitan environment. Natl. Inst. Urban Wildl., Columbia, Md.

SHEDD, D. H. 1990. Aggressive interactions in wintering house finches and purple finches. Wilson Bull. 102:174–178.

SHOESMITH, M. W. 1978. Wildlife management conflicts in urban Winnipeg. Pages 49–57 in C. M. Kirkpatrick, ed. Wildlife and people. The 1978 John S. Wright Forestry Conf., Purdue Univ., West Lafayette, Ind.

———, AND W. H. KOONZ. 1977. The maintenance of an urban deer herd in Winnipeg, Manitoba. Trans. North Am. Wildl. Nat. Resour. Conf. 42:278–285.

SIKOROWSKI, L., AND S. J. BISSELL, EDITORS. 1986. County government and wildlife management: a guide to cooperative habitat development. Colo. Div. Wildl. DOW-R-M-1-86. Var. pagin.

SLATE, D. 1985. Movement, activity and home range patterns among members of a high density suburban raccoon population. Ph.D. Thesis, Rutgers Univ., New Brunswick, N.J. 112pp.

SOLMON, V. E. F., H. BLOKPOEL, W. J. RICHARDSON, AND W. J. LAIDLAW. 1984. Keeping unwanted gulls away—a progress report. Proc. East. Wildl. Damage Control Conf. 1:311.

SOULÉ, M. E., D. T. BOLGER, A. C. ALBERTS, J. WRIGHT, M. SORICE, AND S. HILL. 1988. Reconstructed dynamics of rapid extinctions of chaparral-requiring birds in urban habitat islands. Conserv. Biol. 2:75–92.

SPIRN, A. W. 1984. The granite garden: urban nature and human design. Basic Books, Inc., New York, N.Y. 334pp.

SPITZER, P., AND A. POOLE. 1980. Coastal ospreys between New York City and Boston: a decade of reproductive recovery 1969–1979. Am. Birds 34:234–241.

STENBERG, K., AND W. W. SHAW, EDITORS. 1986. Wildlife conservation and new residential developments. Univ. Arizona School Renewable Nat. Resour., Tucson. 203pp.

STOKES, D., AND L. STOKES. 1987. The bird feeder book: an easy guide to attracting, identifying, and understanding your feeder birds. Little, Brown and Co., Boston, Mass. 86pp.

TALBOT, J. F., AND R. KAPLAN. 1984. Needs and fears: the response to trees and nature in the inner city. J. Arboric. 10:222–228.

TEAFORD, J. W. 1986. Squirrel nest boxes. Sect. 5.1.1, U.S. Army Corps Eng. Tech. Rep. EL 86-11. U.S. Army Waterways Exp. Stn., Vicksburg, Miss. 15pp.

TERRES, J. K. 1968. Songbirds in your garden. Thomas Y. Crowell Co., New York, N.Y. 256pp.

THE WILDLIFE SOCIETY. 1992. Conservation policies of The Wildlife Society. The Wildl. Soc., Bethesda, Md. 20pp.

THOMAS, J. W., AND R. M. DeGRAAF. 1973. Nongame wildlife research in megalopolis: the Forest Service program. U.S. For. Serv. Tech. Rep. NE-4. 12pp.

———, AND ———. 1975. Wildlife habitats in the city. Pages 48–68 *in* D. L. Euler, F. Gilbert, and G. McKeating, eds. Proc. symp. wildlife in urban Canada. Univ. Guelph, Guelph, Ont.

———, ———, AND J. C. MAWSON. 1977. The determination of habitat requirements for birds in suburban areas. U.S. For. Serv. Res. Pap. NE-3557. 15pp.

THOMSEN, L. 1971. Behavior and ecology of burrowing owls on the Oakland Municipal Airport. Condor 73:177–192.

TILGHMAN, N. G. 1987a. Characteristics of urban woodlands affecting breeding bird diversity and abundance. Landscape Urban Plann. 14:481–495.

———. 1987b. Characteristics of urban woodlands affecting winter bird diversity and abundance. For. Ecol. Manage. 21:163–175.

TIMM, R. M., EDITOR. 1983. Prevention and control of wildlife damage. Great Plains Agric. Counc. and Nebraska Coop. Ext. Serv., Lincoln. Var. pagin.

TOMSA, T. N. 1987. An investigation of the factors influencing the frequency and distribution of melanistic gray squirrels (*Sciurus carolinensis*). M.S. Thesis, State Univ. New York, Syracuse. 65pp.

TUFTS, C. 1988. The backyard naturalist. Natl. Wildl. Fed., Washington, D.C. 79pp.

TYLKA, D. L., J. M. SCHAEFER, AND L. W. ADAMS. 1987. Guidelines for implementing urban wildlife programs under state conservation agency administration. Pages 199–205 *in* L. W. Adams and D. L. Leedy, eds. Integrating man and nature in the metropolitan environment. Natl. Inst. Urban Wildl., Columbia, Md.

TZILKOWSKI, W. M., J. S. WAKELEY, AND L. J. MORRIS. 1986. Relative use of municipal street trees by birds during summer in State College, Pennsylvania. Urban Ecol. 9:387–398.

ULRICH, R. S. 1986. Human response to vegetation and landscapes. Landscape Urban Plann. 13:29–44.

U.S. DEPARTMENT OF THE INTERIOR. 1982. The 1980 national survey of fishing, hunting and wildlife-associated recreation. U.S. Fish Wildl. Serv. and Bur. Census, Washington, D.C. 156pp.

URESK, D. W., G. L. SCHENBECK, AND R. CEFKIN, EDITORS. 1988. Proc. Eighth Great Plains Wildlife Damage Control Workshop. U.S. For. Serv. Gen. Tech. Rep. RM-154. 231pp.

VanDRUFF, L. W. 1979. Urban wildlife—neglected resource. Pages 184–190 *in* R. D. Teague and E. Decker, eds. Wildlife conservation: principles and practices. The Wildl. Soc., Washington, D.C.

———, AND R. N. ROWSE. 1986. Habitat association of mammals in Syracuse, New York. Urban Ecol. 9:413–434.

VERNER, J., M. L. MORRISON, AND C. J. RALPH, EDITORS. 1986. Wildlife 2000: modeling habitat relationships of terrestrial vertebrates. Univ. Wisconsin Press, Madison. 470pp.

VINING, J., AND H. W. SCHROEDER. 1989. The effects of perceived conflict, resource scarcity, and information bias on emotions and environmental decisions. Environ. Manage. 13:199–206.

VOGEL, W. O. 1989. Response of deer to density and distribution of housing in Montana. Wildl. Soc. Bull. 17:406–413.

WEBER, W. J. 1979. Health hazards from pigeons, starlings, and English sparrows. Thomson Publ., Fresno, Calif. 138pp.

WEMMER, C., AND M. STÜWE. 1985. Reducing deer populations in large enclosures with drives. Wildl. Soc. Bull. 13:245–248.

WENGER, K. F., EDITOR. 1984. Forestry handbook. John Wiley & Sons, New York, N.Y. 1335pp.

WESEMANN, T., AND M. ROWE. 1987. Factors influencing the distribution and abundance of burrowing owls in Cape Coral, Florida. Pages 129–137 *in* L. W. Adams and D. L. Leedy, eds. Integrating man and nature in the metropolitan environment. Natl. Inst. Urban Wildl., Columbia, Md.

WILEY, M. B. 1986. Eastridge: residential development and the black-tailed gnatcatcher. Pages 77–84 *in* K. Stenberg and W. W. Shaw, eds. Wildlife conservation and new residential developments. Univ. Arizona School Renewable Nat. Resour., Tucson.

WILLIAMSON, R. D. 1983. Identification of urban habitat components which affect eastern gray squirrel abundance. Urban Ecol. 7:345–356.

WITHAM, J. H., AND J. M. JONES. 1987. Deer-human interactions and research in the Chicago metropolitan area. Pages 155–159 *in* L. W. Adams and D. L. Leedy, eds. Integrating man and nature in the metropolitan environment. Natl. Inst. Urban Wildl., Columbia, Md.

———, AND ———. 1990. White-tailed deer abundance on metropolitan forest preserves during winter in northeastern Illinois. Wildl. Soc. Bull. 18:13–16.

WITTER, D. J., D. L. TYLKA, AND J. E. WERNER. 1981. Values of urban wildlife in Missouri. Trans. North Am. Wildl. Nat. Resour. Conf. 46:424–431.

20

RESTORATION AND MANAGEMENT OF ENDANGERED SPECIES

J. Michael Scott, Stanley A. Temple, David L. Harlow, and Mark L. Shaffer

INTRODUCTION

Wildlife managers have been working to maintain or restore rare and declining wildlife populations in North America for nearly a century (Hornaday 1913). But, especially since the adoption of the Endangered Species Act of 1973 in the U.S.A. and other national, state, and provincial legislations modelled after it, more resources and public attention have been focused on species that are legally recognized as endangered or threatened (Bean 1986b, 1987).

A variety of techniques can be used effectively to manage and recover endangered species; some are identical to techniques used with more abundant species, but many others are specially adapted to the needs of rare species. Special approaches are needed because management of endangered species is complicated by their rarity, by legal restrictions intended to protect such species, and by the political and public relations scrutiny under which endangered species management is conducted.

MANDATES FOR ENDANGERED SPECIES CONSERVATION

Not until the passage of the Endangered Species Preservation Act of 1966 did legislation in the United States specifically address society's growing concerns over the increasing rate at which species were becoming extinct. The 1966 Act (80 Stat. 926) directed the Secretary of the Interior to develop a list of endangered species, to further the purposes of the Act, to encourage other agencies to consider protection of endangered species, and to acquire land for the protection of endangered species. The Endan-

gered Species Conservation Act of 1969 (83 Stat. 275) strengthened the 1966 Act by extending protection to invertebrates, developing a list of wildlife threatened with worldwide extinction, and prohibiting their importation, except for specific purposes.

The protection of endangered species continued to be a national priority in the U.S., yet legislation passed in the 1960s did little to reduce the number of species approaching extinction. Thus, the U.S. Congress passed the Endangered Species Act of 1973 (87 Stat. 884), which significantly increased the protective measures available for threatened and endangered species. The 1973 Act clearly mandated that all federal agencies participate in conserving endangered species. In addition, the Act broadened protection to include plant species, added the concept of critical habitat, and added the threatened category in hopes of providing protection to species before they became endangered.

Passage of the Endangered Species Act of 1973 established a major comprehensive program for protecting endangered species in the U.S. Protection of species threatened with extinction was no longer a matter of convenience for federal agencies; it was mandated by law, irrespective of the impact on the mission of a federal agency. The strength of the Act became abundantly clear when the Supreme Court determined, in Hill vs. TVA, that the Tellico Dam could not be completed because its completion would threaten the snail darter with extinction (Bean 1986b). The controversy over this decision led to a Congressional exemption for the Tellico Dam from the provisions of the Act and precipitated substantial revisions of the Act during subsequent amendments (Fitzgerald 1988).

The Endangered Species Act of 1973 also included provisions for increased protection of endangered and threatened species worldwide (Bean 1983). The Endangered Species Act implemented the Convention on International Trade in Endangered Species of Wild Fauna and Flora (CITES) in the U.S. and provided a program to encourage foreign governments to establish programs for the conservation of species (Bean 1986*b*). Species listed under CITES cannot be transported between countries without special permit.

Canada is signatory to the CITES. In addition, representatives of government agencies and conservation groups form the Committee on the Status of Endangered Wildlife in Canada (COSWEC). COSWEC has served to coordinate efforts that identify and manage species at risk since 1977. COSWEC maintains lists of threatened, endangered, vulnerable, extinct, and extirpated species and funds species status surveys and recovery efforts. In 1988, the Council of Canadian Wildlife Ministers signed an agreement establishing a new program called the Recovery of Nationally Endangered Wildlife (RENEW). The objectives of RENEW are to prevent the extinction of species, prevent additional species from becoming threatened or endangered, reintroduce extirpated species, establish recovery programs to remove species from threatened, endangered, or extirpated lists, and prepare recovery plans for all threatened and endangered species (Canadian Wildlife Service 1989).

Mexico is also signatory to the CITES. In addition, the U.S./Mexico Joint Committee on Wildlife Conservation was established in 1974 to help address the conservation of vulnerable species. This agreement provides the basis for cooperation on conservation efforts for species occurring in both nations, such as sharing survey information and cooperation on recovery efforts for species.

ENDANGERED SPECIES DEFINED

One of the most fundamental, yet controversial, aspects of endangered species programs is the determination of which species, or subspecies of populations of a species, should benefit from the special management attention mandated by endangered species legislation (Rojas 1992).

At the international level, registries of taxa considered to be at risk of extinction are maintained by the International Union for the Conservation of Nature and Natural Resources (IUCN). Species or subspecies are the taxonomic units that can be included in the IUCN's registries. The IUCN recognizes several levels of threat in its listings: Endangered, Threatened, Vulnerable, and Rare (International Union for the Conservation of Nature and Natural Resources 1990).

In the U.S., the Endangered Species Act defines ''species'' as ''any subspecies of fish or wildlife or plants, and any distinct population segment of any species of vertebrate fish or wildlife which interbreeds when mature.'' Thus, only subspecies of plants or invertebrates can be placed on the endangered species list provided that they meet the listing criteria; protection of populations is reserved for vertebrate species. What constitutes a population, however, is often defined with nonbiological criteria (O'Brian and Mayr 1991). For example, the bald eagle is listed in the contiguous 48 states but not in Alaska; inland populations of least terns are listed though they may be separated from coastal populations by a distance of <80 km; and gray wolves are listed as threatened in the lower U.S., but not in Alaska or Canada.

Within the Act's definition of a species, an endangered species is in danger of extinction throughout all or a significant portion of its range. A threatened species is likely to become an endangered species within the foreseeable future throughout all or a significant portion of its range.

The U.S. Fish and Wildlife Service also maintains lists of candidate species that are under review (U.S. Fish and Wildlife Service 1990). Three categories of candidates have been established. Category 1 includes species for which currently adequate information is available to support a proposal to formally list the species. Category 2 includes species that may deserve formal listing, but available information will not support a formal listing proposal. Category 3 includes species that may be proposed for removal from the list because the species may have become extinct, taxonomy has been revised such that a formally listed taxon is no longer recognized, or additional biological information has shown that the species does not warrant listing. The list of candidate species serves as an incentive to monitor the status of these taxa and alert conservationists to species of possible concern.

In addition to national lists, states and Canadian provinces maintain lists of endangered, threatened, or rare species within their borders. These lists may or may not include the same species listed by national governments, because the enabling legislation for each state or province usually has different criteria. For instance, many states or provinces list species that have declined within their borders, even though the species may not be nationally listed because it is common throughout most of its range. Some states list only certain taxa, such as vertebrates.

DESIGNATING A SPECIES AS ENDANGERED

Global listings of endangered species originated in the 1960s with the publication of IUCN's *Red Data Books,* the formal international registries of endangered species. In recent years the decisions about which species to include have been consolidated at IUCN's Conservation Data Centre in England (International Union for the Conservation of Nature and Natural Resources 1990). There, teams of taxonomic specialists compile information and make decisions. Listing of a species in a *Red Data Book* is primarily an informational action that facilitates the establishment of priorities for international conservation activities, but it carries no legal authority or mandate to conserve the listed species. This authority typically is vested in sovereign governmental bodies.

The U.S. Endangered Species Act mandates that species be listed solely on the basis of biological information, regardless of social, economic, or political impacts such listing may have. To be listed, a species must be at risk throughout all or a significant portion of its range. The proposed species may be formally listed if it is threatened by any one or a combination of the following factors: present or impending destruction or modification of its habitat; overutilization for commercial, recreational, scientific, or educational purposes; disease or predation; inadequacy of existing regulatory mechanisms; or other natural or human factors affecting its continued existence. Any proposal to remove a species from the list must ensure that none of these factors continues to threaten the species.

A species can be proposed for listing, downlisting (i.e., a status change from endangered to threatened), or delisting (removal from the endangered and threatened list) either by the rule-making process or by petition from any agency, group, or individual (U.S. General Accounting Office 1988). A petition must be responded to in a timely manner by the federal government. Should a change in listing status be warranted, a notice is published in the *Federal Register* proposing the change in listing and providing the justification for the proposal. Within 1 year of the publication of the proposal, the federal government must make a final determination as to the status of the species unless there is substantial disagreement among scientists regarding the accuracy of available information on the species.

IMPLICATIONS OF THE U.S. ENDANGERED SPECIES ACT

The U.S. Endangered Species Act has become one of the most influential legal authorities dealing with endangered species, and many legal activities in other national, state, or provincial governments are, in some way, modelled after it. For these reasons, we will address details of how the Act impacts wildlife management activities.

Implications for the Federal Government

The Endangered Species Act of 1973, as amended, mandates that federal agencies use their authorities to promote the conservation of federally listed species. Section 7 of the Endangered Species Act mandates that each federal agency shall ensure that any action undertaken, funded, or approved not jeopardize the continued existence of a listed species (Bean 1983). The Endangered Species Act also requires that federal agencies confer with the Department of Interior or Department of Commerce (for marine species) on any action that is likely to jeopardize the continued existence of a proposed endangered or threatened species (Yaffee 1988).

A federal agency must request a consultation with the appropriate department when it determines that its action may affect a listed species or jeopardize a proposed species. Often, if a listed species will be affected, minor modifications to the project may, if implemented, avoid impacts on the species in question. If an adverse impact seems unavoidable, formal consultation between the agency and the Fish and Wildlife Service or National Marine Fisheries Service is required. Such consultations result in a biological opinion that determines whether the action is likely to jeopardize the continued existence of a listed species. Opinions concluding that jeopardy to the species exists must offer reasonable and prudent alternatives to the action for consideration by the federal agency or the permit applicant.

The requirements placed on federal agencies by the Endangered Species Act are substantial. Every action they undertake is subject to analysis for impacts on listed species, even those actions that are ongoing.

Implications for States

The Endangered Species Act establishes that the federal government can enter into cooperative agreements with states to further the conservation of species threatened with extinction. To enter into such agreements, a state must develop active programs for the conservation of threatened and endangered species. The state must have the legislative authority to conserve threatened and endangered species, must develop a program that is consistent with the intent of the federal Act, must have the authority to conduct investigations on species, must have the authority to acquire lands for conservation of listed species, and must provide for public participation in the designation of state-listed species. Once an acceptable agreement is developed, the Endangered Species Act returns to the state the authority to regulate listed species, provided that such regulation is not less restrictive than provided by federal law and regulations. The state is also eligible for federal cost-sharing grants to support endangered species programs. Most states have taken advantage of the opportunity to develop their own endangered species programs by entering into cooperative agreements (Swimmer et al. 1992).

Implications for the Private Sector

The Endangered Species Act prohibits the "taking" of federally listed animals. Taking can include habitat destruction when such destruction actually kills or injures wildlife. This provision was completely prohibitive in the 1973 Act.

The 1982 amendments to the Endangered Species Act provide for the issuance of a permit to "take" a federally listed species (Fitzgerald 1988). These amendments provide the opportunity for private individuals, organizations, or local governments to obtain a permit to take listed species incidental to otherwise lawful activities (Murphy and Freas 1988). The federal government can issue such a permit when the applicant submits a "Conservation Plan" that specifies: (1) the impact that likely will result from such taking; (2) what steps the applicant will take to minimize and mitigate such impacts, and the funding that will be available to implement such steps; (3) what alternative actions to such taking the applicant considered, and the reasons why such alternatives are not being used; and (4) such measures that the federal government may require as being necessary or appropriate for purposes of the plan. After a public review period, the federal government can issue a permit authorizing the incidental take of listed species provided it determines that the Conservation Plan provides reasonable reductions in the threats to the species.

Although a permit for incidental take is required only for listed species, the federal government strongly encourages as well the consideration of candidate species. Congress specifically stated that applicants who have adequately considered candidate species in a Conservation Plan may not be required to provide additional investigation should those species be listed subsequently.

RECOVERY PLANNING

The Endangered Species Act directs the Fish and Wildlife Service and National Marine Fisheries Service to prepare and implement recovery plans for the conservation and survival of listed species (U.S. Fish and Wildlife Service 1979, Culbert and Blair 1989). Priority is to be placed on developing plans for those species most likely to benefit from having a plan, particularly those that may be in conflict with impending development (U.S. Fish and Wildlife Service 1990). In March 1993 there were 334 approved recovery plans for 401 of 758 listed species (U.S. Fish and Wildlife Service 1993). The recovery plan approach has been followed in many other countries and by states, provinces, and private conservation organizations (Clark et al. 1989, Clark and Gragun 1991). Recovery planning has become one of the central activities of the U.S. endangered species program (Culbert and Blair 1989).

Recovery teams are appointed to assist in the development of recovery plans. These teams typically comprise experts on the subject species from agencies or the private sector.

Recovery plans are advisory documents that include the best professional recommendations of what is required to recover the species of concern to a nonendangered status. The plans provide guidance to agencies, including budgetary and management advice. These plans are periodically updated when additional information on the species becomes available.

The first step in the recovery planning process is to consolidate all available information on the species and its status and determine what gaps in knowledge must be filled. Frequently, doing this will reveal that basic information on life history and ecology is lacking. Before any effective recovery or management program can be planned or implemented, the distribution, abundance, habitat requirements, and nature of threats to the species must be determined. The ecology of a species must be sufficiently understood to determine what factors limit the species' distribution and abundance. An example of a successful recovery planning effort, in which the limiting factor was properly identified, is the recovery plan for the American alligator. This species was being overharvested, but when protective measures outlined in the recovery plan were implemented, alligator populations recovered.

Having obtained all the relevant biological information, and, most importantly, identified factors responsible for the decline of the species, a recovery team must set goals for population size and distribution. When these goals are achieved the species is no longer in jeopardy and can be proposed for downlisting or delisting.

In setting a recovery goal for any threatened or endangered species, a population viability analysis (PVA) may be useful (Lehmkuhl 1984, Soulé 1987). PVA is a form of risk assessment that predicts the probability of a species (or population) becoming extinct under different scenarios of population size, distribution, and environmental change (Salwasser et al. 1984, Gilpin and Soulé 1986). Extinction is, in part, a chance phenomenon. Stochastic changes in genetics and demography can be devastating to small populations, and virtually no habitat or environment is constant from year to year. "Bad" years, "good" years, and devastating catastrophes can occur, even in protected areas. These natural, year-to-year environmental variations often occur through time in unpredictable, stochastic sequences. The smaller a population, the more easily it can

succumb to such chance events. The more isolated a population, the less likely immigrants from other areas may augment its numbers through a "rescue effect" (Brown and Kodric-Brown 1977). Populations or entire species may go extinct when reduced in size, even if they are protected and their habitats remain in a condition that would sustain them (Shaffer 1981, 1983).

The likelihood of such chance extinctions is directly related to the number, size, and distribution of populations of a species. Thus, it seems reasonable that one criterion to be used in recovery planning is to establish recovery goals that will minimize such chance extinctions over a suitably long time horizon. A PVA is the appropriate technique to employ in determining what pattern of population number, size, and distribution is necessary for any particular recovery effort. PVAs have become central components of recent recovery efforts, e.g., the northern spotted owl, and will become increasingly essential. However, PVA is still a relatively new technique, and there is no standard protocol or "cookbook" for undertaking such an assessment (Shaffer 1990). Developments in this new area are occurring rapidly. Several commercially produced computer programs are now available to recovery teams, e.g., RAMAS-AGE, RAMAS-SPACE, and VORTEX (Akcakaya 1992), and many wildlife biologists are producing their own custom-designed PVAs (Temple 1992). PVAs should soon become essential tools for recovery teams. Recovery plans that set goals on the basis of a comprehensive PVA will be an improvement over plans that set more arbitrary goals.

PROTECTING CRITICAL HABITAT

The purpose of the Endangered Species Act is to conserve species and their habitats. Although occasionally criticized for its perceived focus on individual species (Csuti et al. 1987, Hutto et al. 1987, Rohlf 1991), the Act is, in fact, unambiguous about the need to protect the ecosystems upon which endangered species depend. For most endangered species, protection and management of critical habitat provide the key first steps to improving the chances of long-term survival.

Designating critical habitat is a major responsibility of most recovery teams. Critical habitat is operationally defined as specific areas within or outside the geographic range of the species having features that are essential for the conservation of the species that may require special management or protection (Schreiner 1976). Critical habitat need not be presently occupied to receive protection if it is necessary for eventual recovery (Murphy and Rehm 1990).

It is uncommon for most endangered species to have had their habitat requirements defined specifically enough to guide a recovery effort (Murphy and Noon 1991). Usually critical habitat is characterized through a detailed analysis of potential limiting factors. A variety of approaches that work well for nonendangered species may be used (e.g., Habitat Suitability Index models or various multivariate habitat analyses). Regardless of the specific approach used, the careful definition of critical habitat is important, because it will be used to identify potential conflicts with the provisions of the Endangered Species Act and because protection and restoration of critical habitat often are key components of recovery.

Once defined, critical habitat for an endangered species often needs to be protected (U.S. Fish and Wildlife Service 1980). Federal and state governments and private conservation organizations, notably The Nature Conservancy (TNC), typically take the lead. One way is to purchase land and dedicate it specifically to conservation of an endangered species and its habitat. Another approach is through conservation easements, purchased at a fraction of the commercial value of the land, which protect specific features of the land (e.g., timber rights or water rights) without the need for ownership. Local zoning and land use planning also can play a role (Culbert 1989).

The Nature Conservancy has taken the lead in organizing information on the habitats of endangered species and where they are located. In collaboration with state governments and other organizations with land management objectives, TNC has established a network of Natural Heritage Inventory Programs and Conservation Data Centers (Griffen 1990, Stolzenburg 1992). These computerized databases are invaluable in determining where endangered species and their habitats are located, the land's ownership or management authority, and other useful information about the status of the habitat.

Even after remnant habitat areas are identified for protection, frequently it is necessary to plan for ecological restoration to remove habitat limitations in areas not presently suitable for the species (Howell 1988). In addition to occupying but a fraction of their historical range, many endangered species may occur in relict populations at the extremes of their former range in habitats that may be suboptimal. The whooping crane provides one example (Lewis 1986), as do many of the upland forest birds of Hawaii (Scott et al. 1986). Habitat restoration may be essential to allow an endangered species to expand its size and distributional characteristics to the point where it can be considered recovered.

When areas are set aside to protect critical habitat, managers should consider the potential impact of major habitat perturbations that occur as infrequently as every 100 years. Protected areas must be sufficiently large or a large enough number of smaller areas must be set aside and distributed in such a manner that all habitat is not vulnerable to the same perturbation. A spatially explicit PVA will help to determine what number, size, and distribution of habitat reserves will be adequate.

MANAGING ENDANGERED POPULATIONS

The challenges of managing an endangered population to recovery can best be appreciated by understanding implications of the "population bottleneck" (Fig. 1). A once-viable population declines in response to changes in the environment (the decline phase), lingers for a time at a nonviable population size (the endangered phase), and, perhaps, in response to skillful management, recovers its numbers (the recovery phase).

The essential challenges for an endangered species manager are to prevent the population from becoming too small during the decline phase, to make the endangered phase as brief as possible, and to make population growth during the recovery phase as rapid as possible. These are important objectives for several reasons (Nei et al. 1975). The smaller the population becomes and the longer it remains small in size, the greater will be the loss of genetic

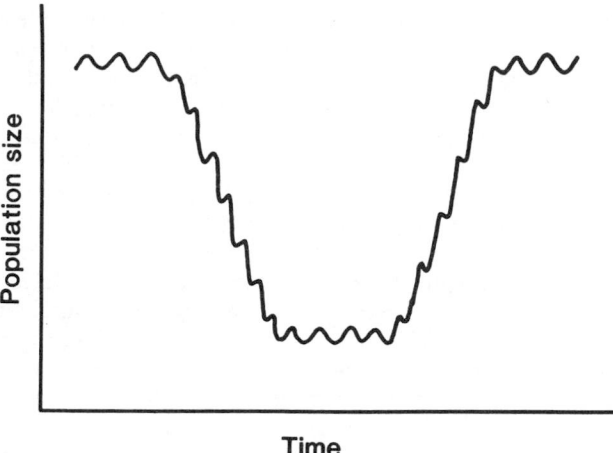

Fig. 1. The population bottleneck.

variation (Nei et al. 1975) and the risks of stochastic demographic or environmental events (Goodman 1987, Shaffer 1987). The more rapidly a population recovers, the sooner the risk of chance extinction can be reduced and the greater the portion of the genetic variation in the endangered-phase population that will be retained.

Ultimate and Proximate Problems

The importance of shortening the endangered phase and accelerating the recovery phase has resulted in a special style of management for some endangered populations (Temple 1977*b*, 1986). The ultimate cause of a species' decline to an endangered status is almost always some extraordinary change in the environment. These changes might be novel events that the species has never experienced before (e.g., the introduction of synthetic toxic chemicals, like DDT, into a species' food chain) or somewhat normal events that suddenly occur in temporal or spatial patterns that are unprecedented (e.g., disturbing a forest through logging more frequently and on a different spatial scale than would occur under natural disturbance regimes, thus eliminating old-growth conditions). Regardless of the ultimate cause, the population fails to cope, and the proximate causes of its decline become reductions in survival and reproduction for which the population cannot compensate through density-dependent responses.

The distinction between ultimate and proximate causes of endangerment can lead to a dichotomy in management approaches (Temple 1977*b*). Clearly, endangered species management must correct the ultimate causes of endangerment by addressing the specific environmental problems that caused the species to decline. These ultimate causes of endangerment all go back to population size and include such widespread problems as habitat loss or deterioration, impacts of exotic organisms, toxic chemicals, and mortality directly related to human activities. Frequently, these ultimate problems are difficult to remedy, and typically a long time is required for corrective measures to improve a species' environment. Time is, of course, a factor working against the survival of most endangered species, particularly if they already have entered the endangered phase of a population bottleneck.

To hasten the recovery of a population, managers often must address proximate factors in a manner that is somewhat unrelated to the ultimate cause of endangerment. Typically, the approach is to increase survival and reproductive rates by whatever means possible to stimulate population growth. Ultimate problems are then addressed somewhat independently of these activities, which often are likened to the treatment of a critically ill patient in a hospital's emergency room: treat the life-threatening symptoms and then remedy underlying problems (Zimmerman 1975).

Two endangered birds, the osprey and peregrine falcon, illustrate well the benefits of addressing proximate problems (Spitzer 1977, Cade et al. 1988). Both species were threatened by the ultimate problem of contamination of their food chains with chlorinated hydrocarbons, especially DDT. In each situation populations declined precipitously after reproduction was impaired when eggs with DDT-thinned shells failed to hatch. Although measures were taken during the late 1960s and early 1970s to address the ultimate problem, by curtailing the use of the offending toxic chemicals, the birds continued to decline and remained at risk because of the persistence of the chemicals already in the environment.

To offset the proximate problem of inadequate recruitment, managers successfully augmented reproduction in wild ospreys and released captive-reared individuals into wild populations of peregrine falcons. Reproduction of ospreys in the wild was enhanced by providing birds with artificial nest platforms on which the birds enjoyed increased success; by reducing nest-site limitations, through erecting artificial nest platforms in areas that lacked natural nesting sites (Postupalsky 1977); and by moving eggs from a population with thick-shelled eggs to the nests of ospreys that were laying thin-shelled eggs (Spitzer 1977). Captive breeding and subsequent reintroduction of peregrine falcons provided an important boost to recruitment to populations that were producing too few young (Cade et al. 1988). In both situations, regional populations were doomed to eventual extinction unless attention to proximate causes improved the status of the species while ultimate problems were being addressed. In the post-DDT era, both ospreys and peregrine falcons recovered impressively after chlorinated hydrocarbon pesticides eventually were purged from the food chain (Spitzer et al. 1983, Cade et al. 1988).

Improving Recruitment

Rapid population growth sometimes can be stimulated by increasing the rate at which new individuals are added to a population through reproduction or immigration. This growth can be particularly dramatic in species that normally reproduce at a slow rate (e.g., K-selected species). For many species, manipulating aspects of reproduction can result in: (1) an increase in natality (i.e., litter size or clutch size); (2) an increase in the frequency of reproduction; (3) an increase in the proportion of individuals in a population that breed; or (4) an increase in the success rate of breeders. For other species, translocations of animals from one population to another and reintroductions of individuals produced in captivity can duplicate the positive contributions of natural immigration toward recruitment in a declining population.

Increasing natality has been a particularly attractive strategy for endangered birds, because many species can be stimulated to lay supernumerary eggs when managers remove eggs from the nest. By artificially incubating the extra eggs or placing them in the nests of foster parents, viability can be multiplied impressively (Cade 1977).

A few species present the opportunity to increase natality. Animals that have long periods of parental care of offspring are inhibited from reproducing while caring for dependent young. Shortening the time to independence can reduce the intervals between breeding. For example, California condors can be induced to breed annually in contrast to the normal biennial rate, thus effectively doubling fecundity (Snyder and Hamber 1985).

In some populations only a proportion of the reproductively competent individuals have the opportunity to breed. Shortages of specific resources essential for breeding typically are responsible for limiting the number of breeders in a population. If these limitations to breeding can be removed, previously excluded individuals can breed. Nesting sites frequently limit the number of breeders, and the provision of artificial nesting sites, such as nest boxes, cavities, and burrows, can result in improved fecundity (Snyder 1977).

Reproductive failures normally reduce recruitment below the potential rate suggested by a species' natality. Losses of young before recruitment into the population sometimes can be substantial. Preventing those losses effectively increases recruitment. Many techniques have been used, including improving the quality of breeding sites, reducing risks (e.g., predation) to young animals, or implementing "head starting" approaches that give young animals improved prospects for survival to breeding age (Temple 1977a).

Translocating individuals from wild populations with positive rates of growth or from captive populations is an effective way to augment a declining population or stimulate growth in a stable but endangered population (Cade 1986, Griffith et al. 1989, Jones 1990). When translocations are undertaken, the choice of a source population can be important. In general, nearby source populations or source populations from areas with similar ecological conditions are likely to have desirable genetic similarities with the recipient population (International Union for the Conservation of Nature and Natural Resources 1987). Source populations also should be selected carefully to avoid translocating pathogens that may be absent in the recipient population (Dobson and Miller 1989). Failure to consider these types of issues can result in translocated individuals having a negative rather than positive impact on the recipient population (Scott and Carpenter 1987).

Enhancing the survival of translocated individuals can improve the effectiveness of this technique. A variety of strategies has been developed, especially to improve the likelihood of a successful transition from captivity to the wild when captive-produced individuals are reintroduced (Chivers 1991, Gipps 1991).

Improving Survival

All else being equal, improved survival of individuals should stimulate population growth. Reducing impacts of predators, pathogens, competitors, and accidents can be effective (Jackson 1977). These manipulations may reduce losses below normally expected levels and enhance survival in a population, even if poor survival is not an important factor in the population's endangerment. For example, protecting an endangered species from human exploitation can improve survival, even if overexploitation is not a cause of endangerment. Improving the quality of a species' habitat also can improve survival.

Avoiding Genetic Problems

Although genetic considerations often are not important in management strategies for nonendangered wildlife, they can be crucial to the success of attempts to recover some endangered species (Schoenwald-Cox et al. 1983, Ballou 1992). Very small populations, as exist for some rare species, can have their continued existence jeopardized by genetic problems that include inbreeding depression and loss of heterozygosity. In most large, panmictic populations, inbreeding rates are low, but, as the size of a population declines, inbreeding rates rise. If deleterious alleles are present in the endangered population, close inbreeding can result in substantial inbreeding depression (i.e., reductions in survival and reproductive performance in inbred individuals). The easiest way to avoid these problems is to keep populations large enough that inbreeding is not likely to be an issue (Soulé 1987). A variety of techniques is available to estimate the effective population size for endangered animals (Reed et al. 1986). If the effective population size is small, it is increasingly important to understand the pedigree of the population and employ strategies to maximize outbreeding or minimize contributions of individuals known to carry deleterious alleles (Templeton 1990).

Losses of heterozygosity in endangered species can result from "founder effects" and from "genetic drifts" (Schoenwald-Cox et al. 1983). When a population passes through a severe bottleneck, the survivors in the endangered phase may not possess a completely representative sample of the genetic diversity in the larger, pre-bottleneck population (Denniston 1977). To prevent losses from this founder effect, managers should keep declining populations from reaching low size during the endangered phase by beginning to manage the population as early in the decline phase as possible.

If a population already has become very small, losses of heterozygosity can result from genetic drift—changes in allele frequencies that are inevitable in small populations with few births per generation. To prevent extreme drift, managers should shorten the endangered phase and accelerate the recovery phase. These activities improve the chances for individuals in the endangered phase to leave behind the best possible sample of the population's genetic diversity.

Loss of genetic diversity, especially the worst cases that result in homozygosity, can seriously compromise the long-term prospects of a population (Frankel and Soulé 1981). Evolutionary change depends on genetic variation upon which natural selection can operate. Post-bottleneck populations that have lost heterozygosity have reduced potential for evolutionary response in a changing environment.

Genetic management of endangered species may be critical in very small populations that can be effectively manipulated, especially captive populations where pedigrees are known and mating schedules can be planned. But even larger populations may benefit from genetic management if they are subdivided into isolated subpopulations. In such metapopulations, managers might consider forcing gene flow between subpopulations by translocating individuals (Temple 1991).

PERMIT REQUIREMENTS FOR ENDANGERED SPECIES PROGRAMS

When a species is listed as endangered, permits are required to undertake all types of "hands-on" management or any activity that may disturb the species. Permits can be granted for scientific research or management activities that enhance the welfare of the species.

Applications for permits to manipulate a listed species must be submitted to the Fish and Wildlife Service or the National Marine Fisheries Service. Application forms are available from national and regional offices of these agencies. Once a permit application has been received and accepted as providing the necessary information, a notice is published in the *Federal Register* to allow for public review and comment on the application. Once the review is completed, a permit is issued if it is determined that the activity is not likely to jeopardize the continued existence of the species.

Scientific permits generally are issued for research or management that will benefit the recovery effort for the species, or for research related to the assessment of threats to the species. In general, permits are not likely to be granted for research that can be accomplished on closely related or surrogate species.

The Endangered Species Act of 1973 authorizes permits for the introduction of endangered or threatened species into areas outside their current range when such introductions will further the conservation of that species. These introduced populations are called "experimental" populations. They are classified as threatened and treated as such, regardless of the status of the species. An experimental population can be designated as nonessential. In such situations, the experimental population is treated as a candidate species, thus receiving only limited protection under the Endangered Species Act.

PUBLIC SCRUTINY OF ENDANGERED SPECIES MANAGEMENT

Although all wildlife management activities are subject to public review, endangered species conservation and management programs can be especially contentious, and endangered species managers must be prepared for the biological, social, economic, and political controversies that seem inevitably to surround most activities (e.g., Snyder and Snyder 1989, Liverman 1990).

Although the recovery process prescribed by the Endangered Species Act specifically avoids issues other than biological ones, endangered species managers frequently are asked to defend their biological proposals in arenas where the challenges are nonbiological (Tilt 1989). Designations of critical habitat and the attendant need to protect or restore specific areas for the benefit of endangered species frequently have social and economic impacts that precipitate acrimonious debates, e.g., the northern spotted owl versus the logging industry in the Pacific Northwest (Doak 1989, Thomas et al. 1990).

SOURCES OF CURRENT INFORMATION

Perhaps the single, best source of information on management of an endangered species will be its recovery plan. These can be obtained from the Fish and Wildlife Reference Service, 5340 Grosvenor Lane, Suite 110, Bethesda, MD 20814.

Two important periodicals, *The Endangered Species Update* and *The Endangered Species Technical Bulletin,* provide the most up-to-date information on recovery efforts for endangered species. They can be obtained from the School of Natural Resources, The University of Michigan, Ann Arbor, MI 48109. The federal list of threatened and endangered species is updated annually and published in the *Federal Register.* Copies may be obtained from offices of the U.S. Fish and Wildlife Service or National Marine Fisheries Service.

LITERATURE CITED

AKCAKAYA, H. R. 1992. Population viability analysis and risk assessment. Pages 148–157 *in* D. R. McCullough and R. H. Barrett, eds. Wildlife 2001: populations. Elsevier Appl. Sci., London, U.K.

BALLOU, J. D. 1992. Genetic and demographic considerations in endangered species captive breeding and reintroduction programs. Pages 262–278 *in* D. R. McCullough and R. H. Barrett, eds. Wildlife 2001: populations. Elsevier Appl. Sci., London, U.K.

BEAN, M. J. 1983. The evolution of national wildlife law. Praeger Publ., New York, N.Y. 449pp.

——. 1986a. International wildlife conservation. Pages 543–578 *in* R. DiSilvestro, ed. Audubon wildlife report 1986. Natl. Audubon Soc., New York, N.Y.

——. 1986b. The endangered species program. Pages 347–371 *in* R. L. DiSilvestro, ed. Audubon wildlife report 1986. Natl. Audubon Soc., New York, N.Y.

——. 1987. The federal endangered species program. Pages 147–160 *in* R. L. DiSilvestro, ed. Audubon wildlife report 1987. Natl. Audubon Soc., New York, N.Y.

BROWN, J. H., AND A. KODRIC-BROWN. 1977. Turnover rates in insular biogeography: effect of immigration on extinction. Ecology 58: 445–449.

CADE, T. J. 1977. Manipulating the nesting biology of endangered birds: a review. Pages 167–170 *in* S. A. Temple, ed. Endangered birds: management techniques for preserving threatened species. Univ. Wisconsin Press, Madison.

——. 1986. Reintroduction as a method of conservation. Raptor Res. 15:72–84.

——, J. ENDERSON, C. THELANDER, AND C. WHITE, EDITORS. 1988. Peregrine falcon populations: their management and recovery. Peregrine Fund, Boise, Id. 949pp.

CANADIAN WILDLIFE SERVICE. 1989. RENEW annual report. Can. Nature Fed, Ottawa, Ont. 15pp.

CHIVERS, D. J. 1991. Guidelines for reintroductions: procedures and problems. Symp. Zool. Soc. Lond. 62:89–99.

CLARK, T., AND J. GRAGUN. 1991. Organization and management of endangered species programs. Endangered Species Update 8:1–4.

——, R. GRETE, AND J. CADA. 1989. Designing and managing successful endangered species recovery programs. Environ. Manage. 13:159–170.

CSUTI, B. A., J. M. SCOTT, AND J. ESTES. 1987. Looking beyond species-oriented conservation. Endangered Species Update 5:4.

CULBERT, R. 1989. Local planning and biological diversity. Endangered Species Update 6:6.

——, AND R. BLAIR, EDITORS. 1989. Recovery planning. Endangered Species Update 6:1–41.

DENNISTON, C. 1977. Small population size and genetic diversity and implications for small populations. Pages 281–289 *in* S. A. Temple, ed. Endangered birds: management techniques for preserving threatened species. Univ. Wisconsin Press, Madison.

DOAK, D. 1989. Spotted owls and old growth logging in the Pacific Northwest. Conserv. Biol. 3:389–396.

DOBSON, D., AND D. MILLER. 1989. Infectious diseases and endangered species management. Endangered Species Update 6:1–4.

FITZGERALD, J. M. 1988. Withering wildlife: wither the Endangered Species Act? A review of amendments to the act. Endangered Species Update 5:27–35.

FRANKEL, O. H., AND M. SOULÉ. 1981. Conservation and evolution. Cambridge Univ. Press, Cambridge, U.K. 327pp.

GILPIN, M. E., AND M. E. SOULÉ. 1986. Minimum viable populations: processes of species extinction. Pages 19–34 *in* M. E. Soulé, ed. Conservation biology. Sinauer Assoc., Sunderland, Mass.

GIPPS, J., EDITOR. 1991. Beyond capture breeding: reintroducing endangered species to the wild. Clarendon Press, Oxford, U.K. 244pp.

GOODMAN, D. 1987. The demography of chance extinction. Pages 11–35 *in* M. Soulé, ed. Viable populations for conservation. Cambridge Univ. Press, Cambridge, U.K.

GRIFFEN, J. 1990. The Nature Conservancy and the Heritage Programs: working together to preserve biodiversity. Endangered Species Update 15:3–5.

GRIFFITH, B., J. SCOTT, J. CARPENTER, AND C. REED. 1989. Translocation as a species conservation tool: status and strategy. Science 245:477–480.

HORNADAY, W. T. 1913. Our vanishing wildlife: its extermination and preservation. New York Zool. Soc., New York, N.Y. 428pp.

HOWELL, E. 1988. The role of restoration in conservation biology. Endangered Species Update 5:1–4.

HUTTO, R., S. REEL, AND P. LANDRES. 1987. A critical evaluation of the species approach to biological conservation. Endangered Species Update 4:1–4.

INTERNATIONAL UNION FOR THE CONSERVATION OF NATURE AND NATURAL RESOURCES. 1987. The IUCN position of translocations of living organisms. Int. Union Conserv. Nat. Nat. Resour., Gland, Switzerland. 20pp.

——. 1990. The 1990 IUCN red list of threatened animals. Int. Union Conserv. Nat. Nat. Resour., Cambridge, U.K. 192pp.

JACKSON, J. 1977. Alleviating problems of competition, predation, parasitism and disease in endangered species. Pages 75–89 *in* S. A. Temple, ed. Endangered birds: management techniques for preserving threatened species. Univ. Wisconsin Press, Madison.

JONES, S. R., EDITOR. 1990. Captive propagation and reintroduction: a strategy for preserving endangered species. Endangered Species Update 8:1–89.

LEHMKUHL, J. F. 1984. Determining size and dispersion of minimum viable populations for land management planning and species conservation. Environ. Manage. 8:167–176.

LEWIS, J. C. 1986. The whooping crane. Pages 659–676 *in* R. L. DiSilvestro, ed. Audubon wildlife report 1986. Natl. Audubon Soc., New York, N.Y.

LIVERMAN, M. C. 1990. The (endangered) Endangered Species Act: political economy of the northern spotted owl. Endangered Species Update 7:1–4.

MURPHY, D., AND B. NOON. 1991. Exercising ambiguity from the Endangered Species Act: critical habitat as an example. Endangered Species Update 8:6.

——, AND K. REHM. 1990. Unoccupied habitats and endangered species protection. Endangered Species Update 7:10.

MURPHY, D. D., AND K. E. FREAS. 1988. Using the Endangered Species Act to resolve conflict between habitat protection and resource development. Endangered Species Update 5:6.

NEI, M., T. MARAYAMA, AND R. CHAKRABORTY. 1975. The bottleneck effect and genetic variability on populations. Evolution 29:1–10.

O'BRIAN, S. J., AND E. MAYR. 1991. Bureaucratic mischief: recognizing endangered species and subspecies. Science 251:1187–1188.

POSTUPALSKY, S. 1977. Artificial nesting platforms for ospreys and bald eagles. Pages 35–45 *in* S. A. Temple, ed. Endangered birds: management techniques for preserving threatened species. Univ. Wisconsin Press, Madison.

REED, J. M., P. D. DOERR, AND J. R. WALTERS. 1986. Determining minimum population sizes for birds and mammals. Wildl. Soc. Bull. 14:255–261.

ROHLF, D. J. 1991. Six reasons why the Endangered Species Act doesn't work and what to do about it. Conserv. Biol. 5:273–282.

ROJAS, J. 1992. The species problem and conservation: what are we protecting? Conserv. Biol. 6:170–178.

SALWASSER, H., S. P. MEALEY, AND K. JOHNSON. 1984. Wildlife population viability: a question of risk. Trans. North Am. Wildl. Nat. Resour. Conf. 49:421–439.

SCHOENWALD-COX, C., S. CHAMBERS, B. MACBRYDE, AND L. THOMAS. 1983. Genetics and conservation: a reference for managing wild animal and plant populations. Benjamin-Cummings Publ., Menlo Park, Calif. 722pp.

SCHREINER, K. M. 1976. Critical habitat: what it is and is not. Endangered Species Tech. Bull. 1:1–4.

SCOTT, J. M., AND J. W. CARPENTER. 1987. Release of captive-reared or translocated endangered birds: what do we need to know? Auk 104:544–545.

———, S. MOUNTAINSPRING, F. RAMSEY, AND C. KEPLER. 1986. Forest bird communities of the Hawaiian Islands: their dynamics, ecology, and conservation. Stud. Avian Biol. 9. 431pp.

SHAFFER, M. L. 1981. Minimum population sizes for species conservation. BioScience 31:131–134.

———. 1983. Determining minimum viable population sizes for the grizzly bear. Int. Conf. Bear Res. Manage. 5:133–139.

———. 1987. Minimum viable populations: coping with uncertainty. Pages 69–87 *in* M. Soulé, ed. Viable populations for conservation. Cambridge Univ. Press, Cambridge, U.K.

———. 1990. Population viability analysis. Conserv. Biol. 4:39–40.

SNYDER, N. F. R. 1977. Increasing reproductive effort and success by reducing nest-site limitations. Pages 27–35 *in* S. A. Temple, ed. Endangered birds: management techniques for preserving threatened species. Univ. Wisconsin Press, Madison.

———, AND J. A. HAMBER. 1985. Replacement-clutching and annual nesting of California condors. Condor 87:374–378.

———, AND H. A. SNYDER. 1989. Biology and conservation of the California condor. Curr. Ornithol. 6:175–263.

SOULÉ, M., EDITOR. 1987. Viable populations for conservation. Cambridge Univ. Press, Cambridge, U.K. 189pp.

SPITZER, P. 1977. Osprey egg and nestling transfers: their value as ecological experiments and as management procedures. Pages 171–187 *in* S. A. Temple, ed. Endangered birds: management techniques for preserving threatened species. Univ. Wisconsin Press, Madison.

———, A. POOLE, AND M. SCHEIBEL. 1983. Initial population recovery of breeding ospreys in the region between New York City and Boston. Pages 231–241 *in* D. Bird, ed. Biology and management of bald eagles and ospreys. Harpell Press, St. Anne de Bellevue, Que.

STOLZENBURG, W. 1992. The heritage network: detectives of diversity. Nat. Conserv. 1992:23–27.

SWIMMER, J. Y., L. MANOR, AND R. L. GOOCH. 1992. Endangered species programs in the 50 states and Puerto Rico. Endangered Species Update 10:8–10.

TEMPLE, S. A. 1977*a*. Endangered birds: management techniques for preserving threatened species. Univ. Wisconsin Press, Madison. 466pp.

———. 1977*b*. The concept of managing endangered birds. Pages 3–8 *in* S. A. Temple, ed. Endangered birds: management techniques for preserving threatened species. Univ. Wisconsin Press, Madison.

———. 1986. The problem of avian extinctions. Curr. Ornithol. 6:453–485.

———. 1991. The role of dispersal in the maintenance of bird populations in a fragmented landscape. Acta Congr. Int. Ornithol. 20: 2298–2305.

———. 1992. Population viability analysis of a sharp-tailed grouse metapopulation in Wisconsin. Pages 730–758 *in* D. R. McCullough and R. H. Barrett, eds. Wildlife 2001: populations. Elsevier Appl. Sci., London, U.K.

TEMPLETON, A. R. 1990. The role of genetics in captive breeding and reintroduction for species conservation. Endangered Species Update 8:14–17.

THOMAS, J. W., E. D. FORSMAN, J. B. LINT, E. C. MESLOW, B. R. NOON, AND J. VERNER. 1990. A conservation strategy for the northern spotted owl. Interagency Scientific Committtee to Address the Conservation of the Northern Spotted Owl, Portland, Oreg. 427pp.

TILT, W. 1989. The biopolitics of endangered species. Endangered Species Update 6:35–40.

U.S. FISH AND WILDLIFE SERVICE. 1979. Service sets guidelines for recovery planning. Endangered Species Tech. Bull. 4:1–7.

———. 1980. Habitat acquisition: costly but necessary to the recovery of many endangered species. Endangered Species Tech. Bull. 5:5–10.

———. 1990. Report to Congress: endangered and threatened species recovery program. U.S. Dep. Inter., Washington, D.C. 406pp.

———. 1993. Box score: listings and recovery plans. Endangered Species Tech. Bull. 17:20.

U.S. GENERAL ACCOUNTING OFFICE. 1988. Endangered species: management improvements could enhance recovery program. U.S. Gen. Accounting Off., Washington, D.C. 100pp.

WOODFORD, M., AND R. KOCK. 1991. Veterinary considerations in reintroduction and translocation projects. Symp. Zool. Soc. Lond. 62: 101–110.

YAFFEE, S. L. 1988. Protecting endangered species through interagency consultation. Endangered Species Update 5:10–19.

ZIMMERMAN, D. 1975. To save a bird in peril. Coward, McCann and Geoghegan, New York, N.Y. 286pp.

21

GEOGRAPHIC INFORMATION SYSTEMS

Gregory T. Koeln, Lewis M. Cowardin, and Laurence L. Strong

INTRODUCTION

Geographic Information Systems (GISs) are a relatively new development in computer technology of particular interest to wildlife and natural resource managers (Peterson and Matney 1986). A GIS consists of software, hardware, and personnel for performing spatial analysis. Many of the things that wildlife managers do with maps, such as calculate areas, measure distances, and calculate amount of edge, now can be automated with GIS technology. The concept of using GISs for resource management can be credited to Ian McHarg. In *Design With Nature,* McHarg (1969) manually overlaid a series of maps on geology, physiography, major aquifers, soils, forest types, wildlife distributions, unique sites, slopes, and other characteristics of the land to generate maps depicting areas most suitable for agriculture, for urban expansion, and for mining, and various other maps depicting land suitability. GISs automate the making of these types of maps and the combining of many maps to create new information.

A study of 61 fish and wildlife agencies in the U.S. revealed that GIS use is widespread and growing rapidly (Rodcay 1991). Thirty-nine percent of the agencies used GISs regularly in their programs, 30.2% used GISs rarely, and 30.5% were not using GISs. Seventy-three percent of those agencies not using GISs plan to use them in the near future. For those agencies using GISs, habitat mapping was the most common use. Other applications included land use inventory, vegetation mapping, species distribution estimation, preferred habitat definition, and land development planning.

Wildlife managers have used GISs for monitoring wetlands for waterfowl habitat (Barnard et al. 1981, Koeln et al. 1988), for mapping Florida scrub jay habitat (Breininger et al. 1991), for evaluating grizzly bear (Craighead et al. 1986, Agee et al. 1989), lesser prairie-chicken (Cannon et al. 1982), and elk habitat (Leckenby et al. 1985), for preserving biological diversity (Davis et al. 1990), for monitoring wood stork foraging habitat (Hodgson et al. 1988), for analyzing radiotelemetry data (Koeln and Cook 1984, Young et al. 1987), for characterizing the spatial structure of habitats (Heinem and Mead 1984, Ripple et al. 1991), for characterizing ecotones (Johnston and Bonde 1989), for predicting wildlife densities (Palmerim 1988, Broschart et al. 1989), for modeling the spatial distribution of species (Palmerim 1987, Walker 1990), for designing reserve systems (Saxon and Dudzinski 1984, Murphy and Noon 1991), for examining the cumulative impacts of habitat loss (Johnston et al. 1988, Gosselink and Lee 1989), for quantifying beaver pond creation (Johnston and Naiman 1990*a,b*), and in many other ways (de Steiguer and Giles 1981, Steenhof 1982, Lyon 1983, Mayer 1984, Peterson and Matney 1986, Ormsby and Lunetta 1987, Scepan et al. 1987, Stenback et al. 1987, Miller and Conroy 1990, Shaw and Atkinson 1990).

The intent of this chapter is to provide wildlife managers and wildlife management students an overview of the technology of GISs. Many universities offer graduate and undergraduate courses in GISs. Various books, journals, and other publications are listed in this chapter for those wanting to further explore the use of GISs in wildlife management and research.

WHAT IS A GIS?

Aronoff (1989) described a GIS as any manual or computer-based set of procedures to store and manipulate geographically referenced data. Geographically referenced data (spatial data) are any data that can be represented on a map as a point, line, or area (polygon) (Fig. 1). Storing and manipulating geographically referenced data quickly become too cumbersome by use of manual procedures. Consequently, Aronoff (1989:39) defined a GIS as "a computer-based system that provides the following four sets of capabilities to handle georeferenced data: 1) input; 2) data management (data storage and retrieval); 3) manipulation and analysis; and 4) output."

Dueker and Kjerne (1989:8–9) defined a GIS as "a system of hardware, software, data, people, organizations, and institutional arrangements for collecting, storing, analyzing, and disseminating information about areas of the earth." GISs differ from general database management systems (DBMS). DBMS do not handle adequately the spatial data requirements of GISs. Spatial data have two components: a geographic reference, and an attribute. A road in a GIS may be represented by a line. The geographic reference of that line would be the coordinates describing its location. The location of the road (line) can be recorded with Universal Transverse Mercator (UTM) coordinates, state plane coordinates, latitude and longitude, or other (including arbitrary) coordinate systems. The attribute component of the spatial data describing the road may include the type of road (gravel), the route number (Interstate 44), or other attributes of the road, such as the average number of cars that use the road in a year. DBMSs often can manage well the attribute component of spatial data, but poorly manage the geographic reference.

A GIS frequently is described in terms of hardware and software, but it should be thought of as a general system with inputs, processes, outputs, and a context. The input component is the most expensive. Capturing, registering, interpreting, and converting spatial data frequently comprise 60–90% of the expense of operating a GIS. Processes of the GIS include efficient and effective means of storing and retrieving both the attribute and geographic reference of spatial data and creating new information derived from spatial data stored in the system. This new information includes such things as the distance to the nearest stream or the size of continuous blocks of forestland.

The outputs of the GIS include hard-copy maps, graphic displays on color or monochrome monitors, and tabular information. Technological advances are vastly improving output capabilities of GISs.

The context of the system includes the organizational and institutional components of the GIS, e.g., staff, funds, and administrative support. Administering the organizational and institutional components of the GIS is frequently much more difficult than selecting, learning, and using the GIS software and hardware (Lauer et al. 1991).

Many types of systems frequently are confused with GISs (Korte 1991). Computer-aided mapping (CAM) systems automate the design, creation, and maintenance of maps. These systems usually are enhancements to computer-aided drafting (CAD) software and provide powerful tools for making and updating maps. CAM/CAD systems handle well the geographic reference of spatial data, but often poorly handle the attribute components of spatial data. The analytical capabilities of CAM/CAD systems

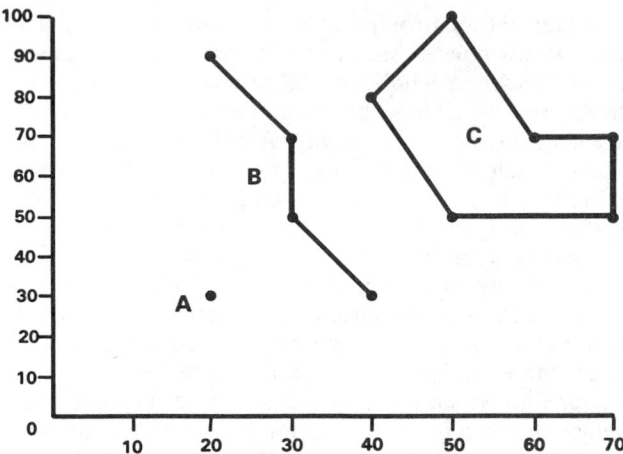

Fig. 1. Geographically referenced data (spatial data) are any data that can be represented as a point (A), line (B), or area (C) (Aronoff 1989).

are not the same as those of a GIS. GISs usually do not perform as well as CAM/CAD for purely cartographic application, but the cartographic capabilities of modern GISs are improving. In addition, many GISs now can use and manipulate data created from CAM/CAD systems.

Automated mapping and facilities management (AM/FM) is the use of CAM or GISs for public works and utility information (Dueker and Kjerne 1989, Vonderohe et al. 1991). Information on telephone lines, electrical lines, water lines, sewers, and other utilities often are managed with AM/FM systems. The AM/FM allows linking attribute data to spatial data, but, like CAM, spatial relationships are not defined and spatial analysis is slow and cumbersome at best.

Cadastral systems are used to manage quantity, value, and ownership of real estate. Multipurpose cadastral systems are parcel-based land information systems (Dueker and Kjerne 1989, Vonderohe et al. 1991). These parcels could be sections of the public land survey, counties, or wildlife management units, for example.

Often, the application of GISs is termed Land Information Systems (LISs). Dueker and Kjerne (1989) described LISs and GISs as containing data primarily describing land records. Vonderohe et al. (1991) described the process of maintaining records on the land as an LIS, which does not necessarily require the use of computers. But once this process is automated by computer, it is a GIS.

The five basic questions that a GIS can be used to answer were described by Walker and Miller (1990) as:

(1) What exists at a particular site or location?
(2) Where are certain conditions met?
(3) What changes have occurred over time and where have these changes occurred?
(4) What are the social, economic, or environmental impacts of a particular change in the use of land?
(5) What will happen if the existing land use for a particular site is altered to another type of use?

The first question is one of the simplest functions of a GIS. The location can be described in many ways and can be defined (1) as a point, line, or area (polygon), (2) by place name (i.e., street address, city, county, or wildlife management area) or post or ZIP code, or (3) by geographic coordinates, such as UTM, state plane, or latitude and longitude. A wildlife manager may use this capability to describe the habitat occurring in a wildlife management area, a study area, or a county. A wildlife researcher, using radiotelemetry techniques, may use a GIS to determine various kinds of information about sites used by the species being studied. For each radiotelemetry location, the habitat type occurring at the location can be ascertained. Many other habitat parameters also can be obtained, such as the distance to the nearest road, stream, or forest edge; the size of a continuous block of habitat being used; the elevation, slope, and aspect of the location; or the area of various habitat types located within a determined distance from the location of the studied animal.

The second question that a GIS can answer is the converse of the first question. Instead of asking what occurs at a particular site, this question asks where do certain situations or conditions occur. The wildlife managers may want to know the location of all red-cockaded woodpecker nesting colonies occurring on private lands, which county sells the most duck stamps, or which county has the most area enrolled in the U.S. Department of Agriculture (USDA) Conservation Reserve Programs (CRP).

The third question addresses changes in time. A waterfowl manager may want to know which wetlands are typically dry during the summer and thus have little value for waterfowl brood habitat. A big-game biologist may want to know which counties have had the greatest reduction in forest cover. The GIS uses two or more inventories acquired at different times to address these types of questions.

The fourth question addresses the social, economic, environmental, or combined impact of an existing change in land use. Ascertaining the benefits of CRP to waterfowl populations requires information on the location of the CRP enrollment areas, the land use that previously occurred in the area, the success of establishment of permanent cover on the CRP areas, and the availability of wetlands within and near the CRP sites.

Walker and Miller (1990) believed that the fifth question may be the most important or highest use of a GIS. Using a GIS, a wildlife manager can answer the "What if?" questions. What will be the impact to waterfowl populations if temporary or seasonal wetlands no longer are protected by federal regulations? Which wetlands are most vulnerable to wetland drainage given that drainage rates are a function of distance to nearest road, wetland type, size of continuous wetland, and surrounding land use?

A GIS is not a computerized system for making maps, even though maps are an important output product of GIS and many GISs now have excellent mapping capabilities. A GIS is not a tool for storing maps or pictures (although many modern GISs can store on one CD-ROM disk images of all topographic maps or aerial photography that are found in stacks of map cases or in rolls in obscure corners of offices). Although maps are an essential source of information for a GIS, the information maintained in the database of the GIS is the central concept, not the maps.

A GIS is an approximate model of the real world that uses computer systems to abstract three key pieces of information about features of the land required for management decisions. For every land feature, the GIS must know (1) what it is, (2) where it is, and (3) how it relates to other features (Walker and Miller 1990). GISs provide a mechanism for maintaining information about the land. Gathering information is the first, and most important, step in developing a GIS, followed closely by updating and maintaining information as features of the land change.

GIS DATA STRUCTURES

Geographic database management systems are more complex than database management systems used for banking, library searches, airline bookings, and medical records. Three general features of the data within a GIS must be maintained: (1) information on the position of the feature being stored, (2) topological information on the spatial relationships of the features (topology is the way in which geographic features are connected and provides a mechanism to identify the positional relationships among features), and (3) attributes of the feature (Burrough 1986). The spatial component of geographic data describes the location of a feature and the possible topological relationships among features. The attribute component describes various attributes of the feature.

The spatial components of geographic data can be represented by three data types: points, lines, and areas (Fig. 1). The spatial data types are referenced to a location by a standard system of coordinates, such as UTM (see Box 1), or by a local coordinate system. Local coordinate systems can be created simply by assigning the southwest corner of a map the X and Y coordinates of 0.0 and 0.0 and measuring the horizontal (X) and vertical (Y) distances from the southwest corner of the map to the feature. The coordinates for the location of a feature, such as a bird's nest, could be ascertained by marking the location of the nest on a map and assigning the X coordinate as the number of centimeters from the west edge of the map to the location marked on the map, and assigning the Y coordinate as the number of centimeters from the south edge of the map to the location marked on the map. Of course, using standard coordinates, such as UTM, state plane, or latitude and longitude, ensures that anyone using the data in the future will know the precise location of the bird's nest. In addition to the coordinates, a label describing which bird's nest is located at the given coordinates will be stored with the coordinates. The attribute record for the bird's nest will be referenced by the label and might include various attributes for the nest, including species of nesting bird, height of the nest, number of eggs laid, and number of eggs hatched. The label links the spatial data with the appropriate attribute record.

Spatial data are represented in GISs in two very different ways. Figure 2 shows the two different ways that a stream could be represented in a GIS. Spatial data can be represented as either rasters or vectors. In raster format, a grid is used to represent the study area. The location of features in the study area is depicted by the values in the cells overlaying the feature. Vector data represent geographic features by coordinates of points, lines, and polygons. Points represent small features such as wells, towers, or nest locations. Linear features such as roads and streams are represented by lines. Areas such as cities, forests, wetlands, and soil units are represented by polygons.

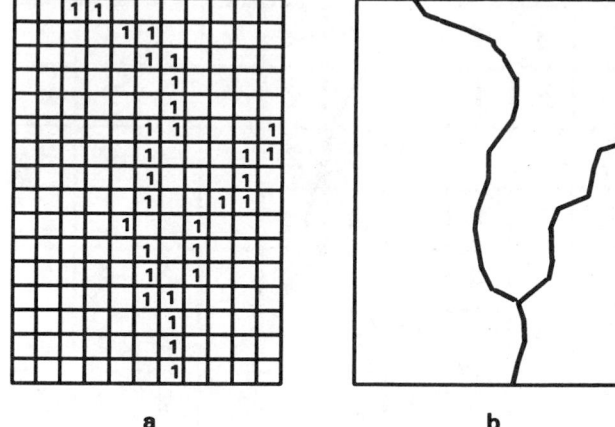

Fig. 2. A stream can be represented in a GIS either by a raster (a) or vector (b) format.

Box 1. Universal Transverse Mercator System.

The Universal Transverse Mercator (UTM) system is frequently used for recording coordinates of features in vector-based GISs, and cells of raster-based GISs are frequently aligned along the UTM grid. UTM coordinates are based upon the UTM map projection. The UTM system consists of 60 east-west zones, each zone measuring 6° wide in longitude. Each of these zones is numbered consecutively, starting with zone 1 between 180° and 174° west longitude and increasing eastward to zone 60 between 174° and 180° east longitude.

Washington, D.C., is located just west of 75° west longitude and is located in UTM zone 18 (72° west longitude to 78° west longitude). St. Louis, Missouri, is located just west of 90° west longitude and is located in UTM zone 15 (90° west longitude to 96° west longitude). San Francisco, California, is located just west of 122° west longitude and is in UTM zone 10 (120° west longitude to 126° west longitude).

UTM coordinates are recorded in meters. The UTM northing coordinate (the Y coordinate) for a feature is the distance in meters north from the equator to the feature. The UTM easting coordinate (the X coordinate) for a feature is the distance in meters east or west from the central meridian of the UTM zone. The central meridian for UTM zone 10 (with boundaries at 120° west longitude and 126° west longitude) is 123° west longitude. UTM eastings are prevented from having both positive values (meters east of the central meridian) and negative values (meters west of the central meridian) by setting the UTM easting for the central meridian at 500,000 m. A feature located 200,000 m west of the central meridian for a UTM zone would have an UTM easting coordinate of 300,000 m. A feature located 200,000 m east of the central meridian for a UTM zone would have a UTM easting coordinate of 700,000 m.

Many maps display a UTM grid as light blue or black lines. The UTM zone for the map will be reported in the map legend. The UTM system is just one example of a map projection. Often data within a GIS will be from various map projections. Snyder (1987) created an excellent reference to map projections.

Fig. 4, cells with value of ''1'' are forests, cells with value of ''2'' are croplands, and cells with value of ''3'' are rangelands.

In more sophisticated raster systems, the cell value is a label that will link to records as an attribute file. In the above example, cells labeled as ''1'' could have many attributes, such as species composition, age of forest stand, and estimated volume of marketable timber.

Because the raster system is strictly a two-dimensional matrix, various types of geographical data are stored as different layers or overlays in the GIS (Fig. 5). One layer may contain land use/land cover, another layer may contain wetland data, and another layer may contain information on the transportation system.

The user of a raster system must determine the size of the cells to be used. This size is referred to as spatial

Polygons are bounded on all sides by a series of straight-line segments.

Raster Data

Raster data are stored in the computer as a matrix. The cells are referenced by lines and elements (Fig. 3). In the simplest form, each line is a computer record. Each record will contain the values for all elements in the line. Any cell not containing a feature would have the value of ''0''. In the simplest raster system, the value stored for each cell is the attribute component of the geographic data. In

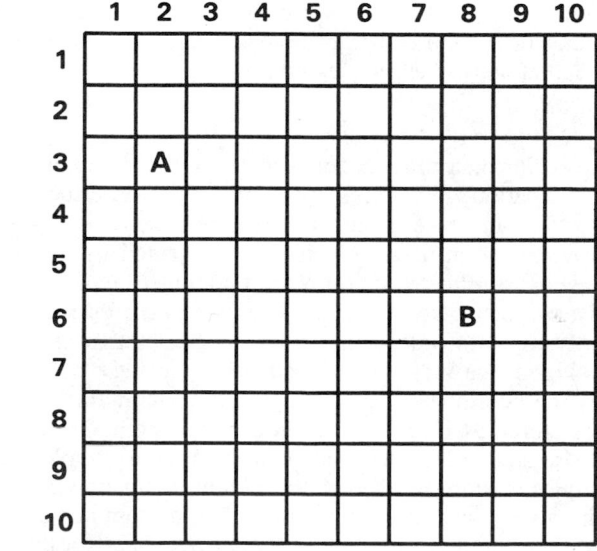

Fig. 3. Raster data are stored in computers as a matrix. Each cell is referenced by its line and element number. The example shown is for a small file with 10 lines and 10 elements. Cell A is located at line 3, element 2. Cell B is located at line 6, element 8.

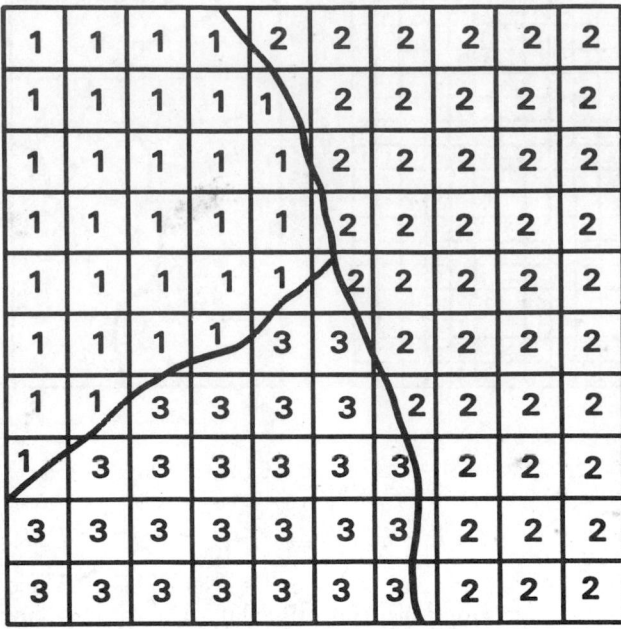

Fig. 4. Land cover as represented in a simple raster system. Cells with a ''1'' are forests, cells with a ''2'' are croplands, and cells with a ''3'' are rangelands.

resolution (see Box 2 for various meanings for the term resolution). The cell size can vary tremendously depending upon the size of the study area and the objectives for the GIS. Cell sizes as large as 20 ha for state or regional planning may be adequate. For a wildlife management area, a cell size of 0.05 ha or smaller might be required, depending upon the application of the GIS and the size of the wildlife management area. Storage requirements increase drastically as the cell size is reduced. Reducing the cell size by one-half will increase the data storage requirements by a factor of four. Conversely, as cell size increases, the precision of the representation of the land feature is reduced. Choosing the appropriate cell size for a particular GIS application is a compromise between cost of data storage and computer time and reliability of the representation of the land feature.

Vector Data

Vector data provide for high precision in representing the location of features. Aronoff (1989) described how vector data can be used to define the location of a point, a line, and an area. A point is represented by a simple pair of coordinates. The line is represented by an ordered list of pairs of coordinates. The area is represented as a polygon with ordered pairs of coordinates that close the polygon (the first and last pair being the same).

The coordinates can be any arbitrary units but usually are stored as UTM, state plane, or latitude and longitude coordinates. The first vector system used simple techniques to store the X and Y coordinates for polygons. In this simple system the coordinates for the common boundary between two areas were stored twice, once for the first area and again for the adjacent area. These duplicate storage techniques simplified computations and plotting but wasted storage space and, more importantly, provided no information as to adjacency or connectivity of geographic

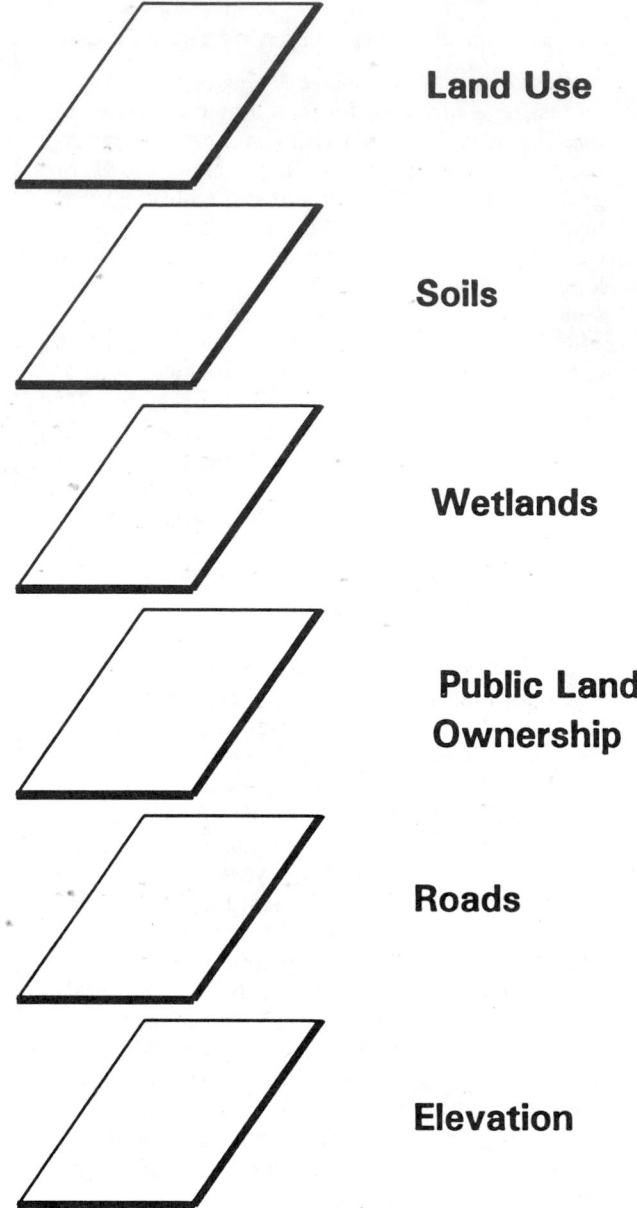

Land Use

Soils

Wetlands

Public Land Ownership

Roads

Elevation

Fig. 5. Various types of geographical data may be stored as different layers or overlays in a GIS.

features (topology). Most vector systems now use topological models (Aronoff 1989) for representing the location of areas.

In topological models (Fig. 6), a polygon is defined by a series of arcs. Arcs begin and end at nodes, which occur wherever two or more arcs meet. Each arc is defined by a series of coordinates, starting with the coordinates for the beginning node and ending with the coordinates for the ending node. Topological relationships are stored in three tables. The polygon topology table describes the arcs that bound each polygon, the node topology table describes the arcs that end at each of the nodes, and the arc topology table describes which end points (nodes) occur on each arc and which polygons are to the left and right of each arc. These three topology tables provide the tools required to efficiently determine the positional relationships of one feature to other features. A coordinate table

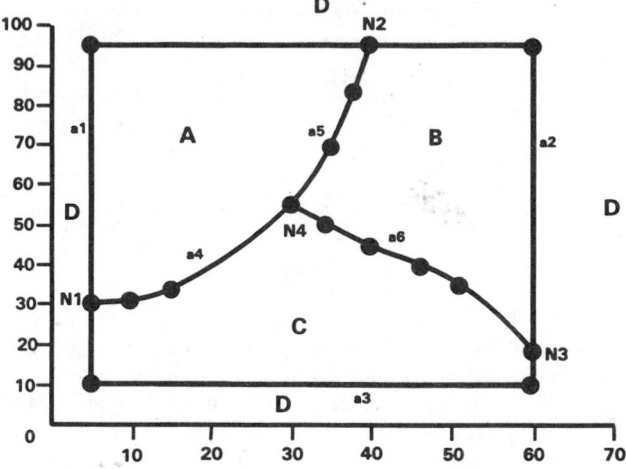

Polygon Topology	
Polygon	Arcs
A	a1, a4, a5
B	a2, a5, a6
C	a3, a4, a6
D	outside study area

Node Topology	
Node	Arcs
N1	a1, a3, a4
N2	a1, a2, a5
N3	a2, a3, a6

Arc Topology				
Arc	Start Node	End Node	Left Polygon	Right Polygon
a1	N1	N2	D	A
a2	N2	N3	D	B
a3	N3	N1	D	C
a4	N1	N4	A	C
a5	N4	N2	A	B
a6	N3	N4	C	B

Fig. 6. In vector systems that use topological models, polygons are represented by a list of arcs (adopted from Aronoff 1989). The arcs required to define each polygon are shown in the polygon topology table. The node topology table defines the arcs associated with each node. The arc topology table describes the starting and end nodes for each arc and defines the polygons to the left and right of each arc.

Box 2. Resolution has various meanings.

In remote sensing and GISs the term resolution has many meanings. The most common use of the term resolution refers to the area on the ground represented by one cell or pixel (picture element) of a raster-based system. For raster data collected with satellite-based digital sensors, the size of the cell is a function of characteristics of the sensor and orbiting altitude of the satellite. Landsat MSS data have a spatial resolution of approximately 0.64 ha (80 × 80 m). Landsat TM data have a spatial resolution of approximately 0.09 ha (30 × 30 m). The panchromatic data from SPOT have a spatial resolution of 0.01 ha (10 × 10 m).

The ability to separate various cover classes or features with data from satellite-based digital sensors is not totally a function of spatial resolution. Spectral resolution often is as important as spatial resolution in selecting a particular satellite for mapping land cover. Spectral resolution refers to the wavelengths of the electromagnetic spectrum selected to be measured by the sensor. Landsat TM data have been able to successfully differentiate among various cover classes that could not be separated by Landsat MSS data. Not only has the improved spatial resolution of TM data accounted for the improved accuracy of classification, but the increase from four to seven spectral bands has had a great impact in the successful use of TM data for mapping cover types. The number of spectral bands and the width of the spectral bands selected make up the spectral resolution of a sensor.

Resolution can also refer to the range of data collected by the satellite. For example, for MSS data collected on Landsats 1-3, the intensity of energy received for each band was measured in the range of 0 to 127. The TM sensors on Landsat 4 and 5 measure the energy received from each band in the range of 0 to 255. The number of computer bits used to store the range sets the maximum limit on this use of the term resolution.

Resolution can also describe the capabilities of some of the hardware used with GISs. When resolution is used to describe display monitors, it refers to size and number of screen pixels. A high-resolution monitor may display 1,024 elements by 1,024 lines, and each pixel is as small as 0.3 mm. A lower resolution monitor may be able to display only 512 elements by 512 lines. The resolution of scanners and plotters is often measured in dots per inch. For plotters, dots per inch is a measure of the number of pixels plotted per linear inch. A high-resolution plotter can display 400 or more pixels per inch, whereas a lower-resolution plotter may be able to display only 200 pixels per inch.

defining the coordinates for each arc also is used in topological models. In addition to these topological databases, the attributes for the features are stored in an attribute database.

Raster Versus Vector Systems

Early GISs were either raster or vector systems. Table 1 lists various advantages and disadvantages of vector and raster data systems (Burrough 1986, Aronoff 1989). Both approaches are equally valid ways of representing spatial data. The advantages and disadvantages of raster and vector systems have been heavily debated. Most modern GISs

Table 1. Comparison of advantages and disadvantages of vector and raster methods as revised from Burrough (1986) and Aronoff (1989).

Raster Method
  Advantages
    Data structure is simple.
    The method is compatible with remotely sensed or scanned data.
    Procedures for spatial analysis are simple.

  Disadvantages
    Greater disk storage is often required.
    Topological relationships are difficult to represent.
    Unless extremely small cell sizes are used, the graphic output is often aesthetically less pleasing.
    Projection transformations are more difficult.

Vector Method
  Advantages
    Compact data structure requires less disk storage.
    Topological relationships are readily maintained.
    Graphic output is aesthetically more pleasing and more closely approximates hand-drawn maps.

  Disadvantages
    Data structures are complex.
    Overlaying multiple vector maps is often time consuming.
    Output graphics may take hours to draw on plotters.
    Some spatial analysis procedures are difficult.
    Software and hardware for vector systems are often more expensive.
    The method is not as compatible with remote sensing data.

---

handle both raster and vector data but usually are designed primarily for one data type. The complete integration of raster and vector data capabilities will be common in GISs of the future (Faust et al. 1991). These new GISs will quickly and efficiently convert among rasters, vectors, and other data structures most appropriate for the application being performed (McKeown 1987, Ripple and Wang 1989, Piwowar and LeDrew 1990). GISs must function equally well with both raster and vector data.

DATA FOR THE GIS

The most expensive component of a GIS is not the software, hardware, or personnel to operate the system, but the cost of acquiring and maintaining the data. It is generally accepted that between 60% and 90% of the investment in a GIS will be the cost of the data (Walklett 1992). The cost is affected by several different factors including geographic coverage, scale, accuracy and reliability of the data, frequency of database updates, efficiency of hardware and software in capturing data from maps, satellite imagery, or aerial photography, and efficiency of hardware and software in reformatting existing available digital data.

Obtaining the data for a GIS is the major bottleneck in implementing a GIS (Aronoff 1989). The creation of an accurate and well-documented database is essential. Information generated from the GIS and resulting decisions made with that information can be accurate only if the initial data are accurate. Accuracy and reliability of all data layers should be documented. Documentation must include such information as date the information was collected, positional accuracy, classification accuracy, completeness, and procedures used to collect and encode the data. The accuracy of spatial databases is a complex issue and was the subject of a recent book (Goodchild and Gopal 1989).

The data to be entered into the GIS include the spatial data (location of the features) and the attribute information (data describing the features). Some data can be captured more readily in vector formats, whereas other data sets are more efficiently extracted from sources by using raster processing techniques.

Because of the expense of data for GISs, all data requirements must be documented before a GIS is initiated. Fortunately, the data required for a wildlife application may be the same data required for a land use planner, soils scientist, geographer, geologist, hydrologist, forester, or direct-marketing expert. Consequently, there is a growing source of existing data in digital format, and sharing or purchasing existing data is much cheaper than digitizing from existing maps or photos or extracting information from satellite data. It is critically important that all available sources of digital data are reviewed and evaluated before new data are acquired.

Various techniques are used to enter data into a GIS. Data can be entered manually from the computer keyboard or a digitizing table. Manual digitizing can be slow and expensive, but at times it may be the most efficient and accurate means of data entry. Scanning or scan digitizing is more automated than manual digitizing. Recent advances in scanning hardware and improvements in software for extracting information from scanned images are making scan digitizing a more attractive option to manual digitizing. For large areas, existing satellite technology and modern digital image-processing techniques can be a cost-effective means of capturing data for a GIS. The most effective and efficient method of capturing data for a GIS is to purchase existing digital data sources. Many federal agencies and a growing number of state and local agencies have digital data available. Many of these agencies are willing to share their data or will provide them at a minimum costs.

Manual Keyboard Data Entry

Often keyboard data entry is used in various digitizing techniques to enter attribute data for a specific feature. Location or the geographic component of features at times can be entered efficiently from the keyboard. This is particularly true for infrequent and widely distributed point data such as the location of cave entrances, nests, or animals that are radio-tracked. The location of points in the field now can be obtained with global positioning systems, termed GPSs (Fig. 7). Hand-held GPSs now can be obtained for <$3,000; they provide locational information in latitude and longitude, UTM, or other coordinates when used in the field. These field locations can be entered manually into the GIS with the computer keyboard, or the coordinates can be stored in the GPS and later downloaded to the GIS. For further information on GPSs, see Box 3.

Fig. 7. A hand-held global positioning system can be used for obtaining accurate locations for features on the ground (photo provided by Trimble Navigation, Sunnyvale, Calif.).

Manual Digitizing

In manual-digitizing techniques, a map or aerial photograph is placed on a digitizing table (Fig. 8) and a pointing device (called a cursor, puck, or mouse) is used to record coordinates of features to be extracted from the map. The digitizing table electronically encodes the position of the cursor. Tracing the map features with the cursor can be time consuming and error prone.

The attribute information about the feature also must be recorded. This frequently is done by labeling each feature with a unique number and building a list of attributes for each uniquely labeled feature. The efficiency of manual digitizing depends on the quality of the digitizing software, the skill of the operator, and the complexity of the map to be digitized. Editing the digitized data and assigning the feature labels or other attributes of the feature may take more time than initially digitizing the map.

Small digitizing tablets (0.3 × 0.3 m) can be purchased for <$100. Large digitizing tables (1.3 × 2 m) that can hold large maps range in cost from $3,000 to $20,000.

Box 3.   Global positioning systems.

Although global positioning systems (GPSs) have been in existence for 2 decades, only recently has this technology become an affordable and effective tool for wildlife managers and researchers. The Navigation Satellite Timing and Ranging (NAVSTAR) GPS developed and operated by the U.S. Department of Defense (DOD) consists of a network of 25 satellites in orbit about the earth (as of October 1993, all 25 satellites were operational), as well as ground operations support. After all the satellites are deployed, GPS will provide all-weather, worldwide, two- and three-dimensional (latitude, longitude, and elevation) positioning capabilities over a 24-hour period.

Wildlife biologists can obtain geographic coordinate position data with GPS receivers that can be hand-held or mounted in vehicles. A receiver manipulates signals emanating from the satellites passing overhead and provides positioning data with accuracies from a few centimeters to nearly 100 m. Currently, the level of accuracy depends on several variables, including the sophistication of the receiver's electronics, number of satellites available for a position fix at the desired time, obstructions to signal reception (e.g., trees), and degradation of the satellite signals by the DOD. Techniques are available to improve position accuracies, but these usually require additional receivers or a more costly individual unit, and they improve only certain problems. Selection of GPS receivers will depend on the desired accuracy levels for the intended applications.

The geographic coordinate data gathered by the GPS receiver can be viewed on the unit's display screen, stored for later analyses, or transmitted to another location for processing. Most receivers require about 2 minutes to obtain a positional fix and then are able to update the location of the receiver every second. Therefore, the GPS receiver maneuvered by a person or vehicle provides an effective method for obtaining data on point locations (e.g., nest site) and the geographic coordinate attributes of land features (e.g., wetland boundary).

The geographic coordinate data collected from the GPS receiver can be an effective means for establishing new layers, or augmenting and updating current data layers, in a GIS. Many receivers come with software that allows the user to directly input the coordinate data into a format usable by the GIS. This allows the data to be quickly verified and used.

Scan Digitizing

Recent advances in scanning hardware and software have made scanning a feasible alternative to manual digitizing for some applications. Continued advancements in this technology are coming and eventually it may replace manual digitizing.

Fig. 8. A digitizing table can be used to record coordinates of features shown on maps or aerial photographs (photo by Altek Corp., Silver Spring, Md.).

Three types of scanners are available. Flat-bed scanners have a flat scanning surface on which a map or a photograph is placed. Small flat-bed scanners (20 × 30 cm) cost <$2,000 and have scanning resolutions of 100–150 dots per centimeter (DPC). The flat-bed scanner that scans a 25 × 25-cm map at 100 DPC will produce a raster data file of 6,250,000 cells (a matrix of 2,500 lines by 2,500 elements). The scanned cells can contain intensity values ranging from 0 for a black object to 255 for a white object (when scanning is done in panchromatic mode with 8-bit data). When scanning is done in color mode, each cell contains the intensity of red, green, and blue light being reflected from the map. These intensities usually are measured in a range from 0 to 255.

Normally when resolutions of >150 DPC are required, or large maps are used, drum scanners are required. The map is mounted on a cylindrical drum, which spins as a detector is moved horizontally across the drum. Black and white intensities are recorded in panchromatic mode, or red, green, and blue intensities are recorded in color mode. The area viewed by the detector is termed the spot size (Aronoff 1989) and can be as small as 20 microns. Scanning a large map at 20 microns will create a large raster file.

For some applications, a video scanner can be used. A video camera is mounted on a copy stand and the map is placed beneath the video camera, which is raised or lowered to include a larger or smaller portion of the map. Video scanning typically produces a raster file with <512 elements and 512 lines. The spatial resolution of the cell depends upon the scale of the map and the distance between the map and the video camera.

Scanning by flat-bed, drum, or video scanners produces raster files. Maps that have been especially prepared for scanning show only the lines between features, and coordinates for these features are extracted readily. Extracted coordinates for features from scanned maps or aerial photographs may be complex and will rely on sophisticated, line-following algorithms or feature classification and extraction algorithms to obtain the desired informa-

tion from the map or photo. As in manual digitizing, much time will be spent editing scanned maps.

GISs are more than simply a warehouse for map information or storage for maps. However, many GISs can effectively store images of maps and aerial photographs obtained from flat-bed or drum scanners. Features on these images are not identified. These high-resolution images often are stored on CD-ROMs. Any of these images can be retrieved by the GIS and viewed on a color monitor. Feature information from these scanned images can be extracted with feature classification and extraction algorithms, or information (such as the distance between two points or areas) can be calculated with available software of the GIS. Many GISs in the future will support scanned-image libraries.

Remote Sensing Techniques

Advances in remote sensing and GIS technology have followed the advances in computer capabilities since the late 1960s. In many situations, remote sensing techniques that use satellite data are the only feasible means for collecting data for GIS applications over large regions. Remote sensing can be defined as any technique by which we gather data about an object without directly touching the object. Remotely sensed data for GIS applications are obtained from satellites or aircraft.

The most effective techniques of remote sensing used for GIS applications are those that provide digital data for the study area. These digital raster data sets can be obtained by satellite sensors, by digital sensors mounted in aircraft, or from scanned aerial photographs.

Remote sensing can employ active or passive systems. Satellite systems, such as Landsat and SPOT, use passive sensors, which measure the intensity of natural radiation. Active systems, such as radar and laser systems, transmit energy to the ground, then measure the energy returned from the ground to the sensor. Photographic cameras, video cameras, and multi-spectral sensors in aircraft or satellite are examples of passive systems. Some satellite and aircraft remote sensing systems use active sensors such as radar. Numerous passive and active remote sensing systems mounted on satellites or aircraft are currently available for acquiring data for GIS applications, and many additional remote sensing systems will become available to GIS users in the near future.

LANDSAT

Landsat, the U.S. land remote sensing satellite system, began as an experimental program conducted by the National Aeronautics and Space Administration (NASA). Landsat 1, launched on 23 July 1972, was expected to function for about 1 year and finally ceased operating in 1978 after nearly 5 years of continuous operation. During that time, it returned digital data for some 300,000 images of the earth's surface. The Landsat system was declared an operational system in 1983 and turned over to the National Oceanic and Atmospheric Administration (NOAA), U.S. Department of Commerce. In 1984, the Land Remote Sensing Commercialization Act (Landsat Act) was established to transfer the commercial operation of the Landsat program to the private sector. Earth Observation Satellite Company (EOSAT) was selected as the commercial operator for the Landsat program.

Landsat 1 through 3 satellites had two sensors. The Return-Beam Vidicom (RBV) sensor, which is similar to the television camera, recorded red, green, and infrared energy reflected from the surface of the earth. The Multi-Spectral Scanner (MSS) was the main instrument carried on these satellites and is still operating in Landsat 4 and 5 satellites. The MSS sensor collects data by scanning the earth from west to east with an oscillating mirror. Radiation from four different spectral bands (green, red, and two in the near infrared) is recorded. The radiation is transferred by fiber optics to filters that permit only certain wavelengths of radiation to strike the sensor's detectors. The picture element (pixel) sampled by the MSS is about 79 × 56 m (the size of a football field in the U.S.). Landsat satellites 2 and 3 ceased operating in 1983. Landsat 1, 2, and 3 satellites orbited the earth at 900 km and provided repeat coverage for any location on earth every 18 days.

Landsat 4 and 5 satellites were launched in 1982 and 1984, respectively. Landsat 4 is used sparingly because of an electrical problem that developed shortly after its launch. As of July 1993, Landsat 5 was still operating. Landsat 4 and 5 satellites circle the earth every 98.9 minutes in a near polar orbit of 705 km. Each satellite provides repeat coverage for any area every 16 days, at the same local time of day. Landsat 4 and 5 satellites weigh nearly 2,000 kg each and carry the MSS sensor and the Thematic Mapper (TM) sensor. The TM sensor has excellent capabilities for meeting the data needs of many GIS applications for large regions. Along each orbital path, the TM and MSS sensors can continually scan a swath 185 km wide. The scanned data are systematically divided into an area termed a "Landsat Scene," which encompasses approximately 185 × 170 km. Each scene covers approximately 3.2 million ha. Users of Landsat data can purchase data from an existing archive maintained by EOSAT or can schedule the collection of data for any site. The images from the TM sensor on Landsats 4 and 5 satellites have significantly better geometric quality than images from sensors on earlier Landsat missions due to engineering enhancements to the spacecraft. This has facilitated geodetic rectifications of the images to the accuracy standards for 1:24,000-scale map products (Welch et al. 1985).

The TM sensor provides significant improvements in spatial, spectral, and radiometric resolution compared to the MSS. The instantaneous field of view (IFOV) of the TM is square and results in a ground-resolution cell and image pixel of approximately 30 m on a side. The TM measures the intensity of reflected radiation in six spectral bands—three in the visible wavelengths, blue (0.45–0.52 μm), green (0.52–0.60 μm), red (0.63–0.69 μm); one in the near infrared (0.76–0.90 μm); and two in the short-wave infrared (1.55–1.75, 2.08–2.35). The TM also measures emitted thermal radiation (10.4–12.5 μm), although the IFOV for this spectral band is 120 m on a side. The greater radiometric resolution is achieved by the analog-to-digital conversion of the electrical signal to 8 bits or 256 gray levels compared to the 127 gray levels of the MSS on the first three Landsat satellites. Figures 9–12 provide examples of the raster data collected by the Landsat TM sensor and types of information that can be extracted for use in GISs. Landsat 6 was launched on 5

October 1993; however, communication with the satellite was not established.

SPOT

The first SPOT (Systeme Pour l'Observation de la Terre) satellite was launched by France in 1986. The SPOT program was designed to be a long-term, operationally commercial program, whereas the Landsat program was designed initially as an experimental system. The SPOT program was established by the French government in 1981 under the French space agency, CNES. France, several European banks, and industries from Belgium and Sweden have invested in this commercial entity. SPOT Images, S.A., which is partly owned by the French government, operates the SPOT system. SPOT Image Corporation was formed to market SPOT data in the U.S., and Radarsat, Inc., markets the data in Canada.

SPOT-1 carries two identical, high-resolution visible (HRV), pushbroom scanners. Each scanner can operate in one of two modes. In the panchromatic mode, 10-m resolution data can be obtained. This single band records visible energy ranging from 0.51 μm to 0.73 μm. In the multi-spectral mode, three bands are recorded at 20-m spatial resolution: green (0.5 μm–0.59 μm), red (0.61 μm–0.73 μm), and near infrared (0.79 μm–0.89 μm).

SPOT orbits at 832 km and repeats the orbit every 26 days. Each of the two sensors images a 60-km-wide swath. Pointed vertically, the two sensors can record a 117-km-wide swath. The sensors can be pointed, which provides two major advantages. First, a particular site can be imaged not only from the path directly over the site but also from adjacent satellite paths. This allows the potential for acquiring data for a site more frequently than every 26 days. Secondly, stereo images can be produced by acquiring scenes for the same area from two widely separated locations.

SPOT's panchromatic band has nine times more spatial detail than Landsat TM data. Combining Landsat's spectral data with the spatial advantages of SPOT's panchromatic data can produce spectacular images. A short-wave infrared spectral band is planned for the fourth satellite of the SPOT series.

COASTAL ZONE COLOR SCANNER

The Coastal Zone Color Scanner (CZCS) was launched on the Nimbus-7 satellite by the U.S. Government in 1978 and operated until June 1986. The CZCS measures ocean color and temperature with six spectral bands, including four bands measuring narrow portions of the visible spectrum, a near-infrared band, and a thermal-infrared band. This sensor provides spatial resolution of 0.825 km^2 at nadir and a scan width of 1,600 km. CZCS data have been used successfully to map suspended sediments and phytoplankton in coastal regions (Clark and Maynard 1986, Tassan and Sturn 1986) and in the detection of acid-waste pollution (Elrod 1988).

AVHRR SENSOR

In 1979 the NOAA-6 satellite and all subsequent satellites in the NOAA series carried the Advanced Very High Resolution Radiometer (AVHRR) sensor. The spatial resolution of the AVHRR varies from 1.1 km^2 at nadir

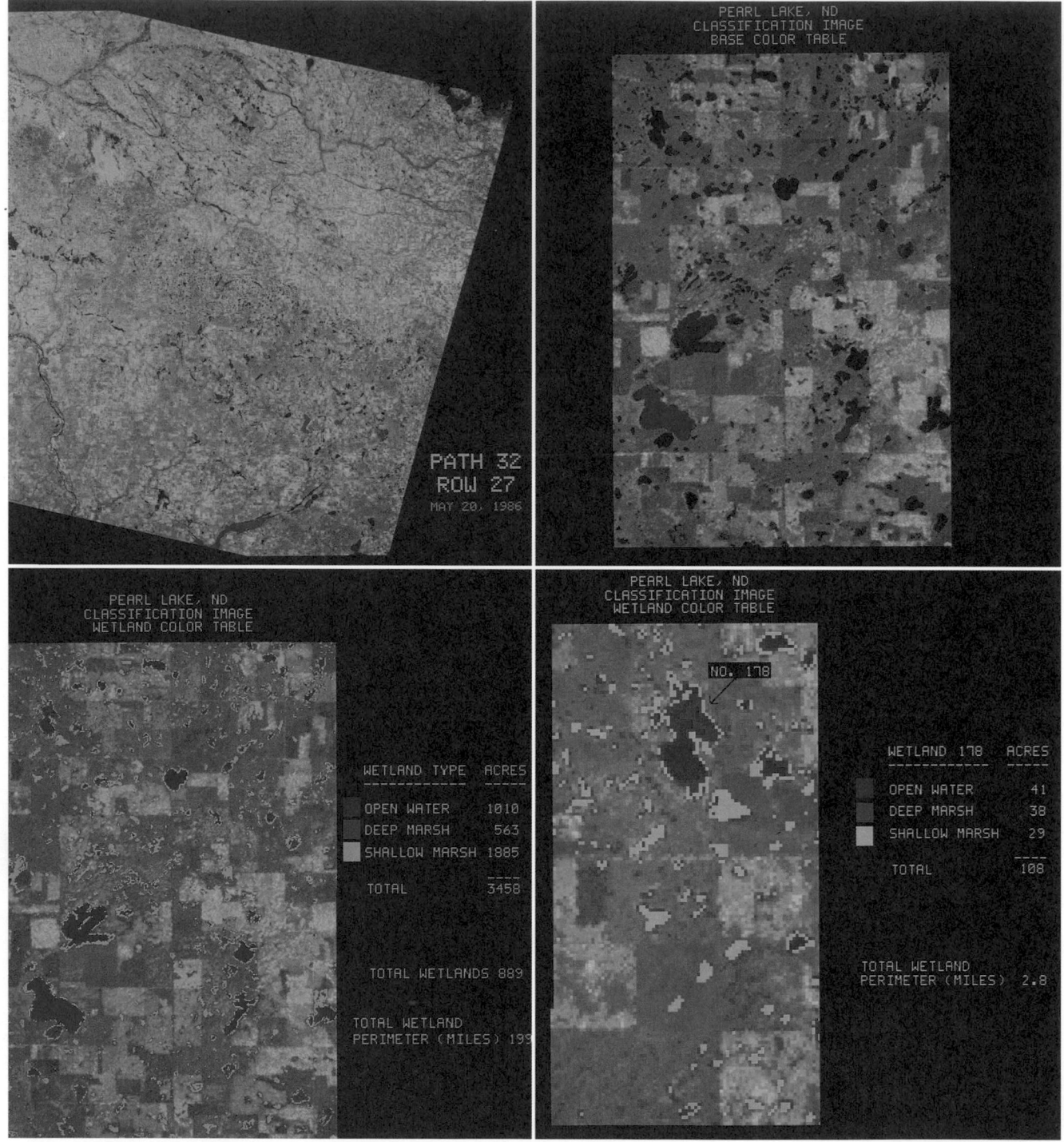

Fig. 9.   (top, left) A TM image for central North Dakota.   Fig. 10.   (top, right) Landsat TM data for the Pearl Lake, North Dakota, map sheet. This is a classified file containing >240 spectral classes. The color assigned to the spectral classes is based upon the mean values from bands 3, 4, and 5. The color table resembles the colors obtained from color-infrared photography.   Fig. 11.   (bottom, left) Digital image-processing techniques were used to extract wetland information from a Landsat TM scene for the Pearl Lake, North Dakota, map sheet.   Fig. 12.   (bottom, right) GIS processing functions can be used to label each wetland basin identified on the Landsat TM classified file and to ascertain the acreage of each wetland type in the basin. The acreages of the various wetland types are shown for the wetland basin labeled number 178.

to 12.6 km² at the end of the scan line. The sensor scans through 110.8° as it examines the earth, producing a scan line of 2,925 km. This wide scan angle, ± 54° of nadir, permits daily views of the earth. The AVHRR measures reflected radiation in the red and near infrared wavelengths and emitted thermal radiation in three spectral

bands. Two data formats are available from NOAA—local area coverage (LAC) at full spatial resolution and reduced resolution global area coverage (GAC) with a spatial resolution of 4 km² at nadir. Originally designed to provide improved determination of hydrologic, oceanographic, and meteorological parameters, AVHRR data, because of their high temporal frequency, also have proven useful for study of the phenology and productivity of terrestrial ecosystems on continental and global scales (Justice et al. 1985).

The digital data and photographic images are used in a variety of time-critical applications over large areas. AVHRR data have been used by the U.S. Fish and Wildlife Service to monitor snow cover in the arctic region of Canada for use in forecasting production of arctic nesting geese. LAC data were used to monitor water distribution for waterfowl wintering in the Central Valley of California (L. Strong, U.S. Fish Wildl. Serv., unpubl. data). Other uses of AVHRR data were described by Lillesand and Kiefer (1987) and Aronoff (1989).

GOES SATELLITE

Geostationary Operational Environmental Satellites (GOES) orbit the earth at an altitude of 36,000 km in the same direction as the earth's rotation. In this orbit, they maintain a stationary position relative to the earth (geostationary orbit). Two GOES satellites are operated by the U.S. and cover the western and eastern parts of North America. Europe and Japan operate additional GOESs. GOESs provide continuous monitoring of temperature, humidity, and cloud cover for weather forecasting. GOES data have been used for some GIS applications over huge regions (Meisner and Arkin 1984).

GOES collects two bands of data, a visible band (0.55–0.75 μm) and a thermal infrared band (10.2–12.5 μm). NOAA can provide data from the visible band at 1-, 2-, 4-, and 8-km resolution and thermal-infrared imagery at 8-km to 14-km resolutions.

MOS-1

The first Japanese remote sensing satellite, the Marine Observation Satellite 1 (MOS-1), was launched in February 1987. It has three sensors: a multi-spectral, self-scanning radiometer (similar to the Landsat MSS), a visible and thermal-infrared radiometer (similar to NOAA AVHRR), and a microwave scanning radiometer. No data tape recorders are on MOS-1, consequently data can be collected only when the satellite is in view of a ground receiving station. The U.S. has no such stations, but data are available from two Canadian receiving stations.

JERS-1

Japan launched JERS-1 (Japanese Earth Remote Sensing Satellite) on 11 February 1992. The three sensors of JERS-1 are an L-band (horizontal polarization synthetic aperture radar system), a visible and near-infrared radiometer, and a short-wave infrared radiometer. All provide 18-m resolution data.

ERS-1

The European Space Agency's first remote sensing satellite, ERS-1, was launched in 1991. ERS-1 carries a C-band, vertical polarization synthetic aperture radar instrument. Both high-resolution (25–35 m) and low-resolution (100 m) data are available in digital and photographic forms. Canada is planning to collect and distribute ERS-1 data for much of North America.

RADARSAT

Canada is developing RADARSAT, a radar remote sensing system to be deployed in 1995, the first Canadian remote sensing satellite. RADARSAT will assume a sun-synchronous orbit at approximately 800 km. The repeat cycle will be every 24 days, but with a change in the look angle, data can be collected for a specific site every 3 days. RADARSAT's synthetic aperture radar (SAR), a C-band with horizontal polarization, is designed to operate in several modes to provide numerous options in terms of swath widths, spatial resolutions, and angles of incidence. The standard beam mode will provide coverage with approximately 100-km-wide swath with a spatial resolution of 28 m. The wide swath beam will collect 28-m data over a 150-km swath. In the fine-resolution beam mode, 10-m data for a 50-km swath will be collected.

In addition to the SAR instrument, RADARSAT will include a scatterometer and two optical instruments. The scatterometer is a microwave sensor that collects data on wind speed and direction for a 600-km swath. One of the optical instruments is a multilinear array sensor, which records four spectral bands at 30-m resolution for a 400-km swath. The other optical instrument is an AVHRR sensor capable of collecting five spectral bands at 1,300-m resolution over a 3,000-km swath.

The radar data collected by RADARSAT could be valuable for mapping and monitoring wetlands, because radar is an active sensor creating its own illumination source, and data can be collected for areas covered with clouds or even at night. Place (1985) reported that the accuracy of mapping forested wetlands was improved by 85%, when radar images collected from SEASAT were used to complement conventional aerial photography used by photo-interpreters for mapping wetlands (SEASAT was launched in 1978, but failed only 99 days after launch).

AIRCRAFT SENSORS

Satellite-based sensors have many advantages for meeting GIS data needs. Satellite data have low cost per hectare of coverage, a geometric fidelity that facilitates registration of images to various maps projections, and freedom from mission planning. However, for some applications, the spatial resolutions of satellite-based sensors may be too gross, and the temporal frequency or clouds prevent data acquisition during optimum times. The time of data acquisition can be critically important for the successful use of the data. For example, temporary wetlands may be inundated for only a few weeks. Acquiring satellite data when the temporary or seasonal wetlands are dry makes detection and identification of temporary or seasonal wetlands difficult. Aerial photography similarly acquired when the wetland basins are dry will not provide acceptable delineation of wetlands.

Aircraft sensors offer great flexibility of spatial resolution, timing, and wavelengths of spectral data. Aircraft sensor data can be scheduled to be collected at optimum time for extracting information from desired features. Spatial resolution can be as fine as 1 m or as coarse as 50 m and is dependent on the aircraft altitude, the optical system, and the size of the sensor's detector elements. However, when compared to satellite data, aircraft sensors provide a relatively narrow swath width. A major problem with aircraft scanner data is the poor geometry of the data. These data are adversely affected by variations in aircraft attitude (roll, pitch, and yaw) and deviations from the flight line. Digital elevation models (DEMs) and GPS can be used to suppress the geometric problems inherent in aircraft multi-spectral data. Lee (1991) provided an excellent review of applications of aircraft multi-spectral data for classifying and mapping wetlands.

VIDEOGRAPHY

Airborne videography recently has been used successfully for assessing wetland and riparian habitats in North Dakota (Cowardin et al. 1988a) and for evaluating rangeland and other vegetation in Texas (Driscoll 1990). Sidle and Ziewitz (1990) described the use of aerial videography for wildlife studies. Lee (1991) described many of the advantages of videography: (1) imagery can be captured by microcomputer for immediate use; (2) in-flight error-proofing can be done; (3) narrow-band filters for fine spectral resolution can be used; (4) data can be acquired in a wide range of atmospheric conditions; (5) data can be acquired any time; 6) cost of videography systems is low; and (7) standard digital image-processing techniques, which are typically used on satellite data, can be used to analyze the video data. Disadvantages of videography include: (1) images provide coverage of only small areas; (2) resolving power is much less than that of aerial photography; (3) geometric distortion from motion in the plane is difficult to correct; (4) spectral resolution of solid-state detectors is limited to visible and near infrared wavelengths; (5) multi-spectral data collection is difficult because single cameras have problems with focus for different wavelengths, and multiple cameras require accurate bore sighting and large camera ports on the aircraft; (6) calibration of video data is difficult because of automatic gain control; and (7) images are vignetted.

Various video systems have been developed (Mausel et al. 1992), including single-band panchromatic systems, single-band color systems, and multi-spectral systems, some of which include near-infrared capabilities. Everitt and Escobar (1989) described many of the available systems. In GIS applications, the best use of videography may be to update existing information layers.

Existing Databases

Because of the expense of acquiring digital data layers for GIS applications by digitizing existing maps or by remote sensing techniques, GIS users always should search for existing digital data sets to meet their data needs, before capturing the data themselves. Sources of existing databases include third-party vendors, federal government agencies, and state and local government agencies.

Before searching for existing databases, one must have a clear vision of what kind of information is required and exactly how the information is to be used. Knowing the exact data requirements is critical to identifying good potential information sources.

The available digital data sets were produced to satisfy a wide range of users. Consequently, the data are not always suitable for a specific GIS application. The cost, accuracy, and currency of the data vary greatly with ex-

isting sources. By the time the data have been collected, reviewed, digitized, edited, and distributed, they may be out of date for some applications. Dulaney (1987) reviewed many of the problems associated with existing databases. Descriptions for some of the more widely used databases follow. The *GIS World Source Book* (Parker 1991), which is published yearly, is an excellent source of information on existing data available for GIS applications.

LAND USE AND LAND COVER AND ASSOCIATED MAPS

The Land Use and Land Cover (LULC) and associated data files are available from the U.S. Geological Survey (USGS) and provide information on five data layers: (1) land use and land cover, (2) political units, (3) hydrologic units, (4) census county subdivisions, and (5) federal land ownership. These files are derived from maps at scales of 1:250,000 and 1:100,000.

Land use and land cover areas are classified into nine major classes: urban or built-up land, agricultural land, rangeland, forestland, water, wetland, barren land, tundra, and perennial snow or ice. Each major class is composed of several minor classes (e.g., forestlands are further classified as deciduous, evergreen, or mixed). This classification system (Anderson et al. 1976) was reviewed by a committee of representatives from the USGS, NASA, Soil Conservation Service (SCS), the Association of American Geographers, and the International Geographical Union. The classification system (Table 2) was designed to be used with data obtained from remote sensors on aircraft and satellites.

The minimum mapping area (smallest area mapped) for all urban areas, bodies of water, surface mines, quarries, gravel pits, and certain agricultural areas is 4 ha. The minimum mapping area for all other categories is 16 ha. Thus, a residential area <4 ha would not be recorded in these files, nor would an area of cropland or pastureland <16 ha. Aerial photographs and satellite data serve as the primary sources used in compiling the LULC maps. Some areas on each map are field checked for accuracy.

The four associated maps are prepared at the same scale as the LULC files. The political units file contains county and state boundaries as shown on USGS maps. The hydrologic-units file was digitized from the 1:500,000-scale state maps delineating hydrologic units, which were compiled by the Water Resources Council and published by USGS's Water Resources Division. The census county subdivisions file shows minor unit divisions or equivalent areas. Census tracts within Standard Metropolitan Statistical Areas (SMSA) are represented in this file. The federal land ownership file delineates surface ownership for all areas >16 ha. Federal subsurface ownerships are not delineated.

The LULC and associated data files are available in vector and raster formats on 9-track tapes or CD-ROMs. The raster format uses a cell size of 4 ha. More information on these files can be obtained from the regional USGS Earth Science Information Centers (ESIC) offices (see Appendix I for the addresses and telephone numbers for USGS ESIC offices).

DIGITAL LINE GRAPHS

Digital line graphs (DLGs) are the digital representation of the planimetric information (line map data) shown on a map. DLGs have been compiled by USGS from 1:2,000,000-scale maps, some 1:250,000- to 1:100,000-scale maps, and some of the 1:24,000- and 1:62,500-scale maps.

DLGs compiled from 1:2,000,000-scale maps are available for three categories: (1) boundaries, which include state and county boundaries and federally administered lands; (2) transportation, which includes roads, railroads, and airports; and (3) hydrography, which includes streams and water boundaries. A CD-ROM that contains data for all 50 states organized into 21 geographic regions can be purchased for only $32 from the ESIC in Reston, Virginia.

The DLGs compiled from 7.5- and 15-minute topographic quadrangles include nine thematic categories: (1) boundaries; (2) transportation; (3) hydrography; (4) U.S. Public Land Survey System (PLSS) (including township, range, and section information); (5) hypsography, including contours and supplemental spot elevation; (6) vegetative surface cover, including woods, scrubs, orchards, vineyards, and marshes and swamps; (7) nonvegetative features including lava, sand, and gravel; (8) survey and control markers, including horizontal and vertical positions of benchmarks; and (9) humanmade features, including cultural features not collected in other major data categories, such as buildings. Any feature shown on a 7.5- or 15-minute topographic map will be delineated on the DLGs. These data are not available currently for many locations. DLG data do not carry quantified accuracy statements. However, the data are inspected for attribute accuracy and topological fidelity.

DIGITAL ELEVATION DATA

Elevation, slope, and aspect can be important information for a variety of wildlife applications of GISs. Digital elevation data, frequently termed digital elevation model (DEM) or digital terrain model (DTM), provide elevation information along a contour or at regularly spaced sample points. Aronoff (1989) described four basic formats for capturing and storing elevation data. These data can be used to derive information about the morphology of the landscape such as slope and aspect, which are important to solar insolation and microclimate. Algorithms have been developed to extract the drainage network from DEMs and to partition the landscape into watersheds, subcatchments, and hillslopes for hydrologic modeling (Jenson and Domingue 1988, Band 1989).

The U.S. Defense Mapping Agency produced the first DEM for the entire U.S. by scanning the contour overlays for all 1:250,000-scale topographic maps. From the scanned contour lines, elevations were sampled every 3 arc-seconds of latitude and longitude (approximately every 90 m). The elevation accuracy of these data ranges from 15-m RMSE (Root Mean Square Error) to 60-m RMSE, depending on the terrain. RMSEs are usually lower in flat terrain and increase in steep terrain. These DEM data are sold by USGS in sections 1 × 1 degree in size.

USGS is compiling elevation data from the 7.5-minute topographic maps. From these maps, elevation is sampled every 30 m. Vertical accuracy varies from 7 m to 15 m.

Table 2. Land use and land cover categories used by the U.S. Geological Survey (Anderson et al. 1976).

1. Urban or built-up land
    - 11 Residential
    - 12 Commercial
    - 13 Industrial
    - 14 Transportation, communications, and utilities
    - 15 Industrial and commercial complexes
    - 16 Mixed urban or built-up land
    - 17 Other urban or built-up land
2. Agriculture land
    - 21 Cropland and pasture
    - 22 Orchards, groves, vineyards, nurseries, and ornamental horticulture areas
    - 23 Confined feeding operations
    - 24 Other agriculture land
3. Rangeland
    - 31 Herbaceous rangeland
    - 32 Shrub and brush rangeland
    - 33 Mixed rangeland
4. Forestland
    - 41 Deciduous forestland
    - 42 Evergreen forestland
    - 43 Mixed forest land
5. Water
    - 51 Streams and canals
    - 52 Lakes
    - 53 Reservoirs
    - 54 Bays and estuaries
6. Wetland
    - 61 Forested wetland
    - 62 Nonforested wetland
7. Barren land
    - 71 Dry salt flats
    - 72 Beaches
    - 73 Sandy areas other than beaches
    - 74 Bare exposed rock
    - 75 Strip mines, quarries, and gravel pits
    - 76 Transitional areas
    - 77 Mixed barren land
8. Tundra
    - 81 Shrub and brush tundra
    - 82 Herbaceous tundra
    - 83 Bare ground tundra
    - 84 Wet tundra
    - 85 Mixed tundra
9. Perennial snow or ice
    - 91 Perennial snowfields
    - 92 Glaciers

DEMs derived from the 7.5-minute topographic maps are available for about 50% of the U.S. as of 1993. Errors in the elevation data can introduce significant errors into calculations of slope and aspect. Errors tend to occur in areas of rapid change in slope and exposure such as along ridges and ravines (Davis and Dozier 1990).

NATIONAL WETLANDS INVENTORY

More than 30,000 detailed wetland maps have been produced by the National Wetland Inventory Program (NWI). NWI maps cover nearly 70% of conterminous

U.S., 21% of Alaska, and all of Hawaii. Most of the maps cover the same area as covered by the 7.5-minute, 1:24,000-scale topographic maps distributed by USGS, but some NWI maps have been produced at scales as small as 1:100,000 (Gravatt 1991).

The maps have excellent consistency, because one classification system (Cowardin et al. 1979), one set of photo-interpretation conventions, and one set of cartographic conventions were used. In April 1991, more than 1,100,000 copies of NWI maps had been distributed (Gravatt 1991). The USFWS is on schedule to complete the mapping of the conterminous U.S. by 1998 as required by the Emergency Wetland Resource Act of 1986. Mapping in Alaska should be completed by 2000.

In 1991, digital data files were available for more than 6,200 maps representing 10.5% of the continental U.S. An index map that shows the current availability of digital NWI data can be obtained from NWI in St. Petersburg, Florida. Digital data are available for sale from the USGS's ESIC offices for $25/map; they are available on magnetic tape in MOSS export, DLG3 optional, or GRASS formats (Gravatt 1991).

DIGITAL SOILS DATA

The SCS has the responsibility for the National Cooperative Soil Survey (NCSS), which includes collecting, storing, maintaining, and distributing soils information for privately owned lands in the U.S. (Nielsen 1991).

The SCS has established three digital geographic databases for soil. Each consists of a spatial component that describes the location of the named soil unit and an attribute component that describes characteristics of the soil unit in detail. These digital data help facilitate the storage, retrieval, analysis, and display of soil data in a highly efficient manner. These data can be integrated readily with other spatial and demographic data in GISs. A soils data layer may be one of the most important GIS data layers for wildlife applications.

The Soil Survey Geographic database (SSURGO) is a vector database describing soil delineation boundaries. The boundaries of the soil units are delineated from aerial photographs ranging in scale from 1:15,840 to 1:31,680 combined with extensive fieldwork. The delineated soil boundaries are transferred to 7.5-minute orthophotoquads or topographic maps before the digitizing proceeds.

State Soil Geographic database (STATSGO) was digitized from 1:250,000-scale topographic maps, on which a generalization of the detailed soil surveys was mapped. For areas for which detailed soil survey maps were not available, the generalized soils information was compiled from existing data on geology, topography, vegetation, and climate. STATSGO data are distributed as complete coverage for a state.

The National Soil Geographic database (NATSGO) was derived from general soil maps for each state. NATSGO data were digitized from a map covering all of the U.S. at a scale of 1:7,500.

The Soil Interpretations Record (SIR) database provides attribute data describing the characteristics for each map-unit component and interpretative data for numerous uses. The accuracy of these maps is not determined. Data standardization between field surveys has been a problem, and as a result soil types and properties at the boundaries of adjacent maps often disagree (Burke et al. 1991). GIS technology not only will revolutionize the way the data are analyzed and displayed, but also the way data are collected. SSURGO, STATSGO, and NATSGO data files and the associated attribute files are available from SCS. NATSGO costs $500 for the entire U.S., STATSGO costs $500/state, and SSRUGO costs $500/county. More information on the availability and distribution of these databases can be obtained from USDA's National Cartographic Center, Fort Worth, TX 76115.

DIME FILES

The U.S. Bureau of the Census created a spatial data set describing street networks, street addresses, political boundaries, and major hydrographic features for approximately 350 major cities and suburbs in the U.S. These files were created with the Dual Independent Map Encoding (DIME) system to automate the processing of the 1970 and 1980 U.S. censuses. DIME files have limited application as a digital map base. For example, streets are represented by straight lines connecting adjacent intersections. Even a curved street is represented by a straight line.

TIGER FILES

To overcome the limitations of the DIME files and to prepare for the 1990 census, the U.S. Bureau of the Census developed the TIGER (Topologically Integrated Geographic Encoding and Referencing) system. The TIGER files provide vector data for hydrography, transportation, political, and statistical areas (such as county, incorporated area, census tract, and census block). Data collected from the 1980 and 1990 censuses, such as population, number of housing units, income, occupation, and housing values, serve as attribute data for these files. Nearly all commercially available GIS software systems have procedures for importing TIGER data. Various companies have developed inexpensive GIS systems strictly for the use of TIGER files and the associated census data. These companies sell hardware, software, TIGER, and census data as a complete package. TIGER data also can be purchased directly from the Bureau of the Census, Washington, DC 20233. With the release of the TIGER files, the Census Bureau no longer supports or sells the DIME files.

The TIGER files comprise one of the most detailed computerized digital map databases ever developed for the U.S. More than 7 years and $200 million dollars were required to complete the TIGER files. The complete TIGER files for the entire U.S. contain nearly 40,000,000 line segments and require more than 15,000 megabytes of storage for the vectors alone (Anonymous 1989).

Appendix II provides the addresses and telephone numbers for many sources of digital data in the U.S. and Canada.

DIGITAL IMAGE PROCESSING

To be effective in management decisions, a geographic information system requires timely, accurate update of many of its spatial data elements. Remote sensing and digital image processing have the potential to meet these needs. During the next 2 decades, there will be an unprecedented availability of digital data from satellite sensors in response to the concerns about human impacts on the earth and global climate change (Ormsby and Soffen 1989). However, Graetz (1990) believed that currently available remote sensing technology far exceeds the scientific capability of interpreting and applying it. If remote sensing data are to be used to their fullest potential, the challenge to ecologists will be to develop realistic spectral, spatial, and temporal models for extracting information from the images. Several excellent books describe remote sensing and digital image processing (Swain and Davis 1978, Estes et al. 1983, Schowengerdt 1983, Curran 1985, Richards 1986).

Understanding of remote sensing models and their interrelationships can benefit from a system view of the image-forming process (Swain and Davis 1978). An important concept is the distinction between the scene, which is real and exists on the earth's surface, and the image, a collection of spatially arranged measurements from the scene (Strahler et al. 1986). The purpose of a remote sensing model is to provide a conceptual and explicit framework for inferring the characteristics of the scene from the image. A remote sensing model may be generalized as having three components: a scene model, an atmospheric model, and a sensor model.

A scene model quantifies the relationships of the objects or targets of interest and their interactions with radiation through the processes of reflectance, transmittance, absorbance, and emittance. Characteristics of the scene objects could include their type, size, number, and spatial and temporal distributions. The model also must consider the background or nontarget components of the scene, including shadow.

An atmospheric model describes the transformation of the radiance due to scattering by molecules and aerosols, and gaseous absorption during the path from the sun to the earth's surface and between the surface and the spacecraft. If an atmospheric model is omitted, the parameters developed to extract information from the image are not transferable and the entire procedure must be repeated for other images. Several methods for the normalization or radiometric calibration of remotely sensed data have been developed (Ahern et al. 1987, Schott et al 1988, Chavez 1989, Tanre et al. 1990).

The sensor model quantifies how the instrument collects the measurements of the scene and includes four key parameters: spectral, spatial, and temporal resolution, and view angle (Duggin 1985). The spectral resolution of the sensor specifies what wavelengths of the electromagnetic spectrum are measured. The spatial resolution specifies the size of the area on the ground from which the measurements that comprise the image are derived. The spatial resolution relative to the spatial structure of the scene objects determines the appropriate analysis methods for scene inference (Woodcock and Strahler 1987). The temporal resolution specifies the frequency with which images are obtained in time. View angle is an important component of the imaging geometry. View angle and illumination geometry (solar zenith and azimuth angles) are important determinants of the measured reflectance since adjustments in observation and illumination geometry result in different sampling of the bidirectional reflectance distribution function, the most fundamental property describing the reflection characteristics of a surface (Silva 1978). Multidirectional observation of this reflectance an-

isotropy will be possible with the new generation of sensors (Ormsby and Soffen 1989).

Digital image processing, the numerical manipulation of digital images, includes procedures for preprocessing, enhancement, and information extraction. Preprocessing involves procedures applied to the original data before enhancement or information extraction. Calibration of image radiometry for atmospheric conditions and illumination and view geometry, the correction of geometric distortions and georegistration of the image, and noise suppression are examples of image-preprocessing procedures (Schowengerdt 1983).

Image enhancement involves the application of procedures designed to facilitate the interpretation of images. These procedures include contrast and color manipulations and spatial-filtering methods (Schowengerdt 1983). The "Tasseled Cap" is a well-known spectral transformation, which derives new variables that allow vegetation and soils information to be extracted, displayed, and understood more easily (Crist et al. 1986). Hodgson et al. (1988) used this transformation with Landsat TM data in a study of wood stork foraging habitat. Jackson (1983) provided a general procedure to develop spectral indices for user-defined features in a scene.

The development of scene models for extracting information from remotely sensed data requires an understanding of the image-forming process. Strahler et al. (1986) provided a framework for identifying appropriate scene models given the characteristics of the image and the scene. The most common information-extraction methods used with remote sensing data are spectral classifiers in which each pixel is processed independently of its neighbors or location in the image. A discrete scene model is appropriate when the scene objects are larger than the spatial resolution of the sensor.

The parameter estimation process for spectral classifiers can be generalized as being supervised or unsupervised (Swain and Davis 1978, Schowengerdt 1983). In supervised classification, a sample of image elements for each land cover class is used to estimate parameters, typically a mean vector and covariance matrix, for input to the classifier. In unsupervised training, a clustering algorithm is used to partition a sample of the data into populations of pixels with similar reflectance, which are referred to as spectral classes and parameters estimated for these spectral classes (Richards and Kelly 1984). In unsupervised training, the analyst then attempts to establish a correspondence among the spectral classes and the land-cover classes. A statistics file consisting of a mean vector and covariance matrix for each land-cover class then is input to a classification algorithm. The output from a maximum likelihood classification, a common method that produces results having the minimum probability of error over the entire set of data classified, is an image in which each pixel is assigned the label of the land-cover class for which the a posteriori probability was the maximum. An enhancement to the standard output from the maximum likelihood classification would be to create a raster for each land-cover class wherein the pixel value would be the a posteriori probabilities of membership for the category. The result is a probabilistic digital map of the geographic distribution for each land-cover class. This would increase the computational and storage requirements, but technological progress in these areas is great (Faust et al. 1991).

In a continuous-scene model, the scene objects are smaller than the resolution element of the sensor. A relationship between the reflectance and a property of a scene, such as canopy coverage, is established and used to estimate the property in each pixel in a continuous fashion. Mixture models are a type of continuous-scene model, in which the objective is to estimate the proportions of scene objects in each pixel. Mixture models have been used for a variety of resource inventories, including waterfowl habitat (Work and Gilmer 1976), rangeland vegetation and soil cover (Pech et al. 1986), and wintering geese (Strong et al. 1991).

Spectral-spatial scene models exploit the spatial structure of images as well as their spectral characteristics to infer the properties and processes at the land surface. A variety of spectral-spatial models is available. Some of these scene models segment the image into contiguous groups of pixels that meet a spectral similarity criterion and perform the classification using all the pixels of the feature (Strahler et al. 1986). Other spectral-spatial models exploit a measure of image texture or the spatial autocorrelation function as an additional feature in the classification process (Shih and Schowengerdt 1983, Pickup and Chewings 1988).

Spectral-temporal models use the change in the spectral properties of images acquired at different times to infer properties or processes at the land surface. The "Tasseled Cap" is an example of a spectral-temporal model of the phenological development of agricultural crops that can be used to identify crops and forecast yields (Kauth and Thomas 1976, Wiegand et al. 1986). Time series of the normalized difference vegetation index (NDVI), calculated from the red and infrared spectral reflectance measurements of the AVHRR sensor, have been used to describe and map the intra- and inter-year phenological dynamics of biomes at regional, continental, and global scales (Justice et al. 1985), to infer net primary productivity (Goward et al. 1985), and to measure the dynamics of vegetation at the transition zones between biomes (Tucker et al. 1991). Various techniques for detecting change (Singh 1989) use images acquired at different times to infer changes in land cover.

The flow of information between remote sensing and GIS should not be one-way. The accuracy of information derived from remote sensing can benefit from access to accurate spatial data within a geographic information system. Integration of the parallel technologies of GIS and RS will be important to the fullest maturation of both areas.

GIS ANALYSIS CAPABILITIES

Much of the initial time and expense in operating a GIS are in developing the various data layers. Once these data layers are in the system, much time is often spent producing maps of the various data layers. These maps are useful but do not represent the real power of the GIS. The unique power of the GIS is in the analysis that it can perform. Unfortunately, because so much effort is devoted to developing the database layers, often the main purpose for developing the GIS (providing aids to the decision-making process) is neglected. It is the analytical functions applied to the spatial and nonspatial attribute data of the GIS that provide the power to aid in the decision-making process.

The analysis capabilities of the GIS provide the answers to the five basic questions defined by Walker and Miller (1990). These five basic questions, which were discussed previously, should be reviewed before the analytical capabilities of GISs are studied.

Aronoff (1989) described four major GIS analysis capabilities: (1) maintenance and analysis of spatial data, (2) maintenance and analysis of attribute data, (3) integrated analysis of spatial and attribute data, and (4) output formatting. The first two functions deal with maintaining the data and the fourth function is for producing output from the system. The real power of the GIS is its ability to integrate the analysis of spatial and attribute data. Spatial data can be stored as raster or vector structures. Some analytical functions can work equally well with raster or vector data, some analytical functions will perform better with raster data, and some functions will perform better with vector data. GIS systems of the future will have improved user interfaces and expert systems to advise the user on how to utilize the existing database and software to obtain the desired resource information (Coulson et al. 1987, Goodenough et al. 1987, McKeown 1987).

Maintenance and Analysis of the Spatial Data

Maintenance and analysis capabilities of the GIS will permit transforming spatial data files, editing spatial data, and assessing the quality and accuracy of the data. Aronoff (1989) described seven maintenance and analysis functions for spatial data: format transformation, geometric transformation, transformation between map projections, conflation, edge matching, editing of graphic elements, and line coordinate thinning.

Because data for the GIS may come in many formats (e.g., DLG, TIGER, MOSS, GRASS, raster run-lengths encoded, and DIME), GISs must be able to transform data among various formats. Conversions from various formats and from raster to vector and vector to raster are examples of format transformation. Because of inaccuracies in source maps, slight differences in map projection, or digitizing errors, data layers may not be precisely registered to a common coordinate system or a standard data layer. Geometric transformations are procedures used to ensure that each data layer precisely overlays the other data layers. Some data may be supplied in geometric coordinates (latitude and longitude), whereas other data may be supplied in UTM or state plane coordinates. Functions to transform from one map projection to another are required to transform all sources of data to a common map projection.

Conflation functions are used to ensure that common boundaries between different data layers share the exact coordinates. For example, the edge of one end of a wetland is adjacent to a gravel road, and the gravel road is the boundary on a wildlife management area. The location of the wetland was digitized for the wetland data layer, the road was digitized for the transportation data layer, and the wildlife management area was digitized for the public lands data layer. When all three layers were plotted as one map, the common boundaries of the wetlands, road, and wildlife management area are similar, but not exactly the same. Conflation procedures ensure that common boundaries are defined with exactly the same coordinates.

Usually all features on one map sheet will be digitized and then all features on the adjacent map will be digitized. Because of slight error in the maps and imprecisions in the digitizing process, the boundary of a feature crossing onto the adjacent map may not match precisely. Edge-matching functions ensure that the edge of a feature on one map perfectly matches the feature edge on the adjacent map.

GISs should have numerous editing functions to aid in adding, deleting, and changing the geographic positions of features. In the digitizing process, more coordinates are collected for a line than are actually required to represent the line. Line-coordinate thinning processes can eliminate coordinates that are not essential to represent the line and can greatly reduce the disk space required to store the coordinates needed to represent the line.

Maintenance and Analysis of Nonspatial Attribute Data

Just as various functions are required to edit, convert, and maintain spatial data, functions for editing, converting, and maintaining the nonspatial attribute data also are required. Attribute editing functions should allow for retrieving, editing, and changing the attributes. Attribute query functions allow records in the attribute databases to be selected. Using the attribute data query functions on the attribute database for the soils layer would allow one to select all soils that are sandy and have a ''T'' value >4 (T values represent the tons of topsoil likely to be eroded per 0.4 ha/year). Attribute query functions should support the use of complex Boolean expressions on the attributes of one or more data layers.

Integrated Analysis of Spatial and Attribute Data

The ability to effectively process spatial and attribute data is what primarily distinguishes GISs from automated mapping and computer-aided drafting systems (Aronoff 1989). The integrated analysis of spatial and attribute data can be divided into four categories (Berry 1987, Aronoff 1989): retrieval/classification/measurement, overlay, neighborhoods, and connectivity of network functions.

The retrieval processes are used to answer the first three basic questions that a GIS can address (Walker and Miller 1990): what exists at a particular location; where are certain conditions met; and what changes have occurred and where have changes occurred over time? Retrieval processes operate on spatial and attribute data. Output of the selective searches can create new layers in the database, tabular reports, interactive displays, or maps.

Classification processes frequently are performed after a retrieval process and are used to assign a new attribute. If deciduous, coniferous, and mixed forest types are retrieved from a vegetational data layer, a classification function could be used to assign a new attribute, forests, to the areas retrieved. Classification functions can be applied on one layer or on multiple data layers.

Measure functions calculate distances between points, lengths of lines, perimeters and areas of polygons, and sizes of continuous areas of the same feature. Selecting

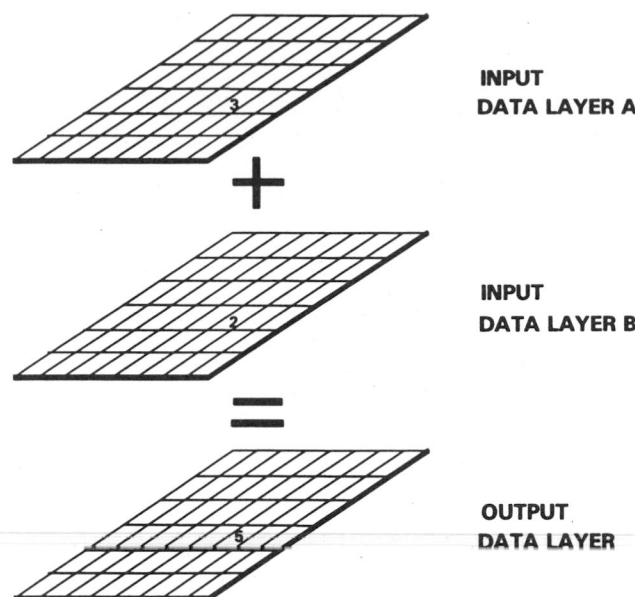

Fig. 13. An arithmetic overlay operation is used to add the value in data layer A to the value in data layer B. The sum of data layers A and B is stored in the output data layer (Arnoff 1989).

all wetlands >8 ha in size from a data layer describing the various wetlands types would require a retrieval process (retrieve all wetlands types), a classification process (classify the various retrieved wetlands types as one category), a measurement process (determine the size of all continuous area of wetlands), and a final retrieval process (select all wetlands >8 ha).

Overlay operators are some of the most fundamental and most frequently used processes in GIS applications. Aronoff (1989) described two types of overlay operators—arithmetic and logical. Arithmetic operators are used to add, subtract, divide, or multiply values in one data layer by a constant or by values in another data layer in corresponding location (Fig. 13). Logical overlays are used to identify areas where one feature or condition exists in one data layer and another feature or condition exists in another data layer. The logical overlays can use various logical Boolean operators such as ''and'' where both condition A and condition B are met, ''or'' where condition A or condition B is met, and ''and not'' to identify areas where condition A is met, but not condition B (Fig. 14).

Neighborhood processes evaluate the characteristics of the area surrounding a specific location (Aronoff 1989). Measuring the length of edge between habitat types within 2 km of the location of a radio-telemetered animal is an example of a neighborhood function. All neighborhood processes require three parameters: the location of one or more target areas, the size of the neighborhood surrounding the targeted area, and a function to perform in the defined neighborhood and assigned to the targeted area. Five numerical functions can be applied in the neighborhood: average, diversity, majority, maximum/minimum, and total (Aronoff 1989).

Point-in-polygon and line-in-polygon operations are fundamental neighborhood process for vector-based GISs (Aronoff 1989). A point-in-polygon function is used to

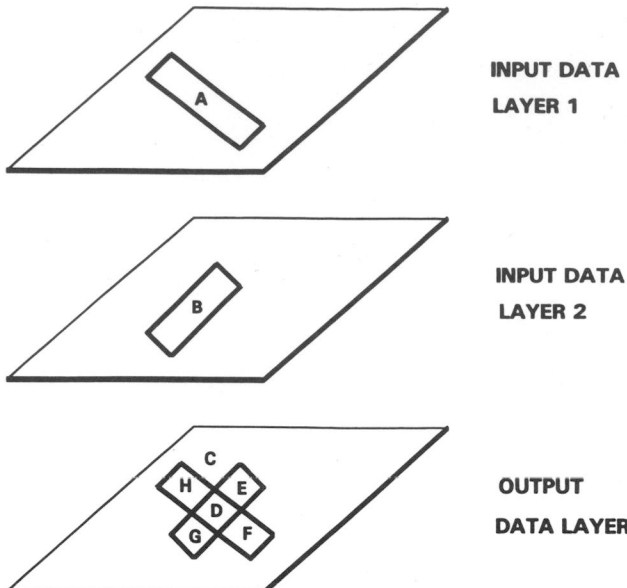

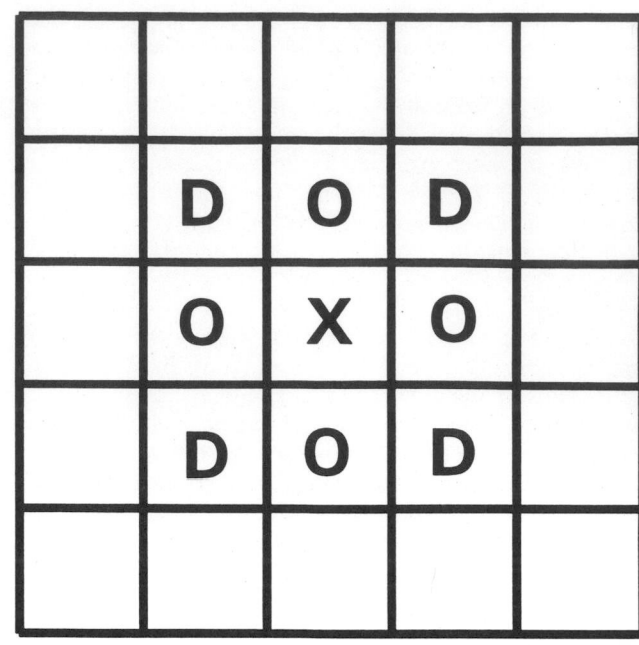

Fig. 14. Logical overlay operation of area A in data layer 1 with area B in data layer 2 can have many results. The resultant area C contains neither A nor B, area D contains A and B, areas H and F contain A but not B, and areas E and G contain B but not A (Aronoff 1989).

Fig. 15. A cell can have orthogonal and diagonal neighbors. Orthogonal neighbors for cell "X" are labeled with an "O", and diagonal neighbors are labeled by a "D." Some neighborhood operations recognize only orthogonal neighbors, whereas other neighborhood operations process orthogonal or diagonal neighbors.

determine which features (as defined by polygons) cover a specific point. In analysis of radiotelemetry data, point-in-polygon routines are used in vector-based GISs to ascertain various habitat parameters occurring at each location made for the animal being studied. In a similar manner, line-in-polygon routines are used to ascertain which polygon contains a specified line.

Various topographic functions can be computed from elevation information provided by DEMs or DTMs. Topographic functions usually are computed from raster data of elevation. The elevation for a particular cell (X), plus the elevation of the eight orthogonal (O) and diagonal (D) neighbors (Fig. 15), can be used to ascertain slope, aspect, and topographic position (ridge, valley, knoll). These topographic parameters frequently are highly correlated with the distribution of plant and animal species, and they frequently can be used in remote sensing applications to distinguish spectrally similar habitats. For example, spectrally, coastal dunes often cannot be separated from sandy flats. Slope and topography position parameters can readily separate these two spectrally similar sandy habitats.

Other topography functions include illumination techniques to enhance display of elevation or relief, viewshed modelling to determine what areas can be seen from a specific point for appraising visual impacts, watershed analysis to determine the extent of the watershed, and perspective views to produce illustrations of the relief as viewed from a given point. These topography functions usually are classified as connectivity functions. Various other neighborhood functions are available for interpolating information from various point locations.

Connectivity functions accumulate values over a connected area (Aronoff 1989). Connectivity functions require (1) a procedure for connecting the areas and (2) a measure for the connected area. Continuity measures are connectivity functions that typically measure the size of a

continuous area. Evaluating habitat for a particular avian species using bottomland hardwood forests will require the locations of all bottomland hardwoods. However, if the species is known to use only bottomland hardwood forests that are at least 50 ha in size, a continuity measure would be used to ascertain the size of each continuous block of bottomland hardwood forests. Continuity measures are a valuable tool for measuring habitat fragmentation.

Proximity to a habitat type also may be valuable in evaluating the availability of habitat to certain species. Proximity analysis is the measure of distance between features. Figure 16 illustrates the use of proximity function to define the area within 250 m of all streams.

Various other connectivity functions are used in GIS applications. Network functions are processes to optimize vehicle routing and to divide areas into service districts for optimizing limited resources (Aronoff 1989). Spread functions can be used to evaluate the cost of transversing an area and often are used to evaluate alternative routes for transportation and utilities. Such routines will find the lowest cost route between two points. Cost can be measured separately for economic, social, or environmental cost, or total costs can be calculated by summing the weighted economic, social, and environmental costs.

Cartographic Modeling

The previous sections described many of the processes of a GIS. Tomlin and Berry (1979) and Berry (1987) developed the term "cartographic modeling" to describe the use of basic GIS processes in a logical sequence to solve complex spatial problems. A series of standard processes, such as reclassifying, overlaying, distance, and neighborhood operations, is used to subdivide the spatial problems into a series of primitive operations. A cartographic model can be depicted with a flowchart. The initial data layers are shown, followed by a standard process performed on the data layer, followed by an output (often a data layer) created by the processing function performed. An output data layer then may be used as input to another process. Various inputs, processes, and outputs are diagrammed until the solution is derived. The cartographic model provides the pathway for solving the specified spatial problem. Prior to implementing the cartographic model with the GIS, the flowchart of the cartographic model can be reviewed and easily revised. Cartographic models are an excellent means to help determine which data layers will be required to solve any spatial problem.

Berry (1987) identified many advantages of the use of cartographic models. These models are capable of dynamic simulations and provide spatial "what if" analysis. Koeln (1980) used a GIS to identify the "best" (or least bad) sites for general aviation airports. Three different sets of weights for economic, social, and environmental impacts were used to simulate the attitude of three different groups of decision makers. One set of simulated decision makers (the misers) was concerned primarily with the economic costs but recognized some social and environmental costs. Another set of simulated decision makers (the environmentalists) weighted the environmental costs of general aviation airport the highest, followed by social costs, then economic costs. The altruist group of decision makers weighted economic, social, and environmental costs nearly equally. Through the cartographic modelling approach, three different sets of "best sites" for general aviation airports were selected based upon weights assigned by the simulated decision makers. Because the choice of location for general aviation airports is quite constrained, many of the best sites as determined from the misers' weights were identical to the best sites chosen with the weights of the environmentalists and altruists.

Another advantage of the cartographic modeling approach is its flexibility. New considerations can be added easily or existing ones can be refined. The cartographic modelling approach also provides an effective means for communicating the process used, including consideration of the specific application and fundamental procedures applied. A flowchart provides an excellent tool for communicating the logic, assumptions, and relationships used in the analysis.

Output Processing

Information from GISs must be reported. The reports can be statistical tabulations, such as reports of various habitat parameters for a study area or the polygon defining the home range of an animal, interactive displays of data on color and monochrome monitors, and hard-copy maps. Both hard-copy and soft-copy (images on a color or monochrome monitor) are essential. The reporting capabilities of GISs vary greatly, depending upon the software used. Modern GISs have interactive capabilities to display images on color or monochrome monitors. Most modern GISs have improved, and are improving, hard-copy output capabilities. Some systems have map automation procedures, text-labeling abilities, and graphic capabilities that nearly match the map-making capabili-

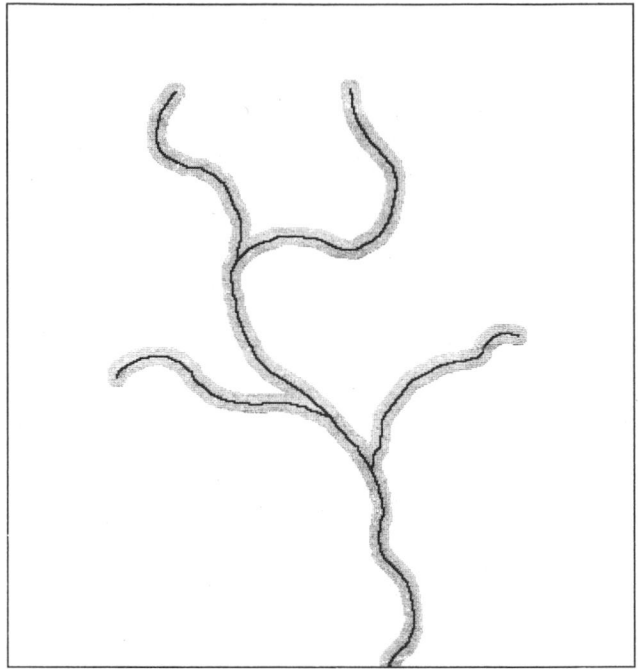

Fig. 16. A proximity function was used to define all areas within 250 m of streams.

ties of automated mapping and drafting systems. The reduction in the cost of image display devices, color plotters, and film recorders allows many GIS users to afford quality output devices. The expansion of software output capabilities seen for most commercially available GISs reflects the availability of an increasing array of quality output devices.

A GIS APPLICATION IN WATERFOWL MANAGEMENT

Management of waterfowl populations involves changing either the recruitment or survival rates of a population. Recruitment and survival are often functions of the amount and type of habitat; therefore, much of waterfowl management consists of preserving, adding, or modifying habitat. Selecting those habitat manipulations that alone or in combination will have the desired effect on a population is a difficult procedure because of the number of variables involved and the complex biological interactions between species biology and habitat.

The same advances in computer technology that have led to recent developments in the use of GISs have led to increases in the use of simulation models that attempt to portray the population biology of waterfowl species. Population data required by these simulation models usually are difficult and expensive to obtain. For many species, habitat availability and quality and various population parameters are correlated. Fortunately, recent advances in remote sensing have made inexpensive estimates of the availability of habitat over vast areas available to waterfowl managers. A marriage between GIS and simulation modeling is the logical outcome of these technological developments. As an example, we will illustrate the use of such a combined GIS and population simulation model for evaluating management practices

Table 3. Availability of nesting habitat for a 10.4-km² area in central North Dakota used to illustrate linkage of a population simulation model to a GIS with habitat data derived from Landsat TM.

Habitat	Percent of area per treatment[a]			
	Control	Impoundment	Conservation reserve	Barrier fence
Autumn-plowed crop	51.86	49.63	28.53	51.86
Grain stubble	1.21	1.02	1.21	1.21
Summer fallow	3.18	3.18	3.18	0.58
No-till winter wheat	11.19	11.19	11.19	11.19
Grassland	13.61	9.72	13.61	13.61
Hayland	2.00	2.00	2.00	2.00
Right-of-way	0.46	0.46	0.46	0.46
Shallow wetland[b]	8.88	7.13	8.88	8.88
Deep wetland[b]	1.42	3.81	1.42	1.42
Permanent water	0.25	0.25	0.25	0.25
Fenced cover	0.00	0.00	0.00	2.61
Conservation reserve	0.00	0.00	23.33	0.00
Barren	5.94	11.24	5.94	5.94

[a]The treatments were: (0) control, no management; (1) creation of an impoundment; (2) half of cropland converted to conservation reserve; (3) addition of a 148-ha predator barrier fence.

[b]Only the portion of the wetland representing nesting cover is shown; the remainder is included in barren.

for mallards in the prairie pothole region of North Dakota.

For the following example, we selected a 10.4-km² plot from typical duck habitat in central North Dakota. The area is glaciated and has numerous small wetland basins. The uplands are used almost exclusively for agriculture. Landsat TM data from May and September of 1986 processed by methods developed by Koeln and Wesley (1987) and Koeln et al. (1988) were used. The digital image processing techniques created a GIS data layer for land use and land cover. The retrieval and reporting functions of a GIS were used to report the areas for three wetland and 10 upland classes. This habitat information was altered to represent the habitats available to birds arriving in early May. For example, areas of growing grain were assumed to have been bare soil (recently plowed ground) when birds arrived in late April. Habitat classes conformed with the classes (Cowardin et al. 1988b) required as data input for a mallard productivity model (Johnson et al. 1987). That model allows the user to vary population parameters and habitat availability to predict the resulting production of young. Cowardin et al. (1988b) illustrated the use of the model in combination with other models that predict the number of breeding birds settling in an area. The same models have been linked with habitat data derived from high-altitude photography and aerial video (Cowardin et al. 1988a).

Population parameters used in the example were based on information from Cowardin et al. (1988b) and Klett et al. (1988). The habitat data (Table 3) can be displayed on the monitor of a microcomputer. These data then were transmitted to the mallard productivity model, and the

model was executed to obtain estimates of expected breeding population and young produced as well as other important population parameters (Table 4). For our demonstration, the control data represent current conditions.

The problem for the waterfowl manager is to select from an array of potential management alternatives the one most likely to maximize production from an area. The simulation models also can be combined with economic models to produce a planning tool (Nelson and Wishart 1988). The system demonstrated here combines GIS and simulation modeling. It cannot tell the manager what management should be applied, but it is a tool to assist evaluation of the outcome of various alternatives. The software has the advantage of allowing this to be done by modifying the habitat displayed on the computer monitor. For illustration, we used the editing functions of the GIS to simulate three management techniques. These techniques were represented by three different land use and land cover data files. In one simulation, we created a large, semipermanent wetland. In another simulation we converted one-half of the cropland to planted cover under the USDA's CRP. In the third simulation, we constructed a predator barrier fence around 148 ha of tall, dense nesting cover (Figs. 17–19). The predicted results are highly dependent on the assumptions made for many population parameters in the model. As a first approximation we used the same assumptions made by Cowardin et al. (1988b) and unpublished nest survival rates prepared for data in Klett et al. (1988).

Results of the simulations are summarized in Table 4. All treatments show some increase in the number of recruits produced. Creation of the large impoundment was the only treatment that increased the amount of wetland used by breeding pairs. The model predicted a corresponding increase in breeding population and recruits produced, but the increase in nest success was negligible. Conservation reserve was modeled under two different nest survival rates for that cover because there were no published data. We therefore used nest survival of 13.2% (unpubl. data, Northern Prairie Wildl. Res. Cent., Jamestown, N.D.) for planted cover in simulation (a) because that cover is similar to what is expected under conservation reserve. Some might argue that nest success in conservation reserve should be higher than planted cover because the large amount of conservation reserve may dilute the effect of predation. In conservation reserve (b), without real data, we chose a nest success rate of 20% which is relatively high but reasonable. The treatment then produced as many recruits as the impoundment from the same number of pairs used in the control. Addition of the predator barrier fence produced the most recruits of all treatments.

Results obtained from the model are extremely sensitive to the input data (Johnson et al. 1987, Cowardin et al. 1988b). It was designed as a tool to assist in comparing the potential outcome to be obtained from treatments. In the example, we used a range of reasonable input data, but there was an absence of data for conservation reserve. Both the input data and the assumptions upon which the model was based must be considered carefully when model results are interpreted.

The model is also sensitive to the amounts of habitat present in the area to be evaluated. Fortunately, these areas can be measured more accurately than some of the mallard population parameters by use of remote sensing techniques. In previous applications of the model, construction and modification of data sets that simulated various treatments were time- and labor-intensive. The combination of GIS technology with the model overcame this problem. Furthermore, the display of a real landscape and the rapid modification of that landscape by means of GIS editing functions are easily understandable by a land manager.

Through the efforts of the North American Waterfowl Management Plan (NAWMP), many waterfowl habitat-enhancement techniques are being applied to the prairie pothole region. By comparing land cover and land use data derived from the 1986 satellite data with data derived from more recent satellite data, we will have an excellent record of the success of the NAWMP in changing the landscape for improving waterfowl habitat. The effects of these actual changes will be evaluated with the GIS and mallard model approach described above.

A CAUTIONARY NOTE

GISs encourage the user to do things that often are not justified by the nature of the data involved (Goodchild and Gopal 1989). The ability to change map scales and to overlay maps can be deceiving; the user must be aware of the imprecision inherent in all cartography and of the ways errors compound when map scales are changed or when maps are merged (Abler 1987).

The presence of error on maps begins with the process of map projection (Vitek et al. 1984). A map projection is a systematic representation of all or part of the three-dimensional earth to a two-dimensional plane. Since this cannot be done without distortion, the user must choose the map property to be shown accurately at the expense of others, or a compromise of several properties (Snyder 1987).

Error is introduced to maps in the process of cartographic abstraction. A map is a model of reality, and map contents are often elegant misrepresentations of changes that are often gradual, vague, or fuzzy (Burrough 1986). Map errors involve both positional and attribute accuracy, which in some instances can be difficult to separate. Furthermore, both position and attribute error are a function of scale.

Errors accumulate during the analysis process in a GIS. Newcomer and Szajgin (1984) demonstrated a method for estimating the error propagation during the map overlay process. They concluded that the accuracy of the final map is a function of the number of map layers, the accuracy of these layers, and the coincidence of errors at the same position from several map layers. The accuracy of spatial databases and the propagation of error during data analysis are complex issues (Newcomer and Szajgin 1984, Vitek et al. 1984, Walsh et al. 1987, Goodchild and Gopal 1989, Lunetta et al. 1991).

The future for the integration of GIS, remote sensing, and expert systems is promising. In 1988, the National Science Foundation announced the formation of the National Center for Geographic Information Analysis, a consortium of universities including the University of California at Santa Barbara, The State University of New York at Buffalo, and the University of Maine at Orono.

Table 4. Estimates of mallard population parameters for a 10.4-ha area produced by using habitat availability data derived from Landsat TM, a GIS, and a population simulation model applied to a control and three habitat enhancement practices.

Treatment[a]	Breeding population (pairs)	Nest success (%)	Hen success (%)	Recruits produced
Control	28	13.4	25.0	26.4
Impoundment	36	14.3	26.3	35.3
Conservation reserve (a)	28	14.9	27.4	28.5
Conservation reserve (b)	28	19.7	33.9	35.6
Barrier fence	28	32.2	50.1	53.0

[a]The treatments were: (0) control, no management; (1) creation of an impoundment; (2) half of cropland converted to conservation reserve, for (a) nest success in conservation reserve = 13.3%, for (b) nest success in conservation reserve = 20.0%; (3) addition of a 148-ha predator barrier fence.

The Center has outlined a program aimed at the systematic removal of perceived impediments to the adoption and use of GIS technology. The program consists of a series of initiatives, several of which already have resulted in publications and symposia.

SUMMARY

This chapter provides wildlife managers and the students of wildlife management with an overview of the capabilities of GISs and possible uses of GISs to aid in land management. The chapter presents the principles behind the uses of GISs, but it is not a technical description of how to operate, design, or select GISs. It is hoped that the information provided in this chapter will encourage future wildlife managers to explore the use of GIS technologies for improving their decision-making abilities. Perhaps with this brief exposure to GISs, future wildlife managers not only will be more receptive to using GISs to help improve the decision-making processes directly affecting wildlife management, but also will try to participate in GIS activities for locating powerline corridors, airports, shopping centers, waste disposal sites, and other development activities. One misplaced development, such as those previously listed, can literally reverse all the wildlife habitat enhancements created by a wildlife manager during his or her entire career (Giles 1991).

At least 100 available GISs of various types for a wide array of computers are available from many government agencies, universities, and commercial vendors (Parker 1991). It is hoped that wildlife managers desiring to use GIS technologies will not get bogged down in the bits, bytes, and ''. . . primordial ooze of system development or primitive promotions'' (Giles 1991:5), but will become excited about the capabilities provided by GISs to gain ''. . . explanatory, descriptive, and predictive control . . .'' (Giles 1991:5) of ecosystems.

COVER TYPES
GRASSLAND
WINTER WHEAT
CROP STUBBLE
SUMMER FALLOW
HAYLAND
CROPLAND
OPEN WATER
DEEP MARSH
SHALLOW MARSH
ROADS
ODD AREAS

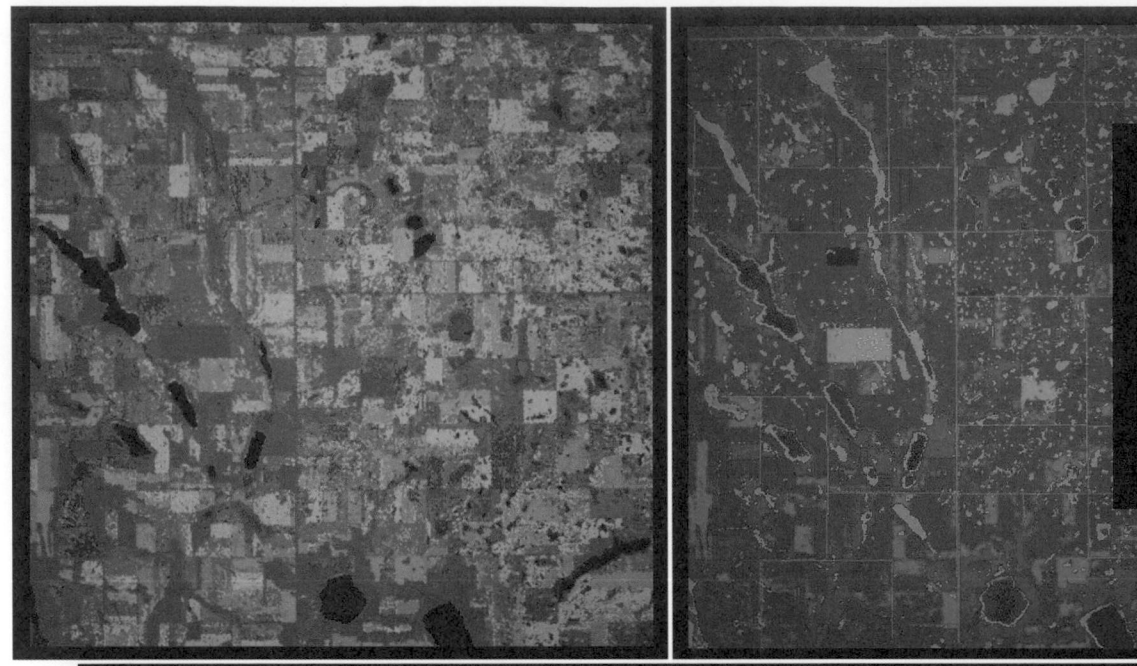

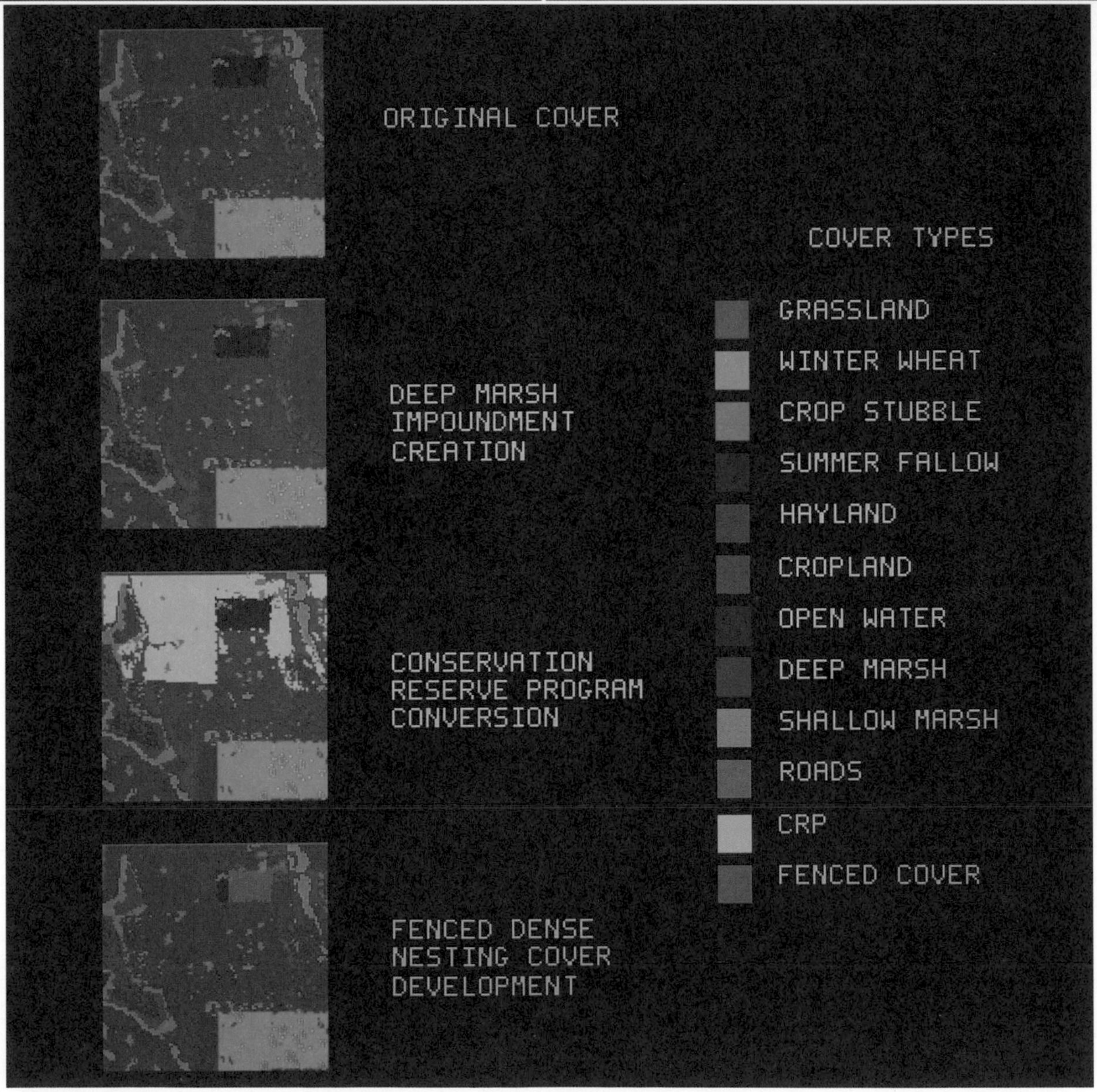

ORIGINAL COVER
DEEP MARSH IMPOUNDMENT CREATION
CONSERVATION RESERVE PROGRAM CONVERSION
FENCED DENSE NESTING COVER DEVELOPMENT
COVER TYPES
GRASSLAND
WINTER WHEAT
CROP STUBBLE
SUMMER FALLOW
HAYLAND
CROPLAND
OPEN WATER
DEEP MARSH
SHALLOW MARSH
ROADS
CRP
FENCED COVER

LITERATURE CITED

ABLER, R. F. 1987. The National Science Foundation National Center for Geographic Information and Analysis. Int. J. Geogr. Inf. Syst. 1:303–326.

AGEE, J. K., S. C. F. STITT, M. NYQUIST, AND R. ROOT. 1989. A geographic analysis of historical grizzly bear sightings in the North Cascades. Photogram. Eng. Remote Sens. 55:1637–1642.

AHERN, F. J., ET AL. 1987. Radiometric correction of visible and infrared remote sensing data at the Canada Centre for Remote Sensing. Int. J. Remote Sens. 98:1349–1376.

ANDERSON, J. R., E. E. HARDY, J. T. ROACH, AND R. E. WITMER. 1976. A land use and land cover classification system for use with remote sensor data. U.S. Geol. Surv. Prof. Pap. 964. 28pp.

ANONYMOUS. 1989. Using the TIGER files. U.S. Stat. Newsl. 5(12): 1–5.

ARONOFF, S. 1989. Geographic information systems: a management perspective. WDL Publ., Ottawa, Ont. 294pp.

BAND, L. E. 1989. A terrain-based watershed information system. Hydrological Processes 3:151–162.

BARNARD, T., R. J. MACFARLANE, T. NERAASEN, R. P. MROCZYNSKI, J. JACOBSON, AND R. SCHMIDT. 1981. Waterfowl habitat inventory of Alberta, Saskatchewan and Manitoba by remote sensing. Proc. Can. Symp. Remote Sens. 7:150–158.

BERRY, J. K. 1987. Fundamental operations in computer-assisted map analysis. Int. J. Geogr. Inf. Syst. 1:119–136.

BREININGER, D. R., M. J. PROVANCHA, AND R. B. SMITH. 1991. Mapping Florida scrub jay habitat for purposes of land-use management. Photogram. Eng. Remote Sens. 57:1467–1474.

BROSCHART, M. R., C. A. JOHNSTON, AND R. J. NAIMAN. 1989. Predicting beaver colony density in boreal landscapes. J. Wildl. Manage. 53:929–934.

BURKE, I. C., T. G. F. KITTEL, W. K. LAUENROTH, P. SNOOK, C. M. YONKER, AND W. J. PARTON. 1991. Regional analysis of the Central Great Plains. BioScience 41:685–692.

BURROUGH, P. A. 1986. Principles of geographical information systems for land resources assessment. Oxford Univ. Press, New York, N.Y. 193pp.

CANNON, R. W., F. L. KNOPF, AND L. R. PETTINGER. 1982. Use of Landsat data to evaluate lesser prairie chicken habitats in western Oklahoma. J. Wildl. Manage. 46:915–922.

CHAVEZ, P. S., JR. 1989. Radiometric calibration of Landsat Thematic Mapper multispectral images. Photogram. Eng. Remote Sens. 55: 1285–1294.

CLARK, D. K., AND N. G. MAYNARD. 1986. Coastal zone color scanner imagery of phytoplankton pigment distribution in Icelandic waters. Pages 350–357 in Proc. SPIE ocean optics VII. Int. Soc. Optical Eng., Billingham, Wash.

COULSON, R. N., L. J. FOLSE, AND D. K. LOH. 1987. Artificial intelligence and natural resource management. Science 237:262–267.

COWARDIN, L. M., P. M. ARNOLD, T. L. SHAFFER, H. R. PYWELL, AND L. D. MILLER. 1988a. Duck numbers estimated from ground counts, MOSS map data, and aerial video. Pages 205–219 in J. D. Scurry, comp. Proc. Natl. MOSS Users' Conf. 5. Louisiana Sea Grant College Program, Baton Rouge, and U.S. Fish Wildl. Serv., Slidell, La.

———, V. CARTER, F. C. GOLET, AND E. T. LAROE. 1979. Classification of wetlands and deep water habitats of the United States. U.S. Fish Wildl. Serv. Rep. FWS/OBS-79/31. 103pp.

———, D. H. JOHNSON, T. L. SHAFFER, AND D. W. SPARLING. 1988b. Applications of a simulation model to decisions in mallard management. U.S. Fish Wildl. Serv. Fish Wildl. Tech. Rep. 17. 28pp.

CRAIGHEAD, J. J., F. L. CRAIGHEAD, AND D. J. CRAIGHEAD. 1986. Using satellites to evaluate ecosystems as grizzly bear habitat. Pages 101–112 in Proc. grizzly bear habitat symposium, Missoula, Mont.

CRIST, E. P., R. LAURIN, AND R. C. CICONE. 1986. Vegetation and soils information contained in transformed thematic mapper data. Pages 1465–1470 in Proc. IGARSS' 86 Symp. ESA SP-254.

CURRAN, P. J. 1985. Principles of remote sensing. Longman Group Limited, London, U.K. 282pp.

DAVIS, F. W., AND J. DOZIER. 1990. Information analysis of a spatial database for ecological land classification. Photogram. Eng. Remote Sens. 56:605–613.

———, D. M. STOMS, J. E. ESTES, J. SCEPAN, AND J. M. SCOTT. 1990. An information systems approach to the preservation of biological diversity. Int. J. Geogr. Inf. Syst. 4:55–78.

DE STEIGUER, J. E., AND R. H. GILES, JR. 1981. Introduction to computerized land-information systems. J. For. 79:734–737.

DRISCOLL, D. 1990. Remote sensing: USFS pest management group. GIS World Mag. 3(5):94–96.

DUEKER, K. J., AND D. KJERNE. 1989. Multipurpose cadastre: terms and definitions. Am. Soc. Photogram. Remote Sens. and Am. Congr. Surv. Mapping, Falls Church, Va. 5:94–103.

DUGGIN, M.J. 1985. Factors limiting the discrimination and quantification of terrestrial features using remotely sensed radiance. Int. J. Remote Sens. 6:3–27.

DULANEY, R. A. 1987. A geographic information system for large area analysis. Pages 206–215 in Proc. of GIS '87. Am. Soc. Photogram. Remote Sens., Falls Church, Va.

ELROD, J. A. 1988. CZCS view of an oceanic acid waste dump. Remote Sens. Environ. 25:245–254.

ESTES, J. A., E. J. HAJIC, AND L. R. TINNEY, EDITORS. 1983. Fundamentals of image analysis: analysis of visible and thermal infrared data. Pages 987–1124 in R. N. Colwell, ed. Manual of remote sensing. Second ed., Vol. 1. Am. Soc. Photogram., Falls Church, Va.

EVERITT, J. H., AND D. E. ESCOBAR. 1989. The status of video systems for remote sensing applications. Proc. Biennial Workshop on Color Aerial Photography and Videography. Am. Soc. Photogram. Remote Sens. 12:6–29.

FAUST, N. L., W. H. ANDERSON, AND J. L. STARR. 1991. Geographic information systems and remote sensing future computing environment. Photogram. Eng. Remote Sens. 57:655–668.

GILES, R. H., JR. 1991. Nine thoughts about geographic information systems. Nat. Resour. Comput. Newsl. 6(4):3–5.

GOODCHILD, M., AND S. GOPAL, EDITORS. 1989. The accuracy of spatial databases. Taylor and Francis, London, U.K. 290pp.

GOODENOUGH, D. G., M. GOLDBERG, G. PLUNKETT, AND J. ZELEK. 1987. An expert system for remote sensing. IEEE Trans. Geoscience Remote Sens. GE-25:349–359.

GOSSELINK, J. G., AND L. C. LEE. 1989. Cumulative impact assessment in bottomland hardwood forests. Wetlands 9. 174pp.

GOWARD, S. N., C. J. TUCKER, AND D. G. DYE. 1985. North American vegetation patterns observed with the NOAA-7 advanced very high resolution radiometer. Vegetatio 64:3–14.

GRAETZ, R.D. 1990. Remote sensing of terrestrial ecosystem structure: an ecologist's pragmatic view. Pages 5–30 in R. J. Hobbs and H. A. Mooney, eds. Remote sensing of biospheric functioning. Springer-Verlag, New York, N.Y.

GRAVATT, G. 1991. National Wetlands Inventory. Pages 29–31 in K. K. Reay, ed. Proc. Natl. Conf. Integrated Water Inf. Manage. Virginia Polytechnic Inst. State Univ., Blacksburg.

HEINEN, J. T. AND R. A. MEAD. 1984. Simulating the effects of clearcuts on deer habitat in the San Juan National Forest, Colorado. Can. J. Remote Sens. 10:17–24.

HODGSON, M. E., J. R. JENSEN, H. E. MACKEY, JR., AND M. C. COULTER. 1988. Monitoring wood stork foraging habitat using remote sensing and geographic information systems. Photogram. Eng. Remote Sens. 54:1601–1607.

JACKSON, R. D. 1983. Spectral indices in n-space. Remote Sens. Environ. 13:409–421.

JENSON, S. K., AND J. O. DOMINGUE. 1988. Extracting topographic structure from digital elevation data for geographic information system analysis. Photogram. Eng. Remote Sens. 54:1593–1600.

JOHNSON, D. H., D. W. SPARLING, AND L. M. COWARDIN. 1987. A

←

Fig. 17. (upper left) Landsat TM data used for estimating current waterfowl production and simulating waterfowl production after various habitat changes were simulated. *Fig. 18.* (upper right) Habitat classes derived from Landsat TM data and used in simulating waterfowl production. *Fig. 19.* Three habitat changes simulated by using editing functions of a GIS. *Fig. 19A* (upper center) depicts the creation of a large impoundment. *Fig. 19B* (lower center) depicts the simulated conversion of 50% of the existing cropland to CRP planted cover. *Fig. 19C* (bottom) shows the location of a simulated predator barrier fence surrounding tall, dense, nesting cover.

model of the productivity of the mallard duck. Ecol. Model. 38: 257–275.

JOHNSTON, C. A., AND J. BONDE. 1989. Quantitative analysis of ecotones using a geographic information system. Photogram. Eng. Remote Sens. 55:1643–1647.

———, N. E. DETENBECK, J. P. BONDE, AND G. J. NIEMI. 1988. Geographic information systems for cumulative impact assessment. Photogram. Eng. Remote Sens. 54:1609–1615.

———, AND R. J. NAIMAN. 1990a. Aquatic patch creation in relation to beaver population trends. Ecology 71:1617–1621.

———, AND ———. 1990b. The use of a geographic information system to analyze long-term landscape alteration by beaver. Landscape Ecol. 4:5–19.

JUSTICE, C. O., J. R. G. TOWNSHEND, B. N. HOLBEN, AND C. J. TUCKER. 1985. Analysis of the phenology of global vegetation using meteorological satellite data. Int. J. Remote Sens. 6:1271–1318.

KAUTH, R. J., AND G. S. THOMAS. 1976. The tasselled cap—a graphic description of the spectral-temporal development of agricultural crops as seen by Landsat. Pages 4B41–4B51 in Proc. symposium machine processing of remotely sensed data. Purdue Univ., West Lafayette, Ind.

KLETT, A. T., T. L. SHAFFER, AND D. H. JOHNSON. 1988. Duck nest success in the prairie pothole region. J. Wildl. Manage. 52:431–440.

KOELN, G. T. 1980. A computer-assisted general aviation airport location and evaluation system for Virginia. Ph.D. Thesis, Virginia Polytechnic Inst. State Univ., Blacksburg. 235pp.

———, AND E. A. COOK. 1984. Applications of geographic information systems for analysis of radio-telemetry data on wildlife. Pecora 9:154–158.

———, J. E. JACOBSON, D. E. WESLEY, AND R. S. REMPLE. 1988. Wetland inventories derived from Landsat data for waterfowl management planning. Trans. North Am. Wildl. Nat. Resour. Conf. 53: 303–310.

———, AND D. E. WESLEY. 1987. Ducks Unlimited's wetland inventory. Pages 225–233 in J. Zelazny and J. S. Feierabend, eds. Proc. increasing our wetland resources. Natl. Wildl. Fed., Washington, D.C.

KORTE, G. B. 1991. How GIS relates to CADD, CAM and AM/FM. Point of Beginning 16:56–66.

LAUER, D. T., J. E. ESTES, J. R. JENSEN, AND D. D. GREENLEE. 1991. Institutional issues affecting the integration and use of remotely sensed data and geographic information systems. Photogram. Eng. Remote Sens. 57:647–654.

LECKENBY, D. A., D. L. ISAACSON, AND S. R. THOMAS. 1985. Landsat application to elk habitat management in northeast Oregon. Wildl. Soc. Bull. 13:130–134.

LEE, K. H. 1991. Wetlands detection methods investigation. U.S. Environ. Prot. Agency Rep. 600/4-91/014. 73pp.

LILLESAND, T. M., AND R. W. KIEFER. 1987. Remote sensing and image interpretation. John Wiley & Sons, New York, N.Y. 721pp.

LUNETTA, R. S., R. G. CONGALTON, L. K. FENSTERMARKER, J. R. JENSEN, K. C. MCGWIRE, AND L. R. TINNEY. 1991. Remote sensing and geographic information system data integration: error sources and research issues. Photogram. Eng. Remote Sens. 57:677–687.

LYON, J. G. 1983. Landsat-derived land-cover classifications for locating potential kestrel nesting habitat. Photogram. Eng. Remote Sens. 49:245–250.

MAUSEL, P. W., J. H. EVERITT, D. E. ESCOBAR, AND D. J. KING. 1992. Airborne videography: current status and future perspectives. Photogram. Eng. Remote Sens. 58:1189–1195.

MAYER, K. E. 1984. A review of selected remote sensing and computer technologies applied to wildlife habitat inventories. Calif. Fish Game 70:101–112.

MCHARG, I. L. 1969. Design with nature. Doubleday and Company, Inc., Garden City, N.J. 197pp.

MCKEOWN, D. M., JR. 1987. The role of artificial intelligence in the integration of remotely sensed data with geographic information systems. IEEE Trans. Geoscience Remote Sens. GE-25:330–348.

MEISNER, B. N., AND P. A. ARKIN. 1984. The GOES precipitation index: large scale tropical rainfall estimates using infrared data. Proc. Conf. Hurricanes Tropical Meteorol. 15:203–206.

MILLER, K. V., AND M. J. CONROY. 1990. SPOT satellite imagery for mapping Kirtland's warbler wintering habitat in the Bahamas. Wildl. Soc. Bull. 18:252–257.

MURPHY, D. D., AND B. D. NOON. 1991. Coping with uncertainty in wildlife biology. J. Wildl. Manage. 55:773–782.

NELSON, J. W., AND R. A. WISHART. 1988. Management of wetland complexes for waterfowl production: planning for the prairie habitat joint venture. Trans. North Am. Wildl. Nat. Resour. Conf. 53: 444–453.

NEWCOMER, J. A., AND J. SZAJGIN. 1984. Accumulation of thematic map errors in digital overlay analysis. Am. Cartographer 11:58–62.

NIELSEN, R. D. 1991. Digital soils data: 1:15,840 to 1:7,500,000 scale digital soils information from SSURGO, STATSGO, and NATSGO data bases. Pages 66–68 in K. K. Reay, ed. Proc. national conference integrated water information management. Virginia Polytechnic Inst. State Univ., Blacksburg.

ORMSBY, J. P., AND R. S. LUNETTA. 1987. Whitetail deer food availability maps from Thematic Mapper data. Photogram. Eng. Remote Sens. 53:1081–1085.

———, AND G. A. SOFFEN. 1989. Foreword: special issue on the Earth Observing System (Eos). Inst. Electrical Electronics Eng. Trans. Geoscience Remote Sens. 27:107–108.

PALMERIM, J. M. 1987. Automatic mapping of avian species habitat using satellite imagery. Oikos 52:59–68.

PARKER, H. D., EDITOR. 1991. GIS world source book. GIS World, Inc., Ft. Collins, Colo. 597pp.

PECH, R. P., R. D. GRAETZ, AND A. W. DAVIS. 1986. Reflectance modelling and the derivation of vegetation indices for an Australian semi-arid shrubland. Int. J. Remote Sens. 7:389–403.

PETERSON, L., AND I. MATNEY. 1986. Data management. Pages 727–740 in A. Y. Cooperider, R. J. Boyd and H. R. Stuart, eds. Inventory and monitoring of wildlife habitat. U.S. Dep. Inter. Bur. Land Manage. Serv. Cent., Denver, Colo.

PICKUP, G., AND V. H. CHEWINGS. 1988. Forecasting patterns of soil erosion in arid lands from Landsat MSS data. Int. J. Remote Sens. 9:69–84.

PIWOWAR, J. M., AND E. F. LEDREW. 1990. Integrating spatial data: a user's perspective. Photogram. Eng. Remote Sens. 56:1497–1502.

PLACE, J. L. 1985. Mapping of forested wetland: use of SEASAT radar images to complement conventional sources. Prof. Geogr. 37:463–469.

RICHARDS, J. A. 1986. Remote sensing digital image analysis. Springer-Verlag, West Berlin. 281pp.

———, AND D. J. KELLY. 1984. On the concept of spectral class. Int. J. Remote Sens. 5:987–991.

RIPPLE, W. J., G. A. BRADSHAW, AND T. A. SPIES. 1991. Measuring forest landscape patterns in the Cascade Range of Oregon, USA. Biol. Conserv. 57:73–88.

———, AND S. WANG. 1989. Quadtree data structures for geographic information systems. Can. J. Remote Sens. 15:172–176.

RODCAY, G. 1991. GIS a "natural" for wildlife management. Pages 365–369 in H. D. Parker ed. GIS world source book. GIS World, Inc., Ft. Collins, Colo.

SAXON, E. C., AND DUDZINSKI, M. L. 1984. Biological survey and reserve design by Landsat mapped ecoclines—a catastrophe theory approach. Aust. J. Ecol. 9:117–123.

SCEPAN, J., F. DAVIS, AND L. L. BLUM. 1987. A geographic information system for managing California condor habitat. Proc. Int. Conf., Exhibits Workshops Geogr. Inf. Syst. 2:276–286.

SCHOTT, J. R., C. SALVAGGIO, AND W. J. VOLCHOK. 1988. Radiometric scene normalization using pseudoinvariant features. Remote Sens. Environ. 26:1–16.

SCHOWENGERDT, R. A. 1983. Techniques for image processing and classification in remote sensing. Academic Press, Inc., New York, N.Y. 249pp.

SHAW, D. M., AND S. F. ATKINSON. 1990. An introduction to the use of geographic information systems for ornithological research. Condor 92:564–570.

SHIH, E. H., AND R. A. SCHOWENGERDT. 1983. Classification of arid geomorphic surfaces using Landsat spectral and textural features. Photogram. Eng. Remote Sens. 49:337–347.

SIDLE, J. G., AND J. W. ZIEWITZ. 1990. Use of aerial videography in wildlife studies. Wildl. Soc. Bull. 18:56–62.

SILVA, L. F. 1978. Radiation and instrumentation in remote sensing. Pages 21–135 in P. H. Swain and S. M. Davis, ed. Remote sensing: the quantitative approach. McGraw-Hill Book Co., New York, N.Y.

SINGH, A. 1989. Digital change detection techniques using remotely-sensed data. Int. J. Remote Sens. 10:989–1003.

SNYDER, J. P. 1987. Map projections—a working manual. U.S. Geol. Surv. Prof. Pap. 1395. 383pp.

STEENHOF, K. 1982. Use of an automated geographic information system by the Snake River Birds of Prey Research Project. Computer-Environ. Urban Syst. 7:245–251.

STENBACK, J. M., C. B. TRAVLOS, R. H. BARRETT, AND R. G. CONGALTON. 1987. Application of remotely sensed digital data and a GIS in evaluating deer habitat suitability on the Tehama Deer winter range. Proc. Int. Conf., Exhibits Workshops Geogr. Inf. Syst. 2:440–445.

STRAHLER, A. H., C. E. WOODCOCK, AND J. A. SMITH. 1986. On the nature of models in remote sensing. Remote Sens. Environ. 20:121–139.

STRONG, L. L., D. S. GILMER, AND J. A. BRASS. 1991. Inventory of wintering geese with a multispectral scanner. J. Wildl. Manage. 55:250–259.

SWAIN, P. H., AND S. M. DAVIS, EDITORS. 1978. Remote sensing: the quantitative approach. McGraw-Hill Book Co., New York, N.Y. 396pp.

TANRE, D., ET AL. 1990. Description of a computer code to simulate the satellite signal in the solar spectrum: the 5S code. Int. J. Remote Sens. 11:659–668.

TASSAN, S., AND B. STURM. 1986. An algorithm for the retrieval of sediment content in turbid coastal waters from CZCS data. Int. J. Remote Sens. 7:643–655.

TOMLIN, C. D., AND J. K. BERRY. 1979. A mathematical structure for cartographic modeling in environmental analysis. Proc. Am Congr. Surv. Mapping 39:269–284.

TUCKER, C. J., H. E. DREGNE, AND W. N. NEWCOMB. 1991. Expansion and contraction of the Sahara Desert from 1980 to 1990. Science 253:299–301.

VITEK, J. D., S. J. WALSH, AND M. S. GREGORY. 1984. Accuracy in geographic information systems: an assessment of inherent and operational errors. Pecora 9:296–302.

VONDEROHE, A. P., R. F. GURDA, S. J. VENTURA, AND P. G. THUM. 1991. Introduction to local land information systems for Wisconsin's future. Wisc. State Cartographic Off., Madison. 59pp.

WALKER, P. A. 1990. Modelling wildlife distributions using a geographic information system: kangaroos in relation to climate. J. Biogeogr. 17:279–289.

WALKER, T. C., AND R. K. MILLER. 1990. Geographic information systems: an assessment of technology, applications, and products. Vol. I. SEAI Tech. Publ., Madison, Ga. 166pp.

WALKLETT, D. C. 1992. Investing in GIS and remote sensing holds the keys to understanding global change. Earth Observation Mag. 1(1):70.

WALSH, S. J., D. R. LIGHTFOOT, AND D. R. BUTLER. 1987. Recognition and assessment of error in geographic information systems. Photogram. Eng. Remote Sens. 53:1423–1430.

WELCH, R., T. R. JORDAN, AND M. EHLERS. 1985. Comparative evaluations of the geodetic accuracy and cartographic potential of Landsat-4 and Landsat-5 Thematic Mapper image data. Photogram. Eng. Remote Sens. 51:1249–1262.

WIEGAND, C. L., ET AL. 1986. Development of agrometeorological crop model inputs from remotely sensed information. IEEE Trans. Geoscience Remote Sens. GE-24:90–98.

WOODCOCK, C. E., AND A. H. STRAHLER. 1987. The factor of scale in remote sensing. Remote Sens. Environ. 21:311–332.

WORK, E. A., JR., AND D. S. GILMER. 1976. Utilization of satellite data for inventorying prairie ponds and lakes. Photogram. Eng. Remote Sens. 42:685–694.

YOUNG, T. N., J. R. EBY, H. L. ALLEN, M. J. HEWITT, III, AND K. R. DIXON. 1987. Wildlife habitat analysis using Landsat and radio-telemetry in a GIS with application to spotted owl preference for old growth. Proc. Int. Conf., Exhibits Workshops Geogr. Inf. Syst. 2:595–600.

Appendix I. USGS information centers

USGS's Earth Science Information Centers (ESIC) are an excellent source for information and advice on the availability of digital data and maps for GIS applications. The ESIC (formerly called the National Cartographic Information Centers) offices have fact sheets, user guides, price lists, order forms, and personnel to assist in selecting the best available information for a GIS application.

Available USGS Circulars include:

895-A, Overview and USGS Activities
895-B, Digital Elevation Models
895-C, Digital Line Graphs from 1:24,000 Maps
895-D, Digital Line Graphs from 1:2,000,000 Maps
895-E, Land Use and Land Cover Digital Data
895-F, Geographic Names Information System
895-G, Digital Line Graph Attribute Coding Standards

Addresses and telephone numbers for the ESIC offices are shown below:

Reston—ESIC
U.S. Geological Survey
507 National Center
Reston, VA 22092
(703) 648-4000

Rolla—ESIC
1400 Independence Road
Rolla, MO 65401
(314) 341-0851

Lakewood—ESIC
Federal Center
Box 25046, MS 504
Denver, CO 80225-0046
(303) 236-5829

Stennis Space Center—ESIC
Building 3101
Stennis Space Center, MS 39529
(601) 688-3544

Salt Lake City—ESIC
8105 Federal Bldg.
1245 S. State St.
Salt Lake City, UT 84138
(801) 524-5652

Menlo Park—ESIC
Building 3, MS 532
345 Middlefield Road
Menlo Park, CA 94025
(415) 329-4309

Anchorage—ESIC
4230 University Drive
Anchorage, AK 99508-4664
(907) 786-7011

Washington, D.C.—ESIC
Dept. of the Interior Bldg.
1849 C St., N.W., Rm. 2650
Washington, DC 20240
(202) 208-4047

Appendix II. Digital data sources in North America

U.S. Sources

Landsat Data
EOSAT
4300 Forbes Blvd.
Lanham, MD 20706
(800) 344-9933

SPOT Data
SPOT Image Corporation
1897 Preston White Drive
Reston, VA 22091
(703) 620-2200

**AVHRR, Coastal Zone Color
Scanner, GOES and SEASAT Data**
National Oceanic and Atmospheric Administration
National Environmental Satellite Data and
    Information Service
National Climatic Center
Satellite Data Services Division
Washington, DC 20233
(301) 763-8399

Landsat and AVHRR Data
U.S. Department of Interior
U.S. Geological Survey
EROS Data Center
Sioux Falls, SD 57198
(605) 594-6511

**DLG Files, Digital Elevation Data,
NWI Data and Land Use/Land Cover Data**
National Cartographic Information Center
U.S. Geological Survey
507 National Center
Reston, VA 22092
(703) 860-6045

**TIGER File Data and Census
Attribute Data Sets**
Customer Services Branch
Data User Services Division
Bureau of the Census
Washington, DC 20233
(301) 763-4100

Soils Data
National Cartographic Center
U.S. Department of Agriculture
Natural Resources Conservation Service
P.O. Box 6567
Fort Worth, TX 76115
(817) 334-5559

Canadian Sources

RadarSat International
275 Slater St., Suite 1203
Ottawa, Ontario K1P 5H9
(613) 238-6413
or
RadarSat International
Satellite Data Distribution Centre
3851 Shell Road, Suite 200
Richmond, British Columbia V6X 2W2
(604) 244-0400

National Digital Topographic Data
Topographic Surveys Division
Surveys and Mapping Branch
Energy, Mines, and Resources Canada
615 Booth Street
Ottawa, Ontario K1A 0E9
(613) 992-0924

**Data From the Canada Land Data
System (includes the Canada
Geographic Information System)**
Environmental Information Systems Division
State of the Environment Reporting Branch
Environment Canada
Ottawa, Ontario K1A 0H3
(613) 997-2800

**Data from the Canada Soils
Information System**
CanSIS Project Leader
Land Resource Research Center
Agriculture Canada, Research Branch
K.W. Neatby Building
Ottawa, Ontario K1A 0C6
(613) 995-5011

22

VEGETATION SAMPLING AND MEASUREMENT

Kenneth F. Higgins, John L. Oldemeyer, Kurt J. Jenkins, Gary K. Clambey, and Richard F. Harlow

INTRODUCTION

What is the utility of vegetation measurements for wildlife managers? In the prairie, savanna, tundra, forest, steppe, and wetland regions of the world, mixtures of plant species provide wildlife with food, cover, and, in some circumstances, water, the three essential habitat elements necessary to sustain viable wildlife populations. In strict definition, the variety of wildlife using plants ranges from snails and voles to bison and elephants in uplands and from mosquitos and ducks to muskrats and manatees in wetlands. Through evolutionary processes, some wildlife species are totally dependent on vegetation for all annual life requirements, whereas other species use vegetation only for cover or food. Regardless of what role vegetation plays in the sustenance of wildlife, any management or research project that requires the evaluation of the wildlife and habitat relationships on a unit of land will necessitate some form of measurement of vegetation.

The term vegetation can refer to a single plant or species on a specific site or to a mixture of plants or species in the landscape. Vegetation may occur naturally or be introduced, and it may be live or dead. The uses of vegetation measurements are many: (1) evaluation of vegetation response to management, (2) estimation of carrying capacity, (3) characterization of cover and habitat components for an endangered species, or (4) long-term monitoring of the general trend of plant vigor or habitat condition.

Surveying and measuring the *quantity* and *quality* of vegetation within various habitats are basic to wildlife research and management. Grassland, shrubland, and woodland habitats are made up of populations of plants in which individual plants are usually too numerous to inventory completely. Consequently, wildlife biologists usually are compelled to use some form of sampling technique to make inferences about the total plant population within a given habitat.

Vegetation sampling methodologies have evolved within several ecological disciplines (e.g., plant ecology, forestry, range science) and for a variety of management or research objectives (e.g., estimating forage for ungulates, describing habitat use by passerine birds). Description of every method that has been used for sampling vegetation is beyond the scope of this chapter. In this chapter, we are concerned about describing how to measure vegetative structure, which Dansereau (1957) described as the spatial organization (distribution) of individuals that form a stand. Thus, we have structured this chapter into a description of basic methods of sampling vegetation and examples of how those methods have been applied or modified in wildlife research and management. We will assume that the investigator has adequate knowledge of the concepts of wildlife ecology, the primary habitat requirements of wildlife species under study, and the ability to systematically identify the species of wildlife and vascular plants within the geographical area of the investigation.

INITIAL STEPS TO SAMPLING VEGETATION
Development of Objectives

The critical element of any project, whether management or research, is to define objectives. If a project has neither an objective for vegetative measurements nor a defined use for each type of measurement, the data should not be collected. Collecting vegetative data is time consuming and often difficult, and that time should be used to meet the needs of the objectives. It is important to have management or research plans reviewed to assure that the information being collected meets the needs of the objectives and that critical information is not being neglected.

Objectives must be specific. They should include what will be sampled, when it will be sampled, and where it will be sampled. Although these factors often are taken for granted, their identification requires that the investigator make a thorough analysis of the biology of the wildlife species to be studied, the factors that relate to the study, and the management or research needs.

General Aspects of Vegetation

After listing the objectives of the study and the primary habitat requirements of the wildlife species under study, one then may determine what aspects of the vegetation to sample. Some or all of the following may be important in describing these primary habitat requirements:

(1) species composition,
(2) spatial distribution, either vertical or horizontal,
(3) temporal variation in structure,
(4) biomass, and
(5) overall stand or landscape structure.

A reconnaissance survey of an area is usually sufficient to provide the investigator with an overview of the vegetative structure. The reconnaissance can be done on the ground or with use of aerial photography. In either instance, the objective of a preliminary survey is to determine *whether* to sample, to identify *what* to sample, and to identify what environmental factors will influence *how* and *when* to sample.

Consider the following example. Suppose the goal of a study is to inventory potential, natural nesting sites for wood ducks. The wood duck nests only in cavities in trees. Because nesting cavities within a reasonable distance of water are a primary habitat requirement of wood ducks, three objectives are:

(1) to quantify the number of wood duck nesting cavities,
(2) to identify the species of trees containing the cavities, and
(3) to determine the age-distribution of trees with cavities.

Assume a reconnaissance survey has revealed that the study area is a riparian system with a permanent stream, riparian vegetation bordering the stream, and farmland bordering the riparian vegetation. Because wood ducks nest in trees, one would not sample the area with crops, but would sample the riparian vegetation. A sample would be designed to randomly select a number of trees for examination. The objectives require identifying the species of trees in which wood duck cavities occur. Because we are interested in determining the number of wood duck cavities, our sample will need to provide an estimate of tree density, one aspect of horizontal spatial distribution. We are not, however, interested in heights of cavities, thus vertical distribution will not be of interest. Cavities often are present in older and larger trees and in dead trees, and we will determine the age distribution. In addition, dead trees are likely to be blown over in windstorms, thus we

may decide to mark cavity trees and follow them over time to determine the rate of loss. Biomass of trees is not of interest; however, if a mast crop is produced by the trees, we would be interested in biomass of mast, a food item of wood ducks.

Study Site Selection

Study site selection is a critical phase of any study and is directly related to the objectives. The first issue to resolve is selection of the habitat type to be sampled. This may require that the project area be mapped so that all habitat types, their location, and their size are enumerated. The objectives may require that samples be taken in all or in only several sites containing a certain habitat type.

The size of the study site must be large enough so that the vegetative characteristics being measured will not be influenced by adjacent habitat types (often called edge effect). Such edge effect may increase the variation in the sample, and unless such variation is explained by the sampling design (see Chapter 1), the results of the sample will be biased with regard to the objectives of the study. For example, if one were sampling browse production in a 100-ha stand of upland willow, one would avoid sampling adjacent to the edge of a habitat type that offered resting cover for moose, because those plants measured within close proximity of resting cover likely would have higher use (and perhaps lower levels of production) than plants measured in the middle of the willow stand.

A variety of other factors influences selection of study sites (e.g., topography, elevation, slope, aspect, soil type, management history, distance to human-caused disturbances). Generally, one is interested in selecting sites that are similar to one another, and care must be taken in selection of sites so that the site-to-site variation is natural and not affected by some factor not accounted for in the objectives and design of the study.

Although sampling design is explained in Chapter 1, visualizing how vegetation sampling plots or plots along a transect appear in field applications may be difficult. Examples of random and transect plot layouts are illustrated in Figs. 1, 2, and 3. Realize, however, many other layout designs are possible, and the final choice of a layout also will depend on the objectives and the requirements of the statistical analysis.

PREPARATIONS AND GETTING STARTED
Leadership

Vegetation sampling is time consuming and demanding (Table 1). Good leadership is essential to maintaining enthusiasm and quality of data collection. The principal investigator can demonstrate leadership by (1) being enthusiastic and knowledgeable about the study area, research design, equipment, plant identification, and data collection, (2) being organized and efficient during all aspects of vegetation sampling, (3) explaining to other team members how the data will be used to make decisions on resource management, and (4) doing his/her share of the data collection. The principal investigator also should listen to input from team members. They often have suggestions that make data collection more efficient. Explaining the entire project, answering questions, and incorporating appropriate

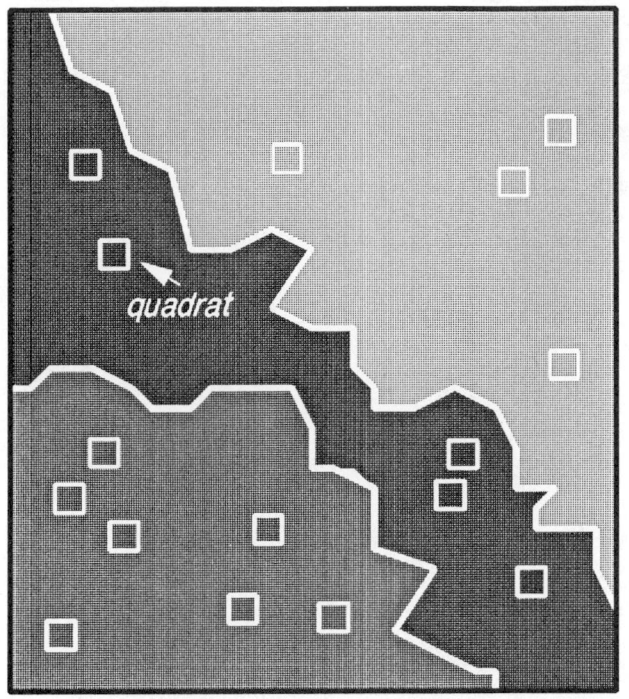 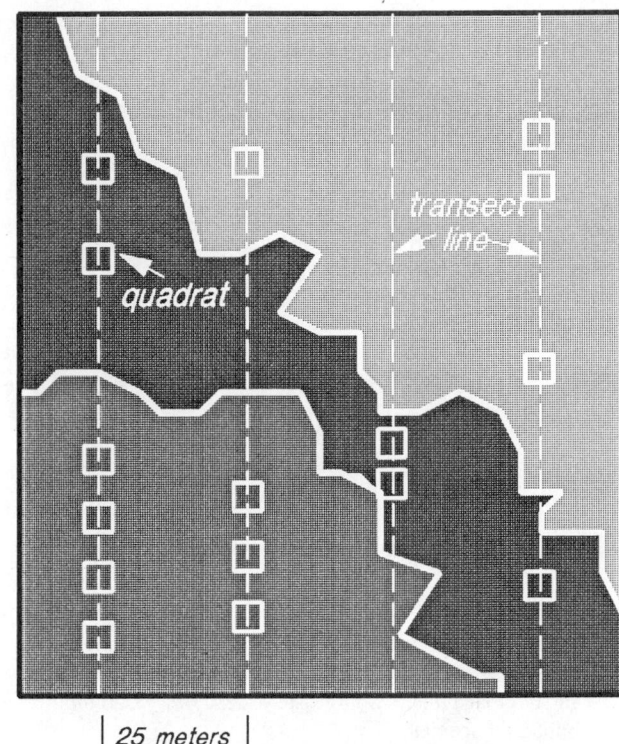

| 25 meters |

Fig. 1. Random (left) and systematic (right) distribution of quadrats with and without the use of transect lines on a site with three different vegetative cover types.

suggestions will make team members feel they are an integral part of the project (and they are!).

Initial Planning and Preparation

Considerable preparation in the office is required before the team goes into the field to conduct a vegetation study. The development of a list of supplies and equipment necessary to complete the task (Table 2) is an important first step. Obviously equipment lists will vary, depending on whether sampling is done in grasslands, wetlands, shrublands, or woodlands. These lists should be all-encompassing and should include everything from the number of pencils, the color of data sheets and plot markers, and

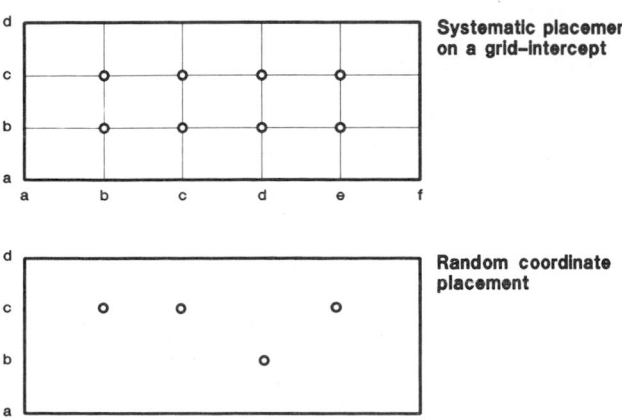

Fig. 2. Systematic and random placement of quadrats with grid coordinates.

the size and shape of the sampling frame, to calipers, photometers, seed traps, and field vehicles.

DATA FORMS

Develop data forms for recording the field data. Even though major advances have been made with entry of field data directly into programmable microcomputers at the time of sampling, most data still are recorded on paper forms. Field data forms can be developed to facilitate simple mathematical analysis with conventional calculators or to facilitate entry onto a personal computer for detailed and complex analysis. In either situation, a set of instruction codes defining what is represented by each numerical or letter entry should accompany each different kind of field data form. Team members must understand the meaning of zeros and blank spaces. Although a blank space usually means no value was available to measure

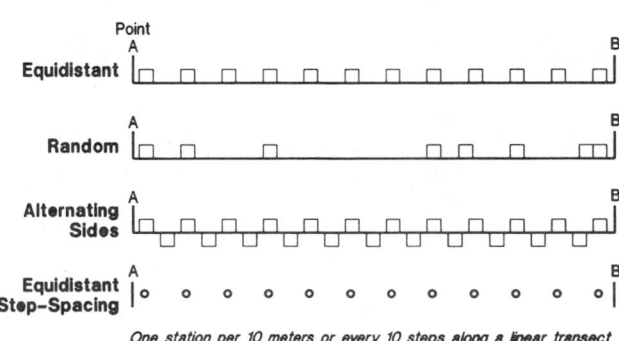

One station per 10 meters or every 10 steps along a linear transect

Fig. 3. Examples of various patterns of quadrat placement along permanent transect lines.

Table 1. Some representative times to complete a transect or a number of plots for different vegetative sampling techniques in various habitats or for different purposes. The numbers are relative, so they may not meet any specific project; however, they should help the investigator during initial planning. The times were derived from published literature, personal communication, and personal experience.

Sampling technique	Habitat type or vegetative component or unit site	Estimated time necessary to complete a plot, a practicable number of sample plots, or a transect by one to three persons	Minimum and range of plots usually necessary to characterize the community vegetative structure	Reference
Grassland				
30.5-m transect, line-intercept method for basal area		1.8–2.5 hr/transect[a]		Johnson 1957
30.5-m transect, point quadrat for basal area		0.5–0.8 hr/transect[a]		Johnson 1957
30.5-m transect, loop method for basal area		0.3–0.4 hr/transect[a]		Johnson 1957
0.30-m^2 clipped plots	California annual grasses	7 min/plot		Reppert et al. 1962
2.9-m^2 circular plots, clipped all species	Southeastern U.S.	32 min/plot, one person		Hilmon 1959
Single-point basal-hit sampling	Tallgrass prairie	7 hr/3,000–4,000 points/three persons	ca. 25 ha	Owensby 1973
Foliage density readings (Robel et al. 1970)	Any grassland	8 hr/1,000 readings/two persons	10 sites	J. M. Callow, pers. commun.
Nudds-board foliage density readings (Nudds 1977)	Any grassland	8 hr/100–200 readings/two persons		L. D. Flake, pers. commun.
10-pin point frame (Smith 1959)	Mixed and tallgrass prairie	8 hr/4,000–6,000 points/two persons		L. L. Manske, pers. commun.
10-pin point frame (Smith 1959)	Wet meadow wetland	8 hr/2,000–3,000 points/two persons		L. L. Manske, pers. commun.
Shrubs				
Shrub dimension/production estimates	Boreal forest	2 hr/25 plants/two persons		Peek 1970
3- × 5-m clipped plots	Boreal forest	24.7 hr/17 plots[a]		Bobek and Bergstrom 1978
Height × diameter measurements in 3- × 5-m plots	Boreal forest	2.3 hr/21 plots[a]		Bobek and Bergstrom 1978
Clipped plots	Southern forests	10–50 plots/2 person-days	28–158/site	Harlow 1977
Twig-length method to measure browse use	Montane shrub	50 min/50 plants/two persons	50/site	Jenson and Scotter 1977
30.5-m^2 plot, weight-estimation method for twig production	Eastern deciduous forest	1.5 hr/41 plots/two persons		Shafer 1963
30.5-m^2 plot, twig-count method for twig production	Eastern deciduous forest	1.5 hr/39 plots/two persons		Shafer 1963
30.5-m^2 plot, clip and weigh for twig production	Eastern deciduous forest	6.5 hr/37 plots/two persons		Shafer 1963
30.5-m line-point transect, sample every 0.30 m for shrub cover	Chaparral	7 min/transect[a]	4–26 transects/site	Heady et al. 1959

Table 1. Continued.

Sampling technique	Habitat type or vegetative component or unit site	Estimated time necessary to complete a plot, a practicable number of sample plots, or a transect by one to three persons	Minimum and range of plots usually necessary to characterize the community vegetative structure	Reference
30.5-m line-intercept for shrub cover	Chaparral	16 min/transect[a]	9–13 transects/site	Heady et al. 1959
0.1- × 0.5-m quadrats for shrub cover	Shrub steppe	15–30 min/80 quadrats/two persons		Hanley 1978
1.2- × 7.6-m plot for shrub cover mapping	Shrub steppe	12 plots/day/two persons		Pickford and Stewart 1935
1- × 5-m quadrats for shrub density	Boreal forest (postburn)	50 quadrats/day[a]		Oldemeyer and Regelin 1980, K. Jenkins, pers. commun.
Trees				
0.1-ha circular plots	Upland forest	10–15/day[a]	5–20 needed per site	Lindsey et al. 1958, James and Shugart 1970
Point-centered quarter method	Upland forest	20–50/day	10–50 needed per site	Lindsey et al. 1958, James and Shugart 1970
Bitterlich variable radius sampling	Upland forest	40–75/day[a]	10–50 needed per site	Lindsey et al. 1958, James and Shugart 1970

[a]One to two persons were used to collect data in specified time.

or no attempt was made to make a measurement, we have found that a hash mark rather than a blank space reduces confusion about whether the blanks were accidental or intentional. To aid organization and recording efficiency of field data, we suggest the use of different color forms for different sampling tasks. For example, one color might be used for sampling shrub density and another for herbaceous cover when both were measured at a site and required the use of two different sampling techniques. White paper is reflective of direct sunlight, and the investigator may want to use colored paper to reduce eyestrain due to reflection. In regions with frequent rainfall or snow, waterproof or water-legible paper is more convenient and reliable than regular bond.

PRELIMINARY FIELD TEST

It is important and useful to conduct a small-scale preliminary field test of a site before initiating full-scale sampling with the entire team. This field test allows the investigator the opportunity to identify and collect plants for field mounts (Burleson 1975) for technician use, evaluate and test equipment and sampling methods, evaluate and adjust experimental design, and make final estimates of the time required to complete the work. Many research projects and surveys have been designed in the office and completely abandoned after the first day of fieldwork, because the investigator failed to test the procedures and equipment under field conditions.

TRAIN THE FIELD CREW

An important step to maximize field efficiency is to properly train field assistants. Field assistants should have a thorough understanding of the safe and proper use of equipment, be familiar with the plants and study area, understand the correct methods for collecting and recording the data, and thoroughly understand the rationale of the study so that, in the principal investigator's absence, they can make an intelligent and informed decision when an unforeseen situation arises. Even when the crew is adequately trained, we have found that several questions and concerns arise during the first week of data collection. We suggest that each day end with a short meeting of the entire field crew to answer questions, inspect data forms for completeness and legibility, and discuss problems encountered in collecting data. If the field crew is divided into smaller teams for collecting data, we suggest that experienced members be teamed with less experienced members and that team memberships rotate daily. We have found that daily rotation of field teams increases the number of questions that arise early in the project, and thus the prompt settlement of problems results in more

Table 2. Supplies and equipment needed in the field for vegetation sampling.

Data forms and notebooks	Camera and film
Pencils and ink pens	Hammer and hatchet
Rulers and tape measures	Transect markers
Plant identification guides	Shovel and hand trowel
Plant press	Knife
Tags and plastic bags	Metal tags and wire
Quadrat frame	Sunscreen lotion
Cover board	Insect repellent
Point frame	Hand gloves
Hand magnifying lens	Backpack on frame
Maps and aerial photos	Compass

uniform collection of data and builds better rapport among crew members.

The principal investigator or field team leader is responsible for the quality control of the project. We recommend that the principal investigator spend at least 1 day working with each crew member early in the field season. This provides the opportunity to discuss the project more fully, to provide assistance and guidance in field technique, to demonstrate enthusiasm about the project, and to learn more about the background and interests of the individual crew members. These all contribute to building a quality field team and improving the quality of data collected.

TECHNIQUES FOR SAMPLING VEGETATION
Frequency of Occurrence

Frequency is the percentage of the sample units in which a species occurs (Bonham 1989). If, for example, 50 small plots were examined in a study site and bitterbrush occurred in 20 of those plots, the frequency of bitterbrush would be 20/50 × 100, or 40%. Frequency is an easy attribute to estimate because the plant either occurs in the sample unit or it does not (Fig. 4). Frequency is a useful characteristic for describing the distribution of plants within a community, and it is useful for monitoring changes in the plant community over time or comparing different communities (Bonham 1989). If frequency is low (<15%), plants have an aggregated distribution (occur in clumps) within the community. When frequency is high (>90%), plants are uniformly distributed. Most statistical procedures rely on plants being randomly distributed, i.e., having a frequency of 63–86% (Bonham 1989:92). In natural plant communities, however, plants generally have an aggregated distribution that is related to the morphological characteristics of the various species in the community, to the degree of sociological interaction among individuals and species, and to environmental patterns (e.g., differences in soil characteristics, fire history, herbivory) (West 1989). Thus, each species may have its own distributional pattern, and the pattern of the plant community may be different from that of the component species. Sampling methods to deal with compound distribution patterns have not been developed adequately (West 1989).

Frequency is dependent on the size and shape of the sample unit; thus when frequency is compared over time or among communities, the sample unit size and shape must remain constant. The size and shape of the sample unit is basically a function of whether one is sampling herbaceous vegetation, shrubs, or trees. Cain and Castro (1959:146) recommended the following sizes for the different forms of vegetation:

Herbaceous vegetation	1–2 m^2
Tall herbs and low shrubs	4 m^2
Tall shrubs and low trees	10 m^2
Trees	100 m^2.

When the total vegetation of a community is sampled, one size of sample unit will not adequately sample frequency for each form of vegetation, because the average frequency of the several species within a given vegetative form should not be <5% or >95% (Hyder et al. 1965). This problem can be solved by nesting plots of different sizes

within each other. Preliminary surveys of vegetation may be made by use of the size recommendations of Cain and Castro (1959); further refinements of sample unit size then may be made by use of the relationship between density and frequency suggested by Hyder et al. (1965).

Plots may be square, rectangular, or round. Ordinarily, plot boundaries are either marked and measured to size with a ruler or tape measure or they are defined by the inside dimensions of a frame. Frames may be of permanent shape and made of welded steel rod or some other rigid material, or they may be collapsible and made with hinged wood products or jointed PVC pipe. Collapsible frames are useful when they enhance efficiency of placement on the ground or of travel to remote areas that are inaccessible to vehicle use. When sampling in shrubby terrain, we have found that frames with one open end are useful for placing the plot around shrubs or other obstructions.

Frequency also may be measured with points. A pin (knitting needle or pointed, small-diameter steel rod) is lowered to the ground over herbaceous cover and will either hit or miss a plant part (Fig. 5). The percentage of hits gives an estimate of the frequency of a species. A single pin may be used to measure frequency (or cover) (Owensby 1973), or, commonly, a frame containing several (usually 10) pins is used. Spacing of the pins within the frame is dependent on the vegetative type, but it is commonly 4–15 cm (Hays et al. 1981). Cook and Stubbendieck (1986) provided useful suggestions for making a 10-pin frame. Along a 10-pin frame (Fig. 6), the same plant may be intercepted more than once in communities with large-sized or clumped plants. This can result in overestimates of cover for those species (Bonham 1989).

Sample size is a consideration when frequency is estimated. Frequency data have a binomial distribution, and confidence limits for small samples are wide. Grieg-Smith (1964:39) recommended that no fewer than 100 sampling units be read to obtain estimates that provide reliable comparisons from one community to another or over time. With a 10-point frame, data from 1,000 (100 frames) points (hits) are usually sufficient to describe grassland vegetation at one location, whereas fewer points (200–500) (20–50 frames) usually will provide data similar to those from a single-point method (Goodall 1952).

Density

Density is the total number of objects (e.g., individual plants, seeds) per unit area. One advantage of the density parameter is that count data are straightforward to obtain and interpret, and results obtained from various methods are directly comparable (Gysel and Lyon 1980). A disadvantage of measuring shrub density is that data are tedious to obtain and often are excessively variable. Such variability requires an often prohibitively large sample size for statistical reliability. For bunchgrasses, annual grasses and forbs, some shrubs, and trees, density is a useful and often important measurement for evaluation of wildlife habitat. However, by itself, density is not an adequate descriptor of a plant community because it does not provide information about how plants are distributed within the plant community. With frequency, density provides a good description of a plant community. With bio-

Fig. 4. Electrician's tape on a rope to aid relocation for initial and subsequent quadrat placement along permanent transects (top). Color demarcation and subplot frame attachments (bottom) are also used to provide quick representation of percentage of frame coverage of vegetation.

Fig. 5. Examples of individual pin hits/misses of vegetation for frequency sampling.

mass of individual plants, density may provide estimates of total biomass within a plant community.

The definition of an individual plant poses a problem when density is sampled. For perennial grasses and forbs, and shrubs that produce several stems from below ground, definition of an individual plant may be impossible and may not be important enough to warrant the effort or the potential error. In such situations, frequency combined with some other measurement, such as cover, may provide more useful descriptions of the plant community. For shrubs, the problem is best resolved either by counting stems at ground level, thereby eliminating the need to define an individual shrub, or by establishing a distance criterion to define individuals arbitrarily. For example, Lyon (1968a) considered stems rooted within 15 cm of each other to represent a single shrub, whereas those sprouting more than 15 cm apart were counted as separate individuals.

QUADRAT METHODS

Density can be determined with either quadrats or plotless methods. If quadrats are used, the investigator must distribute quadrats of uniform size representatively throughout each experimental unit and then must count each individual within each quadrat. Quadrats require that three characteristics be considered (Bonham 1989): (1) distribution of the plants, (2) size and shape of the quadrat, and (3) number of observations needed to obtain an adequate estimate of density.

After defining the individual unit to be counted, the investigator must determine the size and shape of quadrats to be used. Generally sample frames are rectangular, square, or circular. Rectangular plots have the largest circumference per unit area and have the most edge where decisions must be made about including or excluding the plant. Circular quadrats often are more efficient to use than square or rectangular quadrats. Sampling in circular quadrats also is effective for characterizing the vicinity around a point of interest, such as a nest, a den location, or a feeding or resting site. A review of recent wildlife habitat studies (various types of forests and various wildlife species of interest) reveals frequent use of circular quadrats, which are typically in the range of 0.01–0.1 ha (e.g., Hirst 1975, Pierce and Peek 1984, Ratti et al. 1984, Wiggers and Beasom 1986, Degraaf and Chadwick 1987, Edge et al. 1987, Bentz and Woodard 1988). For these areas, the radius of a quadrat would range from 5.6 m to 17.8 m. Increasing the size of a quadrat generally results in a lower variance and reduces the perimeter : area ratio (Bonham 1989). Numerous studies have evaluated quadrat size, and no consistent recommendation has been made about the size to use for herbaceous vegetation, shrubs, or trees.

For herbaceous vegetation, 1- $\times$ 1-m quadrats frequently are used (Bonham 1989). However, in dense vegetation, smaller quadrats such as 20 $\times$ 50 cm may be appropriate. Eddleman et al. (1964) compared quadrats of four sizes and several shapes in alpine vegetation. They recommended against use of 100-cm^2 plots because of the high standard deviations and the fact that the highest frequencies were <50%. The three larger size plots provided similar estimates of density; however, they favored 400-cm^2 rectangular plots because the chance for counting error was reduced and fewer rectangular plots were required (over square plots of the same size) to obtain a 10% standard error of the mean.

In shrub communities, quadrats large enough to contain an average of four individuals have been recommended (Curtis and McIntosh 1950, Cottam and Curtis 1956). Although quadrats as small as 1 m^2 have been used for measuring shrub density (Alaback 1982), 4–10-m^2 plots are more commonly selected (Irwin and Peek 1979). Oldemeyer and Regelin (1980) recommended a 1- × 5-m quadrat over a 2- × 5-m quadrat in an Alaskan shrub community because the smaller quadrat provided nearly the same precision and required only one-half the sampling time as the larger quadrat. Rectangular plots have advantages over square and circular plots in aggregated shrub communities because they have the greatest chance of overlapping individual clusters of shrubs. A rectangular quadrat 1 m wide of any length may be delineated easily by marking one long side of the rectangle with two chaining pins and a chain, and using a meter stick to define the remaining boundaries while one counts shrubs along the strip as the meter stick passes over them.

When trees are sampled, quadrats must be quite large, typically in the range of 0.01–0.1 ha. Curtis (1959), in deciduous and coniferous forests of Wisconsin, used square quadrats 10 m on a side and covering 0.01 ha. Mueller-Dombois and Ellenberg (1974) concluded that forest quadrats typically should be squares of 10 or 20 m on a side (0.01 or 0.04 ha). After sampling points are located, quadrats can be positioned by use of tape measures or other measuring devices and surveyor's pins. In dense vegetation or in some types of terrain, this might require considerable time and effort. To reduce that time, Penfound and Rice (1957) proposed using an elongated 0.0004-ha quadrat established by measuring the width of one's outstretched arms and then, knowing the average pace length, walking the appropriate number of paces along a compass line and recording the trees within reach. It is important to realize that, although this method is faster to implement under natural forest conditions, the area sampled is approximate, and accuracy is sacrificed without careful attention.

The number of samples to take varies from community to community and among the different vegetative forms within a community. Because many species are not distributed randomly, variation normally is quite high and the number of samples required is quite large. To determine sample size, one can use results from the preliminary field test to obtain an estimate of the variance for use in the sample size equation provided in Chapter 1. Frequently, the less common species require a larger number of samples than do the more common species. For example, to obtain a 10% standard error of the mean in a 10- × 40-cm plot, Eddleman et al. (1964) concluded that 816 plots would be required for a species with a density of 0.13 (no area units given), whereas 69 plots would be required for a species with a density of 5.6. Oldemeyer and Regelin (1980) reported that 50 1- × 5-m quadrats

Fig. 6. An inclined 10-pin frame for frequency estimates.

produced estimates of shrub density within two standard errors of actual (counted) shrub densities. Lyon (1968*a*), however, reported that >400 1.5- × 6.1-m quadrat samples would be necessary to obtain an estimate of shrub density within 10% of the true mean 95% of the time. When the equation from Chapter 1 is used, sample sizes in the hundreds are not an uncommon result. As an alternative, one may plot the running mean density against the number of samples taken (Kershaw 1964). When the density of the target, or more abundant, species does not significantly change with additional quadrats, one may stop sampling. Mueller-Dombois and Ellenberg (1974:77) suggested that sampling stop when the running mean of a sample is within 5–10% of a "maximum" sample. Clearly, one must critically evaluate the objectives of a project and the use to be made of the data when designing a study of plant density. One may determine that it is not necessary to have the density (or frequency or cover) estimate be within 5% of the true value; however, it is a waste of time and effort to undersample a community and obtain totally unreliable estimates.

PLOTLESS METHODS

Plotless methods of sampling density have been in use since the 1950s. Plotless methods do not use boundaries and are based on the premise that the density may be

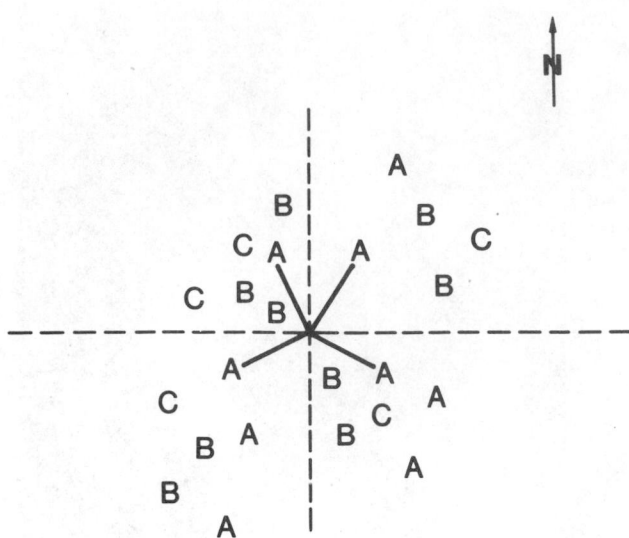

Fig. 7. Point-centered quadrat sampling wherein the point-to-plant distance is measured for the individual of each species nearest the point in each of the 90° quadrants around the point.

estimated from the mean area (i.e., mean distance) between a point and a plant or between two plants. Thus,

$$\text{density} = 1/\text{mean area (distance).}$$

This approach is most reliable when plants are randomly distributed. When they are not, more sophisticated equations must be used.

Cottam and associates (Cottam and Curtis 1949, Cottam et al. 1953, Cottam and Curtis 1956) pioneered research on plotless methods. These included the closest individual method, the nearest-neighbor method, the random-pairs method, and the point-centered-quarter (PCQ) method. Of these, the PCQ method has been widely used in many vegetation types throughout North America. Using the PCQ method, one randomly selects a number of points within a community and measures the distance to the nearest plant within each of four quadrants around the point (Fig. 7). Mean area is calculated by squaring the mean distance between points and individual stems:

$$\text{density} = 1/d^2.$$

This method may be used to calculate density of all species collectively. Or, density of individual species can be estimated by measuring distances to each species in every quadrant around each point. A reliable estimate of an individual species' density cannot be obtained by using only those distances for an individual of the species that was the closest plant in a sample and the distance was measured to the nearest plant regardless of species. That is, if 25 points were sampled, and distances to 100 plants of several species were measured, the density of all plants can be estimated based on the 100 distances. The density of one of the several species from the sample cannot be estimated reliably, because the distance measured to the plant when it was the closest plant in a particular quadrant may not be the least distance when all plants of that species within the entire circle around the point are considered (e.g., examine species A in Fig. 7).

The PCQ method has been criticized because it pro-

vides unreliable density estimates when species have a clumped distribution. Oldemeyer and Regelin (1980), in a stand of known density, concluded that the PCQ method accurately estimated density of white spruce saplings, which were more randomly distributed, but underestimated density of paper birch and aspen saplings, which had clumped distributions. In communities with regularly distributed plants, the PCQ method overestimates density (Mueller-Dombois and Ellenberg 1974). When the total plant density within a community is the only concern, the PCQ method likely provides reliable estimates of density; however, Laycock and Batcheler (1975) reported that determining composition from the proportion of times each species occurred in the total measurements resulted in biased composition estimates.

Methods have been developed to correct for density estimates in nonrandom plant populations (Morisita 1957, Batcheler 1973). The angle-order method (Morisita 1957) measures the distance from the point to the center of the third nearest plant in each quadrant around the point. This method is based on the assumption that the area may be divided into several smaller areas in which the plants will be distributed randomly or uniformly even though they are distributed nonrandomly over the larger area. The method was tested on known populations of grasses, forbs, and shrubs (Laycock and Batcheler 1975, Oldemeyer and Regelin 1980) and provided estimates of the density that were more accurate than those of the PCQ method. Oldemeyer and Regelin (1980) reported that the method provided density estimates closest to the true density in shrub stands and that its coefficient of variation was lower than other accurate estimators. However, because of the time required, Laycock (1965) recommended against its use when density is measured for each species within a community. Bonham (1989) provided a detailed description of the procedures for calculating density and the variance when the angle-order method is used.

The corrected-point-distance (Batcheler 1973) is a modification of the PCQ that uses measurements to the second and third nearest plants to correct for nonrandomness. That is, from a sample point, one measures the distance to the nearest plant, the distance from that plant to its nearest neighbor, and the distance from the nearest neighbor to its nearest neighbor, exclusive of the first plant measured. In aggregated populations, the distance between the nearest plant and its nearest neighbor is generally less than the distance from the point to the nearest plant. Density is determined by the equation:

$$m = \frac{a}{\pi[\Sigma\ r_i^2 + (N - a)R^2]},$$

where

m = density,
R = the maximum distance over which a search is made for a plant at any point,
a = number of points at which a plant is found at a distance $\leq R$, and
r_i = the ith distance measured.

As R decreases, m approaches the true density; however, the variance generally will increase because fewer measurements are included (Bonham 1989). Even though this equation is designed for random as well as nonrandom distributions, densities will be biased in nonrandom populations (Bonham 1989). This problem may be corrected by using a correction factor based on distances from the nearest plant to its nearest neighbor. Laycock and Batcheler (1975) recommended the use of the corrected-point-distance method over other distance methods because the density estimate was within 12% of the true density and because the method is relatively fast and easy to use.

The choice of using a plotless method over a quadrat method will depend on the objectives of the study. If the density of one or two species is required, plotless methods appear to be faster than quadrat methods. If the density of all species in the community is desired, the quadrat method is recommended.

Cover

Cover is defined as the vertical projection of the crown or stem of a plant onto the ground surface. Canopy cover serves as a criterion for relative dominance within a community and is of practical importance because of its influence on interception of light or precipitation and on soil temperature (Hanley 1978). It may be used by plant ecologists to describe total vegetative cover, by range managers to define cover of forage for livestock, or by foresters to describe basal area of merchantable timber. When height structure of a community is known, cover can be an estimator of biomass. Daubenmire (1959) suggested that canopy cover is the surface area over which a plant has influence; thus, cover provided by seedlings and seed stalks might not be measured because they may have little influence in the ecosystem. Although canopy or crown cover may vary within a season or among years, basal cover is relatively stable. Basal cover is a reliable measurement for bunchgrasses, tussocks, and trees. On bunchgrasses and tussocks, it frequently is measured at a height of about 2 cm (Bonham 1989:98), whereas on single-stemmed trees it is measured at 1.5 m above ground (Mueller-Dombois and Ellenberg 1974:88). This latter measurement is referred to as diameter at breast height or DBH. On trees with multiple stems or on trees with buttressed trunks, basal cover is measured at the ground surface. Cover often is expressed as a percentage value, and in a dense or in a multilayered community total vegetative cover may exceed 100%. Cover can be measured directly with a quadrat-charting or a pantographic method (Mueller-Dombois and Ellenberg 1974), an ocular-estimation technique (Daubenmire 1959, Mueller-Dombois and Ellenberg 1974), a line-intercept method (Canfield 1941), or a point-intercept method (Levy and Madden 1933, Owensby 1973).

QUADRAT-CHARTING METHOD

This method has its greatest utility in low, herbaceous vegetation where one can stand and look over the vegetative cover. Cover is mapped to scale on graph paper from a small quadrat (perhaps 1-m² quadrat). The idea is to map the crown area or the basal area onto the graph paper. This may be facilitated by subdividing the larger quadrats into smaller quadrats. Quadrat charting is useful generally only in long-term studies when quadrats are permanently marked at each corner and can be exactly relocated for each measurement. Rather than charting indirectly from what the observer sees on the ground, the observer may use a pantograph (Mueller-Dombois and Ellenberg 1974) or may take photographs (Wimbush et al. 1967).

OCULAR ESTIMATES

Ocular estimates of basal and canopy cover can be obtained with relative ease in grasslands because of their low profile and height. However, the task becomes more difficult in wetland vegetation because of the combination of water depth and plant height, often requiring the use of SCUBA equipment or a ladder.

Cover can be estimated to the nearest percentage point, or to the nearest 5th or 10th percentile; however, most commonly it is estimated according to some form of cover class (Brown 1954, Daubenmire 1959, Braun-Blanquet 1965, Mueller-Dombois and Ellenberg 1974).

The following cover-class scale (Daubenmire 1959) often has been used in grasslands.

Scale of Cover Classes for a 2- × 5-dm Quadrat

Data Integer	Class Range (%)	Midpoint (%)
1	0–5	2.5
2	5–25	15.0
3	25–50	37.5
4	50–75	62.5
5	75–95	85.0
6	95–100	97.5

A variety of plot sizes has been used for estimating shrub cover. Daubenmire recommended using 20- × 50-cm quadrats for both shrubs and herbaceous vegetation because cover is more easily estimated in small quadrats. Meter-square frames also have been used commonly in estimating shrub cover. Cook and Bonham (1977) suggested dividing 1-m² frames into 5- × 5-cm cells, each corresponding to 0.25% cover. Using a gridded quadrat, one may estimate cover by counting the number of grid cells covered by shrubs and adding the number of obstructed cells to determine the total percentage. Although ocular estimation is a rapid method of estimating data on basal or canopy cover, there are drawbacks. Ocular estimates are subject to personal bias, thus estimation error among investigators may add unnecessary variability to the data. Hence, the methods require consistent training and calibration among investigators. Dimensions of plant coverage, even on permanently marked plots, are also subject to the influences of precipitation, heat, and sunlight on plant growth. Consequently, care must be exercised in data interpretation, because a reduction in the coverage of a species on the same plot in different years may be a result of drought as much as of interspecies competition for the same site.

LINE INTERCEPT

Both basal and canopy cover can be measured with the line-intercept method. In this technique, a line or tape measure is stretched between two stakes, and basal width or canopy width of all plants touching the line or tape is measured, even if only a small part of the plant is in contact with the tape. Percent cover is expressed as a percentage of the total length of tape intercepted by vertical projections of the canopy. In tall, dense vegetation, stretching a tape line taut and straight may be difficult.

Canfield (1941) reported that a minimum of 16 15.2–30.4-m-long transects was necessary to describe shrub vegetation adequately in Arizona rangelands. A 15.2-m-long transect was adequate in shrub fields with 5–15% shrub cover, whereas a 30.4-m-long transect was necessary on sites with <5% cover.

The principal advantages of the line-intercept method are the high level of accuracy and precision that are attributed to direct measurement of vegetation rather than estimation (Cook and Stubbendieck 1986). The main limitation of the method is the time required to measure intercepts compared to estimating cover within quadrats. Hanley (1978) reported that the line-intercept method and quadrat sampling produced comparable estimates of shrub cover. Of the two methods, the line-intercept method was the more precise, whereas the quadrat method was quicker. Hanley (1978) concluded that the line-interception method is preferable to 0.1-m^2 quadrats in scientific research when precision of the cover estimate may be more important than cost efficiency. The 0.1-m^2 quadrat method may be preferable when lower levels of statistical confidence are acceptable.

POINT INTERCEPT

Both basal and canopy cover can be measured with a multiple point frame (Levy and Madden 1933) or a single point frame (Owensby 1973). Point frames usually contain 10 pins spaced 5 cm apart, and pins may be positioned vertically or at an inclined angle (Fig. 6). Single point sampling is self-descriptive. Although the point frame can be put in random locations, the pins are spaced systematically. With either method, a single pin is lowered towards the ground. The first strike of any part of the vegetation canopy becomes a canopy cover hit; if it strikes the basal area of a plant it is a basal hit. Often a pin will miss all vegetation in its line of travel. Percent canopy or basal cover is calculated as the total number of hits divided by the total number of pin placements times 100. The diameter of the pin and the point affect the accuracy of cover estimates. Because a point does not have a diameter, and the pin point does have a diameter, cover is generally somewhat overestimated (Winkworth 1955). The point-intercept method frequently is conducted along transect lines. The user should be aware that the line is the sample unit and that it is better to have fewer points per line and more lines than vice versa (Bonham 1989).

Heady et al. (1959) reported that line-transect and point-intercept procedures produced comparable estimates of shrub cover when ground cover was >3%; however, the point-intercept procedure was quicker and thus preferable. Species with ≤3% cover required extremely large samples with the point-intercept method. Thus, the line-interception procedure should be used in sparse shrub communities.

Other modifications of single-point sampling include putting a v-shaped notch in the tip of a boot and using the notch as a single point (Evans and Love 1957) while walking over a tract of grassland. This method offers rapid assessment or survey of cover, but it may have higher observer bias and be less desirable than point sampling when greater precision is necessary.

BITTERLICH VARIABLE RADIUS METHOD

The Bitterlich variable radius method is a modified point-sampling method that was developed for use in forestry (Bitterlich 1948, Grosenbaugh 1952) to measure basal area of trees; it was modified for use in range habitat to measure canopy cover of shrubs (Cooper 1957). Hyder and Sneva (1960) recommended the method for sampling basal cover of bunchgrasses. From randomly located sampling points, shrubs or trees are viewed with one of several types of sighting devices (angle gauges) that delimit a certain sighting angle. The sighting device must be held as nearly horizontal as possible. Wooden sighting sticks (Cooper 1957) have been largely replaced by prisms and other optical devices, but the principle remains the same. Those shrubs whose width or those trees whose trunks are larger in diameter than that specified angle when seen through the sighting device are reported. To be included in the count, small shrubs or trees must be relatively close to the observer, but larger ones can be farther away and yet exceed the viewing angle. The probabilities of species being sampled are proportional to their size, and the correction factor needed to calculate the cover depends on the size of the viewing angle. Percentage cover is defined as:

$$P = \left[(n \times W^2)/L^2\right] \times 25,$$

where

W = the width of the cross-piece of the sighting device,
L = the distance of the cross-piece from the observer's eye, and
n = the number of plants counted.

Using a sighting device with a width : length ratio of 1:50 gives a viewing angle of 1 degree, 10 minutes, and the count of trees within that angle is numerically equal to the tree basal area in square meters per hectare (Mueller-Dombois and Ellenberg 1974). Generally, a ratio of 1:7.07 is most acceptable for shrub communities (Fisser 1961, Cooper 1963), and the average count per plot is divided by the correction factor 2. Correction factors for different width cross-pieces were given by Cooper (1957).

The utility of the variable-plot method for sampling shrub stands is influenced by several factors. The method assumes that the plant is round; thus for species or stands with shrub crowns, particularly of irregular shape, the estimate of cover will be biased (overestimated). In dense stands, individual shrubs or trees may be shielded from view by another plant, resulting in their not being counted. Cooper (1957) reported that the method could be used in desert shrubs stands when cover was <35%, and Fisser (1961) observed that shorter investigators underestimated cover compared to taller investigators. The chief advantage of the variable-plot method is that it is quick and requires counts rather than measurements in the field. Several studies have shown that the variable-plot method produced estimates of cover comparable to those of the line-intercept method in shrub fields with <30% shrub cover (Cooper 1957, Kinsinger et al. 1960, Fisser 1961). Kinsinger et al. (1960) reported that readings from only three to six variable plots were required to produce the same precision as estimates obtained from 20 30-m-long line transects, which required considerably more time to mea-

Fig. 8. A spherical densiometer used to estimate percent overstory cover in woodlands.

sure. Cooper (1963) reasoned that this precision, and the lower coefficients of variation, from the variable-radius method was because of the larger area covered than that covered with point or line-transect methods. Therefore, Kinsinger et al. (1960) concluded that within the stated constraints, the variable-plot method was faster and more precise than the line-intercept method, but it could not be used as effectively to study subtle changes in shrub cover.

Sometimes tree canopy cover is an adequate, perhaps even preferred, measure of overstory structure and composition. In this situation, line or point sampling or ocular estimates within plots can be used to estimate canopy cover. Many workers prefer to use a spherical densiometer (Lemmon 1957) (Fig. 8) for making these estimates. The spherical densiometer uses a curved, gridded mirror that reflects the overstory at a point and provides estimates of relative amounts of the area covered. Lemmon (1957) concluded: (1) that there was no difference in overstory estimates between the spherical densiometer and other instruments used to estimate overstory, (2) that variation among replicated measurements increased with decreases in overstory, and (3) that reliability was greater when the actual grid count was used rather than broader overstory classes obtained from grouping the counts.

Biomass

One of the best indicators of species importance within a plant community is composition based on dry weight (Daubenmire 1968). Wildlife and land managers frequently require data on biomass rather than density or cover because biomass is closely related to forage availability and habitat carrying capacity (Bonham 1989). Woody biomass and size structure are required to estimate fuel loading, a necessity for formulating fire prescriptions and predicting fire behavior on wildlands. Wildlife managers often are interested in measuring biomass of edible components of browse such as current annual growth (CAG), foliage, or twigs. Total biomass and biomass of edible components may be estimated directly by clipping and weighing or indirectly by dimension analyses.

CLIPPING TECHNIQUES

Plant biomass can be measured directly by removing all of the vegetation in a sample plot to ground level and measuring its mass immediately (wet mass) or after air- or oven-drying the sample (dry mass). Clipping, drying, and weighing plant material directly is accomplished with minimal variation in results among investigators; however, proper implementation of methods necessary to obtain good data is both labor and time intensive. For consistency, herbage should be clipped at a specific height or location on the plant and may be separated into live and dead proportions or edible and inedible portions, depending on the objectives. Mean biomass per unit area then may be estimated as the product of mean biomass per plant (e.g., g/plant) and mean density of plants (e.g., plants/m^2). Sample variance may be computed as the variance of a product (Goodman 1960). Because clipping is a destructive sampling method, new plots must be selected in subsequent sampling periods to avoid the effects of previous sampling activities.

In wetlands, biomass samples of macrophytes may be obtained by harvesting all vegetation within a quadrat frame placed above the sediment level (Whigham et al. 1978). Harvesting consists of clipping plants within floating (Tanner and Drummond 1985) or submerged metal rod frames or within an open-ended cylinder or box enclosures (Sefton 1977, Anderson 1978). Water depth also should be measured near the center of each quadrat and recorded. In shallow wetlands (<1 m) clipping can be done easily in conventional waders; however, deeper wetlands (>1 m) may require that sampling be done with the use of specialty gear such as swimmer's goggles, wetsuits, or even SCUBA equipment. Vegetation samples should be dried to a constant weight. Drying temperature is dependent on the purpose of the plant materials; if one is interested only in dry weight, then 80 C may be used. If the plants are to be analyzed for nutritional analysis, lower temperatures (e.g., 60 C) are required to avoid volatilizing nutritional components. If drying and weighing cannot be done onsite, vegetation samples should be frozen or kept at 4 C to stop further respiration activity.

The "clip-and-weigh" method also may be used to estimate twig biomass within plots. Clipping all twigs within plots is a highly accurate yet laborious means of determining browse biomass (Shafer 1963). Several investigations have reported that total browse collection may require 10 to 120 times as long as estimating browse biomass from dimension analysis or twig count methods (Shafer 1963, Uresk et al. 1977, Bobek and Bergstrom 1978).

DIRECT ESTIMATIONS

Herbage biomass also may be determined by direct estimation techniques (Pechanec and Pickford 1937, Ahmed and Bonham 1982, Ahmed et al. 1983). Requirements of biomass estimation techniques include intensive training of investigators. This may be facilitated by incorporating double sampling procedures into the activity. Double sampling requires that ocular biomass estimates be made in each quadrat or for each plant and that a subset of quadrats or plants be clipped and weighed after the estimates are made. Weighing the plants helps the observer develop more accurate ocular estimates. Regression of the estimates and actual weights provides an estimator for the plots or plants for which only estimates were made. Procedures to determine an adequate ratio of clipped to estimated samples were provided by Ahmed and Bonham (1982) and Ahmed et al. (1983).

DIMENSION ANALYSES

Dimension analysis has been used in forestry for timber attributes and in wildlife and range management for estimating shrub biomass. The technique assumes that various plant attributes are related and that one attribute can be predicted from another that is more easily measured (Whittaker 1965). Because clipping, drying, and weighing require so much time, and yet biomass frequently is a critical attribute of a plant community, numerous investigators have developed regression equations of biomass and some more easily measured attribute. Biomass of individual grass plants has been estimated from volume as determined by measurement of height and basal diameter (Johnson et al. 1988). Biomass estimates of individual shrubs have been obtained with, as independent variables, measures of basal stem diameter (Telfer 1969b, Brown 1976), maximum plant height (Ohmann et al. 1976), and various crown dimensions, including diameter, area, volume, and height × circumference (Lyon 1968b, Rittenhouse and Sneva 1977, Uresk et al. 1977, Murray and Jacobson 1982). Common forms of the predictive equations include linear (Y = a + bX) and power (Y = a b^x) curves. Traditionally, researchers have linearized the power curve with logarithmic transformations (ln Y = ln a + b·ln x), but such transformations may introduce bias (Baskerville 1972). With nonlinear regression procedures commonly available in statistical software packages, there is little reason to transform the nonlinear relationships. Several independent variables may provide satisfactory estimates of shrub biomass (Oldemeyer 1982), but care must be taken to select those variables that provide the best predictive accuracy and are not themselves correlated.

In the field, one measures stem and crown dimensions from a sample of individual shrubs. The plant material then is clipped, taken to the laboratory, oven-dried, and weighed. In general, a sample of 25 plants per species is adequate for determining predictive equations for total shrub weight (Peek 1970). Care must be taken in the field to adequately sample the full range of plant sizes present, because one may not estimate biomass of plants that fall outside the size-range of plants used to develop the regression. We believe more reliable regression equations may be developed if one stratifies the plants within the community into size classes, determines the variance of biomass within each size class, and determines the number of plants to measure within each size class on the basis of the variance. For example, if the relative variance of the largest size class was 20% and if 25 plants were to be measured for the regression analysis, then 5 plants (0.2 × 25) would be measured from the largest size class.

Weight-dimension relationships of shrubs vary among sites and years (Oldemeyer 1982), making it necessary to test the influence of various site factors on the regression parameters if predictive relationships are to be applied to a broad area. Developing separate predictive equations for each shrub species in each vegetation community of the study area is often necessary. Once satisfactory predictive equations have been developed, biomass can be estimated from data on shrub density and shrub biomass estimates without destroying shrubs. Dimension analysis represents a substantial savings in time and expenditure over the traditional clip-and-weigh methods when only one, or at most a few, predictive relationships need to be developed for use for a variety of site conditions. Because the method is nondestructive, plants can be measured annually in permanent plots.

Dimension analysis has been used to estimate twig and foliage production of individual shrubs in the same manner as described above for total above-ground standing-crop biomass. Production estimates for individual shrubs are obtained by measuring a sample of shrubs in the field; the shrubs then are harvested, and all CAG of twigs and foliage is clipped, sorted, and dried. Sampling and analytical considerations are the same as for estimating total shrub biomass.

Lyon (1968b) and Peek (1970) reported that total twig production was related linearly to crown volume and crown area, the resulting equation explaining more than 80% of the variation in twig production. Oldemeyer (1982) used multiple regression procedures to estimate twig production as a function of shrub circumference, shrub height, crown length, and number of CAG twigs. Despite the high predictive accuracy of the equations, Lyon (1968b) and Peek (1970) warned that production-dimension relationships of shrubs were influenced strongly by site factors and they varied among species, which necessitated developing unique predictive equations for each shrub species on each distinctive "site type." Once predictive equations are developed for a particular "site type," dimension analysis is a convenient, nondestructive alternative to the traditional clip-and-weigh methods.

The "twig-count" method (Shafer 1963) for measuring browse biomass is based on the simple conversion of twig counts to browse weight by using an average weight per individual twig. In its basic form, an average browsing diameter (DPB) of a particular shrub species is calculated from a random sample of 100 browsed twigs. An average weight per twig then is determined by weighing 50 unbrowsed twigs clipped at the average DPB. Shafer (1963) suggested counting twigs in 9.3-m² circular plots. Twig densities then were converted to biomass estimates from a mean twig weight. Irwin and Peek (1979) observed it was faster and easier to count twigs in 1- × 1-m or 1- × 4-m belt transects. Shafer (1963) reported that the twig-count method was nearly as accurate as the clip-and-weigh method. The twig-count method is also nondestructive, making it suitable for repeated measurement of permanent plots. Additionally, individual twigs are easily counted and tallied in various height categories, permitting easy assessment of influence of snow depth and browsing heights on available browse (Potvin and Huot 1983).

A commonly employed modification of Shafer's (1963) twig-count method involves the development of weight-diameter or weight-length equations to estimate mean twig weights (Basile and Hutchings 1966, Telfer 1969a, Halls and Harlow 1971). The method is based on the principle that average twig weights may be estimated by regressions on twig diameters or twig lengths. Predictive equations relating twig weight to twig diameter or length may be developed by clipping a number of unbrowsed twigs (50 are recommended), measuring twig length and basal diameter, oven-drying, and weighing to the nearest 0.01 g. Care must be taken to collect the full range of twig sizes from several shrubs and to stratify the sample among lower and upper portions of each shrub (Basile and Hutchings 1966). Because twigs are often elliptical in cross section, it may be necessary to estimate twig basal diameter as the average of two perpendicular measurements. Linear regression produces acceptable predictive equations if the range of twig diameters or lengths is not great (Basile and Hutchings 1966, Halls and Harlow 1971); however, curvilinear regression may be required if twig sizes vary widely (Telfer 1969a). Peek et al. (1971) reported that there may be considerable site variation in length-weight and diameter-weight relationships of twigs, which would require developing a separate regression equation for each shrub species and each "site type" under investigation.

Other Attributes

VISUAL OBSCURITY

Visual obstruction caused by vegetation may be functionally important to wildlife both as hiding cover (i.e., cover necessary to escape a sense of danger) and as thermal cover (i.e., cover that creates a beneficial thermal environment). The measurement of horizontal cover of vegetation has been used extensively by wildlife managers and researchers in assessing wildlife habitat suitability, habitat preference, and impacts of land use practices on wildlife habitats. Some measure of horizontal obstruction also has been used by researchers to determine the relative influence of visibility biases associated with wildlife surveys in different vegetation classes.

A variety of devices has been used for measuring the horizontal visual obstruction caused by vegetation (Figs. 9, 10). Wight (1939) first proposed the use of a "density board," a 1.83-m-tall board, each 30.48-cm mark labeled 1 to 6. Horizontal cover is assessed by placing the board in cover, viewing the board from a distance of 20 m, and adding together the numbers that are unobstructed by vegetation. The method produces an index of horizontal cover that ranges from 0 (no obstruction) to 21 (complete obstruction), but it provides no means of describing the vertical distribution of the obstructing vegetation.

Nudds (1977) devised a "vegetation profile board" that enables the investigator to assess visual obstruction of shrub vegetation in five 0.5-m vertical intervals above ground. The board is 2.5 m high and 30.48 cm wide and

Fig. 9. Visual obstruction estimates from a specific height and distance.

Fig. 10. Cover boards used to index or quantify cover or provide visual record to changes in cover when photographed from the same reference point.

is marked in alternate black and white colors at 0.5-m intervals. Horizontal cover is assessed in each interval by viewing the board from 15 m in a randomly chosen direction. The percentage of each interval concealed by vegetation is recorded as a single-digit score, ranging from 1 to 5, corresponding to 0–20, 21–40, 41–60, 61–80, and 81–100% estimated concealment. Although the vegetation profile board has been widely used, its size, weight, and inconvenience associated with using in remote areas are drawbacks of the technique. The board may, however, be reproduced on thin vinyl or nylon material that is easily rolled up and transported in the field, and it can be held in place conveniently by a single pole or by a field assistant.

Robel et al. (1970) used a pole-shaped cover board (3 × 150 cm) that could be read from a standard distance (4 m) and height (1 m) and any direction (Fig. 9). The pole was marked in decimeters and the height of total visual obscurity was recorded; that is, if the pole was not visible until the fifth decimeter, the reading was 4. Additionally, all vegetation was clipped, dried, and weighed from a 2- × 5-dm quadrat next to the pole and regressions were developed from the average obscurity reading and biomass of 30 transects. The $R^2 = 0.95$ indicated that the obscurity reading could be used as a method of estimating biomass in tall grasses to assess prairie-chicken habitat. In short-grass prairie or other habitats of sparse vegetation, poles as cover boards likely do not provide useful information.

Alternatively, Griffith and Youtie (1988) reported that a 2.5- × 200-cm "cover pole," which is easily transported in the field, produced measures of horizontal shrub cover indistinguishable from those produced by the vegetation profile board. The cover pole was painted with alternating 0.1-m black and white bands, and three red bands divided the board into 0.5-m zones. Visual obstruction in each zone is recorded as the number (1–5) of 0.1-m bands that are ≥25% concealed by vegetation in each 0.5-m level.

PLANT HEIGHT

In grasslands, height of herbage is probably the easiest attribute of vegetation to measure but has received little

attention in published information. Herbage height can refer to the tallest portion of a plant or the effective cover height (generally the upper limit of vegetation leafiness), or the area-height of herbage below a specific area such as under a 30-cm-diameter plastic disk. Maximum plant height can be measured readily with a calibrated ruler or tape placed next to a plant. Multiple measurements (≥10) usually are expressed as an average height.

Effective plant height usually is measured as the maximum height of leafy cover for grasses and forbs; however, effective plant height of a forb (e.g., alfalfa) also may be equivalent to its maximum height. Effective herbage height also may be measured by holding a pole or meter stick parallel to the ground and reading the effective height as height at which leafy plant parts touch the horizontal pole in a minimum of three places along its length.

Area-height of herbage is measured with a disk or plate in combination with a ruler (Higgins and Barker 1982). A clear or lightly colored plastic will allow visibility of plant parts below the disk. Maximum area-height measurements are made at the point where the plastic disk is first touched by a plant part. If a weighed disk is used, measurements are made at the lowest point of disk settlement on the vegetation (Bransby et al. 1977).

Plant height can be estimated with a high degree of precision in many grasslands. Plant height also correlates well with other structural attributes of herbage. For example, Higgins and Barker (1982) reported that maximum area-height explained 63% of the foliage density values that were taken concurrently with the use of a modified visual obstruction pole as described by Robel et al. (1970). Herbage height in grassland habitats plays an important role in predator deterrence and prey security.

TREE TRUNK DIMENSIONS

Trunk diameter and cross-sectional area often are taken as measures of tree size. Diameter can be measured with a tape or calipers (Fig. 11), and, by convention, the measurement is made at 1.4 m above ground level, resulting in the DBH mentioned earlier (Spurr 1964) that is above the enlarged base seen with some but not all trees; and, it is also a representative height where measurements can be made consistently and rapidly. Such data often are summarized as numbers of individuals of species per size class per unit of land area. If exact diameter measurements are not needed, a forester's Biltmore stick can be used instead to estimate diameters within size classes.

Trunk cross-sectional area also is determined at breast height, and the results (commonly identified by the misnomer ''basal area'') are given as area units of trunk per unit land area. Individual tree areas can be computed from diameter measurements or determined directly with a tape measure scaled with area equivalent units. Data can be presented as the value just described or as a relative value (percentage of the total contributed by a single species).

TREE AGE

For many wildlife studies it is sufficient to get one or more expressions of tree size, without age, but sometimes the latter is also of value. Age data are beneficial in forest history and dynamics, including predictions of future status. For instance, knowledge of the approximate life-span of a tree species aids in assessment of the current tree population age structure and of regeneration success. Past events influencing the forest and its wildlife inhabitants can be revealed by the presence of fire scars or periods of reduced growth.

Some wildlife species have tree-size and age-specific requirements. For example, in longleaf pine forests of the southeastern U.S., trees >95 years old have been judged important for red-cockaded woodpeckers (Hooper 1988). Ruffed grouse in northern forests do best in a mosaic of aspen stands of various ages (Sharp 1963).

Age determination of trees is possible because trunk lateral growth occurs in annular increments related to the seasonality of temperate zone climates (Raven et al. 1986). In so-called ''ring-porous'' species the increments are especially evident; large-pored vascular tissue is formed early in the growing season, followed by small-pored tissue later, and then termination of growth that year, followed by the onset of obvious spring growth as another growing season begins. Examples of these species include oaks, ashes, and elms. ''Diffuse porous'' angiosperm species, e.g., maples, aspens, and birches, have less apparent growth rings. Conifers, unlike angiosperms, have a somewhat different anatomical structure, yet they too typically have easily recognized growth rings. Extra treatments of the wood, such as applying a light oil, certain stains, or water, or sanding or shaving with a razor blade can help make the growth rings more evident.

Growth rings can be seen on trunk or stump cross sections. Vegetation sampling done in concert with timber harvest or removal of damaged/dead trees is an easy way to collect such data. Where such destructive sampling is not in order, small cylindrical cores can be collected with a wood increment borer (Fig. 10). The cores can be analyzed onsite or stored, for example in soda straws, brought back from the field, and viewed. They also can be affixed to a grooved board and kept for future reference. Together with determining age, tree ring analysis can be used to measure growth rate and to date discernible past events, such as fire resulting in scarred tissue or varied climatic or competitive regimes revealed by varied growth ring widths.

Fig. 11. Calipers used to measure tree diameter at breast height.

PLANT USE

The quantification of plant use and its effect on the ecosystem are important for determining the number of herbivores that can use the land without deterioration of the soil base and plant community (Bonham 1989). Maintenance of adequate plant and litter cover retards water runoff and reduces erosion. Early methods to determine utilization of range grasses were developed from 1930 to 1950 (Stoddart 1935, Pechanec 1936, Lommasson and Jensen 1938, Canfield 1944, Roach 1950) and, with some modification, they are still used today. Many of the methods of estimating shrub utilization are modifications of those used for grasses, and we discuss methods for each in the following paragraphs (see also Chapter 10). To avoid confusion, we use stems to refer to stems of grasses and to twigs of shrubs and saplings.

As with estimation of biomass, plant use may be estimated with ocular methods. These require training with ungrazed plants that are clipped to simulate different intensities of grazing. Such estimates vary by individual investigator and may be inconsistent from year to year. Cole (1959) described an extensive browse survey method that has received wide acceptance in many western states. In Cole's procedure, 50–100 plants of a specific key browse species are marked along a permanently established survey course in selected key winter range areas. For each plant in the survey, the following data are recorded:

(1) Degree of hedging: classified as light, moderate, or severe based on the length and appearance of the previous year's growth below the current leaders.
(2) Availability: classified as available or unavailable based on shrub height and maximum browsing reach of the principal big-game species.
(3) Age/decadence: classified as seedling, young, mature, or decadent based on stem diameter classes. Any living plant with 25% or more of the crown dead is classed as decadent.
(4) Leader use: based on percentage of the total available leaders showing use (recorded as 0 = 0%, 5 = 1–10%, 25 = 10–40%, 50 = 40–60%, 70 = 60–80%, 90 = 80–100%).

Degree of hedging, availability, and age class are summarized as percentages of shrubs in each class, and stem use is summarized as an average. Gysel and Lyon (1980: 321), however, pointed out that "a disadvantage of all extensive surveys, in which the surveyor makes an estimate only, is that they will be no better than the experience, memory, and judgement of the person making the survey." Other commonly accepted methods of measuring use vary from simply counting used or unused stems, to obtaining "before and after" measures of stem lengths, to regression methods.

The stem-count method (Stoddart 1935, G. F. Cole [Mont. P-R Proj. W-37-R-8, 1957]) is a minor modification of the range survey method described above, in which used and unused stems are counted rather than estimated. Stems may be counted in plots or along transect lines. Pechanec (1936) observed that the stem count did not compare favorably with other methods for estimating grass use. Stickney (1966) and Jensen and Scotter (1977) reported that the proportion of shrub stems used correlated well with proportion of lengths removed, but the method was insensitive under heavy use. Stickney (1966) observed that virtually all shrub stems received at least minor use at utilization levels above 55% of length for black chokecherry and 60% of length for Saskatoon serviceberry. Those wishing to compare among sites that receive >50% utilization will need to select a method that remains sensitive under a wider range of use.

Use may be estimated by measuring the height of grass stems or the length of shrub stems before and after use by herbivores. Relationships are determined for height/length removed and biomass used (Lommasson and Jensen 1938, Stickney 1966, Jensen and Scotter 1977). For both grasses and shrubs, the relationship is not linear, and curvilinear relationships must be developed. Jensen and Scotter (1977) reported that the stem-length method provided a sensitive measure of shrub use across a range of levels from 0 to 100%. A primary disadvantage of the method is that it requires two trips to the field, one prior to and one following the browsing season, yet it provides no estimate of production. Curves must be developed individually for each species, site, and year to accurately estimate use (Bonham 1989).

Browse use also may be estimated with dimension analyses of twigs by predicting the prebrowsing lengths or weights of twigs from diameter-weight or diameter-length relationships previously discussed (Basile and Hutchings 1966, Telfer 1969*a,b*, Lyon 1970). Once the diameter-weight or diameter-length equations have been developed, the technique requires three additional types of data:

(1) an estimate of the percentage of twigs browsed,
(2) mean diameters at the point of browsing of a stratified sample of browsed twigs, and
(3) mean lengths or weights of the twig parts remaining after browsing.

Prebrowsing weights or lengths of browsed twigs can be estimated from the regression equations. Postbrowsing weights of browsed twigs can be determined by clipping and weighing the residual twigs. Alternatively, postbrowsing lengths of browsed twigs can be measured directly. The percentage utilization then can be computed from the formula

$$U = B \times [(P - A)/P] \times 100,$$

where

B = percentage of browsed twigs,
P = predicted prebrowsing mean length or weight of browsed twigs, and
A = postbrowsing mean length or weight of browsed twig (Lyon 1970).

As an alternative to the above procedure, several workers have estimated weights of consumed twigs directly using the diameter at point of browsing in weight-diameter equations (Oldemeyer 1982, Rumble 1987). In that instance, utilization may be computed as

$$U = [(B \times C)/P] \times 100,$$

where

B = proportion of browsed twigs,
P = predicted prebrowsing mean weight of browsed twigs (based on diameter of CAG), and
C = predicted mean weight of consumed portions of twigs (based on DPB) of twig consumed.

Several authors (Jensen and Urness 1981, Provenza and Urness 1981) demonstrated that use estimates obtained from twig-diameter measurements are rapid and that they compare favorably with twig-length measurements. Once weight-diameter or length equations have been developed for a site, the method represents a considerable savings in time over the twig-length method because all measurements of use can be obtained during a single trip to the field after use has occurred.

Percentage of plants or stems used by herbivores often is used as an estimator of plant use. This technique requires a combination of techniques discussed previously. For grasses, one determines the percentage of biomass removed, using height-weight relationships, and regresses percentage of plants used on biomass removed from a sample of several sites (Roach 1950). Similar regressions can be developed for shrubs with percentage of plants used and results of dimension analysis (Oldemeyer 1982).

TECHNIQUES FOR SAMPLING FRUITS

When certain species of wildlife become dependent on annual fruit production for their well-being, data on fruit abundance can be quite important (DeGange et al. 1989). However, few habitat analyses include an inventory of fruit production. An enumeration of the number and size of fruiting plants is often as far as managers go to describe fruit-bearing potential and its value to wildlife. The inconsistent and seasonal fruiting tendencies of plants, coupled with their often sporadic distribution in a habitat, minimize the usefulness of simple enumeration.

In studies of wildlife food habits, fruits generally are referred to as mast and are divided into two categories, "hard" and "soft." Consequently, mast can be defined as the fruits and seeds of all plants, both woody and herbaceous, used as food by animals. The importance of fruit as wildlife food is well known; for example, oak mast alone is used by 185 wildlife species and is available for up to 8 months (Van Dersal 1940). Mast is high in food energy, especially carbohydrates and fats (Goodrum et al. 1971).

Soft mast includes fruits with fleshy exteriors such as berries, drupes, and pomes. Hard mast, in contrast, includes fruits with dry or hard exteriors such as achenes, nuts, samaras, cones, pods, seeds, and capsules. Numerous factors affect fruit production, including age of plant, size of plant, genetics of the individual plant, climate, soil, competition for resources, and previous use by animals (Schupp 1990). The annual variation in yield of wild food plants makes it difficult to determine the provision of stable or high fruit production over large land units. Consequently, management practices that provide for the greatest variety of food-producing plants will assure favorable conditions for the greatest variety of wild animals.

Large or Heavy Fruits of Trees

The sampling design necessary for species with large or heavy fruits depends on whether total mast production or an index of annual mast abundance is desired from area to area. Choosing a large number of random points for trap locations may be a practical alternative in large areas. This design avoids intentional bias and allows statistical interpretation of the complete range of trend data. Depending on the objectives of the project, one may want to sample only under the canopies of mast-producing trees. This would be appropriate if one is determining the production per unit area of mast-producing canopy within the forest or obtaining an annual index of mast production.

In forests with well-defined stands of trees, sampling may be random (Thompson 1962) or stratified by vegetation type, stand age classes, or stand location (edge or interior). Sampling methods have been devised to estimate production by small versus large trees (Minckler and McDermott 1960), to compare production of two species of oaks (Tryon and Carvell 1962), and to estimate production in mixed oak stands from 63 to 82 years old (Beck and Olson 1968).

Mast production can be estimated by counts of mast in ground plots (Goodrum et al. 1971), counts of mast on trees (Gysel 1956), or use of seed traps. Counts of mast in plots on the forest floor are generally unreliable estimators of mast production because the mast frequently is taken by wildlife before the counts are made; however, such counts, when used with seed traps, may be a good estimator of wildlife usage of fallen mast. Counts of mast on trees may be quite accurate for small trees, but difficult and time consuming for large trees.

Many kinds of mast traps have been used. Downs and McQuilkin (1944) developed square traps made of hardware cloth on a wood frame. These traps were about 1 m² in size and two were placed under each tree. Since that time, several trap designs have been developed, ranging from makeshift types such as large oil drums to large fruit baskets to those made from wood, cardboard, or polyethylene film and particularly designed for catching acorns. Because squirrels and other wildlife will eat mast in the traps, early traps used predator guards; however, these deflected mast from the trap, and guards are not recommended. A study of eight types of traps comparing catching efficiency, durability, and cost (Thompson and McGinnes 1963) revealed three types to be most suitable: polyethylene film traps; square, wire cage traps; and paperboard seed traps. The polyethylene, conical-shaped, seed trap sampled an area of 0.00004 ha (0.4 m²) and had

Fig. 12. Trap for estimation of fruit production.

an acorn retention efficiency of 99%. Fifty of these traps can be carried by one person a considerable distance without discomfort. The wire cage trap (R. D. Moody, La. Wildl. Fish. Comm., Baton Rouge, P-R Rep. 24-R, 1953) sampled 0.0001 ha (1.0 m²). With a wire cover, it had an acorn-catching efficiency of 87% and durability of 10 years. The design was similar to traps used by Downs and McQuilkin (1944). Of the eight traps compared, the wire cage model was the most expensive to construct. The paperboard seed trap (Klawitter and Stubbs 1961) was a modified version of the pine seed trap (Easley and Chaiken 1951), which has a sampling area of 0.0003 ha (3.2 m²). The paperboard trap had a 96% acorn-retention efficiency and was durable for 2–3 years.

Christisen and Kearby (1984) constructed acorn traps of 8-gauge steel wire formed into a circle 0.73 m in diameter. They attached clear, 4-mil plastic, cut into a semicircle to the wire, forming a cone (Fig. 12). Holes punched in the bottom of the cone allowed for water drainage. The trap was attached to wooden stakes to hold it off the ground. They concluded that the plastic cone was superior to baskets and wire mesh traps, because the soft plastic prevented acorns from bouncing out of the trap, acorn predation was eliminated, and the traps were inexpensive and portable. The primary disadvantage was that the plastic lasted only 1 year.

Mast production varies considerably among tree species, among trees of the same species, and among years (Christisen and Kearby 1984); thus one must design a mast-production study with great care. Traps have been placed under trees at a distance of two-thirds the crown radius from the trunk; however, we are not aware that a consistent distance from the trunk is required. Christisen and Kearby (1984) randomly placed three traps under each sample tree with the stipulations that no two traps lie in the same direction and that no traps be placed under a side of a tree that lacked canopy. Further, they imagined the canopy as consisting of two concentric circles and either placed two traps in the inner circle and one in the outer, or vice versa. Traps should be examined at 1- to 2-week intervals from the time large fruits (e.g., acorns) begin to drop until all have fallen. Fruits removed from traps should be counted and may be placed into such categories as: (1) well developed and sound; (2) well developed but damaged by birds or squirrels; (3) well developed but showing insect emergence holes; and (4) imperfectly developed, deformed, or aborted (Downs and McQuilken 1944, McQuilkin and Musbach 1977).

Gysel (1957) estimated acorn production by multipling the number of acorns collected per trap and species by 1.1 to compensate for losses by deflection. He then multiplied that value (acorns per unit area of trap) by the average weight of sound acorns and total crown area of the stand to derive an estimate of weight of acorn production per unit area.

Small or Light Fruits of Trees

Like large mast, smaller seeds and fruits are important wildlife foods used by many small rodents, tree and ground squirrels, and game and nongame birds (Trousdell 1954, Hooven 1958, Yeatman 1960, Abbott 1961, Abbott and Dodge 1961, Asher 1963, Powell 1965, Landers and Johnson 1976). Abundance of small or light mast (e.g., pine seeds) varies from year to year like all fruiting species; for example, loblolly pine seed varied from nearly 0 to as high as 243,000 seeds/ha (Allen and Trousdell 1961). The two principal techniques of sampling small or light seed production of trees are placing seed traps in a stand (Lotti and LeGrande 1959, Allen and Trousdell 1961, Graber 1970) or counting with binoculars the number of ripening cones on a tree (Wenger 1953). The latter method may be simplified by counting only a portion of the tree (Wenger 1953) or by categorizing the relative abundance of cones on the tree as none, few (1–25 cones), medium (29–90), and heavy (100+).

Fruits of Shrubs

Soft and hard mast of shrubs often is within reach of a biologist and may be counted on or harvested directly from the shrub. In Georgia, Johnson and Landers (1978) collected all fruits, by species, in 4-m² plots on a monthly basis from April through October. Their small sample of five plots per line had such high sampling error that they were not able to compare production among the months sampled. Harlow et al. (1980) counted mast on scrub oaks in Florida in a series of 0.004-ha circular plots to determine mast abundance. Total counts of mast were made for each species within each of 20–40 plots in each stand. Stransky and Halls (1980) counted fruits of shrubs and woody vines in 20 1-m² quadrats within 0.6-ha plots in eastern Texas. They dried fresh fruits of each species to obtain an average weight of each fruit and projected the yield per quadrat based on the quadrat counts. Stransky and Halls (1980) further developed regressions between fruit yield and plant height and density to simplify the sampling effort, similar to regressions of browse production discussed earlier.

Fruits of Herbaceous Vegetation

Herbaceous vegetation provides an abundant supply of seeds for wildlife. Sampling seeds of herbaceous species has not been as well developed as for trees because more plant species are involved, and the wildlife using those seeds generally are less obvious. Nonetheless, sampling for seeds of herbaceous species is merely a miniature of sampling for large mast from trees; samples may be taken from the ground, from traps, or directly from the plant. Ripley and Perkins (1965) sampled ground seed supplies (primarily legumes) for northern bobwhite from soil samples. They removed soil cores (7.6 cm diam × 2.5 cm deep), screened the cores of litter and soil, and counted the number of seeds within each core. Eight soil cores were taken at each of three points along a transect line, and the eight samples were combined to project an estimated seed density and weight. Variation among lines was not greater than variation among points; thus Ripley and Perkins (1965) suggested that random sampling may be as efficient as using lines. They also reported decreased numbers of seeds in the soil cores from autumn to spring, suggesting removal by wildlife. Larger plots and different sampling depths have been used by others. Haugen and Fitch (1955) used 15 30.5- × 30.5-cm plots but took material only to the soil surface when sampling for lespedeza and partridgepea senna seeds. Young et al. (1983) used 32- × 32-cm open-bottom metal boxes driven into the soil 15 cm to estimate abundance of Indian ricegrass seed in Nevada. They further removed the soil in 2.5-cm depth increments to determine where seed reserves occurred.

Seed traps for herbaceous plant seeds are considerably smaller than those used for tree mast. Traps with fine-screen wire for the bottom and 0.64-cm hardware screen for the top have been used for estimating seed yield for game birds (Davison et al. 1955). Traps of this type eliminate predation of seeds by wildlife. Others have used traps with adhesives to hold the seeds. A Petri dish containing filter paper sprayed with Tanglefoot® or other non-drying sticky substance was used by Werner (1975), Rabinowitz and Rapp (1980), and Potvin (1988) to sample seed deposition in prairie grasslands. Rabinowitz and Rapp (1980) believed that seed production was underestimated in tallgrass prairie because leaves closed over the trap and seeds were intercepted by those overhanging leaves. When temperatures dropped below freezing or when traps became covered with snow, they were not effective for catching seed. Huenneke and Graham (1987) used insulation hangers coated with a smooth surface of adhesive to sample seed rain in grasslands. They observed that height of seedfall affected the proportion of seeds adhering to the trap surface; at 60 cm, only about 3% of the seeds adhered, whereas at 10 cm, 65% adhered to the trap surface. Exposure to light, high temperatures, and dust had little effect on capture rates, but shape and form of seed did affect capture rates.

Seed traps also can be used over water to sample seed production and availability in wetlands. Olinde et al. (1985) constructed 12- × 30-cm traps and floated the traps on styrofoam blocks. These blocks were held in place with ropes and stakes driven into the soil, and the blocks could rise and fall with changing water levels.

APPLICATIONS OF VEGETATION MEASUREMENT

To this point, we have presented methods for measuring plants or plant attributes of various life forms of vegetation. We now discuss how some of these methods have been applied to studies of wildlife habitat.

Loft et al. (1987:656) evaluated mule deer habitat during three growing seasons in California. Their objectives were to "determine the effects of cattle stocking rate on hiding cover structure during the summer grazing season" and to determine levels of herbivory on willows and herbaceous meadow vegetation. Estimates of herbaceous forage production, deer hiding cover, and browse utilization were made in 0.1-ha cattle exclosures and in adjacent sites subjected to moderate and heavy levels of cattle grazing. From two to five times each growing season, herbaceous forage was clipped from 0.1-m² plots, oven-dried, and weighed. Hiding cover in aspen and meadow habitats was estimated at eight locations around circular plots of 5.65 m radius with a 1-m² grid subdivided into 100 cells. In patchy willow habitat, where structure of the shrubs precluded use of the larger grid, a narrower 1.0- × 0.4-m grid, similar to that described by Nudds (1977), was placed at 2-m intervals along two 20-m transects and the grids were read from a distance of 5.65 m. The grids were read at three 0.5-m increments to 1.5 m; the percentage obscured by vegetation from ground level to 1 m was considered hiding cover for mule deer fawns and the percentage obscured from 0.5 m to 1.5 m was considered cover for adult deer. To evaluate the browsing level of willows, Loft et al. (1987) tagged willow branches with ≤24 new shoots and determined the percentage of shoots browsed after cattle were removed from the site.

Litvaitis et al. (1985:866) studied understory characteristics of snowshoe hare habitat in Maine. Their objectives were to "examine hare habitat use and density in two areas of Maine with differing forest composition, and determine how those variables were influenced by forest understory characteristics." In each of two sites at each study area, snowshoe hare pellets were counted within 105 circular plots of 1-m radius on seven 700-m transects. Habitat features were measured at each pellet plot. Percent ground (canopy) cover of softwood, hardwood, and herbaceous plants, and of moss was estimated in each circular plot by projecting the plant crown to the ground surface. Understory stem density was estimated by counting the number of hardwood and softwood stems ≤7.5 cm DBH and ≥0.5 m tall in two 15- × 0.5-m quadrats beginning at each pellet plot and running perpendicular to the transect. Visual obscurity at each pellet plot was estimated from a distance of 15 m for three 0.5-m strata 0.5–2.0 m above the plot with profile boards (Nudds 1977). Lastly, overstory canopy closure was estimated with a spherical densiometer (Lemmon 1957) at each pellet plot. Correlation coefficients were calculated between each of the habitat variables and the associated snowshoe hare pellet counts to identify which variables influenced pellet density.

Sedgwick and Knopf (1990:112) studied habitat relationships of cavity-nesting birds along the South Platte River, Colorado. One of their objectives was to "compare nest sites of cavity-nesting birds with available (random) nesting habitat." Each nest tree was characterized by its species, DBH, height (measured with a clinometer), and the estimated length of dead limbs ≥10 cm diameter. Habitat was characterized in a 0.04-ha circle centered at each nest tree and at 31 random points within the cottonwood-dominated riparian habitat. Numbers of snags, trees <23 cm DBH, trees 23–69 cm DBH, and trees >69 cm DBH were counted within each circle to obtain an estimate of density of the four classes. Overstory canopy cover was estimated at four points on the perimeter of each circle with a spherical densiometer. Tree basal area was determined in a circle around each tree and random point with a 10 BAF prism. These data were compared among the species of cavity-nesters using the cavity to characterize habitat use.

Kirsch et al. (1978) studied habitat characteristics of upland nesting birds, particularly of ducks in North Dakota. One of their objectives was to evaluate the height-density (obscurity) relations of residual grassland vegetation structure to success and density of duck nests. Height-density of grassland was measured with a modified version of a visual obscurity pole (Robel et al. 1970) in which readings of 100% obscurity were taken from a distance of 4 m and an eye-level height of 1 m. Results of their study indicated that higher nest density and success for ducks occurred in residual grassland cover with the highest average height-density readings at 100% obscurity.

Gilbert and Allwine (1991) studied relationships between small mammals and habitat characteristics of unmanaged Douglas-fir forests in Oregon. One of their objectives was to determine which environmental factors might be responsible for differences in small-mammal communities among young, mature, and old-growth Douglas-fir stands. They sampled small-mammal abundance and vegetation in 56 young, mature, or old-growth stands in three locations. At each stand, mammals were sampled in a 6 × 6 pitfall grid or 12 × 12 snap trap grid. In the pitfall grids, 9 points were sampled for vegetation; 16 points were sampled in the snap trap grid. Measurements were made in nested circular plots of 5.6-m and 15-m radius. Within the 5.6-m radius plot, cover of logs by decay class, and cover on the ground by bare rock, exposed bare mineral soil, organic litter, moss, and lichen were estimated visually. Cover of foliage to 2-m height and by life form was estimated visually. Number and species of small and medium live trees, snags, and stumps were counted to obtain density. Within the larger circular plot, cover of shrubs and trees >2-m height was estimated in three canopy layers—midstory, main canopy, and super canopy. Number and species of large live trees and snags were counted. Within the larger circle, the presence and type of water and occurrence of rock outcrop and exposed talus were recorded. Number of recent tree-fall mounds with exposed roots and mineral soil was counted. Vegetative components and small-mammal numbers were summarized by stand, and data from the 56 stands were analyzed by detrended correspondence analysis (Hill and Gauch 1980) to explore relationships between species abundance and environmental variables.

Hobbs et al. (1982:12) studied carrying capacity of elk

in Colorado. Their objectives were to "demonstrate that estimates of nutritional carrying capacity are viable habitat-evaluation procedures and to identify sensitive parameters in the range supply-animal demand algorithm." Estimates of biomass of plants comprising >2% of the elk's diet were necessary to develop the carrying capacity model. They obtained biomass estimates from 32 1-ha stands stratified by habitat. In each stand, forbs and grasses were clipped at ground level in 30 0.25-m² plots. Ten 2-m² plots were sampled for shrubs, and current stem growth was collected between ground level and 2.5 m high. Species were individually separated, dried, and weighed. These data were used to develop biomass estimates for habitat types within the winter range of elk and combined with nitrogen concentrations and in vitro dry-matter digestibility to estimate range supply of energy and nitrogen.

Schupp (1990:504) studied seedfall and seedling recruitment on a fruit-producing tree in Panama. Fruits of this tree are eaten by monkeys and birds, and the seeds are eaten by a variety of rodents. Seedlings are eaten by deer and other large browsers. One of the objectives of the study was to determine if there were "extensive year-to-year differences in viable seedfall, postdispersal seed predation, seedling emergence, early seedling mortality, and seedling recruitment." Seedfall was monitored with 84 1.0-m² traps constructed of 1.5-mm mesh plastic window screening in 1- × 1-m frames. Two traps were located randomly in each of 42 adjacent 20- × 20-m plots. Traps were not intentionally located either under or outside the canopy of individual trees, although no traps occurred in large openings. Seeds were counted and removed from traps on a weekly basis. Seedling emergence was studied by scattering a known number of seeds and fruits directly under traps and counting the number of seedlings that emerged. Seedling recruitment was estimated in 3- × 3-m plots that centered at the seed trap. Newly emerged seedlings were counted twice a year and marked with numbered colored plastic bird bands. The number of seedlings marked in a year was an estimate of that year's seedling emergence. From 58% to 74% of the seedlings marked in the first count of the year were present in the second, indicating moderate mortality of newly emerged seedlings. The total number present at the second count represented the year's seedling recruitment. Predation of individual seeds was determined by gluing a 30-cm piece of nylon fishing line to 576 seeds each year, attaching that line to wire-stake flags, and determining unnatural changes in position of the seed or loss of the seed. Schupp's (1990) experiments showed that removal generally indicated loss to vertebrate seed predators. Among-year variation in viable seedfall, seedling emergence, seedling recruitment, and seedling survival was analyzed with parametric and nonparametric analysis of variance methods. An actuarial life-table method was used to analyze seed predation.

LITERATURE CITED

ABBOTT, H. G. 1961. White pine seed consumption by small mammals. J. For. 59:197-201.

———, AND W. E. DODGE. 1961. Photographic observations of white pine seed destruction by birds and mammals. J. For. 59:292-294.

AHMED, J., AND C. D. BONHAM. 1982. Optimum allocation in multivariate double sampling for biomass estimation. J. Range Manage. 35:777-779.

———, ———, AND W. A. LAYCOCK. 1983. Comparison of techniques used for adjusting biomass estimates by double sampling. J. Range Manage. 36:217-221.

ALABACK, P. B. 1982. Dynamics of understory biomass in Sitka spruce-western hemlock forests of southeast Alaska. Ecology 63:1932-1948.

ALLEN, P. H., AND K. B. TROUSDELL. 1961. Loblolly pine seed production in the Virginia-North Carolina Coastal Plain. J. For. 59:187-190.

ANDERSON, M. G. 1978. Distribution and production of sago pondweed (*Potamogeton pectinatus* L.) on a northern prairie marsh. Ecology 59:154-160.

ASHER, W. C. 1963. Squirrels prefer cones from fertilized trees. U.S. For. Serv. Res. Note SE-3. 1p.

BASILE, J. V., AND S. S. HUTCHINGS. 1966. Twig diameter-length-weight relations of bitterbrush. J. Range Manage. 19:34-38.

BASKERVILLE, G. L. 1972. Use of logarithmic regression in the estimation of plant biomass. Can. J. For. Res. 2:49-53.

BATCHELER, C. L. 1973. Estimating density and dispersion from truncated or unrestricted joint point-distance nearest neighbour distances. Proc. N.Z. Ecol. Soc. 20:131-147.

BECK, D. E., AND D. F. OLSON, JR. 1968. Seed production in southern Appalachian oak stands. U.S. For. Serv. Res. Note SE-91. 7pp.

BENTZ, J. A., AND P. M. WOODARD. 1988. Vegetation characteristics and bighorn sheep use on burned and unburned areas in Alberta. Wildl. Soc. Bull. 16:186-193.

BITTERLICH, W. 1948. Die winkelzahlprobe. Allg. Forst. Holzwirtsch. Ztg. 59:4-5.

BOBEK, B., AND R. BERGSTROM. 1978. A rapid method of browse biomass estimation in a forest habitat. J. Range Manage. 31:456-458.

BONHAM, C. D. 1989. Measurements for terrestrial vegetation. John Wiley & Sons, New York, N.Y. 338pp.

BRANSBY, D. I., A. G. MATCHES, AND G. F. KRAUSE. 1977. Disk meter for rapid estimation of herbage yield in grazing trials. Agron. J. 69:393-396.

BRAUN-BLANQUET, J. 1965. Plant sociology: the study of plant communities. Hafner, London, U.K. 439pp.

BROWN, D. 1954. Methods of surveying and measuring vegetation. Commonwealth Agric. Bur. Farnham Royal, Bucks, U.K. 223pp.

BROWN, J. K. 1976. Estimating shrub biomass from basal stem diameters. Can. J. For. Res. 6:153-158.

BURLESON, W. H. 1975. A method of mounting plant specimens in the field. J. Range Manage. 28:240-241.

CAIN, S. A., AND G. M. DE O. CASTRO. 1959. Manual of vegetation analysis. Harper & Brothers Publ., New York, N.Y. 325pp.

CANFIELD, R. H. 1941. Application of the line interception method in sampling range vegetation. J. For. 39:388-394.

———. 1944. Measurement of grazing use by the line interception method. J. For. 42:192-194.

CHRISTISEN, D. M., AND W. H. KEARBY. 1984. Mast measurement and production in Missouri (with special reference to acorns). Missouri Dep. Conserv. Terrestrial Ser. 13. 34pp.

COLE, G. F. 1959. Key browse survey method. Proc. Ann. Conf. West. Assoc. State Fish Game Comm. 39:181-185.

COOK, C. W., AND C. D. BONHAM. 1977. Techniques for vegetation measurements and analysis for a pre- and post-mining inventory. Colorado State Univ. Range Sci. Ser. 28. 82pp.

———, AND J. STUBBENDIECK. 1986. Range research: basic problems and techniques. Soc. Range Manage., Denver, Colo. 317pp.

COOPER, C. F. 1957. The variable plot method for estimating shrub density. J. Range Manage. 10:111-115.

———. 1963. An evaluation of variable plot sampling in shrub and herbaceous vegetation. Ecology 44:565-569.

COTTAM, G., AND J. T. CURTIS. 1949. A method for making rapid surveys of woodlands by means of randomly selected trees. Ecology 30:101-104.

———, AND ———. 1956. The use of distance measures in phytosociological sampling. Ecology 37:451-460.

———, ———, AND B. W. HALE. 1953. Some sampling characteristics of a population of randomly dispersed individuals. Ecology 34:741-757.

CURTIS, J. T. 1959. The vegetation of Wisconsin. Univ. Wisconsin Press, Madison. 657pp.

———, AND R. P. MCINTOSH. 1950. The interrelations of certain analytic and synthetic phytosociological characters. Ecology 31:434–455.

DANSEREAU, P. 1957. Biogeography: an ecological perspective. The Ronald Press, New York, N.Y. 394pp.

DAUBENMIRE, R. F. 1959. A canopy-coverage method of vegetational analysis. Northwest Sci. 33:43–64.

———. 1968. Plant communities: a textbook of plant synecology. Harper & Row Publ., New York, N.Y. 300pp.

DAVISON, V. E., L. M. DICKERSON, K. GRAETZ, W. W. NEELEY, AND L. ROOF. 1955. Measuring the yield and availability of game bird foods. J. Wildl. Manage. 19:302–308.

DEGANGE, A. R., J. W. FITZPATRICK, J. N. LAYNE, AND G. E. WOOLFENDEN. 1989. Acorn harvesting by Florida scrub jays. Ecology 70:348–356.

DEGRAAF, R. M., AND N. L. CHADWICK. 1987. Forest type, timber size class, and New England breeding birds. J. Wildl. Manage. 51:212–217.

DOWNS, A. A., AND W. E. MCQUILKIN. 1944. Seed production of southern Appalachian oaks. J. For. 42:913–920.

EASLEY, L. T., AND L. E. CHAIKEN. 1951. An expendable seed trap. J. For. 49:652–653.

EDDLEMAN, L. E., E. E. REMMENGA, AND R. T. WARD. 1964. An evaluation of plot methods for alpine vegetation. Bull. Torrey Bot. Club 91:439–450.

EDGE, W. D., C. L. MARCUM, AND S. L. OLSON-EDGE. 1987. Summer habitat selection by elk in western Montana: a multivariate approach. J. Wildl. Manage. 51:844–851.

EVANS, R. A., AND R. M. LOVE. 1957. The step-point method of sampling—a practical tool in range research. J. Range Manage. 10:208–212.

FISSER, H. G. 1961. Variable plot, square foot plot, and visual estimate for shrub crown cover measurements. J. Range Manage. 14:202–207.

GILBERT, F. F., AND R. ALLWINE. 1991. Small mammal communities in the Oregon Cascade Range. Pages 257–267 in L. F. Ruggiero, K. B. Aubry, A. B. Carey, and M. H. Huff, tech. coords. Wildlife and vegetation of unmanaged Douglas-fir forests. U.S. For. Serv. Gen. Tech. Rep. PNW-GTR-285.

GOODALL, D. W. 1952. Some considerations in the use of point quadrats for the analysis of vegetation. Aust. J. Sci. Res., Ser. B. 5:1–41.

GOODMAN, L. A. 1960. On the exact variance of products. J. Am. Stat. Assoc. 55:708–713.

GOODRUM, P. D., V. H. REID, AND C. E. BOYD. 1971. Acorn yields, characteristics, and management criteria of oaks for wildlife. J. Wildl. Manage. 35:520–532.

GRABER, R. E. 1970. Natural seed fall in white pine (*Pinus strobus* L.) stands of varying density. U.S. For. Serv. Res. Note NE-119. 6pp.

GRIEG-SMITH, P. 1964. Quantitative plant ecology. Plenum Press, New York, N.Y. 256pp.

GRIFFITH, B., AND B. A. YOUTIE. 1988. Two devices for estimating foliage density and deer hiding cover. Wildl. Soc. Bull. 16:206–210.

GROSENBAUGH, L. R. 1952. Plotless timber estimates, new, fast, easy. J. For. 50:32–37.

GYSEL, L. W. 1956. Measurement of acorn crops. For. Sci. 2:305–313.

———. 1957. Acorn production on good, medium, and poor oak sites in southern Michigan. J. For. 55:570–574.

———, AND L. J. LYON. 1980. Habitat analysis and evaluation. Pages 305–327 in S. D. Schemnitz, ed. Wildlife management techniques manual. Fourth ed. The Wildl. Soc., Washington, D.C.

HALLS, L. K., AND R. F. HARLOW. 1971. Weight-length relations in flowering dogwood twigs. J. Range Manage. 24:236–237.

HANLEY, T. A. 1978. A comparison of the line-interception and quadrat estimation methods of determining shrub canopy coverage. J. Range Manage. 31:60–62.

HARLOW, R. F. 1977. A technique for surveying deer forage in the Southeast. Wildl. Soc. Bull. 5:185–191.

———, B. A. SANDERS, J. B. WHELAN, AND L. C. CHAPPEL. 1980. Deer habitat on the Ocala National Forest: improvement through forage management. South. J. Appl. For. 4:98–102.

HAUGEN, A. O., AND F. W. FITCH, JR. 1955. Seasonal availability of certain bush lespedeza and partridge pea seed as determined from ground samples. J. Wildl. Manage. 19:297–301.

HAYS, R. L., C. SUMMERS, AND W. SEITZ. 1981. Estimating wildlife habitat variables. U.S. Fish Wildl. Serv. FWS/OBS-81/47. 111pp.

HEADY, H. F., R. P. GIBBENS, AND R. W. POWELL. 1959. A comparison of the charting, line intercept, and line point methods of sampling shrub types of vegetation. J. Range Manage. 12:180–188.

HIGGINS, K. F., AND W. T. BARKER. 1982. Changes in vegetation structure in seeded nesting cover in the prairie pothole region. U.S. Fish Wildl. Serv. Spec. Sci. Rep. Wildl. 242. 27pp.

HILL, M. O., AND H. G. GAUCH. 1980. Detrended correspondence analysis: an improved ordination technique. Vegetatio 42:47–58.

HILMON, J. B. 1959. Determination of herbage weight by double-sampling: weight estimate and actual weight. Pages 20–25 in Technique and methods of measuring understory vegetation. U.S. For. Serv., Southern and Southeast. For. Exp. Stns., New Orleans, La.

HIRST, S. M. 1975. Ungulate-habitat relationships in a South African woodland/savanna ecosystem. Wildl. Monogr. 44. 60pp.

HOBBS, N. T., D. L. BAKER, J. E. ELLIS, D. M. SWIFT, AND R. A. GREEN. 1982. Energy- and nitrogen-based estimates of elk winter-range carrying capacity. J. Wildl. Manage. 46:12–21.

HOOPER, R. G. 1988. Longleaf pines used for cavities by red-cockaded woodpeckers. J. Wildl. Manage. 52:392–398.

HOOVEN, E. 1958. Deer mouse and reforestation in the Tillamook burn. Oreg. For. Lands Res. Cent. Res. Note 37. 31pp.

HUENNEKE, L. F., AND C. GRAHAM. 1987. A new sticky trap for monitoring seed rain in grasslands. J. Range Manage. 40:370–372.

HYDER, D. N., R. E. BEMENT, E. E. REMMENGA, AND C. TERWILLIGER, JR. 1965. Frequency sampling of blue grama range. J. Range Manage. 18:90–93.

———, AND F. A. SNEVA. 1960. Bitterlich's plotless method for sampling basal ground cover of bunchgrasses. J. Range Manage. 13:6–9.

IRWIN, L. L., AND J. M. PEEK. 1979. Shrub production and biomass trends following five logging treatments within the cedar-hemlock zone of northern Idaho. For. Sci. 25:415–426.

JAMES, F. C., AND H. H. SHUGART. 1970. A quantitative method of habitat description. Audubon Field Notes 24:727–736.

JENSEN, C. H., AND G. W. SCOTTER. 1977. A comparison of twig-length and browsed-twig methods of determining browse utilization. J. Range Manage. 30:64–67.

———, AND P. J. URNESS. 1981. Establishing browse utilization from twig diameters. J. Range Manage. 34:113–116.

JOHNSON, A. S., AND J. L. LANDERS. 1978. Fruit production in slash pine plantations in Georgia. J. Wildl. Manage. 42:606–613.

JOHNSON, P. S., C. L. JOHNSON, AND N. E. WEST. 1988. Estimation of phytomass for ungrazed crested wheatgrass plants using allometric equations. J. Range Manage. 41:421–425.

JOHNSTON, A. 1957. A comparison of the line interception, vertical point quadrat, and loop methods as used in measuring basal area of grassland vegetation. Can. J. Plant Sci. 37:34–42.

KERSHAW, K. A. 1964. Quantitative and dynamic ecology. Edward Arnold Publ. Co. Ltd., London, U.K. 183pp.

KINSINGER, F. E., R. E. ECKERT, AND P. O. CURRIE. 1960. A comparison of the line-interception, variable-plot, and loop methods as used to measure shrub-crown cover. J. Range Manage. 12:17–21.

KIRSCH, L. M., H. F. DUEBBERT, AND A. D. KRUSE. 1978. Grazing and haying effects on habitats of upland nesting birds. Trans. North Am. Wildl. Nat. Resour. Conf. 43:486–497.

KLAWITTER, R. A., AND J. STUBBS. 1961. A reliable oak seed trap. J. For. 59:291–292.

LANDERS, J. L., AND A. S. JOHNSON. 1976. Bobwhite quail food habits. Tall Timber Res. Stn. Misc. Publ. 4. 90pp.

LAYCOCK, W. A. 1965. Adaptation of distance measurements for range sampling. J. Range Manage. 18:205–211.

———, AND C. L. BATCHELER. 1975. Comparison of distance-measurement techniques for sampling tussock grassland species in New Zealand. J. Range Manage. 28:235–239.

LEMMON, P. E. 1957. A new instrument for measuring forest overstory density. J. For. 55:667–669.

LEVY, E. E., AND E. A. MADDEN. 1933. The point method of pasture analysis. N.Z. Agric. J. 46:267–279.

LITVAITIS, J. A., J. A. SHERBURNE, AND J. A. BISSONETTE. 1985. Influence of understory characteristics on snowshoe hare habitat use and density. J. Wildl. Manage. 49:866–873.

LINDSEY, A. A., J. D. BARTON, AND S. R. MILES. 1958. Field efficiencies of forest sampling methods. Ecology 39:428–444.

LOFT, E. R., J. W. MENKE, J. G. KIE, AND R. C. BERTRAM. 1987. Influence of cattle stocking rate on the structural profile of deer hiding cover. J. Wildl. Manage. 51:655–664.

LOMMASSON, T., AND C. JENSEN. 1938. Grass volume tables for determining range utilization. Science 87:444.

LOTTI, T., AND W. P. LeGRANDE. 1959. Loblolly pine seed production and seedling crops in the lower Coastal Plain of South Carolina. J. For. 57:580–581.

LYON, L. J. 1968a. An evaluation of density sampling methods in a shrub community. J. Range Manage. 21:16–20.

———. 1968b. Estimating twig production of serviceberry from crown volumes. J. Wildl. Manage. 32:115–119.

———. 1970. Length- and weight-diameter relations of serviceberry twigs. J. Wildl. Manage. 34:456–460.

McQUILKIN, R. A., AND R. A. MUSBACH. 1977. Pin oak acorn production on green tree reservoirs in southeastern Missouri. J. Wildl. Manage. 41:218–244.

MINCKLER, L. S., AND R. E. McDERMOTT. 1960. Pin oak acorn production and regeneration as affected by stand density, structure and flooding. Univ. Missouri Agric. Exp. Stn. Res. Bull. 750. 24pp.

MORISITA, M. 1957. A new method for the estimation of density by the spacing method applicable to non-randomly distributed populations. Physiol. Ecol. 7:134–144.

MUELLER-DOMBOIS, D., AND H. ELLENBERG. 1974. Aims and methods of vegetation ecology. John Wiley & Sons, New York, N.Y. 547pp.

MURRAY, R. B., AND M. Q. JACOBSON. 1982. An evaluation of dimension analysis for predicting shrub biomass. J. Range Manage. 35:451–454.

NUDDS, T. D. 1977. Quantifying the vegetative structure of wildlife cover. Wildl. Soc. Bull. 5:113–117.

OHMANN, L. F., D. F. GRIGAL, AND R. B. BRANDER. 1976. Biomass estimation for five shrubs from northeastern Minnesota. U.S. For. Serv. Res. Paper NC-133. 11pp.

OLDEMEYER, J. L. 1982. Estimating production of paper birch and utilization by browsers. Can. J. For. Res. 12:52–57.

———, AND W. L. REGELIN. 1980. Comparison of 9 methods for estimating density of shrubs and saplings in Alaska. J. Wildl. Manage. 44:662–666.

OLINDE, M. W., L. S. PERRIN, F. MONTALBANA, III, L. L. ROWSE, AND M. J. ALLEN. 1985. Smartweed seed production and availability in south-central Florida wetlands. Proc. Annu. Conf. Southeast. Assoc. Fish Wildl. Agencies 39:459–464.

OWENSBY, C. E. 1973. Modified step-point system for botanical composition and basal cover estimates. J. Range Manage. 26:302–303.

PECHANEC, J. F. 1936. Comments on the stem-count method of determining the percentage utilization of range. Ecology 17:329–331.

———, AND G. D. PICKFORD. 1937. A weight method for the determination of range or pasture production. J. Am. Soc. Agron. 29:894–904.

PEEK, J. M. 1970. Relation of canopy area and volume to production of three woody species. Ecology 51:1098–1101.

———, L. W. KREFTING, AND J. C. TAPPEINER. 1971. Variation in twig diameter-weight relationships in northern Minnesota. J. Wildl. Manage. 35:501–507.

PENFOUND, W. T., AND E. L. RICE. 1957. An evaluation of the arms-length rectangle method in forest sampling. Ecology 38:660–661.

PICKFORD, G. D., AND G. STEWART. 1935. The coordinate method of mapping low shrubs. Ecology 16:257–261.

PIERCE, D. J., AND J. M. PEEK. 1984. Moose habitat use and selection patterns in north-central Idaho. J. Wildl. Manage. 48:1335–1343.

POTVIN, F., AND J. HUOT. 1983. Estimating carrying capacity of a white-tailed deer wintering area in Québec. J. Wildl. Manage. 47:463–475.

POTVIN, M. A. 1988. Seed rain on a Nebraska sandhills prairie. Prairie Nat. 20:81–89.

POWELL, J. A. 1965. The Florida wild turkey. Fla. Game Fresh Water Fish Comm. Tech. Bull. 8. 28pp.

PROVENZA, F. D., AND P. J. URNESS. 1981. Diameter-length, weight relations for blackbrush (*Coleogyne ramosissima*) branches. J. Range Manage. 34:215–217.

RABINOWITZ, D., AND J. K. RAPP. 1980. Seed rain in a North American tall grass prairie. J. Appl. Ecol. 17:793–802.

RATTI, J. T., D. L. MACKEY, AND J. R. ALLDREDGE. 1984. Analysis of spruce grouse habitat in north-central Washington. J. Wildl. Manage. 48:1188–1196.

RAVEN, P. H., R. F. EVERT, AND S. E. EICHHORN. 1986. Biology of plants. Fourth ed. Worth Publ., Inc., New York, N.Y. 775pp.

REPPERT, J. N., R. H. HUGHES, AND D. DUNCAN. 1962. Herbage yield and its correlation with other plant measurements. Pages 115–121 *in* Range research methods. U.S. For. Serv. Misc. Publ. 940.

RIPLEY, T. H., AND C. J. PERKINS. 1965. Estimating ground supplies of seed available to bobwhites. J. Wildl. Manage. 29:117–121.

RITTENHOUSE, L. R., AND F. A. SNEVA. 1977. A technique for estimating big sagebrush production. J. Range Manage. 30:68–70.

ROACH, M. E. 1950. Estimating perennial grass utilization on semi-desert cattle range by percentage of ungrazed plants. J. Range Manage. 3:182–185.

ROBEL, R. J., J. N. BRIGGS, A. D. DAYTON, AND L. C. HULBERT. 1970. Relationships between visual obstruction measurements and weight of grassland vegetation. J. Range Manage. 23:295–297.

RUMBLE, M. A. 1987. Using twig diameters to estimate browse utilization on three shrub species in southeastern Montana. Pages 172–175 *in* F. D. Provenza, J. T. Flinders, and E. D. McArthur, eds. Proc. symposium plant-herbivore interactions. U.S. For. Serv. Gen. Tech. Rep. INT-222.

SCHUPP, E. W. 1990. Annual variation in seedfall, postdispersal predation, and recruitment of a neotropical tree. Ecology 71:504–515.

SEDGWICK, J. A., AND F. L. KNOPF. 1990. Habitat relationships and nest site characteristics of cavity-nesting birds in cottonwood floodplains. J. Wildl. Manage. 54:112–124.

SEFTON, D. F. 1977. Productivity and biomass of vascular hydrophytes on the Upper Mississippi. Pages 53–61 *in* C. B. Dewitt and E. Soloway, eds. Wetlands ecology, values and impacts. Proc. Waubesa Conf. Wetlands, Univ. Wisconsin Inst. Environ. Stud., Madison.

SHAFER, E. L., JR. 1963. The twig-count method for measuring hardwood deer browse. J. Wildl. Manage. 27:428–437.

SHARP, W. M. 1963. The effects of habitat manipulation and forest succession on ruffed grouse. J. Wildl. Manage. 27:664–671.

SMITH, J. G. 1959. Additional modifications of the point frame. J. Range Manage. 4:204–205.

SPURR, S. H. 1964. Forest ecology. The Ronald Press Co., New York, N.Y. 352pp.

STICKNEY, P. F. 1966. Browse utilization based on percentage of twig numbers browsed. J. Wildl. Manage. 30:204–206.

STODDART, L. A. 1935. Range capacity determination. Ecology 16:531–533.

STRANSKY, J. J., AND L. K. HALLS. 1980. Fruiting of woody plants affected by site preparation and prior land use. J. Wildl. Manage. 44:258–263.

TANNER, G. W., AND M. E. DRUMMOND. 1985. A floating quadrat. J. Range Manage. 38:287.

TELFER, E. S. 1969a. Twig weight-diameter relationships for browse species. J. Wildl. Manage. 33:917–921.

———. 1969b. Weight-diameter relationships for 22 woody plant species. Can. J. Bot. 47:1851–1855.

THOMPSON, R. L. 1962. An investigation of some techniques for measuring availability of oak mast and deer browse. M.S. Thesis, Virginia Polytechnic Inst. State Univ., Blacksburg. 65pp.

———, AND B. S. McGINNES. 1963. A comparison of eight types of mast traps. J. For. 61:679–680.

TROUSDELL, K. B. 1954. Peak population of seed-eating rodents and shrews occurs 1 year after loblolly stands are cut. U.S. For. Serv., Southeast. For. Exp. Stn. Res. Note 68. 2pp.

TRYON, E. H., AND K. L. CARVELL. 1962. Acorn production and damage. West Virginia Univ. Agric. Exp. Stn. Bull. 466-T. 18pp.

URESK, D. W., R. O. GILBERT, AND W. H. RICKARD. 1977. Sampling big sagebrush for phytomass. J. Range Manage. 30:311–314.

VAN DERSAL, W. R. 1940. Utilization of oaks by birds and mammals. J. Wildl. Manage. 4:404–428.

WENGER, K. F. 1953. The effect of fertilization and injury on the cone and seed production of loblolly pine seed trees. J. For. 51:570–573.

WERNER, P. A. 1975. A seed trap for determining pattern of seed deposition in terrestrial plants. Can. J. Bot. 53:810–813.

WEST, N. E. 1989. Spatial pattern—functional interactions in shrub-dominated plant communities. Pages 283–305 *in* C. M. McKell. The biology and utilization of shrubs. Academic Press, San Diego, Calif.

WHIGHAM, D. F., J. McCORMICK, R. E. GOOD, AND R. L. SIMPSON. 1978. Biomass and primary production in freshwater tidal wetlands of the Middle Atlantic Coast. Pages 3–20 *in* R. E. Good, D. F. Whigham, and R. L. Simpson, eds. Freshwater wetland ecological processes and management potential. Academic Press, New York, N.Y.

WHITTAKER, R. H. 1965. Branch dimensions and estimation of branch production. Ecology 46:365–370.

WIGGERS, E. P., AND S. L. BEASOM. 1986. Characterization of sympatric or adjacent habitats of 2 deer species in west Texas. J. Wildl. Manage. 50:129–134.

WIGHT, H. M. 1939. Field and laboratory technic in wildlife management. Univ. Michigan Press, Ann Arbor. 107pp.

WIMBUSH, D. J., M. D. BARROW, AND A. B. COSTIN. 1967. Color stereo-photography for the measurement of vegetation. Ecology 48: 150–152.

WINKWORTH, R. E. 1955. The use of point quadrats for the analysis of heathland. Aust. J. Bot. 3:68–81.

YEATMAN, H. C. 1960. Population studies of seed-eating mammals. J. Tenn. Acad. Sci. 35:32–48.

YOUNG, J. A., R. A. EVANS, AND B. A. ROUNDY. 1983. Quantity and germinability of *Oryzopsis hymenoides* in seed in Lahontan sands. J. Range Manage. 36:82–86.

23

HABITAT EVALUATION METHODS

Stanley H. Anderson and Kevin J. Gutzwiller

INTRODUCTION

Animals normally are found in areas where their needs for food and shelter are met (Cody 1985). These areas, called habitats, generally are not the same for each species. Some species, such as the tree swallow, are found in a large geographic area, yet very specific features (tree cavities for this species) are needed within this area for nesting. Other species, such as the striped skunk, use a wide variety of areas and probably are limited only by climatic conditions and interactions with other animals. Lyre snakes are found only in the southwestern U.S. where rock slabs are parted enough to provide protection from the heat of the day. Some animals have different seasonal or annual habitat needs, whereas others require different habitats for feeding and nesting during the same season. Typically, wildlife managers have considered wildlife, but not wildlife habitat, to be the primary resource; however, habitat is what enables wildlife to exist (Anderson 1991).

Wildlife managers must evaluate habitat for many reasons. Sometimes management practices need to be instituted for forests, refuges, national parks, or private lands. In such instances, goals are set to increase or decrease numbers of wildlife or to manipulate wildlife diversity. Wildlife managers also are consulted about land use planning, so they need to be able to predict how proposed habitat changes will affect wildlife communities, species (all populations), and populations.

Because managers may want to determine the quantity and quality of available habitat for a particular species, they must be able to measure features of the habitat that relate specifically to the presence, number, or health of the animals. Game managers monitor trends in habitat quality to determine whether a habitat has improved, declined, or remained unchanged and whether hunting quo-

tas thus should be changed. For example, the quality of mule deer winter range influences winter deer survival, so hunting seasons might be lengthened or shortened depending on habitat quality and the number of deer present.

When environmental assessments are made, it may be desirable to know which species are present. In these assessments, biologists determine whether habitat exists for an endangered species before beginning searches for the species. Some forms of habitat evaluation are used to estimate the relative abundance of a species. For example, the size of a marsh will limit the number of pairs of breeding blackbirds. By knowing their approximate territory size, we can predict how many pairs could possibly exist. In addition, habitat assessment procedures can be used along with population monitoring to evaluate habitat improvement efforts. Creating water impoundments or rock piles or leaving timber all are quantified and compared with changes in populations of targeted species (Hoover and Wills 1984).

This chapter demonstrates how and why wildlife fitness, density, and diversity are related to habitat features. It explains the significance of using natural-history data, autecological relations, and knowledge of the temporal and spatial scales of habitat use to assess animal-habitat associations. We provide examples of habitat variables that are related to wildlife species, and we describe how such variables can be measured in aquatic and terrestrial habitats. Standardized and nonstandardized techniques to evaluate habitat are presented. Finally, we discuss the value, interpretation, and application of wildlife-habitat correlations in the context of evaluating habitat quality and quantity.

RELATING ANIMAL FITNESS, DENSITY, AND DIVERSITY TO HABITAT FEATURES

The ability of an organism to survive and reproduce depends in part on the resources available to it. An animal's habitat supplies critical resources, otherwise the animal could not persist there. For example, cavity-nesting birds require natural tree cavities, nest boxes, or other similar human-made structures to reproduce. In the absence of this resource, cavity nesters will not breed. In such an environment, the fitness of an individual—the extent to which its genes are passed on and represented in subsequent generations—would be low. Biologists often are able to demonstrate positive or negative relationships between the number of individuals or productivity of a species and specific habitat features of the area (e.g., Burger 1987, Zwank et al. 1988). Biologists also are interested in the relationship between the number of species (species richness) and habitat characteristics (e.g., Knopf et al. 1988, Soulè et al. 1988). Such wildlife-habitat associations exist because species depend on habitats for resources.

Density, the number of individuals per unit area, is a widely used indicator of habitat quality. We assume that if the density of a species is high, something about the habitat is beneficial to reproduction, survival, or continued high occupation. But Van Horne (1983) emphasized that density may not accurately reflect the influences of habitat quality for three reasons. First, in northern climates, density-habitat associations based on summer studies may be meaningless if winter conditions ultimately dictate survival. Second, densities are subject to fluctuations in many biotic (e.g., food, predators) and abiotic resources. Consequently, current densities may simply mirror recent, short-term changes in environmental conditions, instead of long-term environmental quality. For example, some birds may temporarily use forested areas because of insect outbreaks, not because of present habitat quality. Third, higher-quality habitats may be occupied by dominant individuals, forcing subdominants into lower-quality habitats. So higher densities may be present in poorer, not better, habitats. These types of limitations should be considered when density data are used to evaluate habitat quality.

Correlations between numbers of individuals or species and habitat features are useful to managers for understanding the habitat needs of a species or community. But caution must be exercised in interpreting these animal-habitat relations. Remember, correlation does not imply causation. Many environmental factors typically influence an individual's fitness, a species' abundance, or the diversity of a community. Although researchers measure what they hope are relevant features (see below), there is always the possibility that the factor actually responsible for a particular relationship will not be measured, or that the feature associated with a species is correlated with the causal factor but is not the causal agent itself. For example, suppose we are interested in determining what habitat features influence the number of bird species breeding in separate stands of timber. For each stand, we estimate the number of breeding bird species and several habitat parameters. We find a positive correlation between bird species richness and average tree age. But because the species being studied do not require mature trees, this correlation does not make sense. In fact, each species in the stands is capable of using young, intermediate, and late successional stages. After checking the relationship between bird species richness and stand size, we also find

---

Box 1. Seasonal habitat selection.

A researcher wanted to determine whether salamanders selected summer habitat on the basis of several habitat features. She collected the following data and computed a correlation matrix for all of the variables.

Site	Mean fern cover (%)	Mean distance to water (m)	Mean midday temperature (C)	Mean litter depth (cm)	Mean salamander density (#/ha)
1	80	10	20	4	72
2	83	9	18	4	68
3	81	9	19	4	67
4	79	12	21	4	67
5	40	25	26	2	38
6	22	50	31	1	17
7	90	8	17	5	83
8	54	41	22	3	42
9	67	15	23	3	50
10	38	23	29	2	36

Correlation matrix (significance level)

	Fern cover	Distance to water	Midday temperature	Litter depth
Distance to water	−0.861 (0.0014)			
Midday temperature	−0.964 (0.0001)	0.755 (0.0115)		
Litter depth	0.982 (0.0001)	−0.813 (0.0042)	−0.966 (0.0001)	
Salamander density	0.978 (0.0001)	−0.883 (0.0007)	−0.939 (0.0001)	0.984 (0.0001)

Salamander density was significantly correlated with a number of the habitat features measured, but these same habitat variables were also intercorrelated with one another. Were salamanders selecting habitat on the basis of fern cover, distance to water, or some unmeasured feature that was correlated with these variables? The researcher realized it was impossible to infer causal relations between salamander density and habitat features with this data set. An experiment with controls and replicates would be necessary to avoid interpretation problems associated with correlations.

---

a positive correlation. The larger the stand, the more species. If stand size, but not tree age, is the factor to which birds are actually responding, then the correlation between tree age and number of species is spurious. That is, it does not reflect an actual biological phenomenon. Spurious associations are common, and sometimes difficult to unravel, in wildlife-habitat studies, so we must be careful when interpreting correlations.

HOW DO WE DECIDE WHICH HABITAT FEATURES TO MEASURE?

Time and funding constraints must be considered when objectives are established. Although we all are interested in doing the best possible job, the resources we have may be limited. Investigators should plan carefully for such constraints in their initial study design.

Before fieldwork begins the researcher must decide which habitat features to measure. Several key questions are pertinent: (1) Why is the study being conducted? (2) What is the focus of the study—a population, a species (all populations), or a community? (3) What is the autecology of this species or group of species? (4) During what time of the day or during which season do animals use different parts of the habitat? (5) How do habitat features on different spatial scales influence wildlife? Then one should clearly state the hypothesis(es) to be tested and list the objective(s) of the study. Once the goals are clear, one should make a literature search for clues to habitat factors that may be important to the species. Investigators also should consult experts on the species and spend some time in the field to determine potential habitat correlates.

Determining Which Group of Organisms To Focus On

Often a management or research question results from a problem that appears in the field. For example, we may want to find out why a population of ducks is declining or if an endangered species is present. The questions to be answered then determine if a population (e.g., ducks at one pond) should be studied or if other species (such as an endangered species) should be included in an evaluation. Looking at the community as a whole might be more appropriate. If the proposed work involves managing for a variety of species, try to determine which features of the habitat support the most diverse wildlife community. In this situation, we would not investigate specific features associated with individual species, but instead study features such as canopy volume, shade, and moisture, which affect all species.

If the investigator wants to determine the potential for reintroducing an endangered species such as the black-footed ferret, the main prey (prairie dogs) must be abundant enough to provide an adequate food base. So, physiographic (land features such as slope and elevation) and vegetation features of the community must be capable of supporting prairie dogs. The absence of other prairie-dog predators also is important for suitable reintroduction habitat.

Autecology of the Species

Whenever an investigator is looking at any form of habitat evaluation, it is important to know the autecology, or interrelation, of a species with its environment throughout the annual cycle. For example, migratory waterfowl use a number of different areas: winter habitat must provide adequate food and cover for protection from predators; good migration habitat contains lush food supplies; and breeding habitat must include food, nest and brood cover, and protection from predators. At present, concerns exist over the loss of forest habitat for migratory birds wintering in Central and South America. Although ideal migratory and breeding habitat for these birds may exist in some areas of North America, populations may decline because of habitat losses on the winter range. We know that some migratory birds nesting in the eastern deciduous forest require habitat patches of specific sizes, often larger than those used by nesting resident birds. Thus, a decrease in size of contiguous habitats might be important, and even though specific features within the habitat are correlated with a species, a decline in patch size may be far more influential. Some animals require several different habitats for biannual movement patterns. Others must have feeding, breeding, and winter habitats available in close proximity. In both situations, movement corridors must be considered.

Natural-History Data

The early literature on wildlife and its habitat is dominated by naturalists' accounts obtained subjectively through simple observation and without standard techniques. Early naturalists made little effort to characterize quantitatively specific habitat features that influenced individuals, species, and communities.

Today, natural-history data are still essential for habitat management. There is no point in trying to manage vegetation for a species without knowing when, where, and how that species uses its habitat to survive and reproduce. Detailed natural-history observations thus continue to help us determine which habitat characteristics may be important. Since the late 1960s, however, there has been a definite shift to objective, quantitative analyses of wildlife-habitat relations. We now use a tremendous array of vegetation-, soil-, and water-sampling techniques to describe quantitatively the environments of target organisms. Our goal is to determine more accurately which habitat features are most closely associated with the organism of interest.

Temporal and Spatial Scales

In evaluating habitat for wildlife, we must consider temporal and spatial factors that may affect our assessments. Some species use particular habitats during specific periods of the year. For example, two subspecies of sandhill crane winter in the southern United States and Mexico. They move to the Platte River in southern Nebraska for 6–8 weeks each year between February and April before flying to northern Canada to breed. Habitat measurements at wintering sites taken when the birds are in Nebraska may not reflect conditions that existed when the birds actually used the winter habitat, particularly if the measured conditions change from winter to spring. Evaluations of winter habitat should be made immediately after the cranes leave for Nebraska, or preferably while the birds are still on the wintering grounds, if such activity does not disturb cranes.

If we spend several months measuring habitat features at a series of ruffed grouse nest sites, canopy coverage of herbaceous vegetation height may change. Because of vegetative growth and development, sites analyzed early in the season will not be comparable to those analyzed later. Many biases can be introduced in habitat studies because of the timing of fieldwork. The length of evaluation periods thus should be minimized. Those features that do not change over time can be measured at the investigator's convenience.

Wildlife managers often use published information about wildlife-habitat relations to predict the consequences of habitat alteration or to justify habitat-management decisions. One assumption frequently made is that a relation pertinent on one scale (e.g., habitat immediately around nest sites) can be extrapolated to a larger spatial scale (e.g., entire forests). Such extrapolations may or may not be warranted, and it is not unusual to find that habitat characteristics at various spatial scales, not just one, influence wildlife. Because habitat selection probably involves a series of responses by wildlife to characteristics of the environment related to different spatial scales, this should come as no surprise. We are likely to get a more complete idea of the habitat attributes important to species and communities by analyzing habitat components related to a variety of spatial scales (see Gutzwiller and Anderson 1987*a* and references therein). Patch size, corridors, and degree of isolation may influence population size, the presence or absence of species, and community structure (see Forman and Godron 1986). Thus, the entire landscape, not just site-specific characteristics of vegetation, may be influential (Rodiek and Bolen 1991).

Determining Which Habitat Features Should Be Used To Assess Animal-Habitat Relationships

The literature search at the beginning of the study helps determine which features to study. Also, discussions with experts define and narrow our focus. If there is disagreement about what variables are important for a species, or about what should be measured, such variables might need further study, but if there is agreement, we can avoid unnecessary effort. Careful observation of species in their natural habitats is also valuable. Some investigators tend to overlook time and money constraints, but when we determine which habitat variables are important, we need to keep current and future time and money budgets in mind. This could mean simplifying the project or sampling fewer habitat features.

"Fishing expeditions," in which many habitat features are examined, are inefficient for developing species-habitat relations and can be justified only for general exploratory analyses. A general type of study, in which gross habitat characteristics are measured, can help identify which broad animal taxa (e.g., grassland, shrubland, or forest mammals) might be present, but they frequently do not help identify the key habitat features associated with a particular species.

Examples of Habitat Variables Related to Wildlife

Animals can be classified broadly as generalized and opportunistic, or specialized and competitive. We think of house sparrows, raccoons, and coyotes as being generalized because they can live in a variety of environments. California condors, lyre snakes, and wolves, however, have specific requirements for successful living, and they are more vulnerable to habitat destruction than are less specialized species.

Some important habitat features are vertical and horizontal structure, moisture, sunlight, and temperature. If only one of these conditions is inappropriate, a species may not be able to survive. For example, we may find all the components required by scorpions in the desert except rocks, but without a place to keep cool or hide, this species will die. Evolution plays an important role in how animals adapt to habitat structure. Among Darwin's finches, bills evolved for collecting insects, cracking seeds, and even picking up sticks to use as insect probes. Open-nesting birds begin nesting earlier and the nestling period is shorter, thus the young fledge before they are affected by summer temperatures or detected by predators. Cavity-nesting birds avoid these problems and survive even though they are slow-growing.

Some amphibians spend most of their lives on land, but without water for the egg and larval phases of their lives, they cannot reproduce. All plethodontid salamanders have either reduced lungs or no lungs. Because their skin serves as a respiratory organ, it must be kept moist for oxygen diffusion. Some frogs and toads, however, have thick skin and are able to exist some distance from water.

The effects of sunlight are most noticeable on vegetation, and, in turn, animals are affected by the structure and type of vegetation. Temperature is extremely important to reptiles because they are ectothermic. Without warm temperatures, they remain inactive.

Cooperrider et al. (1986) provided extensive details and examples of habitat variables important in a variety of major habitats. They covered the five vertebrate groups—from fish to mammals.

MEASUREMENT OF HABITAT VARIABLES

Once we have decided which habitat data to collect, we can consider different collection methods. Generally, variables are placed in one of two classes. First, the major or macro features of the habitat, such as size, distance to roadways or water, vegetation cover, and percentage of an area burned, can be measured without actually being in the field. Second, micro variables, which include plant species composition, water chemistry, type of snags, and tree or shrub density, all require work at the site.

Investigators should consider the various scales at which habitat features may influence a given species. Because the influence of vegetation features at different scales is likely to be associated with species' body size, mobility, and life-history requirements, macro and micro scales are relative terms. Scales cannot be defined equally for all species. That is, a macro feature for wide-ranging polar bears may be characterized on a much larger geographic scale than a macro feature for a sessile toad.

Measuring Macro Features

Remote sensing is a time-saving method of collecting wildlife habitat data. It includes satellite imagery (Landsat), infrared aerial photos (Platts et al. 1987), and videography (Sidle and Ziewitz 1990). Habitat size and shape can be determined from satellite digital analyses or interpretation of aerial photographs. Generally, large-scale (1:1,000–1:4,800) aerial photographs are needed to interpret detailed information on streams and vegetation.

Color infrared (CIR) photographs are especially valuable for vegetation analysis. The color tones, along with size, shape, pattern, shadow, and texture, are useful for identifying individual species of trees and shrubs. Color infrared film can be overexposed by 1/2 f-stop to penetrate clear water in lakes and streams (Cuplin 1978). When a fire burns through an area, the degree of burn and levels of recovery can be evaluated from CIR photographs.

Many forms of maps also exist. U.S. Geological Survey topographic maps provide valuable data on the location of physical habitat features such as streams, roads, and powerlines. On recent maps, information on slopes, elevations, aspects, and general habitat shape also can be obtained. Frequently, county maps provide details such as rock quarries, local roads, and buildings. County land-ownership maps, available in the county courthouse, are often helpful before field work begins.

Distance to water, roads, or cliffs is obtainable from maps. Distance to habitat edge, fence posts, buildings, or oil wells can be measured on the ground with a tape or a range finder, or measured from aerial photographs. Such measurements can tell us how important a feature is or how near to a feature an animal can be.

The U.S. Forest Service provides maps that show elevation, landforms, and trail information. Several different maps with varying details are available from some forest supervisors' offices. The U.S. Bureau of Land Management maps, which are available for most western states, provide detail on land ownership and are helpful for locating gross habitat features such as cliffs, ponds, and streams. Other state, provincial, and federal agencies have maps that can be obtained from university libraries or government document repositories. In addition, the highway departments of most states and provinces provide generalized maps that assist in locating study sites and topographic features such as lakes, streams, and mountain ranges.

Measuring Micro Features

AQUATIC HABITATS

Describing the water column, which is the medium of support and movement for fish and aquatic organisms, is often important. Different life stages of fish use water with different flow rates, and wading birds do not stand or feed in fast-flowing or deep water. Stream width is measured as a horizontal transect line from shore to shore along the existing water surface. This measurement can be correlated with the increase or decrease in biomass of some fish. Depth is the vertical height of the water column from the existing water surface level to the channel bottom. It is measured at the same place during the same time of day each month to provide reliable results for comparisons (Platts et al. 1983). Shore depth is critical for young-of-the-year fish, and it is measured at the shoreline or at the edge of a bank overhanging the shoreline. Measurements made on streams and pools include width, depth, distance to riffle, and length of riffles. Pool quality for a species can be estimated by relating these factors to fish abundance, growth, and survival.

Light is important for most fish and wildlife and can be measured with a light meter or with light-intensity scales established by the use of shading devices (Platts et al. 1987) to estimate canopy shading at the stream. The vegetation along a stream can vary from providing almost complete shading to providing none at all. This factor influences water temperature and productivity. Heavy use of stream edges by livestock can reduce shade and increase erosion if the vegetation is trampled and killed.

Substrate measurements must be considered. Silt or sand bottoms can influence the fish species present. Some trout species, for example, commonly occur where rocky bottoms provide habitat for invertebrates, an important food source. Rock bottoms are important spawning areas and are attractive to amphibians because they provide protection from current.

Chemical measurements of lakes and streams provide information on nutrients and harmful chemicals. Basic water chemistry studies can be conducted with kits available from chemical supply stores. When specific pollutants are suspected, testing techniques should be sought from chemical handbooks (e.g., Rand and Petrocelli 1985).

Water turbidity influences the type of fish species present. Most trout require clear water so they can see their prey, whereas carp can live in more turbid water. Turbidity also influences the species of invertebrates present, which in turn influence waterfowl production. Turbidity in lakes often is measured with a Secchi disc (Platts et al. 1987). Conductivity tests in the laboratory also are used to determine turbidity in lakes and streams (Nielsen and Johnson 1983).

In lakes, bottom composition, depth, aquatic vegetation, and debris are related to fish productivity. Water chemistry is an important correlate of invertebrate productivity and species composition. Organic wastes dumped into lakes and streams deplete oxygen because the bacteria degrading such wastes respire aerobically. This decreases overall species diversity because some species cannot tolerate anaerobic conditions. Similarly, decreases in pH resulting from SO_2 and acid rain deposition may reduce the number of fish and invertebrates in lakes and streams.

TERRESTRIAL HABITATS

Many factors, such as community structure, presence or absence of other animals, and diversity of plants and animals, influence terrestrial wildlife. For each organism, we need to determine important features, such as height of the vegetation, number of layers or growth forms, and spatial distribution of vegetation (Anderson 1981, Hays et al. 1981).

Physical Characteristics

Physical features of the terrestrial environment often are correlated with wildlife abundance, diversity, or productivity. Elevation gradients are associated with changes in vegetation and climate, thereby affecting changes in wildlife. For example, Finch (1989) reported a change in bird species composition along an elevation gradient in Wyoming. Generally, elevation data can be taken from maps; then the wildlife correlations with elevation can be assessed. Topographic ruggedness may influence the use of areas by certain wildlife species. Beasom et al. (1983) demonstrated how to estimate land-surface ruggedness from topographic maps.

Types of Disturbance

Rock piles, impoundments, and specific soil types attract wildlife (Dealy et al. 1981) and should be measured. Vegetation can provide shelter for small mammals, which

Box 2.   Importance of cliffs to wildlife.

In south-central Wyoming the arid grasslands are broken by occasional cliffs. Biologists evaluating these cliff sites found that cliffs created an array of vegetative characteristics that appeared to increase wildlife diversity. To evaluate the influence of cliffs on wildlife, they set up 13 cliff study sites and 7 control sites (no cliffs) (Ward and Anderson 1988).

On each study site, birds were sampled within six 400-m transects, and small mammals were sampled within three 210-m^2 square grids. Physical features, including cliff exposure, angle, light, height, surface roughness, talus length, and amount of exposed rock, were measured. Distances to water, roads, and fences were measured from maps. Vegetation was sampled with a 1-m^2 sampling frame placed at each small-mammal trapping station.

Analysis of variance was used to compare abundance (no. of individuals in each species) and species richness (no. of species) on cliff sites to control sites. Biologists ran an all-subsets regression (Dixon 1981) to determine what study-site features were correlated with each measured parameter.

The results showed that small-mammal abundance was greater on cliff sites ($\bar{x} = 97 \pm 5.94$ [SE], $n = 13$) than on control sites ($\bar{x} = 59 \pm 11.03$ [SE], $n = 7$). Abundance of mammals was correlated with talus and topographic roughness. Small-mammal richness was greater on cliff sites ($\bar{x} = 6.0 \pm 0.35$ [SE], $n = 13$) than on control sites ($\bar{x} = 4.0 \pm 0.50$ [SE], $n = 7$). Abundance of male birds was correlated with angle of cliff and distance to water.

Thus, biologists were able to show what features appeared to be associated with wildlife. Their results can be used to improve wildlife habitat in arid, rocky grasslands.

in turn provide food for raptors. Fence posts and buildings are habitat variables related to disturbances that can change species composition. Some birds use fence posts for perching and roosting, and buildings are used for roosting and nesting. For a thorough evaluation of an area's habitat, these remnants of human disturbance should be measured or recorded as present or absent.

Physiographic Features

Two forms of physiographic features generally are described—geomorphic and edaphic (Maser et al. 1979). Geomorphic features are products of the geologic process. They include cliffs, caves, tables, lava flows, sand dunes, and playas. Measurement of geomorphic features generally includes the diameter of the feature; height of cliffs above ground; surface ruggedness; presence of caves; and the depth, width, and parent rock of the feature. Edaphic features are local, distinctive soil characteristics that, along with their vegetation, contrast markedly with the

surroundings. These edaphic areas often provide specific habitat for activities such as reproduction or seclusion (Maser et al. 1979).

Talus is an edaphic feature of accumulated rocks on cliffs or at the base of cliffs. Components of interest include length, depth, width, and rock type of the deposit. Talus can provide protection for some animals, such as the pika, during reproduction and hibernation periods. Other species of wildlife living near talus use the areas for hunting. Lizards, snakes, birds, and mammals are attracted to talus because they find shelter among the rocks.

STANDARDIZED TECHNIQUES OF HABITAT EVALUATION

Generally, managers evaluate an area in terms of its ability to support wildlife populations. Approaches involve spatial diversity, habitat models, and wildlife-habitat correlations.

Spatial Diversity

Spatial diversity is really a measure of the horizontal diversity of habitat present. A technique that combines interspersion (intermixing of units of different habitat types) and juxtaposition (a measure of the proximity of year-round habitat) (Giles 1978) can be used (Heinen and Cross 1983).

Cover types that are critical habitat components for the species under study must be identified. Data then must be collected on this critical cover type. Aerial photos or field sampling can be used. It is best to place these data on a photo or map upon which a grid system can be superimposed. The choice of cell size will depend on home-range size and landscape patchiness.

Interspersion is calculated for a given cell by counting the number of surrounding cells that contain different cover types. Eight cells surround a centroid cell, so the number of cells that differ from the cell of interest is divided by 8. The resultant index value lies between 0 and 1.

An example calculation of interspersion (I_x) is:

A	B	B
B	A	A
A	C	C

where the letters A, B, and C represent different cover-type categories, 5 = total number of cover-type changes from the centroid cell, and 8 = total possible number of cover-type changes. Thus, $I_x = \frac{5}{8} = 0.625$.

Juxtaposition is calculated by first identifying all combinations of edge types around the center cell. A numerical rating is given to each edge type by assigning a value of 1 to diagonal edges and a value of 2 to vertical or horizontal edges. Relative weighting factors ranging from 0 to 1 are assigned to each edge type and represent the quality of different community junctions. The weighting factor is multiplied by the numerical rating of each edge type to give a total value for each edge type. Then all values are totaled and divided by 12 (the maximum total for edge-type ratings) to allow the juxtaposition index for each centroid cell to range from 0 to 1.

An example calculation of juxtaposition (J_x) is:

Edge type	Numerical rating	Quality	Total
A/A	4	0.2	0.8
A/B	5	0.5	2.5
A/C	3	0.6	1.8
			5.1

In this example $J_x = {}^{5.1}/_{12} = 0.425$.

Although the adjacency of two cells of the same cover type does not represent a true edge, it may be given a weighting factor for the juxtaposition index if large stands of that type are important to the species being considered.

The spatial diversity (Sd) index described by Mead et al. (1981) is as follows:

$$Sd_A = ([\sigma_A I_s] + [\alpha_A J_x])(1_A)(2_A)(3_A),$$

where A indicates a particular species, σ_A indicates the importance of interspersion relative to juxtaposition, α_A indicates the importance of juxtaposition relative to interspersion (σ_A and α_A range between 0 and 1 each but must sum to 1), and 1_A, 2_A, and 3_A indicate exclusion factors, which also range between 0 and 1. An exclusion factor is any habitat component with a positive or negative impact on a particular species. For the previous example, the Sd_A index is as follows:

$$Sd_A = (0.5 \times 0.625) + (0.5 \times 0.425) = 0.525.$$

In this example, juxtaposition and interspersion were considered equally important, and no exclusion factors were identified. However, any number of exclusion factors may be used depending on the area and species under consideration. An example of an exclusion factor with a positive impact may be the presence of water within 1.6 km. If it is present, 1_A may be given a value of 1 and thus does not affect the index. If it is absent, a value of 0 may be assigned, thus driving the Sd_A index for that parcel of land to 0. An exclusion factor with a negative impact may be the presence of anything that decreases the habitat suitability of an area for the species in question. An example may be an oil derrick. If it is present, 1_A may be assigned a value of 0, thus driving the index to 0. The numbers assigned for exclusion factors may range between 0 and 1, thus giving the Sd_A index more sensitivity to fine differences in spatial diversity.

Habitat Models

Wildlife-habitat models attempt to describe relationships of change. Often this is done through mathematical equations. It is possible to vary the input on one side of the equation and predict the outcome symbolized by the other side. Thus, this process can be descriptive and predictive. In the past, wildlife managers have not used modeling very much. This approach in wildlife science came into vogue during the 1970s and is used fairly extensively today. However, wildlife managers still are not completely familiar with modeling.

In model building, the natural system must be examined. The various components must be identified and selected so that they can be related to one another. A model can be used to predict how timber cutting might affect forest-bird communities. In such a situation, it would be necessary to know something about the communities and species' feeding and nesting requirements with respect to vegetation. The relationship between the selected components must be understood. Modelers, therefore, must clarify their thinking by being specific. This often is done by writing in long hand the information they want to evaluate in the model. A general description of the problem and a list of input variables with their interrelations often are used. Then the modeler usually develops a box diagram to show how input and output variables interact. Sometimes modelers go no further than this descriptive phase; at other times they develop actual mathematical relationships.

Once a model has been developed, it needs to be evaluated. We must examine the intent of the model and vary the input to determine effects on the output. By doing this, we see what factors are associated with the model and how useful it can be in the field.

Models are used for many different reasons. First, predictions of events can be made. For example, we might want to know how stream channelization affects a trout population. Knowing the habitat requirements for trout populations, we can predict what will occur after channelization. Models are used to predict the number of deer each spring based on population size the preceding autumn, winter weather conditions, and so on. Some models have regional value. For example, knowledge of water flow might be used to predict a general impact on wildlife populations throughout a large region. Other models are useful only in localized areas, sometimes as small as 0.5 ha or less. Thus, geographic scale should be considered when models are developed and applied.

Another important consideration in model development is measurability of the habitat variables in the field. Often models use the subjective terms "more," "less," "little," and "big." These descriptors are not objectively measurable and cannot be used. It is important to select variables, such as the number of trees per hectare or the number of kilometers of streamside habitat, that can be measured and entered into the model. Obviously, these factors should have some probable influence on the species of interest. So, those who develop models must look carefully at how they will use the model and what factors they will consider. Models with variables too difficult or expensive to measure will not be used. Simple, meaningful models are most desirable. Wakeley (1988) described a means of simplifying the application of Habitat Suitability Index (HSI) models (see below) that will enhance their use.

Various models have been developed as a means of standardizing habitat evaluation techniques. We caution that some of the models are oversold and expectations are too high. For example, HSI models sometimes are used to evaluate habitat without looking at the presence of other species that influence the species under consideration. Many other models, including statistical models, are subject to the same problems. Models might, therefore, be useful at a local level, but not applicable on a regional basis.

In 1984, a symposium was held in California on modeling habitat relationships of terrestrial vertebrates (Verner et al. 1986). Much of the symposium discussed models that can be used to standardize techniques. Berry (1986) identified three forms of modeling: single species to multiple species, community models, and habitat analysis models. In this context she (Berry 1986) described HSI models, habitat-capacity models (HC), and pattern-recognition models (PATREC) as correlations between species and different habitat components.

HSI models have been developed by the U.S. Fish and Wildlife Service and have generated a great deal of discussion among wildlife scientists. An HSI model uses the physical and biological attributes of a particular habitat to yield an index of habitat suitability that is assumed to be proportional to the habitat's carrying capacity for a species. HSI models generally are linear models involving a series of habitat variables associated with the different species.

HABITAT SUITABILITY INDEX MODELS

The technique for determining HSI values must be clearly described in a Habitat Evaluation Procedure (HEP) study to establish credibility, to optimize the usefulness of the analysis in decision-making, to provide a permanent record of the basis for a decision, and to make future improvements in HSI models. Studies by Ellis et al. (1979) confirmed that such descriptions increase the repeatability in determining HSI values. Although repeatability does not mean that HSI values will be accurate, repeatability is a prerequisite to improved accuracy.

The recommended method of describing HSI values is through the use of HSI models. An HSI model may be in word or mathematical format but, regardless of the format, it must clearly describe the rules and assumptions used to calculate an HSI. The process of calculating an HSI involves (1) establishing HSI model requirements, (2) developing an HSI model, and (3) determining HSIs for available habitat.

Establishing HSI Model Requirements

Habitat models used in HEP must be in index form. Inhaber (1976) defined an index as a ratio between some value of interest and a standard of comparison. For HEP purposes, the value of interest is an estimate of habitat conditions in the study area, and the standard of comparison is the optimum habitat condition for the species being evaluated. Therefore,

$$\text{Index Value} = \frac{\text{Value of Interest}}{\text{Standard of Comparison}},$$

or

$$\text{HSI} = \frac{\text{Study Area Habitat Conditions}}{\text{Optimum Habitat Conditions}},$$

where the numerator and denominator have the same units of measure. The HSI ranges between 0 and 1.0 and, like any index, is dimensionless.

Developing an HSI Model

The ideal goal of an HSI model is to produce an index with a proven, quantified relationship to carrying capacity (i.e., units of biomass per unit area or units of biomass production per unit area). This goal often will be unobtainable; consequently, a more easily obtainable but acceptable goal must be defined. The minimum acceptable

goal for an HSI model might be, for example, an index that a recognized expert, knowledgeable about the habitat requirements of a species, believes is related to long-term carrying capacity.

Determining HSIs for Available Habitat

As an example, we can show how the growth form of emergent water plants, the canopy cover of aquatic vegetation, and water depth may influence the presence or absence of marsh wrens in different parts of the breeding range (Gutzwiller and Anderson 1987b). HSI models have been developed for more than 100 species (Wakeley 1988), but only a few have been tested or evaluated in the field (e.g., Lancia and Adams 1985).

HABITAT-CAPACITY MODELS

The HC model was developed by the U.S. Forest Service and has been used to describe conditions associated with or necessary to maintain different compositions of species. It uses weighted values based on habitat-capacity ratings at each successive stage of vegetation for reproduction, resting, and feeding by a particular species (Hurley et al. 1982).

A data system in three formats—narratives, habitat-relationships matrices, and status matrices—is generally the beginning of an HC model. Narratives contain life-history data, status (legal and management), distribution by habitat, reproduction data, special habitat requirements, food habits, territory/home-range sizes, references, and other management information.

Habitat-relationships matrices differ for terrestrial species and aquatic species. Matrices for terrestrial species provide information on the use of vegetation types and special habitat needs (e.g., snags and talus slopes). Importance to a terrestrial species of each vegetation type and structural stage is related to the biological functions of reproduction, feeding, and resting. Season of use also is included.

Within each cell of a matrix, a value is assigned for the species association with the particular vegetation type and structural stage for each biological function. This value, referred to as a habitat-capability rating, is based on current literature and professional knowledge. The values range, in whole numbers, from 1 to 3. A habitat-capability rating of 1 indicates the habitat is optimum (it contains all of the required elements, and none is limiting) for that biological function. A habitat-capability rating of 2 is acceptable habitat for a particular biological function, but some elements might be preventing the population from reaching its optimum density. A habitat capability of 3 is marginal; the habitat might be used by the species, but some required elements are missing or limited.

A final value, the habitat-capability coefficient (HCC), is calculated for each vegetation type and structural stage. The HCC is an aggregated, weighted value based on the habitat-capability ratings for reproduction, feeding, and resting. These values can range from 0.00 to 1.00.

Habitat-relationships matrices for aquatic species do not contain habitat capability ratings or coefficients, nor do they denote season of use. These matrices have a variety of aquatic habitat and microhabitat elements, and use of an element is shown as "required for survival," "not required for survival," or "unknown."

Box 3.   Marsh wren HSI.

The habitat suitability index model for marsh wrens (Gutzwiller and Anderson 1987*b*) produces a habitat suitability index (HSI) from 0 (unsuitable habitat) to 1.0 (optimal habitat). The proposed relation between the suitability of wetlands for marsh wrens and wetland features is:

$$HSI = (SIV1 \times SIV2 \times SIV3)^{1/3} \times SIV4,$$

where

    SIV1 = suitability index (SI) for growth form of emergent hydrophytes,

    SIV2 = suitability index for percent canopy cover of emergent herbaceous vegetation,

    SIV3 = suitability index for mean water depth, and

    SIV4 = suitability index for percent canopy cover of woody vegetation.

The overall suitability index (HSI) is thus estimated from the geometric mean and product of individual suitability indices for four habitat variables (V1–V4). This model was developed to estimate the value of wetlands for providing cover and reproductive resources for marsh wrens.

A biologist wanted to determine whether a wetland would support breeding marsh wrens, based on its vegetation and water characteristics. He collected data for the above four habitat variables at randomly selected points. Using the assumed relations (available in model documentation) between the suitability index for each habitat variable and the actual habitat measurement, he converted his field measurements to suitability indices (SIs). The HSI values were then obtained by applying the model above.

Sample	V1[a]	V2 (%)	V3 (cm)	V4 (%)	SIV1	SIV2	SIV3	SIV4	Overall habitat suitability index (HSI)
		Habitat measurements				Suitability indices (SI)			
1	1	72	5	10	1.0	0.63	0.37	0.90	0.55
2	2	83	0	12	0.5	0.90	0.00	0.88	0.00
3	1	77	15	53	1.0	0.79	1.00	0.47	0.43
4	3	64	20	62	0.1	0.43	1.00	0.38	0.13
5	1	81	25	11	1.0	0.89	1.00	0.89	0.86
6	1	92	27	20	1.0	1.00	1.00	0.78	0.78
7	2	91	31	36	0.5	1.00	1.00	0.65	0.52
8	4	80	39	29	0.0	0.88	1.00	0.71	0.00
9	1	40	9	75	1.0	0.08	0.55	0.25	0.09
10	1	52	19	5	1.0	0.13	1.00	0.95	0.48

[a] 1 = cattails, cordgrasses, bulrushes; 2 = reedgrass, canary grass, sedges; 3 = buttonbush, mangrove; 4 = other growth forms.

From these calculations the biologist concluded that the wetland, overall, provided marginal to poor habitat because many sample points had overall suitability values (HSIs) of ≤ 0.55.

The final data format is status matrices. These matrices contain information on life form; federal classification as threatened or endangered; state status as threatened or endangered; protected or unprotected nongame; and hunted, trapped, or fished (Sheppard et al. 1982).

PATTERN-RECOGNITION MODELS

Pattern-recognition or PATREC models are those that use a series of conditional probabilities to assess whether habitat is suitable for a species. To use this approach one must know what constitutes suitable and unsuitable habitat for species (Williams et al. 1978).

Usually a series of habitat attributes (such as habitat size, number of dead trees, and water availability) must be considered. How population density is associated with each habitat factor must be known beforehand. Habitat assessment then can be made from data collected in the field or from aerial photos if appropriate. An "expected habitat suitability" (EHS) then can be calculated as follows:

$$EHS = \frac{P(H) \times P(I/H)}{P(H) \times P(I/H) + P(L) \times P(I/L)},$$

where P(H) = proportion of high-density habitat, P(I/H) = the probability that the area has a high population potential, P(L) = proportion of low-density habitat, and P(I/L) = the probability that the area has a low population potential. Low and high population potentials are determined from survey data.

Some of the multispecies models, such as the life-form system (Thomas 1979) and the community-guild model (Verner et al. 1986), are community models. These models group species with similar habitat requirements for feeding and reproduction, enabling us to describe a general area that would support a specific group of species.

LIFE-FORM MODELS

The life-form models try to include all species found in a community. They create a series of life-form categories for the community based on feeding, reproduction, and other necessary life-history activities. All species then are placed in one of the life-form categories, thereby reducing the number of species that a manager must consider in an area. For example, in developing the life-form concept in the Blue Mountains of Oregon and Washington, Thomas (1979) combined 327 species into 16 life forms (Table 1). The habitat is the basis for grouping animals, so habitat data must be available to develop the model.

Within each life form, more detailed data may be present for individual species (e.g., Table 2). Thus, managers can examine the impact of brush removal and list the life forms affected. How habitat features influence the number of species can be viewed in more detail by examining the detailed needs of each life form. Likewise, the effects of plant-community succession can be examined by looking at the details of change that affect a life form (Table 2).

Thomas (1979) advanced the life-form concept by preparing a summary of information on each species' seasonal use of habitat. He showed on a monthly basis when the animal uses an area for breeding and feeding. He used these data to calculate how susceptible each species is to

Table 1. Life-form descriptions for selected vertebrates (Thomas 1979:246). Table indicates number of organisms that feed and reproduce in each of the community units.

Life form	Reproduces	Feeds	No. of species[1]	Examples
1	in water	in water	1	bullfrog
2	in water	on the ground, in bushes, and/or in trees	9	long-toed salamander, western toad, Pacific treefrog
3	on the ground around water	on the ground, and in bushes, trees, and water	45	common garter snake, killdeer, western jumping mouse
4	in cliffs, caves, rimrock, and/or talus	on the ground or in the air	32	side-blotched lizard, common raven, pika
5	on the ground without specific water, cliff, rimrock, or talus association	on the ground	48	western fence lizard, dark-eyed junco, elk
6	on the ground	in bushes, trees, or the air	7	common nighthawk, Lincoln's sparrow, porcupine
7	in bushes	on the ground, in water, or the air	30	American robin, Swainson's thrush, chipping sparrow
8	in bushes	in trees, bushes, or the air	6	dusky flycatcher, yellow-breasted chat, American goldfinch
9	primarily in deciduous trees	in trees, bushes, or the air	4	cedar waxwing, northern oriole, house finch
10	primarily in conifers	in trees, bushes, or the air	14	golden-crowned kinglet, yellow-rumped warbler, red squirrel
11	in conifers or deciduous trees	in trees, in bushes, on the ground, or in the air	24	goshawk, evening grosbeak, hoary bat
12	on very thick branches	on the ground or in water	7	great blue heron, red-tailed hawk, great horned owl
13	in own hole excavated in tree	in trees, in bushes, on the ground, or in the air	13	common flicker, pileated woodpecker, red-breasted nuthatch
14	in a hole made by another species or in a natural hole	on the ground, in water, or the air	37	wood duck, American kestrel, northern flying squirrel
15	in a burrow underground	on the ground or under it	40	rubber boa, burrowing owl, Columbian ground squirrel
16	in a burrow underground	in the air or in the water	10	bank swallow, muskrat, river otter
		Total:	327	

[1]Species assignment to life form is based on predominant habitat-use patterns.

habitat manipulations. The most versatile species are the least sensitive to habitat manipulation, and the least versatile are the most sensitive.

The versatility (V) score for each species is derived by determining the total number of plant communities and the total number of successional stages to which the species show primary orientation for feeding and reproduction:

$$V = (C_r + S_r) + (C_f + S_f),$$

where C_r is the number of communities used by the species for reproduction, S_r is the number of successional stages used for reproduction, C_f is the number of communities used for feeding, and S_f is the number of successional stages used for feeding.

GUILD MODELS

The guild concept involves a group of species that exploit the same class of environmental resources in a similar way. Each guild is really defined by the investigator. However, statistical procedures such as cluster analysis and principal components analysis also have been used to define guilds.

Verner (1984) used the guild concept to define a management guild. Thus, using birds as an example, he believed that all bird species using a tree canopy for foraging

Habitat Evaluation

Table 2. Wildlife orientation to plant communities by life form (Thomas 1979:247).

Life form 2. Reproduces in water and feeds on the ground, in bushes, and/or in trees (9 species)

Legend: ● Reproduction and feeding   R Reproduction only[a]   F Feeding only[a]

Letter code	Species	Dry meadow	Moist meadow	Other grasses	Sagebrush-bitterbrush	Other shrubs	Curlleaf mountainmahogany	Western juniper	Quaking aspen	Riparian (deciduous)	Ponderosa pine	Mixed conifer	White fir (grand fir)	Lodgepole pine	Subalpine fir	Alpine meadow	Reproduction	Feeding
	AMPHIBIANS																	
AMTI	tiger salamander		●	●	●				●	●							5	5
AMMA	long-toed salamander		●	●	●	●		●		●						●	7	7
ASTR	tailed frog									●	●	●	●	●		●	6	6
SCIN	Great Basin spadefoot toad	●		●	●	●		●		●							6	6
BUBO	western toad		●		●	●	●	●	●	●	●	●	●	●	●		12	12
BUWO	Woodhouse toad	●	●	●	●	●		●		●							7	7
HYRE	Pacific treefrog	●	●	●	●	●	●	●	●	●	●	●	●	●	●		14	14
RAPR	spotted frog				●			●		●	●	●	●	●	●		8	8
RAPI	leopard frog		●							●							2	2
No. of species using each community: Reproduction		3	6	5	7	5	2	6	3	9	4	4	4	4	3	2	Reproduction	Feeding
No. of species using each community: Feeding		3	6	5	7	5	2	6	3	9	4	4	4	4	3	2		

Number of plant communities used by each species for: Reproduction / Feeding

[a]Not present in this example, but see Thomas (1979:248–269).

could be one management guild. Removal of timber would affect all birds that forage there. The guild concept can be extended further to include all animals that forage on the ground of a deciduous forest, or all animals that use snags. This establishes a matrix of units within a plant community, and biologists can indicate which animals would be affected by natural or human changes in habitats.

HABITAT-EVALUATION PROCEDURES

The habitat-evaluation procedure (HEP) is an example of a habitat-analysis model. This procedure, developed by the U.S. Fish and Wildlife Service, uses a series of HSI models to evaluate habitat (Schamberger and Farmer 1978). A suitability index is calculated for a given species by using that species' HSI model. Each HSI model for a species is computed in turn to examine the community as a whole. The process involves developing a habitat-quality index, which is obtained by combining the HSI values for all evaluated species. The "value" of the habitat for the species is expressed as the product of the habitat's area and its habitat-quality index.

The habitat-suitability index values (from 0.0 to 1.0) are multiplied by the area of available habitat to obtain habitat units that are used to compare areas. Thus, an area

that is to be impacted might be the lever a manager uses to improve less suitable habitat elsewhere so it can support the desired species. Similar habitat models are the Wildlife-Fish Relationship Program, simulation models, and economic analysis models (see Verner et al. 1986).

Wildlife-Habitat Correlations

We know that correlations cannot be interpreted as cause-and-effect relationships. Nevertheless, correlations are valuable to managers because they can help them manage habitat for a population, a species (all populations), or a community, even though the exact causal mechanism is not understood. For example, we may find a negative correlation between the number of roads and the number of deer in a series of areas. We cannot demonstrate the underlying cause with such a relationship. Road-kills and hunter success might be expected to increase with increasing numbers of roads, but the real cause might be stress induced by traffic and the presence of people. Knowing the cause would be valuable to managers, but not absolutely necessary in all situations. In this example, managers might use the correlation to decide to keep areas with small deer populations roadless. Alternatively, if populations are too large, building roads may decrease deer populations and minimize habitat degradation that otherwise would result from overpopulation.

Experimentation is the most conclusive means of determining the relevance of specific habitat features for wildlife. Suppose we find a positive relationship between sage grouse numbers and the density of sagebrush 0.3 m in height. Is the density of sagebrush really critical to the birds? We could better determine its significance by setting up a series of control plots where typical sagebrush densities could be maintained, and a series of treatment plots where lower- or higher-than-average densities would be maintained. In addition, we could identify a series of plots that support sage grouse and then remove sagebrush plants on some of them (reduced density) but not others (controls). The objective would be to determine whether sage grouse numbers on treated areas differ from those on control areas. Temporal and spatial controls and adequate replication of each (perhaps at least 5–10) are necessary before techniques such as *t*-tests, analysis of variance, or analysis of covariance will provide meaningful results. In the absence of adequate replication, statistical techniques are not appropriate, but one can graphically examine general trends in the data (see Hurlbert 1984). Basic statistical techniques are described by Zar (1984) and in Chapter 2 of this manual. Wildlife researchers also use more complex techniques, such as canonical correlation analysis, principal components analysis, discriminant function analysis, multivariate analysis of variance, and multiple regression, because these often are more appropriate for examining the multivariate nature of wildlife-habitat relations (Capen 1981, Dubuc et al. 1990, Livingston et al. 1990, Williams et al. 1990). The experimental approach we recommend is often more expensive, sometimes prohibitively so, but if it is executed correctly, the results permit a clearer interpretation of the significance of specific habitat features.

Identifying features that determine whether a habitat is usable or unusable is a central issue in habitat evaluation. In other words, what habitat conditions determine whether a species will be present or absent? The researcher first must define "use" and "nonuse" for a specific sex, age class, place, time, and life-history activity (e.g., nesting, feeding, loafing). Johnson (1981) pointed out that many factors influence whether a habitat is occupied by a species. For example, even if a habitat is usable, it may be vacant when population levels are low and thus appear to be unusable. If a habitat is being used, but the sampling scheme for assessing use is poor, the presence of a species may not be detected. These errors can be minimized by spending more time in the field confirming the species' presence or absence. One then compares habitat characteristics between the two types of areas, assuming that statistically significant differences indicate which habitat conditions make an area usable or unusable (see also Hobbs and Hanley 1990). This assumption is not warranted if sites cannot be correctly identified by the researcher as used or unused.

Regression involving a binary dependent (*Y*) variable (Neter and Wasserman 1974) is especially valuable in these analyses. The dependent variable is coded as 1 or 0 to reflect whether the species was present or absent, respectively. The explanatory variables (*X*s) are habitat characteristics at each site. Habitat features may have linear or nonlinear effects and additive or multiplicative effects on the presence or absence of wildlife. Multiple regression with a binary dependent variable enables the researcher to assess such possibilities simultaneously and include many habitat features in the analysis. Techniques that enable us to examine complex associations are desirable because actual wildlife-habitat relations often have these attributes. In contrast, methods that assume simple linear relationships (e.g., simple correlation) are often inadequate because they are less realistic.

Wildlife scientists spend a great deal of effort trying to identify which habitat features determine whether a habitat will be poor, mediocre, or good for an individual, a species, or a community. Van Horne (1983) and Maurer (1986) pointed out that density may be a poor indicator of habitat quality. Instead, it would be better to consider measuring variables that unequivocally reflect the health, reproduction, or survival of wildlife. For example, suppose we decide to identify which habitat features in an arid environment determine habitat quality for an endangered lizard. Lizard variables to measure could include body weight, growth rate, fat levels, brood size, brood and adult survival rates, maximum longevity, or other physiological measures (see Chapter 11 of this book). Habitat features associated with food, water, loafing, shelter from predators, and temperature extremes, or even the abundance of competitors or predators, should be examined.

We also may assess habitat quality by determining which characteristics are most closely related to the frequency of habitat use for a specific purpose (e.g., Gutzwiller and Wakeley 1982). We assume there is something important about the habitat that repeatedly attracts the species. For example, wintering birds require sources of energy to maintain body heat and weight. Because survival will depend, in part, on calorie intake, sites with food are of higher quality for wintering birds. So, areas with available seeds, insects, or other food sources will be used more frequently (all else being equal) than those without food. Examination of the frequency with which habitats are used, relative to specific habitat characteristics, is applicable for many different wildlife activities.

The concept of preference also can be used to evaluate habitat (e.g., Spencer et al. 1983, Straw et al. 1986). If a particular habitat type constitutes 20% of the available habitat in an area, but a species uses this type more than 20% of the time, we say this habitat type is preferred. Because more than 20% of the species' locations are in this type, it is used more often than we would expect by chance alone. If the species uses it 20% of the time, that species is not associated with the habitat any differently than a random association, and we assume there is no preference. If a species uses the habitat less than 20% of the time, we assume there is habitat avoidance. A safer conclusion in the latter situation may be that there simply is no preference.

Chi-square tests and log-likelihood ratio tests (Zar 1984) typically are used to assess statistical differences between observed and expected frequencies (see also Thomas and Taylor 1990, Alldredge and Ratti 1992). This approach relies on a large random sample of the habitat in an area to determine the availability of resources for use. One also must make extensive observations of the resource levels that actually are used. Inaccurate represen-

Box 4.   Shorebird use of wetlands.

A biologist wanted to identify which of several habitat features was most closely related to the frequency of wetland use by migrant shorebirds. The goal was to find a key characteristic that could be managed for and that could be measured on aerial photographs to assess habitat quality and quantity. The investigator collected the following data.

Wetland	Open water area (ha)	Mean bank slope (%)	Mud-flat area (ha)	Frequency of wetland use (% of 30 visits that migrants were present)
1	4	38	3	15
2	7	45	6	30
3	9	25	7	30
4	11	20	10	45
5	3	17	12	55
6	1	7	16	75
7	1	3	15	75
8	8	9	9	43
9	6	14	4	12
10	2	5	5	15

The following correlation coefficients (significance levels) were computed.

	Open water	Bank slope	Mud-flat area
Wetland use	-0.347 (0.3252)	-0.451 (0.1905)	0.989 (0.0001)

The frequency of wetland use was most closely related to mud-flat area. The relation was positive, suggesting that larger areas of mud flat improved the quality of wetlands for migrant shorebirds, perhaps by supplying more areas for feeding.

---

tation of either of these two categories will lead to inaccurate preference assessments (see Porter and Church 1987). To determine adequate sample sizes, see Chapter 1 of this book. In the example described above, the available resource is a particular habitat type. The general procedure described here can be applied to many other variables, however. Any variable with a range of values that can be categorized into different groups or levels can be subjected to preference analysis. This includes virtually all variables that one might measure for wildlife, such as soil type, tree density, canopy cover, water temperature, oxygen content, pH, and food levels.

Null hypotheses regarding increasing or decreasing relations or specific population means also are used to determine whether wildlife are associated with certain habitat characteristics differently than expected by chance. We make use of random habitat samples and assume no relation between habitat features and wildlife (e.g., Brennan et al. 1986). A random sample of the habitat resource of interest is taken. Features of the same resource known to be used by wildlife are measured, and the two groups of data are compared statistically. If the two groups differ significantly in their means, we conclude that the wildlife are not using the habitat components randomly. In other words, we infer that some important biological reason (habitat characteristic) explains the difference, not chance sampling.

Suppose, for example, we are interested in frog-habitat associations in a large marsh. We characterize the vegetation around 100 randomly selected points in usable habitat. We then find 100 individual frogs and characterize the habitat around their loafing areas. Our null hypothesis predicts there are no habitat differences between randomly selected sites and used sites. Any significant habitat differences would suggest that the frogs are associated with those characteristics nonrandomly. It is important to take a random sample of habitat that is usable. We can better understand important habitat relations by comparing usable and used habitat than by comparing unusable and used habitats. In our example, it would be trivial to obtain habitat data for random points in the marsh that cannot be used by frogs. Although data from such points do not invalidate the comparison, they may only support what is already obvious. The literature and field observations should be used to determine whether an area is usable.

Earlier, we pointed out that many wildlife-habitat models are not verified. The danger, of course, is that a given relationship may hold for one set of conditions but not another (see Maurer 1986). If the results for one area are incorrectly assumed to apply to another, then habitat-management efforts will be misguided and ineffective. Experimental testing of wildlife-habitat relations is the ideal means of verification. But as we emphasized before, this approach is often expensive and sometimes impossible.

Capen et al. (1986) demonstrated that associations can be tested through three other statistical approaches. One is called data splitting or cross validation. Some fraction (e.g., 50–75%) of the total number of observations is randomly selected and used to generate a model. The remaining fraction of known results is compared with the model's predictions for those data points. Large discrepancies between observed and predicted values indicate an inadequate model. A second general analytical method is called the jackknife approach. One observation is temporarily set aside and a model is estimated with the remaining observations. The model is then used to predict the value of the observation that was left out. This is done in turn for each observation (e.g., Montgomery and Peck 1982). The model is assumed to be meaningful and valid when a majority of predicted values agree closely with the actual values. A third technique involves classifying or predicting independent data. Models based on data from one area are tested with comparable data from another area. These three techniques are especially useful for verifying models generated with regression and discriminant-function analysis.

Often, wildlife-habitat models perform poorly when they are tested outside the specific conditions in which they were developed (Maurer 1986, O'Neil et al. 1988). If so, implicit model assumptions regarding ecological principles may not be entirely justified. Flather and Hoekstra (1985) demonstrated how ecological theory could be used as a first step to improve population-habitat models. If the model's assumptions are unreasonable based on accepted ecological theory, alteration is needed. A model also may perform poorly because it applies only to local conditions. In this situation, data from other areas can be combined with the original data to generate a new model or to broaden an existing model's applicability (Maurer 1986, O'Neil et al. 1988). Numerous environmental factors not associated with specific habitat features affect wildlife populations and communities directly, indirectly, or both. Thus, models that incorporate only habitat features are sometimes inadequate (O'Neil and Carey 1986). If more accurate models and improved assessments of habitat quantity and quality are to be obtained, more careful analyses of the many influences (including weather, disease, predators, and competitors) on wildlife are often necessary.

LITERATURE CITED

ALLDREDGE, J. R., AND J. T. RATTI. 1992. Further comparison of some statistical techniques for analysis of resource selection. J. Wildl. Manage. 56:1–9.

ANDERSON, S. H. 1981. Correlating habitat variables and birds. Pages 538–542 in C. J. Ralph and J. M. Scott, eds. Estimating numbers of terrestrial birds. Stud. Avian Biol. 6.

———. 1991. Managing our wildlife resources. Prentice-Hall Inc., Englewood Cliffs, N.J. 492pp.

BEASOM, S. L., E. P. WIGGERS, AND J. R. GIARDINO. 1983. A technique for assessing land surface ruggedness. J. Wildl. Manage. 47:1163–1166.

BERRY, K. H. 1986. Introduction: development, testing, and application of wildlife-habitat models. Pages 3–4 in J. Verner, M. L. Morrison, and C. J. Ralph, eds. Wildlife 2000: modeling habitat relationships of terrestrial vertebrates. Univ. Wisconsin Press, Madison.

BRENNAN, L. A., W. M. BLOCK, AND R. J. GUTIÉRREZ. 1986. The use of multivariate statistics for developing habitat suitability index models. Pages 177–182 in J. Verner, M. L. Morrison, and C. J. Ralph, eds. Wildlife 2000: modeling habitat relationships of terrestrial vertebrates. Univ. Wisconsin Press, Madison.

BURGER, J. 1987. Physical and social determinants of nest-site selection in piping plover in New Jersey. Condor 89:811–818.

CAPEN, D. E., EDITOR. 1981. The use of multivariate statistics in studies of wildlife habitat. U.S. For. Serv. Gen. Tech. Rep. RM-87. 249pp.

———, J. W. FENWICK, D. B. INKLEY, AND A. C. BOYNTON. 1986. Multivariate models of songbird habitat in New England forests. Pages 171–175 in J. Verner, M. L. Morrison, and C. J. Ralph, eds. Wildlife 2000: modeling habitat relationships of terrestrial vertebrates. Univ. Wisconsin Press, Madison.

CODY, M. L., EDITOR. 1985. Habitat selection in birds. Academic Press, New York, N.Y. 558pp.

COOPERRIDER, A. Y., R. J. BOYD, AND H. R. STUART, EDITORS. 1986. Inventory and monitoring of wildlife habitat. U.S. Bur. Land Manage. Serv. Cent., Denver, Colo. 858pp.

CUPLIN, P. 1978. Remote sensing streams. Proc. Int. symp. on remote sensing of observation and inventory of earth resources and the endangered environment. Int. Arch. Photogram., Vol. II. Freiburg, Fed. Republic of Germany.

DEALY, J. E., D. A. LECKENBY, AND D. M. CONCANNON. 1981. Wildlife habitats in managed rangelands—the Great Basin of southeastern Oregon. Plant communities and their importance to wildlife. U.S. For. Serv. Gen. Tech. Rep. PNW-120. 66pp.

DIXON, W. J., EDITOR. 1981. BMDP statistical software 1981. Univ. California Press, Berkeley. 725pp.

DUBUC, L. J., W. B. KROHN, AND R. B. OWEN. 1990. Predicting occurrence of river otters by habitat on Mount Desert Island, Maine. J. Wildl. Manage. 54:594–599.

ELLIS, J. A., J. N. BURROUGHS, M. J. ARMBRUSTER, D. L. HALLET, P. A. KORTE, AND T. S. BASKETT. 1979. Appraising four field methods of terrestrial habitat evaluation. Trans. North Am. Wildl. Nat. Resour. Conf. 44:369–379.

FINCH, D. 1989. Species abundances, guild dominance patterns and community structure of breeding riparian birds. Pages 629–645 in R. Sharitz and J. Gibbons, eds. Freshwater wetlands and wildlife. U.S. Dep. Energy Symp. Ser. 61.

FLATHER, C. H., AND T. W. HOEKSTRA. 1985. Evaluating population-habitat models using ecological theory. Wildl. Soc. Bull. 13:121–130.

FORMAN, R. T. T., AND M. GODRON. 1986. Landscape ecology. John Wiley & Sons, New York, N.Y. 619pp.

GILES, R. H. 1978. Wildlife management. W.H. Freeman, San Francisco, Calif. 416pp.

GUTZWILLER, K. J., AND S. H. ANDERSON. 1987a. Multiscale associations between cavity-nesting birds and features of Wyoming streamside woodlands. Condor 89:534–548.

———, AND ———. 1987b. Habitat suitability index models: marsh wren. U.S. Fish Wildl. Serv. Biol. Rep. 82(10.139). 13pp.

———, AND J. S. WAKELEY. 1982. Differential use of woodcock singing grounds in relation to habitat characteristics. Pages 51–54 in T. J. Dwyer and G. L. Storm, tech. coords. Woodcock ecology and management. U.S. Fish Wildl. Serv. Wildl. Res. Rep. 14.

HAYS, R. L, C. SUMMERS, AND W. SEITZ. 1981. Estimating wildlife habitat variables. U.S. Fish Wildl. Serv., FWS/OBS-81/47. 111pp.

HEINEN, J., AND G. H. CROSS. 1983. An approach to measure interspersion, juxtaposition, and spatial diversity from cover-type maps. Wildl. Soc. Bull. 11:232–237.

HOBBS, N. T., AND T. A. HANLEY. 1990. Habitat evaluation: do use/availability data reflect carrying capacity? J. Wildl. Manage. 54:515–522.

HOOVER, R. L., AND F. L. WILLS, EDITORS. 1984. Managing forested lands for wildlife. Colo. Div. Wildl., Denver. 459pp.

HURLBERT, S. H. 1984. Pseudoreplication and the design of ecological field experiments. Ecol. Monogr. 54:187–211.

HURLEY, J. F., H. SALWASSER, AND K. SHIMAMOTO. 1982. Fish and wildlife habitat capacity models and special habitat criteria. West. Sect., The Wildl. Soc., Cal-Neva Wildl. Trans. 1982:40–48.

INHABER, H. 1976. Environmental indices. John Wiley & Sons, New York, N.Y. 178pp.

JOHNSON, D. H. 1981. The use and misuse of statistics in wildlife habitat studies. Pages 11–19 in D.E. Capen, ed. The use of multivariate statistics in studies of wildlife habitat. U.S. For. Serv. Gen. Tech. Rep. RM-87.

KNOPF, F. L., J. A. SEDGWICK, AND R. W. CANNON. 1988. Guild structure of a riparian avifauna relative to seasonal cattle grazing. J. Wildl. Manage. 52:280–290.

LANCIA, R. A., AND D. A. ADAMS. 1985. A test of habitat suitability index models for five bird species. Proc. Annu. Conf. Southeast. Assoc. Fish Wildl. Agencies 39:412–419.

LIVINGSTON, S. A., C. S. TODD, W. B. KROHN, AND R. B. OWEN, JR. 1990. Habitat models for nesting bald eagles in Maine. J. Wildl. Manage. 54:644–653.

MASER, C., J. M. GEIST, D. M. CONCANNON, R. ANDERSON, AND B. LOVELL. 1979. Wildlife habitats in managed rangelands–the Great Basin of southeastern Oregon. Manmade habitats. U.S. For. Serv. Gen. Tech. Rep. PNW-86. 44pp.

MAURER, B. A. 1986. Predicting habitat quality for grassland birds using density-habitat correlations. J. Wildl. Manage. 50:556–566.

MEAD, R. A., T. L. SHARIK, S. P. PRESLEY, AND J. T. HEINEN. 1981. A computerized spatial analysis system for assessing wildlife habitat from vegetation maps. Can. J. Remote Sensing 7:34–40.

MONTGOMERY, D. C., AND E. A. PECK. 1982. Introduction to linear regression analysis. John Wiley & Sons, New York, N.Y. 504pp.

NIELSEN, L. A., AND D. L. JOHNSON. 1983. Fisheries techniques. Am. Fish. Soc., Bethesda, Md. 468pp.

NETER, J., AND W. WASSERMAN. 1974. Applied linear statistical models. Richard D. Irwin, Inc., Homewood, Ill. 842pp.

O'NEIL, L. J., AND A. B. CAREY. 1986. Introduction: when habitats fail as predictors. Pages 207–208 in J. Verner, M. L. Morrison, and C. J. Ralph, eds. Wildlife 2000: modeling habitat relationships of terrestrial vertebrates. Univ. Wisconsin Press, Madison.

———, T. H. ROBERTS, J. S. WAKELEY, AND J. W. TEAFORD. 1988. A procedure to modify habitat suitability index models. Wildl. Soc. Bull. 16:33–36.

PLATTS, W. S., ET AL. 1987. Methods for evaluating riparian habitats with applications to management. U.S. For. Serv. Gen. Tech. Rep. INT-221. 177pp.

———, W. F. MEGAHAN, AND G. W. MARSHALL. 1983. Methods for evaluating streams, riparian and biotic conditions. U.S. For. Serv. Gen. Tech. Rep. INT-138. 70pp.

PORTER, W. F., AND K. E. CHURCH. 1987. Effects of environmental pattern on habitat preference analysis. J. Wildl. Manage. 51:681–685.

RAND, G. M., AND S. R. PETROCELLI. 1985. Fundamentals of aquatic toxicology. Hemisphere Publ., Washington, D.C. 666pp.

RODIEK, J. E., AND E. G. BOLEN, EDITORS. 1991. Wildlife and habitats in managed landscapes. Island Press, Covelo, Calif. 219pp.

SCHAMBERGER, M., AND A. FARMER. 1978. The habitat evaluation procedures: their applications on project planning and impact evaluation. Trans. North Am. Wildl. Nat. Resour. Conf. 43:274–283.

SHEPPARD, J. L., D. L. WILLS, AND J. L. SIMONSON. 1982. Project applications of the Forest Service Rocky Mountain Region wildlife and fish habitat relationships system. Trans. North Am. Wildl. Nat. Resour. Conf. 47:128–141.

SIDLE, J. G., AND J. W. ZIEWITZ. 1990. Use of aerial videography in wildlife habitat studies. Wildl. Soc. Bull. 18:56–62.

SOULÈ, M. E., D. T. BOLGER, A. C. ALBERTS, J. WRIGHT, M. SORICE, AND S. HILL. 1988. Reconstructed dynamics of rapid extinctions of chaparral-requiring birds in urban habitat islands. Conserv. Biol. 2:75–92.

SPENCER, W. D., R. H. BARRETT, AND W. H. ZIELINSKI. 1983. Marten habitat preferences in the northern Sierra Nevada. J. Wildl. Manage. 47:1181–1186.

STRAW, J. A., JR., J. S. WAKELEY, AND J. E. HUDGINS. 1986. A model for management of diurnal habitat for American woodcock in Pennsylvania. J. Wildl. Manage. 50:378–383.

THOMAS, D. L., AND E. J. TAYLOR. 1990. Study designs and tests for comparing resource use and availability. J. Wildl. Manage. 54:322–330.

THOMAS, J. W., EDITOR. 1979. Wildlife habitat in managed forests, the Blue Mountains of Oregon and Washington. U.S. For. Serv. Agric. Handb. 553. 511pp.

VAN HORNE, B. 1983. Density as a misleading indicator of habitat quality. J. Wildl. Manage. 47:893–901.

VERNER, J. 1984. The guild concept applied to management of bird populations. Environ. Manage. 8:1–14.

———, M. L. MORRISON, AND C. J. RALPH, EDITORS. 1986. Wildlife 2000: modeling habitat relationships of terrestrial vertebrates. Univ. Wisconsin Press, Madison. 470pp.

WAKELEY, J. S. 1988. A method to create simplified versions of existing habitat suitability index (HSI) models. Environ. Manage. 12:79–83.

WARD, J. P., AND S. H. ANDERSON. 1988. Influences of cliffs on wildlife communities in southcentral Wyoming. J. Wildl. Manage. 52:673–678.

WILLIAMS, B. K., K. TITUS, AND J. E. HINES. 1990. Stability and bias of classification rates in biological applications of discriminant analysis. J. Wildl. Manage. 54:331–341.

WILLIAMS, G. L., K. R. RUSSELL, AND W. K. SEITZ. 1978. Pattern recognition as a tool in the ecological analysis of habitat. Pages 521–531 *in* A. Marmelstein, ed. Classification, inventory, and analysis of fish and wildlife habitat. U.S. Fish Wildl. Serv. FWS/OBS-78/76.

ZAR, J. H. 1984. Biostatistical analysis. Second ed. Prentice-Hall Inc., Englewood Cliffs, N.J. 718pp.

ZWANK, P. J., T. H. WHITE, JR., AND F. G. KIMMEL. 1988. Female turkey habitat use in Mississippi River batture. J. Wildl. Manage. 52:253–260.

24

ECOLOGICAL IMPACT ASSESSMENT

Joe C. Truett, Henry L. Short, and Samuel C. Williamson

INTRODUCTION

Human activities encroach on wildlife habitats at a rate that accelerates each year. This encroachment often reduces the ability of waterfowl nesting areas, big-game winter ranges, and other habitats to support wildlife. Long-term population declines in wildlife species frequently result. Most wildlife ecologists agree that measuring or predicting (i.e., assessing) the effects of human activities is a serious challenge. This chapter reviews recent developments in the science of impact assessment and sets forth a common-sense approach for its application by wildlife managers.

The procedure that attempts to identify and predict the effects of anthropogenic activities on the biophysical environment is called environmental impact assessment (EIA) (Munn 1979). An integral part of EIA is mitigation, the process by which managers try to alleviate the expected adverse effects of anthropogenic activities by modifying or abandoning the proposed activities. Such assessments usually address an array of socioeconomic, historical, and natural resource components of the environment (Rosen 1976, Munn 1979, Beanlands and Duinker 1983).

Because this chapter restricts itself to assessing effects on wildlife and wildlife habitat, the term ecological impact assessment is implied by the acronym EIA except when we refer to a wider array of resources. Thus for our purposes EIA estimates how wild vertebrate populations will respond to human-induced changes to their habitats and recommends how to alleviate expected adverse effects.

This chapter begins by briefly reviewing the legal and procedural requirements of EIA in the United States. It then discusses relationships between EIA and traditional wildlife management practices. A review of the historical development of wildlife impact assessment is presented and is followed by an overview of current technology.

Finally, we outline a step-by-step procedure for impact assessment based on current theory and practice.

LEGAL AND PROCEDURAL REQUIREMENTS

The first formal environmental impact assessment process originated in the United States with the enactment of the National Environmental Policy Act (NEPA) of 1969 (Public Law 91-190). NEPA required a detailed statement on ''proposals for legislation and other federal actions significantly affecting the quality of the environment.'' Regulations of the U.S. Council on Environmental Quality (CEQ) (1973, 1978), published under the mandate of NEPA, provided added guidance for the preparation of environmental impact statements. In addition, a series of judicial interpretations of NEPA and CEQ guidelines has strongly influenced the shaping of procedure (Rosen 1976). Many other countries have, since NEPA passage, adopted similar kinds of practices for evaluating the consequences of human activities (Munn 1979, Hollick 1986).

The basic premise of NEPA (and other mandates arising from or patterned after NEPA) is that, to assess impacts, one first must know what activity is proposed (i.e., the action) and in what context this activity will occur (i.e., the environment). Then one can project how components of the environment will change if the action takes place. Assessment philosophies in most countries ascribe to this general line of thinking, though methods vary among countries for describing actions, environments, and interactions between the two (Munn 1979, Beanlands and Duinker 1983, Hollick 1986).

Irrespective of the federal agency, an EIA in the United States follows several sequential steps. First, the party or federal agency that proposes an action or a policy submits a description. Second, specialists describe the existing environment that the action is expected to affect. Third, analysts assess the environmental consequences by identifying the probable changes that the action will cause. Alternatives to the proposed action also must be presented, and the environmental impacts among alternatives must be compared. Fourth, the parties that are involved consider ways in which undesirable impacts can be alleviated, or mitigated. Impact assessments prepared under NEPA guidelines usually address several environmental components—e.g., social, economic, historical—in addition to the ecological ones we discuss in this chapter.

The official documentation of the NEPA-mandated procedure includes draft and subsequent final versions of impact statements. A draft environmental impact statement (DEIS) is prepared and made available to the public for perusal and comment. A final environmental impact statement (FEIS) contains comments on the DEIS submitted by interested parties, responses to those comments, and revisions of the DEIS.

In the United States, numerous documents by various federal agencies provide detailed guidance for preparing environmental impact statements and for conforming to NEPA and associated regulations in other ways (e.g., U.S. Council on Environmental Quality 1973, 1978, Dames and Moore 1981, U.S. Fish and Wildlife Service 1981, U.S. Federal Aviation Administration 1985, U.S. Army Corps of Engineers 1989). This chapter is intended to explain the underlying ecological principles of assessing impacts to wildlife and wildlife habitat, not to elaborate on procedural guidelines. Several considerations support a focus on principles rather than on U.S.-mandated procedure:

(1) This book can be used by individuals in many countries other than the U.S., and no other country except the U.S. is necessarily concerned with NEPA guidelines and procedures.
(2) Many ecologists have found that strict adherence to the NEPA process often hinders rather than helps efficient and effective impact assessment (see Holling 1978, Beanlands and Duinker 1983, Stakhiv 1988).
(3) The required brevity of NEPA documents (U.S. Council on Environmental Quality 1978) may prevent detailed analyses of impacts from being presented in EISs and other mandated documents. Thus impact assessment problems requiring substantial effort for their resolution often are addressed largely outside formal NEPA documents.
(4) Wildlife biologists of agencies that prepare EISs and other mandated documents can readily learn the procedure being used by their organization. What they need prior to preparing NEPA documents is an understanding of the broader principles of impact assessment so that the documents they prepare reflect sound ecological judgement.

COMPARISONS WITH WILDLIFE MANAGEMENT

Though this chapter focuses on assessing impacts to wildlife species and habitats, we must remember that ecosystems have value beyond the wildlife populations they support. The text of NEPA suggests that its authors envisioned an ecosystem-oriented approach to impact assessment. This chapter presents such an approach.

Fortunately for the wildlife ecologist, there is little difference between understanding how to manage habitat for wildlife and knowing how to assess habitat-related impacts to wildlife. Both require being able to predict how animal populations will respond to given habitat changes, and biologists who understand the basics of habitat management will have little trouble grasping the principles of impact assessment.

Wildlife management and impact assessment require, above all, a thorough knowledge of relations between wildlife populations and their habitats. Habitat, defined simply as the place where animals live (Odum 1971, Moen 1973), sustains animals by providing them food and giving them ways to avoid predation, disease, adverse temperatures, and other biophysical hazards.

Interactions between animals and their habitats, often called functional relationships (Moen 1973), regulate, or limit, the abundances of the animals. For example, the welfare of a northern bobwhite population depends on functional relations such as how well the birds can find food and avoid predators. These functional relationships change, resulting in quantitative changes in wildlife populations, as the structure of the habitat (i.e., the juxtaposition of its physical and biological parts) is changed by actions of wildlife managers or developers. Thus, the changes in the ability of northern bobwhites to find food and avoid predators result from changes in such structural factors as the abundance and distribution of selected plant seeds and the stature and distribution of escape cover.

Wildlife biologists often manage habitat by manipulating its structure (Leopold 1933, Moen 1973). Habitat manipulation changes the abundances of wildlife species, generally increasing the abundance of those that are desired by managers. Development actions likewise change the structure of the habitat, but often in ways not beneficial to the desired species.

An important consideration for wildlife management and impact assessment is that managing for higher populations of all species on a given piece of land is impossible (Smith 1974). Wildlife management and development-induced habitat changes bring about declines in populations of some wildlife species and increases in others (Odum et al. 1979). Thus, habitat changes imposed by bobwhite managers are not inherently good for all wildlife, and neither are habitat changes caused by oil-field development necessarily bad for all species. Only in extreme instances of development are habitats changed so drastically that all wildlife populations suffer.

Currently, wildlife management and impact assessment are more complicated than was early-day wildlife management, primarily because public interest has shifted from a few species to many species. Leopold (1933) defined wildlife management as the art of making land produce crops of a few selected species; modern-day wildlife ecologists (e.g., Graul et al. 1976, Westman 1990) stress the need to consider or manage for populations of all species. The general principles for managing many species or assessing impacts on them remain the same as for managing a few species, but the complexities of the task are multiplied. In the next section we examine how methodologies have changed over the past 20 years and how approaches to impact assessment have improved to accommodate the increased complexity.

HISTORICAL DEVELOPMENT

Ecological impact assessment, as a practice separate from wildlife management, began in the United States with the passage of NEPA. It posed a more complex problem than traditional single-species wildlife management, partly because analysts had to consider many species and partly because the managers themselves had little control over the actions that affected the wildlife. Analysts wrestling with these new dimensions progressed through about three sequential approaches over the next score of years: (1) an initial preoccupation with species inventories generally intended to be used as bases against which to measure change, (2) an early shift of focus by some persons to analyses of ecosystem processes (functions) for helping to predict impacts, and (3) an ultimate realization that the complexity introduced by trying to assess impacts to numerous species created a measurement dilemma, and a consequent search for responsible integrative measures that could simplify the measurement problem. The following paragraphs summarize this evolution.

Shortly after the passage of NEPA, the baseline study emerged as the major document supplying information about wildlife and other subjects for the EIA process. Most such studies early in EIA history provided primarily descriptive (and often voluminous) surveys of environmental components (Rosen 1976, Beanlands and Duinker 1983), though the term "baseline study" eventually came to refer to the entire range of predevelopment investigations regardless of their focus (Hirsch 1980). The rationale for conducting baseline studies was to provide a description of existing conditions against which subsequent changes could be detected through monitoring (Hirsch 1980). Some baseline studies additionally purported to enable predictions of impact (Munn 1979).

In the decade after enactment of NEPA, United States scientists and decision makers, in their response to legal mandates, placed heavy and often exclusive reliance on results of baseline studies (Hirsch 1980). But most baseline studies had severe shortcomings as sole providers of information on wildlife. They often produced only taxonomic lists, general descriptions, and relative abundances of wildlife species. They often failed to provide a clear picture of the existing environment against which change could be measured (Hirsch 1980), and most failed completely as a basis for predicting change (Holling 1978, Hirsch 1980).

By the late 1970s, the inability of conventional baseline studies to provide information for predicting impacts caused reactions by rule-makers and scientists. The U.S. Council on Environmental Quality (1978 : Part 1500:4) set forth new guidelines, instructing analysts to prepare "analytic rather than encyclopedic" impact statements. These guidelines directed preparers of NEPA documents to pay more attention to assessing the consequences of actions and less on amassing needless detail and presenting extraneous background data.

Scientists began to call for a narrower focus (Holling 1978, Beanlands and Duinker 1983). Some analysts (Truett 1979, Hirsch 1980, Beanlands and Duinker 1983) thought that understanding ecosystem processes was a desirable approach for predicting the consequences of anthropogenic actions. Studies of processes tried to clarify important functional relationships between species and their biophysical environment, e.g., food-web links, climate effects, and interactions of animals with structural features of their habitat. The so-called ecological characterization (Hirsch 1980), a kind of process study that came into vogue with some federal agencies in the late 1970s, is one example; it attempted to describe not only the important ecosystem components but also the functional relationships among them.

Recall that accurate impact assessment, like successful wildlife management, requires a knowledge of the important functional relationships between wildlife species and their habitats. The process-study approach recognized this. But it failed as an easy solution to assessing impacts to multiple species because quantification of important processes on which each of many species of wildlife depended often proved intractable. To understand why, consider the many years of research that may be required for managers to quantify just the important food-web relationships between white-tailed deer and their habitat at a particular place. The time that would be required to reach this level of understanding for many relatively unstudied species often has stymied accurate and ready predictions of impacts by the process approach.

The late 1970s and the 1980s saw inquiry by many scientists into strategies for assessing impacts to multiple species. Various indices were proposed as measures of the quality of the wildlife community; Schroeder (1987) reviewed and critically analyzed some of these, notably wildlife species richness and diversity, wildlife guilds, species dominance, total wildlife density, wildlife biomass, trophic structure, and community function. Other authors (e.g., Graul and Miller 1984, Kautz 1984, Jarvinen 1985, Duinker and Baskerville 1986, Karr 1987) reviewed some of the existing strategies and proposed indices or conceptual approaches of their own.

The beginning analyst can choose among currently recommended approaches better if he or she understands how they evolved. The progression from single-species management to various multiple-species methodologies has been iterative, and newer methods have responded to past insufficiencies, as illustrated in the following paragraphs.

Single-species management, as noted above, has been for many years standard practice for managing wildlife (Graul and Miller 1984). The accepted procedure has been to identify the factors that regulate or limit a species' abundance or distribution and to modify or control them to affect the abundance of the species (Leopold 1933). Impact assessment for single species requires a slightly different approach using the same information—it predicts how a proposed action will affect a species' abundance by influencing its limiting factors. But because of legal mandates and public concern that now require consideration of many species, a focus on single species sometimes is considered inadequate for EIA (Wagner 1977, Salwasser and Tappeiner 1981).

The featured-species approach, first proposed to manage wildlife on units of National Forest land (Holbrook 1974, Gould 1977), is nearly the same as single-species management. A few featured species that are important to managers or the public are selected for a unit of land, and the effects of management of the unit on these species are evaluated. The manager assumes that other species with incidentally similar needs will respond similarly to given actions (Gould 1977, Salwasser and Tappeiner 1981). This assumption has made some proponents of this approach feel responsive to the multiple-species mandate. They acknowledge that the approach favors some species at the expense of others, but they believe that any management choice does the same and that a practical approach should favor species of concern over others (Gould 1977).

The management-indicator approach evolved from the featured-species concept (Graul and Miller 1984). Management indicators include species that may have the public-interest characteristics of featured species, but also include others that are purposely selected to indicate the welfare of an array of desirable species (Mealey and Horn 1981, Schroeder 1987). Primary considerations in selection of indicator species are the need to maintain viable populations of all vertebrates and to provide wildlife diversity (Mealey and Horn 1981). Thus, in theory, the array of species specifically addressed has habitat requirements that encompass the habitat needs of most or all other species. But, in fact, the needs of each species are unique, and fulfilling the needs of the indicator species often does not provide for the welfare of others (Graul and Miller 1984).

An ecological-indicator approach proposed by Graul et al. (1976) bases management of an area primarily on the needs of species with the most exacting ecological requirements, i.e., stenotopic species. The assumption of this approach is that species with less exacting habitat requirements can readily accommodate to any changes to which the stenotopic species can adjust. Proponents of this approach emphasize its conservative quality of purportedly lessening the risk of losing any species in a given system (Graul and Miller 1984). It has not been widely used, perhaps because the stenotopic species are more sensitive to change than other species, and strict adherence to the approach would severely curtail the extent to which management for other values could change the system.

Indices of wildlife species diversity have been used widely as baseline descriptors (Schroeder 1987). The most widely used index probably is species richness, the total number of species in an area. Equitability, a measure of relative abundance among the species present, also has been considered important. Ratios between species richness and equitability are called species-diversity indices, and their use has been widespread in baseline measures of communities. The advantage of using a species-diversity approach is that it provides a single measure of the wildlife community that is frequently sensitive to habitat change. Critics point out the insensitivity of such indices to changes in wildlife species composition and relative abundance (Schroeder 1987).

The use of wildlife guilds in impact assessment and management has been popular (Severinghaus 1981, Short and Burnham 1982, Landres 1983, Short 1983, Verner 1984). Root (1967) originally defined a guild as a group of species that exploit the same class of environmental resources in a similar way, i.e., species with similar functional relationships with their habitats. Ecologists have proposed guilds based on apparent similarities among species in feeding and breeding strategies, sizes of individuals, habitat preferences, and behavior (Schroeder 1987).

Severinghaus (1981) maintained that, once the impact on any one species in a guild is determined, the impact on every other species in that guild is known. But Landres (1983) and Verner (1984) pointed out that guilds are investigator-defined units that are based on assumed functional relationships between animals and their habitats. The investigator may perceive a group of species to be similarly dependent on a resource and thus include them in the same guild, but each of the species may in fact have different degrees and ways of resource use and thus respond to change differently.

Verner (1984) proposed an operational approach to delineating management guilds based on similar responses of different species to changes caused by development. The rationale for this approach is that the guild members would respond similarly to the same actions. Verner (1984) suggested that structural attributes of habitat (e.g., vertical layering and horizontal spacing of vegetation, substrate type, topographic relief, physical characteristics of water bodies, and presence of special features such as snags or logs) are good criteria for delineating guilds.

Several other habitat-based approaches have emerged. This is not surprising because, as noted earlier, for many years biologists have managed wildlife by managing habitat (Siderits and Radtke 1977). Degree of habitat type interspersion, habitat island size, and extent of connecting corridors between habitat islands are all facets of habitat structure recognized to influence the diversity and abundance of wildlife species (Kautz 1984). Biologists also judge the value of habitat by additional criteria such as vegetation type, vegetation successional stage, and special habitat features with known values to certain species (Graul and Miller 1984).

Some habitat-based approaches ostensibly bypass the needs of individual or multiple wildlife species to concentrate simply on providing diverse environments (Graul and Miller 1984). However, most criteria for describing environmental diversity are in fact based on known or suspected relationships between structural features of habitat and selected wildlife species, species groups, or species-diversity measures.

In the United States, each federal agency responsible for impact assessment usually has adopted one or more formalized procedures based on habitat measures. Two widely applied approaches are the Habitat Evaluation Procedures (HEP) used by the U.S. Fish and Wildlife Service (see Chapter 23) and the Habitat Evaluation System of the U.S. Army Corps of Engineers (1980). Species habitat models drive both of these procedures, which are sophisticated accounting tools for measuring habitat quantity and quality and changes therein. The Wetland Evaluation Technique (Adamus et al. 1987) is applied by some agencies for appraisal of wetland habitats.

PRINCIPLES OF MEASUREMENT AND PROCEDURE

In this section we summarize a few basic principles of measurement and procedure that have evolved over the past few decades and that, despite many differences of opinion about which assessment methods are best, are widely applied today. This overview introduces the theory behind the practical applications presented later.

Assessing and managing ecological impacts invariably require integrating measurement and procedure. Measurement is generally made at one or more of three scales: classificatory (naming), relative-value (ranking), and ratio (zero at origin). Procedure can be viewed as the sequence of actions analysts take; it follows three logical steps that we call scoping (identifying the problem), analysis (assessing its dimensions), and synthesis (exploring solutions).

Among the measurement scales, the classificatory scale is usually the least desirable because it measures objects at the weakest scale—for identification only. The relative-value scale is more desirable; it measures objects that are different and stand in some relative position to each other, but the distances between numbers on the scale are not meaningful. The ratio scale is the most desirable because it quantifies objects—the distances between numbers on the scale are meaningful, and a true zero point exists at the origin. The expense of the measurement is usually directly proportional to its desirability.

The procedural steps require few to many people collaborating over weeks, months, or years. Economic constraints usually determine how many people become involved and how much time is needed. Greater values of proposed projects or affected wildlife resources often are associated with larger numbers of participants and longer time frames in the EIA process; these are usually more desirable than fewer participants and shorter times.

For each of the three procedural steps—scoping, analysis, synthesis—the analyst often considers two options for measurement as follows:

1. Scoping

Classificatory Scale.—Checklists of species and other ecosystem components are useful for organizing information in the early parts of the EIA process (Shopley and Fuggle 1984). Presence/absence matrices are useful for displaying information and judgments about the causes of problems and for identifying interactions between projects or activities and specific components.

Relative-Value Scale.—Mathematical value matrices (Shopley and Fuggle 1984) are useful for displaying value judgments about the relative importance of components, causes, and problems. Networks (Shopley and Fuggle 1984), also known as system diagrams, are useful in classifying, organizing, and displaying problems, processes, and interactions and in producing causal diagrams.

2. Analysis

Relative-Value Scale.—Indices such as the species-diversity index (Schroeder 1987) and habitat-layer index (Short and Williamson 1986) are dimensionless measurement techniques that are useful for comparing the impacts of development alternatives and identifying the preferable alternatives. Mapping land surface cover and use is useful as an inventory and display technique.

Ratio Scale.—Time-series graphs of sampling data (Green 1979) are used to view impacts over time. Time-series maps of remotely sensed data are useful in landscape ecology applications for quantifying habitat fragmentation and for identifying more objectively the preferable alternatives.

3. Synthesis

Relative-Value Scale.—Evaluation techniques are designed for nonquantitative decision making. They are useful for comparing the impacts of development alternatives and for identifying subjectively the more preferable alternatives (Shopley and Fuggle 1984).

Ratio Scale.—Mathematical simulation modeling (Holling 1978, Shopley and Fuggle 1984), usually as time series of ecosystem functions, is most useful for estimating and communicating, in conjunction with other techniques, the long-term, future, and indirect effects. Cartographic modeling (Short and Williamson 1986) is useful for displaying habitat layers and for comparing changes caused by natural succession with those caused by development. Adaptive environmental assessment methods (Holling 1978) combine several techniques, usually emphasizing mathematical simulation modeling, for handling different aspects of impact assessment.

Using the strongest measurement scale possible, given economic constraints, is usually desirable. The recommended combination includes networks and mathematical-value matrices (for relative-value scale in scoping), time-series graphs and time-series maps (for ratio scale in analysis), and mathematical simulation modeling and cartographic modeling (for ratio scale in synthesis). This combination can be used to identify effects produced and mitigation possibilities. The National Research Council of the United States and The Royal Society of Canada (1985) advocated acquisition of four kinds of information to understand the characteristics of the Great Lakes Basin as a system (including human activities taking place): (1) causal models that relate human uses and ecosystem responses, (2) time series of monitored data that provide status and trend, (3) maps that show the ecosystem's key features and human uses, and (4) management case studies that show results in the ecosystem. The innovative use of combinations of assessment techniques (each with distinct purposes and uses) is becoming more common in the ecological impact assessment profession.

WHAT TO MEASURE

How does a biologist armed with the background and theory presented to this point go about measuring the impacts to wildlife when a human action is proposed? Detailed prescriptions are many, and beginning practitioners would benefit from reading a few of the more useful texts (e.g., Holling 1978, Munn 1979, Beanlands and Duinker 1983, Westman 1985). In this section we outline a practical, step-by-step process synthesized from our own experience and that of others over the past 20 years.

To measure impacts of an action to the wildlife community and to plan mitigation for adverse effects, analysts need to take five more or less sequential steps:

1. *Set wildlife management goals.* Early in the assessment process, the biologist describes the wildlife goals (i.e., the species abundances and distributions that are desired in the project area) so that changes expected to be induced by the proposed action can be weighed against these goals.

2. *Describe habitat conditions necessary for reaching goals.* Once wildlife goals have been established, the biologist describes the habitat conditions needed to support

these goals so that predicted habitat changes eventually can be translated into effects on the wildlife goals.

3. *Show how the proposed action will alter habitats.* The biologist portrays and describes projected habitat changes in space and time.

4. *Predict how expected habitat alterations will affect the wildlife community.* The biologist translates habitat changes expected to occur into predictions of changes in the wildlife community.

5. *Plan mitigation for adverse changes.* The biologist and other participants explore options for mitigating impacts of those projected habitat changes expected to cause adverse deviations from wildlife goals.

The following sections provide detail about what to measure at each step. We use five impact assessment projects to illustrate what others have measured under various circumstances; these are:

1. *Resource Management Planning in Northwestern Colorado.*—The U.S. Bureau of Land Management (BLM) (1984) prepared an environmental impact statement to assess the impacts of several alternative resource management plans for BLM-managed lands in the Piceance Basin Planning Area in northwestern Colorado. Because of large deposits of shale oil and other mineral resources in the area, the impact assessment studies providing background for this EIS were numerous.

2. *Oil and Gas Leasing in Northeastern Alaska.*—The U.S. Fish and Wildlife Service (FWS) (1987), in cooperation with the U.S. Geological Survey and BLM, prepared an environmental impact statement relative to a legislative action—the passage of the Alaska National Interest Lands Conservation Act (Public Law 96-487). This piece of legislation required specific study of a coastal region of the Arctic National Wildlife Refuge (NWR) in northeastern Alaska because industry showed interest in leasing it for oil and gas exploration.

3. *Development of an Observatory on Mount Graham, Arizona.*—Proposals for construction of an astrophysics observatory atop Mount Graham in southeastern Arizona required the U.S. Forest Service to study project-related impacts. The presence of the world's only population of the Mount Graham red squirrel in the project area galvanized more impact assessment study than otherwise might have ensued, and much of the assessment effort focused specifically on impacts to this squirrel population (Short and Williamson 1988).

4. *Managing Big Stone National Wildlife Refuge, Minnesota.*—The FWS manages the Big Stone NWR in southwestern Minnesota as habitat for migratory waterbirds and a variety of other species. As part of planning for the refuge, FWS developed a conceptual framework for assessing effects of habitat modification on various species (Short and Williamson 1986).

5. *Forest Conversion in the Tensas River Basin, Louisiana.*—The bottomland hardwood forests of the Tensas River basin in northeastern Louisiana, similarly to many bottomland forests in the eastern United States, are subject to incremental clearing of small tracts. Louisiana State University conducted for the U.S. Environmental Protection Agency an assessment of cumulative effects of those landscape changes on wetlands resources in the Tensas basin (Burdick et al. 1989, Gosselink et al. 1990).

Setting Wildlife Goals

At the beginning of the measurement process, the analyst must set and quantify wildlife management goals for the area to be affected. Biologists usually quantify goals in terms of (1) population levels or, occasionally, recruitment rates of one or more species, or (2) multiple-species measures such as species diversity, wildlife guild diversity, trophic structure, or animal biomass. Change can be assessed without setting goals, but only against a description of society's goals for the wildlife community can managers judge the adversity of change (Odum et al. 1979). Accurate impact assessment requires that goals be few, readily measurable, and unambiguously stated.

Goals in some of our five examples were explicit but in others were ambiguous. In northwestern Colorado, goals focused on desired population levels of game animals (e.g., mule deer, elk, mountain lion, black bear, and sage and blue grouse) and endangered species (bald eagle). On the Arctic NWR, goals were to conserve populations of caribou, muskoxen, wolves, brown bears, wolverines, and migratory birds. An overriding goal at the proposed observatory site on Mount Graham was to attain and maintain a specified population size of the endangered Mount Graham red squirrel. On Big Stone NWR, goals related to maintaining abundances and diversities of selected wildlife populations. The wildlife goals specified in the Tensas River basin were to maintain and enhance balanced indigenous populations of fauna and to conserve existing biota.

Goals were stated more appropriately for impact assessment in three of our examples than they were in the other two examples. In the northwestern Colorado, Mount Graham, and Big Stone NWR examples, goals were sufficiently explicit to determine whether they had been accomplished. But in northeastern Alaska and the Tensas basin, goals as initially stated were too vague to quantify. Biologists on the Tensas basin project ultimately did refine goals to more or less measurable aspects of bird populations. The stated intent to conserve wildlife populations in northeastern Alaska presumably implied maintenance of existing densities.

Describing the Optimum Habitat

Having clarified goals for wildlife management, the analyst must describe the habitat conditions for meeting those goals. The conditions need to be described in terms of those habitat variables that (1) control the abundance and distribution of, or limit, the wildlife populations and (2) will be modified by the proposed action.

How did the analysts in our examples describe the habitat conditions necessary for meeting their wildlife goals? In northwestern Colorado, BLM described desired habitat conditions in terms of (1) animal-unit-months (AUMs) of forage needed for deer and elk (an AUM is the amount required per animal per month), and (2) the presence of crucial habitats believed to limit populations of blue and sage grouse and bald eagles. On the Arctic NWR, FWS desired to maintain the existing availability and quality of habitat necessary for conserving populations of caribou, muskoxen, wolves, bears, wolverines, and migratory birds, but they gave no quantitative ways to measure the habitat's ability to do this. On Mount Graham, analysts proposed measures of the horizontal and vertical structure of the vegetation (e.g., size and spacing of conifer stands, frequency and ages of trees within stands) to describe optimum habitat for red squirrels. For Big Stone NWR, analysts described optimum habitat conditions in terms of a habitat layer index that represented the relative structural diversity of habitats by criteria of layers of vegetation and substrate types present in subunits. Biologists in the Tensas River basin described optimum habitats in terms of sizes and interspersion patterns of bottomland and upland forest tracts, and the presence or absence of connecting corridors among these tracts.

In some of the examples, analysts did not clearly show how changes in habitat could be translated into an assessment of whether wildlife goals could be reached. This inability to quantify connections between the habitat measures and the wildlife populations is the norm rather than the exception (Holling 1978, Beanlands and Duinker 1983). It likely will continue to be one of the major frustrations to accurate impact assessment.

Forecasting Habitat Change

Once analysts have agreed on the habitat conditions necessary for meeting wildlife goals, they must quantify the habitat changes expected to result from the proposed action. This step is invariably more difficult than anticipated because initial descriptions of planned actions are often vague, focused inappropriately for describing habitat changes important to wildlife, and written in the language of policymakers or engineers. Wildlife ecologists must translate these into quantitative descriptions of expected changes in the habitat variables to which wildlife species respond. The advantage of using habitat structure as the basis for describing habitat requirements for wildlife becomes evident at this stage because engineers and policymakers with whom the wildlife biologist must interact usually describe and perceive environmental change in structural terms.

Analysts in the examples quantified forecasts of habitat change in several ways. Managers in northwestern Colorado calculated for each action alternative the AUMs of forage that would remain available for mule deer and elk and the extent to which crucial habitats of other species would be lost. In Alaska's Arctic NWR, impact analysts estimated the total area of tundra (and in some cases the area of special-use habitats) from which caribou, muskoxen, and other animals would be excluded, and reduction in habitat value from industrial activities on remaining areas. On Mount Graham, researchers estimated the area of red squirrel habitat expected to be lost to or degraded by the project and weighed this against the good and excellent habitat presently and potentially available. On Big Stone NWR Refuge, scientists laid the conceptual groundwork for quantifying the spatial changes in habitat layer combinations that would occur under various management scenarios. In the Tensas basin, biologists predicted future cumulative hectares of bottomland hardwood forest lost to agricultural conversion.

Impact analysts generally will find that forecasting the amount of habitat lost (i.e., landscapes made uninhabitable to animals) is easier than measuring a reduction in habitat value. Thus the area of habitat made unusable to terrestrial animals by a reservoir is easy to measure and describe, but estimating the reduction in habitat value caused by selective logging on a given area is much more subjective. This problem is a direct consequence of our inability to quantify connections between the animals and their habitats.

Predicting Wildlife Responses

Once habitat changes are forecast, one must predict how the wildlife community will respond to the changes. Analysts attempting quantitative predictions often find this to be the Achilles' heel of impact assessment. One problem is that habitat lost or changed does not necessarily cause proportionate losses in the wildlife populations of interest because some parts of the habitat may be much more crucial than others. Another problem is that most actions do not totally remove habitat, they simply cause a change in the levels or kinds of wildlife use, and deciding to what extent these changes will affect wildlife populations is difficult.

The five examples illustrate a potentially wide range in analysts' abilities to quantitatively predict wildlife response to habitat change. Predictions seemed particularly difficult to support when multiple species were involved and proposed actions would induce only habitat change and not total habitat loss. In none of the examples have predictions been tested, because the proposed actions have yet to materialize.

Impact analysts' abilities to predict wildlife change on the basis of habitat change will depend on (1) how appropriately the analysts have been trained and (2) how well the species' limiting factors are known. Many wildlife managers accustomed to the concept of manipulating habitat to benefit populations will have gained an intuitive appreciation for how to assess impacts of habitat changes. Wildlife species that have been well studied with respect to habitat requirements are much easier to deal with than those for which habitat needs (i.e., limiting factors) are poorly known. In any event, the precision with which changes can be predicted usually will vary inversely with the number of wildlife species considered.

Planning for Mitigation

Mitigating, or alleviating, adverse impacts to wildlife may range from altering the plans for the proposed action to managing or providing habitat elsewhere to offset losses. It should be evident by now that effective mitigation planning requires a detailed knowledge of how habitats regulate wildlife populations.

In practice, mitigation plans often resemble wildlife management plans. They propose to alter a proposed action, or sometimes to take an action, that will improve habitat for selected species or groups. They describe a course of action that results in wildlife goals being more nearly met than would be possible otherwise.

Mitigation actions themselves, like other human actions, benefit some species to the detriment of others. Here again, the necessity for having specified wildlife goals that give some species and conditions priority over others becomes evident. Mitigation, like impact assessment, must be tailored to meet these goals.

When analysts are preparing formal environmental impact statements, NEPA provides a built-in strategy for mitigation by requiring consideration of several alternatives for each action proposed. This enables agencies protecting wildlife interests to recommend the alternative that poses the fewest obstructions to meeting wildlife goals. For example, the preferred resource management alternative of BLM in northwestern Colorado included a requirement to improve over the long term the forage allocations to mule deer and elk and the habitat conditions for game animals, raptors, and threatened and endangered species. Several alternatives for action on Alaska's Arctic NWR identified specific measures to offset adverse impacts to caribou in their calving areas.

Because most or all action alternatives may cause adverse impacts, mitigation requirements typically extend beyond those accommodated in the alternative plans for action. In northwestern Colorado, proposed mitigation included enhancing forage production by seeding or by manipulating the existing range vegetation, culling wild horse populations to provide additional forage to wildlife, and creating buffer zones of restricted human activity around raptor nest sites and sage grouse leks. On the Arctic NWR, mitigation plans included minimizing vegetation destruction by prohibiting off-road vehicle travel and by requiring drilling in winter from removable pads, and promoting free passage to caribou by elevating or putting ramps over pipelines. On Mount Graham, mitigation measures under consideration focused on maintaining or providing appropriate amounts of forest with optimum vegetation structure for red squirrels. In the Big Stone NWR, analysts proposed no specific actions and thus did not prescribe mitigation measures; mitigation considered necessary for any actions proposed later presumably would be based on altering or maintaining specified habitat layer combinations. Similarly, landscape conversion in the Tensas basin required no specific mitigation plan, but mitigation would entail protecting existing hardwood forest stands and corridors and perhaps, over time, creating others.

Summary

Measurement in impact assessment involves five more or less sequential steps. These steps follow a logical thought process: setting wildlife goals, defining the habitat required to meet these goals, estimating how the proposed action will change that habitat, predicting wildlife responses to the habitat changes, and planning mitigation so that adverse effects can be alleviated.

Accurate and effective impact assessment requires that measuring and describing the wildlife, the habitat characteristics that regulate the wildlife, and the connections between the two be as quantitative as possible. Problems arise when analysts fail to explicitly identify the wildlife goals or to accurately specify the habitat conditions necessary to meet those goals, because all other steps in the assessment process hinge on adequate portrayal of these criteria. Most inefficiencies and inaccuracies in the impact assessment process result from analysts' inabilities to isolate the habitat characteristics that control the wildlife populations of interest. One often finds that structural components of the habitat not only exert major controls on wildlife populations, but they also are relatively easily measured by analysts and readily perceived by lay participants in the process.

HOW TO PROCEED

Unlike conventional wildlife management, impact assessment usually requires extensive interdisciplinary input from decision makers, technical experts other than biologists, and the public. This adds dimensions unfamiliar to many wildlife biologists, making the process of assessing impacts more challenging than simply measuring wildlife and habitat. Several texts provide guidance for how to accommodate this added complexity (e.g., Holling 1978, Beanlands and Duinker 1983, Westman 1985). The following paragraphs give the beginning analyst a few clues to how the orchestration of these added requirements is superimposed on the measurement process. The five steps of measurement outlined above are best contained in a three-step procedural sequence that accommodates conventional team behavior in solving problems: (1) scoping, (2) analysis, and (3) synthesis and quantification.

Scoping

Defining the scope of the problem, or scoping, takes place early in the impact assessment process. It is intended to describe the problem in space and time. All interested parties—technical experts, representatives of the public, proponents of development—typically participate in scoping, usually by attendance at workshops or meetings. Holling (1978) described workshop procedures that are useful. The scoping stage is the appropriate time for the wildlife biologist to set preliminary wildlife goals.

The first available description of the development action is likely to be reviewed at this time, and that description probably will be one prepared solely by project engineers or architects. Invariably, proponents' first descriptions of projects are inadequate bases from which to make predictions of biological impacts, for they emerge from the developer's perspective. During the scoping stage as well as later on, the biologist will need to call upon the project engineer or manager for additional descriptive information.

The scoping exercise must provide an initial view of the proposed project as an agent of habitat change. Determining the area to be affected by the project and the time scale of effects is crucial. For example, construction of a dam will affect a canyon habitat by inundating areas upstream and dewatering floodplain habitats downstream; these changes probably will continue as long as the dam exists.

Spatial and temporal boundaries are set so quantitative predictions can be made (Beanlands and Duinker 1983). Maps of the project area will be required that show the distributions of habitat features and types and the expected distribution and nature of project changes. Remember that formal EIS procedures in the United States require consideration of several alternative actions, not just one, and often the spatial and temporal boundaries are different among alternatives.

The product from the scoping step should be a mutually understandable description of (1) the planned action, (2) the wildlife resources of concern and the goals related to them, and (3) the potential changes to habitats that are likely to influence the wildlife populations. The report should include a spatial depiction (i.e., a map or maps) of areas where wildlife populations are likely to sustain impacts, or at least of areas where habitat structure is likely to be visibly altered. The time scale of impacts may need to be expressed in terms of the estimated life of the project and the time required to replace habitat features expected to be lost.

Analysis

After participants define the scope of the project and set wildlife goals, the biologist must analyze the potential impacts on wildlife in greater detail. This analysis should reveal whether wildlife populations of concern will in fact be affected and whether the effects will be deemed significant.

The biologist will have to conduct an in-depth inquiry into the habitat needs of the wildlife species of interest. A useful approach is to look for relationships between the wildlife populations and the structural features of the habitat. Biologists usually obtain relevant information in one or more of several ways—reviewing the available literature, visiting similarly developed sites, holding meetings with technical experts, or conducting new field studies. As the analysis proceeds, the biologist can enhance his or her own understanding and promote better communication with other participants if he or she builds graphic or word models that show species-habitat relationships. Ways of doing this are described in a later section on synthesis and quantification.

Early in this phase the biologist will need to refine the goal statements developed during scoping. These must be stated as precisely as possible in terms of population measures. Goal statements ideally reflect the public's desires for the composition and abundances of wildlife species; often but not always these equate with the predevelopment condition. Goals that accommodate many species may need to be quantified by integrative measures such as species diversity or richness. Special-interest species may require independent measures.

The biologist must translate the public's apparent desires into descriptors of wildlife populations, but that is not all. He or she also must transform these images of desired populations into descriptions of the habitats required to support them. The biologist thus is the architect of the habitat and analogous to the developer's architect. He or she must portray the habitat conditions that will support the wildlife that the public wants, assess limitations to the supply of habitats, and depict habitat options within the range of potential development designs.

For several reasons, measures of the physical structure of the habitat often provide the best yardstick for the habitat architect. First, alteration of habitat structure is usually the major and most easily perceived channel through which human actions affect the wildlife community. Second, the ways in which wildlife populations will respond to structural change are frequently predictable on the basis of existing or easily gathered information (Willson 1974, Short 1988). Third, plant communities impose much of the structure important to wildlife—e.g., vertical layering of vegetation, horizontal juxtapositioning of life forms, and abundance and distribution of snags and logs (MacArthur and MacArthur 1961, Verner 1986). Fourth, and perhaps most important, using measures of structure enhances communication in the EIA process because a wide range of participants can readily perceive structural char-

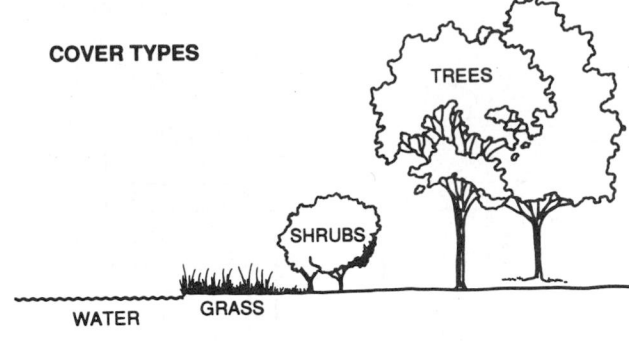

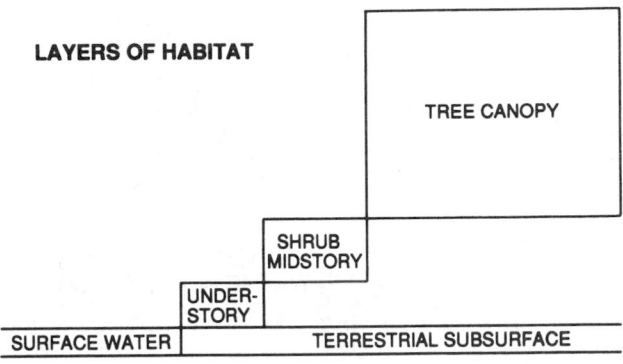

Fig. 1. Abstract representation of vegetation and substrate cover types as layers of habitat (from Short and Williamson 1986).

acteristics and thus comfortably accept them as ''currency'' for measuring losses and gains.

Synthesis and Quantification

Once analysts have clarified relationships among the wildlife populations, the wildlife goals, and the habitat features, projected change needs to be quantified. Making effective assessments of impact and subsequently developing plans for mitigation require that these projections be readily measurable and easily depicted. In this section we show examples of how habitat change can be quantified and trade-offs can be evaluated with simple techniques and tools available to most wildlife biologists. Dealing with single species is relatively simple, so we focus largely on multiple-species problems.

Short (1983) developed a useful technique to simplify the assessment process when numerous species are involved. By reviewing the literature, he observed that the abundance of most species seemed to vary with the extent of habitat structural components that were readily measurable. The important components were various layers of the habitat—trees (overstory), shrubs (midstory), herbs (understory), terrestrial substrates, and water bodies (Fig. 1)—and simple measures could be made of the extent and juxtapositioning of the various layers. That many wildlife populations respond to habitat layers and structural components associated with layers has been well documented (e.g., MacArthur and MacArthur 1961, Willson 1974, Verner 1986).

By using Short's (1983) technique, the biologist can position each species of interest within a species-habitat matrix (Fig. 2). The position or positions of each species in the matrix reflect the layers of habitat on which the

SPECIES - HABITAT MATRIX

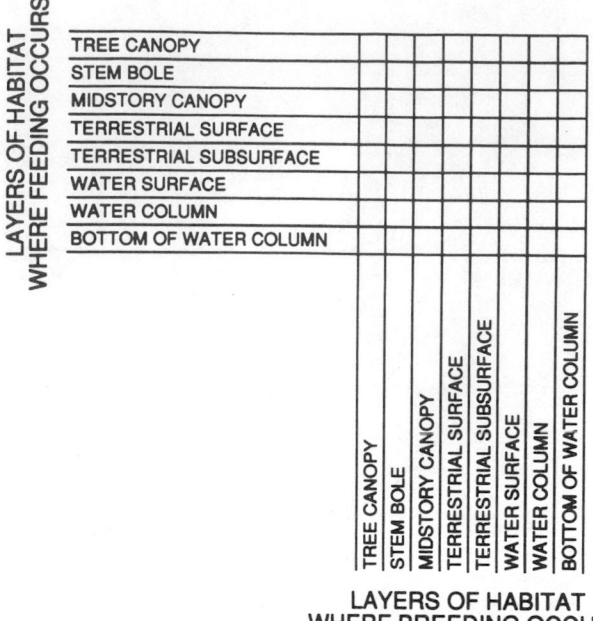

Fig. 2. Species-habitat matrix used to identify habitat guilds. Each wildlife species is positioned in a cell or cells of the matrix by identifying the habitat layer or layers in which it feeds (rows) and those in which it breeds (columns). Guild members share matrix cells (from Short 1989).

species depends for feeding and those on which it depends for breeding (breeding is used here to include other vital, nonfeeding activities such as hiding from predators). Thus a bird species that forages mainly on the ground and nests in tree cavities appears in only one cell (Fig. 2). The biologist thus can develop guilds of species by inspecting

the matrix (or by making a chart as in Table 1) to see which species occur together. Alternatively, guilds can be developed by entering the data into a computer and sorting species into groups by their habitat affiliations.

The major advantage of developing guilds is that it enables the biologist to simplify the analysis. Considering individually the needs of 50 species is clearly intractable, but handling six guilds that contain 50 species may be manageable.

To simplify the analysis further, the biologist should limit the habitat layers in the matrix to those that influence abundances of the species (or guilds) of interest and that are likely to change with development. Inclusion of habitat features that do not influence population levels or will not be affected by development complicates unnecessarily the EIA process and often may lead to misleading impact predictions and pointless mitigation.

Once all the species of interest are positioned in the appropriate matrix cells, displaying the arrangement as a chart or on a computer screen can help the biologist communicate to others the various species-habitat associations. It also helps show the rationale for grouping species into guilds.

The biologist now is ready to begin translating goal statements from population measures into habitat measures. If many species are involved, it is desirable first to restate wildlife goals in terms of guilds rather than individual guild members, if this has not been done already. For example, if species A–F all feed and nest in trees, one may want to shift the focus of the goal from maintaining populations of species A–F to maintaining the "tree-layer" guild. After goal statements are rephrased in terms of guilds, one then can translate them into descriptions of habitat conditions necessary to support the guild. Thus the goal for the tree-layer guild may become to maintain a certain amount and condition of tree-layer habitat.

Table 1. Guild construction based on habitat layers used (x) for feeding and breeding by some species of primary consumers that occur in cottonwood-willow riparian habitats in west-central Arizona (after Short 1983).

Guild	Species	Layers for feeding[a]								Layers for breeding[a]							
		A	B	C	D	E	F	G	H	A	B	C	D	E	F	G	H
1.	American coot				x		x	x	x						x		
2.	Mallard				x		x	x	x				x				
3.	Canada goose				x		x	x	x								
	Gadwall				x		x	x	x								
4.	Longfin dace						x	x	x						x	x	x
	Red shiner						x	x	x						x	x	x
5.	Sonora sucker						x	x	x						x		x
6.	Carp, common						x	x								x	x
	Speckled dace						x	x								x	x
7.	Black bullhead							x									x
	Yellow bullhead							x									x
	Desert sucker							x									x
8.	Roundtail chub						x	x							x	x	
9.	Green sunfish							x							x	x	
10.	Soft-shelled turtle							x						x			
11.	Porcupine	x		x	x	x	x					x	x				
12.	Beaver		x	x	x	x									x	x	

[a]A = tree canopy, B = stem bole, C = midstory, D = terrestrial surface, E = terrestrial subsurface, F = water surface, G = water column, H = bottom water column.

This process of exchanging wildlife currency for habitat currency needs also to consider a special dimension—that of minimum size of habitat blocks. Some guild members may require substantial amounts of contiguous habitat to sustain viable populations. In such instances, translation of wildlife goals into habitat goals must recognize the contiguous area needed. If the intent is to provide habitat for all guild members, the habitat goal must specify a minimum block size that will accommodate the guild member that needs the largest block size. If a special-interest species is the focus of the habitat goal, the goal should state the minimum block size required by that species. Increasing rates of habitat fragmentation in recent years have focused the attention of many impact analysts on the effects of block size (Temple and Wilcox 1986).

The translation requires one final step. The biologist may need to describe, for each habitat layer, special structural features needed by species or guilds that use the layer. Minimum block size is one of these features. There are often other features as well, e.g., minimum snag density, minimum density of hardwood or mast-bearing trees, and maximum water depth. For example, red-cockaded woodpeckers need pine trees with specific characteristics in which to nest, and descriptions of the tree layer that supports them must specify a minimum abundance or density of such trees. Simple word models (Fig. 3), narrative descriptions, and graphic depictions are useful tools here. These descriptions constitute an important standard against which to measure change within layers of habitat. Although one can use maps to depict the distributional status of habitat layers, descriptions often must accompany maps to accurately reflect the quality of each layer.

Impact analysts quantify wildlife goals by using current or desired future conditions as a standard. Measuring or predicting impacts requires estimating deviations from these standards. Thus the biologist usually will want to describe present habitat conditions, future habitat conditions without development, and future conditions with development.

PRESENT HABITAT CONDITIONS

Analysts usually represent present conditions with a combination of maps, tabular data, and word descriptions. Maps showing the distribution of habitat layers can be prepared directly from interpreted aerial photographs (Figs. 4, 5). Coverage by and overlap among layers also can be displayed numerically as area and percentages of total cover (Table 2). Narration or word models (Fig. 3) can describe habitat layers or layer combinations in terms of their block-size distributions and special-feature characteristics.

Biologists will find computerized maps and data sets particularly useful because they expedite the display and analysis of habitat conditions. Adjustment and refinements can be made rapidly by computer, and desired or expected future habitat conditions can be compared with those of the present without extensive manual effort.

FUTURE CONDITIONS WITHOUT DEVELOPMENT

In many ecosystems, representing habitat structure in terms of layers is particularly useful in futures forecasting. This is true because plant succession often involves a pre-

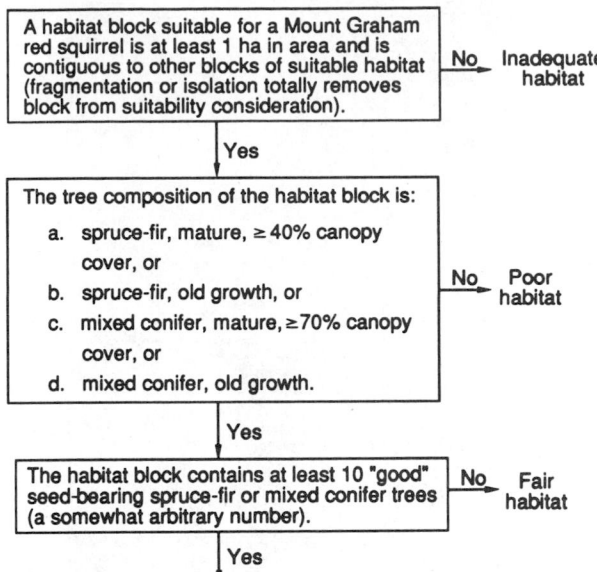

Fig. 3. A word model describing good to excellent habitat condition for a red squirrel (from Short and Williamson 1988).

dictable sequence of change from one life form (i.e., layer type) to another over time. Such natural changes can be superimposed on the changes expected from anthropogenic treatment of the landscape (see below) to project what conditions will prevail at various points in the future.

FUTURE CONDITIONS WITH DEVELOPMENT

To predict the conditions that will prevail after the development project takes place often requires one to consider at least two possible scenarios. One (and often the only one considered) assumes no change other than that caused by the project (Table 3). Another one, generally more realistic, attempts additionally to account for natural changes such as plant succession and for anthropogenic changes not related to the project.

The analysis becomes more complex as one factors in the effects of succession and of other anthropogenic changes. Inclusion of a simple plant-succession model or analysis often is warranted, particularly if major natural changes are likely to occur during the life of the project. Dealing with the effects of other developments is often less manageable, and seldom do analysts attempt detailed analyses of these. We discuss this issue later (see ASSESSING CUMULATIVE IMPACTS [p. 619]).

Overlaying maps of development plans on maps depicting present habitat layer distributions is a first step for projecting future conditions. If different maps are available for each alternative development scenario, relative effects of each option can be visualized.

Many proposed developments have an effective life that can be estimated with reasonable accuracy. The extent and schedule of habitat changes during a project's life vary tremendously with the nature of the project. All habitat layers can disappear entirely with construction of malls or parking lots. Some actions can permanently replace pre-development habitat layers with others; for example, wa-

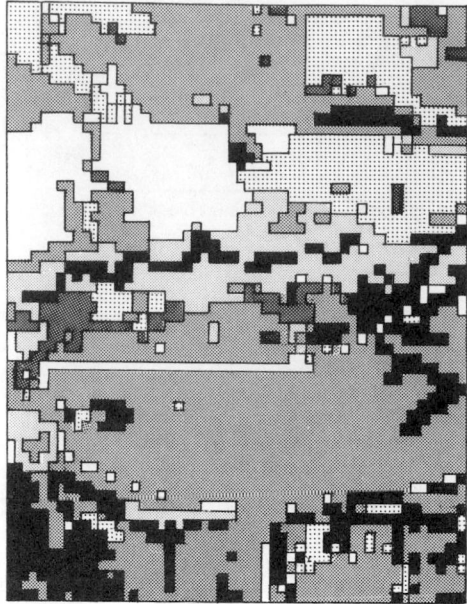

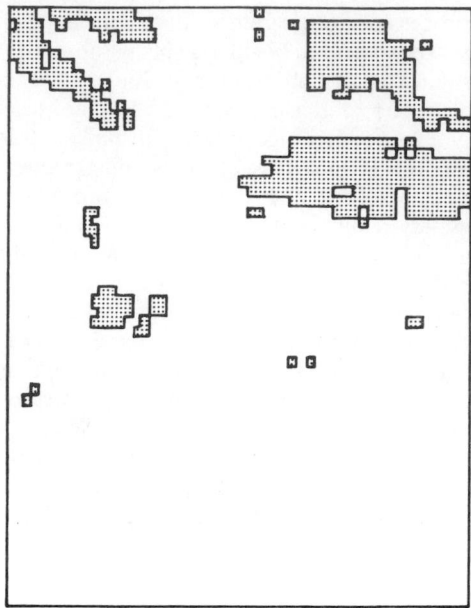

Fig. 5. Distribution of 0.1-ha cells in Fig. 4 that contain three habitat layers: a terrestrial subsurface layer, an understory layer, and a shrub midstory layer (from Short and Williamson 1986).

Fig. 4. Computer-generated map of a portion of the Big Stone National Wildlife Refuge, Minnesota (from Short and Williamson 1986).

ter surface and subsurface layers can replace terrestrial substrate, herb, shrub, and tree layers upstream of a dam. Other actions, such as clear-cutting a forest, temporarily substitute layers associated with early succession for those characteristic of late succession.

The variance between future habitat conditions and those specified as goals can be calculated if analysts can accurately predict the future. But because participants often have difficulty in reaching consensus about the future, they often desire to "game" with various scenarios of change. This is where the value of being able to manipulate by computer the cover-type maps, the development maps, and the data tabulations becomes especially evident.

If analysts project future habitat deficiencies caused by the development, questions arise about potential mitigation. Are the deficiencies tolerable? Should the goal statement be modified as a compromise? Can yet another possible scenario for development be presented to further minimize deficiencies? Is it reasonable or desirable to offset deficiencies by habitat acquisition or management outside the project area?

Planning for Mitigation

How does one go about developing a mitigation plan to offset deficiencies? Ignoring for now the procedural guidelines of some U.S. federal agencies, probably the simplest strategy is to provide habitat layers or layer combinations analogous to those lost to development. Such a mitigation approach is common, partly because it is easily

communicated among participants in the EIA process. It is warranted when predevelopment wildlife conditions are the goal, loss of given structural features is known to cause wildlife loss, and replacement of lost structure is easy.

But this approach is not necessarily the best. First, if provision of habitat layers or other features requires heretofore unaffected (i.e., offsite) habitats to be altered, the mitigation itself can cause deviations from wildlife management goals elsewhere. In such situations, mitigation may "rob Peter to pay Paul." Second, the original distribution and juxtapositioning of habitat layers within the project area may not be the best combination for reaching wildlife goals.

If the original condition is not the best, the manager may be able to make up deficiencies on-site by altering habitats or layers within the project area. Herein lies the challenge for the biological architect of the habitat—to balance habitat needs against habitat possibilities and thereby enhance habitats in situ.

The effectiveness of the mitigation plan in general, and of recommendations for mitigating individual actions, should be monitored as development proceeds. Monitoring is necessary to determine if (1) the recommended mitigation actions were implemented properly, (2) expected results were achieved, and (3) any unexpected consequences to the habitat or to the wildlife community resulted. Formal mitigation evaluation procedures (see U.S. Fish and Wildlife Service 1981, Roelle 1988) can be integrated into the plan. If, over time, mitigation goals clearly are not being achieved as specified, recommendations should be made to modify the mitigation effort or to review and amend the mitigation plan.

Summary

Analysts conducting impact assessments usually employ a team approach that proceeds through three sequen-

Table 2. Numerical representatation of habitat structure from map-based data (after Short and Williamson 1986).

			Habitat layer[a]					
Habitat type	Area (ha)	Area (%)	Water column	Water surface	Terr. surface	Under-story	Mid-story	Over-story
Quarries, roads	37.5	11.7	0	0	0	0	0	0
Lakes, ponds	10.5	3.3	1	1	0	0	0	0
Shallow wetlands w/o emergents	1.1	0.3	0	1	0	0	0	0
Wet meadow	31.4	9.8	0	0	0	1	0	0
Shallow wetlands w/ emergents	32.1	10.0	0	1	0	1	0	0
Grasslands, croplands	113.3	35.3	0	0	1	1	0	0
Lowland shrubs (hydric soils)	7.5	2.3	0	0	0	1	1	0
Upland shrubs	36.1	11.2	0	0	1	1	1	0
Flood-killed trees (hydric soils)	2.1	0.7	0	0	0	1	0	0
Flood-killed trees in water	4.0	1.3	0	1	0	1	0	0
Mature trees (hydric soils), no midstory	3.2	1.0	0	0	0	1	0	1
Mature trees (upland), no midstory	30.7	9.5	0	0	1	1	0	1
Mature trees (upland), with midstory	11.5	3.6	0	0	1	1	1	1
Total %		100.0						
Total ha	321.0		10.5	47.7	191.6	271.9	55.1	45.4

[a] 1 = layer present, 0 = layer absent.

tial steps—scoping, analysis, and synthesis and quantification. Scoping occurs in the initial stages of project planning; the assessment team describes as accurately as they can at this point the spatial and temporal dimensions of the planned action, the wildlife populations potentially affected, and the projected habitat changes that could affect the wildlife populations. Next comes an analysis, based on existing and new data, of the relationships among society's goals for the wildlife populations, the habitat conditions needed for meeting those goals, and the habitat changes likely to be caused by the development. Finally, analysts synthesize relevant information to present a quantitative measure of the projected change in habitats and wildlife. Following the assessment of impacts, biologists recommend mitigative actions to offset expected adverse changes; these recommendations build on the wildlife : habitat relationships that have been defined and quantified during analysis and synthesis.

Measuring habitat structure works well for impact assessment and mitigation planning when numerous species are at issue. The wildlife biologist acts as a kind of habitat architect who measures and displays the status of important structural components of the habitat—e.g., vegetative layers, substrate features, habitat block sizes, special features—and translates potential changes in them to changes in the wildlife community. The biologist's major challenge is to select, from among the variations in habitat condition possible after development, those that most nearly meet society's goals for the wildlife community.

ASSESSING CUMULATIVE IMPACTS

Federal legislative mandates to address ecological cumulative impacts (also called cumulative effects) have existed for more than a decade (U.S. Council on Environmental Quality 1978), but in practice cumulative impacts have received little attention (Muir et al. 1990). Although documentation of research and management dealing specifically with ecological cumulative impacts is not nearly as voluminous as that concerned with traditional EIA, valuable work has occurred that can serve as an introduction to the concepts and issues and as a foundation for the evolution of assessment strategies (Williamson and Hamilton 1989).

Cumulative impacts assessment is the process of situation scoping and problem analysis for the overall impacts of past actions on the ecosystem of concern, not just for a site-specific project development area. Cumulative impacts management planning is the subsequent process of solution synthesis and management direction of the total impacts of present actions and foreseeable future actions on that ecosystem (Williamson 1993). The recommended course of action for dealing with cumulative impacts is to (1) understand and communicate cause-effect relationships, (2) stress measurable action toward progressive goals, (3) use a generation-long, ecosystem-level, problem-solving process, and (4) ratify an interagency collaborative drive toward improving the situation. Natural resource agencies may soon be able to provide ecosystem-level environmental guidance, which should include

Table 3. Numerical representation of habitat structure before and after development. Each habitat type has a unique combination of layers. Development is predicted to remove entirely some habitat types (A, B, D, E, G), to reduce the extent of others (C, F, H), and to create new types (I, J, K) (adapted from Short 1988).

		Habitat layer[a]					
Habitat type	Cover (%)	Tree canopy	Stem bole	Midstory	Terrestrial surface	Terrestrial subsurface	Water surface
Predevelopment structure							
A	5	1	1	0	0	0	1
B	5	1	0	1	0	0	1
C	10	1	1	1	1	1	0
D	20	1	0	0	0	1	0
E	20	0	0	1	1	1	0
F	15	0	0	0	1	1	0
G	5	0	0	1	0	0	1
H	20	0	0	0	0	0	1
Total cover (%)	100	40	15	40	45	65	35
Postdevelopment structure							
I	15	0	1	0	0	0	1
C	5	1	1	1	1	1	0
J	10	1	0	1	1	1	0
H	10	0	0	0	0	0	1
F	10	0	0	0	1	1	0
K	50	0	0	0	0	0	0
Total cover (%)	100	15	20	15	25	25	25

[a]1 = layer present, 0 = layer absent.

promoting positive impacts as well as reducing negative impacts.

The recommended cumulative impacts assessment and management planning process should follow these steps: (1) in the scoping phase, define the ecological situation in specific terms of individual problem statements and select one strategy for each problem; (2) in the problem analysis phase, examine the ecological trends and their causes in detail, using the best available data and analytical tools, and then set several goals; (3) in the synthesis phase, develop and document solutions, estimate changes using mathematical models, and develop an overall plan; and (4) in the direction phase, implement and incrementally improve the management plan and systematically evaluate, improve, and update the problem statements, data, analytical tools, and mathematical models.

At several points during the cumulative impacts assessment process, subjective value judgments must be made with reference to some framework of social values. It has proven essential to deliberate collaboratively on the ramifications of each possible strategy and to gain interagency consensus early in the scoping of the problem. In the United States, strategy selection often is based on the U.S. Fish and Wildlife Service's (1981) five options for mitigation. The option or options selected for mitigation depend on society's "acceptable standards" for ecological resources as follows: (1) if the current ecological condition is below acceptable standards, a restoration strategy is appropriate; (2) if the current condition is about equal to acceptable standards, a strategy of impact avoidance (no net loss of habitat) usually is chosen; and (3) if the current condition is above acceptable standards, a strategy of allowing some decline from current conditions by impact minimization is feasible. Impact minimization is generally the current strategy of the natural resource agencies concerned about cumulative impacts assessment. Simply an agreement on the most desirable strategy for each problem is frequently a major advancement for agencies involved.

The major obstacle in conducting a cumulative impacts assessment is the task of dealing with the total situation in the ecosystem, numerous individual issues, and multiple causes. Nonmathematical models with their wide breadth of consideration are a preferable starting point over more restrictive and less easily explained mathematical models (von Bertalanffy 1968). A cause-effect network analysis produced by a group of technical experts on the local situation can produce a valuable product for better understanding and communicating the ecological situation (Williamson et al. 1987). Historic trend graphs and maps have been valuable for quantifying and displaying the breadth and depth of problems and stimulating a search for specific remedial and research tasks. Landscape structure cartographic modeling has been valuable for quantitatively estimating alternative solutions to problems that involve visible (usually terrestrial) habitat fragmentation and loss (e.g., Gosselink and Lee 1989). Ecosystem function simulation modeling has been used for invisible (usually aquatic) habitat modification and deterioration (e.g., Stone and McHugh 1977). Time series population modeling in combination with cartographic modeling (for terrestrial species) and simulation modeling (for aquatic species) is a promising new capability for modeling interrelations between primary production and vertebrate population declines (e.g., Brinson 1988).

In any event, cumulative impacts assessment is a more complex and less tractable problem than assessing impacts of single projects. This complexity and the rapidly evolving state of the art will discourage most biologists from attempting in-depth analyses of more than one project at a time. Assessing and managing the impacts to wildlife of the total force of human actions, though clearly an urgent priority, probably will be limited to special-interest ecosystems and to experienced teams comprised of biologists, other scientists, and decision makers.

SUMMARY

This chapter reviews the history and current practice of environmental impact assessment as it applies to wildlife, and on this basis recommends an approach for assessing the impacts to wildlife of human-caused habitat change. Its justification lies in the accelerating loss of wildlife habitats worldwide and the consequent need to improve and streamline techniques for assessing losses. It builds on the premise that the principles of ecological impact assessment are the same as those of wildlife management, and that the best assessment approach is to focus on the habitat factors that control the distributions and abundances of wildlife populations. Assessing impacts to single species proceeds directly from identifying population-limiting factors to finding how a given human action will change those factors. Multiple-species assessments require using an integrative method or methods that reflect impacts to many species. Measuring habitat structural features provides a useful basis for predicting impacts to several or many species; it also expedites easy communication during the impact assessment process. We describe a practical approach to impact assessment, explaining what steps to take and how to implement them. Five case histories provide examples. Assessing the cumulative impacts of an array of past actions requires special considerations beyond those used in traditional impact assessment.

LITERATURE CITED

ADAMUS, P. R., E. J. CLAIRAIN, JR., D. R. SMITH, AND R. E. YOUNG. 1987. Wetland evaluation technique (WET). Vol. 2. Operational draft. U.S. Army Corps Eng. Waterways Exp. Stn., Vicksburg, Miss. 200pp.

BEANLANDS, G. E., AND P. N. DUINKER. 1983. An ecological framework for environmental impact assessment in Canada. Inst. Resour. Environ. Stud., Dalhousie Univ., Halifax, N.S., and Fed. Environ. Assessment Rev. Off., Hull, Que. 132pp.

BRINSON, M. M. 1988. Strategies for assessing the cumulative effects of wetland alteration on water quality. Environ. Manage. 12:655–662.

BURDICK, D. M., D. CUSHMAN, R. B. HAMILTON, AND J. G. GOSSELINK. 1989. Faunal changes and bottomland hardwood forest loss in the Tensas Watershed, Louisiana. Conserv. Biol. 3:282–292.

DAMES AND MOORE. 1981. Methodology for the analysis of cumulative impacts of permit activities regulated by the U.S. Army Corps of Engineers. DACW 72-80-C-0012, U.S. Army Corps Eng. Inst. Water Resour., Ft. Belvoir, Va. Var. pagin.

DUINKER, P. N., AND G. L. BASKERVILLE. 1986. A systematic approach to forecasting in environmental impact assessment. J. Environ. Manage. 23:271–290.

GOSSELINK, J. G., AND L. C. LEE. 1989. Cumulative impact assessment in bottomland hardwood forests. Wetlands 9:89–174.

———, ———, AND T. A. MUIR, EDITORS. 1990. Ecological processes and cumulative impacts: illustrated by bottomland hardwood wetland ecosystems. Lewis Publ., Inc., Chelsea, Mich. 708pp.

GOULD, N. E. 1977. Featured species planning for wildlife on southern national forests. Trans. North Am. Wildl. Nat. Resour. Conf. 42:435–437.

GRAUL, W. D., AND G. C. MILLER. 1984. Strengthening ecosystem management approaches. Wildl. Soc. Bull. 12:282–289.

———, J. TORRES, AND R. DENNEY. 1976. A species-ecosystem approach for nongame programs. Wildl. Soc. Bull. 4:79–80.

GREEN, R. H. 1979. Sampling design and statistical methods for environmental biologists. John Wiley & Sons, New York, N.Y. 257pp.

HIRSCH, A. 1980. The baseline study as a tool in environmental impact assessment. Pages 84–93 in Proc. symp. biological evaluation of environmental impacts. U.S. Fish Wildl. Serv. Rep. OBS-80/26.

HOLBROOK, H. L. 1974. A system for wildlife habitat management on southern national forests. Wildl. Soc. Bull. 2:119–123.

HOLLICK, M. 1986. Environmental impact assessment: an international evaluation. Environ. Manage. 10:157–178.

HOLLING, C. S., EDITOR. 1978. Adaptive environmental assessment and management. John Wiley & Sons, New York, N.Y. 377pp.

JARVINEN, O. 1985. Conservation indices in land use planning: dim prospects for a panacea. Ornis Fenn. 62:101–106.

KARR, J. R. 1987. Biological monitoring and environmental assessment: a conceptual framework. Environ. Manage. 11:249–256.

KAUTZ, R. S. 1984. Criteria for evaluating impacts of development on wildlife habitats. Proc. Annu. Conf. Southeast. Assoc. Fish Wildl. Agencies 38:121–136.

LANDRES, P. B. 1983. Use of the guild concept in environmental impact assessment. Environ. Manage. 7:393–398.

LEOPOLD, A. 1933. Game management. Charles Scribner's Sons, New York, N.Y. 481pp.

MACARTHUR, R. H., AND J. W. MACARTHUR. 1961. On bird species diversity. Ecology 42:594–598.

MEALEY, S. P., AND J. R. HORN. 1981. Integrating wildlife habitat objectives into the forest plan. Trans. North Am. Wildl. Nat. Resour. Conf. 46:488–500.

MOEN, A. N. 1973. Wildlife ecology. W. H. Freeman and Co., San Francisco, Calif. 458pp.

MUIR, T. A., C. RHODES, AND J. G. GOSSELINK. 1990. Federal statutes and programs relating to cumulative impacts in wetlands. Pages 223–236 in J. G. Gosselink, L. C. Lee, and T. A. Muir, eds. Ecological processes and cumulative impacts. Lewis Publ., Inc., Chelsea, Mich.

MUNN, R. E., EDITOR. 1979. Environmental impact assessment: principles and procedures. Second ed. John Wiley & Sons, New York, N.Y. 190pp.

NATIONAL RESEARCH COUNCIL OF THE UNITED STATES AND THE ROYAL SOCIETY OF CANADA. 1985. The Great Lakes Water Quality Agreement: an evolving instrument for ecosystem management. Natl. Acad. Press, Washington, D.C. 224pp.

ODUM, E. P. 1971. Fundamentals of ecology. Third ed. W.B. Saunders Co., Philadelphia, Pa. 574pp.

———, J. T. FINN, AND E. H. FRANZ. 1979. Perturbation theory and the subsidy-stress gradient. BioScience 29:349–352.

ROELLE, J.E. 1988. Guidance on formulating and evaluating mitigation recommendations. NERC-88/28, U.S. Fish Wildl. Serv., Natl. Ecol. Res. Cent., Ft. Collins, Colo. Var. pagin.

ROOT, R. B. 1967. The niche exploitation pattern of the blue-gray gnatcatcher. Ecol. Monogr. 37:317–350.

ROSEN, S. J. 1976. Manual for environmental impact evaluation. Prentice-Hall, Inc., Englewood Cliffs, N.J. 232pp.

SALWASSER, H., AND J. C. TAPPEINER, III. 1981. An ecosystem approach to integrated timber and wildlife habitat management. Trans. North Am. Wildl. Nat. Resour. Conf. 46:473–487.

SCHROEDER, R. L. 1987. Community models for wildlife impact assessment: a review of concepts and approaches. U.S. Fish Wildl. Serv. Biol. Rep. 87(2). 41pp.

SEVERINGHAUS, W. D. 1981. Guild theory development as a mechanism for assessing environmental impact. Environ. Manage. 5:187–190.

SHOPLEY, J. B., AND R. F. FUGGLE. 1984. A comprehensive review of current environmental impact assessment methods and techniques. J. Environ. Manage. 18:25–47.

SHORT, H. L. 1983. Wildlife guilds in Arizona desert habitats. U.S. Bur. Land Manage. Tech. Note 352. 258pp.

———. 1988. A habitat structure model for natural resource management. J. Environ. Manage. 27:289–305.

———. 1989. A wildlife habitat model for predicting effects of human activities on nesting birds. Pages 957–973 in R. R. Sharitz and J. W. Gibbons, eds. Freshwater wetlands and wildlife. U.S. Dep. Energy Symp. Ser. 61.

———, AND K. P. BURNHAM. 1982. Techniques for structuring wildlife guilds to evaluate impacts on wildlife communities. U.S. Fish Wildl. Serv. Spec. Sci. Rep. Wildl. 244. 34pp.

———, AND S. C. WILLIAMSON. 1986. Evaluating the structure of habitat for wildlife. Pages 97–104 in J. Verner, M. L. Morrison, and C. J. Ralph, eds. Wildlife 2000: modeling habitat relationships of terrestrial vertebrates. Univ. Wisconsin Press, Madison.

———, AND ———. 1988. An ecological problem-solving process for managing special-interest species. Pages 276–281 in R. C. Szaro, K. E. Severson, and D. R. Patton, tech. coords. Management of amphibians, reptiles, and small mammals in North America. U.S. For. Serv. Gen. Tech. Rep. RM-166.

SIDERITS, K., AND R. E. RADTKE. 1977. Enhancing forest wildlife habitat through diversity. Trans. North Am. Wildl. Nat. Resour. Conf. 42:425–434.

SMITH, R. L. 1974. Ecology and field biology. Second ed. Harper & Row, New York, N.Y. 850pp.

STAKHIV, E. Z. 1988. An evaluation paradigm for cumulative impact analysis. IWR Policy Stud. 88-PS-3. U.S. Army Corps Eng. Inst. Water Resour., Ft. Belvoir, Va. 69pp.

STONE, J. H., AND G. F. McHUGH. 1977. Simulated hydrologic effects of canals in Barataria Basin: a preliminary study of cumulative impacts. Final Rep. for La. State Planning Off., Cent. Wetland Resour., Louisiana State Univ., Baton Rouge. 40pp.

TEMPLE, S. A., AND B. A. WILCOX. 1986. Introduction: predicting effects of habitat patchiness and fragmentation. Pages 261–262 in J. Verner, M. L. Morrison, and C. J. Ralph, eds. Wildlife 2000: modeling habitat relationships of terrestrial vertebrates. Univ. Wisconsin Press, Madison.

TRUETT, J. C. 1979. Pre-impact process analysis: design for mitigation. Pages 355–360 *in* G. A. Swanson, tech. coord. The mitigation symposium: a national workshop on mitigating losses of fish and wildlife habitats. U.S. For. Serv. Gen. Tech. Rep. RM-65.

U.S. ARMY CORPS OF ENGINEERS. 1980. A habitat evaluation system for water resource planning. U.S. Army Corps Eng., Vicksburg, Miss. 88pp.

———. 1989. Procedures for implementing NEPA. 33 Code Fed. Regul. 11, Part 230:329–341.

U.S. BUREAU OF LAND MANAGEMENT. 1984. Draft Piceance Basin resource management plan and environmental impact statement. Vol. 1. U.S. Bur. Land Manage., Colo. State Office, Denver. 243pp.

U.S. COUNCIL ON ENVIRONMENTAL QUALITY. 1973. Guildelines for preparation of environmental impact statements: rules and regulations. Fed. Register 38:20550–20562.

———. 1978. Regulations for implementing the procedural provisions of the National Environmental Policy Act. Fed. Register 43:55978–56007.

U.S. FEDERAL AVIATION ADMINISTRATION. 1985. Airport environmental handbook. U.S. Dep. Transp. Fed. Aviation Adm. Order 5050.4A. 108pp.

U.S. FISH AND WILDLIFE SERVICE. 1981. U.S. Fish and Wildlife Service mitigation policy. Fed. Register 46:7644–7663.

———. 1987. Arctic National Wildlife Refuge, Alaska, coastal plain resource assessment. Report and recommendation to the Congress of the United States and final legislative environmental impact statement. U.S. Fish Wildl. Serv., U.S. Geol. Surv., and U.S. Bur. Land Manage., Washington, D.C. 208pp.

VERNER, J. 1984. The guild concept applied to management of bird populations. Environ. Manage. 8:1–14.

———. 1986. Summary: predicting effects of habitat patchiness and fragmentation—the researcher's viewpoint. Pages 327–329 *in* J. Verner, M. L. Morrison, and C. J. Ralph, eds. Wildlife 2000: modeling habitat relationships of terrestrial vertebrates. Univ. Wisconsin Press, Madison.

VON BERTALANFFY, L. 1968. General system theory: foundations, development, applications. George Brazilier, New York, N.Y. 289pp.

WAGNER, F. H. 1977. Species vs. ecosystem management: concepts and practices. Trans. North Am. Wildl. Nat. Resour. Conf. 42:14–24.

WESTMAN, W. E. 1985. Ecology, impact assessment, and environmental planning. John Wiley & Sons, New York, N.Y. 532pp.

———. 1990. Managing for biodiversity. BioScience 40:26–33.

WILLIAMSON, S. C. 1993. Cumulative impacts assessment and management planning: lessons learned to date. Pages 391–407 *in* S. G. Hildebrand and J. B. Cannon, eds. Environmental analysis: the NEPA experience. Lewis Publ., Boca Raton, Fla..

———, C. L. ARMOUR, G. W. KINSER, S. L. FUNDERBURK, AND T. N. HALL. 1987. Cumulative impacts asessment: an application to Chesapeake Bay. Trans. North Am. Wildl. Nat. Resour. Conf. 52: 377–388.

———, AND K. HAMILTON. 1989. Annotated bibliography of ecological cumulative impacts assessment. U.S. Fish Wildl. Serv. Biol. Rep. 89(11). 80pp.

WILLSON, M. F. 1974. Avian community organization and habitat structure. Ecology 55:1017–1029.

25

MANAGING WETLANDS FOR WILDLIFE

Leigh H. Fredrickson and Murray K. Laubhan

INTRODUCTION

Wetlands were abundant when Europeans first arrived in North America and remained so into the mid-20th century. Wetlands originally composed about 87 million ha of the lower 48 United States (Barton 1986). Although a comprehensive inventory never was conducted, it is likely that wetlands originally encompassed >300 million ha in North America. Unfortunately, nearly all wetlands in Mexico, the 48 conterminous states, and southern Canada have been severely impacted by human activities. Many regional and local wetland areas have been disrupted and now are restricted largely to relatively isolated tracts, except in Alaska and northern Canada. Related impacts on wetland wildlife have not been fully documented, but some studies suggest contiguous habitat parcels ≥4,000 ha in size are required to provide necessary components for greater nest success in prairie wetland habitats (Higgins et al. 1992).

Because of tremendous wetland loss and degradation, resource managers have responsibilities to protect wetlands that retain natural hydroperiods (i.e., seasonal water level pattern) and ecological functions and to manage wetlands disrupted by human activities (Weller 1988). Wetland management should occur at three levels: landscape, regional, and local. Management on a landscape scale is important and essential but is difficult to implement because of environmental differences, intervention of changing political policies in North America, economics, and various constraints of state, provincial, and federal agencies charged with wetland conservation. Thus, we will focus on regional and local scales of wetland management, because these have been studied more extensively and can be influenced to a greater extent because of technological developments, biological/ecological understanding, and administrative opportunities.

Effective wetland management is complex and must simulate short- and long-term natural regimes. Further, seasonal physiological needs of wetland wildlife should be identified to provide required food and habitat in a timely and cost-effective manner. Literature regarding wetland management is increasing rapidly in extent and diversity. Here we identify some of the more comprehensive and important publications, but our literature synthesis is not exhaustive. Our goal is to provide a conceptual overview for effective wetland management. A brief synopsis of wetland types, values, and functions is followed by a general approach to management of selected wetland systems at regional and local scales.

WETLAND CHARACTERIZATION
Definition

Wetlands are transitional areas characteristic of aquatic and terrestrial ecosystems (Mitsch and Gosselink 1986). Therefore, a single definition that accurately describes all wetlands is difficult to formulate because delineating criteria often become arbitrary. For example, hydrology is one criterion commonly used to determine if an area is a wetland (Cowardin et al. 1979). Although frequency of flooding may appear to provide a valid and straightforward approach to wetland determination, this guideline alone can complicate delineation of wetlands. For example, ephemeral pools (i.e., temporarily or intermittently flooded) in the prairies or seasonally flooded oak flats in the South provide important habitats for wetland wildlife, yet they may be dry 11 months of the year and hence not be classified as wetlands by some procedures or experts.

In contrast, a wetland that dries only once in 25 years is readily recognized as a wetland. Such wide variation in flooding has contributed to confusion and inconsistencies in wetland delineation and management.

Wetlands are lands transitional between terrestrial and aquatic systems where the water table is usually at or near the surface or the land is covered by shallow water (Cowardin et al. 1979). "Wetlands must have one or more of the following three attributes: (1) at least periodically, the land supports predominantly hydrophytes; (2) the substrate is predominantly undrained hydric soil; and (3) the substrate is nonsoil and is saturated with water or covered by shallow water at some time during the growing season of each year" (Cowardin et al. 1979:3). This comprehensive definition incorporates plants, soils, and hydrology as important factors in determining classification of a site as a wetland. The definition also is applicable across North America and has a reasonable degree of acceptance by researchers, managers, and regulatory personnel. The U.S. Army Corps of Engineers and the Environmental Protection Agency, the organizations charged with federal regulation of wetlands, also incorporate these attributes into their definition of wetlands. They define wetlands as areas inundated or saturated by surface water or groundwater at a frequency and duration sufficient to support, and that under normal circumstances do support, a prevalence of vegetation typically adapted for life in saturated soil conditions. However, their definition places greater emphasis on plant communities in delineating wetlands. It is imperative that managers consider these factors to ensure effective management of wetlands for wildlife.

Wetland Types

Classification of wetlands into types must incorporate ecological characteristics and functions of wetlands. Because numerous abiotic and biotic elements influence wetland characteristics, a hierarchical approach to separating wetlands is useful. Among the most important elements are hydrological characteristics, water chemistry, substrate type, and vegetation. Knowledge of these factors enables differentiation of wetland types during inventory and evaluation, and also facilitates development of management strategies that maintain wetland integrity and provide resources required by wetland wildlife.

Cowardin et al. (1979) recognized five major wetland types (i.e., marine, estuarine, riverine, lacustrine, and palustrine) that differ with respect to hydrologic, geomorphologic, and chemical factors. Marine wetlands are characterized as occurring in open ocean overlying the continental shelf. Water regimes are determined primarily by oceanic tides, and salinity of water exceeds 30 parts per thousand (ppt). Estuarine wetlands are semienclosed by land but are influenced by open ocean. Salinity from ocean-derived salts must be >0.5 ppt, but an upper salinity limit is not established. Riverine wetlands occur within a channel except in two situations: areas dominated by trees, shrubs, persistent emergents, emergent mosses, or lichens; and habitats with ocean-derived salinities >5 ppt. Lacustrine wetlands exhibit all the following characteristics: (1) sited in a topographic depression or dammed river channel; (2) lack trees, shrubs, persistent emergents, emergent mosses, or lichens with >30% aerial coverage; (3) total area >8 ha; and (4) salinity caused by ocean-derived

salts is <5 ppt. Palustrine wetlands include nontidal areas containing trees, shrubs, persistent emergents, and emergent mosses or lichens, and tidal areas with similar vegetation composition but salinity caused by ocean-derived salts is <5 ppt. The palustrine category also includes wetlands lacking such vegetation, but with all the following characteristics: (1) area <8 ha, (2) wave-formed or bedrock shoreline absent, (3) water depth in the deepest part of the basin is <2 m at low water; and (4) salinities due to ocean-derived salts are <5 ppt.

Within each of these five major types, wetlands can be classified further according to hydrologic regime (e.g., subtidal, intertidal, seasonally flooded), substrate type (e.g., rock, unconsolidated), water chemistry, and vegetation (e.g., aquatic, forested, emergent vegetation). For example, palustrine wetlands can be subdivided into eight types: rock bottom, unconsolidated bottom, aquatic bed, unconsolidated shore, moss/lichen, emergent, scrub/shrub, and forested. Based on this classification scheme, differentiation of wetlands into 55 ecologically distinct types is possible.

DISTRIBUTION AND STATUS

The most recent wetland trend data document that all wetland habitats in the lower 48 states and Hawaii have been greatly reduced or severely degraded. At the time of colonial America, an estimated 89.5 million ha of wetlands existed in the conterminous United States (Dahl 1990). During the past 200 years, wetland loss has been excessive, and <40.1 million ha of wetlands remained by the mid-1970s. However, recent information suggests that the average annual rate of wetland loss has decreased from 185,350 to 117,360 ha (Dahl and Johnson 1991).

Causes of wetland loss are varied; nevertheless, conversion to agricultural land uses has had the greatest impact in the United States and Canada. The reasons for wetland losses in Mexico have been poorly documented (Baldasarre et al. 1989, Kramer and Migoya 1989), but sizeable losses of wetlands in the Colorado River drainage of Mexico are apparent (M. E. Heitmeyer, pers. commun.). From the mid-1950s to the mid-1970s, conversion of wetlands for agricultural purposes was responsible for 87% of wetland losses in the United States (Frayer et al. 1983). Although such conversion has declined subsequently, agriculture remained the primary cause of wetland loss (54%) from the mid-1970s to the mid-1980s. Conversely, wetland destruction caused by "other" land use practices has increased from 8% (mid-1950s–mid-1970s) to 41% (mid-1970s–mid-1980s). A large component of this increase is attributable to wetlands that have been cleared and drained, but a specific use has not been identified (Dahl and Johnson 1991). Urban land uses have accounted for about 5% of wetland losses during the past 30 years (Tiner 1984).

The most recent estimates reveal that freshwater and estuarine wetlands compose 95% (39.6 million ha) and 5% (2.2 million ha), respectively, of the remaining wetland area in the conterminous United States. Of palustrine habitats, 53% are forested, 25% contain emergent vegetation, 16% are scrub/shrub, and 6% are classed as ponds or other wetland types. Estuarine intertidal habitats are composed of emergent (73%), forested/shrub (13%), rock or sandy shoreline (10%), and aquatic bed (4%). Deep-

PHYSICAL FACTORS

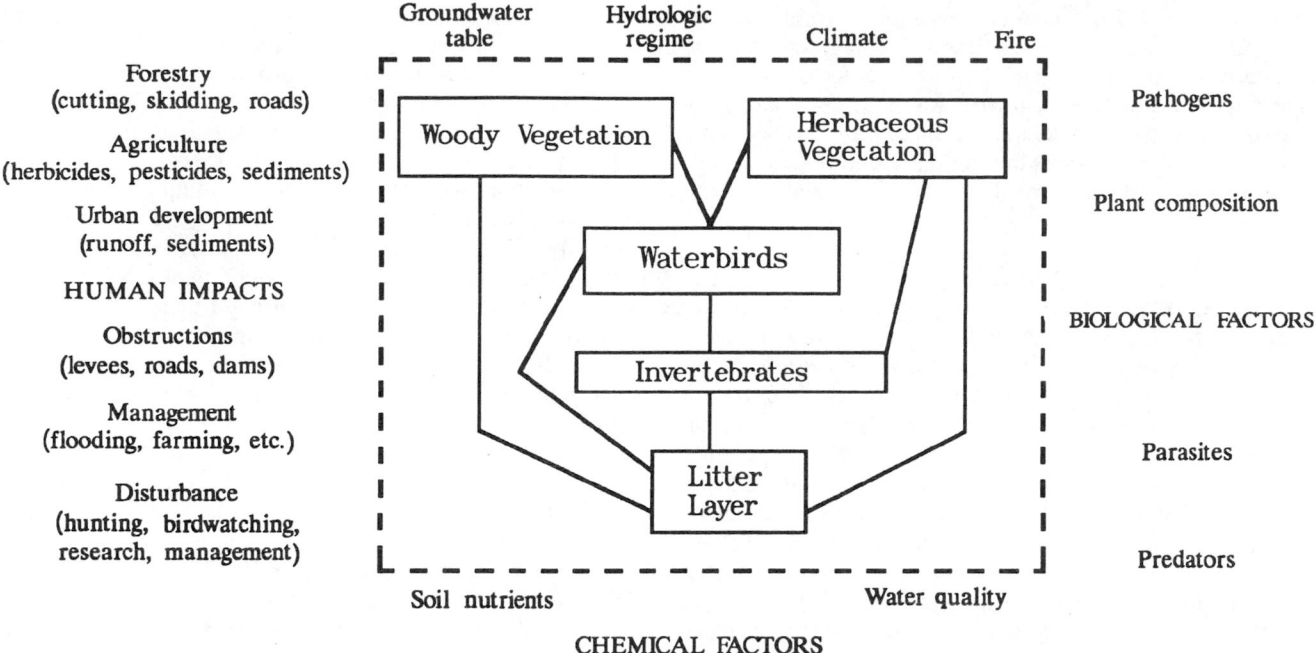

Fig. 1. Chemical, biological, and environmental factors influencing wetland characteristics, functions, and values.

water habitats, composed of lacustrine (71%), riverine (6%), and estuarine subtidal (23%) types, constitute an additional 33.2 million ha of flooded areas (Dahl and Johnson 1991).

Although overall wetland loss has decreased, the destruction of some wetland types continues at an alarming rate. Further, wetland loss in certain regions of the United States and Canada has not been curtailed. For example, forested palustrine wetlands (e.g., swamps, riparian corridors) decreased 6.2% from 22.3 million ha in the mid-1970s to 20.9 million ha in the mid-1980s. Most of this loss has occurred in the southeastern United States (Dahl and Johnson 1991). Although reports documenting the decrease in prairie breeding habitats have been well publicized (Tiner 1984, Pederson et al. 1989), the demise of migration and winter habitats has been equally severe (Korte and Fredrickson 1977, Frayer et al. 1989, Dahl and Johnson 1991). As a result, these figures fail to assess the impact of wetland loss on wildlife populations because we lack information on how major disruptions to entire landscapes influence the ability of individual species to meet life-history requirements.

WETLAND FUNCTIONS

The productivity and characteristics of wetlands constantly are in a state of flux because of dynamic biotic and abiotic factors. As a result, wetlands often are depicted as sieves (van der Valk 1981) (Fig. 1). For example, the type and quantity of nutrients that enter and exit wetlands from adjacent areas vary among locations, seasons, and years (Hammer and Bastian 1989). Such differences, modified and influenced by a myriad of other factors, alter the productivity of the wetland as well as the quality of water that exits the wetland. Other representative abiotic factors influencing wetland characteristics and

productivity include fire, climate, hydroperiod, soil, groundwater, and water quality. Biotic factors can be separated into those that are stationary within a wetland basin (e.g., plants) and those that move among wetlands or use wetlands on a temporary basis (e.g., waterbirds, avian predators, pathogens). A basic understanding of these individual factors and their interrelationships with wetland structure and function is the cornerstone of effective management. Structural components of wetlands include the extent and distribution (horizontal and vertical) of vegetation, which determine the quantity and quality of cover (e.g., nesting, protection) for wildlife. Functional components are less obvious but have important implications for wildlife and human populations and include nutrient cycling, atmospheric stability, and mediation of peak runoff. Moreover, knowledge of these complex processes provides insights to keep natural, human-made, and human-modified wetlands productive. Other factors influencing wetland dynamics related to human activities include: (1) agricultural practices that cause sedimentation, soil subsidence, and herbicide and pesticide accumulations; (2) irrigation and urban water developments; (3) construction of roads, levees, and canals; (4) wetland and wildlife management practices such as flooding, drawdowns, and farming; and (5) urban, industrial, and military developments (Tiner 1984, Grue et al. 1986, Heitmeyer et al. 1989). These factors alter the natural structure and function of wetlands and therefore have important implications for wetland management.

Wildlife use of wetlands largely is determined by the type, quality, and distribution of foods and cover (Weller and Spatcher 1965, Weller and Fredrickson 1973, Kaminski and Prince 1981, Ball and Nudds 1989). Therefore, the vegetation within wetlands is extremely important because plant composition determines (1) type, quantity, and nutritive quality of plant foods available, including seeds, tubers, and browse; (2) distribution, density, and structure of cover; and (3) quantity and type of substrate for invertebrates. In turn, establishment of the plant community is controlled by factors such as seed-bank composition, hydroperiod, soil type, climate, water quality, and past management activities (Simpson et al. 1989). Therefore, knowledge of interactions among these components is essential to understanding wildlife use of wetlands and for enhancing management opportunities.

Vegetation

Although they have been poorly studied, evidence indicates algae likely are an important component of wetland systems because of algal influence on nutrient dynamics (e.g., Murkin 1989, Stevens et al. 1989). For example, algae readily respond to available nutrients in the water column and may form massive blooms, particularly when water temperatures are warm. During slow drainage of a wetland, algae are retained within a basin, thus preventing the export of important nutrients. In addition, epiphytic algae are an important food for many macroinvertebrate grazers (Allanson 1973).

Estuarine intertidal emergent wetlands, palustrine emergent wetlands, and parts of lacustrine and riverine wetlands typically support annual herbaceous vegetation. Annual vegetation communities normally produce abundant seeds that contain important sources of carbohydrates, vitamins, minerals, and essential amino acids (Table 1). After inundation, decomposing vegetation provides substrate for invertebrates, which are important sources of protein for waterbirds. Nutritional components provided by plants and invertebrates are needed by waterbirds to complete annual-cycle events such as molt and egg production (Drobney 1980, Drobney and Fredrickson 1985, Heitmeyer 1988) and are important in the diets of mammals, herptiles, and fish as well. Annual vegetation also provides seasonal cover used by upland as well as wetland wildlife.

Palustrine emergent wetlands and estuarine intertidal emergent wetlands also may contain cattail and bulrush, which are typical examples of perennial vegetation that is ubiquitous in temperate wetlands. These plants are capable of tolerating deep water and propagate by rhizomes or seeds. Although bulrush seeds are consumed by waterbirds, they are of lesser importance than the seeds of many other plants, especially annuals. However, succulent new growth of stems or rhizomes of many perennial plants also serves as browse, and the robust structure also provides vertical and horizontal cover for seclusion of waterfowl pairs, sites for nest attachment, nesting cover for breeding waterbirds, cover from predators, cover for broods, protection from inclement weather, and food for herbivorous mammals (e.g., muskrats). In addition, stems and leaves provide materials for the detrital-based food web that eventually nourishes vertebrates. However, when dense, monotypic stands of robust vegetation develop throughout a basin, plants become a management problem, and waterbird use of the basin may decline (Weller and Spatcher 1965, Weller and Fredrickson 1973). Herbivores such as muskrats and nutria are extremely important in coastal and north temperate wetlands, because their feeding and house-building activities change the structure of robust emergent wetlands, which either improves or degrades the

wetland basin for waterbird use (Weller and Spatcher 1965, Weller and Fredrickson 1973, Chabreck 1988).

Palustrine forested and scrub/shrub wetlands are representative of wetland types containing perennial woody vegetation. Trees and shrubs provide thermal cover important during periods of inclement weather and seclusion for pairs. The mast (e.g., acorns, samaras) produced by woody species also are a good source of carbohydrates, and a distinct macroinvertebrate community often develops after leaf-fall. Vertebrates requiring tree cavities rely heavily on this habitat. Cavities in large trees are particularly important as nest sites for wood ducks, hooded mergansers, and barred owls and provide dens for raccoons, squirrels, and other mammals (Soulliere 1990). Cavities in smaller trees are important for a host of smaller birds and mammals such as prothonotary warblers and golden mice.

Macroinvertebrates

Wetlands provide many habitat niches for invertebrates, which are important foods for waterbirds and fish (Mott et al. 1972, Swanson and Meyer 1973, Weller 1988, Eldridge 1990). Many invertebrates are extremely small (<1 mm), and their use as food is restricted to a few vertebrates with specialized foraging mechanisms or to larger invertebrates. Thus, our focus will be on the macroinvertebrate communities typically associated with different wetland types. Differences in invertebrate composition and distribution among wetland types are driven by hydrologic regimes and vegetation structure (Murkin et al. 1992). The life-history strategies of wetland macroinvertebrates have been shaped by long-term hydrologic cycles, particularly the type of flooding and dynamic water fluctuations. Among the most important are morphological or behavioral adaptations to tolerate or avoid drought. Adaptations that have evolved as a result of long-term hydrologic cycles include at least one of the following: (1) ability to withstand drought in the egg, pupal, or larval state; (2) rapid growth; (3) ability to produce numerous offspring; (4) ability to complete the life cycle within 1 year; and (5) high mobility (Wiggins et al. 1982). Several invertebrate groups, including flatworms; fairy, clam, and seed shrimp; water fleas; mayflies; mosquitoes; and phantom midges, have resistant egg stages that help prevent drought-induced mortality. Aquatic earthworms may use mucosal secretions to survive drought, whereas bloodworm larvae estivate in cocoons and fingernail clams burrow into the wet litter layer and rely on their shell to avoid desiccation. In contrast, aquatic sowbugs and sideswimmers have no morphological adaptations to resist drought, but adults estivate and survive dry seasons by locating suitable conditions in the deeper litter layers.

Because of the dynamic nature of flooding regimes, macroinvertebrates that exhibit rapid growth during periods of adequate water and nutrient availability have an advantage. Furthermore, producing large numbers of offspring and completing the life cycle in 1 year allow for greater success. When water levels decline, species that cannot tolerate drought must avoid dry conditions. The species most successful often are highly mobile and capable of emigrating to suitable sites. Beetles and water boatmen, in particular, respond to drawdowns by aerial dispersal to available wetlands.

Table 1. Chemical composition of selected annual seeds, row crops, and invertebrates.

Food group	Common name	Gross energy (kcal/g)	Fat (%)	Protein (%)	NFE[a] (%)	Ash[b] (%)
Annual seeds	Barnyard grass	3.9	2.4	8.3	40.5	18.0
	Beggarticks	5.2	15.0	25.0	27.5	7.2
	Chufa flatsedge	4.3	6.9	6.7	55.4	2.5
	Fall panicum	4.0	3.1	12.3	50.1	16.1
	Rice cutgrass	4.0	2.0	12.0	57.8	9.5
Row crops	Common sorghum	4.2	3.1	10.2	72.2	3.5
	Corn	4.4	3.8	10.8	79.8	1.5
Invertebrates	Aquatic sowbug	4.0	3.2	42.6	14.9	30.2
	Bloodworm	5.4	17.8	60.1	14.9	7.0
	Pond snail	2.2	1.0	6.7	5.7	57.6
	Sideswimmer (Gammaridae)	3.8	7.6	45.6	14.3	24.1
	Water boatmen	5.2	7.1	71.3	0.8	5.9

[a]Nitrogen-free extract: measure of highly digestible carbohydrates.
[b]Measure of mineral content.

Many macroinvertebrates that exploit wetlands also have adaptations that enable them to thrive in specific habitats or vegetation types. For example, a wetland invertebrate fauna often exists in the dry substrates of an ephemeral or intermittently flooded wetland. When such sites are flooded, invertebrates initiate life-cycle responses that may result in large numbers and biomass of invertebrates in the flooded basin (Batema 1987, Severson 1987, Fredrickson and Reid 1988*d*). Further, the habitat requirements of some invertebrates change, depending on life-cycle stage (Pennak 1978). For example, one habitat may be important for egg laying, whereas a different habitat may be required for feeding. Macroinvertebrate communities can be grouped on the basis of habitat association into those that occur primarily in (1) benthic substrates, (2) submergent vegetation, (3) perennial herbaceous vegetation, (4) annual herbaceous vegetation, and (5) leaf litter. Species that compose each group have life-history strategies that allow them to exploit a particular hydrologic regime (Table 2).

Although long-term hydrologic cycles and habitat type influence adaptive strategies of macroinvertebrates, other factors are involved in determining the occurrence, abundance, growth rate, and reproduction of individual species at any given time. Among the most important are the short-term water regime and physical, chemical, and biological factors (Pennak 1978, Wiggins et al. 1982, Pinder 1986). After senescence of vegetation, plant materials form litter. For herbaceous vegetation, litter includes stems, leaves, and flower structures, whereas leaves are the primary litter input from woody vegetation. Nutrients and organic matter rapidly leach from litter upon initial contact with flood waters, and increases in nutrient concentrations occur in the water column (Peterson and Cummins 1974, Yates and Day 1983, Wylie 1985). The fungi, bacteria, and microinvertebrates associated with litter accelerate the decomposition and release additional energy and nutrients (Fig. 2). Macroinvertebrates also feed on the litter conditioned by the bacteria and fungi, further assisting in litter decomposition and nutrient cycling. Because they are readily consumed as food, macroinvertebrates serve as an important functional link in the transfer of nutrients from detritus to waterbirds, herptiles, and fish (Batema et al. 1985).

The rate of decomposition is governed by several factors including litter type, hydrologic condition, and temperature (Webster and Benfield 1986, Middleton et al. 1992). Under a given set of environmental conditions, herbaceous litter generally decomposes faster than woody vegetation (Fig. 3). Regardless of litter type, decomposition rates generally increase when litter is shallowly flooded and temperatures increase. Deep flooding may result in anaerobic conditions that restrict the faunal community actively associated with the decomposition process (Suthers and Gee 1986). Further, the decomposition process may cause anaerobic conditions, resulting in subsequent elimination of invertebrate communities. Because factors controlling decomposition are changing constantly, peaks in macroinvertebrate abundance often are dramatic and

Table 2. Macroinvertebrates associated with different hydrologic regimes in seasonally flooded impoundments.

Flooding regime	Invertebrates
Winter flooding–late drawdown	Beetle
	Bloodworm
	Mosquito
	Pond snail
	Water boatmen
Winter flooding–early drawdown	Mosquito
	Pond snail
	Water boatmen
Autumn–winter flooding	Bloodworm
	Mosquito
	Water boatmen
Long vernal flooding	Fairy shrimp
	Mosquito
	Water boatmen
Short vernal flooding	Fairy shrimp
	Mosquito

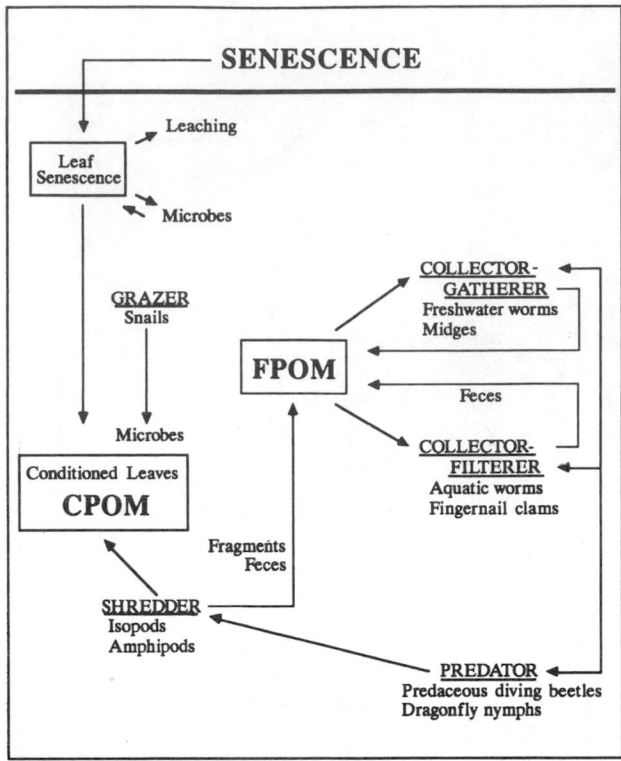

Fig. 2. Litter decomposition pathway (CPOM = coarse particulate organic matter; FPOM = fine particulate organic matter).

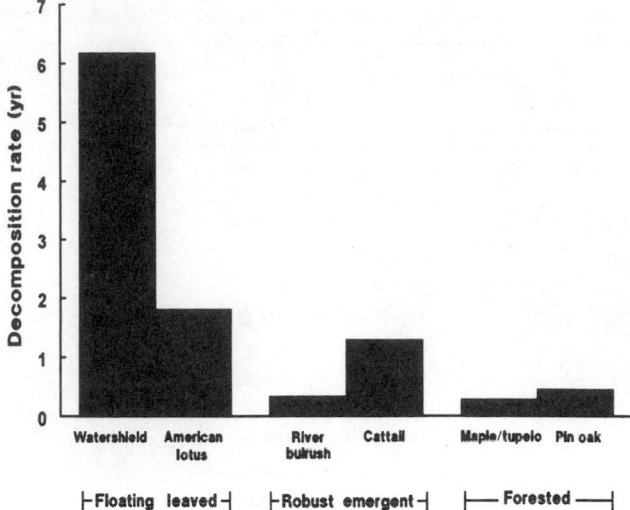

Fig. 3. Decomposition rates of various litter types (data from Boyd 1970, Yates et al. 1983, Wylie 1985).

short-lived. This cycle or "pulsing" of macroinvertebrate populations, although variable among years and habitat types, is typical of invertebrates that exploit nutrient-rich, detrital-based systems that are influenced by constantly fluctuating water levels.

Hydrology

Hydrology probably is the single, most important factor controlling the establishment and maintenance of specific wetlands and wetland processes (Mitsch and Gosselink 1986). In fact, hydrological regime separates wetland systems from true terrestrial and aquatic systems. Hydrology influences chemical and physical properties of wetlands and ultimately determines the biota.

The hydroperiod is unique among different wetlands. Components influencing hydroperiod include those that determine the amount of water entering and leaving a wetland (Marble 1992). Precipitation, groundwater, surface water, and tidal fluctuations represent important inflow components. Possible outflow components include evapotranspiration, groundwater discharge, surface water discharge, and tidal fluctuations. The dominant components change among physiographic regions. For example, tidal fluctuations contribute largely to the hydroperiod of coastal wetlands, whereas changes in surface water are a controlling influence in riparian wetlands. Further, the relative contribution of these components, regardless of locale, is dynamic, changing seasonally and sometimes daily (e.g., storm events, tides). Finally, the balance between inflow and outflow determines the depth, duration, and frequency of flooding of wetlands, and also the amount of water flow through a basin.

Short- and long-term hydrologic conditions affect many abiotic factors. Water fluxes affect nutrient cycling in wetland basins by determining the type and quantity of nutrients that enter and exit wetlands and by influencing decomposition rates (Livingston and Loucks 1979). Hydroperiod also affects water quality (Wharton et al. 1982) and salinity and can alter soil conditions (e.g., oxygen content) that impact chemical processes and determine the availability of nutrients for vegetation growth.

All these abiotic factors in turn influence the distribution, composition, and productivity of vegetation that becomes established (Bedinger 1979). Plant seeds have certain requirements for germination, including appropriate light, pH, redox potential, soil oxygen, moisture, and temperature (Simpson et al. 1989). Hydroperiod alters soil conditions, thereby controlling the plant species capable of growth (Leck 1989). For example, herbaceous annuals such as common barnyard grass require exposed mudflats for germination and growth. In contrast, submerged aquatic species such as American wildcelery require permanent water for growth. Therefore, common barnyard grass is most prevalent in seasonally flooded habitats, whereas American wildcelery is most common in permanently flooded habitats.

Wildlife use of wetlands also is affected by hydroperiod, especially the depth and timing of flooding. Many waterbirds, including dabbling ducks and shorebirds, require shallow water depths to forage efficiently (White and James 1978, Reid et al. 1989) (Fig. 4). Foods and cover in wetlands that are not flooded, or flooded too deeply, largely are unavailable to these species.

WETLAND VALUES

Wetlands have many functions that benefit society and relate to wildlife management. Fur-bearing mammals (e.g., mink, beavers, muskrats) and alligators are commercially harvested from wetlands. Also, about 66% of fish and shellfish species that are harvested commercially are associated with wetlands (Mitsch and Gosselink 1986: 396). In addition, millions of individuals participate in recreational sports (e.g., waterfowl hunting, fishing, bird-watching) that are at least partially associated with wet-lands. These commercial and recreational activities pro-vide an economic foundation for companies that manufacture and sell outdoor equipment and often aug-ment local economies by creating a need for service-ori-ented industries (e.g., motels, restaurants) and jobs.

Wetlands also serve several important environmental functions (Odum 1979, Goldstein 1988, Hammer and Bas-tian 1989). Many function as natural reservoirs that reduce runoff and release water gradually through time. These wetlands can be valuable in reducing flood damage by lessening peak floodwater flow and slowing water dis-charge rates following severe storm events. In addition, coastal wetlands also buffer inland areas from severe storms, thereby reducing damage to physical structures. Wetlands also remove nutrients and toxic substances from waters that flow through them. Thus, wetlands improve water quality. Finally, some wetlands contribute signifi-cantly to the recharge of aquifers, and wetlands may mod-erate extremes in atmospheric conditions (Odum 1979, Mitsch and Gosselink 1986).

WETLAND MANAGEMENT
Design Considerations for Constructed Wetlands

Effective wetland management requires an understand-ing of interrelationships among habitats and resources needed by wetland wildlife to survive and reproduce. Op-timizing value and use of wetlands is possible only if in-formation on hydroperiod and wetland structure and func-tion is integrated with knowledge of requirements and timing of life-cycle events of wildlife. A successfully managed wetland contains foods and cover of a type, quality, and distribution that are the same or functionally similar to those found in natural, unmanaged wetlands. Although these conditions vary within and among wetland types, management should aim to provide resources that meet the physiological and behavioral needs of wildlife.

On the basis of these premises, individual wetlands should be evaluated prior to initiating management (Fig. 5). Pristine wetlands that have retained their inherent hy-drologic characteristics and functions should be protected and passively managed (Errington 1963, Weller 1988, Fredrickson and Reid 1990). Unfortunately, only a small percentage of wetlands in the conterminous United States and Hawaii have remained unaltered by human activity. In contrast, the hydrology of many wetlands in Alaska and northern Canada remains largely unchanged from his-toric conditions. Thus, developing a better understanding and protecting the natural hydrology in these northern ar-eas should be a primary management goal. Conversely, wetlands that have been modified or impacted often must be actively managed to provide resources to wetland wild-life consistently (Fredrickson and Reid 1990). A primary challenge in wetland management is to make hydrological modifications that transform wetlands into suitable habi-tats for wildlife (Weller et al. 1991).

Site characteristics are of primary concern in determin-ing whether the investment in development will be cost effective. For example, sandy soils are unsuitable for lev-ee construction and costs of maintaining water levels would be prohibitive. In riverine areas, porous sandy lay-

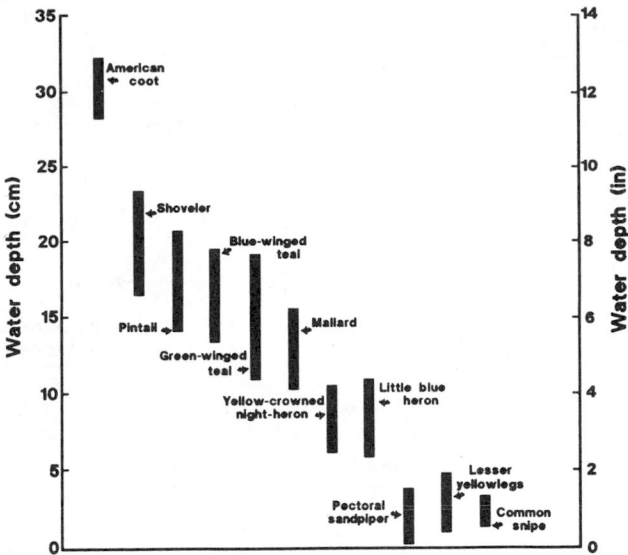

Fig. 4. Foraging depths of selected dabbling ducks, shorebirds, and waders.

ers often are just beneath a shallow clay soil. Natural de-pressions hold water, but they can be destroyed during development if the entire clay profile is removed.

Wetland Design and Construction

Methods vary for ecologically sound management of human-made or human-modified wetlands but should em-phasize creating or restoring natural wetland functions. Implementation of this concept often requires construction and installation of physical structures (e.g., levees, water control structures, water supply and discharge systems, pumping systems) that enable more exact control of water inflow, distribution, and discharge. Water level manage-ment is necessary to create soil and water conditions suit-able for germination and growth of desirable plant com-munities, control problem vegetation, stimulate invertebrate production, and make resources available for target species (Fredrickson 1991). Water control particu-larly is critical for providing habitat conditions required by foraging waterbirds (Fig. 4). For example, of 81 spe-cies of waterbirds using wetlands in the Southwest, only 19 successfully forage in waters >25 cm, but 10 of these readily use water <25 cm. Twice as many species (38) forage in water of <10 cm (Fredrickson and Reid 1986).

The capital investment to rehabilitate or restore a wet-land often is great. However, long-term costs associated with maintenance and operation can be reduced apprecia-bly and wildlife benefits can be maximized if develop-ments are designed to function in an ecologically correct manner. Normally, engineering and management goals can be satisfied if desired requirements are considered in the planning stage. Design engineers must clearly under-stand management goals to construct physical structures compatible with desired natural wetland characteristics and resources. In some situations, rehabilitation efforts have included construction of levees across contours and placement of water control structures at incorrect eleva-tions. Also, ineffective pumping systems have been in-stalled. Although such developments appear as viable methods of restoring wetlands for wildlife from an eco-

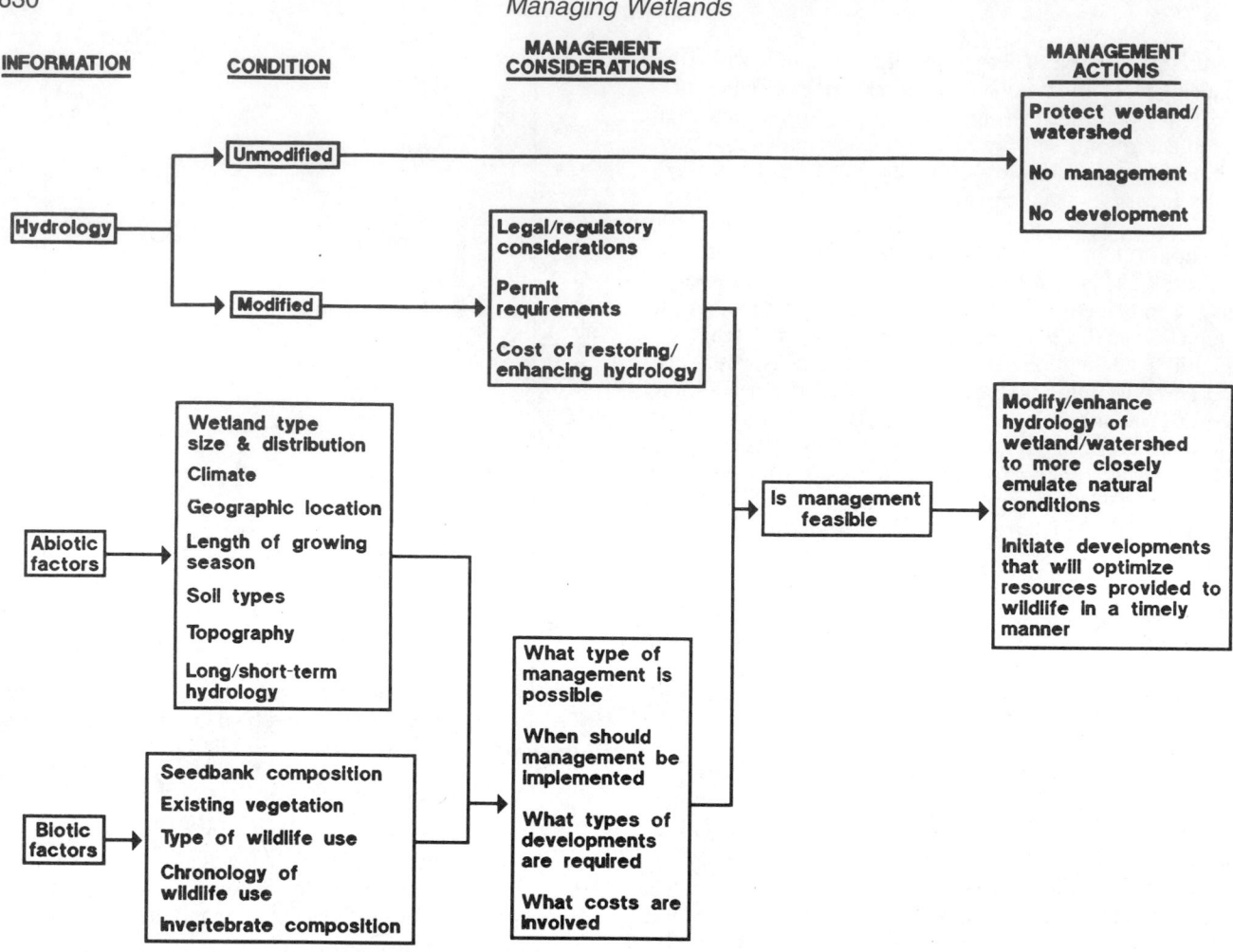

Fig. 5. Factors to consider in determining the type and intensity of wetland management.

nomic or engineering viewpoint, they often result in only superficial improvements in wetland quality and function. In some situations, poorly designed developments may be more deleterious than if no actions were taken. Thus, managers must guard against poorly designed developments and ensure emulation of natural hydrologic cycles and wetland functions.

Wetland Configuration

The capability of flight permits waterbirds to exploit a variety of habitats in close proximity. For example, in the Mississippi Alluvial Valley, dabbling ducks use wetlands within a 16-km radius to meet daily nutritional and physiological requirements (Delnicki and Reinecke 1986). Thus, the extent and type of use a wetland receives by wildlife are influenced not only by habitat conditions in that wetland but also by the condition of adjacent habitats. Generally, provision and management of a complex of different wetland types in a localized area often increase overall diversity and density of wildlife species (Fredrickson and Reid 1988*e*). Further, wetlands ranging in size from 1 ha to 1,400 ha can be successfully managed for waterbirds if they are part of a complex.

Before development, a proposed wetland management area should be assessed to determine potential wildlife use (Fig. 5). Further, food and cover requirements of target species should be considered in relation to existing habitats within the local area to identify habitats in short supply. If possible, the proposed site should be developed to provide the most limiting habitats that historically were present in that physiographic region. Although engineering and biological technology exists for creating or rehabilitating most wetland types, successful, long-term operation and wildlife use of a wetland will occur only if the wetland type is correctly matched with the appropriate region. For example, habitat for breeding birds often is the focus of management, but the site may be best suited for migrating or wintering birds because of its location.

Levees

Levees are an integral component of managed wetlands because they permit control of water levels and determine the maximum water depth that can be maintained. They also may enable water transfer to and from designated wetlands. Although levee dimensions vary depending on wetland type and proposed function, levees should be constructed along natural contours to maximize the area that can be flooded to depths permitting efficient foraging by waterbirds (Fig. 6). Contour levees also assure that the wetland can be completely dewatered if necessary to favor growth of annual plants, control problem vegetation, reduce the incidence of disease outbreaks, and conform to

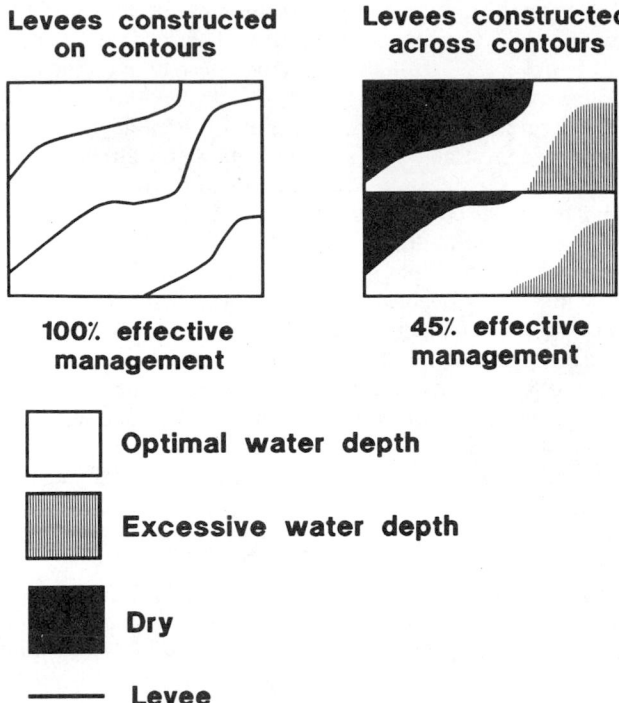

Fig. 6. Schematic representation of the importance of siting levees on contours.

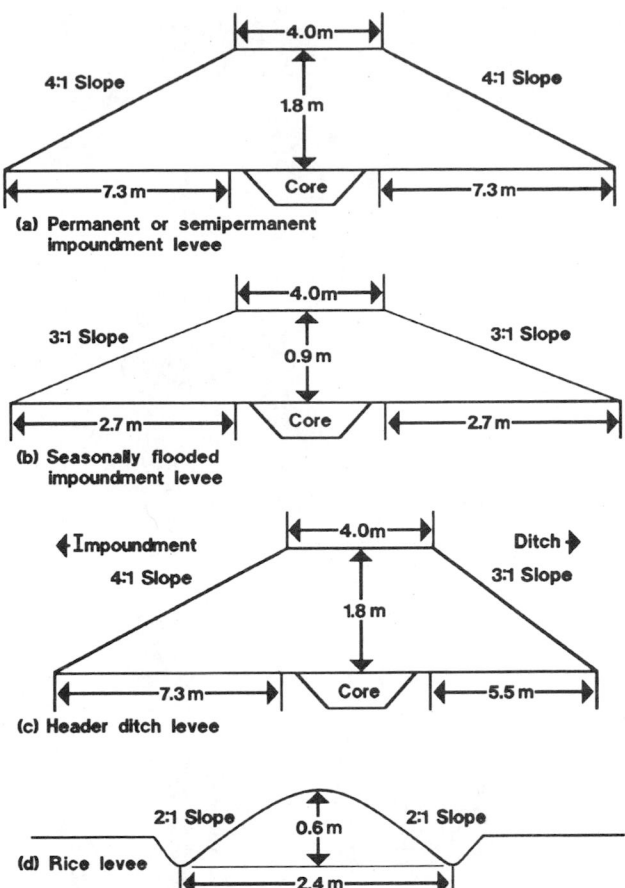

Fig. 7. Base specifications of various levee types necessary to facilitate wetland management activities.

legal statutes (e.g., mosquito abatement policies). In contrast, levees sited perpendicular to natural slopes tend to impede water discharge and may result in surface water that cannot be removed unless lateral ditches are constructed into the interior of the wetland. Although lateral ditches enable complete dewatering of a basin, they increase development costs, require periodic maintenance, and often reduce the area that can be effectively managed. Decisions concerning the contour interval on which to establish levees should be balanced among construction costs, detrimental impacts to existing habitats, and maximizing the area flooded to desired depths. For example, construction of levees on 30-cm contours to create a greentree reservoir may result in optimum water control, but also may require removal of numerous mast-bearing trees. In this situation, the contour interval selected should maximize the area flooded to desired depths while minimizing removal of valuable bottomland hardwoods. Levees can be constructed by varied methods, but the choice often depends on soil type, availability of equipment, and imagination of the technical staff. Commonly used equipment includes bulldozers, motor graders, rice dike plows, and fire plows.

Levee dimensions should be based on engineering criteria. Physical and chemical properties (e.g., organic matter content, texture) of soils influence compactibility and shear strength, and ultimately will dictate the required dimensions necessary to assure the long-term levee stability. However, within acceptable engineering standards, levees also should permit accomplishment of normal management operations. Levees should be capable of supporting machinery (e.g., tractor, mower, disk) necessary to maintain levees and manage vegetation. Further, levee side slopes should be gradual to deter potential damage by

erosion and burrowing mammals (Fig. 7). These objectives normally can be satisfied by constructing levees with 4-m crowns and minimum side slopes of 3:1. Levees with more gradual side slopes require an increased volume of material and also impact more wetland habitat, but they may be needed to satisfy engineering requirements.

Levee height depends on the size of the impoundment and expected depth of flooding. Levees in large wetlands (e.g., >32 ha) or wetlands constructed to simulate permanently flooded habitats are susceptible to wave action and erosion. Consequently, large or deeply flooded wetlands require more substantial levees than seasonally flooded wetlands (Fig. 7). In general, levee height should be a minimum of 1 m above the predicted annual maximum flooding depth. Levee height also should be uniform to minimize damage if the levee is overtopped by floodwaters. In areas subjected to severe deep flooding at regular intervals, a low levee that is submerged quickly and uniformly is damaged less by flooding than a large protective levee of insufficient height. Where flooding is less frequent, emergency spillways can be incorporated into levees to maintain structural integrity.

Water Control Structures

Permanent water control structures that allow relatively precise water level manipulation are essential if natural hydrologic regimes are to be emulated (Fredrickson 1991). Because the hydroperiod of most wetlands is dy-

Stoplog water control structure

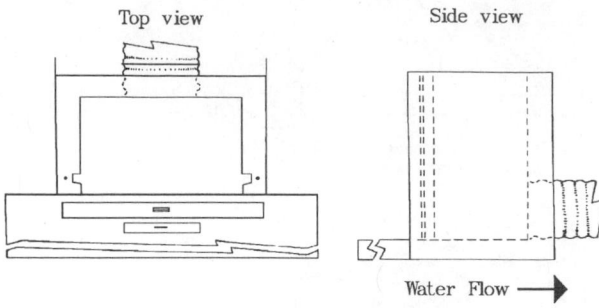

Screwgate water control structure

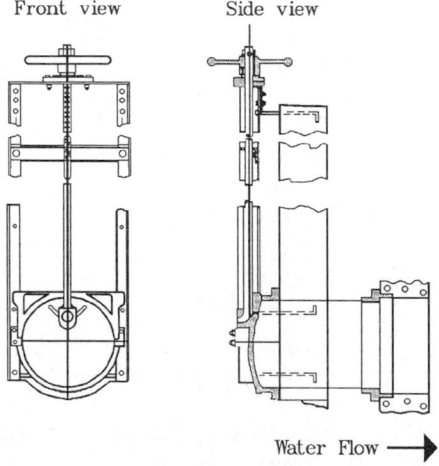

Fig. 8. Diagrams of stoplog and screwgate water control structures.

namic and can involve daily as well as seasonal and annual fluctuations in water levels, the ability to alter water levels slightly and completely remove water from a developed wetland is essential. Therefore, type and location of water control structures are important. In most situations, stoplog water control structures are ideal for controlling water discharge (Fig. 8). Water level adjustments as small as 5 cm can be accomplished with stoplogs. In addition, these structures require less frequent monitoring because stoplogs can be preset at the desired elevation and additional water that enters the wetland is removed from the surface via gravity flow. In contrast, control of water movement with screwgates (i.e., sluice gates) is more difficult because water discharge is accomplished by removing water from the bottom of the wetland (Fig. 8). If screwgates are opened and left unchecked, complete water discharge will occur. Therefore, screwgates may be used as intake structures, but not as discharge structures (Fredrickson 1991). Tidal wetlands have special challenges associated with water inflow and discharge. Not only does water rise and fall daily, but the height of the tides varies during the moon cycle and seasonally (Chabreck 1988). Salinity also varies in relation to storms, season, and upstream flows. Weirs, low dams constructed across channels, often are used to control water movement in tidal systems. In the Southeast, a control structure called a

"rice trunk" also provides opportunities for effective water control (Gordon et al. 1989). These water control structures have excellent longevity, and manipulations require limited manpower.

The proper locations of intake and discharge structures are the highest and lowest elevations within the wetland, respectively. The same structure should never be used for intake and discharge (Fredrickson 1991). In areas with high salinities, multiple inlet structures at the highest elevations and multiple drains at the lowest elevations are essential. This enables complete water discharge necessary to control plant community establishment and permits optimum flooding necessary to make resources available to waterbirds. Additionally, correct siting of water control structures maximizes water circulation within a wetland, which prevents accumulation of soil salts, reduces the risk of disease outbreaks, and facilitates nutrient cycling.

Water Delivery System

Sources of floodwater include precipitation, ditches, rivers, reservoirs, and groundwater (Reid et al. 1989). Each varies in the quantity, quality, timing, predictability, availability, and cost of providing water to flood wetlands. Rainfall is least costly but also least dependable because the amount and timing of precipitation vary spatially and temporally. Similarly, availability of water from ditches and rivers is dependent on characteristics of individual watersheds and precipitation cycles. Historically, floodwater regularly inundated natural wetlands or was available for transfer to managed wetlands. Small, localized precipitation events saturated the soil, whereas larger events resulted in accumulation of surface runoff that filled wetlands connected to natural drainage systems. Although all wetlands were not flooded every year, some habitats were available in most years. Today, however, much of the rainfall in the West is diverted for agricultural or urban uses and never reaches wetland sites.

Channelization of large river systems and associated tributaries also has resulted in serious hydrologic alterations of many wetland systems. Depending on wetland type and location, the impacts of these modifications may be manifest in either severe flooding (Belt 1975) or an almost complete lack of floodwater (Reinecke et al. 1988). This is particularly true in the lower Mississippi Alluvial Valley. Channelization of the upper reaches of the Mississippi River and its tributaries has resulted in severe water fluxes in the lower reaches. Wetlands located near the mouth of the river often are inundated by too much water, whereas wetlands located upstream do not receive sufficient water to enable flooding. In either situation, waterbird use often is limited. Further, the quality of water in many rivers has been degraded by sediments or toxicants (Longcore et al. 1987, Grue et al. 1989). These conditions influence wetland vegetation by restricting germination and growth and often result in monotypic communities.

Groundwater and reservoirs are more reliable sources of good-quality water, but construction of expensive levee or pump systems may be required to accomplish water transport to managed sites. Reservoirs also are disadvantageous because a considerable amount of area often must be converted to deep-water habitats to store sufficient water, and siltation sometimes limits their life-span.

The long- and short-term costs associated with using groundwater will vary, depending on type of pump and power unit and distance to the water table. Pumps powered by diesel engines are cheapest but require frequent maintenance and may be more costly to operate than electric pumps over the long-term (Reid et al. 1989). In contrast, three-phase electric pumps are more costly initially and often require annual line charges to operate. However, electric pumps are less costly to operate, require less maintenance, and operate quietly, causing less disturbance to wildlife compared to diesel-powered pumps. Regardless of power unit, wells should be monitored at least biennially to determine discharge rates. As water output decreases, pumps must be refurbished or replaced.

MANAGEMENT OF PALUSTRINE WETLANDS

The hydrologic cycles influencing palustrine wetlands in North America are diverse, as evidenced by the different structure and function of this group of wetlands, which includes freshwater and brackish wetlands, tidal freshwater wetlands, fens, potholes, playas, wet meadows, seasonally flooded wetlands, and bottomland hardwoods. Management goals for these diverse wetlands vary greatly, depending on location and associated abiotic factors that influence vegetation composition and distribution and the type and timing of wildlife use.

Freshwater Wetlands

Freshwater wetlands are distributed across North America and occur from sea level to >3,000 m in elevation, from the Gulf Coast to the Arctic, and where annual rainfall varies from as little as 5 cm in southwestern deserts to >127 cm in the Pacific rainforest. They also vary greatly in size, ranging from <1 ha to >1,000 ha. Most freshwater wetlands occur as distinct, concave basins (e.g., Prairie Pothole Region), but some occur as extensive perimeter zones associated with lacustrine systems (e.g., Lake Huron). Freshwater wetlands across the continent vary according to formative processes, vegetation composition, number of ice-free days, and long- and short-term fluctuations in water levels.

Classification of freshwater wetlands is based on the permanency of flooding in combination with the type of vegetation (Cowardin et al. 1979). A well-known regional system for management of prairie wetlands was developed by Stewart and Kantrud (1971). They identified seven different wetland types (Table 3) that are all classed as palustrine according to Cowardin et al. (1979). When a group of different wetland types (e.g., seasonal, semipermanent, permanent) is in close juxtaposition, it forms a wetland complex (Fig. 9). Complexes provide many different resources for different species of wildlife or provide resources required for successful completion of certain stages in the life cycle of species (Fredrickson and Reid 1988e, Nelson and Wishart 1988). Water level dynamics are different for each basin type. Some basins maintain water year after year, but others remain flooded only for short periods (Fig. 9). Successful management strategies should focus on providing critical resources for a cross section of wildlife that use a wetland complex (Fredrickson and Reid 1986) (Table 4).

The most common management problems in freshwater wetlands are the development of monotypic communities (i.e., extensive, uninterrupted stands of a single plant species) or nuisance exotics (e.g., purple loosestrife), the loss of all emergent vegetation, and increasing turbidities that destroy submergent plant beds (Robel 1961b, Weller 1988). Monotypic communities often develop because of stabilized water regimes or changes in basin depths caused by siltation. Loss of vegetation cover usually is related to continuous flooding at considerable depths for many years, or muskrats may remove robust emergent vegetation for lodge construction and food (Weller and Fredrickson 1973, Chabreck 1988). Freshwater wetlands contain plants with low salt tolerance, and saltwater intrusion into these wetlands often results in the loss of all emergent vegetation. Turbidity usually is related to land use practices in the drainage basin or from activity of bottom-feeding fish such as carp (Robel 1961a).

BASINS LACKING WATER CONTROL SYSTEMS

Management actions in basins lacking water control are limited, but fire, mechanical treatments, or herbicides can be used to alter the distribution, composition, or density of vegetation. Fire removes excess debris, releases nutrients, exposes soils for new germination, and usually creates a mosaic of vegetation attractive to wildlife. Thus, fire is an effective tool to change the structure, composition, and distribution of vegetation (Kirby et al. 1988). Timing of prescribed burns is dependent on dry conditions, fuel availability, and presence of wildlife within the wetland, especially species listed as threatened or endangered. At some locations, mowing or light discing can be used to create fuel loads sufficient for burning. Because of increasingly stringent air-quality standards, prescribed burns must be carefully planned and executed to meet local, state, or federal regulations.

Mowing, although more restricted in its use than burning, also can modify the distribution and density of vegetation. The technique is best used on wetlands that develop sufficiently thick ice to support a tractor. Robust emergent vegetation that is clipped in winter will have limited regrowth if the stubble is flooded during the growing season (Weller 1975, Kaminski et al. 1985). Because water levels normally are at their lowest level in late autumn when such sites freeze, and highest at about the time of the spring thaw during most years, emergent vegetation can be clipped just above the ice so that spring flooding inundates the cut stems and restricts the oxygen supply to the root zone. Many of the plants do not resprout, thus the distribution and density of robust emergents are modified for as long as several years.

Herbicides have been used for temporary control of dense, monotypic stands of vegetation. However, because herbicides do not modify the ecological conditions that cause the vegetation problem, the continued use of chemicals likely will be necessary until the appropriate causal factors also are addressed. Although the use of herbicides is regulated (i.e., types of chemical and surfactants), caution should be used in application because of potential residual effects of herbicides on algae, a fundamental link in the food chain.

Table 3. Representative cover types and waterbird use of seven prairie palustrine wetland types described by Stewart and Kantrud (1971) and classed as palustrine by Cowardin et al. (1979).

Emergent wetland type[a]	Example	Flood duration (months)	Cover	Wildlife use
Temporarily flooded	Ephemeral ponds	<1	None	Shorebirds Dabbling ducks
Temporarily flooded	Temporary ponds	<1	None	Shorebirds Dabbling ducks Waders
Seasonally flooded	Seasonal ponds and lakes	2–4	Nesting Roosting	Frogs Shorebirds Waders Rails Dabbling ducks Diving ducks
Semipermanently flooded	Semipermanent ponds and lakes	12	Nesting Roosting	Turtles Frogs Waders Grebes Rails Gallinules Terns Dabbling ducks Diving ducks
Permanently flooded	Permanent ponds and lakes	12	Nesting Roosting	Turtles Fish Diving ducks
Permanently flooded—mesosaline	Alkali ponds and lakes	6–12	None	Shorebirds Dabbling ducks Diving ducks
Saturated	Mixosaline	12	Nesting Roosting	Blackbirds

[a]Types based on classification of Stewart and Kantrud (1971).

BASINS WITH WATER CONTROL SYSTEMS

The same techniques described above can be used in basins with water control systems. However, opportunities to consistently create favorable habitat conditions are much greater when control of water supply, water levels, and water discharge is possible (Kadlec and Smith 1992). For example, use of fire or mowing is much more flexible when water can be removed from a basin to create conditions suitable for fire or to enable equipment access.

The use of drawdown techniques to duplicate natural wetland cycles can revitalize wetlands that have reduced values for wildlife. Such drawdowns are most valuable on semipermanent wetlands that have wet-dry periods of 7 years or more. Natural wetland systems undergo a series of vegetation changes that were well described by Weller and Spatcher (1965) and van der Valk and Davis (1978). When mudflats are exposed, germination takes place. As the basin is reflooded in subsequent seasons, the less water-tolerant vegetation disappears, resulting in improved cover-water interspersion (Weller and Spatcher 1965, Weller and Fredrickson 1973). Under optimum conditions the cover:water ratio is about 50:50 across an entire wetland, or a large portion of a wetland (Weller and Spatcher 1965), and vegetation is distributed in patches throughout the basin. During this hemi-marsh stage, the wetland has the greatest diversity and numbers of waterbirds, and muskrat and invertebrate populations often are high (Weller and Fredrickson 1973).

When water level control is possible, drawdowns can be scheduled to produce desired results regardless of natural precipitation or drought cycles. Precise water control also provides the opportunity to control vegetation, stimulate the growth or distribution of certain vegetation, and make resources available for wildlife. For example, recently established cottonwood or willow seedlings (<2 cm) can be controlled (at 40° latitude or farther north) by overtopping the regrowth with as little as 1 cm of water. Careful consideration should be given to timing of drawdowns because of implications for all wildlife populations. Winter drawdowns can help control undesirable vegetation by exposing rootstocks to freezing temperatures, but these same conditions can adversely affect muskrats, herptiles (e.g., turtles), and fish. Delaying drawdown until spring likely would have a lesser impact on the control of some vegetation, but higher water levels in winter would reduce mortality of turtles and other herptile populations.

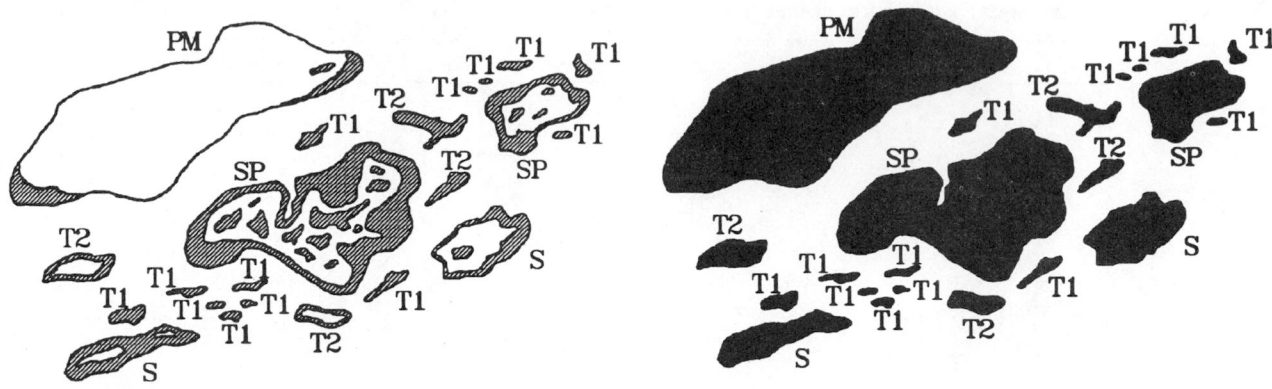

(a) Distribution of robust emergent vegetation and juxtaposition of different wetland types within a palustrine emergent prairie wetland complex.

(b) Surface area flooded at ice-out.

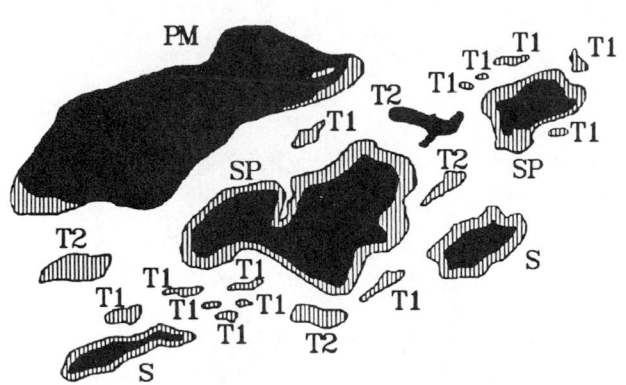
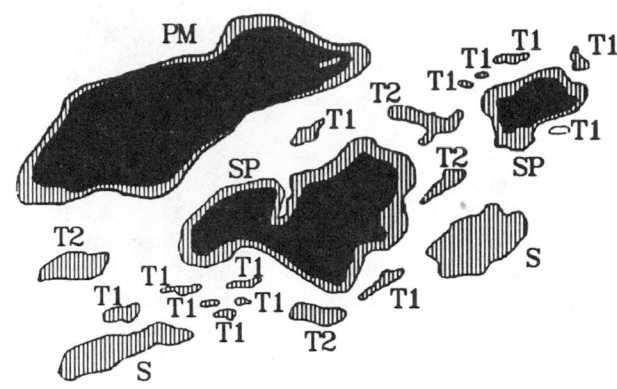

(c) Surface area flooded during waterfowl brood rearing.

(d) Surface area flooded during waterfowl autumn staging.

 Vegetation

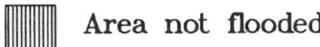 Area not flooded

 Area flooded

T1 – Temporarily flooded <1 month

T2 – Temporarily flooded 1–2 months

S – Seasonally flooded 3–5 months

SP – Semipermanently flooded

PM – Permanently flooded – mesosaline

Fig. 9. Schematic representation of different wetland types composing a prairie wetland complex and seasonal flooding dynamics at ice-out, during brood rearing, and during autumn staging of waterfowl.

Seasonally Flooded Wetlands

Management of seasonally flooded wetlands is best known by wetland managers as moist-soil management. Use of the term moist soil was coined by Low and Bellrose (1944) following their work on the Illinois River in the late 1930s. They associated the term with vegetation that normally germinates on mudflats and produces lush growth after spring or summer drawdowns. Originally, the technique focused on the production of seeds from annual plants such as smartweeds and barnyard grasses. Currently, the technique is used widely on migration and wintering habitats at more southern latitudes (Reid et al. 1989,

Table 4. Example of selected wildlife use of prairie wetland complexes.

Species	Life-history stage	Wetland type[a]
American toad	Breeding/courting	T1, T2, S
Leopard frog	Breeding/courting	T1, T2
	Winter	S, SP
Painted turtle	All phases	SP
Blanding's turtle	All phases	SP
American avocet	Nesting	S, SP
Ring-necked pheasant	Winter	S, SP
Black-crowned night-heron	Nesting	SP
	Foraging	T2, S, SP
Blue-winged teal	Prebreeding	T1, T2, S
	Nesting	Upland near S, SP
	Brood rearing	S, SP, PM
	Autumn staging	SP, PM
Canvasback	Pre-breeding	S, SP, PM
	Nesting	S, SP
	Brood rearing	S, SP, PM
	Autumn staging	SP, PM
Muskrat	Autumn/winter/breeding	SP
Mink	All phases	All
White-tailed deer	Winter	S, SP

[a]T1 = temporarily flooded <1 month, T2 = temporarily flooded 1–2 months, S = seasonally flooded 3–5 months, SP = semipermanently flooded, PM = permanently flooded—mesosaline.

Reinecke et al. 1989, Haukos and Smith [Texas Tech Univ. Dep. Range Wildl. Manage., Manage. Note 14, 1991]), but moist-soil techniques also produce benefits for many different forms of wildlife on sites as far north as Canada and as diverse as the Hawaiian Islands.

The primary strategy in seasonally flooded wetlands is to produce diverse food resources and to make resources readily available to target wildlife. This goal requires an annual drawdown to stimulate the establishment of desirable plant communities, while providing resources for target species from late winter until early summer (Fredrickson and Taylor 1982, Fredrickson 1991). After plant foods are produced, foraging and cover requirements of target species are used to determine the time and type of flooding necessary to make foods available to wildlife from late summer to winter.

Plant response to a drawdown is dependent upon many interacting variables that determine the composition of the plant community (Fig. 10). The seed or propagule bank is a primary factor determining types of vegetation pro-

duced. Soil samples collected in agricultural fields in southeastern Missouri indicate that the number of large moist-soil seeds (i.e., easily separated from soil samples) varies from about 4,000 to >300,000/m^2. This huge number of dormant propagules compared to the number of growing plants is a unique characteristic of the plant kingdom (Harper 1977). Thus, nearly any low-lying site has an adequate seed source to produce an abundance of moist-soil vegetation. The management strategy is to control the conditions associated with germination to produce the desired plant community (Fredrickson and Taylor 1982, Fredrickson 1991). Factors such as time of drawdown, type of drawdown, time since disturbance, topography, soil type, and seasonal variations interact to determine which seeds in the seed bank respond and become established on the exposed mudflat (Fig. 10).

The time and type (i.e., fast, slow) of drawdown influence soil moisture conditions in relation to ambient temperatures and photoperiod (Fredrickson 1991) (Table 5). Drawdowns early in the growing season create conditions suitable for germination of seeds adapted to low soil temperatures and high moisture conditions. Slow drawdowns (i.e., removal of surface water requires >7 days), regardless of their timing, tend to sustain elevated moisture conditions for a longer period. Higher soil moistures tend to have a moderating effect on soil temperatures. Thus, early drawdowns stimulate plant communities that are different from those of late drawdowns. Likewise, fast drawdowns (i.e., removal of surface water requires ≤4 days) may produce plant communities very different from those produced by slow drawdowns (Fredrickson and Taylor 1982). The drawdown type becomes particularly important later in the season, because moist-soil plants with well-established root systems are well adapted to drought conditions. Thus, keeping moisture conditions within a range

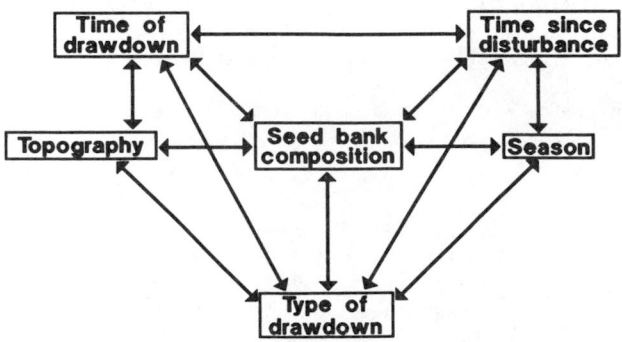

Fig. 10. Factors influencing germination of moist-soil plants.

Table 5. Effect of drawdown date and drawdown rate on the rate of soil drying and germination of selected moist-soil plants on Mingo NWR, Missouri.

Drawdown Date[a]	Rate[b]	Ambient temperature	Photoperiod	Rate of soil drying	Representative plants germinating
Early	Slow	Low	Short	Slow	Smartweed, blunt spikerush, barnyard grass
	Fast	Low	Short	Medium–slow	Barnyard grass, willow, chufa flatsedge
Mid-season	Slow	Intermediate	Long	Medium–slow	Rice cutgrass, common burhead, barnyard grass, chufa flatsedge
	Fast	Intermediate	Long	Medium	Beggarticks, ragweed, panicgrass
Late	Slow	High	Intermediate	Medium	Redroot flatsedge, toothcup
	Fast	High	Intermediate	Fast	Cocklebur, morningglory, aster, crabgrass

[a]Early: prior to 15 May; mid-season: 16 May–1 July; late: after 1 July.
[b]Slow: >14 days; fast: <14 days.

that allows extensive root development is critical to the subsequent survival and seed production of moist-soil plants. For example, in dry seasons, seed production on adjacent moist-soil units of Mingo National Wildlife Refuge, Missouri, varied from <100 kg/ha to >500 kg/ha following a fast and a slow drawdown, respectively. The key difference was that plants germinating during the slow drawdown survived summer drought conditions and responded to late summer rainfall. The greater number of plants surviving and producing seeds resulted in greater total seed production on the sites with slow drawdowns.

Although the relationships among soils and the establishment, growth, and production of specific moist-soil plants is not well studied, factors associated with soil types such as texture, clay pans, pH, cation exchange capacity, available nutrients, and redox potentials are known to influence moist-soil plant communities (Harper 1977). Undoubtedly, different soil properties are related to dissimilar vegetation responses within and among units. Until these factors are understood, managers must rely upon their experience with different units to determine problems and potentials in meeting their vegetation management goals. Similarly, topography within a unit is important, because different elevations influence drawdown date and rate at specific sites. Different microhabitat conditions result in a diverse plant community with a diverse structure. For example, at lower elevations a unit may lack vegetation or have vegetation with a low vertical profile. When these areas are flooded, water is visible and waterbirds are readily attracted to the open site.

The time since soil disturbance or the stage of succession is an important determinant in moist-soil communities. The natural progression is for perennial vegetation to increase over time. Further, most seasonally flooded wetlands occur at more southern latitudes, a region where the rate of woody establishment and growth is rapid. Encroachment of woody and perennial species into moist-soil sites is a widespread management challenge (Fredrickson and Reid 1988b). Mechanical treatments such as discing or plowing are the most common approaches used to reduce such growth and concurrently stimulate the production of annual vegetation. However, these techniques also may stimulate nuisance plants (e.g., cocklebur, hemp

sesbania). During the first growing season following soil disturbance, annual vegetation such as barnyard grasses, smartweeds, sprangletop, and beggarticks are likely to dominate the vegetation community. In some situations mechanical treatment to stimulate annuals is not necessary because grubbing by waterfowl during the winter disturbs the soil enough to stimulate the continued production of annual vegetation.

The structure of the plant community has an important influence on invertebrate communities (Fredrickson and Reid 1988d). The numbers and biomass of invertebrate populations are related to the biomass and structure of the plant community. The most prolific invertebrate groups in moist-soil communities are those with high mobility that can colonize new locations or those with high fecundity that can survive seasonal drought and produce large numbers of eggs quickly (Fredrickson and Reid 1988d). This characteristic of invertebrate communities provides the manager with a wide range of options to produce foods for wildlife. Plants that are not good seed producers but have a complex vegetation structure become an important component in seasonally flooded wetlands because they supply critical habitat to invertebrates. In some instances, plants considered undesirable because they form monocultures can be converted to litter by mechanical treatments and subsequently provide the detrital base for invertebrate populations. By timing such manipulations to coincide with the arrival of certain target species, poor-quality plant foods can be converted into important invertebrate foods, and ideal conditions are created for increased production of plant food in subsequent growing seasons.

Once foods are produced, they must be made available. This requires that habitat conditions, such as vegetation structure and water depth, match the foraging modes of target species (Table 6). Because of constantly changing seasonal needs of wildlife species, foods produced must be made available at the correct time in the annual cycle. The most effective tool is the control of water depths. Stoplog water control structures are the most effective means to achieve this goal because water levels are set at a predetermined elevation, and width of stoplog boards determines precision of water level control possible.

Table 6. Examples of selected wildlife use of seasonally flooded impoundments.

Species	Life-history stage	Habitat condition		Foods
		Flooding[a]	Vegetation	
Bowfin	Spawning	Vernal	Matted	Crayfish, fish
American toad	Breeding/courting	Shallow vernal	Matted	Macroinvertebrates
Cottonmouth	Summer foraging	Shallow/damp	Dense/open	Herps, fish, small mammals
Little blue heron	Summer foraging	Shallow or drawdown	Sparse	Crayfish, fish
Yellow-crowned night-heron	Breeding	Shallow or drawdown	Sparse	Crayfish
	Developing young	Shallow or drawdown	Sparse	Crayfish
Pectoral sandpiper	Migrant/summer	Shallow—mudflats	None	Macroinvertebrates
Common snipe	Migrant/autumn	Shallow—mudflats	Sparse	Macroinvertebrates
Mallard	Migrant/autumn	Shallow	Dense	Seeds, tubers, macroinvertebrates
	Winter	Shallow	Matted/dense	Seeds, tubers
	Spring	Shallow	Matted or dense	Rootlets, seeds, tubers, macroinvertebrates
Canada goose	Autumn migrant	Shallow	Dense but short	Seeds, tubers
	Winter	Shallow	Matted	Seeds, tubers
	Spring	Shallow	Matted	Seeds, tubers
Great horned owl	All year	N/A	N/A	Vertebrate prey
Tree swallow	Spring/summer	N/A	N/A	Aerial insects
Raccoon	Spring	Shallow—mudflats	Any type	Crayfish

[a]Shallow: 0–20 cm.

As the management of native vegetation has become more sophisticated, managers have recognized the value and opportunities to supply cover and food for a variety of species. Although seeds from annual plants were the original focus of such management, tubers, browse, and invertebrates now are recognized as equally important foods produced on these habitats. Likewise, perennial vegetation is recognized as a beneficial component of the plant community. Flooding strategies that are more semipermanent in nature are now part of the diverse approach used in seasonally flooded wetlands to expand the range of target wildlife species from a narrow focus on waterfowl to include rails, bitterns, shorebirds, waders, and passerines (Rundle and Fredrickson 1981, Fredrickson and Reid 1986). In Missouri, >150 species of birds, several mammals, many herptiles, and some fish use moist-soil habitats extensively. Implementing multifaceted management strategies at the correct time to provide benefits for an array of wildlife species is becoming increasingly complex. To assist managers with annual decision making, The U.S. Fish and Wildlife Service (National Ecology Research Center, Ft. Collins, CO 80525) has developed, and distributes, a computerized expert system (i.e., The Moist-Soil Management Advisor).

Forested Wetlands

Forested wetlands are common in the Deciduous Forest Biome of the southern United States. The largest single expanse of forested wetlands originally encompassed 10 million ha within the Mississippi Alluvial Valley (McDonald et al. 1979). Other important forested wetlands occur as riverine swamps from eastern Texas north to Missouri and east to Virginia (Fredrickson 1979). Many famous swamps are associated with riverine systems from North Carolina to Florida, including Congree, Great Dismal, Okeefenokee, and Suwannee River. Other areas of seasonally flooded riparian habitats occur along flowages throughout North America.

Most forested wetlands are seasonally flooded. In the southeastern United States, flooding primarily is during the dormant season but usually extends into early spring (Heitmeyer et al. 1989). The lower elevational zones are flooded longer, especially where common baldcypress, water tupelo, and scrub/shrub swamps with common buttonbush occur. In most other forested sites, flooding is most likely to occur during high winter and spring overflows. However, in some areas, such as along the Mississippi River, an autumn rise may be sufficient to flood some forested habitats.

Southern forested wetlands differ from wetlands in glaciated regions because different wetland types occur along a flooding gradient. From lowest to highest elevations, wetland types are robust emergent, moist-soil, scrub/shrub, cypress-tupelo, overcup oak, water-pin-cherrybark-Nutall oak, and hickory-sugar hackberry. These different wetland types, closely juxtaposed, provide habitats for many different species (Wharton et al. 1982, Fredrickson and Heitmeyer 1988, Fredrickson and Batema 1992) (Table 7).

Management of forested wetlands is practiced widely in the Southeast as greentree reservoir management (Reinecke et al. 1989, Wigley and Filer 1989). The technique refers to flooding live forests during the dormant season by gravity flow from reservoirs or by pumping groundwater or stream water into wetlands. Levees must be constructed to contain water levels, and at times a sophisticated distribution and discharge system is an integral part of development (Fredrickson and Batema 1992). Historically, the usual management strategy was to flood forested wetlands annually to provide foraging and roosting sites before and during waterfowl hunting seasons. As a result, forests regularly were flooded before tree senescence and about the same time, depth, and duration year after year. Each unit usually was flooded to capacity; depth was determined by height of the levee. Because of a concern for tree mortality, most wetlands were drained soon after the waterfowl hunting season. Thus, duration of flooding was similar among years. After 10–15 years of repetitive hydroperiods, waterfowl use declined and damage and mortality of trees were common. Several different management approaches currently are used in forested wetlands, including rehabilitation, reforestation, and development of new sites for intensive management.

REHABILITATION

Rehabilitation of sites usually requires improved water supply and discharge. Inlet structures should be at the highest elevation. Header ditches have been used for water distribution historically, but buried polyvinylchloride (PVC) pipe is now available. PVC pipe is expensive, but it conserves water and eliminates problems associated with control of woody vegetation in ditches. Common problems in greentree reservoirs often result from nonfunctional outlet structures that are of the wrong type, have been clogged, are set at incorrect elevations, or are at the wrong location to allow for total drainage. Replacement with stoplog structures is essential because they can be set at a predetermined water depth, can be adjusted in small increments, and will accommodate storm events without close attendance. A common problem in forested habitats is related to poor control of discharge because of overgrown or silt-filled ditches, beaver dams, or ditches of inadequate size. Thus, ditches carrying discharge water must be large enough to carry the volume of water drained from a unit or from unusual flood events (e.g., overflows during the growing season). Beavers are especially problematic in southern forests where expanding populations have modified drainage, resulting in more extensive and deep flooding. This modified flooding has killed thousands of hectares of valuable oak forests.

REFORESTATION

Reforestation is becoming an increasingly common practice in the Mississippi Alluvial Valley (Haynes and Allen 1988). Seedlings and saplings have been planted, but directly seeding acorns has been the most economical approach to establishing oaks. Acorns must be collected soon after they drop, stored properly (i.e., temperature and humidity), and planted promptly (Allen and Kennedy 1989). The best success occurs when the source of acorns is near the reforestation site. The source of acorns should never be more than 240 km from where planting will occur (Allen and Kennedy 1989). Acorns should not be planted much farther north than where they were collected, but acorns collected in the north can be planted at greater distances to the south.

Table 7. Examples of selected wildlife use of forested wetlands systems.

Species	Life-history stage	Habitat condition			Activity	Foods
		Water depth[a]	Flood duration[b]	Vegetation		
Mole salamander	Autumn egg laying	Shallow	Seasonal	None	Egg laying	Unknown
Green treefrog	Breeding	Shallow	Seasonal	None	Egg laying	Macroinvertebrates
Bullfrog	All phases	Moderate	Semipermanent	Scattered emergent	Foraging	Fish, carrion
Alligator snapping turtle	All phases	Deep	Permanent	Submergent	Foraging	Crayfish
Yellow-crowned night-heron	Breeding	Shallow	Seasonal	Sparse	Foraging	Crayfish
	Subadults	Shallow	Seasonal	Sparse	Foraging	Crayfish
Wood duck	Spring migrant	Shallow–moderate	Semipermanent	Scrub/shrub	Foraging	Acorns, seeds
	Breeding	Shallow	Seasonal	Live forest		Macroinvertebrates
	Brood-rearing	Shallow	Seasonal	Live forest		Macroinvertebrates
	Autumn-staging	Moderate–deep	Semipermanent	Robust emergent	Roosting	None
	Autumn-migrant/post-breeding	Shallow	Seasonal	Live forest	Foraging	Acorns
Mallard	Winter maintenance	Shallow	Seasonal	Live forest	Foraging	Acorns
	Pairing	Shallow	Semipermanent	Scrub/shrub	Courtship	None
	Pre-basic molt (female)	Shallow	Seasonal	Live forest	Foraging	Macroinvertebrates
	Lipid protein/reserve deposition	Shallow	Seasonal	Live forest	Foraging	Macroinvertebrates
River otter	All phases	Moderate–deep	Permanent	Submergent or none	Foraging	Fish, crayfish
Raccoon	All phases	Shallow/damp/dry	Semipermanent–none	All types	Foraging	Fish, crayfish, clams

[a] Shallow: <10 cm; moderate: 10–25 cm; deep: >25 cm.
[b] Seasonal: 3–6 months; semipermanent: 10–12 months; permanent: 12 months.

Careful selection of the tree species must be made in relation to geographic location, flooding regime, and elevation. Species with poor flooding tolerance should be planted only at higher elevations. Careful consideration of the size of area to be planted is essential. Good conditions for germination, growth, and survival do not occur annually. Thus, only part of a site should be planted at one time to avoid duplicating the current problem facing many managers, large expanses of even-aged timber.

Opinions for site preparation vary. With new technology, site preparation depends largely on available equipment and existing vegetation. Dense stands of grasses and forbs should be disrupted before planting, but weed fields with scattered grasses often do not require preparation. Large, multirow planters pulled behind large tractors require good site preparation. Some smaller planters have been designed for use behind all-terrain vehicles. Smaller planters do not require site preparation, although settings for each soil type are necessary. Cultivation of plantings is not recommended because of time and cost. Weedy growth also moderates temperatures and prevents sun scald of the newly established seedlings.

NEW DEVELOPMENTS

New developments should be restricted to sites that receive unreliable flooding, whereas sites with good natural hydrology should be protected. Levees should be on contours, and multiple impoundments should be developed to increase management flexibility during a given year (Fig. 5). The area should be configured so that disturbance to wildlife during critical time periods is limited. Access roads to pumps and water control structures and for public use should not pass through the core of a management area. Further, public use, management, and research activities should be sensitive to critical time periods in the life cycle of target organisms.

WATER LEVEL MANAGEMENT

Flooding regimes in forested wetlands require more careful planning than in other sites because management that damages or kills trees has long-term implications (i.e., herbaceous vegetation can recover in a year but regeneration of a forest requires decades). Natural flooding regimes are highly dynamic within and among years (Heitmeyer et al. 1989). In addition, each wetland type has a flooding regime that varies in timing, depth, and duration each season. Management scenarios should emulate natural variation in these conditions within and among years to maintain productivity and diversify wildlife use (Fredrickson and Batema 1992) (Fig. 11). Not only do water regimes influence establishment and survival of woody vegetation, but they have a unique impact on the production of invertebrate foods and fish (Batema et al. 1985, Finger and Stewart 1987).

Invertebrates, including shredders (e.g., aquatic sowbugs and sideswimmers) and collectors (e.g., fingernail clams), are abundant in the flooded leaf litter of lowland forests (Hubert and Krull 1973, White 1985). Forest invertebrates respond within 2 weeks of flooding and reach peak abundance and biomass by the end of winter. Invertebrate numbers are increased when flooding is slow and water depths do not exceed 25 cm. Rapid flooding when temperatures are high often results in anoxic conditions.

Invertebrates such as rattailed maggots and aquatic earthworms are more common under oxygen-poor conditions. When flooding depths are ≤10 cm or when gradual drawdowns occur over an extended period, invertebrates are more responsive and most available to wildlife.

A major concern in lowland oak forests is the lack of regeneration. Because oaks are shade intolerant, adequate sunlight must reach seedlings if they are to survive. Seedlings also are subject to high mortality during dormant season flooding if water overtops the entire seedling. Variable flooding strategies are the best tools for maintaining tree vigor and productivity and making food available. Impoundments should be flooded at different times, at different depths, or for a different duration among years (Fig. 11). Varying the timing and degree of flooding over a series of years emulates conditions that likely occur in naturally flooded systems (Heitmeyer et al. 1989) and promotes the longevity of the forest.

MANAGEMENT OF TIDAL WETLANDS

Coastal areas have wetlands ranging from fresh to salt marsh (Chabreck 1988). These diverse wetlands provide habitats for rich plant and animal communities. The ecotone between the two types may be divided into brackish (mesosaline) and intermediate (oligosaline) wetlands. Some coastal wetlands may be hypersaline because salt concentrations in the water exceed 36 ppt, which is characteristic of seawater. Tidal wetlands are most prevalent along the Atlantic and Gulf coasts, which account for 98% of all coastal wetlands in the 48 conterminous states (Chabreck 1988:4–6). The inland movement of saltwater and the seaward movement of freshwater govern the distribution of plants and animals in coastal systems. The width of the coastal zone and the size of rivers influence the proportion of these different wetland types in coastal zones.

Tidal influences are a major factor affecting plant and animal communities and thus management practices within this wetland type. The position of the moon in relation to the earth and sun is the primary factor influencing the range and characteristics of tides. Tidal range at each location is variable each day, month, and year. Highest tides occur during full and new moons. Different regions of the United States and Canada have great differences in the daily tidal range that influences vegetation communities and use of these communities by wildlife (Table 8). Tidal ranges in the Gulf Coast are <1 m, whereas in some areas of New England, Maritime Canada, and the Pacific Northwest the range may be >5 m.

The rate of water exchange is determined largely by the size and the length of channels. Tides influence wetlands connected by wide canals more than wetlands connected by small canals. Straight canals allow tidal waters to move farther inland at a faster rate and in greater volumes than through natural, meandering stream channels. The same canals also accelerate the drainage of freshwater from interior wetlands during low tides. Understanding the importance of natural hydrology and the impacts of modified hydrology is essential before management strategies are developed in coastal areas.

Understanding the effects of salinities on plant germination, growth, and production is critical to successful management of coastal wetlands. Plants characteristic of

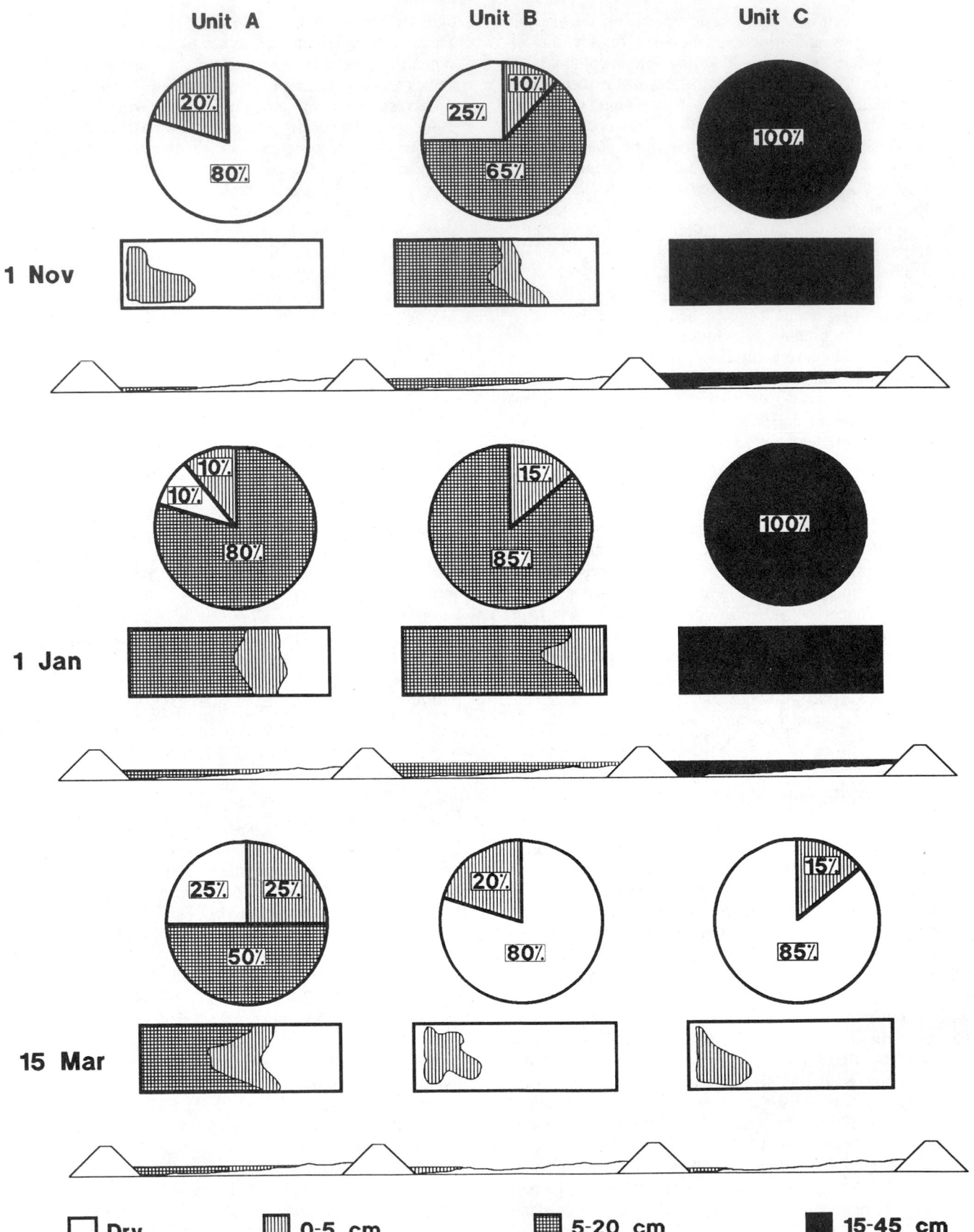

Fig. 11. Planned flooding strategies for three greentree reservoir impoundments during one winter season. The initiation of flooding, the depth of flooding, and the duration of flooding are different for each impoundment. Although two of the three impoundments are intentionally not flooded to capacity, natural events might flood a unit to capacity. Flooding strategies should be varied among years to enhance productivity.

coastal areas occur along a salinity continuum ranging from fresh to saline. The ability to identify wetland plants and an understanding of their tolerance to salinity provide a method of evaluating wetland conditions, drawdown dates, and degree of seasonal flooding.

A rule of thumb is to protect sites with unmodified hydrology. In contrast, sites that have been impacted (e.g., tide gates, ditching for mosquito abatement, gas and oil developments) may require intensive management to restore their natural functions and to provide conditions suitable for wildlife. There is a continuing controversy over the conversion of salt marshes to freshwater wetlands in the Southeast. Both systems have benefits for certain species, but the ecological implications associated with the conversion of salt marsh to freshwater habitats is not understood at a landscape scale. Likewise, permitting a freshwater wetland to convert to salt marsh through lack of management to control saltwater intrusion is a questionable land use.

As in any other wetland type, the key to effective management in coastal areas is the control of water levels, but management efforts also should focus on stabilizing water salinities and minimizing water turbidity (Chabreck 1988). Control of water in tidal areas is more complicated and often more costly than in inland habitats because of tidal fluctuations and because saltwater corrodes most metals commonly used in water distribution and control systems. Unless metals are properly treated, longevity of structures is short or they become inoperative soon after installation.

Weirs commonly are used in Gulf Coast tidal streams to control excessive drainage of wetlands and ponds at low tide (Chabreck 1988). Because weirs reduce the rate of tidal flow, water is impounded behind the structure and cannot recede below the crest. Thus, complete drainage of the basin during low tide is prevented. Turbidity and salinities are reduced in the area behind the weir, and aquatic plant production may increase 400%, which improves the habitat for waterbirds, fur-bearing animals, and alligators. In the Southeast, use of the unique rice trunk water control structure provides an economical and long-lasting approach to the control of tidal waters. These wooden structures can be adjusted to allow water flow in or out and can maintain water depths at a predetermined level.

Burning is used commonly in coastal wetland management. Burning stimulates or controls vegetation for target species such as geese, but concurrently reduces use by other species such as nutria or muskrats. Because spring and summer burns destroy nests and young of many species, controlled burning normally is limited to autumn and winter (Lynch 1941). Along the Gulf Coast, burning brackish wetlands in autumn favors Olney bulrush over marshhay cordgrass. Fresh and moderately saline wetlands should be burned only when woody plant encroachment becomes a problem.

Fresh and brackish water wetlands are common developments in tidal areas. The levees that form the impoundment change the natural hydrology and restrict the dynamic movement of water into and from the site. Changes in hydrology often result in monotypic plant communities that may have limited value for many wildlife species. In the Southeast, soil manipulations by discing or plowing

Table 8. Examples of selected wildlife use of tidal wetlands.

Species	Life-history stage	Wetland type
Loggerhead musk turtle	Egg laying	Barrier beach
Western grebe	Wintering	Intertidal flats
Great egret	All phases	Intertidal flats
		Brackish marsh
		Freshwater marsh
Clapper rail	All phases	Salt marsh
		Brackish marsh
		Freshwater marsh
Snow goose	Wintering	Salt marsh
		Brackish marsh
		Freshwater marsh
American black duck	Wintering	Salt marsh
American wigeon	Wintering	Brackish marsh
Red-breasted merganser	Wintering	Intertidal flats
Black skimmer	All phases	Intertidal flats
Western sand-piper	Migration	Intertidal flats
Muskrat	All phases	Brackish marsh
		Freshwater marsh

with heavy equipment can disrupt monotypic or undesirable vegetation. Impoundments that are high enough can be drawn down and managed as seasonally flooded areas. Many impoundments were built specifically to produce waterfowl foods. Plants commonly associated with waterfowl use are more common in permanent freshwater wetlands, manipulated freshwater wetlands, and permanent brackish water than in natural tidal wetlands. Thus, techniques discussed previously regarding seasonally flooded wetlands may be applicable in tidal areas as well. The primary difference in management from many inland areas is that soil salinities are always of concern in coastal wetlands because plant communities and use by vertebrates are strongly influenced by salinity levels.

Most wetlands along the Atlantic Coast have been modified with grid ditching to control mosquitoes (Bourn and Cottam 1950). The largest unditched area lies within the boundaries of Bombay Hook National Wildlife Refuge (Meredith and Saveikis 1987). A more recent approach to control salt marsh mosquitoes is "open marsh water management" (Meredith et al. 1985). The technique works best on the highest wetlands—those flooded by spring and storm tides and dominated by seashore saltgrass and smooth cordgrass. The rationale is to reduce use of pesticide treatments by eliminating ovipositioning sites for mosquitoes while providing habitats for mosquitofishes, which eat mosquito larvae. The disruption of habitats by earth-moving equipment usually disrupts wildlife use of the manipulated sites for several years. Although there is some controversy concerning bird response to this technique, most studies suggest that bird populations remain unchanged within a few years of treatment. Bird numbers appear to be more closely related to the amount of surface area of open water than to the presence of ponds (Erwin et al. 1991).

MONITORING

Few attempts have been made to assess the impacts of wetland manipulations on soil nutrient condition; vegetation response; production of seeds, tubers, or browse; or wildlife response. The most common monitoring has been to census wildlife present on wetlands. Waterbirds and furbearers are the most common groups that have been monitored, and undoubtedly the best information has been on waterfowl use.

Wildlife

Although bird counts are most common, much of the information has been summarized for reports on peak populations or total use days. Often, important information relating to the response of birds to individual wetlands is combined with other information for the entire area, and determining any relevant management conclusions is impossible. We recommend that standardized census techniques be developed for each area. Records should be kept for each wetland, and specific information on manipulations should be included. Factors such as water depths, duration and dates of flooding, drawdown dates and rates, vegetation conditions or seed production, season, temperature, and weather are pertinent to understanding wildlife response to manipulations. Frequency and timing of counts can be critical to obtain usable assessments of manipulations. For example, sites with early and late nesting species will require more than one census, and diurnal activity patterns may be important in the timing of counts.

Vegetation

Characteristics of plant communities make them difficult to monitor. A few species typically are present in great abundance, whereas many species will be represented by only a few individuals. Different species often occur in a patchy distribution, which makes selection of sample sites and the number of samples collected of great importance. Further, some plants of great interest in moist-soil impoundments are present only early in the growing season, whereas others are present only late in the season (Fredrickson and Reid 1988a). For example, chufa flatsedge is an early-season plant that produces abundant tubers under some conditions (Kelley and Fredrickson 1991). If sampling is done in late August or September, most chufa flatsedge plants have senesced 4–8 weeks earlier and their decomposing stems may not be readily apparent. In contrast, sprangletop is a late-season plant. Sampling scheduled for mid-June would be completed before this plant germinates. Finally, the diverse growth forms of different plants require different sampling strategies. Some important browse species such as blunt spikerush may be abundant beneath the canopy of barnyard grasses or beggarticks. These smaller plants with a low growth form might be completely missed in some sampling schemes.

Seed production is of great interest to managers because it is a means of assessing the rate of succession, the response to manipulations, and the foods available for wildlife. Regression techniques have been developed recently to make satisfactory estimates with minimal time invest-

ment (Laubhan and Fredrickson 1992). Software that will facilitate use of the technique is available from the U.S. Fish and Wildlife Service, Office of Information Transfer, Ft. Collins, CO 80525 (Laubhan 1992).

Invertebrates

Quantitative sampling may be impractical for monitoring wetland invertebrates because invertebrate populations change rapidly and have a patchy distribution (Fredrickson and Reid 1988c, and see Chapter 14). In addition, each wetland habitat and different groups of invertebrates require different sampling techniques. The best management information often can be gained from qualitative samples (e.g., presence/absence, percent occurrence) collected with simplified sampling techniques such as sweep nets, emergence traps, or underwater traps (Merritt and Cummins 1984).

Abiotic Factors

HYDROLOGY

Information on date and rate of drawdown is critical to understanding plant response to management practices. Likewise, information on the time, rate, depth, and duration of flooding is necessary to gain a better understanding of wildlife responses to water level manipulations.

SOIL FERTILITY

Some moist-soil management practices result in denitrification. For example, organic matter and nitrogen continually decrease with each soil disturbance. Monitoring soil nutrients provides opportunities to assess plant responses to fertility and to document changes in fertility in relation to intensive management practices. This can be accomplished by randomly collecting several soil samples annually and submitting them for analysis to a local Cooperative Extension Service or Soil Conservation Service office.

SALINITIES

In arid and coastal areas, soil salinities largely determine the composition of vegetation and provide warnings when certain management practices should be changed or when others should be implemented. Practices that gradually increase salinities might have benefits for certain foraging species, but if such practices continue for an extended period, the site may no longer produce the foods that attract target species. Monitoring soil salinities allows determination of whether saline water can be used for autumn flooding, if units need to be flushed with fresh water, or if some type of soil treatment is necessary. Salinity levels in water also are of great importance. Consistent use of waters with high salinities can reduce productivity of a wetland and reduce wildlife use.

SUMMARY

Effective approaches to wetland management are similar regardless of the wetland type or geographical location. Examination of historical hydrologic regimes and implementation of developments and manipulations to emulate the natural hydrology for a location should be foremost considerations in planning. The ability to identify plants at all stages of their life history and an understanding of their ecology are essential for successful management. Furthermore, a basic understanding of invertebrate ecology is important. Identification of life requisites for target organisms is essential to provide the needed resources in a timely manner. The availability of cover and foods must match the seasonal, social, and biological needs of many species.

Wetland complexes provide the greatest opportunity for success. Pristine systems should be protected, whereas those that have been modified by human activities should be managed to enhance their functional value for wildlife. Because modification of the hydrologic cycle will change the characteristics of a wetland, great care should be taken in the design of wetlands and in water level manipulations on any site. Some general suggestions for successful management include the following:

(1) Wetlands with unmodified hydrology should be protected.

(2) No single wetland or wetland type will provide all the resources needed by a single vertebrate during all of its life-history stages or for all vertebrates adapted to wetlands. Thus, wetland complexes are essential for successful management, and a good mix of different wetland types within a 16-km radius provides optimum conditions for most wetland birds.

(3) Wetlands are highly dynamic within and among years. Constantly fluctuating water levels are necessary to maintain desirable plant communities and associated wildlife. The goal of successful management must include a strategy to change the timing, depth, and duration of water level manipulations within and among years.

(4) The presence of wetland plants and their distributions, growth, and seed production provide important insights into the hydrology of a site and provide cues for ecological and economical approaches to improve management.

(5) Most wetland vertebrates are adapted to use shallow water (<25 cm deep).

(6) Wetland birds are adept at locating food resources quickly and at great distances between wetlands.

(7) Low levees along contours are the most cost effective.

(8) Outlet structures should be of the stoplog type.

(9) The productivity of wetlands varies among years. Therefore, consistent, maximum production from any wetland among years should not be expected.

(10) Monitoring is essential for verifying the effects of manipulations and scheduling treatments to maintain high productivity.

(11) Appropriate agencies should be contacted and appropriate permits must be obtained before certain activities are initiated in wetlands.

LITERATURE CITED

ALLANSON, B. R. 1973. The fine structure of the periphyton of *Chara* sp. and *Potamogeton natans* from Wytham Pond, Oxford, and its significance to the macrophyte-periphyton metabolic model of R. G. Wetzel and H. L. Allen. Freshwater Biol. 3:535–542.

ALLEN, J. A., AND H. E. KENNEDY, JR. 1989. Bottomland hardwood reforestation in the lower Mississippi Valley. U.S. Fish Wildl. Serv. Natl. Wetlands Res. Cent., Slidell, La, and U.S. For. Serv. Southern For. Exp. Stn., Stoneville, Miss. 28pp.

BALDASSARRE, G. A., A. R. BRAZDA, AND E. R. WOODYARD. 1989. The east coast of Mexico. Pages 407–425 in L. M. Smith, R. L. Pederson, and R. M. Kaminski, eds. Habitat management for migrating and wintering waterfowl in North America. Texas Tech Univ. Press, Lubbock.

BALL, J. P., AND T. D. NUDDS. 1989. Mallard habitat selection: an experiment and implications for management. Pages 659–671 in R. R. Sharitz and J. W. Gibbons, eds. Freshwater wetlands and wildlife. DOE Symp. Ser. 61.

BARTON, K. 1986. Federal wetland protection programs. Pages 373–411 in R. L. DiSilvestro, ed. Audubon Wildlife Report 1986. Natl. Audubon Soc., New York, N.Y.

BATEMA, D. L. 1987. Nutrient dynamics in a bottomland hardwood ecosystem. Ph.D. Thesis, Univ. Missouri, Columbia. 191pp.

———, G. S. HENDERSON, AND L. H. FREDRICKSON. 1985. Wetland invertebrate distribution in bottomland hardwoods as influenced by forest type and flooding regime. Proc. Central Hardwoods Forest Conf. 5:196–202.

BEDINGER, M. S. 1979. Relation between forest species and flooding. Pages 427–435 in P. C. Greeson, J. R. Clark, and J. E. Clark, eds. Wetland functions and values: the state of our understanding. Am. Water Resour. Assoc. Tech. Publ. 79-2.

BELT, C. B., JR. 1975. The 1973 flood and man's construction of the Mississippi. River Sci. 189:681–684.

BOURN, W. S., AND C. COTTAM. 1950. Some biological effects of ditching tidewater marshes. U.S. Fish Wildl. Serv. Res. Rep. 19. 30pp.

BOYD, C. E. 1970. Losses of mineral nutrients during decomposition of *Typha latifolia*. Arch. Hydrobiol. 66:511–517.

CHABRECK, R. H. 1988. Coastal marshes: ecology and wildlife management. Univ. Minnesota Press, Minneapolis. 138pp.

COWARDIN, L. M., V. CARTER, F. C. GOLET, AND E. T. LAROE. 1979. Classification of wetlands and deepwater habitats of the United States. U.S. Fish Wildl. Serv. Publ. FWS/OBS-79/31. 103pp.

DAHL, T. E. 1990. Wetlands losses in the United States 1780's to 1980's. U.S. Fish Wildl. Serv., Washington, D.C. 13pp.

———, AND C. E. JOHNSON. 1991. Status and trends of wetlands in the conterminous United States, mid-1970's to mid-1980's. U.S. Fish Wildl. Serv., Washington, D.C. 28pp.

DELNICKI, D., AND K. J. REINECKE. 1986. Mid-winter food use and body weights of mallards and wood ducks in Mississippi. J. Wildl. Manage. 50:43–51.

DROBNEY, R. D. 1980. Reproductive bioenergetics of wood ducks. Auk 97:480–490.

———, AND L. H. FREDRICKSON. 1985. Protein acquisition: a possible proximate factor limiting clutch size in wood ducks. Wildfowl 36: 122–128.

ELDRIDGE, J. 1990. Aquatic invertebrates important for waterfowl production. U.S. Fish Wildl. Serv. Waterfowl Manage. Handb. Leafl. 13.3.3. 7pp.

ERRINGTON, P. L. 1963. The pricelessness of untampered nature. J. Wildl. Manage. 27:313–320.

ERWIN, R. M., D. K. DAWSON, D. B. STOTTS, L. S. MCALLISTER, AND P. H. GEISSLER. 1991. Open marsh water management in the mid-Atlantic region: aerial surveys of waterbird use. Wetlands 11:209–227.

FINGER, T. R., AND E. M. STEWART. 1987. Response of fishes to flooding regimes in lowland hardwood wetlands. Pages 86–92 in W. J. Matthews and D. C. Hines, eds. Evolution and community ecology of North American stream fishes. Univ. Oklahoma Press, Norman.

FRAYER, W. E., T. J. MONAHAN, D. C. BOWDEN, AND F. A. GRAYBILL. 1983. Status and trends of wetlands and deepwater habitats in the conterminous United States, 1950s to 1970s. Dep. For. Wood Sci., Colorado State Univ., Ft. Collins. 32pp.

———, D. D. PETERS, AND H. R. PYWELL. 1989. Wetlands of the California Central Valley: status and trends 1939 to mid-1980's. U.S. Fish Wildl. Serv., Portland, Oreg. 28pp.

FREDRICKSON, L. H. 1979. Lowland hardwood wetlands: current status and habitat values for wildlife. Pages 296–306 in P. E. Greeson, J. R. Clark, and J. E. Clark, eds. Wetland functions and values: the state of our understanding. Am Water Resour. Tech. Publ. 79-2.

———. 1991. Strategies for water level manipulations in moist-soil systems. U.S. Fish Wildl. Serv. Waterfowl Manage. Handb. Leafl. 13.4.6. 8pp.

———, AND D. L. BATEMA. 1992. Greentree reservoir management handbook. Gaylord Mem. Lab., Wetland Manage. Ser. 1. Gaylord Lab., Puxico, Mo. 88pp.

———, AND M. E. HEITMEYER. 1989. Waterfowl use of forested wetlands of the southern United States: an overview. Pages 307–323 *in* M. W. Weller, ed. Waterfowl in winter. Univ. Minnesota Press, Minneapolis.

———, AND F. A. REID. 1986. Wetland and riparian habitats: a nongame management overview. Pages 59–96 *in* J. B. Hale, L. B. Best, and R. L. Clawson, eds. Management of nongame wildlife in the Midwest: a developing art. North-Cent. Sect., The Wildl. Soc., Grand Rapids, Mich.

———, AND ———. 1988*a*. Considerations of community characteristics for sampling vegetation. U.S. Fish Wildl. Serv. Waterfowl Manage. Handb. Leafl. 13.4.1. unnumb.

———, AND ———. 1988*b*. Control of willow and cottonwood seedlings in herbaceous wetlands. U.S. Fish Wildl. Serv. Waterfowl Manage. Handb. Leafl. 13.4.10. unnumb.

———, AND ———. 1988*c*. Initial considerations for sampling wetland invertebrates. U.S. Fish Wildl. Serv. Waterfowl Manage. Handb. Leafl. 13.3.2. unnumb.

———, AND ———. 1988*d*. Invertebrate response to wetland management. U.S. Fish Wildl. Serv. Waterfowl Manage. Handb. Leafl. 13.3.1. unnumb.

———, AND ———. 1988*e*. Waterfowl use of wetland complexes. U.S. Fish Wildl. Serv. Waterfowl Manage. Handb. Leafl. 13.2.1. unnumb.

———, AND ———. 1990. Impacts of hydrologic alteration on management of freshwater wetlands. Pages 72–90 *in* J. M. Sweeney, ed. Management of dynamic ecosystems. North-Cent. Sect., The Wildl. Soc., Springfield, Ill.

———, AND T. S. TAYLOR. 1982. Management of seasonally flooded impoundments for wildlife. U.S. Fish Wildl. Serv. Resour. Publ. 148. 29pp.

GOLDSTEIN, J. H. 1988. The impact of federal programs and subsidies on wetlands. Trans. North Am. Wildl. Nat. Resour. Conf. 53:436–443.

GORDON, D. H., B. T. GRAY, R. D. PERRY, M. B. PREVOST, T. H. STRANGE, AND R. K. WILLIAMS. 1989. South Atlantic coastal wetlands. Pages 57–92 *in* L. M. Smith, R. L. Pederson, and R. M. Kaminski, eds. Habitat management for migrating and wintering waterfowl in North America. Texas Tech Univ. Press, Lubbock.

GRUE, C. E., ET AL. 1986. Potential impacts of agricultural chemicals on waterfowl and other wildlife inhabiting prairie wetlands: an evaluation of research needs and approaches. Trans. North Am. Wildl. Nat. Resour. Conf. 51:357–383.

———, M. W. TOME, T. A. MESSMER, D. B. HENRY, G. A SWANSON, AND L. R. DeWEESE. 1989. Agricultural chemicals and prairie pothole wetlands: meeting the needs of the resource and the farmer—U.S. perspective. Trans. North Am. Wildl. Nat. Resour. Conf. 54:43–58.

HAMMER, D. A., AND R. K. BASTIAN. 1989. Wetlands ecosystems: natural water purifiers? Pages 5–19 *in* D. A. Hammer, ed. Constructed wetlands for wastewater treatment: municipal, industrial, and agricultural. Lewis Publ., Chelsea, Mich.

HARPER, J. L. 1977. Population biology of plants. Academic Press, New York, N.Y. 892pp.

HAYNES, R. J., AND J. A. ALLEN. 1988. Reestablishment of bottomland hardwood forests on disturbed sites: an annotated bibliography. U.S. Fish Wildl. Serv. Biol. Rep. 88(42). 104pp.

HEITMEYER, M. E. 1988. Protein costs of the prebasic molt of female mallards. Condor 90:263–266.

———, L. H. FREDRICKSON, AND G. F. KRAUSE. 1989. Water and habitat dynamics of the Mingo Swamp in southeastern Missouri. U.S. Fish Wildl. Serv., Fish Wildl. Res. 6. 26pp.

HIGGINS, K. F., L. M. KIRSCH, A. T. KLETT, AND H. W. MILLER. 1992. Waterfowl production on the Woodworth Station in southcentral North Dakota, 1965–1981. U.S. Fish Wildl. Serv. Resour. Publ. 180. 79pp.

HUBERT, W. A., AND J. N. KRULL. 1973. Seasonal fluctuations of aquatic macroinvertebrates in Oakwood Bottoms Greentree Reservoir. Am. Midl. Nat. 90:177–185.

KADLEC, J. A., AND L. M. SMITH. 1992. Habitat management for breeding areas. Pages 590–610 *in* B. D. J. Batt, et al., eds. Ecology and management of breeding waterfowl. Univ. Minnesota Press, Minneapolis.

KAMINSKI, R. M., H. R. MURKIN, AND C. E. SMITH. 1985. Control of cattail and bulrush by cutting and flooding. Pages 253–262 *in* H. H. Prince and F. M. D'Itri, eds. Coastal wetlands. Lewis Publ., Chelsea, Mich.

———, AND H. H. PRINCE. 1981. Dabbling duck activity and foraging responses to aquatic macroinvertebrates. Auk 98:115–126.

KELLEY, J. R., JR., AND L. H. FREDRICKSON. 1991. Chufa biology and management. U.S. Fish Wildl. Serv. Waterfowl Manage. Handb. Leafl. 13.4.18. 6pp.

KIRBY, R. E., S. J. LEWIS, AND T. N. SEXSON. 1988. Fire in North American wetland ecosystems and fire wildlife relations. U.S. Fish Wildl. Serv. Biol. Rep. 88. 146pp.

KORTE, P. A., AND L. H. FREDRICKSON. 1977. Loss of Missouri's lowland hardwood ecosystem. Trans. North Am. Wildl. Nat. Resour. Conf. 42:31–41.

KRAMER, G. W., AND R. MIGOYA. 1989. The Pacific Coast of Mexico. Pages 507–528 *in* L. M. Smith, R. L. Pederson, and R. M. Kaminski, eds. Habitat management for migrating and wintering waterfowl in North America. Texas Tech Univ. Press, Lubbock.

LAUBHAN, M. K. 1992. Estimating seed production of common moistsoil plants. U.S. Fish Wildl. Serv. Waterfowl Manage. Handb. Leafl. 13.4.5. 8pp.

———, AND L. H. FREDRICKSON. 1992. Estimating seed production of common plants in seasonally flooded wetlands. J. Wildl. Manage. 56:329–337.

LECK, M. A. 1989. Wetland seed banks. Pages 283–305 *in* M. A. Leck, V. T. Parker, and R. L. Simpson, eds. Ecology of soil seed banks. Academic Press, San Diego, Calif.

LIVINGSTON, R. J., AND O. L. LOUCKS. 1979. Productivity, trophic interactions, and food web relationships in wetlands and associated systems. Pages 101–119 *in* P. E. Greeson, J. R. Clark, and J. E. Clark, eds. Wetland functions and values: the state of our understanding. Am. Water Resour. Tech. Publ. 79-2.

LONGCORE, J. R., R. K. ROSS, AND K. L. FISHER. 1987. Wildlife resources at risk through acidification of wetlands. Trans. North Am. Wildl. Nat. Resour. Conf. 52:608–618.

LOW, J. B., AND F. C. BELLROSE. 1944. The seed and vegetative yield of waterfowl food plants in the Illinois River valley. J. Wildl. Manage. 8:7–22.

LYNCH, J. J. 1941. The place of burning in management of the Gulf Coast wildlife refuges. J. Wildl. Manage. 5:454–457.

MARBLE, A. D. 1992. A guide to wetland functional design. Lewis Publ., Chelsea, Mich. 222pp.

McDONALD, P. O., W. E. FRAYER, AND J. K. CLAUSER. 1979. Documentation, chronology, and future projections of bottomland hardwood habitat losses in the lower Mississippi Alluvial Plain. Vol. I. U.S. Fish Wildl. Serv., Washington, D.C. 133pp.

MEREDITH, W. H., AND D. E. SAVEIKIS. 1987. Effects of open marsh water management (OMWM) on bird populations of a Delaware tidal marsh, and OMWM's use in waterbird habitat restoration and enhancement. Pages 298–321 *in* W. R. Whitman and W. H. Meredith, eds. Waterfowl and wetlands symposium: proceedings of a symposium on waterfowl and wetland management in the Coastal Zone of the Atlantic Flyway. Del. Coastal Manage. Program, Delaware Dep. Nat. Resour. Environ. Control, Dover.

———, ———, AND C. J. STACHECKI. 1985. Guidelines for ''Open marsh water management'' in Delaware's salt marshes—objectives, system design, and installation procedures. Wetlands 5:119–137.

MERRITT, R. W., AND K. W. CUMMINS, EDITORS. 1984. An introduction to the aquatic insects of North America. Second ed. Kendall/Hunt Publ. Co., Dubuque, Ia. 722pp.

MIDDLETON, B. A., A. G. VAN DER VALK, R. L. WILLIAMS, AND D. H. MASON. 1992. Litter decomposition in an Indian monsoonal wetland overgrown with *Paspalum distichum*. Wetlands 12:37–44.

MITSCH, W. J., AND J. G. GOSSELINK. 1986. Wetlands. Van Nostrand Reinhold, New York, N.Y. 537pp.

MOTT, D. F., R. R. WEST, J. W. De GRAZIO, AND J. L. GUARINO. 1972. Foods of the red-winged blackbird in Brown County, South Dakota. J. Wildl. Manage. 36:983–987.

MURKIN, E. J., H. R. MURKIN, AND R. D. TITMAN. 1992. Nektonic invertebrate abundance and distribution at the emergent vegetation-open water interface in the Delta Marsh, Manitoba, Canada. Wetlands 12:45–52.

MURKIN, H. R. 1989. The basis for food chains in prairie wetlands. Pages 316–338 *in* A. van der Valk, ed. Northern prairie wetlands. Iowa State Univ. Press, Ames. 400pp.

NELSON, J. W., AND R. A. WISHART. 1988. Management of wetland complexes for waterfowl production: planning for the Prairie Habitat Joint Venture. Trans. North Am. Wildl. Nat. Resour. Conf. 53:444–453.

ODUM, E. P. 1979. The value of wetlands: a hierarchical approach. Pages 16–25 *in* P. E. Greeson, J. R. Clark, and J. E. Clark, eds.

Wetland functions and values: the state of our understanding. Am. Water Res. Assoc., Minneapolis, Minn.

PEDERSON, R. L., D. G. JORDE, AND S. G. SIMPSON. 1989. Northern Great Plains. Pages 281–310 *in* L. M. Smith, R. L. Pederson, and R. M. Kaminski, eds. Habitat management for migrating and wintering waterfowl in North America. Texas Tech Univ. Press, Lubbock.

PENNAK, R. W. 1978. Freshwater invertebrates of the United States. Second ed. John Wiley & Sons, New York, N.Y. 803pp.

PETERSON, D. L., AND K. W. CUMMINS. 1974. Leaf processing in a woodland stream. Freshwater Biol. 4:343–368.

PINDER, L. C. V. 1986. Biology of freshwater Chironomidae. Annu. Rev. Entomol. 31:1–23.

REID, F. A., J. R. KELLEY, JR., T. S. TAYLOR, AND L. H. FREDRICKSON. 1989. Upper Mississippi Valley wetlands—refuges and moist-soil impoundments. Pages 181–202 *in* L. M. Smith, R. L. Pederson, and R. M. Kaminski, eds. Habitat management for migrating and wintering waterfowl in North America. Texas Tech Univ. Press, Lubbock.

REINECKE, K. J., R. C. BARKLEY, AND C. K. BAXTER. 1988. Potential effects of changing water conditions on mallards wintering in the Mississippi Alluvial Valley. Pages 325–337 *in* M. W. Weller, ed. Waterfowl in winter. Univ. Minnesota Press, Minneapolis.

——, R. M. KAMINSKI, D. J. MOORHEAD, J. D. HODGES, AND J. R. NASSAR. 1989. Mississippi Alluvial Valley. Pages 203–247 *in* L. M. Smith, R. L. Pederson, and R. M. Kaminski, eds. Habitat management for migrating and wintering waterfowl in North America. Texas Tech Univ. Press, Lubbock.

ROBEL, R. J. 1961*a*. The effects of carp populations on the production of waterfowl food plants on a western waterfowl marsh. Trans. North Am. Wildl. Nat. Resour. Conf. 26:147–159.

——. 1961*b*. Water depth and turbidity in relation to growth of sago pondweed. J. Wildl. Manage. 25:436–438.

RUNDLE, W. D., AND L. H. FREDRICKSON. 1981. Managing seasonally flooded impoundments for migrant rails and shorebirds. Wildl. Soc. Bull. 9:80–87.

SEVERSON, D. J. 1987. Macroinvertebrate populations in seasonally flooded marshes in the northern San Joaquin Valley of California. M.S. Thesis, Humbolt State Univ., Arcata, Calif. 113pp.

SIMPSON, R. L., M. A. LECK, AND V. T. PARKER. 1989. Seed banks: general concepts and methodological issues. Pages 3–21 *in* M. A. Leck, V. T. Parker, and R. L. Simpson, eds. Ecology of soil seed banks. Academic Press, San Diego, Calif.

SOULLIERE, G. J. 1990. Review of wood duck nest-cavity characteristics. Pages 153–162 *in* L. H. Fredrickson, G. V. Burger, S. P. Havera, D. A. Graber, R. E. Kirby, and T. S. Taylor, eds. Proc. 1988 North American wood duck symposium, St. Louis, Mo.

STEVENS, S. E., JR., K. DIONIS, AND L. R. STARK. 1989. Manganese and iron encrustation on green algae living in acid mine drainage. Pages 765–773 *in* D. A. Hammer, ed. Constructed wetlands for wastewater treatment: municipal, industrial, and agricultural. Lewis Publ., Chelsea, Mich.

STEWART, R. E., AND H. A. KANTRUD. 1971. Classification of natural

ponds and lakes in the glaciated prairie region. U.S. Fish Wildl. Serv. Resour. Publ. 92. 57pp.

SUTHERS, I. M., AND J. H. GEE. 1986. Role of hypoxia in limiting diel spring and summer distribution of juvenile yellow perch (*Perca flavescens*) in a prairie marsh. Can. J. Fish. Aquat. Sci. 43:1562–1570.

SWANSON, G. A., AND M. I. MEYER. 1973. The role of invertebrates in the feeding ecology of Anatinae during the breeding season. Pages 143–185 *in* Waterfowl habitat management symposium, Moncton, N.B.

TINER, R. W., JR. 1984. Wetlands of the United States: current status and recent trends. U.S. Fish Wildl. Serv., Washington, D.C. 59pp.

VAN DER VALK, A. G. 1981. Succession in wetlands: a gleasonian approach. Ecology 62:688–696.

——, AND C. B. DAVIS. 1978. The role of seed banks in the vegetation dynamics of prairie glacial marshes. Ecology 59:322–335.

WEBSTER, J. R., AND E. F. BENFIELD. 1986. Vascular plant breakdown in freshwater ecosystems. Annu. Rev. Ecol. Syst. 17:567–594.

WELLER, M. W. 1975. Studies of cattail in relation to management for marsh wildlife. Iowa State J. Sci. 49:383–412.

——. 1988. Freshwater marshes: ecology and wildlife management. Second ed. Univ. Minnesota Press, Minneapolis. 150pp.

——, AND L. H. FREDRICKSON. 1973. Avian ecology of a managed glacial marsh. Living Bird 12:269–291.

——, G. W. KAUFMAN, AND P. A. VOHS, JR. 1991. Evaluation of wetland development and waterbird response at Elk Creek Wildlife Management Area, Lake Mills, Iowa, 1961-1990. Wetlands 11: 245–262.

——, AND C. E. SPATCHER. 1965. Role of habitat in the distribution and abundance of marsh birds. Dep. Zool. Entomol. Spec. Rep. 43. Agric. Home Econ. Exp. Stn., Iowa State Univ., Ames.

WHARTON, C. H., W. M. KITCHENS, E. C. PENDLETON, AND T. W. SIPE. 1982. The ecology of bottomland hardwood swamps of the Southeast: a community profile. U.S. Fish Wildl. Serv. Rep. FWS/OBS-81/37. 133pp.

WHITE, D. C. 1985. Lowland hardwood wetland invertebrate community and production in Missouri. Arch. Hydrobiol. 103:509–533.

WHITE, D. H., AND D. JAMES. 1978. Differential use of freshwater environments by wintering waterfowl of coastal Texas. Wilson Bull. 90:99–111.

WIGGINS, G. B., R. J. MACKAY, AND I. M. SMITH. 1982. Evolutionary and ecological strategies of animals in annual temporary pools. Arch. Hydrobiol. (suppl.) 58:97–206.

WIGLEY, T. B., JR., AND T. H. FILER, JR. 1989. Characteristics of green-tree reservoirs: a survey of managers. Wildl. Soc. Bull. 17:136–142.

WYLIE, G. D. 1985. Limnology of lowland hardwood wetlands in southeast Missouri. Ph.D. Thesis, Univ. Missouri, Columbia. 204pp.

YATES, R. F. K., AND F. P. DAY, JR. 1983. Decay rates and nutrient dynamics in confined and unconfined leaf litter in the Great Dismal Swamp. Am. Midl. Nat. 110:37–45.

26

MANAGING FARMLANDS FOR WILDLIFE

Richard E. Warner and Stephen J. Brady

INTRODUCTION

Wildlife management on agricultural lands is at a critical juncture. Many wildlife species that once thrived in farmland settings have declined in association with the pervasive effects of intensified agricultural land use, especially since World War II. However, agriculture is at a turning point. Environmental issues as they pertain to farming have come to the forefront during the 1980s and 1990s, with a strong mandate by society to address these issues. There are movements within agriculture to diversify farm commodities, reduce energy-intensive farming practices, protect natural resources, and in general ensure that agricultural systems are sustainable (National Research Council 1989, Pimentel et al. 1989). These movements are likely to create significant opportunities for wildlife agencies to promote farm practices relatively beneficial to wildlife that are compatible with other goals in agriculture. However, the implication is that successful wildlife management programs often will not stand alone; they must be integrated with other resource-conserving initiatives targeted to farming environments.

Successful farmland habitat programs often are associated with how well managers are able to accommodate the ecological, political, economic, and social contexts in which habitat initiatives must occur (McConnell 1981). Wildlife agencies frequently develop new programs to establish habitat in intensively farmed areas—such attempts have increased in recent years (Vander Zouwen 1990)—and anticipation runs high as plans are made to increase the numbers of target species over large regions. Too often, however, these initiatives dwindle after a few years,

and benefits to wildlife are typically short-lived and highly localized. Agricultural policies and programs, changing farming practices, competition for land use, and the limited resources of wildlife agencies all contribute to the poor track record. Biologists, therefore, must rely on more than ecological theory and habitat development skills; they must be familiar with land use practices, farm policies and programs, the politics of conservation, and rural sociology.

Differing approaches could be taken here for guiding wildlife managers toward habitat development efforts in agricultural environments. One hands-on approach would be to outline specific methods for enhancing vegetation and other habitat components at the patch scale. However, more information appears to be available to wildlife managers about habitat practices at the field/patch scale than how to recognize and accommodate the farm policies, programs, and practices as they affect wildlife habitat at differing spatial scales—and the implications for integrating wildlife conservation practices within agriculture. Underlying factors pertaining to agriculture will amplify, dampen, or negate management efforts by wildlife agencies, which are typically small-scale. Thus, although we do not ignore methods for hands-on habitat development, with agriculture at a crossroads, we consider the integration of wildlife practices with farm policies and practices as the primary focus of this chapter.

Our perspective on habitat management is restricted to arable land east of about 98° longitude, or approximately a line from San Antonio, Texas, to Jamestown, North Dakota. Within that boundary, cropping (tillage of soil) occurs on most of the agricultural land. Although such intensive farming pervades the north-central United States, it is also common in the East. (With irrigation, however, intensively cropped areas are also found in most western states.) The most important commodities associated with intensive cropping are corn, soybeans, wheat, peanuts, tobacco, sorghum, cotton, and forage crops, but additional small grains are also important in localized areas. The agricultural ecosystems of these regions are dynamic, and modern farming presents a complex decision-making environment for farmers. Habitat considerations are not high on their list of priorities as they adopt new technologies associated with intensive agricultural practices. Moreover, they must cope with volatile commodity markets, complicated, ever-changing farm programs, escalating public concern over the effects of pesticide pollution, soil erosion, water quality, and increasing intervention by government, including regulations and incentives.

THE CHALLENGE OF WILDLIFE MANAGEMENT IN AGRICULTURAL ENVIRONMENTS

In recent decades agricultural ecosystems have been tending toward monocultures (National Research Council 1982, Power and Follett 1987) with simplified and unstable pathways for nutrient cycling and the flow of energy (van Emden 1965, Oldfield and Alcorn 1987, Turner 1987, Woolhouse and Harmsen 1987). Broad-scale reviews of changing agricultural land use and responses by wildlife have appeared infrequently in the literature until recently. The interested reader is encouraged to review Baxter and Wolfe (1973), Burger (1978), Taylor et al. (1978), Samson (1980), Edwards et al. (1981), Warner et al. (1984), Warner and Etter (1985), Wooley et al. (1985), Potts (1986), Robbins et al. (1986), Berner (1984, 1988), Brady (1988), and Flather and Hoekstra (1989).

Many upland wildlife species have declined with the loss of grassland, as row crops have supplanted livestock, forage crops, and rotation farming in general (Farris et al. 1977, Etter et al. 1988). For example, during a recent 30-year period dramatic declines in the hunter harvest of ring-necked pheasants, cottontail rabbits, and northern bobwhites in Illinois were highly correlated with increasing amounts of row crops and hay and small grains (Brady 1988). Although our priority here is on species in need of habitat measures, some recent wildlife/agriculture associations have been positive. For example, white-tailed deer and turkeys have increased over much of the Midwest, partly in relation to the juxtaposition of forest land (primarily riparian) in and about agricultural lands. However, as is often true for species experiencing dramatic expansions of range or numerical abundance, or both, the effectiveness of wildlife agencies has been in the regulation of hunting and efforts to recolonize populations (turkeys), not habitat management per se.

Suitability of Agricultural Settings for Wildlife

One way of conceptualizing the impacts of farming on wildlife, and how habitat interventions can mitigate negative impacts, is the "source/sink" model for wildlife habitat. The source/sink paradigm holds that in some patches of habitat, wildlife typically will produce more than enough young to account for average annual mortal-

Fig. 1. Extensive grassed backslope terraces and conservation tillage provide the only elements of habitat in this intensively farmed southwestern Iowa landscape (U.S. Soil Conserv. Serv. photo).

ity—a source. In view of this paradigm, over the past century agricultural landscapes have tended toward fewer habitat patches attracting an abundance and diversity of breeding species, and a diminution in the portion of attractive patches that function as sources (Fig. 1). Other patches, because of factors such as predation, interspecific competition, and farming disturbances, do not allow wildlife to produce enough young to compensate for annual mortality—thus the term habitat sink. The reasons for these changes relate in general to farm policies, programs, and technologies and specifically to the structure of vegetation, the juxtaposition of cover, and the nature and timing of farm disturbances (Best 1986, Brady 1988). In accordance with this paradigm, habitat management on farm mosaics should be directed toward (1) increasing numbers of patches attractive to wildlife, and (2) increasing the portion of attractive patches that afford high rates of reproduction and survival, i.e., source habitats.

Perspectives of Habitat Quality

Farming operations often coincide with critical periods of wildlife reproduction, maturation, and dispersal, ultimately rendering potentially attractive habitats as sinks unless wildlife agencies provide guidance or incentives to encourage farmers to minimize field disturbances during critical periods. Negative disturbances include planting and tillage practices; timing of hay harvest; timing and intensity of livestock grazing; mowing of edge habitats such as road rights-of-way, fencerows, and waterways; and use of pesticides (Best 1986, Brady 1988, Warner and Etter 1989). On the other hand, disturbances are necessary to maintain vegetation in early to mid-successional stages; disturbances are necessary but timing is the key.

Although machinery-related disturbances have become increasingly intensive for more than a century, the use of chemicals on farmland has been limited primarily to the period following World War II. Synthetic fertilizers marketed after the war eliminated the need for crop rotations that include the planting of nitrogen-fixing legumes to maintain soil fertility. Elimination of rotation grasses and legumes also followed the phaseout of cattle operations after the 1950s (Etter et al. 1988) as the production of row crops, especially soybeans, expanded. Row-crop systems are also attractive to farmers because synthetic fertilizers, hybrids, and pesticides have buoyed production per unit of land. Use of insecticides and broad-leaf herbicides rapidly expanded during the 1950s and 1960s, followed by grass herbicides in the 1960s and 1970s (e.g., U.S. Department of Commerce, Bureau of Census 1985).

Soil erosion and sedimentation have further increased farm pesticide loadings in the environment. As pesticides and their metabolites have pervaded aquatic and terrestrial ecosystems, they have reduced target and nontarget plants and insects of value to wildlife. As a result, agricultural sources of pollution adversely affect the fish community in about 29% of the waters of the United States (Auclair 1976, Kroh and Beaver 1978, Judy et al. 1984, Warburton and Klimstra 1984). Sediments containing pollutants also have destroyed many bottomland lakes (Bellrose et al. 1983, Pimentel et al. 1987). Thus, future strategies for enhancing wildlife habitat on farmland should be linked with efforts to improve soil and water quality.

Modern corn- and soybean-head combines came into wide use in the 1960s and left little crop stubble in the field. Prior to 1950, about 20–25% of cropland in the Midwest had been tilled with a moldboard plow after harvest (Warner and Havera 1989). Autumn tillage of cropland increased sharply during the 1960s and 1970s. To reduce soil erosion, several tillage and planting practices that offered less disturbance than the moldboard plow came into use during the 1980s (Gebhardt et al. 1985, Magleby et al. 1985). These practices were accompanied by intensive chemical disturbances (Brady 1985, Castrale 1987) and resulted in high variation from field to field in crop residues available to wildlife on autumn and winter landscapes (Warner et al. 1989).

The spatial separation of where life-sustaining needs are provided for wildlife has been one ultimate effect of modern clean-farming technologies; the number of life-sustaining benefits provided in any given farmland cover type has diminished. The relatively narrow life-sustaining attributes of many patches on farm landscapes of today can lead to relatively extensive movement and the use of numerous cover types by small vertebrates to reproduce and survive. The result can be relatively large expenditures of energy and increased vulnerability to various forms of morbidity (Warner 1984, Basore et al. 1986, Krummel et al. 1987, Stamps et al. 1987). Crop stubble is a good example. Waste corn has been integral to wildlife food webs since settlement agriculture (Warner and Havera 1989), the stubble providing protection from weather and predators, cover for reproduction of small vertebrate animals in early spring, and a diversity of forage in the form of waste grains, other small seeds, and arthropods. However, row-crop stubble left by modern crop harvest machinery has little vertical profile and minimal understory vegetation for protection and diversity of foods. Although waste corn remains an important food source for wildlife, the diverse ecological benefits of stubble that existed over much of the past century have been lost.

Farm Programs

War, variable weather, changing land use practices (increasing productivity through technology), and unpredictable world grain markets have created "boom and bust" agriculture over the past century (Schlebecker 1975). The federal government has repeatedly intervened during times of depressed prices by diverting farmland from production. Most of the diverted land has been seeded with forage grasses and legumes. The Agricultural Conservation Program (ACP) of the 1930s and 1940s, the Soil Bank Program of the 1950s, and the Set-Aside Acres Program of the early 1960s were particularly effective in buffering the impact of increasingly intensive farming on wildlife, by creating grasslands where farm disturbances were curtailed during the breeding season (Joselyn and Warnock 1964, Edwards 1984, Berner 1988).

During the past 20 years, however, most programs diverting land from production (with the notable exception of the Conservation Reserve Program of the 1980s) have been less successful in providing good habitat conditions for wildlife. The year-to-year nature of recent programs has been such that most of the diverted cropland has been tilled in autumn, and fields often were diverted from production for only one growing season. In addition, farmers recently have been allowed (sometimes required) to mow diverted fields during late spring or early summer—the reproductive period for wildlife. By contrast, earlier programs removed land from production through multiple-year contracts, and little or no farm disturbances were allowed during the reproductive season for wildlife (Harmon and Nelson 1973, Berner 1988).

DESIGNING HABITAT PROGRAMS FOR FARMLAND

Wildlife managers should be cognizant of the following factors when devising habitat programs in agricultural settings: (1) the perceived needs of the landholder must be accommodated, (2) management agencies must deliver the technical and material assistance needed to make habitat programs work, (3) a means of evaluating and refining the program should be provided to ensure that management goals are met, (4) management intervention (habitat development and maintenance) must be viewed from a long-term perspective, and (5) the sustainability of agricultural systems (Edwards et al. 1990) must be ensured over the long term—a goal that is compatible with management of wildlife and other natural resources. Although the need for evaluation and refinement seems obvious, these two components often are ignored, and habitat programs fade away without a complete accounting of their costs, benefits, and failures. More effective habitat programs in agricultural settings probably will not be developed until wildlife managers have the benefit of a thorough documentation of past efforts. Thus, we also address evaluating and refining farmland habitat programs that are in place.

Characteristics of Successful Programs

Successful wildlife habitat programs in agricultural environments, i.e., those likely to produce measurable responses, usually include the following elements:

(1) Compatibility with the objectives of the landholder and with the primary use (agricultural production) of the land.
(2) An unshakable resolve by management agencies to succeed; this attitude must be supported by a careful planning phase that sets realistic goals, costs, and strategies for accomplishing tasks.
(3) Cooperation among the principal agencies and

groups—government and private—that influence the decision-making environments of landholders, especially farm operators.

(4) A program of information and education. Such a program should include promotion to capture the interest of landholders and positive reinforcement of desirable farming practices for program participants. In the long run, farming practices essential to the program also must be rewarding to the farmer. Ultimately, concerns over natural resources must be addressed and a land ethic must be built.

(5) A thorough understanding of the ecology and behavioral biology of target species and the impacts of management programs on plant and animal communities.

(6) A habitat development program that has been successful for targeted wildlife species, including the establishment of suitable plant forms and configurations of cover and their maintenance over appropriate physiographic regions.

(7) Scales of time and space large enough to allow for buffering of the changeable physical environment, including weather, chemical and biotic factors, farm policies and programs, and the land use practices of individual landholders.

(8) An ongoing evaluation and refinement process that permits the program to respond to changing circumstances.

The first step in planning a habitat program is to identify target species, evaluate appropriate habitat manipulations, and list important plants and animals that may be affected (positively or negatively) by management efforts. This planning process requires an understanding of plant and animal species endemic to the given physiographic region; information about their current status can be extremely important in developing a successful management plan.

The Community Approach

Early efforts to extend wildlife habitat on farmland focused primarily on game animals, an outgrowth of the initial use of Pittman-Robertson funds for game research and management. Most game species inhabiting farmland thrive in a diversity of cover types that provide edge and associated ecotones (Leopold 1933, Roseberry and Klimstra 1984); however, large, permanent tracts of habitat in predominately agricultural settings are now uncommon. Where large tracts are present, managers should beware of maximizing edge, a practice that can increase penetration of cover by predators and by birds that are parasitic nesters (Brittingham and Temple 1983, Noss 1983, Westemeier 1988).

As a starting point, wildlife biologists should consider the community approach to habitat development, a perspective that implies appraisal of the potential impacts of management on the number and relative abundance of wildlife species at regional levels (Noss 1983, Risser et al. 1984, Klopatek and Kitchings 1985). A community perspective does not preclude emphasizing one or several species in habitat initiatives at localized scales, but such a perspective ensures that the impact of management on the abundance and diversity of wildlife at larger spatial scales is at least considered. In this context, there can be

seemingly contradictory approaches to maximizing diversity of cover and edge for management of game species, while at the same time minimizing fragmentation of key interior habitats for rare and endangered species (Noss 1983). A management program that focuses on relatively common species, for example upland game, should avoid altering prime habitats of uncommon, rare, or endangered species. Thus, priority should be given to larger tracts of natural land forms, areas that might enhance or preserve extant, native plant-animal communities, especially those species that are uncommon or endangered and require interior habitats (Forman and Godron 1981, Brittingham and Temple 1983, Blake and Karr 1987, Dickman 1987, Stamps et al. 1987).

Examine the region to assess significant physiographic features from which to build a plan for habitat management (Jones 1986). Although this often is performed informally by the experienced biologist, it is important to review the process. What are the natural features that give definition to the region? What was the natural vegetation prior to settlement—prairie, savanna, forest? What are the significant habitat features that define the wildlife community, including forests, wetlands, riparian corridors, intensive grain cropping, livestock grazing? Are there remnant patches of natural vegetation or other significant, semipermanent vegetation (such as trees along the stream) from which to start? These are the kinds of factors that generally give definition to the resident fauna and should guide habitat management strategies.

In practice, identifying management priorities for given sites on privately owned farmland is not difficult (Office of Technology Assessment 1985). A first step is to classify sites according to their permanence and their potential for benefitting various species (Anderson and Ohmart 1986). Table 1 considers regional scales of management relative to species richness and abundance and the ensuing priorities for managing cover types on farmland. Such small or linear cover types as corridors, small woodlots near farmsteads, and semipermanent grasses are generally appropriate sites for applying practices to enhance edge-adapted species, e.g., upland game. Regional habitat strategies (Table 1) should permit the development of corridors connecting important land forms and natural areas. Further, the physical dimensions of corridors and how these corridors are managed will affect the reproduction and survival of wildlife at local and regional scales.

River and stream corridors, their riparian vegetation, and the quality of water that flows in them merit special consideration. In addition to upland erosion control, emphasis should be placed on the land-water interface, including near- and in-channel processes (Karr and Schlosser 1978). Streamside vegetative buffers, riparian forests, and wetlands represent excellent opportunities to mitigate nonpoint-source pollution from agricultural lands (Harris 1985). Further, expanded riparian greenbelts provide additional perennial wildlife cover.

Although future biological conservation strategies will emphasize the management of natural (unfarmed) land forms and habitats for rare and endangered plants and animals—typically species requiring interior habitats—many such tracts are primarily edge and therefore will not contribute significantly to such efforts. For example, 75% of 1,089 designated ''natural areas'' in Illinois are <30 ha

Table 1. Priorities for regional habitat management strategies directed to semipermanent cover types in intensively farmed landscapes.[a]

Intervention[b]	Cover type			
	Woodlots	Permanent grasslands	Semipermanent grasslands	Linear corridors
Maximize tract size and interior : edge ratios	High	High	Moderate	Low
Control successional changes in vegetation	Low	High	Low	Moderate
Maximize linkages with similar habitats	High	High	Moderate	Low
Use native plant species	High	High	Low	Moderate
Emphasize native wildlife communities; accommodate uncommon, rare, and endangered plants and animals	High	High	Moderate	Moderate
Emphasize game species and species benefitting from diverse cover types and ecotones	Low	Moderate	High	High

[a]Permanent grasslands include permanent pastures and (large) relict prairie sods; semipermanent grasses include forage grasses and legumes in rotation, and other sods that typically persist 3–5 years; corridors and linear habitats include grassy, shrubby, and woody cover typically <50 m wide, e.g., rights-of-way, drainage ditches, and small tributaries.

[b]Refer to Graber and Graber 1963, MacClintock et al. 1977, Wegner and Merriam 1979, Samson 1980, Forman and Godron 1981, Whitcomb et al. 1981, Yahner 1981, Forman and Baudry 1984, Harris 1984, Forman and Godron 1986, O'Conner and Shrubb 1986, Warner and Joselyn 1986, Baltensperger 1987, Krummel et al. 1987, and Swanson et al. 1988.

in area and typically comprise several smaller tracts with a variety of plant forms. Game and other edge-adapted species are often the primary benefactors of such small natural areas. Nonetheless, attention should be focused on characteristics of the land that can preserve or enhance its biotic integrity.

An appropriate habitat management objective on agricultural lands is to optimize the value to wildlife of each patch of perennial vegetation. Streamside vegetation, woodlots, wetlands, ditchbanks, and other odd areas of permanent vegetation represent habitat patches that collectively contribute to maintain a diversity of wildlife on the agricultural landscape. Complementing these "natural" patches of permanent cover are windbreaks, hedgerows, grassy and woody fencerows, and roadsides that can be managed for wildlife. Some farmers have the interest and resources to manage tracts of land just for wildlife habitat, but in today's economy the frequency of farmers who will forgo a profit to dedicate productive land to wildlife is low.

Scales of Time and Space

Wildlife agencies usually develop habitat at the scale of patch or field (Burgess and Sharpe 1981, Noss 1983).

These sites are readily affected by ecological phenomena occurring in nearby cover types, e.g., interactions at the farm scale (Warner and Etter 1985). Likewise, regional phenomena influence responses by wildlife at farm and field resolutions (Godron and Forman 1983). Regardless of the degree of control over local conditions that is brought about by habitat interventions, the response of wildlife is always affected by habitat conditions at larger (regional) scales; a region is considered to be several contiguous townships or counties where physiographic and land use factors are similar (Crowley 1978, Risser et al. 1984). The importance of spatial scales to resident upland wildlife, for example, is outlined in Table 2.

Diffuse Versus Intensive Programs

Most programs to extend farmland habitat to benefit wildlife can be classified as either diffuse or intensive, depending upon the effort required to benefit target species and the resources available to management agencies. Diffuse programs are typically open-ended initiatives that attract wildlife to managed sites but do not necessarily enhance the reproduction or the survival of target species beyond the farm scale. Further, in diffuse programs there may be no attempt to maximize landholder participation

Table 2. Potential for affecting the relative abundance of resident upland wildlife species into the autumn population by modifying agricultural land use practices compared with habitat interventions by wildlife agencies.[a]

Spatial scale	Impact on target species	Source of habitat improvement	
		Agriculture[b]	Habitat program[c]
Field/farm	Attract to specific sites	Very high	High
Township	Detectable increase in local abundance	Very high	Moderate
Multiple townships/counties	Significant/dramatic increase in local or regional abundance	Very high	Low

[a]Refer to Warner 1988, for example, for a description of spatial scales in relation to ring-necked pheasants.

[b]The modification of agricultural land use practices that are common to most farms in a given region.

[c]Wildlife habitat measures promoted by wildlife agencies that are not typically part of agricultural land use practices in a given farming region.

through incentives and aggressive promotion. Volunteer groups such as Pheasants Forever and Quail Unlimited can play a vital role in implementing diffuse programs as well as more intensive efforts (Wooley et al. 1988).

Intensive habitat interventions, on the other hand, often require incentives and sustained promotional activities to achieve high rates of participation by farmers in the targeted region. The goal of intensive habitat efforts is to improve the reproduction and survival of target species on a regional basis for a minimum of several years. The resources needed to enhance the demographics of target species often are underestimated, and habitat programs may require several years of development (Schwartz and Whitson 1987) before responses by wildlife are apparent. For example, successful intensive pheasant management programs typically require efforts for 5–10 years on a high percentage of farms over a region that is township or larger in size (Warner 1988).

Demonstration Areas and Pilot Programs

Rural sociologists have documented how new practices are evaluated by farm operators, although most studies have been directed to the adoption of commercial farming practices (Fliegel 1956, Rogers and Shoemaker 1971, Pampel and van Es 1977). The process by which farm operators accept or reject innovative practices typically includes awareness, interest, trial, and evaluation. Farmers may rely on such criteria as advantage over present practices, compatibility with existing methods, degree of difficulty in understanding and using a new practice, and opportunity to try a practice—at least on a limited scale—without significant economic or social risk (Warner 1983). Finally, the degree to which the results of a given innovation can be observed is critical. The adoption of a new practice in commercial farming usually produces effects that are apparent relatively quickly, and normally these include economic advantages over previous methods. Plants and animals, however, may require years to respond to new conservation practices, and direct economic return may not necessarily be forthcoming. Wildlife biologists must be prepared to help farm operators and landholders perceive the long-range value of adopting agricultural practices that will benefit wildlife.

Establishing demonstration areas for habitat schemes is a proven way of encouraging participation. A demonstration area enables landholders to see firsthand what the habitat program entails, the techniques for establishing and maintaining vegetation, the integration of the program with existing farm operations, and the anticipated responses of wildlife. Such areas are especially important when participants perceive substantial risk or uncertainty (e.g., demands on time or money, conflicting ethical or social values, or when the program requires practices unfamiliar in a farming community). For example, demonstration areas for the reintroduction of native grasses are emerging in the Midwest because wildlife agencies have learned that farmers are unfamiliar with native plants and the time and care needed for their establishment and maintenance (George et al. 1979, Dumke et al. 1981).

Demonstrations and "field days" are accepted activities in farming communities and are conducted regularly by the Cooperative Extension Service, Soil and Water Conservation Districts, and many agribusinesses. Linking up with these organizations and arranging for one or two stops on a "tour" or 15–30 minutes on the program may mean giving up some of the "ownership" of a fruitful idea but also may mean increased credibility, venerable sources of support, and ultimately more habitat management.

In the developmental stages of a program, the potential benefits to wildlife and the acceptance of the program by landholders are not known with certainty. Pilot projects foster experimentation on a small scale. Further, such projects allow for the development of techniques for manipulating cover, establish requirements for manpower and materials, and document responses by wildlife. In turn, an established pilot project can become a demonstration area.

IMPLEMENTING LARGE-SCALE HABITAT PROGRAMS

Regardless of the approach taken for pilot projects and demonstration areas, wide-scale implementation should not be initiated until (1) methods of establishing and maintaining appropriate plant forms and configurations of cover are proven, (2) benefits to wildlife are documented, (3) a sound basis for promoting the program within target groups is established, and (4) adequate resources are available. The following steps, therefore, should be followed during the implementation phase.

Establishing the Advisory Group

The advisory group is a valuable means for introducing change in agricultural communities (Phipps 1972) and plays a key role in legitimizing the program within the community. Members should represent the targeted groups and include community leaders, personnel from public agencies that influence farm operators and land use practices (e.g., Soil Conservation Service, State Department of Agriculture, Agricultural Stabilization and Conservation Service, Cooperative Extension), and individuals from farm organizations and such private entities as sporting and hunting groups.

The advisory group should be thought of as a dynamic body, its membership subject to change as unforeseen issues arise during the program. The group should be presented with a plan for implementing the habitat program. This plan should list tenets of the program and explain benefits to wildlife. It also should include experimentation and demonstration efforts under way, potential locations for initial habitat efforts, a timetable, and costs. The advisory group should be encouraged to take an active role in proposing modifications for how, when, and where the program is to be implemented. Finally, it should take an active role in evaluating and modifying the program and ensuring that adequate resources are devoted to the project.

Securing Participation of Landholders

Even an ecologically sound habitat management plan based on the latest techniques and most recent information will fail if the varied needs of land users are not met. For this reason, there is often a direct relationship between effective interaction among biologists and landholders during the planning process and the subsequent success of the program. Skills in communicating, marketing, and salesmanship are essential to success but must be backed up with technical know-how that demonstrates to farmers a knowledge of wildlife management and an understanding of agribusiness and the complex world in which a farmer must make a living.

More than a good idea and honorable intentions are needed to make an impact on privately owned land. Biologists should not naively assume that private landowners are willing to manage their land for public benefits at personal expense. Indeed, one of the greatest failures of wildlife professionals has been their inability to translate research findings and management practices into social and economic benefits (Miranowski and Bender 1982, Langner 1985, Gilbert and Dodds 1987). Biologists have little opportunity to develop marketing skills, and a good short course in salesmanship will pay dividends in converting ecological knowledge into changes on the landscape!

The management goals of the landholder—not those of the managing agency—are likely to be the ones applied and maintained over the years. Biologists, therefore, must find out what farm operators want accomplished on their land (Henderson 1984). In addition, they should help the farmer interpret the local landscape, providing information about soils, water, and plant and animal communities that, with management, could be transformed into benefits for the landholder.

Of particular importance is a recognition of the time required for the farm operator to complete the new tasks generated by the habitat plan. Factors other than time that may inhibit the latitude of operators in making farm management decisions include (1) obligations of farm operators to landowners, lending institutions, and farm programs; (2) adherence to other "farm plans" for soil conservation, financial management, marketing, and forestry; and (3) social and economic liabilities perceived by the farmer. A good plan will identify such considerations early and attempt to accommodate them. Wildlife managers also must demonstrate an understanding of such pertinent farming considerations as crop rotation, livestock management, and farm equipment.

When the objectives of land users have been clearly identified, biologists are ready to develop habitat management alternatives. Two or more alternatives should be offered, and estimates for cost and labor should be included as well as responses anticipated from target species. The opportunity to choose among alternatives facilitates negotiation between landholders and biologists. Like any effective salesman, the biologist should conclude the presentation by reviewing the benefits to the landholder and asking him to make the decision. All decisions should be documented in writing, and dates should be assigned to all tasks. If, for example, a landholder has agreed to plant 500 shrub seedlings the following spring, he should know when they will arrive and what to do with them on the day of their arrival.

Follow-up is an important aspect of the landholder/biologist relationship because it gives the biologist opportunity to diagnose and correct such unanticipated events as what to do about "all of the weeds" in the native grass planting. Follow-up also facilitates reevaluating and updating the plan as conditions change on the farm (e.g., government farm programs or finances). Maintaining a close liaison with program participants also cre-

ates opportunities for the biologist to make additional contacts that may make possible the expansion of the habitat program to other farms. If a professional plan was prepared with the farmer's best interest in mind, and follow-up assistance was provided to reinforce pro-conservation attitudes and to correct unforeseen problems, wildlife biologists can be assured that the demand for their assistance in the farm community will grow.

A completed plan also should be professional in appearance, including line drawings, photographs, well-written text, and an attractive cover. By contrast, little habitat management will be achieved and little demand generated if the contact with a farm operator is represented only by a quick trip over his land and a hastily delivered "plan" (one that is likely to be the biologist's plan, not the farmer's).

Selecting the Target Area

The area chosen for initial work should be sufficiently large to demonstrate results but small enough to ensure that proposed habitat development can be completed relatively quickly. As mentioned earlier, the regional context for a given set of habitat interventions should be carefully selected. The more complex and expensive the program, including ecologic, sociologic, and economic factors, the more consideration should be given to how the initial target area can serve as a foundation for demonstration and eventual expansion to a regional scale. Computerized Geographic Information Systems (GIS) are the tools of the future for selecting target areas based on overlays of various factors important to the selection process, including regional land use patterns, connecting corridors, human demography, and climactic factors (Iverson 1988, Iverson et al. 1989, Sidle and Ziewitz 1990).

Assembling Materials and Information

The information required to assemble an attractive and convincing plan comes from a variety of sources and in a variety of forms.

RESOURCE MATERIALS

Government agencies are a primary source of materials and technical advice for habitat development projects. Examples of the types of information available from various agencies are listed in Table 3. Maps provide essential information for habitat programs (Kerr 1986). They identify land tenancy and ownership, geography, and soils. Quadrangle maps available from the U.S. Geological Survey describe topography, identify water surfaces, and locate tributaries, travel corridors, airports, municipalities, residential areas, and other developed land.

Aerial photographs (black and white prints and color slides) maintained by Agricultural Stabilization and Conservation Service (ASCS) offices are useful for mapping vegetation and habitat development. More information pertaining to the use of maps and remote sensing for recording habitat can be found in Chapter 21.

Plat books describe land boundaries and ownership and delineate districts under the jurisdictions of local governments. They can be inspected at libraries and most government offices, including county seats, extension offices, and local branches of government that interact with farm communities. In addition to platting land ownership, the

Table 3. Governmental resource agencies that can provide background information and assistance to wildlife agencies in habitat development programs.

Agency	Organization of offices	Resources available
U.S. Department of Commerce	Regional/national	Human demographic data; county-level agricultural statistics
U.S. Department of Agriculture		
Agricultural Conservation and Stabilization Office	County/state	Landowner/farm records; aerial slides for monitoring annual compliance with farm programs
Soil Conservation Service	County/state	Technical information on soils, plant materials, and soil and water quality programs; technical advice
State Departments of Agriculture	State	Regional farm statistics; guidelines for natural resource and other farm programs available to farmers
U.S. Geological Survey	State/regional/national	Topographic maps; regional geology and geomorphology
University Extension Service	County/regional/state	Technical advice for establishing plant materials and appropriate farm practices
National Oceanographic and Atmospheric Administration	National	Climate data and summaries
Planning Commissions	County or lower	Zoning and land use codes; development plans; local maps

ASCS offices maintain records identifying the farm operator and owner of each land parcel. Private companies in some regions of the country publish directories of rural residents, although the release of such directories has been sporadic in recent years. Local extension and ASCS offices and rural banks will be aware of the existence of such directories.

By analyzing historical trends in land use for target counties, biologists can anticipate to some extent future farm policies and programs that may, for example, divert farmland from production. The Agricultural Census, conducted at 10-year intervals by the U.S. Department of Commerce, provides data for farmland use at the county level extending back to 1850. In recent decades annual statistics describing land use are reported at county or township levels by the Department of Agriculture in each state; these reports describe numbers and sizes of farms and statistics for crop plantings and livestock production.

EQUIPMENT

Most long-term habitat projects require the acquisition of equipment. Purchasing equipment usually requires guidelines established by one or more government agencies. Management biologists should be prepared to describe the availability and costs of suitable machinery and to follow preestablished protocols that culminate in bidding and contracting for equipment. The acquisition process rarely is completed in less than a year. For short-term and small-scale management programs, borrowing or leasing may eliminate the need to purchase equipment.

TIMETABLES AND COST ESTIMATES

Realistic schedules are imperative for a successful habitat program, and deadlines should be set for creating an advisory group, hiring staff, defining job assignments, and

acquiring machinery. When factors of time are uncertain, small-scale pilot projects offer an opportunity to learn about the logistics of implementing a habitat intervention. Further, it is often possible to obtain manpower and borrow equipment from other programs for small start-up projects. Most habitat projects encounter unexpected delays; credibility suffers when schedules are not met, and it is, therefore, better to anticipate possible delays when timetables are prepared.

In addition to equipment costs, estimates should be made for operating, maintaining, and replacing machinery. Costs for vegetation should include seed or nursery stock (including replacement of >25% of seedlings that typically do not survive), and such other essentials as fuel, fertilizer, and chemicals. The cost of maintaining vegetation (e.g., maintaining viability of stands or controlling succession by fertilizing, mowing, grazing, or reseeding) should be anticipated in habitat programs.

Because field offices and staff of wildlife agencies provide much of the personnel for habitat programs, accurate accounting is often useful. The result of inadequate bookkeeping is the inability to accurately project costs and returns of farmland habitat programs. In the absence of records substantiating all costs of pilot endeavors, the resources required to implement wide-scale efforts usually are underestimated. Personnel costs also should be anticipated for evaluating programs; resources for evaluations are a significant cost.

Establishing Habitat and Maintaining Vegetation

Numerous "hands-on" activities are required to implement a habitat management plan. The most common include planting seeds and seedlings or manipulating existing vegetation to achieve a desired condition. A few guiding principles are presented here.

Use native plant material where possible, choosing plants suited to the climatic region, soil, and other site conditions. Seed or seedlings should be obtained from a location where climatic conditions are quite similar to those at the planting site. Switchgrass, for example, a component of the tallgrass prairie, exhibits upland and lowland ecotypes that vary in their tolerance to soil moisture. Some biologists prefer to collect native seed locally, but that option may not be practical on a large scale. Germination tests should be made of all seed stock.

Soil conditions should be evaluated by consulting with a technician of the Soil Conservation Service to identify important variables such as soil texture, organic matter content, moisture holding capacity, depth of water table, drainage class and permeability, degree of past erosion, fertility, and pH. Soil testing is advised where such information is lacking. A suitable match between soil characteristics and plant species (or varieties) will prove economically efficient and reduce the number of seedings lost to disease and environmental stress. In addition, such a match provides competitive advantage in holding the site and may result in a long-lived stand that requires minimal maintenance.

The preparation of seedbeds is a subject for which there is no shortage of advice, and information on suitable times for planting various species can be obtained from seed dealers. Adequate seed-soil contact must be established so that seedlings do not dry out and so that they can take up nutrients and moisture from the soil.

In addition to saving topsoil, erosion control for new seedings saves the cost and effort of replanting the site. All tillage operations should be performed on the contour, and no-till techniques should be used whenever possible. Planting of nurse crops (oats, wheat, rye) with the permanent seeding is a common procedure. The small grains can be rotary mowed after the seed stalks are in the early stages of seed development, but before the seeds mature (sometimes called late boot stage), to reduce competition and prevent them from regrowing. A second pass with the planter should be made on more sloping parts of fields and where water runs off in concentrated flows. On highly erodible sites, the seeding rates should be doubled. Straw also may be needed on slopes, mulched at about 2.2 tonnes/ha (1 ton/acre). Mulch absorbs the kinetic energy of the raindrops before they hit the soil and helps to protect seedlings from desiccation.

The same principles used in planting seeds apply to planting tree and shrub seedlings. The roots of seedlings should be examined when the bundles are unpacked to determine if they will fit in the holes dug or slots prepared by mechanical planters without being turned upward. If needed, roots can be pruned in bundles with a machete. One-third of the top of deciduous seedlings should be removed in the same way. Removing part of the top lowers the transpiration rate and reduces desiccation. The tops will grow back after the roots are established, and the survival rate will be better. Care should be taken to firm the soil around the roots to eliminate air pockets and to ensure that the slots made from mechanical tree planters are fully closed. Mulching around the base of seedlings with wood chips, ground corncobs, or gravel is helpful,

but mulching is not feasible for a large planting. Suitable substitutes for small-scale, labor-intensive projects include newspapers, plastic sheeting, or cardboard (pizza) disks. These materials can be slit to the middle, placed around the seedling, and lightly covered with a shovel of soil. This mulch retards weed growth at the base of the seedling and retains precious moisture while the seedling establishes.

Plant communities are dynamic over successional stages with respect to structural complexity and plant composition. Control of ecological succession may be required to meet species-specific goals. Prescribed burning and grazing are useful for retarding secondary plant succession and stimulating herbaceous growth. Such planned disturbances are widely used in the management of warm-season grasses and prairie forbs. Growth of invading shrubs or cool-season grasses can be inhibited by burning native stands as they begin to emerge in early spring. Burning in early spring favors forbs; later burning favors native grasses. Burning stands of pine (especially in the South) is effective for opening the understory and stimulating the growth of food plants important to northern bobwhites and white-tailed deer. Depending on weather and fuel load, burning may not be controlled easily. Because of liability for out-of-control fires, prescribed burning requires thorough planning, experienced personnel, and conformity with local regulations.

Established stands of grasses need attention annually if they are to remain in optimum condition for wildlife for more than a few years, especially if the nitrogen-fixing legumes are to persist. Mowing brome-alfalfa at the end of the nesting season (early August) controls secondary succession and helps to accumulate plant litter on the soil surface. Through decomposition this litter replenishes the soil with phosphorous and potassium. Alfalfa grows back more vigorously than grass in late summer, thereby allowing the buildup of root reserves before winter dormancy.

Farmers often view whatever was not intentionally planted as weeds. Because these plant forms are important to wildlife, their value should be carefully explained to landholders during the planning process. Maintenance of brush and trees also should be considered. Along hedgerows (Baltensperger 1987), the use of root pruners, the periodic pruning of limbs (for "firewood"), and the planting of a 5-m grass strip mowed once in late summer can provide attractive wildlife cover with minimal maintenance.

EVALUATING AND REFINING PROGRAMS

If possible, document the success or failure of a program, evaluate cost-effectiveness, and consider whether program modifications are appropriate. Evaluation can include (1) adherence by agencies to timetables, costs, and objectives; (2) success rates in establishing and maintaining desired vegetative configurations and structures; (3) participation by landholders, initially and over time (including assessments of opinions, attitudes, knowledge, and conservation behavior); and (4) responses by key species to management. The number of farmland habitat programs that have received rigorous evaluation with well-documented costs and benefits is probably small, and few published accounts are available.

Table 4. Examples of typical sources of errors in planning and assessing the effects of habitat programs for upland wildlife species in intensively farmed areas.

Animal response	Response detected	Sources of errors	Biological explanation
Yes	No	Evaluation period too short	Lag effects Development of vegetation Demographic responses
Yes/No	Yes/No	Measurement techniques	High variability in animal responses and/or census techniques inappropriate
Yes	No	Limited spatial scale	Dispersal

Monitoring Vegetation

The extent to which the monitoring of vegetation is warranted depends on how reliable the planting method is assumed to be. The success of vegetation established and maintained by landholders—when planting methods were proven and widely known—is likely to be more variable than if the task had been undertaken by personnel from wildlife agencies. Monitoring vegetation should minimally include the inspection of a randomly selected sample of treated sites; these should be evaluated with respect to initial and long-term success in achieving desired plant forms. Alternatively, landholders familiar with particular plant forms can provide feedback through questionnaires.

Monitoring Participation by Landholders

Participation by landholders can be evaluated in several ways. Opinion surveys can be taken before promotion of the program is initiated; follow-up surveys can document the effects of participation in the program on attitudes and knowledge. Special recognition of outstanding cooperators can enhance the visibility and importance of wildlife management programs to landholders. Ideally, a rural sociologist or an education specialist should participate in program development and evaluation (Dumke et al. 1981).

Monitoring Responses by Wildlife

Evaluating responses by target species to habitat programs can be accomplished by direct or indirect means, depending in part on the extent to which species-habitat responses are known (Fagen 1988). Direct evaluation involves at a minimum the use of proven census methods to establish trends in relative wildlife abundance (Eng 1986). More involved evaluation of demographic responses may require activities such as nest and radiotelemetry studies. Indirect methods of evaluating habitat programs are gaining wide use by wildlife agencies because they are easy to use and relatively inexpensive. The most widely used method of inferred evaluation is the habitat index in which changes in habitat quality are measured by computing change (Verner et al. 1986). Measuring and computing habitat indices are not, however, the same as measuring the responses of target species to management (O'Neil and Carey 1986). Habitat models are described in Chapter 23.

Evaluating Interactions Between Wildlife and Habitat

If responses by target species to habitat initiatives are a true measure of success, then few farmland habitat programs have been evaluated, let alone proven successful, over a sustained period. Following the concepts of validity in statistical design, faulty evaluations can be categorized as to the type of error. For example, invalid procedures can lead the evaluator to conclude that a positive response by target species to habitat interventions occurred, when in fact the response was negative, negligible, or nonexistent (Table 4). Other common errors are inappropriate evaluation period (typically too short), invalid measurement techniques for evaluating population responses, and management or evaluation over a limited spatial scale. In such instances, underlying biological phenomena remain unmeasured, and some of the same factors that limit success of habitat interventions also lead to failure in evaluation (Table 4).

INTEGRATING WILDLIFE AND AGRICULTURAL PROGRAMS

Many farmers have chosen farming primarily because of the way of life it offers, including a high degree of interaction with nature. Agricultural producers typically have concern for their farmland environment, for the productive capacity of the soil, and for the life it sustains; they ultimately determine how much "country" will persist on their land. Leopold's (1966:177–178) words still ring true:

> There is much confusion between land and country. Land is the place where corn gullies and mortgages grow. Country is the personality of the land, the collective harmony of its soil, life, and weather. . . . Poor land may be rich country and vice versa. Only economists mistake physical opulence for riches. Country may be rich despite a conspicuous poverty of physical endowment, and its quality may not be obvious at first glance, nor at all times. . . . It [wildlife] often represents the difference between rich country and mere land.

Wildlife and agriculture can coexist (Office of Technology Assessment 1985), but wildlife is rarely discussed in agricultural production circles except as a potential secondary source of income (e.g., lease of land for hunting). Many opportunities are available, however, to integrate wildlife management with other farm programs and policies. For example, soil conservation practices can be applied with or without considerations for wildlife; minor changes in a farm plan developed for conserving soil can produce substantial benefits for wildlife. The critical variable for integrating wildlife considerations with other farm management plans is timing; the information has to get to the decision maker before the course of action is set.

The integration of habitat programs with agriculture should take place at several levels of decision making (Jahn 1988). Because regional land use and habitat conditions affect habitat programs of all scales, influence should be brought, where possible, on farm policies and programs so that farmers will be encouraged to provide quality habitat while meeting the guidelines of agricultural programs. Wildlife managers need to understand federal farm programs and the politics of conservation as well as they understand the principles of ecology. As mentioned earlier, the classic example of such integration is the meshing of wildlife habitat with farmland diverted from production. Because these programs affect millions of hectares of cropland in the United States annually, wildlife managers must encourage the management of set-aside lands so that they will continue to benefit wildlife.

These same management goals apply to programs of cropland diversion where cover is established for 5–10 years (e.g., the Conservation Reserve Program [CRP] established in the late 1980s). These more enduring programs also offer an opportunity to establish woodlots and to allow small wetlands to revert to natural plant forms. Further, the CRP allows the establishment of field windbreaks and filter strips along streams, both of which are effective means for integrating soil and water conservation goals with wildlife management (Yahner 1981, see also Berthelsen et al. 1989).

Resource Planning and Interagency Cooperation

Multiple resource management is a concept that is widely ascribed to, but infrequently observed, in habitat programs on private land. One problem is that it requires multidisciplinary expertise—interagency cooperation—a resource often lacking except in management programs on large public lands. In the absence of interagency cooperation, only the most persistent landholders will contact the various agencies responsible for delivering such independently derived plans as wildlife management, soil erosion control, woodland management, and commodity marketing; the process of discovering one program at a time may transpire over several years. Although various efforts are underway around the country to remedy this situation, we are aware of only one manual that outlines the procedures for multiple resource management: *Resource Planning Guidebook* (U.S. Soil Conservation Service 1986).

We recommend resource planning as a method for the comprehensive management of the resource base—soil, water, and related plants and animal resources—within intensively farmed landscapes. Watershed boundaries provide a workable unit, and with input from an advisory group of resident landholders, a plan for the watershed can be prepared with the help of interdisciplinary experts from public agencies. Just as the wildlife biologist would evaluate habitat conditions for selected species and provide technical guidelines for management, other profes-

sionals would provide similar information from their areas of expertise. Technical experts would collectively evaluate and modify individually derived plans to meet multiple resource goals. Local leadership is required to make the system work effectively. Soil and Water Conservation Districts and Extension Councils are examples of groups capable of leading such efforts at a local level.

Several opportunities for integrating wildlife conservation measures with farm practices are emerging. For example, the 1985 Food Security Act and 1990 Agriculture Act mandate that to participate in farm programs, landholders with a predominance of highly erodible soils must (1) file a plan for reducing erosion rates to program-acceptable levels by 1990, and (2) put this plan into effect by the mid-1990s. Wildlife agencies have a unique opportunity to help plan and apply soil conservation practices that also will improve habitat for wildlife.

Farm Programs and Incentives

Incentives can be effective in encouraging farmers to manage set-aside lands for maximum benefits to wildlife (Burger 1978). Modest incentives, such as the provision of seed, can help establish prime cover for wildlife as part of intensive management interventions. Additional incentives may be needed in years when crop prices are relatively strong and farmers are less likely to participate in set-aside programs. A reduction in land taxes for parcels that are managed for wildlife can be another effective incentive.

We caution that where the success of habitat programs is dependent on incentives, how landholders perceive those incentives should be carefully considered before promotion begins on a wide scale. Numerous opinion surveys indicate that although landholders value prime habitats for wildlife—often more than for commercial worth—they will not necessarily accept incentives that will in turn require specific management practices (Dumke et al. 1981). Further, holders of key parcels of habitat may be reluctant to enhance habitat for wildlife if they feel that demand for recreational use of their land will lead to problems of liability and uncontrolled access. The implication is that monetary incentives and land leases are expensive, not easily promoted, and difficult to maintain.

Agricultural Technologies

Agricultural technologies have made farming more profitable and land more productive, and wildlife biologists must pay careful attention to the interface of agricultural technology and ecology. The interest in sustainable or regenerative agriculture is a case in point. The potential that new farm technologies hold for wildlife can be evaluated with Leopold's (1966) criteria: Is soil fertility maintained? Is a diverse flora and fauna sustained?

Where managed parcels are in farmland mosaics, agricultural practices per se should be addressed by managers, even if farm disturbances do not directly affect developed habitats (Brady 1985). For example, the use of pesticides continues to expand; increased autumn tillage since World War II has greatly reduced the availability of stubble foods with suitable forage for wildlife during autumn and winter; and the harvest of forage grasses and legumes (prime nest cover for grassland species) is becoming earlier and progressing more rapidly (Warner and

Fig. 2. No-till corn planted in soybean residue is one of several varieties of conservation tillage providing effective erosion control. This is a dangerous place to nest (Best 1986), but the extensive use of this practice suggests its role will become more important. Pre-emerge banded applications of pesticides would disrupt less of the "habitat" than would post-emerge broad-spectrum sprays (U.S. Soil Conserv. Serv. photo).

Fig. 3. Grassed backslope terraces in western Iowa along with some vegetation in the water course and around the ponds provide perennial cover for wildlife. Management to retain corn and wheat residues on the surface over winter would provide wildlife with a source of waste grain (U.S. Soil Conserv. Serv. photo).

Etter 1989). Future habitat interventions in agricultural landscapes, in addition to establishing and protecting key habitats, must control disturbances at farm and regional scales—a challenging but necessary form of integration.

Thus, agricultural land use affects wildlife habitat at all spatial and temporal scales. Over the long term, successful wildlife habitat interventions on farmland must be compatible with prevailing farm technologies. On the other hand, landholders must embrace the goals and objectives of broadly based natural resource management. Wildlife biologists working on private lands must be diligent in pursuing change on both fronts!

The final increment of habitat to be developed consists of manipulating those practices designed for other purposes (such as erosion control) to secondarily provide habitat for wildlife. Generally, as soil-conserving measures increase, upland wildlife habitat quality also improves (Miranowski and Bender 1982). Although the following practices do not necessarily stand alone in benefitting wildlife, when implemented as one of several components the collective habitat value of such practices may be substantial. A few examples follow to illustrate this point.

Conservation tillage consists of a variety of techniques that leave some or all of the crop residues (cornstalks, wheat stubble) on the soil surface (Fig. 2) rather than turning them under as in conventional tillage with moldboard plows. The residue on the surface substantially reduces soil erosion and also provides some useful habitat. Many forms of intermediate tillage commonly are used that chop or shred the residue to facilitate planting or to incorporate soil amendments or pesticides, all of which reduce the habitat value. The more residue that remains, the better for wildlife. Recent studies in Illinois (Warburton and Klimstra 1984), Indiana (Castrale 1985), and Iowa (Basore et al. 1986) indicated that no-till corn and soybean fields had higher densities and greater varieties of birds during the breeding season than did conventionally tilled fields (see also Rodgers and Wooley 1983, Wooley et al. 1985, Best 1986, Duebbert and Kantrud 1987).

Terraces are a commonly applied erosion control and water management practice. They are ridges of earth about 60–90 cm high constructed across the slope on a gentle grade to remove runoff at a nonerosive velocity (Fig. 3). They may be broad-based and farmed or they may be narrow-based (4–5 m) with grassed ridges or grassed backslopes. Broad-based terraces have no direct benefit to wildlife. Grassed terraces are less expensive to build and increase the diversity and interspersion of vegetative types in cropland settings. These terraces are attractive to many wildlife species. Beck (D.W. Beck, Wildlife use of grassed backslope terraces, abstr., 44th Midwest Fish Wildl. Conf., Des Moines, Ia, 1982) observed that 35 species of vertebrates used grassed backslope terraces in Iowa and reported an average of one successful pheasant nest per 5 ha of grass.

Contour-strip cropping (Fig. 4) is a technique used to control erosion by interspersing strips about 30–40 m wide of close-grown crops (hay and small grains) on the contour between strips of row crops. Alternating strips of corn, oats, and hay can provide the juxtaposition and configuration of cover types necessary to provide for the needs of wildlife during periods of limited mobility, such as when pheasants are tending young broods (Warner et al. 1984, Warner 1988).

Contour-strip cropping frequently results in point rows that are difficult to maneuver with large equipment. Planting grassy strips about 3–4 m wide along field borders can solve this problem, provide a source of hay, and also provide additional cover. Field border strips are a much underutilized practice with multiple benefits. As in the above they may be strips 3–4 m wide to reduce point rows, but they also are useful along end rows where farm traffic is more frequent and where water runs down the rows, ruts form, and erosion is accelerated. They are also useful in the farmer's major struggle—controlling succession. Where crop fields are adjacent to woody cover, there is the continual struggle to keep the woody vegetation from encroaching too far into the field and to avoid hitting tree limbs with the cab of the tractor. From the farmer's viewpoint it costs just as much to plant crops on the edge of the field as it does in the middle, but the crop field

Fig. 4. Coutour-strip cropping in Pennsylvania—an example of country rather than land. Note the diversity, spatial configuration, and interspersion of cover types providing habitat for wildlife while the primary land use is commercial agriculture (U.S. Soil Conserv. Serv. photo).

edge next to the woods, shaded by large trees, frequently does not yield enough to recover costs. Planting the area under the "drip line" of the trees to grassy cover saves the cost of planting where crop yields are below cost. By mowing in late summer each year, the inevitable woody encroachment is controlled and species such as cottontail rabbits and northern bobwhites are benefitted.

Strip cover that occurs along fencerows and roadsides also can provide habitat. Best (1983) reported as many as 30 species of birds using fencerows in Iowa farmlands during the breeding season. Those with greater coverage of trees and shrubs supported a more diverse and abundant avifauna. Best (1983:347) suggests that "providing farm operators with documentation of wildlife's dependence on fencerows may provide the impetus that many farmers need to balance wildlife, soil, and water conservation against possible short-term economic gains." Management of large blocks of rural roadsides for wildlife habitat is a documented technique for providing grassy nesting habitat (Warner and Joselyn 1986, Warner 1992).

Acknowledgments.—We thank the Max McGraw Wildlife Foundation for financial support in the development of this chapter. L. M. David and J. M. Ver Steeg, Illinois Department of Conservation, reviewed a draft of the manuscript. W. R. Edwards, A. S. Hodgins, and G. C. Sanderson, Illinois Natural History Survey, provided technical and editorial support. We also thank G. B. Joselyn, P. A. Vohs, and W. Vander Zouwen for their helpful reviews of the manuscript.

LITERATURE CITED

ANDERSON, B. W., AND R. D. OHMART. 1986. Vegetation. Pages 639–660 *in* A. Y. Cooperrider, R. J. Boyd, and H. R. Stuart, eds. Inventory and monitoring of wildlife habitat. U.S. Dep. Inter. Bur. Land Manage. Serv. Cent., Denver, Colo.

AUCLAIR, A. N. 1976. Ecological factors in the development of intensive-management ecosystems in the midwestern United States. Ecology 57:431–444.

BALTENSPERGER, B. H. 1987. Hedgerow distribution and removal in nonforested regions of the Midwest. J. Soil Water Conserv. 42:60–64.

BASORE, N. S. 1984. Breeding ecology of upland birds in no-tillage and tilled cropland. M.S. Thesis, Iowa State Univ., Ames. 62pp.

———, L. B. BEST, AND J. B. WOOLEY, JR. 1986. Bird nesting in Iowa no-tillage and tilled cropland. J. Wildl. Manage. 50:19–28.

BAXTER, W. L., AND C. W. WOLFE, JR. 1973. Life history and ecology of the ring-necked pheasant in Nebraska. Nebr. Game, Fish, Parks Comm., Lincoln. 58pp.

BELLROSE, F. C., S. P. HAVERA, F. L. PAVEGLIO, AND D. W. STEFFECK. 1983. The fate of lakes in the Illinois River valley. Ill. Nat. Hist. Surv. Biol. Notes 119. 27pp.

BERNER, A. H. 1984. Federal land retirement programs: a land management albatross. Trans. North Am. Wildl. Nat. Resour. Conf. 49:118–131.

———. 1988. Federal pheasants—impact of federal agricultural programs on pheasant habitat. Pages 45–93 *in* D. L. Hallett, W. R. Edwards, and G. V. Burger, eds. Pheasants: symptoms of wildlife problems on agricultural lands. North-Cent. Sect., The Wildl. Soc., Milwaukee, Wis.

BERTHELSEN, P. S., L. M. SMITH, AND C. L. COFFMAN. 1989. CRP land and game bird production in the Texas high plains. J. Soil Water Conserv. 44:504–507.

BEST, L. B. 1983. Bird use of fencerows: implications of contemporary fencerow management practices. Wildl. Soc. Bull. 11:343–347.

———. 1985. Conservation tillage: ecological traps for nesting birds? Wildl. Soc. Bull. 14:308–317.

BLAKE, J. G., AND J. R. KARR. 1987. Breeding birds of isolated woodlots: area and habitat relationships. Ecology 68:1724–1734.

BRADY, S. J. 1985. Important soil conservation techniques that benefit wildlife. Pages 55–62 in Proc. workshop technologies to benefit agriculture and wildlife. U.S. Off. Technol. Assessment OTA-BP-F-34.

———. 1988. Potential implications of sodbuster on wildlife. Trans. North Am. Wildl. Nat. Resour. Conf. 53:239–248.

BRITTINGHAM, M. C., AND S. A. TEMPLE. 1983. Have cowbirds caused forest songbirds to decline? BioScience 33:31–35.

BURGER, G. V. 1978. Agriculture and wildlife. Pages 89–107 in H. P. Brokaw, ed. Wildlife and America. Counc. Environ. Quality, U.S. Gov. Printing Off., Washington, D.C.

BURGESS, R. L., AND D. M. SHARPE, EDITORS. 1981. Forest island dynamics in man-dominated landscapes. Ecol. Stud. 41. Springer-Verlag, New York, N.Y. 310pp.

CASTRALE, J. S. 1985. Responses of wildlife to various tillage conditions. Trans. North Am. Wildl. Nat. Resour. Conf. 50:142–156.

———. 1987. Pesticide use in no-till row-crop fields relative to wildlife. Ind. Acad. Sci. 96:215–222.

CROWLEY, P. H. 1978. Effective size and the persistence of ecosystems. Oceologia 35:185–195.

DICKMAN, C. R. 1987. Habitat fragmentation and vertebrate species richness in an urban environment. J. Appl. Ecol. 24:337–351.

DUEBBERT, H. F., AND H. A. KANTRUD. 1987. Use of no-till winter wheat by nesting ducks in North Dakota. J. Soil Water Conserv. 42:50–53.

DUMKE, R. T., G. V. BURGER, AND J. R. MARCH, EDITORS. 1981. Wildlife management on private lands. Proc. Symp., Wis. Chap., The Wildl. Soc., Madison. 568pp.

EDWARDS, C. A., R. LAL, P. MADDEN, R. H. MILLER, AND G. HOUSE, EDITORS. 1990. Sustainable agricultural systems. Soil Water Conserv. Soc., Ankeny, Ia. 696pp.

EDWARDS, W. R. 1984. Early ACP and pheasant boom and bust! A historic perspective with rationale. Pages 71–83 in R. T. Dumke, R. G. Stiehl, and R. B. Kahl, eds. Perdix III: gray partridge and ring-necked pheasant workshop. Wisconsin Dep. Nat. Resour., Madison.

———, S. P. HAVERA, R. F. LABISKY, J. A. ELLIS, AND R. E. WARNER. 1981. The abundance of cottontails (Sylvilagus floridanus) in relation to agricultural land use in Illinois. Pages 761–789 in K. Myers and C. D. MacInnes, eds. Proc. World Lagomorph Conf., Univ. Guelph, Guelph, Ont.

ENG, R. L. 1986. Upland game birds. Pages 407–428 in A. Y. Cooperrider, R. J. Boyd, and H. R. Stuart, eds. Inventory and monitoring of wildlife habitat. U.S. Dep. Inter. Bur. Land Manage. Serv. Cent., Denver, Colo.

ETTER, S. L., R. E. WARNER, G. B. JOSELYN, AND J. E. WARNOCK. 1988. The dynamics of pheasant abundance during the transition to intensive row-cropping in Illinois. Pages 111–127 in D. L. Hallett, W. R. Edwards, and G. V. Burger, eds. Pheasants: symptoms of wildlife problems on agricultural lands. North-Cent. Sect., The Wildl. Soc., Milwaukee, Wis.

FAGEN, R. 1988. Population effects of habitat change: a quantitative assessment. J. Wildl. Manage. 52:41–46.

FARRIS, A. L., E. D. KLONGLAN, AND R. C. NOMSEN. 1977. The ring-necked pheasant in Iowa. Ia. Conserv. Comm., Des Moines. 147pp.

FLATHER, C. H., AND T. W. HOEKSTRA. 1989. An analysis of the wildlife and fish situation in the United States: 1989–2040. U.S. For. Serv. Gen. Tech. Rep. RM-178. 147pp.

FLIEGEL, F. C. 1956. A multiple correlation analysis of factors associated with adoption of farm practices. Rural Soc. 21:284–292.

FORMAN, R. T. T., AND J. BAUDRY. 1984. Hedgerows and hedgerow networks in landscape ecology. Environ. Manage. 8:495–510.

———, AND M. GODRON. 1981. Patches and structural components for a landscape ecology. BioScience 31:733–740.

———, AND ———. 1986. Landscape ecology. John Wiley & Sons, New York, N.Y. 619pp.

GEBHARDT, M. R., T. C. DANIEL, E. E. SCHWEIZER, AND R. R. ALLMARAS. 1985. Conservation tillage. Science 230:625–630.

GEORGE, R. R., A. L. FARRIS, C. C. SCHWARTZ, D. D. HUMBURG, AND J. C. COFFEY. 1979. Native prairie grass pastures as nest cover for upland birds. Wildl. Soc. Bull. 7:4–9.

GILBERT, F. F., AND D. G. DODDS. 1987. The philosophy and practice of wildlife management. R. E. Krieger Publ. Co., Malabar, Fla. 279pp.

GODRON, M., AND R. T. T. FORMAN. 1983. Landscape modification and changing ecological characteristics. Pages 12–45 in H. A. Mooney and M. Godron, eds. Disturbance and ecosystems. Springer-Verlag, West Berlin.

GRABER, R. R., AND J. W. GRABER. 1963. A comparative study of bird populations in Illinois, 1906–1909 and 1956–1958. Ill. Nat. Hist. Surv. Bull. 28:383–528.

HANSON, L. P., AND C. M. NIXON. 1987. White-tailed deer. Pages 104–105 in R. D. Neely and C. G. Heister, compilers. The natural resources of Illinois. Ill. Nat. Hist. Surv. Spec. Publ. 6.

HARMON, K. W., AND M. M. NELSON. 1973. Wildlife and soil considerations in land retirement programs. Wildl. Soc. Bull. 1:28–38.

HARRIS, L. D. 1984. The fragmented forest: island biogeography and theory and the preservation of biotic diversity. Univ. Chicago Press, Chicago, Ill. 211pp.

———. 1985. Designing landscape mosaics for integrated agricultural and conservation planning in the southeastern United States. Pages 102–111 in Proc. workshop technologies to benefit agriculture and wildlife. U.S. Off. Technol. Assessment OTA-BP-F-34.

HENDERSON, R. F., EDITOR. 1984. Increasing wildlife on farms and ranches. Great Plains Agric. Counc., Wildl. Res. Comm., and Kansas Coop. Ext. Serv., Manhattan.

IVERSON, L. R. 1988. Land-use changes in Illinois, USA: the influence of landscape attributes on current and historic land use. Landscape Ecol. 2:45–61.

———, R. L. OLIVER, D. P. TUCKER, P. G. RISSER, C. D. BURNETT, AND R. G. RAYBURN. 1989. The forest resources of Illinois: an atlas and analysis of spatial and temporal trends. Ill. Nat. Hist. Surv. Spec. Publ. 11. 181pp.

JAHN, L. R. 1988. The potential for wildlife habitat improvements. J. Soil Water Conserv. 43:67–69.

JONES, K. B. 1986. The inventory and monitoring process. Pages 1–28 in A. Y. Cooperrider, R. J. Boyd, and H. R. Stuart, eds. Inventory and monitoring of wildlife habitat. U.S. Dep. Inter. Bur. Land Manage. Serv. Cent., Denver, Colo.

JOSELYN G. B., AND J. E. WARNOCK. 1964. Value of federal feed grain program to production of pheasants in Illinois. J. Wildl. Manage. 28:547–551.

JUDY, R. D., JR., P. N. SEELEY, T. M. MURRAY, S. C. SVIRSKY, M. R. WHITWORTH, AND L. S. ISCHINGER. 1984. 1982 national fisheries survey. Vol. I. Technical report: initial findings. U.S. Fish Wildl. Serv., FWS/OBS-84/06. 140pp.

KARR, J. R., AND I. J. SCHLOSSER. 1978. Water resources and the land-water interface. Science 201:229–234.

KERR, R. M. 1986. Habitat mapping. Pages 49–69 in A. Y. Cooperrider, R. J. Boyd, and H. R. Stuart, eds. Inventory and monitoring of wildlife habitat. U.S. Dep. Inter. Bur. Land Manage. Serv. Cent., Denver, Colo.

KLOPATEK, J. M., AND J. T. KITCHINGS. 1985. A regional technique to address land use changes and animal habitats. Environ. Conserv. 12:343–350.

KROH, G. C., AND D. L. BEAVER. 1978. Insect response to mixture and monoculture patches of Michigan old-field annual herbs. Oceologia 31:269–275.

KRUMMEL, J. R., R. H. GARDNER, G. SUGIHARA, R. V. O'NEILL, AND P. R. COLEMAN. 1987. Landscape patterns in a disturbed environment. Oikos 48:321–324.

LANGNER, L. 1985. An economic perspective on the effects of federal conservation policies on wildlife habitat. Trans. North Am. Wildl. Nat. Resour. Conf. 50:200–209.

LEOPOLD, A. 1933. Game management. Charles Scribner's Sons, New York, N.Y. 481pp.

———. 1966. A Sand County almanac with essays on conservation from Round River. Oxford Univ. Press, New York, N.Y. 295pp.

MacCLINTOCK, L., R. F. WHITCOMB, AND B. L. WHITCOMB. 1977. II. Evidence for the value of corridors and minimization of isolation in preservation of biotic diversity. Am. Birds 31:6–16.

MAGLEBY, R., D. GADSBY, D. COLACICCO, AND J. THIGPEN. 1985. Trends in conservation tillage use. J. Soil Water Conserv. 40:274–276.

McCONNELL, C. A. 1981. Common threads in successful programs benefitting wildlife on private lands. Pages 279–287 in R. T. Dumke, G. V. Burger, and J. R. March, eds. Wildlife management on private lands. Proc. Symp., Wis. Chap., The Wildl. Soc., Madison.

MIRANOWSKI, J. A., AND R. L. BENDER. 1982. Impact of erosion control policies on wildlife habitat on private lands. J. Soil Water Conserv. 37:288–291.

NATIONAL RESEARCH COUNCIL. 1982. Impacts of emerging agricultural trends on fish and wildlife habitat. Natl. Acad. Press, Washington, D.C. 303pp.

———. 1989. Alternative agriculture. Natl. Acad. Press, Washington, D.C. 448pp.

NOSS, R. F. 1983. A regional landscape approach to maintain diversity. BioScience 33:700–706.

O'CONNOR, R. J., AND M. SHRUBB. 1986. Farming & birds. Cambridge Univ. Press, Cambridge, U.K. 290pp.

OFFICE OF TECHNOLOGY ASSESSMENT. 1985. Proc. workshop technologies to benefit agriculture and wildlife. U.S. Off. Technol. Assessment OTA-BP-F-34. 137pp.

OLDFIELD, M. L., AND J. B. ALCORN. 1987. Conservation of traditional agroecosystems. BioScience 37:199–208.

O'NEIL, L. J., AND A. B. CAREY. 1986. Introduction: when habitats fail as predictors. Pages 207–208 *in* J. Verner, M. L. Morrison, and C. J. Ralph, eds. Wildlife 2000: modeling habitat relationships of terrestrial vertebrates. Univ. Wisconsin Press, Madison.

PAMPEL, F., JR., AND J. C. VAN ES. 1977. Environmental quality and issues of adoption research. Rural Soc. 42:57–61.

PHIPPS, L. J. 1972. Handbook on agricultural education in public schools. Third ed. Interstate Printers & Publ. Inc., Danville, Ill. 599pp.

PIMENTEL, D., ET AL. 1987. World agriculture and soil erosion. BioScience 37:277–283.

———, T. W. CULLINEY, I. W. BUTTLER, D. J. REINEMANN, AND K. B. BECKMAN. 1989. Low-input sustainable agriculture using ecological management practices. Agric. Ecosystems Environ. 27:3–24.

POTTS, G. R. 1986. The partridge: pesticides, predation and conservation. Collins, London, U.K. 274pp.

POWER, J. F., AND R. F. FOLLETT. 1987. Monoculture. Sci. Am. 256:79–86.

RISSER, P. G., J. R. KARR, AND R. T. T. FORMAN. 1984. Landscape ecology: directions and approaches. Ill. Nat. Hist. Surv. Spec. Publ. 2. 18pp.

ROBBINS, C. S., D. BYSTRAK, AND P. H. GEISSLER. 1986. The breeding bird survey: its first fifteen years, 1965–1979. U.S. Fish Wildl. Serv. Resour. Publ. 157. 196pp.

RODGERS, R. D., AND J. B. WOOLEY. 1983. Conservation tillage impacts on wildlife. J. Soil Water Conserv. 38:212–213.

ROGERS, E. M., AND F. F. SHOEMAKER. 1971. Communication of innovations: a cross-cultural approach. Second ed. The Free Press, New York, N.Y. 476pp.

ROSEBERRY, J. L., AND W. D. KLIMSTRA. 1984. Population ecology of the bobwhite. Southern Ill. Univ. Press, Carbondale. 259pp.

SAMSON, F. B. 1980. Island biogeography and the conservation of nongame birds. Trans. North Am. Wildl. Nat. Resour. Conf. 45:245–251.

SCHLEBECKER, J. T. 1975. Whereby we thrive: a history of American farming, 1607–1972. Iowa State Univ. Press, Ames. 342pp.

SCHWARTZ, O. A., AND P. D. WHITSON. 1987. A 12-year study of vegetation and mammal succession on a reconstructed tallgrass prairie in Iowa. Am. Midl. Nat. 117:240–249.

SIDLE, J. G., AND J. W. ZIEWITZ. 1990. Use of aerial videography in wildlife habitat studies. Wildl. Soc. Bull. 18:56–62.

STAMPS, J. A., M. BUECHNER, AND V. V. KRISHNAN. 1987. The effects of edge permeability and habitat geometry on emigration from patches of habitat. Am. Nat. 129:533–552.

SWANSON, F. J., T. K. KRATZ, N. CAINE, AND R. G. WOODMANSEE. 1988. Landform effects on ecosystem patterns and processes. BioScience 38:92–98.

TAYLOR, M. W., C. W. WOLFE, AND W. L. BAXTER. 1978. Land-use change and ring-necked pheasants in Nebraska. Wildl. Soc. Bull. 6:226–230.

TURNER, M. G. 1987. Land use changes and net primary production in the Georgia, USA, landscape: 1935–1982. Environ. Manage. 11:237–247.

U.S. DEPARTMENT OF COMMERCE, BUREAU OF CENSUS. 1985. 1982 census of agriculture. Vol. 2. Subject series. Part 1. Graphic summary. AC82-SS-1. Bur. Census, Agric. Div., Washington, D.C. 188pp.

U.S. SOIL CONSERVATION SERVICE. 1986. Resource planning guidebook. U.S. Soil Conserv. Serv., Champaign, Ill. Var. pagin.

VANDER ZOUWEN, W. 1990. State and provincial programs for habitat enhancement on private agricultural lands. Pages 64–83 *in* K. E. Church, R. E. Warner, and S. J. Brady, eds. Perdix V: gray partridge and ring-necked pheasant workshop. Kans. Dep. Wildl. Parks, Emporia.

VAN EMDEN, H. F. 1965. The role of uncultivated land in the biology of crop pests and beneficial insects. Sci. Hort. 17:121–136.

VERNER, J., M. L. MORRISON, AND C. J. RALPH, EDITORS. 1986. Wildlife 2000: modeling habitat relationships of terrestrial vertebrates. Univ. Wisconsin Press, Madison. 470pp.

WARBURTON, D. B., AND W. D. KLIMSTRA. 1984. Wildlife use of no-till and conventionally tilled corn fields. J. Soil Water Conserv. 39:327–330.

WARNER, R. E. 1983. An adoption model for roadside habitat management by Illinois farmers. Wildl. Soc. Bull. 11:238–249.

———. 1984. Effects of changing agriculture on ring-necked pheasant brood movements in Illinois. J. Wildl. Manage. 48:1014–1018.

———. 1988. Habitat management: how well do we understand the pheasant facts of life? Pages 129–146 *in* D. L. Hallett, W. R. Edwards, and G. V. Burger, eds. Pheasants: symptoms of wildlife problems on agricultural lands. North-Cent. Sect., The Wildl. Soc., Milwaukee, Wis.

———. 1992. Nest ecology of grassland passerines on road rights-of-way in central Illinois. Biol. Conserv. 59:1–7.

———, AND S. L. ETTER. 1985. Farm conservation measures to benefit wildlife, especially pheasant populations. Trans. North Am. Wildl. Nat. Resour. Conf. 50:135–141.

———, AND ———. 1989. Hay cutting and the survival of pheasants: a long-term perspective. J. Wildl. Manage. 53:455–461.

———, ———, G. B. JOSELYN, AND J. A. ELLIS. 1984. Declining survival of ring-necked pheasant chicks in Illinois agricultural ecosystems. J. Wildl. Manage. 48:82–88.

———, AND S. P. HAVERA. 1989. Relationships of conservation tillage to the quality of wildlife habitat in row-crop environments of the midwestern United States. J. Environ. Manage. 29:333–343.

———, ———, L. M. DAVID, AND R. J. SIEMERS. 1989. Seasonal abundance of waste corn and soybeans in Illinois. J. Wildl. Manage. 53:142–148.

———, AND G. B. JOSELYN. 1986. Responses of Illinois ring-necked pheasant populations to block roadside management. J. Wildl. Manage. 50:525–532.

WEGNER, J. F., AND G. MERRIAM. 1979. Movements by birds and small mammals between a wood and adjoining farmland habitats. J. Appl. Ecol. 16:349–357.

WESTEMEIER, R. L. 1988. An evaluation of methods for controlling pheasants on Illinois prairie-chicken sanctuaries. Pages 267–288 *in* D. L. Hallett, W. R. Edwards, and G. V. Burger, eds. Pheasants: symptoms of wildlife problems on agricultural lands. North-Cent. Sect., The Wildl. Soc., Milwaukee, Wis.

WHITCOMB, R. F., C. S. ROBBINS, J. F. LYNCH, B. L. WHITCOMB, M. K. KLIMKIEWICZ, AND D. BYSTRAK. 1981. Effects of forest fragmentation on avifauna of the eastern deciduous forest. Pages 125–205 *in* R. L. Burgess and D. M. Sharpe, eds. Forest island dynamics in man-dominated landscapes. Springer-Verlag, New York, N.Y.

WOOLEY, J. B., JR., L. B. BEST, AND W. R. CLARK. 1985. Impacts of no-till row cropping on upland wildlife. Trans. North Am. Wildl. Nat. Resour. Conf. 50:157–168.

———, R. WELLS, AND W. R. EDWARDS. 1988. Pheasants Forever, Quail Unlimited: the role of species constituency groups in upland wildlife management. Pages 111–127 *in* D. L. Hallett, W. R. Edwards, and G. V. Burger, eds. Pheasants: symptoms of wildlife problems on agricultural lands. North-Cent. Sect., The Wildl. Soc., Milwaukee, Wis.

WOOLHOUSE, M. E. J., AND R. HARMSEN. 1987. Just how unstable are agroecosystems? Can. J. Zool. 65:1577–1580.

YAHNER, R. H. 1981. Avian winter abundance patterns in farmstead shelterbelts: weather and temporal effects. J. Field Ornithol. 52:50–56.

27

MANAGING RANGELANDS FOR WILDLIFE

John G. Kie, Vernon C. Bleich, Alvin L. Medina, James D. Yoakum, and Jack Ward Thomas

INTRODUCTION

Rangelands are plant communities dominated by grasses, forbs, and shrubs. The primary use of rangelands by humans worldwide is to support livestock, but those communities also provide habitat for wildlife. Traditionally, wildlife-related concerns of range managers have been directed toward predators of livestock and wildlife species that are hunted. Today, range managers increasingly must consider other species as well. Management of public rangelands in the United States is constrained by both federal and state laws, and those laws require managers to address the impact of livestock grazing on all wildlife.

Land managers are being asked more and tougher questions about how range management activities affect wildlife. Often there are no easy answers. Information may be available on a narrow aspect of a problem based on studies conducted at a single location over a short time. However, habitat change and other impacts of livestock grazing can vary widely from one ecological situation to another. Those effects can be harmful to wildlife in one situation and beneficial in another. Annual variations in weather and resulting vegetation conditions also can be extreme. Therefore, much of the existing literature may seem contradictory.

This chapter summarizes some of that information and provides a framework for its analysis. It discusses range condition concepts and the use of wildlife habitat relationships models, livestock management practices that can

affect wildlife, and technical details of developing water sources on arid rangelands and constructing fences with wildlife considerations in mind.

PLANT SUCCESSION AND WILDLIFE MANAGEMENT GOALS FOR RANGELANDS

Plant succession is the gradual replacement of one assemblage of plant species with others through time until some relatively stable climax community is reached (Clements 1916). As each group of plant species is replaced, the value of the community as habitat to any particular species of wildlife changes. The result of these changes in habitat is a succession of wildlife species (Lyon [paper presented at the Symposium on land classifications based on vegetation: applications for resource management, Univ. Idaho, Moscow, 1987], Kie and Thomas 1988).

For example, as succession in Wisconsin proceeds from open prairie to communities with shrubs and some hardwoods, sharp-tailed grouse replace prairie-chickens. As woodlands and then climax forests replace more open habitats, ruffed grouse and ultimately spruce grouse become dominant (Grange 1948). In the Great Basin of the western United States, livestock grazing affects both habitat structure and rodent populations, and most of the variability in rodent species diversity can be explained by plant diversity (Hanley and Page 1982).

In north-central Colorado, horned larks, McGown's longspurs, and mountain plovers are associated with early-seral plants such as common buffalograss and blue grama

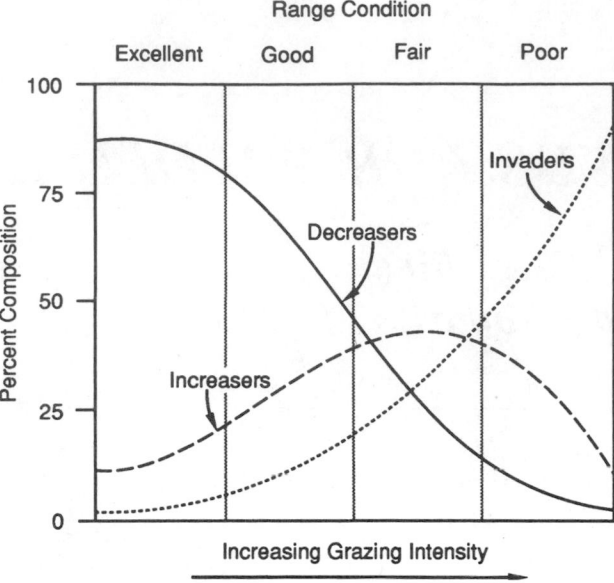

Fig. 1. The relationship of percent cover in decreaser, increaser, and invader plant species, grazing intensity, and traditional definitions of range condition (after Dyksterhuis 1949, Stoddart et al. 1975).

(Ryder 1980). Similarly, western meadowlarks, lark buntings, and Brewer's sparrows are associated with late-stage plants such as western wheatgrass, needlegrass, and fourwing saltbush (Ryder 1980).

As plant communities and populations of primary consumers undergo successional changes, resulting changes can occur upwards through trophic levels. In the late 1950s, Lindheimer pricklypear was common on the Welder Wildlife Refuge in southern Texas, and coyotes ate pricklypear fruit during the summer months (Andelt et al. 1987). Pricklypear also provided cover for southern plains woodrats on which coyotes preyed extensively during winter. Above-average precipitation and reductions in livestock grazing resulted in declines of pricklypear and woodrats by the early 1970s, and both became rare in coyote diets (Andelt et al. 1987).

Numbers of Swainson's hawks have decreased throughout a major portion of their range in the western United States (Sharp 1986). In northeastern California, successional changes in plant communities caused by heavy livestock grazing have resulted in the reduction of native bunchgrasses and the increase of big sagebrush and rabbitbrush (Sharp 1986). Because few microtine rodents are available as prey in these modified habitats, Swainson's hawks now spend most of their foraging time in mowed alfalfa fields, feeding primarily on Belding's ground squirrels (Woodbridge, U.S. For. Serv., Klamath Natl. For., Goosenest Ranger Dist., Macdoel, Calif., 1986).

Range Condition and Wildlife Habitat

Rangelands exist in many different successional stages and structural conditions because of the influences of fire, mechanical disturbance, herbicide treatment, and grazing by wild and domestic herbivores. Plant communities respond to grazing by livestock in a more or less predictable way, depending on the plant species present (Dyksterhuis 1949, Stoddart et al. 1975). Some plant species are dom-

inant in climax communities because they are superior competitors in the absence of disturbance. These species, referred to as decreasers, are often the most palatable to livestock and most susceptible to grazing pressure. They begin to decline in vigor and abundance with increased grazing pressure (Dyksterhuis 1949, Stoddart et al. 1975). As they decline, other less palatable plants present at climax become more abundant with relief from competition. These species are called increasers. If grazing intensity is sufficiently heavy and occurs over a long enough period of time, the increasers also begin to decline, and new plant species called invaders, well-adapted to heavy grazing, appear in the community. The relationship among these three types of plants can be expressed along a successional gradient (Fig. 1).

Traditionally, rangelands have been managed on a concept of range condition based on how closely existing vegetation approximates a climax community (Dyksterhuis 1949). Sites dominated by decreasers are classified in excellent condition, and those made up mostly of invaders are judged to be in poor condition (Fig. 1).

This procedure is used extensively in many areas of the western United States for purposes of livestock management. However, it cannot be used on seeded rangelands or those dominated by introduced, naturalized plant species such as the annual grasslands of California (Smith 1978, 1988). Also, the range condition terms excellent, good, fair, and poor are defined in terms of providing forage for livestock. However, wildlife habitat needs may differ greatly among species. A site rated as poor may provide optimum habitat for a wildlife species dependent on early-seral vegetation, and a site rated excellent may not be used at all. As a first step to correcting some of these problems, the terms excellent, good, fair, and poor should be replaced with climax, late seral, mid-seral, and early seral (Pieper and Beck 1990).

For purposes of managing wildlife habitat, descriptions of range condition must be based on specific management goals for individual or groups of wildlife species (Smith 1978). This might lead to a system of classifying range condition based on management objectives for soil stability and protection, and specific management goals for other resource values such as wildlife habitat, riparian areas, or fish habitat (Schlatterer 1986). In this system, a rating is assigned for each resource value. Overall range condition would be judged satisfactory only if a site was judged satisfactory under all objectives, or, if unsatisfactory, there was a trend towards improving conditions for each objective.

Consider a situation where primary emphasis is placed on the production of forage and cover for mule deer, with secondary emphases on livestock forage production and other resource values. First, a determination is made as to whether soil stability goals are being met. If not, overall condition is unsatisfactory. If soils are stable, mule deer cover and forage production concerns are addressed. If those objectives are not being met and conditions are not improving, the resource value rating for mule deer habitat and overall range condition is unsatisfactory. Management objectives for livestock forage production and other resource values are considered in turn. Overall condition is satisfactory only if the site is judged satisfactory with respect to all objectives, or if the trend among those objec-

Table 1. Generalized structural conditions of rangelands and related environmental variables. Number of asterisks indicates the range of environmental variable from * (least) to ***** (greatest) (from Maser et al. 1984).

Environmental variable	Structural condition				
	Grass–forb	Low–shrub	Tall–shrub	Tree	Tree–shrub
Vegetation height	*	**	***	*****	*****
Canopy closure	*	***	****	***	*****
Canopy volume	*	***	****	***	*****
Plant diversity	**	***	***	*	***
Structural diversity	*	**	***	****	*****
Herbage production	*****	***	****	**	*
Browse production	*	****	*****	**	***
Animal diversity	*	**	***	****	*****

tives rated unsatisfactory is improving. A range condition rating system similar to that described was adopted in 1991 by the U.S. Forest Service (E. R. Schlatterer, pers. commun.).

Models of Rangelands as Wildlife Habitat

The system of classifying wildlife habitats according to potential natural vegetation and seral stage for coniferous forests (Thomas 1979) also was applied to rangeland vegetation in southeastern Oregon (Maser et al. 1984). In that model, habitat data were assembled for 341 species of vertebrates, and the impacts of range management activities on wildlife were weighed. This was done by equating plant communities and their structural conditions with habitat values for wildlife. The structural conditions were grass-forb, low-shrub, tall-shrub, tree, and tree-shrub. As a plant community progresses from grass-forb to tree-shrub conditions as a result of succession, changes occur in various environmental variables important to wildlife (Table 1). Management actions also can result in changes in structural conditions (Table 2) (Maser et al. 1984).

Accounting for large numbers of wildlife species makes land-use planning difficult. The process can be simplified by grouping wildlife species into life-forms based on the relationship of the species to their habitats. In southeastern Oregon, two characteristics of each species (where it feeds and where it reproduces) were used to distinguish 16 life-forms. For example, dark-eyed juncos and mule deer characterize those species that feed and reproduce on the ground. Other examples include long-toed salamanders and western toads, which feed on the ground, in shrubs, or in trees and reproduce in water (Maser et al. 1984). By using information on habitat structural conditions necessary for various wildlife life-forms, coupled with the effects of different management practices on structural conditions, the model predicts which life-forms will benefit and which will not under different land management alternatives (Maser et al. 1984).

Wildlife-habitat relationships models such as the one for southeastern Oregon are potentially useful for managing rangeland wildlife habitats. Similar approaches have been suggested elsewhere (Short 1986); their acceptance and utility will be dependent on extensive field testing and validation of model predictions.

MANAGING RANGELAND LIVESTOCK

Heavy livestock grazing has been detrimental to many wildlife species in western North America (Smith 1977, Gallizioli 1979). Excessive, uncontrolled grazing clearly can dramatically affect the structure and composition of wildlife habitats (Fig. 2). Where such adverse impacts occur, elimination of livestock can improve habitat conditions (Fig. 3), although in many situations changes in livestock management practices can result in similar benefits. When properly managed, livestock grazing can be used to improve habitat for wildlife species dependent on early-seral stage plant communities (Longhurst et al. 1976, Urness 1976, 1990, Kie and Loft 1990). Information on relationships between livestock and wildlife is available in a variety of books, symposium proceedings, and review papers (Smith 1975, Townsend and Smith 1977, Schmidt and Gilbert 1978, DeGraaf 1980, Wallmo 1981, Peek and Dalke 1982, Thomas and Toweill 1982, Menke 1983, Severson and Medina 1983, Halls 1984, Severson 1990).

The relationship between grazing and wildlife habitat is complex. Livestock influence wildlife habitat by modifying plant biomass, species composition, and structural components such as vegetation height and cover (Fig. 4). The impacts of livestock grazing on wild ungulates can be classified as direct negative, indirect negative, operational, and beneficial (Mackie 1978, Wagner 1978). A direct negative impact is competition between cattle and deer for a resource such as food or cover (Mackie 1978, Wagner 1978). Competition occurs when two organisms use a resource in short supply, or when one organism harms another in the process of seeking the resource (Birch 1957, Wagner 1978). Factors affecting impacts of livestock on wildlife include diet similarity, forage availability, animal distribution patterns, season of use, and behavioral interactions (Nelson and Burnell 1975, Severson and Medina 1983).

Indirect negative impacts of cattle grazing include (1) gradual reductions in the vigor of some plants and in the amount and quality of forage produced, (2) elimination or reduction of the ability of forage plants to reproduce, (3) reduction or elimination of locally important cover types and replacement by less favorable types or communities, either by direct actions over time or by changing the rate of natural successional processes, and (4) general alterations and reduction in the kinds, qualities, and amounts

Table 2. Anticipated changes in range-community structural conditions resulting from management actions (from Maser et al. 1984).[a]

Management action	Structural condition				
	Grass-forb	Low-shrub	Tall-shrub	Tree	Tree-shrub
Weed control	<	>	>	0	0
Brush control					
Chemical	—	<	<	—	<
Mechanical	—	<	<	—	<
Biological	—	<	<	—	<
Tree control					
Chain and bull-doze	—	—	—	<	<
Clearcutting	—	—	—	<	<
Shelterwood	—	—	—	<	<
Thin	—	—	—	<	<
Salvage	—	—	—	<	<
Debris disposal	—	—	—	<	<
Prescribed burn					
Cold	<	<	<	0	<
Hot	<	<	<	<	<
Seeding and planting					
Grasses and forbs	>	<	<	<	<
Shrubs	>	0	0	>	0
Trees	>	>	>	0	0
Fertilization	0	<>	<>	<>	<>
Soil treatment					
Pitting	<	<	<	<	<
Contouring	<	<	<	<	<
Water					
Water spreading	<	<	<	<	<
Drainage	>	>	>	>	>
Water development	<	<	<	0	0
Grazing					
Cattle	<>	<>	<>	0	0
Horses	<>	<>	<>	0	0
Sheep	<>	<	<	0	<
Goats	<>	<	<	0	<

[a]> = increases structural diversity, < = decreases structural diversity, <> = could increase or decrease structural diversity, 0 = no effect on structural diversity, — = not a viable practice.

Fig. 2. Heavy livestock grazing can have dramatic effects on the composition and structure of wildlife habitats (photo by J. Kie).

Borchert and Jain 1978) and the degree to which they compete with livestock for forage (Fitch and Bentley 1949, Howard et al. 1959). Because of their size and susceptibility to predation, rodents, rabbits, and other small mammals are highly dependent on the structure of their habitats (Grant et al. 1982, Parmenter and MacMahon 1983, Bock et al. 1984). Grazing by livestock influences vegetation structure in those habitats and can significantly affect small mammal populations (Reynolds and Trost 1980).

Livestock grazing adversely affects many grassland birds, although moderate grazing can be neutral or beneficial to some species (Buttery and Shields 1975). Livestock management practices also can affect birds indirectly. For example, an organophosphate insecticide externally applied to cattle to control warbles may kill black-billed magpies and cause secondary mortality among red-tailed hawks eating the poisoned magpie carcasses (Henny et al. 1985).

Livestock management practices that can affect wildlife habitats and populations include livestock numbers, timing and duration of grazing, animal distribution, livestock types, and specialized grazing systems. These practices

of preferred or otherwise important plants through selective grazing or browsing or other activities (Mackie 1978).

Operational impacts (Mackie 1978) include fence construction, water development (Evans and Kerbs 1977, Wilson 1977, Yoakum 1980), and brush control (Holechek 1981). These impacts can be adverse or beneficial. Disturbance associated with the handling of livestock is another operational impact. For example, deer may temporarily move out of pastures when cattle roundups occur (Hood and Inglis 1974, Rodgers et al. 1978).

Information is also available on the effects of small mammals on rangeland vegetation (Moore and Reid 1951, Wood 1969, Batzli and Pitelka 1970, Turner et al. 1973,

Fig. 3. Changes in livestock management practices can provide wildlife benefits similar to those in this ungrazed meadow (photo by L. Ritter).

Fig. 4. Cattle grazing in aspen habitats can alter understory structure and reduce hiding cover for wildlife (photos by J. Kie).

can be modified to reduce or eliminate adverse effects on wildlife, and sometimes to enhance wildlife habitats (Severson 1990).

Livestock Numbers

Livestock numbers, or stocking rates, usually are specified by animal unit months (AUMs). One AUM is one animal unit (one mature cow with a calf, or equivalent) grazed for 1 month (Heady 1975:117). Livestock effects on wildlife become more pronounced with increasing stocking rates, and the relationship is often nonlinear. A few cattle in a pasture may have no discernible effect on wildlife, but beyond some threshold wildlife response may increase rapidly. Furthermore, a range manager's traditional definition of proper grazing is based on maintaining a mix of plant species valuable as livestock forage and preventing soil erosion. Optimum livestock densities for wildlife may occur at different, and often lower, stocking rates.

Tame mule deer in Utah preferred to forage on areas ungrazed by cattle when first introduced to study enclosures. However, as use levels of deer increased, selectivity for ungrazed areas disappeared. In areas grazed by cattle, deer ate more grass and browse and fewer forbs than in ungrazed areas (Austin and Urness 1986).

White-tailed deer in Texas ate more forbs in ungrazed pastures than in grazed pastures, regardless of livestock stocking rates (McMahan 1964). Mule deer in Montana also ate more forbs when feeding in pastures ungrazed by cattle (Knowles 1975). Under poor range conditions, continued grazing by livestock can adversely affect white-tailed deer through competition for food (McMahan and Ramsey 1965). Conversely, abundant grasses on ranges in excellent condition may relieve livestock pressure on forbs and lessen the effects of competition (Bryant et al. 1979).

On summer ranges in California, cattle stocking rates affected mule deer hiding cover, habitat use, home-range size, and activity patterns. Hiding cover for mule deer declined over the summer as a result of natural processes, even in the absence of cattle grazing (Loft et al. 1987). However, hiding cover is most important early in the summer when fawns are young. Cattle grazing accelerated the decline in hiding cover for deer early in the summer, and the effects were nonlinear, heavy grazing having a more pronounced effect (Fig. 5). Deer home ranges were smallest in the absence of cattle and were centered primarily in creek-bottom, meadow-riparian habitats. Under heavy cattle grazing, home-range locations did not change but became larger and included areas with steeper slopes (Loft 1988). Patterns of habitat use also changed (Loft et al. 1991), and deer spent more time feeding as levels of cattle grazing increased, all factors that may have had adverse

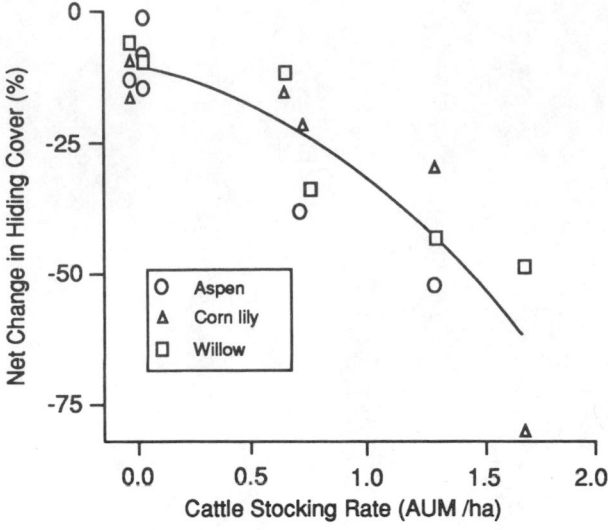

Fig. 5. Net change in mule deer hiding cover between 0 and 1 m in height, from beginning of summer until mid-August, as a function of cattle stocking rate (AUM/ha = animal unit months per hectare) (after Loft et al. 1987).

energetic and population consequences for deer (Kie et al. 1991).

Small mammals also are affected by livestock grazing. More Merriam's kangaroo rats were counted on grazed than on ungrazed grasslands in southern Arizona, and their abundance increased as the native, perennial grass cover declined (Reynolds 1950). White-footed mice were more abundant in grazed than in ungrazed woodlands in Ohio (Dambach 1944), but northern short-tailed shrews were less abundant and woodland voles were absent from grazed areas. Deer mice in Utah preferred pastures from which the least amount of grass cover had been removed by livestock (Frischknecht 1965). In Nevada, small mammal density, species richness, and species diversity were higher in an ungrazed riparian exclosure than in the surrounding area grazed by cattle (Medin and Clary 1989). In Idaho, small mammal density was lower but species richness and species diversity were higher in grazed sites. Deer mice were almost twice as abundant in the grazed area, but montane voles were more common in the ungrazed exclosure (Medin and Clary 1990). These studies emphasize the variety in wildlife responses to grazing reported in the literature. As with most effects of livestock on wildlife, such responses can be difficult to interpret because of inherent site differences (Johnson 1982), as well as differences in grazing intensity, timing, and duration.

Similarly, the response of pocket gophers to livestock grazing depends on range type, intensity and season of grazing, and other factors. Heavy grazing by livestock often results in increased numbers of pocket gophers because of the abundance of deep-rooted and bulbous forbs (Buechner 1942, Ellison 1946, Tevis 1956). However, in Colorado gopher density was twice as high on ungrazed ranges as on grazed areas (Turner et al. 1973).

Changes in livestock grazing practices can affect invertebrate populations that indirectly affect insectivores. As cattle grazing pressure increased in one study, diets of vagrant shrews shifted from flightless invertebrates to fly-

ing insects and caterpillars (Whitaker et al. 1983). The diet shift was related to the decline of flightless invertebrates resulting from cattle trampling and increased soil compaction.

Avian species such as California quail, Gambel's quail, and mountain quail are scrub- and forest-adapted species as a general rule (Brown 1978). Often, their reproductive cycles are related to precipitation, and population sizes often are dependent on annual reproductive success. In such situations, they are not seriously affected by moderate livestock grazing. Conversely, northern bobwhites, Montezuma quail, prairie-chickens, and sharp-tailed grouse are grassland-adapted species. In the southwestern United States, they occur in continental-type climates, reproduction is less dependent on weather, and populations can be more dependent on annual carryover of adult birds. Under such conditions these species are adversely affected by even moderate levels of livestock grazing (Brown 1978).

Livestock grazing does not limit the availability of food for Mearns quail in Arizona, but grazing that results in excessive removal of escape and hiding cover can eliminate breeding populations (Brown 1982). In such situations, grazing should be managed to maintain adequate levels of cover. Livestock grazing also should be carefully controlled or eliminated in Arizona where the endangered masked bobwhite occurs (Goodwin and Hungerford 1977).

Livestock grazing in the arid portions of central and southern California can adversely affect California quail by reducing food supplies and, where shrub cover is lacking, hiding cover (Leopold 1977). However, where rainfall is more plentiful and shrubs are abundant, moderate grazing by livestock can benefit quail by creating open areas and enhancing forb growth (D. A. Duncan, unpubl. abstr., Annu. Meet. Soc. Range Manage., 1980).

In southern Texas, eastern meadowlarks are more abundant under continuous grazing by cattle at moderate rates, but mourning doves are more abundant under heavy grazing (Baker and Guthery 1990). The loss of simulated ground nests in Oklahoma increases exponentially when moderate cattle stocking rates are increased to heavier levels (Jensen et al. 1990). In the southeastern United States, grazing can be detrimental to northern bobwhites because of the removal of food plants (Stoddard 1931, Stoddard and Komarek 1941). However, moderate grazing may benefit bobwhites by reducing heavy grass cover, allowing areas for travel lanes and dusting, and encouraging the growth of forbs (Reid 1954, Moore and Terry 1979).

Because livestock concentrate around sources of water, uncontrolled grazing can be detrimental to nesting waterfowl (Kirsch 1969) and other birds dependent on wetland habitats. Adverse effects can be reduced by limiting cattle stocking rates and by fencing off sections of ponds that are badly overgrazed (Bue et al. 1952). For a thorough review of the effects of livestock on nesting waterfowl, see Kantrud (1990).

Timing and Duration of Grazing

The timing and duration of livestock grazing are factors that influence wildlife and their habitats. Moderate cattle grazing in late autumn in Colorado had no detectable impact on six species of birds dependent on the grass-herb-shrub layer for foraging, nesting, or both (Sedgwick and Knopf 1987). However, summer grazing can eliminate such habitat specialists as willow flycatchers, Lincoln's sparrows, and white-crowned sparrows (Knopf et al. 1988).

The time of year that livestock are present can alter the composition of plant communities. Heavy grazing during a period of rapid growth of one plant species will favor other species that grow more rapidly at other times. For example, spring grazing in annual grasslands in California reduces grass cover and encourages the growth of summer-maturing forbs such as turkeymullein, the seeds of which are readily eaten by mourning doves (Kie 1988). Conversely, many wildlife species are most susceptible to livestock-induced changes in habitat during their reproductive seasons. Birds that nest on the ground or in shrubs can experience reproductive losses if their nests are trampled or otherwise destroyed by cattle.

Willow flycatchers in California breed exclusively in riparian deciduous woodlands and prefer willows as nesting substrate (Valentine et al. 1988). Recent surveys suggest that fewer than 125 breeding pairs are present in California (Serena, Calif. Dep. Fish Game, Wildl. Manage. Div. Adm. Rep. 88-3, 1982). Flycatchers prefer to nest near the edges of willow clumps or along livestock trails (Valentine et al. 1988, Sanders and Flett 1989), where they are susceptible to physical disturbance. In one study, four of 20 willow flycatcher nests found in a 4-year period were destroyed by cattle before the young were fledged, and four other nests were destroyed after the young fledged (Valentine et al. 1988). When cattle stocking levels were reduced and 75% of the remaining cattle were confined to a fenced pasture away from willow flycatcher nest sites until 15 July, no willow flycatcher nests were lost (Valentine et al. 1988).

As previously discussed, loss of hiding cover early in the summer when mule deer fawns are young can be accelerated by excessive grazing (Loft et al. 1987). Such conflicts can be minimized or eliminated by delaying grazing until later in the year (Kie 1991). The timing and duration of grazing, and livestock distribution, are the bases for various specialized grazing systems discussed later.

Livestock Distribution

Livestock congregate around sources of water, supplemental feed, and mineral blocks, and their impacts are most pronounced in those areas. Riparian zones, because of their abundant forage and water, are good examples of livestock concentration areas. Cross-fencing, developing alternative water sources, and providing feeding supplements on upland sites away from riparian areas more evenly distribute livestock. However, in some situations, wildlife species benefit from patchy livestock distribution because some areas remain only lightly grazed.

Livestock Types

Effects of grazing on wildlife depend on the species of livestock. Differences in food habits between cattle and sheep dictate their effects on plant species composition. Also, cattle are usually free-ranging within the confines of a fenced range allotment, but sheep often are herded. Herded bands of sheep may have enhanced some habitats for mule deer in California (Longhurst et al. 1976). How-ever, the transmission of diseases from domestic sheep to wild mountain sheep may have eliminated many populations of the latter from California (Wehausen et al. 1987).

Competition between pronghorns and domestic sheep is greater than between pronghorns and cattle because of greater overlap in forage preferences. On overgrazed sheep ranges, insufficient forb growth was available for pronghorns during the critical midwinter period, and pronghorn die-offs were common (Buechner 1950). In general, domestic sheep are more likely to adversely affect pronghorns than are cattle (Autenrieth 1978, Salwasser 1980, Yoakum 1980, Kindschy et al. 1982), and even moderate use by sheep during the winter dormant period can leave range units unsuitable for pronghorns until plant regrowth in spring (Clary and Beale 1983).

Cows with calves often exhibit grazing patterns different from those of steers, and differences among breeds of cattle and sheep may occur as well. All can be factors in how livestock affect wildlife.

Specialized Grazing Systems

Many specialized grazing systems exist, although most can be classified into a few different types (Heady 1975, Stoddart et al. 1975). *Continuous grazing* allows livestock to graze season-long or year-long. *Deferred grazing* refers to delaying or deferring grazing until after most of the range plants have set seed. Deferred grazing allows plants to grow, store carbohydrates, and reproduce at high rates. *Rotational grazing* involves dividing a range unit and rotating livestock through those different pastures.

Combinations of periodic deferment and rotational grazing are called *deferred-rotation grazing* systems. The most common of these is the *four-pasture deferred-rotation* system, in which four range units or pastures are used, three being grazed year-long and the fourth being deferred for 4 months. The pastures are then rotated each year.

Rest-rotation grazing is similar to a deferred-rotation system, but the period of rest consists of a full year or more. *Short-duration grazing* systems are similar to deferred-rotation systems, except that many small pastures are used (≥8–10 are not uncommon), stocking rates are high in any one pasture, and the livestock are present for short periods of time. Because timing of livestock grazing is critical to most rangeland wildlife species, rotational grazing systems designed with wildlife in mind have the potential to reduce adverse effects.

White-tailed deer in Texas preferred pastures with frequent, periodic deferment from cattle grazing in one study (Reardon et al. 1978). Deer densities were highest in a short-duration grazing system, lower under a four-pasture rotational grazing system, and lowest under heavy, continuous grazing. However, in another study, white-tailed deer tended to avoid concentrations of cattle and traveled more under a short-duration grazing system (Cohen et al. 1989). In addition, water sources in the center of a pasture grazed under a short-duration system may be largely unavailable to white-tailed deer and some other wildlife species because of increased livestock concentrations (Prasad and Guthery 1986).

On the Sheldon National Wildlife Refuge in Nevada, no differences in small mammal abundance or diversity were detected between an ungrazed area and an area man-

Fig. 6. Cattle grazing in late winter and early spring on foothill, annual-grass rangelands in California encourages the growth of forbs that are valuable to many wildlife species (photo by S. Westfall).

aged under a rotational grazing system (Oldemeyer and Allen-Johnson 1988). In this instance, the allotment was grazed from mid-June through early August one year, and from early August through late October the following year.

Short-duration and deferred-rotation cattle grazing systems in Texas provided better cover for northern bobwhites and scaled quail than did continuous grazing (Campbell-Kissock et al. 1984). Both systems resulted in greater grass cover and weight and provided better cover during periods of drought. In a second study in Texas, northern bobwhites were twice as numerous under short-duration grazing than under continuous grazing (Schulz and Guthery 1988). But in a third study, no differences were observed between short-duration and continuous cattle grazing systems in cover, density, or dispersion of nesting cover for northern bobwhites and wild turkeys (Bareiss et al. 1986). For additional information on the effects of short-duration grazing on wildlife, see Guthery et al. (1990).

Loss of artificial nests also does not appear to be affected by type of grazing system (Koerth et al. 1983, Bareiss et al. 1986). However, nest loss among wild turkeys can be higher under deferred-rotation and short-duration grazing than under continuous grazing in some instances (Baker 1979).

Nesting waterfowl densities in Montana were highest on pastures excluded from cattle grazing the previous year under a rest-rotation grazing system and lowest in pastures grazed during late summer and autumn (Gjersing 1975, Mundinger 1976). Early spring grazing during the current year also reduced breeding waterfowl densities (Mundinger 1976). The rest-rotation grazing system allowed an accumulation of plant material used as nesting cover, if rested pastures were not grazed too early the following spring.

Deferred-rotation grazing in Texas was suggested as a method to reduce cattle impacts on shoreline vegetation and nesting waterfowl such as fulvous whistling-ducks, purple gallinules, common moorhens, and American coots (Whyte and Cain 1979, 1981). Fencing one-half of the shoreline of stock-ponds also was recommended. In northeastern California, delaying grazing until after 15 July was

recommended to provide additional residual vegetation for nesting waterfowl (Ruyle et al. 1980).

Rest-rotation grazing may have the most potential to provide benefits to wildlife. Although often economically disruptive in terms of livestock forage foregone, such losses may be compensated for several times over by benefits derived from wildlife-related recreation on public lands. For example, the development of a rest-rotation grazing system in a single deer-hunting zone in California might specify that each range unit would be grazed only 1 of 3 years. The value of livestock forage foregone, calculated on the basis of net economic value at $12.82 per AUM, would equal about $71,000 over each 3-year grazing cycle. However, increased deer populations and additional hunting opportunities would be valued at $6.5 million over the same period (Loomis et al. 1991).

Using Livestock To Manage Wildlife Habitat

In some situations, prescribed livestock grazing can be used to manage wildlife habitat (Longhurst et al. 1976, Holechek 1980, 1982, Longhurst et al. 1982, Urness 1982, 1990, Severson 1990). It has been applied to the management of habitat for species as diverse as mule deer (Smith et al. 1979, Willms et al. 1979, Reiner and Urness 1982), northern bobwhites (Moore and Terry 1979), and Canada geese (Glass 1988). For example, cattle grazing in late winter and spring on foothill, annual grasslands in California encourages the growth of forbs that are valuable to many wildlife species (Fig. 6).

Fire suppression and a century of heavy livestock grazing on foothill rangelands in northern Utah created shrub-dominated communities that supported large herds of mule deer (Urness 1982, 1990). Later reductions in livestock numbers were accompanied by declines in mule deer populations. Prescribed grazing perhaps could be used to increase deer numbers, although the infrastructure (adequate fencing) necessary to manage livestock is no longer in place, and land managers may not be willing to make the commitment necessary to implement such a program (Urness 1990).

In other situations, the application of prescribed grazing has met with mixed results. In some instances, elk and red deer prefer feeding areas that previously were grazed by cattle (Grover and Thompson 1986, Gordon 1988). After establishment of a wildlife management area in northeastern Oregon and the exclusion of cattle in 1961, elk numbers increased from 120 to about 320 animals (Anderson and Scherzinger 1975). A resource management plan implemented in 1964 included a cattle grazing system designed to increase winter forage quality for elk. By 1974, more than 1,100 elk were counted on the area (Anderson and Scherzinger 1975).

Conversely, spring cattle grazing did not increase winter use by elk in southeastern Washington and actually resulted in a decrease in elk use during 1 of 3 years studied (Skovlin et al. 1983). Moreover, although clipping bluebunch wheatgrass at four different phenological stages during the spring resulted in increased crude protein, calcium, and phosphorus in a British Columbia study (Pitt 1986), another study in Montana indicated that summer cattle grazing had no effect on quality of winter forage for elk (Dragt and Havstad 1987).

Too often, the intent of using prescribed livestock grazing has been to manage habitat for a single species, whereas entire communities are affected. Using livestock to maintain a plant community in an early seral stage often will benefit those wildlife species dependent on such habitat, while at the same time adversely affect species associated with climax communities (Kie and Loft 1990).

The prescription for grazing is important. Maximizing benefits to wildlife from prescribed grazing almost always will involve reducing livestock numbers and shortening grazing seasons compared to management plans designed to maximize livestock production. In summary, livestock grazing in itself is neither good nor bad for wildlife, but depends on a variety of factors, including wildlife species of concern, livestock numbers, timing and duration of livestock grazing, livestock distribution, and kinds of livestock (Kie and Loft 1990). Wildlife and range managers should avoid generalizations and evaluate the role of livestock on wildlife and their habitats independently for each species, grazing plan, and management situation.

MANAGING RANGELAND RIPARIAN AREAS

Riparian areas are important habitats for terrestrial and aquatic wildlife (Carothers and Johnson 1975, Thomas et al. 1979*b,c*, Platts and Raleigh 1984, Skovlin 1984, Platts 1990). Their importance is a result of being obligate habitat for many aquatic species, of the uniqueness of their soil and vegetation complexes that produce diverse vegetation structure and concomitant diverse biological communities, and of their limited extent across a diversity of landscapes. Their value for a given species of wildlife is a function of water availability (for example, mule deer in the Sonoran Desert versus those in the pothole region of North America), life stages, animal movements, weather, and other factors.

Riparian vegetation and its structural arrangement have high value for wildlife. Many vertebrate and invertebrate species depend directly or indirectly on riparian vegetation for food, cover, or other life requisites. Wildlife use riparian zones disproportionately more than any other habitat type. For example, of 363 terrestrial species known to occur in the Great Basin of southwestern Oregon, 288 depend directly on riparian zones or use them more than other habitats (Thomas et al. 1979*c*). Herpetofaunas also are strongly associated with riparian areas (Jones 1988). Riparian soils and substrates are also important to amphibians, reptiles, and small mammals because these wildlife forms inhabit subsurface environments. Hence, the temperate microclimate, availability of moisture, and greater biomass production provide for complex food webs of which wildlife is a part.

The value of riparian areas to wildlife is only generally described, owing to the difficulty of long-term observations. Mule deer (Thomas et al. 1979*a*) and white-tailed deer (Compton et al. 1988) select woody riparian vegetation for cover and forage. Selected avifauna have demonstrated an affinity for distinct layers of vegetation (Gutzwiller and Anderson 1986). Riparian zones provide migration routes for birds, bats, deer, and elk (Wauer 1977). Such areas frequently are used by deer and elk as travel corridors between high-elevation summer ranges and low-elevation winter ranges.

Riparian habitats are of further importance because they comprise only about 1% of landscape in the United States (Knopf 1988); moreover, >70% of the original riparian habitats in the United States have been lost through various land use practices (Megahan and King 1985). Barclay (1978) reported natural riparian habitats within the Oklahoma grasslands to have nearly vanished, and channelization was responsible for conversion of 86% of bottomland forests to other land uses. In the Southwest, many historically perennial streams are largely ephemeral watercourses today (Johnson et al. 1989).

Central to development of management strategies for riparian areas are: (1) an understanding of what constitutes a riparian area, (2) their internal functions and processes, (3) the influences on riparian ecosystems, and (4) their importance to wildlife. Elmore (1989) agreed that a fundamental understanding of the functioning of riparian ecosystems was initially necessary to evaluate benefits and incorporate management actions into land use plans.

The primary function of rivers is to transport water and sediments (Jensen and Platts 1987). Riparian habitats are unique products derived from the dynamic processes that a given stream produces and are influenced by the interactions of climate, geology, geomorphology, hydrology, pedogenesis, and chemical and biological processes. No attempt is made to review all literature dealing with riparian interactions; instead, recent reviews are used as a base from which to discuss current advances in management of riparian areas for wildlife.

Little information is available on wildlife/riparian interactions. In general, this results in wildlife management considerations being excluded from land use plans (Dwyer et al. 1984, Dickson and Huntley 1987). However, much work has been done on riverine/riparian dynamics, and those references are used here to illustrate general concepts. The reader is encouraged to read reviews by Curtis and Ripley (1975), Thomas et al. (1979*b,c*), Brinson et al. (1981), Kauffman and Krueger (1984), Platts and Raleigh (1984), Skovlin (1984), Warner and Hendrix (1984), DeBano and Schmidt (1989), and Platts (1990).

Value, Structure, and Function of Riparian Areas

Riparian terminology has been proposed by several authors (Swanson et al. 1982, Johnson and Lowe 1985) who suggested disparity exists among users. Riparian areas are defined herein as the sum of the terrestrial and aquatic components characterized (1) by the presence of permanent or ephemeral surface or subsurface water, (2) by water flowing through channels defined by the local physiography, and (3) by the presence of obligate, occasionally facultative, plants requiring readily available water and rooted in aquatic soils derived from alluvium. Riparian ecosystems usually occur as an ecotone between aquatic and upland ecosystems, but they have distinct and variable vegetation, soil, and water characteristics. Typically, riparian areas are viewed as riverine habitats with perennial surface flows and associated plants and soils. However, surface flows may be ephemeral or periodic, as in desert washes or arroyos of the Southwest. In contrast, riparian areas in the eastern and southern regions of the U.S. are recognized as floodplain and bottomland hardwood forests.

The flow of water may fluctuate severely; hence, these

environments are subject to severe stresses that include lack of water, salinity, anoxia, and oxygen depletion (Kozlowski 1984, Hale and Orcutt 1987). Organisms that persist have biological adaptations to survive in the aquatic-semiaquatic environment (Mitsch and Gosselink 1986). Furthermore, there are distinctions among riparian areas of the northeastern, southern, northwestern, and southwestern United States (Swanson et al. 1982, Johnson and Lowe 1985). Distinctions are largely based on climate, which influences the vegetation and hydrology of these areas.

Riparian areas are valuable wetland habitats capable of cleansing polluted waters (Mitsch and Gosselink 1986), ameliorating the effects of floods (Skinner et al. 1989), and providing a diverse environment for flora and fauna (Thomas et al. 1979c). The associated values derived either directly or indirectly from riparian areas were noted in several symposia and are not recounted here (Johnson and Jones 1977, Swanson 1979, Johnson et al. 1985).

Miller (1987) identified six features of riparian areas with respect to the stream environment—food, large organic debris, solar energy regulation, streambank and bed stability, terrestrial-aquatic buffer, and streamflow regulation. The relevance of these factors to wildlife is through indirect pathways leading from complex food webs emanating from interactions of the physical, chemical, and biological components of the riparian system. In addition, food and cover are products derived and maintained by the diverse vegetation and microclimate. For example, bears or raccoons could forage on insects, fish, or plants produced by the interactions of organic debris, soils, water, and microorganisms while being provided various forms of cover.

In the last 2 decades wetland scientists have provided scientific evidence that riparian areas function as nutrient sinks, sources, filters, or transformers. Definitions of "sink, source, and transformers" as used here are those developed by Mitsch and Gosselink (1986:113). A wetland can be considered a sink if it has a net retention of an element, or specific form thereof, such that the inputs are greater than the outputs. As a sink, the output can be quite significant relative to the input and not simply an input-limited function. A source refers to a wetland that exports more of an element or material to another ecosystem than would occur without that wetland. A wetland is considered a transformer when changes such as chemical to particulate form occur, but the amount exported equals the amount imported. As filters, suspended materials and solids from liquids are removed such that dissolved nutrients readily pass through the ecosystem (Kuenzler 1988, Richardson 1988). Richardson (1988) emphasized the need to view these functions relative to some time frame, because a given wetland may be a sink or source, pending temporal changes. Also, a wetland can function as a combination of the above either simultaneously or singularly.

The ability of a wetland to use greater quantities and different types of nutrients is due partly to the nature of the physical environment itself and the type of vegetation, but more importantly to the individual plant. Vascular aquatic plants have structural and physiologic adaptations to persist in aquatic environments (Mitsch and Gosselink 1986, Hale and Orcutt 1987). Some streamside plants re-

portedly have 50 times greater photosynthetic capacity than marsh plants (Mendelssohn and Postek 1982) and are potentially better biomass producers. The preponderance of information suggests that streamside vegetation has ameliorating effects on nutrient inputs to streams (Lowrance et al. 1983, Gersberg et al. 1986). In addition, riparian plants exhibit varying degrees of tolerance to flooding (Kozlowski 1984) and are subject to distribution patterns dictated by floodwater levels. The buffering function of the riparian area between adjacent uplands and the aquatic zone is not well established for smaller streams but is recognized as important (Miller 1987).

The effects of soil nutrient concentrates on the aquatic and terrestrial systems are complicated and not easily understood. Water quality can be adversely affected by flushes of nutrients such as nitrogen into the aquatic system from terrestrial sources, especially during spring thaws when floodwater exports nutrient-rich detritus. These annual flushes may depauperate some riparian sites while enriching others (Brinson et al. 1981). Peterjohn and Correll (1984) reported an annual loss of 75% of total nitrogen and 41% of phosphorus in groundwater flow from a riparian forest. Surface runoff losses were 22% for nitrogen and 59% for phosphorus. These nutrient losses may have significant ecological effect on the aquatic system (Mitsch and Gosselink 1986, Hale and Orcutt 1987), as well as on the terrestrial system, to the point of inhibiting plant growth.

The influx of essential elements permits development of plant communities that are more diverse (in some instances rare) and more productive than surrounding habitats. This productivity has resulted in development of an equally diverse and productive terrestrial animal community, including many threatened, endangered, and sensitive species. In the Southwest, for example, riparian habitats support higher population densities and a greater number of species than do other forest types (Carothers and Johnson 1975). Indeed, the biodiversity of these areas is great. Furthermore, the condition or health of riparian habitats affects not only wildlife species restricted to its confines, but also faunal composition in contiguous habitats (Szaro and Jakle 1985). In reality, riparian habitats constitute an "edge effect" that frequently increases diversity in animal and plant communities (Campbell 1970). The land/water interface with multiple horizontal and vertical edges, coupled with multiple combinations of successional stages, provides for diverse habitats for wildlife (Thomas et al. 1979c).

The physical environment of riparian areas is dynamic, changing its temporal equilibrium relative to given disturbance factors, such as logging and fire (Heede 1980). When the natural flow and sediment transport regimes are changed for a given stream, the result is an adjustment in the fluvial dynamics within the stream. These adjustments are evidenced by streambank erosion; deposition of sediments that often change the elevation, size, and shape of the floodplain; loss of riparian vegetation; decreased water quality; and many other characteristics (Tiedemann et al. 1979). The stream continually adjusts to accommodate upstream and downstream influences (Heede 1980).

Riparian vegetation functions to allow necessary sediment transport and natural erosional processes but effectively reduces accelerated erosion that could result in loss of riparian habitats (Miller 1987). In addition, large organic debris is supplied by riparian trees and functions to influence the physical (morphology), chemical (nutrient cycling), and biological (flora and fauna) components of the system (Bisson et al. 1987). Changes in stream channel structure and habitat diversity can occur when large organic debris is removed (Bilby 1984). Structural diversity, an important feature of riparian vegetation (Jain 1976, Anderson and Ohmart 1977), is affected by consequences of natural or human-caused habitat disruption.

Management Problems and Recommendations

Management of riparian habitats presently is of concern worldwide because of the acknowledged value of these ecosystems in terms of water quality and nutrient recycling (Stednick 1988). However, riparian habitats have been subjected to a variety of abuses in the past century. Major causes of habitat degradation include logging (Harr and Fredriksen 1988, Hicks et al. 1991), grazing (Kauffman and Krueger 1984, Skovlin 1984, Chaney et al. 1990), channel diversions (Barclay 1978, Rood and Mahoney 1990), fires (Wohl and Pearthree 1991), agriculture (Lowrance et al. 1986, Ritter and Chirnside 1987), urban development (Medina 1990), recreation (Nash 1977), mining (Streeter et al. 1979), and roads (Hill 1974). These activities directly or indirectly affect the aquatic and terrestrial components cumulatively over time. Multiple-use planning related to these activities is crucial (Likens and Bormann 1974, Lewis and Marsh 1977). Not all influences are discussed, because information related to wildlife interactions is scarce.

Riparian zones are easily affected by natural or induced changes on their watersheds. Medina (1990) observed that riparian trees below a water diversion dam were water-stressed during the growing season and that reproduction and establishment of young trees were inhibited; changes in the structural composition would not be obvious immediately but could require >25 years to detect. In other long-term studies, vegetation and stream morphology data showed that protection of riparian areas from grazing failed to prevent channel erosion or deterioration of riparian vegetation; rather, changes were attributed to a wildfire that had occurred >25 years earlier and the subsequent changes in stream dynamics expressed through flooding (Medina and Martin 1988). Thus, problems seemingly related to riparian habitats alone cannot be resolved by considering only that habitat, but must be considered within the context of cumulative effects over the watershed or landscape.

Terrestrial and aquatic ecosystems are intricately interconnected physically, chemically, and biologically, and the riparian habitats are an expression thereof. For example, associated with instream adjustments are changes in the structure, composition, and density of riparian vegetation. These changes may result in loss of habitat for birds (Bull and Skovlin 1982, Gutzwiller and Anderson 1986), small mammals (Geier and Best 1980), deer (Compton et al. 1988), and herpetofauna (Jones 1988). Changes within the stream such as sedimentation, flooding, and streambank erosion more likely will affect aquatic wildlife such as beavers, muskrats, and fisheries. Beavers often abandon dam sites that become laden with sediments (Apple 1985).

Management of riparian areas needs to be considered for two locations: (1) onsite or within the riparian zone; and (2) offsite or outside the riparian zone, which accounts for all adjacent uplands that exert influence over the watershed. Onsite activities such as grazing management and vegetation treatments are performed within riparian habitats; offsite activities include logging, road construction, and slash burning. Management activities outside the riparian zone may change the quantity and quality of water entering the riparian area (Stednick 1988). Various range management options are available for sustaining the health of riparian habitats including complete protection (Stromberg and Patten 1988), multiple-use approaches, and exclusive use.

Platts (1990) detailed the above strategies and provided additional insight to use of grazing options. In essence, at least one grazing strategy is available that would provide riparian areas with the necessary rest or protection needed to restore, maintain, or enhance their productivity. The least accepted option is "no use" by ungulates, but this option is quite attractive in situations where restoration is a major objective of overall riparian management. One recommendation is to fence critical reaches of riparian habitats in an effort to maintain the integrity of the streamside zone.

Livestock grazing is perhaps the greatest biological threat to riparian habitats in the West, given that about 91% of the total rangeland is grazed (Chaney et al. 1990). Improper livestock grazing affects all four components of the riverine/riparian system—channel, streambanks, water column, and vegetation (Platts 1990). Livestock grazing problems likely may be the result of improper distribution of cows, not simply too many (Severson and Medina 1983). Concentrated livestock use results in sparse tree stands of low vigor, generally with much dead material lying on the ground, a tight, sod-bound soil, and a lack of tree reproduction. Damage occurs in several ways. One is compaction of soil, which reduces moisture infiltration and increases runoff. Another is constant removal of herbage, which allows soil temperatures to rise and increases evaporation from the soil surface. A third is physical damage to the trees by rubbing, trampling, and browsing (Severson and Boldt 1978). The primary method for resolving overuse of riparian areas has been modified grazing strategies, which have met with mixed results (Dwyer et al. 1984, Skovlin 1984, Chaney et al. 1990).

Isolated case studies have demonstrated that revised grazing management improved conditions, but the condition of riparian habitats continues to decline (General Accounting Office 1988). Myers (1989) reported 74% of the grazing systems evaluated failed to respond positively within 20 years. He attributed success to provisions for plant phenology, stream function, and livestock-use behavior. Platts (1990) reported that riparian areas are as badly deteriorated today as at any other time in modern history. One reason may be because the number of cattle on western rangelands has steadily increased since 1875. In short, Platts (1990:6) suggested that "the solution is to identify and develop compatible grazing methods," given our state of knowledge of the functions of riparian systems. Fencing of riparian areas to provide protection from livestock is an alternative solution that has gained some support (Platts 1990).

Fig. 7. Water developments for livestock that are designed with wild-life considerations in mind can be used by pronghorns and other species (photo by D. Beale).

Riparian vegetation usually improves from grazing relief within 4–6 years, depending on severity of use (Platts and Nelson 1989). Areas with severe overuse require greater periods of time (>15 years) for native riparian species such as sedges to displace species adapted to overuse (Elmore and Beschta 1987). Conventional grazing systems (Heady 1975) were developed with consideration only for the production and maintenance of forage plants, primarily graminoids. Application of such systems to maintain woody streamside vegetation and streambank integrity likely will not be satisfactory, given the ecophysiology of shrubs and trees. Platts (1990) provided an excellent description of grazing strategies that are designed to complement restoration objectives with livestock management.

Elk, deer, and other wildlife also contribute to overuse of riparian areas. Houston (1982) reported that willow communities of the Yellowstone region were seriously reduced in their extent and stature by wild ungulates. These effects likely impact upon other wildlife such as grizzly bears and beavers (Chadde 1989). Other states, Arizona and New Mexico for example, also are faced with elk herds that exceed the biological carrying capacity of their ranges. Beavers also can have deleterious effects on the structure and composition of riparian vegetation (Barnes and Dibble 1986).

Carothers and Johnson (1975) suggested that riparian vegetation should be managed as the most sensitive and most productive North American wildlife habitat for a variety of reasons. First, many neotropical species of birds, such as the blue-throated hummingbird, violet-crowned hummingbird, and sulphur-bellied flycatcher, reach the northern limits of their range and breed in the southwestern United States only in riparian habitats. Additionally, breeding bird populations can reach close to 1,000 pairs per 40 ha in some Fremont cottonwood stands (Carothers et al. 1974). More than 50% of the species breeding in these homogeneous Fremont cottonwood stands along the Verde River, Arizona, are exclusively dependent on this habitat for reproduction (Carothers and Johnson 1975). Although no other habitat in North America is as important to noncolonial nesting birds, these areas are no less important to other terrestrial vertebrates (Szaro et al. 1985). Indeed, biodiversity is high in these areas.

Recreational use per unit area of the riparian zone is higher than for other habitat types (Lewis and Marsh 1977). The impact on wildlife varies with season, type, intensity, and duration of use (Pfister 1977). Construction of campgrounds in riparian zones enhances the opportunity for human/wildlife contacts but simultaneously decreases the value of the riparian zone as wildlife habitat because of disturbance by humans, trampling, soil erosion, compaction, and loss of vegetation (Settergren 1977).

The best management strategy for sustaining rangeland riparian areas is one that (1) maintains the productivity of the vegetation, e.g., structure, species composition; (2) maintains the integrity of stream dynamics, e.g., channel and bank stability; and (3) recognizes that several factors, e.g., soils, vegetation, hydrology, and animals, interact to maintain a dynamic equilibrium within the riparian zone. Successful biological management in riparian areas is largely dependent on the application of knowledge from the physical sciences, such as hydrology and geomorphology, that describe the structure and function of riparian zones.

DEVELOPING RANGELAND WATER SOURCES

Increasing the amount of water available to wildlife long has been used to enhance habitat for species inhabiting arid rangelands (Nichol 1937, Bond 1943, Glading 1943, 1947, Halloran and Deming 1958). These techniques include the improvement of natural springs, seeps, and waterholes, and the construction of various types of artificial devices to capture and store rainfall (Tsukamoto and Stiver 1990). Water developments designed primarily to benefit livestock also can be of value to many wildlife species if properly constructed (Fig. 7).

Many methods have been used to make subsurface water available to wildlife, including manual techniques, explosives, prescribed fire, and chemicals. Recently, horizontal well technology has been applied to the development of springs and seeps for wildlife (Coombes and Bleich 1979, Bleich 1982, 1990, Bleich et al. 1982a).

Although time-consuming and costly, hand work may be the most practical way to accomplish some types of developments (Weaver et al. 1959). Helicopters can facilitate transport of personnel and hand tools into remote sites, thereby allowing development of those sites (Bleich 1983).

Dynamite can be efficient and useful for providing water (Weaver et al. 1959), but caution is necessary to ensure that the water-yielding subsurface formations are not altered drastically and the flow of water is not interrupted.

When such damage does occur, it is usually the result of a heavy charge opening a crack that allows water to escape. Dynamite should be used only on marginal seeps where sufficient water is not immediately available and where it can be used safely. Explosives also are useful in clearing channels to allow storm flows to bypass a spring or to lay pipe to be used for gravity flow of water to a drinking basin (Weaver et al. 1959).

Prescribed fire can be used to remove phreatophytic vegetation, resulting in a decrease in the transpiration of subsurface water and in increased surface flows (Biswell and Schultz 1958, Weaver et al. 1959). The use of prescribed fire requires extreme caution, and periodic reburning may be necessary to maintain surface flows. However, the importance of small patches of desert riparian vegetation to a multitude of species makes any substantial reduction in the availability of such vegetation undesirable. Where prescribed fire can be used to temporarily clear a spring site or seep so that other development may proceed, its use may be desirable, but its role probably is limited.

Herbicides increase surface flows by eliminating the vegetation responsible for evapotranspiration of subsurface water. They can be particularly useful where the water supply is limited, and the loss of cover or shade may be more than offset by making a permanent water supply available to wildlife (Weaver et al. 1959). The limited distribution of native, riparian vegetation in arid areas, however, makes the widespread use of herbicides undesirable.

Herbicides can be used to control saltcedar tamarisk at desert water sources (Sanchez 1975). Control can be accomplished on a small scale by hand cutting and herbicide application (Sanchez 1975, Neill 1990), although current regulations prevent the use of some herbicides on federal lands.

Development of Springs

Development of springs should: (1) provide at least one escape route for wildlife to and from the water that takes advantage of the natural terrain and vegetation; (2) provide an alternate escape route where feasible; (3) keep water developments from livestock but allow access for wildlife; (4) reduce the possibility of wildlife drowning by providing gentle basin slopes or ramps in tanks; (5) maintain or provide adequate cover around the watering area with natural cover, plantings, or brush piles; (6) provide, where applicable, a sign to inform the public of the purpose of the development; (7) provide water developments of sufficient capacity to supply water whenever it is needed for wild animals; and (8) provide public access to water outside the fenced water development (Yoakum et al. 1980). If shy animals are involved, water for human consumption can be piped some distance from the wildlife water source. For example, sustained camping should be discouraged within a 1-km radius of water used by mountain sheep.

Ramps or walk-in wells offer a simple and inexpensive method of making water available to wildlife. Birds and animals often drown in abandoned wells or flooded mineshafts, but water stored in these can be made available by the construction of a ramp (Weaver et al. 1959). Ramps should extend to the level of the water. Some ramps that extend 4 m below the level of the ground have been con-structed in California with a variety of techniques including hand tools, power tools, and explosives. Unless the ramp is cut through rock, the sides must be boarded to keep material from sloughing into the excavation. Ramps should be a minimum of 1 m wide to allow large animals to enter and exit without becoming trapped. Ramps to allow the escape of trapped animals are also important in other types of water developments such as livestock troughs (Wilson 1977).

The construction of basins or pools at a water source is an effective way to conserve water and make it readily available to wildlife. Basins may be constructed with rock, cement, or masonry, or they may be gouged from solid rock near the source when small seeps originate in a rock stratum. A simple basin, constructed with hand tools, can be chiseled into solid rock and will effectively store water for years. Where appropriate, power tools and explosives may be used to create larger storage basins. When explosives are used, care must be taken not to damage the source of the water or the rock face so that it cannot be modified to store water. A major advantage of this type of development is that such basins are nearly indestructible.

Rock basins can be enlarged with cement and rocks or masonry materials. Similarly, those materials may be used to construct diversions to protect a basin from debris caused by storm flows, or to create an artificial basin at a location where the development of a solid rock basin is impractical. Special masonry techniques may be necessary to ensure a bond between the mortar and the rock (Gray 1974).

Many springs and seeps occur in canyon bottoms. Even when developed, such springs are subject to damage by water from storms. A method of development that often is satisfactory is to bury a length of perforated asphalt or plastic pipe, packed in gravel, at the spring source, and pipe the water to a basin or trough away from the canyon bottom and away from danger of flooding. Placing large rocks over a source after it has been developed and capping the development with cement increase protection. Alternatively, a redwood spring box may be installed at a water source, allowing for access for maintenance purposes, and water can be piped to a trough in a safe location.

Plastic pipe is a good choice for use because it is lightweight, durable, and transportable. Although weaker than galvanized pipe, it is not subject to rust or corrosion and is easy to repair. The pipe should be buried deep enough to prevent freezing, trampling by livestock and wild ungulates, or washing out during floods. A continuous downhill grade will prevent air locks from developing in the pipe and ensure a constant flow of water.

When water is to be piped away from excavated springs, a trough constructed of concrete or masonry is preferred because it will not rust. If the trough poses a potential hazard for small animals and birds, a ramp should be installed to facilitate access to the water (Bond 1943).

Horizontal Wells

The development of springs and seeps has several disadvantages: (1) the flow of water from the source cannot be controlled, (2) a variable flow may be inadequate to

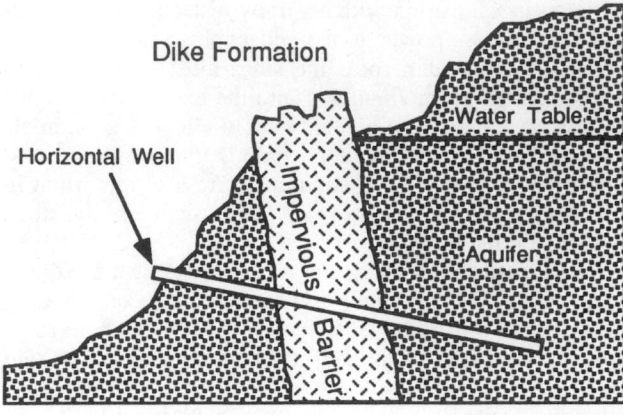

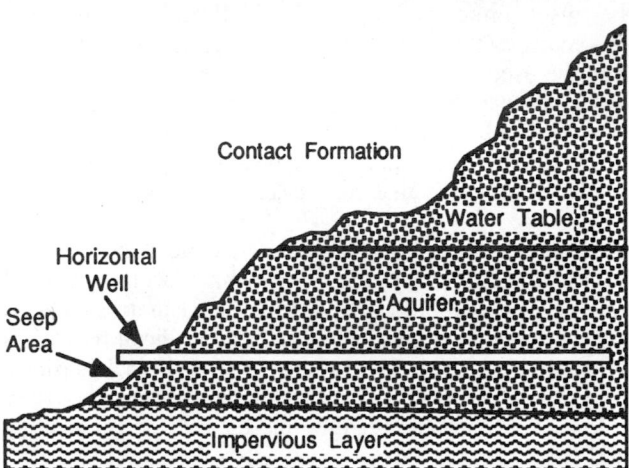

Fig. 8. Horizontal wells developed from dike and contact formations, showing relative positions of the impervious barrier, aquifer, and well casing (after Welchert and Freeman 1973).

generate enough water to create a surface source, and (3) the exposed spring water and the source may be susceptible to contamination (Welchert and Freeman 1973). Horizontal well technology can overcome some of these disadvantages (Coombes and Bleich 1979, Bleich 1982, 1990, Bleich et al. 1982*a*).

Horizontal wells have several advantages: (1) the success rate, particularly in arid regions where historical sources may have dried up, is high; (2) the amount of water can be readily controlled, thus reducing waste; (3) the area is not readily subject to contamination; (4) they are relatively inexpensive to develop; and (5) maintenance requirements are low. Horizontal wells have disadvantages, too: (1) the initial cost of the equipment necessary to construct them can be high (although private contractors who own their equipment can be used to do the work), (2) transporting the necessary equipment to remote sites can be difficult, and (3) some horizontal wells require a vacuum relief valve to prevent air locks from interrupting the flow.

Site selection is the most important and difficult step in the development of a horizontal well. Several factors, including the presence of historical springs and seeps, the distribution of phreatophytes, and the presence of an appropriate geological formation, must be evaluated (Welchert and Freeman 1973). The dike formation (a tilted, impervious formation that forms a natural barrier to an

aquifer) and the contact formation (a perched water table over an impervious material) are both suitable for horizontal well development. Developing a dike formation requires that the impervious barrier be penetrated to tap the stored water (Fig. 8). A contact formation is developed by penetrating at or above a seep area at the boundary of an impervious layer (Fig. 8). Drilling below the seep and into the impervious layer itself will not produce increased flow.

Tenajas

Tenajas are rock tanks created by erosion that hold water. In some desert mountain ranges, tenajas may provide the only source of water for wildlife. The capacity of tenajas can range from a few liters of water to more than 100,000 L.

Several techniques are available to increase the storage capacity of tenajas. Sunshades can be used to reduce evaporation of water from tenajas (Halloran 1949, Halloran and Deming 1956, 1958, Weaver et al. 1959). Such shades can be constructed by anchoring eyebolts into the canyon walls, stretching cables, and attaching shading material such as sheet metal to the cables (Weaver et al. 1959). In Arizona, sunshades have been built with a framework of 5-cm pipe placed into holes drilled into the bedrock, and a wood and pipe framework then is constructed between the uprights. Shading material is placed over the horizontal framework (Werner 1984).

Some tenajas can be deepened or enlarged with explosives (Halloran 1949, Weaver et al. 1959). A safer, more effective method of enlarging the storage capacity of tenajas involves constructing an impervious dam on the downstream side, combined with a pervious structure to divert debris around the tenajas but allowing water to flow into them (Werner 1984).

Development work begins with a thorough cleaning of silt and debris from the tenaja, followed by thorough cleaning with water to ensure that all cracks are clean and ready for sealing (Werner 1984). Next, holes are drilled with a gasoline-powered jackhammer to accommodate steel rebar rods. Rebar is cemented into place with hydraulic cement (Waterplug®, Standard Drywall Products, Newark, CA 94560; product names are for information purposes only, no endorsement is implied) that expands as it sets. The base rock then is treated with an adhesive agent (Acryl 60®, Standard Drywall Products, Newark, CA 94560) to ensure a good joint. Mortar is made from clean sand transported to the site rather than from poor-quality, local sand. Large, flat, square rocks are preferred for dam construction. Once the dam is built, it is sealed with hydraulic cement, Portland cement, lime, and enough Acryl 60 to ensure good adhesion.

A silt diversion dam is built around the tenaja, if such is possible. A silt diversion structure does not require periodic cleaning as does a silt retention dam. The diversion structure can be a rock gabion consisting of V-mesh wire formed between rows of fence posts set in holes drilled into bedrock. The gabion is filled with loose rock and is capped with more V-mesh wire, then wired shut.

Deep, steep-sided tenajas often pose special problems for wildlife, because individuals can become trapped when water levels are low. Pneumatic equipment or explosives can be used to chisel or blast access ramps in such situations (Halloran 1949). Mensch and Weaver (1969) successfully modified such a tenaja, in which 34 mountain sheep had died within a 2-year period.

Sand Dams

Some of the earliest techniques designed to increase water availability in arid regions involved construction of sand dams or sand tanks (Sykes 1937, Halloran 1949, Halloran and Deming l956, 1958). These devices originally were constructed by placing a concrete dam across a narrow canyon. The dam was penetrated by one or more pipes that could be capped to prevent water from draining. The dammed area then filled with sand and gravel washed in by floods. Water soaks into the sand and gravel and is stored, protected from excessive evaporation (National Academy of Sciences 1974).

The dams must be securely anchored in bedrock, and the design and construction of the dam may be the most important aspect of the entire system (Bleich and Weaver 1983). Because seepage at the bedrock interface could be a significant source of water loss, Bleich and Weaver (1983) urged that techniques resulting in an efficient bond between cement and bedrock be used (Gray 1974). These techniques involve the use of chemical compounds such as Acryl 60 to enhance bonding. All areas of bedrock that the dam will contact must be cleaned thoroughly before construction. After construction, the dam and joint should be sealed with a product similar to Thoroseal® (Standard Drywall Products, Newark, CA 94560). Cracks in the bedrock should be sealed with an appropriate material such as Waterplug.

The storage volume of sand dams can be increased in a variety of ways (Sivils and Brock 1981, Bleich and Weaver 1983). Specifically, all culverts must be anchored securely to bedrock, and to each other if more than one layer is used, or damage from storm flows may occur. Additionally, an appropriate filter must be attached on the upstream side of the dam to prevent particles from jamming the float valve if one is used. Heavy-duty, galvanized pipe rather than plastic pipe should be used for the outlet on sand dams because of the potential for the outlet pipe to be damaged by rocks and debris washing over the dam. Water stored behind sand dams can be piped to an appropriately constructed trough located some distance from the dam (Sivils and Brock 1981) or used to flood natural or constructed potholes downstream. The dam should not be too large. If compounds such as calcium aluminate are added to the concrete to decrease set-up time (Gray 1974), such dams should be no more than 12 m long and 3 m high (Halloran and Deming 1956, 1958).

Because precipitation in arid regions often occurs as violent thundershowers, washes and canyons often flow huge amounts of water over a short period of time. Such flows may not allow ample time for storm water to saturate areas behind sand dams, especially if the underground storage capability has been enhanced (Sivils and Brock 1981, Bleich and Weaver 1983). Rock-filled baskets or gabions, anchored into bedrock, can be placed across a wash or canyon perpendicular to the direction of flow to slow water velocity. Such structures also raise and widen the wash.

Reservoirs and Small Ponds

A reservoir consists of open water impounded behind a dam. Yoakum et al. (1980) recommended that the reservoir be formed by building a dam directly across a drainage or by enclosing a depression on one side of a drainage and constructing a diversion ditch into the resulting basin. They also recommended that reservoirs be designed to provide maximum storage with a minimum of surface area to reduce evaporation loss. The following are major points to consider in the selection of reservoir sites: (1) the most suitable soils for dams are clays with a fair proportion of sand and gravel (one part clay to two or three parts grit); (2) the watershed above the dam should be large enough to provide sufficient water to fill the reservoir, but not so large that excessive flows will damage the spillway or wash out the dam; (3) the most economical site is one along a natural drainage where the channel is narrow and relatively deep, the bottom is easily made watertight, and the channel grade immediately above the dam is as flat as possible; (4) wildlife should have easy access to the water; and (5) the dam should be located to take advantage of natural spillway sites, or an adequate spillway must be incorporated into the development.

The dam site should be surveyed and staked before construction proceeds. If the suitability of material for dam construction is questionable, an examination should be made by a soil scientist. Trees and shrubs should be cleared from the dam site and flooded basin. The foundation area of the dam should be plowed or scarified in the direction of the main axis of the dam to produce a good bond between the foundation and the fill material. If stability and permeability of the foundation material are questionable, a narrow core trench should be dug lengthwise to the dam, then refilled and packed with damp clay soil. Suitable material should be obtained on site, so the borrow pit will become part of the reservoir and add depth to the impoundment. The base thickness of the dam must be equal to or greater than 4.5 times the height plus the crest thickness. The slopes of the dam should be 2.5:1 on the upstream face and 2:1 on the downstream face. Minimum width of the top of all dams should be 3 m. Fill of the dam should be carried at least 10% higher than the required height to allow for settling.

Freeboard (depth from the top of the dam to the high-water mark when the spillway is carrying the estimated peak runoff) should not be <60 cm, and the spillway should be designed to handle double the largest known volume of runoff. A natural spillway is preferred, and it should have a broad, relatively flat cross section. Water should be taken out well above the fill and reenter the main channel some distance downstream. Any spillway should be wide, flat-bottomed, and protected from washing by riprapping, or by facing with rocks. The entrance should be wide and smooth, and the grade of the spillway channel should be mild so the water will flow through without cutting (Hamilton and Jepson 1940).

New reservoirs usually do not hold water satisfactorily for several months. Bentonite spread over the bottom and sides of the basin and face of the dam will help seal the impoundment. The basin also can be lined with polyethylene or another appropriate material, and then 15–30 cm

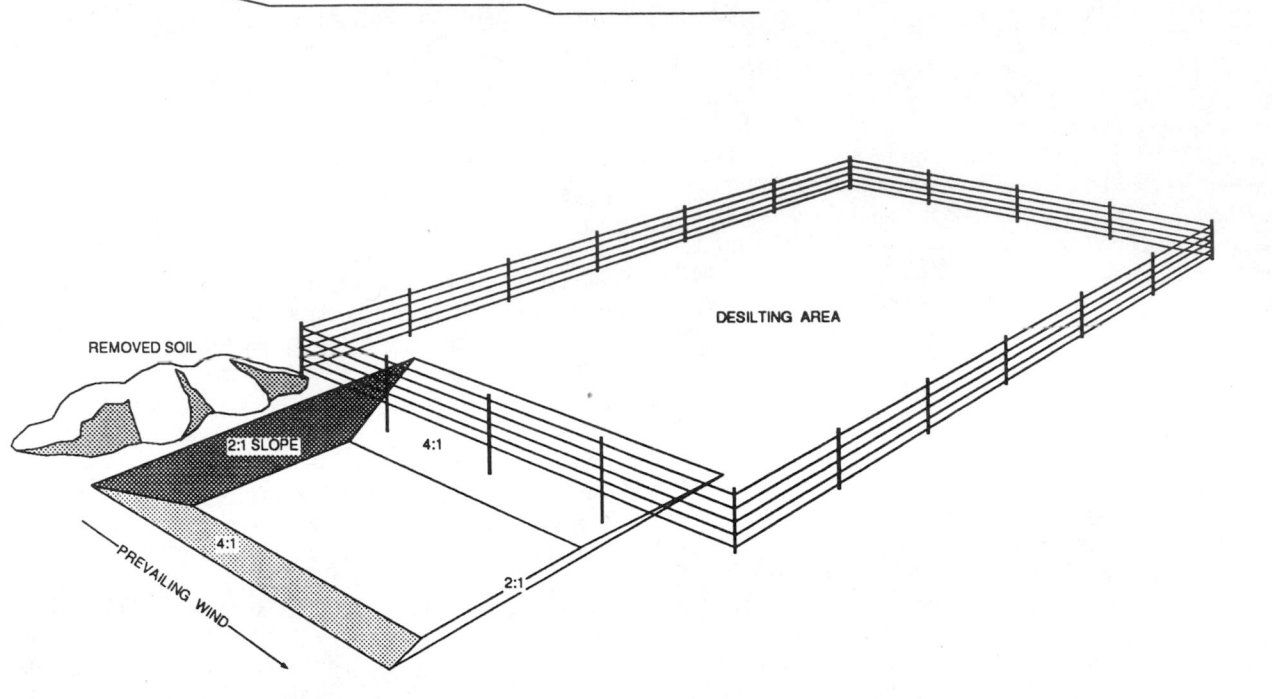

Fig. 9. A dugout or charco can be used to provide water for wildlife and livestock on rangelands (after Yoakum et al. 1980, Kindschy et al. 1982).

of dirt can be rolled evenly over the top (U.S. Bureau of Land Management 1966). Newly available artificial materials such as Hypalon® (Water Saver Company, Denver, CO 80216-0465) are superior to polyethylene because of their strength and resistance to ultraviolet radiation. These liners may be custom-made for reservoirs of various sizes.

Dugouts

Large, earthen catchment basins built to collect water for livestock were commonly called "charcos" by early settlers along the Mexican border and "dugouts" by pioneers in other areas (Yoakum et al. 1980). Dugouts can

Fig. 10. An adit is a tunnel that extends into solid rock, stores water, and provides access for wildlife (photo by B. Garlinger).

be located in almost any type of topography, but they are most common in areas of comparatively flat, well-drained terrain. Flat slopes facilitate maximum storage with minimum excavation. Dugouts can be small, rectangular excavations (Fig. 9). All sides should be sloped sufficiently to prevent sloughing (usually ≤2:1), and one or more relatively flat side slopes (≤4:1) should be provided for livestock or big-game entrances (U.S. Bureau of Land Management 1964).

Adits

Adits are short, dead-end tunnels that extend into solid rock, constructed with a downward sloping floor to allow access by wildlife (Halloran and Deming 1956, 1958) (Fig. 10). Adits have been constructed in Arizona and other western states, primarily to benefit mountain sheep (Parry 1972, Weaver 1973).

Adits are most readily constructed by personnel skilled in hard-rock blasting techniques. Adits should have openings at least 2 × 3 m and should be 4–5 m in length. The water storage depth should be at least 4 m to ensure a dependable water supply (Halloran and Deming 1956, 1958). Commercial masonry sealers should be used to prevent seepage of water through rock fractures (Halloran and Deming 1956, 1958, Gray 1974, Werner 1984).

Because the opening of an adit must be at approximately the same elevation of the wash in which it is situated, it may be necessary to construct a diversion that allows floodwaters to enter, yet causes debris, sand, and boulders to bypass the adit. Boulders placed on the upstream sides of adits can be used for this purpose (Halloran and Deming 1956, 1958). Another effective but simple technique involves the construction of a rock gabion (Werner 1984). Either commercial gabions or those custom-made from heavy wire can be used.

An adit can be designed to store water from a natural source, such as a seasonal or permanent spring (Werner 1984). Occasionally, water can be diverted into adits from natural slick-rock aprons located above. Adits also can be used to store water that normally would be unavailable by pumping water from the adit into a nearby tenaja (Werner 1984). In such instances, the adit should be covered to reduce evaporation. Shade structures have been used to reduce evaporation at adits in which stored water is directly available to wildlife (Halloran and Deming 1956, 1958).

Guzzlers

Guzzlers are permanent, self-filling, constructed devices that collect and store rainwater and make it directly available to wildlife. Guzzlers can be constructed to provide water for small animals only or for animals of all sizes.

Several techniques can be used to collect water for small-game guzzlers. Aprons that collect rainfall can be constructed of manufactured or natural materials. Small-game guzzlers often are made with concrete or sheet metal aprons, but asphalted, oiled, waxed, or otherwise treated soil aprons can be used (Glading 1947, Fink et al. 1973, Rauzi et al. 1973, Myers and Frasier 1974, Frasier et al. 1979, Johnson and Jacobs 1986, Rice 1990).

Small-game guzzlers generally store water in underground tanks, and wildlife walk a ramp to enter the guzzler to drink. Early big-game guzzlers were similar in that water was collected from an artificial apron, and large mammals walked down a ramp into an underground tank to reach water (Halloran and Deming 1956, 1958). However, water can be stored in underground or aboveground concrete, plastic, metal, or fiberglass tanks (Garton 1956a,b, Roberts 1977, Bleich et al. 1982b, Remington et al. 1984, Werner 1984, Bardwell 1990, Bleich and Pauli 1990, deVos and Clarkson 1990, Gunn 1990). Unlike small-game guzzlers, most modern big-game guzzlers have a float-valve system to regulate water at a drinking trough away from the water storage tanks (Roberts 1977, Werner 1984, Bleich and Pauli 1990).

The most important step in the installation of a guzzler is locating an adequate site. A guzzler should not be placed in a wash or gully where it may collect silt or sand or be damaged by floodwaters (Yoakum et al. 1980). Unfortunately, many guzzlers have been installed in areas lacking other critical habitat components (Lewis, Calif. Dep. Fish Game Job Prog. Rep., Proj. W-26-D-29-1, 1973).

Yoakum et al. (1980) provided the following recommendations for the construction of small-game guzzlers. The size of the water-collecting apron should be proportioned so that the storage tank will need no water source other than rainfall to fill it. The cost of digging the hole for the storage tank is one of the largest expenditures, so a site should be chosen where digging is comparatively easy. The tank should be placed with its open end away from the prevailing wind and, if possible, facing in a northerly direction to minimize sunlight entering the tank.

Such placement will reduce water temperature, evaporation, and growth of algae.

The tanks usually are made of concrete or plastic. Occasionally steel tanks are used, as are heavy-equipment tires (Elderkin and Morris 1989, Morris and Elderkin 1990). The plastic guzzler is a prefabricated tank constructed of fiberglass impregnated with plastic resin. Only washed gravel aggregates should be used for construction of concrete tanks, otherwise the concrete may disintegrate in several years. Tanks made of steel are used for guzzlers in some areas and give satisfactory service. Use of tanks constructed of other human-made materials is relatively new.

Water-collecting aprons can be made of many materials. Concrete sealed with bitumol, galvanized metal sheet roofing, glass mat and bitumol, rubber or plastic sheets, asphalt, and plywood have been used successfully. Durable materials such as concrete or metal are least expensive to maintain, although soil cement appears to be a promising material (Rice 1990). Efficiency (percent of water collected) and life-spans (years) vary among materials: steel (98%, 25 years) is best, followed by asphalt roofing (86–92%, 8 years), plastic covered with 2.5 cm of gravel (66–87%, 8–15 years), butyl rubber (98%, 15–20 years), asphalt paving (95%, 15 years), and liquid asphalt soilwater (90%, 5 years) (Fairbourn et al. 1972).

The size of the water-collecting apron needed to fill a guzzler depends on the size of guzzler and the minimum annual rainfall at the construction site. The size of the apron is surprisingly small, because nearly 100% of the rainfall can be collected, depending on material used. Each 10 m^2 in apron surface area will result in the collection of about 1 L of water for each centimeter of rainfall.

Calculations should be made on the basis of minimum precipitation expected, rather than the average or maximum, to prevent guzzler failure during drought years. The size of the apron necessary to fill a guzzler to capacity once each year (an apron yielding a 100% water harvest is assumed) varies as a function of annual precipitation and storage tank capacity (Fig. 11). When different types of aprons are used, required surface area can be calculated from the harvest efficiencies (Fairbourn et al. 1972). Guzzler leakage, evaporation, and heavy use by wildlife also may dictate a larger apron.

Big-game guzzlers have been used for years throughout the southwestern United States (Garton 1956a,b, Bleich et al. 1982b, Gunn 1990). These systems are designed to collect water from either artificial (Gunn 1990) or natural aprons (Stevenson 1990). Using slick-rock catchments to collect runoff from bare rock areas is a common technique (Bleich et al. 1982b, deVos and Clarkson 1990, Stevenson 1990). Such guzzlers take advantage of the fact that rock surfaces yield nearly 100% of the precipitation falling on them as runoff. Several recent studies (Bardwell 1990, Gunn 1990, Stevenson 1990) provided design specifications and other recommendations for the construction of such structures. Bardwell (1990), Bleich and Pauli (1990), deVos and Clarkson (1990), and Gunn (1990) provided information regarding the performance of these units over time. Additionally, those authors evaluated the techniques used in the construction of big-game guzzlers and evaluated the reliability of materials.

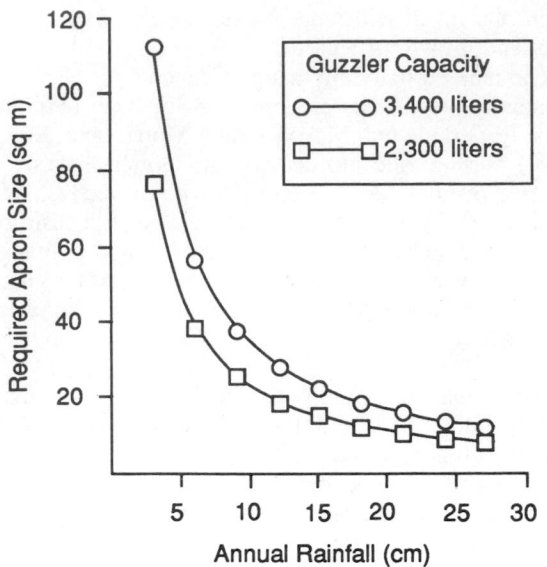

Fig. 11. Recommended apron sizes for 2,300-L and 3,400-L guzzlers as a function of annual rainfall (after Yoakum et al. 1980).

The effectiveness and performance of big-game guzzlers depend on plumbing components. For example, Bleich and Pauli (1990) reported that frozen pipes and fittings accounted for 35 of 98 failures among 22 guzzlers over an 11-year period. Furthermore, of the 98 failures, float-valve failures accounted for 31, design and construction flaws for 9, and natural disasters for 6. Other problems, including rusted tanks, rusted drinker boxes, and vandalism, accounted for 17. Overall, each of the 22 guzzlers evaluated averaged 4.4 mechanical failures over an 11-year period, but each was in service an average of 87% of that time. Mechanical failures did not necessarily lead to inoperative guzzlers but did require effort to repair them.

CONSTRUCTING RANGELAND FENCES

The relationship of fences and wildlife on rangelands in the western United States has been a point of contention over the past century. Fences constructed to control domestic livestock can adversely impact some wildlife species. For example, fences can be major obstacles or traps to pronghorns (Martinka 1967, Spillett et al. 1967, Oakley 1973) and mule deer (Yoakum et al. 1980, Mackie 1981). Proper fence design and use of appropriate construction materials can reduce adverse effects. Details of fence construction on rangelands used by pronghorns, mule deer, elk, bison, and collared peccaries are available elsewhere (U.S. Bureau of Land Management 1985, Karsky 1988). Preventing the movement of some wildlife species may be desirable, and specific fence designs can accomplish that goal (Longhurst et al. 1962, Messner et al. 1973, deCalesta and Cropsey 1978, Jepson et al. 1983, Karsky 1988).

In a recent incident, a rancher in south-central Wyoming erected a net-wire fence around critical winter range that blocked pronghorn access to 3,885 ha of intermingled public and private lands (Moody and Alldredge 1986). The Wyoming Wildlife Federation and the National Wildlife Federation sued to have the fence removed. The District Court judge ruled that the fence violated the Unlawful Enclosures of Public Lands Act of 1885, even though the net-wire fence was entirely on private land. The court ordered the rancher to remove 45 km of fence to permit pronghorns access to critical winter habitat on public lands. The decision was upheld on appeal, setting a major precedent in ensuring wildlife access to public lands.

Fences and Pronghorns

The severity of pronghorn-fence problems varies among areas. Fences are primarily a problem for herds moving seasonally to and from winter grounds on northern rangelands (Oakley 1973). However, seasonal movement problems also were reported for New Mexico (Russell 1964, Howard et al. 1983) and Texas (Buechner 1950, Hailey 1979), especially during drought years.

If fencing is necessary, only that required to provide proper livestock control and minimize hindrance to pronghorns and other wildlife should be used. Unrestricted passage for all age classes during all seasons and all weather conditions should be provided (Yoakum et al. 1980).

Fencing water holes on dry summer rangelands may be as detrimental to pronghorns as fencing migration routes. If a fenced water development is provided specifically for pronghorns, the area should encompass at least 1–2 ha of relatively level terrain (Yoakum et al. 1980).

Pronghorns react to fences differently than do deer, elk, or bison. Pronghorns usually go under a wire fence rather than jump over or negotiate their way through. Some fence lines have well-beaten trails where pronghorns have used the same site to go under a fence. Often such locations are situated at a slight depression in the ground that allows easier access.

Fence specifications to control livestock on pronghorn range have evolved over many years (Spillett et al. 1967, Autenrieth 1978, Salwasser 1980, Yoakum 1980, Kindschy et al. 1982, U.S. Bureau of Land Management 1985). Fences should consist of three strands of wire, the bottom strand being smooth (Fig. 12). Four- to six-strand barbed-wire fences limit pronghorn movements and should not be used. The bottom wire should be at least 41 cm above the ground. Absence of stays between posts will facilitate the occasional movement of pronghorns through the fence (Yoakum et al. 1980, Kindschy et al. 1982, Hall 1985).

New fences should be flagged with pieces of white cloth so pronghorns can become familiar with their locations. By the time a white rag tied to the top of each fence post deteriorates, pronghorns will have become accustomed to the fence (Kindschy et al. 1982). Painting the tops of steel fence posts white also will help to make the fence more visible to pronghorns (Hall 1985).

Where snow accumulation restricts pronghorn movements, let-down or adjustable fences should be used (Yoakum et al. 1980). A let-down fence can consist of a wooden stay at each fence post to which the wires are attached. The stay is secured to the fence post with a wire loop at the top and either a second loop or a pivot bolt at the bottom. Let-down fence sections may be designed to permit pulling the let-down sections back against sections of permanently standing fence.

Let-down fences should provide for adjustments in wire tension. When the wire is so taut that it does not lie flat on the ground or is so loose that wire loops are formed,

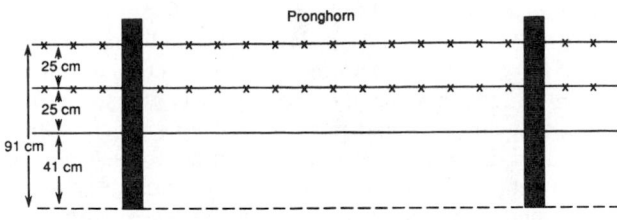

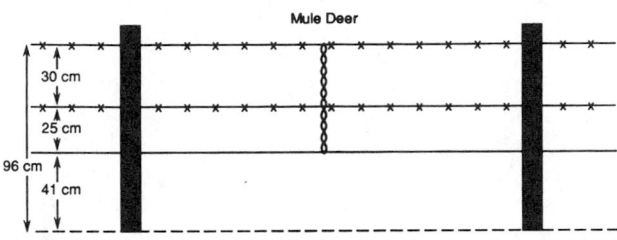

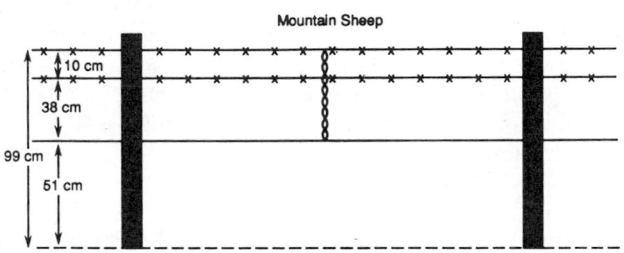

Fig. 12. Recommended specifications for wire fences constructed on ranges used by pronghorns (after Yoakum 1980, Kindschy et al. 1982, U.S. Bureau of Land Management 1985), mule deer (after Jepson et al. 1983, U.S. Bureau of Land Management 1985), and mountain sheep (after Hall 1985, Brigham 1990). Note the use of a smooth bottom wire on all designs and the lack of stays on fences constructed on pronghorn ranges.

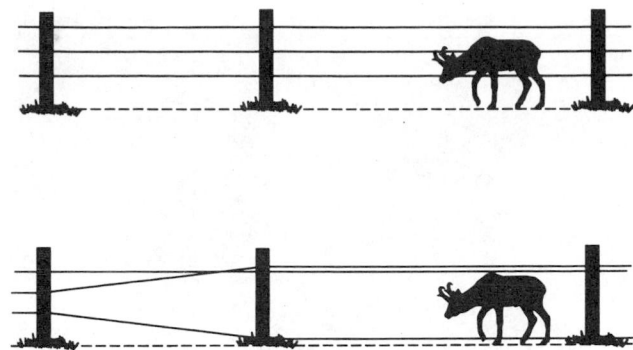

Fig. 13. Adjustable fence modifications to facilitate movement of pronghorns and other ungulates (after Anderson and Denton 1980).

a hazard is created for people and animals (U.S. Bureau of Land Management 1985).

Adjustable fences that allow the movement of one or more wires can allow pronghorn passage during periods when livestock are not present (Anderson and Denton 1980) (Fig. 13). Adjustable fences are particularly useful when winter snow depths exceed 30 cm (Yoakum et al. 1980).

Pronghorn passes are structures that resemble cattle guards intersecting a fence (Spillett et al. 1967, Mapston and ZoBell 1972, Yoakum et al. 1980, Howard et al. 1983). Suitable locations for pronghorn passes make use of the tendency of individuals to parallel a fence, looking for a way to cross. The pass capitalizes on the ability of pronghorns to jump laterally over obstacles.

Pronghorn passes have been built and tested under a variety of conditions (Spillett et al. 1967, Howard et al. 1983). Some adult pronghorns quickly learn to use the facilities, but others do not. Pronghorn fawns often are unable to negotiate the passes. Pronghorn passes are of limited value and should not be used as a panacea for pronghorn access problems (U.S. Bureau of Land Management 1985).

Net-wire fences prevent the movement of pronghorn fawns in particular and should not be used on public rangelands where pronghorns occur (Autenrieth 1978,

Yoakum 1980). However, some adults may become adept at jumping a net-wire fence up to 80 cm high. Higher net-wire fences can be used when the goal is to restrict the movement of animals, such as in live-trapping, control of animals in research projects, decreasing crop depredations, and restricting access to hazardous areas such as highways.

Fences and Mule Deer

The relationship between livestock fences and mule deer has not raised the political furor that it has for pronghorns. However, throughout North America where fences have been built, they have undoubtedly caused far greater mortality to deer than to pronghorns. Deer are more likely to be trapped as individuals, whereas pronghorns at times are caught in large winter concentrations. Also, deer frequently are caught in fences in isolated areas not readily witnessed, whereas pronghorn mortalities in open country are easy to observe.

Deer often crawl under fences when not hurried, but jump them when startled or chased (Mackie 1981). When a deer jumps a fence, its feet can become entangled between the top two wires, resulting in death (Fig. 14). This problem can be reduced by limiting total fence height to 96 cm (U.S. Bureau of Land Management 1985) (Fig. 12). If the top wire is barbed, it should be separated from the next wire by 30 cm. Otherwise, it should be a smooth wire (Jepson et al. 1983). Unlike fences used on pronghorn ranges, wire stays should be placed every 2.5 m between posts to keep the top wires from twisting around the leg of a deer (Yoakum et al. 1980, U.S. Bureau of Land Management 1985).

The effective height of a fence as a barrier to deer is increased on steep slopes. For example, a 110-cm fence on a 20% slope is equivalent to a 140-cm fence on level ground. On a 50% slope, it is equivalent to a 190-cm fence on level ground (Kerr 1979, Anderson and Denton 1980). Height adjustments should be made accordingly.

Let-down fences along seasonal travel routes for deer help ensure free movement. The let-down feature of the fence also helps prevent damage from snow loading during winter. Movements of mule deer also can be aided with an adjustable fence (Fig. 13).

Net-wire fences no higher than 90 cm allow movement of adult deer but prevent passage of fawns. They should

Fig. 14. A mule deer trapped in an improperly constructed barbed wire fence. The fence should consist of only three wires. The top wire should be separated from the next wire by at least 30 cm, the bottom wire should be smooth, and stays should be used between the posts (photo by D. Neal).

not be placed on summer and autumn migration routes used by deer.

Fences and Mountain Sheep

The construction of wire fences on ranges used by mountain sheep (for example, to exclude livestock from water developments) presents particular problems. Mountain sheep are likely to become entangled in a fence when placing their head through the top two wires. This problem is minimized if the two top wires are no more than 10 cm apart (Brigham 1990). A three-wire fence should be used with wires spaced at 51, 38, and 10 cm (Fig. 12), allowing mountain sheep movement under the bottom wire and between it and the middle wire (U.S. Bureau of Land Management 1985, Brigham 1990). Six-wire fence designs (U.S. Bureau of Land Management 1985) are dangerous to mountain sheep and should not be used (Brigham 1990).

Electric Fences

Electric fences can be used to control livestock, and some designs pose little hindrance to movement of wildlife. Electric fences are most effective on moist sites, where two wires may be sufficient to control cattle. On sites with at least 60 cm of rain annually, an electric fence can be made of two smooth wires at heights of 60 cm and 90 cm above ground (U.S. Bureau Land Management 1985, Karsky 1988). The top wire is electrified and the bottom wire serves as the ground. The wires are free-running at all posts and pose little danger of entrapping mule deer. In addition, the fence can be crossed on foot or in a vehicle by standing on it and pushing the wires to the ground. On drier sites, electric fences require more wires to function effectively (Karsky 1988), and the added wires can reduce or eliminate movement of wildlife.

Wood Fences

Fences can be constructed in a variety of designs entirely from wood posts and rails obtained at the site or from manufactured materials (U.S. Bureau of Land Management 1985, Karsky 1988). Wood fences are usually expensive but can be attractive and may require less maintenance than wire fences. Construction options include post and pole, log worm, log and block, and buck and pole designs (Karsky 1988). The same principles apply to wood fences as to wire fences in minimizing hindrance to wildlife movements. The top rail or pole of a wooden fence should be kept low to allow mule deer to jump over and the bottom rail or post kept high enough to allow the movement of fawns.

Rock Jacks

In many areas, soils are too shallow and rocky to allow steel fence posts to be easily driven into the ground (Hall 1985). At such sites, rock jacks are often constructed in the form of wood-rail cribs. The cribs are filled with rocks and serve as anchors to which wire fences can be secured. Cover and dens for small mammals are provided if the bottom rail of a rock jack is kept 10–15 cm off the ground (Hall 1985). Use of rocks at least 30 cm in diameter also will provide crevasses suitable for small mammals (Maser et al. 1979, Hall 1985).

Fences To Exclude Wildlife

Excluding selected wildlife species from certain areas may be desirable. Orchards, vineyards, and other crops often are heavily depredated by mule deer and other ungulates, and appropriate fence designs can alleviate such problems. Highways can be hazardous to mule deer and other ungulates that need to reach critical seasonal ranges. Fences can be used to channel their movement to suitable underpasses and minimize collisions with vehicles. Experimental plots used in research often require the exclusion of one or more species of wildlife. Finally, fencing can be used as an alternative to other control measures in reducing predation on livestock.

A 1.8-m upright net-wire fence, or one slanted at 45 degrees to a total height of about 1.3 m, can be used to exclude mule deer (Longhurst et al. 1962, Messner et al. 1973, Karsky 1988). Electric fences with four to six wires also discourage deer movements (Karsky 1988).

Fences can be used to reduce or eliminate the need for lethal control of coyotes, which can be excluded from pastures either by woven wire (deCalesta and Cropsey 1978, Thompson 1979, Jepson et al. 1983) or electric fences (Gates et al. 1978, Dorrance and Bourne 1980, Karsky 1988, Nass and Theade 1988). To be effective, a woven wire fence must be at least 170 cm high, have mesh openings no larger than 10 × 15 cm, and have an overhang to prevent jumping and an apron to prevent digging, each at least 40 cm wide (Thompson 1979). A seven-wire electric fence (four hot wires alternating with three ground wires) totaling 130 cm in height also can be used (Dorrance and Bourne 1980). Other electric fence designs are available to deter coyotes (Karsky 1988). In general, fencing to control coyotes is expensive and probably justified only to protect small areas of high production capacity, such as irrigated pastures. Also, coyotes can get through a fence, particularly an electric one, and become trapped inside an area that was meant to exclude them (Kie 1977).

Acknowledgments.—We thank Fred Guthery, Marti Kie, and Chris Maser for comments on various drafts of this chapter but accept responsibility for any remaining errors.

LITERATURE CITED

ANDELT, W. F., J. G. KIE, F. F. KNOWLTON, AND K. CARDWELL. 1987. Variation in coyote diets associated with season and successional changes in vegetation. J. Wildl. Manage. 51:273–277.

ANDERSON, B. W., AND R. D. OHMART. 1977. Vegetation structure and bird use in the lower Colorado River valley. Pages 23–34 in R. R. Johnson and D. A. Jones, tech. coords. Importance, preservation, and management of riparian habitat. U.S. For. Serv. Gen. Tech. Rep. RM-43.

ANDERSON, E. W., AND R. J. SCHERZINGER. 1975. Improving quality of winter forage for elk by cattle grazing. J. Range Manage. 28: 120–125.

ANDERSON, L. D., AND J. W. DENTON. 1980. Adjustable wire fences for facilitating big game movement. U.S. Bur. Land Manage. Tech. Note 343. 7pp.

APPLE, L. L. 1985. Riparian habitat restoration and beavers. Pages 489–490 in R. R. Johnson, C. D. Ziebell, D. R. Patton, P. F. Ffolliott, and R. H. Hamre, eds. Riparian ecosystems and their management: reconciling conflicting uses. U.S. For Serv. Gen. Tech. Rep. RM-1.

AUSTIN, D. D., AND P. J. URNESS. 1986. Effects of cattle grazing on mule deer diet and area selection. J. Range Manage. 39:18–21.

AUTENRIETH, R. 1978. Guidelines for the management of pronghorn antelope. Proc. Pronghorn Antelope Workshop 8:473–526.

BAKER, B. W. 1979. Habitat use, productivity, and nest predation of Rio Grande turkeys. Ph.D. Thesis, Texas A&M Univ., College Station. 46pp.

BAKER, D. L., AND F. S. GUTHERY. 1990. Effects of continuous grazing on habitat and density of ground-foraging birds in south Texas. J. Range Manage. 43:2–5.

BARCLAY, J. S. 1978. The effects of channelization on riparian vegetation and wildlife in south central Oklahoma. Pages 129–138 in R. R. Johnson and J. F. McCormick, tech. coords. Strategies for protection and management of floodplain wetlands and other riparian ecosystems. U.S. For. Serv. Gen. Tech. Rep. WO-12.

BARDWELL, P. P. 1990. Artificial water development design, materials, and problems encountered in the BLM, Carson City District, Nevada. Pages 133–139 in G. K. Tsukamoto and S. J. Stiver, eds. Proc. wildlife water development symposium. Nev. Chap. The Wildl. Soc., U.S. Bur. Land Manage., and Nev. Dep. Wildl.

BAREISS, L. J., P. SCHULZ, AND F. S. GUTHERY. 1986. Effects of short-duration and continuous grazing on bobwhite and wild turkey nesting. J. Range Manage. 39:259–260.

BARNES, W. J., AND E. DIBBLE. 1986. The effects of beaver in riverbank forest succession. Can. J. Bot. 66:40–44.

BATZLI, G. O., AND F. A. PITELKA. 1970. Influence of meadow mouse populations on California grassland. Ecology 51:1027–1039.

BILBY, R. E. 1984. Post-logging removal of woody debris affects stream channel stability. J. For. 82:609–613.

BIRCH, L. C. 1957. The meanings of competition. Am. Nat. 91:5–18.

BISSON, P. A., ET AL. 1987. Large woody debris in forested streams in the Pacific Northwest: past, present, and future. Pages 143–190 in E. O. Salo and T. W. Cundy, eds. Streamside management: forestry

and fishery interactions. Univ. Washington Inst. For. Resour. Contrib. 57.

BISWELL, H. H., AND A. M. SCHULTZ. 1958. Effects of vegetation removal on spring flow. Calif. Fish Game 44:211–230.

BLEICH, V. C. 1982. Horizontal wells for mountain sheep: desert bighorn get the shaft. Trans. Desert Bighorn Counc. 26:63–64.

———. 1983. Comments on helicopter use by wildlife agencies. Wildl. Soc. Bull. 11:304–306.

———. 1990. Horizontal wells for wildlife water development. Pages 51–58 in G. K. Tsukamoto and S. J. Stiver, eds. Proc. wildlife water development symposium. Nev. Chap. The Wildl. Soc., U.S. Bur. Land Manage., and Nev. Dep. Wildl.

———, L. J. COOMBES, AND J. H. DAVIS. 1982a. Horizontal wells as a wildlife habitat improvement technique. Wildl. Soc. Bull. 10:324–328.

———, ———, AND G. W. SUDMEIER. 1982b. Volunteer participation in California wildlife habitat improvement projects. Trans. Desert Bighorn Counc. 26:56–58.

———, AND A. M. PAULI. 1990. Mechanical evaluation of artificial watering devices built for mountain sheep in California. Pages 65–72 in G. K. Tsukamoto and S. J. Stiver, eds. Proc. wildlife water development symposium. Nev. Chap. The Wildl. Soc., U.S. Bur. Land Manage., and Nev. Dep. Wildl.

———, AND R. A. WEAVER. 1983. "Improved" sand dams for wildlife habitat management. J. Range Manage. 36:133.

BOCK, C. E., J. H. BOCK, W. R. KENNEY, AND V. M. HAWTHORNE. 1984. Responses of birds, rodents, and vegetation to livestock exclosure in a semidesert grassland site. J. Range Manage. 37:239–242.

BOND, R. M. 1943. Ramps for escape of wildlife from stock-troughs. J. Wildl. Manage. 7:123.

BORCHERT, M. I., AND S. K. JAIN. 1978. The effect of rodent seed predation on four species of California annual grasses. Oecologia 33:101–113.

BRIGHAM, W. R. 1990. Fencing wildlife water developments. Pages 37–43 in G. K. Tsukamoto and S. J. Stiver, eds. Proc. wildlife water development symposium. Nev. Chap. The Wildl. Soc., U.S. Bur. Land Manage., and Nev. Dep. Wildl.

BRINSON, M. M., B. L. SWIFT, R. C. PLANTICO, AND J. S. BARCLAY. 1981. Riparian ecosystems: their ecology and status. U.S. Fish Wildl. Serv. Biol. Serv. Program, FWS/OBS-81. 211pp.

BROWN, D. E. 1978. Grazing, grassland cover, and gamebirds. Trans. North Am. Wildl. Nat. Resour. Conf. 43:477–485.

BROWN, R. L. 1982. Effects of livestock grazing on Mearn's quail in southeastern Arizona. J. Range Manage. 35:727–732.

BRYANT, F. C., M. M. KOTHMANN, AND L. B. MERRILL. 1979. Diets of sheep, angora goats, Spanish goats and white-tailed deer under excellent range conditions. J. Range Manage. 32:412–417.

BUE, I. G., L. BLANKENSHIP, AND W. H. MARSHALL. 1952. The relationship of grazing practices to waterfowl breeding populations and production on stock ponds in western South Dakota. Trans. North Am. Wildl. Conf. 17:396–414.

BUECHNER, H. K. 1942. Interrelationships between the pocket gopher and land use. J. Mammal. 23:346–348.

———. 1950. Life history, ecology, and range use of the pronghorn antelope in Trans-Pecos, Texas. Am. Midl. Nat. 43:257–354.

BULL, E. L., AND J. M. SKOVLIN. 1982. Relationships between avifauna and streamside vegetation. Trans. North Am. Wildl. Nat. Resour. Conf. 47:496–506.

BUTTERY, R. F., AND P. W. SHIELDS. 1975. Range management practices and bird habitat values. Pages 183–189 in D. R. Smith, tech. coord. Proc. symposium on management of forest and range habitats for nongame birds. U.S. For. Serv. Gen. Tech. Rep. WO-1.

CAMPBELL, C. J. 1970. Ecological implications of riparian vegetation management. J. Soil Water Conserv. 25:49–52.

CAMPBELL-KISSOCK, L., L. H. BLANKENSHIP, AND L. D. WHITE. 1984. Grazing management impacts on quail during drought in the northern Rio Grande Plain, Texas. J. Range Manage. 37:442–446.

CAROTHERS, S. W., AND R. R. JOHNSON. 1975. Water management practices and their effects on nongame birds in range habitats. Pages 210–222 in D. R. Smith, tech. coord. Proc. symposium on management of forest and range habitats for nongame birds. U.S. For. Serv. Gen. Tech. Rep. WO-1.

———, ———, AND S. W. AITCHISON. 1974. Population structure and social organization of southwestern riparian birds. Am. Zool. 14: 97–108.

CHADDE, S. 1989. Willows and wildlife of the northern range, Yellowstone National Park. Pages 168–169 *in* R. E. Gresswell, B. A. Barton, and J. L. Kershner, tech. coords. Practical approaches to riparian resource management. U.S. Bur. Land Manage., Billings, Mont.

CHANEY, E., W. ELMORE, AND W. S. PLATTS. 1990. Livestock grazing on western riparian areas. U.S. Environ. Prot. Agency, Denver, Colo. 45pp.

CLARY, W. P., AND D. M. BEALE. 1983. Pronghorn reactions to winter sheep grazing, plant communities, and topography in the Great Basin. J. Range Manage. 36:749–752.

CLEMENTS, F. E. 1916. Plant succession: an analysis of the development of vegetation. Carnegie Inst. Publ. 242. 512pp.

COHEN, W. E., D. L. DRAWE, F. C. BRYANT, AND L. C. BRADLEY. 1989. Observations on white-tailed deer and habitat response to livestock grazing in south Texas. J. Range Manage. 42:361–365.

COMPTON, B. B., R. J. MACKIE, AND G. L. DUSEK. 1988. Factors influencing distribution of white-tailed deer in riparian habitats. J. Wildl. Manage. 52:544–548.

COOMBES, L. J., AND V. C. BLEICH. 1979. Horizontal wells—the DFG's new slant on water for wildlife. Outdoor Calif. 40:10–12.

CURTIS, R. L., AND T. H. RIPLEY. 1975. Water management practices and their effect on nongame bird habit values in a deciduous forest community. Pages 128–141 *in* D. R. Smith, ed. Proc. symposium management of forest and range habitats for nongame birds. U.S. For. Serv. Gen. Tech. Rep. WO-1.

DAMBACH, C. A. 1944. A ten-year ecological study of adjoining grazed and ungrazed woodlands in northeastern Ohio. Ecol. Monogr. 14: 257–270.

DEBANO, L. F., AND L. J. SCHMIDT. 1989. Improving southwestern riparian areas through watershed management. U.S. For. Serv. Gen. Tech. Rep. RM-182. 33pp.

DECALESTA, D. S., AND M. G. CROPSEY. 1978. Field test of a coyote-proof fence. Wildl. Soc. Bull. 6:256–259.

DEGRAAF, R. M., TECHNICAL COORDINATOR. 1980. Workshop proceedings—management of western forests and grasslands for nongame birds. U.S. For. Serv. Gen. Tech. Rep. INT-86. 535pp.

DEVOS, J. C., JR., AND R. W. CLARKSON. 1990. A historic review of Arizona's water developments with discussions on benefits to wildlife, water quality, and design considerations. Pages 157–166 *in* G. K. Tsukamoto and S. J. Stiver, eds. Proc. wildlife water development symposium. Nev. Chap. The Wildl. Soc., U.S. Bur. Land Manage., and Nev. Dep. Wildl.

DICKSON, J. G., AND J. C. HUNTLEY. 1987. Riparian zones and wildlife in southern forests: the problem and squirrel relationships. Pages 37–39 *in* J. G. Dickson and O. E. Maughan, eds. Managing southern forests for wildlife and fish. U.S. For. Serv. Gen. Tech. Rep. SO-65.

DORRANCE, M. J., AND J. BOURNE. 1980. An evaluation of anti-coyote electric fencing. J. Range Manage. 33:385–387.

DRAGT, W. J., AND K. M. HAVSTAD. 1987. Effects of cattle grazing upon chemical constituents within important forages for elk. Northwest Sci. 61:70–73.

DWYER, D. D., J. C. BUCKHOUSE, AND W. S. HUEY. 1984. Impacts of grazing intensity and specialized grazing systems on the use and value of rangeland: summary and recommendations. Pages 867–884 *in* Impacts of grazing intensity and specialized grazing systems on use and values of rangelands. Natl. Acad. Sci., Nat Resour. Counc., and Bur. Land Manage., Washington, D.C. 140pp.

DYKSTERHUIS, E. J. 1949. Condition and management of range land based on quantitative ecology. J. Range Manage. 2:104–115.

ELDERKIN, R. L., AND J. MORRIS. 1989. Design for a durable and inexpensive guzzler. Wildl. Soc. Bull. 17:192–194.

ELLISON, L. 1946. The pocket gopher in relation to soil erosion on mountain range. Ecology 27:101–114.

ELMORE, W. 1989. Rangeland riparian systems. Pages 93–95 *in* A. L. Dana, tech. coord. Proc. California riparian systems conference: protection, management, and restoration for the 1990s. U.S. For. Serv. Gen. Tech. Rep. PSW-110.

———, AND R. L. BESCHTA. 1987. Riparian areas—perceptions in management. Rangelands 9:260–265.

EVANS, K. E., AND R. R. KERBS. 1977. Avian use of livestock watering ponds in western South Dakota. U.S. For. Serv. Gen. Tech. Rep. RM-35. 11pp.

FAIRBOURN, M. L., F. RAUZI, AND H. R. GARDNER. 1972. Harvesting precipitation for a dependable, economical water supply. J. Soil Water Conserv. 27:23–26.

FINK, D. H., K. R. COOLEY, AND G. W. FRASIER. 1973. Wax-treated soils for harvesting water. J. Range Manage. 26:396–398.

FITCH, H. S., AND J. R. BENTLEY. 1949. Use of California annual-plant forage by range rodents. Ecology 30:306–321.

FRASIER, G. W., K. R. COOLEY, AND J. R. GRIGGS. 1979. Performance evaluation of water harvesting catchments. J. Range Manage. 32: 453–456.

FRISCHKNECHT, N. C. 1965. Deer mice on crested wheatgrass range. J. Mammal. 46:529–530.

GALLIZIOLI, S. 1979. Effects of livestock grazing on wildlife. West. Sect., The Wildl. Soc., Cal-Neva Wildl. Trans. 1979:83–87.

GARTON, D. A. 1956a. Experimental big game watering device and detailed information on construction and costs. Calif. Dep. Fish Game, Long Beach. 11pp.

———. 1956b. Experimental big game watering device and information on construction. Calif. Dep. Fish Game, Long Beach. 11pp.

GATES, N. L., J. E. RICH, D. D. GODTEL, AND C. V. HULET. 1978. Development and evaluation of anti-coyote electric fencing. J. Range Manage. 31:151–153.

GEIER, A. R., AND L. B. BEST. 1980. Habitat selection by small mammals of riparian communities: evaluating effects of habitat alterations. J. Wildl. Manage. 44:16–24.

GENERAL ACCOUNTING OFFICE. 1988. Public rangelands: some riparian areas restored, but widespread improvement will be slow. U.S. Gen. Accounting Off., Resour. Community Econ. Dev. Div., Rep. GAO/RCED-88-105. 85pp.

GERSBERG, R. M., B. V. ELKINS, S. R. LYON, AND C. R. GOLDMAN. 1986. Role of aquatic plants in wastewater treatment by artificial wetlands. Water Res. 20:363–368.

GJERSING, F. M. 1975. Waterfowl production in relation to rest-rotation grazing. J. Range Manage. 28:37–42.

GLADING, B. 1943. A self-filling quail watering device. Calif. Fish Game 29:157–164.

———. 1947. Game watering devices for the arid Southwest. Trans. North Am. Wildl. Conf. 12:286–292.

GLASS, R. J. 1988. Habitat improvement costs on state-owned wildlife management areas in New York. U.S. For. Serv. Res. Pap. NE-621. 15pp.

GOODWIN, J. G., JR., AND C. R. HUNGERFORD. 1977. Habitat use by native Gambel's and scaled quail and released masked bobwhite quail in southern Arizona. U.S. For. Serv. Res. Pap. RM-197. 8pp.

GORDON, I. J. 1988. Facilitation of red deer grazing by cattle and its impact on red deer performance. J. Appl. Ecol. 25:1–10.

GRANGE, W. B. 1948. Wisconsin grouse problems. Wis. Conserv. Dep., Madison. 318pp.

GRANT, W. E., E. C. BIRNEY, N. R. FRENCH, AND D. M. SWIFT. 1982. Structure and productivity of grassland small mammal communities related to grazing-induced changes in vegetative cover. J. Mammal. 63:248–260.

GRAY, R. S. 1974. Lasting waters for bighorn. Trans. Desert Bighorn Counc. 18:25–27.

GROVER, K. E., AND M. J. THOMPSON. 1986. Factors influencing spring feeding site selection by elk in the Elkhorn Mountains, Montana. J. Wildl. Manage. 50:466–470.

GUNN, J. 1990. Arizona's standard rainwater catchment. Pages 19–24 *in* G. K. Tsukamoto and S. J. Stiver, eds. Proc. wildlife water development symposium. Nev. Chap. The Wildl. Soc., U.S. Bur. Land Manage., and Nev. Dep. Wildl.

GUTHERY, F. S., C. A. DEYOUNG, F. C. BRYANT, AND D. L. DRAWE. 1990. Using short-duration grazing to accomplish wildlife habitat objectives. Pages 41–55 *in* K. E. Severson, tech. coord. Can livestock be used as a tool to enhance wildife habitat? U.S. For. Serv. Gen. Tech. Rep. RM-194.

GUTZWILLER, K. J., AND S. H. ANDERSON. 1986. Trees used simultaneously and sequentially by breeding cavity-nesting birds. Great Basin Nat. 46:358–360.

HAILEY, T. L. 1979. A handbook on pronghorn antelope management in Texas. Texas Parks Wildl. Dep., Austin. 59pp.

HALE, M. G., AND D. M. ORCUTT. 1987. The physiology of plants under stress. John Wiley & Sons, New York, N.Y. 206pp.

HALL, F. C. 1985. Wildlife habitats in managed rangelands—the Great Basin of southeastern Oregon: management options and practices. U.S. For. Serv. Gen. Tech. Rep. PNW-189. 17pp.

HALLORAN, A. F. 1949. Desert bighorn management. Trans. North Am. Wildl. Conf. 14:527–537.

———, AND O. V. DEMING. 1956. Water development for desert bighorn sheep. U.S. Fish Wildl. Serv. Wildl. Manage. Ser. Leafl. 14. 12pp.

————, AND ————. 1958. Water development for desert bighorn sheep. J. Wildl. Manage. 22:1–9.

HALLS, L. K., EDITOR. 1984. White-tailed deer: ecology and management. Stackpole Books, Harrisburg, Pa. 870pp.

HAMILTON, C. L., AND H. G. JEPSON. 1940. Stock water developments: wells, springs, and ponds. U.S. Dep. Agric. Farmer's Bull. 1859. 70pp.

HANLEY, T. A., AND J. L. PAGE. 1982. Differential effects of livestock use on habitat structure and rodent populations in Great Basin communities. Calif. Fish Game 68:160–174.

HARR, R. D., AND R. L. FREDRIKSEN. 1988. Water quality after logging small watersheds within the Bull Run watershed, Oregon. Water Resour. Bull. 24:1103–1111.

HEADY, H. F. 1975. Rangeland management. McGraw-Hill Inc., New York, N.Y. 460pp.

HEEDE, B. H. 1980. Stream dynamics: an overview for land managers. U.S. For. Serv. Gen. Tech. Rep. RM-72. 26pp.

HENNY, C. J., L. J. BLUS, E. J. KOLBE, AND R. E. FITZNER. 1985. Organophosphate insecticide (Famphur) topically applied to cattle kills magpies and hawks. J. Wildl. Manage. 49:648–658.

HICKS, B. J., R. L. BESCHTA, AND R. D. HARR. 1991. Long-term changes in streamflow following logging in western Oregon and associated fisheries implications. Water Resour. Bull. 27:217–226.

HILL, R. D. 1974. Mining impacts on trout habitat. Pages 47–57 *in* Symp. on trout habitat research and management. U.S. For. Serv. Southeast. For. Exp. Stn., Asheville, N.C.

HOLECHEK, J. 1980. Livestock grazing impacts on rangeland ecosystems. J. Soil Water Conserv. 35:162–164.

————. 1981. Brush control impacts on rangeland wildlife. J. Soil Water Conserv. 36:265–269.

————. 1982. Managing rangelands for mule deer. Rangelands 4:25–28.

HOOD, R. E., AND J. M. INGLIS. 1974. Behavioral responses of white-tailed deer to intensive ranching operations. J. Wildl. Manage. 38:488–498.

HOUSTON, D. 1982. The northern Yellowstone elk: ecology and management. Macmillan, New York, N.Y. 474pp.

HOWARD, V. W., J. L. HOLECHEK, AND R. D. PIEPER. 1983. Roswell pronghorn study. New Mexico State Univ., Las Cruces. 115pp.

HOWARD, W. E., K. A. WAGNON, AND J. R. BENTLEY. 1959. Competition between ground squirrels and cattle for range forage. J. Range Manage. 12:110–115.

JAIN, S., EDITOR. 1976. Vernal pools: their ecology and conservation. Univ. California, Davis, Inst. Ecol. Publ. 9. 93pp.

JENSEN, H. P., D. ROLLINS, AND R. L. GILLEN. 1990. Effects of cattle stock density on trampling loss of simulated ground nests. Wildl. Soc. Bull. 18:71–74.

JENSEN, S., AND W. S. PLATTS. 1987. Processes influencing riparian ecosystems. Proc. Annu. Meet. Soc. Wetland Sci. 8:228–232.

JEPSON, R., R. G. TAYLOR, AND D. W. McKENZIE. 1983. Rangeland fencing systems: state-of-the-art review. U.S. For. Serv. Equipment Dev. Cent., San Dimas, Calif. 23pp.

JOHNSON, M. K. 1982. Response of small mammals to livestock grazing in southcentral Idaho. J. Range Manage. 35:51–53.

JOHNSON, R. R., P. S. BENNETT, AND L. T. HAIGHT. 1989. Southwestern woody riparian vegetation and succession: an evolutionary approach. Pages 135–139 *in* A. L. Dana, tech. coord. Proc. California riparian systems conference: protection, management, and restoration for the 1990s. U.S. For. Serv. Gen. Tech. Rep. PSW-110.

————, AND D. A. JONES, TECHNICAL COORDINATORS. 1977. Importance, preservation, and management of riparian habitat. U.S. For. Serv. Gen. Tech. Rep. RM-43. 217pp.

————, AND C. H. LOWE. 1985. On the development of riparian ecology. Pages 112–116 *in* R. R. Johnson, C. D. Ziebell, D. R. Patton, P. F. Ffolliott, and R. H. Hamre, tech. coords. Riparian ecosystems and their management: reconciling conflicting uses. U.S. For. Serv. Gen. Tech. Rep. RM-120.

————, C. D. ZIEBELL, D. R. PATTON, P. F. FFOLLIOTT, AND R. H. HAMRE, TECHNICAL COORDINATORS. 1985. Riparian ecosystems and their management: reconciling conflicting uses. U.S. For. Serv. Gen. Tech. Rep. RM-120. 523pp.

JOHNSON, T., AND R. A. W. JACOBS. 1986. Gallinaceous guzzlers. U.S. Army Corps Eng., Wildl. Resour. Manage. Manual, Sect. 5.4.1., Tech. Rep. EL-86-8. 20pp.

JONES, K. B. 1988. Comparison of herpetofaunas of a natural and al-

tered riparian ecosystem. Pages 222–227 *in* R. C. Szaro, K. E. Severson, and D. R. Patton, eds. Management of amphibians, reptiles, and small mammals in North America. U.S. For. Serv. Gen. Tech. Rep. RM-166.

KANTRUD, H. A. 1990. Effects of vegetation manipulation on breeding waterfowl in prairie wetlands: a literature review. Pages 93–123 *in* K. E. Severson, tech. coord. Can livestock be used as a tool to enhance wildlife habitat? U.S. For. Serv. Gen. Tech. Rep. RM-194.

KARSKY, R. 1988. Fences. U.S. Bur. Land Manage. and U.S. For. Serv. Technol. Dev. Program, Missoula, Mont. 210pp.

KAUFFMAN, J. B., AND W. C. KRUEGER. 1984. Livestock impacts on riparian ecosystems and streamside management implications—a review. J. Range Manage. 37:430–437.

KERR, R. M. 1979. Mule deer habitat guidelines. U.S. Bur. Land Manage. Tech. Note 336. 61pp.

KIE, J. G. 1977. Effects of coyote predation on population dynamics of white-tailed deer in south Texas. Ph.D. Thesis, Univ. California, Berkeley. 217pp.

————. 1988. Annual grassland. Pages 118–119 *in* K. E. Mayer and W. F. Laudenslayer, Jr., eds. A guide to the wildlife habitats of California. Calif. Dep. For., Sacramento.

————. 1991. Wildlife and livestock grazing alternatives in the Sierra Nevada. Trans. West. Sect., The Wild. Soc. 27:17–29.

————, C. J. EVANS, E. R. LOFT, AND J. W. MENKE. 1991. Foraging behavior by mule deer: the influence of cattle grazing. J. Wildl. Manage. 55:665–674.

————, AND E. R. LOFT. 1990. Using livestock to manage wildlife habitat: some examples from California annual grassland and wet meadow communities. Pages 7–24 *in* K. E. Severson, tech. coord. Can livestock be used as a tool to enhance wildlife habitat? U.S. For. Serv. Gen. Tech. Rep. RM-194.

————, AND J. W. THOMAS. 1988. Rangeland vegetation as wildlife habitat. Pages 585–605 *in* P. T. Tueller, ed. Vegetation science applications for rangeland analysis and management. Kluwer Academic Publ., Dordrecht, The Netherlands.

KINDSCHY, R. R., C. SUNDSTROM, AND J. YOAKUM. 1982. Wildlife habitats in managed rangelands—the Great Basin of southeastern Oregon: pronghorns. U.S. For. Serv. Gen. Tech. Rep. PNW-145. 18pp.

KIRSCH, L. M. 1969. Waterfowl production in relation to grazing. J. Wildl. Manage. 33:821–828.

KNOPF, F. L. 1988. Riparian wildlife habitats: more, worth less, and under invasion. Pages 20–22 *in* K. M. Mutz, D. J. Cooper, M. L. Scott, and L. K. Miller, tech. coords. Restoration, creation and management of wetland and riparian ecosystems in the American West. PIC Technol., Denver, Colo.

————, J. A. SEDGWICK, AND R. W. CANNON. 1988. Guild structure of a riparian avifauna relative to seasonal cattle grazing. J. Wildl. Manage. 52:280–290.

KNOWLES, C. J. 1975. Range relationships of mule deer, elk, and cattle in a rest-rotation grazing system during summer and fall. M.S. Thesis, Montana State Univ., Bozeman. 111pp.

KOERTH, B. H., W. M. WEBB, F. C. BRYANT, AND F. S. GUTHERY. 1983. Cattle trampling of simulated ground nests under short duration and continuous grazing. J. Range Manage. 36:385–386.

KOZLOWSKI, T. T. 1984. Flooding and plant growth. Academic Press, New York, N.Y. 356pp.

KUENZLER, E. J. 1988. Value of forested wetlands as filters for sediments and nutrients. Pages 85–96 *in* D. D. Hook and R. Lea, eds. The forested wetlands of the southern United States. U.S. For. Serv. Gen. Tech. Rep. SE-50.

LEOPOLD, A. S. 1977. The California quail. Univ. California Press, Berkeley. 281pp.

LEWIS D. E., AND G. G. MARSH. 1977. Problems resulting from the increased recreational use of rivers in the West. Pages 27–31 *in* Proc. river recreation management and research symposium. U.S. For. Serv. Gen. Tech. Rep. NC-28.

LIKENS, G. E., AND F. H. BORMANN. 1974. Linkages between terrestrial and aquatic ecosystems. BioScience 24:447–456.

LOFT, E. R. 1988. Habitat and spatial relationships between mule deer and cattle in a Sierra Nevada forest zone. Ph.D. Thesis, Univ. California, Davis. 144pp.

————, J. W. MENKE, AND J. G. KIE. 1991. Habitat shifts by mule deer: the influence of cattle grazing. J. Wildl. Manage. 55:16–26.

————, ————, ————, AND R. C. BERTRAM. 1987. Influence of cattle stocking rate on the structural profile of deer hiding cover. J. Wildl. Manage. 51:655–664.

LONGHURST, W. M., E. O. GARTON, H. F. HEADY, AND G. E. CONNOLLY. 1976. The California deer decline and possibilities for restoration. West. Sect., The Wildl. Soc., Cal-Neva Wildl. Trans. 23: 74–103.

———, R. E. HAFENFELD, AND G. E. CONNOLLY. 1982. Deer-livestock interrelationships in the western states. Proc. wildlife-livestock relationships symposium. 10:409–420.

———, M. B. JONES, R. R. PARKS, L. W. NEUBAUER, AND M. W. CUMMINGS. 1962. Fences for controlling deer damage. Univ. California Agric. Exp. Stn. Circ. 514. 15pp.

LOOMIS, J. B., E. R. LOFT, D. R. UPDIKE, AND J. G. KIE. 1991. Cattle-deer interactions in the Sierra Nevada: a bioeconomic approach. J. Range Manage. 44:395–399.

LOWRANCE, R., J. K. SHARPE, AND J. M. SHERIDAN. 1986. Long-term sediment deposition in the riparian zone of a coastal plain watershed. J. Soil Water Conserv. 41:266–271.

———, R. L. TODD, AND L. E. ASMUSSEN. 1983. Waterborne nutrient budgets for the riparian zone of an agricultural watershed. Agric. Ecosystem Environ. 10:371–384.

MACKIE, R. J. 1978. Impacts of livestock grazing on wild ungulates. Trans. North Am. Wildl. Nat. Resour. Conf. 43:462–476.

———. 1981. Interspecific relationships. Pages 487–507 in O. C. Wallmo, ed. Mule and black-tailed deer of North America. Univ. Nebraska Press, Lincoln.

MAPSTON, R. D., AND R. S. ZOBELL. 1972. Antelope passes: their value and use. U.S. Bur. Land Manage. Tech. Note D-360. 11pp.

MARTINKA, C. J. 1967. Mortality of northern Montana pronghorns in a severe winter. J. Wildl. Manage. 31:159–164.

MASER, C., J. M. GEIST, D. M. CONCANNON, R. ANDERSON, AND B. LOVELL. 1979. Wildlife habitats in managed rangelands—the Great Basin of southeastern Oregon: geomorphic and edaphic habitats. U.S. For. Serv. Gen. Tech. Rep. PNW-99. 84pp.

———, J. W. THOMAS, AND R. G. ANDERSON. 1984. Wildlife habitats in managed rangelands—the Great Basin of southeastern Oregon: the relationship of terrestrial vertebrates to plant communities. U.S. For. Serv. Gen. Tech. Rep. PNW-172. Two parts: 25pp. and 237pp.

MCMAHAN, C. A. 1964. Comparative food habits of deer and three classes of livestock. J. Wildl. Manage. 28:798–808.

———, AND C. RAMSEY. 1965. Response of deer and livestock to controlled grazing in central Texas. J. Range Manage. 18:1–7.

MEDIN, D. E., AND W. C. CLARY. 1989. Small mammal populations in a grazed and ungrazed riparian habitat in Nevada. U.S. For. Serv. Res. Pap. INT-413. 6pp.

———, AND ———. 1990. Bird and small mammal populations in a grazed and ungrazed riparian habitat in Idaho. U.S. For. Serv. Res. Pap. INT-425. 8pp.

MEDINA, A. L. 1990. Possible effects of residential development on streamflow, riparian plant communities, and fisheries on small mountain streams in central Arizona. For. Ecol. Manage. 33/34: 351–361.

———, AND S. C. MARTIN. 1988. Stream channel and vegetation changes in sections of McKnight Creek, New Mexico. Great Basin Nat. 48:373–381.

MEGAHAN, W. F., AND P. N. KING. 1985. Identification of critical areas on forest lands for control of nonpoint sources of pollution. Environ. Manage. 9:7–18.

MENDELSSOHN, I. A., AND M. T. POSTEK. 1982. Elemental analysis of deposits on the roots of *Spartina alterniflora* Loisel. Am. J. Bot. 69:904–912.

MENKE, J. W., EDITOR. 1983. Proceedings of the workshop on livestock and wildlife-fisheries relationships in the Great Basin. Univ. California, Berkeley, Spec. Publ. 3301. 173pp.

MENSCH, J. L., AND R. A. WEAVER. 1969. Desert bighorn (*Ovis canadensis nelsoni*) losses in a natural trap tank. Calif. Fish Game 55: 237–238.

MESSNER, H. E., D. R. DIETZ, AND E. C. GARRETT. 1973. A modification of the slanting deer fence. J. Range Manage. 26:233–235.

MILLER, E. 1987. Effects of forest practices on relationships between riparian area and aquatic ecosystems. Pages 40–47 in J. G. Dickson and O. E. Maughan, eds. Managing southern forests for wildlife and fish. U.S. For. Serv. Gen. Tech. Rep. SO-65.

MITSCH, W. J., AND J. G. GOSSELINK. 1986. Wetlands. Van Nostrand Reinhold Co., New York, N.Y. 539pp.

MOODY, D. S., AND A. W. ALLDREDGE. 1986. Red Rim—mining, fencing, and some decisions. Proc. Pronghorn Antelope Workshop 12: 57.

MOORE, A. W., AND E. H. REID. 1951. The Dalles pocket gopher and its influence on forage production of Oregon mountain meadows. U.S. Dep. Agric. Circ. 884. 36pp.

MOORE, W. H., AND W. S. TERRY. 1979. Short-duration grazing may improve wildlife habitat in southeastern pinelands. Proc. Annual Conf. Southeast. Assoc. Fish Wildl. Agencies 33:279–287.

MORRIS, J. E., AND R. L. ELDERKIN. 1990. A heavy equipment tire guzzler. Pages 49–50 in G. K. Tsukamoto and S. J. Stiver, eds. Proc. wildlife water development symposium. Nev. Chap. The Wildl. Soc., U.S. Bur. Land Manage., and Nev. Dep. Wildl.

MUNDINGER, J. G. 1976. Waterfowl response to rest-rotation grazing. J. Wildl. Manage. 40:60–68.

MYERS, L. 1989. Grazing and riparian management in southwestern Montana. Pages 117–120 in R. E. Gresswell, B. A. Barton, and J. L. Kershner, tech. coords. Practical approaches to riparian resource management. U.S. Bur. Land Manage., Billings, Mont. 193pp.

MYERS, L. E., AND G. W. FRASIER. 1974. Asphalt-fiberglass for precipitation catchments. J. Range Manage. 27:12–15.

NASH, R. 1977. River recreation: history and future. Pages 2–7 in Proc. river recreation management and research symposium. U.S. For. Serv. Gen. Tech. Rep. NC 28.

NASS, R. D., AND J. THEADE. 1988. Electric fences for reducing sheep losses to predators. J. Range Manage. 41:251–252.

NATIONAL ACADEMY OF SCIENCES. 1974. More water for arid lands: a report to the Advisory Committee on Technology Innovation, Board on Science and Technology for International Development, Commission on International Relations. Natl. Acad. Sci., Washington, D.C. 153pp.

NEILL, W. M. 1990. The tamarisk invasion of desert riparian areas. Pages 121–126 in G. K. Tsukamoto and S. J. Stiver, eds. Proc. wildlife water development symposium. Nev. Chap. The Wildl. Soc., U.S. Bur. Land Manage., and Nev. Dep. Wildl.

NELSON, J. R., AND D. G. BURNELL. 1975. Elk-cattle competition in central Washington. Pages 71–83 in Range multiple use management. Coop. Ext. Serv., Washington State Univ., Pullman, Oregon State Univ., Corvallis, and Univ. Idaho, Moscow.

NICHOL, A. A. 1937. Desert bighorn sheep. Ariz. Wildl. Mag. 7(7):9, 16.

OAKLEY, C. 1973. The effects of livestock fencing on antelope. Wyo. Wildl. 37:26–29.

OLDEMEYER, J. L., AND L. R. ALLEN-JOHNSON. 1988. Cattle grazing and small mammals on the Sheldon National Wildlife Refuge, Nevada. Pages 391–402 in R. C. Szaro, K. E. Severson, and D. R. Patton, tech. coords. Management of amphibians, reptiles, and small mammals in North America. U.S. For. Serv. Gen. Tech. Rep. RM-166.

PARMENTER, R. R., AND J. A. MACMAHON. 1983. Factors determining the abundance and distribution of rodents in a shrub-steppe ecosystem: the role of shrubs. Oecologia (Berl.) 59:145–156.

PARRY, P. L. 1972. Development of permanent wildlife water supplies, Joshua Tree National Monument. Trans. Desert Bighorn Counc. 16: 92–96.

PEEK, J. M., AND P. D. DALKE, EDITORS. 1982. Proc. wildlife-livestock relationships symposium, 20–22 April 1981, Coeur d'Alene, Idaho. Proc. 10, Univ. Idaho, Moscow. 614pp.

PETERJOHN, W. T., AND D. L. CORRELL. 1984. Nutrient dynamics in an agricultural watershed: observations on the role of a riparian forest. Ecology 65:1466–1475.

PFISTER, R. E. 1977. Campsite choice behavior in the river setting: a plot study on the Rogue River, Oregon. Pages 351–358 in Proc. river recreation management and research symposium. U.S. For. Serv. Gen. Tech. Rep NC-28.

PIEPER, R. D., AND R. F. BECK. 1990. Range condition from an ecological perspective: modifications to recognize multiple use objectives. J. Range Manage. 43:550–552.

PITT, M. D. 1986. Assessment of spring defoliation to improve fall forage quality of bluebunch wheatgrass (*Agropyron spicatum*). J. Range Manage. 39:175–181.

PLATTS, W. S. 1990. Managing fisheries and wildlife on rangelands grazed by livestock. White Horse Assoc., Smithfield, Ut. 445pp.

———, AND R. L. NELSON. 1989. Characteristics of riparian plant communities and streambanks with respect to grazing in northeastern Utah. Pages 73–81 in R. E. Gresswell, B. A. Barton, and J. L. Kershner, tech. coords. Practical approaches to riparian resource management. U.S. Bur. Land Manage., Billings, Mont.

———, AND R. F. RALEIGH. 1984. Impacts of grazing on wetlands and riparian habitat. Pages 1105–1117 in Developing strategies for rangeland management. Westview Press, Boulder, Colo.

PRASAD, N. L. N. S., AND F. S. GUTHERY. 1986. Wildlife use of livestock water under short duration and continuous grazing. Wildl. Soc. Bull. 14:450–454.

RAUZI, F., M. L. FAIRBOURN, AND L. LANDERS. 1973. Water harvesting efficiencies of four soil surface treatments. J. Range Manage. 26:399–403.

REARDON, P. O., L. B. MERRILL, AND C. A. TAYLOR, JR. 1978. White-tailed deer preferences and hunter success under various grazing systems. J. Range Manage. 31:40–42.

REID, V. H. 1954. Multiple land use: timber, cattle, and bobwhite quail. J. For. 52:575–578.

REINER, R. J., AND P. J. URNESS. 1982. Effect of grazing horses managed as manipulators of big game winter range. J. Range Manage. 35:567–571.

REMINGTON, R., W. E. WERNER, K. R. RAUTENSTRAUCH, AND P. R. KRAUSMAN. 1984. Desert mule deer use of a new permanent water source. Pages 92–94 *in* P. R. Krausman and N. S. Smith, eds. Deer in the Southwest: a symposium. Univ. Arizona School of Renewable Nat. Resour., Tucson.

REYNOLDS, H. G. 1950. Relation of Merriam kangaroo rats to range vegetation in southern Arizona. Ecology 31:456–463.

REYNOLDS, T. D., AND C. H. TROST. 1980. The response of native vertebrate populations to crested wheatgrass planting and grazing by sheep. J. Range Manage. 33:122–125.

RICE, W. E. 1990. Soil cement application for wildlife developments and range developments. Pages 3–10 *in* G. K. Tsukamoto and S. J. Stiver, eds. Proc. wildlife water development symposium. Nev. Chap. The Wildl. Soc., U.S. Bur. Land Manage., and Nev. Dep. Wildl.

RICHARDSON, C. J. 1988. Freshwater wetlands: transformers, filters, or sinks? Forem 11:3–9.

RITTER, W. F., AND A. E. M. CHIRNSIDE. 1987. Influence of agricultural practices on nitrates in the water table aquifer. Biol. Wastes 19:165–178.

ROBERTS, R. F. 1977. Big game guzzlers. Rangeman's J. 4:80–82.

RODGERS, K. J., P. F. FFOLLIOTT, AND D. R. PATTON. 1978. Home range and movement of five mule deer in a semidesert grass-shrub community. U.S. For. Serv. Res. Note RM-355. 6pp.

ROOD, S. B., AND J. M. MAHONEY. 1990. Collapse of riparian poplar forests downstream from dams in western prairies: probable causes and prospects for mitigation. Environ. Manage. 14:451–464.

RUSSELL, T. P. 1964. Antelope of New Mexico. New Mexico Dep. Game and Fish Bull. 12. 103pp.

RUYLE, G. B., J. W. MENKE, AND D. L. LANCASTER. 1980. Delayed grazing may improve upland waterfowl habitat. Calif. Agric. 34:29–31.

RYDER, R. A. 1980. Effects of grazing on bird habitats. Pages 51–66 *in* R. M. DeGraaf, tech. coord. Proc. workshop management of western forests and grasslands for nongame birds. U.S. For. Serv. Gen. Tech. Rep. INT-86.

SALWASSER, H. 1980. Pronghorn antelope population and habitat management in the northwestern Great Basin environments. Interstate Antelope Conf. Guidelines [Fresno, Calif.]. 63pp.

SANCHEZ, P. G. 1975. A tamarisk fact sheet. Trans. Desert Bighorn Counc. 19:12–14.

SANDERS, S. D., AND M. A. FLETT. 1989. Montane riparian habitat and willow flycatchers: threats to a sensitive environment and species. Pages 262–266 *in* D. L. Abell, tech. coord. Proc. California riparian systems conference. U.S. For. Serv. Gen. Tech. Rep. PSW-110.

SCHLATTERER, E. R. 1986. Background, present status, and future of the evaluation of soil condition in rangeland monitoring in the Forest Service. Proc. Annu. Meet. Soc. Range Manage. 39:41–46.

SCHMIDT, J. L., AND D. L. GILBERT, EDITORS. 1978. Big game of North America: ecology and management. Stackpole Books, Harrisburg, Pa. 494pp.

SCHULZ, P. A., AND F. S. GUTHERY. 1988. Effects of short duration grazing on northern bobwhites: a pilot study. Wildl. Soc. Bull. 16:18–24.

SEDGWICK, J. A., AND F. L. KNOPF. 1987. Breeding bird response to cattle grazing of a cottonwood bottomland. J. Wildl. Manage. 51:230–237.

SETTERGREN, C. D. 1977. Impacts of river recreation use on streambank soils and vegetation: state-of-the-knowledge. Pages 55–59 *in* Proc. river recreation management and research symposium. U.S. For. Serv. Gen. Tech. Rep NC-28.

SEVERSON, K. E., TECHNICAL COORDINATOR. 1990. Can livestock be used as a tool to enhance wildlife habitat? U.S. For. Serv. Gen. Tech. Rep. RM-194. 123pp.

———, AND C. E. BOLDT. 1978. Cattle, wildlife and riparian habitats in the western Dakotas. Pages 91–102 *in* Regional rangeland symposium on management and use of northern plains rangeland. North Dakota State Univ., Dickinson.

———, AND A. L. MEDINA. 1983. Deer and elk habitat management in the Southwest. J. Range Manage. Monogr. 2. 64pp.

SHARP, B. 1986. Management guidelines for the Swainson's hawk. U.S. Fish Wildl. Serv., Reg. 1, Portland, Oreg. 28pp.

SHORT, H. L. 1986. Rangelands. Pages 93–122 *in* A. Y. Cooperrider, R. J. Boyd, and H. R. Stuart, eds. Inventory and monitoring of wildlife habitat. U.S. Dep. Inter. Bur. Land Manage. Serv. Cent., Denver, Colo.

SIVILS, B. E., AND J. H. BROCK. 1981. Sand dams as a feasible water development for arid regions. J. Range Manage. 34:238–239.

SKINNER, Q. D., M. A. SKINNER, T. A. WESCHE, AND S. LOWRY. 1989. A survey of values associated with riparian conditions of a stream tributary to the Green/Colorado River. Page 175 *in* R. E. Gresswell, B. A. Barton, and J. L. Kershner, tech. coords. Practical approaches to riparian resource management. U.S. Bur. Land Manage., Billings, Mont.

SKOVLIN, J. M. 1984. Impacts of grazing on wetlands and riparian habitat: a review of our knowledge. Pages 1001–1103 *in* Developing strategies for rangeland management, Westview Press, Boulder, Colo.

———, P. J. EDGERTON, AND B. R. MCCONNELL. 1983. Elk use of winter range as affected by cattle grazing, fertilizing, and burning in southeastern Washington. J. Range Manage. 36:184–189.

SMITH, D. R., EDITOR. 1975. Proc. symposium management of forest and range habitats for nongame birds. U.S. For. Serv. Gen. Tech. Rep. WO-1. 343pp.

SMITH, E. L. 1978. A critical evaluation of the range condition concept. Pages 266–267 *in* International rangelands congress. Soc. Range Manage., Denver, Colo.

———. 1988. Successional concepts in relation to range condition assessment. Pages 113–133 *in* P. T. Tueller, ed. Vegetation science applications for rangeland analysis and management. Kluwer Academic Publ., Dordrecht, The Netherlands.

SMITH, M. A., J. C. MALECHECK, AND K. O. FULGHAM. 1979. Forage selection by mule deer on winter range grazed by sheep in spring. J. Range Manage. 32:40–45.

SMITH, R. J. 1977. Conclusion. Pages 117–118 *in* J. E. Townsend and R. J. Smith, eds. Proc. seminar on improving fish and wildlife benefits in range management. U.S. Fish Wildl. Serv. FSW/OBS-77/1.

SPILLETT, J. J., J. B. LOW, AND D. SILL. 1967. Livestock fences—how they influence pronghorn antelope movements. Ut. Agric. Exp. Stn. Bull. 470. 79pp.

STEDNICK, J. D. 1988. The influence of riparian/wetland systems on surface water quality. Pages 17–19 *in* K. M. Mutz, D. J. Cooper, M. L. Scott, and L. K. Miller, tech. coords. Restoration, creation and management of wetland and riparian ecosystems in the American West. PIC Technol., Denver, Colo.

STEVENSON, C. A. 1990. Identification and construction of slickrock water developments in southern Nevada by Nevada Department of Wildlife. Pages 25–35 *in* G. K. Tsukamoto and S. J. Stiver, eds. Proc. wildlife water development symposium. Nev. Chap. The Wildl. Soc., U.S. Bur. Land Manage., and Nev. Dep. Wildl.

STODDARD, H. L. 1931. The bobwhite quail: its habits, preservation, and increase. Charles Scribner's Sons, New York, N.Y. 559pp.

———, AND E. V. KOMAREK. 1941. The carrying capacity of southeastern quail lands. Trans. North Am. Wildl. Conf. 6:148–155.

STODDART, L. A., A. D. SMITH, AND T. W. BOX. 1975. Range management. Third ed. McGraw-Hill Book Co., New York, N.Y. 532pp.

STREETER, R. G., ET AL. 1979. Energy mining impacts and wildlife management: which way to turn. Trans. North Am. Wildl. Nat. Res. Conf. 44:26–65.

STROMBERG, J. C., AND D. T. PATTEN. 1988. Total protection: one management option. Pages 61–62 *in* K. M. Mutz, D. J. Cooper, M. L. Scott, and L. K. Miller, tech. coords. Restoration, creation and management of wetland and riparian ecosystems in the American West. PIC Technologies, Denver, Colo.

SWANSON, F. J., S. V. GREGORY, J. R. SEDELL, AND A. G. CAMPBELL. 1982. Land-water interactions: the riparian zone. Pages 267–291 *in* L. Rovert, ed. Analyses of coniferous forest ecosystems in the western United States. US/IBP Synthesis Ser. 14. Hutchinson Ross Publ. Co., Stroudsburg, Pa.

SWANSON, G. A., TECHNICAL COORDINATOR. 1979. Mitigating losses of fish and wildlife habitats: a symposium. U.S. For. Serv. Gen. Tech. Rep. RM-65. 684pp.

SYKES, G. 1937. Sand tanks for water storage in desert regions. U.S. For. Serv., Southwestern For. Range Exp. Stn. Res. Note 9. 2pp.

SZARO, R. C., S. C. BELFIT, J. K. AITKIN, AND J. N. RINNE. 1985. Impact of grazing on a riparian garter snake. Pages 359–363 *in* R. R. Johnson, C. D. Ziebell, D. R. Patton, P. F. Ffolliott, and R. H. Hamre, tech. coords. Riparian ecosystems and their management: reconciling conflicting uses. U.S. For. Serv. Gen. Tech. Rep. RM-120.

———, AND M. D. JAKLE. 1985. Avian use of a desert riparian island and its adjacent scrub habitat. Condor 87:511–519.

TEVIS, L., JR. 1956. Pocket gophers and seedlings of red fir. Ecology 37:379–381.

THOMAS, J. W., EDITOR. 1979. Wildlife habitats in managed forests—the Blue Mountains of Oregon and Washington. U.S. For. Serv. Agric. Handb. 553. 512pp.

———, H. C. BLACK, JR., R. J. SCHERZINGER, AND R. J. PEDERSEN. 1979a. Deer and elk. Pages 104–127 *in* J. W. Thomas, ed. Wildlife habitats in managed forests—the Blue Mountains of Oregon and Washington. U.S. For. Serv. Agric. Handb. 553.

———, C. MASER, AND J. E. RODIEK. 1979b. Riparian zones. Pages 40–47 *in* J. W. Thomas, ed. Wildlife habitats in managed forests—the Blue Mountains of Oregon and Washington. U.S. For. Serv. Agric. Handb. 553.

———, ———, AND ———. 1979c. Wildlife habitats in managed rangelands—the Great Basin of southeastern Oregon: riparian zones. U.S. For. Serv. Gen. Tech. Rep. PNW-80. 18pp.

———, AND D. E. TOWEILL, EDITORS. 1982. Elk of North America: ecology and management. Stackpole Books, Harrisburg, Pa. 698pp.

THOMPSON, B. C. 1979. Evaluation of wire fences for coyote control. J. Range Manage. 32:457–461.

TIEDEMANN, A. R., ET AL. 1979. Effects of fire on water, a state-of-knowledge review. U.S. For. Serv. Gen. Tech. Rep. WO-10. 28pp.

TOWNSEND, J. E., AND R. J. SMITH, EDITORS. 1977. Improving fish and wildlife benefits in range management. U.S. Fish Wildl. Serv. FWS/OBS-77/01. 118pp.

TSUKAMOTO, G. K., AND S. J. STIVER, EDITORS. 1990. Wildlife water development. Proc. wildlife water development symposium. Nev. Chap. The Wildl. Soc., U.S. Bur. Land Manage., and Nev. Dep. Wildl. 192pp.

TURNER, G. T., R. M. HANSEN, V. H. REID, H. P. TIETJEN, AND A. L. WARD. 1973. Pocket gophers and Colorado mountain rangeland. Colorado State Univ. Exp. Stn. Bull. 554S. 90pp.

URNESS, P. J. 1976. Mule deer habitat changes resulting from livestock practices. Pages 21–35 *in* G. W. Workman and J. B. Low, eds. Mule deer decline in the West: a symposium. Utah State Univ. Coll. Nat. Resour., Logan.

———. 1982. Livestock as tools for managing big game range in the Intermountain West. Wildl.-Livestock Relationships Symp. 10:20–31.

———. 1990. Livestock as manipulators of mule deer winter habitats in northern Utah. Pages 25–40 *in* K. E. Severson, tech. coord. Can livestock be used as a tool to enhance wildlife habitat? U.S. For. Serv. Gen. Tech. Rep. RM-194.

U.S. BUREAU OF LAND MANAGEMENT. 1964. Water development: range improvements in Nevada for wildlife, livestock, and human use. U.S. Bur. Land Manage., Reno, Nev. 37pp.

———. 1966. Polyethylene liner for pit reservoir including trough and fencing. U.S. Bur. Land Manage., Portland Serv. Cent. Tech. Note P712C. 40pp.

———. 1985. Fencing. U.S. Bur. Land Manage. Manual Handb. H-1741-1. 23pp.

VALENTINE, B. E., T. A. ROBERTS, S. P. BOLAND, AND A. P. WOODMAN. 1988. Livestock management and productivity of willow flycatchers in the central Sierra Nevada. Trans. West. Sect., The Wildl. Soc. 24:105–114.

WAGNER, F. H. 1978. Livestock grazing and the livestock industry. Pages 121–145 *in* H. P. Brokaw, ed. Wildlife and America. Council on Environmental Quality. U.S. Gov. Printing Off., Washington, D.C.

WALLMO, O. C., EDITOR. 1981. Mule and black-tailed deer of North America. Univ. Nebraska Press, Lincoln. 605pp.

WARNER, R. E., AND K. M. HENDRIX. 1984. California riparian systems: ecology, conservation, and productive management. Univ. California Press, Berkeley. 1035pp.

WAUER, R. H. 1977. Significance of Rio Grande riparian systems upon the avifauna. Pages 165–174 *in* R. R. Johnson and D. A. Jones, tech. coords. Importance, preservation, and management of riparian habitat. U.S. For. Serv. Gen. Tech. Rep. RM-43.

WEAVER, R. A. 1973. California's bighorn management plan. Trans. Desert Bighorn Counc. 17:22–42.

———, F. VERNOY, AND B. CRAIG. 1959. Game water development on the desert. Calif. Fish Game 45:333–342.

WEHAUSEN, J. D., V. C. BLEICH, AND R. A. WEAVER. 1987. Mountain sheep in California: a historical perspective on 108 years of full protection. Trans. West. Sect., The Wildl. Soc. 23:65–74.

WELCHERT, W. T., AND B. N. FREEMAN. 1973. "Horizontal" wells. J. Range Manage. 26:253–256.

WERNER, W. E. 1984. Bighorn sheep water development in southwestern Arizona. Trans. Desert Bighorn Counc. 28:12–13.

WHITAKER, J. O., JR., S. P. CROSS, AND C. MASER. 1983. Food of vagrant shrews (*Sorex vagrans*) from Grant County, Oregon, as related to livestock grazing pressures. Northwest Sci. 57:107–111.

WHYTE, R. J., AND B. W. CAIN. 1979. The effect of grazing on nesting marshbird habitat at the Welder Wildlife Refuge, San Patricio County, Texas. Bull. Texas Ornithol. Soc. 12:42–46.

———, AND ———. 1981. Wildlife habitat on grazed or ungrazed small pond shorelines in south Texas. J. Range Manage. 34:64–68.

WILLMS, W., A. MCLEAN, AND R. RITCEY. 1979. Interactions between mule deer and cattle on big sagebrush range in British Columbia. J. Range Manage. 32:299–304.

WILSON, L. O. 1977. Guidelines and recommendations for design and modification of livestock watering developments to facilitate safe use by wildlife. U.S. Bur. Land Manage. T/N 305. 20pp.

WOHL, E. E., AND P. P. PEARTHREE. 1991. Debris flows as geomorphic agents in the Huachuca Mountains of southeastern Arizona. Geomorphology 4:273–292.

WOOD, J. E. 1969. Rodent populations and their impact on desert rangelands. N.M. State Agric. Exp. Stn. Bull. 555. 17pp.

YOAKUM, J. D. 1980. Habitat management guidelines for the American pronghorn antelope. U.S. Bur. Land Manage. Tech. Note 347. 77pp.

———, W. P. DASMANN, H. R. SANDERSON, C. M. NIXON, AND H. S. CRAWFORD. 1980. Habitat improvement techniques. Pages 329–403 *in* S. D. Schemnitz, ed. Wildlife management techniques manual. Fourth ed. The Wildl. Soc. Washington, D.C.

28

MANAGING FORESTLANDS FOR WILDLIFE

R. William Mannan, Richard N. Conner, Bruce Marcot, and James M. Peek

INTRODUCTION

Forestlands often are the focus of wildlife management (Hunter 1990) because they occupy about one-third of the total area of the United States (Haynes 1990), support numerous wildlife species, and are being modified extensively by human activities. Managing forestlands for wildlife primarily involves maintaining or creating forest environments that satisfy the habitat requirements of the species under consideration. Factors that affect the abundance and distribution of forest animals include the age, size, shape, floral composition, and internal structure of forest stands, and the distribution of forest stands across the landscape. Wildlife management in forest ecosystems, therefore, requires an understanding of species-specific habitat requirements, and control of human activities and some of the environmental factors (biological and physical) that modify forest vegetation. It also requires an understanding of the effects and periodicity of environmental factors that cannot be controlled. Decisions about how to use natural resources on forestlands often are influenced greatly by laws, politics, social pressures, and the philosophy of the agency or individual controlling the land. Thus, managing forestlands for wildlife is an activity that frequently involves meshing technical knowledge and skills with societal demands and interests.

Nearly two-thirds of the forestlands in the United States (196 million ha; 66% of total forestlands) are classified as commercial timberland because they can produce ≥ 1.5 m^3 of wood/ha/year and are not reserved for other purposes (e.g., wilderness areas or parks) (Haynes 1990). About 57% of commercial timberland (112 million ha) is privately owned by individuals or companies not primarily interested in timber production, although trees on these holdings frequently are used for industrial wood products or firewood (Haynes 1990). Most of the remaining commercial timberland is managed for commodity values. These timberlands are owned by lumber companies (29 million ha) and private individuals, or managed by federal and state natural resource agencies (55 million ha). The agency that administers the largest area of commercial timberland is the U.S. Forest Service (34 million ha) (Haynes 1990).

The amount and distribution of commercial timberlands that are privately owned versus those that are administered by public agencies vary from region to region across the United States. Forestlands in the western states, for example, generally are managed by public agencies, whereas forestlands in the eastern states generally are privately owned or are a mixture of private and public ownership. Management on a region-wide scale for any renewable natural resource, whether it is timber or wildlife, is easier to coordinate if relatively large tracts of land are administered by a single authority, as among some western forests.

Efforts to manage forest wildlife are perhaps most important on lands where timber harvests and related disturbances dramatically alter forest vegetation. But it is also important to consider the effects on forest wildlife when areas disturbed by other human activities, such as agricultural practices, urbanization, and surface mining, are managed. All of these activities potentially affect the abundance and distribution of indigenous species of animals. Proper management of all forestland habitats is thus critical to ensure the continued existence of viable wildlife populations, viable ecological communities, and high levels of biological diversity (Hunter 1990). We discuss in this chapter concepts and tools for managing forestlands for wildlife.

We first discuss pertinent national laws guiding federal agencies involved with management of forestlands in the U.S. We then introduce the concept of plant succession and its relationship to managing forest wildlife. We follow this section with a review of the types and effects of silvicultural prescriptions on wildlife at the stand level, and management activities for habitat attributes at the landscape scale, including old-growth forests, and riparian zones. Effects of disturbance by humans then are discussed. We end with a review of wildlife-habitat models useful for planning and managing forestlands. Throughout the chapter, we focus on potential strategies and specific activities for managing wildlife in forests where production of timber is a prominent goal. Many of the activities and strategies described, however, should be applicable in all forestlands.

NATIONAL LAWS AND FOREST/WILDLIFE MANAGEMENT

Management strategies for some species of forest wildlife are costly in terms of foregone commodity values and, consequently, are unlikely to be extensively employed on lands whose owner's primary goal is the production of timber. They can be used, however, on forestlands administered by state or federal agencies when laws dictate that these agencies manage forests for more than just wood products. A brief review of the laws that influence forest/wildlife management on federal lands follows; the review summarizes the information provided by Bean (1983).

Laws Governing the U.S. Forest Service

The Forest Reserve Act of 1891 authorized the President to set aside forestlands as federal, public reservations. Timber harvests, mining, and access were restricted in these reservations. The restrictions and the amount of land set aside, especially in the western United States, concerned western interest groups and led to the passage of the "Organic Act" in 1897. The Organic Act mandated that forest reservations could be established only to protect forests, to secure waterflows, or to provide a supply of timber. Despite these restrictions, National Forests were used for a variety of purposes, including grazing and outdoor recreation, during the first decades of their existence.

The unwritten policy of using national forests for a variety of purposes, or "multiple use," functioned adequately in the early 1900s because there were enough public

and private forests to meet the needs of various user groups without serious conflict over how to use individual tracts of land. Demand for timber and outdoor recreational opportunities increased dramatically, however, after World War II, and conflicts between these two disparate uses stimulated questions about how best to manage the demands for national forests. The Multiple Use-Sustained Yield Act, passed in 1960, stipulated that national forests would be administered for outdoor recreation, range, timber, watershed, wildlife, and fish. But the level of consideration given to each of these resources was left to the discretion of the U.S. Forest Service, and harvesting timber remained the dominant activity on many national forests in the southern and western United States.

Several laws passed after 1960 reduced the discretionary power of the Forest Service to emphasize one resource over another. The first of these acts was the Wilderness Act of 1964. Wilderness areas could be established under the Multiple Use-Sustained Yield Act, but there were no provisions to prevent the Secretary of Agriculture from withdrawing a forested area from wilderness status after it had been so designated. The Wilderness Act eliminated this option by giving Congress the authority to identify and permanently set aside wilderness areas.

In 1974, the Sikes Act Extension and the Forest and Rangelands Renewable Resources Planning Act (RPA) were passed. Both laws were designed to facilitate planning for the use and conservation of renewable natural resources on federal lands in general, and the national forest system in particular. The Sikes Act Extension directed the Secretaries of Agriculture and Interior to cooperate with state game and fish agencies in developing comprehensive plans for the conservation and rehabilitation of wildlife, fish, and game. Habitat improvement projects were to be included for threatened and endangered species of animals, and, if a state agreed, permits could be sold to users of federal lands to fund these projects.

The RPA called for administrative units of the national forest system to prepare land management plans for the protection and development of national forests. The plans were to be prepared every 5 years with a multidisciplinary approach to ensure that a variety of resources would be considered. However, neither the RPA nor the Sikes Act Extension specified what to include in the required plans, or how to prepare them. Neither act, therefore, substantially changed the discretion that Forest Service planning teams exercised in fulfilling the mandates of the Multiple Use-Sustained Yield Act. The National Forest Management Act of 1976 (NFMA) modified aspects of the Organic Act to facilitate clearcutting on national forests, but simultaneously imposed standards that limited the discretion of the Forest Service under multiple use management. Among the most important elements of the NFMA was the provision that the Secretary of Agriculture would develop regulations for preparing the land management plans called for in the RPA. The regulations would specify procedures to ensure interdisciplinary resource management and require that the plans comply with the National Environmental Policy Act (NEPA) of 1969 (discussed below). The regulation (36 CFR 219.19) implementing the NFMA further stipulated that management would maintain viable populations of existing native vertebrate species.

Clearcutting or other even-aged management activities were permitted under the NFMA, but only if the potential environmental, biological, esthetic, engineering, and economic impacts had been assessed, and if the cuts did not significantly harm soil, watershed, fish, wildlife, recreation, and esthetic resources. The NFMA also called for public input into the land management plans. Thus, both commercial and noncommercial users of national forests were given equal opportunity to comment on the direction of the plans.

Laws Governing the U.S. Bureau of Land Management

The sequence of laws governing the management of forestlands administered by the Bureau of Land Management was similar to that for the U.S. Forest Service. The Taylor Grazing Act of 1934 was analogous to the Organic Act in seeking to save western grazing lands from abusive treatment so that the cattle industry would not lose forage crops. But the rangelands and forestlands under the jurisdiction of the Bureau of Land Management also were used for other purposes. The concept of managing these lands under a multiple use philosophy was included in the Classification and Multiple Use Act of 1964. Land use plans, similar to those called for by the RPA and the NFMA, were mandated by the Federal Land Policy and Management Act of 1976. For additional details about these laws see Bean (1983).

Other Laws That Influence Management of Federal Forestlands

The National Environmental Policy Act of 1969 (NEPA) and the Endangered Species Act of 1973 also substantially influence how federal forestlands are managed. The NEPA stipulated, among other things, that environmental impact statements would be prepared for any federal project that could significantly affect the quality of the human environment. Wildlife populations or habitats generally have been interpreted by the courts as part of the human environment. Therefore, environmental impact statements or assessments are required for the planning of timber harvests on forestlands under federal jurisdiction. Environmental impact statements promoted consideration of how timber harvests would influence a variety of forest resources, including wildlife populations and habitats. They also provided the public with information about forest management in general and specific issues in particular. The NEPA, therefore, has helped balance the emphasis placed on management of various resources on public forestlands.

The Endangered Species Act is designed to protect species that are near extinction, or those that could be threatened in the foreseeable future (see Chapter 20). It also requires landowners to protect the ecosystems upon which these species depend. Section 7 of the Act stipulates that federal actions must not jeopardize threatened or endangered species or destroy or modify their habitat. Thus, if a proposed action could affect a threatened or endangered species, the agency proposing to carry out the action must consult with the U.S. Fish and Wildlife Service or the National Marine Fisheries Service to determine the effects and identify ways to avoid harmful ones. Under this provision, timber harvests or other management activities on public lands cannot legally destroy the habitat of an endangered species. Management activities on privately owned lands also cannot destroy the habitat of an endangered species because no person under the jurisdiction of the United States may "take" an endangered species in the United States. The term "take" includes harm that would result from a modification or degradation of habitat.

To successfully implement federal laws that influence the availability of forested environments for wildlife, managers need an understanding of ecological dynamics of forests. The next section discusses the concepts of plant succession and its relationship to wildlife management.

PLANT SUCCESSION AND FOREST WILDLIFE

The classical definition of plant succession is the orderly process of one plant community replacing another over time (Clements 1916, 1936). The changes that occur are reasonably directional and predictable and come about because the plants that occupy a site modify the environment to the extent that conditions become suitable for the establishment of other species (Odum 1969). The process equilibrates when a plant community is perpetuated indefinitely. The final community is called a climax community. The sequence of community replacements over time is called a sere.

Classical ideas about succession are based on the assumption that succession is a community characteristic (Noble 1981). Many ecologists, however, believe that changes during plant succession result from interactions among individual species and the environment. As such, they are not a community response per se (e.g., Cooper 1926, Noble 1981). Efforts to explain succession at the species level were prompted, in part, by evidence that the composition of plant species among sites in a given seral stage varied, to some extent. Such variation appeared to depend on type and frequency of disturbance, seed source available after disturbance, and conditions for growth. Other evidence suggested that species in later stages of succession do not always depend on species in earlier stages to create appropriate environments before they can become established (Connell and Slatyer 1977). Thus, the sequence of replacement of plant species in a sere is not always identical nor totally predictable. Current explanations for succession, therefore, generally focus on the adaptations of individual species (e.g., Pickett 1976, Noble 1981).

Succession may begin when a plant community is disturbed by any of a variety of agents or events such as fire, wind, insect outbreaks, pathogens, or timber harvests. Disturbances that create relatively large openings in a forest sometimes are called "coarse-scale" disturbances, whereas those that kill single trees, or small groups of trees, are called "fine-scale" disturbances. The nature, frequency, and pattern of disturbances vary widely among and within forest types (Pickett and Thompson 1978). Fires, for example, may burn a given site in Maine every 250–800 years, but in boreal forests in Minnesota a given site may burn every 5–50 years (Pickett and Thompson 1978). Fires in deciduous woodlands in the northeast and north-central United States tend to be relatively small (3–12 ha) compared to fires in coniferous forests in the western

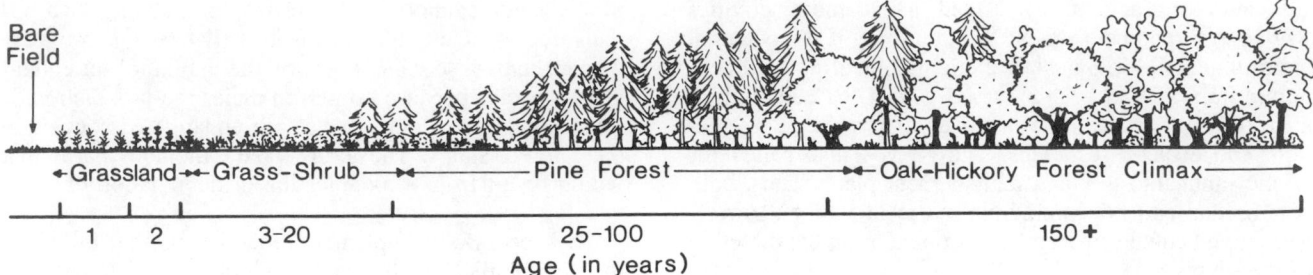

Bare
Field

←Grassland→←Grass-Shrub——→ ←————Pine Forest————→ ←——Oak-Hickory Forest Climax——→

    1   2   3-20   25-100   150 +

                                Age (in years)

Fig. 1.   Typical plant succession in an oak-hickory sere (from Kirk 1975, used with permission from McGraw-Hill, Inc.).

United States (Pickett and Thompson 1978). The disturbance regime dictates how long a forest stand will develop, and the size (area) and distribution of forest stands.

Patterns of stand development have been observed in temperate forests following disturbances. After a severe fire in which all living vegetation is killed and consumed, species of plants that are capable of growing in relatively exposed places invade. These "shade-intolerant" species include annual grasses and forbs, sprout-growth shrubs, and even some trees. They often are eventually replaced by biennial and perennial grasses and shrubs, which in turn yield to trees (Fig. 1). In many seres, different species of trees replace one another in the later stages of succession (Fig. 1). Sometimes seral tree species dominate for long periods. In the Pacific Northwest, for example, a seral species, Douglas-fir, may dominate for 600 years or more before the climax species, western hemlock, replaces it (Franklin and Dyrness 1973:71).

When a forest stand is dominated by climax or long-lived seral species, it is not necessarily static. Fine-scale disturbances may kill single trees, or small groups of trees, and create holes or gaps in the stand. Gaps may be filled in by growth of adjacent trees, if the gaps are small, or by regeneration if the gaps are larger. Primarily shade-tolerant species of trees become established in relatively small gaps, whereas shade-intolerant species become more prevalent in larger gaps. "Gap succession" is considered an important process in mature and old-growth forests (i.e., late seral stages) because it helps maintain much of the biological diversity and variation in structure within these stands and allows them to persist over long periods in the absence of major, coarse-scale disturbances.

Numerous changes in composition and structure take place as forests develop from early to late successional stages. Changes in the composition of plant species are accompanied by structural changes, including (1) density, spacing, and size of living trees; (2) height, profile, and closure of the canopy; (3) density and size of dead trees (standing and on the ground); and (4) density, spacing, and profile of understory vegetation. These changes influence the kinds of food and cover available to animals. Consequently, the composition of the animal community changes as the plant community develops (e.g., Adams 1908, Johnston and Odum 1956, Haapanen 1965, Shugart and James 1973, Meslow and Wight 1975, Repenning and Labisky 1985).

Understanding the relationships among seral stages, the stages of development of trees within a seral stage, and the habitats of forest animals is one of the keys to managing forestlands for wildlife. Some species of animals depend on the resources in one seral or development stage. Other species can live in a wider range of conditions and occupy several different seral or development stages (Fig. 2). Still other species are abundant only in places where different seral or developmental stages are adjacent to each other. The "edges" of stands having different structural characteristics are important to some forest animals because they provide access to different kinds of resources essential to their survival and reproduction (Thomas et al. 1979b).

The nature and size of disturbances are important in shaping the forest environment and thus in determining the animals that live there, but the concepts of "fine-scale" and "coarse-scale" disturbances presented above are relative terms when applied to animals. For example, a 10-ha burn may be a "course-scale" disturbance to a salamander, but a "fine-scale" disturbance to a mountain lion. Silvicultural practices at least partially replace natural disturbances in managed forests and thus provide the means for producing desired changes in stand composition and structure. The next section discusses concepts of silvicultural systems for forest stands.

SILVICULTURAL PRESCRIPTIONS— EFFECTS ON WILDLIFE WITHIN STANDS

Silvicultural practices implemented to provide a sustained yield of wood products largely determine the structure and composition of vegetation in most managed forests. Therefore, wildlife biologists working in forest ecosystems should understand silvicultural operations and must work closely with silviculturists and forest planners to meet objectives associated with wildlife populations and their habitats.

Silviculture is the practice of manipulating forest vegetation to control forest establishment, composition, and growth (modified from Smith 1962). There are some important distinctions among silviculture, forest management, and timber management.

"Forest management is the planning and execution of activities across a large area to meet generally more than one management goal, such as wildlife management, recreation, or timber management. Timber management is the specific objective of producing commercial wood fiber. Silviculture is vegetation management in the broadest sense, the objectives of which may include, but are not restricted to, timber management" (Marcot 1985:102).

Silvicultural practices that remove mature trees for the purpose of establishing a new stand are called reproduc-

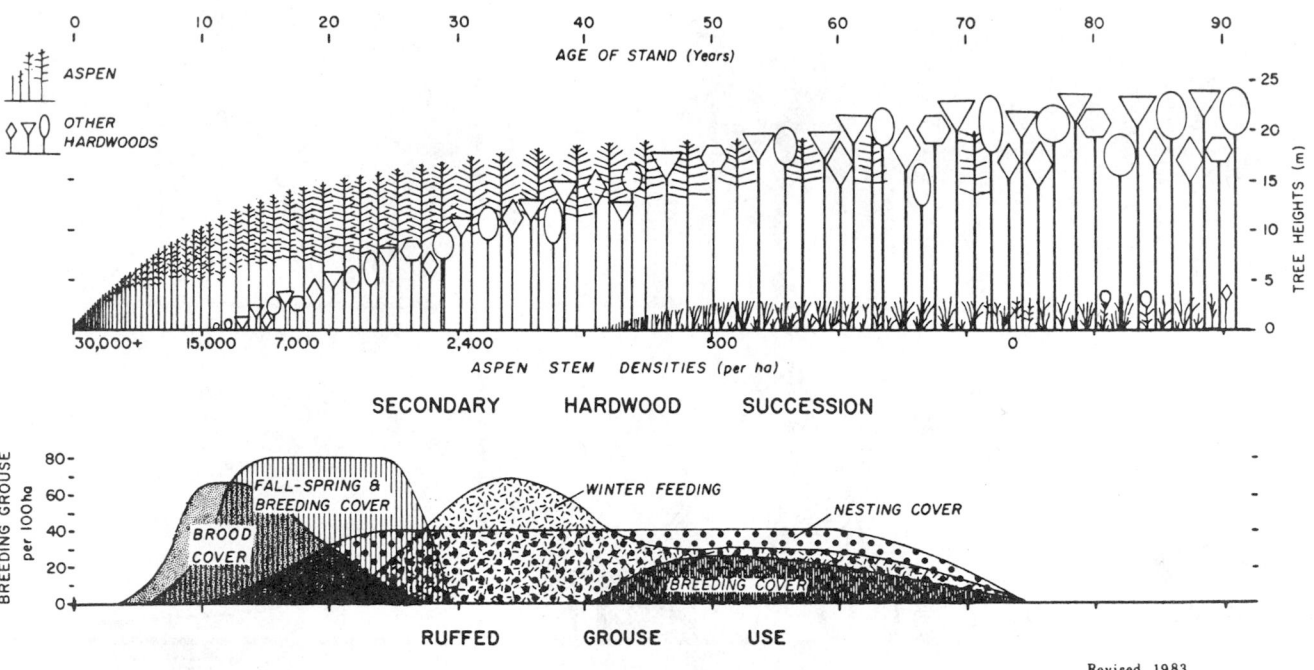

Fig. 2. The relationship between secondary forest succession and habitat use by ruffed grouse (from Gullion 1984, used with permission from The Ruffed Grouse Society).

tion methods. Patterns of disturbance created by reproduction methods generally should approximate disturbances that occur naturally in an area. The underlying assumption, from a silvicultural viewpoint, is that if trees reproduce and grow in an area under certain natural conditions, they will do so in a managed forest as well (Smith 1962). Silvicultural activities that prepare an area for regeneration of young trees are called site preparation methods. Activities that take place during the development of a stand are called intermediate treatments. These treatments usually involve the cutting of selected trees to reduce inter-tree competition and enhance the rate of growth of trees that remain. The collective program of treatments during the life of a stand, including the reproduction method, site preparation methods, and intermediate treatments, is called a silvicultural system (Smith 1962). Below we review common reproduction methods, site preparation methods, and intermediate treatments. We also review the effects of silvicultural activities on forest wildlife to the extent that they are known. The reproduction methods are divided into two groups—those that produce even-aged stands and those that produce uneven-aged stands. The methods and treatments discussed are not intended to span the breadth of possible silvicultural options but are presented to provide an outline of the more commonly used techniques.

Even-Aged Management

Under an even-aged silvicultural system, individual stands are comprised of trees of about the same age (Fig. 3). A stand is considered even-aged if the range of tree age within it does not exceed 20% of the rotation length (Wenger 1984:418). Rotation length is the period of time from stand establishment until financial maturity, when it will be regenerated again. This period varies under different management objectives and for different commu-

nity types. Even-aged stands also usually are made up of trees of about the same size, if they are comprised of shade-intolerant species (Fig. 3). However, stands with trees of different sizes can be grown under an even-aged management system if tree species with different growth rates or shade tolerances are regenerated in the same area. Even-aged forest management begins with the complete, or nearly complete, removal of the existing stand and the establishment of a new cohort of trees. Reproduction methods that produce even-aged stands are the clearcutting, seed-tree, and shelterwood methods.

CLEARCUTTING METHOD

Clearcutting is a widely used method of regenerating even-aged forest stands. The purpose of clearcutting is to remove woody vegetation from the site as a means of initiating a new stand of trees. Reproduction after clearcutting can be induced artificially by seeding or planting seedlings, or can be allowed to begin naturally by seeding from adjacent stands (Wenger 1984). In some hardwood forests, new stands are generated primarily from vegetative reproductions, or coppice sprouting. Among communities of shade-tolerant species, the new stand also may become established from advanced regeneration (i.e., small trees established before the harvest).

Clearcutting a mature forest stand effects changes in the animal species that use the site. Species associated with mature stands generally are replaced by species that live in the early stages of stand development. In eastern North America, big-game species that use clearcuts, primarily as a place to forage, include white-tailed deer (Blymer and Mosby 1977) and moose (Parker and Morton 1978). Some of the small mammals, birds, reptiles, and amphibians that use clearcuts in this region were listed by Conner and Adkisson (1975), Kirkland (1977), Martell and Radvanyi (1977), Webb et al. (1977), Meyers and Johnson (1978),

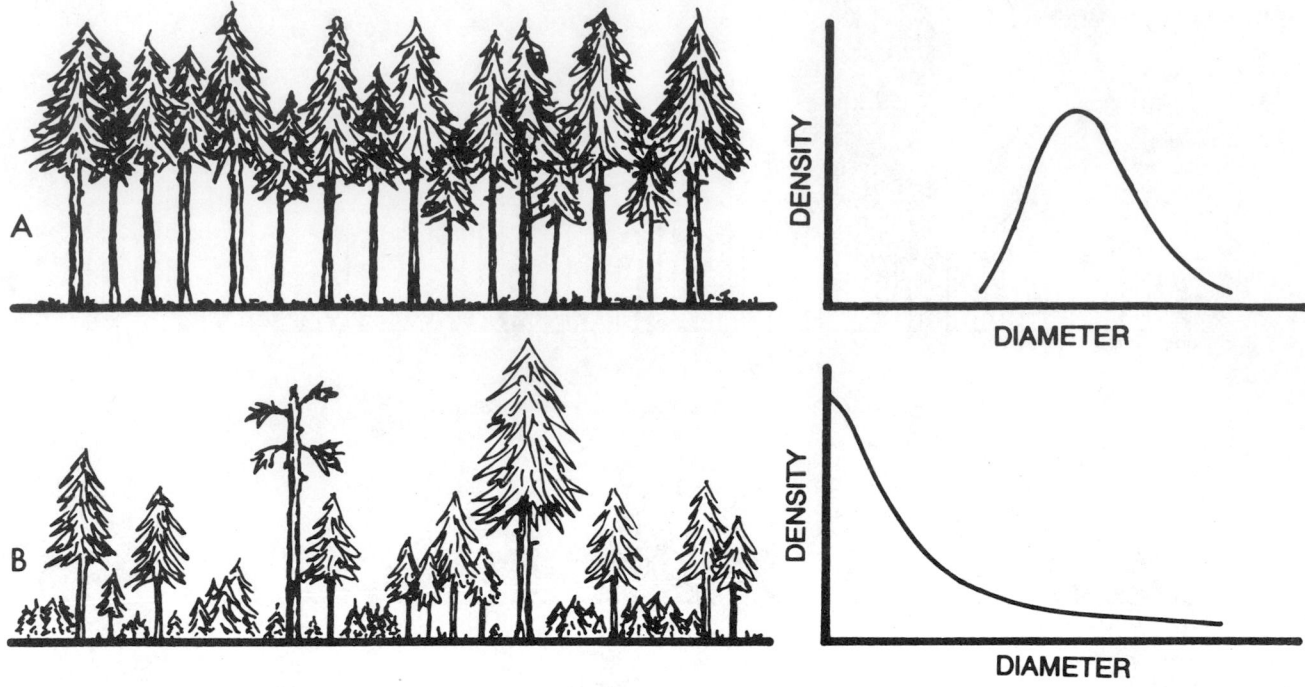

Fig. 3. Characteristic appearances and distribution of tree size in even-aged (A) and uneven-aged (B) forest stands (from Baker 1950, used with permission from McGraw-Hill, Inc.).

Conner et al. (1979), Titterington et al. (1979), Repenning and Labisky (1985), DeGraaf (1987), and Pais et al. (1988). In western North America, elk and mule deer forage in clearcuts (e.g., Regelin and Wallmo 1978, Lyon and Jensen 1980). Gashwiler (1970), Verner and Boss (1980), DeByle (1981), Ramirez and Hornocker (1981), Scott et al. (1982), and Morrison and Meslow (1983) listed some of the small mammals, birds, reptiles, and amphibians that inhabit clearcuts in western forests.

The animals that live or feed in clearcuts sometimes create problems, from the standpoint of timber management, by eating young seedlings. "Wildlife damage" in regenerating stands is a major concern in some areas of the United States. For example, in the Allegheny hardwood forests of Pennsylvania, deer browsing on young trees, and the associated delays in establishment of new stands, costs about $33/ha/year (Marquis 1981). When their abundances are relatively low, deer and elk may be drawn away from regenerating stands of timber by seeding palatable grasses and legumes in other areas (Brown and Mandery 1962, Ramsey and Krueger 1986). When ungulate densities are relatively high, more drastic measures such as woven wire fencing and electric fencing (Brenneman 1982) may be required to regenerate a new stand, but high costs make these techniques impractical for many landowners. Managing ungulate populations via hunting is perhaps the least expensive and most preferable method of controlling ungulate damage (Behrend et al. 1970, Tilghman 1989), but hunting may not be safe in areas of high human populations, or feasible where society inveighs against hunting. Pocket gophers, mountain beavers, and porcupines are among the other animals that can cause damage to regenerating timber; they are controlled with poison, trapping, shooting, and habitat ma-

nipulation (Van Deusen and Meyers 1962, Borrecco et al. 1979, Dodge 1982, Teipner et al. 1983).

Use of clearcuts by forest animals often is influenced by the size and shape of the stands. Distance to protective cover is probably the critical factor for some animals. For example, many forest ungulates feed in clearcuts but retreat to adjacent stands of relatively mature trees for cover. Clearcutting, therefore, is an effective way to increase forage for some big-game animals, if adequate cover is available nearby (Murphy and Ehrenreich 1965, Krefting and Phillips 1970, Monthey 1984).

Recommendations on the most appropriate size of clearcuts vary with the species and areas in question, but most forest ungulates appear to use cuts <50 ha in size. For example, Sweeney et al. (1984) reported that deer used clearcuts to a distance of 208 m from the edge of the forest, indicating that cuts <25 ha received substantial use in southern Arkansas and Louisiana. In western Montana, Lyon and Jensen (1980) concluded that 24 ha was the optimum size of openings for mule deer. Clearcuts between 10 ha and 50 ha appear optimum for elk (Irwin and Peek 1983), but total size may be less important than width and thus distance to cover. Cuts with widths up to 250 m apparently are suitable for elk (Leege 1984), especially if adequate cover is available in stands adjacent to the cut.

The juxtaposition of food and cover for forest ungulates also is an important consideration for regenerating forest stands after any reproduction method, including clearcutting. In forests in the north-central and northeastern United States, dense softwood stands are important to the survival of white-tailed deer in winter. Deer congregate in these stands, called "deeryards," for shelter from low temperatures, high winds, and deep snow (Weber et al.

1983). Not all stands have equal value as deeryards. Weber et al. (1983) observed that stands used as deeryards had high canopy closure of softwoods, some patchiness, and high site indices (i.e., high productivity) and that they were lower in elevation than stands not used as deeryards. They also developed models that could be used to predict which young stands likely would develop into the best deeryards.

In the Upper Peninsula of Michigan and elsewhere, dense pole stands of white cedar not only provide winter cover, but the seedlings and saplings of this tree species are an excellent source of winter food (Verme 1965, Ozoga 1968). Regenerating white cedar to provide future deeryards, or for other reasons, has been difficult partly because deer prevent or suppress the growth of young cedar trees by eating them, when cover is available nearby. Harvest techniques such as clearcutting in strips leave cover next to regenerating stands, and regeneration efforts often have failed. Verme and Johnston (1986) recommended that plans to regenerate stands of white cedar should consider (1) controlling deer density by harvest; (2) drawing deer away from young white cedar stands by cutting in other areas; and (3) cutting small areas adjacent to each other sequentially over 5–10 years until a large block (16–64 ha) is completely cleared, thus precluding use by deer because of lack of shelter.

The size and shape of clearcuts directly affect the amount of edge created. Numerous small cuts with irregular boundaries create more edge than fewer large cuts with straight boundaries. Edges often are considered beneficial to wildlife because many species of birds and mammals, particularly game animals, are abundant along edges (e.g., Leopold 1933, Thomas et al. 1979*b*, Kroodsma 1984, Strelke and Dickson 1980, Williamson and Hirth 1985). However, this phenomenon, termed the "edge effect," is not beneficial for all wildlife species.

In recent years, the negative aspects associated with increasing amounts of edge and fragmentation of the once unbroken blocks of forest have been demonstrated in numerous studies. Birds associated with the interior of forests have been subject to high rates of parasitism, competition, and predation when they nest near edges (Gates and Gysel 1978, Brittingham and Temple 1983, Dueser and Porter 1986, George 1987, Andrén and Angelstam 1988, Martin 1988, Yahner and Scott 1988, Brittingham 1989). Survival of species that require forest environments well away from edges can be problematic where forests have been fragmented into small patches (see Forest Fragmentation [p.49]).

The size, shape, and pattern of clearcuts also dictate the size, shape, and pattern of future stands and influence the species of animals that will inhabit the area in the future. Clearcutting in strips (100 × 400 m; 4 ha) at 5-year intervals in Michigan produced habitat for ruffed grouse when the regenerating aspen stands were 6–15 years of age (Gullion 1989). Clearcutting 15-ha units in a checkerboard arrangement, each with rotation lengths of <100 years, likely would provide habitat in the future for animals that can live in a patchwork of forest stands <100 years of age, but it would not produce habitat for species that live in the interior of large forest stands, or in stands older than 100 years.

SEED-TREE AND SHELTERWOOD METHOD

Other silvicultural methods that produce even-aged stands are seed-tree and shelterwood cuts. In contrast to clearcutting, seed-tree and shelterwood cuts do not remove all mature trees during the initial harvest. Selected trees are left standing to provide a seed source (i.e., seed-cut trees) or a seed source and some shelter for the regenerating stands (i.e., shelterwood cuts) (Smith 1962). Once the regeneration is established, the remaining mature trees are harvested.

The presence of mature, residual trees in seed-tree and shelterwood cuts maintains some of the characteristics of mature stands. Consequently, some species of birds generally associated with mature stands (e.g., brown-headed nuthatches and red-cockaded woodpeckers in longleaf pine forests) can be maintained after the first cutting in the shelterwood sequence (Conner and O'Halloran 1987, O'Halloran and Conner 1987, Conner et al. 1991). Benefits accrued from having mature residual trees can be maintained if the stands are managed by deferment cutting, or the "reserve shelterwood method." In a deferment cut, which resembles a seed-tree cut, "residual trees are not cut when the regenerated stands become established. Instead, residual trees remain until the regenerated stand is at the end of a rotation" (Smith et al. 1989:14). However, deferment cuts are infrequently used in the United States at present, and residual trees in shelterwood and seed-tree cuts typically are removed 6–12 years after the initial harvest.

Seed-tree and shelterwood cuts also provide abundant forage for deer. In northern mixed forests, seed-tree and shelterwood cuts produce more browse for deer than selection cuts (Krefting and Phillips 1970). The value of seed-tree and shelterwood cuts (and other silvicultural treatments) to forest ungulates frequently is evaluated by measuring the response of forage plants (e.g., Behrend and Patric 1969). Changes in the density or distribution of ungulate populations in response to specific silvicultural treatments are not well documented.

Uneven-Aged Management

"Uneven-aged silviculture and management is the manipulation of a forest for a continuous high-forest cover, recurring regeneration of desirable species, and the orderly growth and development of trees through a range of age or diameter classes to provide a sustained yield of forest products" (Gibbs 1978:19). Silvicultural practices designed to produce uneven-aged stands are called selection methods. Forest stands produced and maintained by selection methods contain three or more age classes of trees (Smith 1962, Wenger 1984), and the distribution of tree ages, and often diameters, ideally approximates an inverse "J"-shaped curve (Fig. 3), each age class occupying an equivalent amount of space in the stand. Such a stand is called "balanced." In the selection method of reproduction, single mature trees or small groups of them are removed at relatively short intervals to create space for the regeneration of a new age class (Smith 1962). Residual immature trees are thinned to reduce overcrowding and concentrate the growth potential of the site into trees having characteristics best suited to the objectives of the landowner. Shade-tolerant or moderately shade-intolerant trees normally dominate the regeneration established by selection methods.

SINGLE-TREE SELECTION

One form of uneven-aged management is the selection and cutting of individual trees. This method requires that the stand be entered every 5–30 years, depending on the density of residual trees. During an entry into a stand, single trees from a variety of diameter and age classes and species are cut. Removal of the mature age class creates space for the regeneration of a new one. Individual immature trees are cut (i.e., thinned) to promote the growth of the "best" trees of each age class. Administration and implementation of single-tree selection cuts are complex, because a detailed inventory of all trees by size class is needed before each harvest, and a specified number of trees in each diameter class must be cut while disturbance to other trees is minimized.

Openings or patches produced by cutting single trees are small, and regeneration typically is dominated by shade-tolerant species. In hardwood stands, openings are regenerated by sprouting, seeds stored in the litter and dropped from nearby trees, or seedlings or young trees already established. In conifer stands, seedlings germinate from seeds produced by mature trees adjacent to the openings. Prescribed fire might be used in conjunction with single-tree cuts to prepare the seed bed and reduce competition from hardwoods in some conifer types (e.g., loblolly pine), but only if the youngest age class is old enough to withstand the fire and the interval between harvests is relatively long. Herbicides typically are used to control hardwoods if seedlings and saplings of the desired coniferous tree species are vulnerable to fire.

Information on the effects of single-tree selection cuts on wildlife is scarce. Healy (1989:230) noted that "the initial selection cut in previously unmanaged northern hardwood stands will substantially increase both herbage and browse. Yet repeated treatments at 10- to 15-year intervals will favor the development of tree seedling and shrub understories with a loss of herbs." Overall, however, managed uneven-aged stands provide more herbage, browse, and cover for wildlife than "fully stocked pole and sawtimber-size even-aged" stands (Healy 1987:343).

In contrast, DeGraaf (1987:355) concluded that "even-aged management in northern hardwoods provides habitat for more forest bird species than does uneven-aged management," primarily because young stands following clearcutting supported distinct breeding bird communities. Stands >30 years of age, managed under either even-aged or uneven-aged programs, had similar foliage profiles and diameter distributions and therefore supported similar bird communities (DeGraaf 1987).

Uneven-aged management can change the composition of tree species in mixed hardwood forests (over the course of several entries) by promoting the most shade-tolerant species. In some Appalachian hardwood stands, single-tree selection could favor maples and beech over oaks and would cause a reduction in the production of mast (Healy 1989), and thus a reduction in food for mast-dependent wildlife. Healy (1989) proposed that a mixture of uneven-aged and even-aged management schemes would produce the widest array of habitat features in northern hardwood forests.

GROUP SELECTION

The group selection method also can be used to maintain uneven-aged stands, but it has some features similar to even-aged management. Groups of adjacent, mature trees on small areas (up to 0.1 ha) are cut during group selection harvests. The openings created resemble miniature clearcuts but are too small to be considered even-aged stands. The group selection method is used to regenerate species of trees that are not shade-tolerant enough to regenerate in a single-tree selection system (Leak and Filip 1977). The age of trees regenerated in each opening is relatively uniform, but different age classes are distributed (intermixed) over the entire area of the stand. In stands managed properly with the group selection method, the overall distribution of the ages approximates an inverse "J" (Fig. 3).

Information on the effects of group selection cuts on wildlife is more scarce than information on single-tree selection cuts. Scott and Gottfried (1983) observed that a mix of group and single-tree selection cuts in mixed spruce-fir forests in Arizona decreased total bird abundance by 12% but increased bird species richness by 25%. House wrens, American robins, and pine siskins were among the species that used the area after the selection cuts. The scarcity of information about the effects of uneven-aged management on wildlife should not be interpreted as a negative feature of this management strategy. Uneven-aged management potentially can create and maintain habitat for many wildlife species, and it has been proposed as a way to decrease the negative effects of logging on forest ecosystems (Fritz 1990). One concern about uneven-aged management in general is the relatively high rates of disturbance. Stands are entered relatively frequently (every 5–20 years), and the network of skid trails and roads required is extensive.

Site Preparation and Intermediate Treatments

Methods used to prepare a site for planting trees or for natural regeneration and treatments applied during the development of a stand (i.e., intermediate treatments) may affect the structure and composition of forest vegetation, but only within limits of the species diversity established by the reproduction methods. Habitat components potentially influenced by site preparation and intermediate treatments include the abundance and composition of understory plants, the abundance and thrift of mast-producing trees, the distribution and size of canopy openings, the number of snags, and the density of woody material on the ground. Alteration of these habitat components can significantly influence the current and future value of a forest stand for wildlife.

The primary objectives of site preparation methods are to prepare the site for natural or artificial regeneration, reduce vegetation that could compete with young trees, and dispose of woody debris (often called slash) created by reproduction cuts (Smith 1962). The purpose of intermediate treatments generally is to increase the vigor and growth of selected trees (Smith 1962), but they also can be used to alter the composition of tree species in a stand. Some activities, for example prescribed burning and application of herbicides, can be used to meet the objectives of either site preparation or intermediate treatments. We discuss below some of the more common site preparation and intermediate treatments and their effects on wildlife.

MECHANICAL SITE PREPARATION

Slash remaining after a clearcut harvest frequently is chopped to promote the development of a new crop of seedlings or to reduce fire hazard. Bulldozers with a sharp blade on the front (a K-G blade) sometimes are used to push over residual trees with no commercial value. The layer of organic matter (i.e., leaves, twigs, and other plant remains) sometimes is mixed with the mineral soil by some mechanical action to prepare the soil for planting or improve the seedbed for natural regeneration. This process is called scarification. Typically, intensive mechanical site preparation reduces overall floral diversity by killing the rootstock of some plant species.

Mechanized crushers also have been used to prepare sites for planting. Reduction of woody debris by this method can stimulate resprouting of palatable shrubs, which may increase the quantity and quality of food for big-game animals (e.g., moose; Oldemeyer and Regelin 1980). But because these practices are costly, they seldom are implemented solely for habitat improvement.

One method of mechanical site preparation that adds some diversity to pine plantations in the southern United States is bulldozing dead wood and some soil into long, mounded rows called "windrows." The vegetation that grows in windrows after burning is not typical of the surrounding pine plantations because of the added organic material, "nutrient charge," and residual rootstock. Broadleaf species may become established in these long, narrow brush piles. Windrow strips provide cover for some small mammals, such as woodrats, fulvous harvest mice, and golden mice (Fleet and Dickson 1984), and some species of birds, such as yellow-breasted chats, Carolina wrens, brown thrashers, and mockingbirds (Whiting 1978). Windrowing may, however, reduce the productivity of the surrounding pine forest because of the loss of soil and nutrients (Hunter 1990). Windrowing is not feasible in many forestlands in the western United States, where steep slopes prohibit access by tractors, and trees are removed with overhead cables. Windrowing also is not a practice in northeastern hardwood forests.

Removal of coarse woody debris from cutting units does not benefit all animals. Dead wood (snags and logs) and partially dead trees provide habitat elements for numerous birds and mammals. The role and management of dead wood and defective trees in forest systems are discussed below.

APPLICATION OF HERBICIDES

Herbicides are used in forest management for a variety of purposes. In southern pine and pine-hardwood forests, herbicides are used to kill residual trees after clearcutting, remove plants that compete with young trees in the first years of growth, and kill hardwood tree species in mixed pine-hardwood forests to stimulate the growth of the pines or prepare the site for regenerating new stands (see McComb and Hurst 1986). These practices change the structure and composition of forest vegetation, thereby increasing habitat for some wildlife species and decreasing habitat for others. The last-named practice listed above reduces food for some species of wildlife by killing mast-

producing trees, but the dead trees are used as perches and sites for foraging and nesting by other species (McComb and Hurst 1986).

Similarly, the application of phenoxy herbicides (2,4-D and 2,4,5-T) to young stands of Douglas-fir in Oregon reduced the complexity of vegetation in the stands primarily by killing deciduous trees (Morrison and Meslow 1984). Species of birds that foraged in deciduous trees (e.g., Wilson's warbler) generally decreased in abundance after spraying (also see Osaki 1979), whereas species that used scattered, small conifers (e.g., white-crowned sparrow) generally increased (Morrison and Meslow 1984). In Maine, application of glyphosate in clearcuts of spruce-fir also reduced the structural and floristic complexity of vegetation and caused a reduction in foliage-gleaning birds, but an increase in ground-gleaning birds (Santillo et al. 1989a). Effects of the herbicide treatment on small mammals in the spruce-fir clearcuts were similar: numbers of shrews and red-backed voles decreased on the treated sites, but the abundance of deer mice remained about the same (Santillo et al. 1989b).

The influence of herbicides on vegetation and animals is transitory, particularly in young stands that grow rapidly. For example, application of herbicides to young (4–6 years of age) pine plantations in the southeastern United States may improve conditions for birds that forage on the ground (e.g., mourning doves, northern bobwhites) because openings are created, but releasing the pine seedlings will cause the canopy to close rapidly, so the effects persist only for a few years (McComb and Hurst 1986).

Herbicides also can be used to alter the structure or composition of understory vegetation to increase food for big-game animals. For example, herbicides have been used to kill decadent stems of relatively unpalatable shrubs to induce sprouting of other more palatable species (Krefting and Hansen 1969, Borrecco et al. 1972). Mueggler (1966), however, could not find a time when application of herbicides would benefit all palatable shrubs in a forest in northern Idaho, because one or another of the important plant species was affected adversely by spraying at any given time. Furthermore, Hurst and Warren (1981) reported that application of 2,4,5-T in pine plantations in Mississippi reduced desirable deer foods by promoting unpalatable grasses over shrubs and herbs.

Herbicides potentially are toxic to wildlife and could reduce populations of wild animals to some degree through poisoning. McComb and Hurst (1986:28) reviewed available information on the toxicity of various herbicides to animals and concluded that "Herbicides vary in toxicity to wildlife, but acute toxicity to most species from most herbicides under normal field conditions is unlikely." Assessing toxicity by monitoring changes in animal populations after application of herbicides is ineffective because herbicides also change the structure and composition of vegetation, and separating the toxic effects from the effects of changing the vegetation is difficult.

PRESCRIBED BURNING

Fire played an important role in altering the structure, composition, and growth of vegetation in many forest ecosystems in the past. However, successful human suppression of fire during the last century has reduced this natural phenomenon and its effects in most forest ecosystems (e.g., Taylor 1973). For example, light surface fires once burned the understories of southwestern ponderosa pine forests every 3–10 years. These fires usually did not kill overstory trees, but reduced vegetation in the understory, thereby maintaining open, park-like stands (Cooper 1960). Exclusion of fire from southwestern pine forests allowed dense thickets of young pines to grow, shade-tolerant conifers and some hardwoods to invade the pine woodlands, and large amounts of combustible organic debris to accumulate on the forest floor (Cooper 1960). Changes of this kind are common in many western coniferous forests, where fire has been excluded, and they tend to increase the likelihood or frequency of crown fires and, for some species (e.g., lodgepole pine), insect outbreaks. In other western coniferous forests, aspen woodlands have been declining as a result of fire suppression and subsequent conifer invasion. Fire suppression also affects tree species composition among hardwood forests in the upper Midwest (Niemi and Probst 1990).

"Let burn" policies are not common in forests under management for timber production. Wildfires, therefore, probably will not play a significant role in managed forests, except by accident. Yet, some of the changes brought about by fire suppression can be reversed by applying controlled or prescribed fires (Wagle and Eakle 1979, Komarek 1981). For example, light surface fires applied under specific conditions reduce surface fuel loads (i.e., amounts of combustible woody debris) and thereby reduce the chance of wildfires (Weaver 1951, Arnold 1963) or simply eliminate slash resulting from harvest. Surface fires also prepare the soil for germination of pine seeds, thin pine thickets (Weaver 1947, Wooldridge and Weaver 1965), and control brush and hardwood shrubs. Because of these effects, prescribed fire has gained popularity as a method for preparing a site for planting or natural regeneration (either after a clearcut or before a selection cut) and for meeting a variety of silvicultural objectives during the growth of a stand.

Changes in forest vegetation caused by prescribed fires are beneficial to some species of wildlife. Burning after clearcutting in the western and southern United States often improves the quality of the site for big-game animals by reducing the depth of slash (recommended to be <0.5 m [Lyon et al. 1985]) and by stimulating the growth of palatable forage plants (Garrison and Smith 1974, Leege 1984). In the southeastern United States, sites prepared by burning produced more hardwoods and more soft mast and supported a higher diversity of bird species and more small mammals than sites treated with more intensive mechanical preparation methods (e.g., chopping) (Harris et al. 1974, Stransky and Richardson 1977).

Populations of wild turkeys, northern bobwhites, red-cockaded woodpeckers (Jackson et al. 1986), and Bachman's sparrows (Meanley 1959) generally increase after prescribed fire in pine forests in the southeastern United States. Fire is essential for the perpetuation of intact longleaf pine ecosystems in this area. White-tailed deer and other species that browse also benefit from the conditions established after prescribed fire because woody sprouts and forbs that develop after fires are easier to reach, have higher nutritive qualities, and are more palatable than plants in unburned areas (Stransky and Harlow 1981). Prescribed fire has been important in creating habitat for the Kirtland's warbler, an endangered species (Probst 1988). Burning also enhances berry production for black bears in southern forests (Hamilton 1981) and can help create den sites by scarring the bases of large trees.

The influences of prescribed fires are not always positive. Prescribed fires can reduce the density of snags and defective trees (e.g., Horton and Mannan 1988) and thus potentially reduce the abundance of cavity-nesting birds, although this effect is likely transitory because some trees are killed by prescribed fires. Reduction of understory shrubs, a goal of some prescribed fires, has a negative influence on species associated with this kind of vegetation (Niemi and Probst 1990). For example, Horton (1987) reported that Virginia's warblers were less abundant after prescribed burning in pine forests because substrates they used for nesting and foraging (grasses and understory oaks, respectively) were reduced in the fires. Also, if prescribed fires are too frequent or intense, production of forage and browse plants, usually high after burning, can be reduced.

THINNING

Most timber management programs include thinning densely stocked stands to promote growth of the remaining trees. Thinnings also can reduce the risk of some insect infestations by improving the vigor of residual trees and increasing the distance among trees. Thinnings that cut only small, unmarketable trees are called precommercial thinnings. Thinnings from which some or all of the wood is sold are called commercial thinnings.

Most thinning operations temporarily open the canopy, thereby enhancing the growth of understory plants. Thus, thinnings can increase forage for forest ungulates, but not as effectively as clearcuts or selective cuts. Thinning to provide understory forage plants may be especially important in dense, even-aged stands where understory plants often are suppressed for long periods. Changes in understory vegetation and tree density brought about by thinning also can affect the abundance and composition of forest birds. In general, bird species that use open forests or edges thrive in thinned stands, whereas species that inhabit dense forest generally avoid thinned stands, at least until the canopy closes (Szaro and Balda 1979, Conner et al. 1983a, McComb et al. 1989).

Forest conditions that support a given species of animal often are expressed by biologists in terms of trees per hectare of a given size. Silviculturists frequently use different measures when describing stand density. For example, stand density index (SDI; Reineke 1933) is "a relative measure of stand density that provides a relationship between stand basal area, trees per hectare, average stand diameter, and stocking of a forested stand" (McTague and Patton 1989:59). SDI is commonly used in even-aged stands to help plan thinning schedules. Crown competition factor (Gingrich 1967) and tree area ratio are other measures of relative stand density used most frequently in management of uneven-aged and even-aged stands of eastern hardwoods, respectively. Wildlife biologists should describe some habitat conditions in these terms, to facilitate communication between biologists and silviculturists and to help ensure that appropriate conditions for forest wildlife are maintained in managed forests.

USING SILVICULTURE TO ENHANCE HABITAT FOR WILDLIFE

Mast-Producing Trees

Production of mast is critical for many species of wildlife (Harlow et al. 1975, Nixon et al. 1975, Elowe and Dodge 1989). For example, black bears in northeastern Minnesota heavily use mast produced by two species of oaks as a source of food (Landers et al. 1979, Rogers 1987). These species of oaks make up a relatively small percentage of the trees in the area and should be protected and favored during silvicultural operations. Furthermore, oaks must be of a sufficient diameter and age to maximize acorn production (Huntley 1986). Thus, if oaks are harvested under an even-aged management system, rotation lengths must be at least 70–100 years to ensure mast production for a period of years. Single-tree selection cuts should be avoided if oaks are to be retained over time because other, more shade-tolerant species, if present, will tend to replace oaks after several cutting cycles. Also, retaining oaks in both the ''white oak'' and ''black oak'' groups is important to increase the likelihood of acorn production each year. ''White'' oaks produce seeds each year, whereas ''black'' oaks require 2 years to develop acorns.

Dead Wood in Forest Ecosystems

Dead wood in forest ecosystems frequently is an important element of the habitat of many animals. For example, standing dead trees, often called snags, and live trees with dead tops or branches are important to many species of animals because they provide sites for nesting, roosting, foraging, and other activities (Table 1). In some forests, birds that nest in cavities in snags and in cavities in living trees make up as much as 30–45% of the total avifauna (Raphael and White 1984, Scott et al. 1980) (Table 2). The degree of dependence of animals on cavities varies among species but can be high. In forests of Oregon and Washington, for example, ''snags are used by nearly 100 species of wildlife of which at least 53 (39 birds and 14 mammals) are cavity-dependent'' (Neitro et al. 1985: 130).

SNAGS

Snags are the most important substrate for cavity-nesting animals in western coniferous forests, although trees with dead tops also are important in some areas (e.g., Raphael and White 1984). The number and characteristics of snags vary as the stand matures and are the product of the interaction of tree mortality and snag deterioration. The pattern of snag abundance in Douglas-fir forests in the Pacific Northwest probably typifies many western coniferous forests. Young (35 years old) stands of Douglas-fir produce many small (<19 cm diameter at breast height [dbh]) snags, whereas older stands produce fewer but larger snags (≥48 cm dbh; Fig. 4 [also see Cline et al. 1980]). Large snags remain standing longer than small snags because they deteriorate more slowly (Fig. 5 [also see Raphael and Morrison 1987]). If a large snag (or tree) survives a disturbance that destroys most of its cohort stand, it might become part of a new stand of young trees (Fig. 6). Cline et al. (1980) coined the term ''remnant snag''

to refer to snags that were produced in one stand and survived to occupy a subsequent stand.

The size of snags used most frequently by animals varies among forest types, but snags >38 cm dbh usually are preferred as nest sites (Conner 1978, Evans and Conner 1979, Thomas et al. 1979a, Mannan et al. 1980, Raphael and White 1984, Horton and Mannan 1988). Snags of all sizes show some evidence of use by foraging woodpeckers, although large snags (≥38 cm dbh) are used most frequently (Mannan et al. 1980). Snags in all stages of decay (Fig. 7) are used by animals. Woodpeckers, for example, generally excavate nest holes in snags in the early and intermediate stages of decay (i.e., snags with some decay at the center [Conner et al. 1976]), whereas chestnut-backed chickadees nest in softer snags with more advanced decay (Mannan et al. 1980). Snags in all stages of decay provide sources of food for woodpeckers and other birds because different arthropods occupy different stages of decay.

Silvicultural operations that promote timber production reduce tree mortality by removing trees for commercial purposes before they decline and thus generally reduce the number of snags in a stand. Snags also are felled during clearcutting or thinning to enhance fire control or safety, to make harvesting easier, or to release space that could be used to grow trees. Use of short rotations sometimes means that stands will be harvested before they produce snags large enough for wildlife. The abundance of cavity-nesting birds in a stand, and presumably the abundance of other species that use snags, are at least partly dependent on the abundance of suitable snags (Evans and Conner 1979, Thomas et al. 1979a, Mannan et al. 1980). Furthermore, most species that use snags are most abundant in older stands where large snags are plentiful, but some species are adapted to use snags in open areas or stands in early stages of development (e.g., a clearcut or burn; Mannan et al. 1980, Morrison and Meslow 1983). Clearly, snags of suitable size in various stages of decay should be retained or produced in stands in various stages of development if maintaining a wide spectrum of native species of wildlife is a management objective.

How Many Snags Should Be Retained?

Thomas et al. (1979a) and Neitro et al. (1985) used the following formula (or a derivation of it) to calculate the minimum number of suitable snags (S) per 40 ha needed to support the maximum number of pairs of a primary cavity-nesting species (usually a woodpecker or nuthatch):

$$S = (D) \times (C) \times (X),$$

where D = maximum number of nesting pairs per 40 ha based on average territory size, C = number of cavities excavated per pair per year, and X = the number of snags used, plus the number of suitable snags not used but necessary to support the pair over the planning period. The last variable, X, is a correction factor included to allow some flexibility in selection of nest sites and to mimic empirically derived ratios of used to unused snags. Neitro et al. (1985) suggested a correction factor of 4 unless better site-specific information is available.

If the maximum population of a particular species cannot be maintained (or is not desired), the number of snags

Table 1. Some uses of snags by selected wildlife species (from Neitro et al. 1985).

Use	Pileated woodpecker	Red-breasted sapsucker	Acorn woodpecker	Turkey vulture	Owls and raptors	Osprey	Bald eagle	Fly-catchers	Brown creeper	Bats	Raccoon and black bear	Small mammals
Cavity nest sites	x	x	x		x		x					x
Nesting platforms						x	x					
Feeding substrate	x	x	x						x			
Plucking posts					x							
Singing or drumming (communication)	x	x	x									
Food cache or granary			x									x
Location of courtship	x	x	x									
Overwintering sites	x	x	x		x					x	x	x
Roosting	x	x	x	x	x	x	x			x		
Lookout posts				x	x	x	x	x				
Hunting and hawking perch					x	x	x	x				
Fledging site						x	x					
Dwelling or dens											x	x
Loafing sites				x		x	x					
Nesting under bark									x			
Communal nesting or nursery colonies			x							x		
Anvil sites (substrate upon which to hammer)			x									
Thermally regulated habitat	x	x	x		x					x		x

needed to maintain some percentage of the maximum is calculated by multiplying S by the desired population level (e.g., 75% of maximum [Thomas et al. 1979*a*, Neitro et al. 1985]). The total number of snags per 40 ha needed to support all primary cavity-nesting species is calculated by summing the requirements of individual species.

What Kind of Snag Should Be Retained?

Ideally, information on the species, size, and characteristics of snags needed by each animal species should be used in conjunction with the above formula to identify goals for snag management (Conner 1978, Evans and Conner 1979). Different tree species remain standing for different lengths of time after they die. Snags of species that remain standing the longest should be retained, if they are expected to be used by the animals under consideration. If information of this kind is not available, Neitro et al. (1985:163) recommended three general rules:

1. "Leave all hard snags [i.e., those in the early stages of decay], damaged and dying trees, and defective (cull) trees . . . , except those considered safety hazards. Hard snags or cull trees should be left for recruitment of future soft snags."

2. "Select snags and defective (cull) trees for retention that meet or exceed the minimum size requirements for nesting [usually woodpeckers]. Place emphasis on larger diameter trees because the larger trees remain standing longer, retain bark longer, and support a larger variety of wildlife."

3. "If a tradeoff must be made, retain hard snags in favor of soft snags, large diameter [>38 cm dbh] snags in favor of small diameter snags, tall [>18 m] snags in favor of short snags, and snags with greater bark cover in favor of snags with little bark cover."

Where To Leave Snags

The most appropriate area in which to leave snags or defective trees, from the standpoint of safety, is around the edges of cutting units if stands are even-aged. Snags left standing in the interior of cutting units can be of great benefit to some cavity-nesters (Dickson et al. 1983), but they should be <18.3 m tall to reduce potential interference with aerial application of herbicides and other logging operations (Neitro et al. 1985). If the cutting unit will be burned, snags inside the fireline and closest to the edge should be protected so they do not ignite and start fires in adjacent stands and so they remain suitable for wildlife. Neitro et al. (1985) discussed other criteria for snag retention in the Pacific Northwest. Snags could be retained throughout stands under uneven-aged management because the logging is selective and trees (and presumably some snags) of a variety of sizes should be present.

How Can Snags Be Maintained Over Time?

Retaining snags during clearcutting or thinning will provide habitat for snag-dependent wildlife during some, but not all, of the rotation. Eventually the retained snags will fall or deteriorate and should be replaced. Replacement snags can be provided by intentionally killing trees, if the stand does not yet have suitable snags and live trees of suitable size are present.

The most effective ways to create snags with appropri-

Table 2. Cavity-nesting birds commonly censused in temperate forest ecosystems of North America (from Harmon et al. 1986, used with permission by Academic Press).

Common name	Type of cavity use[a]
Northern flicker	P (L)[b]
Pileated woodpecker	P (L)
Red-bellied woodpecker	P
Gila woodpecker	P
Red-headed woodpecker	P
Acorn woodpecker	P (L)
Lewis' woodpecker	P (L)
Yellow-bellied sapsucker	P
Williamson's sapsucker	P
Hairy woodpecker	P (L)
Downy woodpecker	P
Ladder-backed woodpecker	P
Nuttall's woodpecker	P
Strickland's woodpecker	P
White-headed woodpecker	P (L)
Black-backed woodpecker	P (L)
Three-toed woodpecker	P (L)
Wied's crested flycatcher	S
Brown-crested flycatcher	S (L)
Dusky-capped flycatcher	S
Western flycatcher	S (L)
Violet-green swallow	S
Tree swallow	S
Black-capped chickadee	P (L)
Carolina chickadee	P
Mountain chickadee	S (L)
Boreal chickadee	P
Chestnut-backed chickadee	P
Tufted titmouse	S
Plain titmouse	S
Bridled titmouse	S
White-breasted nuthatch	S (L)
Red-breasted nuthatch	S (L)
Pygmy nuthatch	S (L)
Brown creeper	S (L)
House wren	S (L)
Winter wren	S (L)
Bewick's wren	S (L)
Carolina wren	S (L)
Eastern bluebird	S (L)
Western bluebird	S (L)
Mountain bluebird	S

[a] P = primary excavator; S = secondary nonexcavator.
[b] (L) = also uses logs.

ate decay characteristics (e.g., heart rot) are by blasting the tops of trees with dynamite (Bull et al. 1981), by injecting herbicides (Conner et al. 1981, McComb and Rumsey 1983), and by inoculating them with suitable fungi (Conner et al. 1983*b*). Also, individual trees or small groups of trees can be retained during clearcutting and then killed at designated intervals during the following rotation. This strategy is useful when replacement snags are needed in stands in early stages of development and when rotation lengths are so short that the new stand will never contain large trees. In the latter situation, trees left

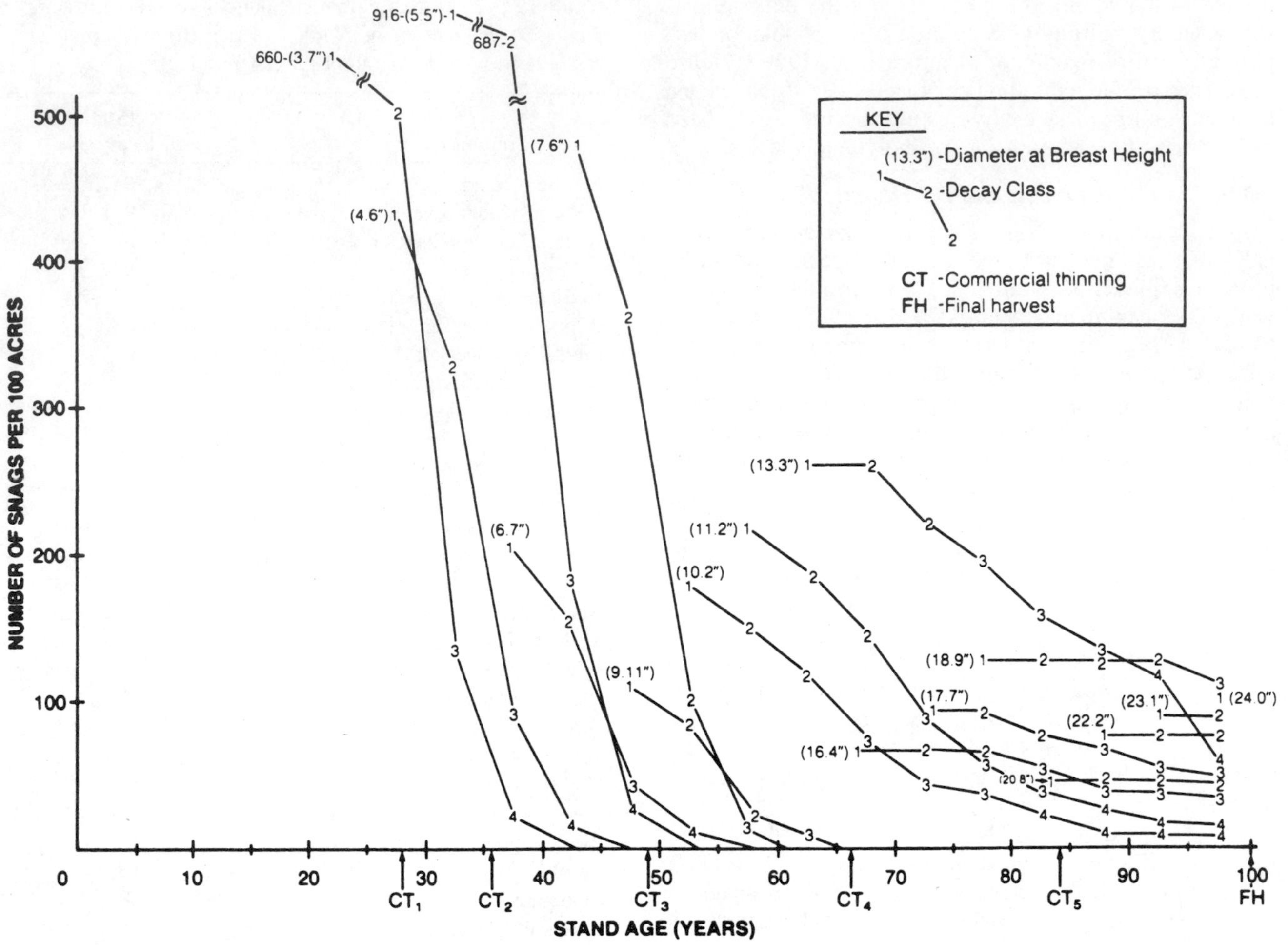

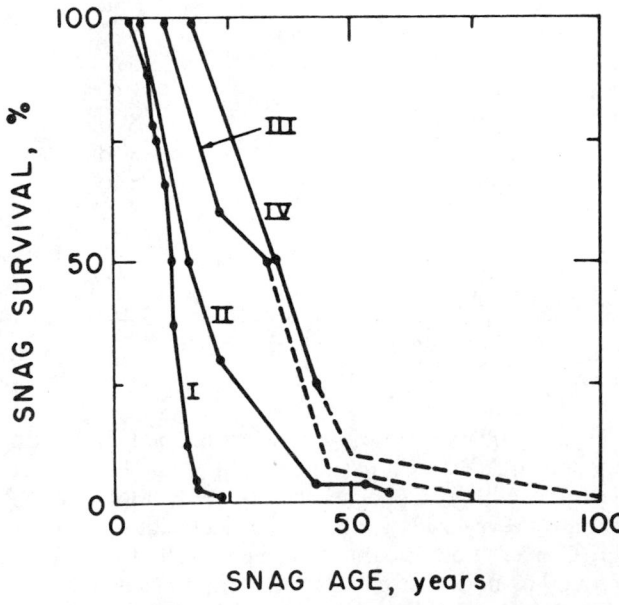

KEY

(13.3") -Diameter at Breast Height
1
  ↘2 -Decay Class
      ↘2

CT -Commercial thinning
FH -Final harvest

Fig. 4. Number of snags per 100 acres (40.5 ha) in an intensively managed, even-aged stand of Douglas-fir. Each curve traces the fate of a snag "cohort." CT = commercial thinning, FH = final harvest (regeneration cut) (from Neitro et al. 1985).

Fig. 5. Survival curves (I, 10–18 cm dbh; II, 29–31 cm dbh; III, 32–46 cm dbh; IV, 47–71 cm dbh) of Douglas-fir snags on unmanaged permanent plots, western Oregon. Dashed lines indicate estimations from survival of remnant snags (from Cline et al. 1980, used with permission from The Wildlife Society).

standing at the end of a rotation (e.g., reserve shelter-wood) should be allowed to grow until they are large enough to produce suitable snags.

Maintaining snags in stands under uneven-aged management should be relatively easy if large trees are present in the stand. Some of these can be retained until they decline and become defective, or they can be killed at designated intervals to ensure an adequate number of snags through time.

Distribution of Snags Across the Landscape

Every hectare of managed forest probably will not have large snags. The distribution of snags on a local scale must be wide enough to serve the needs of territorial species that use them, however. For example, if 2,000 snags were to be retained in a 500-ha tract of forest (i.e., 4 snags/ha), clumping them in a 100-ha patch (i.e., a burn) would be inappropriate. Similarly, the distribution of snags on a region-wide or forest-wide basis also must be wide enough that forest stands with snags are not isolated from one another. Thomas et al. (1979a) recommended that snags of appropriate size and number be retained on more than 60% of the land base. However, because tree size is, in part, a function of site quality, not every forest

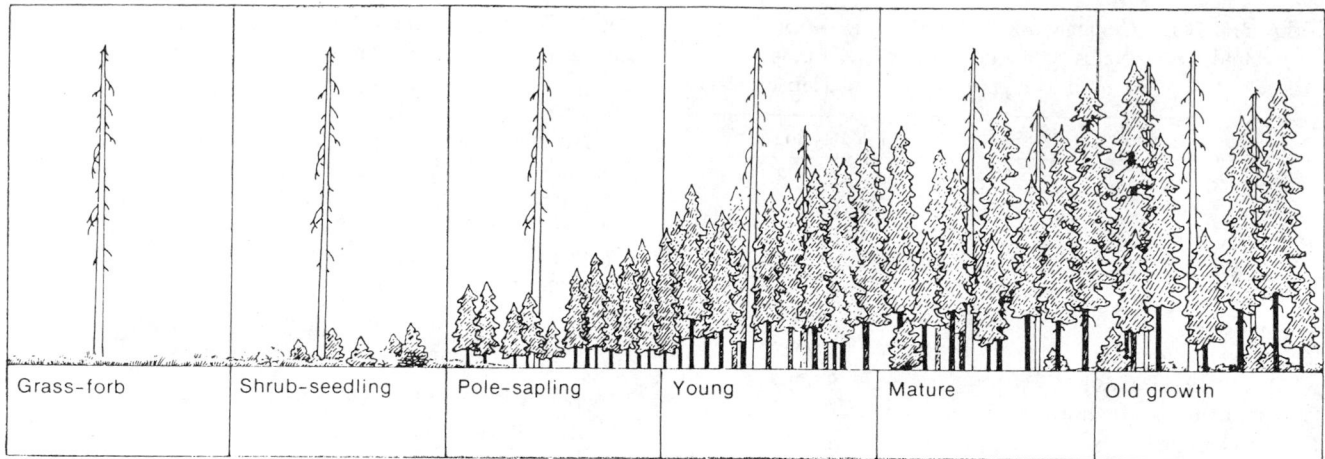

Fig. 6. Snags that survive to occupy a new stand of trees are sometimes called remnant snags. Different species of animals may use the snag as the stand develops (from Thomas et al. 1979a).

stand is capable of growing trees that will provide homes for the largest cavity-using species.

Den Trees

Snags are not as important to cavity-using animals in many deciduous forests as they are in western coniferous forests, because many cavities in deciduous forests are excavated or occur naturally in live trees, or dead portions of live trees. For example, 84% of trees with cavities in oak-hickory forests in West Virginia were partially alive (Carey 1983). Similarly, Sedgwick and Knopf (1986) reported that 94% of the nest substrate for cavity-nesting birds was in large, dead limbs in live trees in mature cottonwood stands in northeastern Colorado. Management for cavity-using animals should focus on the retention of live trees with cavities, sometimes called "den trees" (McComb et al. 1986), in forests where snags do not provide adequate habitat.

McComb et al. (1986) examined the abundance of den trees in different forest types in Florida and South Carolina. They observed that hardwood stands (e.g., oak-tupelo-baldcypress) produced more dens and den trees per 40 ha than did pine stands (e.g., longleaf pine and slash pine). Also, in both hardwood and pine stands the density of dens increased with stand age. Pine stands under intensive management (including artificial regeneration, intermediate treatments to remove hardwoods, and rotations of 30–60 years) produced three to four times fewer dens than natural pine stands.

Harlow and Guynn (1983) estimated the number of cavities needed to support average secondary cavity-nesting bird populations in southeastern forests. Management activities in these forests that help maintain the appropriate density of dens include extending the rotation age beyond 60 years for extensive tracts of forestland, promoting management of pine species prone to den formation (e.g., loblolly and pond pine), increasing the hardwood component of pine stands, and maintaining den trees through two or more rotations (McComb et al. 1986). Similarly, Sedgwick and Knopf (1986) recommended that large trees (≥55 cm dbh) with large limbs be maintained in bottomland cottonwood stands in Colorado to retain habitat for secondary cavity-nesting birds.

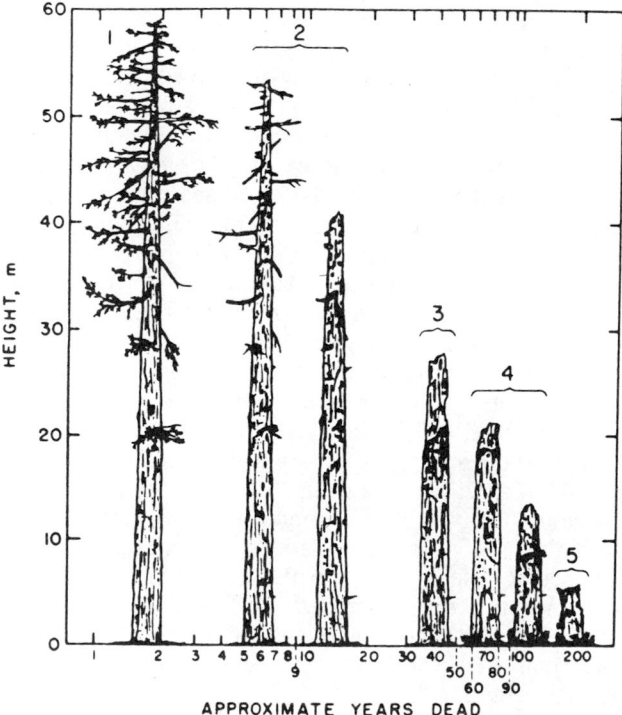

Fig. 7. Stages of deterioration (1–5) of large Douglas-fir snags, western Oregon. Diameters are drawn 2× the height scale (from Cline et al. 1980, used with permission from The Wildlife Society).

Nest Boxes

Placing nest boxes in forest stands can satisfy the nest hole requirements of some secondary cavity-nesting animals. Some species of birds and mammals actually use nest boxes more frequently than natural cavities when boxes are available (McComb and Noble 1981). Nest boxes can be used, therefore, in intensive management of one or a few species (e.g., American kestrels, eastern bluebirds, wood ducks). Nest boxes do not, however, substitute for snags in all ways (e.g., providing habitat for the

Table 3. Small mammals using coarse woody debris in temperate forest ecosystems of North America and Europe (from Harmon et al. 1986, used with permission by Academic Press).

Common name	Type of log use[a]
Short-tailed shrew	P
Masked shrew	P
Smoky shrew	P
Dusky shrew	P
Trowbridge's shrew	P
Vagrant shrew	P
Yellow-pine chipmunk	P (C)[b]
Northern flying squirrel	S (C)
California ground squirrel	S
Eastern chipmunk	S
Red squirrel	S (C)
Bushy-tailed woodrat	S (C)
White-footed mouse	P
Deer mouse	P (C)
Southern red-backed vole	P
Common red-backed vole	P (C)
Yellow-necked field mouse	P (C)
Meadow jumping mouse	S
Western jumping mouse	S
Pacific jumping mouse	S
Ermine	P (C)

[a]P = primary, use logs to fulfill the three major life-history functions: reproduction, feeding, and cover. S = secondary, use logs to fulfill only one or two of the major life-history functions.

[b](C) = also use snag or tree cavities or nest boxes.

same kinds of insects and arthropods). Also, erecting and maintaining them on large areas is impractical.

LOGS

Dead woody materials on the ground, especially large logs, store energy and some nutrients, serve as sites for nitrogen fixation, provide favorable moisture conditions for the growth of young trees, and protect the soil from surface erosion (Harmon et al. 1986). Large logs also are important to many species of forest animals. Woodpeckers eat insects that inhabit logs, and numerous species of mammals use logs as sites for reproduction, foraging, and cover (Table 3). Forest-dwelling amphibians also often are found in association with decaying logs on the forest floor (DeGraaf and Rudis 1990).

Retaining some dead and downed wood in managed forests is important because dead wood plays a number of important roles in forest ecosystems. But the slash created by logging operations often is excessive and can become a fire hazard or interfere with forest regeneration. Excessive slash also can impede the movements of some big-game animals (e.g., Dimock 1974, Garrison and Smith 1974, Pierovich et al. 1975). Removal of some slash, therefore, may be necessary after logging in some regions. Maser et al. (1979) and Bartels et al. (1985) recommended that for forests in the Pacific Northwest slash be reduced to a depth of ≤20 cm on at least 75% of an area, if the area is important for production of big game. They also recommended that some woody debris be retained after logging. They placed particular importance on retaining large logs—those >30 cm in diameter at the large end

and >6 m long—and recommended that at least five large, uncharred logs per hectare be retained as wildlife habitat. Large logs can be retained during prescribed fires by burning when smaller fuels are dry and larger logs are wet. In the Northeast, where fire hazards are low and natural regeneration of forests is common, efforts to control slash are primarily for aesthetic purposes. In areas where deer populations are high, however, slash may improve the chances of stand regeneration by preventing deer from browsing on the seedlings (Grisez 1960).

MANAGING FOREST WILDLIFE— LANDSCAPE CONCERNS

Most of the previous discussion focused on the effects of silvicultural practices on plants and animals within stands. Yet, many of the goals associated with wildlife in managed forest ecosystems concern the size and distribution of animal populations over larger areas and the health of entire species complexes. Management for the benefit of one or several species with similar habitat requirements can negatively affect other species. Wildlife biologists must be concerned, therefore, about the quantity and distribution of all seral and development stages across a forest or region.

Forest Fragmentation

Forestlands often are cleared for agricultural purposes, urban development, and other land uses. Patches of forest that remain after extensive clearing may be small and widely separated (Fig. 8). Fragmentation of forestlands is widespread in midwestern and eastern deciduous woodlands (e.g., Whitcomb et al. 1981), although some areas in the Northeast, formerly fragmented by agriculture, now support forests again because the farmlands were abandoned. Tracts of the oldest age classes of timber in western coniferous forests and elsewhere also are being reduced and fragmented rapidly by timber management (Harris 1984, also see Old Growth, below).

Fragmentation of forestlands into small patches changes the quality of the forest environment for wildlife in both obvious and subtle ways. Fragmentation reduces the size of forest patches, changes the types and quality of food and cover, alters temperature and moisture regimes, and potentially exposes animals to increased predation, competition, parasitism, and exploitation by humans (Morrison et al. 1992). Because of these changes, small and isolated patches of forest generally support fewer animal species than do large forest tracts (Whitcomb et al. 1981).

Vertebrate species frequently absent or in low abundance in small and isolated patches often are those with large territories (e.g., pileated woodpecker [Whitcomb et al. 1981]) and those that cannot easily disperse across expanses of unfavorable environments. Also, some species of birds that live in the interior of forests are reduced in number or absent in the smallest forest patches because of competition with, or predation or parasitism by, species associated with edges (Gates and Gysel 1978, Brittingham and Temple 1983, Dueser and Porter 1986, George 1987, Andrén and Angelstam 1988, Martin 1988, Yahner and Scott 1988, Brittingham 1989). Even those species that can inhabit small forest tracts may not survive for long periods because small and isolated populations are at greater risk of local extinction from catastrophes, demo-

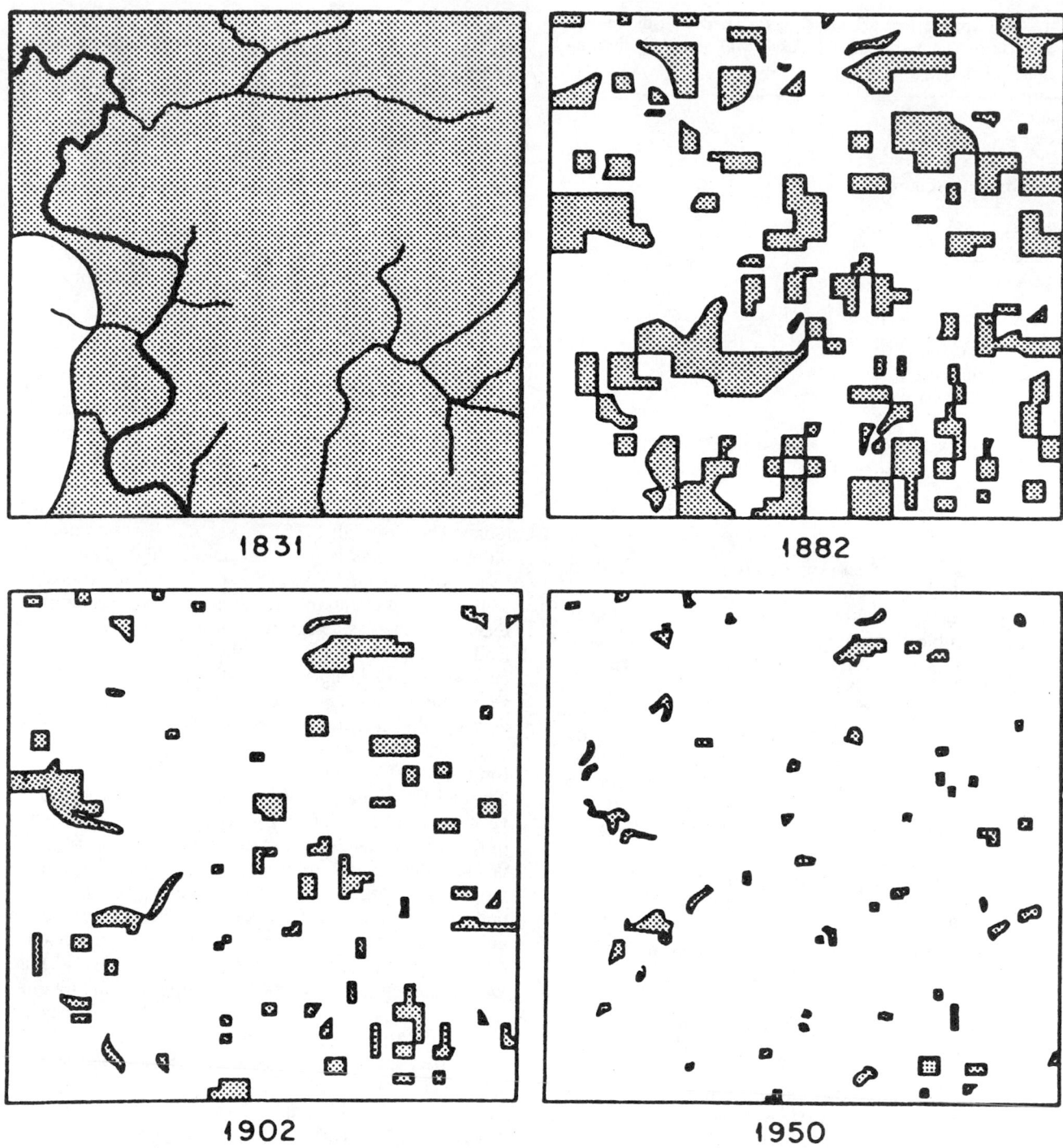

1831   1882

1902   1950

Fig. 8. Fragmentation of forestlands in an area of Cadiz Township, Green County, Wisconsin, from 1831 to 1950. Shaded areas represent land remaining in, or reverting to, forest (from Curtis 1956, composed and printed by The University of Chicago Press, Chicago, Illinois, U.S.A., used with permission from The University of Chicago Press).

graphic variability, or genetic deterioration (Karr 1982, Wilcove 1987).

Forest patches of appropriate size, stage of development, and distribution must be maintained on a landscape scale if the full complement of native animal species is to survive in a given region or forest type. Ideally, decisions about patch size, age, structure, shape, and distribution should be based on the needs of individual species and communities (see example below under Old

Growth) and desired ecosystem conditions. But, frequently, details about the habitat requirements and population dynamics of animals that are sensitive to forest fragmentation, and the significance of patch size and distribution to other ecosystem functions, are not known. Thus, specific guidelines for managing fragmented forests for wildlife are difficult to develop, and they likely would differ among forest types and regions. Some general guidelines are possible, however, and several rec-

Table 4. Species that are closely associated with old-growth forests (from Thomas et al. 1988, used with permission by The Wildlife Society).

Marten
Fisher
Red tree vole
Northern flying squirrel
Elk
Mule deer
Mountain goat
Brown bear
Several species of bats
Bald eagle
Northern spotted owl
Pileated woodpecker
Vancouver Canada goose
Marbled murrelet
Red-cockaded woodpecker
A variety of cavity-nesting birds
Several species of amphibians

---

ommendations have been proposed (e.g., Diamond 1975, Franklin and Forman 1987):

1. Maintain or plan for the largest contiguous blocks of forest possible. The number of bird species that occupy forest patches is positively correlated with patch size (e.g., Lynch and Whigham 1981, Freemark and Merriam 1986). Robbins (1979) recommended patches of 1,000 ha in shapes that minimize edge to maintain most or all of the avian species pool in eastern forestlands. Ideally, the area maintained should be large enough so that adequate forestland remains for wildlife after natural disturbances (e.g., fires, storms, floods) (Pickett and Thompson 1978).

2. Minimize distances among patches of forest in the landscape to facilitate movement of animals among patches. The animals that recolonize forest "islands" must come from within the "island" or from other forest patches. For this reason, small patches are likely to be more valuable as habitat for forest animals if they are close to a large patch.

3. Provide corridors to facilitate movement of animals among patches when patches are separated by considerable distances (MacClintock et al. 1977, Harris and Gallagher 1989). Corridors could operate at a several different scales and could link, for example, national forests and national parks within a geographic province, relatively large patches of a particular type of forest environment within a national forest, or relatively small patches of forest that collectively constitute the habitat needs of a single animal or pair of animals (Morrison et al. 1992).

The utility of corridors in a landscape design, although intuitively attractive, has not been firmly established with empirical data, and studies are needed to determine if corridors actually function to facilitate use of patchy forest environments by individual animals or to enhance colonization between isolated patches (Morrison et al. 1992). An alternative to corridors is to manage stands or open lands between patches so that they promote movement of animals among patches. For example, Thomas et al. (1990:309–310) observed that corridors between patches of habitat for spotted owls might be unnecessary provid-

ing that the forest structure in more than 50% of the forestlands outside the patches did not discourage movement by owls.

4. Retain or manage for patches that maximize the ratio of forest interior to forest edge. The portions of a forest patch that are most useful to animals that depend on forest interiors might be >300 m from an edge. Long, narrow patches (e.g., <600 m wide) might provide no habitat for these species.

Other strategies for managing forestlands at the landscape scale are presented below in the discussion of management practices for important forest environments or special habitat components.

Old Growth

Stands of predominantly old trees throughout North America are being harvested at a rapid rate primarily because they contain large quantities of valuable wood. The oldest of these stands are frequently referred to as "old growth." Only 2–15% of the old-growth forests that were present when Europeans arrived in North America exist today (Thomas et al. 1988). Pressure to harvest remaining old-growth stands is great because of high demands for wood products and because the stability of many local and some regional economies depends on wood products and related businesses. Regrowth of old-growth stands on lands devoted primarily to timber production is unlikely because rotation lengths generally vary from 25 to 120 years, and most forest types require at least 200 years to develop old-growth characteristics (Thomas et al. 1988).

Concern exists about the loss of old-growth stands from forest ecosystems, because old stands often are structurally and functionally more complex than young, intensively managed stands (Thomas et al. 1988). The structure of old-growth stands varies with forest type but frequently is characterized by multilayered canopies, large trees, large snags, large logs, and patchiness of overstory and understory vegetation. Functions performed by some old-growth stands that are not performed or are performed less effectively by younger, managed stands include: "(1) provide important pathways for fixing nitrogen, and retaining and recycling nutrients (Franklin et al. 1981, Maser and Trappe 1984); (2) produce high-quality water; (3) support specialized species of plants and animals (Franklin et al. 1981, Schoen et al. 1981, Matthews and McKnight 1982, Meehan et al. 1984, Raphael and Barrett 1984, Sigman 1985); and (4) represent a unique ecosystem with intrinsic values and extraordinary aesthetic qualities" (Thomas et al. 1988:253).

Several wildlife species are closely associated with old-growth forests in some regions (Table 4). These species tend to use old growth more than expected based on its relative occurrence in their home ranges. The degree of dependence of these species on old growth is not completely understood, however. Species that appear to be the most highly dependent on old-growth forests are the northern spotted owl in Douglas-fir forests of the Pacific Northwest (Forsman et al. 1984, Thomas et al. 1990), the red-cockaded woodpecker in the pine forests of the Southeast (U.S. Fish and Wildlife Service 1985), and the Sitka black-tailed deer in western hemlock and Sitka spruce forests of southeastern Alaska (Wallmo and Schoen 1980).

DEFINING OLD GROWTH

Efforts to retain old-growth stands in managed forests must begin with a definition of old growth. Definitions should be specific to a particular forest type and include "(1) composition of plant and animal species; (2) vegetative structure, including sizes [ages] and densities of living and dead trees (standing and fallen), and the number and nature of canopy layers; and (3) minimum stand size (area) as related to specific ecological functions, particularly those concerning wildlife and fish habitat" (Thomas et al. 1988:253). Working definitions of old growth have been developed for Douglas-fir forests in the Pacific Northwest (Society of American Foresters 1984), hemlock-spruce forests in southeastern Alaska, and several forest types in the Blue Mountains of Oregon and Washington (Thomas 1979).

MANAGEMENT OF OLD GROWTH

Plans for maintaining old growth in a given region must be based, at least initially, on existing old-growth stands, because developing old-growth stands from younger stands requires long periods of time (Thomas et al. 1988). Information on the amounts, locations, and sizes of remaining old-growth stands is necessary before management plans can be developed. Old-growth forests, however, are rare or do not exist in many regions. Furthermore, those stands that are reserved today eventually will succumb to various destructive agents (e.g., disease, fire) (Harris 1984). Therefore, strategies for replacing existing old-growth stands must be developed.

Decisions about how much old growth to retain, and the size and spatial arrangement of old-growth stands, should be based in part on the needs of animals that depend on old-growth stands for survival. For example, current efforts to protect old-growth forests in the Pacific Northwest are interrelated with efforts to maintain viable populations of the northern spotted owl. A proposed strategy for conservation of the northern spotted owl calls for the creation of a system of large blocks of mature and old-growth forests on public lands in Washington, Oregon, and northern California (Thomas et al. 1990). The size of those blocks is based on the area required to support ≥20 pairs of owls. Distance between blocks is based on the dispersal capabilities of young spotted owls. Distribution of blocks across the landscape is set to ensure an adequate distribution of owls and to facilitate reoccupancy of vacated habitat.

Relying on the needs of one or a few species to develop plans for retaining old-growth stands is a logical result of the laws regulating forest management and those requiring maintenance of viable wildlife populations. But Thomas et al. (1988) cautioned that the issue of old-growth management is broader than the maintenance of wildlife habitat and urged that, whenever possible, information on other values of old growth be incorporated into management plans.

Options for retaining old-growth stands vary among regions. In southeastern Alaska, vast tracts of old growth remain intact, and preserving entire watersheds (>2,000 ha) of primarily old-growth stands may be possible. Considerable amounts of old growth also exist in the Pacific Northwest. Harris (1984) developed a plan for preserving old-growth Douglas-fir forests in this region. His plan was based on managing islands of old growth of various size on extremely long rotation lengths (≥320 years). He also proposed that the distribution of old-growth islands be associated with the dendritic pattern of water courses. Another plan for managing old-growth stands in this region is based on the needs of the spotted owl (Thomas et al. 1990).

Other regions of North America, particularly the eastern U.S., have little old growth remaining. Management plans for old growth in these regions must focus on setting aside forested areas to develop into old growth. Thomas et al. (1988) emphasized that whatever strategy is adopted, management plans for old-growth forests should be conservative to preserve the greatest number of options for future management efforts (also see Conner 1979).

Riparian Zones

Riparian zones are areas immediately adjacent to streams, rivers, ponds, lakes, and wetlands. Vegetation in riparian zones is usually different from that in upland areas primarily because of the greater availability of water. Disturbance regimes in riparian zones also are different than in upland areas. Floods and other fluvial disturbances tend to be more frequent and severe than disturbances in upland zones. Many forest animals use riparian vegetation as an important habitat feature, and some animals, such as the beaver, profoundly influence the nature of streams and nearby vegetation (Naiman et al. 1988). In some regions, particularly the arid Southwest, animals associated with riparian zones add significantly to the total diversity of forest ecosystems (e.g., Davis 1977). Even animals that do not depend on riparian zones for survival might use them secondarily, or as travel corridors (Harris 1984).

The importance of riparian zones in forested ecosystems, and the relatively small area that they occupy, suggest that the best way to manage them is preservation and protection. The initial step is to define the riparian zone. Gregory and Ashkenas (unpubl. U.S. For. Serv. Rep., Willamette Natl. For., 1990) suggested that the riparian management zone include the entire floodplain because removal of vegetation in any part of the floodplain could lead to erosion in subsequent floods. They recommended delineating riparian zones at ecological boundaries. Riparian zones would be defined, therefore, on the basis of topography, geology, groundwater, and plant communities (Hunter 1990). On large streams, the riparian zone could range up to 400 m on either side of the stream (Hunter 1990).

Recommendations for managing riparian zones are relatively simple. First, construction of landings, roads, campgrounds, and other facilities in riparian vegetation should be avoided. Second, logging, grazing, or other activities in the riparian zone that would disrupt the integrity of the riparian vegetation and the aquatic system should not be permitted. Third, erosion within upland areas should be controlled.

Guidelines for managing riparian zones developed by many public land management agencies do not include protection of areas in and adjacent to intermittent streams. The upper reaches of some streams may flow only in early spring, yet these areas may provide key habitat to the life cycle of some wildlife species, especially amphibians (Welsh 1990). It is important, therefore, to consider protecting forests adjacent to all parts of streams and forested environments around other potentially important aquatic environments (e.g., ponds, seeps, springs).

CONTROL OF HUMAN ACCESS

Control of human access to forests is important in the management of some wild animals, especially big-game species. Habitat management programs for big-game animals can be ineffective if many kinds of human activities are not controlled. In most instances, frequent disturbance by humans alters only the distribution of big-game animals and not total density or survival. However, increased access can affect survival of big-game animals if legal hunters, or poachers, are allowed to move through an area more easily (Thiessen 1976). Also, in extreme situations, human activities may permanently prevent use of an area, thereby reducing the overall carrying capacity.

The primary way that human activity increases in an area is by the construction of roads. A road eliminates more than 1.2 ha of habitat/km (two-lane highway, 12 m wide [Rost and Bailey 1979]) and may be a barrier to the movement of some animals (Hunter 1990). However, concerns about the effects of roads on big-game animals generally are not associated with the road itself. Responses of caribou and moose to the Alaskan pipeline illustrate that even the presence of major obstacles can be tolerated if human activity is not common (Cameron et al. 1979, Curatolo and Murphy 1986, Eide et al. 1986, Sopuck and Vernam 1986). The negative influence of roads on big-game animals is associated with the humans that use the roads. Frequent disturbance by cars, trucks, and ORVs, or humans on skis, horseback, and foot can frighten animals away from the area near a road (e.g., 0.6–1.3 km from the road [Perry and Overly 1977]). The potential effectiveness of otherwise appropriate habitat for big-game animals thus is diminished as road density and traffic levels increase.

Some animals may habituate to the presence of humans on roads, especially if they are not hunted (Geist 1970, Schultz and Bailey 1978). Others animals avoid human activity. For example, brown bears in Norway reportedly avoid forest roads and logging operations (Elgmork 1978), and densities of wolves tend to be highest where road density is lowest (Thiel 1985, Mech et al. 1988). Control of human activities, especially those associated with roads, is therefore of critical importance for the management of some animals.

Black et al. (1976), Lyon (1983), and Jageman (1984) recommended the following activities to reduce the effects of roads on big-game animals: (1) screen meadows and other openings (including clearcuts) from roads with vegetation or topography; (2) avoid constructing roads in saddles, meadows, riparian zones, ridges, and other areas used as travel corridors by big-game animals; (3) locate road cuts and fills so that movements of big-game animals are not restricted; (4) lay out roads to facilitate closure after the need for the road has ended; (5) upon closure, revegetate the driving surface as well as cut and fill slopes; (6) retain vegetation along roads to reduce sight distance (<0.4 km when possible); (7) restrict access to specific areas during periods important to big-game animals (e.g., areas for calving, rutting, and wintering); (8) if access must occur in an area important to big-game animals, be sure traffic will move at a constant pace; and (9) inform the public about the objectives of road management. The last recommendation is particularly important. People generally tolerate road management policies if they understand the reasons.

ROLE OF MODELING IN MANAGING FORESTLANDS FOR WILDLIFE

A model is valuable for managing forestlands for wildlife. Models help explain observed patterns of abundance of a species, patterns of distribution of future species, and how a biological system functions. Each purpose might require models of different structures and incorporation of different types of data. Most models in wildlife management ultimately are used for prediction.

A "model"—from the Latin *modus*, meaning "mode" or "measure"—is essentially any depiction of the real world. A model can be conceptual or can take the form of a diagram, a mathematical formula, or a computer program (Hall and Day 1977). Developing conceptual and diagrammatic models is usually the first and the most valuable step in the model-building process. These models are also the most difficult forms to develop, however, because they force the builder to articulate how a particular biological system works and the key assumptions underlying the vision of that system.

Conceptual and diagrammatic models help explain and predict, and help a biologist understand, species' responses to changes in forest conditions. Too often, model construction seems to begin at the mathematical or computer programming levels, leaping ahead of the critical assumptions about how a biological system functions. For example, using linear equations in a population response model assumes that the population will respond to environmental changes in a linear manner. Whether this assumption is at all useful or appropriate for a particular modeling exercise might be overlooked unless the conceptual model is developed first.

As with any technique, models should be used only in appropriate contexts. This means that models should be used for the species and the geographic range and forest types for which the model originally was developed or intended. Restricting the use of models to appropriate conditions helps to safeguard against inappropriate uses of a model's depictions and predictions.

Appropriate context also refers to the administrative setting in which a model is used. The degree to which models are accepted by managers and decision makers is an important and often overlooked facet of model validity. A related aspect of model validity is how well it can be used in established evaluation and decision-making processes as, for example, in environmental impact statements required by NEPA. Attending to how well the use of a model fits with existing procedures for evaluating habitat conditions, predicting species' responses, and making resource allocation decisions are essential to help ensure the success of any model use. Even the most accurate and reliable model can be useless if administrative conditions do not allow users to apply it in existing evaluation and decision-making processes.

MODELING AND THE PLANNING PROCESS

The following discussion focuses on the planning process used in some government agencies charged with managing wildlife habitat in a multiple resource context. However, the process is pertinent to developing a wildlife habitat plan on most any land base, including planning for habitat improvement on small woodlots.

Models are used to predict the likely distribution and abundance of wildlife species following proposed or actual forest management activities, or to help select from the possible alternate prescriptions for forest management to achieve a desired objective for wildlife distribution and abundance. The form of the model—its specific parameters and characteristics, including its accuracy and reliability—will vary with the objectives and scale (stand or landscape).

The process of planning for wildlife habitat in any given area can be depicted as follows. First, *goals and objectives* are established. Objectives for larger areas often are set by agency policy, legislated mandates, and social interest. For specific tracts, such as at the scale of a drainage or watershed, goals and objectives might derive from a broader land management plan. Setting exact objectives identifies the species of interest and their desired distributions and abundances in the planning area. A useful concept for setting specific objectives, to borrow from the NEPA terminology, is to explicitly describe the "desired future condition" for species and habitat in the planning area. Questions that help define wildlife habitat goals include: (1) what particular forest structures and patterns are desired, and (2) what are the desired patterns of species distribution and abundance?

Second, *alternatives* are developed depicting alternative approaches or conditions for habitat in the area specified. Next, the alternatives are *analyzed* for their effect on species' distribution and abundance. Then, after *decision criteria* to evaluate the alternatives are developed and applied, a particular alternative is selected. Next, the alternative is *implemented*, given appropriate funding and administrative authority.

Implementation is then evaluated by *monitoring* the results to determine if the alternative is carried out as intended and if species and habitats respond as predicted. Any of the planning steps, including establishing new goals and objectives, can be modified after the results of monitoring are evaluated. This feedback process is called *adaptive management*.

Modeling can help with developing planning alternatives, analyzing alternatives, and monitoring results. Decision-aiding models, discussed below, also can help planners to analyze the alternatives, select an appropriate alternative, and implement the adaptive management process. Throughout, however, it must be remembered that models are tools to aid in the planning process. They are best used to inform, not rule.

A REVIEW OF FOREST HABITAT-WILDLIFE MODELS

Several models depicting forest-wildlife relationships are available (Table 5). An initial step in deciding which model to use should be to describe what is expected from the model.

Model Selection

Models are tools to use in planning. To this end, the following steps should be taken in selecting a model and assessing if a mathematical or computer modeling approach is appropriate.

1. *Specify the step in the planning process when the model will be used and how it will be employed.* For example, if the model is to aid in evaluating effects of different planning alternatives on species distribution and abundance, it should predict species responses to various habitat conditions. Some model forms, such as simple correlation models (discussed below), might be inappropriate for such predictive purposes.

2. *List the parameters of interest.* Describe the desired prediction and response variables. If the management would alter specific features of the landscape or forest habitat, such as tree density, the model should use these habitat parameters as prediction variables. For example, if the objective for the planning effort is to provide seasonal elk habitat, response variables may include habitat capability, presence or absence of elk, or herd density, depending on the level of detail required for decision making.

3. *Specify the acceptable degree of reliability.* The levels of precision and accuracy should be specified for the models that predict future conditions. Identification of acceptable levels of confidence (error of estimates) is especially important for models associated with monitoring. Monitoring population trends, for example, can be expensive and labor-intensive (Verner 1984). Sample sizes can be reduced, thus saving cost and labor, by designing the monitoring program to: (1) detect only population declines (entailing a one-tailed test) at a moderate level of assurance (e.g., at 80% confidence level), or (2) detect population declines at the desired confidence level over a broader geographic area.

Models of Forest Stand Structure

Several models display and predict forest composition and structure at the *stand* scale (Table 5). These include silvicultural *stand growth* and *yield models*, such as: CLIMACS (Dale and Hemstrom 1984) and Douglas-Fir Simulator (DFSIM [Curtis et al. 1981, Fight et al. 1984]); FORCYTE (Kimmins 1987); FOREST (Ek and Monserud 1974); FREP (U.S. Forest Service 1979); Prognosis (Wykoff et al. 1982); Stand Projection System (SPS [Arney 1985]); STEMS (Belcher et al. 1982); VARP (Tappeiner et al. 1985); and WOODPLAN (Williamson 1983). Growth and yield information generally is available for a variety of forest types on commercial forestland. For example, Oliver and Powers (1978) presented growth models for unthinned ponderosa pine plantations in northern California, and Ramm and Miner (1986) reviewed 14 growth and yield programs in the north-central region of the U.S. Meldahl (1986) compared and critiqued alternative modeling methodologies for growth and yield prediction models, including yield tables, multiple regression models, diameter distribution models, differential/difference equation models, and individual tree models.

Many stand growth and yield models assume even-age silviculture for a stand. The stand typically is depicted as

Table 5. Summary of models of forest habitat–wildlife relationships.

Model name	Description	Reference
Stand growth and yield models		
CLIMACS	Projects long-term effects of disturbances	Dale and Hemstrom 1984
Douglas-fir Similator (DFSIM)	Even-age Douglas-fir yield	Curtis et al. 1981, Fight et al. 1984
FORCYTE		Kimmins 1987
FOREST		Ek and Monserud 1974
FREP		U.S. Forest Service 1979
Prognosis	Managed stand yield model	Wykoff et al. 1982
Stand Projection System (SPS)	Even-aged managed stand yield model	Arney 1985
STEMS		Belcher et al. 1982
VARP	Assesses stand cruise data on HP-41C	Tappeiner et al. 1985
WOODPLAN		Williamson 1983
Forest succession models		
DYNAST	Projects multiple stand growth	Boyce 1980
FORPLAN	Linear programming of stand growth	U.S. Forest Service 1979
Statistical models		
CORRELATION MODELS	Associate habitat parameters with species parameters	Many studies
MULTIVARIATE MODELS	Associate multiple habitat parameters with single or multiple species parameters	e.g., Capen 1981
Species-habitat models		
HABITAT SUITABILITY (HSI) MODELS	Relate three habitat variables to species parameters	Schamberger et al. 1982
HABITAT CAPABILITY MODELS	Relate habitat variables to capability of habitat to support a species; similar to HSI models	e.g., Wisdom et al. 1986
HABITAT EVALUATION PROCE-DURES (HEP)	Assess habitat condition	U.S. Fish and Wildlife Service 1980
PATTERN RECOGNITION (PATREC)	Predicts probabilities of wildlife effects from habitat conditions	Williams et al. 1977
SPECIES-HABITAT MATRICES	List wildlife species by habitat types and conditions	e.g., Thomas 1979, Verner and Boss 1980
GUILD AND LIFE	Denotes response of guilds to habitat conditions	e.g., Severinghaus 1981, Short 1983, De-Graaf et al. 1985
COMMUNITY STRUCTURE	Denotes wildlife species distribution, abundance, and diversity as a function of vegetation structure	Raphael and Barrett 1981, Schroeder 1987
Forest landscape models		
HABITAT DISTURBANCE	Simulates effects of disturbance on habitat composition and structure	Shugart 1984, Pickett and White 1985, Shugart and Seagle 1985
FOREST FRAGMENTATION	Displays effects of forest stand fragmentation on species' distribution and abundance	Dueser and Porter 1986, Askins et al. 1987, Bock 1987, Stamps et al. 1987, many others (see text)
Cumulative effects models		
DYNAST	Can integrate species habitat relationships	Benson and Laudenslayer 1986, Holthausen 1986, Sweeney 1986
ECOSYM	Tracks multiple stand development	Henderson et al. 1978, Davis 1980, Davis and DeLain 1986
FORHAB	Projects stand with species data	Smith et al. 1981, Smith 1986
FORPLAN	Integrates species needs as constraints	Davis and DeLain 1986, Holthausen 1986
FSSIM	Projects multiple stand growth and species requirements	Holthausen and Dobbs 1985
HABSIM	Habitat capability model	Raedeke and Lehmkuhl 1986
STEMS		Belcher et al. 1982
TWIGS		Belcher 1982, Brand et al. 1986
Grizzly Bear	Assesses spatial effects from management activities on occurrence and viability of grizzly bears	Weaver et al. 1985

Table 5.   Continued.

Model name	Description	Reference
Decision-aiding models		
DECISION SUPPORT	Helps advise, weigh, and prioritize management decisions; includes expert systems	Marcot 1986, 1988
Monitoring models		
ADAPTIVE MANAGEMENT	Allows amending management direction from monitoring results	Walters and Hilborn 1978, Walters 1986

beginning at final harvest, such as clearcutting, although data are available for other even-aged reproduction methods, such as shelterwood cutting. The user may specify such parameters as stand origin (such as artificial planting or natural seeding), seedling spacing, fertilization, presence and degree of pre-commercial thinning, number and intensity of commercial thinnings, sanitation entries, other intermediate treatments, and timing of final harvest. Such models typically describe the expected structure of a forest stand in terms of stem density, stem volume, stem basal area, quadratic mean diameter at breast height (dbh), canopy characteristics, and tree heights. Some stand growth models, such as SPS, provide stem dbh distributions as well as quadratic mean dbh.

Most forest growth models are more accurate within stand ages typical of rotations scheduled for maximum timber production because they are based on empirical data. Such models may produce unreliable predictions of stand conditions for extended rotations, such as describing old-forest characteristics. One exception is CLIMACS, which was designed to simulate long-term effects of disturbance on forest stands (Hemstrom and Adams 1982).

Some stand growth and yield models provide estimates of suppression mortality, which may be useful for predicting the density and size of future snags or down wood in the stand. For example, Neitro et al. (1985) used DFSIM to model the occurrence of snags in even-age stands of Douglas-fir in western Washington and Oregon. SPS also predicts the rate distribution of trees that likely will die during some period of time. Thus, this model may be especially useful to predict when snags of a certain size will appear in a stand.

Another class of forest stand growth models tracks changes in the number of hectares of *successional* or *structural forest stages*. Examples are the DYNamically Analytic Silviculture Technique (DYNAST [Boyce 1980]) and FORPLAN, which are used in the U.S. Forest Service for habitat and general resource planning (e.g., Benson and Laudenslayer 1986, Kirkman et al. 1986, Sweeney 1986). These models are based on current number of hectares of various forest types and their growth stages, and rates of succession and development for each type. The output describes the numbers of hectares of each forest type and growth stage over time. Such models often are used to calculate habitat capability for a variety of wildlife species, as discussed below. One shortcoming of such models is that they typically lack sensitivity to spatial patterns of forest stages; the same array of areas of various forest developmental stages will produce the same habitat

capability estimates regardless of habitat patch sizes or arrangements. This might be of little consequence to wildlife species that use early- to mid-successional stages, but such models might produce great errors in predicting responses of species requiring older forests or forest interior conditions.

Models of Species' Responses to Changing Forest Structure

Modeling species' responses to changing forest structure ultimately entails linking models of stand growth, structure, and succession with predictions of occurrence and abundance of wildlife species. A variety of model forms serves this purpose. Such models may be categorized generally as single species models and multiple species models.

SINGLE SPECIES MODELS
Correlation Models

Correlation models display the degree to which species variables are explained by habitat parameters. Correlations are based on empirical data and are best used for explaining patterns in an existing data set, rather than for forecasting species' responses. Too often, however, unvalidated correlation models are taken as predictive models. Garsd (1984) reviewed various pitfalls of correlation models, including the common mistake of interpreting spurious correlation as a causal relation.

Multivariate Statistical Models

Modeling species-forest habitat relationships with multivariate statistics is common (e.g., see Folse 1979 and papers in Capen 1981). Multivariate approaches include multiple regression, various forms of principle components analysis, discriminant function analysis, and canonical correlation. Recently, logistic regression has come into favor, because it accounts for some of the nonlinearity between predictor and response variables and can include ordinal data (e.g., Hassler et al. 1986). In general, multivariate models help identify significant forest habitat parameters that account for observed variation in the distribution and abundance of wildlife species. One shortcoming of a multivariate approach is that results might be difficult to interpret. Mathematically, many multivariate statistical techniques combine several habitat parameters into one collapsed function, which then is correlated with species' distribution and abundance. Biologically, however, what the functions or the correlations mean is not always clear. Nevertheless, multivariate models are indispensable for discovering patterns (i.e., for hindcasting) in large empirical data sets and for identifying possible relationships among wildlife species and forest habitat variables. These relationships then can be examined in future studies. Examples of multivariate approaches to modeling forest-species relationships are assessments of nest site selection by kingbirds (MacKenzie and Sealy 1981) and habitat selection by songbirds (Conner et al. 1983*a*). Multivariate statistics can be used on habitat information alone; Radloff and Betters (1978) used a multivariate approach to classify physical site data for wildland management.

Habitat Suitability Index (HSI) Models

One of the more popular approaches to modeling the response of species to habitat conditions is the use of habitat suitability index (HSI) models. HSI models are used extensively by U.S. Fish and Wildlife Service (Schamberger et al. 1982) and other federal land management agencies. These models typically denote habitat suitability for a particular species as the geometric mean of *n* habitat variables deemed to most affect species presence, distribution, or abundance. The general model form of the HSI is:

$$HSI = (V_1 \times V_2 \cdots V_n)^{1/n},$$

where V_1, V_2, and V_n are key habitat variables. Each variable and the resulting HSI values are scaled from 0 to 1.

For example, the habitat variables denoted in an HSI model for yellow warblers (Schroeder 1982) are percent deciduous shrub crown cover, average height of deciduous shrub canopy, and percent of shrub canopy comprised of hydrophytic shrubs. The resulting suitability index in the yellow warbler model represents relative habitat values for reproduction. HSI models have been used to assess habitat conditions for a variety of species, including fish (Terrell 1984).

HSI models are best viewed as hypotheses of species-habitat relationships rather than causal functions (Schamberger et al. 1982). Their values lie in documenting a repeatable assessment procedure and providing an index to particular habitat characteristics that can be compared among alternative management plans (see Chapter 23).

Habitat Capability (HC) Models

Closely allied to habitat suitability index models are habitat capability models. HC models essentially perform the same function as HSI models but may vary slightly in structure. An example is the HC model for assessing habitat effectiveness for Rocky Mountain elk winter range in the Blue Mountains of eastern Oregon and Washington (Wisdom et al. 1986). This model calculates an elk habitat effectiveness index as the geometric mean of four habitat variables. The model is currently undergoing field testing by the U.S. Forest Service.

For both HC and HSI models, whether the resulting index value represents habitat conditions or potentials for population response is unclear. Also, in both model forms, the sensitivity of the resulting habitat index values to any one habitat variable is diminished as more variables are added to the model. This is a function of the mathematics of a geometric mean model and might not accurately reflect the relative importance of variables in the model. Finally, HC models should be used to represent relative habitat conditions and as a means of generating hypotheses about species-habitat relationships rather than as definitive statements of cause-and-effect relations and reliable predictions of species response.

Habitat Evaluation Procedures (HEP)

Habitat Evaluation Procedures are used extensively by the U.S. Fish and Wildlife Service to assess habitat conditions at the species level (U.S. Fish and Wildlife Service 1980). The procedure is based on habitat units (HUs), which are defined as the product of habitat quality (on a 0 to 1 index) and habitat quantity. HEP models might require much field data on specific habitat attributes, such as forage quality or quantity. However, the procedure provides a structured way to document a repeatable assessment of habitat conditions. HEP often is used to evaluate impacts of, and mitigations for, proposed projects on habitat conditions for species of special interest (see Chapter 23).

Bayesian and Pattern Recognition (PATREC) Models

Pattern recognition models are useful for predicting effects on forest wildlife from changes in habitat conditions in the form of a risk analysis. PATREC is used to help judge how likely that specific wildlife populations will achieve various densities, given habitat conditions. The general form of the model is:

$$P(S \mid H) = P(H \mid S) \, P(S)/P(H),$$

where $P(S \mid H)$ = the probability of a population's density, given that specific habitat conditions are present; $P(H \mid S)$ = the probability of habitat conditions being present, given a specific species' density; $P(S)$ = the unconditional probability of the species having a specific density; and $P(H)$ = the unconditional probability of the habitat being in specific conditions. $P(S \mid H)$ is calculated by the model and is used for resource decision making. $P(H \mid S)$ usually is based on field estimates of population densities in known habitat conditions. $P(H)$ and $P(S)$ often are estimated through professional judgments or from field studies. PATREC models have been used in forest planning by integrating them with vegetation response models (Kirkman et al. 1986).

MULTIPLE SPECIES MODELS

Several approaches have been taken to assess habitat conditions for multiple forest species. Principal methods include use of species-habitat matrices, guild and lifeform models, and community structure models. In general, multiple species models have the advantage over single species models in simultaneously assessing potentially conflicting species' requirements. Schroeder (1987) reviewed many ecological community models and discussed their structures and utility.

Species-Habitat Matrices

One simple form of representing relationships between wildlife species and forest habitats is species-habitat matrices. These are tables listing vegetation types and habitat conditions with which wildlife species are associated. Often the data are qualitative and derived from a combination of field studies and professional judgment. Examples from the Wildlife Habitat Relationships Program of the U.S. Forest Service include species-habitat matrices for amphibians, reptiles, birds, and mammals in the Pacific Northwest (Thomas 1979), California (Marcot 1979, Verner and Boss 1980), Colorado (Hoover and Wills 1984), New England (DeGraaf and Rudis 1986, DeGraaf and Chadwick 1987), and the Southeast (Hamel et al. 1982). Such information bases might be useful for predicting sets of wildlife species associated with specific habitat conditions, such as old-growth forests in the Pacific Northwest (Marcot 1980). They also can be useful for assessing optimal patterns of habitat to meet requirements of many species simultaneously (Toth et al. 1986).

Validation of a wildlife-habitat relationships model by Raphael and Marcot (1986) suggested that such information bases probably are used best to predict the occurrence of species in general forest types and habitat conditions across broad regions rather than in individual stands. Such models do not quantify population responses. Thus, they cannot be used to gauge population density or to quantify population trends.

Guild and Life-Form Models

Guild or life-form models denote the response of a set of species with similar characteristics to changes in habitat conditions (e.g., Thomas 1979, Severinghaus 1981, Short and Burnham 1982, Verner 1984, DeGraaf et al. 1985). Such models simplify the assessment of many species by referring to fewer sets of species. Guilds or life forms may be defined a priori as sets of species with common attributes, as illustrated by the models of Thomas (1979), Short (1983), and Verner (1984). They also may be defined by multivariate analysis of empirical data on species abundance and distribution, as with the assessment of Hubbard Brook bird guilds by Holmes et al. (1979).

The guild approach might be useful when habitat conditions and target species are well defined (e.g., Landres and MacMahon 1980, Maurer et al. 1981, Block et al. 1987, Knopf et al. 1988). However, individual species of a guild might vary disparately in response to habitat conditions, whereas the guild as a whole shows little or no variation (Hairston 1981, Mannan et al. 1984). In this situation, it might be more appropriate to individually model each species comprising the guild and combine results.

Community Structure Models

Community structure models describe distribution, abundance, and diversity of sets of wildlife species as a function of forest structure. Multivariate statistics are commonly used to assess these relationships (e.g., Erdelen 1984, Swift et al. 1984, Scott et al. 1987). A shortcoming of this approach, as with the guild approach, is that individual species' responses can vary considerably while the composite measures of community structure remain more or less constant. Also, forest management objectives seldom are expressed in terms of indices of wildlife species diversity.

Models of Forest Landscapes

Included in this category are models that display habitat spacing and patterns explicitly at the *landscape scale.*

Models of forest landscapes include habitat disturbance models, forest fragmentation models, and cumulative effects assessments.

HABITAT DISTURBANCE MODELS

Habitat disturbance models simulate the extent and distribution of various forest stages across a landscape given the frequency and intensity of disturbances such as fire and timber harvesting. Such models (e.g., Shugart 1984, Pickett and White 1985, Shugart and Seagle 1985) can be used to plan habitat management over space and time (Smith et al. 1981, Karr and Freemark 1985).

FOREST FRAGMENTATION MODELS

Models displaying species response to forest fragmentation and isolation of forest patches include an assessment of the probability of a species occurring in a particular habitat patch given its size. These assessments, called "incidence functions," generally suggest that the greater the degree of forest fragmentation, the lower the likelihood of occurrence of species associated with forest interiors (Wilcove 1987). Increasing forest fragmentation on a continental or global scale likely will increase the rate of extinction of these species.

Forest fragmentation models address a variety of factors that can influence species persistence in a forested landscape: habitat size (Cole 1981, Lynch and Whigham 1984, Askins et al. 1987, Blake and Karr 1987), habitat isolation (Faanes 1984, Fahrig and Merriam 1985, Bock 1987, Fahrig and Paloheimo 1988), habitat patch patterns (Lynch and Whigham 1981, Toth et al. 1986, Stamps et al. 1987), competition and habitat structure (Dueser and Porter 1986), and the interaction of habitat fragmentation with parasites (Dobson 1988) and predators (George 1987, Savidge 1987, Martin 1988, Yahner and Scott 1988) as it affects population regulation. The use of island biogeography theory for predicting effects of forest fragmentation on population persistence and quality of forest environments has served as the basis for management guidelines (Franklin and Forman 1987). Noss and Harris (1986), Simberloff and Cox (1987), and others have addressed the utility of planning for habitat corridors in forested landscapes.

CUMULATIVE EFFECTS ASSESSMENTS

Cumulative effects models are used to gauge how landscape conditions affect the distribution and abundance of wildlife species. The term cumulative effects model is a generic one. It can be applied to any assessment of the effects on wildlife species from management activities or natural disturbances across a geographic area or over time. Salwasser and Samson (1985) discussed development of cumulative effects models, noting that the major steps are stating management goals and standards, representing major habitat factors, projecting changes in habitats, and estimating wildlife effects.

Weaver et al. (1985) presented a cumulative effects simulation model for assessing grizzly bear habitat in the Yellowstone ecosystem. Their model consisted of submodels to assess habitat, bear displacement from human activities, and bear mortality.

Habitat capability models also can be used for assessing cumulative effects when used with models that predict future amounts of forest habitats in sub-drainage scale areas. Examples are habitat capability models incorporated into FORHAB (Smith et al. 1981, Smith 1986), HABSIM (Raedeke and Lehmkuhl 1986), ECOSYM (Henderson et al. [ECOSYM, Dep. For. Resour., Utah State Univ., Logan, 1978], Davis 1980, Davis and DeLain 1986), DYNAST (Benson and Laudenslayer 1986, Holthausen 1986, Sweeney 1986), FORPLAN (Davis and DeLain 1986, Holthausen 1986), TWIGS (Belcher 1982, Brand et al. 1986), and STEMS (Belcher et al. 1982). Powers (1979) used linear programming to develop an optimal mix of agricultural and wildlife land use.

Holthausen and Dobbs (paper presented at Soc. Am. For. Conf., Ft. Collins, Colo., 1985) presented a cumulative effects model, FSSIM, that was used to assess management activities in 400-ha to 8,100-ha management units and to project the consequences of those actions over time. The model presented by Salwasser and Tappeiner (1981) was designed to assess effects of timber harvest scheduling on spatial patterns of habitats.

Models That Aid Evaluation and Decision Making

Another group of models aids in habitat evaluation and decision making. Such models help organize and document factors associated with habitat evaluation and planning. Included are models for decision support and for monitoring species and habitats.

Decision support models help weigh and prioritize decisions for habitat planning. These models can use optimization algorithms, as with FORPLAN, or they can be expert systems that capture the expertise of a specialist or group of specialists. Some expert models provide decision support for integration of wildlife habitat evaluation and management with other resource users. For example, the NE Model (Marquis 1990) provides site-specific evaluations and management recommendations for wildlife, timber, and aesthetics for forests in the Northeast at the stand and management unit levels.

Related are advisory models that offer expertise in habitat evaluation. Such models include expert systems that capture the expertise of a specialist in "If . . . , then . . ."-type rules. Expert systems are a growing area of computer modeling (Marcot 1986, Marcot et al. 1989). Marcot (1986, 1988) discussed expert systems that aided evaluation of forest habitat conditions for birds and identification of wildlife species.

Models for Monitoring Species and Habitats

Models of forest habitat-species relationships might become useful for the expensive task of monitoring species and habitats over time. Single-species models are probably best for monitoring management indicator species and other species of singular concern. Habitat association models, such as HSI, HC, PATREC, and HEP, are useful for species that are too expensive to monitor directly. However, it is vital to first demonstrate the degree of reliability and validity of the model. Also, population trends of species of high concern, especially state- and federal-listed threatened or endangered species, are better monitored directly in the field rather than inferred through habitat relationships models.

A useful model in monitoring is the adaptive management paradigm (Walters and Hilborn 1978, Walters 1986). Adaptive management entails viewing a forest management plan and its expected effects on wildlife species as a hypothesis; monitoring habitats and species to ascertain how they respond to the habitat management; and revising the direction of management if results so warrant. Monitoring is an essential step in the adaptive management feedback process and helps managers deal directly with limitations of biological uncertainty (Lee and Lawrence 1986).

The Right Model for the Right Job

Selecting the right model first entails clearly stating objectives for the task at hand (Lipscomb et al. 1984) and then selecting the tool that addresses the correct spatial scale, time scale, and set of biological parameters used as predictor and dependent variables. Two examples illustrate.

The first example is that of developing a general information base for depicting and predicting presence and abundance of wildlife species in various forest conditions. This might be called a wildlife-habitat relationships information system (WHRIS). WHRISs have been developed by many federal and state wildlife agencies, such as U.S. Forest Service and many state heritage programs. A WHRIS depicts a set of generalized relationships between presence or relative abundance of wildlife species and forest environments and is useful for producing lists of species potentially present in forest types and their developmental stages.

It is most efficacious to build a WHRIS program in several stages, ranging from compiling general information on wildlife communities to developing species-specific prediction models. Specific steps in building a WHRIS program and developing specific models depend on existing inventory data and on program objectives. An initial step likely would entail a broad-scale inventory of habitat conditions and species presence. Examples include the procedure by Scott et al. (1987) for depicting species presence and richness with GIS, and the regional landscape wildlife evaluation procedure by Hawes and Hudson (1976).

The next step in a WHRIS program might be to identify associations of individual wildlife species with categories of potential natural vegetation (plant associations or climax vegetation) and structural and successional variants thereof. An example is the species-habitat matrices by Verner and Boss (1980) for Sierra Nevada forests. The guild and life-form models of Short (1983) and Thomas (1979) are other examples. A next step might entail quantifying the structural and functional relationships between forest habitat conditions and species' presence and abundance. This might include field studies and development of appropriate statistical empirical models (e.g., see Sparrowe and Sparrowe 1977, Grue et al. 1981, Hays et al. 1981). Finally, models predicting wildlife population responses to occurrence of and changes in specific forest habitat conditions can be constructed. These models can include HSI, HC, HEP, pattern recognition, and statistical regression and correlation models. Validation of such quantitative relationships is an important step in reliably using such models.

The second example is that of a biological evaluation

(BE). A BE typically is conducted by wildlife biologists on private, state, or federal lands to evaluate potential impacts to wildlife from a proposed forest management project. A BE is commonly part of an environmental impact assessment and helps identify changes or mitigations to a proposed activity to help create or maintain desired wildlife populations or habitats. Although BEs vary widely in scope and objectives, they typically are used to assess the immediate effects of a forest management project on wildlife or habitats within the planned activity area, and the long-term effects of how changes in forest composition and structure influence the associated wildlife community. BEs often include assessing how cumulative effects of other nearby projects influence on-site objectives and also how the proposed project influences conditions off-site.

Which models are useful for conducting BEs? Initially, a WHRIS information base would provide a starting point for predicting potential species' presence on the project site and in the general project area. Field reconnaissance and inventory of species and habitats are usually necessary early steps in the BE process. Multiresource inventories (e.g., Chalk et al. 1984), such as simultaneous assessments of timber and habitat conditions, are a particularly efficient use of available labor. Intensive sampling of wildlife populations (e.g., Raphael and Barrett 1981) is desirable where information on species' abundance and distribution is lacking, but it is expensive and often infeasible for use in BEs. However, directly sampling or censusing populations and habitats of threatened, endangered, or some sensitive (TE&S) species is frequently necessary. Use of models alone to predict presence of and population response by TE&S species is not recommended.

The next step in a BE might entail using species-habitat models to predict current presence and relative abundance of wildlife species and species groups in the project site and the surrounding area (Schroeder 1987). Next, models of stand growth and forest succession are useful for projecting changes in habitat conditions on site under the proposed management activity. Coupled with the species-habitat models, these tools comprise a powerful means of depicting potential changes in wildlife communities on the site over time (e.g., Benson and Laudenslayer 1986, Holthausen 1986). The spatial context of such projections then can be broadened to include cumulative effects from off-site conditions and activities, by using cumulative effects models and forest landscape models that account for habitat disturbance and fragmentation. Decision-aiding models can be helpful in focusing management objectives (Marcot et al. 1989) and potential mitigation measures. Finally, monitoring models and procedures can aid in tracking effects of implementing the project (Marcot et al. 1983) and in adapting management to new information over time (Salwasser and Samson 1985, Kirkman et al. 1986, Lee and Lawrence 1986).

CONCLUSIONS

Maintaining appropriate conditions for wildlife in forestlands is complex and difficult, especially in areas where timber harvests and silvicultural practices change the structure and composition of forest vegetation. Perhaps the most critical element in the management of forestlands for wildlife is the need to maintain in a given region the full spectrum of seral and developmental stages for each forest type. A variety of silvicultural techniques and models of forest growth and wildlife habitat relationships is available to help develop and maintain desired conditions. Unfortunately, we do not know currently how large individual forest stands should be and how the stands should be distributed. Certainly some of the patches will need to be very large (on the order of 2,000–4,000 ha) and some, perhaps many, will need to be maintained in older seral or developmental stages. Specific answers to questions about the appropriate configuration of stand age, size, and distribution will be forthcoming as we learn more about the habitat requirements of the animal species we seek to maintain. Only when armed with information about how individual species interact with each other and respond to various changes in the forest environment can we hope to maintain natural levels of biotic diversity in managed forest landscapes.

Acknowledgments.—We thank W. Knapp, W. McComb, R. Nyland, N. Tilghman, and an anonymous reviewer for reviewing drafts of this chapter. Their comments and suggestions were helpful. We also thank V. Catt for typing the chapter.

LITERATURE CITED

ADAMS, C. C. 1908. The ecological succession of birds. Auk 25:109–153.

ANDRÉN, H., AND P. ANGELSTAM. 1988. Elevated predation rates as an edge effect in habitat islands: experimental evidence. Ecology 69:544–547.

ARNEY, J. D. 1985. User's guide for the Stand Projection System (SPS). Rep. 1. Appl. Biometrics, Spokane Wash. 9pp.

ARNOLD, J. F. 1963. Uses of fire in the management of Arizona watersheds. Proc. Tall Timbers Fire Ecol. Conf. 2:99–111.

ASKINS, R. A., M. J. PHILBRICK, AND D. S. SUGENO. 1987. Relationship between the regional abundance of forest and the composition of forest bird communities. Biol. Conserv. 39:129–152.

BAKER, F. S. 1950. The principles of silviculture. McGraw-Hill, New York, N.Y. 414pp.

BARTELS, R., J. D. DELL, R. L. KNIGHT, AND G. SCHAEFER. 1985. Dead and down woody material. Pages 171–186 *in* E. R. Brown, tech. ed. Management of wildlife and fish habitats in forests of western Oregon and Washington. U.S. For. Serv. PNW Publ. R6-F&WL-192-1985.

BEAN, M. J. 1983. The evolution of national wildlife law. Environmental Defense Fund, Inc. Praeger Publ., New York, N.Y. 449pp.

BEHREND, D. F., G. F. MATTFELD, W. C. TIERSON, AND J. E. WILEY, III. 1970. Deer density control for comprehensive forest management. J. For. 68:695–700.

―――, AND E. F. PATRIC. 1969. Influence of site disturbance and removal of shade on regeneration of deer browse. J. Wildl. Manage. 33:394–398.

BELCHER, D. M. 1982. TWIGS: the woodsman's ideal growth projection system. Pages 70–95 *in* J. W. Moser, Jr., ed. Microcomputers: a new tool for foresters. Purdue Univ. Press, West Lafayette, Ind.

―――, M. R. HOLDAWAY, AND G. J. BRAND. 1982. A description of STEMS—the stand and tree evaluation modeling system. U.S. For. Serv. Gen. Tech. Rep. NC-79. 18pp.

BENSON, G. L., AND W. F. LAUDENSLAYER, JR. 1986. DYNAST: simulating wildlife responses to forest-management strategies. Pages 351–355 *in* J. Verner, M. L. Morrison, and C. J. Ralph, eds. Wildlife 2000: modeling habitat relationships of terrestrial vertebrates. Univ. Wisconsin Press, Madison.

BLACK, H., R. SCHERZINGER, AND J. W. THOMAS. 1976. Relationships of Rocky Mountain elk and Rocky Mountain mule deer to timber management in the Blue Mountains of Oregon and Washington. Pages 11–31 *in* S. R. Hieb, ed. Proc. elk-logging-roads symposium. Univ. Idaho, Moscow.

BLAKE, J. G., AND J. R. KARR. 1987. Breeding birds of isolated woodlots: area and habitat relationships. Ecology 68:1724–1734.

BLOCK, W. M., L. A. BRENNAN, AND R. J. GUTIÉRREZ. 1987. Evaluation of guild-indicator species for use in resource management. Environ. Manage. 11:265–269.

BLYMER, J. J., AND H. S. MOSBY. 1977. Deer utilization of clearcuts in southwestern Virginia. South. J. Appl. For. 1:10–13.

BOCK, C. E. 1987. Distribution-abundance relationships of some Arizona landbirds: a matter of scale? Ecology 68:124–129.

BORRECCO, J. E., ET AL. 1979. Survey of mountain beaver damage to forests in the Pacific Northwest, 1977. Washington Dep. Nat. Resour. Note 26. 16pp.

———, H. C. BLACK, AND E. F. HOOVEN. 1972. Response of black-tailed deer to herbicide-induced habitat changes. Proc. Annu. Conf. West. Assoc. State Game and Fish Comm. 52:437–451.

BOYCE, S. 1980. Management of forests for optimal benefits (DYNAST-OB). U.S. For. Serv. Res. Pap. SE-204. 92pp.

BRAND, G. J., S. R. SHIFLEY, AND L. F. OHMANN. 1986. Linking wildlife and vegetation models to forecast the effects of management. Pages 383–387 in J. Verner, M. L. Morrison, and C. J. Ralph, eds. Wildlife 2000: modeling habitat relationships of terrestrial vertebrates. Univ. Wisconsin Press, Madison.

BRENNEMAN, R. 1982. Electric fencing to prevent deer browsing on hardwood clearcuts. J. For. 80:660–661.

BRITTINGHAM, M. C. 1989. Effects of timber management practices on forest interior birds. Pages 162–170 in J. C. Finley and M. C. Brittingham, eds. Timber management and its effects on wildlife. Proc. Pa. State For. Resour. Issues Conf., Pa. State Coop. Ext., University Park.

———, AND S. A. TEMPLE. 1983. Have cowbirds caused forest songbirds to decline? BioScience 33:31–35.

BROWN, E. R., AND J. H. MANDERY. 1962. Planting and fertilization as a possible means of controlling distribution of big game animals. J. For. 60:33–35.

BULL, E. L, A. D. PARTRIDGE, AND W. G. WILLIAMS. 1981. Creating snags with explosives. U.S. For. Serv. Res. Note PNW-393. 4pp.

CAMERON, R. D., K. R. WHITTEN, W. T. SMITH, AND D. D. ROBY. 1979. Caribou distribution and group composition associated with construction of the Trans-Alaska pipeline. Can. Field-Nat. 93:155–162.

CAPEN, D. E., EDITOR. 1981. The use of multivariate statistics in studies of wildlife habitat. U.S. For. Serv. Gen. Tech. Rep. RM-87. 249pp.

CAREY, A. B. 1983. Cavities in trees in hardwood forests. Pages 167–184 in J. W. Davis, G. A. Goodwin, and R. A. Okenfels, tech. coords. Snag habitat management: proceedings of the symposium. U.S. For. Serv. Gen. Tech. Rep. RM-99.

CHALK, D. E., S. A. MILLER, AND T. W. HOEKSTRA. 1984. Multiresource inventories: integrating information on wildlife resources. Wildl. Soc. Bull. 12:357–364.

CLEMENTS, F. E. 1916. Plant succession—an analysis of the development of vegetation. Carnegie Inst. Publ. 242. 512pp.

———. 1936. Nature and structure of the climax. J. Ecol. 24:252–284.

CLINE, S. P., A. B. BERG, AND H. M. WIGHT. 1980. Snag characteristics and dynamics in Douglas-fir forests, western Oregon. J. Wildl. Manage. 44:773–786.

COLE, B. J. 1981. Colonizing abilities, island size, and the number of species on archipelagoes. Am. Nat. 117:629–638.

CONNELL, J. H., AND R. O. SLATYER. 1977. Mechanisms of succession in natural communities and their role in community stability and organization. Am. Nat. 111:1119–1144.

CONNER, R. N. 1978. Snag management for cavity nesting birds. Pages 120–128 in R. M. DeGraaf, tech. coord. Proc. workshop management of southern forests for nongame birds. U.S. For. Serv. Gen. Tech. Rep. SE-14.

———. 1979. Minimum standards and forest wildlife management. Wildl. Soc. Bull. 7:293–296.

———, AND C. S. ADKISSON. 1975. Effects of clearcutting on the diversity of breeding birds. J. For. 73:781–785.

———, J. G. DICKSON, AND B. A. LOCKE. 1981. Herbicide-killed trees infected by fungi: potential cavity sites for woodpeckers. Wildl. Soc. Bull. 9:308–310.

———, ———, ———, AND C. A. SEGELQUIST. 1983a. Vegetation characteristics important to common songbirds in east Texas. Wilson Bull. 95:349–361.

———, AND J. H. WILLIAMSON. 1983b. Potential woodpecker nest trees through artificial inoculation of heart rots. Pages 68–72 in J. W. Davis, G. A. Goodwin, and R. A. Ockenfels, tech. coords. Proc. symp. snag habitat management. U.S. For. Serv., Gen. Tech. Rep. RM-99.

———, O. K. MILLER, JR., AND C. S. ADKISSON. 1976. Woodpecker dependence on trees infected by fungal heart rots. Wilson Bull. 88:575–581.

———, AND K. A. O'HALLORAN. 1987. Cavity-tree selection by red-cockaded woodpeckers as related to growth dynamics of southern pines. Wilson Bull. 99:398–412.

———, A. E. SNOW, AND K. A. O'HALLORAN. 1991. Red-cockaded woodpecker use of seed-tree/shelterwood cuts in eastern Texas. Wildl. Soc. Bull. 19:67–73.

———, J. W. VIA, AND I. D. PRATHER. 1979. Effects of pine-oak clearcutting on winter and breeding birds in southwestern Virginia. Wilson Bull. 91:301–316.

COOPER, C. F. 1960. Changes in vegetation, structure, and growth of southwestern pine forests since white settlement. Ecol. Monogr. 30:129–164.

COOPER, W. S. 1926. The fundamentals of vegetational change. Ecology 7:391–413.

CURATOLO, J. A., AND S. M. MURPHY. 1986. The effects of pipelines, roads, and traffic on the movements of caribou, *Rangifer tarandus*. Can. Field-Nat. 100:218–224.

CURTIS, J. T. 1956. The modification of mid-latitude grasslands and forests by man. Pages 721–736 in W. L. Thomas, ed. Man's role in changing the face of the earth. Univ. Chicago Press, Chicago, Ill.

CURTIS, R. O., G. W. CLENDENEN, AND D. J. DeMARS. 1981. A new stand simulator for coast Douglas-fir: DFSIM user's guide. U.S. For. Serv. Gen. Tech. Rep. PNW-128. 79pp.

DALE, V. H., AND M. HEMSTROM. 1984. CLIMACS: a computer model of forest stand development for western Oregon and Washington. U.S. For. Serv. Res. Pap. PNW-327. 60pp.

DAVIS, G. A. 1977. Management alternatives for riparian habitat in the southwest. Pages 59–67 in R. R. Johnson and D. A. Jones, tech. coords. Importance, preservation and management of riparian habitat: a symposium. U.S. For. Serv. Gen. Tech. Rep. RM-43.

DAVIS, L. S. 1980. Strategy for building a location-specific, multi-purpose information system for wildland management. J. For. 78:402–408.

———, AND L. I. DeLAIN. 1986. Linking wildlife-habitat analysis to forest planning with ECOSYM. Pages 361–369 in J. Verner, M. L. Morrison, and C. J. Ralph, eds. Wildlife 2000: modeling habitat relationships of terrestrial vertebrates. Univ. Wisconsin Press, Madison.

DeBYLE, N. V. 1981. Songbird populations and clearcut harvesting of aspen in northern Utah. U.S. For. Serv. Res. Note INT-302. 7pp.

DeGRAAF, R. M. 1987. Managing northern hardwoods for breeding birds. Pages 348–362 in R. D. Nyland, ed. Managing northern hardwoods. Proc. Silvicultural Symp. SUNY Coll. Environ. Sci. For., Faculty For., Misc. Publ. 13 (ESF 87-002).

———, AND N. L. CHADWICK. 1987. Forest type, timber size class, and New England breeding birds. J. Wildl. Manage. 51:212–217.

———, AND D. D. RUDIS. 1986. New England wildlife: habitat, natural history, and distribution. U.S. For. Serv. Gen. Tech. Rep. NE-108. 491pp.

———, AND ———. 1990. Herpetofaunal species composition and relative abundance among three New England forest types. For. Ecol. Manage. 32:155–165.

———, N. G. TILGHMAN, AND S. H. ANDERSON. 1985. Foraging guilds of North American birds. Environ. Manage. 9:493–536.

DIAMOND, J. M. 1975. The island dilemma: lessons of modern geographic studies for the design of natural reserves. Biol. Conserv. 7:129–146.

DICKSON, J. G., R. N. CONNER, AND J. H. WILLIAMSON. 1983. Snag retention increases bird use of a clear-cut. J. Wildl. Manage. 47:799–804.

DIMOCK, E. J., II. 1974. Animal populations and damage. Pages 0-1–0-28 in O. P. Cramer, ed. Environmental effects of forest residues management in the Pacific Northwest—a state-of-the-knowledge compendium. U.S. For. Serv. Gen. Tech. Rep. PNW-24.

DOBSON, A. P. 1988. Restoring island ecosystems: the potential of parasites to control introduced mammals. Conserv. Biol. 2:31–39.

DODGE, W. E. 1982. Porcupine (*Erethizon dorsatum*). Pages 355–366 in J. A. Chapman and G. A. Feldhamer, eds. Wild mammals of North America: biology, management, and economics. Johns Hopkins Univ. Press, Baltimore, Md.

DUESER, R. D., AND J. H. PORTER. 1986. Habitat use by insular small mammals: relative effects of competition and habitat structure. Ecology 67:195–201.

EIDE, S. H., S. D. MILLER, AND M. A. CHIHULY. 1986. Oil pipeline crossing sites utilized in winter by moose, *Alces alces*, and caribou, *Rangifer tarandus*, in southcentral Alaska. Can. Field-Nat. 100: 197–207.

EK, A. R., AND R. A. MONSERUD. 1974. FOREST: a computer model for simulating the growth and reproduction of mixed species forest stands. Univ. Wisconsin, School Nat. Resour. Res. Rep. R2635. 85pp.

ELGMORK, K. 1978. Human impact on a brown bear population (*Ursus arctos* L.). Biol. Conserv. 13:81–103.

ELOWE, K. D., AND W. E. DODGE. 1989. Factors affecting black bear reproductive success and cub survival. J. Wildl. Manage. 53:962–968.

ERDELEN, M. 1984. Bird communities and vegetation structure: I. Correlations and comparisons of simple and diversity indices. Oecologia 61:277–284.

EVANS, K. E., AND R. N. CONNER. 1979. Snag management. Pages 214–225 *in* R. M. DeGraaf, tech. coord. Proc. workshop management of north central and northeastern forests for nongame birds. U.S. For. Serv. Gen. Tech. Rep. NC-51.

FAANES, C. A. 1984. Wooded islands in a sea of prairie. Am. Birds 38:3–6.

FAHRIG, L., AND G. MERRIAM. 1985. Habitat patch connectivity and population survival. Ecology 66:1762–1768.

———, AND J. PALOHEIMO. 1988. Effect of spatial arrangement of habitat patches on local population size. Ecology 69:468–475.

FIGHT, R. D., J. M. CHITTESTER, AND G. W. CLENDENEN. 1984. DFSIM with economics: a financial analysis option for the DFSIM Douglas-fir simulator. U.S. For. Serv. Gen. Tech. Rep. PNW-175. 22pp.

FLEET, R. R., AND J. G. DICKSON. 1984. Small mammals in two adjacent forests stands in east Texas. Pages 264–269 *in* W. C. McComb, ed. Proc. workshop management of nongame species and ecological communities. Univ. Kentucky, Lexington.

FOLSE, L. J., JR. 1979. Analysis of community census data: a multivariate approach. Pages 9–22 *in* J. G. Dickson, R. N. Conner, R. R. Fleet, J. C. Kroll, and J. A. Jackson, eds. The role of insectivorous birds in forest ecosystems. Academic Press, New York, N.Y.

FORSMAN, E. D., E. C. MESLOW, AND H. M. WIGHT. 1984. Distribution and biology of the spotted owl in Oregon. Wildl. Monogr. 87. 64pp.

FRANKLIN, J. F., AND C. T. DYRNESS. 1973. Natural vegetation of Oregon and Washington. U.S. For. Serv. Gen. Tech. Rep. PNW-8. 417pp.

———, ET AL. 1981. Ecological characteristics of old-growth Douglas-fir forests. U.S. For. Serv. Gen. Tech. Rep. PNW-118. 48pp.

———, AND R. T. T. FORMAN. 1987. Creating landscape patterns by forest cutting: ecological consequences and principles. Landscape Ecol. 1:5–18.

FREEMARK, K. E., AND H. G. MERRIAM. 1986. Importance of area and habitat heterogeneity to bird assemblages in temperate forest fragments. Biol. Conserv. 36:115–141.

FRITZ, E. C. 1990. Whats all this about "new forestry." For. Watch 11(7):7–14.

GARRISON, G. A., AND J. G. SMITH. 1974. Habitat of grazing animals. Pages P-1–P-10 *in* O. P. Cramer, ed. Environmental effects of forest residues management in the Pacific Northwest—a state-of-the-knowledge compendium. U.S. For. Serv. Gen. Tech. Rep. PNW-24.

GARSD, A. 1984. Spurious correlation in ecological modelling. Ecol. Model. 23:191–201.

GASHWILER, J. S. 1970. Plant and mammal changes on a clearcut in west-central Oregon. Ecology 51:1018–1026.

GATES, J. E., AND L. W. GYSEL. 1978. Avian nest dispersion and fledging success in field-forest ecotones. Ecology 59:871–883.

GEIST, V. 1970. A behavioural approach to the management of wild ungulates. Symp. Br. Ecol. Soc. 11:413–424.

GEORGE, T. L. 1987. Greater land bird densities on island vs. mainland: relation to nest predation level. Ecology 68:1393–1400.

GIBBS, C. B. 1978. Uneven-aged silviculture and management? Even-aged silviculture management? Definitions and differences. Pages 18–24 *in* Uneven-aged silviculture and management in the United States. U.S. For. Serv. Gen. Tech. Rep. WO-24.

GINGRICH, S. F. 1967. Measuring and evaluating stock and stand density in upland hardwood forests in the central states. For. Sci. 13: 38–53.

GRISEZ, T. J. 1960. Slash helps protect seedlings from deer browsing. J. For. 58:385–387.

GRUE, C. E., R. R. REID, AND N. J. SILVY. 1981. A windshield and multivariate approach to the classification, inventory and evaluation

of wildlife habitat: an exploratory study. Pages 124–140 *in* D. E. Capen, ed. The use of multivariate statistics in studies of wildlife habitat. U.S. For. Serv. Gen. Tech. Rep. RM-87.

GULLION, G. W. 1984. Managing northern forests for wildlife. The Ruffed Grouse Soc., Coraopolis, Pa., Publ. 13,442, Misc. J. Ser., Minn. Agric. Exp. Stn., St. Paul. 72pp.

———. 1989. Managing the woods for the birds' sake. Pages 334–349 *in* S. Atwater and J. Schnell, eds. The wildlife series: ruffed grouse. Stackpole Books, Harrisburg, Pa.

HAAPANEN, A. 1965. Bird fauna of the Finnish forests in relation to forest succession. I. Ann. Zool. Fenn. 2:153–196.

HAIRSTON, N. G. 1981. An experimental test of a guild: salamander competition. Ecology 62:65–72.

HALL, C. A. S., AND J. W. DAY. 1977. Systems and models: terms and basic principles. Pages 5–36 *in* C. A. S. Hall and J. W. Day, Jr., eds. Ecosystem modeling in theory and practice: an introduction with case histories. John Wiley & Sons, New York, N.Y.

HAMEL, P. B., H. E. LEGRAND, JR., M. R. LENNARTZ, AND S. A. GAUTHREAUX, JR. 1982. Bird-habitat relationships on southeastern forest lands. U.S. For. Serv. Gen. Tech. Rep. SE-22. 417pp.

HAMILTON, R. J. 1981. Effects of prescribed fire on black bear populations in southern forests. Pages 129–134 *in* G. W. Wood, ed. Prescribed fire and wildlife in southern forests. Belle W. Baruch For. Sci. Inst. Clemson Univ., Georgetown, S.C.

HARLOW, R. F., AND D. C. GUYNN, JR. 1983. Snag densities in managed stands of the South Carolina coastal plain. South. J. Appl. For. 7:224–229.

———, J. B. WHELAN, H. S. CRAWFORD, AND J. E. SKEEN. 1975. Deer foods during years of oak mast abundance and scarcity. J. Wildl. Manage. 39:330–336.

HARMON, M. E., ET AL. 1986. Ecology of coarse woody debris in temperate ecosystems. Adv. Ecol. Res. 15:133–302.

HARRIS, L. D. 1984. The fragmented forest. Univ. Chicago Press, Chicago, Ill. 211pp.

———, AND P. B. GALLAGHER. 1989. New initiatives for wildlife conservation: the need for movement corridors. Pages 11–34 *in* G. Mackintosh, ed. Preserving communities and corridors. Defenders of Wildlife, Washington, D.C.

———, L. D. WHITE, J. E. JOHNSTON, AND D. G. MILCHUNAS. 1974. Impact of forest plantations on north Florida wildlife and habitat. Proc. Annu. Conf. Southeast. Assoc. Game and Fish Comm. 28: 659–667.

HASSLER, C. C., S. A. SINCLAIR, AND E. KALLIO. 1986. Logistic regression: a potentially useful tool for researchers. For. Prod. J. 36(9):16–18.

HAWES, A. R., AND R. J. HUDSON. 1976. A method of regional landscape evaluation for wildlife. J. Soil Water Conserv. 31:210–211.

HAYNES, R. W., COORDINATOR. 1990. An analysis of the timber situation in the United States: 1989–2040. U.S. For. Serv. Gen. Tech. Rep. RM-199. 268pp.

HAYS, R. L., C. SUMMERS, AND W. SEITZ. 1981. Estimating wildlife habitat variables. U.S. Fish Wildl. Serv. FWS/OBS-81/47. 111pp.

HEALY, W. M. 1987. Habitat characteristics of uneven-aged stands. Pages 338–347 *in* R. D. Nyland, ed. Managing northern hardwoods. Proc. Silvicultural Symp. SUNY Coll. Environ. Sci. For., Faculty For., Misc. Publ. 13 (ESF 87-002).

———. 1989. Uneven-aged silviculture and wildlife habitat. Pages 225–237 *in* J. C. Finley and M. C. Brittingham, eds. Timber management and its effects on wildlife. Proc. Pa. State For. Resour. Issues Conf., Pa. State Coop. Ext., University Park.

HEMSTROM, M., AND V. D. ADAMS. 1982. Modeling long-term forest succession in the Pacific Northwest. Pages 14–23 *in* J. E. Means, ed. Forest succession and stand development research in the northwest. For. Res. Lab., Oregon State Univ., Corvallis.

HOLMES, R. T., R. E. BONNEY, JR., AND S. W. PACALA. 1979. Guild structure of the Hubbard Brook bird community: a multivariate approach. Ecology 60:512–520.

HOLTHAUSEN, R. S. 1986. Use of vegetation projection models for management problems. Pages 371–375 *in* J. Verner, M. L. Morrison, and C. J. Ralph, eds. Wildlife 2000: modeling habitat relationships of terrestrial vertebrates. Univ. Wisconsin Press, Madison.

HOOVER, R. L., AND D. L. WILLS, EDITORS. 1984. Managing forested lands for wildlife. Colo. Div. Wildl., U.S. For. Serv., Rocky Mt. Reg., Denver. 459pp.

HORTON, S. P. 1987. Effects of prescribed burning on breeding birds in a ponderosa pine forest, southeastern Arizona. M.S. Thesis, Univ. Arizona, Tucson. 75pp.

———, AND R. W. MANNAN. 1988. Effects of prescribed fire on snags and cavity-nesting birds in southeastern Arizona pine forests. Wildl. Soc. Bull. 16:37–44.

HUNTER, M. L., JR. 1990. Wildlife, forests, and forestry. Prentice Hall, Englewood Cliffs, N.J. 370pp.

HUNTLEY, J. C. 1986. Wilderness areas: impact on gray and fox squirrels. Pages 54–61 *in* D. L. Kulhavy and R. N. Conner, eds. Wilderness and natural areas in the eastern United States: a management challenge. Cent. Appl. Stud., School For., Stephen F. Austin State Univ., Nacogdoches, Tex.

HURST, G. A., AND R. C. WARREN. 1981. Enhancing white-tailed deer habitat of pine plantations by intensive management. Miss. Agric. For. Exp. Stn. Tech. Bull. 107. 8pp.

IRWIN, L. L., AND J. M. PEEK. 1983. Elk habitat use relative to forest succession in Idaho. J. Wildl. Manage. 47:664–672.

JACKSON, J. A., R. N. CONNER, AND B. J. S. JACKSON. 1986. The effects of wilderness on the endangered red-cockaded woodpecker. Pages 71–78 *in* D. L. Kulhavy and R. N. Conner, eds. Wilderness and natural areas in the eastern United States: a management challenge. Cent. Appl. Stud., School For., Stephen F. Austin State Univ., Nacogdoches, Tex.

JAGEMAN, H. 1984. White-tailed deer habitat management guidelines. Univ. Idaho For. Wildl. Range Exp. Stn. Bull. 37. 14pp.

JOHNSTON, D. W., AND E. P. ODUM. 1956. Breeding bird populations in relation to plant succession on the Piedmont of Georgia. Ecology 37:50–62.

KARR, J. R. 1982. Population variability and extinction in the avifauna of a tropical land bridge island. Ecology 63:1975–1978.

———, AND K. E. FREEMARK. 1985. Disturbance and vertebrates: an integrative perspective. Pages 153–168 *in* S. T. A. Pickett and P. S. White, eds. The ecology of natural disturbance and patch dynamics. Academic Press, Orlando Fla.

KIMMINS, J. P. 1987. Forest ecology. Macmillan, New York, N.Y. 531pp.

KIRK, D. 1975. Biology today. Second ed. Random House, Inc., New York, N.Y. 847pp.

KIRKLAND, G. L., JR. 1977. Responses of small mammals to the clearcutting of northern Appalachian forests. J. Mammal. 58:600–609.

KIRKMAN, R. L., J. A. EBERLY, W. R. PORATH, AND R. R. TITUS. 1986. A process for integrating wildlife needs into forest management planning. Pages 347–350 *in* J. Verner, M. L. Morrison, and C. J. Ralph, eds. Wildlife 2000: modeling habitat relationships of terrestrial vertebrates. Univ. Wisconsin Press, Madison.

KNOPF, F. L., J. A. SEDGWICK, AND R. W. CANNON. 1988. Guild structure of a riparian avifauna relative to seasonal cattle grazing. J. Wildl. Manage. 52:280–290.

KOMAREK, E. V. 1981. History of prescribed fire and controlled burning in wildlife management in the south. Pages 1–14 *in* G. W. Wood, ed. Prescribed fire and wildlife in southern forests. Belle W. Baruch For. Sci. Inst. Clemson Univ., Georgetown, S.C.

KREFTING, L. W., AND H. L. HANSEN. 1969. Increasing browse for deer by aerial applications of 2,4-D. J. Wildl. Manage. 33:784–790.

———, AND R. L. PHILLIPS. 1970. Improving deer habitat in Upper Michigan by cutting mixed-conifer swamps. J. For. 68:701–704.

KROODSMA, R. L. 1984. Effect of edge on breeding forest bird species. Wilson Bull. 96:426–436.

LANDERS, J. L., R. J. HAMILTON, A. S. JOHNSON, AND R. L. MARCHINTON. 1979. Foods and habitat of black bears in southeastern North Carolina. J. Wildl. Manage. 43:143–153.

LANDRES, P. B., AND J. A. MACMAHON. 1980. Guilds and community organization: analysis of an oak woodland avifauna in Sonora, Mexico. Auk 97:351–365.

LEAK, W. B., AND S. M. FILIP. 1977. Thirty-eight years of group selection in New England northern hardwoods. J. For. 75:641–643.

LEE, K. N., AND J. LAWRENCE. 1986. Adaptive management: learning from the Columbia River basin fish and wildlife program. Environ. Law 16:431–460.

LEEGE, T. A. 1984. Guidelines for evaluating and managing summer elk habitat in northern Idaho. Idaho Dep. Fish Game Wildl. Bull. 11. 37pp.

LEOPOLD, A. 1933. Game management. Charles Scribner's Sons, New York, N.Y. 481pp.

LIPSCOMB, J. F., J. C. CAPP, S. P. MEALEY, AND W. W. SANDFORT. 1984. Establishing wildlife goals and objectives. Pages 305–321 *in* R. L. Hoover and D. L. Wills, eds. Managing forested lands for wildlife. Colo. Div. Wildl., U.S. For. Serv., Rocky Mt. Reg., Denver.

LYNCH, J. F., AND D. F. WHIGHAM. 1981. Configuration of forest patches necessary to maintain bird and plant communities. Smithsonian Inst. Feb. 1982, from U.S. Gov. Rep. 82(18):3546.

———, AND ———. 1984. Effects of forest fragmentation on breeding bird communities in Maryland, USA. Biol. Conserv. 28:287–324.

LYON, L. J. 1983. Road density models describing habitat effectiveness for elk. J. For. 81:592–595.

———, AND C. E. JENSEN. 1980. Management implications of elk and deer use of clear-cuts in Montana. J. Wildl. Manage. 44:352–362.

———, ET AL. 1985. Coordinating elk and timber management. Mont. Dep. Fish, Wildl., Parks, Bozeman. 53pp.

MACCLINTOCK, L., R. F. WHITCOMB, AND B. L. WHITCOMB. 1977. Island biogeography and ''habitat islands'' of eastern forest. II. Evidence for the value of corridors and minimization of isolation in preservation of biotic diversity. Am. Birds 31:6–12.

MACKENZIE, D. I., AND S. G. SEALY. 1981. Nest site selection in eastern and western kingbirds: a multivariate approach. Condor 83: 310–321.

MANNAN, R. W., E. C. MESLOW, AND H. M. WIGHT. 1980. Use of snags by birds in Douglas-fir forests, western Oregon. J. Wildl. Manage. 44:787–797.

———, M. L. MORRISON, AND E. C. MESLOW. 1984. Comment: the use of guilds in forest bird management. Wildl. Soc. Bull. 12:426–430.

MARCOT, B. G. 1979. California wildlife/habitat relationships program, North Coast/Cascades zone. Five vols. U.S. For. Serv., Eureka, Calif. Var. pagin.

———. 1980. Use of a habitat/niche model for old growth management: a preliminary discussion. Pages 390–402 *in* R. M. DeGraaf, tech. coord. Proc. workshop management of western forests and grasslands for nongame birds. U.S. For. Serv. Gen. Tech. Rep. INT-86.

———. 1985. Habitat relationships of birds and young-growth Douglas-fir in northwestern California. Ph.D. Thesis, Oregon State Univ., Corvallis. 282pp.

———. 1986. Use of expert systems in wildlife-habitat modeling. Pages 145–150 *in* J. Verner, M. L. Morrison, and C. J. Ralph, eds. Wildlife 2000: modeling habitat relationships of terrestrial vertebrates. Univ. Wisconsin Press, Madison.

———. 1988. 1st-class expert systems: 1st-class. AI Expert 3:77–80.

———, R. S. MCNAY, AND R. E. PAGE. 1989. Use of microcomputers for planning and managing silviculture-habitat relationships. U.S. For. Serv. Gen. Tech. Rep. PNW-GTR-228. 19pp.

———, M. G. RAPHAEL, AND K. H. BERRY. 1983. Monitoring wildlife habitat and validation of wildlife-habitat relationships models. Trans. North Am. Wildl. Nat. Resour. Conf. 48:315–329.

MARQUIS, D. A. 1981. Effect of deer browsing on timber production in Allegheny hardwood forests of northwestern Pennsylvania. U.S. For. Serv. Res. Paper NE-475. 10pp.

———. 1990. A multi-resource silviculture decision model for forests of the northeastern United States. Proc. Int. Union For. Res. Organ. World Congr. 19:419–430.

MARTELL, A. M., AND A. RADVANYI. 1977. Changes in small mammal populations after clearcutting of northern Ontario black spruce forest. Can. Field–Nat. 91:41–46.

MARTIN, T. E. 1988. Habitat and area effects on forest bird assemblages: is nest predation an influence? Ecology 69:74–84.

MASER, C., R. G. ANDERSON, K. CROMACK, JR., J. T. WILLIAMS, AND R. E. MARTIN. 1979. Dead and down wood material. Pages 78–95 *in* J. W. Thomas, tech. ed. Wildlife habitats in managed forests—the Blue Mountains of Oregon and Washington. U.S. For. Serv. Handb. 553.

———, AND J. M. TRAPPE, EDITORS. 1984. The seen and unseen world of the fallen tree. U.S. For. Serv. Gen. Rep. PNW-164. 56pp.

MATTHEWS, J. W., AND D. E. MCKNIGHT. 1982. Renewable resource commitments and conflicts in southeast Alaska. Trans. North Am. Wildl. Nat. Resour. Conf. 47:573–582.

MAURER, B. A., L. B. MCARTHUR, AND R. C. WHITMORE. 1981. Effects of logging on guild structure of a forest bird community in West Virginia. Am. Birds 35:11–13.

MCCOMB, W. C., S. A. BONNEY, R. M. SHEFFIELD, AND N. D. COST. 1986. Den tree characteristics and abundance in Florida and South Carolina. J. Wildl. Manage. 50:584–591.

———, P. L. GROETSCH, G. E. JACOBY, AND G. A. MCPEEK. 1989. Response of forest birds to an improvement cut in Kentucky. Proc. Annu. Conf. Southeast. Assoc. Fish Wildl. Agencies 43:313–325.

———, AND G. A. HURST. 1986. Herbicides and wildlife in southern forests. Pages 28–36 *in* J. G. Dickson and O. E. Maughan, eds. Managing southern forests for wildlife and fish. U.S. For. Serv. Gen. Tech. Rep. SO-65.

———, AND R. E. NOBLE. 1981. Nest-box and natural-cavity use in three mid-south forest habitats. J. Wildl. Manage. 45:93–101.

———, AND R. L. RUMSEY. 1983. Characteristics and cavity-nesting bird use of picloram-created snags in the central Appalachians. South. J. Appl. For. 7:34–37.

MCTAGUE, J. P., AND D. R. PATTON. 1989. Stand density index and its application in describing wildlife habitat. Wildl. Soc. Bull. 17: 58–62.

MEANLEY, B. 1959. Notes on Bachman's sparrow in central Louisiana. Auk 76:232–234.

MECH, L. D., S. H. FRITTS, G. L. RADDE, AND W. J. PAUL. 1988. Wolf distribution and road density in Minnesota. Wildl. Soc. Bull. 16: 85–87.

MEEHAN, W. R., T. R. MERRELL, JR., AND T. A. HANLEY, EDITORS. 1984. Fish and wildlife habitat relationships in old-growth forests. Am. Inst. Fish. Res. Biol., Juneau, Alas. 425pp.

MELDAHL, R. 1986. Alternative modeling methodologies for growth and yield projection systems. Pages 27–31 *in* Data management issues in forestry. Proc. Computer Conf.; April 7–9, 1986, Atlanta Ga. For. Resour. Systems Inst., Florence Ala.

MESLOW, E. C., AND H. M. WIGHT. 1975. Avifauna and succession in Douglas-fir forests of the Pacific Northwest. Pages 266–271 *in* D. R. Smith, tech. coord. Symp. management of forest and range habitats for nongame birds. U.S. For. Serv. Gen. Tech. Rep. WO-1.

MEYERS, J. M., AND A. S. JOHNSON. 1978. Bird communities associated with succession and management of loblolly-shortleaf pine forests. Pages 50–65 *in* R. M. DeGraaf, tech. coord. Proc. workshop management of southern forests for nongame birds. U.S. For. Serv. Gen. Tech. Rep. SE-14.

MONTHEY, R. W. 1984. Effects of timber harvesting on ungulates in northern Maine. J. Wildl. Manage. 48:279–285.

MORRISON, M. L., B. G. MARCOT, AND R. W. MANNAN. 1992. Wildlife-habitat relationships: concepts and applications. Univ. Wisconsin Press, Madison. 343pp.

———, AND E. C. MESLOW. 1983. Avifauna associated with early growth vegetation on clearcuts in the Oregon coast ranges. U.S. For. Serv. Res. Pap. PNW-305. 12pp.

———, AND ———. 1984. Response of avian communities to herbicide-induced vegetation changes. J. Wildl. Manage. 48:14–22.

MUEGGLER, W. F. 1966. Herbicide treatment of browse on a big-game winter range in northern Idaho. J. Wildl. Manage. 30:141–151.

MURPHY, D. A., AND J. H. EHRENREICH. 1965. Effects of timber harvest and stand improvement on forage production. J. Wildl. Manage. 29:734–739.

NAIMAN, R. J., C. A. JOHNSTON, AND J. C. KELLEY. 1988. Alteration of North American streams by beaver. BioScience 38:753–762.

NEITRO, W. A., ET. AL. 1985. Snags (wildlife trees). Pages 129–169 *in* E. R. Brown, tech. ed. Management of wildlife and fish habitats in forests of western Oregon and Washington. U.S. For. Serv. PNW Publ. R6-F&WL-192-1985.

NIEMI, G. J., AND J. R. PROBST. 1990. Wildlife and fire in the Upper Midwest. Pages 35–49 *in* J. M. Sweeney, ed. Management of dynamic ecosystems. North–Cent. Sect., The Wildl. Soc., Springfield, Ill.

NIXON, C. M., M. W. MCCLAIN, AND R. W. DONOHOE. 1975. Effects of hunting and mast crops on a squirrel population. J. Wildl. Manage. 39:1–25.

NOBLE, I. R. 1981. Predicting successional change. Pages 278–300 *in* Fire regimes and ecosystem properties. U.S. For. Serv. Gen. Tech. Rep. WO-26.

NOSS, R. F., AND L. D. HARRIS. 1986. Nodes, networks, and MUMs: preserving diversity at all scales. Environ. Manage. 10:299–309.

ODUM, E. P. 1969. The strategy of ecosystem development. Science 164:262–270.

O'HALLORAN, K. A., AND R. N. CONNER. 1987. Habitat used by brown-headed nuthatches. Bull. Tex. Ornithol. Soc. 20:7–13.

OLDEMEYER, J. L., AND W. L. REGELIN. 1980. Response of vegetation to tree crushing in Alaska. Proc. North Am. Moose Conf. Workshop 16:429–443.

OLIVER, W. W., AND R. F. POWERS. 1978. Growth models for ponderosa pine: I. Yield of unthinned plantations in northern California. U.S. For. Serv. Res. Pap. PSW-133. 21pp.

OSAKI, S. K. 1979. An assessment of wildlife populations and habitat in herbicide-treated Jeffrey pine plantations. M.S. Thesis, Univ. California, Berkeley. 83pp.

OZOGA, J. J. 1968. Variations in microclimate in a conifer swamp deeryard in northern Michigan. J. Wildl. Manage. 32:574–585.

PAIS, R. C., S. A. BONNEY, AND W. C. MCCOMB. 1988. Herpetofaunal species richness and habitat associations in an eastern Kentucky forest. Proc. Annu. Conf. Southeast. Assoc. Fish Wildl. Agencies 42:448–455.

PARKER, G. R., AND L. D. MORTON. 1978. The estimation of winter forage and its use by moose on clearcuts in northcentral Newfoundland. J. Range Manage. 31:300–304.

PERRY, C., AND R. OVERLY. 1977. Impact of roads on big game distribution in portions of the Blue Mountains of Washington, 1972–1973. Wash. Game Dep. Appl. Res. Sect. Bull. 11. 39pp.

PICKETT, S. T. A. 1976. Succession: an evolutionary interpretation. Am. Nat. 110:107–119.

———, AND J. N. THOMPSON. 1978. Patch dynamics and the design of nature reserves. Biol. Conserv. 36:27–37.

———, AND P. S. WHITE. 1985. The ecology of natural disturbance and patch dynamics. Academic Press, Orlando, Fla. 472pp.

PIEROVICH, J. M., E. H. CLARKE, S. G. PICKFORD, AND F. R. WARD. 1975. Forest residues management guidelines for the Pacific Northwest. U.S. For. Serv. Gen. Tech. Rep. PNW-33. 281pp.

POWERS, J. E. 1979. Planning for an optimal mix of agricultural and wildlife land use. J. Wildl. Manage. 43:493–502.

PROBST, J. R. 1988. Kirtland's warbler breeding biology and habitat management. Pages 28–35 *in* T. W. Hoekstra and J. Capp, comps. Integrating forest management for wildlife and fish. U.S. For. Serv. Gen. Tech. Rep. NC-122.

RADLOFF, D. L., AND D. R. BETTERS. 1978. Multivariate analysis of physical site data for wildland classification. For. Sci. 24:2–10.

RAEDEKE, K. J., AND J. F. LEHMKUHL. 1986. A simulation procedure for modeling the relationships between wildlife and forest management. Pages 377–381 *in* J. Verner, M. L. Morrison, and C. J. Ralph, eds. Wildlife 2000: modeling habitat relationships of terrestrial vertebrates. Univ. Wisconsin Press, Madison.

RAMIREZ, P., JR., AND M. HORNOCKER. 1981. Small mammal populations in different-aged clearcuts in northwestern Montana. J. Mammal. 62:400–403.

RAMM, C. W., AND C. L. MINER. 1986. Growth and yield programs used on microcomputers in the North Central Region. North. J. Appl. For. 3:44–45, 79.

RAMSEY, K. J., AND W. C. KRUEGER. 1986. Grass-legume seeding to improve winter forage for Roosevelt elk: a literature review. Oregon State Univ. Agric. Exp. Stn. Spec. Rep. 763. 31pp.

RAPHAEL, M. G., AND R. H. BARRETT. 1981. Methodologies for a comprehensive wildlife survey and habitat analysis in old-growth Douglas-fir forests. Cal-Neva Wildl. Trans. 1981:106–121.

———, AND ———. 1984. Diversity and abundance of wildlife in late successional Douglas-fir forests. Pages 34–43 *in* New forests for a changing world. Proc. 1983 Annu. Meet. Soc. Am. For., Portland, Oreg. Soc. Am. For., Bethesda, Md.

———, AND B. G. MARCOT. 1986. Validation of a wildlife-habitat-relationships model: vertebrates in a Douglas-fir sere. Pages 129–138 *in* J. Verner, M. L. Morrison, and C. J. Ralph, eds. Wildlife 2000: modeling habitat relationships of terrestrial vertebrates. Univ. Wisconsin Press, Madison.

———, AND M. L. MORRISON. 1987. Decay and dynamics of snags in the Sierra Nevada, California. For. Sci. 33:774–783.

———, AND M. WHITE. 1984. Use of snags by cavity-nesting birds in the Sierra Nevada. Wildl. Monogr. 86. 66pp.

REGELIN, W. L., AND O. C. WALLMO. 1978. Duration of deer forage benefits after clearcut logging of subalpine forest in Colorado. U.S. For. Serv. Res. Note RM-356. 4pp.

REINEKE, L. H. 1933. Perfecting a stand density index for even-aged forests. J. Agric. Res. 46:627–638.

REPENNING, R. W., AND R. F. LABISKY. 1985. Effects of even-age timber management on bird communities of the longleaf pine forest in northern Florida. J. Wildl. Manage. 49:1088–1098.

ROBBINS, C. S. 1979. Effect of forest fragmentation on bird populations. Pages 198–212 *in* R. DeGraaf, tech. coord. Proc. workshop management of north central and northeastern forests for nongame birds. U.S. For. Serv. Gen. Tech. Rep. NC-51.

ROGERS, L. L. 1987. Effects of food supply and kinship on social behavior, movements, and population growth of black bears in northeastern Minnesota. Wildl. Monogr. 97. 72pp.

ROST, G. R., AND J. A. BAILEY. 1979. Distribution of mule deer and elk in relation to roads. J. Wildl. Manage. 43:634–641.

SALWASSER, H., AND F. B. SAMSON. 1985. Cumulative effects analysis: an advance in wildlife planning and management. Trans. North Am. Wildl. Nat. Resour. Conf. 50:313–321.

———, AND J. C. TAPPEINER, II. 1981. An ecosystem approach to integrated timber and wildlife habitat management. Trans. North Am. Wildl. Nat. Resour. Conf. 46:473–487.

SANTILLO, D. J., P. W. BROWN, AND D. M. LESLIE, JR. 1989a. Response of songbirds to glyphosate-induced habitat changes on clearcuts. J. Wildl. Manage. 53:64–71.

———, D. M. LESLIE, JR., AND P. W. BROWN. 1989b. Responses of small mammals and habitat to glyphosate application on clearcuts. J. Wildl. Manage. 53:164–172.

SAVIDGE, J. A. 1987. Extinction of an island forest avifauna by an introduced snake. Ecology 68:660–668.

SCHAMBERGER, M., A. H. FARMER, AND J. W. TERRELL. 1982. Habitat suitability index models: introduction. U.S. Fish Wildl. Serv. FWS/OBS-82/10. 2pp.

SCHOEN, J. W., O. C. WALLMO, AND M. D. KIRCHOFF. 1981. Wildlife-forest relationships: is a reevaluation of old growth necessary? Trans. North Am. Wildl. Nat. Resour. Conf. 46:531–544.

SCHROEDER, R. L. 1982. Habitat suitability index models: yellow warbler. U.S. Fish Wildl. Serv. FWS/OBS-82/10.27. 8pp.

———. 1987. Community models for wildlife impact assessment: a review of concepts and approaches. U.S. Fish Wildl. Serv. Biol. Rep. 87(2). 41pp.

SCHULTZ, R. D., AND J. A. BAILEY. 1978. Responses of national park elk to human activity. J. Wildl. Manage. 42:91–100.

SCOTT, J. M., B. CSUTI, J. D. JACOBI, AND J. E. ESTES. 1987. Species richness: a geographic approach to protecting future biological diversity. BioScience 37:782–788.

———, G. L. CROUCH, AND J. A. WHELAN. 1982. Responses of birds and small mammals to clearcutting in a subalpine forest in central Colorado. U.S. For. Serv. Res. Note RM-422. 6pp.

———, AND G. J. GOTTFRIED. 1983. Bird response to timber harvest in a mixed conifer forest in Arizona. U.S. For. Serv. Res. Pap. RM-245. 8pp.

———, J. A. WHELAN, AND P. L. SVOBODA. 1980. Cavity-nesting birds and forest management. Pages 311–324 in R. M. DeGraaf, tech. coord. Proc. workshop management of western forests and grasslands for nongame birds. U.S. For. Serv. Gen. Tech. Rep. INT-86.

SEDGWICK, J. A., AND F. L. KNOPF. 1986. Cavity-nesting birds and the cavity-tree resource in plains cottonwood bottomlands. J. Wildl. Manage. 50:247–252.

SEVERINGHAUS, W. D. 1981. Guild theory development as a mechanism for assessing environmental impact. Environ. Manage. 5:187–190.

SHORT, H. L. 1983. Wildlife guilds in Arizona desert habitats. U.S. Bur. Land Manage. Tech. Note 362. 258pp.

———, AND K. P. BURNHAM. 1982. Technique for structuring wildlife guilds to evaluate impacts on wildlife communities. U.S. Fish Wildl. Serv. Spec. Sci. Rep. Wildl. 234. 34pp.

SHUGART, H. H. 1984. A theory of forest dynamics. Springer-Verlag, New York, N.Y. 278pp.

———, AND D. JAMES. 1973. Ecological succession of breeding bird populations in northwestern Arkansas. Auk 90:62–77.

———, AND S. W. SEAGLE. 1985. Modeling forest landscapes and the role of disturbance in ecosystems and communities. Pages 353–368 in The ecology of natural disturbance and patch dynamics. Academic Press, New York, N.Y.

SIGMAN, M. 1985. Impacts of clear-cut logging in the fish and wildlife resources of southeast Alaska. Alas. Dep. Fish Game Tech. Rep. 85-3. 95pp.

SIMBERLOFF, D., AND J. COX. 1987. Consequences and costs of conservation corridors. Conserv. Biol. 1:63–71.

SMITH, H. C., N. I. LAMSON, AND G. W. MILLER. 1989. An esthetic alternative to clearcutting? Deferment cutting in eastern hardwoods. J. For. 87:14–18.

SMITH, D. M. 1962. The practice of silviculture. John Wiley & Sons, Inc., New York, N.Y. 578pp.

SMITH, T. M. 1986. Habitat-simulation models: integrating habitat-classification and forest-simulation models. Pages 389–393 in J. Verner, M. L. Morrison, and C. J. Ralph, eds. Wildlife 2000: modeling habitat relationships of terrestrial vertebrates. Univ. Wisconsin Press, Madison.

———, H. H. SHUGART, AND D. C. WEST. 1981. Use of forest simulation models to integrate timber harvest and nongame bird management. Trans. North Am. Wildl. Nat. Res. Conf. 46:501–510.

SOCIETY OF AMERICAN FORESTERS. 1984. Scheduling the harvest of old growth. Soc. Am. For., Bethesda, Md. 44pp.

SOPUCK, L. G., AND D. J. VERNAM. 1986. Distribution and movements of moose (Alces alces) in relation to the trans-Alaska oil pipeline. Arctic 39:138–144.

SPARROWE, R. D., AND B. F. SPARROWE. 1977. Use of critical parameters for evaluating wildlife habitat. Pages 385–405 in Classification, inventory, and analysis of fish and wildlife habitat. U.S. Fish Wildl. Serv. FWS/OBS-78/76-604.

STAMPS, J. A., M. BUECHNER, AND V. V. KRISHNAN. 1987. The effects of edge permeability and habitat geometry on emigration from patches of habitat. Am. Nat. 129:533–552.

STRANSKY, J. J., AND R. F. HARLOW. 1981. Effects of fire on deer habitat in the Southeast. Pages 135–142 in G. W. Wood, ed. Prescribed fire and wildlife in southern forests. Belle W. Baruch For. Sci. Inst. Clemson Univ., Georgetown, S.C.

———, AND D. RICHARDSON. 1977. Fruiting of browse plants affected by pine site preparation in east Texas. Proc. Annu. Conf. Southeast. Assoc. Fish Wildl. Agencies 31:5–7.

STRELKE, W. K., AND J. G. DICKSON. 1980. Effect of forest clear-cut edge on breeding birds in east Texas. J. Wildl. Manage. 44:559–567.

SWEENEY, J. M. 1986. Refinement of DYNAST's forest structure simulation. Pages 357–360 in J. Verner, M. L. Morrison, and C. J. Ralph, eds. Wildlife 2000: modeling habitat relationships of terrestrial vertebrates. Univ. Wisconsin Press, Madison.

———, M. E. GARNER, AND R. P. BURKERT. 1984. Analysis of white-tailed deer use of forest clear-cuts. J. Wildl. Manage. 48:652–655.

SWIFT, B. L., J. S. LARSON, AND R. M. DEGRAAF. 1984. Relationship of breeding bird density and diversity to habitat variables in forested wetlands. Wilson Bull. 96:48–59.

SZARO, R. C., AND R. P. BALDA. 1979. Bird community dynamics in a ponderosa pine forest. Stud. Avian Biol. 3. 66pp.

TAPPEINER, J. C., J. C. GOURLEY, AND W. H. EMMINGHAM. 1985. A user's guide for on-site determinations of stand density and growth with a programmable calculator. Oregon State Univ., For. Res. Lab., Spec. Publ. 11. 18pp.

TAYLOR, D. L. 1973. Some ecological implications of forest fire control in Yellowstone National Park, Wyoming. Ecology 54:1394–1396.

TEIPNER, C. L., E. O. GARTON, AND L. NELSON, JR. 1983. Pocket gophers in forest ecosystems. U.S. For. Serv. Gen. Tech. Rep. INT-154. 53pp.

TERRELL, J. W., EDITOR. 1984. Proceedings of a workshop on fish habitat suitability index models. U.S. Fish Wildl. Serv. Biol. Rep. 85(6). 393pp.

THIEL, R. P. 1985. Relationship between road densities and wolf habitat suitability in Wisconsin. Am. Midl. Nat. 113:404–407.

THIESSEN, J. L. 1976. Some relations of elk to logging, roading and hunting in Idaho's Game Management Unit 39. Pages 3–5 in S. R. Hieb, ed. Proc. elk-logging–roads symposium. Univ. Idaho, Moscow.

THOMAS, J. W., TECHNICAL EDITOR. 1979. Wildlife habitats in managed forests—the Blue Mountains of Oregon and Washington. U.S. For. Serv. Agric. Handb. 553. 512pp.

———, R. G. ANDERSON, C. MASER, AND E. L. BULL. 1979a. Snags. Pages 60–77 in J. W. Thomas, tech. ed. Wildlife habitats in managed forests—the Blue Mountains of Oregon and Washington. U.S. For. Serv. Agric. Handbk. 553.

———, E. D. FORSMAN, J. B. LINT, E. C. MESLOW, B. R. NOON, AND J. VERNER. 1990. A conservation strategy for the northern spotted owl. U.S. For. Serv. Interagency Sci. Comm. Rep. 427pp.

———, C. MASER, AND J. E. RODIEK. 1979b. Edges. Pages 48–59 in J. W. Thomas, tech. ed. Wildlife habitats in managed forests—the Blue Mountains of Oregon and Washington. U.S. For. Serv. Agric. Handbk. 553.

———, L. F. RUGGIERO, R. W. MANNAN, J. W. SCHOEN, AND R. A. LANCIA. 1988. Management and conservation of old-growth forests in the United States. Wildl. Soc. Bull. 16:252–262.

TILGHMAN, N. C. 1989. Impacts of white-tailed deer on forest regeneration in northwestern Pennsylvania. J. Wildl. Manage. 53:524–532.

TITTERINGTON, R. W., H. S. CRAWFORD, AND B. N. BURGASON. 1979. Songbird responses to commercial clear-cutting in Maine spruce-fir forests. J. Wildl. Manage. 43:602–609.

Toth, E. F., D. M. Solis, and B. G. Marcot. 1986. A management strategy for habitat diversity: using models of wildlife-habitat relationships. Pages 139–144 *in* J. Verner, M. L. Morrison, and C. J. Ralph, eds. Wildlife 2000: modeling habitat relationships of terrestrial vertebrates. Univ. Wisconsin Press, Madison.

U.S. Fish and Wildlife Service. 1980. Habitat evaluation procedures (HEP). Div. Ecol. Serv. ESM 102. 123pp.

———. 1985. Endangered species recovery plan: red-cockaded woodpecker (*Picoides borealis*). U.S. Fish Wildl. Serv., Atlanta, Ga. 88pp.

U.S. Forest Service. 1979. A generalized forest growth projection system applied to the Lakes States region. U.S. For. Serv. Gen. Tech. Rep. NC-49. 96pp.

Van Deusen, J. L., and C. A. Meyers. 1962. Porcupine damage in immature stands of ponderosa pine in the Black Hills. J. For. 60: 811–813.

Verme, L. J. 1965. Swamp conifer deeryards in northern Michigan—their ecology and management. J. For. 63:523–529.

———, and W. F. Johnston. 1986. Regeneration of northern white cedar deeryards in Upper Michigan. J. Wildl. Manage. 50:307–313.

Verner, J. 1984. The guild concept applied to management of bird populations. Environ. Manage. 8:1–14.

———, and A. S. Boss, technical coordinators. 1980. California wildlife and their habitats: western Sierra Nevada. U.S. For. Serv. Gen. Tech. Rep. PSW-37. 439pp.

Wagle, R. F., and T. W. Eakle. 1979. A controlled burn reduces the impact of a subsequent wildfire in a ponderosa pine vegetation type. For. Sci. 25:123–129.

Wallmo, O. C., and J. W. Schoen. 1980. Response of deer to secondary forest succession in southeast Alaska. For. Sci. 26:448–462.

Walters, C. J. 1986. Adaptive management of renewable resources. Macmillan Publ. Co., New York, N.Y. 374pp.

———, and R. Hilborn. 1978. Ecological optimization and adaptive management. Annu. Rev. Ecol. Syst. 9:157–188.

Weaver, H. 1947. Fire—nature's thinning agent in ponderosa pine stands. J. For. 45:437–444.

———. 1951. Fire as an ecological factor in the southwestern ponderosa pine forests. J. For. 49:93–98.

Weaver, J., R. Escano, D. Mattson, T. Puchlerz, and D. Despain. 1985. A cumulative effects model for grizzly bear management in the Yellowstone ecosystem. Pages 234–246 *in* G. P. Contreras and K. E. Evans, comp. Proc. grizzly bear habitat symposium. U.S. For. Serv. Gen. Tech. Rep. INT-207.

Webb, W. L., D. F. Behrend, and B. Saisorn. 1977. Effect of logging on songbird populations in a northern hardwood forest. Wildl. Monogr. 55. 35pp.

Weber, S. J., W. W. Mautz, J. W. Lanier, and J. E. Wiley, III. 1983. Predictive equations for deeryards in northern New Hampshire. Wildl. Soc. Bull. 11:331–338.

Welsh, H. H., Jr. 1990. Relictual amphibians and old-growth forests. Conserv. Biol. 4:309–319.

Wenger, K. F., editor. 1984. Forestry handbook. Second ed. John Wiley & Sons, New York, N.Y. 1335pp.

Whitcomb, R. F., C. S. Robbins, J. F. Lynch, B. L. Whitcomb, M. K. Klimkiewicz, and D. Bystrak. 1981. Effects of forest fragmentation on avifauna of the eastern deciduous forest. Pages 125–205 *in* R. L. Burgess, and D. M. Sharpe, eds. Forest island dynamics in man-dominated landscapes. Springer-Verlag, New York, N.Y.

Whiting, R. M. 1978. Avian diversity in various age pine forests in east Texas. Ph.D. Thesis, Texas A&M Univ., College Station. 160pp.

Wilcove, D. S. 1987. From fragmentation to extinction. Nat. Areas J. 7:23–29.

Williams, G. L., K. R. Russell, and W. K. Seitz. 1977. Pattern recognition as a tool in the ecological analysis of habitat. Pages 521–531 *in* Classification, inventory, and analysis of fish and wildlife habitat. U.S. Fish Wildl. Serv., FWS/OBS-78/76.

Williamson, J. F. 1983. Woodplan: microcomputer programs for forest management. Pages 128–130 *in* Proc. national workshop on computer uses in fish and wildlife programs. Virginia Polytechnic Inst. State Univ., Blacksburg.

Williamson, S. J., and D. H. Hirth. 1985. An evaluation of edge use by white-tailed deer. Wildl. Soc. Bull. 13:252–257.

Wisdom, M. J., et al. 1986. A model to evaluate elk habitat in western Oregon. U.S. For. Serv. Publ. R6-F&WL-216-1986. 36pp.

Wooldridge, D. D., and H. Weaver. 1965. Some effects of thinning a ponderosa pine thicket with a prescribed fire, II. J. For. 63:92–95.

Wykoff, W. R., N. L. Crookston, and A. R. Stage. 1982. User's guide to the stand prognosis model. U.S. For. Serv. Gen. Tech. Rep. INT-133. 112pp.

Yahner, R. H., and D. P. Scott. 1988. Effects of forest fragmentation on depredation of artificial nests. J. Wildl. Manage. 52:158–161.

APPENDICES

Appendix A. Common and scientific names of birds mentioned in text. Authority for scientific names of North American birds is Banks et al. 1987 (Checklist of vertebrates of the United States, U.S. Territories, and Canada. U.S. Fish Wildl. Serv. Resour. Publ. 166). Authority for scientific names of non-North American birds is Sibley, C.G., and B. L. Monroe, Jr. 1990 (Distribution and taxonomy of birds of the world. Yale Univ. Press, New Haven, Conn.).

Common name	Scientific name	Common name	Scientific name
Albatross	*Diomedea* spp.	mottled	*Anas fulvigula*
Avocet, American	*Recurvirostra americana*	ring-necked	*Aythya collaris*
Bananaquit	*Coereba flaveola*	ruddy	*Oxyura jamaicensis*
Bittern, American	*Botaurus lentiginosus*	wood	*Aix sponsa*
Blackbird, Brewer's	*Euphagus cyanocephalus*	Eagle, bald	*Haliaeetus leucocephalus*
red-winged	*Agelaius phoeniceus*	golden	*Aquila chrysaetos*
rusty	*Euphagus carolinus*	Egret, cattle	*Bubulcus ibis*
tricolored	*Agelaius tricolor*	great	*Casmerodius albus*
yellow-headed	*Xanthocephalus xanthocephalus*	Eider, common	*Somateria mollissima*
		Falcon, brown	*Falco berigora*
Bluebird, eastern	*Sialia sialis*	peregrine	*Falco peregrinus*
mountain	*Sialia currucoides*	Finch, Darwin's	Geospizinae
western	*Sialia mexicana*	house	*Carpodacus mexicanus*
Bobwhite, masked	*Colinus virginianus*	purple	*Carpodacus purpureus*
northern	*Colinus virginianus*	zebra	*Poephila guttata*
Brant, black	*Branta bernicla*	Flicker, northern	*Colaptes auratus*
Bullfinch, Puerto Rican	*Loxigilla portoricensis*	Flycatcher, brown-crested	*Myiarchus tyrannulus*
Bunting, indigo	*Passerina cyanea*	dusky	*Empidonax oberholseri*
lark	*Calamospiza melanocorys*	dusky-capped	*Myiarchus tuberculifer*
Canvasback	*Aythya valisineria*	great crested	*Myiarchus crinitus*
Cardinal, northern	*Cardinalis cardinalis*	least	*Empidonax minimus*
Catbird, gray	*Dumetella carolinensis*	sulphur-bellied	*Myiodynastes luteiventris*
Chat, yellow-breasted	*Icteria virens*	western	*Empidonax difficilis*
Chickadee, black-capped	*Parus atricapillus*	Wied's crested	*Myiarchus tyrannulus*
boreal	*Parus hudsonicus*	willow	*Empidonax traillii*
Carolina	*Parus carolinensis*	Frigate birds	*Fregata* spp.
chestnut-backed	*Parus rufescens*	Gadwall	*Anas strepera*
mountain	*Parus gambeli*	Gallinule, purple	*Porphyrula martinica*
Chukar	*Alectoris chukar*	Gnatcatcher, blue-gray	*Polioptila caerulea*
Condor, California	*Gymnogyps californianus*	Goldeneye, common	*Bucephala clangula*
Coot, American	*Fulica americana*	Goldeneye, Barrow's	*Bucephala islandica*
Cormorant, double-crested	*Phalacrocorax auritus*	Goldfinch, American	*Carduelis tristis*
Coturnix, common	*Coturnix coturnix*	Goose, barnacle	*Branta leucopsis*
Cowbird, brown-headed	*Molothrus ater*	cackling	*Branta canadensis*
Crane, sandhill	*Grus canadensis*	Canada	*Branta canadensis*
white-naped	*Grus vipio*	dusky Canada	*Branta canadensis*
whooping	*Grus americana*	Egyptian	*Alopochen aegyptiaca*
Creeper, brown	*Certhia americana*	emperor	*Chen canagica*
Crow, American	*Corvus brachyrhynchos*	greater white-fronted	*Anser albifrons*
Cuckoo, Puerto Rican lizard-	*Saurothera vieilloti*	lesser snow	*Chen caerulescens*
yellow-billed	*Coccyzus americanus*	Pacific white-fronted	*Anser albifrons*
Dove, mourning	*Zenaida macroura*	Ross'	*Chen rossii*
rock	*Columba livia*	snow	*Chen caerulescens*
white-winged	*Zenaida asiatica*	Vancouver Canada	*Branta canadensis*
Duck, American black	*Anas rubripes*	Goshawk, brown	*Accipiter fasciatus*
black-bellied whistling-	*Dendrocygna autumnalis*	northern	*Accipiter gentilis*
fulvous whistling-	*Dendrocygna bicolor*	Grackle, common	*Quiscalus quiscula*

722

Appendix A. Continued.

Common name	Scientific name	Common name	Scientific name
Grebe	Podicipedidae	common barn-	*Tyto alba*
western	*Aechmophorus occidentalis*	eastern screech-	*Otus asio*
Grosbeak, evening	*Coccothraustes vespertinus*	great horned	*Bubo virginianus*
rose-breasted	*Pheucticus ludovicianus*	long-eared	*Asio otus*
Grouse, blue	*Dendragapus obscurus*	northern saw-whet	*Aegolius acadicus*
forest	*Dendragapus* spp.	northern spotted	*Strix occidentalis*
red	*Dendragapus obscurus*	short-eared	*Asio flammeus*
ruffed	*Bonasa umbellus*	spotted	*Strix occidentalis*
sage	*Centrocercus urophasianus*	Parakeet, monk	*Myiopsitta monachus*
sharp-tailed	*Tympanuchus phasianellus*	Partridge, gray	*Perdix perdix*
spruce	*Dendragapus canadensis*	Pelican, brown	*Pelecanus occidentalis*
Gull, glaucous-winged	*Larus glaucescens*	Penguin, king	*Aptenodytes pataganicus*
herring	*Larus argentatus*	Petrel	Procellariidae
ring-billed	*Larus delawarensis*	Phalarope, Wilson's	*Phalaropus tricolor*
Harrier, northern	*Circus cyaneus*	Pheasant, ring-necked	*Phasianus colchicus*
Hawk, Cooper's	*Accipiter cooperii*	Phoebe, eastern	*Sayornis phoebe*
ferruginous	*Buteo regalis*	Pigeon, band-tailed	*Columba fasciata*
red-tailed	*Buteo jamaicensis*	tippler	*Columba* spp.
Swainson's	*Buteo swainsoni*	Pintail, northern	*Anas acuta*
Heron, black-crowned night-	*Nycticorax nycticorax*	Plover, mountain	*Charadrius montanus*
great blue	*Ardea herodias*	semipalmated	*Charadrius semipalmatus*
little blue	*Egretta caerulea*	Prairie-chicken, greater	*Tympanuchus cupido*
yellow-crowned night-	*Nycticorax violaceus*	lesser	*Tympanuchus pallidicinctus*
Hummingbird, blue-throated	*Lampornis clemenciae*	Ptarmigan, rock	*Lagopus mutus*
emerald	*Chlorostilbon maugaeus*	white-tailed	*Lagopus leucurus*
ruby-throated	*Archilochus colubris*	willow	*Lagopus lagopus*
violet-crowned	*Amazilia violiceps*	Quail, California	*Callipepla californica*
Ibis, white	*Eudocimus albus*	Gambel's	*Callipepla gambelii*
Jay, blue	*Cyanocitta cristata*	harlequin	*Cyrtonyx montezumae*
scrub	*Aphelocoma coerulescens*	Japanese	*Coturnix coturnix*
Junco, dark-eyed	*Junco hyemalis*	Mearns	*Cyrtonyx montezumae*
Kestrel, American	*Falco sparverius*	Montezuma	*Cyrtonyx montezumae*
Killdeer	*Charadrius vociferus*	mountain	*Oreortyx pictus*
Kingbird, eastern	*Tyrannus tyrannus*	scaled	*Callipepla squamata*
Kinglet, golden-crowned	*Regulus satrapa*	Rail, clapper	*Rallus longirostris*
Lark, horned	*Eremophila alpestris*	Virginia	*Rallus limicola*
Longspur, McCown's	*Calcarius mccownii*	Raven, common	*Corvus corax*
Magpie, black-billed	*Pica pica*	white-necked	*Corvus cryptoleucus*
Mallard	*Anas platyrhynchos*	Redhead	*Aythya americana*
Meadowlark, eastern	*Sturnella magna*	Redstart, American	*Setophaga ruticilla*
western	*Sturnella neglecta*	Robin, American	*Turdus migratorius*
Merganser, common	*Mergus merganser*	Sandpiper, pectoral	*Calidris melanotos*
hooded	*Lophodytes cucullatus*	upland	*Bartramia longicauda*
red-breasted	*Mergus serrator*	western	*Caladris mauri*
Mockingbird, northern	*Mimus polyglottos*	Sapsucker, red-breasted	*Sphyrapicus ruber*
Moorhen, common	*Gallinula chloropus*	Williamson's	*Sphyrapicus thyroideus*
Murrelet, marbled	*Brachyramphus marmoratus*	yellow-bellied	*Sphyrapicus varius*
Nighthawk, common	*Chordeiles minor*	Scoter, surf	*Melanitta perspicillata*
Nuthatch, brown-headed	*Sitta pusilla*	Shoveler, northern	*Anas clypeata*
pygmy	*Sitta pygmaea*	Shrike, loggerhead	*Lanius ludovicianus*
red-breasted	*Sitta canadensis*	Siskin, pine	*Carduelis pinus*
white-breasted	*Sitta carolinensis*	Skimmer, black	*Rynchops niger*
Oldsquaw	*Clangula hyemalis*	Snipe, common	*Gallinago gallinago*
Oriole, northern	*Icterus galbula*	Sora	*Porzana carolina*
Osprey	*Pandion haliaetus*	Sparrow, American tree	*Spizella arborea*
Ovenbird	*Seiurus aurocapillus*	Bachman's	*Aimophila aestivalis*
Owl, barred	*Strix varia*	Brewer's	*Spizella breweri*
boreal	*Aegolius funereus*	chipping	*Spizella passerina*
burrowing	*Athene cunicularia*	field	*Spizella pusilla*

Appendix A. Continued.

Common name	Scientific name	Common name	Scientific name
Gambel's	*Zonotrichia leucophrys*	solitary	*Vireo solitarius*
house	*Passer domesticus*	warbling	*Vireo gilvus*
Lincoln's	*Melospiza lincolnii*	Vulture, black	*Coragyps atratus*
song	*Melospiza melodia*	turkey	*Cathartes aura*
swamp	*Melospiza georgiana*	Warbler, blackpoll	*Dendroica striata*
white-crowned	*Zonotrichia leucophrys*	black-and-white	*Mniotilta varia*
white-throated	*Zonotrichia albicollis*	blue-winged	*Vermivora pinus*
Starling, European	*Sturnus vulgaris*	Kirtland's	*Dendroica kirtlandii*
Stork, wood	*Mycteria americana*	prothonotary	*Protonotaria citrea*
Swallow, bank	*Riparia riparia*	Virginia's	*Vermivora virginiae*
barn	*Hirundo rustica*	Wilson's	*Wilsonia pusilla*
tree	*Tachycineta bicolor*	yellow	*Dendroica petechia*
violet-green	*Tachycineta thalassina*	yellow-rumped	*Dendroica coronata*
welcome	*Hirundo neoxena*	yellow-throated	*Dendroica dominica*
Swan, Bewick's	*Cygnus columbianus*	Waxwing, cedar	*Bombycilla cedrorum*
mute	*Cygnus olor*	Wigeon, American	*Anas americana*
trumpeter	*Cygnus buccinator*	Woodcock, American	*Scolopex minor*
tundra	*Cygnus columbianus*	Woodpecker, acorn	*Melanerpes formicivorus*
Swift	Apodidae	black-backed	*Picoides arcticus*
Tanager, Puerto Rican	*Nesospingus speculiferus*	downy	*Picoides pubescens*
scarlet	*Piranga olivacea*	gila	*Melanerpes uropygialis*
striped-headed	*Spindalis zena*	hairy	*Picoides villosus*
Teal, blue-winged	*Anas discors*	ladder-backed	*Picoides scalaris*
green-winged	*Anas crecca*	Lewis'	*Melanerpes lewis*
red-billed	*Anas erythrorhyncha*	Nuttall's	*Picoides nuttallii*
Tern, least	*Sterna antillarum*	pileated	*Dryocopus pileatus*
Thrasher, brown	*Toxostoma rufum*	Puerto Rican	*Melanerpes portoricensis*
pearly-eyed	*Margarops fuscatus*	red-bellied	*Melanerpes carolinus*
Thrush, hermit	*Catharus guttatus*	red-cockaded	*Picoides borealis*
red-legged	*Turdus plumbeus*	red-headed	*Melanerpes erythrocephalus*
Swainson's	*Catharus ustulatus*	Strickland's	*Picoides stricklandi*
wood	*Hylocichla mustelina*	three-toed	*Picoides tridactylus*
Titmouse, bridled	*Parus wollweberi*	white-headed	*Picoides albolarvatus*
plain	*Parus inornatus*	Wood-pewee, eastern	*Contopus virens*
tufted	*Parus bicolor*	Wren, Bewick's	*Thryomanes bewickii*
Tody, Puerto Rican	*Todus mexicanus*	Carolina	*Thryothorus ludovicianus*
Towhee, rufous-sided	*Pipilo erythrophthalmus*	house	*Troglodytes aedon*
Tropicbird, white-tailed	*Phaethon lepturus*	marsh	*Cistothorus palustris*
Turkey, wild	*Meleagris gallopavo*	winter	*Troglodytes troglodytes*
Vireo, black-whiskered	*Vireo altiloquus*	Yellowlegs, lesser	*Tringa flavipes*
red-eyed	*Vireo olivaceus*	Yellowthroat, common	*Geothlypis trichas*

Appendix B. Common and scientific names of mammals mentioned in text. Authority for scientific names of North American mammals is Banks et al. 1987 (Checklist of vertebrates of the United States, the U.S. Territories, and Canada. U.S. Fish Wildl. Serv. Resour. Publ. 166). Authority for scientific names of non-North American mammals is Grzimek 1990 (Grzimek's encyclopedia of mammals. McGraw-Hill Publ. Co., New York, N.Y.).

Common name	Scientific name	Common name	Scientific name
Armadillo, nine-banded	*Dasypus novemcinctus*	Guar	*Bos gaurus*
Badger	*Taxidea taxus*	Hare, snowshoe	*Lepus americanus*
European	*Meles meles*	Hog, feral	*Sus scrofa*
Bat, big brown	*Eptesicus fuscus*	Impala	*Aepyceros melampus*
Brazilian free-tailed	*Tadarida brasiliensis*	Jack rabbit	*Lepus* spp.
hoary	*Lasiurus cinereus*	black-tailed	*Lepus californicus*
little brown	*Myotis lucifugus*	Jaguar	*Panthera onca*
pallid	*Antrozous pallidus*	Kangaroo, gray	*Macropus* spp.
Bear, black	*Ursus americanus*	Lemmings	Arvicolinae
brown	*Ursus arctos*	Lion	*Panthera leo*
grizzly	*Ursus arctos*	Llama	*Lama guanicoe*
polar	*Ursus maritimus*	Lynx	*Lynx canadensis*
Beaver	*Castor canadensis*	Macaque, pig-tailed	*Macaca nemestrina*
mountain	*Aplodontia rufa*	Manatee	*Trichechus manatus*
Beluga	*Delphinapterus leucas*	Marmot	*Marmota* spp.
Bison	*Bison bison*	yellow-bellied	*Marmota flaviventris*
Bobcat	*Lynx rufus*	Marten	*Martes americana*
Buffalo, black	*Syncerus caffer*	pine	*Martes martes*
Caribou	*Rangifer tarandus*	Mink	*Mustela vison*
Chipmunk, eastern	*Tamias striatus*	Mole	Talpidae
red-tailed	*Tamias ruficaudus*	Mongoose	*Herpestes* spp.
Townsend's	*Tamias townsendii*	Monkey, howler	*Alouatta palliata*
yellow-pine	*Tamias amoenus*	vervet	*Cercopithecus pygerythus*
Cottontail, desert	*Sylvilagus auduboni*	Moose	*Alces alces*
eastern	*Sylvilagus floridanus*	Mountain lion	*Felis concolor*
Nuttall's	*Sylvilagus nuttallii*	Mouse, brush	*Peromyscus boylii*
Cougar	*Felis concolor*	deer	*Peromyscus maniculatus*
Coyote	*Canis latrans*	fulvous harvest	*Reithrodontomys fulvescens*
Deer, black-tailed	*Odocoileus hemionus*	golden	*Ochrotomys nuttalli*
Columbian black-tailed	*Odocoileus hemionus*	harvest	*Reithrodontomys* spp.
fallow	*Dama dama*	house	*Mus musculus*
mule	*Odocoileus hemionus*	meadow jumping	*Zapus hudsonius*
musk	*Moschus moschiferus*	Pacific jumping	*Zapus trinotatus*
Père David's	*Elaphurus davidanus*	western jumping	*Zapus princeps*
red	*Cervus elaphus*	white-footed	*Peromyscus leucopus*
Sitka black-tailed	*Odocoileus hemionus*	wood	*Apodemus sylvaticus*
white-tailed	*Odocoileus virginianus*	yellow-necked field	*Apodemus flavicollis*
Dolphin, bottle-nosed	*Tursiops truncatus*	Muskox	*Ovibos moschatus*
Elephant, African	*Loxodonta africanus*	Muskrat	*Ondatra zibethicus*
Asian	*Elephas maximus*	round-tailed	*Neofiber alleni*
Elk	*Cervus elaphus*	Myotis, Yuma	*Myotis yumanensis*
Rocky Mountain	*Cervus elaphus*	Nilgai	*Boselaphus tragocamelus*
Roosevelt	*Cervus elaphus*	Nilgiri tahr	*Hemitragus hylocrius*
Ermine	*Mustela erminea*	Nutria	*Myocastor coypus*
Ferret, black-footed	*Mustela nigripes*	Okapi	*Okapia johnstoni*
Fisher	*Martes pennanti*	Opossum, Virginia	*Didelphis virginiana*
Fox, arctic	*Alopex lagopus*	Otter, river	*Lutra canadensis*
kit	*Vulpes macrotis*	sea	*Enhydra lutris*
gray	*Urocyon cinereoargenteus*	Panda, giant	*Ailuropoda melanoleuca*
red	*Vulpes vulpes*	Panther	*Felis concolor*
Gazelle, Thomson's	*Gazella thomsoni*	Peccary, collared	*Tayassu tajacu*
Giraffe	*Giraffa camelopardalis*	Pika	*Ochotona princeps*
Goat, mountain	*Oreamnos americanus*	Porcupine	*Erethizon dorsatum*
Rocky Mountain	*Oreamnos americanus*	Possum, brush-tailed	*Trichosurus vulpecula*
Gopher, pocket	*Thomomys* spp.	Prairie dog, black-tailed	*Cynomys ludovicianus*
Groundhog	*Marmota monax*	Utah	*Cynomys parvidens*

Appendix B. Continued.

Common name	Scientific name	Common name	Scientific name
Pronghorn	*Antilocapra americana*	Columbian ground	*Spermophilus columbianus*
Puma	*Felis concolor*	Douglas'	*Tamiasciurus douglasii*
Rabbit, European wild	*Oryctolagus cuniculus*	fox	*Sciurus niger*
swamp	*Sylvilagus palustris*	gray (grey)	*Sciurus carolinensis*
Raccoon	*Procyon lotor*	ground	*Spermophilus* spp.
Rat, black	*Rattus rattus*	Mount Graham red	*Tamiasciurus hudsonicus*
cotton	*Sigmodon* spp.	northern flying	*Glaucomys sabrinus*
hispid cotton	*Sigmodon hispidus*	red	*Tamiasciurus hudsonicus*
kangaroo	*Dipodomys* spp.	southern flying	*Glaucomys volans*
Norway	*Rattus norvegicus*	tassel-eared	*Sciurus aberti*
Reindeer	*Rangifer tarandus*	tree	*Sciurus* spp.
Rhinoceros, black	*Diceros bicornis*	Uinta ground	*Spermophilus armatus*
Indian	*Rhinoceros unicornis*	Vole, creeping	*Microtus oregoni*
Seal, gray	*Halichoerus grypus*	field	*Microtus agrestis*
harbor	*Phoca vitulina*	meadow	*Microtus pennsylvanicus*
Hawaiian monk	*Monachus schauinslandi*	montane	*Microtus montanus*
northern elephant	*Mirounga angustirostris*	pine	*Microtus pinetorum*
northern fur	*Callorhinus ursinus*	prairie	*Microtus ochrogaster*
Sea lion, northern	*Eumetopias jubatus*	red-backed	*Clethrionomys glareolus*
Steller's	*Eumetopias jubatus*	red tree	*Aborimus longicaudus*
Sheep, bighorn	*Ovis canadensis*	Townsend's	*Microtus townsendii*
Dall	*Ovis dalli*	Wallaby	*Petrogale* spp.
desert bighorn	*Ovis canadensis*	Wapiti	*Cervus elaphus*
mountain	*Ovis canadensis*	Weasel, least	*Mustela nivalis*
Stone	*Ovis stonei*	long-tailed	*Mustela frenata*
Shrew, dusky	*Sorex obscurus*	Whale, killer	*Orcinus orca*
masked	*Sorex cinereus*	Minke	*Balaenoptera acutorostrata*
short-tailed	*Blarina brevicauda*	sperm	*Physeter catodon*
smoky	*Sorex fumeus*	Wildebeest	*Connochaetes* spp.
Trowbridge's	*Sorex trowbridgii*	Wolf, gray	*Canis lupus*
vagrant	*Sorex vagrans*	Wolverine	*Gulo gulo*
Skunk, spotted	*Spilogale putorius*	Wombat	*Vombatidae* spp.
striped	*Mephitis mephitis*	Woodchuck	*Marmota monax*
Squirrel, arctic ground	*Spermophilus parryii*	Woodrat	*Neotoma* spp.
Belding's ground	*Spermophilus beldingi*	bushy-tailed	*Neotoma cinerea*
California ground	*Spermophilus beecheyi*	southern plains	*Neotoma micropus*

Appendix C. Common and scientific names of reptiles and amphibians mentioned in text. Authority for scientific names of North American reptiles and amphibians is Banks et al. 1987 (Checklist of vertebrates of the United States, the U.S. Territories, and Canada. U.S. Fish Wildl. Serv. Resour. Publ. 166). Authority for non-North American reptiles and amphibians is Sokolov 1988 (Dictionary of animal names in five languages: amphibians and reptiles. Russky Yazyk Publ., Moscow).

Common name	Scientific name	Common name	Scientific name
Alligator, American	*Alligator mississippiensis*	mountain dusky	*Desmognathus ochrophaeus*
Anole	*Anolis nebulosus*	plethodontid	Plethodontidae
green	*Anolis carolinensis*	ringed	*Ambystoma annulatum*
Boa, rubber	*Charina bottae*	slender	*Batrochoseps* spp.
Bullfrog	*Rana catesbeiana*	slimy	*Plethodon glutinosus*
Caiman, spectacled	*Caiman crocodilus*	tiger	*Ambystoma tigrinum*
Cottonmouth	*Agkistrodon piscivorus*	Wehrle's	*Plethodon wehrlei*
Crocodile, American	*Crocodylus acutus*	Skink, five-lined	*Eumeces fasciatus*
Frog, cascades	*Rana cascadae*	ground	*Scincella laterale*
green	*Rana clamitans*	New Guinea	*Egernia physicae*
harlequin	*Atelopus oxyrhynchus*	Slider, common	*Trachemys scripta*
Kenyan reed	*Hyperolius viridiflavus*	Snake, Dekay's brown	*Storeria dekayi*
leopard	*Rana pipiens*	common garter	*Thamnophis sirtalis*
northern leopard	*Rana pipiens*	grass	*Natrix natrix*
red-eared	*Rana erythraea*	king	*Lampropeltis* spp.
spotted	*Rana pretiosa*	lyre	*Trimorphodon biscutatus*
tailed	*Ascaphus truei*	pine	*Pituophis melanoleucus*
wood	*Rana sylvatica*	rat	*Elaphe obsoleta*
Hellbender	*Cryptobranchus alleganiensis*	redbelly	*Storeria occipitomaculata*
Iguana, desert	*Dipsosaurus dorsalis*	ring-necked	*Diadophis punctatus*
green	*Iguana iguana*	worm	*Carphophis amoenus*
Lizard, bloodsucker	*Calotes nemoricola*	Spadefoot, Great Basin	*Scaphiopus intermontanus*
eastern fence	*Sceloporus undulatus*	Terrapin, diamondback	*Malaclemys terrapin*
fringe-toed	*Uma notata*	Toad, American	*Bufo americanus*
keeled earless	*Holbrookia propinqua*	Fowler's	*Bufo woodhousii*
northern fence	*Sceloporus undulatus*	Gulf Coast	*Bufo valliceps*
side-blotched	*Uta stansburiana*	western	*Bufo boreas*
Sita's	*Sitana ponticeriana*	Woodhouse's	*Bufo woodhousii*
slender glass	*Ophisaurus attenuatus*	Yosemite	*Bufo canorus*
slow worm	*Anguis fragilis*	Treefrog, green	*Hyla cinerea*
Texas horned	*Phrynosoma cornutum*	Pacific	*Hyla regilla*
viviparous	*Lacerta vivipara*	Tortoise, Aldabra	*Geochelone gigantea*
western fence	*Sceloporus occidentalis*	desert	*Gopherus agassizii*
Newt, alpine	*Triturus alpestris*	Turtle, alligator snapping	*Macroclemys temminckii*
eastern	*Notophthalmus viridescens*	Blanding's	*Emydoidea blandingii*
roughskin	*Taricha granulosa*	chicken	*Deirochelys reticularia*
smooth	*Triturus vulgaris*	common box	*Terrapene carolina*
warty	*Triturus cristatus*	common mud	*Kinosternon subrubrum*
Racer	*Coluber constrictor*	green sea	*Chelonia mydas*
Racerunner, six-lined	*Cnemidophorus sexlineatus*	leatherback sea	*Dermochelys coriacea*
Rattlesnake, western	*Crotalus viridis*	loggerhead	*Caretta caretta*
Salamander, Appalachian	*Plethodon jordani*	loggerhead musk	*Sternotherus minor*
cave	*Eurycea lucifuga*	olive ridley	*Lepidochelys olivacea*
dusky	*Desmognathus fuscus*	painted	*Chrysemys picta*
eastern red-backed	*Plethodon cinereus*	snapping	*Chelydra serpentina*
long-tailed	*Eurycea longicauda*	spiny softshell	*Trionyx spiniferus*
long-toed	*Ambystoma macrodactylum*	spotted	*Clemmys guttata*
many-ribbed	*Eurycea multiplicata*	Whipsnake, striped	*Masticophis taeniatus*
mole	*Ambystoma talpoideum*		

Appendix D. Common and scientific names of fishes mentioned in text. Authority for scientific names of North American fishes is Robins et al. 1991 (Common and scientific names of fishes from the United States and Canada, fifth ed. Am. Fish. Soc. Spec. Publ. 20).

Common name	Scientific name	Common name	Scientific name
Bowfin	*Amia calva*	Mosquitofish, western	*Gambusia affinis*
Bullhead, black	*Ameiurus melas*	Sardine	Clupeidae
yellow	*Ameiurus natalis*	Shiner, red	*Cyprinella lutrensis*
Carp, common	*Cyprinus carpio*	Sucker, Sonora	*Catostomus insignis*
Chub, roundtail	*Gila robusta*	desert	*Catostomus clarki*
Dace, longfin	*Agosia chrysogaster*	Sunfish, green	*Lepomis cyanellus*
speckled	*Rhinichthys osculus*	Trout	Salmonidae
Darter, snail	*Percina tanasi*		

Appendix E. Common and scientific names of invertebrates mentioned in text. Authorities for scientific names are Pennak 1989 (Fresh-water invertebrates of the United States: protozoa to mollusca. Third ed. John Wiley & Sons, New York, N.Y.) for aquatic insects and Borror et al. 1989 (An introduction to the study of insects. Sixth ed. Saunders College Publ., Philadelphia, Pa.).

Common name	Scientific name	Common name	Scientific name
Boatman, water	Corixidae	Maggot, rattailed	*Eristalis* spp.
Beetle	Coleoptera	Mayfly	Ephemeroptera
Black fly	Simuliidae	Midge, phantom	Chaoboridae
Bloodworm	Chironomidae	Mosquito	Culicidae
Budworm, western spruce	*Choristoneura occidentalis*	Scorpion	Scorpiones
Clam, fingernail	Pelecypoda	Shrimp, clam	Conchostraca
Crab, ghost	*Ocypode albicas*	fairy	Anostraca
Crayfish	Decapoda	seed	Ostracoda
Dragonfly	Anisoptera	Sideswimmer	Amphipoda
Earthworm, aquatic	Oligochaeta	Snail, pond	Gastropoda
Flatworm	Turbellaria	Sowbug, aquatic	Isopoda
Flea, water	Cladocera	Spider	Araneae
Grasshopper	Orthoptera	Springtail	Collembola
Horsefly	Tabanidae	Water strider	Gerridae
Leech, nasal	*Theromyzon* spp.		

Appendix F. Common and scientific names of plants mentioned in text. Authority for scientific names is U.S. Soil Conservation Service 1982 (National list of scientific plant names. U.S. Dep. Agric. Soil Conserv. Serv. SCS-TP-159).

Common name	Scientific name	Common name	Scientific name
Acacia, catclaw	*Acacia greggii*	yellow	*Betula alleghaniensis*
sweet	*Acacia farnesiana*	Bitterbrush	*Purshia tridentata*
Alder, hazel	*Alnus rugosa*	Bittersweet	*Celastrus* spp.
Anise	*Pimpinella* spp.	Blackberry	*Rubus alleghaniensis*
Arrowwood	*Viburnum dentatum*	Blackgum	*Nyssa sylvatica*
Asafetida	*Ferula* spp.	Blackhaw	*Viburnum prunifolium*
Ash, green	*Fraxinus pennsylvanica*	Blueberry, highbush	*Vaccinium corymbosum*
Aspen, bigtooth	*Populus grandidentata*	Boxelder	*Acer negundo*
quaking	*Populus tremuloides*	Bristlegrass	*Setaria* spp.
Aster	*Aster* spp.	Buckthorn	*Rhamnus* spp.
Baldcypress	*Taxodium distichum*	Bulrush, Olney	*Scirpus olneyi*
Barnyard grass, common	*Echinochloa crusgalli*	Burhead, common	*Echinodorus cordifolius*
Bayberry	*Myrica* spp.	Buttonbush, common	*Cephalanthus occidentalis*
Beech, American	*Fagus grandifolia*	Canary grass	*Phalaris arundinacea*
Beggarticks	*Bidens frondosa*	Cattail	*Typha* spp.
Birch, gray	*Betula populifolia*	Cedar, white	*Thuja occidentalis*
paper	*Betula papyrifera*	Cherry, black	*Prunus serotina*
river	*Betula nigra*	pin	*Prunus pensylvanica*
sweet	*Betula lenta*	Cholla (cactus)	*Opuntia* spp.

Appendix F. Continued.

Common name	Scientific name	Common name	Scientific name
Chokecherry, black	*Aronia melanocarpa*	striped	*Acer pensylvanicum*
common	*Aronia melanocarpa*	sugar	*Acer saccharum*
Cocklebur	*Xanthium* spp.	Mesquite	*Prosopis* spp.
Coralberry	*Symphoricarpos orbiculatus*	Morningglory	*Ipomoea* spp.
Cordgrass, marshhay	*Spartina patens*	Mountain ash, American	*Sorbus americana*
Cottonwood, eastern	*Populus deltoides*	Mountain-mahogany	*Cercocarpus betuloides*
Crabapple, prairie	*Malus ioensis*	curlleaf	*Cercocarpus ledifolius*
Crabgrass	*Digitaria* spp.	Mulberry, red	*Morus rubra*
Cranberry, highbush	*Viburnum trilobum*	Nannyberry	*Viburnum lentago*
Current, golden	*Ribes aureum*	Oak, black	*Quercus velutina*
Cutgrass, rice	*Leersia oryzoides*	blackjack	*Quercus marilandica*
Desert ironwood	*Olneya tesota*	bur	*Quercus macrocarpa*
Devil's walkingstick	*Aralia spinosa*	California black	*Quercus kelloggi*
Dogwood, alternate leaf	*Cornus alternifolia*	cherrybark	*Quercus falcata*
flowering	*Cornus florida*	chestnut	*Quercus prinus*
red-osier	*Cornus stolonifera*	chinkapin	*Quercus muhlenbergii*
Douglas-fir	*Pseudotsuga menziesii*	live	*Quercus virginiana*
Elderberry	*Sambucus* spp.	northern red	*Quercus rubra*
Elm	*Ulmus* spp.	Nutall	*Quercus nuttallii*
Fern	Polypodiaceae	overcup	*Quercus lyrata*
Filaree	*Erodium cicutarium*	pin	*Quercus palustris*
Fir, balsam	*Abies balsamea*	post	*Quercus stellata*
subalpine	*Abies lasiocarpa*	scarlet	*Quercus coccinea*
white (grand fir)	*Abies concolor*	shingle	*Quercus imbricaria*
Firethorn	*Cotoneaster pyracantha*	swamp white	*Quercus bicolor*
Flatsedge, chufa	*Cyperus esculentus*	water	*Quercus nigra*
redroot	*Cyperus erythrorhizos*	white	*Quercus alba*
Grape, Oregon	*Berberis aquifolium*	Olive, autumn	*Elaeagnus umbellata*
Gooseberry	*Ribes* spp.	Russian	*Elaeagnus angustifolia*
Greenbrier	*Smilax* spp.	Palmetto	*Sabal* spp.
Hackberry, common	*Celtis occidentalis*	Panicgrass	*Panicum* spp.
sugar	*Celtis laevigata*	Panicum, fall	*Panicum dichotomiflorum*
Hawthorn, cockspur	*Crataegus crus-galli*	Partridgepea senna	*Cassia fasciculata*
dotted	*Crataegus punctata*	Pecan	*Carya illinoensis*
downy	*Crataegus mollis*	Persimmon, common	*Diospyros virginiana*
frosted	*Crataegus pruinosa*	Pine, eastern white	*Pinus strobus*
glossy	*Crataegus nitida*	jack	*Pinus banksiana*
Washington	*Crataegus phaenopyrum*	loblolly	*Pinus taeda*
Hemlock, eastern	*Tsuga canadensis*	lodgepole	*Pinus contorta*
western	*Tsuga heterophylla*	longleaf	*Pinus palustris*
Hickory, mockernut	*Carya tomentosa*	pinyon	*Pinus edulis*
pignut	*Carya glabra*	pitch	*Pinus rigida*
shagbark	*Carya ovata*	ponderosa	*Pinus ponderosa*
Holly	*Ilex* spp.	slash	*Pinus elliottii*
Honeysuckle, Japanese	*Lonicera japonica*	white	*Pinus strobus*
tartarian	*Lonicera tatarica*	Plum	*Prunus* spp.
Indian ricegrass	*Oryzopsis hymenoides*	Pokeweed	*Phytolacca americana*
Juniper	*Juniperus* spp.	Pricklyash, common	*Zanthoxylum americanum*
western	*Juniperus occidentalis*	Pricklypear, Lindheimer	*Opuntia lindheimeri*
Knotweed	*Polygonum* spp.	Rabbitbrush	*Chrysothamnus* spp.
Larch, eastern	*Larix laricina*	Ragweed	*Ambrosia* spp.
Lespedeza	*Lespedeza* spp.	Redcedar, eastern	*Juniperus virginiana*
Loosestrife, purple	*Lythrum salicaria*	Reedgrass	*Calamagrostis canadensis*
Lotus, American	*Nelumbo lutea*	Rose, multiflora	*Rosa multiflora*
Mangrove	*Rhizophora* spp.	Sagebrush, big	*Artemisia tridendata*
Manzanita	*Arctostaphylos* spp.	Saguaro (cactus)	*Cereus giganteus*
Maple, black	*Acer saccharum*	Saltbush, furrowing	*Avicennia germinans*
mountain	*Acer spicatum*	Saltgrass, seashore	*Distichlis spicata*
red	*Acer rubrum*	Sedges	*Carex* spp.

Appendix F. Continued.

Common name	Scientific name	Common name	Scientific name
Serviceberry, Allegany	*Amelanchier laevis*	Sweetgum, American	*Liquidambar styraciflua*
Saskatoon	*Amelanchier alnifolia*	Switchgrass	*Panicum virgatum*
shadblow	*Amelanchier canadensis*	Tarweed	*Madia* spp.
Sesbania, hemp	*Sesbania exaltata*	Thorn apple	*Crataegus columbiana*
Smartweed	*Polygonum* spp.	Timothy	*Phleum* spp.
Snowberry, common	*Symphoricarpos albus*	Toothcup	*Ammannia coccinea*
Sorghum, common	*Sorghum vulgare*	Tupelo, black	*Nyssa sylvatica*
Spicebush	*Lindera benzoin*	water	*Nyssa aquatica*
Spikerush, blunt	*Eleocharis obtusa*	Turkeymullein	*Eremocarpus* spp.
Sprangletop	*Leptochloa* spp.	Valerian	*Valeriana* sp.
Spruce, Colorado	*Picea pungens*	Viburnum, mapleleaf	*Viburnum acerifolium*
Engelmann	*Picea engelmannii*	Virginia creeper	*Parthenocissus quinquefolia*
Sitka	*Picea sitchensis*	Watershield	*Brasenia* spp.
white	*Picea glauca*	Wheatgrass,bluebunch	*Agropyron spicatum*
Squaw apple	*Peraphyllum ramosissimum*	Wildcelery, American	*Vallisneria americana*
Sumac, flameleaf	*Rhus copallinum*	Willow, black	*Salix nigra*
smooth	*Rhus glabra*	pussy	*Salix discolor*
staghorn	*Rhus typhina*	Winterberry	*Ilex* spp.
Sunflower	*Helianthus* spp.		

INDEX

A

2, 4-D, 697
2,4, 5-T, 697
abundance
  absolute, 216
  definition, 216
  proportional, 224
  relative, 216
  true, 219
accipiter(s), 185
acepromazine, 128–130
acid rain, 596
acidosis, 135
ACTH, 300
adits, 678, 679
aflatoxin, 329
after hatching year (AHY), 173
age distribution, 433
age ratio, 426
age structure, 187, 419, 433, 435, 437, 438
aggregate percentage, 269
aggregate volume, 269
Agriculture Act of 1990, 658
agricultural technology, 658
albatross, 347
algorithm, 79, 83
alkaloid, 307, 311
alligator, 118, 224, 249, 426, 628
alligator, American, 156–158
alpha-chloralose, 111, 118, 478, 480
American black duck, 227
American coot, 466, 670
American robin, 696
American Society of Ichthyologists and Herpetologists (ASIH), 97
amino acid, 626
analgesia, 127–129
analgesic, 127, 129, 131
analysis
  canonical correlation, 58
  cluster, 59, 60
  correlation, 4
  crop, 6
  data, 4–6
  discriminant, 57, 58
  discriminant function, 603
  factor, 56, 57
  failure-time, 82
  fecal, 6, 265, 267, 268
  food-habits, 268
  life-table, 226
  microscopic, 267
  multiple regression, 222
  multivariate, 2, 4, 19, 24, 60
  principal components, 56, 57, 601, 603
  quantitative, 28
  single-factor, 19
  statistical, 19, 21, 75, 87, 90
analysis of variance, 80
analysis of variance, multivariate, 603
Anectine, 131

anesthetic, 127, 131
animal (see also wildlife)
  collection, 98
  confinement during shipping, 100, 102
  density, 592, 593
  disposition of, 103
  diversity, 592, 596
  fitness, 592, 593
  housing, 100, 101, 104
  marking, 100
  restraint, 99–102
  transportation, 101
Animal Care and Use Committees (ACUC), 97, 98
Animal Welfare Act, 96, 97, 100, 101
anole, 155
ANOVA, 18, 19
ANOVA, multivariate, 80, 90
ANOVA, univariate, 80
antagonist, 126, 128–130, 135
antelope, 147
antenna
  Adcock, 383, 392
  dipole, 381, 382, 392
  H, 383
  loop, 375, 382, 386, 392
  mounting on aircraft, 384
  omni-directional, 383
  receiving, 379–383, 390, 392
  transmitting, 373–376, 381, 389, 390
  whip, 375–377, 383, 386
  Yagi, 383, 384, 392, 393
antibiotic, 135
antidote, 126, 130
antler beam diameter, 277
aphid, 419
apnea, 128
aquatic invertebrates, artificial substrates for sampling, 354
aquatic invertebrates, sampling, 352, 353
aquatic invertebrates, sampling vegetation for, 354
aquifer, 629
Arachnida, 358
archery, 453, 455
armadillo, 315, 484, 490
ASCII, 81, 85, 87
ash, 519, 583
aspen, 485, 576, 583, 587, 698
ataxia, 129
atropine, 129, 130
autecology, 594
autocorrelation, 262
avian cholera, 325, 326, 329
Avitrol, 480

B

bacteria, 325, 329, 334, 346, 627
baculum, 196, 198, 201, 204
badger, 131, 146, 292, 294, 312, 313, 488, 494, 500
badger, American, 201
bait, 106, 118

bait consumption, 147
baldcypress, 703
band
  arm-, 143
  bird, 143
  butt-end, 150, 157
  close-ring, 150
  colored, 150
  leg, 149, 150, 157
  lock-on, 150
  rivet, 143, 150
  split-ring, 150
band analysis, 82
band recovery, 239, 466, 468, 470
band return, 82
barnyard grass, 628, 635, 637, 644
basal metabolic rate, 315
bat, 339, 671
  big brown, 484
  Brazilian free-tailed, 484
  frugivore, 146
  little brown, 484
  pallid, 484
  Yuma myotis, 484
bear, 111, 672
  black, 32, 44, 126, 142, 148, 197, 284, 386, 500, 612, 698, 699
  brown, 197, 198, 294, 387, 612, 708
  grizzly, 126, 284, 441, 445, 500, 514, 540, 674, 713
  polar, 127, 132, 142, 144, 197, 198, 595
beaver, 145, 146, 186, 188, 204, 223, 249, 288, 292, 294, 387, 441, 483, 484, 490, 492, 493, 500, 502, 540, 639, 673, 674, 707
beaver, mountain, 147, 484, 487, 489, 492, 500, 694
bee balm, 519
beech, 696
beetles, 626
behavior, 3, 6, 21, 141, 150
behavior, animal, 84
behavior, reproductive, 141
benthos, 352
Berlese-Tullgren funnel, 358, 361
Betalight, 149, 154
bias, 5, 9, 10, 15, 18, 20, 29, 30, 32, 217, 223, 240, 242, 269, 270
big game
  harvest, 446
  hunting regulations, 454
  inventory, 446
  population and management objectives, 446, 447, 450, 454
biodiversity, 672, 674
biological diversity, 510, 511, 540, 690, 692
biomass, 627, 637, 641
  fish, 596

vegetative, 268
wildlife, 609
biome, 255
biotelemetry, 82, 84
biotic factors, 625
birch, 518, 576, 583
bird damage control
  chemical agents, 477, 478, 480
  cultural practices, 477, 479
  frightening devices, 477–480
  habitat modification, 477–479
  methods of assessment, 476
  proofing and screening, 477–479
  repellents, 477, 478, 480
  shooting, 477, 478, 481
  toxicants, 477, 478, 481, 482
  traps, 478, 480
birds
  gallinaceous, 175, 178
  game, 106, 111
  migrating, 630
  upland game, 113, 115
  wintering, 630
birth rate, 81
  age-dependent, 431
  estimation, 425
  instantaneous, 425
bison, 129, 133, 190, 193, 421–424, 426, 567, 680
bite-count, 263, 264, 266
bittern, 478
bivariate normal ellipse estimator, 83
blackbird, 114, 592
  Brewer's, 476
  red-winged, 4, 151, 476, 477
  rusty, 476
  tri-colored, 476
  yellow-headed, 476
blood
  collection, 97, 98
  sample, 334
  sampling, 298
  serum, 334
  smear, 333
blood urea nitrogen (BUN), 279, 282, 285
bloodworm, 626
blowgun, 132, 133
bluebird, 519
bobcat, 30, 132, 133, 148, 202, 224, 312, 315, 441, 494, 495, 497, 499, 500
bobwhite, masked, 668
bobwhite, northern, 10, 19, 172, 176, 283, 285, 312, 347, 441, 458–460, 462, 608, 668, 670, 697, 698
bomb calorimetry, 311, 312, 362
Boston ivy, 518
botulism, 326, 334, 346
bradycardia, 129
brain cholinesterase, 328
branding, 141, 145, 153, 155, 156